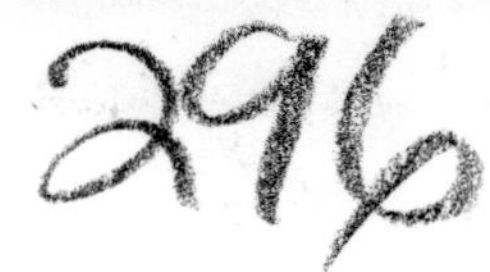

PROBABILITY AND STATISTICS

PSYCHOLOGY

TECHNOLOGY AND ENGINEERING

TRIGONOMETRIC APPLICATIONS

College Algebra and Trigonometry

FOURTH EDITION

College Algebra and Trigonometry

Richard N. Aufmann

Vernon C. Barker

Richard D. Nation

Palomar College

Houghton Mifflin Company

Boston New York

Editor-in-Chief: *Jack Shira*
Senior Sponsoring Editor: *Lynn Cox*
Senior Development Editor: *Dawn Nuttall*
Editorial Assistant: *Melissa Parkin*
Senior Project Editor: *Maria Morelli*
Editorial Assistant: *Tanius Stamper*
Senior Production/Design Coordinator: *Carol Merrigan*
Senior Manufacturing Coordinator: *Sally Culler*
Senior Marketing Manager: *Michael Busnach*

Cover Photo and page 413: FPG 2001

PHOTO CREDITS

Page 1 top: The Granger Collection; page 1 bottom: Francesco Regineto/The Image Bank; page 63 top: The Granger Collection; page 63 bottom: CORBIS; page 133 top and bottom: Bill Gallery/Stock Boston; page 237 top: Bob Rowen/Progressive Image/CORBIS; page 261: The Granger Collection; page 269: The Granger Collection; page 299 right: CORBIS/Bettmann: page 299 top: Tony Craddock/STONE Images; page 299 bottom: Edward Miller/Stock Boston; page 371: Richard T. Nowitz/CORBIS; page 387 top: Don & Liysa Kaing/The Image Bank; page 387 bottom: Peter Menzel/Stock Boston; page 400: Courtesy of NASA/SPScI; page 431: Reuters New Media Inc./CORBIS; page 454: The Granger Collection; page 503: The photo used on page is provided courtesy of Texas Instruments Incorporate; page 535: Courtesy of NASA/JPL; page 549: Macduff Everton/CORBIS; page 576: The Granger Collection; page 657 top: Michael Grecco/Stock Boston; page 657 bottom: Miro Vintoniv/Stock Boston; page 713 top: STONE Images; page 713 bottom: The Image Bank; page 771 top and bottom: Stephen Johnson/STONE Images.

Printed in the U.S.A.

Library of Congress Control Number: 2001131469

ISBNs:
Student's Edition: 0-618-13068-3
Instructor's Annotated Edition: 0-618-13069-1

123456789-VH-05 04 03 02 01

CONTENTS

2 FUNCTIONS AND GRAPHS 133

3 POLYNOMIAL AND RATIONAL FUNCTIONS 237

4 EXPONENTIAL AND LOGARITHMIC FUNCTIONS 299

PREFACE

With each successive edition of *College Algebra and Trigonometry* we strive to enhance and refine our instructional materials. In this edition we have continued our emphasis on *doing* mathematics rather than duplicating mathematics through extensive drill. Students are encouraged to investigate concepts, apply those concepts, and then present their findings. We are ever cognizant of the motivating influence contemporary, relevant applications have on students. As a result, we have added many new application problems and deleted those that have little instructional value.

Technology is introduced very naturally in the text to illustrate or enhance a concept. Our intention is to demonstrate appropriate technology when necessary to support a concept. We attempt to foster the idea that the concept motivates the use of technology and, therefore, don't introduce technology for technology's sake. The optional Graphing Calculator Exercises are designed to develop a student's appreciation for both the power and the limitations of technology.

In this edition, we have retained our basic philosophy to deliver a comprehensive and mathematically sound treatment of the topics considered essential for a college algebra course. To help students master these concepts, we have tried to maintain a balance among theory, application, modeling and drill. Carefully developed mathematics is complemented by abundant, creative applications that are both contemporary and representative of a wide range of disciplines. Many application problems are accompanied by a diagram that helps the student visualize the mathematics of the application.

NEW! CONTENT CHANGES TO THIS EDITION

Besides adding many contemporary application problems, we have made the following organizational and topical changes.

The first chapter has been renamed *Preliminary Concepts* and has been labeled Chapter P. This chapter is a review of many topics that are a prerequisite to success in college algebra. Students who have successfully completed an intermediate algebra course may not need to cover this chapter.

In Chapter 1, *Equations and Inequalities,* absolute value equations and inequalities have been integrated within the chapter rather than existing as a separate section. Complex numbers have been moved to Section 1.3, Quadratic Equations. This allowed us to more naturally motivate complex numbers.

Chapter 2, *Functions and Graphs,* now includes a section on linear and quadratic regression, *Modeling Data Using Regression.* Actual data sets are given in the exercises so that the relationship between concept and model is set as a real circumstance. If you wish to omit the sections that involve mathematical modeling, then this section, which requires a graphing utility, may be omitted without disturbing the flow of the text.

In Chapter 3, *Polynomial and Rational Functions,* many new applications have been integrated into each section. In section 3.3 we have added a "guidelines for finding the zeros of a polynomial with integer coefficients" feature and examples that illustrate the steps in the guidelines. Section 3.4 has an expanded coverage of the process of finding the complex zeros of a polynomial.

Chapter 4, *Exponential and Logarithmic Functions,* now begins with Inverse Functions. Moving inverse functions into this chapter enabled us to better connect a logarithmic function as the inverse of an exponential function. Many application problems have been added to this chapter. The section on *Properties of Logarithms* includes an expanded coverage on the topic of logarithmic scales. The last section of this chapter is new to this edition. The title of the section *is Modeling Data with Exponential and Logarithmic Functions* and it includes many real world applications. The process of using empirical data to determine exponential, logarithmic, and logistic models is fully illustrated.

In Chapter 6, *Trigonometric Identities and Equations,* we have included several examples that illustrate both an algebraic solution and a graphical solution of trigonometric equations. These examples are in a side-by-side format that helps the student visualize the solution. Section 6.6 now includes the topic of modeling sinusoidal data and includes several real world exercises that illustrate the relevance of this process.

Chapter 7, *Applications of Trigonometry,* includes several new applications along with a guideline to help students determine whether to solve a triangle by using the Law of Sines or the Law of Cosines. A chart listing the properties of the nth root of a complex number has been inserted into Section 7.5.

In Chapter 8, *Topics in Analytic Geometry,* we have included several new applications of a more contemporary nature. Also, we have expanded the use of graphing utilities to illustrate important concepts.

Chapter 10, *Matrices,* now includes interpolating polynomials as an application of solving systems of equations by using matrices.

Chapter Opener

Each chapter presents a motivating illustrated application of the mathematical content of the chapter.

CHAPTER

5

TOPICS IN ANALYTIC GEOMETRY

Radio Telescopes, Conic Sections, and the Search for Extraterrestrial Intelligence

The movie *Contact* was based on the novel by astronomer Carl Sagan. In the movie, Jodie Foster plays an astronomer who is searching for extraterrestrial intelligence. One scene from the movie takes place at the Very Large Array (VLA) in New Mexico. The VLA consists of 27 large radio telescopes that are electronically connected. Each of the telescopes has a dish measuring 81 feet across. A reflective property of each dish is such that electronic signals from space are collected by the surface of the dish and reflected to a receiver located at a point called the focus. The signals collected by each telescope are then sent to a comp… The s… as th… this c…

◆ The moment of contact with an alien life form. *Contact*, Warner Bros., 1997.

◆ Radio Telescopes in the Very

NEW! Section Opener

Under the title of each section there is a list of the major topics that are covered in the section. The  next to this list indicates which video from the text-specific video set contains further review for the content in this section. The indicates that there is a tutorial (available on CD and on-line) to cover the content in this section (and *every* section) of the text.

Math Matters

These margin notes contain interesting sidelights about mathematics, its history, or its application.

Take Note

These notes alert students to a point requiring special attention or are used to amplify the concept under discussion. NEW! Special *Take Notes*, identified by , appear next to a concept for which an interactive example or a downloadable file has been created for the web-e.g. applets and TI™ graphing calculator programs. These resources can be used for presentation purposes or homework and can help students visualize a concept.

Key Terms and Concepts

Definitions, properties, and theorems (in blue boxes) highlight important concepts for easy reference. Key terms (in blue bold) emphasize important terms. The key terms are also provided in a Glossary at the back of the text.

SECTION **5.1** PARABOLAS

- PARABOLAS WITH VERTEX AT (0, 0)
- PARABOLAS WITH VERTEX AT (h, k)
- APPLICATIONS

MATH MATTERS

Appollonius (262–200 B.C.) wrote an eight-volume treatise entitled *On Conic Sections* in which he derived the formulas for all the conic sections. He was the first to use the words *parabola*, *ellipse*, and *hyperbola*.

take note

If the intersection of a plane and a cone is a point, a line, or two intersecting lines, then the intersection is called a *degenerate conic section*.

take note

A web applet is available to experiment with parabolas by manipulating the focus, directrix, and vertex. This applet, Parabola with Horizontal Directrix, can be found on our web site at http://college.hmco.com

The graph of a parabola, circle, ellipse, or hyperbola can be formed by the intersection of a plane and a cone. Hence these figures are referred to as conic sections. See **Figure 5.1**.

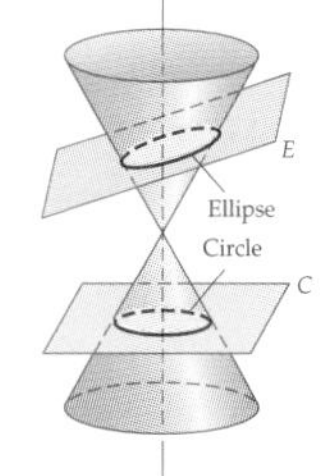

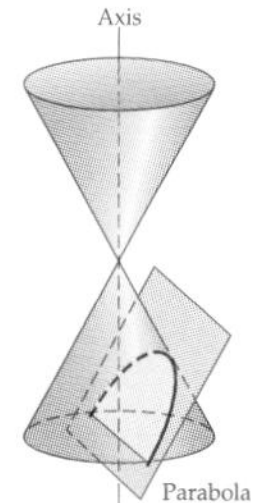

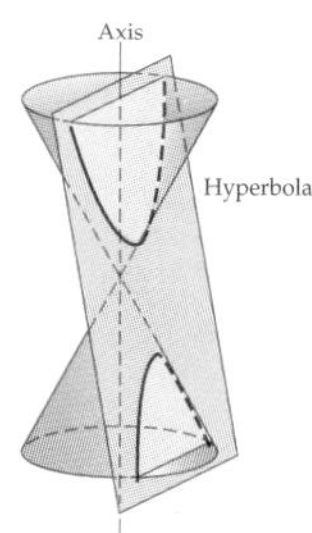

Figure 5.1
Cones intersected by planes

A plane perpendicular to the axis of the cone intersects the cone in a circle (plane C). The plane E, tilted so that it is not perpendicular to the axis, intersects the cone in an ellipse. When the plane is parallel to a line on the surface of the cone, the plane intersects the cone in a parabola. When the plane intersects both portions of the cone, a hyperbola is formed.

◆ PARABOLAS WITH VERTEX AT (0, 0)

Besides the geometric description of a conic section just given, a conic section can be defined as a set of points. This method uses some specified conditions about the curve to determine which points in a coordinate system are points of the graph. For example, a parabola can be defined by the following set of points.

Definition of a Parabola

A **parabola** is the set of points in the plane that are equidistant from a fixed line (the **directrix**) and a fixed point (the **focus**) not on the directrix.

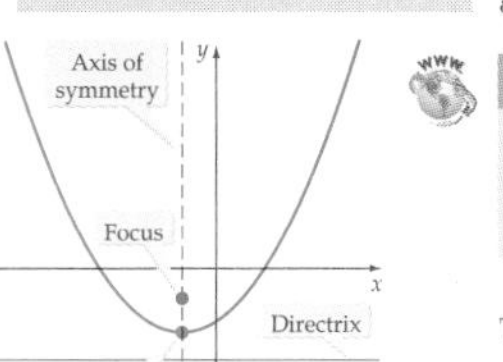

Figure 5.2

The line that passes through the focus and is perpendicular to the directrix is called the **axis of symmetry** of the parabola. The midpoint of the line segment between the focus and directrix on the axis of symmetry is the **vertex** of the parabola, as shown in **Figure 5.2**.

SECTION P.6 RATIONAL EXPRESSIONS 55

◆ APPLICATION OF RATIONAL EXPRESSIONS

EXAMPLE 6 Solve an Application

The *average speed* for a round trip is given by the complex fraction

$$\frac{2}{\frac{1}{v_1}+\frac{1}{v_2}}$$

where v_1 is the average speed on the way to your destination and v_2 is the average speed on your return trip. Find the average speed for a round trip if $v_1 = 50$ mph and $v_2 = 40$ mph.

Solution

Evaluate the complex fraction with $v_1 = 50$ and $v_2 = 40$.

$$\frac{2}{\frac{1}{v_1}+\frac{1}{v_2}} = \frac{2}{\frac{1}{50}+\frac{1}{40}} = \frac{2}{\frac{1\cdot 4}{50\cdot 4}+\frac{1\cdot 5}{40\cdot 5}}$$

• Substitute and simplify the denominator.

$$= \frac{2}{\frac{4}{200}+\frac{5}{200}} = \frac{2}{\frac{9}{200}}$$

$$= 2\cdot\frac{200}{9} = \frac{400}{9} = 44\frac{4}{9}$$

The average speed of the round trip is $44\frac{4}{9}$ mph.

TRY EXERCISE 64, EXERCISE SET P.6, PAGE 57

QUESTION In Example 6, why is the speed of the round trip *not* the average of v_1 and v_2?

TOPICS FOR DISCUSSION

1. Discuss the meaning of the phrase *rational expression*. Is a rational expression the same as a fraction? If not, give some examples of a fraction that is not a rational expression.
2. What is the domain of a rational expression?
3. Explain why the following is *not* correct.

$$\frac{2x^2+5}{x^2} = 2+5 = 7$$

ANSWER Because you were traveling slower on the return trip, the return trip took longer than the time spent going to your destination. More time was spent traveling at the slower speed. Thus the average speed is less than the average of v_1 and v_2.

Interactive Presentation

College Algebra and Trigonometry is written in a style that encourages the student to interact with the textbook. At various places throughout the text, we pose a **Question** to the student about the material being read. This question encourages the reader to pause and think about the current discussion and to answer the question. To make sure the student does not miss important information, the **Answer** to the question is provided as a footnote on the same page.

Each section contains a variety of worked examples. Each example is given a name so that the student can see at a glance the type of problem being illustrated. Most examples are accompanied by annotations that assist the student in moving from step to step, and the final answer is in color in order to be readily identifiable.

Following the worked example is a suggested exercise, "Try Exercise 64, Exercise Set P.6, page 57" from that section's exercise set for the student to work. The exercises are color coded by number in the exercise set and the *complete solution* of that exercise can be found in an appendix to the text.

SECTION P.6 RATIONAL EXPRESSIONS 57

...mplify each complex fraction.

$$\frac{3-\frac{2}{a}}{5+\frac{3}{a}}$$

43. $$\frac{\frac{x}{y}-2}{y-x}$$

$$\frac{5-\frac{1}{x+2}}{1+\frac{3}{1+\frac{3}{x}}}$$

46. $$\frac{\frac{1}{(x+h)^2}-1}{h}$$

48. $$r-\frac{r}{r+\frac{1}{3}}$$

$$1+\frac{1}{x}$$

50. $$\frac{1}{\frac{1}{a}+\frac{1}{b}}$$

a. Find the average speed for a round trip by helicopter with $v_1 = 180$ mph and $v_2 = 110$ mph.

b. Simplify the complex fraction.

64. **RELATIVITY THEORY** Using Einstein's theory of relativity, the "sum" of the two speeds v_1 and v_2 is given by the complex fraction

$$\frac{v_1+v_2}{1+\frac{v_1v_2}{c^2}}$$

where c is the speed of light.

a. Evaluate this expression with $v_1 = 1.2\times 10^8$ mph, $v_2 = 2.4\times 10^8$ mph, and $c = 6.7\times 10^8$ mph.

b. Simplify the complex fraction.

65. Find the rational expression in simplest form that represents the sum of the reciprocals of the consecutive integers x and $x+1$.

66. Find the rational expression in simplest form that represents the positive difference between the reciprocals of the consecutive even integers x and $x+2$.

67. Find the rational expression in simplest form that represents the sum of the reciprocals of the consecutive even integers $x-2$, x, and $x+2$.

68. Find the rational expression in simplest form that represents the sum of the reciprocals of the squares of the consecutive even integers $x-2$, x, and $x+2$.

SUPPLEMENTAL EXERCISES

In Exercises 69 to 72, simplify each algebraic fraction.

69. $$\frac{(x+5)-x(x+5)^{-1}}{x+5}$$

70. $$\frac{(y+2)+y^2(y+2)^{-1}}{y+2}$$

S2 SOLUTIONS TO SELECTED EXERCISES

Exercise Set P.6, page 56

2. $$\frac{2x^2-5x-12}{2x^2+5x+3} = \frac{(2x+3)(x-4)}{(2x+3)(x+1)} = \frac{x-4}{x+1}$$

16. $$\frac{x^2-16}{x^2+7x+12}\cdot\frac{x^2-4x-21}{x^2-4x}$$

$$= \frac{(x-4)(x+4)(x+3)(x-7)}{\ldots-4)} = \frac{x-7}{x}$$

$$= \frac{f-e^2}{e^2f}\cdot\frac{1}{ef} = \frac{f-e^2}{e^3f^2}$$

64. a. $$\frac{v_1+v_2}{1+\frac{v_1v_2}{c^2}} = \frac{1.2\times 10^8+2.4\times 10^8}{1+\frac{(1.2\times 10^8)(2.4\times 10^8)}{(6.7\times 10^8)^2}} \approx 3.4\times 10^8 \text{ mph}$$

b. $$\frac{v_1+v_2}{1+\frac{v_1\cdot v_2}{c^2}} = \frac{c^2(v_1+v_2)}{c^2\left(1+\frac{v_1\cdot v_2}{c^2}\right)} = \frac{c^2(v_1+v_2)}{c^2+v_1\cdot v_2}$$

Simple interest problems can be solved by using the formula $I = Prt$, where I is the interest, P is the principal, r is the simple interest rate per period, and t is the number of periods.

EXAMPLE 5 Solve a Simple Interest Problem

An accountant invests part of a \$6000 bonus in a 5% simple interest account and the remainder of the money is invested at 8.5% simple interest. Together the investments earn \$370 per year. Find the amount invested at each rate.

Solution

Let x be the amount invested at 5%. The remainder of the money is \$6000 $- x$, which will be the amount invested at 8.5%. Using $I = Prt$, with $t = 1$ year, yields

$$\text{Interest at } 5\% = x \cdot 0.05 = 0.05x$$
$$\text{Interest at } 8.5\% = (6000 - x) \cdot (0.085) = 510 - 0.085x$$

The interest earned on the two accounts equals \$370.

$$\begin{aligned} 0.05x + (510 - 0.085x) &= 370 \\ -0.035x + 510 &= 370 \\ -0.035x &= -140 \\ x &= 4000 \end{aligned}$$

Therefore, the accountant invested \$4000 at 5% and the remaining \$2000 at 8.5%. Check as before.

TRY EXERCISE 40, EXERCISE SET 1.2, PAGE 80

Percent mixture problems involve combining solutions or alloys that have different concentrations of a common substance. Percent mixture problems can be solved by using the formula $pA = Q$, where p is the percent of concentration, A is the amount of the solution or alloy, and Q is the quantity of a substance in the solution or alloy. For example, in 4 liters of a 25% acid solution, p is the percent of acid (25%), A is the amount of solution (4 liters), and Q is the amount of acid in the solution, which equals $(0.25) \cdot (4)$ liters = 1 liter.

EXAMPLE 6 Solve a Percent Mixture Problem

A chemist mixes an 11% hydrochloric acid solution with a 6% hydrochloric acid solution. How many milliliters (ml) of each solution should the chemist use to make a 600-milliliter solution that is 8% hydrochloric acid?

Solution

Let x be the number of milliliters of the 11% solution. Because the final solution will have a total of 600 milliliters of fluid, $600 - x$ is the number of milliliters of the 6% solution. See **Figure 1.5**.

Applications

One way to motivate an interest in mathematics is through applications. The applications in *College Algebra and Trigonometry* have been taken from many disciplines including agriculture, architecture, biology, business, chemistry, earth science, economics, engineering, medicine, and physics. Besides providing motivation to study mathematics, the applications assist students in developing good problem-solving skills.

...n has been increasing over ...owing table shows the per ...per day, generated in the

...0	1980	1990	2000
...7	3.61	4.00	4.30

...= 60.

...find a linear model and a ...data. Use t as the independ-...p as the dependent variable

...efficients of the two regres-...which model provides the

...d in part **b** to predict the ...l be generated per capita in ...0.01 pound.

...H FUNCTION The scientists ...determined that the pH of ...blood is a function of the ratio q of the blood's bicarbonate and carbonic acid.

a. Use a graphing utility and the data in the following table to determine a linear model and a logarithmic model for the data. Use q as the independent variable (domain) and pH as the dependent variable (range). Which model provides the best fit?

q	7.9	12.6	31.6	50.1	79.4
pH	7.0	7.2	7.6	7.8	8.0

b. A blood pH of 9.0 results in death. Use the model you chose in part **a** to find, to the nearest tenth, the q-value associated with a pH of 9.0.

27. WORLD POPULATION The following table lists the years in which the world's population first reached 3, 4, 5, and 6 billion. (*Source: Time Almanac 2000*)

World Population Milestones

1960	3 billion
1974	4 billion
1987	5 billion
1999	6 billion

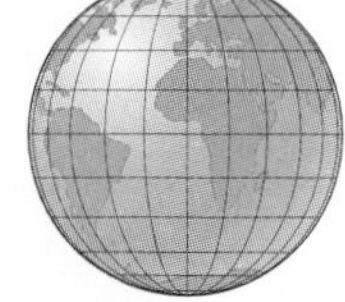

Some scientists think that the Earth has only enough resources to support 12 billion people.

a. Find an *exponential model* for the data in the table and use the model to predict in what year the world's population will reach 12 billion. (Use 60 to represent the year 1960.)

b. Find a *logistic model* for the data in the table. According to the logistic model, what will the world's population approach as $t \to \infty$?

c. Do you think the exponential model or the logistic model is the most realistic model for predicting the world's future population? Explain.

28. INTEREST RATES The following table shows the interest rates paid on jumbo CDs of various terms in April of 2000.

CD Term, t years	0.25	0.5	1.0	1.5	2.0	3.0	5.0
Rate	6.50%	6.80%	7.00%	7.10%	7.15%	7.20%	7.25%

a. Find a *logarithmic model* for the data in the table and use the model to predict, to the nearest 0.01%, the interest rate on a jumbo CD that is invested for a term of 4.0 years.

b. According to your model in part **a**, for how long, to the nearest 0.1 year, would you need to invest to earn 7.5% on your money?

29. PANDA POPULATION One estimate gives the panda population as 3200 in 1980 and as 590 in 2000.

a. Find an exponential model for the data and use the model to predict the year in which the panda population will be reduced to 200. (Use 0 to represent the year 1980.)

b. The exponential model in part **a** fits the data perfectly. Does this mean that the model will accurately predict future panda populations? Explain.

30. NUMBER OF AUTOMOBILES The number of automobiles in the United States in 1900 was around 8000. In the year 2000, the number of automobiles in the United States reached 200 million.

a. Find an exponential model for the data and use the model to predict, to the nearest 100,000, the number of automobiles in the United States in 2010. Use $t = 0$ to represent the year 1900.

b. According to the model, in what year will the number of automobiles first reach 300 million?

Real Data

Real data examples and exercises, identified by [icon], ask students to analyze and solve problems taken from actual situations. Students are often required to work with tables, graphs, and charts drawn from a variety of disciplines.

Technology

Optional graphing calculator discussions, examples, and exercises, identified by , are presented throughout the text. These can be used to further explore a topic or concept or to introduce technology as an alternative way to solve certain problems.

When a graphing utility is used to draw a graph, it is important to look at the graph and ask yourself, "Does this graph have the characteristics of the function I intended to graph?" What you must guard against is incorrectly entering the function. The graphing utility will graph only what you enter—not what you intended to enter.

For instance, suppose we want to draw the graph of $f(x) = 3(2^{-x^2})$ and produce the graph shown in **Figure 4.20**. First note that

$$f(-x) = 3(2^{-(-x)^2}) = 3(2^{-x^2}) = f(x)$$

Thus f is an even function, and its graph should be symmetric with respect to the y-axis. This appears to be the case from our graph. Next, if we write f as

$$f(x) = \frac{3}{2^{x^2}}$$

we observe that as $|x|$ increases without bound, the denominator increases without bound. Therefore, $f(x)$ is approaching zero. This is also consistent with our graph. For another observation, recall that the value of a fraction with a constant numerator is as large as possible when the denominator is as small as possible. For our function, the smallest denominator occurs when $x = 0$. In that case $f(0) = 3$, which is again consistent with our graph. It appears that we have entered the function correctly and produced the desired graph. As a final check, we can evaluate the function for various values of x and compare those values to values found by using the TRACE feature of the graphing utility.

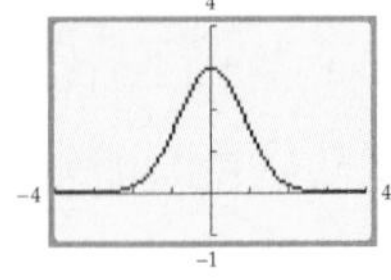

$f(x) = 3 \cdot 2^{-x^2}$

Figure 4.20

EXAMPLE 2 Graph a Function of the Form $f(x) = b^{p(x)}$

Use a graphing utility to graph $f(x) = 2^{|x|}$. Then use your knowledge of functions to verify the accuracy of the graph.

Solution

The graph as drawn with a graphing utility is shown in **Figure 4.21**. The minimum value of f occurs when $|x| = 0$, which means $x = 0$. At $x = 0$, $f(x) = 2^{|0|} = 1$. This is consistent with the graph. As shown below, f is an even function.

$$f(x) = 2^{|x|}$$
$$f(-x) = 2^{|-x|} = 2^{|x|} = f(x)$$

Because f is an even function, the graph of f should be symmetric with respect to the y-axis. This is consistent with the graph of f in **Figure 4.21**. As a final check, evaluate f at a few values of x and compare them to the corresponding values you find by tracing along the curve.

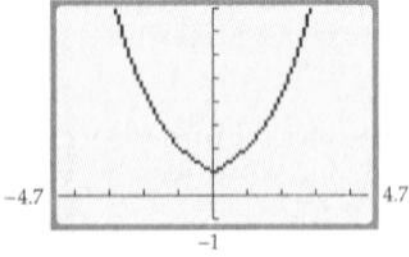

$f(x) = 2^{|x|}$

Figure 4.21

TRY EXERCISE 40, EXERCISE SET 4.2, PAGE 317

NEW! Modeling

Special modeling sections, which rely heavily on the graphing calculator, have been added to this edition. These special sections introduce the idea of mathematical modeling of data through linear, quadratic, exponential, logarithmic, and logistic regression.

◆ APPLICATIONS

The methods used to model data with an exponential, a logarithmic, or a logistic function are similar to the methods used in Chapter 2 to model data with a linear or a quadratic function. Here is a summary of the modeling process.

The Modeling Process

Use a graphing utility to

1. **Construct a *scatter plot* of the data** to determine which type of function will best model the data.
2. **Find the *regression equation*** of the modeling function and the correlation coefficient for the regression.
3. **Examine the *correlation coefficient*** and *view a graph* that displays both the modeling function and the scatter plot to determine how well your function fits the data.

In the following example, we use this modeling process to find a function that closely models the value of a diamond as a function of its weight.

EXAMPLE 2 Model an Application with an Exponential Function

A diamond merchant has determined the value of several white diamonds that have different weights (measured in carats) but are *similar in quality*. See Table 4.12.

Table 4.12

4.00 ct	3.00 ct	2.00 ct	1.75 ct	1.50 ct	1.25 ct	1.00 ct	0.75 ct	0.50 ct
$14,500	$10,700	$7,900	$7,300	$6,700	$6,200	$5,800	$5,000	$4,600

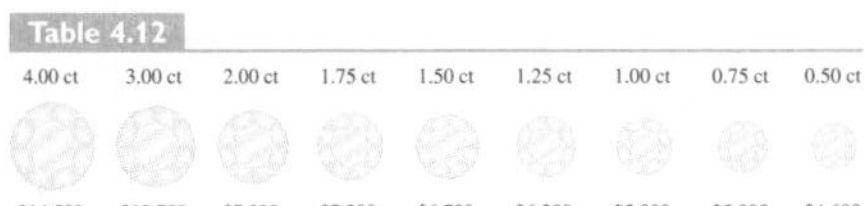

Find a function that models the value of the diamonds as a function of their weight, and use the function to predict the value of a 3.5-carat diamond of similar quality.

Solution

1. **Construct a scatter plot of the data.** See **Figure 4.53**.

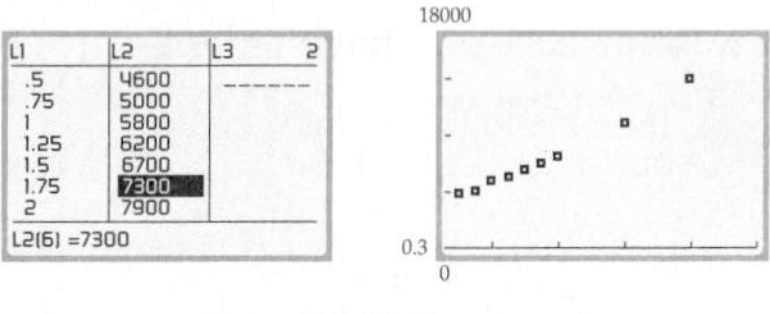

Figure 4.53

MATH MATTERS

The value of a diamond is generally

EXPLORING CONCEPTS WITH TECHNOLOGY

Finding Zeros of a Polynomial Using *Mathematica*

Computer algebra systems (CAS) are computer programs that are used to solve equations, graph functions, simplify algebraic expressions, and help us perform many other mathematical tasks. In this exploration, we will demonstrate how to use one of these programs, *Mathematica*, to find zeros of a polynomial.

Recall that a zero of a function P is a number, x, for which $P(x) = 0$. The idea behind finding a zero of a polynomial by using a CAS is to solve the polynomial equation $P(x) = 0$ for x.

Two commands in *Mathematica* that can be used to solve an equation are Solve and NSolve. (*Mathematica* is sensitive about syntax (the way in which an expression is typed.) You *must* use upper-case and lower-case letters as we indicate.) Solve will attempt to find an *exact* solution of the equation; NSolve attempts to find *approximate* solutions. Here are some examples.

To find the exact values of the zeros of $P(x) = x^3 + 5x^2 + 11x + 15$, input the following. *Note:* The two equals signs are necessary.

```
Solve[x^3+5x^2+11x+15==0]
```

Press Enter. The result should be

```
{{x->-3}, {x->-1-2 I}, {x->-1+2 I}}
```

Thus the three zeros of P are -3, $-1 - 2i$, and $-1 + 2i$.

To find the approximate values of the zeros of $P(x) = x^4 - 3x^3 + 4x^2 + x - 4$, input the following.

```
NSolve[x^4-3x^3+4x^2+x-4==0]
```

Press Enter. The result should be

```
{{x->-0.821746}, {x->1.2326}, {x->1.29457-1.50771 I},
{x->1.29457+1.50771 I}}
```

The four zeros are (approximately) -0.821746, 1.2326, $1.29457 - 1.50771i$, and $1.29457 + 1.50771i$.

Exploring Concepts with Technology

A special end of chapter feature, *Exploring Concepts with Technology*, extends ideas introduced in the text by using technology (graphing calculator, CAS, etc.) to investigate extended applications or mathematical topics. These explorations can serve as group projects, class discussions, or extra-credit assignments.

254 CHAPTER 3 POLYNOMIAL AND RATIONAL FUNCTIONS

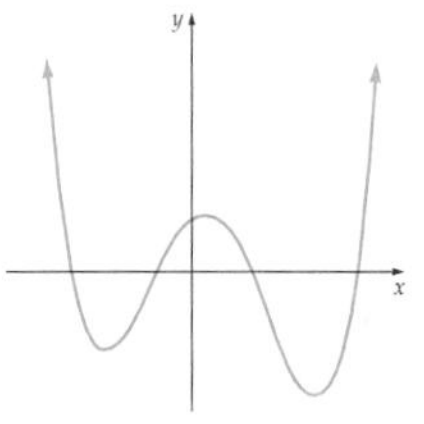

Figure 3.19

TOPICS FOR DISCUSSION

1. Discuss the meaning of the phrase *polynomial function*. Give examples of polynomial functions and of functions that are not polynomials.
2. Is it possible for the graph of the polynomial shown in Figure 3.19 to have a degree that is an odd number? If so, explain how. If not, explain why not.
3. Explain the difference between a relative minimum and an absolute minimum and the difference between a relative maximum and an absolute maximum.
4. Discuss how the Zero Location Theorem can be used to find a real zero of a polynomial.
5. A complex number may be a zero of a polynomial. For instance, i is a zero of $P(x) = x^2 + 1$ because $P(i) = 0$. Explain why the Zero Location Theorem cannot be used to find the complex number zeros of a polynomial.
6. Let $P(x)$ be a polynomial with real coefficients. Explain the relationship among a real zero of a polynomial, the x-coordinate of the x-intercept of the graph of the polynomial, and the solution of the equation $P(x) = 0$.

Topics for Discussion

Discussion topics are found at the end of each section. These conceptual questions can form the basis for a group or class discussion or serve as writing assignments.

316 CHAPTER 4 EXPONENTIAL AND LOGARITHMIC FUNCTION

EXAMPLE 5 Determine the Zero of a Fun

Use a graphing utility to determine, to the nearest

the zero of $f(x) = -\frac{1}{3}e^x + 2$.

Solution

Graph f and use the features of your graphing utility to find the x-intercept to the nearest thousandth. The x-coordinate of the x-intercept in Figure 4.25 is approximately 1.792.

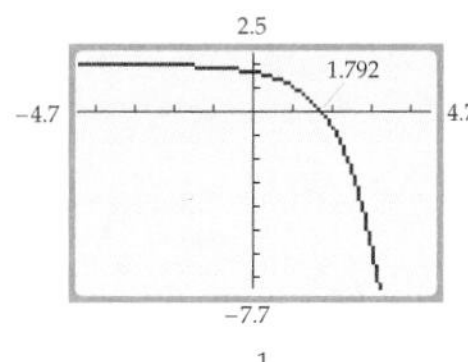

$f(x) = -\frac{1}{3}e^x + 2$

Figure 4.25

Verify the Solution

Evaluating $f(x)$ when $x = 1.792$, shows that $f(x) \approx 0$. Thus 1.792 approximates a zero of f.

$$f(x) = -\frac{1}{3}e^x + 2$$
$$\approx -\frac{1}{3}e^{1.792} + 2$$
$$\approx -0.00048112$$
$$\approx 0$$

The zero of f to the nearest thousandth is 1.792.

TRY EXERCISE 50, EXERCISE SET 4.2, PAGE 317

◆ AN APPLICATI

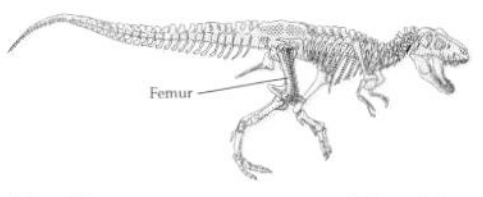

The *Tyrannosaurus rex* named Sue. The Field Museum, Chicago.

Paleontologists estimat of its femur. The weigh where C is the circumf

EXAMPLE 6

Estimate the weight cumference of the fe

Solution

SECTION 6.1 SYSTEMS OF LINEAR EQUATIONS IN TWO VARIABLES 427

◆ SUBSTITUTION METHOD FOR SOLVING A SYSTEM OF LINEAR EQUATIONS

The substitution method is one procedure for solving a system of equations. This method is illustrated in Example 1.

EXAMPLE 1 Solve a System of Equations by the Substitution Method

Solve: $\begin{cases} 3x - 5y = 7 & (1) \\ y = 2x & (2) \end{cases}$

Solution

The solutions of $y = 2x$ are the ordered pairs $(x, 2x)$. For the system of equations to have a solution, ordered pairs of the form $(x, 2x)$ must also be solutions of $3x - 5y = 7$. To determine whether the ordered pairs $(x, 2x)$ are solutions of Equation (1), substitute $(x, 2x)$ into Equation (1) and solve for x. Think of this as *substituting* $2x$ for y.

$3x - 5y = 7$ • Equation (1)
$3x - 5(2x) = 7$ • Substitute $2x$ for y.
$3x - 10x = 7$
$-7x = 7$
$x = -1$
$y = 2x$ • Equation (2)
$= 2(-1) = -2$ • Substitute -1 for x in Equation 2.

The only ordered-pair solution of the system of equations is $(-1, -2)$. When a system of equations has a unique solution, the system of equations is independent.

Visualize the Solution

Graphing $3x - 5y = 7$ and $y = 2x$ shows that the ordered pair $(-1, -2)$ belongs to both lines. Therefore, $(-1, -2)$ is a solution of the system of equations. See Figure 6.4.

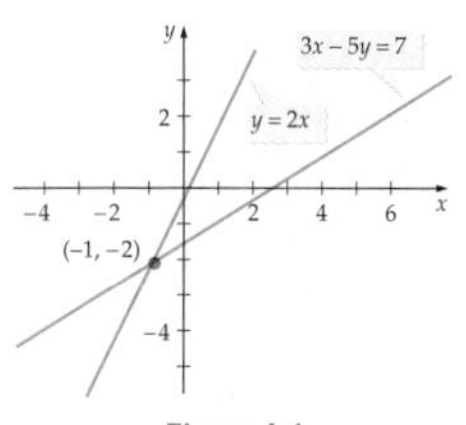

Figure 6.4
An independent system of equations

TRY EXERCISE 6, EXERCISE SET 6.1, PAGE 433

NEW! Visualize the Solution

For appropriate examples within the text, we have provided both an algebraic solution and a graphical representation (either a coordinate grid graph or a graphing calculator screen) of the solution. This allows the student to visualize the algebraic solution. For other optional graphing calculator examples, an algebraic verification of a graphing calculator solution is presented. The intent is to create a link between the algebraic and visual components of a solution.

Projects

Projects at the end of exercise sets encourage students to research and write about math and its applications. In the Instructor's Resource Manual additional Projects (with solutions) may be assigned. Additional projects can be found at *http://college.hmco.com.* Follow the links to these projects and bookmark the location to make it easy to return.

44. Find the value of x in the domain of $f(x) = 4x - 3$ for which $f(x) = -2$.

In Exercises 45 to 48 find the solution $f(x) = 0$, verify that the solution of $f(x) = 0$ is the same as the x-coordinate of the x-intercept of the graph of $y = f(x)$.

45. $f(x) = 3x - 12$ 46. $f(x) = -2x - 4$

47. $f(x) = \frac{1}{4}x + 5$ 48. $f(x) = -\frac{1}{3}x + 2$

In Exercises 49 to 52, solve $f_1(x) = f_2(x)$ by an algebraic method and by graphing.

49. $f_1(x) = 4x + 5$ $f_2(x) = x + 6$

50. $f_1(x) = -2x - 11$ $f_2(x) = 3x + 7$

51. $f_1(x) = 2x - 4$ $f_2(x) = -x + 12$

52. $f_1(x) = \frac{1}{2}x + 5$ $f_2(x) = \frac{2}{3}x - 7$

53. AUTOMOTIVE TECHNOLOGY The table below shows the EPA estimates for city and highway driving for ten selected luxury cars. (Source: **www.money.com**, May 26, 2000)

EPA miles per gallon estimates for city and highway driving for selected luxury cars.

Car	City mpg	Highway mpg
Acura RL	18	24
Audi A8, 4.2L	17	24
BMW 528i	21	29
Cadillac Deville	17	27
Infiniti Q45	18	23
Jaguar XJ8	17	24
Lexus LS400	18	25
Lincoln Continental	17	25
Mercedes S500	16	23
Saab	18	24

A linear function that approximates this data is given by $f(x) = 0.95652x + 7.86957$, where $f(x)$ is the highway mpg and x is the city mpg.

a. Use the function to interpolate the estimated highway mpg for a car whose city mpg is 19 mpg.

b. Use the function to extrapolate the estimated highway mpg for a car whose city mpg is 24 mpg.

54. AVIATION The table in the next column is based on data from the Federal Aviation Administration for planes flown in 1998. The table shows, for selected airlines, the total number of hours an airline's planes operated and the total number of miles flown by those planes.

Total hours of operation of planes and the total number of miles flown by those planes.

Airline	Hours flown (in thousands)	Miles flown (in thousands)
Alaska	295	126
American	2054	945
Continental	1054	476
Delta	1787	793
Frontier	43	17
Midwest Express	72	31
Northwest		
Southwest		
United		
US Air		

A linear function tha... $f(x) = 0.46175x - 17$... miles flown and x is...

a. Use the function... an airline whos... would fly.

PROJECTS

1. A MODELING PROJECT The purpose of this project is for you to find data that can be modeled by an exponential, a logarithmic, or a logistic function. Choose data from a *life-like* situation that you find interesting. Search for the data in a magazine, a newspaper, or an almanac or on the Internet. If you wish, you can collect your data by performing an experiment. Use the following steps to complete this project.

a. List the source of your data. Include the date, page number, and any other specifics about the source. If your data was collected by performing an experiment, then provide all the details about the experiment.

b. Explain what you have chosen as your variables. Which variable is the dependent variable and which the independent variable?

c. Use the three-step modeling process to find a regression equation that models the data.

d. Graph the regression equation on the scatter plot of the data. What is the regression coefficient for the model? Do you think that your regression equation accurately models your data? Explain.

e. Use the regression equation to predict the value of (1) the dependent variable for a specific value of the independent variable and (2) the independent variable for a specific value of the dependent variable.

f. Write a few comments about what you have learned from this project.

43. PREDICTING A... quacy of a city... eled by a *gamma den*... from this function e... probability that cert... Suppose a city has ... being able to provid... per day is given by

P

a. Use a graphing utility to graph $P(x)$ for $x = 0$.

b. Determine the probability that a city can supply more than 5 million liters of water per day.

c. The city manager wants to determine the minimum water supply in a reservoir that the city can maintain so that there is less than a 0.25 chance that the city will not be able to meet demand. What must the capacity of the water supply be to meet the goal of the manager?

d. As x 1 ∞, P 1 0. Explain why this makes sense in the context of this application.

44. PREDICTING ADEQUACY OF RESOURCES The probability (see Exercise 43) that an electric company can supply more than x million kilowatt-hours of electricity per day is given by

$$P = \left(\frac{1}{4}x + 1\right)e^{-x/4}$$

a. Use a graphing utility to graph $P(x)$ for $x \ge 0$.

b. Determine, to the nearest 0.001, the probability that this electric company can supply more than 8 million kilowatt-hours of electricity per day.

c. The electric company wants to determine what capacity it must have so that there is less than a 0.50 chance that the company will not be able to meet demand. What, to the nearest 0.1 million, must the capacity of the electric company supply be to meet the goal of the company?

d. As x 1 ∞, P 1 0. Explain why this makes sense in the context of this application.

45. PHYSICS If air resistance is proportional to velocity, then the time t in seconds for a particular object to reach a velocity of v feet per second is given by

$$t = 3.125 \ln \frac{100}{100 - v}$$

a. How long, to the nearest 0.1 second, is required before the velocity is 50 feet per second?

b. There is a vertical asymptote when $v = 100$. Describe the meaning of this asymptote in the context of the application.

46. PHYSICS If air resistance is proportional to velocity, then the time t in seconds for a particular object to reach a

...d before ... 75. De- ... the con-

SUPPLEMENTAL EXERCISES

47. MEDICATION LEVEL A patient is given three dosages of aspirin. Each dosage contains 1 gram of aspirin. The second and third dosages are each taken 3 hours after the previous dosage is administered. The half-life of the aspirin is 2 hours. The amount of aspirin, A, in the patient's body t hours after the first dosage is administered is

$$A(t) = \begin{cases} 0.5^{t/2} & 0 \le t < 3 \\ 0.5^{t/2} + 0.5^{(t-3)/2} & 3 \le t < 6 \\ 0.5^{t/2} + 0.5^{(t-3)/2} + 0.5^{(t-6)/2} & t \ge 6 \end{cases}$$

Find, to the nearest 0.01 gram, the amount of aspirin in the patient's body when

a. $t = 1$ b. $t = 4$ c. $t = 9$

48. MEDICATION LEVEL Use a graphing calculator and the dosage formula in Exercise 47 to determine when, to the nearest 0.1 hour, the amount of aspirin in the patient's body first reaches 0.25 gram.

Exercises 49 to 51 make use of the factorial function which is defined as follows. For whole numbers n, the number $n!$, (which is read "n factorial"), is given by

$$n! = \begin{cases} n(n-1)(n-2)\cdots 1, & \text{if } n \ge 1 \\ 1, & \text{if } n = 0 \end{cases}$$

Thus, $0! = 1$ and $4! = 4 \cdot 3 \cdot 2 \cdot 1 = 24$.

49. QUEUEING THEORY A study shows that the number of people who arrive at a bank teller's window averages 4.1 people every 10 minutes. The probability P that exactly x people will arrive at the teller's window in a given 10-minute period is

$$P(x) = \frac{4.1^x e^{-4.1}}{x!}$$

Find, to the nearest 0.1%, the probability that in a given 10-minute period, exactly

a. 0 people arrive at the window.

b. 2 people arrive at the window.

c. 3 people arrive at the window.

Exercises

The exercise sets of *College Algebra and Trigonometry* were carefully developed to provide a wide variety of exercises. The exercises range from drill and practice to interesting challenges and were chosen to illustrate the many facets of topics discussed in the text. Exercise sets emphasizes skill building, skill maintenance, and, as appropriate, applications.

Included in each exercise set are **Supplemental Exercises** that include material from previous chapters, present extensions of topics, require data analysis, or offer challenge problems, or problems of the form "prove or disprove."

To identify the various types of exercises, we use the following symbols: writing, data analysis, group

activity , 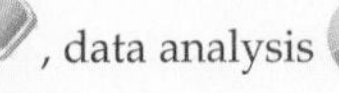 and graphing calculator .

Chapter Summary

The chapter summary provides a convenient place where students can review the essential concepts of the chapter.

CHAPTER 5 SUMMARY

5.1 Parabolas

- A parabola is the set of points in the plane that are equidistant from a fixed line (the directrix) and a fixed point (the focus) not on the directrix.
- The equations of a parabola with vertex at (h, k) and axis of symmetry parallel to a coordinate axis are given by

$(x - h)^2 = 4p(y - k)$; focus $(h, k + p)$; directrix $y = k - p$

$(y - k)^2 = 4p(x - h)$; focus $(h + p, k)$; directrix $x = h - p$

5.2 Ellipses

- An ellipse is the set of all points in the plane, the sum of whose distances from two fixed points (foci) is a positive constant.
- The equations of an ellipse with center at (h, k) and major axis parallel to a coordinate axis are given by

$\frac{(x-h)^2}{a^2} + \frac{(y-k)^2}{b^2} = 1$; foci $(h \pm c, k)$; vertices $(h \pm a, k)$

$\frac{(x-h)^2}{b^2} + \frac{(y-k)^2}{a^2} = 1$; foci $(h, k \pm c)$; vertices $(h, k \pm a)$

For each equation, $a > b$ and $c^2 = a^2 - b^2$.

- The eccentricity e of an ellipse is given by $e = c/a$.

5.3 Hyperbolas

- A hyperbola is the set of all points in the plane, the difference of whose distances from two fixed points (foci) is a positive constant.
- The equations of a hyperbola with center at (h, k) and transverse axis parallel to a coordinate axis are given by

$\frac{(x-h)^2}{a^2} - \frac{(y-k)^2}{b^2} = 1$; foci $(h \pm c, k)$; vertices $(h \pm a, k)$

$\frac{(y-k)^2}{a^2} - \frac{(x-h)^2}{b^2} = 1$; foci $(h, k \pm c)$; vertices $(h, k \pm a)$

For each equation, $c^2 = a^2 + b^2$.

- The eccentricity e of a hyperbola is given by $e = c/a$.

True/False Exercises

Following each chapter summary are true/false exercises. These exercises are intended to help students understand concepts and can be used to initiate class discussions.

CHAPTER 5 TRUE/FALSE EXERCISES

In Exercises 1 to 9, answer true or false. If the statement is false, give an example to show that the statement is false.

1. The graph of a parabola is the same shape as that of one branch of a hyperbola.
2. For the two axes of an ellipse, the major axis and the minor axis, the major axis is always the longer axis.
3. For the two axes of a hyperbola, the transverse axis and the conjugate axis, the transverse axis is always the longer axis.
4. If two ellipses have the same foci, they have the same graph.
5. A hyperbola is similar to a parabola in that both curves have asymptotes.
6. If a hyperbola with center at the origin and a parabola with vertex at the origin have the same focus, $(0, c)$, then the two graphs always intersect.
7. The graphs of all the conic sections are not the graphs of functions.
8. If F_1 and F_2 are the two foci of an ellipse and P is a point on the ellipse, then $d(P, F_1) + d(P, F_2) = 2a$, where a is the length of the semimajor axis of the ellipse.
9. The eccentricity of a hyperbola is always greater than 1.

CHAPTER 5 REVIEW EXERCISES

In Exercises 1 to 12, find the foci and the vertices of each conic. If the conic is a hyperbola, find the asymptotes. Graph each equation.

1. $x^2 - y^2 = 4$
2. $y^2 = 16x$
3. $x^2 + 4y^2 - 6x + 8y - 3 = 0$
4. $3x^2 - 4y^2 + 12x - 24y - 36 = 0$
5. $3x - 4y^2 + 8y + 2 = 0$
6. $3x + 2y^2 - 4y - 7 = 0$

17. Parabola with vertex $(0, -2)$ and passing through the point $(3, 4)$.
18. Ellipse with eccentricity 2/3 and foci $(-4, -1)$ and $(0, -1)$.
19. Hyperbola with vertices $(\pm 6, 0)$ and asymptotes whose equations are $y = \pm\frac{1}{9}x$.
20. Parabola passing through the points $(1, 0)$, $(2, 1)$, and $(0, 1)$ with axis of symmetry parallel to the y-axis.
21. Find the equation of the parabola traced by a point $P(x, y)$ that moves in such a way that the distance between $P(x, y)$ and the line $x = 2$ equals the distance between $P(x, y)$ and the point $(-2, 3)$.
22. Find the equation of the parabola traced by a point $P(x, y)$ that moves in such a way that the distance between $P(x, y)$ and the line $y = 1$ equals the distance between $P(x, y)$ and the point $(-1, 2)$.
23. Find the equation of the ellipse traced by a point $P(x, y)$ that moves in such a way that the sum of its distances to $(-3, 1)$ and $(5, 1)$ is 10.
24. Find the equation of the ellipse traced by a point $P(x, y)$ that moves in such a way that the sum of its distances to $(3, 5)$ and $(3, -1)$ is 8.

Chapter Review Exercises

Review exercises can be used to assist students in assessing the understanding of the concepts in the chapter.

CHAPTER 5 TEST

1. Find the vertex, focus, and directrix of the parabola given by the equation $y = \frac{1}{8}x^2$.
2. Find the vertex, focus, and directrix of the parabola given by the equation $x^2 + 4x - 12y + 16 = 0$.
3. Find the equation in standard form of the parabola with directrix $x = 3$ and focus $(-1, -2)$.
4. Graph the parabola with focus $(0, -1)$ and directrix $y = -5$.
5. Find the vertices and foci of the ellipse given by the equation $\frac{x^2}{9} + \frac{y^2}{64} = 1$.
6. Graph: $\frac{x^2}{16} + \frac{y^2}{1} = 1$
7. Find the vertices and foci of the ellipse given by the equation $25x^2 - 150x + 9y^2 + 18y + 9 = 0$.
8. Find the equation in standard form of the ellipse with center $(0, -3)$, foci $(-6, -3)$ and $(6, -3)$, and minor axis of length 6.
9. Find the eccentricity of the ellipse given by the equation $9x^2 + 25y^2 = 81$.
10. Graph: $\frac{y^2}{25} - \frac{x^2}{16} = 1$
11. Find the vertices, foci, and asymptotes of the hyperbola given by the equation $\frac{x^2}{36} - \frac{y^2}{64} = 1$.
12. Graph: $16y^2 + 32y - 4x^2 - 24x = 84$
13. Find the vertices and foci of the hyperbola given by the equation $\frac{(y-4)^2}{36} - \frac{(x+5)^2}{9} = 1$.
14. Find the equation in standard form of the hyperbola with vertices at $(-2, -3)$ and $(-6, -3)$ and foci $(-4 + \sqrt{34}, -3)$ and $(-4 - \sqrt{34}, -3)$.
15. Find the equation in standard form of the parabola with focus $(-2, 4)$ and directrix $x = 6$.

Chapter Test

The chapter test is a self-assessment guide for the student. Answers to all the questions on the chapter test are in an appendix.

Supplements for the Instructor

College Algebra and Trigonometry has a complete set of teaching aids for the instructor.

Instructor's Annotated Edition This edition contains a replica of the student text and additional items just for the instructor. These include: *Instructor Notes, Transparency Master icons, Alternates to Examples, Concept Checks, Discuss the Concepts, New Vocabulary, Challenge Problems, Special Symbols, Quizzes,* and *Suggested Assignments.* Answers to all exercises are also provided.

Instructor's Solutions Manual The *Instructor's Solutions Manual* contains worked-out solutions for all end-of-section, supplemental, and review exercises.

Instructor's Resource Manual with Chapter Tests The *Instructor's Resource Manual* contains ready-to-use printed Chapter Tests, which is the first of three sources of testing material. Six printed tests (in two formats - free response and multiple choice) are provided for each chapter. These tests are available on the *Class Prep* CD or can be downloaded from our web site at *http://college.hmco.com.* The tests are in Microsoft Word format and can be edited to suit the needs of the instructor. The *Instructor's Resource Manual* also includes transparency masters and outlines for solutions of the Projects in the text. In addition, there are suggestions and solutions for additional Projects that can be assigned as group activities or extra-credit.

NEW! ***HM Testing*** *HM Testing*, our computerized test generator, is our second source of testing material. The database contains more than 3000 test items-many of which are algorithmic. These questions are unique to *HM Testing* and do not repeat items provided in the Chapter Tests of the *Instructor's Resource Manual. HM Testing* is designed to produce an unlimited number of tests for each chapter of the text, including cumulative tests and final exams. It is available for Microsoft Windows® and the Macintosh. Both versions provide **algorithms**, **on-line testing** and **gradebook** functions.

Printed Test Bank The *Printed Test Bank*, the third component of the testing material, is a printout of the items in *HM Testing*. Items that are algorithmic in *HM Testing* are identified with an asterisk. Instructors can use the test bank to select specific items from the database. Instructors who do not have access to a computer can use the *Printed Test Bank* to create a test being prepared by hand.

NEW! ***WebCT Courselets*** *WebCT Courselets* provide instructors with a flexible, Internet-based education platform providing multiple ways to present learning materials. The *WebCT Courselets* come with a full array of features to enrich the online learning experience.

NEW! ***BlackBoard Course Cartridges*** The *Houghton Mifflin Blackboard course cartridge* allows flexible, efficient, and creative ways to present learning materials and opportunities. In addition to course management benefits, instructors may make use of an electronic grade book, receive papers from students enrolled in the course via the Internet, and track student use of the communication and collaboration functions.

NEW! ***HMClassPrep*** These CD-ROMs contain a multitude of text specific resources for instructors to use to enhance the classroom experience. The resources (or 'assets') are available as pdf and/or customizable Microsoft Word® files and include: transparency masters, Chapter Tests from the IRM, and Quizzes from the IAE to name only a few. These resources can be accessed from the CD-ROM easily by chapter or asset type. The CD can also link you the text's web site.

NEW! ***Text-specific web site-instructor*** The assets available on the *Class Prep CD* are also available on the instructor web site at http://college.hmco.com. In addition, a syllabus builder, the web resources referenced in the text, the desktop version of the *Computer Tutor* with management system, and additional Projects with answers, are also offered on the web site. Appropriate items will be password protected. Instructors also have access to the student part of the text's web site.

Supplements for the Student

Student Study Guide The *Student Study Guide* contains complete solutions to all odd-numbered problems in the text as well as study tips and a practice test for each chapter.

NEW! ***Algorithmic Computer Tutor - Web Version*** The *Computer Tutor* is an interactive tutorial containing lessons and exercises for every section of the text as indicated by at each section title. However, the web version also offers quizzing and has a management system that can be used in both Macintosh and Windows environments. The web version can be accessed from our web site at *http://college.hmco.com.*

Algorithmic Computer Tutor - Desktop Version The *Computer Tutor* is an interactive tutorial containing lessons and exercises for every section of the text. The lessons provide additional instruction and practice and can be used in several ways: (1) to cover material that was missed because of absence from class; (2) to reinforce instruction on a concept that has not yet mastered; (3) to review material in preparation for an examination. Following each lesson there are exercises for the student to try. These exercises are created by carefully constructed algorithms that allow a student to practice a variety of exercise types. Because the exercises are created algorithmically, each time the student uses the tutorial, the student is presented with different problems. This tutorial is available for both Windows and Macintosh operating systems. It is offered on CD. For the Window's operating system, the Instructor's version of the Tutor also contains a management system that can be used, in conjunction with a computer network, to monitor student use of the Tutor. The management system records how long a student has been using the Tutor, the number of problems attempted, the number answered correctly, and the percent correct.

Algorithmic Review Computer Tutor The *Review Computer Tutor* is a self-paced, interactive tutorial in the same style as the desktop version of the Algorithmic Computer Tutor mentioned above. This tutorial covers all of the necessary prerequisite material that would be found in an intermediate algebra class. This tutorial is available for Macintosh and Windows operating systems. It is offered on CD.

NEW! ***SMARTHINKING™ live, on-line tutoring*** Houghton Mifflin has partnered with SMARTHINKING to provide an easy-to-use and effective on-line tutorial service. A **Graphing Calculator** function enables students and e-structors to collaborate in drawing graphs using a whiteboard feature. **Whiteboard Simulations** and **Practice Area** further promote real-time visual interaction.

Three levels of service are offered.

- **Prescheduled Text-specific Tutoring** provides real-time, one-on-one instruction with a specially qualified 'e-structor.'
- **Questions Any Time** allows students to submit questions to the tutor outside the scheduled hours and receive a reply within 24 hours.

- **Independent Study Resources** connect students with around-the-clock access to additional educational services, ranging from interactive web sites and on-line textbooks to diagnostic tests and Frequently Asked Questions posed to SMARTHINKING e-structors.

NEW! ***Videos*** This edition offers brand new text-specific videos, hosted by Dana Mosely, covering all sections of the text. At each section title is a video icon identifying the specific tape for that section. These videos, professionally produced specifically for the text, offer a valuable resource for further instruction and review.

NEW! ***Graphing Calculator Instructional Video*** This new two-video set, hosted by Dana Mosely, uses the TI-83+™ calculator to demonstrate the benefits of using a graphing calculator to illustrate the major mathematical concepts from college algebra through calculus.

NEW! ***Real Deal UpGrade CD*** These CD-ROMs have been carefully tailored to supplement and enhance the content of each textbook. Features are designed to help students improve their understanding of the textbook material and the course for which they are using the book. Resources, or assets, include: Study Tips, Learning Tips from Houghton Mifflin's best-selling *Becoming a Master Student* text, chapter summaries, and ACE self-quizzes with answers to name only a few. These resources can be accessed from the CD-ROM easily by chapter or asset type.

NEW! ***Text-specific web site-student*** The assets available on the *Real Deal CD* are also available on the student web site at *http://college.hmco.com.*

Acknowledgments

The authors would like to thank the people who have reviewed this manuscript and provided many valuable suggestions.

Ebrahim Ahmadizadeh, *Northampton Community College, PA*
James Alsobrook, *Southern Union State Community College, AL*
Danny T. Barnes, *University of Maryland University College, MD*
John J. Bray, *Broward Community College, FL*
Lawrence M. Clar, *Monroe Community College, NY*
Jeanne M. Draper, *Solano Community College*, CA
Matthew Frueh
Anne Haney
John Mark Henry, *Lincoln Land Community College, IL*
William C. Hoston
Barbara Krueger, *Cochise Community College, AZ*
Linda Kuroski, *Erie Community College–City, NY*
Linda Marable, *Nashville State Technical Institute, TN*
Lauri Semarne
Patricia G. Shelton, *North Carolina A & T State University, NC*

Special thanks to Sandy Doerfel, *Palomar College*, for her assistance in preparing material for the *Instructor's Annotated Edition*.

C H A P T E R P

PRELIMINARY CONCEPTS

◆ Georg Cantor

◆ The "infinity" of space.

How Large Is Infinity?

The German mathematician Georg Cantor (1845–1918) developed the idea of the cardinality of a set. The cardinality of a finite set is the number of elements in the set. For example, the set $\{5, 7, 11\}$ has a cardinality of 3. The set of natural numbers $\{1, 2, 3, 4, 5, 6, 7, \ldots\}$ is an infinite set. Cantor denoted its cardinality by the symbol $\aleph_0$, which is read "aleph null."

The set of whole numbers consists of all the elements of the set of natural numbers and the number 0. The following display shows a one-to-one correspondence between the set of natural numbers and the set of whole numbers.

$$\begin{array}{cccccccc} \{1, & 2, & 3, & 4, & 5, & \ldots, & n, & \ldots\} \\ \updownarrow & \updownarrow & \updownarrow & \updownarrow & \updownarrow & & \updownarrow & \\ \{0, & 1, & 2, & 3, & 4, & \ldots, & n-1, & \ldots\} \end{array}$$

Cantor reasoned that because of this one-to-one correspondence, the set of natural numbers and the set of whole numbers both have the same cardinality, namely $\aleph_0$.

Cantor was also able to show that the set of irrational numbers has a cardinality that is different from $\aleph_0$. The idea that some infinite sets have more elements than other infinite sets was not readily accepted. A century before Cantor's work, the philosopher Voltaire (1694–1778) had expressed the following opinion:

We admit, in geometry, not only infinite magnitudes, that is to say magnitudes greater than any assignable magnitude, but infinite magnitudes infinitely greater, the one than the other. This astonishes our dimension of brains, which is only about six inches long, five broad, and six in depth, in the largest of heads.

SECTION P.1 THE REAL NUMBER SYSTEM

- SETS
- PROPERTIES OF REAL NUMBERS
- PROPERTIES OF FRACTIONS

PA

◆ SETS

Human beings share the desire to organize and classify. Ancient astronomers classified stars into groups called constellations. Modern astronomers continue to classify stars by such characteristics as color, mass, size, temperature, and distance from earth. In mathematics it is useful to place numbers with similar characteristics into **sets.** The following sets of numbers are used extensively in the study of algebra:

Integers	$\{\ldots, -3, -2, -1, 0, 1, 2, 3, \ldots\}$
Rational numbers	{all terminating or repeating decimals}
Irrational numbers	{all nonterminating, nonrepeating decimals}
Real numbers	{all rational or irrational numbers}

If a number in decimal form terminates or repeats a block of digits, then the number is a rational number. Rational numbers can also be written in the form p/q, where p and q are integers and $q \neq 0$. For example,

$$\frac{3}{4} = 0.75 \quad \text{and} \quad \frac{5}{11} = 0.\overline{45}$$

are rational numbers. The bar over the 45 means that the block repeats without end; that is, $0.\overline{45} = 0.454545\ldots$.

In its decimal form, an irrational number neither terminates nor repeats. For example, $0.272272227\ldots$ is a nonterminating, nonrepeating decimal and thus is an irrational number. One of the best-known irrational numbers is pi, denoted by the Greek symbol π. The number π is defined as the ratio of the circumference of a circle to its diameter. Often in applications, the rational number 3.14 or the rational number 22/7 is used as an approximation of the irrational number π.

Every real number is either a rational number or an irrational number. If a real number is written in decimal form, it is a terminating decimal, a repeating decimal, or a nonterminating and nonrepeating decimal.

Each member of a set is called an **element** of the set. For instance, if $C = \{2, 3, 5\}$, then the elements of C are 2, 3, and 5. The notation $2 \in C$ is read "2 is an element of C." Set A is a **subset** of set B if every element of A is also an element of B, and we write $A \subseteq B$. For instance, the set of **negative integers** $\{-1, -2, -3, -4, \ldots\}$ is a subset of the set of integers. The set of **positive integers** $\{1, 2, 3, 4, \ldots\}$ (also known as the set of **natural numbers**) is also a subset of the set of integers. **Figure P.1** illustrates the subset relationships among the sets defined above.

Prime numbers and *composite numbers* play an important role in almost every branch of mathematics. A **prime number** is a positive integer other than 1 that has no positive-integer factors[1] other than itself and 1. The ten smallest prime

MATH MATTERS

Archimedes (c. 287–212 B.C.) was the first to calculate π with any degree of precision. He was able to show that

$$3\frac{10}{71} < \pi < 3\frac{1}{7}$$

from which we get the approximation $3\frac{1}{7} = \frac{22}{7} \approx \pi$. The use of the symbol π for this quantity was introduced by Leonhard Euler (1707–1783) in 1739, approximately 2000 years after Archimedes.

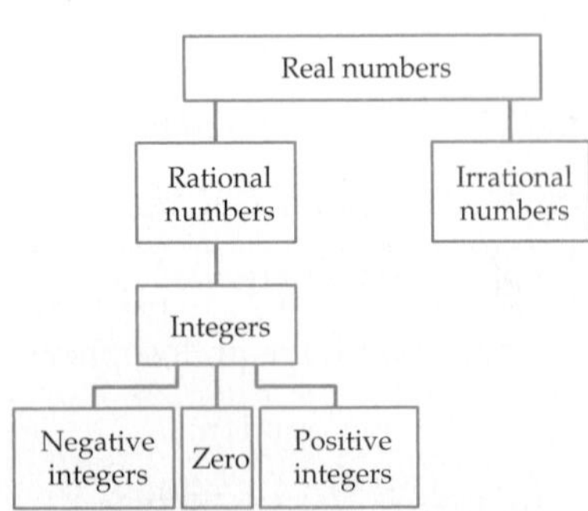

Figure P.1

[1] Recall that a factor of a number divides the number evenly. For instance, 3 and 7 are factors of 21; 5 is not a factor of 21.

numbers are 2, 3, 5, 7, 11, 13, 17, 19, 23, and 29. Each of these numbers has only itself and 1 as factors.

A **composite number** is a positive integer greater than 1 that is not a prime number. For example, 10 is a composite number because 10 has both 2 and 5 as factors. The ten smallest composite numbers are 4, 6, 8, 9, 10, 12, 14, 15, 16, and 18.

MATH MATTERS

Sophie Germain, (1776–1831) was born in Paris, France. Because enrollment in the university she wanted to attend was only available to men, Germain attended under the name of Antoine-August Le Blanc. Eventually her ruse was discovered but not before she came to the attention of Pierre Lagrange, one of the best mathematicians of the time. He encouraged her work and became a mentor to her. A certain type of prime number is named after her called a Germain prime number. It is one such that p and $2p + 1$ are both prime. For instance, 23 is a Germain prime because $23 = 2(11) + 1$ and 23 and 11 are prime numbers. Germain primes are used in public key cryptography, a method used to send secure communications over the internet.

EXAMPLE 1 Classify Real Numbers

Determine which of the following numbers are

a. integers **b.** rational numbers **c.** irrational numbers
d. real numbers **e.** prime numbers **f.** composite numbers

$$-0.2, \quad 0, \quad 0.\overline{3}, \quad \pi, \quad 6, \quad 7, \quad 41, \quad 51, \quad 0.71771777177771\ldots$$

Solution

a. Integers: 0, 6, 7, 41, 51

b. Rational numbers: $-0.2, 0, 0.\overline{3}, 6, 7, 41, 51$

c. Irrational numbers: $0.71771777177771\ldots, \pi$

d. Real numbers: $-0.2, 0, 0.\overline{3}, \pi, 6, 7, 41, 51, 0.71771777177771\ldots$

e. Prime numbers: 7, 41

f. Composite numbers: 6, 51

TRY EXERCISE 2, EXERCISE SET P.1, PAGE 8

MATH MATTERS

Prime numbers are used to encrypt messages on the Internet via a coding system called PGP (Pretty Good Privacy). This is part of an encryption scheme called Public Key Cryptography.

Sets are often written using **set-builder notation,** which makes use of a variable and a characteristic property that the elements of the set alone possess. This notation is especially useful to describe infinite sets. The set-builder notation

$$\{x^2 \mid x \text{ is an integer}\}$$

is read as "the set of all elements x^2 such that x is an integer." This is the infinite set of **perfect squares:** $\{0, 1, 4, 9, 16, 25, 36, 49, \ldots\}$.

The **empty set** or **null set** is a set without any elements. The set of numbers that are both prime and also composite is an example of the null set. The null set is denoted by the symbol $\varnothing$.

Just as addition and subtraction are operations performed on real numbers, there are operations performed on sets. Two of these set operations are called *intersection* and *union.* The **intersection** of sets A and B, denoted by $A \cap B$, is the set of all elements belonging to both set A and set B. The **union** of sets A and B, denoted by $A \cup B$, is the set of all elements belonging to set A, to set B, or to both.

EXAMPLE 2 Find the Intersection and the Union of Two Sets

Find each intersection or union, given $A = \{0, 1, 4, 6, 9\}$, $B = \{1, 3, 5, 7, 9\}$ and $P = \{x \mid x \text{ is a prime number} < 10\}$.

a. $A \cap B$ **b.** $A \cap P$ **c.** $A \cup B$ **d.** $A \cup P$

Continued

Solution

a. $A \cap B = \{0, 1, 4, 6, 9\} \cap \{1, 3, 5, 7, 9\}$
$= \{1, 9\}$ • **Only 1 and 9 belong to both sets.**

b. First determine that $P = \{2, 3, 5, 7\}$. Therefore,

$A \cap P = \{0, 1, 4, 6, 9\} \cap \{2, 3, 5, 7\}$
$= \varnothing$ • **There are no common elements.**

c. $A \cup B = \{0, 1, 4, 6, 9\} \cup \{1, 3, 5, 7, 9\}$
$= \{0, 1, 3, 4, 5, 6, 7, 9\}$ • **List the elements of the first set. Include elements from the second set that are not already listed.**

d. $A \cup P = \{0, 1, 4, 6, 9\} \cup \{2, 3, 5, 7\} = \{0, 1, 2, 3, 4, 5, 6, 7, 9\}$

TRY EXERCISE 14, EXERCISE SET P.1, PAGE 8

◆ PROPERTIES OF REAL NUMBERS

Addition, multiplication, subtraction, and *division* are the operations of arithmetic. **Addition** of the two real numbers a and b is designated by $a + b$. If $a + b = c$, then c is the **sum** and the real numbers a and b are called **terms.**

Multiplication of the real numbers a and b is designated by ab or $a \cdot b$. If $ab = c$, then c is the **product** and the real numbers a and b are called **factors** of c.

The number $-b$ is referred to as the **additive inverse** of b. **Subtraction** of the real numbers a and b is designated by $a - b$ and is defined as the sum of a and the additive inverse of b. That is,

$$a - b = a + (-b)$$

If $a - b = c$, then c is called the **difference** of a and b.

The **multiplicative inverse** or **reciprocal** of the nonzero number b is $1/b$. The **division** of a and b, designated by $a \div b$ with $b \neq 0$, is defined as the product of a and the reciprocal of b. That is,

$$a \div b = a\left(\frac{1}{b}\right) \quad \text{provided that } b \neq 0$$

If $a \div b = c$, then c is called the **quotient** of a and b.

The notation $a \div b$ is often represented by the fractional notation a/b or $\frac{a}{b}$.

The real number a is the **numerator,** and the nonzero real number b is the **denominator** of the fraction.

Properties of Real Numbers

Let a, b, and c be real numbers.

	Addition Properties	Multiplication Properties
Closure	$a + b$ is a unique real number.	ab is a unique real number.
Commutative	$a + b = b + a$	$ab = ba$
Associative	$(a + b) + c = a + (b + c)$	$(ab)c = a(bc)$
Identity	There exists a unique real number 0 such that $a + 0 = 0 + a = a$.	There exists a unique real number 1 such that $a \cdot 1 = 1 \cdot a = a$.
Inverse	For each real number a, there is a unique real number $-a$ such that $a + (-a) = (-a) + a = 0$.	For each *nonzero* real number a, there is a unique real number $1/a$ such that $a \cdot \dfrac{1}{a} = \dfrac{1}{a} \cdot a = 1$.
Distributive	$a(b + c) = ab + ac$	

We can identify which property of real numbers has been used to rewrite expressions by closely comparing the expressions and noting any changes.

EXAMPLE 3 Identify Properties of Real Numbers

Identify the property of real numbers illustrated in each statement.

a. $(2a)b = 2(ab)$

b. $\left(\dfrac{1}{5}\right)11$ is a real number.

c. $4(x + 3) = 4x + 12$

d. $(a + 5b) + 7c = (5b + a) + 7c$

e. $\left(\dfrac{1}{2} \cdot 2\right)a = 1 \cdot a$

f. $1 \cdot a = a$

Solution

a. Associative property of multiplication

b. Closure property of multiplication of real numbers

c. Distributive property

d. Commutative property of addition

e. Inverse property of multiplication

f. Identity property of multiplication

TRY EXERCISE 26, EXERCISE SET P.1, PAGE 8

An **equation** is a statement of equality between two numbers or two expressions. There are four basic properties of equality that relate to equations.

Properties of Equality

Let a, b, and c be real numbers.

Reflexive	$a = a$
Symmetric	If $a = b$, then $b = a$.
Transitive	If $a = b$ and $b = c$, then $a = c$.
Substitution	If $a = b$, then a may be replaced by b in any expression that involves a.

EXAMPLE 4 Identify Properties of Equality

Identify the property of equality illustrated in each statement.

a. If $3a + b = c$, then $c = 3a + b$. **b.** $5(x + y) = 5(x + y)$

c. If $4a - 1 = 7b$ and $7b = 5c + 2$, then $4a - 1 = 5c + 2$.

d. If $a = 5$ and $b(a + c) = 72$, then $b(5 + c) = 72$.

Solution

a. Symmetric **b.** Reflexive **c.** Transitive **d.** Substitution

TRY EXERCISE 28, EXERCISE SET P.1, PAGE 8

◆ PROPERTIES OF FRACTIONS

The following properties of fractions will be used throughout this text.

Properties of Fractions

For all fractions a/b and c/d, where $b \neq 0$ and $d \neq 0$:

Equality	$\frac{a}{b} = \frac{c}{d}$ if and only if $ad = bc$
Equivalent fractions	$\frac{a}{b} = \frac{ac}{bc}$, $c \neq 0$
Addition	$\frac{a}{b} + \frac{c}{b} = \frac{a + c}{b}$
Subtraction	$\frac{a}{b} - \frac{c}{b} = \frac{a - c}{b}$
Multiplication	$\frac{a}{b} \cdot \frac{c}{d} = \frac{ac}{bd}$
Division	$\frac{a}{b} \div \frac{c}{d} = \frac{a}{b} \cdot \frac{d}{c} = \frac{ad}{bc}$, $c \neq 0$
Sign	$-\frac{a}{b} = \frac{-a}{b} = \frac{a}{-b}$

The equality property of fractions contains the terminology "if and only if," which implies each of the following:

$$\text{If } \frac{a}{b} = \frac{c}{d}, \qquad \text{then } ad = bc.$$

$$\text{If } ad = bc, \qquad \text{then } \frac{a}{b} = \frac{c}{d}.$$

The number zero has many special properties. The following division properties of zero play an important role in this text.

Division Properties of Zero

1. For $a \neq 0, \dfrac{0}{a} = 0$. (Zero divided by any nonzero number is zero.)

2. $\dfrac{a}{0}$ is undefined. (Division by zero is undefined.)

The properties of fractions can be used to find the sum, difference, product, or quotient of fractions.

EXAMPLE 5 Compute with Fractions

Use the properties of fractions to perform the indicated operations. Assume that $a \neq 0$.

a. $\dfrac{2a}{3} - \dfrac{a}{5}$ b. $\dfrac{2a}{5} \cdot \dfrac{3a}{4}$ c. $\dfrac{5a}{6} \div \dfrac{3a}{4}$ d. $\dfrac{0}{3a}$

Solution

a. Rewrite each fraction as an equivalent fraction with a common denominator of 15 by multiplying both the numerator and the denominator of $2a/3$ by 5 and by multiplying both the numerator and the denominator of $a/5$ by 3.

$$\frac{2a}{3} - \frac{a}{5} = \frac{2a(5)}{3(5)} - \frac{a(3)}{5(3)} = \frac{10a}{15} - \frac{3a}{15} = \frac{10a - 3a}{15} = \frac{7a}{15}$$

b. $\dfrac{2a}{5} \cdot \dfrac{3a}{4} = \dfrac{(2a)(3a)}{(5)(4)} = \dfrac{6a^2}{20} = \dfrac{3a^2}{10}$

c. $\dfrac{5a}{6} \div \dfrac{3a}{4} = \dfrac{5a}{6} \cdot \dfrac{4}{3a} = \dfrac{20a}{18a} = \dfrac{10}{9}$

d. $\dfrac{0}{3a} = 0$ • **Zero divided by any nonzero number is zero.**

TRY EXERCISE 40, EXERCISE SET P.1, PAGE 8

TOPICS FOR DISCUSSION

1. Archimedes determined that $223/71 < \pi < 22/7$. Is it possible to find an exact expression for π of the form a/b, where a and b are integers?
2. Is the intersection of two infinite sets always an infinite set? Is the union of two infinite sets always an infinite set?
3. Explain why division by zero is not allowed.
4. Explain the similarities and differences between rational and irrational numbers.

EXERCISE SET P.1

In Exercises 1 and 2, determine which of the numbers are a. integers, b. rational numbers, c. irrational numbers, d. real numbers, e. prime numbers, f. composite numbers.

1. $-3 \quad 4 \quad \frac{1}{5} \quad 11 \quad 3.14 \quad 57 \quad 0.252252225\ldots$

2. $5.\overline{17} \quad -4.25 \quad \frac{1}{4} \quad \pi \quad 21 \quad 53 \quad 0.45454545\ldots$

In Exercises 3 to 6, list the elements of the set.

3. $A = \{x \mid x \text{ is a composite number less than } 11\}$
4. $B = \{x \mid x \text{ is an even prime number}\}$
5. $C = \{x \mid 50 < x < 60 \text{ and } x \text{ is a prime number}\}$
6. $D = \{x \mid x \text{ is the smallest odd composite number}\}$

For Exercises 7 to 12, list the 4 smallest elements of each infinite set.

7. $\{2x \mid x \text{ is a positive integer}\}$
8. $\{|x| \mid x \text{ is an integer}\}$
9. $\{y \mid y = 2x + 1, x \text{ is a natural number}\}$
10. $\{y \mid y = x^2 - 1, x \text{ is an integer}\}$
11. $\{z \mid z = |x|, x \text{ is an integer}\}$
12. $\{z \mid z = |x| - x, x \text{ is a negative integer}\}$

In Exercises 13 to 24, use $A = \{0, 1, 2, 3, 4\}$, $B = \{1, 3, 5, 11\}$, $C = \{1, 3, 6, 10\}$, and $D = \{0, 2, 4, 6, 8, 10\}$ to find the indicated intersection or union.

13. $A \cap B$
14. $A \cap C$
15. $B \cap C$
16. $B \cap D$
17. $A \cap D$
18. $C \cap D$
19. $A \cup B$
20. $A \cup C$
21. $A \cap (B \cup C)$
22. $A \cup (B \cap C)$
23. $(B \cap C) \cap D$
24. $A \cup (B \cup C)$

In Exercises 25 to 38, identify the property of real numbers or the property of equality that is illustrated.

25. $3 + (2 + 5) = (3 + 2) + 5$
26. $6 + (2 + 7) = 6 + (7 + 2)$
27. $1 \cdot a = a$
28. If $a + b = 2$, then $2 = a + b$.
29. $a(bx) = a(bx)$
30. If $x + 2y = 7$ and $7 = y$, then $x + 2(7) = 7$.
31. If $x = 2(y + z)$ and $2(y + z) = 5w$, then $x = 5w$.
32. $p(q + r) = pq + pr$
33. $m + (-m) = 0$
34. $t\left(\frac{1}{t}\right) = 1, t \neq 0$
35. $7(a + b) = 7(b + a)$
36. $8(gh + 5) = 8(hg + 5)$
37. If $x + 2y = 7$ and $w = 7$, then $x + 2y = w$.
38. $5[x + (y + z)] = 5x + 5(y + z)$

In Exercises 39 to 48, use the properties of fractions to perform the indicated operations. State each answer in lowest terms. Assume a is a nonzero real number.

39. $\frac{2a}{7} - \frac{5a}{7}$
40. $\frac{2a}{5} + \frac{3a}{7}$
41. $\frac{-3a}{5} + \frac{a}{4}$
42. $\frac{7}{8}a - \frac{13}{5}a$
43. $\frac{-5}{7} \cdot \frac{2}{3}$
44. $\frac{7}{11} \cdot \frac{-22}{21}$
45. $\frac{12a}{5} \div \frac{-2a}{3}$
46. $\frac{2}{5} \div 3\frac{2}{3}$

47. $\dfrac{2a}{3} - \dfrac{4a}{5}$

48. $\dfrac{1}{2a} - \dfrac{3}{a}$

49. **POOL MAINTENANCE** One pipe can fill a pool in 11 hours. A second pipe can fill the same pool in 15 hours. Assume the first pipe fills 1/11 of the pool every hour and the second pipe fills 1/15 of the pool every hour.

a. Find the amount of the pool the two pipes together fill in 3 hours.

b. Find the amount of the pool they fill together in x hours.

50. **FOCAL LENGTH OF A MIRROR** The relationship between the distance of an object d_0 from a curved mirror, the distance of its image d_i from the mirror, and the focal length f of the mirror is given by the **mirror equation:**

$$\frac{1}{f} = \frac{1}{d_0} + \frac{1}{d_i}$$

What is the focal length f of a mirror[2] for which $d_0 = 25$ centimeters and $d_i = -5$ centimeters?

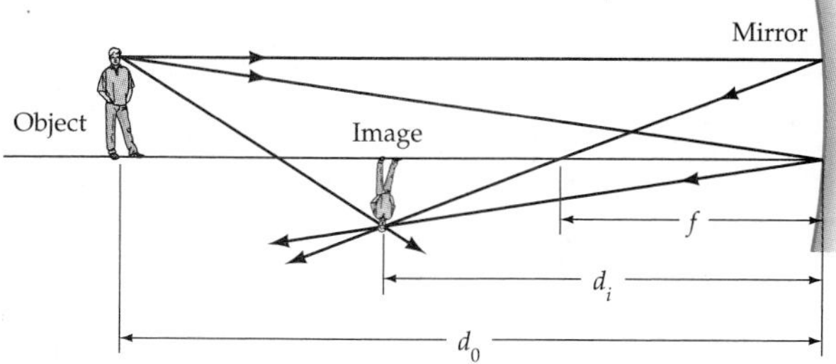

51. State the multiplicative inverse of $7\frac{3}{8}$.

52. State the multiplicative inverse of $-4\frac{2}{5}$.

53. Show by an example that the operation of subtraction of real numbers is not a commutative operation.

54. Show by an example that the operation of division of nonzero real numbers is not a commutative operation.

55. Show by an example that the operation of subtraction of real numbers is not an associative operation.

56. Show by an example that the operation of division of real numbers is not an associative operation.

In Exercises 57 to 66, classify each statement as true or false.

57. $a/0$ is the multiplicative inverse of $0/a$.

58. $(-1/\pi)$ is the multiplicative inverse of $-\pi$.

59. If $p = q + \dfrac{t}{2}$ and $q + \dfrac{t}{2} = \dfrac{1}{2}s$, then $\dfrac{1}{2}s = p$.

60. If $a - b = 7$, then $7 = b - a$.

61. The sum of two composite numbers is a composite number.

62. All integers are natural numbers.

63. Every real number is either a rational or an irrational number.

64. Every rational number is either even or odd.

65. 1 is the only positive integer that is not prime and not composite.

66. All repeating decimals are rational numbers.

67. Use a calculator to write each of the following rational numbers as a decimal. If the number is represented by a nonterminating decimal, then use a *bar* over the repeating portion of the decimal.

a. $\dfrac{8}{11}$ b. $\dfrac{33}{40}$ c. $\dfrac{2}{7}$ d. $\dfrac{5}{37}$

68. Use a calculator to determine whether 3.14 or 22/7 is a closer approximation to π.

SUPPLEMENTAL EXERCISES

69. Use a calculator to complete the following table.

x	0.1	0.01	0.001	0.0000001
$\dfrac{\sqrt{x+9}-3}{x}$				

Now make a guess as to the number the fraction seems to be approaching as x assumes the values of real numbers that are closer and closer to zero.

70. Use a calculator to complete the following table.

x	0.1	0.01	0.001	0.0000001
$\dfrac{\frac{1}{2}-\frac{1}{x+2}}{x}$				

Now make a guess as to the number the fraction seems to be approaching as x assumes the values of real numbers that are closer and closer to zero.

71. Which of the properties of real numbers are satisfied by the set of positive integers?

72. Which of the properties of real numbers are satisfied by the set of integers?

[2] For convex mirrors, both the focal length f and the image distance d_i are *negative* quantities.

73. Which of the properties of real numbers are satisfied by the set of rational numbers?

74. Which of the properties of real numbers are satisfied by the set of irrational numbers?

75. **GOLDBACH'S CONJECTURE** In 1742 Christian Goldbach conjectured that every even number greater than 2 can be written as the sum of two prime numbers. Many mathematicians have tried to prove or disprove this conjecture without succeeding. Show that Goldbach's conjecture is true for the following even numbers.

a. 12 b. 30

76. **TWIN PRIMES** If the natural numbers n and $n + 2$ are both prime numbers, then they are said to be twin primes. For example, 11 and 13 are twin primes. It is not known whether the set of twin primes is an infinite set or a finite set. List all the twin primes less than 50.

PROJECTS

1.  **NUMBER THEORY** *Theorem:* If a number of the form 111...1 is a prime number, then the number of 1's is a prime number. For instance, the numbers

11 and 1111111111111111111

are prime numbers, and the number of 1's in each number is a prime number (2 in the first number and 19 in the second number).

a. The number $111 = 3 \cdot 37$, so 111 is not a prime number. Explain why this does not contradict the above theorem.

b. What is the converse of a theorem? State the converse of the theorem above.

c. If a theorem is true, is the converse of a theorem also true? Explain your answer.

2. **PERFECT SQUARES** Explain why a perfect-square integer must have an odd number of distinct natural-number divisors.

SECTION P.2 INTERVALS, ABSOLUTE VALUE, AND DISTANCE

- INTERVAL NOTATION
- ABSOLUTE VALUE
- DISTANCE BETWEEN TWO POINTS ON A NUMBER LINE

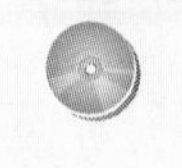

The real numbers can be represented geometrically by a **coordinate axis** called a **real number line.** **Figure P.2** shows a portion of a real number line. The number associated with a particular point on a real number line is called the **coordinate** of the point. It is customary to label those points whose coordinates are integers. The point corresponding to zero is called the **origin,** denoted 0. Numbers to the right of the origin are **positive real numbers;** numbers to the left of the origin are **negative real numbers.**

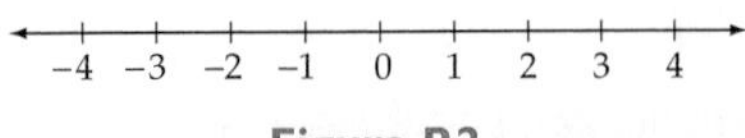

Figure P.2

A real number line provides a picture of the real numbers. That is, each real number corresponds to one and only one point on the real number line, and each point on a real number line corresponds to one and only one real number. This type of correspondence is referred to as a **one-to-one correspondence.** The real numbers -3, $-1/2$, and 1.75 are graphed in **Figure P.3.**

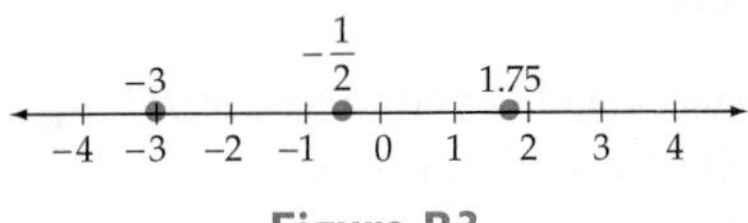

Figure P.3

Certain order relationships exist between real numbers. For example, if a and b are real numbers, then

a **equals** b (denoted by $a = b$) if $a - b = 0$.

a is **greater than** b (denoted by $a > b$) if $a - b$ is positive.

a is **less than** b (denoted by $a < b$) if $b - a$ is positive.

On a horizontal number line, the notation

$a = b$ implies that the point with coordinate a is the *same* point as the point with coordinate b.

$a > b$ implies that the point with coordinate a is to the *right* of the point with coordinate b.

$a < b$ implies that the point with coordinate a is to the *left* of the point with coordinate b.

The **inequality** symbols $<$ and $>$ are sometimes combined with the equality symbol in the following manner:

$a \geq b$ This is read "a is greater than or equal to b," which means $a > b$ or $a = b$.

$a \leq b$ This is read "a is less than or equal to b," which means $a < b$ or $a = b$.

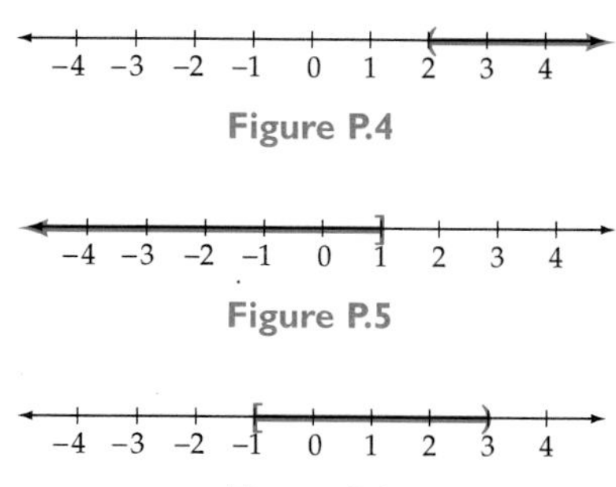

Figure P.4

Figure P.5

Figure P.6

Inequalities can be used to represent subsets of real numbers. For example, the inequality $x > 2$ represents all real numbers greater than 2; **Figure P.4** shows its graph. The parenthesis at 2 means that 2 is not part of the graph.

The inequality $x \leq 1$ represents all real numbers less than or equal to 1; **Figure P.5** shows its graph. The bracket at 1 means that 1 is part of the graph.

The inequality $-1 \leq x < 3$ represents all real numbers between -1 and 3, including -1 but not including 3. **Figure P.6** shows its graph.

◆ INTERVAL NOTATION

Subsets of real numbers can also be represented by a compact form of notation called **interval notation.** For example, $[-1, 3)$ is the interval notation for the subset of real numbers in **Figure P.6.**

In general, the interval notation

(a, b) represents all real numbers between a and b, not including a and not including b. This is an **open interval.** Using inequalities, this is written $a < x < b$.

$[a, b]$ represents all real numbers between a and b, including a and including b. This is a **closed interval.** Using inequalities, this is written $a \leq x \leq b$.

$(a, b]$ represents all real numbers between a and b, not including a but including b. This is a **half-open interval.** Using inequalities, this is written $a < x \leq b$.

$[a, b)$ represents all real numbers between a and b, including a but not including b. This is a **half-open interval.** Using inequalities, this is written $a \le x < b$.

Figure P.7 shows the four subsets of real numbers that are associated with the four interval notations (a, b), $[a, b]$, $(a, b]$, and $[a, b)$.

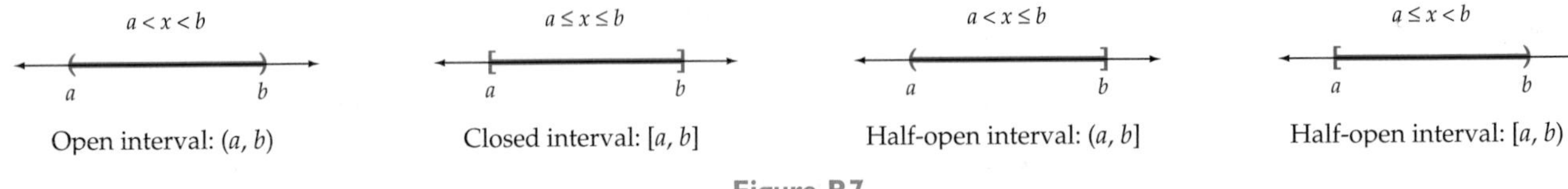

Open interval: (a, b) Closed interval: $[a, b]$ Half-open interval: $(a, b]$ Half-open interval: $[a, b)$

Figure P.7
Finite intervals

Subsets of the real numbers whose graphs extend forever in one or both directions can be represented by interval notation using the **infinity symbol** ∞ or the **negative infinity symbol** $-\infty$.

As Figure P.8 shows, the interval notation

$(-\infty, a)$	represents all real numbers less than a.
(b, ∞)	represents all real numbers greater than b.
$(-\infty, a]$	represents all real numbers less than or equal to a.
$[b, \infty)$	represents all real numbers greater than or equal to b.
$(-\infty, \infty)$	represents all real numbers.

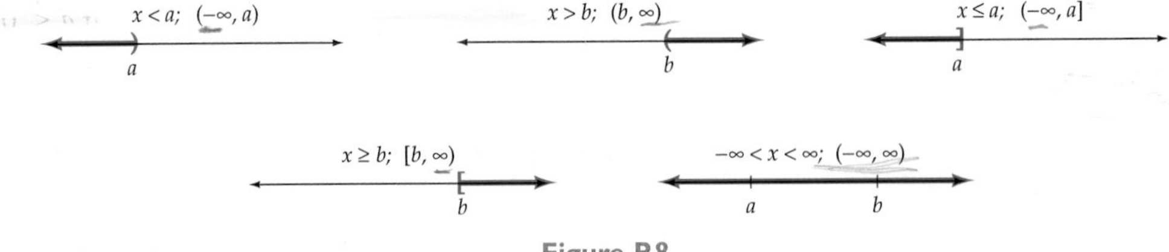

Figure P.8
Infinite intervals

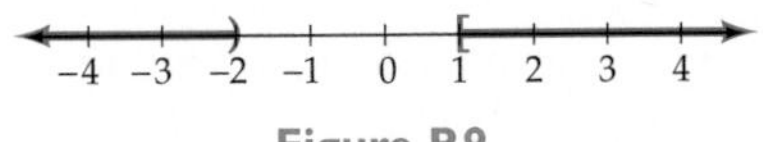

Figure P.9

Some graphs consist of more than one interval of the real number line. Figure P.9 is a graph of the interval $(-\infty, -2)$, along with the interval $[1, \infty)$.

The word *or* is used to denote the union of two sets. The word *and* is used to denote intersection. Thus the graph in Figure P.9 is denoted by the inequality notation

$$x < -2 \quad \text{or} \quad x \ge 1$$

To represent this graph using interval notation, use the union symbol $\cup$ and write $(-\infty, -2) \cup [1, \infty)$.

EXAMPLE 1 Graph Intervals and Inequalities

Graph the following. Also write **a.** and **b.** using interval notation, and write **c.** and **d.** using inequality notation.

a. $-2 \le x < 3$ **b.** $x \ge -3$ **c.** $[-4, -2] \cup [0, \infty)$ **d.** $(-\infty, 2)$

Solution

a. 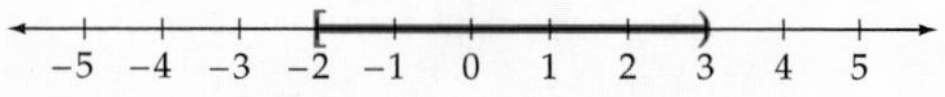 $[-2, 3)$

b. 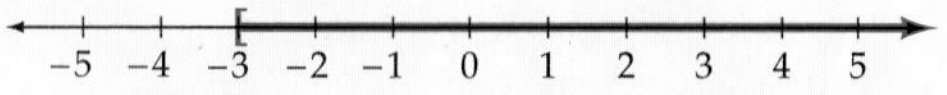 $[-3, \infty)$

c. 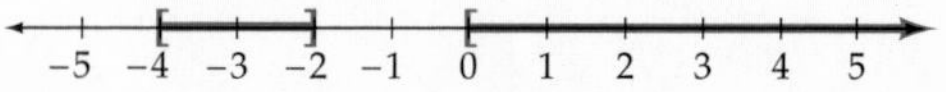 $-4 \leq x \leq -2$ or $x \geq 0$

d. 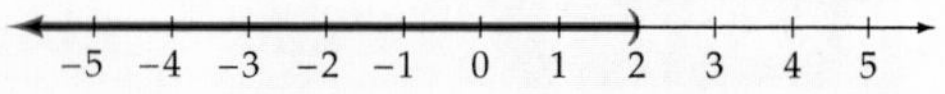 $x < 2$

TRY EXERCISE 16, EXERCISE SET P.2, PAGE 16

◆ ABSOLUTE VALUE

The *absolute value* of the real number a, denoted $|a|$, is the distance between a and 0 on the number line. For example, $|2| = 2$ and $|-2| = 2$. In general, if $a \geq 0$, then $|a| = a$; however, if $a < 0$, then $|a| = -a$ because $-a$ is positive when $a < 0$. This leads us to the following definition.

take note

The second part of the definition of absolute value states that if $a < 0$, then $|a| = -a$. For instance, if $a = -4$, then

$$|a| = |-4| = -(-4) = 4$$

Definition of Absolute Value

The **absolute value** of the real number a is defined by

$$|a| = \begin{cases} a & \text{if } a \geq 0 \\ -a & \text{if } a < 0 \end{cases}$$

The following theorems can be derived by using the definition of absolute value.

take note

Note the term *nonnegative* that is used at the right. Nonnegative means *greater than or equal to zero*. *Positive* means *greater than zero*.

Absolute Value Theorems

For all real numbers a and b,

Nonnegative	$\lvert a\rvert \geq 0$ greater than or equal to zero
Product	$\lvert ab\rvert = \lvert a\rvert \lvert b\rvert$
Quotient	$\left\lvert \frac{a}{b} \right\rvert = \frac{\lvert a\rvert}{\lvert b\rvert}, \quad b \neq 0$
Triangle inequality	$\lvert a + b\rvert \leq \lvert a\rvert + \lvert b\rvert$
Difference	$\lvert a - b\rvert = \lvert b - a\rvert$

The definition of absolute value and the absolute value theorems can be used to write some expressions without absolute value symbols. For instance, because $1 - \pi < 0$,

$$|1 - \pi| = -(1 - \pi) = \pi - 1$$

More complicated expressions can also be simplified by using these theorems. For example, given $-1 < x < 1$,

$$\begin{aligned} |x+3| - |x-2| &= (x+3) - [-(x-2)] \\ &= (x+3) + (x-2) \\ &= 2x + 1 \end{aligned}$$

• $-1 < x < 1$. Thus $|x - 2| = -(x - 2)$.

EXAMPLE 2 Evaluate Absolute Value Expressions

Write $\left|\dfrac{2x}{|x| + |x-2|}\right|$, given $0 < x < 2$, without absolute value symbols.

Solution

Use the quotient theorem to write the expression as a quotient of absolute values.

$$\left|\frac{2x}{|x| + |x-2|}\right| = \frac{|2x|}{||x| + |x-2||}$$

Because $0 < x < 2$, $|2x| = 2x$, $|x| = x$, and $|x - 2| = -x + 2$. Substituting yields

$$\frac{|2x|}{||x| + |x-2||} = \frac{2x}{|x + (-x + 2)|} = \frac{2x}{|2|} = \frac{2x}{2} = x$$

TRY EXERCISE 56, EXERCISE SET P.2, PAGE 16

◆ DISTANCE BETWEEN TWO POINTS ON A NUMBER LINE

The definition of *distance* between any two points on a real number line makes use of absolute value.

Distance Between Points on a Real Number Line

For any real numbers a and b, the **distance** between the graph of a and the graph of b is denoted by $d(a, b)$, where

$$d(a, b) = |a - b|$$

EXAMPLE 3 Find the Distance Between Points

Find the distance between the points whose coordinates are given.

a. $5, -2$ **b.** $-\pi, -2$

Solution

a. $d(5, -2) = |5 - (-2)| = |5 + 2| = |7| = 7$

b. $d(-\pi, -2) = |-\pi - (-2)| = |-\pi + 2|$
$= -(-\pi + 2) = \pi - 2$ • $-\pi + 2 < 0$. Thus $|-\pi + 2| = -(-\pi + 2)$.

TRY EXERCISE 64, EXERCISE SET P.2, PAGE 16

Absolute value notation and the notion of distance can also be used to describe intervals.

EXAMPLE 4 Use Absolute Value Notation

Express "the distance between a real number x and 7 is less than 2" using absolute value notation.

Solution

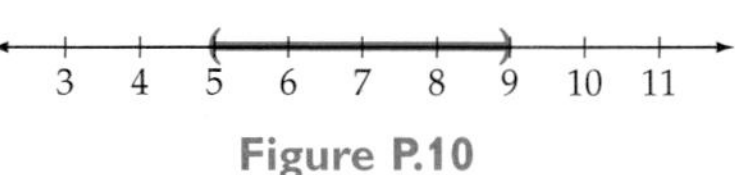

Figure P.10

The distance between x and 7 is $|x - 7|$. To express that this distance is less than 2, we write $|x - 7| < 2$. See Figure P.10.

TRY EXERCISE 80, EXERCISE SET P.2, PAGE 16

TOPICS FOR DISCUSSION

1. Explain the similarities and differences between open intervals and closed intervals.
2. Discuss why it is *not* correct to write intervals of real numbers such as $(a, \infty]$ or $[-\infty, \infty]$.
3. Discuss the correctness of the statement "If x is a real number, then $|x| = x$."
4. What is an order relation? Can all things be ordered? For instance, can colors such as blue, brown, aqua, purple, and yellow be put in order? Are the letters of the alphabet ordered?

EXERCISE SET P.2

In Exercises 1 and 2, graph each number on a real number line.

1. $-4; -2; \frac{7}{4}; 2.5$

2. $-3.5; 0; 3; \frac{9}{4}$

In Exercises 3 to 14, replace the □ with the appropriate symbol (<, =, or >).

3. $\frac{5}{2} \square 4$

4. $-\frac{3}{2} \square -3$

5. $\frac{2}{3} \square 0.6666$

6. $\frac{1}{5} \square 0.2$

7. $1.75 \square 2.23$

8. $1.25 \square 1.3$

9. $0.\overline{36} \square \frac{4}{11}$

10. $0.4 \square \frac{4}{9}$

11. $\frac{10}{5} \square 2$

12. $\frac{0}{2} \square -\frac{0}{5}$

13. $\pi \square 3.14159$

14. $\frac{22}{7} \square \pi$

In Exercises 15 to 26, graph each inequality and write the inequality using interval notation.

15. $3 < x < 5$
16. $-2 \le x < 1$
17. $x < 3$
18. $x \ge 4$
19. $x \ge 0$ and $x < 3$
20. $x > -4$ and $x \le 4$
21. $x < -3$ or $x \ge 2$
22. $x \le 2$ or $x > 3$
23. $x > 3$ and $x < 4$
24. $x > -5$ or $x < 1$
25. $x \le 3$ and $x > -1$
26. $x < 5$ and $x \le 2$

In Exercises 27 to 38, graph each interval and write each interval as an inequality.

27. $[-4, 1]$
28. $[-2, 3)$
29. $(1, 5)$
30. $(1, 4]$
31. $[2.5, \infty)$
32. $(-\infty, 3]$
33. $(-\infty, 2)$
34. (π, ∞)
35. $(-\infty, 2] \cup (3, \infty)$
36. $(-\infty, 1) \cup (4, \infty)$
37. $(-\infty, 3) \cup (3, \infty)$
38. $(-\infty, 1) \cup [2, \infty)$

In Exercises 39 to 46, use the given notation or graph to supply the notation or graph that is marked with a question mark.

	Inequality Notation	Interval Notation	Graph
39.	$x \le 3$	?	?
40.	?	$(-2, \infty)$	?
41.	?	?	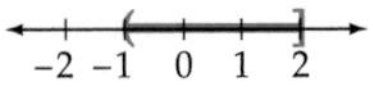 $-2\ \ -1\ \ 0\ \ 1\ \ 2$
42.	$-3 \le x < -1$	?	?
43.	?	$[1, 4]$	?
44.	?	?	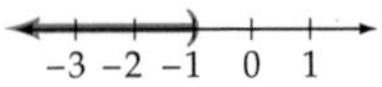 $-3\ \ -2\ \ -1\ \ 0\ \ 1$
45.	?	$[-2, \pi)$	?
46.	$x < 2$ or $x \ge 4$	?	?

In Exercises 47 to 60, write each expression without absolute value symbols.

47. $|4|$
48. $|-8|$
49. $|-27.4|$
50. $|3| - |-7|$
51. $-|-3| - |8|$
52. $|4||-8|$
53. $|y^2 + 10|$
54. $|x^2 + 1|$
55. $|-1 - \pi|$
56. $|x + 6| + |x - 2|$, given $0 < x < 1$
57. $|x - 4| + |x + 5|$, given $2 < x < 3$
58. $|x + 1| + |x - 3|$, given $x > 5$
59. $\left|\dfrac{x + 7}{|x| + |x - 1|}\right|$, given $0 < x < 1$
60. $\left|\dfrac{x + 3}{\left|x - \frac{1}{2}\right| + \left|x + \frac{1}{2}\right|}\right|$, given $0 < x < 0.2$

In Exercises 61 to 72, find the distance between the points whose coordinates are given.

61. 8, 1
62. $-2, -7$
63. $-3, 5$
64. $-5, 8$
65. $16, -34$
66. $-108, 22$
67. $-38, -5$
68. $\pi, 3$
69. $-\pi, 3$
70. $\frac{1}{7}, -\frac{1}{2}$
71. $\frac{1}{3}, \frac{3}{4}$
72. $0, -8$

In Exercises 73 to 80, use absolute value notation to describe the given expression.

73. Distance between a and 2
74. Distance between b and -7
75. $d(m, n)$
76. $d(p, -8)$
77. The distance between a and 4 is less than z.
78. The distance between z and 5 is greater than 4.
79. The distance between x and -2 is less than 7.
80. The distance between y and -3 is greater than 6.

In Exercises 81 to 84, write interval notation for the given expression.

81. x is a real number and $x \ne 3$.
82. x is a real number whose square is nonnegative.
83. x is a real number whose absolute value is less than 3.
84. x is a real number whose absolute value is greater than 2.

In Exercises 85 to 87, determine whether each statement is true or false.

85. $|x|$ is a positive number.
86. $|-y| = y$
87. If $m < 0$, then $|m| = -m$.
88. For any two different real numbers x and y, the smaller of the two numbers is given by

$$\frac{1}{2}(x + y - |x - y|)$$

Verify the statement given for

a. $x = 5$ and $y = 8$
b. $x = -2$ and $y = 7$
c. $x = -4$ and $y = -7$

89. Prove that the expression in Exercise 88 yields the smaller of the numbers x and y. *Hint:* Evaluate the expression for the two cases

$$x > y \quad \text{and} \quad x < y$$

90. The inequality $|a + b| \le |a| + |b|$ is called the triangle inequality. For what values of a and b does

$$|a + b| = |a| + |b|?$$

SUPPLEMENTAL EXERCISES

In Exercises 91 to 94, use inequalities to describe the given statement.

91. The interest I is not greater than $120.

92. The rent R will be at least $650 a month.

93. The property has an area A that is at least 2 acres but less than 3 acres.

94. The distance D is greater than 7 miles, and it is not more than 8 miles.

In Exercises 95 to 102, use absolute value notation to describe the given statement.

95. x is closer to 2 than it is to 6.

96. x is closer to a than it is to b.

97. x is farther from 3 than it is from -7.

98. x is farther from 0 than it is from 5.

99. x is more than 2 units from 4 but less than 7 units from 4.

100. x is more than b units from a but less than c units from a.

101. x is within δ units of a.

102. x is not equal to a, but it is within δ units of a.

103. Prove the product theorem: $|ab| = |a|\,|b|$

104. Prove the quotient theorem: $\left|\dfrac{a}{b}\right| = \dfrac{|a|}{|b|}$

PROJECTS

1.  **INFINITE SETS** Explain how Georg Cantor (see page 1) was able to prove that the set of irrational numbers has a cardinality that is larger than the set of rational numbers. One source of information is *From Zero to Infinity* by Constance Reid (New York: Thomas Y. Crowell, 1964).

SECTION P.3 INTEGER AND RATIONAL NUMBER EXPONENTS

- PROPERTIES OF EXPONENTS
- SCIENTIFIC NOTATION
- RATIONAL EXPONENTS AND RADICALS
- SIMPLIFY RADICAL EXPRESSIONS

◆ PROPERTIES OF EXPONENTS

A compact method of writing $5 \cdot 5 \cdot 5 \cdot 5$ is 5^4. The expression 5^4 is written in **exponential notation.** Similarly, we can write

$$\frac{2x}{3} \cdot \frac{2x}{3} \cdot \frac{2x}{3} \quad \text{as} \quad \left(\frac{2x}{3}\right)^3$$

Exponential notation can be used to express the product of any expression that is used repeatedly as a factor.

MATH MATTERS

The expression 10^{100} is called a *googol.* The term was coined by the nine-year-old nephew of the American mathematician Edward Kasner. Many calculators do not provide for numbers of this magnitude, but it is no serious loss. To appreciate the magnitude of a googol, consider that if all the atoms in the known universe were counted, the number would not be even close to a googol. But if a googol is too small for you, try 10^{googol} which is called a *googolplex.*

Definition of Natural Number Exponents

If b is any real number and n is any natural number, then

$$b^n = \underbrace{b \cdot b \cdot b \cdot \cdots \cdot b}_{n \text{ factors of } b}$$

In the expression b^n, b is the **base,** n is the **exponent,** and b^n is the **nth power of b.**

For instance,

$$(-5)^4 = (-5)(-5)(-5)(-5) = 625$$
$$-5^4 = -(5 \cdot 5 \cdot 5 \cdot 5) = -625$$

Note the difference between $(-5)^4 = 625$ and $-5^4 = -625$. The parentheses in $(-5)^4$ indicate that the base is -5; however, the expression -5^4 means $-(5^4)$. This time the base is 5.

Definition of b^0

For any nonzero real number b, $b^0 = 1$.

Any nonzero real number raised to the zero power equals 1. For example,

$$7^0 = 1 \qquad \left(\frac{1}{2}\right)^0 = 1 \qquad (-3)^0 = 1 \qquad \pi^0 = 1 \qquad (a^2 + 1)^0 = 1$$

Definition of b^{-n}

If $b \neq 0$ and n is any natural number, then $b^{-n} = \dfrac{1}{b^n}$ and $\dfrac{1}{b^{-n}} = b^n$.

Here are some examples of this definition.

$$3^{-2} = \frac{1}{3^2} = \frac{1}{9} \qquad \frac{1}{4^{-3}} = 4^3 = 64 \qquad \frac{5^{-2}}{7^{-1}} = \frac{7}{5^2} = \frac{7}{25}$$

Restriction Agreement

The expressions 0^0, 0^n where n is a negative integer, and $x/0$ are all undefined expressions. Therefore, all values of variables in this text are restricted to avoid any one of these expressions.

For instance, in the expression

$$\frac{x^0 y^{-3}}{z - 4}$$

we assume that $x \neq 0$, $y \neq 0$, and $z \neq 4$.

Simplifying exponential expressions requires use of the following properties of exponents.

Properties of Exponents

If m, n, and p are integers and a and b are real numbers, then

Product $\quad b^m \cdot b^n = b^{m+n}$

Quotient $\quad \dfrac{b^m}{b^n} = b^{m-n}, \quad b \neq 0$

Power $\quad (b^m)^n = b^{mn} \qquad (a^m b^n)^p = a^{mp} b^{np}$

$$\left(\frac{a^m}{b^n}\right)^p = \frac{a^{mp}}{b^{np}}, \quad b \neq 0$$

Exponential expressions such as a^{b^c} can be confusing. The generally accepted meaning of a^{b^c} is $a^{(b^c)}$. However, some graphing calculators do not evaluate exponential expressions in this way. Enter 2^3^4 in a graphing calculator. If the result is approximately 2.42×10^{24}, then the calculator evaluated $2^{(3^4)}$. If the result is 4096, then the calculator evaluated $(2^3)^4$. To ensure that you calculate the value you intend, we strongly urge you to use parentheses. For instance, entering 2^(3^4) will produce 2.42×10^{24} and entering (2^3)^4 will produce 4096.

To simplify an expression involving exponents, write the expression in a form in which *each base appears at most once* and *no powers of powers or negative exponents appear.*

EXAMPLE 1 Simplify Exponential Expressions

Simplify. a. $\left(\dfrac{2abc^2}{5a^2b}\right)^3$ b. $\dfrac{x^n y^{2n}}{x^{n-1}y^n}$

Solution

a. $\left(\dfrac{2abc^2}{5a^2b}\right)^3 = \left(\dfrac{2c^2}{5a}\right)^3$ • **The quotient property**

$= \dfrac{8c^6}{125a^3}$ • **A power property**

b. $\dfrac{x^n y^{2n}}{x^{n-1}y^n} = x^{n-(n-1)}y^{2n-n}$ • **The quotient property**

$= xy^n$

TRY EXERCISE 24, EXERCISE SET P.3, PAGE 28

◆ Scientific Notation

MATH MATTERS

- Approximately 3.1×10^6 orchid seeds weigh 1 ounce.
- Computer scientists measure an operation in nanoseconds. 1 nanosecond $= 1 \times 10^{-9}$ seconds.
- If a space ship traveled 25,000 mph, it would require approximately 2.7×10^9 years to travel from one end of the universe to the other.

The exponent theorems provide a compact method of writing very large or very small numbers. The method is called *scientific notation.* A number written in **scientific notation** has the form $a \cdot 10^n$, where n is an integer and $1 \le a < 10$. The following procedure is used to change a number from its decimal form to scientific notation.

For numbers greater than 10, move the decimal point to the position to the right of the first digit. The exponent n will equal the number of places the decimal point has been moved. For example,

$$7{,}430{,}000 = 7.43 \times 10^6$$

6 places

For numbers less than 1, move the decimal point to the right of the first nonzero digit. The exponent n will be negative, and its absolute value will equal the number of places the decimal point has been moved. For example,

$$0.00000078 = 7.8 \times 10^{-7}$$

7 places

To change a number from scientific notation to its decimal form, reverse the procedure. That is, if the exponent is positive, move the decimal point to the right the same number of places as the exponent. For example,

$$3.5 \times 10^5 = 350{,}000$$

5 places

If the exponent is negative, move the decimal point to the left the same number of places as the absolute value of the exponent. For example,

$$2.51 \times 10^{-8} = 0.0000000251$$

8 places

Most scientific calculators display very large and very small numbers in scientific notation. The number $450{,}000^2$ is displayed as 2.025 11. This means $450{,}000^2 = 2.025 \times 10^{11}$.

◆ Rational Exponents and Radicals

To this point, the expression b^n has been defined for real numbers b and integers n. Now we wish to extend the definition of exponents to include rational numbers so that expressions such as $2^{1/2}$ will be meaningful. Not just any definition will do. We want a definition of rational exponents for which the properties of integer exponents are true. The following example shows the direction we can take to accomplish our goal.

If the product property for exponential expressions is to hold for rational exponents, then for rational numbers p and q, $b^p b^q = b^{p+q}$. For example,

$$9^{1/2} \cdot 9^{1/2} \quad \text{must equal} \quad 9^{1/2+1/2} = 9^1 = 9$$

Thus $9^{1/2}$ must be a square root of 9. That is, $9^{1/2} = 3$.

The example suggests that $b^{1/n}$ can be defined in terms of roots according to the following definition.

Definition of $b^{1/n}$

If n is an even positive integer and $b \geq 0$, then $b^{1/n}$ is the nonnegative real number such that $(b^{1/n})^n = b$.

If n is an odd positive integer, then $b^{1/n}$ is the real number such that $(b^{1/n})^n = b$.

As examples,

- $25^{1/2} = 5$ because $5^2 = 25$.
- $(-64)^{1/3} = -4$ because $(-4)^3 = -64$.
- $16^{1/2} = 4$ because $4^2 = 16$.
- $-16^{1/2} = -(16^{1/2}) = -4$.
- $(-16)^{1/2}$ is not a real number.
- $(-32)^{1/5} = -2$ because $(-2)^5 = -32$.

If n is an even positive integer and $b < 0$, then $b^{1/n}$ is a *complex number.* Complex numbers are discussed in Chapter 1.

To define expressions such as $8^{2/3}$, we will extend our definition of exponents even further. Because we want the power property $(b^p)^q = b^{pq}$ to be true for rational exponents also, we must have $(b^{1/n})^m = b^{m/n}$. With this in mind, we make the following definition.

Definition of $b^{m/n}$

For all positive integers m and n such that m/n is in simplest form, and for all real numbers b for which $b^{1/n}$ is a real number,

$$b^{m/n} = (b^{1/n})^m = (b^m)^{1/n}$$

take note

Some graphing calculators do not evaluate $b^{m/n}$ when $b < 0$. Try entering **(-8)^(2/3)**. The answer should be 4, but some calculators display an error message for this expression. You can still use your calculator to evaluate this expression, but you must use parentheses. You can enter **((-8)^(1/3))^2** to evaluate $(-8)^{2/3}$.

Because $b^{m/n}$ is defined as $(b^{1/n})^m$ and also as $(b^m)^{1/n}$, we can evaluate expressions such as $8^{4/3}$ in more than one way. For example, because $8^{1/3}$ is a real number, $8^{4/3}$ can be evaluated in either of the following ways:

$$8^{4/3} = (8^{1/3})^4 = 2^4 = 16$$
$$8^{4/3} = (8^4)^{1/3} = 4096^{1/3} = 16$$

Of the two methods, the $b^{m/n} = (b^{1/n})^m$ method is usually easier to apply, provided you can evaluate $b^{1/n}$.

The following exponent properties were stated earlier, but they are restated here to remind you that they have now been extended to apply to rational exponents.

Properties of Rational Exponents

If p, q, and r represent rational numbers and a and b are positive real numbers, then

Product $\quad b^p \cdot b^q = b^{p+q}$

Quotient $\quad \dfrac{b^p}{b^q} = b^{p-q}$

Power $\quad (b^p)^q = b^{pq} \qquad (a^p b^q)^r = a^{pr} b^{qr}$

$$\left(\frac{a^p}{b^q}\right)^r = \frac{a^{pr}}{b^{qr}} \qquad b^{-p} = \frac{1}{b^p}$$

Recall that an exponential expression is in simplest form when no powers of powers or negative exponents appear and each base occurs at most once.

EXAMPLE 2 Simplify Exponential Expressions

Simplify: $\left(\dfrac{x^2y^3}{x^{-3}y^5}\right)^{1/2}$ (Assume $x > 0$, $y > 0$.)

Solution

$$\left(\frac{x^2y^3}{x^{-3}y^5}\right)^{1/2} = (x^{2-(-3)}y^{3-5})^{1/2} = (x^5y^{-2})^{1/2} = (x^{5/2}y^{-1})^{1/2} = \frac{x^{5/2}}{y}$$

TRY EXERCISE 36, EXERCISE SET P.3, PAGE 28

◆ SIMPLIFY RADICAL EXPRESSIONS

MATH MATTERS

The formula for kinetic energy (energy of motion) that is used in Einstein's Theory of Relativity involves a radical.

$$\text{K.E.}_r = mc^2\left[\frac{1}{\sqrt{1-\frac{v^2}{c^2}}} - 1\right]$$

where m is the mass of the object at rest, v is the speed of the object, and c is the speed of light.

Radicals, expressed by the notation $\sqrt[n]{b}$, are also used to denote roots. The number b is the **radicand,** and the positive integer n is the **index** of the radical.

Definition of $\sqrt[n]{b}$

If n is a positive integer and b is a real number such that $b^{1/n}$ is a real number, then $\sqrt[n]{b} = b^{1/n}$.

If the index n equals 2, then the radical $\sqrt[2]{b}$ is written as simply $\sqrt{b}$, and it is referred to as the **principal square root of b** or simply the **square root of b.**

The symbol $\sqrt{b}$ is reserved to represent the nonnegative square root of b. To represent the negative square root of b, write $-\sqrt{b}$. For example, $\sqrt{25} = 5$, whereas $-\sqrt{25} = -5$.

Definition of $(\sqrt[n]{b})^m$

For all positive integers n, all integers m, and all real numbers b such that $\sqrt[n]{b}$ is a real number, then $(\sqrt[n]{b})^m = \sqrt[n]{b^m} = b^{m/n}$.

When $\sqrt[n]{b}$ is a real number, the equations

$$b^{m/n} = \sqrt[n]{b^m} \qquad \text{and} \qquad b^{m/n} = (\sqrt[n]{b})^m$$

can be used to write exponential expressions such as $b^{m/n}$ in radical form. Use the denominator n as the index of the radical and the numerator m as the power of the radicand or as the power of the radical. For example,

$(5xy)^{2/3} = (\sqrt[3]{5xy})^2 = \sqrt[3]{25x^2y^2}$ • **Use the denominator 3 as the index of the radical and the numerator 2 as the power of the radical.**

The equations

$$b^{m/n} = \sqrt[n]{b^m} \qquad \text{and} \qquad b^{m/n} = (\sqrt[n]{b})^m$$

can also be used to write radical expressions in exponential form. For example,

$\sqrt{(2ab)^3} = (2ab)^{3/2}$ • **Use the index 2 as the denominator of the power and the exponent 3 as the numerator of the power.**

The definition of $\sqrt[n]{b^m}$ can often be used to evaluate radical expressions. For instance,

$$(\sqrt[3]{8})^4 = 8^{4/3} = (8^{1/3})^4 = 2^4 = 16$$

Care must be exercised when simplifying even roots (square roots, fourth roots, sixth roots, . . .) of variable expressions. Consider $\sqrt{x^2}$ when $x = 5$ and when $x = -5$.

Case 1 If $x = 5$, then $\sqrt{x^2} = \sqrt{5^2} = \sqrt{25} = 5 = x$.

Case 2 If $x = -5$, then $\sqrt{x^2} = \sqrt{(-5)^2} = \sqrt{25} = 5 = -x$.

These two cases suggest that

$$\sqrt{x^2} = \begin{cases} x, & \text{if } x \geq 0 \\ -x, & \text{if } x < 0 \end{cases}$$

Recalling the definition of absolute value, we can write this more compactly as $\sqrt{x^2} = |x|$.

Simplifying odd roots of a variable expression does not require using the absolute value symbol. Consider $\sqrt[3]{x^3}$ when $x = 5$ and when $x = -5$.

Case 1 If $x = 5$, then $\sqrt[3]{x^3} = \sqrt[3]{5^3} = \sqrt[3]{125} = 5 = x$.

Case 2 If $x = -5$, then $\sqrt[3]{x^3} = \sqrt[3]{(-5)^3} = \sqrt[3]{-125} = -5 = x$.

Thus $\sqrt[3]{x^3} = x$.

Although we have illustrated this principle only for square roots and cube roots, the same reasoning can be applied to other cases. The general result is given below.

Definition of $\sqrt[n]{b^n}$

If n is an even natural number and b is a real number, then

$$\sqrt[n]{b^n} = |b|$$

If n is an odd natural number and b is a real number, then

$$\sqrt[n]{b^n} = b$$

Here are some examples of these properties.

$$\sqrt[4]{16z^4} = 2|z| \qquad \sqrt[5]{32a^5} = 2a$$

Because radicals are defined in terms of rational powers, the properties of radicals are similar to those of exponential expressions.

Properties of Radicals

If m and n are natural numbers and a and b are nonnegative real numbers, then

Product $\quad \sqrt[n]{a} \cdot \sqrt[n]{b} = \sqrt[n]{ab}$

Quotient $\quad \dfrac{\sqrt[n]{a}}{\sqrt[n]{b}} = \sqrt[n]{\dfrac{a}{b}}$

Index $\quad \sqrt[m]{\sqrt[n]{a}} = \sqrt[mn]{a}$

A radical is in **simplest form** if it meets all of the following criteria.

1. The radicand contains only powers less than the index. ($\sqrt{x^5}$ does not satisfy this requirement because 5, the exponent, is greater than 2, the index.)
2. The index of the radical is as small as possible. ($\sqrt[9]{x^3}$ does not satisfy this requirement because $\sqrt[9]{x^3} = x^{3/9} = x^{1/3} = \sqrt[3]{x}$.)
3. The denominator has been rationalized. That is, no radicals appear in the denominator. ($1/\sqrt{2}$ does not satisfy this requirement.)
4. No fractions appear under the radical sign. ($\sqrt[4]{2/x^3}$ does not satisfy this requirement.)

Radical expressions are simplified by using the properties of radicals. Here are some examples.

EXAMPLE 3 Simplify Radical Expressions

Simplify.

a. $\sqrt[4]{32x^3y^4}$ b. $\sqrt[3]{162x^4y^6}$

Solution

a. $\sqrt[4]{32x^3y^4} = \sqrt[4]{2^5x^3y^4} = \sqrt[4]{(2^4y^4) \cdot (2x^3)}$ • **Factor and group factors that can be written as a power of the index.**

$= \sqrt[4]{2^4y^4} \cdot \sqrt[4]{2x^3}$ • **Use the product property of radicals.**

$= 2|y|\sqrt[4]{2x^3}$ • **Recall that for n even, $\sqrt[n]{b^n} = |b|$.**

b. $\sqrt[3]{162x^4y^6} = \sqrt[3]{(2 \cdot 3^4)x^4y^6}$ • **Factor and group factors that can be written as a power of the index.**

$= \sqrt[3]{(3xy^2)^3 \cdot (2 \cdot 3x)}$

$= \sqrt[3]{(3xy^2)^3} \cdot \sqrt[3]{6x}$ • **Use the product property of radicals.**

$= 3xy^2\sqrt[3]{6x}$ • **Recall that for n odd, $\sqrt[n]{b^n} = b$.**

TRY EXERCISE 76, EXERCISE SET P.3, PAGE 28

Like radicals have the same radicand and the same index. For instance,

$$3\sqrt[3]{5xy^2} \quad \text{and} \quad -4\sqrt[3]{5xy^2}$$

are like radicals. Addition and subtraction of like radicals are accomplished by using the distributive property. For example,

$$4\sqrt{3x} - 9\sqrt{3x} = (4 - 9)\sqrt{3x} = -5\sqrt{3x}$$
$$2\sqrt[3]{y^2} + 4\sqrt[3]{y^2} - \sqrt[3]{y^2} = (2 + 4 - 1)\sqrt[3]{y^2} = 5\sqrt[3]{y^2}$$

The sum $2\sqrt{3} + 6\sqrt{5}$ cannot be simplified further because the radicands are not the same. The sum $3\sqrt[3]{x} + 5\sqrt[4]{x}$ cannot be simplified because the indices are not the same.

Sometimes it is possible to simplify radical expressions that do not appear to be like radicals by simplifying each radical expression.

EXAMPLE 4 Combine Radical Expressions

Simplify: $5x\sqrt[3]{16x^4} - \sqrt[3]{128x^7}$

Solution

$5x\sqrt[3]{16x^4} - \sqrt[3]{128x^7}$

$= 5x\sqrt[3]{2^4x^4} - \sqrt[3]{2^7x^7}$ • **Factor.**

$= 5x\sqrt[3]{2^3x^3} \cdot \sqrt[3]{2x} - \sqrt[3]{2^6x^6} \cdot \sqrt[3]{2x}$ • **Group factors that can be written as a power of the index.**

$= 5x(2x\sqrt[3]{2x}) - 2^2x^2 \cdot \sqrt[3]{2x}$ • **Use the product property of radicals.**

$= 10x^2\sqrt[3]{2x} - 4x^2\sqrt[3]{2x}$ • **Simplify.**

$= 6x^2\sqrt[3]{2x}$

TRY EXERCISE 84, EXERCISE SET P.3, PAGE 28

Multiplication of radical expressions is accomplished by using the distributive property. For instance,

$$\sqrt{5}(\sqrt{20} - 3\sqrt{15}) = \sqrt{5}(\sqrt{20}) - \sqrt{5}(3\sqrt{15})$$ • Use the distributive property.

$$= \sqrt{100} - 3\sqrt{75}$$ • Multiply the radicals.

$$= 10 - 3 \cdot 5\sqrt{3}$$ • Simplify.

$$= 10 - 15\sqrt{3}$$

The product of more complicated radical expressions may require repeated use of the distributive property.

EXAMPLE 5 Multiply Radical Expressions

Perform the indicated operation: $(\sqrt{3} + 5)(\sqrt{3} - 2)$

Solution

$$(\sqrt{3} + 5)(\sqrt{3} - 2)$$

$$= (\sqrt{3} + 5)\sqrt{3} - (\sqrt{3} + 5)2$$ • Use the distributive property.

$$= (\sqrt{3}\sqrt{3} + 5\sqrt{3}) - (2\sqrt{3} + 2 \cdot 5)$$ • Use the distributive property.

$$= 3 + 5\sqrt{3} - 2\sqrt{3} - 10$$

$$= -7 + 3\sqrt{3}$$

TRY EXERCISE 92, EXERCISE SET P.3, PAGE 28

To **rationalize the denominator** of a fraction means to write it in an equivalent form that does not involve any radicals in its denominator.

EXAMPLE 6 Rationalize the Denominator

Rationalize the denominator. **a.** $\dfrac{5}{\sqrt[3]{a}}$ **b.** $\sqrt{\dfrac{3}{32y}}$

Solution

a. $$\frac{5}{\sqrt[3]{a}} = \frac{5}{\sqrt[3]{a}} \cdot \frac{\sqrt[3]{a^2}}{\sqrt[3]{a^2}} = \frac{5\sqrt[3]{a^2}}{\sqrt[3]{a^3}} = \frac{5\sqrt[3]{a^2}}{a}$$ • Use $\sqrt[3]{a} \cdot \sqrt[3]{a^2} = \sqrt[3]{a^3} = a$.

b. $$\sqrt{\frac{3}{32y}} = \frac{\sqrt{3}}{\sqrt{32y}} = \frac{\sqrt{3}}{4\sqrt{2y}} = \frac{\sqrt{3}}{4\sqrt{2y}} \cdot \frac{\sqrt{2y}}{\sqrt{2y}} = \frac{\sqrt{6y}}{8y}$$

TRY EXERCISE 104, EXERCISE SET P.3, PAGE 28

To rationalize the denominator of a fractional expression such as

$$\frac{1}{\sqrt{m} + \sqrt{n}}$$

we make use of the conjugate of $\sqrt{m} + \sqrt{n}$, which is $\sqrt{m} - \sqrt{n}$. The product of these conjugate pairs does not involve a radical.

$$(\sqrt{m} + \sqrt{n})(\sqrt{m} - \sqrt{n}) = m - n$$

In Example 7 we use the conjugate of the denominator to rationalize the denominator.

EXAMPLE 7 Rationalize the Denominator

Rationalize the denominator.

a. $\dfrac{2}{\sqrt{3} + \sqrt{2}}$ **b.** $\dfrac{a + \sqrt{5}}{a - \sqrt{5}}$

Solution

a. $\dfrac{2}{\sqrt{3} + \sqrt{2}} = \dfrac{2}{\sqrt{3} + \sqrt{2}} \cdot \dfrac{\sqrt{3} - \sqrt{2}}{\sqrt{3} - \sqrt{2}} = \dfrac{2\sqrt{3} - 2\sqrt{2}}{3 - 2} = 2\sqrt{3} - 2\sqrt{2}$

b. $\dfrac{a + \sqrt{5}}{a - \sqrt{5}} = \dfrac{a + \sqrt{5}}{a - \sqrt{5}} \cdot \dfrac{a + \sqrt{5}}{a + \sqrt{5}} = \dfrac{a^2 + 2a\sqrt{5} + 5}{a^2 - 5}$

TRY EXERCISE 110, EXERCISE SET P.3, PAGE 29

TOPICS FOR DISCUSSION

1. Given that a is a real number, discuss when the expression $a^{p/q}$ represents a real number.

2. The expressions $-a^n$ and $(-a)^n$ do not always represent the same number. Discuss the situations in which the two expressions are equal and those in which they are not equal.

3. Most calculators will automatically convert a number to scientific notation whenever the calculation produces a number that cannot be represented exactly. However, there are limits as to how large or small a number can be and still have a calculator representation. Discuss the limits of your calculator and the practical effects they would have on the most demanding calculations. For instance, can you find the distance (in miles) to the Orion nebula, which is 1600 light-years away? Can your calculator represent the mass of a quark (a subatomic particle)?

4. If you enter the expression for $\sqrt{5}$ on your calculator, the calculator will respond with 2.236067977 or some number close to that. Is this the exact value of $\sqrt{5}$? Is it possible to find the exact decimal value of $\sqrt{5}$ with a calculator? with a computer?

EXERCISE SET P.3

In Exercises 1 to 22, evaluate each expression.

1. -4^4
2. $(-4)^4$
3. $\left(\frac{2^2 \cdot 3^{-5}}{2^{-3} \cdot 5^4}\right)^0$
4. -7^0
5. $\left(\frac{4}{9}\right)^{-2}$
6. $\left(\frac{5^{-3} \cdot 7}{3^{-2}}\right)^{-1}$
7. $4^{3/2}$
8. $16^{3/2}$
9. $-9^{1/2}$
10. $-25^{1/2}$
11. $-64^{2/3}$
12. $-125^{2/3}$
13. $(-64)^{2/3}$
14. $(-125)^{2/3}$
15. $9^{-1/2}$
16. $16^{-1/2}$
17. $27^{-2/3}$
18. $4^{-3/2}$
19. $\left(\frac{9}{16}\right)^{1/2}$
20. $\left(\frac{4}{25}\right)^{3/2}$
21. $\left(\frac{8}{27}\right)^{-2/3}$
22. $\left(\frac{4}{9}\right)^{-3/2}$

In Exercises 23 to 46, simplify each exponential expression.

23. $(2x^2y^3)(3x^5y)$
24. $\left(\frac{2ab^2c^3}{5ab^2}\right)^3$
25. $\frac{(3xy^{-3})^2}{(2xy)^{-2}}$
26. $(2x^{-3}y^0)(3^{-1}xy)^2$
27. $\left(\frac{3x}{y}\right)^{-1}$
28. $(x^2y^{-3})^{-2}$
29. $a^{-1} + b^{-1}$
30. $\frac{4a^2(bc)^{-1}}{(-2)^2a^3b^{-2}c}$
31. $(2ab^{-3})^2(-2a^{-1}b^2)^2$
32. $\left[\left(\frac{b^{-3}}{a^2}\right)^2\left(\frac{a^{-2}}{ab}\right)^{-1}\right]^0$
33. $(81x^4y^{12})^{1/4}$
34. $(625a^8b^4)^{1/4}$
35. $\frac{a^{3/4}b^{1/2}}{a^{1/4}b^{1/5}}$
36. $\frac{x^{1/3}y^{5/6}}{x^{3/2}y^{1/6}}$
37. $a^{1/3}(a^{5/3} + 7a^{2/3})$
38. $m^{3/4}(m^{1/4} - 8m^{5/4})$
39. $(p^{1/2} + q^{1/2})(p^{1/2} - q^{1/2})$
40. $(c + d^{1/3})(c - d^{1/3})$
41. $\left(\frac{m^2n^4}{m^{-2}n}\right)^{1/2}$
42. $\left(\frac{r^3s^{-2}}{rs^4}\right)^{1/2}$
43. $\frac{x^{n+1/2} \cdot x^{-n}}{x^{1/2}}$
44. $\frac{r^{n/2} \cdot r^{2n}}{r^{-n}}$
45. $\frac{r^{1/n}}{r^{1/m}}$
46. $\frac{s^{2/n}}{s^{-2/n}}$

In Exercises 47 to 50, write each number in scientific notation.

47. 21,000,000
48. 163,000,000
49. 0.00095
50. 0.0000000821

In Exercises 51 to 54, change each number from scientific notation to decimal notation.

51. 6.5×10^3
52. 6.86×10^{-9}
53. 2.17×10^{-4}
54. 3.75×10^0

In Exercises 55 to 60, write each exponential expression in radical form.

55. $(3x)^{1/2}$
56. $(6y)^{1/3}$
57. $5(xy)^{1/4}$
58. $2a(bc)^{1/5}$
59. $(5w)^{2/3}$
60. $(a + b)^{3/4}$

In Exercises 61 to 66, write each radical in exponential form.

61. $\sqrt[3]{17k}$
62. $4\sqrt{3m}$
63. $\sqrt[5]{a^2}$
64. $3\sqrt[4]{5n}$
65. $\sqrt{\frac{7a}{3}}$
66. $\sqrt[3]{\frac{5b^2}{7}}$

In Exercises 67 to 78, simplify each radical.

67. $\sqrt{45}$
68. $\sqrt{75}$
69. $\sqrt[3]{24}$
70. $\sqrt[3]{135}$
71. $\sqrt[3]{-81}$
72. $\sqrt[3]{-250}$
73. $-\sqrt[3]{32}$
74. $-\sqrt[3]{243}$
75. $\sqrt{24x^3y^2}$
76. $\sqrt{18x^2y^5}$
77. $-\sqrt[3]{16a^3y^7}$
78. $-\sqrt[3]{54c^2d^5}$

In Exercises 79 to 86, simplify each radical and then combine like radicals.

79. $2\sqrt{32} - 3\sqrt{98}$
80. $5\sqrt[3]{32} + 2\sqrt[3]{108}$
81. $-8\sqrt[4]{48} + 2\sqrt[4]{243}$
82. $2\sqrt[3]{40} - 3\sqrt[3]{135}$
83. $4\sqrt[3]{32y^4} + 3y\sqrt[3]{108y}$
84. $-3x\sqrt[3]{54x^4} + 2\sqrt[3]{16x^7}$
85. $3x\sqrt[3]{8x^3y^4} + 4y\sqrt[3]{64x^6y}$
86. $4\sqrt{a^5b} - a^2\sqrt{ab}$

In Exercises 87 to 96, find the indicated product of the radical expressions. Express each result in simplest form.

87. $(\sqrt{5} + 8)(\sqrt{5} + 3)$
88. $(\sqrt{7} + 4)(\sqrt{7} - 1)$
89. $(\sqrt{2x} + 3)(\sqrt{2x} - 3)$
90. $(7 - \sqrt{3a})(7 + \sqrt{3a})$
91. $(5\sqrt{2y} + \sqrt{3z})^2$
92. $(3\sqrt{5y} - 4)^2$
93. $(\sqrt{x - 3} + 5)^2$
94. $(\sqrt{x + 7} - 3)^2$
95. $(\sqrt{2x + 5} + 7)^2$
96. $(\sqrt{9x - 2} + 11)^2$

In Exercises 97 to 112, simplify each expression by rationalizing the denominator. Write the result in simplest form.

97. $\frac{2}{\sqrt{2}}$
98. $\frac{3x}{\sqrt{3}}$
99. $\sqrt{\frac{5}{18}}$
100. $\sqrt{\frac{7}{40}}$
101. $\frac{3}{\sqrt[3]{2}}$
102. $\frac{2}{\sqrt[3]{4}}$
103. $\frac{4}{\sqrt[3]{8x^2}}$
104. $\frac{2}{\sqrt[4]{4y}}$

105. $\sqrt{\dfrac{10}{18}}$ **106.** $\sqrt{\dfrac{14}{40}}$ **107.** $\sqrt{\dfrac{2x}{27y}}$ **108.** $\sqrt{\dfrac{4c}{50d}}$

109. $\dfrac{3}{\sqrt{5}+\sqrt{x}}$ **110.** $\dfrac{5}{\sqrt{y}-\sqrt{3}}$

111. $\dfrac{\sqrt{7}}{2-\sqrt{7}}$ **112.** $\dfrac{6\sqrt{6}}{5+\sqrt{6}}$

113. **COLOR MONITORS** The number of colors that a computer with a color monitor can display is sometimes indicated in *bits* (a bit is a *b*inary dig*it*). For example, a computer with an 8-bit color interface can display $2^8 = 256$ colors. Determine how many colors each of the following can display.

a. A computer with a 16-bit interface.

b. A computer with a 32-bit interface.

114. **PHYSIOLOGY** It has been estimated that the human eye can detect 36,000 different colors. How many colors (to the nearest 1000) would go undetected by a human using a computer with a 24-bit color interface? (See Exercise 113.)

115. **ASTRONOMY** Pluto is 5.91×10^{12} meters from the Sun. The speed of light is 3.00×10^8 meters per second. Find the time it takes light from the Sun to reach Pluto.

116. **ASTRONOMY** The Earth's mean distance from the Sun is 9.3×10^7 miles. The distance is called the *astronomical unit* (AU). Jupiter is 5.2 AU from the Sun. Find the distance in miles from the Sun to Jupiter.

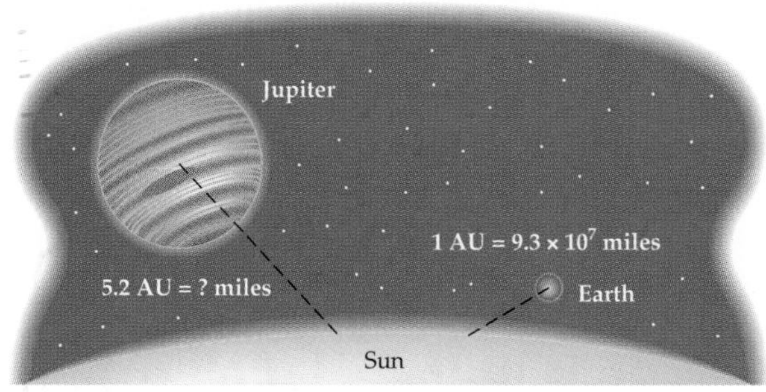

117. **FINANCE** A principal P invested at an annual interest rate r compounded n times per year yields a balance A given by the formula

$$A = P\left(1 + \frac{r}{n}\right)^n$$

Find the balance after 1 year when \$4500 is deposited in an account with an annual interest rate of 8% that is compounded monthly.

118. **FINANCE** You plan to save 1¢ the first day of a month, 2¢ the second day, and 4¢ the third day and to continue this pattern of saving twice what you saved on the previous day for every day in a month that has 30 days.

a. How much money will you need to save on the 30th day?

b. How much money will you have after 30 days?

(*Hint:* Note that after 2 days you will have saved $2^2 - 1 = 3$¢ and that after 3 days you will have saved $2^3 - 1 = 7$¢.)

119. **OPTICS** The percent P of light that will pass through a frosted glass is given by the equation $P = 10^{-kd}$, where d is the thickness of the glass in centimeters and k is a constant that depends on the glass. Find, to the nearest percent, the amount of light that will pass through the frosted glass for which

a. $k = 0.15$, $d = 0.6$ centimeters.

b. $k = 0.15$, $d = 1.2$ centimeters.

120. **FOOD SCIENCE** The number of hours h needed to cook a pot roast that weighs p pounds can be approximated by using the formula $h = 0.9p^{0.6}$.

a. Find the time (to the nearest hundredth of an hour) required to cook a 12-pound pot roast.

b. If pot roast A weighs twice as much as pot roast B, then roast A should be cooked for a period of time that is how many times longer than the time required for roast B to cook?

SUPPLEMENTAL EXERCISES

In Exercises 121 to 124, write each expression as an equivalent expression in which the variables x and y occur only once.

121. $\dfrac{x^n y^{n+2}}{x^{n-3} y}$ **122.** $\left(\dfrac{x^n y^{2n}}{y^{3-n}}\right)^{-2}$

123. $\left(\dfrac{x^{3n} y^{2n}}{x^{-2n} y^{3n+1}}\right)^{-1}$ **124.** $\left(\dfrac{x^{4-n} y^{n+4}}{x y^{n-4}}\right)^2$

In Exercises 125 to 128, find the value of p for which the statement is true.

125. $a^{2/5} a^p = a^2$ **126.** $b^{-3/4} b^{2p} = b^3$

127. $\dfrac{x^{-3/4}}{x^{3p}} = x^4$ **128.** $(x^4 x^{2p})^{1/2} = x$

129. Which is larger, $3^{(3^3)}$ or $(3^3)^3$?

130. If $2^x = y$, then find 2^{x-3} in terms of y.

131. Prove: $\sqrt{a^2 + b^2} \neq a + b$.
(*Hint:* Find a counter-example.)

132. When does $\sqrt[3]{a^3 + b^3} = a + b$?
(*Hint:* Cube each side of the equation.)

In Exercises 133 to 138, rationalize the numerator.

133. $\dfrac{\sqrt{4+h}-2}{h}$

134. $\dfrac{\sqrt{9+h}-3}{h}$

135. $\dfrac{\sqrt{a+h}-\sqrt{a}}{h}$

136. $\dfrac{\sqrt{2x+2h}-\sqrt{2x}}{h}$

137. $\sqrt{n^2+1}-n$ $\left(\textit{Hint: } \sqrt{n^2+1}-n=\dfrac{\sqrt{n^2+1}-n}{1}\right)$

138. $\sqrt{n^2+n}-n$ $\left(\textit{Hint: } \sqrt{n^2+n}-n=\dfrac{\sqrt{n^2+n}-n}{1}\right)$

PROJECTS

1.  **RELATIVITY THEORY** A moving object has energy, called kinetic energy, by virtue of its motion. As mentioned earlier in this chapter, the theory of relativity uses

$$\text{K.E}_r = mc^2\left[\frac{1}{\sqrt{1-\dfrac{v^2}{c^2}}}-1\right]$$

as the formula for kinetic energy. When the speed of an object is much less than the speed of light (3.0×10^8 meters per second) the formula

$$\text{K.E}_n = \frac{1}{2}mv^2$$

is used. In each formula, v is the velocity of the object in meters per second, m is its rest mass in kilograms, and c is the speed of light given above. Calculate the percent error (in **a.** through **e.**) for each of the given velocities. The formula for percent error is

$$\%\text{ error} = \frac{|\text{K.E}_r - K.E_n|}{\text{K.E}_r} \times 100$$

a. $v = 30$ meters per second (speeding car on an expressway)

b. $v = 240$ meters per second (speed of a commercial jet)

c. $v = 3.0 \times 10^7$ meters per second (10% of the speed of light)

d. $v = 1.5 \times 10^8$ meters per second (50% of the speed of light)

e. $v = 2.7 \times 10^8$ meters per second (90% of the speed of light)

f. Use your answers from **a.** through **e.** to give a reason why the formula for kinetic energy given by K.E_n is adequate for most of our common experiences involving motion (walking, running, bicycle, car, plane).

g. According to relativity theory, the mass, m, of an object changes as its velocity according to

$$m = \frac{m_0}{\sqrt{1-\dfrac{v^2}{c_2}}}$$

where m_0 is the rest mass of the object. The approximate rest mass of an electron is 9.11×10^{-31} kilogram. What is the percent change, from its rest mass, in the mass of an electron that is traveling at $0.99c$ (99% of the speed of light)?

h. According to the theory of relativity, a particle (such as an electron or a space craft) cannot exceed the speed of light. Explain why the equation for K.E_r suggests that conclusion.

SECTION P.4 POLYNOMIALS

- OPERATIONS ON POLYNOMIALS
- APPLICATION OF POLYNOMIALS

◆ OPERATIONS ON POLYNOMIALS

A **monomial** is a constant, a variable, or a product of a constant and one or more variables, with the variables having only nonnegative integer exponents. The constant is called the **numerical coefficient** or simply the **coefficient** of the monomial. The **degree of a monomial** is the sum of the exponents of the variables. For example, $-5xy^2$ is a monomial with coefficient -5 and degree 3.

The algebraic expression $3x^{-2}$ is not a monomial because it cannot be written as a product of a constant and a variable with a *nonnegative* integer exponent.

A sum of a finite number of monomials is called a **polynomial.** Each monomial is called a **term** of the polynomial. The **degree of a polynomial** is the largest degree of the terms in the polynomial.

Terms that have exactly the same variables raised to the same powers are called **like terms.** For example, $14x^2$ and $-31x^2$ are like terms; however, $2x^3y$ and $7xy$ are not like terms because x^3y and xy are not identical.

A polynomial is said to be simplified if all its like terms have been combined. For example, the simplified form of $4x^2 + 3x + 5x$ is $4x^2 + 8x$. A simplified polynomial that has two terms is a **binomial,** and a simplified polynomial that has three terms is a **trinomial.** For example, $4x + 7$ is a binomial, and $2x^3 - 7x^2 + 11$ is a trinomial.

A nonzero constant, such as 5, is called a **constant polynomial.** It has degree zero because $5 = 5x^0$. The number 0 is defined to be a polynomial with no degree.

General Form of a Polynomial

The **general form of a polynomial** of degree n in the variable x is

$$a_nx^n + a_{n-1}x^{n-1} + \cdots + a_2x^2 + a_1x + a_0$$

where $a_n \neq 0$ and n is a nonnegative integer. The coefficient a_n is the **leading coefficient,** and a_0 is the **constant term.**

If a polynomial in the variable x is written with decreasing powers of x, then it is in **standard form.** For example, the polynomial

$$3x^2 - 4x^3 + 7x^4 - 1$$

is written in standard form as

$$7x^4 - 4x^3 + 3x^2 - 1$$

The following table shows the leading coefficient, degree, terms, and coefficients of the given polynomials.

Polynomial	Leading Coefficient	Degree	Terms	Coefficients
$9x^2 - x + 5$	9	2	$9x^2, -x, 5$	$9, -1, 5$
$11 - 2x$	-2	1	$-2x, 11$	$-2, 11$
$x^3 + 5x - 3$	1	3	$x^3, 5x, -3$	$1, 5, -3$

To add polynomials, we combine like terms.

EXAMPLE 1 Add Polynomials

Simplify: $(3x^2 + 7x - 5) + (4x^2 - 2x + 1)$

Solution

$$\begin{aligned}(3x^2 + 7x - 5) + (4x^2 - 2x + 1) &= (3x^2 + 4x^2) + (7x - 2x) + [(-5) + 1] \\ &= 7x^2 + 5x - 4\end{aligned}$$

TRY EXERCISE 24, EXERCISE SET P.4, PAGE 36

The **additive inverse of the polynomial** $3x - 7$ is

$$-(3x - 7) = -3x + 7$$

To subtract a polynomial, we add its additive inverse. For example,

$$\begin{aligned}(2x - 5) - (3x - 7) &= (2x - 5) + (-3x + 7) \\ &= [2x + (-3x)] + [(-5) + 7] \\ &= -x + 2\end{aligned}$$

The distributive property is used to find the product of polynomials. For instance, to find the product of $(3x - 4)$ and $(2x^2 + 5x + 1)$, we treat $3x - 4$ as a *single* quantity and *distribute it* over the trinomial $2x^2 + 5x + 1$, as shown in Example 2.

EXAMPLE 2 Multiply Polynomials

Simplify: $(3x - 4)(2x^2 + 5x + 1)$

Solution

$$\begin{aligned}&(3x - 4)(2x^2 + 5x + 1) \\ &\quad = (3x - 4)(2x^2) + (3x - 4)(5x) + (3x - 4)(1) \\ &\quad = (3x)(2x^2) - 4(2x^2) + (3x)(5x) - 4(5x) + (3x)(1) - 4(1) \\ &\quad = 6x^3 - 8x^2 + 15x^2 - 20x + 3x - 4 \\ &\quad = 6x^3 + 7x^2 - 17x - 4\end{aligned}$$

TRY EXERCISE 32, EXERCISE SET P.4, PAGE 36

In the following calculation, a vertical format has been used to find the product of $(x^2 + 6x - 7)$ and $(5x - 2)$. Note that like terms are arranged in the same vertical column.

$$
\begin{array}{r}
x^2 + 6x - 7 \\
5x - 2 \\
\hline
-2x^2 - 12x + 14 \\
5x^3 + 30x^2 - 35x \quad\;\; \\
\hline
5x^3 + 28x^2 - 47x + 14
\end{array}
$$

If the terms of the binomials $(a + b)$ and $(c + d)$ are labeled as shown below, then the product of the two binomials can be computed mentally by the **FOIL method.**

First, Last, Inner, Outer

$$(a + b) \cdot (c + d) = \underset{\text{First}}{ac} + \underset{\text{Outer}}{ad} + \underset{\text{Inner}}{bc} + \underset{\text{Last}}{bd}$$

In the following illustration, we find the product of $(7x - 2)$ and $(5x + 4)$ by the FOIL method.

$$
\begin{aligned}
(7x - 2)(5x + 4) &= \underset{\text{First}}{(7x)(5x)} + \underset{\text{Outer}}{(7x)(4)} + \underset{\text{Inner}}{(-2)(5x)} + \underset{\text{Last}}{(-2)(4)} \\
&= 35x^2 + 28x - 10x - 8 \\
&= 35x^2 + 18x - 8
\end{aligned}
$$

Certain products occur so frequently in algebra that they deserve special attention.

Special Product Formulas

Special Form	Formula(s)
(Sum)(Difference)	$(x + y)(x - y) = x^2 - y^2$
(Binomial)^2	$(x + y)^2 = x^2 + 2xy + y^2$ $(x - y)^2 = x^2 - 2xy + y^2$

The variables x and y in these special product formulas can be replaced by other algebraic expressions, as shown in Example 3.

EXAMPLE 3 Use the Special Product Formulas

Find each special product. **a.** $(7x + 10)(7x - 10)$ **b.** $(2y^2 + 11z)^2$

Solution

a. $(7x + 10)(7x - 10) = (7x)^2 - (10)^2 = 49x^2 - 100$

b. $(2y^2 + 11z)^2 = (2y^2)^2 + 2[(2y^2)(11z)] + (11z)^2 = 4y^4 + 44y^2z + 121z^2$

TRY EXERCISE 56, EXERCISE SET P.4, PAGE 37

Many application problems require you to *evaluate polynomials*. To **evaluate a polynomial,** substitute the given value(s) for the variable(s) and then perform the indicated operations using the **Order of Operations Agreement.**

The Order of Operations Agreement

If grouping symbols are present, evaluate by performing the operations within the grouping symbols, innermost grouping symbol first, while observing the order given in steps 1 to 3.

1. First, evaluate each power.
2. Next, do all multiplications and divisions, working from left to right.
3. Last, do all additions and subtractions, working from left to right.

EXAMPLE 4 Evaluate a Polynomial

Evaluate the polynomial $2x^3 - 6x^2 + 7$ for $x = -4$.

Solution

$2x^3 - 6x^2 + 7$

$2(-4)^3 - 6(-4)^2 + 7 = 2(-64) - 6(16) + 7$ • **Substitute -4 for x. Evaluate the powers.**

$= -128 - 96 + 7$ • **Perform the multiplications.**

$= -217$ • **Perform the additions and subtractions.**

TRY EXERCISE 66, EXERCISE SET P.4, PAGE 37

◆ APPLICATION OF POLYNOMIALS

EXAMPLE 5 Solve an Application

The number of singles tennis matches that can be played between n tennis players is given by the polynomial $\frac{1}{2}n^2 - \frac{1}{2}n$. Find the number of singles tennis matches that can be played among 4 tennis players.

Figure P.11
4 tennis players can play a total of 6 singles matches.

Solution

$$\frac{1}{2}n^2 - \frac{1}{2}n$$

$$\frac{1}{2}(4)^2 - \frac{1}{2}(4) = \frac{1}{2}(16) - \frac{1}{2}(4) = 8 - 2 = 6$$ • **Substitute 4 for *n*. Then simplify.**

Therefore, 4 tennis players can play a total of 6 singles matches. See **Figure P.11.**

TRY EXERCISE 76, EXERCISE SET P.4, PAGE 37

MATH MATTERS

The procedure used by the computer to determine whether a number is a prime or a composite is a *polynomial time algorithm,* because the time required can be estimated via a polynomial. The procedure used to factor a number is an *exponential time algorithm.* In the field of *computational complexity,* it is important to distinguish between polynomial time algorithms and exponential time algorithms. Example 6 illustrates that the polynomial time algorithm can be run in about 2 seconds, whereas the exponential time algorithm requires about 44 minutes!

EXAMPLE 6 Solve an Application

A scientist determines that the average time in seconds that it takes a particular computer to determine whether an n-digit natural number is prime or composite is given by

$$0.002n^2 + 0.002n + 0.009, \quad 20 \le n \le 40$$

The average time in seconds that it takes the computer to factor an n-digit number is given by

$$0.00032(1.7)^n, \quad 20 \le n \le 40$$

Estimate the average time it takes the computer to

a. determine whether a 30-digit number is a prime or a composite.

b. factor a 30-digit number.

Solution

a. $0.002n^2 + 0.002n + 0.009$

$0.002(30)^2 + 0.002(30) + 0.009 = 1.8 + 0.06 + 0.009 = 1.869 \approx 2$ seconds

b. $0.00032(1.7)^n$

$0.0032(1.7)^{30} \approx 0.00032(8{,}193{,}465.726)$

≈ 2600 seconds

TRY EXERCISE 78, EXERCISE SET P.4, PAGE 37

TOPICS FOR DISCUSSION

1. Discuss the definition of the term *polynomial.* Give some examples of expressions that are polynomials and some examples of expressions that are not polynomials.

2. Suppose that P and Q are both polynomials of degree n. Discuss the degrees of $P + Q$, $P - Q$, PQ, $P + P$, and $P - P$.

3. Suppose that you evaluate a polynomial P of degree n for larger and larger values of x (for instance, when $x = 1, 2, 3, 4, \ldots$). Discuss whether the value of the polynomial would eventually (for very large values of x) continually increase, decrease, or fluctuate between increasing and decreasing.

4. Discuss the similarities and differences among monomials, binomials, trinomials, and polynomials.

EXERCISE SET P.4

In Exercises 1 to 10, match the descriptions, labeled A, B, C, …., J, with the appropriate examples.

A. $x^3y + xy$ B. $7x^2 + 5x - 11$

C. $\frac{1}{2}x^2 + xy + y^2$ D. $4xy$

E. $8x^3 - 1$ F. $3 - 4x^2$

G. 8 H. $3x^5 - 4x^2 + 7x - 11$

I. $8x^4 - \sqrt{5}x^3 + 7$ J. 0

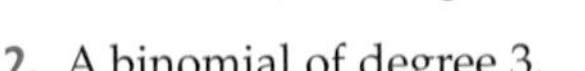

1. A monomial of degree 2.
2. A binomial of degree 3.
3. A polynomial of degree 5.
4. A binomial with leading coefficient of -4.
5. A zero-degree polynomial.

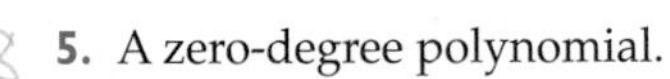

6. A fourth-degree polynomial that has a third-degree term.
7. A trinomial with integer coefficients.
8. A trinomial in x and y.
9. A polynomial with no degree.
10. A fourth-degree binomial.

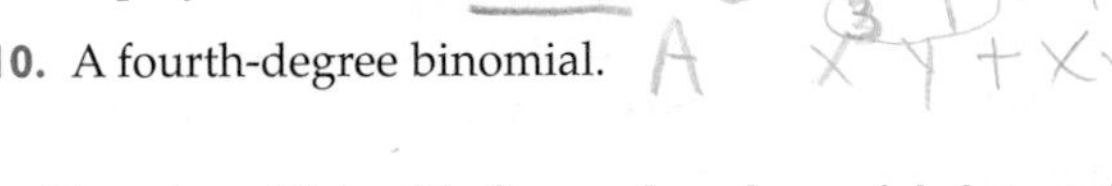

In Exercises 11 to 16, for each polynomial determine its a. standard form, b. degree, c. coefficients, d. leading coefficient, e. terms.

11. $2x + x^2 - 7$
12. $-3x^2 - 11 - 12x^4$
13. $x^3 - 1$
14. $4x^2 - 2x + 7$
15. $2x^4 + 3x^3 + 5 + 4x^2$
16. $3x^2 - 5x^3 + 7x - 1$

In Exercises 17 to 22, determine the degree of the given polynomial.

17. $3xy^2 - 2xy + 7x$
18. $x^3 + 3x^2y + 3xy^2 + y^3$
19. $4x^2y^2 - 5x^3y^2 + 17xy^3$
20. $-9x^5y + 10xy^4 - 11x^2y^2$
21. xy
22. $5x^2y - y^4 + 6xy$

In Exercises 23 to 34, perform the indicated operations and simplify if possible by combining like terms. Write the result in standard form.

23. $(3x^2 + 4x + 5) + (2x^2 + 7x - 2)$
24. $(5y^2 - 7y + 3) + (2y^2 + 8y + 1)$
25. $(4w^3 - 2w + 7) + (5w^3 + 8w^2 - 1)$
26. $(5x^4 - 3x^2 + 9) + (3x^3 - 2x^2 - 7x + 3)$
27. $(r^2 - 2r - 5) - (3r^2 - 5r + 7)$
28. $(7s^2 - 4s + 11) - (-2s^2 + 11s - 9)$
29. $(u^3 - 3u^2 - 4u + 8) - (u^3 - 2u + 4)$
30. $(5v^4 - 3v^2 + 9) - (6v^4 + 11v^2 - 10)$
31. $(4x - 5)(2x^2 + 7x - 8)$
32. $(5x - 7)(3x^2 - 8x - 5)$
33. $(3x^2 - 2x + 5)(2x^2 - 5x + 2)$
34. $(2y^3 - 3y + 4)(2y^2 - 5y + 7)$

In Exercises 35 to 48, use the FOIL method to find the indicated product.

35. $(2x + 4)(5x + 1)$
36. $(5x - 3)(2x + 7)$
37. $(y + 2)(y + 1)$
38. $(y + 5)(y + 3)$
39. $(4z - 3)(z - 4)$
40. $(5z - 6)(z - 1)$
41. $(a + 6)(a - 3)$
42. $(a - 10)(a + 4)$
43. $(5x - 11y)(2x - 7y)$
44. $(3a - 5b)(4a - 7b)$
45. $(9x + 5y)(2x + 5y)$
46. $(3x - 7z)(5x - 7z)$
47. $(3p + 5q)(2p - 7q)$
48. $(2r - 11s)(5r + 8s)$

In Exercises 49 to 54, perform the indicated operations and simplify.

49. $(4d - 1)^2 - (2d - 3)^2$
50. $(5c - 8)^2 - (2c - 5)^2$
51. $(r + s)(r^2 - rs + s^2)$
52. $(r - s)(r^2 + rs + s^2)$

53. $(3c - 2)(4c + 1)(5c - 2)$

54. $(4d - 5)(2d - 1)(3d - 4)$

In Exercises 55 to 62, use the special product formulas to perform the indicated operation.

55. $(3x + 5)(3x - 5)$

56. $(4x^2 - 3y)(4x^2 + 3y)$

57. $(3x^2 - y)^2$

58. $(6x + 7y)^2$

59. $(4w + z)^2$

60. $(3x - 5y^2)^2$

61. $[(x + 5) + y][(x + 5) - y]$

62. $[(x - 2y) + 7][(x - 2y) - 7]$

In Exercises 63 to 70, evaluate the given polynomial for the indicated value of the variable.

63. $x^2 + 7x - 1$, for $x = 3$

64. $x^2 - 8x + 2$, for $x = 4$

65. $-x^2 + 5x - 3$, for $x = -2$

66. $-x^2 - 5x + 4$, for $x = -5$

67. $3x^3 - 2x^2 - x + 3$, for $x = -1$

68. $5x^3 - x^2 + 5x - 3$, for $x = -1$

69. $1 - x^5$, for $x = -2$

70. $1 - x^3 - x^5$, for $x = 2$

71. **RECREATION** The air resistance (in pounds) on a cyclist riding a bicycle in an upright position can be given by $0.016v^2$, where v is the speed of the cyclist in miles per hour. Find the air resistance on a cyclist when

a. $v = 10$ mph **b.** $v = 15$ mph

72. **HIGHWAY ENGINEERING** On an expressway, the recommended *safe distance* between cars in feet is given by $0.015v^2 + v + 10$, where v is the speed of the car in miles per hour. Find the safe distance when

a. $v = 30$ mph **b.** $v = 55$ mph

73. **GEOMETRY** The volume of a right circular cylinder (as shown below) is given by $\pi r^2 h$, where r is the radius of the base and h is the height of the cylinder. Find the volume when

a. $r = 3$ inches, $h = 8$ inches

b. $r = 5$ cm, $h = 12$ cm

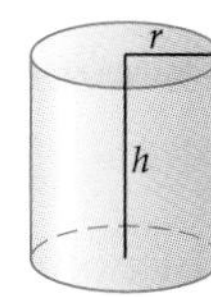

74. **AUTOMOTIVE ENGINEERING** The fuel efficiency (in miles per gallon of gas) of a car is given by the expression $-0.02v^2 + 1.5v + 2$, where v is the speed of the car in miles per hour. Find the fuel efficiency when

a. $v = 45$ mph **b.** $v = 60$ mph

75. **PSYCHOLOGY** Based on data from one experiment, the reaction time, in hundredths of a second, of a person to visual stimulus varies according to age and is given by $0.005x^2 - 0.32x + 12$, where x is the age of the person. Find the reaction time to the stimulus for the person when

a. $x = 20$ years old **b.** $x = 50$ years old

76. The number of committees consisting of exactly 3 people that can be formed from a group of n people is given by the polynomial

$$\frac{1}{6}n^3 - \frac{1}{2}n^2 + \frac{1}{3}n$$

Find the number of committees consisting of exactly 3 people that can be formed from a group of 21 people.

77. Find the number of chess matches that can be played between the members of a group of 150 people. Use the formula from Example 5.

78. **COMPUTER SCIENCE** A computer scientist determines that the time in seconds it takes a particular computer to calculate n digits of π is given by the polynomial

$$4.3 \times 10^{-6}n^2 - 2.1 \times 10^{-4}n$$

where $1000 \le n \le 10{,}000$. Estimate the time it takes the computer to calculate π to

a. 1000 digits **b.** 5000 digits **c.** 10,000 digits

79. **COMPUTER SCIENCE** If n is a positive integer, then $n!$, which is read "n factorial," is given by

$$n(n - 1)(n - 2) \cdots 2 \cdot 1$$

For example, $4! = 4 \cdot 3 \cdot 2 \cdot 1 = 24$. A computer scientist determines that each time a program is run on a particular computer, the time in seconds required to compute $n!$ is given by the polynomial

$$1.9 \times 10^{-6}n^2 - 3.9 \times 10^{-3}n$$

where $1000 \le n \le 10{,}000$. Using this polynomial, estimate the time it takes this computer to calculate 4000! and 8000!.

SUPPLEMENTAL EXERCISES

The following special product formulas can be used to find the cube of a binomial.

$$(x + y)^3 = x^3 + 3x^2y + 3xy^2 + y^3$$
$$(x - y)^3 = x^3 - 3x^2y + 3xy^2 - y^3$$

In Exercises 80 to 85, make use of the above special product formulas to find the indicated products.

80. $(a + b)^3$

81. $(a - b)^3$

82. $(x - 1)^3$

83. $(y + 2)^3$

84. $(2x - 3y)^3$

85. $(3x + 5y)^3$

PROJECTS

1. **ODD NUMBERS** Every odd number can be written in the form $2n - 1$, and every even number can be expressed as $2n$, where n is a natural number. Explain, by writing a few paragraphs and giving the supporting mathematics, why the product of two odd numbers is an odd number, the product of two even numbers is an even number, and the product of an even number and an odd number is an even number.

2. **PRIME NUMBERS** Fermat's Little Theorem states, "If n is a prime number and a is *any* natural number, then $a^n - a$ is divisible by n." For instance, for $n = 11$ and $a = 14$, $\frac{14^{11} - 14}{11} = 368{,}142{,}288{,}150$. The important aspect of this theorem is that no matter what natural number is chosen for a, $a^{11} - a$ is evenly divisible by 11.

Knowing whether a number is prime plays a central role in the security of computer systems. A restatement (called the *contrapositive*) of Fermat's Little Theorem as "If n is a number and a is some number for which $a^n - a$ is *not* divisible by n, then n is *not* a prime number" is used to determine when a number is *not* prime. For example, if $n = 14$, then $\frac{2^{14} - 2}{14} = \frac{8191}{7}$, and thus there is some number ($a = 2$) for which $2^{14} - 2$ is not evenly divisible by 14. Therefore, 14 is not prime.

a. Explain the meaning of the *contrapositive* (used above) of a theorem. Use your explanation to write the contrapositive of "If two triangles are congruent, then they are similar."

b. $7^{14} - 7$ is divisible by 14. Explain why this does not contradict the fact that 14 is not a prime.

c. Explain the meaning of the *converse* of a theorem. State the converse of Fermat's Little Theorem.

d. The number 561 has the property that $a^{561} - a$ is divisible by 561 for all natural numbers a. Can you use Fermat's Little Theorem to conclude that 561 is a prime number? Explain.

e. Suppose that $a^n - a$ is divisible by n for all values of a. Can you conclude that n is a prime number? Explain your answer.

f. Find a definition of a Carmichael number. What do Carmichael numbers have to do with the information in part **e.**?

SECTION P.5 FACTORING

- GREATEST COMMON FACTOR
- FACTORING TRINOMIALS
- SPECIAL FACTORING
- FACTOR BY GROUPING
- GENERAL FACTORING

PB

Writing a polynomial as a product of polynomials of lower degree is called **factoring.** Factoring is an important procedure that is often used to simplify fractional expressions and to solve equations.

In this section we consider only the factorization of polynomials that have integer coefficients. Also, we are concerned only with **factoring over the integers.** That is, we search only for polynomial factors that have integer coefficients.

◆ GREATEST COMMON FACTOR

The first step in any factorization of a polynomial is to use the distributive property to factor out the **greatest common factor (GCF)** of the terms of the polynomial. Given two or more exponential expressions with the same prime number base or the same variable base, the GCF is the exponential expression with the smallest exponent. For example,

2^3 is the GCF of 2^3, 2^5, and 2^8, and a is the GCF of a^4 and a.

The GCF of two or more monomials is the product of the GCFs of all of the *common* bases. For example, to find the GCF of $27a^3b^4$ and $18b^3c$, factor the coefficients into prime factors and then write each common base with its smallest exponent.

$$27a^3b^4 = 3^3 \cdot a^3 \cdot b^4 \qquad 18b^3c = 2 \cdot 3^2 \cdot b^3 \cdot c$$

The only common bases are 3 and b. The product of these common bases with their smallest exponents is 3^2b^3. The GCF of $27a^3b^4$ and $18b^3c$ is $9b^3$.

The expressions $3x(2x + 5)$ and $4(2x + 5)$ have a common *binomial* factor which is $2x + 5$. Thus, the GCF of $3x(2x + 5)$ and $4(2x + 5)$ is $2x + 5$.

EXAMPLE 1 Factor Out the Greatest Common Factor

Factor out the GCF.

a. $10x^3 + 6x$ **b.** $15x^{2n} + 9x^{n+1} - 3x^n$ (where n is a positive integer)
c. $(m + 5)(x + 3) + (m + 5)(x - 10)$

Solution

a. $10x^3 + 6x = (2x)(5x^2) + (2x)(3)$ • **The GCF is 2x.**

$= 2x(5x^2 + 3)$ • **Factor out the GCF.**

b. $15x^{2n} + 9x^{n+1} - 3x^n$

$= (3x^n)(5x^n) + (3x^n)(3x) - (3x^n)(1)$ • **The GCF is $3x^n$.**

$= 3x^n(5x^n + 3x - 1)$ • **Factor out the GCF.**

c. $(m + 5)(x + 3) + (m + 5)(x - 10)$

$= (m + 5)[(x + 3) + (x - 10)]$ • **Use the distributive property to factor out (m + 5).**

$= (m + 5)(2x - 7)$ • **Simplify.**

TRY EXERCISE 6, EXERCISE SET P.5, PAGE 47

◆ FACTORING TRINOMIALS

Some trinomials of the form $x^2 + bx + c$ can be factored by a trial procedure. This method makes use of the FOIL method in reverse. For example, consider the following products:

$$(x + 3)(x + 5) = x^2 + 5x + 3x + (3)(5) = x^2 + 8x + 15$$
$$(x - 2)(x - 7) = x^2 - 7x - 2x + (-2)(-7) = x^2 - 9x + 14$$
$$(x + 4)(x - 9) = x^2 - 9x + 4x + (4)(-9) = x^2 - 5x - 36$$

The coefficient of x is the sum of the constant terms of the binomials.

The constant term of the trinomial is the product of the constant terms of the binomials.

Points to Remember to Factor $x^2 + bx + c$

1. The constant term c of the trinomial is the product of the constant terms of the binomials.
2. The coefficient b in the trinomial is the sum of the constant terms of the binomials.
3. If the constant term c of the trinomial is positive, the constant terms of the binomials have the same sign as the coefficient b of the trinomial.
4. If the constant term c of the trinomial is negative, the constant terms of the binomials have opposite signs.

EXAMPLE 2 Factor a Trinomial of the Form $x^2 + bx + c$

Factor: $x^2 + 7x - 18$

Solution

We must find two binomials whose first terms have a product of x^2 and whose last terms have a product of -18; also, the sum of the product of the outer terms and the product of the inner terms must be $7x$. Begin by listing the possible integer factorizations of -18.

Factors of −18	Sum of the Factors
$1 \cdot (-18)$	$1 + (-18) = -17$
$(-1) \cdot 18$	$(-1) + 18 = 17$
$2 \cdot (-9)$	$2 + (-9) = -7$
$(-2) \cdot 9$	$(-2) + 9 = 7$

• **Stop. This is the desired sum.**

Thus -2 and 9 are the numbers whose sum is 7 and whose product is -18. Therefore,

$$x^2 + 7x - 18 = (x - 2)(x + 9)$$

multiply for last # add for middle

The FOIL method can be used to verify that the factorization is correct.

TRY EXERCISE 12, EXERCISE SET P.5, PAGE 47

The trial method can sometimes be used to factor trinomials of the form $ax^2 + bx + c$, which do not have a leading coefficient of 1. We use the factors of a and c to form trial binomial factors. Factoring trinomials of this type may require testing many factors. To reduce the number of trial factors, make use of the following points.

Points to Remember to Factor $ax^2 + bx + c$, $a > 0$

1. If the constant term of the trinomial is positive, the constant terms of the binomials have the same sign as the coefficient b in the trinomial.
2. If the constant term of the trinomial is negative, the constant terms of the binomials have opposite signs.
3. If the terms of the trinomial do not have a common factor, then neither binomial will have a common factor.

EXAMPLE 3 Factor a Trinomial of the Form $ax^2 + bx + c$

Factor: $6x^2 - 11x + 4$

Solution

Because the constant term of the trinomial is positive and the coefficient of the x term is negative, the constant terms of the binomials will both be negative. This time we find factors of the first term as well as factors of the constant term.

Factors of $6x^2$	Factors of 4 (both negative)
x, $6x$	-1, -4
$2x$, $3x$	-2, -2

Use these factors to write trial factors. Use the FOIL method to see whether any of the trial factors produce the correct middle term. If the terms of a

Continued

trinomial do not have a common factor, then a binomial factor cannot have a common factor (point 3). Such trial factors need not be checked.

Trial Factors	Middle Term	
$(x - 1)(6x - 4)$	Common factor	• **6x and 4 have a common factor.**
$(x - 4)(6x - 1)$	$-1x - 24x = -25x$	
$(x - 2)(6x - 2)$	Common factor	• **6x and 2 have a common factor.**
$(2x - 1)(3x - 4)$	$-8x - 3x = -11x$	• **This is the correct middle term.**

Thus $6x^2 - 11x + 4 = (2x - 1)(3x - 4)$.

TRY EXERCISE 16, EXERCISE SET P.5, PAGE 47

Sometimes it is impossible to factor a polynomial into the product of two polynomials having integer coefficients. Such polynomials are said to be **nonfactorable over the integers.** For example, $x^2 + 3x + 7$ is nonfactorable over the integers because there are no integers whose product is 7 and whose sum or difference is 3.

If you have difficulty factoring a trinomial, you may wish to use the following theorem. It will indicate whether the trinomial is factorable over the integers.

Factorization Theorem

The trinomial $ax^2 + bx + c$, with integer coefficients a, b, and c, can be factored as the product of two binomials with integer coefficients if and only if $b^2 - 4ac$ is a perfect square.

EXAMPLE 4 Apply the Factorization Theorem

Determine whether each trinomial is factorable over the integers.

a. $4x^2 + 8x - 7$ **b.** $6x^2 - 5x - 4$

Solution

a. The coefficients of $4x^2 + 8x - 7$ are $a = 4$, $b = 8$, and $c = -7$. Applying the factorization theorem yields

$$b^2 - 4ac = 8^2 - 4(4)(-7) = 176$$

Because 176 is not a perfect square, the trinomial is nonfactorable over the integers.

b. The coefficients of $6x^2 - 5x - 4$ are $a = 6$, $b = -5$, and $c = -4$. Thus

$$b^2 - 4ac = (-5)^2 - 4(6)(-4) = 121$$

Because 121 is a perfect square, the trinomial is factorable over the integers. Using the methods we have developed, we find

$$6x^2 - 5x - 4 = (3x - 4)(2x + 1)$$

TRY EXERCISE 24, EXERCISE SET P.5, PAGE 47

◆ SPECIAL FACTORING

Some polynomials of degree greater than 2 can be factored by the trial procedure. Consider $2x^6 + 9x^3 + 9$. Because all the signs of the trinomial are positive, the coefficients of all the terms in the binomial factors must be positive.

Factors of $2x^6$	Factors of 9 (both positive)
$x^3, 2x^3$	1, 9 3, 3

The factors $(x^3 + 3)$ and $(2x^3 + 3)$ are the only trial factors whose product has the correct middle term, $9x^3$. Thus $2x^6 + 9x^3 + 9 = (x^3 + 3)(2x^3 + 3)$.

Some polynomials can be factored by making use of the following factoring formulas.

Factoring Formulas

Special Form	Formula(s)
Difference of two squares	$x^2 - y^2 = (x + y)(x - y)$
Perfect-square trinomials	$x^2 + 2xy + y^2 = (x + y)^2$ $x^2 - 2xy + y^2 = (x - y)^2$
Sum of cubes	$x^3 + y^3 = (x + y)(x^2 - xy + y^2)$
Difference of cubes	$x^3 - y^3 = (x - y)(x^2 + xy + y^2)$

take note

The polynomial $x^2 + y^2$ is the sum of two squares. You may be tempted to factor it in a manner similar to the method used on the difference of two squares; however, $x^2 + y^2$ is nonfactorable over the integers.

The monomial a^2 is a square of a, and a is called a **square root** of a^2. The factoring formula

$$x^2 - y^2 = (x + y)(x - y)$$

indicates that the **difference of two squares** can be written as the product of the sum and the difference of the square roots of the squares.

EXAMPLE 5 Factor the Difference of Squares

Factor: $49x^2 - 144$

Solution

$49x^2 - 144 = (7x)^2 - (12)^2$

• Recognize the difference-of-squares form.

$= (7x + 12)(7x - 12)$

• The binomial factors are the sum and the difference of the square roots of the squares.

TRY EXERCISE 32, EXERCISE SET P.5, PAGE 48

A **perfect-square trinomial** is a trinomial that is the square of a binomial. For example, $x^2 + 6x + 9$ is a perfect-square trinomial because

$$(x + 3)^2 = x^2 + 6x + 9$$

Every perfect-square trinomial can be factored by the trial method, but it generally is faster to factor perfect-square trinomials by using the factoring formulas.

EXAMPLE 6 Factor a Perfect-Square Trinomial

Factor: $16m^2 - 40mn + 25n^2$

Solution

$16m^2 - 40mn + 25n^2$

$= (4m)^2 - 2(4m)(5n) + (5n)^2$

• Recognize the perfect-square trinomial form.

$= (4m - 5n)^2$

TRY EXERCISE 42, EXERCISE SET P.5, PAGE 48

The product of the same three factors is called a **cube.** For example, $8a^3$ is a cube because $8a^3 = (2a)^3$. The **cube root** of a cube is one of the three equal factors. To factor the sum or the difference of two cubes, use the appropriate factoring formula. It helps to use the following patterns, which involve the signs of the terms.

Same sign

$$x^3 + y^3 = (x + y)(x^2 - xy + y^2)$$

Opposite signs

Same sign

$$x^3 - y^3 = (x - y)(x^2 + xy + y^2)$$

Opposite signs

In the factorization of the sum or difference of two cubes, the terms of the binomial factor are the cube roots of the cubes. For example,

$$8a^3 - 27b^3 = (2a)^3 - (3b)^3 = (2a - 3b)(4a^2 + 6ab + 9b^2)$$

EXAMPLE 7 Factor the Sum or Difference of Cubes

Factor: **a.** $8a^3 + b^3$ **b.** $a^3 - 64$

Solution

a. $8a^3 + b^3 = (2a)^3 + b^3$ • **Recognize the sum-of-cubes form.**

$= (2a + b)(4a^2 - 2ab + b^2)$ • **Factor**

b. $a^3 - 64 = a^3 - 4^3$ • **Recognize the difference-of-cubes form.**

$= (a - 4)(a^2 + 4a + 16)$ • **Factor**

TRY EXERCISE 48, EXERCISE SET P.5, PAGE 48

◆ FACTOR BY GROUPING

> **take note**
>
> $-a + b = -(a - b)$. Thus,
>
> $-4y + 14 = -(4y - 14)$.

Some polynomials can be **factored by grouping.** Pairs of terms that have a common factor are first grouped together. The process makes repeated use of the distributive property, as shown in the following factorization of $6y^3 - 21y^2 - 4y + 14$.

$6y^3 - 21y^2 - 4y + 14$

$= (6y^3 - 21y^2) - (4y - 14)$ • **Group the first two terms and the last two terms.**

$= 3y^2(2y - 7) - 2(2y - 7)$ • **Factor out the GCF from each of the groups.**

$= (2y - 7)(3y^2 - 2)$ • **Factor out the common binomial factor.**

When you factor by grouping, some experimentation may be necessary to find a grouping that is of the form of one of the special factoring formulas.

EXAMPLE 8 Factor by Grouping

Factor by grouping. **a.** $a^2 + 10ab + 25b^2 - c^2$ **b.** $p^2 + p - q - q^2$

Solution

a. $a^2 + 10ab + 25b^2 - c^2$

$= (a^2 + 10ab + 25b^2) - c^2$ • **Group the terms of the perfect-square trinomial.**

$= (a + 5b)^2 - c^2$ • **Factor the trinomial.**

$= [(a + 5b) + c][(a + 5b) - c]$ • **Factor the difference of squares.**

$= (a + 5b + c)(a + 5b - c)$ • **Simplify.**

Continued •►

b. $p^2 + p - q - q^2$

$= p^2 - q^2 + p - q$ • **Rearrange the terms.**

$= (p^2 - q^2) + (p - q)$ • **Regroup.**

$= (p + q)(p - q) + (p - q)$ • **Factor the difference of squares.**

$= (p - q)(p + q + 1)$ • **Factor out the common factor $(p - q)$.**

TRY EXERCISE 58, EXERCISE SET P.5, PAGE 48

◆ GENERAL FACTORING

Here is a general factoring strategy for polynomials:

General Factoring Strategy

1. Factor out the GCF of all terms.
2. Try to factor a binomial as
   a. the difference of two squares.
   b. the sum or difference of two cubes.
3. Try to factor a trinomial
   a. as a perfect-square trinomial.
   b. using the trial method
4. Try to factor a polynomial with more than three terms by grouping.
5. After each factorization, examine the new factors to see whether they can be factored.

EXAMPLE 9 Factor Using the General Factoring Strategy

Completely factor: $x^6 + 7x^3 - 8$

Solution

Factor $x^6 + 7x^3 - 8$ as the product of two binomials.

$$x^6 + 7x^3 - 8 = (x^3 + 8)(x^3 - 1)$$

Now factor $x^3 + 8$, which is the sum of two cubes, and factor $x^3 - 1$, which is the difference of two cubes.

$$x^6 + 7x^3 - 8 = (x + 2)(x^2 - 2x + 4)(x - 1)(x^2 + x + 1)$$

TRY EXERCISE 64, EXERCISE SET P.5, PAGE 48

TOPICS FOR DISCUSSION

1. Discuss the meaning of the phrase *nonfactorable over the integers.*

2. You know that if $ab = 0$, then $a = 0$ or $b = 0$. Suppose a polynomial is written in factored form and then set equal to zero. For instance, suppose
$$x^2 - 2x - 15 = (x - 5)(x + 3) = 0$$
Discuss what implications this has for the values of x. Do not only answer this question for the polynomial above, but also discuss the implications for the values of x for any polynomial written as a product of linear factors and then set equal to zero.

3. Let P be a polynomial of degree n. Discuss the number of possible distinct linear polynomials that can be a factor of P.

4. A method of evaluating polynomials, sometimes called Horner's method, involves factoring a polynomial in a certain manner. For instance,
$$4x^3 - 2x^2 + 5x - 3 = [(4x - 2)x + 5]x - 3$$
$$5x^4 - 2x^3 + 4x^2 + x - 6 = \{[(5x - 2)x + 4]x + 1\}x - 6$$
To evaluate the polynomial, the factored form is evaluated. Discuss the advantages and disadvantages of using this method to evaluate a polynomial.

5. Recall that if n is a natural number, $n! = n(n - 1)(n - 2)\cdots 3 \cdot 2 \cdot 1$. Explain why none of the following consecutive integers is a prime number.
$$5! + 2 \qquad 5! + 3 \qquad 5! + 4 \qquad 5! + 5$$
How many numbers are in the following list of consecutive integers? How many of those numbers are prime numbers?
$$k! + 2, k! + 3, k! + 4, k! + 5, \ldots, k! + k$$
Explain why this result means that there are arbitrarily long sequences of consecutive natural numbers that do not contain a prime number.

EXERCISE SET P.5

In Exercises 1 to 8, factor out the GCF from each polynomial.

1. $5x + 20$
2. $8x^2 + 12x - 40$
3. $-15x^2 - 12x$
4. $-6y^2 - 54y$
5. $10x^2y + 6xy - 14xy^2$
6. $6a^3b^2 - 12a^2b + 72ab^3$
7. $(x - 3)(a + b) + (x - 3)(a + 3b)$
8. $(x - 4)(2a - b) + (x + 4)(2a - b)$

In Exercises 9 to 22, factor each trinomial.

9. $x^2 + 7x + 12$
10. $x^2 + 9x + 20$
11. $a^2 - 10a - 24$
12. $b^2 + 12b - 28$
13. $6x^2 + 25x + 4$
14. $8a^2 - 26a + 15$
15. $51x^2 - 5x - 4$
16. $57y^2 + y - 6$
17. $6x^2 + xy - 40y^2$
18. $8x^2 + 10xy - 25y^2$
19. $x^4 + 6x^2 + 5$
20. $x^4 + 11x^2 + 18$
21. $6x^4 + 23x^2 + 15$
22. $9x^4 + 10x^2 + 1$

In Exercises 23 to 28, use the factorization theorem to determine whether each trinomial is factorable over the integers.

23. $8x^2 + 26x + 15$
24. $16x^2 + 8x - 35$
25. $4x^2 - 5x + 6$
26. $6x^2 + 8x - 3$
27. $6x^2 - 14x + 5$
28. $10x^2 - 4x - 5$

In Exercises 29 to 38, factor each difference of squares.

29. $x^2 - 9$ 30. $x^2 - 64$ 31. $4a^2 - 49$

32. $81b^2 - 16c^2$ 33. $1 - 100x^2$ 34. $1 - 121y^2$

35. $x^4 - 9$ 36. $y^4 - 196$

37. $(x + 5)^2 - 4$ 38. $(x - 3)^2 - 16$

In Exercises 39 to 46, factor each perfect-square trinomial.

39. $x^2 + 10x + 25$ 40. $y^2 + 6y + 9$

41. $a^2 - 14a + 49$ 42. $b^2 - 24b + 144$

43. $4x^2 + 12x + 9$ 44. $25y^2 + 40y + 16$

45. $z^4 + 4z^2w^2 + 4w^4$ 46. $9x^4 - 30x^2y^2 + 25y^4$

In Exercises 47 to 54, factor each sum or difference of cubes.

47. $x^3 - 8$ 48. $b^3 + 64$ 49. $8x^3 - 27y^3$

50. $64u^3 - 27v^3$ 51. $8 - x^6$ 52. $1 + y^{12}$

53. $(x - 2)^3 - 1$ 54. $(y + 3)^3 + 8$

In Exercises 55 to 60, factor by grouping in pairs.

55. $3x^3 + x^2 + 6x + 2$

56. $18w^3 + 15w^2 + 12w + 10$

57. $ax^2 - ax + bx - b$ 58. $a^2y^2 - ay^3 + ac - cy$

59. $6w^3 + 4w^2 - 15w - 10$ 60. $10z^3 - 15z^2 - 4z + 6$

In Exercises 61 to 80, use the general factoring strategy to completely factor each polynomial. If the polynomial does not factor, then state that it is nonfactorable over the integers.

61. $18x^2 - 2$ 62. $4bx^3 + 32b$

63. $16x^4 - 1$ 64. $81y^4 - 16$

65. $12ax^2 - 23axy + 10ay^2$ 66. $6ax^2 - 19axy - 20ay^2$

67. $3bx^3 + 4bx^2 - 3bx - 4b$ 68. $2x^6 - 2$

69. $72bx^2 + 24bxy + 2by^2$ 70. $64y^3 - 16y^2z + yz^2$

71. $(w - 5)^3 + 8$ 72. $5xy + 20y - 15x - 60$

73. $x^2 + 6xy + 9y^2 - 1$ 74. $4y^2 - 4yz + z^2 - 9$

75. $8x^2 + 3x - 4$ 76. $16x^2 + 81$

77. $5x(2x - 5)^2 - (2x - 5)^3$ 78. $6x(3x + 1)^3 - (3x + 1)^4$

79. $4x^2 + 2x - y - y^2$ 80. $a^2 + a + b - b^2$

SUPPLEMENTAL EXERCISES

In Exercises 81 and 82, find all positive values of k such that the trinomial is a perfect-square trinomial.

81. $x^2 + kx + 16$ 82. $36x^2 + kxy + 100$

In Exercises 83 and 84, find k such that the trinomial is a perfect-square trinomial.

83. $x^2 + 16x + k$ 84. $x^2 - 14xy + ky^2$

In Exercises 85 and 86, use the general strategy to completely factor each polynomial. In each exercise n represents a positive integer.

85. $x^{4n} - 1$ 86. $x^{4n} - 2x^{2n} + 1$

In Exercises 87 to 90, write, in its factored form, the area of the shaded portion of each geometric figure.

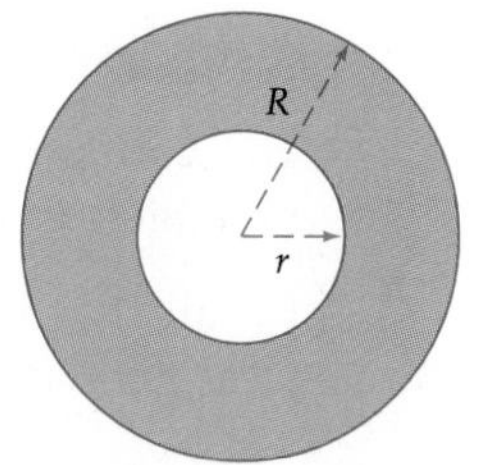

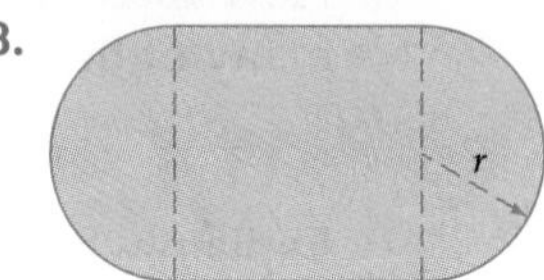

89.

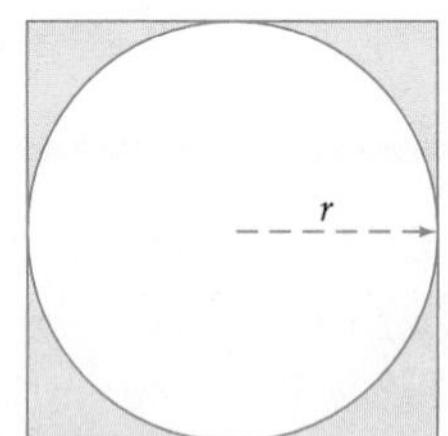

90.

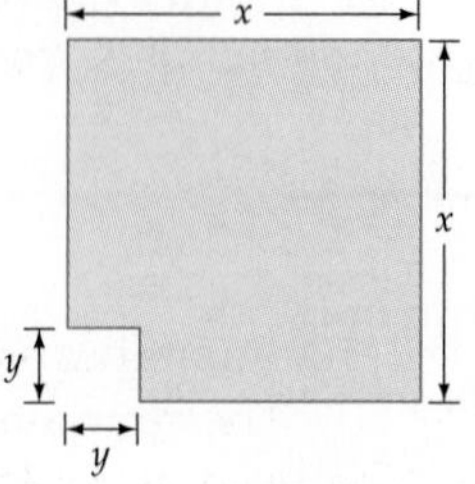

PROJECTS

1. **GEOMETRY** The ancient Greeks used geometric figures and the concept of area to illustrate many algebraic concepts. The factoring formula $x^2 - y^2 = (x + y)(x - y)$ can be illustrated by the figure at the left below.
   a. Which regions are represented by $(x + y)(x - y)$?
   b. Which regions are represented by $x^2 - y^2$?
   c. Explain why the area of the regions listed in **a.** must equal the area of the regions listed in **b.**

2. **GEOMETRY** What algebraic formula does the geometric figure in the middle below illustrate?

3. **GEOMETRY** Show how the figure at the right below can be used to illustrate the factoring formula for the difference of two cubes.

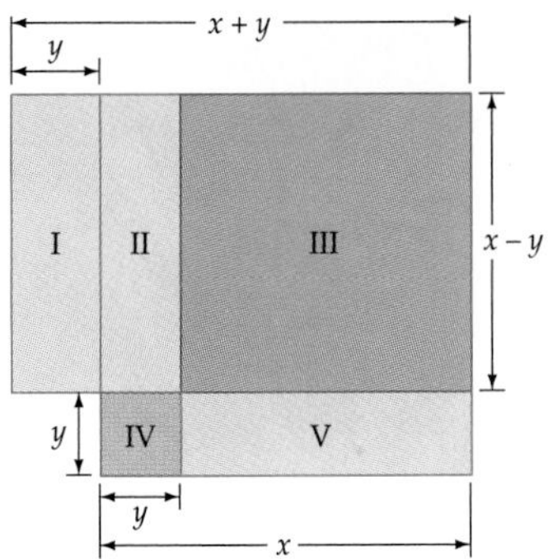

Figure for Project 1

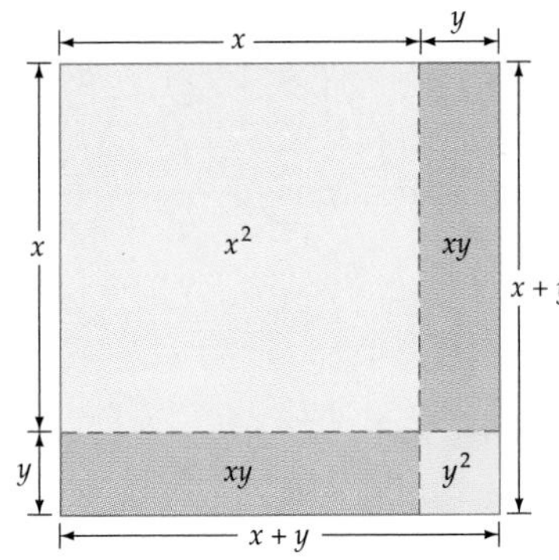

Figure for Project 2

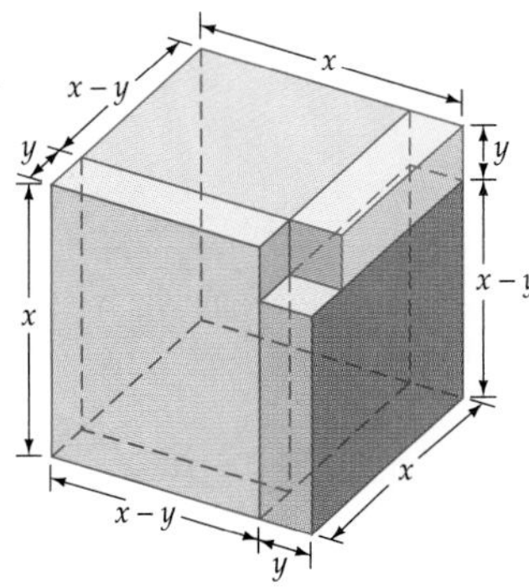

Figure for Project 3

SECTION P.6 RATIONAL EXPRESSIONS

- SIMPLIFY A RATIONAL EXPRESSION
- OPERATIONS ON RATIONAL EXPRESSIONS
- DETERMINING THE LCD OF RATIONAL EXPRESSIONS
- COMPLEX FRACTIONS
- APPLICATION OF RATIONAL EXPRESSIONS

A **rational expression** is a fraction in which the numerator and denominator are polynomials. For example,

$$\frac{3}{x + 1} \quad \text{and} \quad \frac{x^2 - 4x - 21}{x^2 - 9}$$

are rational expressions.

The **domain of a rational expression** is the set of all real numbers that can be used as replacements for the variable. Any value of the variable that causes division by zero is excluded from the domain of the rational expression. For example, the domain of

$$\frac{7x}{x^2 - 5x} \quad x \neq 0, x \neq 5$$

is the set of all real numbers except 0 and 5. Both 0 and 5 are excluded values because the denominator $x^2 - 5x$ equals zero when $x = 0$ and also when $x = 5$. Sometimes the excluded values are specified to the right of a rational expression, as shown here. However, a rational expression is meaningful only for those real numbers that are not excluded values, regardless of whether the excluded values are specifically stated.

Rational expressions have properties similar to the properties of rational numbers.

MATH MATTERS

Evidence from work left by early Egyptians more than 3600 years ago shows that they used, with one exception, unit fractions–that is, fractions whose numerators are 1. The one exception was 2/3. A unit fraction was represented by placing an oval over the symbol for the number in the denominator. For instance, 1/4 = ⬭ over IIII.

Properties of Rational Expressions

For all rational expressions P/Q and R/S where $Q \neq 0$ and $S \neq 0$,

Equality	$\frac{P}{Q} = \frac{R}{S}$	if and only if $PS = QR$
Equivalent expressions	$\frac{P}{Q} = \frac{PR}{QR}$,	$R \neq 0$
Sign	$-\frac{P}{Q} = \frac{-P}{Q} = \frac{P}{-Q}$	

◆ SIMPLIFY A RATIONAL EXPRESSION

To **simplify a rational expression**, factor the numerator and the denominator. Then use the equivalent expressions property to eliminate factors common to both the numerator and the denominator. A rational expression is *simplified* when 1 is the only common factor of both the numerator and the denominator.

EXAMPLE 1 Simplify a Rational Expression

Simplify: $\dfrac{7 + 20x - 3x^2}{2x^2 - 11x - 21}$

Solution

$$\frac{7 + 20x - 3x^2}{2x^2 - 11x - 21} = \frac{(7 - x)(1 + 3x)}{(x - 7)(2x + 3)}$$ • **Factor.**

$$= \frac{-(x - 7)(1 + 3x)}{(x - 7)(2x + 3)}$$ • **Use $(7 - x) = -(x - 7)$.**

$$= \frac{-\cancel{(x - 7)}(1 + 3x)}{\cancel{(x - 7)}(2x + 3)}$$

$$= \frac{-(1 + 3x)}{2x + 3} = -\frac{3x + 1}{2x + 3} \quad x \neq 7, x \neq -\frac{3}{2}$$

take note

A rational expression like $(x + 3)/3$ does not simplify to $x + 1$ because

$$\frac{x + 3}{3} = \frac{x}{3} + \frac{3}{3} = \frac{x}{3} + 1$$

Rational expressions can be simplified by dividing nonzero factors common to the numerator and the denominator, but not terms.

TRY EXERCISE 2, EXERCISE SET P.6, PAGE 56

◆ OPERATIONS ON RATIONAL EXPRESSIONS

Arithmetic operations are defined on rational expressions just as they are on rational numbers.

Arithmetic Operations Defined on Rational Expressions

For all rational expressions P/Q, R/Q, and R/S where $Q \neq 0$ and $S \neq 0$,

Addition $$\frac{P}{Q} + \frac{R}{Q} = \frac{P + R}{Q}$$

Subtraction $$\frac{P}{Q} - \frac{R}{Q} = \frac{P - R}{Q}$$

Multiplication $$\frac{P}{Q} \cdot \frac{R}{S} = \frac{PR}{QS}$$

Division $$\frac{P}{Q} \div \frac{R}{S} = \frac{P}{Q} \cdot \frac{S}{R} = \frac{PS}{QR} \quad R \neq 0$$

Factoring and the equivalent expressions property of rational expressions are used in the multiplication and division of rational expressions.

EXAMPLE 2 Divide a Rational Expression

Simplify: $\dfrac{x^2 + 6x + 9}{x^3 + 27} \div \dfrac{x^2 + 7x + 12}{x^3 - 3x^2 + 9x}$

Solution

$$\frac{x^2 + 6x + 9}{x^3 + 27} \div \frac{x^2 + 7x + 12}{x^3 - 3x^2 + 9x}$$

$$= \frac{(x + 3)^2}{(x + 3)(x^2 - 3x + 9)} \div \frac{(x + 4)(x + 3)}{x(x^2 - 3x + 9)}$$ • **Factor.**

$$= \frac{(x + 3)^2}{(x + 3)(x^2 - 3x + 9)} \cdot \frac{x(x^2 - 3x + 9)}{(x + 4)(x + 3)}$$ • **Multiply by the reciprocal.**

$$= \frac{\cancel{(x + 3)^2}x\cancel{(x^2 - 3x + 9)}}{\cancel{(x + 3)}\cancel{(x^2 - 3x + 9)}(x + 4)\cancel{(x + 3)}}$$ • **Simplify.**

$$= \frac{x}{x + 4}$$

TRY EXERCISE 16, EXERCISE SET P.6, PAGE 56

Addition of rational expressions with a **common denominator** is accomplished by writing the sum of the numerators over the common denominator. For example,

$$\frac{5x}{18} + \frac{x}{18} = \frac{5x + x}{18} = \frac{6x}{18} = \frac{x}{3}$$

If the rational expressions do not have a common denominator, then they can be written as equivalent rational expressions that have a common denominator by multiplying the numerator and denominator of each of the rational expressions by the required polynomials. The following procedure can be used to determine the least common denominator (LCD) of rational expressions. It is similar to the process used to find the LCD of rational numbers.

◆ Determining the LCD of Rational Expressions

1. Factor each denominator completely and express repeated factors using exponential notation.
2. Identify the largest power of each factor in any single factorization. The LCD is the product of each factor raised to its largest power.

For example,

$$\frac{1}{x + 3} \quad \text{and} \quad \frac{5}{2x - 1}$$

have an LCD of $(x + 3)(2x - 1)$. The rational expressions

$$\frac{5x}{(x + 5)(x - 7)^3} \quad \text{and} \quad \frac{7}{x(x + 5)^2(x - 7)}$$

have an LCD of $x(x + 5)^2(x - 7)^3$.

EXAMPLE 3 Add and Subtract Rational Expressions

Perform the indicated operation and then simplify if possible.

a. $\frac{5x}{48} + \frac{x}{15}$ b. $\frac{x}{x^2 - 4} - \frac{2x - 1}{x^2 - 3x - 10}$

Solution

a. Determine the prime factorization of the denominators.

$$48 = 2^4 \cdot 3 \quad \text{and} \quad 15 = 3 \cdot 5$$

The desired common denominator is the product of each of the prime factors raised to its largest power. Thus the common denominator is $2^4 \cdot 3 \cdot 5 = 240$. Write each rational expression as an equivalent rational expression with a denominator of 240.

$$\frac{5x}{48} + \frac{x}{15} = \frac{5x \cdot 5}{48 \cdot 5} + \frac{x \cdot 16}{15 \cdot 16} = \frac{25x}{240} + \frac{16x}{240} = \frac{41x}{240}$$

b. Factor each denominator to determine the LCD of the rational expressions.

$$x^2 - 4 = (x + 2)(x - 2)$$
$$x^2 - 3x - 10 = (x + 2)(x - 5)$$

The LCD is $(x + 2)(x - 2)(x - 5)$. Forming equivalent rational expressions that have the LCD, we have

$$\begin{aligned}
&\frac{x}{x^2 - 4} - \frac{2x - 1}{x^2 - 3x - 10} \\
&= \frac{x(x - 5)}{(x + 2)(x - 2)(x - 5)} - \frac{(2x - 1)(x - 2)}{(x + 2)(x - 5)(x - 2)} \\
&= \frac{x^2 - 5x - (2x^2 - 5x + 2)}{(x + 2)(x - 2)(x - 5)} = \frac{x^2 - 5x - 2x^2 + 5x - 2}{(x + 2)(x - 2)(x - 5)} \\
&= \frac{-x^2 - 2}{(x + 2)(x - 2)(x - 5)} = -\frac{x^2 + 2}{(x + 2)(x - 2)(x - 5)}
\end{aligned}$$

TRY EXERCISE 30, EXERCISE SET P.6, PAGE 56

◆ COMPLEX FRACTIONS

A **complex fraction** is a fraction whose numerator or denominator contains one or more fractions. Complex fractions can be simplified by using one of the following two methods.

Methods for Simplifying Complex Fractions

Method 1: Multiply by the LCD

1. Determine the LCD of all the fractions in the complex fraction.
2. Multiply both the numerator and the denominator of the complex fraction by the LCD.
3. If possible, simplify the resulting rational expression.

Method 2: Multiply by the reciprocal of the denominator

1. Simplify the numerator to a single fraction and the denominator to a single fraction.
2. Multiply the numerator by the reciprocal of the denominator.
3. If possible, simplify the resulting rational expression.

EXAMPLE 4 Simplify Complex Fractions

Simplify: $\dfrac{\dfrac{2}{x-2}+\dfrac{1}{x}}{\dfrac{3x}{x-5}-\dfrac{2}{x-5}}$

Solution

First simplify the numerator to a single fraction and then simplify the denominator to a single fraction.

$$\frac{\dfrac{2}{x-2}+\dfrac{1}{x}}{\dfrac{3x}{x-5}-\dfrac{2}{x-5}} = \frac{\dfrac{2\cdot x}{(x-2)\cdot x}+\dfrac{1\cdot(x-2)}{x\cdot(x-2)}}{\dfrac{3x-2}{x-5}}$$

- **Simplify numerator and denomimator.**

$$= \frac{\dfrac{2x+(x-2)}{x(x-2)}}{\dfrac{3x-2}{x-5}} = \frac{\dfrac{3x-2}{x(x-2)}}{\dfrac{3x-2}{x-5}}$$

$$= \frac{\cancel{3x-2}}{x(x-2)}\cdot\frac{x-5}{\cancel{3x-2}}$$

- **Multiply by the reciprocal of the denominator.**

$$= \frac{x-5}{x(x-2)}$$

TRY EXERCISE 42, EXERCISE SET P.6, PAGE 57

EXAMPLE 5 Simplify a Fraction

Simplify the fraction $\dfrac{c^{-1}}{a^{-1}+b^{-1}}$.

Solution

The fraction written without negative exponents becomes

$$\frac{c^{-1}}{a^{-1}+b^{-1}} = \frac{\dfrac{1}{c}}{\dfrac{1}{a}+\dfrac{1}{b}}$$

- **Using $x^{-n} = \dfrac{1}{x^n}$.**

$$= \frac{\dfrac{1}{c}\cdot abc}{\left(\dfrac{1}{a}+\dfrac{1}{b}\right)abc}$$

- **Multiply the numerator and the denominator by abc, which is the LCD of the fraction in the numerator and the fraction in the denominator.**

$$= \frac{ab}{bc+ac}$$

TRY EXERCISE 60, EXERCISE SET P.6, PAGE 57

take note

It is a mistake to write

$\dfrac{c^{-1}}{a^{-1}+b^{-1}}$ as $\dfrac{a+b}{c}$

because a^{-1} and b^{-1} are terms and cannot be treated as factors.

◆ APPLICATION OF RATIONAL EXPRESSIONS

EXAMPLE 6 Solve an Application

The *average speed* for a round trip is given by the complex fraction

$$\frac{2}{\dfrac{1}{v_1}+\dfrac{1}{v_2}}$$

where v_1 is the average speed on the way to your destination and v_2 is the average speed on your return trip. Find the average speed for a round trip if $v_1 = 50$ mph and $v_2 = 40$ mph.

Solution

Evaluate the complex fraction with $v_1 = 50$ and $v_2 = 40$.

$$\frac{2}{\dfrac{1}{v_1}+\dfrac{1}{v_2}} = \frac{2}{\dfrac{1}{50}+\dfrac{1}{40}} = \frac{2}{\dfrac{1\cdot 4}{50\cdot 4}+\dfrac{1\cdot 5}{40\cdot 5}}$$

• **Substitute and simplify the denominator.**

$$= \frac{2}{\dfrac{4}{200}+\dfrac{5}{200}} = \frac{2}{\dfrac{9}{200}}$$

$$= 2\cdot\frac{200}{9} = \frac{400}{9} = 44\frac{4}{9}$$

The average speed of the round trip is $44\frac{4}{9}$ mph.

TRY EXERCISE 64, EXERCISE SET P.6, PAGE 57

QUESTION In Example 6, why is the speed of the round trip *not* the average of v_1 and v_2?

TOPICS FOR DISCUSSION

1. Discuss the meaning of the phrase *rational expression.* Is a rational expression the same as a fraction? If not, give some examples of a fraction that is not a rational expression.
2. What is the domain of a rational expression?
3. Explain why the following is *not* correct.

$$\frac{2x^2+5}{x^2} = 2+5 = 7$$

ANSWER Because you were traveling slower on the return trip, the return trip took longer than the time spent going to your destination. More time was spent traveling at the slower speed. Thus the average speed is less than the average of v_1 and v_2.

4. Consider the rational expression $\frac{x^2 - 3x - 10}{x^2 + x - 30}$. By simplifying this expression, we have

$$\frac{x^2 - 3x - 10}{x^2 + x - 30} = \frac{(x-5)(x+2)}{(x-5)(x+6)} = \frac{x+2}{x+6}$$

Does this really mean that $\frac{x^2 - 3x - 10}{x^2 + x - 30} = \frac{x+2}{x+6}$ for every value of x? If not, for what values of x are the two expressions equal?

EXERCISE SET P.6

In Exercises 1 to 10, simplify each rational expression.

1. $\frac{x^2 - x - 20}{3x - 15}$
2. $\frac{2x^2 - 5x - 12}{2x^2 + 5x + 3}$
3. $\frac{x^3 - 9x}{x^3 + x^2 - 6x}$
4. $\frac{x^3 + 125}{2x^3 - 50x}$
5. $\frac{a^3 + 8}{a^2 - 4}$
6. $\frac{y^3 - 27}{-y^2 + 11y - 24}$
7. $\frac{x^2 + 3x - 40}{-x^2 + 3x + 10}$
8. $\frac{2x^3 - 6x^2 + 5x - 15}{9 - x^2}$
9. $\frac{4y^3 - 8y^2 + 7y - 14}{-y^2 - 5y + 14}$
10. $\frac{x^3 - x^2 + x}{x^3 + 1}$

In Exercises 11 to 40, simplify each expression.

11. $\left(-\frac{4a}{3b^2}\right)\left(\frac{6b}{a^4}\right)$
12. $\left(\frac{12x^2y}{5z^4}\right)\left(-\frac{25x^2z^3}{15y^2}\right)$
13. $\left(\frac{6p^2}{5q^2}\right)^{-1}\left(\frac{2p}{3q^2}\right)^2$
14. $\left(\frac{4r^2s}{3t^3}\right)^{-1}\left(\frac{6rs^3}{5t^2}\right)$
15. $\frac{x^2 + x}{2x + 3} \cdot \frac{3x^2 + 19x + 28}{x^2 + 5x + 4}$
16. $\frac{x^2 - 16}{x^2 + 7x + 12} \cdot \frac{x^2 - 4x - 21}{x^2 - 4x}$
17. $\frac{3x - 15}{2x^2 - 50} \cdot \frac{2x^2 + 16x + 30}{6x + 9}$
18. $\frac{y^3 - 8}{y^2 + y - 6} \cdot \frac{y^2 + 3y}{y^3 + 2y^2 + 4y}$
19. $\frac{12y^2 + 28y + 15}{6y^2 + 35y + 25} \div \frac{2y^2 - y - 3}{3y^2 + 11y - 20}$
20. $\frac{z^2 - 81}{z^2 - 16} \div \frac{z^2 - z - 20}{z^2 + 5z - 36}$
21. $\frac{a^2 + 9}{a^2 - 64} \div \frac{a^3 - 3a^2 + 9a - 27}{a^2 + 5a - 24}$
22. $\frac{6x^2 + 13xy + 6y^2}{4x^2 - 9y^2} \div \frac{3x^2 - xy - 2y^2}{2x^2 + xy - 3y^2}$
23. $\frac{p+5}{r} + \frac{2p-7}{r}$
24. $\frac{2s + 5t}{4t} + \frac{-2s + 3t}{4t}$
25. $\frac{x}{x-5} + \frac{7x}{x+3}$
26. $\frac{2x}{3x+1} + \frac{5x}{x-7}$
27. $\frac{5y-7}{y+4} - \frac{2y-3}{y+4}$
28. $\frac{6x-5}{x-3} - \frac{3x-8}{x-3}$
29. $\frac{4z}{2z-3} + \frac{5z}{z-5}$
30. $\frac{3y-1}{3y+1} - \frac{2y-5}{y-3}$
31. $\frac{x}{x^2-9} - \frac{3x-1}{x^2+7x+12}$
32. $\frac{m-n}{m^2 - mn - 6n^2} + \frac{3m-5n}{m^2 + mn - 2n^2}$
33. $\frac{1}{x} + \frac{2}{3x-1} \cdot \frac{3x^2 + 11x - 4}{x-5}$
34. $\frac{2}{y} - \frac{3}{y+1} \cdot \frac{y^2-1}{y+4}$
35. $\frac{q+1}{q-3} - \frac{2q}{q-3} \div \frac{q+5}{q-3}$
36. $\frac{p}{p+5} + \frac{p}{p-4} \div \frac{p+2}{p^2 - p - 12}$
37. $\frac{1}{x^2 + 7x + 12} + \frac{1}{x^2 - 9} + \frac{1}{x^2 - 16}$
38. $\frac{2}{a^2 - 3a + 2} + \frac{3}{a^2 - 1} - \frac{5}{a^2 + 3a - 10}$
39. $\left(1 + \frac{2}{x}\right)\left(3 - \frac{1}{x}\right)$
40. $\left(4 - \frac{1}{z}\right)\left(4 + \frac{2}{z}\right)$

In Exercises 41 to 58, simplify each complex fraction.

41. $\dfrac{4+\dfrac{1}{x}}{1-\dfrac{1}{x}}$ **42.** $\dfrac{3-\dfrac{2}{a}}{5+\dfrac{3}{a}}$ **43.** $\dfrac{\dfrac{x}{y}-2}{y-x}$

44. $\dfrac{3+\dfrac{2}{x-3}}{4+\dfrac{1}{2+\dfrac{1}{x}}}$ **45.** $\dfrac{5-\dfrac{1}{x+2}}{1+\dfrac{3}{1+\dfrac{3}{x}}}$ **46.** $\dfrac{\dfrac{1}{(x+h)^2}-1}{h}$

47. $\dfrac{1+\dfrac{1}{b-2}}{1-\dfrac{1}{b+3}}$ **48.** $r-\dfrac{r}{r+\dfrac{1}{3}}$

49. $\dfrac{1-\dfrac{1}{x^2}}{1+\dfrac{1}{x}}$ **50.** $\dfrac{1}{\dfrac{1}{a}+\dfrac{1}{b}}$

51. $2-\dfrac{m}{1-\dfrac{1-m}{-m}}$ **52.** $\dfrac{\dfrac{x+h+1}{x+h}-\dfrac{x}{x+1}}{h}$

53. $\dfrac{\dfrac{1}{x}-\dfrac{x-4}{x+1}}{\dfrac{x}{x+1}}$ **54.** $\dfrac{\dfrac{2}{y}-\dfrac{3y-2}{y-1}}{\dfrac{y}{y-1}}$

55. $\dfrac{\dfrac{1}{x+3}-\dfrac{2}{x-1}}{\dfrac{x}{x-1}+\dfrac{3}{x+3}}$ **56.** $\dfrac{\dfrac{x+2}{x^2-1}+\dfrac{1}{x+1}}{\dfrac{x}{2x^2-x-1}+\dfrac{1}{x-1}}$

57. $\dfrac{\dfrac{x^2+3x-10}{x^2+x-6}}{\dfrac{x^2-x-30}{2x^2-15x+18}}$ **58.** $\dfrac{\dfrac{2y^2+11y+15}{y^2-4y-21}}{\dfrac{6y^2+11y-10}{3y^2-23y+14}}$

In Exercises 59 to 62, simplify each algebraic fraction. Write all answers with positive exponents.

59. $\dfrac{a^{-1}+b^{-1}}{a-b}$ **60.** $\dfrac{e^{-2}-f^{-1}}{ef}$

61. $\dfrac{a^{-1}b-ab^{-1}}{a^2+b^2}$ **62.** $(a+b^{-2})^{-1}$

63. AVERAGE SPEED According to Example 6, the average speed for a round trip in which the average speed on the way to your destination was v_1 and the average speed on your return was v_2 is given by the complex fraction

$$\frac{2}{\dfrac{1}{v_1}+\dfrac{1}{v_2}}$$

a. Find the average speed for a round trip by helicopter with $v_1 = 180$ mph and $v_2 = 110$ mph.

b. Simplify the complex fraction.

64. RELATIVITY THEORY Using Einstein's theory of relativity, the "sum" of the two speeds v_1 and v_2 is given by the complex fraction

$$\frac{v_1+v_2}{1+\dfrac{v_1v_2}{c^2}}$$

where c is the speed of light.

a. Evaluate this expression with $v_1 = 1.2 \times 10^8$ mph, $v_2 = 2.4 \times 10^8$ mph, and $c = 6.7 \times 10^8$ mph.

b. Simplify the complex fraction.

65. Find the rational expression in simplest form that represents the sum of the reciprocals of the consecutive integers x and $x + 1$.

66. Find the rational expression in simplest form that represents the positive difference between the reciprocals of the consecutive even integers x and $x + 2$.

67. Find the rational expression in simplest form that represents the sum of the reciprocals of the consecutive even integers $x - 2$, x, and $x + 2$.

68. Find the rational expression in simplest form that represents the sum of the reciprocals of the squares of the consecutive even integers $x - 2$, x, and $x + 2$.

SUPPLEMENTAL EXERCISES

In Exercises 69 to 72, simplify each algebraic fraction.

69. $\dfrac{(x+5)-x(x+5)^{-1}}{x+5}$ **70.** $\dfrac{(y+2)+y^2(y+2)^{-1}}{y+2}$

71. $\dfrac{x^{-1}-4y}{(x^{-1}-2y)(x^{-1}+2y)}$ **72.** $\dfrac{x+y}{x-y}\cdot\dfrac{x^{-1}-y^{-1}}{x^{-1}+y^{-1}}$

73. FINANCE The **present value** of an ordinary annuity is given by

$$R\left[\frac{1-\dfrac{1}{(1+i)^n}}{i}\right]$$

where n is the number of payments of R dollars each invested at an interest rate of i per conversion period. Simplify the complex fraction.

74. ELECTRICITY The total resistance of the three resistances R_1, R_2, and R_3 in parallel is given by

$$\frac{1}{\dfrac{1}{R_1}+\dfrac{1}{R_2}+\dfrac{1}{R_3}}$$

Simplify the complex fraction.

PROJECTS

$$\cfrac{1}{1+\cfrac{1}{1+\cfrac{1}{1+\cfrac{1}{1+\cdots}}}}$$

1. **CONTINUED FRACTIONS** The complex fraction shown at the right is called a **continued fraction.** The three dots in $\dfrac{1}{1+\cdots}$ indicate that the pattern continues in the same manner. A **convergent** of a complex fraction is an approximation of the continued fraction that is found by stopping the process at some point.

   a. Calculate the convergent $C_2 = \cfrac{1}{1+\cfrac{1}{1+1}}$.

   b. Calculate the convergent $C_3 = \cfrac{1}{1+\cfrac{1}{1+\cfrac{1}{1+1}}}$.

   c. Calculate the convergent $C_5 = \cfrac{1}{1+\cfrac{1}{1+\cfrac{1}{1+\cfrac{1}{1+\cfrac{1}{1+1}}}}}$.

   d. Show that $C_5 \approx \dfrac{-1+\sqrt{5}}{2}$. Using some techniques from more advanced math courses, it can be shown that the convergents of the continued fraction become closer and closer to $\dfrac{-1+\sqrt{5}}{2}$.

2. **REPRESENTATION OF π** There are a few continued-fraction representations for π. Find two of these representations. Compute the value of π accurate to 4 decimal places using a convergent from each of the continued fractions you found.

EXPLORING CONCEPTS WITH TECHNOLOGY

Can You Trust Your Calculator?

You may think that your calculator always produces correct results in a *predictable* manner. However, the following experiment may change your opinion.

First note that the algebraic expression

$$p + 3p(1 - p)$$

is equal to the expression

$$4p - 3p^2$$

Use a graphing calculator to evaluate both of these expressions with $p = 0.05$. You should find that both expressions equal 0.1925. So far we do not observe any unexpected results. Now replace p in each expression with the current value of that expression (0.1925 in this case). This is called *feedback* because we are feeding our outputs back into each expression as inputs. Each new evaluation is referred to as an *iteration*. This time each expression takes on the value

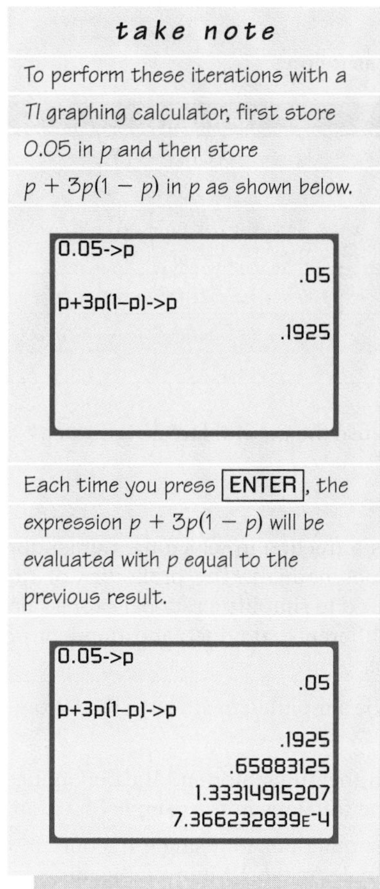

take note

To perform these iterations with a TI graphing calculator, first store 0.05 in p and then store $p + 3p(1 - p)$ in p as shown below.

Each time you press ENTER, the expression $p + 3p(1 - p)$ will be evaluated with p equal to the previous result.

0.65883125. Still no surprises. Continue the feedback process. That is, replace p in each expression *with the current value* of that expression. Now each expression takes on the value 1.33314915207, as shown in the following table. The iterations were performed on a *TI-85* calculator.

Iteration	$p + 3p(1 - p)$	$4p - 3p^2$
1	0.1925	0.1925
2	0.65883125	0.65883125
3	1.33314915207	1.33314915207

The following table shows that if we continue this feedback process on a calculator, the expressions $p + 3p(1 - p)$ and $4p - 3p^2$ will start to take on different values starting with the fourth iteration. By the 37th iteration, the values do not even agree to two decimal places.

Iteration	$p + 3p(1 - p)$	$4p - 3p^2$
4	7.366232839E–4	7.366232838E–4
5	0.002944865294	0.002944865294
6	0.011753444481	0.0117534448
7	0.046599347553	0.046599347547
20	1.12135618652	1.12135608405
30	0.947163304835	0.947033128433
37	0.285727963839	0.300943417861

1. Use a calculator to find the first 20 iterations of $p + 3p(1 - p)$ and $4p - 3p^2$, with the initial value of $p = 0.5$.

2. Write a report on chaos and fractals. Include information on the "butterfly effect." An excellent source is *Chaos and Fractals, New Frontiers of Science* by Heinz-Otto Peitgen, Hartmut Jurgens, and Dietmar Saupe (New York: Springer-Verlag, 1992).

3. Equations of the form $p_{n+1} = p_n + rp_n(1 - p_n)$ are called Verhulst population models. Write a report on Verhulst population models.

CHAPTER P SUMMARY

P.1 The Real Number System

- The following sets of numbers are used extensively in the study of algebra:

Integers	$\{\ldots, -3, -2, -1, 0, 1, 2, 3, \ldots\}$
Rational numbers	{all terminating or repeating decimals}
Irrational numbers	{all nonterminating, nonrepeating decimals}
Real numbers	{all rational or irrational numbers}

P.2 Intervals, Absolute Value, and Distance

- The absolute value of the real number a is defined by

$$|a| = \begin{cases} a & \text{if } a \geq 0 \\ -a & \text{if } a < 0 \end{cases}$$

- For any real numbers a and b, the distance between the graph of a and the graph of b is denoted by $d(a, b)$, where $d(a, b) = |a - b|$.

P.3 Integer and Rational Number Exponents

- If b is any real number and n is any natural number, then

$$b^n = \underbrace{b \cdot b \cdot b \cdot \cdots \cdot b}_{n \text{ factors of } b}$$

- For any nonzero real number b, $b^0 = 1$.
- If $b \neq 0$ and n is any natural number, then $b^{-n} = \dfrac{1}{b^n}$ and $\dfrac{1}{b^{-n}} = b^n$.
- **Properties of Rational Exponents**
  If p, q, and r represent rational numbers, and a and b are positive real numbers, then

  Product $b^p \cdot b^q = b^{p+q}$

  Quotient $\dfrac{b^p}{b^q} = b^{p-q}$

  Power $(b^p)^q = b^{pq}$ $\quad (a^p b^q)^r = a^{pr} b^{qr}$

  $\left(\dfrac{a^p}{b^q}\right)^r = \dfrac{a^{pr}}{b^{qr}}$ $\quad b^{-p} = \dfrac{1}{b^p}$

- **Properties of Radicals**
  If m and n are natural numbers and a and b are positive real numbers, then

  Product $\sqrt[n]{a} \cdot \sqrt[n]{b} = \sqrt[n]{ab}$

  Quotient $\dfrac{\sqrt[n]{a}}{\sqrt[n]{b}} = \sqrt[n]{\dfrac{a}{b}}$

  Index $\sqrt[m]{\sqrt[n]{b}} = \sqrt[mn]{b}$

P.4 Polynomials

- A polynomial is an expression of the form

$$a_n x^n + a_{n-1} x^{n-1} + \cdots + a_2 x^2 + a_1 x + a_0$$

- Special product formulas are as follows:

Special Form	Formula(s)
(Sum)(Difference)	$(x + y)(x - y) = x^2 - y^2$
(Binomial)^2	$(x + y)^2 = x^2 + 2xy + y^2$ $(x - y)^2 = x^2 - 2xy + y^2$

P.5 Factoring

- Factoring formulas are as follows:

Special Form	Formula(s)
Difference of two squares	$x^2 - y^2 = (x + y)(x - y)$
Perfect-square trinomials	$x^2 + 2xy + y^2 = (x + y)^2$ $x^2 - 2xy + y^2 = (x - y)^2$
Sum of cubes	$x^3 + y^3 = (x + y)(x^2 - xy + y^2)$
Difference of cubes	$x^3 - y^3 = (x - y)(x^2 + xy + y^2)$

- To factor a polynomial, use the general factoring strategy.

P.6 Rational Expressions

- A rational expression is a fraction in which the numerator and denominator are polynomials. The properties of rational expressions are used to simplify a rational expression and to find the sum, difference, product, and quotient of two rational expressions.
- Complex fractions can be simplified in either of the following ways:

  Method 1: Multiply both the numerator and the denominator by the LCD of all the fractions in the complex fraction.

  Method 2: Simplify the numerator to a single fraction and the denominator to a single fraction. Multiply the numerator by the reciprocal of the denominator.

CHAPTER P TRUE/FALSE EXERCISES

In Exercises 1 to 9, answer true or false. If the statement is false, give an example to show that the statement is false.

1. If a and b are real numbers, then $|a - b| = |b - a|$.
2. If a is a real number, then $a^2 \geq a$.
3. The set of rational numbers is closed under the operation of addition.
4. The set of irrational numbers is closed under the operation of addition.

5. Let $x \oplus y$ denote the average of the two real numbers x and y. That is,

$$x \oplus y = \frac{x + y}{2}$$

The operation $\oplus$ is an associative operation because $(x \oplus y) \oplus z = x \oplus (y \oplus z)$ for all real numbers x, y, and z.

6. Using interval notation, we write the inequality $x > a$ as $[a, \infty)$.

7. If n is a real number, then $\sqrt{n^2} = n$.

8. $(a + b)^2 = a^2 + b^2$

9. $\sqrt[3]{a^3 + b^3} = a + b$

CHAPTER P REVIEW EXERCISES

In Exercises 1 to 4, classify each number as one or more of the following: integer, rational number, irrational number, real number, prime number, composite number.

1. 3
2. $\sqrt{7}$
3. $-\frac{1}{2}$
4. $0.\overline{5}$

In Exercises 5 and 6, use $A = \{1, 5, 7\}$ and $B = \{2, 3, 5, 11\}$ to find the indicated intersection or union.

5. $A \cup B$
6. $A \cap B$

In Exercises 7 to 14, identify the real number property or the property of equality that is illustrated.

7. $5(x + 3) = 5x + 15$
8. $a(3 + b) = a(b + 3)$
9. $(6c)d = 6(cd)$
10. $\sqrt{2} + 3$ is a real number.
11. $7 + 0 = 7$
12. $1x = x$
13. If $7 = x$, then $x = 7$.
14. If $3x + 4 = y$, and $y = 5z$, then $3x + 4 = 5z$.

In Exercises 15 and 16, graph each inequality and write the inequality using interval notation.

15. $-4 < x \le 2$
16. $x \le -1$ or $x > 3$

In Exercises 17 and 18, graph each interval and write each interval as an inequality.

17. $[-3, 2)$
18. $(-1, \infty)$

In Exercises 19 to 22, write each real number without absolute value symbols.

19. $|7|$
20. $|2 - \pi|$
21. $|4 - \pi|$
22. $|-11|$

In Exercises 23 and 24, find the distance on the real number line between the points whose coordinates are given.

23. $-3, 14$
24. $\sqrt{5}, -\sqrt{2}$

In Exercises 25 and 26, evaluate each expression.

25. $-5^2 + (-11)$
26. $\dfrac{(2^2 \cdot 3^{-2})^2}{3^{-1} \cdot 2^3}$

In Exercises 27 and 28, simplify each expression.

27. $(3x^2y)(2x^3y)^2$
28. $\left(\dfrac{2a^2b^3c^{-2}}{3ab^{-1}}\right)^2$

In Exercises 29 and 30, write each number in scientific notation.

29. 620,000
30. 0.0000017

In Exercises 31 and 32, change each number from scientific notation to decimal form.

31. 3.5×10^4
32. 4.31×10^{-7}

In Exercises 33 to 36, perform the indicated operation and express each result as a polynomial in standard form.

33. $(2a^2 + 3a - 7) + (-3a^2 - 5a + 6)$
34. $(5b^2 - 11) - (3b^2 - 8b - 3)$
35. $(2x^2 + 3x - 5)(3x^2 - 2x + 4)$
36. $(3y - 5)^3$

In Exercises 37 to 40, completely factor each polynomial.

37. $3x^2 + 30x + 75$
38. $25x^2 - 30xy + 9y^2$
39. $20a^2 - 4b^2$
40. $16a^3 + 250$

In Exercises 41 and 42, simplify each rational expression.

41. $\dfrac{6x^2 - 19x + 10}{2x^2 + 3x - 20}$
42. $\dfrac{4x^3 - 25x}{8x^4 + 125x}$

In Exercises 43 to 46, perform the indicated operation and simplify if possible.

43. $\dfrac{10x^2 + 13x - 3}{6x^2 - 13x - 5} \cdot \dfrac{6x^2 + 5x + 1}{10x^2 + 3x - 1}$

44. $\dfrac{15x^2 + 11x - 12}{25x^2 - 9} \div \dfrac{3x^2 + 13x + 12}{10x^2 + 11x + 3}$

45. $\dfrac{x}{x^2 - 9} + \dfrac{2x}{x^2 + x - 12}$

46. $\dfrac{3x}{x^2 + 7x + 12} - \dfrac{x}{2x^2 + 5x - 3}$

In Exercises 47 and 48, simplify each complex fraction.

47. $\dfrac{2 + \dfrac{1}{x - 5}}{3 - \dfrac{2}{x - 5}}$

48. $\dfrac{1}{2 + \dfrac{3}{1 + \dfrac{4}{x}}}$

In Exercises 49 and 50, evaluate each exponential expression.

49. $25^{1/2}$

50. $-27^{2/3}$

In Exercises 51 to 54, simplify each expression.

51. $x^{2/3} \cdot x^{3/4}$

52. $\left(\dfrac{8x^{5/4}}{x^{1/2}}\right)^{2/3}$

53. $\left(\dfrac{x^2y}{x^{1/2}y^{-3}}\right)^{1/2}$

54. $(x^{1/2} - y^{1/2})(x^{1/2} + y^{1/2})$

In Exercises 55 to 64, simplify each radical expression. Assume the variables are positive real numbers.

55. $\sqrt{48a^2b^7}$

56. $\sqrt{12a^3b}$

57. $\sqrt{72x^2y}$

58. $\sqrt{18x^3y^5}$

59. $\sqrt{\dfrac{54xy^3}{10x}}$

60. $-\sqrt{\dfrac{24xyz^3}{15z^6}}$

61. $\dfrac{7x}{\sqrt[3]{2x^2}}$

62. $\dfrac{5y}{\sqrt[3]{9y}}$

63. $\sqrt[3]{-135x^2y^7}$

64. $\sqrt[3]{-250xy^6}$

CHAPTER P TEST

1. For real numbers a, b, and c, identify the property that is illustrated by $(a + b)c = ac + bc$.
2. Given $A = \{0, 2, 4, 6, 8\}$ and $B = \{1, 3, 5, 7, 9\}$, find $A \cup B$.
3. Find the distance between the points -12 and -5 on the number line.
4. Simplify: $(-2x^0y^{-2})^2(-3x^2y^{-1})^{-2}$
5. Simplify: $\dfrac{(2a^{-1}bc^{-2})^2}{(3^{-1}b)(2^{-1}ac^{-2})^3}$
6. Write 0.00137 in scientific notation.
7. Simplify: $(x - 2y)(x^2 - 2x + y)$
8. Evaluate the polynomial $3y^3 - 2y^2 - y + 2$ for $y = -3$.
9. Factor: $7x^2 + 34x - 5$
10. Factor: $3ax - 12bx - 2a + 8b$
11. Factor: $16x^4 - 2xy^3$
12. Simplify: $\dfrac{x^2 - 2x - 15}{25 - x^2}$
13. Simplify: $\dfrac{x}{x^2 + x - 6} - \dfrac{2}{x^2 - 5x + 6}$
14. Simplify: $\dfrac{2x^2 + 3x - 2}{x^2 - 3x} \div \dfrac{2x^2 - 7x + 3}{x^3 - 3x^2}$
15. Simplify: $\dfrac{3}{a + b} \cdot \dfrac{a^2 - b^2}{2a - b} - \dfrac{5}{a}$
16. Simplify: $x - \dfrac{x}{x + \dfrac{1}{2}}$
17. Simplify: $\dfrac{x^{1/3}y^{-3/4}}{x^{-1/2}y^{3/2}}$
18. Simplify: $3x\sqrt[3]{81xy^4} - 2y\sqrt[3]{3x^4y}$
19. Simplify: $\dfrac{x}{\sqrt[4]{2x^3}}$
20. Simplify: $\dfrac{3}{\sqrt{x} + 2}$

CHAPTER 1

EQUATIONS AND INEQUALITIES

◆ Pythagoras was the first to notice that vibrating strings produce harmonius tones when the ratios of the lengths of the strings are whole numbers.

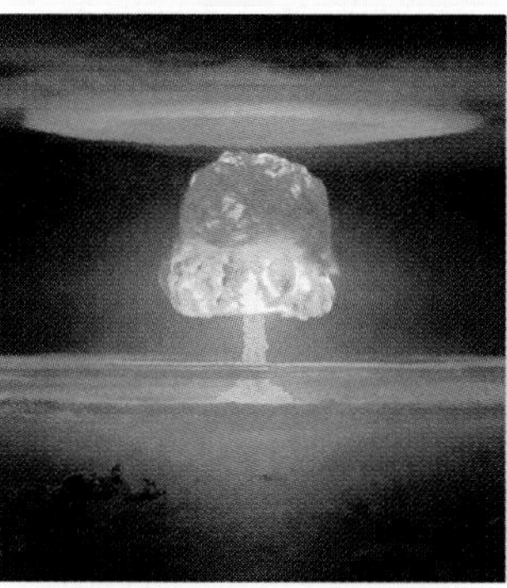

◆ In a hydrogen bomb, the fusion of two isotopes of hydrogen results in a release of energy.

Famous Equations[1]

Equations, the subject of this chapter, have played a fundamental role in the development of many disciplines. Some of these equations are quite famous because they significantly advanced the state of human knowledge. Here are some of those equations.

$a^2 + b^2 = c^2$ This is the Pythagorean Theorem which states that the sum of the squares of the sides of a right triangle equal the square of the hypotenuse. Although this theorem is credited to Pythagoras (circa 550 B.C.), it was known to the Babylonians at least 1500 years before Pythagoras was born. Roman builders used the principle of this equation to create right angles in buildings.

$F = \dfrac{GMm}{r^2}$ This equation is called Newton's Universal Law of Gravitation and was developed by Isaac Newton around 1660. This equation shows how to calculate the force between two bodies (like Earth and a person, or the sun and Venus). In 1682, Edmund Halley used this equation to calculate the equation of a comet's path and used that equation to predict the next time the comet would appear. The comet is known as Halley's comet.

$E = mc^2$ This equation derived by Albert Einstein in 1905 states that any mass (atom, golf ball, car) can be converted into energy. The startling fact about this equation is how much energy results when mass is converted to energy. The constant c in this equation is the speed of light, which is approximately 3.00×10^8 meters per second. Multiplying even a small mass by the *square* of the speed of light produces a significant amount of energy. This is evident any time a nuclear bomb explodes.

[1] See "Five Equations that Changed the World" by Michael Guillen, published by Hyperion, copyright 1995, for a discussion of some of these equations and others.

SECTION 1.1 LINEAR EQUATIONS

- SOLVE BY PRODUCING EQUIVALENT EQUATIONS
- CONTRADICTIONS, CONDITIONAL EQUATIONS, AND IDENTITIES
- ABSOLUTE VALUE EQUATIONS
- APPLICATIONS

◆ SOLVE BY PRODUCING EQUIVALENT EQUATIONS

An **equation** is a statement about the equality of two expressions. If either of the expressions contains a variable, the equation may be a true statement for some values of the variable and a false statement for other values. For example, the equation $2x + 1 = 7$ is a true statement for $x = 3$, but it is false for any number except 3. The number 3 is said to **satisfy** the equation $2x + 1 = 7$, because substituting 3 for x produces $2(3) + 1 = 7$, which is a true statement.

To **solve** an equation means to find all values of the variable that satisfy the equation. The values that satisfy an equation are called **solutions** or **roots** of the equation. For instance, 2 and 3 are both solutions of $x^2 - 5x + 6 = 0$.

Equivalent equations are equations that have exactly the same solution(s). The process of solving an equation involving the variable x is often accomplished by producing a sequence of equivalent equations until we produce an equation or equations of the form

$$x = \text{a constant}$$

To produce these equivalent equations that lead us to the solution(s), we often perform one or more of the following procedures.

Procedures That Produce Equivalent Equations

1. Simplification of an expression on either side of the equation by such procedures as (i) combining like terms and (ii) applying the properties explained in Chapter P, such as the commutative, associative, and distributive properties.

   $2x + 3 + 5x = -11$ and $7x + 3 = -11$ are equivalent equations.

2. Addition or subtraction of the same quantity on both sides of an equation.

   $3x - 7 = 2$ and $3x = 9$ are equivalent equations.

3. Multiplication or division by the same nonzero quantity on both sides of an equation.

   $\frac{5}{6}x = 10$ and $x = 12$ are equivalent equations.

Many applications can be modeled by *linear equations.*

Definition of a Linear Equation

A **linear equation** in the single variable x is an equation that can be written in the form

$$ax + b = 0$$

where a and b are real numbers, with $a \neq 0$.

Linear equations are generally solved by applying the procedures that produce equivalent equations.

EXAMPLE 1 Solve a Linear Equation

Solve: $\frac{3}{4}x - 6 = 0$

Solution

$$\frac{3}{4}x - 6 = 0$$

$$\frac{3}{4}x - 6 + 6 = 0 + 6 \qquad \text{• Add 6 to each side.}$$

$$\frac{3}{4}x = 6$$

$$\left(\frac{4}{3}\right)\left(\frac{3}{4}x\right) = \left(\frac{4}{3}\right)(6) \qquad \text{• Multiply each side by } \frac{4}{3}.$$

$$x = 8$$

Because 8 satisfies the original equation (see the *Take Note*), 8 is the solution.

take note

Check the proposed solution by substituting 8 for x in the original equation.

$$\frac{3}{4}x - 6 = 0$$

$$\frac{3}{4}(8) - 6 \stackrel{?}{=} 0$$

$$0 = 0 \quad \text{True}$$

TRY EXERCISE 2, EXERCISE SET 1.1, PAGE 69

If an equation involves fractions, it is helpful to multiply each side of the equation by the LCD of all the denominators to produce an equivalent equation that does not contain fractions.

EXAMPLE 2 Solve by Clearing Fractions

Solve: $\frac{2}{3}x + 10 - \frac{x}{5} = \frac{36}{5}$

Continued •➤

Solution

$$\frac{2}{3}x + 10 - \frac{x}{5} = \frac{36}{5}$$

$$15\left(\frac{2}{3}x + 10 - \frac{x}{5}\right) = 15\left(\frac{36}{5}\right)$$ • Multiply each side of the equation by 15, the LCD of the denominators.

$$10x + 150 - 3x = 108$$ • Simplify.

$$7x + 150 = 108$$

$$7x + 150 - 150 = 108 - 150$$ • Subtract 150 from each side.

$$7x = -42$$

$$\frac{7x}{7} = \frac{-42}{7}$$ • Divide each side by 7.

$$x = -6$$ • Check as before.

TRY EXERCISE 12, EXERCISE SET 1.1, PAGE 70

EXAMPLE 3 Solve an Equation by Applying Properties

Solve: $(x + 2)(5x + 1) = 5x(x + 1)$

Solution

$$(x + 2)(5x + 1) = 5x(x + 1)$$

$$5x^2 + 11x + 2 = 5x^2 + 5x$$ • Simplify each product.

$$11x + 2 = 5x$$ • Subtract $5x^2$ from each side.

$$6x + 2 = 0$$ • Subtract $5x$ from each side.

$$6x = -2$$ • Subtract 2 from each side.

$$x = -\frac{1}{3}$$ • Divide each side of the equation by 6.

TRY EXERCISE 18, EXERCISE SET 1.1, PAGE 70

◆ CONTRADICTIONS, CONDITIONAL EQUATIONS, AND IDENTITIES

An equation that has no solutions is called a **contradiction.** The equation $x = x + 1$ is a contradiction. No number is equal to itself increased by 1.

An equation that is true for some values of the variable but not true for other values of the variable is called a **conditional equation.** For example, $x + 2 = 8$ is a conditional equation because it is true for $x = 6$ and false for any number not equal to 6.

An **identity** is an equation that is true for *every* real number for which all terms of the equation are defined. Examples of identities include the equations $x + x = 2x$ and $(x + 3)^2 = x^2 + 6x + 9$.

EXAMPLE 4 Verify an Identity

Verify the identity $\dfrac{3(x^3 - 8)}{x - 2} = 3x^2 + 6x + 12, x \neq 2.$

Solution

Simplify the left side of the equation.

$$\begin{aligned}\frac{3(x^3 - 8)}{x - 2} &= \frac{3(x - 2)(x^2 + 2x + 4)}{x - 2} \\ &= 3(x^2 + 2x + 4) \\ &= 3x^2 + 6x + 12\end{aligned}$$

• **Factor the difference of cubes and simplify.**

Because we have shown that it is possible to write the left side of the equation exactly as the right side is written, we have verified the identity.

TRY EXERCISE 24, EXERCISE SET 1.1, PAGE 70

Multiplying each side of an equation by the same *nonzero* number always yields an equivalent equation. If each side of an equation is multiplied by an expression that involves a variable, then we restrict the variable so that the expression is not equal to zero. Example 5b illustrates the fact that you may produce incorrect results if you fail to restrict the variable.

EXAMPLE 5 Solve Equations That Have Restrictions

Solve each equation.

a. $\dfrac{x}{x - 3} = \dfrac{9}{x - 3} - 5$ b. $1 + \dfrac{x}{x - 5} = \dfrac{5}{x - 5}$

Solution

take note

When we multiply both sides of an equation by $x - a$, we assume that $x \neq a$.

a. First, note that the denominator $x - 3$ would equal zero if x were 3. To produce a simpler equivalent equation, multiply each side by $x - 3$, with the restriction that $x \neq 3$.

$$\begin{aligned}(x - 3)\left(\frac{x}{x - 3}\right) &= (x - 3)\left(\frac{9}{x - 3} - 5\right) \\ x &= (x - 3)\left(\frac{9}{x - 3}\right) - (x - 3)5 \\ x &= 9 - 5x + 15 \\ 6x &= 24 \\ x &= 4\end{aligned}$$

Substituting 4 for x in the original equation establishes that 4 is the solution.

Continued •➤

MATH MATTERS

Determine the incorrect step in the following "proof" that $2 = 1$.

$a = b$	Given.
$a^2 = ab$	Multiply by a.
$a^2 - b^2 = ab - b^2$	Subtract b^2.
$(a + b)(a - b) = b(a - b)$	Factor.
$\frac{(a + b)(a - b)}{(a - b)} = \frac{b(a - b)}{(a - b)}$	Divide by $a - b$.
$a + b = b$	Simplify.
$b + b = b$	Substitute b for a.
$2b = b$	Simplify.
$2 = 1$	Divide by b.

b. To produce a simpler equivalent equation, multiply each side of the equation by $x - 5$, with the restriction that $x \neq 5$.

$$(x - 5)\left(1 + \frac{x}{x - 5}\right) = (x - 5)\left(\frac{5}{x - 5}\right)$$

$$(x - 5)1 + (x - 5)\left(\frac{x}{x - 5}\right) = 5$$

$$x - 5 + x = 5$$

$$2x = 10$$

$$x = 5$$

Although we have obtained 5 as a proposed solution, 5 is *not* a solution of the original equation because it contradicts our restriction $x \neq 5$. Substitution of 5 for x in the original equation results in denominators of 0. In this case the original equation has no solution.

TRY EXERCISE 30, EXERCISE SET 1.1, PAGE 70

◆ ABSOLUTE VALUE EQUATIONS

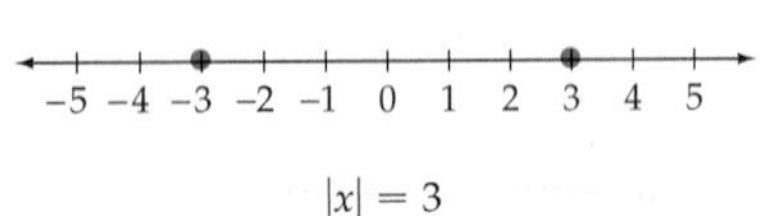

$|x| = 3$

Figure 1.1

The absolute value of a real number x is the distance between the number x and 0 on the real number line. For example, the solution set of $|x| = 3$ is the set of all real numbers that are 3 units from 0. Therefore, the solution set of $|x| = 3$ is $x = 3$ or $x = -3$. See **Figure 1.1**.

The following property is used to solve absolute value equations.

A Property of Absolute Value Equations

For any variable expression E and any nonnegative real number k,

$$|E| = k \quad \text{if and only if} \quad E = k \quad \text{or} \quad E = -k$$

take note

Some absolute value equations have an empty solution set. For example, $|x + 2| = -5$ is false for all values of x. Because an absolute value is always nonnegative, the equation is never true.

When solving an absolute value equation such as $|3x - 7| = 5$, remember to keep the $3x - 7$ the same when writing the two resulting equations, $3x - 7 = 5$ and $3x - 7 = -5$. Do not change the sign inside the absolute value bars.

EXAMPLE 6 Solve an Absolute Value Equation

Solve: $|2x - 5| = 21$

Solution

$|2x - 5| = 21$ implies $2x - 5 = 21$ or $2x - 5 = -21$. Solving each of these equations produces

$$2x - 5 = 21 \quad \text{or} \quad 2x - 5 = -21$$

$$2x = 26 \qquad\qquad 2x = -16$$

$$x = 13 \qquad\qquad x = -8$$

Therefore, the solutions of $|2x - 5| = 21$ are -8 and 13.

TRY EXERCISE 50, EXERCISE SET 1.1, PAGE 70

◆ APPLICATIONS

Linear equations can often be used to model real-world data.

EXAMPLE 7 An Application to Amusement Park Attendance

According to *Amusement Business* magazine, the attendance at amusement parks has been increasing steadily since 1994. An equation that approximates this increase is given by

$$\text{Attendance} = 10x + 260$$

where Attendance is in millions of people and x is the number of years *after* 1994. (This means that $x = 0$ corresponds to 1994.) Using this equation, determine in what year attendance first reached 300 million people.

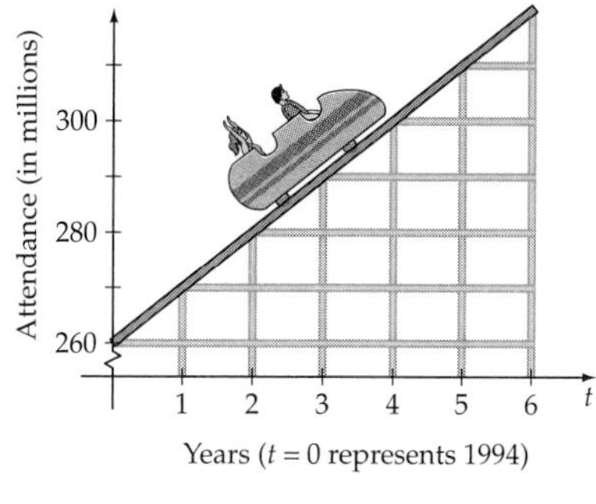

Figure 1.2

Solution

Replace Attendance by 300 and solve for x.

$$\text{Attendance} = 10x + 260$$

$$300 = 10x + 260 \quad \text{• Replace Attendance by 300.}$$

$$40 = 10x \quad \text{• Solve for } x.$$

$$4 = x$$

The year that corresponds to $x = 4$ is 1998. Thus the attendance first reached 300 million people in 1998. See **Figure 1.2.**

TRY EXERCISE 62, EXERCISE SET 1.1, PAGE 70

TOPICS FOR DISCUSSION

1. A student multiplies each side of the equation $\frac{1}{2}x + 3 = 4$ by 2 to produce the equation $x + 3 = 8$. Has the student produced an equivalent equation? Explain.
2. Suppose we attempt to solve the equation $4x = 7x$ by first dividing each side of the equation by x. The result is $4 = 7$, which is surely not true. Explain what happened.
3. If $P = Q$, is it also true that $Q = P$?
4. If $y = 5$ and $x + 1 = y$, does $x = 4$? Explain what property of equality is being used to reach the conclusion.
5. Consider the equation $|x + y| = |x| + |y|$. Is this equation true for all values of x and y, true for some values of x and y, or never true?

EXERCISE SET 1.1

In Exercises 1 to 28, solve and check each equation.

1. $2x + 10 = 40$
2. $-3y + 20 = 2$
3. $5x + 2 = 2x - 10$
4. $4x - 11 = 7x + 20$
5. $2(x - 3) - 5 = 4(x - 5)$

6. $5(x-4)-7=-2(x-3)$

7. $4(2r-17)+5(3r-8)=0$

8. $6(5s-11)-12(2s+5)=0$

9. $\frac{3}{4}x+\frac{1}{2}=\frac{2}{3}$

10. $\frac{x}{4}-5=\frac{1}{2}$

11. $\frac{2}{3}x-5=\frac{1}{2}x-3$

12. $\frac{1}{2}x+7-\frac{1}{4}x=\frac{19}{2}$

13. $0.2x+0.4=3.6$

14. $0.04x-0.2=0.07$

15. $x+0.08(60)=0.20(60+x)$

16. $6(t+1.5)=12t$

17. $3(x+5)(x-1)=(3x+4)(x-2)$

18. $5(x+4)(x-4)=(x-3)(5x+4)$

19. $5[x-(4x-5)]=3-2x$

20. $6[3y-2(y-1)]-2+7y=0$

21. $\frac{40-3x}{5}=\frac{6x+7}{8}$

22. $\frac{12+x}{-4}=\frac{5x-7}{3}+2$

In Exercises 23 to 28, determine whether the equation is an identity, a conditional equation, or a contradiction.

23. $-3(x-5)=-3x+15$

24. $2x+\frac{1}{3}=\frac{6x+1}{3}$

25. $2y+7=3(y-1)$

26. $x^2+10x=x(x+10)$

27. $\frac{4y+7}{4}=y+7$

28. $(x+3)^2=x^2+9$

In Exercises 29 to 44, solve and check each equation.

29. $\frac{3}{x+2}=\frac{5}{2x-7}$

30. $\frac{4}{y+2}=\frac{7}{y-4}$

31. $\frac{30}{10+x}=\frac{20}{10-x}$

32. $\frac{6}{8+x}=\frac{4}{8-x}$

33. $\frac{3x}{x+4}=2-\frac{12}{x+4}$

34. $\frac{8}{2m+1}-\frac{1}{m-2}=\frac{5}{2m+1}$

35. $2+\frac{9}{r-3}=\frac{3r}{r-3}$

36. $\frac{t}{t-4}+3=\frac{4}{t-4}$

37. $\frac{5}{x-3}-\frac{3}{x-2}=\frac{4}{x-3}$

38. $\frac{4}{x-1}+\frac{7}{x+7}=\frac{5}{x-1}$

39. $\frac{2x+5}{3x-1}=1$

40. $\frac{4x-1}{3x+2}=\frac{5}{6}$

41. $\frac{x}{x-3}=\frac{x+4}{x+2}$

42. $\frac{x}{x-5}=\frac{x+7}{x+1}$

43. $\frac{x+3}{x+5}=\frac{x-3}{x-4}$

44. $\frac{x-6}{x+4}=\frac{x-1}{x+2}$

In Exercises 45 to 60, solve each absolute value equation for x.

45. $|x|=4$

46. $|x|=7$

47. $|x-5|=2$

48. $|x-8|=3$

49. $|2x-5|=11$

50. $|2x-3|=21$

51. $|2x+6|=10$

52. $|2x+14|=60$

53. $\left|\frac{x-4}{2}\right|=8$

54. $\left|\frac{x+3}{4}\right|=6$

55. $|2x+5|=-8$

56. $|4x-1|=-17$

57. $2|x+3|+4=34$

58. $3|x-5|-16=2$

59. $|2x-a|=b \quad (b>0)$

60. $3|x-d|=c \quad (c>0)$

61. **RECREATION** The revenues of all the amusement and theme parks in the United States have been increasing since 1990. An equation that approximates the total revenues of all parks is given by the equation

$$\text{Revenues (in billions)} = 0.35x + 5.7$$

where x is the number of years after 1990. Using this equation, determine between what two years the revenues for all amusement and theme parks in the U.S. first exceeded \$10 billion. (Source: *Amusement Business* magazine as reported in the San Diego *Union-Tribune*, March 19, 2000)

62. **PATENTS** Data from the U.S. Patent and Trademark Office suggest that the number of patents that have been issued each year in the U.S. since 1993 can be approximated by the equation

$$\text{Number of patents (in thousands)} = 5.4x + 110$$

where x is the number of years after 1993. Using this equation, determine in what year the number of patents will first exceed 150,000 patents.

63. **COMPUTER SCIENCE** The percent of a file that remains to be downloaded using a dialup Internet connection for a certain modem is given by the equation

$$\text{Percent remaining} = 100 - \frac{42{,}000}{N}t$$

where N is the size of the file in bytes and t is the number of seconds since the download began. In how many minutes will 25% of a 500,000-byte file remain to be downloaded? Round to the nearest minute.

64. **AVIATION** The number of miles that remain to be flown by a commercial jet traveling from Boston to Los Angeles can be approximated by the equation

$$\text{Miles remaining} = 2650 - 475t$$

where t is the number of hours since leaving Boston. In how many hours will the plane be 1000 miles from Los Angeles. Round to the nearest tenth.

To benefit from an aerobic exercise program, many experts recommend that you exercise three to five times a week for 20 minutes to an hour. It is also important that your heart rate be in the *training zone*, which is defined by the following linear equations, where *a* is your age in years and the heart rate is in beats per minute.[1]

65. **MAXIMUM EXERCISE HEART RATE** Find the maximum exercise heart rate and the minimum exercise heart rate for a person who is 25 years of age. (Round to the nearest beat per minute.)

66. **MAXIMUM EXERCISE HEART RATE** How old is a person who has a maximum exercise heart rate of 153 beats per minute?

SUPPLEMENTAL EXERCISES

In Exercises 67 to 70, determine whether the given pair of equations are equivalent.

67. $3x - 11 = -5, \quad \dfrac{3x - 11}{x - 2} = \dfrac{-5}{x - 2}$

68. $3x - 9 = x - 3, \quad \dfrac{3x - 9}{x - 3} = \dfrac{x - 3}{x - 3}$

69. $\dfrac{1}{t} = \dfrac{1}{a} + \dfrac{1}{b}, \quad t = \dfrac{ab}{a + b}$, where t is a variable and a and b are nonzero constants, $a \neq -b$.

70. $\dfrac{2}{x} = \dfrac{1}{x - 1}, \quad 2(x - 1) = x$

71. Let a, b, and c be real constants. Show that an equation of the form $ax + b = c$ has $x = \dfrac{c - b}{a}$ $(a \neq 0)$ as its solution.

72. Let a, b, c and d be real constants. Show that an equation of the form $ax + b = cx + d$ has $x = \dfrac{d - b}{a - c}$ $(a - c \neq 0)$ as its solution.

In Exercises 73 to 80, find the values of *x* that make each equation true.

73. $|x + 4| = x + 4$

74. $|x - 1| = x - 1$

75. $|x + 7| = -(x + 7)$

76. $|x - 3| = -(x - 3)$

77. $|2x + 7| = 2x + 7$

78. $|3x - 11| = -3x + 11$

79. $|x - 2| + |x + 4| = 8$

80. $|x + 1| - |x + 3| = 4$

PROJECTS

1. **PERFECT GAMES** In baseball, a **perfect game** is a game in which one of the teams gives up no hits, no walks, and no errors. Statistics show that a batter will get on base roughly 30% of the time. Thus the probability that a pitcher will retire the batter is 70%, or 0.7 as a decimal. The probability that a pitcher will retire two batters in a row is $0.7^2 = 0.49$. The probability is 0.7^{27} that a pitcher will retire 27 batters in succession and thus pitch a perfect game.[2]

   a. Explain why the linear equation

   $$p = 2(0.7^{27})x$$

   provides a good estimate of the number of perfect games p we can expect after x games are completed.

[1] "The Heart of the Matter," *American Health*, September 1995.

[2] *A Mathematician Reads the Newspaper* by John Allen Paulos (New York: BasicBooks, A Division of HarperCollins Publishers, Inc., 1995).

b. Check a major league baseball almanac to determine how many perfect games have been played in the last 40 years and how many games have been played in the last 40 years.

c. Use the linear equation in **a.** to estimate how many perfect games we should expect to have been pitched over the last 40 years of major league baseball. How does this result compare with the actual result found in **b.**?

SECTION 1.2 FORMULAS AND APPLICATIONS

- FORMULAS
- APPLICATIONS

◆ FORMULAS

A **formula** is an equation that expresses known relationships between two or more variables. Table 1.1 lists several formulas from geometry that are used in this text. The variable P represents perimeter, C represents circumference of a circle, A represents area, S represents surface area of an enclosed solid, and V represents volume.

Table 1.1 Formulas from Geometry

Rectangle	Square	Triangle	Circle	Parallelogram
$P = 2l + 2w$	$P = 4s$	$P = a + b + c$	$C = \pi d = 2\pi r$	$P = 2b + 2s$
$A = lw$	$A = s^2$	$A = \frac{1}{2}bh$	$A = \pi r^2$	$A = bh$
w, l	s, s	c, h, a, b	r	h, s, b
Rectangular Solid	**Right Circular Cone**	**Sphere**	**Right Circular Cylinder**	**Frustum of a Cone**
$S = 2(wh + lw + hl)$	$S = \pi r\sqrt{r^2 + h^2} + \pi r^2$	$S = 4\pi r^2$	$S = 2\pi rh + 2\pi r^2$	$S = \pi(R + r)\sqrt{h^2 + (R - r)^2} + \pi r^2 + \pi R^2$
$V = lwh$	$V = \frac{1}{3}\pi r^2 h$	$V = \frac{4}{3}\pi r^3$	$V = \pi r^2 h$	$V = \frac{1}{3}\pi h(r^2 + rR + R^2)$
h, l, w	h, r	r	h, r	r, h, R

It is often necessary to solve a formula for a specified variable. Begin the process by isolating all terms that contain the specified variable on one side of the equation and all terms that do not contain the specified variable on the other side.

MATH MATTERS

- In Example 1a, the solution $l = \frac{P - 2w}{2}$ can also be written as $l = \frac{P}{2} - w$.
- In Example 1b, the restriction $x - z \neq 0$ is necessary to ensure that each side is divided by a *nonzero* expression.

EXAMPLE 1 Solve a Formula for a Specified Variable

a. Solve $2l + 2w = P$ for l. **b.** Solve $xy - z = yz$ for y.

Solution

a. $2l + 2w = P$

$2l = P - 2w$ • **Subtract 2*w* from each side to isolate the 2*l* term.**

$l = \frac{P - 2w}{2}$ • **Divide each side by 2.**

b. To solve for y, first isolate the terms that involve the variable y on the left side of the equation.

$xy - z = yz$

$xy - yz - z = 0$ • **Subtract yz from each side so that all terms that contain y are on the same side of the equation.**

$xy - yz = z$ • **Add z to each side to isolate the terms that contain y.**

$y(x - z) = z$ • **Factor y from each term on the left side of the equation.**

$y = \frac{z}{x - z}$ • **Divide each side of the equation by $x - z$, $x - z \neq 0$.**

TRY EXERCISE 4, EXERCISE SET 1.2, PAGE 79

◆ APPLICATIONS

People with good problem-solving skills generally work application problems by applying specific techniques in a series of small steps.

Guidelines for Solving Application Problems

1. Read the problem carefully. If necessary, reread the problem several times.
2. When appropriate, draw a sketch and label parts of the drawing with the specific information given in the problem.
3. Determine the unknown quantities, and label them with variables. Write down any equation that relates the variables.
4. Use the information from step 3, along with a known formula or some additional information given in the problem, to write an equation.
5. Solve the equation obtained in step 4, and check to see whether these results satisfy all the conditions of the original problem.

EXAMPLE 2 Solve an Application

The length of a rectangular garden is 2 feet greater than three times its width. If the perimeter of the garden is 92 feet, find the width and the length of the garden.

Solution

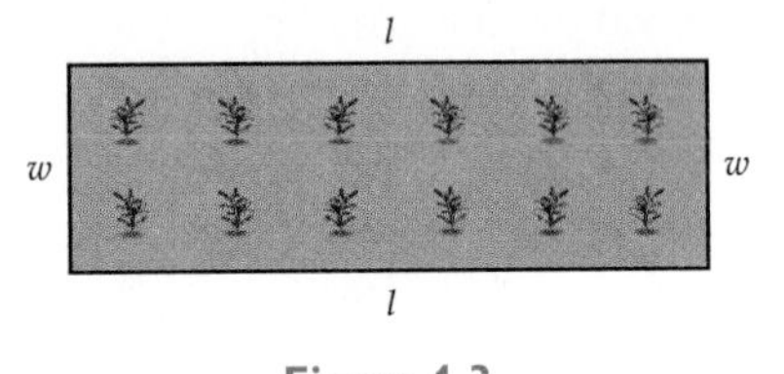

Figure 1.3

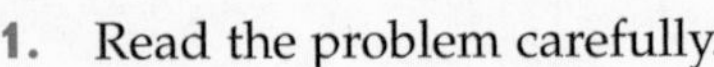

1. Read the problem carefully.

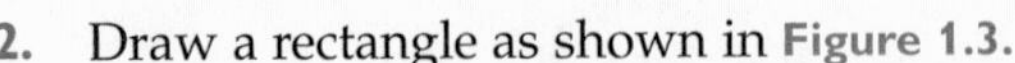

2. Draw a rectangle as shown in Figure 1.3.

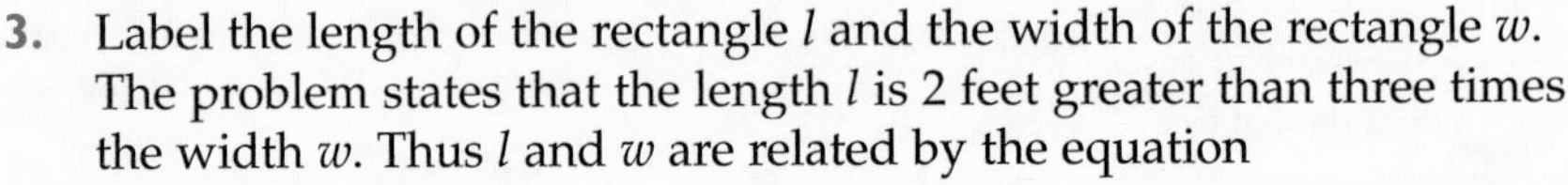

3. Label the length of the rectangle l and the width of the rectangle w. The problem states that the length l is 2 feet greater than three times the width w. Thus l and w are related by the equation

$$l = 3w + 2$$

4. Because the problem involves the length, width, and perimeter of a rectangle, we use the geometric formula $2l + 2w = P$. To write an equation that involves only constants and a single variable (say, w), substitute 92 for P and $3w + 2$ for l.

$$\begin{aligned} 2l + 2w &= P \\ 2(3w + 2) + 2w &= 92 \end{aligned}$$

5. Solve for the unknown w.

$$\begin{aligned} 6w + 4 + 2w &= 92 \\ 8w + 4 &= 92 \\ 8w &= 88 \\ w &= 11 \end{aligned}$$

Because the length l is two more than three times the width,

$$l = 3(11) + 2 = 35$$

A check verifies that 35 is two more than three times 11. Also, twice the length (70) plus twice the width (22) gives the perimeter (92). The width of the rectangle is 11 feet, and its length is 35 feet.

TRY EXERCISE 22, EXERCISE SET 1.2, PAGE 79

Many *uniform motion* problems can be solved by using the formula $d = rt$, where d is the distance traveled, r is the rate of speed, and t is the time.

EXAMPLE 3 Solve a Uniform Motion Problem

A runner runs a course at a constant speed of 6 mph. One hour after the runner begins, a cyclist starts on the same course at a constant speed of 15 mph. How long after the runner starts does the cyclist overtake the runner?

Solution

If we represent the time the runner has spent on the course by t, then the time the cyclist takes to overtake the runner is $t - 1$. The following table organizes the information and helps us determine how to write the distances each person travels.

	rate r	$\cdot$	time t	$=$	distance d
Runner	6	$\cdot$	t	$=$	$6t$
Cyclist	15	$\cdot$	$t - 1$	$=$	$15(t - 1)$

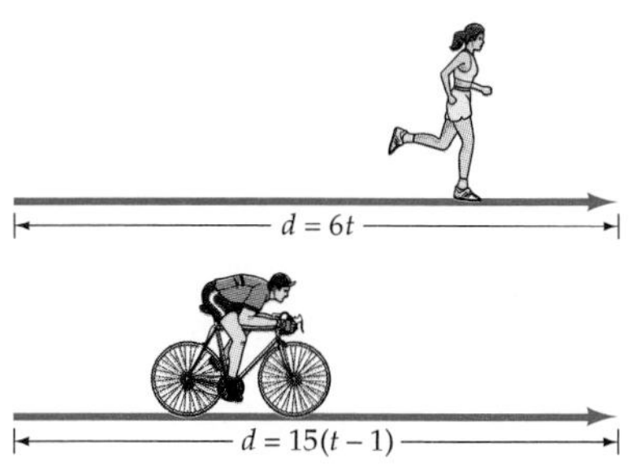

Figure 1.4

Figure 1.4 indicates that the runner and the cyclist cover the same distance. Thus

$$\begin{aligned} 6t &= 15(t - 1) \\ 6t &= 15t - 15 \\ -9t &= -15 \\ t &= 1\frac{2}{3} \end{aligned}$$

A check will verify that the cyclist does overtake the runner $1\frac{2}{3}$ hours after the runner starts.

TRY EXERCISE 28, EXERCISE SET 1.2, PAGE 79

Many business applications can be solved by using the equation

$$\text{Profit} = \text{revenue} - \text{cost}$$

EXAMPLE 4 Solve a Business Application

It costs a tennis shoe manufacturer \$26.55 to produce a pair of tennis shoes that sells for \$49.95. How many pairs of tennis shoes must the manufacturer sell to make a profit of \$14,274.00?

Solution

The *profit* is equal to the *revenue* minus the *cost*. If x equals the number of pairs of tennis shoes to be sold, then the revenue will be $\$49.95x$ and the cost will be $\$26.55x$. Therefore,

$$\begin{aligned} \text{Profit} &= \text{revenue} - \text{cost} \\ 14{,}274.00 &= 49.95x - 26.55x \\ 14{,}274.00 &= 23.40x \\ 610 &= x \end{aligned}$$

The manufacturer must sell 610 pairs of tennis shoes to make the desired profit.

TRY EXERCISE 36, EXERCISE SET 1.2, PAGE 80

Simple interest problems can be solved by using the formula $I = Prt$, where I is the interest, P is the principal, r is the simple interest rate per period, and t is the number of periods.

EXAMPLE 5 Solve a Simple Interest Problem

An accountant invests part of a \$6000 bonus in a 5% simple interest account and the remainder of the money is invested at 8.5% simple interest. Together the investments earn \$370 per year. Find the amount invested at each rate.

Solution

Let x be the amount invested at 5%. The remainder of the money is \$6000 − x, which will be the amount invested at 8.5%. Using $I = Prt$, with $t = 1$ year, yields

$$\text{Interest at } 5\% = x \cdot 0.05 = 0.05x$$
$$\text{Interest at } 8.5\% = (6000 - x) \cdot (0.085) = 510 - 0.085x$$

The interest earned on the two accounts equals \$370.

$$\begin{aligned} 0.05x + (510 - 0.085x) &= 370 \\ -0.035x + 510 &= 370 \\ -0.035x &= -140 \\ x &= 4000 \end{aligned}$$

Therefore, the accountant invested \$4000 at 5% and the remaining \$2000 at 8.5%. Check as before.

TRY EXERCISE 40, EXERCISE SET 1.2, PAGE 80

Percent mixture problems involve combining solutions or alloys that have different concentrations of a common substance. Percent mixture problems can be solved by using the formula $pA = Q$, where p is the percent of concentration, A is the amount of the solution or alloy, and Q is the quantity of a substance in the solution or alloy. For example, in 4 liters of a 25% acid solution, p is the percent of acid (25%), A is the amount of solution (4 liters), and Q is the amount of acid in the solution, which equals $(0.25) \cdot (4)$ liters $= 1$ liter.

EXAMPLE 6 Solve a Percent Mixture Problem

A chemist mixes an 11% hydrochloric acid solution with a 6% hydrochloric acid solution. How many milliliters (ml) of each solution should the chemist use to make a 600-milliliter solution that is 8% hydrochloric acid?

Solution

Let x be the number of milliliters of the 11% solution. Because the final solution will have a total of 600 milliliters of fluid, $600 - x$ is the number of milliliters of the 6% solution. See **Figure 1.5.**

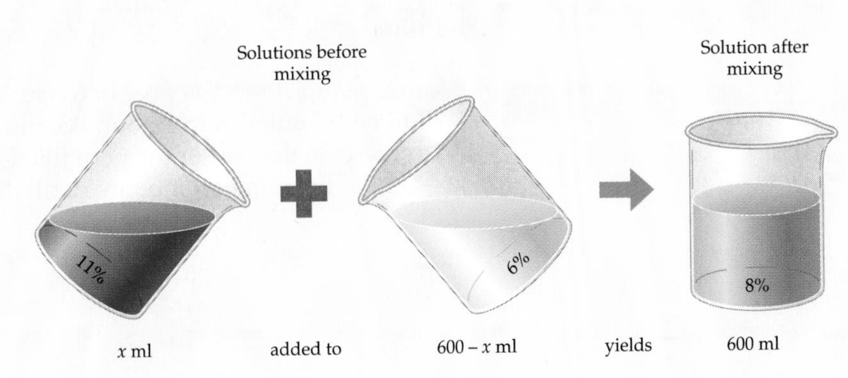

Figure 1.5

Because all the hydrochloric acid in the final solution comes from either the 11% solution or the 6% solution, the number of milliliters of hydrochloric acid in the 11% solution added to the number of milliliters of hydrochloric acid in the 6% solution must equal the number of milliliters of hydrochloric acid in the 8% solution.

$$\begin{pmatrix}\text{ml of acid in}\\ \text{11\% solution}\end{pmatrix} + \begin{pmatrix}\text{ml of acid in}\\ \text{6\% solution}\end{pmatrix} = \begin{pmatrix}\text{ml of acid in}\\ \text{8\% solution}\end{pmatrix}$$

$$0.11x \quad + \quad 0.06(600 - x) \quad = \quad 0.08(600)$$

$$0.11x + 36 - 0.06x = 48$$

$$0.05x + 36 = 48$$

$$0.05x = 12$$

$$x = 240$$

Therefore, the chemist should use 240 milliliters of the 11% solution and 360 milliliters of the 6% solution to make a 600-milliliter solution that is 8% hydrochloric acid.

TRY EXERCISE 44, EXERCISE SET 1.2, PAGE 80

To solve a *work problem,* use the equation

Rate of work × time worked = part of task completed

For example, if a painter can paint a wall in 15 minutes, then the painter can paint 1/15 of the wall in 1 minute. The painter's *rate of work* is 1/15 of the wall each minute. In general, if a task can be completed in x minutes, then the rate of work is $1/x$ of the task each minute.

EXAMPLE 7 Solve a Work Problem

Pump A can fill a pool in 6 hours, and pump B can fill the same pool in 3 hours. How long will it take to fill the pool if both pumps are used?

Continued •

Solution

Because pump A fills the pool in 6 hours, 1/6 represents the part of the pool filled by pump A in 1 hour. Because pump B fills the pool in 3 hours, 1/3 represents the part of the pool filled by pump B in 1 hour.

Let $t =$ the number of hours to fill the pool together. Then

$$t \cdot \frac{1}{6} = \frac{t}{6}$$ • **Part of the pool filled by pump A**

$$t \cdot \frac{1}{3} = \frac{t}{3}$$ • **Part of the pool filled by pump B**

$$\begin{pmatrix}\text{Part filled}\\ \text{by pump } A\end{pmatrix} + \begin{pmatrix}\text{Part filled}\\ \text{by pump } B\end{pmatrix} = \begin{pmatrix}\text{1 filled}\\ \text{pool}\end{pmatrix}$$

$$\frac{t}{6} \quad + \quad \frac{t}{3} \quad = \quad 1$$

Multiplying each side of the equation by 6 produces

$$t + 2t = 6$$
$$3t = 6$$
$$t = 2$$

Check: Pump A fills 2/6, or 1/3, of the pool in 2 hours and pump B fills 2/3 of the pool in 2 hours, so 2 hours is the time required to fill the pool if both pumps are used.

TRY EXERCISE 54, EXERCISE SET 1.2, PAGE 81

TOPICS FOR DISCUSSION

1. A student solves the formula $A = P + Prt$ for the variable P. The student's answer is $P = A - Prt$. Is this a correct response? Explain.
2. A student takes reciprocals of each term to write the formula
$$\frac{1}{f} = \frac{1}{d_0} + \frac{1}{d_i}$$
as $f = d_0 + d_i$. Did this technique produce a valid formula? Explain.
3. In the formula $S = a_1/(1 - r)$, what restrictions are placed on the variable r?
4. A tutor states that the formula $A = \frac{1}{2}bh$ can also be expressed as $A = \frac{bh}{2}$. Do you agree?
5. A tutor claims that a runner who runs the length of a track at 8 yards per second and then jogs back at 2 yards per second will take the same amount of time as a second runner who runs the length of the same track and back at a rate of 5 yards per second. Do you agree?

EXERCISE SET 1.2

In Exercises 1 to 18, solve the formula for the specified variable.

1. $V = \frac{1}{3}\pi r^2 h$; h (geometry)

2. $P = S - Sdt$; t (business)

3. $I = Prt$; t (business)

4. $A = P + Prt$; P (business)

5. $F = \frac{Gm_1m_2}{d^2}$; m_1 (physics)

6. $A = \frac{1}{2}h(b_1 + b_2)$; b_1 (geometry)

7. $s = v_0t - 16t^2$; v_0 (physics)

8. $\frac{1}{f} = \frac{1}{d_0} + \frac{1}{d_i}$; f (astronomy)

9. $Q_w = m_wc_w(T_f - T_w)$; T_w (physics)

10. $T\Delta t = Iw_f - Iw_i$; I (physics)

11. $a_n = a_1 + (n-1)d$; d (mathematics)

12. $y - y_1 = m(x - x_1)$; x (mathematics)

13. $S = \frac{a_1}{1-r}$; r (mathematics)

14. $\frac{P_1V_1}{T_1} = \frac{P_2V_2}{T_2}$; V_2 (chemistry)

15. $\frac{w_1}{w_2} = \frac{f_2 - f}{f - f_1}$; f_1 (hydrostatics)

16. $v = \frac{v_1 + v_2}{1 + \frac{v_1v_2}{c^2}}$; v_1 (physics)

17. $f_{LC} = f_v\frac{v + v_{LC}}{v}$; v_{LC} (physics)

18. $F_1d_1 + F_2d_2 = F_3d_3 + F_4d_4$; F_3 (physics)

In Exercises 19 to 58, solve by using the Guidelines for Solving Application Problems (see page 73).

19. One-fifth of a number plus one-fourth of the number is five less than one-half the number. What is the number?

20. The numerator of a fraction is 4 less than the denominator. If the numerator is increased by 14 and the denominator is decreased by 10, the resulting number is 5. What is the original fraction?

21. **GEOMETRY** The length of a rectangle is 3 feet less than twice the width of the rectangle. If the perimeter of the rectangle is 174 feet, find the width and the length.

22. **GEOMETRY** The width of a rectangle is 1 meter more than half the length of the rectangle. If the perimeter of the rectangle is 110 meters, find the width and the length.

23. **GEOMETRY** A triangle has a perimeter of 84 centimeters. Each of the two longer sides of the triangle is three times as long as the shortest side. Find the length of each side of the triangle.

24. **GEOMETRY** A triangle has a perimeter of 161 miles. Each of the two smaller sides of the triangle is two-thirds the length of the longest side. Find the length of each side of the triangle.

25. **NUMBER SENSE** If $5k - 3 = 2k + 9$, factor $x^2 + kx - 5$.

26. **NUMBER SENSE** Consider the arrangement of numbers below. In which column is the number 2002?

a	b	c	d	e
2	3	4	5	
	9	8	7	6
10	11	12	13	
	17	16	15	14

27. **UNIFORM MOTION** Running at an average rate of 6 meters per second, a sprinter ran to the end of a track and then jogged back to the starting point at an average rate of 2 meters per second. The total time for the sprint and the jog back was 2 minutes 40 seconds. Find the length of the track.

28. **UNIFORM MOTION** A motorboat left a harbor and traveled to an island at an average rate of 15 knots. The average speed on the return trip was 10 knots. If the total trip took 7.5 hours, how far is the harbor from the island?

29. **UNIFORM MOTION** A plane leaves an airport traveling at an average speed of 240 kilometers per hour. How long will it take a second plane traveling the same route at an average speed of 600 kilometers per hour to catch up with the first plane if it leaves 3 hours later?

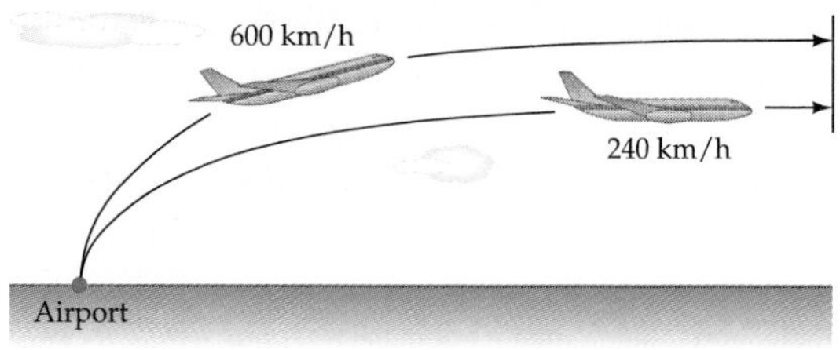

30. **UNIFORM MOTION** A plane leaves Chicago headed for Los Angeles at 540 mph. One hour later, a second plane leaves Los Angeles headed for Chicago at 660 mph. If the air route from Chicago to Los Angeles is 1800 miles, how long will it take for the first plane to pass the second plane? How far from Chicago will they be at that time?

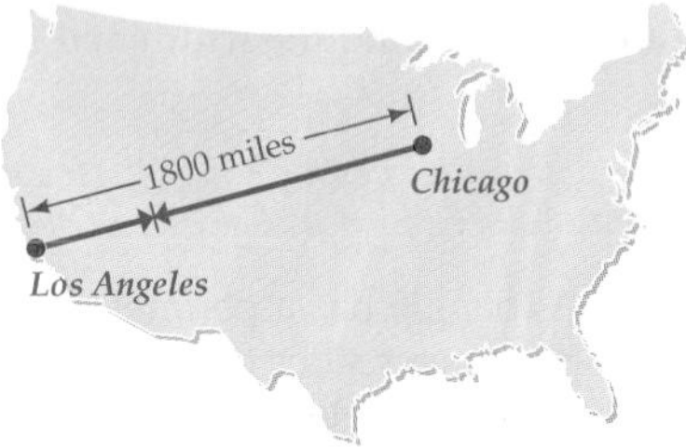

31. **UNIFORM MOTION** Marlene rides her bicycle to her friend Jon's house and returns home by the same route. Marlene rides her bike at constant speeds of 6 mph on level ground, 4 mph when going uphill, and 12 mph when going down hill. If her total time riding was 1 hour, how far is it to Jon's house?

32. **UNIFORM MOTION** A car traveling at 80 km/h is passed by a second car going in the same direction at a constant speed. After 30 seconds, the two cars are 500 meters apart. Find the speed of the second car.

33. **FINDING AN AVERAGE** A student has test scores of 80, 82, 94, and 71. What score does the student need on the next test to produce an average score of 85?

34. **FINDING AN AVERAGE** A student has test scores of 90, 74, 82, and 90. The next examination is the final examination, which will count as two tests. What score does the student need on the final examination to produce an average score of 85?

35. **BUSINESS** It costs a manufacturer of sunglasses \$8.95 to produce sunglasses that sell for \$29.99. How many sunglasses must the manufacturer sell to make a profit of \$17,884?

36. **BUSINESS** It costs a restaurant owner 18 cents per glass for orange juice, which is sold for 75 cents per glass. How many glasses of orange juice must the restaurant owner sell to make a profit of \$2337?

37. **BUSINESS** The price of a computer fell 20% this year. If the computer now costs \$750, how much did it cost last year?

38. **BUSINESS** The price of a magazine subscription rose 4% this year. If the subscription now costs \$26, how much did it cost last year?

39. **INVESTMENT** An investment adviser invested \$14,000 in two accounts. One investment earned 8% annual simple interest, and the other investment earned 6.5% annual simple interest. The amount of interest earned for 1 year was \$1024. How much was invested in each account?

40. **INVESTMENT** A total of \$7500 is deposited into two simple interest accounts. On one account the annual simple interest rate is 5%, and on the second account the annual simple interest rate is 7%. The amount of interest earned for 1 year was \$405. How much was invested in each account?

41. **INVESTMENT** An investment of \$2500 is made at an annual simple interest rate of 5.5%. How much additional money must be invested at an annual simple interest rate of 8% so that the total interest earned is 7% of the total investment?

42. **INVESTMENT** An investment of \$4600 is made at an annual simple interest rate of 6.8%. How much additional money must be invested at an annual simple interest rate of 9% so that the total interest earned is 8% of the total investment?

43. **METALLURGY** How many grams of pure silver must a silversmith mix with a 45% silver alloy to produce 200 grams of a 50% alloy?

44. **CHEMISTRY** How many liters of a 40% sulfuric acid solution should be mixed with 4 liters of a 24% sulfuric acid solution to produce a 30% solution?

45. **NURSING** How many liters of water should be evaporated from 160 liters of a 12% saline solution so that the solution that remains is a 20% saline solution?

46. **AUTOMOTIVE** A radiator contains 6 liters of a 25% antifreeze solution. How much should be drained and replaced with pure antifreeze to produce a 33% antifreeze solution?

47. **COMMERCE** A ballet performance brought in \$61,800 on the sale of 3000 tickets. If the tickets sold for \$14 and \$25, how many of each were sold?

48. **COMMERCE** A vending machine contains \$41.25. The machine contains 255 coins, which consist only of nickels,

dimes, and quarters. If the machine contains twice as many dimes as nickels, how many of each type of coin does the machine contain?

49. **COMMERCE** A coffee shop decides to blend a coffee that sells for \$12 per pound with a coffee that sells for \$9 per pound to produce a blend that will sell for \$10 per pound. How much of each should be used to yield 20 pounds of the new blend?

50. **DETERMINE NUMBER OF COINS** A bag contains 42 coins, with a total weight of 246 grams. If the bag contains only gold coins that weigh 8 grams each and silver coins that weigh 5 grams each, how many gold and how many silver coins are in the bag?

51. **METALLURGY** How much pure gold should be melted with 15 grams of 14-karat gold to produce 18-karat gold? *Hint:* A karat is a measure of the purity of gold in an alloy. Pure gold measures 24 karats. An alloy that measures x karats is $x/24$ gold. For example, 18-karat gold is $18/24 = 3/4$ gold.

52. **METALLURGY** How much 14-karat gold should be melted with 4 ounces of pure gold to produce 18-karat gold? (*Hint:* See Exercise 51.)

53. **INSTALL ELECTRICAL WIRES** An electrician can install the electric wires in a house in 14 hours. A second electrician requires 18 hours. How long would it take both electricians, working together, to install the wires?

54. **PRINT A REPORT** Printer A can print a report in 3 hours. Printer B can print the same report in 4 hours. How long would it take both printers, working together, to print the report?

55. **BUILD A FENCE** A worker can build a fence in 8 hours. With the help of an assistant, the fence can be built in 5 hours. How long would it take the assistant to build the fence alone?

56. **REPAIR A ROOF** A roofer and an assistant can repair a roof together in 6 hours. The assistant can complete the repair alone in 14 hours. If both the roofer and the assistant work together for 2 hours and then the assistant is left alone to finish the job, how much longer will the assistant need to finish the repairs?

57. **DETERMINE INDIVIDUAL PRICES** A book and a bookmark together sell for \$10.10. If the price of the book is \$10.00 more than the price of the bookmark, find the price of the book and the price of the bookmark.

58. **SHARE AN EXPENSE** Three people decide to share the cost of a yacht. By bringing in an additional partner, they can reduce the cost for each by \$4000. What is the total cost of the yacht?

SUPPLEMENTAL EXERCISES

The *Archimedean law of the lever* states that for a lever to be in a state of balance with respect to a point called the fulcrum, the sum of the downward forces times their respective distances from the fulcrum on one side of the fulcrum must equal the sum of the downward forces times their respective distances from the fulcrum on the other side of the fulcrum. The accompanying figure shows this relationship.

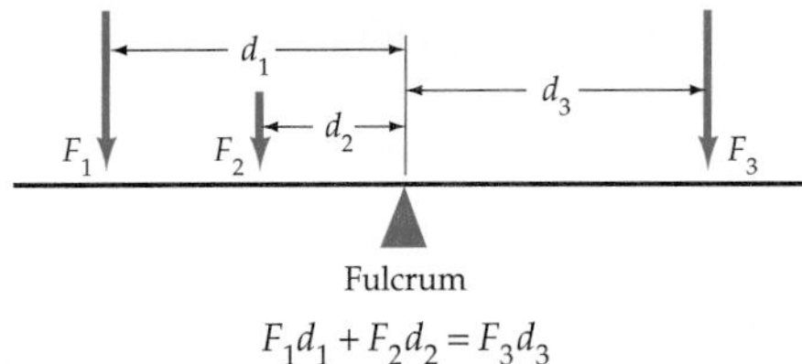

$$F_1d_1 + F_2d_2 = F_3d_3$$

59. **LOCATE THE FULCRUM** A 100-pound person 8 feet to the left of the fulcrum and a 40-pound person 5 feet to the left of the fulcrum balance with a 160-pound person on a teeter-totter. How far from the fulcrum is the 160-pound person?

60. **LOCATE THE FULCRUM** A lever 21 feet long has a force of 117 pounds applied to one end of the lever and a force of 156 pounds applied to the other end. Where should the fulcrum be located to produce a state of balance?

61. **DETERMINE A FORCE** How much force applied 5 feet from the fulcrum is needed to lift a 400-pound weight that is located on the other side, 0.5 foot from the fulcrum?

62. **DETERMINE A FORCE** Two workers need to lift a 1440-pound rock. They use a 6-foot steel bar with the fulcrum 1 foot from the rock, as the accompanying figure shows. One worker applies 180 pounds of force to the other end of the lever. How much force will the second worker need to apply 1 foot from that end to lift the rock?

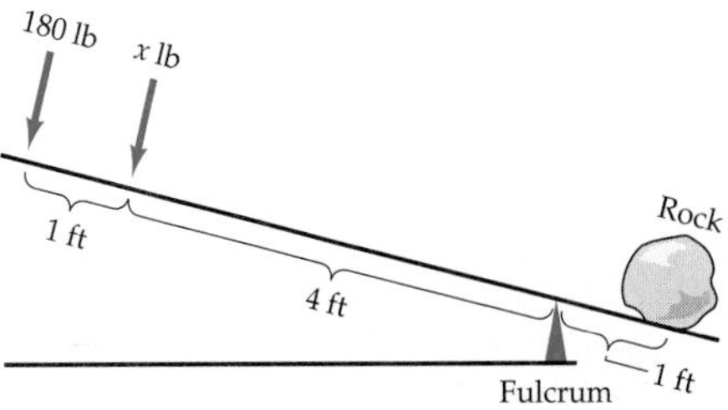

63. **SPEED OF SOUND IN AIR** Two seconds after firing a rifle at a target, the shooter hears the impact of the bullet. Sound travels at 1100 feet per second and the bullet at 1865 feet per second. Determine the distance to the target (to the nearest foot).

64. SPEED OF SOUND IN WATER Sound travels through sea water 4.62 times faster than through air. The sound of an exploding mine on the surface of the water and partially submerged reaches a ship through the water 4 seconds before it reaches the ship through the air. How far is the ship from the explosion? Use 1100 feet per second as the speed of sound through the air.

65. AGE OF DIOPHANTUS The work of the ancient Greek mathematician Diophantus had great influence on later European number theorists. Nothing is known about his personal life except for the information given in the following epigram. "Diophantus passed 1/6 of his life in childhood, 1/12 in youth, and 1/7 more as a bachelor. Five years after his marriage was born a son who died four years before his father, at 1/2 his father's (final) age." How old was Diophantus when he died?

66. EQUIVALENT TEMPERATURES The relationship between the Fahrenheit temperature (F) and the Celsius temperature (C) is given by the formula

$$F = \frac{9}{5}C + 32$$

At what temperature will a Fahrenheit thermometer and a Celsius thermometer read the same?

PROJECTS

1. **A WORK PROBLEM AND ITS EXTENSIONS** If a pump can fill a pool in A hours, and a second pump can fill the same pool in B hours, then the total time T in hours to fill the pool with both pumps working together is given by

$$T = \frac{AB}{A + B}$$

a. Verify this formula.

b. Consider the case where a pool is to be filled by three pumps. One can fill the pool in A hours, a second in B hours, and a third in C hours. Derive a formula in terms of A, B, and C for the total time T needed to fill the pool.

c. Consider the case where a pool is to be filled by n pumps. One pump can fill the pool in A_1 hours, a second in A_2 hours, a third in A_3 hours, ..., and the nth pump can fill the pool in A_n hours. Write a formula in terms of $A_1, A_2, A_3, \ldots, A_n$ for the total time T needed to fill the pool.

The chart at the right is called an *alignment chart* or a *nomogram*. If you know any two of the values A, B, and T, then you can use the alignment chart to determine the unknown value. For example, the straight line segment that connects 3 on the A-axis with 6 on the B-axis crosses the T-axis at 2. Thus the total time required for a pump that takes 3 hours to fill the pool and a pump that takes 6 hours to fill the pool is 2 hours when they work together.

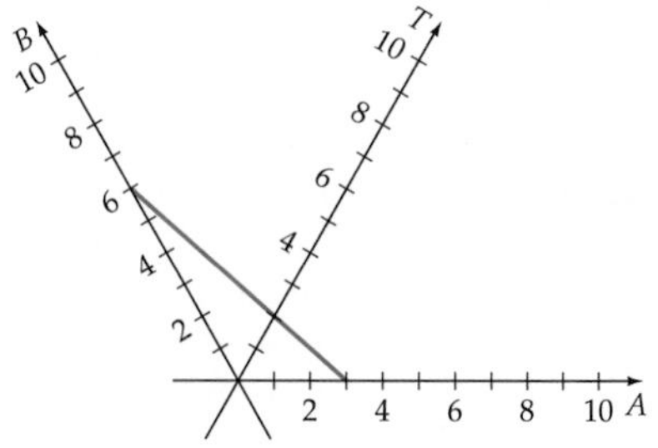

Alignment Chart for $T = \dfrac{AB}{A+B}$

d. Consider the case where a pool is to be filled by three pumps. One can fill the pool in $A = 6$ hours, a second in $B = 8$ hours, and a third in $C = 12$ hours. Write a few sentences explaining how you could make use of the alignment chart at the right to show that it takes about 2.7 hours for the three pumps to fill the pool when they work together.

2. **RESISTANCE OF PARALLEL CIRCUITS** The alignment chart shown in Project 1 can be used to solve some problems in electronics that concern the total resistance of a *parallel* circuit. Read an electronics text, and write a paragraph or two that explain this problem and how it is related to the problem of filling a pool with two pumps.

SECTION 1.3 QUADRATIC EQUATIONS

- SOLVING QUADRATIC EQUATIONS BY FACTORING
- SOLVING QUADRATIC EQUATIONS BY TAKING SQUARE ROOTS
- COMPLEX NUMBERS
- SOLVING QUADRATIC EQUATIONS BY COMPLETING THE SQUARE
- SOLVING QUADRATIC EQUATIONS BY USING THE QUADRATIC FORMULA
- THE DISCRIMINANT
- APPLICATIONS

1B

MATH MATTERS

The term *quadratic* is derived from the Latin word *quadrāre*, which means "to make square." Because the area of a square that measures x units on each side is x^2, we refer to equations that can be written in the form $ax^2 + bx + c = 0$ as equations that are quadratic in x.

◆ SOLVING QUADRATIC EQUATIONS BY FACTORING

A **quadratic equation** in x is an equation that can be written in the **standard quadratic form** $ax^2 + bx + c = 0, a \neq 0$.

Several methods can be used to solve quadratic equations. If the quadratic polynomial $ax^2 + bx + c$ can be factored over the integers, then the equation can be solved by factoring and using the **zero product property.**

Zero Product Property

If A and B are algebraic expressions, then

$$AB = 0 \quad \text{if and only if} \quad A = 0 \text{ or } B = 0$$

This property states that when the product of two factors equals zero, then at least one of the factors is zero.

EXAMPLE 1 Solve by Using the Zero Product Property

Solve each quadratic equation.

a. $3x^2 + 10x = 8$ **b.** $x^2 + 10x + 25 = 0$

Solution

a.

$$3x^2 + 10x = 8$$

$$3x^2 + 10x - 8 = 0$$ • **Write in standard quadratic form.**

$$(3x - 2)(x + 4) = 0$$ • **Factor.**

$$3x - 2 = 0 \quad \text{or} \quad x + 4 = 0$$ • **Apply the zero product property.**

$$3x = 2 \quad \text{or} \quad x = -4$$

$$x = \frac{2}{3} \quad \text{or} \quad x = -4$$ • **Check as before.**

The solutions are -4 and $\frac{2}{3}$.

Continued

b. $x^2 + 10x + 25 = 0$

$(x + 5)^2 = 0$ • **Factor.**

$x + 5 = 0$ or $x + 5 = 0$ • **Apply the zero product property.**

$x = -5$ or $x = -5$ • **Check as before.**

The only solution is -5.

TRY EXERCISE 6, EXERCISE SET 1.3, PAGE 96

In Example 1b, the solution or root -5 is called a **double root** of the equation because the application of the zero product property produced the two identical equations $x + 5 = 0$, both of which have a root of -5.

◆ SOLVING QUADRATIC EQUATIONS BY TAKING SQUARE ROOTS

The quadratic equation $x^2 = c$ can be solved by taking the square root of each side of the equation. Before we do this, however, recall that $\sqrt{x^2} = |x|$. This fact is an important part of what follows.

$x^2 = c$

$\sqrt{x^2} = \sqrt{c}$ • **Take the square root of each side of the equation.**

$|x| = \sqrt{c}$ • $\sqrt{x^2} = |x|$

$x = \pm\sqrt{c}$

This result is known as the square root theorem, which we will use to solve quadratic equations that can be written in the form $A^2 = B$.

The Square Root Theorem

If A and B are algebraic expressions such that

$$A^2 = B, \quad \text{then} \quad A = \pm\sqrt{B}$$

EXAMPLE 2 Solve by Using the Square Root Theorem

Use the square root theorem to solve $(x + 1)^2 = 49$.

Solution

$(x + 1)^2 = 49$

$x + 1 = \pm\sqrt{49}$ • **Apply the square root theorem.**

$x + 1 = \pm 7$ • **Solve for x.**

$x = -1 \pm 7$

Thus $x = -1 - 7 = -8$ or $x = -1 + 7 = 6$.

The solutions of $(x + 1)^2 = 49$ are -8 and 6.

TRY EXERCISE 14, EXERCISE SET 1.3, PAGE 96

◆ COMPLEX NUMBERS

Now consider the equation $x^2 + 1 = 0$. If we attempt to solve this equation using the square root theorem, we have

$$x^2 + 1 = 0$$
$$x^2 = -1$$
$$x = \pm\sqrt{-1}$$

To find $\sqrt{-1}$ we would need a number whose square is -1. However, the square of any nonzero real number is a positive number. Thus, during the seventeenth century, a new number, called an **imaginary number,** was defined. The letter i was chosen to represent this number.

MATH MATTERS

Electrical engineers use i to represent the current in a circuit. To avoid confusion between i for current and the imaginary unit i, these engineers use j as the imaginary unit.

Definition of i

The number i, called the **imaginary unit,** is the number such that

$$i^2 = -1$$

Many of the solutions to equations in the remainder of this text will involve radicals such as $\sqrt{-a}$, where a is a positive real number. The expression $\sqrt{-a}$, with $a > 0$, is defined as follows:

Definition of $\sqrt{-a}$

For any positive real number a,

$$\sqrt{-a} = i\sqrt{a}$$

This definition with $a = 1$ implies that $\sqrt{-1} = i$. It is often used to write the square root of a negative real number as the product of the imaginary unit i and a positive real number. For example,

$$\sqrt{-4} = i\sqrt{4} = 2i \quad \text{and} \quad \sqrt{-7} = i\sqrt{7}$$

take note

Even though b is a real number, it is called the imaginary part of the complex number $a + bi$. For example, the complex number $3 + 8i$ has the real number 8 as its imaginary part.

Definition of a Complex Number

If a and b are real numbers and i is the imaginary unit, then $a + bi$ is called a **complex number.** The real number a is called the **real part** and the real number b is called the **imaginary part** of the complex number.

The real numbers are a subset of the complex numbers. This can be observed by letting $b = 0$. Then $a + bi = a + 0i = a$, which is a real number. It can be shown that the associative, commutative, distributive, and identity properties also apply to complex numbers. Any number that can be written in the form $0 + bi = bi$, where b is a nonzero real number, is an **imaginary number** (or a pure imaginary number). For example, i, $3i$, and $-0.5i$ are all imaginary numbers.

take note

The expression $\sqrt{a}\,i$ is often written $i\sqrt{a}$ so that it is not mistaken for $\sqrt{ai}$.

A complex number is in **standard form** when it is written in the form $a + bi$. For instance,

$3 + \sqrt{-4} = 3 + i\sqrt{4} = 3 + 2i$ • $a + bi$ form with $a = 3$ and $b = 2$.

$\sqrt{-37} - 3 = i\sqrt{37} - 3 = -3 + i\sqrt{37}$

EXAMPLE 3 Solve an Equation with Complex Number Solutions

Solve: $(2x + 4)^2 + 20 = 0$

Solution

$(2x + 4)^2 + 20 = 0$

$(2x + 4)^2 = -20$ • Solve for $(2x + 4)^2$.

$\sqrt{(2x + 4)^2} = \sqrt{-20}$ • Take the square root of each side of the equation.

$2x + 4 = \pm 2i\sqrt{5}$ • Solve for x.

$2x = -4 \pm 2i\sqrt{5}$

$x = -2 \pm i\sqrt{5}$

The solutions are $-2 + i\sqrt{5}$ and $-2 - i\sqrt{5}$.

TRY EXERCISE 22, EXERCISE SET 1.3, PAGE 96

If we were to check the solutions to Example 3, it would be necessary to perform arithmetic operations such as addition and multiplication of complex numbers.

Definition of Addition and Subtraction of Complex Numbers

If $a + bi$ and $c + di$ are complex numbers, then

Addition $(a + bi) + (c + di) = (a + c) + (b + d)i$

Subtraction $(a + bi) - (c + di) = (a - c) + (b - d)i$

To add two complex numbers, add their real parts to produce the real part of the sum and add their imaginary parts to produce the imaginary part of the sum. For instance,

$(4 + 2i) + (3 + 7i) = (4 + 3) + (2 + 7)i = 7 + 9i$

$i - (3 - 4i) = (0 + 1i) - (3 - 4i)$ • $i = 0 + 1i$

$= (0 - 3) + [1 - (-4)]i = -3 + 5i$

Definition of Multiplication of Complex Numbers

If $a + bi$ and $c + di$ are complex numbers, then

$$(a + bi)(c + di) = (ac - bd) + (ad + bc)i$$

Because every complex number can be written as a sum of two terms, it is natural to perform multiplication on complex numbers in a manner consistent with the operation of multiplication defined on binomials and the definition $i^2 = -1$. Thus to multiply complex numbers, it is not necessary to memorize the definition of multiplication.

EXAMPLE 4 Multiply Complex Numbers

Simplify: $(3 + 5i)(2 - 4i)$

Solution

$$\begin{aligned}(3 + 5i)(2 - 4i) &= 6 - 12i + 10i - 20i^2 \\ &= 6 - 12i + 10i - 20(-1) && \text{• Substitute } -1 \text{ for } i^2. \\ &= 6 - 12i + 10i + 20 && \text{• Simplify.} \\ &= 26 - 2i\end{aligned}$$

TRY EXERCISE 32, EXERCISE SET 1.3, PAGE 96

Caution To compute $\sqrt{a}\sqrt{b}$ when both a and b are negative numbers, write each radical in terms of i before multiplying. For example,

Correct method $\quad \sqrt{-4}\sqrt{-9} = i\sqrt{4} \cdot i\sqrt{9} = 2i \cdot 3i = 6i^2 = -6$

Here are two other examples.

$$\sqrt{-20}\sqrt{-5} = (i\sqrt{20})(i\sqrt{5}) = i^2\sqrt{100} = -1(10) = -10$$

$$\begin{aligned}(2 + 5\sqrt{-3})(1 - 2\sqrt{-3}) &= (2 + 5i\sqrt{3})(1 - 2i\sqrt{3}) \\ &= 2 - 4i\sqrt{3} + 5i\sqrt{3} - 10i^2\sqrt{9} \\ &= 2 + i\sqrt{3} + 30 = 32 + i\sqrt{3}\end{aligned}$$

The complex numbers $a + bi$ and $a - bi$ are called **complex conjugates** or **conjugates** of each other. The conjugate of the complex number z is denoted by $\overline{z}$. For example,

$$\overline{3 + 2i} = 3 - 2i \qquad \text{and} \qquad \overline{7 - 11i} = 7 + 11i$$

Consider the product of a complex number and its conjugate. For instance,

$$\begin{aligned}(3 + 2i)(3 - 2i) &= 9 - 6i + 6i - 4i^2 && \text{• } 3 + 2i \text{ and } 3 - 2i \text{ are conjugates.} \\ &= 9 - 4(-1) \\ &= 13\end{aligned}$$

Note that the product is a *real* number. This is aways true.

Product of Conjugates

The product of a complex number and its conjugate is a real number. That is, if $a + bi$ and $a - bi$ are complex conjugates, then

$$(a + bi)(a - bi) = a^2 + b^2$$

The fact that the product of a complex number and its conjugate is a real number can be used to find the quotient of two complex numbers. For instance, to find the quotient $\frac{a + bi}{c + di}$, multiply the numerator and denominator by the conjugate of the denominator.

EXAMPLE 5 Divide Complex Numbers

Find the quotient: $\frac{3 + 2i}{5 - i}$

Solution

$$\frac{3 + 2i}{5 - i} = \frac{(3 + 2i)(5 + i)}{(5 - i)(5 + i)}$$

• **Multiply numerator and denominator by 5 + i, which is the conjugate of the denominator.**

$$= \frac{15 + 3i + 10i + 2i^2}{25 + 1}$$

$$= \frac{13 + 13i}{26} = \frac{1}{2} + \frac{1}{2}i$$

• **Write in standard form.**

TRY EXERCISE 40, EXERCISE SET 1.3, PAGE 96

The following powers of i illustrate a pattern:

$$i^1 = i \qquad i^5 = i^4 \cdot i = (1)i = i$$
$$i^2 = -1 \qquad i^6 = i^4 \cdot i^2 = (1)(-1) = -1$$
$$i^3 = i^2 \cdot i = (-1)i = -i \qquad i^7 = i^4 \cdot i^3 = (1)(-i) = -i$$
$$i^4 = i^2 \cdot i^2 = (-1)(-1) = 1 \qquad i^8 = (i^4)^2 = 1^2 = 1$$

Because $i^4 = 1$, $(i^4)^n = 1$ for any integer n. Thus it is possible to evaluate powers of i by factoring out powers of i^4, as shown in the following example:

$$i^{25} = (i^4)^6(i) = 1^6(i) = i$$

The following theorem can be used to evaluate powers of i. Essentially, it makes use of division to eliminate powers of i^4.

Powers of i

If n is a positive integer, then $i^n = i^r$, where r is the remainder of the division of n by 4.

For an example of this theorem, consider the evaluation of i^{543}. Divide 543 (the exponent) by 4 and determine the remainder.

$$i^{543} = i^3 = -i$$

• **$4\overline{)543}$ = 135 r3**

◆ SOLVING QUADRATIC EQUATIONS BY COMPLETING THE SQUARE

Consider the following binomial squares and their perfect-square trinomial products.

Square of a Binomial		Perfect-Square Trinomial
$(x + 6)^2$	=	$x^2 + 12x + 36$
$(x - 3)^2$	=	$x^2 - 6x + 9$

In each perfect-square trinomial, the coefficient of x^2 is 1, and the constant term of the perfect-square trinomial is the square of half the coefficient of its x term.

$$x^2 + 12x + 36, \qquad \left(\frac{1}{2} \cdot 12\right)^2 = 36$$

$$x^2 - 6x + 9, \qquad \left(\frac{1}{2}(-6)\right)^2 = 9$$

Adding, to a binomial of the form $x^2 + bx$, the constant that makes that binomial a perfect-square trinomial is called **completing the square.** For example, to complete the square of $x^2 + 8x$, add

$$\left(\frac{1}{2} \cdot 8\right)^2 = 16$$

to produce the perfect-square trinomial $x^2 + 8x + 16$.

Completing the square is a powerful method because it can be used to solve any quadratic equation.

To solve $x^2 = 6x - 13$ by completing the square, begin by writing the variable terms on one side of the equals sign and the constant term on the other side.

$x^2 - 6x = -13$ • **Subtract $6x$ from each side of the equation.**

$x^2 - 6x + 9 = -13 + 9$ • **Complete the square. $\left[\frac{1}{2}(-6)\right]^2 = 9$.**

$(x - 3)^2 = -4$ • **Factor the left side of the equation.**

$\sqrt{(x - 3)^2} = \sqrt{-4}$ • **Take the square root of each side.**

$x - 3 = \pm i\sqrt{4} = \pm 2i$ • **Solve for x.**

$x = 3 \pm 2i$

The solutions are $3 - 2i$ and $3 + 2i$.

By using the operations on complex numbers, we can check these solutions. We will show the check for $3 - 2i$.

x^2	$= 6x - 13$
$(3 - 2i)^2$	$6(3 - 2i) - 13$
$(3 - 2i)(3 - 2i)$	$18 - 12i - 13$
$9 - 6i - 6i + 4i^2$	$5 - 12i$
$9 - 12i + 4(-1)$	$5 - 12i$
$5 - 12i$	$= 5 - 12i$

The solution checks.

MATH MATTERS

Ancient mathematicians thought of "completing the square" in a geometrical manner. For instance, to complete the square of $x^2 + 8x$, draw a square that measures x units on each side, and add four rectangles that measure 1 unit by x units to the right side and the bottom of the square.

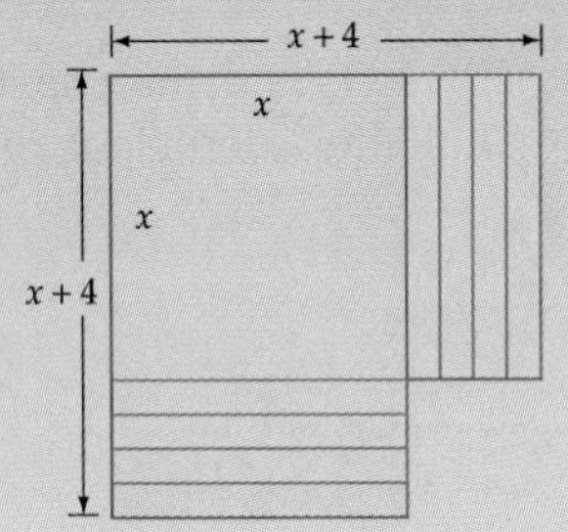

Each of the rectangles has an area of x square units, so the total area of the figure is $x^2 + 8x$. To make this figure a complete square, we must add 16 squares that measure 1 unit by 1 unit as shown below.

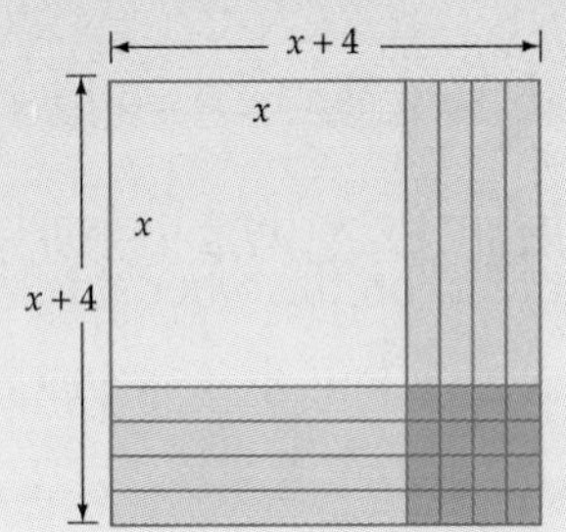

This figure is a *complete square* whose area is

$$(x + 4)^2 = x^2 + 8x + 16$$

EXAMPLE 6 Solve by Completing the Square

Solve: $x^2 = 2x - 6$

Solution

$$x^2 = 2x - 6$$

$$x^2 - 2x = -6 \qquad \text{• Isolate the constant term.}$$

$$x^2 - 2x + 1 = -6 + 1 \qquad \text{• Complete the square.}$$

$$(x - 1)^2 = -5 \qquad \text{• Factor and simplify.}$$

$$x - 1 = \pm\sqrt{-5} \qquad \text{• Apply the square root theorem.}$$

$$x = 1 \pm i\sqrt{5} \qquad \text{• Solve for } x.$$

The solutions are $1 - i\sqrt{5}$ and $1 + i\sqrt{5}$.

TRY EXERCISE 56, EXERCISE SET 1.3, PAGE 96

Completing the square by adding the square of half the coefficient of the x term requires that the coefficient of the x^2 term be 1. If the coefficient of the x^2 term is not 1, multiply each term on each side of the equation by the reciprocal of the coefficient of x^2.

EXAMPLE 7 Solve by Completing the Square

Solve: $2x^2 + 8x - 15 = 0$

Solution

$$2x^2 + 8x - 15 = 0$$

$$2x^2 + 8x = 15 \qquad \text{• Isolate the constant term.}$$

$$\frac{1}{2}(2x^2 + 8x) = \frac{1}{2}(15) \qquad \text{• Multiply both sides of the equation by the reciprocal of the leading coefficient.}$$

$$x^2 + 4x = \frac{15}{2}$$

$$x^2 + 4x + 4 = \frac{15}{2} + 4 \qquad \text{• Add the square of half the } x \text{ coefficient to both sides.}$$

$$(x + 2)^2 = \frac{23}{2} \qquad \text{• Factor and simplify.}$$

$$x + 2 = \pm\sqrt{\frac{23}{2}} \qquad \text{• Apply the square root theorem.}$$

Continued ➤

$$x = -2 \pm \frac{\sqrt{46}}{2}$$ • **Add −2 to each side of the equation, and rationalize the denominator.**

$$x = \frac{-4 \pm \sqrt{46}}{2}$$

The solutions are $\frac{-4 - \sqrt{46}}{2}$ and $\frac{-4 + \sqrt{46}}{2}$.

TRY EXERCISE 60, EXERCISE SET 1.3, PAGE 96

◆ SOLVING QUADRATIC EQUATIONS BY USING THE QUADRATIC FORMULA

Completing the square on $ax^2 + bx + c = 0$ $(a \neq 0)$ produces a formula for x in terms of the coefficients a, b, and c. The formula is known as the **quadratic formula,** and applying it is another way to solve quadratic equations.

MATH MATTERS

There is a general procedure to solve "by radicals" the general cubic

$$ax^3 + bx^2 + cx + d = 0$$

and the general quartic

$$ax^4 + bx^3 + cx^2 + dx + e = 0$$

However, it has been proved that there are no general procedures that can solve "by radicals" general equations of degree 5 or larger.

The Quadratic Formula

If $ax^2 + bx + c = 0$, $a \neq 0$, then

$$x = \frac{-b \pm \sqrt{b^2 - 4ac}}{2a}$$

Proof We assume a is a positive real number. If a were a negative real number, then we could multiply each side of the equation by -1 to make it positive.

$$ax^2 + bx + c = 0 \quad (a \neq 0)$$ • **Given.**

$$ax^2 + bx = -c$$ • **Isolate the constant term.**

$$x^2 + \frac{b}{a}x = -\frac{c}{a}$$ • **Multiply each term on each side of the equation by 1/*a*.**

$$x^2 + \frac{b}{a}x + \left(\frac{b}{2a}\right)^2 = \left(\frac{b}{2a}\right)^2 - \frac{c}{a}$$ • **Complete the square.**

$$\left(x + \frac{b}{2a}\right)^2 = \frac{b^2}{4a^2} - \frac{c}{a}$$ • **Factor the left side. Simplify the powers on the right side.**

$$\left(x + \frac{b}{2a}\right)^2 = \frac{b^2}{4a^2} - \frac{4a}{4a} \cdot \frac{c}{a}$$ • **Use a common denominator to simplify the right side.**

$$x + \frac{b}{2a} = \pm\sqrt{\frac{b^2 - 4ac}{4a^2}}$$ • **Apply the square root theorem.**

$$x + \frac{b}{2a} = \pm\frac{\sqrt{b^2 - 4ac}}{2a}$$ • **Because $a > 0$, $\sqrt{4a^2} = 2a$.**

Continued •➤

$$x = -\frac{b}{2a} \pm \frac{\sqrt{b^2 - 4ac}}{2a}$$ • Add $-\frac{b}{2a}$ to each side.

$$x = \frac{-b \pm \sqrt{b^2 - 4ac}}{2a}$$

◆

As a general rule, you should first try to solve quadratic equations by factoring. If the factoring process proves difficult, then solve either by using the quadratic formula or by completing the square.

take note

You can visit college.hmco.com for a graphing calculator program that can be used to check your answer to quadratic equations with real number solutions.

EXAMPLE 8 Solve by Using the Quadratic Formula

Solve:

a. $4x^2 - 8x + 1 = 0$ **b.** $x^2 = 3x - 5$

Solution

Use the quadratic formula $x = \frac{-b \pm \sqrt{b^2 - 4ac}}{2a}$.

a. $a = 4, b = -8, c = 1$

$$x = \frac{-(-8) \pm \sqrt{(-8)^2 - 4(4)(1)}}{2(4)} = \frac{8 \pm \sqrt{48}}{8} = \frac{8 \pm 4\sqrt{3}}{8} = \frac{2 \pm \sqrt{3}}{2}$$

The solutions are $\frac{2 - \sqrt{3}}{2}$ and $\frac{2 + \sqrt{3}}{2}$.

b.

$$x^2 = 3x - 5$$

$$x^2 - 3x + 5 = 0$$ • Write the equation in standard form. $a = 1, b = -3, c = 5$.

$$x = \frac{-(-3) \pm \sqrt{(-3)^2 - 4(1)(5)}}{2(1)}$$

$$= \frac{3 \pm \sqrt{-11}}{2} = \frac{3 \pm i\sqrt{11}}{2}$$

The solutions are $\frac{3}{2} - \frac{i\sqrt{11}}{2}$ and $\frac{3}{2} + \frac{i\sqrt{11}}{2}$.

TRY EXERCISE 70, EXERCISE SET 1.3, PAGE 96

◆ THE DISCRIMINANT

The solutions of $ax^2 + bx + c = 0, a \neq 0$, are given by

$$x = \frac{-b \pm \sqrt{b^2 - 4ac}}{2a}$$

The expression $b^2 - 4ac$ is called the **discriminant** of the equation $ax^2 + bx + c = 0$. If $b^2 - 4ac \geq 0$, then $\sqrt{b^2 - 4ac}$ is a real number; if $b^2 - 4ac < 0$, then $\sqrt{b^2 - 4ac}$ is not a real number. Thus the sign of the discriminant determines whether the roots of a quadratic equation are real numbers or nonreal complex numbers.

The Discriminant and Roots of a Quadratic Equation

The quadratic equation $ax^2 + bx + c = 0$, with real coefficients and $a \neq 0$, has discriminant $b^2 - 4ac$.

If $b^2 - 4ac > 0$, then the quadratic equation has *two distinct real roots.*

If $b^2 - 4ac = 0$, then the quadratic equation has *a real root* that is a double root.

If $b^2 - 4ac < 0$, then the quadratic equation has *two distinct complex roots* that are not real. These roots are conjugates of each other.

By examining the discriminant, it is possible to determine whether the roots of a quadratic equation are real numbers without actually finding the roots.

EXAMPLE 9 Use the Discriminant to Classify Roots

Classify the roots of each quadratic equation as real numbers or nonreal complex numbers.

a. $2x^2 - 5x + 1 = 0$ **b.** $3x^2 + 6x + 7 = 0$ **c.** $x^2 + 6x + 9 = 0$

Solution

a. $2x^2 - 5x + 1 = 0$ has coefficients $a = 2$, $b = -5$, and $c = 1$.

$$b^2 - 4ac = (-5)^2 - 4(2)(1) = 25 - 8 = 17$$

Because the discriminant 17 is *positive,* $2x^2 - 5x + 1 = 0$ has *two distinct real roots.*

b. $3x^2 + 6x + 7 = 0$ has coefficients $a = 3$, $b = 6$, and $c = 7$.

$$b^2 - 4ac = 6^2 - 4(3)(7) = 36 - 84 = -48$$

Because the discriminant -48 is *negative,* $3x^2 + 6x + 7 = 0$ has *two distinct nonreal complex roots.*

c. $x^2 + 6x + 9 = 0$ has coefficients $a = 1$, $b = 6$, and $c = 9$.

$$b^2 - 4ac = 6^2 - 4(1)(9) = 36 - 36 = 0$$

Because the discriminant is 0, $x^2 + 6x + 9 = 0$ has *a real root.* The root is a double root.

take note

Some students state, "If the discriminant is negative, then the roots of the quadratic equation are complex numbers." Although this statement is true, it is not precise, because the real numbers are a subset of the complex numbers.

TRY EXERCISE 78, EXERCISE SET 1.3, PAGE 96

◆ APPLICATIONS

A **right triangle** contains one 90° angle. The sidε opposite the 90° angle is called the **hypotenuse.** The two other sides are called **legs.** See **Figure 1.6.**

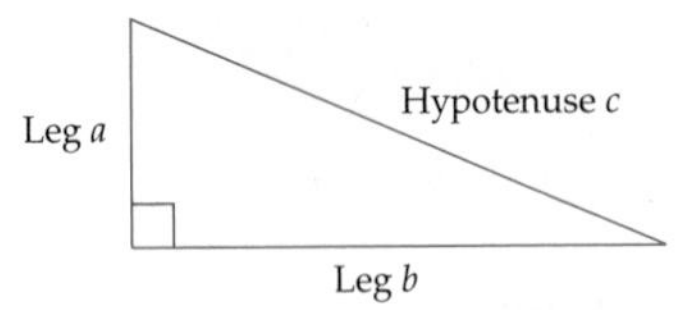

Figure 1.6

The Pythagorean Theorem

If a and b denote the lengths of the legs of a right triangle and c the length of the hypotenuse, then

$$c^2 = a^2 + b^2$$

The Pythagorean Theorem states that the square of the length of the hypotenuse of a right triangle is equal to the sum of the squares of the lengths of the two legs. This theorem is often used to solve applications that involve right triangles.

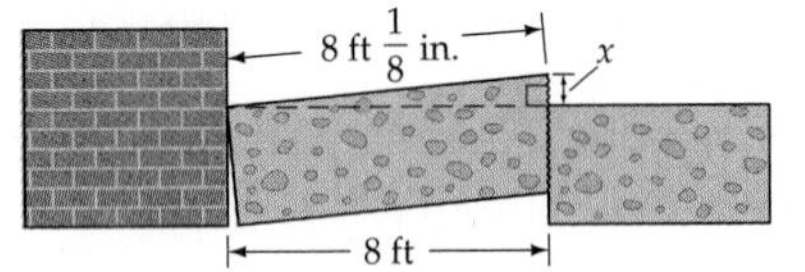

Figure 1.7

EXAMPLE 10 Solve a Construction Application

Concrete slabs often crack and buckle if proper expansion joints are not installed. Suppose a concrete slab expands as a result of an increase in temperature, as shown in **Figure 1.7**. Determine the height x, to the nearest inch, to which the concrete will rise as a consequence of this expansion.

Solution

Use the Pythagorean Theorem.

$$\left(8 \text{ feet} + \frac{1}{8} \text{ inch}\right)^2 = x^2 + (8 \text{ feet})^2$$

$$(96.125)^2 = x^2 + (96)^2$$ • **Change units to inches.**

$$(96.125)^2 - (96)^2 = x^2$$

$$\sqrt{(96.125)^2 - (96)^2} = x$$ • **Only the positive root is taken because $x > 0$.**

$$4.9 \approx x$$

Thus, to the nearest inch, the concrete will rise 5 inches.

TRY EXERCISE 84, EXERCISE SET 1.3, PAGE 96

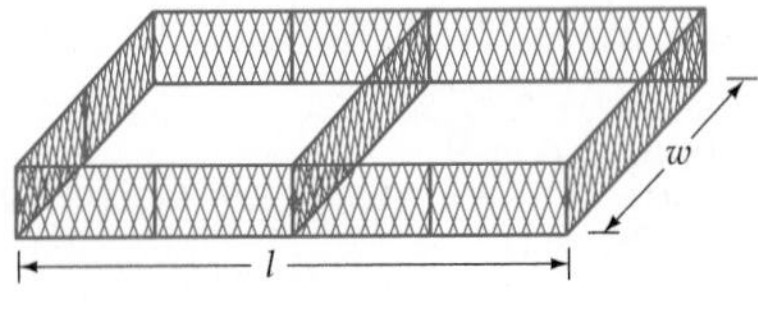

Figure 1.8

EXAMPLE 11 Solve a Geometric Application

A veterinarian wishes to use 132 feet of chain-link fencing to enclose a rectangular region and subdivide the region into two smaller rectangles, as shown in **Figure 1.8**. If the total enclosed area is 576 square feet, find the dimensions of the enclosed region.

Solution

Let w be the width of the enclosed region. Then $3w$ represents the amount of fencing used to construct the three widths. The amount of fencing left for the two lengths is $132 - 3w$. Thus each length must be half of the remaining fencing, or $(132 - 3w)/2$.

Now we have variable expressions in w for both the width and the length. Substituting these into the area formula $lw = A$ produces

$$\left(\frac{132 - 3w}{2}\right)w = 576 \qquad \bullet \textbf{ Substitute } \frac{132 - 3w}{2} \textbf{ for } l.$$

$$132w - 3w^2 = 1152 \qquad \bullet \textbf{ Simplify.}$$

$$-3w^2 + 132w - 1152 = 0$$

$$w^2 - 44w + 384 = 0 \qquad \bullet \textbf{ Divide each term by } -3.$$

Although this quadratic equation can be solved by factoring, the following solution makes use of the quadratic formula.

$$w = \frac{-(-44) \pm \sqrt{(-44)^2 - 4(1)(384)}}{2(1)} \qquad \bullet \textbf{ Apply the quadratic formula.}$$

$$w = \frac{44 \pm \sqrt{400}}{2} = \frac{44 \pm 20}{2} = 12 \text{ or } 32$$

Thus there are two solutions to the problem:

1. If the width $w = 12$ feet, then the length is $\frac{132 - 3(12)}{2} = 48$ feet.
2. If the width $w = 32$ feet, then the length is $\frac{132 - 3(32)}{2} = 18$ feet.

TRY EXERCISE 90, EXERCISE SET 1.3, PAGE 97

QUESTION In Example 11, what reason can you give for using the quadratic formula, rather than factoring, to solve $w^2 - 44w + 384 = 0$?

TOPICS FOR DISCUSSION

1. If A and B are algebraic expressions and the product AB equals 0, then $A = 0$ or $B = 0$. Do you agree with this statement? Explain.
2. If A and B are algebraic expressions and the product AB equals 10, then $A = 10$ or $B = 10$. Do you agree with this statement? Explain.
3. Every quadratic equation has two distinct solutions. Do you agree with this statement? Explain.
4. Every quadratic equation $ax^2 + bx + c = 0$ (with real coefficients and $a \neq 0$) can be solved by applying the quadratic formula. Do you agree? Explain.
5. Every quadratic equation of the form $ax^2 + bx = 0$ $(a \neq 0, b \neq 0)$ has two distinct real solutions. Do you agree? Explain.

ANSWER Factoring $w^2 - 44w + 384$ may be time-consuming because 384 has several integer factors.

EXERCISE SET 1.3

In Exercises 1 to 12, solve each quadratic equation by factoring and applying the zero product property.

1. $x^2 - 2x - 15 = 0$
2. $y^2 + 3y - 10 = 0$
3. $2x^2 - x = 1$
4. $2x^2 + 5x = 3$
5. $8y^2 + 189y - 72 = 0$
6. $12w^2 - 41w + 24 = 0$
7. $3x^2 - 7x = 0$
8. $5x^2 = -8x$
9. $(x - 5)^2 - 9 = 0$
10. $(3x + 4)^2 - 16 = 0$
11. $(2x - 5)^2 - (4x - 11)^2 = 0$
12. $(5x + 3)^2 - (x + 7)^2 = 0$

In Exercises 13 to 22, use the square root theorem to solve each quadratic equation.

13. $x^2 = 81$
14. $y^2 = 225$
15. $2x^2 = 48$
16. $3x^2 = 144$
17. $3x^2 + 12 = 0$
18. $4y^2 + 20 = 0$
19. $(x - 5)^2 = 36$
20. $(x + 4)^2 = 121$
21. $(x - 3)^2 + 16 = 0$
22. $(x + 2)^2 + 28 = 0$

In Exercises 23 to 42, simplify and write the complex number in standard form.

23. $(2 + 5i) + (3 + 7i)$
24. $(1 - 3i) + (6 + 2i)$
25. $(8 - 6i) - (10 - i)$
26. $(-3 + i) - (-8 + 2i)$
27. $(7 - 3i) - (-5 - i)$
28. $7 - (3 - 2i)$
29. $3(2 + 7i) + 5(2 - i)$
30. $8(4 - i) - (4 - 3i)$
31. $(2 + 3i)(4 - 5i)$
32. $(5 + 3i)(-2 - 4i)$
33. $(5 + 7i)(5 - 7i)$
34. $(-3 - 5i)(-3 + 5i)$
35. $\dfrac{4 + i}{3 + 5i}$
36. $\dfrac{5 - i}{4 + 5i}$
37. $\dfrac{1}{7 - 3i}$
38. $\dfrac{1}{-8 + i}$
39. $\dfrac{3 + 2i}{3 - 2i}$
40. $\dfrac{5 - 7i}{5 + 7i}$
41. $(3 - 5i)^2$
42. $(-5 + 7i)^2$

In Exercises 43 to 50, simplify and write the complex number as i, $-i$, 1, or -1.

43. i^{10}
44. i^{28}
45. $-i^{40}$
46. i^{40}
47. i^{223}
48. i^{553}
49. i^{2001}
50. i^{5000}

In Exercises 51 to 64, solve by completing the square.

51. $x^2 + 6x + 1 = 0$
52. $x^2 + 8x - 10 = 0$
53. $x^2 - 2x - 15 = 0$
54. $x^2 + 2x - 8 = 0$
55. $x^2 + 4x + 5 = 0$
56. $x^2 - 6x + 10 = 0$
57. $x^2 + 3x - 1 = 0$
58. $x^2 + 7x - 2 = 0$
59. $2x^2 + 4x - 1 = 0$
60. $2x^2 + 10x - 3 = 0$
61. $3x^2 - 8x + 1 = 0$
62. $4x^2 - 4x + 15 = 0$
63. $33 - 6x + x^2 = 0$
64. $10 + 4x + x^2 = 0$

In Exercises 65 to 76, solve by using the quadratic formula.

65. $x^2 - 2x - 15 = 0$
66. $x^2 - 5x - 24 = 0$
67. $x^2 + x - 1 = 0$
68. $x^2 + x + 1 = 0$
69. $2x^2 + 4x + 1 = 0$
70. $2x^2 + 4x - 1 = 0$
71. $3x^2 - 5x + 3 = 0$
72. $3x^2 - 5x + 4 = 0$
73. $\dfrac{1}{2}x^2 + \dfrac{3}{4}x - 1 = 0$
74. $\dfrac{2}{3}x^2 - 5x + \dfrac{1}{2} = 0$
75. $\sqrt{2}x^2 + 3x + \sqrt{2} = 0$
76. $2x^2 + \sqrt{5}x - 3 = 0$

In Exercises 77 to 82, determine the discriminant of the quadratic equation, and then classify the roots of the equation as *a.* two distinct real numbers, *b.* one real number (which is a double root), or *c.* two distinct nonreal complex numbers. Do not solve the equations.

77. $2x^2 - 5x - 7 = 0$
78. $x^2 + 3x - 11 = 0$
79. $3x^2 - 2x + 10 = 0$
80. $x^2 + 3x + 3 = 0$
81. $x^2 - 20x + 100 = 0$
82. $4x^2 + 12x + 9 = 0$

83. **GEOMETRY** The length of each side of a square is 54 inches. Find the length of the diagonal of the square. Round to the nearest tenth of an inch.

84. **CONSTRUCTION** A concrete slab cracks and expands as a result of an increase in temperature, as shown in the following figure. Determine the height x, to the nearest inch, to which the concrete will rise as a consequence of this expansion.

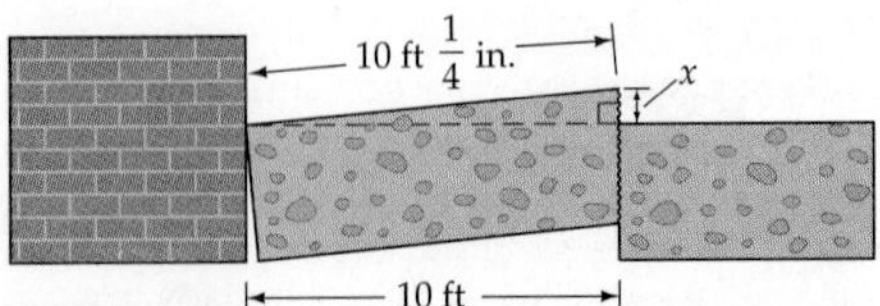

85. **GEOMETRY** The length of each side of an equilateral triangle is 31 centimeters. Find the altitude of the triangle. Round to the nearest tenth of a centimeter.

86. **BASEBALL** How far, to the nearest foot, is it from home-plate to second base on a square baseball diamond? *Hint:* The distance between home plate and first base is 90 feet.

87. **GEOMETRY** The perimeter of a rectangle is 27 centimeters and its area is 35 square centimeters. Find the length and the width of the rectangle.

88. **GEOMETRY** The perimeter of a rectangle is 34 feet and its area is 60 square feet. Find the length and the width of the rectangle.

89. **RECTANGULAR ENCLOSURE** A gardener wishes to use 600 feet of fencing to enclose a rectangular region and subdivide the region into two smaller rectangles. The total enclosed area is 15,000 square feet. Find the dimensions of the enclosed region.

90. **RECTANGULAR ENCLOSURE** A farmer wishes to use 400 yards of fencing to enclose a rectangular region and subdivide the region into three smaller rectangles. The total enclosed area is 4800 square yards. Find the dimensions of the enclosed region.

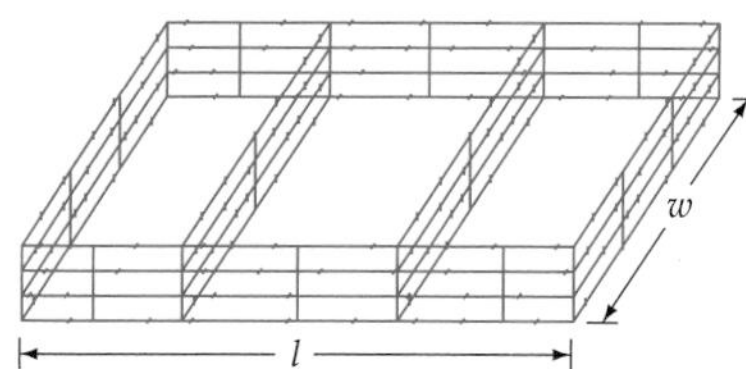

91. **TRAFFIC CONTROL** Traffic engineers have installed "flow lights" at the entrances of freeways to control the number of cars entering a freeway each minute. This results in a backup of cars at the flow lights. The number N of cars waiting to enter the freeway at a certain flow light is given by

$$N = -100x^2 + 230x + 20$$

where x is the number of hours after 6:30 A.M. and $x \le 2.25$. At what times, to the nearest minute, will there be 100 cars waiting to enter the freeway?

92. **AUTOMOTIVE ENGINEERING** The number N of feet that it will take a car to stop on a certain road surface is given by

$$N = -0.015v^2 + 3v$$

where v is the speed of the car in miles per hour when the driver applies the breaks. At what speed, to the nearest whole number, can a motorist be traveling and stop the car in 100 feet?

93. **WATER TECHNOLOGY** The number N of gallons (in millions) of water used by a small community during the hours from 11:00 P.M. to 8:00 A.M. is given by

$$N = 0.34x^2 - 2.66x + 6$$

where x is the number of hours after 11:00 P.M. At what time, to the nearest minute, will the community be using 7 million gallons of water?

94. **ENTERTAINMENT** The number N of guests that enter an amusement park each hour can be approximated by

$$N = -80t^2 + 440t + 200$$

where t is the number of hours after 9:00 A.M. At what times, to the nearest minute, are 700 guests per hour entering the amusement park?

95. **AVERAGE RATE** A salesperson drove the first 105 miles of a trip in 1 hour more than it took to drive the last 90 miles. The average rate during the last 90 miles was 10 mph faster than the average rate during the first 105 miles. Find the average rate for each portion of the trip.

96. **AVERAGE RATE** A car and a bus both completed a 240-mile trip. The car averaged 10 mph faster than the bus and completed the trip in 48 minutes less time than the bus. Find the average rate, in miles per hour, of the bus.

97. **INDIVIDUAL TIME** A mason can build a wall in 6 hours less than an apprentice. Together they can build the wall in 4 hours. How long would it take the apprentice, working alone, to build the wall?

98. **INDIVIDUAL TIME** Pump A can fill a pool in 2 hours less time than pump B. Together the pumps can fill the pool in 2 hours 24 minutes. Find how long it takes pump A to fill the pool.

99. **SPORTS** Michael Jordan was known for his "hang time," the time he was in the air when he jumped. An equation that approximates the height, s, of his jump is given by $s = -16t^2 + 26.6t$. Using this equation, determine Michael Jordan's hang time.

100. **FOOTBALL** A football player kicks a football downfield. The height s in feet of the football t seconds after it leaves the kicker's foot is given by

$$s = -16t^2 + 88t + 2$$

Find the "hang time."

SUPPLEMENTAL EXERCISES

The following theorem is known as the *sum and product of the roots theorem.*

Let $ax^2 + bx + c = 0$, $a \neq 0$, be a quadratic equation. Then r_1 and r_2 are roots of the equation if and only if

$$r_1 + r_2 = -\frac{b}{a} \text{ and } r_1 r_2 = \frac{c}{a}.$$

In Exercises 101 to 106, use the sum and product of the roots theorem given in the preceding column to determine whether the given numbers are roots of the quadratic equation.

101. $x^2 - 5x - 24 = 0; \quad -3, 8$

102. $x^2 + 4x - 21 = 0; \quad -7, 3$

103. $2x^2 - 7x - 30 = 0;\quad -5/2, 6$

104. $9x^2 - 12x - 1 = 0;\quad (2 + \sqrt{5})/3, (2 - \sqrt{5})/3$

105. $x^2 - 2x + 2 = 0;\quad 1 + i, 1 - i$

106. $x^2 - 4x + 12 = 0;\quad 2 + 3i, 2 - 3i$

In Exercises 107 to 112, use the quadratic formula to solve each equation for the indicated variable in terms of the other variables. Assume that none of the denominators is zero.

107. $0 = -\frac{1}{2}gt^2 + v_0t + s_0,\quad$ for t

108. $3x^2 + xy + 4y^2 = 0,\quad$ for y

109. $x = y^2 + y - 8,\quad$ for y

110. $3x^2 + xy - 4y^2 = 0,\quad$ for x

111. $(x - 8)^2 = (x + 1)^2$

112. $(x + 5)^2 = (2x + 1)^2$

113. Simplify: $i + i^2 + i^3 + i^4 + \cdots + i^{28}$

114. Simplify: $i + i^2 + i^3 + i^4 + \cdots + i^{100}$

115. When we say $\sqrt[3]{8} = 2$, we are really stating that the *real* cube root of 8 is 2. There are, however, two other cube roots of 8, both of which are complex numbers. Show that

$$(-1 + i\sqrt{3})^3 = 8 \quad \text{and} \quad (-1 - i\sqrt{3})^3 = 8$$

Explain how this shows that $-1 + i\sqrt{3}$ and $-1 - i\sqrt{3}$ are cube roots of 8.

116. It is possible, although we will not show the technique here, to take the square root of a complex number. Verify that

$$\sqrt{i} = \frac{\sqrt{2}}{2}(1 + i)$$

by showing that

$$\left[\frac{\sqrt{2}}{2}(1 + i)\right]^2 = i$$

The ***absolute value of the complex number*** $a + bi$ **is denoted by** $|a + bi|$ **and defined as the real number** $\sqrt{a^2 + b^2}$**. In Exercises 117 to 124, find the indicated absolute value of each complex number.**

117. $|3 + 4i|$

118. $|5 + 12i|$

119. $|2 - 5i|$

120. $|4 - 4i|$

121. $|7 - 4i|$

122. $|11 - 2i|$

123. $|-3i|$

124. $|18i|$

PROJECTS

1. **A GOLDEN RECTANGLE** A rectangle is a "golden rectangle" if its length l and its width w satisfy the equation

$$\frac{l}{w} = \frac{w}{l - w}$$

   a. Solve this formula for w.

   b. If the length l of a golden rectangle measures 101 feet, what is the width of the rectangle?

2. **THE SUM AND PRODUCT OF THE ROOTS THEOREM** Use the quadratic formula to prove the sum and product of the roots theorem stated just before Exercise 101.

3. **ATTRACTORS** Consider the quadratic equation

$$y = 2x(1 - x)$$

Let x be any number between 0 and 1. Evaluate y for that value of x, substitute that value of y back in for x, and evaluate for the next y value. Continuing this process over and over will produce a sequence of numbers that are *attracted* to 0.5. Verify the above statements for

   a. $x = 0.2$ b. $x = 0.713$

   c. Write an essay on *attractors*. Explain how the topic of attractors is related to *chaos*. An excellent source of information on attractors is *The Mathematical Tourist* by Ivars Peterson (New York: Freeman, 1988).

4. VISUAL INSIGHT

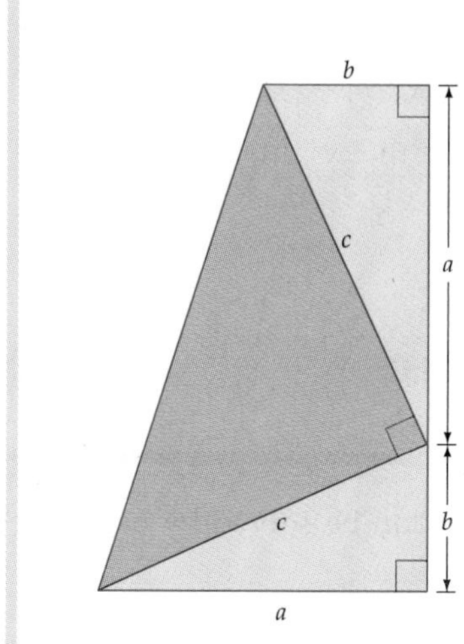

$$\text{Area} = \frac{1}{2}c^2 + 2\left(\frac{1}{2}ab\right) = \frac{1}{2}(\text{height})(\text{sum of bases})$$

$$\frac{1}{2}c^2 + ab = \frac{1}{2}(a + b)(a + b)$$

$$\frac{1}{2}c^2 = \frac{1}{2}a^2 + \frac{1}{2}b^2$$

$$c^2 = a^2 + b^2$$

President James A. Garfield is credited with the above proof of the Pythagorean Theorem. Write the supporting reasons for each of the steps in this proof.

SECTION 1.4 OTHER TYPES OF EQUATIONS

- SOLVING HIGHER-DEGREE EQUATIONS BY FACTORING
- SOLVING EQUATIONS BY USING THE POWER PRINCIPLE
- SOLVING EQUATIONS THAT ARE QUADRATIC IN FORM

1B

◆ SOLVING HIGHER-DEGREE EQUATIONS BY FACTORING

Some equations that are neither linear nor quadratic can be solved by the various techniques presented in this section. For instance, the **third-degree equation,** or **cubic equation,** in Example 1 can be solved by factoring the polynomial on the left side of the equation and using the zero product property.

EXAMPLE 1 Solve an Equation by Factoring

Solve: $x^3 - 16x = 0$

Solution

$$x^3 - 16x = 0$$

$$x(x^2 - 16) = 0 \quad \text{• Factor out the GCF, } x.$$

$$x(x + 4)(x - 4) = 0 \quad \text{• Factor the difference of squares.}$$

take note

If you attempt to solve Example 1 by dividing each side by x, you will produce the equation $x^2 - 16 = 0$, which has roots of only −4 and 4. In this case the division of each side of the equation by the variable x does not produce an equivalent equation. To avoid this common mistake, factor out any variable factors that are common to each term instead of dividing each side of the equation by the factor.

Set each factor equal to zero.

$$x = 0 \quad \text{or} \quad x + 4 = 0 \quad \text{or} \quad x - 4 = 0$$
$$x = 0 \quad \text{or} \quad x = -4 \quad \text{or} \quad x = 4$$

A check will show that −4, 0, and 4 are roots of the original equation.

TRY EXERCISE 6, EXERCISE SET 1.4, PAGE 105

◆ SOLVING EQUATIONS BY USING THE POWER PRINCIPLE

Some equations that involve radical expressions can be solved by using the following result.

The Power Principle

If P and Q are algebraic expressions and n is a positive integer, then every solution of $P = Q$ is a solution of $P^n = Q^n$.

EXAMPLE 2 Solve a Radical Equation

Use the power principle to solve $\sqrt{x + 4} = 3$.

Solution

$$\sqrt{x + 4} = 3$$
$$\left(\sqrt{x + 4}\right)^2 = 3^2$$ • **Apply the power principle with $n = 2$.**
$$x + 4 = 9$$
$$x = 5$$

Check: $\sqrt{x + 4} = 3$
$$\sqrt{5 + 4} \stackrel{?}{=} 3$$ • **Substitute 5 for x.**
$$\sqrt{9} \stackrel{?}{=} 3$$
$$3 = 3$$ • **5 checks.**

The only solution is 5.

TRY EXERCISE 14, EXERCISE SET 1.4, PAGE 105

Some care must be taken when using the power principle, because the equation $P^n = Q^n$ may have more solutions than the original equation $P = Q$. As an example, consider $x = 3$. The only solution is the real number 3. Square each side of the equation to produce $x^2 = 9$, which has both 3 and −3 as solutions. The −3 is called an *extraneous solution* because it is not a solution of the original equation $x = 3$.

Extraneous Solutions

Any solution of $P^n = Q^n$ that is not a solution of $P = Q$ is called an **extraneous solution.** Extraneous solutions *may* be introduced whenever we raise each side of an equation to an *even* power.

EXAMPLE 3 Solve a Radical Equation

Solve $x = 2 + \sqrt{2 - x}$. Check all proposed solutions.

Solution

$$
\begin{aligned}
x &= 2 + \sqrt{2 - x} \\
x - 2 &= \sqrt{2 - x} && \text{• Isolate the radical.} \\
(x - 2)^2 &= (\sqrt{2 - x})^2 && \text{• Square each side of the equation.} \\
x^2 - 4x + 4 &= 2 - x \\
x^2 - 3x + 2 &= 0 && \text{• Collect and combine like terms.} \\
(x - 2)(x - 1) &= 0 && \text{• Factor.}
\end{aligned}
$$

$$x - 2 = 0 \quad \text{or} \quad x - 1 = 0$$

$$x = 2 \quad \text{or} \quad x = 1 \qquad \text{• Proposed solutions}$$

Check for $x = 2$: $x = 2 + \sqrt{2 - x}$

$$
\begin{aligned}
2 &\stackrel{?}{=} 2 + \sqrt{2 - (2)} && \text{• Substitute 2 for } x. \\
2 &\stackrel{?}{=} 2 + \sqrt{0} \\
2 &= 2 && \text{• 2 is a solution.}
\end{aligned}
$$

Check for $x = 1$: $x = 2 + \sqrt{2 - x}$

$$
\begin{aligned}
1 &\stackrel{?}{=} 2 + \sqrt{2 - (1)} && \text{• Substitute 1 for } x. \\
1 &\stackrel{?}{=} 2 + \sqrt{1} \\
1 &\neq 3 && \text{• 1 is not a solution.}
\end{aligned}
$$

The check shows that 1 is not a solution. It is an extraneous solution that we created by squaring each side of the equation. The only solution is 2.

TRY EXERCISE 16, EXERCISE SET 1.4, PAGE 105

In Example 4 it will be necessary to square $(1 + \sqrt{2x - 5})$. Recall the special product formula $(x + y)^2 = x^2 + 2xy + y^2$. Using this special product formula to square $(1 + \sqrt{2x - 5})$ produces

$$(1 + \sqrt{2x - 5})^2 = 1 + 2\sqrt{2x - 5} + (2x - 5)$$

EXAMPLE 4 Solve a Radical Equation

Solve $\sqrt{x + 1} - \sqrt{2x - 5} = 1$. Check all proposed solutions.

Continued

Solution

First write an equivalent equation in which one radical is isolated on one side of the equation.

$$\sqrt{x+1} - \sqrt{2x-5} = 1$$
$$\sqrt{x+1} = 1 + \sqrt{2x-5}$$

The next step is to square each side. Using the result from the discussion preceding this example, we have

$$(\sqrt{x+1})^2 = (1 + \sqrt{2x-5})^2$$
$$x + 1 = 1 + 2\sqrt{2x-5} + (2x - 5)$$
$$-x + 5 = 2\sqrt{2x-5}$$ • **Isolate the remaining radical.**

The right side still contains a radical, so we square each side again.

$$(-x + 5)^2 = (2\sqrt{2x+5})^2$$
$$x^2 - 10x + 25 = 4(2x - 5)$$
$$x^2 - 10x + 25 = 8x - 20$$
$$x^2 - 18x + 45 = 0$$
$$(x - 3)(x - 15) = 0$$
$$x = 3 \quad \text{or} \quad x = 15$$ • **Proposed solutions**

3 checks as a solution, but 15 does not. Therefore, 3 is the only solution.

TRY EXERCISE 20, EXERCISE SET 1.4, PAGE 105

Some equations that involve fractional exponents can be solved by raising each side to a reciprocal power. For example, to solve $x^{1/3} = 4$, raise each side to the third power to find that $x = 64$. Be sure to check all proposed solutions to determine whether they are actual solutions or extraneous solutions.

EXAMPLE 5 Solve Equations That Involve Fractional Exponents

Solve: $(x^2 + 4x + 52)^{3/2} = 512$

Solution

Because the equation involves a three-halves power, start by raising each side of the equation to the two-thirds power.

$$[(x^2 + 4x + 52)^{3/2}]^{2/3} = 512^{2/3}$$ • **The reciprocal of 3/2 is 2/3.**
$$x^2 + 4x + 52 = 64$$ • **Think: $512^{2/3} = (\sqrt[3]{512})^2 = 8^2 = 64$**
$$x^2 + 4x - 12 = 0$$ • **Subtract 64 from each side.**
$$(x - 2)(x + 6) = 0$$ • **Factor.**

Continued •►

$$x - 2 = 0 \quad \text{or} \quad x + 6 = 0$$
$$x = 2 \quad \text{or} \quad x = -6$$

A check will verify that 2 and -6 are both solutions of the original equation.

TRY EXERCISE 32, EXERCISE SET 1.4, PAGE 105

◆ SOLVING EQUATIONS THAT ARE QUADRATIC IN FORM

The equation $4x^4 - 25x^2 + 36 = 0$ is said to be **quadratic in form,** which means it can be written in the form

$$au^2 + bu + c = 0 \qquad a \neq 0$$

where u is an algebraic expression involving x. For example, if we make the substitution $u = x^2$ (which implies $u^2 = x^4$), then our original equation can be written as

$$4u^2 - 25u + 36 = 0$$

This quadratic equation can be solved for u, and then, using the relationship $u = x^2$, we can find the solutions of the original equation.

EXAMPLE 6 Solve an Equation That Is Quadratic in Form

Solve: $4x^4 - 25x^2 + 36 = 0$

Solution

Make the substitutions $u = x^2$ and $u^2 = x^4$ to produce the quadratic equation $4u^2 - 25u + 36 = 0$. Factor the quadratic polynomial on the left side of the equation.

$$(4u - 9)(u - 4) = 0$$
$$4u - 9 = 0 \quad \text{or} \quad u - 4 = 0$$
$$u = \frac{9}{4} \quad \text{or} \quad u = 4$$

Substitute x^2 for u to produce

$$x^2 = \frac{9}{4} \quad \text{or} \quad x^2 = 4$$
$$x = \pm\sqrt{\frac{9}{4}} \quad \text{or} \quad x = \pm\sqrt{4}$$
$$x = \pm\frac{3}{2} \quad \text{or} \quad x = \pm 2$$

• Check as before.

The solutions are $-2, -\frac{3}{2}, \frac{3}{2}$, and 2.

TRY EXERCISE 42, EXERCISE SET 1.4, PAGE 105

Following is a table of equations that are quadratic in form. Each is accompanied by an appropriate substitution that will enable it to be written in the form $au^2 + bu + c = 0$.

Equations That Are Quadratic in Form

Original Equation	Substitution	$au^2 + bu + c = 0$ Form
$x^4 - 8x^2 + 15 = 0$	$u = x^2$	$u^2 - 8u + 15 = 0$
$x^6 + x^3 - 12 = 0$	$u = x^3$	$u^2 + u - 12 = 0$
$x^{1/2} - 9x^{1/4} + 20 = 0$	$u = x^{1/4}$	$u^2 - 9u + 20 = 0$
$2x^{2/3} + 7x^{1/3} - 4 = 0$	$u = x^{1/3}$	$2u^2 + 7u - 4 = 0$
$15x^{-2} + 7x^{-1} - 2 = 0$	$u = x^{-1}$	$15u^2 + 7u - 2 = 0$

EXAMPLE 7 Solve an Equation That is Quadratic in Form

Solve: $3x^{2/3} - 5x^{1/3} - 2 = 0$

Solution

Substituting u for $x^{1/3}$ gives us

$$3u^2 - 5u - 2 = 0$$

$$(3u + 1)(u - 2) = 0 \qquad \text{• Factor.}$$

$$3u + 1 = 0 \quad \text{or} \quad u - 2 = 0$$

$$u = -\frac{1}{3} \quad \text{or} \quad u = 2$$

$$x^{1/3} = -\frac{1}{3} \quad \text{or} \quad x^{1/3} = 2 \qquad \text{• Replace } u \text{ with } x^{1/3}.$$

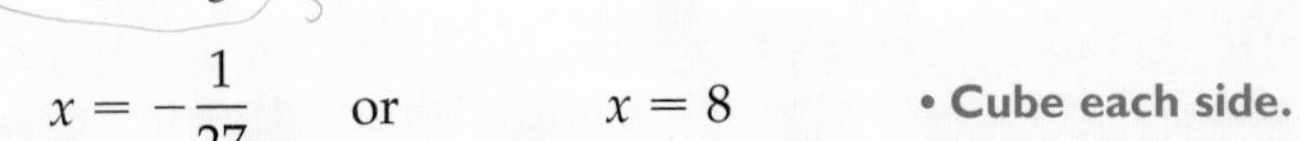

$$x = -\frac{1}{27} \quad \text{or} \quad x = 8 \qquad \text{• Cube each side.}$$

A check will verify that both $-1/27$ and 8 are solutions.

TRY EXERCISE 52, EXERCISE SET 1.4, PAGE 105

It is possible to solve equations that are quadratic in form without making a formal substitution. For example, to solve $x^4 + 5x^2 - 36 = 0$, factor the equation and apply the zero product property.

$$x^4 + 5x^2 - 36 = 0$$

$$(x^2 + 9)(x^2 - 4) = 0$$

$$x^2 + 9 = 0 \quad \text{or} \quad x^2 - 4 = 0$$

$$x^2 = -9 \quad \text{or} \quad x^2 = 4$$

$$x = \pm 3i \quad \text{or} \quad x = \pm 2$$

TOPICS FOR DISCUSSION

1. If P and Q are algebraic expressions and n is a positive integer, then the equation $P^n = Q^n$ is equivalent to the equation $P = Q$. Do you agree? Explain.
2. Consider the equation $(x^2 - 1)(x - 2) = 3(x - 2)$. Dividing each side of the equation by $x - 2$ yields $x^2 - 1 = 3$. Is this second equation equivalent to the first equation?
3. A tutor claims that cubing each side of $(4x - 1)^{1/3} = -2$ will not introduce any extraneous solutions. Do you agree?
4. A tutor claims that the equation $x^{-2} - 2/x = 15$ is quadratic in form. Do you agree? If so, what would be an appropriate substitution that would enable you to write the equation as a quadratic?
5. A classmate solves the equation $x^2 + y^2 = 25$ for y and produces the equation $y = \sqrt{25 - x^2}$. Do you agree with this result?

EXERCISE SET 1.4

In Exercises 1 to 10, factor to solve each equation.

1. $x^3 - 25x = 0$
2. $x^3 - x = 0$
3. $x^3 - 2x^2 - x + 2 = 0$
4. $x^3 - 4x^2 - 2x + 8 = 0$
5. $2x^5 - 18x^3 = 0$
6. $x^4 - 36x^2 = 0$
7. $x^4 - 3x^3 - 40x^2 = 0$
8. $x^4 + 3x^3 - 8x - 24 = 0$
9. $x^4 - 16x^2 = 0$
10. $x^4 - 16 = 0$

In Exercises 11 and 12, solve each equation by factoring and by using the quadratic formula.

11. $x^3 - 8 = 0$
12. $x^3 + 8 = 0$

In Exercises 13 to 30, use the power principle to solve each radical equation. Check all proposed solutions.

13. $\sqrt{x - 4} - 6 = 0$
14. $\sqrt{10 - x} = 4$
15. $x = 3 + \sqrt{3 - x}$
16. $x = \sqrt{5 - x} + 5$
17. $\sqrt{3x - 5} - \sqrt{x + 2} = 1$
18. $\sqrt{6 - x} + \sqrt{5x + 6} = 6$
19. $\sqrt{2x + 11} - \sqrt{2x - 5} = 2$
20. $\sqrt{x + 7} - 2 = \sqrt{x - 9}$
21. $\sqrt{x + 7} + \sqrt{x - 5} = 6$
22. $x = \sqrt{12x - 35}$
23. $2x = \sqrt{4x + 15}$
24. $\sqrt[3]{7x - 3} = \sqrt[3]{2x + 7}$
25. $\sqrt[3]{2x^2 + 5x - 3} = \sqrt[3]{x^2 + 3}$
26. $\sqrt[4]{x^2 + 20} = \sqrt[4]{9x}$
27. $\sqrt{3\sqrt{5x + 16}} = \sqrt{5x - 2}$
28. $\sqrt{4\sqrt{2x - 5}} = \sqrt{x + 5}$
29. $\sqrt{3x + 1} + \sqrt{2x - 1} = \sqrt{10x - 1}$
30. $\sqrt{x - 3} + \sqrt{x + 3} = \sqrt{9 - x}$

In Exercises 31 to 40, solve each equation that involves fractional exponents. Check all proposed solutions.

31. $(3x + 5)^{1/3} = (-2x + 15)^{1/3}$
32. $(4z + 7)^{1/3} = 2$
33. $(x + 4)^{2/3} = 9$
34. $(x - 5)^{3/2} = 125$
35. $(4x)^{2/3} = (30x + 4)^{1/3}$
36. $z^{2/3} = (3z - 2)^{1/3}$
37. $4x^{3/4} = x^{1/2}$
38. $x^{3/5} = 2x^{1/5}$
39. $(3x - 5)^{2/3} + 6(3x - 5)^{1/3} = -8$
40. $2(x + 1)^{1/2} - 11(x + 1)^{1/4} + 12 = 0$

In Exercises 41 to 60, find all the real solutions of each equation by first rewriting each equation as a quadratic equation.

41. $x^4 - 9x^2 + 14 = 0$
42. $x^4 - 10x^2 + 9 = 0$
43. $2x^4 - 11x^2 + 12 = 0$
44. $6x^4 - 7x^2 + 2 = 0$
45. $x^6 + x^3 - 6 = 0$
46. $6x^6 + x^3 - 15 = 0$
47. $21x^6 + 22x^3 = 8$
48. $-3x^6 + 377x^3 - 250 = 0$
49. $x^{1/2} - 3x^{1/4} + 2 = 0$
50. $2x^{1/2} - 5x^{1/4} - 3 = 0$
51. $3x^{2/3} - 11x^{1/3} - 4 = 0$
52. $6x^{2/3} - 7x^{1/3} - 20 = 0$
53. $9x^4 = 30x^2 - 25$
54. $4x^4 - 28x^2 = -49$

55. $x^{2/5} - 1 = 0$

56. $2x^{2/5} - x^{1/5} = 6$

57. $\frac{1}{x^2} + \frac{3}{x} - 10 = 0$

58. $10\left(\frac{x-2}{x}\right)^2 + 9\left(\frac{x-2}{x}\right) - 9 = 0$

59. $9x - 52\sqrt{x} + 64 = 0$

60. $8x - 38\sqrt{x} + 9 = 0$

In Exercises 61 to 64, solve each equation. Round each solution to the nearest hundredth.

61. $x^4 - 3x^2 + 1 = 0$

62. $x - 4\sqrt{x} + 1 = 0$

63. $x^2 - \sqrt{9x^2 - 1} = 0$

64. $2x^2 = \sqrt{10x^2 - 3}$

SUPPLEMENTAL EXERCISES

In Exercises 65 to 70, solve for x in terms of the other variables.

65. $x^2 + y^2 = 9$

66. $\frac{x^2}{a^2} + \frac{y^2}{b^2} = 1$

67. $\sqrt{x} - \sqrt{y} = \sqrt{z}$

68. $x - y = \sqrt{x^2 + y^2 + 5}$

69. $x + y = \sqrt{x^2 - y^2 + 7}$

70. $x + \sqrt{x} = -y$

71. Solve $(\sqrt{x} - 2)^2 - 5\sqrt{x} + 14 = 0$ for x. (*Hint:* Use the substitution $u = \sqrt{x} - 2$, and then rewrite so that the equation is quadratic in terms of the variable u.)

72. Solve $(\sqrt[3]{x} + 3)^2 - 8\sqrt[3]{x} = 12$ for x. (*Hint:* Use the substitution $u = \sqrt[3]{x} + 3$, and then rewrite so that the equation is quadratic in terms of the variable u.)

73. **RADIUS OF A CONE** A conical funnel has a height h of 4 inches and a lateral surface area L of 15π square inches. Find the radius r of the cone. (*Hint:* Use the formula $L = \pi r\sqrt{r^2 + h^2}$.)

74. **DIAMETER OF A CONE** As flour is poured onto a table, it forms a right circular cone whose height is one-third the diameter of the base. What is the diameter of the base when the cone has a volume of 192 cubic inches?

75. **SPHERES** A solid silver sphere has a diameter of 8 millimeters, and a second silver sphere has a diameter of 12 millimeters. The spheres are melted down and recast to form a single cube. What is the length s of each edge of the cube? Round your answer to the nearest tenth of a millimeter.

76. **PENDULUM** The period of a pendulum T is the time it takes the pendulum to complete one swing from left to right and back. For a pendulum near the surface of the earth,

$$T = 2\pi\sqrt{\frac{L}{32}}$$

where T is measured in seconds and L is the length of the pendulum in feet. Find the length of a pendulum that has a period of 4 seconds. Round to the nearest tenth of a foot.

77. **DISTANCE TO THE HORIZON** On a ship, the distance d that you can see to the horizon is given by $d = 1.5\sqrt{h}$, where h is the height of your eye measured in feet above sea level and d is measured in miles. How high is the eye level of a navigator who can see 14 miles to the horizon? Round to the nearest foot.

78. **RADIUS OF A CIRCLE** The radius r of a circle inscribed in a triangle with sides of length a, b, and c is given by

$$r = \sqrt{\frac{(s-a)(s-b)(s-c)}{s}}$$

where $s = \frac{1}{2}(a + b + c)$.

a. Find the length of the radius of a circle inscribed in a triangle with sides of 5 inches, 6 inches, and 7 inches.

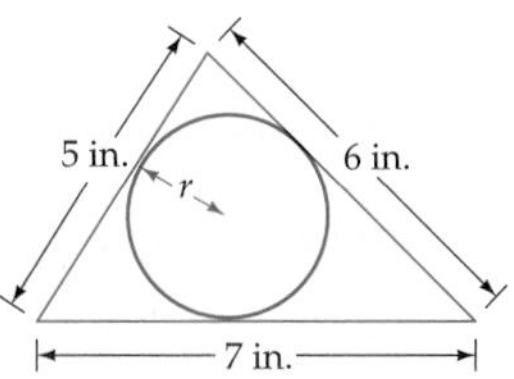

b. The radius of a circle inscribed in an equilateral triangle measures 2 inches. What is the length of each side of the equilateral triangle?

79. **RADIUS OF A CIRCLE** The radius r of a circle that is circumscribed about a triangle with sides a, b, and c is given by

$$r = \frac{abc}{4\sqrt{s(s-a)(s-b)(s-c)}}$$

where $s = \frac{1}{2}(a + b + c)$.

a. Find the radius of a circle that is circumscribed about a triangle with sides of 7 inches, 10 inches, and 15 inches.

b. A circle with radius 5 inches is circumscribed about an equilateral triangle. What is the length of each side of the equilateral triangle?

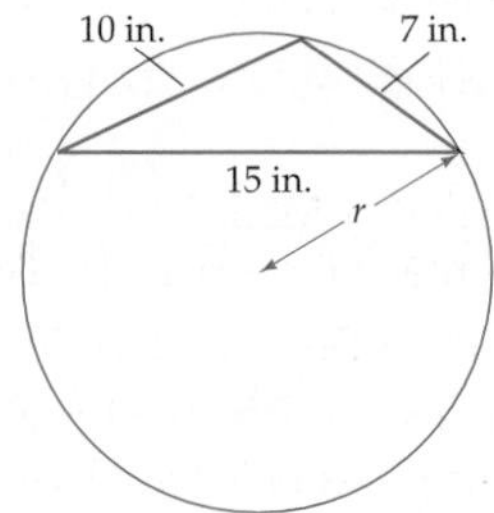

In Exercises 80 and 81, consider the depth s from the opening of a well to the water can be determined by measuring the total time between the instant you drop a stone and the time you hear it hit the water. The time (in seconds) it takes the stone to hit the water is given by $\sqrt{s}/4$, where s is measured in feet. The time (also in seconds) required for the sound of the impact to travel up to your ears is given by $s/1100$. Thus the total time T (in seconds) between the instant you drop a stone and the moment you hear its impact is

$$T = \frac{\sqrt{s}}{4} + \frac{s}{1100}$$

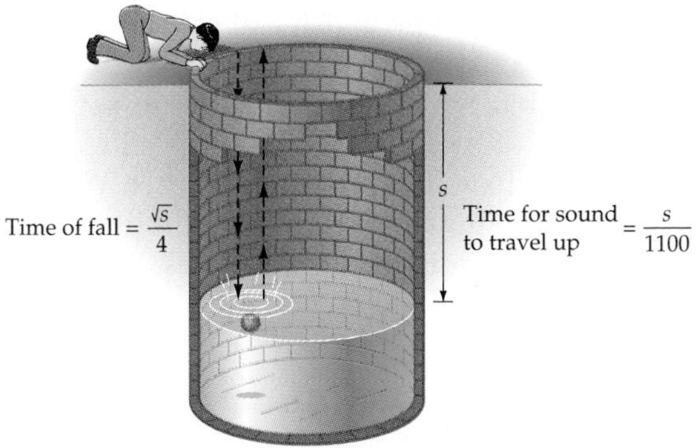

80. **Time of Fall** One of the world's deepest water wells is 7320 feet deep. Find the time between the instant you drop a stone and the time you hear it hit the water if the surface of the water is 7100 feet below the opening of the well. Round your answer to the nearest tenth of a second.

81. Solve $T = \dfrac{\sqrt{s}}{4} + \dfrac{s}{1100}$ for s.

82. **Depth of a Well** Use the result of Exercise 81 to determine the depth from the opening of a well to the water level if the time between the instant you drop a stone and the moment you hear its impact is 3 seconds. Round your answer to the nearest foot.

Projects

1. **The Reduced Cubic** The mathematician Francois Vieta knew a method of solving the "reduced cubic" $x^3 + mx + n = 0$ by using the substitution $x = m/(3z) - z$.
   a. Show that this substitution results in the equation $z^6 - nz^3 - m^3/27 = 0$.
   b. Show that the equation in **a.** is quadratic in form.
   c. Solve the equation in **a.** for z.
   d. Use your solution from **c.** to find the real solution of the equation $x^3 + 3x = 14$.

2. **Fermat's Last Theorem** One of the most famous theorems is known as *Fermat's Last Theorem*. Write an essay on Fermat's Last Theorem. Include information about
   - the history of Fermat's Last Theorem.
   - the relationship between Fermat's Last Theorem and the Pythagorean Theorem.
   - Dr. Andrew Wiles's proof of Fermat's Last Theorem.

   The following list includes a few of the sources you may wish to consult.
   - *The Last Problem* by Eric Temple Bell. The Mathematical Association of America, 1990.
   - "Andrew Wiles: A Math Whiz Battles 350-Year-Old Puzzle" by Gina Kolata, *Math Horizons*, Winter 1993, pp. 8–11. The Mathematical Association of America.
   - "Introduction to Fermat's Last Theorem," by David A. Cox, *The American Mathematical Monthly*, vol. 101, no. 1 (January 1994), pp. 3–14.
   - "The Evidence: Fermat's Last Theorem," by S. Wagon, *The Mathematical Intelligencer*, vol. 8, no. 1, 1986, pp. 59–61.

SECTION 1.5 INEQUALITIES

- SOLVING FIRST-DEGREE INEQUALITIES
- COMPOUND INEQUALITIES
- ABSOLUTE VALUE INEQUALITIES
- SOLVING INEQUALITIES BY THE CRITICAL VALUE METHOD
- RATIONAL INEQUALITIES
- APPLICATIONS

◆ SOLVING FIRST-DEGREE INEQUALITIES

In Section P.2 we used the concept of an inequality to describe the order of real numbers on the real number line, and we also used inequalities to represent subsets of real numbers. In this section we consider inequalities that involve a variable. In particular, we consider how to determine which real values of the variable make the inequality a true statement.

The set of all solutions of an inequality is called the **solution set of the inequality.** For example, the solution set of $x + 1 > 4$ is the set of all real numbers greater than 3. **Equivalent inequalities** have the same solution set. We can solve an inequality by producing *simpler* but equivalent inequalities until the solutions are found. To produce these *simpler* but equivalent inequalities, we often apply one or more of the following properties.

Properties of Inequalities

Let a, b, and c be real numbers.

1. ***Addition Property*** Adding the same real number to each side of an inequality preserves the direction of the inequality symbol.

   $a < b$ and $a + c < b + c$ are equivalent inequalities.

2. ***Multiplication Properties***
   - **a.** Multiplying each side of an inequality by the same *positive* real number *preserves* the direction of the inequality symbol.

     If $c > 0$, then $a < b$ and $ac < bc$ are equivalent inequalities.
   - **b.** Multiplying each side of an inequality by the same *negative* real number *changes* the direction of the inequality symbol.

     If $c < 0$, then $a < b$ and $ac > bc$ are equivalent inequalities.

Note the difference between Property 2a and Property 2b. Property 2a states that an equivalent inequality is produced when each side of a given inequality is multiplied by the same *positive* real number and that the direction of the inequality symbol is *not* changed. By contrast, Property 2b states that when each side of a given inequality is multiplied by a *negative* real number, we must *reverse* the direction of the inequality to produce an equivalent inequality.

For instance, $-2b < 6$ and $b > -3$ are equivalent inequalities. (We multiplied each side of the first inequality by $-1/2$, and we changed the "less than" symbol to a "greater than" symbol.)

Because subtraction is defined in terms of addition, subtracting the same real number from each side of an inequality does not change the direction of the inequality symbol.

Because division is defined in terms of multiplication, dividing each side of the inequality by the same *positive* real number does *not* change the direction of the inequality symbol, and dividing each side of an inequality by a *negative* real number *changes* the direction of the inequality symbol.

take note

Solutions of inequalities are often stated using set notation or interval notation. For instance, the real numbers that are solutions of the inequality in Example 1 can be written in set notation as $\{x \mid x > -2\}$ or in interval notation as $(-2, \infty)$.

EXAMPLE 1 Solve an Inequality

Solve $2(x + 3) < 4x + 10$. Write the solution set in set notation.

Solution

$2(x + 3) < 4x + 10$

$2x + 6 < 4x + 10$ • **Use the distributive property.**

$-2x < 4$ • **Subtract 4x and 6 from each side of the inequality.**

$x > -2$ • **Divide each side by −2 and reverse the inequality symbol.**

Thus the original inequality is true for all real numbers greater than -2. The solution set is $\{x \mid x > -2\}$.

TRY EXERCISE 8, EXERCISE SET 1.5, PAGE 117

◆ COMPOUND INEQUALITIES

A **compound inequality** is formed by joining two inequalities with the connective word *and* or *or.* The inequalities shown below are compound inequalities.

$$x + 1 > 3 \quad \text{and} \quad 2x - 11 < 7$$
$$x + 3 > 5 \quad \text{or} \quad x - 1 < 9$$

The solution set of a compound inequality with the connective word *or* is the *union* of the solution sets of the two inequalities. The solution set of a compound inequality with the connective word *and* is the *intersection* of the solution sets of the two inequalities.

EXAMPLE 2 Solve Compound Inequalities

Solve each compound inequality. Write each solution in set-builder notation.

a. $2x < 10$ or $x + 1 > 9$ **b.** $x + 3 > 4$ and $2x + 1 > 15$

Continued •➤

Solution

a. $2x < 10$ or $x + 1 > 9$

$x < 5$ $\quad x > 8$ • Solve each inequality.

$\{x \mid x < 5\}$ $\quad \{x \mid x > 8\}$ • Write each solution as a set.

$\{x \mid x < 5\} \cup \{x \mid x > 8\} = \{x \mid x < 5 \text{ or } x > 8\}$ • Write the union of the solution sets.

b. $x + 3 > 4$ and $2x + 1 > 15$

$x > 1$ $\quad 2x > 14$ • Solve each inequality.

$x > 7$

$\{x \mid x > 1\}$ $\quad \{x \mid x > 7\}$ • Write each solution as a set.

$\{x \mid x > 1\} \cap \{x \mid x > 7\} = \{x \mid x > 7\}$ • Write the intersection of the solution sets.

TRY EXERCISE 12, EXERCISE SET 1.5, PAGE 117

take note

We reserve the notation $a < b < c$ to mean $a < b$ and $b < c$. Thus the solution set of $2 > x > 5$ is the empty set, because there are no numbers less than 2 and greater than 5.

take note

The compound inequality $a < b$ and $b < c$ can be written in the compact form $a < b < c$. However, the compound inequality $a < b$ or $b > c$ cannot be expressed in a compact form.

The inequality given by

$$12 < x + 5 < 19$$

is equivalent to the compound inequality $12 < x + 5$ *and* $x + 5 < 19$. You can solve $12 < x + 5 < 19$ by either of the following methods.

Method 1 Find the intersection of the solution sets of the inequalities $12 < x + 5$ and $x + 5 < 19$.

$$12 < x + 5 \quad \text{and} \quad x + 5 < 19$$
$$7 < x \quad \text{and} \quad x < 14$$

The solution set is $\{x \mid x > 7\} \cap \{x \mid x < 14\} = \{x \mid 7 < x < 14\}$.

Method 2 Subtract 5 from each of the three parts of the inequality.

$$12 < x + 5 < 19$$
$$12 - 5 < x + 5 - 5 < 19 - 5$$
$$7 < x < 14$$

The solution set is $\{x \mid 7 < x < 14\}$.

◆ ABSOLUTE VALUE INEQUALITIES

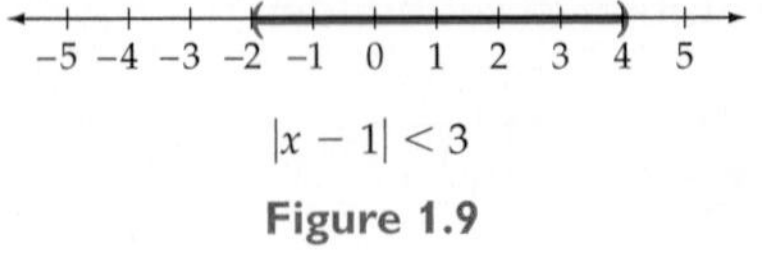

$|x - 1| < 3$

Figure 1.9

The solution set of the absolute value inequality $|x - 1| < 3$ is the set of all real numbers whose distance from 1 is *less than* 3. Therefore, the solution set consists of all numbers between -2 and 4. See **Figure 1.9.** In interval notation, the solution set is $(-2, 4)$.

The solution set of the absolute value inequality $|x - 1| > 3$ is the set of all real numbers whose distance from 1 is *greater than* 3. Therefore, the solution set

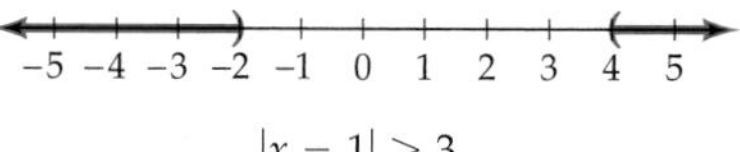

$|x - 1| > 3$

Figure 1.10

consists of all real numbers less than -2 *or* greater than 4. See **Figure 1.10.** In interval notation, the solution set is $(-\infty, -2) \cup (4, \infty)$.

The following properties are used to solve absolute value inequalities.

Properties of Absolute Value Inequalities

For any variable expression E and any nonnegative real number k,

$|E| \le k$ if and only if $-k \le E \le k$

$|E| \ge k$ if and only if $E \le -k$ or $E \ge k$

EXAMPLE 3 Solve an Absolute Value Inequality

Solve: $|2 - 3x| < 7$

Solution

$|2 - 3x| < 7$ implies $-7 < 2 - 3x < 7$. Solve this compound inequality.

$-7 < 2 - 3x < 7$

$-9 < -3x < 5$ • **Subtract 2 from each of the three parts of the inequality.**

$3 > x > -\frac{5}{3}$ • **Multiply each part of the inequality by $-1/3$ and reverse the inequality symbols.**

In interval notation, the solution set is given by $(-5/3, 3)$. See **Figure 1.11.**

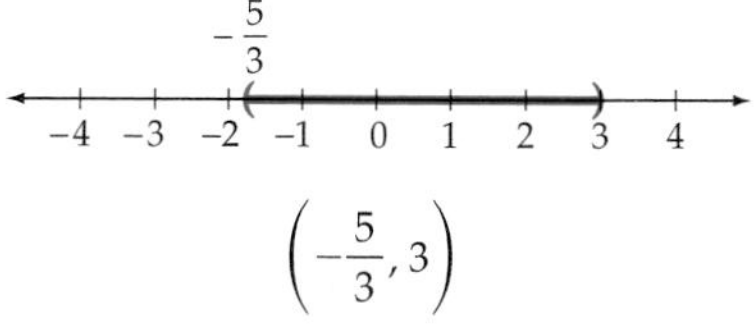

$\left(-\frac{5}{3}, 3\right)$

Figure 1.11

TRY EXERCISE 22, EXERCISE SET 1.5, PAGE 117

take note

Some inequalities have a solution set that consists of all real numbers. For example, $|x + 9| \ge 0$ is true for all values of x. Because an absolute value is always nonnegative, the equation is always true.

EXAMPLE 4 Solve an Absolute Value Inequality

Solve: $|4x - 3| \ge 5$

Solution

$|4x - 3| \ge 5$ implies $4x - 3 \le -5$ or $4x - 3 \ge 5$. Solving each of these inequalities produces

$$
\begin{array}{lcl}
4x - 3 \le -5 & \text{or} & 4x - 3 \ge 5 \\
4x \le -2 & & 4x \ge 8 \\
x \le -\frac{1}{2} & & x \ge 2
\end{array}
$$

Therefore, the solution set is $(-\infty, -1/2] \cup [2, \infty)$. See **Figure 1.12.**

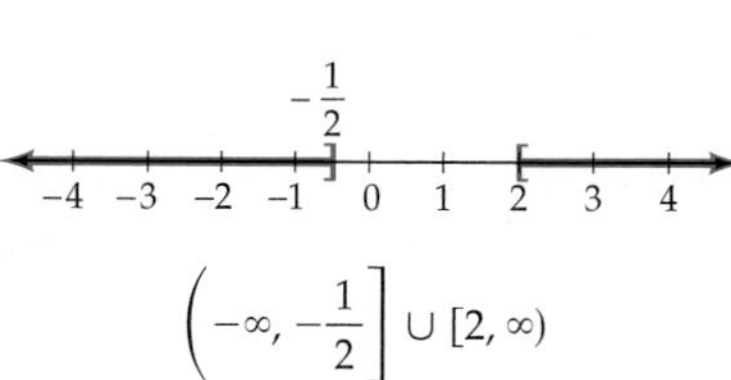

$\left(-\infty, -\frac{1}{2}\right] \cup [2, \infty)$

Figure 1.12

TRY EXERCISE 26, EXERCISE SET 1.5, PAGE 117

◆ SOLVING INEQUALITIES BY THE CRITICAL VALUE METHOD

Any value of x that causes a polynomial in x to equal zero is called a **zero of the polynomial.** For example, -4 and 1 are both zeros of the polynomial $x^2 + 3x - 4$, because $(-4)^2 + 3(-4) - 4 = 0$ and $1^2 + 3 \cdot 1 - 4 = 0$.

A Sign Property of Polynomials

Nonzero polynomials in x have the property that for any value of x between two consecutive real zeros, either all values of the polynomial are positive or all values of the polynomial are negative.

In our work with inequalities that involve polynomials, the real zeros of the polynomial are also referred to as **critical values of the inequality,** because on a number line they separate the real numbers that make an inequality involving a polynomial true from those that make it false. In Example 5, we use critical values and the sign property of polynomials to solve an inequality.

EXAMPLE 5 Solve a Polynomial Inequality

Solve: $x^2 + 3x - 4 < 0$

Solution

Factoring the polynomial $x^2 + 3x - 4$ produces the equivalent inequality

$$(x + 4)(x - 1) < 0$$

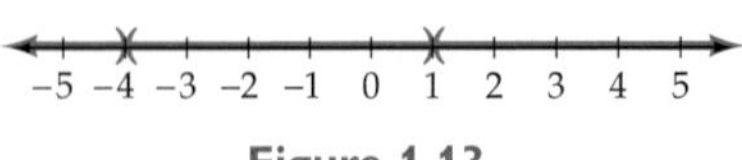

Figure 1.13

Thus the zeros of the polynomial $x^2 + 3x - 4$ are -4 and 1. They are the critical values of the inequality $x^2 + 3x - 4 < 0$. They separate the real number line into the three intervals shown in **Figure 1.13.**

To determine the intervals on which $x^2 + 3x - 4 < 0$, pick a number called a **test value** from each of the three intervals and then determine whether $x^2 + 3x - 4 < 0$ for each of these test values. For example, in the interval $(-\infty, -4)$, pick a test value of, say, -5. Then

$$x^2 + 3x - 4 = (-5)^2 + 3(-5) - 4 = 6$$

Because 6 is not less than 0, by the sign property of polynomials, no number in the interval $(-\infty, -4)$ makes $x^2 + 3x - 4 < 0$.

Now pick a test value from the interval $(-4, 1)$, say, 0. When $x = 0$,

$$x^2 + 3x - 4 = 0^2 + 3(0) - 4 = -4$$

Because -4 is less than 0, by the sign property of polynomials, all numbers in the interval $(-4, 1)$ make $x^2 + 3x - 4 < 0$.

If we pick a test value of 2 from the interval $(1, \infty)$, then

$$x^2 + 3x - 4 = (2)^2 + 3(2) - 4 = 6$$

Because 6 is not less than 0, by the sign property of polynomials, no number in the interval $(1, \infty)$ makes $x^2 + 3x - 4 < 0$.

The following table is a summary of our work.

Interval	Test Value x	$x^2 + 3x - 4 \overset{?}{<} 0$
$(-\infty, -4)$	-5	$(-5)^2 + 3(-5) - 4 < 0$ $6 < 0$ False
$(-4, 1)$	0	$(0)^2 + 3(0) - 4 < 0$ $-4 < 0$ True
$(1, \infty)$	2	$(2)^2 + 3(2) - 4 < 0$ $6 < 0$ False

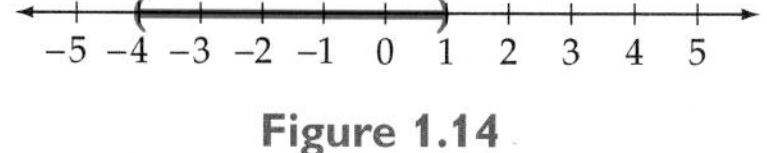

Figure 1.14

In interval notation, the solution set of $x^2 + 3x - 4 < 0$ is $(-4, 1)$. The solution set is graphed in **Figure 1.14**. Note that in this case the critical values -4 and 1 are not included in the solution set because they do not make $x^2 + 3x - 4$ less than 0.

TRY EXERCISE 42, EXERCISE SET 1.5, PAGE 117

To avoid the arithmetic in Example 5, we often use a *sign diagram*. For example, note that the factor $(x + 4)$ is negative for all $x < -4$ and positive for all $x > -4$. The factor $(x - 1)$ is negative for all $x < 1$ and positive for all $x > 1$. These results are shown in **Figure 1.15**.

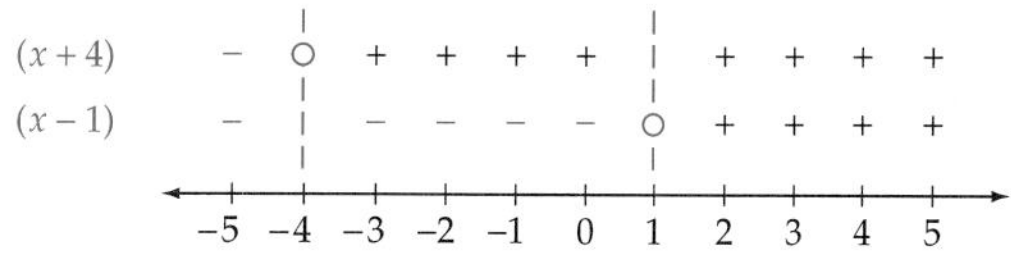

Figure 1.15

To determine on which intervals the product $(x + 4)(x - 1)$ is negative, we examine the sign diagram to see where the factors have opposite signs. This occurs only on the interval $(-4, 1)$, where $(x + 4)$ is positive and $(x - 1)$ is negative, so the original equality is true only on the interval $(-4, 1)$.

Following is a summary of the steps used to solve polynomial inequalities by the critical value method.

Solving a Polynomial Inequality by the Critical Value Method

1. Write the inequality so that one side of the inequality is a nonzero polynomial and the other side is 0.
2. Find the real zeros of the polynomial.[3] They are the critical values of the original inequality.
3. Use test values to determine which of the intervals formed by the critical values are to be included in the solution set.

[3]In Chapter 3, additional ways to find the zeros of a polynomial are developed. For the present, however, we will find the zeros by factoring or by using the quadratic formula.

◆ RATIONAL INEQUALITIES

A rational expression is the quotient of two polynomials. **Rational inequalities** involve rational expressions, and they can be solved by an extension of the critical value method.

Critical Values of a Rational Expression

The **critical values of a rational expression** are the numbers that cause the numerator of the rational expression to equal zero or the denominator of the rational expression to equal zero.

Rational expressions also have the property that they remain either positive for all values of the variable between consecutive critical values or negative for all values of the variable between consecutive critical values.

Following is a summary of the steps used to solve rational inequalities by the critical value method.

Solving a Rational Inequality by the Critical Value Method

1. Write the inequality so that one side of the inequality is a rational expression and the other side is 0.
2. Find the real zeros of the numerator of the rational expression and the real zeros of its denominator. They are the critical values of the inequality.
3. Use test values to determine which of the intervals formed by the critical values are to be included in the solution set.

EXAMPLE 6 Solve a Rational Inequality

Solve: $\dfrac{3x + 4}{x + 1} \le 2$

Solution

Write the inequality so that 0 appears on the right side of the inequality.

$$\frac{3x + 4}{x + 1} \le 2$$

$$\frac{3x + 4}{x + 1} - 2 \le 0$$

Write the left side as a rational expression.

$$\frac{3x + 4}{x + 1} - \frac{2(x + 1)}{x + 1} \leq 0 \qquad \text{• The LCD is } x + 1.$$

$$\frac{3x + 4 - 2x - 2}{x + 1} \leq 0 \qquad \text{• Simplify.}$$

$$\frac{x + 2}{x + 1} \leq 0$$

The critical values of this inequality are -2 and -1 because the numerator $x + 2$ is equal to zero when $x = -2$, and the denominator $x + 1$ is equal to zero when $x = -1$. The critical values -2 and -1 separate the real number line into the three intervals $(-\infty, -2)$, $(-2, -1)$, and $(-1, \infty)$.

All values of x on the interval $(-2, -1)$ make $(x + 2)/(x + 1)$ negative, as desired. On the other intervals, the quotient $(x + 2)/(x + 1)$ is positive. See the sign diagram in **Figure 1.16.**

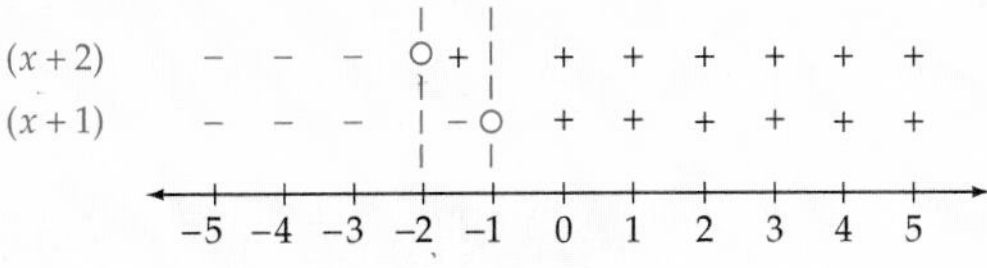

Figure 1.16

The solution set is $[-2, -1)$. The graph of the solution set is shown in **Figure 1.17.** Note that -2 is included in the solution set because $(x + 2)/(x + 1) = 0$ when $x = -2$. However, -1 is not included in the solution set because the denominator $(x + 1)$ is zero when $x = -1$.

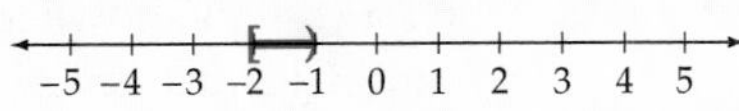

Figure 1.17

TRY EXERCISE 56, EXERCISE SET 1.5, PAGE 117

◆ APPLICATIONS

EXAMPLE 7 Solve an Application That Involves an Inequality

You can rent a car from Company A for \$46 per day plus \$0.09 a mile. Company B charges \$32 per day plus \$0.14 a mile. Find the number of miles for which it is cheaper to rent from Company A if you rent a car for 1 day.

Continued •➤

Solution

Let m equal the number of miles the car is to be driven. Then the cost of renting the car will be

$$\$46 + \$0.09m \quad \text{from Company A}$$
$$\$32 + \$0.14m \quad \text{from Company B}$$

If renting from Company A is to be cheaper than renting from Company B, then we must have

$$46 + 0.09m < 32 + 0.14m$$

Solving for m produces

$$14 < 0.05m$$
$$\frac{14}{0.05} < m$$
$$280 < m$$

Renting from Company A is cheaper if you drive over 280 miles per day.

TRY EXERCISE 74, EXERCISE SET 1.5, PAGE 118

EXAMPLE 8 Solve an Application That Involves a Compound Inequality

A photographic developer needs to be kept at a temperature between 15°C and 25°C. What is that temperature range in degrees Fahrenheit (°F)?

Solution

Figure 1.18 shows how some temperatures are related on the Fahrenheit and Celsius scales. The formula that relates the Celsius temperature (C) to the Fahrenheit temperature (F) is

$$C = \frac{5}{9}(F - 32)$$

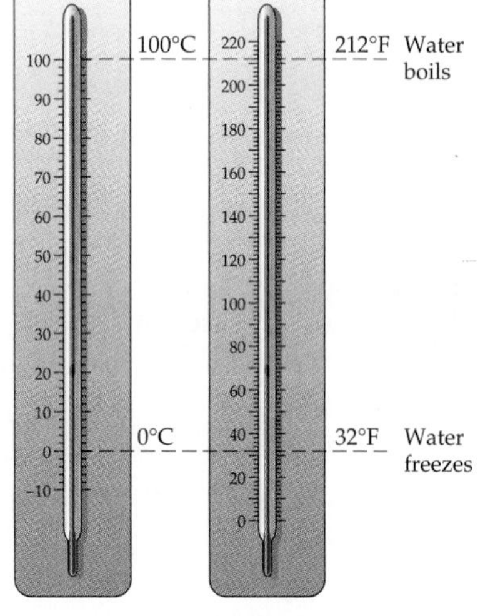

Figure 1.18

We are given that

$$15 < C < 25$$

Substituting $\frac{5}{9}(F - 32)$ for C yields

$$15 < \frac{5}{9}(F - 32) < 25$$

$$27 < F - 32 < 45$$ • **Multiply each of the three parts of the inequality by 9/5.**

$$59 < F < 77$$ • **Add 32 to each of the three parts of the inequality.**

Thus the developer needs to be kept between 59°F and 77°F.

TRY EXERCISE 78, EXERCISE SET 1.5, PAGE 118

SUPPLEMENTAL EXERCISES

In Exercises 85 to 88, use the critical value method to solve each inequality. Use interval notation to write each solution set.

85. $\dfrac{(x-3)^2}{(x-6)^2} > 0$

86. $\dfrac{(x-1)^2}{(x-4)^4} \ge 0$

87. $\dfrac{(x-4)^2}{(x+3)^3} \ge 0$

88. $\dfrac{(2x-7)}{(x-1)^2(x+2)^2} \ge 0$

In Exercises 89 to 92, determine the set of all real numbers *x* such that *y* will be a real number.

89. $y = \sqrt[4]{x^3 - 3x}$

90. $y = \sqrt[4]{x^4 - 4x^3 + 4x^2}$

91. $y = \sqrt[6]{5 + x^2}$

92. $y = \sqrt[6]{(x+3)^6}$

In Exercises 93 to 96, find the values of *k* such that the given equation will have at least one real solution.

93. $x^2 + kx + 6 = 0$

94. $x^2 + kx + 11 = 0$

95. $2x^2 + kx + 7 = 0$

96. $-3x^2 + kx - 4 = 0$

In Exercises 97 to 108, use interval notation to express the solution set of each inequality.

97. $1 < |x| < 5$

98. $2 < |x| < 3$

99. $3 \le |x| < 7$

100. $0 < |x| \le 3$

101. $0 < |x - a| < \delta \quad (\delta > 0)$

102. $0 < |x - 5| < 2$

103. $2 < |x - 6| < 4$

104. $1 \le |x - 3| < 5$

105. $|x| > |x - 1|$

106. $|x - 2| \le |x + 4|$

107. $|x^2 - 10| < 6$

108. $|x^2 - 1| > 1$

109. **HEIGHT OF A PROJECTILE** The equation

$$s = -16t^2 + v_0 t + s_0$$

gives the height s in feet above ground level, at the time t seconds, of an object thrown directly upward from a height s_0 feet above the ground and with an initial velocity of v_0 feet per second. A ball is thrown directly upward from ground level with an initial velocity of 64 feet per second. Find the time interval for which the ball has a height of more than 48 feet.

110. **HEIGHT OF A PROJECTILE** A ball is thrown directly upward from a height of 32 feet above the ground with an initial velocity of 80 feet per second. Find the time interval for which the ball will be more than 96 feet above the ground. (*Hint:* See Exercise 109.)

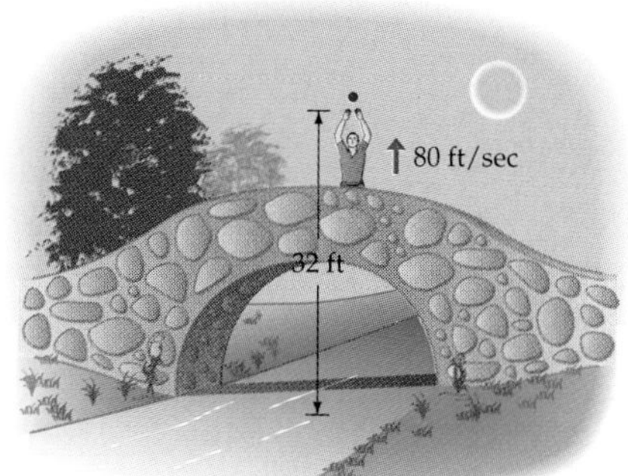

PROJECTS

1. **TRIANGLES** In any triangle, the sum of the lengths of the two shorter sides must be greater than the length of the longest side. Find all possible values of x if a triangle has sides of length

   a. x, $x + 5$, and $x + 9$ b. x, $x^2 + x$, and $2x^2 + x$ c. $\dfrac{1}{x+2}, \dfrac{1}{x+1}$, and $\dfrac{1}{x}$

2. **FAIR COINS** A coin is considered a **fair** coin if it has an equal chance of landing heads up or tails up. To decide whether a coin is a fair coin, a statistician tosses it 1000 times and records the number of tails t. The statistician is prepared to state that the coin is a fair coin if

$$\left|\frac{t - 500}{15.81}\right| \le 2.33$$

   a. Determine what values of t will cause the statistician to state that the coin is a fair coin.

   b. Pick a coin and test it according to the criteria above to see whether it is a fair coin.

SECTION 1.6 VARIATION AND APPLICATIONS

- DIRECT VARIATION
- INVERSE VARIATION
- JOINT VARIATION AND COMBINED VARIATION

◆ DIRECT VARIATION

Many real-life situations involve variables that are related by a type of equation called a **variation.** For example, a stone thrown into a pond generates circular ripples whose circumference and diameter are increasing. The equation $C = \pi d$ expresses the relationship between the circumference C of a circle and its diameter d. If d increases, then C increases. In fact, if d doubles in size, then C also doubles in size. The circumference C is said to *vary directly* as the diameter d.

Definition of Direct Variation

The variable y **varies directly** as the variable x, or y is **directly proportional** to x, if and only if

$$y = kx$$

where k is a constant called the **constant of proportionality** or the **variation constant.**

Direct variations occur in many daily applications. For example, the cost of a newspaper is 35 cents. The cost C to purchase n newspapers is directly proportional to the number n. That is, $C = 35n$. In this example the variation constant is 35.

To solve a problem that involves a variation, we typically write a general equation that relates the variables and then use given information to solve for the variation constant.

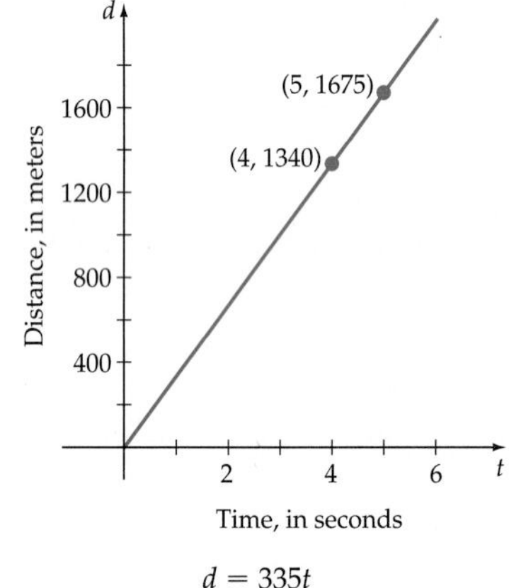

$d = 335t$

Figure 1.19

EXAMPLE 1 Solve a Direct Variation

The distance sound travels varies directly as the time it travels. If sound travels 1340 meters in 4 seconds, find the distance sound will travel in 5 seconds.

Solution

Write an equation that relates the distance d to the time t. Because d varies directly as t, our equation is $d = kt$. Because $d = 1340$ when $t = 4$, we obtain

$$1340 = k \cdot 4 \quad \text{which implies} \quad k = \frac{1340}{4} = 335$$

Therefore, the specific equation that relates the distance d sound travels in t seconds is $d = 335t$. To find the distance sound travels in 5 seconds, replace t with 5 to produce

$$d = 335(5) = 1675$$

Under the same conditions, sound will travel 1675 meters in 5 seconds. See **Figure 1.19.**

TRY EXERCISE 22, EXERCISE SET 1.6, PAGE 126

Direct Variation as the *n*th Power

If y **varies directly as the *n*th power,** of x, then

$$y = kx^n$$

where k is a constant

EXAMPLE 2 Solve a Variation of the Form $y = kx^2$

The distance s that an object falls from rest (neglecting air resistance) varies directly as the square of the time t that it has been falling. If an object falls 64 feet in 2 seconds, how far will it fall in 10 seconds?

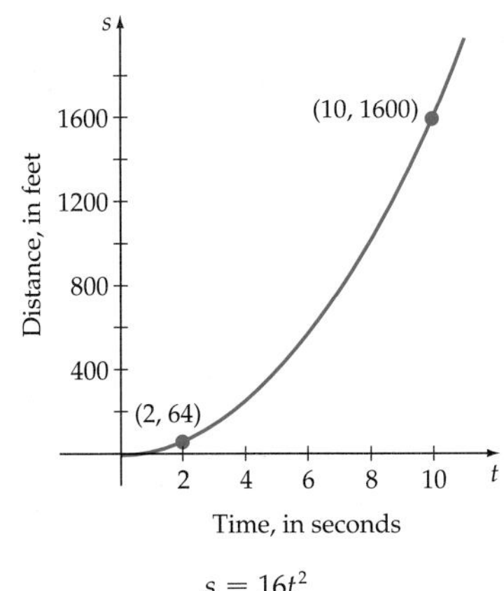

$s = 16t^2$

Figure 1.20

Solution

Because s varies directly as the square of t, $s = kt^2$. The variable s is 64 when t is 2, so

$$64 = k \cdot 2^2 \qquad \text{which implies} \qquad k = \frac{64}{4} = 16$$

The specific equation that relates the distance s an object falls in t seconds is $s = 16t^2$. Letting $t = 10$ yields

$$s = 16(10^2) = 16(100) = 1600$$

Under the same conditions, the object will fall 1600 feet in 10 seconds. See **Figure 1.20.**

TRY EXERCISE 24, EXERCISE SET 1.6, PAGE 126

◆ INVERSE VARIATION

Two variables can also vary *inversely.*

Definition of Inverse Variation

The variable y **varies inversely** as the variable x, or y is **inversely proportional** to x, if and only if

$$y = \frac{k}{x}$$

where k is the variation constant.

In 1661, Robert Boyle made a study of the *compressibility* of gases. **Figure 1.21** shows that he used a J-shaped tube to demonstrate the inverse relationship between the volume of a gas at a given temperature and the applied pressure. The J-shaped tube on the left shows that the volume of a gas at normal atmospheric pressure is 60 milliliters. If the pressure is doubled by adding mercury (Hg), as shown in the middle tube, the volume of the gas is halved to 30 milliliters. Tripling the pressure decreases the volume of the gas to 20 milliliters, as shown in the tube at the right.

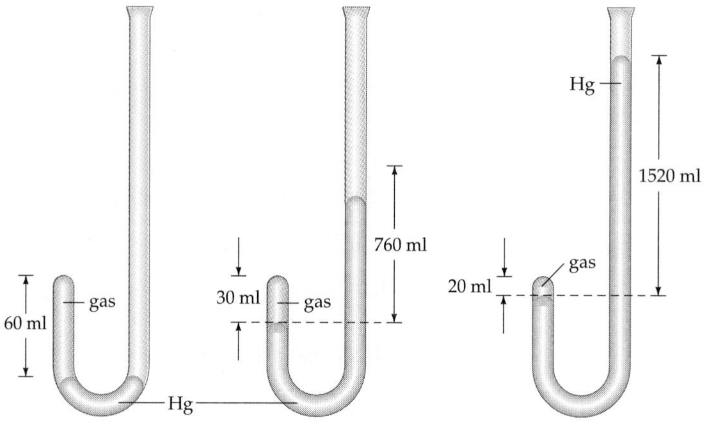

Figure 1.21

EXAMPLE 3 Solve an Inverse Variation

Boyle's Law states that the volume V of a sample of gas (at a constant temperature) varies inversely as the pressure P. The volume of a gas in a J-shaped tube is 75 milliliters when the pressure is 1.5 atmospheres. Find the volume of the gas when the pressure is increased to 2.5 atmospheres.

Solution

The volume V varies inversely as the pressure P, so $V = k/P$. The volume V is 75 milliliters when the pressure is 1.5 atmospheres, so

$$75 = \frac{k}{1.5} \quad \text{and} \quad k = (75)(1.5) = 112.5$$

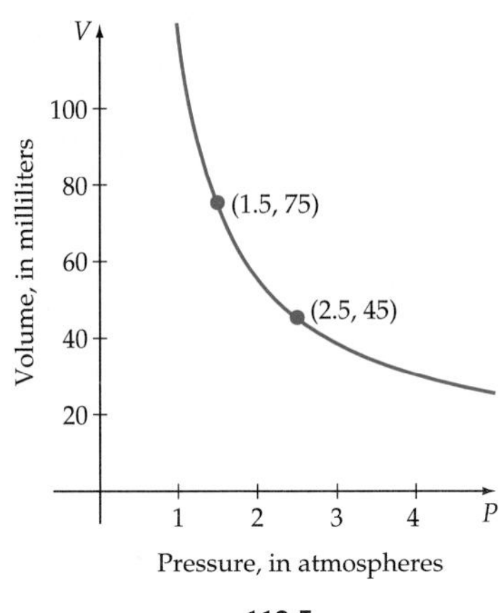

$V = \frac{112.5}{P}$

Figure 1.22

take note

Because the volume V varies inversely as the pressure P, the function $V = 112.5/P$ is a decreasing function, as shown in Figure 1.22.

Thus $V = 112.5/P$. When the pressure is 2.5 atmospheres, we have

$$V = \frac{112.5}{2.5} = 45 \text{ milliliters}$$

See **Figure 1.22.**

TRY EXERCISE 28, EXERCISE SET 1.6, PAGE 126

Inverse Variation as the nth Power

If y **varies inversely as the nth power** of x, then

$$y = \frac{k}{x^n}$$

where k is a constant.

◆ JOINT VARIATION AND COMBINED VARIATION

Some variations involve more than two variables.

Definition of Joint Variation

The variable z **varies jointly** as the variables x and y if and only if

$$z = kxy$$

where k is a constant.

EXAMPLE 4 Solve a Joint Variation

The cost of insulating the ceiling of a house varies jointly with the thickness of the insulation and the area of the ceiling. It costs $175 to insulate a 2100-square-foot ceiling with insulation 4 inches thick. Find the cost of insulating a 2400-square-foot ceiling with insulation that is 6 inches thick.

Solution

Because the cost C varies jointly as the area A of the ceiling and the thickness T of the insulation, we know $C = kAT$. Using the fact that $C = 175$ when $A = 2100$ and $T = 4$ gives us

$$175 = k(2100)(4) \qquad \text{which implies} \qquad k = \frac{175}{(2100)(4)} = \frac{1}{48}$$

Continued ▸

Consequently, the specific formula for C is $C = \frac{1}{48}AT$. Now when $A = 2400$ and $T = 6$, we have

$$C = \frac{1}{48}(2400)(6) = 300$$

Thus the cost of insulating the 2400-square-foot ceiling with 6-inch insulation is \$300.

TRY EXERCISE 30, EXERCISE SET 1.6, PAGE 126

Combined variations involve more than one type of variation.

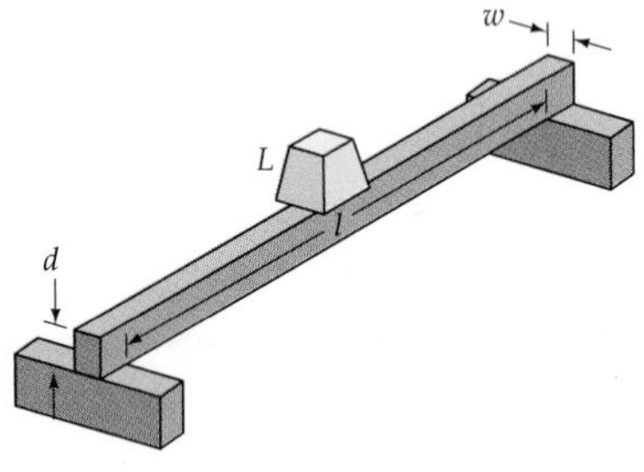

Figure 1.23

EXAMPLE 5 Solve a Combined Variation

The weight that a horizontal beam with a rectangular cross section can safely support varies jointly as the width and square of the depth of the cross section and inversely as the length of the beam. See **Figure 1.23.** If a 4-inch by 4-inch beam 10 feet long safely supports a load of 256 pounds, what load L can be safely supported by a beam made of the same material and with a width w of 4 inches, a depth d of 6 inches, and a length l of 16 feet?

Solution

The general variation equation is $L = k\frac{wd^2}{l}$. Using the given data yields

$$256 = k\frac{4(4^2)}{10}$$

Solving for k produces $k = 40$, so the specific formula for L is

$$L = 40\frac{wd^2}{l}$$

Substituting 4 for w, 6 for d, and 16 for l gives

$$L = 40\frac{4(6^2)}{16} = 360 \text{ pounds}$$

TRY EXERCISE 34, EXERCISE SET 1.6, PAGE 126

TOPICS FOR DISCUSSION

1. The area A of a trapezoid varies jointly as the product of its height h and the sum of its bases b and B. State an equation that represents this variation. Given that $A = 15$ square inches when $h = 6$ inches, $b = 2$ inches, and $B = 3$ inches, explain how you would determine the value of the variation constant.

2. Given that the variation constant $k > 0$ and that A varies directly as b, then A ______ when b increases, and A ______ when b decreases.

3. Given that the variation constant $k > 0$ and that S varies inversely as d, then S ______ when d increases, and S ______ when d decreases.

4. All direct variations can be written in the form $y =$ ______, and all inverse variations can be written in the form $y =$ ______.

5. The volume V of a right circular cylinder varies jointly as the square of the radius r and the height h. Tell what happens to V when

   a. h is tripled b. r is tripled

   c. r is doubled and h is decreased to $\frac{1}{2}h$

6. Give some examples of real situations where one quantity varies inversely as a second quantity.

EXERCISE SET 1.6

In Exercises 1 to 12, write an equation that represents the relationship between the given variables. Use k as the variation constant.

1. d varies directly as t.
2. r varies directly as the square of s.
3. y varies inversely as x.
4. p is inversely proportional to q.
5. m varies jointly as n and p.
6. t varies jointly as r and the cube of s.
7. V varies jointly as l, w, and h.
8. u varies directly as v and inversely as the square of w.
9. A is directly proportional to the square of s.
10. A varies jointly as h and the square of r.
11. F varies jointly as m_1 and m_2 and inversely as the square of d.
12. T varies jointly as t and r and the square of a.

In Exercises 13 to 20, write the equation that expresses the relationship between the variables, and then use the given data to solve for the variation constant.

13. y varies directly as x, and $y = 64$ when $x = 48$.
14. m is directly proportional to n, and $m = 92$ when $n = 23$.
15. r is directly proportional to the square of t, and $r = 144$ when $t = 108$.
16. C varies directly as r, and $C = 94.2$ when $r = 15$.
17. T varies jointly as r and the square of s, and $T = 210$ when $r = 30$ and $s = 5$.
18. u varies directly as v and inversely as the square root of w, and $u = 0.04$ when $v = 8$ and $w = 0.04$.
19. V varies jointly as l, w, and h, and $V = 240$ when $l = 8$ and $w = 6$ and $h = 5$.
20. t varies directly as the cube of r and inversely as the square root of s, and $t = 10$ when $r = 5$ and $s = 0.09$.
21. **CHARLES'S LAW** *Charles's Law* states that the volume V occupied by a gas (at a constant pressure) is directly proportional to its absolute temperature T. An experiment with a balloon shows that the volume of the balloon is 0.85 liter at 270 K (absolute temperature).[4] What will the volume of the balloon be when its temperature is 324 K?

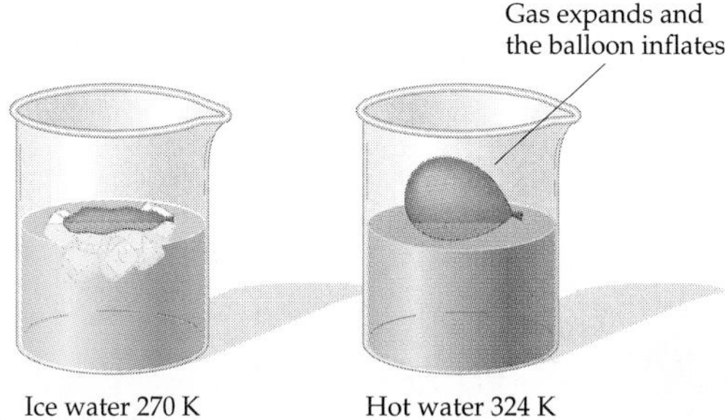

[4] Absolute temperature is measured on the Kelvin scale. A unit (called a kelvin) on the Kelvin scale is the same measure as a degree on the Celsius scale; however, 0 on the Kelvin scale corresponds to −273°C on the Celsius scale.

22. **HOOKE'S LAW** *Hooke's Law* states that the distance a spring stretches varies directly as the weight on the spring. A weight of 80 pounds stretches a spring 6 inches. How far will a weight of 100 pounds stretch the spring?

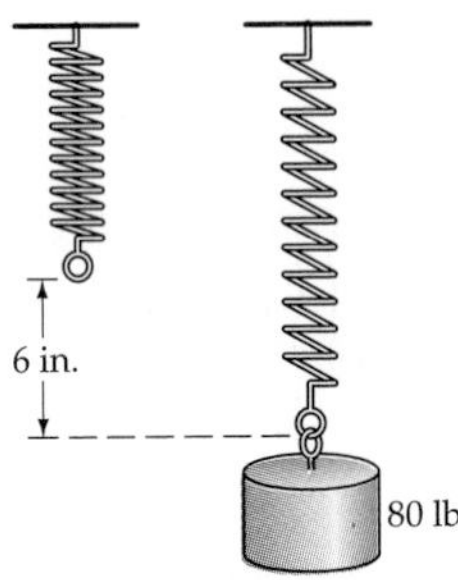

23. **PRESSURE AND DEPTH** The pressure a liquid exerts at a given point on a submarine is directly proportional to the depth of the point below the surface of the liquid. If the pressure at a depth of 3 feet is 187.5 pounds per square foot, find the pressure at a depth of 7 feet.

24. **MOTORCYCLE JUMP** The range of a projectile is directly proportional to the square of its velocity. If a motorcyclist can make a jump of 140 feet by coming off a ramp at 60 mph, find the distance the motorcyclist could expect to jump if the speed coming off the ramp were increased to 65 mph.

25. **PERIOD OF A PENDULUM** The period T (the time it takes a pendulum to make one complete oscillation) varies directly as the square root of its length L. A pendulum 3 feet long has a period of 1.8 seconds.

    a. Find the period of a pendulum 10 feet long.

    b. What is the length of a pendulum that *beats seconds* (that is, it has a 2-second period)?

26. **AREA OF A PROJECTED PICTURE** The area of a projected picture on a movie screen varies directly as the square of the distance from the projector to the screen. If a distance of 20 feet produces a picture with an area of 64 square feet, what distance produces an area of 100 square feet?

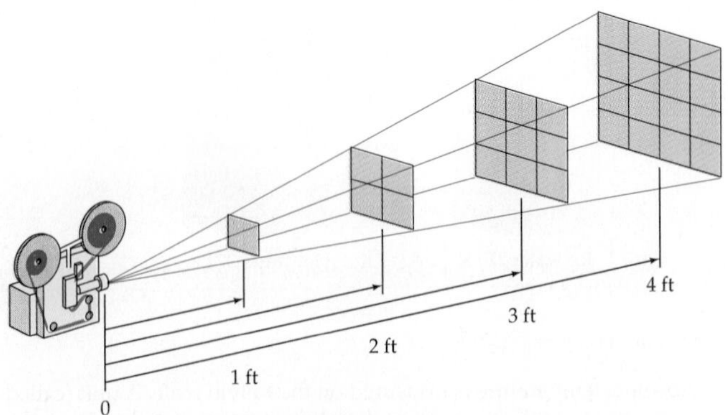

27. **DECIBELS** The loudness, measured in decibels, of a stereo speaker is inversely proportional to the square of the distance of the listener from the speaker. The loudness is 28 decibels at a distance of 8 feet. What is the loudness when the listener is 4 feet from the speaker?

28. **ILLUMINATION** The illumination a source of light provides is inversely proportional to the square of the distance from the source. If the illumination at a distance of 10 feet from the source is 50 footcandles, what is the illumination at a distance of 15 feet from the source?

29. **VOLUME RELATIONSHIPS** The volume V of a right circular cone varies jointly as the square of the radius r and the height h. Tell what happens to V when

    a. r is tripled

    b. h is tripled

    c. both r and h are tripled

30. **SAFE LOAD** The load L that a horizontal beam can safely support varies jointly as the width w and the square of the depth d. If a beam with width 2 inches and depth 6 inches safely supports up to 200 pounds, how many pounds can a beam of the same length that has width 4 inches and depth 4 inches be expected to support?

31. **IDEAL GAS LAW** The *Ideal Gas Law* states that the volume V of a gas varies jointly as the number of moles of gas n and the absolute temperature T and inversely as the pressure P. What happens to V when n is tripled and P is reduced by a factor of one-half?

32. **MAXIMUM LOAD** The maximum load of a cylindrical column of circular cross section can support varies directly as the fourth power of the diameter and inversely as the square of the height. If a column 2 feet in diameter and 10 feet high supports up to 6 tons, how much of a load does a column 3 feet in diameter and 14 feet high support?

33. **ASTRONOMY** A meteorite approaching the earth has a velocity that varies inversely as the square root of the distance from the center of the earth. The meteorite has a velocity of 3 miles per second at 4900 miles from the center of the earth. Find the velocity of the meteorite when it is 4225 miles from the center of the earth.

34. **SAFE LOAD** The load L a horizontal beam can safely support varies jointly as the width w and the square of the depth d and inversely as the length l. If a 12-foot beam with width 4 inches and depth 8 inches safely supports 800 pounds, how many pounds can a 16-foot beam that has width 3.5 inches and depth 6 inches be expected to support?

35. **FORCE, SPEED, AND RADIUS RELATIONSHIPS** The force needed to keep a car from skidding on a curve varies jointly as the weight of the car and the square of the speed and inversely as the radius of the curve. It takes 2800 pounds of force to keep an 1800-pound car from skidding on a curve with radius 425 feet at 45 mph. What force is needed to keep the same car from skidding when it takes a similar curve with radius 450 feet at 55 mph?

36. **STIFFNESS OF A BEAM** A cylindrical log is to be cut so that it will yield a beam that has a rectangular cross section of depth d and width w. The stiffness of a beam of given length is directly proportional to the width and the cube of the depth. The diameter of the log is 18 inches.

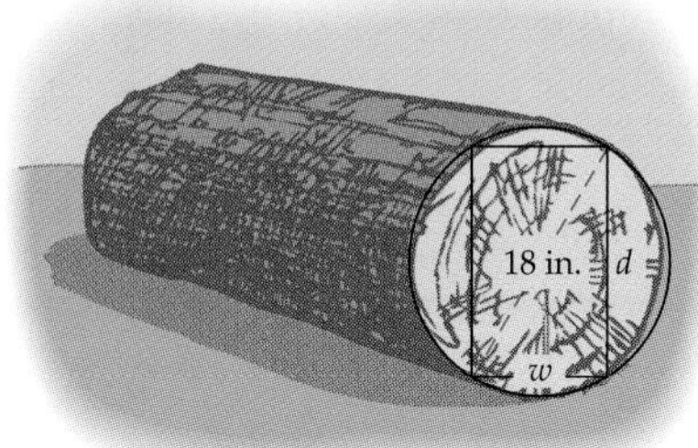

What depth will yield the "stiffest" beam, $d = 10$ inches, $d = 12$ inches, $d = 14$ inches, or $d = 16$ inches?

SUPPLEMENTAL EXERCISES

37. **KEPLER'S THIRD LAW** *Kepler's Third Law* states that the time T needed for a planet to make one complete revolution about the sun is directly proportional to the 3/2 power of the average distance d between the planet and the sun. The earth, which averages 93 million miles from the sun, completes one revolution in 365 days. Find the average distance from the sun to Mars if Mars completes one revolution about the sun in 686 days.

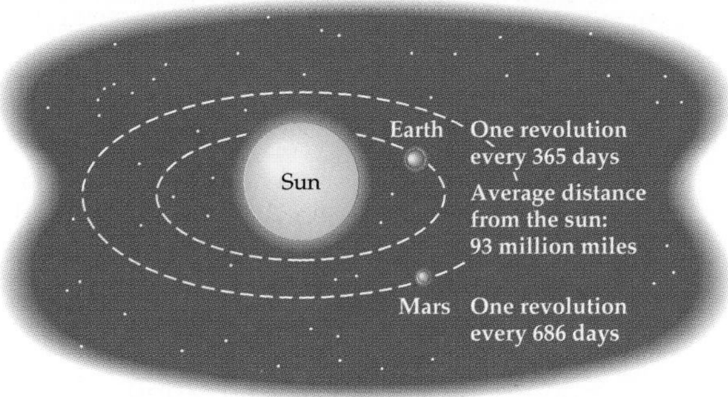

38. **ILLUMINATION** The illumination a light source provides is directly proportional to the strength of the source and inversely proportional to the square of the distance from the source. Two light sources are 10 feet apart. The strength of the light source at point B is 8 times the strength of the light source at point A.

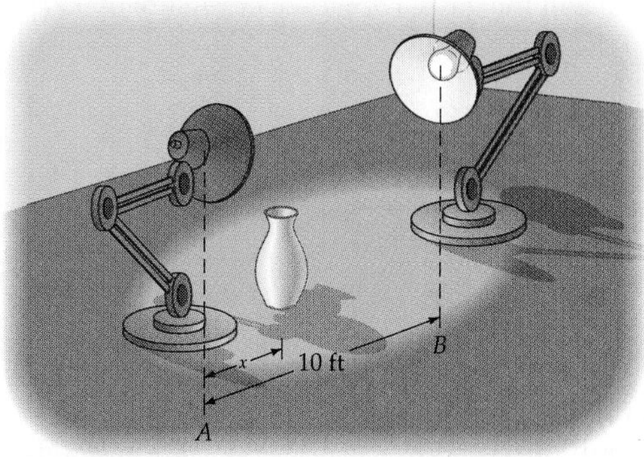

Which position will receive the least amount of illumination, the point on line segment AB where $x = 2.5$ feet, where $x = 3$ feet, where $x = 3.3$ feet, or where $x = 3.5$ feet?

PROJECTS

1. **A DIRECT VARIATION FORMULA** If $f(x)$ varies directly as x, prove that $f(x_2) = f(x_1)\dfrac{x_2}{x_1}$.

   Use this formula to solve the following direct variation without solving for the variation constant. The distance a spring stretches varies directly as the force applied. An experiment shows that a force of 17 kilograms stretches the spring 8.5 centimeters. How far will a 22-kilogram force stretch the spring?

2. **AN INVERSE VARIATION FORMULA** Given that $f(x)$ varies inversely as x, prove that $f(x_2) = f(x_1)\dfrac{x_1}{x_2}$. Use this formula to solve the following inverse variation *without* solving for the variation constant. The volume of a gas varies inversely as pressure (assuming the temperature remains constant). An experiment shows that a particular gas has a volume of 2.4 liters under a pressure of 280 grams per square centimeter. What volume will the gas have when a pressure of 330 grams per square centimeter is applied?

EXPLORING CONCEPTS WITH TECHNOLOGY

The Mandelbrot Replacement Procedure

The following procedure is called the **Mandelbrot replacement procedure.**

Mandelbrot Replacement Procedure

Pick a complex number s.

1. Square s and add the result to s.
2. Square the last result and add it to s.
3. Repeat step 2.

The number s is referred to as the seed of the procedure. The number s is a seed in the sense that each seed produces a different sequence of numbers. Some seeds produce sequences that grow without bound. Some seeds produce sequences that grow toward some constant. Still other seeds yield sequences that are cyclic. Consider the following illustrations.

- Let the seed $s = 1$.

$$1^2 + 1 = 2, \qquad 2^2 + 1 = 5, \qquad 5^2 + 1 = 26, \qquad 26^2 + 1 = 677$$

  As the replacement procedure continues, we get larger and larger numbers.

- Let the seed $s = -1$.

$$(-1)^2 + (-1) = 0, \qquad 0^2 + (-1) = -1, \qquad (-1)^2 + (-1) = 0, \ldots$$

  As the replacement procedure continues, the results *cycle:* 0, −1, 0, −1, 0, ...

- Let the seed $s = 0.25$.

$$(0.25)^2 + 0.25 = 0.3125, \qquad (0.3125)^2 + 0.25 = 0.34765625,$$
$$(0.34765625)^2 + 0.25 \approx 0.3708648682, \ldots$$

1. Use your calculator to continue the above Mandelbrot replacement procedure with seed $s = 0.25$. What number do you have after
   a. 25 applications of step 2?
   b. 50 applications of step 2?
   c. 75 applications of step 2?
   d. What constant do you think the sequence of numbers is approaching?

- Let the seed $s = i$.

$$i^2 + 1 = -1 + i, \qquad (-1 + i)^2 + i = -i, \ldots$$

2. **a.** What is the next number produced by the Mandelbrot replacement procedure?

b. What happens as the procedure is continued?

The Mandelbrot replacement procedure can be used to determine special kinds of numbers called **attractors.** The attractors produced by the Mandelbrot replacement procedure are an essential part of the Mandelbrot set, which is shown in black in **Figure 1.27.**

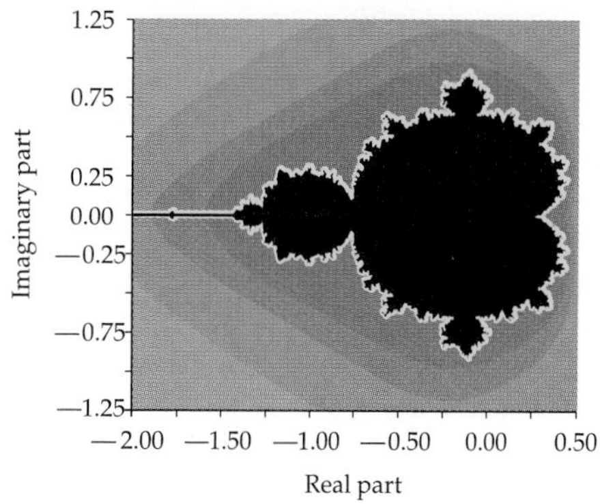

Figure 1.27

CHAPTER 1 SUMMARY

1.1 Linear Equations

- A number is said to satisfy an equation if substituting the number for the variable results in an equation that is a true statement. To solve an equation means to find all values of the variable that satisfy the equation. These values that make the equation true are called solutions or roots of the equation. Equivalent equations have the same solution(s).
- A linear equation in the single variable x is an equation that can be written in the form $ax + b = 0$, where a and b are real numbers, with $a \neq 0$.
- An equation of the form $|ax + b| = c$, $a \neq 0$, is an absolute value equation.

1.2 Formulas and Applications

- A formula is an equation that expresses known relationships between two or more variables. Application problems are best solved by using the guidelines developed in this section.

1.3 Quadratic Equations

- A quadratic equation in x is an equation that can be written in the form $ax^2 + bx + c = 0$, where $a \neq 0$. If the quadratic polynomial in a quadratic equation is factorable over the set of integers, then the equation can be solved by factoring and using the zero product property (see page 83). Every quadratic equation can be solved by completing the square or by using the quadratic formula.
- A number of the form $a + bi$ is a complex number, where a and b are real numbers and i is defined so that $i^2 = -1$. The number a is called the real part of the imaginary number, and the number b is called the imaginary part of the complex number. The operations on complex numbers can be reviewed on pages 85 to 88.
- **The Quadratic Formula**
  If $ax^2 + bx + c = 0$, $a \neq 0$, then $x = \dfrac{-b \pm \sqrt{b^2 - 4ac}}{2a}$.

1.4 Other Types of Equations

- **The Power Principle**
  If P and Q are algebraic expressions and n is a positive integer, then every solution of $P = Q$ is a solution of $P^n = Q^n$.
- An equation is said to be quadratic in form if it can be written in the form $au^2 + bu + c = 0$, where $a \neq 0$ and u is an algebraic expression.

1.5 Inequalities

- The set of all solutions of an inequality is the solution set of the inequality. Equivalent inequalities have the same solution set. To solve an inequality, use the properties of inequalities or the critical value method.
- An inequality of the form $|ax + b| > c$, $a \neq 0$, is an absolute value inequality. The inequality symbol $>$ can be replaced by $<, \leq$, or $\geq$.

1.6 Variation and Applications

- The variable y varies directly as the variable x if and only if $y = kx$, where k is a constant called the variation constant.
- The variable y varies inversely as the variable x if and only if $y = k/x$, where k is the variation constant.
- The variable z varies jointly as the variables x and y if and only if $z = kxy$, where k is the variation constant.

CHAPTER 1 TRUE/FALSE EXERCISES

In Exercises 1 to 10, answer true or false. If the statement is false, give an example to show that the statement is false.

1. If $x^2 = 9$, then $x = 3$.
2. The equations
$$x = \sqrt{12 - x} \quad \text{and} \quad x^2 = 12 - x$$
are equivalent equations.
3. Adding the same constant to each side of a given equation produces an equation that is equivalent to the given equation.
4. If $a > b$, then $-a < -b$.
5. If $a \neq 0$, $b \neq 0$, and $a > b$, then $1/a > 1/b$.
6. The discriminant of $ax^2 + bx + c = 0$ is $\sqrt{b^2 - 4ac}$.
7. If $\sqrt{a} + \sqrt{b} = c$, then $a + b = c^2$.
8. The solution set of $|x - a| < b$ with $b > 0$ is given by the interval $(a - b, a + b)$.
9. The only quadratic equation that has roots of 4 and -4 is $x^2 - 16 = 0$.
10. Every quadratic equation $ax^2 + bx + c = 0$ with real coefficients such that $ac < 0$ has two distinct real roots.

CHAPTER 1 REVIEW EXERCISES

In Exercises 1 to 30, solve each equation.

1. $x - 2(5x - 3) = -3(-x + 4)$
2. $3x - 5(2x - 7) = -4(5 - 2x)$
3. $\dfrac{4x}{3} - \dfrac{4x - 1}{6} = \dfrac{1}{2}$
4. $\dfrac{3x}{4} - \dfrac{2x - 1}{8} = \dfrac{3}{2}$
5. $\dfrac{x}{x + 2} + \dfrac{1}{4} = 5$
6. $\dfrac{y - 1}{y + 1} - 1 = \dfrac{2}{y}$
7. $x^2 - 5x + 6 = 0$
8. $6x^2 + x - 12 = 0$
9. $3x^2 - x - 1 = 0$
10. $x^2 - x + 1 = 0$
11. $3x^3 - 5x^2 = 0$
12. $2x^3 - 8x = 0$
13. $6x^4 - 23x^2 + 20 = 0$
14. $3x + 16\sqrt{x} - 12 = 0$
15. $\sqrt{x^2 - 15} = \sqrt{-2x}$
16. $\sqrt{x^2 - 24} = \sqrt{2x}$
17. $\sqrt{3x + 4} + \sqrt{x - 3} = 5$
18. $\sqrt{2x + 2} - \sqrt{x + 2} = \sqrt{x - 6}$
19. $\sqrt{4 - 3x} - \sqrt{5 - x} = \sqrt{5 + x}$
20. $\sqrt{3x + 9} - \sqrt{2x + 4} = \sqrt{x + 1}$
21. $\dfrac{1}{(y + 3)^2} = 1$
22. $\dfrac{1}{(2s - 5)^2} = 4$
23. $|x - 3| = 2$
24. $|x + 5| = 4$
25. $|2x + 1| = 5$
26. $|3x - 7| = 8$
27. $(x + 2)^{1/2} + x(x + 2)^{3/2} = 0$
28. $x^2(3x - 4)^{1/4} + (3x - 4)^{5/4} = 0$
29. $(2x - 1)^{2/3} + (2x - 1)^{1/3} = 12$
30. $6(x + 1)^{1/2} - 7(x + 1)^{1/4} - 3 = 0$

In Exercises 31 to 48, solve each inequality. Express your solution sets by using interval notation.

31. $-3x + 4 \geq -2$
32. $-2x + 7 \leq 5x + 1$
33. $x^2 + 3x - 10 \leq 0$
34. $x^2 - 2x - 3 > 0$
35. $61 \leq \dfrac{9}{5}C + 32 \leq 95$
36. $30 < \dfrac{5}{9}(F - 32) < 65$
37. $x^3 - 7x^2 + 12x \leq 0$
38. $x^3 + 4x^2 - 21x > 0$
39. $\dfrac{x + 3}{x - 4} > 0$
40. $\dfrac{x(x - 5)}{x + 7} \leq 0$

41. $\dfrac{2x}{3-x} \le 10$

42. $\dfrac{x}{5-x} \ge 1$

43. $|3x-4| < 2$

44. $|2x-3| \ge 1$

45. $0 < |x| < 2$

46. $0 < |x| \le 1$

47. $0 < |x-2| < 1$

48. $0 < |x-a| < b \quad (b > 0)$

In Exercises 49 to 54, solve each equation for the indicated unknown.

49. $V = \pi r^2 h$, for h

50. $P = \dfrac{A}{1+rt}$, for t

51. $A = \dfrac{h}{2}(b_1 + b_2)$, for b_1

52. $P = 2(l + w)$, for w

53. $e = mc^2$, for m

54. $F = G\dfrac{m_1 m_2}{s^2}$, for m_1

In Exercises 55 and 56, write the complex number in standard form and give its conjugate.

55. $3 - \sqrt{-64}$

56. $\sqrt{-4} + 6$

In Exercises 57 to 60, simplify and write the complex number in standard form.

57. $(3 + 7i) + (2 - 5i)$

58. $(6 - 8i) - (9 - 11i)$

59. $(5 + 3i)(2 - 5i)$

60. $\dfrac{4+i}{7-2i}$

In Exercises 61 to 64, simplify and write each complex number as i, $-i$, 1, or -1.

61. i^{20}

62. i^{57}

63. $\dfrac{1}{i^{28}}$

64. i^{-200}

65. **UNKNOWN NUMBER** One-half of a number minus one-fourth of the number is four more than one-fifth of the number. What is the number?

66. **RECTANGULAR REGION** The length of a rectangle is 9 feet less than twice the width of the rectangle. The perimeter of the rectangle is 54 feet. Find the width and the length.

67. **DISTANCE TO AN ISLAND** A motorboat left a harbor and traveled to an island at an average rate of 8 knots. The average speed on the return trip was 6 knots. If the total trip took 7 hours, how far is it from the harbor to the island?

68. **PRICE OF SUBSCRIPTION** The price of a magazine subscription rose 5% this year. If the subscription now costs \$21, how much did the subscription cost last year?

69. **INVESTMENT** A total of \$5500 was deposited into two simple interest accounts. On one account the annual simple interest rate is 4%, and on the second account the annual simple interest rate is 6%. The amount of interest earned for 1 year was \$295. How much was invested in each account?

70. **INDIVIDUAL PRICE** A calculator and a battery together sell for \$21. The price of the calculator is \$20 more than the price of the battery. Find the price of the calculator and the price of the battery.

71. **MAINTENANCE COST** Eighteen owners share the maintenance cost of a condominium complex. If six more units are sold, the maintenance cost will be reduced by \$12 per month for each of the present owners. What is the total monthly maintenance cost for the condominium complex?

72. **RECTANGULAR REGION** The perimeter of a rectangle is 40 inches and its area is 96 square inches. Find the length and the width of the rectangle.

73. **BUILD A WALL** A mason can build a wall in 9 hours less than an apprentice. Together they can build the wall in 6 hours. How long would it take the apprentice, working alone, to build the wall?

74. **COMMERCE** An art show brought in \$33,196 on the sale of 4526 tickets. The adult tickets sold for \$8 and the student tickets sold for \$2. How many of each type of ticket were sold?

75. **DIAMETER OF A CONE** As sand is poured from a chute, it forms a right circular cone whose height is one-fourth the diameter of the base. What is the diameter of the base when the cone has a volume of 144 cubic feet?

76. **REVENUE** A manufacturer of calculators finds that the monthly revenue R from a particular style of calculator is given by $R = 72x - 2x^2$, where x is the price in dollars of each calculator. Find the interval, in terms of x, for which the monthly revenue is greater than \$576.

77. The acceleration due to gravity on the surface of a planetary body is directly proportional to the mass of the body and inversely proportional to the square of its radius. If the acceleration due to gravity is 9.8 meters/sec^2 on Earth, whose radius is 6,370,000 meters and whose mass

is 5.98×10^{26} grams, find the acceleration due to gravity on the moon, whose radius is 1,740,000 meters and whose mass is 7.46×10^{24} grams. Round to the nearest hundredth.

78. The maximum load that a cylindrical column of circular cross section can support varies as the fourth power of the diameter and inversely as the square of the height. If a column 1.5 feet in diameter and 8 feet high supports up to 4 tons, how much of a load does a column 4 feet in diameter and 12 feet high support? Round to the nearest integer.

CHAPTER 1 TEST

1. Solve: $3 - \frac{x}{4} = \frac{3}{5}$

2. Solve: $\frac{3}{x+2} - \frac{3}{4} = \frac{5}{x+2}$

3. Solve $ax - c = c(x - d)$ for x.

4. Solve $x^2 + 4x - 1 = 0$ by completing the square.

5. Solve: $3x^2 + 2x - 9 = 0$

6. Solve: $x^4 + 4x^3 - x - 4 = 0$

7. Solve: $\sqrt{x-2} - 1 = \sqrt{3-x}$

8. Solve: $3x^{2/3} + 10x^{1/3} - 8 = 0$

9. Solve: $(x - 3)^{2/3} = 16$

10. Solve: $-3(x + 2) \le 4 - 7x$

11. Solve: $\frac{x^2 + x - 12}{x + 1} \ge 0$

12. Solve: $|2x + 7| = 5$

13. Solve: $|x + 4| < 3$

14. Solve: $|3x - 2| \ge 7$

15. Simplify: $\frac{2 + i}{3 - 2i}$

16. A boat has a speed of 5 mph in still water. The boat can travel 21 miles with the current in the same time in which it can travel 9 miles against the current. Find the rate of the current.

17. A radiator contains 6 liters of a 20% antifreeze solution. How much should be drained and replaced with pure antifreeze to produce a 50% antifreeze solution?

18. A worker can cover a parking lot with asphalt in 10 hours. With the help of an assistant, the work can be done in 6 hours. How long would it take the assistant, working alone, to cover the parking lot with asphalt?

19. You can rent a car for the day from Company A for \$28 plus \$0.10 a mile. Company B charges \$20 plus \$0.18 a mile. At what point, in terms of miles driven per day, is it cheaper to rent from Company A?

20. A meteorite approaching the moon has a velocity that varies inversely as the square root of its distance from the center of the moon. If the meteorite has a velocity of 4 miles per second at 3000 miles from the center of the moon, find the velocity of the meteorite when it is 2500 miles from the center of the moon. Round to the nearest tenth.

2

FUNCTIONS AND GRAPHS

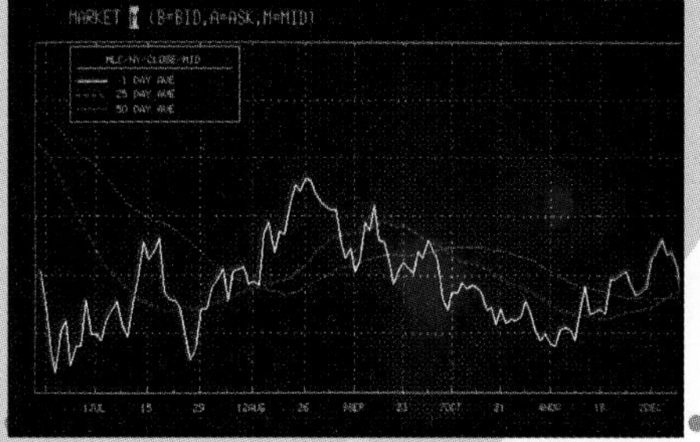

◆ Stockbrokers display market trends using Cartesian graphs.

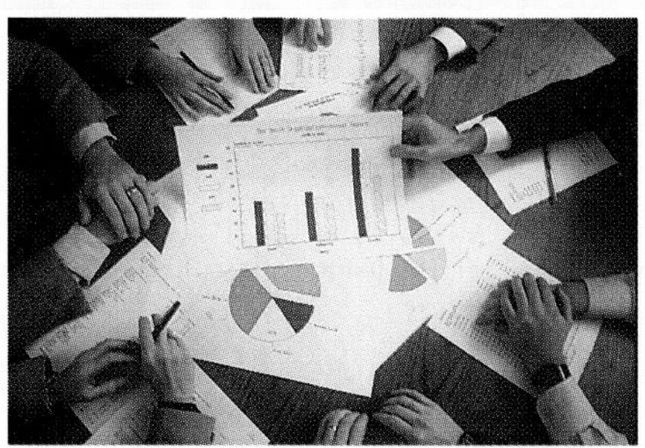

◆ At a business meeting, different types of graphs prove useful in displaying a range of information.

A Simple But Powerful Concept

This chapter is concerned with the concepts of a function, the graph of a function, and elementary analytic geometry. The following two quotes lend support to the power of these ideas.

> *As long as algebra and geometry proceeded along separate paths, their advance was slow and their applications limited. But when the sciences joined company, they drew from each others' vitality and thence forward marched on at a rapid pace toward perfection.*
>
> —*Joseph Louis Lagrange*

> *[Analytic geometry], far more than any of his metaphysical speculations, immortalized the name of Descartes, and constitutes the greatest single step ever made in the progress of the exact sciences.*
>
> —*John Stuart Mill*

Two types of graphs that are used in this chapter are shown below.

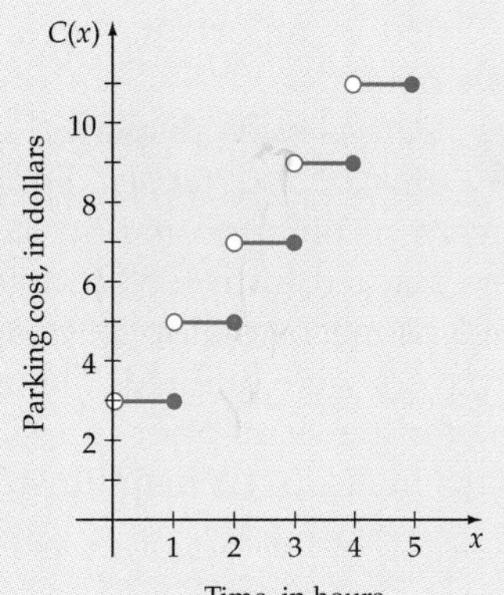

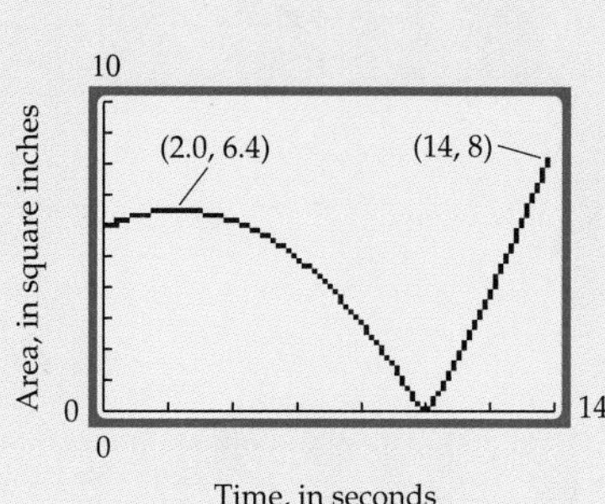

SECTION 2.1 A TWO-DIMENSIONAL COORDINATE SYSTEM AND GRAPHS

- CARTESIAN COORDINATE SYSTEMS
- THE DISTANCE AND MIDPOINT FORMULAS
- GRAPH OF AN EQUATION
- INTERCEPTS
- CIRCLES, THEIR EQUATIONS, AND THEIR GRAPHS

take note

Abscissa comes from the same root word as scissors. An open pair of scissors looks like an x.

MATH MATTERS

The concepts of *analytic geometry* developed over an extended period of time culminating in 1637 with the publication of two works: *Discourse on the Method for Rightly Directing One's Reason and Searching for Truth in the Sciences* by René Descartes (1596–1650) and *Introduction to Plane and Solid Loci* by Pierre de Fermat. Each of these works was an attempt to integrate the study of geometry with the study of algebra. Of the two mathematicians, Descartes is usually given most of the credit for developing analytic geometry. In fact, Descartes became so famous in La Haye, the city in which he was born, that it was renamed La Haye-Descartes.

◆ CARTESIAN COORDINATE SYSTEMS

Each point on a coordinate axis is associated with a number called its coordinate. Each point on a flat, two-dimensional surface, called a **coordinate plane** or *xy*-plane, is associated with an **ordered pair** of numbers called **coordinates** of the point. Ordered pairs are denoted by (a, b), where the real number a is the **x-coordinate** or **abscissa** and the real number b is the **y-coordinate** or **ordinate.**

The coordinates of a point are determined by the point's position relative to a horizontal coordinate axis called the **x-axis** and a vertical coordinate axis called the **y-axis.** The axes intersect at the point (0, 0), called the **origin.** In **Figure 2.1**, the axes are labeled such that positive numbers appear to the right of the origin on the x-axis and above the origin on the y-axis. The four regions formed by the axes are called **quadrants** and are numbered counterclockwise. This two-dimensional coordinate system is referred to as a **Cartesian coordinate system** in honor of René Descartes (1596–1650).

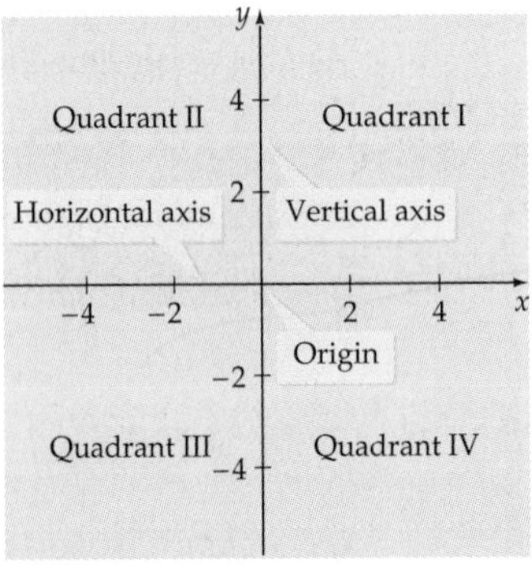

Figure 2.1

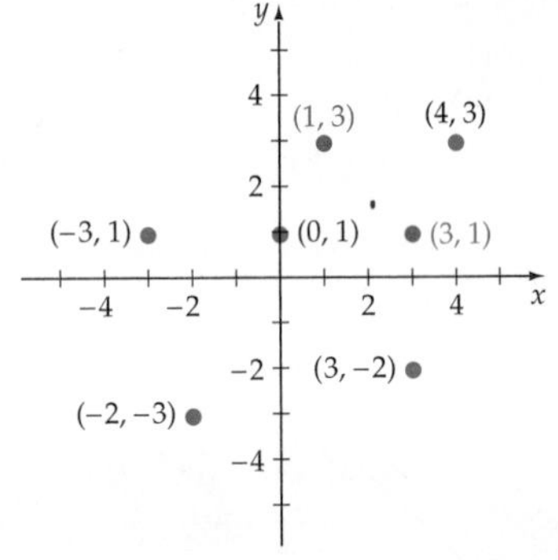

Figure 2.2

To **plot a point** $P(a, b)$ means to draw a dot at its location in the coordinate plane. In **Figure 2.2** we have plotted the points (4, 3), (−3, 1), (−2, −3), (3, −2), (0, 1), (1, 3), and (3, 1). The order in which the coordinates of an ordered pair are listed is important. **Figure 2.2** shows that (1, 3) and (3, 1) do not denote the same point.

Data is often displayed in visual form as a set of points called a *scatter diagram* or a *scatter plot*. For instance, the scatter diagram in **Figure 2.3** shows the percent of high-income taxpayers—those making over $100,000 per year—that were audited from 1988 to 1999. The point whose coordinates are approximately (1993, 4) means that in 1993, approximately 4% of the tax returns of high-income taxpayers were audited. The zigzag line on the horizontal axis indicates that no data for the years prior to 1988 was available. The line segments that connect the points in **Figure 2.3** help illustrate trends.

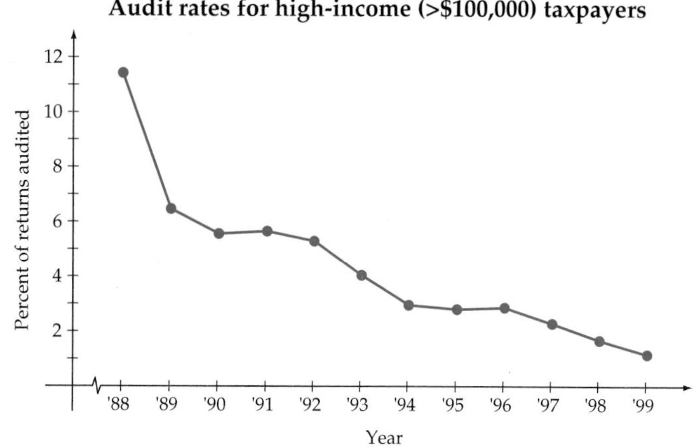

Figure 2.3

take note

The notation (a, b) was used earlier to denote an interval on a one-dimensional number line. In this section, (a, b) denotes an ordered pair in a two-dimensional plane. This should not cause confusion in future sections, because as each mathematical topic is introduced, it will be clear whether a one-dimensional or a two-dimensional coordinate system is involved.

QUESTION According to the graph in **Figure 2.3**, which is based on data supplied by Syracuse University's Transactional Records Access Clearinghouse, has the percent of audits of high-income taxpayers been increasing, remaining about the same, or decreasing?

In some instances, it is important to know when two ordered pairs are equal.

Equality of Ordered Pairs

The ordered pairs (a, b) and (c, d) are equal if and only if $a = c$ and $b = d$.

For instance, if $(3, y) = (x, -2)$, then $x = 3$ and $y = -2$.

◆ THE DISTANCE AND MIDPOINT FORMULAS

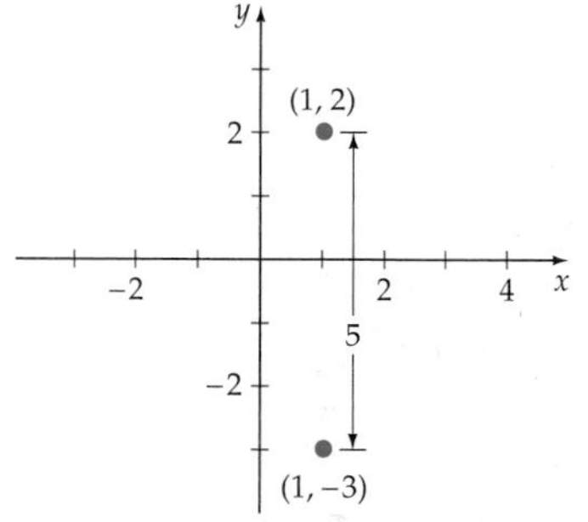

Figure 2.4

The Cartesian coordinate system makes it possible to combine the concepts of algebra and geometry into a branch of mathematics called *analytic geometry.*

The distance between two points on a horizontal line is the absolute value of the difference between the x-coordinates of the two points. The distance between two points on a vertical line is the absolute value of the difference between the y-coordinates of the two points. For example, as shown in **Figure 2.4**, the distance d between the points with coordinates $(1, 2)$ and $(1, -3)$ is $d = |2 - (-3)| = 5$.

If two points are not on a horizontal or vertical line, then a *distance formula* for the distance between the two points can be developed as follows.

The distance between the points $P_1(x_1, y_1)$ and $P_2(x_2, y_2)$ in **Figure 2.5** is the length of the hypotenuse of a right triangle whose sides are horizontal and

ANSWER As a general trend, the percent of audits of high-income taxpayers has been decreasing.

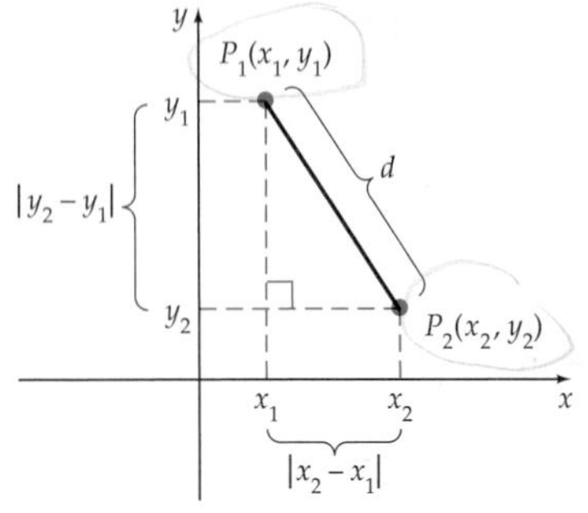

Figure 2.5

vertical line segments that measure $|x_2 - x_1|$ and $|y_2 - y_1|$, respectively. Applying the Pythagorean Theorem to this triangle produces

$$d^2 = |x_2 - x_1|^2 + |y_2 - y_1|^2$$

$$d = \sqrt{|x_2 - x_1|^2 + |y_2 - y_1|^2}$$

- **The square root theorem. Because d is nonnegative, the negative root is not listed.**

$$= \sqrt{(x_2 - x_1)^2 + (y_2 - y_1)^2}$$

- **Because $|x_2 - x_1|^2 = (x_2 - x_1)^2$ and $|y_2 - y_1|^2 = (y_2 - y_1)^2$**

Thus we have established the following theorem.

The Distance Formula

The distance d between the points $P_1(x_1, y_1)$ and $P_2(x_2, y_2)$ is

$$d = \sqrt{(x_2 - x_1)^2 + (y_2 - y_1)^2}$$

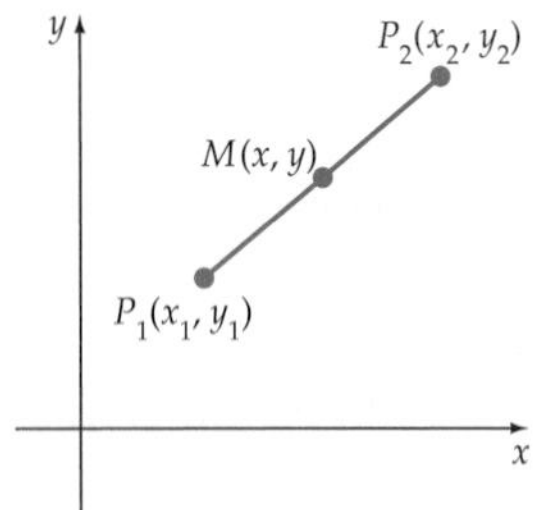

Figure 2.6

The distance d between the points whose coordinates are $P_1(x_1, y_1)$ and $P_2(x_2, y_2)$ is denoted by $d(P_1, P_2)$. To find the distance $d(P_1, P_2)$ between the points $P_1(-3, 4)$ and $P_2(7, 2)$, we apply the distance formula with $x_1 = -3$, $y_1 = 4$, $x_2 = 7$, and $y_2 = 2$.

$$\begin{aligned} d(P_1, P_2) &= \sqrt{(x_2 - x_1)^2 + (y_2 - y_1)^2} \\ &= \sqrt{[7 - (-3)]^2 + (2 - 4)^2} \\ &= \sqrt{104} = 2\sqrt{26} \approx 10.2 \end{aligned}$$

The **midpoint** M of a line segment is the point on the line segment that is equidistant from the endpoints $P_1(x_1, y_1)$ and $P_2(x_2, y_2)$ of the segment. See **Figure 2.6.**

> **MATH MATTERS**
>
> The midpoint formula is a special case of the following theorem. The coordinates (x, y) of the point P dividing the line segment from $P_1(x_1, y_1)$ to $P_2(x_2, y_2)$ into the ratio
>
> $$\frac{d(P_1, P)}{d(P, P_2)} = r$$
>
> are given by the formulas
>
> $$x = \frac{x_1 + rx_2}{1 + r} \quad \text{and}$$
>
> $$y = \frac{y_1 + ry_2}{1 + r}$$
>
> See Exercises 93 to 96, Exercise Set 2.1.

The Midpoint Formula

The midpoint M of the line segment from $P_1(x_1, y_1)$ to $P_2(x_2, y_2)$ is given by

$$\left(\frac{x_1 + x_2}{2}, \frac{y_1 + y_2}{2}\right)$$

The midpoint formula states that the x-coordinate of the midpoint of a line segment is the *average* of the x-coordinates of the endpoints of the line segment and that the y-coordinate of the midpoint of a line segment is the *average* of the y-coordinates of the endpoints of the line segment.

The midpoint M of the line segment connecting $P_1(-2, 6)$ and $P_2(3, 4)$ is

$$M = \left(\frac{x_1 + x_2}{2}, \frac{y_1 + y_2}{2}\right) = \left(\frac{(-2) + 3}{2}, \frac{6 + 4}{2}\right) = \left(\frac{1}{2}, 5\right)$$

◆ GRAPH OF AN EQUATION

The equations below are equations in two variables.

$$y = 3x^3 - 4x + 2 \qquad x^2 + y^2 = 25 \qquad y = \frac{x}{x + 1}$$

The solution of an equation in two variables is an ordered pair (x, y) whose coordinates satisfy the equation. For instance, the ordered pairs $(3, 4)$, $(4, -3)$, and $(0, 5)$ are some of the solutions of $x^2 + y^2 = 25$. Generally, there are an infinite number of solutions of an equation in two variables. These solutions can be displayed in a *graph.*

MATH MATTERS

Maria Agnesi (1718–1799) wrote *Foundations of Analysis for the Use of Italian Youth,* one of the most successful textbooks of the 18th century. The French Academy authorized a translation into French in 1749, noting that "there is no other book, in any language, which would enable a reader to penetrate as deeply, or as rapidly, into the fundamental concepts of analysis." A curve that she discusses in her text is given by the equation $y = \frac{a\sqrt{a - x}}{\sqrt{x}}$. Unfortunately, due to a translation error from Italian to English, the curve became known as the "witch of Agnesi."

Graph of an Equation

The **graph of an equation** in the two variables x and y is the set of all points whose coordinates satisfy the equation.

Consider $y = 2x - 1$. Substituting various values of x into the equation and solving for y produces some of the ordered pairs of the equation. It is convenient to record the results in a table similar to the one shown below. The graph of the ordered pairs is shown in **Figure 2.7.**

x	$y = 2x - 1$	y	(x, y)
−2	2(−2) − 1	−5	(−2, −5)
−1	2(−1) − 1	−3	(−1, −3)
0	2(0) − 1	−1	(0, −1)
1	2(1) − 1	1	(1, 1)
2	2(2) − 1	3	(2, 3)

Choosing some noninteger values of x produces more ordered pairs to graph, such as $(-3/2, -4)$ and $(5/2, 4)$, as shown in **Figure 2.8.** Using still other values of x would result in more and more ordered pairs being graphed. The result would be so many dots that the graph would appear as the straight line shown in **Figure 2.9,** which is the graph of $y = 2x - 1$.

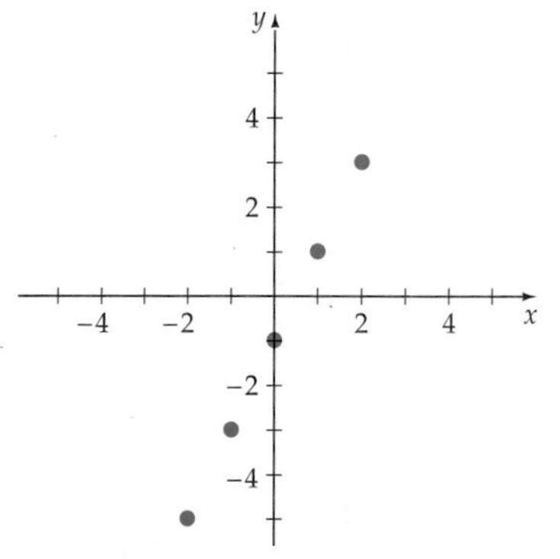

Figure 2.7

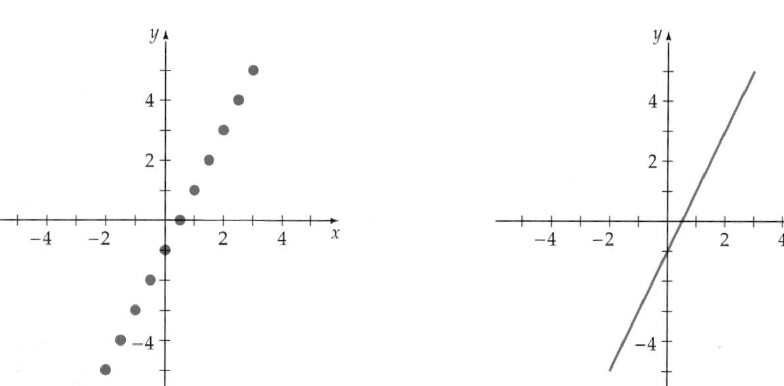

Figure 2.8

Figure 2.9

EXAMPLE 1 Draw a Graph by Plotting Points

Graph: $-x^2 + y = 1$

Continued

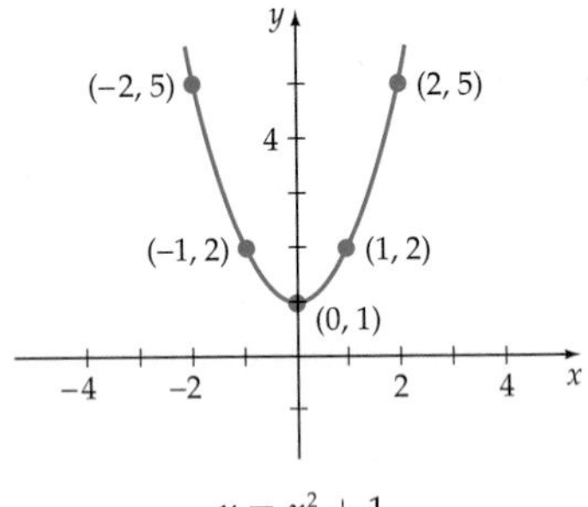

$y = x^2 + 1$

Figure 2.10

Solution

Solve the equation for y.

$$y = x^2 + 1$$

Select values of x and use the equation to calculate y. Choose enough values of x so that an accurate graph can be drawn. Plot the points and draw a curve through them. See **Figure 2.10.**

x	$y = x^2 + 1$	y	(x, y)
−2	$(-2)^2 + 1$	5	(−2, 5)
−1	$(-1)^2 + 1$	2	(−1, 2)
0	$(0)^2 + 1$	1	(0, 1)
1	$(1)^2 + 1$	2	(1, 2)
2	$(2)^2 + 1$	5	(2, 5)

TRY EXERCISE 26, EXERCISE SET 2.1, PAGE 145

Some graphing calculators, such as the *TI-83*, have a TABLE feature that allows you to create a table similar to the one shown in Example 1. Enter the equation to be graphed, the first value for x, and the increment (the difference between successive values of x). For instance, entering $y_1 = x^2 + 1$, an initial value of x as -2, and an increment of 1 yields a display similar to the one in **Figure 2.11.** Changing the initial value to -6 and the increment to 2 gives the table in **Figure 2.12.**

X	Y1	
-2	5	
-1	2	
0	1	
1	2	
2	5	
3	10	
4	17	

X=-2

Figure 2.11

X	Y1	
-6	37	
-4	17	
-2	5	
0	1	
2	5	
4	17	
6	37	

X=-6

Figure 2.12

With some calculators, you may scroll through the table by using the up- or down-arrow keys. In this way, you can determine many more ordered pairs of the graph.

EXAMPLE 2 Graph by Plotting Points

Graph: $y = |x - 2|$

Solution

This equation is already solved for y, so start by choosing an x value and using the equation to determine the corresponding y value. For example, if

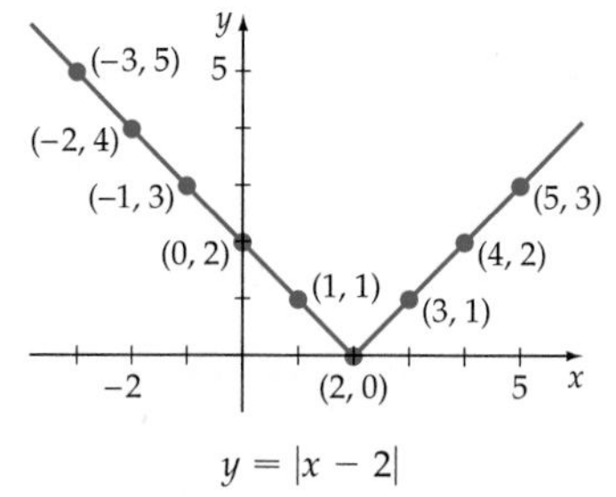

$y = |x - 2|$

Figure 2.13

$x = -3$, then $y = |(-3) - 2| = |-5| = 5$. Continuing in this manner produces the following table:

When x is	−3	−2	−1	0	1	2	3	4	5
y is	5	4	3	2	1	0	1	2	3

Now plot the points listed in the table. Connecting the points forms a V shape, as shown in **Figure 2.13.**

TRY EXERCISE 30, EXERCISE SET 2.1, PAGE 145

EXAMPLE 3 Graph by Plotting Points

Graph: $y^2 = x$

Solution

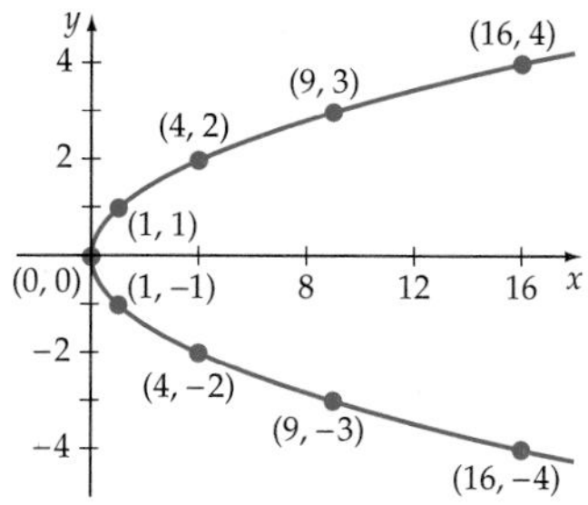

Solving this equation for y yields

$$y = \pm\sqrt{x}$$

Choose several x-values and use the equation to determine the corresponding y values.

When x is	0	1	4	9	16
y is	0	±1	±2	±3	±4

$y^2 = x$

Figure 2.14

Plot the points as shown in **Figure 2.14.** The graph is a *parabola.*

TRY EXERCISE 32, EXERCISE SET 2.1, PAGE 145

A graphing calculator or computer graphing software can be used to draw the graphs in Example 2 and Example 3. These graphing utilities graph a curve in much the same way as you would, by selecting values of x and calculating the corresponding values of y. A curve is then drawn through the points.

If you use a graphing utility to graph $y = |x - 2|$, you will need to use the *absolute value* function that is built-in to the utility. The equation you enter will look similar to Y1=abs(X−2).

To graph the equation in Example 3, you will enter two equations. The equations you enter will be similar to

Y1=√(X)
Y2=−√(X)

The first equation will graph the top half of the parabola, and the second equation will graph the bottom half.

◆ INTERCEPTS

Any point that has an x- or a y-coordinate of zero is called an **intercept** of the graph of an equation, because it is at these points that the graph intersects the x- or the y-axis.

Definition of x-Intercepts and y-Intercepts

If $(x_1, 0)$ satisfies an equation, then the point $(x_1, 0)$ is called an **x-intercept** of the graph of the equation.

If $(0, y_1)$ satisfies an equation, then the point $(0, y_1)$ is called a **y-intercept** of the graph of the equation.

To find the x-intercepts of the graph of an equation, let $y = 0$ and solve the equation for x. To find the y-intercepts of the graph of an equation, let $x = 0$ and solve the equation for y.

EXAMPLE 4 Find x- and y-intercepts

Find the x- and y-intercepts of the graph of $y = x^2 - 2x - 3$.

Solution

To find the y-intercept, let $x = 0$ and solve for y.

$$y = 0^2 - 2(0) - 3 = -3$$

To find the x-intercepts, let $y = 0$ and solve for x.

$$0 = x^2 - 2x - 3$$
$$0 = (x - 3)(x + 1)$$
$$(x - 3) = 0 \text{ or } (x + 1) = 0$$
$$x = 3 \text{ or } \quad x = -1$$

Because $y = -3$ when $x = 0$, $(0, -3)$ is a y-intercept. Because $x = 3$ or -1 when $y = 0$, $(3, 0)$ and $(-1, 0)$ are x-intercepts. **Figure 2.15** confirms that these three points are intercepts.

Visualize the Solution

The graph of $y = x^2 - 2x - 3$ is shown below. Observe that the graph intersects the x-axis at $(-1, 0)$ and $(3, 0)$, the x-intercepts. The graph also intersects the y-axis at $(0, -3)$, the y-intercept.

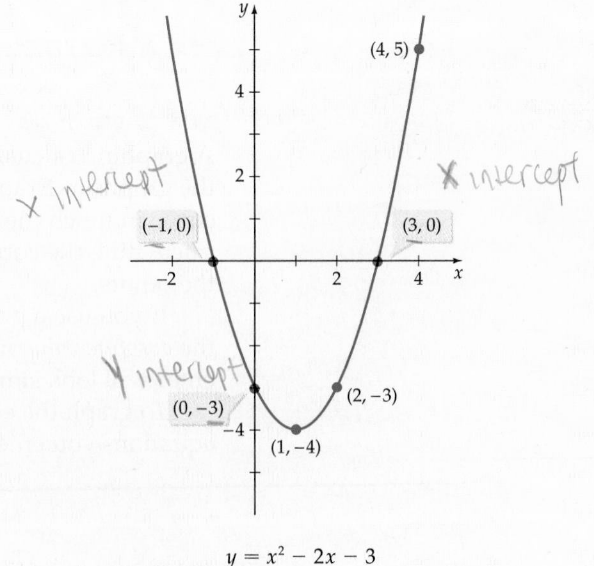

$y = x^2 - 2x - 3$

Figure 2.15

TRY EXERCISE 40, EXERCISE SET 2.1, PAGE 145

In Example 4, it was possible to find the x-intercepts by solving a quadratic equation. In some instances, however, solving an equation to find the intercepts may be very difficult. In these cases, a graphing calculator can be used to find the x-intercepts.

The x-intercepts for the graph of $y = x^3 + x + 4$ were found using the INTERCEPT feature of a TI-83 calculator. The keystrokes and some sample screens for this procedure are shown below.

Press [Y=] Now enter X^3+X+4. Press [ZOOM] and select the standard viewing window.

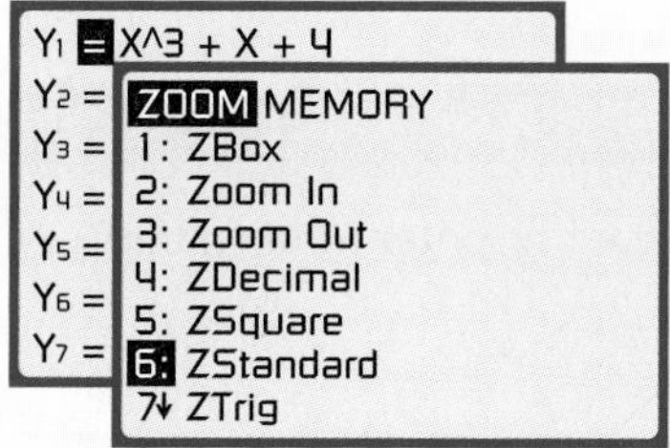

Press [2nd] CALC to access the CALCULATE menu. The x-coordinate of an x-interceptis zero. Therefore, select 2:zero. Press [ENTER]

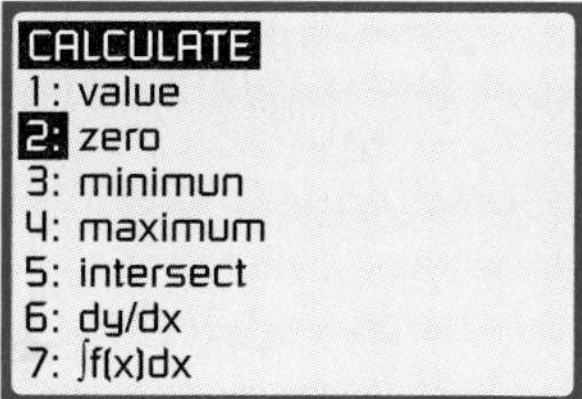

The "Left Bound?" shown on the bottom of the screen means to move an cursor until it is to the left of an x-intercept. Press [ENTER]

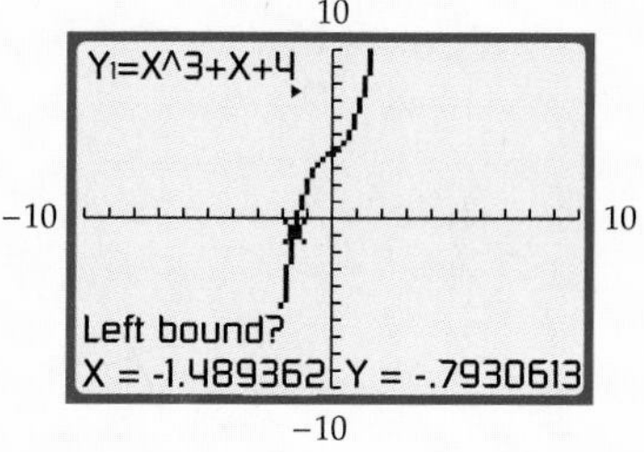

The "Right Bound?" shown on the bottom of the screen means to move the cursor until it is to the right of the desired x-intercept. Press [ENTER]

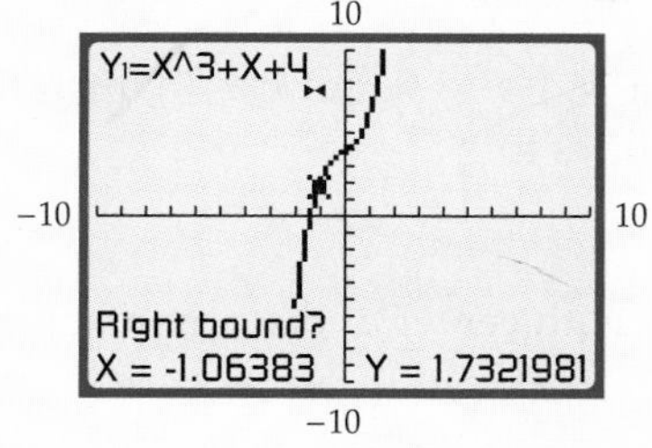

"Guess?" is shown on the bottom of the screen. Move the cursor until it is approximately on the x-intercept. Press [ENTER]

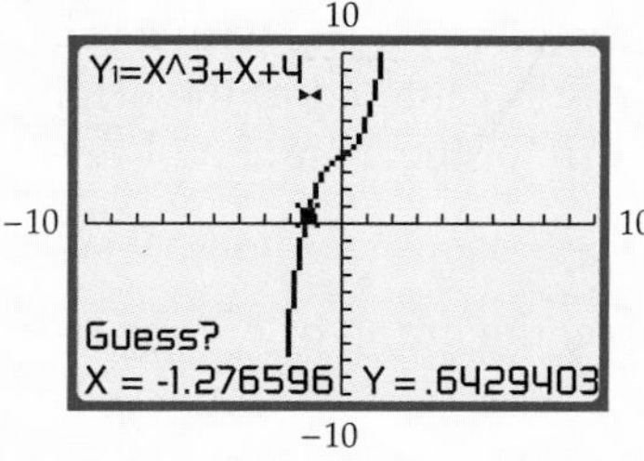

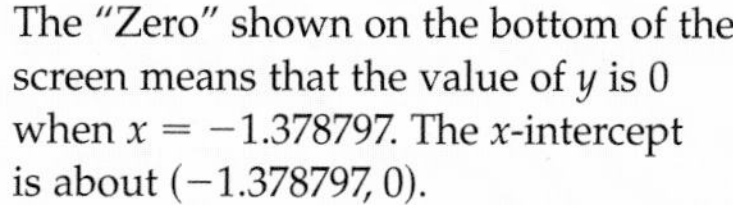

The "Zero" shown on the bottom of the screen means that the value of y is 0 when $x = -1.378797$. The x-intercept is about $(-1.378797, 0)$.

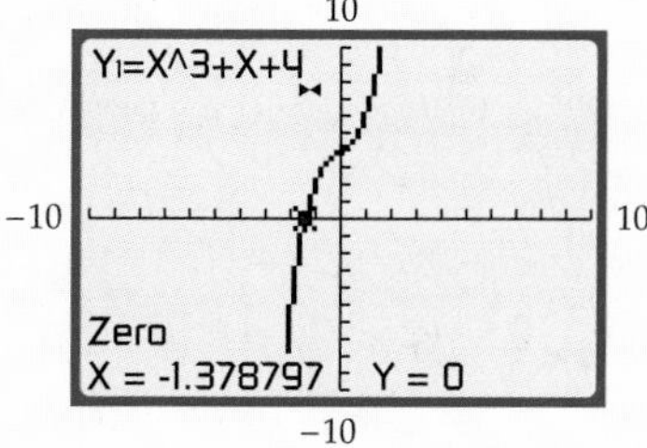

◆ CIRCLES, THEIR EQUATIONS, AND THEIR GRAPHS

Frequently you will sketch graphs by plotting points. However, some graphs can be sketched by merely recognizing the form of the equation. A *circle* is an example of a curve whose graph you can sketch after you have inspected its equation.

Definition of a Circle

A **circle** is the set of points in a plane that are a fixed distance from a specified point. The distance is the **radius** of the circle, and the specified point is the **center** of the circle.

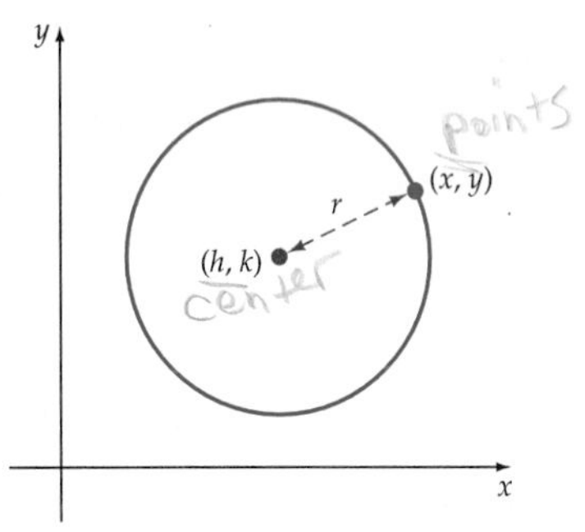

Figure 2.16

The standard form of the equation of a circle is derived by using this definition. To derive the standard form, we use the distance formula. **Figure 2.16** is a circle with center (h, k) and radius r. The point (x, y) is on the circle if and only if it is a distance of r units from the center (h, k). Thus (x, y) is on the circle if and only if

$$\sqrt{(x-h)^2 + (y-k)^2} = r$$

$$(x-h)^2 + (y-k)^2 = r^2 \quad \text{• Square each side.}$$

Standard Form of the Equation of a Circle

The **standard form of the equation of a circle** with center at (h, k) and radius r is

$$(x-h)^2 + (y-k)^2 = r^2$$

For example, the equation $(x-3)^2 + (y+1)^2 = 4$ is the equation of a circle. The standard form of the equation is

$$(x-3)^2 + (y-(-1))^2 = 2^2$$

from which it can be determined that $h = 3$, $k = -1$, and $r = 2$. Thus the graph is a circle centered at $(3, -1)$ with a radius of 2.

If a circle is centered at the origin $(0, 0)$ (that is, if $h = 0$ and $k = 0$), then the standard form of the equation of the circle simplifies to

$$x^2 + y^2 = r^2$$

For example, the graph of $x^2 + y^2 = 9$ is a circle with center at the origin and radius of 3.

QUESTION What are the radius and the coordinates of the center of the circle with equation $x^2 + (y-2)^2 = 10$?

EXAMPLE 5 Find the Standard Form of the Equation of a Circle

Find the standard form of the equation of a circle that has center $C(-4, -2)$ and contains the point $P(-1, 2)$.

ANSWER The radius is $\sqrt{10}$ and the coordinates of the center are $(0, 2)$.

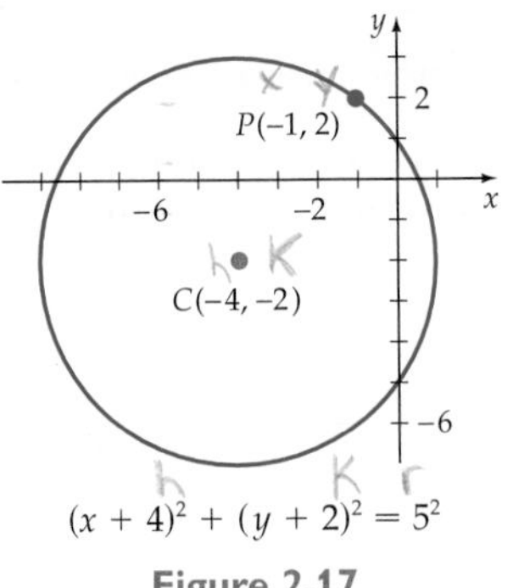

$(x + 4)^2 + (y + 2)^2 = 5^2$

Figure 2.17

Solution

See the graph of the circle in **Figure 2.17**. Because the point P is on the circle, the radius r of the circle must equal the distance from C to P. Thus

$$r = \sqrt{(-1 - (-4))^2 + (2 - (-2))^2}$$
$$= \sqrt{9 + 16} = \sqrt{25} = 5$$

Using the standard form with $h = -4$, $k = -2$, and $r = 5$, we obtain

$$(x + 4)^2 + (y + 2)^2 = 5^2$$

TRY EXERCISE 66, EXERCISE SET 2.1, PAGE 145

If we rewrite $(x + 4)^2 + (y + 2)^2 = 5^2$ by squaring and combining like terms, we produce

$$x^2 + 8x + 16 + y^2 + 4y + 4 = 25$$
$$x^2 + y^2 + 8x + 4y - 5 = 0$$

This form of the equation is known as the **general form of the equation of a circle.** By completing the square, it is always possible to write the general form $x^2 + y^2 + Ax + By + C = 0$ in the standard form

$$(x - h)^2 + (y - k)^2 = s$$

for some number s. If $s > 0$, the graph is a circle with radius $r = \sqrt{s}$. If $s = 0$, the graph is the point (h, k), and if $s < 0$, the equation has no real solutions and there is no graph.

EXAMPLE 6 Find the Center and Radius of a Circle by Completing the Square

Find the center and the radius of the circle that is given by

$$x^2 + y^2 - 6x + 4y - 3 = 0$$

Solution

First rearrange and group the terms as shown.

$$(x^2 - 6x) + (y^2 + 4y) = 3$$

Now complete the square of $(x^2 - 6x)$ and $(y^2 + 4y)$.

$$(x^2 - 6x + 9) + (y^2 + 4y + 4) = 3 + 9 + 4$$
$$(x - 3)^2 + (y + 2)^2 = 16$$
$$(x - 3)^2 + (y - (-2))^2 = 4^2$$

• **Add 9 and 4 to each side of the equation.**

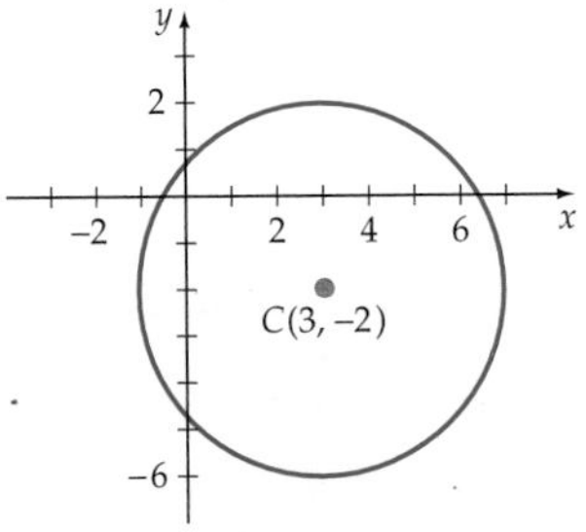

This equation is the standard form of the equation of a circle and indicates that the graph of the original equation is a circle centered at $(3, -2)$ with radius 4. See **Figure 2.18**.

$x^2 + y^2 - 6x + 4y - 3 = 0$

Figure 2.18

TRY EXERCISE 68, EXERCISE SET 2.1, PAGE 145

TOPICS FOR DISCUSSION

1. The distance formula states that the distance d between the points $P_1(x_1, y_1)$ and $P_2(x_2, y_2)$ is $d = \sqrt{(x_2 - x_1)^2 + (y_2 - y_1)^2}$. Can the distance formula also be written as follows? Explain.

$$d = \sqrt{(x_1 - x_2)^2 + (y_1 - y_2)^2}$$

2. Does the equation $(x - 3)^2 + (y + 4)^2 = -6$ have a graph that is a circle? Explain.
3. Explain why the graph of $|x| + |y| = 1$ does not contain any points that have
   a. a y-coordinate that is greater than 1 or less than -1.
   b. an x-coordinate that is greater than 1 or less than -1.
4. Discuss the graph of $xy = 0$.
5. Explain how to determine the x- and y-intercepts of a graph defined by an equation (without using the graph).

EXERCISE SET 2.1

In Exercises 1 to 2, plot the points whose coordinates are given on a Cartesian coordinate system.

1. $(2, 4), (0, -3), (-2, 1), (-5, -3)$
2. $(-3, -5), (-4, 3), (0, 2), (-2, 0)$

3. HEALTH A study at the Ohio State University measured the changes in heart rates of students doing stepping exercises. Students stepped onto a platform that was approximately 11 inches high at a rate of 14 steps per minute. The heart rate, in beats per minute, before and after the exercise is given in the table below.

Before	After	Before	After
63	84	96	141
72	99	69	93
87	111	81	96
90	129	75	90
90	108	84	90

   a. Draw a scatter diagram for this data.
   b. For these students, what is the average increase in heart rate?

4. HEALTH The table is based on data from the Bureau of Labor Statistics that shows the number of people in the U.S. who are covered by pharmacy-benefit companies. These companies can purchase prescription drugs at a discount and pass the savings along to the members.

Year	Number of People Covered, in millions
1996	150
1997	160
1998	168
1999	192
2000	200
2001	210

   a. Draw a scatter diagram for this data using $t = 0$ for 1996.
   b. From the data, it appears that the number of people covered by these companies is increasing. If the number of people covered by these companies in 2002 increases by the same percent as the percent increase between 2000 and 2001, how many people will be covered by these companies in 2002?

In Exercises 5 to 16, find the distance between the points whose coordinates are given.

5. $(6, 4), (-8, 11)$
6. $(-5, 8), (-10, 14)$
7. $(-4, -20), (-10, 15)$
8. $(40, 32), (36, 20)$

9. $(5, -8), (0, 0)$

10. $(0, 0), (5, 13)$

11. $(\sqrt{3}, \sqrt{8}), (\sqrt{12}, \sqrt{27})$

12. $(\sqrt{125}, \sqrt{20}), (6, 2\sqrt{5})$

13. $(a, b), (-a, -b)$

14. $(a - b, b), (a, a + b)$

15. $(x, 4x), (-2x, 3x)$ given that $x < 0$

16. $(x, 4x), (-2x, 3x)$ given that $x > 0$

17. Find all points on the x-axis that are 10 units from $(4, 6)$. (*Hint:* First write the distance formula with $(4, 6)$ as one of the points and $(x, 0)$ as the other point.)

18. Find all points on the y-axis that are 12 units from $(5, -3)$.

In Exercises 19 to 24, find the midpoint of the line segment with the following endpoints.

19. $(1, -1), (5, 5)$

20. $(-5, -2), (6, 10)$

21. $(6, -3), (6, 11)$

22. $(4, 7), (-10, 7)$

23. $(1.75, 2.25), (-3.5, 5.57)$

24. $(-8.2, 10.1), (-2.4, -5.7)$

In Exercises 25 to 38, graph each equation by plotting points that satisfy the equation.

25. $x - y = 4$

26. $2x + y = -1$

27. $y = 0.25x^2$

28. $3x^2 + 2y = -4$

29. $y = -2|x - 3|$

30. $y = |x + 3| - 2$

31. $y = x^2 - 3$

32. $y = x^2 + 1$

33. $y = \frac{1}{2}(x - 1)^2$

34. $y = 2(x + 2)^2$

35. $y = x^2 + 2x - 8$

36. $y = x^2 - 2x - 8$

37. $y = -x^2 + 2$

38. $y = -x^2 - 1$

In Exercises 39 to 48, find the x- and y-intercepts of the graph of each equation. Use the intercepts and additional points as needed, to draw the graph of the equation.

39. $2x + 5y = 12$

40. $3x - 4y = 15$

41. $x = -y^2 + 5$

42. $x = y^2 - 6$

43. $x = |y| - 4$

44. $x = y^3 - 2$

45. $x^2 + y^2 = 4$

46. $x^2 = y^2$

47. $|x| + |y| = 4$

48. $|x - 4y| = 8$

In Exercises 49 to 58, determine the center and radius of the circle with the given equation.

49. $x^2 + y^2 = 36$

50. $x^2 + y^2 = 49$

51. $x^2 + y^2 = 10^2$

52. $x^2 + y^2 = 4^2$

53. $(x - 1)^2 + (y - 3)^2 = 7^2$

54. $(x - 2)^2 + (y - 4)^2 = 5^2$

55. $(x + 2)^2 + (y + 5)^2 = 25$

56. $(x + 3)^2 + (y + 5)^2 = 121$

57. $(x - 8)^2 + y^2 = \frac{1}{4}$

58. $x^2 + (y - 12)^2 = 1$

In Exercises 59 to 66, find an equation of a circle that satisfies the given conditions. Write your answer in standard form.

59. Center $(4, 1)$, radius $r = 2$

60. Center $(5, -3)$, radius $r = 4$

61. Center $(1/2, 1/4)$, radius $r = \sqrt{5}$

62. Center $(0, 2/3)$, radius $r = \sqrt{11}$

63. Center $(0, 0)$, passing through $(-3, 4)$

64. Center $(0, 0)$, passing through $(5, 12)$

65. Center $(1, 3)$, passing through $(4, -1)$

66. Center $(-2, 5)$, passing through $(1, 7)$

In Exercises 67 to 76, find the center and the radius of each circle. The equations of the circles are written in the general form.

67. $x^2 + y^2 - 6x + 5 = 0$

68. $x^2 + y^2 - 6x - 4y + 12 = 0$

69. $x^2 + y^2 - 4x - 10y + 20 = 0$

70. $x^2 + y^2 + 4x - 2y - 11 = 0$

71. $x^2 + y^2 - 14x + 8y + 56 = 0$

72. $x^2 + y^2 - 10x + 2y + 25 = 0$

73. $4x^2 + 4y^2 + 4x - 63 = 0$

74. $9x^2 + 9y^2 - 6y - 17 = 0$

75. $x^2 + y^2 - x + \frac{2}{3}y + \frac{1}{3} = 0$

76. $x^2 + y^2 - 2x + 2y + \frac{7}{4} = 0$

SUPPLEMENTAL EXERCISES

In Exercises 77 to 88, graph the set of all points whose x- and y-coordinates satisfy the given conditions.

77. $x = 3$

78. $y = 2$

79. $x = 1, y \geq 1$

80. $y = -3, x \geq -2$

81. $y \leq 3$

82. $x \geq 2$

83. $xy \geq 0$

84. $|y| \geq 1, \frac{x}{y} \leq 0$

85. $|x| = 2, |y| = 3$

86. $|x| = 4, |y| = 1$

87. $|x| \leq 2, y \geq 2$

88. $x \geq 1, |y| \leq 3$

In Exercises 89 to 92, find the other endpoint of the line segment that has the given endpoint and midpoint.

89. Endpoint (5, 1), midpoint (9, 3)

90. Endpoint (4, −6), midpoint (−2, 11)

91. Endpoint (−3, −8), midpoint (2, −7)

92. Endpoint (5, −4), midpoint (0, 0)

The coordinates (x, y) of the point P that divides the line segment from $P_1(x_1, y_1)$ to $P_2(x_2, y_2)$ into the ratio

$$\frac{d(P_1, P)}{d(P, P_2)} = r$$

are given by the formulas

$$x = \frac{x_1 + rx_2}{1 + r} \quad \text{and} \quad y = \frac{y_1 + ry_2}{1 + r}$$

In Exercises 93 to 96 use these formulas to find the indicated point on the line segment with endpoints P_1 and P_2.

93. Find the point three-fourths of the way from $P_1(-8, 11)$ to $P_2(6, 20)$.

94. Find the point two-fifths of the way from $P_1(-9, 6)$ to $P_2(2, 4)$.

95. Find the point seven-eighths of the way from $P_1(6, 10)$ to $P_2(2, 1)$.

96. Find the point nine-tenths of the way from $P_1(-1, 8)$ to $P_2(0, 3)$.

97. Use the distance formula to determine whether the points given by (1, −4), (3, 2), (−3, 4), and (−5, −2) are the vertices of a square.

98. Use the distance formula to determine whether the points given by (2, −1), (5, 0), (6, 3), and (3, 2) are the vertices of a parallelogram, a rhombus, or a square.

99. Find a formula for the set of all points (x, y) for which the distance from (x, y) to (3, 4) is 5.

100. Find a formula for the set of all points (x, y) for which the distance from (x, y) to (−5, 12) is 13.

101. Find a formula for the set of all points (x, y) for which the sum of the distances from (x, y) to (4, 0) and from (x, y) to (−4, 0) is 10.

102. Find a formula for the set of all points for which the absolute value of the difference of the distances from (x, y) to (0, 4) and from (x, y) to (0, −4) is 6.

103. Find an equation of a circle that has a diameter with endpoints (2, 3) and (−4, 11). Write your answer in standard form.

104. Find an equation of a circle that has a diameter with endpoints (7, −2) and (−3, 5). Write your answer in standard form.

105. Find an equation of a circle that has its center at (7, 11) and is tangent to the x-axis. Write your answer in standard form.

106. Find an equation of a circle that has its center at (−2, 3) and is tangent to the y-axis. Write your answer in standard form.

107. Find an equation of a circle that is tangent to both axes, has its center in the second quadrant, and has a radius of 3.

108. Find an equation of a circle that is tangent to both axes, has its center in the third quadrant, and has a diameter of $\sqrt{5}$.

PROJECTS

1. **VERIFY A GEOMETRIC THEOREM** Use the midpoint formula and the distance formula to prove that the midpoint M of the hypotenuse of a right triangle is equidistant from each of the vertices of the triangle. (*Hint:* Label the vertices of the triangle as shown in the figure at the right.)

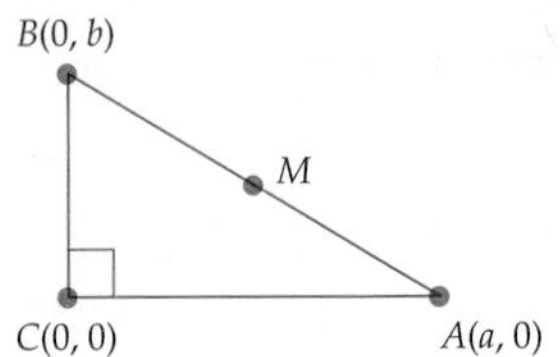

2. **SOLVE A QUADRATIC EQUATION GEOMETRICALLY** In the 17th century, Descartes (and others) solved equations by using both algebra and geometry. This project outlines the method Descartes used to solve certain quadratic equations.

a. Consider the equation $x^2 = 2ax + b^2$. Construct a right triangle ABC with $d(A, C) = a$ and $d(C, B) = b$. Now draw a circle with center at A and radius a. Let P be the point at which the circle intersects the hypotenuse of the right triangle and Q the point where an extension of the hypotenuse intersects the circle. Your drawing should be similar to the one at the right.

b. Show that a solution of the equation $x^2 = 2ax + b^2$ is $d(Q, B)$.

c. Show that $d(P, B)$ is a solution of the equation $x^2 = -2ax + b^2$.

d. Construct a line parallel to AC and passing through B. Let S and T be the points at which the line intersects the circle. Show that $d(S, B)$ and $d(T, B)$ are solutions of the equation $x^2 = 2ax - b^2$.

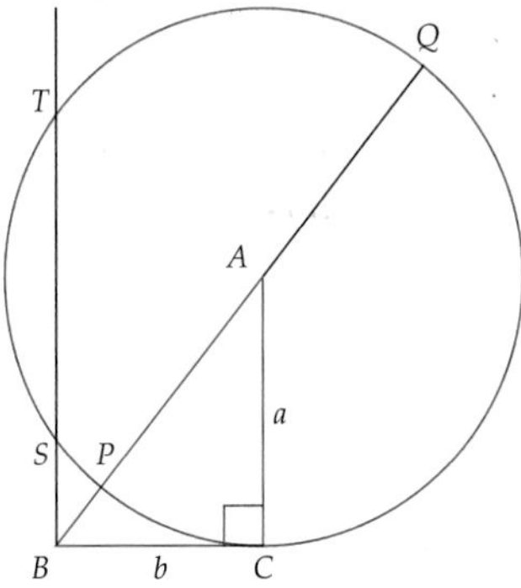

SECTION 2.2 INTRODUCTION TO FUNCTIONS

- RELATIONS
- FUNCTIONS
- FUNCTIONAL NOTATION
- IDENTIFYING FUNCTIONS
- GRAPHS OF FUNCTIONS
- THE GREATEST INTEGER FUNCTION
- APPLICATIONS

◆ RELATIONS

In many situations in science, business, and mathematics, a correspondence exists between two sets. The correspondence is often defined by a *table*, an *equation*, or a *graph*, each of which can be viewed from a mathematical perspective as a set of ordered pairs. In mathematics, any set of ordered pairs is called a **relation.**

Table 2.1 defines a correspondence between a set of percent scores and a set of letter grades. For each score from 0 to 100, there corresponds only one letter grade. The score 94% corresponds to the letter grade of A. Using ordered-pair notation we record this correspondence as (94, A).

Table 2.1

Score	Grade
[90, 100]	A
[80, 90)	B
[70, 80)	C
[60, 70)	D
[0, 60)	F

The *equation* $d = 16t^2$ indicates the distance d that a rock falls (neglecting air resistance) corresponds to the time t that it has been falling. For each non-negative value t, the equation assigns only one value for the distance d. According to this equation, in 3 seconds a rock will fall 144 feet, which we record as (3, 144). Some of the other ordered pairs determined by $d = 16t^2$ are (0, 0), (1, 16), (2, 64), and (2.5, 100).

Equation: $d = 16t^2$

If $t = 3$, then $d = 16(3)^2 = 144$

The *graph* in **Figure 2.19** defines a correspondence between the length of a pendulum and the time it takes the pendulum to complete one oscillation. For each nonnegative pendulum length, the graph yields only one time. According to the graph, a pendulum length of 2 feet yields an oscillation time of 1.6 seconds, and a length of 4 feet yields an oscillation time of 2.2 seconds, where the time is measured to the nearest tenth of a second. These results can be recorded as the ordered pairs (2, 1.6) and (4, 2.2).

Graph: A pendulum's oscillation time

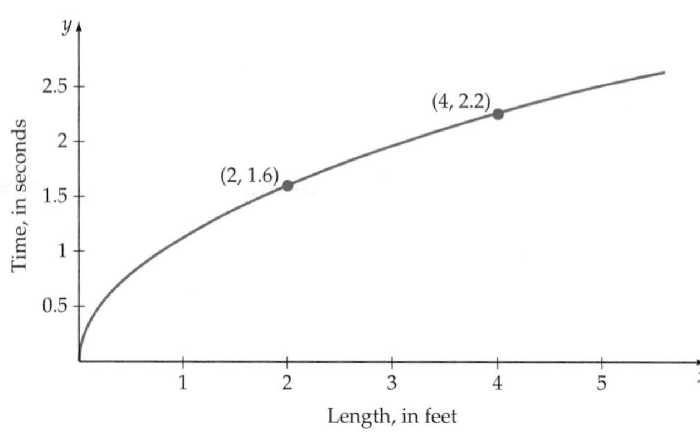

Figure 2.19

◆ FUNCTIONS

> **MATH MATTERS**
>
> It is generally agreed among historians that Leonard Euler (1707–1783) was the first person to use the word *function.* His definition of function occurs in his book *Introduction to Analysis of the Infinite*, published in 1748. Euler contributed to many areas of mathematics and was one of the most prolific expositors of mathematics.

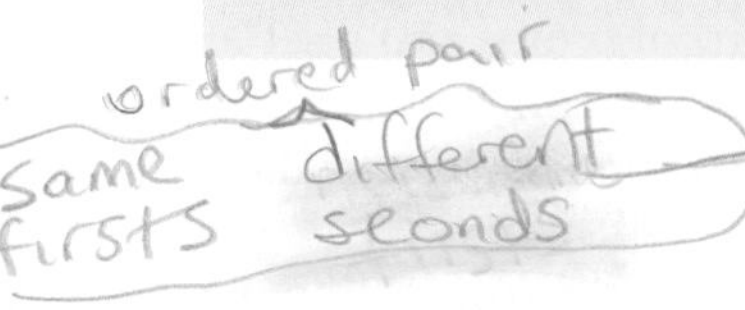

The preceding table, equation, and graph each determines a special type of relation called a *function.*

Definition of a Function

A **function** is a set of ordered pairs in which no two ordered pairs have the same first coordinate and different second coordinates.

Although every function is a relation, not every relation is a function. For instance, consider (94, A) from the grading correspondence. The first coordinate, 94, is paired with a second coordinate of A. It would not make sense to have 94 paired with A, (94, A), and 94 paired with B, (94, B). The same first coordinate would be paired with two different second coordinates. This would mean that two students with the same score received different grades, one student an A and the other a B!

Functions may have ordered pairs with the same second coordinate. For instance, (94, A) and (95, A) are both ordered pairs that belong to the function defined by Table 2.1. A function may have different first coordinates and the same second coordinate.

The equation $d = 16t^2$ represents a function because for each value of t there is only one value of d. Not every equation, however, represents a function. For instance, $y^2 = 25 - x^2$ does not represent a function. The ordered pairs $(-3, 4)$ and $(-3, -4)$ are both solutions of the equation. But these ordered pairs do not satisfy the definition of a function; there are two ordered pairs with the same first coordinate but *different* second coordinates.

QUESTION Does the set $\{(0, 0), (1, 0), (2, 0), (3, 0), (4, 0)\}$ define a function?

ANSWER Yes. There are no two ordered pairs with the same first coordinate that have different second coordinates.

The **domain** of a function is the set of all the first coordinates of the ordered pairs. The **range** of a function is the set of all the second coordinates. In the function determined by the grading correspondence in Table 2.1, the domain is the interval [0, 100]. The range is {A, B, C, D, F}. In a function, each domain element is paired with one and only one range element.

If a function is defined by an equation, the variable that represents elements of the domain is the **independent variable.** The variable that represents elements of the range is the **dependent variable.** In the free-fall experiment, we used the equation $d = 16t^2$. The elements of the domain represented the time the rock fell, and the elements of the range represented the distance the rock fell. Thus in $d = 16t^2$, the independent variable is t and the dependent variable is d.

The specific letters used for the independent and the dependent variable are not important. For example, $y = 16x^2$ represents the same function as $d = 16t^2$. Traditionally, x is used for the independent variable, and y for the dependent variable. Anytime we use the phrase "y is a function of x" or a similar phrase with different letters, the variable that follows "function of" is the independent variable.

◆ FUNCTIONAL NOTATION

Functions can be named by using a letter or a combination of letters, such as f, g, A, log, or tan. If x is an element of the domain of f, then $f(x)$, which is read "f of x" or "the value of f at x," is the element in the range of f that corresponds with the domain element x. The notation "f" and the notation "$f(x)$" mean different things. "f" is the name of the function, whereas "$f(x)$" is the value of the function at x. Finding the value of $f(x)$ is referred to as *evaluating* f at x. To evaluate $f(x)$ at $x = a$, substitute a for x, and simplify.

EXAMPLE 1 Evaluate Functions

Let $f(x) = x^2 - 1$, and evaluate.

a. $f(-5)$ **b.** $f(3b)$ **c.** $3f(b)$ **d.** $f(a + 3)$ **e.** $f(a) + f(3)$

Solution

a. $f(-5) = (-5)^2 - 1 = 25 - 1 = 24$ • **Substitute −5 for x, and simplify.**

b. $f(3b) = (3b)^2 - 1 = 9b^2 - 1$ • **Substitute 3b for x, and simplify.**

c. $3f(b) = 3(b^2 - 1) = 3b^2 - 3$ • **Substitute b for x, and simplify.**

d. $f(a + 3) = (a + 3)^2 - 1$ • **Substitute a + 3 for x.**

$= a^2 + 6a + 8$ • **Simplify.**

e. $f(a) + f(3) = (a^2 - 1) + (3^2 - 1)$ • **Substitute a for x; substitute 3 for x.**

$= a^2 + 7$ • **Simplify.**

take note

In Example 1, observe that

$f(3b) \neq 3f(b)$

and that

$f(a + 3) \neq f(a) + f(3)$

TRY EXERCISE 2, EXERCISE SET 2.2, PAGE 158

Piecewise-defined functions are functions represented by more than one expression. The function shown below is an example of a piecewise-defined function.

$$f(x) = \begin{cases} 2x, & x < -2 \\ x^2, & -2 \le x < 1 \\ 4 - x, & x \ge 1 \end{cases}$$

• This function is made up of different *pieces*, $2x$, x^2, and $4 - x$, depending on the value of x.

The expression that is used to evaluate this function depends on the value of x. For instance, to find $f(-3)$ we note that $-3 < -2$ and therefore use the expression $2x$ to evaluate the function.

$$f(-3) = 2(-3) = -6$$ **• When $x < -2$, use the expression $2x$.**

Here are some additional instances of evaluating this function:

$$f(-1) = (-1)^2 = 1$$ **• When x satisfies $-2 \le x < 1$, use the expression x^2.**

$$f(4) = 4 - 4 = 0$$ **• When $x \ge 1$, use the expression $4 - x$.**

The graph of this function is shown in **Figure 2.20**. Note the use of the open and closed circles at the endpoints of the intervals. These circles are used to show the evaluation of the function at the endpoints of each interval. For instance, because -2 is in the interval $-2 \le x < 1$, the value of the function at -2 is 4 $[f(-2) = (-2)^2 = 4]$. Therefore a closed dot is placed at $(-2, 4)$. Similarly, when $x = 1$, because 1 is in the interval $x \ge 1$, the value of the function at 1 is 3 $(f(1) = 4 - 1 = 3)$.

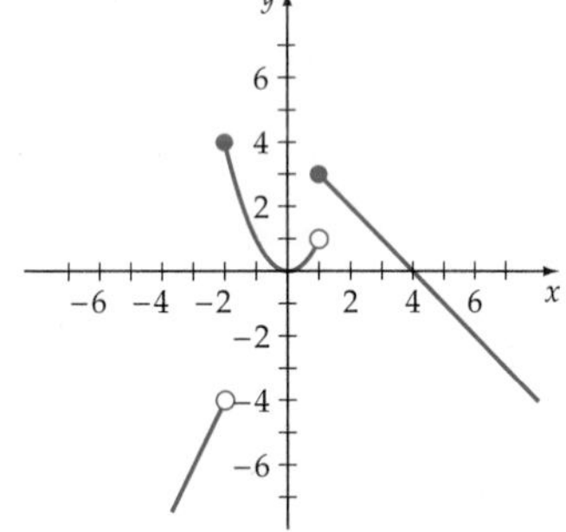

Figure 2.20

QUESTION Evaluate the function given above when $x = 0.5$.

The absolute value function is another example of a piecewise-defined function. Below is the definition of this function, which is sometimes abbreviated abs(x). Its graph **(Figure 2.21)** is shown at the left.

$$\text{abs}(x) = \begin{cases} -x, & x < 0 \\ x, & x \ge 0 \end{cases}$$

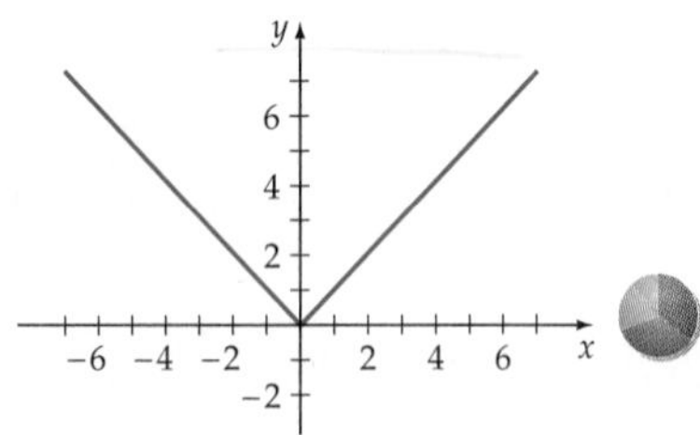

Figure 2.21

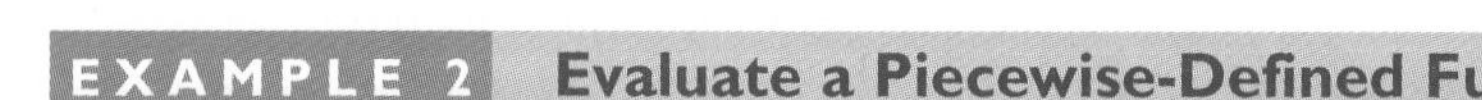

EXAMPLE 2 Evaluate a Piecewise-Defined Function

Figure 2.22, which is based on data from the Worldwatch Institute, shows the number of cigarettes smoked per person annually in the United States.

The data in the graph can be approximated by

$$N(t) = \begin{cases} -74t + 2810, & 0 \le t < 6 \\ -7.5t^2 + 157t + 1531, & 6 \le t \le 20 \end{cases}$$

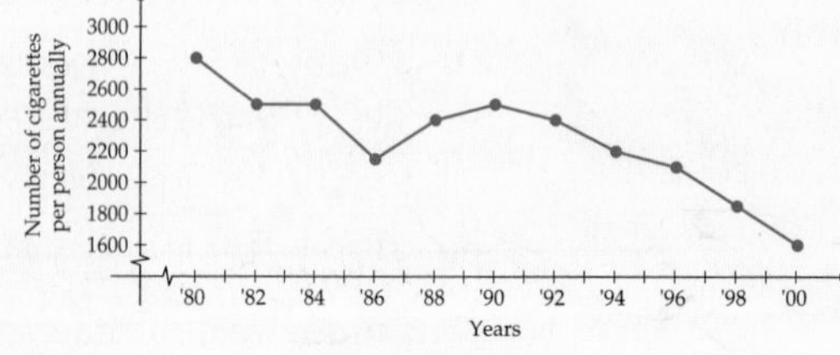

Figure 2.22

ANSWER 0.5 is in the interval $-2 \le x < 1$. Therefore, $f(0.5) = 0.5^2 = 0.25$.

where $N(t)$ is the number of cigarettes smoked per person annually in the United States for year t, and $t = 0$ corresponds to 1980. Use this function to estimate the number of cigarettes smoked per person per year in the following years.

a. 1985 **b.** 1998

Solution

a. The year 1985 corresponds to a value of $t = 5$. Because 5 is in the interval $0 \le t < 6$, evaluate $-74t + 2810$ at 5.

$$-74(5) + 2810 = 2440$$

The model suggests that 2440 cigarettes per person were smoked annually in 1985.

b. The year 1998 corresponds to a value of $t = 18$. Because 18 is in the interval $6 \le t \le 20$, evaluate $-7.5t^2 + 157t + 1531$ at 18.

$$-7.5(18)^2 + 157(18) + 1531 = 1927$$

The model suggests that 1927 cigarettes were smoked annually in 1998.

TRY EXERCISE 10, EXERCISE SET 2.2, PAGE 159

◆ IDENTIFYING FUNCTIONS

Recall that although every function is a relation, not every relation is a function. In the next example, we examine four relations to determine which are functions.

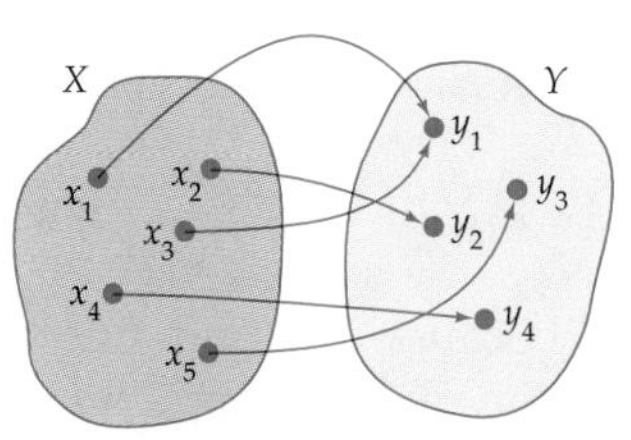

Figure 2.23

EXAMPLE 3 Identify Functions

Which relations define y as a function of x?

a. $\{(2, 3), (4, 1), (4, 5)\}$ **b.** $3x + y = 1$ **c.** $-4x^2 + y^2 = 9$

d. The correspondence between the x values and the y values in **Figure 2.23.**

Solution

a. There are two ordered pairs, (4, 1) and (4, 5), with the same first coordinate and different second coordinates. This set does not define y as a function of x.

b. Solving $3x + y = 1$ for y yields $y = -3x + 1$. Because $-3x + 1$ is a unique real number for each x, this equation defines y as a function of x.

Continued ➤

c. Solving $-4x^2 + y^2 = 9$ for y yields $y = \pm\sqrt{4x^2 + 9}$. The right side $\pm\sqrt{4x^2 + 9}$ produces two values of y for each value of x. For example, when $x = 0$, $y = 3$ or $y = -3$. Thus $-4x^2 + y^2 = 9$ does not define y as a function of x.

d. Each x is paired with one and only one y. The correspondence in **Figure 2.23** defines y as a function of x.

TRY EXERCISE 14, EXERCISE SET 2.2, PAGE 159

take note

If a function is defined by an equation, then it may be difficult to determine the range of the function by examining its equation. To determine the range of a function generally requires a graph of the function or the application of a theorem such as the theorem on horizontal asymptotes that is stated in Section 3.5. Thus for the present we will concentrate on using an equation only to determine the domain of a function.

Sometimes the domain of a function is explicitly stated. For example, each of f, g, and h below is given by an equation, followed by a statement that indicates the domain of the function.

$$f(x) = x^2, x > 0; \qquad g(t) = \frac{1}{t^2 + 4}, 0 \le t \le 5; \qquad h(x) = x^2, x = 1, 2, 3$$

Although f and h have the same equation, they are different functions because they have different domains. If the domain of a function is not explicitly stated, then its domain is determined by the following convention.

Domain of a Function

Unless otherwise stated, the domain of a function is the set of all real numbers for which the function makes sense and yields real numbers.

take note

You may indicate the domain of a function using set notation, interval notation, a graph, or words. For instance, the domain of $f(x) = \sqrt{x - 3}$ may be given in each of the following ways.

Set notation: $\{x \mid x \ge 3\}$

Interval notation: $[3, \infty)$

A graph: 

Words: All real numbers x, where x is greater than or equal to 3.

EXAMPLE 4 Determine the Domain of a Function

Determine the domain of each function.

a. $G(t) = \dfrac{1}{t - 4}$ b. $f(x) = \sqrt{x + 1}$

c. $A(s) = s^2$, where $A(s)$ is the area of a square whose sides are s units.

Solution

a. The number 4 is not an element of the domain because G is undefined when the denominator $t - 4$ equals 0. The domain of G is all real numbers except 4. In interval notation the domain is $(-\infty, 4) \cup (4, \infty)$.

b. The radical $\sqrt{x + 1}$ is a real number only when $x + 1 \ge 0$ or when $x \ge -1$. Thus, in set notation, the domain of f is $\{x \mid x \ge -1\}$.

c. Because s represents the length of the side of a square, s must be positive. In set notation the domain of A is $\{s \mid s > 0\}$.

TRY EXERCISE 28, EXERCISE SET 2.2, PAGE 159

◆ GRAPHS OF FUNCTIONS

If a is an element of the domain of a function, then $(a, f(a))$ is an ordered pair that belongs to the function.

Graph of a Function

The **graph of a function** is the graph of all the ordered pairs that belong to the function.

take note

Knowing how many ordered pairs to plot and how to connect the points requires specific knowledge about the function. Much of the rest of this chapter will be concerned with this topic.

EXAMPLE 5 Graph a Function by Plotting Points

Graph each function. State the domain and the range of each function.

a. $f(x) = |x - 1|$ **b.** $n(x) = \begin{cases} 2, & \text{if } x \leq 1 \\ x, & \text{if } x > 1 \end{cases}$

Solution

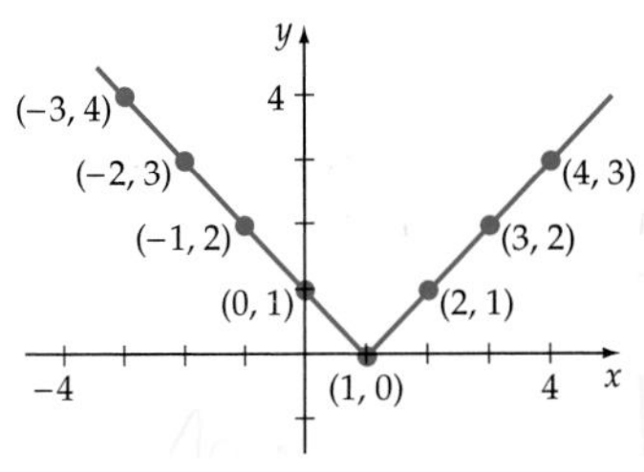

$f(x) = |x - 1|$

Figure 2.24

a. The domain of f is the set of all real numbers. Write the function as $y = |x - 1|$. Evaluate the function for several domain values. We have used $x = -3, -2, -1, 0, 1, 2, 3$, and 4.

x	−3	−2	−1	0	1	2	3	4
y = |x − 1|	4	3	2	1	0	1	2	3

Plot the points determined by the ordered pairs. Connect the points to form the graph in **Figure 2.24.**

Because $|x - 1| \geq 0$, we can conclude that the graph of f extends from a height of 0 upward, so the range is $\{y \mid y \geq 0\}$.

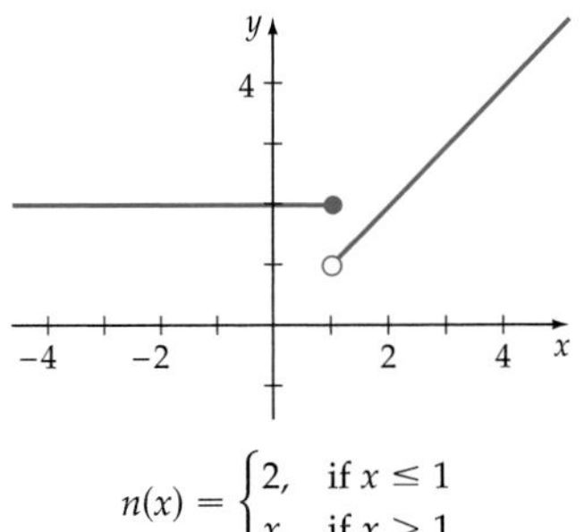

$n(x) = \begin{cases} 2, & \text{if } x \leq 1 \\ x, & \text{if } x > 1 \end{cases}$

Figure 2.25

b. The domain is the union of the inequalities $x \leq 1$ and $x > 1$. Thus the domain of n is the set of all real numbers. For $x \leq 1$, graph $n(x) = 2$. This results in the horizontal ray in **Figure 2.25.** The solid circle indicates that the point (1, 2) *is* part of the graph. For $x > 1$, graph $n(x) = x$. This produces the second ray in **Figure 2.26.** The open circle indicates that the point (1, 1) *is not* part of the graph.

Examination of the graph shows that it includes only points whose y values are greater than 1. Thus the range of n is $\{y \mid y > 1\}$.

TRY EXERCISE 40, EXERCISE SET 2.2, PAGE 159

take note

A web applet is available to explore the properties of the graph of the absolute value. This applet, Absolute Value of a Linear Function, can be found on our web site at http://college.hmco.com

A graphing utility can also be used to draw the graph of a function. For instance, to graph $f(x) = x^2 - 1$, you will enter an equation similar to Y1=X²−1.The graph is shown in **Figure 2.26.**

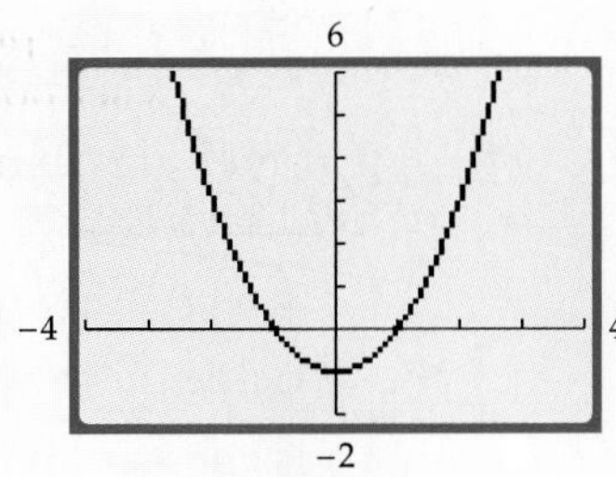

Figure 2.26

The definition that a function is a set of ordered pairs in which no two ordered pairs that have the same first coordinate have different second coordinates implies that any vertical line intersects the graph of a function at no more than one point. This is known as the *vertical line test.*

The Vertical Line Test for Functions

A graph is the graph of a function if and only if no vertical line intersects the graph at more than one point.

EXAMPLE 6 Apply the Vertical Line Test

Which of the following graphs are graphs of functions?

a.

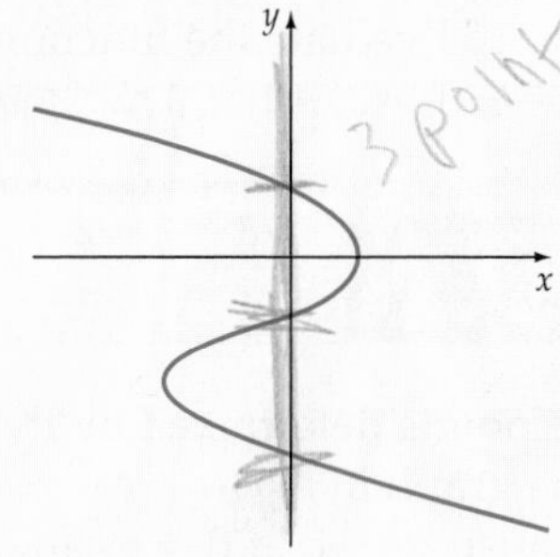

b.

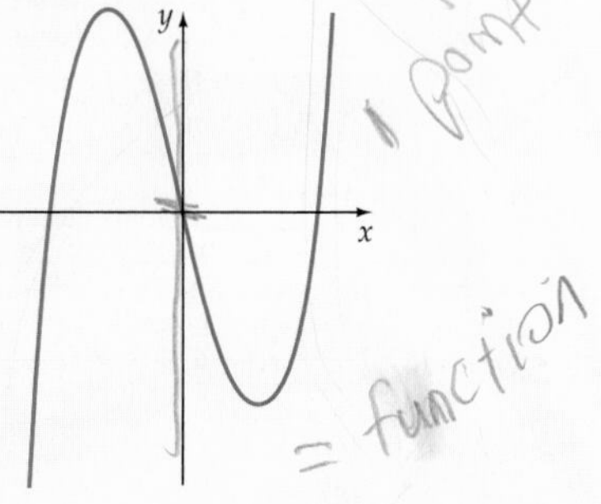

Solution

a. This graph *is not* the graph of a function because some vertical lines intersect the graph in more than one point.

b. This graph *is* the graph of a function because every vertical line intersects the graph in at most one point.

TRY EXERCISE 50, EXERCISE SET 2.2, PAGE 160

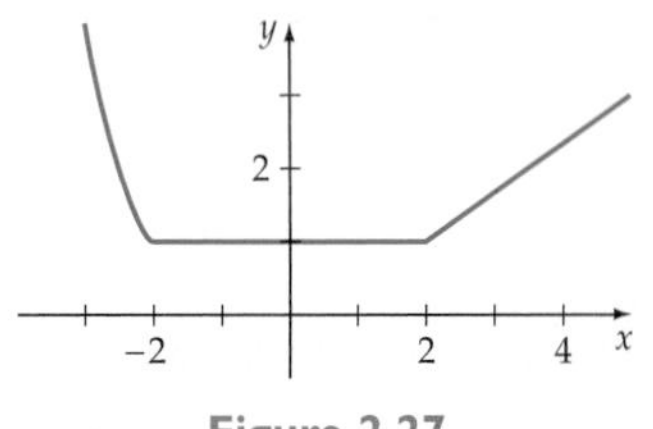

Figure 2.27

Consider the graph in Figure 2.27. As a point on the graph moves from left to right, this graph falls for values of $x \leq -2$, it remains the same height from $x = -2$ to $x = 2$, and it rises for $x \geq 2$. The function represented by the graph is said to be *decreasing* on the interval $(-\infty, -2]$, *constant* on the interval $[-2, 2]$, and *increasing* on the interval $[2, \infty)$.

Definition of Increasing, Decreasing, and Constant Functions

If a and b are elements of an interval I that is a subset of the domain of a function f, then

- f is **increasing** on I if $f(a) < f(b)$ whenever $a < b$.
- f is **decreasing** on I if $f(a) > f(b)$ whenever $a < b$.
- f is **constant** on I if $f(a) = f(b)$ for all a and b.

Recall that a function is a relation in which no two ordered pairs that have the same first coordinate have different second coordinates. This means that given any x, there is only one y that can be paired with that x. A **one-to-one function** satisfies the additional condition that given any y, there is only one x that can be paired with that given y. In a manner similar to applying the vertical line test, we can apply a *horizontal line test* to identify one-to-one functions.

Horizontal Line Test for a One-To-One Function

If every horizontal line intersects the graph of a function at most once, then the graph is the graph of a one-to-one function.

For example, some horizontal lines intersect the graph in **Figure 2.28** at more than one point. It is *not* the graph of a one-to-one function. Every horizontal line intersects the graph in **Figure 2.29** at most once. This is the graph of a one-to-one function.

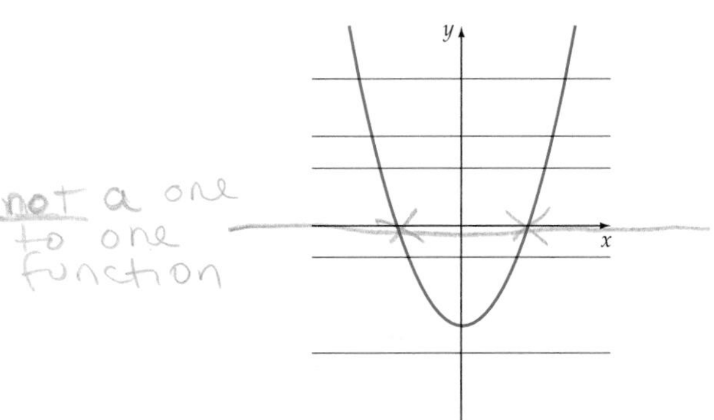

Figure 2.28
Some horizontal lines intersect this graph at more than one point. It is *not* the graph of a one-to-one function.

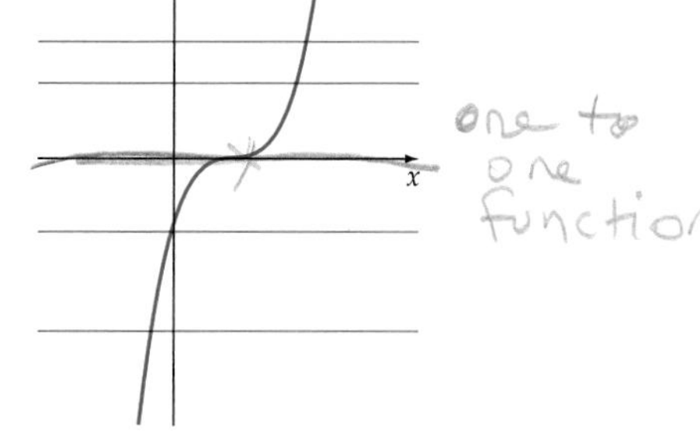

Figure 2.29
Every horizontal line intersects this graph at most once. It is the graph of a one-to-one function.

take note

The greatest integer function is an important function that is often used in advanced mathematics and also in computer science.

◆ THE GREATEST INTEGER FUNCTION

The notation $[\![x]\!]$ or $\text{int}(x)$ is defined to be the greatest integer less than or equal to x. For example:

$$\text{int}(-3) = -3 \qquad \text{int}(-7.4) = -8 \qquad \text{int}(\pi) = 3$$

The function defined by $y = [\![x]\!]$ or $y = \text{int}(x)$ is known as the **greatest integer function.** The greatest integer function is a **step function. Figure 2.30** shows that its graph resembles a series of steps. The graph of $y = \text{int}(x)$ has a break or *discontinuity* whenever x is an integer. The domain of $y = \text{int}(x)$ is the set of all real numbers, and its range is the set of integers.

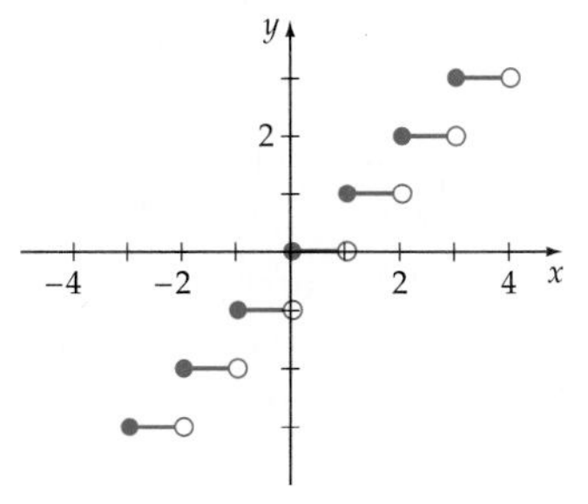

$f(x) = [\![x]\!]$

Figure 2.30

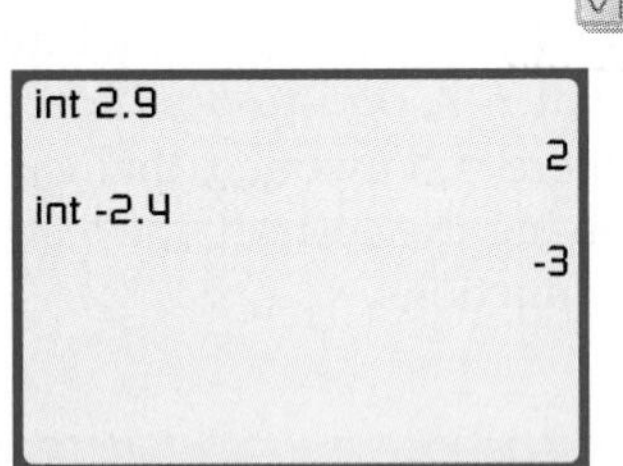

Figure 2.31

The greatest integer function is programmed into many graphing utilities. A typical display might be as shown in Figure 2.31.

By entering Y1=int(X), a graph of the greatest integer function can also be produced.

take note

It may seem that the equation $y = 3 + 2\,\text{int}(x)$ would model the parking cost in Example 7; however, this is not the case. The graph of $y = 3 + 2\,\text{int}(x)$ and the graph of $y = C(x)$ are the same for noninteger values of x, but for integer values they produce different results. For example, evaluating $y = 3 + 2\,\text{int}(x)$ for $x = 1$ yields $y = 5$, whereas $C(1) = 3$.

EXAMPLE 7 Use the Greatest Integer Function to Model Expenses

The cost of parking in a garage is \$3 for the first hour or any part of the hour and \$2 for each additional hour or any part of the hour thereafter. If x is the time in hours that you park your car, then the cost is given by

$$C(x) = 3 - 2\,\text{int}(1 - x), \quad x > 0$$

a. Evaluate $C(2)$ and $C(2.5)$. **b.** Graph $y = C(x)$ for $0 < x \leq 5$.

Solution

a.
$$\begin{aligned} C(2) &= 3 - 2\,\text{int}(1 - 2) \\ &= 3 - 2\,\text{int}(-1) \\ &= 3 - 2(-1) \\ &= \$5 \end{aligned} \qquad \begin{aligned} C(2.5) &= 3 - 2\,\text{int}(1 - 2.5) \\ &= 3 - 2\,\text{int}(-1.5) \\ &= 3 - 2(-2) \\ &= \$7 \end{aligned}$$

b. You could construct the graph by plotting several points, but the graph in Figure 2.32 was constructed by using a graphing utility. Because the graphing utility was in "connected" mode, the graph does not show the discontinuities that occur whenever x is an integer.

The graph in Figure 2.33 was constructed by graphing $y = C(x)$ in "dot" mode. This graph is a better representation of C because it shows the discontinuities at $x = 1$, $x = 2$, $x = 3$, and $x = 4$.

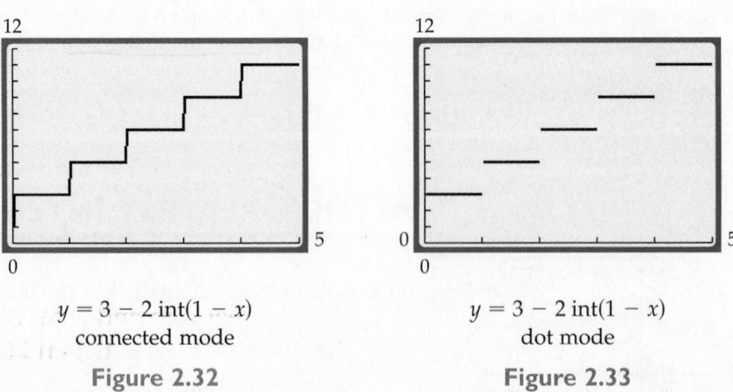

$y = 3 - 2\,\text{int}(1 - x)$
connected mode

Figure 2.32

$y = 3 - 2\,\text{int}(1 - x)$
dot mode

Figure 2.33

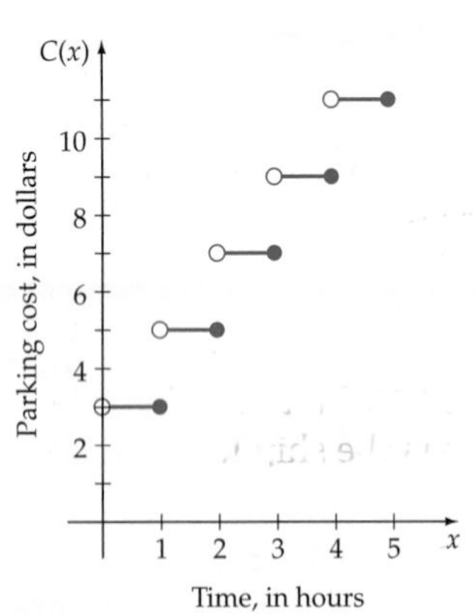

$C(x) = 3 - 2\,\text{int}(1 - x)$

Figure 2.34

Because $C(1) = 3$, $C(2) = 5$, $C(3) = 7$, $C(4) = 9$, and $C(5) = 11$, we can use a solid circle at the right endpoint of each "step" and an open circle at the left endpoint of each "step" to better indicate the true nature of the function C as shown in Figure 2.34.

TRY EXERCISE 48, EXERCISE SET 2.2, PAGE 160

Example 7 illustrates that a graphing calculator may not produce a graph that is a good representation of a function. You may be required to *make adjustments* in the MODE, SET UP, or WINDOW of the graphing calculator so that it will produce a better representation of the function. Some graphs may also require some *fine tuning,* such as open or solid circles at particular points, to accurately represent the function.

◆ APPLICATIONS

EXAMPLE 8 Solve an Application

A car was purchased for \$16,500. Assuming the car depreciates at a constant rate of \$2200 per year (*straight-line depreciation*) for the first 7 years, write the value v of the car as a function of time, and calculate the value of the car 3 years after purchase.

Solution

Let t represent the number of years that have passed since the car was purchased. Then $2200t$ is the amount that the car has depreciated after t years. The value of the car at time t is given by

$$v(t) = 16{,}500 - 2200t, \quad 0 \le t \le 7$$

When $t = 3$, the value of the car is

$$v(3) = 16{,}500 - 2200(3) = 16{,}500 - 6600 = \$9900$$

TRY EXERCISE 66, EXERCISE SET 2.2, PAGE 161

Often in applied mathematics, formulas are used to determine the functional relationship that exists between two variables.

EXAMPLE 9 Solve an Application

A lighthouse is 2 miles south of a port. A ship leaves port and sails east at a rate of 7 mph. Express the distance d between the ship and the lighthouse as a function of time, given that the ship has been sailing for t hours.

Solution

Draw a diagram and label it as shown in **Figure 2.35.** Note that because distance = (rate)(time) and the rate is 7, in t hours the ship has sailed a distance of $7t$.

$$[d(t)]^2 = (7t)^2 + 2^2 \qquad \text{• The Pythagorean Theorem}$$

$$[d(t)]^2 = 49t^2 + 4$$

$$d(t) = \sqrt{49t^2 + 4} \qquad \text{• The } \pm \text{ sign is not used because } d \text{ must be nonnegative.}$$

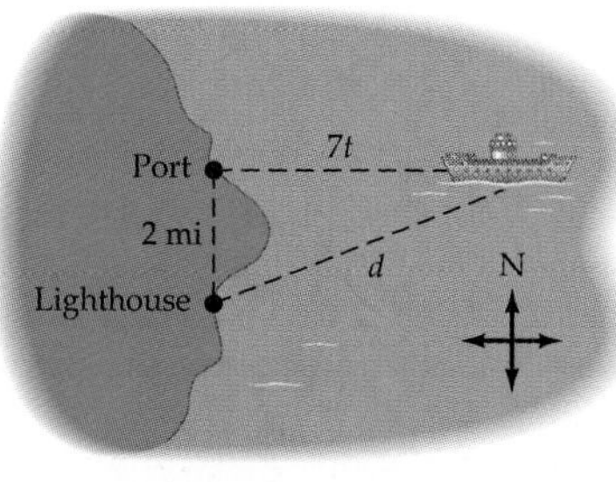

Figure 2.35

TRY EXERCISE 72, EXERCISE SET 2.2, PAGE 162

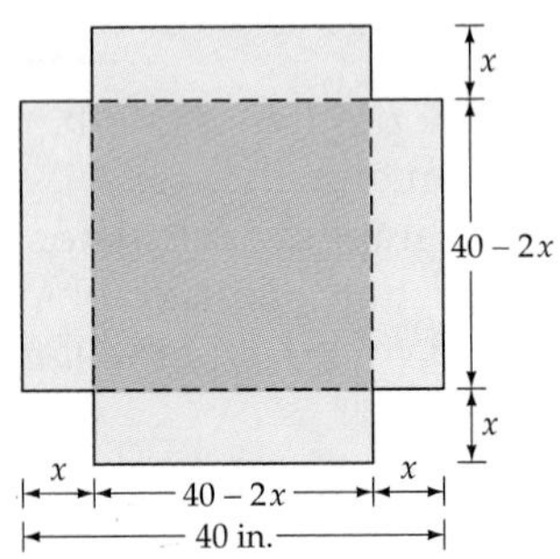

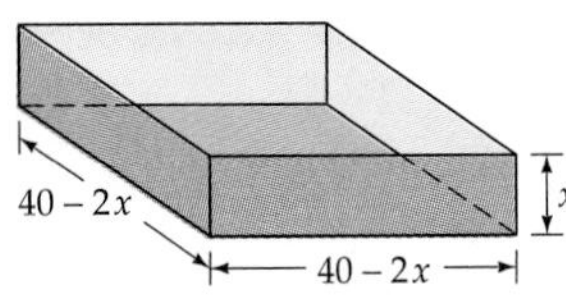

Figure 2.36

EXAMPLE 10 Solve an Application

An open box is to be made from a square piece of cardboard that measures 40 inches on each side. To construct the box, squares that measure x inches on each side are cut from each corner of the cardboard as shown in **Figure 2.36.**

a. Express the volume V of the box as a function of x.

b. Determine the domain of V.

Solution

a. The length l of the box is $40 - 2x$. The width w is also $40 - 2x$. The height of the box is x. The volume V of a box is a product of its length, its width, and its height. Thus

$$V = (40 - 2x)^2 x$$

b. The squares that are cut from each corner require x to be larger than 0 inches but less than 20 inches. Thus the domain is $\{x \mid 0 < x < 20\}$.

TRY EXERCISE 68, EXERCISE SET 2.2, PAGE 161

TOPICS FOR DISCUSSION

1. Discuss the definition of *function*. Give some examples of relationships that are functions and some that are not functions.
2. What is the difference between the domain and range of a function?
3. How many y-intercepts can a function have? How many x-intercepts can a function have?
4. Discuss how the vertical line test is used to determine whether or not a graph is the graph of a function. Explain why the vertical line test works.
5. What is the domain of $f(x) = \dfrac{\sqrt{1 - x}}{x^2 - 9}$? Explain.
6. Is 2 in the range of $g(x) = \dfrac{6x - 5}{3x + 1}$? Explain the process you used to make your decision.
7. Suppose that f is a function and that $f(a) = f(b)$. Does this imply that $a = b$? Explain your answer.

EXERCISE SET 2.2

In Exercises 1 to 8, evaluate each function.

1. Given $f(x) = 3x - 1$, find
   a. $f(2)$ b. $f(-1)$ c. $f(0)$
   d. $f\left(\dfrac{2}{3}\right)$ e. $f(k)$ f. $f(k + 2)$

2. Given $g(x) = 2x^2 + 3$, find
   a. $g(3)$ b. $g(-1)$ c. $g(0)$
   d. $g\left(\dfrac{1}{2}\right)$ e. $g(c)$ f. $g(c + 5)$

3. Given $A(w) = \sqrt{w^2 + 5}$, find
   a. $A(0)$ b. $A(2)$ c. $A(-2)$
   d. $A(4)$ e. $A(r + 1)$ f. $A(-c)$

4. Given $J(t) = 3t^2 - t$, find
   a. $J(-4)$ b. $J(0)$ c. $J\left(\frac{1}{3}\right)$
   d. $J(-c)$ e. $J(x + 1)$ f. $J(x + h)$

5. Given $f(x) = \frac{1}{|x|}$, find
   a. $f(2)$ b. $f(-2)$ c. $f\left(\frac{-3}{5}\right)$
   d. $f(2) + f(-2)$ e. $f(c^2 + 4)$ f. $f(2 + h)$

6. Given $T(x) = 5$, find
   a. $T(-3)$ b. $T(0)$ c. $T\left(\frac{2}{7}\right)$
   d. $T(3) + T(1)$ e. $T(x + h)$ f. $T(3k + 5)$

7. Given $s(x) = \frac{x}{|x|}$, find
   a. $s(4)$ b. $s(5)$ c. $s(-2)$
   d. $s(-3)$ e. $s(t), t > 0$ f. $s(t), t < 0$

8. Given $r(x) = \frac{x}{x + 4}$, find
   a. $r(0)$ b. $r(-1)$ c. $r(-3)$
   d. $r\left(\frac{1}{2}\right)$ e. $r(0.1)$ f. $r(10{,}000)$

In Exercises 9 and 10, evaluate each piecewise-defined function for the indicated values.

9. $P(x) = \begin{cases} 3x + 1, & \text{if } x < 2 \\ -x^2 + 11, & \text{if } x \geq 2 \end{cases}$
   a. $P(-4)$ b. $P(\sqrt{5})$
   c. $P(c),\ \ c < 2$ d. $P(k + 1),\ \ k \geq 1$

10. $Q(t) = \begin{cases} 4, & \text{if } 0 \leq t \leq 5 \\ -t + 9, & \text{if } 5 < t \leq 8 \\ \sqrt{t - 7}, & \text{if } 8 < t \leq 11 \end{cases}$
   a. $Q(0)$ b. $Q(e),\ \ 6 < e < 7$
   c. $Q(n),\ \ 1 < n < 2$ d. $Q(m^2 + 7),\ \ 1 < m \leq 2$

In Exercises 11 to 20, identify the equations that define y as a function of x.

11. $2x + 3y = 7$ 12. $5x + y = 8$
13. $-x + y^2 = 2$ 14. $x^2 - 2y = 2$
15. $y = 4 \pm \sqrt{x}$ 16. $x^2 + y^2 = 9$
17. $y = \sqrt[3]{x}$ 18. $y = |x| + 5$
19. $y^2 = x^2$ 20. $y^3 = x^3$

In Exercises 21 to 26, identify the sets of the ordered pairs (x, y) that define y as a function of x.

21. $\{(2, 3), (5, 1), (-4, 3), (7, 11)\}$
22. $\{(5, 10), (3, -2), (4, 7), (5, 8)\}$
23. $\{(4, 4), (6, 1), (5, -3)\}$
24. $\{(2, 2), (3, 3), (7, 7)\}$
25. $\{(1, 0), (2, 0), (3, 0)\}$
26. $\left\{\left(-\frac{1}{3}, \frac{1}{4}\right), \left(-\frac{1}{4}, \frac{1}{3}\right), \left(\frac{1}{4}, \frac{2}{3}\right)\right\}$

In Exercises 27 to 38, determine the domain of the function represented by the given equation.

27. $f(x) = 3x - 4$ 28. $f(x) = -2x + 1$
29. $f(x) = x^2 + 2$ 30. $f(x) = 3x^2 + 1$
31. $f(x) = \frac{4}{x + 2}$ 32. $f(x) = \frac{6}{x - 5}$
33. $f(x) = \sqrt{7 + x}$ 34. $f(x) = \sqrt{4 - x}$
35. $f(x) = \sqrt{4 - x^2}$ 36. $f(x) = \sqrt{12 - x^2}$
37. $f(x) = \frac{1}{\sqrt{x + 4}}$ 38. $f(x) = \frac{1}{\sqrt{5 - x}}$

In Exercises 39 to 46, graph each function. State the domain of each function. Insert solid circles or hollow circles where necessary to indicate the true nature of the function.

39. $f(x) = \begin{cases} |x|, & \text{if } x \leq 1 \\ 2, & \text{if } x > 1 \end{cases}$

40. $g(x) = \begin{cases} -4, & \text{if } x \leq 0 \\ x^2 - 4, & \text{if } 0 < x \leq 1 \\ -x, & \text{if } x > 1 \end{cases}$

41. $J(x) = \begin{cases} 4, & \text{if } x = -4, -3, \text{ or } -2 \\ x^2, & \text{if } x = -1, 0, \text{ or } 1 \\ -x + 6, & \text{if } x = 2, 3, \text{ or } 4 \end{cases}$

42. $K(x) = \begin{cases} 1, & \text{if } x = -5, -4, -3, \text{ or } -2 \\ x^2 - 3, & \text{if } x = -1, 0, 1, \text{ or } 2 \\ \frac{1}{2}x, & \text{if } x = 3, 4, 5, \text{ or } 6 \end{cases}$

43. $L(x) = \left[\!\left[\frac{1}{3}x\right]\!\right]$ for $-6 \leq x \leq 6$

44. $M(x) = [\![x]\!] + 2$ for $0 \leq x \leq 4$

45. $N(x) = \text{int}(-x)$ for $-3 \leq x \leq 3$

46. $P(x) = \text{int}(x) + x$ for $0 \leq x \leq 4$

47. **MAILING CHARGES** The cost of mailing a parcel is given by

$$C(w) = 0.29 - 0.29\,\text{int}(1 - w), \quad w > 0$$

where C is in dollars and w is the weight of the parcel in ounces.

a. Evaluate $C(3.97)$.

b. Graph $y = C(w)$ for $0 < w \leq 5$.

48. **TELEPHONE CHARGES** The cost of a long-distance telephone call is given by

$$C(t) = 0.85 - 0.50\,\text{int}(1 - t), \quad t > 0$$

where C is in dollars and t is the length of the call in minutes.

a. Evaluate $C(4.75)$.

b. Graph $y = C(t)$ for $0 < t \leq 6$.

49. Use the vertical line test to determine which of the following graphs are graphs of functions.

a.

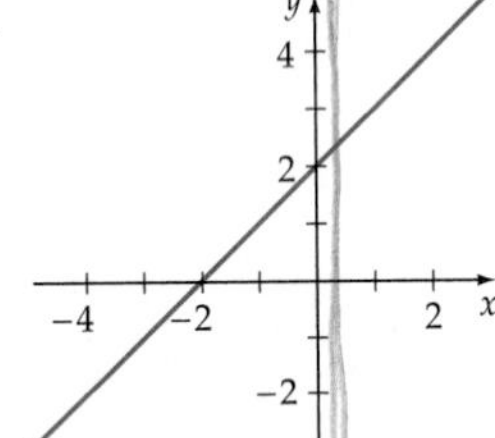

b.

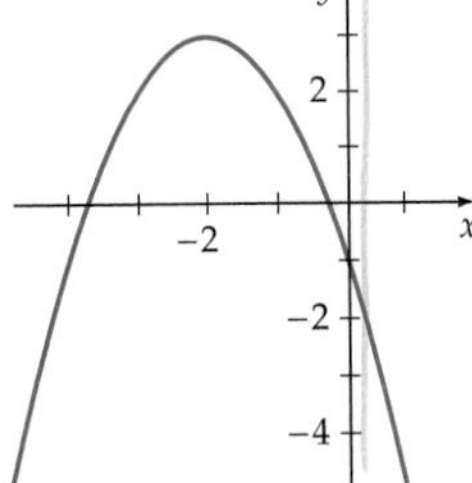

c. d.

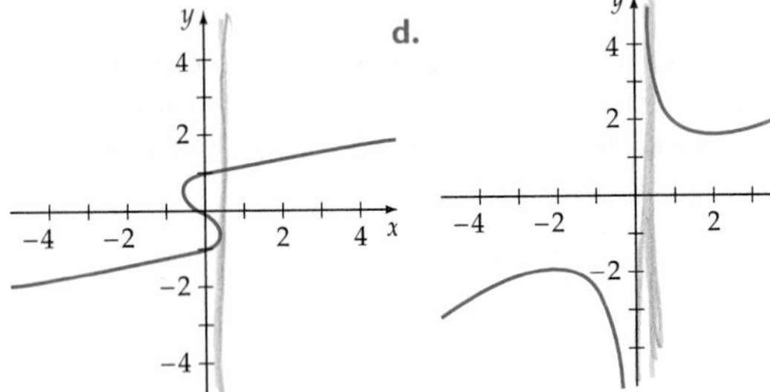

50. Use the vertical line test to determine which of the following graphs are graphs of functions.

a. b.

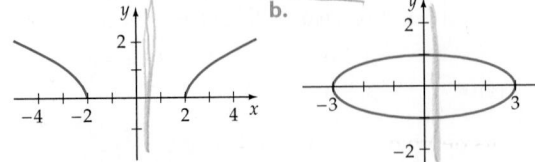

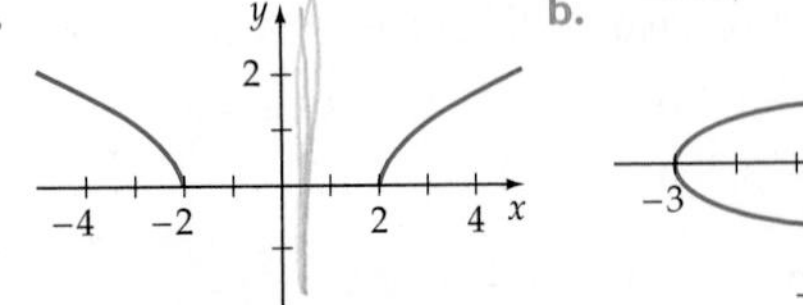

c. d.

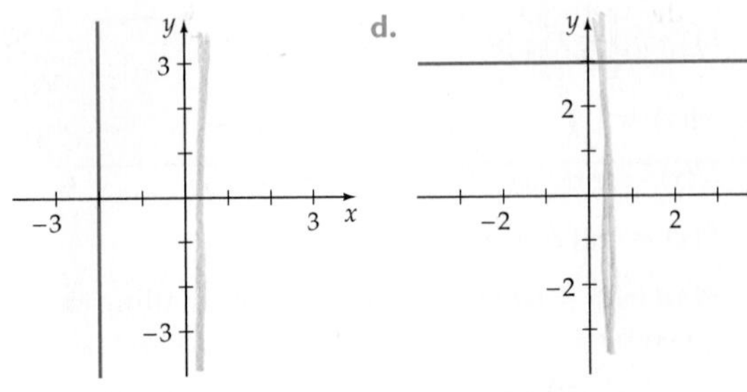

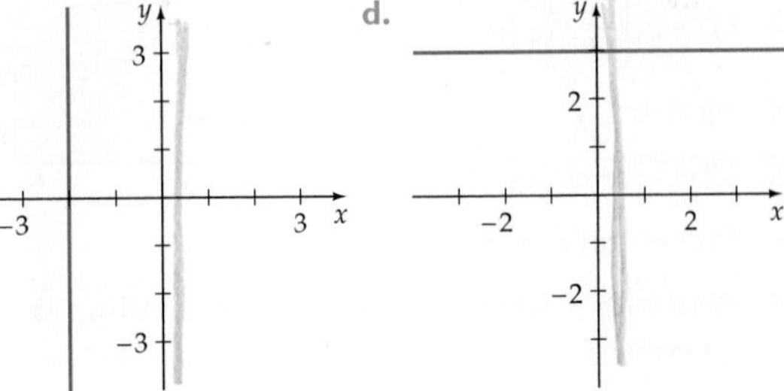

In Exercises 51 to 60, use the indicated graph to identify the intervals over which the function is increasing, constant, or decreasing.

51.

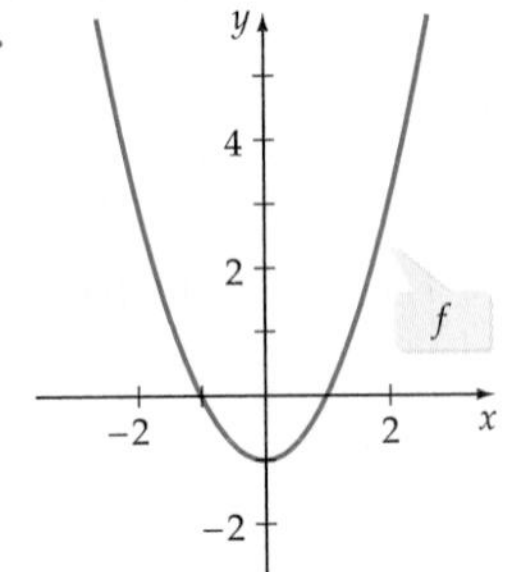

52.

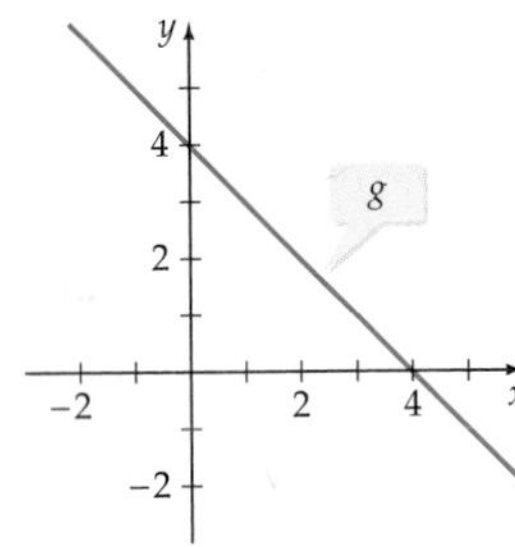

53.

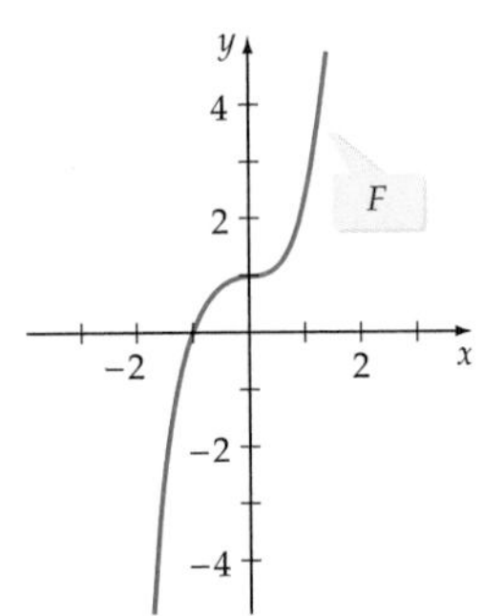

54.

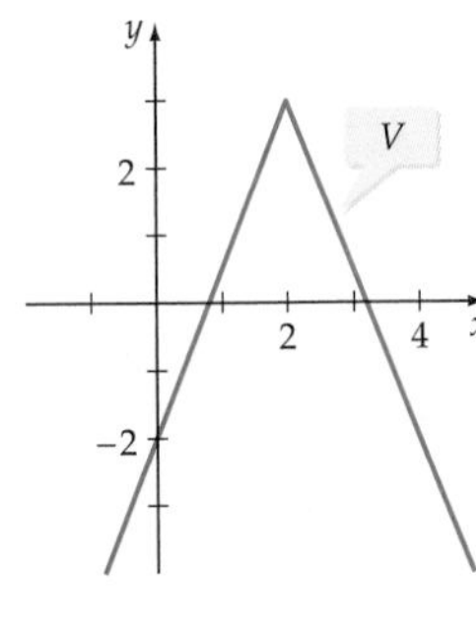

55.

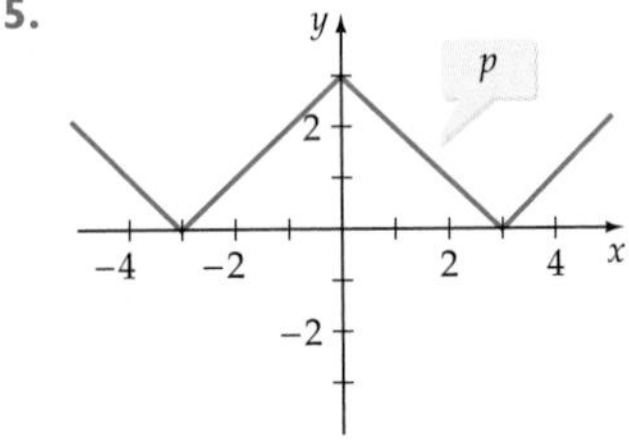

56.

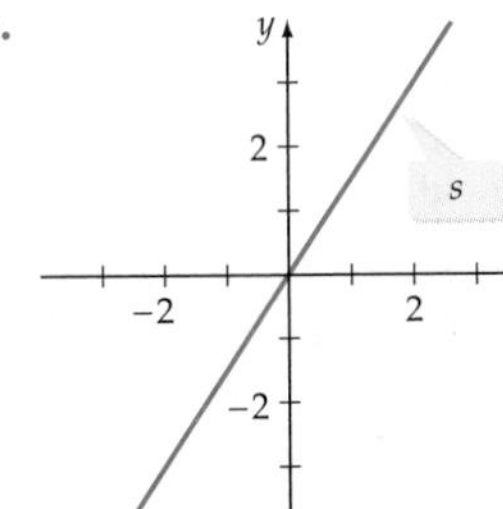

57.

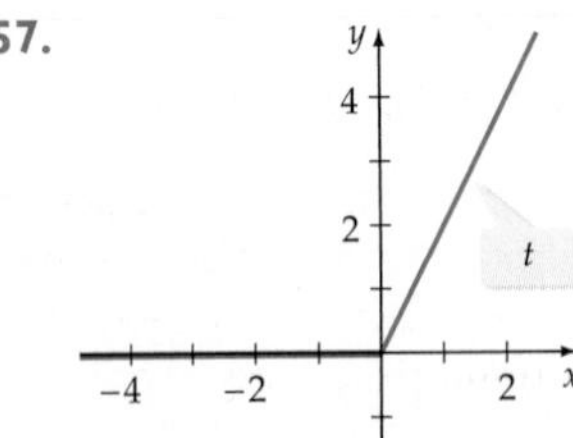

58.

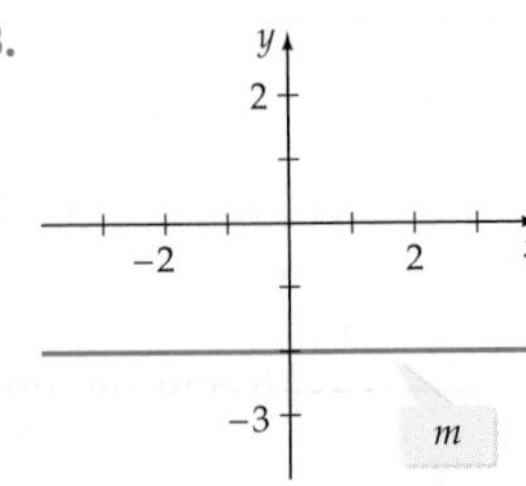

59.

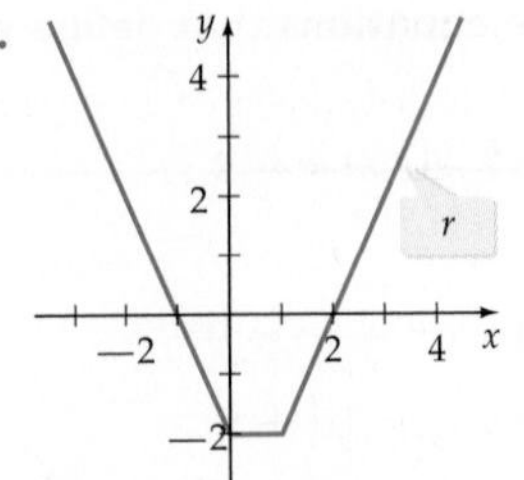

60.

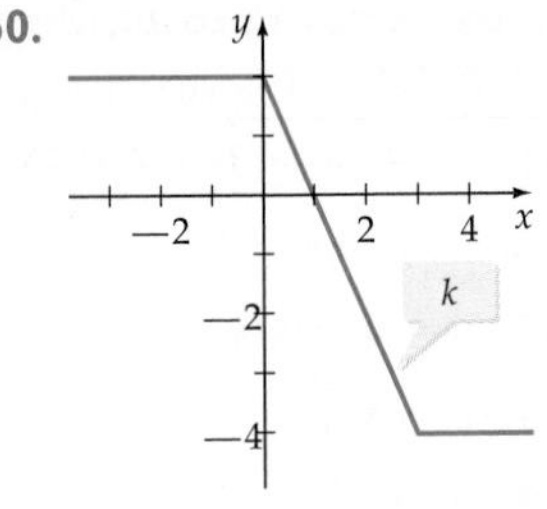

61. Use the horizontal line test to determine which of the following functions are one-to-one.

f as shown in Exercise 51
g as shown in Exercise 52
F as shown in Exercise 53
V as shown in Exercise 54
p as shown in Exercise 55

62. Use the horizontal line test to determine which of the following functions are one-to-one.

s as shown in Exercise 56
t as shown in Exercise 57
m as shown in Exercise 58
r as shown in Exercise 59
k as shown in Exercise 60

63. A rectangle has a length of l feet and a perimeter of 50 feet.
   a. Write the width w of the rectangle as a function of its length.
   b. Write the area A of the rectangle as a function of its length.

64. The sum of two numbers is 20. Let x represent one of the numbers.
   a. Write the second number y as a function of x.
   b. Write the product P of the two numbers as a function of x.

65. **DEPRECIATION** A bus was purchased for \$80,000. Assuming the bus depreciates at a rate of \$6500 per year (*straight-line depreciation*) for the first 10 years, write the value v of the bus as a function of the time t (measured in years) for $0 \le t \le 10$.

66. **DEPRECIATION** A boat was purchased for \$44,000. Assuming the boat depreciates at a rate of \$4200 per year (*straight-line depreciation*) for the first 8 years, write the value v of the boat as a function of the time t (measured in years) for $0 \le t \le 8$.

67. **COST, REVENUE, AND PROFIT** A manufacturer produces a product at a cost of \$22.80 per unit. The manufacturer has a fixed cost of \$400.00 per day. Each unit retails for \$37.00. Let x represent the number of units produced in a 5-day period.
   a. Write the total cost C as a function of x.
   b. Write the revenue R as a function of x.
   c. Write the profit P as a function of x. (*Hint:* The profit function is given by $P(x) = R(x) - C(x)$.)

68. **VOLUME OF A BOX** An open box is to be made from a square piece of cardboard having dimensions 30 inches by 30 inches by cutting out squares of area x^2 from each corner, as shown in the figure.

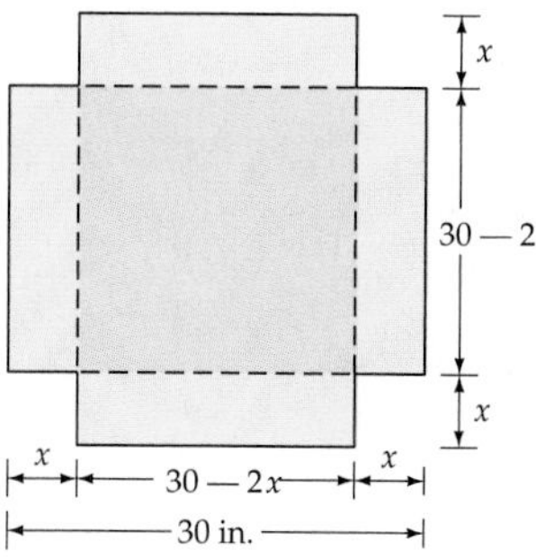

   a. Express the volume V of the box as a function of x.
   b. State the domain of V.

69. **HEIGHT OF AN INSCRIBED CYLINDER** A cone has an altitude of 15 centimeters and a radius of 3 centimeters. A right circular cylinder of radius r and height h is inscribed in the cone as shown in the figure. Use similar triangles to write h as a function of r.

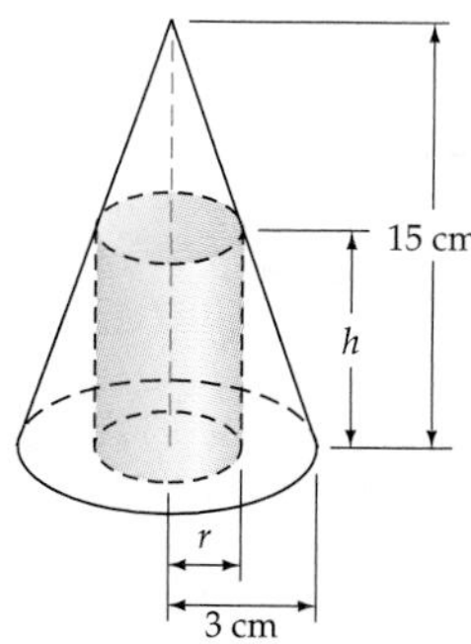

70. **VOLUME OF WATER** Water is running out of a conical funnel that has an altitude of 20 inches and a radius of 10 inches, as shown in the figure.
   a. Write the radius r of the water as a function of its depth h.
   b. Write the volume V of the water as a function of its depth h.

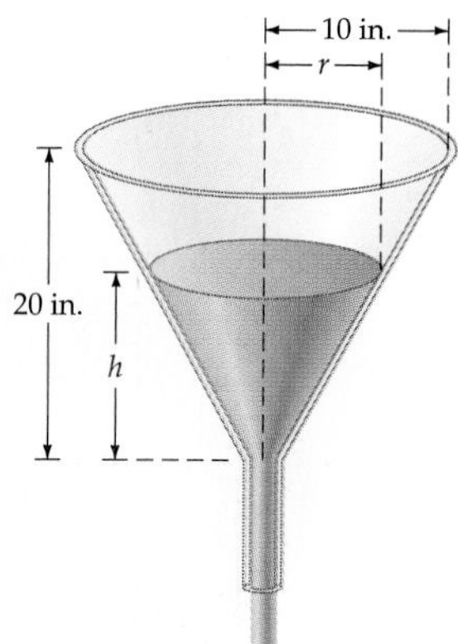

71. DISTANCE FROM A BALLOON For the first minute of flight, a hot air balloon rises vertically at a rate of 3 meters per second. If t is the time in seconds that the balloon has been airborne, write the distance d between the balloon and a point on the ground 50 meters from the point of lift-off as a function of t.

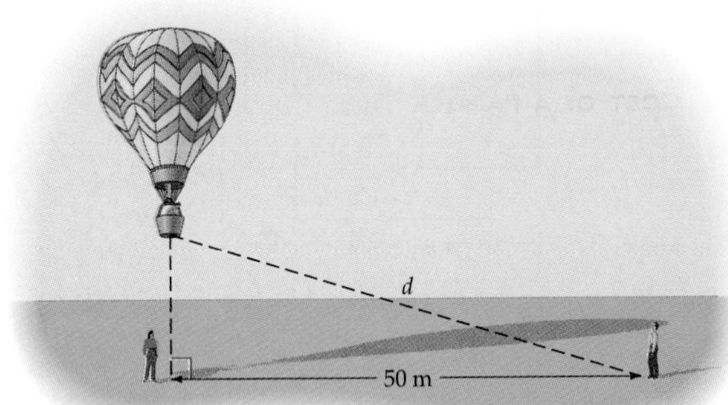

72. TIME FOR A SWIMMER An athlete swims from point A to point B at the rate of 2 mph and runs from point B to point C at a rate of 8 mph. Use the dimensions in the figure to write the time t required to reach point C as a function of x.

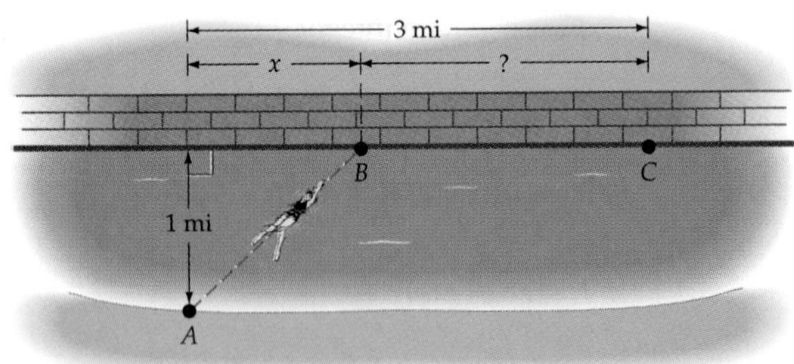

73. DISTANCE BETWEEN SHIPS At 12:00 noon Ship A is 45 miles due south of ship B and is sailing north at a rate of 8 mph. Ship B is sailing east at a rate of 6 mph. Write the distance d between the ships as a function of the time t where $t = 0$ represents 12:00 noon.

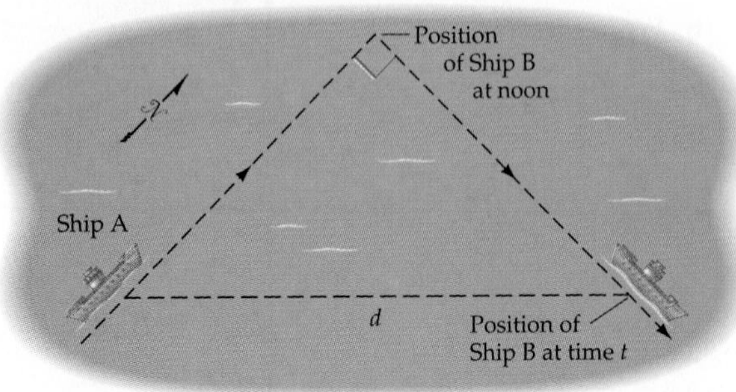

74. SALES VS. PRICE A business finds that the number of feet f of pipe it can sell per week is a function of the price p in cents per foot as given by

$$f(p) = \frac{320{,}000}{p + 25}, \quad 40 \le p \le 90$$

Complete the following table by evaluating f (to the nearest 100 feet) for the indicated values of p.

p	40	50	60	75	90
$f(p)$					

75. MODEL YIELD The yield Y of apples per tree is related to the amount x of a particular type of fertilizer applied (in pounds per year) by the function

$$Y(x) = 400[1 - 5(x - 1)^{-2}], \quad 5 \le x \le 20$$

Complete the following table by evaluating Y (to the nearest apple) for the indicated applications.

x	5	10	12.5	15	20
$Y(x)$					

76. MODEL COST A manufacturer finds that the cost C in dollars of producing x items of a product is given by

$$C(x) = \left(225 + 1.4\sqrt{x}\right)^2, \quad 100 \le x \le 1000$$

Complete the following table by evaluating C (to the nearest dollar) for the indicated numbers of items.

x	100	200	500	750	1000
$C(x)$					

77. If $f(x) = x^2 - x - 5$ and $f(c) = 1$, find c.

78. If $g(x) = -2x^2 + 4x - 1$ and $g(c) = -4$, find c.

79. Determine whether 1 is in the range of $f(x) = \dfrac{x - 1}{x + 1}$.

80. Determine whether 0 is in the range of $g(x) = \dfrac{1}{x - 3}$.

 **In Exercises 81 to 86, use a graphing utility.**

81. Graph $f(x) = \dfrac{[\![x]\!]}{|x|}$ for $-4 \le x \le 4$ and $x \ne 0$.

82. Graph $f(x) = \dfrac{[\![2x]\!]}{|x|}$ for $-4 \le x \le 4$ and $x \ne 0$.

83. Graph: $f(x) = x^2 - 2|x| - 3$

84. Graph: $f(x) = x^2 - |2x - 3|$

85. Graph: $f(x) = |x^2 - 1| - |x - 2|$

86. Graph: $f(x) = |x^2 - 2x| - 3$

SUPPLEMENTAL EXERCISES

The notation $f(x)|_a^b$ is used to denote the difference $f(b) - f(a)$. That is,

$$f(x)|_a^b = f(b) - f(a)$$

In Exercises 87 to 90, evaluate $f(x)|_a^b$ for the given function f and the indicated values of a and b.

87. $f(x) = x^2 - x; f(x)|_2^3$

88. $f(x) = -3x + 2; f(x)|_4^7$

89. $f(x) = 2x^3 - 3x^2 - x; f(x)|_0^2$

90. $f(x) = \sqrt{8 - x}; f(x)|_0^8$

In Exercises 91 to 94, each function has two or more independent variables.

91. Given $f(x, y) = 3x + 5y - 2$, find

a. $f(1, 7)$ b. $f(0, 3)$ c. $f(-2, 4)$

d. $f(4, 4)$ e. $f(k, 2k)$ f. $f(k + 2, k - 3)$

92. Given $g(x, y) = 2x^2 - |y| + 3$, find

a. $g(3, -4)$ b. $g(-1, 2)$

c. $g(0, -5)$ d. $g\left(\frac{1}{2}, -\frac{1}{4}\right)$

e. $g(c, 3c), c > 0$ f. $g(c + 5, c - 2), c < 0$

93. **AREA OF A TRIANGLE** The area of a triangle with sides a, b, and c is given by the function

$$A(a, b, c) = \sqrt{s(s - a)(s - b)(s - c)}$$

where s is the semiperimeter

$$s = \frac{a + b + c}{2}$$

Find $A(5, 8, 11)$.

94. **COST OF A PAINTER** The cost in dollars to hire a house painter is given by the function

$$C(h, g) = 15h + 14g$$

where h is the number of hours it takes to paint the house and g is the number of gallons of paint required to paint the house. Find $C(18, 11)$.

A *fixed point* of a function is a number a such that $f(a) = a$. In Exercises 95 and 96, find all fixed points for the given function.

95. $f(x) = x^2 + 3x - 3$

96. $g(x) = \dfrac{x}{x + 5}$

In Exercises 97 and 98, sketch the graph of the piecewise-defined function.

97. $s(x) = \begin{cases} 1 & \text{if } x \text{ is an integer} \\ 2 & \text{if } x \text{ is not an integer} \end{cases}$

98. $v(x) = \begin{cases} 2x - 2 & \text{if } x \neq 3 \\ 1 & \text{if } x = 3 \end{cases}$

PROJECTS

1. **DAY OF THE WEEK** A formula known as Zeller's Congruence makes use of the greatest integer function $[\![x]\!]$ to determine the day of the week on which a given day fell or will fall. To use Zeller's Congruence, we first compute the integer z given by

$$z = \left[\!\left[\frac{13m - 1}{5}\right]\!\right] + \left[\!\left[\frac{y}{4}\right]\!\right] + \left[\!\left[\frac{c}{4}\right]\!\right] + d + y - 2c$$

The variables c, y, d, and m are defined as follows:

c = the century

y = the year of the century

d = the day of the month

m = the month, using 1 for March, 2 for April, . . . , 10 for December. January and February are assigned the values 11 and 12 of the previous year.

For example, for the date September 12, 2001, we use $c = 20$, $y = 1$, $d = 12$, and $m = 7$. The remainder of z divided by 7 gives the day of the week. A remainder of 0 represents a Sunday, a remainder of 1 a Monday, . . . , a remainder of 6 a Saturday.

a. Verify that December 7, 1941 was a Sunday.
b. Verify that January 1, 2010 will fall on a Friday.
c. Determine on what day of the week Independence Day (July 4, 1776) fell.
d. Determine on what day of the week you were born.

SECTION 2.3 LINEAR FUNCTIONS

- SLOPES OF LINES
- APPLICATIONS
- PARALLEL AND PERPENDICULAR LINES

The following function has many applications.

Definition of a Linear Function

A function of the form

$$f(x) = mx + b, \quad m \neq 0$$

where m and b are real numbers, is a **linear function** of x.

◆ SLOPES OF LINES

The graph of $f(x) = mx + b$, or $y = mx + b$, is a nonvertical straight line.

The graphs shown in **Figure 2.37** are the graphs of $f(x) = mx + b$ for various values of m. The graphs intersect at the point $(-2, -1)$, but they differ in *steepness*. The steepness of a line is called the *slope* of the line and is denoted by the symbol m. The slope of a line is the ratio of the change in the y values of any two points on the line to the change in the x values of the same two points. For example, the graph of the line L_1 in **Figure 2.37** passes through the points $(-2, -1)$ and $(3, 5)$. The change in the y values is determined by subtracting the two y-coordinates.

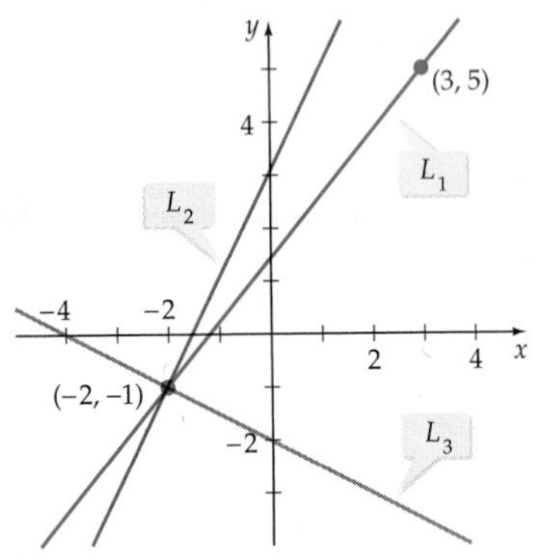

Figure 2.37

$$\text{Change in } y = 5 - (-1) = 6$$

The change in the x values is determined by subtracting the two x-coordinates.

$$\text{Change in } x = 3 - (-2) = 5$$

The slope m of L_1 is the ratio of the change in the y values of the two points to the change in the x values of the two points. That is,

$$m = \frac{\text{change in } y}{\text{change in } x} = \frac{6}{5}$$

Because the slope of a nonvertical line can be calculated by using any two arbitrary points on the line, we have the following formula.

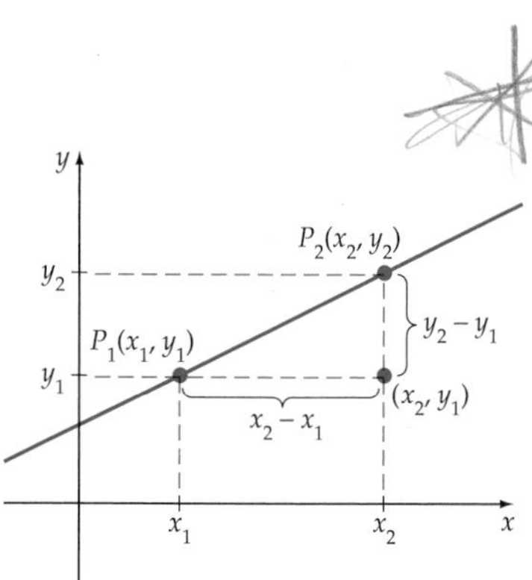

Figure 2.38

Slope of a Nonvertical Line

The **slope** m of the line passing through the points $P_1(x_1, y_1)$ and $P_2(x_2, y_2)$ with $x_1 \neq x_2$ is given by

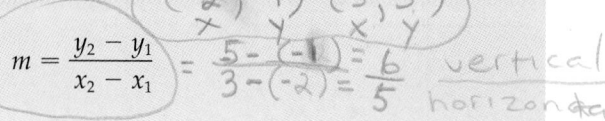

$$m = \frac{y_2 - y_1}{x_2 - x_1}$$

Because the numerator $y_2 - y_1$ is the vertical **rise** and the denominator $x_2 - x_1$ is the horizontal **run** from P_1 to P_2, slope is often referred to as the *rise over the run* or the *change in y divided by the change in x*. See **Figure 2.38.** Lines that have a positive slope slant upward from left to right. Lines that have a negative slope slant downward from left to right.

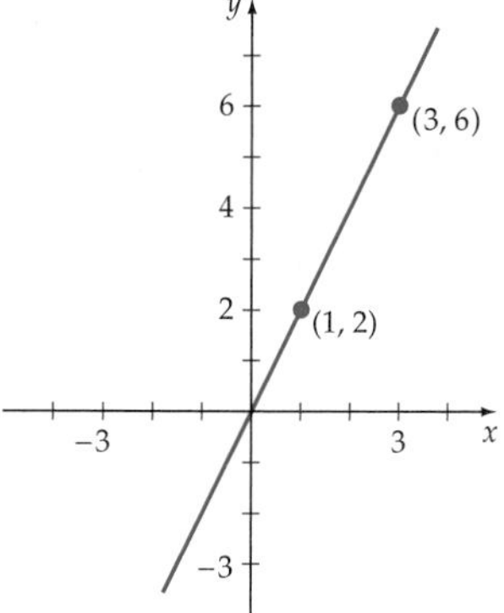

Figure 2.39

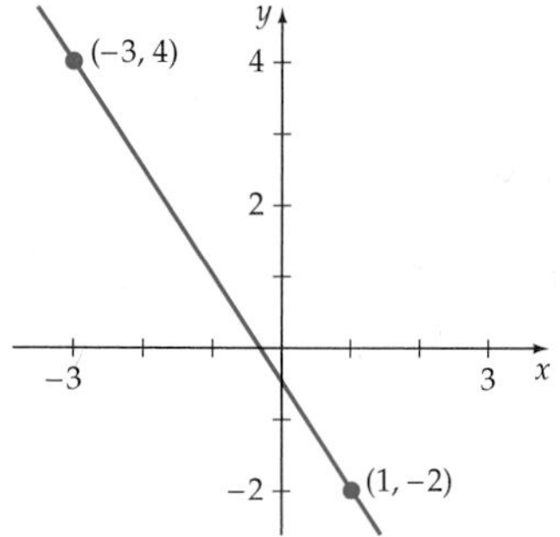

Figure 2.40

EXAMPLE 1 Find the Slope of a Line

Find the slope of the line passing through the points whose coordinates are given.

a. (1, 2) and (3, 6) **b.** (−3, 4) and (1, −2)

Solution

a. The slope of the line passing through (1, 2) and (3, 6) is

$$m = \frac{y_2 - y_1}{x_2 - x_1} = \frac{6 - 2}{3 - 1} = \frac{4}{2} = 2$$

Because $m > 0$, the line slants upward from left to right. See **Figure 2.39.**

b. The slope of the line passing through (−3, 4) and (1, −2) is

$$m = \frac{y_2 - y_1}{x_2 - x_1} = \frac{-2 - 4}{1 - (-3)} = \frac{-6}{4} = -\frac{3}{2}$$

Because $m < 0$, the line slants downward from left to right. See **Figure 2.40.**

TRY EXERCISE 2, EXERCISE SET 2.3, PAGE 175

take note

A web applet is available to explore the calculation of slope. This applet, Slope Calculation, can be found on our web site at

http://college.hmco.com

The definition of slope does not apply to vertical lines. Consider, for example, the points (2, 1) and (2, 3) on the vertical line l_1 in **Figure 2.41.** Applying the definition of slope to this line produces

$$m = \frac{3 - 1}{2 - 2}$$

which is undefined because it requires division by zero. Because division by zero is undefined, we say that the slope of any vertical line is undefined.

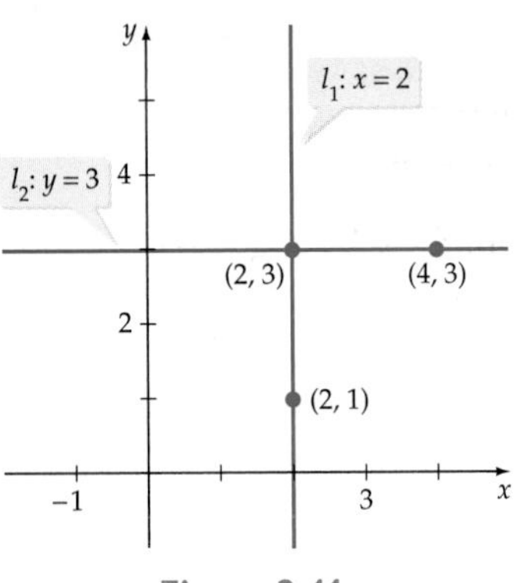

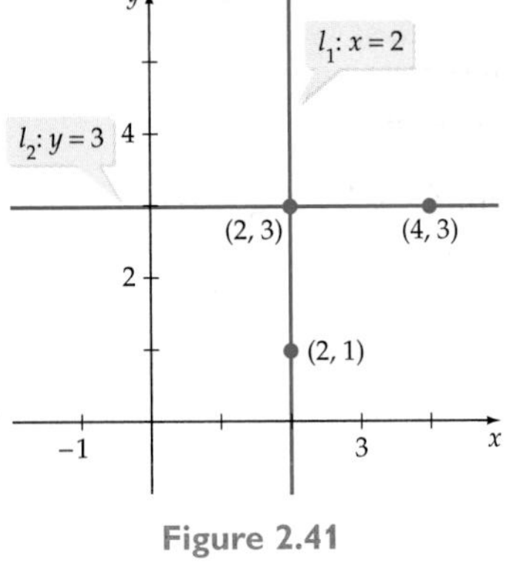

Figure 2.41

QUESTION Is the graph of a vertical line the graph of a function?

All horizontal lines have 0 slope. For example, the line l_2 through (2, 3) and (4, 3) in **Figure 2.41** is a horizontal line. Its slope is given by

$$m = \frac{3-3}{4-2} = \frac{0}{2} = 0$$

When computing the slope of a line, it does not matter which point we label P_1 and which P_2 because

$$\frac{y_2 - y_1}{x_2 - x_1} = \frac{y_1 - y_2}{x_1 - x_2}$$

In functional notation, the points P_1 and P_2 can be represented by

$$(x_1, f(x_1)) \qquad \text{and} \qquad (x_2, f(x_2))$$

In this notation, the slope formula

$$m = \frac{y_2 - y_1}{x_2 - x_1} \quad \text{is expressed as} \quad m = \frac{f(x_2) - f(x_1)}{x_2 - x_1} \qquad (1)$$

Figure 2.42

If $m = 0$, then $f(x) = mx + b$ can be written as $f(x) = b$, or $y = b$. The graph of $y = b$ is the horizontal line through $(0, b)$. See **Figure 2.42.** Because every point on the graph of $y = b$ has a y-coordinate of b, the function $y = b$ is called a **constant function.**

Every point on the vertical line through $(a, 0)$ has an x-coordinate of a. The equation of the vertical line through $(a, 0)$ is $x = a$. See **Figure 2.42.**

Horizontal Lines and Vertical Lines

The graph of $y = b$ is a horizontal line through $(0, b)$.

The graph of $x = a$ is a vertical line through $(a, 0)$.

take note

A web applet is available to explore the slope-intercept form of a straight line. This applet, Slope-Intercept Form, can be found on our web site at

http://college.hmco.com

The equation $f(x) = mx + b$ is called the **slope-intercept form** of the equation of a line because of the following theorem.

Slope-Intercept Form

The graph of $f(x) = mx + b$ is a line with slope m and y-intercept $(0, b)$.

ANSWER No. For example, the vertical line passing through $x = 2$ contains the ordered pairs (2, 3) and (2, −5). Thus, there are two ordered pairs with the same first coordinate but different second coordinates.

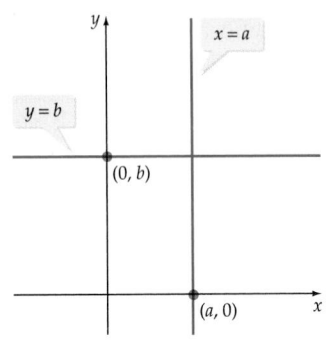

Proof The slope of the graph of $f(x) = mx + b$ is given by Equation (1).

$$\frac{f(x_2) - f(x_1)}{x_2 - x_1} = \frac{(mx_2 + b) - (mx_1 + b)}{x_2 - x_1} = \frac{m(x_2 - x_1)}{x_2 - x_1} = m, \quad x_1 \neq x_2$$

The y-intercept of the graph of $f(x) = mx + b$ is found by letting $x = 0$.

$$f(0) = m(0) + b = b$$

Thus $(0, b)$ is the y-intercept, and m is the slope, of the graph of $f(x) = mx + b$ ◆

If a function is written in the form $f(x) = mx + b$, then its graph can be drawn by first plotting the y-intercept $(0, b)$ and then using its slope m to determine another point on the line.

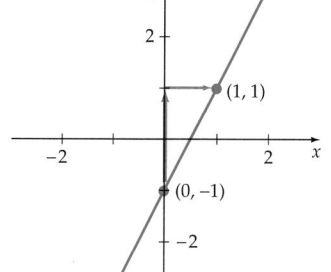

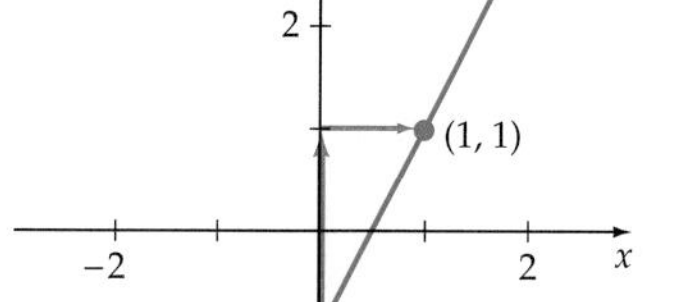

$y = 2x - 1$

Figure 2.43

EXAMPLE 2 Graph a Linear Function

Graph: $f(x) = 2x - 1$

Solution

The equation $y = 2x - 1$ is in slope-intercept form, with $b = -1$ and $m = 2$ or $2/1$. To graph the equation, first plot the y-intercept $(0, -1)$ and then use the slope to plot a second point, which is two units up and one unit to the right of the y-intercept. See **Figure 2.43**.

TRY EXERCISE 16, EXERCISE SET 2.3, PAGE 175

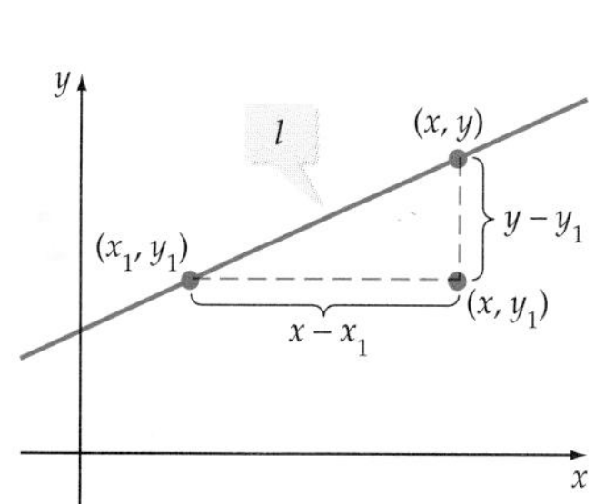

The slope of line l is $m = \dfrac{y - y_1}{x - x_1}$.

Figure 2.44

We can find an equation of a line, provided we know its slope and at least one point on the line. **Figure 2.44** suggests that if (x_1, y_1) is a point on a line l of slope m, and (x, y) is *any other* point on the line, then

$$\frac{y - y_1}{x - x_1} = m, \quad x \neq x_1$$

Multiplying each side by $x - x_1$ produces $y - y_1 = m(x - x_1)$. This equation is called the **point-slope form** of the equation of line l.

Point-Slope Form

The graph of

$$y - y_1 = m(x - x_1)$$

is a line that has slope m and passes through (x_1, y_1).

take note

A web applet is available to explore the equation of a line using the point-slope formula. This applet, Point-Slope Form, can be found on our web site at

http://college.hmco.com

EXAMPLE 3 Use the Point-Slope Form

Find an equation of a line with slope -3 that passes through $(-1, 4)$.

Continued ►

take note

To determine an equation of a nonvertical line that passes through two points, first determine the slope of the line and then use the coordinates of either one of the points in the point-slope form.

Solution

Use the point-slope form with $m = -3$, $x_1 = -1$, and $y_1 = 4$.

$$y - y_1 = m(x - x_1)$$
$$y - 4 = -3[x - (-1)] \quad \text{• Substitute.}$$
$$y - 4 = -3x - 3 \quad \text{• Solve for } y.$$
$$y = -3x + 1 \quad \text{• Slope-intercept form}$$

TRY EXERCISE 28, EXERCISE SET 2.3, PAGE 175

An equation of the form $Ax + By + C = 0$, where A, B, and C are real numbers and both A and B are not zero, is called the **general form of the equation of a line.** For example, the equation $y = -3x + 1$ in Example 3 can be written in general form as $3x + y - 1 = 0$.

By solving a first-degree equation, the specific relationship between an element of the domain and an element of the range of a linear function can be determined.

EXAMPLE 4 Find the Value in the Domain of f for which $f(x) = b$

Find the value x in the domain of $f(x) = 3x - 4$ for which $f(x) = 5$.

Solution

$$f(x) = 3x - 4$$
$$5 = 3x - 4 \quad \text{• Replace } f(x) \text{ by 5 and solve for } x.$$
$$9 = 3x$$
$$3 = x$$

When $x = 3$, $f(x) = 5$. This means that 3, in the domain of f, is paired with 5, in the range of f. Another way of stating this is that the ordered pair (3, 5) is an element of f.

Visualize the Solution

By graphing $y = 5$ and $f(x) = 3x - 4$, we can see that $f(x) = 5$ when $x = 3$.

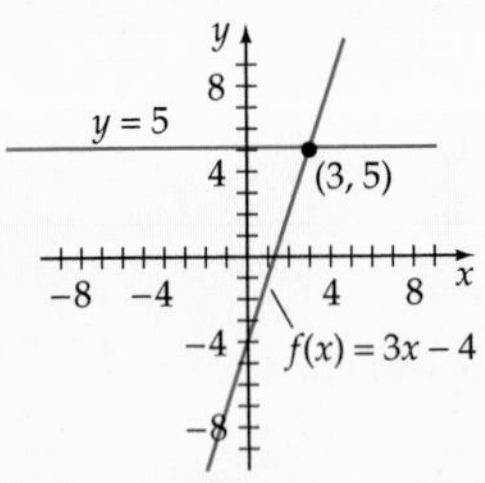

TRY EXERCISE 42, EXERCISE SET 2.3, PAGE 175

Although we are mainly concerned with linear functions in this section, the following theorem applies to all functions. It illustrates a powerful relationship between the real solutions of $f(x) = 0$ and the x-intercepts of the graph of $y = f(x)$.

Real Solutions and x-Intercepts Theorem

For every function f, the real number c is a solution of $f(x) = 0$ if and only if $(c, 0)$ is an x-intercept of the graph of $y = f(x)$.

The real solutions and x-intercepts theorem tells us that we can find real solutions of $f(x) = 0$ by graphing. The following example illustrates the theorem for a linear function of x.

EXAMPLE 5 Verify the Real Solutions and x-Intercepts Theorem

Let $f(x) = -2x + 6$. Find the real solution of $f(x) = 0$ and then graph $f(x)$. Compare the solution of $f(x) = 0$ with the x-intercept of the graph of f.

Solution

To find the real solution of $f(x) = 0$, replace $f(x)$ by $-2x + 6$ and solve for x.

$$f(x) = 0$$
$$-2x + 6 = 0$$
$$-2x = -6$$
$$x = 3$$

The x-coordinate of the x-intercept is 3. The real solution of $f(x) = 0$ is 3.

Visualize the Solution

Graph $f(x) = -2x + 6$ (see Figure 2.45).

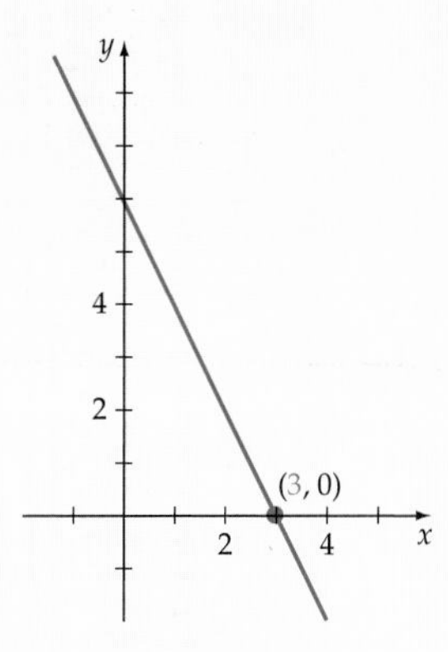

Figure 2.45

The x-intercept is $(3, 0)$.

TRY EXERCISE 46, EXERCISE SET 2.3, PAGE 176

Let $f_1(x) = ax + b$ and $f_2(x) = cx + d$, with the restriction that $a \neq c$. The solution of $f_1(x) = f_2(x)$ can be determined by solving $ax + b = cx + d$ for x. Graphically, the solution of this equation is the x-coordinate of the intersection of the graphs of $y = f_1(x)$ and $y = f_2(x)$.

EXAMPLE 6 Solve $ax + b = cx + d$

Let $f_1(x) = 2x - 1$ and $f_2(x) = -x + 11$. Find the values x for which $f_1(x) = f_2(x)$.

Continued

Solution

$$f_1(x) = f_2(x)$$
$$2x - 1 = -x + 11$$
$$3x = 12$$
$$x = 4$$

When $x = 4$, $f_1(x) = f_2(x)$.

Visualize the Solution

The graph of $y = f_1(x)$ and $y = f_2(x)$ are shown on the same coordinate axes (see Figure 2.46). Note that the point of intersection is (4, 7).

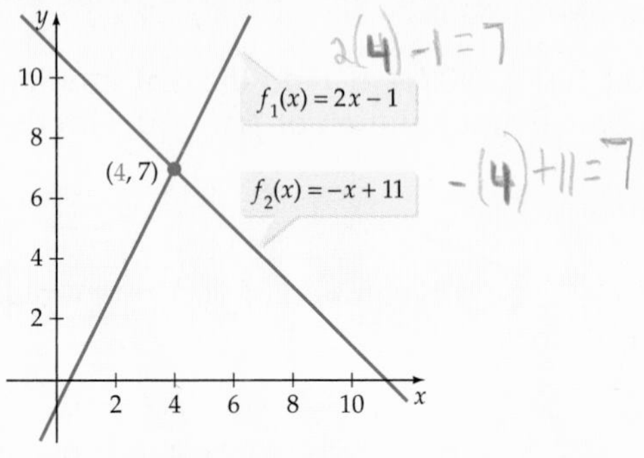

Figure 2.46

TRY EXERCISE 50, EXERCISE SET 2.3, PAGE 176

◆ APPLICATIONS

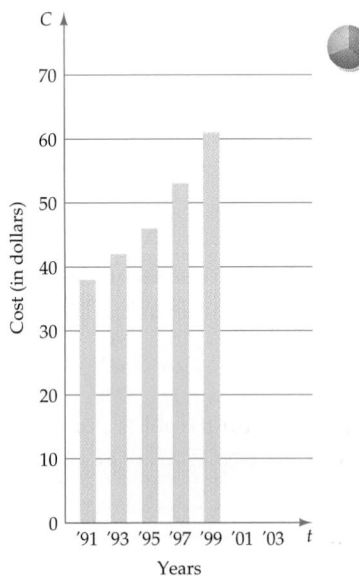

EXAMPLE 7 Find a Linear Model of Data

The bar graph in Figure 2.47 is based on data from the Pharmaceutical Research and Manufacturer's of America. The graph illustrates the average cost of filling a prescription drug order in the United States since 1991. Let $t = 1$ represent the year 1991. The ordered pair (1, 38) indicates that in 1991, the average cost to fill a prescription drug order was \$38. The ordered pair (9, 61) indicates that in 1999, the average cost to fill a prescription drug order was \$61.

a. Use the ordered pairs (1, 38) and (9, 61) to find a linear function that models the data in the graph.

b. What average cost for a prescription drug does the model predict for 1996? Round to the nearest dollar.

c. Use the model to predict in what year will the average cost of a prescription drug first exceed \$70?

Figure 2.47

take note

In Example 7 and Example 9 on the next page, you must find equation of a line between two points. You can visit college.hmco.com for a graphing calculator program that can be used to check your answer to these types of problems.

Solution

a. First calculate the slope of the line and then use the point-slope formula to find the equation of the line.

$$m = \frac{C_2 - C_1}{t_2 - t_1} = \frac{61 - 38}{9 - 1} = \frac{23}{8} = 2.875$$ • **Calculate the slope.**

$$C - C_1 = m(t - t_1)$$ • **Use the point-slope formula.**

$$C - 38 = 2.875(t - 1)$$ • **Substitute 38 for C_1, 2.875 for m, and 1 for t.**

$$C = 2.875t + 35.125$$

In functional notation, the linear model is $C(t) = 2.875t + 35.125$.

b. To find the average cost of a prescription drug in 1996, evaluate $C(t)$ when $t = 6$.

$$C(t) = 2.875t + 35.125$$
$$C(6) = 2.875(6) + 35.125$$
$$= 52.375 \approx 52$$

The model predicts that the average cost of a prescription drug was \$52 in 1996.

take note

The model in Example 7 is based on using t as the number of years after 1990. You can think of a year as $1990 + t$. In part c, $t = 12$ represents the year $1990 + 12 = 2002$.

c. To predict the year in which the average cost of a prescription drug will first exceed \$70, let $C(t) = 70$ and solve for t.

$$C(t) = 2.875t + 35.125$$
$$70 = 2.875t + 35.125$$
$$34.875 = 2.875t$$
$$12.130435 = t$$

The model predicts that the average price of a prescription drug will first exceed \$70 in the thirteenth year after 1990, which is 2003.

TRY EXERCISE 54, EXERCISE SET 2.3, PAGE 176

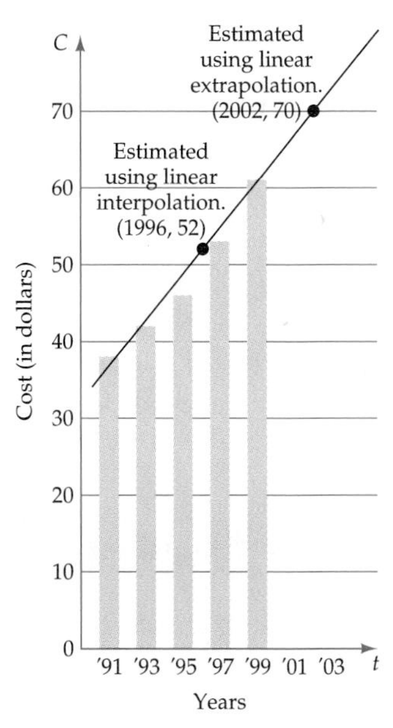

Figure 2.48

The procedure in part **b** of Example 7 used a linear function to determine a point between two known points. This procedure is referred to as **linear interpolation.** In part **c** of Example 7, the linear function was used to estimate a value in the *future*. This is called **linear extrapolation.** See Figure 2.48.

If a manufacturer produces x units of a product that sells for p dollars per unit, then the **cost function** C, the **revenue function** R, and the **profit function** P are defined as follows:

$C(x)$ = cost of producing and selling x units

$R(x) = xp$ = revenue from the sale of x units at p dollars each

$P(x)$ = profit from selling x units

Because profit equals the revenue less the cost, we have

$$P(x) = R(x) - C(x)$$

The value of x for which $R(x) = C(x)$ is called the **break-even point.** At the break-even point, $P(x) = 0$.

EXAMPLE 8 Find the Profit Function and the Break-even Point

A manufacturer finds that the costs incurred in the manufacture and sale of a particular type of calculator are \$180,000 plus \$27 per calculator.

a. Determine the profit function P, given that x calculators are manufactured and sold at \$59 each.

b. Determine the break-even point.

Solution

a. The cost function is $C(x) = 27x + 180{,}000$. The revenue function is $R(x) = 59x$. Thus the profit function is

$$\begin{aligned} P(x) &= R(x) - C(x) \\ &= 59x - (27x + 180{,}000) \\ &= 32x - 180{,}000, \quad x \geq 0 \text{ and } x \text{ is an integer} \end{aligned}$$

b. At the break-even point, $R(x) = C(x)$.

$$\begin{aligned} 59x &= 27x + 180{,}000 \\ 32x &= 180{,}000 \\ x &= 5625 \end{aligned}$$

The manufacturer will break even when 5625 calculators are sold.

TRY EXERCISE 56, EXERCISE SET 2.3, PAGE 176

take note

The graphs of C, R, and P are shown below. Observe that the graphs of C and R intersect at the break-even point, where $x = 5625$ and $P(5625) = 0$.

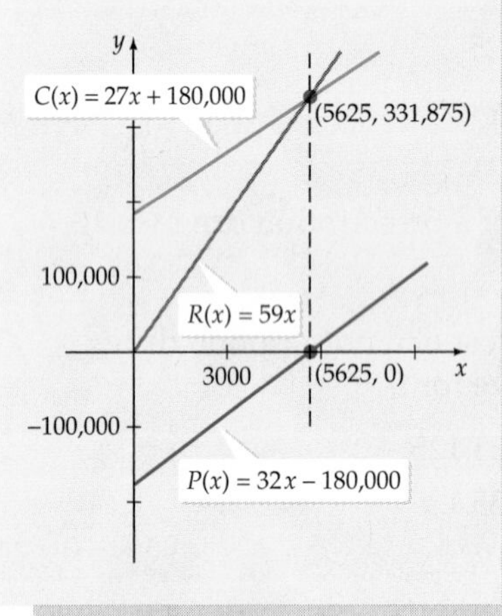

The scatter plot in the next example appears to indicate a linear relationship between the variables t and E. But what linear function best models the relationship? The last section of this chapter illustrates an analytic method that can be used to find a linear function that models data. However, in the next example, we find a linear model by sketching the line that appears to fit the data better than any other line.

Acre-feet of Rain for Unseeded versus Seeded Clouds

Unseeded, u	Seeded, s
24	47
56	104
33	68
82	172
66	130
80	169

EXAMPLE 9 Approximate a Linear Model of Data

Cloud seeding is a technique of depositing silver nitrate into clouds to increase the amount of rain from a cloud formation. The table at the left shows the number of acre-feet of rain for unseeded and seeded cloud formations.

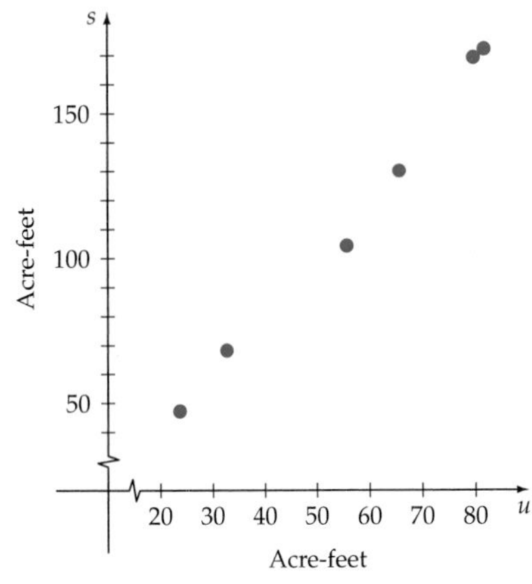

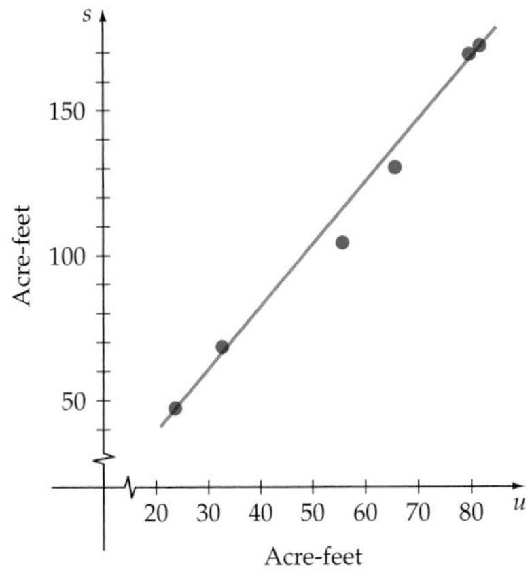

Figure 2.49

a. Construct a scatter diagram for the data. Use the variable u to represent the acre-feet of rain from unseeded clouds, and let s be the acre-feet of rain from seeded clouds.

b. Using the points $P_1(24, 47)$ and $P_2(82, 172)$, find the equation of a linear model for this data. Round the slope to the nearest hundredth.

Solution

a. See Figure 2.49.

b. $m = \dfrac{s_2 - s_1}{u_2 - u_1} = \dfrac{172 - 47}{82 - 24} \approx 2.16$ • **Find the slope of the line.**

$s - s_1 = m(u - u_1)$ • **Use the point-slope formula to find the equation of the line.**

$s - 47 = 2.16(u - 24)$

$s = 2.16u - 4.84$

In functional notation, a linear model for the data is $s(u) = 2.16u - 4.84$.

TRY EXERCISE 66, EXERCISE SET 2.3, PAGE 177

◆ PARALLEL AND PERPENDICULAR LINES

Two nonintersecting lines in a plane are **parallel.** All vertical lines are parallel to each other. All horizontal lines are parallel to each other.

Two lines are **perpendicular** if and only if they intersect and form adjacent angles each of which measures 90°. In a plane, vertical and horizontal lines are perpendicular to one another.

Parallel and Perpendicular Lines

Let l_1 be the graph of $f_1(x) = m_1x + b$ and l_2 be the graph of $f_2(x) = m_2x + b$. Then

- l_1 and l_2 are parallel if and only if $m_1 = m_2$.
- l_1 and l_2 are perpendicular if and only if $m_1 = -\dfrac{1}{m_2}$.

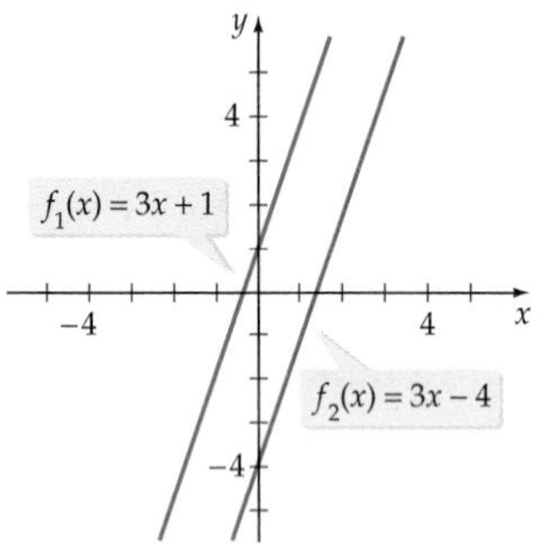

Figure 2.50

The graphs of $f_1(x) = 3x + 1$ and $f_2(x) = 3x - 4$ are shown in **Figure 2.50**. Because $m_1 = m_2 = 3$, the lines are parallel.

If $m_1 = -\dfrac{1}{m_2}$, then m_1 and m_2 are negative reciprocals of each other. The graphs of $g_1(x) = 2x + 3$ and $g_2(x) = -\dfrac{1}{2}x + 1$ are shown in **Figure 2.51**. Because 2 and $-\dfrac{1}{2}$ are negative reciprocals of each other, the lines are perpendicular. The symbol ⌝ indicates an angle of 90°. In **Figure 2.51** it is used to indicate that the lines are perpendicular.

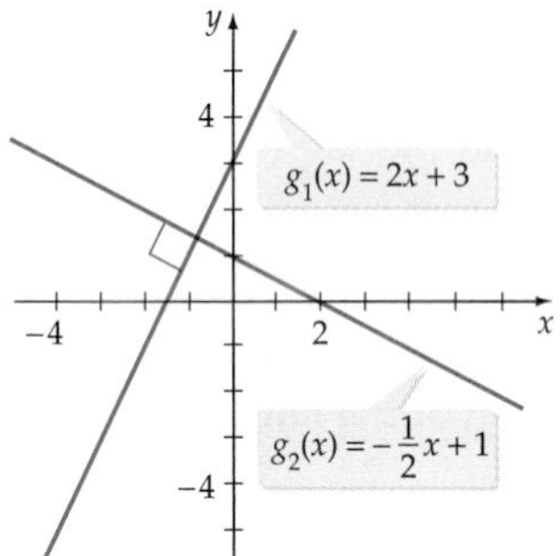

Figure 2.51

EXAMPLE 10 Determine a Point of Impact

A rock is whirled horizontally in a circular counterclockwise path about the origin. See **Figures 2.52** and **2.53**. When the string breaks, the rock travels on a linear path perpendicular to the radius $\overline{OP}$ and hits a wall located at

$$y = x + 12 \tag{2}$$

If the string breaks when the rock is at $P(4, 3)$, determine the point where the rock hits the wall.

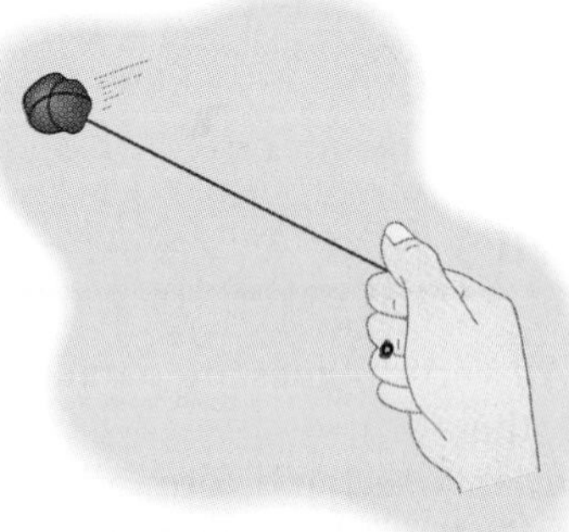
Figure 2.52

Solution

The slope of the radius from (0, 0) to (4, 3) is 3/4. The negative reciprocal of 3/4 is −4/3. Therefore, the linear path of the rock is given by

$$y - 3 = -\frac{4}{3}(x - 4)$$

$$y = -\frac{4}{3}x + \frac{25}{3} \tag{3}$$

To find the point where the rock hits the wall, set the right side of Equation (2) equal to the right side of Equation (3) and solve for x. This is the procedure explained in Example 6.

$$-\frac{4}{3}x + \frac{25}{3} = x + 12$$

$$-4x + 25 = 3x + 36 \quad \text{• Multiply all terms by 3.}$$

$$-7x = 11$$

$$x = -\frac{11}{7}$$

For every point on the wall, x and y are related by $y = x + 12$. Therefore, substituting −11/7 for x in $y = x + 12$ yields $y = -11/7 + 12 = 73/7$, and the rock hits the wall at $(-11/7, 73/7)$.

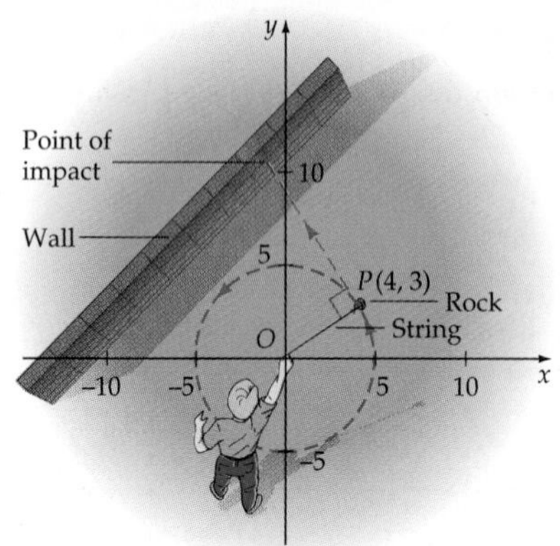

Figure 2.53

TRY EXERCISE 72, EXERCISE SET 2.3, PAGE 178

TOPICS FOR DISCUSSION

1. If a linear function has a negative y-intercept and a negative slope, then its graph does not contain any points in Quadrant I. Do you agree? Explain.
2. Is a "break-even point" a point or a number? Explain.
3. A tutor states that some perpendicular lines do not have the property that their slopes are negative reciprocals of each other. Explain why the tutor is correct.
4. The real solutions and x-intercepts theorem applies only to linear functions. Do you agree?
5. Explain why the function $f(x) = x$ is referred to as the identity function.

EXERCISE SET 2.3

In Exercises 1 to 10, find the slope of the line that passes through the given points.

1. $(3, 4)$ and $(1, 7)$
2. $(-2, 4)$ and $(5, 1)$
3. $(4, 0)$ and $(0, 2)$
4. $(-3, 4)$ and $(2, 4)$
5. $(0, 0)$ and $(0, 4)$
6. $(0, 0)$ and $(3, 0)$
7. $(-3, 4)$ and $(-4, -2)$
8. $(-5, -1)$ and $(-3, 4)$
9. $\left(-4, \frac{1}{2}\right)$ and $\left(\frac{7}{3}, \frac{7}{2}\right)$
10. $\left(\frac{1}{2}, 4\right)$ and $\left(\frac{7}{4}, 2\right)$

In Exercises 11 to 14, find the slope of the line that passes through the given points.

11. $(3, f(3))$ and $(3 + h, f(3 + h))$
12. $(-2, f(-2 + h))$ and $(-2 + h, f(-2 + h))$
13. $(0, f(0))$ and $(h, f(h))$
14. $(a, f(a))$ and $(a + h, f(a + h))$

In Exercises 15 to 26, graph y as a function of x by finding the slope and y-intercept of each.

15. $y = 2x - 4$
16. $y = -x + 1$
17. $y = -\frac{1}{3}x + 4$
18. $y = \frac{2}{3}x - 2$
19. $y = 3$
20. $y = x$
21. $y = 2x$
22. $y = -3x$
23. $2x + y = 5$
24. $x - y = 4$
25. $4x + 3y - 12 = 0$
26. $2x + 3y + 6 = 0$

In Exercises 27 to 38, find the equation of the indicated line. Write the equation in the form $y = mx + b$.

27. y-intercept $(0, 3)$, slope 1
28. y-intercept $(0, 5)$, slope -2
29. y-intercept $\left(0, \frac{1}{2}\right)$, slope $\frac{3}{4}$
30. y-intercept $\left(0, \frac{3}{4}\right)$, slope $-\frac{2}{3}$
31. y-intercept $(0, 4)$, slope 0
32. y-intercept $(0, -1)$, slope $\frac{1}{2}$
33. Through $(-3, 2)$, slope -4
34. Through $(-5, -1)$, slope -3
35. Through $(3, 1)$ and $(-1, 4)$
36. Through $(5, -6)$ and $(2, -8)$
37. Through $(7, 11)$ and $(2, -1)$
38. Through $(-5, 6)$ and $(-3, -4)$
39. Find the value of x in the domain of $f(x) = 2x + 3$ for which $f(x) = -1$.
40. Find the value of x in the domain of $f(x) = 4 - 3x$ for which $f(x) = 7$.
41. Find the value of x in the domain of $f(x) = 1 - 4x$ for which $f(x) = 3$.
42. Find the value of x in the domain of $f(x) = \frac{2x}{3} + 2$ for which $f(x) = 4$.
43. Find the value of x in the domain of $f(x) = 3 - \frac{x}{2}$ for which $f(x) = 5$.

44. Find the value of x in the domain of $f(x) = 4x - 3$ for which $f(x) = -2$.

In Exercises 45 to 48 find the solution $f(x) = 0$, verify that the solution of $f(x) = 0$ is the same as the x-coordinate of the x-intercept of the graph of $y = f(x)$.

45. $f(x) = 3x - 12$

46. $f(x) = -2x - 4$

47. $f(x) = \frac{1}{4}x + 5$

48. $f(x) = -\frac{1}{3}x + 2$

In Exercises 49 to 52, solve $f_1(x) = f_2(x)$ by an algebraic method and by graphing.

49. $f_1(x) = 4x + 5$ $\quad f_2(x) = x + 6$

50. $f_1(x) = -2x - 11$ $\quad f_2(x) = 3x + 7$

51. $f_1(x) = 2x - 4$ $\quad f_2(x) = -x + 12$

52. $f_1(x) = \frac{1}{2}x + 5$ $\quad f_2(x) = \frac{2}{3}x - 7$

53. **AUTOMOTIVE TECHNOLOGY** The table below shows the EPA estimates for city and highway driving for ten selected luxury cars. (Source: **www.money.com**, May 26, 2000)

EPA miles per gallon estimates for city and highway driving for selected luxury cars.

Car	City mpg	Highway mpg
Acura RL	18	24
Audi A8, 4.2L	17	24
BMW 528i	21	29
Cadillac Deville	17	27
Infiniti Q45	18	23
Jaguar XJ8	17	24
Lexus LS400	18	25
Lincoln Continental	17	25
Mercedes S500	16	23
Saab	18	24

A linear function that approximates this data is given by $f(x) = 0.95652x + 7.86957$, where $f(x)$ is the highway mpg and x is the city mpg.

a. Use the function to interpolate the estimated highway mpg for a car whose city mpg is 19 mpg.

b. Use the function to extrapolate the estimated highway mpg for a car whose city mpg is 24 mpg.

54. **AVIATION** The table in the next column is based on data from the Federal Aviation Administration for planes flown in 1998. The table shows, for selected airlines, the total number of hours an airline's planes operated and the total number of miles flown by those planes.

Total hours of operation of planes and the total number of miles flown by those planes.

Airline	Hours flown (in thousands)	Miles flown (in thousands)
Alaska	295	126
American	2054	945
Continental	1054	476
Delta	1787	793
Frontier	43	17
Midwest Express	72	31
Northwest	1089	485
Southwest	898	357
United	1912	895
US Air	1045	423

A linear function that approximates this data is given by $f(x) = 0.46175x - 17.84843$, where $f(x)$ is the number of miles flown and x is the number of hours flown.

a. Use the function to interpolate the number of miles an airline whose planes operated for 1025 hours would fly.

b. Use the function to extrapolate the number of miles an airline whose planes operated for 2100 hours would fly.

In Exercises 55 to 58, determine the profit function for the given revenue function and cost function. Also determine the break-even point.

55. $R(x) = 92.50x$; $C(x) = 52x + 1782$

56. $R(x) = 124x$; $C(x) = 78.5x + 5005$

57. $R(x) = 259x$; $C(x) = 180x + 10,270$

58. $R(x) = 14,220x$; $C(x) = 8010x + 1,602,180$

59. **MARGINAL COST** In business, *marginal cost* is a phrase used to represent the rate of change or slope of a cost function that relates the cost C to the number of units x produced. If a cost function is given by $C(x) = 8x + 275$, find

a. $C(0)$ b. $C(1)$ c. $C(10)$ d. marginal cost

60. **MARGINAL REVENUE** In business, *marginal revenue* is a phrase used to represent the rate of change or slope of a revenue function that relates the revenue R to the number of units x sold. If a revenue function is given by the function $R(x) = 210x$, find

a. $R(0)$ b. $R(1)$ c. $R(10)$ d. marginal revenue

61. **BREAK-EVEN POINT FOR A RENTAL TRUCK** A rental company purchases a truck for \$19,500. The truck requires an average of \$6.75 per day in maintenance.

a. Find the linear function that expresses the total cost C of owning the truck after t days.

b. The truck rents for \$55.00 a day. Find the linear function that expresses the revenue R when the truck has been rented for t days.

c. The profit after t days, $P(t)$, is given by the function $P(t) = R(t) - C(t)$. Find the linear function $P(t)$.

d. Use the function $P(t)$ that you obtained in **c.** to determine how many days it will take the company to break even on the purchase of the truck. Assume that the truck is always in use.

62. **BREAK-EVEN POINT FOR A PUBLISHER** A magazine company had a profit of \$98,000 per year when it had 32,000 subscribers. When it obtained 35,000 subscribers, it had a profit of \$117,500. Assume that the profit P is a linear function of the number of subscribers s.

a. Find the function P.

b. What will the profit be if the company obtains 50,000 subscribers?

c. What is the number of subscribers needed to break even?

63. **LINEAR MODEL IN SPORTS STATISTICS** The table below shows the number of high school girls participating in sports. (Source: Information Please web site)

High School Girls Sports Participation

Year	Number, in millions
1991	1.86
1992	1.89
1993	1.94
1994	2.00
1995	2.13
1996	2.24
1997	2.37
1998	2.47

a. Using $t = 1$ for 1991 and $t = 8$ for 1998, find the equation of a linear model of this data, given that the graph of this line passes through $(1, 1.86)$ and $(8, 2.47)$.

b. How many high school girls does the model you found in part **a.** predict for the year 2002?

64. **LINEAR MODEL IN SPORTS STATISTICS** The table below shows the number of home runs hit and the number of runs batted in (RBI) by all National League baseball teams in 1999. (Source: CBS Sportsline web site)

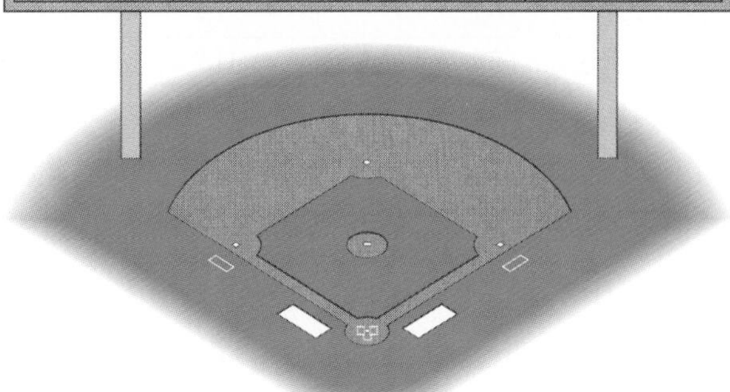

Baseball Statistics

Home runs	Runs batted in	Home runs	Runs batted in
216	865	165	777
197	791	163	680
189	717	181	814
209	820	161	797
223	863	170	717
128	655	153	671
168	784	188	828
187	761	194	763

a. Find the equation of a linear model of this data, given that the graph of the line passes through $P_1(128, 655)$ and $P_2(223, 863)$.

b. How many runs batted in does the model you found in part **a.** predict for a team that has 175 home runs?

65. **HEALTH** Framingham Heart Study is an ongoing research project that is attempting to identify risk factors associated with heart disease. Selected blood pressure data from that study is shown in the table below.

Selected Framingham Blood Pressure Statistics

Diastolic	Systolic
100	135
88	154
80	110
70	110
80	114
108	180
85	135
75	115

a. Find the equation of a linear model of this data, given that the graph of the line passes through $P_1(70, 110)$ and $P_2(108, 180)$.

b. What systolic blood pressure does the model you found in part **a.** predict for a diastolic pressure of 90?

66. **HEALTH** The table below shows average remaining lifetime, by age, of all people in the United States in 1997. (Source: National Institutes of Health)

Average Remaining Lifetime by Age in the United States

Current Age	Remaining Years
0	76.5
15	62.3
35	43.4
65	17.7
75	11.2

a. Find the equation of a linear model of this data, given that the graph of the line passes through (0, 76.5) and (75, 11.2).

b. Based on your model, what is the average remaining lifetime of a person whose current age is 25?

In Exercises 67 to 70, find the equation of the indicated line. Write the equation in the form $y = mx + b$.

67. Through (1, 3) and parallel to $3x + 4y = -24$

68. Through (2, −1) and parallel to $x + y = 10$

69. Through (1, 2) and perpendicular to $x + y = 4$

70. Through (−3, 4) and perpendicular to $2x - y = 7$

71. **POINT OF IMPACT** A rock is whirled horizontally, in a counterclockwise circular path with radius 5 feet, about the origin. When the string breaks, the rock travels on a linear path perpendicular to the radius $\overline{OP}$ and hits a wall located at $y = 10$ feet.

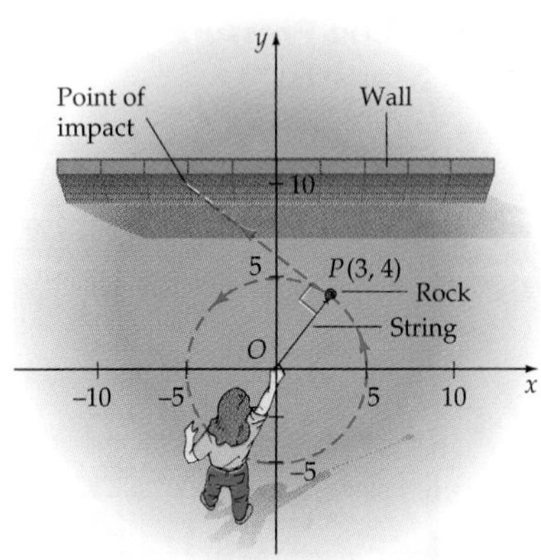

If the string breaks when the rock is at P(3 feet, 4 feet), find the x-coordinate of the point where the rock hits the wall.

72. **POINT OF IMPACT** A rock is whirled horizontally, in a counterclockwise circular path with radius 4 feet, about the origin. When the string breaks, the rock travels on a linear path perpendicular to the radius $\overline{OP}$ and hits a wall located at $y = 14$ feet. If the string breaks when the rock is at $P(\sqrt{15}$ feet, 1 foot$)$, find the x-coordinate of the point where the rock hits the wall.

73. **SLOPE OF A SECANT LINE** The graph of $y = x^2 + 1$ is

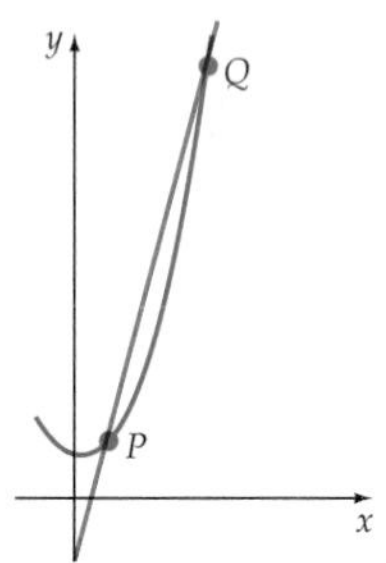

$P(2, 5)$ and $Q(2 + h, [2 + h]^2 + 1)$ are points on the graph.

a. If $h = 1$, determine the coordinates of Q and the slope of the line PQ.

b. If $h = 0.1$, determine the coordinates of Q and the slope of the line PQ.

c. If $h = 0.01$, determine the coordinates of Q and the slope of the line PQ.

d. As h approaches 0, what value does the slope of the line PQ seem to be approaching?

e. Verify that the slope of the line passing through (2, 5) and $(2 + h, [2 + h]^2 + 1)$ is $4 + h$.

74. **SLOPE OF A SECANT LINE** The graph of $y = 3x^2$ is shown

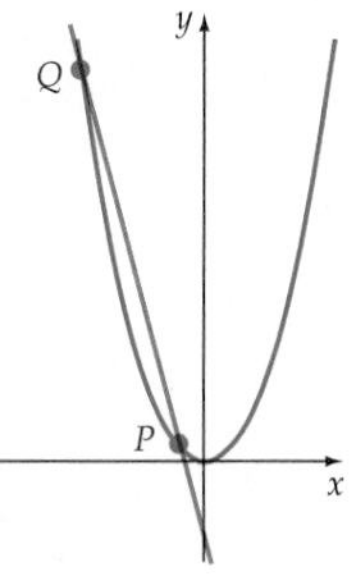

$P(-1, 3)$ and $Q(-1 + h, 3[-1 + h]^2)$ are points on the graph.

a. If $h = 1$, determine the coordinates of Q and the slope of the line PQ.

b. If $h = 0.1$, determine the coordinates of Q and the slope of the line PQ.

c. If $h = 0.01$, determine the coordinates of Q and the slope of the line PQ.

d. As h approaches 0, what value does the slope of the line PQ seem to be approaching?

e. Verify that the slope of the line passing through (−1, 3) and $(-1 + h, 3[-1 + h]^2)$ is $-6 + 3h$.

75. Verify that the slope of the line passing through (x, x^2) and $(x + h, [x + h]^2)$ is $2x + h$.

76. Verify that the slope of the line passing through $(x, 4x^2)$ and $(x + h, 4[x + h]^2)$ is $8x + 4h$.

SUPPLEMENTAL EXERCISES

77. The Two-Point Form Use the point-slope form to derive the following equation, which is called the two-point form.

$$y - y_1 = \left(\frac{y_2 - y_1}{x_2 - x_1}\right)(x - x_1)$$

78. The Intercept Form Use the two-point form from Exercise 77 to show that the line with intercepts $(a, 0)$ and $(0, b)$, $a \neq 0$ and $b \neq 0$, has the equation

$$\frac{x}{a} + \frac{y}{b} = 1$$

In Exercises 79 and 80, use the two-point form to find an equation of the line that passes through the indicated points. Write your answers in slope-intercept form.

79. $(5, 1)$, $(4, 3)$

80. $(2, 7)$, $(-1, 6)$

In Exercises 81 to 84, use the equation from Exercise 78 (called the intercept form) to write an equation of a line with the indicated intercepts.

81. x-intercept $(3, 0)$, y-intercept $(0, 5)$

82. x-intercept $(-2, 0)$, y-intercept $(0, 7)$

83. x-intercept $(a, 0)$, y-intercept $(0, 3a)$, point on the line $(5, 2)$, $a \neq 0$

84. x-intercept $(-b, 0)$, y-intercept $(0, 2b)$, point on the line $(-3, 10)$, $b \neq 0$

85. Verify that the slope of the line passing through $(1, 3)$ and $(1 + h, 3[1 + h]^3)$ is $9 + 9h + 3h^2$.

86. Find the two points on the circle given by $x^2 + y^2 = 25$ such that the slope of the radius from $(0, 0)$ to each point is 0.5.

87. Find a point $P(x, y)$ on the graph of the equation $y = x^2$ such that the slope of the line through the point $(3, 9)$ and P is $15/2$.

88. Determine whether there is a point $P(x, y)$ on the graph of the equation $y = \sqrt{x + 1}$ such that the slope of the line through the point $(3, 2)$ and P is $3/8$.

PROJECTS

1. VISUAL INSIGHT

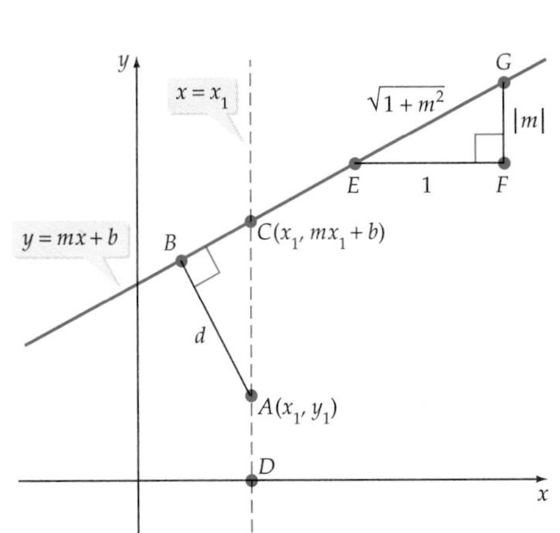

The distance d between the point $A(x_1, y_1)$ and the line $y = mx + b$ is

$$d = \frac{|mx_1 + b - y_1|}{\sqrt{1 + m^2}}$$

Write a paragraph that explains how to make use of the figure at the bottom of page 179 to verify the formula for the distance d.

2. **VERIFY GEOMETRIC THEOREMS**

   a. Prove that in any triangle, the line segment that joins the midpoints of two sides of the triangle is parallel to the third side. (*Hint:* Assign coordinates to the vertices of the triangle as shown in the figure at the left below.)

   b. Prove that in any square, the diagonals are perpendicular bisectors of each other. (*Hint:* Assign coordinates to the vertices of the square as shown in the figure at the right below.)

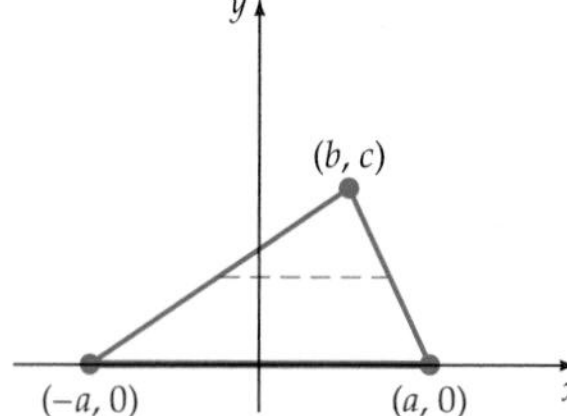

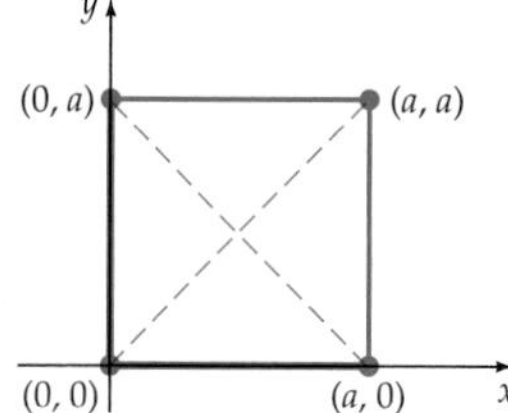

SECTION 2.4 QUADRATIC FUNCTIONS

- VERTEX OF A PARABOLA
- APPLICATIONS

Some applications can be modeled by a *quadratic function.*

Definition of a Quadratic Function

A **quadratic function** of x is a function that can be represented by an equation of the form

$$f(x) = ax^2 + bx + c$$

where a, b, and c are real numbers and $a \neq 0$.

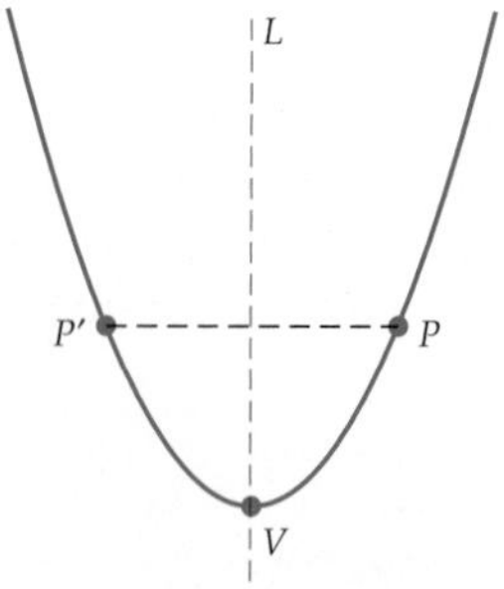

Figure 2.54

The graph of $f(x) = ax^2 + bx + c$ is a *parabola*. The graph opens up if $a > 0$, and it opens down if $a < 0$. The **vertex of a parabola** is the lowest point on a parabola that opens up or the highest point on a parabola that opens down. Point V is the vertex of the parabola in Figure 2.54.

The graph of $f(x) = ax^2 + bx + c$ is *symmetric* with respect to a vertical line through its vertex.

Definition of Symmetry with Respect to a Line

A graph is **symmetric with respect to a line** L if for each point P on the graph there is a point P' on the graph such that the line L is the perpendicular bisector of the line segment PP'.

take note

The axis of symmetry is a line. When asked to determine the axis of symmetry, the answer is an equation, not just a number.

In Figure 2.54, the parabola is symmetric with respect to the line L. The line L is called the **axis of symmetry.** The points P and P' are reflections or images of each other with respect to the axis of symmetry.

If $b = 0$ and $c = 0$, then $f(x) = ax^2 + bx + c$ simplifies to $f(x) = ax^2$. The graph of $f(x) = ax^2$ ($a \neq 0$) is a parabola with its vertex at the origin, and the y-axis is its axis of symmetry. The graph of $f(x) = ax^2$ can be constructed by plotting a few points and drawing a smooth curve that passes through these points with the origin as the vertex and the y-axis as its axis of symmetry. The graphs of $f(x) = x^2$, $g(x) = 2x^2$, and $h(x) = -\dfrac{1}{2}x^2$ are shown in Figure 2.55.

take note

The equation $z = x^2 - y^2$ defines z as a quadratic function of x and y. You might think that the graph of every quadratic function is a parabola. However, the graph of $z = x^2 - y^2$ is the *saddle* shown in the figure below. You will study quadratic functions involving two or more independent variables in calculus.

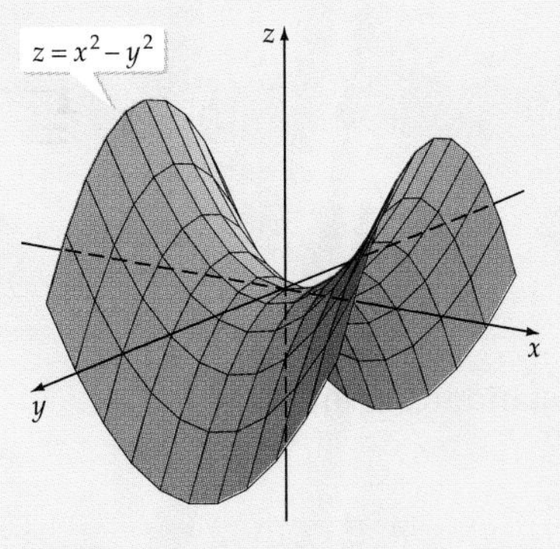

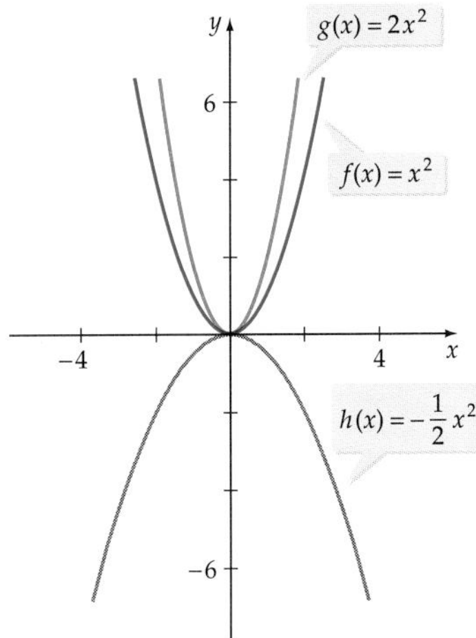

Figure 2.55

Standard Form of Quadratic Functions

Every quadratic function f given by $f(x) = ax^2 + bx + c$ can be written in the **standard form of a quadratic function:**

$$f(x) = a(x - h)^2 + k, \quad a \neq 0$$

The graph of f is a parabola with vertex (h, k). The parabola opens up if $a > 0$, and it opens down if $a < 0$. The vertical line $x = h$ is the axis of symmetry of the parabola.

The standard form is useful because it readily gives information about the vertex of the parabola and its axis of symmetry. For example, note the graph of $f(x) = 2(x - 4)^2 - 3$ is a parabola. The coordinates of the vertex are $(4, -3)$, and the line $x = 4$ is its axis of symmetry. Because a is the positive number 2, the parabola opens upward.

EXAMPLE 1 Find the Standard Form of a Quadratic Function

Use the technique of completing the square to find the standard form of $g(x) = 2x^2 - 12x + 19$. Sketch the graph.

Continued

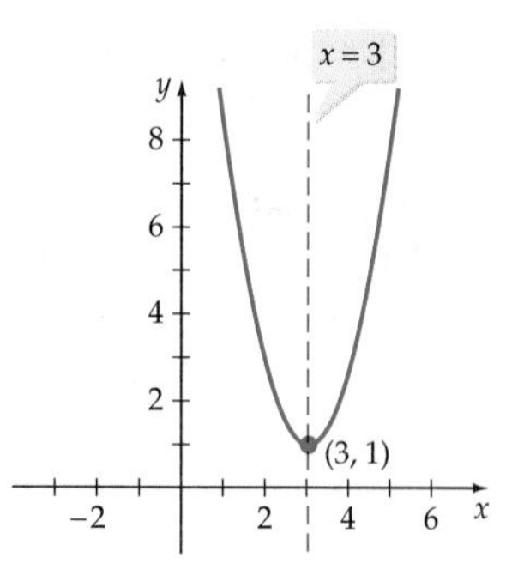

$g(x) = 2x^2 - 12x + 19$

Figure 2.56

Solution

$$g(x) = 2x^2 - 12x + 19$$
$$= 2(x^2 - 6x) + 19 \qquad \text{• Factor 2 from the variable terms.}$$
$$= 2(x^2 - 6x + 9 - 9) + 19 \qquad \text{• Complete the square.}$$
$$= 2(x^2 - 6x + 9) - 2(9) + 19 \qquad \text{• Regroup.}$$
$$= 2(x - 3)^2 - 18 + 19 \qquad \text{• Factor and simplify.}$$
$$= 2(x - 3)^2 + 1 \qquad \text{• Standard form.}$$

The vertex is (3, 1). The axis of symmetry is $x = 3$. Because $a > 0$, the parabola opens up. See **Figure 2.56**.

TRY EXERCISE 10, EXERCISE SET 2.4, PAGE 188

take note

A web applet is available to explore the graph a vertex of a parabola. This applet, Quadratics: Vertex Form, can be found on our web site at

http://college.hmco.com

◆ VERTEX OF A PARABOLA

By completing the square of $ax^2 + bx + c$, the x-coordinate of the vertex of the graph of $f(x) = ax^2 + bx + c$ can be shown to be $-\frac{b}{2a}$. The y-coordinate of the vertex is $f\left(-\frac{b}{2a}\right)$. This result is summarized by the following formula.

Vertex Formula

The vertex of the graph of $f(x) = ax^2 + bx + c$ is $\left(-\frac{b}{2a}, f\left(-\frac{b}{2a}\right)\right)$.

take note

The vertex formula can be used to write the standard form of a quadratic function. Use the formulas

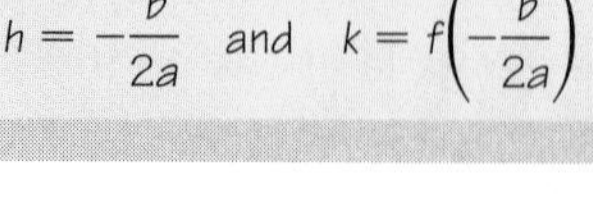

$$h = -\frac{b}{2a} \quad \text{and} \quad k = f\left(-\frac{b}{2a}\right)$$

EXAMPLE 2 Find the Vertex and Standard Form of a Quadratic Function

Use the vertex formula to find the vertex and standard form of $f(x) = 2x^2 - 8x + 3$. See **Figure 2.57**.

Solution

$$f(x) = 2x^2 - 8x + 3 \qquad \text{• } a = 2, b = -8, c = 3$$
$$h = -\frac{b}{2a} = -\frac{-8}{2(2)} = 2 \qquad \text{• } x\text{-coordinate of the vertex}$$
$$k = f\left(-\frac{b}{2a}\right) = 2(2)^2 - 8(2) + 3 = -5 \qquad \text{• } y\text{-coordinate of the vertex}$$

The vertex is (2, −5). Substituting into the standard form $f(x) = a(x - h)^2 + k$ yields the standard form $f(x) = 2(x - 2)^2 - 5$.

$f(x) = 2x^2 - 8x + 3$

Figure 2.57

TRY EXERCISE 20, EXERCISE SET 2.4, PAGE 189

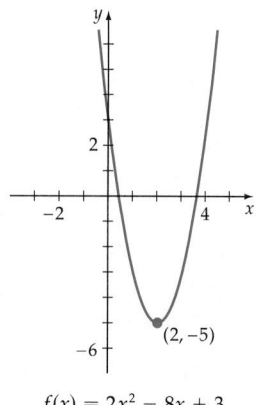

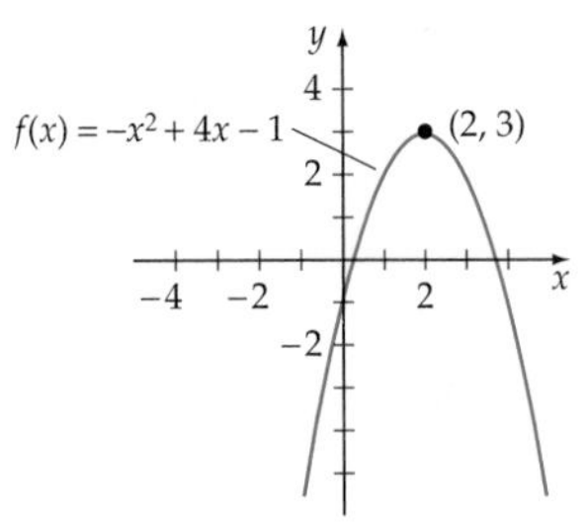

Note from Example 2 that graph of the parabola opens up and the vertex is the *lowest* point on the graph of the parabola. Therefore, the y-coordinate of the vertex is the *minimum* value of that function. This information can be used to determine the range of $f(x) = 2x^2 - 8x + 3$. The range is $\{y \mid y \geq -5\}$. Similarly, if the graph of a parabola opened down, the vertex would be the *highest* point on the graph and the y-coordinate of the vertex would be the *maximum* value of the function. For instance, the maximum value of the $f(x) = -x^2 + 4x - 1$, graphed at the left is 3, the y-coordinate of the vertex. The range of the function is $\{y \mid y \leq 3\}$. For the function in Example 2 and the function whose graph is shown at the left, the domain is the set of real numbers.

EXAMPLE 3 Find the Range of $f(x) = ax^2 + bx + c$

Find the range of $f(x) = -2x^2 - 6x - 1$. Determine the values of x for which $f(x) = 3$.

Solution

To find the range of f, determine the y-coordinate of the vertex of the graph of f.

$f(x) = -2x^2 - 6x - 1$ • $a = -2, b = -6, c = -1$

$h = -\frac{b}{2a} = -\frac{-6}{2(-2)} = -\frac{3}{2}$ • Find the x-coordinate of the vertex.

$k = f\left(-\frac{3}{2}\right) = -2\left(-\frac{3}{2}\right)^2 - 6\left(\frac{-3}{2}\right) - 1 = \frac{7}{2}$ • Find the y-coordinate of the vertex.

The vertex is $\left(-\frac{3}{2}, \frac{7}{2}\right)$. Because the parabola opens down, $\frac{7}{2}$ is the maximum value of f. Therefore, the range of f is $\left\{y \mid y \leq \frac{7}{2}\right\}$.

To determine the values of x for which $f(x) = 3$, replace $f(x)$ by $-2x^2 - 6x - 1$ and solve for x.

$f(x) = 3$

$-2x^2 - 6x - 1 = 3$ • Replace $f(x)$ by $-2x^2 - 6x - 1$.

$-2x^2 - 6x - 4 = 0$ • Solve for x.

$-2(x + 1)(x + 2) = 0$ • Factor

$x + 1 = 0$ or $x + 2 = 0$ • Use the Principle of Zero Products to solve for x.

$x = -1 \qquad x = -2$

The values of x for which $f(x) = 3$ are -1 and -2.

Visualize the Solution

The graph of f is shown below. The vertex of the graph is $\left(-\frac{3}{2}, \frac{7}{2}\right)$. Note that the line $y = 3$ intersects the graph of f when $x = -2$ and when $x = -1$.

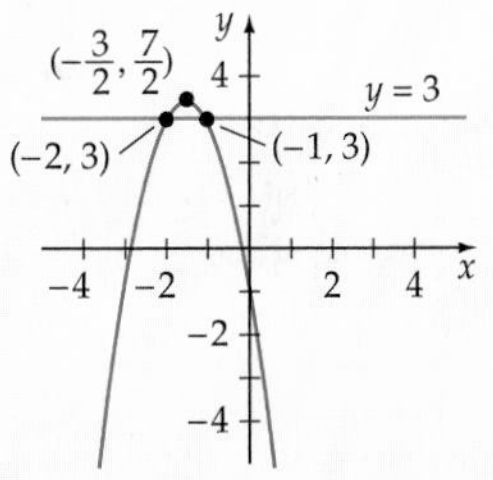

TRY EXERCISE 32, EXERCISE SET 2.4, PAGE 189

The following theorem can be used to determine the maximum value or the minimum value of a quadratic function.

take note

A web applet is available to explore the maximum and minimum of a quadratic function. These applets, Maximize Area and Minimize Perimeter, can be found on our web site at

http://college.hmco.com

Maximum or Minimum Value of a Quadratic Function

If $a > 0$, then the vertex (h, k) is the lowest point on the graph of $f(x) = a(x - h)^2 + k$, and the y-coordinate k of the vertex is the **minimum value** of the function f. See **Figure 2.58a.**

If $a < 0$, then the vertex (h, k) is the highest point on the graph of $f(x) = a(x - h)^2 + k$, and the y-coordinate k is the **maximum value** of the function f. See **Figure 2.58b.**

In either case, the maximum or minimum is achieved when $x = h$.

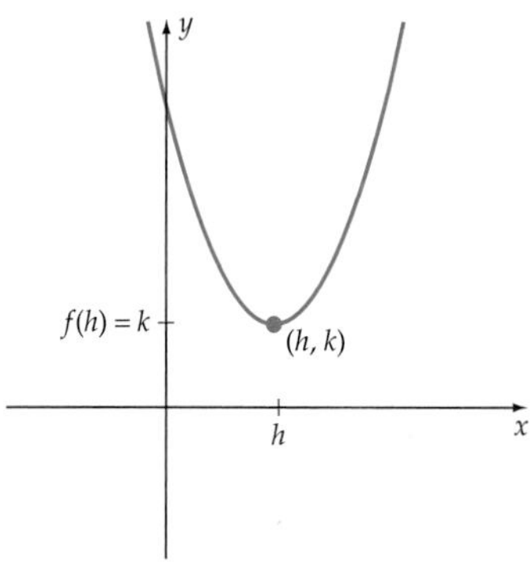

a. k is the minimum value of f.

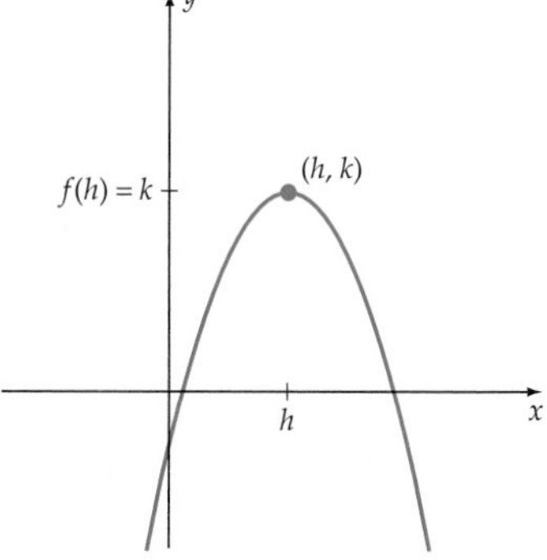

b. k is the maximum value of f.

Figure 2.58

EXAMPLE 4 Find the Maximum or Minimum of a Quadratic Function

Find the maximum or minimum value of each quadratic function. State whether the value is a maximum or a minimum.

a. $F(x) = -2x^2 + 8x - 1$ **b.** $G(x) = x^2 - 3x + 1$

Solution

The maximum or minimum value of a quadratic function is the y-coordinate of the vertex of the graph of the function.

a. $h = -\dfrac{b}{2a} = -\dfrac{8}{2(-2)} = 2$ • **x-coordinate of the vertex**

$k = F\left(-\dfrac{b}{2a}\right) = -2(2)^2 + 8(2) - 1 = 7$ • **y-coordinate of the vertex**

Because $a < 0$, the function has a maximum value. The maximum value is 7. See **Figure 2.59.**

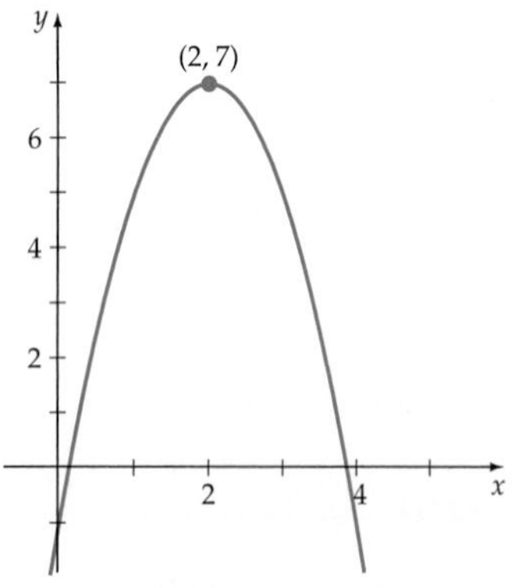

$F(x) = -2x^2 + 8x - 1$

Figure 2.59

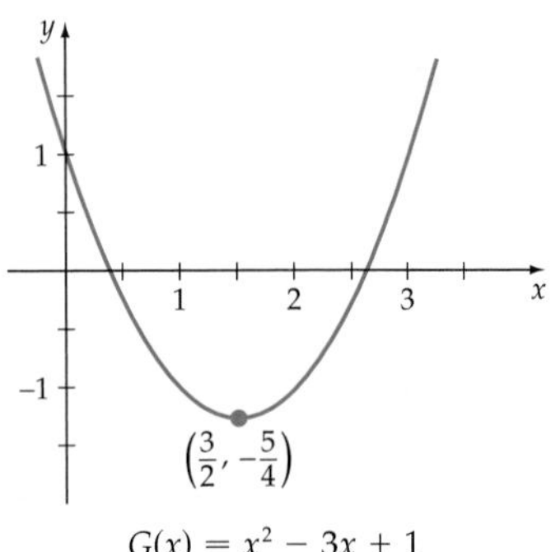

$G(x) = x^2 - 3x + 1$

Figure 2.60

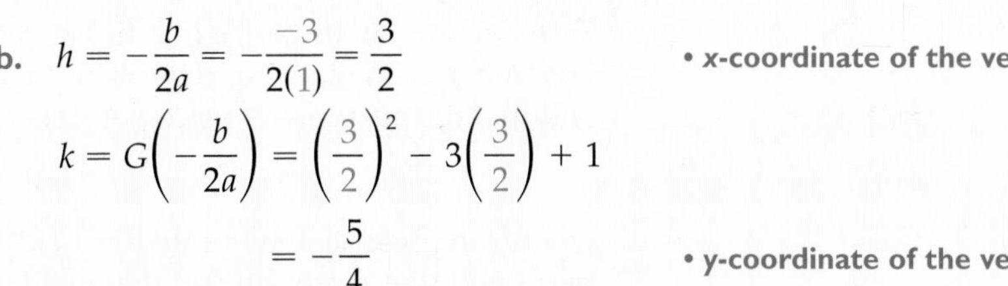

b. $h = -\dfrac{b}{2a} = -\dfrac{-3}{2(1)} = \dfrac{3}{2}$ • *x*-coordinate of the vertex

$$k = G\left(-\frac{b}{2a}\right) = \left(\frac{3}{2}\right)^2 - 3\left(\frac{3}{2}\right) + 1$$

$$= -\frac{5}{4}$$ • *y*-coordinate of the vertex

Because $a > 0$, the function has a minimum value. The minimum value is $-5/4$. See **Figure 2.60.**

TRY EXERCISE 36, EXERCISE SET 2.4, PAGE 189

◆ APPLICATIONS

EXAMPLE 5 Find the Maximum of a Quadratic Function

A long sheet of tin 20 inches wide is to be made into a trough by bending up two sides until they are perpendicular to the bottom. How many inches should be turned up so that the trough will achieve its maximum carrying capacity?

Solution

Figure 2.61

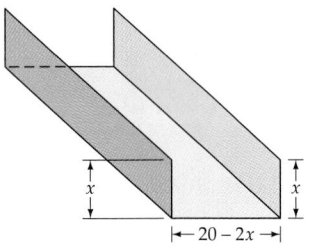

The trough is shown in **Figure 2.61.** If x is the number of inches to be turned up on each side, then the width of the base is $20 - 2x$ inches. The maximum carrying capacity of the trough will occur when the cross-sectional area is a maximum. The cross-sectional area $A(x)$ is given by

$$A(x) = x(20 - 2x)$$ • Area = (length)(width)

$$= -2x^2 + 20x$$

To find when A obtains its maximum value, find the x-coordinate of the vertex of the graph of A. Using the vertex formula with $a = -2$ and $b = 20$, we have

$$x = -\frac{b}{2a} = -\frac{20}{2(-2)} = 5$$

Therefore, the maximum carrying capacity will be achieved when 5 inches are turned up. See **Figure 2.62.**

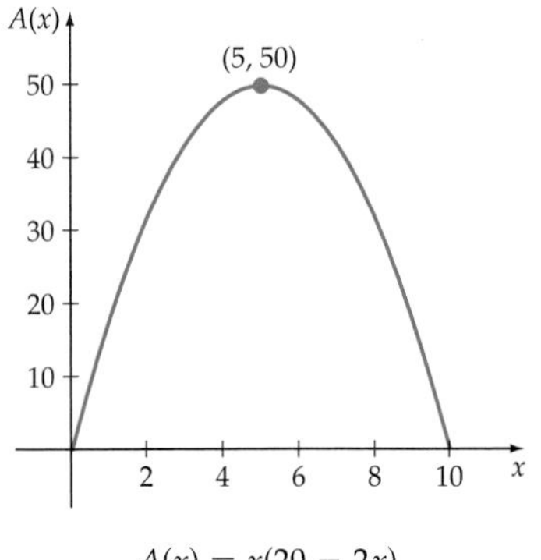

$A(x) = x(20 - 2x)$

Figure 2.62

TRY EXERCISE 46, EXERCISE SET 2.4, PAGE 189

EXAMPLE 6 Solve a Business Application

The owners of a travel agency have determined that they can sell all 160 tickets for a tour if they charge \$8 (their cost) for each ticket. For each

Continued ➤

\$0.25 increase in the price of a ticket, they estimate they will sell 1 ticket less. A business manager determines that their cost function is $C(x) = 8x$ and that the customer's price per ticket is

$$p(x) = 8 + 0.25(160 - x) = 48 - 0.25x$$

where x represents the number of tickets sold. Determine the maximum profit and the cost per ticket that yields the maximum profit.

Solution

The profit from selling x tickets is $P(x) = R(x) - C(x)$, where P, R, and C are the profit function, the revenue function, and the cost function as defined in Section 2.3. Thus

$$\begin{aligned} P(x) &= R(x) - C(x) \\ &= x[p(x)] - C(x) \\ &= x(48 - 0.25x) - 8x \\ &= 40x - 0.25x^2 \end{aligned}$$

$P(x) = 40x - 0.25x^2$

Figure 2.63

The graph of the profit function is a parabola that opens down. Thus the maximum profit occurs when

$$x = -\frac{b}{2a} = -\frac{40}{2(-0.25)} = 80$$

The maximum profit is determined by evaluating $P(x)$ with $x = 80$.

$$P(80) = 40(80) - 0.25(80)^2 = 1600$$

The maximum profit is \$1600.
To find the price per ticket that yields the maximum profit, we evaluate $p(x)$ with $x = 80$.

$$p(80) = 48 - 0.25(80) = 28$$

Thus the travel agency can expect a maximum profit of \$1600 when 80 people take the tour at a ticket price of \$28 per person. The graph of the profit function is shown in **Figure 2.63.**

QUESTION In **Figure 2.63,** why have we shown only the portion of the graph that lies in Quadrant I?

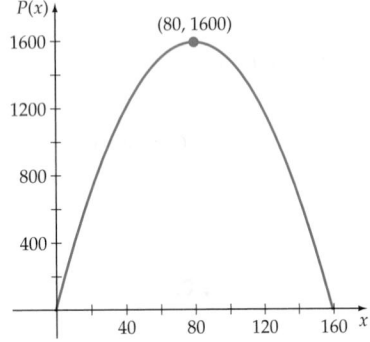

TRY EXERCISE 64, EXERCISE SET 2.4, PAGE 190

EXAMPLE 7 Solve a Projectile Application

In **Figure 2.64,** a ball is thrown vertically upward with an initial velocity of 48 feet per second. If the ball started its flight at a height of 8 feet, then its height s at time t can be determined by $s(t) = -16t^2 + 48t + 8$, where $s(t)$ is

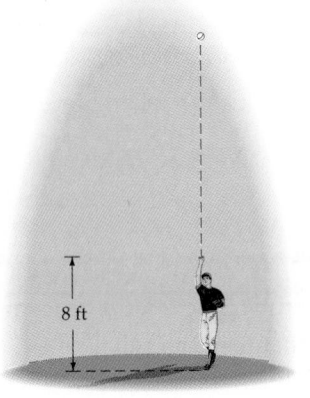

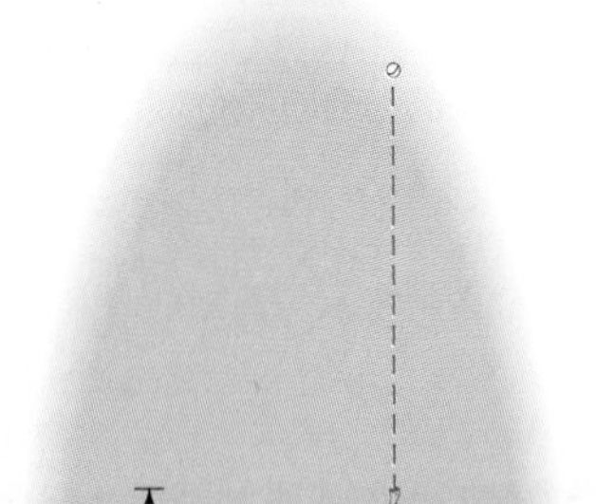

Figure 2.64

ANSWER Since x represents the number of tickets sold, x must be greater than or equal to zero but less than or equal to 160. $P(x)$ is nonnegative for $0 \le x \le 160$.

measured in feet above ground level and t is the number of seconds of flight.

a. Determine the time it takes the ball to attain its maximum height.

b. Determine the maximum height the ball attains.

c. Determine the time it takes the ball to hit the ground.

Solution

a. The graph of $s(t) = -16t^2 + 48t + 8$ is a parabola that opens downward. See Figure 2.65. Therefore, s will attain its maximum value at the vertex of its graph. Using the vertex formula with $a = -16$ and $b = 48$, we get

$$t = -\frac{b}{2a} = -\frac{48}{2(-16)} = \frac{3}{2}$$

Therefore, the ball attains its maximum height $1\frac{1}{2}$ seconds into its flight.

b. When $t = 3/2$, the height of the ball is

$$s\left(\frac{3}{2}\right) = -16\left(\frac{3}{2}\right)^2 + 48\left(\frac{3}{2}\right) + 8 = 44 \text{ feet}$$

c. The ball will hit the ground when its height $s(t) = 0$. Therefore, solve $-16t^2 + 48t + 8 = 0$ for t.

$$-16t^2 + 48t + 8 = 0$$

$$-2t^2 + 6t + 1 = 0$$

• **Divide each side by 8.**

$$t = \frac{-(6) \pm \sqrt{6^2 - 4(-2)(1)}}{2(-2)}$$

• **Use the quadratic formula.**

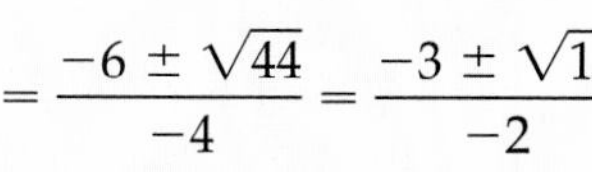

$$= \frac{-6 \pm \sqrt{44}}{-4} = \frac{-3 \pm \sqrt{11}}{-2}$$

Using a calculator to approximate the positive root, we find that the ball will hit the ground in approximately $t \approx 3.16$ seconds. This is also the value of the t-coordinate of the t-intercept in Figure 2.65.

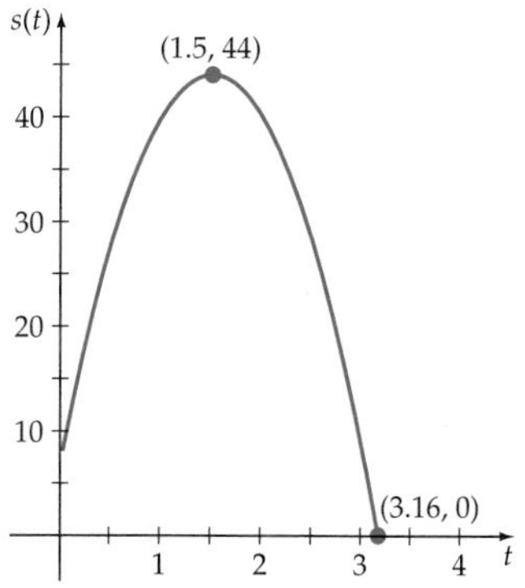

$s(t) = -16t^2 + 48t + 8$

Figure 2.65

take note

The graph of s in Figure 2.65 is a graph of the height of the ball at time t. It is *not* a graph of the path of the ball.

TRY EXERCISE 66, EXERCISE SET 2.4, PAGE 191

TOPICS FOR DISCUSSION

1. A classmate states that the graph of every quadratic function of the form

$$f(x) = ax^2 + bx + c$$

has a y-intercept. Do you agree? Explain.

2. The graph of $f(x) = -x^2 + 6x + 11$ has a vertex of $(3, 20)$. Is this vertex point the highest point or the lowest point on the graph of f?

3. A tutor states that the graph of $f(x) = ax^2 + bx + c$ $(a \neq 0)$ is a parabola and that its axis of symmetry is $y = -\dfrac{b}{2a}$. Do you agree?

4. Every quadratic function of the form $f(x) = ax^2 + bx + c$ has a domain of all real numbers. Do you agree?

5. A classmate states that the graph of every quadratic function of the form

$$f(x) = ax^2 + bx + c$$

must contain points from at least two quadrants. Do you agree?

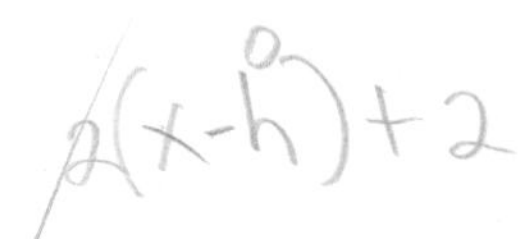

EXERCISE SET 2.4

In Exercises 1 to 8, match each graph in *a.* through *h.* with the proper quadratic function.

1. $f(x) = x^2 - 3$
2. $f(x) = x^2 + 2$
3. $f(x) = (x - 4)^2$
4. $f(x) = (x + 3)^2$
5. $f(x) = -2x^2 + 2$
6. $f(x) = -\frac{1}{2}x^2 + 3$
7. $f(x) = (x + 1)^2 + 3$
8. $f(x) = -2(x - 2)^2 + 2$

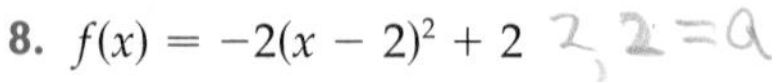

a.

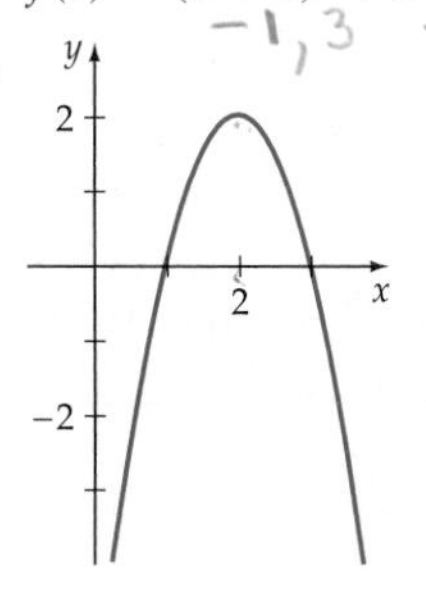

b.

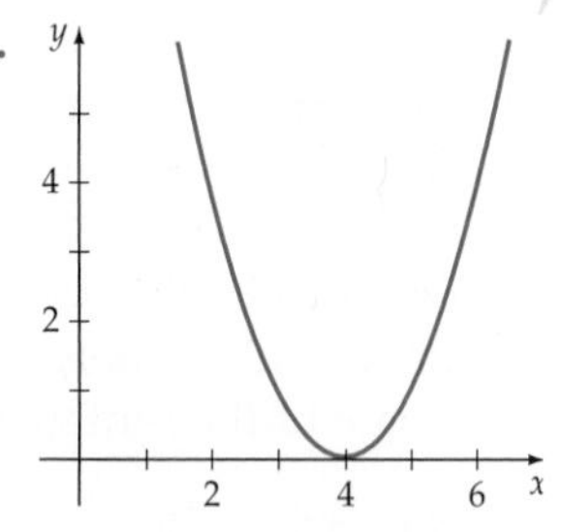

c.

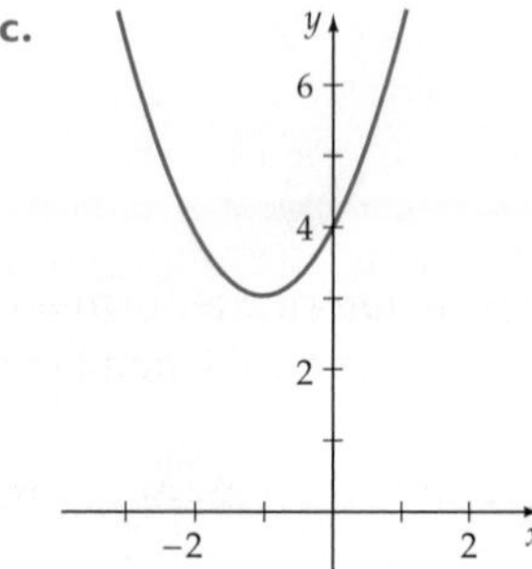

d.

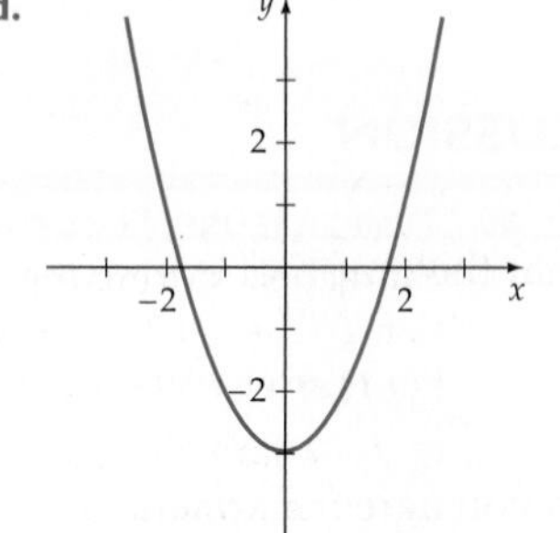

e.

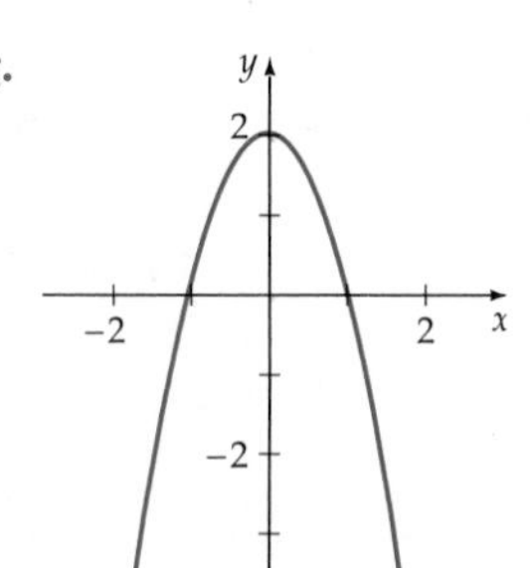

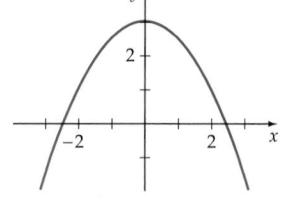

f.

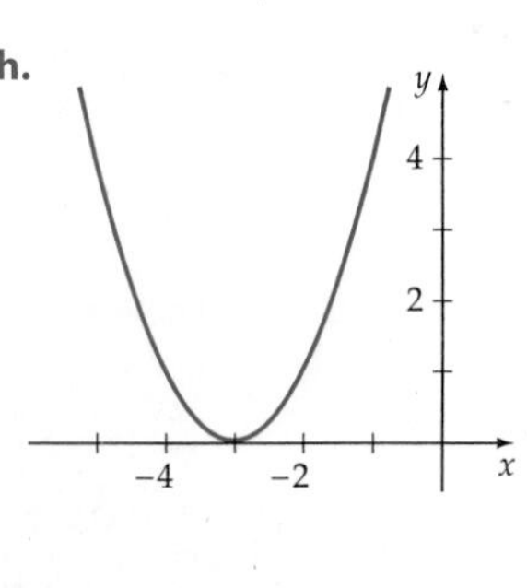

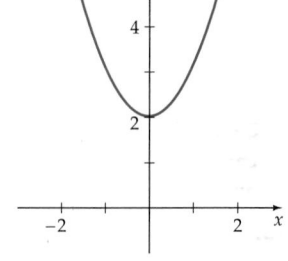

g.

h.

In Exercises 9 to 18, use the method of completing the square to find the standard form of the quadratic function, and then sketch its graph. Label its vertex and axis of symmetry.

9. $f(x) = x^2 + 4x + 1$
10. $f(x) = x^2 + 6x - 1$
11. $f(x) = x^2 - 8x + 5$
12. $f(x) = x^2 - 10x + 3$
13. $f(x) = x^2 + 3x + 1$
14. $f(x) = x^2 + 7x + 2$

15. $f(x) = -x^2 + 4x + 2$

16. $f(x) = -x^2 - 2x + 5$

17. $f(x) = -3x^2 + 3x + 7$

18. $f(x) = -2x^2 - 4x + 5$

In Exercises 19 to 28, use the vertex formula to determine the vertex of the graph of the function and write the function in standard form.

19. $f(x) = x^2 - 10x$

20. $f(x) = x^2 - 6x$

21. $f(x) = x^2 - 10$

22. $f(x) = x^2 - 4$

23. $f(x) = -x^2 + 6x + 1$

24. $f(x) = -x^2 + 4x + 1$

25. $f(x) = 2x^2 - 3x + 7$

26. $f(x) = 3x^2 - 10x + 2$

27. $f(x) = -4x^2 + x + 1$

28. $f(x) = -5x^2 - 6x + 3$

29. Find the range of $f(x) = x^2 - 2x - 1$. Determine the values of x in the domain of f for which $f(x) = 2$.

30. Find the range of $f(x) = -x^2 - 6x - 2$. Determine the values of x in the domain of f for which $f(x) = 3$.

31. Find the range of $f(x) = -2x^2 + 5x - 1$. Determine the values of x in the domain of f for which $f(x) = 2$.

32. Find the range of $f(x) = 2x^2 + 6x - 5$. Determine the values of x in the domain of f for which $f(x) = 15$.

33. Is 3 in the range of $f(x) = x^2 + 3x + 6$? Explain your answer.

34. Is -2 in the range of $f(x) = -2x^2 - x + 1$? Explain your answer.

In Exercises 35 to 44, find the maximum or minimum value of the function. State whether this value is a maximum or a minimum.

35. $f(x) = x^2 + 8x$

36. $f(x) = -x^2 - 6x$

37. $f(x) = -x^2 + 6x + 2$

38. $f(x) = -x^2 + 10x - 3$

39. $f(x) = 2x^2 + 3x + 1$

40. $f(x) = 3x^2 + x - 1$

41. $f(x) = 5x^2 - 11$

42. $f(x) = 3x^2 - 41$

43. $f(x) = -\frac{1}{2}x^2 + 6x + 17$

44. $f(x) = -\frac{3}{4}x^2 - \frac{2}{5}x + 7$

45. HEIGHT OF AN ARCH The height of an arch is given by the equation

$$h(x) = -\frac{3}{64}x^2 + 27, \quad -24 \le x \le 24$$

where $|x|$ is the horizontal distance in feet from the center of the arch.

a. What is the maximum height of the arch?

b. What is the height of the arch 10 feet to the right of center?

c. How far from the center is the arch 8 feet tall?

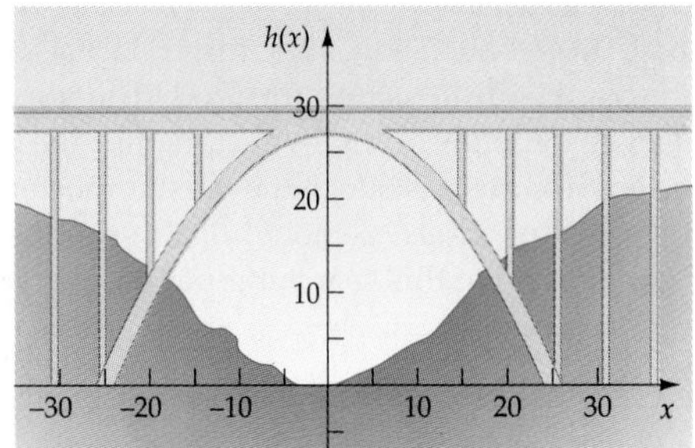

46. The sum of the length l and the width w of a rectangular area is 240 meters.

a. Write w as a function of l.

b. Write the area A as a function of l.

c. Find the dimensions that produce the greatest area.

47. RECTANGULAR ENCLOSURE A veterinarian uses 600 feet of chain-link fencing to enclose a rectangular region and also to subdivide the region into two smaller rectangular regions by placing a fence parallel to one of the sides, as shown in the figure.

a. Write the width w as a function of the length l.

b. Write the total area A as a function of l.

c. Find the dimensions that produce the greatest enclosed area.

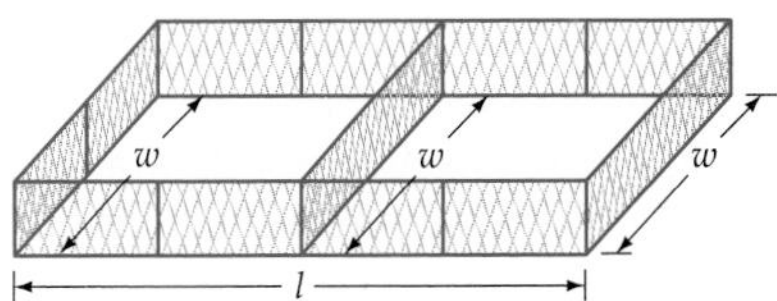

48. RECTANGULAR ENCLOSURE A farmer uses 1200 feet of fence to enclose a rectangular region and also to subdivide the region into three smaller rectangular regions by placing the fences parallel to one of the sides. Find the dimensions that produce the greatest enclosed area.

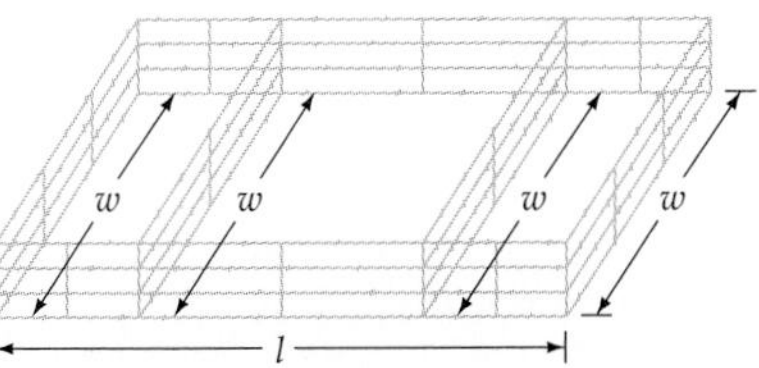

49. TEMPERATURE FLUCTUATIONS The temperature, $T(t)$ in Fahrenheit, during the day can be modeled by the equation $T(t) = -0.7t^2 + 9.4t + 59.3$, where t is the number of hours after 6:00 A.M.

a. At what time is the temperature a maximum? Round to the nearest minute.

b. What is the maximum temperature? Round to the nearest degree.

50. **LARVAE SURVIVAL** Soon after insect larvae are hatched, they must begin to search for food. The survival rate of the larvae depends on many factors, but the temperature of the environment is one of the most important. For a certain species of insect, a model of the number of larvae, $N(T)$, that survive this searching period is given by

$$N(T) = -0.6T^2 + 32.1T - 350$$

where T is the temperature in degrees Celsius.

a. At what temperature will the maximum number of larvae survive? Round to the nearest degree.

b. What is the maximum number of surviving larvae? Round to the nearest whole number.

c. Find the x-intercepts, to the nearest whole number, for the graph of this function.

d. Write a sentence that describes the meaning of the x-intercepts in the context of this problem.

51. **AUTOMOTIVE ENGINEERING** The fuel efficiency for a certain midsize car is given by

$$E(v) = -0.018v^2 + 1.476v + 3.4$$

where $E(v)$ is the fuel efficiency in miles per gallon for a car traveling v miles per hour.

a. What speed will yield the maximum fuel efficiency? Round to the nearest mile per hour.

b. What is the maximum fuel efficiency for this car? Round to the nearest mile per gallon.

52. **SPORTS** Some football fields are built in a parabolic mound shape so that water will drain off the field. A model for the shape of a certain field is given by

$$h(x) = -0.0002348x^2 + 0.0375x$$

where $h(x)$ is the height, in feet, of the field at a distance of x feet from one sideline. Find the maximum height of the field. Round to the nearest tenth of a foot.

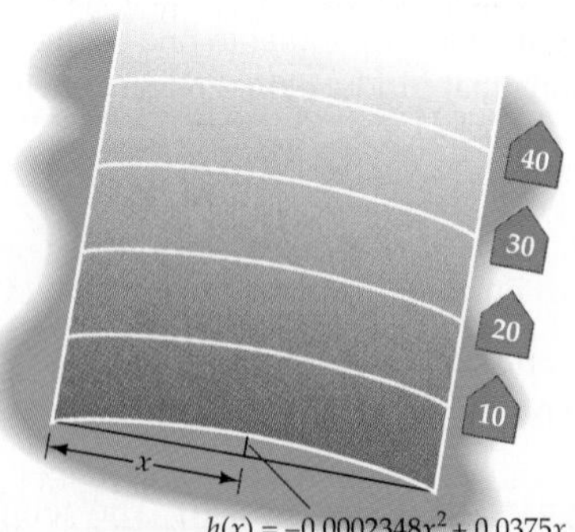

In Exercises 53 to 56, determine the y- and x-intercepts (if any) of the quadratic function.

53. $f(x) = x^2 + 6x$

54. $f(x) = -x^2 + 4x$

55. $f(x) = -3x^2 + 5x - 6$

56. $f(x) = 2x^2 + 3x + 4$

In Exercises 57 and 58, determine the number of units x that produce a maximum revenue for the given revenue function. Also determine the maximum revenue.

57. $R(x) = 296x - 0.2x^2$

58. $R(x) = 810x - 0.6x^2$

In Exercises 59 and 60, determine the number of units x that produce a maximum profit for the given profit function. Also determine the maximum profit.

59. $P(x) = -0.01x^2 + 1.7x - 48$

60. $P(x) = -\dfrac{x^2}{14{,}000} + 1.68x - 4000$

In Exercises 61 and 62, determine the profit function for the given revenue function and cost function. Also determine the break-even point(s).

61. $R(x) = x(102.50 - 0.1x)$; $C(x) = 52.50x + 1840$

62. $R(x) = x(210 - 0.25x)$; $C(x) = 78x + 6399$

63. **TOUR COST** A charter bus company has determined that its cost of providing x people a tour is

$$C(x) = 180 + 2.50x$$

A full tour consists of 60 people. The ticket price per person is \$15 plus \$0.25 for each unsold ticket. Determine

a. the revenue function

b. the profit function

c. the company's maximum profit

d the number of ticket sales that yields the maximum profit

64. **DELIVERY COST** An air freight company has determined that its cost, in dollars, of delivering x parcels per flight is

$$C(x) = 2025 + 7x$$

The price, in dollars, per parcel the company charges to send x parcels is

$$p(x) = 22 - 0.01x$$

Determine

a. the revenue function

b. the profit function

c. the company's maximum profit

d the price per parcel that yields the maximum profit

e. the minimum number of parcels the air freight company must ship to break even

65. **PROJECTILE** If the initial velocity of a projectile is 128 feet per second, then its height h in feet is a function of time t in seconds, given by the equation $h(t) = -16t^2 + 128t$.

a. Find the time t when the projectile achieves its maximum height.

b. Find the maximum height of the projectile.

c. Find the time t when the projectile hits the ground.

66. **PROJECTILE** The height in feet of a projectile with an initial velocity of 64 feet per second and an initial height of 80 feet is a function of time t in seconds, given by

$$h(t) = -16t^2 + 64t + 80$$

a. Find the maximum height of the projectile.

b. Find the time t when the projectile achieves its maximum height.

c. Find the time t when the projectile has a height of 0 feet.

67. **FIRE MANAGEMENT** The height of a stream of water from the nozzle of the fire hose can be modeled by

$$y(x) = -0.014x^2 + 1.19x + 5$$

where $y(x)$ is the height of the stream x feet from the firefighter. What is the maximum height that the stream of water from this nozzle can reach? Round to the nearest foot.

68. **OLYMPIC SPORTS** In 1988, Louise Ritter of the United States set the women's Olympic record for the high jump. A mathematical model that approximates her jump is given by

$$h(t) = -204.8t^2 + 256t$$

where $h(t)$ is her height in inches t seconds after beginning her jump. Find the maximum height of her jump.

69. **NORMAN WINDOW** A Norman window has the shape of a rectangle surmounted by a semicircle. The exterior perimeter of the window shown in the figure is 48 feet. Find the height h and the radius r that will allow the maximum amount of light to enter the window. (*Hint:* Write the area of the window as a quadratic function of the radius r.)

70. **NORMAN WINDOW** Assume the semicircle in Exercise 69 permits only 1/3 as much light per square unit to enter as does the rectangular portion of the window. Find the height h and the radius r that will allow the maximum amount of light to enter the window.

SUPPLEMENTAL EXERCISES

71. Let $f(x) = x^2 - (a + b)x + ab$, where a and b are real numbers.

a. Show that the x-intercepts are $(a, 0)$ and $(b, 0)$.

b. Show that the minimum value of the function occurs at the x-value of the midpoint of the line segment defined by the x-intercepts.

72. Let $f(x) = ax^2 + bx + c$, where a, b, and c are real numbers.

a. What conditions must be imposed on the coefficients so that f has a maximum?

b. What conditions must be imposed on the coefficients so that f has a minimum?

c. What conditions must be imposed on the coefficients so that the graph of f intersects the x-axis?

73. Find the quadratic function of x whose graph has a minimum at $(2, 1)$ and passes through $(0, 4)$.

74. Find the quadratic function of x whose graph has a maximum at $(-3, 2)$ and passes through $(0, -5)$.

75. **AREA OF A RECTANGLE** A wire 32 inches long is bent so that it has the shape of a rectangle. The length of the rectangle is x and the width is w.

a. Write w as a function of x.

b. Write the area A of the rectangle as a function of x.

76. **MAXIMIZE AREA** Use the function A from **b.** in Exercise 75 to prove that the area A is greatest if the rectangle is a square.

77. Show that the function $f(x) = x^2 + bx - 1$ has a real zero for any value b.

78. Show that the function $g(x) = -x^2 + bx + 1$ has a real zero for any value b.

79. What effect does increasing the constant c have on the graph of $f(x) = ax^2 + bx + c$?

80. If $a > 0$, what effect does decreasing the coefficient a have on the graph of $f(x) = ax^2 + bx + c$?

81. Find two numbers whose sum is 8 and whose product is a maximum.

82. Find two numbers whose difference is 12 and whose product is a minimum.

83. Verify that the slope of the line passing through (x, x^3) and $(x + h, [x + h]^3)$ is $3x^2 + 3xh + h^2$.

84. Verify that the slope of the line passing through $(x, 4x^3 + x)$ and $(x + h, 4[x + h]^3 + [x + h])$ is given by $12x^2 + 12xh + 4h^2 + 1$.

PROJECTS

1. **THE CUBIC FORMULA** Write an essay on the development of the cubic formula. An excellent source of information is the chapter "Cardano and the Solution of the Cubic" in *Journey Through Genius*, by William Dunham (New York: Wiley, 1990).

2. **SIMPSON'S RULE** In calculus a procedure known as *Simpson's Rule* is often used to approximate the area under a curve. The figure at the right shows the graph of a parabola that passes through $P_0(-h, y_0)$, $P_1(0, y_1)$, and $P_2(h, y_2)$. The equation of the parabola is of the form $y = Ax^2 + Bx + C$. Using calculus procedures, we can show that the area bounded by the parabola, the x-axis, and the vertical lines $x = -h$ and $x = h$ is

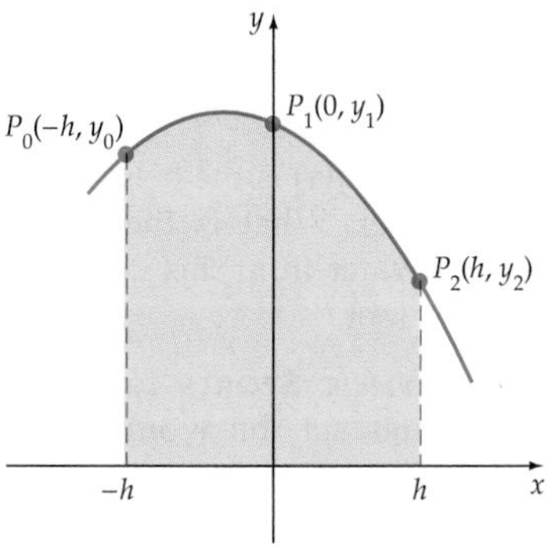

$$\frac{h}{3}(2Ah^2 + 6C)$$

Use algebra to show that $y_0 + 4y_1 + y_2 = 2Ah^2 + 6C$, from which we can deduce that the area of the bounded region can also be written as

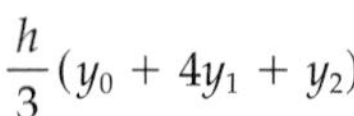

$$\frac{h}{3}(y_0 + 4y_1 + y_2)$$

(*Hint:* Evaluate $Ax^2 + Bx + C$ at $x = -h$, $x = 0$, and $x = h$ to determine values of y_0, y_1, and y_2, respectively. Then compute $y_0 + 4y_1 + y_2$.)

SECTION 2.5 PROPERTIES OF GRAPHS

- SYMMETRY
- EVEN AND ODD FUNCTIONS
- TRANSLATIONS OF GRAPHS
- REFLECTIONS OF GRAPHS
- SHRINKING AND STRETCHING OF GRAPHS

◆ SYMMETRY

The graph in **Figure 2.66** is symmetric with respect to the line l. Note that the graph has the property that if the paper is folded along the dotted line l, the point A' will coincide with the point A, the point B' will coincide with the point B, and the point C' will coincide with the point C. One part of the graph is a *mirror image* of the rest of the graph across the line l.

A graph is **symmetric with respect to the y-axis** if, whenever the point given by (x, y) is on the graph, then $(-x, y)$ is also on the graph. The graph in **Figure 2.67** is symmetric with respect to the y-axis. A graph is **symmetric with respect to the x-axis** if, whenever the point given by (x, y) is on the graph, then $(x, -y)$ is also on the graph. The graph in **Figure 2.68** is symmetric with respect to the x-axis.

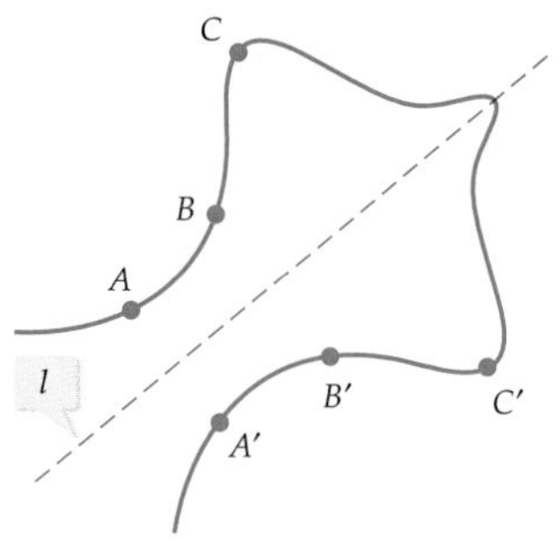

Figure 2.66

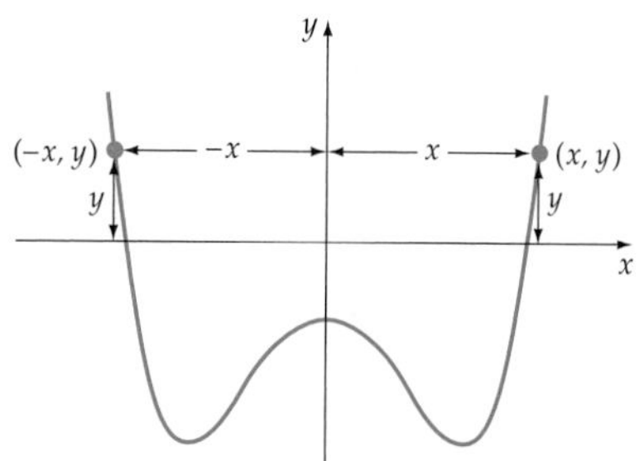

Figure 2.67
Symmetry with respect to the y-axis

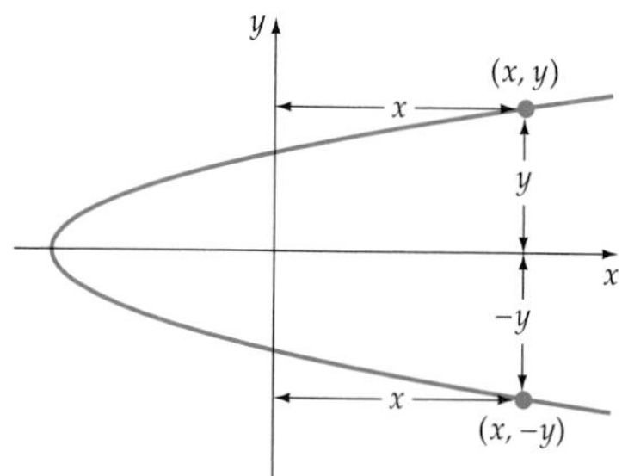

Figure 2.68
Symmetry with respect to the x-axis

Tests for Symmetry with Respect to a Coordinate Axis

The graph of an equation is symmetric with respect to

- the y-axis if the replacement of x with $-x$ leaves the equation unaltered.
- the x-axis if the replacement of y with $-y$ leaves the equation unaltered.

EXAMPLE 1 Determine Symmetries of a Graph

Determine whether the graph of the given equations has symmetry with respect to either the x- or the y-axis.

a. $y = x^2 + 2$ b. $x = |y| - 2$

Solution

a. The equation $y = x^2 + 2$ is unaltered by the replacement of x with $-x$. That is, the simplification of $y = (-x)^2 + 2$ yields the original equation $y = x^2 + 2$. Thus the graph of $y = x^2 + 2$ is symmetric with respect to the y-axis. However, the equation $y = x^2 + 2$ *is altered* by the replacement of y with $-y$. That is, the simplification of $-y = x^2 + 2$, which is $y = -x^2 - 2$, *does not* yield the original equation $y = x^2 + 2$. The graph of $y = x^2 + 2$ is not symmetric with respect to the x-axis. See **Figure 2.69.**

$y = x^2 + 2$

Figure 2.69

b. The equation $x = |y| - 2$ *is altered* by the replacement of x with $-x$. That is, the simplification of $-x = |y| - 2$, which is $x = -|y| + 2$, *does not* yield the original equation $x = |y| - 2$. This implies that the graph of $x = |y| - 2$ is not symmetric with respect to the y-axis. However, the equation $x = |y| - 2$ is unaltered by the replacement of y with $-y$. That is, the simplification of $x = |-y| - 2$ yields the original equation $x = |y| - 2$. The graph of $x = |y| - 2$ is symmetric with respect to the x-axis. See **Figure 2.70.**

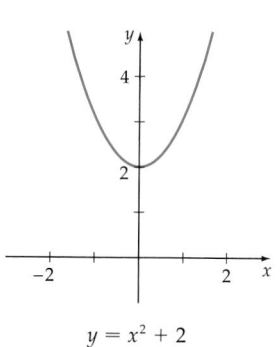

$x = |y| - 2$

Figure 2.70

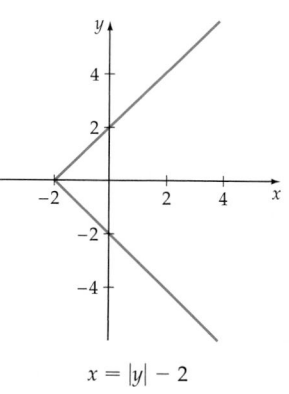

TRY EXERCISE 14, EXERCISE SET 2.5, PAGE 202

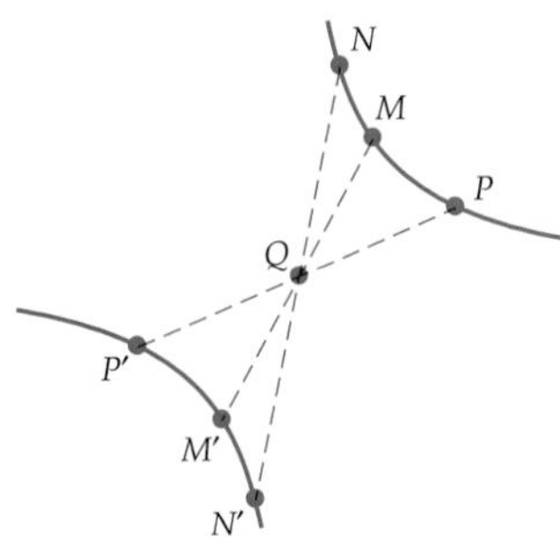

Figure 2.71

Symmetry with Respect to a Point

A graph is **symmetric with respect to a point** Q if for each point P on the graph there is a point P' on the graph such that Q is the midpoint of the line segment PP'.

The graph in **Figure 2.71** is symmetric with respect to the point Q. For any point P on the graph, there exists a point P' on the graph such that Q is the midpoint of $P'P$.

When we discuss symmetry with respect to a point, we frequently use the origin. A graph is symmetric with respect to the origin if, whenever the point given by (x, y) is on the graph, then $(-x, -y)$ is also on the graph. The graph in **Figure 2.72** is symmetric with respect to the origin.

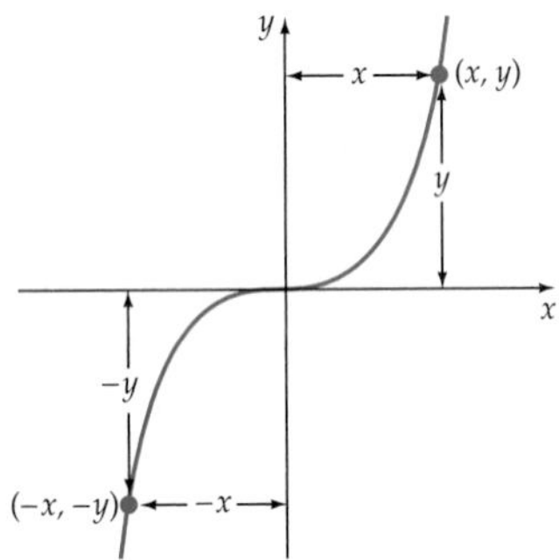

Symmetry with respect to the origin

Figure 2.72

Test for Symmetry with Respect to the Origin

The graph of an equation is symmetric with respect to the origin if the replacement of x with $-x$ and of y with $-y$ leaves the equation unaltered.

EXAMPLE 2 Determine Symmetry with Respect to the Origin

Determine whether the graph of each equation has symmetry with respect to the origin.

a. $xy = 4$ **b.** $y = x^3 + 1$

Solution

a. The equation $xy = 4$ is unaltered by the replacement of x with $-x$ and of y with $-y$. That is, the simplification of $(-x)(-y) = 4$ yields the original equation $xy = 4$. Thus the graph of $xy = 4$ is symmetric with respect to the origin. See **Figure 2.73**.

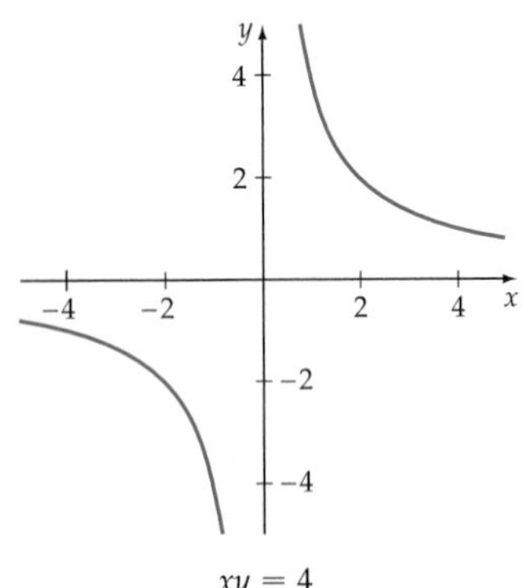

$xy = 4$

Figure 2.73

b. The equation $y = x^3 + 1$ *is altered* by the replacement of x with $-x$ and of y with $-y$. That is, the simplification of $-y = (-x)^3 + 1$, which is $y = x^3 - 1$, *does not* yield the original equation $y = x^3 + 1$. Thus the graph $y = x^3 + 1$ is not symmetric with respect to the origin. See **Figure 2.74**.

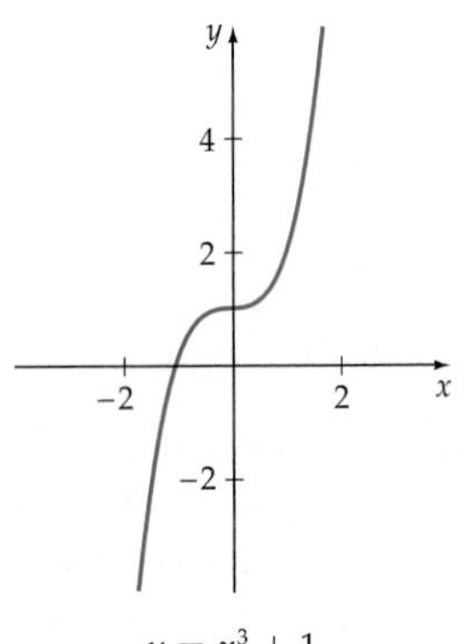

$y = x^3 + 1$

Figure 2.74

TRY EXERCISE 24, EXERCISE SET 2.5, PAGE 202

Some graphs have more than one symmetry. For example, the graph of $|x| + |y| = 2$ has symmetry with respect to the x-axis, the y-axis, and the origin. **Figure 2.75** is the graph of $|x| + |y| = 2$.

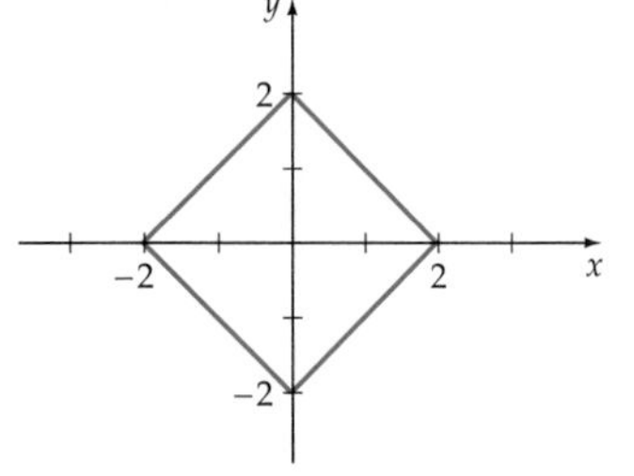

$|x| + |y| = 2$

Figure 2.75

◆ EVEN AND ODD FUNCTIONS

Some functions are classified as either *even* or *odd*.

Definition of Even and Odd Functions

The function f is an **even function** if

$$f(-x) = f(x) \quad \text{for all } x \text{ in the domain of } f$$

The function f is an **odd function** if

$$f(-x) = -f(x) \quad \text{for all } x \text{ in the domain of } f$$

EXAMPLE 3 Identify Even or Odd Functions

Determine whether each function is even, odd, or neither.

a. $f(x) = x^3$ **b.** $F(x) = |x|$ **c.** $h(x) = x^4 + 2x$

Solution

Replace x with $-x$ and simplify.

a. $f(-x) = (-x)^3 = -x^3 = -(x^3) = -f(x)$
Because $f(-x) = -f(x)$, this function is an odd function.

b. $F(-x) = |-x| = |x| = F(x)$
Because $F(-x) = F(x)$, this function is an even function.

c. $h(-x) = (-x)^4 + 2(-x) = x^4 - 2x$
This function is neither an even nor an odd function because

$$h(-x) = x^4 - 2x,$$

which is not equal to either $h(x)$ or $-h(x)$.

TRY EXERCISE 44, EXERCISE SET 2.5, PAGE 202

The following properties are a result of the tests for symmetry.

- The graph of an even function is symmetric with respect to the y-axis.
- The graph of an odd function is symmetric with respect to the origin.

The graph of f in **Figure 2.76** is symmetric with respect to the y-axis. It is the graph of an even function. The graph of g in **Figure 2.77** is symmetric with respect to the origin. It is the graph of an odd function. The graph of h in **Figure 2.78** is not symmetric to the y-axis and is not symmetric to the origin. It is neither an even nor an odd function.

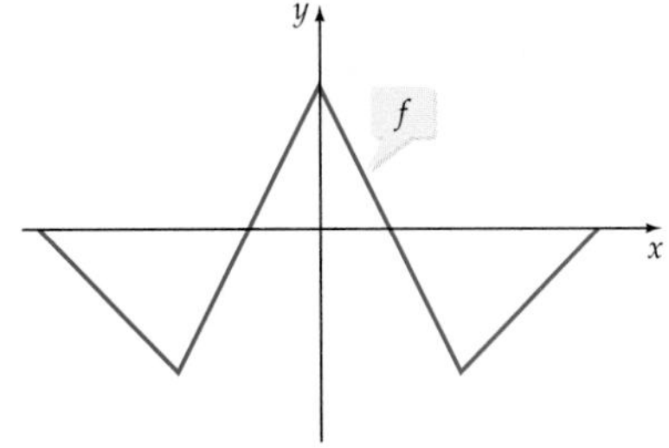

Figure 2.76
The graph of an even function is symmetric with respect to the y-axis.

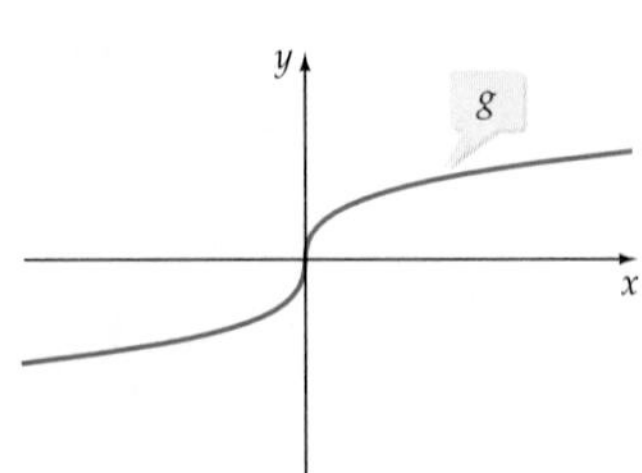

Figure 2.77
The graph of an odd function is symmetric with respect to the origin.

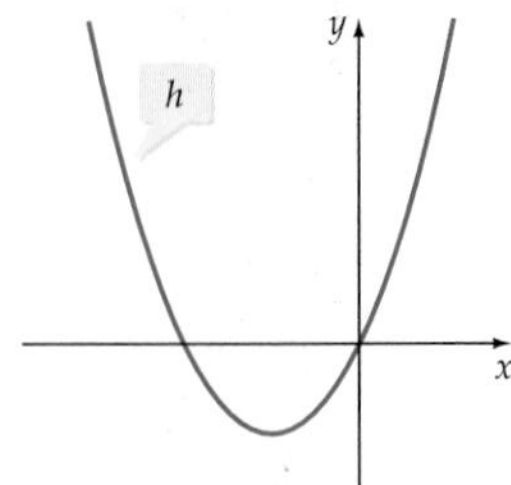

Figure 2.78
If the graph of a function is not symmetric to the y-axis or to the origin then the function is neither even nor odd.

Figure 2.79

◆ TRANSLATIONS OF GRAPHS

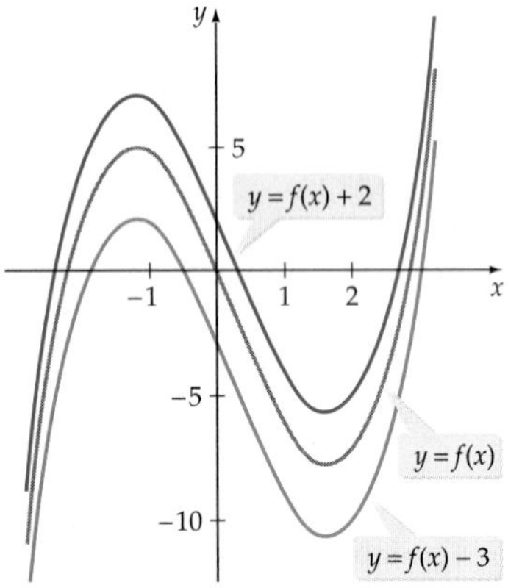

The shape of a graph may be exactly the same as the shape of another graph; only their position in the xy-plane may differ. For example, the graph of $y = f(x) + 2$ is the graph of $y = f(x)$ with each point moved up vertically 2 units. The graph of $y = f(x) - 3$ is the graph of $y = f(x)$ with each point moved down vertically 3 units. See **Figure 2.79.**

The graphs of $y = f(x) + 2$ and $y = f(x) - 3$ in **Figure 2.79** are called *vertical translations* of the graph of $y = f(x)$.

Vertical Translations

If f is a function and c is a positive constant, then the graph of

- $y = f(x) + c$ is the graph of $y = f(x)$ shifted up *vertically* c units.
- $y = f(x) - c$ is the graph of $y = f(x)$ shifted down *vertically* c units.

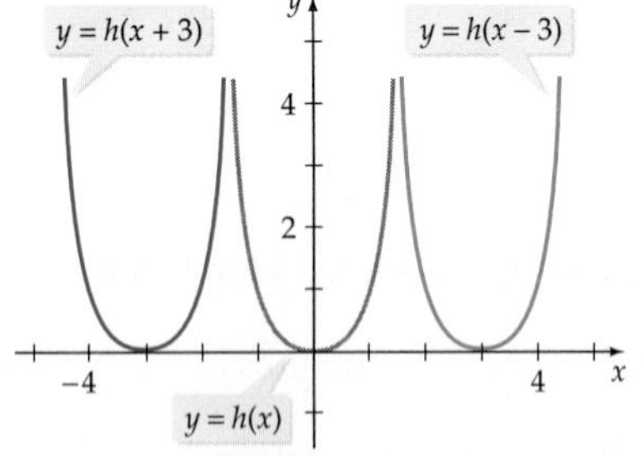

Figure 2.80

In **Figure 2.80,** the graph of $y = h(x + 3)$ is the graph of $y = h(x)$ with each point shifted to the left horizontally 3 units. Similarly, the graph of $y = h(x - 3)$ is the graph of $y = h(x)$ with each point shifted to the right horizontally 3 units.

The graphs of $y = h(x + 3)$ and $y = h(x - 3)$ in **Figure 2.80** are called *horizontal translations* of the graph of $y = h(x)$.

Horizontal Translations

If f is a function and c is a positive constant, then the graph of

- $y = f(x + c)$ is the graph of $y = f(x)$ shifted left *horizontally* c units.
- $y = f(x - c)$ is the graph of $y = f(x)$ shifted right *horizontally* c units.

take note

A TI graphing calculator program is available to explore the translation of the graph of a parabola. This program, TRANSQUAD, can be found on our web site at

http://college.hmco.com

A graphing calculator can be used to draw the graphs of a *family* of functions. The word *family* refers to a function one of the constants of which may change. That constant is called a **parameter.** For instance, $f(x) = x^2 + c$ constitutes a family of functions with parameter c. The only feature of the graph that changes is the value of c.

A graphing calculator can be used to produce the graphs of a family of curves for specific values of the parameter. The LIST feature of the calculator can be used. For instance, to graph $f(x) = x^2 + c$ for $c = -2, 0,$ and 1, we will create a list and use that list to produce the family of curves. The keystrokes for a TI-83 calculator are given below.

Here are the keystrokes to create the list:

2nd { -2 , 0 , 1 2nd } STO 2nd L1

Now use the Y= key to enter

Y= X x² + 2nd L1 ZOOM 6

Sample screens for the keystrokes and graphs are shown here. You can use similar keystrokes for Exercises 67–74 of this section.

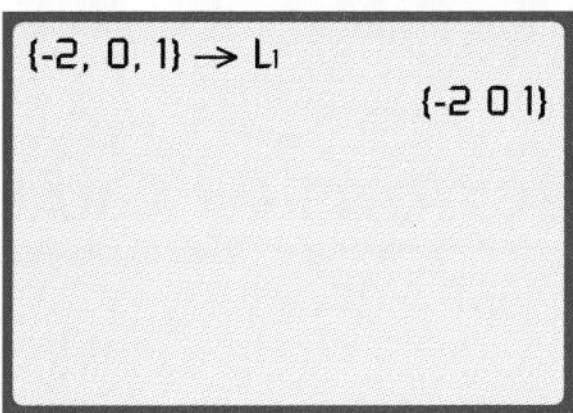

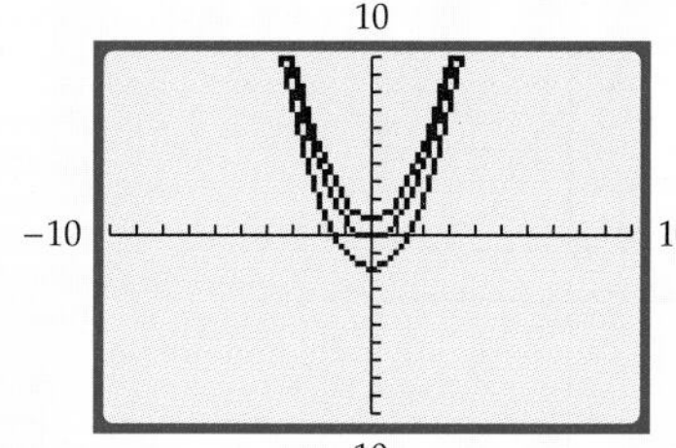

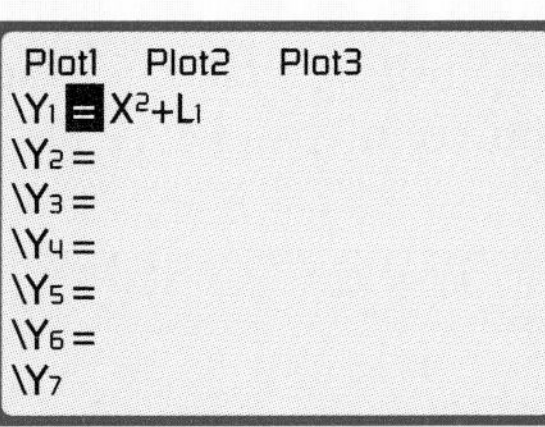

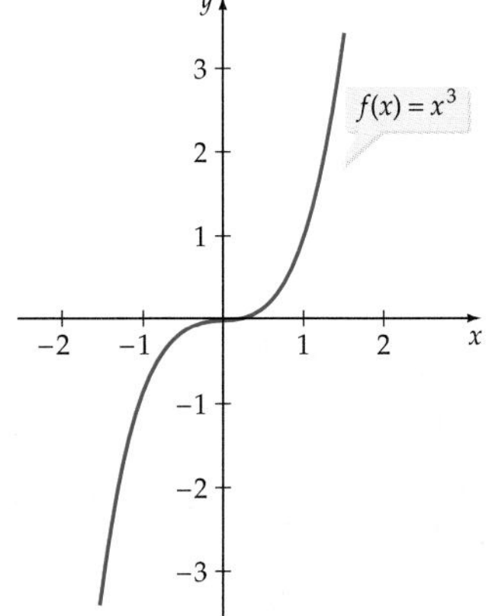

Figure 2.81

EXAMPLE 4 Graph by Using Translations

Use vertical and horizontal translations of the graph of $f(x) = x^3$, shown in Figure 2.81, to graph

a. $g(x) = x^3 - 2$ b. $h(x) = (x + 1)^3$

Solution

a. The graph of $g(x) = x^3 - 2$ is the graph of $f(x) = x^3$ shifted down vertically 2 units. See Figure 2.82.

Continued

b. The graph of $h(x) = (x + 1)^3$ is the graph of $f(x) = x^3$ shifted to the left horizontally 1 unit. See **Figure 2.83.**

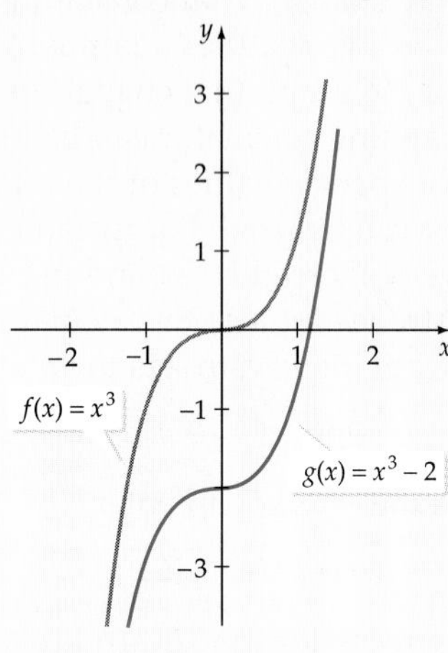

Figure 2.82

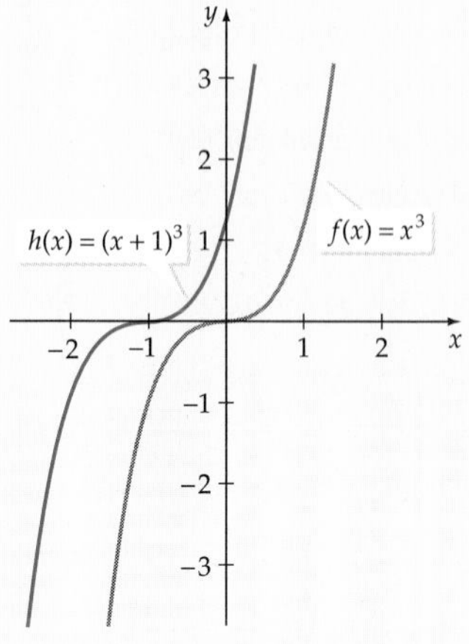

Figure 2.83

TRY EXERCISE 58, EXERCISE SET 2.5, PAGE 202

◆ REFLECTIONS OF GRAPHS

The graph of $y = -f(x)$ cannot be obtained from the graph of $y = f(x)$ by a combination of vertical and/or horizontal shifts. **Figure 2.84** illustrates that the graph of $y = -f(x)$ is the reflection of the graph of $y = f(x)$ across the x-axis.

The graph of $y = f(-x)$ is the reflection of the graph of $y = f(x)$ across the y-axis as shown in **Figure 2.85.**

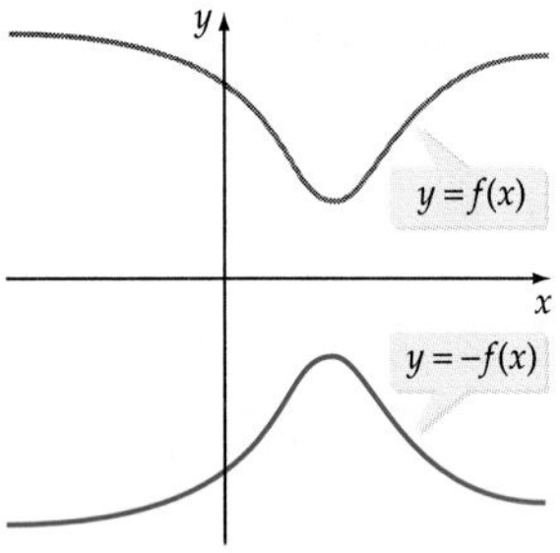

Figure 2.84

Reflections

The graph of

- $y = -f(x)$ is the graph of $y = f(x)$ reflected across the x-axis.
- $y = f(-x)$ is the graph of $y = f(x)$ reflected across the y-axis.

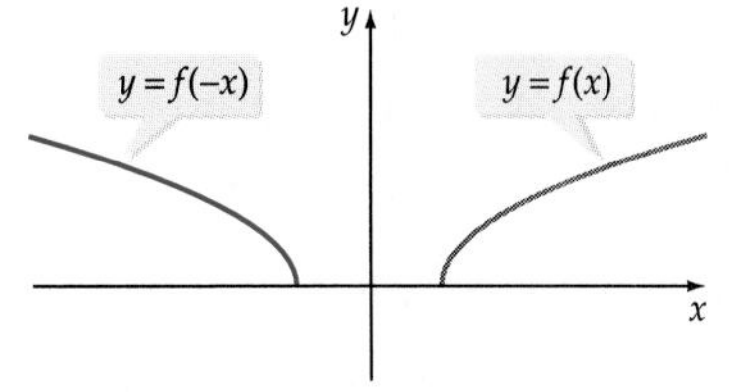

Figure 2.85

EXAMPLE 5 Graph by Using Reflections

Use reflections of the graph of $f(x) = \sqrt{x - 1} + 1$, shown in **Figure 2.86,** to graph

a. $g(x) = -\left(\sqrt{x - 1} + 1\right)$ **b.** $h(x) = \sqrt{-x - 1} + 1$

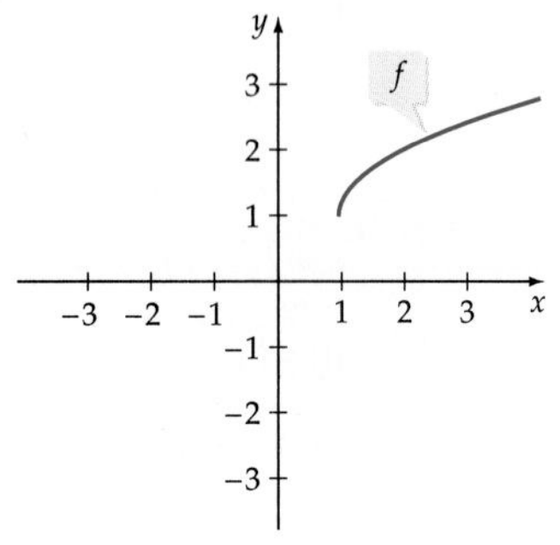

Figure 2.86

Solution

a. Because $g(x) = -f(x)$, the graph of g is the graph of f reflected across the x-axis. See **Figure 2.87.**

b. Because $h(x) = f(-x)$, the graph of h is the graph of f reflected across the y-axis. See **Figure 2.88.**

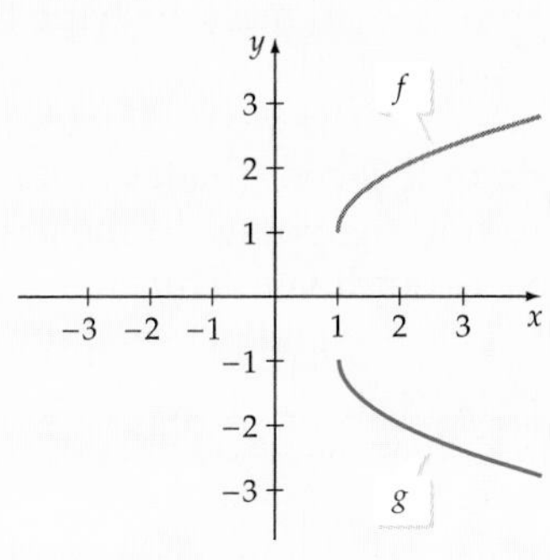

Figure 2.87

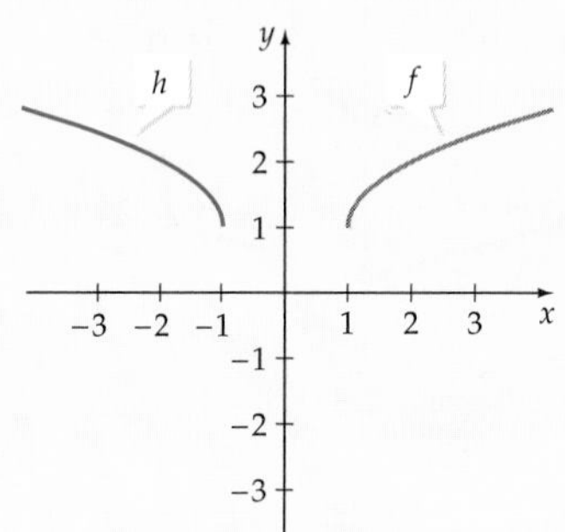

Figure 2.88

TRY EXERCISE 60, EXERCISE SET 2.5, PAGE 203

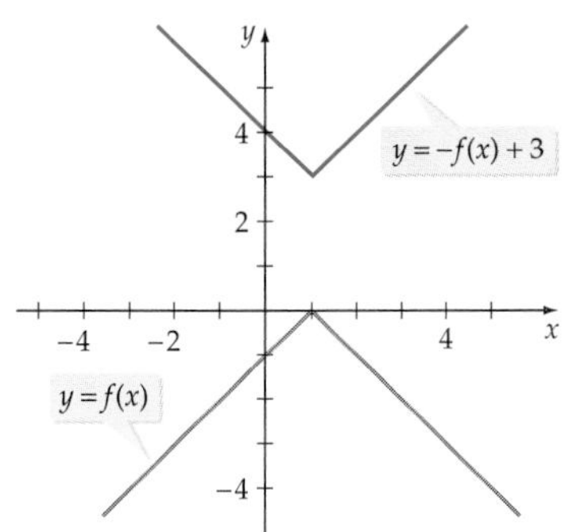

Figure 2.89

Some graphs of functions can be constructed by using a combination of translations and reflections. For instance, the graph of $y = -f(x) + 3$ in **Figure 2.89** was obtained by reflecting the graph of $y = f(x)$ in **Figure 2.89** with respect to the x-axis and then shifting that graph up vertically 3 units.

◆ SHRINKING AND STRETCHING OF GRAPHS

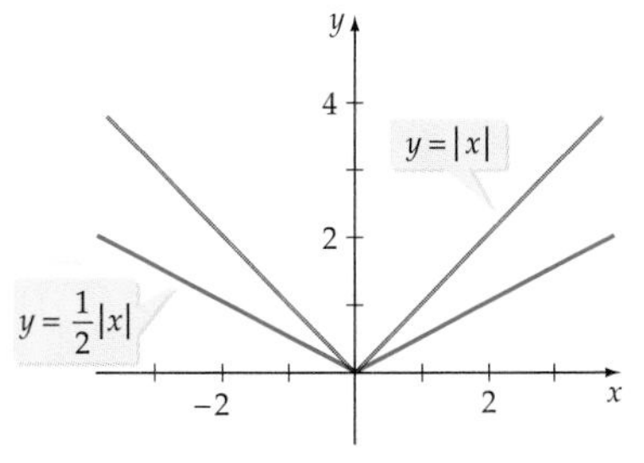

Figure 2.90

The graph of the equation $y = c \cdot f(x)$ for $c \neq 1$ vertically shrinks or stretches the graph of $y = f(x)$. To determine the points on the graph of $y = c \cdot f(x)$, multiply each y-coordinate of the points on the graph of $y = f(x)$ by c. For example, **Figure 2.90** shows that the graph of $y = \frac{1}{2}|x|$ can be obtained by plotting points that have a y-coordinate that is one-half of the y-coordinate of those found on the graph of $y = |x|$.

If $0 < c < 1$, then the graph of $y = c \cdot f(x)$ is obtained by *shrinking* the graph of $y = f(x)$. **Figure 2.90** illustrates the vertical shrinking of the graph of $y = |x|$ toward the x-axis to form the graph of $y = \frac{1}{2}|x|$.

If $c > 1$, then the graph of $y = c \cdot f(x)$ is obtained by *stretching* the graph of $y = f(x)$. For example, if $f(x) = |x|$, then we obtain the graph of

$$y = 2f(x) = 2|x|$$

by stretching the graph of f away from the x-axis. See **Figure 2.91.**

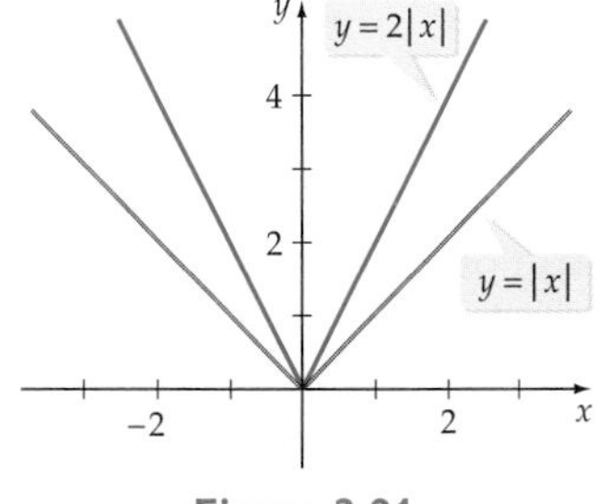

Figure 2.91

EXAMPLE 6 **Graph by Using Vertical Shrinking and Shifting**

Graph: $H(x) = \dfrac{1}{4}|x| - 3$

Continued ►

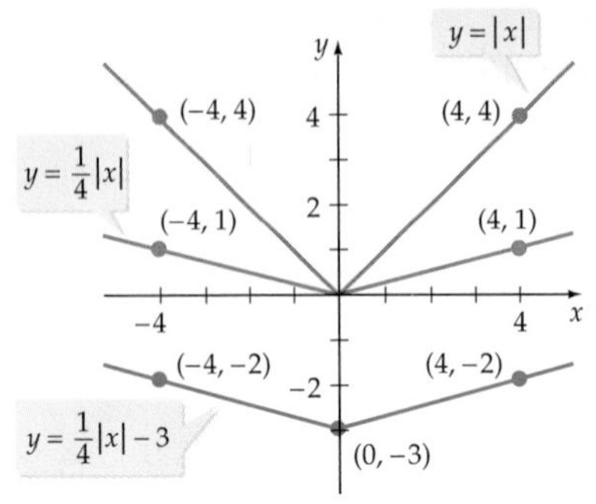

Figure 2.92

Solution

The graph of $y = |x|$ has a V shape that has its lowest point at (0, 0) and passes through (4, 4) and (−4, 4). The graph of $y = \frac{1}{4}|x|$ is a shrinking of the graph of $y = |x|$. The y-coordinates (0, 0), (4, 1), and (−4, 1) are obtained by multiplying the y-coordinates of the ordered pairs (0, 0), (4, 4), and (4, −4) by 1/4. To find the points on the graph of H, we still need to subtract 3 from each y-coordinate. Thus the graph of H is a V shape that has its lowest point at (0, −3) and passes through (4, −2) and (−4, −2). See **Figure 2.92.**

TRY EXERCISE 62, EXERCISE SET 2.5, PAGE 203

Some functions can be graphed by using a horizontal shrinking or stretching of a given graph. The procedure makes use of the following concept.

Horizontal Shrinking and Stretching

If $a > 0$ and the graph of $y = f(x)$ contains the point (x, y), then the graph of $y = f(ax)$ contains the point $\left(\frac{1}{a}x, y\right)$.

If $a > 1$, then the graph of $y = f(ax)$ is a *horizontal shrinking* of the graph of $y = f(x)$. If $0 < a < 1$, then the graph of $y = f(ax)$ is a *horizontal stretching* of the graph of $y = f(x)$.

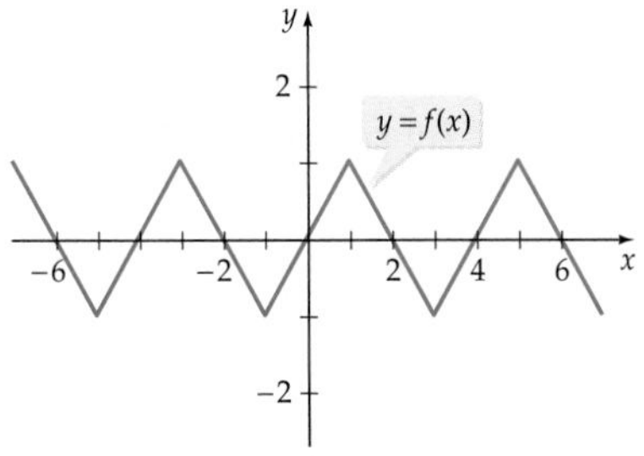

Figure 2.93

EXAMPLE 7 Graph by Using Horizontal Shrinking and Stretching

Use the graph of $y = f(x)$ shown in **Figure 2.93** to graph

a. $y = f(2x)$ **b.** $y = f\left(\frac{1}{3}x\right)$

Solution

a. Because $2 > 1$, the graph of $y = f(2x)$ is a horizontal contraction (shrinking) of the graph of $y = f(x)$. The graph of $y = f(2x)$ can be constructed by contracting each point on the graph of $y = f(x)$ toward the y-axis by a factor of 1/2. For example, the point (2, 0) on the graph of $y = f(x)$ becomes the point (1, 0) on the graph of $y = f(2x)$. See **Figure 2.94.**

b. Since $0 < \frac{1}{3} < 1$, the graph of $y = f\left(\frac{1}{3}x\right)$ is a horizontal dilation (stretching) of the graph of $y = f(x)$. The graph of $y = f\left(\frac{1}{3}x\right)$ can be constructed by moving each point on the graph of $y = f(x)$ away from

take note

A web applet is available to explore stretching, shrinking, and reflection of a quadratic function. This applet, Quadratic: Polynomial Form, can be found on our web site at

http://college.hmco.com

the y-axis by a factor of 3. For example, the point (1, 1) on the graph of $y = f(x)$ becomes the point (3, 1) on the graph of $y = f\left(\frac{1}{3}x\right)$. See **Figure 2.95.**

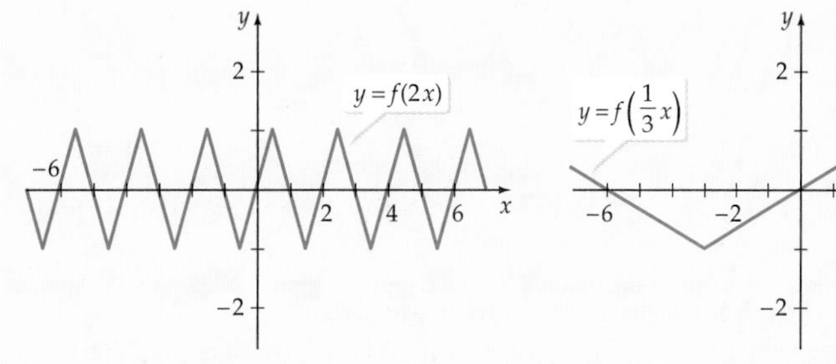

Figure 2.94 **Figure 2.95**

TRY EXERCISE 64, EXERCISE SET 2.5, PAGE 203

TOPICS FOR DISCUSSION

1. Discuss the meaning of symmetry of a graph with respect to a line. How do you determine whether a graph has symmetry with respect to the x-axis? with respect to the y-axis?
2. Discuss the meaning of symmetry of a graph with respect to a point. How do you determine whether a graph has symmetry with respect to the origin?
3. What does it mean to reflect a graph across the x-axis or across the y-axis?
4. Explain how the graphs of $y_1 = 2x^3 - x^2$ and $y_2 = 2(-x)^3 - (-x)^2$ are related.
5. Given the graph of $y_3 = f(x)$, explain how to obtain the graph of $y_4 = f(x - 3) + 1$.
6. The graph of the *step function* $y_5 = [\![x]\!]$ has steps that are 1 unit wide. Determine how wide the steps are in the graph of $y_6 = \left[\!\!\left[\frac{1}{3}x\right]\!\!\right]$.

EXERCISE SET 2.5

In Exercises 1 to 6, plot the image of the given point with respect to

a. **the y-axis. Label this point *A*.**

b. **the x-axis. Label this point *B*.**

c. **the origin. Label this point *C*.**

1. $P(5, -3)$
2. $Q(-4, 1)$
3. $R(-2, 3)$
4. $S(-5, 3)$
5. $T(-4, -5)$
6. $U(5, 1)$

In Exercises 7 and 8, sketch a graph that is symmetric to the given graph with respect to the *x*-axis.

7.

8.

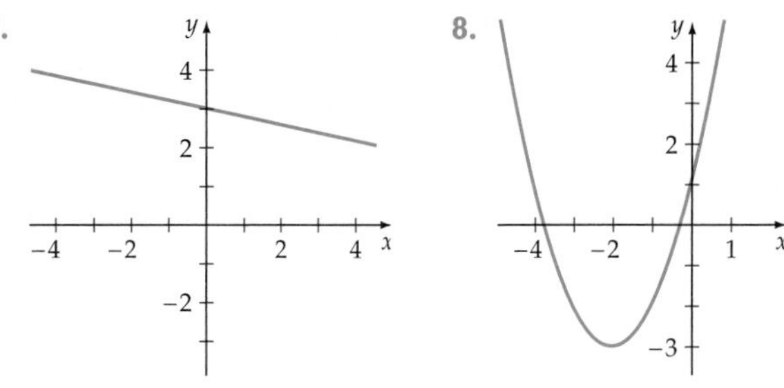

In Exercises 9 and 10, sketch a graph that is symmetric to the given graph with respect to the y-axis.

9. 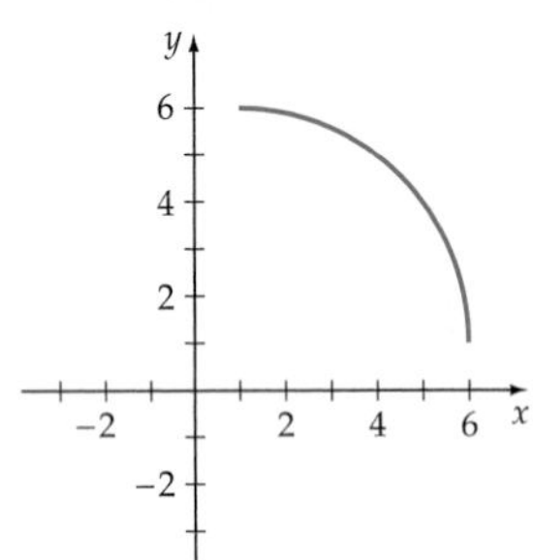

10. 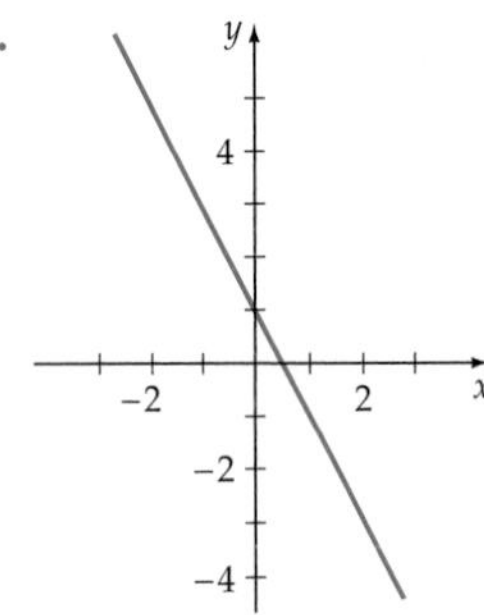

In Exercises 11 and 12, sketch a graph that is symmetric to the given graph with respect to the origin.

11. 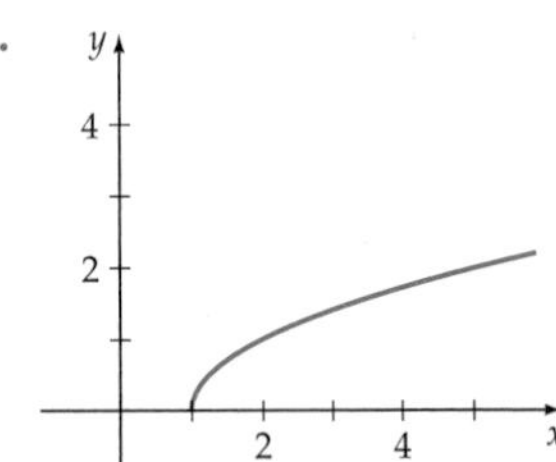

12. 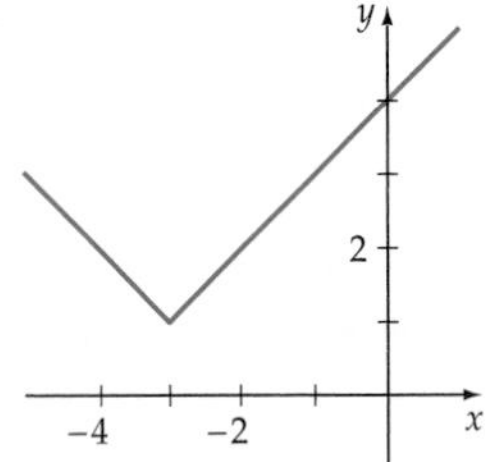

In Exercises 13 to 21, determine whether the graph of each equation is symmetric with respect to the *a.* x-axis, *b.* y-axis.

13. $y = 2x^2 - 5$
14. $x = 3y^2 - 7$
15. $y = x^3 + 2$
16. $y = x^5 - 3x$
17. $x^2 + y^2 = 9$
18. $x^2 - y^2 = 10$
19. $x^2 = y^4$
20. $xy = 8$
21. $|x| - |y| = 6$

In Exercises 22 to 30, determine whether the graph of each equation is symmetric with respect to the origin.

22. $y = x + 1$
23. $y = 3x - 2$
24. $y = x^3 - x$
25. $y = -x^3$
26. $y = \dfrac{9}{x}$
27. $x^2 + y^2 = 10$
28. $x^2 - y^2 = 4$
29. $y = \dfrac{x}{|x|}$
30. $|y| = |x|$

In Exercises 31 to 42, graph the given equations. Label each intercept. Use the concept of symmetry to confirm that the graph is correct.

31. $y = x^2 - 1$
32. $x = y^2 - 1$
33. $y = x^3 - x$
34. $y = -x^3$
35. $xy = 4$
36. $xy = -8$
37. $y = 2|x - 4|$
38. $y = |x - 2| - 1$
39. $y = (x - 2)^2 - 4$
40. $y = (x - 1)^2 - 4$
41. $y = x - |x|$
42. $|y| = |x|$

In Exercises 43 to 56, identify whether the given function is an even function, an odd function, or neither.

43. $g(x) = x^2 - 7$
44. $h(x) = x^2 + 1$
45. $F(x) = x^5 + x^3$
46. $G(x) = 2x^5 - 10$
47. $H(x) = 3|x|$
48. $T(x) = |x| + 2$
49. $f(x) = 1$
50. $k(x) = 2 + x + x^2$
51. $r(x) = \sqrt{x^2 + 4}$
52. $u(x) = \sqrt{3 - x^2}$
53. $s(x) = 16x^2$
54. $v(x) = 16x^2 + x$
55. $w(x) = 4 + \sqrt[3]{x}$
56. $z(x) = \dfrac{x^3}{x^2 + 1}$

57. Use the graph of $f(x) = \sqrt{4 - x^2}$ to sketch the graph of
    a. $y = f(x) + 3$ b. $y = f(x - 3)$

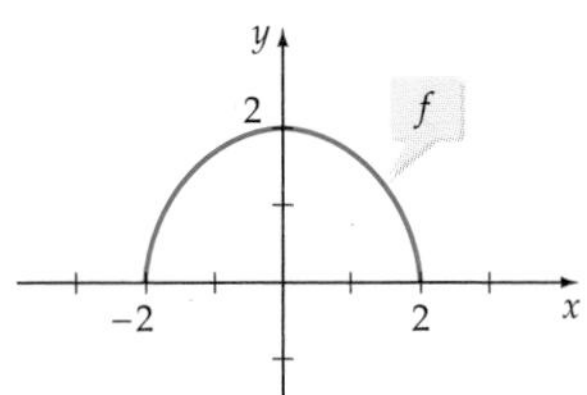

58. Use the graph of $g(x) = |x|$ to sketch the graph of
    a. $y = g(x) - 2$ b. $y = g(x - 3)$

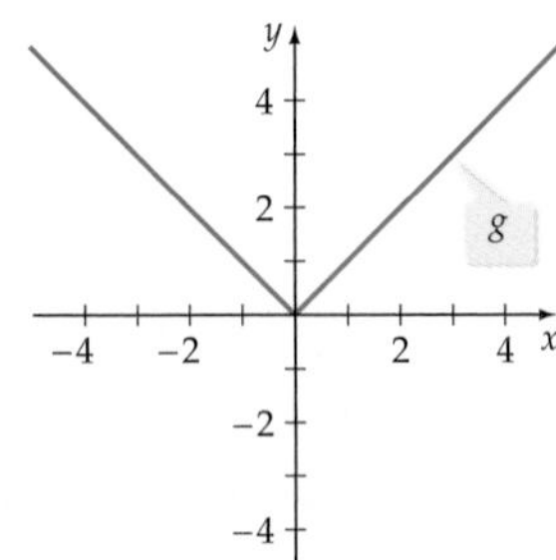

59. Use the graph of $F(x) = (x - 1)^{2/3}$ to sketch the graph of
    a. $y = -F(x)$ b. $y = F(-x)$

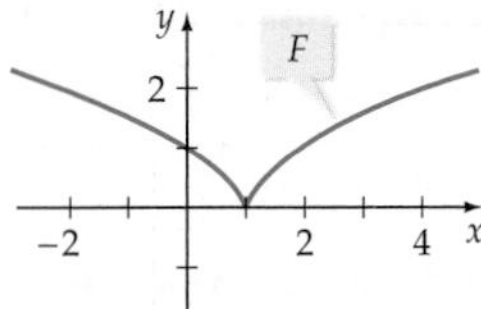

60. Use the graph of $E(x) = |x - 1| + 1$ to sketch the graph of

a. $y = -E(x)$ **b.** $y = E(-x)$

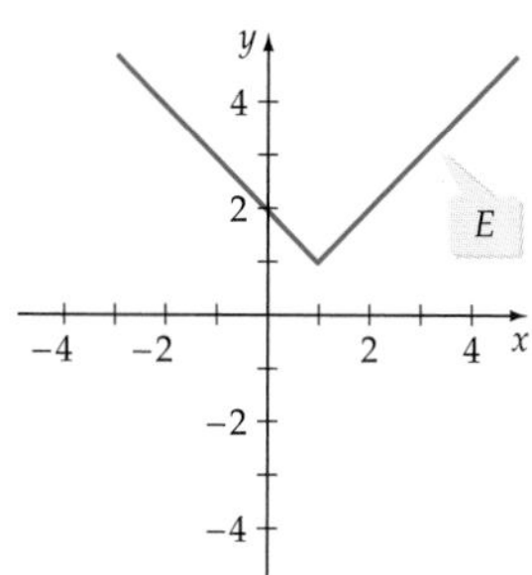

61. Use the graph of $m(x) = x^2 - 2x - 3$ to sketch the graph of $y = -\frac{1}{2}m(x) + 3$.

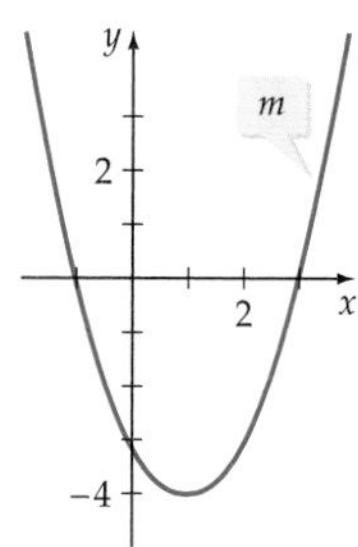

62. Use the graph of $n(x) = -x^2 - 2x + 8$ to sketch the graph of $y = \frac{1}{2}n(x) + 1$.

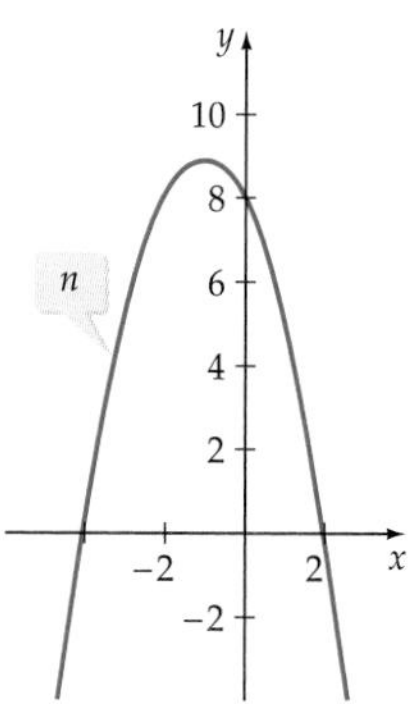

63. Use the graph of $y = f(x)$ to sketch the graph of

a. $y = f(2x)$ **b.** $y = f\left(\frac{1}{3}x\right)$

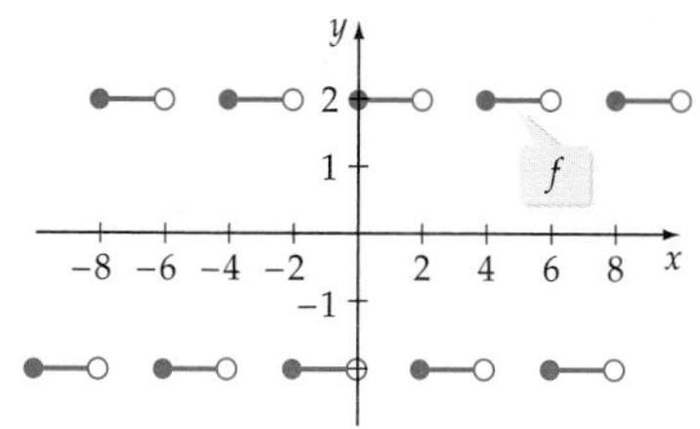

64. Use the graph of $y = g(x)$ to sketch the graph of

a. $y = g(2x)$ **b.** $y = g\left(\frac{1}{2}x\right)$

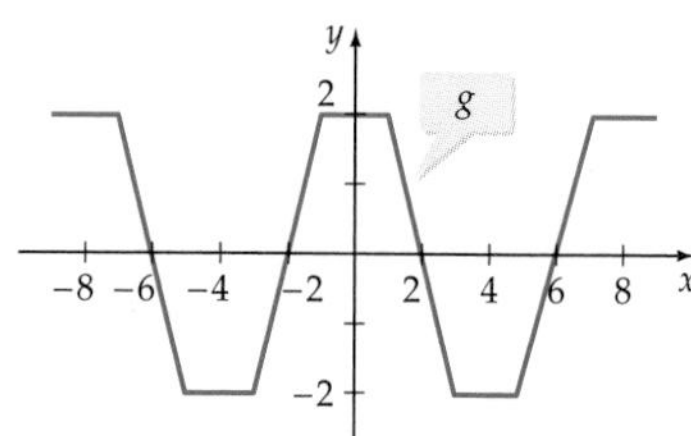

65. Use the graph of $y = h(x)$ to sketch the graph of

a. $y = h(2x)$ **b.** $y = h\left(\frac{1}{2}x\right)$

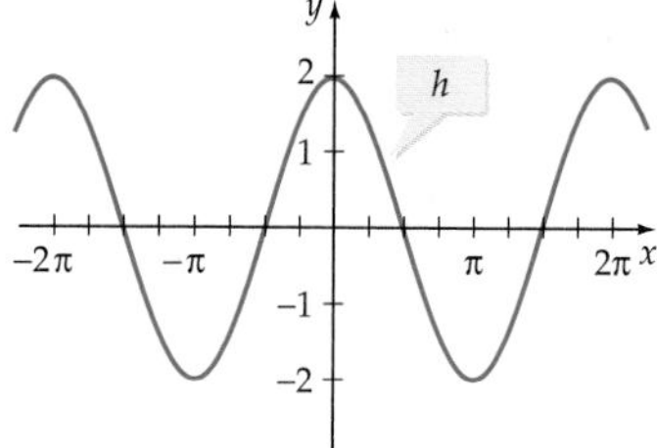

66. Use the graph of $y = j(x)$ to sketch the graph of

a. $y = j(2x)$ **b.** $y = j\left(\frac{1}{3}x\right)$

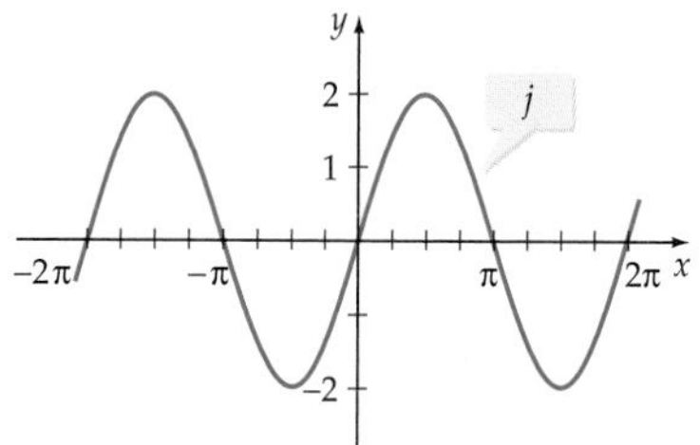

In Exercises 67 to 74, use a graphing utility.

67. On the same coordinate axes, graph

$$G(x) = \sqrt[3]{x} + c$$

for $c = 0, -1$, and 3.

68. On the same coordinate axes, graph

$$H(x) = \sqrt[3]{x + c}$$

for $c = 0, -1$, and 3.

69. On the same coordinate axes, graph

$$J(x) = |2(x + c) - 3| - |x + c|$$

for $c = 0, -1$, and 2.

70. On the same coordinate axes, graph

$$K(x) = |x - 1| - |x| + c$$

for $c = 0, -1$, and 2.

71. On the same coordinate axes, graph

$$L(x) = cx^2$$

for $c = 1, 1/2$, and 2.

72. On the same coordinate axes, graph

$$M(x) = c\sqrt{x^2 - 4}$$

for $c = 1, 1/3$, and 3.

73. On the same coordinate axes, graph

$$S(x) = c(|x - 1| - |x|)$$

for $c = 1, 1/4$, and 4.

74. On the same coordinate axes, graph

$$T(x) = c\left(\frac{x}{|x|}\right)$$

for $c = 1, 2/3$, and $3/2$.

75. Graph $V(x) = [\![cx]\!]$, $0 \le x \le 6$, for each value of c.

a. $c = 1$

b. $c = 1/2$

c. $c = 2$

76. Graph $W(x) = [\![cx]\!] - cx$, $0 \le x \le 6$, for each value of c.

a. $c = 1$

b. $c = 1/3$

c. $c = 3$

SUPPLEMENTAL EXERCISES

77. Use the graph of $f(x) = 2/(x^2 + 1)$ to determine an equation for the graphs shown in **a.** and **b.**

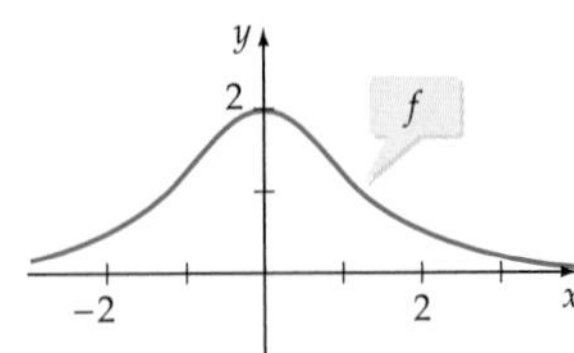

a.

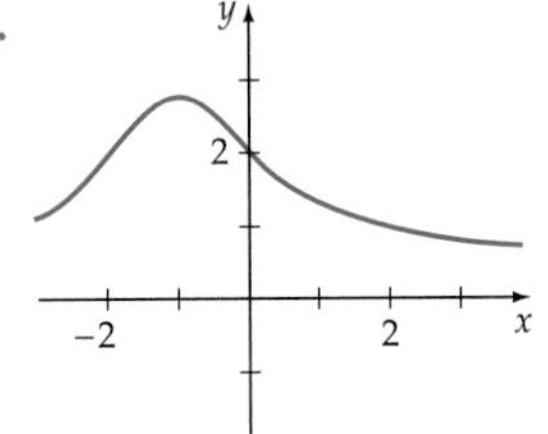

b.

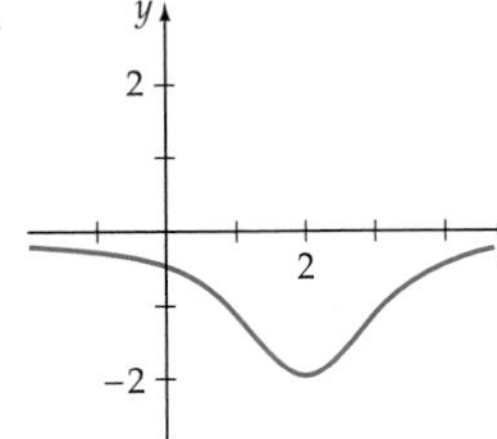

78. Use the graph of $f(x) = x\sqrt{2 + x}$ to determine an equation for the graphs shown in **a.** and **b.**

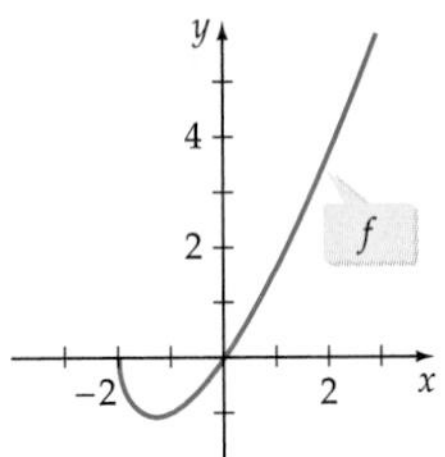

a.

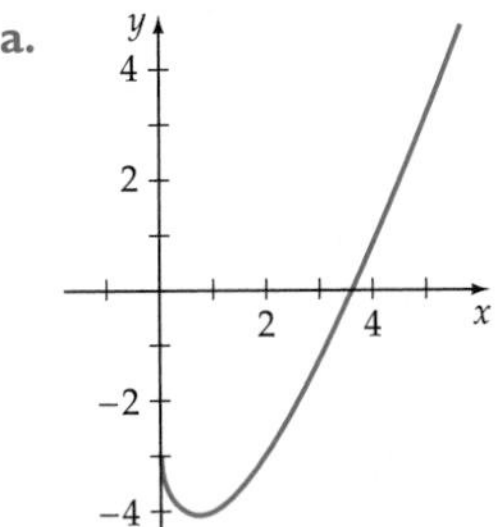

b.

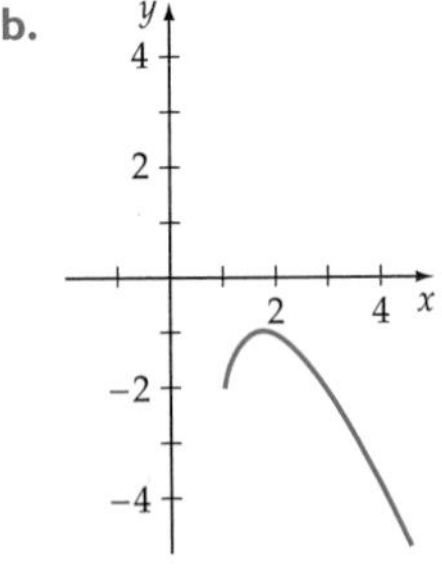

PROJECTS

1. DIRICHLET FUNCTION We owe our present-day definition of a function to the German mathematician Peter Gustav Dirichlet (1805–1859). He created the following unusual function, which is now known as the *Dirichlet function.*

$$f(x) = \begin{cases} 0, & \text{if } x \text{ is a rational number} \\ 1, & \text{if } x \text{ is an irrational number} \end{cases}$$

Answer the following questions about the Dirichlet function.

a. What is its domain?

b. What is its range?

c. What are its x-intercepts? **d.** What is its y-intercept?

e. Is it an even or an odd function?

f.  Explain why a graphing calculator cannot be used to produce an accurate graph of the function.

g.  Write a sentence or two that describes its graph.

2. 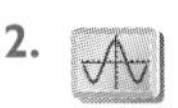  **ISOLATED POINT** Consider the function given by

$$y = \sqrt{(x-1)^2(x-2)} + 1$$

Verify that the point (1, 1) is a solution of the equation. Now use a graphing utility to graph the function. Does your graph include the isolated point at (1, 1), as shown at the right? If the graphing utility you used failed to include the point (1, 1), explain at least one reason for the omission of this isolated point.

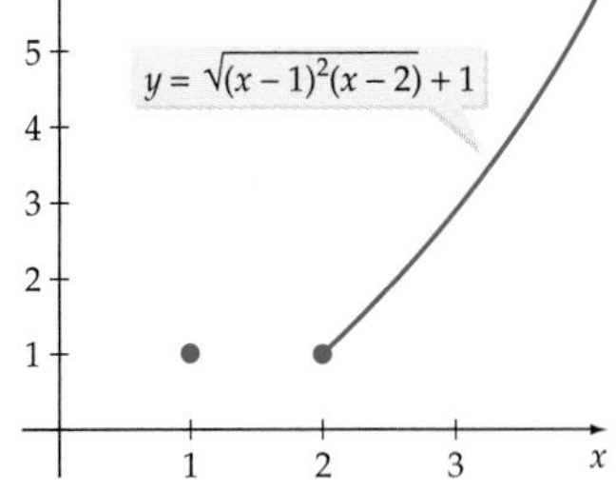

3.  **A LINE WITH A HOLE** The function

$$f(x) = \frac{(x-2)(x+1)}{(x-2)}$$

graphs as a line with a y-intercept of 1, a slope of 1, and a hole at (2, 3). Use a graphing utility to graph f. Explain why a graphing utility might not show the hole at (2, 3).

4. **FINDING A COMPLETE GRAPH** Use a graphing utility to graph the function $f(x) = 3x^{5/3} - 6x^{4/3} + 2$ for $-2 \le x \le 10$. Compare your graph with the graph at the right. Does your graph include the part to the left of the y-axis? If not, how might you enter the function in such a way that the graphing utility you used would include this part?

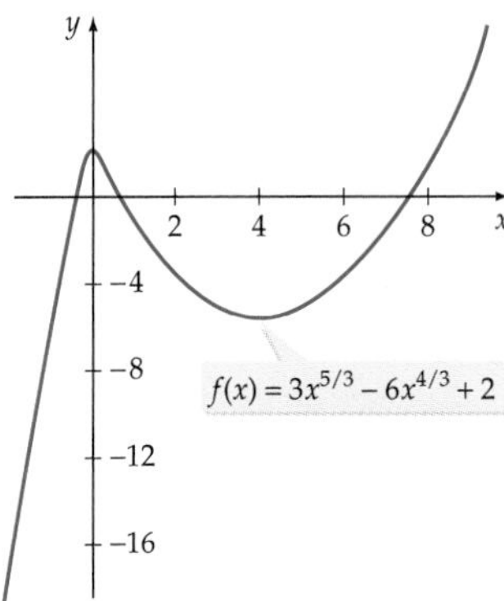

SECTION 2.6 THE ALGEBRA OF FUNCTIONS

- THE DIFFERENCE QUOTIENT
- COMPOSITION OF FUNCTIONS

Functions can be defined in terms of other functions. For example, the function defined by $h(x) = x^2 + 8x$ is the sum of

$$f(x) = x^2 \quad \text{and} \quad g(x) = 8x$$

Thus if we are given any two functions, f and g, we can define the four new functions $f + g$, $f - g$, fg, and f/g as follows.

Operations on Functions

For all values of x for which both $f(x)$ and $g(x)$ are defined, we define the following functions.

Sum	$(f + g)(x) = f(x) + g(x)$
Difference	$(f - g)(x) = f(x) - g(x)$
Product	$(fg)(x) = f(x) \cdot g(x)$
Quotient	$\left(\dfrac{f}{g}\right)(x) = \dfrac{f(x)}{g(x)}, \quad g(x) \neq 0$

Domain of $f + g, f - g, fg, f/g$

For the given functions f and g, the domains of $f + g$, $f - g$, and $f \cdot g$ consist of all real numbers formed by the intersection of the domains of f and g. The domain of f/g is the set of all real numbers formed by the intersection of the domains of f and g, except for those real numbers x such that $g(x) = 0$.

EXAMPLE 1 Determine the Domain of a Function

If $f(x) = \sqrt{x - 1}$ and $g(x) = x^2 - 4$, find the domain of $f + g$, of $f - g$, of fg, and of f/g.

Solution

Note that f has the domain $\{x \mid x \geq 1\}$ and g has the domain of all real numbers. Therefore, the domain of $f + g$, $f - g$, and fg is $\{x \mid x \geq 1\}$. Because $g(x) = 0$ when $x = -2$ or $x = 2$, neither -2 nor 2 is in the domain of f/g. The domain of f/g is $\{x \mid x \geq 1 \text{ and } x \neq 2\}$.

TRY EXERCISE 10, EXERCISE SET 2.6, PAGE 213

EXAMPLE 2 Evaluate Functions

Let $f(x) = x^2 - 9$ and $g(x) = 2x + 6$. Find

a. $(f + g)(5)$ **b.** $(fg)(-1)$ **c.** $\left(\dfrac{f}{g}\right)(4)$

Solution

a. $(f + g)(x) = f(x) + g(x) = (x^2 - 9) + (2x + 6) = x^2 + 2x - 3$
Therefore, $(f + g)(5) = (5)^2 + 2(5) - 3 = 25 + 10 - 3 = 32$.

b. $(fg)(x) = f(x) \cdot g(x) = (x^2 - 9)(2x + 6) = 2x^3 + 6x^2 - 18x - 54$

Therefore, $(fg)(-1) = 2(-1)^3 + 6(-1)^2 - 18(-1) - 54$

$= -2 + 6 + 18 - 54 = -32.$

c. $\left(\dfrac{f}{g}\right)(x) = \dfrac{f(x)}{g(x)} = \dfrac{x^2 - 9}{2x + 6} = \dfrac{\cancel{(x+3)}(x-3)}{2\cancel{(x+3)}} = \dfrac{x-3}{2}, \quad x \neq -3$

Therefore, $\left(\dfrac{f}{g}\right)(4) = \dfrac{4-3}{2} = \dfrac{1}{2}.$

TRY EXERCISE 14, EXERCISE SET 2.6, PAGE 213

◆ THE DIFFERENCE QUOTIENT

> **take note**
>
> The difference quotient is an important concept that plays a fundamental role in calculus.

The expression

$$\frac{f(x+h) - f(x)}{h}, \quad h \neq 0$$

is called the **difference quotient** of f. It enables us to study the manner in which a function changes in value as the independent variable changes.

EXAMPLE 3 Determine a Difference Quotient

Determine the difference quotient of $f(x) = x^2 + 7$.

Solution

$$\frac{f(x+h) - f(x)}{h} = \frac{[(x+h)^2 + 7] - [x^2 + 7]}{h}$$

• **Apply the difference quotient.**

$$= \frac{[x^2 + 2xh + h^2 + 7] - [x^2 + 7]}{h}$$

$$= \frac{x^2 + 2xh + h^2 + 7 - x^2 - 7}{h}$$

$$= \frac{2xh + h^2}{h} = \frac{\cancel{h}(2x + h)}{\cancel{h}} = 2x + h$$

TRY EXERCISE 30, EXERCISE SET 2.6, PAGE 213

The difference quotient $2x + h$ of $f(x) = x^2 + 7$ from Example 3 is the slope of the secant line through the points

$$(x, f(x)) \quad \text{and} \quad (x + h, f(x + h))$$

For instance, let $x = 1$ and $h = 1$. Then the difference quotient is

$$2x + h = 2(1) + 1 = 3$$

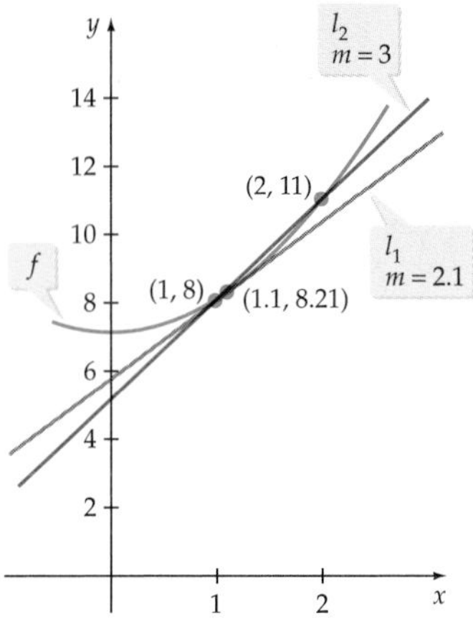

Figure 2.96

Figure 2.97

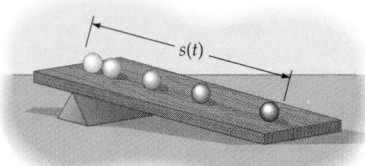

This is the slope of the secant line l_2 through $(1, 8)$ and $(2, 11)$, as shown in **Figure 2.96.** If we let $x = 1$ and $h = 0.1$, then the difference quotient is

$$2x + h = 2(1) + 0.1 = 2.1$$

This is the slope of the secant line l_1 through $(1, 8)$ and $(1.1, 8.21)$.

The difference quotient

$$\frac{f(x + h) - f(x)}{h}$$

can be used to compute *average velocities.* In such cases it is traditional to replace f with s (for distance), the variable x with the varibale a (for the time at the start of an observed interval of time), and the variable h with Δt, where Δt is the difference between the time at the end of an interval and the time at the start of the interval. For example, if an experiment is observed over the time interval from $t = 3$ seconds to $t = 5$ seconds, then the time interval is denoted as $[3, 5]$ with $a = 3$, and $\Delta t = 5 - 3 = 2$. Thus if the distance traveled by a ball that rolls down a ramp is given by $s(t)$, where t is the time in seconds after the ball is released (see **Figure 2.97**), then the **average velocity** of the ball over the interval $t = a$ to $t = a + \Delta t$ is the difference quotient

$$\frac{s(a + \Delta t) - s(a)}{\Delta t}$$

take note

Δt is read "delta t." Delta is a Greek letter often used to indicate difference.

The average velocity of the ball in Example 4 over the interval $[3, 3 + \Delta t]$ is

$$\frac{s(3 + \Delta t) - s(3)}{\Delta t}$$
$$= \frac{4(3 + \Delta t)^2 - 4(3)^2}{\Delta t}$$
$$= \frac{4(9 + 6(\Delta t) + (\Delta t)^2) - 36}{\Delta t}$$
$$= \frac{36 + 24(\Delta t) + 4(\Delta t)^2 - 36}{\Delta t}$$
$$= \frac{24(\Delta t) + 4(\Delta t)^2}{\Delta t}$$
$$= 24 + 4\Delta t \text{ feet per second}$$

Observe that as Δt approaches 0, the average velocity approaches 24 feet per second.

EXAMPLE 4 Evaluate Average Velocities

The distance traveled by a ball rolling down a ramp is given by $s(t) = 4t^2$, where t is the time in seconds after the ball is released, and $s(t)$ is measured in feet. Evaluate the average velocity of the ball for each time interval.

a. $[3, 5]$ **b.** $[3, 4]$ **c.** $[3, 3.5]$ **d.** $[3, 3.01]$

Solution

a. In this case, $a = 3$ and $\Delta t = 2$. Thus the average velocity over this interval is

$$\frac{s(a + \Delta t) - s(a)}{\Delta t} = \frac{s(3 + 2) - s(3)}{2} = \frac{s(5) - s(3)}{2} = \frac{100 - 36}{2}$$
$$= 32 \text{ feet per second}$$

b. Let $a = 3$ and $\Delta t = 4 - 3 = 1$.

$$\frac{s(a + \Delta t) - s(a)}{\Delta t} = \frac{s(3 + 1) - s(3)}{1} = \frac{s(4) - s(3)}{1} = \frac{64 - 36}{1}$$
$$= 28 \text{ feet per second}$$

c. Let $a = 3$ and $\Delta t = 3.5 - 3 = 0.5$.

$$\frac{s(a + \Delta t) - s(a)}{\Delta t} = \frac{s(3 + 0.5) - s(3)}{0.5} = \frac{49 - 36}{0.5} = 26 \text{ feet per second}$$

d. Let $a = 3$ and $\Delta t = 3.01 - 3 = 0.01$.

$$\frac{s(a + \Delta t) - s(a)}{\Delta t} = \frac{s(3 + 0.01) - s(3)}{0.01} = \frac{36.2404 - 36}{0.01}$$
$$= 24.04 \text{ feet per second}$$

TRY EXERCISE 72, EXERCISE SET 2.6, PAGE 214

◆ COMPOSITION OF FUNCTIONS

Composition of functions is yet another method of constructing a function from two given functions. The process consists of using the range element of one function as the domain element of another function.

Composite functions occur in many situations. For example, suppose the manufacturing cost (in dollars) per compact disc player is given by

$$m(x) = \frac{180x + 2600}{x}$$

where x is the number of compact disc players to be manufactured. An electronics outlet agrees to sell the compact discs by marking up the manufacturing cost per player $m(x)$ by 30%. Note that the selling price s will be a function of $m(x)$. More specifically,

$$s[m(x)] = 1.30[m(x)]$$

Simplifying $s[m(x)]$ produces

$$s[m(x)] = 1.30\left(\frac{180x + 2600}{x}\right) = 1.30(180) + 1.30\frac{2600}{x} = 234 + \frac{3380}{x}$$

The function produced in this manner is referred to as the composition of m by s. The notaiton $s \circ m$ is used to denote this composition function. That is,

$$(s \circ m)(x) = 234 + \frac{3380}{x}$$

Composition of Functions

For the functions f and g, the **composite function** or **composition** of f by g is given by

$$(g \circ f)(x) = g[f(x)]$$

for all x in the domain of f such that $f(x)$ is in the domain of g.

If f and g are specified by equations, you can use substitution to find equations that specify $(g \circ f)$ and $(f \circ g)$.

EXAMPLE 5 Form Composite Functions

If $f(x) = x^2 - 3x$ and $g(x) = 2x + 1$, find

a. $(g \circ f)$ b. $(f \circ g)$

Solution

a. $(g \circ f) = g[f(x)] = 2(f(x)) + 1$ • Substitute $f(x)$ for x in g.

$= 2(x^2 - 3x) + 1$ • $f(x) = x^2 - 3x$

$= 2x^2 - 6x + 1$

b. $(f \circ g) = f[g(x)] = (g(x))^2 - 3(g(x))$ • Substitute $g(x)$ for x in f.

$= (2x + 1)^2 - 3(2x + 1)$ • $g(x) = 2x + 1$

$= 4x^2 - 2x - 2$

TRY EXERCISE 38, EXERCISE SET 2.6, PAGE 213

Note that in this example $(f \circ g) \neq (g \circ f)$. In general, the composition of functions is not a commutative operation.

Caution Some care must be used when forming the composition of functions. For instance, if $f(x) = x + 1$ and $g(x) = \sqrt{x - 4}$, then

$$(g \circ f)(2) = g[f(2)] = g(3) = \sqrt{3 - 4} = \sqrt{-1}$$

which is not a real number. We can avoid this problem by imposing suitable restrictions on the domain of f so that the range of f is part of the domain of g. If the domain of f is restricted to $[3, \infty)$, then the range of f is $[4, \infty)$. But this is precisely the domain of g. Note that $2 \notin [3, \infty)$, and thus we avoid the problem of $(g \circ f)(2)$ not being a real number.

To evaluate $(f \circ g)(c)$ for some constant c, you can use either of the following methods.

Method 1 First evaluate $g(c)$. Then substitute this result for x in $f(x)$.

Method 2 First determine $f[g(x)]$ and then substitute c for x.

EXAMPLE 6 Evaluate a Composite Function

take note

In Example 6, both Method 1 and Method 2 produce the same result. Although Method 2 is longer, it is the better method if you must evaluate $(f \circ g)(x)$ for several values of x.

Evaluate $(f \circ g)(3)$, where $f(x) = 2x - 7$ and $g(x) = x^2 + 4$.

Solution

Method 1 $(f \circ g)(3) = f[g(3)]$

$= f[(3)^2 + 4]$ • Evaluate $g(3)$.

$= f(13)$

$= 2(13) - 7 = 19$ • Substitute 13 for x in f.

Method 2 $(f \circ g)(x) = 2[g(x)] - 7$ • **Form $f[g(x)]$.**

$$= 2[x^2 + 4] - 7$$

$$= 2x^2 + 1$$

$(f \circ g)(3) = 2(3)^2 + 1 = 19$ • **Substitute 3 for x.**

TRY EXERCISE 50, EXERCISE SET 2.6, PAGE 213

Figures 2.98 and **2.99** graphically illustrate the difference between Method 1 and Method 2.

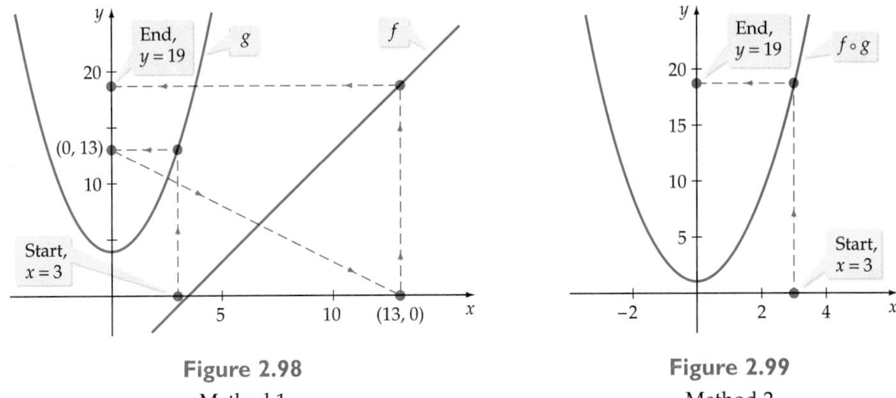

Figure 2.98
Method 1

Figure 2.99
Method 2

EXAMPLE 7 Use a Composite Function to Solve an Application

A graphic artist has drawn a 3-inch by 2-inch rectangle on a computer screen. The artist has been scaling the size of the rectangle for t seconds in such a way that the upper right corner of the original rectangle is moving to the right at the rate of 0.5 inch per second and downward at the rate of 0.2 inch per second. See **Figure 2.100**.

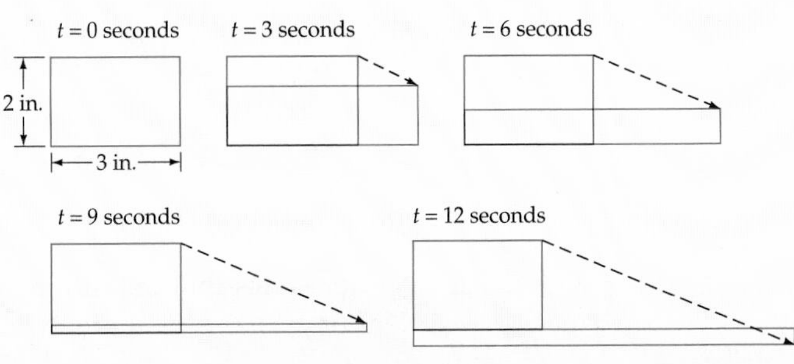

Figure 2.100

a. Write the length l and the width w of the scaled rectangles as functions of t.

Continued •➤

b. Write the area A of the scaled rectangle as a function of t.

c. Find the intervals for which A is an increasing function on $0 \le t \le 14$. Also find the intervals where A is a decreasing function.

d. Find the value of t (where $0 \le t \le 14$) that maximizes $A(t)$.

Solution

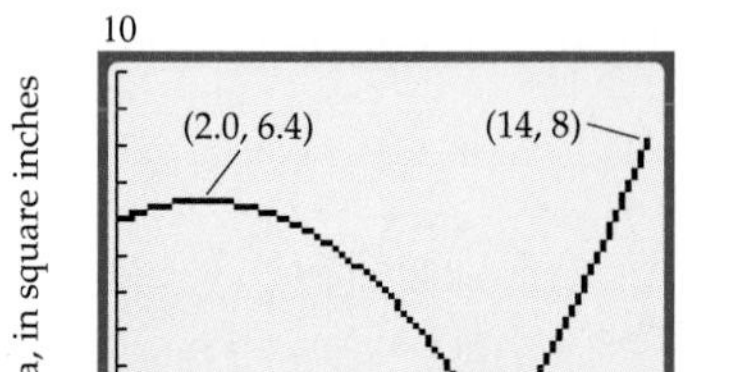

$A = (3 + 0.5t)|2 - 0.2t|$

Figure 2.101

a. Because *distance* = *rate* · *time*, we see that the change in l is given by $0.5t$. Therefore, the length at any time t is $l = 3 + 0.5t$. For $0 \le t \le 10$, the width is given by $w = 2 - 0.2t$. For $10 \le t \le 14$, the width is $w = -2 + 0.2t$. In either case the width can be determined by finding $w = |2 - 0.2t|$. (The absolute value symbol is needed to keep the width positive for $10 < t \le 14$.)

b. $A = lw = (3 + 0.5t)|2 - 0.2t|$

c. Use a graphing utility to determine that A is increasing on $[0, 2]$ and on $[10, 14]$ and that A is decreasing on $[2, 10]$. See **Figure 2.101.**

d. The highest point on the graph of A occurs when $t = 14$ seconds. See **Figure 2.101.**

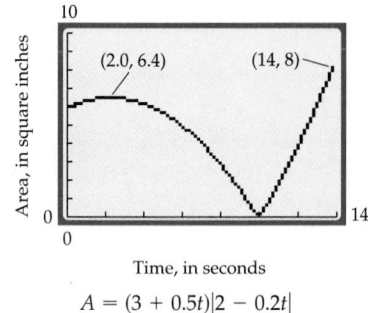

TRY EXERCISE 66, EXERCISE SET 2.6, PAGE 214

You may be inclined to think that if the area of a rectangle is decreasing, then its perimeter is also decreasing, but this is not always the case. For example, the area of the scaled rectangle in Example 7 was shown to decrease on [2, 10] even though its perimeter is always increasing. See Exercise 68 in Exercise Set 2.6.

TOPICS FOR DISCUSSION

1. The domain of $f + g$ consists of all real numbers formed by the *union* of the domain of f and the domain of g. Do you agree?

2. Given $f(x) = 3x - 2$ and $g(x) = \frac{1}{3}x + \frac{2}{3}$, determine $f \circ g$ and $g \circ f$. Does this show that composition of functions is a commutative operation?

3. A tutor states that the difference quotient of $f(x) = x^2$ and the difference quotient of $g(x) = x^2 + 4$ are the same. Do you agree?

4. A classmate states that the difference quotient of any linear function $f(x) = mx + b$ is always m. Do you agree?

5. When we use a difference quotient to determine an average velocity, we generally replace the variable h with the variable Δt. What does Δt represent?

EXERCISE SET 2.6

In Exercises 1 to 12, use the given functions f and g to find $f + g$, $f - g$, fg, and f/g. State the domain of each.

1. $f(x) = x^2 - 2x - 15, \quad g(x) = x + 3$
2. $f(x) = x^2 - 25, \quad g(x) = x - 5$
3. $f(x) = 2x + 8, \quad g(x) = x + 4$
4. $f(x) = 5x - 15, \quad g(x) = x - 3$
5. $f(x) = x^3 - 2x^2 + 7x, \quad g(x) = x$
6. $f(x) = x^2 - 5x - 8, \quad g(x) = -x$
7. $f(x) = 2x^2 + 4x - 7, \quad g(x) = 2x^2 + 3x - 5$
8. $f(x) = 6x^2 + 10, \quad g(x) = 3x^2 + x - 10$
9. $f(x) = \sqrt{x - 3}, \quad g(x) = x$
10. $f(x) = \sqrt{x - 4}, \quad g(x) = -x$
11. $f(x) = \sqrt{4 - x^2}, \quad g(x) = 2 + x$
12. $f(x) = \sqrt{x^2 - 9}, \quad g(x) = x - 3$

In Exercises 13 to 28, evaluate the indicated function, where $f(x) = x^2 - 3x + 2$ and $g(x) = 2x - 4$.

13. $(f + g)(5)$
14. $(f + g)(-7)$
15. $(f + g)\left(\frac{1}{2}\right)$
16. $(f + g)\left(\frac{2}{3}\right)$
17. $(f - g)(-3)$
18. $(f - g)(24)$
19. $(f - g)(-1)$
20. $(f - g)(0)$
21. $(fg)(7)$
22. $(fg)(-3)$
23. $(fg)\left(\frac{2}{5}\right)$
24. $(fg)(-100)$
25. $\left(\frac{f}{g}\right)(-4)$
26. $\left(\frac{f}{g}\right)(11)$
27. $\left(\frac{f}{g}\right)\left(\frac{1}{2}\right)$
28. $\left(\frac{f}{g}\right)\left(\frac{1}{4}\right)$

In Exercises 29 to 36, find the difference quotient of the given function.

29. $f(x) = 2x + 4$
30. $f(x) = 4x - 5$
31. $f(x) = x^2 - 6$
32. $f(x) = x^2 + 11$
33. $f(x) = 2x^2 + 4x - 3$
34. $f(x) = 2x^2 - 5x + 7$
35. $f(x) = -4x^2 + 6$
36. $f(x) = -5x^2 - 4x$

In Exercises 37 to 48, find $g \circ f$ and $f \circ g$ for the given functions f and g.

37. $f(x) = 3x + 5, \quad g(x) = 2x - 7$
38. $f(x) = 2x - 7, \quad g(x) = 3x + 2$
39. $f(x) = x^2 + 4x - 1, \quad g(x) = x + 2$
40. $f(x) = x^2 - 11x, \quad g(x) = 2x + 3$
41. $f(x) = x^3 + 2x, \quad g(x) = -5x$
42. $f(x) = -x^3 - 7, \quad g(x) = x + 1$
43. $f(x) = \dfrac{2}{x + 1}, \quad g(x) = 3x - 5$
44. $f(x) = \sqrt{x + 4}, \quad g(x) = \dfrac{1}{x}$
45. $f(x) = \dfrac{1}{x^2}, \quad g(x) = \sqrt{x - 1}$
46. $f(x) = \dfrac{6}{x - 2}, \quad g(x) = \dfrac{3}{5x}$
47. $f(x) = \dfrac{3}{|5 - x|}, \quad g(x) = -\dfrac{2}{x}$
48. $f(x) = |2x + 1|, \quad g(x) = 3x^2 - 1$

In Exercises 49 to 64, evaluate each composite function, where $f(x) = 2x + 3$, $g(x) = x^2 - 5x$, and $h(x) = 4 - 3x^2$.

49. $(g \circ f)(4)$
50. $(f \circ g)(4)$
51. $(f \circ g)(-3)$
52. $(g \circ f)(-1)$
53. $(g \circ h)(0)$
54. $(h \circ g)(0)$
55. $(f \circ f)(8)$
56. $(f \circ f)(-8)$
57. $(h \circ g)\left(\frac{2}{5}\right)$
58. $(g \circ h)\left(-\frac{1}{3}\right)$
59. $(g \circ f)(\sqrt{3})$
60. $(f \circ g)(\sqrt{2})$
61. $(g \circ f)(2c)$
62. $(f \circ g)(3k)$
63. $(g \circ h)(k + 1)$
64. $(h \circ g)(k - 1)$

65. **WATER TANK** A water tank has the shape of a right circular cone, with height 16 feet and radius 8 feet. Water is

running into the tank so that the radius r (in feet) of the surface of the water is given by $r = 1.5t$, where t is the time (in minutes) that the water has been running.

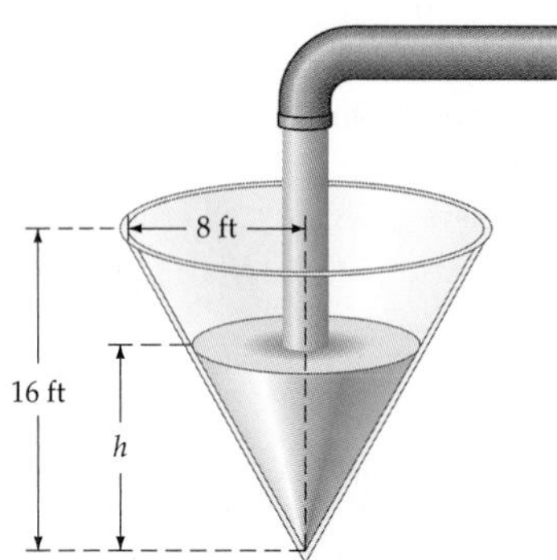

a. The area A of the surface of the water is $A = \pi r^2$. Find $A(t)$ and use it to determine the area of the surface of the water when $t = 2$ minutes.

b. The volume V of the water is given by $V = \dfrac{1}{3}\pi r^2 h$.

Find $V(t)$ and use it to determine the volume of the water when $t = 3$ minutes. (*Hint:* The height of the water in the cone is always twice the radius of the water.

66. **SCALING A RECTANGLE** Work Example 7 of this section with the scaling as follows. The upper right corner of the original rectangle is pulled to the *left* at 0.5 inch per second and downward at 0.2 inch per second.

67. **TOWING A BOAT** A boat is towed by a rope that runs through a pulley that is 4 feet above the point where the rope is tied to the boat. The length (in feet) of the rope from the boat to the pulley is given by $s = 48 - t$, where t is the time in seconds that the boat has been in tow. The horizontal distance from the pulley to the boat is d.

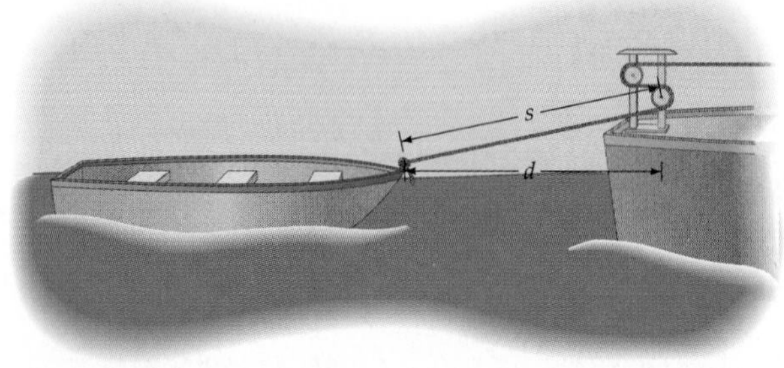

a. Find $d(t)$. b. Evaluate $s(35)$ and $d(35)$.

68. **PERIMETER OF A SCALED RECTANGLE** Show by a graph that the perimeter

$$P = 2(3 + 0.5t) + 2|2 - 0.2t|$$

of the scaled rectangle in Example 7 of this section is an increasing function over $0 \le t \le 14$.

69. **CONVERSION FUNCTIONS** The function $F(x) = x/12$ converts x inches to feet. The function $Y(x) = x/3$ converts x feet to yards. Explain the meaning of $(Y \circ F)(x)$.

70. **CONVERSION FUNCTIONS** The function $F(x) = 3x$ converts x yards to feet. The function $I(x) = 12x$ converts x feet to inches. Explain the meaning of $(I \circ F)(x)$.

71. **CONCENTRATION OF A MEDICATION** The concentration $C(t)$ (in milligrams per liter) of a medication in a patient's blood is given by the data in the following table.

Concentration of Medication in Patient's Blood

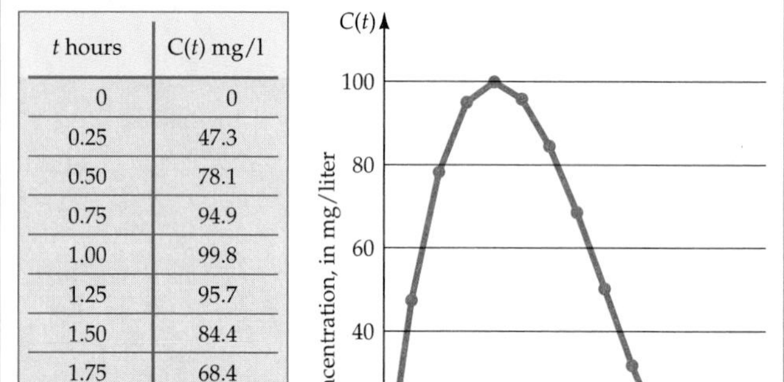

t hours	C(t) mg/l
0	0
0.25	47.3
0.50	78.1
0.75	94.9
1.00	99.8
1.25	95.7
1.50	84.4
1.75	68.4
2.00	50.1
2.25	31.6
2.50	15.6
2.75	4.3

The **average rate of change** of the concentration over the time interval from $t = a$ to $t = a + \Delta t$ is

$$\frac{C(a + \Delta t) - C(a)}{\Delta t}$$

Use the data in the table to evaluate the average rate of change for each of the following time intervals.

a. $[0, 1]$ (*Hint:* In this case, $a = 0$ and $\Delta t = 1$.) Compare this result to the slope of the line through $(0, C(0))$ and $(1, C(1))$.

b. $[0, 0.5]$ c. $[1, 2]$ d. $[1, 1.5]$ e. $[1, 1.25]$

f. The data in the table can be modeled by the function $Con(t) = 25t^3 - 150t^2 + 225t$. Use $Con(t)$ to verify that the average rate of change over $[1, 1 + \Delta t]$ is $-75(\Delta t) + 25(\Delta t)^2$. What does the average rate of change over $[1, 1 + \Delta t]$ seem to approach as Δt approaches 0?

72. **BALL ROLLING ON A RAMP** The distance traveled by a ball rolling down a ramp is given by $s(t) = 6t^2$, where t is the time in seconds after the ball is released, and $s(t)$ is measured in feet. The ball travels 6 feet in 1 second and it travels 24 feet in 2 seconds. Use the difference quotient for average velocity given on page 208 to evaluate the average velocity for each of the following time intervals.

a. $[2, 3]$ (*Hint:* In this case, $a = 2$ and $\Delta t = 1$.) Compare this result to the slope of the line through $(2, f(2))$ and $(3, f(3))$.

b. $[2, 2.5]$ **c.** $[2, 2.1]$ **d.** $[2, 2.01]$ **e.** $[2, 2.001]$

f. Verify that the average velocity over $[2, 2 + \Delta t]$ is $24 + 6(\Delta t)$. What does the average velocity seem to approach as Δt approaches 0?

SUPPLEMENTAL EXERCISES

In Exercises 73 to 78, show that

$$(g \circ f)(x) = x \quad \text{and} \quad (f \circ g)(x) = x$$

73. $f(x) = 2x + 3, \quad g(x) = \dfrac{x - 3}{2}$

74. $f(x) = 4x - 5, \quad g(x) = \dfrac{x + 5}{4}$

75. $f(x) = \dfrac{4}{x + 1}, \quad g(x) = \dfrac{4 - x}{x}$

76. $f(x) = \dfrac{2}{1 - x}, \quad g(x) = \dfrac{x - 2}{x}$

77. $f(x) = x^3 - 1, \quad g(x) = \sqrt[3]{x + 1}$

78. $f(x) = -x^3 + 2, \quad g(x) = \sqrt[3]{2 - x}$

79. Let x be the number of computer monitors to be manufactured. The manufacturing cost (in dollars) per computer monitor is given by the function

$$m(x) = \frac{60x + 34{,}000}{x}$$

A computer store will sell the monitors by marking up the manufacturing cost per monitor $m(x)$ by 45%. Thus the selling price s is a function of $m(x)$ given by the equation

$$s[m(x)] = 1.45[m(x)]$$

a. Express the selling price as a function of the number of monitors to be manufactured. That is, find $(s \circ m)(x)$.

b. Find $(s \circ m)(24{,}650)$.

PROJECTS

1. 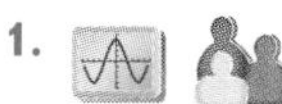 **A GRAPHING UTILITY PROJECT** For any two different real numbers x and y, the larger of the two numbers is given by

$$\text{Maximum}(x, y) = \frac{x + y}{2} + \frac{|x - y|}{2} \qquad (1)$$

a. Verify Equation (1) for $x = 5$ and $y = 9$.

b. Verify Equation (1) for $x = 201$ and $y = 80$.

For any two different functional values $f(x)$ and $g(x)$, the larger of the two is given by

$$\text{Maximum}(f(x), g(x)) = \frac{f(x) + g(x)}{2} + \frac{|f(x) - g(x)|}{2} \qquad (2)$$

To illustrate how we might make use of Equation (2), consider the functions $y_1 = x^2$ and $y_2 = \sqrt{x}$ on the interval from Xmin $= -1$ to Xmax $= 6$. The graphs of y_1 and y_2 are shown at the left and in the middle below.

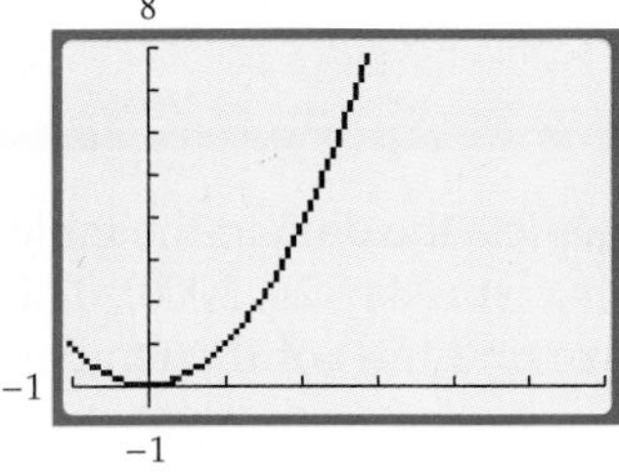

$y_1 = x^2$

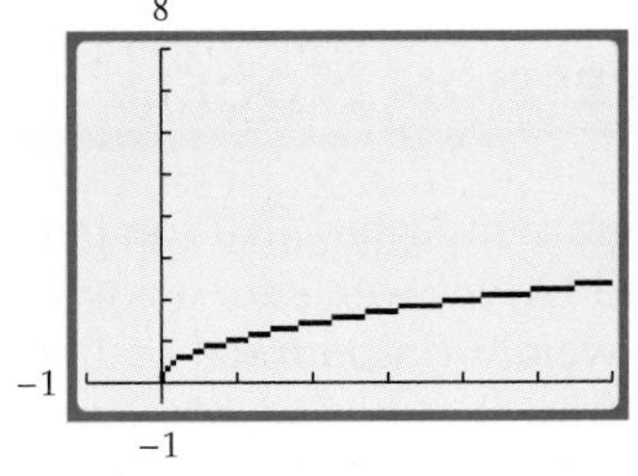

$y_2 = \sqrt{x}$

8

-1 6

-1

$y_3 = (y_1 + y_2)/2 + (\text{abs}\,(y_1 - y_2))/2$

Now consider the function $y_3 = (y_1 + y_2)/2 + (\text{abs}(y_1 - y_2))/2$, where "abs" represents the absolute value function. The graph of y_3 is shown at the right on previous page.

c. Write a sentence or two that explains why the graph of y_3 is as shown.

d. What is the domain of y_1? of y_2? of y_3? Write a sentence that explains how to determine the domain of y_3, given the domain of y_1 and the domain of y_2.

e. Determine a formula for the function Minimum$(f(x), g(x))$.

2. **THE NEVER-NEGATIVE FUNCTION** The author J. D. Murray describes a function f_+ that is defined in the following manner.[1]

$$f_+ = \begin{cases} f & \text{if } f \geq 0 \\ 0 & \text{if } f < 0 \end{cases}$$

We will refer to this function as a **never-negative** function. Never-negative functions can be graphed by using Equation (2) in Exercise 1. For example, if we let $g(x) = 0$, then Equation (2) simplifies to

$$\text{Maximum}(f(x), 0) = \frac{f(x)}{2} + \frac{|f(x)|}{2} \tag{3}$$

The graph of $y = \text{Maximum}(f(x), 0)$ is the graph of $y = f(x)$ provided that $f(x) \geq 0$, and it is the graph of $y = 0$ provided that $f(x) < 0$.

An application: The mosquito population per area of a large resort is controlled by spraying on a monthly basis. A biologist has determined that the mosquito population can be approximated by the never-negative function M_+ with

$$M(t) = -35{,}400(t - \text{int}(t))^2 + 35{,}400(t - \text{int}(t)) - 4000$$

Here t represents the month, and $t = 0$ corresponds to June 1, 2001.

a. Use a graphing utility to graph M for $0 \leq t \leq 3$.

b. Use a graphing utility to graph M_+ for $0 \leq t \leq 3$.

c. Write a sentence or two that explains how the graph of M_+ differs from the graph of M.

d. What is the maximum mosquito population per acre for $0 \leq t \leq 3$? When does this maximum mosquito population occur?

e. Explain when would be the best time to visit the resort, provided that you wished to minimize your exposure to mosquitos.

SECTION 2.7 MODELING DATA USING REGRESSION

- LINEAR REGRESSION MODELS
- CORRELATION COEFFICIENT AND COEFFICIENT OF DETERMINATION
- QUADRATIC REGRESSION MODELS

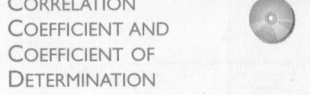

◆ LINEAR REGRESSION MODELS

The scatter diagram in **Figure 2.102** on page 217 depicts the data in the table. (Source: Car and Driver web site, www.10bestcars.com; May 28, 2000). This data shows the curb weight (in pounds) and engine size (in liters) of the ten best cars of 1999 as ranked by *Car and Driver* magazine.

[1] *Mathematical Biology* (New York: Springer-Verlag, 1989), p. 101.

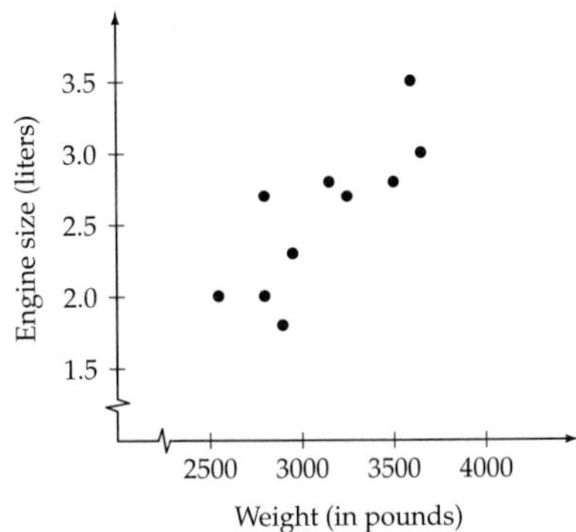

Figure 2.102

take note

A web applet is available to explore linear regression lines. This applet, Least Squares Fit Lines, can be found on our web site at

http://college.hmco.com

Curb Weight and Engine Size of Ten Selected Cars

Car Model	Weight, in pounds	Engine, in liters	Car Model	Weight, in pounds	Engine, in liters
Audi A6	3250	2.7	Ford Focus	2550	2.0
Audi TT	2900	1.8	Honda Accord	2950	2.3
BMW 3 Series	3150	2.8	Honda S2000	2800	2.0
BMW 5 Series	3500	2.8	Lexus GS300	3650	3.0
Chrysler 300M	3600	3.5	Porsche Boxster	2800	2.7

Although there is no one line that passes through every point, we could find an approximate linear model of this data. For instance, the line shown in **Figure 2.103** in blue approximates the data better than the line shown in red. However, as **Figure 2.104** shows, there are many other lines we could have drawn that seem to approximate the data.

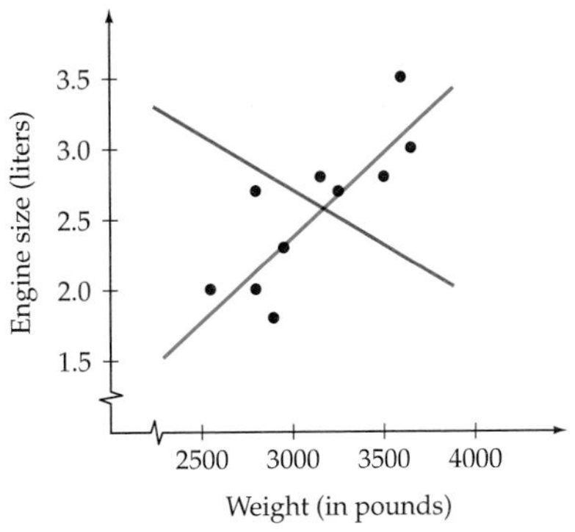

Figure 2.103

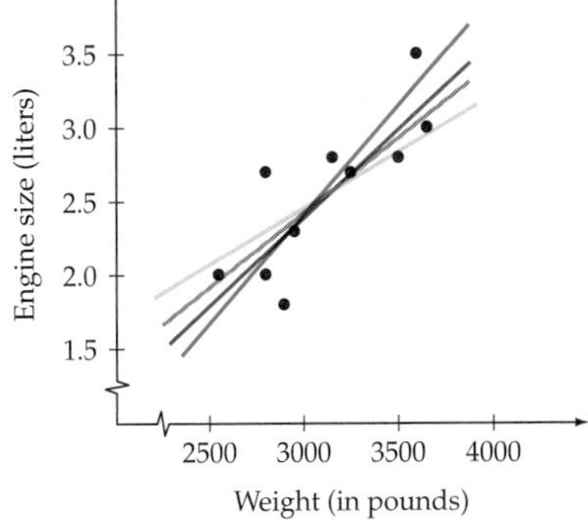

Figure 2.104

To find the line that "best" approximates the data, **regression analysis** is used. This analysis produces the linear function whose graph is called the **line of best fit** or the **least-squares regression line.**[2]

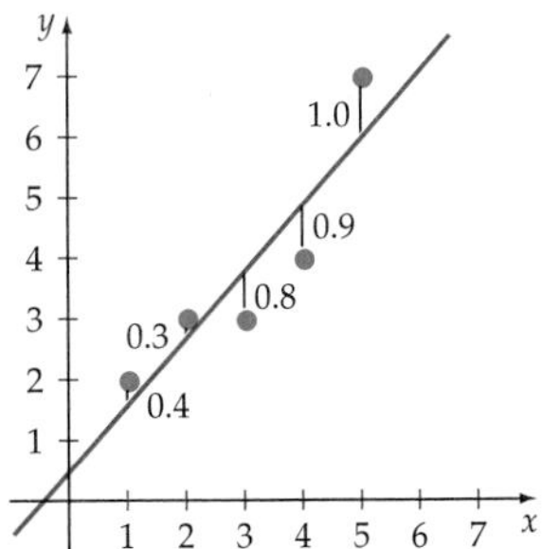

Figure 2.105

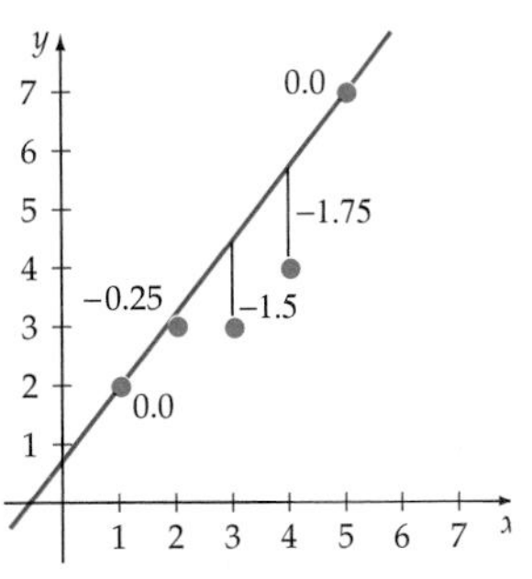

Figure 2.106

Definition of the Least-Squares Regression Line

The **least-squares regression line** is the line that minimizes the sum of the squares of the vertical deviations of all data points from the line.

To help understand this definition, consider the data set $S = \{(1, 2), (2, 3), (3, 3), (4, 4), (5, 7)\}$. As we will show later, the least-squares line for this data is $y = 1.1x + 0.5$. If we evaluate this function at the x-coordinates of the data set S, we obtain the set of ordered pairs $T = \{(1, 1.6), (2, 2.7), (3, 3.8), (4, 4.9), (5, 6)\}$. The vertical deviations are the differences between the y-coordinates in S and the y-coordinates in T. From the definition, we must calculate the sum of the squares of these deviations.

$$(2 - 1.6)^2 + (3 - 2.7)^2 + (3 - 3.8)^2 + (4 - 4.9)^2 + (7 - 6)^2 = 2.7$$

Because $y = 1.1x + 0.5$ is the least squares regression line, for no other line is the sum of the squares of the deviations less than 2.7. For instance, if we consider the

[2] The least-squares regression line is also called the least-squares line and the regression line.

equation $y = 1.25x + 0.75$, which is the equation of the line through the two points $P_1(1, 2)$ and $P_2(5, 7)$ of the data set, the sum of the square deviations is larger than 2.7.

$$(2 - 2)^2 + (3 - 3.25)^2 + (3 - 4.5)^2 + (4 - 5.75)^2 + (7 - 7)^2 = 5.375$$

The equations used to calculate a regression line are somewhat cumbersome. Fortunately, these equations are preprogrammed into most graphing calculators. We will now illustrate the technique for a TI-83 calculator using data set S given previously.

Press STAT. Select EDIT. Press ENTER.

```
EDIT CALC TESTS
1:Edit...
2:SortA(
3:SortD(
4:ClrList
5:SetUpEditor
```

Enter the data.

```
L1     L2     L3   2
1      2      ------
2      3
3      3
4      4
5      7
------
L2(6) =
```

Press STAT. Select 4, LinReg(ax+b). Press ENTER

```
EDIT CALC TESTS
1:1-Var Stats
2:1-Var Stats
3:Med-Med
4:LinReg(ax+b)
5:QuadReg
6:CubicReg
7↓QuartReg
```

Press VARS.

```
LinReg(ax+b)
```

Press Y-VARS. Press ENTER.

```
VARS Y-VARS
1:Function...
2:Parametric...
3:Polar...
4:On/Off...
```

Select 1. Press ENTER.

```
FUNCTION
1:Y1
2:Y2
3:Y3
4:Y4
5:Y5
6:Y6
7↓Y7
```

Press ENTER

```
LinReg(ax+b) Y1
```

View the results.

```
LinReg
 y=ax+b
 a=1.1
 b=.5
 r²=.8175675676
 r=.9041944302
```

From the last screen, the equation of the regression line is $y = 1.1x + 0.5$. The last screen may not look exactly like ours. The information provided on our screen requires that DiagnosticsOn be enabled. This is accomplished using the following keystrokes:

2nd CATALOG (Scroll to DiagnosticsOn) ENTER

With DiagnosticsOn enabled, besides the values for the regression equation, two other values are given. We will discuss these values later in this section.

If you used the keystrokes we have shown above, the regression line will be stored in Y1. This is helpful if you wish to graph the regression line. However, if it is not necessary to graph the regression line, then instead of pressing VARS at step 4, just press ENTER. The result will be the last screen showing the results of the regression calculations.

```
LinReg
y=ax+b
a=.0011450888
b=-1.006951767
r²=.6704074496
r=.8187841288
```

Figure 2.107

EXAMPLE 1 Find a Linear Regression Equation

Find the regression equation for the data on curb weight and engine size given at the beginning of this section. What size engine does the regression equation predict for a car whose curb weight is 2700 pounds? Round to the nearest tenth.

Solution

Using your calculator, enter the data from the table. Then have the calculator produce the values for the regression equation. Your results should be similar to those shown in **Figure 2.107.** The equation of the regression line is

$$y = 0.0011450888x - 1.006951767$$

To find what size engine the regression equation predicts for a car whose curb weight is 2700 pounds, evaluate the regression equation for $x = 2700$.

$$\begin{aligned} y &= 0.001145088x - 1.006951767 \\ &= 0.001145088(2700) - 1.006951767 \approx 2.0847858 \end{aligned}$$

This predicts that a car that weighs 2700 pounds has a 2.1-liter engine.

TRY EXERCISE 18, EXERCISE SET 2.7, PAGE 224

take note

If you followed the steps we gave earlier and stored the regression equation in Y1, then you can evaluate the regression equation using the following keystrokes:

[VARS] ▶ [ENTER] [ENTER] [(] 2700 [)] [ENTER]

◆ CORRELATION COEFFICIENT AND COEFFICIENT OF DETERMINATION

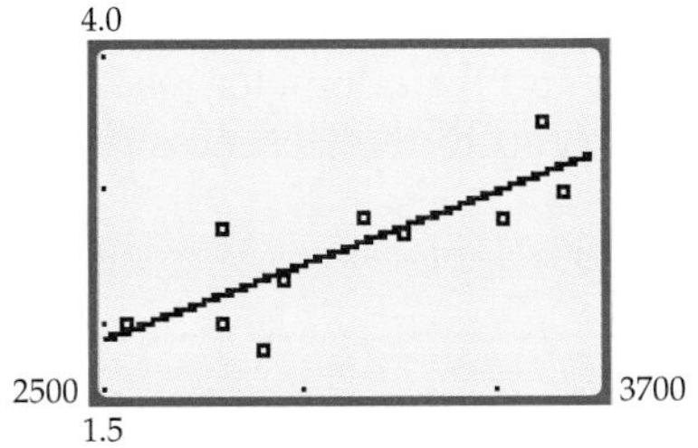

The scatter plot of curb weight versus engine size is shown in the figure at the left, along with the graph of the regression line. Note that the slope of the regression line is positive. This indicates that as the weight of a car increases, the size of the engine increases. Note also that for that data, the value of r on the regression calculation screen was positive, $r \approx 0.904$.

Now consider the data in the table below (Source: Kelley Blue Book web site, May 29, 2000), which shows the trade-in value of a 1998 Porsche Boxster for various odometer readings.

The scatter diagram below is based on the table at the left. The graph of the regression line is also shown. The details for the calculations are shown in "Take Note".

take note

The data for the Porsche Boxster was created assuming that the condition of the car was excellent. The only variable that changed was the odometer reading.

```
LinReg
y=ax+b
a=-228.9189189
b=43261.48649
r²=.9890296836
r=.9944997152
```

Trade-in Value of 1998 Porsche Boxster, May 2000

Odometer Reading, in thousands	Trade-in value, in $
20	38,550
25	37,450
30	36,500
35	35,600
45	32,725

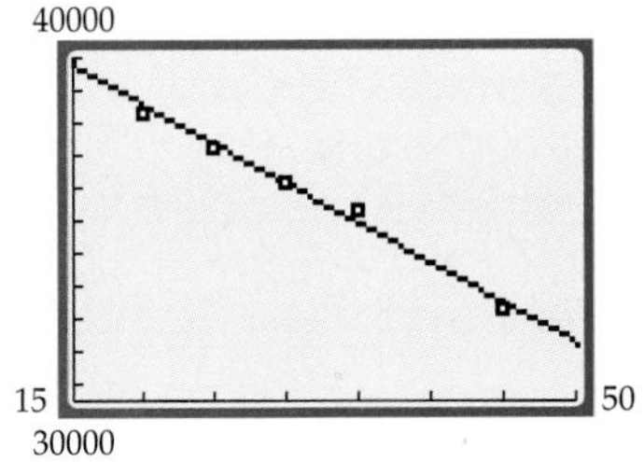

In this case the slope of the regression line is negative. This means that as the odometer reading increases, the trade-in value of the car decreases. Note also that the value of r is negative, $r \approx -0.994$.

Linear Correlation Coefficient

The **linear correlation coefficient r** is a measure of how close the points of a data set can be modeled by a straight line. If $r = -1$, then the points of the data set can be modeled *exactly* by a straight line with negative slope. If $r = 1$, then the data set can be modeled *exactly* by a straight line with positive slope. For all data sets, $-1 \le r \le 1$.

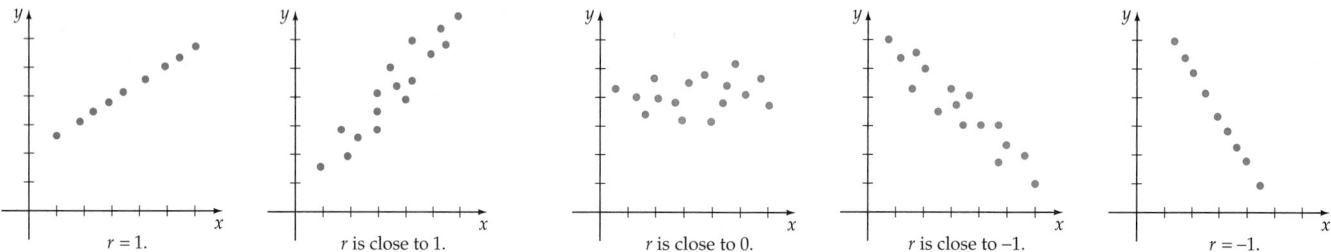

If $r \neq 1$ or $r \neq -1$, then the data set *cannot* be modeled exactly by a straight line. The further the value of r is from 1 or -1, (or in other words, the closer the value of r to zero) the more the ordered pairs of the data set deviate from a straight line.

The graphs that follow show the points of the data sets and the graphs of the regression lines for the curb weight/engine size data and the odometer reading/trade-in data. Note the values of r and their relationship to the closeness of the data points to the regression line.

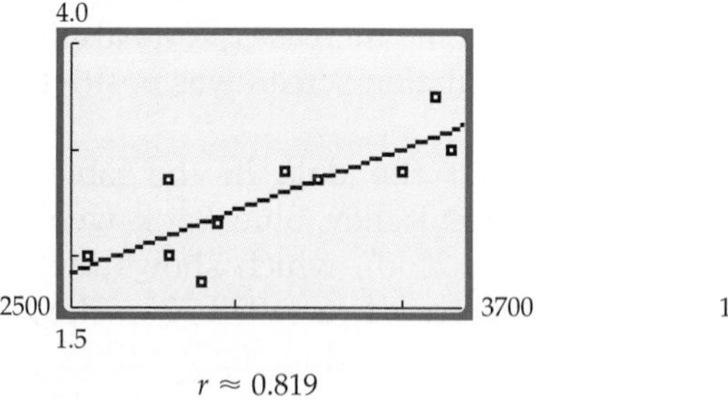

$r \approx 0.819$

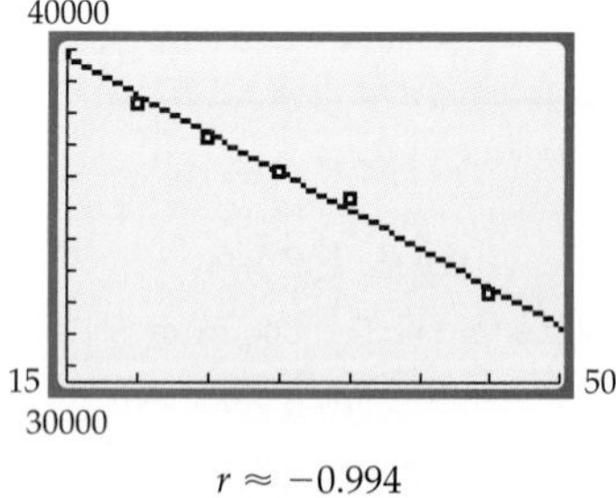

$r \approx -0.994$

Researchers calculate a regression line to determine a relationship between two variables. The researcher wants to know whether the change in one variable produces a predictable change in a second variable. The value of r^2 tells the researcher the extent of that relationship.

Coefficient of Determination

The **coefficient of determination** is r^2. It measures the percent of the total variation in the dependent variable that is explained by the regression equation.

For the curb weight/engine size data, $r^2 \approx 0.670$. This means that approximately 67% of the total variation in the dependent variable (engine size) can be attributed to the regression equation. It also means that car weight alone does not predict with certainty the size of an engine. Other factors, such as desire to promote fuel efficiency, are also involved in the engine size of a car.

QUESTION What is the coefficient of determination for the odometer reading/trade-in value data and what is its significance?

◆ QUADRATIC REGRESSION MODELS

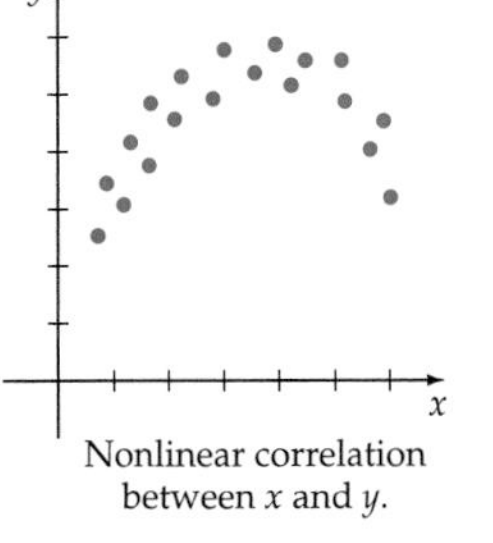

Nonlinear correlation between x and y.

To this point, our focus has been *linear* regression equations. However, there may be a nonlinear relationship between two quantities. The accompanying scatter diagram suggests that a quadratic function might be a better model of the data than a linear model. As we proceed through this text, various functional models will be discussed.

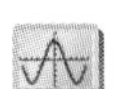

EXAMPLE 2 Find a Quadratic Regression Model

The data in the table below was collected on five successive Saturdays. It shows the average number of cars entering a shopping center parking lot. The value of t is the number of minutes after 9:00 A.M. The value of N is the number of cars that entered the parking lot in the 10 minutes prior to the value of t. Find a regression model for this data.

Average Number of Cars Entering a Shopping Center Parking Lot

t	N	t	N
20	70	140	301
40	135	160	298
60	178	180	284
80	210	200	286
100	260	220	260
120	280	240	195

Continued ▸

ANSWER $r^2 \approx 0.989$. This means that 98.9% of the total variation in trade-in value can be attributed to the odometer reading.

Solution

1. **Construct a scatter diagram for this data.** Enter the data into your calculator as shown earlier.

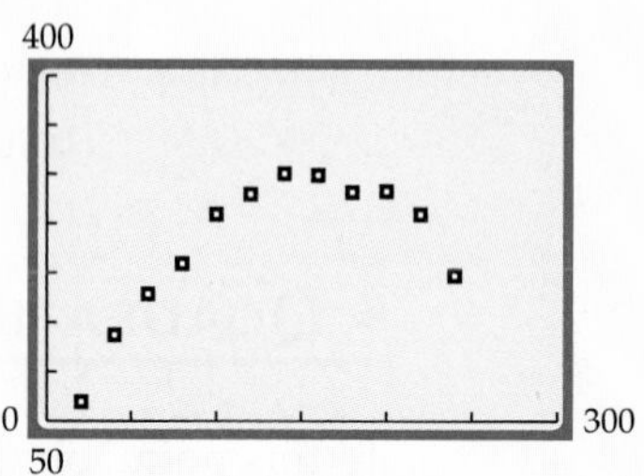

From the scatter diagram, it appears that there is a nonlinear relationship between the variables.

2. **Find the regression equation.** Try a quadratic regression model. For a TI-83 calculator, press STAT ▶ 5 ENTER.

```
QuadReg
 y=ax²+bx+c
 a=.0124881369
 b=3.904433067
 c=-7.25
 R²=.9840995401
```

take note

In the case of nonlinear regression calculations, the value of r cannot be computed. In these cases, the coefficient of determination is used to determine how well the data fits the model.

3. **Examine the coefficient of determination.** The coefficient of determination is approximately 0.984. Because this number is fairly close to 1, the regression equation $y = -0.0124881369x^2 + 3.904433067x - 7.25$ provides a good model of the data.

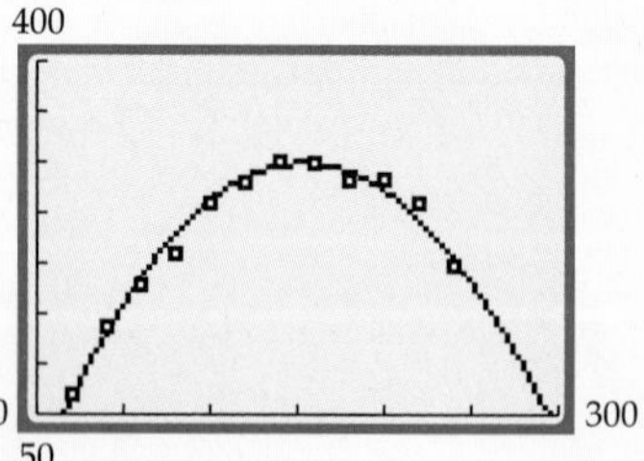

TRY EXERCISE 32, EXERCISE SET 2.7, PAGE 227

```
LinReg
 y=ax+b
 a=.6575174825
 b=144.2727273
 r²=.4193509866
 r=.6475731515
```

For Example 2, we could have calculated the *linear* regression line for the data. The results are shown here. Note that the coefficient of determination for this calculation is approximately 0.419. Because this number is less than the coefficient of determination for the quadratic model, we choose a quadratic model of the data rather than a linear model.

Now for a final note: The regression line equation does not *prove* that the changes in the dependent variable are *caused* by the independent variable. For instance, suppose various cities throughout the United States were randomly selected and the numbers of gas stations (independent variable) and restaurants (dependent variable) were recorded in a table. If we calculated the regression equation for this data, we would find that r would be close to 1. However, this does not mean that gas stations *cause* restaurants to be built. The primary cause is that there are fewer gas stations and restaurants in cities with small populations and greater numbers of gas stations and restaurants in cities with large populations.

TOPICS FOR DISCUSSION

1. What is the purpose of calculating the equation of a regression line?
2. Discuss the implications of the following correlation coefficients: $r = -1$, $r = 0$, and $r = 1$.
3. Discuss the coefficient of determination and what its value says about a data set.
4. What are the implications of $r^2 = 1$ for a nonlinear regression equation?

EXERCISE SET 2.7

Use a graphing calculator for this Exercise Set.

For Exercises 1 to 4, determine whether the scatter diagram suggests a linear relationship between x and y, a nonlinear relationship between x and y, or no relationship between x and y.

1. 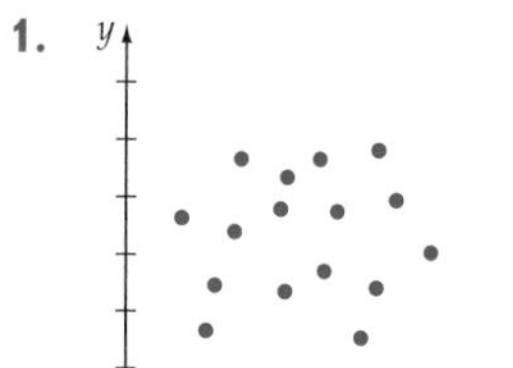

2.

3. 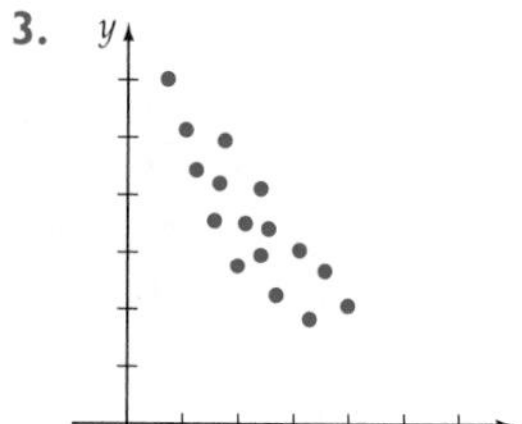

4. 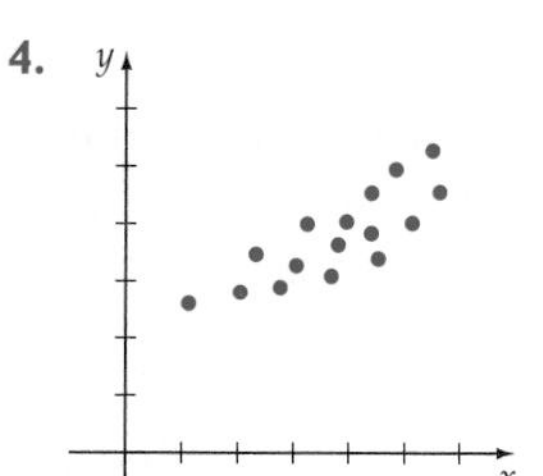

For Exercises 5 and 6, determine for which scatter diagram, A or B, the coefficient of determination is closer to 1.

5. 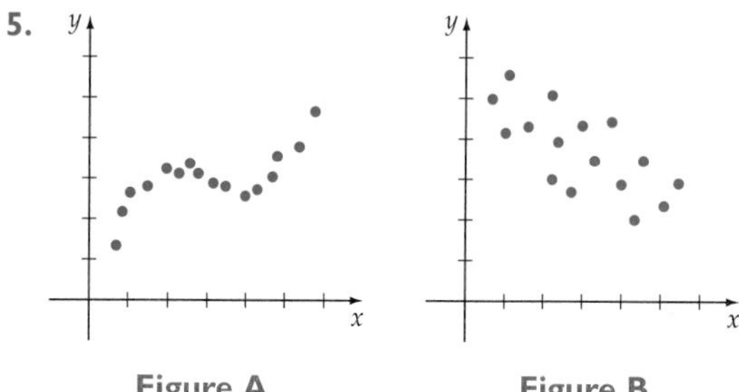

Figure A Figure B

6. 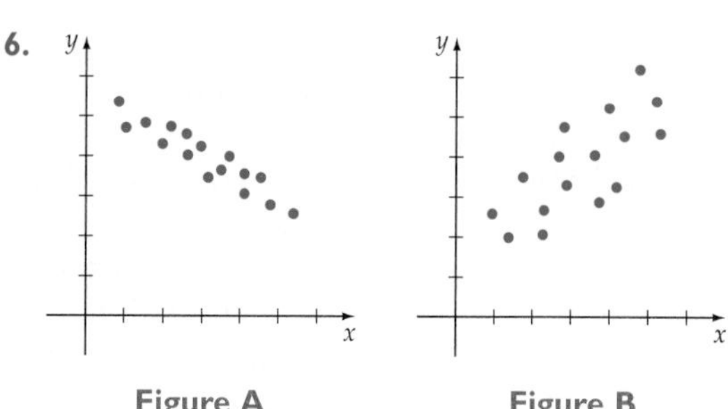

Figure A Figure B

For Exercises 7 to 12, find the linear regression equation for the given set.

7. $\{(2, 6), (3, 6), (4, 8), (6, 11), (8, 18)\}$

8. $\{(2, -3), (3, -4), (4, -9), (5, -10), (7, -12)\}$

9. $\{(-3, 11.8), (-1, 9.5), (0, 8.6), (2, 8.7), (5, 5.4)\}$

10. $\{(-7, -11.7), (-5, -9.8), (-3, -8.1), (1, -5.9), (2, -5.7)\}$

11. $\{(1.3, -4.1), (2.6, -0.9), (5.4, 1.2), (6.2, 7.6), (7.5, 10.5)\}$

12. $\{(-1.5, 8.1), (-0.5, 6.2), (3.0, -2.3), (5.4, -7.1), (6.1, -9.6)\}$

For Exercises 13 to 16, find a quadratic model of the given data.

13. $\{(1, -1), (2, 1), (4, 8), (5, 14), (6, 25)\}$

14. $\{(-2, -5), (-1, 0), (0, 1), (1, 4), (2, 4)\}$

15. $\{(1.5, -2.2), (2.2, -4.8), (3.4, -11.2), (5.1, -20.6), (6.3, -28.7)\}$

16. $\{(-2, -1), (-1, -3.1), (0, -2.9), (1, 0.8), (2, 6.8), (3, 15.9)\}$

For Exercises 17 to 32, determine a regression model of the data.

17. ARCHEOLOGY The data below shows the length, in centimeters, of the humerous and the total wingspan, in centimeters, of several pterosaurs, which are extinct flying reptiles of the order Pterosauria. (Source: Southwest Educational Development Laboratory)

Pterosaur Data

Humerous	Wingspan	Humerous	Wingspan
24	600	20	500
32	750	27	570
22	430	15	300
17	370	15	310
13	270	9	240
4.4	68	4.4	55
3.2	53	2.9	50
1.5	24		

a. Compute the linear regression equation for this data.

b. On the basis of this model, what is the projected wingspan of the pterosaur *Quetzalcoatlus northropi*, which is thought to have been largest of the prehistoric birds, if its humerous is 54 centimeters?

18. CONSUMER SCIENCE The table in the next column shows the trade-in value for a 2-door, 1996 Ford Explorer in excellent condition for various odometer readings in thousands of miles. (Source: Kelley Blue Book web site, May–June, 2000, Edition)

Trade-in Value of 1996 Ford Explorer, May 2000

Odometer	Trade-in	Odometer	Trade-in
45	11,635	70	9,710
50	11,435	75	9,460
60	10,735	80	8,985
68	10,060	95	8,260

a. Compute the linear regression equation for this data.

b. On the basis of this model, what is the expected trade-in value of a similar Ford Explorer with 55,000 miles on the odometer?

19. BOTANY The data in the table below is based on a study by R. A. Fisher of various flowers of the iris family. The width, in centimeters, and length, in centimeters, of the petal for selected flowers are shown in the table.

Iris Petal Data

Width	Length	Width	Length
2	14	24	56
23	51	10	36
20	52	19	51
13	45	16	47
17	45	14	47
16	31	17	45
14	47	16	31

a. Compute the linear regression equation for this data.

b. On the basis of this model, what is the estimated length of an iris petal if the iris has a petal width of 18 centimeters?

20. BOTANY The study by R. A. Fisher (see Exercise 19) also included the width, in centimeters, and length, in centimeters, of the sepal for these flowers. Some of the data is shown below.

Iris Sepal Data

Width	Length	Width	Length
33	50	31	67
31	69	36	46
30	65	27	58
28	57	33	63
25	49	32	70
31	48	25	63
32	70	25	63

a. Compute the linear correlation coefficient for this data.

b. On the basis of the value of the linear correlation coefficient, is a linear model of the data reasonable?

21. **HEALTH** The body mass index (BMI) of a person is a measure of the person's ideal body weight. The table below shows the BMI for different weights for a person 5 feet 6 inches tall. (Source: San Diego *Union-Tribune*, May 31, 2000)

BMI Data for Person 5' 6" Tall

Weight (lb)	BMI	Weight (lb)	BMI
110	17	160	25
120	19	170	27
125	20	180	29
135	21	190	30
140	22	200	32
145	23	205	33
150	24	215	34

a. Compute the linear regression equation for this data.

b. On the basis of the model, what is the estimated BMI for a person 5 feet 6 inches tall whose weight is 158 pounds?

22. **HEALTH** The BMI (see Exercise 21) of a person depends on height as well as weight. The table below shows the changes in BMI for a 150-pound person as height (in inches) changes. (Source: San Diego *Union-Tribune*, May 31, 2000)

BMI Data for 150-Pound Person

Height (in.)	BMI	Height (in.)	BMI
60	29	71	21
62	27	72	20
64	25	73	19
66	24	74	19
67	23	75	18
68	23	76	18
70	21		

a. Compute the linear regression equation for this data.

b. On the basis of the model, what is the estimated BMI for a 150 pound-person who is 5 feet 8 inches tall?

23. **INDUSTRIAL ENGINEERING** Permanent-magnet direct-current motors are used in a variety of industrial applications. For these motors to be effective, there must be a strong linear relationship between the current (in amps, A) supplied to the motor and the resulting torque (in newton-centimeters, N-cm) produced by the motor. A randomly selected motor is chosen from a production line and tested, with the following results.

Direct-Current Motor Data at 12 Volts

Current, in A	Torque, in N-cm	Current, in A	Torque, in N-cm
7.3	9.4	8.5	8.6
11.9	2.8	7.9	4.3
5.6	5.6	14.5	9.5
14.2	4.9	12.7	8.3
7.9	7.0	10.6	4.7

Based on the data in this table, is the chosen motor effective? Explain.

24. **HEALTH SCIENCES** The average remaining lifetime for men in the United States is given in the table below (Source: National Institutes of Health).

Average Remaining Lifetime for Men

Age	Years	Age	Years
0	73.6	65	15.9
15	59.4	75	9.9
35	40.8		

Based on the data in this table, is there a strong correlation between a man's age and the average remaining lifetime for that man? Explain.

25. **HEALTH SCIENCES** The average remaining lifetime for women in the United States is given in the table below (Source: National Institutes of Health).

Average Remaining Lifetime for Women

Age	Years	Age	Years
0	79.4	65	19.2
15	65.1	75	12.1
35	45.7		

Based on the data in this table, is there a strong correlation between a woman's age and the average remaining lifetime for that woman?

a. Compute the linear regression equation for this data.

b. On the basis of the model, what is the estimated remaining lifetime of a woman of age 25?

26. HEALTH SCIENCES The infant mortality rate for respiratory disease syndrome (RDS) in the United States has been declining since 1975. The data in the table below show the number of RDS deaths per 100,000 live births for various years (Source: National Institutes of Health).

Infant Mortality for RDS in U.S.

Year	Deaths	Year	Deaths
1975	248	1988	103
1978	180	1990	75
1980	149	1998	40
1985	105		

a. Based on this data, what was the expected infant mortality rate from RDS in 2001?

b. Does the answer to part **a** make sense? Explain.

27. AUTOMOTIVE TECHNOLOGY The table below shows the EPA estimates for city and highway driving for ten selected luxury cars (Source: www.money.com, May 26, 2000).

EPA Miles-per-Gallon Estimates for City and Highway Driving for Selected Luxury Cars

Car	City mpg	Highway mpg
Acura RL	18	24
Audi A8, 4.2L	17	24
BMW 528i	21	29
Cadillac Deville	17	27
Infiniti Q45	18	23
Jaguar XJ8	17	24
Lexus LS400	18	25
Lincoln Continental	17	25
Mercedes S500	16	23
Saab	18	24

Is there a strong linear relationship between city mpg and highway mpg for these cars. Explain.

28. AVIATION The table in the next column is based on data from the Federal Aviation Administration for planes flown in 1998. The table shows, for selected airlines, the total number of hours an airline's planes operated and the total number of miles flown by those planes (Source: Federal Aviation Administration).

Total Hours of Operation of Planes and the Total Number of Miles Flown by Those Planes

Airline	Hours Flown, in thousands	Miles Flown, in millions
Alaska	295	126
American	2054	945
Continental	1054	476
Delta	1787	793
Frontier	43	17
Midwest Express	72	31
Northwest	1089	485
Southwest	898	357
United	1912	895
US Air	1045	423

Is there a strong linear relationship between hours flown and miles flown for these airlines? Explain.

29. BIOLOGY The survival of certain larvae after hatching depends on the temperature (in degrees Celsius) of the surrounding environment. The table below shows the number of larvae that survive at various temperatures. Find a quadratic model of this data.

Larvae Surviving for Various Temperatures

Temperature	Number Surviving	Temperature	Number Surviving
20	40	26	68
21	47	27	67
22	52	28	64
23	61	29	62
24	64	30	61
25	64		

30. METEOROLOGY The temperature at various times of a summer day at a resort in Southern California is given in the following table. The variable t is the number of minutes after 6:00 A.M., and the variable T is the temperature in degrees Fahrenheit.

Temperatures at a Resort

Time, t	Temperature, T	Time, t	Temperature, T
20	59	240	86
40	65	280	88
80	71	320	86
120	78	360	85
160	81	400	80
200	83		

a. Find a quadratic model for this data.

b. Use the model to predict the temperature at 1:00 P.M.

31. **AUTOMOTIVE ENGINEERING** The fuel efficiency, in miles per gallon, for a certain midsize car at various speeds, in miles per hour, is given in the table below.

Fuel Efficiency of a Midsize Car

mph	mpg	mph	mpg
25	29	55	31
30	32	60	28
35	33	65	24
40	35	70	19
45	34	75	17
50	33		

a. Find a quadratic model for this data.

b. Use the model to predict the fuel efficiency of this car when it is traveling at a speed of 50 mph.

32. **BIOLOGY** The data in the table below shows the oxygen consumption in milliliters per minute of a bird flying level at various speeds in kilometers per hour.

Oxygen Consumption

Speed	Consumption
20	32
25	27
28	22
35	21
42	26
50	34

a. Find a quadratic model for this data.

b. Use the model to determine the speed at which the bird has minimum oxygen consumption.

SUPPLEMENTAL EXERCISES

33. **PHYSICS** Galileo (1564–1642) studied the acceleration due to gravity by allowing balls of various weights to roll down an incline. This allowed him to time the descent of a ball more accurately than by just dropping the ball. The data in the table show some possible results of such an experiment using balls of different masses. Time, t, is measured in seconds; distance, s, is measured in centimeters.

Distance Traveled for Balls of Various Weights

5-Pound Ball		10-Pound Ball		15-Pound Ball	
t	s	t	s	t	s
2	2	3	5	3	5
4	10	6	22	5	15
6	22	9	49	7	30
8	39	12	87	9	49
10	61	15	137	11	75
12	86	18	197	13	103
14	120			15	137
16	156				

a. Find a quadratic model for each of the balls.

b. On the bases of a similar experiment, Galileo concluded that if air resistance is excluded, all falling objects fall with the same acceleration. Explain how one could conclude that from the regression equations.

34. **ASTRONOMY** In 1929, Edwin Hubble published a paper that revolutionized astronomy ("A Relationship Between Distance and Radial Velocity Among Extra-Galactic Nebulae," *Proceedings of the National Academy of Science,* 168). His paper dealt with the distance an extra-galactic nebula was from the Milky Way galaxy and the nebula's velocity with respect to the Milky Way. The data is given in the table below. Distance is measured in megaparsecs (1 megaparsec equals 1.918×10^{19} miles), and velocity (called the *recession velocity*) is measured in kilometers per second. A negative velocity means the nebula is moving toward the Milky Way; a positive velocity means the nebula is moving away from the Milky Way.

Recession Velocities

Distance	Velocity	Distance	Velocity
0.032	170	0.9	650
0.034	290	0.9	150
0.214	−130	0.9	500
0.263	−70	1.0	920
0.275	−185	1.1	450
0.275	−220	1.1	500
0.45	200	1.4	500
0.5	290	1.7	960
0.5	270	2.0	500
0.63	200	2.0	850
0.8	300	2.0	800
0.9	−30	2.0	1090

a. Find the linear regression model for this data.

b. On the basis of this model, what is the recession velocity of a nebula that is 1.5 megaparsecs from the Milky Way?

35. The data in the table at the right was collected on five successive Saturdays. It shows the average number of cars entering a shopping center parking lot. The value of t is the number of minutes after 9:00 A.M. The value of N is the number of cars that entered the parking lot in the 10 minutes prior to the value of t. Does a linear or quadratic regression model better fit this data? Explain.

Average Number of Cars Entering a Parking Lot

t	N	t	N
20	70	140	301
40	135	160	298
60	178	180	284
80	210	200	286
100	260	220	260
120	280	240	195

PROJECTS

Another linear model of data is called the **median–median line**. This line employs *summary points* calculated using the medians of subsets of the independent and dependent variables. The **median** of a data set is the middle number or the average of the two middle numbers for a data set arranged in numerical order. For instance, to find the median of $\{8, 12, 6, 7, 9\}$, first arrange the data in numerical order.

$$6, 7, 8, 9, 12$$

The median is 8, the number in the middle. To find the median of $\{15, 12, 20, 9, 13, 10\}$, arrange the numbers in numerical order.

$$9, 10, 12, 13, 15, 20$$

The median is 12.5, the average of the two middle numbers.

$$\text{Median} = \frac{12 + 13}{2} = 12.5$$

The median–median line is determined by dividing a data set into 3 equal groups. (If the set cannot be divided into 3 equal groups, the first and third groups should be equal. For instance, if there are 11 data points, divide the set into groups of 4, 3, and 4.) The slope of the median–median line is the slope of the line through the x-medians and y-medians of the first and third set of points. The median–median line passes through the average of the x- and y-medians of all three sets.

A graphing calculator can be used to find the median–median line. This line, along with the linear regression line, is shown below for the data in the accompanying table.

x	y
2	3
3	5
4	4
5	7
6	8
7	9
8	12
9	12
10	14
11	15
12	14

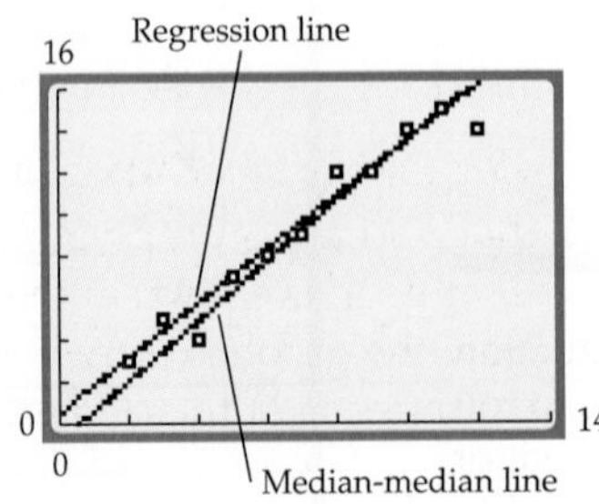

1. Find the median–median line for the data in Exercise 17.
2. Find the median–median line for the data in Exercise 18.
3. Consider the data set $\{(1,3), (2,5), (3,7), (4,9), (5,11), (6,13), (7,15), (8,17)\}$.
   a. Find the linear regression line for this data.
   b. Find the median–median line for this data.
   c. What conclusion might you draw from the answers to parts **a** and **b**?
4. For this exercise, use the data in the table in Project 1.
   a. Calculate the median–median line and the linear regression line.
   b. Change the entry (12, 14) to (12, 1) and then recalculate the median–median line and the linear regression line.
   c. Explain why there is more change in the linear regression line than in the median–median line.

EXPLORING CONCEPTS WITH TECHNOLOGY

Graphing Piecewise Functions with a Graphing Calculator

A graphing calculator can be used to graph piecewise functions by including as part of the function the interval on which each piece of the function is defined. The method is based on the fact that a graphing calculator "evaluates" inequalities. For purposes of this Exploration, we will use keystrokes for a TI-83 calculator.

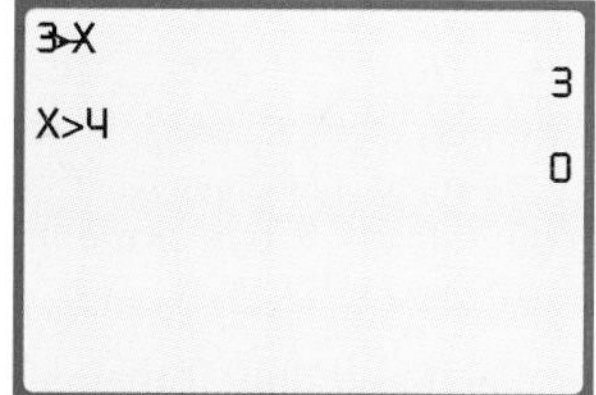

For instance, store 3 in X by pressing 3 [STO▶] [X,T,Θ,n] [ENTER]. Now enter the inequality $x > 4$ by pressing [X,T,Θ,n] [2nd] TEST 3 4 [ENTER]. Your screen should look something like the one at the left. Note that the value of the inequality is 0. This occurs because the calculator replaced X by 3 and then determined whether the inequality $3 > 4$ was true or false. The calculator expresses the fact that the inequality is false by placing a zero on the screen. If we repeat the sequence of steps above, except that we store 5 in X instead of 3, the calculator will determine that the inequality is true and place a 1 on the screen.

This property of calculators is used to graph piecewise functions. Graphs of these functions work best when the Dot mode rather than Connected mode is used. To switch to Dot mode, select [MODE], use the arrow keys to highlight [DOT], and then press [ENTER].

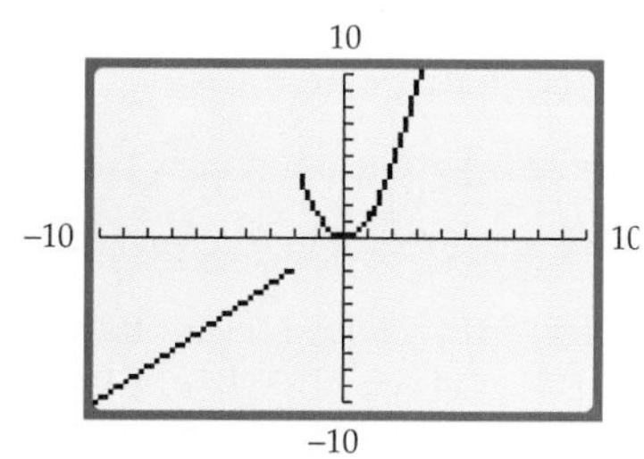

Now we will graph the piecewise function defined by $f(x) = \begin{cases} x, & x \le -2 \\ x^2, & x > -2 \end{cases}$.

Enter the function[3] as Y1=X*(X≤-2)+X²*(X>-2) and graph this in the standard viewing window. Note that you are multiplying each piece of the function by its domain. The graph will appear as shown at the left.

To understand how the graph is drawn, we will consider two values of x, -8 and 2, and evaluate Y1 for each of those values.

[3] Note that pressing [2nd] TEST will display the inequality menu.

Y1=X*(X≤-2)+X²*(X>-2)

$= -8(-8 \le -2) + (-8)^2(-8 > -2)$

$= -8(1) + 64(0) = -8$

• When $x = -8$, the value assigned to $-8 \le -2$ is 1; the value assigned to $-8 > -2$ is 0.

Y1=X*(X≤-2)+X²*(X>-2)

$= 2(2 \le -2) + 2^2(2 > -2)$

$= 2(0) + 4(1) = 4$

• When $x = 2$, the value assigned to $2 \le -2$ is 0; the value assigned to $2 > -2$ is 1.

In a similar manner, for any value of x for which $x \le -2$, the value assigned to (X≤−2) is 1 and the value assigned to (X>−2) is 0. Thus Y1=X*1+X²*0=X on that interval. This means that only the $f(x) = x$ piece of the function is graphed. When $x > -2$, the value assigned to (X≤−2) is 0 and the value assigned to (X>−2) is 1. Thus Y1=X*0+X²*1=X² on that interval. This means that only the $f(x) = x^2$ piece of the function is graphed on that interval.

1. Graph: $f(x) = \begin{cases} x^2, & x < 2 \\ -x, & x \ge 2 \end{cases}$

2. Graph: $f(x) = \begin{cases} x^2 - x, & x < 2 \\ -x + 4, & x \ge 2 \end{cases}$

3. Graph: $f(x) = \begin{cases} -x^2 + 1, & x < 0 \\ x^2 - 1, & x \ge 0 \end{cases}$

4. Graph: $f(x) = \begin{cases} x^3 - 4x, & x < 1 \\ x^2 - x + 2, & x \ge 1 \end{cases}$

CHAPTER 2 SUMMARY

2.1 A Two-Dimensional Coordinate System and Graphs

- *The Distance Formula* The distance d between the points represented by (x_1, y_1) and (x_2, y_2) is

$$d = \sqrt{(x_2 - x_1)^2 + (y_2 - y_1)^2}$$

- The midpoint of the line segment from $P_1(x_1, y_1)$ to $P_2(x_2, y_2)$ is

$$\left(\frac{x_1 + x_2}{2}, \frac{y_1 + y_2}{2}\right)$$

- The standard form of the equation of a circle with center at (h, k) and radius r is $(x - h)^2 + (y - k)^2 = r^2$.

2.2 Introduction to Functions

- *Definition of a Function* A function is a set of ordered pairs in which no two ordered pairs that have the same first coordinate have different second coordinates.
- A graph is the graph of a function if and only if no vertical line intersects the graph at more than one point. If every horizontal line intersects the graph of a function at most once, then the graph is the graph of a one-to-one function.

2.3 Linear Functions

- A function is a linear function of x if it can be written in the form $f(x) = mx + b$, where m and b are real numbers and $m \ne 0$.
- The slope m of the line passing through the points $P_1(x_1, y_1)$ and $P_2(x_2, y_2)$ with $x_1 \ne x_2$ is given by

$$m = \frac{y_2 - y_1}{x_2 - x_1}$$

- The graph of the equation $f(x) = mx + b$ has slope m and y intercept $(0, b)$.
- Two nonvertical lines are parallel if and only if their slopes are equal. Two lines with slopes m_1 and m_2 are perpendicular if and only if $m_1 = -\dfrac{1}{m_2}$.

2.4 Quadratic Functions

- A quadratic function of x is a function that can be represented by an equation of the form $f(x) = ax^2 + bx + c$, where a, b, and c are real numbers and $a \ne 0$.
- The vertex of the graph of $f(x) = ax^2 + bx + c$ is

$$\left(-\frac{b}{2a}, f\left(-\frac{b}{2a}\right)\right)$$

- Every quadratic function $f(x) = ax^2 + bx + c$ can be written in the standard form $f(x) = a(x - h)^2 + k$, $a \ne 0$. The graph of f is a parabola with vertex (h, k). The parabola is symmetric with respect to the vertical line $x = h$, which is called the axis of symmetry of the parabola. The parabola opens up if $a > 0$; it opens down if $a < 0$.

2.5 Properties of Graphs

- The graph of an equation is symmetric with respect to

  the y-axis if the replacement of x with $-x$ leaves the equation unaltered.

  the x-axis if the replacement of y with $-y$ leaves the equation unaltered.

  the origin if the replacement of x with $-x$ and y with $-y$ leaves the equation unaltered.

- If f is a function and c is a positive constant, then

  $y = f(x) + c$ is the graph of $y = f(x)$ shifted up *vertically* c units

  $y = f(x) - c$ is the graph of $y = f(x)$ shifted down *vertically* c units

  $y = f(x + c)$ is the graph of $y = f(x)$ shifted left *horizontally* c units

  $y = f(x - c)$ is the graph of $y = f(x)$ shifted right *horizontally* c units

- The graph of

  $y = -f(x)$ is the graph of $y = f(x)$ reflected across the x-axis.

  $y = f(-x)$ is the graph of $y = f(x)$ reflected across the y-axis.

2.6 The Algebra of Functions

- For all values of x for which both $f(x)$ and $g(x)$ are defined, we define the following functions.

  Sum $(f + g)(x) = f(x) + g(x)$

  Difference $(f - g)(x) = f(x) - g(x)$

  Product $(fg)(x) = f(x) \cdot g(x)$

  Quotient $\left(\frac{f}{g}\right)(x) = \frac{f(x)}{g(x)}, \quad g(x) \neq 0$

- The expression

$$\frac{f(x + h) - f(x)}{h}, \quad h \neq 0$$

  is called the difference quotient of f. The difference quotient is an important function because it can be used to compute the *average rate of change* of f over the time interval $[x, x + h]$.

- For the functions f and g, the composite function, or composition, of f by g is given by $(g \circ f)(x) = g[f(x)]$ for all x in the domain of f such that $f(x)$ is in the domain of g.

2.7 Modeling Data Using Regression

- Regression analysis is used to find a mathematical model of collected data.
- The least-squares regression line is the line that minimizes the sum of the squares of the vertical deviations of all data points from the line.
- The linear correlation coefficient r is a measure of how close the points of a data set can be modeled by a straight line. If $r = -1$, then the points of the data set can be modeled *exactly* by a straight line with negative slope. If $r = 1$, then the data set can be modeled *exactly* by a straight line with positive slope. For all data sets $-1 \leq r \leq 1$.
- The coefficient of determination is r^2. It measures the percent of the total variation in the dependent variable that is explained by the regression line.
- It is possible to find both linear and nonlinear mathematical models of data.

CHAPTER 2 TRUE/FALSE EXERCISES

In Exercises 1 to 12, answer true or false. If the statement is false, give an example to show that the statement is false.

1. Let f be any function. Then $f(a) = f(b)$ implies that $a = b$.
2. If f and g are two functions, then $(f \circ g)(x) = (g \circ f)(x)$.
3. If f is not a one-to-one function, then there are at least two numbers, u and v, in the domain of f, for which $f(u) = f(v)$.
4. Let f be a function such that $f(x) = f(x + 4)$ for all real numbers x. If $f(2) = 3$, then $f(18) = 3$.
5. For all functions, f, $[f(x)]^2 = f[f(x)]$.
6. Let f be any function. Then for all a and b in the domain of f such that $f(b) \neq 0$ and $b \neq 0$,
$$\frac{f(a)}{f(b)} = \frac{a}{b}$$
7. The **identity function** $f(x) = x$ is its own inverse.
8. If f is a function, then $f(a + b) = f(a) + f(b)$ for all real numbers a and b in the domain of f.
9. If f is defined by $f(x) = |x|$, then $f(ab) = f(a)f(b)$ for all real numbers a and b.
10. If f is a one-to-one function and a and b are real numbers in the domain of f with $a < b$, then $f(a) \neq f(b)$.

11. The coordinates of a point on the graph of $y = f(x)$ are (a, b). If k is a positive constant, then (a, kb) are the coordinates of a point on the graph of $y = kf(x)$.

12. For every function f, the real number c is a solution of $f(x) = 0$ if and only if $(c, 0)$ is an x-intercept of the graph of $y = f(x)$.

13. The domain of every polynomial function is the real numbers.

14. If the linear coefficient of determination is 0.8, then the slope of the regression line is positive.

CHAPTER 2 REVIEW EXERCISES

In Exercises 1 and 2, find the distance between the points whose coordinates are given.

1. $(-3, 2)$ $(7, 11)$
2. $(5, -4)$ $(-3, -8)$

In Exercises 3 and 4, find the midpoint of the line segment with the given endpoints.

3. $(2, 8)$ $(-3, 12)$
4. $(-4, 7)$ $(8, -11)$

In Exercises 5 and 6, determine the center and radius of the circle with the given equation.

5. $(x - 3)^2 + (y + 4)^2 = 81$
6. $x^2 + y^2 + 10x + 4y + 20 = 0$

In Exercises 7 and 8, find the equation in standard form of a circle that satisfies the given conditions.

7. Center $C = (2, -3)$, radius $r = 5$
8. Center $C = (-5, 1)$, passing through $(3, 1)$
9. If $f(x) = 3x^2 + 4x - 5$, find
   a. $f(1)$ b. $f(-3)$ c. $f(t)$
   d. $f(x + h)$ e. $3f(t)$ f. $f(3t)$
10. If $g(x) = \sqrt{64 - x^2}$, find
   a. $g(3)$ b. $g(-5)$ c. $g(8)$
   d. $g(-x)$ e. $2g(t)$ f. $g(2t)$
11. If $f(x) = x^2 + 4x$ and $g(x) = x - 8$, find
   a. $(f \circ g)(3)$ b. $(g \circ f)(-3)$
   c. $(f \circ g)(x)$ d. $(g \circ f)(x)$
12. If $f(x) = 2x^2 + 7$ and $g(x) = |x - 1|$, find
   a. $(f \circ g)(-5)$ b. $(g \circ f)(-5)$
   c. $(f \circ g)(x)$ d. $(g \circ f)(x)$
13. If $f(x) = 4x^2 - 3x - 1$, find the difference quotient
$$\frac{f(x + h) - f(x)}{h}$$
14. If $g(x) = x^3 - x$, find the difference quotient
$$\frac{g(x + h) - g(x)}{h}$$

In Exercises 15 to 20, sketch the graph of *f*. Find the interval(s) in which *f* is *a.* increasing, *b.* constant, *c.* decreasing.

15. $f(x) = |x - 3| - 2$
16. $f(x) = x^2 - 5$
17. $f(x) = |x + 2| - |x - 2|$
18. $f(x) = [\![x + 3]\!]$
19. $f(x) = \frac{1}{2}x - 3$
20. $f(x) = \sqrt[3]{x}$

In Exercises 21 to 24, determine the domain of the function represented by the given equation.

21. $f(x) = -2x^2 + 3$
22. $f(x) = \sqrt{6 - x}$
23. $f(x) = \sqrt{25 - x^2}$
24. $f(x) = \frac{3}{x^2 - 2x - 15}$

In Exercises 25 and 26, find the slope-intercept form of the equation of the line through the two points.

25. $(-1, 3)$ $(4, -7)$
26. $(0, 0)$ $(7, 11)$
27. Find the slope-intercept form of the equation of the line that is parallel to the graph of $3x - 4y = 8$ and passes through $(2, 11)$.
28. Find the slope-intercept form of the equation of the line that is perpendicular to the graph of $2x = -5y + 10$ and passes through $(-3, -7)$.

In Exercises 29 to 34, use the method of completing the square to write each quadratic equation in its standard form.

29. $f(x) = x^2 + 6x + 10$
30. $f(x) = 2x^2 + 4x + 5$
31. $f(x) = -x^2 - 8x + 3$
32. $f(x) = 4x^2 - 6x + 1$
33. $f(x) = -3x^2 + 4x - 5$
34. $f(x) = x^2 - 6x + 9$

In Exercises 35 to 38, find the vertex of the graph of the quadratic function.

35. $f(x) = 3x^2 - 6x + 11$

36. $h(x) = 4x^2 - 10$

37. $k(x) = -6x^2 + 60x + 11$

38. $m(x) = 14 - 8x - x^2$

39. Use the formula

$$d = \frac{|mx_1 + b - y_1|}{\sqrt{1 + m^2}}$$

to find the distance from the point $(1, 3)$ to the line given by $y = 2x - 3$.

40. A freight company has determined that its cost of delivering x parcels per delivery is

$$C(x) = 1050 + 0.5x$$

The price it charges to send a parcel is \$13.00 per parcel. Determine

a. the revenue function

b. the profit function

c. the minimum number of parcels the company must ship to break even

In Exercises 41 and 42, sketch a graph that is symmetric to the given graph with respect to the *a.* x-axis, *b.* y-axis, *c.* origin.

41. 42.

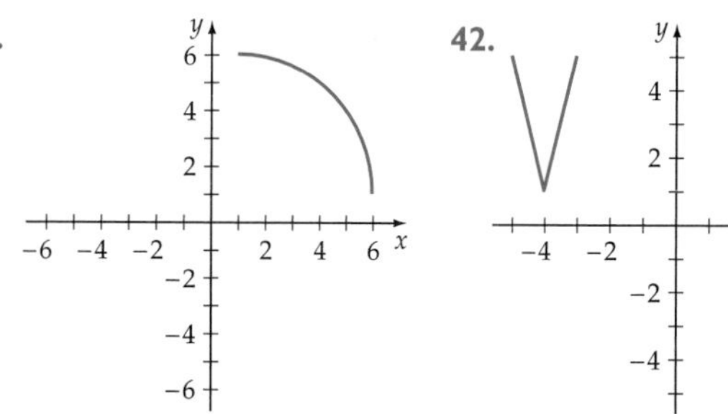

In Exercises 43 to 50, determine whether the graph of each equation is symmetric with respect to the *a.* x-axis, *b.* y-axis, *c.* origin.

43. $y = x^2 - 7$

44. $x = y^2 + 3$

45. $y = x^3 - 4x$

46. $y^2 = x^2 + 4$

47. $\dfrac{x^2}{3^2} + \dfrac{y^2}{4^2} = 1$

48. $xy = 8$

49. $|y| = |x|$

50. $|x + y| = 4$

In Exercises 51 to 56, sketch the graph of g. *a.* Find the domain and the range of g. *b.* State whether g is even, odd, or neither even nor odd.

51. $g(x) = -x^2 + 4$

52. $g(x) = -2x - 4$

53. $g(x) = |x - 2| + |x + 2|$

54. $g(x) = \sqrt{16 - x^2}$

55. $g(x) = x^3 - x$

56. $g(x) = 2[\![x]\!]$

In Exercises 57 to 62, first write the function in standard form, and then make use of translations to graph the function.

57. $F(x) = x^2 + 4x - 7$

58. $A(x) = x^2 - 6x - 5$

59. $P(x) = 3x^2 - 4$

60. $G(x) = 2x^2 - 8x + 3$

61. $W(x) = -4x^2 - 6x + 6$

62. $T(x) = -2x^2 - 10x$

63. On the same set of coordinate axes, sketch the graph of $p(x) = \sqrt{x} + c$ for $c = 0, -1$, and 2.

64. On the same set of coordinate axes, sketch the graph of $q(x) = \sqrt{x + c}$ for $c = 0, -1$, and 2.

65. On the same set of coordinate axes, sketch the graph of $r(x) = c\sqrt{9 - x^2}$ for $c = 1, 1/2$, and -2.

66. On the same set of coordinate axes, sketch the graph of $s(t) = [\![cx]\!]$ for $c = 1, 1/4$, and 4.

In Exercises 67 and 68, graph each piecewise-defined function.

67. $f(x) = \begin{cases} x, & \text{if } x \le 0 \\ \dfrac{1}{2}x, & \text{if } x > 0 \end{cases}$

68. $g(x) = \begin{cases} -2, & \text{if } x < -3 \\ \dfrac{2}{3}x, & \text{if } -3 \le x \le 3 \\ 2, & \text{if } x > 3 \end{cases}$

In Exercises 69 and 70, use the given functions f and g to find $f + g, f - g, fg$, and f/g. State the domain of each.

69. $f(x) = x^2 - 9, \quad g(x) = x + 3$

70. $f(x) = x^3 + 8, \quad g(x) = x^2 - 2x + 4$

71. Find two numbers whose sum is 50 and whose product is a maximum.

72. Find two numbers whose difference is 10 and the sum of whose squares is a minimum.

73. The distance traveled by a ball rolling down a ramp is given by $s(t) = 3t^2$, where t is the time in seconds after the ball is released and $s(t)$ is measured in feet. Evaluate the average velocity of the ball for each of the following time intervals.

a. $[2, 4]$ b. $[2, 3]$ c. $[2, 2.5]$ d. $[2, 2.01]$

e. What appears to be the average velocity of the ball for the time interval $[2, 2 + \Delta t]$ if Δt approaches 0?

74. The distance traveled by a ball that is pushed down a ramp is given by $s(t) = 2t^2 + t$, where t is the time in seconds after the ball is released and $s(t)$ is measured in feet. Evaluate the average velocity of the ball for each of the following time intervals.

a. $[3, 5]$ b. $[3, 4]$ c. $[3, 3.5]$ d. $[3, 3.01]$

e. What appears to be the average velocity of the ball for the time interval $[3, 3 + \Delta t]$ if Δt approaches 0?

75. **COMPUTER SCIENCE** A test of an Internet service provider showed the following download times (in seconds) for files of various sizes (in kilobytes).

a. Find a linear regression model for this data.

b. Judging on the basis of the value of r, is a linear model of this data a reasonable model? Explain.

c. On the basis of the model, what is the expected download time of a file that is 100 kilobytes in size?

76. **PHYSICS** The rate at which water will escape from the bottom of a can depends on a number of factors, including the height of the water, the size of the hole, and the diameter of the can. The table below shows the height (in millimeters) of water in a can after t seconds.

Water Escaping a Ruptured Can

Height	Time	Height	Time
0	180	60	93
10	163	70	81
20	147	80	70
30	133	90	60
40	118	100	50
50	105	110	48

a. Find the quadratic regression model for this data.

b. On the basis of this model, will the can ever empty?

c. Explain why there seems to be a contradiction between the model and reality, in that we know the can will eventually run out of water.

CHAPTER 2 TEST

1. Find the midpoint and the length of the line segment with endpoints $(-2, 3)$ and $(4, -1)$.

2. Determine the x- and y-intercepts, and then graph the equation $x = 2y^2 - 4$.

3. Graph the equation $y = |x + 2| + 1$.

4. Find the center and radius of the circle that has the general form $x^2 - 4x + y^2 + 2y - 4 = 0$.

5. Determine the domain of the function

$$f(x) = -\sqrt{x^2 - 16}$$

6. Use the formula

$$d = \frac{|mx_1 + b - y_1|}{\sqrt{1 + m^2}}$$

to find the distance from the point $(3, 4)$ to the line given by $y = 3x + 1$.

7. Graph $f(x) = -2|x - 2| + 1$. Identify the intervals over which the function is

a. increasing

b. constant

c. decreasing

8. Graph the function $f(x) = x^2 + 2$. From the graph, find the domain and range of the function.

9. An air freight company has determined that its cost of delivering x parcels per flight is

$$C(x) = 875 + 0.75x$$

The price it charges to send a parcel is \$12.00 per parcel. Determine

a. the revenue function

b. the profit function

c. the minimum number of parcels the company must ship to break even

10. Use the graph of $f(x) = |x|$ to graph $y = -f(x + 2) - 1$.

11. Classify each of the following as either an even function, an odd function, or neither an even nor an odd function.

a. $f(x) = x^4 - x^2$ b. $f(x) = x^3 - x$

c. $f(x) = x - 1$

12. Find the slope-intercept form of the equation of the line that passes through $(4, -2)$ and is perpendicular to the graph of $3x - 2y = 4$.

13. Find the maximum or minimum value of the function $f(x) = x^2 - 4x - 8$. State whether this value is a maximum or a minimum value.

14. Let $f(x) = x^2 - 1$ and $g(x) = x - 2$. Find $(f + g)$ and (f/g).

15. Find the difference quotient of the function

$$f(x) = x^2 + 1$$

16. Evaluate $(f \circ g)$, where

$$f(x) = x^2 - 2x + 1 \quad \text{and} \quad g(x) = \sqrt{x - 2}$$

17. The distance traveled by a ball rolling down a ramp is given by $s(t) = 5t^2$, where t is the time in seconds after the ball is released and $s(t)$ is measured in feet. Evaluate the average velocity of the ball for each of the following time intervals.

a. $[2, 3]$ b. $[2, 2.5]$ c. $[2, 2.01]$

18. The table below shows the percent of water and the number of calories in various canned soups to which 100 grams of water are added.

Percent Water in Soups

% Water	Calories
93.2	28
92.3	26
91.9	39
89.5	56
89.6	56
90.5	36
91.9	32
91.7	32

a. Find the linear regression line for this data.

b. Using the linear model from part **a**, find the expected number of calories in a soup that is 89% water. Round to the nearest whole number.

CHAPTER 3

POLYNOMIAL AND RATIONAL FUNCTIONS

Bézier Curves: An Application of Polynomials

Many computer-generated figures are drawn by using Bézier (pronounced *bay-zee-ay*) curves. These curves were first created by the mathematician Pierre Bézier (1910–1999) when he was employed by the car company Renault. Up until this time, designers drew curves by using a French curve.

◆ Using a French curve

Bézier's work dramatically changed the design process, because he discovered a way to define the shape of a free-form curve using polynomials. On a computer screen, Bézier curves are displayed as curves with anchor points and handles. The following figure shows two Bézier curves in red. The endpoints of the blue line segments are the handles. You can change the shape of a Bézier curve by moving one of its handles.

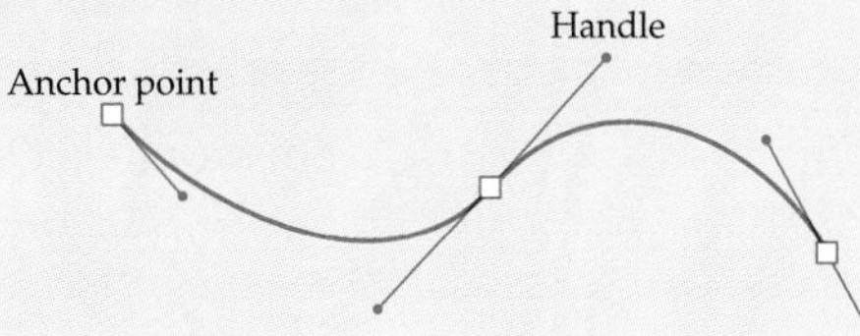

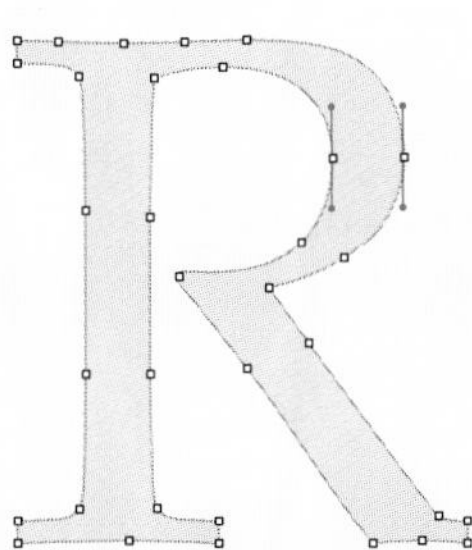

◆ A character composed of 32 Bézier curves

Bézier curves are used in the design of fonts. For instance, the "R" shown to the left is composed of approximately 32 Bézier curves. Each Bézier curve is defined by a pair of polynomials. Fonts and graphics that are created with Bézier curves can be enlarged without loss of resolution, and they do not require a large amount of memory.

SECTION 3.1 POLYNOMIAL DIVISION AND SYNTHETIC DIVISION

- POLYNOMIAL DIVISION
- SYNTHETIC DIVISION
- THE REMAINDER THEOREM
- THE FACTOR THEOREM
- REDUCED POLYNOMIALS

◆ POLYNOMIAL DIVISION

If $P(x)$ is a polynomial, then the values of x for which $P(x)$ is equal to 0 are called the **zeros** of $P(x)$ or the **roots** of $P(x) = 0$. Much of the work in this chapter concerns finding the zeros of a polynomial. Sometimes the zeros of a polynomial can be determined by dividing the polynomial by another polynomial. Dividing a polynomial by another polynomial is similar to the long-division process used for dividing positive integers. For example, to divide $(x^2 + 9x - 16)$ by $(x - 3)$, we use the following procedure:

$$\begin{array}{r}
x + 12 \\
x - 3 \overline{) x^2 + 9x - 16} \\
\underline{x^2 - 3x} \\
12x - 16 \\
\underline{12x - 36} \\
20
\end{array}$$

Thus $(x^2 + 9x - 16) \div (x - 3) = x + 12$, with a remainder of 20.

In this example, $x^2 + 9x - 16$ is called the **dividend,** $x - 3$ is the **divisor,** $x + 12$ is the **quotient,** and 20 is the **remainder**. The dividend is equal to the product of the divisor and the quotient, plus the remainder. That is,

$$\underbrace{x^2 + 9x - 16}_{\textbf{Dividend}} = \underbrace{(x - 3)}_{\textbf{Divisor}} \cdot \underbrace{(x + 12)}_{\textbf{Quotient}} + \underbrace{20}_{\textbf{Remainder}}$$

The above result is a special case of a theorem known as the *Division Algorithm for Polynomials.*

The Division Algorithm for Polynomials

If $P(x)$ and $D(x)$ are polynomials such that $D(x) \neq 0$, then there exist unique polynomials $Q(x)$ and $R(x)$ such that $P(x) = D(x)Q(x) + R(x)$, where either $R(x) = 0$ or the degree of $R(x)$ is less than the degree of $D(x)$.

take note

If $R(x) = 0$, then $D(x)$ is a factor of $P(x)$. If the degree of $D(x)$ is greater than the degree of $P(x)$, then $Q(x) = 0$ and $R(x) = P(x)$.

The polynomial $P(x)$ is the dividend, $D(x)$ is the divisor, $Q(x)$ is the quotient, and the polynomial $R(x)$ is the remainder.

$$\underbrace{P(x)}_{\textbf{Dividend}} = \underbrace{D(x)}_{\textbf{Divisor}} \cdot \underbrace{Q(x)}_{\textbf{Quotient}} + \underbrace{R(x)}_{\textbf{Remainder}}$$

Multiplying both sides of $P(x) = D(x)Q(x) + R(x)$ by $1/D(x)$ produces the fractional form

$$\frac{P(x)}{D(x)} = Q(x) + \frac{R(x)}{D(x)}$$

EXAMPLE 1 Divide Polynomials

Perform the indicated division.

$$\frac{x^4 + 3x^2 - 6x - 10}{x^2 + 3x - 5}$$

Solution

$$\begin{array}{r}
x^2 - 3x + 17 \\
x^2 + 3x - 5\overline{)x^4 + 0x^3 + 3x^2 - 6x - 10} \\
\underline{x^4 + 3x^3 - 5x^2} \\
-3x^3 + 8x^2 - 6x \\
\underline{-3x^3 - 9x^2 + 15x} \\
17x^2 - 21x - 10 \\
\underline{17x^2 + 51x - 85} \\
-72x + 75
\end{array}$$

• Writing $0x^3$ for the missing term helps us align like terms in the same column.

Thus $\dfrac{x^4 + 3x^2 - 6x - 10}{x^2 + 3x - 5} = x^2 - 3x + 17 + \dfrac{-72x + 75}{x^2 + 3x - 5}$.

TRY EXERCISE 6, EXERCISE SET 3.1, PAGE 245

◆ SYNTHETIC DIVISION

The procedure for dividing a polynomial by a binomial of the form $x - c$ can be condensed by a method called **synthetic division.** To understand the synthetic division method, consider the following division:

$$\begin{array}{r}
3x^2 - 2x + 3 \\
x - 2\overline{)3x^3 - 8x^2 + 7x + 2} \\
\underline{3x^3 - 6x^2} \\
-2x^2 + 7x \\
\underline{-2x^2 + 4x} \\
3x + 2 \\
\underline{3x - 6} \\
8
\end{array}$$

No essential data are lost when we omit the variables, because the position of a term indicates the power of the term.

$$\begin{array}{rrrrr}
 & 3 & -2 & 3 & \\
-2) & 3 & -8 & 7 & 2 \\
 & \underline{3} & \underline{-6} & & \\
 & & -2 & 7 & \\
 & & \underline{-2} & \underline{4} & \\
 & & & 3 & 2 \\
 & & & \underline{3} & \underline{-6} \\
 & & & & 8
\end{array}$$

The coefficients shown in red are duplicates of those directly above them. Omitting these repeated coefficients (in red) enables us to condense the vertical spacing.

$$\begin{array}{r|rrr} & 3 & -2 & 3 \\ \hline -2) & 3 \quad -8 & 7 & 2 \\ & -6 & 4 & -6 \\ \hline & -2 & 3 & 8 \end{array}$$

The coefficients in blue in the top row can be omitted because they are duplicates of those in the bottom row. The leading coefficient of the quotient (top row) can be written in the bottom row with the coefficients of the other terms in order to condense the vertical spacing even more.

$$\begin{array}{r|rrrr} -2 & 3 & -8 & 7 & 2 \\ & & -6 & 4 & -6 \\ \hline & 3 & -2 & 3 & 8 \end{array}$$

So that we may add the numbers in each column instead of subtracting them, we change the sign of the divisor. This changes the sign of each number in the second row.

$$\begin{array}{r|rrrr} 2 & 3 & -8 & 7 & 2 \\ & & 6 & -4 & 6 \\ \hline & 3 & -2 & 3 & 8 \end{array}$$

Coefficients of the quotient (3, −2, 3) **Remainder** (8)

The following procedure illustrates how to use the synthetic-division method, used here to find the quotient and the remainder of

$$\frac{2x^3 - 9x^2 + 5}{x - 3}$$

take note

The synthetic-division method shown at the right is used only to divide by a polynomial of the form $x - c$, where the coefficient of x is 1. To divide a polynomial by a polynomial that is not of the form $x - c$, use the long-division method shown in Example 1.

Coefficients of the dividend

$$\begin{array}{r|rrrr} 3 & 2 & -9 & 0 & 5 \\ & & & & \\ \hline \end{array}$$

- **Synthetic-division form with 0 inserted for the missing x term.**

$$\begin{array}{r|rrrr} 3 & 2 & -9 & 0 & 5 \\ & & & & \\ \hline & 2 & & & \end{array}$$

- **Bring down the leading coefficient 2.**

$$\begin{array}{r|rrrr} 3 & 2 & -9 & 0 & 5 \\ & & 6 & & \\ \hline & 2 & & & \end{array}$$

- **Multiply $3 \cdot 2$ and place the product (6) in the middle row and in the next column to the right.**

$$\begin{array}{r|rrrr} 3 & 2 & -9 & 0 & 5 \\ & & 6 & & \\ \hline & 2 & -3 & & \end{array}$$

- **Add −9 and 6 and place the sum (−3) in the bottom row.**

$$\begin{array}{r|rrrr} 3 & 2 & -9 & 0 & 5 \\ & & 6 & -9 & -27 \\ \hline & 2 & -3 & -9 & -22 \end{array}$$

Coefficients of the quotient (2, −3, −9) — **Remainder** (−22)

• **Repeat the previous steps for columns 3 and 4.**

$$\frac{2x^3 - 9x^2 + 5}{x - 3} = 2x^2 - 3x - 9 + \frac{-22}{x - 3}$$

EXAMPLE 2 Use Synthetic Division to Divide Polynomials

Use synthetic division to perform the indicated division.

$$\frac{x^4 - 4x^2 + 7x + 15}{x + 4}$$

Solution

Because the divisor is $x + 4$, we perform the synthetic division with $c = -4$.

$$\begin{array}{r|rrrrr} -4 & 1 & 0 & -4 & 7 & 15 \\ & & -4 & 16 & -48 & 164 \\ \hline & 1 & -4 & 12 & -41 & 179 \end{array}$$

The quotient is $x^3 - 4x^2 + 12x - 41$ and the remainder is 179.

$$\frac{x^4 - 4x^2 + 7x + 15}{x + 4} = x^3 - 4x^2 + 12x - 41 + \frac{179}{x + 4}$$

TRY EXERCISE 12, EXERCISE SET 3.1, PAGE 245

COEF OF QUOTIENT
1
-4
12
-41

Figure 3.1

A TI-82/83 synthetic-division program SYDIV is available on the Internet at

http://college.hmco.com

The SYDIV program prompts you to enter the degree of the dividend, the coefficients of the dividend, and the constant c, from the divisor $x - c$. For instance, to perform the synthetic division in Example 2, enter 4 for the degree of the dividend, followed by the coefficients 1, 0, −4, 7, and 15, followed by −4, which is the constant c. The calculator display appears as in Figure 3.1. Press ENTER to produce the display in Figure 3.2.

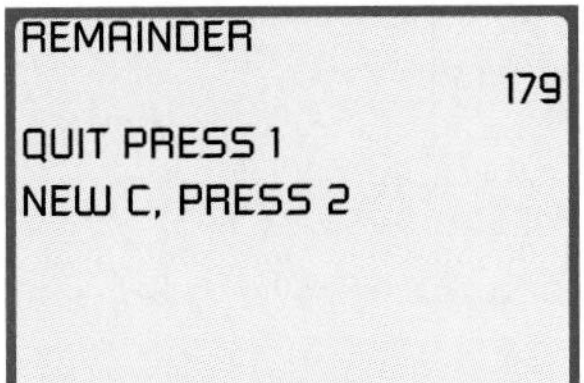

Figure 3.2

◆ THE REMAINDER THEOREM

The following theorem shows that synthetic division can be used to find the value $P(c)$ for any polynomial function P and constant c.

The Remainder Theorem

If a polynomial $P(x)$ is divided by $x - c$, then the remainder is $P(c)$.

Proof The Division Algorithm states that

$$P(x) = (x - c)Q(x) + R(x)$$

where $R(x)$ is zero or the degree of $R(x)$ is less than the degree of $x - c$. Because the degree of $x - c$ is 1, the remainder $R(x)$ must be some constant—say, r. Therefore,

$$P(x) = (x - c)Q(x) + r$$

This equality evaluated at $x = c$ produces

$$\begin{aligned} P(c) &= (c - c)Q(c) + r \\ &= (0)Q(c) + r \\ &= r \end{aligned}$$

◆

EXAMPLE 3 Use the Remainder Theorem to Evaluate a Polynomial

Evaluate $P(x) = 2x^3 + 3x^2 + 2x - 2$ when $x = -2$ and $x = 1/2$.

Solution

Perform the synthetic division for $x = -2$ and $x = 1/2$ and examine the remainders.

$$\begin{array}{r|rrrr} -2 & 2 & 3 & 2 & -2 \\ & & -4 & 2 & -8 \\ \hline & 2 & -1 & 4 & -10 \end{array}$$

The remainder is -10. Therefore $P(-2) = -10$.

$$\begin{array}{r|rrrr} \frac{1}{2} & 2 & 3 & 2 & -2 \\ & & 1 & 2 & 2 \\ \hline & 2 & 4 & 4 & 0 \end{array}$$

The remainder is 0. Therefore $P\left(\frac{1}{2}\right) = 0$.

Visualize the Solution

A graph of P shows that points $(-2, -10)$ and $(1/2, 0)$ are on the graph.

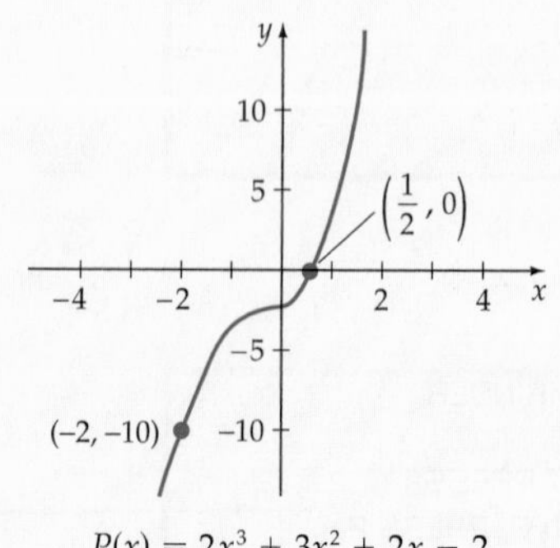

$P(x) = 2x^3 + 3x^2 + 2x - 2$

Figure 3.3

TRY EXERCISE 30, EXERCISE SET 3.1, PAGE 245

◆ THE FACTOR THEOREM

Note from Example 3 that $P(1/2) = 0$. The number $1/2$ is called a zero of the polynomial because the value of the polynomial is 0 when $x = 1/2$.

Zero of a Polynomial

If $P(x)$ is a polynomial and a is a number for which $P(a) = 0$, then a is a **zero** of $P(x)$.

The following theorem is a result of the Remainder Theorem.

The Factor Theorem

A polynomial $P(x)$ has a factor $(x - c)$ if and only if $P(c) = 0$. That is, $(x - c)$ is a factor of $P(x)$ if and only if c is a zero of P.

Proof **Part 1:** Given that $P(x)$ has a factor of $(x - c)$, show that $P(c) = 0$. If $(x - c)$ is a factor of $P(x)$, then $P(x) = (x - c) \cdot Q(x)$ for some $Q(x)$. Thus the division of $P(x)$ by $(x - c)$ has a remainder of zero, and the Remainder Theorem implies that $P(c) = 0$.

Part 2: Given $P(c) = 0$, we need to show that $(x - c)$ is a factor of $P(x)$. The division algorithm applied to the polynomial $P(x)$ with divisor $(x - c)$ produces

$$P(x) = (x - c)Q(x) + R(x)$$

Because $P(c) = 0$, the Remainder Theorem implies that $R(x) = 0$. Thus

$$P(x) = (x - c)Q(x)$$

which shows that $(x - c)$ is a factor of $P(x)$. ◆

EXAMPLE 4 Find a Factor of a Polynomial

Determine whether $(x + 5)$ is a factor of

$$P(x) = x^4 + x^3 - 21x^2 - x + 20$$

Solution

$$\begin{array}{r|rrrrr} -5 & 1 & 1 & -21 & -1 & 20 \\ & & -5 & 20 & 5 & -20 \\ \hline & 1 & -4 & -1 & 4 & 0 \end{array}$$

The remainder 0 implies that $(x + 5)$ is a factor of $P(x)$.

TRY EXERCISE 40, EXERCISE SET 3.1, PAGE 245

QUESTION From the result of Example 4, is -5 a zero of $P(x)$?

◆ REDUCED POLYNOMIALS

From Example 4, $(x + 5)$ is a factor of $P(x) = x^4 + x^3 - 21x^2 - x + 20$, and the quotient is $Q(x) = x^3 - 4x^2 - x + 4$ (from the last line of the synthetic division). Thus

$$P(x) = (x + 5)(x^3 - 4x^2 - x + 4)$$

The polynomial $Q(x) = x^3 - 4x^2 - x + 4$ is called a **reduced polynomial** or a **depressed polynomial** because it is 1 degree less than the degree of $P(x)$. Reduced polynomials play an important role in our work in Section 3.3.

EXAMPLE 5 Find a Reduced Polynomial

Verify that $(x - 3)$ is a factor of $P(x) = 2x^3 - 3x^2 - 4x - 15$, and write $P(x)$ as the product of $(x - 3)$ and the reduced polynomial $Q(x)$.

Solution

3	2	−3	−4	−15
		6	9	15
	2	3	5	0

Coefficients of the reduced polynomial $Q(x)$

Thus $(x - 3)$ and the reduced polynomial $2x^2 + 3x + 5$ are both factors of $P(x)$. That is,

$$2x^3 - 3x^2 - 4x - 15 = (x - 3)(2x^2 + 3x + 5)$$

TRY EXERCISE 62, EXERCISE SET 3.1, PAGE 246

TOPICS FOR DISCUSSION

1. Explain the meaning of the phrase *zero of a polynomial.*
2. If $P(x)$ is a polynomial of degree $n \geq 2$, what is the degree of the quotient $\dfrac{P(x)}{x - a}$?
3. Discuss how the Factor Theorem can be used to determine whether a number is a zero of a polynomial.

ANSWER Yes. Because $(x + 5)$ is a factor of $P(x)$, the Factor Theorem states that $P(-5) = 0$, and thus -5 is a zero of P.

4. A zero of $P(x) = x^3 - x^2 - 14x + 24$ is -4. Discuss how this information and the Factor Theorem can be used to solve $x^3 - x^2 - 14x + 24 = 0$.

5. Discuss the advantages and disadvantages of using synthetic division rather than substitution to evaluate a polynomial at $x = c$.

EXERCISE SET 3.1

In Exercises 1 to 10, use long division to divide the first polynomial by the second.

1. $5x^3 + 6x^2 - 17x + 20, \quad x + 3$
2. $6x^3 + 15x^2 - 8x + 2, \quad x + 4$
3. $2x^4 + 15x^3 + 7x^2 - 135x - 225, \quad 2x + 5$
4. $6x^4 + 3x^3 - 11x^2 - 3x + 9, \quad 2x - 3$
5. $3x^4 + x^3 - 98x^2 - 30, \quad 3x^2 + x + 1$
6. $2x^4 - x^3 - 23x^2 + 9x + 45, \quad 2x^2 - x - 5$
7. $20x^4 - 3x^2 + 9, \quad 5x^2 - 2$
8. $24x^5 + 20x^3 - 16x^2 - 15, \quad 6x^2 + 5$
9. $x^3 + 5x^2 + 6x - 19, \quad x^2 + x - 4$
10. $2x^4 + 3x^3 - 7x - 10, \quad x^2 - 2x - 5$

In Exercises 11 to 28, use synthetic division to divide the first polynomial by the second.

11. $4x^3 - 5x^2 + 6x - 7, \quad x - 2$
12. $5x^3 + 6x^2 - 8x + 1, \quad x - 5$
13. $4x^3 - 2x + 3, \quad x + 1$
14. $6x^3 - 4x^2 + 17, \quad x + 3$
15. $x^5 - 10x^3 + 5x - 1, \quad x - 4$
16. $6x^4 - 2x^3 - 3x^2 - x, \quad x - 5$
17. $x^5 - 1, \quad x - 1$
18. $x^4 + 1, \quad x + 1$
19. $8x^3 - 4x^2 + 6x - 3, \quad x - \dfrac{1}{2}$
20. $12x^3 + 5x^2 + 5x + 6, \quad x + \dfrac{3}{4}$
21. $x^8 + x^6 + x^4 + x^2 + 4, \quad x - 2$
22. $-x^7 - x^5 - x^3 - x - 5, \quad x + 1$
23. $x^6 + x - 10, \quad x + 3$
24. $2x^5 - 3x^4 - 5x^2 - 10, \quad x - 4$
25. $3x^2 - 4x + 5, \quad x - 0.3$
26. $2x^2 - 12x + 1, \quad x + 0.4$
27. $2x^3 - 11x^2 - 17x + 3, \quad x$
28. $5x^4 - 2x^2 + 6x - 1, \quad x$

In Exercises 29 to 38, use the Remainder Theorem to find $P(c)$.

29. $P(x) = 3x^3 + x^2 + x - 5, c = 2$
30. $P(x) = 2x^3 - x^2 + 3x - 1, c = 3$
31. $P(x) = 4x^4 - 6x^2 + 5, c = -2$
32. $P(x) = 6x^3 - x^2 + 4x, c = -3$
33. $P(x) = -2x^3 - 2x^2 - x - 20, c = 10$
34. $P(x) = -x^3 + 3x^2 + 5x + 30, c = 8$
35. $P(x) = -x^4 + 1, c = 3$
36. $P(x) = x^5 - 1, c = 1$
37. $P(x) = x^4 - 10x^3 + 2, c = 3$
38. $P(x) = x^5 + 20x^2 - 1, c = -5$

In Exercises 39 to 50, use synthetic division and the Factor Theorem to determine whether the given binomial is a factor of $P(x)$.

39. $P(x) = x^3 + 2x^2 - 5x - 6, x - 2$
40. $P(x) = x^3 + 4x^2 - 27x - 90, x + 6$
41. $P(x) = 2x^3 + x^2 - 2x - 1, x + 1$
42. $P(x) = 3x^3 + 4x^2 - 27x - 36, x - 4$
43. $P(x) = x^4 - 25x^2 + 144, x + 3$
44. $P(x) = x^4 - 25x^2 + 144, x - 3$
45. $P(x) = x^5 + 2x^4 - 22x^3 - 50x^2 - 75x, x - 5$
46. $P(x) = 9x^4 - 6x^3 - 23x^2 - 4x + 4, x + 1$

47. $P(x) = 16x^4 - 8x^3 + 9x^2 + 14x - 4, x - \frac{1}{4}$

48. $P(x) = 10x^4 + 9x^3 - 4x^2 + 9x + 6, x + \frac{1}{2}$

49. $P(x) = x^2 - 4x - 1, x - (2 + \sqrt{5})$

50. $P(x) = x^2 - 4x - 1, x - (2 - \sqrt{5})$

In Exercises 51 to 60, use synthetic division to show that c is a zero of $P(x)$.

51. $P(x) = 3x^3 - 8x^2 - 10x + 28, c = 2$

52. $P(x) = 4x^3 - 10x^2 - 8x + 6, c = 3$

53. $P(x) = x^4 - 1, c = 1$

54. $P(x) = x^3 + 8, c = -2$

55. $P(x) = 3x^4 + 8x^3 + 10x^2 + 2x - 20, c = -2$

56. $P(x) = x^4 - 2x^2 - 100x - 75, c = 5$

57. $P(x) = 2x^3 - 18x^2 - 50x + 66, c = 11$

58. $P(x) = 2x^4 - 34x^3 + 70x^2 - 153x + 45, c = 15$

59. $P(x) = 3x^2 - 8x + 4, c = \frac{2}{3}$

60. $P(x) = 5x^2 + 12x + 4, c = -\frac{2}{5}$

In Exercises 61 to 64, verify that the given binomial is a factor of $P(x)$, and write $P(x)$ as the product of the binomial and its reduced polynomial $Q(x)$.

61. $P(x) = x^3 + x^2 + x - 14, x - 2$

62. $P(x) = x^4 + 5x^3 + 3x^2 - 5x - 4, x + 1$

63. $P(x) = x^4 - x^3 - 9x^2 - 11x - 4, x - 4$

64. $P(x) = 2x^5 - x^4 - 7x^3 + x^2 + 7x - 10, x - 2$

You can use a graph to factor some polynomials. For example, the graph of $y = x^2 - x - 12$ intersects the x-axis at $x = -3$ and $x = 4$. Thus $x^2 - x - 12$ has -3 and 4 as zeros. Hence $x^2 - x - 12$ has factors of $(x + 3)$ and $(x - 4)$. Use a graphing utility to factor each polynomial in Exercises 65 to 68.

65. $x^3 - 7x + 6$

66. $x^3 + 6x^2 + 3x - 10$

67. $x^4 + 2x^3 - 13x^2 - 38x - 24$

68. $x^4 + 2x^3 - 7x^2 - 8x + 12$

SUPPLEMENTAL EXERCISES

69. Use the Factor Theorem to prove that for any positive odd integer n, $x^n + 1$ has $x + 1$ as a factor.

70. Use the Factor Theorem to prove that for any positive integer n, $x^n - 1$ has $x - 1$ as a factor.

71. Find the remainder of $5x^{48} + 6x^{10} - 5x + 7$ divided by $x - 1$.

72. Find the remainder of $18x^{80} - 6x^{50} + 4x^{20} - 2$ divided by $x + 1$.

73. Prove that $P(x) = 4x^4 + 7x^2 + 12$ has no factor of the form $x - c$, where c is a real number.

74. Prove that $P(x) = -5x^6 - 4x^2 - 10$ has no factor of the form $x - c$, where c is a real number.

75. Use synthetic division to show that $(x - i)$ is a factor of $x^3 - 3x^2 + x - 3$.

76. Use synthetic division to show that $(x + 2i)$ is a factor of $x^4 - 2x^3 + x^2 - 8x - 12$.

PROJECTS

1. **HORNER'S METHOD AND SYNTHETIC DIVISION** A method of factoring a polynomial, called Horner's Method, was described in the Topics for Discussion in Section P.6. Show that this method is essentially synthetic division.

SECTION 3.2 POLYNOMIAL FUNCTIONS

- FAR-LEFT AND FAR-RIGHT BEHAVIOR
- MAXIMUM AND MINIMUM VALUES
- REAL ZEROS OF A POLYNOMIAL
- THE EVEN AND ODD POWERS OF $(x - c)$ THEOREM

3A

Table 3.1 summarizes information developed in Chapter 2 about graphs of polynomial functions of degree 0, 1, or 2. Polynomial functions of degree 3 or higher can be graphed by the technique of plotting points. However, some additional knowledge about polynomial functions will make graphing easier.

Table 3.1

Polynomial Function $P(x)$	Graph
$P(x) = a$ (degree 0), $a \neq 0$	Horizontal line through $(0, a)$
$P(x) = ax + b$ (degree 1)	Line with y-intercept $(0, b)$ and slope a
$P(x) = ax^2 + bx + c$ (degree 2)	Parabola with vertex $\left(-\frac{b}{2a}, P\left(-\frac{b}{2a}\right)\right)$

All polynomial functions have graphs that are **smooth continuous curves.** The terms *smooth* and *continuous* are defined rigorously in calculus, but for the present, a smooth curve is a curve that does not have sharp corners such as that shown in **Figure 3.4a.** A continuous curve does not have a break or hole such as those shown in **Figure 3.4b.**

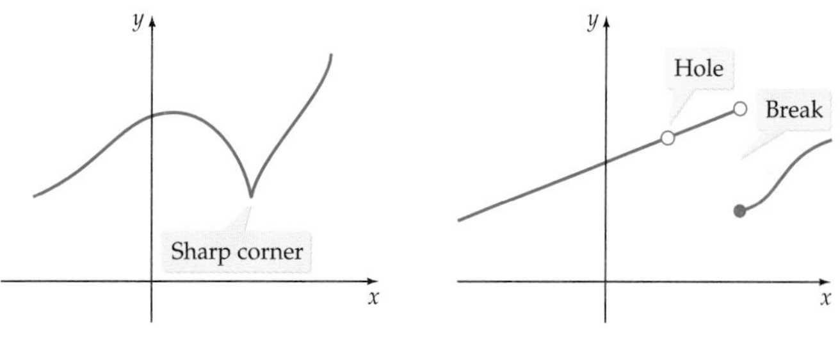

a. Continuous, but not smooth b. Not continuous

Figure 3.4

◆ FAR-LEFT AND FAR-RIGHT BEHAVIOR

The graph of a polynomial function may have several up and down fluctuations; however, the graph of every polynomial function will eventually increase or decrease without bound as the graph moves far to the left or far to the right. The **leading term** a_nx^n is said to **dominate** the polynomial function $P(x) = a_nx^n + a_{n-1}x^{n-1} + \cdots + a_1x + a_0$ as $|x|$ becomes large, because the absolute value of a_nx^n will be much larger than the absolute value of any of the other terms. Because of this condition, you can determine the far-left and far-right behavior of the polynomial by examining the leading coefficient a_n and the degree n of the polynomial.

Table 3.2 indicates the far-left and the far-right behavior of a polynomial function $P(x)$ with leading term a_nx^n.

Table 3.2 Far-Right and Far-Left Behavior of a Polynomial with Leading Term $a_n x^n$

	n is even	n is odd
$a_n > 0$	Up to left and up to right	Down to left and up to right
$a_n < 0$	Down to left and down to right	Up to left and down to right

EXAMPLE 1 Determine the Far-Left and Far-Right Behavior of a Polynomial Function

Examine the leading term to determine the far-left and the far-right behavior of the graph of each polynomial function.

a. $P(x) = x^3 - x$ **b.** $S(x) = \dfrac{1}{2}x^4 - \dfrac{5}{2}x^2 + 2$

c. $T(x) = -2x^3 + x^2 + 7x - 6$ **d.** $U(x) = -x^4 + 8x^2 + 9$

Solution

a. Because $a_n = 1$ is *positive* and $n = 3$ is *odd*, the graph of P goes down to its far left and up to its far right. See **Figure 3.5.**

b. Because $a_n = \frac{1}{2}$ is *positive* and $n = 4$ is *even*, the graph of S goes up to its far left and up to its far right. See **Figure 3.6.**

c. Because $a_n = -2$ is *negative* and $n = 3$ is *odd*, the graph of T goes up to its far left and down to its far right. See **Figure 3.7.**

d. Because $a_n = -1$ is *negative* and $n = 4$ is *even*, the graph of U goes down to its far left and down to its far right. See **Figure 3.8.**

TRY EXERCISE 2, EXERCISE SET 3.2, PAGE 254

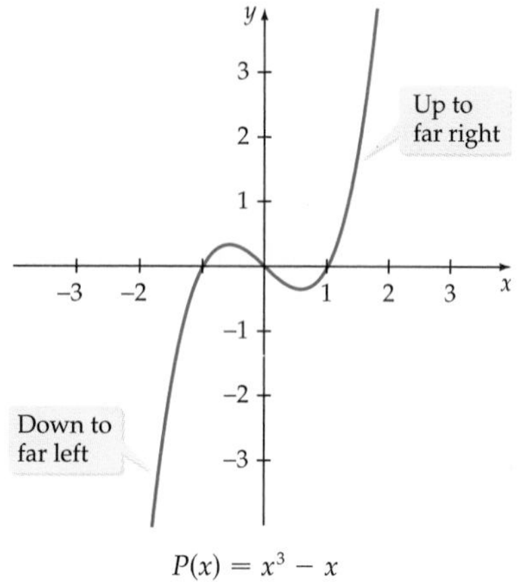

$P(x) = x^3 - x$

Figure 3.5

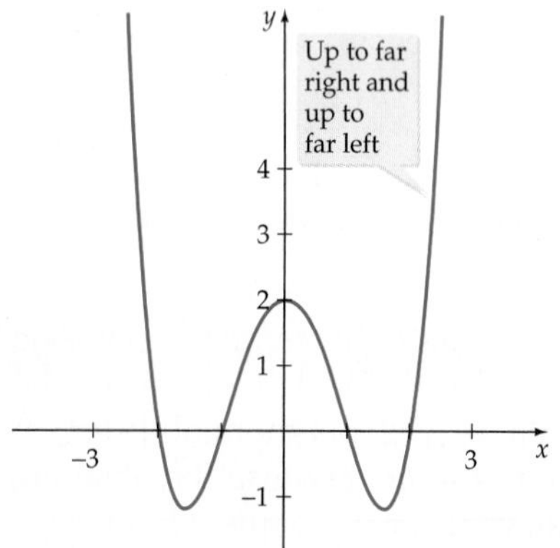

$S(x) = \dfrac{1}{2}x^4 - \dfrac{5}{2}x^2 + 2$

Figure 3.6

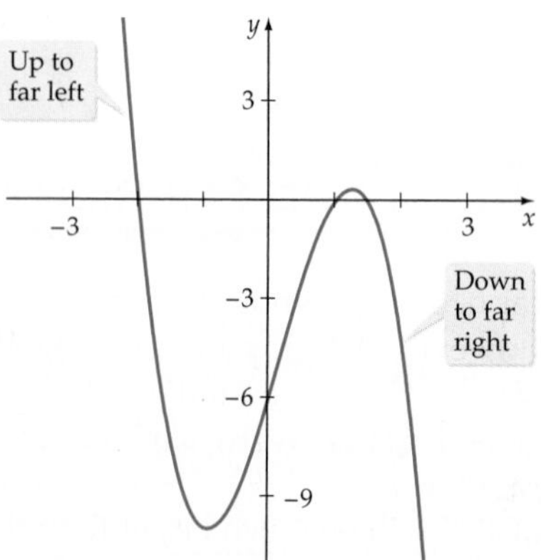

$T(x) = -2x^3 + x^2 + 7x - 6$

Figure 3.7

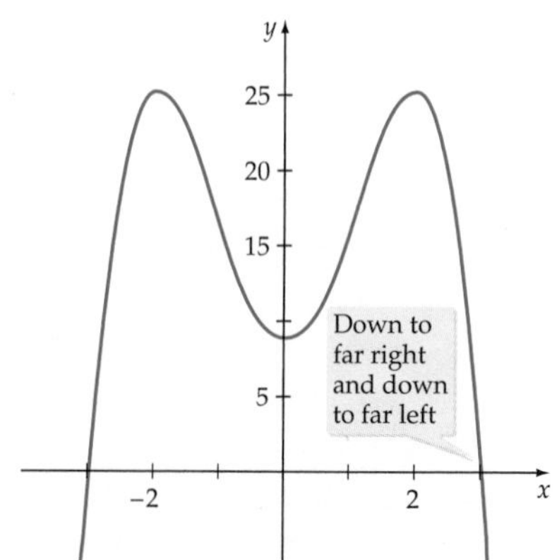

$U(x) = -x^4 + 8x^2 + 9$

Figure 3.8

◆ Maximum and Minimum Values

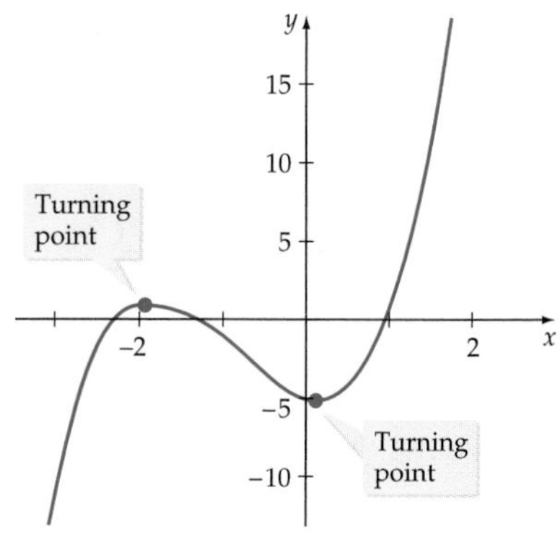

$P(x) = 2x^3 + 5x^2 - x - 5$

Figure 3.9

Figure 3.9 illustrates the graph of a polynomial function of degree 3 with two **turning points,** points where the function changes from an increasing function to a decreasing function, or vice versa. In general, the graph of a polynomial function of degree n has at most $n - 1$ turning points.

Turning points can be related to the concepts of maximum and minimum value of a function. These concepts were introduced in the discussion of graphs of second-degree equations in two variables earlier in the text. Recall that the minimum value of a function f is the smallest range value of f. It is often called the **absolute minimum.** For the function whose graph is shown in Figure 3.10, the y value of point E is the absolute minimum. There are no y values less than y_5.

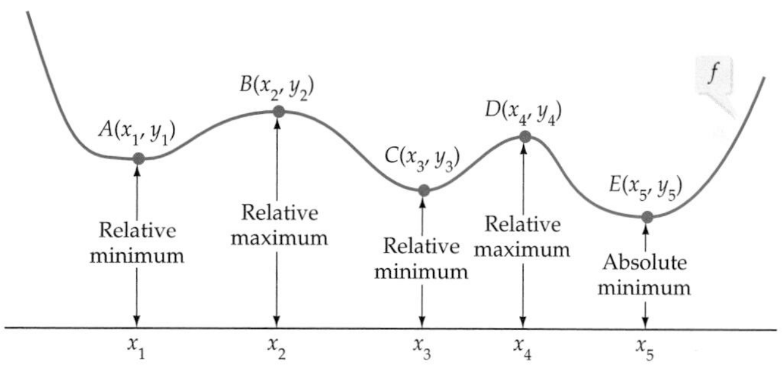

Figure 3.10

take note

A web applet is available to explore the relationships between the graph of a polynomial and its coefficients. This applet, Fourth Degree Polynomial, can be found on our web site at

http://college.hmco.com

Now consider y_1, the y value of turning point A in Figure 3.10. It is not the smallest y value of every point on the graph of f; however, it is the smallest y value if we *localize* our field of view to a small neighborhood or open interval containing x_1. It is for this reason that we refer to y_1 as a *local minimum,* or *relative minimum,* of f. The y value of point C is also a relative minimum of f.

The function does not have an absolute maximum because it goes up both to its far left and to its far right.

The y value of the point B is a relative maximum, as is the y value of point D. The formal definitions of *relative maximum* and *relative minimum* are presented below.

Relative Maximum and Relative Minimum

Let f be a function defined on the open interval I, and let c be an element of I. Then

- $f(c)$ is a **relative minimum** of f on I if $f(c) \leq f(x)$ for all x in I.
- $f(c)$ is a **relative maximum** of f on I if $f(c) \geq f(x)$ for all x in I.

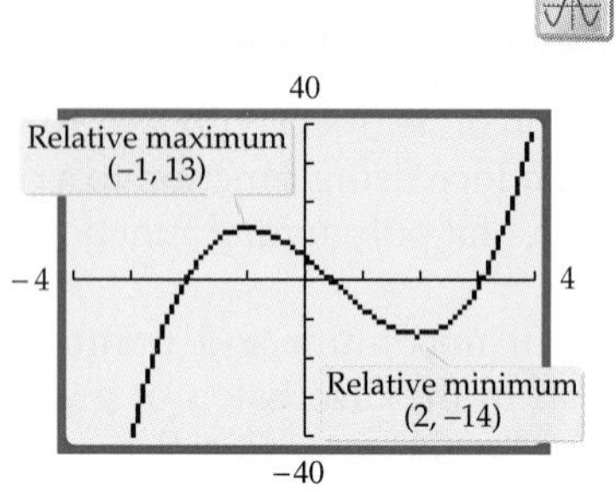

$P(x) = 2x^3 - 3x^2 - 12x + 6$

Figure 3.11

A graphing utility can be used to estimate minimum and maximum values of a polynomial.

Figure 3.11 shows the graph of $P(x) = 2x^3 - 3x^2 - 12x + 6$. The relative maximum of the function is $y = 13$, and it occurs at $x = -1$. The relative minimum of the function is $y = -14$, and it occurs at $x = 2$. Because of the far-left and far-right behavior of the cubic function, the function does not have an absolute maximum or an absolute minimum.

The following example illustrates the role a relative maximum may play in an application.

EXAMPLE 2 Solve a Maximization Application Problem

A rectangular piece of cardboard measures 12 inches by 16 inches. An open box is formed by cutting congruent squares that measure x by x from each of the corners of the cardboard and folding as shown in **Figure 3.12**.

a. Express the volume V of the box as a function of x.

b. Determine (to the nearest 0.1 inch) the x value that maximizes the volume.

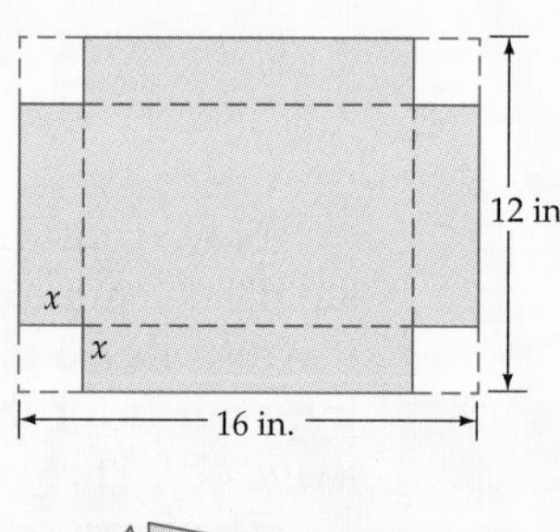

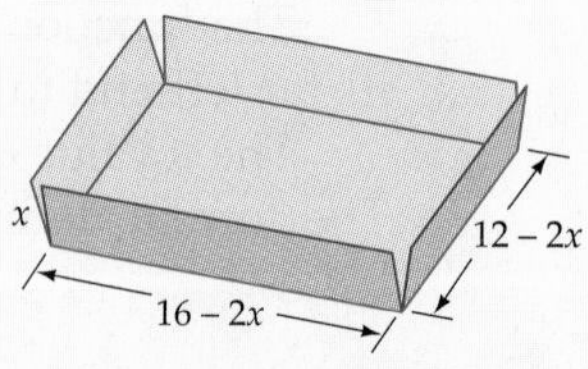

Figure 3.12

Solution

a. The lengths of the sides of the open box are x, $12 - 2x$, and $16 - 2x$. The volume is given by

$$\begin{aligned} V(x) &= x(12 - 2x)(16 - 2x) \\ &= 4x^3 - 56x^2 + 192x \end{aligned}$$

b. Use a graphing utility to graph $y = V(x)$. The graph is shown in **Figure 3.13**. Note that we are interested only in the part of the graph for

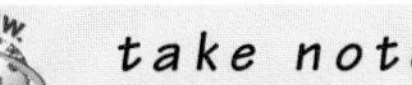

take note

A TI graphing calculator program is available to simulate the construction of a box by cutting out squares from each corner of a rectangular piece of cardboard. This program, CUTOUT, can be found on our web site at

http://college.hmco.com

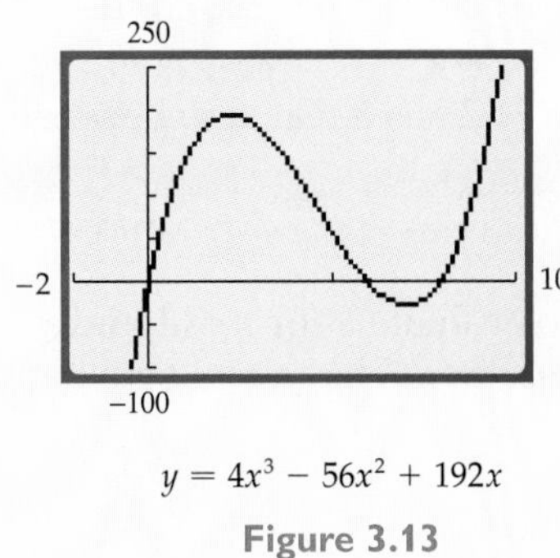

$y = 4x^3 - 56x^2 + 192x$

Figure 3.13

which $0 < x < 6$. This conclusion is a result of the following. The length of each side of the box must be positive. Hence,

$$x > 0, \quad 12 - 2x > 0, \quad \text{and} \quad 16 - 2x > 0$$

The domain of V is the intersection of the solution sets of the three inequalities. Thus the domain is $\{x \mid 0 < x < 6\}$.

Now use a graphing utility to find that V attains its maximum of 194.068 when $x \approx 2.3$. See **Figures 3.14 and 3.15.**

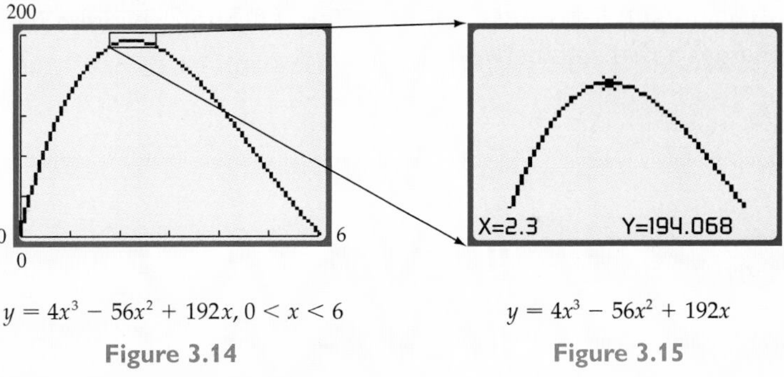

$y = 4x^3 - 56x^2 + 192x, 0 < x < 6$

Figure 3.14

$y = 4x^3 - 56x^2 + 192x$

Figure 3.15

TRY EXERCISE 26, EXERCISE SET 3.2, PAGE 255

◆ REAL ZEROS OF A POLYNOMIAL

The graph of every polynomial P is a smooth continuous curve, and if the value of P changes sign on an interval, then $P(c)$ must equal zero for at least one c in the interval. This result is known as the *Zero Location Theorem.*

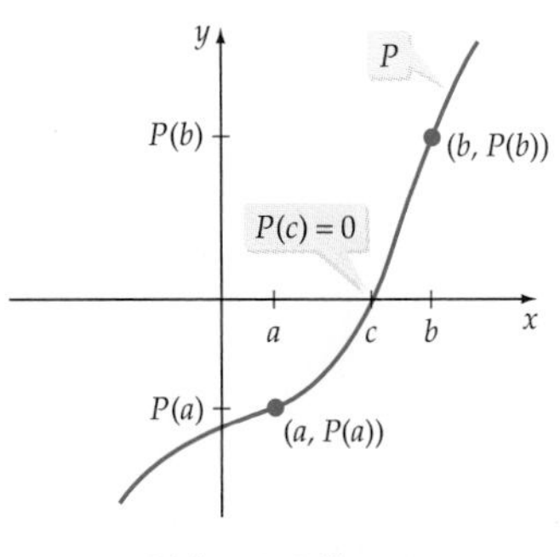

$P(a) < 0, P(b) > 0$

Figure 3.16

The Zero Location Theorem

Let $P(x)$ be a polynomial. If $a < b$, and if $P(a)$ and $P(b)$ have opposite signs, then there is at least one value c between a and b such that $P(c) = 0$.

If, for instance, the value of a polynomial P is negative at $x = a$ and positive at $x = b$, then there is at least one real number c between a and b such that $P(c) = 0$. See **Figure 3.16.**

EXAMPLE 3 Apply the Zero Location Theorem

Verify that $S(x) = x^3 - x - 2$ has a real zero between 1 and 2.

Solution

Use synthetic division to evaluate S for $x = 1$ and $x = 2$. If S changes sign between these two values, then S has a real zero between 1 and 2.

$$\begin{array}{r|rrrr} 1 & 1 & 0 & -1 & -2 \\ & & 1 & 1 & 0 \\ \hline & 1 & 1 & 0 & -2 \end{array}$$

• $S(1)$ is negative.

$$\begin{array}{r|rrrr} 2 & 1 & 0 & -1 & -2 \\ & & 2 & 4 & 6 \\ \hline & 1 & 2 & 3 & 4 \end{array}$$

• $S(2)$ is positive.

The graph of S is continuous because S is a polynomial. Also $S(1)$ is negative and $S(2)$ is positive. Thus the Zero Location Theorem indicates that there is a real zero between 1 and 2.

Visualize the Solution

The graph of S crosses the x-axis between $x = 1$ and $x = 2$. Thus S has a real zero between 1 and 2.

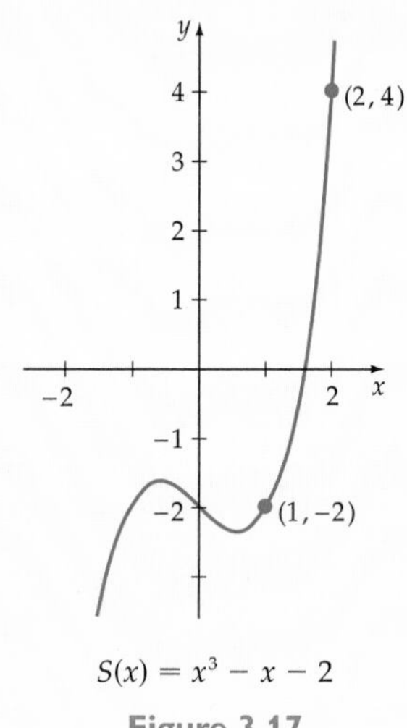

$S(x) = x^3 - x - 2$

Figure 3.17

TRY EXERCISE 34, EXERCISE SET 3.2, PAGE 256

The following theorem expresses important relationships among the real zeros of a polynomial, the x-intercepts of its graph, and its factors that can be written in the form $(x - c)$, where c is a real number.

Polynomials, Real Zeros, Graphs, and Factors $(x - c)$

If P is a polynomial and c is a real number, then all of the following statements are equivalent in the sense that if any one statement is true, then they are all true, and if any one statement is false, then they are all false.

- $(x - c)$ is a factor of P.
- $x = c$ is a real solution of $P(x) = 0$.
- $x = c$ is a real zero of P.
- $(c, 0)$ is an x-intercept of the graph of $y = P(x)$.

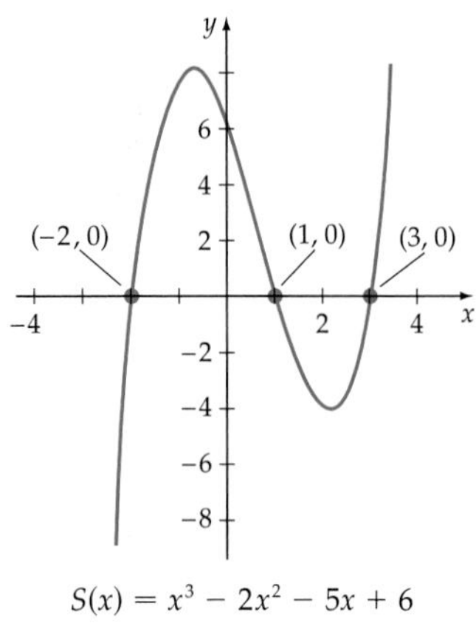

$S(x) = x^3 - 2x^2 - 5x + 6$

Figure 3.18

Sometimes it is possible to make use of the preceding theorem and a graph of a polynomial to factor the polynomial. For example, the graph of

$$S(x) = x^3 - 2x^2 - 5x + 6$$

is shown in **Figure 3.18.** The x-intercepts are $(-2, 0)$, $(1, 0)$, and $(3, 0)$. Hence -2, 1, and 3 are zeros of S, and $[x - (-2)]$, $(x - 1)$, and $(x - 3)$ are all factors of S. Thus, in factored form,

$$S(x) = (x + 2)(x - 1)(x - 3)$$

◆ THE EVEN AND ODD POWERS OF $(x - c)$ THEOREM

Use a graphing utility to graph $P(x) = (x + 3)(x - 4)^2$. Examine the graph near the x-intercepts $(-3, 0)$ and $(4, 0)$. Observe that the graph of P

- crosses the x-axis at $(-3, 0)$.
- intersects the x-axis but does not cross the x-axis at $(4, 0)$.

The following theorem can be used to determine at which x-intercepts the graph of a polynomial will cross the x-axis and at which x-intercepts the graph will intersect but not cross the x-axis.

Even and Odd Powers of $(x - c)$

If c is a real number and the polynomial $P(x)$ has $(x - c)$ as a factor exactly k times, then the graph of P will

- intersect but not cross the x-axis at $(c, 0)$, provided that k is an even positive integer.
- cross the x-axis at $(c, 0)$, provided that k is an odd positive integer.

EXAMPLE 4 Apply the Even and Odd Powers of $(x - c)$ Theorem

Determine where the graph of $P(x) = (x + 3)(x - 2)^2(x - 4)^3$ will cross the x-axis and where the graph will intersect but not cross the x-axis.

Solution

The exponents of the factors $(x + 3)$ and $(x - 4)$ are odd integers. Therefore, the graph of P will cross the x-axis at the x-intercepts $(-3, 0)$ and $(4, 0)$.

The exponent of the factor $(x - 2)$ is an even integer. Therefore, the graph of P will intersect but not cross the axis at $(2, 0)$.

Use a graphing utility to check these results.

TRY EXERCISE 52, EXERCISE SET 3.2, PAGE 254

TOPICS FOR DISCUSSION

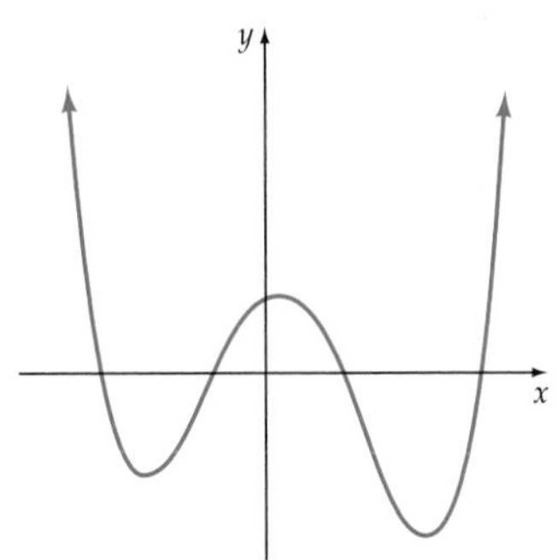

Figure 3.19

1. Discuss the meaning of the phrase *polynomial function*. Give examples of polynomial functions and of functions that are not polynomials.
2. Is it possible for the graph of the polynomial shown in **Figure 3.19** to have a degree that is an odd number? If so, explain how. If not, explain why not.
3. Explain the difference between a relative minimum and an absolute minimum and the difference between a relative maximum and an absolute maximum.
4. Discuss how the Zero Location Theorem can be used to find a real zero of a polynomial.
5. A complex number may be a zero of a polynomial. For instance, i is a zero of $P(x) = x^2 + 1$ because $P(i) = 0$. Explain why the Zero Location Theorem cannot be used to find the complex number zeros of a polynomial.
6. Let $P(x)$ be a polynomial with real coefficients. Explain the relationship among a real zero of a polynomial, the x-coordinate of the x-intercept of the graph of the polynomial, and the solution of the equation $P(x) = 0$.

EXERCISE SET 3.2

In Exercises 1 to 10, examine the leading term and determine the far-left and far-right behavior of the graph of the polynomial function.

1. $P(x) = 3x^4 - 2x^2 - 7x + 1$
2. $P(x) = -2x^3 - 6x^2 + 5x - 1$
3. $P(x) = 5x^5 - 4x^3 - 17x^2 + 2$
4. $P(x) = -6x^4 - 3x^3 + 5x^2 - 2x + 5$
5. $P(x) = 2 - 3x - 4x^2$
6. $P(x) = -16 + x^4$
7. $P(x) = \frac{1}{2}(x^3 + 5x^2 - 2)$
8. $P(x) = -\frac{1}{4}(x^4 + 3x^2 - 2x + 6)$
9. $P(x) = -\frac{2}{3}(x + 1)^3$
10. $P(x) = \frac{1}{5}(x - 1)^4$

In Exercises 11 to 14, use your knowledge of the vertex of a parabola to find the maximum or minimum of each quadratic function.

11. $P(x) = x^2 + 4x - 1$
12. $P(x) = x^2 + 6x + 1$
13. $P(x) = -x^2 - 8x + 1$
14. $P(x) = -2x^2 + 8x - 1$
15. **MAXIMIZE AN AREA FUNCTION** A farmer has \$1000 to spend to fence a rectangular corral. Because extra reinforcing is needed for one side, the corral cost \$6 per foot along that side. It costs \$2 per foot to fence the remaining three sides. What dimensions of the corral will maximize the area of the corral?
16. **MAXIMIZE A PROFIT** A manufacturer has determined that the revenue received from selling x items of a product is given by $R(x) = -\frac{1}{10}x^2 + 90x$ and the cost to produce x items of the product is given by $C(x) = 40x + 8000$. Assuming that all of the products produced can be sold, how many should be produced to maximize profit? *Suggestion:* profit = revenue − cost
17. **MINIMIZE A SUM** Let $S = (3 - x)^2 + (7 - x)^2 + (8 - x)^2$. Find the number x for which the value of S is a minimum. Now show that this number is the average of 3, 7, and 8.
18. **MINIMIZE A DISTANCE** Find the coordinates (x, y) of the point on the graph of $y = -x + 4$ that is closest to the origin. *Suggestion:* The *square* of the distance from the point is $D^2 = x^2 + y^2$. Replace y by $-x + 4$ and find the point that minimizes the square of the distance. Then use the fact that the point that minimizes the *square* of the distance also minimizes the distance.

In Exercises 19 to 24, use a graphing utility to graph each polynomial. Now use the maximum and minimum features of the graphing utility to estimate, to the nearest tenth, the coordinates of the relative maximum and relative minimum. The number in parentheses to the right of the polynomial is the total number of relative maxima and minima.

19. $P(x) = x^3 + x^2 - 9x - 9$ (2)

20. $P(x) = x^3 + 4x^2 - 4x - 16$ (2)

21. $P(x) = x^3 - 3x^2 - 24x + 3$ (2)

22. $P(x) = -2x^3 - 3x^2 + 12x + 1$ (2)

23. $P(x) = x^4 - 4x^3 - 2x^2 + 12x - 5$ (3)

24. $P(x) = x^4 - 10x^2 + 9$ (3)

25. **MAXIMIZING VOLUME** An open box is to be constructed from a rectangular sheet of cardboard that measures 16 inches by 22 inches. To assemble the box make the four cuts shown in the figure below and then fold on the dashed lines. What value of x (to the nearest 0.001 inch) will produce a box with maximum volume? What is the maximum volume (to the nearest 0.1 cubic inch)?

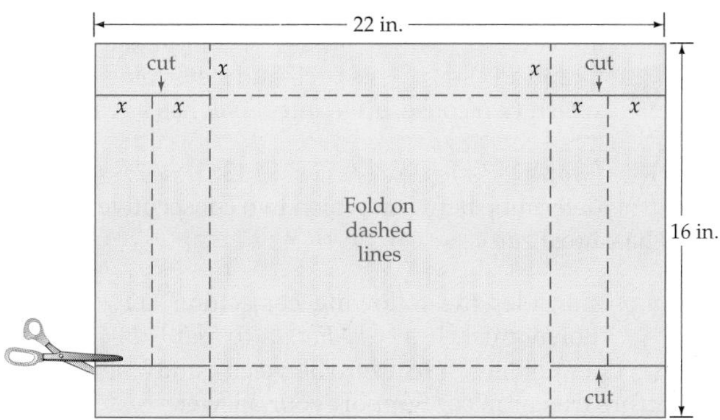

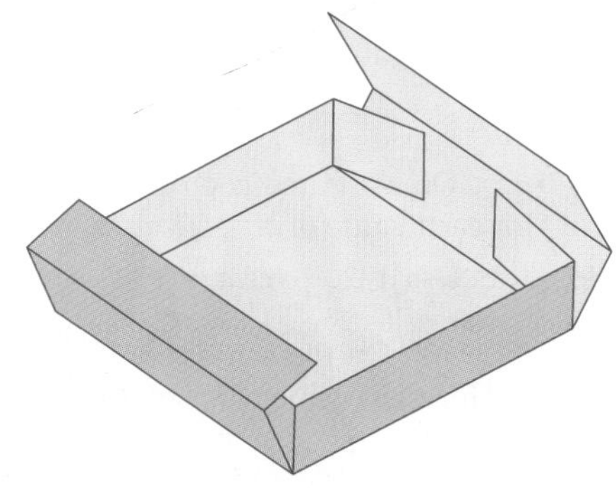

26. **MAXIMIZING VOLUME** A closed box is to be constructed from a rectangular sheet of cardboard that measures 18 inches by 42 inches. The box is made by cutting rectangles that measure x inches by $2x$ inches from two of the corners and by cutting two squares that measure x inches by x inches from the top and from the bottom of the rectangle, as shown in the following figure. What value of x (to the nearest 0.001 inch) will produce a box with maximum volume? What is the maximum volume (to the nearest 0.1 cubic inch)?

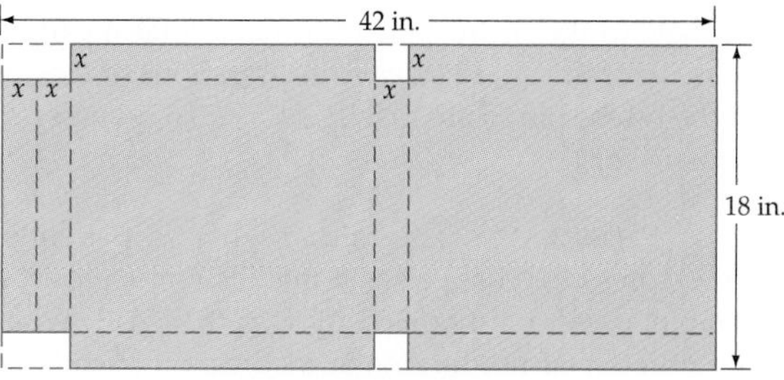

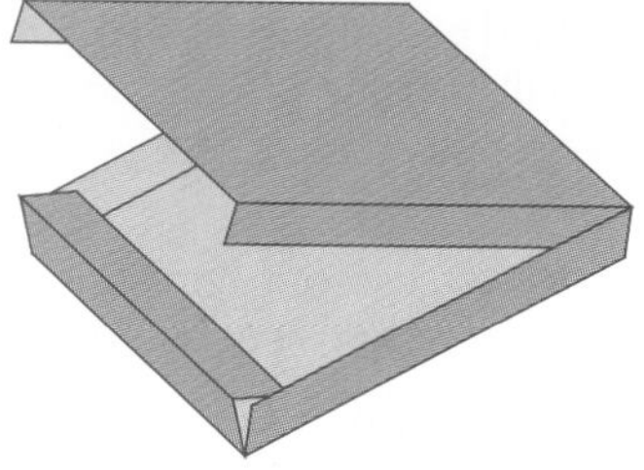

27. **MAXIMIZING VELOCITY** A car traveled at a velocity (kilometers per hour) of

$$v(t) = 0.038(t + 2)(t - 18)(t - 21) + 48$$

where t is the time in minutes after 1 P.M. What was the maximum velocity (to the nearest 0.1 kilometer per hour) that the car attained during the period $0 \le t \le 24$? At what time (to the nearest 0.01 minute) did it reach this maximum velocity?

28. **MINIMIZING VELOCITY** A plane traveled at a velocity (mph) of

$$v(t) = 0.00182(t + 2)(t - 48)(t - 31)(t + 9) + 378$$

where t is the time in minutes after 8 A.M. What was the minimum velocity (to the nearest mph) that the plane attained during the period $0 \le t \le 50$? At what time (to the nearest 0.1 minute) did it reach this minimum velocity?

Exercises 29 and 30 refer to the following mathematical model. The Fahrenheit temperature T in a city during a particular 12-hour period can be modeled by

$$T(t) = 0.051(t - 1)(t - 11)(t - 14) + 44$$

where t is the time in hours after 9 A.M.

29. **MAXIMIZING TEMPERATURE** What was the maximum temperature (to the nearest degree), and at what time (to the nearest minute) was it achieved?

30. **MINIMIZING TEMPERATURE** What was the minimum temperature (to the nearest degree), and at what time (to the nearest minute) was it achieved?

31. **PHYSIOLOGY** The velocity of the air that is expelled during a cough can be modeled by $v = 6r^2 - 10r^3$, where v is measured in meters per second and r is the radius of the trachea in centimeters. Find the radius of the trachea, to the nearest 0.1 centimeter, that maximizes the velocity of the air. Physical considerations of the structure of the trachea indicate that $0.3 \le r \le 0.6$ is an appropriate domain for r.

32. **PHYSIOLOGY** The rate at which a patient's blood pressure changes as a medication is absorbed into the blood can be modeled by $R = (8 - A)A$, where A is the number of milligrams of medication absorbed and R is the rate, in millimeters of mercury per minute, at which a patient's blood pressure is changing. At what amount of absorption will the rate of the patient's blood pressure change most rapidly?

In Exercises 33 to 38, use the Zero Location Theorem to verify that *P* has a zero between *a* and *b*.

33. $P(x) = 2x^3 + 3x^2 - 23x - 42;\quad a = 3, b = 4$

34. $P(x) = 4x^3 - x^2 - 6x + 1;\quad a = 0, b = 1$

35. $P(x) = 3x^3 + 7x^2 + 3x + 7;\quad a = -3, b = -2$

36. $P(x) = 2x^3 - 21x^2 - 2x + 21;\quad a = 10, b = 11$

37. $P(x) = 4x^4 + 7x^3 - 11x^2 + 7x - 15;\quad a = 1, b = 1\frac{1}{2}$

38. $P(x) = 5x^3 - 16x^2 - 20x + 64;\quad a = 3, b = 3\frac{1}{2}$

In Exercises 39 to 48, find the *x*-intercepts of the graph of *P*.

39. $P(x) = (x + 2)(x - 3)(2x + 7)$

40. $P(x) = (x - 5)(x - 1)(4x + 1)$

41. $P(x) = x(x - 1)(5x - 2)$

42. $P(x) = x(x - 4)(x - 1)(x + 7)$

43. $P(x) = (3x + 7)(2x - 11)(x + 5)^2$

44. $P(x) = (x - 3)^2(2x - 1)$

45. $P(x) = x^3 + x^2 - 6x$

46. $P(x) = 2x^3 - 7x^2 - 15x$

47. $P(x) = x^3 - 1$

48. $P(x) = x^3 + 8$

In Exercises 49 to 58, determine the *x*-intercepts of the graph of *P*. For each *x*-intercept, use the Even and Odd Powers of $(x - c)$ Theorem to determine whether the graph of *P* crosses the *x*-axis or intersects but does not cross the *x*-axis.

49. $P(x) = (x - 1)(x + 1)(x - 3)$

50. $P(x) = (x - 2)(x + 3)(x + 1)$

51. $P(x) = -(x - 3)^2(x - 7)^5$

52. $P(x) = (x + 2)^3(x - 6)^{10}$

53. $P(x) = (2x - 3)^4(x - 1)^{15}$

54. $P(x) = (5x + 10)^6(x - 2.7)^5$

55. $P(x) = x^4(x - 201)^2(3x - 2)$

56. $P(x) = x^5(x + 11)(3x - 7)^3$

57. $P(x) = x^3 - 6x^2 + 9x$

58. $P(x) = x^4 + 3x^3 + 4x^2$

SUPPLEMENTAL EXERCISES

59. Graph $P(x) = x^3 - x - 25$ and determine between which two consecutive integers P has a real zero.

60. Graph $P(x) = 4x^4 - 12x^3 + 13x^2 - 12x + 9$ and determine between which two consecutive integers P has a real zero.

61. Consider the following conjecture. Let $P(x)$ be a polynomial. If $a < b$, $P(a) > 0$, and $P(b) > 0$, then $P(x)$ does not have a real zero between a and b. Is this conjecture true or false? Support your answer.

62. Let $f(x) = x^3 + c$. On the same coordinate axes, sketch the graph of f for each value of c.
    a. $c = 0$ b. $c = 2$ c. $c = -3$

63. Let $f(x) = ax^3$. On the same coordinate axes, sketch the graph of f for each value of a.
    a. $a = 2$ b. $a = 1/2$ c. $a = -1$

64. Let $f(x) = (x - h)^3$. On the same coordinate axes, sketch the graph of f for each value of h.
    a. $h = 2$ b. $h = -1$ c. $h = -5$

65. Make use of the graphs in Exercises 62 through 64 to explain how the graph of the equation $f(x) = a(x - h)^3 + c$ compares with the graph of $g(x) = x^3$.

PROJECTS

1.  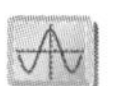 **REAL ZEROS AND THE DEGREE OF A POLYNOMIAL** This project examines the connection between the number of real zeros of a polynomial and its degree.

   a. Graph the polynomials

   $$P(x) = x^4 + 1, Q(x) = x^4 - x^3 - 11x^2 - x - 12$$

   and

   $$R(x) = x^4 - 2x^3 - 13x^2 + 14x + 24$$

   How many real zeros does each of the polynomials $P(x)$, $Q(x)$, and $R(x)$ have? Graph some other polynomials of degree 4. On the basis of your graphs, make a conjecture about the number of real zeros a polynomial of degree 4 can have.

   b. Graph the polynomials

   $$P(x) = x^5, Q(x) = x^5 - 3x^4 - 11x^3 + 27x^2 + 10x - 24$$

   and

   $$R(x) = x^5 - 2x^4 - 10x^3 + 10x^2 - 11x + 12$$

   How many real zeros do the polynomials $P(x)$, $Q(x)$, and $R(x)$ have? Graph some other polynomials of degree 5. On the basis of your graphs, make a conjecture about the number of real zeros a polynomial of degree 5 can have.

   c. Graph some second-degree and third-degree polynomials, and note the number of real zeros of the polynomials. On the basis of all your graphs and the graphs in **a.** and **b.**, make a conjecture about the number of real zeros a polynomial of degree n can have.

SECTION 3.3 ZEROS OF POLYNOMIAL FUNCTIONS

- MULTIPLE ZEROS OF A POLYNOMIAL
- THE RATIONAL ZERO THEOREM
- UPPER AND LOWER BOUNDS FOR REAL ZEROS
- DESCARTES' RULE OF SIGNS
- FIND ZEROS OF A POLYNOMIAL
- APPLICATIONS

3A

◆ MULTIPLE ZEROS OF A POLYNOMIAL

Recall that if $P(x)$ is a polynomial function, then the values of x for which $P(x)$ is equal to 0 are called the *zeros* of $P(x)$ or the *roots* of the equation $P(x) = 0$. A zero of a polynomial may be a **multiple zero.** For example, the polynomial $x^2 + 6x + 9$ can be expressed in factored form as $(x + 3)(x + 3)$. Setting each factor equal to zero yields $x = -3$ in both cases. Thus $x^2 + 6x + 9$ has a zero of -3 that occurs twice. The following definition will be most useful when we are discussing multiple zeros.

Definition of Multiple Zeros of a Polynomial

If a polynomial $P(x)$ has $(x - r)$ as a factor exactly k times, then r is a **zero of multiplicity k** of the polynomial $P(x)$.

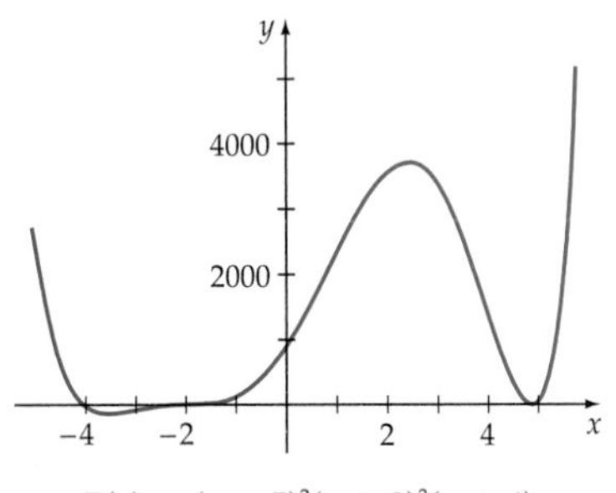

$P(x) = (x - 5)^2(x + 2)^3(x + 4)$

Figure 3.20

The graph of the polynomial

$$P(x) = (x - 5)^2(x + 2)^3(x + 4)$$

is shown in **Figure 3.20.** This polynomial has

- 5 as a zero of multiplicity 2
- -2 as a zero of multiplicity 3
- -4 as a zero of multiplicity 1

A zero of multiplicity 1 is generally referred to as a **simple zero.**

When searching for the zeros of a polynomial function, it is important that we know how many zeros to expect. This question is answered completely in Section 3.4. For the work in this section, the following result is valuable.

Number of Zeros of a Polynomial Function

A polynomial function P of degree n has at most n zeros, where each zero of multiplicity k is counted k times.

◆ THE RATIONAL ZERO THEOREM

The rational zeros of polynomials with integer coefficients can be found with the aid of the following theorem.

The Rational Zero Theorem

If $P(x) = a_nx^n + a_{n-1}x^{n-1} + \cdots + a_1x + a_0$ has integer coefficients, and p/q (where p and q have no common factors) is a rational zero of $P(x)$, then p is a factor of a_0 and q is a factor of a_n.

The Rational Zero Theorem often is used to make a list of all possible rational zeros of a polynomial. The list consists of all rational numbers of the form p/q, where p is an integer factor of the constant term a_0, and q is an integer factor of the leading coefficient a_n.

EXAMPLE 1 Apply the Rational Zero Theorem

Use the Rational Zero Theorem to list all possible rational zeros of

$$P(x) = 4x^4 + x^3 - 40x^2 + 38x + 12$$

take note

The Rational Zero Theorem gives the *possible* rational zeros of a polynomial. That is, if P has a rational zero, then it must be one indicated by the theorem. However, P may not have any rational zeros. In the case of the polynomial in Example 1, the only rational zeros are $-1/4$ and 2. The remaining rational numbers in the list are not zeros of P.

Solution

List all integers p that are factors of 12 and all integers q that are factors of 4.

$$p: \quad \pm1, \pm2, \pm3, \pm4, \pm6, \pm12$$
$$q: \quad \pm1, \pm2, \pm4$$

Form all possible rational numbers using ±1, ±2, ±3, ±4, ±6, and ±12 for the numerator and ±1, ±2, and ±4 for the denominator. By the Rational Zero Theorem, the possible rational zeros are

$$\pm1, \pm\frac{1}{2}, \pm\frac{1}{4}, \pm2, \pm3, \pm\frac{3}{2}, \pm\frac{3}{4}, \pm4, \pm6, \pm12$$

TRY EXERCISE 14, EXERCISE SET 3.3, PAGE 268

It is not necessary to list a factor that is already listed in reduced form. For example, $\pm6/4$ is not listed because it is equal to $\pm3/2$.

◆ UPPER AND LOWER BOUNDS FOR REAL ZEROS

A real number b is called an **upper bound** of the zeros of the polynomial function P if no zero is greater than b. A real number a is called a **lower bound** of the zeros of P if no zero is less than a. The following theorem is often used to find positive upper bounds and negative lower bounds for the real zeros of a polynomial function.

Upper- and Lower-Bound Theorem

Upper bound If $b > 0$ and all the numbers in the bottom row of the synthetic division of P by $x - b$ are either positive or zero, then b is an upper bound for the real zeros of P.

Lower bound If $a < 0$ and the numbers in the bottom row of the synthetic division of P by $x - a$ alternate in sign (the number zero can be considered positive or negative), then a is a lower bound for the real zeros of P.

Upper and lower bounds are not unique. For example, if b is an upper bound for the real zeros of P, then any number greater than b is also an upper bound. Also, if a is a lower bound for the real zeros of P, then any number less than a is also a lower bound.

EXAMPLE 2 Find Upper and Lower Bounds

According to the Upper- and Lower-Bound Theorem, what is the smallest positive integer that is an upper bound and the largest negative integer that is a lower bound of the zeros of $P(x) = 2x^3 + 7x^2 - 4x - 14$?

Continued ▶

take note

When you check for bounds, you do not need to limit your choices to the possible zeros given by the Rational Zero Theorem. For instance, in Example 2 the integer -4 is a lower bound; however, -4 is not one of the possible zeros of P as given by the Rational Zero Theorem.

Solution

To find the smallest positive-integer upper bound, use synthetic division with $1, 2, \ldots,$ as test values.

$$\begin{array}{r|rrrr} 1 & 2 & 7 & -4 & -14 \\ & & 2 & 9 & 5 \\ \hline & 2 & 9 & 5 & -9 \end{array} \qquad \begin{array}{r|rrrr} 2 & 2 & 7 & -4 & -14 \\ & & 4 & 22 & 36 \\ \hline & 2 & 11 & 18 & 22 \end{array}$$

• **All positive signs**

Thus 2 is the smallest positive-integer upper bound.

Now find the largest negative-integer lower bound.

$$\begin{array}{r|rrrr} -1 & 2 & 7 & -4 & -14 \\ & & -2 & -5 & 9 \\ \hline & 2 & 5 & -9 & -5 \end{array} \qquad \begin{array}{r|rrrr} -2 & 2 & 7 & -4 & -14 \\ & & -4 & -6 & 20 \\ \hline & 2 & 3 & -10 & 6 \end{array}$$

$$\begin{array}{r|rrrr} -3 & 2 & 7 & -4 & -14 \\ & & -6 & -3 & 21 \\ \hline & 2 & 1 & -7 & 7 \end{array} \qquad \begin{array}{r|rrrr} -4 & 2 & 7 & -4 & -14 \\ & & -8 & 4 & 0 \\ \hline & 2 & -1 & 0 & -14 \end{array}$$

• **Alternating signs**

Thus -4 is the largest negative-integer lower bound.

TRY EXERCISE 22, EXERCISE SET 3.3, PAGE 268

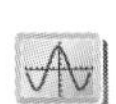

You can use the Upper- and Lower-Bound Theorem in conjunction with a graphing utility to determine Xmin (the lower bound) and Xmax (the upper bound) for the viewing window. This will ensure that all the real zeros, which are the x-coordinates of the x-intercepts of the polynomial, will be shown. Note in **Figure 3.21** that the zeros of $P(x) = 2x^3 + 7x^2 - 4x - 14$ are between -4 (the lower bound) and 2 (the upper bound).

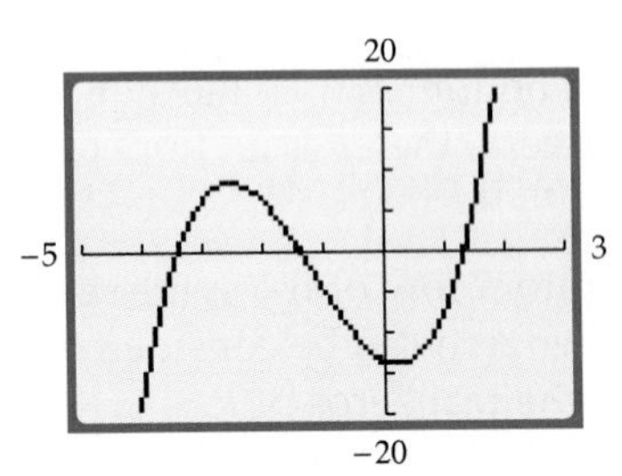

$P(x) = 2x^3 + 7x^2 - 4x - 14$

Figure 3.21

◆ DESCARTES' RULE OF SIGNS

Descartes' Rule of Signs is another theorem often used to obtain information about the zeros of a polynomial. In Descartes' Rule of Signs, the number of *variations in sign* of the coefficients of a polynomial $P(x)$ or $P(-x)$ refers to sign changes of the coefficients from positive to negative or from negative to positive that we find when we examine successive terms of the polynomial. The terms of the polynomial are assumed to appear in the order of descending powers of x. For example, the polynomial

$$P(x) = \underbrace{+3x^4 - 5x^3}_{1} \underbrace{- 7x^2 + x}_{2} \underbrace{+ x - 7}_{3}$$

has three variations of sign. The polynomial

$$\begin{aligned} P(-x) &= +3(-x)^4 - 5(-x)^3 - 7(-x)^2 + (-x) - 7 \\ &= +\ 3x^4 \ + \ \underbrace{5x^3 \ - \ 7x^2}_{1} \ - \ x \ - 7 \end{aligned}$$

has one variation in sign.

Terms that have a coefficient of 0 are not counted as a variation of sign and may be ignored. For example,

$$P(x) = \underbrace{-x^5 + 4x^2}_{1} + 1$$

has one variation in sign.

MATH MATTERS

Descartes' Rule of Signs first appeared in his *La Géométrie* (1673). Although a proof of Descartes' Rule of Signs is beyond the scope of this course, we can see that a polynomial with no variations in sign cannot have a positive zero. For instance, consider $P(x) = x^3 + x^2 + x + 1$. Each term of P is positive for any positive value of x. Thus P is never zero for $x > 0$.

Descartes' Rule of Signs

Let $P(x)$ be a polynomial with real coefficients and with the terms arranged in the order of decreasing powers of x.

1. The number of positive real zeros of $P(x)$ is equal to the number of variations in sign of $P(x)$ or is equal to that number decreased by an even integer.
2. The number of negative real zeros of $P(x)$ is equal to the number of variations in sign of $P(-x)$ or is equal to that number decreased by an even integer.

EXAMPLE 3 Apply Descartes' Rule of Signs

Determine both the number of possible positive and the number of possible negative real zeros of each polynomial.

a. $x^4 - 5x^3 + 5x^2 + 5x - 6$ **b.** $2x^5 + 3x^3 + 5x^2 + 8x + 7$

Solution

a. $$P(x) = \underbrace{+x^4 - 5x^3}_{1}\underbrace{\ +\ 5x^2}_{2} + \underbrace{5x - 6}_{3}$$

There are three variations of sign. By Descartes' Rule of Signs, there are either three or one positive real zeros. Now examine the variations of sign of $P(-x)$.

$$P(-x) = x^4 + 5x^3 + \underbrace{5x^2 - 5x}_{1} - 6$$

There is one variation of sign of $P(-x)$. By Descartes' Rule of Signs, there is one negative real zero.

b. $P(x) = 2x^5 + 3x^3 + 5x^2 + 8x + 7$ has no variation of sign, so there are no positive real zeros.

$$P(-x) = -2x^5 - 3x^3 + 5x^2 - 8x + 7$$

1 2 3

$P(-x)$ has three variations of sign, so there are either three or one negative real zeros.

TRY EXERCISE 32, EXERCISE SET 3.3, PAGE 268

In applying Descartes' Rule of Signs, we count each zero of multiplicity k as k zeros. For instance, the polynomial

$$P(x) = x^2 - 10x + 25$$

has two variations in sign. Thus by Descartes' Rule of Signs it must have either two or zero positive real zeros. Factoring the polynomial produces $(x - 5)^2$, from which it can be observed that 5 is a positive zero of multiplicity 2.

◆ FIND ZEROS OF A POLYNOMIAL

Guidelines for Finding the Zeros of a Polynomial with Integer Coefficients

1. ***Gather general information.*** Determine the degree n of the polynomial. The number of zeros of the polynomial is at most n. Apply Descartes' Rule of Signs to find the possible number of positive zeros and also the possible number of negative zeros.

2. ***Check suspects.*** Apply the Rational Zero Theorem to list rational numbers that are possible zeros. Use synthetic division to test numbers in your list. If you find an upper or a lower bound, then eliminate from your list any number that is greater than an upper bound or less than a lower bound.

3. ***Work with the reduced polynomials.*** Each time a zero is found, you obtain a reduced polynomial.

   - If a reduced polynomial is of degree 2, find its zeros either by factoring or by applying the quadratic formula.

   - If the degree of a reduced polynomial is 3 or greater, repeat the above steps for this polynomial.

Example 4 illustrates the procedures discussed in the above guidelines.

take note

If you have a graphing utility, you can produce a graph similar to the one below. By looking at the x-intercepts of the graph, you can reject as possible zeros some of the values suggested by the Rational Zero Theorem. This will reduce the amount of work that is necessary to find the zeros of the polynomial.

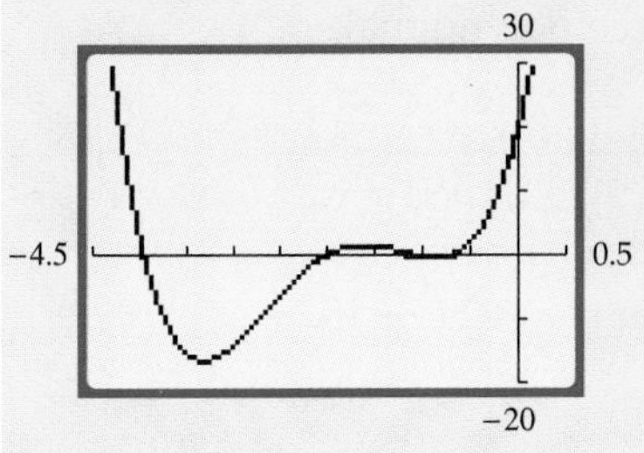

$P(x) = 3x^4 + 23x^3 + 56x^2 + 52x + 16$

EXAMPLE 4 Find the Zeros of a Polynomial

Find the zeros of $P(x) = 3x^4 + 23x^3 + 56x^2 + 52x + 16$.

Solution

1. *Gather general information.* The degree of P is 4. Thus the number of zeros of the polynomial is at most 4. By Descartes' Rule of Signs, there are no positive zeros, and there are either four, two, or no negative zeros.

2. *Check suspects.* By the Rational Zero theorem, the possible negative rational zeros of P are

$$\frac{p}{q}: \quad -1, -2, -4, -8, -16, -\frac{1}{3}, -\frac{2}{3}, -\frac{4}{3}, -\frac{8}{3}, -\frac{16}{3}$$

Use synthetic division to test the possible rational zeros. The following work shows that -4 is a zero of P.

$$\begin{array}{r|rrrrr} -4 & 3 & 23 & 56 & 52 & 16 \\ & & -12 & -44 & -48 & -16 \\ \hline & 3 & 11 & 12 & 4 & 0 \end{array}$$

Coefficients of the first reduced polynomial

3. *Work with the reduced polynomials.* Because -4 is a zero, the factors of P are $(x + 4)$ and the first reduced polynomial $(3x^3 + 11x^2 + 12x + 4)$. Thus

$$P(x) = (x + 4)(3x^3 + 11x^2 + 12x + 4)$$

All remaining zeros of P must be zeros of $3x^3 + 11x^2 + 12x + 4$. The Rational Zero Theorem indicates that the only possible negative rational zeros of $3x^3 + 11x^2 + 12x + 4$ are

$$\frac{p}{q}: \quad -1, -2, -4, -\frac{1}{3}, -\frac{2}{3}, -\frac{4}{3}$$

Synthetic division is again use to test possible zeros.

$$\begin{array}{r|rrrr} -2 & 3 & 11 & 12 & 4 \\ & & -6 & -10 & -4 \\ \hline & 3 & 5 & 2 & 0 \end{array}$$

Coefficients of the second reduced polynomial

Because -2 is a zero, $(x + 2)$ is also a factor of P. Thus

$$P(x) = (x + 4)(x + 2)(3x^2 + 5x + 2)$$

Continued

The remaining zeros of P must be zeros of $3x^2 + 5x + 2$.

$$3x^2 + 5x + 2 = 0$$
$$(3x + 2)(x + 1) = 0$$
$$x = -\frac{2}{3} \quad \text{and} \quad x = -1$$

The zeros of $P(x) = 3x^4 + 23x^3 + 56x^2 + 52x + 16$ are -4, -2, $-\frac{2}{3}$ and -1.

TRY EXERCISE 42, EXERCISE SET 3.3, PAGE 268

In Example 5 we find the zeros of a fifth-degree polynomial.

EXAMPLE 5 Find the Zeros of a Polynomial

Find the zeros of $P(x) = 2x^5 - 23x^4 + 80x^3 - 58x^2 - 82x - 15$.

Solution

1. *Gather general information.* The degree of P is 5. The number of zeros of the polynomial is at most 5. By Descartes' Rule of Signs, there are three or one positive zeros, and two, or no negative zeros.

2. *Check suspects.* By the Rational Zero Theorem, the possible rational zeros are

$$\frac{p}{q}: \quad \pm1, \pm3, \pm5, \pm15, \pm\frac{1}{2}, \pm\frac{3}{2}, \pm\frac{5}{2}, \pm\frac{15}{2}$$

The following synthetic division shows that 3 is a zero of P.

3	2	−23	80	−58	−82	−15
		6	−51	87	87	15
	2	−17	29	29	5	0

Coefficients of the first reduced polynomial

3. *Work with the reduced polynomials.* Now look for zeros of the reduced polynomial $2x^4 - 17x^3 + 29x^2 + 29x + 5$. The number of zeros of this polynomial is at most 4. By Descartes' Rule of Signs, there are two or no positive zeros, and two or no negative zeros. By the Rational Zero Theorem, the possible rational zeros are

$$\frac{p}{q}: \quad \pm1, \pm5, \pm\frac{1}{2}, \pm\frac{5}{2}$$

The following division shows that 5 is a zero of $2x^4 - 17x^3 + 29x^2 + 29x + 5$.

$$\begin{array}{r|rrrrr} 5 & 2 & -17 & 29 & 29 & 5 \\ & & 10 & -35 & -30 & -5 \\ \hline & 2 & -7 & -6 & -1 & 0 \end{array}$$

Coefficients of the second reduced polynomial

The possible rational zeros of the reduced polynomial $2x^3 - 7x^2 - 6x - 1$ are

$$\frac{p}{q}: \qquad \pm 1, \pm\frac{1}{2}$$

Synthetic division shows that $-\frac{1}{2}$ is a zero of $2x^3 - 7x^2 - 6x - 1$.

$$\begin{array}{r|rrrr} -\frac{1}{2} & 2 & -7 & -6 & -1 \\ & & -1 & 4 & 1 \\ \hline & 2 & -8 & -2 & 0 \end{array}$$

Coefficients of the third reduced polynomial

The third reduced polynomial is $2x^2 - 8x - 2$ or $2(x^2 - 4x - 1)$. Solving $x^2 - 4x - 1 = 0$ by the quadratic formula yields

$$x = \frac{-(-4) \pm \sqrt{(-4)^2 - 4(1)(-1)}}{2(1)} = \frac{4 \pm \sqrt{20}}{2} = \frac{4}{2} \pm \frac{2\sqrt{5}}{2} = 2 \pm \sqrt{5}$$

Thus the zeros of $2x^5 - 23x^4 + 80x^3 - 58x^2 - 82x - 15$ are $3, 5, -\frac{1}{2}$, $2 + \sqrt{5}$, and $2 - \sqrt{5}$.

TRY EXERCISE 54, EXERCISE SET 3.3, PAGE 268

◆ APPLICATIONS

In the following application we make use of an upper bound to eliminate most of the possible zeros that are given by the Rational Zero Theorem.

EXAMPLE 6 Solve a Geometric Application

The dimensions of a rectangular box are consecutive natural numbers. The volume of the box is 2184 cubic inches. Find the dimensions of the box.

Continued

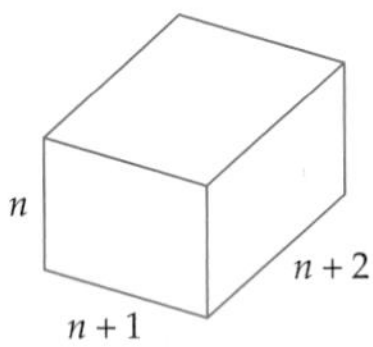

Figure 3.22

Solution

Label the dimensions of the box as n, $n + 1$, and $n + 2$ as shown in **Figure 3.22.** The volume of any box equals the product of its length, width, and height. Thus $n(n + 1)(n + 2) = 2184$, which can be rewritten as

$$n^3 + 3n^2 + 2n - 2184 = 0$$

We need to find the natural-number zeros of $n^3 + 3n^2 + 2n - 2184$. The constant 2184 has many divisors; however, the following synthetic division shows that 14 is an upper bound.

$$\begin{array}{r|rrrr} 14 & 1 & 3 & 2 & -2184 \\ & & 14 & 238 & 3360 \\ \hline & 1 & 17 & 240 & 1176 \end{array}$$

Each of these numbers is positive. Thus 14 is an upper bound.

The divisors of 2184 that are less than 14 are 1, 2, 3, 4, 6, 7, 8, 12, and 13. The following division shows that 12 is a zero of $n^3 + 3n^2 + 2n - 2184$.

$$\begin{array}{r|rrrr} 12 & 1 & 3 & 2 & -2184 \\ & & 12 & 180 & 2184 \\ \hline & 1 & 15 & 182 & 0 \end{array}$$

If $n = 12$, then the dimensions of the box are 12 in. by 13 in. by 14 in. There is no need to seek additional solutions, because any increase (decrease) in n produces a corresponding increase (decrease) in the volume.

TRY EXERCISE 64, EXERCISE SET 3.3, PAGE 268

The procedures developed in this section will not find all the real zeros of every polynomial. However, a graphing utility can be used to estimate the real zeros of a polynomial. In Example 7 we rely on a graphing utility to estimate a zero that is the solution to an application.

EXAMPLE 7 Solve an Engineering Application

A beam extends 54 feet beyond a support as shown in **Figure 3.23.** An engineer has determined that a load of 500 pounds placed x feet from the support causes the beam to deflect d feet, where

$$d = -0.000004x^3 + 0.0014x^2 \quad \text{for } 0 \le x \le 54$$

How far (to the nearest 0.1 foot) from the support should the load be placed to make the deflection equal to 2 feet?

Figure 3.23

Solution

We need to solve

$$-0.000004x^3 + 0.0014x^2 = 2 \qquad \text{(I)}$$

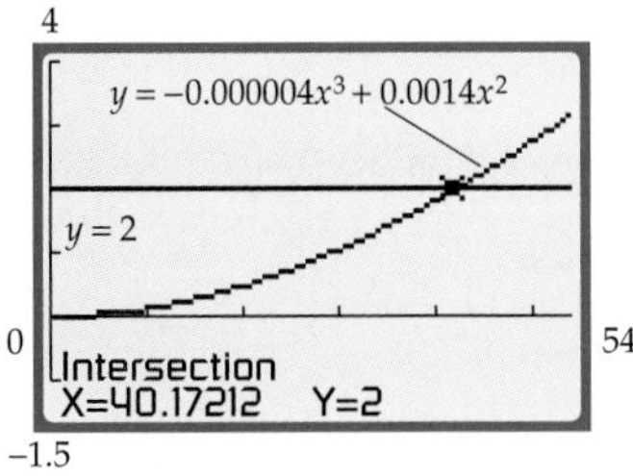

Figure 3.24

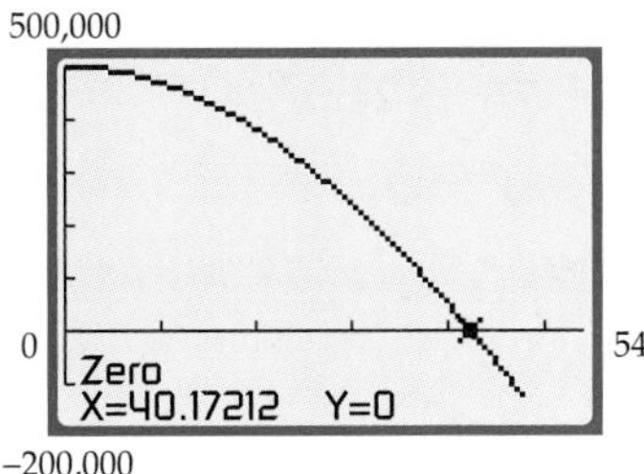

$y = x^3 - 350x^2 + 500{,}000$

Figure 3.25

for x, with the requirement that $0 \le x \le 54$. The following shows two methods of solving for x.

Method 1 Use a graphing utility to graph $y = -0.000004x^3 + 0.0014x^2$ and $y = 2$ on the same screen for $0 \le x \le 54$. The x-coordinate of the point of intersection of the two graphs is the desired solution. Figure 3.24 shows that the graphs intersect at $x = 40.2$ (rounded to the nearest 0.1).

Method 2 Rewrite Equation (I) as a polynomial with integer coefficients.

$$\begin{aligned} -0.000004x^3 + 0.0014x^2 &= 2 \\ -4x^3 + 1400x^2 &= 2{,}000{,}000 \\ 4x^3 - 1400x^2 + 2{,}000{,}000 &= 0 \\ x^3 - 350x^2 + 500{,}000 &= 0 \end{aligned}$$

Graph $y = x^3 - 350x^2 + 500{,}000$. Examine the graph for an x-intercept between $x = 0$ and $x = 54$. The graph of y is shown in Figure 3.25. The x-intercept of the graph is located at $x = 40.2$ (nearest 0.1).

TRY EXERCISE 66, EXERCISE SET 3.3, PAGE 269

You could try to solve $x^3 - 350x^2 + 500{,}000 = 0$ by using the Rational Zero Theorem. However, the constant 500,000 has many integer factors, and the results of Example 7 show that the desired zero is not an integer.

TOPICS FOR DISCUSSION

1. What is a multiple zero of a polynomial? Give an example of a polynomial that has -2 as a multiple zero.
2. Discuss how the Rational Zero Theorem is used.
3. In Topics 4 and 5, we talk about polynomials with integer coefficients and those with real coefficients. Discuss the similarities and differences between these two types of polynomials.
4. Let $P(x)$ be a polynomial with real coefficients. Explain why $(a, 0)$ is an x-intercept of the graph of $P(x)$ if a is a real zero of $P(x)$.
5. Let $P(x)$ be a polynomial with integer coefficients. Suppose that the Rational Zero Theorem is applied to $P(x)$ and that after testing each possible rational zero it is determined that the polynomial has no rational zero. Does this mean that all of the zeros of the polynomial are irrational numbers?

EXERCISE SET 3.3

In Exercises 1 to 10, find the zeros of the polynomial and state the multiplicity of each zero.

1. $P(x) = (x - 3)^2(x + 5)$
2. $P(x) = (x + 4)^3(x - 1)^2$
3. $P(x) = x^2(3x + 5)^2$
4. $P(x) = x^3(2x + 1)(3x - 12)^2$
5. $P(x) = (x^2 - 4)(x + 3)^2$

6. $P(x) = (x + 4)^3(x^2 - 9)^2$
7. $P(x) = (x^2 - 3x - 10)^2$
8. $P(x) = (x^3 - 4x)(2x - 7)^2$
9. $P(x) = x^4 - 10x^2 + 9$
10. $P(x) = x^4 - 12x^2 + 32$

In Exercises 11 to 20, use the Rational Zero Theorem to list possible rational zeros for each polynomial.

11. $x^3 + 3x^2 - 6x - 8$
12. $x^3 - 19x - 30$
13. $2x^3 + x^2 - 25x + 12$
14. $3x^3 + 11x^2 - 6x - 8$
15. $6x^4 + 23x^3 + 19x^2 - 8x - 4$
16. $2x^3 + 9x^2 - 2x - 9$
17. $4x^4 - 12x^3 - 3x^2 + 12x - 7$
18. $x^5 - x^4 - 7x^3 + 7x^2 - 12x - 12$
19. $x^5 - 32$
20. $x^4 - 1$

In Exercises 21 to 30, find the smallest positive integer and the largest negative integer that, by the Upper- and Lower-Bound Theorem, are upper and lower bounds for the real zeros of each polynomial.

21. $x^3 + 3x^2 - 6x - 6$
22. $x^3 - 19x - 28$
23. $2x^3 + x^2 - 25x + 10$
24. $3x^3 + 11x^2 - 6x - 9$
25. $6x^4 + 23x^3 + 19x^2 - 8x - 4$
26. $2x^3 + 9x^2 - 2x - 9$
27. $4x^4 - 12x^3 - 3x^2 + 12x - 7$
28. $x^5 - x^4 - 7x^3 + 7x^2 - 12x - 12$
29. $x^5 - 32$
30. $x^4 - 1$

In Exercises 31 to 40, use Descartes' Rules of Signs to state the number of possible positive and negative real zeros of each polynomial.

31. $x^3 + 3x^2 - 6x - 8$
32. $x^3 - 19x - 30$
33. $2x^3 + x^2 - 25x + 12$
34. $3x^3 + 11x^2 - 6x - 8$
35. $6x^4 + 23x^3 + 19x^2 - 8x - 4$
36. $2x^3 + 9x^2 - 2x - 9$
37. $4x^4 - 12x^3 - 3x^2 + 12x - 7$
38. $x^5 - x^4 - 7x^3 + 7x^2 - 12x - 12$
39. $x^5 - 32$
40. $x^4 - 1$

In Exercises 41 to 62, find the zeros of each polynomial. If a zero is a multiple zero, state its multiplicity.

41. $x^3 + 3x^2 - 6x - 8$
42. $x^3 - 19x - 30$
43. $2x^3 + x^2 - 25x + 12$
44. $3x^3 + 11x^2 - 6x - 8$
45. $6x^4 + 23x^3 + 19x^2 - 8x - 4$
46. $2x^3 + 9x^2 - 2x - 9$
47. $2x^4 - 9x^3 - 2x^2 + 27x - 12$
48. $3x^3 - x^2 - 6x + 2$
49. $x^3 - 8x^2 + 8x + 24$
50. $x^3 - 7x^2 - 7x + 69$
51. $2x^4 - 19x^3 + 51x^2 - 31x + 5$
52. $4x^4 - 35x^3 + 71x^2 - 4x - 6$
53. $3x^6 - 10x^5 - 29x^4 + 34x^3 + 50x^2 - 24x - 24$
54. $3x^5 + 16x^4 + 2x^3 - 58x^2 - 61x - 14$
55. $x^3 - 3x - 2$
56. $3x^4 - 4x^3 - 11x^2 + 16x - 4$
57. $x^4 - 5x^2 - 2x$
58. $x^3 - 2x + 1$
59. $x^4 + x^3 - 3x^2 - 5x - 2$
60. $6x^4 - 17x^3 - 11x^2 + 42x$
61. $2x^4 - 17x^3 + 4x^2 + 35x - 24$
62. $x^5 + 5x^4 + 10x^3 + 10x^2 + 5x + 1$

63. **FIND THE DIMENSIONS** A cube measures n inches on each edge. If a slice 2 inches thick is cut from one face of the cube, the resulting solid has a volume of 567 cubic inches. Find n.

64. **FIND THE DIMENSIONS** A cube measures n units on each edge. If a slice 1 inch thick is cut from one face of the cube, and then a slice 3 inches thick is cut from another face of the cube as shown, the resulting solid has a volume of 1560 cubic inches. Find the dimensions of the original cube.

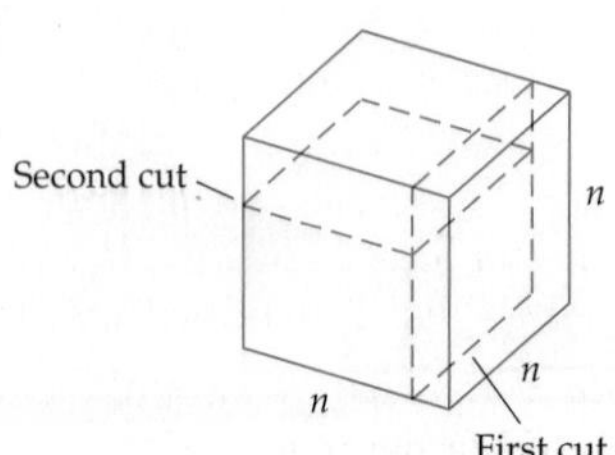

65. **FIND THE DIMENSIONS** A hollow cube has a thickness of 1 cm. The volume of the space inside the cube is one-eighth the volume of the material that was

used to construct the cube. Determine x (nearest hundredth), the outer dimension of the cube.

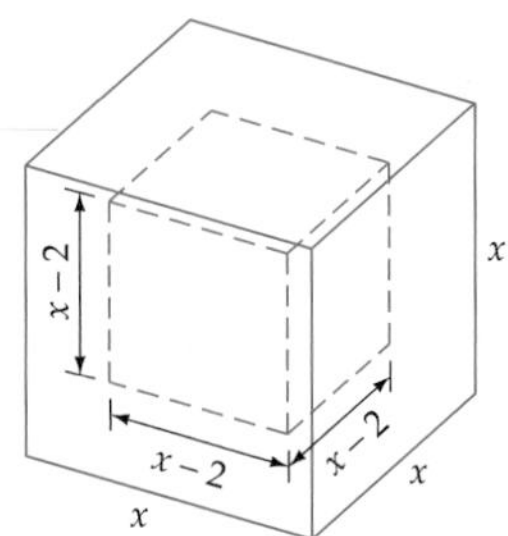

66. **FIND THE RADIUS** A spherical shell has a thickness of 8 mm. The volume of the space inside the sphere is one-tenth the volume of the material that was used to construct the shell. Find the outer radius r of the shell.

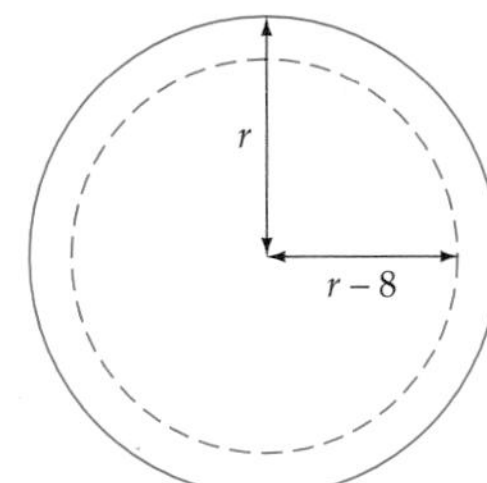

Recall that a real zero of a polynomial is the x-coordinate of the x-intercept of the graph of the polynomial. In Exercises 67 to 72 use this fact and a graphing utility to find, to the nearest tenth, the real zeros of each polynomial. The number in parentheses is the number of real zeros.

67. $P(x) = x^4 + x^3 - 21x^2 - x + 20$; (4)

68. $P(x) = x^4 - x^3 - 16x^2 + 4x + 48$; (4)

69. $P(x) = 4x^4 - 8x^3 - 39x^2 + 43x + 70$; (4)

70. $P(x) = 4x^5 - 28x^4 - 3x^3 + 280x^2 - 283x - 210$; (5)

71. $P(x) = 2x^4 - 3x^3 + 6x^2 - 12x - 8$; (2)

72. $P(x) = 2x^4 + 3x^3 + 2x^2 - x - 6$; (2)

SUPPLEMENTAL EXERCISES

In Exercises 73 to 78, verify that each polynomial has no rational zeros.

73. $x^4 - 2x^3 + 11x^2 - 2x + 10$

74. $x^4 - 2x^3 + 21x^2 - 2x + 20$

75. $2x^4 + x^2 + 5$

76. $4x^4 + 14x^2 + 5$

77. $x^4 - 4x^3 + 14x^2 - 4x + 13$

78. $x^6 + 3x^4 + 3x^2 + 1$

In Exercises 79 to 82, determine whether the given polynomial satisfies the following theorem.

Theorem **Let $P(x) = a_nx^n + a_{n-1}x^{n-1} + \cdots + a_1x + a_0$ be a polynomial with integer coefficients and $n \geq 2$. If a_n, a_0, and $P(1)$ are all odd, then $P(x)$ has no rational zeros.**

79. $P(x) = x^5 + 2x^4 + x^3 - x^2 + x + 945$

80. $P(x) = 5x^3 - 2x^2 - x + 1815$

81. $P(x) = 3x^4 - 5x^3 + 6x^2 - 2x + 9009$

82. $P(x) = 15x^7 - 4x^3 + x^2 - 6075$

PROJECTS

1. **WRITE A PROOF** Use the Rational Zero Theorem to prove that $\sqrt{p}$ is an irrational number, where p is a prime number.

2. **A BRILLIANT BUT TRAGIC LIFE** Many of the concepts that have been presented in this section were further developed by Evariste Galois. The mathematician Felix Klein said of Galois, "In France, about 1830, a new star of unimaginable brightness appeared in the heavens of pure mathematics…."[1] Unfortunately, Galois's talent did not shine for long as he died in a duel at the age of 21. Write an essay about the life and death of Galois. The following list includes two sources you may wish to consult.

   - *Whom the Gods Love* by Leopold Infeld (Reston, Virginia: The National Council of Teachers of Mathematics, 1975)
   - *Men of Mathematics* by Eric Temple Bell (Touchstone Books, 1986)

[1] From *Whom the God Love*, page xi.

SECTION 3.4 THE FUNDAMENTAL THEOREM OF ALGEBRA

- THE FUNDAMENTAL THEOREM OF ALGEBRA
- THE NUMBER OF ZEROS OF A POLYNOMIAL
- THE CONJUGATE PAIR THEOREM
- FACTORS OF A POLYNOMIAL
- FIND A POLYNOMIAL WITH GIVEN ZEROS

◆ THE FUNDAMENTAL THEOREM OF ALGEBRA

The German mathematician Carl Friedrich Gauss (1777–1855) was the first to prove that every polynomial has at least one complex zero. This concept is so basic to the study of algebra that it is called the **Fundamental Theorem of Algebra.** The proof of the Fundamental Theorem is beyond the scope of this text; however, it is important to understand the theorem and its consequences. As you consider each of the following theorems, keep in mind that the terms *complex coefficients* and *complex zeros* include real coefficients and real zeros, because the set of real numbers is a subset of the set of complex numbers.

The Fundamental Theorem of Algebra

If $P(x)$ is a polynomial of degree $n \geq 1$ with complex coefficients, then $P(x)$ has at least one complex zero.

◆ THE NUMBER OF ZEROS OF A POLYNOMIAL

Let $P(x)$ be a polynomial of degree $n \geq 1$ with complex coefficients. The Fundamental Theorem implies that $P(x)$ has a complex zero—say, c_1. The Factor Theorem implies that

$$P(x) = (x - c_1)Q(x)$$

where $Q(x)$ is a polynomial of degree one less than the degree of $P(x)$. Recall that the polynomial $Q(x)$ is called a reduced polynomial. Assuming that the degree of $Q(x)$ is 1 or more, the Fundamental Theorem implies that it must also have a zero. A continuation of this reasoning process leads to the following theorem, which is a corollary of the Fundamental Theorem.

The Number of Zeros of a Polynomial

If $P(x)$ is a polynomial of degree $n \geq 1$ with complex coefficients, then $P(x)$ has exactly n complex zeros, provided that each zero is counted according to its multiplicity.

Even though every polynomial of nth degree has exactly n zeros, the zeros may not be distinct. For example, the third-degree polynomial

$$x^3 - 5x^2 + 3x + 9$$

factors into

$$(x + 1)(x - 3)(x - 3)$$

which has zeros -1, 3, and 3. The zero 3 is a zero of multiplicity 2.

◆ THE CONJUGATE PAIR THEOREM

Although the Fundamental Theorem and its corollary give information about the existence and the number of zeros of a polynomial, they do not provide a method of actually finding the zeros. If a polynomial has real coefficients, then the following theorem can help us determine the zeros of the polynomial.

The Conjugate Pair Theorem

If $a + bi$ $(b \neq 0)$ is a complex zero of the polynomial $P(z)$, *with real coefficients,* then the conjugate $a - bi$ is also a complex zero of the polynomial.

EXAMPLE 1 Use the Conjugate Pair Theorem to Find Zeros

Find all the zeros of $x^4 - 4x^3 + 14x^2 - 36x + 45$ given that $2 + i$ is a zero.

Solution

Because the coefficients are real numbers and $2 + i$ is a zero, the Conjugate Pair Theorem implies that $2 - i$ must also be a zero. Using synthetic division with $2 + i$ and then $2 - i$, we have

$$\begin{array}{r|rrrrr} 2+i & 1 & -4 & 14 & -36 & 45 \\ & & 2+i & -5 & 18+9i & -45 \\ \hline & 1 & -2+i & 9 & -18+9i & 0 \end{array}$$

$$\begin{array}{r|rrrr} 2-i & 1 & -2+i & 9 & -18+9i \\ & & 2-i & 0 & 18-9i \\ \hline & 1 & 0 & 9 & 0 \end{array}$$

• The coefficients of the reduced polynomial

• The coefficients of the next reduced polynomial

The resulting reduced polynomial is $x^2 + 9$, which has $3i$ and $-3i$ as zeros. Therefore, the four zeros of $x^4 - 4x^3 + 14x^2 - 36x + 45$ are $2 + i$, $2 - i$, $3i$, and $-3i$.

TRY EXERCISE 2, EXERCISE SET 3.4, PAGE 276

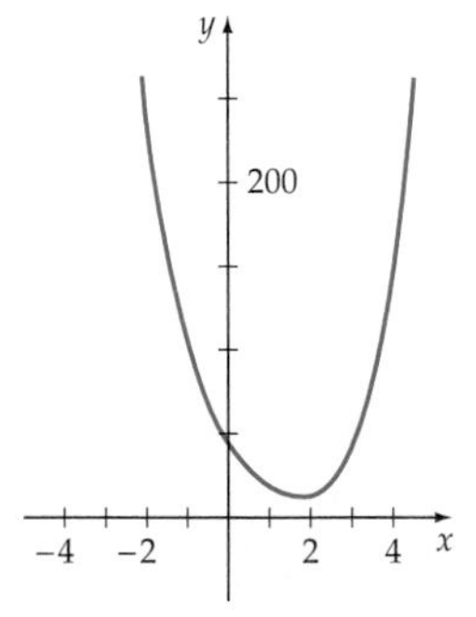

$y = x^4 - 4x^3 + 14x^2 - 36x + 45$

Figure 3.26

A graph of $y = x^4 - 4x^3 + 14x^2 - 36x + 45$ is shown in **Figure 3.26**. Since the polynomial in Example 1 is a fourth-degree polynomial and since we have verified that it has four complex solutions, it comes as no surprise that the graph does not intersect the x-axis.

When performing synthetic division with complex numbers, it is helpful to write the coefficients of the given polynomial as complex coefficients. For instance, -10 can be written as $-10 + 0i$. This technique is illustrated in the next example.

take note

Many graphing calculators can be used to do computations with complex numbers. The following TI-83 screen display shows that the product of $3 - 5i$ and $-7 - 5i$ is $-46 + 20i$.

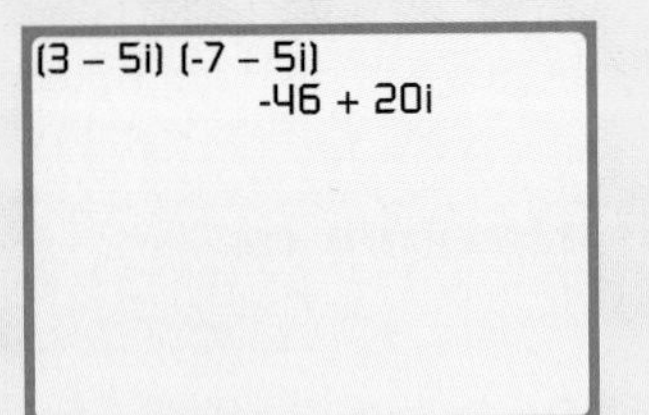

EXAMPLE 2 Use the Conjugate Pair Theorem to Find Zeros

Find all the zeros of $x^5 - 10x^4 + 65x^3 - 184x^2 + 274x - 204$ given that $3 - 5i$ is a zero.

Solution

Because the coefficients are real numbers and $3 - 5i$ is a zero, $3 + 5i$ must also be a zero. Use synthetic division to produce

$$\begin{array}{r|rrrrrr} 3-5i & 1 & -10+0i & 65+0i & -184+0i & 274+0i & -204 \\ & & 3-5i & -46+20i & 157-35i & -256+30i & 204 \\ \hline 3+5i & 1 & -7-5i & 19+20i & -27-35i & 18+30i & 0 \\ & & 3+5i & -12-20i & 21+35i & -18-30i & \\ \hline & 1 & -4 & 7 & -6 & 0 & \end{array}$$

The reduced polynomial $x^3 - 4x^2 + 7x - 6$ has three or one positive zeros and no negative zeros.

$$\frac{p}{q} = 1, 2, 3, 6$$

$$\begin{array}{r|rrrr} 2 & 1 & -4 & 7 & -6 \\ & & 2 & -4 & 6 \\ \hline & 1 & -2 & 3 & 0 \end{array}$$

Use the quadratic formula to solve $x^2 - 2x + 3 = 0$.

$$x = \frac{-(-2) \pm \sqrt{(-2)^2 - 4(1)(3)}}{2(1)} = \frac{2 \pm \sqrt{-8}}{2} = \frac{2 \pm 2\sqrt{2}i}{2} = 1 \pm \sqrt{2}i$$

The zeros of $x^5 - 10x^4 + 65x^3 - 184x^2 + 274x - 204$ are $3 - 5i$, $3 + 5i$, 2, $1 + \sqrt{2}i$, and $1 - \sqrt{2}i$.

TRY EXERCISE 14, EXERCISE SET 3.4, PAGE 276

Recall that the real zeros of a polynomial P are the x-coordinates of the x-intercepts of the graph of P. This important connection between real zeros of a polynomial and x-intercepts of the graph of the polynomial is the basis for using a graphing utility to solve equations. Careful analysis of the graph of a polynomial and your knowledge of the properties of polynomials can be used to solve many polynomial equations.

EXAMPLE 3 Solve a Polynomial Equation

Solve: $x^4 - 5x^3 + 4x^2 + 3x + 9 = 0$

Solution

Let $P(x) = x^4 - 5x^3 + 4x^2 + 3x + 9$. The x-intercepts of the graph of P are the real solutions of the equation. Use a graphing utility to graph P. See Figure 3.27.

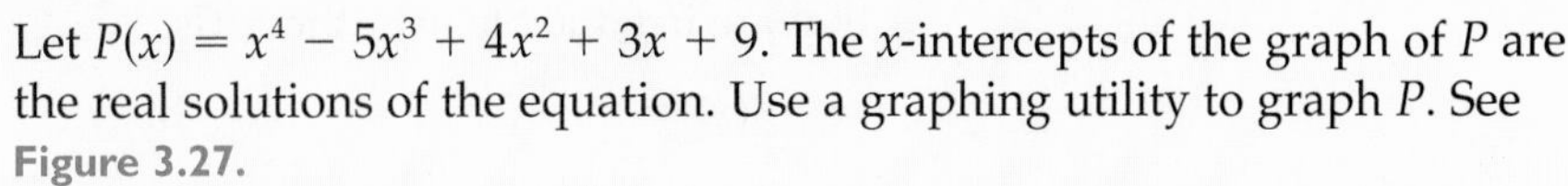

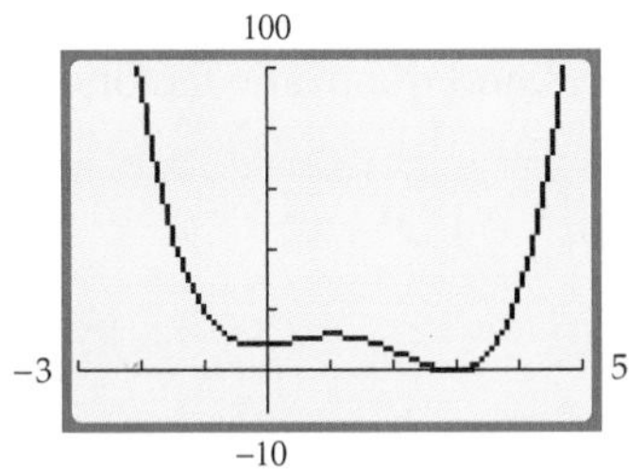

$P(x) = x^4 - 5x^3 + 4x^2 + 3x + 9$

Figure 3.27

From the graph, it appears that (3, 0) is an x-intercept and the only x-intercept. Because the graph of P intersects but does not cross the x-axis at (3, 0), we know that 3 is a multiple zero of P with an even multiplicity.

$$\begin{array}{r|rrrrr} 3 & 1 & -5 & 4 & 3 & 9 \\ & & 3 & -6 & -6 & -9 \\ \hline & 1 & -2 & -2 & -3 & 0 \end{array}$$

• Coefficients of P

• Remainder is zero. Thus 3 is a zero.

By the Number of Zeros Theorem, there are three more zeros of P. Use synthetic division to show that 3 is also a zero of the reduced polynomial $x^3 - 2x^2 - 2x - 3$.

$$\begin{array}{r|rrrr} 3 & 1 & -2 & -2 & -3 \\ & & 3 & 3 & 3 \\ \hline & 1 & 1 & 1 & 0 \end{array}$$

• Coefficients of reduced polynomial

• Remainder is zero. Thus 3 is a zero of multiplicity 2.

We now have 3 as a double root of the original equation, and from the last line of the preceding synthetic division, the remaining solutions must be solutions of $x^2 + x + 1 = 0$. Use the quadratic formula to solve this equation.

$$x = \frac{-1 \pm \sqrt{1^2 - 4(1)(1)}}{2(1)} = \frac{-1 \pm \sqrt{-3}}{2} = \frac{-1 \pm i\sqrt{3}}{2}$$

The solutions of $x^4 - 5x^3 + 4x^2 + 3x + 9 = 0$ are 3, 3, $\frac{-1 + i\sqrt{3}}{2}$, and $\frac{-1 - i\sqrt{3}}{2}$.

TRY EXERCISE 28, EXERCISE SET 3.4, PAGE 276

◆ FACTORS OF A POLYNOMIAL

The following theorem is a result of the Conjugate Pair Theorem.

Linear and Quadratic Factors of a Polynomial

Every polynomial with real coefficients and positive degree n can be written as the product of linear and quadratic factors with real coefficients, where the quadratic factors have no real zeros.

A quadratic factor with no real zeros is said to be **irreducible over the reals.**

EXAMPLE 4 Factor a Polynomial into Linear and Quadratic Factors

Write each polynomial as a product of linear factors and quadratic factors that are irreducible over the reals.

a. $P(x) = x^3 - 3x^2 + x - 3$ **b.** $P(x) = x^3 - 6x^2 + 13x - 10$

Solution

a. Factoring by grouping produces

$$\begin{aligned} P(x) &= x^3 - 3x^2 + x - 3 = (x^3 - 3x^2) + (x - 3) \\ &= x^2(x - 3) + 1(x - 3) = (x - 3)(x^2 + 1) \end{aligned}$$

Because each binomial factor is irreducible over the reals, the factorization is complete.

b. Because $x^3 - 6x^2 + 13x - 10$ cannot be factored by grouping, synthetic division is used to determine zeros that, by the Factor Theorem, also determine the factors of P. By the Rational Zero Theorem, we know that ± 1, ± 2, ± 5, and ± 10 are possible rational zeros. Testing each of these, we find

$$\begin{array}{r|rrrr} 2 & 1 & -6 & 13 & -10 \\ & & 2 & -8 & 10 \\ \hline & 1 & -4 & 5 & 0 \end{array}$$

• **2 is a zero**

Using the quadratic formula, we find that the reduced polynomial $x^2 - 4x + 5$ has zeros of $2 \pm i$, so it cannot be factored using real numbers. Thus $x^3 - 6x^2 + 13x - 10$ factors into

$$(x - 2)(x^2 - 4x + 5)$$

which is a product of a linear factor and a quadratic factor that is irreducible over the reals.

TRY EXERCISE 36, EXERCISE SET 3.4, PAGE 276

◆ FIND A POLYNOMIAL WITH GIVEN ZEROS

Many of the problems in this section and in Section 3.3 dealt with the process of finding the zeros of a given polynomial. Example 5 considers the reverse process, finding a polynomial when the zeros are given.

EXAMPLE 5 Determine a Polynomial Given Its Zeros

Find each polynomial.

a. A polynomial of degree 3 that has 1, 2, and -3 as zeros

b. A polynomial of degree 4 that has real coefficients and zeros $2i$ and $3 - 7i$

Solution

a. Because 1, 2, and -3 are zeros, $(x - 1)$, $(x - 2)$, and $(x + 3)$ are factors. The product of these factors produces a polynomial that has the indicated zeros.

$$(x - 1)(x - 2)(x + 3) = (x^2 - 3x + 2)(x + 3) = x^3 - 7x + 6$$

b. By the Conjugate Pair Theorem, the polynomial also must have $-2i$ and $3 + 7i$ as zeros. The product of the factors $x - 2i$, $x - (-2i)$, $x - (3 - 7i)$, and $x - (3 + 7i)$ produces the desired polynomial.

$$\begin{aligned}(x - 2i)(x + 2i)[x - (3 - 7i)][x - (3 + 7i)]\\ &= (x^2 + 4)(x^2 - 6x + 58)\\ &= x^4 - 6x^3 + 62x^2 - 24x + 232\end{aligned}$$

TRY EXERCISE 56, EXERCISE SET 3.4, PAGE 276

A polynomial that has a given set of zeros is not unique. For example, $x^3 - 7x + 6$ has zeros 1, 2, and -3, but so does any nonzero multiple of that polynomial, such as $2x^3 - 14x + 12$. This concept is illustrated in Figure 3.28. The graphs of the two polynomials are different, but they have the same x-intercepts.

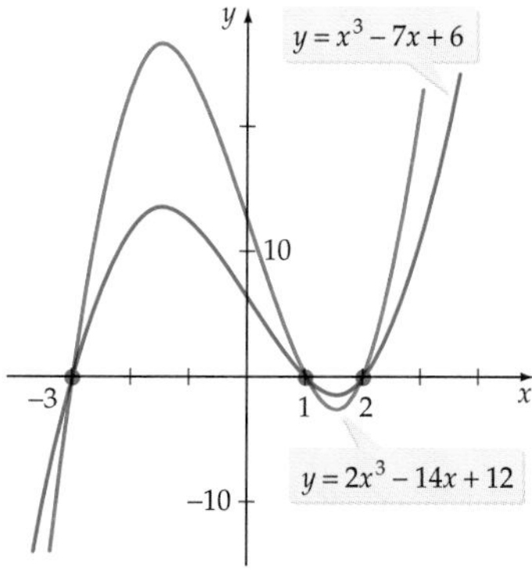

Figure 3.28

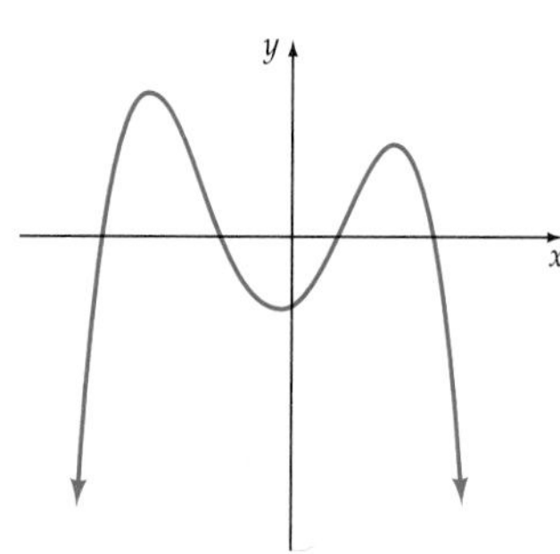

Figure 3.29

TOPICS FOR DISCUSSION

1. What is the Fundamental Theorem of Algebra and why is this theorem so important?
2. Let $P(x)$ be a polynomial of degree n with real coefficients. Discuss the number of *possible* real zeros of this polynomial. Include in your discussion the cases when n is even and when n is odd.
3. Consider the graph of a polynomial in Figure 3.29. Is it possible that the degree of the polynomial is 3? Explain.

4. If two polynomials have exactly the same zeros, do the graphs of the polynomials look exactly the same?

5. Does the graph of every polynomial have at least one x-intercept?

EXERCISE SET 3.4

In Exercises 1 to 16, use the given zero to find the remaining zeros of each polynomial.

1. $2x^3 - 5x^2 + 6x - 2;\quad 1 + i$
2. $3x^3 - 29x^2 + 92x + 34;\quad 5 + 3i$
3. $x^3 + 3x^2 + x + 3;\quad -i$
4. $x^4 - 6x^3 + 71x^2 - 146x + 530;\quad 2 + 7i$
5. $x^5 - x^4 - 3x^3 + 3x^2 - 10x + 10;\quad i\sqrt{2}$
6. $x^4 - 4x^3 + 14x^2 - 4x + 13;\quad 2 - 3i$
7. $12x^3 - 28x^2 + 23x - 5;\quad \frac{1}{3}$
8. $8x^4 - 2x^3 + 199x^2 - 50x - 25;\quad -5i$
9. $x^4 - 4x^3 + 19x^2 - 30x + 50;\quad 1 + 3i$
10. $12x^4 - 52x^3 + 19x^2 - 13x + 4;\quad \frac{1}{2}i$
11. $x^5 - x^4 - 4x^3 - 4x^2 - 5x - 3;\quad i$
12. $x^5 - 3x^4 + 7x^3 - 13x^2 + 12x - 4;\quad -2i$
13. $x^4 - 8x^3 + 18x^2 - 8x + 17;\quad i$
14. $x^5 - 6x^4 + 22x^3 - 64x^2 + 117x - 90;\quad 3i$
15. $x^4 - 17x^3 + 112x^2 - 333x + 377;\quad 5 + 2i$
16. $2x^5 - 8x^4 + 61x^3 - 99x^2 + 12x + 182;\quad 1 - 5i$

In Exercises 17 to 26, find all the zeros of the polynomial. (*Hint:* First determine the rational zeros.)

17. $x^4 + x^3 - 2x^2 + 4x - 24$
18. $x^4 - 3x^3 + 5x^2 - 27x - 36$
19. $2x^4 + x^3 + 39x^2 + 136x - 78$
20. $x^3 - 13x^2 + 65x - 125$
21. $x^5 - 9x^4 + 34x^3 - 58x^2 + 45x - 13$
22. $x^4 - 4x^3 + 53x^2 - 196x + 196$
23. $2x^4 - x^3 - 15x^2 + 23x + 15$
24. $3x^4 - 17x^3 - 39x^2 + 337x + 116$
25. $2x^4 - 14x^3 + 33x^2 - 46x + 40$
26. $3x^4 - 10x^3 + 15x^2 + 20x - 8$

In Exercises 27 to 34, use a graph and your knowledge of the zeros of polynomial functions to determine the *exact* value of all the solutions of each equation.

27. $2x^3 - x^2 + x - 6 = 0$
28. $4x^3 + 3x^2 + 16x + 12 = 0$
29. $24x^3 - 62x^2 - 7x + 30 = 0$
30. $12x^3 - 52x^2 + 27x + 28 = 0$
31. $x^4 - 4x^3 + 5x^2 - 4x + 4 = 0$
32. $x^4 + 4x^3 + 8x^2 + 16x + 16 = 0$
33. $x^4 + 4x^3 - 2x^2 - 12x + 9 = 0$
34. $x^4 + 3x^3 - 6x^2 - 28x - 24 = 0$

In Exercises 35 to 44, factor each polynomial into linear factors and/or quadratic factors that are irreducible over the reals.

35. $x^3 - x^2 - 2x$
36. $6x^3 - 23x^2 - 4x$
37. $x^3 + 9x$
38. $x^3 + 10x$
39. $x^4 + 2x^2 - 24$
40. $x^4 - 8x^2 - 20$
41. $x^4 + 3x^2 + 2$
42. $x^5 + 11x^3 + 18x$
43. $x^4 - 2x^3 + x^2 - 8x - 12$
44. $x^4 + 2x^3 + 6x^2 + 32x + 40$

In Exercises 45 to 54, find a polynomial of lowest degree that has the given zeros.

45. $4, -3, 2$
46. $-1, 1, -5$
47. $3, 2i, -2i$
48. $0, i, -i$
49. $3 + i, 3 - i, 2 + 5i, 2 - 5i$
50. $2 + 3i, 2 - 3i, -5, 2$
51. $6 + 5i, 6 - 5i, 2, 3, 5$
52. $\frac{1}{2}, 4 - i, 4 + i$
53. $\frac{3}{4}, 2 + 7i, 2 - 7i$
54. $\frac{1}{4}, -\frac{1}{5}, i, -i$

In Exercises 55 to 62, find a polynomial $P(x)$ with real coefficients that has the indicated zeros and satisfies the given conditions.

55. Zeros: $2 - 5i, -4$, degree 3
56. Zeros: $3 + 2i, 7$, degree 3
57. Zeros: $4 + 3i, 5 - i$, degree 4
58. Zeros: $i, 3 - 5i$, degree 4
59. Zeros: $-1, 2, 3$, degree 3, $P(1) = 12$

60. Zeros: $3i, 2$, degree 3, $P(3) = 27$

61. Zeros: $3, -5, 2 + i$, degree 4, $P(1) = 48$

62. Zeros: $\frac{1}{2}, 1 - i$, degree 3, $P(4) = 140$

SUPPLEMENTAL EXERCISES

63. Verify that $x^3 - x^2 - ix^2 - 9x + 9 + 9i$ has $1 + i$ as a zero and that its conjugate $1 - i$ is not a zero. Explain why this does not contradict the Conjugate Pair Theorem.

64. Verify that $x^3 - x^2 - ix^2 - 20x + ix + 20i$ has a zero of i and that its conjugate $-i$ is not a zero. Explain why this does not contradict the Conjugate Pair Theorem.

65. Show that 2 is a zero of multiplicity 3 of

$$P(x) = x^5 - 6x^4 + 21x^3 - 62x^2 + 108x - 72$$

and express $P(x)$ as a product of linear factors and/or quadratic factors that are irreducible over the reals.

66. Show that -1 is a zero of multiplicity 4 of

$$P(x) = x^6 + 5x^5 + 11x^4 + 14x^3 + 11x^2 + 5x + 1$$

and express $P(x)$ as a product of linear factors and/or quadratic factors that are irreducible over the reals.

67. Find a polynomial $P(x)$ of degree 5 such that 1 is a zero of multiplicity 2, 2 is a zero of multiplicity 3, and $P(-1) = -54$.

68. Find a polynomial $P(x)$ of degree 5 such that -4 is a zero of multiplicity 4, $1/2$ is a zero of multiplicity 1, and $P(1) = 125$.

PROJECTS

1. INVESTIGATE THE ROOTS OF A CUBIC EQUATION Hieronimo Cardano, using a technique he stole from Nicolo Tartaglia, was able so solve some cubic equations.

a. Show that the cubic equation $x^3 + bx^2 + cx + d = 0$ can be transformed into the "reduced" cubic $y^3 + my = n$, where m and n are constants, depending on b, c, and d, by using the substitution $x = y - b/3$.

b. Cardano then showed that a solution of the reduced cubic is given by

$$\sqrt[3]{\frac{n}{2} + \sqrt{\frac{n^2}{4} + \frac{m^3}{27}}} - \sqrt[3]{-\frac{n}{2} + \sqrt{\frac{n^2}{4} + \frac{m^3}{27}}}$$

Use Cardano's procedure to solve the equation $x^3 - 6x^2 + 20x - 33 = 0$.

SECTION 3.5 RATIONAL FUNCTIONS AND THEIR GRAPHS

- ASYMPTOTES
- A SIGN PROPERTY OF RATIONAL FUNCTIONS
- GENERAL GRAPHING PROCEDURE
- SLANT ASYMPTOTES
- GRAPH RATIONAL FUNCTIONS THAT HAVE A COMMON FACTOR
- APPLICATION

3B

If $P(x)$ and $Q(x)$ are polynomials, then the function F given by

$$F(x) = \frac{P(x)}{Q(x)}$$

is called a **rational function.** The domain of F is the set of all real numbers except those for which $Q(x) = 0$. For example, the domain of

$$F(x) = \frac{x^2 - x - 5}{x(2x - 5)(x + 3)}$$

is the set of all real numbers except 0, $5/2$, and -3.

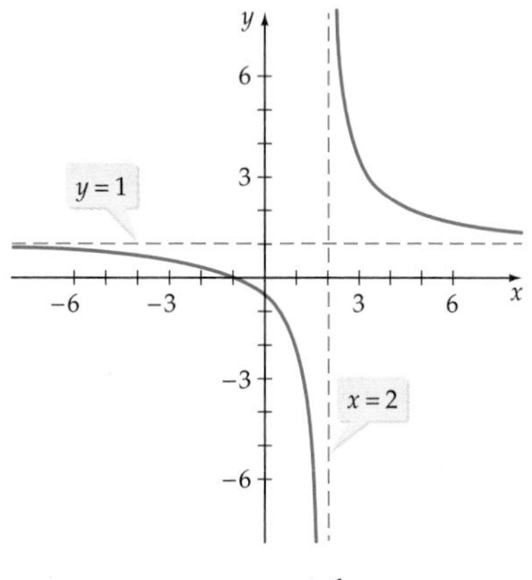

$G(x) = \dfrac{x + 1}{x - 2}$

Figure 3.30

The graph of $G(x) = \dfrac{x + 1}{x - 2}$ is given in **Figure 3.30**. The graph shows that G has the following properties:

- The graph has an x-intercept at $(-1, 0)$ and a y-intercept at $(0, -1/2)$.
- The graph does not exist when $x = 2$.

Note the behavior of the graph as x takes on values that are close to 2 but *less* than 2. Mathematically, we say that "x approaches 2 from the left."

x	1.9	1.95	1.99	1.995	1.999
$G(x)$	−29	−59	−299	−599	−2999

From this table and the graph, it appears that as x approaches 2 from the left, the functional values $G(x)$ decrease without bound.

- We say that "$G(x)$ approaches negative infinity."

Now observe the behavior of the graph as x takes on values that are close to 2 but *greater* than 2. Mathematically, we say that "x approaches 2 from the right."

x	2.1	2.05	2.01	2.005	2.001
$G(x)$	31	61	301	601	3001

From this table and the graph, it appears that as x approaches 2 from the right, the functional values $G(x)$ increase without bound.

- In this case, we say that "$G(x)$ approaches positive infinity."

Now consider the values of $G(x)$ as x *increases* without bound. The table below indicates this for selected values of x.

x	1000	5000	10,000	50,000	100,000
$G(x)$	1.00301	1.00060	1.00030	1.00006	1.00003

- As x increases without bound, the values of $G(x)$ are becoming closer to 1.

Now let the values of x *decrease* without bound. The table below gives the value of $G(x)$ for selected values of x.

x	−1000	−5000	−10,000	−50,000	−100,000
$G(x)$	0.997006	0.999400	0.997001	0.999940	0.999970

- As x decreases without bound, the values of $G(x)$ are becoming closer to 1.

When we are discussing graphs that increase or decrease without bound, it is convenient to use mathematical notation. The notation

$$f(x) \to \infty \text{ as } x \to a^+$$

means that the functional values $f(x)$ increase without bound as x approaches a from the right. Recall that the symbol ∞ does not represent a real number but is used merely to describe the concept of a variable taking on larger and larger values without bound. See **Figure 3.31a**.

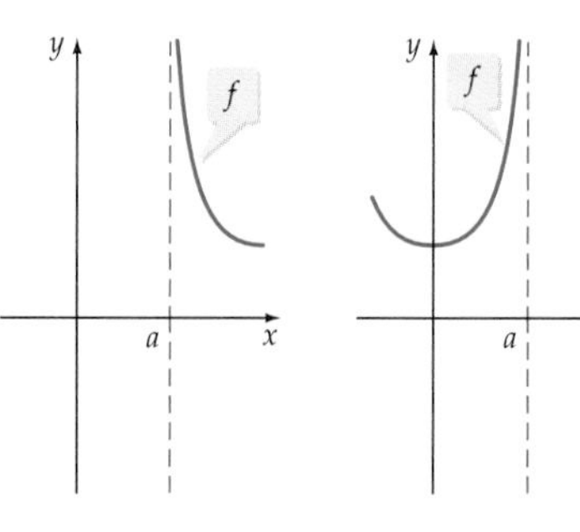

a. $f(x) \to \infty$ as $x \to a^+$

b. $f(x) \to \infty$ as $x \to a^-$

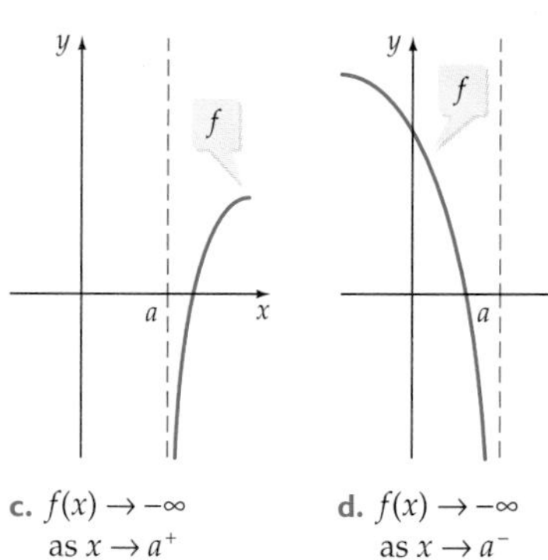

c. $f(x) \to -\infty$ as $x \to a^+$

d. $f(x) \to -\infty$ as $x \to a^-$

Figure 3.31

The notation

$$f(x) \to \infty \text{ as } x \to a^-$$

means that the functional values $f(x)$ increase without bound as x approaches a from the left. See **Figure 3.31b.**

The notation

$$f(x) \to -\infty \text{ as } x \to a^+$$

means that the functional values $f(x)$ decrease without bound as x approaches a from the right. See **Figure 3.31c.**

The notation

$$f(x) \to -\infty \text{ as } x \to a^-$$

means that the functional values $f(x)$ decrease without bound as x approaches a from the left. See **Figure 3.31d.**

◆ ASYMPTOTES

Each graph in **Figure 3.31** approaches a vertical line through $(a, 0)$ as $x \to a^+$ or a^-. The line is said to be a *vertical asymptote* to the graph.

Definition of a Vertical Asymptote

The line $x = a$ is a **vertical asymptote** of the graph of a function F provided that

$$F(x) \to \infty \qquad \text{or} \qquad F(x) \to -\infty$$

as x approaches a from either left of right.

In **Figure 3.30**, the line $x = 2$ is a vertical asymptote of the graph of G. Note that the graph of G in **Figure 3.30** also approaches the horizontal line $y = 1$ as $x \to \infty$ and as $x \to -\infty$. The line $y = 1$ is a *horizontal asymptote* of the graph of G.

Definition of a Horizontal Asymptote

The line $y = b$ is a **horizontal asymptote** of the graph of a function F provided that

$$F(x) \to b \quad \text{as} \quad x \to \infty \quad \text{or} \quad x \to -\infty$$

Figure 3.32 illustrates some of the ways in which the graph of a rational function may approach its horizontal asymptote. It is common practice to display the asymptotes of the graph of a rational function by using dashed lines. Although a rational function may have several vertical asymptotes, it can have at most one horizontal asymptote. The graph may intersect its horizontal asymptote.

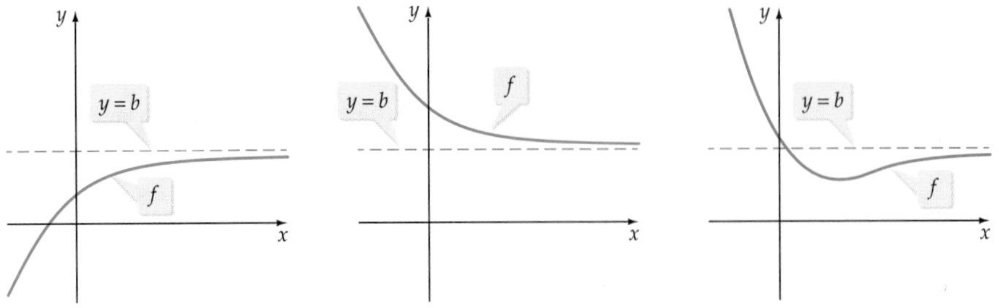

$f(x) \to b$ as $x \to \infty$

Figure 3.32

QUESTION Can a graph of a rational function cross its vertical asymptote? Why or why not?

Geometrically, a line is an asymptote to a curve if the distance between the line and a point $P(x, y)$ on the curve approaches zero as the distance between the origin and the point P increases without bound.

Vertical asymptotes of the graph of a rational function can be found by using the following theorem.

Theorem on Vertical Asymptotes

If the real number a is a zero of the denominator $Q(x)$, then the graph of $F(x) = P(x)/Q(x)$, where $P(x)$ and $Q(x)$ have no common factors, has the vertical asymptote $x = a$.

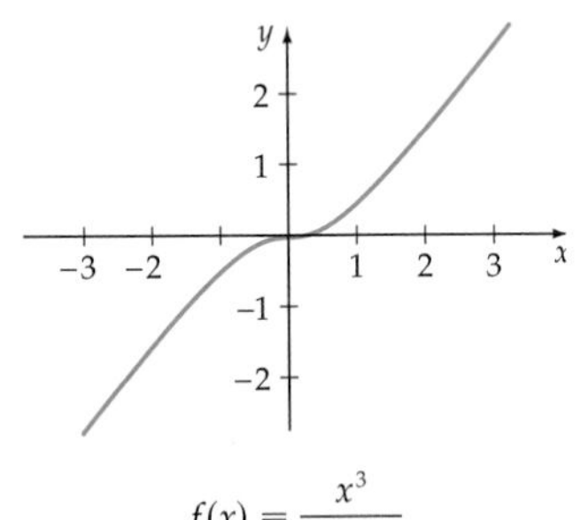

$f(x) = \dfrac{x^3}{x^2 + 1}$

Figure 3.33

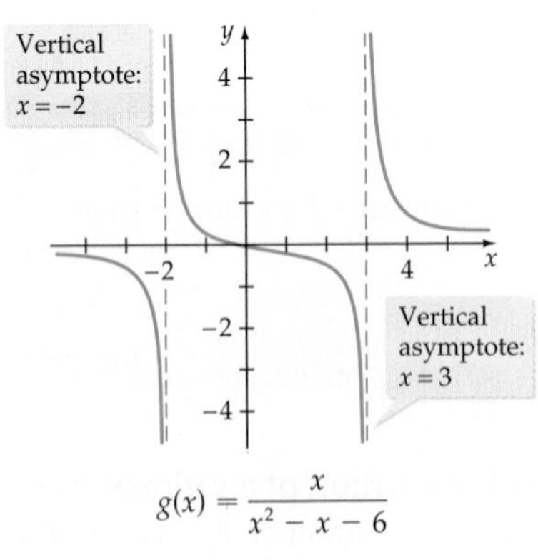

$g(x) = \dfrac{x}{x^2 - x - 6}$

Figure 3.34

EXAMPLE 1 Find the Vertical Asymptotes of a Rational Function

Find the vertical asymptotes of each rational function.

a. $f(x) = \dfrac{x^3}{x^2 + 1}$ b. $g(x) = \dfrac{x}{x^2 - x - 6}$

Solution

a. To find the vertical asymptotes, set the denominator equal to zero. The denominator $x^2 + 1$ has no real zeros, so the graph of f has no vertical asymptotes. See **Figure 3.33**.

b. The denominator $x^2 - x - 6 = (x - 3)(x + 2)$ has zeros of 3 and -2. The numerator has no common factors with the denominator, so $x = 3$ and $x = -2$ are both vertical asymptotes of the graph of g, as shown in **Figure 3.34**.

TRY EXERCISE 2, EXERCISE SET 3.5, PAGE 291

ANSWER No. If $x = a$ is a vertical asymptote of a rational function R, then $R(a)$ is undefined.

The following theorem states that a horizontal asymptote can be determined by examining the leading terms of the numerator and the denominator of a rational function.

Theorem on Horizontal Asymptotes

Let
$$F(x) = \frac{a_n x^n + a_{n-1}x^{n-1} + \cdots + a_1 x + a_0}{b_m x^m + b_{m-1}x^{m-1} + \cdots + b_1 x + b_0}$$

be a rational function with numerator of degree n and denominator of degree m.

1. If $n < m$, then the x-axis is the horizontal asymptote of the graph of F.
2. If $n = m$, then the line $y = a_n/b_m$ is the horizontal asymptote of the graph of F.
3. If $n > m$, the graph of F has no horizontal asymptote.

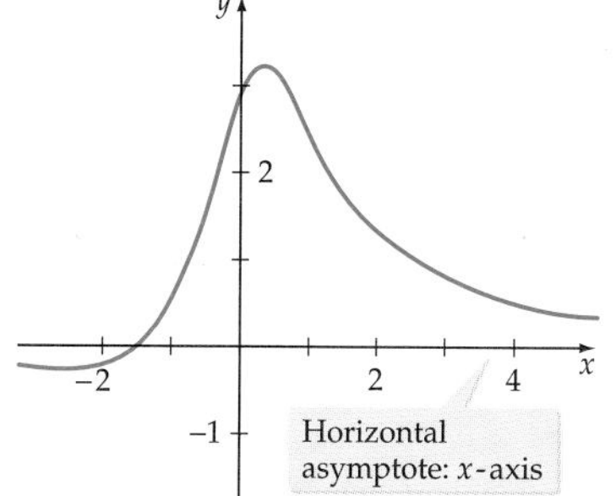

$f(x) = \dfrac{2x + 3}{x^2 + 1}$

Figure 3.35

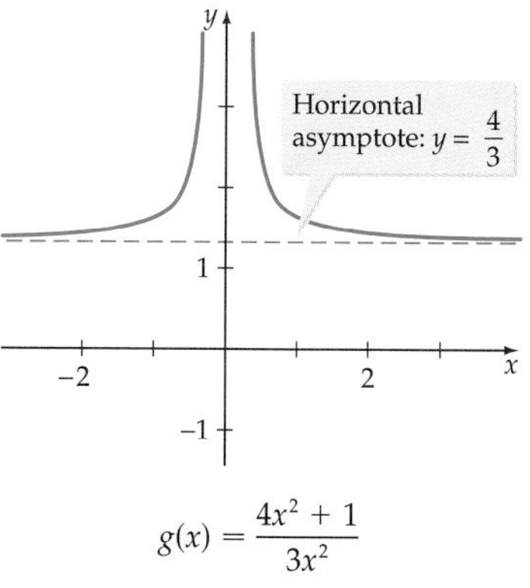

$g(x) = \dfrac{4x^2 + 1}{3x^2}$

Figure 3.36

EXAMPLE 2 Find the Horizontal Asymptote of a Rational Function

Find the horizontal asymptote of each rational function.

a. $f(x) = \dfrac{2x + 3}{x^2 + 1}$ b. $g(x) = \dfrac{4x^2 + 1}{3x^2}$ c. $h(x) = \dfrac{x^3 + 1}{x - 2}$

Solution

a. The degree of the numerator $2x + 3$ is less than the degree of the denominator $x^2 + 1$. By the Theorem on Horizontal Asymptotes, the x-axis is the horizontal asymptote of f. See the graph of f in **Figure 3.35.**

b. The numerator $4x^2 + 1$ and the denominator $3x^2$ of g are both of degree 2. By the Theorem on Horizontal Asymptotes, the line $y = 4/3$ is the horizontal asymptote of g. See the graph of g in **Figure 3.36.**

c. The degree of the numerator $x^3 + 1$ is larger than the degree of the denominator $x - 2$, so by the Theorem on Horizontal Asymptotes, the graph of h has no horizontal asymptotes.

TRY EXERCISE 6, EXERCISE SET 3.5, PAGE 291

The proof of the Theorem on Horizontal Asymptotes makes use of the technique employed in the following verification. To verify that

$$y = \frac{5x^2 + 4}{3x^2 + 8x + 7}$$

has a horizontal asymptote of $y = 5/3$, divide the numerator and the denominator by the largest power of the variable x (x^2 in this case).

$$y = \frac{\dfrac{5x^2 + 4}{x^2}}{\dfrac{3x^2 + 8x + 7}{x^2}} = \frac{5 + \dfrac{4}{x^2}}{3 + \dfrac{8}{x} + \dfrac{7}{x^2}}, \quad x \neq 0$$

As x increases without bound or decreases without bound, the fractions $4/x^2$, $8/x$, and $7/x^2$ approach zero. Thus

$$y \to \frac{5 + 0}{3 + 0 + 0} = \frac{5}{3} \quad \text{as} \quad x \to \pm\infty$$

and hence the line $y = 5/3$ is a horizontal asymptote of the graph.

◆ A Sign Property of Rational Functions

The zeros and vertical asymptotes of a rational function F divide the x-axis into intervals. In each interval,

- $F(x)$ is positive for all x in the interval, or
- $F(x)$ is negative for all x in the interval.

For example, consider the rational function

$$g(x) = \frac{x + 1}{x^2 + 2x - 3}$$

which has vertical asymptotes of $x = -3$ and $x = 1$ and a zero of -1. These three numbers divide the x-axis into the four intervals $(-\infty, -3)$, $(-3, -1)$, $(-1, 1)$, and $(1, \infty)$. Note in **Figure 3.37** that the graph of g is

- negative for all x such that $x < -3$.
- positive for all x such that $-3 < x < -1$.
- negative for all x such that $-1 < x < 1$.
- positive for all x such that $x > 1$.

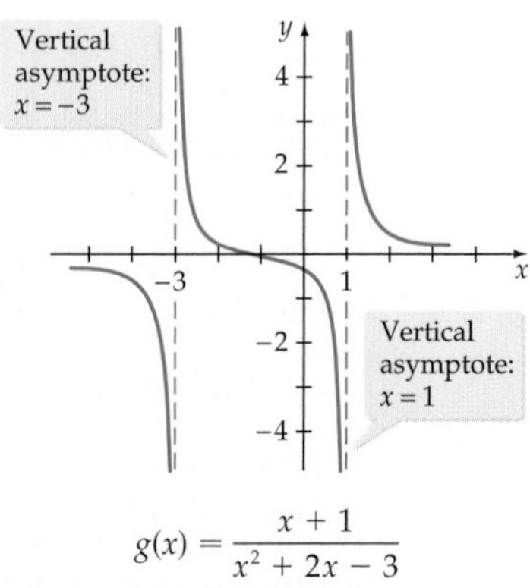

$$g(x) = \frac{x + 1}{x^2 + 2x - 3}$$

Figure 3.37

◆ GENERAL GRAPHING PROCEDURE

If $F(x) = P(x)/Q(x)$, where $P(x)$ and $Q(x)$ are polynomials that have no common factor, then the following general procedure offers useful guidelines for graphing F.

General Procedure for Graphing Rational Functions That Have No Common Factors

1. ***Asymptotes*** Find the real zeros of the denominator $Q(x)$. For each zero a, draw the dashed line $x = a$. Each line is a vertical asymptote of the graph of F. Graph any horizontal asymptotes. These can be found by using the Theorem on Horizontal Asymptotes. If the degree of the numerator $P(x)$ is larger than the degree of the denominator $Q(x)$, then the graph of F does not have a horizontal asymptote.
2. ***Intercepts*** Find the real zeros of the numerator $P(x)$. For each zero a, plot the point $(a, 0)$. Each such point is an x-intercept of the graph of F. Evaluate $F(0)$. Plot $(0, F(0))$, the y-intercept of the graph of F.
3. ***Symmetry*** Use the tests for symmetry to determine whether the graph of the function has symmetry with respect to the y-axis or symmetry with respect to the origin.
4. ***Addition points*** Plot at least two points that lie in the intervals between and beyond the vertical asymptotes and the x-intercepts.
5. ***Behavior near asymptotes*** If $x = a$ is a vertical asymptote, determine whether $F(x) \to \infty$ or $F(x) \to -\infty$ as $x \to a^-$ and also as $x \to a^+$.
6. ***Complete the sketch*** Use all the information obtained above to sketch the graph of F. Plot additional points if necessary to gain additional knowledge about the function.

EXAMPLE 3 Graph a Rational Function

Sketch a graph of $f(x) = \dfrac{2x^2 - 18}{x^2 + 3}$.

Solution

Asymptotes The denominator $x^2 + 3$ has no real zeros, so the graph of f has no vertical asymptotes. The numerator and denominator both have degree 2. The leading coefficients are 2 and 1, respectively. By the Theorem on Horizontal Asymptotes, the graph of f has a horizontal asymptote $y = 2/1 = 2$.

Intercepts The zeros of the numerator occur when $2x^2 - 18 = 0$ or, solving for x, when $x = -3$ and $x = 3$. Therefore, the x-intercepts are $(-3, 0)$

Continued ➤

and (3, 0). To find the y-intercept, evaluate f when $x = 0$. This gives $y = -6$. Therefore, the y-intercept is $(0, -6)$.

Symmetry Below we show that $f(-x) = f(x)$, which means that f is an even function and therefore its graph is symmetric with respect to the y-axis.

$$f(-x) = \frac{2(-x)^2 - 18}{(-x)^2 + 3} = \frac{2x^2 - 18}{x^2 + 3} = f(x)$$

Because $f(x) = f(-x)$, f is an even function.

Additional points The intervals determined by the x-intercepts are $x < -3$, $-3 < x < 3$, and $x > 3$. Generally, it is necessary to determine points in all intervals. However, because f is an even function, its graph is symmetric with respect to the y-axis. The following table lists a few points for $x > 0$. Symmetry can be used to locate corresponding points for $x < 0$.

x	1	2	6
$f(x)$	-4	$-\frac{10}{7} \approx -1.43$	$\frac{18}{13} \approx 1.38$

Behavior near asymptotes As x increases or decreases without bound, $f(x)$ approaches the horizontal asymptote $y = 2$.

To determine whether the graph of f intersects the horizontal asymptote at any point, solve the equation $f(x) = 2$.

There are no solutions of $f(x) = 2$ because

$$\frac{2x^2 - 18}{x^2 + 3} = 2 \quad \text{implies} \quad 2x^2 - 18 = 2x^2 + 6$$

This is not possible. Thus the graph of f does not intersect the horizontal asymptote but approaches it from below as x increases or decreases without bound.

Complete the sketch Use the summary in Table 3.3 to finish the sketch. The completed graph is shown in **Figure 3.38.**

Table 3.3

Vertical Asymptote	None
Horizontal Asymptote	$y = 2$
x-Intercepts	$(-3, 0), (3, 0)$
y-Intercept	$(0, -6)$
Additional Points	$(1, -4), (2, -1.43), (6, 1.38)$

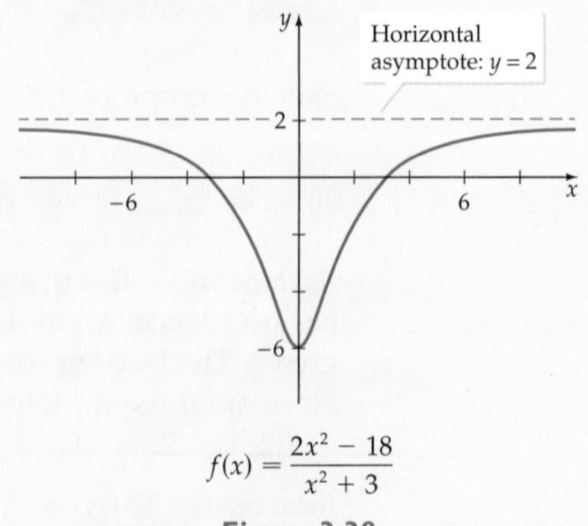

Figure 3.38

TRY EXERCISE 10, EXERCISE SET 3.5, PAGE 291

EXAMPLE 4 Graph a Rational Function

Sketch a graph of $h(x) = \dfrac{x^2 + 1}{x^2 + x - 2}$.

Solution

Asymptotes The denominator $x^2 + x - 2 = (x + 2)(x - 1)$ has zeros -2 and 1; because there are no common factors of the numerator and the denominator, the lines $x = -2$ and $x = 1$ are vertical asymptotes.

The numerator and denominator both have degree 2. The leading coefficients of the numerator and denominator are both 1. Thus h has the horizontal asymptote $y = 1/1 = 1$.

Intercept(s) The numerator $x^2 + 1$ has no real zeros, so the graph of h has no x-intercepts. Because $h(0) = -0.5$, h has the y-intercept $(0, -0.5)$.

Symmetry By applying the tests for symmetry, we can determine that the graph of h is not symmetric with respect to the origin or to the y-axis.

Additional points The intervals determined by the vertical asymptotes are $(-\infty, -2)$, $(-2, 1)$, and $(1, \infty)$. Plot a few points from each interval:

x	-5	-3	-1	0.5	2	3	4
$h(x)$	$\frac{13}{9}$	2.5	-1	-1	$\frac{5}{4}$	1	$\frac{17}{18}$

The graph of h will intersect the horizontal asymptote $y = 1$ exactly once. This can be determined by solving the equation $h(x) = 1$.

$$\frac{x^2 + 1}{x^2 + x - 2} = 1$$

$$x^2 + 1 = x^2 + x - 2 \qquad \bullet \text{ Multiply both sides by } x^2 + x - 2.$$

$$1 = x - 2$$

$$3 = x$$

The only solution is $x = 3$. Therefore, the graph of h intersects the horizontal asymptote at $(3, 1)$.

Behavior near asymptotes As x approaches -2 from the left, the denominator $(x + 2)(x - 1)$ approaches 0 but remains positive. The numerator $x^2 + 1$ approaches 5, which is positive, so the quotient $h(x)$ increases without bound. Stated in mathematical notation,

$$h(x) \to \infty \quad \text{as} \quad x \to -2^-$$

Similarly, it can be determined that

$$h(x) \to -\infty \quad \text{as} \quad x \to -2^+$$

$$h(x) \to -\infty \quad \text{as} \quad x \to 1^-$$

$$h(x) \to \infty \quad \text{as} \quad x \to 1^+$$

Table 3.4

Vertical Asymptote	$x = -2, x = 1$
Horizontal Asymptote	$y = 1$
x-Intercepts	None
y-Intercept	$(0, -0.5)$
Additional Points	$(-5, 1.\overline{4}), (-3, 2.5), (-1, -1), (0.5, -1), (2, 1.25), (3, 1), (4, 0.9\overline{4})$

Complete the sketch Use the summary in Table 3.4 to obtain the graph sketched in **Figure 3.39.**

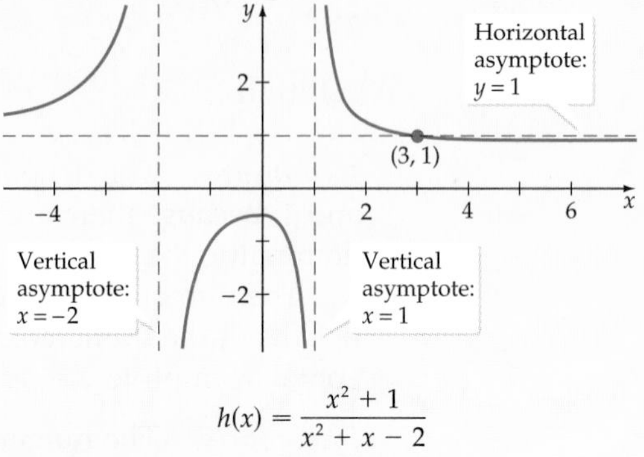

$$h(x) = \frac{x^2 + 1}{x^2 + x - 2}$$

Figure 3.39

TRY EXERCISE 26, EXERCISE SET 3.5, PAGE 291

◆ SLANT ASYMPTOTES

Some rational functions have an asymptote that is neither vertical nor horizontal but slanted.

Theorem on Slant Asymptotes

The rational function given by $F(x) = P(x)/Q(x)$, where $P(x)$ and $Q(x)$ have no common factors, has a **slant asymptote** if the degree of the polynomial $P(x)$ in the numerator is one greater than the degree of the polynomial $Q(x)$ in the denominator.

To find the slant asymptote, use division to express $F(x)$ in the form

$$F(x) = \frac{P(x)}{Q(x)} = (mx + b) + \frac{r(x)}{Q(x)}$$

where the degree of $r(x)$ is less than the degree of $Q(x)$. Because

$$\frac{r(x)}{Q(x)} \to 0 \quad \text{as} \quad x \to \pm\infty$$

we know that $F(x) \to mx + b$ as $x \to \pm\infty$.

The line represented by $y = mx + b$ is called the slant asymptote of the graph of F.

EXAMPLE 5 **Find the Slant Asymptote of a Rational Function**

Find the slant asymptote of $f(x) = \dfrac{2x^3 + 5x^2 + 1}{x^2 + x + 3}$.

Solution

Because the degree of the numerator $2x^3 + 5x^2 + 1$ is exactly one larger than the degree of the denominator $x^2 + x + 3$ and f is in simplest form, f has a slant asymptote. To find the asymptote, divide $2x^3 + 5x^2 + 1$ by $x^2 + x + 3$.

$$\begin{array}{r}
2x + 3 \\
x^2 + x + 3 \overline{)\, 2x^3 + 5x^2 + 0x + 1} \\
\underline{2x^3 + 2x^2 + 6x } \\
3x^2 - 6x + 1 \\
\underline{3x^2 + 3x + 9} \\
-9x - 8
\end{array}$$

Therefore,

$$f(x) = \frac{2x^3 + 5x^2 + 1}{x^2 + x + 3} = (2x + 3) + \frac{-9x - 8}{x^2 + x + 3}$$

and the line given by $y = 2x + 3$ is the slant asymptote for the graph of f. **Figure 3.40** shows the graph of f and its slant asymptote.

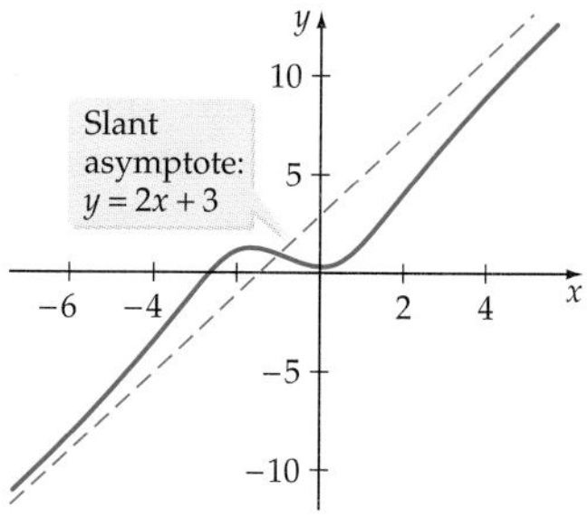

Slant asymptote: $y = 2x + 3$

$f(x) = \dfrac{2x^3 + 5x^2 + 1}{x^2 + x + 3}$

Figure 3.40

TRY EXERCISE 34, EXERCISE SET 3.5, PAGE 291

The function f in Example 5 does not have a vertical asymptote because the denominator $x^2 + x + 3$ does not have any real zeros. However, the function

$$g(x) = \frac{2x^2 - 4x + 5}{3 - x}$$

has both a slant asymptote and a vertical asymptote. The vertical asymptote is $x = 3$, and the slant asymptote is $y = -2x - 2$. **Figure 3.41** shows the graph of g and its asymptotes.

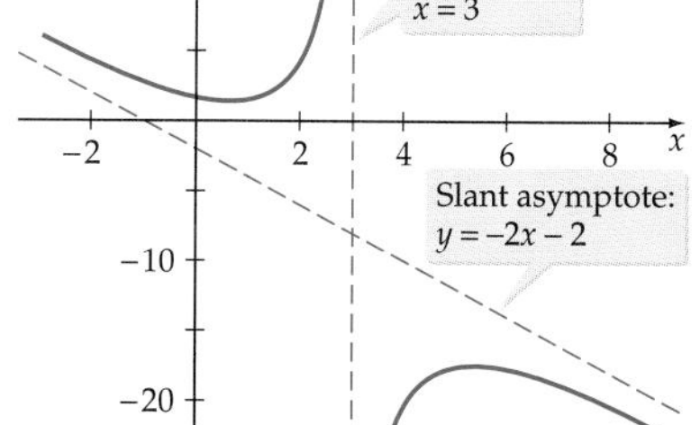

$g(x) = \dfrac{2x^2 - 4x + 5}{3 - x}$

Figure 3.41

EXAMPLE 6 Graph a Rational Function That Has a Slant Asymptote

Sketch the graph of $j(x) = \dfrac{x^2 - 1}{x}$.

Solution

Asymptotes The denominator x has 0 as its only zero. Because there are no common factors of the numerator and the denominator, the y-axis is the vertical asymptote of the graph of j.

The degree of the numerator $x^2 - 1$ is exactly one more than the degree of the denominator x, so j has a slant asymptote. Dividing $x^2 - 1$ by x shows that j can be expressed as

$$j(x) = \frac{x^2 - 1}{x} = \frac{x^2}{x} - \frac{1}{x} = x - \frac{1}{x}$$

From this we can conclude that $j(x) \to x$ as $x \to \pm\infty$. Therefore, the graph of j has a slant asymptote of $y = x$.

Intercepts By setting the numerator of j to zero, we can determine that the zeros of j are $x = -1$ and $x = 1$.

Symmetry As shown below, $j(-x) = -j(x)$, and therefore j is an odd function. The graph of an odd function is symmetric with respect to the origin.

$$j(-x) = \frac{(-x)^2 - 1}{-x} = \frac{x^2 - 1}{-x} = -\left(\frac{x^2 - 1}{x}\right) = -j(x)$$

Additional points The intervals determined by the x-intercepts and the vertical asymptotes are $x < -1$, $-1 < x < 0$, $0 < x < 1$, and $x > 1$. The following table lists some points from each interval.

x	-5	-2	-0.5	0.5	2	5
$j(x)$	-4.8	-1.5	1.5	-1.5	1.5	4.8

Complete the sketch Use all the previous information to complete the sketch of j as shown in **Figure 3.42.**

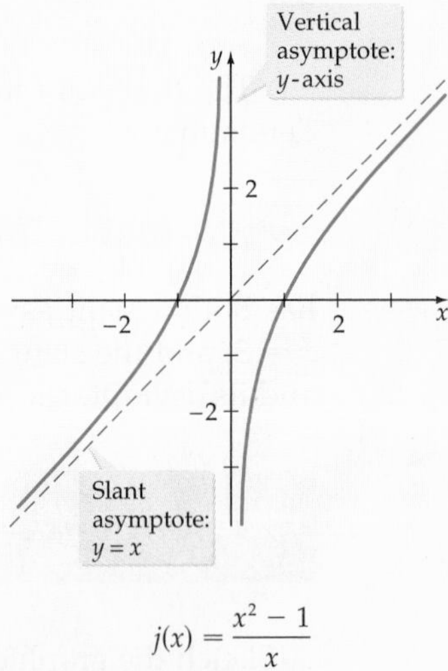

$$j(x) = \frac{x^2 - 1}{x}$$

Figure 3.42

TRY EXERCISE 40, EXERCISE SET 3.5, PAGE 291

◆ GRAPH RATIONAL FUNCTIONS THAT HAVE A COMMON FACTOR

If a rational function has a numerator and denominator that have a common factor, then you should reduce the rational function to lowest terms before you apply the general procedure for sketching the graph of a rational function.

EXAMPLE 7 Graph a Rational Function that has a Common Factor

Sketch the graph of $f(x) = \dfrac{x^2 - 3x - 4}{x^2 - 6x + 8}$.

Solution

Factor the numerator and denominator to obtain

$$f(x) = \frac{x^2 - 3x - 4}{x^2 - 6x + 8} = \frac{(x + 1)(x - 4)}{(x - 2)(x - 4)}, \quad x \neq 2, x \neq 4$$

Thus for all x values other than $x = 4$, the graph of f is the same as the graph of

$$G(x) = \frac{x + 1}{x - 2}$$

Figure 3.30, on page 278, shows a graph of G. The graph of f will be the same as this graph, except that it will have an open circle at (4, 2.5) to indicate that it is undefined for $x = 4$. See the graph of f in **Figure 3.43**.

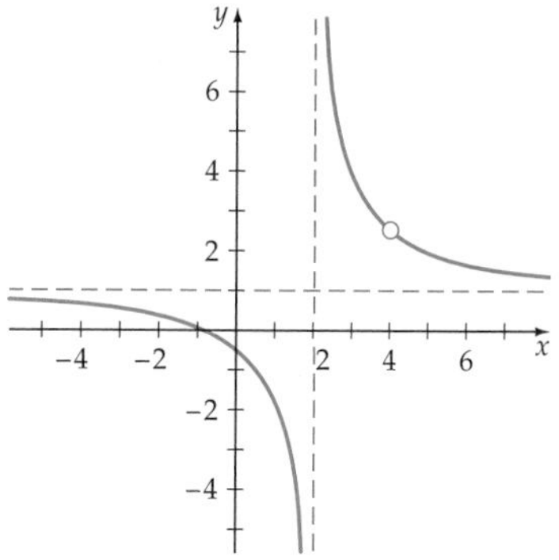

$f(x) = \dfrac{x^2 - 3x - 4}{x^2 - 6x + 8}$

Figure 3.43

TRY EXERCISE 50, EXERCISE SET 3.5, PAGE 291

QUESTION Does $F(x) = \dfrac{x^2 - x - 6}{x^2 - 9}$ have a vertical asymptote when $x = 3$?

◆ APPLICATION

EXAMPLE 8 Solve an Application

A cylindrical soft drink can is to be constructed so that it will have a volume of 21.6 cubic inches. See **Figure 3.44**.

a. Write the total surface area A of the can as a function of r, where r is the radius of the can in inches.

b. Use a graphing utility to estimate the value of r (to the nearest 0.1 inch) that produces the minimum surface area.

Figure 3.44

Continued

ANSWER No. $F(x) = \dfrac{x^2 - x - 6}{x^2 - 9} = \dfrac{(x - 3)(x + 2)}{(x - 3)(x + 3)} = \dfrac{x + 2}{x + 3}$. As x approaches 3, $F(x)$ approaches $\dfrac{5}{6}$.

take note

A web applet is available to explore the relationship between the radius of a cylinder, with a given volume, and the surface area of the cylinder. This applet, CYLINDER, can be found on our web site at

http://college.hmco.com

Solution

a. The formula for the volume of a cylinder is $V = \pi r^2 h$, where r is the radius and h is the height. Because we are given that the volume is 21.6 cubic inches, we have

$$21.6 = \pi r^2 h$$

$$\frac{21.6}{\pi r^2} = h \qquad \text{• Solve for } h.$$

The surface area of the cylinder is given by

$$A = 2\pi r^2 + 2\pi r h$$

$$A = 2\pi r^2 + 2\pi r\left(\frac{21.6}{\pi r^2}\right) \qquad \text{• Substitute for } h.$$

$$A = 2\pi r^2 + \frac{2(21.6)}{r} \qquad \text{• Simplify.}$$

$$A = \frac{2\pi r^3 + 43.2}{r} \qquad (1)$$

b. Use Equation (1) with $y = A$ and $x = r$ and a graphing utility to determine that A is a minimum when $r \approx 1.5$ inches. See **Figure 3.45.**

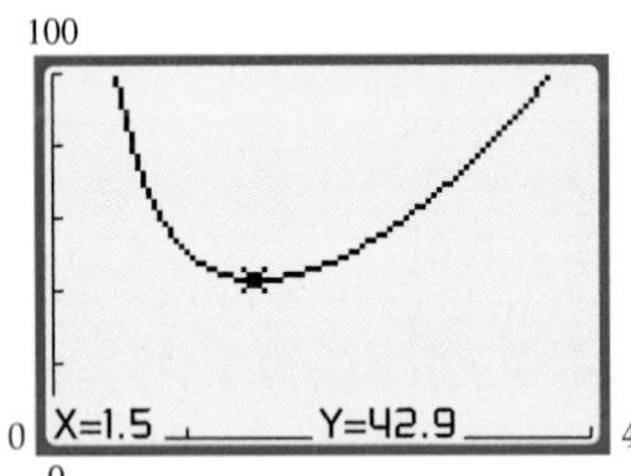

$$y = \frac{2\pi x^3 + 43.2}{x}$$

Figure 3.45

TRY EXERCISE 64, EXERCISE SET 3.5, PAGE 292

TOPICS FOR DISCUSSION

1. Discuss the meaning of a rational function. Give examples of functions that are rational functions and examples of functions that are not rational functions.

2. Does the graph of every rational function have at least one vertical asymptote? If so, explain why. If not, give an example of a rational function without a vertical asymptote.

3. Does the graph of every rational function have a horizontal asymptote? If so, explain why. If not, give an example of a rational function without a horizontal asymptote.

4. What conditions must exist to ensure that a rational function has a slant asymptote? Give an example of a rational function that has a slant asymptote.

5. Can the graph of a polynomial function have a vertical asymptote? a horizontal asymptote?

EXERCISE SET 3.5

In Exercises 1 to 4, find all vertical asymptotes of each rational function.

1. $F(x) = \dfrac{2x - 1}{x^2 + 3x}$

2. $F(x) = \dfrac{3x^2 + 5}{x^2 - 4}$

3. $F(x) = \dfrac{x^2 + 11}{6x^2 - 5x - 4}$

4. $F(x) = \dfrac{3x - 5}{x^3 - 8}$

In Exercises 5 to 8, find the horizontal asymptote of each rational function.

5. $F(x) = \dfrac{4x^2 + 1}{x^2 + x + 1}$

6. $F(x) = \dfrac{3x^3 - 27x^2 + 5x - 11}{x^5 - 2x^3 + 7}$

7. $F(x) = \dfrac{15{,}000x^3 + 500x - 2000}{700 + 500x^3}$

8. $F(x) = 6000\left(1 - \dfrac{25}{(x + 5)^2}\right)$

In Exercises 9 to 32, determine the vertical and horizontal asymptotes and sketch the graph of the rational function *F*. Label all intercepts and asymptotes.

9. $F(x) = \dfrac{1}{x + 4}$

10. $F(x) = \dfrac{1}{x - 2}$

11. $F(x) = \dfrac{-4}{x - 3}$

12. $F(x) = \dfrac{-3}{x + 2}$

13. $F(x) = \dfrac{4}{x}$

14. $F(x) = \dfrac{-4}{x}$

15. $F(x) = \dfrac{x}{x + 4}$

16. $F(x) = \dfrac{x}{x - 2}$

17. $F(x) = \dfrac{x + 4}{2 - x}$

18. $F(x) = \dfrac{x + 3}{1 - x}$

19. $F(x) = \dfrac{1}{x^2 - 9}$

20. $F(x) = \dfrac{-2}{x^2 - 4}$

21. $F(x) = \dfrac{1}{x^2 + 2x - 3}$

22. $F(x) = \dfrac{1}{x^2 - 2x - 8}$

23. $F(x) = \dfrac{x}{9 - x^2}$

24. $F(x) = \dfrac{x}{x^2 - 16}$

25. $F(x) = \dfrac{x^2}{x^2 + 4x + 4}$

26. $F(x) = \dfrac{x^2}{x^2 - 6x + 9}$

27. $F(x) = \dfrac{10}{x^2 + 2}$

28. $F(x) = \dfrac{-20}{x^2 + 4}$

29. $F(x) = \dfrac{2x^2 - 2}{x^2 - 9}$

30. $F(x) = \dfrac{6x^2 - 5}{2x^2 + 6}$

31. $F(x) = \dfrac{x^2 + x + 4}{x^2 + 2x - 1}$

32. $F(x) = \dfrac{2x^2 - 14}{x^2 - 6x + 5}$

In Exercises 33 to 36, find the slant asymptote of each rational function.

33. $F(x) = \dfrac{3x^2 + 5x - 1}{x + 4}$

34. $F(x) = \dfrac{x^3 - 2x^2 + 3x + 4}{x^2 - 3x + 5}$

35. $F(x) = \dfrac{x^3 - 1}{x^2}$

36. $F(x) = \dfrac{4000 + 20x + 0.0001x^2}{x}$

In Exercises 37 to 46, determine the vertical and slant asymptotes and sketch the graph of the rational function *F*.

37. $F(x) = \dfrac{x^2 - 4}{x}$

38. $F(x) = \dfrac{x^2 + 10}{2x}$

39. $F(x) = \dfrac{x^2 - 3x - 4}{x + 3}$

40. $F(x) = \dfrac{x^2 - 4x - 5}{2x + 5}$

41. $F(x) = \dfrac{2x^2 + 5x + 3}{x - 4}$

42. $F(x) = \dfrac{4x^2 - 9}{x + 3}$

43. $F(x) = \dfrac{x^2 - x}{x + 2}$

44. $F(x) = \dfrac{x^2 + x}{x - 1}$

45. $F(x) = \dfrac{x^3 + 1}{x^2 - 4}$

46. $F(x) = \dfrac{x^3 - 1}{3x^2}$

In Exercises 47 to 56, sketch the graph of the rational function *F*. (*Hint:* First examine the numerator and denominator to determine whether there are any common factors.)

47. $F(x) = \dfrac{x^2 + x}{x + 1}$

48. $F(x) = \dfrac{x^2 - 3x}{x - 3}$

49. $F(x) = \dfrac{2x^3 + 4x^2}{2x + 4}$

50. $F(x) = \dfrac{x^2 - x - 12}{x^2 - 2x - 8}$

51. $F(x) = \dfrac{-2x^3 + 6x}{2x^2 - 6x}$

52. $F(x) = \dfrac{x^3 + 3x^2}{x(x + 3)(x - 1)}$

53. $F(x) = \dfrac{x^2 - 3x - 10}{x^2 + 4x + 4}$

54. $F(x) = \dfrac{2x^2 + x - 3}{x^2 - 2x + 1}$

55. $F(x) = \dfrac{x^3 + x^2 - 14x - 24}{x + 2}$

56. $F(x) = \dfrac{2x^3 + 5x^2 - 4x - 3}{x - 1}$

In Exercises 57 to 60, use a graphing utility to graph each function and determine, to the nearest 0.1, the equations of the vertical asymptotes.

57. $R(x) = \dfrac{x^2 + 4}{x^2 - x - 3}$

58. $R(x) = \dfrac{2x^2 - x}{x^2 + x - 4}$

59. $P(x) = \dfrac{x^3 - x - 3}{x^3 - 2x^2 - 5x + 6}$

60. $V(x) = \dfrac{2x^3 + x + 1}{x^3 - 2x^2 - 11x + 12}$

61. **DESALINIZATION** The cost C in dollars to remove $p\%$ of the salt in a tank of sea water is given by

$$C(p) = \frac{2000p}{100 - p}, \quad 0 \le p < 100$$

a. Find the cost of removing 40% of the salt.

b. Find the cost of removing 80% of the salt.

c. Sketch the graph of C.

62. **FOOD SCIENCE** The temperature F (measured in degrees Fahrenheit) of a dessert placed in a freezer for t hours is given by the rational function

$$F(t) = \frac{60}{t^2 + 2t + 1}, \quad t \ge 0$$

a. Find the temperature of the dessert after it has been in the freezer for 1 hour.

b. Find the temperature of the dessert after 4 hours.

c. Sketch the graph of F.

63. **MANUFACTURING** A large electronics firm finds that the number of computers it can produce per week after t weeks of production is approximated by

$$C(t) = \frac{2000t^2 + 20{,}000t}{t^2 + 10t + 25}, \quad 0 \le t \le 50$$

a. Find the number of computers it produced during the first week.

b. Find the number of computers it produced during the tenth week.

c. What is the equation of the horizontal asymptote of the graph of C?

d. Sketch the graph of C, and then use the graph to estimate how many weeks pass until the firm can produce 1900 computers in a single week.

64. **PRODUCTION COSTS** The cost, in dollars, of publishing x books is

$$C(x) = 40{,}000 + 20x + 0.0001x^2$$

The average cost, in dollars, per book is given by

$$A(x) = \frac{C(x)}{x} = \frac{40{,}000 + 20x + 0.0001x^2}{x}$$

where $1000 \le x \le 100{,}000$.

a. What is the average cost per book if 5000 books are published?

b. What is the average cost per book if 10,000 books are published?

c. What is the equation of the slant asymptote of the graph of the average cost of function?

d. Graph A, and estimate the number of books that should be published to minimize the average cost per book.

65. **PHYSIOLOGY** One of Poiseuille's Laws states that the resistance R encountered by blood flowing through a blood vessel is given by the rational function

$$R(r) = C\frac{L}{r^4}$$

where C is a positive constant determined by the viscosity of the blood, L is the length of the blood vessel, and r is the radius.

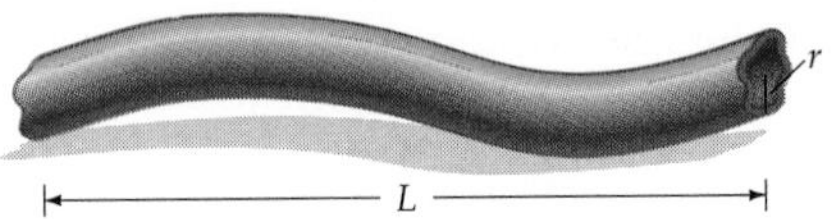

a. Explain the meaning of $R(r) \to \infty$ as $r \to 0$.

b. Explain the meaning of $R(r) \to 0$ as $r \to \infty$.

c. Graph R for $0 < r \le 4$ millimeters, given that $C = 1$ and $L = 100$ millimeters.

66. **MINIMIZING A CYLINDRICAL CONTAINER'S SURFACE AREA** A cylindrical soft drink can is to be made so that it will have a volume of 354 milliliters. If r is the radius of the can in centimeters, then the total surface area A of the can is given by the rational function

$$A(r) = \frac{2\pi r^3 + 708}{r}$$

a. Use the graph of A to estimate (to the nearest 0.1 cm) the value of r that produces the minimum value of A.

b. Does the graph of A have a slant asymptote?

c. 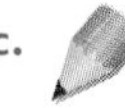 Explain the meaning of of the following statement as it applies to the graph of A.

$$\text{As } r \to \infty, A \to 2\pi r^2.$$

SUPPLEMENTAL EXERCISES

67. Determine the point where the graph of

$$F(x) = \frac{2x^2 + 3x + 4}{x^2 + 4x + 7}$$

intersects its horizontal asymptote.

68. Determine the point where the graph of

$$F(x) = \frac{3x^3 + 2x^2 - 8x - 12}{x^2 + 4}$$

intersects its slant asymptote.

69. Determine the two points where the graph of

$$F(x) = \frac{x^3 + x^2 + 4x + 1}{x^3 + 1}$$

intersects its horizontal asymptote.

70. Give an example of a rational function that intersects its slant asymptote at two points.

PROJECTS

1. **INVESTIGATE THE VERTICAL ASYMPTOTES** The Theorem on Vertical Asymptotes given in this section requires that the numerator and denominator of a rational function have no common factors. Note from the graphs of rational functions in this section that in the case of all vertical asymptotes, the values of the function approach plus or minus infinity as x approaches the vertical asymptote.

a. Because $Q(x) = \dfrac{x + 1}{x - 2}$ has no common factors, by the Theorem on Vertical Asymptotes the graph of Q has a vertical asymptote when $x = 2$. Show that $Q(x)$ approaches plus or minus infinity as x approaches 2 by completing the following tables.

x	2.1	2.01	2.001	2.0001	2.00001
Q(x)					

On the basis of the table results, complete the following sentence. "As x approaches 2 from the right, $Q(x)$ approaches ________ infinity." Now complete the table below.

x	1.9	1.99	1.999	1.9999	1.99999
Q(x)					

On the basis of the table results, complete the following sentence. "As x approaches 2 from the left, $Q(x)$ approaches ________ infinity."

b. Now consider the rational function $R(x) = \dfrac{x^2 + 2x - 8}{x^2 + x - 6}$. In this case, $x - 2$ is a common factor. Verify this! Complete the table below.

x	2.1	2.01	2.001	2.0001	2.00001
R(x)					

On the basis of the table results, complete the following sentence. "As x approaches 2 from the right, $R(x)$ approaches ________." Now complete the table below.

x	1.9	1.99	1.999	1.9999	1.99999
$R(x)$					

On the basis of the table results, complete the following sentence. "As x approaches 2 from the left, $R(x)$ approaches ________." From your work on the rational function R, does the graph of R have a vertical asymptote at $x = 2$?

c. Explain the conditions under which a rational function will be undefined at $x = a$ but whose graph will not have an asymptote at $x = a$.

d. Explain the conditions under which a rational function will be undefined at $x = a$ and whose graph will have an asymptote at $x = a$.

EXPLORING CONCEPTS WITH TECHNOLOGY

Finding Zeros of a Polynomial Using *Mathematica*

Computer algebra systems (CAS) are computer programs that are used to solve equations, graph functions, simplify algebraic expressions, and help us perform many other mathematical tasks. In this exploration, we will demonstrate how to use one of these programs, *Mathematica,* to find zeros of a polynomial.

Recall that a zero of a function P is a number, x, for which $P(x) = 0$. The idea behind finding a zero of a polynomial by using a CAS is to solve the polynomial equation $P(x) = 0$ for x.

Two commands in *Mathematica* that can be used to solve an equation are `Solve` and `NSolve`. (*Mathematica* is sensitive about syntax (the way in which an expression is typed.) You *must* use upper-case and lower-case letters as we indicate.) `Solve` will attempt to find an *exact* solution of the equation; `NSolve` attempts to find *approximate* solutions. Here are some examples.

To find the exact values of the zeros of $P(x) = x^3 + 5x^2 + 11x + 15$, input the following. *Note:* The two equals signs are necessary.

```
Solve[x^3+5x^2+11x+15==0]
```

Press Enter. The result should be

```
{{x->-3}, {x->-1-2 I}, {x->-1+2 I}}
```

Thus the three zeros of P are -3, $-1 - 2i$, and $-1 + 2i$.

To find the approximate values of the zeros of $P(x) = x^4 - 3x^3 + 4x^2 + x - 4$, input the following.

```
NSolve[x^4-3x^3+4x^2+x-4==0]
```

Press Enter. The result should be

```
{{x->-0.821746}, {x->1.2326}, {x->1.29457-1.50771 I},
{x->1.29457+1.50771 I}}
```

The four zeros are (approximately) -0.821746, 1.2326, $1.29457 - 1.50771i$, and $1.29457 + 1.50771i$.

Not all polynomial equations can be solved exactly. This means that **Solve** will not always give solutions with *Mathematica*. Consider the two examples below.

Input `NSolve[x^5–3x^3+2x^2–5==0]`

Output `{{x->-1.80492}, {x->-1.12491}, {x->0.620319-1.03589 I}, {x->0.620319+1.03589 I}, {x->1.68919}}`

These are the approximate zeros of the polynomial.

Input `Solve[x^5–3x^3+2x^2–5==0]`

Output `{ToRules[Roots[2x²–3x³+x⁵==5]]}`

In this case, no exact solution could be found. In general, there are no formulas like the quadratic formula, for instance, that yield exact solutions for fifth- or higher-degree polynomial equations.

Use *Mathematica* (or another CAS) to find the zeros of each polynomial.

1. $P(x) = x^4 - 3x^3 + x - 5$
2. $P(x) = 3x^3 - 4x^2 + x - 3$
3. $P(x) = 4x^5 - 3x^3 + 2x^2 - x + 2$
4. $P(x) = -3x^4 - 6x^3 + 2x - 8$

CHAPTER 3 SUMMARY

3.1 Polynomial Division and Synthetic Division

- *The Remainder Theorem* If a polynomial $P(x)$ is divided by $(x - c)$, then the remainder is $P(c)$.
- *The Factor Theorem* A polynomial $P(x)$ has a factor $(x - c)$ if and only if $P(c) = 0$.

3.2 Polynomial Functions

- Characteristics and properties used in graphing polynomial functions:
  1. Continuity—Polynomial functions are smooth continuous curves.
  2. Leading term test—Determines the behavior of the graph of a polynomial function at the far right or the far left.
  3. Zeros of the function determine the x-intercepts.
- Let f be a function defined on the open interval I, and let c be an element of I. Then

$f(c)$ is a relative minimum of f on I if $f(c) \le f(x)$ for all x in I

$f(c)$ is a relative maximum of f on I if $f(c) \ge f(x)$ for all x in I

- *The Zero Location Theorem* Let $P(x)$ be a polynomial with real coefficients. If $a < b$, and if $P(a)$ and $P(b)$ have opposite signs, then there is at least one value c between a and b such that $P(c) = 0$.

3.3 Zeros of Polynomial Functions

- Values of x that satisfy $P(x) = 0$ are called zeros of P.
- Definition of Multiple Zeros of a Polynomial If a polynomial $P(x)$ has $(x - r)$ as a factor exactly k times, then r is said to be a zero of multiplicity k of the polynomial $P(x)$.
- *The Rational Zero Theorem* If

$$P(x) = a_n x^n + a_{n-1}x^{n-1} + \cdots + a_1 x + a_0$$

has integer coefficients, and p/q (where p and q have no common factors) is a rational zero of $P(x)$, then p is a factor of a_0 and q is a factor of a_n.

- *Upper- and Lower-Bound Theorem*

Upper Bound If $b > 0$ and all the numbers in the bottom row of the synthetic division of P by $x - b$ are either positive or zero, then b is an upper bound for the real zeros of P.

Lower Bound If $a < 0$ and the numbers in the bottom row of the synthetic division of P by $x - a$ alternate in sign, then a is a lower bound for the real zeros of P.

- *Descartes' Rule of Signs* Let $P(x)$ be a polynomial with real coefficients and with terms arranged in order of decreasing powers of x.
  1. The number of positive real zeros of $P(x)$ is equal to the number of variations in sign of $P(x)$ or is equal to that number decreased by an even integer.

2. The number of negative real zeros of $P(x)$ is equal to the number of variations in sign of $P(-x)$ or is equal to that number decreased by an even integer.

- The zeros of some polynomials with integer coefficients can be found by using the guidelines stated on page 262.

3.4 The Fundamental Theorem of Algebra

- *The Fundamental Theorem of Algebra* If $P(x)$ is a polynomial of degree $n \geq 1$ with complex coefficients, then $P(x)$ has at least one complex zero.
- *The Number of Zeros of a Polynomial* If $P(x)$ is a polynomial of degree $n \geq 1$ with complex coefficients, then $P(x)$ has exactly n complex zeros, provided that each zero is counted according to its multiplicity.
- *The Conjugate Pair Theorem* If $a + bi$ $(b \neq 0)$ is a complex zero of the polynomial $P(x)$, with real coefficients, then the conjugate $a - bi$ is also a complex zero of the polynomial.

3.5 Rational Functions and Their Graphs

- If $P(x)$ and $Q(x)$ are polynomials, then the function F given by

$$F(x) = \frac{P(x)}{Q(x)}$$

is called a rational function.

- *General Procedure for Graphing Rational Functions That Have No Common Factors*

1. Find the real zeros of the denominator. For each zero a the vertical line $x = a$ will be a vertical asymptote. The vertical asymptotes will occur at these points. Use the Theorem on Horizontal Asymptotes to determine the equation of any horizontal asymptote. Graph any horizontal asymptotes.
2. Find the real zeros of the numerator. For each zero a, plot $(a, 0)$. These points are the x-intercepts.
3. Use the tests for symmetry to determine whether the graph has symmetry with respect to the x-axis or to the origin.
4. Find additional points that lie in the intervals between the x-intercepts and the vertical asymptotes.
5. Determine the behavior of the graph near the asymptotes.
6. Use the information obtained in the above steps to sketch graph.

- *Theorem on Slant Asymptotes* The rational function given by $F(x) = P(x)/Q(x)$, where $P(x)$ and $Q(x)$ have no common factors, has a slant asymptote if the degree of the polynomial $P(x)$ in the numerator is one greater than the degree of the polynomial $Q(x)$ in the denominator.

CHAPTER 3 TRUE/FALSE EXERCISES

In Exercises 1 to 14, answer true or false. If the statement is false, give an example.

1. The complex zeros of a polynomial with complex coefficients always occur in conjugate pairs.
2. Descartes' Rule of Signs indicates that $x^3 - x^2 + x - 1$ must have three positive zeros.
3. The polynomial $2x^5 + x^4 - 7x^3 - 5x^2 + 4x + 10$ has two variations in sign.
4. If 4 is an upper bound of the zeros of the polynomial P, then 5 is also an upper bound of the zeros of P.
5. The graph of every rational function has a vertical asymptote.
6. The graph of the rational function

$$F(x) = \frac{x^2 - 4x + 4}{x^2 - 5x + 6}$$

has a vertical asymptote of $x = 2$.

7. If 7 is a zero of the polynomial P, then $x - 7$ is a factor of P.
8. According to the Zero Location Theorem, the polynomial function $P(x) = x^3 + 6x - 2$ has a real zero between 0 and 1.
9. Synthetic division can be used to show that $3i$ is a zero of $x^3 - 2x^2 + 9x - 18$.
10. Every fourth-degree polynomial with complex coefficients has exactly four complex zeros, provided that each zero is counted according to its multiplicity.
11. The graph of a rational function never intersects any of its vertical asymptotes.
12. The graph of a rational function can have at most one horizontal asymptote.
13. Descartes' Rule of Signs indicates that the polynomial function $P(x) = x^3 + 2x^2 + 4x - 7$ does have a positive zero.
14. Every polynomial has at least one real zero.

False; $x^2+1=0$ does not have real zero
$x^2 = -1$
$x = \sqrt{-1}$

CHAPTER 3 REVIEW EXERCISES

In Exercises 1 to 6, use long division to divide the first polynomial by the second.

1. $x^3 + 5x^2 + 2x - 17, x^2 + x + 3$
2. $2x^3 - 5x + 1, x^2 + 4$
3. $-x^4 + 2x^2 - 12x - 3, x^3 + x$
4. $x^3 - 5x^2 - 6x - 11, x^2 - 6x - 1$
5. $6x^4 + 8x^3 - 47x^2 + 19x + 5, 2x^2 + 6x - 5$
6. $x^4 + 3x^3 - 6x^2 - 13x + 15, x^2 + 2x - 3$

In Exercises 7 to 12, use synthetic division to divide the first polynomial by the second.

7. $4x^3 - 11x^2 + 5x - 2, x - 3$
8. $5x^3 - 18x + 2, x - 1$
9. $3x^3 - 5x + 1, x + 2$
10. $2x^3 + 7x^2 + 16x - 10, x - \frac{1}{2}$
11. $3x^3 - 10x^2 - 36x + 55, x - 5$
12. $x^4 + 9x^3 + 6x^2 - 65x - 63, x + 7$

In Exercises 13 to 16, use the Remainder Theorem to find $P(c)$.

13. $P(x) = x^3 + 2x^2 - 5x + 1, c = 4$
14. $P(x) = -4x^3 - 10x + 8, c = -1$
15. $P(x) = 6x^4 - 12x^2 + 8x + 1, c = -2$
16. $P(x) = 5x^5 - 8x^4 + 2x^3 - 6x^2 - 9, c = 3$

In Exercises 17 to 20, use synthetic division to show that c is a zero of the given polynomial.

17. $x^3 + 2x^2 - 26x + 33, c = 3$
18. $2x^4 + 8x^3 - 8x^2 - 31x + 4, c = -4$
19. $x^5 - x^4 - 2x^2 + x + 1, c = 1$
20. $2x^3 + 3x^2 - 8x + 3, c = \frac{1}{2}$

In Exercises 21 to 26, graph the polynomial function.

21. $P(x) = x^3 - x$
22. $P(x) = -x^3 - x^2 + 8x + 12$
23. $P(x) = x^4 - 6$
24. $P(x) = x^5 - x$
25. $P(x) = x^4 - 10x^2 + 9$
26. $P(x) = x^5 - 5x^3$

In Exercises 27 to 32, use the Rational Zero Theorem to list all possible rational zeros for each polynomial.

27. $x^3 - 7x - 6$
28. $2x^3 + 3x^2 - 29x - 30$
29. $15x^3 - 91x^2 + 4x + 12$
30. $x^4 - 12x^3 + 52x^2 - 96x + 64$
31. $x^3 + x^2 - x - 1$
32. $6x^5 + 3x - 2$

In Exercises 33 to 36, use Descartes' Rule of Signs to state the number of possible positive and negative real zeros of each polynomial.

33. $x^3 + 3x^2 + x + 3$
34. $x^4 - 6x^3 - 5x^2 + 74x - 120$
35. $x^4 - x - 1$
36. $x^5 - 4x^4 + 2x^3 - x^2 + x - 8$

In Exercises 37 to 42, find the zeros of the polynomial.

37. $x^3 + 6x^2 + 3x - 10$
38. $x^3 - 10x^2 + 31x - 30$
39. $6x^4 + 35x^3 + 72x^2 + 60x + 16$
40. $2x^4 + 7x^3 + 5x^2 + 7x + 3$
41. $x^4 - 4x^3 + 6x^2 - 4x + 1$
42. $2x^3 - 7x^2 + 22x + 13$
43. Find a third-degree polynomial with zeros of 4, -3, and $1/2$.
44. Find a fourth-degree polynomial with zeros of 2, -3, i, and $-i$.
45. Find a fourth-degree polynomial with real coefficients that has zeros of 1, 2, and $5i$.
46. Find a fourth-degree polynomial with real coefficients that has -2 as a zero of multiplicity 2 and also has $1 + 3i$ as a zero.

In Exercises 47 to 50, find the vertical, horizontal, and slant asymptotes for each rational function.

47. $f(x) = \frac{3x + 5}{x + 2}$
48. $f(x) = \frac{2x^2 + 12x + 2}{x^2 + 2x - 3}$
49. $f(x) = \frac{2x^2 + 5x + 11}{x + 1}$
50. $f(x) = \frac{6x^2 - 1}{2x^2 + x + 7}$

In Exercises 51 to 58, graph each rational function.

51. $f(x) = \frac{3x - 2}{x}$
52. $f(x) = \frac{x + 4}{x - 2}$
53. $f(x) = \frac{6}{x^2 + 2}$
54. $f(x) = \frac{4x^2}{x^2 + 1}$

55. $f(x) = \dfrac{2x^3 - 4x + 6}{x^2 - 4}$

56. $f(x) = \dfrac{x}{x^3 - 1}$

57. $f(x) = \dfrac{3x^2 - 6}{x^2 - 9}$

58. $f(x) = \dfrac{-x^3 + 6}{x^2}$

In Exercises 59 to 62, the given polynomials have one zero that satisfies the given condition. Use a graphing utility and the Zero Location Theorem to approximate the zero to the nearest 0.001.

59. $x^3 - x - 1 = 0, \quad x > 0$

60. $x^3 - 3x - 6 = 0, \quad x > 0$

61. $x^4 + x^2 - 1 = 0, \quad x > 0$

62. $x^4 - 2x^2 - 2 = 0, \quad x > 0$

CHAPTER 3 TEST

1. Use synthetic division to divide:

$$(3x^3 + 5x^2 + 4x - 1) \div (x + 2)$$

2. Use the Remainder Theorem to find $P(-2)$ if

$$P(x) = -3x^3 + 7x^2 + 2x - 5$$

3. Show that $x - 1$ is a factor of

$$x^4 - 4x^3 + 7x^2 - 6x + 2$$

4. Examine the leading term of the function given by the equation $P(x) = -3x^3 + 2x^2 - 5x + 2$ and determine the far-left and far-right behavior of the graph of P.

5. Find the real solutions of $3x^3 + 7x^2 - 6x = 0$.

6. Use the Zero Location Theorem to verify that

$$P(x) = 2x^3 - 3x^2 - x + 1$$

has a zero between 1 and 2.

7. Find the zeros of the polynomial

$$P(x) = (x^2 - 4)^2(2x - 3)(x + 1)^3$$

and state the multiplicity of each.

8. Use the Rational Zero Theorem to list the possible rational zeros for the polynomial

$$P(x) = 6x^3 - 3x^2 + 2x - 3$$

9. Find, by using the Upper- and Lower-Bound Theorem, the smallest positive integer and the largest negative integer that are upper and lower bounds for the polynomial

$$P(x) = 2x^4 + 5x^3 - 23x^2 - 38x + 24$$

10. Use Descartes' Rule of Signs to state the number of possible positive and negative real zeros of

$$P(x) = x^4 - 3x^3 + 2x^2 - 5x + 1$$

11. Find the zeros of the polynomial

$$P(x) = 2x^3 - 3x^2 - 11x + 6$$

12. Given that $2 + 3i$ is a zero of

$$P(x) = 6x^4 - 5x^3 + 12x^2 + 207x + 130,$$

find the remaining zeros.

13. Find all the zeros of the polynomial

$$P(x) = x^5 - 6x^4 + 14x^3 - 14x^2 + 5x$$

14. Find a polynomial of lowest degree that has real coefficients and zeros $1 + i$, 3, and 0.

15. Find all vertical asymptotes of the graph of

$$f(x) = \frac{3x^2 - 2x + 1}{x^2 - 5x + 6}$$

16. Find the horizontal asymptote of the graph of

$$f(x) = \frac{3x^2 - 2x + 1}{2x^2 - 1}$$

17. Graph $f(x) = \dfrac{x^2 - 1}{x^2 - 2x - 3}$ and label the open circle that appears on the graph with its coordinates.

18. Graph $f(x) = \dfrac{2x^2 + 2x + 1}{x + 1}$ and label the slant asymptote with its equation.

19. Graph $f(x) = \dfrac{x}{x^2 + 1}$

20. Use a graphing utility to approximate the zero in the interval $1 < x < 2$ of the polynomial $P(x) = x^3 - 5x + 3$ to within one-tenth of a unit.

CHAPTER

4 EXPONENTIAL AND LOGARITHMIC FUNCTIONS

◆ The shape of the Eiffel Tower can be modeled by an equation involving the number e.

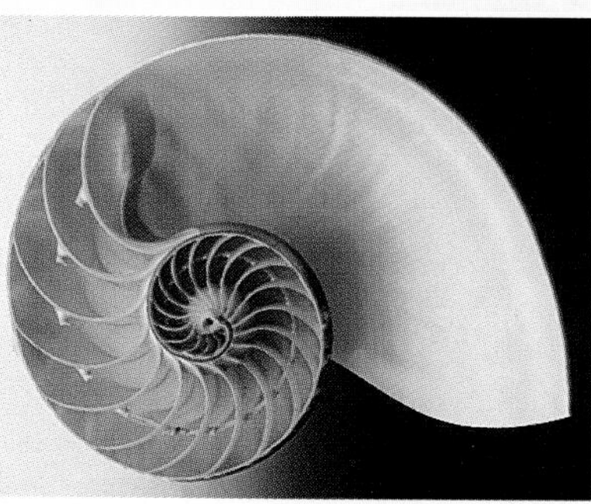

◆ The spiral in a nautilus shell can be modeled by an equation involving logarithms.

An Optical Illusion

The St. Louis Gateway Arch is one of the largest optical illusions ever created. As you look at the arch, it seems to be much taller than it is wide. However, the two distances are exactly the same, 630 feet. You can measure the arch on this page and see for yourself that its width and height are equal.

The equation of the curve described by the Gateway Arch is approximated by

$$y = 693.8597 - 68.7672\left(\frac{e^{0.0100333x} + e^{-0.0100333x}}{2}\right)$$

This equation involves *exponential functions*, one of the topics of this chapter.

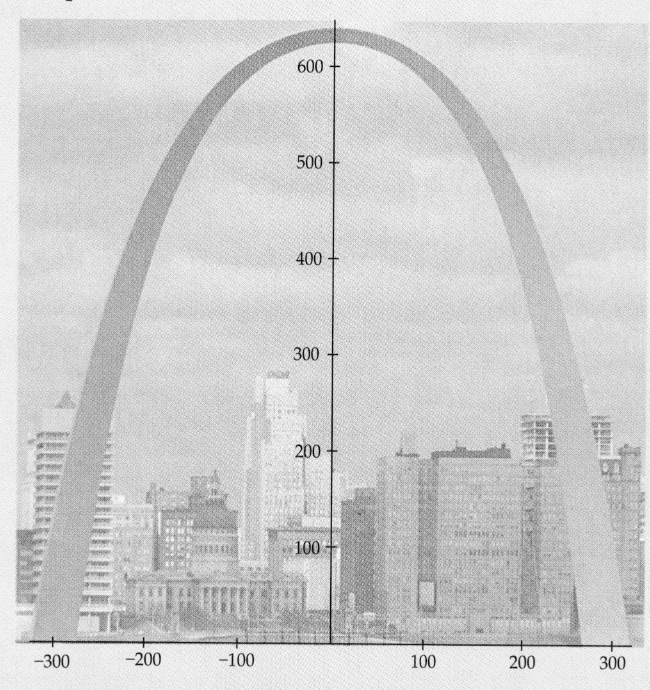

SECTION 4.1 INVERSE FUNCTIONS

- INVERSE FUNCTIONS DEFINED
- ONE-TO-ONE FUNCTIONS AND INVERSES
- FIND AN INVERSE FUNCTION
- PLACE A RESTRICTION ON A DOMAIN
- GRAPHS OF INVERSE FUNCTIONS
- AN APPLICATION OF INVERSE FUNCTIONS

◆ INVERSE FUNCTIONS DEFINED

Some functions are inverses of each other in the sense that one undoes the other. The function $g(x) = \frac{1}{2}x - 4$ undoes the function $f(x) = 2x + 8$. To illustrate, let $x = 10$. Then

$$f(10) = 2(10) + 8 = 28$$

Now evaluate $g[f(10)] = g(28)$.

$$g[f(10)] = g(28) = \frac{1}{2}(28) - 4 = 14 - 4 = 10$$

Thus we started with $x = 10$, we evaluated $f(10)$, we evaluated $g[f(10)]$, and the end result was the number that we started with, 10. This was not a coincidence. See **Figure 4.1**.

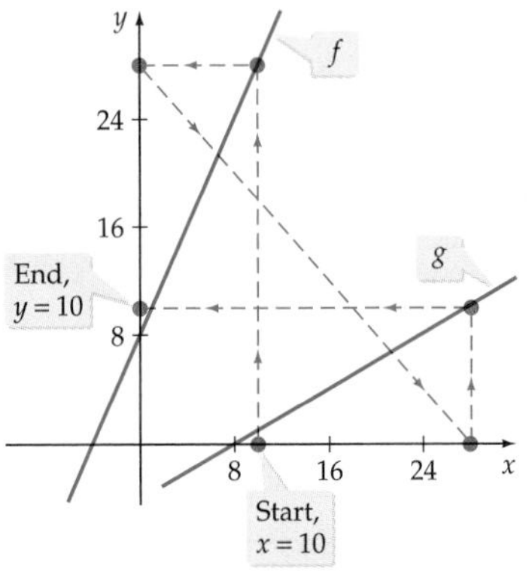

Figure 4.1

Definition of an Inverse Function

If f is a one-to-one function with domain X and range Y, and g is a function with domain Y and range X, then g is the **inverse function** of f if and only if

$(f \circ g)(x) = x$ for all x in the domain of g

and $(g \circ f)(x) = x$ for all x in the domain of f

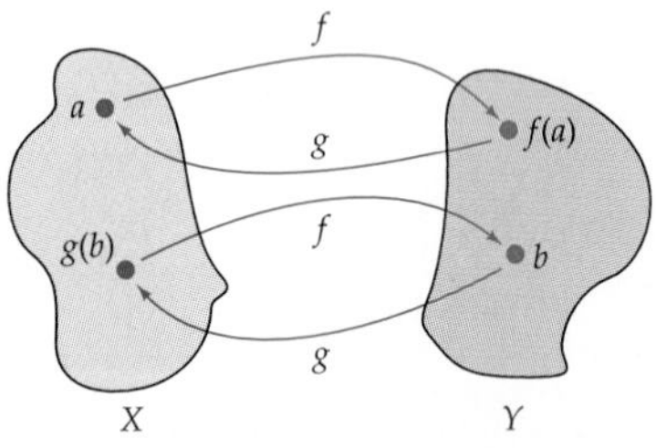

Figure 4.2

Figure 4.2 illustrates a function f and its inverse g. The set X is both the domain of f and the range of g. The set Y is both the domain of g and the range of f.

EXAMPLE 1 Verify that Functions are Inverse Functions

Verify that $g(x) = \frac{1}{2}x - 4$ is the inverse function of $f(x) = 2x + 8$.

Solution

We need to show that $(f \circ g)(x) = x$ and that $(g \circ f)(x) = x$.

$$(f \circ g)(x) = f[g(x)] = f\left[\frac{1}{2}x - 4\right] = 2\left[\frac{1}{2}x - 4\right] + 8 = x - 8 + 8 = x$$

$$(g \circ f)(x) = g[f(x)] = g[2x + 8] = \frac{1}{2}[2x + 8] - 4 = x + 4 - 4 = x$$

Therefore, the function g is the inverse function of f. This work also shows that f is the inverse function of g.

TRY EXERCISE 2, EXERCISE SET 4.1, PAGE 306

take note

Recall from Chapter 2 that a function is a one-to-one function provided each range element corresponds with one and only one domain element.

◆ ONE-TO-ONE FUNCTIONS AND INVERSES

The definition of an inverse function requires f to be a one-to-one function. The reason for this restriction is now explained.

If a one-to-one function is given as a set of ordered pairs, then its inverse is the set of ordered pairs with their components interchanged. For example, the inverse of

$$\{(4, 7), (5, 2), (6, 11)\} \quad \text{is} \quad \{(7, 4), (2, 5), (11, 6)\}$$

Now consider the function j defined by $j(x) = x^2 - 1$. Some of the ordered pairs of j are

$$(-2, 3), \quad (-1, 0), \quad (0, -1), \quad (1, 0), \quad \text{and} \quad (2, 3)$$

The inverse of j contains the ordered pairs

$$(3, -2), \quad (0, -1), \quad (-1, 0), \quad (0, 1), \quad \text{and} \quad (3, 2)$$

This set of ordered pairs does *not* satisfy the definition of a function, because there are ordered pairs with the same first component and *different* second components. For example, the ordered pairs $(3, -2)$ and $(3, 2)$ both have 3 as their first component, but they have different second components. This example illustrates that not all functions have inverses that are functions.

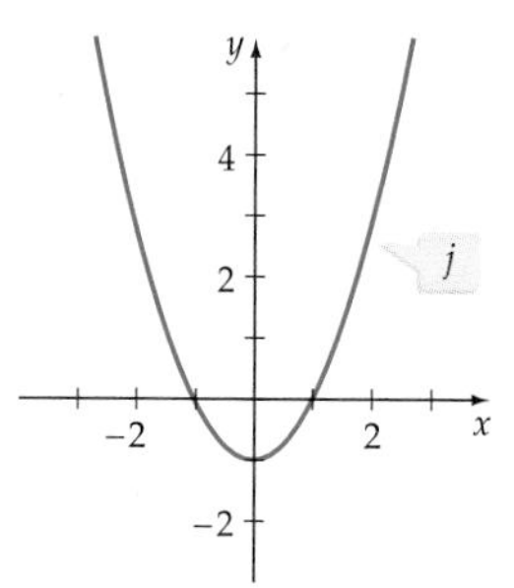

$j(x) = x^2 - 1$

Figure 4.3

Figure 4.3 is the graph of the function j. The horizontal line test indicates that j is *not* the graph of a *one-to-one* function. The horizontal line test can be used to show that the function $h(x) = \frac{1}{2}x^3$ is a one-to-one function. See **Figure 4.4.** Some of the ordered pairs of h are

$$(-2, -4), \quad \left(-1, -\frac{1}{2}\right), \quad (0, 0), \quad \left(1, \frac{1}{2}\right), \quad \text{and} \quad (2, 4)$$

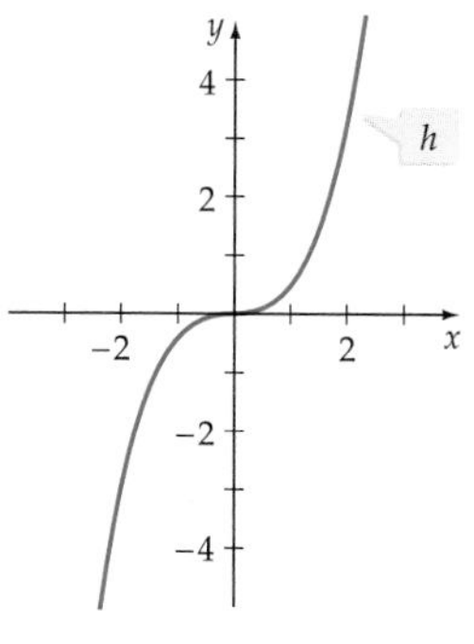

$h(x) = \frac{1}{2}x^3$

Figure 4.4

Because h is a one-to-one function, given any y in the range of h, there corresponds exactly one x in the domain of h. Thus interchanging the coordinates of each ordered pair defined by h yields a set of ordered pairs that is a function. This function with the coordinates interchanged is the inverse function of h. The one-to-one property is exactly what is required for a function to have an inverse function.

Condition for a Function to Have an Inverse Function

A function f has an inverse function if and only if it is a one-to-one function.

take note

The notation f^{-1} for an inverse function does not mean $1/f$. The function denoted by $1/f$ is called the *reciprocal function* and is an entirely different function from f^{-1}. For instance, in Example 1 we showed that

$$f^{-1}(x) = g(x) = \frac{1}{2}x - 4,$$

whereas $1/f(x) = \dfrac{1}{2x + 8}$.

◆ Find an Inverse Function

The inverse of the function f is often denoted by f^{-1}. In Example 1, we verified that g was the inverse of f, so in this case the function g could be written as f^{-1}.

If a one-to-one function f is defined by an equation, then we use the following method to find the equation of the inverse f^{-1}.

Find the Equation for f^{-1}

To find the inverse f^{-1} of the one-to-one function f:

- Substitute y for $f(x)$.
- Interchange x and y.
- Solve, if possible, for y in terms of x.
- Substitute $f^{-1}(x)$ for y.
- Verify that the domain of f is the range of f^{-1} and that the range of f is the domain of f^{-1}.

take note

We use different procedures to verify that two functions are inverse functions than we use to find an inverse function of a given function.

EXAMPLE 2 Find the Inverse of a One-to-One Function

Find the inverse of the one-to-one function $f(x) = 2x - 6$.

Solution

Begin by substituting y for $f(x)$.

$$y = 2x - 6$$

$$x = 2y - 6 \qquad \text{• Interchange } x \text{ and } y.$$

$$x + 6 = 2y \qquad \text{• Solve for } y.$$

$$\frac{x + 6}{2} = y$$

This equation can be written as

$$y = \frac{1}{2}x + 3$$

In inverse notation,

$$f^{-1}(x) = \frac{1}{2}x + 3$$

In this example, the function f has a domain of all real numbers and a range of all real numbers, so the inverse f^{-1} also has a domain of all real numbers and a range of all real numbers.

TRY EXERCISE 14, EXERCISE SET 4.1, PAGE 307

EXAMPLE 3 Find the Inverse of a One-to-One Function

Find the inverse of the function defined by $g(x) = \dfrac{2x}{x+3}$.

Solution

$$y = \frac{2x}{x+3}, x \neq -3$$ • Replace g(x) with y.

$$x = \frac{2y}{y+3}$$ • Interchange x and y.

$$x(y+3) = 2y$$ • Multiply each side by (y + 3).

$$xy + 3x = 2y$$

$$xy - 2y = -3x$$ • Collect on one side the terms that contain a factor of y.

$$y(x-2) = -3x$$ • Factor out the y.

$$y = \frac{-3x}{x-2}$$ • Solve for y.

$$g^{-1}(x) = \frac{-3x}{x-2}, x \neq 2$$

take note

Because $\dfrac{-3x}{x-2} = \dfrac{3x}{2-x}$, we can also write the inverse of g as $g^{-1}(x) = \dfrac{3x}{2-x}$.

TRY EXERCISE 22, EXERCISE SET 4.1, PAGE 307

◆ PLACE A RESTRICTION ON A DOMAIN

The graph of the function defined by $f(x) = x^2 - 4x$ is a parabola that opens upward. It is not a one-to-one function and therefore does not have an inverse function. However, the function $G(x) = x^2 - 4x$ with domain restricted to $\{x \mid x \geq 2\}$ is a one-to-one function. It has an inverse function denoted by G^{-1}.

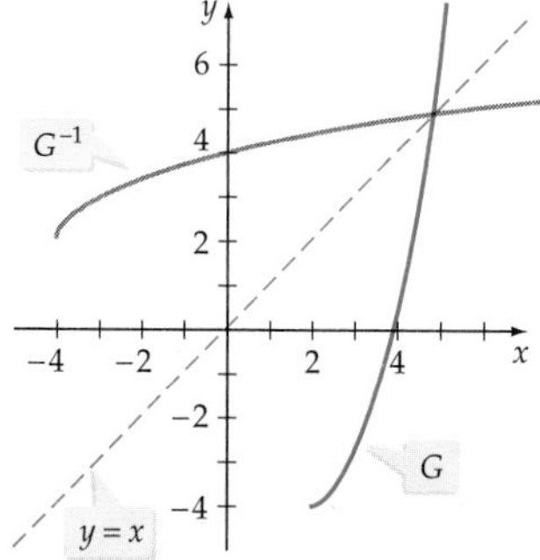

$G(x) = x^2 - 4x, \quad x \geq 2$
$G^{-1}(x) = 2 + \sqrt{x+4}$

Figure 4.5

EXAMPLE 4 Find the Inverse Function and State Its Domain and Range

Find the inverse G^{-1} of the function $G(x) = x^2 - 4x$, for $x \geq 2$. State the domain and range of both G and G^{-1}.

Solution

First note that the domain of G is given as $\{x \mid x \geq 2\}$. The graph of G in **Figure 4.5** shows that G has the range $\{y \mid y \geq -4\}$. Because the domain of G^{-1} is the range of G and the range of G^{-1} is the domain of G, G^{-1} has the domain $\{x \mid x \geq -4\}$ and the range $\{y \mid y \geq 2\}$.

Continued ►

Now we proceed to find G^{-1}. The method shown uses the technique of completing the square.

$$G(x) = x^2 - 4x \quad \text{for } x \geq 2$$
$$y = x^2 - 4x$$
$$x = y^2 - 4y$$
• **Interchange x and y.**
$$x + 4 = y^2 - 4y + 4$$
• **Solve for y by completing the square of $y^2 - 4y$.**
$$x + 4 = (y - 2)^2$$
• **Factor.**
$$\pm\sqrt{x + 4} = y - 2$$
• **Apply the Square Root Theorem.**
$$2 \pm \sqrt{x + 4} = y$$

The range of G^{-1} is $\{y \mid y \geq 2\}$. Recall that the radical $\sqrt{x + 4}$ is a nonnegative number. Therefore, to make $G^{-1}(x) = 2 \pm \sqrt{x + 4}$ a real number greater than or equal to 2 requires that we consider only the nonnegative square root. Thus G^{-1} is given by

$$G^{-1}(x) = 2 + \sqrt{x + 4}$$

TRY EXERCISE 36, EXERCISE SET 4.1, PAGE 307

◆ GRAPHS OF INVERSE FUNCTIONS

The graphs of G and G^{-1} from Example 4 are shown in **Figure 4.5** on page 303. The graphs are symmetric with respect to the line $y = x$. This is always the case for the graph of a function and its inverse.

Symmetry Property of f and f^{-1}

The graph of a function f and the graph of the inverse function f^{-1} are symmetric with respect to the line given by $y = x$.

The symmetry property of f and f^{-1} can be used to graph the inverse of a one-to-one function.

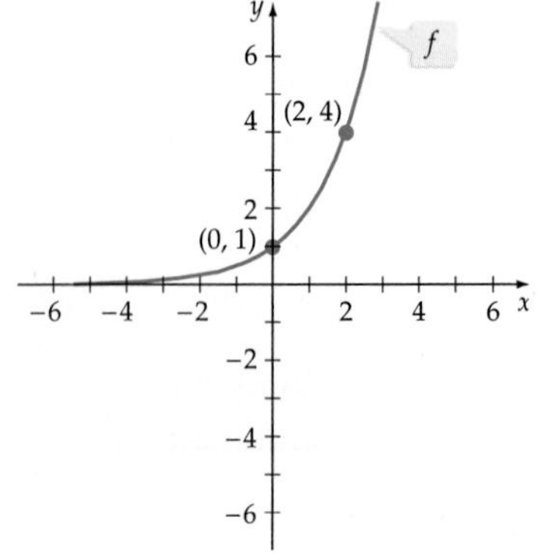

Figure 4.6

EXAMPLE 5 Graph the Inverse of a Function

Sketch the graph of f^{-1}, given that f is the function shown in **Figure 4.6**.

Solution

Because the graph of f passes through (0, 1) and (2, 4), the graph of f^{-1} must pass through (1, 0) and (4, 2). Draw a smooth curve through (1, 0) and

(4, 2) that is the reflection of f with respect to the graph of $y = x$. See Figure 4.7.

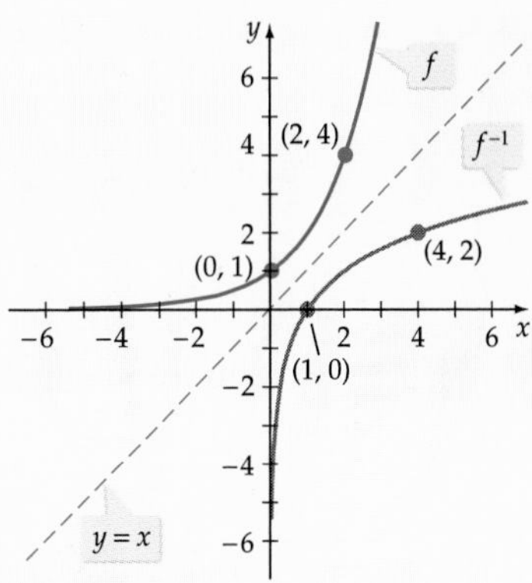

Figure 4.7

TRY EXERCISE 40, EXERCISE SET 4.1, PAGE 307

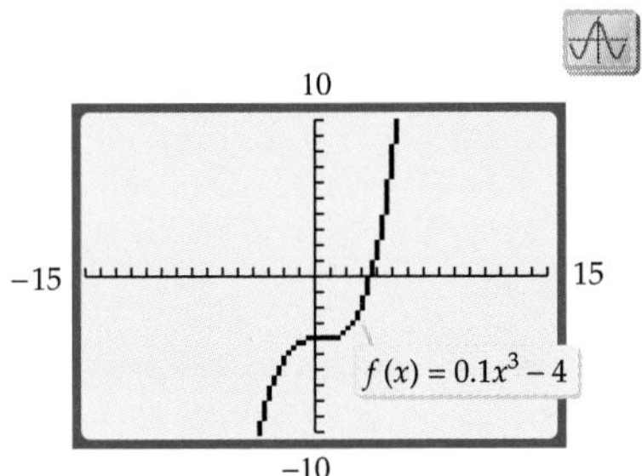

Figure 4.8

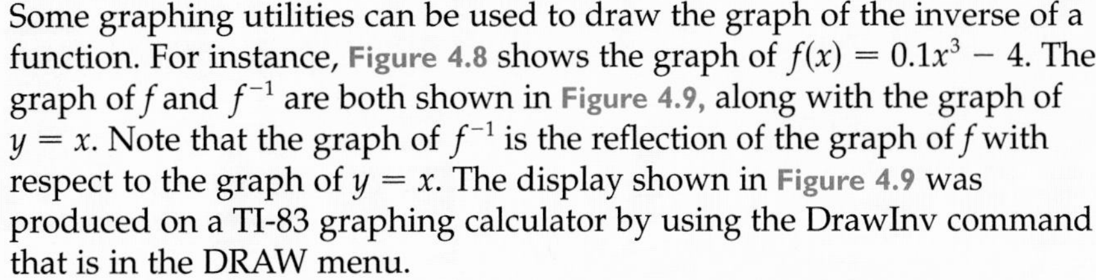

Some graphing utilities can be used to draw the graph of the inverse of a function. For instance, Figure 4.8 shows the graph of $f(x) = 0.1x^3 - 4$. The graph of f and f^{-1} are both shown in Figure 4.9, along with the graph of $y = x$. Note that the graph of f^{-1} is the reflection of the graph of f with respect to the graph of $y = x$. The display shown in Figure 4.9 was produced on a TI-83 graphing calculator by using the DrawInv command that is in the DRAW menu.

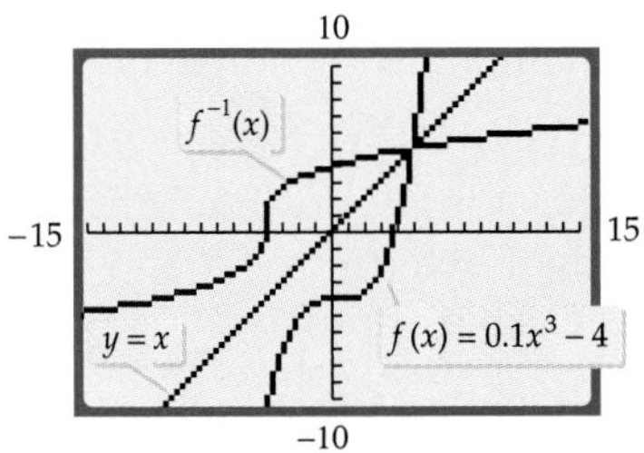

Figure 4.9

◆ AN APPLICATION OF INVERSE FUNCTIONS

EXAMPLE 6 An Application Concerning Men's Fashions

The function $IT(x) = 2x + 8$ converts a man's shirt size x in the United States to the equivalent shirt size in Italy.

a. Use IT to determine the equivalent Italian shirt size for a size 16.5 U.S. shirt.

b. Find IT^{-1} and use IT^{-1} to determine the U.S. men's shirt size that is equivalent to an Italian shirt size of 36.

Solution

a. $IT(16.5) = 2(16.5) + 8 = 33 + 8 = 41$

A size 16.5 U.S. shirt is equivalent to a size 41 Italian shirt.

Continued ➤

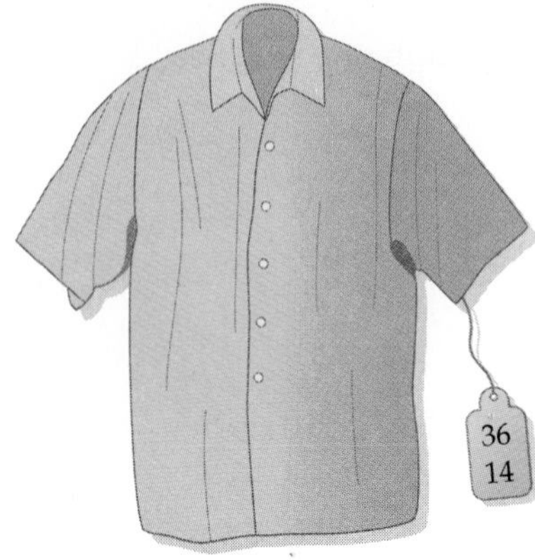

b. To find the inverse function, begin by substituting y for $IT(x)$.

$$IT(x) = 2x + 8$$
$$y = 2x + 8$$
$$x = 2y + 8 \qquad \text{• Interchange } x \text{ and } y.$$
$$x - 8 = 2y \qquad \text{• Solve for } y.$$
$$\frac{x-8}{2} = y$$

In inverse notation, the above equation can be written as

$$IT^{-1}(x) = \frac{x-8}{2} \quad \text{or} \quad IT^{-1}(x) = \frac{1}{2}x - 4$$

Substitute the Italian shirt size 36 for x to find the equivalent U.S. shirt size.

$$IT^{-1}(36) = \frac{1}{2}(36) - 4 = 18 - 4 = 14$$

A size 36 Italian shirt is equivalent to a size 14 U.S. shirt.

TRY EXERCISE 46, EXERCISE SET 4.1, PAGE 308

TOPICS FOR DISCUSSION

1. "The notation f^{-1} for an inverse function is also written as $\frac{1}{f}$." Do you agree?
2. "Every odd function has an inverse function." Do you agree?
3. "The domain of f^{-1} is the range of f, and the domain of f is the range of f^{-1}." Do you agree?
4. A student finds it difficult to determine the range of the function $f(x) = 2x/(5x - 3)$. A tutor suggests that the student determine f^{-1} and then use the domain of f^{-1} as the range of f. Is this a feasible approach?
5. The function $y = -x$ is its own inverse. Determine at least three other functions that are their own inverses.

EXERCISE SET 4.1

In Exercises 1 to 8, verify that f and g are inverse functions by showing that $(f \circ g)(x) = x$ and $(g \circ f)(x) = x$.

1. $f(x) = 2x + 1, g(x) = \dfrac{x-1}{2}$
2. $f(x) = \dfrac{1}{2}x - 3, g(x) = 2x + 6$
3. $f(x) = 3x - 5, g(x) = \dfrac{x+5}{3}$
4. $f(x) = -2x + 1, g(x) = -\dfrac{1}{2}x + \dfrac{1}{2}$
5. $f(x) = \dfrac{1}{x+1}, g(x) = \dfrac{1-x}{x}$

6. $f(x) = \frac{1}{x} + 1, g(x) = \frac{1}{x-1}$

7. $f(x) = \sqrt[3]{x-1}, g(x) = x^3 + 1$

8. $f(x) = x^3 - 2, g(x) = \sqrt[3]{x+2}$

In Exercises 9 to 12, find the inverse of the function defined by the given set of ordered pairs.

9. $\{(-3, 1), (-2, 2), (1, 5), (4, -7)\}$

10. $\{(-5, 4), (-2, 3), (0, 1), (3, 2), (7, 11)\}$

11. $\{(0, 1), (1, 2), (2, 4), (3, 8), (4, 16)\}$

12. $\{(1, 0), (10, 1), (100, 2), (1000, 3), (10000, 4)\}$

In Exercises 13 to 28, find the inverse of the given function.

13. $f(x) = 4x + 1$

14. $g(x) = \frac{2}{3}x + 4$

15. $F(x) = -6x + 1$

16. $h(x) = -3x - 2$

17. $j(t) = 2t + 1$

18. $m(s) = -3s + 8$

19. $f(v) = 1 - v^3$

20. $u(t) = 2t^3 + 5$

21. $f(x) = \frac{-3x}{x+4}$

22. $G(x) = \frac{3x}{x-5}$

23. $M(t) = \frac{t-5}{t}$

24. $P(v) = \frac{2v}{v+1}$

25. $r(t) = \frac{1}{t^2}, t < 0$

26. $F(x) = \frac{1}{x}, x > 0$

27. $J(x) = x^2 + 4, x \geq 0$

28. $N(x) = 2x^2 + 1, x \leq 0$

In Exercises 29 to 38, find the inverse of *f*. State the domain and range of both *f* and f^{-1}.

29. $f(x) = x^2 + 3, x \geq 0$

30. $f(x) = x^2 - 4, x \geq 0$

31. $f(x) = \sqrt{x}, x \geq 0$

32. $f(x) = \sqrt{16 - x}, x \leq 16$

33. $f(x) = \sqrt{9 - x^2}, 0 \leq x \leq 3$

34. $f(x) = \sqrt{16 - x^2}, -4 \leq x \leq 0$

35. $f(x) = x^2 - 4x + 1, x \geq 2$

36. $f(x) = x^2 + 6x - 6, x \geq -3$

37. $f(x) = x^2 + 8x - 9, x \leq -4$

38. $f(x) = x^2 - 2x - 2, x \leq 1$

In Exercises 39 to 44, graph f^{-1} if *f* is the function defined by the graph.

39.

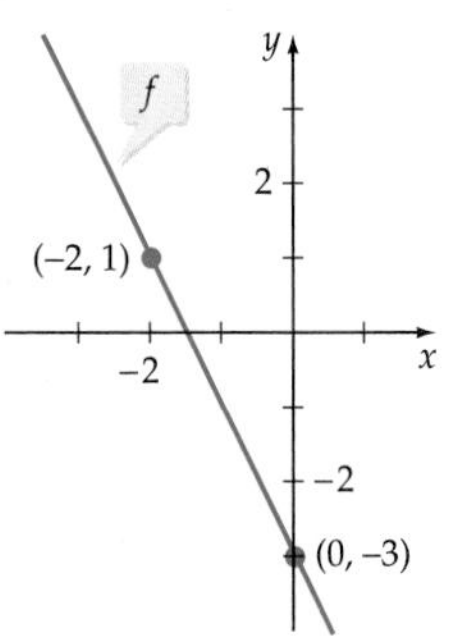

40.

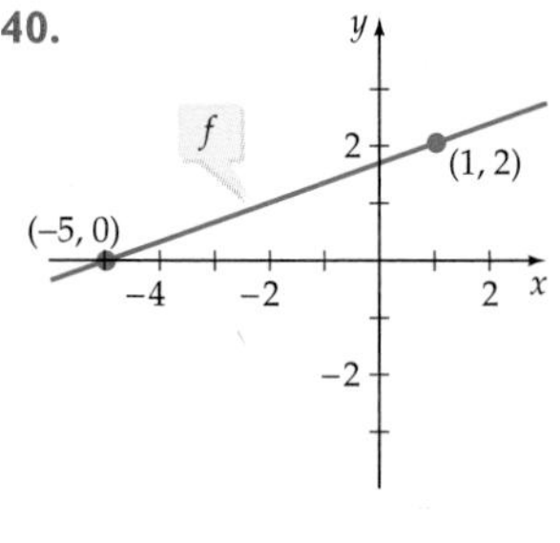

41.

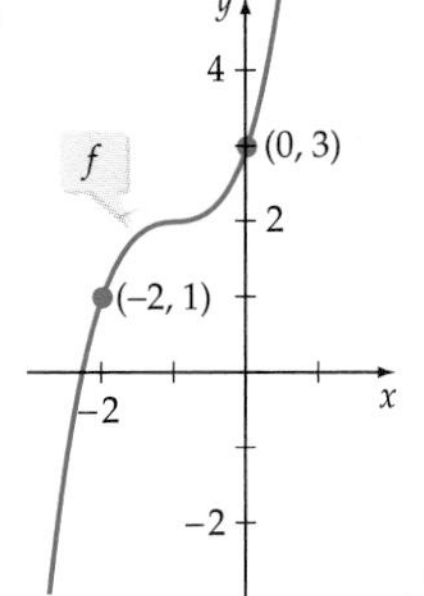

42.

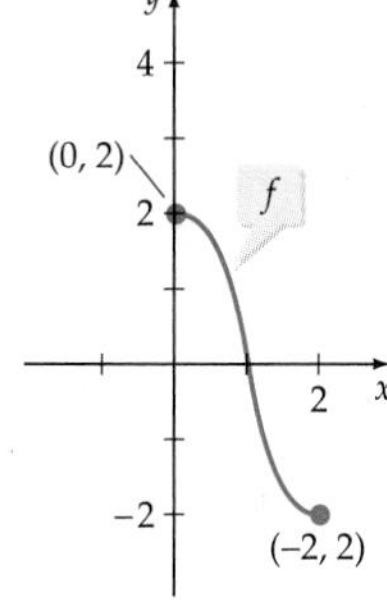

43.

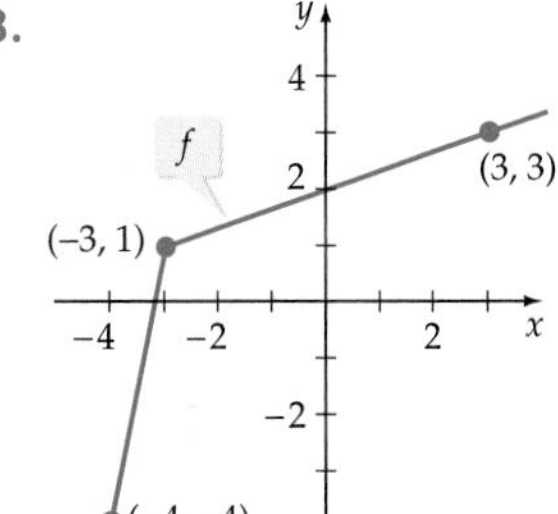

44.

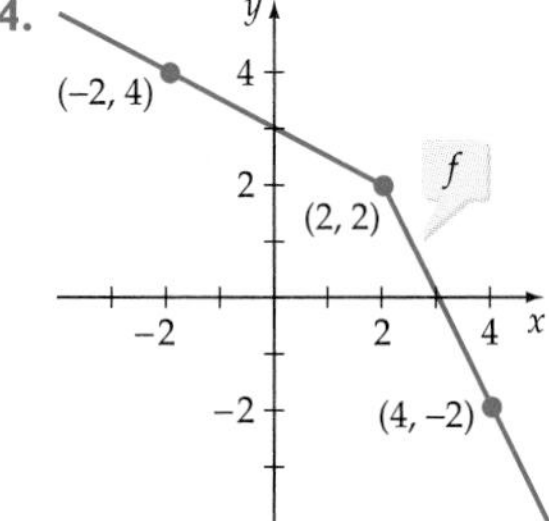

45. **WOMEN'S SHOE SIZES** The following table lists equivalent shoe sizes for women's shoes in the United States and Italy.

Women's Shoe Sizes

U.S.	4	5	6	7	8	9
Italy	36	37	38	39	40	41

a. Determine a linear function s that can be used to convert a U.S. women's shoe size x to its equivalent Italian shoe size $s(x)$.

b. Find s^{-1} and explain how it can be used.

46. **Men's Shoe Sizes** The function $UK(x) = 1.3x - 4.7$ converts a man's shoe size in the United States to the equivalent shoe size in the United Kingdom. For instance, a man who wears a U.S. size 12 shoe would wear a U.K. shoe of size $UK(12) = 1.3(12) - 4.7 = 10.9 \approx 11$

   a. Determine the U.K. shoe size for a man who wears a size 9 U.S. shoe.

   b. Find UK^{-1}. Use UK^{-1} to determine the men's U.S. shoe size that corresponds to a U.K. shoe size of 5.

In Exercises 47 to 54, graph each function f and its inverse f^{-1} on the same coordinate plane. Note that the graphs are symmetric with respect to the line $y = x$.

47. $f(x) = 3x + 3, f^{-1}(x) = \frac{1}{3}x - 1$

48. $f(x) = x - 4, f^{-1}(x) = x + 4$

49. $f(x) = \frac{1}{2}x, f^{-1}(x) = 2x$

50. $f(x) = 2x - 4, f^{-1}(x) = \frac{1}{2}x + 2$

51. $f(x) = x^2 + 2, x \geq 0, f^{-1}(x) = \sqrt{x - 2}, x \geq 2$

52. $f(x) = x^2 - 3, x \geq 0, f^{-1}(x) = \sqrt{x + 3}, x \geq -3$

53. $f(x) = (x - 2)^2, x \leq 2, f^{-1}(x) = 2 - \sqrt{x}, x \geq 0$

54. $f(x) = (x + 3)^2, x \geq -3, f^{-1}(x) = \sqrt{x} - 3, x \geq 0$

SUPPLEMENTAL EXERCISES

In Exercises 55 to 58, find the inverse of the given function.

55. $f(x) = ax + b, a \neq 0$

56. $f(x) = ax^2 + bx + c; a \neq 0, x > -\frac{b}{2a}$

57. $f(x) = \frac{x - 1}{x + 1}, x \neq -1$

58. $f(x) = \frac{2 - x}{x + 2}, x \neq -2$

Only one-to-one functions have inverses that are functions. In Exercises 59 to 66, determine whether or not the given function is a one-to-one function.

59. $f(x) = x^2 + 8$

60. $v(s) = s^2 - 4$

61. $p(t) = \sqrt{9 - t}$

62. $v(t) = \sqrt{16 + t}$

63. $G(x) = -\sqrt{x}$

64. $K(x) = 1 - \sqrt{x - 5}$

65. $F(x) = |x| + x$

66. $T(x) = |x| - x$

In Exercises 67 to 70, assume that the given function has an inverse function.

67. If $f(5) = 2$, find $f^{-1}(2)$.

68. If $v(3) = 11$, find $v^{-1}(11)$.

69. If $s(4) = 60$, find $s^{-1}(60)$.

70. If $F(-8) = 5$, find $F^{-1}(5)$.

71. Graph $f(x) = -x + 3$. Use the graph to explain why f is its own inverse.

PROJECT

1. **SYMMETRY WITH RESPECT TO $y = x$** If the ordered pair (a, b) belongs to the graph of the function f, then (b, a) belongs to the graph of f^{-1}. Prove that the points $P(a, b)$ and $Q(b, a)$ are symmetric with respect to the graph of the line $y = x$ by using the definition of symmetry with respect to a line.

SECTION 4.2 EXPONENTIAL FUNCTIONS AND THEIR GRAPHS

- EXPONENTIAL FUNCTIONS DEFINED 4A
- GRAPHS OF EXPONENTIAL FUNCTIONS
- THE NATURAL EXPONENTIAL FUNCTION
- AN APPLICATION

◆ EXPONENTIAL FUNCTIONS DEFINED

In 1965 Gordon Moore, one of the cofounders of Intel Corporation, observed that the number of transistors on a chip seemed to be doubling every 18 to 24 months. Table 4.1 shows how the number of transistors on various Intel processors has changed over time. (Source: Intel Museum Home Page)

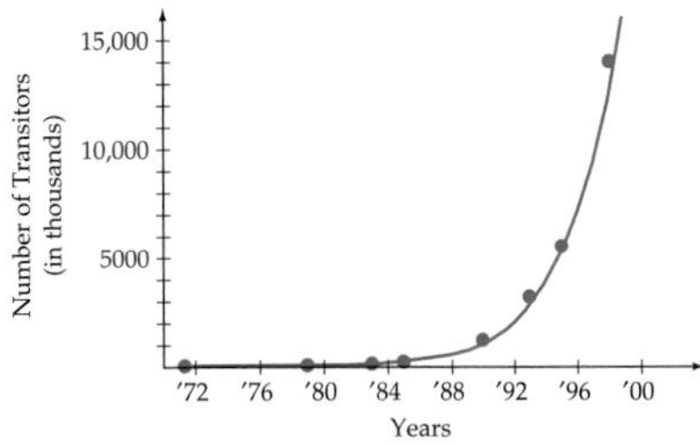

Figure 4.10

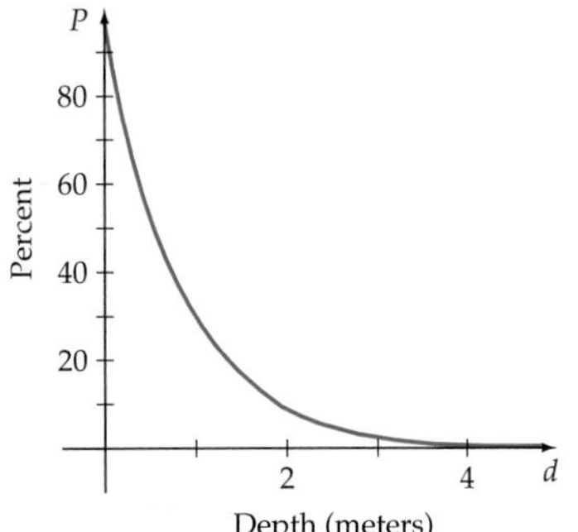

Figure 4.11

Table 4.1

Chip model	4004	8086	80286	80386	80486	P5	P6	P7
Year	1971	1979	1983	1985	1990	1993	1995	1998
Number of transistors, in thousands	2.3	31	110	280	1200	3100	5500	14,000

The curve that approximately passes through the points is a mathematical model of the data (see **Figure 4.10**). The model is based on an *exponential function.*

When light enters water, the intensity of the light decreases with the depth of the water. The graph in **Figure 4.11** shows a model of the decrease in the percent of light available as the depth of the water increases. This model is also based on an exponential function.

Definition of an Exponential Function

The **exponential function** f with base b is defined by

$$f(x) = b^x$$

where $b > 0$, $b \neq 1$, and x is any real number.

In the definition of an exponential function, b, the base, is required to be positive. If the base of an exponential function were a negative number, the value of the function would be a complex number for some values of x. For instance, the value of $f(x) = (-4)^x$ when $x = 1/2$ is

$$f\left(\frac{1}{2}\right) = (-4)^{1/2} = \sqrt{-4} = 2i$$

For this reason, the base of an exponential function is defined to be a positive number. If $b = 1$, then $1^x = 1$ for all values of x. If $b = 0$, then $0^x = 0$ for $x > 0$ and is undefined for $x \leq 0$.

Examples of exponential functions are

$$f(x) = 2^x, \qquad g(x) = \left(\frac{2}{3}\right)^x, \qquad \text{and} \qquad h(x) = \pi^x$$

The value of $f(x) = 2^x$ when $x = 3$ is $f(3) = 2^3 = 8$.

The value of $g(x) = \left(\frac{2}{3}\right)^x$ when $x = -2$ is $g(-2) = \left(\frac{2}{3}\right)^{-2} = \left(\frac{3}{2}\right)^2 = \frac{9}{4}$.

To evaluate an exponential function for an irrational number such as $\sqrt{3}$ or π, an approximation to the value of the function can be obtained by approximating the irrational number. For instance, the value of $f(x) = 4^x$ when $x = \sqrt{5}$ can be approximated by using an approximation of $\sqrt{5}$.

$$f(\sqrt{5}) = 4^{\sqrt{5}} \approx 4^{2.236068} \approx 22.194587$$

Because $f(x) = b^x$ is a real number for both rational and irrational numbers x, the domain of f is all real numbers. Because $b > 0$, $b^x > 0$ for all values of x; therefore, the range of f is the set of positive real numbers. Other properties of the exponential function can be determined from its graph.

◆ GRAPHS OF EXPONENTIAL FUNCTIONS

The graph of $f(x) = 2^x$ is shown in **Figure 4.12.** The coordinates of some of the points on the curve are given in Table 4.2.

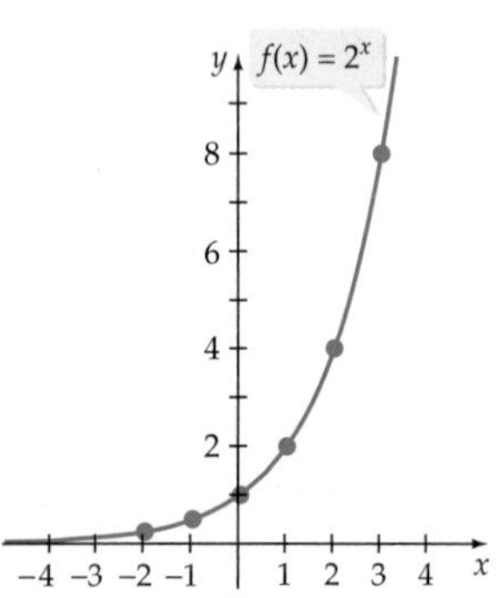

Figure 4.12

Table 4.2

x	$y = f(x) = 2^x$	(x, y)
−2	$f(-2) = 2^{-2} = \frac{1}{4}$	$\left(-2, \frac{1}{4}\right)$
−1	$f(-1) = 2^{-1} = \frac{1}{2}$	$\left(-1, \frac{1}{2}\right)$
0	$f(0) = 2^0 = 1$	$(0, 1)$
1	$f(1) = 2^1 = 2$	$(1, 2)$
2	$f(2) = 2^2 = 4$	$(2, 4)$
3	$f(3) = 2^3 = 8$	$(3, 8)$

> **take note**
>
> A web applet is available to experiment with exponential functions. This applet, Exponential Functions, can be found on our web site at
>
> http://college.hmco.com

Observe the following properties of the graph of an exponential function for which the base, b, is greater than 1. (In **Figure 4.12,** $b = 2 > 1$.)

- The y-intercept is $(0, 1)$.
- As x decreases without bound, that is, as $x \to -\infty$, $f(x) \to 0$.
- The graph of f is the graph of an increasing function. By the horizontal line test, f is a one-to-one function.

Now consider the graph of an exponential function for which the base is between 0 and 1. The graph of $f(x) = \left(\frac{1}{2}\right)^x$ is shown in **Figure 4.13.** The coordinates of some of the points on the curve are given in Table 4.3.

Table 4.3

x	$y = f(x) = \left(\frac{1}{2}\right)^x$	(x, y)
−3	$f(-3) = \left(\frac{1}{2}\right)^{-3} = 8$	$(-3, 8)$
−2	$f(-2) = \left(\frac{1}{2}\right)^{-2} = 4$	$(-2, 4)$
−1	$f(-1) = \left(\frac{1}{2}\right)^{-1} = 2$	$(-1, 2)$
0	$f(0) = \left(\frac{1}{2}\right)^{0} = 1$	$(0, 1)$
1	$f(1) = \left(\frac{1}{2}\right)^{1} = \frac{1}{2}$	$\left(1, \frac{1}{2}\right)$
2	$f(2) = \left(\frac{1}{2}\right)^{2} = \frac{1}{4}$	$\left(2, \frac{1}{4}\right)$

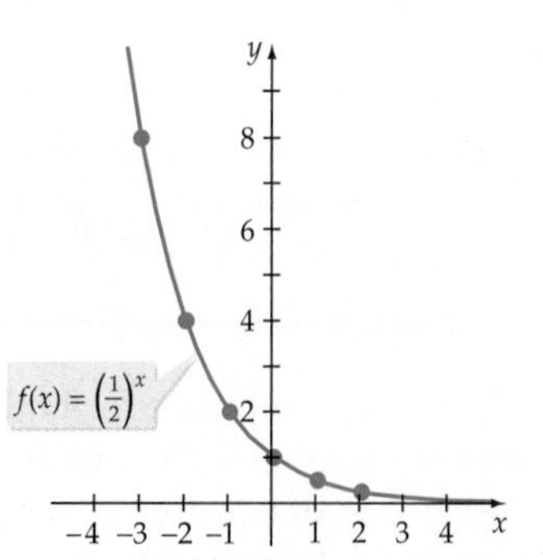

Figure 4.13

Note the following properties of the graph in Figure 4.13:

- The y-intercept is $(0, 1)$.
- As x increases without bound, that is, as $x \to \infty$, $f(x) \to 0$.
- The graph of f is the graph of a decreasing function. By the horizontal line test, f is a one-to-one function.

When using a graphing utility to produce the graph of an exponential function, be sure to observe the Order of Operations Agreement. For instance, to graph $f(x) = \left(\frac{3}{2}\right)^x$, use parentheses around $\frac{3}{2}$. Entering 3/2^X will result in $\frac{3}{2^x}$, which is not the same as $\left(\frac{3}{2}\right)^x$. The graphs of $f(x) = \left(\frac{3}{2}\right)^x$ and $g(x) = \frac{3}{2^x}$ are shown in Figures 4.14 and 4.15.

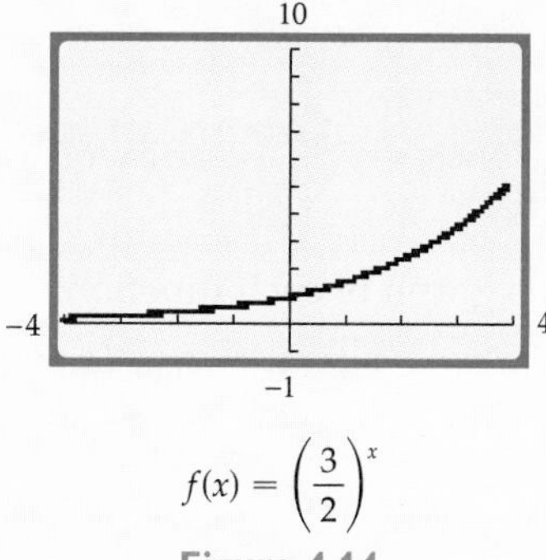

$f(x) = \left(\frac{3}{2}\right)^x$

Figure 4.14

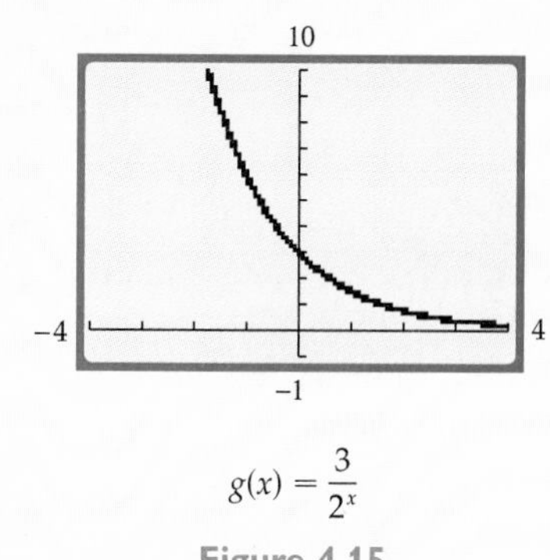

$g(x) = \frac{3}{2^x}$

Figure 4.15

To graph an exponential function over a certain portion of its domain, you may need to use different scales on the x- and y-axes. For example, a graph of $f(x) = 10^x$, for $-2 \le x \le 1$, is shown in Figure 4.16. Observe that each space between tick marks on the y-axis represents a distance of 5 units and that each space between tick marks on the x-axis represents 1 unit.

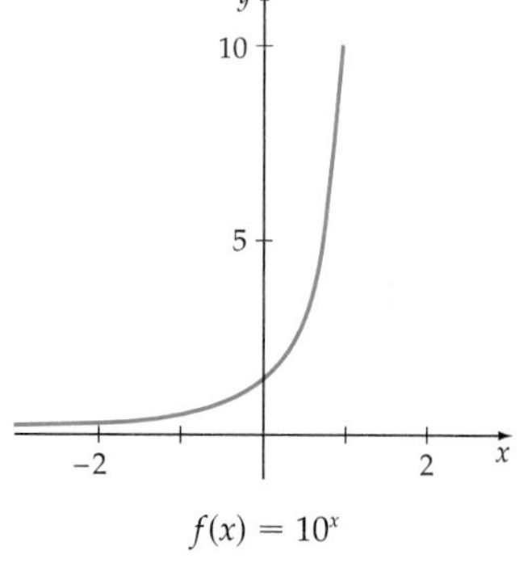

$f(x) = 10^x$

Figure 4.16

Properties of $f(x) = b^x$

For positive real numbers b, $b \neq 1$, the exponential function defined by $f(x) = b^x$ has the following properties:

1. f has the set of real numbers as its domain.
2. f has the set of positive real numbers as its range.
3. f has a graph with a y-intercept of $(0, 1)$.
4. f is a one-to-one function.
5. f has a graph asymptotic to the x-axis. If $b > 1$, $f(x) \to 0$ as $x \to -\infty$. If $0 < b < 1$, $f(x) \to 0$ as $x \to \infty$.
6. f is an increasing function if $b > 1$. See Figure 4.17a. f is a decreasing function if $0 < b < 1$. See Figure 4.17b.

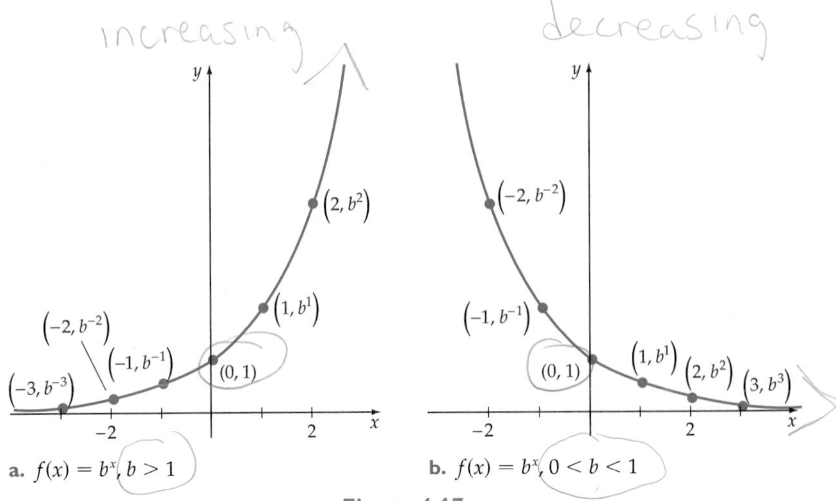

Figure 4.17

EXAMPLE 1 Graph an Exponential Function Using Translations

Sketch the graph of each function.

a. $F(x) = 2^x - 3$ b. $G(x) = 2^{x-3}$

Solution

a. Let $f(x) = 2^x$. Then $F(x) = 2^x - 3 = f(x) - 3$. From Section 2.5, the graph of F is a vertical translation of f down 3 units, as shown in Figure 4.18.

b. Let $f(x) = 2^x$. Then $G(x) = 2^{x-3} = f(x - 3)$. From Section 2.5, the graph of G is a horizontal translation of f to the right 3 units, as shown in Figure 4.19.

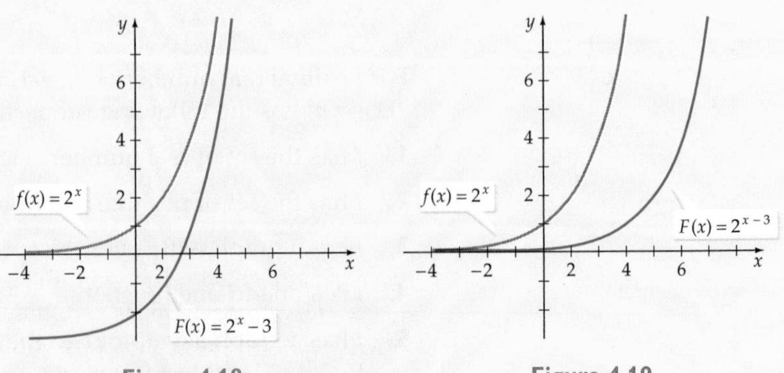

Figure 4.18 Figure 4.19

TRY EXERCISE 36, EXERCISE SET 4.2, PAGE 317

When a graphing utility is used to draw a graph, it is important to look at the graph and ask yourself, "Does this graph have the characteristics of the function I intended to graph?" What you must guard against is incorrectly entering the function. The graphing utility will graph only what you enter—not what you intended to enter.

For instance, suppose we want to draw the graph of $f(x) = 3(2^{-x^2})$ and produce the graph shown in **Figure 4.20.** First note that

$$f(-x) = 3(2^{-(-x)^2}) = 3(2^{-x^2}) = f(x)$$

Thus f is an even function, and its graph should be symmetric with respect to the y-axis. This appears to be the case from our graph. Next, if we write f as

$$f(x) = \frac{3}{2^{x^2}}$$

we observe that as $|x|$ increases without bound, the denominator increases without bound. Therefore, $f(x)$ is approaching zero. This is also consistent with our graph. For another observation, recall that the value of a fraction with a constant numerator is as large as possible when the denominator is as small as possible. For our function, the smallest denominator occurs when $x = 0$. In that case $f(0) = 3$, which is again consistent with our graph. It appears that we have entered the function correctly and produced the desired graph. As a final check, we can evaluate the function for various values of x and compare those values to values found by using the TRACE feature of the graphing utility.

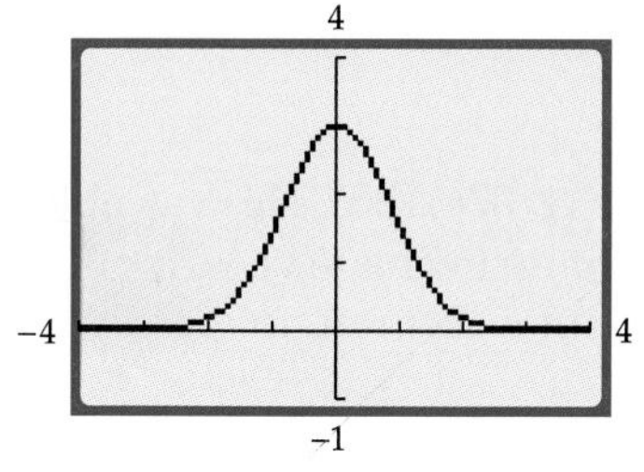

$f(x) = 3 \cdot 2^{-x^2}$

Figure 4.20

EXAMPLE 2 Graph a Function of the Form $f(x) = b^{p(x)}$

Use a graphing utility to graph $f(x) = 2^{|x|}$. Then use your knowledge of functions to verify the accuracy of the graph.

Solution

The graph as drawn with a graphing utility is shown in **Figure 4.21.** The minimum value of f occurs when $|x| = 0$, which means $x = 0$. At $x = 0$, $f(x) = 2^{|0|} = 1$. This is consistent with the graph. As shown below, f is an even function.

$$f(x) = 2^{|x|}$$
$$f(-x) = 2^{|-x|} = 2^{|x|} = f(x)$$

Because f is an even function, the graph of f should be symmetric with respect to the y-axis. This is consistent with the graph of f in **Figure 4.21.** As a final check, evaluate f at a few values of x and compare them to the corresponding values you find by tracing along the curve.

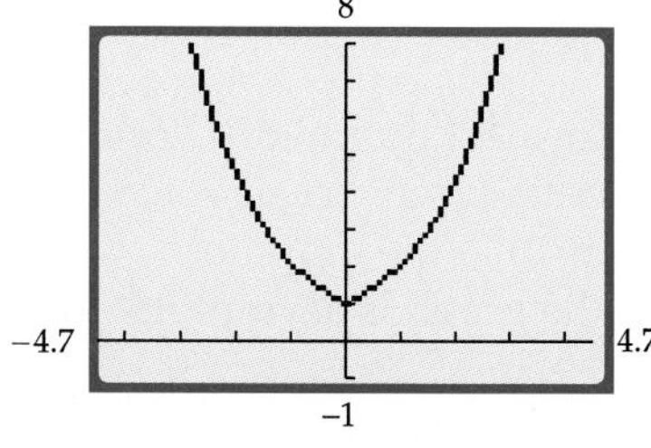

$f(x) = 2^{|x|}$

Figure 4.21

TRY EXERCISE 40, EXERCISE SET 4.2, PAGE 317

◆ THE NATURAL EXPONENTIAL FUNCTION

Table 4.4

Value of n	Value of $\left(1+\frac{1}{n}\right)^n$
1	2
10	2.59374246
100	2.704813829
1000	2.716923932
10,000	2.718145927
100,000	2.718268237
1,000,000	2.718280469
10,000,000	2.718281693

The irrational number π is often used in applications that involve circles. Another irrational number called e is useful for applications that involve the growth of a population or radioactive decay. Using techniques developed in calculus, it can be shown that as $n \to \infty$,

$$\left(1+\frac{1}{n}\right)^n \to e$$

The letter e was chosen in honor of Leonhard Euler (1707–1783). He was able to compute e to several places by using large values of n to evaluate $(1 + 1/n)^n$. The entries in Table 4.4 illustrate the process. The value of e accurate to eight places is 2.71828183.

The Natural Exponential Function

For all real numbers x, the function defined by

$$f(x) = e^x$$

is called the **natural exponential function.**

To evaluate e^x for specific values of x, you use a calculator with an $\boxed{e^x}$ key. For example,

$$e^2 \approx 7.389056099$$
$$e^{4.21} \approx 67.35653981$$
$$e^{-1.8} \approx 0.165298888$$

To graph the natural exponential function, use a calculator to approximate e^x for the desired domain values. Then plot the resulting points and connect them with a smooth curve.

EXAMPLE 3 Graph the Natural Exponential Function

Graph: $f(x) = e^x$

Solution

The values in Table 4.5 have been rounded to the nearest hundredth. Plot the points and then connect the points with a smooth curve. Because $e > 1$, we know by the properties of exponential functions that the graph of $f(x) = e^x$ is an increasing function. To the far left the graph is asymptotic to the x-axis. The y-intercept is $(0, 1)$. See **Figure 4.22.**

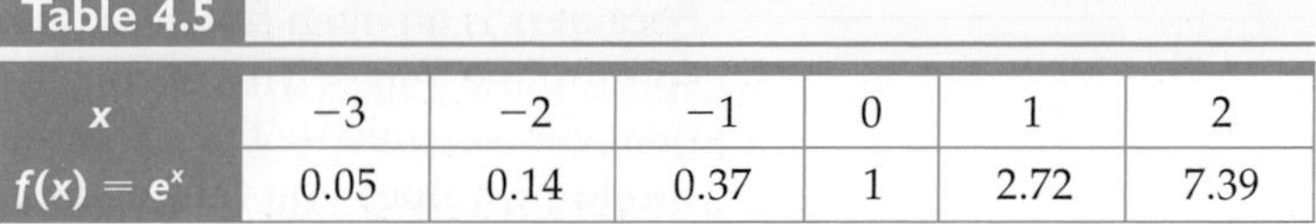

Table 4.5

x	−3	−2	−1	0	1	2
$f(x) = e^x$	0.05	0.14	0.37	1	2.72	7.39

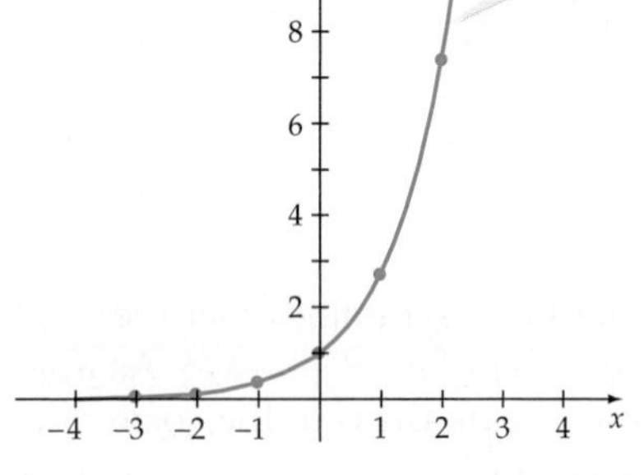

$f(x) = e^x$

Figure 4.22

TRY EXERCISE 44, EXERCISE SET 4.2, PAGE 317

Note in Figure 4.23 how the graph of $f(x) = e^x$ compares with the graphs of $g(x) = 2^x$ and $h(x) = 3^x$. You may have anticipated that the graph of f would be between the graph of g and that of h because e is between 2 and 3.

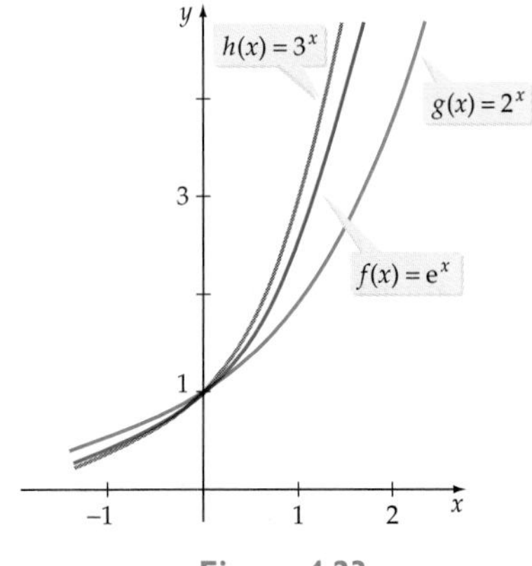

Figure 4.23

EXAMPLE 4 Graph a Combination of Exponential Functions

Use a graphing utility to graph $S(x) = \dfrac{e^x - e^{-x}}{2}$.

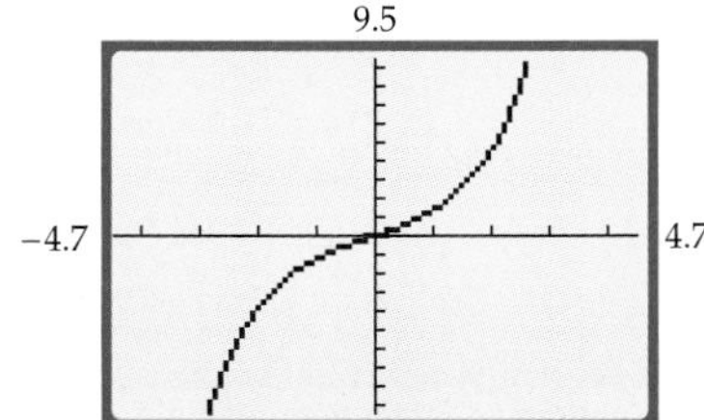

$S(x) = \dfrac{e^x - e^{-x}}{2}$

Figure 4.24

Solution

The graph as drawn with a graphing utility is shown in Figure 4.24. The graph is symmetric about the origin because S, as shown below, is an odd function.

$$S(x) = \frac{e^x - e^{-x}}{2}$$

$$S(-x) = \frac{e^{-x} - e^{-(-x)}}{2} = \frac{e^{-x} - e^x}{2} = -\left(\frac{e^x - e^{-x}}{2}\right) = -S(x)$$

As another check, evaluate S for a few values of x and compare them to the corresponding values you find by tracing along the curve. For instance, $S(0) = \dfrac{e^0 - e^{-0}}{2} = \dfrac{1 - 1}{2} = 0$. Thus, $(0, 0)$ is on the graph. This result also is consistent with the graph.

TRY EXERCISE 48, EXERCISE SET 4.2, PAGE 317

Recall that a real zero of a function is a real number x for which $f(x) = 0$, and when $f(x) = 0$, the graph of f intersects the x-axis. Thus the real number zeros of a function f can be approximated by using a graphing calculator to determine the x-intercepts of its graph. This technique will be applied to an exponential function in Example 5. An algebraic method of finding real zeros of exponential functions is given later in this chapter.

EXAMPLE 5 Determine the Zero of a Function

Use a graphing utility to determine, to the nearest thousandth, the zero of $f(x) = -\frac{1}{3}e^x + 2$.

Solution

Graph f and use the features of your graphing utility to find the x-intercept to the nearest thousandth. The x-coordinate of the x-intercept in Figure 4.25 is approximately 1.792.

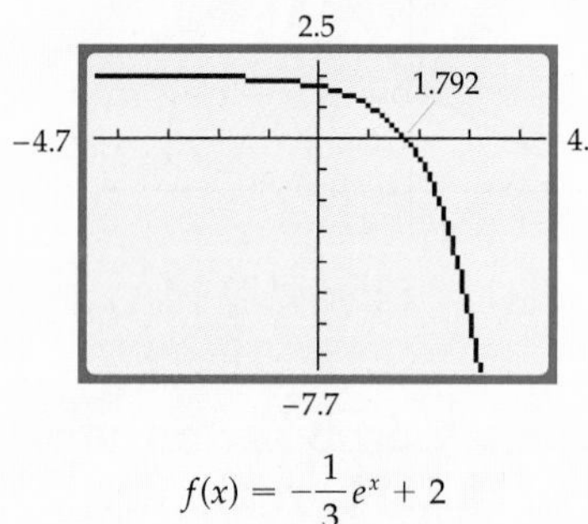

$f(x) = -\frac{1}{3}e^x + 2$

Figure 4.25

The zero of f to the nearest thousandth is 1.792.

Verify the Solution

Evaluating $f(x)$ when $x = 1.792$, shows that $f(x) \approx 0$. Thus 1.792 approximates a zero of f.

$$\begin{aligned} f(x) &= -\frac{1}{3}e^x + 2 \\ &\approx -\frac{1}{3}e^{1.792} + 2 \\ &\approx -0.00048112 \\ &\approx 0 \end{aligned}$$

TRY EXERCISE 50, EXERCISE SET 4.2, PAGE 317

◆ AN APPLICATION

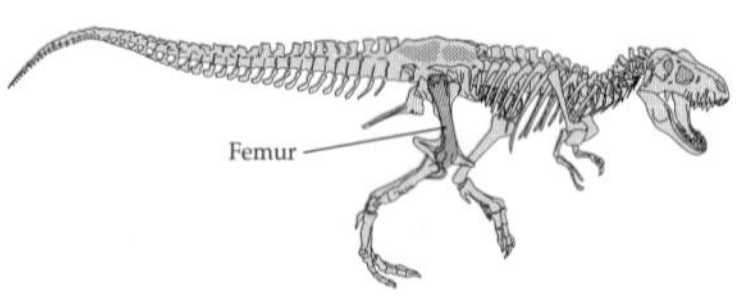

The *Tyrannosaurus rex* named Sue. The Field Museum, Chicago.

Paleontologists estimate the weight of a dinosaur by measuring the circumference of its femur. The weight, in kilograms, of the dinosaur is given by $W = 0.00031C^{2.9}$, where C is the circumference of the femur in millimeters.

EXAMPLE 6 Estimate the Weight of a Dinosaur

Estimate the weight of the *Tyrannosaurus rex* named Sue, for which the circumference of the femur is approximately 332 millimeters.

Solution

$W = 0.00031C^{2.9}$

$W = 0.00031(332^{2.9}) \approx 6348$

The estimated weight of the dinosaur named Sue is 6348 kilograms.

TRY EXERCISE 58, EXERCISE SET 4.2, PAGE 317

TOPICS FOR DISCUSSION

1. The definition of an exponential function, as $f(x) = b^x$, requires that $b > 0$ and $b \neq 1$. Discuss why these conditions are imposed.

2. Discuss the properties of the graph of $f(x) = b^x$ when $b > 1$.
3. Discuss the properties of the graph of $f(x) = b^x$ when $0 < b < 1$.
4. What is the base of the natural exponential function? How is it calculated? What is its approximate value?

EXERCISE SET 4.2

In Exercises 1 to 12, evaluate each power. Round to the nearest ten-thousandth.

1. $3^{\sqrt{2}}$
2. $5^{\sqrt{3}}$
3. $10^{\sqrt{7}}$
4. $10^{\sqrt{11}}$
5. $\sqrt{3}^{\sqrt{2}}$
6. $\sqrt{5}^{\sqrt{7}}$
7. $e^{5.1}$
8. $e^{-3.2}$
9. $e^{\sqrt{3}}$
10. $e^{\sqrt{5}}$
11. $e^{-0.031}$
12. $e^{-0.42}$

In Exercises 13 to 24, evaluate each functional value given that $f(x) = 3^x$ and $g(x) = e^x$. Round to the nearest ten-thousandth.

13. $f(\sqrt{15})$
14. $f(\pi)$
15. $f(e)$
16. $f(-\sqrt{15})$
17. $g(\sqrt{7})$
18. $g(\pi)$
19. $g(e)$
20. $g(-3.4)$
21. $f[g(2)]$
22. $f[g(-1)]$
23. $g[f(2)]$
24. $g[f(-1)]$

In Exercises 25 to 48, sketch the graph of each function.

25. $f(x) = 3^x$
26. $f(x) = 4^x$
27. $f(x) = \left(\frac{3}{2}\right)^x$
28. $f(x) = \left(\frac{4}{3}\right)^x$
29. $f(x) = \left(\frac{1}{3}\right)^x$
30. $f(x) = \left(\frac{2}{3}\right)^x$
31. $f(x) = \left(\frac{1}{2}\right)^{-x}$
32. $f(x) = \left(\frac{1}{3}\right)^{-x}$
33. $f(x) = \frac{5^x}{2}$
34. $f(x) = \frac{10^x}{10}$
35. $f(x) = 2^{x+2}$
36. $f(x) = 2^{x+3}$
37. $f(x) = 3^x - 1$
38. $f(x) = 3^x + 1$
39. $f(x) = 3^{x^2}$
40. $f(x) = 2^{-|x|}$
41. $f(x) = \frac{3^x + 3^{-x}}{2}$
42. $f(x) = 4 \cdot 3^{-x^2}$
43. $f(x) = e^{-x}$
44. $f(x) = 2e^x$
45. $f(x) = -e^x$
46. $f(x) = 0.5e^{-x}$
47. $f(x) = e^{-x^2}$
48. $f(x) = \frac{e^x + e^{-x}}{2}$

In Exercises 49 to 56, use a graphing utility to determine the zero of f to the nearest hundredth.

49. $f(x) = 2^x - 3$
50. $f(x) = 3^{-x} - 4$
51. $f(x) = 1 - 2e^{-x}$
52. $f(x) = 3 - 4e^x$
53. $f(x) = e^x + x - 3$
54. $f(x) = 2x - 2^{-\frac{1}{2}x}$
55. $f(x) = 3^x + 2x - 4$
56. $f(x) = e^x + x - 5$

57. **INTENSITY OF LIGHT** The percent, $I(x)$, of the original intensity of light striking the surface of a lake that is available x feet below the surface of the lake is given by $I(x) = 100e^{-0.95x}$. What percent of the light, to the nearest tenth of a percent, is available 2 feet below the surface of the lake?

58. **RADIATION** Lead shielding is used to contain radiation. The percent of a certain radiation that can penetrate x millimeters of lead shielding is given by $I(x) = 100e^{-1.5x}$. What percent of radiation, to the nearest tenth of a percent, will penetrate a lead shield that is 1 millimeter thick?

59. **INTERNET CONNECTIONS** Data from Forrester Research suggest that the number of broadband [cable and digital subscriber line (DSL)] connections to the Internet can be modeled by $f(x) = 1.353(1.9025)^x$, where x is the number of years after 1998 and $f(x)$ is the number of connections in millions. How many broadband Internet connections, to the nearest million, does this equation predict will exist in 2004?

60. **TEMPERATURE OF SOUP** Soup that is at a temperature of 170°F is poured into a bowl and placed in a room with a constant temperature. The temperature of the soup decreases according to the model given by $T(t) = 75 + 95e^{-0.25t}$.
   a. What is the temperature, to the nearest tenth of a degree, of the soup after 2 minutes?
   b. What is the temperature of the room?

61. Graph $f(x) = x^x$ on $(0, 3]$. Estimate
   a. the minimum value of f to the nearest ten thousandth on this interval
   b. the behavior of f as x approaches 0 from the right

62. **THE NORMAL DISTRIBUTION CURVE** Graph the **normal distribution curve** defined by

$$f(x) = \frac{1}{\sqrt{2\pi}}e^{-x^2/2}$$

Estimate the maximum value of f (to the nearest tenth).

63. **BACTERIAL GROWTH** The number of bacteria present in a culture is given by $N(t) = 10{,}000(2^t)$, where $N(t)$ is the

number of bacteria present after t hours. Find the number of bacteria present for each value of t.

a. $t = 1$ hour b. $t = 2$ hours c. $t = 5$ hours

64. **OIL PRODUCTION** The production function for an oil well is given by the function $B(t) = 100{,}000(e^{-0.2t})$, where $B(t)$ is the number of barrels of oil the well can produce per month after t years. Find the number of barrels of oil the well can produce per month for each value of t.

a. $t = 1$ year b. $t = 2$ years c. $t = 5$ years

65. **CALCULATING MONTHLY CAR PAYMENTS** A formula to determine the monthly payment (PMT) for a car loan, home mortgage, or other installment loan is given by

$$\text{PMT} = P\left(\frac{i/12}{1 - (1 + i/12)^{-n}}\right)$$

where P (called the present value) is the amount borrowed, i is the annual interest rate, and n is the total number of payments.

a. An accountant purchases a car and secures a loan for \$9000 at an annual interest rate of 10% for a term of 4 years. Find the monthly car payment.

b. If the accountant makes all 48 payments, how much money will be repaid?

c. How much of the total repaid is interest?

66. **CALCULATING MONTHLY CAR PAYMENTS** Using the formula in Exercise 65, answer the following questions.

a. A nurse purchases a car and secures a loan for \$15,000 at an annual interest rate of 9% for a term of 5 years. Find the monthly car payment.

b. If the nurse makes all 60 payments, how much money will be repaid?

c. How much of the total repaid is interest?

67. **CALCULATING THE PAYOFF OF A CAR LOAN** The formula in Exercise 65 can be solved for P with the following result:

$$P = \text{PMT}\left(\frac{1 - (1 + i/12)^{-n}}{i/12}\right)$$

This formula gives the present value (amount owed on a loan) for the *remaining* n payments of an installment loan.

a. Suppose a person has a loan that has a monthly payment of \$258, has an annual interest rate of 9%, and runs for 5 years. What is the loan amount?

b. How much is owed after the borrower has made 12 payments?

c. After how many months will the present value first be less than one-half the original present value?

68. **MUSICAL SCALES** Starting on the left of a standard 88-key piano, the frequency of the nth note is given by $f(n) = (27.5)2^{(n-1)/12}$.

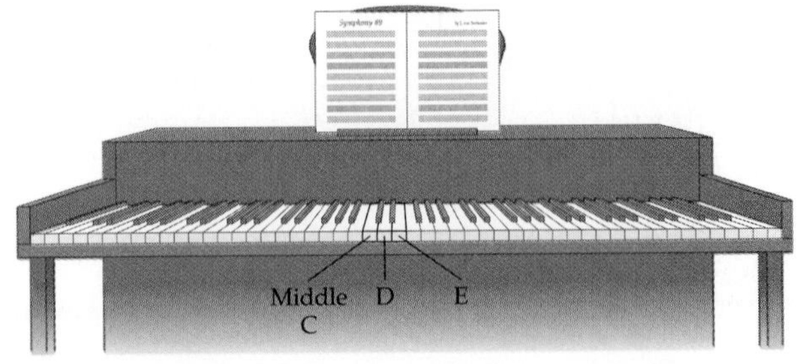

a. Using this formula, determine the frequency of middle C, key number 40 on an 88-key piano.

b. Is the difference in frequency between middle C and the next note, D, the same as the difference in frequency between D and the next note, E? Explain why this is so.

c. Some animals make a sound by blowing air through a nasal passage. The frequency f of the sound is approximately given by $f = 170/x$, where x is the length of the nasal passage in meters. Estimate the length of the nasal passage of an animal that emits a sound that has the frequency of the 70th key (from the left) on a piano.

SUPPLEMENTAL EXERCISES

In Exercises 69 to 76, graph f. State the domain and range of f using interval notation. When necessary, estimate values to the nearest tenth. Also state whether f is an even function, an odd function, or neither an even nor an odd function.

69. $f(x) = \dfrac{e^x - e^{-x}}{e^x + e^{-x}}$ 70. $f(x) = x^2e^x$

71. $f(x) = \dfrac{4x^2}{e^{|x|}}$ 72. $f(x) = \sqrt{\dfrac{|x|}{1 + e^x}}$

73. $f(x) = \dfrac{e^{|x|}}{1 + e^x}$ 74. $f(x) = 1 + e^{(x^3 - x^2 - 2x)}$

75. $f(x) = \sqrt{1 - e^x}$ 76. $f(x) = \sqrt{e^x - e^{-x}}$

In Exercises 77 and 78, use $f(x) = e^x$ and $g(x) = e^{-x}$ to graph the given equations. State the domain and range of y using interval notation.

77. a. $y = (f + g)(x)$ b. $y = (f - g)(x)$

78. a. $y = (f \cdot g)(x)$ b. $y = (f/g)(x)$

79. Evaluate $h(x) = (-4)^x$ for each value of x.

a. $x = 1$ b. $x = 2$ c. $x = \dfrac{3}{2}$

d. Explain what is meant by the statement that h is not a *real-valued* exponential function.

80. Graph $j(x) = 1^x$. Explain why j is not an exponential function.

81. Graph $f(x) = e^x$, and then sketch the graph of f reflected about the graph of the line given by $y = x$.

82. Graph $g(x) = 10^x$, and then sketch the graph of g reflected about the graph of the line given by $y = x$.

83. Prove that the hyperbolic sine function

$$\sinh(x) = \frac{e^x - e^{-x}}{2}$$

is an odd function.

84. Prove that $G(x) = e^x$ is neither an odd function nor an even function.

85. Which of the numbers e^π and π^e is larger?

86. Let $f(x) = x^{(x^x)}$ and $g(x) = (x^x)^x$. Which is larger, $f(3)$ or $g(3)$?

In Exercises 87 to 92, determine the *a.* vertical and horizontal asymptotes, *b.* x-intercepts, and *c.* graph of *f*.

87. $f(x) = \dfrac{2}{1 + e^x}$

88. $f(x) = \dfrac{4}{1 + 2^x}$

89. $f(x) = \dfrac{x + 1}{1 - 2^x}$

90. $f(x) = \dfrac{x - 1}{2 - 2e^x}$

91. $f(x) = \dfrac{1 - 2x}{1 - e^x}$

92. $f(x) = \dfrac{1 - x}{1 - 2^x}$

PROJECTS

1. **PROPERTIES OF AN EXPONENTIAL FUNCTION** Graph $f(x) = b^x$ for $b = 2, 3, 4$, and 10. On the basis of the graphs, complete the following statements.

   a. As $x \to \infty$, $f(x) \to$ ___?___. b. As $x \to -\infty$, $f(x) \to$ ___?___.

   c. Let $x_1 = 1$ and $x_2 = 3$. For each value of b, calculate the slope of the line through $P_1(x_1, f(x_1))$ and $P_2(x_2, f(x_2))$. On the basis of your answers, complete the following statement: As b increases, the slope of the line through P_1 and P_2 ________.

   d. Calculate the slope of the line through $(x_1, f(x_1))$ and $(x_2, f(x_2))$, where x_1 and x_2 are consecutive integers. Is the slope of each line the same? If not, what relationship exists between the successive slopes?

SECTION 4.3 LOGARITHMIC FUNCTIONS AND THEIR GRAPHS

- LOGARITHMIC FUNCTIONS DEFINED 4A
- GRAPHS OF LOGARITHMIC FUNCTIONS 
- DOMAINS OF LOGARITHMIC FUNCTIONS
- COMMON AND NATURAL LOGARITHMS
- APPLICATIONS OF LOGARITHMIC FUNCTIONS

◆ LOGARITHMIC FUNCTIONS DEFINED

Every exponential function of the form $g(x) = b^x$ is a one-to-one function and therefore has an inverse function. Sometimes we can determine the inverse of a function represented by an equation by interchanging the variables and then solving for the dependent variable. If we attempt to use this procedure for $g(x) = b^x$, we obtain

$$g(x) = b^x$$
$$y = b^x$$
$$x = b^y \qquad \text{• Interchange the variables.}$$

None of our previous methods can be used to solve the equation $x = b^y$ for the exponent y. Thus we need to develop a new procedure. One method would be merely to write

$$y = \text{the power of } b \text{ that produces } x$$

Although this would work, it is not very concise. We need a compact notation to represent "y is the power of b that produces x." This more compact notation is given in the following definition.

MATH MATTERS

Logarithms were developed independently by Jobst Burgi (1552–1632) and John Napier (1550–1617) as a means of simplifying the calculations of astronomers. The idea was to devise a method by which two numbers could be multiplied by performing additions. Napier is usually given credit for devising logarithms, because he published his result first.

Definition of Logarithm

If $x > 0$ and b is a positive constant ($b \neq 1$), then

$$y = \log_b x \qquad \text{if and only if} \qquad b^y = x$$

The notation $\log_b x$ is read "the logarithm (or log) base b of x."

The function defined by $f(x) = \log_b x$ is a **logarithmic function** with base b.

This function is the inverse of the exponential function $g(x) = b^x$.

take note

The notation $\log_b x$ replaces the phrase "the power of b that produces x." For instance, "3 is the power of 2 that produces 8" is abbreviated $3 = \log_2 8$. In your work with logarithms, remember that a logarithm is an *exponent*.

It is essential to remember that $f(x) = \log_b x$ is the inverse function of $g(x) = b^x$. Because these functions are inverses of one another, and because functions that are inverse of one another have the property that $f(g(x)) = x$ and $g(f(x)) = x$, we have the following important relationships.

Inverse of Logarithmic and Exponential Functions

Let $g(x) = b^x$ and $f(x) = \log_b x$ ($x > 0, b > 0, b \neq 1$). Then

$$g(f(x)) = b^{\log_b x} = x \qquad \text{and} \qquad f(g(x)) = \log_b b^x = x$$

As an example of these relationships, let $g(x) = 2^x$ and $f(x) = \log_2 x$. Then

$$2^{\log_2 x} = x \qquad \text{and} \qquad \log_2 2^x = x$$

The equations

$$y = \log_b x \qquad \text{and} \qquad b^y = x$$

are different ways of expressing the same concept.

Exponential Form and Logarithmic Form

The exponential form of $y = \log_b x$ is $b^y = x$.

The logarithmic form of $b^y = x$ is $y = \log_b x$.

These concepts are illustrated in the next two examples.

EXAMPLE 1 Change from Logarithmic to Exponential Form

Write each equation in its exponential form.

a. $2 = \log_7 x$ b. $3 = \log_{10}(x + 8)$ c. $\log_5 125 = x$

Solution

Use the definition $y = \log_b x$ if and only if $b^y = x$.

a.

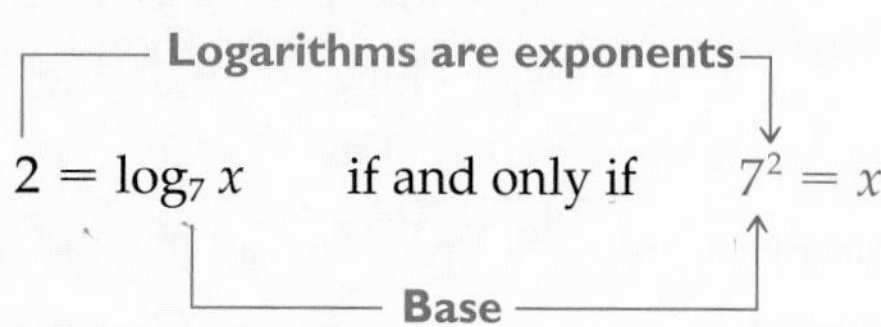

b. $3 = \log_{10}(x + 8)$ if and only if $10^3 = (x + 8)$.

c. $\log_5 125 = x$ if and only if $5^x = 125$.

TRY EXERCISE 2, EXERCISE SET 4.3, PAGE 327

EXAMPLE 2 Change from Exponential to Logarithmic Form

Write each equation in its logarithmic form.

a. $3^2 = 9$ b. $5^3 = x$ c. $a^b = c$

Solution

The logarithmic form of $b^y = x$ is $y = \log_b x$.

a.

b. $5^3 = x$ if and only if $3 = \log_5 x$.

c. $a^b = c$ if and only if $b = \log_a c$.

TRY EXERCISE 12, EXERCISE SET 4.3, PAGE 327

Basic Logarithmic Properties

1. $\log_b b = 1$ 2. $\log_b 1 = 0$ 3. $\log_b (b^p) = p$

A proof of each of the basic logarithmic properties follows directly from the definition of a logarithm.

Proofs:

- $\log_b b = 1$ because $b = b^1$
- $\log_b 1 = 0$ because $1 = b^0$
- $\log_b (b^p) = p$ because $b^p = b^p$

◆

EXAMPLE 3 Apply the Basic Logarithmic Properties

Evaluate each of the following logarithms.

a. $\log_8 1$ b. $\log_3 3$ c. $\log_2 (2^4)$

Solution

a. Property 2 indicates that $\log_8 1 = 0$.

b. Property 1 indicates that $\log_3 3 = 1$.

c. Property 3 indicates that $\log_2 (2^4) = 4$.

TRY EXERCISE 22, EXERCISE SET 4.3, PAGE 327

Some logarithms can be evaluated by remembering that a logarithm is an exponent. For instance, $\log_5 25 = 2$ because $5^2 = 25$.

EXAMPLE 4 Evaluate Logarithms

Evaluate each of the following:

a. $\log_{10} 100$ b. $\log_4 64$ c. $\log_7 \dfrac{1}{49}$

Solution

a. $\log_{10} 100 = 2$ because $10^2 = 100$

b. $\log_4 64 = 3$ because $4^3 = 64$

c. $\log_7 \dfrac{1}{49} = -2$ because $7^{-2} = \dfrac{1}{7^2} = \dfrac{1}{49}$

TRY EXERCISE 30, EXERCISE SET 4.3, PAGE 327

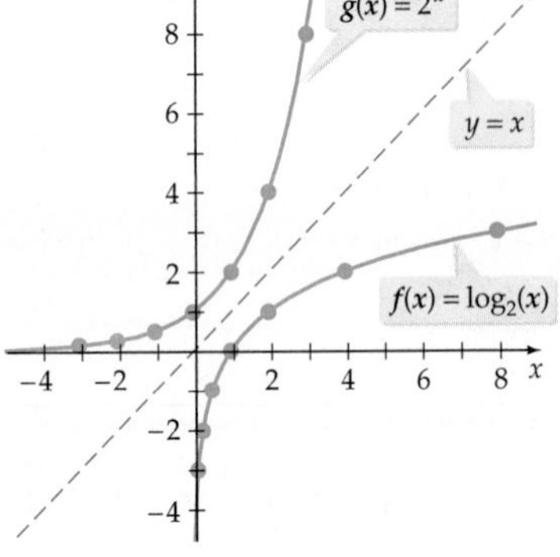

◆ GRAPHS OF LOGARITHMIC FUNCTIONS

Because $f(x) = \log_b x$ is the inverse function of $g(x) = b^x$, the graph of f is a reflection with respect to the line $y = x$. The graph of $g(x) = 2^x$ is shown in **Figure 4.26**, and Table 4.6 gives some of the ordered pairs that belong to the function.

Figure 4.26

Table 4.6

x	-3	-2	-1	0	1	2	3
$g(x) = 2^x$	$\frac{1}{8}$	$\frac{1}{4}$	$\frac{1}{2}$	1	2	4	8

The graph of the inverse of g, which is $f(x) = \log_2 x$, is also shown in **Figure 4.26.** Some of the ordered pairs of f are given in Table 4.7. Note that the x and y coordinates of the ordered pairs are the reverse of those for g and that the graph of f is a reflection of the graph of g with respect to the line $y = x$.

Table 4.7

x	$\frac{1}{8}$	$\frac{1}{4}$	$\frac{1}{2}$	1	2	4	8
$f(x) = \log_2 x$	-3	-2	-1	0	1	2	3

The graph of a logarithmic function can be drawn by first rewriting the function in its exponential form. This procedure is illustrated in Example 5.

EXAMPLE 5 Graph a Logarithmic Function

Graph $f(x) = \log_3 x$.

Solution

To graph $f(x) = \log_3 x$, consider the equivalent exponential equation $x = 3^y$. Because this equation is solved for x, choose values of y and calculate the corresponding values of x, as shown in Table 4.8.

Table 4.8

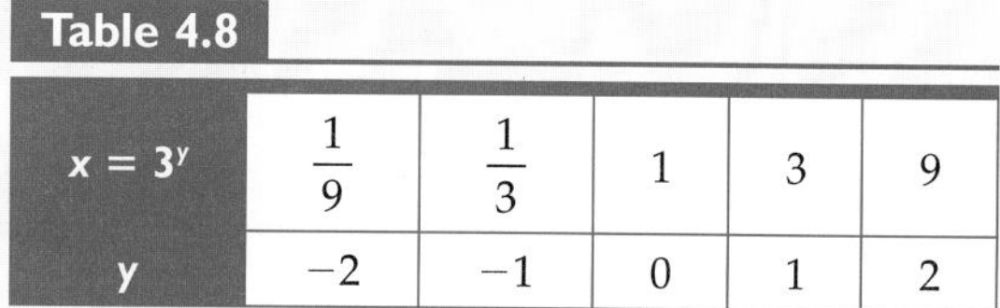

$x = 3^y$	$\frac{1}{9}$	$\frac{1}{3}$	1	3	9
y	-2	-1	0	1	2

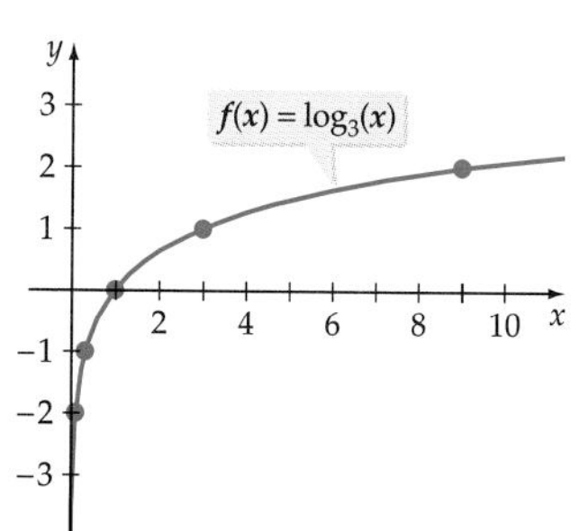

Figure 4.27

Now plot the ordered pairs and connect the points with a smooth curve, as shown in **Figure 4.27.**

TRY EXERCISE 32, EXERCISE SET 4.3, PAGE 327

In a similar manner, we can draw the graph of a logarithmic function with a fractional base. For instance, consider $y = \log_{2/3} x$. Rewriting this in exponential form yields $(2/3)^y = x$. Choose values of y and calculate the corresponding values of x (see Table 4.9), plot the points corresponding to the ordered pairs, and then draw a smooth graph through the points, as shown in **Figure 4.28.**

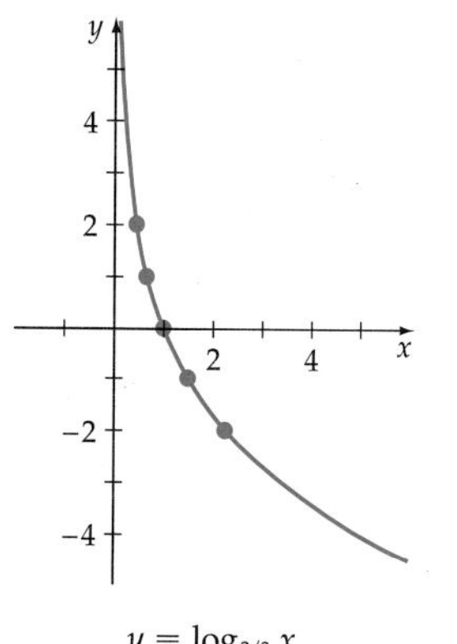

$y = \log_{2/3} x$

Figure 4.28

Table 4.9

$x = \left(\frac{2}{3}\right)^y$	$\left(\frac{2}{3}\right)^{-2} = \frac{9}{4}$	$\left(\frac{2}{3}\right)^{-1} = \frac{3}{2}$	$\left(\frac{2}{3}\right)^{0} = 1$	$\left(\frac{2}{3}\right)^{1} = \frac{2}{3}$	$\left(\frac{2}{3}\right)^{2} = \frac{4}{9}$
y	-2	-1	0	1	2

take note

A web applet is available to experiment with logarithmic functions. This applet, Logarithm, can be found on our web site at http://college.hmco.com

Properties of $f(x) = \log_b x$

For all positive real numbers b, $b \neq 1$, the function defined by $f(x) = \log_b x$ has the following properties:

1. f has the set of positive real numbers as its domain.
2. f has the set of real numbers as its range.
3. f has a graph with an x-intercept of $(1, 0)$.
4. f is a one-to-one function.
5. f has a graph asymptotic to the y-axis. If $b > 1$, $f(x) \to -\infty$ as $x \to 0^+$. If $0 < b < 1$, $f(x) \to \infty$, as $x \to 0^+$.
6. f is an increasing function if $b > 1$. See **Figure 4.29a.** f is a decreasing function if $0 < b < 1$. See **Figure 4.29b.**

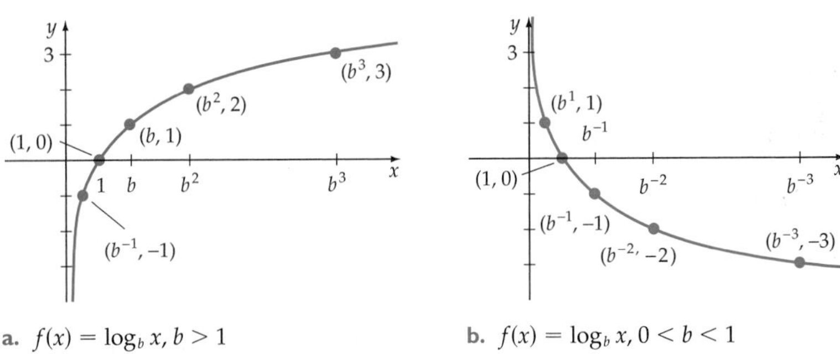

a. $f(x) = \log_b x, b > 1$ b. $f(x) = \log_b x, 0 < b < 1$

Figure 4.29

◆ Domains of Logarithmic Functions

The function $f(x) = \log_b x$ has the set of positive real numbers as its domain. The function $f(x) = \log_b g(x)$ has as its domain the set of all x for which $g(x) > 0$.

EXAMPLE 6 Find the Domain of a Logarithmic Function

Find the domain of each of the following logarithmic functions.

a. $f(x) = \log_6 (x - 3)$ b. $F(x) = \log_2 |x + 2|$

c. $R(x) = \log_5 \left(\dfrac{x}{8 - x}\right)$

Solution

a. Solving $(x - 3) > 0$ gives us $x > 3$. Thus the domain of f consists of all real numbers greater than 3. In interval notation the domain is $(3, \infty)$.

b. The solution set of $|x + 2| > 0$ consists of all real numbers except -2. Thus the domain of F consists of all real numbers $x \neq -2$. In interval notation the domain is $(-\infty, -2) \cup (-2, \infty)$.

c. Solving $\left(\frac{x}{8 - x}\right) > 0$ yields the set of all real numbers x between 0 and 8. Thus the domain of R is all real numbers x such that $0 < x < 8$. In interval notation the domain is $(0, 8)$.

TRY EXERCISE 42, EXERCISE SET 4.3, PAGE 328

Translations of the graph of the logarithmic function $f(x) = \log_b x$ sometimes can be used to obtain the graph of functions that involve logarithms.

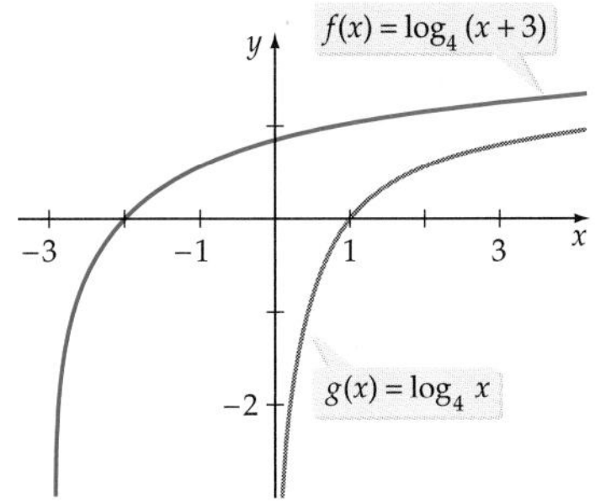

Figure 4.30

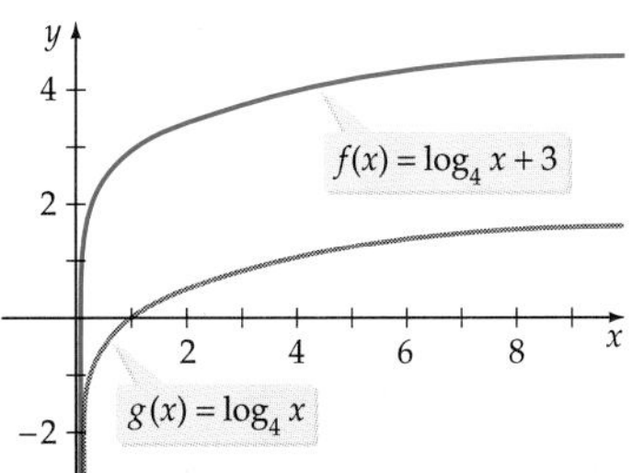

Figure 4.31

EXAMPLE 7 Use Translations to Graph

Graph.

a. $f(x) = \log_4 (x + 3)$ b. $f(x) = \log_4 x + 3$

Solution

a. The graph of $f(x) = \log_4 (x + 3)$ can be obtained by shifting the graph of $g(x) = \log_4 x$ three units to the left. Figure 4.30 shows the graph of $g(x) = \log_4 x$ and the graph of $f(x) = \log_4 (x + 3)$. Note that the domain of $f(x) = \log_4 (x + 3)$ consists of all real numbers x such that $(x + 3) > 0$. That is all real numbers $x > -3$.

b. The graph of $f(x) = \log_4 x + 3$ can be obtained by shifting the graph of $g(x) = \log_4 x$ three units upward. Figure 4.31 shows the graph of $g(x) = \log_4 x$ and the graph of $f(x) = \log_4 x + 3$.

TRY EXERCISE 52, EXERCISE SET 4.3, PAGE 328

◆ COMMON AND NATURAL LOGARITHMS

Two of the most frequently used logarithmic functions are *common logarithms,* which have base 10, and *natural logarithms,* which have base e (the base of the natural exponential function).

Definition of Common and Natural Logarithms

The function defined by $f(x) = \log_{10} x$ is called the **common logarithmic function.** It is customarily written without the base as $f(x) = \log x$.

The function defined by $f(x) = \log_e x$ is called the **natural logarithmic function.** It is customarily written as $f(x) = \ln x$.

Most scientific or graphing calculators have a log key for evaluating common logarithms and a ln key to evaluate natural logarithms. For instance,

$$\log 24 \approx 1.3802112$$
$$\ln 81 \approx 4.3944492$$
$$\log 0.58 \approx -0.23657201$$

The graphs of $f(x) = \log x$ and $f(x) = \ln x$ can be drawn by using the same techniques we used to draw the graphs in the preceding examples. However, these graphs can also be produced with a graphing calculator by using the log key or the ln key on the calculator. See Figures 4.32 and 4.33.

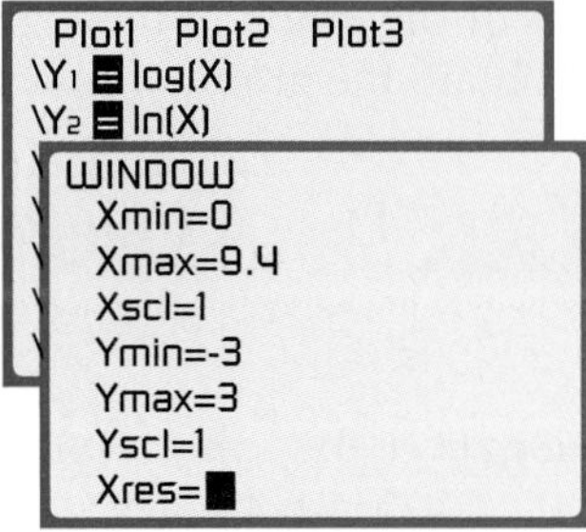

Figure 4.32

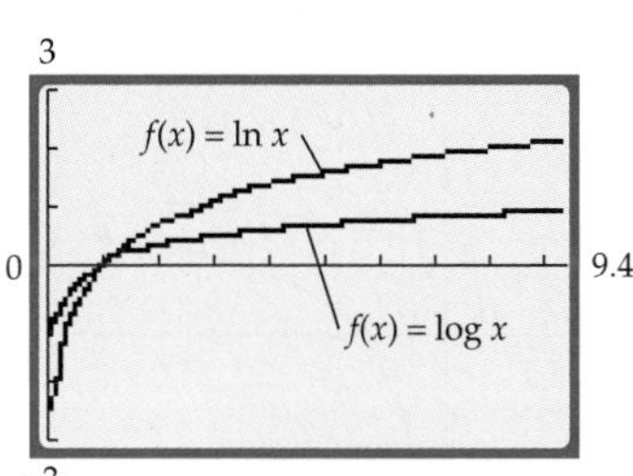

Figure 4.33

Observe that each graph passes through $(1, 0)$. Also note that as $x \to 0^+$, $y \to -\infty$; thus the y-axis is a vertical asymptote. From the definition of $y = \log_b x$, $x > 0$. Thus the domain of $f(x) = \log x$ and $f(x) = \ln x$ is the set of positive real numbers; the range is the set of real numbers.

◆ APPLICATIONS OF LOGARITHMIC FUNCTIONS

EXAMPLE 8 A Logarithmic Model of a Skill Level

The following logarithmic equation models the average typing speed S (in words per minute) for a student who has been typing for t months.

$$S(t) = 5 + 29 \ln (t + 1), \quad 0 \le t \le 9$$

a. What was the student's average typing speed when the student first started to type?

b. What was the student's average typing speed after 3 months?

c. How long will it take the student to achieve an average typing speed of 65 words per minute?

MATH MATTERS

Although logarithms were originally developed to assist with computations, logarithmic functions have a much broader use today. These functions occur in geology, acoustics, chemistry, and economics, for example.

Solution

a. $S(0) = 5 + 29 \ln (0 + 1) = 5 + 0 = 5$

At the start, the student had an average typing speed of 5 words per minute.

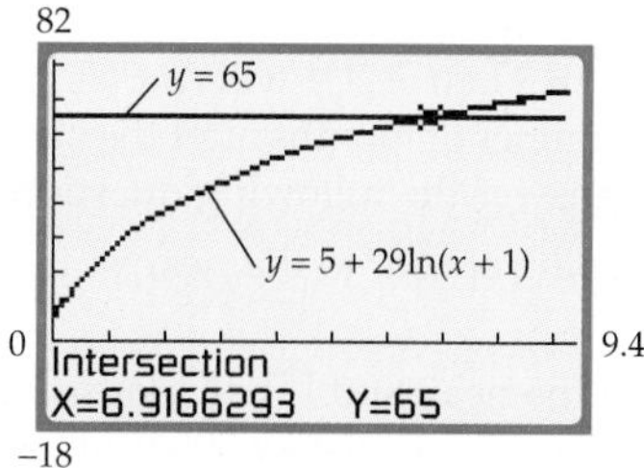

Figure 4.34

b. $S(3) = 5 + 29 \ln(3 + 1) \approx 5 + 40.2 = 45.2$

After 3 months, the student's average typing speed was about 45 words per minute.

c. Graph $y = 5 + 29 \ln(x + 1)$ and $y = 65$ on the same viewing window. Figure 4.34 shows that these graphs intersect at approximately (6.9, 65).

Thus it will take the student about 6.9 months to achieve an average typing speed of 65 words per minute.

TRY EXERCISE 80, EXERCISE SET 4.3, PAGE 328

TOPICS FOR DISCUSSION

1. The definition of a logarithm as $f(x) = \log_b x$ requires that $x > 0$ and that $b \neq 1$. Discuss why these conditions are imposed.
2. Discuss the characteristics of the graph of $f(x) = \log_b x$ if $b > 0$.
3. Let b be a positive real number and let a be its reciprocal. Discuss the relationship between the graphs of $f(x) = \log_b x$ and $g(x) = \log_a x$.
4. Explain why the y-axis is a vertical asymptote of the graph of $f(x) = \log_b x$. Does the graph of f have a horizontal asymptote?

EXERCISE SET 4.3

In Exercises 1 to 10, change each equation to its exponential form.

1. $\log_{10} 100 = 2$
2. $\log_{10} 1000 = 3$
3. $\log_5 125 = 3$
4. $\log_5 \frac{1}{25} = -2$
5. $\log_3 81 = 4$
6. $\log_3 1 = 0$
7. $\log_b r = t$
8. $\log_b (s + t) = r$
9. $-3 = \log_3 \frac{1}{27}$
10. $-1 = \log_7 \frac{1}{7}$

In Exercises 11 to 20, change each equation to its logarithmic form.

11. $2^4 = 16$
12. $3^5 = 243$
13. $7^3 = 343$
14. $7^{-4} = \frac{1}{2401}$
15. $10{,}000 = 10^4$
16. $\frac{1}{1000} = 10^{-3}$
17. $b^k = j$
18. $p = m^n$
19. $b^1 = b$
20. $b^0 = 1$

In Exercises 21 to 30, evaluate each logarithm. Do not use a calculator.

21. $\log_{10} 1{,}000{,}000$
22. $\log_b b$
23. $\log_2 32$
24. $\log_3 243$
25. $\log_{3/2} \frac{27}{8}$
26. $\log_{0.5} 16$
27. $\log_5 \frac{1}{25}$
28. $\log_{0.3} \frac{100}{9}$
29. $\log_b 1$
30. $\log_{10} \frac{1}{1000}$

In Exercises 31 to 40, graph each function by using its exponential form.

31. $f(x) = \log_4 x$
32. $f(x) = \log_5 x$
33. $f(x) = \log_{12} x$
34. $f(x) = \log_8 x$
35. $f(x) = \log_{1/2} x$
36. $f(x) = \log_{1/4} x$
37. $f(x) = \log_{5/2} x$
38. $f(x) = \log_{7/3} x$
39. $f(x) = -\log_6 x$
40. $f(x) = -\log_{1/5} x$

In Exercises 41 to 50, find the domain of the function.

41. $f(x) = \log_5 (x - 3)$
42. $g(x) = \log_2 (3x - 1)$
43. $k(x) = \log_{2/3} (11 - x)$
44. $H(x) = \log_{1/4} (x^2 + 1)$
45. $P(x) = \ln (x^2 - 4)$
46. $J(x) = \ln \left(\frac{x - 3}{x}\right)$
47. $h(x) = \ln \left(\frac{x^2}{x - 4}\right)$
48. $R(x) = \ln (x^4 - x^2)$
49. $N(x) = \log_2 (x^3 - x)$
50. $s(x) = \log_7 (x^2 + 7x + 10)$

In Exercises 51 to 60, use translations of the graphs from Exercises 31 to 40 to produce the graph of the given function.

51. $f(x) = \log_4 (x - 3)$
52. $f(x) = \log_5 (x + 3)$
53. $f(x) = \log_{12} x + 2$
54. $f(x) = \log_8 x - 4$
55. $f(x) = 3 + \log_{1/2} x$
56. $f(x) = 2 + \log_{1/4} x$
57. $f(x) = 1 + \log_{5/2} (x - 4)$
58. $f(x) = \log_{7/3} (x - 3) - 4$
59. $f(x) = -\log_6 (x + 1) + 2$
60. $f(x) = 4 - \log_{1/5} x$

In Exercises 61 to 74, use a graphing utility to graph the function.

61. $f(x) = -2 \ln x$
62. $f(x) = -\log x$
63. $f(x) = |\ln x|$
64. $f(x) = \ln |x|$
65. $f(x) = \log \sqrt[3]{x}$
66. $f(x) = \ln \sqrt{x}$
67. $f(x) = \log (x + 10)$
68. $f(x) = \ln (x + 3)$
69. $f(x) = \ln (x - 5)$
70. $f(x) = \log (x^2)$
71. $f(x) = 3 + \ln (x + 2)$
72. $f(x) = 2 + \log (x + 5)$
73. $f(x) = 3 \log |2x + 10|$
74. $f(x) = \frac{1}{2} \ln |x - 4|$

In Exercises 75 to 78, use a graphing utility.

75. Graph $f(x) = \frac{\log x}{x}$. What does $f(x)$ approach as $x \to \infty$?

76. Graph $f(x) = x \log x$. Is $x = 0$ a vertical asymptote for the graph of f? Explain your answer.

77. Graph $f(x) = \ln (e^x)$. Is the graph the same as the graph of $y = x$? Explain your answer.

78. Graph $f(x) = e^{\ln x}$. Is the graph the same as the graph of $y = x$? Explain your answer.

79. **MAINTAINING A MANUAL SKILL** If a skill is not practiced, the proficiency of the person performing the skill diminishes over time. In one study, the average typing speed for students after a 6-week class was 58 words per minute. Every month after the class, the people who did not practice were tested. The results can be modeled by the equation $S = 58 - 6.8 \ln (t + 1)$, where S is the typing speed and t is the number of months after the class. Use a graphing utility to answer the following questions.
   a. What is the average typing speed, to the nearest integer, after 3 months?
   b. How many months, to the nearest 0.1, will elapse before the average typing speed falls below 50 words per minute?

80. **ADVERTISING EXPENDITURES** An advertising agency estimates that the number of sales, N (in thousands), of a certain product is related to the amount A spent on advertising (in thousands of dollars) by the equation $N = 1.6 + 2.3 \ln A$. Graph this equation and use the graph to estimate the amount spent on advertising when 6000 units of the product were sold.

81. **MEDICINE** In anesthesiology it is necessary to estimate accurately the body surface area of a patient. One formula for estimating body surface area (BSA) was developed by Edith Boyd (University of Minnesota Press, 1935). Her formula for BSA (m^2) of a patient with a height H (cm), and weight W (g) is

$$\text{BSA} = 0.0003207 \cdot H^{0.3} \cdot W^{(0.7285 - 0.0188 \log W)}$$

Use Boyd's formula to estimate the body surface area for patients whose weight and height are as follows.
   a. $W = 110$ lb (49,886.6 g); $H = 5'4''$ (162.56 cm)
   b. $W = 180$ lb (81,632.7 g); $H = 6'1''$ (185.42 cm)
   c. $W = 40$ lb (18,140.6 g); $H = 29''$ (73.66 cm)

82. **SALARY GROWTH** The inflation-adjusted salary S (in thousands of dollars) of a computer programmer for a corporation can be approximated by the equation $S = 10 \ln (y + 1) + 26$, where y is the number of years the programmer has worked for the corporation. Graph this equation and use the graph to estimate the number of years, to the nearest tenth of a year, an employee must work to reach an inflation-adjusted salary of \$52,000.

83. **NUMBER OF DIGITS IN b^x** The number of digits N in the expansion of b^x, with both b and x positive integers, is

$$N = \text{int}(x \log b) + 1$$

where int $(x \log b)$ denotes the greatest integer of $x \log b$.
   a. Because $2^{10} = 1024$, we know that 2^{10} has 4 digits. Use the equation $N = \text{int}(x \log b) + 1$ to verify this result.
   b. Find the number of digits in 3^{200}.
   c. Find the number of digits in 7^{4005}.
   d. The largest known prime number as of June 1999 was the number $2^{6972593} - 1$. Find the number of digits in this prime number. Do $2^{6972593} - 1$ and $2^{6972593}$ both have the same number of digits? Explain.

84. **NUMBER OF DIGITS IN $9^{(9^9)}$** A mathematics teacher has offered 10 points extra credit to any student who will write out all of the digits in the expansion of $9^{(9^9)}$.

a. Use the formula from Exercise 83 to determine the number of digits in this number.

b. If you can write 1000 digits per page and 500 pages of paper constitute a ream of paper, how many reams of paper will you need to write out the expansion of the number? Assume that you write only on one side of the page.

SUPPLEMENTAL EXERCISES

In Exercises 85 to 88, use a graphing utility.

85. Graph $f(x) = \dfrac{e^x - e^{-x}}{2}$ and $g(x) = \ln\left(x + \sqrt{x^2 + 1}\right)$ on the same coordinate axes. Use the same scale on both the x- and the y-axis. What appears to be the relationship between f and g?

86. On the same coordinate axes, graph $f(x) = \dfrac{e^x + e^{-x}}{2}$ for $x \geq 0$ and $g(x) = \ln\left(x + \sqrt{x^2 - 1}\right)$ for $x \geq 1$. Use the same scale on both the x- and y-axis. What appears to be the relationship between f and g?

87. Graph $f(x) = e^{-x}(\ln x)$ for $1 \leq x \leq e^2$.

88. Graph $g(x) = \log [\![x]\!]$ for $1 \leq x \leq 10$. Recall that $[\![x]\!]$ represents the greatest integer function.

In Exercises 89 to 94, use a graphing utility to graph each function. Determine the domain and range of the function.

89. $f(x) = \sqrt{\log x}$

90. $f(x) = \sqrt{\ln x^3}$

91. $f(x) = 100 - \ln \sqrt{1 - x^2}$

92. $f(x) = 10 + |\ln (x - e)|$

93. $f(x) = \log (\log x)$

94. $f(x) = |\ln (-\ln x)|$

PROJECTS

1. **THE HARMONIC SERIES** The sum

$$S_n = 1 + \frac{1}{2} + \frac{1}{3} + \frac{1}{4} + \frac{1}{5} + \cdots + \frac{1}{n}$$

is called the *harmonic series*. As the natural number n increases, S_n increases, but it increases very slowly, as shown by the following examples.

$$S_1 = 1, \quad S_2 = 1 + \frac{1}{2} = 1.5, \quad S_3 = 1 + \frac{1}{2} + \frac{1}{3} = 1.8\overline{3}, \quad S_4 = 1 + \frac{1}{2} + \frac{1}{3} + \frac{1}{4} = 2.08\overline{3},$$

$$\text{and} \quad S_{10} = 1 + \frac{1}{2} + \frac{1}{3} + \frac{1}{4} + \cdots + \frac{1}{10} = \frac{7381}{2520} \approx 2.93$$

The sum S_n can be approximated by the equation $y = 0.5772 + \ln (n + 1)$. A graph of this equation can be used to approximate the smallest value of n for which S_n is greater than a given constant. For instance, to estimate the smallest value of n for which $S_n > 5$, graph $y = 5$ and $y = 0.5772 + \ln (n + 1)$ on the same viewing window. The following figure shows that the intersection of the graphs is about (82.3, 5). Thus the equation $S_n \approx 0.5772 + \ln (n + 1)$ predicts that S_n first exceeds 5 for $n = 83$.

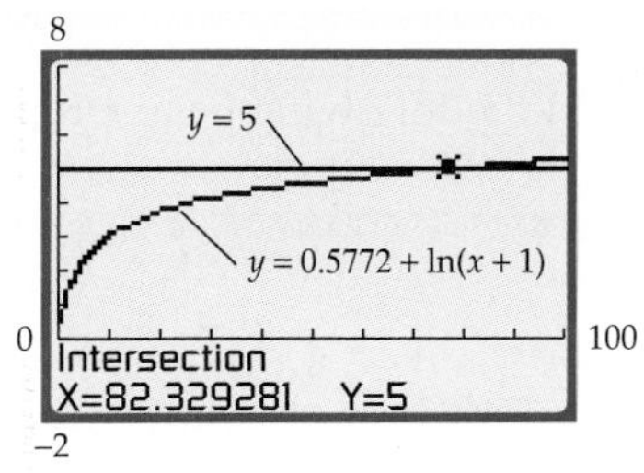

Use a graph of $S_n \approx 0.5772 + \ln(n+1)$ to estimate the smallest natural number n for which S_n first exceeds

a. 10 **b.** 100 **c.** 120

d. Explain why most graphing calculators cannot be used to find the smallest n such that $S_n > 1000$.

e. Search the Internet to find why $S_n = 1 + \frac{1}{2} + \frac{1}{3} + \frac{1}{4} + \frac{1}{5} + \cdots + \frac{1}{n}$ is called the harmonic series.

2. **PRIME NUMBERS AND LOGARITHMS** The *Prime Number Theorem* states that if $\pi(n)$ is the number of prime numbers less than the natural number n, then the ratio of $\pi(n)$ to $P(n) = \frac{n}{\ln n}$ approaches 1 as $n \to \infty$. To better understand this idea, use a graphing utility to complete the following table and then answer the questions below.

n	$\pi(n)$	$P(n) = \frac{n}{\ln n}$	*Difference* $\pi(n) - P(n)$	*Ratio* $\pi(n)/P(n)$	*Relative error* $[\pi(n) - P(n)]/\pi(n)$
1000	168	145	23	1.16	0.137
10000	1229	1086	143	1.13	0.116
100000	9592				
1000000	78498				
10000000	664579				
100000000	5761455				

a. Does the difference $\pi(n) - P(n)$ tend to increase or decrease as $n \to \infty$?

b. What happens to the ratio $\pi(n)/P(n)$ as $n \to \infty$?

c. The relative error between $\pi(n)$ and $P(n)$ is defined as $[\pi(n) - P(n)]/\pi(n)$. What happens to this relative error as $n \to \infty$?

d. Explain how to argue why $P(n)$ becomes a better approximation to $\pi(n)$ as $n \to \infty$, even though the difference between $\pi(n)$ and $P(n)$ is increasing.

SECTION 4.4 PROPERTIES OF LOGARITHMS

- PROPERTIES OF LOGARITHMS
- CHANGE-OF-BASE FORMULA
- LOGARITHMIC SCALES

4A

◆ PROPERTIES OF LOGARITHMS

In Section 4.3 we introduced the following basic properties of logarithms.

Basic Properties of Logarithms

1. $\log_b b = 1$ **2.** $\log_b 1 = 0$ **3.** $\log_b (b^p) = p$

In addition to these basic properties, we can use the properties of exponents to establish the following additional logarithmic properties.

Properties of Logarithms

In the following properties, b, M, and N are positive real numbers ($b \neq 1$), and p is any real number.

Product property	$\log_b (MN) = \log_b M + \log_b N$
Quotient property	$\log_b \frac{M}{N} = \log_b M - \log_b N$
Power property	$\log_b (M^p) = p \log_b M$
One-to-one property	$\log_b M = \log_b N$ implies $M = N$
Logarithm-of-each-side property	$M = N$ implies $\log_b M = \log_b N$
Inverse property	$b^{\log_b p} = p$ (for $p > 0$)

The above properties of logarithms are often used to rewrite logarithmic expressions in an equivalent form.

EXAMPLE 1 Rewrite Logarithmic Expressions

Use the properties of logarithms to express the following logarithms in terms of logarithms of x, y, and z.

a. $\log_b (3y^2)$ b. $\log_b \frac{x^2\sqrt{y}}{z^5}$

Solution

a. $\log_b (3y^2) = \log_b 3 + \log_b y^2$ • Product property

$= \log_b 3 + 2 \log_b y$ • Power property

b. $\log_b \frac{x^2\sqrt{y}}{z^5} = \log_b (x^2\sqrt{y}) - \log_b z^5$ • Quotient property

$= \log_b x^2 + \log_b \sqrt{y} - \log_b z^5$ • Product property

$= 2 \log_b x + \frac{1}{2} \log_b y - 5 \log_b z$ • Power property

TRY EXERCISE 2, EXERCISE SET 4.4, PAGE 339

The properties of logarithms are also used to rewrite expressions that involve logarithms as a single logarithm.

take note

Pay close attention to the product property of logarithms. It states that

$$\log_b (MN) = \log_b M + \log_b N$$

It does not state any relationship that involves $\log_b (M + N)$. **The logarithm of a sum cannot be simplified.**

EXAMPLE 2 Rewrite Logarithmic Expressions

Use the properties of logarithms to rewrite each expression as a single logarithm with a coefficient of 1.

a. $2 \log_b x + \dfrac{1}{2} \log_b (x + 4)$ **b.** $4 \log_b (x + 2) - 3 \log_b (x - 5)$

Solution

a. $2 \log_b x + \dfrac{1}{2} \log_b (x + 4)$

$= \log_b x^2 + \log_b (x + 4)^{1/2}$ • Power property

$= \log_b [x^2(x + 4)^{1/2}]$ • Product property

b. $4 \log_b (x + 2) - 3 \log_b (x - 5)$

$= \log_b (x + 2)^4 - \log_b (x - 5)^3$ • Power property

$= \log_b \dfrac{(x + 2)^4}{(x - 5)^3}$ • Quotient property

TRY EXERCISE 12, EXERCISE SET 4.4, PAGE 339

Sometimes it is possible to use known logarithmic values and the properties of logarithms to evaluate logarithms.

EXAMPLE 3 Evaluate Logarithms

Given $\log_8 2 \approx 0.3333$, $\log_8 3 \approx 0.5283$, and $\log_8 5 \approx 0.7740$, evaluate the following:

a. $\log_8 15$ **b.** $\log_8 \dfrac{5}{2}$ **c.** $\log_8 \sqrt[3]{9}$

Solution

a. $\log_8 15 = \log_8 (3 \cdot 5) = \log_8 3 + \log_8 5$ • Product property

$\approx 0.5283 + 0.7740 = 1.3023$

b. $\log_8 \dfrac{5}{2} = \log_8 5 - \log_8 2$ • Quotient property

$\approx 0.7740 - 0.3333 = 0.4407$

c. $\log_8 \sqrt[3]{9} = \log_8 3^{2/3} = \dfrac{2}{3} \log_8 3$ • Power property

$\approx \dfrac{2}{3}(0.5283) = 0.3522$

TRY EXERCISE 22, EXERCISE SET 4.4, PAGE 340

◆ CHANGE-OF-BASE FORMULA

Recall that to determine the value of y in $\log_3 81 = y$, we are basically asking "What power of 3 is 81?" Because $3^4 = 81$, we have $\log_3 81 = 4$. Now suppose that we need to determine the value of $\log_3 50$. In this case we need to determine the power of 3 that produces 50. Because $3^3 = 27$ and $3^4 = 81$, the value we are seeking is somewhere between 3 and 4. The exponential form of $\log_3 50 = y$ is $3^y = 50$. Applying logarithmic properties gives us

$$\ln 3^y = \ln 50$$ • **Logarithm-of-each-side property**

$$y \ln 3 = \ln 50$$ • **Power property**

$$y = \frac{\ln 50}{\ln 3}$$ • **Solve for *y*.**

$$\approx \frac{3.91202}{1.09861}$$ • **Use a calculator to evaluate ln 50 and ln 3.**

$$\approx 3.56088$$

Thus $\log_3 50 \approx 3.56088$. The above procedure can be used to verify the following formula.

Change-of-Base Formula

If x, a, and b are positive real numbers with $a \neq 1$ and $b \neq 1$, then

$$\log_b x = \frac{\log_a x}{\log_a b}$$

Because most calculators use only common logarithms ($a = 10$) or natural logarithms ($a = e$), the change-of-base formula is often written in the following form.

If x and b are positive real numbers and $b \neq 1$, then

$$\log_b x = \frac{\log x}{\log b} = \frac{\ln x}{\ln b}$$

EXAMPLE 4 Use the Change-of-Base Formula

Evaluate each logarithm.

a. $\log_3 18$ **b.** $\log_{12} 400$

Continued •➤

take note

Common logarithms could have been used for the calculation in Example 4a.

$$\log_3 18 = \frac{\log 18}{\log 3} \approx 2.63093$$

Solution

In each case we will use the change-of-base formula with $a = e$. That is, we will evaluate these logarithms by using the ln key on a scientific or graphing calculator.

a. $\log_3 18 = \dfrac{\ln 18}{\ln 3} \approx 2.63093$

b. $\log_{12} 400 = \dfrac{\ln 400}{\ln 12} \approx 2.41114$

TRY EXERCISE 32, EXERCISE SET 4.4, PAGE 340

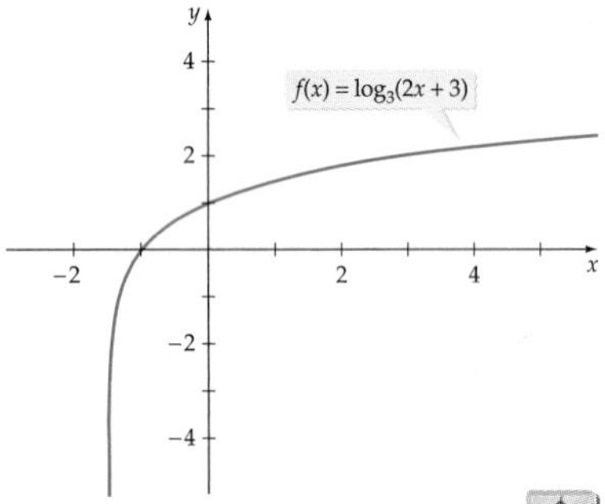

Figure 4.35

The change-of-base formula and a graphing calculator can be used to graph logarithmic functions that have a base other than 10 or e. For instance, to graph $f(x) = \log_3 (2x + 3)$, we rewrite the function in terms of base 10 or base e. Using logarithms base 10, we have $f(x) = \log_3 (2x + 3) = \dfrac{\log (2x + 3)}{\log 3}$. The graph is shown in **Figure 4.35.**

EXAMPLE 5 Use the Change-of-Base Formula to Graph a Logarithmic Function

Graph $f(x) = \log_2 |x - 3|$.

Solution

Rewrite f using the change-of-base formula. We will use the natural logarithm function; however, the common logarithm function could have been used instead.

$$f(x) = \log_2 |x - 3| = \frac{\ln |x - 3|}{\ln 2}$$

Enter $\dfrac{\ln |x - 3|}{\ln 2}$ into Y1. The graph is shown in **Figure 4.36.** Note that the domain of $f(x) = \log_2 |x - 3|$ is all real numbers except 3, because $|x - 3| = 0$ when $x = 3$.

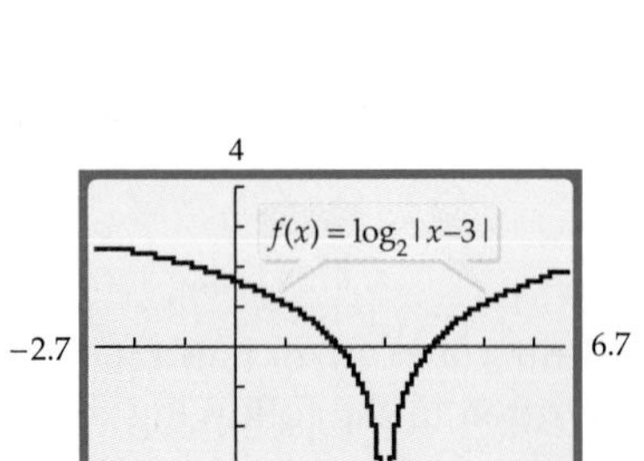

Figure 4.36

TRY EXERCISE 42, EXERCISE SET 4.4, PAGE 340

◆ LOGARITHMIC SCALES

Logarithmic functions are often used to scale very large (or very small) numbers into numbers that are easier to comprehend. For instance, the *Richter scale* magnitude of an earthquake uses a logarithmic function to convert the intensity of its shock waves I into a number M, which for most earthquakes is in the range of 0 to 10. The intensity I of an earthquake is often given in terms of the constant I_0, where I_0 is the intensity of the smallest earthquake (called a **zero-level earthquake**) that can be measured on a seismograph near the earthquake's epicenter. The following formula is used to compute the Richter scale magnitude of an earthquake.

The Richter Scale Magnitude of an Earthquake

An earthquake with an intensity of I has a Richter scale magnitude of

$$M = \log\left(\frac{I}{I_0}\right)$$

where I_0 is the measure of a zero-level earthquake.

EXAMPLE 6 Determine the Magnitude of an Earthquake

Find, to the nearest 0.1, the Richter scale magnitude of the 1999 Joshua Tree, California, earthquake that had an intensity of $I = 12{,}589{,}254I_0$.

Solution

$$M = \log\left(\frac{I}{I_0}\right) = \log\left(\frac{12{,}589{,}254I_0}{I_0}\right) = \log(12{,}589{,}254) \approx 7.1$$

The 1999 Joshua Tree earthquake had a Richter scale magnitude of 7.1.

TRY EXERCISE 54, EXERCISE SET 4.4, PAGE 340

If you know the Richter scale magnitude of an earthquake, then you can also determine the intensity of the earthquake.

EXAMPLE 7 Determine the Intensity of an Earthquake

Find the intensity of the 1999 Taiwan earthquake, which measured 7.6 on the Richter scale.

Solution

$$\log\left(\frac{I}{I_0}\right) = 7.6$$

$$\frac{I}{I_0} = 10^{7.6} \qquad \text{• Write in exponential form.}$$

$$I = 10^{7.6}I_0 \qquad \text{• Solve for } I.$$

$$I \approx 39{,}810{,}717I_0$$

The 1999 Taiwan earthquake had an intensity that was approximately 39,811,000 times the intensity of a zero-level earthquake.

TRY EXERCISE 56, EXERCISE SET 4.4, PAGE 340

MATH MATTERS

The Richter scale was created by the seismologist Charles F. Richter in 1935. Note that a tenfold increase in the intensity level of an earthquake increases the Richter scale magnitude of the earthquake by only 1.

In Example 8 we make use of the Richter scale magnitudes of two earthquakes to compare the intensities of the earthquakes.

EXAMPLE 8 Compare Earthquakes

The 1960 Chile earthquake had a Richter scale magnitude of 9.5. The 1989 San Francisco earthquake had a Richter scale magnitude of 7.1. Compare the intensities of the earthquakes.

Solution

Let I_1 be the intensity of the Chilean earthquake and I_2 be the intensity of the San Francisco earthquake. Thus

$$\log\left(\frac{I_1}{I_0}\right) = 9.5 \qquad \text{and} \qquad \log\left(\frac{I_2}{I_0}\right) = 7.1$$

$$\frac{I_1}{I_0} = 10^{9.5} \qquad\qquad \frac{I_2}{I_0} = 10^{7.1}$$

$$I_1 = 10^{9.5}I_0 \qquad\qquad I_2 = 10^{7.1}I_0$$

To compare the intensities of the earthquakes, we compute the ratio I_1/I_2.

$$\frac{I_1}{I_2} = \frac{10^{9.5}I_0}{10^{7.1}I_0} = \frac{10^{9.5}}{10^{7.1}} = 10^{9.5-7.1} = 10^{2.4} \approx 251$$

The earthquake in Chile was approximately 251 times as intense as the San Francisco earthquake.

TRY EXERCISE 58, EXERCISE SET 4.4, PAGE 340

take note

The results of Example 8 show that if an earthquake has a Richter scale magnitude of M_1 and a smaller earthquake has a Richter scale magnitude of M_2, then the first earthquake is $10^{M_1-M_2}$ times as intense as the smaller earthquake.

Seismologists generally determine the Richter scale magnitude of an earthquake by examining a *seismogram*. See **Figure 4.37.**

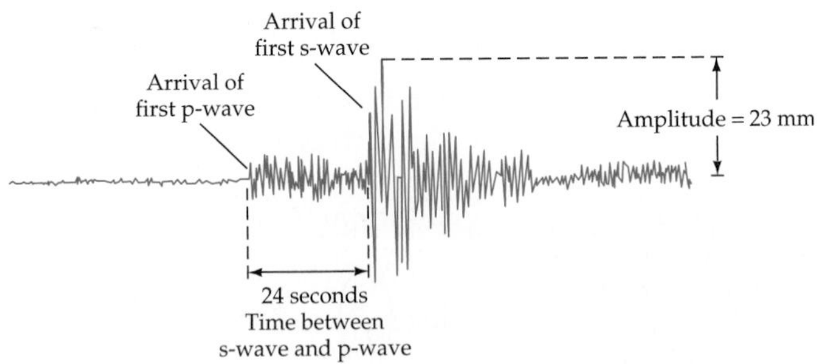

Figure 4.37

The magnitude of an earthquake cannot be determined just by examining the amplitude of a seismogram, because this amplitude decreases as the distance between the epicenter of an earthquake and the observation station is increased. To account for the distance between the epicenter and the observation station, a seismologist examines a seismogram for both small waves called *p-waves* and larger waves called *s-waves*. The Richter scale magnitude, M, of the earthquake is a function of both the amplitude, A, of the s-waves and the difference in time, t, between the occurrence of the first s-wave and that of the first p-wave. In the 1950s Charles Richter developed the following formula to determine the magnitude, M, of an earthquake from the data in a seismogram.

The Amplitude–Time-Difference Formula

The Richter scale magnitude of an earthquake is given by

$$M = \log A + 3\log 8t - 2.92$$

where A is the amplitude of the s-waves on a seismogram, in millimeters, and t is the difference in time between the first s-wave and the first p-wave, in seconds.

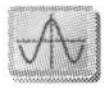

EXAMPLE 9 Determine the Magnitude of an Earthquake from Its Seismogram

Find the magnitude of the earthquake that produced the seismogram in **Figure 4.37.**

Solution

$$\begin{aligned} M &= \log A + 3\log 8t - 2.92 \\ &= \log 23 + 3\log[8 \cdot 24] - 2.92 \\ &\approx 1.36173 + 6.84990 - 2.92 \\ &\approx 5.3 \end{aligned}$$

• **Substitute 23 for *A* and 24 for *t*.**

The earthquake had a magnitude of 5.3 on the Richter scale.

TRY EXERCISE 62, EXERCISE SET 4.4, PAGE 340

Logarithmic scales are also used in chemistry. As an example, we consider the acidity of a solution, which is a function of the hydronium-ion concentration of the solution. Because the hydronium-ion concentration of a solution can be very small, it is convenient to measure the acidity of a solution in terms of pH, which is defined as follows.

The pH of a Solution

The pH of a solution with a hydronium-ion concentration of H^+ moles per liter is given by

$$\text{pH} = -\log[H^+]$$

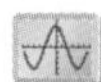

EXAMPLE 10 Find the pH of a Solution

MATH MATTERS

The pH scale was created by the Danish biochemist Søren Sørensen in 1909.

Find the pH of each liquid.

a. Orange juice with $[H^+] = 2.8 \times 10^{-4}$ mole per liter

b. Milk with $[H^+] = 3.97 \times 10^{-7}$ mole per liter

c. Rain water with $[H^+] = 6.31 \times 10^{-5}$ mole per liter

d. A baking soda solution with $[H^+] = 3.98 \times 10^{-9}$ mole per liter

Solution

a. $\text{pH} = -\log[H^+] = -\log(2.8 \times 10^{-4}) \approx 3.6$

The orange juice has a pH of 3.6.

b. $\text{pH} = -\log[H^+] = -\log(3.97 \times 10^{-7}) \approx 6.4$

The milk has a pH of 6.4.

c. $\text{pH} = -\log[H^+] = -\log(6.31 \times 10^{-5}) \approx 4.2$

The rain water has a pH of 4.2.

d. $\text{pH} = -\log[H^+] = -\log(3.98 \times 10^{-9}) \approx 8.4$

The baking soda solution has a pH of 8.4.

TRY EXERCISE 64, EXERCISE SET 4.4, PAGE 340

Figure 4.38 illustrates that the pH scale ranges from 0 to 14, distilled water having a pH of 7.0. A solution with a pH of less than 7 is an **acid**, and a solution with a pH greater than 7 is an **alkaline solution** or a **base**. A solution with a pH of 5 is ten times more acidic than a solution with a pH of 6. From Example 10 we see that the orange juice, the rain water, and the milk are acids, whereas the baking soda solution is a base.

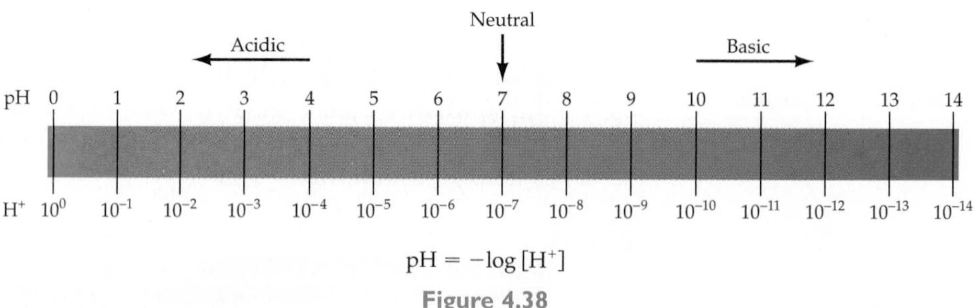

$\text{pH} = -\log[H^+]$

Figure 4.38

In Example 10, the hydronium-ion concentrations of the orange juice and the milk were given as 2.8×10^{-4} and 3.97×10^{-7} respectively. The difference between 2.8×10^{-4} and 3.97×10^{-7} is very small.

$$0.00028 - 0.000000397 \approx 0.0002796$$

Figure 4.38 shows how the pH function scales small hydronium-ion concentration numbers that are relatively close together on the H^+ axis into larger (more convenient) numbers that are farther apart on the pH axis.

EXAMPLE 11 Find the Hydronium-Ion Concentration

A sample of blood has a pH of 7.3. Find the hydronium-ion concentration of the blood.

Solution

$$\text{pH} = -\log[\text{H}^+]$$
$$7.3 = -\log[\text{H}^+]$$ • **Substitute 7.3 for pH.**
$$-7.3 = \log[\text{H}^+]$$ • **Multiply both sides by −1.**
$$10^{-7.3} = \text{H}^+$$ • **Change to exponential form.**
$$5.0 \times 10^{-8} \approx \text{H}^+$$

The hydronium-ion concentration of the blood is 5.0×10^{-8} mole per liter.

TRY EXERCISE 66, EXERCISE SET 4.4, PAGE 341

TOPICS FOR DISCUSSION

1. The function $f(x) = \log_b x$ is defined only for $x > 0$. Explain why this condition is imposed.
2. What are the product, quotient, and power properties for logarithms?
3. If $f(x) = \log_b x$ and $f(c) = f(d)$, can we conclude that $c = d$?
4. Explain a method that could be used to show that $\log_b (M - N) \neq \log_b \left(\frac{M}{N}\right)$. Assume $M > 0$ and $N > 0$.

EXERCISE SET 4.4

In Exercises 1 to 10, write the given logarithm in terms of logarithms of x, y, and z.

1. $\log_b (xyz)$
2. $\log_b (x^2y^3)$
3. $\log_3 \frac{x}{z^4}$
4. $\log_5 \frac{x^2}{yz^3}$
5. $\log_b \frac{\sqrt{x}}{y^3}$
6. $\log_b \frac{\sqrt{x}}{\sqrt[3]{z}}$
7. $\log_b x\sqrt[3]{\frac{y^2}{z}}$
8. $\log_b \sqrt[3]{x^2z\sqrt{y}}$
9. $\log_7 \frac{\sqrt{xz}}{y^2}$
10. $\log_5 x^2\left(\frac{y}{z^3}\right)^2$

In Exercises 11 to 20, write each logarithmic expression as a single logarithm with a coefficient of 1.

11. $\log_{10} (x + 5) + 2\log_{10} x$
12. $5\log_3 x - 4\log_3 y + 2\log_3 z$
13. $\frac{1}{2}[3\log_b (x - y) + \log_b (x + y) - \log_b z]$
14. $\log_b (y^3z^2) - 3\log_b (x\sqrt{y}) + 2\log_b \frac{x}{z}$
15. $\log_8 (x^2 - y^2) - \log_8 (x - y)$
16. $\log_4 (x^3 - y^3) - \log_4 (x - y)$

17. $4 \ln (x - 3) + 2 \ln x$

18. $3 \ln z - 2 \ln (z + 1)$

19. $\ln x - \ln y + \ln z$

20. $\frac{1}{2} \log x + 2 \log y$

In Exercises 21 to 30, evaluate the logarithm using the values $\log_7 2 \approx 0.3562$, $\log_7 3 \approx 0.5646$, and $\log_7 5 \approx 0.8271$ and the properties of logarithms. Do not use a calculator.

21. $\log_7 6$

22. $\log_7 20$

23. $\log_7 9$

24. $\log_7 4$

25. $\log_7 \frac{2}{5}$

26. $\log_7 \frac{3}{2}$

27. $\log_7 30$

28. $\log_7 45$

29. $\log_7 14$

30. $\log_7 \frac{7^2}{3}$

In Exercises 31 to 40, use the change-of-base formula to approximate the logarithm accurate to five significant digits.

31. $\log_7 20$

32. $\log_5 37$

33. $\log_{11} 8$

34. $\log_{50} 22$

35. $\log_6 0.045$

36. $\log_4 \sqrt{7}$

37. $\log_{0.5} 5$

38. $\log_{0.2} 17$

39. $\log_\pi e$

40. $\log_\pi \sqrt{15}$

In Exercises 41 to 52, use a graphing utility and the change-of-base formula to graph the logarithmic function.

41. $f(x) = \log_4 x$

42. $s(x) = \log_7 x$

43. $g(x) = \log_8 (x - 3)$

44. $t(x) = \log_9 (5 - x)$

45. $h(x) = \log_3 (x - 3)^2$

46. $v(x) = \log_2 (x + 1)^2$

47. $F(x) = -\log_\pi |x - 2|$

48. $J(x) = \log_{12} (-x)$

49. $K(x) = |\log_6 (2x - 1)|$

50. $L(x) = \log_{3.5} (5 - x)$

51. $m(x) = -\log_{5.5} (x^2)$

52. $n(x) = \log_2 \sqrt{x - 8}$

53. EARTHQUAKE MAGNITUDE What is the Richter scale magnitude of an earthquake with an intensity of $I = 100{,}000I_0$?

54. EARTHQUAKE MAGNITUDE The Colombia earthquake of 1906 had an intensity of $I = 398{,}107{,}000I_0$. What did it measure on the Richter scale?

55. EARTHQUAKE INTENSITY The Coalinga, California, earthquake of 1983 had a Richter scale magnitude of 6.5. Find the intensity of this earthquake.

56. EARTHQUAKE INTENSITY The earthquake that occurred just south of Concepción, Chile, in 1960 had a Richter scale magnitude of 9.5. Find the intensity of this earthquake.

57. COMPARISON OF EARTHQUAKES Compare the intensity of an earthquake that measures 5 on the Richter scale to the intensity of an earthquake that measures 3 on the Richter scale by finding the ratio of the larger intensity to the smaller intensity.

58. COMPARISON OF EARTHQUAKES How many times more intense was the 1960 earthquake in Chile, which measured 9.5 on the Richter scale, than the San Francisco earthquake of 1906, which measured 8.3 on the Richter scale?

59. COMPARISON OF EARTHQUAKES On March 2, 1933, an earthquake of magnitude 8.9 on the Richter scale struck Japan. In October 1989, an earthquake of magnitude 7.1 on the Richter scale struck San Francisco, California. Compare the intensity of the larger earthquake to the intensity of the smaller earthquake by finding the ratio of the larger intensity to the smaller intensity.

60. COMPARISON OF EARTHQUAKES An earthquake that occurred in China in 1978 measured 8.2 on the Richter scale. In 1988, an earthquake in California measured 6.9 on the Richter scale. Compare the intensity of the larger earthquake to the intensity of the smaller earthquake by finding the ratio of the larger intensity to the smaller intensity.

61. EARTHQUAKE MAGNITUDE Find the magnitude of the earthquake that produced the seismogram in the following figure.

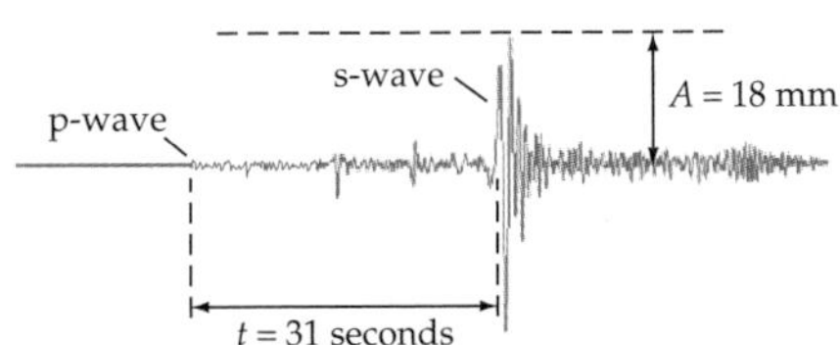

62. EARTHQUAKE MAGNITUDE Find the magnitude of the earthquake that produced the seismogram in the following figure.

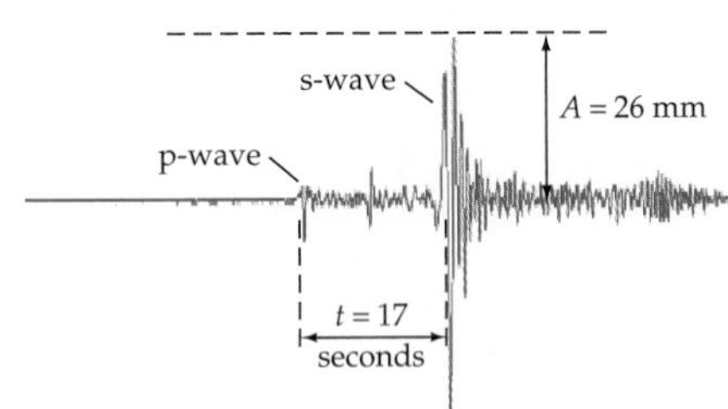

63. PH Household ammonia has a hydronium-ion concentration of 1.26×10^{-12} mole per liter. Determine the pH of the ammonia and state whether it is an acid or a base.

64. PH Vinegar has a hydronium-ion concentration of 1.26×10^{-3} mole per liter. Determine the pH of the ammonia and state whether it is an acid or a base.

65. HYDRONIUM-ION CONCENTRATION A morphine solution has a pH of 9.5. Determine the hydronium-ion concentration of the morphine solution.

66. HYDRONIUM-ION CONCENTRATION A rain storm in New York City produced rainwater with a pH of 5.6. Determine the hydronium-ion concentration of the rainwater.

67. DECIBEL LEVEL The range of sound intensities that the human ear can detect is so large that a special decibel scale (named after Alexander Graham Bell) is used to measure and compare sound intensities. The level in decibels, dB, of a sound is given by

$$dB(I) = 10 \log \left(\frac{I}{I_0}\right)$$

where I_0 is the intensity of sound that is barely audible to the human ear. Find the decibel level for each of the following sounds.

Sound	Intensity
a. Automobile traffic	$I = 1.58 \times 10^8 \cdot I_0$
b. Quiet conversation	$I = 10{,}800 \cdot I_0$
c. Fender guitar	$I = 3.16 \times 10^{11} \cdot I_0$
d. Jet engine	$I = 1.58 \times 10^{15} \cdot I_0$

68. EQUIVALENT FORMULAS Show that the Richter scale magnitude formula

$$M = \log \left(\frac{I}{I_0}\right) \quad \text{can be written as} \quad M = \frac{\ln I - \ln I_0}{\ln 10}.$$

69. COMPARISON OF INTENSITIES How many times more intense is a sound that measures 120 decibels than a sound that measures 110 decibels?

70. DECIBEL LEVEL If the intensity of a sound is doubled, what is the increase in the decibel level? *Hint:* Find $dB(2I) - dB(I)$.

SUPPLEMENTAL EXERCISES

71. Evaluate $\log_3 5 \cdot \log_5 7 \cdot \log_7 9$. Do not use a calculator.

72. Evaluate $\log_5 20 \cdot \log_{20} 60 \cdot \log_{60} 100 \cdot \log_{100} 125$. Do not use a calculator.

73. Supply the reason or property for each step in the following proof of the product property.

Prove: $\log_b (MN) = \log_b M + \log_b N$

Proof:

Let $\log_b M = x$ and $\log_b N = y$. Then

$M = b^x$ and $N = b^y$ ________

$MN = b^x b^y$ ________

$MN = b^{(x+y)}$ ________

$\log_b (MN) = \log_b b^{(x+y)}$ ________

$\log_b (MN) = x + y$ ________

$\log_b (MN) = \log_b M + \log_b N$ ________

74. Supply the reason or property for each step in the following proof of the power property.

Prove: $\log_b (M^p) = p \log_b M$

Proof:

Let $\log_b M = x$. Then

$M = b^x$ ________

$M^p = (b^x)^p$ ________

$M^p = b^{px}$ ________

$\log_b (M^p) = \log_b (b^{px})$ ________

$\log_b (M^p) = px$ ________

$\log_b (M^p) = p \log_b M$ ________

In Exercises 75 to 80, find all the real numbers that are solutions of the given inequality. Use interval notation to write your answers.

75. $0 \le \log x \le 1000$

76. $-3 \le \log x \le -2$

77. $e \le \ln x \le e^3$

78. $-2 \le \ln x \le 3$

79. $-\log x > 0$

80. $100 - 10 \log (x + 1) > 0$

81. NOMOGRAMS AND LOGARITHMIC SCALES A **nomogram** is a diagram used to determine a numerical result by drawing a line across numerical scales. The following nomogram, used by Richter, determines the magnitude of an earthquake from its seismogram. To use the nomogram, mark the amplitude of a seismogram on the amplitude scale, and mark the time between the s-wave and the p-wave on the S-P scale. Draw a line between these marks. The magnitude of the earthquake that produced the seismogram is shown by the intersection of this line and the center scale. Our example shows that an earthquake with a seismogram amplitude of 23 mm and an S-P time of 24 seconds has a Richter magnitude of about 5.

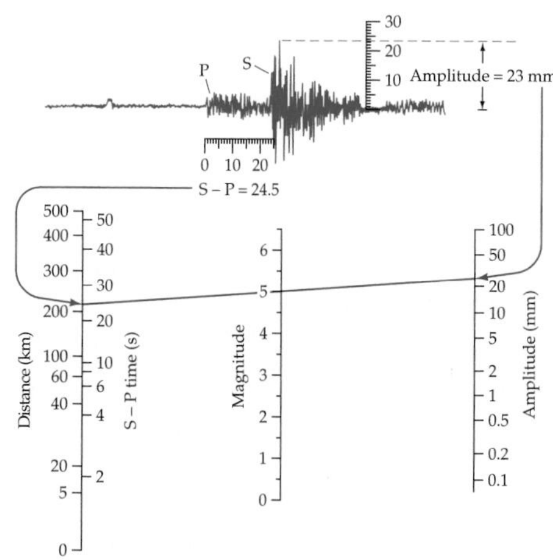

Richter's Earthquake Nomogram

The amplitude and the S-P time are shown on logarithmic scales. On the amplitude scale, the distance from 1 to 10 is the same as the distance from 10 to 100, because $\log 100 - \log 10 = \log 10 - \log 1$.

Use the nomogram on page 341 to determine the Richter magnitude of an earthquake with each of the following seismogram readings.

a. amplitude of 50 mm and S-P time of 40 seconds.

b. amplitude of 1 mm and S-P time of 30 seconds.

How do the results in parts **a** and **b** compare with the Richter magnitude produced by using the amplitude–time-difference formula?

PROJECTS

1. **BIOLOGICAL DIVERSITY** To discuss the variety of species that live in a certain environment, a biologist needs a precise definition of *diversity*. Let $p_1, p_2, \ldots, p_n$ be the proportion of n species that live in an environment. The biological diversity, D, of this system is

$$D = -(p_1 \log_2 p_1 + p_2 \log_2 p_2 + \cdots + p_n \log_2 p_n)$$

Suppose an ecosystem has exactly five different varieties of grass: rye (R), bermuda (B), blue (L), fescue (F), and St. Augustine (A). The various proportions of these grasses are as shown in the tables at the right.

a. Calculate the diversity of this ecosystem if the proportions of these grasses are as shown in Table 1.

b. Because bermuda and St. Augustine are virulent grasses, after a time the proportions will be as shown in Table 2. Does this system have more or less diversity than the system given in Table 1?

c. After an even longer time period, the bermuda and St. Augustine grasses completely overrun the environment and the proportions are as shown in Table 3. Calculate the diversity of this system. (*Note:* Although the equation is not technically correct, for purposes of the diversity definition, we may say that $0 \log_2 0 = 0$. By using very small values of p_i, we can demonstrate that this definition makes sense.) Does this system have more or less diversity than the system given in Table 2?

d. Finally, the St. Augustine grasses overruns the bermuda grasses and the proportions are as shown in Table 4. Calculate the diversity of this system. Write a sentence that explains the meaning of the value you obtained.

Table 1

R	B	L	F	A
$\frac{1}{5}$	$\frac{1}{5}$	$\frac{1}{5}$	$\frac{1}{5}$	$\frac{1}{5}$

Table 2

R	B	L	F	A
$\frac{1}{8}$	$\frac{3}{8}$	$\frac{1}{16}$	$\frac{1}{8}$	$\frac{5}{16}$

Table 3

R	B	L	F	A
0	$\frac{1}{4}$	0	0	$\frac{3}{4}$

Table 4

R	B	L	F	A
0	0	0	0	1

SECTION 4.5 EXPONENTIAL AND LOGARITHMIC EQUATIONS

- SOLVE EXPONENTIAL EQUATIONS
- SOLVE LOGARITHMIC EQUATIONS
- SOLVE AN APPLICATION

4B

◆ SOLVE EXPONENTIAL EQUATIONS

If a variable appears in an exponent of a term of an equation, such as $2^{x+1} = 32$, then the equation is called an **exponential equation**. Example 1 uses the following equality-of-exponents theorem to solve $2^{x+1} = 32$.

Equality-of-Exponents Theorem

If $b^x = b^y$, then $x = y$, provided that $b > 0$ and $b \neq 1$.

EXAMPLE 1 Solve an Exponential Equation

Use the equality-of-exponents theorem to solve $2^{x+1} = 32$.

Solution

$$2^{x+1} = 32$$

$$2^{x+1} = 2^5 \qquad \text{• Write each side as a power of 2.}$$

$$x + 1 = 5 \qquad \text{• Equate the exponents.}$$

$$x = 4$$

Check Let $x = 4$, then $2^{x+1} = 2^{4+1} = 2^5 = 32$

TRY EXERCISE 2, EXERCISE SET 4.5, PAGE 348

A graphing utility can also be used to find solutions of an equation of the form $f(x) = g(x)$. Either of the following two methods can be employed.

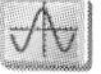

Using a Graphing Utility to Find the Solutions of $f(x) = g(x)$

Intersection Method Graph $y_1 = f(x)$ and $y_2 = g(x)$ on the same screen. The solutions of $f(x) = g(x)$ are the x-coordinates of the points of intersection of the graphs.

Intercept Method The solutions of $f(x) = g(x)$ are the x-coordinates of the x-intercepts of the graph of $y = f(x) - g(x)$.

Figures 4.39 and 4.40 illustrate the graphical methods for solving $2^{x+1} = 32$.

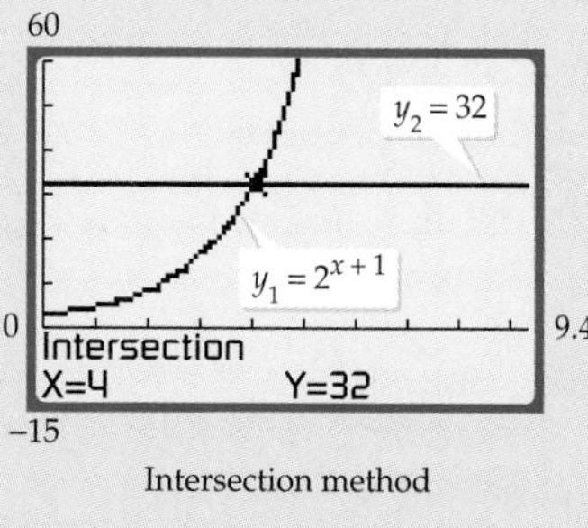

Intersection method

Figure 4.39

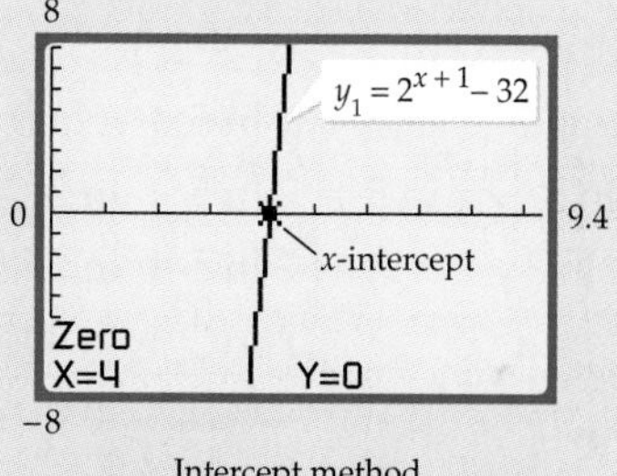

Intercept method

Figure 4.40

In Example 1, we were able to write both sides of the equation as a power of the same base. If you find it difficult to write both sides of an exponential equation in terms of the same base, then try the procedure of taking the logarithm of each side of the equation. This procedure is used in Example 2.

EXAMPLE 2 Solve an Exponential Equation

Solve: $5^x = 40$

Solution

$$5^x = 40$$

$$\log(5^x) = \log 40 \quad \text{• Take the logarithm of each side.}$$

$$x \log 5 = \log 40 \quad \text{• Power property}$$

$$x = \frac{\log 40}{\log 5} \quad \text{• Exact solution}$$

$$x \approx 2.3 \quad \text{• Decimal approximation}$$

To the nearest 0.1, the solution is 2.3.

Visualize the Solution

Intersection Method The solution of $5^x = 40$ is the x-coordinate of the point of intersection of $y = 5^x$ and $y = 40$ (see **Figure 4.41**).

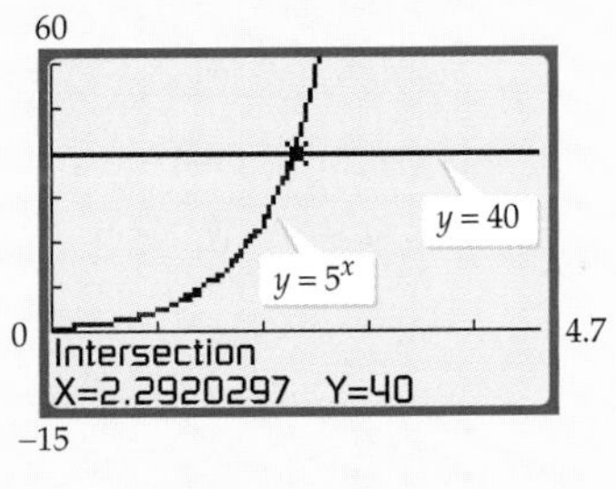

Figure 4.41

TRY EXERCISE 10, EXERCISE SET 4.5, PAGE 348

EXAMPLE 3 Solve an Exponential Equation

Solve: $3^{2x-1} = 5^{x+2}$

Solution

$$3^{2x-1} = 5^{x+2}$$

$$\ln 3^{2x-1} = \ln 5^{x+2} \quad \text{• Take the natural logarithm of each side.}$$

$$(2x - 1)\ln 3 = (x + 2)\ln 5 \quad \text{• Power property}$$

$$2x \ln 3 - \ln 3 = x \ln 5 + 2 \ln 5 \quad \text{• Distributive property}$$

$$2x \ln 3 - x \ln 5 = 2 \ln 5 + \ln 3 \quad \text{• Solve for } x.$$

$$x(2 \ln 3 - \ln 5) = 2 \ln 5 + \ln 3$$

$$x = \frac{2 \ln 5 + \ln 3}{2 \ln 3 - \ln 5} \quad \text{• Exact solution}$$

$$x \approx 7.3 \quad \text{• Decimal approximation}$$

To the nearest 0.1, the solution is 7.3.

Visualize the Solution

Intercept Method The solution of $3^{2x-1} = 5^{x+2}$ is the x-coordinate of the x-intercept of $y = 3^{2x-1} - 5^{x+2}$ (see **Figure 4.42**).

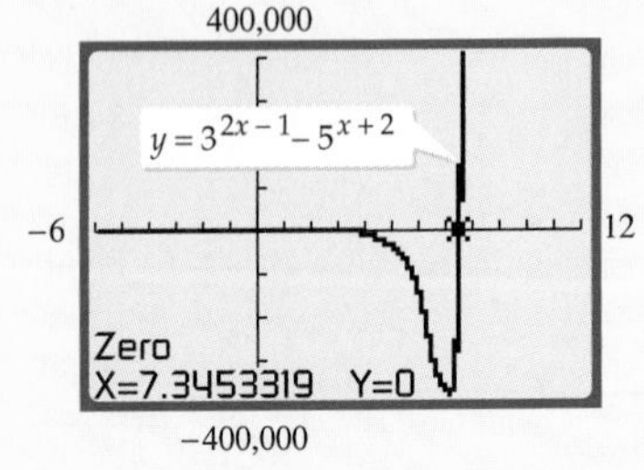

Figure 4.42

TRY EXERCISE 18, EXERCISE SET 4.5, PAGE 348

In Example 4 we solve an exponential equation that has two solutions.

EXAMPLE 4 Solve an Exponential Equation Involving $b^x + b^{-x}$

Solve: $\dfrac{2^x + 2^{-x}}{2} = 3$

Solution

Multiplying each side by 2 produces

$$2^x + 2^{-x} = 6$$

$$2^{2x} + 2^0 = 6(2^x) \qquad \text{• Multiply each side by } 2^x \text{ to clear negative exponents.}$$

$$(2^x)^2 - 6(2^x) + 1 = 0 \qquad \text{• Write in quadratic form.}$$

$$(u)^2 - 6(u) + 1 = 0 \qquad \text{• Substitute } u \text{ for } 2^x.$$

By the quadratic formula,

$$u = \frac{6 \pm \sqrt{36 - 4}}{2} = \frac{6 \pm 4\sqrt{2}}{2} = 3 \pm 2\sqrt{2}$$

$$2^x = 3 \pm 2\sqrt{2} \qquad \text{• Replace } u \text{ with } 2^x.$$

$$\log 2^x = \log\left(3 \pm 2\sqrt{2}\right) \qquad \text{• Take the common logarithm of each side.}$$

$$x \log 2 = \log\left(3 \pm 2\sqrt{2}\right) \qquad \text{• Power property}$$

$$x = \frac{\log\left(3 \pm 2\sqrt{2}\right)}{\log 2} \approx \pm 2.54$$

The approximate solutions are -2.54 and 2.54.

Visualize the Solution

Intersection Method The solutions of $\dfrac{2^x + 2^{-x}}{2} = 3$ are the x-coordinates of the points of intersection of $y = \dfrac{2^x + 2^{-x}}{2}$ and $y = 3$ (see **Figure 4.43**).

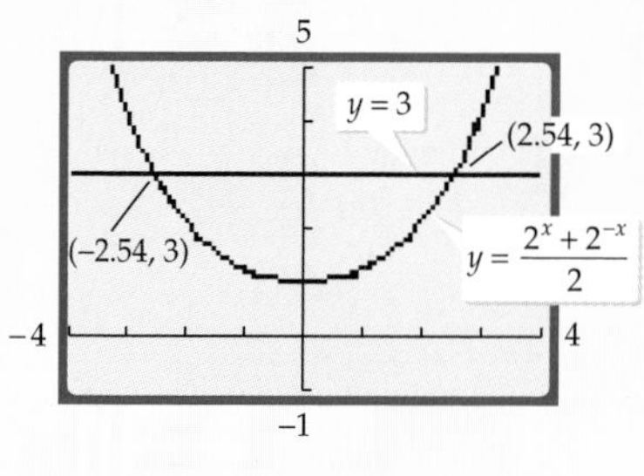

Figure 4.43

TRY EXERCISE 40, EXERCISE SET 4.5, PAGE 349

◆ SOLVE LOGARITHMIC EQUATIONS

Equations that involve logarithms are called **logarithmic equations.** The properties of logarithms, along with the definition of a logarithm, are often used to find the solutions of a logarithmic equation.

EXAMPLE 5 Solve a Logarithmic Equation

Solve: $\log(3x - 5) = 2$

Solution

$$\log(3x - 5) = 2$$

$$3x - 5 = 10^2 \qquad \text{• Definition of a logarithm}$$

Continued •➤

$$3x = 105 \quad \text{• Solve for } x.$$
$$x = 35$$

Check: $\log[3(35) - 5] = \log 100 = 2$

TRY EXERCISE 22, EXERCISE SET 4.5, PAGE 348

QUESTION Can a negative number be a solution of a logarithmic equation?

EXAMPLE 6 Solve a Logarithmic Equation

Solve: $\log 2x - \log(x - 3) = 1$

Solution

$$\log 2x - \log(x - 3) = 1$$
$$\log \frac{2x}{x - 3} = 1 \quad \text{• Quotient property}$$
$$\frac{2x}{x - 3} = 10^1 \quad \text{• Definition of logarithm}$$
$$2x = 10x - 30 \quad \text{• Solve for } x.$$
$$-8x = -30$$
$$x = \frac{15}{4}$$

Check the solution by substituting 15/4 into the original equation.

TRY EXERCISE 26, EXERCISE SET 4.5, PAGE 349

In Example 7 we make use of the one-to-one property of logarithms to find the solution of a logarithmic equation. This example illustrates that the process of solving a logarithmic equation by using logarithmic properties may introduce an extraneous solution.

EXAMPLE 7 Solve a Logarithmic Equation

Solve: $\ln(3x + 8) = \ln(2x + 2) + \ln(x - 2)$

Solution

$$\ln(3x + 8) = \ln(2x + 2) + \ln(x - 2)$$
$$\ln(3x + 8) = \ln[(2x + 2)(x - 2)] \quad \text{• Product property}$$
$$\ln(3x + 8) = \ln(2x^2 - 2x - 4)$$

Visualize the Solution

The graph of $y = \ln(3x + 8) - \ln(2x + 2) - \ln(x - 2)$ has only one x-intercept (see

Continued •

ANSWER Yes. For instance, -10 is a solution of $\log(-x) = 1$.

$$3x + 8 = 2x^2 - 2x - 4$$ • One-to-one property of logarithms

$$0 = 2x^2 - 5x - 12$$

$$0 = (2x + 3)(x - 4)$$ • Solve for x.

$$x = -\frac{3}{2} \quad \text{or} \quad x = 4$$

Thus $-3/2$ and 4 are possible solutions. It can be shown that 4 checks as a solution of the equation but that $-3/2$ does not check. The only solution is 4.

Figure 4.44). Thus there is only one real solution.

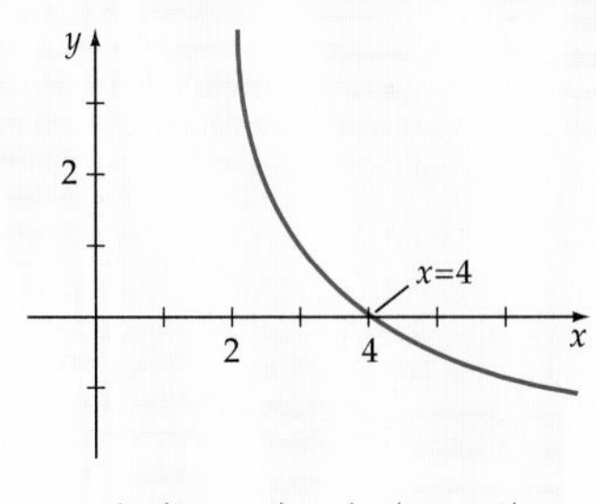

$y = \ln(3x + 8) - \ln(2x + 2) - \ln(x - 2)$

Figure 4.44

TRY EXERCISE 36, EXERCISE SET 4.5, PAGE 349

QUESTION Why does $x = -3/2$ not check in Example 7?

◆ SOLVE AN APPLICATION

EXAMPLE 8 Velocity of an Object Experiencing Air Resistance

The time t in seconds required for an object that is dropped to reach a velocity v (in feet per second) is given by $t = v/32$ when air resistance is not considered. If air resistance is considered, then one possible model is given by $t = 2.43 \ln \frac{150 + v}{150 - v}$ for $0 \leq v < 150$. Use this equation to determine, to the nearest tenth, the velocity of an object that has been falling for 5 seconds. The graph of the equation has a vertical asymptote at $v = 150$. Explain the meaning of that asymptote in the context of this problem.

Solution

Substitute 5 for t and solve for v.

$$t = 2.43 \ln \frac{150 + v}{150 - v}$$

$$5 = 2.43 \ln \frac{150 + v}{150 - v}$$ • Replace t by 5.

$$2.0576132 = \ln \frac{150 + v}{150 - v}$$ • Divide by 2.43.

Continued •

ANSWER If $x = -3/2$, the original equation becomes $\ln(7/2) = \ln(-1) + \ln(-7/2)$. This cannot be true, because the function $f(x) = \ln x$ is not defined for negative values of x.

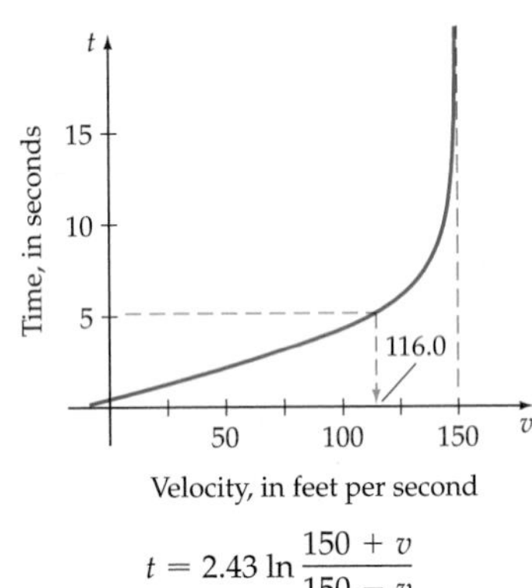

$t = 2.43 \ln \dfrac{150 + v}{150 - v}$

Figure 4.45

$$e^{2.0576132} = \frac{150 + v}{150 - v}$$ • If $a = \ln b$, then $e^a = b$.

$$e^{2.0576132}(150 - v) = 150 + v$$ • Solve for v.

$$150e^{2.0576132} - e^{2.0576132}v = 150 + v$$

$$-v - e^{2.0576132}v = 150 - 150e^{2.0576132}$$

$$v(-e^{2.0576132} - 1) = 150 - 150e^{2.0576132}$$

$$v \approx 116.014$$

After 5 seconds, the velocity of the object will be approximately 116 feet per second. See Figure 4.45.

The vertical asymptote $v = 150$ indicates that the object cannot attain a speed greater than 150 feet per second. In Figure 4.45, note that as $v \to 150$, $t \to \infty$.

TRY EXERCISE 70, EXERCISE SET 4.5, PAGE 351

TOPICS FOR DISCUSSION

1. Discuss how to solve the equation $a = \log_b x$ for x.
2. What is the domain of $y = \log_4 (2x - 5)$? Explain why this means that the equation $\log_4 (x - 3) = \log_4 (2x - 5)$ has no real number solution.
3. -8 is not a solution of the equation $\log_2 x + \log_2 (x + 6) = 4$. Discuss at which step in the following solution the extraneous solution -8 was introduced.

$$\begin{aligned} \log_2 x + \log_2 (x + 6) &= 4 \\ \log_2 x(x + 6) &= 4 \\ x(x + 6) &= 2^4 \\ x^2 + 6x &= 16 \\ x^2 + 6x - 16 &= 0 \\ (x + 8)(x - 2) &= 0 \\ x = -8 \quad \text{or} \quad x &= 2 \end{aligned}$$

EXERCISE SET 4.5

In Exercises 1 to 46, solve for x algebraically.

1. $2^x = 64$
2. $3^x = 243$
3. $49^x = \dfrac{1}{343}$
4. $9^x = \dfrac{1}{243}$
5. $2^{5x+3} = \dfrac{1}{8}$
6. $3^{4x-7} = \dfrac{1}{9}$
7. $\left(\dfrac{2}{5}\right)^x = \dfrac{8}{125}$
8. $\left(\dfrac{2}{5}\right)^x = \dfrac{25}{4}$
9. $5^x = 70$
10. $6^x = 50$
11. $3^{-x} = 120$
12. $7^{-x} = 63$
13. $10^{2x+3} = 315$
14. $10^{6-x} = 550$
15. $e^x = 10$
16. $e^{x+1} = 20$
17. $2^{1-x} = 3^{x+1}$
18. $3^{x-2} = 4^{2x+1}$
19. $2^{2x-3} = 5^{-x-1}$
20. $5^{3x} = 3^{x+4}$
21. $\log (4x - 18) = 1$
22. $\log (x^2 + 19) = 2$

23. $\ln(x^2 - 12) = \ln x$

24. $\log(2x^2 + 3x) = \log(10x + 30)$

25. $\log_2 x + \log_2(x - 4) = 2$

26. $\log_3 x + \log_3(x + 6) = 3$

27. $\log(5x - 1) = 2 + \log(x - 2)$

28. $1 + \log(3x - 1) = \log(2x + 1)$

29. $\ln(1 - x) + \ln(3 - x) = \ln 8$

30. $\log(4 - x) = \log(x + 8) + \log(2x + 13)$

31. $\log\sqrt{x^3 - 17} = \dfrac{1}{2}$

32. $\log(x^3) = (\log x)^2$

33. $\log(\log x) = 1$

34. $\ln(\ln x) = 2$

35. $\ln(e^{3x}) = 6$

36. $\ln x = \dfrac{1}{2}\ln\left(2x + \dfrac{5}{2}\right) + \dfrac{1}{2}\ln 2$

37. $e^{\ln(x-1)} = 4$

38. $10^{\log(2x+7)} = 8$

39. $\dfrac{10^x - 10^{-x}}{2} = 20$

40. $\dfrac{10^x + 10^{-x}}{2} = 8$

41. $\dfrac{10^x + 10^{-x}}{10^x - 10^{-x}} = 5$

42. $\dfrac{10^x - 10^{-x}}{10^x + 10^{-x}} = \dfrac{1}{2}$

43. $\dfrac{e^x + e^{-x}}{2} = 15$

44. $\dfrac{e^x - e^{-x}}{2} = 15$

45. $\dfrac{1}{e^x - e^{-x}} = 4$

46. $\dfrac{e^x + e^{-x}}{e^x - e^{-x}} = 3$

In Exercises 47 to 56, use a graphing utility to approximate the solutions of the equation to the nearest hundredth.

47. $2^{-x+3} = x + 1$

48. $3^{x-2} = -2x - 1$

49. $e^{3-2x} - 2x = 1$

50. $2e^{x+2} + 3x = 2$

51. $3\log_2(x - 1) = -x + 3$

52. $2\log_3(2 - 3x) = 2x - 1$

53. $\ln(2x + 4) + \dfrac{1}{2}x = -3$

54. $2\ln(3 - x) + 3x = 4$

55. $2^{x+1} = x^2 - 1$

56. $\ln x = -x^2 + 4$

57. **POPULATION GROWTH** The population P of a city grows exponentially according to the function

$$P(t) = 8500(1.1)^t, \quad 0 \le t \le 8$$

where t is measured in years.

a. Find the population at time $t = 0$ and also at time $t = 2$.

b. When, to the nearest year, will the population reach 15,000?

58. **PHYSICAL FITNESS** After a race, a runner's pulse rate R in beats per minute decreases according to the function

$$R(t) = 145e^{-0.092t}, \quad 0 \le t \le 15$$

where t is measured in minutes.

a. Find the runner's pulse rate at the end of the race and also 1 minute after the end of the race.

b. How long, to the nearest minute, after the end of the race will the runner's pulse rate be 80 beats per minute?

59. **RATE OF COOLING** A can of soda at 79°F is placed in a refrigerator that maintains a constant temperature of 36°F. The temperature T of the soda t minutes after it is placed in the refrigerator is given by

$$T(t) = 36 + 43e^{-0.058t}$$

a. Find the temperature, to the nearest degree, of the soda 10 minutes after it is placed in the refrigerator.

b. When, to the nearest minute, will the temperature of the soda be 45°F?

60. **MEDICINE** During surgery, a patient's circulatory system requires at least 50 milligrams of an anesthetic. The amount of anesthetic present t hours after 80 milligrams of anesthetic is administered is given by

$$T(t) = 80(0.727)^t$$

a. How much, to the nearest milligram, of the anesthetic is present in the patient's circulatory system 30 minutes after the anesthetic is administered?

b. How long, to the nearest minute, can the operation last if the patient does not receive additional anesthetic?

61. **PSYCHOLOGY** Industrial psychologists study employee training programs to assess the effectiveness of the instruction. In one study, the percent score P on a test for a person who has completed t hours of training was given by

$$P = \frac{100}{1 + 30e^{-0.088t}}$$

a. Use a graphing utility to graph the equation for $t \ge 0$.

b. Use the graph to estimate (to the nearest hour) the number of hours of training necessary to achieve a 70% score on the test.

c. From the graph, determine the horizontal asymptote.

d. Write a sentence that explains the meaning of the horizontal asymptote.

62. **PSYCHOLOGY** An industrial psychologist has determined that the average percent score for an employee on a test of the employee's knowledge of the company's product is given by

$$P = \frac{100}{1 + 40e^{-0.1t}}$$

where t is the number of weeks on the job and P is the percent score.

a. Use a graphing utility to graph the equation for $t \geq 0$.

b. Use the graph to estimate (to the nearest week) the number of weeks of employment that are necessary for the average employee to earn a 70% score on the test.

c. Determine the horizontal asymptote of the graph.

d. Write a sentence that explains the meaning of the horizontal asymptote.

63. **ECOLOGY** A herd of bison was placed in a wildlife preserve that can support a maximum of 1000 bison. A population model for the bison is given by

$$B = \frac{1000}{1 + 30e^{-0.127t}}$$

where B is the number of bisons in the preserve and t is in years, with the year 1999 represented by $t = 0$.

a. Use a graphing utility to graph the equation for $t \geq 0$.

b. Use the graph to estimate (to the nearest year) the number of years before the bison population reaches 500.

c. Determine the horizontal asymptote of the graph.

d. Write a sentence that explains the meaning of the horizontal asymptote.

64. **POPULATION GROWTH** A yeast culture grows according to the equation

$$Y = \frac{50{,}000}{1 + 250e^{-0.305t}}$$

where Y is the number of yeast and t is in hours.

a. Use a graphing utility to graph the equation for $t \geq 0$.

b. Use the graph to estimate (to the nearest hour) the number of hours before the yeast population reaches 35,000.

c. From the graph, estimate the horizontal asymptote.

d. Write a sentence that explains the meaning of the horizontal asymptote.

65. **CONSUMPTION OF NATURAL RESOURCES** A model for how long our coal resources will last is approximated by

$$T = \frac{1}{r}\ln(300r + 1)$$

where r is the percent increase in consumption from current levels of use and T is the time (in years) before the resource is depleted.

a. Use a graphing utility to graph this equation. (*Hint:* r is a percent, so use a domain from 0 to 1.)

b. If our consumption of coal increases by 3%, in how many years (to the nearest year) will we deplete our coal resources?

c. What percent increase in consumption of coal will deplete the resource in 100 years? Round to the nearest tenth of a percent.

66. **CONSUMPTION OF NATURAL RESOURCES** A model for how long our aluminum resources will last is approximated by

$$T = \frac{1}{r}\ln(20{,}500r + 1)$$

where r is the percent increase of consumption from current levels of use and T is the time (in years) before the resource is depleted.

a. Use a graphing utility to graph this equation. (*Hint:* r is a percent, so use a domain from 0 to 1.)

b. If our consumption of aluminum increases by 5% per year, in how many years (to the nearest year) will we deplete our aluminum resources?

c. What percent increase in consumption of aluminum will deplete the resource in 100 years? Round to the nearest tenth of a percent.

67. **CONSUMPTION OF NATURAL RESOURCES** A more accurate model for how long our coal resources will last (see Exercise 65) is given by

$$T = \frac{\ln(300r + 1)}{\ln(r + 1)}$$

where r is the percent increase in consumption from current levels of use and T is the time (in years) before the resource is depleted.

a. Use a graphing utility to graph this equation.

b. If our consumption of coal increases by 3%, in how many years will we deplete our coal resources?

c. What percent increase in consumption of coal will deplete the resource in 100 years? Round to the nearest tenth of a percent.

d. If our consumption of coal increases by r% per year, does this model predict more or less time than the model in Exercise 65 before the resource is depleted? Explain.

68. **CONSUMPTION OF NATURAL RESOURCES** A more accurate model for how long our aluminum resources

will last (see Exercise 66) is given by

$$T = \frac{\ln(20{,}500r + 1)}{\ln(r + 1)}$$

where r is the percent increase in consumption from current levels of use and T is the time (in years) before the resource is depleted.

a. Use a graphing utility to graph this equation.

b. If our consumption of aluminum increases by 5% per year, in how many years (to the nearest year) will we deplete our aluminum resources?

c. What percent increase in consumption of aluminum will deplete the resource in 100 years? Round to the nearest tenth of a percent.

d. If our consumption of aluminum resources increases $r\%$ per year, does this model predict more or less time than the model in Exercise 66 before the resource is depleted? Explain.

69. **VELOCITY OF A SKY DIVER** The time t in seconds required for a sky diver to reach a velocity v in feet per second is given by

$$t = -\frac{175}{32}\ln\left(1 - \frac{v}{175}\right)$$

a. Determine the velocity of the sky diver after 10 seconds.

b. Determine the vertical asymptote for the graph of this function.

c. Write a sentence that describes the meaning of the vertical asymptote in the context of this problem.

70. **EFFECTS OF AIR RESISTANCE ON VELOCITY** If we assume that air resistance is proportional to the square of the velocity, then the time t in seconds required for an object to reach a velocity v in feet per second is given by

$$t = \frac{9}{24}\ln\frac{24 + v}{24 - v}$$

a. Determine the velocity, to the nearest 0.01 foot/second, of the object after 1.5 seconds.

b. Determine the vertical asymptote for the graph of this function.

c. Write a sentence that describes the meaning of the vertical asymptote in the context of this problem.

71. **TERMINAL VELOCITY WITH AIR RESISTANCE** The velocity v of an object t seconds after it's been dropped from a height above the surface of the earth is given by the equation $v = 32t$ feet/second, assuming no air resistance. If we assume that air resistance is proportional to the square of the velocity, then the velocity after t seconds is given by

$$v = 100\left(\frac{e^{0.64t} - 1}{e^{0.64t} + 1}\right)$$

a. In how many seconds will the velocity be 50 feet per second?

b. Determine the horizontal asymptote for the graph of this function.

c. Write a sentence that describes the meaning of the horizontal asymptote in the context of this problem.

72. **TERMINAL VELOCITY WITH AIR RESISTANCE** If we assume that air resistance is proportional to the square of the velocity, then the velocity v in feet per second of an object t seconds after it has been dropped is given by

$$v = 50\left(\frac{e^{1.6t} - 1}{e^{1.6t} + 1}\right)$$

(See Exercise 71. The reason for the difference in the equations is that the proportionality constants are different.)

a. In how many seconds will the velocity be 20 feet per second?

b. Determine the horizontal asymptote for the graph of this function.

c. Write a sentence that describes the meaning of the horizontal asymptote in the context of this problem.

73. **EFFECTS OF AIR RESISTANCE ON DISTANCE** The distance s in feet that the object in Exercise 71 will fall in t seconds is given by

$$s = \frac{100^2}{32}\ln\left(\frac{e^{0.32t} + e^{-0.32t}}{2}\right)$$

a. Use a graphing utility to graph this equation for $t \ge 0$.

b. How long does it take for the object to fall 100 ft? Round to the nearest tenth of a second.

74. **EFFECTS OF AIR RESISTANCE ON DISTANCE** The distance s in feet that the object in Exercise 72 will fall in t seconds is given by

$$s = \frac{50^2}{40}\ln\left(\frac{e^{0.8t} + e^{-0.8t}}{2}\right)$$

a. Use a graphing utility to graph this equation for $t \ge 0$.

b. How long does it take for the object to fall 100 ft? Round to the nearest tenth of a second.

75. **RETIREMENT PLANNING** The retirement account for a graphic designer contains \$250,000 on January 1, 1999, and earns interest at a rate of 0.5% per month. On February 1, 1999, the designer withdraws \$2000 and plans

to continue these withdrawals as retirement income each month. The value V in the account after x months is

$$V = 400{,}000 - 150{,}000(1.005)^x$$

a. What is the domain for this equation?

b. If the designer wishes to leave \$100,000 to a scholarship foundation, what is the maximum number of withdrawals (to the nearest month) the designer can make from this account and still have \$100,000 to donate?

76. **RETIREMENT PLANNING** The retirement account for an assembly line shift manager contains \$300,000 on January 1, 2000, and earns interest at a rate of 0.75% per month. On February 1, 2000, the manager withdraws \$3000 from the account, and these withdrawals continue each month thereafter. The value V in the account after x months is given by the function

$$V = 700{,}000 - 400{,}000(1.0075)^x$$

a. What is the domain for this function?

b. What is the maximum number of monthly withdrawals (to the nearest month) possible at \$3000 per month that the manager can make and still have a balance of more than \$3000?

SUPPLEMENTAL EXERCISES

77. The following argument seems to indicate that $0.125 > 0.25$. Find the first incorrect statement in the argument.

$$3 > 2$$
$$3(\log 0.5) > 2(\log 0.5)$$
$$\log 0.5^3 > \log 0.5^2$$
$$0.5^3 > 0.5^2$$
$$0.125 > 0.25$$

78. The following argument seems to indicate that $4 = 6$. Find the first incorrect statement in the argument.

$$4 = \log_2 16$$
$$4 = \log_2 (8 + 8)$$
$$4 = \log_2 8 + \log_2 8$$
$$4 = 3 + 3$$
$$4 = 6$$

79. A common mistake that students make is to write $\log(x + y)$ as $\log x + \log y$. For what values of x and y does $\log(x + y) = \log x + \log y$? (*Hint:* Solve for x in terms of y.)

80. Which is larger, 500^{501} or 506^{500}? (*Hint:* Let $x = 500^{501}$ and $y = 506^{500}$ and then compare $\ln x$ with $\ln y$.)

81. Explain why the functions $F(x) = 1.4^x$ and $G(x) = e^{0.336x}$ represent essentially the same function.

82. Find the constant k that will make $f(t) = 2.2^t$ and $g(t) = e^{-kt}$ represent essentially the same function.

83. Solve $e^{1/x} > 2$. Write the solution in interval notation.

84. Solve $\log(x^2) > (\log x)^2$. Write the solution in interval notation.

PROJECTS

1. **NAVIGATING** The pilot of a boat is trying to cross a river to a point O two miles due west of the boat's starting position by always pointing the nose of the boat toward O. Suppose the speed of the current is w miles per hour and the speed of the boat is v miles per hour. If point O is the origin and the boat's starting position is $(2, 0)$ (see the diagram at the right), then the equation of the boat's path is given by

$$y = \left(\frac{x}{2}\right)^{1-(w/v)} - \left(\frac{x}{2}\right)^{1+(w/v)}$$

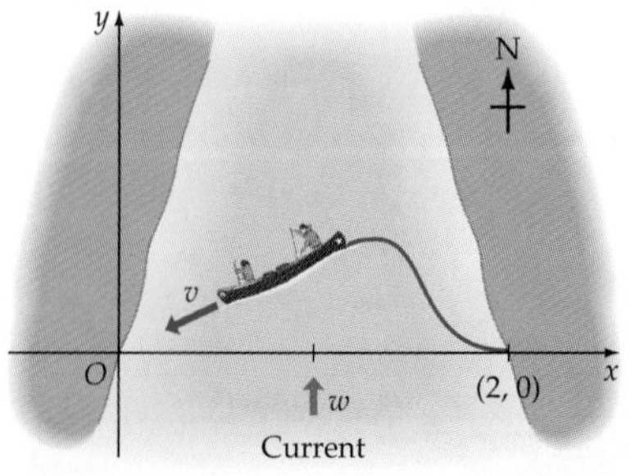

a. If the speed of the current and the speed of the boat are the same, can the pilot reach point O by always having the nose of the boat pointed toward O? If not, at what point will the pilot arrive? Explain your answer.

b. If the speed of the current is greater than the speed of the boat, can the pilot reach point O by always pointing the nose of the boat toward point O? If not, where will the pilot arrive? Explain.

c. If the speed of the current is less than the speed of the boat, can the pilot reach point O by always pointing the nose of the boat toward point O? If not, where will the pilot arrive? Explain.

SECTION 4.6 APPLICATIONS OF EXPONENTIAL AND LOGARITHMIC FUNCTIONS

- COMPOUND INTEREST
- EXPONENTIAL GROWTH
- EXPONENTIAL DECAY
- CARBON DATING
- THE LOGISTIC MODEL

In many applications, a quantity N grows or decays according to the function $N(t) = N_0e^{kt}$. In this function, N is a function of time t, and N_0 is the value of N at time $t = 0$. If k is a *positive* constant, then $N(t) = N_0e^{kt}$ is called an **exponential growth function.** If k is a *negative* constant, then $N(t) = N_0e^{kt}$ is called an **exponential decay function.** The following examples illustrate how growth and decay functions arise naturally in the investigation of certain phenomena.

Interest is money paid for the use of money. The interest I is called **simple interest** if it is a fixed percent r, per time period t, of the amount of money invested. The amount of money invested is called the **principal** P. Simple interest is computed using the formula $I = Prt$. For example, if \$1000 is invested at 12% for 3 years, the simple interest is

$$I = Prt = \$1000(0.12)(3) = \$360$$

The balance after t years is $A = P + I = P + Prt$. In the previous example, the \$1000 invested for 3 years produced \$360 interest. Thus the balance after 3 years is \$1360.

◆ COMPOUND INTEREST

In many financial transactions, interest is added to the principal at regular intervals so that interest is paid on interest as well as on the principal. Interest earned in this manner is called **compound interest.** For example, if \$1000 is invested at 12% annual interest compounded annually for 3 years, then the total interest after 3 years is

First-year interest	\$1000(0.12) = \$120.00	
Second-year interest	\$1120(0.12) = \$134.40	
Third-year interest	\$1254.40(0.12) ≈ \$150.53	
	\$404.93	• **Total interest**

This method of computing the balance can be tedious and time-consuming. A *compound interest formula* that can be used to determine the balance due after t years of compounding can be developed as follows.

Note that if P dollars is invested at an interest rate of r per year, then the balance after one year is $A_1 = P + Pr = P(1 + r)$, where Pr represents the interest earned for the year. Observe that A_1 is the product of the original principal P and $(1 + r)$. If the amount A_1 is reinvested for another year, then the balance after the second year is

$$A_2 = (A_1)(1 + r) = P(1 + r)(1 + r) = P(1 + r)^2$$

Successive reinvestments lead to the results shown in Table 4.10. The equation $A_t = P(1 + r)^t$ is valid if r is the interest rate paid during each of the t years.

Table 4.10

Number of Years	Balance
3	$A_3 = P(1 + r)^3$
4	$A_4 = P(1 + r)^4$
⋮	⋮
n	$A_n = P(1 + r)^n$

If r is an annual interest rate and n is the number of compounding periods per year, then the interest rate each period is r/n and the number of compounding periods after t years is nt. Thus the compound interest formula is expressed as follows:

The Compound Interest Formula

A principal P invested at an annual interest rate r, expressed as a decimal and compounded n times per year for t years, produces the balance

$$A = P\left(1 + \frac{r}{n}\right)^{nt}$$

EXAMPLE 1 Solve a Compound Interest Application

Find the balance if \$1000 is invested at an annual interest rate of 10% for 2 years compounded

a. annually **b.** monthly **c.** daily

Solution

a. Use the compound interest formula, with $P = 1000$, $r = 0.1$, $t = 2$, and $n = 1$.

$$A = \$1000\left(1 + \frac{0.1}{1}\right)^{1\cdot 2} = \$1000(1.1)^2 = \$1210.00$$

b. Because there are 12 months in a year, use $n = 12$.

$$A = \$1000\left(1 + \frac{0.1}{12}\right)^{12\cdot 2} \approx \$1000(1.008333333)^{24} \approx \$1220.39$$

c. Because there are 365 days in a year, use $n = 365$.

$$A = \$1000\left(1 + \frac{0.1}{365}\right)^{365\cdot 2} \approx \$1000(1.000273973)^{730} \approx \$1221.37$$

TRY EXERCISE 4, EXERCISE SET 4.6, PAGE 361

To **compound continuously** means to increase the number of compounding periods without bound.

To derive a continuous compounding interest formula, substitute $1/m$ for r/n in the compound interest formula

$$A = P\left(1 + \frac{r}{n}\right)^{nt} \qquad (1)$$

to produce

$$A = P\left(1 + \frac{1}{m}\right)^{nt} \qquad (2)$$

This substitution is motivated by the desire to express $(1 + r/n)^n$ as $[(1 + 1/m)^m]^r$, which approaches e^r as m gets larger without bound.

Solving the equation $1/m = r/n$ for n yields $n = mr$, so the exponent nt can be written as mrt. Therefore Equation (2) can be expressed as

$$A = P\left(1 + \frac{1}{m}\right)^{mrt} = P\left[\left(1 + \frac{1}{m}\right)^{m}\right]^{rt} \qquad (3)$$

By the definition of e, we know that as m increases without bound,

$$\left(1 + \frac{1}{m}\right)^{m} \quad \text{approaches} \quad e$$

Thus, using continuous compounding, Equation (3) simplifies to $A = Pe^{rt}$.

Continuous Compounding Interest Formula

If an account with principal P and annual interest rate r is compounded continuously for t years, then the balance is $A = Pe^{rt}$.

EXAMPLE 2 Solve a Continuous Compound Interest Application

Find the balance after 4 years on $800 invested at an annual rate of 6% compounded continuously.

Solution

Use the continuous compounding formula.

$$\begin{aligned} A &= Pe^{rt} \\ &= 800e^{0.06(4)} \\ &= 800e^{0.24} \\ &\approx 800(1.27124915) \\ &\approx 1017.00 \end{aligned}$$

• To the nearest cent

The balance after 4 years will be $1017.00

Visualize the Solution

Figure 4.46, a graph of $A = 800e^{0.06t}$, shows that the balance is about $1017.00 when $t = 4$.

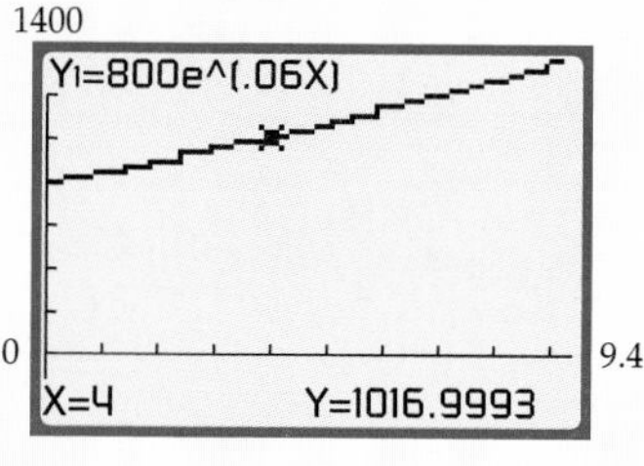

Figure 4.46

TRY EXERCISE 6, EXERCISE SET 4.6, PAGE 361

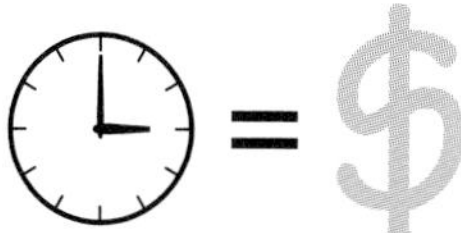

You have probably heard it said that time is money. In fact, many investors ask the question "How long will it take to double my money?" The following example answers this question.

EXAMPLE 3 Doubling a Sum of Money

Find the time it takes for money invested at an annual rate of r to double if the money is compounded continuously.

Continued •►

Solution

Use $A = Pe^{rt}$ with $A = 2P$, twice the principal P.

$$2P = Pe^{rt}$$

$$2 = e^{rt} \qquad \text{• Divide each side by } P.$$

$$\ln 2 = rt \qquad \text{• Take the natural logarithm of each side.}$$

$$\frac{\ln 2}{r} = t \qquad \text{• Solve for } t.$$

The time it takes for money to double when interest is compounded continuously at an annual rate of r is $t = (\ln 2)/r$.

TRY EXERCISE 10, EXERCISE SET 4.6, PAGE 362

◆ EXPONENTIAL GROWTH

Given any two points on the graph of $N(t) = N_0e^{kt}$, you can use the given data to solve for the constants N_0 and k.

EXAMPLE 4 Find the Exponential Growth Function That Models Given Data

a. Find the exponential growth function for a town whose population was 16,400 in 1990 and 20,200 in 2000.

b. Use the function from part **a** to predict, to the nearest 100, the population of the town in 2005.

Solution

a. We need to determine N_0 and k in $N(t) = N_0e^{kt}$. If we represent the year 1990 by $t = 0$, then our given data are $N(0) = 16{,}400$ and $N(10) = 20{,}200$. Because N_0 is defined to be $N(0)$, we know $N_0 = 16{,}400$. To determine k, substitute $t = 10$ and $N_0 = 16{,}400$ into $N(t) = N_0e^{kt}$ to produce

$$N(10) = 16{,}400e^{k \cdot 10}$$

$$20{,}200 = 16{,}400e^{10k} \qquad \text{• Substitute 20,200 for } N(10).$$

$$\frac{20{,}200}{16{,}400} = e^{10k} \qquad \text{• Solve for } e^{10k}.$$

$$\ln \frac{20{,}200}{16{,}400} = \ln e^{10k} \qquad \text{• Take the natural logarithm of each side.}$$

$$\ln \frac{20{,}200}{16{,}400} = 10k \qquad \text{• Use } \log_b (b^p) = p.$$

take note

Because $e^{0.0208} \approx 1.021$, the growth equation can also be written as

$N(t) \approx 16{,}400(1.021)^t$

In this form we see that the population is growing by 2.1% $(1.021 - 1 = 0.021 = 2.1\%)$ per year.

$$\frac{1}{10}\ln\frac{20{,}200}{16{,}400} = k \qquad \text{• Solve for } k.$$

$$0.0208 \approx k$$

The exponential growth function is $N(t) \approx 16{,}400e^{0.0208t}$.

b. The year 1990 was represented by $t = 0$, so we will use $t = 15$ to represent the year 2005.

$$N(t) \approx 16{,}400e^{0.0208t}$$

$$N(15) \approx 16{,}400e^{0.0208 \cdot 15}$$

$$\approx 22{,}400 \quad \text{(nearest 100)}$$

The exponential growth function yields 22,400 as the approximate population of the town in 2005.

TRY EXERCISE 18, EXERCISE SET 4.6, PAGE 362

◆ EXPONENTIAL DECAY

Many radioactive materials *decrease* in mass exponentially over time. This decrease, called radioactive decay, is measured in terms of **half-life,** which is defined as the time required for the disintegration of half the atoms in a sample of a radioactive substance. Table 4.11 shows the half-lives of selected radioactive isotopes.

Table 4.11

Isotope	Half-Life
Carbon (^{14}C)	5730 years
Radium (^{226}Ra)	1660 years
Polonium (^{210}Po)	138 days
Phosphorus (^{32}P)	14 days
Polonium (^{214}Po)	1/10,000th of a second

EXAMPLE 5 Find the Exponential Decay Function That Models Given Data

Find the exponential decay function for the amount of phosphorus (^{32}P) that remains in a sample after t days.

Solution

When $t = 0$, $N(0) = N_0e^{k(0)} = N_0$. Thus $N(0) = N_0$. Also, because the phosphorus has a half-life of 14 days (from Table 4.11) $N(14) = 0.5N_0$. To find k, substitute $t = 14$ into $N(t) = N_0e^{kt}$ and solve for k.

$$N(14) = N_0 \cdot e^{k \cdot 14}$$

$$0.5N_0 = N_0e^{14k} \qquad \text{• Substitute } 0.5N_0 \text{ for } N(14).$$

$$0.5 = e^{14k} \qquad \text{• Divide each side by } N_0.$$

$$\ln 0.5 = 14k \qquad \text{• Take the natural logarithm of each side.}$$

$$\frac{1}{14}\ln 0.5 = k \qquad \text{• Solve for } k.$$

$$-0.0495 \approx k$$

The exponential decay function is $N(t) = N_0e^{-0.0495t}$.

TRY EXERCISE 20, EXERCISE SET 4.6, PAGE 362

take note

Because $e^{-0.0495} \approx (0.5)^{1/14}$, the decay function $N(t) = N_0e^{-0.0495t}$ can also be written as $N(t) = N_0(0.5)^{t/14}$. In this form it is easy to see that if t is increased by 14, then N will decrease by a factor of 0.5.

EXAMPLE 6 Application to Air Resistance

Assuming that air resistance is proportional to the velocity of a falling object, the velocity (in feet per second) of the object t seconds after it has been dropped is given by $v = 82(1 - e^{-0.39t})$.

a. Determine when the velocity will be 70 feet per second.

b. Write a sentence that explains the meaning of the horizontal asymptote, which is $v = 82$, in the context of this problem.

Solution

a.

$$v = 82(1 - e^{-0.39t})$$

$$70 = 82(1 - e^{-0.39t}) \quad \text{• Replace } v \text{ by 70.}$$

$$\frac{70}{82} = 1 - e^{-0.39t} \quad \text{• Divide each side by 82.}$$

$$e^{-0.39t} = 1 - \frac{70}{82} \quad \text{• Solve for } e^{-0.39t}.$$

$$\ln e^{-0.39t} = \ln \frac{6}{41} \quad \text{• Take the natural logarithm of each side.}$$

$$-0.39t = \ln \frac{6}{41} \quad \text{• Solve for } t.$$

$$t = \frac{\ln(6/41)}{-0.39} \approx 4.9277246$$

The time is approximately 4.9 seconds.

b. The horizontal asymptote $v = 82$ means that as time increases, the velocity of the object will approach but never exceed 82 feet per second.

Visualize the Solution

a. A graph of $y = 82(1 - e^{-0.39x})$ and $y = 70$ shows that the x-coordinate of the point of intersection is about 4.9.

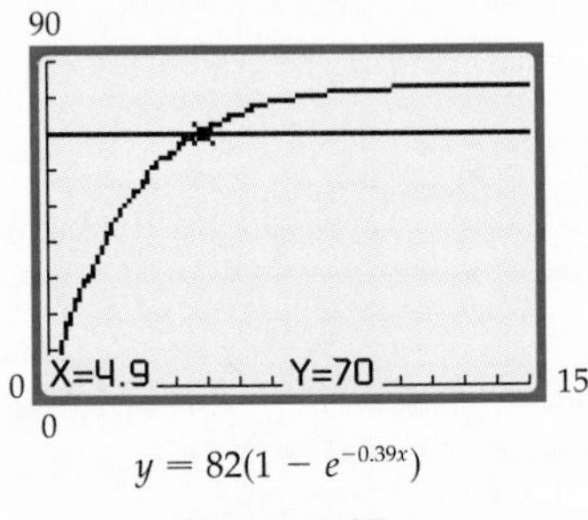

$y = 82(1 - e^{-0.39x})$

Figure 4.47

Note: The x value shown is rounded to the nearest tenth.

TRY EXERCISE 32, EXERCISE SET 4.6, PAGE 363

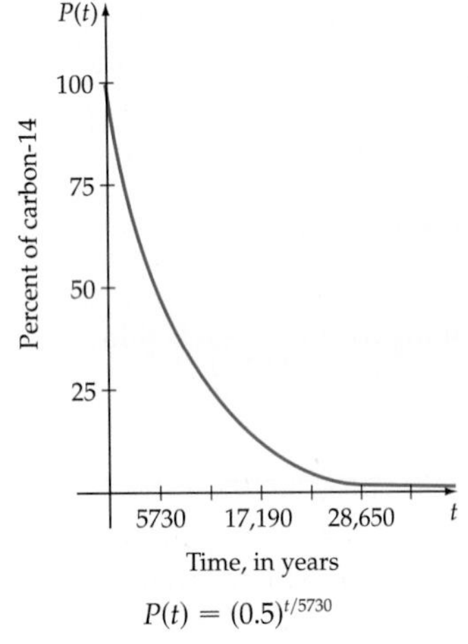

$P(t) = (0.5)^{t/5730}$

Figure 4.48

◆ CARBON DATING

The bone tissue in all living animals contains both carbon-12, which is nonradioactive, and carbon-14, which is radioactive with a half-life of approximately 5730 years. See **Figure 4.48.** As long as the animal is alive, the ratio of carbon-14 to carbon-12 remains constant. When the animal dies ($t = 0$), the carbon-14 begins to decay. Thus a bone that has a smaller ratio of carbon-14 to carbon-12 is older than a bone that has a larger ratio. The percent of carbon-14 present at time t is

$$P(t) = 0.5^{t/5730}$$

EXAMPLE 7 Application to Archeology

Find the age of a bone if it now has 85% of the carbon-14 it had when $t = 0$.

MATH MATTERS

The chemist Willard Frank Libby developed the carbon-14 dating technique in 1947. In 1960 he was awarded the Nobel Prize in chemistry for this achievement.

Solution

Let t be the time at which $P(t) = 0.85$

$$0.85 = 0.5^{t/5730}$$

$$\ln 0.85 = \ln 0.5^{t/5730}$$ • Take the natural logarithm of each side.

$$\ln 0.85 = \frac{t}{5730} \ln 0.5$$ • Power property

$$5730\left(\frac{\ln 0.85}{\ln 0.5}\right) = t$$ • Solve for t.

$$1340 \approx t$$ • To the nearest 10 years

The bone is about 1340 years old.

TRY EXERCISE 24, EXERCISE SET 4.6, PAGE 362

◆ THE LOGISTIC MODEL

The population growth model $P(t) = P_0e^{kt}$ is called the **Malthusian model** after Robert Malthus (1766–1834), who wrote about population growth in *An Essay on the Principle of Population Growth,* which was published in 1798. The Malthusian model does not consider that there are limited resources (for instance, food) and that this limitation will eventually curb population growth.

The *logistic model* is a growth model that does consider limited resources.

The Logistic Model

The magnitude of a population at time t is given by

$$P(t) = \frac{mP_0}{P_0 + (m - P_0)e^{-kt}}$$

where $P_0 = P(0)$ is the population at time $t = 0$, m is the maximum population that can be supported, and k is a constant called the **growth rate constant.**

EXAMPLE 8 Use the Logistic Model to Estimate the World's Population

In 1987 the world's population first reached 5 billion, and in 1999 it reached 6 billion. Assume that the maximum possible world population is 14 billion and that its population growth satisfies the logistic model.

Continued •➤

a. Use the data given to determine the growth rate constant for a logistic model of the world's population.

b. Use the logistic model from part **a,** to predict the year in which the world's population will reach 8 billion.

Solution

a. If we represent the year 1987 by $t = 0$, then the year 1999 will be represented by $t = 12$. In the logistic model we make the following substitutions.

$$P_0 = 5, \qquad m = 14, \qquad t = 12, \qquad P(12) = 6$$

$$P(t) = \frac{mP_0}{P_0 + (m - P_0)e^{-kt}}$$

$$P(12) = \frac{14 \cdot 5}{5 + (14 - 5)e^{-k \cdot 12}}$$

$$6 = \frac{70}{5 + 9e^{-k \cdot 12}}$$

$$6(5 + 9e^{-k \cdot 12}) = 70$$

$$30 + 54e^{-k \cdot 12} = 70$$

$$54e^{-k \cdot 12} = 40$$

$$e^{-k \cdot 12} = \frac{40}{54}$$

$$-k \cdot 12 = \ln\left(\frac{40}{54}\right)$$

$$k = -\frac{1}{12}\ln\left(\frac{40}{54}\right)$$

$$k \approx 0.0250087$$

With $k = 0.0250087$, the logistic model is $P(t) = \dfrac{70}{5 + 9e^{-0.0250087t}}$.

b. To determine in what year the logistic model predicts that the world's population will first reach 8 billion, replace $P(t)$ with 8 and solve for t.

$$8 = \frac{70}{5 + 9e^{-0.0250087t}}$$

$$8 \cdot (5 + 9e^{-0.0250087t}) = 70$$

$$5 + 9e^{-0.0250087t} = \frac{70}{8}$$

$$9e^{-0.0250087t} = 8.75 - 5$$

$$e^{-0.0250087t} = \frac{3.75}{9}$$

$$-0.0250087t = \ln\left(\frac{3.75}{9}\right)$$

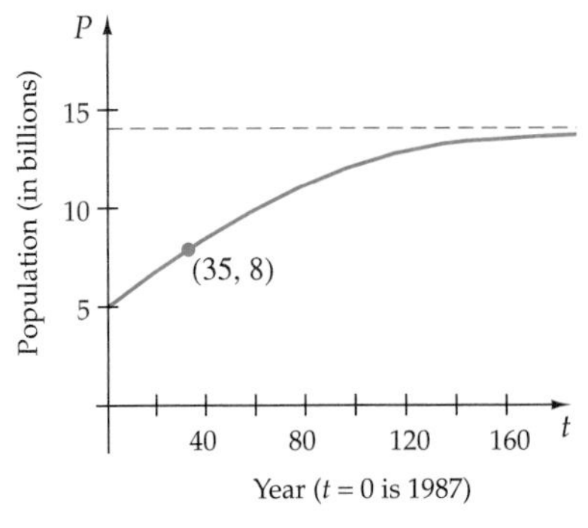

$$P(t) = \frac{70}{5 + 9e^{-0.0250087t}}$$

Figure 4.49

$$t = \frac{1}{-0.0250087} \ln\left(\frac{3.75}{9}\right)$$

$$t \approx 35.007$$

Thus, according to the logistic model, the world's population will reach 8 billion people a little more than 35 years after 1987, in 2022.

TRY EXERCISE 36, EXERCISE SET 4.6, PAGE 364

The graph of $P(t) = 70/(5 + 9e^{-0.0250087t})$ from Example 8 is shown in **Figure 4.49.** Note that when $t = 35$, $P \approx 8$ billion. Also note that as $t \to \infty$, the graph approaches the horizontal asymptote $P = 14$.

TOPICS FOR DISCUSSION

1. Explain the difference between compound interest and simple interest.
2. What is an exponential growth model? Give an example of an application for which the exponential growth model might be appropriate.
3. What is an exponential decay model? Give an example of an application for which the exponential decay model might be appropriate.
4. Consider the exponential model $P(t) = P_0e^{kt}$, and the logistic model $P(t) = \dfrac{mP_0}{P_0 + (m - P_0)e^{-kt}}$. Explain the similarities and differences between the two models.

EXERCISE SET 4.6

1. **COMPOUND INTEREST** If \$8000 is invested at an annual interest rate of 5% and compounded annually, find the balance after

   a. 4 years b. 7 years

2. **COMPOUND INTEREST** If \$22,000 is invested at an annual interest rate of 4.5% and compounded annually, find the balance after

   a. 2 years b. 10 years

3. **COMPOUND INTEREST** If \$38,000 is invested at an annual interest rate of 6.5% for 4 years, find the balance if the interest is compounded

   a. annually b. daily c. hourly

4. **COMPOUND INTEREST** If \$12,500 is invested at an annual interest rate of 8% for 10 years, find the balance if the interest is compounded

   a. annually b. daily c. hourly

5. **COMPOUND INTEREST** Find the balance if \$15,000 is invested at an annual rate of 10% for 5 years, compounded continuously.
6. **COMPOUND INTEREST** Find the balance if \$32,000 is invested at an annual rate of 8% for 3 years, compounded continuously.
7. **COMPOUND INTEREST** How long will it take \$4000 to double if it is invested in a certificate of deposit that pays 7.84% annual interest compounded continuously? Round to the nearest tenth of a year.
8. **COMPOUND INTEREST** How long will it take \$25,000 to double if it is invested in a savings account that pays 5.88% annual interest compounded continuously? Round to the nearest tenth of a year.
9. **CONTINUOUS COMPOUNDING INTEREST** Use the Continuous Compounding Interest Formula to derive an expression for the time it will take money to triple when

invested at an annual interest rate of r compounded continuously.

10. **CONTINUOUS COMPOUNDING INTEREST** How long will it take \$1000 to triple if it is invested at an annual interest rate of 5.5% compounded continuously? Round to the nearest year.

11. **CONTINUOUS COMPOUNDING INTEREST** How long will it take \$6000 to triple if it is invested in a savings account that pays 7.6% annual interest compounded continuously? Round to the nearest year.

12. **CONTINUOUS COMPOUNDING INTEREST** How long will it take \$10,000 to triple if it is invested in a savings account that pays 5.5% annual interest compounded continuously? Round to the nearest year.

13. **POPULATION GROWTH** The number of bacteria $N(t)$ present in a culture at time t hours is given by

$$N(t) = 2200(2)^t$$

Find the number of bacteria present when

a. $t = 0$ hours b. $t = 3$ hours

14. **POPULATION GROWTH** The population of a town grows exponentially according to the function

$$f(t) = 12{,}400(1.14)^t$$

for $0 \le t \le 5$ years. Find, to the nearest hundred, the population of the town when t is

a. 3 years b. 4.25 years

15. **POPULATION GROWTH** A town had a population of 22,600 in 1990 and a population of 24,200 in 1995.

a. Find the exponential growth function for the town. Use $t = 0$ to represent the year 1990.

b. Use the growth function to predict the population of the town in 2005. Round to the nearest hundred.

16. **POPULATION GROWTH** A town had a population of 53,700 in 1996 and a population of 58,100 in 2000.

a. Find the exponential growth function for the town. Use $t = 0$ to represent the year 1996.

b. Use the growth function to predict the population of the town in 2008. Round to the nearest hundred.

17. **POPULATION GROWTH** The growth of the population of Los Angeles, California, for the years 1992 through 1996 can be approximated by the equation

$$P = 10{,}130(1.005)^t$$

where $t = 0$ corresponds to January 1, 1992 and P is in thousands.

a. Assuming this growth rate continues, what will be the population of Los Angeles on January 1 in the year 2004?

b. In what year will the population of Los Angeles first exceed 13,000,000?

18. **POPULATION GROWTH** The growth of the population of Mexico City, Mexico, for the years 1991 through 1998 can be approximated by the equation

$$P = 20{,}899(1.027)^t$$

where $t = 0$ corresponds to 1991 and P is in thousands.

a. Assuming this growth rate continues, what will be the population of Mexico City in the year 2003?

b. Assuming this growth rate continues, in what year will the population of Mexico City first exceed 35,000,000?

19. **MEDICINE** Sodium-24 is a radioactive isotope of sodium that is used to study circulatory dysfunction. Assuming that 4 micrograms of sodium-24 is injected into a person, the amount A in micrograms remaining in that person after t hours is given by the equation $A = 4e^{-0.046t}$.

a. Graph this equation.

b. What amount of sodium-24 remains after 5 hours?

c. What is the half-life of sodium-24?

d. In how many hours will the amount of sodium-24 be 1 microgram?

20. **RADIOACTIVE DECAY** Polonium (^{210}Po) has a half-life of 138 days. Find the decay function for the amount of polonium (^{210}Po) that remains in a sample after t days.

21. **GEOLOGY** Geologists have determined that Crater Lake in Oregon was formed by a volcanic eruption. Chemical analysis of a wood chip that is assumed to be from a tree that died during the eruption has shown that it contains approximately 45% of its original carbon-14. Determine how long ago the volcanic eruption occurred. Use 5730 years as the half-life of carbon-14.

22. **RADIOACTIVE DECAY** Use $N(t) = N_0(0.5)^{t/138}$, where t is measured in days, to estimate the percentage of polonium (^{210}Po) that remains in a sample after 2 years.

23. **ARCHEOLOGY** The Rhind papyrus, named after A. Henry Rhind, contains most of what we know today of ancient Egyptian mathematics. A chemical analysis of a sample from the papyrus has shown that it contains approximately 75% of its original carbon-14. What is the age of the Rhind papyrus? Use 5730 years as the half-life of carbon-14.

24. **ARCHEOLOGY** Determine the age of a bone if it now contains 65% of its original amount of carbon-14. Round to the nearest 100 years.

25. **PHYSICS** Newton's Law of Cooling states that if an object at temperature T_0 is placed into an environment at constant temperature A, then the temperature of the object, $T(t)$, after t minutes is given by $T(t) = A + (T_0 - A)e^{-kt}$, where k is a constant that depends on the object.

a. Determine the constant k (to the nearest thousandth) for a canned soda drink that takes 5 minutes to cool from 75°F to 65°F after being placed in a refrigerator that maintains a constant temperature of 34°F.

b. What will be the temperature (to the nearest degree) of the soda drink after 30 minutes?

c. When (to the nearest minute) will the temperature of the soda drink be 36°F?

26. **PSYCHOLOGY** According to a software company, the users of its typing tutorial can expect to type $N(t)$ words per minute after t hours of practice with the product, according to the function $N(t) = 100(1.04 - 0.99^t)$.

a. How many words per minute can a student expect to type after 2 hours of practice?

b. How many words per minute can a student expect to type after 40 hours of practice?

c. According to the function N, how many hours (to the nearest hour) of practice will be required before a student can expect to type 60 words per minute?

27. **PSYCHOLOGY** In the city of Whispering Palms, which has a population of 80,000 people, the number of people $P(t)$ exposed to a rumor in t hours is given by the function $P(t) = 80{,}000(1 - e^{-0.0005t})$.

a. Find the number of hours until 10% of the population have heard the rumor.

b. Find the number of hours until 50% of the population have heard the rumor.

28. **LAW** A lawyer has determined that the number of people $P(t)$ in a city of 1,200,000 people who have been exposed to a news item after t days is given by the function

$$P(t) = 1{,}200{,}000(1 - e^{-0.03t})$$

a. How many days after a major crime has been reported have 40% of the population heard of the crime?

b. A defense lawyer knows it will be very difficult to pick an unbiased jury after 80% of the population have heard of the crime. After how many days will 80% of the population have heard of the crime?

29. **DEPRECIATION** An automobile depreciates according to the function $V(t) = V_0(1 - r)^t$, where $V(t)$ is the value in dollars after t years, V_0 is the original value, and r is the yearly depreciation rate. A car has a yearly depreciation rate of 20%. Determine, to the nearest 0.1 year, in how many years the car will depreciate to half its original value.

30. **PHYSICS** The current $I(t)$ (measured in amperes) of a circuit is given by the function $I(t) = 6(1 - e^{-2.5t})$, where t is the number of seconds after the switch is closed.

a. Find the current when $t = 0$.

b. Find the current when $t = 0.5$.

c. Solve the equation for t.

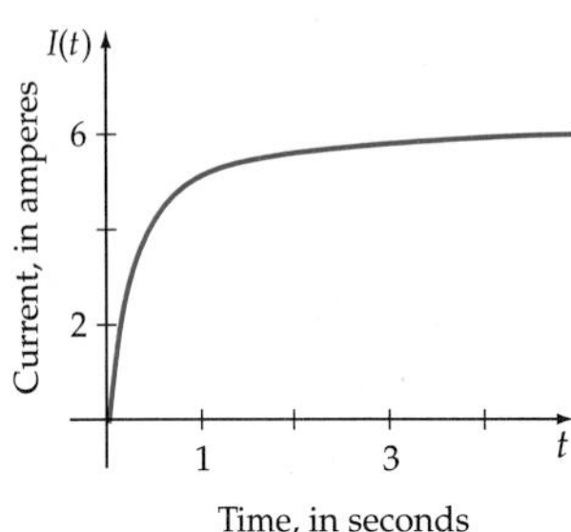

31. **AIR RESISTANCE** Assuming that air resistance is proportional to velocity, the velocity v, in feet per second, of a certain object after t seconds is given by $v = 32(1 - e^{-t})$.

a. Graph this equation for $t \geq 0$.

b. Determine algebraically, to the nearest 0.01 second, when the velocity is 20 feet per second.

c. Determine the horizontal asymptote of the graph of v.

d. Write a sentence that explains the meaning of the horizontal asymptote in the context of this application.

32. **AIR RESISTANCE** Assuming that air resistance is proportional to velocity, the velocity v, in feet per second, of a certain object after t seconds is given by $v = 64(1 - e^{-t/2})$.

a. Graph this equation for $t \geq 0$.

b. Determine algebraically, to the nearest 0.1 second, when the velocity is 50 feet per second.

c. Determine the horizontal asymptote of the graph of v.

d. Write a sentence that explains the meaning of the horizontal asymptote in the context of this application.

33. The distance s (in feet) that the object in Exercise 31 will fall in t seconds is given by $s = 32t + 32(e^{-t} - 1)$.

a. Use a graphing utility to graph this equation for $t \geq 0$.

b. Determine, to the nearest 0.1 second, the time it takes the object to fall 50 feet.

c. Calculate the slope of the secant line through $(1, s(1))$ and $(2, s(2))$.

d. Write a sentence that explains the meaning of the slope of the secant line you calculated in **c**.

34. The distance s (in feet) that the object in Exercise 32 will fall in t seconds is given by $s = 64t + 128(e^{-t/2} - 1)$.

a. Use a graphing utility to graph this equation for $t \geq 0$.

b. Determine, to the nearest 0.1 second, the time it takes the object to fall 50 feet.

c. Calculate the slope of the secant line through $(1, s(1))$ and $(2, s(2))$.

d. Write a sentence that explains the meaning of the slope of the secant line you calculated in **c**.

35. **POPULATION GROWTH** The population of squirrels in a nature preserve satisfies the logistic model with $P_0 = 1500$ in 1995, $m = 16{,}500$, and $P(2) = 2500$ (the population in 1997).

a. Determine k to the nearest hundredth.

b. Determine the year in which the population first exceeds 10,000.

36. **POPULATION GROWTH** The population of walruses in a colony satisfies the logistic model with $P_0 = 800$ in 1998, $m = 5500$, and $P(1) = 900$ (the population in 1999).

a. Determine k to the nearest hundredth.

b. Determine the year in which the population first exceeds 2000.

37. **RATE OF GROWTH** The *rate of growth* of the squirrel population in Exercise 35 is given by

$$R = \frac{47{,}850e^{0.29t}}{(10 + e^{0.29t})^2}$$

where R is the rate of growth; that is, the units of R are squirrels per year.

a. Use a graphing utility to graph this equation.

b. What is the maximum value, to the nearest whole number, of this function?

c. What is the t-coordinate, to the nearest hundredth, associated with the maximum value of the function?

d. Write a sentence that explains the meaning of the maximum value of this function in the context of the application.

38. **RATE OF GROWTH** The *rate of growth* of the walrus population in Exercise 36 is given by

$$R = \frac{289{,}520e^{0.14t}}{(47 + 8e^{0.14t})^2}$$

where R is the rate of growth; that is, the units of R are walruses per year.

a. Use a graphing utility to graph this equation.

b. What, to the nearest whole number, is the maximum value of this function?

c. What, to the nearest 0.1 year, is the t-coordinate associated with the maximum value of the function?

d. Write a sentence that explains the meaning of the maximum value of this function in the context of the application.

39. **RATE OF GROWTH** The equation for the *rate of growth* of the bison population in Exercise 63 of Section 4.5 is given by

$$R = \frac{3810e^{0.127t}}{(30 + e^{0.127t})^2}$$

where R is the rate of growth; that is, the units of R are bison per year.

a. Use a graphing utility to graph this equation using the domain you used in Exercise 63 of Section 4.5.

b. What is the maximum value of this function?

c. What, to the nearest hundredth, is the t-coordinate associated with the maximum value of the function?

d. Write a sentence that explains the meaning of the maximum value of this function in the context of the application.

40. **RATE OF GROWTH** The equation for the *rate of growth* of the yeast population in Exercise 64 of Section 4.5 is given by

$$R = \frac{3.8125 \times 10^6 e^{0.305t}}{(250 + e^{0.305t})^2}$$

where R is the rate of growth; that is, the units of R are numbers of yeast per hour.

a. Use a graphing utility to graph this equation using the domain you used in Exercise 64 of Section 4.5.

b. What, to the nearest 0.1, is the maximum value of this function?

c. What, to the nearest 0.1, is the t-coordinate associated with the maximum value of the function?

d. Write a sentence that explains the meaning of the maximum value of this function in the context of the application.

41. **LEARNING THEORY** The logistic model is also used in learning theory. Suppose that historical records from employee training at a company show that the percent score on a product information test is given by

$$P = \frac{100}{1 + 25e^{-0.095t}}$$

where t is the number of hours of training. What is the number of hours (to the nearest hour) of training needed before a new employee will answer 75% of the questions correctly?

42. **LEARNING THEORY** A company provides training in the assembly of a computer circuit to new employees. Past experience has shown that the number of correctly assembled circuits per week can be modeled by

$$N = \frac{250}{1 + 249e^{-0.503t}}$$

where t is the number of weeks of training. What is the number of weeks (to the nearest week) of training needed before a new employee will correctly make 140 circuits?

43. **PREDICTING ADEQUACY OF RESOURCES** The adequacy of a city's resources can sometimes be modeled by a *gamma density function*. The distribution created from this function enables city planners to determine the probability that certain city services can be maintained. Suppose a city has determined that the probability of its being able to provide more than x million liters of water per day is given by

$$P = \left(\frac{1}{3}x + 1\right)e^{-x/3}$$

a. Use a graphing utility to graph $P(x)$ for $x \geq 0$.

b. Determine the probability that a city can supply more than 5 million liters of water per day.

c. The city manager wants to determine the minimum water supply in a reservoir that the city can maintain so that there is less than a 0.25 chance that the city will not be able to meet demand. What must the capacity of the water supply be to meet the goal of the manager?

d. As $x \to \infty$, $P \to 0$. Explain why this makes sense in the context of this application.

44. **PREDICTING ADEQUACY OF RESOURCES** The probability (see Exercise 43) that an electric company can supply more than x million kilowatt-hours of electricity per day is given by

$$P = \left(\frac{1}{4}x + 1\right)e^{-x/4}$$

a. Use a graphing utility to graph $P(x)$ for $x \geq 0$.

b. Determine, to the nearest 0.001, the probability that this electric company can supply more than 8 million kilowatt-hours of electricity per day.

c. The electric company wants to determine what capacity it must have so that there is less than a 0.50 chance that the company will not be able to meet demand. What, to the nearest 0.1 million, must the capacity of the electric company supply be to meet the goal of the company?

d. As $x \to \infty$, $P \to 0$. Explain why this makes sense in the context of this application.

45. **PHYSICS** If air resistance is proportional to velocity, then the time t in seconds for a particular object to reach a velocity of v feet per second is given by

$$t = 3.125 \ln \frac{100}{100 - v}$$

a. How long, to the nearest 0.1 second, is required before the velocity is 50 feet per second?

b. There is a vertical asymptote when $v = 100$. Describe the meaning of this asymptote in the context of the application.

46. **PHYSICS** If air resistance is proportional to velocity, then the time t in seconds for a particular object to reach a velocity of v feet per second is given by

$$t = 2.34 \ln \frac{75}{75 - v}$$

a. How long, to the nearest 0.1 second, is required before the velocity is 30 feet per second?

b. There is a vertical asymptote when $v = 75$. Describe the meaning of this asymptote in the context of the application.

SUPPLEMENTAL EXERCISES

47. **MEDICATION LEVEL** A patient is given three dosages of aspirin. Each dosage contains 1 gram of aspirin. The second and third dosages are each taken 3 hours after the previous dosage is administered. The half-life of the aspirin is 2 hours. The amount of aspirin, A, in the patient's body t hours after the first dosage is administered is

$$A(t) = \begin{cases} 0.5^{t/2} & 0 \leq t < 3 \\ 0.5^{t/2} + 0.5^{(t-3)/2} & 3 \leq t < 6 \\ 0.5^{t/2} + 0.5^{(t-3)/2} + 0.5^{(t-6)/2} & t \geq 6 \end{cases}$$

Find, to the nearest 0.01 gram, the amount of aspirin in the patient's body when

a. $t = 1$ b. $t = 4$ c. $t = 9$

48. **MEDICATION LEVEL** Use a graphing calculator and the dosage formula in Exercise 47 to determine when, to the nearest 0.1 hour, the amount of aspirin in the patient's body first reaches 0.25 gram.

Exercises 49 to 51 make use of the factorial function which is defined as follows. For whole numbers *n*, the number *n*!, (which is read "*n* factorial"), is given by

$$n! = \begin{cases} n(n-1)(n-2)\cdots 1, & \text{if } n \geq 1 \\ 1, & \text{if } n = 0 \end{cases}$$

Thus, 0! = 1 and 4! = 4 · 3 · 2 · 1 = 24.

49. **QUEUEING THEORY** A study shows that the number of people who arrive at a bank teller's window averages 4.1 people every 10 minutes. The probability P that exactly x people will arrive at the teller's window in a given 10-minute period is

$$P(x) = \frac{4.1^x e^{-4.1}}{x!}$$

Find, to the nearest 0.1%, the probability that in a given 10-minute period, exactly

a. 0 people arrive at the window.

b. 2 people arrive at the window.

c. 3 people arrive at the window.

d. 4 people arrive at the window.

e. 9 people arrive at the window.

As $x \to \infty$, what does P approach?

50. STIRLING'S FORMULA *Stirling's Formula* (after James Stirling, 1692–1770),

$$n! \approx \left(\frac{n}{e}\right)^n \sqrt{2\pi n}$$

is often used to approximate very large factorials. Use Stirling's Formula to approximate 10!, and then compute the ratio of Stirling's approximation of 10! divided by the actual value of 10!, which is 3,628,800.

51. RUBIK'S CUBE The Rubik's cube shown here was invented by Erno Rubik in 1975. The small outer cubes are held together in such a way that they can be rotated around three axes. The total number of positions in which the Rubik's cube can be arranged is

$$\frac{3^8 2^{12} 8!\, 12!}{2 \cdot 3 \cdot 2}$$

If you can arrange a Rubik's cube into a new arrangement every second, how many centuries would it take to place the cube into each of its arrangements? Assume that there are 365 days in a year.

52. AGRICULTURE A farmer knows that planting the same crop in the same field year after year reduces the yield. If the yield on each succeeding year's crop is 90% of the preceding year's yield, then the yield $Y(t)$ at any time t is given by the function $Y(t) = Y_0(0.90)^t$, where Y_0 is the yield when $t = 0$. In how many years (to the nearest year) will the yield be 60% of Y_0?

53. OIL SPILLS Crude oil leaks from a tank at a rate that depends on the amount of oil that remains in the tank. Because 1/8 of the oil in the tank leaks out every 2 hours, the volume of oil $V(t)$ in the tank at t hours is given by $V(t) = V_0(0.875)^{t/2}$, where $V_0 = 350{,}000$ gallons is the number of gallons in the tank at the time the tank started to leak ($t = 0$).

a. How many gallons does the tank hold after 3 hours?

b. How many gallons does the tank hold after 5 hours?

c. How long, to the nearest hour, will it take until 90% of the oil has leaked from the tank?

54. HANGING CABLE The height h in feet of any point P on the cable shown is given by

$$h(x) = 10(e^{x/20} + e^{-x/20}), \quad -15 \le x \le 15$$

where x is the horizontal distance in feet between P and the y-axis.

a. What is the lowest height of the cable?

b. What is the height of the cable 10 feet to the right of the y-axis? Round to the nearest 0.1 foot.

c. How far to the right of the y-axis is the cable 24 feet in height? Round to the nearest 0.1 foot.

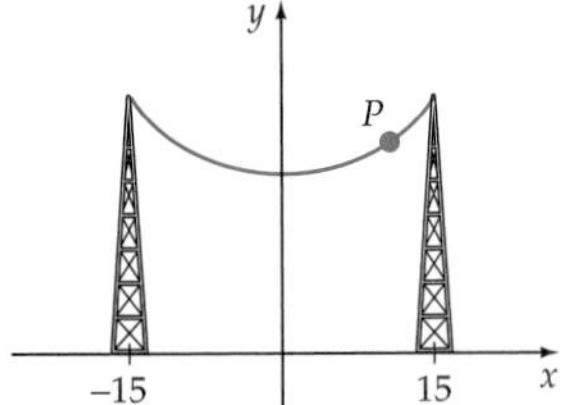

55. AIR POLLUTION MODEL The following diagram shows a plume of smoke from a smokestack.

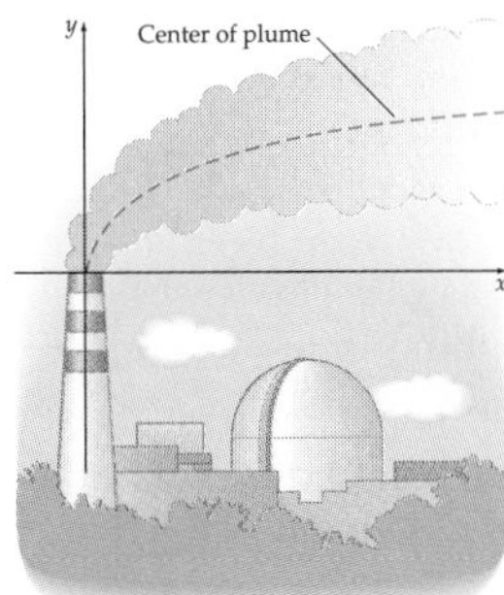

For the plume in the diagram, the concentration, C, of a pollutant (in grams per cubic meter) at a point along the center of the plume, x kilometers downwind, is

$$C = \left(\frac{E}{w(5000x^2 + 350x + 6)\pi}\right) e^{(-1.4^2/(50x+2)^2)}$$

The variable E is the emission rate of the pollutant at the source (in grams per second), and w is the speed of the wind in meters per second. Assume that $E = 4$ g/s of nitrogen oxides and $w = 5$ m/s. Find, to 5 significant digits, the concentration of nitrogen oxides in the center of the plume, at a distance of

a. $x = 0.5$ km b. $x = 1$ km c. $x = 1.5$ km

How far downwind, to the nearest 0.01 km, will the concentration of nitrogen oxides, in the center of the plume, reach 0.001 gram per cubic meter?

Additional air pollution models can be found on the Internet. One source is the Environmental Protection Agency (EPA).

56. Use the concentration formula in Exercise 55 to answer the following questions.
   a. Determine whether C increases or decreases if both E and x remain constant, and w is increased.
   b. Determine whether C increases or decreases if both E and w remain constant, and x is increased.
   c. Determine whether C increases or decreases if both x and w remain constant, and E is increased.

PROJECTS

When you are trying to determine an equation that models certain data, it is sometimes convenient to use a logarithmic scale. The following project uses this technique.

1. **AVIATION** A pressure altimeter is used to determine the height of a plane above sea level. The table to the right gives values for the pressure p and altitude h for an altimeter.

p, in pounds per square inch	h, in feet
14.1	1100
13.8	1700
13.5	2300
12.8	3700
12.1	5200
11.0	7700

   a. Make a scatter plot of these data with pressure along the horizontal axis and height on the vertical axis. *Suggestion:* Choose a scale that will enable you to plot these points fairly accurately.
   b. Draw a smooth curve that goes through the points. *Note:* This curve does not have to pass through each point. The idea is to draw a *smooth* curve that approximates the data. Your task is to determine the equation of the curve you have drawn by completing the remaining parts of this project.
   c. Make a new graph in which you graph the natural logarithm of each pressure and the corresponding height. That is, plot $(\ln p, h)$ for each value of p.
   d. Draw a smooth curve through the new data points. This curve should be very close to a straight line.
   e. Assuming that the points in **d.** are approximately on a straight line, find the equation of the line by using the points $(\ln 14.1, 1100)$ and $(\ln 11.0, 7700)$. *Suggestion:* Because the values along the horizontal axis are natural logarithms, the point-slope formula becomes $h - h_1 = m[\ln p - \ln p_1]$.
   f. Evaluate the function you found in part **e.** for the pressures given in the table above and verify that the values are approximately the heights of the plane for the given pressures.
   g. Explain why the function you found in **e.** is the model for the pressure altimeter.
   h. Use your model to predict the height of the plane when the pressure is 11.5 pounds per square inch.

SECTION 4.7 MODELING DATA WITH EXPONENTIAL AND LOGARITHMIC FUNCTIONS

- ANAYLIZE SCATTER PLOTS
- APPLICATIONS

◆ ANALYZE SCATTER PLOTS

In Chapter 2 we used linear and quadratic functions to model several data sets. However, in some applications the data can be modeled more closely by using exponential or logarithmic functions. For instance, **Figure 4.50** illustrates some scatter plots that can best be modeled by exponential and logarithmic functions.

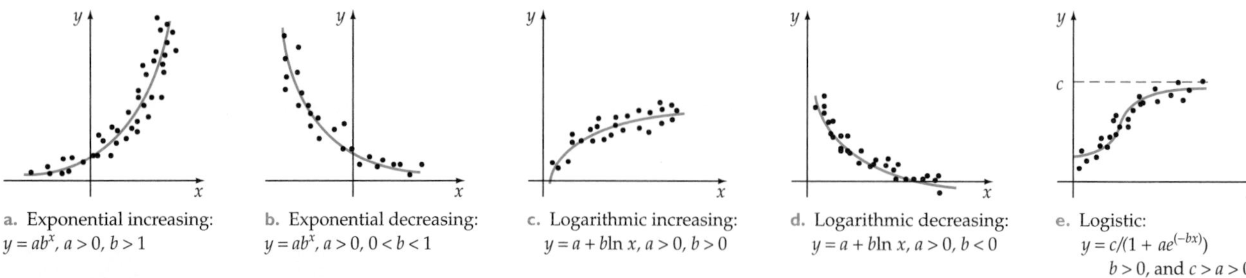

Figure 4.50
Exponential and Logarithmic Models

The terms *concave upward* and *concave downward* are often used to describe a graph. For instance, **Figures 4.51a** and **4.51c** show the graphs of two increasing functions that join the points P and Q. The graph of f and the graph of g differ in that they bend in different directions. We can distinguish between these two types of "bending" by examining the position of lines tangent to the graph. In **Figures 4.51b** and **4.51d,** lines tangent to the graphs of f and g have been drawn in black. The graph of f lies above its tangent lines and the graph of g lies below its tangent lines. The function f is said to be concave upward, and g is concave downward.

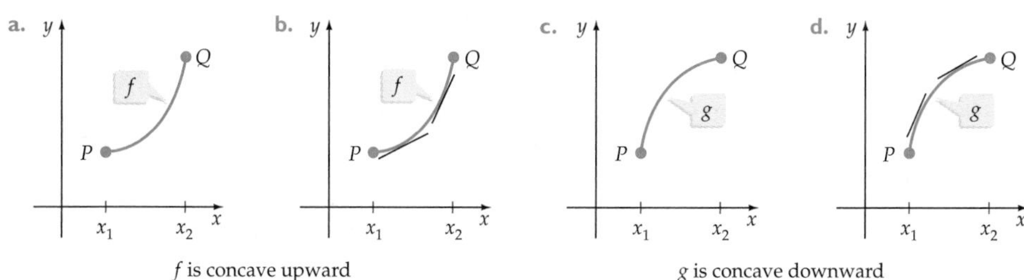

Figure 4.51

Definition of Concavity

If the graph of f lies above all of its tangents on an interval $[x_1, x_2]$, then f is **concave upward** on $[x_1, x_2]$.

If the graph of f lies below all of its tangents on an interval $[x_1, x_2]$, then f is **concave downward** on $[x_1, x_2]$.

An examination of the graphs in **Figure 4.50** shows that the graphs of all exponential functions (of the form $y = ab^x$, $a > 0$, $b > 0$, $b \neq 1$) are concave upward.

The graphs of increasing logarithmic functions are concave downward, and the graphs of decreasing logarithmic functions are concave upward. The graph of the logistic function shown in **Figure 4.50e** is concave upward at first and then become concave downward. The point where a logistic function changes from concave upward to concave downward is called its **point of inflection.**

In Example 1 we analyze scatter plots by determining whether the shape of the scatter plot can be approximated by an increasing or decreasing function and also by a function that is concave upward and/or concave downward.

EXAMPLE 1 Analyze Scatter Plots

For each of the following sets of data, determine whether a suitable model of the data would be a linear function, a quadratic function, an exponential function, a logarithmic function, or a logistic function.

$$A = \{(1, 0.6), (2, 0.7), (2.8, 0.8), (4, 1.3), (6, 1.5), (6.5, 1.6), (8, 2.1), (11.2, 4.1), (12, 4.6), (15, 8.2)\}$$

$$B = \{(0, 1.8), (0.6, 2.2), (1.2, 2.4), (2, 3.1), (2.5, 4.1), (3.1, 4.6), (4, 5.1), (4.4, 5.3), (4.9, 5.4), (5.5, 5.6), (6.0, 5.5), (6.3, 5.7), (6.9, 5.8), (7.5, 5.9)\}$$

$$C = \{(1.5, 2.8), (2, 3.5), (4.1, 5.1), (5, 5.5), (5.5, 5.7), (7, 6.1), (7.2, 6.4), (8, 6.6), (9, 6.9), (11.6, 7.4), (12.3, 7.5), (14.7, 7.9)\}$$

Solution

For each set, construct a scatter plot of the data. See **Figure 4.52.**

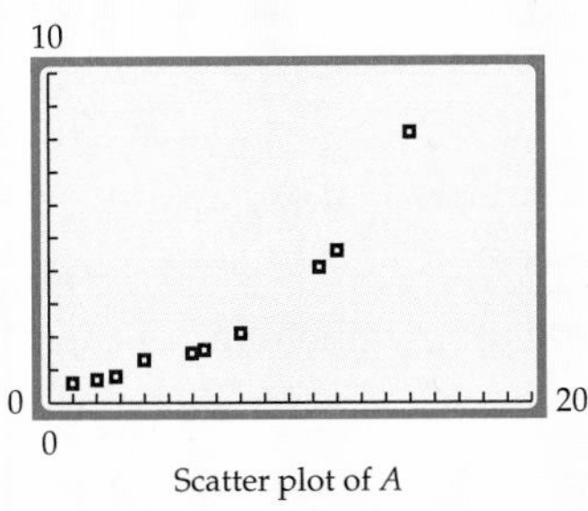

Scatter plot of *A*

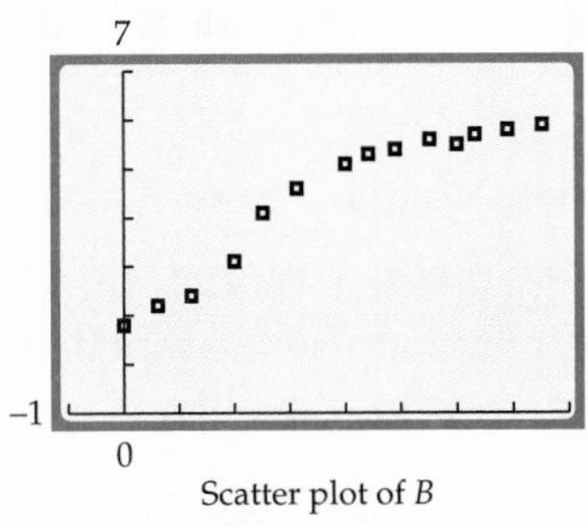

Scatter plot of *B*

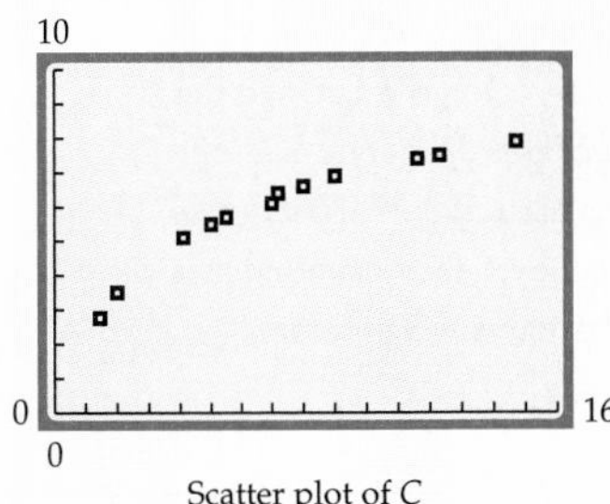

Scatter plot of *C*

Figure 4.52

The scatter plot of *A* suggests that *A* is an increasing function that is concave upward. Thus *A* can be effectively modeled by an increasing exponential function and also by a quadratic function.

The scatter plot of *B* suggests that *B* is an increasing function that is concave upward on [0, 2] and concave downward on [2, 8]. Thus *B* can be effectively modeled by a logistic function.

The scatter plot of *C* suggests that *C* is an increasing function that is concave downward. Thus *C* can be effectively modeled by an increasing logarithmic function and also by a quadratic function.

TRY EXERCISE 2, EXERCISE SET 4.7, PAGE 375

◆ APPLICATIONS

The methods used to model data with an exponential, a logarithmic, or a logistic function are similar to the methods used in Chapter 2 to model data with a linear or a quadratic function. Here is a summary of the modeling process.

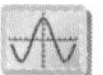

The Modeling Process

Use a graphing utility to

1. **Construct a** ***scatter plot*** **of the data** to determine which type of function will best model the data.
2. **Find the** ***regression equation*** of the modeling function and the correlation coefficient for the regression.
3. **Examine the** ***correlation coefficient*** and *view a graph* that displays both the modeling function and the scatter plot to determine how well your function fits the data.

In the following example, we use this modeling process to find a function that closely models the value of a diamond as a function of its weight.

EXAMPLE 2 Model an Application with an Exponential Function

A diamond merchant has determined the value of several white diamonds that have different weights (measured in carats) but are *similar in quality.* See Table 4.12.

Table 4.12

4.00 ct	3.00 ct	2.00 ct	1.75 ct	1.50 ct	1.25 ct	1.00 ct	0.75 ct	0.50 ct
$14,500	$10,700	$7,900	$7,300	$6,700	$6,200	$5,800	$5,000	$4,600

Find a function that models the value of the diamonds as a function of their weight, and use the function to predict the value of a 3.5-carat diamond of similar quality.

MATH MATTERS

The value of a diamond is generally determined by its color, cut, clarity, and carat weight. These characteristics of a diamond are known as the four C's. In Example 2 we have assumed that the color, the cut, and the clarity of all of the diamonds are similar. Using this assumption allows us to model the value of each diamond as a function of its carat weight alone.

Solution

1. **Construct a scatter plot of the data.** See **Figure 4.53.**

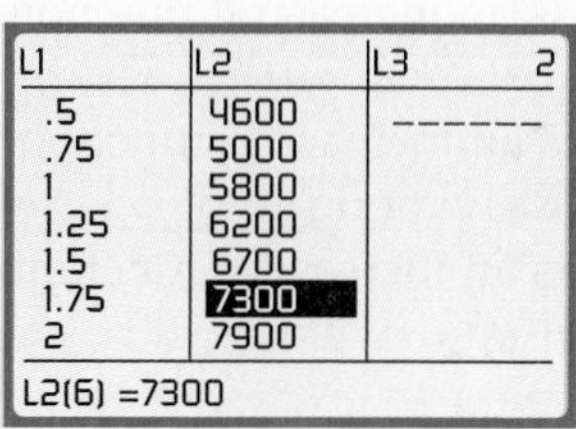

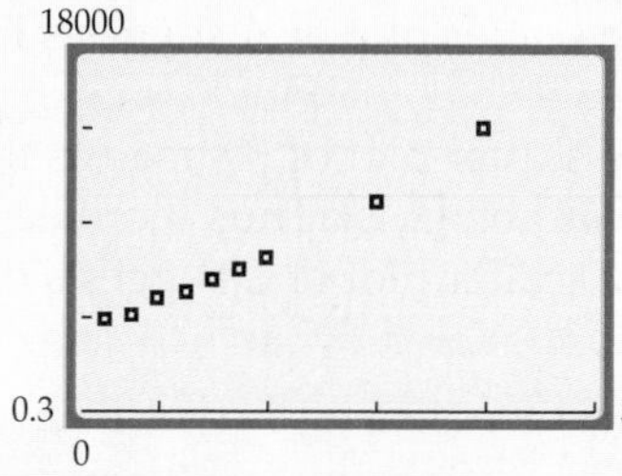

Figure 4.53

take note

Most graphing utilities have built-in routines that can be used to determine the exponential, logarithmic, or logistic regression equation that models the data. The TI-83 does not show the value of the regression coefficient r unless the "DiagnosticOn" command has been entered. The "DiagnosticOn" command is in the "CATALOG" menu.

From the above scatter plot it appears that the data can be closely modeled by an exponential function of the form $y = ab^x$ with $b > 1$.

2. **Find the regression equation.** The calculator display in Figure 4.54 shows that the exponential regression equation is $y \approx 4067.6 \cdot 1.3816^x$, where x is the carat weight of the diamond and y is the value of the diamond.

```
ExpReg
 y=a*b^x
 a=4067.641145
 b=1.381644186
 r²=.994881215
 r=.9974373238
```

Figure 4.54
ExpReg display (DiagnosticOn)

3. **Examine the correlation coefficient.** The correlation coefficient $r \approx 0.9974$ is close to 1. This indicates that the function $y \approx 4067.6 \cdot 1.3816^x$ provides a good fit to the data. The graph of $y \approx 4067.6 \cdot 1.3816^x$ (see Figure 4.55) also shows that the regression function provides a good model of the data.

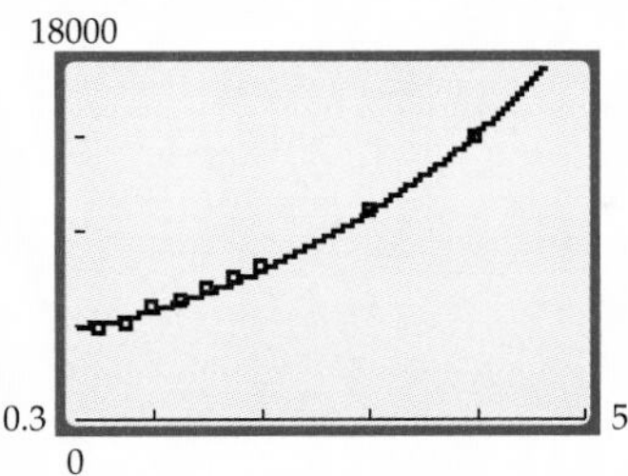

Figure 4.55

According to the modeling function, the value of a 3.5-carat diamond of similar quality is

$$y \approx 4067.6 \cdot 1.3816^{3.5} \approx \$12{,}610$$

TRY EXERCISE 8, EXERCISE SET 4.7, PAGE 375

MATH MATTERS

The Hope diamond is the world's largest deep blue diamond. It has a weight of 45.52 carats. We should not expect the function $y \approx 4067.6 \cdot 1.3816^x$ in Example 2 to yield an accurate value of the Hope diamond, because it is not the same type of diamond as the diamonds in Table 4.12 and because its weight is so much greater.

The Hope Diamond is on display at the Smithsonian Museum of Natural History in Washington D.C.

EXAMPLE 3 Model an Application with a Logistic Function

Table 4.13 shows the population of deer in an animal preserve over the years from 1985 to 1999.

Find a function that effectively models the deer population as a function of the year. Use the function to predict the deer population for the year 2005.

Continued

Table 4.13 Deer Population at the Wild West Animal Preserve

Year	Population	Year	Population
1985	320	1993	2040
1986	410	1994	2310
1987	560	1995	2620
1988	730	1996	2940
1989	940	1997	3100
1990	1150	1998	3300
1991	1410	1999	3460
1992	1760		

Solution

1. **Construct a scatter plot of the data.** In this example we have represented the year 1985 by $x = 0$, the year 1999 by $x = 14$, and the deer population by y.

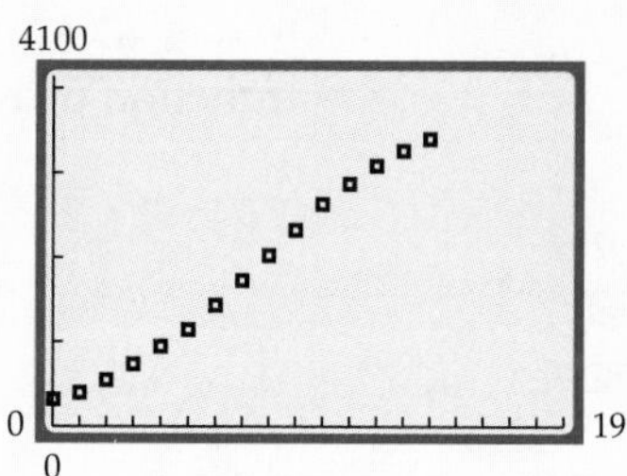

Figure 4.56

Figure 4.56 shows that the data can be effectively modeled by a logistic function.

2. **Find the regression equation.** **Figure 4.57** shows that a logistic regression equation for the data is

$$y \approx \frac{3965.3}{1 + 11.445e^{-0.31152x}}$$

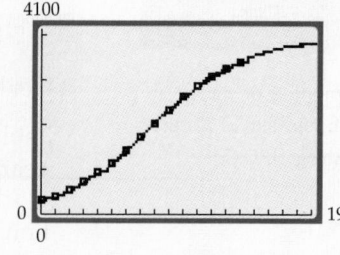

Figure 4.57

Figure 4.58

take note

In Section 4.6 logistic functions were written in the form

$$P(x) = \frac{mP_0}{P_0 + (m - P_0)e^{-kx}}$$

Many graphing utilities express logistic functions in the form

$$y = \frac{c}{1 + ae^{-bx}}$$

To convert between the two forms, use the following relationships.

$$c = m;\ b = k;\ \text{and}\ a = \frac{m - P_0}{P_0}$$

Exercise 35 page 378 explains a procedure that can be used to establish the above relationships.

3. **Examine the correlation coefficient.** A TI-83 calculator does not give the correlation coefficient for a logistic regression. However, **Figure 4.58** shows that the graph of

$$y \approx \frac{3965.3}{1 + 11.445e^{-0.31152x}}$$

provides a good fit to the data.

The model predicts that in the year 2005 ($x = 20$), the deer population will be

$$y \approx \frac{3965.3}{1 + 11.445e^{-0.31152(20)}} \approx 3878$$

TRY EXERCISE 16, EXERCISE SET 4.7, PAGE 375

In the next example we consider a data set that can be effectively modeled by more than one type of function.

EXAMPLE 4 Choosing the Best Model

Table 4.14 shows the winning times in the women's Olympic 100-meter freestyle for the years from 1968 to 1996. (*Source: 1998 Information Please Almanac*)

Table 4.14 Women's Olympic 100-Meter Freestyle, 1968–1996

Year	Time	Year	Time
1968	60.0 s	1984	55.92 s
1972	58.59 s	1988	54.93 s
1976	55.65 s	1992	54.64 s
1980	54.79 s	1996	54.50 s

Find a function that models the winning time in the women's Olympic 100-meter freestyle as a function of the year in which the winning time was produced. Then use the function to predict the winning time of the women's Olympic 100-meter freestyle for the year 2004.

Solution

1. **Construct a scatter plot of the data.** In this example we have represented the year 1968 by $x = 68$, the year 1996 by $x = 96$, and the winning time by y.

From **Figure 4.59**, it appears that the data can be effectively modeled by a decreasing exponential function and a decreasing logarithmic function.

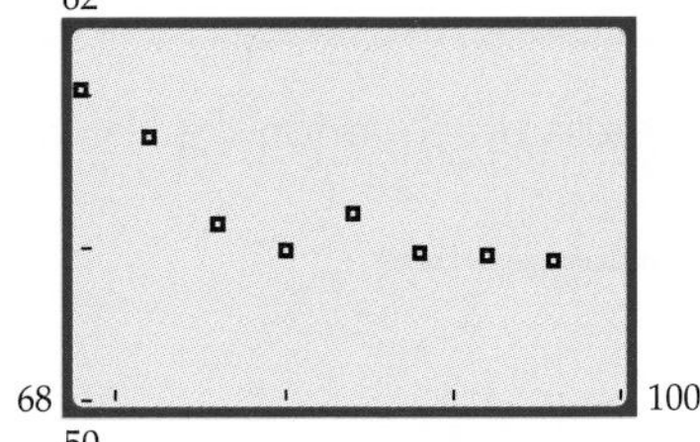

Figure 4.59

Continued

2. **Find the regression equations.** Use a graphing utility to determine both an exponential regression equation and a logarithmic regression equation for the data. Figure 4.60 shows the exponential regression equation, and Figure 4.61 shows the logarithmic regression equation.

```
ExpReg
 y=a*b^x
 a=72.31443696
 b=.9969076216
 r²=.7151837083
 r=-.8456853483
```

Figure 4.60

```
LnReg
 y=a+blnx
 a=120.7473143
 b=-14.68497634
 r²=.7454398922
 r=-.8633886102
```

Figure 4.61

3. **Examine the correlation coefficients.** In this example, the regression coefficients are both negative. In such cases, the regression that has a regression coefficient closest to -1 provides the best fit for the data. Thus the logarithmic model provides a slightly better fit for the data in this example.

The logarithmic regression equation is $y \approx 120.75 - 14.685 \ln x$. The graph of $y \approx 120.75 - 14.685 \ln x$, along with a scatter plot of the data, is shown in Figure 4.62.

take note

Although the logarithmic regression model provides a better fit for the data in Example 4, it may not be the best regression model for predicting future times in the women's Olympic 100-meter freestyle. In athletic events, future results are not dependent on past results. Thus a function that provides a good model of present data may not provide a good model of future results.

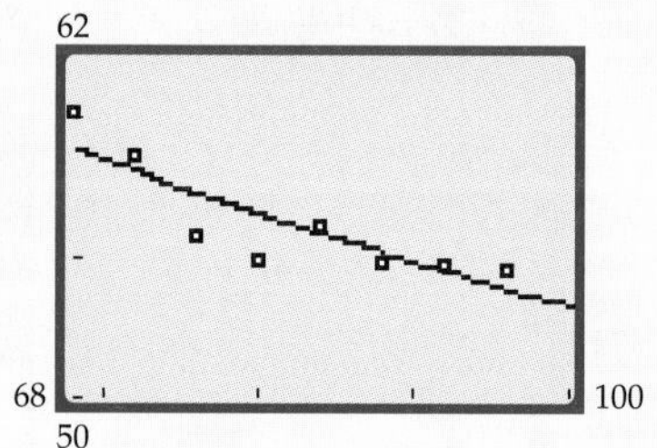

Figure 4.62

The logarithmic model predicts that in the year 2004, the winning time for the women's Olympic 100-meter freestyle will be

$$y \approx 120.75 - 14.685 \ln (104) \approx 52.55 \text{ s}$$

TRY EXERCISE 20, EXERCISE SET 4.7, PAGE 376

TOPICS FOR DISCUSSION

1. A student tries to determine the exponential regression equation for the following data.

x	1	2	3	4	5
y	8	2	0	−1.5	−2

The student's calculator displays an ERROR message. Explain why the calculator was unable to determine the exponential regression equation for the data.

2. Consider the following logistic model.

$$g(x) = \frac{14}{1 + 0.5e^{-0.1x}}$$

a. Find $g(0)$.

b. What does $g(x)$ approach as $x \to \infty$?

3. Consider the logarithmic model $h(x) = 6 - 2 \ln x$.

a. Is h an increasing or a decreasing function?

b. Is h concave up or concave down on the interval $(0, \infty)$?

c. Find (if possible) $h(0)$ and $h(e)$.

d. Does h have a horizontal asymptote? Explain.

EXERCISE SET 4.7

 **Use a graphing utility for this exercise set.**

In Exercises 1 to 6, use a scatter plot to determine which of the following types of functions would provide a suitable model of the data: *a.* a linear function, *b.* a quadratic function, *c.* an exponential function, *d.* a logarithmic function, *e.* a logistic function. *Note:* In some exercises, the data can be closely modeled by more than one type of function.

1. $\{(1, 3), (1.5, 4), (2, 6), (3, 13), (3.5, 19), (4, 27)\}$

2. $\{(1.0, 1.12), (2.1, 0.87), (3.2, 0.68), (3.5, 0.63), (4.4, 0.52)\}$

3. $\{(-2, 11), (-1, 5), (0, 1), (1.5, -1.3), (3, 1), (5, 11), (6, 19)\}$

4. $\{(0, 1.3), (0.5, 1.7), (1, 2.16), (2, 3.2), (3, 4.4), (4.6, 6.1)\}$

5. $\{(1, 2.5), (1.5, 1.7), (2, 0.7), (3, -0.5), (3.5, -1.3), (4, -1.5)\}$

6. $\{(1, 3), (1.5, 3.8), (2, 4.4), (3, 5.2), (4, 5.8), (6, 6.6)\}$

In Exercises 7 to 10, use a graphing utility to find the exponential regression function for the data. State the correlation coefficient *r*.

7. $\{(10, 6.8), (12, 6.9), (14, 15.0), (16, 16.1), (18, 50.0), (19, 20.0)\}$

8. $\{(2.6, 16.2), (3.8, 48.8), (5.1, 160.1), (6.5, 590.2), (7, 911.2)\}$

9. $\{(0, 1.83), (1, 0.92), (2, 0.51), (3, 0.25), (4, 0.13), (5, 0.07)\}$

10. $\{(4.5, 1.92), (6.0, 1.48), (7.5, 1.14), (10.2, 0.71), (12.3, 0.49)\}$

In Exercises 11 to 14, use a graphing utility to find the logarithmic regression function for the data. State the correlation coefficient *r*.

11. $\{(5, 2.7), (6, 2.5), (7.2, 2.2), (9.3, 1.9), (11.4, 1.6), (14.2, 1.3)\}$

12. $\{(11, 15.75), (14, 15.52), (17, 15.34), (20, 15.18), (23, 15.05)\}$

13. $\{(3, 16.0), (4, 16.5), (5, 16.9), (7, 17.5), (8, 17.7), (9.8, 18.1)\}$

14. $\{(8, 67.1), (10, 67.8), (12, 68.4), (14, 69.0), (16, 69.4)\}$

In Exercises 15 to 18, use a graphing utility to find the logistic regression function for the data.

15. $\{(0, 6.5), (0.6, 9.1), (1, 11.1), (1.8, 15.2), (2.2, 17.1), (3, 20.4), (4, 23.2), (5, 24.6), (8, 25.9)\}$

16. $\{(0, 25.22), (1, 47.81), (2, 65.55), (3, 73.84), (4, 76.71), (5, 77.63), (6, 77.92), (7, 77.96), (8, 77.99)\}$

17. $\{(1.6, 151.2), (2, 191.9), (2.5, 251.8), (3.4, 382.5), (4, 468.6), (6.1, 700.2), (8, 772.7), (10, 793.8), (12, 798.6)\}$

18. $\{(0.5, 466.7), (1, 583.2), (2, 872.8), (3, 1225.8), (3.5, 1410.4), (4, 1595.6), (6, 2191.2), (8, 2514.3), (10, 2626.1)\}$

19. **OLYMPIC RECORDS** The following table shows the Olympic gold medal records for the women's high jump from 1968 to 2000. (*Source: 2001 ESPN Information Please Sports Almanac*)

Women's Olympic High Jump, 1968–2000

Year	Height	Year	Height
1968	5 ft 11 3/4 in	1984	6 ft 7 1/2 in
1972	6 ft 3 5/8 in	1988	6 ft 8 in
1976	6 ft 4 in	1992	6 ft 7 1/2 in
1980	6 ft 5 1/2 in	1996	6 ft 8 3/4 in
		2000	6 ft 7 in

Represent the year 1968 by 68.

a. Use a graphing utility to determine a linear model and a logarithmic model for the data with the height measured in inches.

b. Use the regression coefficient for each of the models in part **a** to determine which model provides the best fit for the data.

c. Use the model you selected in part **b** to predict the women's Olympic gold medal high jump record for the year 2012. Round to the nearest 0.1 inch.

20. **OLYMPIC RECORDS** The following table shows the Olympic records for the men's shot put for 1948–2000. (*Source: 2001 ESPN Information Please Sports Almanac*)

Men's Olympic Shot Put, 1948–2000

Year	Distance	Year	Distance
1948	56 ft 2 in	1976	69 ft 3/4 in
1952	57 ft 1 1/2 in	1980	70 ft 1/2 in
1956	60 ft 11 in	1984	69 ft 9 in
1960	64 ft 6 3/4 in	1988	73 ft 8 3/4 in
1964	66 ft 8 1/4 in	1992	71 ft 2 1/2 in
1968	67 ft 4 3/4 in	1996	70 ft 11 1/4 in
1972	69 ft 6 in	2000	69 ft 10 1/4 in

Represent the year 1948 by 48.

a. Use a graphing utility to determine a logistic model and a logarithmic model for the data with the distance measured in feet.

b. Use a graph of the models in part **a** to determine which model provides the best fit for the data.

c. Use the model you selected in part **b** to predict the men's shot put record for the year 2008. Round to the nearest 0.01 foot.

21. **BASEBALL STATISTICS** The following table shows the average times of major-league nine-inning baseball games from 1981 to 1999. (*Source:* Elias Sports Bureau, reported in the San Diego *Union Tribune*)

Average Time of Nine-Inning Major League Baseball Games, 1981–1999

Year	Time	Year	Time
1981	2:33	1991	2:49
1982	2:34	1992	2:49
1983	2:36	1993	2:48
1984	2:35	1994	2:54
1985	2:40	1995	2:50
1986	2:44	1996	2:51
1987	2:48	1997	2:52
1988	2:45	1998	2:47
1989	2:46	1999	2:53
1990	2:48		

a. Use a graphing utility to find a linear model and a logarithmic model for the data with the time measured in minutes. Represent the year 1981 by $t = 1$.

b. Use the regression coefficient for each of the models in part **a** to determine which model provides the best fit for the data.

c. Use the model you selected in part **b** to predict in which year, to the nearest year, the average time of a nine-inning game will reach 2 hours 55 minutes.

22. **BASEBALL STATISTICS**

a. Use a graphing utility to determine a logarithmic model for the data in Exercise 21, with the year 1981 represented by $t = 81$.

b. Does the logarithmic model from part **a** provide a better fit to the data than the logarithmic model from part **b** of Exercise 21?

c. Explain why the two logarithmic models do not have the same correlation coefficient.

23. **TEMPERATURE OF COFFEE** A cup of coffee is placed in a room that is maintained at a constant temperature of 70°F. The following table shows both the coffee temperature T after t minutes and the difference between the coffee temperature and the room temperature after t minutes.

Time (minutes), t	0	5	10	15	20	25
Coffee temp., T	165°	140°	121°	107°	97°	89°
$T - 70°$	95°	70°	51°	37°	27°	19°

a. Use a graphing utility to find an exponential model for the difference $T - 70$°F as a function of t.

b. Use the model to predict how long, to the nearest minute, it will take for the coffee to cool to 80°F.

24. **DESALINATION** The following table shows the amount of fresh water w (in cubic yards) produced from salt water after t hours of a desalination process.

t	1	2.5	3.5	4.0	5.1	6.5
w	18.2	46.6	57.4	61.5	68.7	76.2

a. Use a graphing utility to find a linear model and a logarithmic model for the data.

b. Examine the correlation coefficients of the two regression models to determine which model provides the best fit.

c. Use the model you selected in part **b** to predict the amount of fresh water that will be produced after 10 hours of the desalination process. Round to the nearest 0.1 cubic yard.

25. **GENERATION OF GARBAGE** According to the U.S. Environmental Protection Agency, the amount of

garbage generated per person has been increasing over the last few decades. The following table shows the per capita garbage, in pounds per day, generated in the United States.

Year, t	1960	1970	1980	1990	2000
Pounds per day, p	2.66	3.27	3.61	4.00	4.30

Represent the year 1960 by $t = 60$.

a. Use a graphing utility to find a linear model and a logarithmic model for the data. Use t as the independent variable (domain) and p as the dependent variable (range).

b. Examine the correlation coefficients of the two regression models to determine which model provides the best fit.

c. Use the model you selected in part **b** to predict the amount of garbage that will be generated per capita in 2005. Round to the nearest 0.01 pound.

26. THE HENDERSON-HASSELBACH FUNCTION The scientists Henderson and Hasselbach determined that the pH of blood is a function of the ratio q of the blood's bicarbonate and carbonic acid.

a. Use a graphing utility and the data in the following table to determine a linear model and a logarithmic model for the data. Use q as the independent variable (domain) and pH as the dependent variable (range). Which model provides the best fit?

q	7.9	12.6	31.6	50.1	79.4
pH	7.0	7.2	7.6	7.8	8.0

b. A blood pH of 9.0 results in death. Use the model you chose in part **a** to find, to the nearest tenth, the q-value associated with a pH of 9.0.

27. WORLD POPULATION The following table lists the years in which the world's population first reached 3, 4, 5, and 6 billion. (*Source: Time Almanac 2000*)

World Population Milestones

1960	3 billion
1974	4 billion
1987	5 billion
1999	6 billion

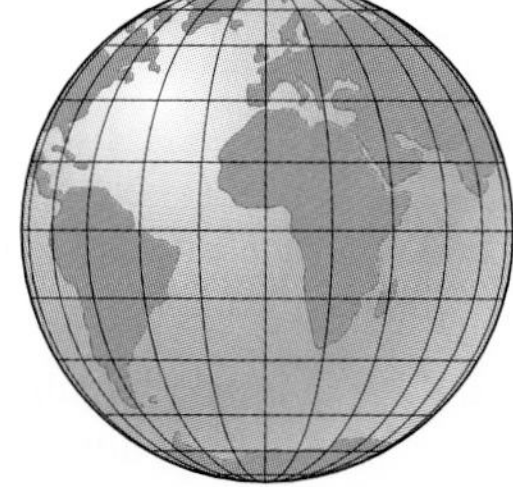

Some scientists think that the Earth has only enough resources to support 12 billion people.

a. Find an *exponential model* for the data in the table and use the model to predict in what year the world's population will reach 12 billion. (Use 60 to represent the year 1960.)

b. Find a *logistic model* for the data in the table. According to the logistic model, what will the world's population approach as $t \to \infty$?

c. Do you think the exponential model or the logistic model is the most realistic model for predicting the world's future population? Explain.

28. INTEREST RATES The following table shows the interest rates paid on jumbo CDs of various terms in April of 2000.

CD Term, t years	0.25	0.5	1.0	1.5	2.0	3.0	5.0
Rate	6.50%	6.80%	7.00%	7.10%	7.15%	7.20%	7.25%

a. Find a *logarithmic model* for the data in the table and use the model to predict, to the nearest 0.01%, the interest rate on a jumbo CD that is invested for a term of 4.0 years.

b. According to your model in part **a**, for how long, to the nearest 0.1 year, would you need to invest to earn 7.5% on your money?

29. PANDA POPULATION One estimate gives the panda population as 3200 in 1980 and as 590 in 2000.

a. Find an exponential model for the data and use the model to predict the year in which the panda population will be reduced to 200. (Use 0 to represent the year 1980.)

b. The exponential model in part **a** fits the data perfectly. Does this mean that the model will accurately predict future panda populations? Explain.

30. NUMBER OF AUTOMOBILES The number of automobiles in the United States in 1900 was around 8000. In the year 2000, the number of automobiles in the United States reached 200 million.

a. Find an exponential model for the data and use the model to predict, to the nearest 100,000, the number of automobiles in the United States in 2010. Use $t = 0$ to represent the year 1900.

b. According to the model, in what year will the number of automobiles first reach 300 million?

POINT OF INFLECTION **A point P on a graph is called a point of inflection if the graph changes from concave upward to concave downward or from concave downward to concave upward at P. It can be shown that the point of inflection of the logistic function $y = \dfrac{c}{1 + ae^{-bx}}$ occurs at $P = \left(\dfrac{\ln a}{b}, \dfrac{c}{2}\right)$. For a logistic function, the point of inflection is also the point at which the largest population growth rate occurs.**

31. Find the point of inflection of the logistic model for the data in Exercise 15. Round the coordinates to the nearest 0.1.

32. Find the point of inflection of the logistic model for the data in Exercise 17. Round the coordinates to the nearest 0.1.

33. Find the point of inflection of the logistic model for the deer population in Example 3. Round the coordinates to the nearest 0.1.

34. Find the point of inflection of the logistic model for the data in Exercise 18. Round the coordinates to the nearest 0.1.

35. **RELATIONSHIPS BETWEEN MODELS** In Section 4.6 the logistic model was given as

$$P(x) = \frac{mP_0}{P_0 + (m - P_0)e^{-kx}} \qquad \text{I}$$

Most graphing utilities use

$$y = \frac{c}{1 + ae^{-bx}} \qquad \text{II}$$

as the logistic model. To determine a relationship between the constants used in the two equations, divide the numerator and denominator of Equation I by P_0. Compare your result with Equation II.

In the two equations, what

a. is the relationship between c and m?

b. is the relationship between b and k?

c. does a equal in terms of m and P_0?

36. **DUPLICATE DATA POINTS** A student needs to model the data in set A with an exponential function.

$$A = \{(2, 5), (3, 10), (4, 17), (4, 17), (5, 28)\}$$

Because the ordered pair (4, 17) is listed twice, the student decides to eliminate one of these ordered pairs and model the data in set B.

$$B = \{(2, 5), (3, 10), (4, 17), (5, 28)\}$$

Determine whether sets A and B have the same exponential regression function.

SUPPLEMENTAL EXERCISES

37. **DOMAIN ERROR** A student needs to model the data in set A.

$$A = \{(0, 1.2), (1, 2.3), (2, 2.8), (3, 3.1), (4, 3.3), (5, 3.4)\}$$

The student views a scatter plot of the data and decides to model the data with a logarithmic function of the form $y = a + b \ln x$.

a. After attempting to use a graphing calculator to determine the logarithmic regression equation, the calculator displays the message "ERR:DOMAIN". Explain why the calculator was unable to determine the logarithmic regression equation for the data.

b. Explain what the student could do so that the data in set A could be modeled by a logarithmic function of the form $y = a + b \ln x$.

38. **A CORRELATION COEFFICIENT OF 1** A scientist uses a graphing utility to model the data set {(2, 5), (4, 6)} with a logarithmic function. The following display shows the results. What is the significance of the fact that the correlation coefficient for this regression is $r = 1$?

```
LnReg
 y=a+blnx
 a=4
 b=1.442695041
 r²=1
 r=1
```

POWER FUNCTIONS **A function that can be written in the form $y = ax^b$ is said to be a power function. Some data sets can best be modeled by a power function. On a TI-83, the PwrReg instruction is used to produce a power regression function for a set of data.**

39. a. Use a graphing utility to find an exponential regression function and a power regression function for the following data.

x	1	2	3	4	5	6
y	2.1	5.5	9.8	14.6	20.1	25.8

b. Which of the two regression functions produces the best fit?

40. a. Use a graphing utility to find an exponential regression function and a power regression function for the following data.

x	6.0	8.0	10.0	12.0	14.0	16.0
y	38.4	54.5	72.4	90.5	105.2	123.3

b. Which of the two regression functions produces the best fit?

41. **PERIOD OF A PENDULUM** The following table shows the time t (in seconds) of the period (the time it takes the pendulum to complete a swing to the left and back) of a pendulum of length l (in feet).

a. Use a graphing utility to determine the equation of the best model of the data. Your model must be a power, exponential, or logarithmic model.

Length, l	1	2	3	4	6	8
Time, t	1.11	1.57	1.92	2.25	2.72	3.14

b. What, to the nearest 0.1 foot, is the length of a pendulum that has a period of 12 seconds?

PROJECTS

1. **A MODELING PROJECT** The purpose of this project is for you to find data that can be modeled by an exponential, a logarithmic, or a logistic function Choose data from a *life-like* situation that you find interesting. Search for the data in a magazine, a newspaper, or an almanac or on the Internet. If you wish, you can collect your data by performing an experiment. Use the following steps to complete this project.

a. List the source of your data. Include the date, page number, and any other specifics about the source. If your data was collected by performing an experiment, then provide all the details about the experiment.

b. Explain what you have chosen as your variables. Which variable is the dependent variable and which the independent variable?

c. Use the three-step modeling process to find a regression equation that models the data.

d. Graph the regression equation on the scatter plot of the data. What is the regression coefficient for the model? Do you think that your regression equation accurately models your data? Explain.

e. Use the regression equation to predict the value of (1) the dependent variable for a specific value of the independent variable and (2) the independent variable for a specific value of the dependent variable.

f. Write a few comments about what you have learned from this project.

EXPLORING CONCEPTS WITH TECHNOLOGY

Table 4.15

T	V
90	700
100	500
110	350
120	250
130	190
140	150
150	120

Using a Semi-Log Graph to Model Exponential Decay

Consider the data in Table 4.15, which shows the viscosity V of SAE 40 motor oil at various temperatures T. The graph of this data is shown below, along with a curve that passes through the points. The graph in **Figure 4.63** appears to have an exponential decay shape.

One way to determine whether the graph in **Figure 4.63** is the graph of an exponential function is to plot the data on *semi-log* graph paper. On this graph paper, the horizontal axis remains the same, but the vertical axis uses a logarithmic scale.

The data in Table 4.15 are graphed again in **Figure 4.64**, but this time the vertical axis is a natural logarithm axis. This graph is approximately a straight line.

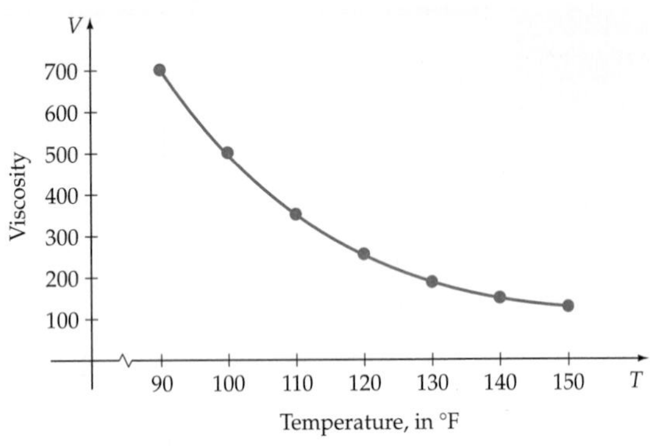

Figure 4.63

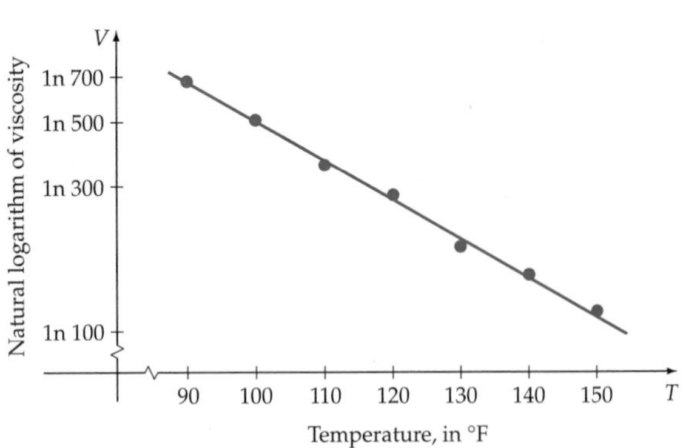

Figure 4.64

The slope of the line, to the nearest ten-thousandth, is

$$m = \frac{\ln 500 - \ln 120}{100 - 150} \approx -0.0285$$

Using this slope and the point-slope formula with V replaced by $\ln V$, we have

$$\ln V - \ln 120 = -0.0285(T - 150)$$

$$\ln V \approx -0.0285T + 9.062 \qquad (1)$$

Equation (1) is the equation of the line on a semi-log coordinate grid.

Now solve Equation (1) for V.

$$e^{\ln V} = e^{-0.0285T+9.062}$$

$$V = e^{-0.0285T}e^{9.062}$$

$$V \approx 8621e^{-0.0285T} \qquad (2)$$

Equation (2) is a model of the data in the rectangular coordinate system shown in **Figure 4.63.**

Table 4.16

t	A
1	91.77
4	70.92
8	50.30
15	27.57
20	17.95
30	7.60

1. A chemist wishes to determine the decay characteristics of iodine-131. A 100-mg sample of iodine-131 is observed over a 30-day period. Table 4.16 shows the amount A (in milligrams) of iodine-131 remaining after t days.

a. Graph the ordered pairs (t, A) on semi-log paper. (*Note:* Semi-log paper comes in different varieties. Our calculations are based on semi-log paper that has a natural logarithm scale on the vertical axis.)

b. Use the points $(4, 4.3)$ and $(15, 3.3)$ to approximate the slope of the line that passes through the points.

c. Using the slope calculated in part **b** and the point $(4, 4.3)$, determine the equation of the line.

d. Solve the equation you derived in part **c** for A.

e. Graph the equation you derived in **d** in a rectangular coordinate system.

f. What is the half-life of iodine-131?

Table 4.17

t	B
0	15.5
1	15.7
2	15.9
3	16.2
4	16.7

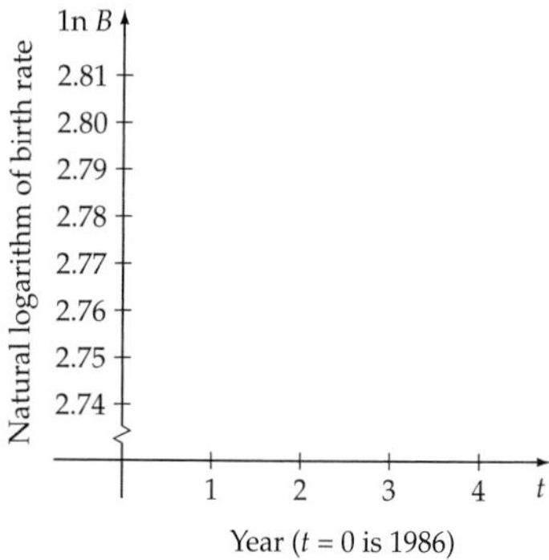

Figure 4.65

2. The live birth rates B per thousand births in the United States are given in Table 4.17 for the years 1986 through 1990 ($t = 0$ corresponds to 1986).

a. Graph the ordered pairs $(t, \ln B)$. (You will need to adjust the scale so that you can discriminate between plotted points. A suggestion is given in **Figure 4.65.**)

b. Use the points $(1, 2.754)$ and $(3, 2.785)$ to approximate the slope of the line that passes through the points.

c. Using the slope calculated in part **b** and the point $(1, 2.754)$, determine the equation of the line.

d. Solve the equation you derived in part **c** for B.

e. Graph the equation you derived in part **d** in a rectangular coordinate system.

f. If the birth rate continues as predicted by your model, in what year will the birth rate be 17.5 per thousand?

The difference in graphing strategies between Exercise 1 and Exercise 2 is that in Exercise 1, semi-log paper was used. When a point is graphed on this coordinate paper, the y-coordinate is $\ln y$. In Exercise 2, graphing a point at $(x, \ln y)$ in a rectangular coordinate system has the same effect as graphing (x, y) in a semi-log coordinate system.

CHAPTER 4 SUMMARY

4.1 Inverse Functions

- If f is a one-to-one function with domain X and range Y, and g is a function with domain Y and range X, then g is the inverse function of f if and only if $(f \circ g)(x) = x$ for all x in the domain of g and $(g \circ f)(x) = x$ for all x in the domain of f.
- A function f has an inverse function if and only if it is a one-to-one function. The graph of a function f and the graph of the inverse function f^{-1} are symmetric with respect to the line given by $y = x$.

4.2 Exponential Functions and Their Graphs

- For all positive real numbers b, $b \neq 1$, the exponential function defined by $f(x) = b^x$ has the following properties:

  1. f has the set of real numbers as its domain.
  2. f has the set of positive real numbers as its range.
  3. f has a graph with a y-intercept of $(0, 1)$.
  4. f has a graph asymptotic to the x-axis.
  5. f is a one-to-one function.
  6. f is an increasing function if $b > 1$.
  7. f is a decreasing function if $0 < b < 1$.

- As n increases without bound, $(1 + 1/n)^n$ approaches an irrational number denoted by e. The value of e accurate to eight decimal places is 2.71828183.
- The function defined by $f(x) = e^x$ is called the natural exponential function.

4.3 Logarithmic Functions and Their Graphs

- *Definition of a Logarithm* If $x > 0$ and b is a positive constant ($b \neq 1$), then

$$y = \log_b x \quad \text{if and only if} \quad b^y = x$$

- For all positive real numbers b, $b \neq 1$, the function defined by $f(x) = \log_b x$ has the following properties:

  1. f has the set of positive real numbers as its domain.
  2. f has the set of real numbers as its range.
  3. f has a graph with an x-intercept of $(1, 0)$.
  4. f has a graph asymptotic to the y-axis.
  5. f is a one-to-one function.
  6. f is an increasing function if $b > 1$.
  7. f is a decreasing function if $0 < b < 1$.

- The exponential form of $y = \log_b x$ is $b^y = x$.

- The logarithmic form of $b^y = x$ is $y = \log_b x$.
- *Basic Logarithmic Properties*

  1. $\log_b b = 1$ **2.** $\log_b 1 = 0$ **3.** $\log_b (b^p) = p$
- The function $f(x) = \log_{10} x$ is the common logarithmic function. It is customarily written as $f(x) = \log x$.
- The function $f(x) = \log_e x$ is the natural logarithmic function. It is customarily written as $f(x) = \ln x$.

4.4 Properties of Logarithms

- If b, M, and N are positive real numbers ($b \neq 1$), and p is any real number, then

$$\log_b (MN) = \log_b M + \log_b N$$
$$\log_b \frac{M}{N} = \log_b M - \log_b N$$
$$\log_b (M^p) = p \log_b M$$
$$\log_b M = \log_b N \quad \text{implies} \quad M = N$$
$$M = N \quad \text{implies} \quad \log_b M = \log_b N$$
$$b^{\log_b p} = p \quad (\text{for } p > 0)$$

- *Change-of-Base Formula* If x, a, and b are positive real numbers with $a \neq 1$ and $b \neq 1$, then

$$\log_b x = \frac{\log_a x}{\log_a b}$$

- An earthquake with an intensity of I has a Richter scale magnitude of $M = \log\left(\frac{I}{I_0}\right)$, where I_0 is the measure of a zero-level earthquake.
- The pH of a solution with a hydronium-ion concentration of H^+ mole per liter is given by $\text{pH} = -\log[H^+]$.

4.5 Exponential and Logarithmic Equations

- *Equality-of-Exponents Theorem* If b is a positive real number ($b \neq 1$) such that $b^x = b^y$, then $x = y$.
- Exponential equations of the form $b^x = b^y$ can be solved by using the equality-of-exponents theorem.
- Exponential equations of the form $b^x = c$ can be solved by taking either the common logarithm or the natural logarithm of each side of the equation.
- Logarithmic equations can often be solved by using the properties of logarithms and the definition of a logarithm.

4.6 Applications of Exponential and Logarithmic Functions

- The function defined by $N(t) = N_0 e^{kt}$ is called an exponential growth function if k is a positive constant, and it is called an exponential decay function if k is a negative constant.
- *The Compound Interest Formula* A principal P invested at an annual interest rate r, expressed as a decimal and compounded n times per year for t years, produces the balance

$$A = P\left(1 + \frac{r}{n}\right)^{nt}$$

- *Continuous Compounding Interest Formula* If an account with principal P and annual interest rate r is compounded continuously for t years, then the balance is $A = Pe^{rt}$.
- *The Logistic Model* The magnitude of a population at time t is given by

$$P(t) = \frac{mP_0}{P_0 + (m - P_0)e^{-kt}}$$

  where $P_0 = P(0)$ is the population at time $t = 0$, m is the maximum population that can be supported, and k is a constant called the growth rate constant.

4.7 Modeling Data with Exponential and Logarithmic Functions

- If the graph of f lies above all of its tangents on $[x_1, x_2]$, then f is concave upward on $[x_1, x_2]$.
- If the graph of f lies below all of its tangents on $[x_1, x_2]$, then f is concave downward on $[x_1, x_2]$.
- *The Modeling Process* Use a graphing utility to
  1. Construct a scatter plot of the data to determine which type of function will best model the data.
  2. Find the regression equation of the modeling function and the correlation coefficient for the regression.
  3. Examine the correlation coefficient and view a graph that displays both the function and the scatter plot to determine how well the function fits the data.

CHAPTER 4 TRUE/FALSE EXERCISES

In Exercises 1 to 16, answer true or false. If the statement is false, give an example to show that the statement is false.

1. Every function has an inverse function.
2. If $(f \circ g)(a) = a$ and $(g \circ f)(a) = a$ for some constant a, then f and g are inverse functions.
3. If $7^x = 40$, then $\log_7 40 = x$. True

4. If $\log_4 x = 3.1$, then $4^{3.1} = x$.

5. If $f(x) = \log x$ and $g(x) = 10^x$, then $f[g(x)] = x$ for all real numbers x.

6. If $f(x) = \log x$ and $g(x) = 10^x$, then $g[f(x)] = x$ for all real numbers x.

7. The exponential function $h(x) = b^x$ is an increasing function.

8. The logarithmic function $j(x) = \log_b x$ is an increasing function.

9. The exponential function $h(x) = b^x$ is a one-to-one function.

10. The logarithmic function $j(x) = \log_b x$ is a one-to-one function.

11. The graph of $f(x) = \dfrac{2^x + 2^{-x}}{2}$ is symmetric with respect to the y-axis.

12. The graph of $f(x) = \dfrac{2^x - 2^{-x}}{2}$ is symmetric with respect to the origin.

13. If $x > 0$ and $y > 0$, then $\log(x + y) = \log x + \log y$.

14. If $x > 0$, then $\log x^2 = 2 \log x$.

15. If M and N are positive real numbers, then

$$\ln \frac{M}{N} = \ln M - \ln N$$

16. For all $p > 0$, $e^{\ln p} = p$.

CHAPTER 4 REVIEW EXERCISES

In Exercises 1 to 4, determine whether the given functions are inverses.

1. $F(x) = 2x - 5 \qquad G(x) = \dfrac{x+5}{2}$

2. $h(x) = \sqrt{x} \qquad k(x) = x^2, \quad x \geq 0$

3. $l(x) = \dfrac{x+3}{x} \qquad m(x) = \dfrac{3}{x-1}$

4. $p(x) = \dfrac{x-5}{2x} \qquad q(x) = \dfrac{2x}{x-5}$

In Exercises 5 to 8, find the inverse of the function. Sketch the graph of the function and its inverse on the same set of coordinates axes.

5. $f(x) = 3x - 4$

6. $g(x) = -2x + 3$

7. $h(x) = -\dfrac{1}{2}x - 2$

8. $k(x) = \dfrac{1}{x}$

In Exercises 9 to 20, solve each equation. Do not use a calculator.

9. $\log_5 25 = x$

10. $\log_3 81 = x$

11. $\ln e^3 = x$

12. $\ln e^{\pi} = x$

13. $3^{2x+7} = 27$

14. $5^{x-4} = 625$

15. $2^x = \dfrac{1}{8}$

16. $27(3^x) = 3^{-1}$

17. $\log x^2 = 6$

18. $\dfrac{1}{2}\log|x| = 5$

19. $10^{\log 2x} = 14$

20. $e^{\ln x^2} = 64$

In Exercises 21 to 34, sketch the graph of each function.

21. $f(x) = (2.5)^x$

22. $f(x) = \left(\dfrac{1}{4}\right)^x$

23. $f(x) = 3^{|x|}$

24. $f(x) = 4^{-|x|}$

25. $f(x) = 2^x - 3$

26. $f(x) = 2^{(x-3)}$

27. $f(x) = \dfrac{4^x + 4^{-x}}{2}$

28. $f(x) = \dfrac{3^x - 3^{-x}}{2}$

29. $f(x) = \dfrac{1}{3}\log x$

30. $f(x) = 3\log x^{1/3}$

31. $f(x) = -x + \log x$

32. $f(x) = 2^{-x}\log x$

33. $f(x) = -\dfrac{1}{2}\ln x$

34. $f(x) = -\ln|x|$

In Exercises 35 to 38, change each logarithmic equation to its exponential form.

35. $\log_4 64 = 3$

36. $\log_{1/2} 8 = -3$

37. $\log_{\sqrt{2}} 4 = 4$

38. $\ln 1 = 0$

In Exercises 39 to 42, change each exponential equation to its logarithmic form.

39. $5^3 = 125$

40. $2^{10} = 1024$

41. $10^0 = 1$

42. $8^{1/2} = 2\sqrt{2}$

In Exercises 43 to 46, write the given logarithm in terms of logarithms of x, y, and z.

43. $\log_b \dfrac{x^2y^3}{z}$

44. $\log_b \dfrac{\sqrt{x}}{y^2z}$

45. $\ln xy^3$

46. $\ln \dfrac{\sqrt{xy}}{z^4}$

In Exercises 47 to 50, write each logarithmic expression as a single logarithm with a coefficient of 1.

47. $2 \log x + \frac{1}{3} \log (x + 1)$

48. $5 \log x - 2 \log (x + 5)$

49. $\frac{1}{2} \ln 2xy - 3 \ln z$

50. $\ln x - (\ln y - \ln z)$

In Exercises 51 to 54, use the change-of-base formula and a calculator to approximate each logarithm accurate to six significant digits.

51. $\log_5 101$

52. $\log_3 40$

53. $\log_4 0.85$

54. $\log_8 0.3$

In Exercises 55 to 70, solve each equation for x. Give exact answers. Do not use a calculator.

55. $4^x = 30$

56. $5^{x+1} = 41$

57. $\ln 3x - \ln (x - 1) = \ln 4$

58. $\ln 3x + \ln 2 = 1$

59. $e^{\ln (x+2)} = 6$

60. $10^{\log (2x+1)} = 31$

61. $\frac{4^x + 4^{-x}}{4^x - 4^{-x}} = 2$

62. $\frac{5^x + 5^{-x}}{2} = 8$

63. $\log (\log x) = 3$

64. $\ln (\ln x) = 2$

65. $\log \sqrt{x - 5} = 3$

66. $\log x + \log (x - 15) = 1$

67. $\log_4 (\log_3 x) = 1$

68. $\log_7 (\log_5 x^2) = 0$

69. $\log_5 x^3 = \log_5 16x$

70. $25 = 16^{\log_4 x}$

71. **EARTHQUAKE MAGNITUDE** Determine, to the nearest 0.1, the Richter scale magnitude of an earthquake with an intensity of $I = 51{,}782{,}000 I_0$.

72. **EARTHQUAKE MAGNITUDE** A seismogram has an amplitude of 18 millimeters and a time delay of 21 seconds. Find, to the nearest 0.1, the Richter scale magnitude of the earthquake that produced the seismogram.

73. **COMPARISON OF EARTHQUAKES** An earthquake had a Richter scale magnitude of 7.2. Its aftershock had a Richter scale magnitude of 3.7. Compare the intensity of the earthquake to the intensity of the aftershock by finding, to the nearest unit, the ratio of the larger intensity to the smaller intensity.

74. **COMPARISON OF EARTHQUAKES** An earthquake has an intensity 600 times the intensity of a second earthquake. Find, to the nearest 0.1, the difference between the Richter scale magnitudes of the earthquakes.

75. **CHEMISTRY** Find the pH of tomatoes that have a hydronium-ion concentration of 6.28×10^{-5}.

76. **CHEMISTRY** Find the hydronium-ion concentration of rainwater that has a pH of 5.4.

77. **COMPOUND INTEREST** Find the balance when \$16,000 is invested at an annual rate of 8% for 3 years if the interest is compounded

a. monthly b. continuously

78. **COMPOUND INTEREST** Find the balance when \$19,000 is invested at an annual rate of 6% for 5 years if the interest is compounded

a. daily b. continuously

79. **DEPRECIATION** The scrap value S of a product with an expected life span of n years is given by $S(n) = P(1 - r)^n$, where P is the original purchase price of the product and r is the annual rate of depreciation. A taxicab is purchased for \$12,400 and is expected to last 3 years. What is its scrap value if it depreciates at a rate of 29% per year?

80. **MEDICINE** A skin wound heals according to the function given by $N(t) = N_0 e^{-0.12t}$, where N is the number of square centimeters of unhealed skin t days after the injury, and N_0 is the number of square centimeters covered by the original wound.

a. What percentage of the wound will be healed after 10 days?

b. How many days, to the nearest day, will it take for 50% of the wound to heal?

c. How long, to the nearest day, will it take for 90% of the wound to heal?

In Exercises 81 to 84, find the exponential growth/decay function $N(t) = N_0 e^{kt}$ that satisfies the given conditions.

81. $N(0) = 1, N(2) = 5$

82. $N(0) = 2, N(3) = 11$

83. $N(1) = 4, N(5) = 5$

84. $N(-1) = 2, N(0) = 1$

85. **POPULATION GROWTH**

a. Find the exponential growth function for a city whose population was 25,200 in 1998 and 26,800 in 1999. Use $t = 0$ to represent the year 1998.

b. Use the growth function to predict, to the nearest 100, the population of the city in 2005.

86. **CARBON DATING** Determine, to the nearest 10 years, the age of a bone if it now contains 96% of its original amount of carbon-14. The half-life of carbon-14 is 5730 years.

87. **ACTIVE MILITARY DUTY PERSONNEL** The following table at the top of page 385 shows the number of U.S. military personnel on active duty for each year from 1990 to 1999. (*Source: Time Almanac 2000 with Information Please*)

Active Military Duty Personnel, 1990–1999

1990	2,043,705	1995	1,518,224
1991	1,985,555	1996	1,471,722
1992	1,807,177	1997	1,438,562
1993	1,705,103	1998	1,406,830
1994	1,610,490	1999	1,379,756

a. Use a graphing utility to find a linear model, an exponential model, and a logarithmic model for the number of active-duty personnel, P, as a function of the year. Represent the year 1990 by $t = 90$.

b. Examine the correlation coefficients of the three regression models to determine which model provides the best fit.

c. Use the model you selected in part **b** to predict, to the nearest 10,000, the number of active-duty military personnel for the year 2004.

88. MORTALITY RATE The following table shows the infant mortality rate in the United States each year from 1940 to 1997. (*Source:* U.S. Dept. Health and Human Services, National Center for Health Statistics, *National Vital Statistics Report,* Oct. 7, 1998)

U.S. Infant Mortality Rate, 1940–1997 (per 1000 live births)

Year	Rate
1940	47.0
1950	29.2
1960	26.0
1970	20.0
1980	12.6
1990	9.2
1996	7.2
1997	7.1

a. Use a graphing utility to find a linear model, an exponential model, and a logarithmic model for the infant mortality rate, R, as a function of the year. Represent the year 1940 by $t = 40$.

b. Examine the correlation coefficients of the three regression models to determine which model provides the best fit.

c. Use the model you selected in part **b** to predict, to the nearest 0.1, the infant mortality rate in 2008.

89. LOGISTIC GROWTH The population of coyotes in a national park satisfies the logistic model with $P_0 = 210$ in 1992, $m = 1400$, and $P(3) = 360$ (the population in 1995).

a. Determine the logistic model.

b. Use the model to predict, to the nearest 10, the coyote population in 2005.

90. Consider the logistic function

$$P(t) = \frac{128}{1 + 5e^{-0.27t}}$$

a. Find P_0.

b. Find the coordinates of the point of inflection for $P(t)$. Round the coordinates to the nearest 0.01.

c. What does $P(t)$ approach as $t \to \infty$?

CHAPTER 4 TEST

1. Find the inverse of $f(x) = 2x - 3$. Graph f and f^{-1} on the same coordinate axes.

2. Find the inverse of $f(x) = \dfrac{x}{4x - 8}$. State the domain and the range of f^{-1}.

3. Graph: $f(x) = 3^{-x/2}$

4. Graph: $f(x) = e^{x/2}$

5. Write $\log_b (5x - 3) = c$ in exponential form.

6. Write $3^{x/2} = y$ in logarithmic form.

7. Write $\log_b \frac{z^2}{y^3\sqrt{x}}$ in terms of logarithms x, y, and z.

8. Write $\log(2x+3) - 3\log(x-2)$ as a single logarithm with a coefficient of 1.

9. Use the change-of-base formula and a calculator to approximate $\log_4 12$. Round your result to the nearest 0.0001.

10. Graph: $f(x) = -\ln(x+1)$

11. Solve: $5^x = 22$ Round your solution to the nearest 0.0001.

12. Find the *exact* solution(s) of

$$\frac{3^x + 3^{-x}}{2} = 21$$

13. Solve: $\log(x+99) - \log(3x-2) = 2$

14. Solve: $\ln(2-x) + \ln(5-x) = \ln(37-x)$

15. Find the balance on \$20,000 invested at an annual interest rate of 7.8% for 5 years:

   a. compounded monthly.

   b. compounded continuously.

16. a. What, to the nearest 0.1, will an earthquake measure on the Richter scale if it has an intensity of $I = 42{,}304{,}000I_0$?

   b. Compare the intensity of an earthquake that measures 6.3 on the Richter scale to an intensity of an earthquake that measures 4.5 on the Richter scale by finding the ratio of the larger intensity to the smaller intensity. Round to the nearest whole number.

17. a. Find the exponential growth function for a city whose population was 34,600 in 1996 and 39,800 in 1999. Use $t = 0$ to represent the year 1996.

   b. Use the growth function to predict the population of the city in 2006. Round to the nearest 1000.

18. Determine, to the nearest 10 years, the age of a bone if it now contains 92% of its original amount of carbon-14. The half-life of carbon-14 is 5730 years.

19. a. Use a graphing utility to find the exponential regression function for the following data.

$$\{(2.5, 16), (3.7, 48), (5.0, 155), (6.5, 571), (6.9, 896)\}$$

   b. Use the function to predict, to the nearest whole number, the y-value associated with $x = 7.8$.

20. The following table shows the interest rates paid by a bank on CDs of different terms in May of 2000.

CD Term, t years	0.25	0.5	1.0	1.5	2.0	3.0	5.0
Rate	3.50%	3.80%	4.00%	4.10%	4.15%	4.20%	4.25%

   a. Find the *logarithmic regression function* for the data and use the function to predict, to the nearest 0.01%, the interest rate on a CD invested for 3.5 years.

   b. According to your function, for how long (to the nearest 0.1 year) would you need to invest to receive a 4.4% interest rate?

CHAPTER

5

TRIGONOMETRIC FUNCTIONS

Tsunamis and Earthquakes

Imagine an undersea earthquake. The energy from the earthquake would be translated to the water as water waves. These water waves are called tsunamis. Although the phrase *tidal waves* is still used to describe them, *tsunami* is the preferred term because the waves have nothing to do with the tides.

In the open ocean, the distance between crests of a tsunami may be as great as 60 miles and the height of the wave no more than 2 feet. As the depth of the ocean decreases, however, the water wave slows down. As it slows, the height of the wave increases. When a tsunami reaches the shore, it may include crests 100 feet above the normal tide level.

◆ Nature's beauty can possess terrific force and power.

The earthquake that generated the tsunamis also creates waves within the earth. Two of the wave types created are the *primary wave* or *P-wave* and the *secondary wave* or *S-wave*. These two waves are quite different. The P-wave is very much like a sound wave. It alternately compresses and dilates the substances within the earth. These waves can travel through solid rock and water.

◆ A geologist determines the magnitude of an earthquake by examining a seisomogram.

S-waves are slower than P-waves and are more like water waves. As an S-wave travels through the earth, it shears the rock sideways at right angles to the direction of travel. The S-wave causes much of the structural damage associated with earthquakes. Wave phenomena exhibited by tsunami and earthquake waves can be described by trigonometric functions.

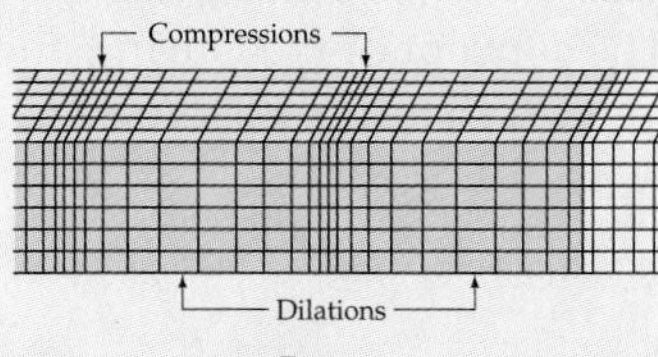

P-wave

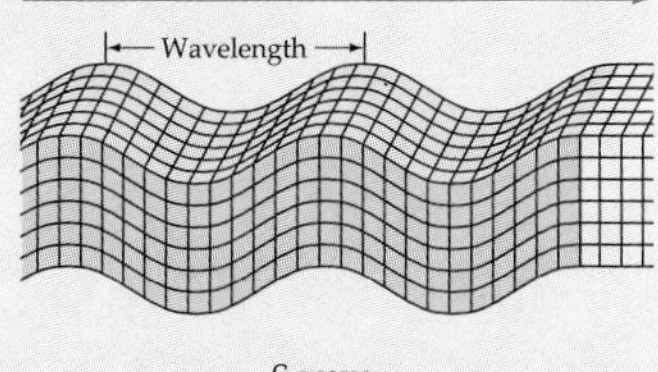

S-wave

SECTION

5.1 ANGLES AND ARCS

- DEGREE MEASURE
- CLASSIFICATION OF ANGLES
- CONVERSION BETWEEN UNITS
- RADIAN MEASURE
- ARCS AND ARC LENGTH
- LINEAR AND ANGULAR SPEED

A point P on a line separates the line into two parts, each of which is called a **half-line.** The union of point P and the half-line formed by P that includes point A is called a **ray,** and it is represented as $\overrightarrow{PA}$. The point P is the **endpoint** of ray $\overrightarrow{PA}$. **Figure 5.1** shows the ray $\overrightarrow{PA}$ and a second ray $\overrightarrow{QR}$.

In geometry an *angle* is defined simply as the union of two rays that have a common endpoint. In trigonometry and many advanced mathematics courses, it is beneficial to define an angle in terms of a rotation.

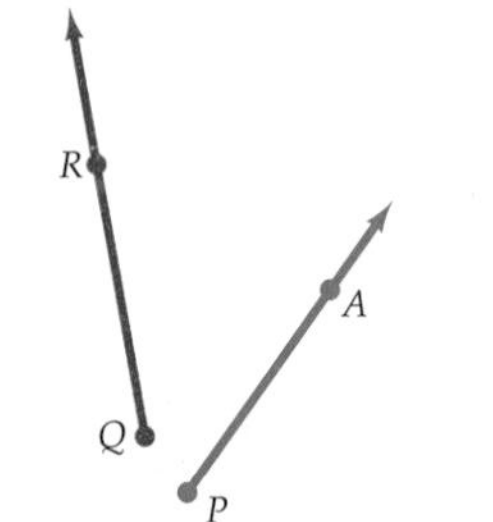

Figure 5.1

Definition of an Angle

An **angle** is formed by rotating a given ray about its endpoint to some terminal position. The original ray is the **initial side** of the angle, and the second ray is the **terminal side** of the angle. The common endpoint is the **vertex** of the angle.

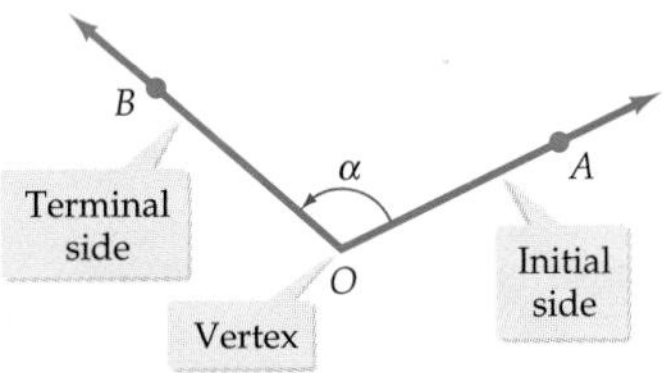

Figure 5.2

There are several methods used to name an angle. One way is to employ Greek letters. For example, the angle shown in **Figure 5.2** can be designated as α or as $\angle\alpha$. It can also be named $\angle O$, $\angle AOB$, or $\angle BOA$. If you name an angle by using three points, such as $\angle AOB$, it is traditional to list the vertex point between the other two points.

Angles formed by a counterclockwise rotation are considered **positive angles,** and angles formed by a clockwise rotation are considered **negative angles.** See **Figure 5.3.**

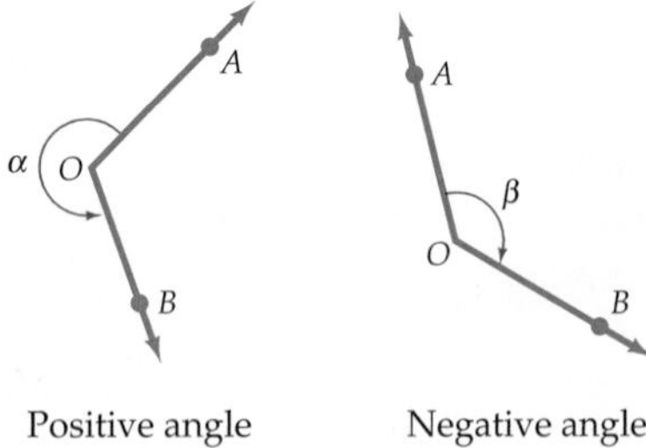

Figure 5.3

◆ DEGREE MEASURE

The **measure** of an angle is determined by the amount of rotation of the initial ray. The concept of measuring angles in *degrees* grew out of the belief of the early Babylonians that the seasons repeated every 360 days.

> **Definition of Degree**
>
> One **degree** is the measure of an angle formed by rotating a ray $\frac{1}{360}$ of a complete revolution. The symbol for degree is °.

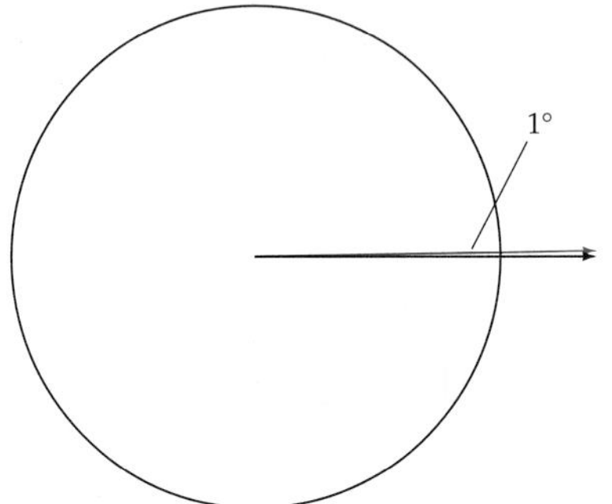

Figure 5.4
$1° = 1/360$ of a revolution

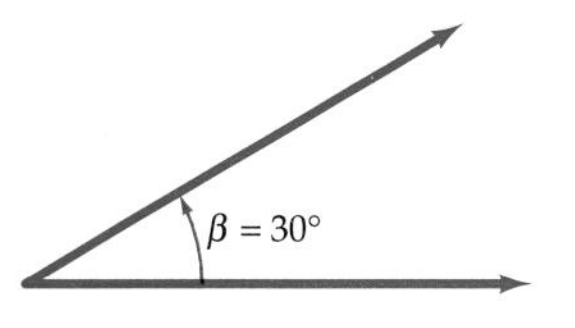

Figure 5.5

The angle shown in **Figure 5.4** has a measure of 1°. The angle β shown in **Figure 5.5** has a measure of 30°. We will use the notation $\beta = 30°$ to denote that the measure of angle β is 30°. The protractor shown in **Figure 5.6** can be used to measure an angle in degrees or to draw an angle with a given degree measure.

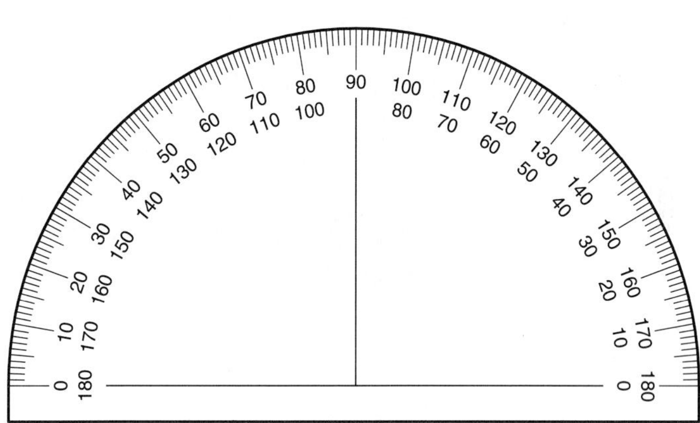

Figure 5.6
Protractor for measuring angles in degrees.

◆ CLASSIFICATION OF ANGLES

Angles are often classified according to their measure.

- 180° angles are **straight angles.** See **Figure 5.7a.**
- 90° angles are **right angles**. See **Figure 5.7b.**
- Angles that have a measure greater than 0° but less than 90° are **acute angles.** See **Figure 5.7c.**
- Angles that have a measure greater than 90° but less than 180° are **obtuse angles.** See **Figure 5.7d.**

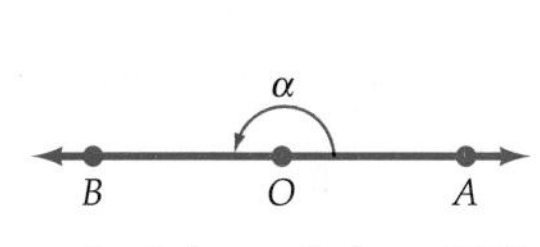

a. Straight angle ($\alpha = 180°$)

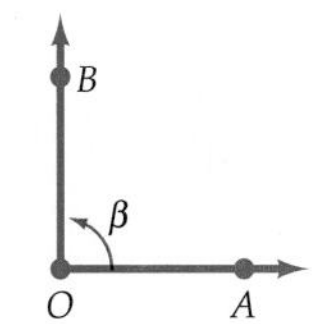

b. Right angle ($\beta = 90°$)

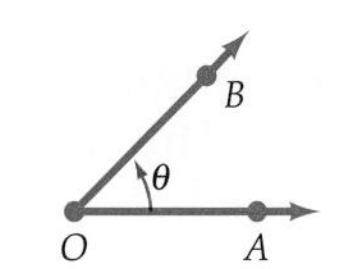

c. Acute angle ($0° < \theta < 90°$)

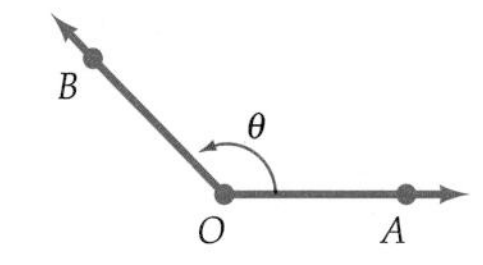

d. Obtuse angle ($90° < \theta < 180°$)

Figure 5.7

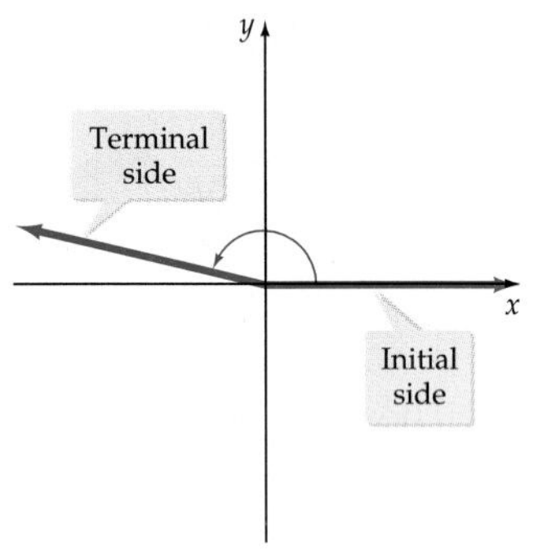

Figure 5.8
An angle in standard position

An angle superimposed in a Cartesian coordinate system is in **standard position** if its vertex is at the origin and its initial side is on the positive x-axis. See **Figure 5.8.**

Two positive angles are **complementary angles (Figure 5.9a)** if the sum of the measures of the angles is 90°. Each angle is the *complement* of the other angle. Two positive angles are **supplementary angles (Figure 5.9b)** if the sum of the measures of the angles is 180°. Each angle is the *supplement* of the other angle.

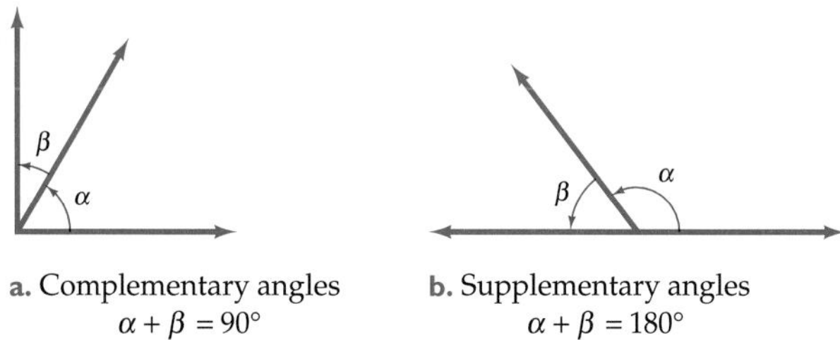

a. Complementary angles $\alpha + \beta = 90°$ **b.** Supplementary angles $\alpha + \beta = 180°$

Figure 5.9

EXAMPLE 1 Find the Measure of the Complement and the Supplement of an Angle

For each angle, find the measure (if possible) of its complement and of its supplement.

a. $\theta = 40°$ **b.** $\theta = 125°$

Solution

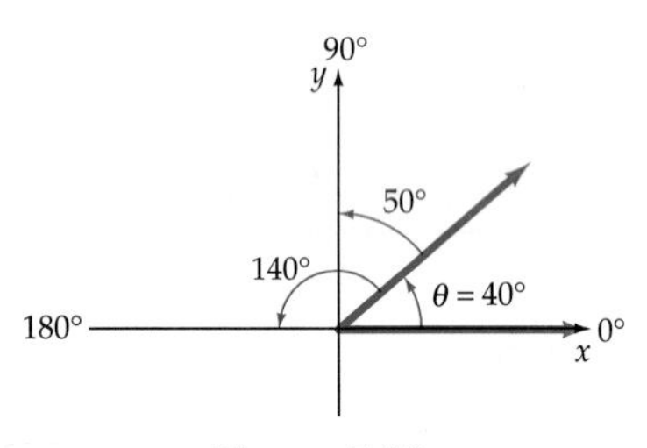

Figure 5.10

a. **Figure 5.10** shows $\angle\theta = 40°$ in standard position. The measure of its complement is $90° - 40° = 50°$. The measure of its supplement is $180° - 40° = 140°$.

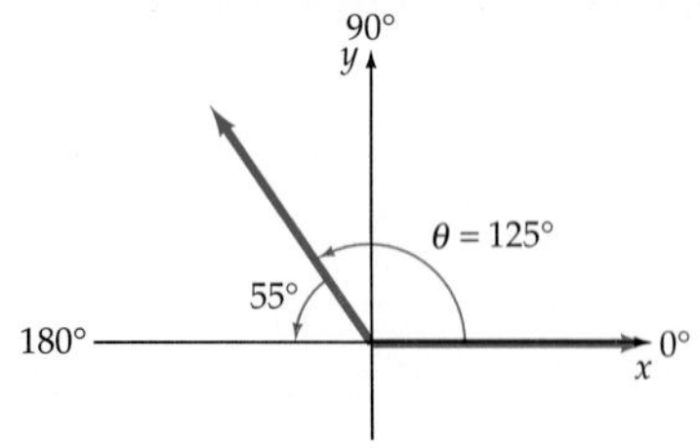

b. **Figure 5.11** shows $\angle\theta = 125°$ in standard position. Angle θ does not have a complement because there is no positive number x such that

$$x° + 125° = 90°$$

The measure of its supplement is $180° - 125° = 55°$.

Figure 5.11

TRY EXERCISE 2, EXERCISE SET 5.1, PAGE 399

QUESTION Are the two acute angles of any right triangle complementary angles? Explain.

Some angles have a measure greater than 360°. See **Figure 5.12a** and **Figure 5.12b.** The angle shown in **Figure 5.12c** has a measure less than −360°,

ANSWER Yes. The sum of the measures of the angles of any triangle is 180°. The right angle has a measure of 90°. Thus the measure of the sum of the two acute angles must be $180° - 90° = 90°$.

because it is formed by a clockwise rotation of more than one revolution of the initial ray.

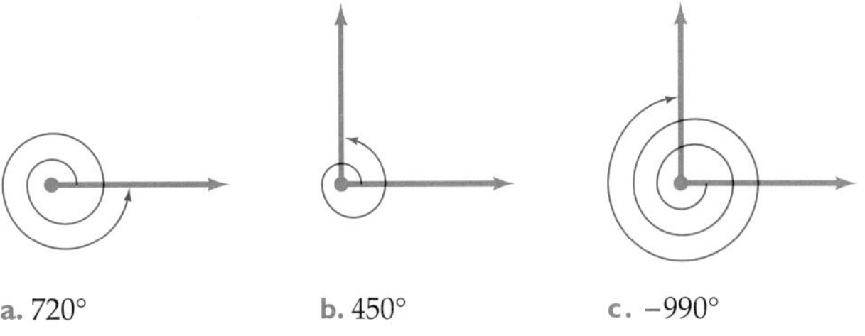
a. 720°   b. 450°   c. −990°

Figure 5.12

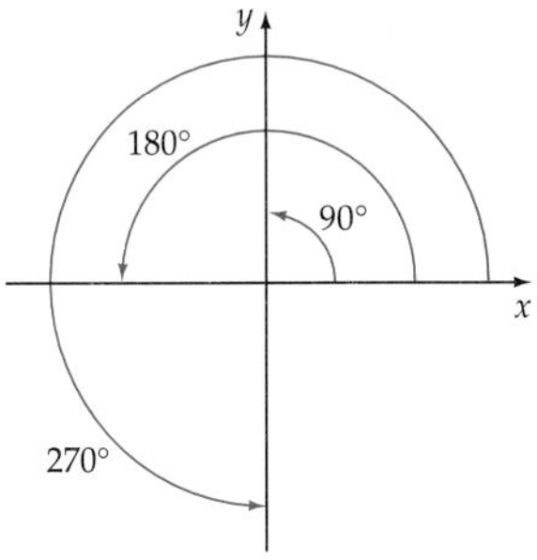

180° 90° 270° x y

Figure 5.13

If the terminal side of an angle in standard position lies on a coordinate axis, then the angle is classified as a **quadrantal angle.** For example, the 90° angle, the 180° angle, and the 270° angle shown in **Figure 5.13** are all quadrantal angles.

If the terminal side of an angle in standard position does not lie on a coordinate axis, then the angle is classified according to the quadrant that contains the terminal side. For example, $\angle\beta$ in **Figure 5.14** is a Quadrant III angle.

Angles in standard position that have the same terminal sides are **coterminal angles.** Every angle has an unlimited number of coterminal angles. **Figure 5.15** shows $\angle\theta$ and two of its coterminal angles, labeled $\angle 1$ and $\angle 2$.

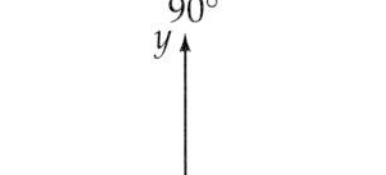
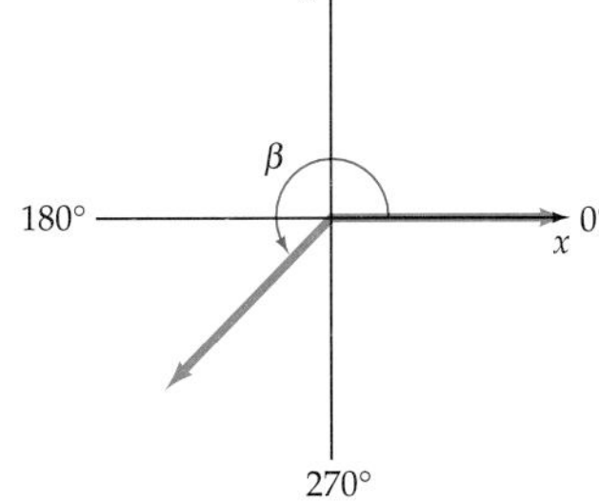

Figure 5.14

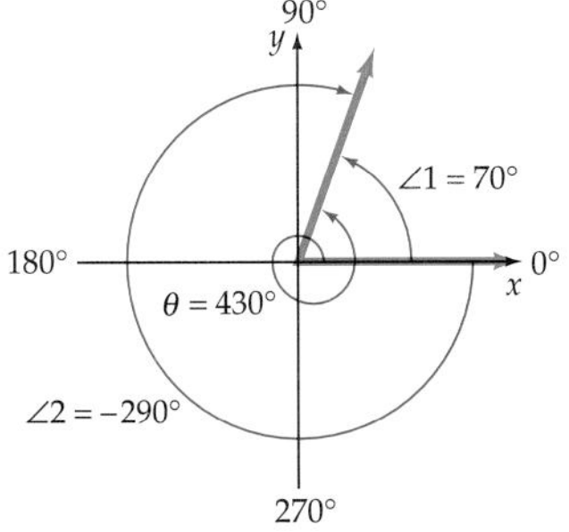

Figure 5.15

Measures of Coterminal Angles

Given $\angle\theta$ in standard position with measure $x°$, then the measures of the angles that are coterminal with $\angle\theta$ are given by

$$x° + k \cdot 360°$$

where k is an integer.

This theorem states that the measures of any two coterminal angles differ by an integer multiple of 360°. For instance, in **Figure 5.15,** $\theta = 430°$,

$$\angle 1 = 430° + (-1) \cdot 360° = 70°, \quad \text{and}$$
$$\angle 2 = 430° + (-2) \cdot 360° = -290°$$

If we add positive multiples of 360° to 430°, we find that the angles with measures 790°, 1150°, 1510°, ... are also coterminal with $\angle\theta$.

EXAMPLE 2 Classify by Quadrant and Find a Coterminal Angle

Assume the following angles are in standard position. Classify each angle by quadrant, and then determine the measure of the positive angle with measure less than 360° that is coterminal with the given angle.

a. $\alpha = 550°$ **b.** $\beta = -225°$ **c.** $\gamma = 1105°$

Continued ►

a.

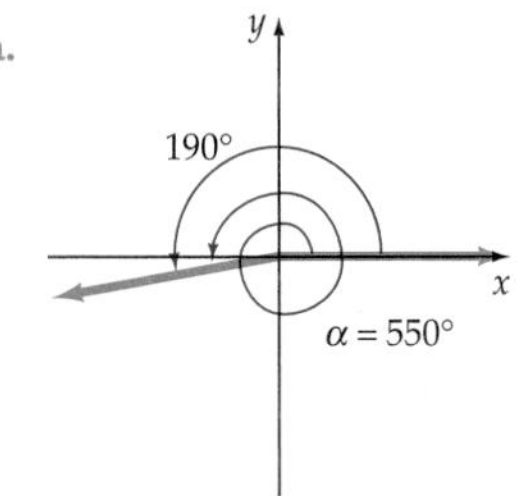

b.

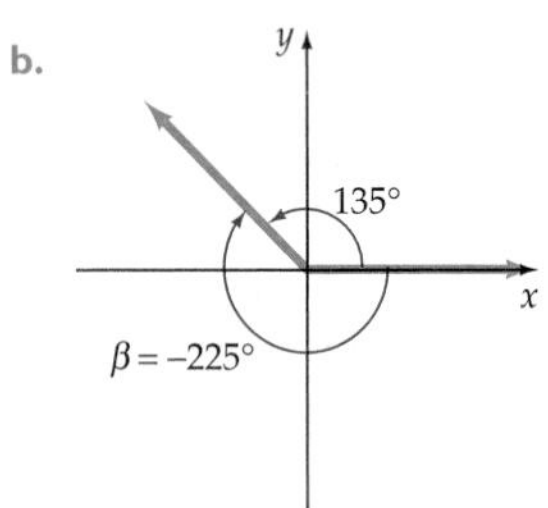

c.

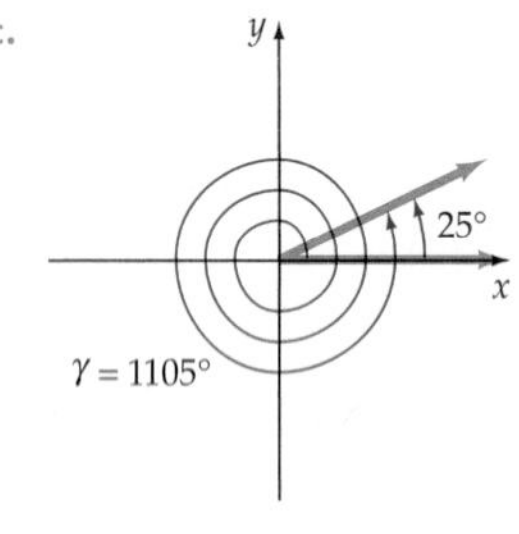

Figure 5.16

Solution

a. Because $550° = 190° + 360°$, $\angle\alpha$ is coterminal with an angle that has measure of 190°. $\angle\alpha$ is a Quadrant III angle. See **Figure 5.16a.**

b. Because $-225° = 135° + (-1) \cdot 360°$, $\angle\beta$ is coterminal with an angle that has measure of 135°. $\angle\beta$ is a Quadrant II angle. See **Figure 5.16b.**

c. $1105° \div 360° = 3.069\overline{4}$. Thus $\angle\gamma$ is an angle formed by 3 complete counterclockwise rotations, plus $0.069\overline{4}$ of a rotation. To convert $0.069\overline{4}$ of a rotation to degrees, multiply $0.069\overline{4}$ times 360°.

$$0.069\overline{4} \cdot 360° = 25°$$

Thus $1105° = 25° + 3 \cdot 360°$. Hence $\angle\gamma$ is coterminal with an angle that has a measure of 25°. $\angle\gamma$ is a Quadrant I angle. See **Figure 5.16c.**

TRY EXERCISE 14, EXERCISE SET 5.1, PAGE 399

◆ CONVERSION BETWEEN UNITS

There are two popular methods for representing a fractional part of a degree. One is the decimal degree method. For example, the measure 29.76° is a decimal degree. It means

29° plus 76 hundredths of 1°

A second method of measurement is known as the DMS (**D**egree, **M**inute, **S**econd) method. In the DMS method a degree is subdivided into 60 equal parts, each of which is called a minute, denoted by ′. Thus $1° = 60'$. Furthermore, a minute is subdivided into 60 equal parts, each of which is called a second, denoted by ″. Thus $1' = 60''$ and $1° = 3600''$. The fractions

$$\frac{1°}{60'} = 1, \quad \frac{1'}{60''} = 1, \quad \text{and} \quad \frac{1°}{3600''} = 1$$

are another way of expressing the relationship among degrees, minutes, and seconds. Each of the fractions is known as a **unit fraction** or a **conversion factor.** Because all conversion factors are equal to 1, you can multiply a magnitude by a conversion factor and not change the magnitude, even though you change the units used to express the magnitude. The following illustrates the process of multiplying by conversion factors to write 126°12′27″ as a decimal degree.

$$\begin{aligned} 126°12'27'' &= 126° + 12' + 27'' \\ &= 126° + 12'\left(\frac{1°}{60'}\right) + 27''\left(\frac{1°}{3600''}\right) \\ &= 126° + 0.2° + 0.0075° = 126.2075° \end{aligned}$$

Many graphing calculators can be used to convert a decimal degree measure to its equivalent DMS measure, and vice versa. For instance, Figure 5.17 shows that 31.57° is equivalent to 31°34′12″. On a TI-83 graphing calculator, the degree symbol, °, and the DMS function are in the ANGLE menu.

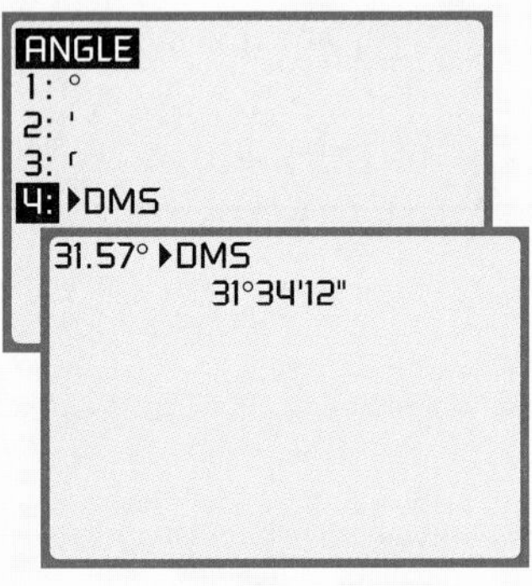

Figure 5.17

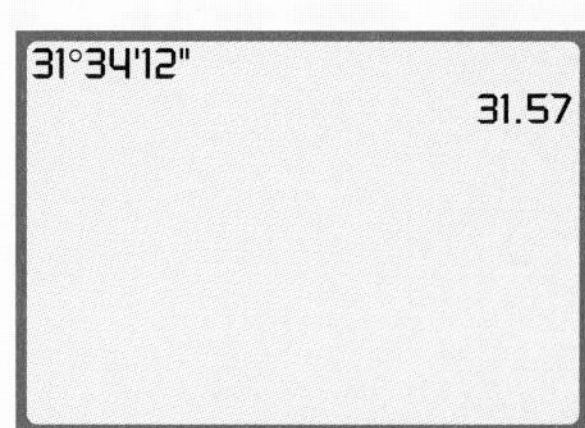

Figure 5.18

To convert a DMS measure to its equivalent decimal degree measure, enter the DMS measure and press ENTER. The calculator screen in Figure 5.18 shows that 31°34′12″ is equivalent to 31.57°. A TI-83 needs to be in degree mode to produce the results displayed in Figures 5.17 and 5.18. On a TI-83, the degree symbol, °, and the minute symbol, ′, are both in the ANGLE menu, however, the second symbol, ″, is entered by pressing ALPHA ["].

◆ RADIAN MEASURE

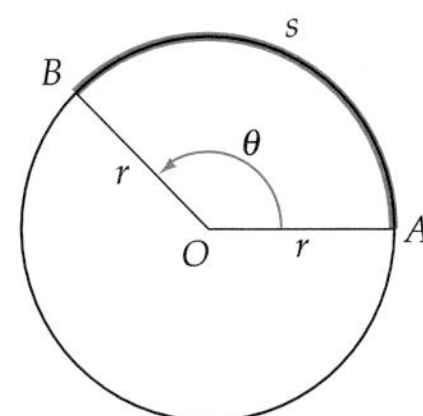

Figure 5.19

Another commonly used angle measurement is the *radian*. To define a radian, first consider a circle of radius r and two radii $\overline{OA}$ and $\overline{OB}$. The angle θ formed by the two radii is a **central angle**. The portion of the circle between A and B is an **arc** of the circle and is written $\overset{\frown}{AB}$. We say that $\overset{\frown}{AB}$ *subtends* the angle θ. The length of $\overset{\frown}{AB}$ is s (see Figure 5.19).

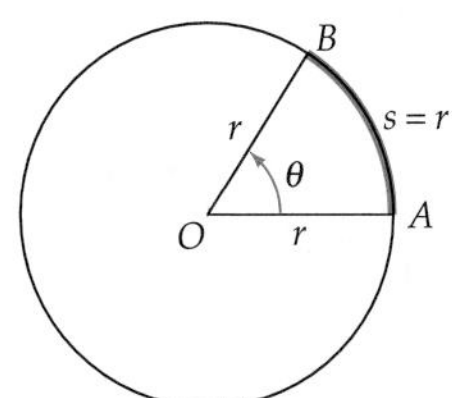

Figure 5.20
Central angle θ has a measure of 1 radian.

Definition of Radian

One **radian** is the measure of the central angle subtended by an arc of length r on a circle of radius r. See Figure 5.20.

Figure 5.21 shows a protractor that can be used to measure angles in radians or to construct angles given in radian measure.

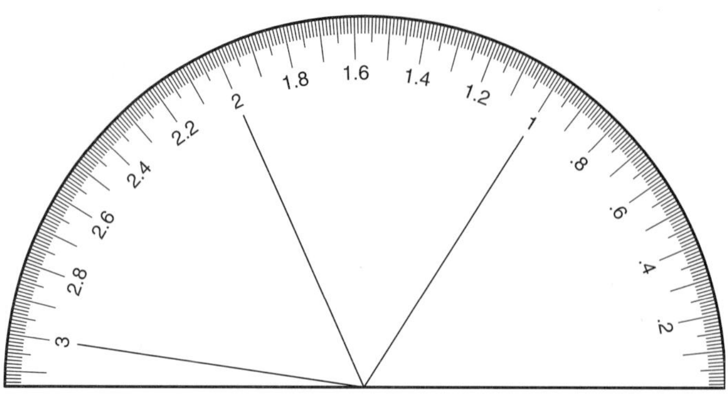

Figure 5.21
Protractor for measuring angles in radians.

Radian Measure

Given an arc of length s on a circle of radius r, the measure of the central angle subtended by the arc is $\theta = \dfrac{s}{r}$ radians.

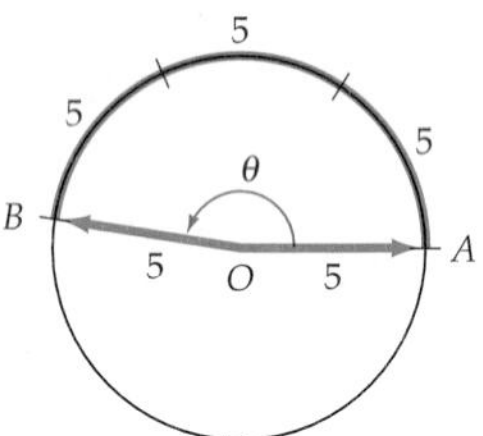

Figure 5.22
Central angle θ has a measure of 3 radians.

As an example, consider that an arc of length 15 centimeters on a circle with a radius of 5 centimeters subtends an angle of 3 radians, as shown in **Figure 5.22.** The same result can be found by dividing 15 centimeters by 5 centimeters.

To find the measure in radians of any central angle θ, divide the length s of the arc that subtends θ by the length of the radius of the circle. Using the formula for radian measure, we find that an arc of length 12 centimeters on a circle of radius 8 centimeters subtends a central angle θ whose measure is

$$\theta = \frac{s}{r} \text{ radians} = \frac{12 \text{ centimeters}}{8 \text{ centimeters}} \text{ radians} = \frac{3}{2} \text{ radians}$$

Note that the centimeter units are *not* part of the final result. The radian measure of a central angle formed by an arc of length 12 miles on a circle of radius 8 miles would be the same, 3/2 radians. We say that radian is a *dimensionless* quantity because there are no units of measurement associated with a radian.

Recall that the circumference of a circle is given by the equation $C = 2\pi r$. The radian measure of the central angle θ subtended by the circumference is $\theta = \dfrac{2\pi r}{r} = 2\pi$. In degree measure, the central angle θ subtended by the circumference is 360°. Thus we have the relationship $360° = 2\pi$ radians. Dividing each side of the equation by 2 gives $180° = \pi$ radians. From this last equation, we can establish the following conversion factors.

take note

A calculator shows that

1 radian ≈ 57.29577951°

and

1° ≈ 0.017453293 radian

Radian-Degree Conversion

- To convert from radians to degrees, multiply by $\left(\dfrac{180°}{\pi \text{ radians}}\right)$.
- To convert from degrees to radians, multiply by $\left(\dfrac{\pi \text{ radians}}{180°}\right)$.

EXAMPLE 3 Convert Degree Measure to Radian Measure

Convert 300° to radians.

Solution

$$300° = 300°\left(\frac{\pi \text{ radians}}{180°}\right)$$

$$= \frac{5}{3}\pi \text{ radians} \qquad \text{• Exact answer}$$

$$\approx 5.23598776 \text{ radians} \qquad \text{• Approximate answer}$$

TRY EXERCISE 32, EXERCISE SET 5.1, PAGE 399

In Example 3, note that $\frac{5}{3}\pi$ radians is an exact result. Many times it will be convenient to leave the measure of an angle in terms of π and not change it to a decimal approximation.

EXAMPLE 4 Convert Radian Measure to Degree Measure

Convert $-\frac{3}{4}\pi$ radians to degrees.

Solution

$$-\frac{3}{4}\pi \text{ radians} = -\frac{3}{4}\pi \text{ radians}\left(\frac{180°}{\pi \text{ radians}}\right) = -135°$$

TRY EXERCISE 40, EXERCISE SET 5.1, PAGE 399

Table 5.1

Degrees	Radians
0	0
30	$\pi/6$
45	$\pi/4$
60	$\pi/3$
90	$\pi/2$
120	$2\pi/3$
135	$3\pi/4$
150	$5\pi/6$
180	π
210	$7\pi/6$
225	$5\pi/4$
240	$4\pi/3$
270	$3\pi/2$
300	$5\pi/3$
315	$7\pi/4$
330	$11\pi/6$
360	2π

Table 5.1 lists the degree and radian measures of selected angles. Figure 5.23 illustrates each angle as measured from the positive x-axis.

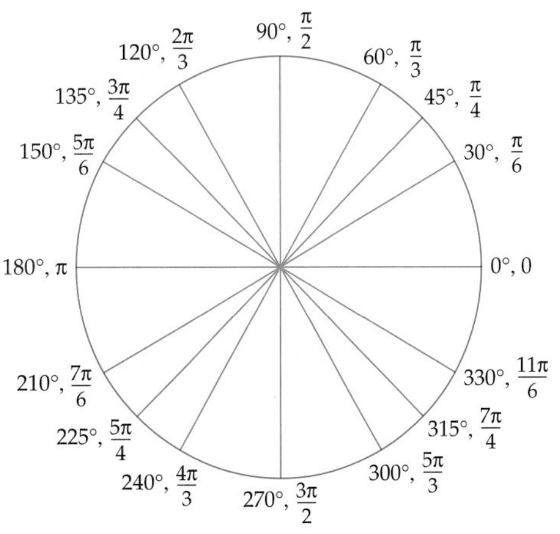

Figure 5.23
Degree and radian measures of selected angles

Figure 5.24

Figure 5.25

A graphing calculator can convert degree measure to radian measure, and vice versa. For example, the calculator display in **Figure 5.24** shows that 100° is approximately 1.74533 radians. The calculator must be in radian mode to convert from degrees to radians. The display in **Figure 5.25** shows that 2.2 radians is approximately 126.051°. The calculator must be in degree mode to convert from radians to degrees.

On a TI-83, the symbol for radian measure is r, and it is in the ANGLE menu.

◆ Arcs and Arc Length

Consider a circle of radius r. By solving the formula $\theta = s/r$ for s, we have an equation for arc length.

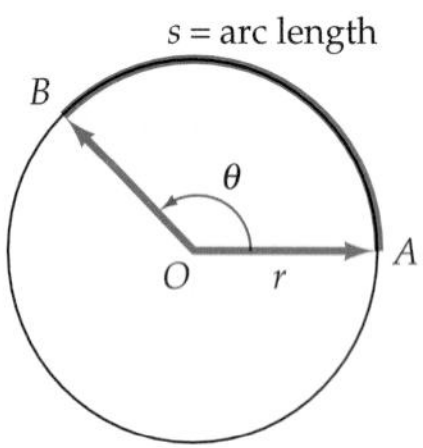

Figure 5.26
$s = r\theta$

Arc Length Formula

Let r be the length of the radius of a circle and θ the nonnegative radian measure of a central angle of the circle. Then the length of the arc s that subtends the central angle is $s = r\theta$. See **Figure 5.26.**

take note

The formula $s = r\theta$ is valid only when θ is expressed in radians.

EXAMPLE 5 Find the Length of an Arc

Find the length of an arc that subtends a central angle of 120° in a circle of radius 10 cm.

Solution

The formula $s = r\theta$ requires that θ be expressed in radians. We first convert 120° to radian measure and then use the formula $s = r\theta$.

$$\theta = 120^\circ = 120^\circ\left(\frac{\pi \text{ radians}}{180^\circ}\right) = \frac{2\pi}{3} \text{ radians} = \frac{2\pi}{3}$$

$$s = r\theta = (10 \text{ cm})\left(\frac{2\pi}{3}\right) = \frac{20\pi}{3} \text{ cm}$$

TRY EXERCISE 62, EXERCISE SET 5.1, PAGE 399

EXAMPLE 6 Solve an Application

A pulley with a radius of 10 inches uses a belt to drive a pulley with a radius of 6 inches. Find the angle through which the smaller pulley turns as the 10-inch pulley makes one revolution. State your answer in radians and also in degrees.

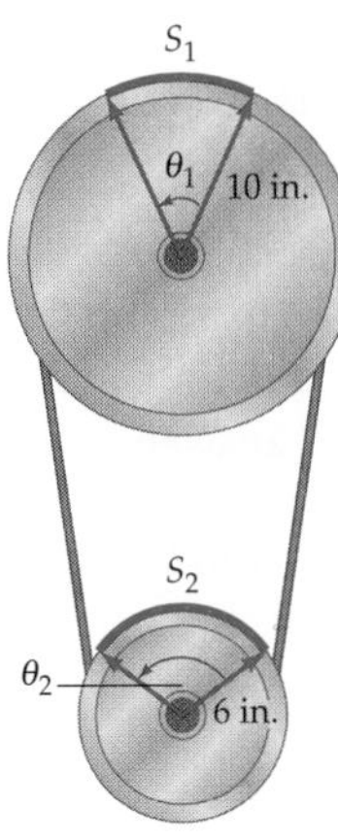

Figure 5.27

Solution

Use the formula $s = r\theta$. As the 10-inch pulley turns through an angle θ_1, a point on that pulley moves s_1 inches, where $s_1 = 10\theta_1$. See **Figure 5.27.** At the same time, the 6-inch pulley turns through an angle of θ_2 and a point on that pulley moves s_2 inches, where $s_2 = 6\theta_2$. Assuming that the belt does not slip on the pulleys, we have $s_1 = s_2$. Thus

$$10\theta_1 = 6\theta_2$$

$$10(2\pi) = 6\theta_2 \qquad \text{• Solve for } \theta_2\text{, when } \theta_1 = 2\pi \text{ radians.}$$

$$\frac{10}{3}\pi = \theta_2$$

The 6-inch pulley turns through an angle of $\frac{10}{3}\pi$ radians, or 600°.

TRY EXERCISE 66, EXERCISE SET 5.1, PAGE 400

◆ LINEAR AND ANGULAR SPEED

A car traveling at a speed of 55 miles per hour covers a distance of 55 miles in 1 hour. **Linear speed** v is *distance* traveled per unit time. In equation form,

$$v = \frac{s}{t}$$

where v is the linear speed, s is the distance traveled, and t is the time.

The floppy disk in a computer disk drive revolving at 300 revolutions per minute makes 300 complete revolutions in 1 minute. **Angular speed** ω is the *angle* through which a point on a circle moves per unit time. In equation form,

$$\omega = \frac{\theta}{t}$$

where ω is the angular speed, θ is the measure (in radians) of the angle through which a point has moved, and t is the time. Some common units of angular speed are revolutions per second, revolutions per minute, radians per second, and radians per minute.

EXAMPLE 7 Convert an Angular Speed

A hard disk in a computer rotates at 3600 revolutions per minute. Find the angular speed of the disk in radians per second.

Solution

As a point on the disk rotates 1 revolution (rev), the angle through which the point moves is 2π radians. Thus (2π radians/1 rev) will be the unit

Continued ►

fraction we will use to convert from revolutions to radians. To convert from minutes to seconds, use the unit fraction (1 minute/60 seconds).

$$3600 \text{ rev/minute} = \frac{3600 \text{ rev}}{1 \text{ minute}}\left(\frac{2\pi \text{ radians}}{1 \text{ rev}}\right)\left(\frac{1 \text{ minute}}{60 \text{ seconds}}\right)$$

$$= \frac{120\pi \text{ radians}}{1 \text{ second}} \qquad \text{• Exact answer}$$

$$\approx 377 \text{ radians/second} \qquad \text{• Approximate answer}$$

TRY EXERCISE 68, EXERCISE SET 5.1, PAGE 400

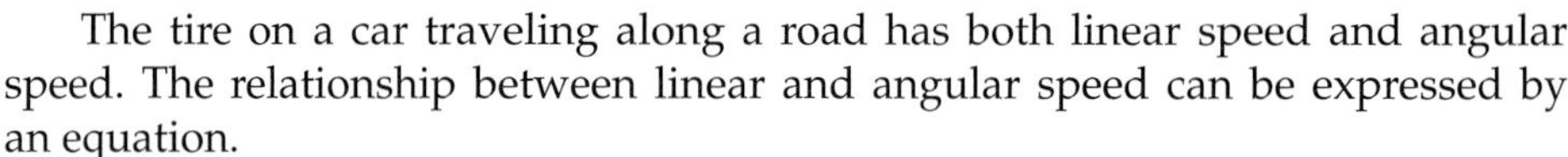

The tire on a car traveling along a road has both linear speed and angular speed. The relationship between linear and angular speed can be expressed by an equation.

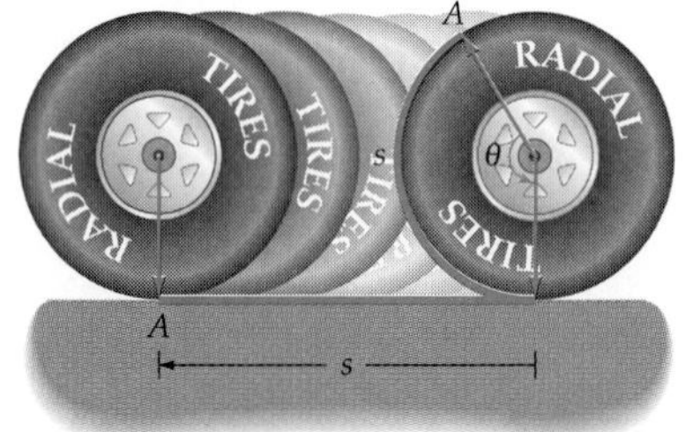

Figure 5.28

Assume that the wheel in **Figure 5.28** is rolling without slipping. As the wheel moves a distance s, point A moves through an angle θ. The arc length subtending angle θ is also s, the distance traveled by the wheel. From the equations for linear and angular speed, we have

$$v = \frac{s}{t} = \frac{r\theta}{t} = r\frac{\theta}{t} \qquad \text{• } s = r\theta$$

$$v = r\omega \qquad \text{• } \omega = \theta/t$$

The equation $v = r\omega$ gives the linear speed of a point on a rotating body in terms of distance r from the axis of rotation and the angular speed ω, provided that ω is in radians per unit of time.

EXAMPLE 8 Find Linear Speed

A wind machine is used to generate electricity. The wind machine has propeller blades that are 12 feet in length (see **Figure 5.29**). If the propeller is rotating at 3 revolutions per second, what is the linear speed in feet per second of the tips of the blades?

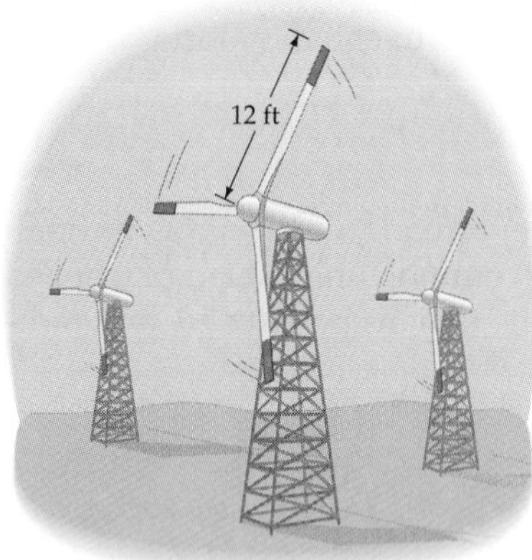

Figure 5.29

Solution

Convert the angular speed $\omega = 3$ revolutions per second into radians per second, and then use the formula $v = r\omega$.

$$\omega = \frac{3 \text{ revolutions}}{1 \text{ second}} = \left(\frac{3 \text{ revolutions}}{1 \text{ second}}\right)\left(\frac{2\pi \text{ radians}}{1 \text{ revolution}}\right) = \frac{6\pi \text{ radians}}{1 \text{ second}}$$

Thus

$$v = r\omega = (12 \text{ ft})\left(\frac{6\pi \text{ radians}}{1 \text{ second}}\right)$$

$$= 72\pi \text{ feet per second} \approx 226 \text{ feet per second}$$

TRY EXERCISE 74, EXERCISE SET 5.1, PAGE 400

TOPICS FOR DISCUSSION

1. The measure of a radian differs depending on the length of the radius of the circle used. Do you agree? Explain.
2. The measure of 1 radian is over 100 times larger than the measure of 1 degree. Do you agree? Explain.
3. What are the necessary conditions for an angle to be in standard position?
4. Is the supplement of an obtuse angle an acute angle?
5. Do all acute angles have a positive measure?

EXERCISE SET 5.1

In Exercises 1 to 12, find the measure (if possible) of the complement and the supplement of each angle.

1. $15°$ 2. $87°$ 3. $70°15'$
4. $22°43'$ 5. $56°33'15''$ 6. $19°42'05''$
7. 1 8. 0.5 9. $\pi/4$
10. $\pi/3$ 11. $\pi/2$ 12. $\pi/6$

In Exercises 13 to 18, classify each angle by quadrant, and state the measure of the positive angle with measure less than 360° that is coterminal with the given angle.

13. $\alpha = 610°$ 14. $\alpha = 765°$ 15. $\alpha = -975°$
16. $\alpha = -872°$ 17. $\alpha = 2456°$ 18. $\alpha = -3789°$

In Exercises 19 to 24, use a calculator to convert each decimal degree measure to its equivalent DMS measure.

19. $24.56°$ 20. $110.24°$ 21. $64.158°$
22. $18.96°$ 23. $3.402°$ 24. $224.282°$

In Exercises 25 to 30, use a calculator to convert each DMS measure to its equivalent decimal measure.

25. $25°25'12''$ 26. $63°29'42''$ 27. $183°33'36''$
28. $141°6'9''$ 29. $211°46'48''$ 30. $19°12'18''$

In Exercises 31 to 39, convert the degree measure to exact radian measure.

31. $30°$ 32. $-45°$ 33. $90°$
34. $15°$ 35. $165°$ 36. $315°$
37. $420°$ 38. $630°$ 39. $585°$

In Exercises 40 to 48, convert the radian measure to exact degree measure.

40. $\frac{\pi}{4}$ 41. $\frac{\pi}{5}$ 42. $-\frac{2\pi}{3}$
43. $\frac{\pi}{6}$ 44. $\frac{\pi}{9}$ 45. $\frac{3\pi}{8}$
46. $\frac{11\pi}{18}$ 47. $\frac{11\pi}{3}$ 48. $\frac{6\pi}{5}$

In Exercises 49 to 54, convert radians to degrees or degrees to radians. Round answers to the nearest hundredth.

49. 1.5 50. -2.3 51. $133°$
52. $427°$ 53. 8.25 54. $-90°$

In Exercises 55 to 58, find the measure in radians and degrees of the central angle of a circle subtended by the given arc. Round answers to the nearest hundredth.

55. $r = 2$ inches, $s = 8$ inches
56. $r = 7$ feet, $s = 4$ feet
57. $r = 5.2$ centimeters, $s = 12.4$ centimeters
58. $r = 35.8$ meters, $s = 84.3$ meters

In Exercises 59 to 62, find the measure of the intercepted arc of a circle with the given radius and central angle. Round answers to the nearest hundredth.

59. $r = 8$ inches, $\theta = \pi/4$ 60. $r = 3$ feet, $\theta = 7\pi/2$
61. $r = 25$ centimeters, $\theta = 42°$
62. $r = 5$ meters, $\theta = 144°$

63. Find the number of radians in $1\frac{1}{2}$ revolutions.

64. Find the number of radians in 3/8 revolution.

65. **ANGULAR ROTATION OF TWO PULLEYS** A pulley with a radius of 14 inches uses a belt to drive a pulley with a radius of 28 inches. The 14-inch pulley turns through an angle of 150°. Find the angle through which the 28-inch pulley turns.

66. **ANGULAR ROTATION OF TWO PULLEYS** A pulley with a diameter of 1.2 meters uses a belt to drive a pulley with a diameter of 0.8 meter. The 1.2-meter pulley turns through an angle of 240°. Find the angle through which the 0.8-meter pulley turns.

67. **ANGULAR SPEED** Find the angular speed, in radians per second, of the second hand on a clock.

68. **ANGULAR SPEED** Find the angular speed, in radians per second, of a point on the equator of the earth.

69. **ANGULAR SPEED** A wheel is rotating at 50 revolutions per minute. Find the angular speed in radians per second.

70. **ANGULAR SPEED** A wheel is rotating at 200 revolutions per minute. Find the angular speed in radians per second.

71. **ANGULAR SPEED** The turntable of a record player turns at $33\frac{1}{3}$ revolutions per minute. Find the angular speed in radians per second.

72. **ANGULAR SPEED** A car with a wheel of radius 14 inches is moving with a speed of 55 mph. Find the angular speed of the wheel in radians per second.

73. **LINEAR SPEED OF A CAR** Each tire on a car has a radius of 15 inches. The tires are rotating at 450 revolutions per minute. Find the speed of the automobile to the nearest mile per hour.

74. **LINEAR SPEED OF A TRUCK** Each tire on a truck has a radius of 18 inches. The tires are rotating at 500 revolutions per minute. Find the speed of the truck to the nearest mile per hour.

75. **ASTRONOMY** At a time when the earth was 93,000,000 miles from the sun, you observed through a tinted glass that the diameter of the sun occupied an arc of 31′. Determine, to the nearest ten thousand miles, the diameter of the sun.

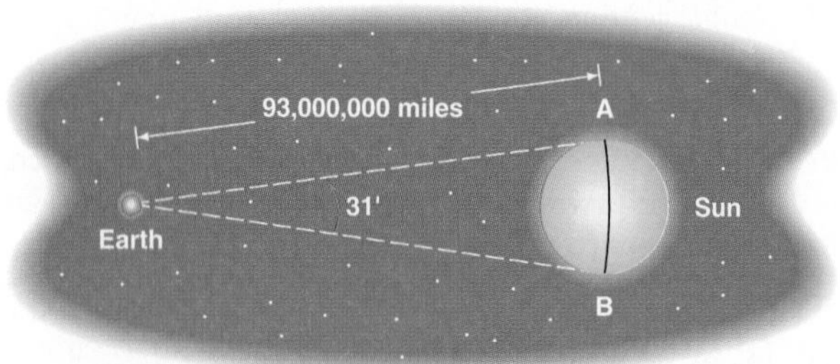

(*Hint:* Because the radius of arc *AB* is large and its central angle is small, the length of the diameter of the sun is approximately the length of the arc *AB*.)

76. **ANGLE OF ROTATION AND DISTANCE** The minute hand on the clock atop city hall measures 6 feet 3 inches from the tip to its axle.
    a. Through what angle (in radians) does the minute hand pass between 9:12 A.M. and 9:48 A.M.?
    b. What distance, to the nearest tenth of a foot, does the tip of the minute hand travel during this period?

77. **VELOCITY OF THE HUBBLE SPACE TELESCOPE** On April 25, 1990, the Hubble space telescope (HST) was deployed into a circular orbit 625 kilometers above the surface of the earth. The HST completes an earth orbit every 1.61 hours.

    a. Find the angular velocity, with respect to the center of the earth, of the HST. Round your answer to the nearest 0.1 radian per hour.
    b. Find the linear velocity of the HST. *Hint:* The radius of the earth is about 6370 kilometers. Round your answer to the nearest 100 kilometers per hour.

78. **ESTIMATING THE RADIUS OF THE EARTH** Eratosthenes, the fifth librarian of Alexandria (230 B.C.), was able to estimate the radius of the earth from the following data: The distance between the Egyptian cities of Alexandria and Syrene was 5000 stadia (520 miles). Syrene was located directly south of Alexandria. One summer, at noon, the sun was directly overhead at Syrene, whereas, at the same time in Alexandria the sun was at a 7.5° angle from the zenith.

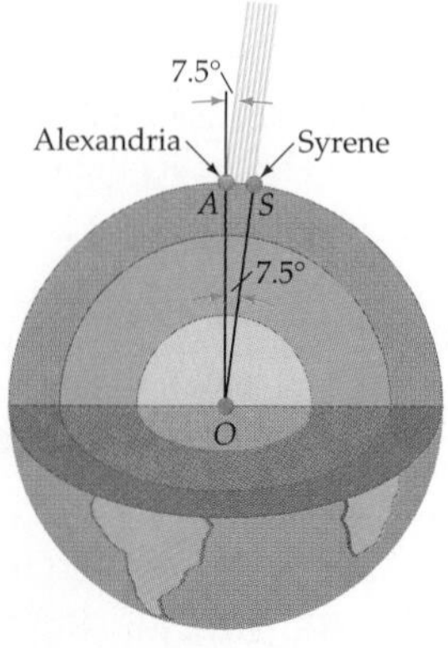

Eratosthenes reasoned that because the sun is far away, the rays of sunlight that reach the earth must be nearly parallel. From this assumption he concluded that the measure of $\angle AOS$, in the accompanying figure, must be 7.5°. Use this information to estimate the radius (to the nearest 10 miles) of the earth.

79. **VELOCITY COMPARISONS** Assume that the bicycle in the figure is moving forward at a constant rate. Point A is on the edge of the 30-inch rear tire, and point B is on the edge of the 20-inch front tire.

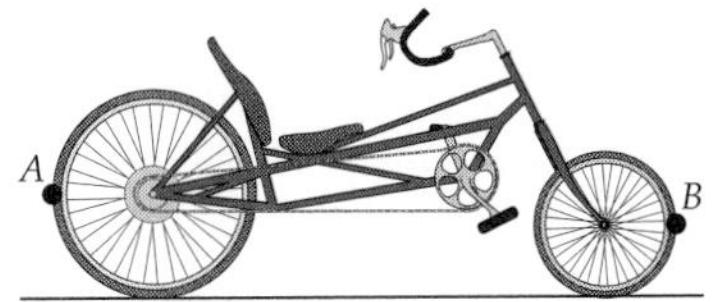

a. Which point (A or B) has the greater angular velocity? Explain.

b. Which point (A or B) has the greater linear velocity? Explain.

80. Given that s, r, θ, t, v, and ω are as defined in Section 5.1, determine which of the following formulas are valid.

$$s = r\theta \qquad r = \frac{s}{\theta} \qquad v = \frac{r\theta}{t}$$

$$v = r\omega \qquad v = \frac{s}{t} \qquad \omega = \frac{\theta}{t}$$

81. **NAUTICAL MILES AND STATUTE MILES** A **nautical mile** is the length of an arc, on the earth's equator, that subtends a 1′ central angle. The equatorial radius of the earth is about 3960 **statute miles.**

a. Convert 1 nautical mile to statute miles. Round to the nearest hundredth of a statute mile.

b. Determine what percent (to the nearest 1 percent) of the earth's circumference is covered by a trip from Los Angeles, California, to Honolulu, Hawaii (a distance of 2217 nautical miles).

82. **PHOTOGRAPHY** The field of view for a camera with a 200-millimeter lens is 12°. A photographer takes a photograph of a large building that is 485 feet in front of the camera. What is the approximate width, to the nearest foot, of the building that will appear in the photograph? (*Hint:* If the radius of an arc AB is large and its central angle is small, then the length of the chord AB is approximately the length of the arc AB.)

SUPPLEMENTAL EXERCISES

A *sector* of a circle is the region bounded by radii OA and OB and the intercepted arc AB. See the figure at the top of the next column. The area of the sector is given by

$$A = \frac{1}{2}r^2\theta,$$

where r is the radius of the circle and θ is the measure of the central angle in radians.

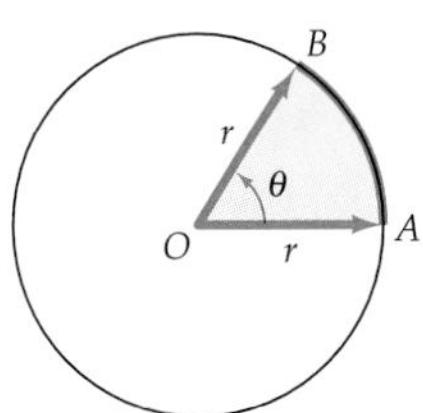

In Exercises 83 to 88, find the area, to the nearest square unit, of the sector of a circle with the given radius and central angle.

83. $r = 5$ inches, $\theta = \pi/3$ radians

84. $r = 2.8$ feet, $\theta = 5\pi/2$ radians

85. $r = 120$ centimeters, $\theta = 0.65$ radian

86. $r = 30$ feet, $\theta = 62°$

87. $r = 20$ meters, $\theta = 125°$

88. $r = 25$ centimeters, $\theta = 220°$

89. **DISTANCE AND LINEAR SPEED** A merry-go-round horse is 11.6 meters from the center. The merry-go-round makes $14\frac{1}{4}$ revolutions per ride in 5 minutes.

a. How many meters, to the nearest meter, does the horse travel?

b. How fast is it moving in meters per second?

90. **TIRE SIZE AND REVOLUTIONS**

a. A car with 13-inch-radius tires makes an 8-mile trip. Find the number of revolutions a tire makes on the 8-mile trip. Round to nearest 100 revolutions.

b. A car with 15-inch-radius tires makes an 8-mile trip. Find the number of revolutions a tire makes on the 8-mile trip. Round to the nearest 100 revolutions.

91. **SPEED OF A RIVER** A water wheel has a 10-foot radius. When the wheel makes 18 revolutions per minute, what is the speed of the river? Round to the nearest tenth of a foot per second.

92. **LINEAR SPEED OF A PULLEY** A pulley with a 50-centimeter diameter drives a pulley with a 20-centimeter diameter. The larger pulley makes 30 revolutions per minute. What is the linear speed of a point on the circumference of the smaller pulley in centimeters per second?

93. **AREA OF A CIRCULAR SEGMENT** Find the area of the shaded portion of the circle. The radius of the circle is 9 inches.

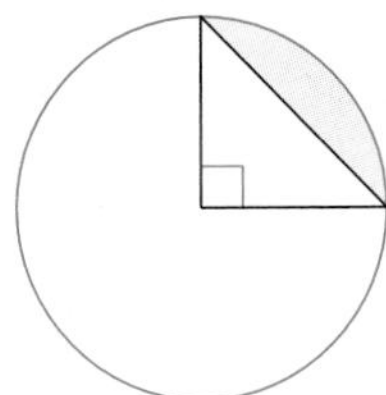

Latitude describes the position of a point on the earth's surface in relation to the equator. A point on the equator has a latitude of 0°. The north pole has a latitude of 90°. The radius of the earth is approximately 3960 miles.

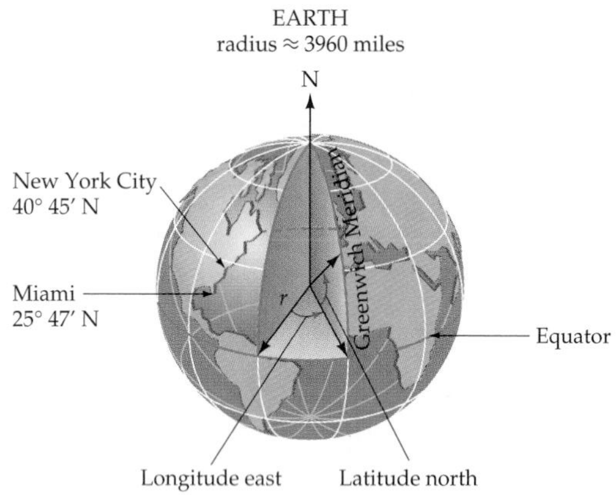

94. **GEOGRAPHY** The city of New York has a latitude of 40°45′N. How far north, to the nearest 10 miles, is it from the equator? Use 3960 miles as the radius of the earth.

95. **GEOGRAPHY** The city of Miami has a latitude of 25°47′N. How far north, to the nearest 10 miles, is it from the equator? Use 3960 miles as the radius of the earth.

96. **GEOGRAPHY** Assuming that the earth is a perfect sphere, and expressing your answer to three significant digits, find the distance along the earth's surface (in miles) that subtends a central angle of

a. 1° b. 1′ c. 1″

PROJECTS

1. **CONVERSION OF UNITS** You are traveling in a foreign country. You discover that the currency used consists of lollars, mollars, nollars, and tollars.

$$5 \text{ lollars} = 1 \text{ mollar}$$
$$3 \text{ mollars} = 5 \text{ nollars}$$
$$4 \text{ nollars} = 7 \text{ tollars}$$

The fare for a taxi is 14 tollars.

a. How much is the fare in mollars? (*Hint:* Make use of unit fractions to convert from tollars to mollars.)

b. If you have only mollars and lollars, how many of each could you use to pay the fare?

c. Explain to a classmate the concept of unit fractions. Also explain how you knew which of the unit fractions

$$\left(\frac{4 \text{ nollars}}{7 \text{ tollars}}\right) \qquad \left(\frac{7 \text{ tollars}}{4 \text{ nollars}}\right)$$

should be used in the conversion in **a.**

d. If you wish to convert x degrees to radians, you need to multiply x by which of the following unit fractions?

$$\left(\frac{\pi}{180°}\right) \qquad \left(\frac{180°}{\pi}\right)$$

2. **SPACE SHUTTLE** The rotational period of the earth is 23.933 hours. A space shuttle revolves around the earth's equator every 2.231 hours. Both are rotating in the same direction. At the present time, the space shuttle is directly above the Galápagos Islands. How long will it take for the space shuttle to circle the earth and return to a position directly above the Galápagos Islands?

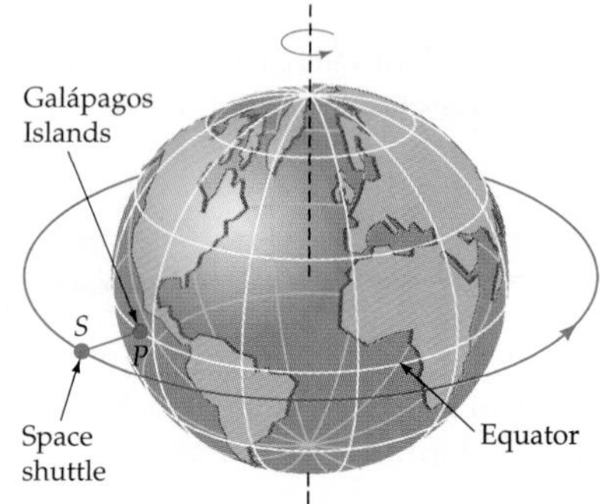

SECTION 5.2 TRIGONOMETRIC FUNCTIONS OF ACUTE ANGLES

- TRIGONOMETRIC FUNCTIONS OF ACUTE ANGLES
- TRIGONOMETRIC FUNCTIONS OF SPECIAL ANGLES
- APPLICATIONS INVOLVING RIGHT TRIANGLES

◆ TRIGONOMETRIC FUNCTIONS OF ACUTE ANGLES

The study of trigonometry, which means "triangle measurement," began more than 2000 years ago, partially as a means of solving surveying problems. Early trigonometry used the length of a chord of a circle as the value of a *trigonometric function.* In the sixteenth century, right triangles were used to define a trigonometric function. We will use a modification of this approach.

When working with right triangles, it is convenient to refer to the side *opposite* an angle or the side *adjacent* to (next to) an angle. **Figure 5.30a** shows the sides opposite and adjacent to the angle α. For angle β, the opposite and adjacent sides are shown as in **Figure 5.30b.** In both cases, the hypotenuse remains the same.

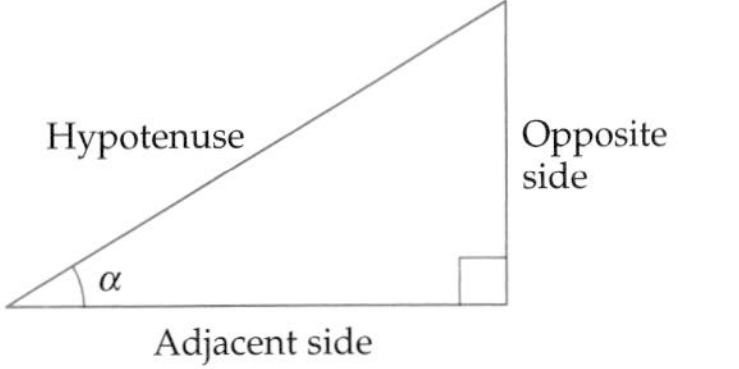

a. Adjacent and opposite sides of $\angle\alpha$

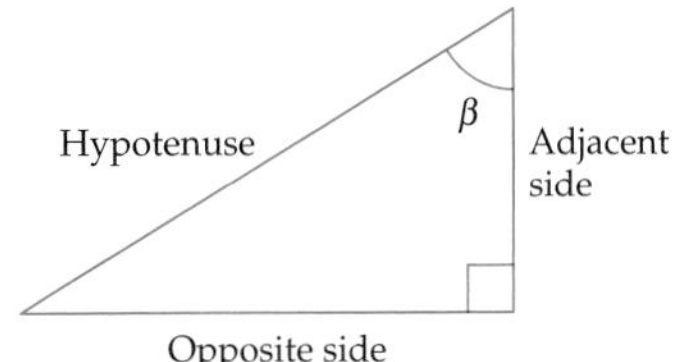

b. Adjacent and opposite sides of $\angle\beta$

Figure 5.30

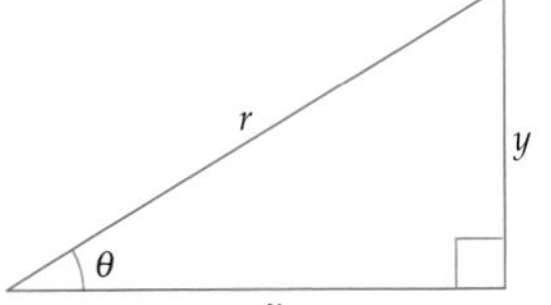

Figure 5.31

Consider an angle θ in the right triangle shown in **Figure 5.31.** Let x and y represent the lengths, respectively, of the adjacent and opposite sides of the triangle, and let r be the length of the hypotenuse. Six possible ratios can be formed:

$$\frac{y}{r} \quad \frac{x}{r} \quad \frac{y}{x} \quad \frac{r}{y} \quad \frac{r}{x} \quad \frac{x}{y}$$

Each ratio defines a value of a trigonometric function of the acute angle θ. The functions are **sine** (sin), **cosine** (cos), **tangent** (tan), **cosecant** (csc), **secant** (sec), and **cotangent** (cot).

Trigonometric Functions of an Acute Angle

Let θ be an acute angle of a right triangle. The values of the six trigonometric functions of θ are

$$\sin\theta = \frac{\text{length of opposite side}}{\text{length of hypotenuse}} = \frac{y}{r} \qquad \cos\theta = \frac{\text{length of adjacent side}}{\text{length of hypotenuse}} = \frac{x}{r}$$

$$\tan\theta = \frac{\text{length of opposite side}}{\text{length of adjacent side}} = \frac{y}{x} \qquad \cot\theta = \frac{\text{length of adjacent side}}{\text{length of opposite side}} = \frac{x}{y}$$

$$\sec\theta = \frac{\text{length of hypotenuse}}{\text{length of adjacent side}} = \frac{r}{x} \qquad \csc\theta = \frac{\text{length of hypotenuse}}{\text{length of opposite side}} = \frac{r}{y}$$

We will write opp, adj, and hyp as abbreviations for *the length of the* opposite side, adjacent side, and hypotenuse, respectively.

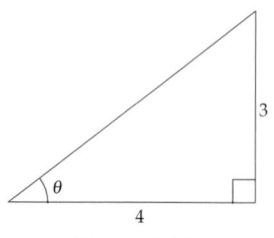

EXAMPLE 1 Evaluate Trigonometric Functions

Find the values of the six trigonometric functions of θ for the triangle given in **Figure 5.32.**

Figure 5.32

Solution

Use the Pythagorean Theorem to find the length of the hypotenuse.

$$r = \sqrt{3^2 + 4^2} = \sqrt{25} = 5$$

From the definitions of the trigonometric functions,

$$\sin\theta = \frac{\text{opp}}{\text{hyp}} = \frac{3}{5} \qquad \cos\theta = \frac{\text{adj}}{\text{hyp}} = \frac{4}{5} \qquad \tan\theta = \frac{\text{opp}}{\text{adj}} = \frac{3}{4}$$

$$\cot\theta = \frac{\text{adj}}{\text{opp}} = \frac{4}{3} \qquad \sec\theta = \frac{\text{hyp}}{\text{adj}} = \frac{5}{4} \qquad \csc\theta = \frac{\text{hyp}}{\text{opp}} = \frac{5}{3}$$

TRY EXERCISE 6, EXERCISE SET 5.2, PAGE 410

Given the value of one trigonometric function of θ, it is possible to find the value of any of the remaining trigonometric functions of θ.

EXAMPLE 2 Find the Value of a Trigonometric Function

Given $\cos\theta = \frac{5}{8}$, find $\tan\theta$.

Solution

$$\cos\theta = \frac{5}{8} = \frac{\text{adj}}{\text{hyp}}$$

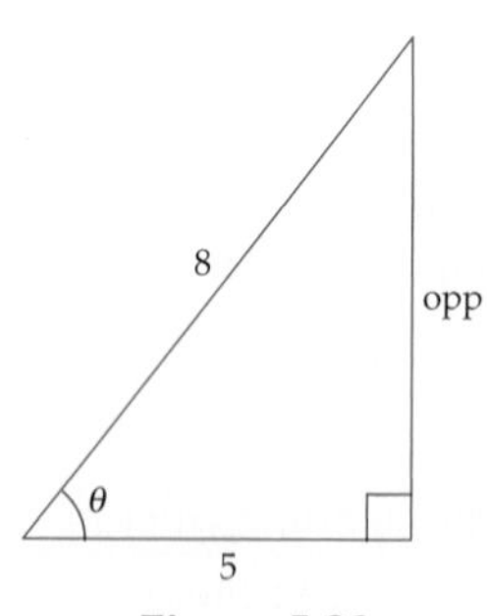

Figure 5.33

Sketch a right triangle with one leg of length 5 units and a hypotenuse of length 8 units. Label as θ the acute angle that has the leg of length 5 units as its adjacent side (see **Figure 5.33**). Use the Pythagorean Theorem to find the length of the opposite side.

$$8^2 = (\text{opp})^2 + 5^2$$
$$(\text{opp})^2 = 39$$
$$\text{opp} = \sqrt{39}$$

Therefore, $\tan\theta = \frac{\text{opp}}{\text{adj}} = \frac{\sqrt{39}}{5}$.

TRY EXERCISE 20, EXERCISE SET 5.2, PAGE 411

◆ TRIGONOMETRIC FUNCTIONS OF SPECIAL ANGLES

In Example 1, the lengths of the legs of the triangle were given, and you were asked to find the values of the six trigonometric functions of the angle θ. Often we will want to find the value of a trigonometric function when we are given *the measure of an angle* rather than the measure of the sides of a triangle. For most angles, advanced mathematical methods are required to evaluate a trigonometric function. For some *special angles,* however, the value of a trigonometric function can be found by geometric methods. These special acute angles are 30°, 45°, and 60°.

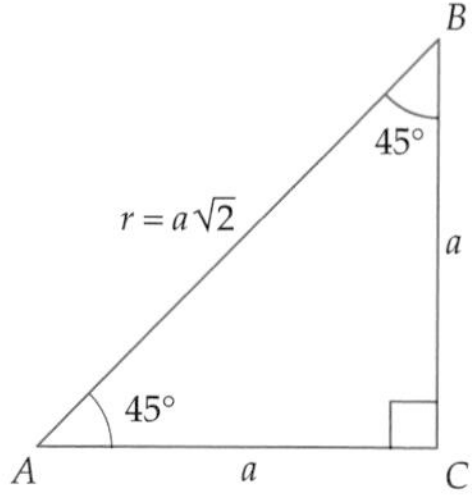

Figure 5.34

First we will find the values of the six trigonometric functions of 45°. (This discussion is based on angles measured in degrees. Radian measure could have been used without changing the results.) **Figure 5.34** shows a right triangle with angles 45°, 45°, and 90°. Because $\angle A = \angle B$, the lengths of the sides opposite these angles are equal. Let the length of each equal side be denoted by a. From the Pythagorean Theorem,

$$r^2 = a^2 + a^2 = 2a^2$$
$$r = \sqrt{2a^2} = a\sqrt{2}$$

The values of the six trigonometric functions of 45° are

$$\sin 45^\circ = \frac{a}{a\sqrt{2}} = \frac{1}{\sqrt{2}} = \frac{\sqrt{2}}{2} \qquad \cos 45^\circ = \frac{a}{a\sqrt{2}} = \frac{1}{\sqrt{2}} = \frac{\sqrt{2}}{2}$$

$$\tan 45^\circ = \frac{a}{a} = 1 \qquad \cot 45^\circ = \frac{a}{a} = 1$$

$$\sec 45^\circ = \frac{a\sqrt{2}}{a} = \sqrt{2} \qquad \csc 45^\circ = \frac{a\sqrt{2}}{a} = \sqrt{2}$$

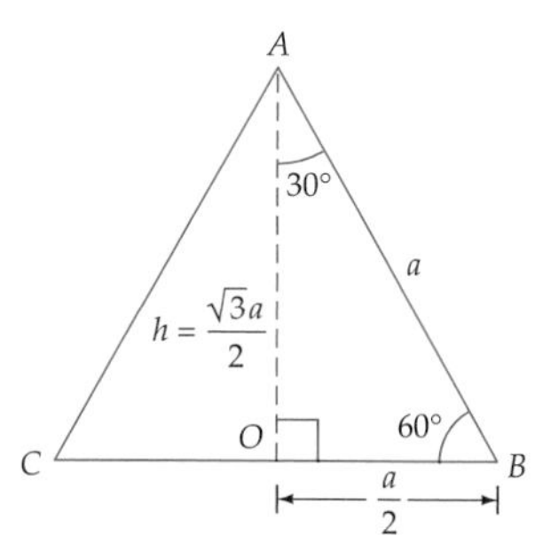

Figure 5.35

The values of the trigonometric functions of the special angles 30° and 60° can be found by drawing an equilateral triangle and bisecting one of the angles, as **Figure 5.35** shows. The angle bisector also bisects one of the sides. Thus the length of the side opposite the 30° angle is one-half the length of the hypotenuse of triangle OAB.

Let a denote the length of the hypotenuse. Then the length of the side opposite the 30° angle is $a/2$. The length of the side adjacent to the 30° angle, h, is found by using the Pythagorean Theorem.

$$a^2 = \left(\frac{a}{2}\right)^2 + h^2$$

$$a^2 = \frac{a^2}{4} + h^2$$

$$\frac{3a^2}{4} = h^2 \qquad \text{• Subtract } \frac{a^2}{4} \text{ from each side.}$$

$$h = \frac{\sqrt{3}a}{2} \qquad \text{• Solve for } h.$$

The values of the six trigonometric functions of 30° are

$$\sin 30^\circ = \frac{a/2}{a} = \frac{1}{2} \qquad \cos 30^\circ = \frac{\sqrt{3}a/2}{a} = \frac{\sqrt{3}}{2}$$

$$\tan 30^\circ = \frac{a/2}{\sqrt{3}a/2} = \frac{1}{\sqrt{3}} = \frac{\sqrt{3}}{3} \qquad \cot 30^\circ = \frac{\sqrt{3}a/2}{a/2} = \sqrt{3}$$

$$\sec 30^\circ = \frac{a}{\sqrt{3}a/2} = \frac{2}{\sqrt{3}} = \frac{2\sqrt{3}}{3} \qquad \csc 30^\circ = \frac{a}{a/2} = 2$$

The values of the trigonometric functions of 60° can be found by again using Figure 5.35. The length of the side opposite the 60° angle is $\sqrt{3}a/2$, and the length of the side adjacent to the 60° angle is $a/2$. The values of the trigonometric functions of 60° are

$$\sin 60^\circ = \frac{\sqrt{3}a/2}{a} = \frac{\sqrt{3}}{2} \qquad \cos 60^\circ = \frac{a/2}{a} = \frac{1}{2}$$

$$\tan 60^\circ = \frac{\sqrt{3}a/2}{a/2} = \sqrt{3} \qquad \cot 60^\circ = \frac{a/2}{\sqrt{3}a/2} = \frac{1}{\sqrt{3}} = \frac{\sqrt{3}}{3}$$

$$\sec 60^\circ = \frac{a}{a/2} = 2 \qquad \csc 60^\circ = \frac{a}{\sqrt{3}a/2} = \frac{2}{\sqrt{3}} = \frac{2\sqrt{3}}{3}$$

Table 5.2 summarizes the values of the trigonometric functions for the special angles 30° ($\pi/6$), 45° ($\pi/4$), and 60° ($\pi/3$).

take note

Memorizing the values given in Table 5.2 will prove to be extremely useful in the remaining trigonometric sections.

Table 5.2 Trigonometric Functions of Special Angles

θ	$\sin\theta$	$\cos\theta$	$\tan\theta$	$\csc\theta$	$\sec\theta$	$\cot\theta$
$30°; \frac{\pi}{6}$	$\frac{1}{2}$	$\frac{\sqrt{3}}{2}$	$\frac{\sqrt{3}}{3}$	2	$\frac{2\sqrt{3}}{3}$	$\sqrt{3}$
$45°; \frac{\pi}{4}$	$\frac{\sqrt{2}}{2}$	$\frac{\sqrt{2}}{2}$	1	$\sqrt{2}$	$\sqrt{2}$	1
$60°; \frac{\pi}{3}$	$\frac{\sqrt{3}}{2}$	$\frac{1}{2}$	$\sqrt{3}$	$\frac{2\sqrt{3}}{3}$	2	$\frac{\sqrt{3}}{3}$

take note

The patterns in the following chart can be used to memorize the sine and cosine of 30°, 45°, and 60°.

$$\sin 30^\circ = \frac{\sqrt{1}}{2} \qquad \cos 30^\circ = \frac{\sqrt{3}}{2}$$

$$\sin 45^\circ = \frac{\sqrt{2}}{2} \qquad \cos 45^\circ = \frac{\sqrt{2}}{2}$$

$$\sin 60^\circ = \frac{\sqrt{3}}{2} \qquad \cos 60^\circ = \frac{\sqrt{1}}{2}$$

EXAMPLE 3 Evaluate a Trigonometric Expression

Find the *exact* value of $\sin^2 45^\circ + \cos^2 60^\circ$.

Solution

Substitute the values of sin 45° and cos 60° and simplify.

Note: $\sin^2\theta = (\sin\theta)(\sin\theta) = (\sin\theta)^2$ and $\cos^2\theta = (\cos\theta)(\cos\theta) = (\cos\theta)^2$.

$$\sin^2 45^\circ + \cos^2 60^\circ = \left(\frac{\sqrt{2}}{2}\right)^2 + \left(\frac{1}{2}\right)^2 = \frac{2}{4} + \frac{1}{4} = \frac{3}{4}$$

TRY EXERCISE 38, EXERCISE SET 5.2, PAGE 411

From the definition of the sine and cosecant functions,

$$(\sin\theta)(\csc\theta) = \frac{y}{r}\cdot\frac{r}{y} = 1 \quad \text{or} \quad (\sin\theta)(\csc\theta) = 1$$

By rewriting the last equation, we find

$$\sin\theta = \frac{1}{\csc\theta} \quad \text{and} \quad \csc\theta = \frac{1}{\sin\theta}, \text{ provided } \sin\theta \neq 0$$

The sine and cosecant functions are called **reciprocal functions.** The cosine and secant are also reciprocal functions, as are the tangent and cotangent functions. Table 5.3 shows each trigonometric function and its reciprocal. These relationships hold for all values of θ for which both of the functions are defined.

Table 5.3 Trigonometric Functions and Their Reciprocals

$\sin\theta = \frac{1}{\csc\theta}$	$\cos\theta = \frac{1}{\sec\theta}$	$\tan\theta = \frac{1}{\cot\theta}$
$\csc\theta = \frac{1}{\sin\theta}$	$\sec\theta = \frac{1}{\cos\theta}$	$\cot\theta = \frac{1}{\tan\theta}$

take note

Some graphing calculators will allow you to display the degree symbol. For example, the *TI-83* display

was produced by entering

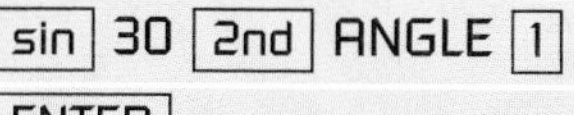

ENTER

Displaying the degree symbol will cause the calculator to evaluate a trigonometric function using degree mode even if the calculator is in radian mode.

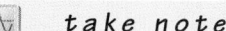

When evaluating a trigonometric function by using a graphing calculator, be sure the calculator is in the correct mode. If the measure of an angle is written with the degree symbol, then make sure the calculator is in degree mode. If the measure of an angle is given in radians (no degree symbol is used), then make sure the calculator is in radian mode. *Many errors are made because the correct mode is not selected.*

Some graphing calculators can be used to construct a table of functional values. For instance, the *TI-83* keystrokes shown below generate the table in **Figure 5.36**, in which the first column lists the domain values

$$0°, 1°, 2°, 3°, \ldots$$

and the second column lists the range values

$$\sin 0°, \sin 1°, \sin 2°, \sin 3°, \ldots$$

TI-83 Keystrokes

2nd TBLSET 0

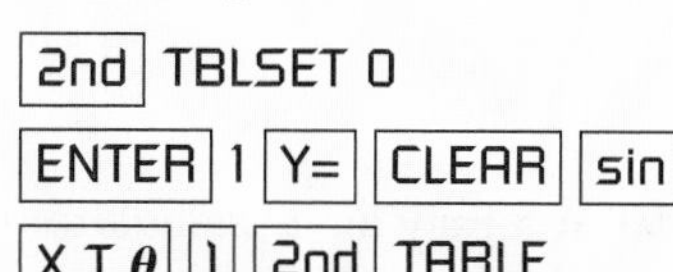

X	Y1
0	0
1	.01745
2	.0349
3	.05234
4	.06976
5	.08716
6	.10453

Y1 = .017452406437

Figure 5.36

The graphing calculator must be in degree mode to produce this table.

◆ APPLICATIONS INVOLVING RIGHT TRIANGLES

One of the major reasons for the development of trigonometry was to solve application problems. In this section we will consider some applications involving right triangles. In some application problems, a horizontal line of sight is used as a reference line. An angle measured above the line of sight is called an **angle of elevation,** and an angle measured below the line of sight is called an **angle of depression.** See **Figure 5.37.**

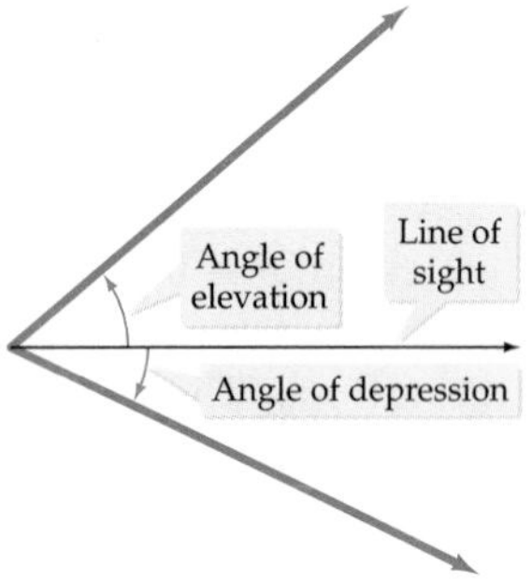

Angle of elevation
Line of sight
Angle of depression

Figure 5.37

EXAMPLE 4 Solve an Angle-of-Elevation Problem

From a point 115 feet from the base of a redwood tree, the angle of elevation to the top of the tree is 64.3°. Find the height of the tree to the nearest foot.

Solution

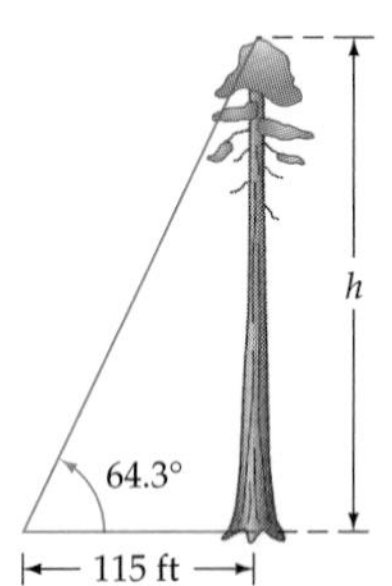

Figure 5.38

From **Figure 5.38,** the length of the adjacent side of the angle is known (115 feet). Because we need to determine the height of the tree (length of the opposite side), we use the tangent function. Let h represent the length of the opposite side.

$$\tan 64.3° = \frac{\text{opp}}{\text{adj}} = \frac{h}{115}$$

$$h = 115 \tan 64.3° \approx 238.952$$

The height of the redwood tree is approximately 239 feet.

TRY EXERCISE 66, EXERCISE SET 5.2, PAGE 412

Because the cotangent function involves the sides adjacent to and opposite an angle, we could have solved Example 4 by using the cotangent function. The solution would have been

$$\cot 64.3° = \frac{\text{adj}}{\text{opp}} = \frac{115}{h}$$

$$h = \frac{115}{\cot 64.3°} \approx 238.952$$

The accuracy of a calculator is sometimes beyond the limits of measurement. In the last example, the distance from the base of the tree was given as 115 feet (three significant digits), whereas the height of the tree was shown as 238.952 ft (six significant digits). When using approximate numbers, we will use the conventions given at the top of the next page for calculating with trigonometric functions.

take note

This rounding convention for approximate trigonometric calculations will be used throughout the remaining trigonometric sections.

A Rounding Convention:
Significant Digits for Trigonometric Calculations

Angle Measure to the Nearest	Significant Digits of the Lengths
Degree	Two
Tenth of a degree	Three
Hundredth of a degree	Four

EXAMPLE 5 Solve an Angle-of-Depression Problem

DME (Distance Measuring Equipment) is standard avionic equipment on a commercial airplane. This equipment measures the distance from a plane to a radar station. If the distance from a plane to a radar station is 160 miles and the angle of depression is 33°, find the number of ground miles from a point directly below the plane to the radar station.

Solution

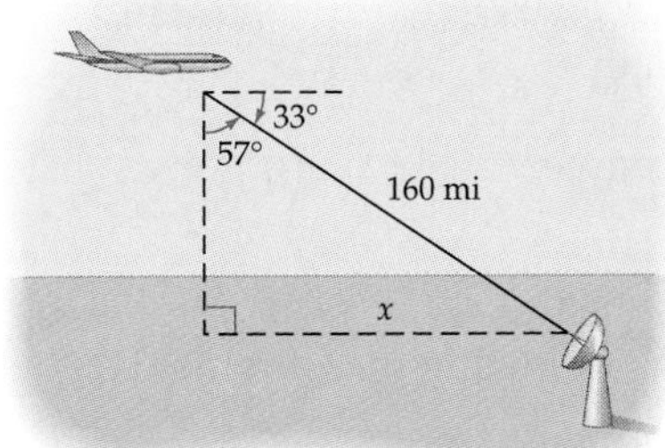

Figure 5.39

From **Figure 5.39**, the length of the hypotenuse is known (160 miles). The length of the side opposite the angle of 57° is unknown. The sine function involves the hypotenuse and the opposite side, x, of the 57°angle.

$$\sin 57° = \frac{x}{160}$$

$$x = 160 \sin 57° \approx 134.1873$$

The plane is 130, rounded to two significant digits, ground miles from the radar station.

TRY EXERCISE 68, EXERCISE SET 5.2, PAGE 412

EXAMPLE 6 Solve an Angle-of-Elevation Problem

An observer notes that the angle of elevation from point A to the top of a space shuttle is 27.2°. From a point 17.5 meters further from the space shuttle, the angle of elevation is 23.9°. Find the height of the space shuttle.

Solution

Figure 5.40

From **Figure 5.40**, let x denote the distance from point A to the base of the space shuttle, and let y denote the height of the space shuttle. Then

$$(1)\quad \tan 27.2° = \frac{y}{x} \qquad \text{and} \qquad (2)\quad \tan 23.9° = \frac{y}{x + 17.5}$$

Continued

take note

The intermediate calculations in Example 6 were not rounded off. This ensures better accuracy for the final result. Using the rounding convention stated on page 409, we round off only the last result.

Solving Equation (1) for x, $\left(x = \dfrac{y}{\tan 27.2°} = y \cot 27.2°\right)$, and substituting into Equation (2), we have

$$\tan 23.9° = \frac{y}{y \cot 27.2° + 17.5}$$

$$y = (\tan 23.9°)(y \cot 27.2° + 17.5) \quad \text{• Solve for } y.$$

$$y - y \tan 23.9° \cot 27.2° = (\tan 23.9°)(17.5)$$

$$y = \frac{(\tan 23.9°)(17.5)}{1 - \tan 23.9° \cot 27.2°} \approx 56.2993$$

To three significant digits, the height of the space shuttle is 56.3 meters.

TRY EXERCISE 72, EXERCISE SET 5.2, PAGE 413

TOPICS FOR DISCUSSION

1. If θ is an acute angle of a right triangle for which $\cos \theta = 3/8$, then it must be the case that $\sin \theta = 5/8$. Do you agree? Explain.
2. A tutor claims that $\tan 30° = \cot 60°$. Do you agree?
3. Does $\sin 2\theta = 2 \sin \theta$? Explain.
4. How many significant digits are in each of the following measurements?
   a. 0.0042 inches b. 5.03 inches c. 62.00 inches
5. A student claims that $\sin^2 30° = (\sin 30°)^2$. Do you agree? Explain.

EXERCISE SET 5.2

In Exercises 1 to 14, find the values of the six trigonometric functions of θ for the right triangle with the given sides.

1.
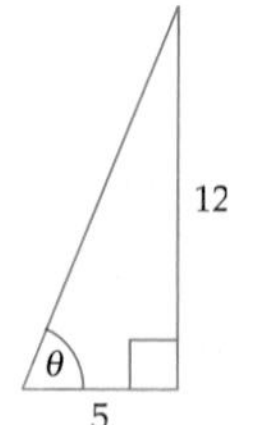

2.
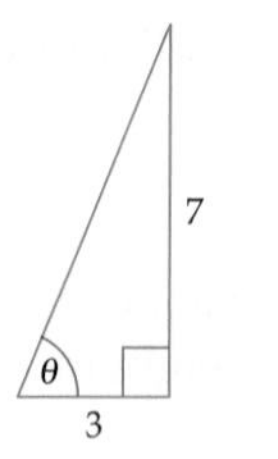

3.

4.
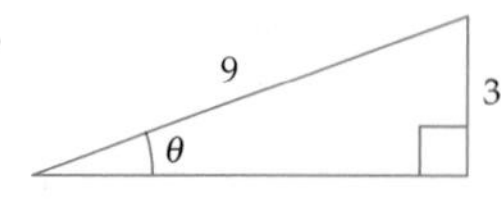

5.
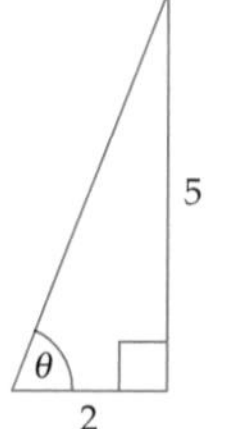

6.
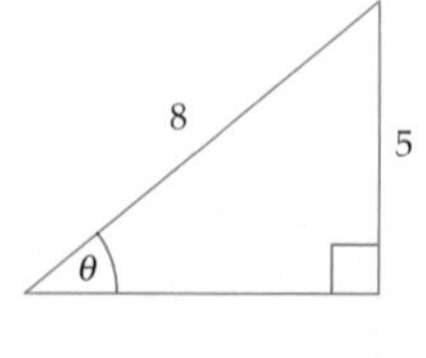

7.
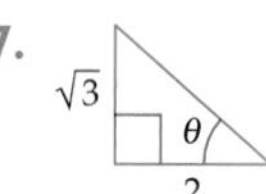
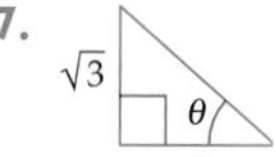

8.
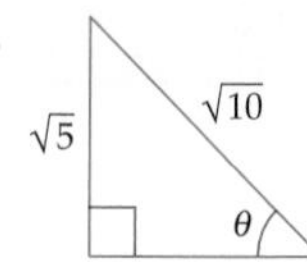

9. 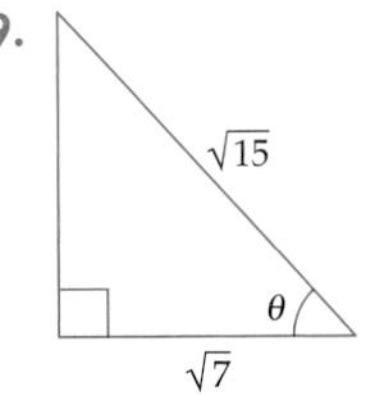

10. 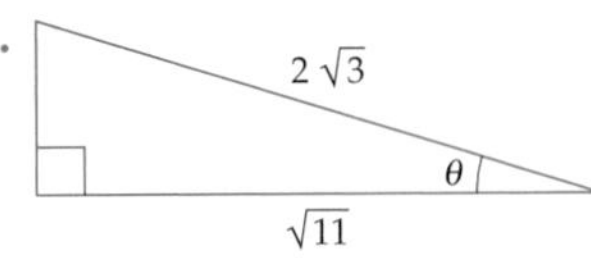

11. 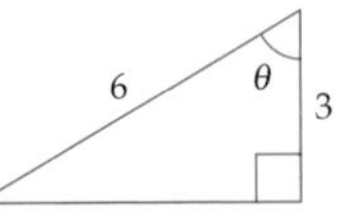

12. 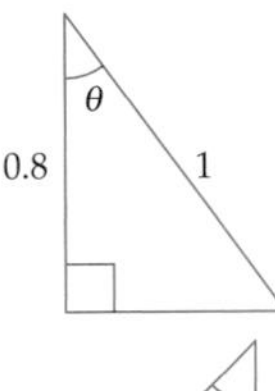

13. 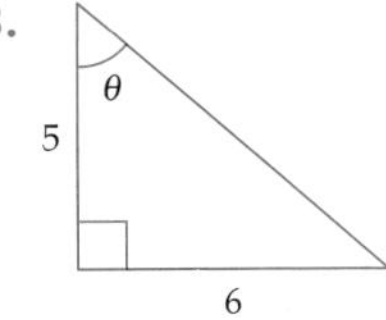

14. 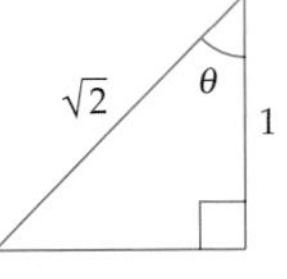

For Exercises 15 to 17, let θ be an acute angle of a right triangle and $\sin \theta = 3/5$. Find

15. $\tan \theta$ 16. $\sec \theta$ 17. $\cos \theta$

For Exercises 18 to 20, let θ be an acute angle of a right triangle and $\tan \theta = 4/3$. Find

18. $\sin \theta$ 19. $\cot \theta$ 20. $\sec \theta$

For Exercises 21 to 23, let β be an acute angle of a right triangle and $\sec \beta = 13/12$. Find

21. $\cos \beta$ 22. $\cot \beta$ 23. $\csc \beta$

For Exercises 24 to 26, let θ be an acute angle of a right triangle and $\cos \theta = 2/3$. Find

24. $\sin \theta$ 25. $\sec \theta$ 26. $\tan \theta$

For Exercises 27 to 42, find the *exact* value of each expression.

27. $\sin 45^\circ + \cos 45^\circ$
28. $\csc 45^\circ - \sec 45^\circ$
29. $\sin 30^\circ \cos 60^\circ - \tan 45^\circ$
30. $\csc 60^\circ \sec 30^\circ + \cot 45^\circ$
31. $\sin 30^\circ \cos 60^\circ + \tan 45^\circ$
32. $\sec 30^\circ \cos 30^\circ - \tan 60^\circ \cot 60^\circ$
33. $2 \sin 60^\circ - \sec 45^\circ \tan 60^\circ$
34. $\sec 45^\circ \cot 30^\circ + 3 \tan 60^\circ$
35. $\sin \frac{\pi}{3} + \cos \frac{\pi}{6}$
36. $\csc \frac{\pi}{6} - \sec \frac{\pi}{3}$
37. $\sin \frac{\pi}{4} + \tan \frac{\pi}{6}$
38. $\sin \frac{\pi}{3} \cos \frac{\pi}{4} - \tan \frac{\pi}{4}$
39. $\sec \frac{\pi}{3} \cos \frac{\pi}{3} - \tan \frac{\pi}{6}$
40. $\cos \frac{\pi}{4} \tan \frac{\pi}{6} + 2 \tan \frac{\pi}{3}$
41. $2 \csc \frac{\pi}{4} - \sec \frac{\pi}{3} \cos \frac{\pi}{6}$
42. $3 \tan \frac{\pi}{4} + \sec \frac{\pi}{6} \sin \frac{\pi}{3}$

In Exercises 43 to 58, use a calculator to find the value of the trigonometric function to four decimal places.

43. $\tan 32^\circ$ 44. $\sec 88^\circ$ 45. $\cos 63^\circ 20'$
46. $\cot 55^\circ 50'$ 47. $\cos 34.7^\circ$ 48. $\tan 81.3^\circ$
49. $\sec 5.9^\circ$ 50. $\sin \frac{\pi}{5}$ 51. $\tan \frac{\pi}{7}$
52. $\sec \frac{3\pi}{8}$ 53. $\csc 1.2$ 54. $\sin 0.45$
55. $\cos 1.25$ 56. $\tan \frac{3}{4}$ 57. $\sec \frac{5}{8}$
58. $\cot \frac{3}{5}$

59. **VERTICAL HEIGHT FROM SLANT HEIGHT** A 12-foot ladder is resting against a wall and makes an angle of 52° with the ground. Find the height to which the ladder will reach on the wall.

60. **DISTANCE ACROSS A MARSH** Find the distance AB across the marsh shown in the accompanying figure.

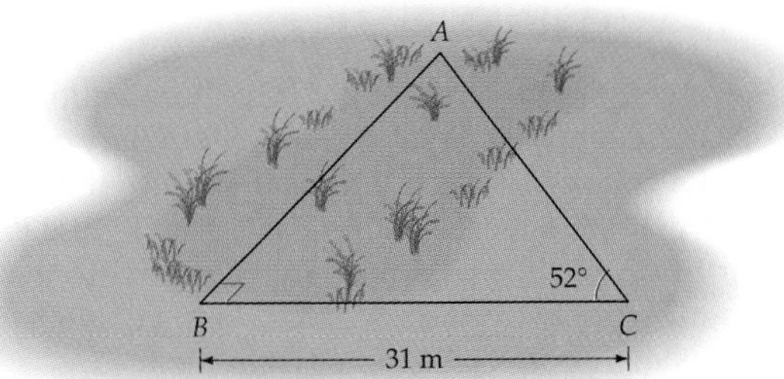

61. **SLOPE OF A LINE** Show that the slope of a line that makes an angle θ with the positive x-axis equals $\tan \theta$.

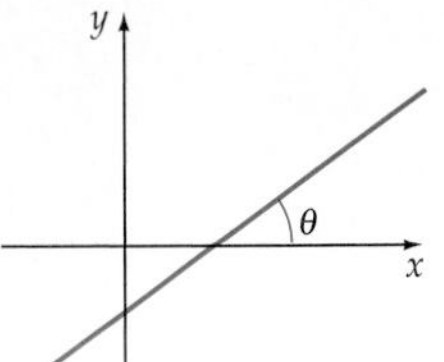

62. **WIDTH OF A SCREEN** Television screens are measured by the length of the diagonal of the screen. Find the width of a 19-inch television screen if the diagonal makes an angle of 38° with the base of the screen.

63. **TIME OF CLOSEST APPROACH** At 1:00 P.M., a boat is 40 kilometers due east of a lighthouse and traveling at a rate of 10 kilometers per hour in a direction that is 30° south of an east-west line. At what time will the boat be closest to the lighthouse?

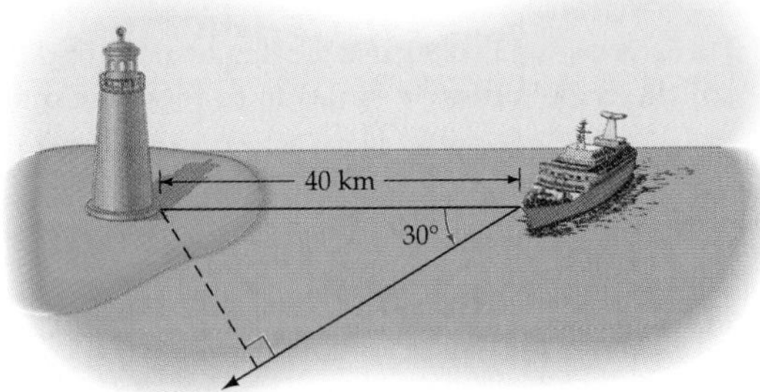

64. **TIME OF CLOSEST APPROACH** At 3:00 P.M., a boat is 12.5 miles due west of a radar station and traveling at 11 mph in a direction that is 57.3° south of an east-west line. At what time will the boat be closest to the radar station?

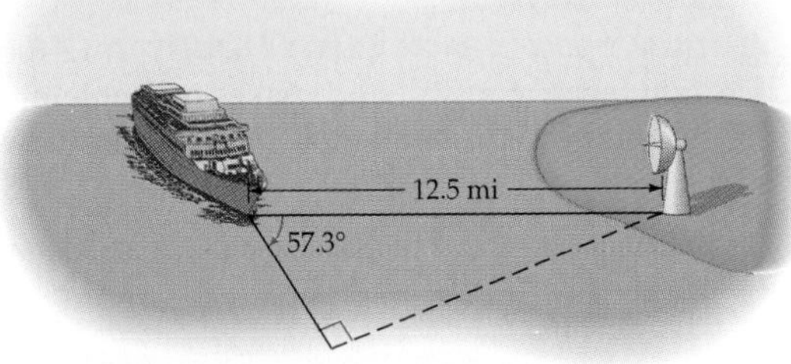

65. **PLACEMENT OF A LIGHT** For best illumination of a piece of art, a lighting specialist for an art gallery recommends that a ceiling-mounted light be 6 feet from a piece of art and that the angle of depression of the light be 38°. How far from a wall should the light be placed so that the recommendations of the specialist are met? Notice that the art extends outward 4 inches from the wall.

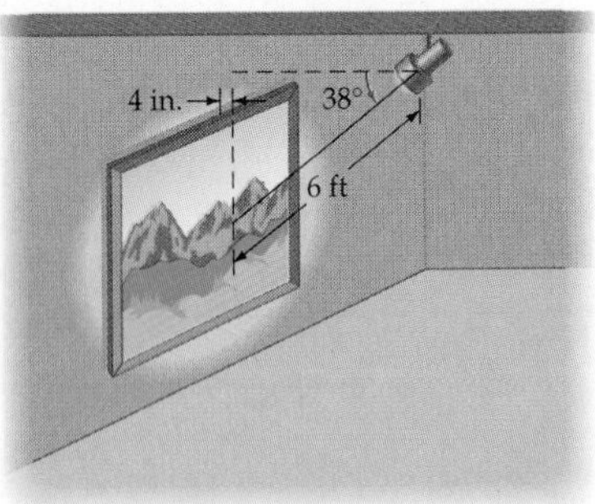

66. **HEIGHT OF THE EIFFEL TOWER** The angle of elevation from a point 116 meters from the base of the Eiffel Tower to the top of the tower is 68.9°. Find the approximate height of the tower.

67. **DISTANCE OF A DESCENT** An airplane traveling 240 mph is descending at an angle of depression of 6°. How many miles will the plane descend in 4 minutes?

68. **TIME OF A DESCENT** A submarine traveling 9.0 mph is descending at an angle of depression of 5°. How many minutes does it take the submarine to reach a depth of 80 feet?

69. **HEIGHT OF AN AQUEDUCT** From a point 300 feet from the base of a Roman aqueduct in southern France, the angle of elevation to the top of the aqueduct is 78°. Find the height of the aqueduct.

70. **WIDTH OF A LAKE** The angle of depression of one side of a lake, measured from a balloon 2500 feet above the lake as shown in the accompanying figure is 43°. The angle of depression to the opposite side of the lake is 27°. Find the width of the lake.

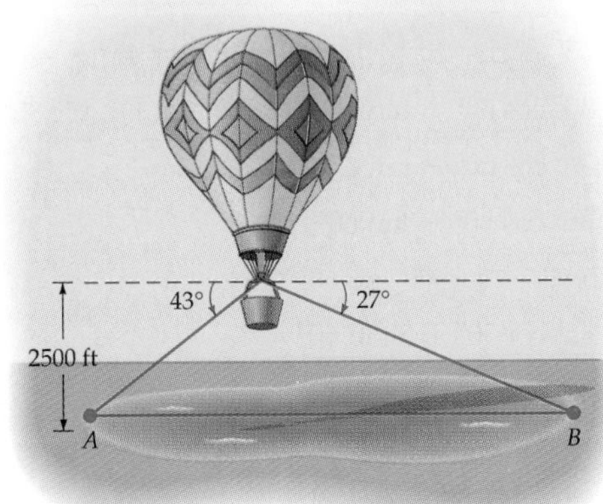

71. **HEIGHT OF A PYRAMID** The angle of elevation to the top of the Egyptian pyramid Cheops is 36.4°, measured from a point 350 feet from the base of the pyramid. The angle of elevation of a face of the pyramid is 51.9°. Find the height of Cheops.

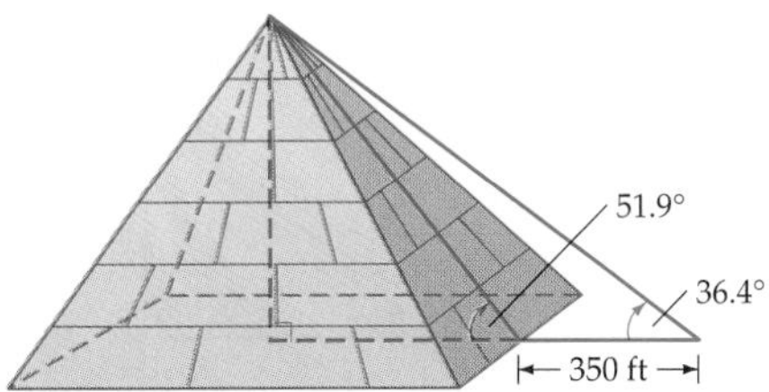

72. **HEIGHT OF A BUILDING** Two buildings are 240 feet apart. The angle of elevation from the top of the shorter building to the top of the other building is 22°. If the shorter building is 80 feet high, how high is the taller building?

73. **HEIGHT OF THE WASHINGTON MONUMENT** From a point A on a line from the base of the Washington Monument, the angle of elevation to the top of the monument is 42.0°. From a point 100 feet away and on the same line, the angle to the top is 37.8°. Find the approximate height of the Washington Monument.

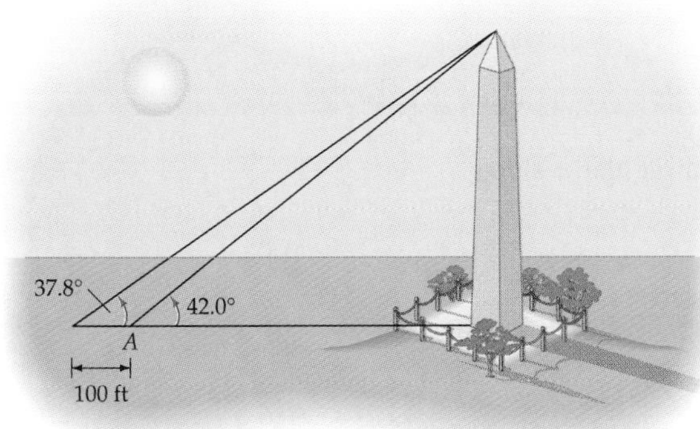

74. **HEIGHT OF A BUILDING** The angle of elevation to the top of a radio antenna on the top of a building is 53.4°. After moving 200 feet closer to the building, the angle of elevation is 64.3°. Find the height of the building if the height of the antenna is 180 feet.

75. **THE PETRONAS TOWERS** The Petronas Towers in Kuala Lumpur, Malaysia, are the world's tallest buildings. Each tower is 1483 feet in height. The towers are connected by a skybridge at the forty-first floor. Note the information given in the accompanying figure.

   a. Determine the height of the skybridge.

   b. Determine the length of the skybridge.

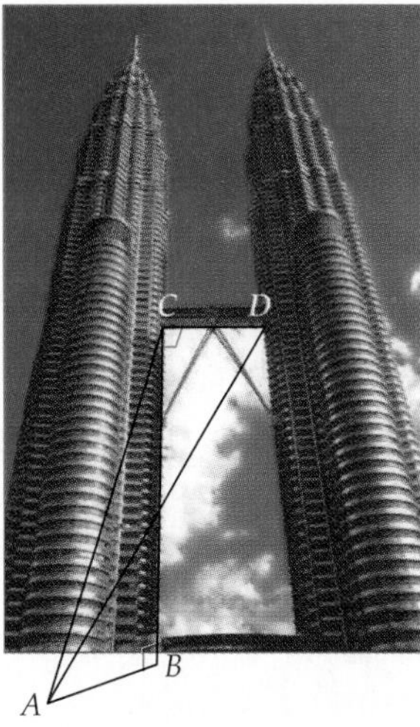

$AB = 412$ feet
$\angle CAB = 53.6^\circ$
$\overline{AB}$ is at ground level
$\angle CAD = 15.5^\circ$

76. **AN EIFFEL TOWER REPLICA** Use the information in the accompanying figure to estimate the height of the Eiffel Tower replica that stands in front of the Paris Las Vegas Hotel in Las Vegas, Nevada.

SUPPLEMENTAL EXERCISES

77. A circle is inscribed in a regular hexagon with each side 6.0 meters long. Find the radius of the circle.

78. Show that the area A of the triangle given in the figure is
$$A = \frac{1}{2}ab \sin \theta.$$

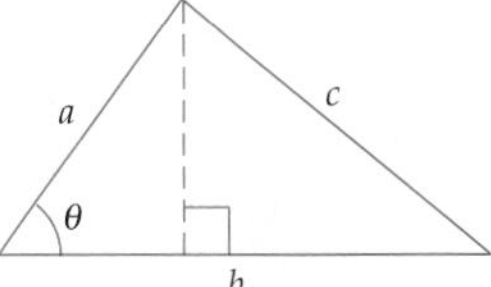

79. **DETERMINE A RANGE OF HEIGHTS** If an angle of 27° has been measured to the nearest degree, then the actual

measure of the angle θ is such that $26.5° \leq \theta < 27.5$. From a distance of exactly 100 meters from the base of a tree, the angle of elevation is measured as 27° to the nearest degree. Find the range in which the height of the tree must fall.

80. **HEIGHT OF A TOWER** Let B denote the base of a clock tower. The angle of elevation from a point A, on the ground, to the top of the clock tower is 56.3°. On a line on the ground that is perpendicular to AB and 25 feet from A, the angle of elevation is 53.3°. Find the height of the clock tower.

81. **FIND A MAXIMUM LENGTH** Find the length of the longest piece of wood that can be slid around the corner of the hallway in the figure.

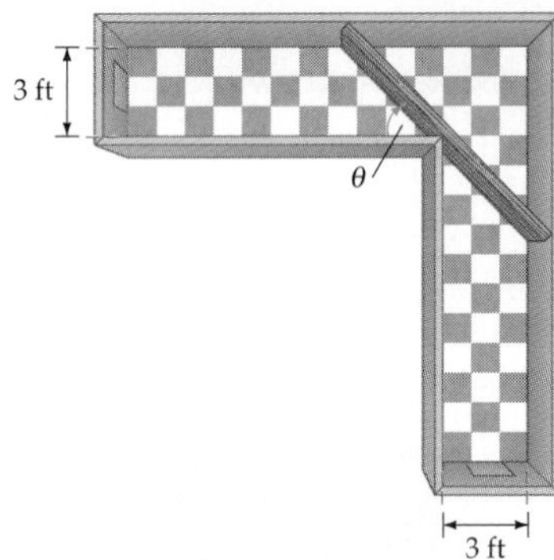

82. **FIND A MAXIMUM LENGTH** In Exercise 81, suppose that the hall is 8 feet high. Find the length of the longest piece of wood that can be taken around the corner. Round to the nearest 0.1 foot.

83. Using the triangle given in the figure, show that

a. $\sin^2 \theta + \cos^2 \theta = 1$ b. $\tan \theta = \dfrac{\sin \theta}{\cos \theta}$

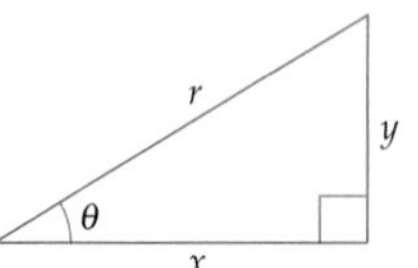

PROJECTS

1. a. **PERIMETER OF A REGULAR n-GON** Show that the perimeter P of a regular n-sided polygon (n-gon) inscribed in a circle of radius 1 is $P = 2n \sin \dfrac{180°}{n}$.

b. Let P_n denote the perimeter of a regular n-gon inscribed in a circle of radius 1. Use the result from **a.** to complete the following table.

n	10	50	100	1000	10,000
P_n					

Write a few sentences explaining why P_n approaches 2π as n increases without bound.

2. a. **AREA OF A REGULAR n-GON** Show that the area A of a regular n-gon inscribed in a circle of radius 1 is $A = \dfrac{n}{2} \sin \dfrac{360°}{n}$.

b. Let A_n denote the area of a regular n-gon inscribed in a circle of radius 1. Use the result from **a.** to complete the following table.

n	10	50	100	1000	10,000
A_n					

Write a few sentences explaining why A_n approaches π as n increases without bound.

SECTION 5.3 TRIGONOMETRIC FUNCTIONS OF ANY ANGLE

- TRIGONOMETRIC FUNCTIONS OF QUADRANTAL ANGLES
- SIGNS OF TRIGONOMETRIC FUNCTIONS
- THE REFERENCE ANGLE

5B

The application of trigonometry would be quite limited if all angles had to be acute angles. Fortunately, this is not the case. In this section we extend the definition of a trigonometric function to include any angle.

Consider angle θ in **Figure 5.41** in standard position and a point $P(x, y)$ on the terminal side of the angle. We define the trigonometric functions of any angle according to the following definitions.

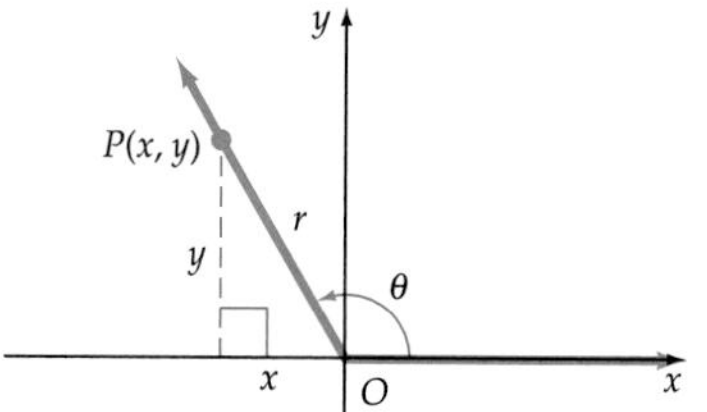

Figure 5.41

The Trigonometric Functions of Any Angle

Let $P(x, y)$ be any point, except the origin, on the terminal side of an angle θ in standard position. Let $r = d(O, P)$, the distance from the origin to P. The six trigonometric functions of θ are

$$\sin\theta = \frac{y}{r} \qquad \cos\theta = \frac{x}{r} \qquad \tan\theta = \frac{y}{x},\quad x \neq 0$$

$$\csc\theta = \frac{r}{y},\quad y \neq 0 \qquad \sec\theta = \frac{r}{x},\quad x \neq 0 \qquad \cot\theta = \frac{x}{y},\quad y \neq 0$$

where $r = \sqrt{x^2 + y^2}$.

take note

The measure of angle θ can be positive or negative. Note from the following figure that $\sin 120^\circ = \sin(-240^\circ)$ because $P(x, y)$ is on the terminal side of each angle. In a similar way, the value of any trigonometric function of 120° is equal to the value of that function of -240°.

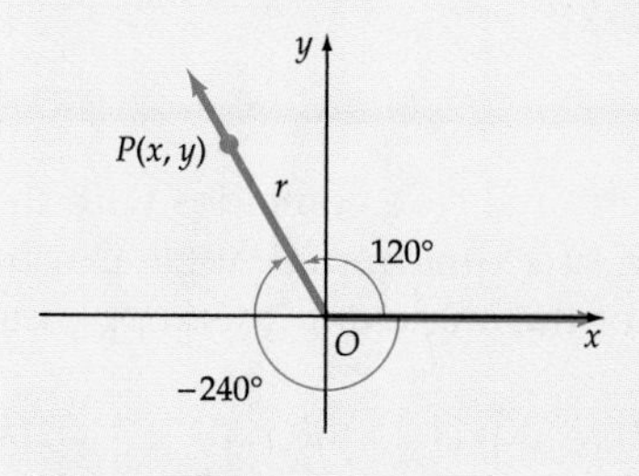

The value of a trigonometric function is independent of the point chosen on the terminal side of the angle. Consider any two points on the terminal side of an angle θ in standard position, as shown in **Figure 5.42**. The right triangles formed are similar triangles, so the ratios of the corresponding sides are equal. Thus, for example, $\frac{b}{a} = \frac{b'}{a'}$. Because $\tan\theta = \frac{b}{a} = \frac{b'}{a'}$, we have $\tan\theta = \frac{b'}{a'}$. Therefore, the value of the tangent function is independent of the point chosen on the terminal side of the angle. By a similar argument, we can show that the value of any trigonometric function is independent of the point chosen on the terminal side of the angle.

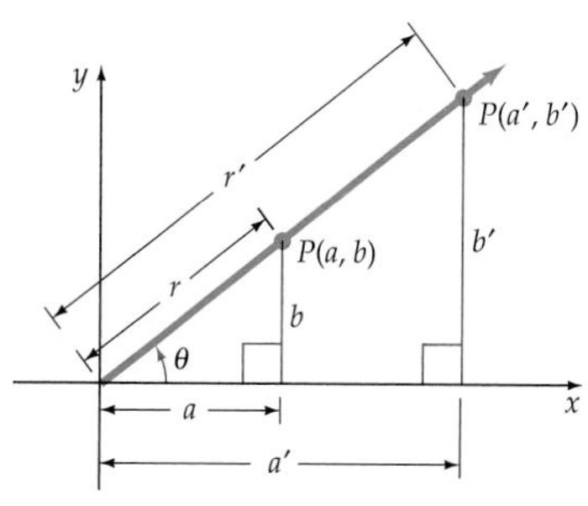

Figure 5.42

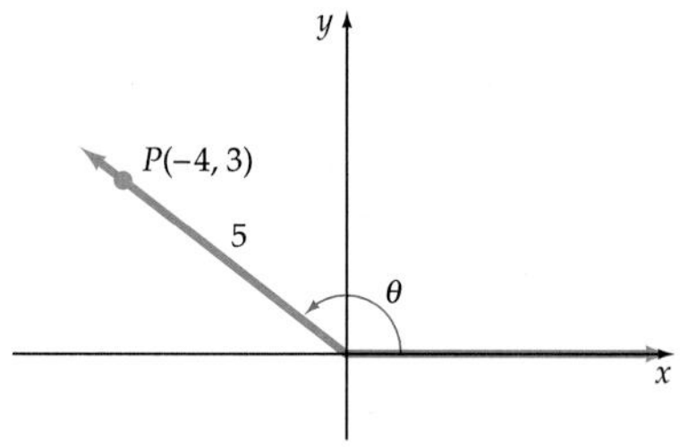

Any point in a rectangular coordinate system (except the origin) can determine an angle in standard position. For example, $P(-4, 3)$ in **Figure 5.43** is a point in the second quadrant and determines an angle θ in standard position with $r = \sqrt{(-4)^2 + 3^2} = 5$. The values of the trigonometric functions of θ are

$$\sin\theta = \frac{3}{5} \qquad \cos\theta = \frac{-4}{5} = -\frac{4}{5} \qquad \tan\theta = \frac{3}{-4} = -\frac{3}{4}$$

$$\csc\theta = \frac{5}{3} \qquad \sec\theta = \frac{5}{-4} = -\frac{5}{4} \qquad \cot\theta = \frac{-4}{3} = -\frac{4}{3}$$

Figure 5.43

EXAMPLE 1 Evaluate Trigonometric Functions

Find the value of each of the six trigonometric functions of an angle θ whose terminal side contains the point $P(-3, -2)$.

Solution

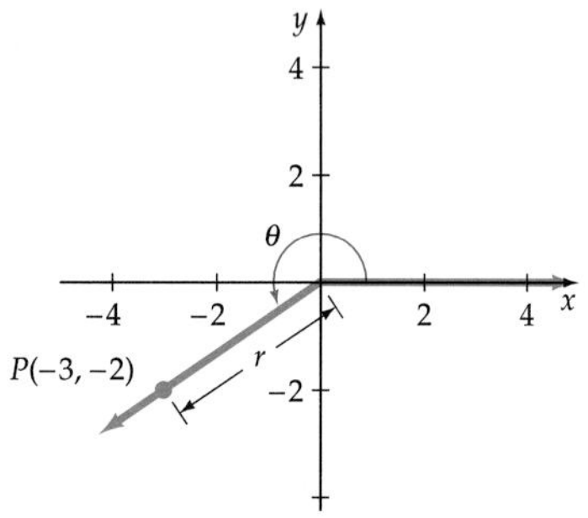

The angle is sketched in **Figure 5.44**. Find r by using the equation $r = \sqrt{x^2 + y^2}$, where $x = -3$ and $y = -2$.

$$r = \sqrt{(-3)^2 + (-2)^2} = \sqrt{9 + 4} = \sqrt{13}$$

Now use the definitions for trigonometric functions.

$$\sin\theta = \frac{-2}{\sqrt{13}} = -\frac{2\sqrt{13}}{13} \qquad \cos\theta = \frac{-3}{\sqrt{13}} = -\frac{3\sqrt{13}}{13} \qquad \tan\theta = \frac{-2}{-3} = \frac{2}{3}$$

$$\csc\theta = \frac{\sqrt{13}}{-2} = -\frac{\sqrt{13}}{2} \qquad \sec\theta = \frac{\sqrt{13}}{-3} = -\frac{\sqrt{13}}{3} \qquad \cot\theta = \frac{-3}{-2} = \frac{3}{2}$$

Figure 5.44

TRY EXERCISE 6, EXERCISE SET 5.3, PAGE 421

◆ TRIGONOMETRIC FUNCTIONS OF QUADRANTAL ANGLES

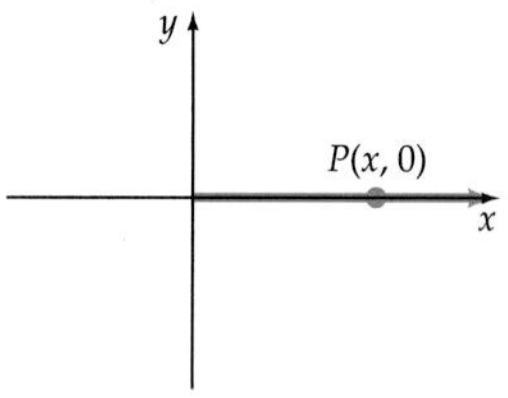

Recall that a quadrantal angle is an angle whose terminal side coincides with the x- or y-axis. The value of a trigonometric function of a quadrantal angle can be found by choosing any point on the terminal side of the angle and then applying the definition of that trigonometric function.

The terminal side of 0° coincides with the positive x-axis. Let $P(x, 0)$, $x > 0$, be any point on the x-axis as shown in **Figure 5.45**. Then $y = 0$ and $r = x$. The values of the six trigonometric functions of 0° are

Figure 5.45

$$\sin 0^\circ = \frac{0}{r} = 0 \qquad \cos 0^\circ = \frac{x}{r} = \frac{x}{x} = 1 \qquad \tan 0^\circ = \frac{0}{x} = 0$$

$$\csc 0^\circ \text{ is undefined} \qquad \sec 0^\circ = \frac{r}{x} = \frac{x}{x} = 1 \qquad \cot 0^\circ \text{ is undefined}$$

QUESTION Why are csc 0° and cot 0° undefined?

In like manner, the values of the trigonometric functions of the other quadrantal angles can be found. The results are shown in Table 5.4.

Table 5.4 Values of Trigonometric Functions for Quadrantal Angles

θ	$\sin\theta$	$\cos\theta$	$\tan\theta$	$\csc\theta$	$\sec\theta$	$\cot\theta$
0°	0	1	0	undefined	1	undefined
90°	1	0	undefined	1	undefined	0
180°	0	−1	0	undefined	−1	undefined
270°	−1	0	undefined	−1	undefined	0

◆ SIGNS OF TRIGONOMETRIC FUNCTIONS

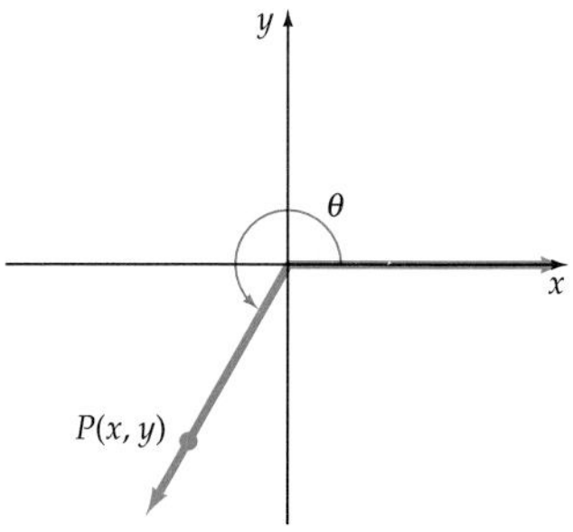

The sign of a trigonometric function depends on the quadrant in which the terminal side of the angle lies. For example, if θ is an angle whose terminal side lies in Quadrant III and $P(x, y)$ is on the terminal side of θ, then both x and y are negative and therefore y/x and x/y are positive. See **Figure 5.46**. Because $\tan\theta = y/x$ and $\cot\theta = x/y$, the values of the tangent and cotangent functions are positive for any Quadrant III angle. The values of the other four trigonometric functions of any Quadrant III angle are all negative.

Table 5.5 lists the signs of the six trigonometric functions in each quadrant. **Figure 5.47** is a graphical display of the contents of Table 5.5.

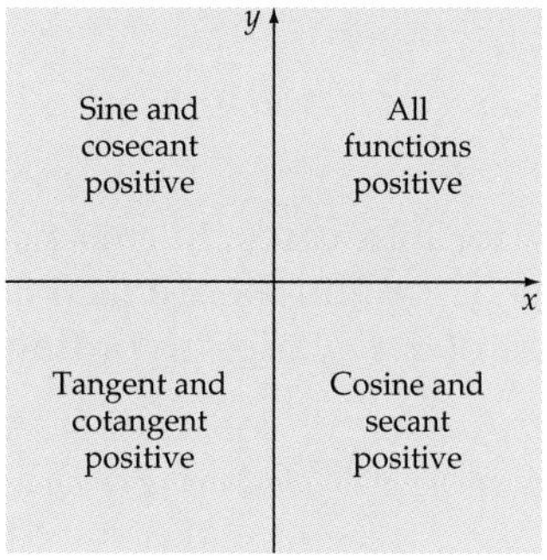

y
θ
x
P(x, y)

Figure 5.46

y
Sine and cosecant positive
All functions positive
x
Tangent and cotangent positive
Cosine and secant positive

Figure 5.47

Table 5.5 Signs of the Trigonometric Functions

Sign of	Terminal Side of θ in Quadrant I	II	III	IV
$\sin\theta$ and $\csc\theta$	positive	positive	negative	negative
$\cos\theta$ and $\sec\theta$	positive	negative	negative	positive
$\tan\theta$ and $\cot\theta$	positive	negative	positive	negative

In the next example we are asked to evaluate two trigonometric functions of the angle θ. A key step is to use our knowledge about trigonometric functions and their signs to determine that θ is a Quadrant IV angle.

EXAMPLE 2 Evaluate Trigonometric Functions

Given $\tan\theta = -\dfrac{7}{5}$ and $\sin\theta < 0$, find $\cos\theta$ and $\csc\theta$.

Continued •➤

ANSWER $P(x, 0)$ is a point on the terminal side of 0°. Thus $\csc 0° = r/0$, which is undefined. Similarly, $\cot 0° = x/0$, which is undefined.

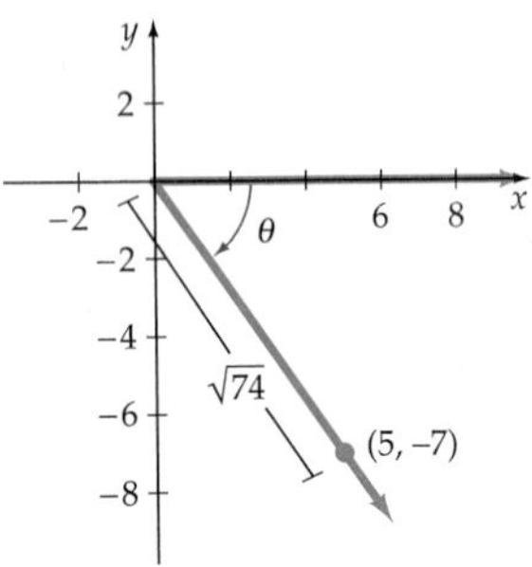

Figure 5.48

take note

In this section r is a distance, and hence it is nonnegative.

Solution

The terminal side of angle θ must lie in Quadrant IV; that is the only quadrant for which $\sin\theta$ and $\tan\theta$ are both negative. Because

$$\tan\theta = -\frac{7}{5} = \frac{y}{x} \tag{1}$$

and the terminal side of θ is in Quadrant IV, we know that y must be negative and x must be positive. Thus Equation (1) is true for $y = -7$ and $x = 5$. Now $r = \sqrt{5^2 + (-7)^2} = \sqrt{74}$. Hence

$$\cos\theta = \frac{x}{r} = \frac{5}{\sqrt{74}} = \frac{5\sqrt{74}}{74} \quad \text{and} \quad \csc\theta = \frac{r}{y} = \frac{\sqrt{74}}{-7} = -\frac{\sqrt{74}}{7}$$

See **Figure 5.48.**

TRY EXERCISE 18, EXERCISE SET 5.3, PAGE 422

◆ THE REFERENCE ANGLE

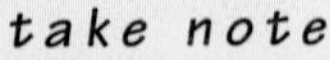

take note

The reference angle is a very important concept that will be used time and time again in the remaining trigonometry sections.

We will often find it convenient to evaluate trigonometric functions by making use of the concept of a *reference angle.*

Reference Angle

Given $\angle\theta$ in standard position, its **reference angle** θ' is the smallest positive angle formed by the terminal side of $\angle\theta$ and the x-axis.

Figure 5.49 shows $\angle\theta$ and its reference angle θ', for four cases. In every case the reference angle θ' is formed by the terminal side of $\angle\theta$ and the x-axis (never the y-axis). The process of determining the measure of $\angle\theta'$ varies according to what quadrant contains the terminal side of $\angle\theta$.

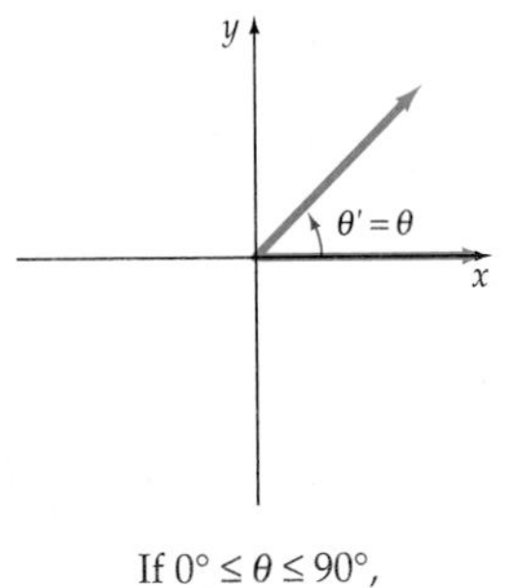

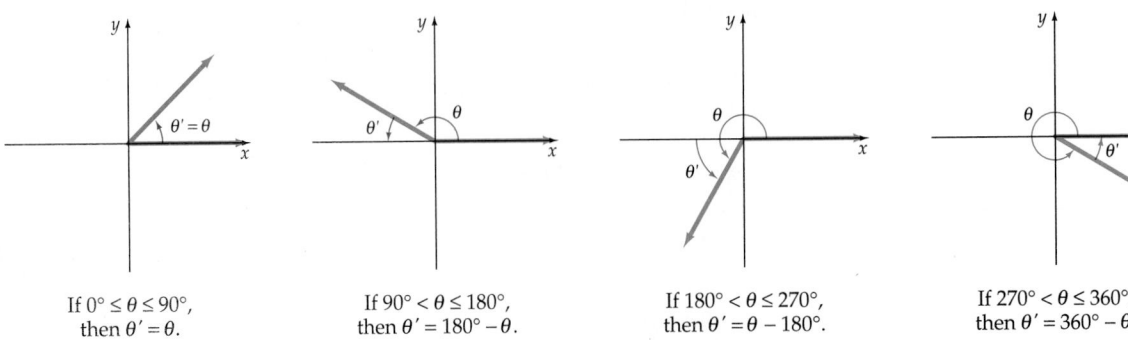

If $0° \le \theta \le 90°$, then $\theta' = \theta$.

If $90° < \theta \le 180°$, then $\theta' = 180° - \theta$.

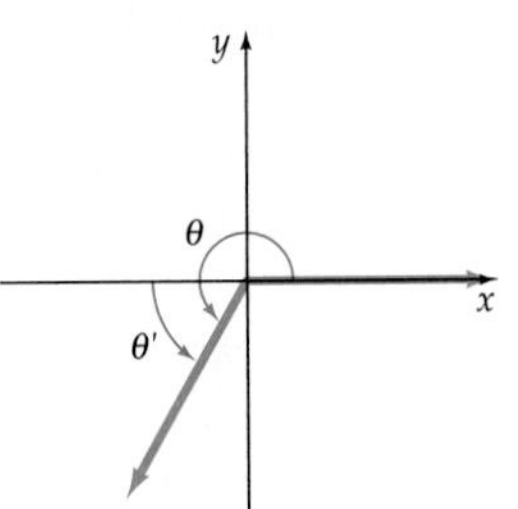

If $180° < \theta \le 270°$, then $\theta' = \theta - 180°$.

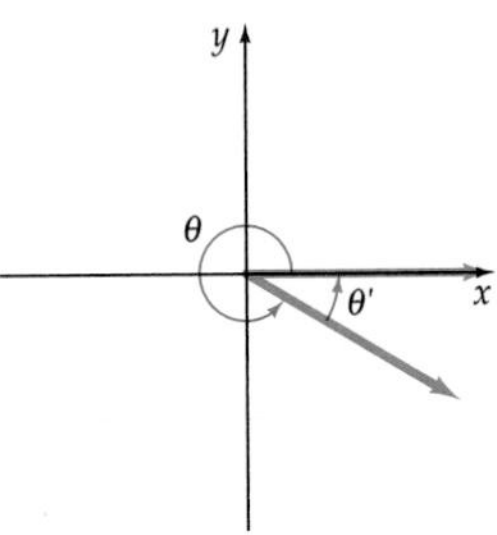

If $270° < \theta \le 360°$, then $\theta' = 360° - \theta$.

Figure 5.49

take note

A TI graphing calculator program is available to compute the measure of the reference angle for a given angle. This program, REFANG, can be found on our web site at

http://college.hmco.com

EXAMPLE 3 Find the Measure of the Reference Angle

For each of the following, sketch the given angle θ (in standard position) and its reference angle θ'. Then determine the measure of θ'.

a. $\theta = 120°$ b. $\theta = 345°$ c. $\theta = 924°$

d. $\theta = \dfrac{9}{5}\pi$ e. $\theta = -4$ f. $\theta = 17$

Solution

a.

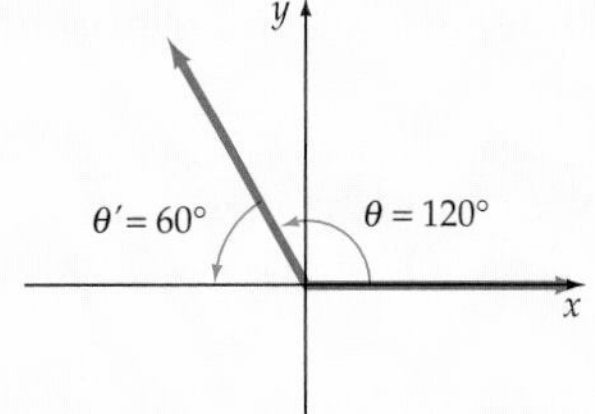

$\theta' = 180° - 120° = 60°$

b.

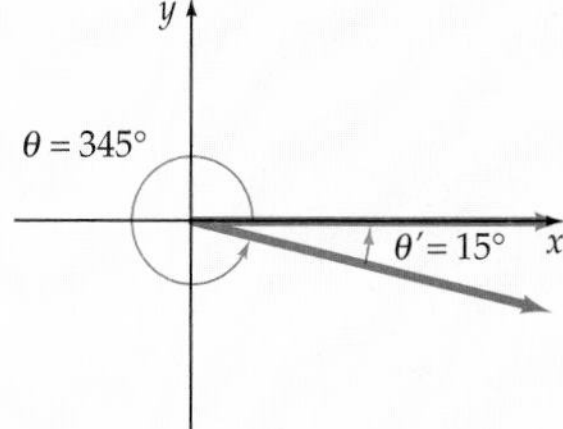

$\theta' = 360° - 345° = 15°$

c.

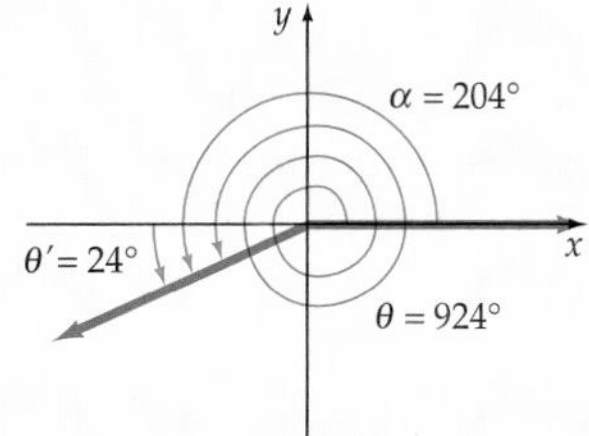

Because $\theta = 924° > 360°$, we first determine the coterminal angle, $\alpha = 204°$.

$\theta' = 204° - 180° = 24°$

d.

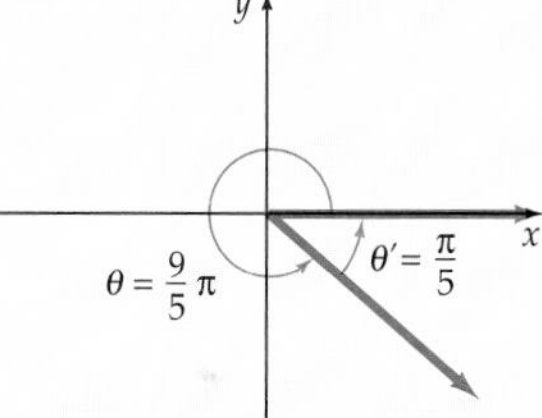

$$\theta' = 2\pi - \frac{9}{5}\pi$$

$$= \frac{10\pi}{5} - \frac{9\pi}{5} = \frac{\pi}{5} \approx 0.628$$

e.

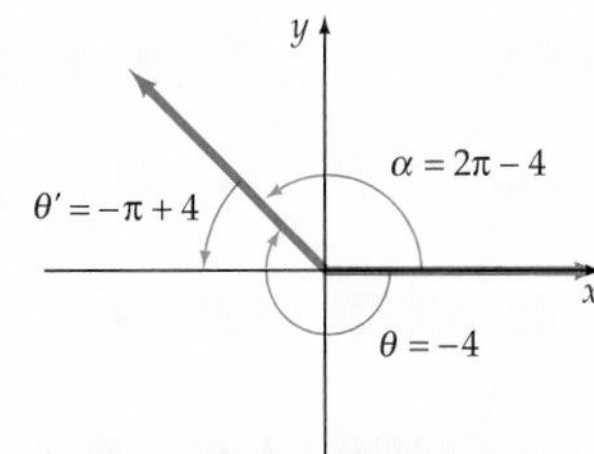

-4 radians is coterminal with $\alpha = 2\pi - 4$.

$$\theta' = \pi - (2\pi - 4)$$
$$= -\pi + 4 \approx 0.8584$$

f.

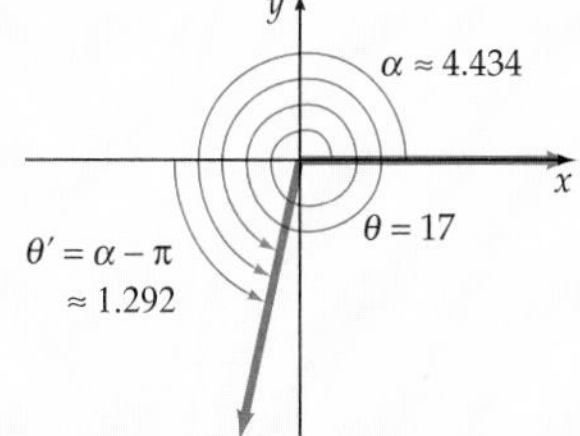

Because $17 \div (2\pi) \approx 2.70563$, a coterminal angle is $\alpha = 0.70563(2\pi) \approx 4.434$.

Because α is a Quadrant III angle, $\theta' = \alpha - \pi \approx 1.292$

TRY EXERCISE 26, EXERCISE SET 5.3, PAGE 422

take note

The Reference Angle Theorem is also valid if the sine function is replaced by any other trigonometric function.

The following theorem states an important relationship that exists between $\sin\theta$ and $\sin\theta'$, where θ' is the reference angle for angle θ.

Reference Angle Theorem

To evaluate $\sin\theta$, determine $\sin\theta'$. Then use either $\sin\theta'$ or its opposite as the answer, depending on which has the correct sign.

In the following example, we illustrate how to evaluate a trigonometric function of θ by first evaluating the trigonometric function of θ'.

EXAMPLE 4 Use the Reference Angle Theorem to Evaluate Trigonometric Functions

Evaluate each function.

a. $\sin 210^\circ$ **b.** $\cos 405^\circ$ **c.** $\tan 300^\circ$

Solution

a. We know that $\sin 210^\circ$ is negative (the sign chart is given in Table 5.5). The reference angle for $\theta = 210^\circ$ is $\theta' = 30^\circ$. By the Reference Angle Theorem, we know that $\sin 210^\circ$ equals either

$$\sin 30^\circ = \frac{1}{2} \qquad \text{or} \qquad -\sin 30^\circ = -\frac{1}{2}$$

Thus $\sin 210^\circ = -\frac{1}{2}$.

b. Because $\theta = 405^\circ$ is a Quadrant I angle, we know $\cos 405^\circ > 0$. The reference angle for $\theta = 405^\circ$ is $\theta' = 45^\circ$. By the Reference Angle Theorem, $\cos 405^\circ$ equals either

$$\cos 45^\circ = \frac{\sqrt{2}}{2} \qquad \text{or} \qquad -\cos 45^\circ = -\frac{\sqrt{2}}{2}$$

Thus $\cos 405^\circ = \frac{\sqrt{2}}{2}$.

c. Because $\theta = 300^\circ$ is a Quadrant IV angle, $\tan 300^\circ < 0$. The reference angle for $\theta = 300^\circ$ is $\theta' = 60^\circ$. Hence, $\tan 300^\circ$ equals either

$$\tan 60^\circ = \sqrt{3} \qquad \text{or} \qquad -\tan 60^\circ = -\sqrt{3}$$

Thus $\tan 300^\circ = -\sqrt{3}$.

TRY EXERCISE 38, EXERCISE SET 5.3, PAGE 422

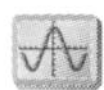

In many applications, a calculator is used to evaluate a trigonometric function of the angle. Use *degree mode* to evaluate a trigonometric function of an angle given in degrees. Use *radian mode* to evaluate a trigonometric function of an angle given in radians. **Figure 5.50** shows the sine of 137.4 degrees, and **Figure 5.51** shows the cotangent of 3.4 radians.

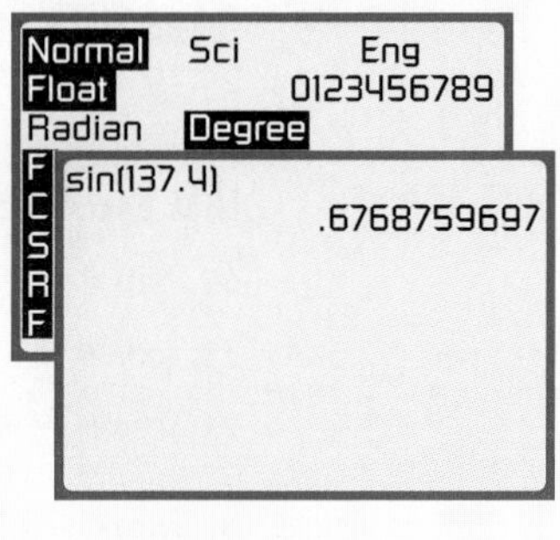

Figure 5.50

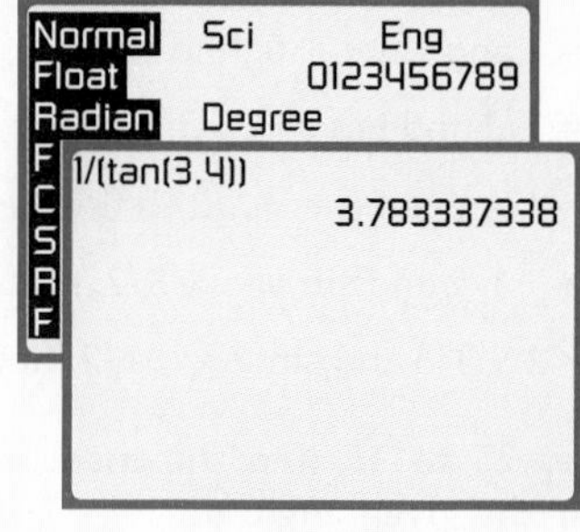

Figure 5.51

TOPICS FOR DISCUSSION

1. Is every reference angle an acute angle? Explain.
2. If θ' is the reference angle for the angle θ, then $\sin\theta = \sin\theta'$. Do you agree? Explain.
3. If $\sin\theta < 0$, and $\cos\theta > 0$, then the terminal side of the angle θ lies in which quadrant?
4. A student claims that if $\theta = 160$, then $\theta' = 20$. Explain why the student is not correct.
5. Explain how to find the measure of the reference angle θ' for the angle $\theta = \dfrac{19}{5}\pi$.

EXERCISE SET 5.3

In Exercises 1 to 8, find the value of each of the six trigonometric functions for the angle whose terminal side passes through the given point.

1. $P(2, 3)$
2. $P(3, 7)$
3. $P(-2, 3)$
4. $P(-3, 5)$
5. $P(-8, -5)$
6. $P(-6, -9)$
7. $P(-5, 0)$
8. $P(0, 2)$

In Exercises 9 to 14, let θ be an angle in standard position. State the quadrant in which the terminal side of θ lies.

9. $\sin\theta > 0,\quad \cos\theta > 0$
10. $\tan\theta < 0,\quad \sin\theta < 0$
11. $\cos\theta > 0,\quad \tan\theta < 0$
12. $\sin\theta < 0,\quad \cos\theta > 0$
13. $\sin\theta < 0,\quad \cos\theta < 0$
14. $\tan\theta < 0,\quad \cos\theta < 0$

In Exercises 15 to 24, find the value of each expression.

15. $\sin\theta = -1/2$, $180° < \theta < 270°$; find $\tan\theta$.

16. $\cot\theta = -1$, $90° < \theta < 180°$; find $\cos\theta$.

17. $\csc\theta = \sqrt{2}$, $\pi/2 < \theta < \pi$; find $\cot\theta$.

18. $\sec\theta = 2\sqrt{3}/3$, $3\pi/2 < \theta < 2\pi$; find $\sin\theta$.

19. $\sin\theta = -1/2$ and $\cos\theta > 0$; find $\tan\theta$.

20. $\tan\theta = 1$ and $\sin\theta < 0$; find $\cos\theta$.

21. $\cos\theta = 1/2$ and $\tan\theta = \sqrt{3}$; find $\csc\theta$.

22. $\tan\theta = 1$ and $\sin\theta = -\sqrt{2}/2$; find $\sec\theta$.

23. $\cos\theta = -1/2$ and $\sin\theta = \sqrt{3}/2$; find $\cot\theta$.

24. $\sec\theta = 2\sqrt{3}/3$ and $\sin\theta = -1/2$; find $\cot\theta$.

In Exercises 25 to 36, find the measure of the reference angle θ' for the given angle θ.

25. $\theta = 160°$ 26. $\theta = 255°$ 27. $\theta = 351°$

28. $\theta = 48°$ 29. $\theta = \frac{11}{5}\pi$ 30. $\theta = -6$

31. $\theta = \frac{8}{3}$ 32. $\theta = \frac{18}{7}\pi$ 33. $\theta = 1406°$

34. $\theta = 840°$ 35. $\theta = -475°$ 36. $\theta = -650°$

In Exercises 37 to 48, use the Reference Angle Theorem to find the exact value of each trigonometric function.

37. $\sin 225°$ 38. $\cos 300°$ 39. $\tan 405°$

40. $\sec 150°$ 41. $\csc\frac{4}{3}\pi$ 42. $\cot\frac{7}{6}\pi$

43. $\cos\frac{17\pi}{4}$ 44. $\tan\left(-\frac{\pi}{3}\right)$ 45. $\sec 765°$

46. $\csc(-510°)$ 47. $\cot 540°$ 48. $\cos 570°$

In Exercises 49 to 60, use a calculator to approximate the given trigonometric functions to six significant digits.

49. $\sin 127°$ 50. $\sin(-257°)$ 51. $\cos(-116°)$

52. $\cot 398°$ 53. $\sec 578°$ 54. $\sec 740°$

55. $\sin(-\pi/5)$ 56. $\cos(3\pi/7)$ 57. $\csc(9\pi/5)$

58. $\tan(-4.12)$ 59. $\sec(-4.45)$ 60. $\csc 0.34$

In Exercises 61 to 68, find (without using a calculator) the exact value of each expression.

61. $\sin 210° - \cos 330° \tan 330°$

62. $\tan 225° + \sin 240° \cos 60°$

63. $\sin^2 30° + \cos^2 30°$

64. $\cos\pi \sin(7\pi/4) - \tan(11\pi/6)$

65. $\sin(3\pi/2)\tan(\pi/4) - \cos(\pi/3)$

66. $\cos(7\pi/4)\tan(4\pi/3) + \cos(7\pi/6)$

67. $\sin^2(5\pi/4) + \cos^2(5\pi/4)$

68. $\tan^2(7\pi/4) - \sec^2(7\pi/4)$

SUPPLEMENTAL EXERCISES

In Exercises 69 to 74, find two values of θ, $0° \le \theta < 360°$, that satisfy the given trigonometric equation.

69. $\sin\theta = 1/2$ 70. $\tan\theta = -\sqrt{3}$

71. $\cos\theta = -\sqrt{3}/2$ 72. $\tan\theta = 1$

73. $\csc\theta = -\sqrt{2}$ 74. $\cot\theta = -1$

In Exercises 75 to 80, find two values of θ, $0 \le \theta < 2\pi$, that satisfy the given trigonometric equation.

75. $\tan\theta = -1$ 76. $\cos\theta = 1/2$

77. $\tan\theta = -\sqrt{3}/3$ 78. $\sec\theta = -2\sqrt{3}/3$

79. $\sin\theta = \sqrt{3}/2$ 80. $\cos\theta = -1/2$

If $P(x, y)$ is a point on the terminal side of an acute angle θ in standard position and $r = \sqrt{x^2 + y^2}$, then $\sin\theta = \frac{y}{r}$ and $\cos\theta = \frac{x}{r}$. Using these definitions, we find that

$$\cos^2\theta + \sin^2\theta = \left(\frac{x}{r}\right)^2 + \left(\frac{y}{r}\right)^2 = \frac{x^2}{r^2} + \frac{y^2}{r^2} = \frac{x^2 + y^2}{r^2}$$
$$= \frac{r^2}{r^2} = 1$$

Hence $\cos^2\theta + \sin^2\theta = 1$ for all acute angles θ. This important identity is actually true for all angles θ. We will show this later. In the meantime, use the definitions of the trigonometric functions to prove the identities in Exercises 81 to 90 for the acute angle θ.

81. $1 + \tan^2\theta = \sec^2\theta$ 82. $\cot^2\theta + 1 = \csc^2\theta$

83. $\tan\theta = \frac{\sin\theta}{\cos\theta}$ 84. $\cot\theta = \frac{\cos\theta}{\sin\theta}$

85. $\cos(90° - \theta) = \sin\theta$ 86. $\sin(90° - \theta) = \cos\theta$

87. $\tan(90° - \theta) = \cot\theta$ 88. $\cot(90° - \theta) = \tan\theta$

89. $\sin(\theta + \pi) = -\sin\theta$ 90. $\cos(\theta + \pi) = -\cos\theta$

For each of the angles given in Exercises 91 to 96, find the coordinates of a point on the terminal side of angle θ in standard position to the nearest ten thousandth.

91. $\theta = 78°$ 92. $\theta = 165°$ 93. $\theta = 3$

94. $\theta = 2$ 95. $\theta = -68°$ 96. $\theta = -1$

97. Complete the table below.

θ	0°	15°	30°	45°	60°	75°	90°	105°	120°	135°	150°	165°	180°
sin θ													

Use this table to complete the following sentences where $0° \le \theta \le 180°$.

a. The maximum value of sin θ is ________.

b. For $0° \le \theta \le 90°$, the value of sin θ is ________.

c. For $90° \le \theta \le 180°$, the value of sin θ is ________.

PROJECTS

1. **FIND SUMS OR PRODUCTS** Determine the following sums or products. Do not use a calculator. (*Hint:* The Reference Angle Theorem may be helpful.) Explain to a classmate how you know you are correct.

a. $\cos 0° + \cos 1° + \cos 2° + \cdots + \cos 178° + \cos 179° + \cos 180°$

b. $\sin 0° + \sin 1° + \sin 2° + \cdots + \sin 358° + \sin 359° + \sin 360°$

c. $\cot 1° + \cot 2° + \cot 3° + \cdots + \cot 177° + \cot 178° + \cot 179°$

d. $(\cos 1°)(\cos 2°)(\cos 3°) \cdots (\cos 177°)(\cos 178°)(\cos 179°)$

e. $\cos^2 1° + \cos^2 2° + \cos^2 3° + \cdots + \cos^2 357° + \cos^2 358° + \cos^2 359°$

SECTION 5.4 TRIGONOMETRIC FUNCTIONS OF REAL NUMBERS

- THE WRAPPING FUNCTION
- PROPERTIES OF TRIGONOMETRIC FUNCTIONS OF REAL NUMBERS
- TRIGONOMETRIC IDENTITIES
- AN APPLICATION INVOLVING A TRIGONOMETRIC FUNCTION OF A REAL NUMBER

◆ THE WRAPPING FUNCTION

In the seventeenth century, applications of trigonometry were extended to problems in physics and engineering. These kinds of problems required trigonometric functions whose domains were sets of real numbers rather than sets of angles. During this time, the definitions of trigonometric functions were extended to real numbers by using a correspondence between an angle and a number.

Consider a circle given by the equation $x^2 + y^2 = 1$, called a **unit circle,** and a vertical coordinate line l tangent to the unit circle at (1, 0). We define a function W that pairs a real number t on the coordinate line with a point $P(x, y)$ on the unit circle. This function is called the *wrapping function* because it is analogous to wrapping a line around a circle.

As shown in **Figure 5.52,** the positive part of the coordinate line is wrapped around the unit circle in a counterclockwise direction. The negative part of the coordinate line is wrapped around the circle in a clockwise direction. The wrapping function is defined by the equation $W(t) = P(x, y)$, where t is a real number and $P(x, y)$ is the point on the unit circle that corresponds to t.

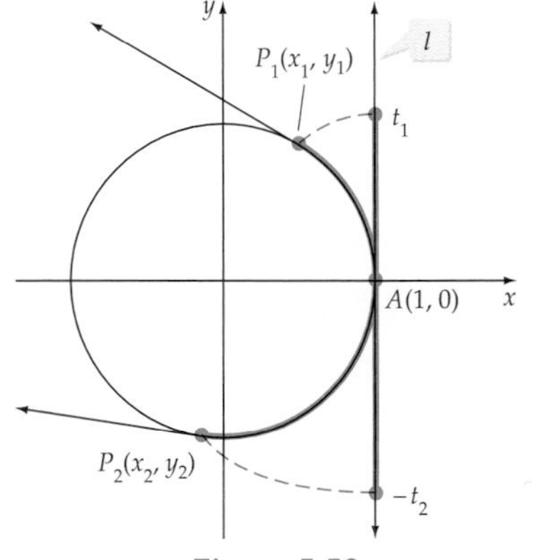

Figure 5.52

Through the wrapping function, each real number t defines an arc $\widehat{AP}$ that subtends a central angle with a measure of θ radians. The length of the arc $\widehat{AP}$ is t (see **Figure 5.53**). From the equation $s = r\theta$ for the arc length of a circle, we have (with $t = s$) $t = r\theta$. For a unit circle, $r = 1$ and the equation becomes $t = \theta$. Thus, on a unit circle, *the measure of a central angle and the length of its arc can be represented by the same real number t.*

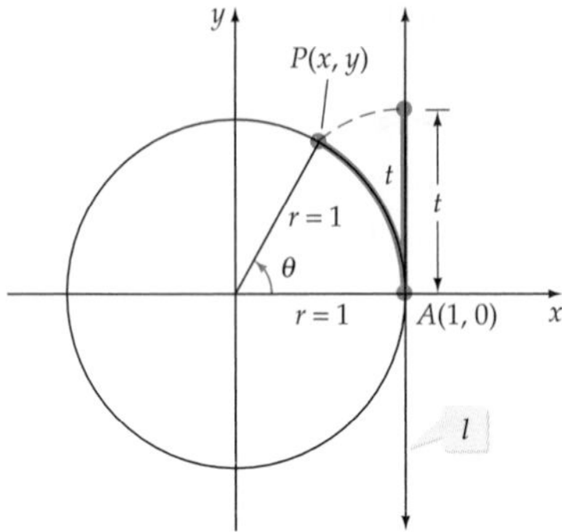

Figure 5.53

EXAMPLE 1 Evaluate the Wrapping Function

Find the values of x and y such that $W\left(\frac{\pi}{3}\right) = P(x, y)$.

Solution

The point $\pi/3$ on line l is shown in Figure 5.54. From the wrapping function, $W(\pi/3)$ is the point P on the unit circle for which arc $\widehat{AP}$ subtends an angle θ, the measure of which is $\pi/3$ radians. The coordinates of P can be determined from the definitions of $\cos\theta$ and $\sin\theta$ given in Section 5.3 and from Table 5.2.

$$\cos\theta = \frac{x}{r} \qquad \sin\theta = \frac{y}{r}$$

$$\cos\frac{\pi}{3} = \frac{x}{1} = x \qquad \sin\frac{\pi}{3} = \frac{y}{r} = y \qquad \bullet\ \theta = \frac{\pi}{3};\ r = 1$$

$$\frac{1}{2} = x \qquad \frac{\sqrt{3}}{2} = y \qquad \bullet\ \cos\frac{\pi}{3} = \frac{1}{2};\ \sin\frac{\pi}{3} = \frac{\sqrt{3}}{2}$$

From these equations, $x = 1/2$ and $y = \sqrt{3}/2$. Therefore, $W(\pi/3) = P(1/2, \sqrt{3}/2)$.

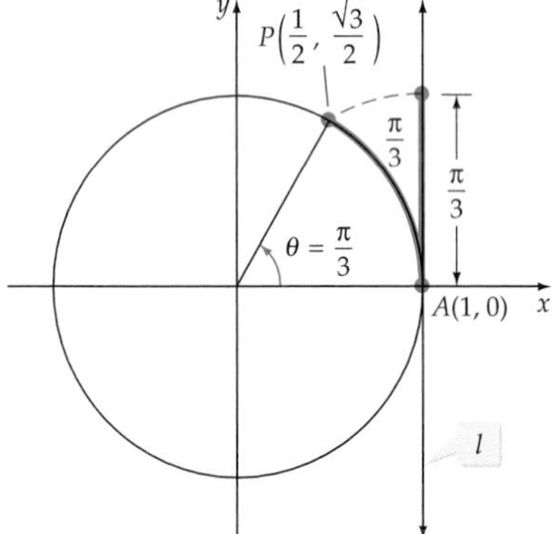

Figure 5.54

TRY EXERCISE 10, EXERCISE SET 5.4, PAGE 431

To determine $W(\pi/2)$, recall that the circumference of a unit circle is 2π. One-fourth the circumference is $\frac{1}{4}(2\pi) = \frac{\pi}{2}$ (see Figure 5.55). Thus $W(\pi/2) = P(0, 1)$.

Note from the last two examples that for the given real number t, $\cos t = x$ and $\sin t = y$. That is, for a real number t and $W(t) = P(x, y)$, the value of the cosine of t is the x-coordinate of P, and the value of the sine of t is the y-coordinate of P. See Figure 5.56.

The following definition makes use of the wrapping function $W(t)$ to define trigonometric functions of real numbers.

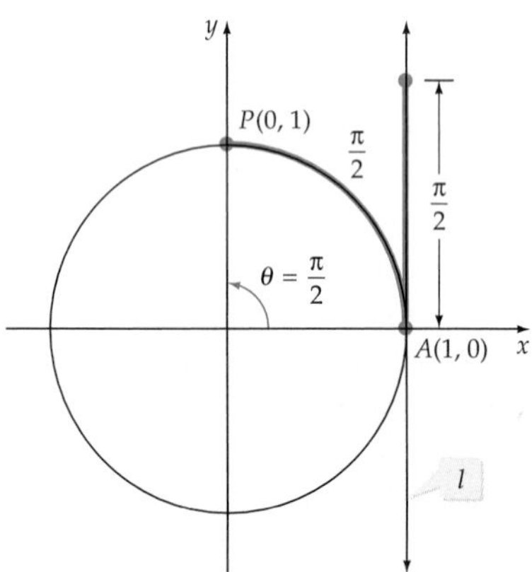

Figure 5.55

Definition of the Trigonometric Functions of Real Numbers

Let W be the wrapping function, t be a real number, and $W(t) = P(x, y)$. Then

$$\sin t = y \qquad \cos t = x \qquad \tan t = \frac{y}{x},\ x \neq 0$$

$$\csc t = \frac{1}{y},\ y \neq 0 \qquad \sec t = \frac{1}{x},\ x \neq 0 \qquad \cot t = \frac{x}{y},\ y \neq 0$$

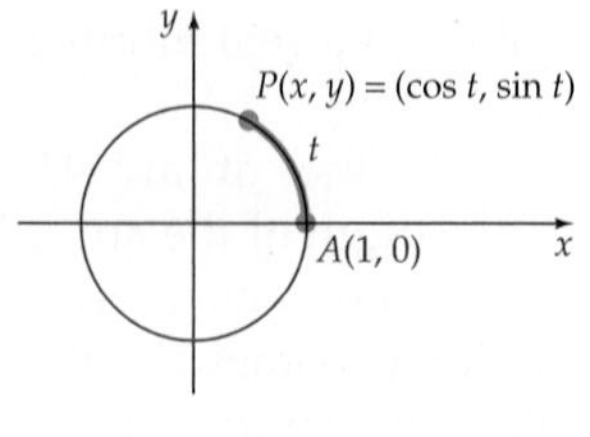

$x = \cos t$, $y = \sin t$

Figure 5.56

Trigonometric functions of real numbers are frequently called *circular functions* to distinguish them from trigonometric functions of angles.

The *trigonometric functions of real numbers* (or circular functions) look remarkably like the trigonometric functions defined in the last section. The difference

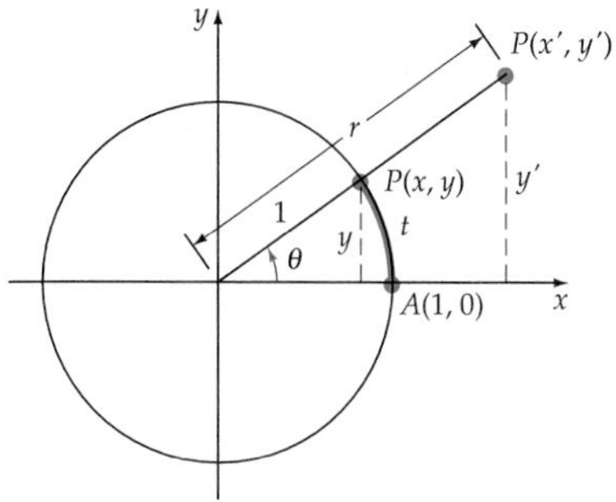

Figure 5.57

between the two is that of domain: In one case, the domains are sets of *real numbers;* in the other case, the domains are sets of *angles.* However, there are similarities between the two functions.

Consider an angle θ (in radians) in standard position as shown in **Figure 5.57.** Let $P(x, y)$ and $P'(x', y')$ be two points on the terminal side of θ, where $x^2 + y^2 = 1$ and $(x')^2 + (y')^2 = r^2$. Let t be the length of the arc from $A(1, 0)$ to $P(x, y)$. Then

$$\sin\theta = \frac{y'}{r} = \frac{y}{1} = \sin t$$

Thus the value of the sine function of θ, measured in radians, is equal to the value of the sine of the real number t. Similar arguments can be given to show corresponding results for the other five trigonometric functions. With this in mind, we can assert that *the value of a trigonometric function at the real number t is its value at an angle of t radians.*

EXAMPLE 2 Evaluate Trigonometric Functions of Real Numbers

Find the exact value of each function.

a. $\cos\frac{\pi}{4}$ b. $\sin\left(-\frac{7\pi}{6}\right)$ c. $\tan\left(-\frac{5\pi}{4}\right)$ d. $\sec\frac{5\pi}{3}$

Solution

The value of a trigonometric function at the real number t is its value at an angle of t radians. Using Table 5.2, we have

a. $\cos\frac{\pi}{4} = \frac{\sqrt{2}}{2}$

b. $\sin\left(-\frac{7\pi}{6}\right) = \sin\frac{\pi}{6} = \frac{1}{2}$
- **Reference angle for $-7\pi/6$ is $\pi/6$ and $\sin t > 0$ in Quadrant II.**

c. $\tan\left(-\frac{5\pi}{4}\right) = -\tan\frac{\pi}{4} = -1$
- **Reference angle for $-5\pi/4$ is $\pi/4$ and $\tan t < 0$ in Quadrant II.**

d. $\sec\frac{5\pi}{3} = \sec\frac{\pi}{3} = 2$
- **Reference angle for $5\pi/3$ is $\pi/3$ and $\sec t > 0$ in Quadrant IV.**

TRY EXERCISE 16, EXERCISE SET 5.4, PAGE 431

◆ PROPERTIES OF TRIGONOMETRIC FUNCTIONS OF REAL NUMBERS

The domain and range of the trigonometric functions can be found from the definition of these functions. If t is any real number and $P(x, y)$ is the point corresponding to $W(t)$, then by definition $\cos t = x$ and $\sin t = y$. Thus the domain of the sine and cosine functions is the set of real numbers.

Because the radius of the unit circle is 1, we have

$$-1 \le x \le 1 \qquad \text{and} \qquad -1 \le y \le 1$$

Therefore, with $x = \cos t$ and $y = \sin t$, we have

$$-1 \le \cos t \le 1 \qquad \text{and} \qquad -1 \le \sin t \le 1$$

The range of the cosine and sine functions is $[-1, 1]$.

Using the definitions of tangent and secant,

$$\tan t = \frac{y}{x} \qquad \text{and} \qquad \sec t = \frac{1}{x}$$

The domain of the tangent function is all real numbers t except those for which the x-coordinate of $W(t)$ is zero. The x-coordinate is zero when $t = \pm\pi/2$, $t = \pm 3\pi/2$, and $t = \pm 5\pi/2$ and in general when $t = (2n + 1)\pi/2$, where n is an integer. Thus the domain of the tangent function is the set of all real numbers t except $t = (2n + 1)\pi/2$, where n is an integer. The range of the tangent function is all real numbers.

Similar methods can be used to find the domain and range of the cotangent, secant, and cosecant functions. The results are summarized in Table 5.6.

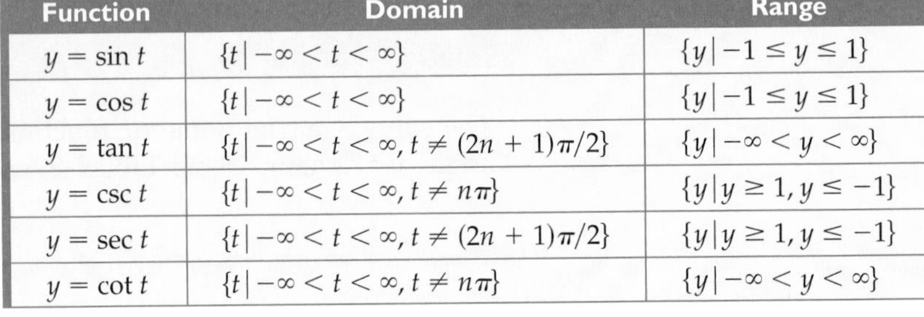

Table 5.6 Domain and Range of the Trigonometric Functions (n is an integer)

Function	Domain	Range
$y = \sin t$	$\{t \mid -\infty < t < \infty\}$	$\{y \mid -1 \le y \le 1\}$
$y = \cos t$	$\{t \mid -\infty < t < \infty\}$	$\{y \mid -1 \le y \le 1\}$
$y = \tan t$	$\{t \mid -\infty < t < \infty, t \ne (2n + 1)\pi/2\}$	$\{y \mid -\infty < y < \infty\}$
$y = \csc t$	$\{t \mid -\infty < t < \infty, t \ne n\pi\}$	$\{y \mid y \ge 1, y \le -1\}$
$y = \sec t$	$\{t \mid -\infty < t < \infty, t \ne (2n + 1)\pi/2\}$	$\{y \mid y \ge 1, y \le -1\}$
$y = \cot t$	$\{t \mid -\infty < t < \infty, t \ne n\pi\}$	$\{y \mid -\infty < y < \infty\}$

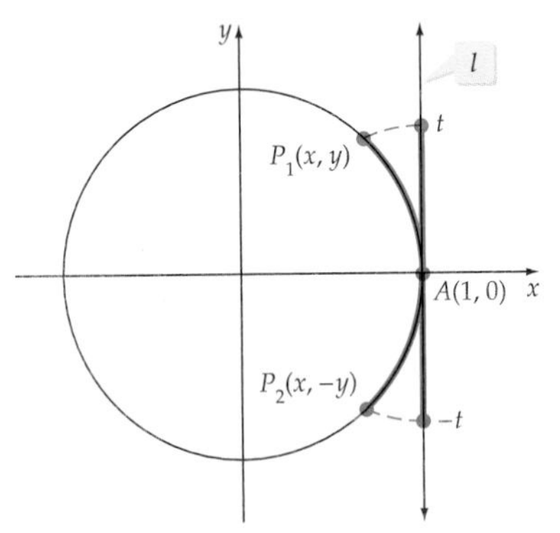

Figure 5.58

Consider the points t and $-t$ on the coordinate line l tangent to the unit circle at the point $(1, 0)$. The points $W(t)$ and $W(-t)$ are symmetric with respect to the x-axis. Therefore, if $P_1(x, y)$ are the coordinates of $W(t)$, then $P_2(x, -y)$ are the coordinates of $W(-t)$. See **Figure 5.58.**

From the definitions of the trigonometric functions, we have

$$\sin t = y \quad \text{and} \quad \sin(-t) = -y \qquad \text{and} \qquad \cos t = x \quad \text{and} \quad \cos(-t) = x$$

Substituting $\sin t$ for y and $\cos t$ for x yields

$$\sin(-t) = -\sin t \qquad \text{and} \qquad \cos(-t) = \cos t$$

Thus the sine is an odd function and the cosine is an even function. Because $\csc t = 1/\sin t$ and $\sec t = 1/\cos t$, it follows that

$$\csc(-t) = -\csc t \qquad \text{and} \qquad \sec(-t) = \sec t$$

These equations state that the cosecant is an odd function and the secant an even function.

From the definition of the tangent function, we have $\tan t = y/x$ and $\tan(-t) = -y/x$. Substituting $\tan t$ for y/x yields $\tan(-t) = -\tan t$. Because

$\cot t = 1/\tan t$, it follows that $\cot(-t) = -\cot t$. Thus the tangent and cotangent are odd functions.

Even and Odd Trigonometric Functions

The odd trigonometric functions are $y = \sin t$, $y = \csc t$, $y = \tan t$, and $y = \cot t$. The even trigonometric functions are $y = \cos t$ and $y = \sec t$.

Thus for all t in their domain,

$$\sin(-t) = -\sin t \qquad \cos(-t) = \cos t \qquad \tan(-t) = -\tan t$$
$$\csc(-t) = -\csc t \qquad \sec(-t) = \sec t \qquad \cot(-t) = -\cot t$$

EXAMPLE 3 Determine Whether a Function Is Even, Odd, or Neither

Is the function defined by $f(x) = x - \tan x$ even, odd, or neither?

Solution

Find $f(-x)$ and compare it to $f(x)$.

$$\begin{aligned} f(-x) &= (-x) - \tan(-x) = -x + \tan x \\ &= -(x - \tan x) \\ &= -f(x) \end{aligned}$$

• $\tan(-x) = -\tan x$

The function defined by $f(x) = x - \tan x$ is an odd function.

TRY EXERCISE 36, EXERCISE SET 5.4, PAGE 432

Let W be the wrapping function, t be a point on the coordinate line tangent to the unit circle at $(1, 0)$, and $W(t) = P(x, y)$. Because the circumference of the unit circle is 2π, $W(t + 2\pi) = W(t) = P(x, y)$. Thus the value of the wrapping function repeats itself in 2π units. *The wrapping function is periodic and the period is 2π.*

Recall the definitions of $\cos t$ and $\sin t$:

$$\cos t = x \qquad \text{and} \qquad \sin t = y$$

where $W(t) = P(x, y)$. Because $W(t + 2\pi) = W(t) = P(x, y)$ for all t,

$$\cos(t + 2\pi) = x \qquad \text{and} \qquad \sin(t + 2\pi) = y$$

Thus $\cos t$ and $\sin t$ have period 2π. Because

$$\sec t = \frac{1}{\cos t} = \frac{1}{\cos(t + 2\pi)} = \sec(t + 2\pi) \qquad \text{and}$$

$$\csc t = \frac{1}{\sin t} = \frac{1}{\sin(t + 2\pi)} = \csc(t + 2\pi)$$

$\sec t$ and $\csc t$ have a period of 2π.

Period of cos t, sin t, sec t, and csc t

The period of $\cos t$, $\sin t$, $\sec t$, and $\csc t$ is 2π.

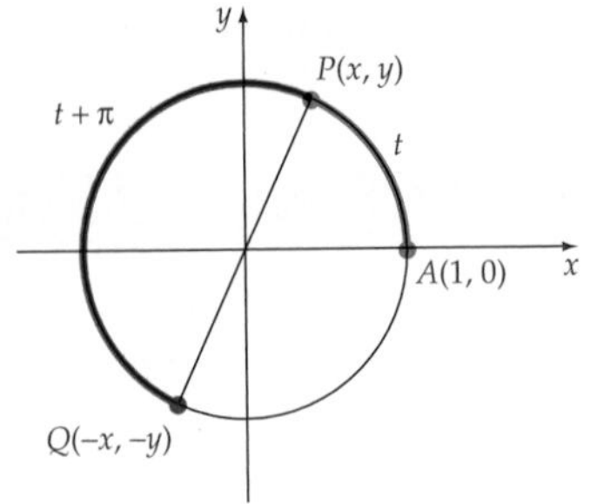

Figure 5.59

Although it is true that $\tan t = \tan(t + 2\pi)$, the period of $\tan t$ is not 2π. Recall that the period of a function is the *smallest* value of p for which $f(t) = f(t + p)$.

If W is the wrapping function (see **Figure 5.59**) and $W(t) = P(x, y)$, then $W(t + \pi) = P(-x, -y)$. Because

$$\tan t = \frac{y}{x} \qquad \text{and} \qquad \tan(t + \pi) = \frac{-y}{-x} = \frac{y}{x} = \tan t$$

we have $\tan(t + \pi) = \tan t$ for all t. A similar argument applies to $\cot t$.

Period of tan t and cot t

The period of $\tan t$ and $\cot t$ is π.

◆ TRIGONOMETRIC IDENTITIES

Recall that any equation that is true for every number in the domain of the equation is an identity. The statement

$$\csc t = \frac{1}{\sin t}, \quad \sin t \neq 0$$

is an identity because the two expressions produce the same result for all values of t for which both the functions are defined.

The **ratio identities** are obtained by writing the tangent and cotangent functions in terms of the sine and cosine functions.

$$\tan t = \frac{y}{x} = \frac{\sin t}{\cos t} \qquad \text{and} \qquad \cot t = \frac{x}{y} = \frac{\cos t}{\sin t}$$

• $x = \cos t$ and $y = \sin t$

The **Pythagorean identities** are based on the equation of a unit circle, $x^2 + y^2 = 1$, and on the definitions of the sine and cosine functions.

$$x^2 + y^2 = 1$$

$$\cos^2 t + \sin^2 t = 1$$

• Replace x by $\cos t$ and y by $\sin t$

Dividing each term of $\cos^2 t + \sin^2 t = 1$ by $\cos^2 t$, we have

$$\frac{\cos^2 t}{\cos^2 t} + \frac{\sin^2 t}{\cos^2 t} = \frac{1}{\cos^2 t}$$

• $\cos t \neq 0$

$$1 + \tan^2 t = \sec^2 t$$

• $\frac{\sin t}{\cos t} = \tan t$

Dividing each term of $\cos^2 t + \sin^2 t = 1$ by $\sin^2 t$, we have

$$\frac{\cos^2 t}{\sin^2 t} + \frac{\sin^2 t}{\sin^2 t} = \frac{1}{\sin^2 t}$$

• $\sin t \neq 0$

$$\cot^2 t + 1 = \csc^2 t$$

• $\frac{\cos t}{\sin t} = \cot t$

Here is a summary of the Fundamental Trigonometric Identities:

Fundamental Trigonometric Identities

The reciprocal identities are

$$\sin t = \frac{1}{\csc t} \qquad \cos t = \frac{1}{\sec t} \qquad \tan t = \frac{1}{\cot t}$$

The ratio identities are

$$\tan t = \frac{\sin t}{\cos t} \qquad \cot t = \frac{\cos t}{\sin t}$$

The Pythagorean identities are

$$\cos^2 t + \sin^2 t = 1 \qquad 1 + \tan^2 t = \sec^2 t \qquad 1 + \cot^2 t = \csc^2 t$$

EXAMPLE 4 Use the Unit Circle to Verify an Identity

By using the unit circle and the definitions of the trigonometric functions, show that $\sin(t + \pi) = -\sin t$.

Solution

Sketch a unit circle, and let P be the point on the unit circle such that $W(t) = P(x, y)$, as shown in **Figure 5.60.** Draw a diameter from P and label the endpoint Q. For any line through the origin, if $P(x, y)$ is a point on the line, then $Q(-x, -y)$ is also a point on the line. Because PQ is a diameter, the length of arc PQ is π. Thus the length of arc AQ is $t + \pi$. Therefore, $W(t + \pi) = Q(-x, -y)$. From the definition of $\sin t$, we have

$$\sin t = y \qquad \text{and} \qquad \sin(t + \pi) = -y$$

Thus $\sin(t + \pi) = -\sin t$.

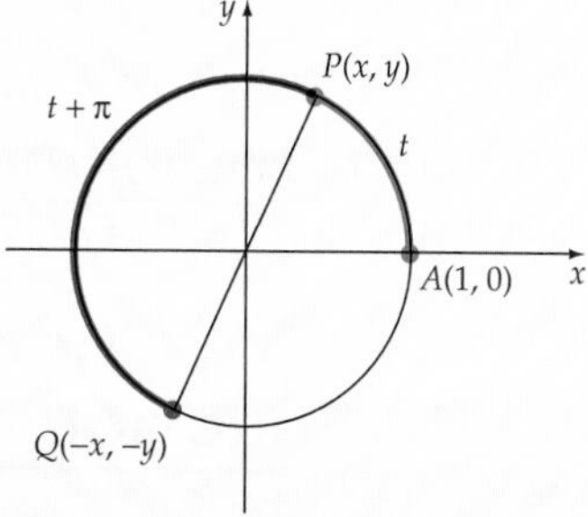

Figure 5.60

TRY EXERCISE 42, EXERCISE SET 5.4, PAGE 432

Using identities and basic algebra concepts, we can rewrite trigonometric expressions in different forms.

EXAMPLE 5 Simplify a Trigonometric Expression

Write the expression $\frac{1}{\sin^2 t} + \frac{1}{\cos^2 t}$ as a single term.

Solution

Express each fraction in terms of a common denominator. The common denominator is $\sin^2 t \cos^2 t$.

$$\frac{1}{\sin^2 t} + \frac{1}{\cos^2 t} = \frac{1}{\sin^2 t}\frac{\cos^2 t}{\cos^2 t} + \frac{1}{\cos^2 t}\frac{\sin^2 t}{\sin^2 t}$$

$$= \frac{\cos^2 t + \sin^2 t}{\sin^2 t \cos^2 t} = \frac{1}{\sin^2 t \cos^2 t} \qquad \text{• } \cos^2 t + \sin^2 t = 1$$

take note

Because

$$\frac{1}{\sin^2 t \cos^2 t} = \frac{1}{\sin^2 t} \cdot \frac{1}{\cos^2 t} = (\csc^2 t)(\sec^2 t)$$

we could have written the answer to Example 5 in terms of the cosecant and secant functions.

TRY EXERCISE 60, EXERCISE SET 5.4, PAGE 432

EXAMPLE 6 Write a Trigonometric Expression in Terms of a Given Function

For $\pi/2 < t < \pi$, write $\tan t$ in terms of $\sin t$.

Solution

Write $\tan t = \frac{\sin t}{\cos t}$. Now solve $\cos^2 t + \sin^2 t = 1$ for $\cos t$.

$$\cos^2 t + \sin^2 t = 1$$

$$\cos^2 t = 1 - \sin^2 t$$

$$\cos t = \pm\sqrt{1 - \sin^2 t}$$

Because $\pi/2 < t < \pi$, $\cos t$ is negative. Therefore, $\cos t = -\sqrt{1 - \sin^2 t}$. Thus

$$\tan t = -\frac{\sin t}{\sqrt{1 - \sin^2 t}} \qquad \text{• } \frac{\pi}{2} < t < \pi$$

TRY EXERCISE 66, EXERCISE SET 5.4, PAGE 432

◆ AN APPLICATION INVOLVING A TRIGONOMETRIC FUNCTION OF A REAL NUMBER

EXAMPLE 7 Determine a Height as a Function of Time

The Millennium Wheel, in London, is the world's largest Ferris wheel. It has a diameter of 450 feet. When the Millennium Wheel is in uniform motion, it completes 1 revolution every 30 minutes. The height h, in feet

The Millennium Wheel, on the banks of the Thames river, London.

above the Thames river, of a person riding on the Millennium Wheel can be estimated by

$$h(t) = 255 - 225 \cos\left(\frac{\pi}{15}t\right)$$

where t is the time in minutes since the person started a ride.

a. How high is the person at the start of a ride ($t = 0$)?

b. How high is the person after 18.0 minutes?

Solution

a. $h(0) = 255 - 225 \cos\left(\frac{\pi}{15}0\right)$
$= 255 - 225$
$= 30$

At the start of the ride, the person is 30 feet above the Thames.

b. $h(18) = 255 - 225 \cos\left(\frac{\pi}{15} \cdot 18.0\right)$
$\approx 255 - (-182)$
≈ 437

After 18.0 minutes, the person is about 437 feet above the Thames.

TRY EXERCISE 70, EXERCISE SET 5.4, PAGE 432

TOPICS FOR DISCUSSION

1. Is $W(t)$ a number? Explain.
2. Explain how to find the exact value of $\cos(13\pi/6)$.
3. Explain why the equation $\cos^2 t + \sin^2 t = 1$ is called a Pythagorean Identity.
4. Is $f(x) = \cos^3 x$ an even function or an odd function? Explain how you made your decision.
5. Explain how to make use of a unit circle to show that

$$\sin(-t) = -\sin t$$

EXERCISE SET 5.4

In Exercises 1 to 12, find $W(t)$ for each given t.

1. $t = \pi/6$
2. $t = \pi/4$
3. $t = 7\pi/6$
4. $t = 4\pi/3$
5. $t = 5\pi/3$
6. $t = -\pi/6$
7. $t = 11\pi/6$
8. $t = 0$
9. $t = \pi$
10. $t = -7\pi/4$
11. $t = -2\pi/3$
12. $t = -\pi$

In Exercises 13 to 22, find the exact value of each function.

13. $\tan(11\pi/6)$
14. $\cot(2\pi/3)$
15. $\cos(-2\pi/3)$
16. $\sec(-5\pi/6)$
17. $\csc(-\pi/3)$
18. $\tan(12\pi)$
19. $\sin(3\pi/2)$
20. $\cos(7\pi/3)$
21. $\sec(-7\pi/6)$
22. $\sin(-5\pi/3)$

In Exercises 23 to 32, use a calculator to find an approximate value of each function. Round your answers to the nearest ten-thousandth.

23. sin 1.22
24. cos 4.22
25. csc (−1.05)
26. sin (−0.55)
27. tan $(11\pi/12)$
28. cos $(2\pi/5)$
29. cos $(-\pi/5)$
30. csc 8.2
31. sec 1.55
32. cot 2.11

In Exercises 33 to 40, determine whether the function defined by each equation is even, odd, or neither.

33. $f(x) = -4 \sin x$
34. $f(x) = -2 \cos x$
35. $G(x) = \sin x + \cos x$
36. $F(x) = \tan x + \sin x$
37. $S(x) = \dfrac{\sin x}{x}, x \neq 0$
38. $C(x) = \dfrac{\cos x}{x}, x \neq 0$
39. $v(x) = 2 \sin x \cos x$
40. $w(x) = x \tan x$

In Exercises 41 to 48, use the unit circle to verify each identity.

41. $\cos(-t) = \cos t$
42. $\tan(t - \pi) = \tan t$
43. $\cos(t + \pi) = -\cos t$
44. $\sin(-t) = -\sin t$
45. $\sin(t - \pi) = -\sin t$
46. $\sec(-t) = \sec t$
47. $\csc(-t) = -\csc t$
48. $\tan(-t) = -\tan t$

In Exercises 49 to 64, use the trigonometric identities to write each expression in terms of a single trigonometric function or a constant. Answers may vary.

49. $\tan t \cos t$
50. $\cot t \sin t$
51. $\dfrac{\csc t}{\cot t}$
52. $\dfrac{\sec t}{\tan t}$
53. $1 - \sec^2 t$
54. $1 - \csc^2 t$
55. $\tan t - \dfrac{\sec^2 t}{\tan t}$
56. $\dfrac{\csc^2 t}{\cot t} - \cot t$
57. $\dfrac{1 - \cos^2 t}{\tan^2 t}$
58. $\dfrac{1 - \sin^2 t}{\cot^2 t}$
59. $\dfrac{1}{1 - \cos t} + \dfrac{1}{1 + \cos t}$
60. $\dfrac{1}{1 - \sin t} + \dfrac{1}{1 + \sin t}$
61. $\dfrac{\tan t + \cot t}{\tan t}$
62. $\dfrac{\csc t - \sin t}{\csc t}$
63. $\sin^2 t(1 + \cot^2 t)$
64. $\cos^2 t(1 + \tan^2 t)$

65. Write $\sin t$ in terms of $\cos t$, $0 < t < \pi/2$.

66. Write $\tan t$ in terms of $\sec t$, $3\pi/2 < t < 2\pi$.

67. Write $\csc t$ in terms of $\cot t$, $\pi/2 < t < \pi$.

68. Write $\sec t$ in terms of $\tan t$, $\pi < t < 3\pi/2$.

69. **PATH OF A SATELLITE** A satellite is launched into space from Cape Canaveral. The distance, in miles, that the satellite is north of the equator is

$$d(t) = 1970 \cos\left(\frac{\pi}{64}t\right)$$

where t is the number of minutes since liftoff.

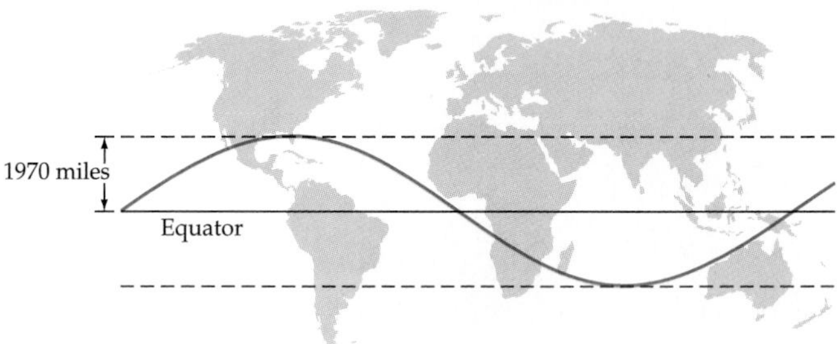

What distance, to the nearest 10 miles, is the satellite north of the equator 24 minutes after liftoff?

70. **AVERAGE HIGH TEMPERATURE** The average high temperature T, in degrees Fahrenheit, for Fairbanks, Alaska, is given by

$$T(t) = -41 \cos\left(\frac{\pi}{6}t\right) + 36$$

where t is the number of months after January 5. Use the formula to estimate (to the nearest 0.1 degree Fahrenheit) the average high temperature in Fairbanks for March 5 and July 20.

In Exercises 71 to 82, perform the indicated operation and simplify.

71. $\cos t - \dfrac{1}{\cos t}$
72. $\tan t + \dfrac{1}{\tan t}$
73. $\cot t + \dfrac{1}{\cot t}$
74. $\sin t - \dfrac{1}{\sin t}$
75. $(1 - \sin t)^2$
76. $(1 - \cos t)^2$
77. $(\sin t - \cos t)^2$
78. $(\sin t + \cos t)^2$
79. $(1 - \sin t)(1 + \sin t)$
80. $(1 - \cos t)(1 + \cos t)$
81. $\dfrac{\sin t}{1 + \cos t} + \dfrac{1 + \cos t}{\sin t}$
82. $\dfrac{1 - \sin t}{\cos t} - \dfrac{1}{\tan t + \sec t}$

In Exercises 83 to 90, factor the expression.

83. $\cos^2 t - \sin^2 t$
84. $\sec^2 t - \csc^2 t$
85. $\tan^2 t - \tan t - 6$
86. $\cos^2 t + 3 \cos t - 4$

87. $2 \sin^2 t - \sin t - 1$

88. $4 \cos^2 t + 4 \cos t + 1$

89. $\cos^4 t - \sin^4 t$

90. $\sec^4 t - \csc^4 t$

SUPPLEMENTAL EXERCISES

In Exercises 91 to 94, use the trigonometric identities to find the value of the function.

91. Given $\csc t = \sqrt{2}$, $0 < t < \pi/2$, find $\cos t$.

92. Given $\cos t = 1/2$, $3\pi/2 < t < 2\pi$, find $\sin t$.

93. Given $\sin t = 1/2$, $\pi/2 < t < \pi$, find $\tan t$.

94. Given $\cot t = \sqrt{3}/3$, $\pi < t < 3\pi/2$, find $\cos t$.

In Exercises 95 to 98, simplify the first expression to the second expression.

95. $\dfrac{\sin^2 t + \cos^2 t}{\sin^2 t}$; $\csc^2 t$

96. $\dfrac{\sin^2 t + \cos^2 t}{\cos^2 t}$; $\sec^2 t$

97. $(\cos t - 1)(\cos t + 1)$; $-\sin^2 t$

98. $(\sec t - 1)(\sec t + 1)$; $\tan^2 t$

PROJECTS

1. **Visual Insight**

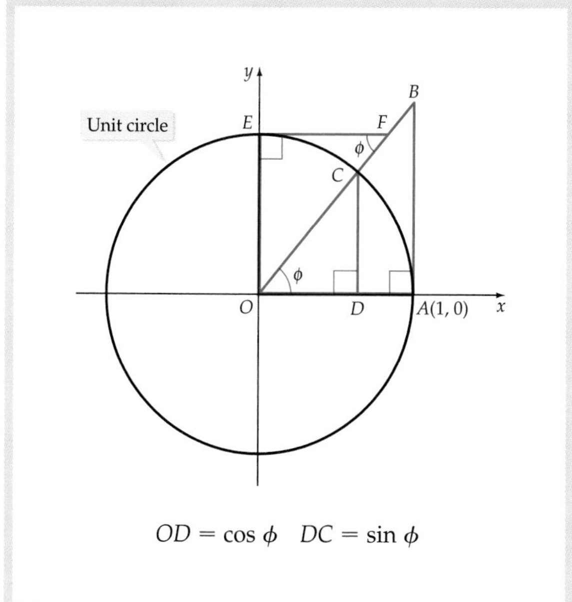

Make use of the above circle and similar triangles to explain why the length of line segment

a. AB is equal to $\tan \phi$

b. EF is equal to $\cot \phi$

c. OB is equal to $\sec \phi$

d. OF is equal to $\csc \phi$

2. **PERIODIC FUNCTIONS**

a. A function f is periodic with a period of 3. If $f(2) = -1$, determine $f(14)$.

b. If g is a periodic function with period 2 and h is a periodic function with period 3, determine the period of $f + g$.

c. Find two periodic functions f and g such that the function $f + g$ is not a periodic function.

3. Consider a square as shown. Start at the point $(1, 0)$ and travel counterclockwise around the square for a distance t ($t \geq 0$).

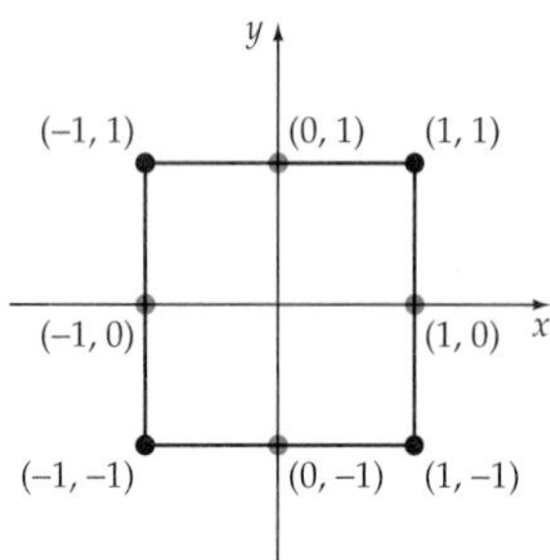

Let $WSQ(t) = P(x, y)$ be the point on the square determined by traveling counterclockwise a distance of t units from $(1, 0)$. For instance,

$$WSQ(0.5) = (1, 0.5)$$
$$WSQ(1.75) = (0.25, 1).$$

Find $WSQ(4.2)$ and $WSQ(6.4)$. We define the square sine of t, denoted by ssin t, to be the y-value of point P. The square cosine of t, denoted by scos t, is defined to be the x-value of point P. For example,

$$\text{ssin } 0.4 = 0.4 \qquad \text{scos } 0.4 = 1$$
$$\text{scos } 1.2 = 0.8 \qquad \text{scos } 5.3 = -0.7$$

The square tangent of t, denoted by stan t is defined as

$$\text{stan } t = \frac{\text{ssin } t}{\text{scos } t} \qquad \text{scos } t \neq 0$$

Find each of the following.

a. ssin 3.2 b. scos 4.4 c. stan 5.5

d. ssin 11.2 e. scos -5.2 f. stan -6.5

SECTION 5.5 GRAPHS OF THE SINE AND COSINE FUNCTIONS

- THE GRAPH OF THE SINE FUNCTION (5B)
- THE GRAPH OF THE COSINE FUNCTION

take note

A web applet is available to explore the relationships between the unit circle and the graphs of the sine, and cosine functions. This applet, Unit Circle, can be found on our web site at

http://college.hmco.com

◆ THE GRAPH OF THE SINE FUNCTION

The trigonometric functions can be graphed on a rectangular coordinate system by plotting the points whose coordinates belong to the function. We begin with the graph of the sine function.

Table 5.7 lists some ordered pairs (x, y), where $y = \sin x$, $0 \leq x \leq 2\pi$. In **Figure 5.61**, the points are plotted and a smooth curve is drawn through the points.

Table 5.7

x	$y = \sin x$
0	0
$\pi/4$	0.7
$\pi/2$	1
$3\pi/4$	0.7
π	0
$5\pi/4$	-0.7
$3\pi/2$	-1
$7\pi/4$	-0.7
2π	0

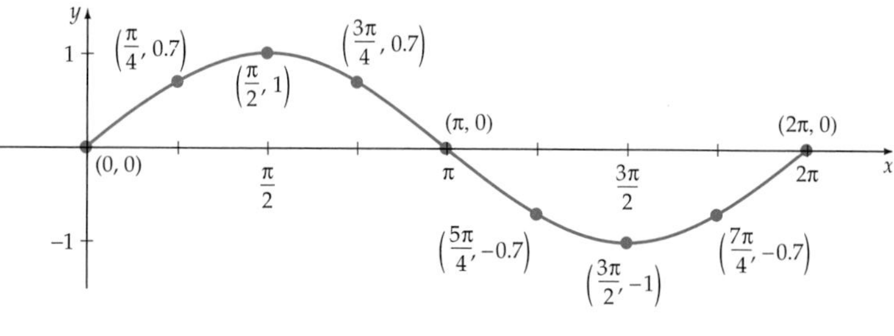

$y = \sin x, 0 \leq x \leq 2\pi$

Figure 5.61

Because the domain of the sine function is the real numbers and the period is 2π, the graph of $y = \sin x$ is drawn by repeating the portion shown in **Figure 5.61**. Any part of the graph that corresponds to one period (2π) is one cycle of the graph of $y = \sin x$ (see **Figure 5.62**).

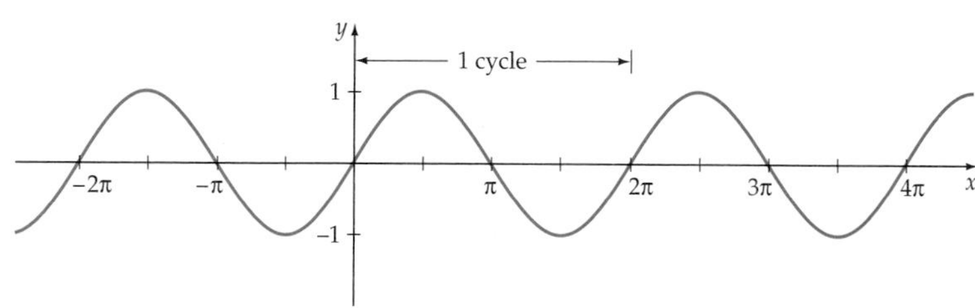

$y = \sin x$

Figure 5.62

> **take note**
>
> The amplitude is defined to be half the difference between the maximum height and the minimum height. It may not be equal to the maximum height. For example, the graph of $y = 4 + \sin x$ has a maximum height of 5 and an amplitude of 1.

The maximum value M reached by $\sin x$ is 1 and the minimum value m is -1. The amplitude of the graph of $y = \sin x$ is given by

$$\text{Amplitude} = \frac{1}{2}(M - m)$$

QUESTION What is the amplitude of $y = \sin x$?

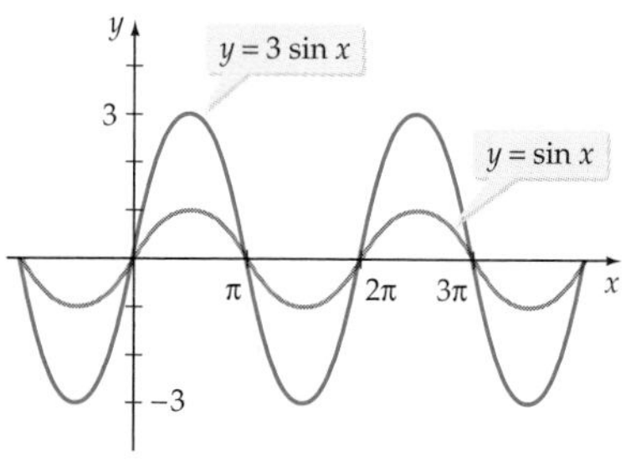

Figure 5.63

Recall that the graph of $y = a \cdot f(x)$ is obtained by *stretching* ($|a| > 1$) or *shrinking* ($0 < |a| < 1$) the graph of $y = f(x)$. **Figure 5.63** shows the graph of $y = 3 \sin x$ that was drawn by stretching the graph of $y = \sin x$. The amplitude of $y = 3 \sin x$ is 3 because

$$\text{Amplitude} = \frac{1}{2}(M - m) = \frac{1}{2}[3 - (-3)] = 3$$

Note that for $y = \sin x$ and $y = 3 \sin x$, the amplitude of the graph was the coefficient of $\sin x$. This suggests the theorem at the top of the next page.

ANSWER Amplitude $\frac{1}{2}(M - m) = \frac{1}{2}[1 - (-1)] = \frac{1}{2}(2) = 1$

Amplitude of $y = a \sin x$

The amplitude of $y = a \sin x$ is $|a|$.

EXAMPLE 1 Graph $y = a \sin x$

Graph: $y = -2 \sin x$

Solution

The amplitude of $y = -2 \sin x$ is 2. The graph of $y = -f(x)$ is a *reflection* across the x-axis of $y = f(x)$. Thus the graph of $y = -2 \sin x$ is a reflection across the x-axis of $y = 2 \sin x$. See **Figure 5.64.**

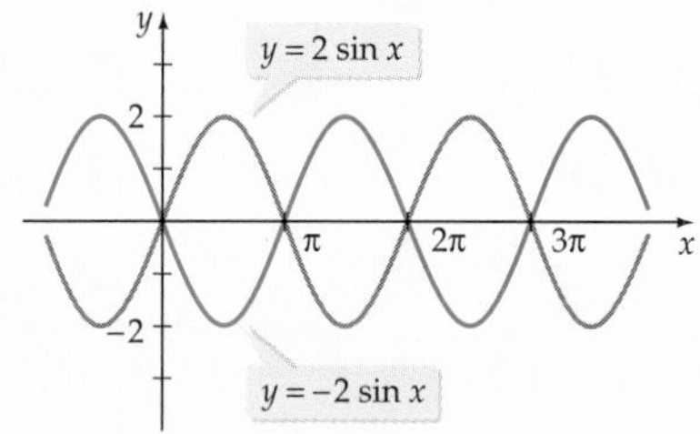

Figure 5.64

TRY EXERCISE 20, EXERCISE SET 5.5, PAGE 441

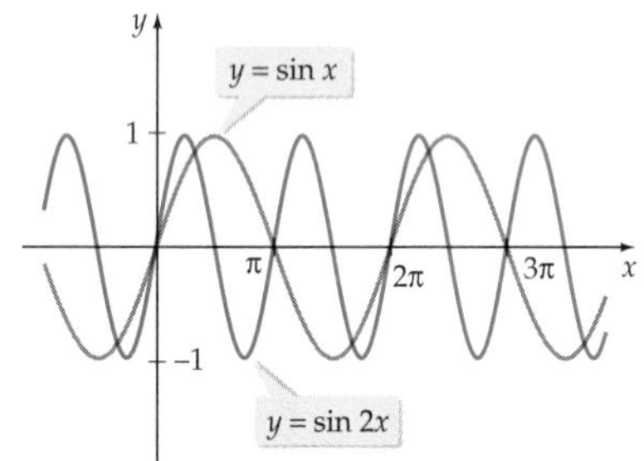

Figure 5.65

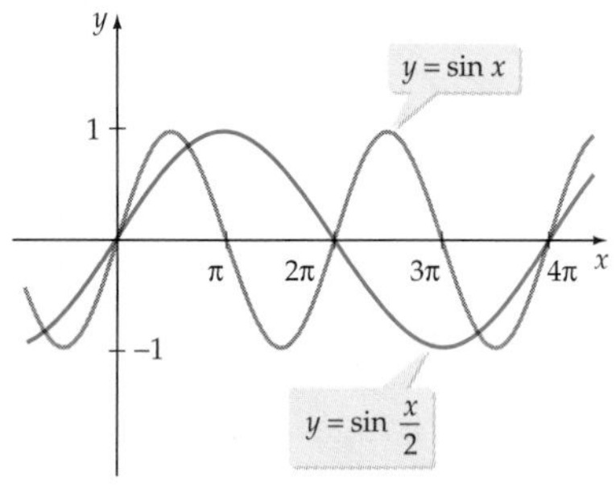

Figure 5.66

The graphs of $y = \sin x$ and $y = \sin 2x$ are shown in **Figure 5.65.** Because one cycle of the graph of $y = \sin 2x$ is completed in an interval of length π, the period of $y = \sin 2x$ is π.

The graphs of $y = \sin x$ and $y = \sin (x/2)$ are shown in **Figure 5.66.** Because one cycle of the graph of $y = \sin (x/2)$ is completed in an interval of length 4π, the period of $y = \sin (x/2)$ is 4π.

Generalizing the last two examples, one cycle of $y = \sin bx$, $b > 0$, is completed as bx varies from 0 to 2π. Algebraically, one cycle of $y = \sin bx$ is completed as bx varies from 0 to 2π. Therefore,

$$0 \le bx \le 2\pi$$

$$0 \le x \le \frac{2\pi}{b}$$

The length of the interval, $2\pi/b$, is the period of $y = \sin bx$. Now we consider the case when the coefficient of x is negative. If $b > 0$, then using the fact that the sine function is an odd function, we have $y = \sin (-bx) = -\sin bx$ and thus the period is still $2\pi/b$. This gives the following theorem.

Period of $y = \sin bx$

The period of $y = \sin bx$ is $\dfrac{2\pi}{|b|}$.

Table 5.8 gives the amplitude and period of several sine functions.

Table 5.8

Function	$y = a \sin bx$	$y = 3 \sin(-2x)$	$y = -\sin \frac{x}{3}$	$y = -2 \sin \frac{3x}{4}$
Amplitude	$\lvert a \rvert$	$\lvert 3 \rvert = 3$	$\lvert -1 \rvert = 1$	$\lvert -2 \rvert = 2$
Period	$\frac{2\pi}{\lvert b \rvert}$	$\frac{2\pi}{2} = \pi$	$\frac{2\pi}{1/3} = 6\pi$	$\frac{2\pi}{3/4} = \frac{8\pi}{3}$

EXAMPLE 2 Graph y = sin bx

Graph: $y = \sin \pi x$

Solution

$$\text{Amplitude} = 1 \qquad \text{Period} = \frac{2\pi}{b} = \frac{2\pi}{\pi} = 2 \qquad \bullet\, b = \pi$$

The graph is sketched in **Figure 5.67**.

$y = \sin \pi x$

Figure 5.67

TRY EXERCISE 30, EXERCISE SET 5.5, PAGE 441

Figure 5.68 shows the graph of $y = a \sin bx$ for both a and b positive. Note from the graph that

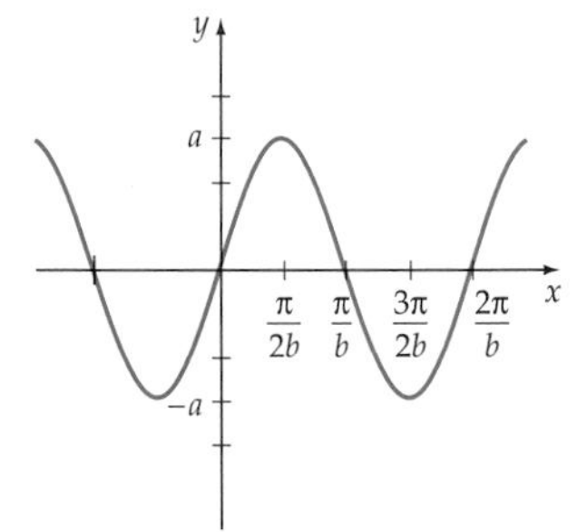

$y = a \sin bx$

Figure 5.68

- The amplitude is a.
- The period is $\frac{2\pi}{b}$.
- For $0 \le x \le \frac{2\pi}{b}$, the zeros are 0, $\frac{\pi}{b}$, and $\frac{2\pi}{b}$.
- The maximum value is a when $x = \frac{\pi}{2b}$, and the minimum value is $-a$ when $x = \frac{3\pi}{2b}$.
- If $a < 0$, the graph is reflected across the x-axis.

EXAMPLE 3 Graph y = a sin bx

Graph: $y = -\frac{1}{2} \sin \frac{x}{3}$

Solution

$$\text{Amplitude} = \left| -\frac{1}{2} \right| = \frac{1}{2} \qquad \text{Period} = \frac{2\pi}{1/3} = 6\pi \qquad \bullet\, b = \frac{1}{3}$$

Continued •➤

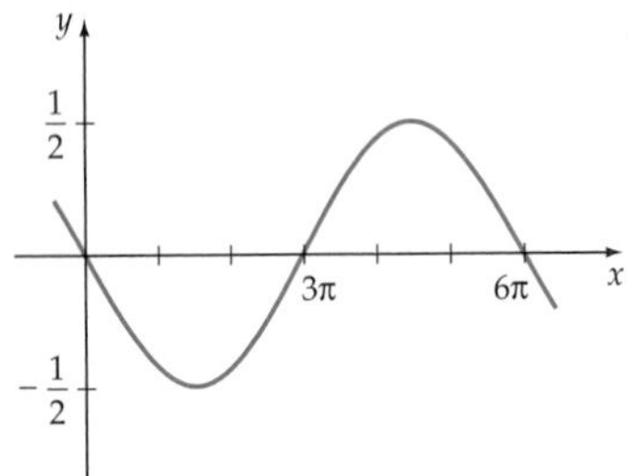

$y = -\frac{1}{2}\sin\frac{x}{3}$

Figure 5.69

The zeros in the interval $0 \le x \le 6\pi$ are 0, $\frac{\pi}{1/3} = 3\pi$, and $\frac{2\pi}{1/3} = 6\pi$.

Because $-\frac{1}{2} < 0$, the graph is the graph of $y = \frac{1}{2}\sin\frac{x}{3}$ reflected across the x-axis as shown in **Figure 5.69.**

TRY EXERCISE 38, EXERCISE SET 5.5, PAGE 441

◆ THE GRAPH OF THE COSINE FUNCTION

Table 5.9

x	y = cos x
0	1
$\pi/4$	0.7
$\pi/2$	0
$3\pi/4$	−0.7
π	−1
$5\pi/4$	−0.7
$3\pi/2$	0
$7\pi/4$	0.7
2π	1

Table 5.9 lists some of the ordered pairs of $y = \cos x$, $0 \le x \le 2\pi$. In **Figure 5.70,** the points are plotted and a smooth curve is drawn through the points.

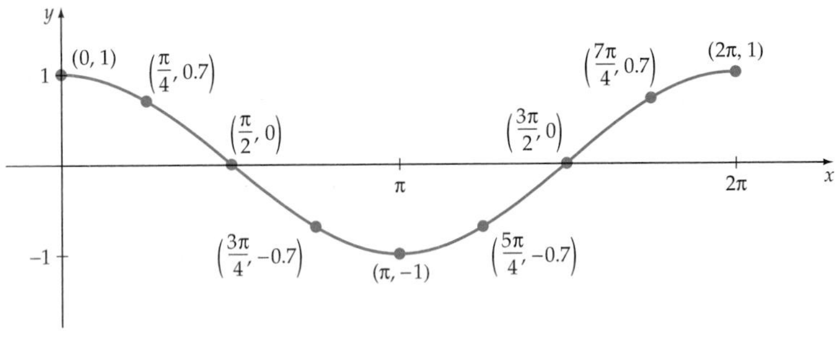

$y = \cos x, 0 \le x \le 2\pi$

Figure 5.70

Because the domain of $y = \cos x$ is the real numbers and the period is 2π, the graph of $y = \cos x$ is drawn by repeating the portion shown in **Figure 5.70.** Any part of the graph corresponding to one period (2π) is one cycle of $y = \cos x$ (see **Figure 5.71**).

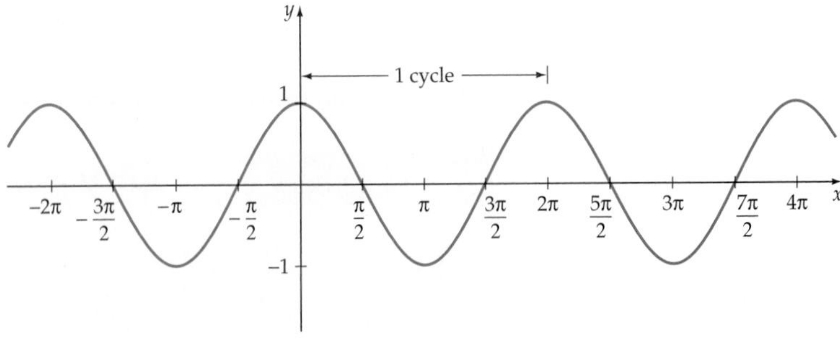

$y = \cos x$

Figure 5.71

The following two theorems concerning cosine functions can be developed using methods that are analogous to those we used to determine the amplitude and period of a sine function.

Amplitude of $y = a \cos x$

The amplitude of $y = a \cos x$ is $|a|$.

Period of $y = \cos bx$

The period of $y = \cos bx$ is $\dfrac{2\pi}{|b|}$.

Table 5.10 gives the amplitude and period of some cosine functions.

Table 5.10

Function	$y = a \cos bx$	$y = 2 \cos 3x$	$y = -3 \cos \dfrac{2x}{3}$
Amplitude	$\lvert a\rvert$	$\lvert 2\rvert = 2$	$\lvert -3\rvert = 3$
Period	$\dfrac{2\pi}{\lvert b\rvert}$	$\dfrac{2\pi}{3}$	$\dfrac{2\pi}{2/3} = 3\pi$

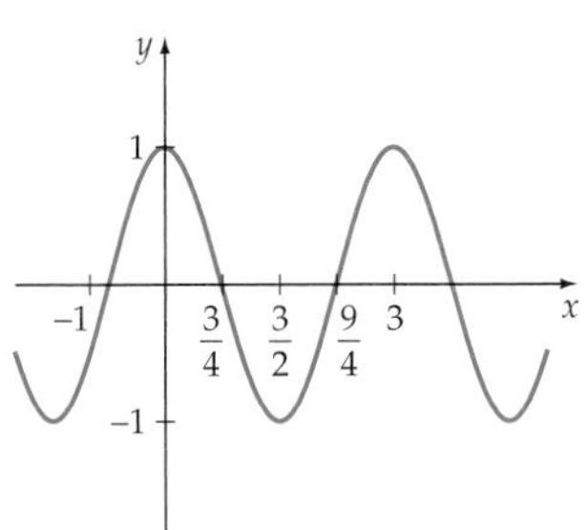

$y = \cos \dfrac{2\pi}{3} x$

Figure 5.72

EXAMPLE 4 Graph $y = \cos bx$

Graph: $y = \cos \dfrac{2\pi}{3} x$

Solution

$$\text{Amplitude} = 1 \qquad \text{Period} = \frac{2\pi}{b} = \frac{2\pi}{2\pi/3} = 3 \qquad \bullet\ b = \frac{2\pi}{3}$$

The graph is shown in **Figure 5.72**.

TRY EXERCISE 32, EXERCISE SET 5.5, PAGE 441

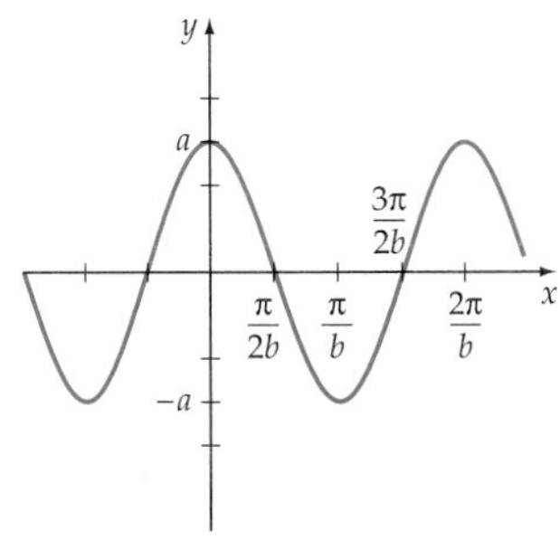

$y = a \cos bx$

Figure 5.73

Figure 5.73 shows the graph of $y = a \cos bx$ for both a and b positive. Note from the graph that

- The amplitude is a.
- The period is $\dfrac{2\pi}{b}$.
- For $0 \le x \le \dfrac{2\pi}{b}$, the zeros are $\dfrac{\pi}{2b}$ and $\dfrac{3\pi}{2b}$.
- The maximum value is a when $x = 0$, and the minimum value is $-a$ when $x = \dfrac{\pi}{b}$.
- If $a < 0$, then the graph is reflected across the x-axis.

EXAMPLE 5 Graph a Cosine Function

Graph: $y = -2\cos\dfrac{\pi x}{4}$

Solution

$$\text{Amplitude} = |-2| = 2 \qquad \text{Period} = \frac{2\pi}{\pi/4} = 8 \qquad \bullet\ b = \pi/4$$

The zeros in the interval $0 \le x \le 8$ are $\dfrac{\pi}{2\pi/4} = 2$ and $\dfrac{3\pi}{2\pi/4} = 6$. Because $-2 < 0$, the graph is the graph of $y = 2\cos\dfrac{\pi x}{4}$ reflected across the x-axis as shown in **Figure 5.74.**

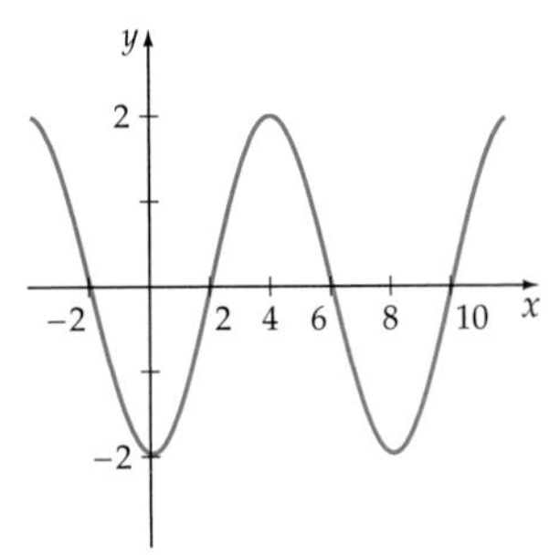

$y = -2\cos\dfrac{\pi x}{4}$

Figure 5.74

TRY EXERCISE 46, EXERCISE SET 5.5, PAGE 441

EXAMPLE 6 Graph the Absolute Value of the Cosine Function

Graph $y = |\cos x|$, where $0 \le x \le 2\pi$.

Solution

Because $|\cos x| \ge 0$, the graph of $y = |\cos x|$ is drawn by reflecting the negative portion of the graph of $y = \cos x$ across the x-axis. The graph is the one shown in dark blue and light blue in **Figure 5.75.**

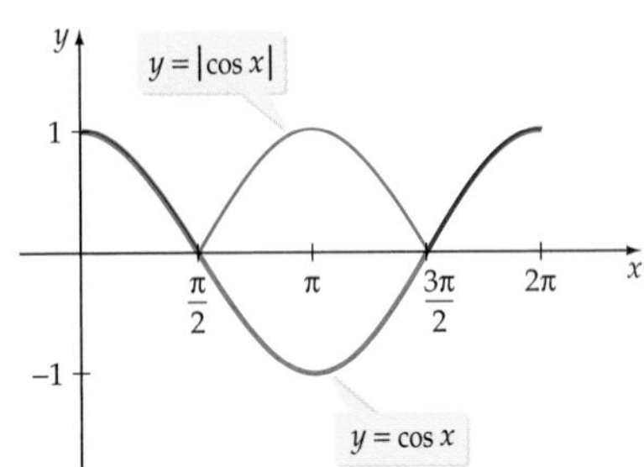

Figure 5.75

TRY EXERCISE 52, EXERCISE SET 5.5, PAGE 441

TOPICS FOR DISCUSSION

1. Is the graph of $f(x) = |\sin x|$ the same as the graph of $y = \sin|x|$? Explain.
2. Explain how the graph of $y = \cos 2x$ differs from the graph of $y = \cos x$.
3. Does the graph of $y = \sin(-2x)$ have the same period as the graph of $y = \sin 2x$? Explain.
4. The function $h(x) = a\sin bt$ has an amplitude of 3 and a period of 4. What are the possible values of a? What are the possible values of b?

EXERCISE SET 5.5

In Exercises 1 to 16, state the amplitude and period of the function defined by each equation.

1. $y = 2\sin x$
2. $y = -\dfrac{1}{2}\sin x$
3. $y = \sin 2x$
4. $y = \sin\dfrac{2x}{3}$
5. $y = \dfrac{1}{2}\sin 2\pi x$
6. $y = 2\sin\dfrac{\pi x}{3}$

7. $y = -2 \sin \frac{x}{2}$

8. $y = -\frac{1}{2} \sin \frac{x}{2}$

9. $y = \frac{1}{2} \cos x$

10. $y = -3 \cos x$

11. $y = \cos \frac{x}{4}$

12. $y = \cos 3x$

13. $y = 2 \cos \frac{\pi x}{3}$

14. $y = \frac{1}{2} \cos 2\pi x$

15. $y = -3 \cos \frac{2x}{3}$

16. $y = \frac{3}{4} \cos 4x$

In Exercises 17 to 54, graph at least one full period of the function defined by each equation.

17. $y = \frac{1}{2} \sin x$

18. $y = \frac{3}{2} \cos x$

19. $y = 3 \cos x$

20. $y = -\frac{3}{2} \sin x$

21. $y = -\frac{7}{2} \cos x$

22. $y = 3 \sin x$

23. $y = -4 \sin x$

24. $y = -5 \cos x$

25. $y = \cos 3x$

26. $y = \sin 4x$

27. $y = \sin \frac{3x}{2}$

28. $y = \cos \pi x$

29. $y = \cos \frac{\pi}{2} x$

30. $y = \sin \frac{3\pi}{4} x$

31. $y = \sin 2\pi x$

32. $y = \cos 3\pi x$

33. $y = 4 \cos \frac{x}{2}$

34. $y = 2 \cos \frac{3x}{4}$

35. $y = -2 \cos \frac{x}{3}$

36. $y = -\frac{4}{3} \cos 3x$

37. $y = 2 \sin \pi x$

38. $y = \frac{1}{2} \sin \frac{\pi x}{3}$

39. $y = \frac{3}{2} \cos \frac{\pi x}{2}$

40. $y = \cos \frac{\pi x}{3}$

41. $y = 4 \sin \frac{2\pi x}{3}$

42. $y = 3 \cos \frac{3\pi x}{2}$

43. $y = 2 \cos 2x$

44. $y = \frac{1}{2} \sin 2.5x$

45. $y = -2 \sin 1.5x$

46. $y = -\frac{3}{4} \cos 5x$

47. $y = \left|2 \sin \frac{x}{2}\right|$

48. $y = \left|\frac{1}{2} \sin 3x\right|$

49. $y = |-2 \cos 3x|$

50. $y = \left|-\frac{1}{2} \cos \frac{x}{2}\right|$

51. $y = -\left|2 \sin \frac{x}{3}\right|$

52. $y = -\left|3 \sin \frac{2x}{3}\right|$

53. $y = -|3 \cos \pi x|$

54. $y = -\left|2 \cos \frac{\pi x}{2}\right|$

In Exercises 55 to 60, find an equation of each graph.

55.

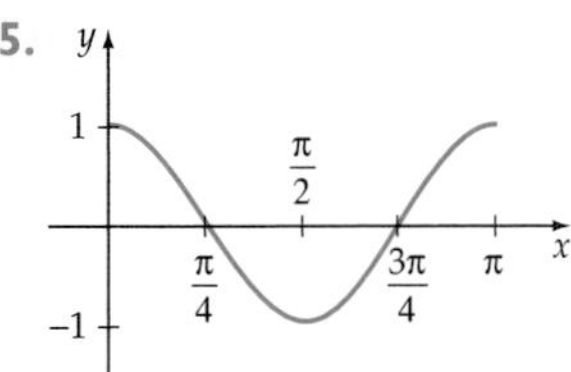

56.

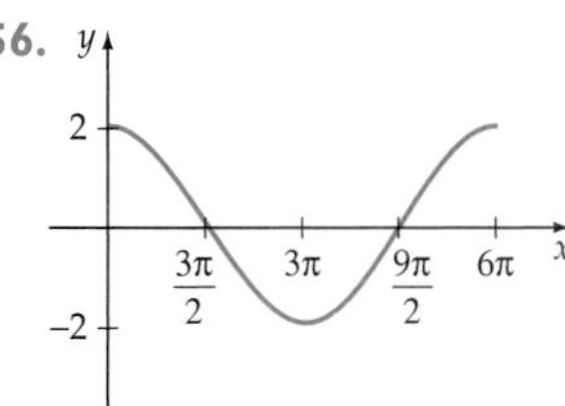

57.

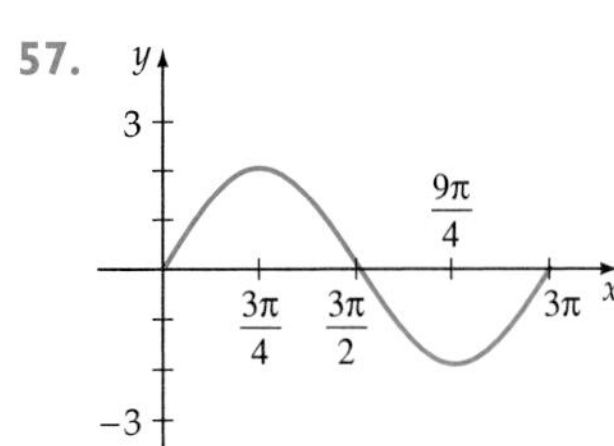

58.

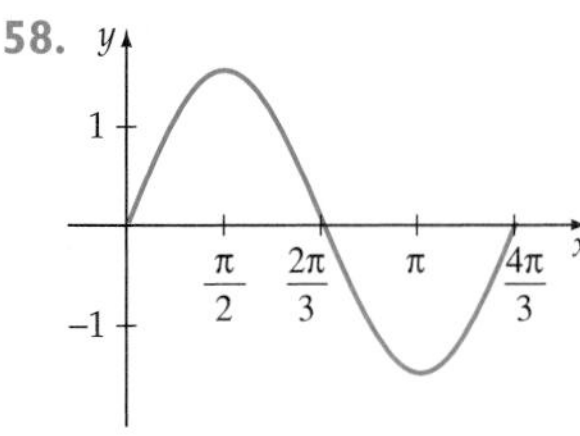

59.

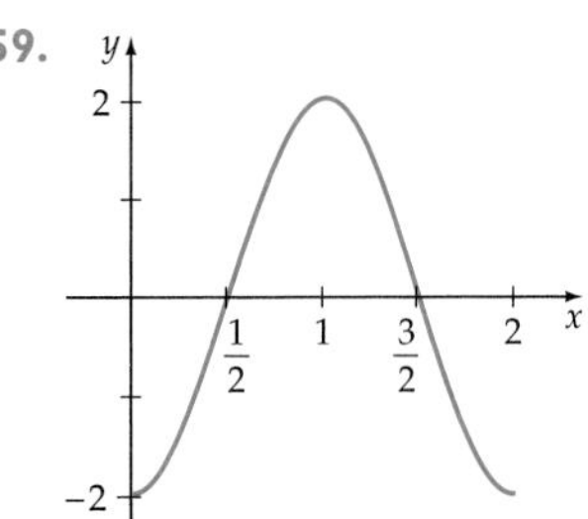

60.

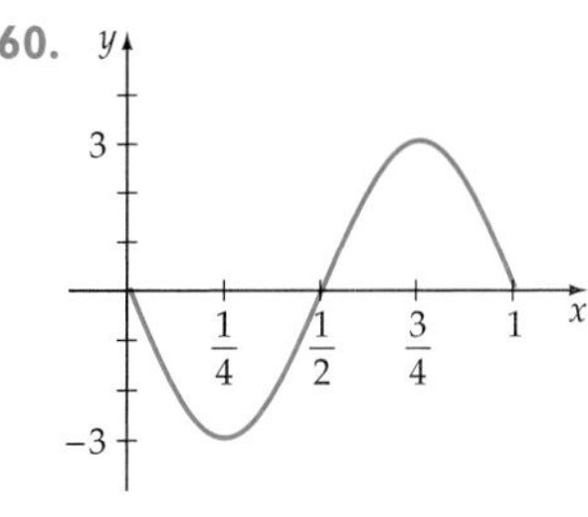

61. Sketch the graph of $y = 2 \sin \frac{2x}{3}$, $-3\pi \le x \le 6\pi$.

62. Sketch the graph of $y = -3 \cos \frac{3x}{4}$, $-2\pi \le x \le 4\pi$.

63. Sketch the graphs of

$$y_1 = 2 \cos \frac{x}{2} \quad \text{and} \quad y_2 = 2 \cos x$$

on the same set of axes for $-2\pi \le x \le 4\pi$.

64. Sketch the graphs of

$$y_1 = \sin 3\pi x \quad \text{and} \quad y_2 = \sin \frac{\pi x}{3}$$

on the same set of axes for $-2 \le x \le 4$.

In Exercises 65 to 72, use a graphing utility to graph each function.

65. $y = \cos^2 x$

66. $y = 3^{\cos^2 x} \cdot 3^{\sin^2 x}$

67. $y = \cos |x|$

68. $y = \sin |x|$

69. $y = \frac{1}{2}x \sin x$

70. $y = \frac{1}{2}x + \sin x$

71. $y = -x \cos x$

72. $y = -x + \cos x$

73. Complete the table for $f(x) = \frac{\sin x}{x}$.

x	−0.1	−0.05	−0.01	−0.001	0.001	0.01	0.05	0.1
$\frac{\sin x}{x}$								

What conclusion might you draw about the value of $f(x)$ as $x \to 0$? Does the graph of f have a vertical asymptote as $x = 0$?

74. Complete the table for $f(x) = \frac{\cos x}{x}$.

x	−0.1	−0.05	−0.01	−0.001	0.001	0.01	0.05	0.1
$\frac{\cos x}{x}$								

What conclusion might you draw about the value of $f(x)$ as $x \to 0$? Does the graph of f have a vertical asymptote at $x = 0$?

75. Graph $y = e^{\sin x}$. What is the maximum value of $e^{\sin x}$? What is the minimum value of $e^{\sin x}$? Is the function defined by $y = e^{\sin x}$ a periodic function? If so, what is the period?

76. Graph $y = e^{\cos x}$. What is the maximum value of $e^{\cos x}$? What is the minimum value of $e^{\cos x}$? Is the function defined by $y = e^{\cos x}$ a periodic function? If so, what is the period?

SUPPLEMENTAL EXERCISES

In Exercises 77 to 80, write an equation for a sine function with the given information.

77. Amplitude = 2; period = 3π

78. Amplitude = 5; period = $2\pi/3$

79. Amplitude = 4; period = 2

80. Amplitude = 2.5; period = 3.2

In Exercises 81 to 84, write an equation for a cosine function with the given information.

81. Amplitude = 3; period = $\pi/2$

82. Amplitude = 0.8; period = 4π

83. Amplitude = 3; period = 2.5

84. Amplitude = 4.2; period = 1

85. **EQUATION OF A WAVE** A tidal wave that is caused by an earthquake under the ocean is called a **tsunami.** A model of a tsunami is given by $f(t) = A \cos Bt$. Find the equation of a tsunami that has an amplitude of 60 feet and a period of 20 seconds.

86. **EQUATION OF HOUSEHOLD CURRENT** The electricity supplied to your home, called *alternating current,* can be modeled by $I = A \sin \omega t$, where I is the number of amperes of current at time t seconds. Write the equation of household current whose graph is given in the figure below. Calculate I when $t = 0.5$ second.

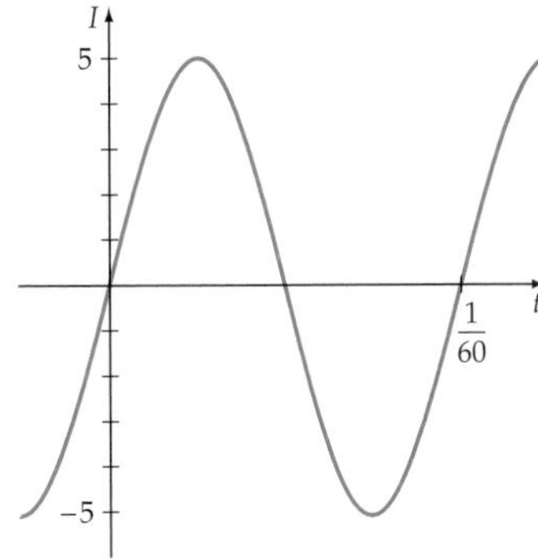

87. **EQUATION OF A SECONDARY WAVE** The secondary wave or S-wave of an earthquake can be approximated by a sine curve. Assuming that an S-wave has an amplitude of 4 feet and a period of $2\pi/3$ seconds, write the equation for the S-wave. What is the displacement of the S-wave when $t = 0.75$ second?

PROJECTS

1. **A TRIGONOMETRIC POWER FUNCTION**

a. Determine the domain and the range of $y = (\sin x)^{\cos x}$. Explain.

b. What is the amplitude of the function? Explain.

SECTION 5.6 GRAPHS OF THE OTHER TRIGONOMETRIC FUNCTIONS

- THE GRAPH OF THE TANGENT FUNCTION
- THE GRAPH OF THE COTANGENT FUNCTION
- THE GRAPH OF THE COSECANT FUNCTION
- THE GRAPH OF THE SECANT FUNCTION

◆ THE GRAPH OF THE TANGENT FUNCTION

Figure 5.76 shows the graph of $y = \tan x$ for $-\pi/2 < x < \pi/2$. The lines $x = \pi/2$ and $x = -\pi/2$ are vertical asymptotes for the graph of $y = \tan x$. From Section 5.4, the period of $y = \tan x$ is π. Therefore, the portion of the graph shown in **Figure 5.76** is repeated along the x-axis as shown in **Figure 5.77.**

Because the tangent function is unbounded, there is no amplitude for the tangent function. The graph of $y = a \tan x$ is drawn by stretching ($|a| > 1$) or shrinking ($|a| < 1$) the graph of $y = \tan x$. If $a < 0$, then the graph is reflected across the x-axis. **Figure 5.78** shows the graph of three tangent functions. Because $\tan \pi/4 = 1$, the point $(\pi/4, a)$ is convenient to plot as a guide for the graph of $y = a \tan x$.

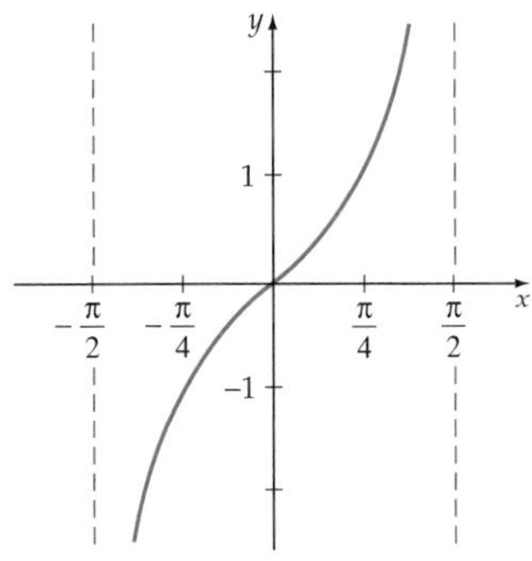

$y = \tan x, -\frac{\pi}{2} < x < \frac{\pi}{2}$

Figure 5.76

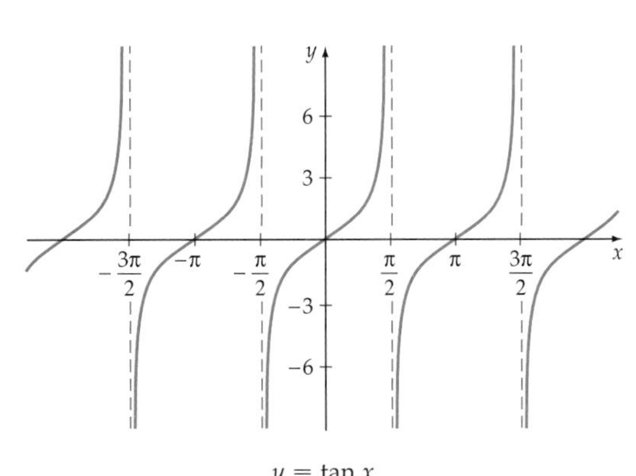

$y = \tan x$

Figure 5.77

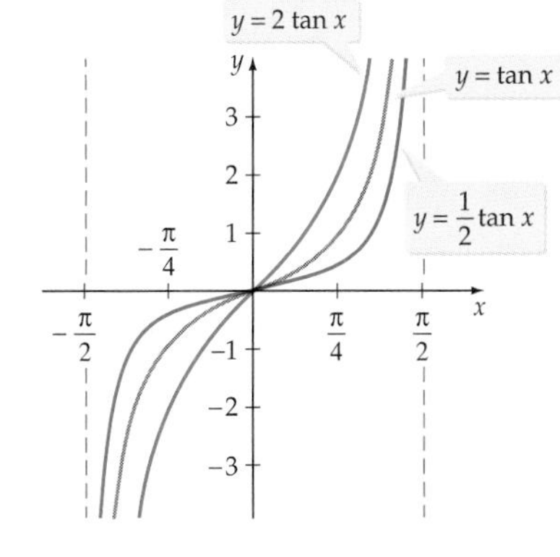

Figure 5.78

EXAMPLE 1 Graph $y = a \tan x$

Graph: $y = -3 \tan x$

Solution

The graph of $y = -3 \tan x$ is the reflection across the x-axis of the graph of $y = 3 \tan x$ as shown in **Figure 5.79.**

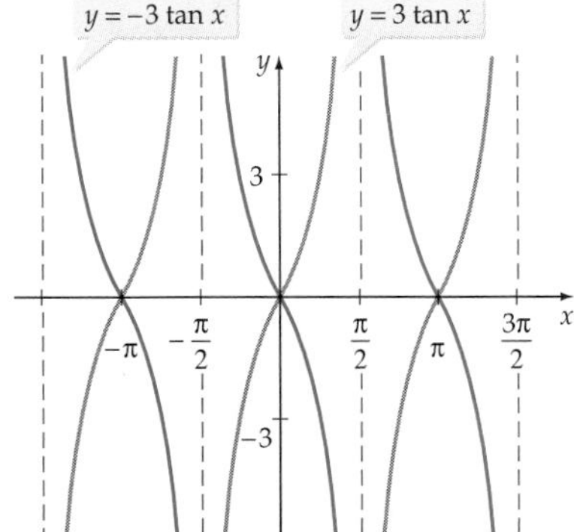

Figure 5.79

TRY EXERCISE 22, EXERCISE SET 5.6, PAGE 448

The period of $y = \tan x$ is π and the graph completes one cycle on the interval $-\pi/2 < x < \pi/2$. The period of $y = \tan bx$ $(b > 0)$ is π/b. The graph of $y = \tan bx$ completes one cycle on the interval $(-\pi/(2b), \pi/(2b))$.

Period of $y = \tan bx$

The period of $y = \tan bx$ is $\dfrac{\pi}{|b|}$.

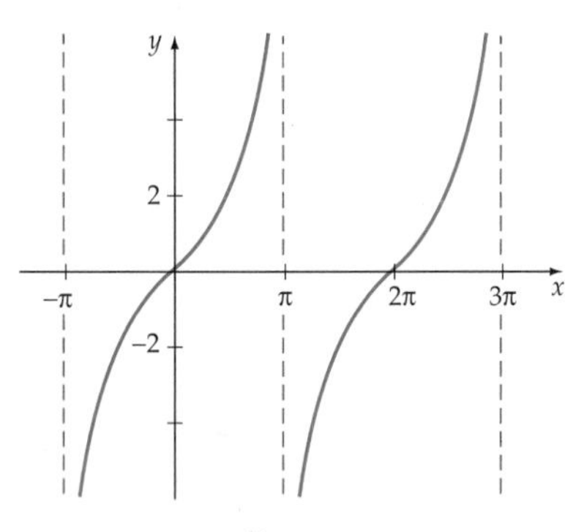

$y = 2\tan\dfrac{x}{2}, -\pi < x < 3\pi$

Figure 5.80

EXAMPLE 2 Graph $y = a \tan bx$

Graph: $y = 2\tan\dfrac{x}{2}$

Solution

Period $= \dfrac{\pi}{b} = \dfrac{\pi}{1/2} = 2\pi$. Graph one cycle for values of x such that $-\pi < x < \pi$. This curve is repeated along the x-axis as shown in **Figure 5.80.**

TRY EXERCISE 30, EXERCISE SET 5.6, PAGE 448

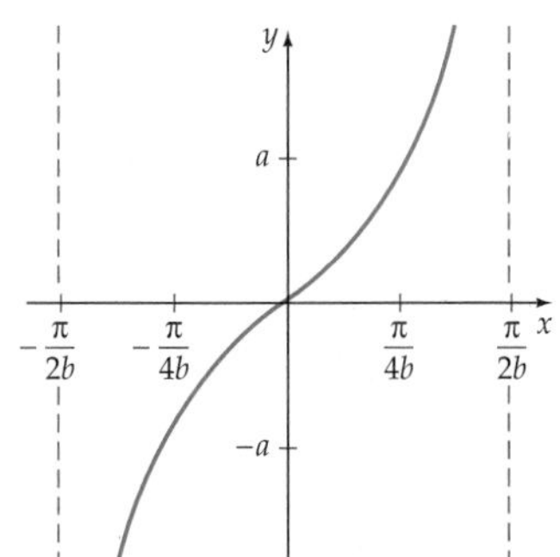

$y = a\tan bx, -\dfrac{\pi}{2b} < x < \dfrac{\pi}{2b}$

Figure 5.81

Figure 5.81 shows one cycle of the graph of $y = a \tan bx$ for both a and b positive. Note from the graph that

- The period is $\dfrac{\pi}{b}$.
- $x = 0$ is a zero.
- The graph passes through $\left(-\dfrac{\pi}{4b}, -a\right)$ and $\left(\dfrac{\pi}{4b}, a\right)$.
- If $a < 0$, the graph is reflected across the x-axis.

◆ THE GRAPH OF THE COTANGENT FUNCTION

Figure 5.82 shows the graph of $y = \cot x$ for $0 < x < \pi$. The lines $x = 0$ and $x = \pi$ are vertical asymptotes for the graph of $y = \cot x$. From Section 5.4, the period of $y = \cot x$ is π. Therefore, the graph cycle shown in **Figure 5.82** is repeated along the x-axis as shown in **Figure 5.83.** As with the graph of $y = \tan x$, the graph of $y = \cot x$ is unbounded and there is no amplitude.

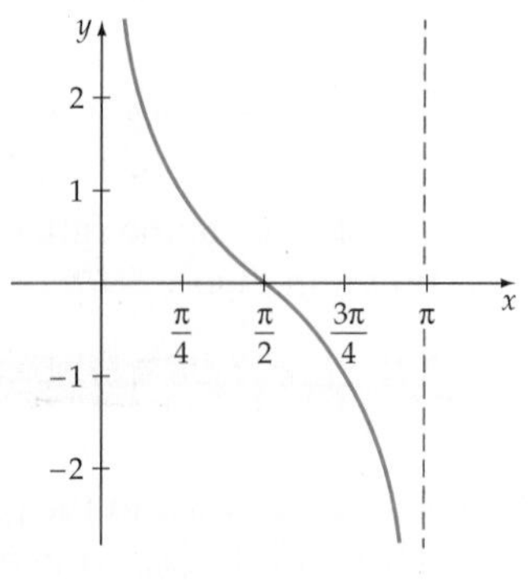

$y = \cot x, 0 < x < \pi$

Figure 5.82

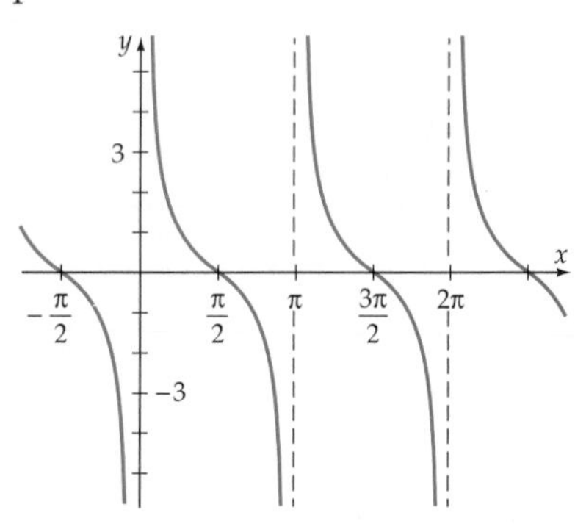

$y = \cot x$

Figure 5.83

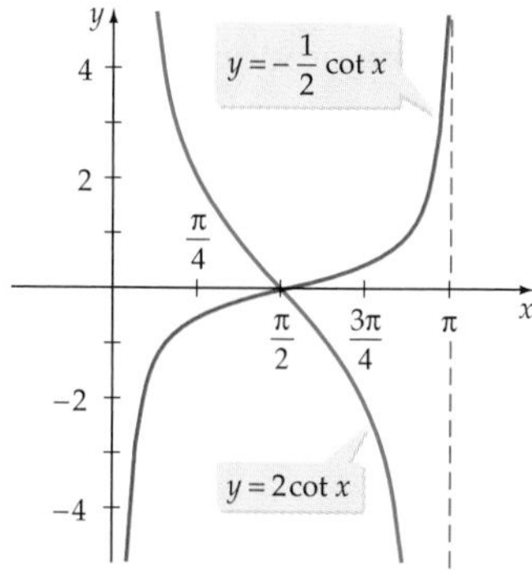

Figure 5.84

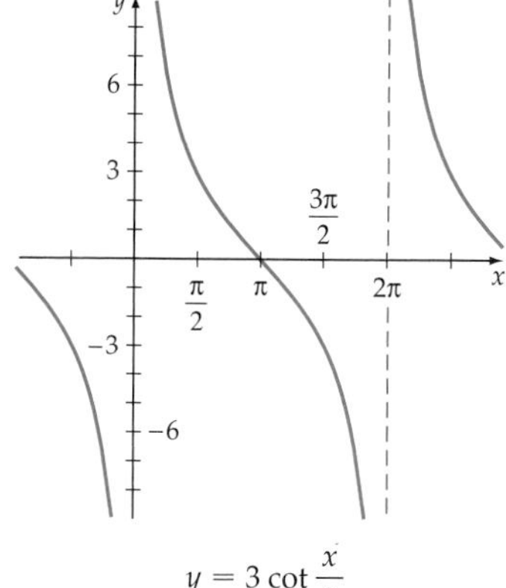

$y = 3\cot\frac{x}{2}$

Figure 5.85

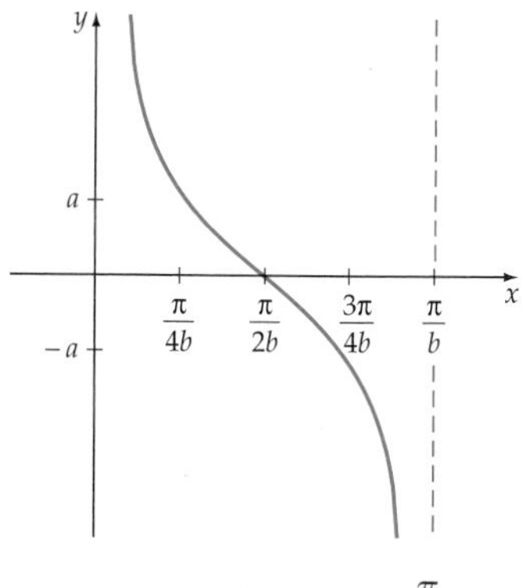

$y = a\cot bx, 0 < x < \frac{\pi}{b}$

Figure 5.86

The graph of $y = a\cot x$ is drawn by stretching ($|a| > 1$) or shrinking ($|a| < 1$) the graph of $y = \cot x$. The graph is reflected across the x-axis when $a < 0$. **Figure 5.84** shows the graphs of two cotangent functions.

The period of $y = \cot x$ is π, the period of $y = \cot bx$ is $\pi/|b|$. One cycle of the graph of $y = \cot bx$ is completed on the interval $(0, \pi/b)$.

Period of $y = \cot bx$

The period of $y = \cot bx$ is $\frac{\pi}{|b|}$.

EXAMPLE 3 Graph $y = a\cot bx$

Graph: $y = 3\cot\frac{x}{2}$

Solution

Period $= \frac{\pi}{b} = \frac{\pi}{1/2} = 2\pi$. Sketch the graph for values of x where $0 < x < 2\pi$. This curve is repeated along the x-axis as shown in **Figure 5.85.**

TRY EXERCISE 32, EXERCISE SET 5.6, PAGE 448

Figure 5.86 shows one cycle of the graph of $y = a\cot bx$ for both a and b positive. Note from the graph that

- The period is $\frac{\pi}{b}$.
- $x = \frac{\pi}{2b}$ is a zero.
- The graph passes through $\left(\frac{\pi}{4b}, a\right)$ and $\left(\frac{3\pi}{4b}, -a\right)$.
- If $a < 0$, the graph is reflected across the x-axis.

◆ THE GRAPH OF THE COSECANT FUNCTION

Because $\csc x = \frac{1}{\sin x}$, the value of $\csc x$ is the reciprocal of the value of $\sin x$. Therefore, $\csc x$ is undefined when $\sin x = 0$ or when $x = n\pi$, where n is an integer. The graph of $y = \csc x$ has vertical asymptotes at $n\pi$. Because $y = \csc x$ has period 2π, the graph will be repeated along the x-axis every 2π units. A graph of $y = \csc x$ is shown in **Figure 5.87.**

The graph of $y = \sin x$ is also shown in **Figure 5.87.** Note the relationships among the zeros of $y = \sin x$ and the asymptotes of $y = \csc x$. Also note that because $|\sin x| \le 1$, $\frac{1}{|\sin x|} \ge 1$. Thus the range of $y = \csc x$ is $|y| \ge 1$. The general procedure for graphing $y = a \csc bx$ is first to graph $y = a \sin bx$. Then sketch the graph of the cosecant function by using y values equal to the product of a and the reciprocal values of $\sin bx$.

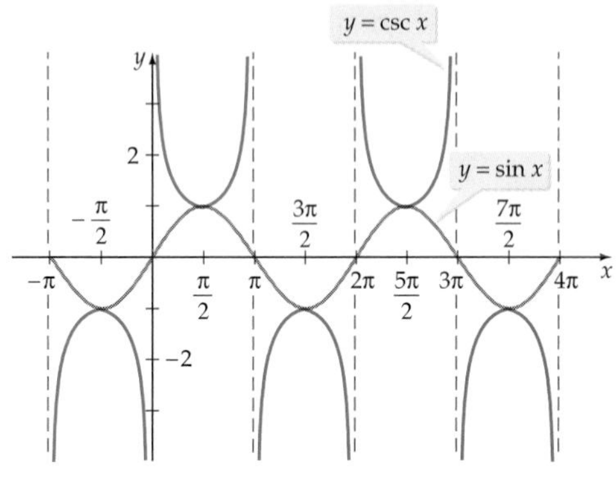

Figure 5.87

EXAMPLE 4 Graph $y = a \csc bx$

Graph: $y = 2 \csc \dfrac{\pi x}{2}$

Solution

First sketch the graph of $y = 2 \sin \dfrac{\pi x}{2}$ and draw vertical asymptotes through the zeros. Now sketch the graph of $y = 2 \csc \dfrac{\pi x}{2}$, using the asymptotes as guides for the graph as shown in **Figure 5.88.**

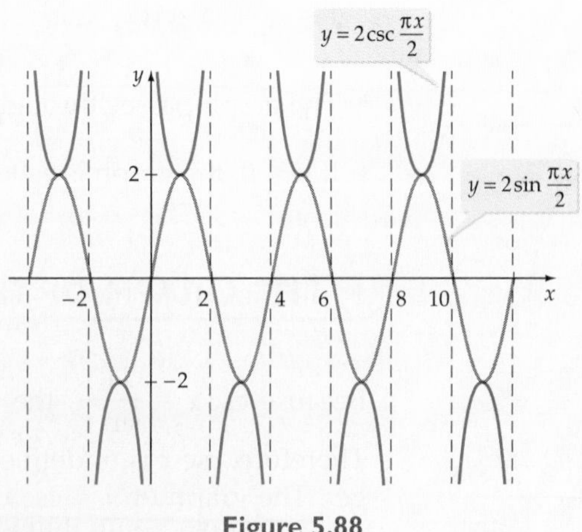

Figure 5.88

TRY EXERCISE 38, EXERCISE SET 5.6, PAGE 448

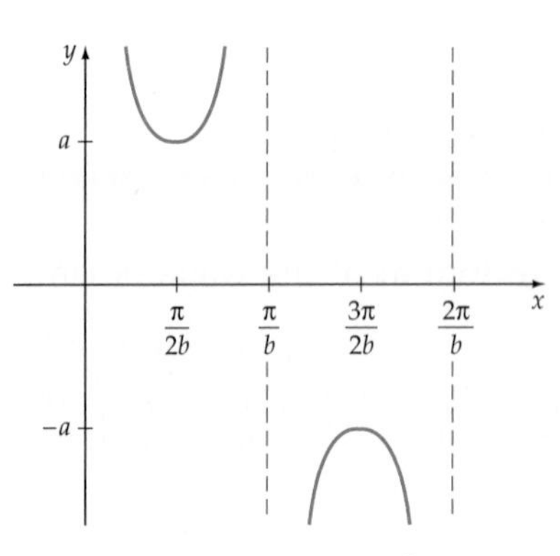

$y = a \csc bx, 0 < x < \dfrac{2\pi}{b}$

Figure 5.89

Figure 5.89 shows one cycle of the graph of $y = a \csc bx$ for both a and b positive. Note from the graph that

- The period is $\frac{2\pi}{b}$.
- The vertical asymptotes of $y = a \csc bx$ are the zeros of $y = a \sin bx$.
- The graph passes through $\left(\frac{\pi}{2b}, a\right)$ and $\left(\frac{3\pi}{2b}, -a\right)$.
- If $a < 0$, then the graph is reflected across the x-axis.

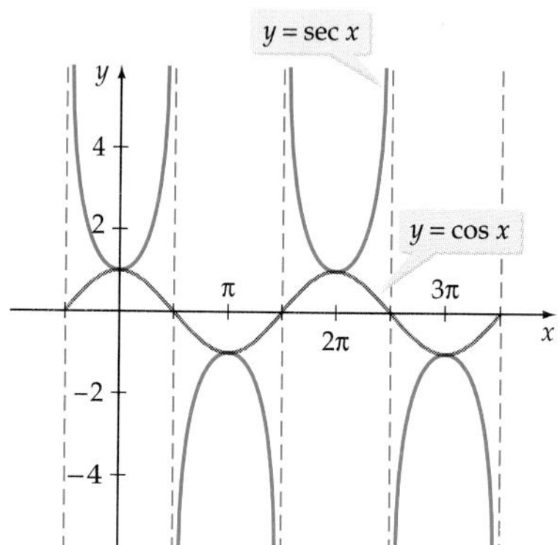

Figure 5.90

◆ THE GRAPH OF THE SECANT FUNCTION

Because $\sec x = \frac{1}{\cos x}$, the value of $\sec x$ is the reciprocal of the value of $\cos x$. Therefore, $\sec x$ is undefined when $\cos x = 0$ or when $x = \frac{\pi}{2} + n\pi$, n an integer. The graph of $y = \sec x$ has vertical asymptotes at $\frac{\pi}{2} + n\pi$. Because $y = \sec x$ has period 2π, the graph will be replicated along the x-axis every 2π units. A graph of $y = \sec x$ is shown in **Figure 5.90.**

The graph of $y = \cos x$ is also shown in **Figure 5.90.** Note the relationships among the zeros of $y = \cos x$ and the asymptotes of $y = \sec x$. Also note that because $|\cos x| \le 1$, $\frac{1}{|\cos x|} \ge 1$. Thus the range of $y = \sec x$ is $|y| \ge 1$. The general procedure for graphing $y = a \sec bx$ is first to graph $y = a \cos bx$. Then sketch the graph of the secant function by using y values equal to the product of a and the reciprocal values of $\cos bx$.

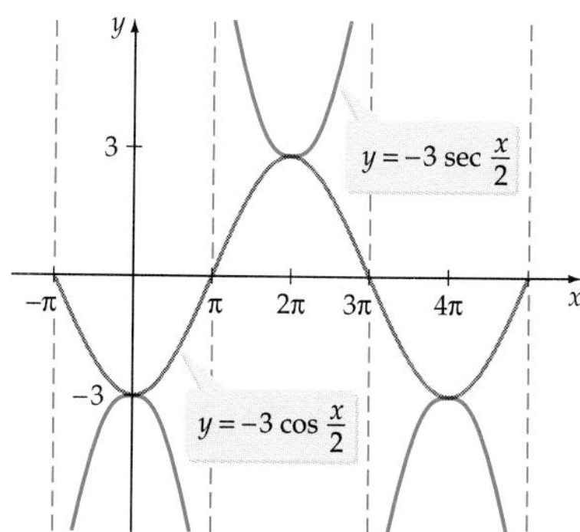

Figure 5.91

EXAMPLE 5 Graph $y = a \sec bx$

Graph: $y = -3 \sec \frac{x}{2}$

Solution

First sketch the graph of $y = -3 \cos \frac{x}{2}$ and draw vertical asymptotes through the zeros. Now sketch the graph of $y = -3 \sec \frac{x}{2}$, using the asymptotes as guides for the graph as shown in **Figure 5.91.**

TRY EXERCISE 42, EXERCISE SET 5.6, PAGE 448

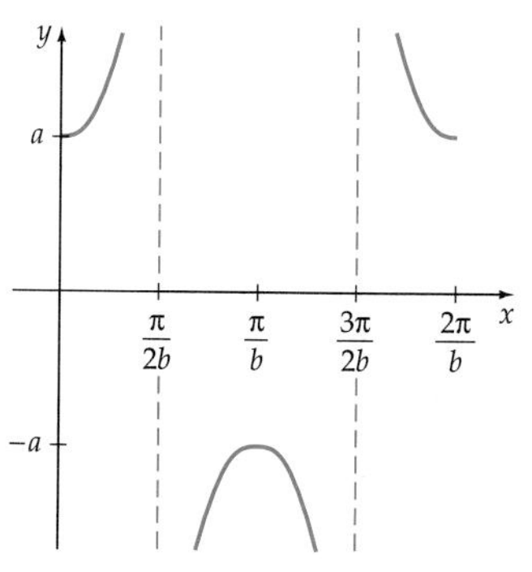

$y = a \sec bx, 0 < x < \frac{2\pi}{b}$

Figure 5.92

Figure 5.92 shows one cycle of the graph of $y = a \sec bx$ for both a and b positive. Note from the graph that

- The period is $\frac{2\pi}{b}$.
- The vertical asymptotes of $y = a \sec bx$ are the zeros of $y = a \cos bx$.

- The graph passes through $(0, a)$, $\left(\frac{\pi}{b}, -a\right)$, and $\left(\frac{2\pi}{b}, a\right)$.
- If $a < 0$, then the graph is reflected across the x-axis.

TOPICS FOR DISCUSSION

1. What are the zeros of $y = \tan x$? Explain.
2. What are the zeros of $y = \sec x$? Explain.
3. The functions $f(x) = \tan x$ and $g(x) = \tan(-x)$ both have a period of π. Do you agree?
4. What is the amplitude of the function $k(x) = 4 \cot(\pi x)$? Explain.

EXERCISE SET 5.6

1. For what values of x is $y = \tan x$ undefined?
2. For what values of x is $y = \cot x$ undefined?
3. For what values of x is $y = \sec x$ undefined?
4. For what values of x is $y = \csc x$ undefined?

In Exercises 5 to 20, state the period of each function.

5. $y = \sec x$
6. $y = \cot x$
7. $y = \tan x$
8. $y = \csc x$
9. $y = 2 \tan \frac{x}{2}$
10. $y = \frac{1}{2} \cot 2x$
11. $y = \csc 3x$
12. $y = \csc \frac{x}{2}$
13. $y = -\tan 3x$
14. $y = -3 \cot \frac{2x}{3}$
15. $y = -3 \sec \frac{x}{4}$
16. $y = -\frac{1}{2} \csc 2x$
17. $y = \cot \pi x$
18. $y = \cot \frac{\pi x}{3}$
19. $y = 2 \csc \frac{\pi x}{2}$
20. $y = -3 \cot \pi x$

In Exercises 21 to 40, sketch one full period of the graph of each function.

21. $y = 3 \tan x$
22. $y = \frac{1}{3} \tan x$
23. $y = \frac{3}{2} \cot x$
24. $y = 4 \cot x$
25. $y = 2 \sec x$
26. $y = \frac{3}{4} \sec x$
27. $y = \frac{1}{2} \csc x$
28. $y = 2 \csc x$
29. $y = 2 \tan \frac{x}{2}$
30. $y = -3 \tan 3x$
31. $y = -3 \cot \frac{x}{2}$
32. $y = \frac{1}{2} \cot 2x$
33. $y = -2 \csc \frac{x}{3}$
34. $y = \frac{3}{2} \csc 3x$
35. $y = \frac{1}{2} \sec 2x$
36. $y = -3 \sec \frac{2x}{3}$
37. $y = -2 \sec \pi x$
38. $y = 3 \csc \frac{\pi x}{2}$
39. $y = 3 \tan 2\pi x$
40. $y = -\frac{1}{2} \cot \frac{\pi x}{2}$
41. Graph $y = 2 \csc 3x$ from -2π to 2π.
42. Graph $y = \sec \frac{x}{2}$ from -4π to 4π.
43. Graph $y = 3 \sec \pi x$ from -2 to 4.
44. Graph $y = \csc \frac{\pi x}{2}$ from -4 to 4.

45. Graph $y = 2 \cot 2x$ from $-\pi$ to π.

46. Graph $y = \dfrac{1}{2} \tan \dfrac{x}{2}$ from -4π to 4π.

47. Graph $y = 3 \tan \pi x$ from -2 to 2.

48. Graph $y = \cot \dfrac{\pi x}{2}$ from -4 to 4.

In Exercises 49 to 54, find an equation of each blue graph.

49.

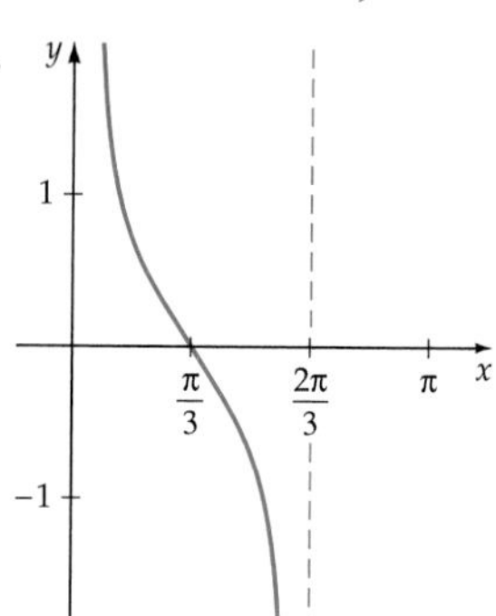

50.

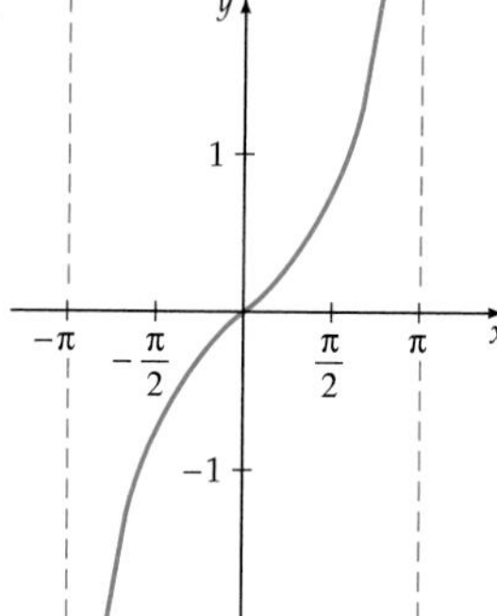

51.

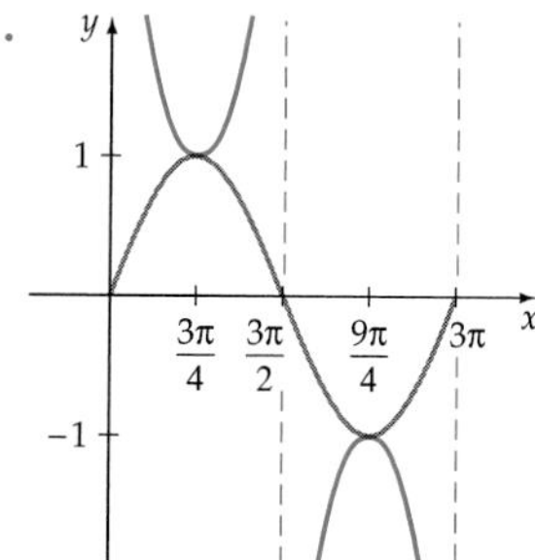

52.

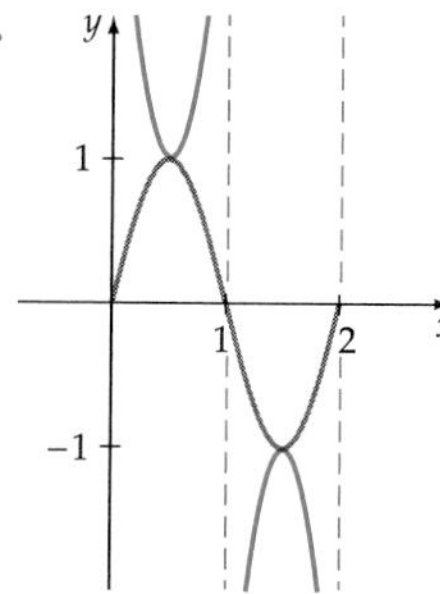

53.

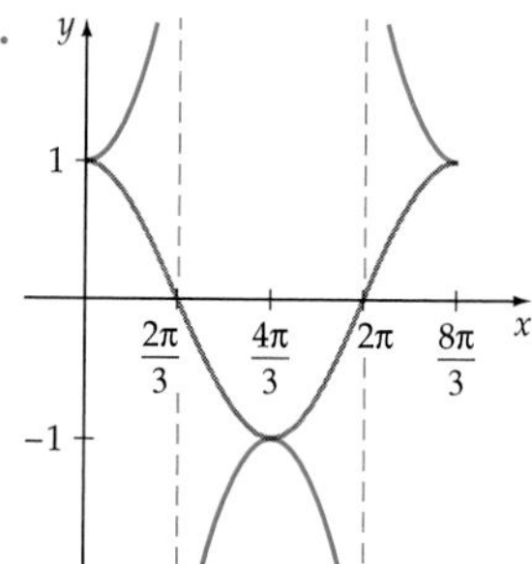

54.

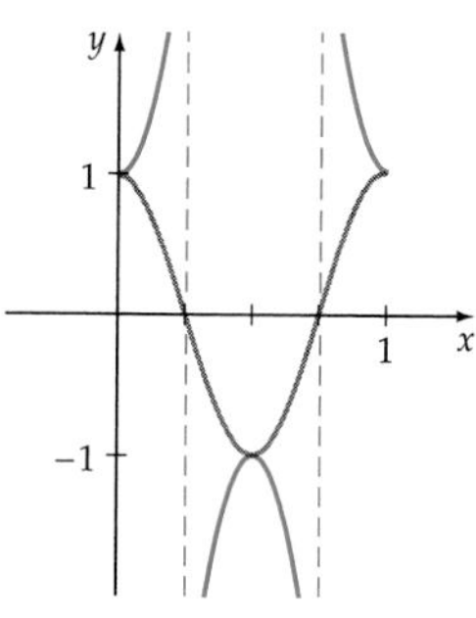

In Exercises 55 to 60, use a graphing utility to graph each equation. If needed, use open circles so that your graph is accurate.

55. $y = \tan |x|$

56. $y = \sec |x|$

57. $y = |\csc x|$

58. $y = |\cot x|$

59. $y = \tan x \cos x$

60. $y = \cot x \sin x$

61. Graph $y = \tan x$ and $x = \tan y$ on the same coordinate axes.

62. Graph $y = \sin x$ and $x = \sin y$ on the same coordinate axes.

SUPPLEMENTAL EXERCISES

In Exercises 63 to 70, write an equation that is of the form $y = \tan bx$, $y = \cot bx$, $y = \sec bx$, or $y = \csc bx$ and satisfies the given conditions.

63. Tangent, period: $\pi/3$

64. Cotangent, period: $\pi/2$

65. Secant, period: $3\pi/4$

66. Cosecant, period: $5\pi/2$

67. Cotangent, period: 2

68. Tangent, period: 0.5

69. Cosecant, period: 1.5

70. Secant, period: 3

PROJECTS

1. **A TECHNOLOGY QUESTION** A student's calculator shows the display at the right. Note that the domain values 4.7123 and 4.7124 are close together, but the range values are over 100,000 units apart. Explain how this is possible.

```
tan 4.7123
                11238.43194
tan 4.7124
               -90747.26955
```

2. **SOLUTIONS OF A TRIGONOMETRIC EQUATION** How many solutions does $\tan(1/x) = 0$ have on the interval $-1 \le x \le 1$? Explain.

SECTION

5.7 GRAPHING TECHNIQUES

- TRANSLATION OF TRIGONOMETRIC FUNCTIONS
- ADDITION OF ORDINATES
- DAMPING FACTOR

◆ TRANSLATION OF TRIGONOMETRIC FUNCTIONS

Recall that the graph of $y = f(x) \pm c$ is a *vertical translation* of the graph of $y = f(x)$. For $c > 0$, the graph of $y = f(x) - c$ is shifted c units down; the graph of $y = f(x) + c$ is shifted c units up. The graph in **Figure 5.93** is a graph of the equation $y = 2 \sin \pi x - 3$, which is a vertical translation of $y = 2 \sin \pi x$ down 3 units. Note that subtracting 3 from $2 \sin \pi x$ changes neither its amplitude nor its period.

Also, the graph of $y = f(x \pm c)$ is a *horizontal translation* of the graph of $y = f(x)$. For $c > 0$, the graph of $y = f(x - c)$ is shifted c units to the right; the graph of $y = f(x + c)$ is shifted c units to the left. The graph in **Figure 5.94** is a graph of the equation $y = 2 \sin\left(x - \frac{\pi}{4}\right)$, which is the graph of $y = 2 \sin x$ translated $\pi/4$ units to the right. Note that neither the period nor the amplitude was affected. The horizontal shift of the graph of a trigonometric function is called its **phase shift.**

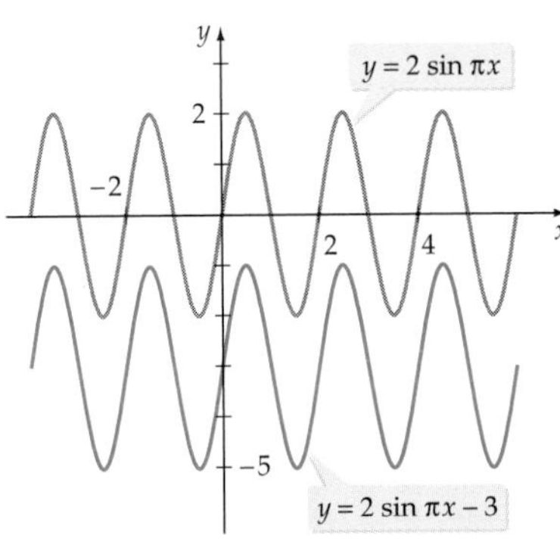

Figure 5.93

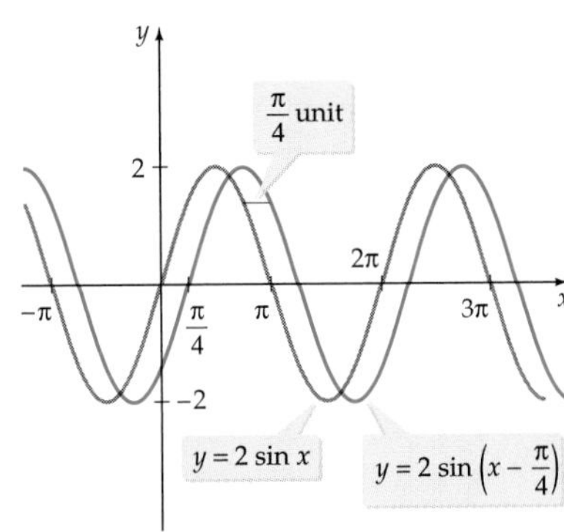

Figure 5.94

Because one cycle of $y = a \sin x$ is completed for $0 \le x \le 2\pi$, one cycle of the graph of $y = a \sin(bx + c)$, where $b > 0$, is completed for $0 \le bx + c \le 2\pi$. Solving this inequality for x, we have

$$0 \le bx + c \le 2\pi$$
$$-c \le bx \le -c + 2\pi$$
$$-\frac{c}{b} \le x \le -\frac{c}{b} + \frac{2\pi}{b}$$

The number $-\frac{c}{b}$ is the phase shift for $y = a \sin(bx + c)$. The graph of the equation $y = a \sin(bx + c)$ is the graph of $y = a \sin bx$ shifted $-\frac{c}{b}$ units horizontally. Similar arguments apply to the remaining trigonometric functions.

take note

A web applet is available to explore the graphs of sine and cosine functions. This applet, Shifting and Scaling Sine and Cosine Curves, can be found on our web site at

http://college.hmco.com

The Graphs of $y = a \sin (bx + c)$ and $y = a \cos (bx + c)$

The graphs of $y = a \sin (bx + c)$ and $y = a \cos (bx + c)$, with $b > 0$ have

$$\text{Amplitude: } |a| \qquad \text{Period: } \frac{2\pi}{b} \qquad \text{Phase shift: } -\frac{c}{b}$$

One cycle of each graph is completed on the interval

$$-\frac{c}{b} \le x \le -\frac{c}{b} + \frac{2\pi}{b}$$

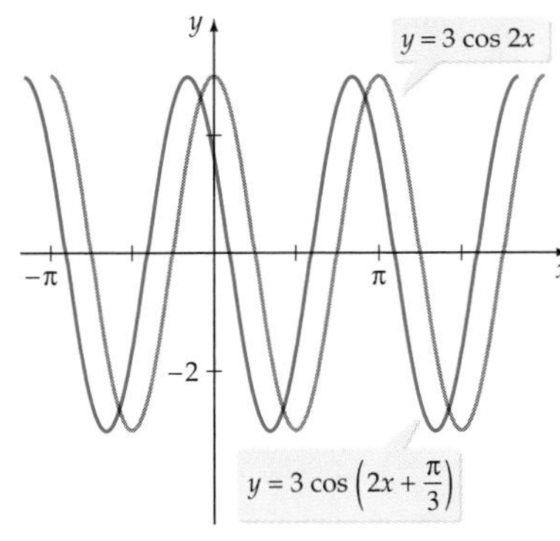

Figure 5.95

EXAMPLE 1 Graph $y = a \cos (bx + c)$

Graph: $y = 3 \cos \left(2x + \frac{\pi}{3}\right)$

Solution

The phase shift is $-\frac{c}{b} = -\frac{\pi/3}{2} = -\frac{\pi}{6}$. The graph of the equation $y = 3 \cos \left(2x + \frac{\pi}{3}\right)$ is the graph of $y = 3 \cos 2x$ shifted $\frac{\pi}{6}$ units to the left as shown in **Figure 5.95**.

TRY EXERCISE 20, EXERCISE SET 5.7, PAGE 456

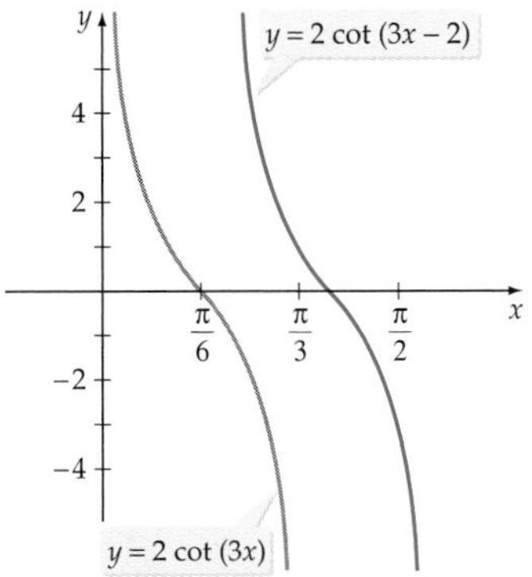

Figure 5.96

EXAMPLE 2 Graph $y = a \cot (bx + c)$

Graph: $y = 2 \cot (3x - 2)$

Solution

The phase shift is

$$-\frac{c}{b} = -\frac{-2}{3} = \frac{2}{3} \qquad \bullet\ 3x - 2 = 3x + (-2)$$

The graph of $y = 2 \cot (3x - 2)$ is the graph of $y = 2 \cot (3x)$ shifted 2/3 unit to the right as shown in **Figure 5.96**.

TRY EXERCISE 22, EXERCISE SET 5.7, PAGE 456

The graph of a trigonometric function may be the combination of a vertical translation and a phase shift.

EXAMPLE 3 Graph $y = a \sin (bx + c) + d$

Graph: $y = \frac{1}{2} \sin \left(x - \frac{\pi}{4}\right) - 2$

Continued

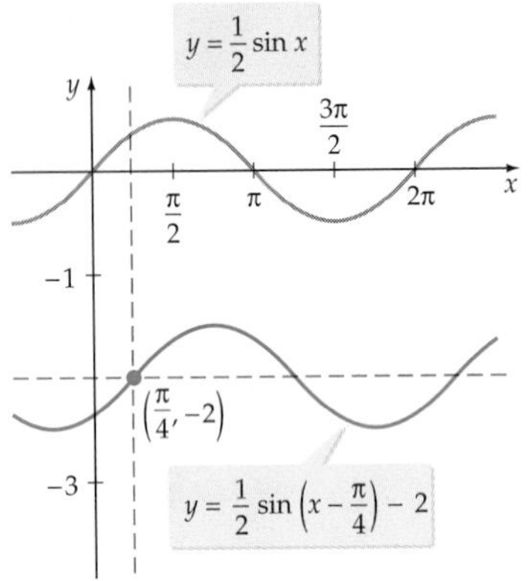

Figure 5.97

Solution

The phase shift is $-\frac{c}{b} = -\frac{-\pi/4}{1} = \frac{\pi}{4}$. The vertical shift is 2 units down.

The graph of $y = \frac{1}{2}\sin\left(x - \frac{\pi}{4}\right) - 2$ is the graph of $y = \frac{1}{2}\sin x$ shifted $\pi/4$ units to the right and 2 units down as shown in **Figure 5.97.**

TRY EXERCISE 40, EXERCISE SET 5.7, PAGE 456

EXAMPLE 4 Graph $y = a\cos(bx + c) + d$

Graph: $y = -2\cos\left(\pi x + \frac{\pi}{2}\right) + 1$

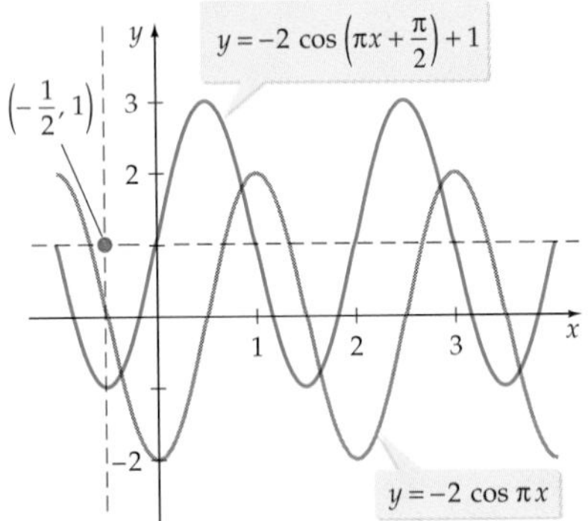

Figure 5.98

Solution

The phase shift is $-\frac{c}{b} = -\frac{\pi/2}{\pi} = -\frac{1}{2}$. The vertical shift is 1 unit up. The graph of $y = -2\cos\left(\pi x + \frac{\pi}{2}\right) + 1$ is the graph of $y = -2\cos \pi x$ shifted 1/2 unit to the left and 1 unit up as shown in **Figure 5.98.**

TRY EXERCISE 42, EXERCISE SET 5.7, PAGE 456

The following example involves a function of the form $y = \cos(bx + c) + d$.

EXAMPLE 5 A Mathematical Model of a Patient's Blood Pressure

The function $bp(t) = 32\cos\left(\frac{10\pi}{3}t - \frac{\pi}{3}\right) + 112, 0 \le t \le 20$ gives the blood pressure, in millimeters of mercury (mm Hg), of a patient during a 20-second interval.

a. Find the phase shift and the period of *bp*.

b. Graph one period of *bp*.

c. What are the patient's maximum (*systolic*) and minimum (*diastolic*) blood pressure during the given time interval?

d. What is the patient's pulse rate, in beats per minute?

Solution

a. Phase shift $= -\frac{c}{b} = -\frac{\left(-\frac{\pi}{3}\right)}{\left(\frac{10\pi}{3}\right)} = 0.1$

$$\text{Period} = \frac{2\pi}{b} = \frac{2\pi}{\left(\frac{10\pi}{3}\right)} = 0.6 \text{ second}$$

b. The graph of bp is the graph of $y_1 = 32\cos\left(\frac{10\pi}{3}t\right)$ shifted 0.1 unit to the right, shown by $y_2 = 32\cos\left(\frac{10\pi}{3}t - \frac{\pi}{3}\right)$ in **Figure 5.99**, and upward 112 units.

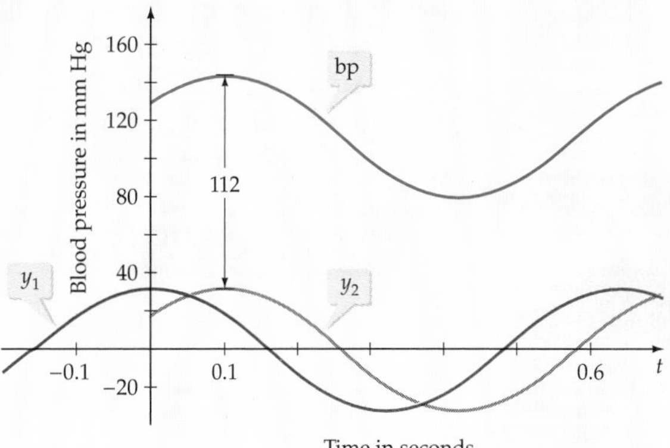

Figure 5.99

c. The function $y_2 = 32\cos\left(\frac{10\pi}{3}t - \frac{\pi}{3}\right)$ has a maximum of 32 and a minimum of −32. Thus the patient's maximum blood pressure is $32 + 112 = 144$ mm Hg and the patient's minimum blood pressure is $-32 + 112 = 80$ mm Hg.

d. From part **a** we know that the patient has 1 heartbeat every 0.6 second. Therefore, during the given time interval, the patient has a pulse rate of

$$\left(\frac{1 \text{ heartbeat}}{0.6 \text{ second}}\right)\left(\frac{60 \text{ seconds}}{1 \text{ minute}}\right) = 100 \text{ heartbeats per minute}$$

TRY EXERCISE 52, EXERCISE SET 5.7, PAGE 457

Translation techniques can also be used to graph secant and cosecant functions.

EXAMPLE 6 Graph a Cosecant Function

Graph: $y = 2\csc(2x - \pi)$

Solution

The phase shift is $-\frac{c}{b} = -\frac{-\pi}{2} = \frac{\pi}{2}$. The graph of the equation $y = 2\csc(2x - \pi)$ is the graph of $y = 2\csc 2x$ shifted $\pi/2$ units to the

Continued

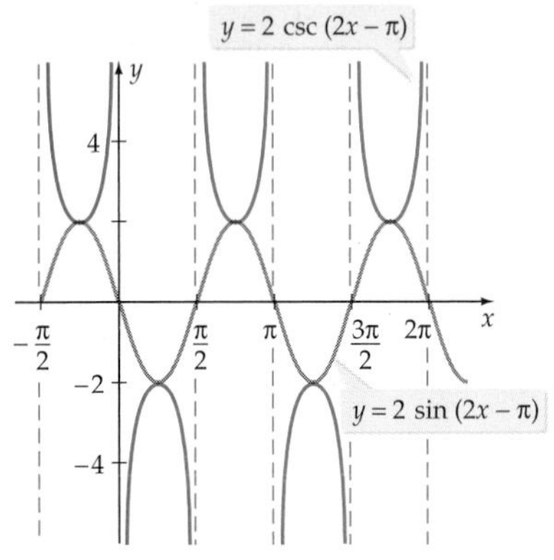

Figure 5.100

right. Sketch the graph of the equation $y = 2\sin 2x$ shifted $\pi/2$ units to the right. Use this graph to draw the graph of $y = 2\csc(2x - \pi)$ as shown in **Figure 5.100**.

TRY EXERCISE 48, EXERCISE SET 5.7, PAGE 456

◆ ADDITION OF ORDINATES

Given two functions g and h, the sum of the functions is the function f defined by $f(x) = g(x) + h(x)$. The graph of the sum f can be obtained by graphing g and h separately and then geometrically adding the y-coordinates of each function for a given value of x. It is convenient, when we are drawing the graph of the sum of two functions, to pick zeros of the function.

EXAMPLE 7 Graph the Sum of Two Functions

Graph: $y = x + \cos x$

Solution

Graph $g(x) = x$ and $h(x) = \cos x$ on the same coordinate grid. Then add the y-coordinates geometrically point by point. **Figure 5.101** shows the results of adding, by using a ruler, the y-coordinates of the two functions for selected values of x.

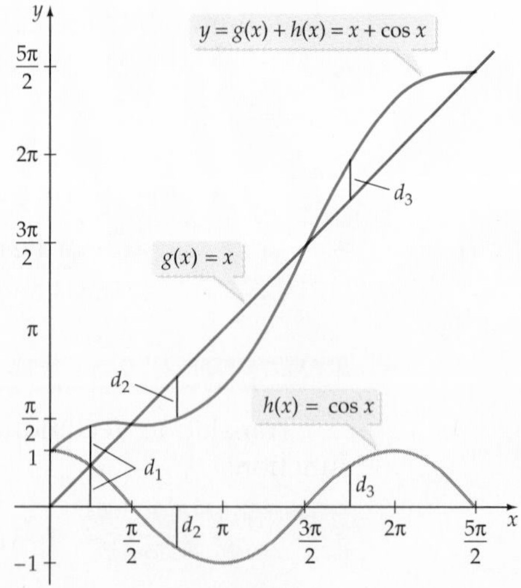

Figure 5.101

TRY EXERCISE 54, EXERCISE SET 5.7, PAGE 457

MATH MATTERS

In 1798 the mathematician Joseph Fourier served as Napoleon's scientific adviser during the French invasion of Egypt. In Egypt he was in charge of many archaeological explorations and served as the secretary of the Cairo Institute.

After returning to France, Fourier began a study of the conduction of heat in solid objects. In 1822 he published his famous work *The Analytical Theory of Heat.* In this paper, Fourier represented many functions that occur in physical applications by a series of sine and cosine functions. Today these series are known as Fourier series, and they are often used in physics and advanced mathematics.

EXAMPLE 8 Graph the Difference of Two Functions

Graph: $y = \sin x - \cos x$ for $0 \le x \le 2\pi$.

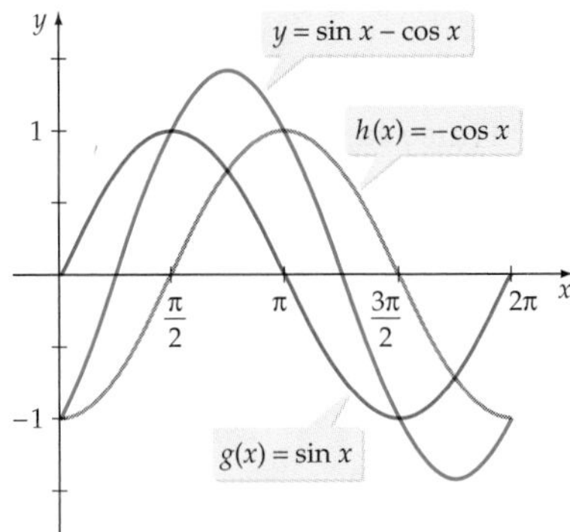

Figure 5.102

Solution

Graph $g(x) = \sin x$ and $h(x) = -\cos x$ on the same coordinate grid. For selected values of x, add $g(x)$ and $h(x)$ geometrically. Now draw a smooth curve through the points. See **Figure 5.102.**

TRY EXERCISE 58, EXERCISE SET 5.7, PAGE 457

◆ DAMPING FACTOR

The factor $\frac{1}{4}x$ in $f(x) = \frac{1}{4}x \cos x$ is referred to as the *damping factor.* In the next example, we analyze the role of the damping factor.

EXAMPLE 9 Graph the Product of Two Functions

Use a graphing utility to graph $f(x) = \frac{1}{4}x \cos x$, $x \geq 0$, and analyze the role of the damping factor.

Solution

Figure 5.103 shows that the graph of f intersects

- the graph of $y = \frac{1}{4}x$ for $x = 0, 2\pi, 4\pi, \ldots$ • **Because $\cos x = 1$ for $x = 2n\pi$**
- the graph of $y = -\frac{1}{4}x$ for $x = \pi, 3\pi, 5\pi, \ldots$ • **Because $\cos x = -1$ for $x = (2n - 1)\pi$**
- the x-axis for $x = \frac{1}{2}\pi, \frac{3}{2}\pi, \frac{5}{2}\pi, \ldots$ • **Because $\cos x = 0$ for $x = \frac{2n-1}{2}\pi$**

Figure 5.103 also shows that the graph of f lies on or between the lines

$$y = \frac{1}{4}x \quad \text{and} \quad y = -\frac{1}{4}x$$ • **Because $|\cos x| \leq 1$ for all x**

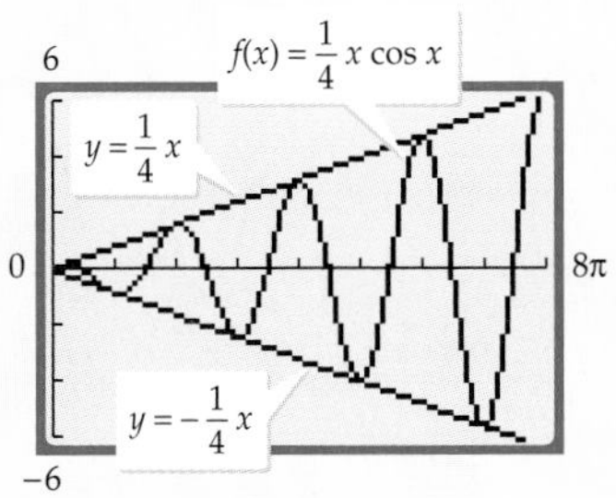

Figure 5.103

TRY EXERCISE 78, EXERCISE SET 5.7, PAGE 459

take note

Replacing the damping factor can make a dramatic change in the graph of a function. For instance, the graph of $f(x) = 2^{-0.1x} \cos x$ approaches 0 as x approaches ∞.

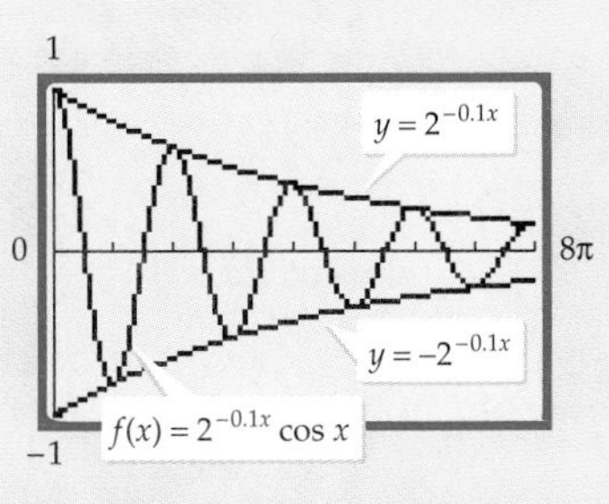

TOPICS FOR DISCUSSION

1. The maximum value of $f(x) = (3 \sin x) + 4$ is 7. Thus f has an amplitude of 7. Do you agree? Explain.
2. The graph of $y = \sec x$ has a period of 2π. What is the period of the graph of $y = |\sec x|$?
3. The zeros of $y = \sin x$ are the same as the zeros of $y = x \sin x$. Do you agree? Explain.
4. What is the phase shift of the graph of $y = \tan\left(3x - \frac{\pi}{6}\right)$?

EXERCISE SET 5.7

In Exercises 1 to 8, find the amplitude, phase shift, and period for the graph of each function.

1. $y = 2 \sin\left(x - \frac{\pi}{2}\right)$
2. $y = -3 \sin(x + \pi)$
3. $y = \cos\left(2x - \frac{\pi}{4}\right)$
4. $y = \frac{3}{4} \cos\left(\frac{x}{2} + \frac{\pi}{3}\right)$
5. $y = -4 \sin\left(\frac{2x}{3} + \frac{\pi}{6}\right)$
6. $y = \frac{3}{2} \sin\left(\frac{x}{4} - \frac{3\pi}{4}\right)$
7. $y = \frac{5}{4} \cos(3x - 2\pi)$
8. $y = 6 \cos\left(\frac{x}{3} - \frac{\pi}{6}\right)$

In Exercises 9 to 16, find the phase shift and the period for the graph of each function.

9. $y = 2 \tan\left(2x - \frac{\pi}{4}\right)$
10. $y = \frac{1}{2} \tan\left(\frac{x}{2} - \pi\right)$
11. $y = -3 \csc\left(\frac{x}{3} + \pi\right)$
12. $y = -4 \csc\left(3x - \frac{\pi}{6}\right)$
13. $y = 2 \sec\left(2x - \frac{\pi}{8}\right)$
14. $y = 3 \sec\left(\frac{x}{4} - \frac{\pi}{2}\right)$
15. $y = -3 \cot\left(\frac{x}{4} + 3\pi\right)$
16. $y = \frac{3}{2} \cot\left(2x - \frac{\pi}{4}\right)$

In Exercises 17 to 32, graph one full period of each function.

17. $y = \sin\left(x - \frac{\pi}{2}\right)$
18. $y = \sin\left(x + \frac{\pi}{6}\right)$
19. $y = \cos\left(\frac{x}{2} + \frac{\pi}{3}\right)$
20. $y = \cos\left(2x - \frac{\pi}{3}\right)$
21. $y = \tan\left(x + \frac{\pi}{4}\right)$
22. $y = \tan(x - \pi)$
23. $y = 2 \cot\left(\frac{x}{2} - \frac{\pi}{8}\right)$
24. $y = \frac{3}{2} \cot\left(3x + \frac{\pi}{4}\right)$
25. $y = \sec\left(x + \frac{\pi}{4}\right)$
26. $y = \csc(2x + \pi)$
27. $y = \csc\left(\frac{x}{3} - \frac{\pi}{2}\right)$
28. $y = \sec\left(2x + \frac{\pi}{6}\right)$
29. $y = -2 \sin\left(\frac{x}{3} - \frac{2\pi}{3}\right)$
30. $y = -\frac{3}{2} \sin\left(2x + \frac{\pi}{4}\right)$
31. $y = -3 \cos\left(3x + \frac{\pi}{4}\right)$
32. $y = -4 \cos\left(\frac{3x}{2} + 2\pi\right)$

In Exercises 33 to 50, graph each function using translations.

33. $y = \sin x + 1$
34. $y = -\sin x + 1$
35. $y = -\cos x - 2$
36. $y = 2 \sin x + 3$
37. $y = \sin 2x - 2$
38. $y = -\cos \frac{x}{2} + 2$
39. $y = 4 \cos(\pi x - 2) + 1$
40. $y = 2 \sin\left(\frac{\pi x}{2} + 1\right) - 2$
41. $y = -\sin(\pi x + 1) - 2$
42. $y = -3 \cos(2\pi x - 3) + 1$
43. $y = \sin\left(x - \frac{\pi}{2}\right) - \frac{1}{2}$
44. $y = -2 \cos\left(x + \frac{\pi}{3}\right) + 3$
45. $y = \tan \frac{x}{2} - 4$
46. $y = \cot 2x + 3$
47. $y = \sec 2x - 2$
48. $y = \csc \frac{x}{3} + 4$
49. $y = \csc \frac{x}{2} - 1$
50. $y = \sec\left(x - \frac{\pi}{2}\right) + 1$

51. **RETAIL SALES** The manager of a major department store finds that the number of men's suits S, in hundreds, that it sells is given by

$$S = 4.1 \cos\left(\frac{\pi}{6}t - 1.25\pi\right) + 7$$

where t is time measured in months, with $t = 0$ representing January 1.

a. Find the phase shift and the period of S.

b. Graph one period of S.

c. Use the graph to determine in which month the store sells the most suits.

52. **RETAIL SALES** The owner of a shoe store finds that the number of pairs of shoes S, in hundreds, that it sells can be modeled by the function

$$S = 2.7 \cos\left(\frac{\pi}{6}t - \frac{7}{12}\pi\right) + 4$$

where t is time measured in months, with $t = 0$ representing January 1.

a. Find the phase shift and the period of S.

b. Graph one period of S.

c. Use the graph to determine in which month the store sells the most shoes.

In Exercises 53 to 58, graph the given function by using the addition-of-ordinates method.

53. $y = x - \sin x$

54. $y = \dfrac{x}{2} + \cos x$

55. $y = x + \sin 2x$

56. $y = \dfrac{2x}{3} - \sin x$

57. $y = \sin x + \cos x$

58. $y = -\sin x + \cos x$

In Exercises 59 to 64, find an equation of each blue graph.

59.

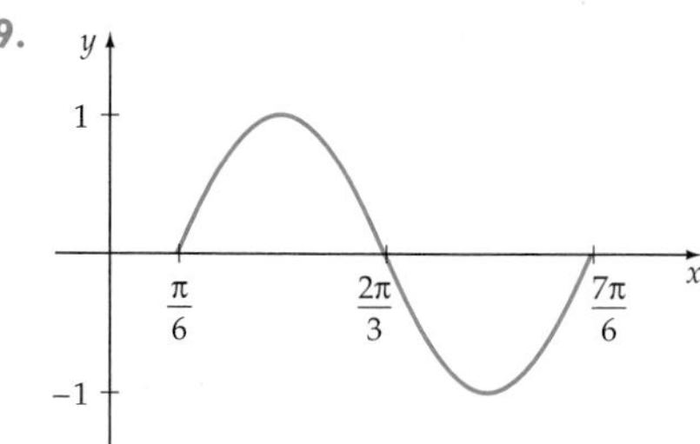

60.

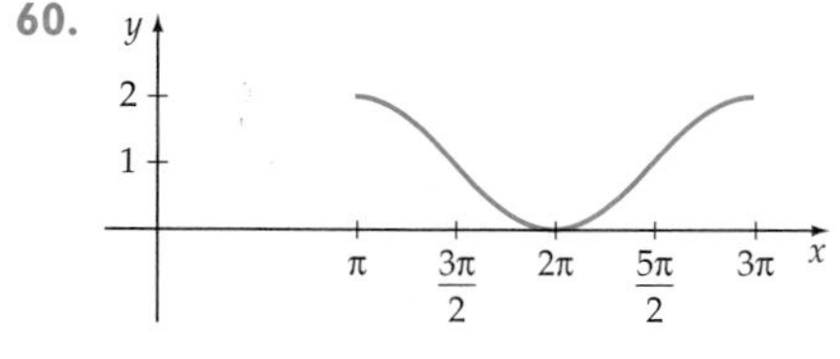

61.

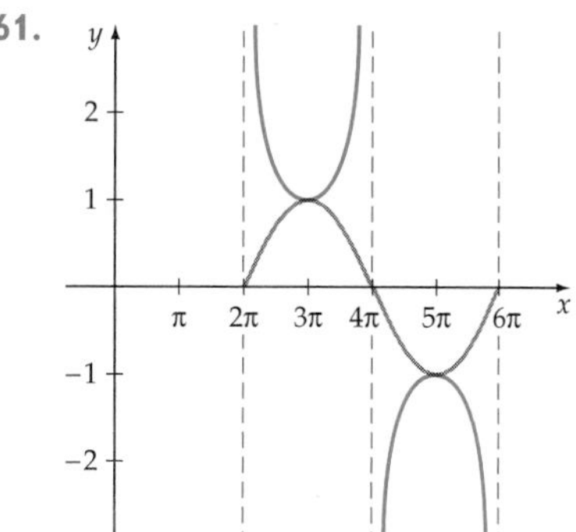

62.

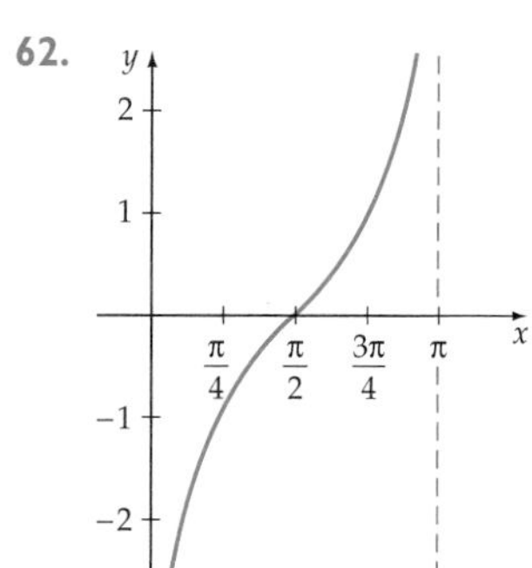

63.

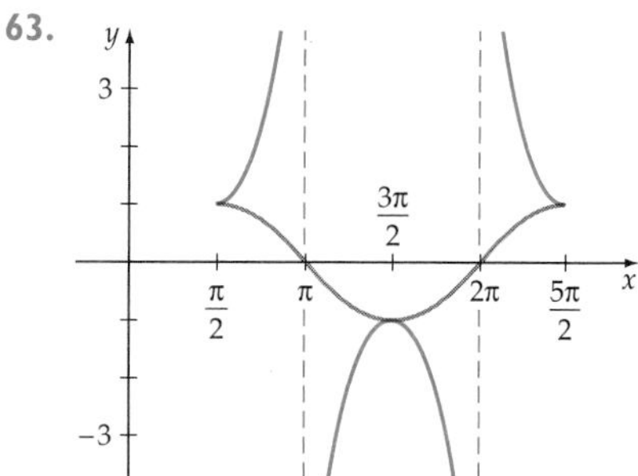

64.

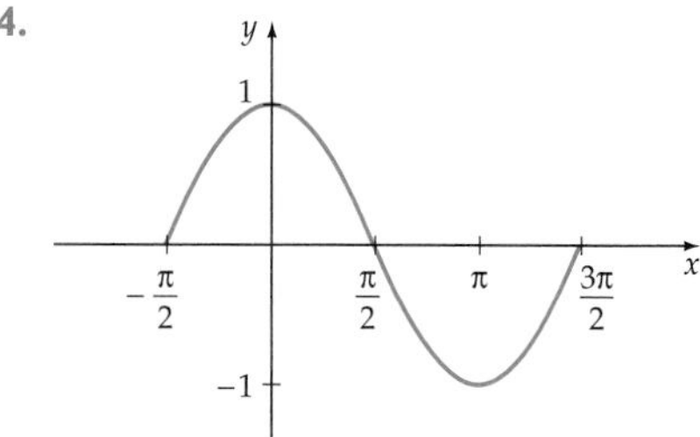

65. **CARBON DIOXIDE LEVELS** Because of seasonal changes in vegetation, carbon dioxide (CO_2) levels, as a product of photosynthesis, rise and fall during the year. Besides the naturally occurring CO_2 from plants, additional CO_2 is given off as a pollutant. A reasonable model of CO_2 levels in a city for the years 1976–1996 is given by $y = 2.3 \sin 2\pi t + 1.25t + 315$, where t is the number of years since 1976 and y is the concentration of CO_2 in parts per million (ppm). Find the difference in CO_2 levels between the beginning of 1976 and the beginning of 1996.

66. **ENVIRONMENTAL SCIENCE** Some environmentalists contend that the rate of growth of atmospheric CO_2 in parts per million is given by the equation $y = 2.54e^{0.112t} + \sin 2\pi t + 315$. See Exercise 65. Use this model to find the difference between CO_2 levels from the beginning of 1976 to the beginning of 1996.

67. **HEIGHT OF A PADDLE** The paddle wheel on a river boat is shown in the accompanying figures. Write an equation for the height of a paddle relative to the water at time t. The radius of the paddle wheel is 7 feet, and the distance from the center of the paddle wheel to the water is 5 feet. Assume that the paddle wheel rotates at 5 revolutions per minute and that the paddle is at its highest point at $t = 0$. Graph the equation for $0 \le t \le 0.20$ minute.

68. **VOLTAGE AND AMPERAGE** The graphs of the voltage and the amperage of an alternating household circuit are shown in the accompanying figures. Note that there is a phase shift between the graph of the voltage and the graph of the current. The current is said to *lag* the voltage. Write an equation for the voltage and an equation for the current.

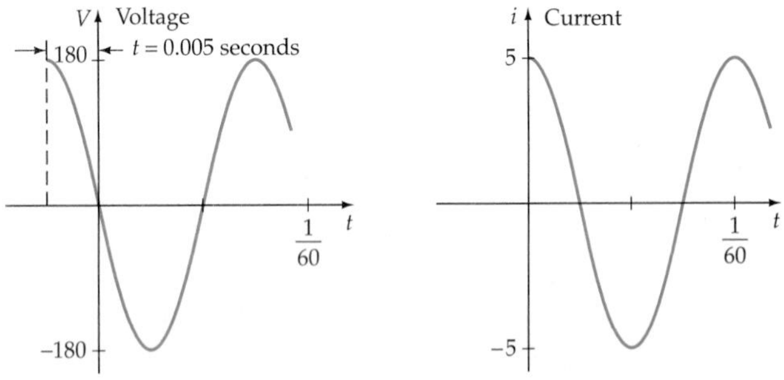

69. **A LIGHTHOUSE BEACON** The beacon of a lighthouse 400 meters from a straight sea wall rotates at 6 revolutions per minute. Using the accompanying figures, write an equation expressing the distance s measured in meters in terms of time t. Assume that when $t = 0$, the beam is perpendicular to the sea wall. Sketch a graph of the equation for $0 \le t \le 10$ seconds.

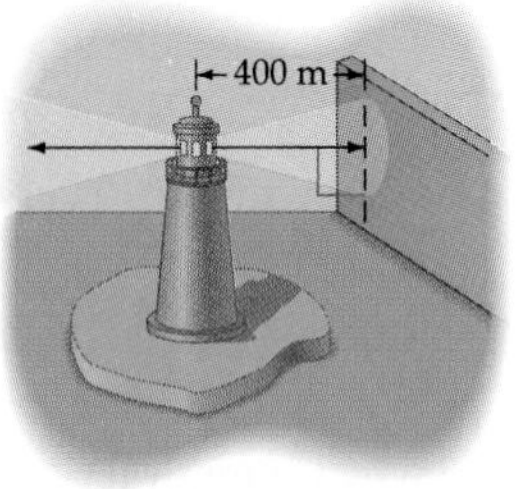

Side view

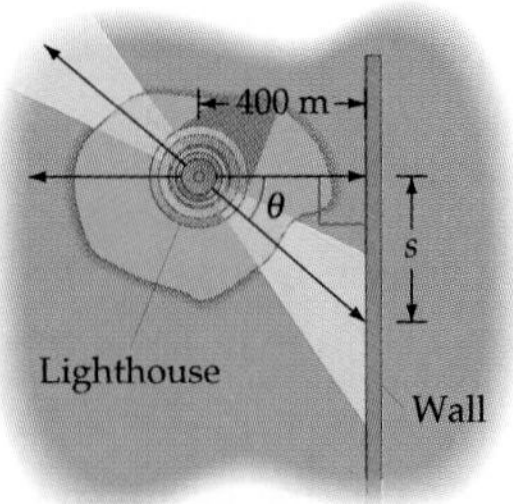

Top view

70. **HOURS OF DAYLIGHT** The duration of daylight for a region is dependent not only on the time of year but also on the latitude of the region. The graph gives the daylight hours for a one-year period at various latitudes. Assuming that a sine function can model these curves, write an equation for each curve.

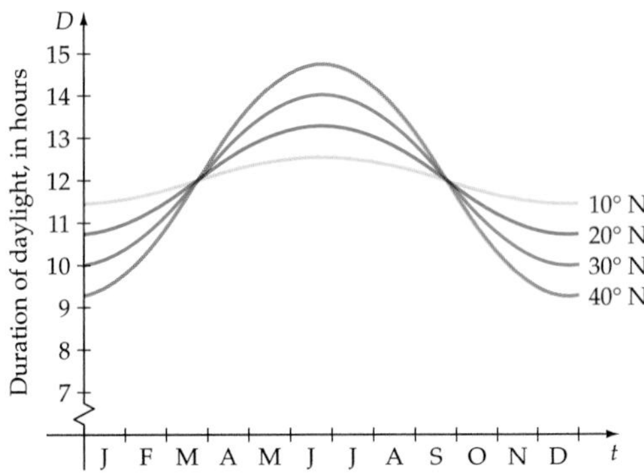

71. **TIDES** During a 24-hour day, the tides raise and lower the depth of water at a pier as shown in the figure below. Write an equation in the form $f(t) = A \cos Bt + k$, and find the depth of the water at 6 P.M.

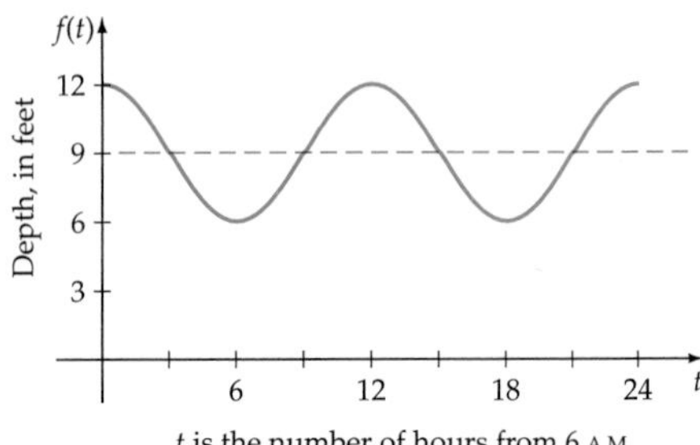

t is the number of hours from 6 A.M.

72. **TEMPERATURE** During a summer day, the ground temperature at a desert location was recorded and graphed as a function of time as shown in the figure on next page. The graph can be approximated by $f(t) = A \cos(bt + c) + k$.

Find the equation, and approximate the temperature (to the nearest degree) at 1: 00 P.M.

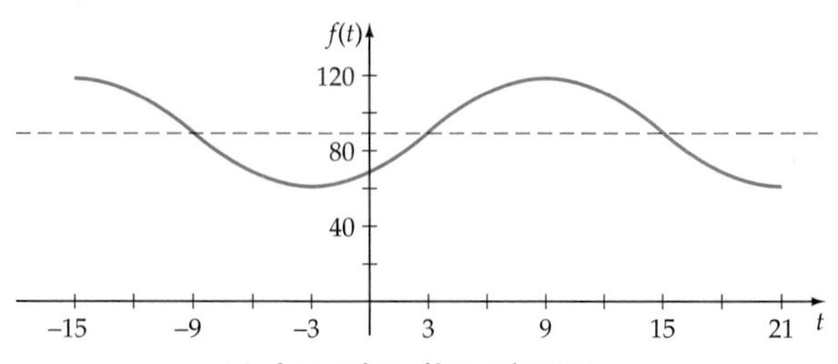

t is the number of hours from 6 A.M.

In Exercises 73 to 82, use a graphing utility to graph each function.

73. $y = \sin x - \cos \dfrac{x}{2}$

74. $y = 2 \sin 2x - \cos x$

75. $y = 2 \cos x + \sin \dfrac{x}{2}$

76. $y = -\dfrac{1}{2} \cos 2x + \sin \dfrac{x}{2}$

77. $y = \dfrac{x}{2} \sin x$

78. $y = x \cos x$

79. $y = x \sin \dfrac{x}{2}$

80. $y = \dfrac{x}{2} \cos \dfrac{x}{2}$

81. $y = x \sin \left(x + \dfrac{\pi}{2}\right)$

82. $y = x \cos \left(x - \dfrac{\pi}{2}\right)$

BEATS When two sound waves have approximately the same frequency, the sound waves interfere with one another and produce phenomena called *beats*, which are heard as variations in the loudness of the sound. A piano tuner can use these phenomena to tune a piano. By striking a tuning fork and then tapping the corresponding key on a piano, the piano tuner listens for beats and adjusts the tension in the string until the beats disappear. Use a graphing utility to graph the functions in Exercises 83 to 86, which are based on beats.

83. $y = \sin (5\pi x) \cdot \sin \left(-\dfrac{\pi}{2}x\right)$

84. $y = \sin (9\pi x) \cdot \sin \left(-\dfrac{\pi}{2}x\right)$

85. $y = \sin (13\pi x) \cdot \sin \left(-\dfrac{\pi}{2}x\right)$

86. $y = \sin (17\pi x) \cdot \sin \left(-\dfrac{\pi}{2}x\right)$

SUPPLEMENTAL EXERCISES

87. Find an equation of the sine function with amplitude 2, period π, and phase shift $\pi/3$.

88. Find an equation of the cosine function with amplitude 3, period 3π, and phase shift $-\pi/4$.

89. Find an equation of the tangent function with period 2π and phase shift $\pi/2$.

90. Find an equation of the cotangent function with period $\pi/2$ and phase shift $-\pi/4$.

91. Find an equation of the secant function with period 4π and phase shift $3\pi/4$.

92. Find an equation of the cosecant function with period $3\pi/2$ and phase shift $\pi/4$.

93. If $g(x) = \sin^2 x$ and $h(x) = \cos^2 x$, find $g(x) + h(x)$.

94. If $g(x) = 2 \sin x - 3$ and $h(x) = 4 \cos x + 2$, find the sum $g(x) + h(x)$.

95. If $g(x) = x^2 + 2$ and $h(x) = \cos x$, find $g[h(x)]$.

96. If $g(x) = \sin x$ and $h(x) = x^2 + 2x + 1$, find $h[g(x)]$.

In Exercises 97 to 100, use a graphing utility to graph each function.

97. $y = \dfrac{\sin x}{x}$

98. $y = 2 + \sec \dfrac{x}{2}$

99. $y = |x| \sin x$

100. $y = |x| \cos x$

PROJECTS

1. **PREDATOR-PREY RELATIONSHIP** Predator-prey interactions can produce cyclic population growth for both the predator population and the prey population. Consider an animal reserve where the rabbit population r is given by

$$r(t) = 850 + 210 \sin \left(\frac{\pi}{6}t\right)$$

and the wolf population w is given by

$$w(t) = 120 + 30 \sin\left(\frac{\pi}{6}t - 2\right)$$

where t is the number of months after March 1, 2000. Graph $r(t)$ and $w(t)$ on the same coordinate system for $0 \leq t \leq 24$. Write an essay that explains a possible relationship between the two populations.

Write an equation that could be used to model the rabbit population shown by the graph at the right where t is measured in months.

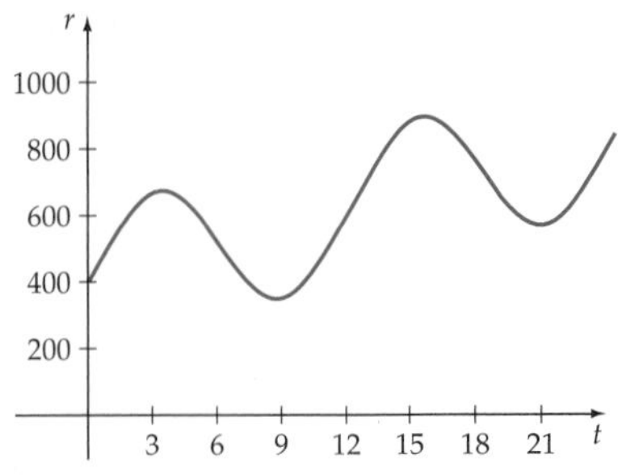

SECTION 5.8 HARMONIC MOTION—AN APPLICATION OF THE SINE AND COSINE FUNCTIONS

- SIMPLE HARMONIC MOTION
- DAMPED HARMONIC MOTION

Many phenomena that occur in nature can be modeled by periodic functions, including vibrations of a swing or in a spring. These phenomena can be described by the *sinusoidal* functions, which are the sine and cosine functions or a sum of these two functions.

◆ SIMPLE HARMONIC MOTION

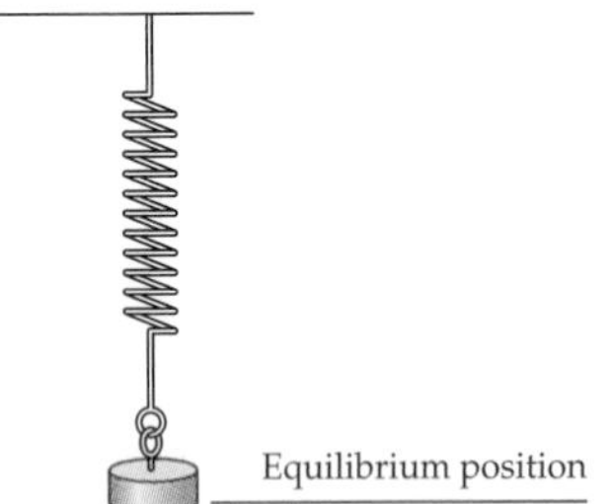

Figure 5.104

We will consider a mass on a spring to illustrate vibratory motion. Assume that we have placed a mass on a spring and allowed the spring to come to rest, as shown in **Figure 5.104.** The system is said to be in equilibrium when the mass is at rest. The point of rest is called the origin of the system. We consider the distance above the equilibrium point as positive and the distance below the equilibrium point as negative.

If the mass is now lifted a distance a and released, the mass will oscillate up and down in periodic motion. If there is no friction, the motion repeats itself in a certain period of time. The distance a is called the displacement from the origin. The number of times the mass oscillates in 1 unit of time is called the frequency f of the motion, and the time one oscillation takes is the period p of the motion. The motion is referred to as *simple harmonic motion.* **Figure 5.105** shows the displacement y of the mass for one oscillation for $t = 0, p/4, p/2, 3p/4,$ and p.

take note

The graph of the displacement y as a function of the time t is a cosine curve.

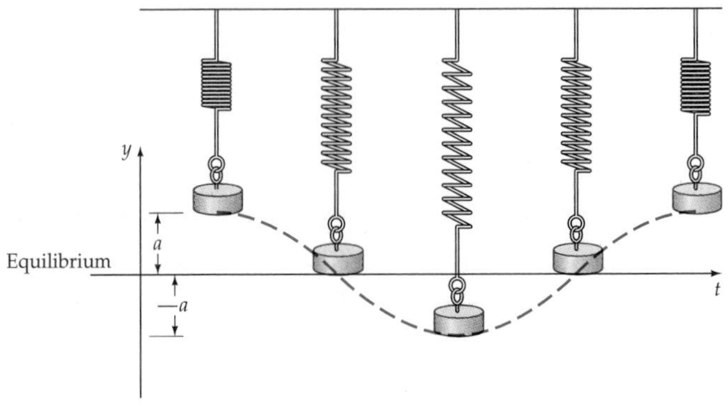

Figure 5.105

The frequency and the period are related by the formulas

$$f = \frac{1}{p} \quad \text{and} \quad p = \frac{1}{f}$$

The maximum displacement from the equilibrium position is called the *amplitude of the motion.* Vibratory motion can be quite complicated. However, the simple harmonic motion of the mass on the spring can be described by one of the following equations.

> **take note**
>
> Function (1) is used if the displacement from the origin is at a maximum at time $t = 0$. Function (2) is used if the displacement at time $t = 0$ is zero.

Simple Harmonic Motion

Simple harmonic motion can be modeled by one of the following functions:

$$y = a \cos 2\pi ft \quad (1) \qquad \text{or} \qquad y = a \sin 2\pi ft \quad (2)$$

where $|a|$ is the amplitude (maximum displacement), f is the frequency, $1/f$ is the period, y is the displacement, and t is the time.

EXAMPLE 1 Find the Equation of Motion of a Mass on a Spring

A mass on a spring has been displaced 4 centimeters above the equilibrium point and released. The mass is vibrating with a frequency of 1/2 cycle per second. Write the equation of simple harmonic motion, and graph three cycles of the displacement as a function of time.

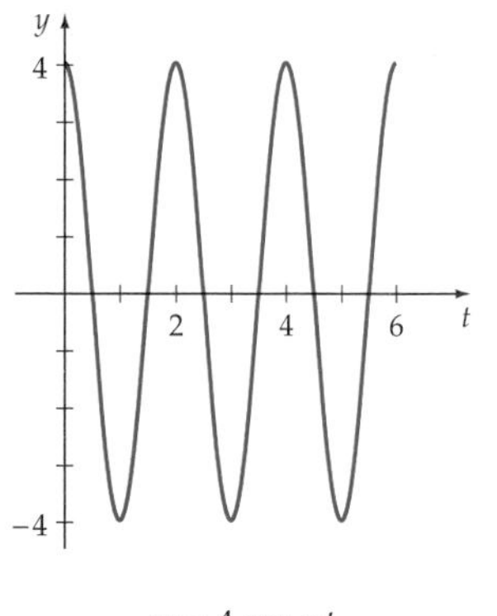

$y = 4 \cos \pi t$

Figure 5.106

Solution

Because the maximum displacement is 4 centimeters when $t = 0$, use $y = a \cos 2\pi ft$. See **Figure 5.106.**

$$\begin{aligned} y &= a \cos 2\pi ft && \text{• Equation for simple harmonic motion} \\ &= 4 \cos 2\pi\left(\frac{1}{2}\right)t && \text{• } a = 4, f = \frac{1}{2} \\ &= 4 \cos \pi t \end{aligned}$$

TRY EXERCISE 20, EXERCISE SET 5.8, PAGE 464

From physical laws determined by experiment, the frequency of oscillation of a mass on a spring is given by

$$f = \frac{1}{2\pi}\sqrt{\frac{k}{m}}$$

where k is a spring constant determined by experiment and m is the mass. The simple harmonic motion of the mass on the spring (with maximum displacement at $t = 0$) can then be described by

$$\begin{aligned} y &= a \cos 2\,\pi ft = a \cos 2\pi\left(\frac{1}{2\pi}\sqrt{\frac{k}{m}}\right)t \\ &= a \cos \sqrt{\frac{k}{m}}\,t \qquad (3) \end{aligned}$$

The equation of the simple harmonic motion for zero displacement at $t = 0$ is

$$y = a \sin \sqrt{\frac{k}{m}}\, t \qquad (4)$$

EXAMPLE 2 Find the Equation of Motion of a Mass on a Spring

A mass of 2 units is in equilibrium suspended from a spring with a spring constant of $k = 18$. The mass is pulled down 0.5 unit and released. Find the period, frequency, and amplitude of the resulting simple harmonic motion. Write the equation of the motion, and graph two cycles of the displacement as a function of time.

Solution

At the start of the motion, the displacement is at a maximum but in the negative direction. The resulting motion is described by Equation (3), using $a = -0.5$, $k = 18$, and $m = 2$.

$$y = a \cos \sqrt{\frac{k}{m}}\, t = -0.5 \cos \sqrt{\frac{18}{2}}\, t$$ • Substitute for *a*, *k*, and *m*.

$$= -0.5 \cos 3t$$ • Equation of motion

Period: $\dfrac{2\pi}{|b|} = \dfrac{2\pi}{3}$

Frequency: $\dfrac{1}{\text{period}} = \dfrac{3}{2\pi}$

Amplitude: $|a| = |-0.5| = 0.5$

See **Figure 5.107**.

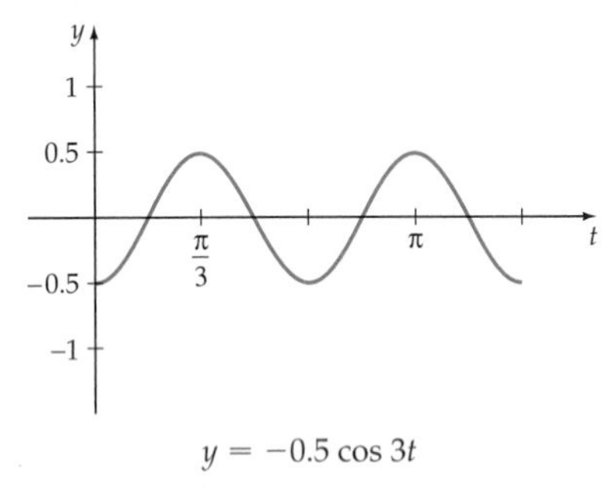

$y = -0.5 \cos 3t$

Figure 5.107

TRY EXERCISE 28, EXERCISE SET 5.8, PAGE 464

◆ DAMPED HARMONIC MOTION

The previous examples have assumed that there is no friction within the spring and no air resistance. If we consider friction and air resistance, then the motion of the mass tends to decrease as t increases. The motion is called **damped harmonic motion.**

EXAMPLE 3 Model Damped Harmonic Motion

A mass on a spring has been displaced 14 inches below the equilibrium point and released. The damped harmonic motion of the mass is given by

$$f(t) = -14e^{-0.4t} \cos 2t, \quad t \ge 0$$

where $f(t)$ is measured in inches and t is the time in seconds.

take note

- The motion in Example 3 is not periodic in a strict sense, because the motion does not repeat exactly but tends to diminish as t increases. However, because the motion cycles every π seconds, we call π the *pseudoperiod* of the motion. In general, the damped harmonic motion modeled by $f(t) = ae^{-kt}\cos\omega t$ has pseudoperiod $p = (2\pi)/\omega$ and frequency $f = \omega/(2\pi)$.

a. Find the values of t for which $f(t) = 0$.

b. Use a graphing utility to determine how long it will be until the absolute value of the displacement of the mass is always less than 0.01 inch.

Solution

a. $f(t) = 0$ if and only if $\cos 2t = 0$, and $\cos 2t = 0$ when $2t = \frac{\pi}{2} + n\pi$.

Therefore, $f(t) = 0$ when $t = \frac{\pi}{4} + n\left(\frac{\pi}{2}\right) \approx 0.79 + n(1.57)$. The first time the displacement $f(t) = 0$ occurs when $t \approx 0.79$ second. The graph of f in **Figure 5.108** confirms these results. **Figure 5.108** also shows that the graph of f lies on or between the graphs of $y = -14e^{-0.4t}$ and $y = 14e^{-0.4t}$. Recall from Section 5.7 that $-14e^{0.4t}$ is the damping factor.

b. We need to find the smallest value of t for which the absolute value of the displacement of the mass is always less than 0.01. **Figure 5.109** shows a graph of f using a viewing window of $0 \le t \le 20$ and $-0.01 \le f(t) \le 0.01$. Use the TRACE or the ISECT feature to determine that $t \approx 17.59$ seconds.

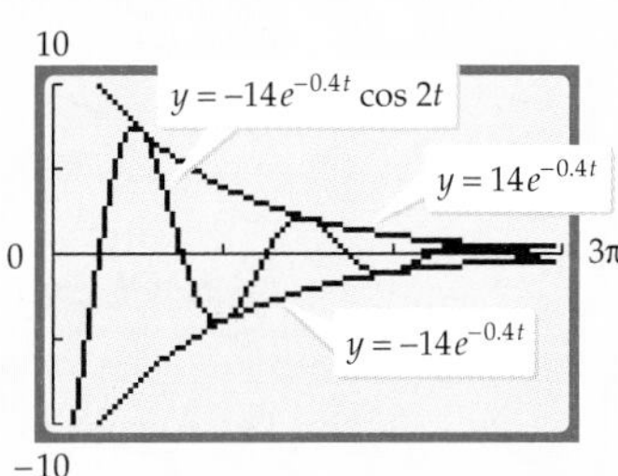

Figure 5.108

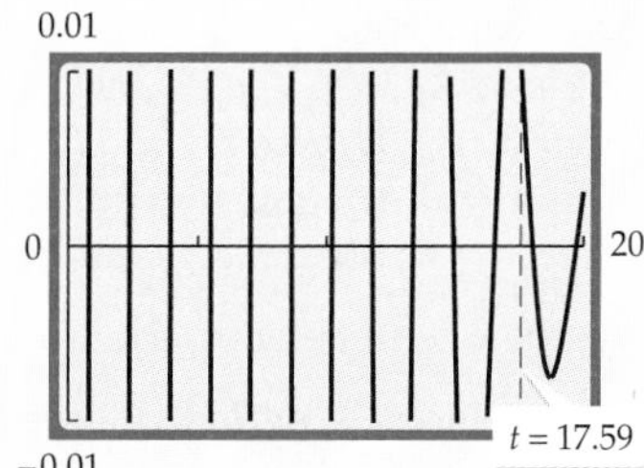

$f(t) = -14e^{-0.4t}\cos 2t$

Figure 5.109

TRY EXERCISE 30, EXERCISE SET 5.8, PAGE 464

TOPICS FOR DISCUSSION

1. The period of a simple harmonic motion is the same as the frequency of the motion. Do you agree? Explain.

2. In the simple harmonic motion modeled by $y = 3\cos 2\pi t$, does the displacement y approach 0 as t increases without bound? Explain.

3. Explain how you know whether to use

$$y = a\cos 2\pi ft \qquad \text{or} \qquad y = a\sin 2\pi ft$$

to model a particular harmonic motion.

4. If the mass on a spring is increased from m to $4m$, what effect will this have on the frequency of the simple harmonic motion of the mass?

EXERCISE SET 5.8

In Exercises 1 to 8, find the amplitude, period, and frequency of the simple harmonic motion.

1. $y = 2\sin 2t$
2. $y = \frac{2}{3}\cos\frac{t}{3}$
3. $y = 3\cos\frac{2t}{3}$
4. $y = 4\sin 3t$
5. $y = 4\cos \pi t$
6. $y = 2\sin\frac{\pi t}{3}$
7. $y = \frac{3}{4}\sin\frac{\pi t}{2}$
8. $y = 5\cos 2\pi t$

In Exercises 9 to 12, write an equation for the simple harmonic motion that satisfies the given conditions. Assume that the maximum displacement occurs at t = 0. Sketch a graph of the equation.

9. Frequency = 1.5 cycles per second, $a = 4$ inches
10. Frequency = 0.8 cycle per second, $a = 4$ centimeters
11. Period = 1.5 seconds, $a = 3/2$ feet
12. Period = 0.6 second, $a = 1$ meter

In Exercises 13 to 18, write an equation for the simple harmonic motion with the given conditions. Assume zero displacement at t = 0. Sketch a graph of the equation.

13. Amplitude 2 centimeters, period π seconds
14. Amplitude 4 inches, period $\pi/2$ seconds
15. Amplitude 1 inch, period 2 seconds
16. Amplitude 3 centimeters, period 1 second
17. Amplitude 2 centimeters, frequency 1 second
18. Amplitude 4 inches, frequency 4 seconds

In Exercises 19 to 26, write an equation for simple harmonic motion. Assume that the maximum displacement occurs when t = 0.

19. Amplitude 1/2 centimeter, frequency $2/\pi$ cycles per second
20. Amplitude 3 inches, frequency $1/\pi$ cycles per second
21. Amplitude 2.5 inches, frequency 0.5 cycle per second
22. Amplitude 5 inches, frequency 1/8 cycle per second
23. Amplitude 1/2 inch, period 3 seconds
24. Amplitude 5 centimeters, period 5 seconds
25. Amplitude 4 inches, period $\pi/2$ seconds
26. Amplitude 2 centimeters, period π seconds

27. A mass of 32 units is in equilibrium suspended from a spring. The mass is pulled down 2 feet and released. Find the period, frequency, and amplitude of the resulting simple harmonic motion. Write an equation of the motion. Assume a spring constant of $k = 8$.

28. A mass of 27 units is in equilibrium suspended from a spring. The mass is pulled down 1.5 feet and released. Find the period, frequency, and amplitude of the resulting simple harmonic motion. Write an equation of the motion. Assume a spring constant of $k = 3$.

In Exercises 29 to 36, each of the equations models a damped harmonic motion.

a. Find the number of complete oscillations that occur during the time interval $0 \le t \le 10$ seconds.

b. Use a graph to determine how long it will be (to the nearest 0.1 second) until the absolute value of the displacement of the mass is always less than 0.01.

29. $f(t) = 4e^{-0.1t}\cos 2t$
30. $f(t) = 12e^{-0.6t}\cos t$
31. $f(t) = -6e^{-0.09t}\cos 2\pi t$
32. $f(t) = -11e^{-0.4t}\cos \pi t$
33. $f(t) = e^{-0.5t}\cos 2\pi t$
34. $f(t) = e^{-0.2t}\cos 3\pi t$
35. $f(t) = e^{-0.75t}\cos 2\pi t$
36. $f(t) = e^{-t}\cos 2\pi t$

SUPPLEMENTAL EXERCISES

37. Assuming that a mass of m pounds on the end of a spring is oscillating in simple harmonic motion. What effect will there be on the period of the motion if the mass is increased to $9m$?

38. A mass on a spring is displaced 9 inches below its equilibrium position and then released. The weight oscillates in simple harmonic motion with a frequency of 2 cycles per second. Find the period and the equation of the motion.

In Exercises 39 to 42, use a graphing utility to determine whether both of the given functions model the same damped harmonic motion.

39. $f(t) = \sqrt{2}e^{-0.2t}\sin\left(t + \frac{\pi}{4}\right)$
$g(t) = e^{-0.2t}(\cos t + \sin t)$

40. $f(t) = 5e^{-0.3t}\cos(t - 0.927295)$
$g(t) = e^{-0.2t}(3\cos t + 4\sin t)$

41. $f(t) = 13e^{-0.4t}\cos(t - 1.176005)$
$g(t) = e^{-0.4t}(5\cos t + 12\sin t)$

42. $f(t) = \sqrt{2}e^{-0.2t}\cos(t + 0.785398)$
$g(t) = e^{-0.2t}(\cos t - \sin t)$

PROJECTS

1. **THREE TYPES OF DAMPED HARMONIC MOTION** In some cases the damped harmonic motion of a mass on the end of a spring does not cycle about the equilibrium point. The following three functions illustrate different types of damped harmonic motion. Use a graphing utility to graph each function, and then write a few sentences that explain the major differences among the motions.

$$f(t) = (0.5t + 1)e^{-0.5t} \qquad g(t) = -2e^{-0.4t} + 5e^{-t} \qquad h(t) = 4e^{-0.2t}\cos 2\pi t$$

2. **LOGARITHMIC DECREMENT** If a damped harmonic motion is modeled by

$$f(t) = ae^{-\alpha t}\cos \omega t$$

then the ratio of any two consecutive relative maxima of the motion is a constant γ.

a. Use a graphing utility to determine γ for the damped harmonic motion modeled by

$$f(t) = -14e^{-0.4t}\cos 2t, \quad t \geq 0$$

b. The constant $\Delta = 2\pi\alpha/\omega$ is called the **logarithmic decrement** of the motion. Compute Δ for the damped harmonic motion in the equation in **a.** How does $\ln \gamma$ compare with Δ?

EXPLORING CONCEPTS WITH TECHNOLOGY

Sinusoidal Families

Some graphing calculators have a feature that allows you to graph a family of functions easily. For instance, entering `Y1={2,4,6}sin(X)` in the `Y=` menu and pressing the `GRAPH` key on a TI-83 calculator produces a graph of the three functions $y = 2\sin x$, $y = 4\sin x$, and $y = 6\sin x$, all displayed in the same window.

1. Use a graphing calculator to graph `Y1={2,4,6}sin(X)`. Write a sentence that indicates the similarities and the differences among the three graphs.

2. Use a graphing calculator to graph `Y1=sin({π,2π,4π}X)`. Write a sentence that indicates the similarities and the differences among the three graphs.

3. Use a graphing calculator to graph `Y1=sin(X+{π/4,π/6,π/12})`. Write a sentence that indicates the similarities and the differences among the three graphs.

4. A student has used a graphing calculator to graph `Y1=sin(X+{π,3π,5π})` and expects to see three graphs. However, the student sees only one graph displayed on the graph window. Has the calculator displayed all three graphs? Explain.

CHAPTER 5 SUMMARY

5.1 Angles and Arcs

- An angle is in standard position when its initial side is along the positive x-axis and its vertex is at the origin of the coordinate axes.
- Angle α is an acute angle when $0° < \alpha < 90°$; it is an obtuse angle when $90° < \alpha < 180°$.
- α and β are complementary angles when $\alpha + \beta = 90°$; they are supplementary angles when $\alpha + \beta = 180°$.

- The length of the arc s that subtends the central angle θ (in radians) on a circle of radius r is given by $s = r\theta$.
- Angular speed is given by $\omega = \dfrac{\theta}{t}$.

5.2 Trigonometric Functions of Acute Angles

- Let θ be an acute angle of a right triangle. The six trigonometric functions of θ are given by

$$\sin\theta = \frac{\text{opp}}{\text{hyp}} \qquad \csc\theta = \frac{\text{hyp}}{\text{opp}}$$
$$\cos\theta = \frac{\text{adj}}{\text{hyp}} \qquad \sec\theta = \frac{\text{hyp}}{\text{adj}}$$
$$\tan\theta = \frac{\text{opp}}{\text{adj}} \qquad \cot\theta = \frac{\text{adj}}{\text{opp}}$$

5.3 Trigonometric Functions of Any Angle

- Let $P(x, y)$ be a point, except the origin, on the terminal side of an angle θ in standard position. The six trigonometric functions of θ are

$$\sin\theta = \frac{y}{r} \qquad \csc\theta = \frac{r}{y},\quad y \neq 0$$
$$\cos\theta = \frac{x}{r} \qquad \sec\theta = \frac{r}{x},\quad x \neq 0$$
$$\tan\theta = \frac{y}{x},\quad x \neq 0 \qquad \cot\theta = \frac{x}{y},\quad y \neq 0$$

5.4 Trigonometric Functions of Real Numbers

- The wrapping function pairs a real number with a point on the unit circle.
- Let W be the wrapping function, t be a real number, and $W(t) = P(x, y)$. Then the trigonometric functions of the real number t are defined as follows:

$$\sin t = y \qquad \csc t = \frac{1}{y},\quad y \neq 0$$
$$\cos t = x \qquad \sec t = \frac{1}{x},\quad x \neq 0$$
$$\tan t = \frac{y}{x},\quad x \neq 0 \qquad \cot t = \frac{x}{y},\quad y \neq 0$$

- $\sin t$, $\csc t$, $\tan t$, and $\cot t$ are odd functions.
- $\cos t$ and $\sec t$ are even functions.
- $\sin t$, $\cos t$, $\sec t$, and $\csc t$ have period 2π.
- $\tan t$ and $\cot t$ have period π.

Domain and Range of Each Trigonometric Function (n is an integer)

Function	Domain	Range
$\sin t$	$\{t \mid -\infty < t < \infty\}$	$\{y \mid -1 \le y \le 1\}$
$\cos t$	$\{t \mid -\infty < t < \infty\}$	$\{y \mid -1 \le y \le 1\}$
$\tan t$	$\{t \mid -\infty < t < \infty, t \neq (2n+1)\pi/2\}$	$\{y \mid -\infty < y < \infty\}$
$\csc t$	$\{t \mid -\infty < t < \infty, t \neq n\pi\}$	$\{y \mid y \ge 1, y \le -1\}$
$\sec t$	$\{t \mid -\infty < t < \infty, t \neq (2n+1)\pi/2\}$	$\{y \mid y \ge 1, y \le -1\}$
$\cot t$	$\{t \mid -\infty < t < \infty, t \neq n\pi\}$	$\{y \mid -\infty < y < \infty\}$

5.5 Graphs of the Sine and Cosine Functions

- The graph of $y = a\sin bx$ and that of $y = a\cos bx$ both have an amplitude of $|a|$ and a period of $\dfrac{2\pi}{|b|}$. The graph of each for $a > 0$ and $b > 0$ is given below.

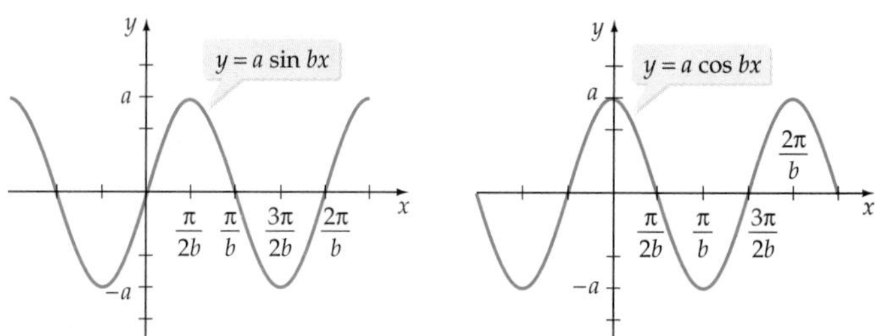

5.6 Graphs of the Other Trigonometric Functions

- The period of $y = a\tan bx$ and $y = a\cot bx$ is $\dfrac{\pi}{|b|}$.
- The period of $y = a\sec bx$ and $y = a\csc bx$ is $\dfrac{2\pi}{|b|}$.

5.7 Graphing Techniques

- Phase shift is a horizontal translation of the graph of a trigonometric function. If $y = f(bx + c)$, where f is a trigonometric function, then the phase shift is $-c/b$.
- The graphs of $y = a\sin(bx + c)$ and $y = a\cos(bx + c)$, $b > 0$, have amplitude $|a|$, period $\dfrac{2\pi}{b}$, and phase shift $-\dfrac{c}{b}$. One cycle of each graph is completed on the interval $-\dfrac{c}{b} \le x \le -\dfrac{c}{b} + \dfrac{2\pi}{b}$.

- Addition of ordinates is a method of graphing the sum of two functions by graphically adding the values of their y-coordinates.

- The factor $g(x)$ in $f(x) = g(x) \cos x$ is called a damping factor. The graph of f lies on or between the graphs of the equations $y = g(x)$ and $y = -g(x)$.

5.8 Harmonic Motion—An Application of the Sine and Cosine Functions

- The equations of the simple harmonic motion are
$$y = a \cos 2\pi ft \qquad \text{and} \qquad y = a \sin 2\pi ft$$
where a is the amplitude and f is the frequency.
- Functions of form $f(t) = ae^{-at} \cos \omega t$ are often used to model some forms of damped harmonic motion.

CHAPTER 5 TRUE/FALSE EXERCISES

In Exercises 1 to 16, answer true or false. If the statement is false, give an example to show that the statement is false.

1. An angle is in standard position when the vertex is at the origin of a coordinate system.
2. The angle θ in radians is in standard position with the terminal side in the second quadrant. The reference angle of θ is $\pi - \theta$.
3. In the formula $s = r\theta$, the angle θ must be measured in radians.
4. If $\tan \theta < 0$ and $\cos \theta > 0$, then the terminal side of θ is in Quadrant III.
5. $\sec^2 \theta + \tan^2 \theta = 1$ is an identity.
6. The amplitude of the graph of $y = 2 \tan x$ is 2.
7. The period of $y = \cos x$ is π.
8. The graph of $y = \sin x$ is symmetric to the origin.
9. For any acute angle θ, $\sin \theta + \cos (90° - \theta) = 1$.
10. $\sin (x + y) = \sin x + \sin y$.
11. $\sin^2 x = \sin x^2$.
12. The phase shift of $f(x) = 2 \sin \left(2x - \dfrac{\pi}{3}\right)$ is $\dfrac{\pi}{3}$.
13. The measure of one radian is more than 50 times the measure of one degree.
14. The measure of one radian differs depending on the radius of the circle used.
15. The graph of $y = 2^{-x} \cos x$ lies on or between the graphs of $y = 2^{-x}$ and $y = 2^x$.
16. The function $f(t) = e^{-0.1t} \cos t$, $t > 0$, models damped harmonic motion in which $|f(t)| \to 0$ as $t \to 0$.

CHAPTER 5 REVIEW EXERCISES

1. Find the complement and supplement of the angle θ whose measure is 65°.
2. Find the measure of the reference angle θ' for the angle θ whose measure is 980°.
3. Convert 2 radians to the nearest hundredth of a degree.
4. Convert 315° to radian measure.
5. Find the length (to the nearest 0.01 meter) of the arc on a circle of radius 3 meters that subtends an angle of 75°.
6. Find the radian measure of the angle subtended by an arc of length 12 centimeters on a circle whose radius is 40 centimeters.
7. A car with a 16-inch-radius wheel is moving with a speed of 50 mph. Find the angular speed (to the nearest radian per second) of the wheel in radians per second.

In Exercises 8 to 11, let θ be an acute angle of a right triangle and $\csc \theta = \dfrac{3}{2}$. Evaluate each function.

8. $\cos \theta$
9. $\cot \theta$
10. $\sin \theta$
11. $\sec \theta$
12. Find the values of the six trigonometric functions of an angle in standard position with the point $P(1, -3)$ on the terminal side of the angle.
13. Find the exact value of
    a. $\sec 150°$
    b. $\tan (-3\pi/4)$
    c. $\cot (-225°)$
    d. $\cos 2\pi/3$
14. Find the value of each of the following to the nearest ten-thousandth.
    a. $\cos 123°$
    b. $\cot 4.22$
    c. $\sec 612°$
    d. $\tan 2\pi/5$

15. Given $\cos\phi = -\sqrt{3}/2$, $180° < \phi < 270°$, find the exact value of

a. $\sin\phi$ b. $\tan\phi$

16. Given $\tan\phi = -\sqrt{3}/3$, $90° < \phi < 180°$, find the exact value of

a. $\sec\phi$ b. $\csc\phi$

17. Given $\sin\phi = -\sqrt{2}/2$, $270° < \phi < 360°$, find the exact value of

a. $\cos\phi$ b. $\cot\phi$

18. Let W be the wrapping function. Evaluate

a. $W(\pi)$ b. $W\left(-\frac{\pi}{3}\right)$ c. $W\left(\frac{5\pi}{4}\right)$ d. $W(28\pi)$

19. Is the function defined by $f(x) = \sin(x)\tan(x)$ even, odd, or neither?

In Exercises 20 to 21, use the unit circle to show that each equation is an identity.

20. $\cos(\pi + t) = -\cos t$ **21.** $\tan(-t) = -\tan t$

In Exercises 22 to 27, use trigonometric identities to write each expression in terms of a single trigonometric function.

22. $1 + \dfrac{\sin^2\phi}{\cos^2\phi}$ **23.** $\dfrac{\tan\phi + 1}{\cot\phi + 1}$

24. $\dfrac{\cos^2\phi + \sin^2\phi}{\csc\phi}$ **25.** $\sin^2\phi(\tan^2\phi + 1)$

26. $1 + \dfrac{1}{\tan^2\phi}$ **27.** $\dfrac{\cos^2\phi}{1 - \sin^2\phi} - 1$

In Exercises 28 to 33, state the amplitude (if there is one), period, and phase shift of the graph of each function.

28. $y = 3\cos(2x - \pi)$ **29.** $y = 2\tan 3x$

30. $y = -2\sin\left(3x + \frac{\pi}{3}\right)$ **31.** $y = \cos\left(2x - \frac{2\pi}{3}\right) + 2$

32. $y = -4\sec\left(4x - \frac{3\pi}{2}\right)$ **33.** $y = 2\csc\left(x - \frac{\pi}{4}\right) - 3$

In Exercises 34 to 51, graph each function.

34. $y = 2\cos\pi x$ **35.** $y = -\sin\dfrac{2x}{3}$

36. $y = 2\sin\dfrac{3x}{2}$ **37.** $y = \cos\left(x - \frac{\pi}{2}\right)$

38. $y = \dfrac{1}{2}\sin\left(2x + \frac{\pi}{4}\right)$ **39.** $y = 3\cos 3(x - \pi)$

40. $y = -\tan\dfrac{x}{2}$ **41.** $y = 2\cot 2x$

42. $y = \tan\left(x - \frac{\pi}{2}\right)$ **43.** $y = -\cot\left(2x + \frac{\pi}{4}\right)$

44. $y = -2\csc\left(2x - \frac{\pi}{3}\right)$ **45.** $y = 3\sec\left(x + \frac{\pi}{4}\right)$

46. $y = 3\sin 2x - 3$ **47.** $y = 2\cos 3x + 3$

48. $y = -\cos\left(3x + \frac{\pi}{2}\right) + 2$

49. $y = 3\sin\left(4x - \frac{2\pi}{3}\right) - 3$

50. $y = 2 - \sin 2x$

51. $y = \sin x - \sqrt{3}\cos x$

52. A car climbs a hill that has a constant angle of 4.5° for a distance of 1.14 miles. What is the car's increase in altitude?

53. A tree casts a shadow of 8.55 feet when the angle of elevation of the sun is 55.3°. Find the height of the tree.

54. Find the sine of the angle α formed by the intersection of a diagonal of a face of a cube and the diagonal of the cube originating from the same vertex.

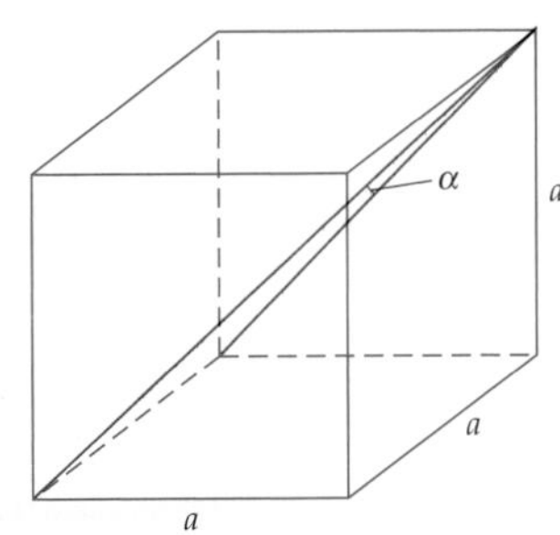

55. Find the height of a building if the angle of elevation to the top of the building changes from 18° to 37° as the observer moves a distance of 80 feet toward the building.

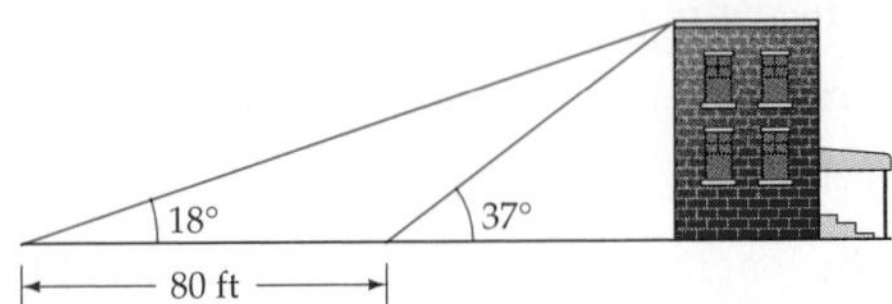

56. Find the amplitude, period, and frequency of the simple harmonic motion given by $y = 2.5\sin 50t$.

57. A mass of 5 kilograms is in equilibrium suspended from a spring. The mass is pulled down 0.5 foot and released. Find the period, frequency, and amplitude of the motion, assuming the mass oscillates in simple harmonic motion. Write an equation of motion. Assume $k = 20$.

58. Use a graphing utility to graph the damped harmonic motion that is modeled by

$$f(t) = 3e^{-0.75t} \cos \pi t$$

where t is in seconds. Use the graph to determine, to the nearest tenth, how long (to the nearest 0.1 second) it will be until the absolute value of the displacement of the mass is always less than 0.01.

CHAPTER 5 TEST

1. Convert 150° to exact radian measure.
2. Find the supplement of the angle whose radian measure is $\frac{11}{12}\pi$. Express your answer in terms of π.
3. Find the length (to the nearest 0.1 centimeter) of an arc that subtends a central angle of 75° in a circle of radius 10 centimeters.
4. A wheel is rotating at 6 revolutions per second. Find the angular speed in radians per second.
5. A wheel with a diameter of 16 centimeters is rotating at 10 radians per second. Find the linear speed (in centimeters per second) of a point on the edge of the wheel.
6. If θ is an acute angle of a right triangle and $\tan \theta = \frac{3}{7}$, find $\sec \theta$.
7. Use a calculator to find the value of csc 67° to the nearest ten-thousandth.
8. Find the exact value of $\tan \frac{\pi}{6} \cos \frac{\pi}{3} - \sin \frac{\pi}{2}$.
9. Find the exact coordinates of $W\left(\frac{11\pi}{6}\right)$.
10. Express $\frac{\sec^2 t - 1}{\sec^2 t}$ in terms of a single trigonometric function.
11. State the period of $y = -4 \tan 3x$.
12. State the amplitude, period, and phase shift for the function $y = -3 \cos\left(2x + \frac{\pi}{2}\right)$.
13. State the period and phase shift for the function $y = 2 \cot\left(\frac{\pi}{3}x + \frac{\pi}{6}\right)$.
14. Graph one full period of $y = 3 \cos \frac{1}{2}x$.
15. Graph one full period of $y = -2 \sec \frac{1}{2}x$.
16. Write a sentence that explains how to obtain the graph of $y = 2 \sin\left(2x - \frac{\pi}{2}\right) - 1$ from the graph of $y = 2 \sin 2x$.
17. Graph one full period of $y = 2 - \sin \frac{x}{2}$
18. Graph one full period of $y = \sin x - \cos 2x$.
19. The angle of elevation from point A to the top of a tree is 42.2°. At point B, 5.24 meters from A and on a line through the base of the tree and A, the angle of elevation is 37.4°. Find the height of the tree.
20. Write the equation for simple harmonic motion, given that the amplitude is 13 feet, the period is 5 seconds, and the displacement is zero when $t = 0$.

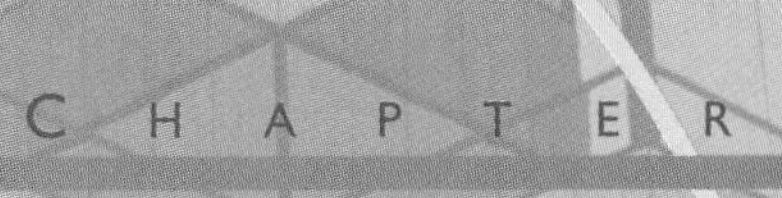

6

Trigonometric Identities and Equations

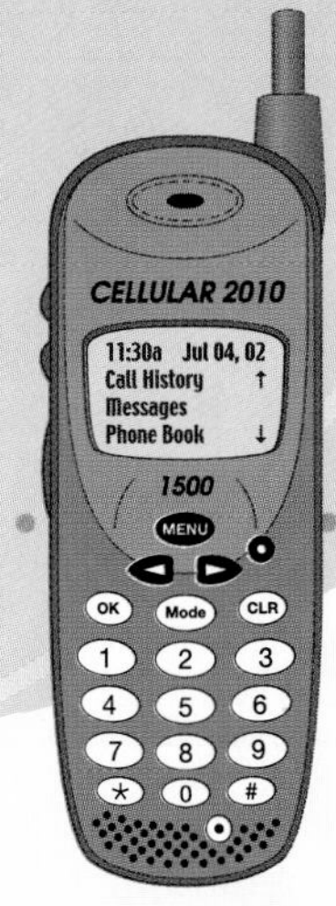

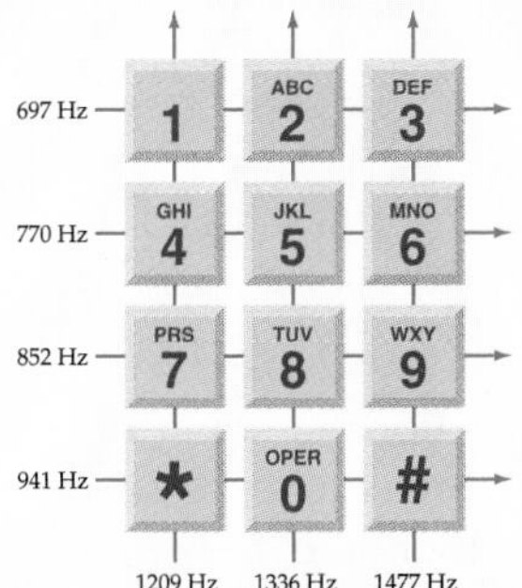

◆ *Source:* Data in chart from **http://www.howstuffworks.com/telephone2.htm** and also found at **http://hyperarchive.lcs.mit.edu/telecom-archives/tribute/touch_tone_info.html**

Touch-Tone Phones and Trigonometry

The dial tone emitted by a telephone is produced by adding a 350-hertz sound to a 440-hertz sound. An equation that models the dial tone is

$$v_1(t) = \cos(2\pi \cdot 440t) + \cos(2\pi \cdot 350t)$$

where t is in seconds. Concepts from this chapter can be used to show that the dial tone can also be modeled by

$$v_2(t) = 2\cos(790\pi t)\cos(90\pi t)$$

The equation $v_1(t) = v_2(t)$ is called an *identity* because the left side of the equation equals the right side for all domain values t. Use a graphing utility to graph v_1 and v_2 on the interval $[0, 0.1]$ to see that they appear to represent the same function.

Every dial sound made on a touch-tone phone is produced by adding a pair of sounds. The chart to the left shows the frequencies used for each key. For instance, the sound emitted by pressing 3 on the keypad is produced by adding a 1477-hertz sound to a 697-hertz sound. An equation that models this tone is

$$v(t) = \cos(2\pi \cdot 1477t) + \cos(2\pi \cdot 697t)$$

QUESTION What is an equation for the tone emitted by pressing 4 on the keypad?

ANSWER $v(t) = \cos(2\pi \cdot 1209t) + \cos(2\pi \cdot 770t)$

SECTION 6.1 VERIFICATION OF TRIGONOMETRIC IDENTITIES

- FUNDAMENTAL TRIGONOMETRIC IDENTITIES
- VERIFICATION OF TRIGONOMETRIC IDENTITIES

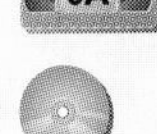

◆ FUNDAMENTAL TRIGONOMETRIC IDENTITIES

The domain of an equation consists of all values of the variable for which every term is defined. For example, the domain of

$$\frac{\sin x \cos x}{\sin x} = \cos x \tag{1}$$

includes all real numbers x except $x = n\pi$ where n is an integer, because $\sin x = 0$ for $x = n\pi$, and division by 0 is undefined. An **identity** is an equation that is true for all its domain values. Table 6.1 lists identities that were introduced earlier.

Table 6.1 Fundamental Trigonometric Identities

Reciprocal identities	$\sin x = \dfrac{1}{\csc x}$	$\cos x = \dfrac{1}{\sec x}$	$\tan x = \dfrac{1}{\cot x}$
Ratio identities	$\tan x = \dfrac{\sin x}{\cos x}$	$\cot x = \dfrac{\cos x}{\sin x}$	
Pythagorean identities	$\sin^2 x + \cos^2 x = 1$	$\tan^2 x + 1 = \sec^2 x$	$1 + \cot^2 x = \csc^2 x$
Odd-even identities	$\sin(-x) = -\sin x$ $\cos(-x) = \cos x$	$\tan(-x) = -\tan x$ $\cot(-x) = -\cot x$	$\sec(-x) = \sec x$ $\csc(-x) = -\csc x$

◆ VERIFICATION OF TRIGONOMETRIC IDENTITIES

To verify an identity, we show that one side of the identity can be rewritten in a form that is identical to the other side. There is no one method that can be used to verify every identity; however, the following guidelines should prove useful.

Guidelines for Verifying Trigonometric Identities

- If one side of the identity is more complex than the other, then it is generally best to try first to simplify the more complex side until it becomes identical to the other side.
- Perform indicated operations such as adding fractions or squaring a binomial. Also be aware of any factorization that may help you achieve your goal of producing the expressions on the other side.
- Make use of previously established identities that enable you to rewrite one side of the identity in an equivalent form.
- Rewrite one side of the identity so that it involves only sines and/or cosines.
- Rewrite one side of the identity in terms of a single trigonometric function.

- Multiplying both the numerator and the denominator of a fraction by the same factor (such as the conjugate of the denominator or the conjugate of the numerator) may get you closer to your goal.
- Keep your goal in mind. Does it involve products, quotients, sums, radicals, or powers? Knowing exactly what your goal is may provide the insight you need to verify the identity.

EXAMPLE 1 **Determine Whether an Equation is an Identity**

Determine whether each equation is an identity. If the equation is an identity, then verify the identity. If the equation is not an identity, then find a value of the domain for which the left side of the equation is not equal to the right side.

a. $\sin\left(x + \dfrac{\pi}{6}\right) = \sin x + \sin\dfrac{\pi}{6}$

b. $(\sin x + \cos x)^2 = 2\sin x\cos x + 1$

Solution

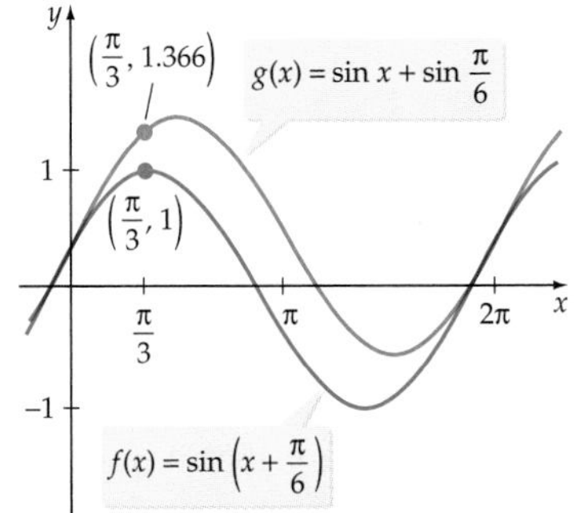

Figure 6.1

a. The graphs of $f(x) = \sin\left(x + \dfrac{\pi}{6}\right)$ and $g(x) = \sin x + \sin\dfrac{\pi}{6}$ are shown in Figure 6.1. Because the graphs are not identical, we know that the equation is not an identity. This can be further confirmed by letting $x = \pi/3$ and observing that

$$f\left(\frac{\pi}{3}\right) = \sin\left(\frac{\pi}{3} + \frac{\pi}{6}\right) = \sin\frac{\pi}{2} = 1$$

whereas

$$g\left(\frac{\pi}{3}\right) = \sin\frac{\pi}{3} + \sin\frac{\pi}{6} = \frac{\sqrt{3}}{2} + \frac{1}{2} \approx 1.366$$

Thus $\sin\left(x + \dfrac{\pi}{6}\right) \neq \sin x + \sin\dfrac{\pi}{6}$ for $x = \dfrac{\pi}{3}$. Therefore, the equation is not an identity.

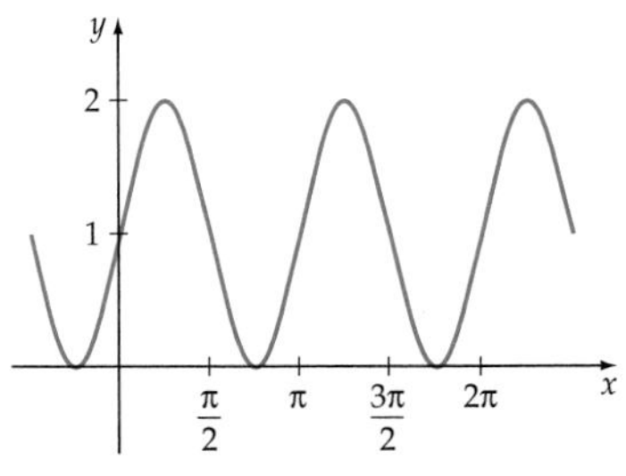

$f(x) = (\sin x + \cos x)^2$
$g(x) = 2\sin x\cos x + 1$

Figure 6.2

b. The graphs of $f(x) = (\sin x + \cos x)^2$ and $g(x) = 2\sin x\cos x + 1$ are shown in Figure 6.2. The graphs appear to be identical. To verify that $(\sin x + \cos x)^2 = 2\sin x\cos x + 1$ is an identity, we expand $(\sin x + \cos x)^2$ as shown below.

$$\begin{aligned}(\sin x + \cos x)^2 &= \sin^2 x + 2\sin x\cos x + \cos^2 x \\ &= 2\sin x\cos x + (\sin^2 x + \cos^2 x) \\ &= 2\sin x\cos x + 1 \qquad \bullet\ \sin^2 x + \cos^2 x = 1\end{aligned}$$

We have rewritten the left side in an equivalent form that is identical to the right side. Thus $(\sin x + \cos x)^2 = 2\sin x\cos x + 1$ is an identity.

TRY EXERCISE 2, EXERCISE SET 6.1, PAGE 476

take note

Each of the Pythagorean identities can be written in several different forms. For instance,

$$\sin^2 x + \cos^2 x = 1$$

can also be written as

$$\sin^2 x = 1 - \cos^2 x$$

and as

$$\cos^2 x = 1 - \sin^2 x$$

EXAMPLE 2 Verify an Identity

Verify the identity $1 - 2\sin^2 x = 2\cos^2 x - 1$.

Solution

Rewrite the right side of the equation.

$$\begin{aligned} 2\cos^2 x - 1 &= 2(1 - \sin^2 x) - 1 && \bullet\ \cos^2 x = 1 - \sin^2 x \\ &= 2 - 2\sin^2 x - 1 \\ &= 1 - 2\sin^2 x \end{aligned}$$

TRY EXERCISE 24, EXERCISE SET 6.1, PAGE 476

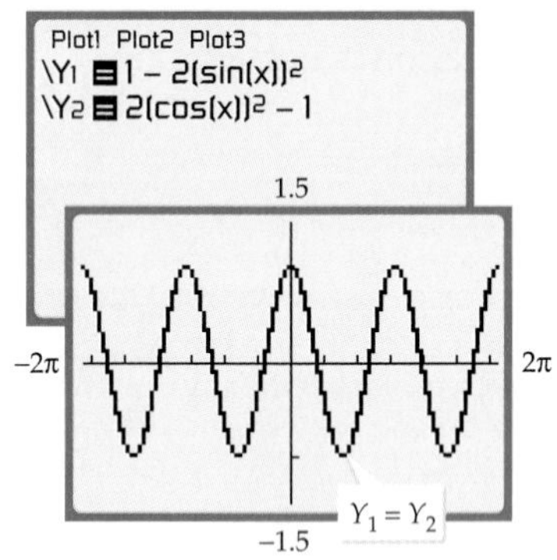

Figure 6.3

Figure 6.3 shows the graph of $f(x) = 1 - 2\sin^2 x$ and the graph of $g(x) = 2\cos^2 x - 1$ on the same coordinate axes. The fact that the graphs appear to be identical on the domain $[-2\pi, 2\pi]$ supports the verification in Example 2.

EXAMPLE 3 Factor to Verify an Identity

Verify the identity $\csc^2 x - \cos^2 x \csc^2 x = 1$.

Solution

Simplify the left side of the equation.

$$\begin{aligned} \csc^2 x - \cos^2 x \csc^2 x &= \csc^2 x(1 - \cos^2 x) && \bullet\ \text{Factor out } \csc^2 x. \\ &= \csc^2 x \sin^2 x \\ &= \frac{1}{\sin^2 x} \cdot \sin^2 x = 1 && \bullet\ \csc^2 x = \frac{1}{\sin^2 x} \end{aligned}$$

TRY EXERCISE 36, EXERCISE SET 6.1, PAGE 476

In the next example we make use of the guideline that indicates that it may be useful to multiply both the numerator and the denominator of a fraction by the same factor.

EXAMPLE 4 Multiply by a Conjugate to Verify an Identity

Verify the identity $\dfrac{\sin x}{1 + \cos x} = \dfrac{1 - \cos x}{\sin x}$.

Solution

Multiply the numerator and denominator of the left side of the identity by the conjugate of $1 + \cos x$ which is $1 - \cos x$.

$$\begin{aligned} \frac{\sin x}{1 + \cos x} &= \frac{\sin x}{1 + \cos x} \cdot \frac{1 - \cos x}{1 - \cos x} = \frac{\sin x(1 - \cos x)}{1 - \cos^2 x} \\ &= \frac{\sin x(1 - \cos x)}{\sin^2 x} = \frac{1 - \cos x}{\sin x} \end{aligned}$$

take note

The sum $a + b$ and the difference $a - b$ are called conjugates of each other.

TRY EXERCISE 46, EXERCISE SET 6.1, PAGE 477

EXAMPLE 5 Change to Sines and Cosines to Verify an Identity

Verify the identity $\dfrac{\sin x + \tan x}{1 + \cos x} = \tan x$.

Solution

Rewrite the left side of the identity in terms of sines and cosines.

$$\frac{\sin x + \tan x}{1 + \cos x} = \frac{\sin x + \dfrac{\sin x}{\cos x}}{1 + \cos x} \qquad \bullet\ \tan x = \frac{\sin x}{\cos x}$$

$$= \frac{\dfrac{\sin x \cos x + \sin x}{\cos x}}{1 + \cos x} \qquad \bullet\ \text{Write the terms in the numerator with a common denominator.}$$

$$= \frac{\sin x \cos x + \sin x}{\cos x(1 + \cos x)} \qquad \bullet\ \text{Simplify.}$$

$$= \frac{\sin x\cancel{(1 + \cos x)}}{\cos x\cancel{(1 + \cos x)}}$$

$$= \tan x$$

TRY EXERCISE 56, EXERCISE SET 6.1, PAGE 477

TOPICS FOR DISCUSSION

1. Explain why tan = sin/cos is not an identity.

2. Is $\cos |x| = |\cos x|$ an identity? Explain. What about $\cos |x| = \cos x$? Explain.

3. The identity $\sin^2 x + \cos^2 x = 1$ is one of the Pythagorean Identities. What are the other two Pythagorean Identities and how are they derived?

4. The graph of $y = \sin \dfrac{x}{2}$ for $0 \le x \le 2\pi$ is shown on the left. The graph of $y = \sqrt{\dfrac{1 - \cos x}{2}}$ for $0 \le x \le 2\pi$ is shown on the right.

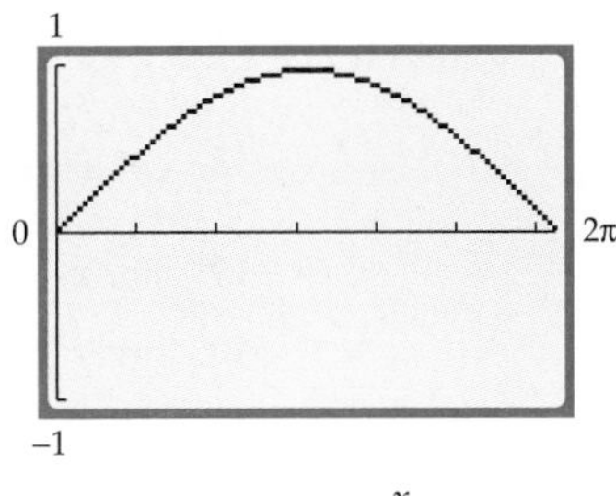

$y = \sin \dfrac{x}{2}$

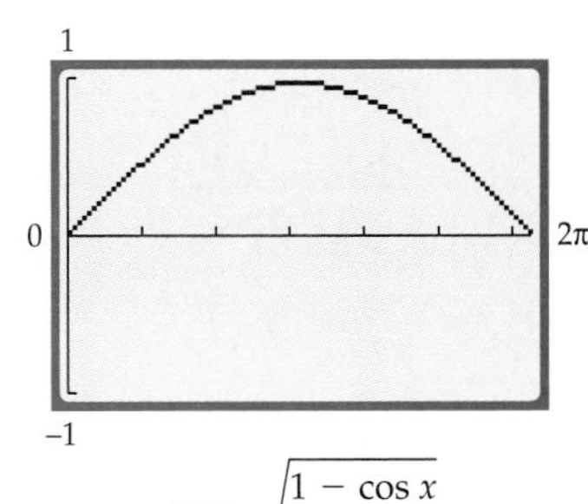

$y = \sqrt{\dfrac{1 - \cos x}{2}}$

The graphs appear identical, but the equation

$$\sin \frac{x}{2} = \sqrt{\frac{1 - \cos x}{2}}$$

is not an identity. Explain.

EXERCISE SET 6.1

In Exercises 1 to 12, decide whether each equation is an identity. If the equation is an identity, then verify the identity. If the equation is not an identity, then find a value in the domain for which the left side of the equation is not equal to the right side.

1. $(\sin x + \cos x)^2 = \sin^2 x + \cos^2 x$
2. $\tan 2x = 2 \tan x$
3. $\cos (x + 30°) = \cos x + \cos 30°$
4. $\sqrt{1 - \sin^2 x} = \cos x$
5. $\tan^4 x - \sec^4 x = \tan^2 x + \sec^2 x$
6. $\sqrt{1 + \tan^2 x} = |\sec x|$
7. $\tan^4 x - 1 = \sec^2 x$
8. $\sin^3 x + \cos^3 x = (\sin x + \cos x)(1 + \sin x \cos x)$
9. $2 \sin 30° = \sin 60°$
10. $\cot x \csc x \sec x = 1$
11. $\sin^4 x + \cos^2 x = 1 + \sin^2 x$
12. $\sec^2 x + \csc^2 x = 2 + \cot^2 x + \tan^2 x$

In Exercises 13 to 66, verify each identity.

13. $\tan x \csc x \cos x = 1$
14. $\sin x \cot x \sec x = 1$
15. $\dfrac{4 \sin^2 x - 1}{2 \sin x + 1} = 2 \sin x - 1$
16. $\dfrac{\sin^2 x - 2 \sin x + 1}{\sin x - 1} = \sin x - 1$
17. $(\sin x - \cos x)(\sin x + \cos x) = 1 - 2 \cos^2 x$
18. $(\tan x)(1 - \cot x) = \tan x - 1$
19. $\dfrac{1}{\sin x} - \dfrac{1}{\cos x} = \dfrac{\cos x - \sin x}{\sin x \cos x}$
20. $\dfrac{1}{\sin x} + \dfrac{3}{\cos x} = \dfrac{\cos x + 3 \sin x}{\sin x \cos x}$
21. $\dfrac{\cos x}{1 - \sin x} = \sec x + \tan x$
22. $\dfrac{\sin x}{1 - \cos x} = \csc x + \cot x$
23. $\dfrac{1 - \tan^4 x}{\sec^2 x} = 1 - \tan^2 x$
24. $\sin^4 x - \cos^4 x = \sin^2 x - \cos^2 x$
25. $\dfrac{1 + \tan^3 x}{1 + \tan x} = 1 - \tan x + \tan^2 x$
26. $\dfrac{\cos x \tan x - \sin x}{\cot x} = 0$
27. $\dfrac{\sin x - 2 + \dfrac{1}{\sin x}}{\sin x - \dfrac{1}{\sin x}} = \dfrac{\sin x - 1}{\sin x + 1}$
28. $\dfrac{\sin x}{1 - \cos x} - \dfrac{\sin x}{1 + \cos x} = 2 \cot x$
29. $(\sin x + \cos x)^2 = 1 + 2 \sin x \cos x$
30. $(\tan x + 1)^2 = \sec^2 x + 2 \tan x$
31. $\dfrac{\cos x}{1 + \sin x} = \sec x - \tan x$
32. $\dfrac{\sin x}{1 + \cos x} = \csc x - \cot x$
33. $\csc x = \dfrac{\cot x + \tan x}{\sec x}$
34. $\sec x = \dfrac{\cot x + \tan x}{\csc x}$
35. $\dfrac{\cos x \tan x + 2 \cos x - \tan x - 2}{\tan x + 2} = \cos x - 1$
36. $\dfrac{2 \sin x \cot x + \sin x - 4\cot x - 2}{2 \cot x + 1} = \sin x - 2$
37. $\sec x - \tan x = \dfrac{1 - \sin x}{\cos x}$
38. $\cot x - \csc x = \dfrac{\cos x - 1}{\sin x}$
39. $\sin^2 x - \cos^2 x = 2 \sin^2 x - 1$
40. $\sin^2 x - \cos^2 x = 1 - 2 \cos^2 x$
41. $\dfrac{1}{\sin^2 x} + \dfrac{1}{\cos^2 x} = \csc^2 x \sec^2 x$

42. $\dfrac{1}{\tan^2 x} - \dfrac{1}{\cot^2 x} = \csc^2 x - \sec^2 x$

43. $\sec x - \cos x = \sin x \tan x$

44. $\tan x + \cot x = \sec x \csc x$

45. $\dfrac{\dfrac{1}{\sin x} + 1}{\dfrac{1}{\sin x} - 1} = \tan^2 x + 2\tan x \sec x + \sec^2 x$

46. $\dfrac{\dfrac{1}{\sin x} + \dfrac{1}{\cos x}}{\dfrac{1}{\sin x} - \dfrac{1}{\cos x}} = \dfrac{\cos^2 x - \sin^2 x}{1 - 2\cos x \sin x}$

47. $\sin^4 x - \cos^4 x = 2\sin^2 x - 1$

48. $\sin^6 x + \cos^6 x = \sin^4 x - \sin^2 x \cos^2 x + \cos^4 x$

49. $\dfrac{1}{1 - \cos x} = \dfrac{1 + \cos x}{\sin^2 x}$

50. $1 + \sin x = \dfrac{\cos^2 x}{1 - \sin x}$

51. $\dfrac{\sin x}{1 - \sin x} - \dfrac{\cos x}{1 - \sin x} = \dfrac{1 - \cot x}{\csc x - 1}$

52. $\dfrac{\tan x}{1 + \tan x} - \dfrac{\cot x}{1 + \tan x} = 1 - \cot x$

53. $\dfrac{1}{1 + \cos x} - \dfrac{1}{1 - \cos x} = -2\cot x \csc x$

54. $\dfrac{1}{1 - \sin x} - \dfrac{1}{1 + \sin x} = 2\tan x \sec x$

55. $\dfrac{\dfrac{1}{\sin x} + \csc x}{\dfrac{1}{\sin x} - \sin x} = \dfrac{2}{\cos^2 x}$

56. $\dfrac{\dfrac{1}{\tan x} + \cot x}{\dfrac{1}{\tan x} + \tan x} = \dfrac{2}{\sec^2 x}$

57. $\sqrt{\dfrac{1 + \sin x}{1 - \sin x}} = \dfrac{1 + \sin x}{\cos x}, \quad \cos x > 0$

58. $\dfrac{\cos x + \cot x \sin x}{\cot x} = 2\sin x$

59. $\dfrac{\sin^3 x + \cos^3 x}{\sin x + \cos x} = 1 - \sin x \cos x$

60. $\dfrac{1 - \sin x}{1 + \sin x} - \dfrac{1 + \sin x}{1 - \sin x} = -4\sec x \tan x$

61. $\dfrac{\sec x - 1}{\sec x + 1} - \dfrac{\sec x + 1}{\sec x - 1} = -4\csc x \cot x$

62. $\dfrac{1}{1 - \cos x} - \dfrac{\cos x}{1 + \cos x} = 2\csc^2 x - 1$

63. $\dfrac{1 + \sin x}{\cos x} - \dfrac{\cos x}{1 - \sin x} = 0$

64. $(\sin x + \cos x + 1)^2 = 2(\sin x + 1)(\cos x + 1)$

65. $\dfrac{\sec x + \tan x}{\sec x - \tan x} = \dfrac{(\sin x + 1)^2}{\cos^2 x}$

66. $\dfrac{\sin^3 x - \cos^3 x}{\sin x + \cos x} = \dfrac{\csc^2 x - \cot x - 2\cos^2 x}{1 - \cot^2 x}$

67. Express $\cos x$ in terms of $\sin x$.

68. Express $\tan x$ in terms of $\cos x$.

69. Express $\sec x$ in terms of $\sin x$.

70. Express $\csc x$ in terms of $\sec x$.

In Exercises 71 to 78, use a graphing utility to graph each side of the equation to suggest that the equation is an identity.

71. $\sin 2x = 2\sin x \cos x$

72. $\sin^2 x + \cos^2 x = 1$

73. $\sin x + \cos x = \sqrt{2}\sin\left(x + \dfrac{\pi}{4}\right)$

74. $\cos 2x = 2\cos^2 x - 1$

75. $\cos\left(x + \dfrac{\pi}{3}\right) = \cos x \cos\dfrac{\pi}{3} - \sin x \sin\dfrac{\pi}{3}$

76. $\cos\left(x - \dfrac{\pi}{4}\right) = \cos x \cos\dfrac{\pi}{4} + \sin x \sin\dfrac{\pi}{4}$

77. $\sin\left(x + \dfrac{\pi}{6}\right) = \sin x \cos\dfrac{\pi}{6} + \cos x \sin\dfrac{\pi}{6}$

78. $\sin\left(x - \dfrac{\pi}{3}\right) = \sin x \cos\dfrac{\pi}{3} - \cos x \sin\dfrac{\pi}{3}$

SUPPLEMENTAL EXERCISES

In Exercises 79 to 84, verify the identity.

79. $\dfrac{1 - \sin x + \cos x}{1 + \sin x + \cos x} = \dfrac{\cos x}{\sin x + 1}$

80. $\dfrac{1 - \tan x + \sec x}{1 + \tan x - \sec x} = \dfrac{1 + \sec x}{\tan x}$

81. $\dfrac{2\sin^4 x + 2\sin^2 x \cos^2 x - 3\sin^2 x - 3\cos^2 x}{2\sin^2 x}$

$= 1 - \dfrac{3}{2}\csc^2 x$

82. $\dfrac{4\tan x \sec^2 x - 4\tan x - \sec^2 x + 1}{4\tan^3 x - \tan^2 x} = 1$

83. $\dfrac{\sin x(\tan x + 1) - 2\tan x \cos x}{\sin x - \cos x} = \tan x$

84. $\dfrac{\sin^2 x \cos x + \cos^3 x - \sin^3 x \cos x - \sin x \cos^3 x}{1 - \sin^2 x} = \dfrac{\cos x}{1 + \sin x}$

85. Verify the identity $\sin^4 x + \cos^4 x = 1 - 2\sin^2 x \cos^2 x$ by completing the square of the left side of the identity.

86. Verify the identity $\tan^4 x + \sec^4 x = 1 + 2\tan^2 x \sec^2 x$ by completing the square of the left side of the identity.

PROJECTS

1.  **GRADING A QUIZ** Suppose you are a teacher's assistant. You are to assist the teacher of a trigonometry class by grading a four-question quiz. Each question asks the student to find a trigonometric expression that models a given application. The teacher has prepared an answer key. These answers are shown on the left below. A student gives as answers the expressions shown on the right. Determine for which problems the student has given a correct response.

Answer Key	Student's Response
1. $\csc x \sec x$	1. $\cot x + \tan x$
2. $\cos^2 x$	2. $(1 + \sin x)(1 - \sin x)$
3. $\cos x \cot x$	3. $\csc x - \sec x$
4. $\csc x \cot x$	4. $\sin x(\cot x + \cot^3 x)$

SECTION 6.2 SUM, DIFFERENCE, AND COFUNCTION IDENTITIES

- IDENTITIES THAT INVOLVE $(\alpha \pm \beta)$
- COFUNCTIONS
- ADDITIONAL SUM AND DIFFERENCE IDENTITIES
- REDUCTION FORMULAS

6A

◆ IDENTITIES THAT INVOLVE $(\alpha \pm \beta)$

Each identity in Section 6.1 involved only one variable. We now consider identities that involve a trigonometric function of the sum or difference of two variables.

Sum and Difference Identities

$$\cos(\alpha - \beta) = \cos\alpha\cos\beta + \sin\alpha\sin\beta$$

$$\cos(\alpha + \beta) = \cos\alpha\cos\beta - \sin\alpha\sin\beta$$

$$\sin(\alpha - \beta) = \sin\alpha\cos\beta - \cos\alpha\sin\beta$$

$$\sin(\alpha + \beta) = \sin\alpha\cos\beta + \cos\alpha\sin\beta$$

$$\tan(\alpha + \beta) = \frac{\tan\alpha + \tan\beta}{1 - \tan\alpha\tan\beta}$$

$$\tan(\alpha - \beta) = \frac{\tan\alpha - \tan\beta}{1 + \tan\alpha\tan\beta}$$

To establish the identity for $\cos(\alpha - \beta)$, we make use of the unit circle shown in **Figure 6.4.** The angles α and β are drawn in standard position, with OA and OB as the terminal sides of α and β, respectively. The coordinates of A are $(\cos\alpha, \sin\alpha)$, and the coordinates of B are $(\cos\beta, \sin\beta)$. The angle $(\alpha - \beta)$ is formed by the terminal sides of the angles α and β (angle AOB).

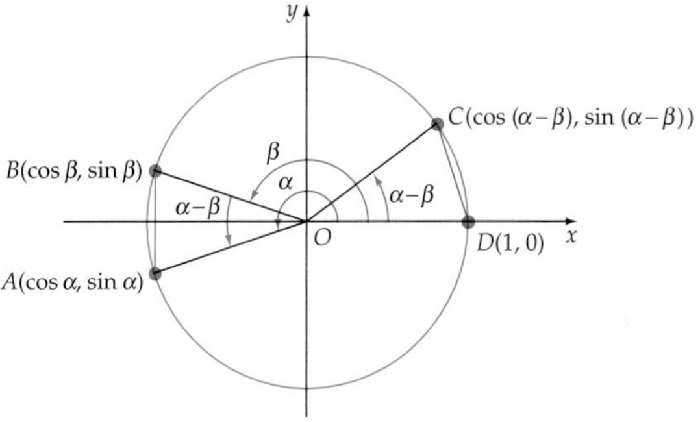

Figure 6.4

An angle equal in measure to angle $(\alpha - \beta)$ is placed in standard position in the same figure (angle COD). From geometry, if two central angles of a circle have the same measure, then their chords are also equal in measure. Thus the chords AB and CD are equal in length. Using the distance formula, we can calculate the lengths of the chords AB and CD.

$$d(A, B) = \sqrt{(\cos\alpha - \cos\beta)^2 + (\sin\alpha - \sin\beta)^2}$$
$$d(C, D) = \sqrt{[\cos(\alpha - \beta) - 1]^2 + [\sin(\alpha - \beta) - 0]^2}$$

Because $d(A, B) = d(C, D)$, we have

$$\sqrt{(\cos\alpha - \cos\beta)^2 + (\sin\alpha - \sin\beta)^2} = \sqrt{[\cos(\alpha - \beta) - 1]^2 + [\sin(\alpha - \beta)]^2}$$

Squaring each side of the equation and simplifying, we obtain

$$(\cos\alpha - \cos\beta)^2 + (\sin\alpha - \sin\beta)^2 = [\cos(\alpha - \beta) - 1]^2 + [\sin(\alpha - \beta)]^2$$

$$\cos^2\alpha - 2\cos\alpha\cos\beta + \cos^2\beta + \sin^2\alpha - 2\sin\alpha\sin\beta + \sin^2\beta$$
$$= \cos^2(\alpha - \beta) - 2\cos(\alpha - \beta) + 1 + \sin^2(\alpha - \beta)$$

$$\cos^2\alpha + \sin^2\alpha + \cos^2\beta + \sin^2\beta - 2\cos\alpha\cos\beta - 2\sin\alpha\sin\beta$$
$$= \cos^2(\alpha - \beta) + \sin^2(\alpha - \beta) + 1 - 2\cos(\alpha - \beta)$$

Simplifying by using $\sin^2\theta + \cos^2\theta = 1$, we have

$$2 - 2\sin\alpha\sin\beta - 2\cos\alpha\cos\beta = 2 - 2\cos(\alpha - \beta)$$

Solving for $\cos(\alpha - \beta)$ gives us

$$\cos(\alpha - \beta) = \cos\alpha\cos\beta + \sin\alpha\sin\beta$$

To derive an identity for $\cos(\alpha + \beta)$, write $\cos(\alpha + \beta)$ as $\cos[\alpha - (-\beta)]$.

$$\cos(\alpha + \beta) = \cos[\alpha - (-\beta)] = \cos\alpha\cos(-\beta) + \sin\alpha\sin(-\beta)$$

Recall that $\cos(-\beta) = \cos\beta$ and $\sin(-\beta) = -\sin\beta$. Substituting into the previous equation, we obtain the identity

$$\cos(\alpha + \beta) = \cos\alpha\cos\beta - \sin\alpha\sin\beta$$

◆ COFUNCTIONS

take note

To visualize the cofunction identities, consider the right triangle shown in the following figure.

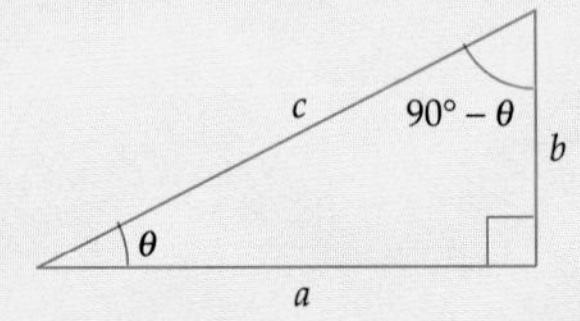

If θ is the degree measure of one of the acute angles, then the degree measure of the other acute angle is $(90° - \theta)$. Using the definitions of the trigonometric functions gives us

$$\sin\theta = \frac{b}{c} = \cos(90° - \theta)$$

$$\tan\theta = \frac{b}{a} = \cot(90° - \theta)$$

$$\sec\theta = \frac{c}{a} = \csc(90° - \theta)$$

These identities state that the value of a trigonometric function of θ is equal to the cofunction of the complement of θ.

Any pair of trigonometric functions f and g for which

$$f(x) = g(90° - x) \quad \text{and} \quad g(x) = f(90° - x)$$

are said to be **cofunctions.**

Cofunction Identities

$$\sin(90° - \theta) = \cos\theta \qquad \cos(90° - \theta) = \sin\theta$$

$$\tan(90° - \theta) = \cot\theta \qquad \cot(90° - \theta) = \tan\theta$$

$$\sec(90° - \theta) = \csc\theta \qquad \csc(90° - \theta) = \sec\theta$$

If θ is in radian measure, replace 90° with $\pi/2$.

To verify that the sine function and the cosine function are cofunctions, we make use of the identity for $\cos(\alpha - \beta)$.

$$\begin{aligned}\cos(90° - \beta) &= \cos 90° \cos\beta + \sin 90° \sin\beta \\ &= 0 \cdot \cos\beta + 1 \cdot \sin\beta\end{aligned}$$

which gives

$$\cos(90° - \beta) = \sin\beta$$

Thus the sine of an angle is equal to the cosine of its complement. Using $\cos(90° - \beta) = \sin\beta$ with $\beta = 90° - \alpha$, we have

$$\cos\alpha = \cos[90° - (90° - \alpha)] = \sin(90° - \alpha)$$

Therefore,

$$\cos\alpha = \sin(90° - \alpha)$$

We can use the ratio identities to show that the tangent and cotangent functions are cofunctions.

$$\tan(90° - \theta) = \frac{\sin(90° - \theta)}{\cos(90° - \theta)} = \frac{\cos\theta}{\sin\theta} = \cot\theta$$

$$\cot(90° - \theta) = \frac{\cos(90° - \theta)}{\sin(90° - \theta)} = \frac{\sin\theta}{\cos\theta} = \tan\theta$$

The secant and cosecant functions are also cofunctions.

◆ ADDITIONAL SUM AND DIFFERENCE IDENTITIES

We can use the cofunction identities to verify the remaining sum and difference identities. To derive an identity for $\sin(\alpha + \beta)$, substitute $\alpha + \beta$ for θ in the cofunction identity $\sin\theta = \cos(90° - \theta)$.

$$\begin{aligned}\sin\theta &= \cos(90^\circ - \theta)\\ \sin(\alpha+\beta) &= \cos[90^\circ - (\alpha+\beta)]\\ &= \cos[(90^\circ-\alpha)-\beta] && \text{• Rewrite as the difference of two angles.}\\ &= \cos(90^\circ-\alpha)\cos\beta + \sin(90^\circ-\alpha)\sin\beta\\ &= \sin\alpha\cos\beta + \cos\alpha\sin\beta\end{aligned}$$

Therefore,

$$\sin(\alpha+\beta) = \sin\alpha\cos\beta + \cos\alpha\sin\beta$$

We can also derive an identity for $\sin(\alpha-\beta)$ by rewriting $(\alpha-\beta)$ as $[\alpha+(-\beta)]$.

$$\begin{aligned}\sin(\alpha-\beta) &= \sin[\alpha+(-\beta)]\\ &= \sin\alpha\cos(-\beta) + \cos\alpha\sin(-\beta)\\ &= \sin\alpha\cos\beta - \cos\alpha\sin\beta && \text{• } \cos(-\beta)=\cos\beta,\ \sin(-\beta)=-\sin\beta\end{aligned}$$

Thus

$$\sin(\alpha-\beta) = \sin\alpha\cos\beta - \cos\alpha\sin\beta$$

The identity for $\tan(\alpha+\beta)$ is a result of the identity $\tan\theta = \dfrac{\sin\theta}{\cos\theta}$ and the identities for $\sin(\alpha+\beta)$ and $\cos(\alpha+\beta)$.

$$\begin{aligned}\tan(\alpha+\beta) &= \frac{\sin(\alpha+\beta)}{\cos(\alpha+\beta)} = \frac{\sin\alpha\cos\beta+\cos\alpha\sin\beta}{\cos\alpha\cos\beta-\sin\alpha\sin\beta}\\ &= \frac{\dfrac{\sin\alpha\cos\beta}{\cos\alpha\cos\beta}+\dfrac{\cos\alpha\sin\beta}{\cos\alpha\cos\beta}}{\dfrac{\cos\alpha\cos\beta}{\cos\alpha\cos\beta}-\dfrac{\sin\alpha\sin\beta}{\cos\alpha\cos\beta}} && \text{• Multiply both the numerator and the denominator by } \frac{1}{\cos\alpha\cos\beta} \text{ and simplify.}\end{aligned}$$

Therefore,

$$\tan(\alpha+\beta) = \frac{\tan\alpha+\tan\beta}{1-\tan\alpha\tan\beta}$$

The tangent function is an odd function, so $\tan(-\theta) = -\tan\theta$. Rewriting $(\alpha-\beta)$ as $[\alpha+(-\beta)]$ enables us to derive an identity for $\tan(\alpha-\beta)$.

$$\tan(\alpha-\beta) = \tan[\alpha+(-\beta)] = \frac{\tan\alpha+\tan(-\beta)}{1-\tan\alpha\tan(-\beta)}$$

Therefore,

$$\tan(\alpha-\beta) = \frac{\tan a-\tan\beta}{1+\tan\alpha\tan\beta}$$

The sum and difference identities can be used to simplify some trigonometric expressions.

EXAMPLE 1 Simplify Trigonometric Expressions

Write each expression in terms of a single trigonometric function.

a. $\sin 5x \cos 3x - \cos 5x \sin 3x$ b. $\dfrac{\tan 4\alpha + \tan \alpha}{1 - \tan 4\alpha \tan \alpha}$

Solution

a. $\sin 5x \cos 3x - \cos 5x \sin 3x = \sin (5x - 3x) = \sin 2x$

b. $\dfrac{\tan 4\alpha + \tan \alpha}{1 - \tan 4\alpha \tan \alpha} = \tan (4\alpha + \alpha) = \tan 5\alpha$

TRY EXERCISE 20, EXERCISE SET 6.2, PAGE 484

EXAMPLE 2 Evaluate a Trigonometric Function

Given $\tan \alpha = -4/3$ for α in Quadrant II and $\tan \beta = -5/12$ for β in Quadrant IV, find $\sin (\alpha + \beta)$

Solution

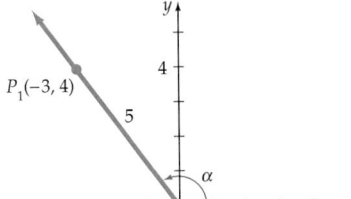

See Figure 6.5. Because $\tan \alpha = y/x = -4/3$ and the terminal side of α is in Quadrant II, $P_1(-3, 4)$ is a point on the terminal side of α. Similarly, $P_2(12, -5)$ is a point on the terminal side of β. Using the Pythagorean Theorem, we find that the length of the line segment OP_1 is 5 and the length of OP_2 is 13.

$$\sin (\alpha + \beta) = \sin \alpha \cos \beta + \cos \alpha \sin \beta$$
$$= \frac{4}{5} \cdot \frac{12}{13} + \frac{-3}{5} \cdot \frac{-5}{13} = \frac{48}{65} + \frac{15}{65} = \frac{63}{65}$$

Figure 6.5

TRY EXERCISE 32, EXERCISE SET 6.2, PAGE 484

EXAMPLE 3 Verify an Identity

Verify the identity $\cos (\pi - \theta) = -\cos \theta$.

Solution

Use the identity for $\cos (\alpha - \beta)$.

$$\cos (\pi - \theta) = \cos \pi \cos \theta + \sin \pi \sin \theta = -1 \cdot \cos \theta + 0 \cdot \sin \theta = -\cos \theta$$

TRY EXERCISE 44, EXERCISE SET 6.2, PAGE 485

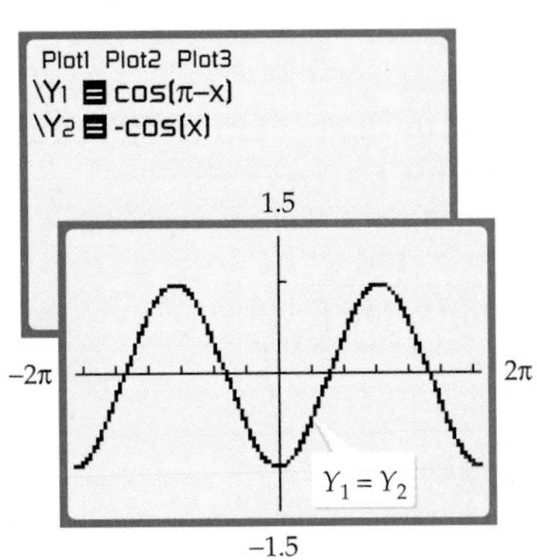

Figure 6.6

Figure 6.6 shows the graphs of $f(\theta) = \cos (\pi - \theta)$ and $g(\theta) = -\cos \theta$ on the same coordinate axes. The fact that the graphs appear to be identical supports the verification in Example 3.

EXAMPLE 4 Verify an Identity

Verify the identity $\dfrac{\cos 4\theta}{\sin \theta} - \dfrac{\sin 4\theta}{\cos \theta} = \dfrac{\cos 5\theta}{\sin \theta \cos \theta}$.

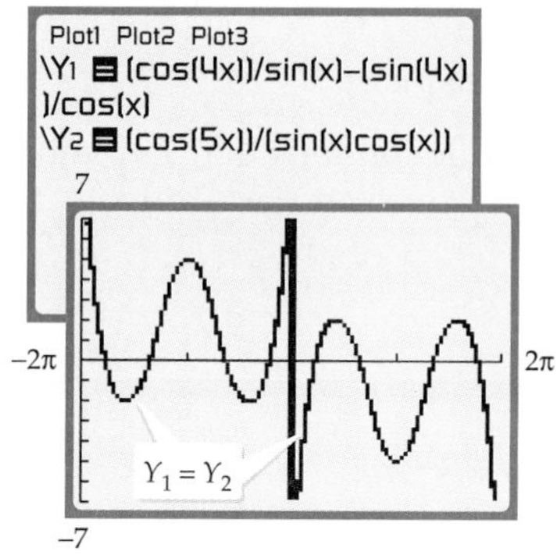

Plot1 Plot2 Plot3
\Y1 ■ (cos(4x))/sin(x)−(sin(4x))/cos(x)
\Y2 ■ (cos(5x))/(sin(x)cos(x))

7, −2π, 2π, −7, $Y_1 = Y_2$

Figure 6.7

Solution

Subtract the fractions on the left side of the equation.

$$\frac{\cos 4\theta}{\sin \theta} - \frac{\sin 4\theta}{\cos \theta} = \frac{\cos 4\theta \cos \theta - \sin 4\theta \sin \theta}{\sin \theta \cos \theta}$$

$$= \frac{\cos(4\theta + \theta)}{\sin \theta \cos \theta} = \frac{\cos 5\theta}{\sin \theta \cos \theta}$$

• Use the identity for $\cos(\alpha + \beta)$.

TRY EXERCISE 56, EXERCISE SET 6.2, PAGE 485

Figure 6.7 shows the graph of $f(\theta) = \dfrac{\cos 4\theta}{\sin \theta} - \dfrac{\sin 4\theta}{\cos \theta}$ and the graph of $g(\theta) = \dfrac{\cos 5\theta}{\sin \theta \cos \theta}$ on the same coordinate axes. The fact that the graphs appear to be identical supports the verification in Example 4.

◆ REDUCTION FORMULAS

The sum or difference identities can be used to write expressions such as

$$\sin(\theta + k\pi) \qquad \sin(\theta + 2k\pi) \qquad \text{and} \qquad \cos[\theta + (2k + 1)\pi]$$

where k is an integer, as expressions involving only $\sin \theta$ or $\cos \theta$. The resulting formulas are called **reduction formulas.**

EXAMPLE 5 Find Reduction Formulas

Write as a function involving only $\sin \theta$.

$$\sin[\theta + (2k + 1)\pi], \quad \text{where } k \text{ is an integer.}$$

Solution

Applying the identity $\sin(\alpha + \beta) = \sin \alpha \cos \beta + \cos \alpha \sin \beta$ yields

$$\sin[\theta + (2k + 1)\pi] = \sin \theta \cos[(2k + 1)\pi] + \cos \theta \sin[(2k + 1)\pi]$$

If k is an integer, then $2k + 1$ is an odd integer. The cosine of any odd multiple of π equals -1, and the sine of any odd multiple of π is 0. This gives us

$$\sin[\theta + (2k + 1)\pi] = (\sin \theta)(-1) + (\cos \theta)(0) = -\sin \theta$$

Thus $\sin[\theta + (2k + 1)\pi] = -\sin \theta$, for any integer k.

TRY EXERCISE 68, EXERCISE SET 6.2, PAGE 485

TOPICS FOR DISCUSSION

1. Does $\sin(\alpha + \beta) = \sin \alpha + \sin \beta$ for all values of α and β? If not, find nonzero values of α and β for which $\sin(\alpha + \beta) \neq \sin \alpha + \sin \beta$.
2. If k is an integer, then $2k + 1$ is an odd integer. Do you agree? Explain.
3. What are the trigonometric cofunction identities? Explain.
4. Is $\tan(\theta + k\pi) = \tan \theta$, where k is an integer, a reduction formula? Explain.

EXERCISE SET 6.2

In Exercises 1 to 18, find the exact value of the expression.

1. $\sin(45° + 30°)$
2. $\sin(330° + 45°)$
3. $\cos(45° - 30°)$
4. $\cos(120° - 45°)$
5. $\tan(45° - 30°)$
6. $\tan(240° - 45°)$
7. $\sin\left(\frac{5\pi}{4} - \frac{\pi}{6}\right)$
8. $\sin\left(\frac{4\pi}{3} + \frac{\pi}{4}\right)$
9. $\cos\left(\frac{3\pi}{4} + \frac{\pi}{6}\right)$
10. $\cos\left(\frac{\pi}{4} - \frac{\pi}{3}\right)$
11. $\tan\left(\frac{\pi}{6} + \frac{\pi}{4}\right)$
12. $\tan\left(\frac{11\pi}{6} - \frac{\pi}{4}\right)$
13. $\cos 212° \cos 122° + \sin 212° \sin 122°$
14. $\sin 167° \cos 107° - \cos 167° \sin 107°$
15. $\sin\frac{5\pi}{12}\cos\frac{\pi}{4} - \cos\frac{5\pi}{12}\sin\frac{\pi}{4}$
16. $\cos\frac{\pi}{12}\cos\frac{\pi}{4} - \sin\frac{\pi}{12}\sin\frac{\pi}{4}$
17. $\dfrac{\tan 7\pi/12 - \tan \pi/4}{1 + \tan 7\pi/12 \tan \pi/4}$
18. $\dfrac{\tan \pi/6 + \tan \pi/3}{1 - \tan \pi/6 \tan \pi/3}$

In Exercises 19 to 30, write each expression in terms of a single trigonometric function.

19. $\sin 7x \cos 2x - \cos 7x \sin 2x$
20. $\sin x \cos 3x + \cos x \sin 3x$
21. $\cos x \cos 2x + \sin x \sin 2x$
22. $\cos 4x \cos 2x - \sin 4x \sin 2x$
23. $\sin 7x \cos 3x - \cos 7x \sin 3x$
24. $\cos x \cos 5x - \sin x \sin 5x$
25. $\cos 4x \cos(-2x) - \sin 4x \sin(-2x)$
26. $\sin(-x)\cos 3x - \cos(-x)\sin 3x$
27. $\sin\frac{x}{3}\cos\frac{2x}{3} + \cos\frac{x}{3}\sin\frac{2x}{3}$
28. $\cos\frac{3x}{4}\cos\frac{x}{4} + \sin\frac{3x}{4}\sin\frac{x}{4}$
29. $\dfrac{\tan 3x + \tan 4x}{1 - \tan 3x \tan 4x}$
30. $\dfrac{\tan 2x - \tan 3x}{1 + \tan 2x \tan 3x}$

In Exercises 31 to 42, find the exact value of the given functions.

31. Given $\tan \alpha = -4/3$, α in Quadrant II, and $\tan \beta = 15/8$, β in Quadrant III, find
    a. $\sin(\alpha - \beta)$ b. $\cos(\alpha + \beta)$ c. $\tan(\alpha - \beta)$
32. Given $\tan \alpha = 24/7$, α in Quadrant I, and $\sin \beta = -8/17$, β in Quadrant III, find
    a. $\sin(\alpha + \beta)$ b. $\cos(\alpha + \beta)$ c. $\tan(\alpha - \beta)$
33. Given $\sin \alpha = 3/5$, α in Quadrant I, and $\cos \beta = -5/13$, β in Quadrant II, find
    a. $\sin(\alpha - \beta)$ b. $\cos(\alpha + \beta)$ c. $\tan(\alpha - \beta)$
34. Given $\sin \alpha = 24/25$, α in Quadrant II, and $\cos \beta = -4/5$, β in Quadrant III, find
    a. $\cos(\beta - \alpha)$ b. $\sin(\alpha + \beta)$ c. $\tan(\alpha + \beta)$
35. Given $\sin \alpha = -4/5$, α in Quadrant III, and $\cos \beta = -12/13$, β in Quadrant II, find
    a. $\sin(\alpha - \beta)$ b. $\cos(\alpha + \beta)$ c. $\tan(\alpha + \beta)$
36. Given $\sin \alpha = -7/25$, α in Quadrant IV, and $\cos \beta = 8/17$, β in Quadrant IV, find
    a. $\sin(\alpha + \beta)$ b. $\cos(\alpha - \beta)$ c. $\tan(\alpha + \beta)$

37. Given $\cos\alpha = 15/17$, α in Quadrant I, and $\sin\beta = -3/5$, β in Quadrant III, find

a. $\sin(\alpha+\beta)$ **b.** $\cos(\alpha-\beta)$ **c.** $\tan(\alpha-\beta)$

38. Given $\cos\alpha = -7/25$, α in Quadrant II, and $\sin\beta = -12/13$, β in Quadrant IV, find

a. $\sin(\alpha+\beta)$ **b.** $\cos(\alpha+\beta)$ **c.** $\tan(\alpha-\beta)$

39. Given $\cos\alpha = -3/5$, α in Quadrant III, and $\sin\beta = 5/13$, β in Quadrant I, find

a. $\sin(\alpha-\beta)$ **b.** $\cos(\alpha+\beta)$ **c.** $\tan(\alpha+\beta)$

40. Given $\cos\alpha = 8/17$, α in Quadrant IV, and $\sin\beta = -24/25$, β in Quadrant III, find

a. $\sin(\alpha-\beta)$ **b.** $\cos(\alpha+\beta)$ **c.** $\tan(\alpha+\beta)$

41. Given $\sin\alpha = 3/5$, α in Quadrant I, and $\tan\beta = 5/12$, β in Quadrant III, find

a. $\sin(\alpha+\beta)$ **b.** $\cos(\alpha-\beta)$ **c.** $\tan(\alpha-\beta)$

42. Given $\tan\alpha = 15/8$, α in Quadrant I, and $\tan\beta = -7/24$, β in Quadrant IV, find

a. $\sin(\alpha-\beta)$ **b.** $\cos(\alpha-\beta)$ **c.** $\tan(\alpha+\beta)$

In Exercises 43 to 66, verify the identity.

43. $\cos\left(\frac{\pi}{2}-\theta\right) = \sin\theta$

44. $\cos(\theta+\pi) = -\cos\theta$

45. $\sin\left(\theta+\frac{\pi}{2}\right) = \cos\theta$

46. $\sin(\theta+\pi) = -\sin\theta$

47. $\tan\left(\theta+\frac{\pi}{4}\right) = \frac{\tan\theta+1}{1-\tan\theta}$

48. $\tan 2\theta = \frac{2\tan\theta}{1-\tan^2\theta}$

49. $\cos\left(\frac{3\pi}{2}-\theta\right) = -\sin\theta$

50. $\sin\left(\frac{3\pi}{2}+\theta\right) = -\cos\theta$

51. $\cot\left(\frac{\pi}{2}-\theta\right) = \tan\theta$

52. $\cot(\pi+\theta) = \cot\theta$

53. $\csc(\pi-\theta) = \csc\theta$

54. $\sec\left(\frac{\pi}{2}-\theta\right) = \csc\theta$

55. $\sin 6x\cos 2x - \cos 6x\sin 2x = 2\sin 2x\cos 2x$

56. $\cos 5x\cos 3x + \sin 5x\sin 3x = \cos^2 x - \sin^2 x$

57. $\cos(\alpha+\beta) + \cos(\alpha-\beta) = 2\cos\alpha\cos\beta$

58. $\cos(\alpha-\beta) - \cos(\alpha+\beta) = 2\sin\alpha\sin\beta$

59. $\sin(\alpha+\beta) + \sin(\alpha-\beta) = 2\sin\alpha\cos\beta$

60. $\sin(\alpha-\beta) - \sin(\alpha+\beta) = -2\cos\alpha\sin\beta$

61. $\frac{\cos(\alpha-\beta)}{\sin(\alpha+\beta)} = \frac{\cot\alpha+\tan\beta}{1+\cot\alpha\tan\beta}$

62. $\frac{\sin(\alpha+\beta)}{\sin(\alpha-\beta)} = \frac{1+\cot\alpha\tan\beta}{1-\cot\alpha\tan\beta}$

63. $\sin\left(\frac{\pi}{2}+\alpha-\beta\right) = \cos\alpha\cos\beta + \sin\alpha\sin\beta$

64. $\cos\left(\frac{\pi}{2}+\alpha+\beta\right) = -(\sin\alpha\cos\beta + \cos\alpha\sin\beta)$

65. $\sin 3x = 3\sin x - 4\sin^3 x$

66. $\cos 3x = 4\cos^3 x - 3\cos x$

In Exercises 67 to 72, write the given expression as a function that involves only $\sin\theta$, $\cos\theta$, or $\tan\theta$. (In Exercises 70, 71 and 72, assume k is an integer.)

67. $\cos(\theta+3\pi)$

68. $\sin(\theta+2\pi)$

69. $\tan(\theta+\pi)$

70. $\cos[\theta+(2k+1)\pi]$

71. $\sin(\theta+2k\pi)$

72. $\sin(\theta-k\pi)$

In Exercises 73 to 76, use a graphing utility to graph the function on each side of the equation to suggest that the equation is an identity.

73. $\sin\left(\frac{\pi}{2}-x\right) = \cos x$

74. $\cos(x+\pi) = -\cos x$

75. $\sin 7x\cos 2x - \cos 7x\sin 2x = \sin 5x$

76. $\sin 3x = 3\sin x - 4\sin^3 x$

SUPPLEMENTAL EXERCISES

In Exercises 77 to 83, verify the identity.

77. $\sin(x-y)\cdot\sin(x+y) = \sin^2 x\cos^2 y - \cos^2 x\sin^2 y$

78. $\sin(x+y+z) = \sin x\cos y\cos z + \cos x\sin y\cos z + \cos x\cos y\sin z - \sin x\sin y\sin z$

79. $\cos(x+y+z) = \cos x\cos y\cos z - \sin x\sin y\cos z - \sin x\cos y\sin z - \cos x\sin y\sin z$

80. $\frac{\sin(x+y)}{\sin x\sin y} = \cot x + \cot y$

81. $\frac{\cos(x-y)}{\cos x\sin y} = \cot y + \tan x$

82. $\frac{\sin(x+h)-\sin x}{h} = \cos x\frac{\sin h}{h} + \sin x\frac{(\cos h-1)}{h}$

83. $\frac{\cos(x+h)-\cos x}{h} = \cos x\frac{(\cos h-1)}{h} - \sin x\frac{\sin h}{h}$

84. **MODEL RESISTANCE** The drag (resistance) on a fish when it is swimming is 2 to 3 times the drag when it is gliding. To compensate for this, some fish swim in a sawtooth pattern as shown in the accompanying figure. The ratio of the amount of energy the fish expends when swimming upward at angle β and then gliding down at angle α to the energy it expends swimming horizontally is given by

$$E_R = \frac{k \sin \alpha + \sin \beta}{k \sin (\alpha + \beta)}$$

where k is a value such that $2 \le k \le 3$, and k depends on the assumptions we make about the amount of drag experienced by the fish. Find E_R for $k = 2$, $\alpha = 10°$, and $\beta = 20°$.

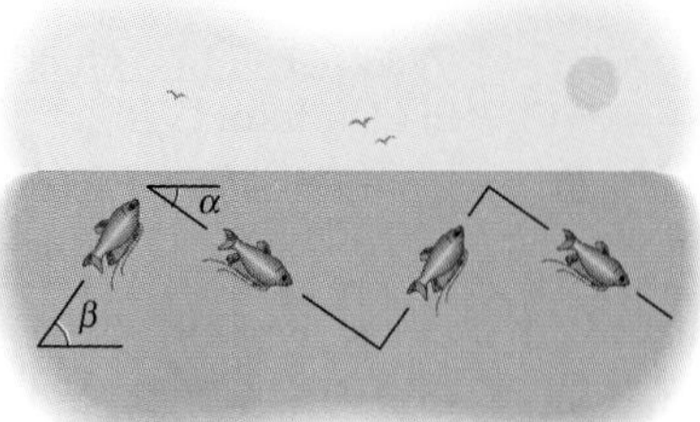

PROJECTS

1. **INTERSECTING LINES** In the figure shown at the right, two nonvertical lines intersect in a plane. The slope of line l_1 is m_1 and the slope of line l_2 is m_2.
   a. Show that the tangent of the smallest positive angle γ from l_1 to l_2 is given by

   $$\tan \gamma = \frac{m_2 - m_1}{1 + m_1 m_2}$$

   b. Use this equation to find the measure of the angle (to the nearest 0.1°) from the line $y = x + 5$ to the line $y = 3x - 4$.
   c. Two nonvertical lines intersect at the point (1, 5). The measure of the smallest positive angle between the lines is $\gamma = 60°$. The first line is given by $y = 0.5x + 4.5$. What is the equation (in slope-intercept form) of the second line?

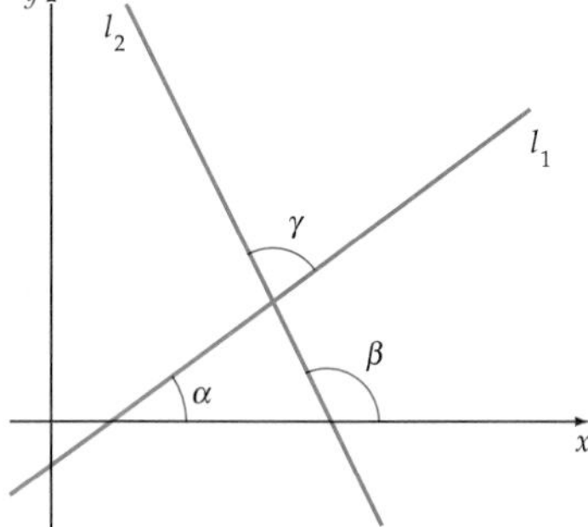

SECTION 6.3 DOUBLE- AND HALF-ANGLE IDENTITIES

- DOUBLE-ANGLE IDENTITIES
- HALF-ANGLE IDENTITIES

◆ DOUBLE-ANGLE IDENTITIES

By using the sum identities, we can derive identities for $f(2\alpha)$, where f is a trigonometric function. These are called the *double-angle identities*. To find the sine of a double angle, substitute α for β in the identity for $\sin(\alpha + \beta)$.

$$\sin(\alpha + \beta) = \sin \alpha \cos \beta + \cos \alpha \sin \beta$$

$$\sin(\alpha + \alpha) = \sin \alpha \cos \alpha + \cos \alpha \sin \alpha \qquad \text{• Let } \beta = \alpha.$$

$$\sin 2\alpha = 2 \sin \alpha \cos \alpha$$

take note

One way of showing that $\sin 2x \neq 2 \sin x$ is by graphing $y = \sin 2x$ and $y = 2 \sin x$ and observing that the graphs are not the same.

A double-angle identity for cosine is derived in a similar manner.

$$\cos(\alpha + \beta) = \cos\alpha\cos\beta - \sin\alpha\sin\beta$$
$$\cos(\alpha + \alpha) = \cos\alpha\cos\alpha - \sin\alpha\sin\alpha \qquad \text{• Let } \beta = \alpha.$$
$$\cos 2\alpha = \cos^2\alpha - \sin^2\alpha$$

There are two alternative forms of the double-angle identity for $\cos 2\alpha$. Using $\cos^2\alpha = 1 - \sin^2\alpha$, we can rewrite the identity for $\cos 2\alpha$ as follows:

$$\cos 2\alpha = \cos^2\alpha - \sin^2\alpha$$
$$\cos 2\alpha = (1 - \sin^2\alpha) - \sin^2\alpha \qquad \text{• } \cos^2\alpha = 1 - \sin^2\alpha$$
$$\cos 2\alpha = 1 - 2\sin^2\alpha$$

We can also rewrite $\cos 2\alpha$ as

$$\cos 2\alpha = \cos^2\alpha - \sin^2\alpha$$
$$\cos 2\alpha = \cos^2\alpha - (1 - \cos^2\alpha) \qquad \text{• } \sin^2\alpha = 1 - \cos^2\alpha$$
$$\cos 2\alpha = 2\cos^2\alpha - 1$$

The double-angle identity for the tangent function is derived from the identity for $\tan(\alpha + \beta)$ with $\beta = \alpha$.

$$\tan(\alpha + \beta) = \frac{\tan\alpha + \tan\beta}{1 - \tan\alpha\tan\beta}$$
$$\tan(\alpha + \alpha) = \frac{\tan\alpha + \tan\alpha}{1 - \tan\alpha\tan\alpha} \qquad \text{• Let } \beta = \alpha.$$
$$\tan 2\alpha = \frac{2\tan\alpha}{1 - \tan^2\alpha}$$

The double-angle identities are often used to write a trigonometric expression in terms of a single trigonometric function.

EXAMPLE 1 Simplify a Trigonometric Expression

Write $4 \sin 5\theta \cos 5\theta$ as a single trigonometric function.

Solution

$$4\sin 5\theta\cos 5\theta = 2(2\sin 5\theta\cos 5\theta) \qquad \text{• Use } 2\sin\alpha\cos\alpha = \sin 2\alpha, \text{ with } \alpha = 5\theta.$$
$$= 2(\sin 10\theta) = 2\sin 10\theta$$

TRY EXERCISE 2, EXERCISE SET 6.3, PAGE 491

The double-angle identities can also be used to evaluate some trigonometric expressions.

EXAMPLE 2 Evaluate a Trigonometric Function

For an angle α in Quadrant I, $\sin\alpha = 4/5$. Find $\sin 2\alpha$.

Continued •➤

Solution

Use the identity $\sin 2\alpha = 2 \sin \alpha \cos \alpha$. Find $\cos \alpha$ by substituting for $\sin \alpha$ in $\sin^2 \alpha + \cos^2 \alpha = 1$ and solving for $\cos \alpha$.

$$\cos \alpha = \sqrt{1 - \sin^2 \alpha} = \sqrt{1 - \left(\frac{4}{5}\right)^2} = \frac{3}{5}$$

• $\cos \alpha > 0$ if α is in Quadrant I.

Substitute the values of $\sin \alpha$ and $\cos \alpha$ in the double-angle formula for $\sin 2\alpha$.

$$\sin 2\alpha = 2 \sin \alpha \cos \alpha = 2\left(\frac{4}{5}\right)\left(\frac{3}{5}\right) = \frac{24}{25}$$

TRY EXERCISE 26, EXERCISE SET 6.3, PAGE 492

EXAMPLE 3 Verify a Double-Angle Identity

Verify the identity $\csc 2\alpha = \frac{1}{2}(\tan \alpha + \cot \alpha)$.

Solution

Work on the right-hand side of the equation.

$$\frac{1}{2}(\tan \alpha + \cot \alpha) = \frac{1}{2}\left(\frac{\sin \alpha}{\cos \alpha} + \frac{\cos \alpha}{\sin \alpha}\right) = \frac{1}{2}\left(\frac{\sin^2 \alpha + \cos^2 \alpha}{\cos \alpha \sin \alpha}\right)$$

$$= \frac{1}{2 \cos \alpha \sin \alpha} = \frac{1}{\sin 2\alpha} = \csc 2\alpha$$

TRY EXERCISE 54, EXERCISE SET 6.3, PAGE 492

◆ HALF-ANGLE IDENTITIES

An identity for one-half an angle, $\alpha/2$, is called a *half-angle identity*. To derive a half-angle identity for $\sin \alpha/2$, we solve for $\sin^2 \theta$ in the following double-angle identity for $\cos 2\theta$.

$$\cos 2\theta = 1 - 2 \sin^2 \theta$$

$$\sin^2 \theta = \frac{1 - \cos 2\theta}{2}$$

Substitute $\alpha/2$ for θ and take the square root of each side of the equation.

$$\sin^2 \frac{\alpha}{2} = \frac{1 - \cos 2\left(\frac{\alpha}{2}\right)}{2}$$

$$\sin \frac{\alpha}{2} = \pm\sqrt{\frac{1 - \cos \alpha}{2}}$$

take note

The sign of $\sqrt{\dfrac{1-\cos\alpha}{2}}$ depends on the quadrant in which the terminal side of $\alpha/2$ lies, not that of α.

The sign of the radical is determined by the quadrant in which the terminal side of angle $\alpha/2$ lies.

In a similar manner, we derive an identity for $\cos \alpha/2$.

$$\cos 2\theta = 2\cos^2\theta - 1$$

$$\cos^2\theta = \frac{1+\cos 2\theta}{2}$$

Substitute $\alpha/2$ for θ and take the square root of each side of the equation.

$$\cos^2\frac{\alpha}{2} = \frac{1+\cos 2\left(\dfrac{\alpha}{2}\right)}{2}$$

$$\cos\frac{\alpha}{2} = \pm\sqrt{\frac{1+\cos\alpha}{2}}$$

Two different identities for $\tan \alpha/2$ are possible.

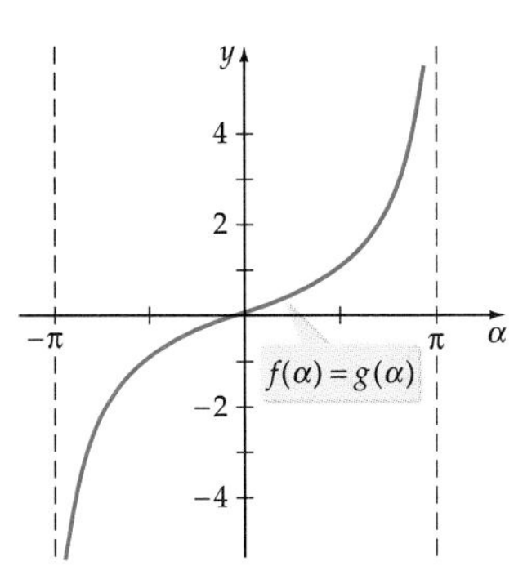

$f(\alpha) = \tan\dfrac{\alpha}{2}$

$g(\alpha) = \dfrac{\sin\alpha}{1+\cos\alpha}$

Figure 6.8

$$\tan\frac{\alpha}{2} = \frac{\sin\dfrac{\alpha}{2}}{\cos\dfrac{\alpha}{2}} = \frac{\sin\dfrac{\alpha}{2}}{\cos\dfrac{\alpha}{2}}\cdot\frac{2\cos\dfrac{\alpha}{2}}{2\cos\dfrac{\alpha}{2}}$$

$$= \frac{2\sin\dfrac{\alpha}{2}\cos\dfrac{\alpha}{2}}{2\cos^2\dfrac{\alpha}{2}} = \frac{\sin 2\left(\dfrac{\alpha}{2}\right)}{2\left(\pm\sqrt{\dfrac{1+\cos\alpha}{2}}\right)^2}$$

• $\cos\dfrac{\alpha}{2} = \pm\sqrt{\dfrac{1+\cos\alpha}{2}}$

$$\tan\frac{\alpha}{2} = \frac{\sin\alpha}{1+\cos\alpha}$$

• See Figure 6.8.

To obtain another identity for $\tan \alpha/2$, multiply by the conjugate of the denominator.

$$\tan\frac{\alpha}{2} = \frac{\sin\alpha}{1+\cos\alpha}\cdot\frac{1-\cos\alpha}{1-\cos\alpha}$$

• $\alpha \neq 2k\pi$, where k is an integer

$$= \frac{\sin\alpha(1-\cos\alpha)}{1-\cos^2\alpha}$$

$$= \frac{\sin\alpha(1-\cos\alpha)}{\sin^2\alpha}$$

$$\tan\frac{\alpha}{2} = \frac{1-\cos\alpha}{\sin\alpha}$$

EXAMPLE 4 Verify a Half-Angle Identity

Verify the identity $2\csc x\cos^2\dfrac{x}{2} = \dfrac{\sin x}{1-\cos x}$.

Continued

Solution

Work on the left side of the identity.

$$2\csc x\cos^2\frac{x}{2} = 2\csc x\left(\frac{1+\cos x}{2}\right)$$

• $\cos^2\frac{x}{2} = \frac{1+\cos x}{2}$

$$= \frac{1+\cos x}{\sin x}$$

• $\csc x = \frac{1}{\sin x}$

$$= \frac{1+\cos x}{\sin x}\cdot\frac{1-\cos x}{1-\cos x}$$

• **Multiply the numerator and denominator by the conjugate of the numerator.**

$$= \frac{1-\cos^2 x}{\sin x(1-\cos x)}$$

$$= \frac{\sin^2 x}{\sin x(1-\cos x)}$$

• $1 - \cos^2 x = \sin^2 x$

$$= \frac{\sin x}{1-\cos x}$$

TRY EXERCISE 72, EXERCISE SET 6.3, PAGE 492

EXAMPLE 5 Verify a Half-Angle Identity

Verify the identity $\tan\frac{\alpha}{2} = \sin\alpha + \cos\alpha\cot\alpha - \cot\alpha$.

Solution

Work on the left side of the identity.

$$\tan\frac{\alpha}{2} = \frac{1-\cos\alpha}{\sin\alpha} = \frac{\sin^2\alpha+\cos^2\alpha-\cos\alpha}{\sin\alpha}$$

• $1 = \sin^2\alpha + \cos^2\alpha$

$$= \frac{\sin^2\alpha}{\sin\alpha} + \frac{\cos^2\alpha}{\sin\alpha} - \frac{\cos\alpha}{\sin\alpha}$$

• **Write each numerator over the common denominator.**

$$= \sin\alpha + \cos\alpha\cot\alpha - \cot\alpha$$

TRY EXERCISE 78, EXERCISE SET 6.3, PAGE 493

Here is a summary of the double-angle and half-angle identities.

Double-Angle Identities

$$\sin 2\alpha = 2\sin\alpha\cos\alpha$$

$$\cos 2\alpha = \cos^2\alpha - \sin^2\alpha = 1 - 2\sin^2\alpha = 2\cos^2\alpha - 1$$

$$\tan 2\alpha = \frac{2\tan\alpha}{1-\tan^2\alpha}$$

Half-Angle Identities

$$\sin\frac{\alpha}{2} = \pm\sqrt{\frac{1-\cos\alpha}{2}}$$

$$\cos\frac{\alpha}{2} = \pm\sqrt{\frac{1+\cos\alpha}{2}}$$

$$\tan\frac{\alpha}{2} = \frac{\sin\alpha}{1+\cos\alpha} = \frac{1-\cos\alpha}{\sin\alpha}$$

TOPICS FOR DISCUSSION

1. True or False: If $\sin\alpha = \sin\beta$, then $\alpha = \beta$. Why?
2. Does $\sin 2\alpha = 2\sin\alpha$ for all values of α? If not, find a value of α for which $\sin 2\alpha \neq 2\sin\alpha$.
3. Because

$$\tan\frac{\alpha}{2} = \frac{\sin\alpha}{1+\cos\alpha} \quad \text{and} \quad \tan\frac{\alpha}{2} = \frac{1-\cos\alpha}{\sin\alpha}$$

are both identities, it follows that

$$\frac{\sin\alpha}{1+\cos\alpha} = \frac{1-\cos\alpha}{\sin\alpha}$$

is also an identity. Do you agree? Explain.

4. Is $\sin 10x = 2\sin 5x\cos 5x$ an identity? Explain.
5. Is $\sin\alpha/2 = \cos\alpha/2$ an identity? Explain.

EXERCISE SET 6.3

In Exercises 1 to 8, write each trigonometric expression in terms of a single trigonometric function.

1. $2\sin 2\alpha\cos 2\alpha$
2. $2\sin 3\theta\cos 3\theta$
3. $1 - 2\sin^2 5\beta$
4. $2\cos^2 2\beta - 1$
5. $\cos^2 3\alpha - \sin^2 3\alpha$
6. $\cos^2 6\alpha - \sin^2 6\alpha$
7. $\dfrac{2\tan 3\alpha}{1-\tan^2 3\alpha}$
8. $\dfrac{2\tan 4\theta}{1-\tan^2 4\theta}$

In Exercises 9 to 24, use the half-angle identities to find the exact value of each trigonometric expression.

9. $\sin 75°$
10. $\cos 105°$
11. $\tan 67.5°$
12. $\tan 165°$
13. $\cos 157.5°$
14. $\sin 112.5°$
15. $\sin 22.5°$
16. $\cos 67.5°$
17. $\sin\dfrac{7\pi}{8}$
18. $\cos\dfrac{5\pi}{8}$
19. $\cos\dfrac{5\pi}{12}$
20. $\sin\dfrac{3\pi}{8}$
21. $\tan\dfrac{7\pi}{12}$
22. $\tan\dfrac{3\pi}{8}$
23. $\cos\dfrac{\pi}{12}$
24. $\sin\dfrac{\pi}{8}$

In Exercises 25 to 36, find the exact value of $\sin 2\theta$, $\cos 2\theta$, and $\tan 2\theta$ given the following information.

25. $\cos\theta = -\dfrac{4}{5}$ $\quad\theta$ is in Quadrant II.

26. $\cos\theta = \frac{24}{25}$ θ is in Quadrant IV.

27. $\sin\theta = \frac{8}{17}$ θ is in Quadrant II.

28. $\sin\theta = -\frac{9}{41}$ θ is in Quadrant III.

29. $\tan\theta = -\frac{24}{7}$ θ is in Quadrant IV.

30. $\tan\theta = \frac{4}{3}$ θ is in Quadrant I.

31. $\sin\theta = \frac{15}{17}$ θ is in Quadrant I.

32. $\sin\theta = -\frac{3}{5}$ θ is in Quadrant III.

33. $\cos\theta = \frac{40}{41}$ θ is in Quadrant IV.

34. $\cos\theta = \frac{4}{5}$ θ is in Quadrant IV.

35. $\tan\theta = \frac{15}{8}$ θ is in Quadrant III.

36. $\tan\theta = -\frac{40}{9}$ θ is in Quadrant II.

In Exercises 37 to 48, find the exact value of the sine, cosine, and tangent of $\alpha/2$ given the following information.

37. $\sin\alpha = \frac{5}{13}$ α is in Quadrant II.

38. $\sin\alpha = -\frac{7}{25}$ α is in Quadrant III.

39. $\cos\alpha = -\frac{8}{17}$ α is in Quadrant III.

40. $\cos\alpha = \frac{12}{13}$ α is in Quadrant I.

41. $\tan\alpha = \frac{4}{3}$ α is in Quadrant I.

42. $\tan\alpha = -\frac{8}{15}$ α is in Quadrant II.

43. $\cos\alpha = \frac{24}{25}$ α is in Quadrant IV.

44. $\sin\alpha = -\frac{9}{41}$ α is in Quadrant IV.

45. $\sec\alpha = \frac{17}{15}$ α is in Quadrant I.

46. $\csc\alpha = -\frac{5}{3}$ α is in Quadrant IV.

47. $\cot\alpha = \frac{8}{15}$ α is in Quadrant III.

48. $\sec\alpha = -\frac{13}{5}$ α is in Quadrant II.

In Exercises 49 to 94, verify the given identity.

49. $\sin 3x \cos 3x = \frac{1}{2}\sin 6x$

50. $\cos 8x = \cos^2 4x - \sin^2 4x$

51. $\sin^2 x + \cos 2x = \cos^2 x$

52. $\frac{\cos 2x}{\sin^2 x} = \cot^2 x - 1$

53. $\frac{1 + \cos 2x}{\sin 2x} = \cot x$

54. $\frac{1}{1 - \cos 2x} = \frac{1}{2}\csc^2 x$

55. $\frac{\sin 2x}{1 - \sin^2 x} = 2\tan x$

56. $\frac{\cos^2 x - \sin^2 x}{2\sin x\cos x} = \cot 2x$

57. $1 - \tan^2 x = \frac{\cos 2x}{\cos^2 x}$

58. $\tan 2x = \frac{2\sin x\cos x}{\cos^2 x - \sin^2 x}$

59. $\sin 2x - \tan x = \tan x\cos 2x$

60. $\sin 2x - \cot x = -\cot x\cos 2x$

61. $\cos^4 x - \sin^4 x = \cos 2x$

62. $\sin 4x = 4\sin x\cos^3 x - 4\cos x\sin^3 x$

63. $\cos^2 x - 2\sin^2 x\cos^2 x - \sin^2 x + 2\sin^4 x = \cos^2 2x$

64. $2\cos^4 x - \cos^2 x - 2\sin^2 x\cos^2 x + \sin^2 x = \cos^2 2x$

65. $\cos 4x = 1 - 8\cos^2 x + 8\cos^4 x$

66. $\sin 4x = 4\sin x\cos x - 8\cos x\sin^3 x$

67. $\cos 3x - \cos x = 4\cos^3 x - 4\cos x$

68. $\sin 3x + \sin x = 4\sin x - 4\sin^3 x$

69. $\sin^3 x + \cos^3 x = (\sin x + \cos x)\left(1 - \frac{1}{2}\sin 2x\right)$

70. $\cos^3 x - \sin^3 x = (\cos x - \sin x)\left(1 + \frac{1}{2}\sin 2x\right)$

71. $\sin^2\frac{x}{2} = \frac{\sec x - 1}{2\sec x}$

72. $\cos^2\frac{x}{2} = \frac{\sec x + 1}{2\sec x}$

73. $\tan\frac{x}{2} = \csc x - \cot x$

74. $\tan\frac{x}{2} = \frac{\tan x}{\sec x + 1}$

75. $2\sin\frac{x}{2}\cos\frac{x}{2} = \sin x$

76. $\cos^2\frac{x}{2} - \sin^2\frac{x}{2} = \cos x$

77. $\left(\cos\frac{x}{2} + \sin\frac{x}{2}\right)^2 = 1 + \sin x$

78. $\tan^2\frac{x}{2} = \frac{\sec x - 1}{\sec x + 1}$

79. $\sin^2\frac{x}{2}\sec x = \frac{1}{2}(\sec x - 1)$

80. $\cos^2\frac{x}{2}\sec x = \frac{1}{2}(\sec x + 1)$

81. $\cos^2\frac{x}{2} - \cos x = \sin^2\frac{x}{2}$

82. $\sin^2\frac{x}{2} + \cos x = \cos^2\frac{x}{2}$

83. $\sin^2\frac{x}{2} - \cos^2\frac{x}{2} = -\cos x$

84. $\cos^2\frac{x}{2} - \sin^2\frac{x}{2} = \frac{1}{2}\csc x \sin 2x$

85. $\sin 2x - \cos x = (\cos x)(2\sin x - 1)$

86. $\frac{\cos 2x}{\sin^2 x} = \csc^2 x - 2$

87. $\tan 2x = \frac{2}{\cot x - \tan x}$

88. $\frac{2\cos 2x}{\sin 2x} = \cot x - \tan x$

89. $2\tan\frac{x}{2} = \frac{\sin^2 x + 1 - \cos^2 x}{(\sin x)(1 + \cos x)}$

90. $\frac{1}{2}\csc^2\frac{x}{2} = \csc^2 x + \cot x \csc x$

91. $\csc 2x = \frac{1}{2}\csc x \sec x$

92. $\sec 2x = \frac{\sec^2 x}{2 - \sec^2 x}$

93. $\cos\frac{x}{5} = 1 - 2\sin^2\frac{x}{10}$

94. $\sec^2\frac{x}{2} = \frac{2}{1 + \cos x}$

SUPPLEMENTAL EXERCISES

In Exercises 95 to 98, use a graphing utility to graph the function on each side of the equation to suggest that the equation is an identity.

95. $\sin^2 x + \cos 2x = \cos^2 x$

96. $\frac{\sin 2x}{1 - \sin^2 x} = 2\tan x$

97. $2\sin\frac{x}{2}\cos\frac{x}{2} = \sin x$

98. $\left(\cos\frac{x}{2} + \sin\frac{x}{2}\right)^2 = 1 + \sin x$

In Exercises 99 to 102, verify the identity.

99. $\frac{\sin^3 x + \cos^3 x}{\sin x + \cos x} = 1 - \frac{1}{2}\sin 2x$

100. $\cos^4 x = \frac{1}{8}\cos 4x + \frac{1}{2}\cos 2x + \frac{3}{8}$

101. $\sin\frac{x}{2} - \cos\frac{x}{2} = \sqrt{1 - \sin x}$, $0° \le x \le 90°$

102. $\frac{\sin x - \sin 2x}{\cos x + \cos 2x} = -\tan\frac{x}{2}$

103. If $x + y = 90°$, verify that $\sin(x - y) = -\cos 2x$.

104. If $x + y = 90°$, verify that $\sin(x - y) = \cos 2y$.

105. If $x + y = 180°$, verify that $\sin(x - y) = -\sin 2x$.

106. If $x + y = 180°$, verify that $\cos(x - y) = -\cos 2x$.

PROJECTS

1. **MACH NUMBERS** Earnst Mach (1838–1916) was an Austrian physicist. He made a study of the motion of objects at high speeds. Today we often state the speed of aircraft in terms of a *Mach number*. A **Mach number** is the speed of an object divided by the speed of sound. For example, a plane flying at the speed of sound is said to have a speed M of Mach 1. Mach 2 is twice the speed of sound. An airplane that travels faster than the speed of sound creates a sonic boom. This sonic boom emanates from the airplane in the shape of a cone.

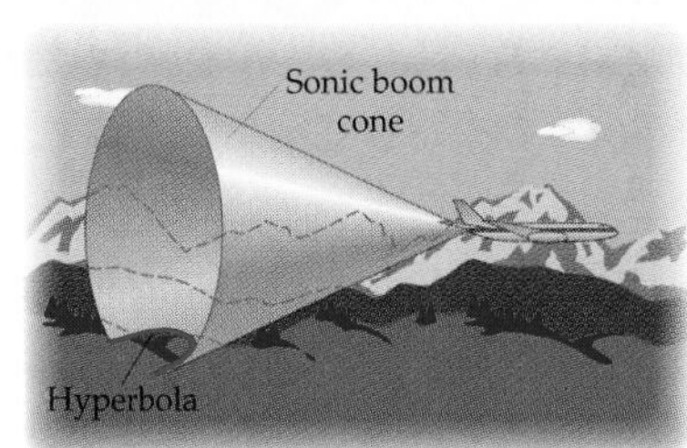

The following equation shows the relationship between the measure of the cone's vertex angle α and the Mach speed M of an aircraft that is flying faster than the speed of sound.

$$M \sin \frac{\alpha}{2} = 1$$

a. If $\alpha = \pi/4$, determine the Mach speed M of the airplane. State your answer as an *exact* value and as a decimal accurate to the nearest hundredth.

b. How fast is Mach 1 in miles per hour?

c. 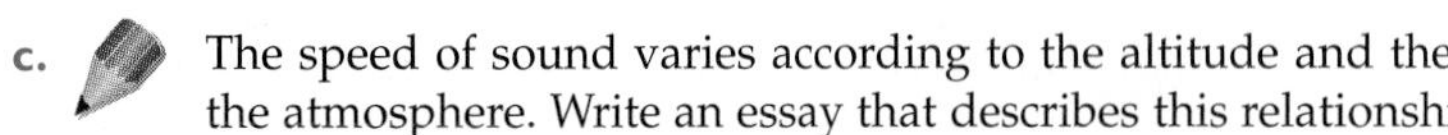
The speed of sound varies according to the altitude and the temperature of the atmosphere. Write an essay that describes this relationship.

2. **VISUAL INSIGHT**

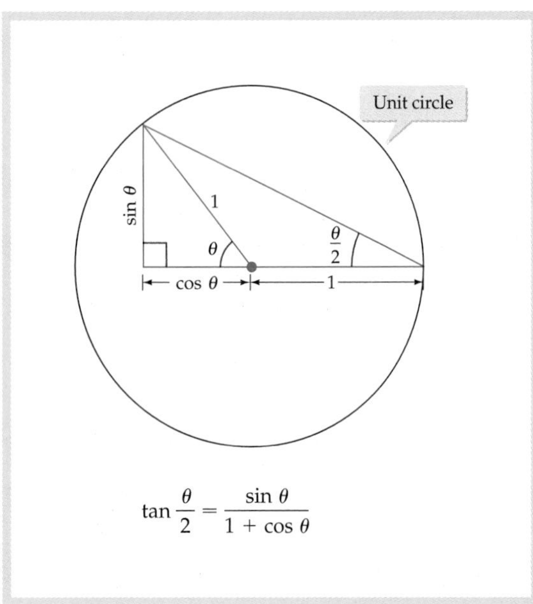

Explain how the figure above can be used to verify the half-angle identity shown.

SECTION 6.4 IDENTITIES INVOLVING THE SUM OF TRIGONOMETRIC FUNCTIONS

- THE PRODUCT-TO-SUM IDENTITIES
- THE SUM-TO-PRODUCT IDENTITIES
- FUNCTIONS OF THE FORM $f(x) = a \sin x + b \cos x$

6A

Some applications require that a product of trigonometric functions be written as a sum or difference of these functions. Other applications require that the sum or difference of trigonometric functions be represented as a product of these functions. The *product-to-sum identities* are particularly useful in these types of problems.

◆ THE PRODUCT-TO-SUM IDENTITIES

The product-to-sum identities can be derived by using the sum or difference identities. Adding the identities for $\sin(\alpha + \beta)$ and $\sin(\alpha - \beta)$, we have

$$\sin(\alpha + \beta) = \sin\alpha\cos\beta + \cos\alpha\sin\beta$$

$$\underline{\sin(\alpha - \beta)} = \underline{\sin\alpha\cos\beta - \cos\alpha\sin\beta}$$

$$\sin(\alpha + \beta) + \sin(\alpha - \beta) = 2\sin\alpha\cos\beta$$ • **Add the identities.**

Solving for $\sin\alpha\cos\beta$, we obtain the first product-to-sum identity:

$$\sin\alpha\cos\beta = \frac{1}{2}[\sin(\alpha + \beta) + \sin(\alpha - \beta)]$$

The identity for $\cos\alpha\sin\beta$ is obtained when $\sin(\alpha - \beta)$ is subtracted from $\sin(\alpha + \beta)$. The result is

$$\cos\alpha\sin\beta = \frac{1}{2}[\sin(\alpha + \beta) - \sin(\alpha - \beta)]$$

In like manner, the identities for $\cos(\alpha + \beta)$ and $\cos(\alpha - \beta)$ are used to derive the identities for $\cos\alpha\cos\beta$ and $\sin\alpha\sin\beta$.

$$\cos\alpha\cos\beta = \frac{1}{2}[\cos(\alpha + \beta) + \cos(\alpha - \beta)]$$

$$\sin\alpha\sin\beta = \frac{1}{2}[\cos(\alpha - \beta) - \cos(\alpha + \beta)]$$

The product-to-sum identities can be used to verify some identities.

EXAMPLE 1 Verify an Identity

Verify the identity $\cos 2x \sin 5x = \frac{1}{2}(\sin 7x + \sin 3x)$.

Solution

$$\cos 2x \sin 5x = \frac{1}{2}[\sin(2x + 5x) - \sin(2x - 5x)]$$ • **Use the product-to-sum identity: $\cos\alpha\sin\beta$.**

$$= \frac{1}{2}[\sin 7x - \sin(-3x)]$$

$$= \frac{1}{2}(\sin 7x + \sin 3x)$$ • **$\sin(-3x) = -\sin 3x$**

TRY EXERCISE 36, EXERCISE SET 6.4, PAGE 500

◆ THE SUM-TO-PRODUCT IDENTITIES

The *sum-to-product identities* can be derived from the product-to-sum identities. To derive the sum-to-product identity for $\sin x + \sin y$, we first let $x = \alpha + \beta$ and $y = \alpha - \beta$. Then

$$x + y = \alpha + \beta + \alpha - \beta \quad \text{and} \quad x - y = \alpha + \beta - (\alpha - \beta)$$

$$x + y = 2\alpha \qquad\qquad x - y = 2\beta$$

$$\alpha = \frac{x + y}{2} \qquad\qquad \beta = \frac{x - y}{2}$$

Substituting these expressions for α and β into the product-to-sum identity

$$\frac{1}{2}[\sin (\alpha + \beta) + \sin (\alpha - \beta)] = \sin \alpha \cos \beta$$

yields

$$\sin\left(\frac{x + y}{2} + \frac{x - y}{2}\right) + \sin\left(\frac{x + y}{2} - \frac{x - y}{2}\right) = 2 \sin \frac{x + y}{2} \cos \frac{x - y}{2}$$

Simplifying the left side, we have a sum-to-product identity.

$$\sin x + \sin y = 2 \sin \frac{x + y}{2} \cos \frac{x - y}{2}$$

In like manner, three other sum-to-product identities can be derived from the other product-to-sum identities. The proofs of these identities are left as exercises.

$$\sin x - \sin y = 2 \cos \frac{x + y}{2} \sin \frac{x - y}{2}$$

$$\cos x + \cos y = 2 \cos \frac{x + y}{2} \cos \frac{x - y}{2}$$

$$\cos x - \cos y = -2 \sin \frac{x + y}{2} \sin \frac{x - y}{2}$$

EXAMPLE 2 Write the Difference of Trigonometric Expressions as a Product

Write $\sin 4\theta - \sin \theta$ as the product of two functions.

Solution

$$\sin 4\theta - \sin \theta = 2 \cos \frac{4\theta + \theta}{2} \sin \frac{4\theta - \theta}{2} = 2 \cos \frac{5\theta}{2} \sin \frac{3\theta}{2}$$

TRY EXERCISE 22, EXERCISE SET 6.4, PAGE 500

EXAMPLE 3 Verify a Sum-to-Product Identity

Verify the identity $\dfrac{\sin 6x + \sin 2x}{\sin 6x - \sin 2x} = \tan 4x \cot 2x$.

Solution

$$\frac{\sin 6x + \sin 2x}{\sin 6x - \sin 2x} = \frac{2 \sin \dfrac{6x + 2x}{2} \cos \dfrac{6x - 2x}{2}}{2 \cos \dfrac{6x + 2x}{2} \sin \dfrac{6x - 2x}{2}} = \frac{\sin 4x \cos 2x}{\cos 4x \sin 2x}$$

$$= \tan 4x \cot 2x$$

TRY EXERCISE 44, EXERCISE SET 6.4, PAGE 500

◆ FUNCTIONS OF THE FORM $f(x) = a \sin x + b \cos x$

The function given by $f(x) = a \sin x + b \cos x$ can be written in the form $f(x) = k \sin (x + \alpha)$. This form of the function is useful in graphing and engineering applications because the amplitude, period, and phase shift can be readily calculated.

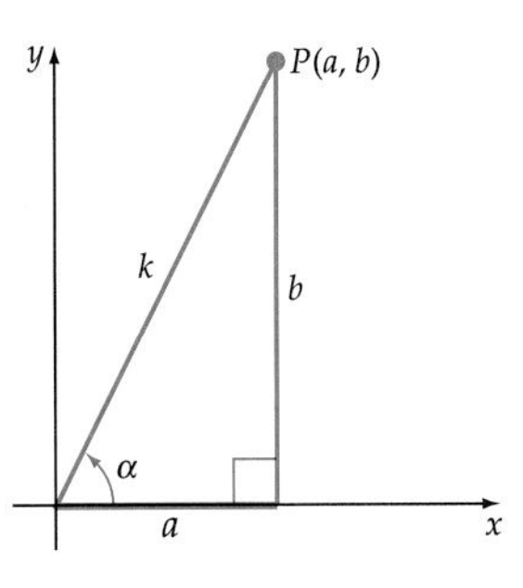

Figure 6.9

Let $P(a, b)$ be a point on a coordinate plane, and let α represent an angle in standard position. See Figure 6.9. To rewrite $y = a \sin x + b \cos x$, multiply and divide the expression $a \sin x + b \cos x$ by $\sqrt{a^2 + b^2}$.

$$a \sin x + b \cos x = \frac{\sqrt{a^2 + b^2}}{\sqrt{a^2 + b^2}}(a \sin x + b \cos x)$$

$$= \sqrt{a^2 + b^2}\left(\frac{a}{\sqrt{a^2 + b^2}} \sin x + \frac{b}{\sqrt{a^2 + b^2}} \cos x\right) \qquad (1)$$

From the definition of the sine and cosine of an angle in standard position, let

$$k = \sqrt{a^2 + b^2}, \quad \cos \alpha = \frac{a}{\sqrt{a^2 + b^2}}, \quad \text{and} \quad \sin \alpha = \frac{b}{\sqrt{a^2 + b^2}}$$

Substituting these expressions into Equation (1) yields

$$a \sin x + b \cos x = k(\cos \alpha \sin x + \sin \alpha \cos x)$$

Now, using the identity for the sine of the sum of two angles, we have

$$a \sin x + b \cos x = k \sin (x + \alpha)$$

Thus $a \sin x + b \cos x = k \sin (x + \alpha)$, where $k = \sqrt{a^2 + b^2}$ and α is the angle for which $\sin \alpha = b/\sqrt{a^2 + b^2}$ and $\cos \alpha = a/\sqrt{a^2 + b^2}$.

EXAMPLE 4 Rewrite $a \sin x + b \cos x$

Rewrite $\sin x + \cos x$ in the form $k \sin (x + \alpha)$.

Continued

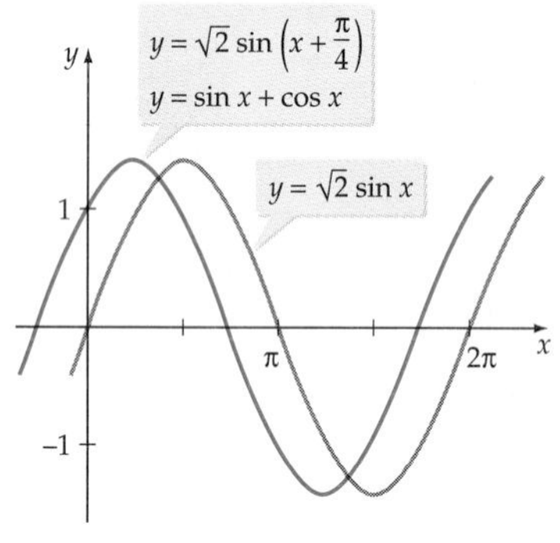

Figure 6.10

Solution

Comparing $\sin x + \cos x$ to $a \sin x + b \cos x$, $a = 1$ and $b = 1$. Thus $k = \sqrt{1^2 + 1^2} = \sqrt{2}$, $\sin \alpha = 1/\sqrt{2}$, and $\cos \alpha = 1/\sqrt{2}$. Thus $\alpha = \pi/4$.

$$\sin x + \cos x = k \sin (x + \alpha) = \sqrt{2} \sin\left(x + \frac{\pi}{4}\right)$$

TRY EXERCISE 62, EXERCISE SET 6.4, PAGE 501

The graphs of $y = \sin x + \cos x$ and $y = \sqrt{2}\sin\left(x + \frac{\pi}{4}\right)$ are both the graph of $y = \sqrt{2}\sin x$ shifted $\frac{\pi}{4}$ units to the left. See **Figure 6.10**.

EXAMPLE 5 Use an Identity to Graph a Trigonometric Function

Graph $f(x) = -\sin x + \sqrt{3}\cos x$.

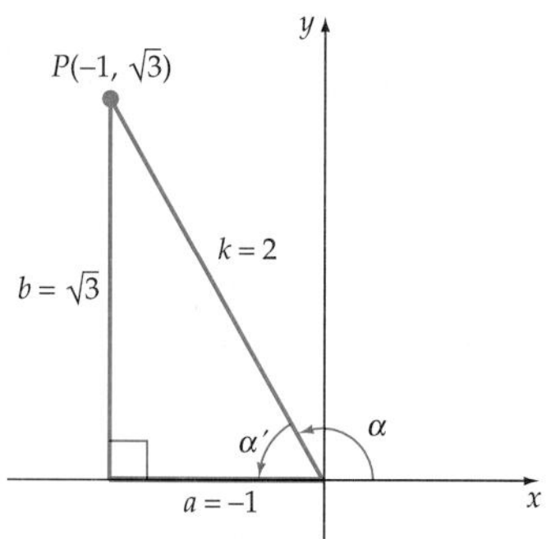

Figure 6.11

Solution

First we write $f(x)$ as $k \sin (x + \alpha)$. Let $a = -1$ and $b = \sqrt{3}$; then $k = \sqrt{(-1)^2 + (\sqrt{3})^2} = 2$. The point $P(-1, \sqrt{3})$ is in the second quadrant (see **Figure 6.11**). Let α be an angle with P on its terminal side. Let α' be the reference angle for α. Then

$$\sin \alpha' = \frac{\sqrt{3}}{2}$$

$$\alpha' = \frac{\pi}{3}$$

$$\alpha = \pi - \alpha' = \pi - \frac{\pi}{3} = \frac{2\pi}{3}$$

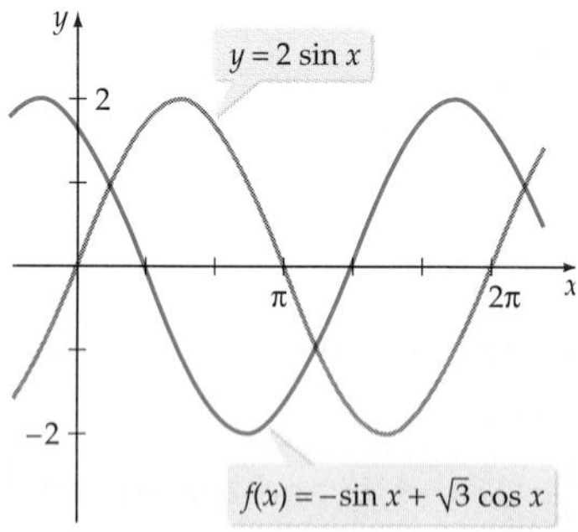

Figure 6.12

Substituting 2 for k and $2\pi/3$ for α in $y = k \sin (x + \alpha)$, we have

$$y = 2 \sin\left(x + \frac{2\pi}{3}\right)$$

The phase shift is $-c/b = -2\pi/3$. Thus the graph of the equation $f(x) = -\sin x + \sqrt{3}\cos x$ is the graph of $y = 2 \sin x$ shifted $2\pi/3$ units to the left. See **Figure 6.12**.

TRY EXERCISE 70, EXERCISE SET 6.4, PAGE 501

We now list the identities that have been discussed.

Product-to-Sum Identities

$$\sin\alpha\cos\beta = \frac{1}{2}[\sin(\alpha+\beta) + \sin(\alpha-\beta)]$$

$$\cos\alpha\sin\beta = \frac{1}{2}[\sin(\alpha+\beta) - \sin(\alpha-\beta)]$$

$$\cos\alpha\cos\beta = \frac{1}{2}[\cos(\alpha+\beta) + \cos(\alpha-\beta)]$$

$$\sin\alpha\sin\beta = \frac{1}{2}[\cos(\alpha-\beta) - \cos(\alpha+\beta)]$$

Sum-to-Product Identities

$$\sin x + \sin y = 2\sin\frac{x+y}{2}\cos\frac{x-y}{2}$$

$$\cos x + \cos y = 2\cos\frac{x+y}{2}\cos\frac{x-y}{2}$$

$$\sin x - \sin y = 2\cos\frac{x+y}{2}\sin\frac{x-y}{2}$$

$$\cos x - \cos y = -2\sin\frac{x+y}{2}\sin\frac{x-y}{2}$$

Sums of the Form $a \sin x + b \cos x$

$$a\sin x + b\cos x = k\sin(x+\alpha)$$

where $k = \sqrt{a^2+b^2}$, $\sin\alpha = \dfrac{b}{\sqrt{a^2+b^2}}$, and $\cos\alpha = \dfrac{a}{\sqrt{a^2+b^2}}$.

TOPICS FOR DISCUSSION

1. A student claims that the *exact* value of $\sin 75^\circ \cos 15^\circ$ is $\dfrac{2+\sqrt{3}}{4}$. Do you agree? Explain.

2. Do you agree with the following work? Explain.

$$\cos 195^\circ + \cos 105^\circ = \cos(195^\circ + 105^\circ) = \cos 300^\circ = -\frac{1}{2}$$

3. The graphs of $y_1 = \sin x + \cos x$ and $y_2 = \sqrt{2}\sin\left(x + \frac{\pi}{4}\right)$ are identical. Do you agree? Explain.

4. Explain to a classmate how to determine the amplitude of the graph of $y = a\sin x + b\cos x$.

EXERCISE SET 6.4

In Exercises 1 to 8, write each expression as the sum or difference of two functions.

1. $2\sin x\cos 2x$
2. $2\sin 4x\sin 2x$
3. $\cos 6x\sin 2x$
4. $\cos 3x\cos 5x$
5. $2\sin 5x\cos 3x$
6. $2\sin 2x\cos 6x$
7. $\sin x\sin 5x$
8. $\cos 3x\sin x$

In Exercises 9 to 16, find the exact value of each expression. Do not use a calculator.

9. $\cos 75^\circ\cos 15^\circ$
10. $\sin 105^\circ\cos 15^\circ$
11. $\cos 157.5^\circ\sin 22.5^\circ$
12. $\sin 195^\circ\cos 15^\circ$
13. $\sin\frac{13\pi}{12}\cos\frac{\pi}{12}$
14. $\sin\frac{11\pi}{12}\sin\frac{7\pi}{12}$
15. $\sin\frac{\pi}{12}\cos\frac{7\pi}{12}$
16. $\cos\frac{17\pi}{12}\sin\frac{7\pi}{12}$

In Exercises 17 to 32, write each expression as the product of two functions.

17. $\sin 4\theta + \sin 2\theta$
18. $\cos 5\theta - \cos 3\theta$
19. $\cos 3\theta + \cos\theta$
20. $\sin 7\theta - \sin 3\theta$
21. $\cos 6\theta - \cos 2\theta$
22. $\cos 3\theta + \cos 5\theta$
23. $\cos\theta + \cos 7\theta$
24. $\sin 3\theta + \sin 7\theta$
25. $\sin 5\theta + \sin 9\theta$
26. $\cos 5\theta - \cos\theta$
27. $\cos 2\theta - \cos\theta$
28. $\sin 2\theta + \sin 6\theta$
29. $\cos\frac{\theta}{2} - \cos\theta$
30. $\sin\frac{3\theta}{4} + \sin\frac{\theta}{2}$
31. $\sin\frac{\theta}{2} - \sin\frac{\theta}{3}$
32. $\cos\theta + \cos\frac{\theta}{2}$

In Exercises 33 to 48, verify the identity.

33. $2\cos\alpha\cos\beta = \cos(\alpha + \beta) + \cos(\alpha - \beta)$
34. $2\sin\alpha\sin\beta = \cos(\alpha - \beta) - \cos(\alpha + \beta)$
35. $2\cos 3x\sin x = 2\sin x\cos x - 8\cos x\sin^3 x$
36. $\sin 5x\cos 3x = \sin 4x\cos 4x + \sin x\cos x$
37. $2\cos 5x\cos 7x = \cos^2 6x - \sin^2 6x + 2\cos^2 x - 1$
38. $\sin 3x\cos x = \sin x\cos x(3 - 4\sin^2 x)$
39. $\sin 3x - \sin x = 2\sin x - 4\sin^3 x$
40. $\cos 5x - \cos 3x = -8\sin^2 x(2\cos^3 x - \cos x)$
41. $\sin 2x + \sin 4x = 2\sin x\cos x(4\cos^2 x - 1)$
42. $\cos 3x + \cos x = 4\cos^3 x - 2\cos x$
43. $\dfrac{\sin 3x - \sin x}{\cos 3x - \cos x} = -\cot 2x$
44. $\dfrac{\cos 5x - \cos 3x}{\sin 5x + \sin 3x} = -\tan x$
45. $\dfrac{\sin 5x + \sin 3x}{4\sin x\cos^3 x - 4\sin^3 x\cos x} = 2\cos x$
46. $\dfrac{\cos 4x - \cos 2x}{\sin 2x - \sin 4x} = \tan 3x$
47. $\sin(x + y)\cos(x - y) = \sin x\cos x + \sin y\cos y$
48. $\sin(x + y)\sin(x - y) = \sin^2 x - \sin^2 y$

In Exercises 49 to 58, write the given equation in the form $y = k\sin(x + \alpha)$, where the measure of α is in degrees.

49. $y = -\sin x - \cos x$
50. $y = \sqrt{3}\sin x - \cos x$
51. $y = \frac{1}{2}\sin x - \frac{\sqrt{3}}{2}\cos x$
52. $y = \frac{\sqrt{3}}{2}\sin x - \frac{1}{2}\cos x$
53. $y = \frac{1}{2}\sin x - \frac{1}{2}\cos x$
54. $y = -\frac{\sqrt{3}}{2}\sin x - \frac{1}{2}\cos x$
55. $y = -3\sin x + 3\cos x$
56. $y = \frac{\sqrt{2}}{2}\sin x + \frac{\sqrt{2}}{2}\cos x$
57. $y = \pi\sin x - \pi\cos x$
58. $y = -0.4\sin x + 0.4\cos x$

In Exercises 59 to 66, write the given equation in the form $y = k\sin(x + \alpha)$, where the measure of α is in radians.

59. $y = -\sin x + \cos x$
60. $y = -\sqrt{3}\sin x - \cos x$

61. $y = \frac{\sqrt{3}}{2}\sin x + \frac{1}{2}\cos x$ **62.** $y = \sin x + \sqrt{3}\cos x$

63. $y = -10\sin x + 10\sqrt{3}\cos x$

64. $y = 3\sin x - 3\sqrt{3}\cos x$

65. $y = -5\sin x + 5\cos x$ **66.** $y = 3\sin x - 3\cos x$

In Exercises 67 to 76, graph one cycle of each equation.

67. $y = -\sin x - \sqrt{3}\cos x$ **68.** $y = -\sqrt{3}\sin x + \cos x$

69. $y = 2\sin x + 2\cos x$ **70.** $y = \sin x + \sqrt{3}\cos x$

71. $y = -\sqrt{3}\sin x - \cos x$ **72.** $y = -\sin x + \cos x$

73. $y = -5\sin x + 5\sqrt{3}\cos x$

74. $y = -\sqrt{2}\sin x + \sqrt{2}\cos x$

75. $y = 6\sqrt{3}\sin x - 6\cos x$

76. $y = 5\sqrt{2}\sin x + 5\sqrt{2}\cos x$

In Exercises 77 to 82, use a graphing utility to graph the function on each side of the equation to suggest that the equation is an identity.

77. $\sin 3x - \sin x = 2\sin x - 4\sin^3 x$

78. $\frac{\sin 3x - \sin x}{\cos 3x - \cos x} = -\frac{1}{\tan 2x}$

79. $-\sqrt{3}\sin x - \cos x = 2\sin\left(x - \frac{5\pi}{6}\right)$

80. $-\sqrt{3}\sin x + \cos x = 2\sin\left(x + \frac{5\pi}{6}\right)$

81. $\frac{1}{2}\sin x - \frac{\sqrt{3}}{2}\cos x = \sin\left(x - \frac{\pi}{3}\right)$

82. $\frac{\sqrt{3}}{2}\sin x + \frac{1}{2}\cos x = \sin\left(x + \frac{\pi}{6}\right)$

SUPPLEMENTAL EXERCISES

83. Derive the sum-to-product identity

$$\cos x + \cos y = 2\cos\frac{x+y}{2}\cos\frac{x-y}{2}$$

84. Derive the product-to-sum identity

$$\sin x \sin y = \frac{1}{2}[\cos(x - y) - \cos(x + y)]$$

85. If $x + y = 180°$, show that $\sin x + \sin y = 2\sin x$.

86. If $x + y = 360°$, show that $\cos x + \cos y = 2\cos x$.

In Exercises 87 to 92, verify the identity.

87. $\sin 2x + \sin 4x + \sin 6x = 4\sin 3x\cos 2x\cos x$

88. $\sin 4x - \sin 2x + \sin 6x = 4\cos 3x\sin 2x\cos x$

89. $\frac{\cos 10x + \cos 8x}{\sin 10x - \sin 8x} = \cot x$

90. $\frac{\sin 10x + \sin 2x}{\cos 10x + \cos 2x} = \frac{2\tan 3x}{1 - \tan^2 3x}$

91. $\frac{\sin 2x + \sin 4x + \sin 6x}{\cos 2x + \cos 4x + \cos 6x} = \tan 4x$

92. $\frac{\sin 2x + \sin 6x}{\cos 6x - \cos 2x} = -\cot 2x$

93. Verify that $\cos^2 x - \sin^2 x = \cos 2x$ by using a product-to-sum identity.

94. Verify that $2\sin x\cos x = \sin 2x$ by using a product-to-sum identity.

95. Verify that $a\sin x + b\cos x = k\cos(x - \alpha)$, where $k = \sqrt{a^2 + b^2}$ and $\tan\alpha = a/b$.

96. Verify that $a\sin cx + b\cos cx = k\sin(cx + \alpha)$, where $k = \sqrt{a^2 + b^2}$ and $\tan\alpha = b/a$.

In Exercises 97 to 102, find the amplitude, phase shift, and period, and then graph the equation.

97. $y = \sin\frac{x}{2} - \cos\frac{x}{2}$

98. $y = -\sqrt{3}\sin\frac{x}{2} + \cos\frac{x}{2}$

99. $y = \sqrt{3}\sin 2x - \cos 2x$

100. $y = -\sin 2x + \cos 2x$

101. $y = \sin\pi x + \sqrt{3}\cos\pi x$

102. $y = \sin 2\pi x + \cos 2\pi x$

PROJECTS

1. **INTERFERENCE OF SOUND WAVES** The following figure on page 502 shows the waveforms of two tuning forks. The frequency of one of the tuning forks is 10 cycles per second, and the frequency of the other tuning fork is 8 cycles per second. Each of the sound waves can be modeled by an equation of the form

$$p(t) = A\cos 2\pi ft$$

where p is the pressure produced on the eardrum at time t, A is the amplitude of the sound wave, and f is the frequency of the sound.

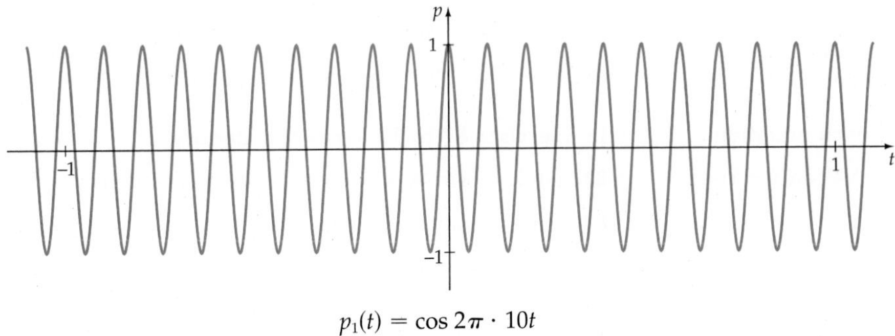

$p_1(t) = \cos 2\pi \cdot 10t$

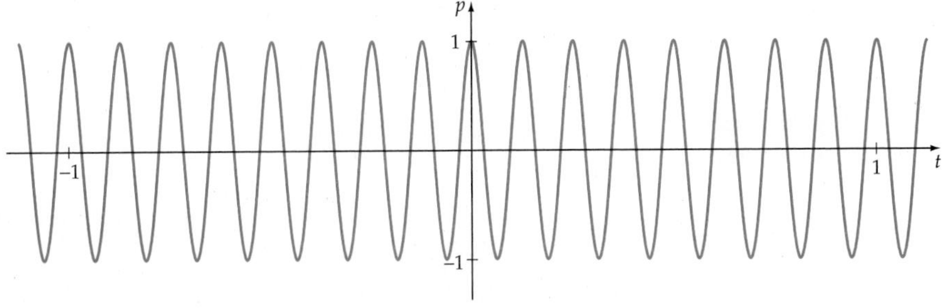

$p_2(t) = \cos 2\pi \cdot 8t$

If the two tuning forks are struck at the same time, with the same force, the sound that we hear fluctuates between a loud tone and silence. These regular fluctuations are called **beats.** The loud periods occur when the sound waves reinforce (interfere constructively with) one another, and the nearly silent periods occur when the waves interfere destructively with each other. The pressure p produced on the eardrum from the combined sound waves is given by

$$p(t) = p_1 + p_2 = \cos 2\pi \cdot 10t + \cos 2\pi \cdot 8t$$

The following figure shows the graph of p.

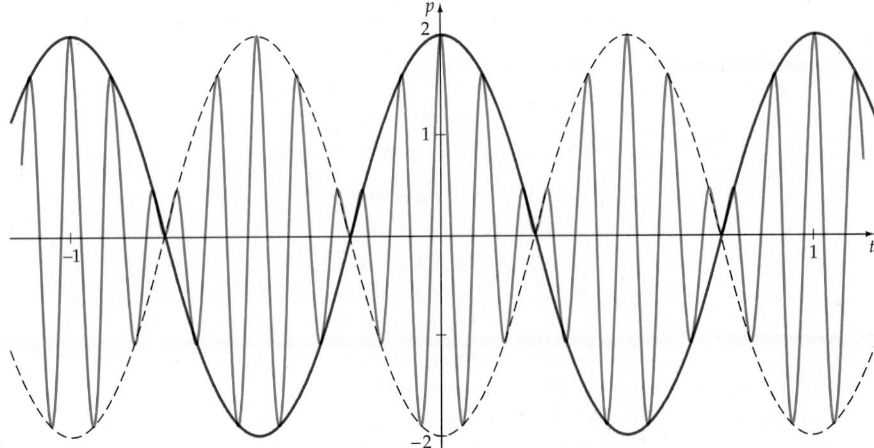

Graph of $p = p_1 + p_2 = \cos 2\pi \cdot 10t + \cos 2\pi \cdot 8t$
showing the beats in the combined sounds

a. Use the sum-to-product identity for $\cos x + \cos y$ to write p as a product.

b. Explain why the graph of $p = A\cos 2\pi f_1 t + A\cos 2\pi f_2 t$ can be thought of as a cosine curve with period $2/(f_1 + f_2)$ and a *variable* amplitude of

$$2A\cos\left[2\pi\left(\frac{f_1 - f_2}{2}\right)t\right]$$

which changes with frequency $(f_1 - f_2)/2$. The factor $2A\cos\left[2\pi\left(\frac{f_1 - f_2}{2}\right)t\right]$ is the *beat* factor. Because $f_1 = 10$ cycles per second and $f_2 = 8$ cycles per second, the frequency of the beats is $(f_1 - f_2)/2 = (10 - 8)/2 = 1$ cycle per second. The frequency f of a sound wave and the period λ of a sound wave are related by the formula $f = 1/\lambda$.

c. Two tuning forks are struck at the same time with the same force and held on a sounding board. The tuning forks have frequencies of 560 and 568 cycles per second, respectively. How many beats will be heard each second?

d. A piano tuner strikes a tuning fork and a key on a piano that is supposed to have the same frequency as the tuning fork. The piano tuner notices that the sound produced by the piano is lower than that produced by the tuning fork. The piano tuner also notes that the combined sound of the piano and the tuning fork has 2 beats per second. How much lower is the frequency of the piano than the frequency of the tuning fork?

2. Use some tuning forks to demonstrate to the class the phenomenon of beats (see Project 1).

3. Texas Instruments makes a Calculator-Based Laboratory™ System (CBL™) that can be connected to a microphone (sound sensor) that can be used to demonstrate the phenomenon of beats in a visual manner on a TI graphing calculator. Use some tuning forks and a CBL™ unit to demonstrate "beats" to your class.

Calculator-Based Laboratory and CBL are trademarks of Texas Instruments Incorporated.

SECTION 6.5 INVERSE TRIGONOMETRIC FUNCTIONS

- INVERSE TRIGONOMETRIC FUNCTIONS
- COMPOSITION OF TRIGONOMETRIC FUNCTIONS AND THEIR INVERSES
- GRAPHS OF INVERSE TRIGONOMETRIC FUNCTIONS
- AN APPLICATION INVOLVING AN INVERSE TRIGONOMETRIC FUNCTION

◆ INVERSE TRIGONOMETRIC FUNCTIONS

Because the graph of $y = \sin x$ fails the horizontal line test, it is not the graph of a one-to-one function. Therefore, it does not have an inverse function. **Figure 6.13** (on page 504) shows the graph of $y = \sin x$ on the interval $-2\pi \le x \le 2\pi$ and the graph of the inverse relation $x = \sin y$. Note that the graph of $x = \sin y$ does not satisfy the vertical line test and therefore is not the graph of a function.

If the domain of $y = \sin x$ is restricted to $-\pi/2 \le x \le \pi/2$, the graph of $y = \sin x$ satisfies the horizontal line test and therefore it has an inverse function. The graphs of $y = \sin x$ for $-\pi/2 \le x \le \pi/2$ and its inverse are shown in **Figure 6.14.**

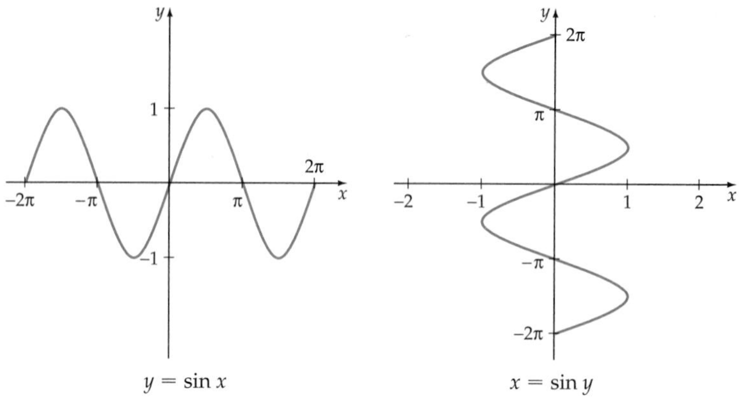

Figure 6.13

To find the inverse of the function defined by $y = \sin x$, with $-\pi/2 \le x \le \pi/2$, interchange x and y. Then solve for y.

$$y = \sin x \qquad \bullet\ -\pi/2 \le x \le \pi/2$$

$$x = \sin y \qquad \bullet\ \text{Interchange } x \text{ and } y.$$

$$y = ? \qquad \bullet\ \text{Solve for } y.$$

$y = \sin x$: $-\dfrac{\pi}{2} \le x \le \dfrac{\pi}{2}$

Unfortunately, there is no algebraic solution for y. Thus we establish new notation and write

$$y = \sin^{-1} x$$

which is read "y is the inverse sine of x." Some textbooks use the notation $\arcsin x$ instead of $\sin^{-1} x$.

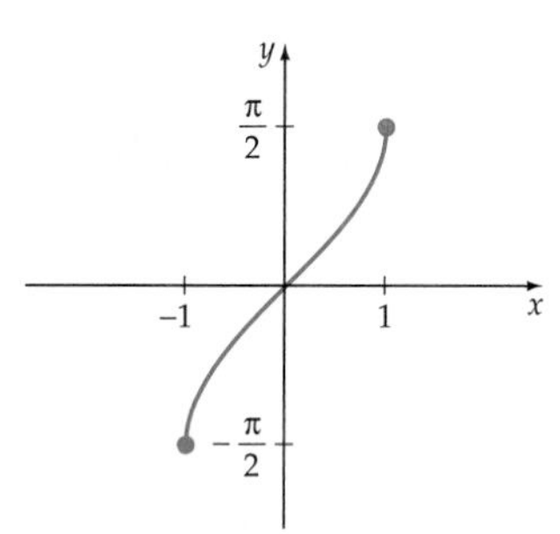

$y = \sin^{-1} x$: $-1 \le x \le 1$

Figure 6.14

Definition of $\sin^{-1} x$

$$y = \sin^{-1} x \quad \text{if and only if} \quad x = \sin y$$

where $-1 \le x \le 1$ and $-\dfrac{\pi}{2} \le y \le \dfrac{\pi}{2}$.

take note

The −1 in $\sin^{-1} x$ is not an exponent. The −1 is used to denote the inverse function. To use −1 as an exponent for a sine function, enclose the function in parentheses.

$$(\sin x)^{-1} = \frac{1}{\sin x} = \csc x$$

$$\sin^{-1} x \ne \frac{1}{\sin x}$$

It is convenient to think of the value of an inverse trigonometric function as an angle. For instance, if $y = \sin^{-1}(1/2)$, then y is the angle in the interval $[-\pi/2, \pi/2]$ whose sine is $1/2$. Thus $y = \pi/6$.

Because the graph of $y = \cos x$ fails the horizontal line test, it is not the graph of a one-to-one function. Therefore, it does not have an inverse function. Figure 6.15 shows the graph of $y = \cos x$ on the interval $-2\pi \le x \le 2\pi$ and the graph of the inverse relation $x = \cos y$. Note that the graph of $x = \cos y$ does not satisfy the vertical line test and therefore is not the graph of a function.

If the domain of $y = \cos x$ is restricted to $0 \le x \le \pi$, the graph of $y = \cos x$ satisfies the horizontal line test and therefore is the graph of a one-to-one function. The graph of $y = \cos x$ for $0 \le x \le \pi$ and that of $x = \cos y$ are shown in Figure 6.16.

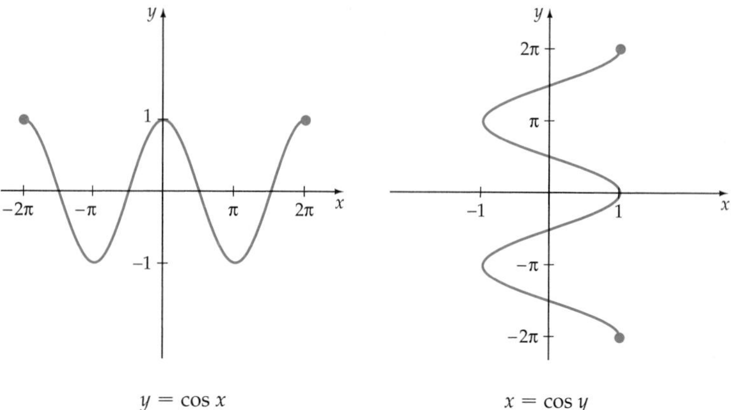

$y = \cos x$ $x = \cos y$

Figure 6.15

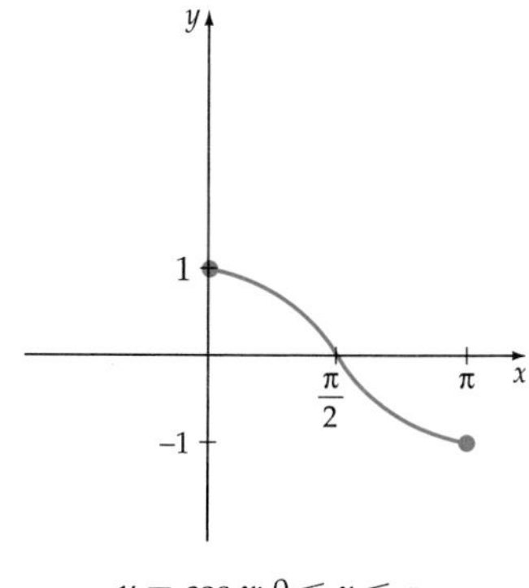

$y = \cos x$: $0 \le x \le \pi$

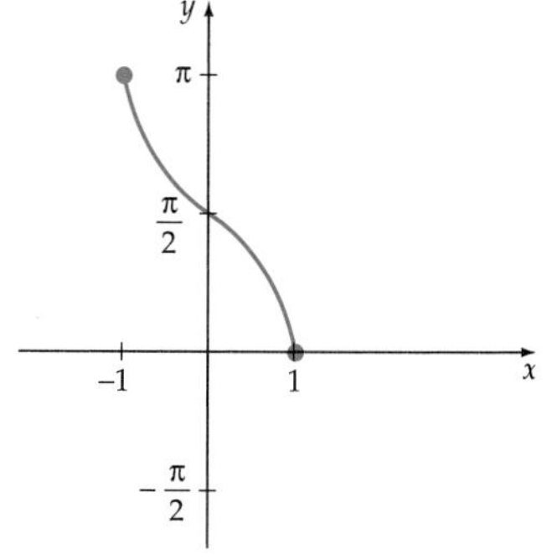

$y = \cos^{-1} x$: $-1 \le x \le 1$

Figure 6.16

To find the inverse of the function defined by $y = \cos x$, with $0 \le x \le \pi$, interchange x and y. Then solve for y.

$y = \cos x$ • $0 \le x \le \pi$

$x = \cos y$ • **Interchange x and y.**

$y = ?$ • **Solve for y.**

As in the case for the inverse sine function, there is no algebraic solution for y. Thus the notation for the inverse cosine function becomes $y = \cos^{-1} x$. We can write the following definition of the inverse cosine function.

Definition of $\cos^{-1} x$

$$y = \cos^{-1} x \quad \text{if and only if} \quad x = \cos y$$

where $-1 \le x \le 1$ and $0 \le y \le \pi$.

Because the graphs of $y = \tan x$, $y = \csc x$, $y = \sec x$, and $y = \cot x$ fail the horizontal line test, these functions are not one-to-one functions. Therefore, these functions do not have inverse functions. If the domains of all of these functions are restricted in a certain way, however, the graphs satisfy the horizontal line test. Thus each of these functions has an inverse function over a restricted domain. Table 6.2 on page 506 shows the restricted function and the inverse function for $\tan x$, $\csc x$, $\sec x$, and $\cot x$.

The choice of the ranges for $y = \sec^{-1} x$ and $y = \csc^{-1} x$ is not universally accepted. For example, some calculus texts use $[0, \pi/2) \cup [\pi, 3\pi/2)$ as the range of $y = \sec^{-1} x$. This definition has some advantages and some disadvantages that are explained in more advanced mathematics courses.

Table 6.2

	$y = \tan x$	$y = \tan^{-1} x$	$y = \csc x$	$y = \csc^{-1} x$
Domain	$-\frac{\pi}{2} < x < \frac{\pi}{2}$	$-\infty < x < \infty$	$-\frac{\pi}{2} \le x \le \frac{\pi}{2}, x \ne 0$	$x \le -1$ or $x \ge 1$
Range	$-\infty < y < \infty$	$-\frac{\pi}{2} < y < \frac{\pi}{2}$	$y \le -1$ or $y \ge 1$	$-\frac{\pi}{2} \le y \le \frac{\pi}{2}, y \ne 0$
Asymptotes	$x = -\frac{\pi}{2}, x = \frac{\pi}{2}$	$y = -\frac{\pi}{2}, y = \frac{\pi}{2}$	$x = 0$	$y = 0$
Graph				

	$y = \sec x$	$y = \sec^{-1} x$	$y = \cot x$	$y = \cot^{-1} x$
Domain	$0 \le x \le \pi, x \ne \frac{\pi}{2}$	$x \le -1$ or $x \ge 1$	$0 < x < \pi$	$-\infty < x < \infty$
Range	$y \le -1$ or $y \ge 1$	$0 \le y \le \pi, y \ne \frac{\pi}{2}$	$-\infty < y < \infty$	$0 < y < \pi$
Asymptotes	$x = \frac{\pi}{2}$	$y = \frac{\pi}{2}$	$x = 0, x = \pi$	$y = 0, y = \pi$
Graph				

EXAMPLE 1 Evaluate Inverse Functions

Find the exact value of each inverse function.

a. $y = \tan^{-1} \frac{\sqrt{3}}{3}$ **b.** $y = \cos^{-1}\left(-\frac{\sqrt{2}}{2}\right)$

Solution

a. Because $y = \tan^{-1} \frac{\sqrt{3}}{3}$, y is the angle whose measure is in the interval $(-\pi/2, \pi/2)$, and $\tan y = \sqrt{3}/3$. Therefore, $y = \pi/6$.

b. Because $y = \cos^{-1}(-\sqrt{2}/2)$, y is the angle whose measure is in the interval $[0, \pi]$, and $\cos y = -\sqrt{2}/2$. Therefore, $y = 3/4\pi$.

TRY EXERCISE 2, EXERCISE SET 6.5, PAGE 513

A calculator may not have keys for the inverse secant, cosecant, and cotangent functions. The following procedure shows an identity for the inverse cosecant function in terms of the inverse sine function. If we need to determine y, which is the angle whose cosecant is x, we can rewrite $y = \csc^{-1} x$ as follows.

$$y = \csc^{-1} x$$
• **Domain: $x \le -1$ or $x \ge 1$**
Range: $-\pi/2 \le y \le \pi/2$, $y \ne 0$

$$\csc y = x$$
• **Definition of inverse function**

$$\frac{1}{\sin y} = x$$
• **Substitute $\frac{1}{\sin x}$ for csc y.**

$$\sin y = \frac{1}{x}$$
• **Solve for sin y.**

$$y = \sin^{-1}\frac{1}{x}$$

$$\csc^{-1} x = \sin^{-1}\frac{1}{x}$$

Thus $\csc^{-1} x$ is the same as $\sin^{-1}(1/x)$. There are similar identities for $\sec^{-1} x$ and $\cot^{-1} x$.

Identities for $\csc^{-1} x$, $\sec^{-1} x$, and $\cot^{-1} x$

If $x \le -1$ or $x \ge 1$, then

$$\csc^{-1} x = \sin^{-1}\frac{1}{x} \quad \text{and} \quad \sec^{-1} x = \cos^{-1}\frac{1}{x}$$

If x is a real number, then

$$\cot^{-1} x = \begin{cases} \tan^{-1}\dfrac{1}{x}, & \text{for } x > 0 \\ \tan^{-1}\dfrac{1}{x} + \pi, & \text{for } x < 0 \\ \dfrac{\pi}{2}, & \text{for } x = 0 \end{cases}$$

◆ COMPOSITION OF TRIGONOMETRIC FUNCTIONS AND THEIR INVERSES

Recall that a function f and its inverse f^{-1} have the property that $f[f^{-1}(x)] = x$ for all x in the domain of f^{-1} and that $f^{-1}[f(x)] = x$ for all x in the domain of f. Applying this property to the functions $\sin x$, $\cos x$, and $\tan x$ and their inverse functions produces the following theorems on page 508.

Composition of Trigonometric Functions and Their Inverses

- If $-1 \le x \le 1$, then $\sin(\sin^{-1} x) = x$, and $\cos(\cos^{-1} x) = x$.
- If x is any real number, then $\tan(\tan^{-1} x) = x$.
- If $-\frac{\pi}{2} \le x \le \frac{\pi}{2}$, then $\sin^{-1}(\sin x) = x$.
- If $0 \le x \le \pi$, then $\cos^{-1}(\cos x) = x$.
- If $-\frac{\pi}{2} < x < \frac{\pi}{2}$, then $\tan^{-1}(\tan x) = x$.

In the next example, we make use of some of the composition theorems to evaluate trigonometric expressions.

EXAMPLE 2 Evaluate the Composition of a Function and its Inverse

Find the exact value of each composition of functions.

a. $\sin(\sin^{-1} 0.357)$ **b.** $\cos^{-1}(\cos 3)$ **c.** $\tan[\tan^{-1}(-11.27)]$

d. $\sin(\sin^{-1} \pi)$ **e.** $\cos(\cos^{-1} 0.277)$ **f.** $\tan^{-1}\left(\tan \frac{4\pi}{3}\right)$

Solution

a. Because 0.357 is in the interval $[-1, 1]$, $\sin(\sin^{-1} 0.357) = 0.357$.

b. Because 3 is in the interval $[0, \pi]$, $\cos^{-1}(\cos 3) = 3$.

c. Because -11.27 is a real number, $\tan[\tan^{-1}(-11.27)] = -11.27$.

d. Because π is not in the domain of the inverse sine function, $\sin(\sin^{-1} \pi)$ is undefined.

e. Because 0.277 is in the interval $[-1, 1]$, $\cos(\cos^{-1} 0.277) = 0.277$.

f. $\frac{4\pi}{3}$ is not in the interval $\left(-\frac{\pi}{2}, \frac{\pi}{2}\right)$; however, the reference angle for $\theta = \frac{4\pi}{3}$ is $\theta' = \frac{\pi}{3}$. Thus $\tan^{-1}\left(\tan \frac{4\pi}{3}\right) = \tan^{-1}\left(\tan \frac{\pi}{3}\right)$. Because $\frac{\pi}{3}$ is in the interval $\left(-\frac{\pi}{2}, \frac{\pi}{2}\right)$, $\tan^{-1}\left(\tan \frac{\pi}{3}\right) = \frac{\pi}{3}$. Hence $\tan^{-1}\left(\tan \frac{4\pi}{3}\right) = \frac{\pi}{3}$.

TRY EXERCISE 24, EXERCISE SET 6.5, PAGE 513

It is often easy to evaluate a trigonometric expression by referring to a sketch of a right triangle that satisfies given conditions. In Example 3 we make use of this technique.

EXAMPLE 3 Evaluate a Trigonometric Expression

Find the exact value of $\sin\left(\cos^{-1}\dfrac{2}{5}\right)$.

Solution

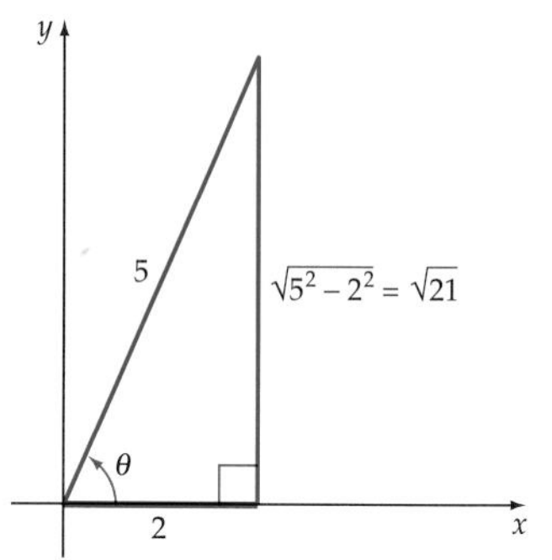

Figure 6.17

Let $\theta = \cos^{-1}\dfrac{2}{5}$, which implies $\cos\theta = \dfrac{2}{5}$. Because $\cos\theta$ is positive, θ is a first-quadrant angle. We draw a right triangle with base 2 and hypotenuse 5 so that we can view θ, as shown in **Figure 6.17.** The height of the triangle is $\sqrt{5^2 - 2^2} = \sqrt{21}$. Our goal is to find $\sin\theta$, which by definition is opp/hyp $= \sqrt{21}/5$. Thus

$$\sin\left(\cos^{-1}\frac{2}{5}\right) = \sin(\theta) = \frac{\sqrt{21}}{5}$$

In Example 4, we sketch two right triangles to evaluate the given expression.

EXAMPLE 4 Evaluate a Trigonometric Expression

Find the exact value of $\sin\left[\sin^{-1}\dfrac{3}{5} + \cos^{-1}\left(-\dfrac{5}{13}\right)\right]$.

Solution

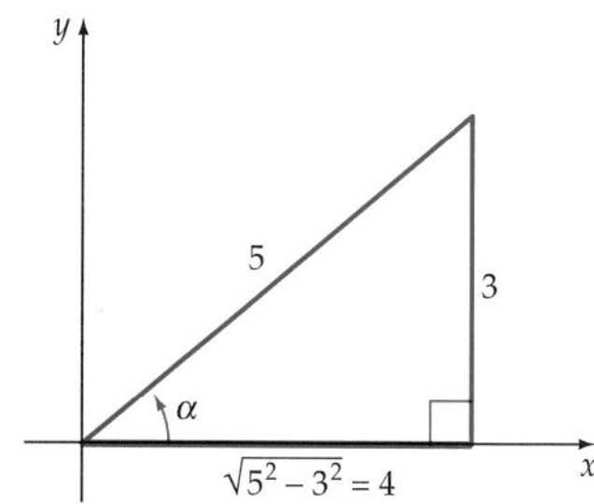

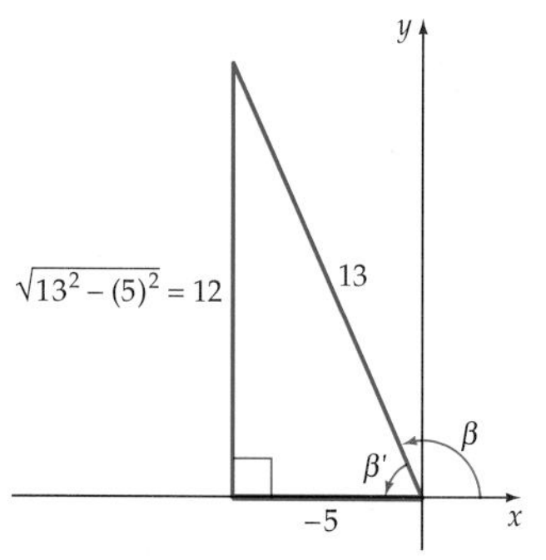

Figure 6.18

Let $\alpha = \sin^{-1}\dfrac{3}{5}$. Thus $\sin\alpha = \dfrac{3}{5}$. Let $\beta = \cos^{-1}\left(-\dfrac{5}{13}\right)$, which implies that $\cos\beta = -\dfrac{5}{13}$. Sketch angles α and β as shown in **Figure 6.18.** We wish to evaluate

$$\begin{aligned}\sin\left[\sin^{-1}\frac{3}{5} + \cos^{-1}\left(-\frac{5}{13}\right)\right] &= \sin(\alpha + \beta)\\ &= \sin\alpha\cos\beta + \cos\alpha\sin\beta \qquad (1)\end{aligned}$$

A close look at the triangles in **Figure 6.18** shows us that

$$\cos\alpha = \frac{4}{5} \quad \text{and} \quad \sin\beta = \frac{12}{13}$$

Substituting in Equation (1) gives us our desired result.

$$\begin{aligned}\sin\left[\sin^{-1}\frac{3}{5} + \cos^{-1}\left(-\frac{5}{13}\right)\right] &= \sin\alpha\cos\beta + \cos\alpha\sin\beta\\ &= \left(\frac{3}{5}\right)\left(-\frac{5}{13}\right) + \left(\frac{4}{5}\right)\left(\frac{12}{13}\right) = \frac{33}{65}\end{aligned}$$

TRY EXERCISE 54, EXERCISE SET 6.5, PAGE 514

In Example 5, we make use of the identity $\cos(\cos^{-1} x) = x$, where $-1 \le x \le 1$, to solve an equation.

EXAMPLE 5 Solve an Inverse Trigonometric Equation

Solve the inverse trigonometric equation $\sin^{-1}\frac{3}{5} + \cos^{-1} x = \pi$.

Solution

Solve for $\cos^{-1} x$, and then take the cosine of both sides of the equation.

$$\sin^{-1}\frac{3}{5} + \cos^{-1} x = \pi$$

$$\cos^{-1} x = \pi - \sin^{-1}\frac{3}{5}$$

$$\cos(\cos^{-1} x) = \cos\left(\pi - \sin^{-1}\frac{3}{5}\right)$$

$$x = \cos(\pi - \alpha)$$

• **Let $\alpha = \sin^{-1}(3/5)$. Note that α is the angle whose sine is $3/5$. (See Figure 6.19.)**

$$= \cos\pi\cos\alpha + \sin\pi\sin\alpha$$

$$= (-1)\cos\alpha + (0)\sin\alpha$$

$$= -\cos\alpha$$

$$= -\frac{4}{5}$$

• **$\cos\alpha = 4/5$ (See Figure 6.19.)**

Figure 6.19

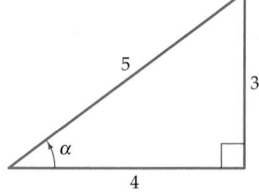

TRY EXERCISE 64, EXERCISE SET 6.5, PAGE 514

EXAMPLE 6 Verify a Trigonometric Identity That Involves Inverses

Verify the identity $\sin^{-1} x + \cos^{-1} x = \frac{\pi}{2}$.

Solution

Let $\alpha = \sin^{-1} x$ and $\beta = \cos^{-1} x$. These equations imply that $\sin\alpha = x$ and $\cos\beta = x$. From the right triangles in **Figure 6.20,**

$$\cos\alpha = \sqrt{1 - x^2} \qquad \text{and} \qquad \sin\beta = \sqrt{1 - x^2}$$

Our goal is to show that the left side of the equation equals $\pi/2$.

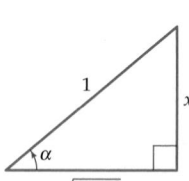

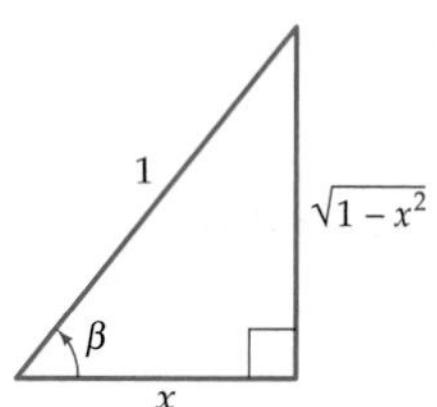

Figure 6.20

$$\begin{aligned}\sin^{-1} x + \cos^{-1} x &= \alpha + \beta \\ &= \cos^{-1}[\cos(\alpha + \beta)] \quad &&\text{• Because } 0 \le \alpha + \beta \le \pi, \text{ we can apply } \alpha + \beta = \cos^{-1}[\cos(\alpha+\beta)]. \\ &= \cos^{-1}[\cos\alpha\cos\beta - \sin\alpha\sin\beta] &&\text{• Addition formula for cosine} \\ &= \cos^{-1}\left[\left(\sqrt{1-x^2}\right)(x) - (x)\left(\sqrt{1-x^2}\right)\right] \\ &= \cos^{-1} 0 = \frac{\pi}{2}\end{aligned}$$

TRY EXERCISE 72, EXERCISE SET 6.5, PAGE 514

$y = \sin^{-1} x$

$y = \sin^{-1}(x - 2)$

Figure 6.21

◆ GRAPHS OF INVERSE TRIGONOMETRIC FUNCTIONS

The inverse trigonometric functions can be graphed by using the procedures of stretching, shrinking, and translation that were discussed earlier in the text. For instance, the graph of $y = \sin^{-1}(x - 2)$ is a horizontal shift 2 units to the right of the graph of $y = \sin^{-1} x$, as shown in **Figure 6.21**.

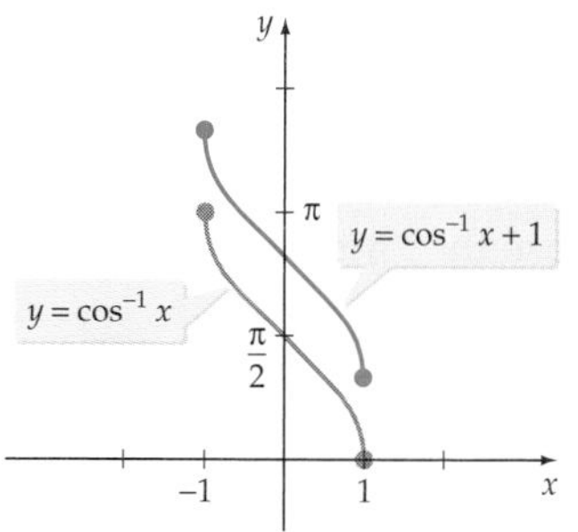

Figure 6.22

EXAMPLE 7 Graph an Inverse Cosine Function

Graph: $y = \cos^{-1} x + 1$

Solution

Recall that the graph of $y = f(x) + c$ is a vertical translation of the graph of f. Because $c = 1$, a positive number, the graph of $y = \cos^{-1} x + 1$ is the graph of $y = \cos^{-1} x$ shifted 1 unit up. See **Figure 6.22**.

TRY EXERCISE 76, EXERCISE SET 6.5, PAGE 514

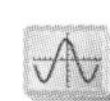

When you use a graphing utility to draw the graph of an inverse trigonometric function, use the properties of these functions to verify the correctness of your graph. For instance, the graph of $y = 3\sin^{-1} 0.5x$ is shown in **Figure 6.23**. The domain of $y = \sin^{-1} x$ is $-1 \le x \le 1$. Therefore the domain of $y = 3\sin^{-1} 0.5x$ is $-1 \le 0.5x \le 1$, or, multiplying the inequality by 2, $-2 \le x \le 2$. This is consistent with the graph in **Figure 6.23**.

The range of $y = \sin^{-1} x$ is $-\pi/2 \le y \le \pi/2$. Thus the range of $y = 3\sin^{-1} 0.5x$ is $-3\pi/2 \le y \le 3\pi/2$. This is also consistent with the graph. Verifying some of the properties of $y = \sin^{-1} x$ serves as a check that you have correctly entered the equation for the graph.

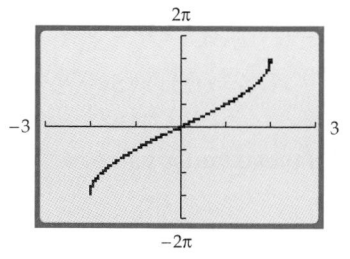

$y = 3\sin^{-1} 0.5x$

Figure 6.23

◆ AN APPLICATION INVOLVING AN INVERSE TRIGONOMETRIC FUNCTION

EXAMPLE 8 Solve an Application

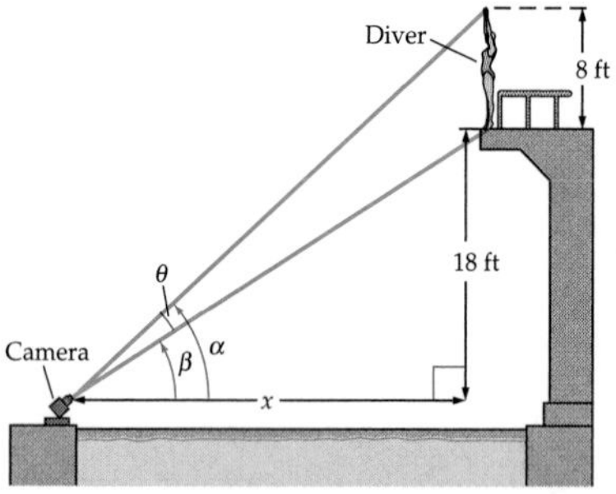

A camera is placed on a deck of a pool as shown in **Figure 6.24.** A diver is 18 feet above the camera lens. The extended length of the diver is 8 feet.

Figure 6.24

a. Show that the angle θ subtended at the lens by the diver is

$$\theta = \tan^{-1}\frac{26}{x} - \tan^{-1}\frac{18}{x}$$

b. For what values of x will $\theta = 9°$?

c. What value of x maximizes θ?

Solution

a. From **Figure 6.24,** we see that $\alpha = \tan^{-1}\frac{26}{x}$ and $\beta = \tan^{-1}\frac{18}{x}$. Because $\theta = \alpha - \beta$, we have $\theta = \tan^{-1}\frac{26}{x} - \tan^{-1}\frac{18}{x}$.

b. Use a graphing utility to graph $\theta = \tan^{-1}\frac{26}{x} - \tan^{-1}\frac{18}{x}$ and $\theta = \frac{\pi}{20}$ ($9° = \pi/20$ radians). See **Figure 6.25.** Use the "intersect" command to show that θ is 9° for $x \approx 12.22$ feet and $x \approx 38.29$ feet.

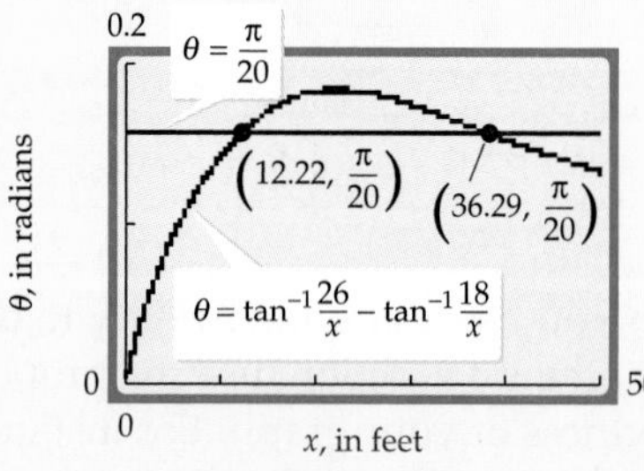

Figure 6.25

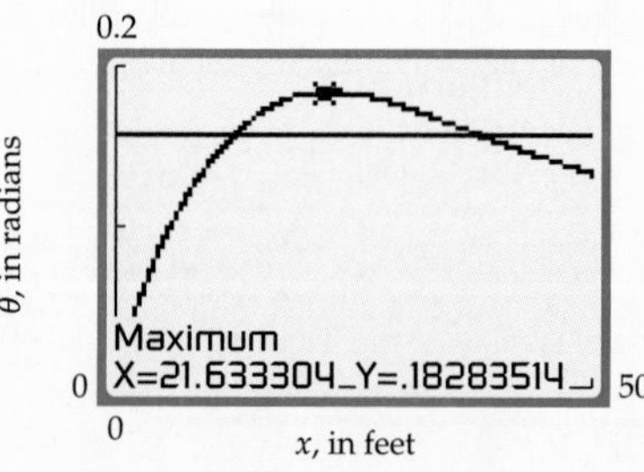

Figure 6.26

c. Use the "maximum" command to show that the maximum value of $\theta = \tan^{-1}\frac{26}{x} - \tan^{-1}\frac{18}{x}$ occurs when $x \approx 21.63$ feet. See **Figure 6.26.**

TRY EXERCISE 84, EXERCISE SET 6.5, PAGE 514

TOPICS FOR DISCUSSION

1. Is the equation

$$\tan^{-1} x = \frac{1}{\tan x}$$

true for all values of x, true for some values of x, or false for all values of x?

2. Are there real numbers x for which the following is true? Explain.

$$\sin(\sin^{-1} x) \neq \sin^{-1}(\sin x)$$

3. Explain how to find the value of $\sec^{-1} 3$ by using a scientific calculator.

4. Explain how you can determine the range of $y = (2\cos^{-1} x) - 1$ using

a. algebra b. a graph

EXERCISE SET 6.5

In Exercises 1 to 20, find the exact radian value.

1. $\sin^{-1} 1$
2. $\sin^{-1} \frac{\sqrt{2}}{2}$
3. $\cos^{-1}\left(-\frac{\sqrt{3}}{2}\right)$
4. $\cos^{-1}\left(-\frac{1}{2}\right)$
5. $\tan^{-1}(-1)$
6. $\tan^{-1} \sqrt{3}$
7. $\cot^{-1} \frac{\sqrt{3}}{3}$
8. $\cot^{-1} 1$
9. $\sec^{-1} 2$
10. $\sec^{-1} \frac{2\sqrt{3}}{3}$
11. $\csc^{-1}(-\sqrt{2})$
12. $\csc^{-1}(-2)$
13. $\sin^{-1}\left(-\frac{\sqrt{3}}{2}\right)$
14. $\sin^{-1} \frac{1}{2}$
15. $\cos^{-1}\left(-\frac{1}{2}\right)$
16. $\cos^{-1} \frac{\sqrt{3}}{2}$
17. $\tan^{-1} \frac{\sqrt{3}}{3}$
18. $\tan^{-1} 1$
19. $\cot^{-1} \sqrt{3}$
20. $\cot^{-1}(-1)$

In Exercises 21 to 56, find the exact value of the given expression. If an exact value cannot be given, give the value to the nearest ten-thousandth.

21. $\cos\left(\cos^{-1} \frac{1}{2}\right)$
22. $\cos(\cos^{-1} 2)$
23. $\tan(\tan^{-1} 2)$
24. $\tan\left(\tan^{-1} \frac{1}{2}\right)$
25. $\sin\left(\tan^{-1} \frac{3}{4}\right)$
26. $\cos\left(\sin^{-1} \frac{5}{13}\right)$
27. $\tan\left(\sin^{-1} \frac{\sqrt{2}}{2}\right)$
28. $\sin\left[\cos^{-1}\left(-\frac{\sqrt{3}}{2}\right)\right]$
29. $\cos(\sec^{-1} 2)$
30. $\sin^{-1}(\sin 2)$
31. $\sin^{-1}\left(\sin \frac{\pi}{6}\right)$
32. $\sin^{-1}\left(\sin \frac{5\pi}{6}\right)$
33. $\cos^{-1}\left(\sin \frac{\pi}{4}\right)$
34. $\cos^{-1}\left(\cos \frac{5\pi}{4}\right)$
35. $\sin^{-1}\left(\tan \frac{\pi}{3}\right)$
36. $\cos^{-1}\left(\tan \frac{2\pi}{3}\right)$
37. $\tan^{-1}\left(\sin \frac{\pi}{6}\right)$
38. $\cot^{-1}\left(\cos \frac{2\pi}{3}\right)$
39. $\sin^{-1}\left[\cos\left(-\frac{2\pi}{3}\right)\right]$
40. $\cos^{-1}\left[\tan\left(-\frac{\pi}{3}\right)\right]$
41. $\tan\left(\sin^{-1} \frac{1}{2}\right)$
42. $\cot(\csc^{-1} 2)$
43. $\sec\left(\sin^{-1} \frac{1}{4}\right)$
44. $\csc\left(\cos^{-1} \frac{3}{4}\right)$
45. $\cos\left(\sin^{-1} \frac{7}{25}\right)$
46. $\tan\left(\cos^{-1} \frac{3}{5}\right)$
47. $\sec\left(\tan^{-1} \frac{12}{5}\right)$
48. $\csc\left(\sin^{-1} \frac{12}{13}\right)$
49. $\cos\left(2\sin^{-1} \frac{\sqrt{2}}{2}\right)$
50. $\tan\left(2\sin^{-1} \frac{\sqrt{3}}{2}\right)$
51. $\sin\left(2\sin^{-1} \frac{4}{5}\right)$
52. $\cos(2\tan^{-1} 1)$
53. $\sin\left(\sin^{-1} \frac{2}{3} + \cos^{-1} \frac{1}{2}\right)$

54. $\cos\left(\sin^{-1}\frac{3}{4} + \cos^{-1}\frac{5}{13}\right)$

55. $\tan\left(\cos^{-1}\frac{1}{2} - \sin^{-1}\frac{3}{4}\right)$

56. $\sec\left(\cos^{-1}\frac{2}{3} + \sin^{-1}\frac{2}{3}\right)$

In Exercises 57 to 66, solve the equation for x algebraically.

57. $\sin^{-1} x = \cos^{-1}\frac{5}{13}$

58. $\tan^{-1} x = \sin^{-1}\frac{24}{25}$

59. $\sin^{-1}(x - 1) = \frac{\pi}{2}$

60. $\cos^{-1}\left(x - \frac{1}{2}\right) = \frac{\pi}{3}$

61. $\tan^{-1}\left(x + \frac{\sqrt{2}}{2}\right) = \frac{\pi}{4}$

62. $\sin^{-1}(x - 2) = -\frac{\pi}{6}$

63. $\sin^{-1}\frac{3}{5} + \cos^{-1} x = \frac{\pi}{4}$

64. $\sin^{-1} x + \cos^{-1}\frac{4}{5} = \frac{\pi}{6}$

65. $\sin^{-1}\frac{\sqrt{2}}{2} + \cos^{-1} x = \frac{2\pi}{3}$

66. $\cos^{-1} x + \sin^{-1}\frac{\sqrt{3}}{2} = \frac{\pi}{2}$

In Exercises 67 to 70, evaluate each expression.

67. $y = \cos(\sin^{-1} x)$

68. $y = \tan(\cos^{-1} x)$

69. $y = \sin(\sec^{-1} x)$

70. $y = \sec(\sin^{-1} x)$

In Exercises 71 to 74, verify the identity.

71. $\sin^{-1} x + \sin^{-1}(-x) = 0$

72. $\cos^{-1} x + \cos^{-1}(-x) = \pi$

73. $\tan^{-1} x + \tan^{-1}\frac{1}{x} = \frac{\pi}{2}, x > 0$

74. $\sec^{-1}\frac{1}{x} + \csc^{-1}\frac{1}{x} = \frac{\pi}{2}$

In Exercises 75 to 82, use stretching, shrinking, and translation procedures to graph each equation.

75. $y = \sin^{-1} x + 2$

76. $y = \cos^{-1}(x - 1)$

77. $y = \sin^{-1}(x + 1) - 2$

78. $y = \tan^{-1}(x - 1) + 2$

79. $y = 2\cos^{-1} x$

80. $y = -2\tan^{-1} x$

81. $y = \tan^{-1}(x + 1) - 2$

82. $y = \sin^{-1}(x - 2) + 1$

83. DOT-MATRIX PRINTING In dot-matrix printing, the *blank-area factor* is the ratio of the blank area (unprinted area) to the total area of the line. If circular dots are used to print, then the blank-area factor is given by

$$\frac{A}{(S)(D)} = 1 - \frac{1}{2}\left[1 - \left(\frac{S}{D}\right)^2 + \frac{D}{S}\sin^{-1}\left(\frac{S}{D}\right)\right]$$

where $A = A_1 + A_2$, A_1 and A_2 are the areas of the regions shown in the figure, S is the distance between centers of overlapping dots, and D is the diameter of a dot.

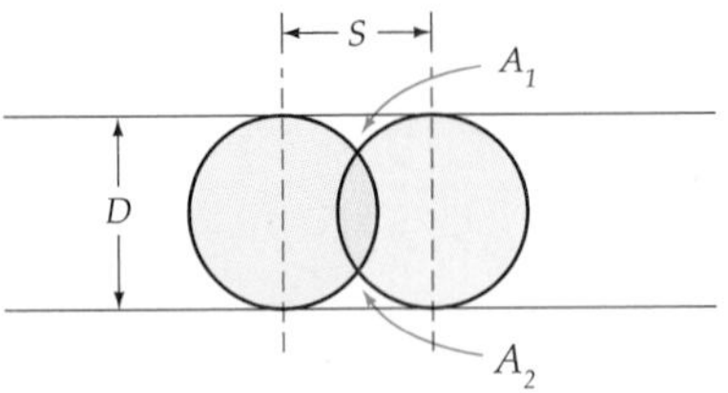

Calculate the blank-area factor where

a. $D = 0.2$ millimeter and $S = 0.1$ millimeter

b. $D = 0.16$ millimeter and $S = 0.1$ millimeter

84. VOLUME IN A WATER TANK The volume V of water (measured in cubic feet) in a horizontal cylindrical tank of radius 5 feet and length 12 feet is given by

$$V(x) = 12\left[25\cos^{-1}\left(\frac{5 - x}{5}\right) - (5 - x)\sqrt{10x - x^2}\right]$$

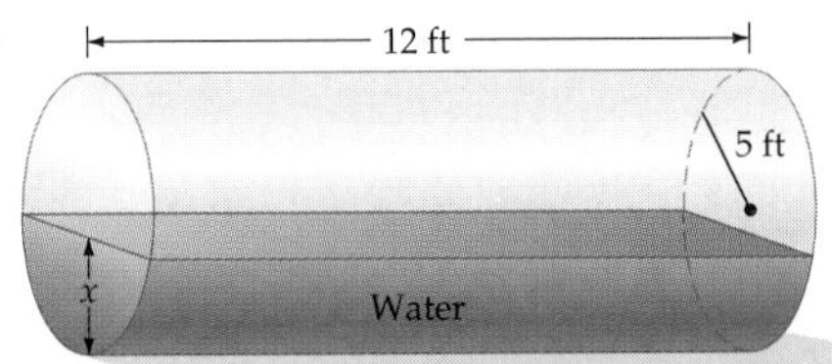

where x is the depth of the water in feet.

a. Graph V over its domain $0 \le x \le 10$.

b. Write a sentence that explains why the graph of V increases more rapidly when x increases from 4.9 feet to 5 feet than it does when x increases from 0.1 foot to 0.2 foot.

c. If $x = 4$ feet, find the volume (to the nearest 0.01 cubic foot) of the water in the tank.

d. Find the depth x (to the nearest 0.01 foot) if there are 288 cubic feet of water in tank.

85. Graph $f(x) = \cos^{-1} x$ and $g(x) = \sin^{-1}\sqrt{1 - x^2}$ on the same coordinate axes. Does $f(x) = g(x)$ on the interval $[-1, 1]$?

86. Graph $y = \cos(\cos^{-1} x)$ on $[-1, 1]$. Graph $y = \cos^{-1}(\cos x)$ on $[-2\pi, 2\pi]$.

 **In Exercises 87 to 94, use a graphing utility to graph each equation.**

87. $y = \csc^{-1} 2x$

88. $y = 0.5\sec^{-1}\frac{x}{2}$

89. $y = \sec^{-1}(x - 1)$

90. $y = \sec^{-1}(x + \pi)$

91. $y = 2\tan^{-1} 2x$

92. $y = \tan^{-1}(x - 1)$

93. $y = \cot^{-1}\dfrac{x}{3}$

94. $y = 2\cot^{-1}(x - 1)$

SUPPLEMENTAL EXERCISES

In Exercises 95 to 98, verify the identity.

95. $\cos(\sin^{-1} x) = \sqrt{1 - x^2}$

96. $\sec(\sin^{-1} x) = \dfrac{\sqrt{1 - x^2}}{1 - x^2}$

97. $\tan(\csc^{-1} x) = \dfrac{\sqrt{x^2 - 1}}{x^2 - 1}, x > 1$

98. $\sin(\cot^{-1} x) = \dfrac{\sqrt{x^2 + 1}}{x^2 + 1}$

In Exercises 99 to 102, solve for y in terms of x.

99. $5x = \tan^{-1} 3y$

100. $2x = \dfrac{1}{2}\sin^{-1} 2y$

101. $x - \dfrac{\pi}{3} = \cos^{-1}(y - 3)$

102. $x + \dfrac{\pi}{2} = \tan^{-1}(2y - 1)$

PROJECTS

1. VISUAL INSIGHT

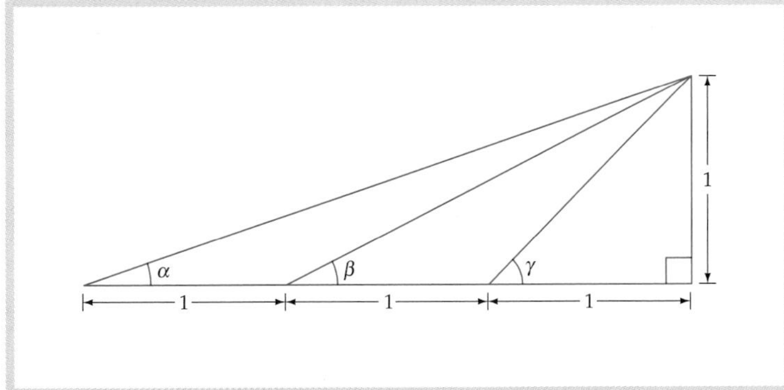

Explain how the figure above can be used to verify each identity.

a. $\tan^{-1}\dfrac{1}{3} + \tan^{-1}\dfrac{1}{2} = \dfrac{\pi}{4}$ **b.** $\alpha + \beta = \gamma$

SECTION 6.6 TRIGONOMETRIC EQUATIONS

- SOLVE TRIGONOMETRIC EQUATIONS
- AN APPLICATION INVOLVING A TRIGONOMETRIC EQUATION
- MODEL SINUSOIDAL DATA

6B

◆ SOLVE TRIGONOMETRIC EQUATIONS

Consider the equation $\sin x = 1/2$. The graph of $y = \sin x$, along with the line $y = 1/2$, is shown in **Figure 6.27**. The x values of the intersections of the two graphs are the solutions of $\sin x = 1/2$. The solutions in the interval $0 \le x < 2\pi$ are $x = \pi/6$ and $5\pi/6$.

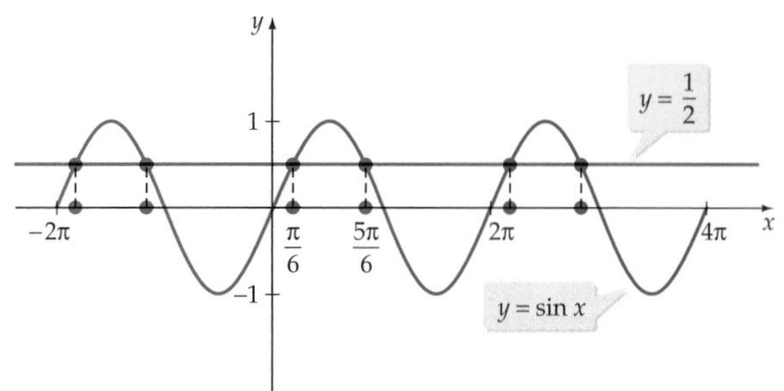

Figure 6.27

If we remove the restriction $0 \le x < 2\pi$, there are many more solutions. Because the sine function is periodic with a period of 2π, other solutions are obtained by adding $2k\pi$, k an integer, to either of the previous solutions. Thus the solutions of $\sin x = 1/2$ are

$$x = \frac{\pi}{6} + 2k\pi, \quad k \text{ an integer}$$

$$x = \frac{5\pi}{6} + 2k\pi, \quad k \text{ an integer}$$

Algebraic methods and trigonometric identities are used frequently to find the solutions of trigonometric equations. Algebraic methods that are often employed include solving by factoring, solving by using the quadratic formula, and squaring each side of the equation.

EXAMPLE 1 Solve a Trigonometric Equation by Factoring

Solve $2\sin^2 x \cos x - \cos x = 0$, where $0 \le x < 2\pi$.

Solution

$2\sin^2 x \cos x - \cos x = 0$

$\cos x(2\sin^2 x - 1) = 0$ • **Factor cos x from each term.**

$\cos x = 0 \quad \text{or} \quad 2\sin^2 x - 1 = 0$ • **Use the Principle of Zero Products.**

$x = \frac{\pi}{2}, \frac{3\pi}{2} \qquad \sin^2 x = \frac{1}{2}$ • **Solve each equation for x with $0 \le x < 2\pi$.**

$$\sin x = \pm\frac{\sqrt{2}}{2}$$

$$x = \frac{\pi}{4}, \frac{3\pi}{4}, \frac{5\pi}{4}, \frac{7\pi}{4}$$

The solutions in the interval $0 \le x < 2\pi$ are $\pi/4, \pi/2, 3\pi/4, 5\pi/4, 3\pi/2$, and $7\pi/4$.

Visualize the Solution

The solutions are the x-coordinates of the x-intercepts of $y = 2\sin^2 x \cos x - \cos x$ on the interval $[0, 2\pi)$. See Figure 6.28.

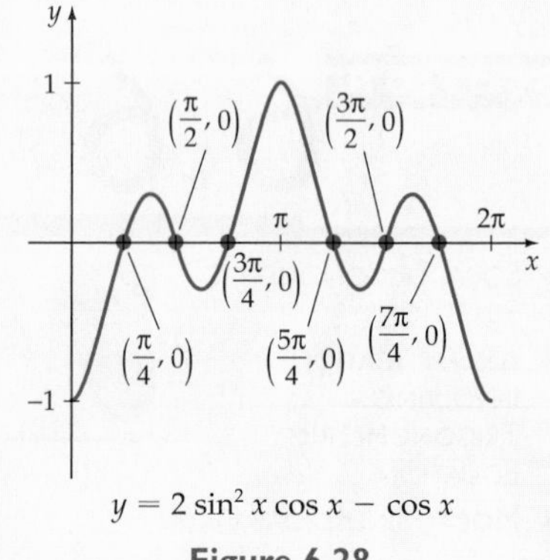

Figure 6.28

TRY EXERCISE 14, EXERCISE SET 6.6, PAGE 524

Squaring both sides of an equation may not produce an equivalent equation. Thus, when this method is used, the proposed solutions must be checked to eliminate any extraneous solutions.

EXAMPLE 2 Solve a Trigonometric Equation by Squaring Each Side of an Equation

Solve $\sin x + \cos x = 1$, where $0 \le x < 2\pi$.

Solution

$$\sin x + \cos x = 1$$

$$\sin x = 1 - \cos x \qquad \text{• Solve for } \sin x.$$

$$\sin^2 x = (1 - \cos x)^2 \qquad \text{• Square each side.}$$

$$\sin^2 x = 1 - 2\cos x + \cos^2 x$$

$$1 - \cos^2 x = 1 - 2\cos x + \cos^2 x \qquad \text{• } \sin^2 x = 1 - \cos^2 x$$

$$2\cos^2 x - 2\cos x = 0$$

$$2\cos x(\cos x - 1) = 0$$

$$2\cos x = 0 \quad \text{or} \quad \cos x = 1 \qquad \text{• Factor.}$$

$$x = \frac{\pi}{2}, \frac{3\pi}{2} \qquad x = 0 \qquad \text{• Solve each equation for } x \text{ with } 0 \le x < 2\pi.$$

Squaring each side of an equation may introduce extraneous solutions. Therefore, we must check the solutions. A check will show that 0 and $\pi/2$ are solutions but $3\pi/2$ is not a solution.

Visualize the Solution

The solutions are the x-coordinates of the points of intersection of $y = \sin x + \cos x$ and $y = 1$ on the interval $[0, 2\pi)$. See **Figure 6.29.**

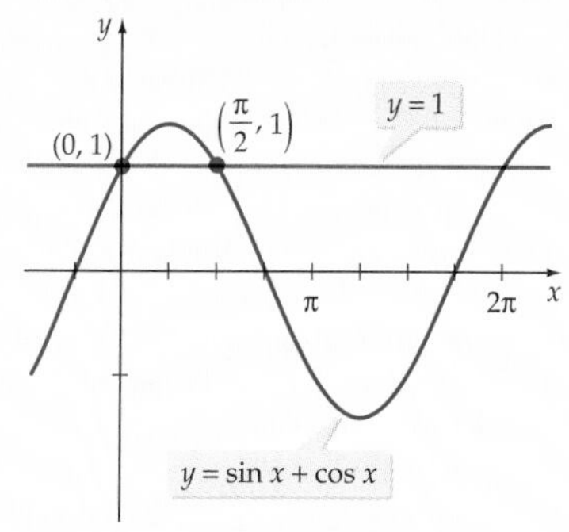

Figure 6.29

TRY EXERCISE 52, EXERCISE SET 6.6, PAGE 524

EXAMPLE 3 Solve a Trigonometric Equation by Using the Quadratic Formula

Solve $3\cos^2 x - 5\cos x - 4 = 0$, where $0 \le x < 2\pi$.

Solution

The given equation is quadratic in form and cannot be factored easily. However, we can use the quadratic formula to solve for $\cos x$.

$$3\cos^2 x - 5\cos x - 4 = 0 \qquad \text{• } a = 3, b = -5, c = -4$$

$$\cos x = \frac{-(-5) \pm \sqrt{(-5)^2 - 4(3)(-4)}}{(2)(3)} = \frac{5 \pm \sqrt{73}}{6}$$

The equation $\cos x = \dfrac{5 + \sqrt{73}}{6}$ does not have a solution because $\dfrac{5 + \sqrt{73}}{6} > 2$ and for any x the maximum value of $\cos x$ is 1. Thus

Visualize the Solution

The solutions are the x-coordinates of the x-intercepts of $y = 3\cos^2 x - 5\cos x - 4$ on the interval $[0, 2\pi)$. See **Figure 6.30.**

Continued ►

$\cos x = \dfrac{5 - \sqrt{73}}{6}$, and because $(5 - \sqrt{73})/6$ is a negative number (about -0.59), the equation $\cos x = (5 - \sqrt{73})/6$ will have two solutions on the interval $[0, 2\pi)$. Thus

$$x = \cos^{-1}\left(\frac{5 - \sqrt{73}}{6}\right) \approx 2.2027 \quad \text{or}$$

$$x = 2\pi - \cos^{-1}\left(\frac{5 - \sqrt{73}}{6}\right) \approx 4.0805$$

To the nearest 0.0001 the solutions on the inteval $[0, 2\pi)$ are 2.2027 and 4.0805.

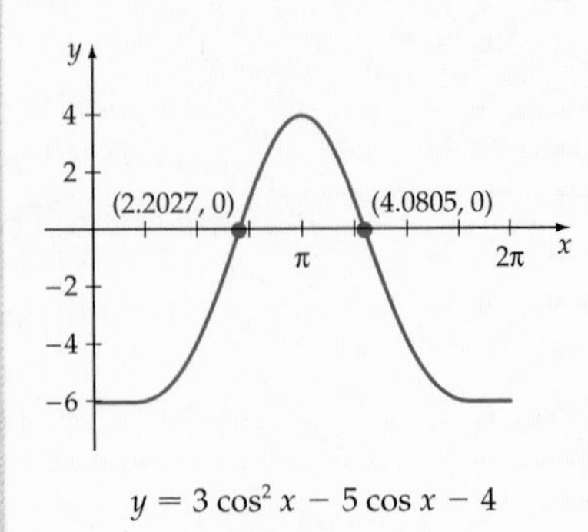

$y = 3\cos^2 x - 5\cos x - 4$

Figure 6.30

TRY EXERCISE 56, EXERCISE SET 6.6, PAGE 524

When solving equations that contain multiple angles, we must be sure we find all the solutions of the equation for the given interval. For example, to find all solutions of $\sin 2x = 1/2$, where $0 \le x < 2\pi$, we first solve for $2x$.

$$\sin 2x = \frac{1}{2}$$

$$2x = \frac{\pi}{6} + 2k\pi \quad \text{or} \quad 2x = \frac{5\pi}{6} + 2k\pi \qquad \bullet\ k \text{ is an integer.}$$

Solving for x, we have $x = \pi/12 + k\pi$ or $x = 5\pi/12 + k\pi$. Substituting integers for k, we obtain

$$k = 0: \quad x = \frac{\pi}{12} \quad \text{or} \quad x = \frac{5\pi}{12}$$

$$k = 1: \quad x = \frac{13\pi}{12} \quad \text{or} \quad x = \frac{17\pi}{12}$$

$$k = 2: \quad x = \frac{25\pi}{12} \quad \text{or} \quad x = \frac{29\pi}{12}$$

Note that for $k \ge 2$, $x \ge 2\pi$ and the solutions to $\sin 2x = 1/2$ are not in the interval $0 \le x < 2\pi$. Thus for $0 \le x < 2\pi$, the solutions are $\pi/12$, $5\pi/12$, $13\pi/12$, and $17\pi/12$. See **Figure 6.31.**

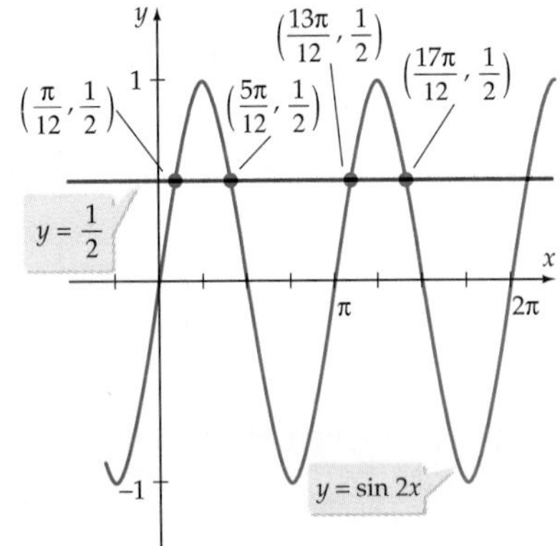

Figure 6.31

EXAMPLE 4 Solve a Trigonometric Equation

Solve: $\sin 3x = 1$

Solution

The equation $\sin 3x = 1$ implies

$$3x = \frac{\pi}{2} + 2k\pi, \quad k \text{ an integer}$$

$$x = \frac{\pi}{6} + \frac{2k\pi}{3}, \quad k \text{ an integer} \qquad \bullet\ \text{Divide each side by 3.}$$

Visualize the Solution

The solutions are the x-coordinates of the points of intersection of $y = \sin 3x$ and $y = 1$. **Figure 6.32** shows eight of the points of intersection.

Because x is not restricted to a finite interval, the given equation has an infinite number of solutions. All of the solutions are represented by the equation

$$x = \frac{\pi}{6} + \frac{2k\pi}{3}, \quad \text{where } k \text{ is an integer}$$

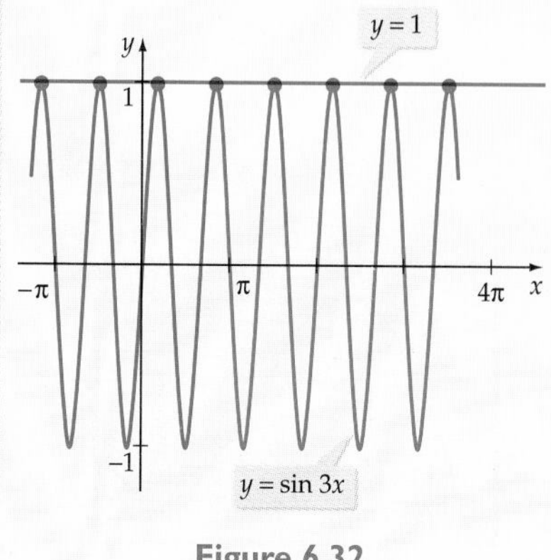

Figure 6.32

TRY EXERCISE 66, EXERCISE SET 6.6, PAGE 525

EXAMPLE 5 Solve a Trigonometric Equation

Solve $\sin^2 2x - \frac{\sqrt{3}}{2}\sin 2x + \sin 2x - \frac{\sqrt{3}}{2} = 0$, where $0° \le x < 360°$.

Solution

Factor the left side of the equation by grouping, and then set each factor equal to zero.

$$\sin^2 2x - \frac{\sqrt{3}}{2}\sin 2x + \sin 2x - \frac{\sqrt{3}}{2} = 0$$

$$\sin 2x\left(\sin 2x - \frac{\sqrt{3}}{2}\right) + \left(\sin 2x - \frac{\sqrt{3}}{2}\right) = 0$$

$$(\sin 2x + 1)\left(\sin 2x - \frac{\sqrt{3}}{2}\right) = 0$$

$$\sin 2x + 1 = 0 \quad \text{or} \quad \sin 2x - \frac{\sqrt{3}}{2} = 0$$

$$\sin 2x = -1 \qquad\qquad \sin 2x = \frac{\sqrt{3}}{2}$$

The equation $\sin 2x = -1$ implies that $2x = 270° + 360° \cdot k$, k an integer. Thus $x = 135° + 180° \cdot k$. The solutions of this equation with $0° \le x < 360°$ are 135° and 315°. Similarly, the equation $\sin 2x = \sqrt{3}/2$ implies

$$2x = 60° + 360° \cdot k \quad \text{or} \quad 2x = 120° + 360° \cdot k$$

$$x = 30° + 180° \cdot k \qquad\qquad x = 60° + 180° \cdot k$$

The solutions with $0° \le x < 360°$ are 30°, 60°, 210°, and 240°. Combining the solutions from each equation, we have 30°, 60°, 135°, 210°, 240°, and 315° as our solutions.

Visualize the Solution

The solutions are the x-coordinates of the x-intercepts of

$$y = \sin^2 2x - \frac{\sqrt{3}}{2}\sin 2x + \sin 2x - \frac{\sqrt{3}}{2}$$

on the interval $[0, 2\pi)$. See Figure 6.33.

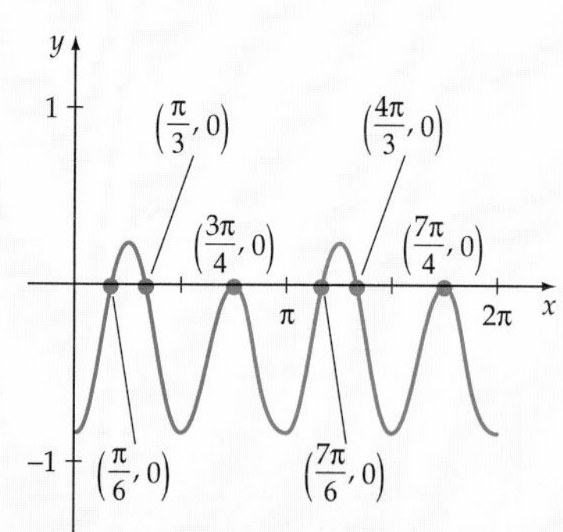

$$y = \sin^2 2x - \frac{\sqrt{3}}{2}\sin 2x + \sin 2x - \frac{\sqrt{3}}{2}$$

Figure 6.33

TRY EXERCISE 84, EXERCISE SET 6.6, PAGE 525

In Example 6, algebraic methods do not provide the solutions, so we rely on a graph.

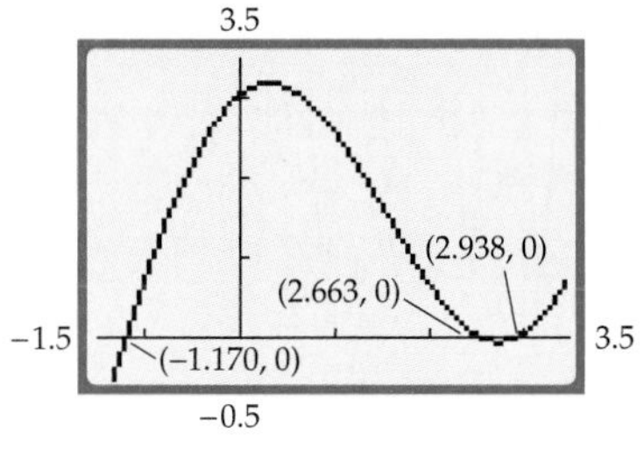

EXAMPLE 6 Approximate Solutions Graphically

Use a graphing utility to approximate the solutions of $x + 3\cos x = 0$.

Solution

The solutions are the x-intercepts of $y = x + 3\cos x$. See **Figure 6.34.** A close-up view of the graph of $y = x + 3\cos x$ shows that, to the nearest thousandth, the solutions are

$$x_1 = -1.170, \quad x_2 = 2.663, \quad \text{and} \quad x_3 = 2.938$$

3.5
(2.938, 0)
(2.663, 0)
−1.5
(−1.170, 0)
3.5
−0.5

$y = x + 3\cos x$

Figure 6.34

TRY EXERCISE 86, EXERCISE SET 6.6, PAGE 525

◆ AN APPLICATION INVOLVING A TRIGONOMETRIC EQUATION

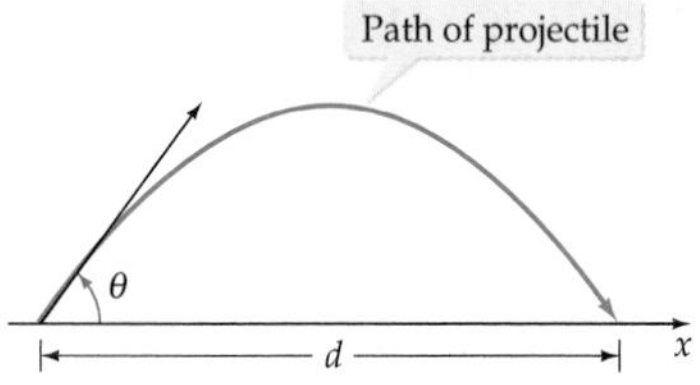

EXAMPLE 7 Solve a Projectile Application

A projectile is fired at an angle of inclination θ from the horizon with an initial velocity v_0. Its range d (neglecting air resistance) is given by

$$d = \frac{v_0^2}{16} \sin\theta\cos\theta$$

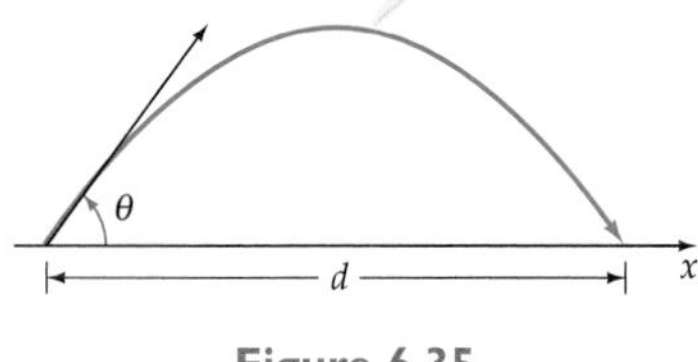

Figure 6.35

where v_0 is measured in feet per second and d is measured in feet. See **Figure 6.35.**

a. If $v_0 = 325$ feet per second, find the angles θ (in degrees) for which the projectile will hit a target 2295 feet downrange.

b. What is the maximum horizontal range for a projectile that has an initial velocity of 474 feet per second?

c. Determine the angle of inclination that produces the maximum range.

Solution

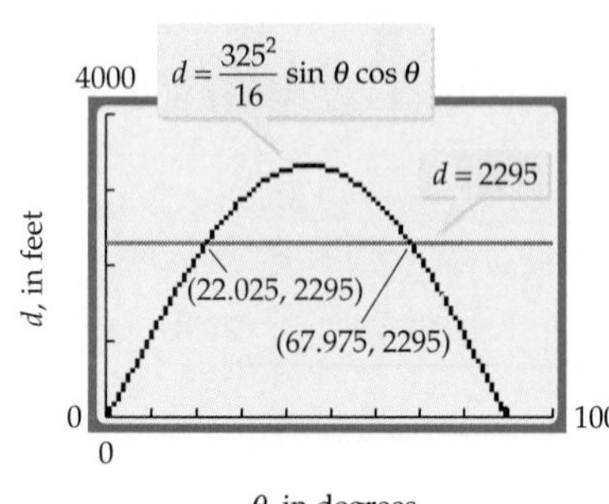

Figure 6.36

a. We need to solve

$$2295 = \frac{325^2}{16} \sin\theta\cos\theta \tag{1}$$

for θ, where $0° < \theta < 90°$.

Method 1 The following solutions were obtained by using a graphing utility to graph $d = 2295$ and $d = \frac{325^2}{16} \sin\theta\cos\theta$. See **Figure 6.36.** Thus there are two angles for which the projectile will hit the target. To the nearest thousandth of a degree, they are

$$\theta = 22.025° \quad \text{and} \quad \theta = 67.975°$$

It should be noted that the graph in **Figure 6.36** is *not* a graph of the path of the projectile. It is a graph of the distance d as a function of the angle θ.

Method 2 To solve algebraically, we proceed as follows. Multiply each side of Equation (1) by 16 and divide by 325^2 to produce

$$\sin\theta\cos\theta = \frac{(16)(2295)}{325^2}$$

The identity $2\sin\theta\cos\theta = \sin 2\theta$ gives us $\sin\theta\cos\theta = \dfrac{\sin 2\theta}{2}$.

Hence
$$\frac{\sin 2\theta}{2} = \frac{(16)(2295)}{325^2}$$

$$\sin 2\theta = 2\frac{(16)(2295)}{325^2} \approx 0.69529$$

There are two angles in the interval $[0°, 180°]$ whose sine is 0.69529. One is $\sin^{-1} 0.69529$, and the other one is the *reference angle* for $\sin^{-1} 0.69529$. Therefore,

$$2\theta \approx \sin^{-1} 0.69529 \quad \text{or} \quad 2\theta \approx 180° - \sin^{-1} 0.69529$$

$$\theta \approx \frac{1}{2}\sin^{-1} 0.69529 \quad \text{or} \quad \theta \approx \frac{1}{2}(180° - \sin^{-1} 0.69529)$$

$$\theta \approx 22.025° \quad \text{or} \quad \theta \approx 67.975°$$

These are the same angles that we obtained in Method 1.

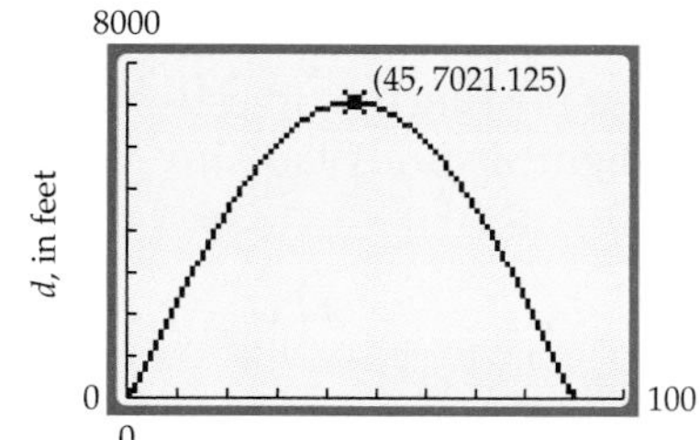

$d = \dfrac{474^2}{16}\sin\theta\cos\theta$

Figure 6.37

b. Use a graphing utility to find that the graph of $d = \dfrac{474^2}{16}\sin\theta\cos\theta$ has a maximum value of $d = 7021.125$ feet. See **Figure 6.37.**

c. In part b, the maximum value is attained for $\theta = 45°$. To prove that this is true in general, we use $2\sin\theta\cos\theta = \sin 2\theta$ to write

$$d = \frac{v_0^2}{16}\sin\theta\cos\theta \quad \text{as} \quad d = \frac{v_0^2}{32}\sin 2\theta$$

This equation enables us to determine that d will attain its maximum when $\sin 2\theta$ attains its maximum—that is, when $2\theta = 90°$, or $\theta = 45°$.

TRY EXERCISE 92, EXERCISE SET 6.6, PAGE 525

◆ MODEL SINUSOIDAL DATA

Data that can be closely modeled by a function of the form $y = a\sin(bx + c) + d$ is called **sinusoidal data.** Many graphing utilities are designed to perform a **sine**

regression to find the sine function that provides the best least-squares fit for sinusoidal data. For instance, the TI-83 uses the command **SinReg** and the TI-86 uses **SinR** to perform a sine regression. The process generally works best for those sets of data for which we have a good estimate of the period.

In Example 8, we use a sine regression to model the percent of illumination of the moon, as seen from earth. See **Figure 6.38.**

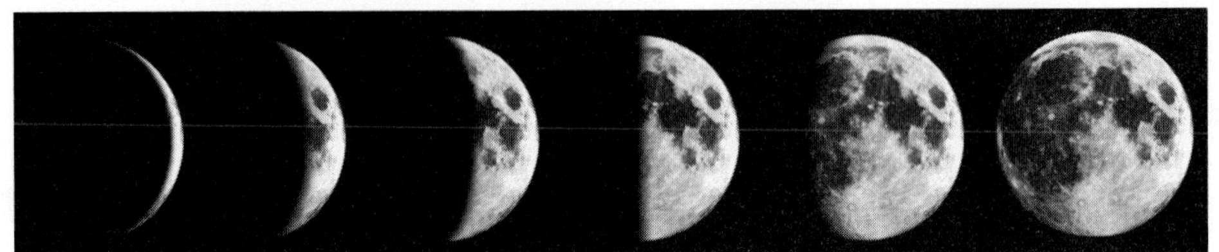

Figure 6.38

EXAMPLE 8 Use a Sine Regression to Model Data

Table 6.3 shows the percent of the moon illuminated, at midnight Central Standard Time, for selected days of January 2003. Find the regression function that models the percent of the moon illuminated as a function of the day, with January 1, 2003 represented by $x = 1$. Use the function to estimate the percent of the moon illuminated at midnight Central Standard Time, on January 21, 2003.

Table 6.3

Midnight of: Date in 2003 (CST)	Day Number	% of Moon Illuminated
Jan. 3	3	0
Jan. 7	7	20
Jan. 11	11	57
Jan. 15	15	89
Jan. 19	19	99
Jan. 23	23	73
Jan. 27	27	29
Jan. 31	31	2

The percent of the moon illuminated is nearly the same regardless of one's position on earth.

Source: The Astronomical Applications Department of the U.S. Naval Observatory.

Solution

1. **Construct a scatter plot of the data.** Enter the data from Table 6.3 into your graphing utility. See **Figure 6.39.** The sinusoidal nature of

L1	L2	L3 1
3	0	------
7	20	
11	57	
15	89	
19	99	
23	73	
27	29	

L1(1) = 3

Figure 6.39

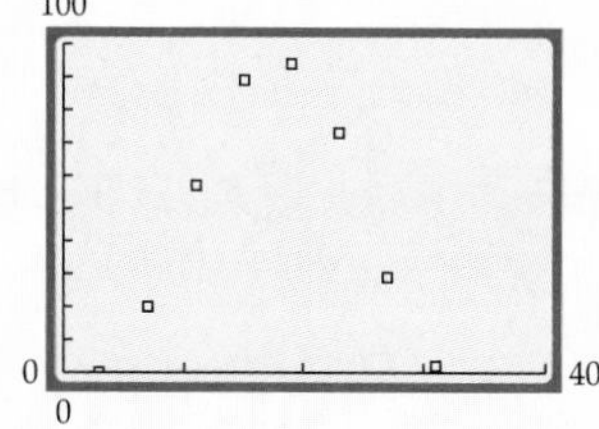

Figure 6.40

the scatter plot in **Figure 6.40**, indicates that the data can be effectively modeled by a sine function.

> **take note**
>
> A sine regression requires at least four data points. The iterations number 16 in Figure 6.41 is the maximum number of times the SinReg command will iterate to find a regression equation. Any integer from 1 to 16 can be used; however, the integer 16 generally produces more accurate results than smaller integers.

2. **Find the regression equation.** On a TI-83 graphing calculator the input shown in **Figure 6.41**, produces the results in **Figure 6.42**.

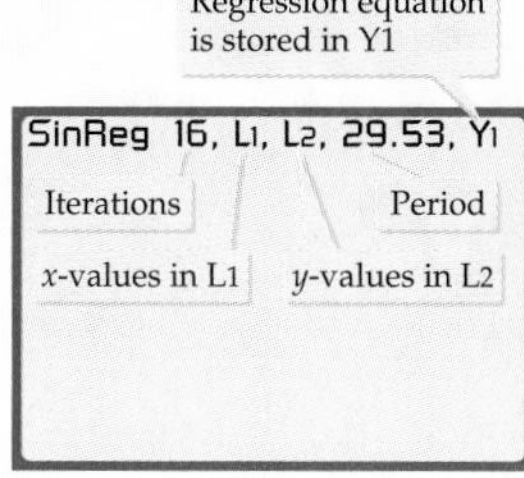

Figure 6.41

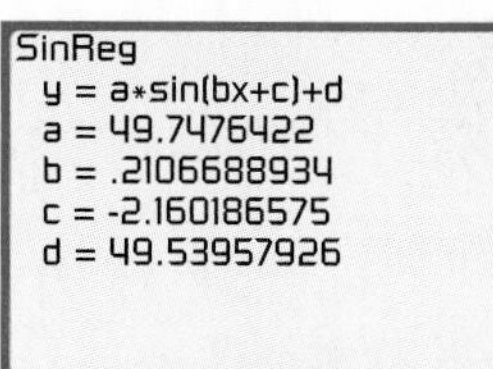

Figure 6.42

The regression equation is $y \approx 49.75 \sin(0.2107x - 2.160) + 49.54$.

3. **Examine the fit.** The SinReg command does not yield a correlation coefficient. However, a graph of the regression equation and the scatter plot of the data shows that the regression equation provides a good model. See **Figure 6.43**.

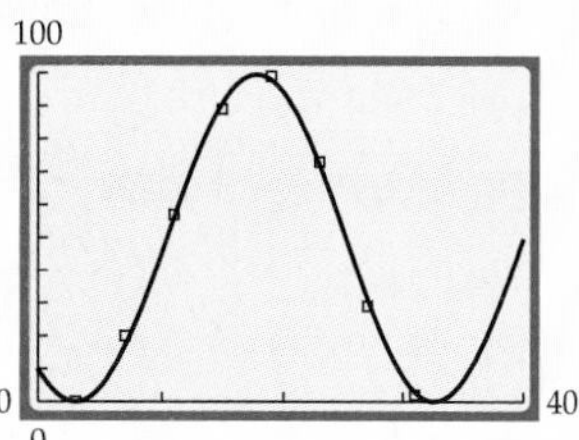

Figure 6.43

According to the following result, the percent of the moon illuminated at midnight (CST) on January 21, 2003, ($x = 21$), will be about 88%.

$$y \approx 49.75 \sin(0.2107(21) - 2.160) + 49.54 \approx 88$$

TRY EXERCISE 94, EXERCISE SET 6.6, PAGE 525

TOPICS FOR DISCUSSION

1. Explain why it is not necessary for graphing utilities to have a cosine regression.

2. A student finds that $x = 0$ is a solution of $\sin x = x$. Because the function $y = \sin x$ has a period of 2π, the student reasons that $\pm 2\pi, \pm 4\pi, \pm 6\pi, \ldots$ are also solutions. Explain why the student is not correct.

3. How many solutions does $2 \sin\left(x - \frac{\pi}{2}\right) = 5$ have on the interval $0 \le x < 2\pi$? Explain.

4. How many solutions does $\sin(1/x) = 0$ have on the interval $0 < x < \pi/2$? Explain.

5. On the interval $0 \le x < 2\pi$, the equation $\sin x = 1/2$ has solutions of $x = \pi/6$ and $x = 5\pi/6$. How would you write the solutions of $\sin x = 1/2$, if the real number x is not restricted to the interval $[0, 2\pi)$?

EXERCISE SET 6.6

In Exercises 1 to 22, solve each equation for exact solutions in the interval $0 \le x < 2\pi$.

1. $\sec x - \sqrt{2} = 0$
2. $2 \sin x = \sqrt{3}$
3. $\tan x - \sqrt{3} = 0$
4. $\cos x - 1 = 0$
5. $2 \sin x \cos x = \sqrt{2} \cos x$
6. $2 \sin x \cos x = \sqrt{3} \sin x$
7. $\sin^2 x - 1 = 0$
8. $\cos^2 x - 1 = 0$
9. $4 \sin x \cos x - 2\sqrt{3} \sin x - 2\sqrt{2} \cos x + \sqrt{6} = 0$
10. $\sec^2 x + \sqrt{3} \sec x - \sqrt{2} \sec x - \sqrt{6} = 0$
11. $\csc x - \sqrt{2} = 0$
12. $3 \cot x + \sqrt{3} = 0$
13. $2 \sin^2 x + 1 = 3 \sin x$
14. $2 \cos^2 x + 1 = -3 \cos x$
15. $4 \cos^2 x - 3 = 0$
16. $2 \sin^2 x - 1 = 0$
17. $2 \sin^3 x = \sin x$
18. $4 \cos^3 x = 3 \cos x$
19. $4 \sin^2 x + 2\sqrt{3} \sin x - \sqrt{3} = 2 \sin x$
20. $\tan^2 x + \tan x - \sqrt{3} = \sqrt{3} \tan x$
21. $\sin^4 x = \sin^2 x$
22. $\cos^4 x = \cos^2 x$

In Exercises 23 to 60, solve each equation, where $0° \le x < 360°$. Round approximate solutions to the nearest tenth of a degree.

23. $\cos x - 0.75 = 0$
24. $\sin x + 0.432 = 0$
25. $3 \sin x - 5 = 0$
26. $4 \cos x - 1 = 0$
27. $3 \sec x - 8 = 0$
28. $4 \csc x + 9 = 0$
29. $\cos x + 3 = 0$
30. $\sin x - 4 = 0$
31. $3 - 5 \sin x = 4 \sin x + 1$
32. $4 \cos x - 5 = \cos x - 3$
33. $\frac{1}{2} \sin x + \frac{2}{3} = \frac{3}{4} \sin x + \frac{3}{5}$
34. $\frac{2}{5} \cos x - \frac{1}{2} = \frac{1}{3} - \frac{1}{2} \cos x$
35. $3 \tan^2 x - 2 \tan x = 0$
36. $4 \cot^2 x + 3 \cot x = 0$
37. $3 \cos x + \sec x = 0$
38. $5 \sin x - \csc x = 0$
39. $\tan^2 x = 3 \sec^2 x - 2$
40. $\csc^2 x - 1 = 3 \cot^2 x + 2$
41. $2 \sin^2 x = 1 - \cos x$
42. $\cos^2 x + 4 = 2 \sin x - 3$
43. $3 \cos^2 x + 5 \cos x - 2 = 0$
44. $2 \sin^2 x + 5 \sin x + 3 = 0$
45. $2 \tan^2 x - \tan x - 10 = 0$
46. $2 \cot^2 x - 7 \cot x + 3 = 0$
47. $3 \sin x \cos x - \cos x = 0$
48. $\tan x \sin x - \sin x = 0$
49. $2 \sin x \cos x - \sin x - 2 \cos x + 1 = 0$
50. $6 \cos x \sin x - 3 \cos x - 4 \sin x + 2 = 0$
51. $2 \sin x - \cos x = 1$
52. $\sin x + 2 \cos x = 1$
53. $2 \sin x - 3 \cos x = 1$
54. $\sqrt{3} \sin x + \cos x = 1$
55. $3 \sin^2 x - \sin x - 1 = 0$
56. $2 \cos^2 x - 5 \cos x - 5 = 0$
57. $2 \cos x - 1 + 3 \sec x = 0$
58. $3 \sin x - 5 + \csc x = 0$
59. $\cos^2 x - 3 \sin x + 2 \sin^2 x = 0$
60. $\sin^2 x = 2 \cos x + 3 \cos^2 x$

In Exercises 61 to 70, find the exact solutions, in radians, of each trigonometric equation.

61. $\tan 2x - 1 = 0$

62. $\sec 3x - \frac{2\sqrt{3}}{3} = 0$

63. $\sin 5x = 1$

64. $\cos 4x = -\frac{\sqrt{2}}{2}$

65. $\sin 2x - \sin x = 0$

66. $\cos 2x = -\frac{\sqrt{3}}{2}$

67. $\sin\left(2x + \frac{\pi}{6}\right) = -\frac{1}{2}$

68. $\cos\left(2x - \frac{\pi}{4}\right) = -\frac{\sqrt{2}}{2}$

69. $\sin^2 \frac{x}{2} + \cos x = 1$

70. $\cos^2 \frac{x}{2} - \cos x = 1$

In Exercises 71 to 84, find exact solutions where $0 \le x < 2\pi$.

71. $\cos 2x = 1 - 3 \sin x$

72. $\cos 2x = 2 \cos x - 1$

73. $\sin 4x - \sin 2x = 0$

74. $\sin 4x - \cos 2x = 0$

75. $\tan \frac{x}{2} = \sin x$

76. $\tan \frac{x}{2} = 1 - \cos x$

77. $\sin 2x \cos x + \cos 2x \sin x = 0$

78. $\cos 2x \cos x - \sin 2x \sin x = 0$

79. $\sin x \cos 2x - \cos x \sin 2x = \frac{\sqrt{3}}{2}$

80. $\cos 2x \cos x + \sin 2x \sin x = -1$

81. $\sin 3x - \sin x = 0$

82. $\cos 3x + \cos x = 0$

83. $2 \sin x \cos x + 2 \sin x - \cos x - 1 = 0$

84. $2 \sin x \cos x - 2\sqrt{2} \sin x - \sqrt{3} \cos x + \sqrt{6} = 0$

In Exercises 85 to 88, use a graphing utility to solve the equation. State each solution accurate to the nearest ten-thousandth.

85. $\cos x = x$, where $0 \le x < 2\pi$

86. $2 \sin x = x$, where $0 \le x < 2\pi$

87. $\sin 2x = \frac{1}{x}$, where $-4 \le x \le 4$

88. $\cos x = \frac{1}{x}$, where $0 \le x \le 5$

89. Use a graphing utility to solve $\cos x = x^3 - x$ by graphing each side and finding all points of intersection.

90. Approximate the largest value of k for which the equation $\sin x \cos x = k$ has a solution.

PROJECTILES Exercises 91 and 92 make use of the following. A projectile is fired at an angle of inclination θ from the horizon with an initial velocity v_0. Its range d (neglecting air resistance) is given by

$$d = \frac{v_0^2}{16} \sin \theta \cos \theta$$

where v_0 is measured in feet per second and d is measured in feet.

91. If $v_0 = 288$ feet per second, use a graphing utility to find the angles θ (to the nearest hundredth of a degree) for which the projectile will hit a target 1295 feet downrange.

92. Use a graphing utility to find the maximum horizontal range, to the nearest tenth of a foot, for a projectile that has an initial velocity of 375 feet per second. What value of θ produces this maximum horizontal range?

93. **SUNRISE TIME** The table shows the sunrise time for Atlanta, Georgia, for selected days in 2004.

Date	Day of the Year, x	Sunrise Time
Jan. 1	1	7:42
Feb. 1	32	7:35
Mar. 1	61	7:06
April 1	92	6:25
May 1	122	5:48
June 1	153	5:28
July 1	183	5:31
Aug. 1	214	5:50
Sept. 1	245	6:12
Oct. 1	275	6:32
Nov. 1	306	6:57
Dec. 1	336	7:25

Source: The U.S. Naval Observatory. *Note:* The times do not reflect daylight savings time.

a. Use a graphing utility to find the sine regression function that models the sunrise time, in hours, as a function of the day of the year. Let $x = 1$ represent January 1, 2004. Assume that the sunrise times have a period of 365.25 days.

b. Use the regression function to estimate the sunrise time (to the nearest minute) for March 11, 2004 ($x = 71$).

94. **SUNSET TIME** The table shows the sunset time for Sioux City, Iowa, for selected days in 2003.

a. Use a graphing utility to find the sine regression function that models the sunset time, in hours, as a function of the day of the year. Let $x = 1$ represent January 1, 2003.

Date	Day of the Year, x	Sunset Time
Jan. 1	1	17:03
Feb. 1	32	17:39
Mar. 1	60	18:15
April 1	91	18:51
May 1	121	19:25
June 1	152	19:56
July 1	182	20:07
Aug. 1	213	19:46
Sept. 1	244	19:00
Oct. 1	274	18:08
Nov. 1	305	17:19
Dec. 1	335	16:54

Source: The U.S. Naval Observatory. *Note:* The times do not reflect daylight savings time.

Assume that the sunset times have a period of 365.25 days.

b. Use the regression function to estimate the sunset time (to the nearest minute) for May 21, 2003 ($x = 141$).

95. **PERCENT OF THE MOON ILLUMINATED** The table shows the percent of the moon illuminated at 10 P.M., Central Standard Time, for selected days in October and November of 2002.

10:00 P.M. of: Date in 2002	Day Number	% of Moon Illuminated
Oct. 1	1	35
Oct. 5	5	3
Oct. 9	9	11
Oct. 13	13	50
Oct. 17	17	85
Oct. 21	21	100
Oct. 25	25	87
Oct. 29	29	50
Nov. 2	33	10
Nov. 6	37	3
Nov. 10	41	33
Nov. 14	45	72
Nov. 18	49	97
Nov. 22	53	96
Nov. 26	57	66
Nov. 30	61	22

Source: The U.S. Naval Observatory.

a. Use a graphing utility to find the sine regression function that models the percent of the moon illuminated as a function of the day of the year. Let $x = 1$ represent October 1, 2002. Use 29.53 for the period of the data.

b. Use the regression function to estimate the percent of the moon illuminated (to the nearest 1 percent) at 10 P.M. Central Standard Time on October 31, 2002.

96. **HOURS OF DAYLIGHT** The table shows the hours of daylight for Houston, Texas, for selected days in 2003.

Date	Day of the Year, x	Hours of Daylight (hours:minutes)
Jan. 1	1	10:16
Feb. 1	32	10:47
Mar. 1	60	11:33
April 1	91	12:29
May 1	121	13:19
June 1	152	13:55
July 1	182	14:01
Aug. 1	213	13:34
Sept. 1	244	12:45
Oct. 1	274	11:52
Nov. 1	305	11:00
Dec. 1	335	10:23

Source: Data extracted from sunrise-sunset times given on the World Wide Web by the U.S. Naval Observatory.

a. Use a graphing utility to find the sine regression function that models the hours of daylight as a function of the day of the year. Let $x = 1$ represent January 1, 2003. Use 365.25 for the period of the data.

b. Use the regression function to estimate the hours of daylight (stated in hours and minutes, with the minutes rounded to the nearest minute) for Houston on May 12, 2003.

97. **ALTITUDE OF THE SUN** The table, on page 527 shows the altitude of the sun for Detroit, Michigan, at selected times during October 19, 2004.

a. Use a graphing utility to find the sine regression function that models the altitude, in degrees, of the sun as a function of the time of the day. Use 24.03 hours (the time from sunrise October 19 to sunrise October 20) for the period.

b. Use the regression function to estimate the altitude of the sun (to the nearest 0.1 of a degree) on October 19, 2004, at 9:25.

Time of day	Altitude (degrees)
6:00	−10.0
7:00	1.3
8:00	11.4
9:00	20.8
10:00	28.7
11:00	34.5
12:00	37.2
13:00	36.4
14:00	32.3
15:00	25.5
16:00	16.8
17:00	7.0
18:00	−3.8

Source: The U.S. Naval Observatory.

98. **RAINFALL TOTALS FOR SAN FRANCISCO** The table shows some average monthly rainfall totals for San Francisco, California.

Month	Month Number	Average Rainfall (inches)
January	1	4.48
March	3	2.58
May	5	0.35
July	7	0.04
September	9	0.24
November	11	2.49

a. Use a graphing utility to find the sine regression function that models the average monthly rainfall totals as a function of the month number. Use 12 months for the period.

b. Use the regression function to estimate San Francisco's average rainfall total (to the nearest 0.01 inch) for the month of April. How does this result compare with the recorded value of 1.48 inches?

c. Use the regression function to estimate San Francisco's average rainfall total for the month of June. Explain how you know that this result is incorrect.

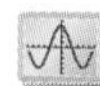

In Exercises 99 and 100, use a graphing utility.

99. **MODEL THE DAYLIGHT HOURS** For a particular day of the year t, the number of daylight hours in Mexico City can be approximated by

$$d(t) = 1.208 \sin\left(\frac{2\pi(t - 80)}{365}\right) + 12.133$$

where t is an integer and $t = 1$ corresponds to January 1. According to d, how many days per year will Mexico City have at least 12 hours of daylight?

100. **MODEL THE DAYLIGHT HOURS** For a particular day of the year t, the number of daylight hours in New Orleans can be approximated by

$$d(t) = 1.792 \sin\left(\frac{2\pi(t - 80)}{365}\right) + 12.145$$

where t is an integer and $t = 1$ corresponds to January 1. According to d, how many days per year will New Orleans have at least 10.75 hours of daylight?

SUPPLEMENTAL EXERCISES

In Exercises 101 to 110, solve each equation for exact solutions in the interval $0 \le x < 2\pi$.

101. $\sqrt{3}\sin x + \cos x = \sqrt{3}$

102. $\sin x - \cos x = 1$

103. $-\sin x + \sqrt{3}\cos x = \sqrt{3}$

104. $-\sqrt{3}\sin x - \cos x = 1$

105. $\cos 5x - \cos 3x = 0$

106. $\cos 5x - \cos x - \sin 3x = 0$

107. $\sin 3x + \sin x = 0$

108. $\sin 3x + \sin x - \sin 2x = 0$

109. $\cos 4x + \cos 2x = 0$

110. $\cos 4x + \cos 2x - \cos 3x = 0$

111. **MODEL THE MOVEMENT OF A BUS** As bus A_1 makes a left turn, the back B of the bus moves to the right. If bus A_2 were waiting at a stoplight while A_1 turned left, as shown in the figure, there is a chance the two buses would scrape against one another. For a bus 28 feet long and 8 feet wide, the movement of the back of the bus to the right can be approximated by

$$x = \sqrt{(4 + 18\cot\theta)^2 + 100} - (4 + 18\cot\theta)$$

where θ is the angle the bus driver has turned the front of the bus. Find the value of x for $\theta = 20°$ and $\theta = 30°$.

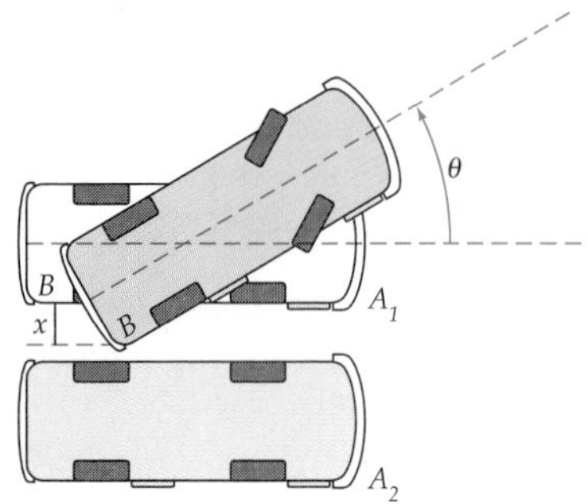

112. **OPTIMAL BRANCHING OF BLOOD VESSELS** It is hypothesized that the system of blood vessels in primates has evolved so that it has an optimal structure. In the case of a blood vessel splitting into two vessels, as shown in the accompanying figure, we assume that both new branches carry equal amounts of blood. A model of the angle θ is given by the equation $\cos \theta = 2^{(x-4)/(x+4)}$. The value of x is such that $1 \le x \le 2$ and depends on assumptions about the thickness of the blood vessels. Assuming this is an accurate model, find the values of the angle θ.

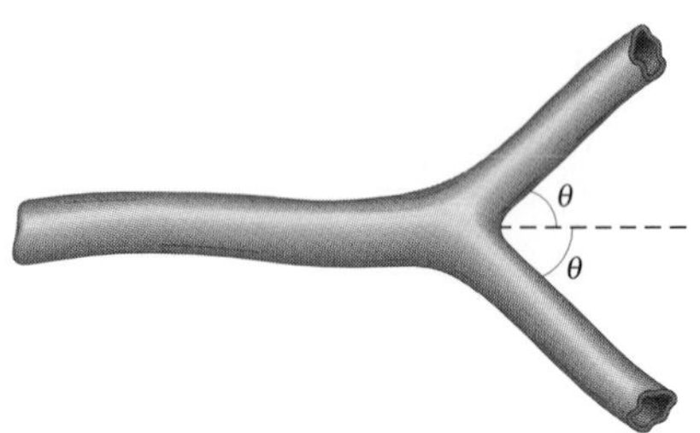

PROJECTS

1. **THE MOONS OF SATURN** The accompanying figure shows the east-west displacement of five moons of Saturn. The figure shows that the period of the sine curve that models the displacement of the moon Titan is about 15.9 Earth days. The period of the sine curve that models the displacement of the moon Rhea is about 4.5 Earth days.

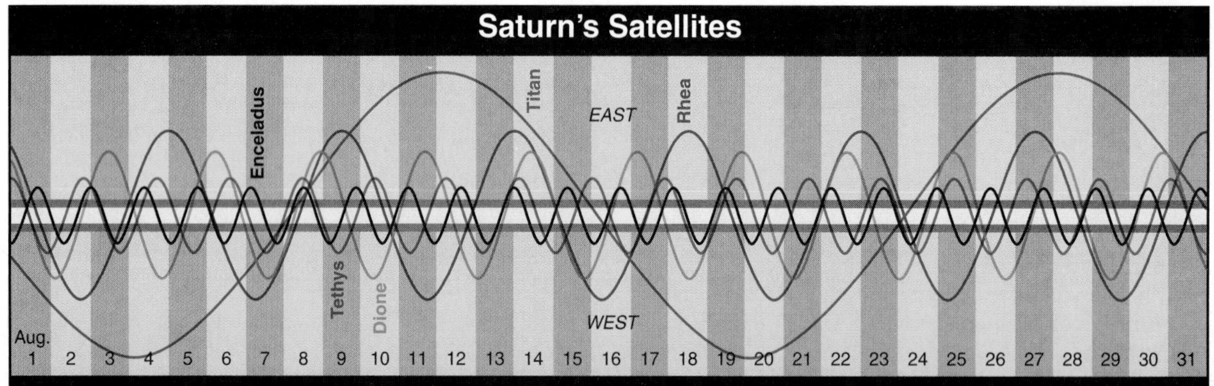

"Saturn's Satellites" from *Sky & Telescope*, August 2000, p. 102, Copyright 2000 by Sky Publishing Company. Reprinted with permission.

a. Write an equation of the form

$$d(t) = A \sin (Bt + C)$$

that models the displacement of the moon Titan. Use $t = 0$ to represent the beginning of August 1, 2000, and think of "east" as a positive displacement and of "west" as a negative displacement.

b. Work part **a** for the moon Rhea.

c. An astronomer viewed the moons Titan and Rhea at 10 P.M. on September 10, 2000. Were these moons on opposite sides of Saturn or on the same side?

EXPLORING CONCEPTS WITH TECHNOLOGY

Approximate an Inverse Trigonometric Function with Polynomials

The function $y = \sin^{-1} x$ can be approximated by polynomials. For example, consider the following:

$$f_1(x) = x + \frac{x^3}{2 \cdot 3} \qquad \text{where } -1 \le x \le 1$$

$$f_2(x) = x + \frac{x^3}{2 \cdot 3} + \frac{1 \cdot 3x^5}{2 \cdot 4 \cdot 5} \qquad \text{where } -1 \le x \le 1$$

$$f_3(x) = x + \frac{x^3}{2 \cdot 3} + \frac{1 \cdot 3x^5}{2 \cdot 4 \cdot 5} + \frac{1 \cdot 3 \cdot 5x^7}{2 \cdot 4 \cdot 6 \cdot 7} \qquad \text{where } -1 \le x \le 1$$

$$f_4(x) = x + \frac{x^3}{2 \cdot 3} + \frac{1 \cdot 3x^5}{2 \cdot 4 \cdot 5} + \frac{1 \cdot 3 \cdot 5x^7}{2 \cdot 4 \cdot 6 \cdot 7} + \frac{1 \cdot 3 \cdot 5 \cdot 7x^9}{2 \cdot 4 \cdot 6 \cdot 8 \cdot 9} \qquad \text{where } -1 \le x \le 1$$

$$\vdots$$

$$f_n(x) = x + \frac{x^3}{2 \cdot 3} + \frac{1 \cdot 3x^5}{2 \cdot 4 \cdot 5} + \frac{1 \cdot 3 \cdot 5x^7}{2 \cdot 4 \cdot 6 \cdot 7} + \cdots + \frac{(2n)!\,x^{2n+1}}{(2^n n!)^2(2n+1)}$$

where $-1 \le x \le 1,\ n! = 1 \cdot 2 \cdot 3 \cdots (n-1)n$

and $(2n)! = 1 \cdot 2 \cdot 3 \cdots (2n-1)(2n)$

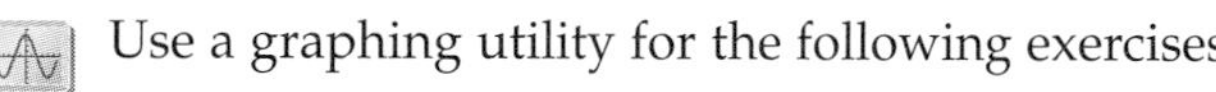

Use a graphing utility for the following exercises.

1. Graph $y = f_1(x)$, $y = f_2(x)$, $y = f_3(x)$, and $y = f_4(x)$ on the viewing window Xmin $= -1$, Xmax $= 1$, Ymin $= -1.5708$, Ymax $= 1.5708$.

2. Determine the values of x for which $f_3(x)$ and $\sin^{-1} x$ differ by less than 0.001. That is, determine the values of x for which

$$|f_3(x) - \sin^{-1} x| < 0.001$$

3. Determine the values of x for which

$$|f_4(x) - \sin^{-1} x| < 0.001$$

4. Write all seven terms of $f_6(x)$. Graph $y = f_6(x)$ and $y = \sin^{-1} x$ on the viewing window Xmin $= -1$, Xmax $= 1$, Ymin $= -\pi/2$, Ymax $= \pi/2$.

5. Write all seven terms of $f_6(1)$. What do you notice about the size of a term compared to that of the previous term?

6. What is the largest-degree term in $f_{10}(x)$?

CHAPTER 6 SUMMARY

6.1 Verification of Trigonometric Identities

- Trigonometric identities are verified by using algebraic methods and previously proved identities.

$$\sin x = \frac{1}{\csc x} \qquad \cos x = \frac{1}{\sec x} \qquad \tan x = \frac{1}{\cot x}$$

$$\tan x = \frac{\sin x}{\cos x} \qquad \cot x = \frac{\cos x}{\sin x}$$

$$\sin^2 x + \cos^2 x = 1; \tan^2 x + 1 = \sec^2 x;$$

$$1 + \cot^2 x = \csc^2 x$$

6.2 Sum, Difference, and Cofunction Identities

- Sum and difference identities for the cosine function are

$$\cos(\alpha - \beta) = \cos\alpha\cos\beta + \sin\alpha\sin\beta$$
$$\cos(\alpha + \beta) = \cos\alpha\cos\beta - \sin\alpha\sin\beta$$

- Sum and difference identities for the sine function are

$$\sin(\alpha - \beta) = \sin\alpha\cos\beta - \cos\alpha\sin\beta$$
$$\sin(\alpha + \beta) = \sin\alpha\cos\beta + \cos\alpha\sin\beta$$

- Sum and difference identities for the tangent function are

$$\tan(\alpha + \beta) = \frac{\tan\alpha + \tan\beta}{1 - \tan\alpha\tan\beta}$$

$$\tan(\alpha - \beta) = \frac{\tan\alpha - \tan\beta}{1 + \tan\alpha\tan\beta}$$

- The cofunction identities are

$$\sin(90^\circ - \theta) = \cos\theta \qquad \cos(90^\circ - \theta) = \sin\theta$$
$$\tan(90^\circ - \theta) = \cot\theta \qquad \cot(90^\circ - \theta) = \tan\theta$$
$$\sec(90^\circ - \theta) = \csc\theta \qquad \csc(90^\circ - \theta) = \sec\theta$$

where θ is in degrees. If θ is in radian measure, replace 90° with $\pi/2$.

6.3 Double- and Half-Angle Identities

- The double-angle identities are

$$\sin 2\alpha = 2\sin\alpha\cos\alpha$$
$$\cos 2\alpha = \cos^2\alpha - \sin^2\alpha$$
$$= 1 - 2\sin^2\alpha$$
$$= 2\cos^2\alpha - 1$$
$$\tan 2\alpha = \frac{2\tan\alpha}{1 - \tan^2\alpha}$$

- The half-angle identities are

$$\sin\frac{\alpha}{2} = \pm\sqrt{\frac{1 - \cos\alpha}{2}}$$

$$\cos\frac{\alpha}{2} = \pm\sqrt{\frac{1 + \cos\alpha}{2}}$$

$$\tan\frac{\alpha}{2} = \frac{\sin\alpha}{1 + \cos\alpha} = \frac{1 - \cos\alpha}{\sin\alpha}$$

6.4 Identities Involving the Sum of Trigonometric Functions

- The product-to-sum identities are

$$\sin\alpha\cos\beta = \frac{1}{2}[\sin(\alpha + \beta) + \sin(\alpha - \beta)]$$

$$\cos\alpha\sin\beta = \frac{1}{2}[\sin(\alpha + \beta) - \sin(\alpha - \beta)]$$

$$\cos\alpha\cos\beta = \frac{1}{2}[\cos(\alpha + \beta) + \cos(\alpha - \beta)]$$

$$\sin\alpha\sin\beta = \frac{1}{2}[\cos(\alpha - \beta) - \cos(\alpha + \beta)]$$

- The sum-to-product identities are

$$\sin x + \sin y = 2\sin\frac{x + y}{2}\cos\frac{x - y}{2}$$

$$\cos x - \cos y = -2\sin\frac{x + y}{2}\sin\frac{x - y}{2}$$

$$\sin x - \sin y = 2\cos\frac{x + y}{2}\sin\frac{x - y}{2}$$

$$\cos x + \cos y = 2\cos\frac{x + y}{2}\cos\frac{x - y}{2}$$

- For sums of the form $a\sin x + b\cos x$,

$$a\sin x + b\cos x = k\sin(x + \alpha)$$

where $k = \sqrt{a^2 + b^2}$, $\sin\alpha = \dfrac{b}{\sqrt{a^2 + b^2}}$, and

$\cos\alpha = \dfrac{a}{\sqrt{a^2 + b^2}}$.

6.5 Inverse Trigonometric Functions

- The inverse of $y = \sin x$ is $y = \sin^{-1} x$, with $-1 \le x \le 1$ and $-\pi/2 \le y \le \pi/2$.
- The inverse of $y = \cos x$ is $y = \cos^{-1} x$, with $-1 \le x \le 1$ and $0 \le y \le \pi$.
- The inverse of $y = \tan x$ is $y = \tan^{-1} x$, with $-\infty < x < \infty$ and $-\pi/2 < y < \pi/2$.

- The inverse of $y = \cot x$ is $y = \cot^{-1} x$, with $-\infty < x < \infty$ and $0 < y < \pi$.
- The inverse of $y = \csc x$ is $y = \csc^{-1} x$, with $x \le -1$ or $x \ge 1$ and $-\pi/2 \le y \le \pi/2$, $y \ne 0$.
- The inverse of $y = \sec x$ is $y = \sec^{-1} x$, with $x \le -1$ or $x \ge 1$ and $0 \le y \le \pi$, $y \ne \pi/2$.

6.6 Trigonometric Equations

- Algebraic methods and identities are used to solve trigonometric equations. Because the trigonometric functions are periodic, there may be an infinite number of solutions. If solutions cannot be found by algebraic methods, then we often use a graph to find approximate solutions.

CHAPTER 6 TRUE/FALSE EXERCISES

In Exercises 1 to 12, answer true or false. If the statement is false, give an example to show that the statement is false.

1. $\dfrac{\tan \alpha}{\tan \beta} = \dfrac{\alpha}{\beta}$

2. $\dfrac{\sin x}{\cos y} = \tan \dfrac{x}{y}$

3. $\sin^{-1} x = \csc x^{-1}$

4. $\sin 2\alpha = 2 \sin \alpha$ for all α

5. $\sin(\alpha + \beta) = \sin \alpha + \sin \beta$

6. An equation that has an infinite number of solutions is an identity.

7. If $\tan \alpha = \tan \beta$, then $\alpha = \beta$.

8. $\cos^{-1}(\cos x) = x$

9. $\cos(\cos^{-1} x) = x$

10. $\csc^{-1} \dfrac{1}{\alpha} = \dfrac{1}{\csc \alpha}$

11. If $0^\circ \le \theta \le 90^\circ$, then $\cos \theta = \sin(180^\circ - \theta)$.

12. $\sin^2 \theta = \sin \theta^2$

CHAPTER 6 REVIEW EXERCISES

In Exercises 1 to 10, find the exact value.

1. $\cos(45^\circ + 30^\circ)$

2. $\tan(210^\circ - 45^\circ)$

3. $\sin\left(\dfrac{2\pi}{3} + \dfrac{\pi}{4}\right)$

4. $\sec\left(\dfrac{4\pi}{3} - \dfrac{\pi}{4}\right)$

5. $\sin(60^\circ - 135^\circ)$

6. $\cos\left(\dfrac{5\pi}{3} - \dfrac{7\pi}{4}\right)$

7. $\sin\left(22\dfrac{1}{2}\right)^\circ$

8. $\cos 105^\circ$

9. $\tan\left(67\dfrac{1}{2}\right)^\circ$

10. $\sin 112.5^\circ$

In Exercises 11 to 14, find the exact value of the given functions.

11. Given $\sin \alpha = 1/2$, α in Quadrant I, and $\cos \beta = 1/2$, β in Quadrant IV, find

    a. $\cos(\alpha - \beta)$ b. $\tan 2\alpha$ c. $\sin(\beta/2)$

12. Given $\sin \alpha = \sqrt{3}/2$, α in Quadrant II, and $\cos \beta = -1/2$, β in Quadrant III, find

    a. $\sin(\alpha + \beta)$ b. $\sec 2\beta$ c. $\cos(\alpha/2)$

13. Given $\sin \alpha = -1/2$, α in Quadrant IV, and $\cos \beta = -\sqrt{3}/2$, β in Quadrant III, find

    a. $\sin(\alpha - \beta)$ b. $\tan 2\alpha$ c. $\cos(\beta/2)$

14. Given $\sin \alpha = \sqrt{2}/2$, α in Quadrant I, and $\cos \beta = \sqrt{3}/2$, β in Quadrant IV, find

    a. $\cos(\alpha - \beta)$ b. $\tan 2\beta$ c. $\sin 2\alpha$

In Exercises 15 to 20, write the given expression as a single trigonometric function.

15. $2 \sin 3x \cos 3x$

16. $\dfrac{\tan 2x + \tan x}{1 - \tan 2x \tan x}$

17. $\sin 4x \cos x - \cos 4x \sin x$

18. $\cos^2 2\theta - \sin^2 2\theta$

19. $1 - 2\sin^2 \dfrac{\beta}{2}$

20. $\pm\sqrt{\dfrac{1 - \cos 4\theta}{2}}$

In Exercises 21 to 24, evaluate each expression.

21. $\sin 47^\circ \sin 22^\circ$

22. $\cos 14^\circ \cos 92^\circ$

23. $2 \sin \dfrac{\pi}{3} \cos \dfrac{2\pi}{3}$

24. $2 \cos \dfrac{\pi}{4} \cos \dfrac{3\pi}{2}$

In Exercises 25 to 28, write each expression as the product of two functions.

25. $\cos 2\theta - \cos 4\theta$

26. $\sin 3\theta - \sin 5\theta$

27. $\sin 6\theta + \sin 2\theta$

28. $\sin 5\theta - \sin \theta$

In Exercises 29 to 46, verify the identity.

29. $\dfrac{1}{\sin x - 1} + \dfrac{1}{\sin x + 1} = -2 \tan x \sec x$

30. $\dfrac{\sin x}{1 - \cos x} = \csc x + \cot x, \quad 0 < x < \dfrac{\pi}{2}$

31. $\dfrac{1 + \sin x}{\cos^2 x} = \tan^2 x + 1 + \tan x \sec x$

32. $\cos^2 x - \sin^2 x - \sin 2x = \dfrac{\cos^2 2x - \sin^2 2x}{\cos 2x + \sin 2x}$

33. $\dfrac{1}{\cos x} - \cos x = \tan x \sin x$

34. $\sin(270° - \theta) - \cos(270° - \theta) = \sin\theta - \cos\theta$

35. $\sin\left(\dfrac{\pi}{4} - \alpha\right) = \dfrac{\sqrt{2}}{2}(\cos\alpha - \sin\alpha)$

36. $\sin(180° - \alpha + \beta) = \sin\alpha\cos\beta - \cos\alpha\sin\beta$

37. $\dfrac{\sin 4x - \sin 2x}{\cos 4x - \cos 2x} = -\cot 3x$

38. $2\sin x \sin 3x = (1 - \cos 2x)(1 + 2\cos 2x)$

39. $\sin x - \cos 2x = (2\sin x - 1)(\sin x + 1)$

40. $\cos 4x = 1 - 8\sin^2 x + 8\sin^4 x$

41. $\tan 4x = \dfrac{4\tan x - 4\tan^3 x}{1 - 6\tan^2 x + \tan^4 x}$

42. $\dfrac{\sin 2x - \sin x}{\cos 2x + \cos x} = \dfrac{1 - \cos x}{\sin x}$

43. $2\cos 4x \sin 2x = 2\sin 3x \cos 3x - 2\sin x \cos x$

44. $2\sin x \sin 2x = 4\cos x \sin^2 x$

45. $\cos(x + y)\cos(x - y) = \cos^2 x + \cos^2 y - 1$

46. $\cos(x + y)\sin(x - y) = \sin x \cos x - \sin y \cos y$

In Exercises 47 to 52, solve the equation.

47. $y = \sec\left(\sin^{-1}\dfrac{12}{13}\right)$

48. $y = \cos\left(\sin^{-1}\dfrac{3}{5}\right)$

49. $2\sin^{-1}(x - 1) = \dfrac{\pi}{3}$

50. $y = \cos\left[\sin^{-1}\left(-\dfrac{3}{5}\right) + \cos^{-1}\dfrac{5}{13}\right]$

51. $\sin^{-1} x + \cos^{-1}\dfrac{4}{5} = \dfrac{\pi}{2}$

52. $y = \cos\left(2\sin^{-1}\dfrac{3}{5}\right)$

In Exercises 53 and 54, solve each equation on $0° \le x < 360°$.

53. $4\sin^2 x + 2\sqrt{3}\sin x - 2\sin x - \sqrt{3} = 0$

54. $2\sin x \cos x - \sqrt{2}\cos x - 2\sin x + \sqrt{2} = 0$

In Exercises 55 and 56, solve the trigonometric equation.

55. $3\cos^2 x + \sin x = 1$

56. $\tan^2 x - 2\tan x - 3 = 0$

In Exercises 57 and 58, solve each equation on $0 \le x < 2\pi$.

57. $\sin 3x \cos x - \cos 3x \sin x = \dfrac{1}{2}$

58. $\cos\left(2x - \dfrac{\pi}{3}\right) = -\dfrac{\sqrt{3}}{2}$

In Exercises 59 to 62, find the amplitude and phase shift of each function. Graph each function.

59. $f(x) = \sqrt{3}\sin x + \cos x$

60. $f(x) = -2\sin x - 2\cos x$

61. $f(x) = -\sin x - \sqrt{3}\cos x$

62. $f(x) = \dfrac{\sqrt{3}}{2}\sin x - \dfrac{1}{2}\cos x$

In Exercises 63 to 66, graph each function.

63. $f(x) = 2\cos^{-1} x$

64. $f(x) = \sin^{-1}(x - 1)$

65. $f(x) = \sin^{-1}\dfrac{x}{2}$

66. $f(x) = \sec^{-1} 2x$

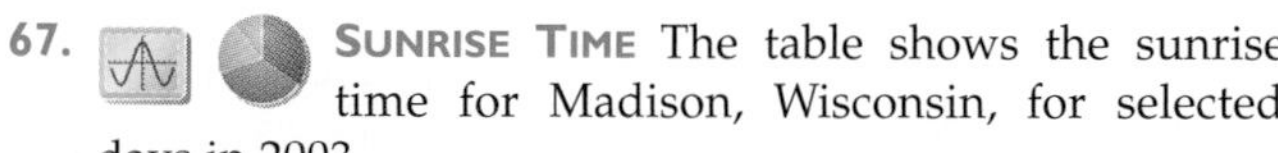

67. **SUNRISE TIME** The table shows the sunrise time for Madison, Wisconsin, for selected days in 2003.

Date	Day of the Year, x	Sunrise Time (hours:minutes)
Jan. 1	1	7:29
Feb. 1	32	7:13
Mar. 1	60	6:34
April 1	91	5:40
May 1	121	4:52
June 1	152	4:21
July 1	182	4:22
Aug. 1	213	4:48
Sept. 1	244	5:22
Oct. 1	274	5:55
Nov. 1	305	6:32
Dec. 1	335	7:09

Source: The U.S. Naval Observatory. *Note:* The times are Central Standard Times. The times do not reflect daylight savings time.

a. Use a graphing utility to find the sine regression function that models the sunrise time, in hours, as a function of the day of the year. Let $x = 1$ represent January 1, 2003. Assume that the sunrise times have a period of 365.25 days.

b. Use the regression function to estimate the sunrise time (to the nearest minute) for April 14, 2003 ($x = 104$).

CHAPTER 6 TEST

1. Verify the identity $1 + \sin^2 x \sec^2 x = \sec^2 x$.

2. Verify the identity

$$\frac{1}{\sec x - \tan x} - \frac{1}{\sec x + \tan x} = 2 \tan x$$

3. Verify the identity $\cos^3 x + \cos x \sin^2 x = \cos x$.

4. Verify the identity $\csc x - \cot x = \dfrac{1 - \cos x}{\sin x}$.

5. Find the exact value of $\sin 195°$.

6. Given $\sin \alpha = -3/5$, α in Quadrant III, and $\cos \beta = -\sqrt{2}/2$, β in Quadrant II, find $\sin(\alpha + \beta)$.

7. Verify the identity $\sin\left(\theta - \dfrac{3\pi}{2}\right) = \cos \theta$.

8. Write $\cos 6x \sin 3x + \sin 6x \cos 3x$ in terms of a single trigonometric function.

9. Find the exact value of $\cos 2\theta$ given that $\sin \theta = 4/5$ and θ is in Quadrant II.

10. Verify the identity $\tan \dfrac{\theta}{2} + \dfrac{\cos \theta}{\sin \theta} = \csc \theta$.

11. Verify the identity $\sin^2 2x + 4 \cos^4 x = 4 \cos^2 x$.

12. Find the exact value of $\sin 15° \cos 75°$.

13. Write $y = -\sqrt{3}/2 \sin x + 1/2 \cos x$ in the form $y = k \sin(x + \alpha)$, where α is measured in radians.

14. Use a calculator to approximate the radian measure of $\cos^{-1} 0.7644$ to the nearest thousandth.

15. Find the exact value of $\sin(\cos^{-1} 12/13)$.

16. Graph: $y = \sin^{-1}(x + 2)$

17. Solve $3 \sin x - 2 = 0$, where $0° \le x < 360°$. (State solutions to the nearest 0.1°.)

18. Solve $\sin x \cos x - \dfrac{\sqrt{3}}{2} \sin x = 0$, where $0 \le x < 2\pi$.

19. Find the exact solutions of $\sin 2x + \sin x - 2 \cos x - 1 = 0$, where $0 \le x < 2\pi$.

20. The table shows the altitude of the sun for Fort Lauderdale, Florida, at selected times during April 20, 2003.

Time of day	Altitude (degrees)
6:00	1.1
7:00	14.0
8:00	27.5
9:00	40.9
10:00	54.0
11:00	66.2
12:00	74.7
13:00	72.6
14:00	62.2
15:00	49.6
16:00	36.3
17:00	22.9
18:00	9.5
19:00	−3.7

Source: The U.S. Naval Observatory.

a. Use a graphing utility to find the sine regression function that models the altitude of the sun as a function of the time of the day. Use 23.983 hours (the time from sunrise April 20 to sunrise April 21) for the period.

b. Use the regression function to estimate the altitude of the sun (to the nearest 0.1 of a degree) on April 20, 2003, at 10:40.

CHAPTER 7

APPLICATIONS OF TRIGONOMETRY

Trigonometry and Cartography

In February 2000, NASA launched its Shuttle Radar Topography Mission (SRTM). During this 10-day mission, the space shuttle bounced radar beams off the Earth's surface and used two types of radar antennas to collect data that can be used to produce high-resolution three-dimensional images of the Earth's topography. Many of the theorems from this chapter were used in the design and implementation of this mission.

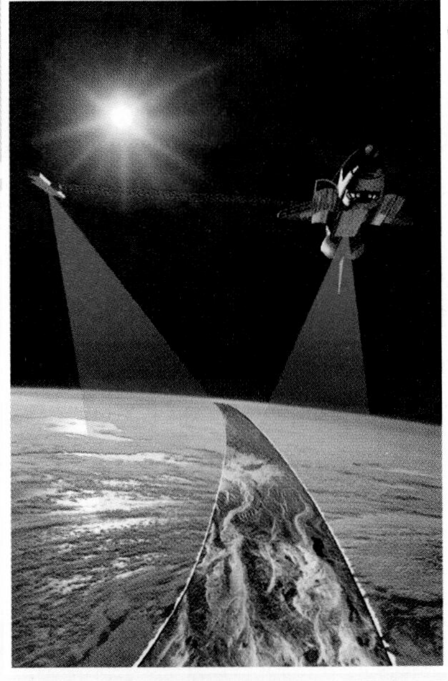

◆ The space shuttle with its 197 foot radar mast deployed during NASA's Shuttle Radar Topography Mission.

SECTION 7.1 THE LAW OF SINES

- THE LAW OF SINES
- THE AMBIGUOUS CASE (SSA)
- APPLICATIONS OF THE LAW OF SINES

◆ THE LAW OF SINES

Solving a triangle involves finding the lengths of all sides and the measure of all angles in the triangle. In this section and the next, we develop formulas for solving an **oblique triangle,** which is a triangle that does not contain a right angle. The *Law of Sines* can be used to solve oblique triangles in which either two angles and a side (AAS) or two sides and an angle opposite one of the sides (SSA) are known. In **Figure 7.1**, altitude CD is drawn from C. The length of the altitude is h. Triangles ACD and BCD are right triangles.

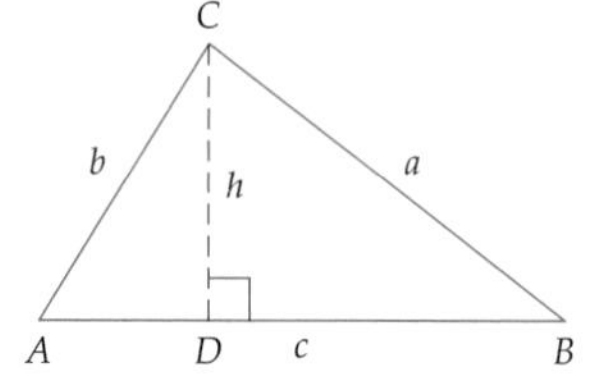

Figure 7.1

Using the definition of the sine of an angle of a right triangle, we have from **Figure 7.1**

$$\sin B = \frac{h}{a} \qquad\qquad \sin A = \frac{h}{b}$$

$$h = a \sin B \quad (1) \qquad\qquad h = b \sin A \quad (2)$$

Equating the values of h in Equations (1) and (2), we obtain

$$a \sin B = b \sin A$$

Dividing each side of the equation by $\sin A \sin B$, we obtain

$$\frac{a}{\sin A} = \frac{b}{\sin B}$$

Similarly, when an altitude is drawn to a different side, the following formulas result:

$$\frac{c}{\sin C} = \frac{b}{\sin B} \quad \text{and} \quad \frac{c}{\sin C} = \frac{a}{\sin A}$$

take note

The Law of Sines may also be written as

$$\frac{\sin A}{a} = \frac{\sin B}{b} = \frac{\sin C}{c}$$

The Law of Sines

If A, B, and C are the measures of the angles of a triangle, and a, b, and c are the lengths of the sides opposite these angles, then

$$\frac{a}{\sin A} = \frac{b}{\sin B} = \frac{c}{\sin C}$$

EXAMPLE 1 Solve a Triangle Using the Law of Sines (AAS)

Solve triangle ABC if $A = 42°$, $B = 63°$, and $c = 18$ centimeters.

Solution

Find C by using the fact that the sum of the interior angles of a triangle is 180°.

$$\begin{aligned} A + B + C &= 180^\circ \\ 42^\circ + 63^\circ + C &= 180^\circ \\ C &= 75^\circ \end{aligned}$$

Use the Law of Sines to find a.

$$\frac{a}{\sin A} = \frac{c}{\sin C}$$

$$\frac{a}{\sin 42^\circ} = \frac{18}{\sin 75^\circ} \qquad \bullet\ A = 42^\circ, c = 18, C = 75^\circ$$

$$a = \frac{18 \sin 42^\circ}{\sin 75^\circ} \approx 12 \text{ centimeters}$$

Use the Law of Sines again, this time to find b.

$$\frac{b}{\sin B} = \frac{c}{\sin C}$$

$$\frac{b}{\sin 63^\circ} = \frac{18}{\sin 75^\circ} \qquad \bullet\ B = 63^\circ, c = 18, C = 75^\circ$$

$$b = \frac{18 \sin 63^\circ}{\sin 75^\circ} \approx 17 \text{ centimeters}$$

The solution is $C = 75^\circ$, $a \approx 12$ centimeters, and $b \approx 17$ centimeters.

TRY EXERCISE 4, EXERCISE SET 7.1, PAGE 541

◆ THE AMBIGUOUS CASE (SSA)

When you are given two sides of a triangle and an angle opposite one of them, you sometimes find that the triangle is not unique. Some information may result in two triangles, and some may result in no triangle at all. It is because of this that the case of knowing two sides and an angle opposite one of them (SSA) is called the *ambiguous case* of the Law of Sines.

Suppose sides a and c and the nonincluded angle A of a triangle are known and we are then asked to solve triangle ABC. The relationships among h, the height of the triangle, a (the side opposite $\angle A$), and c determine whether there are no, one, or two triangles.

Case 1 First consider the case in which $\angle A$ is an acute angle (see **Figure 7.2**). There are four possible situations.

1. $a < h$; there is no possible triangle.
2. $a = h$; there is one triangle, a right triangle.
3. $h < a < c$; there are two possible triangles. One has all acute angles, and the second has one obtuse angle.
4. $a \geq c$; there is one triangle, which is not a right triangle.

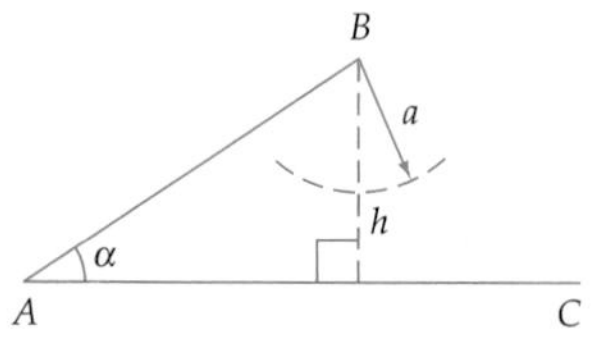

1. $a < h$; no triangle

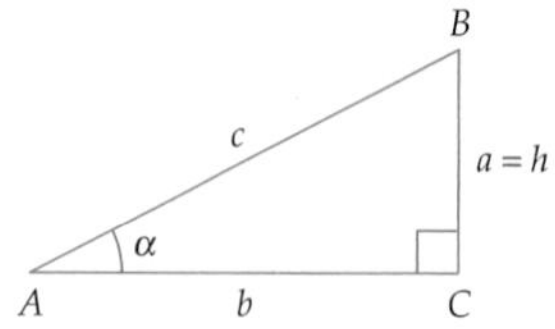

2. $a = h$; one triangle

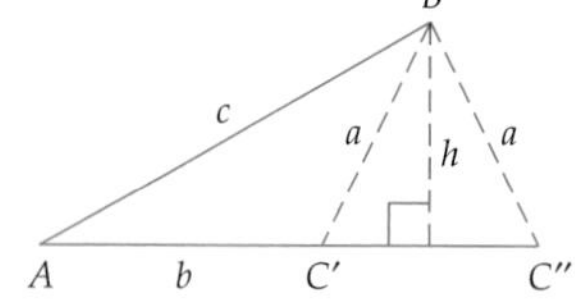

3. $h < a < c$; two triangles

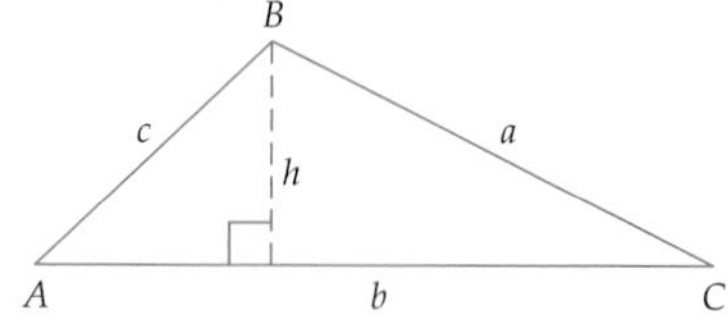

4. $a \geq c$; one triangle

Figure 7.2
Case 1: A is an acute angle

Case 2 Now consider the case in which $\angle A$ is an obtuse angle (see **Figure 7.3**). Here, there are two possible situations.

1. $a \leq c$; there is no triangle.
2. $a > c$; there is one triangle.

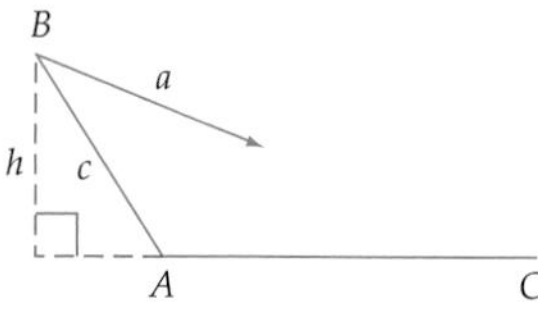

1. $a \leq c$; no triangle

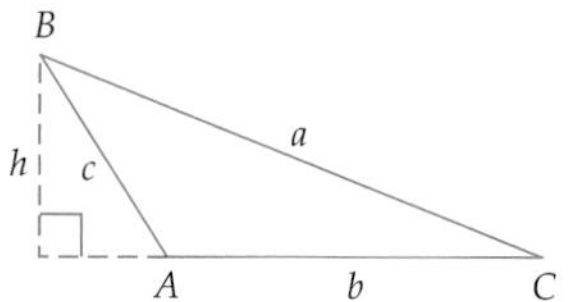

2. $a > c$; one triangle

Figure 7.3
Case 2: A is an obtuse angle

Figure 7.4

take note

When you solve for an angle of a triangle by using the Law of Sines, be aware that the value of the inverse sine function will give the measure of an acute angle. If the situation is the ambiguous case (SSA), you must consider a second, obtuse, angle by using the supplement of the angle. You can use a scale drawing to see whether your results are reasonable.

EXAMPLE 2 Solve a Triangle Using the Law of Sines (SSA)

a. Find A, given triangle ABC with $B = 32°$, $a = 42$, and $b = 30$.

b. Find C, given triangle ABC with $A = 57°$, $a = 15$ feet, and $c = 20$ feet.

Solution

a.
$$\frac{b}{\sin B} = \frac{a}{\sin A}$$

$$\frac{30}{\sin 32°} = \frac{42}{\sin A} \qquad \text{• } B = 32°, a = 42, b = 30$$

$$\sin A = \frac{42 \sin 32°}{30} \approx 0.7419$$

$$A \approx 48° \text{ or } 132°$$

• The two angles with measure between 0° and 180° that have a sine of 0.7419 are approximately 48° and 132°.

To check that $A \approx 132°$ is a valid result, add 132° to the measure of the given angle B (32°). Because $132° + 32° < 180°$, we know that $A \approx 132°$ is a valid result. Thus angle $A \approx 48°$, or $A \approx 132°$ ($\angle BAC$ in **Figure 7.4**).

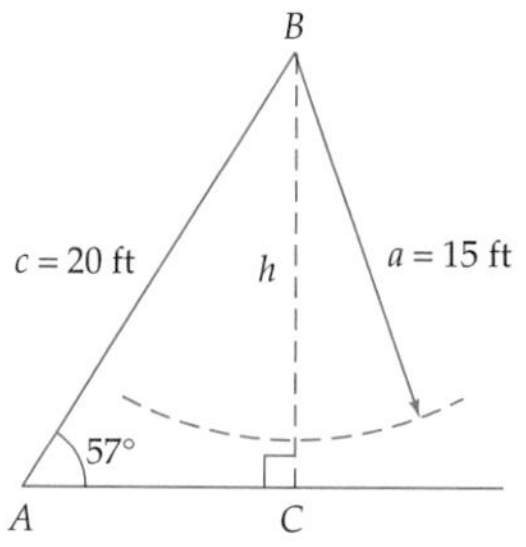

Figure 7.5

b. $$\frac{a}{\sin A} = \frac{c}{\sin C}$$

$$\frac{15}{\sin 57°} = \frac{20}{\sin C}$$ • $A = 57°, a = 15, c = 20$

$$\sin C = \frac{20 \sin 57°}{15} \approx 1.1182$$

Because 1.1182 is not in the range of the sine function, there is no solution of the equation. Thus there is no triangle for these values of A, a, and c. See **Figure 7.5.**

TRY EXERCISE 18, EXERCISE SET 7.1, PAGE 541

◆ APPLICATIONS OF THE LAW OF SINES

EXAMPLE 3 Solve an Application Using the Law of Sines

A radio antenna 85 feet high is located on top of an office building. At a distance AD from the base of the building, the angle of elevation to the top of the antenna is 26°, and the angle of elevation to the bottom of the antenna is 16°. Find the height of the building.

Solution

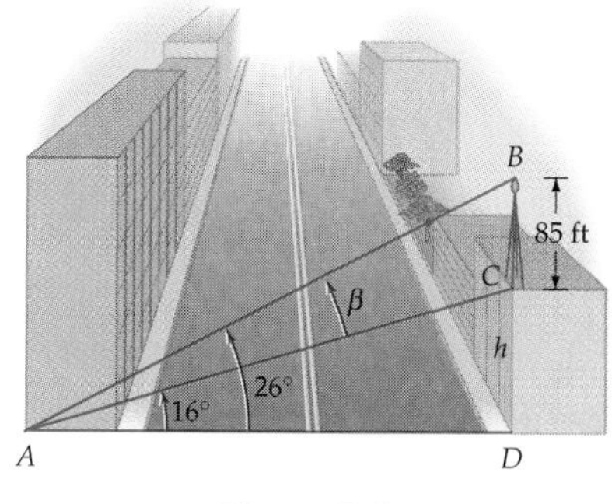

Figure 7.6

Sketch the diagram. See **Figure 7.6.** Find B and β.

$$B = 90° - 26° = 64°$$
$$\beta = 26° - 16° = 10°$$

Because we know the length BC and the measure of β, we can use triangle ABC and the Law of Sines to find length AC.

$$\frac{BC}{\sin \beta} = \frac{AC}{\sin B}$$

$$\frac{85}{\sin 10°} = \frac{AC}{\sin 64°}$$ • $BC = 85, \beta = 10°, B = 64°$

$$AC = \frac{85 \sin 64°}{\sin 10°}$$

Having found AC, we can now find the height of the building.

$$\sin 16° = \frac{h}{AC}$$

$$h = AC \sin 16°$$

$$= \frac{85 \sin 64°}{\sin 10°} \sin 16° \approx 121 \text{ feet}$$ • **Substitute for AC.**

The height of the building to two significant digits is 120 feet.

TRY EXERCISE 26, EXERCISE SET 7.1, PAGE 541

take note

In Example 3 we rounded the height of the building to two significant digits to comply with the rounding convention given on page 409.

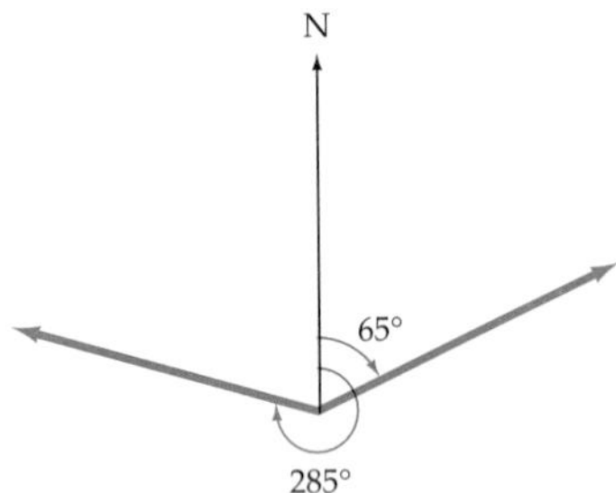

Figure 7.7

In navigation and surveying problems, there are two commonly used methods for specifying direction. The angular direction in which a craft is pointed is called the **heading.** Heading is expressed in terms of an angle measured clockwise from north. Figure 7.7 shows a heading of 65° and a heading of 285°.

The angular direction used to locate one object in relation to another object is called the **bearing.** Bearing is expressed in terms of the acute angle formed by a north–south line and the line of direction. Figure 7.8 shows a bearing of N38°W and a bearing of S15°E.

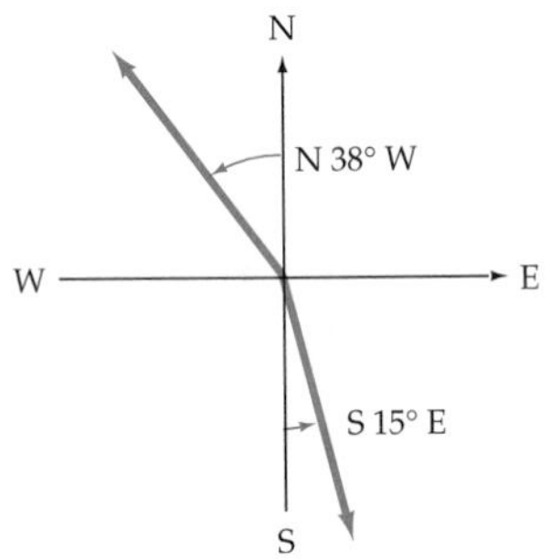

Figure 7.8

EXAMPLE 4 Solve an Application

A ship with a heading of 330° first sighted a lighthouse (point B) at a bearing of N65°E. After traveling 8.5 miles, the ship observed the lighthouse at a bearing of S50°E. Find the distance from the ship to the lighthouse when the first sighting was made.

Solution

From Figure 7.9 we see that the measure of $\angle CAB = 65° + 30° = 95°$, the measure of $\angle BCA = 50° - 30° = 20°$, and $B = 180° - 95° - 20° = 65°$. Use triangle ABC and the Law of Sines to find c.

$$\frac{b}{\sin B} = \frac{c}{\sin C}$$

$$\frac{8.5}{\sin 65°} = \frac{c}{\sin 20°} \qquad \bullet\ b = 8.5, B = 65°, C = 20°$$

$$c = \frac{8.5 \sin 20°}{\sin 65°} \approx 3.2$$

The lighthouse was 3.2 miles (two significant digits) from the ship when the first sighting was made.

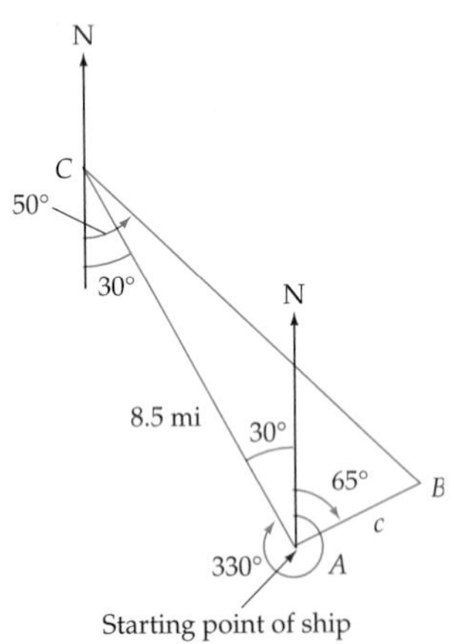

Figure 7.9

TRY EXERCISE 34, EXERCISE SET 7.1, PAGE 542

TOPICS FOR DISCUSSION

1. Is it possible to solve a triangle if the only given information consists of the measures of the three angles of the triangle? Explain.
2. Explain why it is not possible (in general) to use the Law of Sines to solve a triangle for which we are given the lengths of all of the sides.
3. Draw a triangle with dimensions $A = 30°$, $c = 3$ inches, and $a = 2.5$ inches. Is your answer unique? That is, can more than one triangle with the given dimensions be drawn?
4. Argue for or against the following proposition: In a scalene triangle (a triangle with no congruent sides), the largest angle is always opposite the longest side and the smallest angle is always opposite the shortest side.

EXERCISE SET 7.1

In Exercises 1 to 12, solve the triangles.

1. $A = 42°, B = 61°, a = 12$
2. $B = 25°, C = 125°, b = 5.0$
3. $A = 110°, C = 32°, b = 12$
4. $B = 28°, C = 78°, c = 44$
5. $A = 132°, a = 22, b = 16$
6. $B = 82.0°, b = 6.0, c = 3.0$
7. $A = 82.0°, B = 65.4°, b = 36.5$
8. $B = 54.8°, C = 72.6°, a = 14.4$
9. $A = 33.8°, C = 98.5°, c = 102$
10. $B = 36.9°, C = 69.2°, a = 166$
11. $C = 114.2°, c = 87.2, b = 12.1$
12. $A = 54.32°, a = 24.42, c = 16.92$

In Exercises 13 to 24, solve the triangles that exist.

13. $A = 37°, c = 40, a = 28$
14. $B = 32°, c = 14, b = 9.0$
15. $C = 65°, b = 10, c = 8.0$
16. $A = 42°, a = 12, c = 18$
17. $A = 30°, a = 1.0, b = 2.4$
18. $B = 22.6°, b = 5.55, a = 13.8$
19. $A = 14.8°, c = 6.35, a = 4.80$
20. $C = 37.9°, b = 3.50, c = 2.84$
21. $C = 47.2°, a = 8.25, c = 5.80$
22. $B = 52.7°, b = 12.3, c = 16.3$
23. $B = 117.32°, b = 67.25, a = 15.05$
24. $A = 49.22°, a = 16.92, c = 24.62$

25. **HURRICANE WATCH** A satellite weather map shows a hurricane off the coast of North Carolina. Use the information in the map to find the distance from the hurricane to Nags Head.

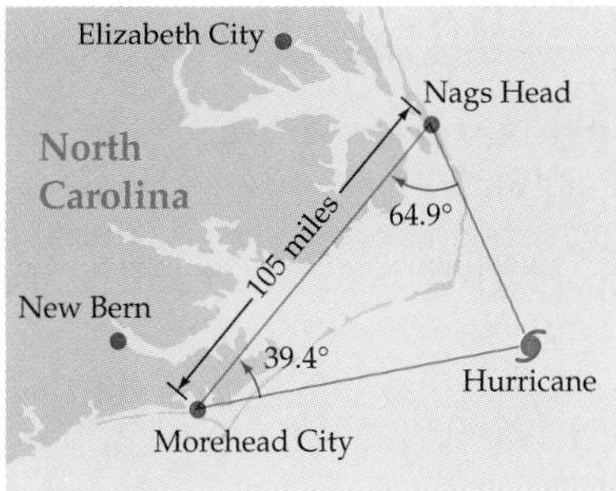

26. **NAVAL MANEUVERS** The distance between an aircraft carrier and a Navy destroyer is 7620 feet. The angle of elevation from the destroyer to a helicopter is 77.2°, and the angle of elevation from the aircraft carrier to the helicopter is 59.0°. The helicopter is in the same vertical plane as the two ships, as shown in the following figure. Use this data to determine the distance x from the helicopter to the aircraft carrier.

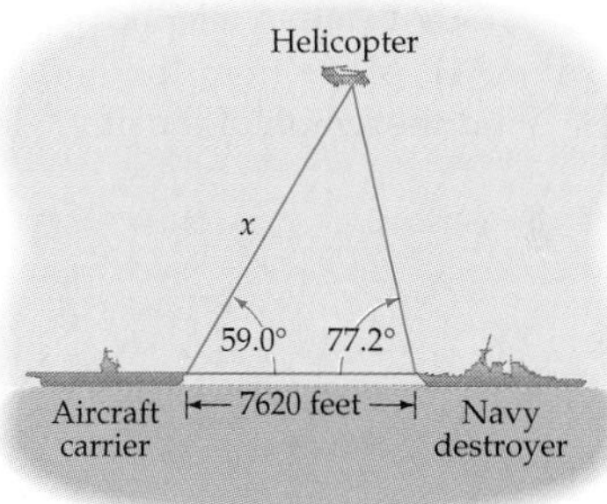

27. **DISTANCE TO A HOT AIR BALLOON** The angle of elevation of a balloon from one observer is 67°, and the angle of elevation from another observer, 220 feet away, is 31°. If the balloon is in the same vertical plane as the two observers and in between them, find the distance of the balloon from the first observer.

28. **LENGTH OF A DIAGONAL** The longer side of a parallelogram is 6.0 meters. The measure of $\angle BAD$ is 56° and α is 35°. Find the length of the longer diagonal.

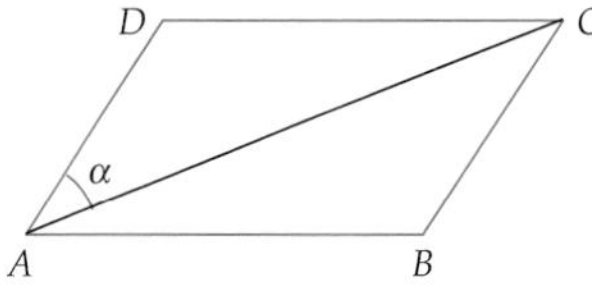

29. **DISTANCE ACROSS A CANYON** To find the distance across a canyon, a surveying team locates points A and B on one side of the canyon and point C on the other side of the canyon. The distance between A and B is 85 yards. The measure of $\angle CAB$ is 68°, and the measure of $\angle CBA$ is 75°. Find the distance across the canyon.

30. **HEIGHT OF A KITE** Two observers, in the same vertical plane as a kite and at a distance of 30 feet apart, observe the kite at an angle of 62° and 78°, as shown in the diagram on page 542. Find the height of the kite.

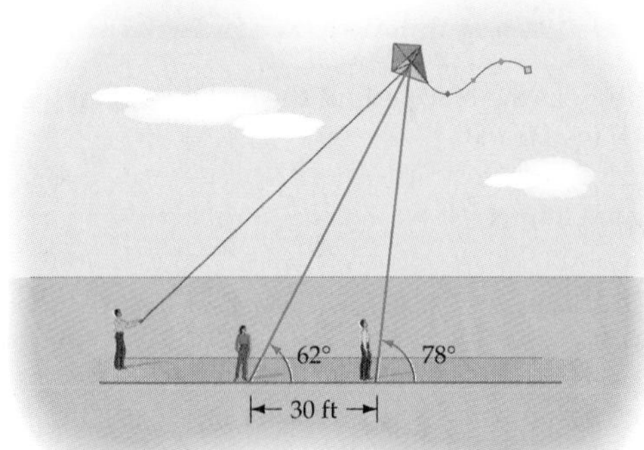

31. **LENGTH OF A GUY WIRE** A telephone pole 35 feet high is situated on an 11° slope from the horizontal. The angle CAB is 21°. Find the length of the guy wire AC.

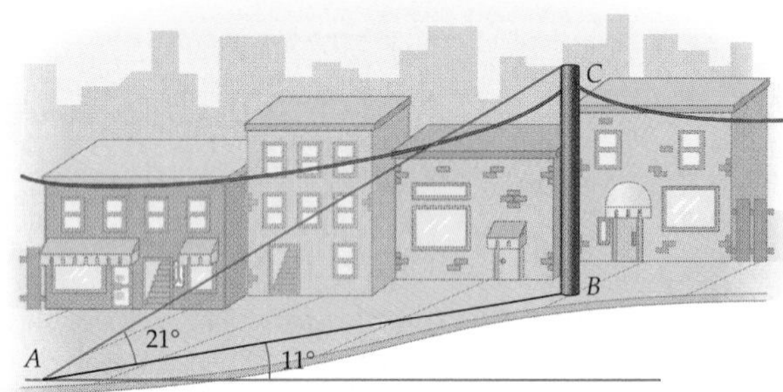

32. **DIMENSIONS OF A PLOT OF LAND** Three roads intersect in such a way as to form a triangular piece of land. See the accompanying figure. Find the lengths of the other two sides of the land.

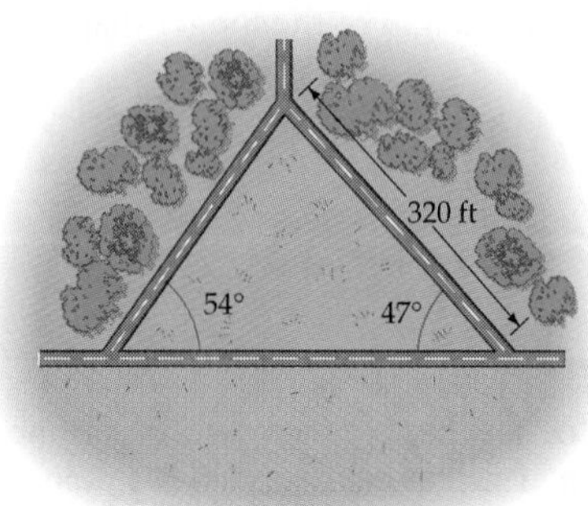

33. **HEIGHT OF A HILL** A surveying team determines the height of a hill by placing a 12-foot pole at the top of the hill and measuring the angles of elevation to the bottom and the top of the pole. They find the angles of elevation shown in the figure at the top of next column. Find the height of the hill.

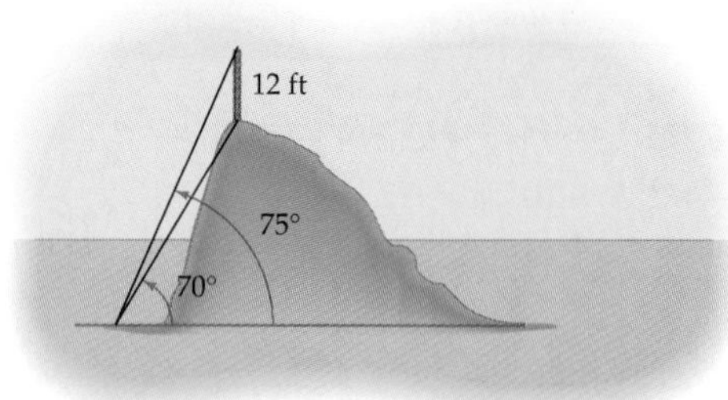

34. **DISTANCE TO A FIRE** Two fire lookouts are located on mountains 20 miles apart. Lookout B is at a bearing of S65°E from A. A fire was sighted at a bearing of N50°E from A and at a bearing of N8°E from B. Find the distance of the fire from lookout A.

35. **DISTANCE TO A LIGHTHOUSE** A navigator on a ship sights a lighthouse at a bearing of N36°E. After traveling 8.0 miles at a heading of 332°, the ship sights the lighthouse at a bearing of S82°E. How far is the ship from the lighthouse at the second sighting?

36. **MINIMUM DISTANCE** The navigator on a ship traveling due east at 8 mph sights a lighthouse at a bearing of S55°E. One hour later it is sighted at a bearing of S25°W. Find the closest the ship came to the lighthouse.

37. **DISTANCE BETWEEN AIRPORTS** An airplane flew 450 miles at a bearing of N65°E from airport A to airport B. The plane then flew at a bearing of S38°E to airport C. Find the distance from A to C if the bearing from airport A to airport C is S60°E.

38. **LENGTH OF A BRACE** A 12-foot solar panel is to be installed on a roof with a 15° pitch. Find the length of the vertical brace d if the panel must be installed to make a 40° angle with the horizontal.

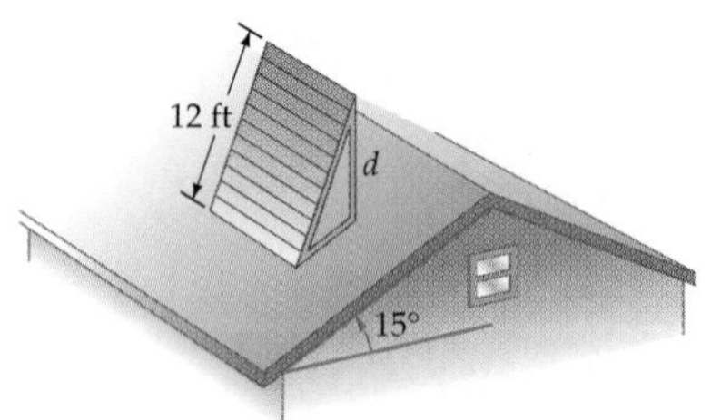

SUPPLEMENTAL EXERCISES

39. **DISTANCES BETWEEN HOUSES** House B is located at a bearing of N67°E from house A. House C is 300 meters at a bearing of S68°E from house A. House B is located at a bearing of N11°W from house C. Find the distance from house A to house B.

40. Show that for any triangle ABC, $\dfrac{a - b}{b} = \dfrac{\sin A - \sin B}{\sin B}$.

41. Show that for any triangle ABC, $\dfrac{a+b}{b} = \dfrac{\sin A + \sin B}{\sin B}$.

42. Show that for any triangle ABC, $\dfrac{a-b}{a+b} = \dfrac{\sin A - \sin B}{\sin A + \sin B}$.

43. **MAXIMUM LENGTH OF A ROD** The longest rod that can be carried horizontally around a corner from a hall 3 meters wide into one that is 5 meters wide is the minimum of the length L of the dashed line shown in the figure below.

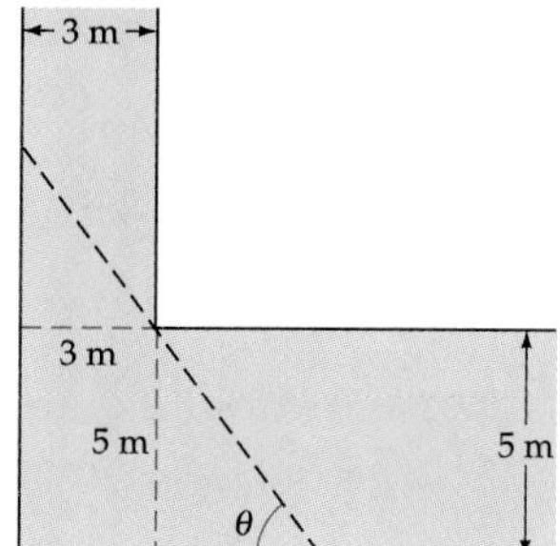

Use similar triangles to show that the length L is a function of the angle θ, given by

$$L(\theta) = \frac{5}{\sin \theta} + \frac{3}{\cos \theta}$$

Use a graphing utility to graph L and estimate the minimum value of L.

PROJECT

1. **FERMAT'S PRINCIPLE AND SNELL'S LAW** State Fermat's Principle and Snell's Law. The refractive index of a glass ring is found to be 1.82. The refractive index of a particular diamond ring is 2.38. Use Snell's Law to explain what this means in terms of the reflective properties of the two rings.

SECTION 7.2 THE LAW OF COSINES AND AREA

- THE LAW OF COSINES
- AN APPLICATION OF THE LAW OF COSINES
- AREA OF A TRIANGLE
- HERON'S FORMULA

7A

◆ THE LAW OF COSINES

The *Law of Cosines* can be used to solve triangles in which two sides and the included angle (SAS) are known or in which three sides (SSS) are known. Consider the triangle in **Figure 7.10.** The height BD is drawn from B perpendicular to the x-axis. The triangle BDA is a right triangle, and the coordinates of B are $(a \cos C, a \sin C)$. The coordinates of A are $(b, 0)$. Using the distance formula, we can find the distance c.

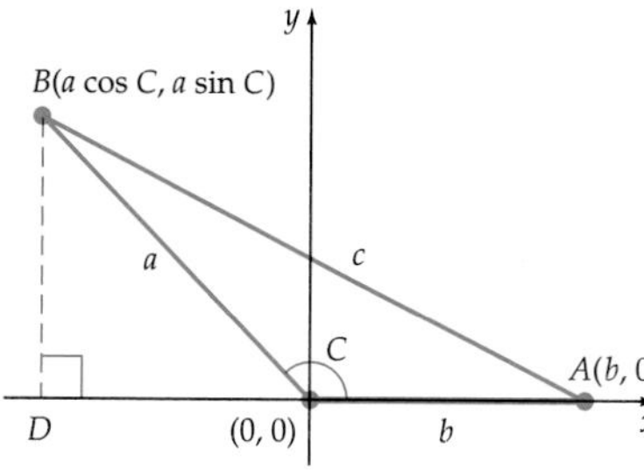

Figure 7.10

$$c = \sqrt{(a \cos C - b)^2 + (a \sin C - 0)^2}$$
$$c^2 = a^2 \cos^2 C - 2ab \cos C + b^2 + a^2 \sin^2 C$$
$$c^2 = a^2(\cos^2 C + \sin^2 C) + b^2 - 2ab \cos C$$
$$c^2 = a^2 + b^2 - 2ab \cos C$$

The Law of Cosines

If A, B, and C are the measures of the angles of a triangle, and a, b, and c are the lengths of the sides opposite these angles, then

$$c^2 = a^2 + b^2 - 2ab \cos C$$

$$a^2 = b^2 + c^2 - 2bc \cos A$$

$$b^2 = a^2 + c^2 - 2ac \cos B$$

EXAMPLE 1 Use the Law of Cosines (SAS)

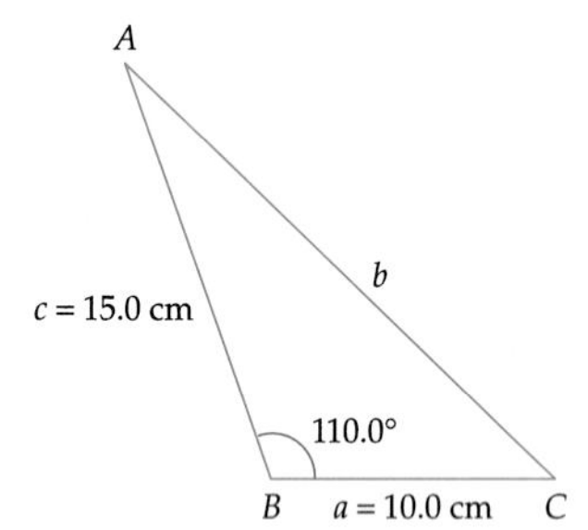

Figure 7.11

In triangle ABC, $B = 110.0°$, $a = 10.0$ centimeters, and $c = 15.0$ centimeters. See Figure 7.11. Find b.

Solution

The Law of Cosines can be used because two sides and the included angle are known.

$$b^2 = a^2 + c^2 - 2ac \cos B$$

$$= 10.0^2 + 15.0^2 - 2(10.0)(15.0) \cos 110.0°$$

$$b = \sqrt{10.0^2 + 15.0^2 - 2(10.0)(15.0) \cos 110.0°}$$

$$b \approx 20.7 \text{ centimeters}$$

TRY EXERCISE 12, EXERCISE SET 7.2, PAGE 550

In the next example we know the length of each side, but we do not know the measure of any of the angles.

EXAMPLE 2 Use the Law of Cosines (SSS)

In triangle ABC, $a = 32$ feet, $b = 20$ feet, and $c = 40$ feet. Find B. This is the SSS case.

Solution

$$b^2 = a^2 + c^2 - 2ac \cos B$$

$$\cos B = \frac{a^2 + c^2 - b^2}{2ac}$$ • Solve for $\cos B$.

$$= \frac{32^2 + 40^2 - 20^2}{2(32)(40)}$$ • Substitute for a, b, and c and solve for angle B.

$$B = \cos^{-1}\left(\frac{32^2 + 40^2 - 20^2}{2(32)(40)}\right)$$

$B \approx 30°$ • To the nearest degree

TRY EXERCISE 18, EXERCISE SET 7.2, PAGE 550

◆ AN APPLICATION OF THE LAW OF COSINES

EXAMPLE 3 Solve an Application Using the Law of Cosines

A car traveled 3.0 miles at a heading of 78°. The road turned and the car traveled another 4.3 miles at a heading of 138°. Find the distance and the bearing of the car from the starting point.

Solution

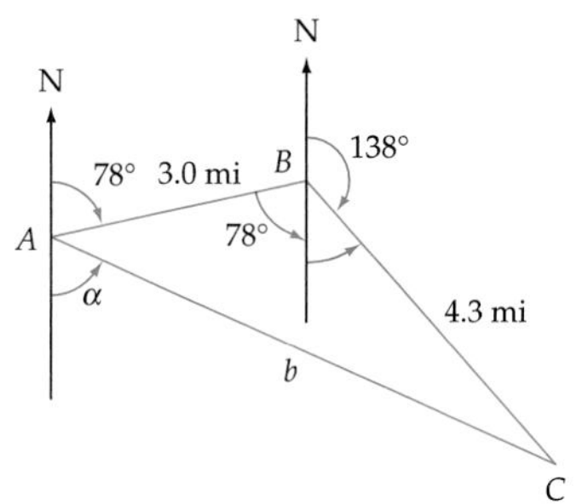

Figure 7.12

Sketch a diagram (see Figure 7.12). First find B.

$$B = 78° + (180° - 138°) = 120°$$

Use the Law of Cosines to find b.

$$b^2 = a^2 + c^2 - 2ac \cos B$$
$$= 4.3^2 + 3.0^2 - 2(4.3)(3.0) \cos 120°$$
$$b = \sqrt{4.3^2 + 3.0^2 - 2(4.3)(3.0) \cos 120°}$$
$$b \approx 6.4 \text{ miles}$$

take note

The measure of A in Example 3 can also be determined by using the Law of Sines.

Find A.

$$\cos A = \frac{b^2 + c^2 - a^2}{2bc}$$

$$A = \cos^{-1}\left(\frac{b^2 + c^2 - a^2}{2bc}\right) \approx \cos^{-1}\left(\frac{6.4^2 + 3.0^2 - 4.3^2}{(2)(6.4)(3.0)}\right)$$

$$A \approx 35°$$

The bearing of the present position of the car from the starting point A can be determined by calculating the measure of angle α in Figure 7.12.

$$\alpha = 180° - (78° + 35°) = 67°$$

The distance is approximately 6.4 miles, and the bearing (to the nearest degree) is S67°E.

TRY EXERCISE 52, EXERCISE SET 7.2, PAGE 552

There are five different cases that we may encounter when solving an oblique triangle. They are listed in the following guideline, along with the law that can be used to solve the triangle.

A Guideline for Choosing Between the Law of Sines and the Law of Cosines

Apply the Law of Sines to solve an oblique triangle for each of the following cases.

ASA The measures of two angles of the triangle and the length of the included side are known.

AAS The measures of two angles of the triangle and the length of a side opposite one of these angles are known.

SSA The lengths of two sides of the triangle and the measure of an angle opposite one of these sides are known. This case is called the ambiguous case. It may yield one solution, two solutions, or no solution.

Apply the Law of Cosines to solve an oblique triangle for each of the following cases.

SSS The lengths of all three sides of the triangle are known. After finding the measure of an angle, you can complete your solution by using the Law of Sines.

SAS The lengths of two sides of the triangle and the measure of the included angle are known. After finding the measure of the third side, you can complete your solution by using the Law of Sines.

◆ Area of a Triangle

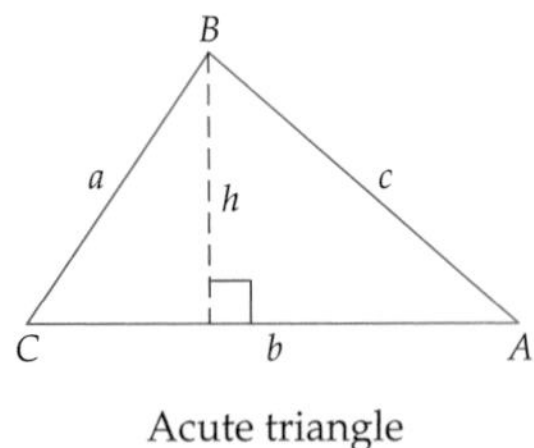

Acute triangle

B
h
a
c
C
b
A

Obtuse triangle

Figure 7.13

The formula $A = \frac{1}{2}bh$ is for the area of a triangle when the base and height are given. In this section, we will find the area of triangles when the height is not given. We will use K for the area of a triangle because A is often used to represent the measure of an angle.

Consider the areas of the acute and obtuse triangles in **Figure 7.13**.

Height of each triangle: $h = c \sin A$

Area of each triangle: $K = \frac{1}{2}bh$

$$K = \frac{1}{2}bc \sin A \qquad \text{• Substitute for } h.$$

Thus we have established the following theorem.

Area of a Triangle

The area K of triangle ABC is one-half the product of the lengths of any two sides and the sine of the included angle. Thus

$$K = \frac{1}{2}bc \sin A$$

$$K = \frac{1}{2}ab \sin C$$

$$K = \frac{1}{2}ac \sin B$$

take note

Because each formula requires two sides and the included angle, it is necessary to learn only one formula.

EXAMPLE 4 Find the Area of a Triangle

Given angle $A = 62°$, $b = 12$ meters, and $c = 5.0$ meters, find the area of triangle ABC.

Solution

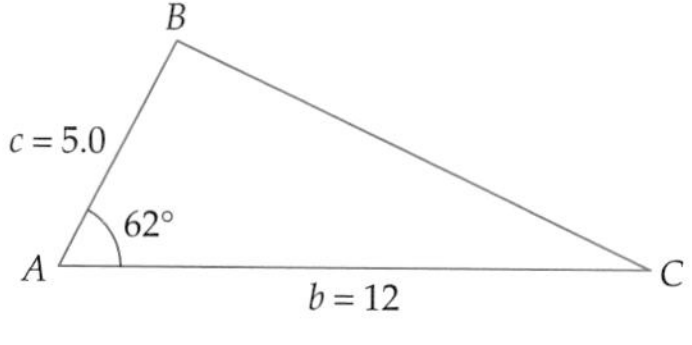

Figure 7.14

In **Figure 7.14,** two sides and the included angle of the triangle are given. Using the formula for area, we have

$$K = \frac{1}{2}bc \sin A = \frac{1}{2}(12)(5.0)(\sin 62°) \approx 26 \text{ square meters}$$

TRY EXERCISE 26, EXERCISE SET 7.2, PAGE 550

When two angles and an included side are given, the Law of Sines is used to derive a formula for the area of a triangle. First, solve for c in the Law of Sines.

$$\frac{c}{\sin C} = \frac{b}{\sin B}$$

$$c = \frac{b \sin C}{\sin B}$$

Substitute for c in the formula $K = \frac{1}{2}bc \sin A$.

$$K = \frac{1}{2}bc \sin A = \frac{1}{2}b\left(\frac{b \sin C}{\sin B}\right) \sin A$$

$$K = \frac{b^2 \sin C \sin A}{2 \sin B}$$

In like manner, the following two alternative formulas can be derived for the area of a triangle.

$$K = \frac{a^2 \sin B \sin C}{2 \sin A} \quad \text{and} \quad K = \frac{c^2 \sin A \sin B}{2 \sin C}$$

EXAMPLE 5 Find the Area of a Triangle

Given $A = 32°$, $C = 77°$, and $a = 14$ inches, find the area of triangle ABC.

Solution

To use the area formula, we need to know two angles and the included side. Therefore, we need to determine the measure of angle B.

$$B = 180° - 32° - 77° = 71°$$

Thus

$$K = \frac{a^2 \sin B \sin C}{2 \sin A} = \frac{14^2 \sin 71° \sin 77°}{2 \sin 32°} \approx 170 \text{ square inches}$$

TRY EXERCISE 28, EXERCISE SET 7.2, PAGE 550

MATH MATTERS

Recent findings indicate that Heron's formula for finding the area of a triangle was first discovered by Archimedes. However, the formula is called Heron's formula in honor of the geometer Heron of Alexandria (A.D. 50), who gave an ingenious proof of the theorem in his work *Metrica*. Because Heron of Alexandria was also know as Hero, some texts refer to Heron's formula as Hero's formula.

◆ HERON'S FORMULA

The Law of Cosines can be used to derive *Heron's formula* for the area of a triangle in which three sides of the triangle are given.

Heron's Formula for Finding the Area of a Triangle

If a, b, and c are the lengths of the sides of a triangle, then the area K of the triangle is

$$K = \sqrt{s(s-a)(s-b)(s-c)}, \quad \text{where } s = \frac{1}{2}(a+b+c)$$

Because s is one-half of the perimeter of the triangle, it is called the **semi-perimeter.**

EXAMPLE 6 Find an Area by Heron's Formula

Find the area of the triangle with $a = 7.0$ meters, $b = 15$ meters, and $c = 12$ meters.

Solution

Calculate the semi-perimeter s.

$$s = \frac{a+b+c}{2} = \frac{7.0 + 15 + 12}{2} = 17$$

Use Heron's formula.

$$\begin{aligned} K &= \sqrt{s(s-a)(s-b)(s-c)} \\ &= \sqrt{17(17-7.0)(17-15)(17-12)} \\ &= \sqrt{1700} \approx 41 \text{ square meters} \end{aligned}$$

TRY EXERCISE 36, EXERCISE SET 7.2, PAGE 550

EXAMPLE 7 **Use Heron's Formula to Solve an Application**

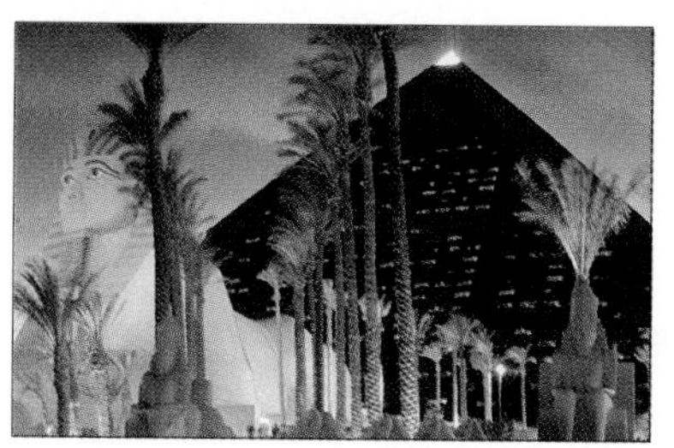

The pyramid portion of the Luxor Hotel, Las Vegas, Nevada.

The original portion of the Luxor Hotel in Las Vegas has the shape of a square pyramid. Each face of the pyramid is an isosceles triangle with a base of 646 feet and sides of length 576 feet. Assuming that the glass on the exterior of the Luxor Hotel costs \$35 per square foot, determine the cost of the glass, to the nearest \$10,000, for one of the triangular faces of the hotel.

Solution

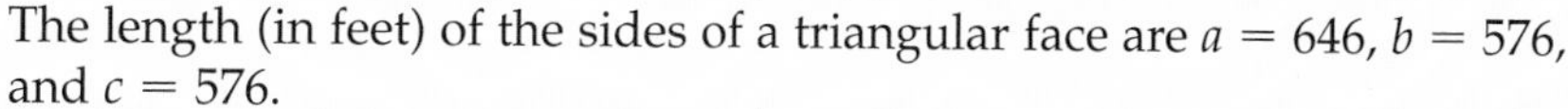

The length (in feet) of the sides of a triangular face are $a = 646$, $b = 576$, and $c = 576$.

$$\begin{aligned} s &= \frac{a+b+c}{2} = \frac{646+576+576}{2} = 899 \text{ feet} \\ K &= \sqrt{s(s-a)(s-b)(s-c)} \\ &= \sqrt{899(899-646)(899-576)(899-576)} \\ &= \sqrt{23{,}729{,}318{,}063} \\ &\approx 154{,}043 \text{ square feet} \end{aligned}$$

The cost C of the glass is the product of the cost per square foot and the area.

$$C = 35 \cdot 154{,}043 = 5{,}391{,}505$$

The approximate cost of the glass for one face of the Luxor Hotel is \$5,390,000.

TRY EXERCISE 60, EXERCISE SET 7.2, PAGE 552

TOPICS FOR DISCUSSION

1. Explain why there is no triangle that has sides of length $a = 2$ inches, $b = 11$ inches, and $c = 3$ inches.
2. The Pythagorean Theorem is a special case of the Law of Cosines. Explain.
3. To solve a triangle in which the lengths of the three sides are given (SSS), a mathematics professor recommends the following procedure.
   (i) Use the Law of Cosines to find the measure of the largest angle.
   (ii) Use the Law of Sines to find the measure of a second angle.

(iii) Find the measure of the third angle by using the formula $A + B + C = 180°$.

Explain why this procedure is easier than using the Law of Cosines to find the measure of all three angles.

4. To solve a triangle in which the lengths of two sides and the measure of the included angle are given (SAS), a tutor recommends the following procedure.

(i) Use the Law of Cosines to find the length of the third side.

(ii) Use the Law of Sines to find the smaller of the unknown angles.

(iii) Find the measure of the third angle by using the formula $A + B + C = 180°$.

Explain why this procedure is easier than using the Law of Cosines three times to find each of the unknowns.

EXERCISE SET 7.2

In Exercises 1 to 14, find the third side of the triangle.

1. $a = 12, b = 18, C = 44°$
2. $b = 30, c = 24, A = 120°$
3. $a = 120, c = 180, B = 56°$
4. $a = 400, b = 620, C = 116°$
5. $b = 60, c = 84, A = 13°$
6. $a = 122, c = 144, B = 48°$
7. $a = 9.0, b = 7.0, C = 72°$
8. $b = 12, c = 22, A = 55°$
9. $a = 4.6, b = 7.2, C = 124°$
10. $b = 12.3, c = 14.5, A = 6.5°$
11. $a = 25.9, c = 33.4, B = 84.0°$
12. $a = 14.2, b = 9.30, C = 9.20°$
13. $a = 122, c = 55.9, B = 44.2°$
14. $b = 444.8, c = 389.6, A = 78.44°$

In Exercises 15 to 24, given three sides of a triangle, find the specified angle.

15. $a = 25, b = 32, c = 40$; find A.
16. $a = 60, b = 88, c = 120$; find B.
17. $a = 8.0, b = 9.0, c = 12$; find C.
18. $a = 108, b = 132, c = 160$; find A.
19. $a = 80.0, b = 92.0, c = 124$; find B.
20. $a = 166, b = 124, c = 139$; find B.
21. $a = 1025, b = 625.0, c = 1420$; find C.
22. $a = 4.7, b = 3.2, c = 5.9$; find A.
23. $a = 32.5, b = 40.1, c = 29.6$; find B.
24. $a = 112.4, b = 96.80, c = 129.2$; find C.

In Exercises 25 to 36, find the area of the given triangle. Round each area to the same number of significant digits as are in each of the given sides.

25. $A = 105°, b = 12, c = 24$
26. $B = 127°, a = 32, c = 25$
27. $A = 42°, B = 76°, c = 12$
28. $B = 102°, C = 27°, a = 8.5$
29. $a = 16, b = 12, c = 14$
30. $a = 32, b = 24, c = 36$
31. $B = 54.3°, a = 22.4, b = 26.9$
32. $C = 18.2°, b = 13.4, a = 9.84$
33. $A = 116°, B = 34°, c = 8.5$
34. $B = 42.8°, C = 76.3°, c = 17.9$
35. $a = 3.6, b = 4.2, c = 4.8$
36. $a = 13.3, b = 15.4, c = 10.2$

37. **DISTANCE BETWEEN AIRPORTS** A plane leaves airport A and travels 560 miles to airport B at a bearing of N32°E. The plane leaves airport B and travels to airport C 320 miles away at a bearing of S72°E. Find the distance from airport A to airport C.

38. **LENGTH OF A STREET** A developer has a triangular lot at the intersection of two streets. The streets meet at an angle of 72°, and the lot has 300 feet of frontage along one street

and 416 feet of frontage along the other street. Find the length of the third side of the lot.

39. **BASEBALL** In a baseball game, a batter hits a ground ball 26 feet in the direction of the pitcher's mound. The pitcher runs forward and reaches for the ball. At that moment, how far is the ball from first base? (*Note:* A baseball infield is a square that measures 90 feet on each side.)

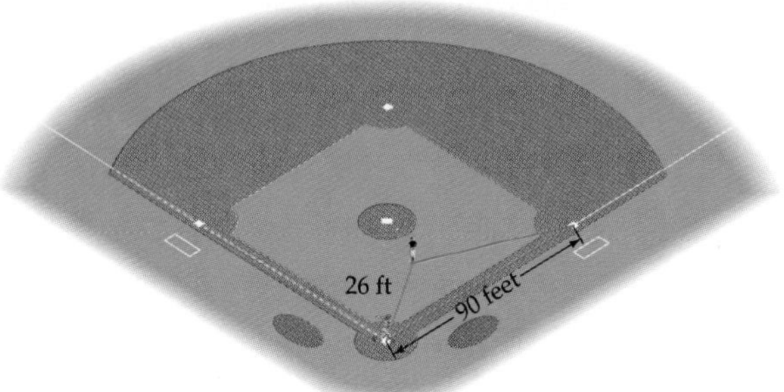

40. **B-2 BOMBER** The leading edge of each wing of the B-2 Stealth Bomber measures 105.6 feet in length. The angle between the wing's leading edges ($\angle ABC$) is 109.05°. What is the wing span (the distance from A to C) of the B-2 Bomber?

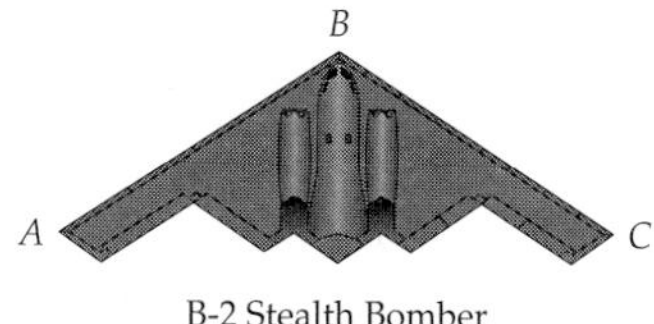

B-2 Stealth Bomber

41. **ANGLE BETWEEN THE DIAGONALS OF A BOX** The rectangular box in the figure measures 6.50 feet by 3.25 feet by 4.75 feet. Find the measure of the angle θ that is formed by the union of the diagonal shown on the front of the box and the diagonal shown on the right side of the box.

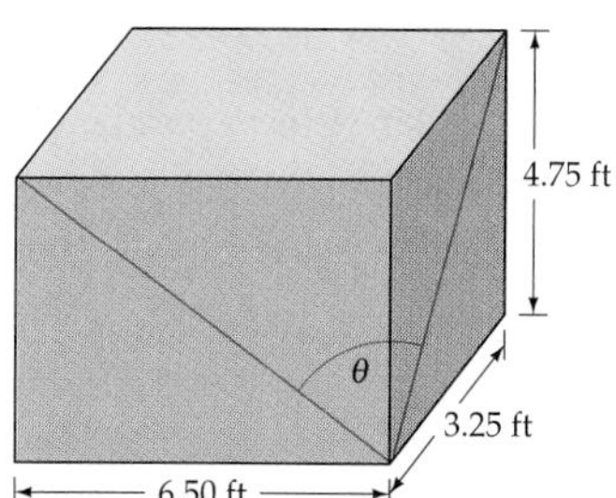

42. **SUBMARINE RESCUE MISSION** The surface ships shown in the figure at the top of the next column have determined the indicated distances. Use this data to determine the depth of the submarine below the surface of the water. Assume that the line segment between the surface ships is directly above the submarine.

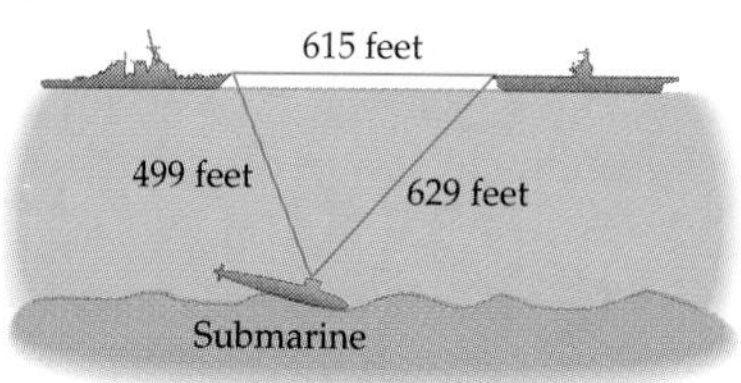

43. **DISTANCE BETWEEN SHIPS** Two ships left a port at the same time. One ship traveled at a speed of 18 mph at a heading of 318°. The other ship traveled at a speed of 22 mph at a heading of 198°. Find the distance between the two ships after 10 hours of travel.

44. **DISTANCE ACROSS A LAKE** Find the distance across a lake, using the measurements shown in the figure.

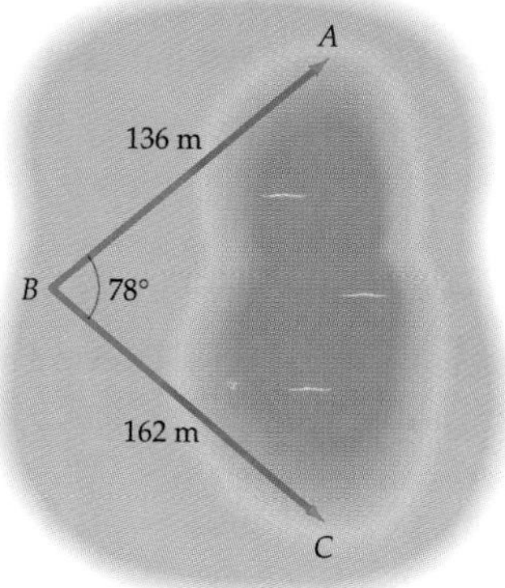

45. A regular hexagon is inscribed in a circle with a radius of 40 centimeters. Find the length of one side of the hexagon.

46. A regular pentagon is inscribed in a circle with a radius of 25 inches. Find the length of one side of the pentagon.

47. The lengths of the diagonals of a parallelogram are 20 inches and 32 inches. The diagonals intersect at an angle of 35°. Find the lengths of the sides of the parallelogram. (*Hint:* The diagonals of a parallelogram bisect one another.)

48. The sides of a parallelogram are 10 feet and 14 feet. The longer diagonal of the parallelogram is 18 feet. Find the length of the shorter diagonal of the parallelogram. (*Hint:* The diagonals of a parallelogram bisect one another.)

49. The sides of a parallelogram are 30 centimeters and 40 centimeters. The shorter diagonal of the parallelogram is 44 centimeters. Find the length of the longer diagonal of the parallelogram. (*Hint:* The diagonals of a parallelogram bisect one another.)

50. **ANGLE BETWEEN BOUNDARIES OF A LOT** A triangular city lot has sides of 224 feet, 182 feet, and 165 feet. Find the angle between the longer two sides of the lot.

51. **DISTANCE TO A PLANE** A plane traveling at 180 mph passes 400 feet directly over an observer. The plane is traveling along a straight path with an angle of elevation of 14°. Find the distance of the plane from the observer 10 seconds after the plane has passed directly overhead.

52. **DISTANCE BETWEEN SHIPS** A ship leaves a port at a speed of 16 mph at a heading of 32°. One hour later another ship leaves the port at a speed of 22 mph at a heading of 254°. Find the distance between the ships 4 hours after the first ship leaves the port.

53. **AREA OF A TRIANGULAR LOT** Find the area of a triangular piece of land that is bounded by sides of 236 meters, 620 meters, and 814 meters.

54. Find the area of a parallelogram whose diagonals are 24 inches and 32 inches and intersect at an angle of 40°.

55. Find the area of a parallelogram with sides of 12 meters and 18 meters and with one angle of 70°.

56. Find the area of a parallelogram with sides of 8 feet and 12 feet. The shorter diagonal is 10 feet.

57. Find the area of a square inscribed in a circle with a radius of 9 inches.

58. Find the area of a regular hexagon inscribed in a circle with a radius of 24 centimeters.

59. **COST OF A LOT** A commercial piece of real estate is priced at $2.20 per square foot. Find, to the nearest $1000, the cost of a triangular lot measuring 212 feet by 185 feet by 240 feet.

60. **COST OF A LOT** An industrial piece of real estate is priced at $4.15 per square foot. Find, to the nearest $1000, the cost of a triangular lot measuring 324 feet by 516 feet by 412 feet.

61. **AREA OF A PASTURE** Find the number of acres in a pasture whose shape is a triangle measuring 800 feet by 1020 feet by 680 feet. (An acre is 43,560 square feet.)

62. **AREA OF A HOUSING TRACT** Find the number of acres in a housing tract whose shape is a triangle measuring 420 yards by 540 yards by 500 yards. (An acre is 4840 square yards.)

SUPPLEMENTAL EXERCISES

HERON'S FORMULA FOR QUADRILATERALS The following formula is a generalization of Heron's formula. It gives the area K of a convex quadrilateral with sides of length a, b, c, and d.

$$K = \sqrt{(s-a)(s-b)(s-c)(s-d)}$$

where

$$s = \frac{a+b+c+d}{2}$$

63. Verify that Heron's formula for quadrilaterals produces the correct area for each of the following:
    a. A square with sides of length 8.
    b. A rectangle with sides of length 5 and 11.

64. Use Heron's formula for quadrilaterals to find the area (to the nearest 0.01 square unit) of the following convex quadrilateral.

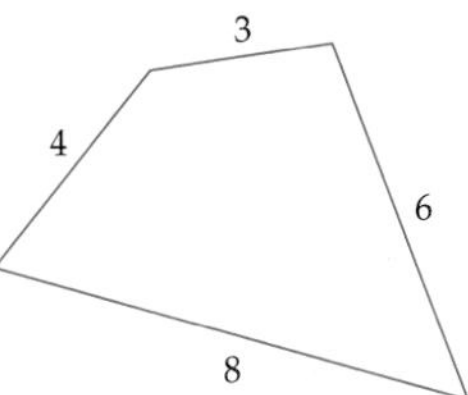

65. Find the measure of the angle formed by the sides P_1P_2 and P_1P_3 of a triangle with vertices at $P_1(-2, 4)$, $P_2(2, 1)$, and $P_3(4, -3)$.

66. A regular pentagon is inscribed in a circle with a radius of 4 inches. Find the perimeter of the pentagon.

67. An equilateral triangle is inscribed in a circle with a radius of 10 centimeters. Find the perimeter of the triangle.

68. Given a triangle ABC, prove that

$$a^2 = b^2 + c^2 - 2bc\cos A$$

69. Use the Law of Cosines to show that

$$\cos A = \frac{(b+c-a)(b+c+a)}{2bc} - 1$$

70. Prove that $K = xy\sin A$ for a parallelogram, where x and y are the lengths of adjacent sides, A is the measure of the angle between side x and side y, and K is the area of the parallelogram.

71. Show that the area of the parallelogram in the figure is $K = 2ab\sin C$.

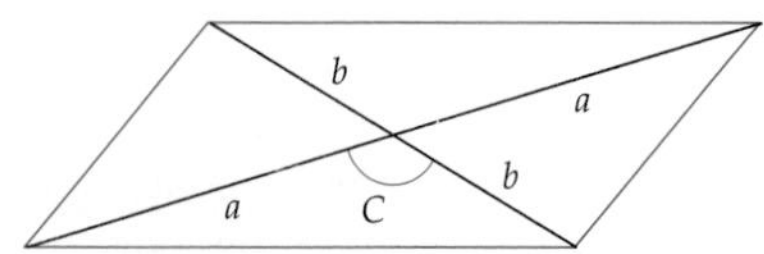

72. Given a regular hexagon inscribed in a circle with a radius of 10 inches, find the area of a segment of the circle (see the dark-shaded region in the figure at the top of the next page).

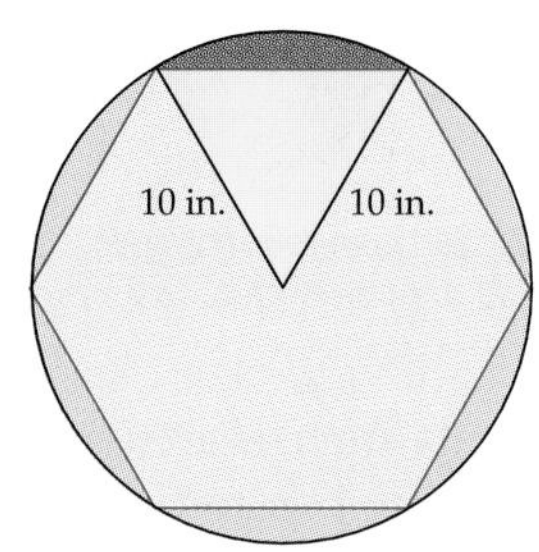

73. Find the volume of the triangular prism shown in the figure.

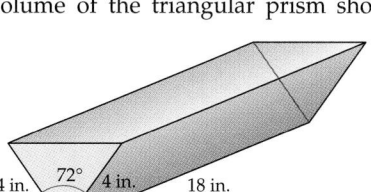

74. Show that the area of the circumscribed triangle in the figure is $K = rs$, where $s = \dfrac{a + b + c}{2}$.

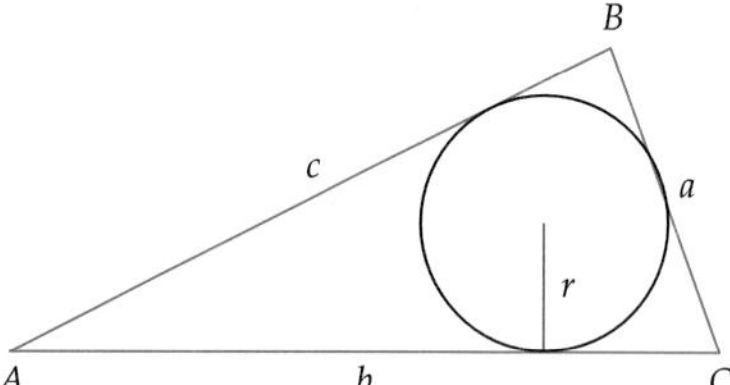

PROJECTS

1. **CLOCK WATCHING** The minute hand and the hour hand of a clock are shown in the figure at the right. The minute hand is 6 inches in length, and the hour hand is 4 inches in length.

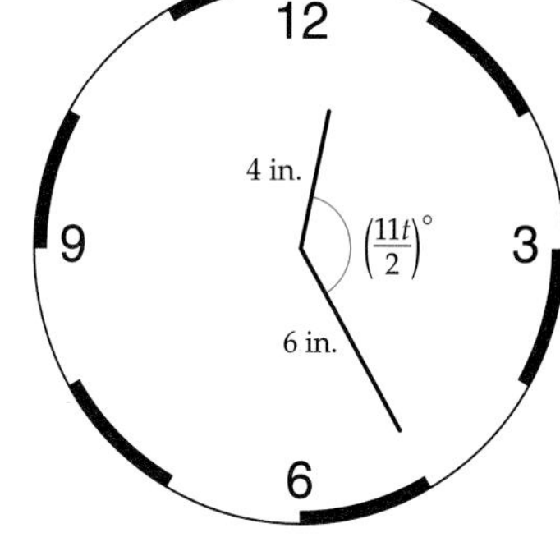

   a. What is the distance d (to the nearest 0.1 inch) between the tips of the hands at 12:12? *Hint:* Use the Law of Cosines. For every minute t after 12:00, the minute hand rotates $6t°$, and the hour hand rotates 1/12 of the rotation made by the minute hand. Thus at time t, the angle between the hands is $[6t - (1/12)6t]° = (11t/2)°$.

   b. The maximum distance between the tips of the hands is 10 inches. What is the first time after 12:00 (to the nearest 0.1 minute) when the tips of the hands will be 10 inches apart? *Hint:* Use the Law of Cosines to find a function for the distance d between the hands at any time t (where t is in minutes and $t = 0$ represents 12:00). Use a graphing utility to graph the function.

   c. The minimum distance between the tips of the hands is 2 inches. What is the first time after 12:00 when the tips of the hands will be 2 inches apart?

2. **VISUAL INSIGHT**

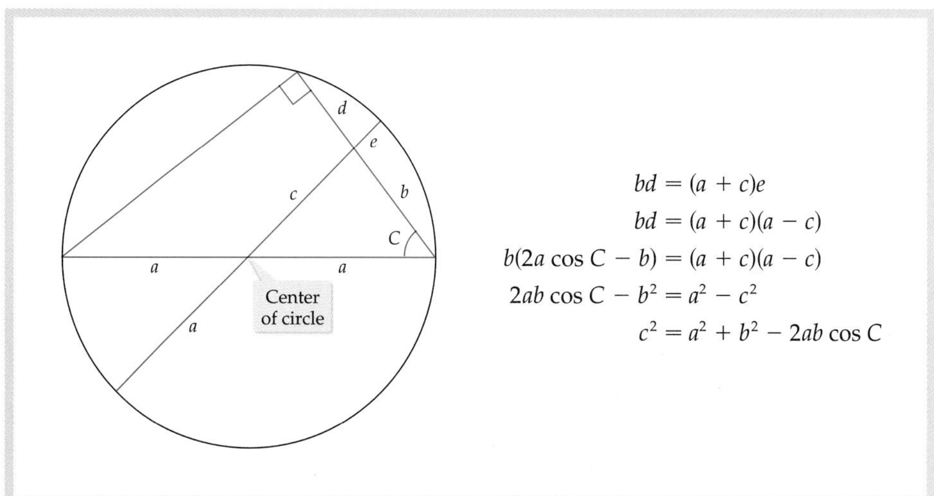

Give the rule or reason that justifies each step in the above proof of the Law of Cosines.

SECTION 7.3 VECTORS

- VECTORS
- VECTORS IN A COORDINATE PLANE
- UNIT VECTORS
- APPLICATION PROBLEMS USING VECTORS
- DOT PRODUCT
- SCALAR PROJECTION
- PARALLEL AND PERPENDICULAR VECTORS
- WORK: AN APPLICATION OF THE DOT PRODUCT

7B

In scientific applications, some measurements such as area, mass, distance, speed, and time are completely described by a real number and a unit. Examples include 30 square feet (area), 25 meters/second (speed), and 5 hours (time). These measurements are **scalar quantities,** and the number used to indicate the magnitude of the measurement is called a **scalar.** Two other examples of scalar quantities are volume and temperature.

For other quantities, besides the numerical and unit description, it is also necessary to include a *direction* to describe the quantity completely. For example, applying a force of 25 pounds at various angles to a small metal box will influence how the box moves. In **Figure 7.15,** applying the 25-pound force straight down (A) will not move the box to the left. However, applying a 25-pound force (C) parallel to the floor will move the box along the floor.

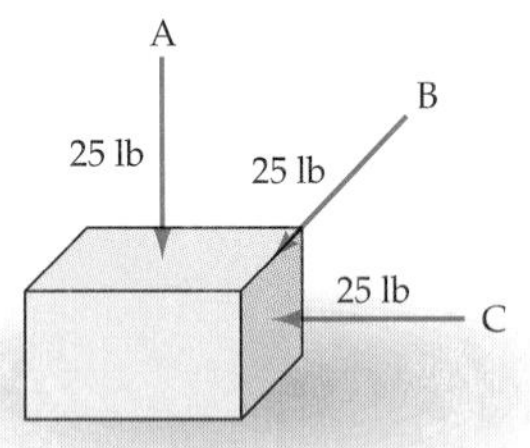

Figure 7.15

◆ VECTORS

Vector quantities have a *magnitude* (numerical and unit description) and a *direction*. Force is a vector quantity. Velocity is another. Velocity includes the speed (magnitude) and a direction. A velocity of 40 mph east is different from a velocity of 40 mph north. Displacement is another vector quantity; it consists of distance (a scalar) moved in a certain direction; for example, we might speak of a displacement of 13 centimeters at an angle of 15° from the positive x-axis.

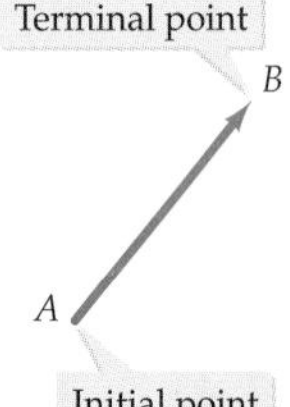

Figure 7.16

Definition of a Vector

A **vector** is a directed line segment. The length of the line segment is the magnitude of the vector, and the direction of the vector is measured by an angle.

The point A for the vector in **Figure 7.16** is called the **initial point** (or tail) of the vector, and the point B is the **terminal point** (or head) of the vector. An arrow over the letters ($\overrightarrow{AB}$), an arrow over a single letter ($\vec{V}$), or boldface type (**AB** or **V**) is used to denote a vector. The magnitude of the vector is the length of the line segment and is denoted by $\|\overrightarrow{AB}\|$, $\|\vec{V}\|$, $\|\mathbf{AB}\|$, or $\|\mathbf{V}\|$.

Equivalent vectors have the same magnitude and the same direction. The vectors in **Figure 7.17** are equivalent. They have the same magnitude and direction.

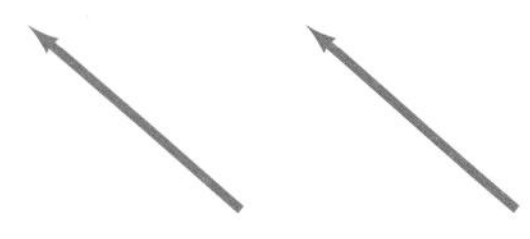

Figure 7.17

Multiplying a vector by a positive real number (other than 1) changes the magnitude of the vector but not its direction. If **v** is any vector, then 2**v** is the vector that has the same direction as **v** but is twice the magnitude of **v**. The multiplication of 2 and **v** is called the **scalar multiplication** of the vector **v** and the scalar 2. Multiplying a vector by a negative number a reverses the direction of the vector and multiplies the magnitude of the vector by $|a|$. See **Figure 7.18.**

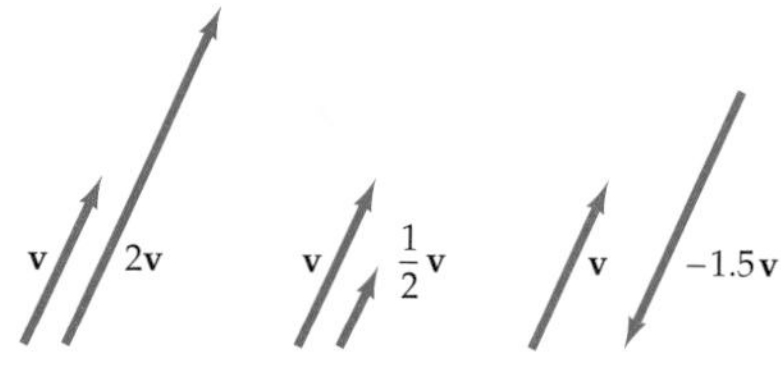

Figure 7.18

The sum of two vectors, called the **resultant vector,** or the **resultant,** is the single equivalent vector that will have the same effect as the application of those two vectors. For example, a displacement 40 meters along the positive x-axis and then 30 meters in the positive y direction is equivalent to a vector of magnitude 50 meters at an angle of approximately 37° to the positive x-axis. See Figure 7.19.

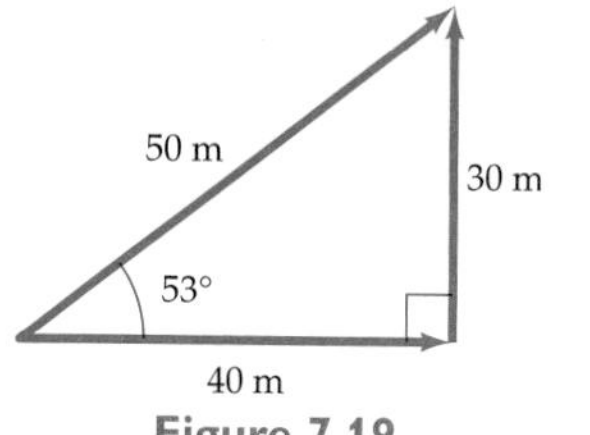

Figure 7.19

Vectors can be added graphically by using the *triangle method* or the *parallelogram method*. In the triangle method shown in Figure 7.20, the tail of **V** is placed at the head of **U**. The vector connecting the tail of **U** with the head of **V** is the sum **U** + **V**.

The parallelogram method of adding two vectors graphically places the tails of the two vectors **U** and **V** together, as in Figure 7.21. Complete the parallelogram so that **U** and **V** are sides of the parallelogram. The diagonal beginning at the tails of the two vectors is **U** + **V**.

To find the difference between two vectors, first rewrite the expression as **V** − **U** = **V** + (−**U**). The difference is shown geometrically in Figure 7.22.

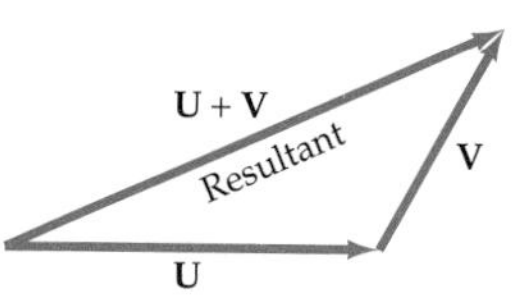

Figure 7.20

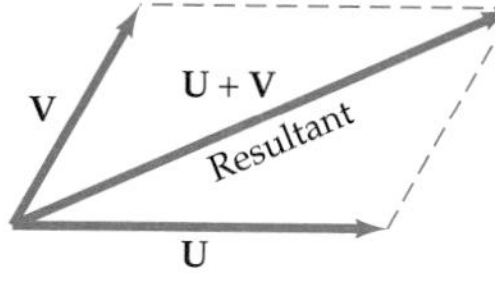

Figure 7.21

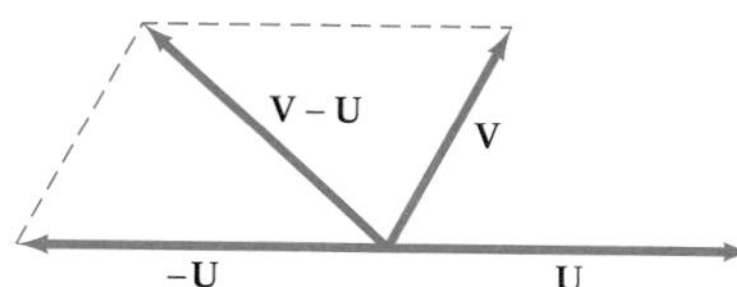

Figure 7.22

◆ VECTORS IN A COORDINATE PLANE

By introducing a coordinate plane, it is possible to develop an analytic approach to vectors. Recall from our discussion about equivalent vectors that a vector can be moved in the plane as long as *the magnitude and direction* are not changed.

With this in mind, consider **AB**, whose initial point is $A(2, -1)$ and whose terminal point is $B(-3, 4)$. If this vector is moved so that the initial point is at the origin O, the terminal point becomes $P(-5, 5)$ as shown in Figure 7.23. The vector **OP** is equivalent to the vector **AB**.

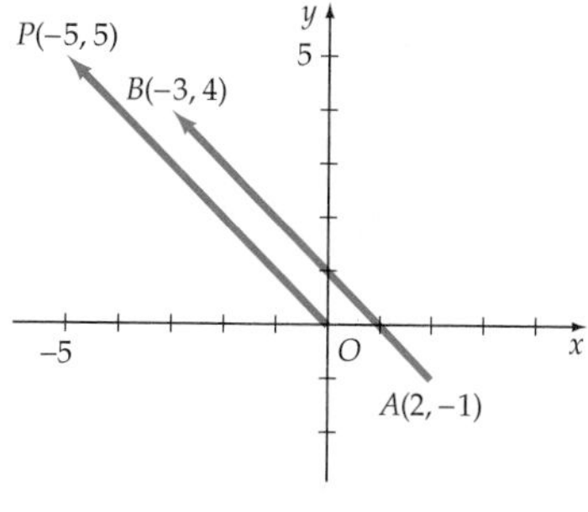

Figure 7.23

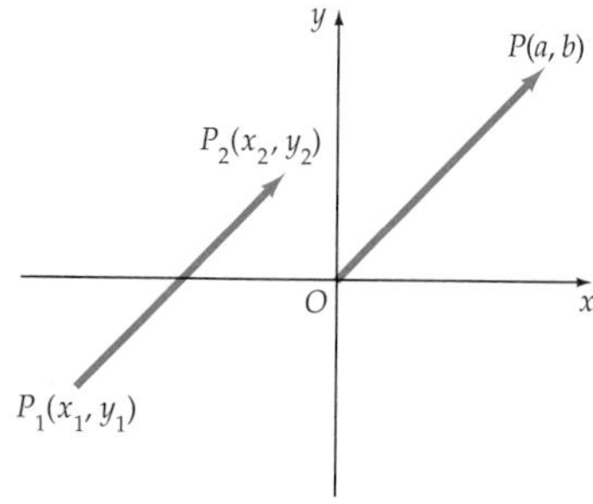

Figure 7.24

In Figure 7.24, let $P_1(x_1, y_1)$ be the initial point of a vector and $P_2(x_2, y_2)$ its terminal point. Then an equivalent vector **OP** has its initial point at the origin and terminal point at $P(a, b)$, where $a = x_2 - x_1$ and $b = y_2 - y_1$. The vector **OP** can be denoted by $\mathbf{v} = \langle a, b \rangle$; a and b are called the **components** of the vector.

EXAMPLE 1 Find the Components of a Vector

Find the components of a vector **AB** whose tail is the point $A(2, -1)$ and whose head is the point $B(-2, 6)$. Determine a vector **v** that is equivalent to **AB** and has an initial point at the origin.

Solution

The components of **AB** are $\langle a, b\rangle$ where

$$a = x_2 - x_1 = -2 - 2 = -4 \quad \text{and} \quad b = y_2 - y_1 = 6 - (-1) = 7$$

Thus $\mathbf{v} = \langle -4, 7\rangle$.

Visualize the Solution

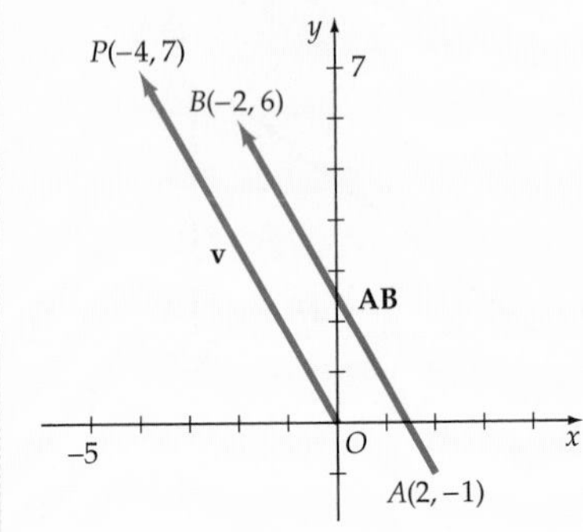

Figure 7.25

TRY EXERCISE 6, EXERCISE SET 7.3, PAGE 566

The magnitude and direction of a vector can be found from its components. For instance, the head of vector **v** sketched in Figure 7.25 is the ordered pair $(-4, 7)$. Applying the Pythagorean Theorem, we find

$$\|\mathbf{v}\| = \sqrt{(-4)^2 + 7^2} = \sqrt{16 + 49} = \sqrt{65}$$

Let θ be the angle made by the positive x-axis and **v**. Let α be the reference angle for θ. Then

$$\tan \alpha = \left|\frac{b}{a}\right| = \left|\frac{7}{-4}\right| = \frac{7}{4}$$

$$\alpha = \tan^{-1}\frac{7}{4} \approx 60^\circ$$ • α **is the reference angle.**

$$\theta = 180^\circ - 60^\circ = 120^\circ$$ • θ **is the angle made by the vector and the positive x-axis.**

The magnitude of **v** is $\sqrt{65}$, and its direction is 120° as measured from the positive x-axis. The angle between a vector and the positive x-axis is called the **direction angle** of the vector. Because **AB** in Figure 7.25 is equivalent to **v**, $\|\mathbf{AB}\| = \sqrt{65}$ and the direction angle of **AB** is also 120°.

Expressing vectors in terms of components provides a convenient method for performing operations on vectors.

Fundamental Vector Operations

If $\mathbf{v} = \langle a, b\rangle$ and $\mathbf{w} = \langle c, d\rangle$ are two vectors and k is a real number, then

1. $\|\mathbf{v}\| = \sqrt{a^2 + b^2}$
2. $\mathbf{v} + \mathbf{w} = \langle a, b\rangle + \langle c, d\rangle = \langle a + c, b + d\rangle$
3. $k\mathbf{v} = k\langle a, b\rangle = \langle ka, kb\rangle$

In terms of components, the zero vector $\mathbf{0} = \langle 0, 0\rangle$. The additive inverse of a vector $\mathbf{v} = \langle a, b\rangle$ is given by $-\mathbf{v} = \langle -a, -b\rangle$.

EXAMPLE 2 Perform Operations on Vectors

Given $\mathbf{v} = \langle -2, 3\rangle$ and $\mathbf{w} = \langle 4, -1\rangle$, find

a. $\|\mathbf{w}\|$ b. $\mathbf{v} + \mathbf{w}$ c. $-3\mathbf{v}$ d. $2\mathbf{v} - 3\mathbf{w}$

Solution

a. $\|\mathbf{w}\| = \sqrt{4^2 + (-1)^2} = \sqrt{17}$

b. $\mathbf{v} + \mathbf{w} = \langle -2, 3\rangle + \langle 4, -1\rangle = \langle -2 + 4, 3 + (-1)\rangle = \langle 2, 2\rangle$

c. $-3\mathbf{v} = -3\langle -2, 3\rangle = \langle 6, -9\rangle$

d. $2\mathbf{v} - 3\mathbf{w} = 2\langle -2, 3\rangle - 3\langle 4, -1\rangle = \langle -4, 6\rangle - \langle 12, -3\rangle = \langle -16, 9\rangle$

TRY EXERCISE 20, EXERCISE SET 7.3, PAGE 566

◆ UNIT VECTORS

A **unit vector** is a vector whose magnitude is 1. For example, the vector $\mathbf{v} = \left\langle \frac{3}{5}, -\frac{4}{5}\right\rangle$ is a unit vector because

$$\|\mathbf{v}\| = \sqrt{\left(\frac{3}{5}\right)^2 + \left(-\frac{4}{5}\right)^2} = \sqrt{\frac{9}{25} + \frac{16}{25}} = \sqrt{\frac{25}{25}} = 1$$

Given any nonzero vector $\mathbf{v}$, we can obtain a unit vector in the direction of $\mathbf{v}$ by dividing each component of $\mathbf{v}$ by the magnitude of $\mathbf{v}$, $\|\mathbf{v}\|$.

EXAMPLE 3 Find a Unit Vector

Find a unit vector $\mathbf{u}$ in the direction of $\mathbf{v} = \langle -4, 2\rangle$.

Solution

Find the magnitude of $\mathbf{v}$.

$$\|\mathbf{v}\| = \sqrt{(-4)^2 + 2^2} = \sqrt{16 + 4} = \sqrt{20} = 2\sqrt{5}$$

Divide each component of $\mathbf{v}$ by $\|\mathbf{v}\|$.

$$\mathbf{u} = \left\langle \frac{-4}{2\sqrt{5}}, \frac{2}{2\sqrt{5}}\right\rangle = \left\langle \frac{-2}{\sqrt{5}}, \frac{1}{\sqrt{5}}\right\rangle = \left\langle -\frac{2\sqrt{5}}{5}, \frac{\sqrt{5}}{5}\right\rangle.$$

A unit vector in the direction of $\mathbf{v}$ is $\mathbf{u}$.

TRY EXERCISE 8, EXERCISE SET 7.3, PAGE 566

Two unit vectors, one parallel to the x-axis and one parallel to the y-axis, are of special importance. See **Figure 7.26**.

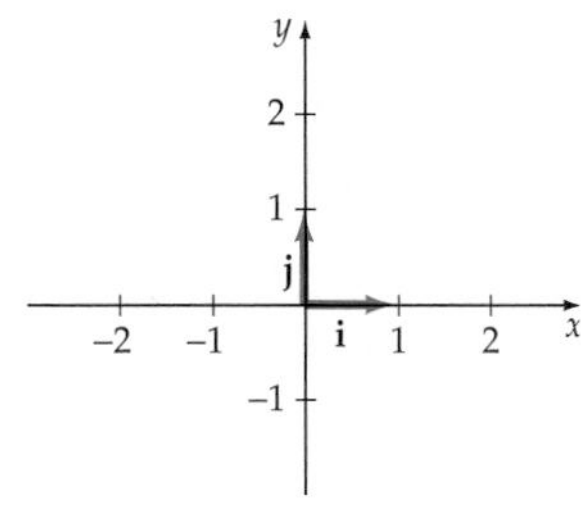

Figure 7.26

Definition of Unit Vectors i and j

$$\mathbf{i} = \langle 1, 0 \rangle \qquad \mathbf{j} = \langle 0, 1 \rangle$$

Figure 7.27

The vector $\mathbf{v} = \langle 3, 4 \rangle$ can be written in terms of the unit vectors $\mathbf{i}$ and $\mathbf{j}$ as shown in Figure 7.27.

$$\begin{aligned} \langle 3, 4 \rangle &= \langle 3, 0 \rangle + \langle 0, 4 \rangle && \text{• Vector Addition Property} \\ &= 3\langle 1, 0 \rangle + 4\langle 0, 1 \rangle && \text{• Scalar multiplication of a vector} \\ &= 3\mathbf{i} + 4\mathbf{j} && \text{• Definition of i and j} \end{aligned}$$

By means of scalar multiplication and addition of vectors, any vector can be expressed in terms of the unit vectors $\mathbf{i}$ and $\mathbf{j}$. Let $\mathbf{v} = \langle a_1, a_2 \rangle$. Then

$$\mathbf{v} = \langle a_1, a_2 \rangle = a_1\langle 1, 0 \rangle + a_2\langle 0, 1 \rangle = a_1\mathbf{i} + a_2\mathbf{j}$$

This gives the following result.

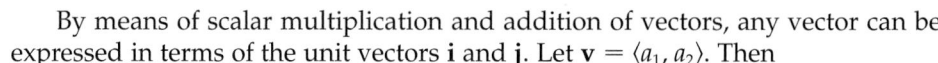

Representation of a Vector in Terms of i and j

If $\mathbf{v}$ is a vector and $\mathbf{v} = \langle a_1, a_2 \rangle$, then $\mathbf{v} = a_1\mathbf{i} + a_2\mathbf{j}$.

The rules for addition and scalar multiplication of vectors can be restated in terms of $\mathbf{i}$ and $\mathbf{j}$. If $\mathbf{v} = a_1\mathbf{i} + a_2\mathbf{j}$ and $\mathbf{w} = b_1\mathbf{i} + b_2\mathbf{j}$, then

$$\mathbf{v} + \mathbf{w} = (a_1\mathbf{i} + a_2\mathbf{j}) + (b_1\mathbf{i} + b_2\mathbf{j}) = (a_1 + b_1)\mathbf{i} + (a_2 + b_2)\mathbf{j}$$
$$k\mathbf{v} = k(a_1\mathbf{i} + a_2\mathbf{j}) = ka_1\mathbf{i} + ka_2\mathbf{j}$$

EXAMPLE 4 Operate on Vectors Written in Terms of i and j

Given $\mathbf{v} = 3\mathbf{i} - 4\mathbf{j}$ and $\mathbf{w} = 5\mathbf{i} + 3\mathbf{j}$, find $3\mathbf{v} - 2\mathbf{w}$.

Solution

$$\begin{aligned} 3\mathbf{v} - 2\mathbf{w} &= 3(3\mathbf{i} - 4\mathbf{j}) - 2(5\mathbf{i} + 3\mathbf{j}) \\ &= (9\mathbf{i} - 12\mathbf{j}) - (10\mathbf{i} + 6\mathbf{j}) \\ &= (9 - 10)\mathbf{i} + (-12 - 6)\mathbf{j} \\ &= -\mathbf{i} - 18\mathbf{j} \end{aligned}$$

TRY EXERCISE 26, EXERCISE SET 7.3, PAGE 566

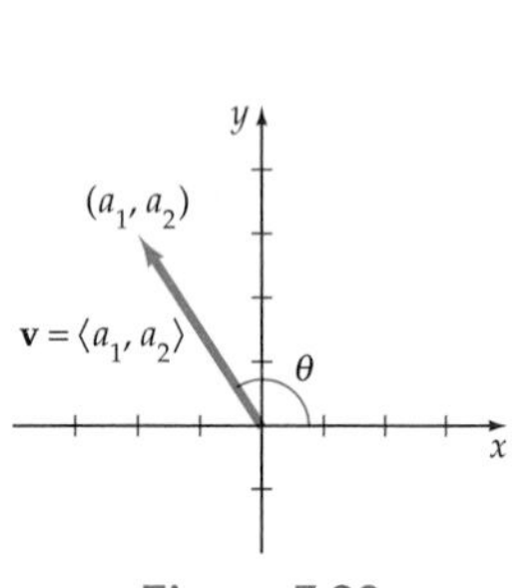

Figure 7.28

The components a_1 and a_2 of the vector $\mathbf{v} = \langle a_1, a_2 \rangle$ can be expressed in terms of the magnitude of $\mathbf{v}$ and the direction angle of $\mathbf{v}$ (the angle that $\mathbf{v}$ makes with the positive x-axis). Consider the vector $\mathbf{v}$ in Figure 7.28. Then

$$\|\mathbf{v}\| = \sqrt{(a_1)^2 + (a_2)^2}$$

From the definitions of sine and cosine, we have

$$\cos\theta = \frac{a_1}{\|\mathbf{v}\|} \quad \text{and} \quad \sin\theta = \frac{a_2}{\|\mathbf{v}\|}$$

Rewriting the last two equations, we find that the components of **v** are

$$a_1 = \|\mathbf{v}\|\cos\theta \quad \text{and} \quad a_2 = \|\mathbf{v}\|\sin\theta$$

Horizontal and Vertical Components of a Vector

Let $\mathbf{v} = \langle a_1, a_2 \rangle$, where $\mathbf{v} \neq \mathbf{0}$, the zero vector. Then

$$a_1 = \|\mathbf{v}\|\cos\theta \quad \text{and} \quad a_2 = \|\mathbf{v}\|\sin\theta$$

where θ is the angle between the positive x-axis and **v**.

The **horizontal component** of **v** is $\|\mathbf{v}\|\cos\theta$. The **vertical component** of **v** is $\|\mathbf{v}\|\sin\theta$.

QUESTION Is $\mathbf{u} = \cos\theta\mathbf{i} + \sin\theta\mathbf{j}$ a unit vector?

Any nonzero vector can be written in terms of its horizontal and vertical components. Let $\mathbf{v} = a_1\mathbf{i} + a_2\mathbf{j}$. Then

$$\begin{aligned}\mathbf{v} &= a_1\mathbf{i} + a_2\mathbf{j}\\ &= (\|\mathbf{v}\|\cos\theta)\mathbf{i} + (\|\mathbf{v}\|\sin\theta)\mathbf{j}\\ &= \|\mathbf{v}\|(\cos\theta\mathbf{i} + \sin\theta\mathbf{j})\end{aligned}$$

$\|\mathbf{v}\|$ is the magnitude of **v**, and the vector $\cos\theta\mathbf{i} + \sin\theta\mathbf{j}$ is a unit vector. The last equation shows that any vector **v** can be written as the product of its magnitude and a unit vector in the direction of **v**.

EXAMPLE 5 Find the Horizontal and Vertical Components of a Vector

Find the approximate horizontal and vertical components of a vector **v** of magnitude 10 meters with direction angle 228°. Write the vector in the form $\mathbf{v} = a_1\mathbf{i} + a_2\mathbf{j}$.

Solution

$$a_1 = 10\cos 228^\circ \approx -6.7$$
$$a_2 = 10\sin 228^\circ \approx -7.4$$

The approximate horizontal and vertical components are -6.7 and -7.4, respectively.

$$\mathbf{v} \approx -6.7\mathbf{i} - 7.4\mathbf{j}$$

TRY EXERCISE 36, EXERCISE SET 7.3, PAGE 566

ANSWER Yes, because $\|\cos\theta\mathbf{i} + \sin\theta\mathbf{j}\| = \sqrt{\cos^2\theta + \sin^2\theta} = \sqrt{1} = 1$

◆ APPLICATION PROBLEMS USING VECTORS

take note

Ground speed is the magnitude of the resultant of the plane's velocity vector and the wind velocity vector.

Consider an object on which two vectors are acting simultaneously. This occurs when a boat is moving in a current or an airplane is flying in a wind. The **airspeed** of a plane is the speed at which the plane would be moving if there were no wind. The actual velocity of a plane is the velocity relative to the ground. The magnitude of the actual velocity is the **ground speed.**

EXAMPLE 6 Solve an Application Involving Airspeed

An airplane is traveling with an airspeed of 320 mph and a heading of 62°. A wind of 42 mph is blowing at a heading of 125°. Find the ground speed and the course of the airplane.

Solution

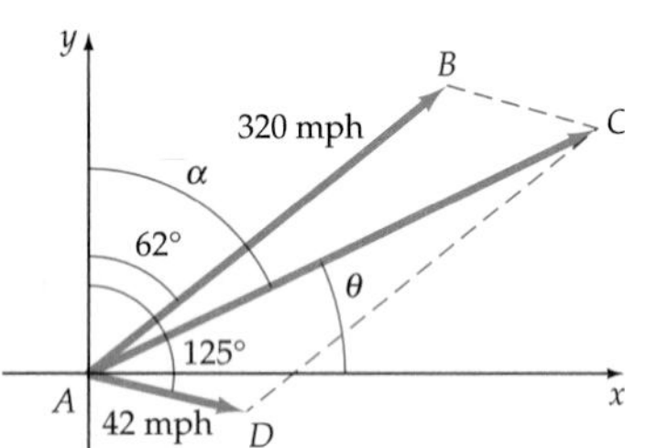

Figure 7.29

Sketch a diagram similar to Figure 7.29 showing the relevant vectors. **AB** represents the heading and the airspeed, **AD** represents the wind velocity, and **AC** represents the course and the ground speed. By vector addition, **AC** = **AB** + **AD**. From the figure,

$$\begin{aligned}\mathbf{AB} &= 320(\cos 28°\mathbf{i} + \sin 28°\mathbf{j})\\ \mathbf{AD} &= 42[\cos(-35°)\mathbf{i} + \sin(-35°)\mathbf{j}]\\ \mathbf{AC} &= 320(\cos 28°\mathbf{i} + \sin 28°\mathbf{j}) + 42[\cos(-35°)\mathbf{i} + \sin(-35°)\mathbf{j}]\\ &\approx (282.5\mathbf{i} + 150.2\mathbf{j}) + (34.4\mathbf{i} - 24.1\mathbf{j})\\ &= 316.9\mathbf{i} + 126.1\mathbf{j}\end{aligned}$$

AC is the course of the plane. The ground speed is $\|\mathbf{AC}\|$. The heading is $\alpha = 90° - \theta$.

$$\|\mathbf{AC}\| = \sqrt{(316.9)^2 + (126.1)^2} \approx 340$$

$$\alpha = 90° - \theta = 90° - \tan^{-1}\left(\frac{126.1}{316.9}\right) \approx 68°$$

The ground speed is approximately 340 mph at a heading of 68°.

TRY EXERCISE 40, EXERCISE SET 7.3, PAGE 566

There are numerous problems involving force that can be solved by using vectors. One type involves objects that are resting on a ramp. For these problems, we frequently try to find the components of a force vector relative to the ramp rather than to the x-axis.

EXAMPLE 7 Solve an Application Involving Force

A 110-pound box is on a 24° ramp. Find the component of the force that is parallel to the ramp.

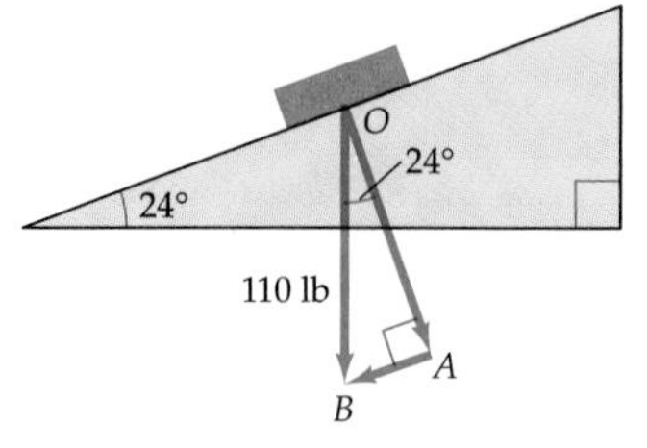

Figure 7.30

Solution

The force-of-gravity vector (110 pounds) is the sum of two components, one parallel to the ramp and the other (called the *normal component*) perpendicular to the ramp. (See Figure 7.30). **AB** is the vector that represents the force tending to move the box down the ramp. Because triangle OAB is a right triangle and $\angle AOB$ is 24°,

$$\sin 24° = \frac{\|\mathbf{AB}\|}{110}$$

$$\|\mathbf{AB}\| = 110 \sin 24° \approx 45$$

The component of force parallel to the ramp is approximately 45 pounds.

TRY EXERCISE 42, EXERCISE SET 7.3, PAGE 567

◆ DOT PRODUCT

We have considered the product of a real number (scalar) and a vector. We now turn our attention to the product of two vectors. Finding the *dot product* of two vectors is one way to multiply a vector by a vector. The dot product of two vectors is a real number and *not* a vector. The dot product is also called the *inner product* or the *scalar product*. This product is useful in engineering and physics.

Definition of Dot Product

Given $\mathbf{v} = \langle a, b\rangle$ and $\mathbf{w} = \langle c, d\rangle$, the **dot product** of **v** and **w** is given by

$$\mathbf{v} \cdot \mathbf{w} = ac + bd$$

take note

As illustrated in Example 8, the dot product of two vectors is a real number, not a vector.

EXAMPLE 8 Find the Dot Product of Two Vectors

Find the dot product of $\mathbf{v} = \langle 6, -2\rangle$ and $\mathbf{w} = \langle -2, 4\rangle$.

Solution

$\mathbf{v} \cdot \mathbf{w} = 6(-2) + (-2)4 = -12 - 8 = -20$

TRY EXERCISE 50, EXERCISE SET 7.3, PAGE 567

If the vectors in Example 8 were given in terms of the vectors **i** and **j**, then $\mathbf{v} = 6\mathbf{i} - 2\mathbf{j}$ and $\mathbf{w} = -2\mathbf{i} + 4\mathbf{j}$. In this case,

$$\mathbf{v} \cdot \mathbf{w} = (6\mathbf{i} - 2\mathbf{j}) \cdot (-2\mathbf{i} + 4\mathbf{j}) = 6(-2) + (-2)4 = -20$$

Properties of the Dot Product

In the following properties, $\mathbf{u}$, $\mathbf{v}$, and $\mathbf{w}$ are vectors and a is a scalar.

1. $\mathbf{v} \cdot \mathbf{w} = \mathbf{w} \cdot \mathbf{v}$
2. $\mathbf{u} \cdot (\mathbf{v} + \mathbf{w}) = \mathbf{u} \cdot \mathbf{v} + \mathbf{u} \cdot \mathbf{w}$
3. $a(\mathbf{u} \cdot \mathbf{v}) = (a\mathbf{u}) \cdot \mathbf{v} = \mathbf{u} \cdot (a\mathbf{v})$
4. $\mathbf{v} \cdot \mathbf{v} = \|\mathbf{v}\|^2$
5. $\mathbf{0} \cdot \mathbf{v} = 0$
6. $\mathbf{i} \cdot \mathbf{i} = \mathbf{j} \cdot \mathbf{j} = 1$
7. $\mathbf{i} \cdot \mathbf{j} = \mathbf{j} \cdot \mathbf{i} = 0$

The proofs of these properties follow from the definition of dot product. Here is the proof of the fourth property. Let $\mathbf{v} = a\mathbf{i} + b\mathbf{j}$.

$$\mathbf{v} \cdot \mathbf{v} = (a\mathbf{i} + b\mathbf{j}) \cdot (a\mathbf{i} + b\mathbf{j}) = a^2 + b^2 = \|\mathbf{v}\|^2$$

Rewriting the fourth property of the dot product yields an alternative way of expressing the magnitude of a vector.

Magnitude of a Vector in Terms of the Dot Product

If $\mathbf{v} = \langle a, b \rangle$. then $\|\mathbf{v}\| = \sqrt{\mathbf{v} \cdot \mathbf{v}}$.

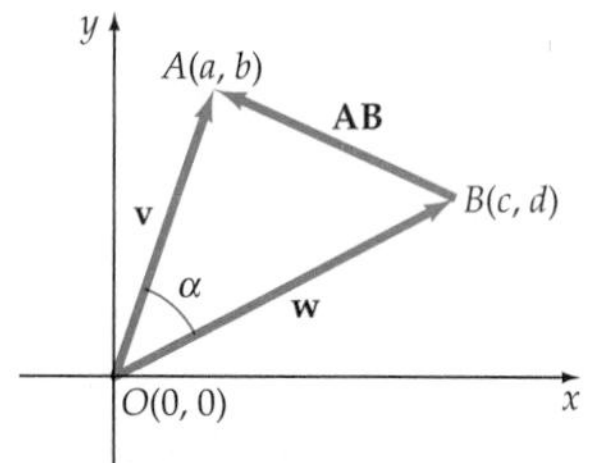

Figure 7.31

The Law of Cosines can be used to derive an alternative formula for the dot product. Consider the vectors $\mathbf{v} = \langle a, b \rangle$ and $\mathbf{w} = \langle c, d \rangle$ as shown in Figure 7.31. Using the Law of Cosines for triangle OAB, we have

$$\|\mathbf{AB}\|^2 = \|\mathbf{v}\|^2 + \|\mathbf{w}\|^2 - 2\|\mathbf{v}\|\,\|\mathbf{w}\| \cos \alpha$$

By the distance formula, $\|\mathbf{AB}\|^2 = (a - c)^2 + (b - d)^2$, $\|\mathbf{v}\|^2 = a^2 + b^2$, and $\|\mathbf{w}\|^2 = c^2 + d^2$. Thus

$$\begin{aligned}
(a - c)^2 + (b - d)^2 &= (a^2 + b^2) + (c^2 + d^2) - 2\|\mathbf{v}\|\,\|\mathbf{w}\|\cos \alpha \\
a^2 - 2ac + c^2 + b^2 - 2bd + d^2 &= a^2 + b^2 + c^2 + d^2 - 2\|\mathbf{v}\|\,\|\mathbf{w}\| \cos \alpha \\
-2ac - 2bc &= -2\|\mathbf{v}\|\,\|\mathbf{w}\| \cos \alpha \\
ac + bd &= \|\mathbf{v}\|\,\|\mathbf{w}\| \cos \alpha \\
\mathbf{v} \cdot \mathbf{w} &= \|\mathbf{v}\|\,\|\mathbf{w}\| \cos \alpha \qquad \bullet\ \mathbf{v} \cdot \mathbf{w} = ac + bd
\end{aligned}$$

Alternative Formula for the Dot Product

If $\mathbf{v}$ and $\mathbf{w}$ are two nonzero vectors and α is the smallest non-negative angle between $\mathbf{v}$ and $\mathbf{w}$, then $\mathbf{v} \cdot \mathbf{w} = \|\mathbf{v}\|\,\|\mathbf{w}\| \cos \alpha$.

Solving the alternative formula for the dot product for $\cos \alpha$, we have a formula for the cosine of the angle between two vectors.

Angle Between Two Vectors

If $\mathbf{v}$ and $\mathbf{w}$ are two nonzero vectors and α is the smallest non-negative angle between $\mathbf{v}$ and $\mathbf{w}$, then $\cos \alpha = \dfrac{\mathbf{v} \cdot \mathbf{w}}{\|\mathbf{v}\|\,\|\mathbf{w}\|}$ and $\alpha = \cos^{-1}\left(\dfrac{\mathbf{v} \cdot \mathbf{w}}{\|\mathbf{v}\|\,\|\mathbf{w}\|}\right)$.

EXAMPLE 9 Find the Angle Between Two Vectors

Find the measure of the smallest positive angle between the vectors $\mathbf{v} = 2\mathbf{i} - 3\mathbf{j}$ and $\mathbf{w} = -\mathbf{i} + 5\mathbf{j}$ as shown in Figure 7.32.

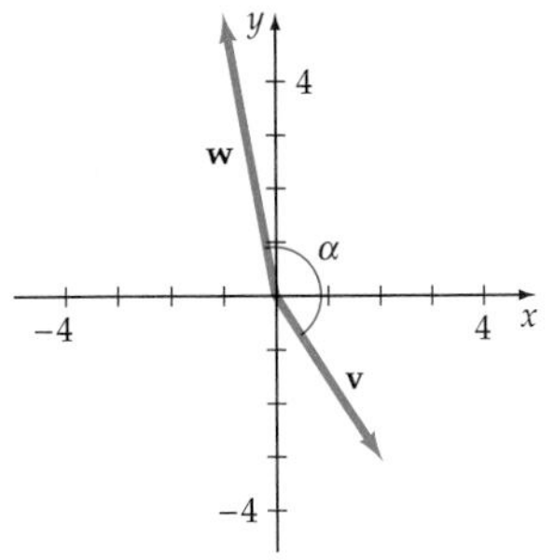

Figure 7.32

Solution

Use the equation for the angle between two vectors.

$$\cos \alpha = \frac{\mathbf{v} \cdot \mathbf{w}}{\|\mathbf{v}\|\|\mathbf{w}\|} = \frac{(2\mathbf{i} - 3\mathbf{j}) \cdot (-\mathbf{i} + 5\mathbf{j})}{(\sqrt{2^2 + (-3)^2})(\sqrt{(-1)^2 + 5^2})}$$

$$= \frac{-2 - 15}{\sqrt{13}\,\sqrt{26}} = \frac{-17}{\sqrt{338}}$$

$$\alpha = \cos^{-1}\left(\frac{-17}{\sqrt{338}}\right) \approx 157.6^\circ$$

The angle between the two vectors is approximately 157.6°.

TRY EXERCISE 60, EXERCISE SET 7.3, PAGE 567

◆ SCALAR PROJECTION

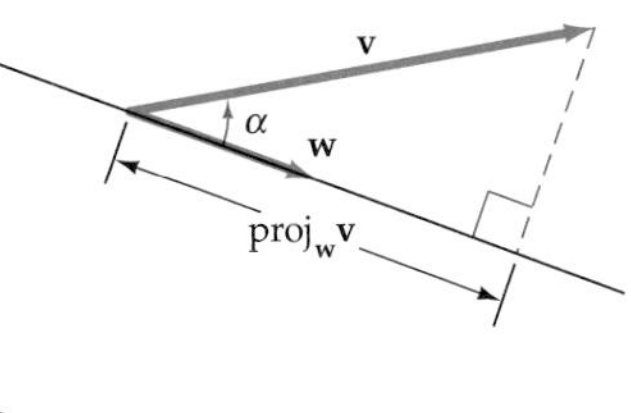

Figure 7.33

Let $\mathbf{v} = \langle a_1, a_2 \rangle$ and $\mathbf{w} = \langle b_1, b_2 \rangle$ be two nonzero vectors and let α be the angle between the vectors. Two possible configurations, one for which α is an acute angle and one for which α is an obtuse angle, are shown in Figure 7.33. In each case, a right triangle is formed by drawing a line segment from the head of $\mathbf{v}$ to a line through $\mathbf{w}$.

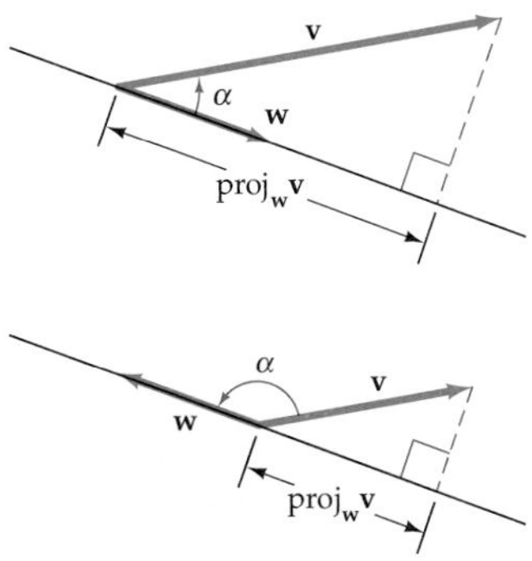

Definition of the Scalar Projection of v on w

If $\mathbf{v}$ and $\mathbf{w}$ are two nonzero vectors and α is the smallest positive angle between $\mathbf{v}$ and $\mathbf{w}$, then the scalar projection of $\mathbf{v}$ on $\mathbf{w}$, $\text{proj}_{\mathbf{w}}\mathbf{v}$, is given by

$$\text{proj}_{\mathbf{w}}\mathbf{v} = \|\mathbf{v}\| \cos \alpha$$

To derive an alternate formula for $\text{proj}_{\mathbf{w}}\mathbf{v}$, consider the dot product, $\mathbf{v} \cdot \mathbf{w} = \|\mathbf{v}\| \|\mathbf{w}\| \cos \alpha$. Solving for $\|\mathbf{v}\| \cos \alpha$, which is $\text{proj}_{\mathbf{w}}\mathbf{v}$, we have

$$\text{proj}_{\mathbf{w}}\mathbf{v} = \frac{\mathbf{v} \cdot \mathbf{w}}{\|\mathbf{w}\|}$$

When the angle α between the two vectors is an acute angle, $\text{proj}_{\mathbf{w}}\mathbf{v}$ is positive. When α is an obtuse angle, $\text{proj}_{\mathbf{w}}\mathbf{v}$ is negative.

EXAMPLE 10 Find the Projection of v on w

Given $\mathbf{v} = 2\mathbf{i} + 4\mathbf{j}$ and $\mathbf{w} = -2\mathbf{i} + 8\mathbf{j}$ as shown in Figure 7.34, find $\text{proj}_{\mathbf{w}}\mathbf{v}$.

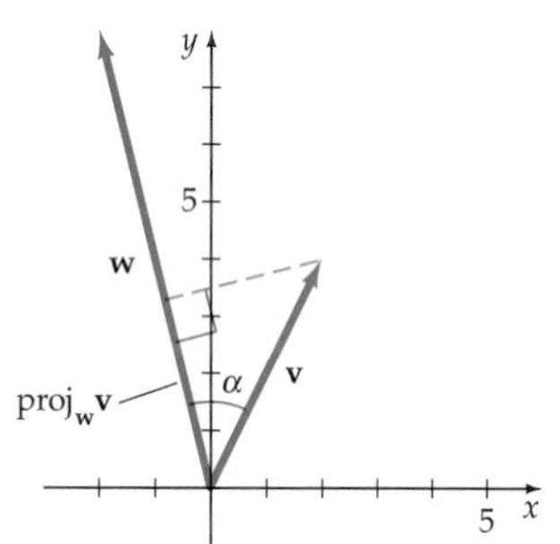

Figure 7.34

Solution

Use the equation $\text{proj}_{\mathbf{w}}\mathbf{v} = \dfrac{\mathbf{v} \cdot \mathbf{w}}{\|\mathbf{w}\|}$.

$$\text{proj}_{\mathbf{w}}\mathbf{v} = \frac{(2\mathbf{i} + 4\mathbf{j}) \cdot (-2\mathbf{i} + 8\mathbf{j})}{\sqrt{(-2)^2 + 8^2}} = \frac{28}{\sqrt{68}} = \frac{14\sqrt{17}}{17} \approx 3.4$$

TRY EXERCISE 62, EXERCISE SET 7.3, PAGE 567

◆ PARALLEL AND PERPENDICULAR VECTORS

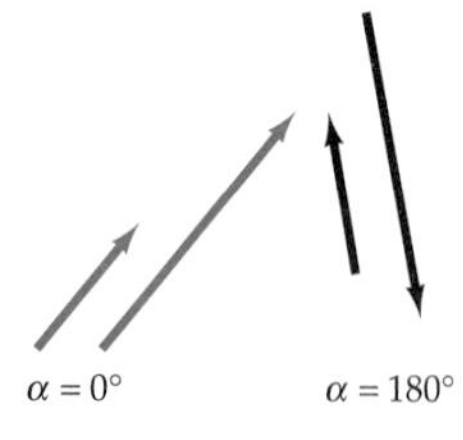

Figure 7.35

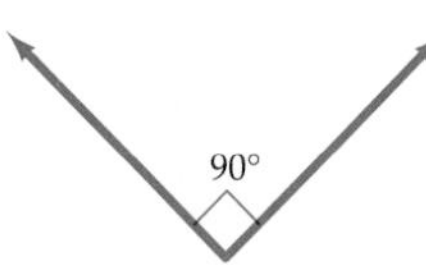

Figure 7.36

Two vectors are *parallel* when the angle α between the vectors is 0° or 180°, as shown in Figure 7.35. When the angle α is 0°, the vectors point in the same direction; the vectors point in opposite directions when α is 180°.

Let $\mathbf{v} = a_1\mathbf{i} + b_1\mathbf{j}$, let c be a real number, and let $\mathbf{w} = c\mathbf{v}$. Because $\mathbf{w}$ is a constant multiple of $\mathbf{v}$, $\mathbf{w}$ and $\mathbf{v}$ are parallel vectors. When $c > 0$, the vectors point in the same direction. When $c < 0$, the vectors point in opposite directions.

Two vectors are *perpendicular* when the angle between the vectors is 90°. See Figure 7.36. Perpendicular vectors are referred to as **orthogonal vectors.** If $\mathbf{v}$ and $\mathbf{w}$ are two nonzero orthogonal vectors, then from the formula for the angle between two vectors and the fact that $\cos \alpha = 0$, we have

$$0 = \frac{\mathbf{v} \cdot \mathbf{w}}{\|\mathbf{v}\| \|\mathbf{w}\|}$$

If a fraction equals zero, the numerator must be zero. Thus, for orthogonal vectors $\mathbf{v}$ and $\mathbf{w}$, $\mathbf{v} \cdot \mathbf{w} = 0$. This gives the following result.

Condition for Perpendicular Vectors

Two nonzero vectors $\mathbf{v}$ and $\mathbf{w}$ are orthogonal if and only if $\mathbf{v} \cdot \mathbf{w} = 0$.

◆ WORK: AN APPLICATION OF THE DOT PRODUCT

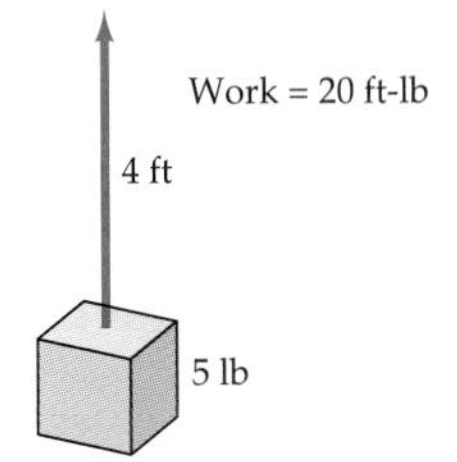

Figure 7.37

When a 5-pound force is used to lift a box from the ground a distance of 4 feet, *work* is done. The amount of **work** is the product of the force on the box and the distance the box is moved. In this case, the work is 20 foot-pounds. When the box is lifted, the force and the displacement vector (the direction and the distance in which the box was moved) are in the same direction. (See **Figure 7.37.**)

Now consider a sled being pulled by a child along the ground by a rope attached to the sled, as shown in **Figure 7.38.** The force vector (along the rope) is *not* in the same direction as the displacement vector (parallel to the ground). In this case, the dot product is used to determine the work done by the force.

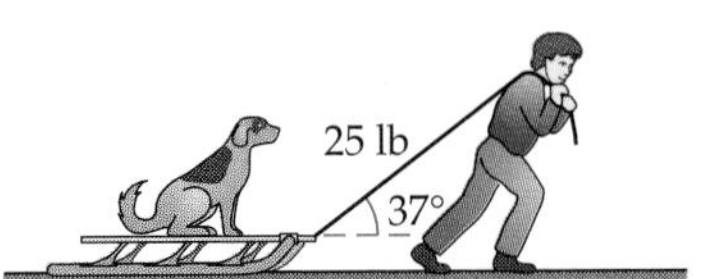

Figure 7.38

Definition of Work

The work W done by a force $\mathbf{F}$ applied along a displacement $\mathbf{s}$ is

$$W = \mathbf{F} \cdot \mathbf{s}$$

In the case of the child pulling the sled 7 feet, the work done is

$$\begin{aligned} W &= \mathbf{F} \cdot \mathbf{s} \\ &= \|\mathbf{F}\|\,\|\mathbf{s}\| \cos \alpha \qquad \bullet\ \alpha \text{ is the angle between } \mathbf{F} \text{ and } \mathbf{s}. \\ &= (25)(7) \cos 37^\circ \approx 140 \text{ foot-pounds} \end{aligned}$$

EXAMPLE 11 Solve a Work Problem

A force of 50 pounds on a rope is used to drag a box up a ramp that is inclined 10°. If the rope makes an angle of 37° with the ground, find the work done in moving the box 15 feet along the ramp. See **Figure 7.39.**

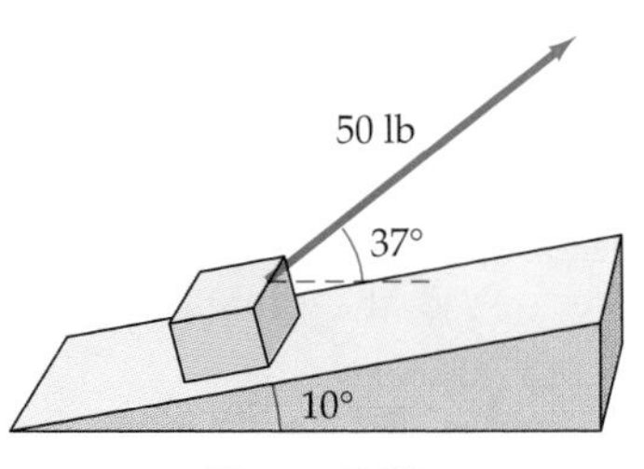

Figure 7.39

Solution

We will provide two solutions to this example.

Method 1 From the last section, $\mathbf{u} = \cos 10^\circ\mathbf{i} + \sin 10^\circ\mathbf{j}$ is a unit vector parallel to the ramp. Multiplying $\mathbf{u}$ by 15 (the magnitude of the displacement vector) gives $\mathbf{s} = 15(\cos 10^\circ\mathbf{i} + \sin 10^\circ\mathbf{j})$. Similarly, the force vector is $\mathbf{F} = 50(\cos 37^\circ\mathbf{i} + \sin 37^\circ\mathbf{j})$. The work done is given by the dot product.

$$\begin{aligned} W = \mathbf{F} \cdot \mathbf{s} &= 50(\cos 37^\circ\mathbf{i} + \sin 37^\circ\mathbf{j}) \cdot 15(\cos 10^\circ\mathbf{i} + \sin 10^\circ\mathbf{j}) \\ &= [50 \cdot 15](\cos 37^\circ \cos 10^\circ + \sin 37^\circ \sin 10^\circ) \approx 668.3 \text{ foot-pounds} \end{aligned}$$

Method 2 When we write the work equation as $W = \|\mathbf{F}\|\|\mathbf{s}\| \cos \alpha$, α is the angle between the force and the displacement. Thus $\alpha = 37^\circ - 10^\circ = 27^\circ$. The work done is

$$W = \|\mathbf{F}\|\|\mathbf{s}\| \cos \alpha = 50 \cdot 15 \cdot \cos 27^\circ \approx 668.3 \text{ foot-pounds}$$

TRY EXERCISE 70, EXERCISE SET 7.3, PAGE 567

TOPICS FOR DISCUSSION

1. Is the dot product of two vectors a vector or a scalar? Explain.
2. Is the projection of **v** on **w** a vector or a scalar? Explain.
3. Is the nonzero vector $\langle a, b\rangle$ perpendicular to the vector $\langle -b, a\rangle$? Explain.
4. Explain how to determine the angle between the vector $\langle 3, 4\rangle$ and the vector $\langle 5, -1\rangle$.
5. Consider the nonzero vector $\mathbf{u} = \langle a, b\rangle$ and the vector
$$\mathbf{v} = \left\langle \frac{a}{\sqrt{a^2+b^2}}, \frac{a}{\sqrt{a^2+b^2}} \right\rangle.$$
   a. Are the vectors parallel? Explain.
   b. Which one of the vectors is a unit vector?
   c. Which vector has the larger magnitude? Explain.

EXERCISE SET 7.3

In Exercises 1 to 6, find the components of a vector with the given initial and terminal points. Write an equivalent vector in terms of its components.

1. $P_1(-3, 0)$; $P_2(4, -1)$
2. $P_1(5, -1)$; $P_2(3, 1)$
3. $P_1(4, 2)$; $P_2(-3, -3)$
4. $P_1(0, -3)$; $P_2(0, 4)$
5. $P_1(2, -5)$; $P_2(2, 3)$
6. $P_1(3, -2)$; $P_2(3, 0)$

In Exercises 7 to 14, find the magnitude and direction of each vector. Find the unit vector in the direction of the given vector.

7. $\mathbf{v} = \langle -3, 4\rangle$
8. $\mathbf{v} = \langle 6, 10\rangle$
9. $\mathbf{v} = \langle 20, -40\rangle$
10. $\mathbf{v} = \langle -50, 30\rangle$
11. $\mathbf{v} = 2\mathbf{i} - 4\mathbf{j}$
12. $\mathbf{v} = -5\mathbf{i} + 6\mathbf{j}$
13. $\mathbf{v} = 42\mathbf{i} - 18\mathbf{j}$
14. $\mathbf{v} = -22\mathbf{i} - 32\mathbf{j}$

In Exercises 15 to 23, perform the indicated operations where u = ⟨−2, 4⟩ and v = ⟨−3, −2⟩.

15. $3\mathbf{u}$
16. $-4\mathbf{v}$
17. $2\mathbf{u} - \mathbf{v}$
18. $4\mathbf{v} - 2\mathbf{u}$
19. $\frac{2}{3}\mathbf{u} + \frac{1}{6}\mathbf{v}$
20. $\frac{3}{4}\mathbf{u} - 2\mathbf{v}$
21. $\|\mathbf{u}\|$
22. $\|\mathbf{v} + 2\mathbf{u}\|$
23. $\|3\mathbf{u} - 4\mathbf{v}\|$

In Exercises 24 to 32, perform the indicated operations where u = 3i − 2j and v = −2i + 3j.

24. $-2\mathbf{u}$
25. $4\mathbf{v}$
26. $3\mathbf{u} + 2\mathbf{v}$
27. $6\mathbf{u} + 2\mathbf{v}$
28. $\frac{1}{2}\mathbf{u} - \frac{3}{4}\mathbf{v}$
29. $\frac{2}{3}\mathbf{v} + \frac{3}{4}\mathbf{u}$
30. $\|\mathbf{v}\|$
31. $\|\mathbf{u} - 2\mathbf{v}\|$
32. $\|2\mathbf{v} + 3\mathbf{u}\|$

In Exercises 33 to 36, find the horizontal and vertical components of each vector. Write an equivalent vector in the form $\mathbf{v} = a_1\mathbf{i} + a_2\mathbf{j}$.

33. Magnitude = 5, direction angle = 27°
34. Magnitude = 4, direction angle = 127°
35. Magnitude = 4, direction angle = $\pi/4$
36. Magnitude = 2, direction angle = $8\pi/7$
37. **GROUND SPEED OF A PLANE** A plane is flying at an airspeed of 340 mph at a heading of 124°. A wind of 45 mph is blowing from the west. Find the ground speed of the plane.
38. **HEADING OF A BOAT** A person who can row 2.6 mph in still water wants to row due east across a river. The river is flowing from the north at a rate of 0.8 mph. Determine the heading of the boat that will be required for it to travel due east across the river.
39. **GROUND SPEED AND COURSE OF A PLANE** A pilot is flying at a heading of 96° at 225 mph. A 50-mph wind is blowing from the southwest at a heading of 37°. Find the ground speed and course of the plane.
40. **COURSE OF A BOAT** The captain of a boat is steering at a heading of 327° at 18 mph. The current is flowing at 4 mph at a heading of 60°. Find the course (to the nearest degree) of the boat.
41. **MAGNITUDE OF A FORCE** Find the magnitude of force necessary to keep a 3000-pound car from sliding down a ramp inclined at an angle of 5.6°.

42. **ANGLE OF A RAMP** A 120-pound force keeps an 800-pound object from sliding down an inclined ramp. Find the angle of the ramp.

43. **MAGNITUDE OF THE NORMAL COMPONENT** A 25-pound box is resting on a ramp that is inclined 9.0°. Find the magnitude of the normal component of force.

44. **MAGNITUDE OF THE NORMAL COMPONENT** Find the magnitude of the normal component of force for a 50-pound crate that is resting on a ramp that is inclined 12°.

In Exercises 45 to 52, find the dot product of the vectors.

45. $\mathbf{v} = \langle 3, -2 \rangle$; $\mathbf{w} = \langle 1, 3 \rangle$

46. $\mathbf{v} = \langle 2, 4 \rangle$; $\mathbf{w} = \langle 0, 2 \rangle$

47. $\mathbf{v} = \langle 4, 1 \rangle$; $\mathbf{w} = \langle -1, 4 \rangle$

48. $\mathbf{v} = \langle 2, -3 \rangle$; $\mathbf{w} = \langle 3, 2 \rangle$

49. $\mathbf{v} = \mathbf{i} + 2\mathbf{j}$; $\mathbf{w} = -\mathbf{i} + \mathbf{j}$

50. $\mathbf{v} = 5\mathbf{i} + 3\mathbf{j}$; $\mathbf{w} = 4\mathbf{i} - 2\mathbf{j}$

51. $\mathbf{v} = 6\mathbf{i} - 4\mathbf{j}$; $\mathbf{w} = -2\mathbf{i} - 3\mathbf{j}$

52. $\mathbf{v} = -4\mathbf{i} + 2\mathbf{j}$; $\mathbf{w} = -2\mathbf{i} - 4\mathbf{j}$

In Exercises 53 to 60, find the angle between the two vectors. State which pair of vectors is orthogonal.

53. $\mathbf{v} = \langle 2, -1 \rangle$; $\mathbf{w} = \langle 3, 4 \rangle$

54. $\mathbf{v} = \langle 1, -5 \rangle$; $\mathbf{w} = \langle -2, 3 \rangle$

55. $\mathbf{v} = \langle 0, 3 \rangle$; $\mathbf{w} = \langle 2, 2 \rangle$

56. $\mathbf{v} = \langle -1, 7 \rangle$; $\mathbf{w} = \langle 3, -2 \rangle$

57. $\mathbf{v} = 5\mathbf{i} - 2\mathbf{j}$; $\mathbf{w} = 2\mathbf{i} + 5\mathbf{j}$

58. $\mathbf{v} = 8\mathbf{i} + \mathbf{j}$; $\mathbf{w} = -\mathbf{i} + 8\mathbf{j}$

59. $\mathbf{v} = 5\mathbf{i} + 2\mathbf{j}$; $\mathbf{w} = -5\mathbf{i} - 2\mathbf{j}$

60. $\mathbf{v} = 3\mathbf{i} - 4\mathbf{j}$; $\mathbf{w} = 6\mathbf{i} - 12\mathbf{j}$

In Exercises 61 to 68, find $\text{proj}_{\mathbf{w}}\mathbf{v}$.

61. $\mathbf{v} = \langle 6, 7 \rangle$; $\mathbf{w} = \langle 3, 4 \rangle$

62. $\mathbf{v} = \langle -7, 5 \rangle$; $\mathbf{w} = \langle -4, 1 \rangle$

63. $\mathbf{v} = \langle -3, 4 \rangle$; $\mathbf{w} = \langle 2, 5 \rangle$

64. $\mathbf{v} = \langle 2, 4 \rangle$; $\mathbf{w} = \langle -1, 5 \rangle$

65. $\mathbf{v} = 2\mathbf{i} + \mathbf{j}$; $\mathbf{w} = 6\mathbf{i} + 3\mathbf{j}$

66. $\mathbf{v} = 5\mathbf{i} + 2\mathbf{j}$; $\mathbf{w} = -5\mathbf{i} - 2\mathbf{j}$

67. $\mathbf{v} = 3\mathbf{i} - 4\mathbf{j}$; $\mathbf{w} = -6\mathbf{i} + 12\mathbf{j}$

68. $\mathbf{v} = 2\mathbf{i} + 2\mathbf{j}$; $\mathbf{w} = -4\mathbf{i} - 2\mathbf{j}$

69. **WORK** A 150-pound box is dragged 15 feet along a level floor. Find the work done if a force of 75 pounds at an angle of 32° is used.

70. **WORK** A 100-pound force is pulling a sled loaded with bricks that weighs 400 pounds. The force is at an angle of 42° with the displacement. Find the work done in moving the sled 25 feet.

71. **WORK** A rope is being used to pull a box up a ramp that is inclined at 15°. The rope exerts a force of 75 pounds on the box, and it makes an angle of 30° with the plane of the ramp. Find the work done in moving the box 12 feet.

72. **WORK** A dock worker exerts a force on a box sliding down the ramp of a truck. The ramp makes an angle of 48° with the road, and the worker exerts a 50-pound force parallel to the road. Find the work done in sliding the box 6 feet.

SUPPLEMENTAL EXERCISES

73. For $\mathbf{u} = \langle -1, 1 \rangle$, $\mathbf{v} = \langle 2, 3 \rangle$, and $\mathbf{w} = \langle 5, 5 \rangle$, find the sum of the three vectors geometrically by using the triangle method of adding vectors.

74. For $\mathbf{u} = \langle 1, 2 \rangle$, $\mathbf{v} = \langle 3, -2 \rangle$, and $\mathbf{w} = \langle -1, 4 \rangle$, find $\mathbf{u} + \mathbf{v} - \mathbf{w}$ geometrically by using the triangle method of adding vectors.

75. Find a vector that has the initial point $(3, -1)$ and is equivalent to $\mathbf{v} = 2\mathbf{i} - 3\mathbf{j}$.

76. Find a vector that has the initial point $(-2, 4)$ and is equivalent to $\mathbf{v} = \langle -1, 3 \rangle$.

77. If $\mathbf{v} = 2\mathbf{i} - 5\mathbf{j}$ and $\mathbf{w} = 5\mathbf{i} + 2\mathbf{j}$ have the same initial point, is $\mathbf{v}$ perpendicular to $\mathbf{w}$? Why or why not?

78. If $\mathbf{v} = \langle 5, 6 \rangle$ and $\mathbf{w} = \langle 6, 5 \rangle$ have the same initial point, is $\mathbf{v}$ perpendicular to $\mathbf{w}$? Why or why not?

79. Let $\mathbf{v} = \langle -2, 7 \rangle$. Find a vector perpendicular to $\mathbf{v}$.

80. Let $\mathbf{w} = 4\mathbf{i} + \mathbf{j}$. Find a vector perpendicular to $\mathbf{w}$.

In Example 7, of this section, if the box were to be kept from sliding down the ramp, it would be necessary to provide a force of 45 pounds parallel to the ramp but pointed *up* the ramp. Some of this force would be provided by a frictional force between the box and the ramp. The force of friction is $\mathbf{F}_\mu = \mu\mathbf{N}$, where N is the normal component of the force of gravity. In Exercises 81 and 82, find the frictional force.

81. **FRICTIONAL FORCE** A 50-pound box is resting on a ramp inclined at 12°. Find the force of friction if the coefficient of friction, μ, is 0.13.

82. **FRICTIONAL FORCE** A car weighing 2500 pounds is resting on a ramp inclined at 15°. Find the frictional force if the coefficient of friction, μ, is 0.21.

83. Is the dot product an associative operation? That is, given nonzero vectors $\mathbf{u}$, $\mathbf{v}$, and $\mathbf{w}$, does

$$(\mathbf{u} \cdot \mathbf{v}) \cdot \mathbf{w} = \mathbf{u} \cdot (\mathbf{v} \cdot \mathbf{w})?$$

84. Prove that $\mathbf{v} \cdot \mathbf{w} = \mathbf{w} \cdot \mathbf{v}$.

85. Prove that $c(\mathbf{v} \cdot \mathbf{w}) = (c\mathbf{v}) \cdot \mathbf{w}$.

86. Show that the dot product of two nonzero vectors is positive if the angle between the vectors is an acute angle and that the dot product is negative if the angle between the two vectors is an obtuse angle.

87. **COMPARISON OF WORK DONE** Consider the following two situations. (1) A rope is being used to pull a box up a ramp inclined at an angle α. The rope exerts a force $\mathbf{F}$ on the box, and the rope makes an angle θ with the ramp. The box is pulled s feet. (2) A rope is being used to pull a box along a level floor. The rope exerts the same force $\mathbf{F}$ on the box. The box is pulled the same s feet. In which case is more work done?

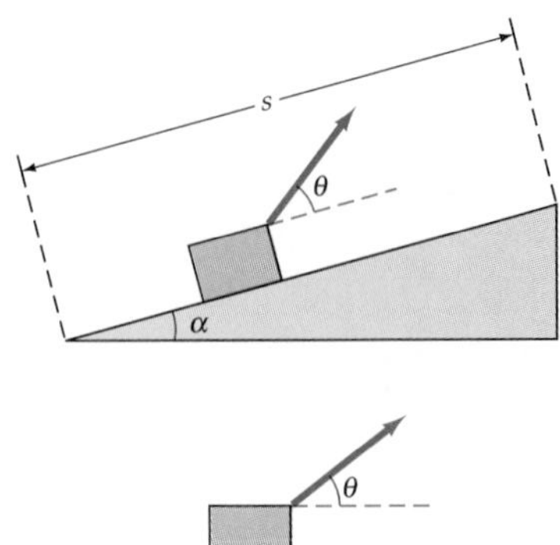

PROJECTS

1. **SAME DIRECTION OR OPPOSITE DIRECTIONS** Let $\mathbf{v} = c\mathbf{w}$, where c is a nonzero real number and $\mathbf{w}$ is a nonzero vector. Show that $\dfrac{\mathbf{v} \cdot \mathbf{w}}{\|\mathbf{v}\|\,\|\mathbf{w}\|} = \pm 1$ and that the result is 1 when $c > 0$ and -1 when $c < 0$.
2. **THE LAW OF COSINES AND VECTORS** Prove that $\|\mathbf{v} - \mathbf{w}\|^2 = \|\mathbf{v}\|^2 + \|\mathbf{w}\|^2 - 2\mathbf{v} \cdot \mathbf{w}$.
3. **PROJECTION RELATIONSHIPS** What is the relationship between $\mathbf{v}$ and $\mathbf{w}$ if

   a. $\text{proj}_{\mathbf{w}}\mathbf{v} = 0$? b. $\text{proj}_{\mathbf{w}}\mathbf{v} = \|\mathbf{v}\|$?

SECTION 7.4 TRIGONOMETRIC FORM OF COMPLEX NUMBERS

- GRAPHICAL REPRESENTATION OF A COMPLEX NUMBER
- ABSOLUTE VALUE OF A COMPLEX NUMBER
- TRIGONOMETRIC FORM OF A COMPLEX NUMBER
- THE PRODUCT AND THE QUOTIENT OF COMPLEX NUMBERS WRITTEN IN TRIGONOMETRIC FORM

7C

◆ GRAPHICAL REPRESENTATION OF A COMPLEX NUMBER

Real numbers are graphed as points on a number line. Complex numbers can be graphed in a coordinate plane called the **complex plane.** The horizontal axis of the complex plane is called the **real axis;** the vertical axis is called the **imaginary axis.**

A complex number written in the form $z = a + bi$ is written in **standard form** or **rectangular form.** The graph of $a + bi$ is associated with the point $P(a, b)$ in the complex plane. **Figure 7.40** shows the graphs of several complex numbers.

◆ ABSOLUTE VALUE OF A COMPLEX NUMBER

The length of the line segment from the origin to the point $(-3, 4)$ in the complex plane is the *absolute value* of $z = -3 + 4i$. See **Figure 7.41.** From the Pythagorean

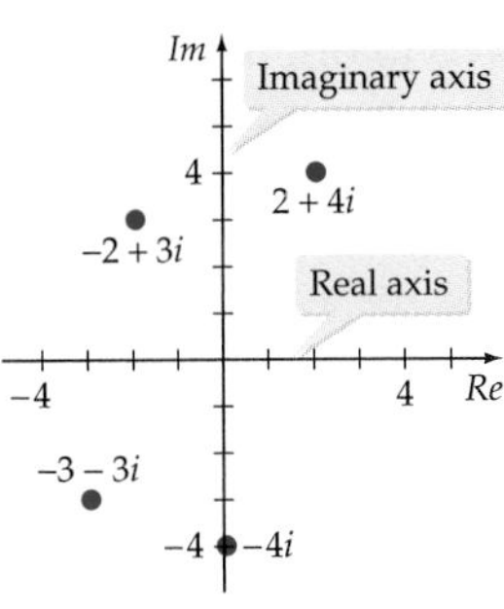

Figure 7.40

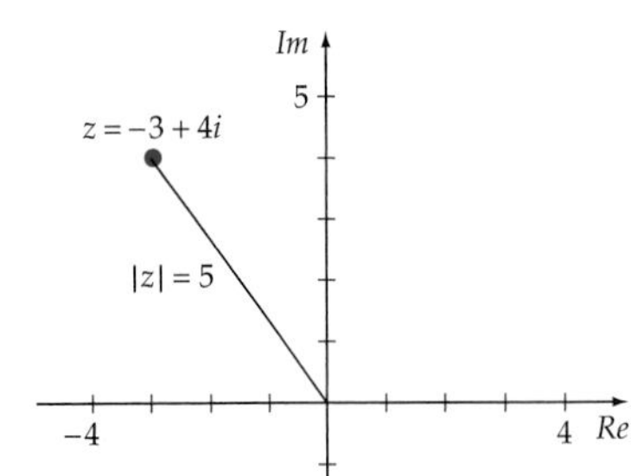

Figure 7.41

Theorem, the absolute value of $z = -3 + 4i$ is

$$\sqrt{(-3)^2 + 4^2} = \sqrt{25} = 5$$

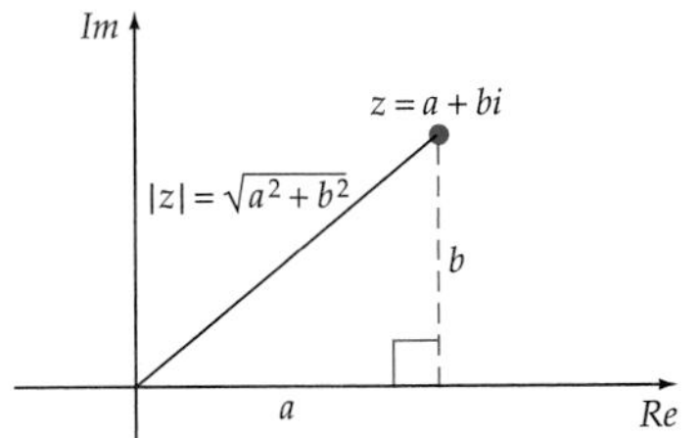

Figure 7.42

Definition of the Absolute Value of a Complex Number

The absolute value of the complex number $z = a + bi$, denoted by $|z|$, is

$$|z| = |a + bi| = \sqrt{a^2 + b^2}$$

Thus $|z|$ is the distance from the origin to z (see **Figure 7.42**).

◆ TRIGONOMETRIC FORM OF A COMPLEX NUMBER

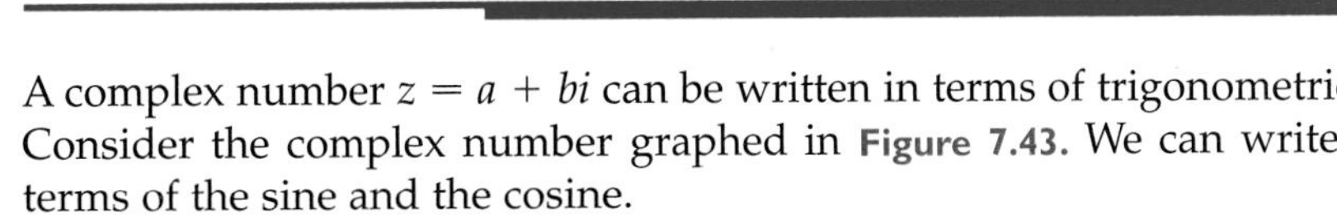

A complex number $z = a + bi$ can be written in terms of trigonometric functions. Consider the complex number graphed in **Figure 7.43.** We can write a and b in terms of the sine and the cosine.

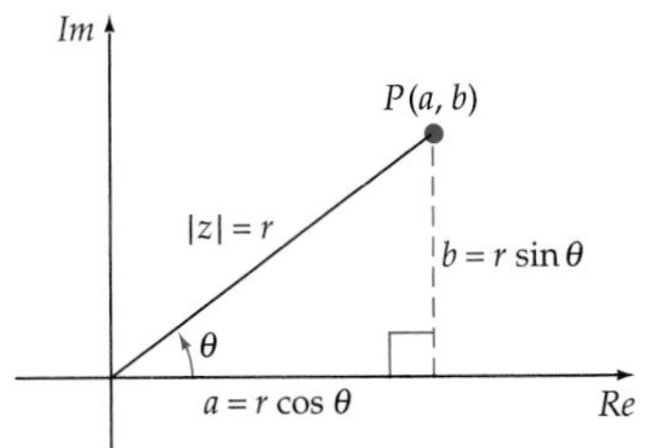

Figure 7.43

$$\cos\theta = \frac{a}{r} \qquad \sin\theta = \frac{b}{r}$$

$$a = r\cos\theta \qquad b = r\sin\theta$$

where $r = |z| = \sqrt{a^2 + b^2}$. Substituting for a and b in $z = a + bi$, we obtain

$$z = r\cos\theta + ir\sin\theta = r(\cos\theta + i\sin\theta)$$

The expression $z = r(\cos\theta + i\sin\theta)$ is known as the **trigonometric form** of a complex number. The trigonometric form of a complex number is also called the **polar form** of the complex number. The notation $\cos\theta + i\sin\theta$ is often abbreviated as cis θ using the c from cos θ, the imaginary unit i, and the s from sin θ.

Trigonometric Form of a Complex Number

The complex number $z = a + bi$ can be written in trigonometric form as

$$z = r(\cos\theta + i\sin\theta) = r \operatorname{cis}\theta$$

where $a = r\cos\theta$, $b = r\sin\theta$, $r = \sqrt{a^2 + b^2}$, and $\tan\theta = \dfrac{b}{a}$.

In this text, we will often write the trigonometric form of a complex number in its abbreviated form $z = r \text{ cis } \theta$. The value of r is called the **modulus** of the complex number z, and the angle θ is called the **argument** of the complex number z. The modulus r and the argument θ of a complex number $z = a + bi$ are given by

$$r = \sqrt{a^2 + b^2} \qquad \text{and} \qquad \cos\theta = \frac{a}{r}, \quad \sin\theta = \frac{b}{r}$$

We can also write $\alpha = \tan^{-1}\left|\frac{b}{a}\right|$, where α is the reference angle for θ. As a result of the periodic nature of the sine and cosine functions, the trigonometric form of a complex number is not unique. Because $\cos\theta = \cos(\theta + 2k\pi)$ and $\sin\theta = \sin(\theta + 2k\pi)$, where k is an integer, the following complex numbers are equal.

$$r \text{ cis } \theta = r \text{ cis } (\theta + 2k\pi) \quad \text{for } k \text{ an integer}$$

For example, $2 \text{ cis } \frac{\pi}{6} = 2 \text{ cis } \left(\frac{\pi}{6} + 2\pi\right)$.

EXAMPLE 1 Write a Complex Number in Trigonometric Form

Write $z = -2 - 2i$ in trigonometric form.

Solution

Find the modulus and the argument of z. Then substitute these values in the trigonometric form of z.

$$r = \sqrt{(-2)^2 + (-2)^2} = \sqrt{8} = 2\sqrt{2}$$

To determine θ, we first determine α. See **Figure 7.44.**

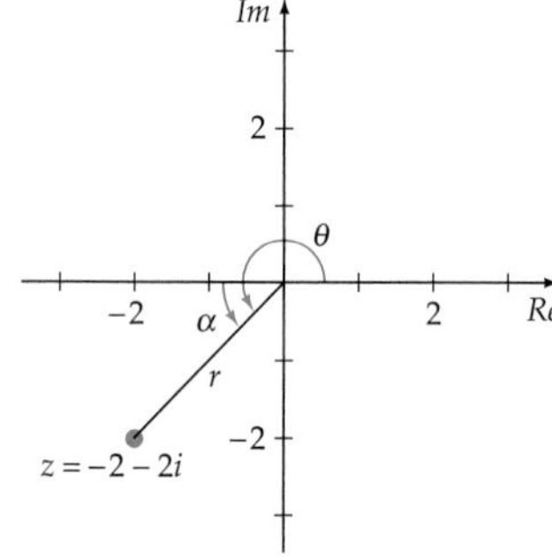

Figure 7.44

$$\alpha = \tan^{-1}\left|\frac{b}{a}\right|$$ • **α is the reference angle of angle θ.**

$$\alpha = \tan^{-1}\left|\frac{-2}{-2}\right| = \tan^{-1}1 = 45^\circ$$

$$\theta = 180^\circ + 45^\circ = 225^\circ$$ • **Because z is in the third quadrant, $180^\circ < \theta < 270^\circ$.**

The trigonometric form is

$$z = r \text{ cis } \theta = 2\sqrt{2} \text{ cis } 225^\circ$$ • **$r = 2\sqrt{2}$, $\theta = 225^\circ$**

TRY EXERCISE 12, EXERCISE SET 7.4, PAGE 574

EXAMPLE 2 Write a Complex Number in Standard Form

Write $z = 2 \text{ cis } 120^\circ$ in standard form.

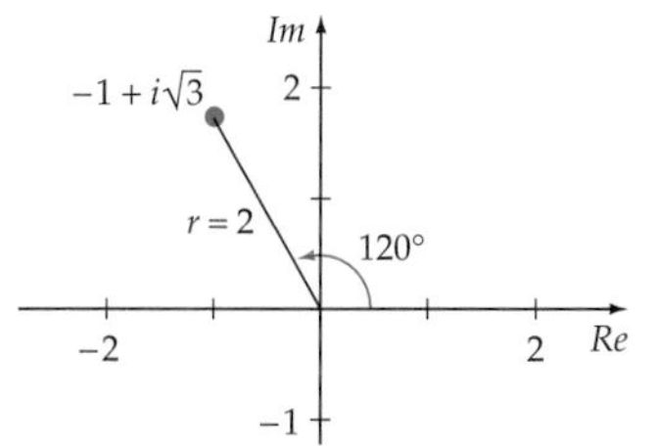

Figure 7.45

Solution

Write z in the form $r(\cos\theta + i\sin\theta)$ and then evaluate $\cos\theta$ and $\sin\theta$. See **Figure 7.45.**

$$z = 2\text{ cis } 120° = 2(\cos 120° + i\sin 120°) = 2\left(-\frac{1}{2} + \frac{\sqrt{3}}{2}i\right) = -1 + i\sqrt{3}$$

TRY EXERCISE 26, EXERCISE SET 7.4, PAGE 574

◆ THE PRODUCT AND THE QUOTIENT OF COMPLEX NUMBERS WRITTEN IN TRIGONOMETRIC FORM

Let z_1 and z_2 be two complex numbers written in trigonometric form. The product of z_1 and z_2 can be found by using trigonometric identities. If $z_1 = r_1(\cos\theta_1 + i\sin\theta_1)$ and $z_2 = r_2(\cos\theta_2 + i\sin\theta_2)$, then

$$\begin{aligned} z_1z_2 &= r_1(\cos\theta_1 + i\sin\theta_1)\cdot r_2(\cos\theta_2 + i\sin\theta_2)\\ &= r_1r_2(\cos\theta_1\cos\theta_2 + i\cos\theta_1\sin\theta_2 + i\sin\theta_1\cos\theta_2 + i^2\sin\theta_1\sin\theta_2)\\ &= r_1r_2[(\cos\theta_1\cos\theta_2 - \sin\theta_1\sin\theta_2) + i(\sin\theta_1\cos\theta_2 + \cos\theta_1\sin\theta_2)]\\ &= r_1r_2[\cos(\theta_1+\theta_2) + i\sin(\theta_1+\theta_2)] \end{aligned}$$

• **Identities for $\cos(\theta_1+\theta_2)$ and $\sin(\theta_1+\theta_2)$**

$$z_1z_2 = r_1r_2\text{ cis }(\theta_1+\theta_2)$$

• **The Product Property of Complex Numbers**

Thus the modulus for the product of two complex numbers in trigonometric form is the product of the moduli of the two numbers, and the argument of the product is the sum of the arguments of the two numbers.

EXAMPLE 3 Find the Product of Two Complex Numbers

Find the product of $z_1 = -1 + i\sqrt{3}$ and $z_2 = -\sqrt{3} + i$ by using the trigonometric form of the complex numbers. Write the answer in standard form.

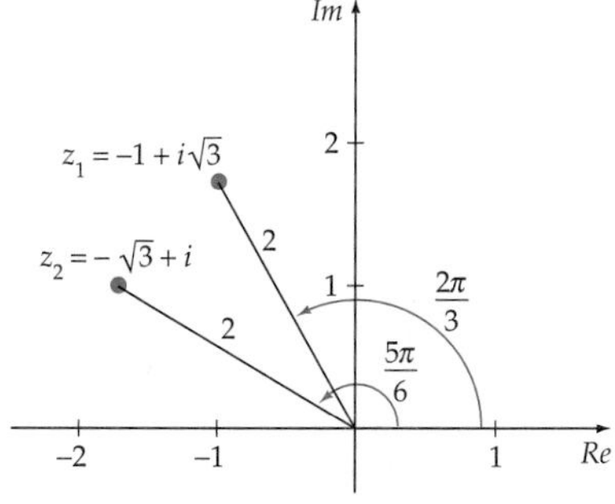

Figure 7.46

Solution

Write each complex number in trigonometric form. Then use the Product Property of complex numbers. See **Figure 7.46.**

$$z_1 = -1 + i\sqrt{3} = 2\text{ cis }\frac{2\pi}{3}$$ • $r_1 = 2,\ \theta_1 = \frac{2\pi}{3}$

$$z_2 = -\sqrt{3} + i = 2\text{ cis }\frac{5\pi}{6}$$ • $r_2 = 2,\ \theta_2 = \frac{5\pi}{6}$

Continued

$$\begin{aligned} z_1z_2 &= 2 \text{ cis } \frac{2\pi}{3} \cdot 2 \text{ cis } \frac{5\pi}{6} \\ &= 4 \text{ cis } \left(\frac{2\pi}{3} + \frac{5\pi}{6}\right) \\ &= 4 \text{ cis } \frac{3\pi}{2} \\ &= 4\left(\cos \frac{3\pi}{2} + i \sin \frac{3\pi}{2}\right) \\ &= 4(0 - i) \\ &= -4i \end{aligned}$$

TRY EXERCISE 52, EXERCISE SET 7.4, PAGE 574

Similarly, the quotient of z_1 and z_2 can be found by using trigonometric identities. If $z_1 = r_1(\cos \theta_1 + i \sin \theta_1)$ and $z_2 = r_2(\cos \theta_2 + i \sin \theta_2)$, then

$$\begin{aligned} \frac{z_1}{z_2} &= \frac{r_1(\cos \theta_1 + i \sin \theta_1)}{r_2(\cos \theta_2 + i \sin \theta_2)} \\ &= \frac{r_1(\cos \theta_1 + i \sin \theta_1)(\cos \theta_2 - i \sin \theta_2)}{r_2(\cos \theta_2 + i \sin \theta_2)(\cos \theta_2 - i \sin \theta_2)} \\ &= \frac{r_1(\cos \theta_1 \cos \theta_2 - i \cos \theta_1 \sin \theta_2 + i \sin \theta_1 \cos \theta_2 - i^2 \sin \theta_1 \sin \theta_2)}{r_2(\cos^2 \theta_2 - i^2 \sin^2 \theta_2)} \\ &= \frac{r_1[(\cos \theta_1 \cos \theta_2 + \sin \theta_1 \sin \theta_2) + i(\sin \theta_1 \cos \theta_2 - \cos \theta_1 \sin \theta_2)]}{r_2(\cos^2 \theta_2 + \sin^2 \theta_2)} \\ &= \frac{r_1}{r_2}[\cos (\theta_1 - \theta_2) + i \sin (\theta_1 - \theta_2)] \end{aligned}$$

• **Identities for $\cos (\theta_1 - \theta_2)$, $\sin (\theta_1 - \theta_2)$, and $\cos^2 \theta_2 + \sin^2 \theta_2$**

$$\frac{z_1}{z_2} = \frac{r_1}{r_2} \text{ cis } (\theta_1 - \theta_2)$$

• **The Quotient Property of Complex Numbers**

Thus the modulus for the quotient of two complex numbers in trigonometric form is the quotient of the moduli of the two numbers, and the argument of the quotient is the difference of the arguments of the two numbers.

EXAMPLE 4 Find the Quotient of Two Complex Numbers

Find the quotient of $z_1 = -1 + i$ and $z_2 = \sqrt{3} - i$ by using the trigonometric form of the complex numbers. Write the answer in standard form.

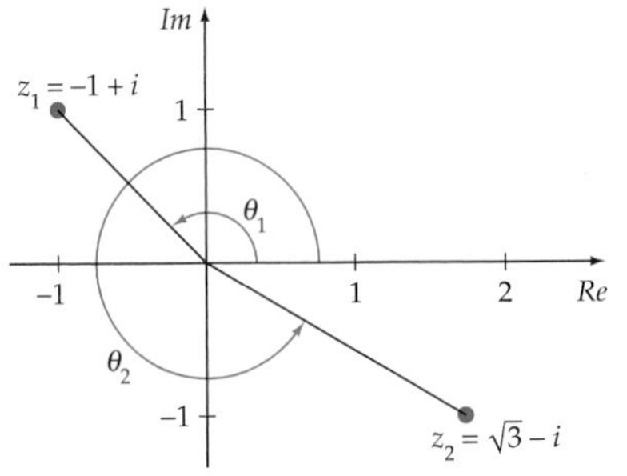

Figure 7.47

Solution

Write the numbers in trigonometric form. Then use the Quotient Property of complex numbers. See **Figure 7.47.**

$$z_1 = -1 + i = \sqrt{2} \text{ cis } 135^\circ \qquad \bullet\ r_1 = \sqrt{2},\ \theta_1 = 135^\circ$$

$$z_2 = \sqrt{3} - i = 2 \text{ cis } 330^\circ \qquad \bullet\ r_2 = 2,\ \theta_2 = 330^\circ$$

$$\begin{aligned}\frac{z_1}{z_2} = \frac{-1 + i}{\sqrt{3} - i} &= \frac{\sqrt{2} \text{ cis } 135^\circ}{2 \text{ cis } 330^\circ} \\ &= \frac{\sqrt{2}}{2} \text{ cis } (135^\circ - 330^\circ) \\ &= \frac{\sqrt{2}}{2} \text{ cis } (-195^\circ) \\ &= \frac{\sqrt{2}}{2}[\cos(-195^\circ) + i \sin(-195^\circ)] \\ &= \frac{\sqrt{2}}{2}(\cos 195^\circ - i \sin 195^\circ) \\ &\approx -0.6830 + 0.1830i\end{aligned}$$

TRY EXERCISE 56, EXERCISE SET 7.4, PAGE 575

Here is a summary of the product and quotient theorems.

Product and Quotient of Complex Numbers Written in Trigonometric Form

Let $z_1 = r_1(\cos \theta_1 + i \sin \theta_1)$ and $z_2 = r_2(\cos \theta_2 + i \sin \theta_2)$. Then

$$z_1 z_2 = r_1 r_2[\cos(\theta_1 + \theta_2) + i \sin(\theta_1 + \theta_2)]$$

$$z_1 z_2 = r_1 r_2 \text{ cis } (\theta_1 + \theta_2) \qquad \bullet \text{ Using cis notation}$$

and

$$\frac{z_1}{z_2} = \frac{r_1}{r_2}[\cos(\theta_1 - \theta_2) + i \sin(\theta_1 - \theta_2)]$$

$$\frac{z_1}{z_2} = \frac{r_1}{r_2} \text{ cis } (\theta_1 - \theta_2) \qquad \bullet \text{ Using cis notation}$$

TOPICS FOR DISCUSSION

1. Explain why the absolute value of $a + bi$ is equal to the absolute value of $-a + bi$.
2. Describe the graph of all complex numbers with an absolute value of 5.

3. Explain two different ways to find $4/i$.

4. The complex numbers z_1 and z_2 both have an absolute value of 1. What is the absolute value of the product z_1z_2? Explain.

5. Explain how to use the Product Property to prove that the product of a complex number and its conjugate is a real number.

EXERCISE SET 7.4

In Exercises 1 to 8, graph each complex number. Find the absolute value of each complex number.

1. $z = -2 - 2i$
2. $z = 4 - 4i$
3. $z = \sqrt{3} - i$
4. $z = 1 + i\sqrt{3}$
5. $z = -2i$
6. $z = -5$
7. $z = 3 - 5i$
8. $z = -5 - 4i$

In Exercises 9 to 16, write each complex number in trigonometric form.

9. $z = 1 - i$
10. $z = -4 - 4i$
11. $z = \sqrt{3} - i$
12. $z = 1 + i\sqrt{3}$
13. $z = 3i$
14. $z = -2i$
15. $z = -5$
16. $z = 3$

In Exercises 17 to 34, write each complex number in standard form.

17. $z = 2(\cos 45° + i \sin 45°)$
18. $z = 3(\cos 240° + i \sin 240°)$
19. $z = (\cos 315° + i \sin 315°)$
20. $z = 5(\cos 120° + i \sin 120°)$
21. $z = 6 \text{ cis } 135°$
22. $z = \text{cis } 315°$
23. $z = 8 \text{ cis } 0°$
24. $z = 5 \text{ cis } 90°$
25. $z = 2\left(\cos \frac{5\pi}{6} + i \sin \frac{5\pi}{6}\right)$
26. $z = 4\left(\cos \frac{5\pi}{3} + i \sin \frac{5\pi}{3}\right)$
27. $z = 3\left(\cos \frac{3\pi}{2} + i \sin \frac{3\pi}{2}\right)$
28. $z = 5(\cos \pi + i \sin \pi)$
29. $z = 8 \text{ cis } \frac{3\pi}{4}$
30. $z = 9 \text{ cis } \frac{4\pi}{3}$
31. $z = 9 \text{ cis } \frac{11\pi}{6}$
32. $z = \text{cis } \frac{3\pi}{2}$
33. $z = 2 \text{ cis } 2$
34. $z = 5 \text{ cis } 4$

In Exercises 35 to 42, multiply the complex numbers. Write the answer in trigonometric form.

35. $2 \text{ cis } 30° \cdot 3 \text{ cis } 225°$
36. $4 \text{ cis } 120° \cdot 6 \text{ cis } 315°$
37. $3(\cos 122° + i \sin 122°) \cdot 4(\cos 213° + i \sin 213°)$
38. $8(\cos 88° + i \sin 88°) \cdot 12(\cos 112° + i \sin 112°)$
39. $5\left(\cos \frac{2\pi}{3} + i \sin \frac{2\pi}{3}\right) \cdot 2\left(\cos \frac{2\pi}{5} + i \sin \frac{2\pi}{5}\right)$
40. $5 \text{ cis } \frac{11\pi}{12} \cdot 3 \text{ cis } \frac{4\pi}{3}$
41. $4 \text{ cis } 2.4 \cdot 6 \text{ cis } 4.1$
42. $7 \text{ cis } 0.88 \cdot 5 \text{ cis } 1.32$

In Exercises 43 to 50, divide the complex numbers. Write the answers in standard form.

43. $\dfrac{32 \text{ cis } 30°}{4 \text{ cis } 150°}$
44. $\dfrac{15 \text{ cis } 240°}{3 \text{ cis } 135°}$
45. $\dfrac{27(\cos 315° + i \sin 315°)}{9(\cos 225° + i \sin 225°)}$
46. $\dfrac{9(\cos 25° + i \sin 25°)}{3(\cos 175° + i \sin 175°)}$
47. $\dfrac{12 \text{ cis } \frac{2\pi}{3}}{4 \text{ cis } \frac{11\pi}{6}}$
48. $\dfrac{10 \text{ cis } \frac{\pi}{3}}{5 \text{ cis } \frac{\pi}{4}}$
49. $\dfrac{25(\cos 3.5 + i \sin 3.5)}{5(\cos 1.5 + i \sin 1.5)}$
50. $\dfrac{18(\cos 0.56 + i \sin 0.56)}{6(\cos 1.22 + i \sin 1.22)}$

In Exercises 51 to 58, perform the indicated operation in trigonometric form. Write the solution in standard form.

51. $(1 - i\sqrt{3})(1 + i)$
52. $(\sqrt{3} - i)(1 + i\sqrt{3})$

53. $(3 - 3i)(1 + i)$

54. $(2 + 2i)(\sqrt{3} - i)$

55. $\dfrac{1 + i\sqrt{3}}{1 - i\sqrt{3}}$

56. $\dfrac{1 + i}{1 - i}$

57. $\dfrac{\sqrt{2} - i\sqrt{2}}{1 + i}$

58. $\dfrac{1 + i\sqrt{3}}{4 - 4i}$

SUPPLEMENTAL EXERCISES

In Exercises 59 to 64, perform the indicated operation in trigonometric form. Write the solution in standard form.

59. $(\sqrt{3} - i)(2 + 2i)(2 - 2i\sqrt{3})$

60. $(1 - i)(1 + i\sqrt{3})(\sqrt{3} - i)$

61. $\dfrac{\sqrt{3} + i\sqrt{3}}{(1 - i\sqrt{3})(2 - 2i)}$

62. $\dfrac{(2 - 2i\sqrt{3})(1 - i\sqrt{3})}{4\sqrt{3} + 4i}$

63. $(1 - 3i)(2 + 3i)(4 + 5i)$

64. $\dfrac{(2 - 5i)(1 - 6i)}{3 + 4i}$

65. Use the trigonometric forms of the complex numbers z and $\bar{z}$ to find $z \cdot \bar{z}$. (Note $\bar{z}$ is the conjugate of z.)

66. Use the trigonometric forms of z and $\bar{z}$ to find $\bar{z}/\bar{z}$.

PROJECTS

1. **A GEOMETRICAL INTERPRETATION** Multiplying a real number by a number greater than 1 increases the magnitude of that number. For example, multiplying 2 by 3 triples the magnitude of 2. Explain the effect of multiplying a real number by i, the imaginary unit. Now multiply a complex number by i and note the effect. The use of a complex plane may be helpful in your explanation.

2. **HISTORICAL PERSPECTIVE** The complex number system was not widely accepted in the mathematical community until the 1600s. Other kinds of numbers, such as zero and negative numbers, also did not gain immediate acceptance when they were first introduced. Write a brief history of negative numbers. Be sure to include the contributions of early Chinese and Middle-Eastern mathematicians.

SECTION 7.5 DE MOIVRE'S THEOREM

- DE MOIVRE'S THEOREM
- DE MOIVRE'S THEOREM FOR FINDING ROOTS

◆ DE MOIVRE'S THEOREM

De Moivre's Theorem is a procedure for finding powers and roots of complex numbers when the complex numbers are expressed in trigonometric form. This theorem can be illustrated by repeated multiplication of a complex number.

Let $z = r \text{ cis } \theta$. Then z^2 can be written as

$$z \cdot z = r \text{ cis } \theta \cdot r \text{ cis } \theta$$

$$z^2 = r^2 \text{ cis } 2\theta$$

The product $z^2 \cdot z$ is

$$z^2 \cdot z = r^2 \operatorname{cis} 2\theta \cdot r \operatorname{cis} \theta$$
$$z^3 = r^3 \operatorname{cis} 3\theta$$

If we continue this process, the results suggest a formula for the nth power of a complex number that is known as De Moivre's Theorem.

De Moivre's Theorem

If $z = r \operatorname{cis} \theta$ and n is a positive integer, then

$$z^n = r^n \operatorname{cis} n\theta$$

MATH MATTERS

Moivre.

Abraham de Moivre (1667–1754) was a French mathematician who fled to England during the expulsion of the Huguenots in 1685.

De Moivre made important contributions in probability theory and analytic geometry. In 1718 he published *The Doctrine of Chance,* in which he developed the theory of annuities, mortality statistics, and the concept of statistical independence. In 1730 de Moivre stated the theorem to the right, which we now call De Moivre's Theorem. This theorem is significant because it provides a connection between trigonometry and mathematical analysis.

Although de Moivre was well respected in the mathematical community and was elected to the Royal Society of England, he never was able to secure a university teaching position. His income came mainly from tutoring, and he died in poverty.

De Moivre is often remembered for predicting the day of his own death. At one point in his life, he noticed that he was sleeping a few minutes longer each night. Thus he calculated that he would die when he needed 24 hours of sleep. As it turned out, his calculation was correct.

EXAMPLE 1 Find the Power of a Complex Number

Find $(2 \operatorname{cis} 30°)^5$. Write the answer in standard form.

Solution

By De Moivre's Theorem,

$$\begin{aligned}(2 \operatorname{cis} 30°)^5 &= 2^5 \operatorname{cis} (5 \cdot 30°) \\ &= 2^5[\cos (5 \cdot 30°) + i \sin (5 \cdot 30°)] \\ &= 32(\cos 150° + i \sin 150°) \\ &= 32\left(-\frac{\sqrt{3}}{2} + \frac{1}{2}i\right) = -16\sqrt{3} + 16i\end{aligned}$$

TRY EXERCISE 6, EXERCISE SET 7.5, PAGE 579

EXAMPLE 2 Use De Moivre's Theorem

Find $(1 + i)^8$ using De Moivre's Theorem. Write the answer in standard form.

Solution

Convert $1 + i$ to trigonometric form and then use De Moivre's Theorem.

$$\begin{aligned}(1 + i)^8 &= (\sqrt{2} \operatorname{cis} 45°)^8 = (\sqrt{2})^8 \operatorname{cis} 8(45°) = 16 \operatorname{cis} 360° \\ &= 16(\cos 360° + i \sin 360°) = 16(1 + 0i) = 16\end{aligned}$$

TRY EXERCISE 14, EXERCISE SET 7.5, PAGE 579

◆ DE MOIVRE'S THEOREM FOR FINDING ROOTS

De Moivre's Theorem can be used to find the nth roots of any number.

De Moivre's Theorem for Finding Roots

If $z = r \operatorname{cis} \theta$ is a complex number, then there are n distinct nth roots of z given by

$$w_k = r^{1/n} \operatorname{cis} \frac{\theta + 360^\circ k}{n} \quad \text{for } k = 0, 1, 2, \ldots, n-1, \text{ and } n \geq 1.$$

EXAMPLE 3 Find Cube Roots by De Moivre's Theorem

Find the three cube roots of 27.

Solution

Write 27 in trigonometric form: $27 = 27 \operatorname{cis} 0^\circ$. Then, from De Moivre's Theorem, the cube roots of 27 are

$$w_k = 27^{1/3} \operatorname{cis} \frac{0^\circ + 360^\circ k}{3} \quad \text{for } k = 0, 1, 2$$

Substitute for k to find the cube roots of 27.

$$w_0 = 27^{1/3} \operatorname{cis} 0^\circ \qquad \bullet\ k = 0;\ \frac{0^\circ + 360^\circ(0)}{3} = 0^\circ$$

$$= 3(\cos 0^\circ + i \sin 0^\circ)$$

$$= 3$$

$$w_1 = 27^{1/3} \operatorname{cis} 120^\circ \qquad \bullet\ k = 1;\ \frac{0^\circ + 360^\circ(1)}{3} = 120^\circ$$

$$= 3(\cos 120^\circ + i \sin 120^\circ)$$

$$= -\frac{3}{2} + \frac{3\sqrt{3}}{2}i$$

$$w_2 = 27^{1/3} \operatorname{cis} 240^\circ \qquad \bullet\ k = 2;\ \frac{0^\circ + 360^\circ(2)}{3} = 240^\circ$$

$$= 3(\cos 240^\circ + i \sin 240^\circ)$$

$$= -\frac{3}{2} - \frac{3\sqrt{3}}{2}i$$

For $k = 3$, $\frac{0^\circ + 1080^\circ}{3} = 360^\circ$. The angles start repeating; thus there are only three cube roots of 27. The three cube roots are graphed in **Figure 7.48.**

Visualize the Solution

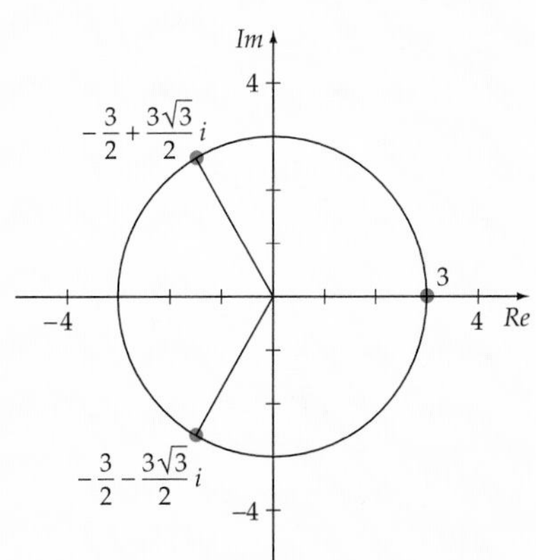

Figure 7.48

Note that the arguments of the three cube roots of 27 are 0°, 120°, and 240° and that $|w_0| = |w_1| = |w_2| = 3$. In geometric terms, this means that the three cube roots of 27 are equally spaced on a circle centered at the origin with a radius of 3.

TRY EXERCISE 26, EXERCISE SET 7.5, PAGE 580

EXAMPLE 4 Find the Fifth Roots of a Complex Number

Find the fifth roots of $z = 1 + i\sqrt{3}$.

Solution

Write z in trigonometric form: $z = r \text{ cis } \theta$.

$$r = \sqrt{1^2 + (\sqrt{3})^2} = 2$$

$$z = 2 \text{ cis } 60^\circ \qquad \bullet\ \theta = \tan^{-1}\frac{\sqrt{3}}{1} = 60^\circ$$

From De Moivre's Theorem, the modulus of each root is $\sqrt[5]{2}$, and the arguments are determined by $\frac{60^\circ + 360^\circ k}{5}$, $k = 0, 1, 2, 3, 4$.

$$w_k = \sqrt[5]{2} \text{ cis } \frac{60^\circ + 360^\circ k}{5} \qquad \bullet\ k = 0, 1, 2, 3, 4$$

Substitute for k to find the five fifth roots of z.

$$w_0 = \sqrt[5]{2} \text{ cis } 12^\circ \qquad \bullet\ k = 0; \frac{60^\circ + 360^\circ(0)}{5} = 12^\circ$$

$$w_1 = \sqrt[5]{2} \text{ cis } 84^\circ \qquad \bullet\ k = 1; \frac{60^\circ + 360^\circ(1)}{5} = 84^\circ$$

$$w_2 = \sqrt[5]{2} \text{ cis } 156^\circ \qquad \bullet\ k = 2; \frac{60^\circ + 360^\circ(2)}{5} = 156^\circ$$

$$w_3 = \sqrt[5]{2} \text{ cis } 228^\circ \qquad \bullet\ k = 3; \frac{60^\circ + 360^\circ(3)}{5} = 228^\circ$$

$$w_4 = \sqrt[5]{2} \text{ cis } 300^\circ \qquad \bullet\ k = 4; \frac{60^\circ + 360^\circ(4)}{5} = 300^\circ$$

Visualize the Solution

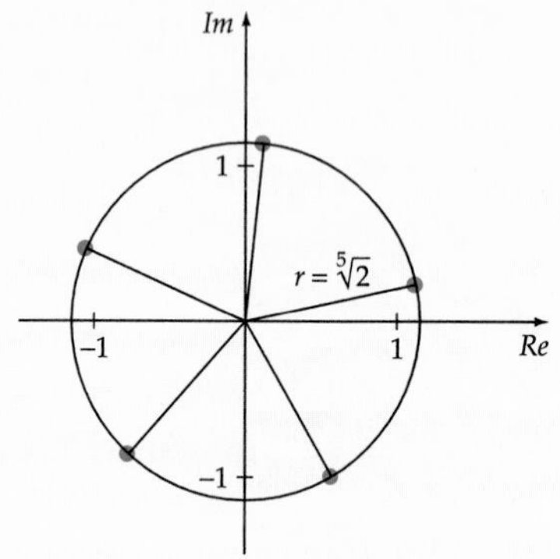

Figure 7.49

The five fifth roots of $1 + i\sqrt{3}$ are graphed in Figure 7.49. Note that the roots are equally spaced on a circle with center (0, 0) and a radius of $\sqrt[5]{2} \approx 1.15$.

TRY EXERCISE 28, EXERCISE SET 7.5, PAGE 580

Keep the following properties in mind as you compute the n distinct nth roots of the complex number z.

Properties of the nth Roots of z

Geometric Property
All nth roots of z are equally spaced on a circle centered at the origin with radius $|z|^{1/n}$.

Absolute Value Properties

1. If $|z| = 1$, then each nth root of z has an absolute value of 1.
2. If $|z| > 1$, then each nth root of z has an absolute value of $|z|^{1/n}$, and $|z|^{1/n}$ is greater than 1 but less than $|z|$.
3. If $|z| < 1$, then each nth root of z has an absolute value of $|z|^{1/n}$, and $|z|^{1/n}$ is less than 1 but greater than $|z|$.

take note

A web applet is available to show the nth roots of a complex number. This applet, Nth Roots, can be found on our web site at http://college.hmco.com.

Argument Property

Given that the argument of z is θ, then the argument of w_0 is $\frac{\theta}{n}$ and the arguments of the remaining nth roots can be determined by adding multiples of $\frac{360^\circ}{n}$ (or $\frac{2\pi}{n}$ if you are using radians) to $\frac{\theta}{n}$.

TOPICS FOR DISCUSSION

1. How many solutions are there for $z = (a + bi)^8$? How many solutions are there for $z^8 = a + bi$? Explain.
2. A tutor claims that $z^3 = 8$ has solutions of
$$2, \frac{-1 + i\sqrt{3}}{2} \quad \text{and} \quad \frac{-1 - i\sqrt{3}}{2}$$
Cube each of these to see whether the tutor is correct.
3. To solve $z^4 = 16$, a student first observes that $z = 2$ is one solution. The other solutions are equally spaced on a circle with center (0, 0) and radius 2, so the student reasons that $z = 2i$, $z = -2$, and $z = -2i$ are the other three solutions. Do you agree? Explain.
4. If $|z| = 1$, then the n solutions of $w^n = z$ all have an absolute value of 1. Explain.
5. If z is a solution of $z^2 = c + di$, then the conjugate of z is also a solution. Do you agree? Explain.

EXERCISE SET 7.5

In Exercises 1 to 14, find the indicated power. Write the answers in standard form.

1. $[2(\cos 30^\circ + i \sin 30^\circ)]^8$
2. $(\cos 240^\circ + i \sin 240^\circ)^{12}$
3. $[2(\cos 240^\circ + i \sin 240^\circ)]^5$
4. $[2(\cos 45^\circ + i \sin 45^\circ)]^{10}$
5. $(2 \text{ cis } 225^\circ)^5$
6. $(2 \text{ cis } 330^\circ)^4$
7. $\left(2 \text{ cis } \frac{2\pi}{3}\right)^6$
8. $\left(4 \text{ cis } \frac{5\pi}{6}\right)^3$
9. $(1 - i)^{10}$
10. $(1 + i\sqrt{3})^8$
11. $(2 + 2i)^7$
12. $(2\sqrt{3} - 2i)^5$
13. $\left(\frac{\sqrt{2}}{2} + i\frac{\sqrt{2}}{2}\right)^6$
14. $\left(-\frac{\sqrt{2}}{2} + i\frac{\sqrt{2}}{2}\right)^{12}$

In Exercises 15 to 28, find all of the indicated roots. Write all answers in standard form.

15. The two square roots of 9
16. The two square roots of 16
17. The six sixth roots of 64
18. The five fifth roots of 32
19. The five fifth roots of -1
20. The four fourth roots of -16
21. The three cube roots of 1
22. The three cube roots of i
23. The four fourth roots of $1 + i$
24. The five fifth roots of $-1 + i$

25. The three cube roots of $2 - 2i\sqrt{3}$

26. The three cube roots of $-2 + 2i\sqrt{3}$

27. The two square roots of $-16 + 16i\sqrt{3}$

28. The two square roots of $-1 + i\sqrt{3}$

In Exercises 29 to 40, find all roots of the equation. Write the answers in trigonometric form.

29. $x^3 + 8 = 0$

30. $x^5 - 32 = 0$

31. $x^4 + i = 0$

32. $x^3 - 2i = 0$

33. $x^3 - 27 = 0$

34. $x^5 + 32i = 0$

35. $x^4 + 81 = 0$

36. $x^3 - 64i = 0$

37. $x^4 - (1 - i\sqrt{3}) = 0$

38. $x^3 + (2\sqrt{3} - 2i) = 0$

39. $x^3 + (1 + i\sqrt{3}) = 0$

40. $x^6 - (4 - 4i) = 0$

SUPPLEMENTAL EXERCISES

41. Show that the conjugate of $z = r(\cos\theta + i\sin\theta)$ is equal to $\bar{z} = r(\cos\theta - i\sin\theta)$.

42. Show that if $z = r(\cos\theta + i\sin\theta)$, then

$$z^{-1} = r^{-1}(\cos\theta - i\sin\theta)$$

43. Show that if $z = r(\cos\theta + i\sin\theta)$, then

$$z^{-2} = r^{-2}(\cos 2\theta - i\sin 2\theta)$$

Note that Exercises 42 and 43 suggest that if $z = r(\cos\theta + i\sin\theta)$, then the general expression is

$$z^{-n} = r^{-n}(\cos n\theta - i\sin n\theta)$$

44. Use the results of Exercise 43 to find z^{-4} for $z = 1 - i\sqrt{3}$.

PROJECTS

1. **VERIFY IDENTITIES** Raise $(\cos\theta + i\sin\theta)$ to the second power by using De Moivre's Theorem. Now square $(\cos\theta + i\sin\theta)$ as a binomial. Equate the real and imaginary parts of the two complex numbers and show that

   a. $\cos 2\theta = \cos^2\theta - \sin^2\theta$ b. $\sin 2\theta = 2\sin\theta\cos\theta$

2. **DISCOVER IDENTITIES** Raise $(\cos\theta + i\sin\theta)$ to the fourth power by using De Moivre's Theorem. Now find the fourth power of the binomial $(\cos\theta + i\sin\theta)$ by multiplying. Equate the real and imaginary parts of the two complex numbers.

   a. What identity have you discovered for $\cos 4\theta$?

   b. What identity have you discovered for $\sin 4\theta$?

EXPLORING CONCEPTS WITH TECHNOLOGY

Optimal Branching of Arteries

The physiologist Jean Louis Poiseuille (1799–1869) developed several laws concerning the flow of blood. One of his laws states that the resistance R of a blood vessel of length l and radius r is given by

$$R = k\frac{l}{r^4} \tag{1}$$

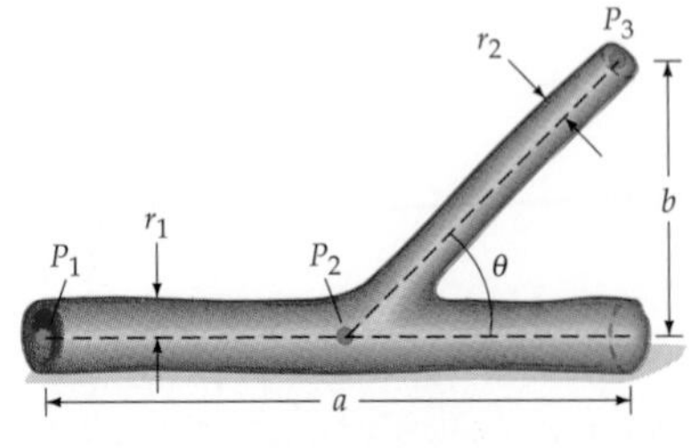

Figure 7.50

The number k is a variation constant that depends on the viscosity of the blood. Figure 7.50 shows a large artery with radius r_1 and a smaller artery with radius r_2. The branching angle between the arteries is θ. Make use of Poiseuille's Law, Equation (1), to show that the resistance R of the blood along the path $P_1P_2P_3$ is

$$R = k\left(\frac{a - b\cot\theta}{r_1^4} + \frac{b\csc\theta}{r_2^4}\right) \tag{2}$$

Use a graphing utility to graph R with $k = 0.0563$, $a = 8$ centimeters, $b = 4$ centimeters, $r_1 = 0.4$ centimeter, and $r_2 = (3/4)r_1 = 0.3$ centimeter.

Then estimate (to the nearest degree) the angle θ that minimizes R. By using calculus, it can be demonstrated that R is minimized when

$$\cos\theta = \left(\frac{r_2}{r_1}\right)^4 \qquad (3)$$

This equation is remarkable because it is much simpler than Equation (2) and because it does not involve the distance a or b. Solve Equation (3) for θ, with $r_2 = (3/4)r_1$. How does this value of θ compare with the value of θ you obtained by graphing?

CHAPTER 7 SUMMARY

7.1 The Law of Sines

- The Law of Sines is used to solve triangles when two angles and a side are given (AAS) or when two sides and an angle opposite one of them are given (SSA).

$$\frac{a}{\sin A} = \frac{b}{\sin B} = \frac{c}{\sin C}$$

7.2 The Law of Cosines and Area

- The Law of Cosines, $a^2 = b^2 + c^2 - 2bc\cos A$, is used to solve general triangles when two sides and the included angle (SAS) or three sides (SSS) of the triangle are given.
- Area K of a triangle ABC is

$$K = \frac{1}{2}bc\sin A = \frac{b^2 \sin C \sin A}{2\sin B}$$

- Area for a triangle in which three sides are given (Heron's formula):

$$K = \sqrt{s(s-a)(s-b)(s-c)}, \quad \text{where } s = \frac{1}{2}(a+b+c).$$

7.3 Vectors

- A vector is a quantity with magnitude and direction. Two vectors are equivalent if they have the same magnitude and the same direction. The resultant of two or more vectors is the sum of the vectors.
- Vectors can be added by the parallelogram method, the triangle method, or addition of the x- and y-components.
- If $\mathbf{v} = \langle a, b\rangle$ and k is a real number, then $k\mathbf{v} = \langle ka, kb\rangle$.
- The dot product of $\mathbf{v} = \langle a, b\rangle$ and $\mathbf{w} = \langle c, d\rangle$ is given by

$$\mathbf{v}\cdot\mathbf{w} = ac + bd$$

- If $\mathbf{v}$ and $\mathbf{w}$ are two nonzero vectors and α is the smallest positive angle between $\mathbf{v}$ and $\mathbf{w}$, then $\cos\alpha = \dfrac{\mathbf{v}\cdot\mathbf{w}}{\|\mathbf{v}\|\,\|\mathbf{w}\|}$.

7.4 Trigonometric Form of Complex Numbers

- The complex number $z = a + bi$ can be written in trigonometric form as

$$z = r(\cos\theta + i\sin\theta) = r\operatorname{cis}\theta$$

where $a = r\cos\theta$, $b = r\sin\theta$, $r = \sqrt{a^2+b^2}$, and $\tan\theta = \dfrac{b}{a}$.

- If $z_1 = r_1(\cos\theta_1 + i\sin\theta_1)$ and $z_2 = r_2(\cos\theta_2 + i\sin\theta_2)$, then

$$z_1z_2 = r_1r_2\operatorname{cis}(\theta_1+\theta_2)$$

and

$$\frac{z_1}{z_2} = \frac{r_1}{r_2}\operatorname{cis}(\theta_1-\theta_2)$$

7.5 De Moivre's Theorem

- **De Moivre's Theorem**
If $z = r\operatorname{cis}\theta$ and n is a positive integer, then

$$z^n = r^n\operatorname{cis} n\theta$$

- If $z = r\operatorname{cis}\theta$, then the n distinct roots of z are given by

$$w_k = r^{1/n}\operatorname{cis}\frac{\theta + 360^\circ k}{n} \quad \text{for } k = 0, 1, 2, \ldots, n-1$$

CHAPTER 7 TRUE/FALSE EXERCISES

For Exercises 1 to 16, answer true or false. If the statement is false, give an example to show that the statement is false.

1. The Law of Cosines can be used to solve any triangle given two sides and an angle.
2. The Law of Sines can be used to solve any triangle given two angles and any side.
3. In any triangle, the longest side is opposite the largest angle.

4. If two vectors have the same magnitude, then they are equal.

5. It is possible for the sum of two nonzero vectors to equal zero.

6. The expression $a^2 = b^2 + c^2 + 2bc \cos D$ is true for triangle ABC, in which angle D is the supplement of angle A.

7. The measure of angle α formed by two vectors is greater than or equal to $0°$ and less than or equal to $180°$.

8. If A, B, and C are the angles of a triangle, then

$$\sin (A + B + C) = 0$$

9. Real numbers are complex numbers.

10. Let $\mathbf{v} = a\mathbf{i} + b\mathbf{j}$. Then $\mathbf{v} \cdot \mathbf{v} = a^2\mathbf{i} + b^2\mathbf{j}$.

11. If $\mathbf{v}$ and $\mathbf{w}$ are vectors with $\mathbf{v} \cdot \mathbf{w} = 0$, then $\mathbf{v} = \mathbf{0}$ or $\mathbf{w} = \mathbf{0}$.

12. The n roots of a complex number can be graphed on a circle and are equally spaced on the circle.

13. Let $z = r(\cos \theta + i \sin \theta)$. Then $z^2 = r^2(\cos^2 \theta + i \sin^2 \theta)$.

14. $|a + bi| = \sqrt{a^2 + b^2}$

15. $i = \cos \pi + i \sin \pi$

16. $z = \cos 45° + i \sin 45°$ is a square root of i.

CHAPTER 7 REVIEW EXERCISES

In Exercises 1 to 10, solve each triangle.

1. $A = 37°, b = 14, C = 90°$

2. $B = 77.4°, c = 11.8, C = 90°$

3. $a = 12, b = 15, c = 20$

4. $a = 24, b = 32, c = 28$

5. $a = 18, b = 22, C = 35°$

6. $b = 102, c = 150, A = 82°$

7. $A = 105°, a = 8, c = 10$

8. $C = 55°, c = 80, b = 110$

9. $A = 55°, B = 80°, c = 25$

10. $B = 25°, C = 40°, c = 40$

In Exercises 11 to 18, find the area of each triangle. Round each area accurate to two significant digits.

11. $a = 24, b = 30, c = 36$

12. $a = 9.0, b = 7.0, c = 12$

13. $a = 60, b = 44, C = 44°$

14. $b = 8.0, c = 12, A = 75°$

15. $b = 50, c = 75, C = 15°$

16. $b = 18, a = 25, A = 68°$

17. $A = 110°, a = 32, b = 15$

18. $C = 45°, c = 22, b = 18$

In Exercises 19 and 20, find the components of each vector with the given initial and terminal points. Write an equivalent vector in terms of its components.

19. $P_1(-2, 4); P_2(3, 7)$

20. $P_1(-4, 0); P_2(-3, 6)$

In Exercises 21 to 24, find the magnitude and the direction angle of each vector.

21. $\mathbf{v} = \langle -4, 2 \rangle$

22. $\mathbf{v} = \langle 6, -3 \rangle$

23. $\mathbf{u} = -2\mathbf{i} + 3\mathbf{j}$

24. $\mathbf{u} = -4\mathbf{i} - 7\mathbf{j}$

In Exercises 25 to 28, find a unit vector in the direction of the given vector.

25. $\mathbf{w} = \langle -8, 5 \rangle$

26. $\mathbf{w} = \langle 7, -12 \rangle$

27. $\mathbf{v} = 5\mathbf{i} + \mathbf{j}$

28. $\mathbf{v} = 3\mathbf{i} - 5\mathbf{j}$

In Exercises 29 and 30, perform the indicated operation where $\mathbf{u} = \langle 3, 2 \rangle$, and $\mathbf{v} = \langle -4, -1 \rangle$.

29. $\mathbf{v} - \mathbf{u}$

30. $2\mathbf{u} - 3\mathbf{v}$

In Exercises 31 and 32, perform the indicated operation where $\mathbf{u} = 10\mathbf{i} + 6\mathbf{j}$, and $\mathbf{v} = 8\mathbf{i} - 5\mathbf{j}$.

31. $-\mathbf{u} + \frac{1}{2}\mathbf{v}$

32. $\frac{2}{3}\mathbf{v} - \frac{3}{4}\mathbf{u}$

33. **GROUND SPEED OF A PLANE** A plane is flying at an airspeed of 400 mph at a heading of 204°. A wind of 45 mph is blowing from the east. Find the ground speed of the plane.

34. **ANGLE OF A RAMP** A 40-pound force keeps a 320-pound object from sliding down an inclined ramp. Find the angle of the ramp.

In Exercises 35 to 38, find the dot product of the vectors.

35. $\mathbf{u} = \langle 3, 7 \rangle; \mathbf{v} = \langle -1, 3 \rangle$

36. $\mathbf{v} = \langle -8, 5 \rangle; \mathbf{u} = \langle 2, -1 \rangle$

37. $\mathbf{v} = -4\mathbf{i} - \mathbf{j}; \mathbf{u} = 2\mathbf{i} + \mathbf{j}$

38. $\mathbf{u} = -3\mathbf{i} + 7\mathbf{j}; \mathbf{v} = -2\mathbf{i} + 2\mathbf{j}$

In Exercises 39 to 42, find the angle between the vectors.

39. $\mathbf{u} = \langle 7, -4 \rangle; \mathbf{v} = \langle 2, 3 \rangle$

40. $\mathbf{v} = \langle -5, 2 \rangle; \mathbf{u} = \langle 2, -4 \rangle$

41. $\mathbf{v} = 6\mathbf{i} - 11\mathbf{j}$; $\mathbf{u} = 2\mathbf{i} + 4\mathbf{j}$

42. $\mathbf{u} = \mathbf{i} - 5\mathbf{j}$; $\mathbf{v} = \mathbf{i} + 5\mathbf{j}$

In Exercises 43 and 44, find $\text{proj}_{\mathbf{w}}\mathbf{v}$.

43. $\mathbf{v} = \langle -2, 5\rangle$; $\mathbf{w} = \langle 5, 4\rangle$

44. $\mathbf{v} = 4\mathbf{i} - 7\mathbf{j}$; $\mathbf{w} = -2\mathbf{i} - 5\mathbf{j}$

45. WORK A 120-pound box is dragged 14 feet along a level floor. Find the work done if a force of 60 pounds at an angle of 38° is used.

In Exercises 46 and 47, find the modulus and the argument, and graph the complex number.

46. $z = 2 - 3i$

47. $z = -5 + i\sqrt{3}$

In Exercises 48 and 49, write the complex number in trigonometric form.

48. $z = 2 - 2i$

49. $z = -\sqrt{3} + 3i$

In Exercises 50 and 51, write the complex number in standard form.

50. $z = 5 \text{ cis } 315°$

51. $z = 6 \text{ cis } \dfrac{4\pi}{3}$

In Exercises 52 to 55, multiply the complex numbers. Write the answers in standard form.

52. $(5 \text{ cis } 162°)(2 \text{ cis } 63°)$

53. $(3 \text{ cis } 12°)(4 \text{ cis } 126°)$

54. $\left(7 \text{ cis } \dfrac{2\pi}{3}\right)\left(4 \text{ cis } \dfrac{\pi}{4}\right)$

55. $(3 \text{ cis } 1.8)(5 \text{ cis } 2.5)$

In Exercises 56 to 59, divide the complex numbers. Write the answers in trigonometric form.

56. $\dfrac{6 \text{ cis } 50°}{2 \text{ cis } 150°}$

57. $\dfrac{30 \text{ cis } 165°}{10 \text{ cis } 55°}$

58. $\dfrac{40 \text{ cis } 66°}{8 \text{ cis } 125°}$

59. $\dfrac{\sqrt{3} - i}{1 + i}$

In Exercises 60 to 63, find the indicated power. Write the answers in standard form.

60. $(3 \text{ cis } 45°)^5$

61. $\left(\text{cis } \dfrac{11\pi}{6}\right)^8$

62. $\left(1 - i\sqrt{3}\right)^7$

63. $(-2 - 2i)^{10}$

In Exercises 64 to 67, find all of the indicated roots. Write the answers in trigonometric form.

64. cube roots of $27i$

65. fourth roots of $8i$

66. fourth roots of $256 \text{ cis } 120°$

67. fifth roots of $-1 - i$

CHAPTER 7 TEST

1. Solve triangle ABC if $A = 70°$, $C = 16°$, and $c = 14$.

2. Find B in triangle ABC if $A = 140°$, $b = 13$, and $a = 45$.

3. In triangle ABC, $C = 42°$, $a = 20$, and $b = 12$. Find side c.

4. In triangle ABC, $a = 32$, $b = 24$, and $c = 18$. Find angle B.

In Exercises 5 to 7, round your answers to two significant digits.

5. Given angle $C = 110°$, side $a = 7.0$, and side $b = 12$, find the area of triangle ABC.

6. Given angle $B = 42°$, angle $C = 75°$, and side $b = 12$, find the area of triangle ABC.

7. Given side $a = 17$, side $b = 55$, and side $c = 42$, find the area of triangle ABC.

8. A vector has a magnitude of 12 and direction 220°. Write an equivalent vector in the form $\mathbf{v} = a_1\mathbf{i} + a_2\mathbf{j}$.

9. Find $3\mathbf{u} - 5\mathbf{v}$ given the vectors $\mathbf{u} = 2\mathbf{i} - 3\mathbf{j}$ and $\mathbf{v} = 5\mathbf{i} + 4\mathbf{j}$.

10. Find the dot product of $\mathbf{u} = -2\mathbf{i} + 3\mathbf{j}$ and $\mathbf{v} = 5\mathbf{i} + 3\mathbf{j}$.

11. Find the smallest positive angle to the nearest degree between the vectors $\mathbf{u} = \langle 3, 5\rangle$ and $\mathbf{v} = \langle -6, 2\rangle$.

12. Write $z = -3\sqrt{2} + 3i$ in trigonometric form.

13. Write $z = 5 \text{ cis } 315°$ in standard form.

14. Multiply $3 \text{ cis } 80° \cdot 8 \text{ cis } 210°$. Write the answer in standard form.

15. Divide $\dfrac{25 \text{ cis } 115°}{10 \text{ cis } 210°}$. Write the answer in trigonometric form.

16. Use De Moivre's theorem to find $\left(\sqrt{2} - i\right)^5$. Write the answer in standard form.

17. Find the three cube roots of $27i$. Write your answer in standard form.

18. One ship leaves a port at 1:00 P.M. traveling at 12 mph at a heading of 65°. At 2:00 P.M. another ship leaves the port traveling at 18 mph at a heading of 142°. Find the distance between the ships at 3:00 P.M.

19. Two fire lookouts are located 12 miles apart. Lookout *A* is at a bearing of N32°W of lookout *B*. A fire was sighted at a bearing of S82°E from *A* and N72°E from *B*. Find the distance of the fire from lookout *B*.

20. A triangular commercial piece of real estate is priced at $8.50 per square foot. Find, to the nearest $1000, the cost of the lot, which measures 112 feet by 165 feet by 140 feet.

C H A P T E R

8

Topics in Analytic Geometry

Radio Telescopes, Conic Sections, and the Search for Extraterrestrial Intelligence

◆ The moment of contact with an alien life form. *Contact*, Warner Bros., 1997.

◆ Radio Telescopes in the Very Large Array.

The movie *Contact* was based on the novel by astronomer Carl Sagan. In the movie, Jodie Foster plays an astronomer who is searching for extraterrestrial intelligence. One scene from the movie takes place at the Very Large Array (VLA) in New Mexico. The VLA consists of 27 large radio telescopes that are electronically connected. Each of the telescopes has a dish measuring 81 feet across. A reflective property of each dish is such that electronic signals from space are collected by the surface of the dish and reflected to a receiver located at a point called the focus. The signals collected by each telescope are then sent to a computer where they are combined and analyzed.

The shape and the reflective properties of parabolic dishes, such as the radio telescopes in the VLA, are two of the topics covered in this chapter.

SECTION

8.1 PARABOLAS

- PARABOLAS WITH VERTEX AT (0, 0)
- PARABOLAS WITH VERTEX AT (h, k)
- APPLICATIONS

The graph of a parabola, circle, ellipse, or hyperbola can be formed by the intersection of a plane and a cone. Hence these figures are referred to as conic sections. See **Figure 8.1**.

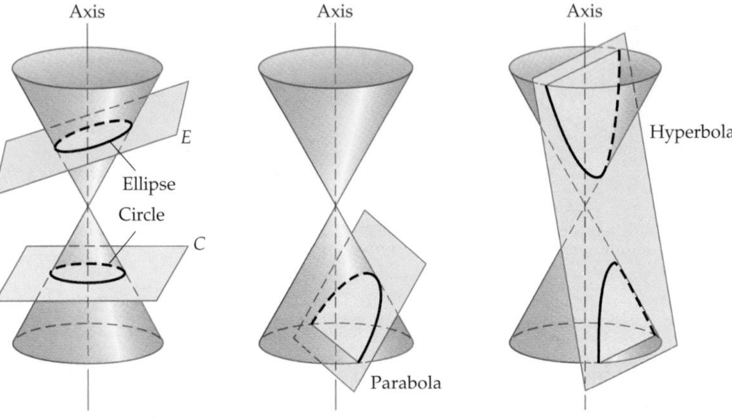

Figure 8.1
Cones intersected by planes

MATH MATTERS

Appollonius (262–200 B.C.) wrote an eight-volume treatise entitled *On Conic Sections* in which he derived the formulas for all the conic sections. He was the first to use the words *parabola*, *ellipse*, and *hyperbola*.

take note

If the intersection of a plane and a cone is a point, a line, or two intersecting lines, then the intersection is called a *degenerate conic section.*

take note

A web applet is available to experiment with parabolas by manipulating the focus, directrix, and vertex. This applet, Parabola with Horizontal Directrix, can be found on our web site at
http://college.hmco.com

A plane perpendicular to the axis of the cone intersects the cone in a circle (plane C). The plane E, tilted so that it is not perpendicular to the axis, intersects the cone in an ellipse. When the plane is parallel to a line on the surface of the cone, the plane intersects the cone in a parabola. When the plane intersects both portions of the cone, a hyperbola is formed.

PARABOLAS WITH VERTEX AT (0, 0)

Besides the geometric description of a conic section just given, a conic section can be defined as a set of points. This method uses some specified conditions about the curve to determine which points in a coordinate system are points of the graph. For example, a parabola can be defined by the following set of points.

Definition of a Parabola

A **parabola** is the set of points in the plane that are equidistant from a fixed line (the **directrix**) and a fixed point (the **focus**) not on the directrix.

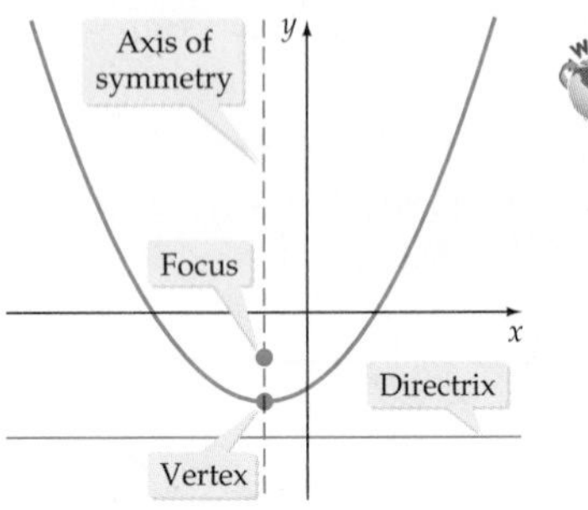

Figure 8.2

The line that passes through the focus and is perpendicular to the directrix is called the **axis of symmetry** of the parabola. The midpoint of the line segment between the focus and directrix on the axis of symmetry is the **vertex** of the parabola, as shown in **Figure 8.2**.

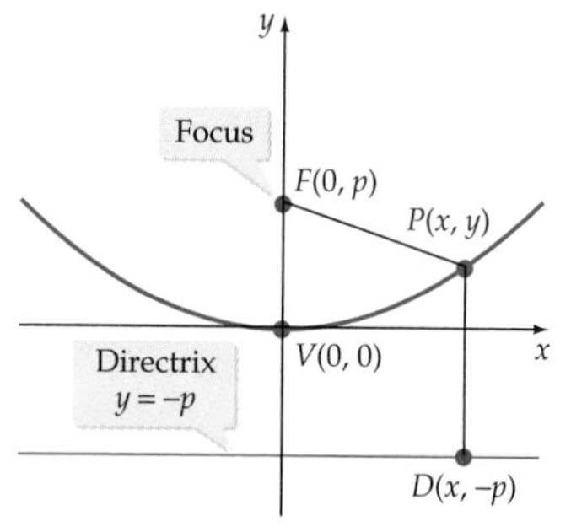

Figure 8.3

Using this definition of a parabola, we can determine an equation of a parabola. Suppose that the coordinates of the vertex of a parabola are $V(0, 0)$ and the axis of symmetry is the y-axis. The equation of the directrix is $y = -p, p > 0$. The focus lies on the axis of symmetry and is the same distance from the vertex as the vertex is from the directrix. Thus the coordinates of the focus are $F(0, p)$, as shown in **Figure 8.3.**

Let $P(x, y)$ be any point P on the parabola. Then, using the distance formula and the fact that the distance between any point P on the parabola and the focus is equal to the distance from the point P to the directrix, we can write the equation

$$d(P, F) = d(P, D)$$

By the distance formula,

$$\sqrt{(x - 0)^2 + (y - p)^2} = y + p$$

Now, squaring each side and simplifying, we get

$$\left(\sqrt{(x - 0)^2 + (y - p)^2}\right)^2 = (y + p)^2$$
$$x^2 + y^2 - 2py + p^2 = y^2 + 2py + p^2$$
$$x^2 = 4py$$

This is an equation of a parabola with vertex at the origin and the y-axis as its axis of symmetry. The equation of a parabola with vertex at the origin and the x-axis as its axis of symmetry is derived in a similar manner.

take note

The tests for y-axis and x-axis symmetry can be used to verify these statements and provide connections to earlier topics on symmetry.

Standard Forms of the Equation of a Parabola with Vertex at the Origin

Axis of Symmetry is the y-axis
The standard form of the equation of a parabola with vertex (0, 0) and the y-axis as its axis of symmetry is $x^2 = 4py$. The focus is $(0, p)$, and the equation of the directrix is $y = -p$.

Axis of Symmetry is the x-axis
The standard form of the equation of a parabola with vertex (0, 0) and the x-axis as its axis of symmetry is $y^2 = 4px$. The focus is $(p, 0)$ and the equation of the directrix is $x = -p$.

In the equation $x^2 = 4py$, $x^2 \geq 0$. Therefore, $4py \geq 0$. Thus if $p > 0$, then $y \geq 0$ and the parabola opens up. If $p < 0$, then $y \leq 0$ and the parabola opens down. A similar analysis shows that for $y^2 = 4px$, the parabola opens to the right when $p > 0$ and opens to the left when $p < 0$.

EXAMPLE 1 Find the Focus and Directrix of a Parabola

Find the focus and directrix of the parabola given by the equation $y = -\frac{1}{2}x^2$.

Continued ►

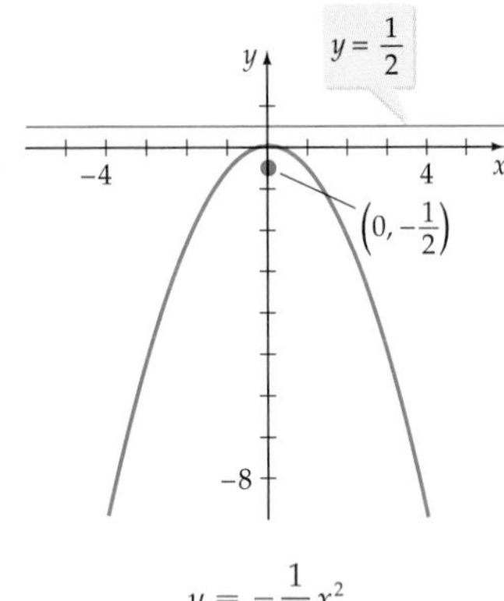

$y = -\frac{1}{2}x^2$

Figure 8.4

Solution

Because the x term is squared, the standard form of the equation is $x^2 = 4py$.

$$y = -\frac{1}{2}x^2$$

$$x^2 = -2y \qquad \text{• Write the given equation in standard form.}$$

Comparing this equation with $x^2 = 4py$ gives

$$4p = -2$$

$$p = -\frac{1}{2}$$

Because p is negative, the parabola opens down and the focus is below the vertex (0, 0), as shown in **Figure 8.4.** The coordinates of the focus are $(0, -1/2)$. The equation of the directrix is $y = 1/2$.

TRY EXERCISE 4, EXERCISE SET 8.1, PAGE 592

EXAMPLE 2 Find the Equation of a Parabola in Standard Form

Find the equation of the parabola in standard form with vertex at the origin and focus at $(-2, 0)$.

Solution

Because the vertex is (0, 0) and the focus is at $(-2, 0)$, $p = -2$. The graph of the parabola opens toward the focus, so in this case, the parabola opens to the left. The equation of the parabola in standard form that opens to the left is $y^2 = 4px$. Substitute -2 for p in this equation and simplify.

$$y^2 = 4(-2)x = -8x$$

The equation of the parabola is $y^2 = -8x$.

TRY EXERCISE 28, EXERCISE SET 8.1, PAGE 592

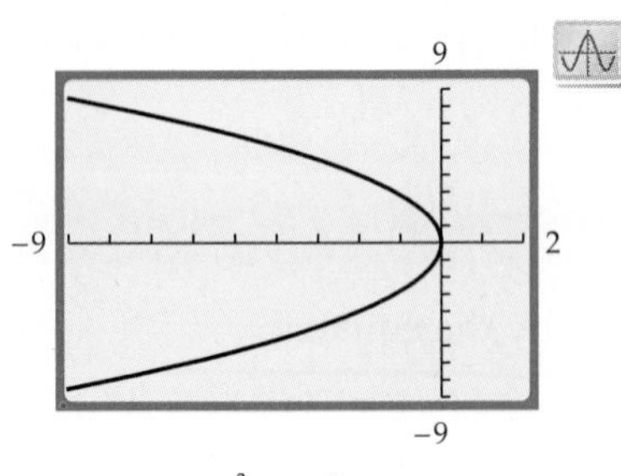

$y^2 = -8x$

Figure 8.5

The graph of $y^2 = -8x$ is shown in **Figure 8.5.** Note that the graph is not the graph of a function. To graph $y^2 = -8x$ with a graphing utility, we first solve for y to produce $y = \pm\sqrt{-8x}$. From this equation we can see that for any $x < 0$, there are two values of y. For example, when $x = -2$,

$$y = \pm\sqrt{(-8)(-2)} = \pm\sqrt{16} = \pm 4$$

The graph of $y^2 = -8x$ in **Figure 8.5** was drawn by graphing both $y_1 = \sqrt{-8x}$ and $y_2 = -\sqrt{-8x}$ on the same window.

◆ PARABOLAS WITH VERTEX AT (h, k)

The equation of a parabola with a vertical or horizontal axis of symmetry and with the vertex at a point (h, k) can be found by using the translations discussed previously. Consider a coordinate system with coordinate axes labeled x' and y' placed so that its origin is at (h, k) of the xy-coordinate system.

The relationship between an ordered pair in the $x'y'$-coordinate system and in the xy-coordinate system is given by the transformation equations

$$\begin{aligned} x' &= x - h \\ y' &= y - k \end{aligned} \qquad (1)$$

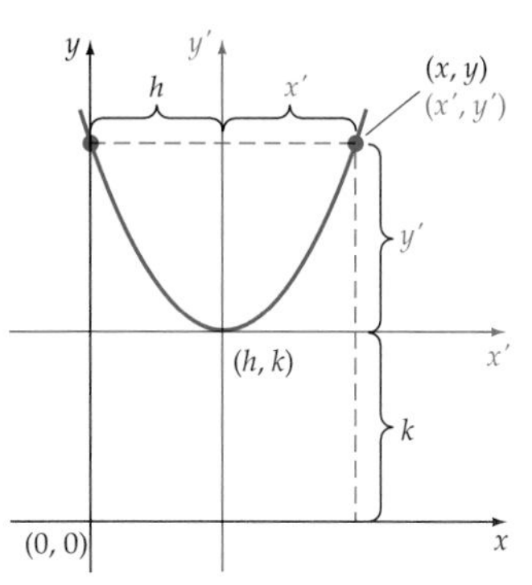

Figure 8.6

Now consider a parabola with vertex at (h, k) as shown in **Figure 8.6.** Place a new coordinate system labeled x' and y' with its origin at (h, k). The equation of a parabola in the $x'y'$-coordinate system is

$$(x')^2 = 4py' \qquad (2)$$

Using the transformation Equations (1), we can substitute the expressions for x' and y' into Equation (2). The standard form of the equation of the parabola with vertex (h, k) and a vertical axis of symmetry is

$$(x - h)^2 = 4p(y - k)$$

Similarly, we can derive the standard form of the equation of the parabola with vertex (h, k) and a horizontal axis of symmetry.

Standard Forms of the Equation of a Parabola with Vertex at (h, k)

Vertical Axis of Symmetry

The standard form of the equation of the parabola with vertex $V(h, k)$ and a vertical axis of symmetry is

$$(x - h)^2 = 4p(y - k)$$

The focus is $(h, k + p)$, and the equation of the directrix is $y = k - p$. See **Figure 8.7.**

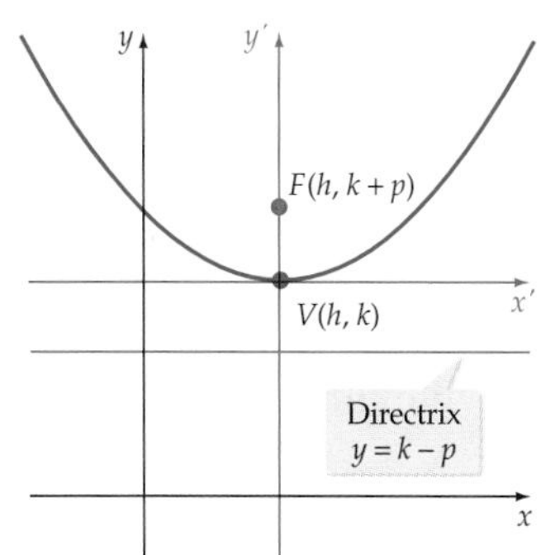

Figure 8.7

Horizontal Axis of Symmetry

The standard form of the equation of the parabola with vertex (h, k) and a horizontal axis of symmetry is

$$(y - k)^2 = 4p(x - h)$$

The focus is $(h + p, k)$, and the equation of the directrix is $x = h - p$.

EXAMPLE 3 Find the Focus and Directrix of a Parabola

Find the equation of the directrix and the coordinates of the vertex and focus of the parabola given by the equation $3x + 2y^2 + 8y - 4 = 0$.

Continued ·➤

Solution

Rewrite the equation so that the y terms are on one side of the equation, and then complete the square on y.

$$3x + 2y^2 + 8y - 4 = 0$$

$$2y^2 + 8y = -3x + 4$$

$$2(y^2 + 4y) = -3x + 4$$

$$2(y^2 + 4y + 4) = -3x + 4 + 8$$

• **Complete the square. Note that $2 \cdot 4 = 8$ is added to each side.**

$$2(y + 2)^2 = -3(x - 4)$$

• **Simplify and then factor.**

$$(y + 2)^2 = -\frac{3}{2}(x - 4)$$

• **Write the equation in standard form.**

Comparing this equation to $(y - k)^2 = 4p(x - h)$, we have a parabola that opens to the left with vertex $(4, -2)$ and $4p = -3/2$. Thus $p = -3/8$.

The coordinates of the focus are

$$\left(4 + \left(-\frac{3}{8}\right), -2\right) = \left(\frac{29}{8}, -2\right)$$

The equation of the directrix is

$$x = 4 - \left(-\frac{3}{8}\right) = \frac{35}{8}$$

Choosing some values for y and finding the corresponding values for x, we plot a few points. Because the line $y = -2$ is the axis of symmetry, for each point on one side of the axis of symmetry, there is a corresponding point on the other side. Two points are $(-2, 1)$ and $(-2, -5)$. See **Figure 8.8.**

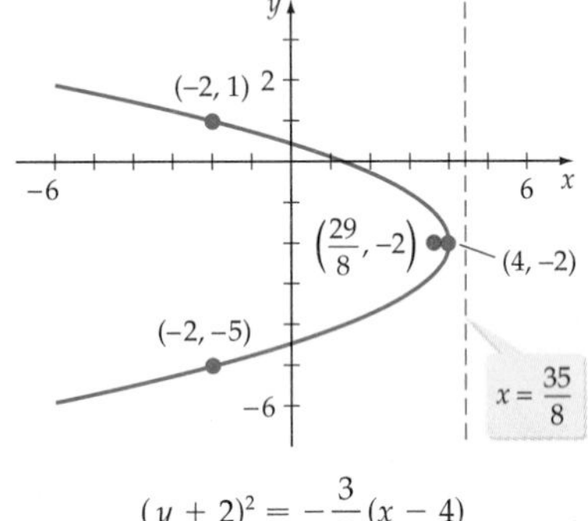

Figure 8.8

TRY EXERCISE 20, EXERCISE SET 8.1, PAGE 592

EXAMPLE 4 Find the Equation in Standard Form of a Parabola

Find the equation in standard form of the parabola with directrix $x = -1$ and focus $(3, 2)$.

Solution

The vertex is the midpoint of the line segment joining $(3, 2)$ and the point $(-1, 2)$ on the directrix.

$$(h, k) = \left(\frac{-1 + 3}{2}, \frac{2 + 2}{2}\right) = (1, 2)$$

The standard form of the equation is $(y - k)^2 = 4p(x - h)$. The distance from the vertex to the focus is 2. Thus $4p = 4(2) = 8$, and the equation of the parabola in standard form is $(y - 2)^2 = 8(x - 1)$. See **Figure 8.9.**

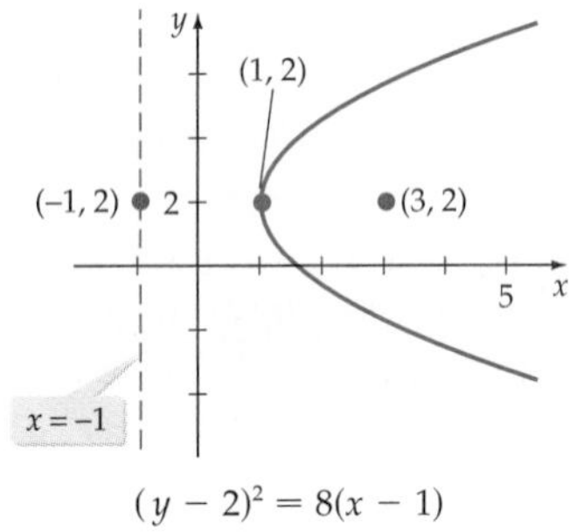

Figure 8.9

TRY EXERCISE 30, EXERCISE SET 8.1, PAGE 592

APPLICATIONS

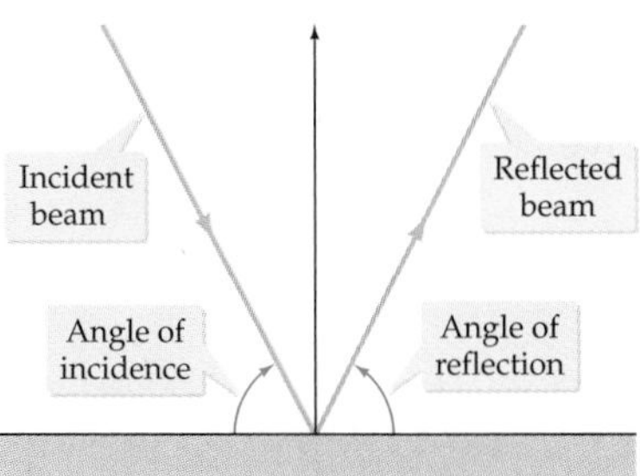

Figure 8.10

A principle of physics states that when light is reflected from a point P on a surface, the angle of incidence (that of the incoming ray) equals the angle of reflection (that of the outgoing ray). See **Figure 8.10.** This principle applied to parabolas has some useful consequences.

Optical Property of a Parabola

The line tangent to a parabola at a point P makes equal angles with the line through P and parallel to the axis of symmetry and the line through P and the focus of the parabola (see **Figure 8.11**).

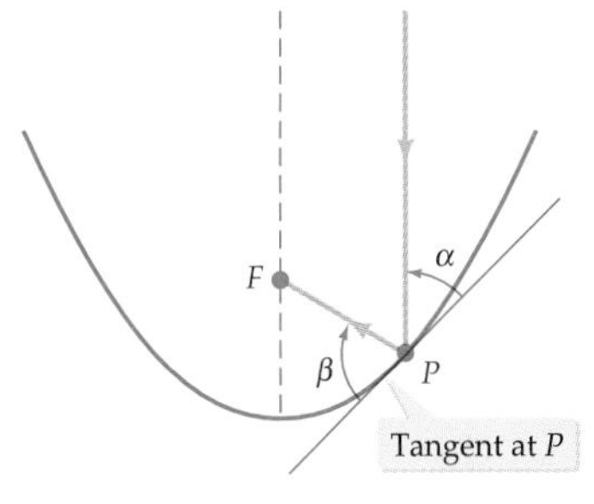

$\alpha = \beta$

Figure 8.11

A cross section of the reflecting mirror of a telescope has the shape of a parabola. The incoming parallel rays of light are reflected from the surface of the mirror and to the focus. See **Figure 8.12.**

Flashlights and car headlights also make use of this property. The light bulb is positioned at the focus of the parabolic reflector, which causes the reflected light to be reflected outward in parallel rays. See **Figure 8.13.**

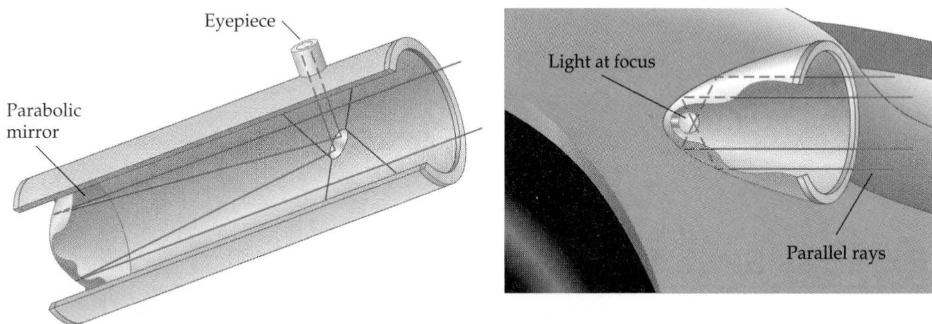

Figure 8.12

Figure 8.13

EXAMPLE 5 Find the Focus of a Satellite Dish

A satellite dish has the shape of a paraboloid. The signals that it receives are reflected to a receiver that is located at the focus of the paraboloid. If the dish is 8 feet across at its opening and $1\frac{1}{4}$ feet deep at its center, determine the location of its focus.

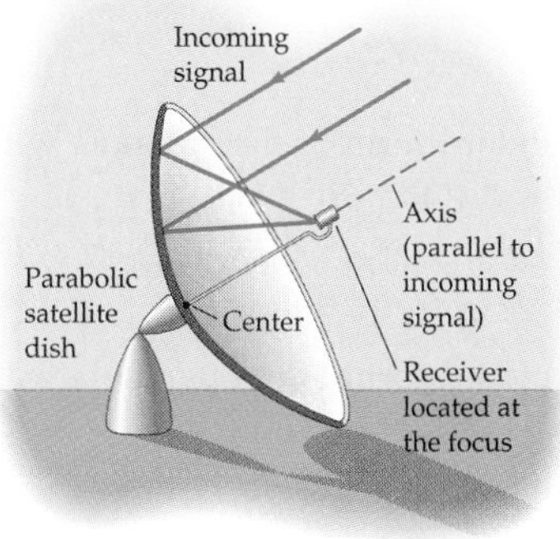

Figure 8.14

Solution

Figure 8.14 shows that a cross section of the paraboloid along its axis of symmetry is a parabola. **Figure 8.15** shows this cross section placed in a rectangular coordinate system with the vertex of the parabola at (0, 0) and the axis of symmetry of the parabola on the y-axis. The parabola has an equation of the form

$$4py = x^2$$

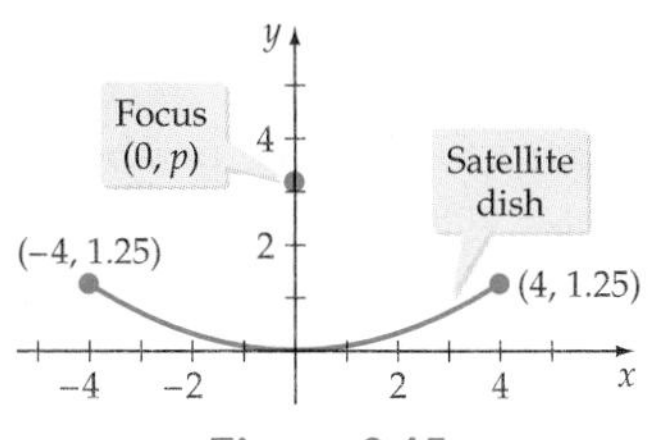

Figure 8.15

Continued

Because the parabola contains the point $\left(4, 1\frac{1}{4}\right)$, this equation is satisfied by the substitutions $x = 4$ and $y = 1\frac{1}{4}$. Thus we have

$$4p\left(1\frac{1}{4}\right) = 4^2$$
$$5p = 16$$
$$p = \frac{16}{5}$$

The focus of the satellite dish is on the axis of symmetry of the dish, and it is $3\frac{1}{5}$ feet above the vertex of the dish.

TRY EXERCISE 36, EXERCISE SET 8.1, PAGE 593

TOPICS FOR DISCUSSION

1. Do the graphs of the parabola given by $y = x^2$ and the vertical line given by $x = 10{,}000$ intersect? Explain.
2. A student claims that the focus of the parabola given by $y = 8x^2$ is at (0, 2) because $4p = 8$ implies that $p = 2$. Explain the error in the student's reasoning.
3. "The vertex of a parabola is always halfway between its focus and its directrix." Do you agree? Explain.
4. A tutor claims that the graph of $(x - h)^2 = 4p(y - k)$ has a y-intercept of $(0, h^2/(4p) + k)$. Explain why the tutor is correct.

EXERCISE SET 8.1

In Exercises 1 to 26, find the vertex, focus, and directrix of the parabola given by each equation. Sketch the graph.

1. $x^2 = -4y$
2. $2y^2 = x$
3. $y^2 = \frac{1}{3}x$
4. $x^2 = -\frac{1}{4}y$
5. $(x - 2)^2 = 8(y + 3)$
6. $(y + 1)^2 = 6(x - 1)$
7. $(y + 4)^2 = -4(x - 2)$
8. $(x - 3)^2 = -(y + 2)$
9. $(y - 1)^2 = 2x + 8$
10. $(x + 2)^2 = 3y - 6$
11. $(2x - 4)^2 = 8y - 16$
12. $(3x + 6)^2 = 18y - 36$
13. $x^2 + 8x - y + 6 = 0$
14. $x^2 - 6x + y + 10 = 0$
15. $x + y^2 - 3y + 4 = 0$
16. $x - y^2 - 4y + 9 = 0$
17. $2x - y^2 - 6y + 1 = 0$
18. $3x + y^2 + 8y + 4 = 0$
19. $x^2 + 3x + 3y - 1 = 0$
20. $x^2 + 5x - 4y - 1 = 0$
21. $2x^2 - 8x - 4y + 3 = 0$
22. $6x - 3y^2 - 12y + 4 = 0$
23. $2x + 4y^2 + 8y - 5 = 0$
24. $4x^2 - 12x + 12y + 7 = 0$
25. $3x^2 - 6x - 9y + 4 = 0$
26. $2x - 3y^2 + 9y + 5 = 0$
27. Find the equation in standard form of the parabola with vertex at the origin and focus (0, −4).
28. Find the equation in standard form of the parabola with vertex at the origin and focus (5, 0).
29. Find the equation in standard form of the parabola with vertex at (−1, 2) and focus (−1, 3).
30. Find the equation in standard form of the parabola with vertex at (2, −3) and focus (0, −3).
31. Find the equation in standard form of the parabola with focus (3, −3) and directrix $y = -5$.
32. Find the equation in standard form of the parabola with focus (−2, 4) and directrix $x = 4$.
33. Find the equation in standard form of the parabola that has vertex (−4, 1), has its axis of symmetry parallel to the y-axis, and passes through the point (−2, 2).
34. Find the equation in standard form of the parabola that has vertex (3, −5), has its axis of symmetry parallel to the x-axis, and passes through the point (4, 3).

35. **SATELLITE DISH** A satellite dish has the shape of a paraboloid. The signals that it receives are reflected to a receiver that is located at the focus of the paraboloid. If the dish is 8 feet across at its opening and 1 foot deep at its vertex, determine the location (distance above the vertex of the dish) of its focus.

36. **THE VERY LARGE ARRAY** Each of the antennas of the radio telescopes in the Very Large Array (see the first page of this chapter) is a paraboloid measuring 81 feet across with a depth of 16 feet. Determine, to the nearest 0.1 of a foot, the distance from the vertex to the focus of one of these antennas.

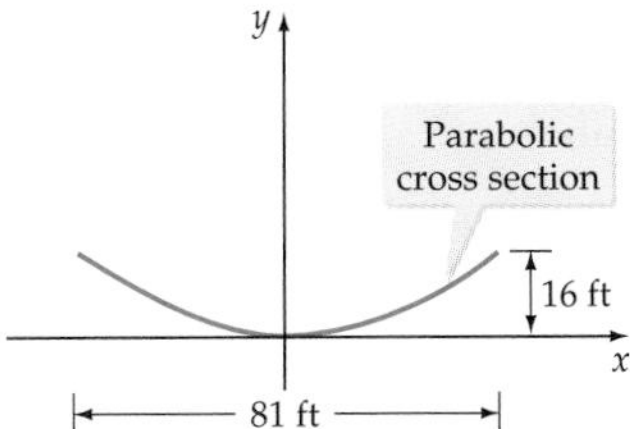

37. **CAPTURING THE SOUND** During televised football games, a parabolic microphone is used to capture sounds. The shield of the microphone is a paraboloid with a diameter 18.75 inches and a depth of 3.66 inches. To pick up the sounds, a microphone is placed at the focus of the paraboloid. How far (to the nearest 0.1 of an inch) from the vertex of the paraboloid should the microphone be placed?

38. **THE LOVELL TELESCOPE** The Lovell Telescope is a radio telescope located at the Jodrell Bank Observatory in Cheshire, England. The dish of the telescope has the shape of a paraboloid with a diameter of 250 feet and a focal length of 75 feet.

a. Find an equation of a cross section of the paraboloid that passes through the vertex of the paraboloid. Assume that the dish has its vertex at (0, 0) and a vertical axis of symmetry.

b. Find the depth of the dish. Round to the nearest foot.

39. The surface area of a paraboloid with radius r and depth d is given by $S = \dfrac{\pi r}{6d^2}[(r^2 + 4d^2)^{3/2} - r^3]$. Approximate (to the nearest 100 square feet) the surface area of:

a. One of the radio telescopes in the Very Large Array (see Exercise 36).

b. The Lovell Telescope (see Exercise 38).

40. **THE HALE TELESCOPE** The parabolic mirror in the Hale telescope at the Palomar Observatory in southern California has a diameter of 200 inches, and it has a concave depth of 3.75375 inches. Determine the location of its focus (to the nearest inch).

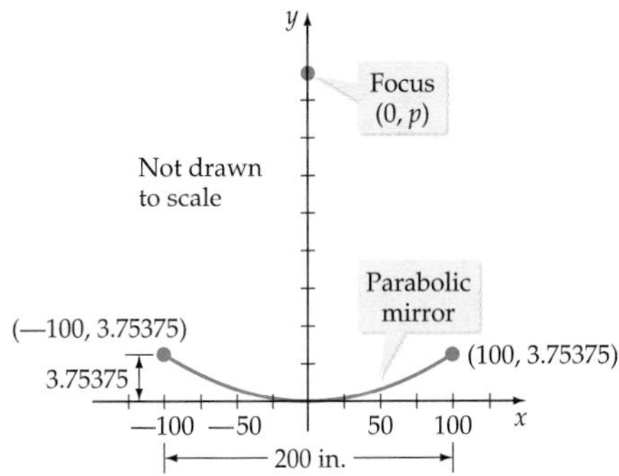

Mirror in the Hale Telescope

41. **THE LICK TELESCOPE** The parabolic mirror in the Lick telescope at the Lick Observatory on Mount Hamilton has a diameter of 120 inches, and it has a focal length of 600 inches. In the construction of the mirror, workers ground the mirror as shown in the following diagram. Determine the dimension a, which is the concave depth of the mirror.

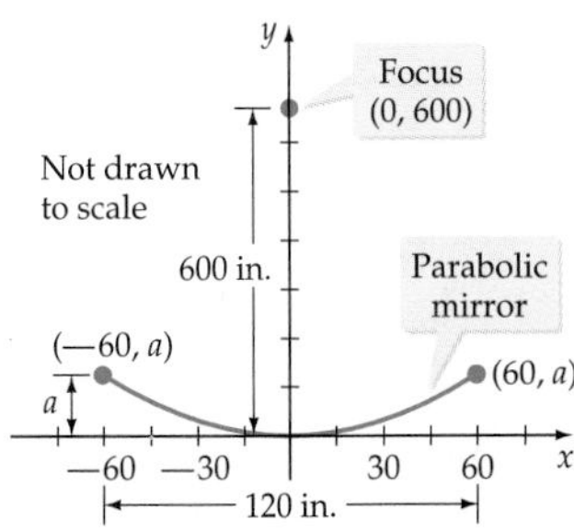

Mirror in the Lick Telescope

42. **HEADLIGHT DESIGN** A light source is to be placed on the axis of symmetry of the parabolic reflector shown in the figure at the top of the next column. How far to the right

of the vertex point should the light source be located if the designer wishes the reflected light rays to form a beam of parallel rays?

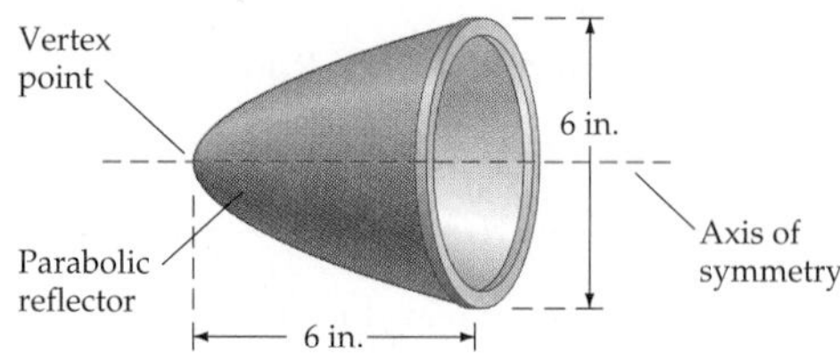

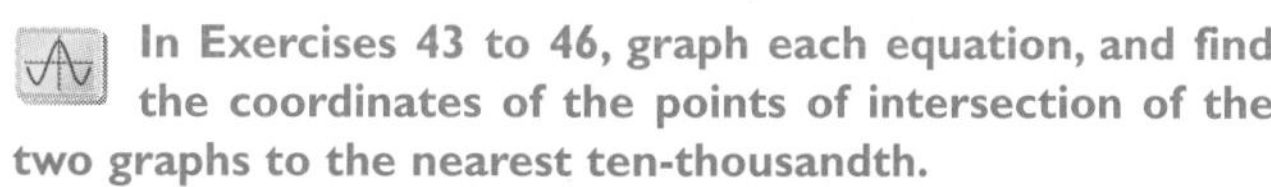

In Exercises 43 to 46, graph each equation, and find the coordinates of the points of intersection of the two graphs to the nearest ten-thousandth.

43. $y = 2x^2 - x - 1$
$y = x$

44. $y = x^2 + 2x - 4$
$y = x - 1$

45. $y = 2x^2 - 1$
$y = x^2 + x + 3$

46. $y = 2x^2 - x - 1$
$y = x^2 - 4$

SUPPLEMENTAL EXERCISES

In Exercises 47 to 49, use the following definition of latus rectum: The line segment that has endpoints on a parabola, passes through the focus of the parabola, and is perpendicular to the axis of symmetry is called the *latus rectum* of the parabola.

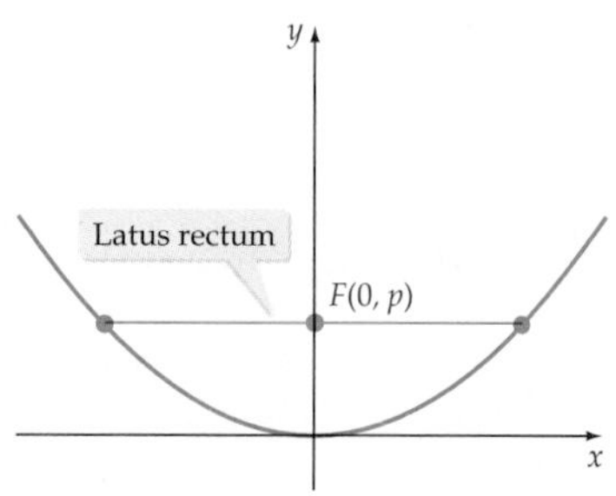

47. Find the length of the latus rectum for the parabola $x^2 = 4y$.

48. Find the length of the latus rectum for the parabola $y^2 = -8x$.

49. Find the length of the latus rectum for any parabola in terms of $|p|$, the distance from the vertex of the parabola to the focus.

The result of Exercise 49 can be stated as the following theorem: Two points on a parabola will be $2|p|$ units on each side of the axis of symmetry on the line through the focus and perpendicular to that axis.

50. Use the theorem to sketch a graph of the parabola given by the equation $(x - 3)^2 = 2(y + 1)$.

51. Use the theorem to sketch a graph of the parabola given by the equation $(y + 4)^2 = -(x - 1)$.

52. By using the definition of a parabola, find the equation in standard form of the parabola with $V(0, 0)$, $F(-c, 0)$, and directrix $x = c$.

53. Sketch a graph of $4(y - 2) = x|x| - 1$.

54. Find the equation of the directrix of the parabola with vertex at the origin and focus at the point $(1, 1)$.

55. Find the equation of the parabola with vertex at the origin and focus at the point $(1, 1)$. (*Hint:* You will need the answer to Exercise 54 and the definition of a parabola.)

PROJECTS

1. **PARABOLAS AND TANGENTS** Calculus procedures can be used to show that the equation of a tangent line to the parabola $4py = x^2$ at the point (x_0, y_0) is given by

$$y - y_0 = \left(\frac{1}{2p}x_0\right)(x - x_0)$$

Use this equation to verify each of the following statements.

a. If two tangent lines to a parabola intersect at right angles, then the point of intersection of the tangent lines is on the directrix of the parabola.

b. If two tangent lines to a parabola intersect at right angles, then the focus of the parabola is located on the line segment that connects the two points of tangency.

c. The tangent line to the parabola $4py = x^2$ at the point (x_0, y_0) intersects the y-axis at the point $(0, -y_0)$.

SECTION 8.2 ELLIPSES

- ELLIPSES WITH CENTER AT (0, 0)
- ELLIPSES WITH CENTER AT (h, k)
- ECCENTRICITY OF AN ELLIPSE
- APPLICATIONS
- ACOUSTIC PROPERTY OF AN ELLIPSE

An ellipse is another of the conic sections formed when a plane intersects a right circular cone. If β is the angle at which the plane intersects the axis of the cone and α is the angle shown in **Figure 8.16**, an ellipse is formed when $\alpha < \beta < 90°$. If $\beta = 90°$, then a circle is formed.

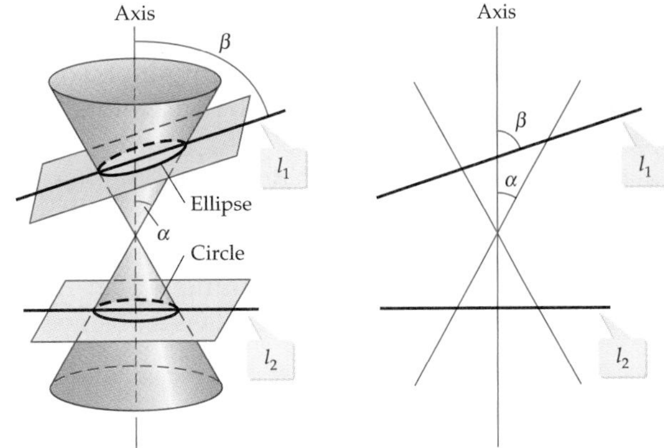

Figure 8.16

take note

If the plane intersects the cone at the vertex of the cone so that the resulting figure is a point, the point is a *degenerate ellipse*. See the accompanying figure.

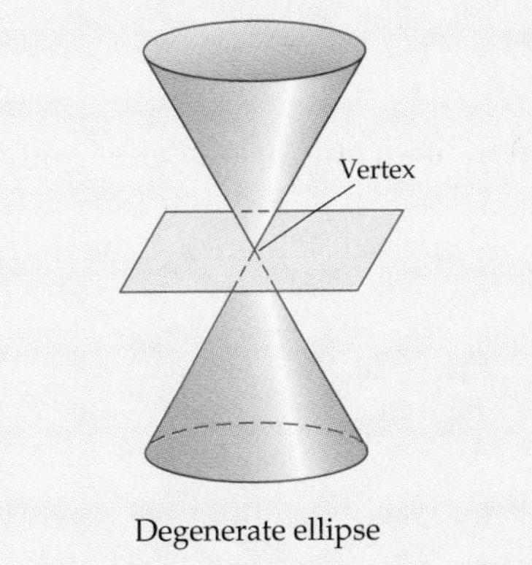

Degenerate ellipse

As was the case for a parabola, there is a definition for an ellipse in terms of a certain set of points in the plane.

Definition of an Ellipse

An **ellipse** is the set of all points in the plane, the sum of whose distances from two fixed points (**foci**) is a positive constant.

We can use this definition to draw an ellipse, equipped only with a piece of string and two tacks (see **Figure 8.17**). Tack the ends of the string to the foci, and trace a curve with a pencil held tight against the string. The resulting curve is an ellipse. The positive constant mentioned in the definition of an ellipse is the length of the string.

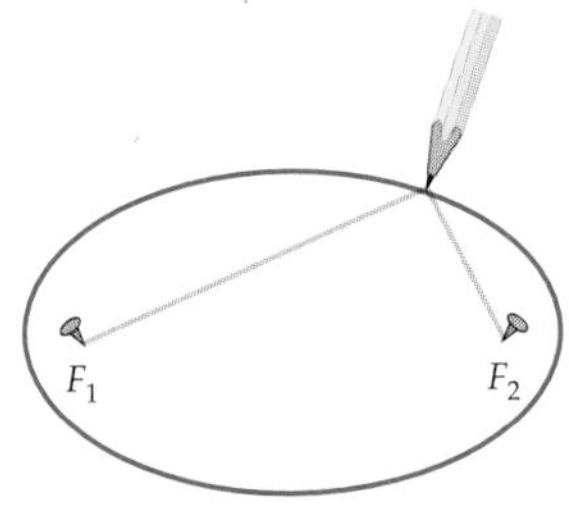

Figure 8.17

◆ ELLIPSES WITH CENTER AT (0, 0)

The graph of an ellipse has two axes of symmetry (see **Figure 8.18**). The longer axis is called the **major axis.** The foci of the ellipse are on the major axis. The shorter axis is called the **minor axis.** It is customary to denote the length of the major axis as $2a$ and the length of the minor axis as $2b$. The **semiaxes** are one-half the axes in length. Thus the length of the semimajor axis is denoted by a and the

take note

A web applet is available to experiment with ellipses by manipulating the foci and vertices. This applet, Ellipse, can be found on our web site at

http://college.hmco.com

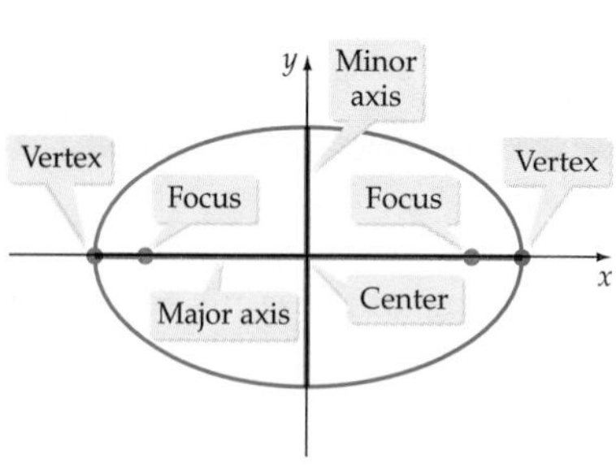

Figure 8.18

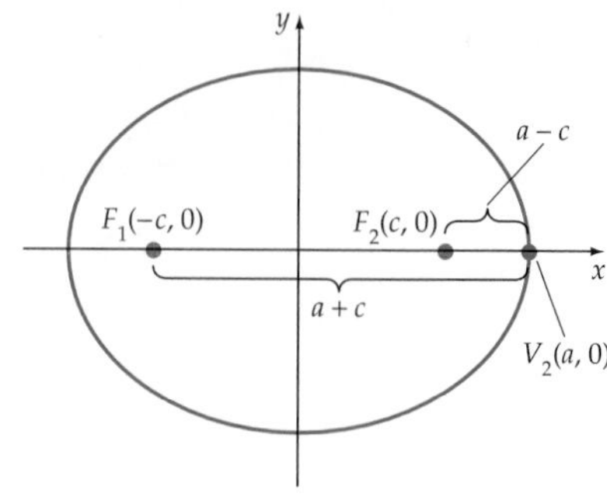

Figure 8.19

length of the semiminor axis by b. The **center** of the ellipse is the midpoint of the major axis. The endpoints of the major axis are the **vertices** (plural of *vertex*) of the ellipse.

Consider the point $V_2(a, 0)$, which is one vertex on an ellipse, and the points $F_2(c, 0)$ and $F_1(-c, 0)$, which are the foci of the ellipse shown in **Figure 8.19**. The distance from V_2 to F_1 is $a + c$. Similarly, the distance from V_2 to F_2 is $a - c$. From the definition of an ellipse, the sum of the distances from any point on the ellipse to the foci is a positive constant. By adding the expressions $a + c$ and $a - c$, we have

$$(a + c) + (a - c) = 2a$$

Thus the positive constant referred to in the definition of an ellipse is $2a$, the length of the major axis.

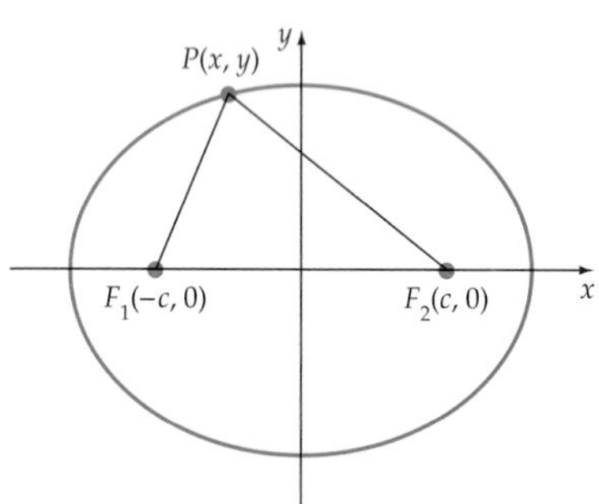

Figure 8.20

Now let $P(x, y)$ be any point on the ellipse (see **Figure 8.20**). By using the definition of an ellipse, we have

$$d(P, F_1) + d(P, F_2) = 2a$$
$$\sqrt{(x + c)^2 + y^2} + \sqrt{(x - c)^2 + y^2} = 2a$$

Subtract the second radical from each side of the equation, and then square each side.

$$\left[\sqrt{(x + c)^2 + y^2}\right]^2 = \left[2a - \sqrt{(x - c)^2 + y^2}\right]^2$$
$$(x + c)^2 + y^2 = 4a^2 - 4a\sqrt{(x - c)^2 + y^2} + (x - c)^2 + y^2$$
$$x^2 + 2cx + c^2 + y^2 = 4a^2 - 4a\sqrt{(x - c)^2 + y^2} + x^2 - 2cx + c^2 + y^2$$
$$4cx - 4a^2 = -4a\sqrt{(x - c)^2 + y^2}$$
$$[-cx + a^2]^2 = \left[a\sqrt{(x - c)^2 + y^2}\right]^2$$

• **Divide by −4, and then square each side.**

$$c^2x^2 - 2cxa^2 + a^4 = a^2x^2 - 2cxa^2 + a^2c^2 + a^2y^2$$
$$-a^2x^2 + c^2x^2 - a^2y^2 = -a^4 + a^2c^2$$

• **Rewrite with *x* and *y* terms on the left side.**

$$-(a^2 - c^2)x^2 - a^2y^2 = -a^2(a^2 - c^2)$$

• **Factor and let $b^2 = a^2 - c^2$.**

$$-b^2x^2 - a^2y^2 = -a^2b^2$$

• **Divide each side by $-a^2b^2$.**

$$\frac{x^2}{a^2} + \frac{y^2}{b^2} = 1$$

• **An equation of an ellipse with center at (0, 0).**

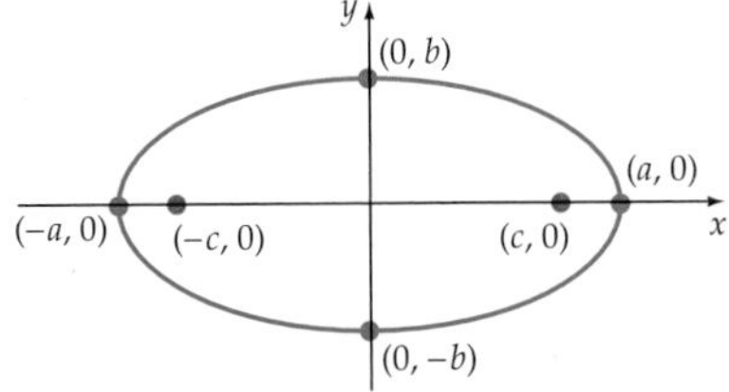

a. Major axis on x-axis

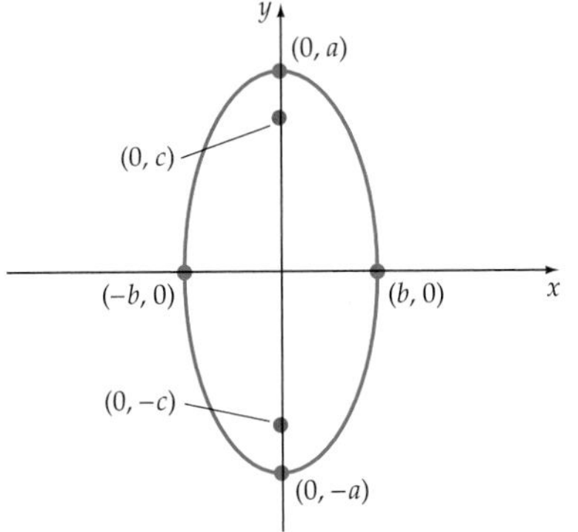

b. Major axis on y-axis

Figure 8.21

Standard Forms of the Equation of an Ellipse with Center at the Origin

Major Axis on the x-axis

The standard form of the equation of an ellipse with the center at the origin and major axis on the x-axis (see **Figure 8.21a**) is given by

$$\frac{x^2}{a^2} + \frac{y^2}{b^2} = 1, \quad a > b$$

The length of the major axis is $2a$. The length of the minor axis is $2b$. The coordinates of the vertices are $(a, 0)$ and $(-a, 0)$, and the coordinates of the foci are $(c, 0)$ and $(-c, 0)$, where $c^2 = a^2 - b^2$.

Major Axis on the y-axis

The standard form of the equation of an ellipse with the center at the origin and major axis on the y-axis (see **Figure 8.21b**) is given by

$$\frac{x^2}{b^2} + \frac{y^2}{a^2} = 1, \quad a > b$$

The length of the major axis is $2a$. The length of the minor axis is $2b$. The coordinates of the vertices are $(0, a)$ and $(0, -a)$, and the coordinates of the foci are $(0, c)$ and $(0, -c)$, where $c^2 = a^2 - b^2$.

EXAMPLE 1 Find the Vertices and Foci of an Ellipse

Find the vertices and foci of the ellipse given by the equation $\frac{x^2}{25} + \frac{y^2}{49} = 1$.

Sketch the graph.

Solution

Because the y^2 term has the larger denominator, the major axis is on the y-axis.

$$a^2 = 49 \qquad b^2 = 25 \qquad c^2 = a^2 - b^2$$
$$a = 7 \qquad b = 5 \qquad = 49 - 25 = 24$$
$$c = \sqrt{24} = 2\sqrt{6}$$

The vertices are $(0, 7)$ and $(0, -7)$. The foci are $(0, 2\sqrt{6})$ and $(0, -2\sqrt{6})$. See **Figure 8.22**.

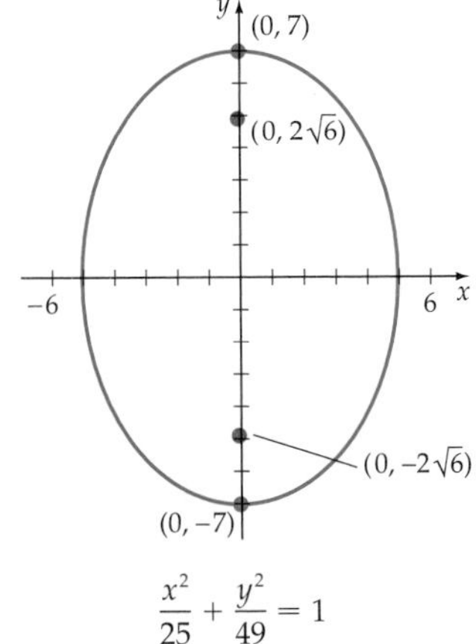

$\frac{x^2}{25} + \frac{y^2}{49} = 1$

Figure 8.22

TRY EXERCISE 20, EXERCISE SET 8.2, PAGE 605

An ellipse with foci $(3, 0)$ and $(-3, 0)$ and major axis of length 10 is shown in **Figure 8.23**. To find the equation of the ellipse in standard form, we must find a^2 and b^2. Because the foci are on the major axis, the major axis is on the x-axis. The length of the major axis is $2a$. Thus $2a = 10$. Solving for a, we have $a = 5$ and $a^2 = 25$.

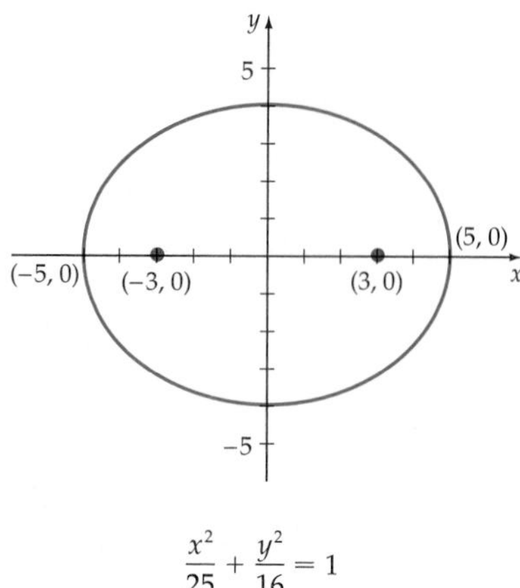

$$\frac{x^2}{25} + \frac{y^2}{16} = 1$$

Figure 8.23

Because the foci are $(3, 0)$ and $(-3, 0)$ and the center of the ellipse is the midpoint between the two foci, the distance from the center of the ellipse to a focus is 3. Therefore, $c = 3$. To find b^2, use the equation

$$c^2 = a^2 - b^2$$
$$9 = 25 - b^2$$
$$b^2 = 16$$

The equation of the ellipse in standard form is $\dfrac{x^2}{25} + \dfrac{y^2}{16} = 1$.

◆ ELLIPSES WITH CENTER AT (h, k)

The equation of an ellipse with center (h, k) and with horizontal or vertical major axes can be found by using a translation of coordinates. On a coordinate system with axes labeled x' and y', the standard form of the equation of an ellipse with center at the origin of the $x'y'$-coordinate system is

$$\frac{(x')^2}{a^2} + \frac{(y')^2}{b^2} = 1$$

Now place the origin of the $x'y'$-coordinate system at (h, k) in an xy-coordinate system. See **Figure 8.24.**

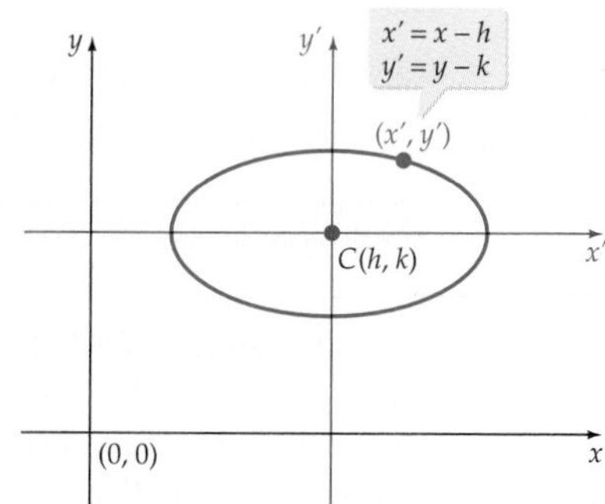

Figure 8.24

The relationship between an ordered pair in the $x'y'$-coordinate system and the xy-coordinate system is given by the transformation equations

$$x' = x - h$$
$$y' = y - k$$

Substitute the expressions for x' and y' into the equation of an ellipse. The equation of the ellipse with center at (h, k) is

$$\frac{(x-h)^2}{a^2} + \frac{(y-k)^2}{b^2} = 1$$

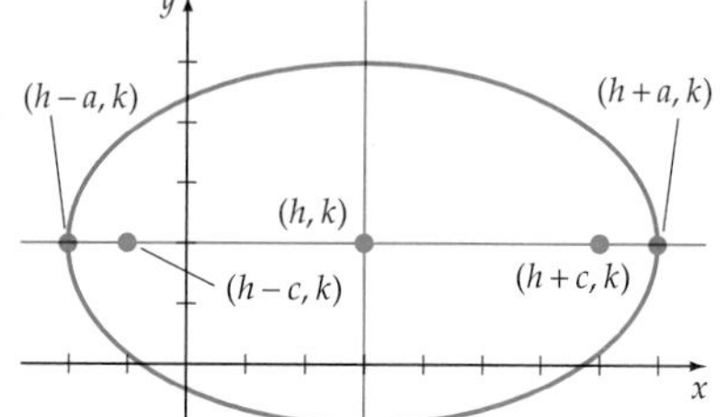

a. Major axis parallel to x-axis

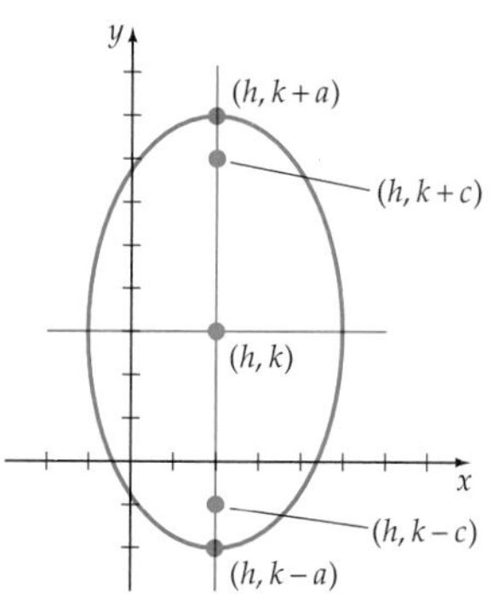

b. Major axis parallel to y-axis

Figure 8.25

Standard Forms of the Equation of an Ellipse with Center at (h, k)

Major Axis Parallel to the x-axis
The standard form of the equation of an ellipse with the center at (h, k) and major axis parallel to the x-axis (see **Figure 8.25a**) is given by

$$\frac{(x-h)^2}{a^2} + \frac{(y-k)^2}{b^2} = 1 \quad a > b$$

The length of the major axis is $2a$. The length of the minor axis is $2b$. The coordinates of the vertices are $(h + a, k)$ and $(h - a, k)$, and the coordinates of the foci are $(h + c, k)$ and $(h - c, k)$, where $c^2 = a^2 - b^2$.

Major Axis Parallel to the y-axis
The standard form of the equation of an ellipse with the center at (h, k) and major axis parallel to the y-axis (see **Figure 8.25b**) is given by

$$\frac{(x-h)^2}{b^2} + \frac{(y-k)^2}{a^2} = 1 \quad a > b$$

The length of the major axis is $2a$. The length of the minor axis is $2b$. The coordinates of the vertices are $(h, k + a)$ and $(h, k - a)$, and the coordinates of the foci are $(h, k + c)$ and $(h, k - c)$, where $c^2 = a^2 - b^2$.

EXAMPLE 2 Find the Vertices and Foci of an Ellipse

Find the vertices and foci of the ellipse $4x^2 + 9y^2 - 8x + 36y + 4 = 0$. Sketch the graph.

Solution

Write the equation of the ellipse in standard form by completing the square.

$$4x^2 + 9y^2 - 8x + 36y + 4 = 0$$
$$4x^2 - 8x + 9y^2 + 36y = -4 \qquad \text{• Rearrange terms.}$$
$$4(x^2 - 2x) + 9(y^2 + 4y) = -4 \qquad \text{• Factor.}$$

Continued •►

$$4(x^2 - 2x + 1) + 9(y^2 + 4y + 4) = -4 + 4 + 36$$ • **Complete the square.**

$$4(x - 1)^2 + 9(y + 2)^2 = 36$$ • **Factor.**

$$\frac{(x - 1)^2}{9} + \frac{(y + 2)^2}{4} = 1$$ • **Divide by 36.**

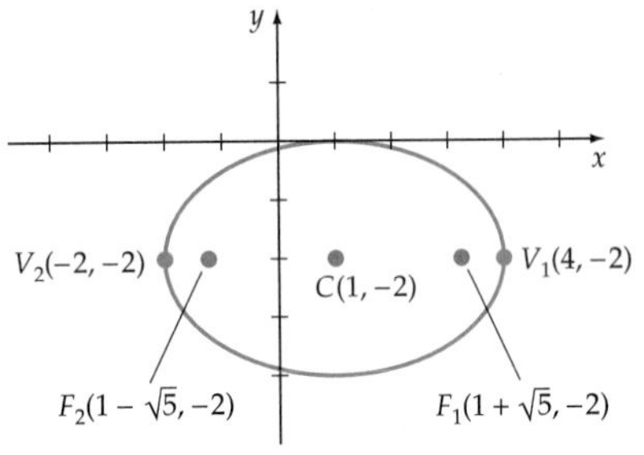

From the equation of the ellipse in standard form, the coordinates of the center of the ellipse are $(1, -2)$. Because the larger denominator is 9, the major axis is parallel to the x-axis and $a^2 = 9$. Thus $a = 3$. The vertices are $(4, -2)$ and $(-2, -2)$.

To find the coordinates of the foci, we find c.

$$c^2 = a^2 - b^2 = 9 - 4 = 5$$
$$c = \sqrt{5}$$

The foci are $\left(1 + \sqrt{5}, -2\right)$ and $\left(1 - \sqrt{5}, -2\right)$. See **Figure 8.26**.

$V_2(-2, -2)$, $C(1, -2)$, $V_1(4, -2)$, $F_2(1 - \sqrt{5}, -2)$, $F_1(1 + \sqrt{5}, -2)$

$$\frac{(x - 1)^2}{9} + \frac{(y + 2)^2}{4} = 1$$

Figure 8.26

TRY EXERCISE 26, EXERCISE SET 8.2, PAGE 605

A graphing utility can be used to graph an ellipse. For instance, consider the equation $4x^2 + 9y^2 - 8x + 36y + 4 = 0$ from Example 2. Rewrite the equation as

$$9y^2 + 36y + (4x^2 - 8x + 4) = 0$$

In this form, the equation is a quadratic equation in terms of the variable y with

$$A = 9, B = 36, \text{ and } C = 4x^2 - 8x + 4$$

Apply the quadratic formula to produce

$$y = \frac{-36 \pm \sqrt{1296 - 36(4x^2 - 8x + 4)}}{18}$$

The graph of $Y_1 = \dfrac{-36 + \sqrt{1296 - 36(4x^2 - 8x + 4)}}{18}$ is the part of the ellipse on or above the line $y = -2$ (see **Figure 8.27**).

The graph of $Y_2 = \dfrac{-36 - \sqrt{1296 - 36(4x^2 - 8x + 4)}}{18}$ is the part of the ellipse on or below the line $y = -2$, as shown in **Figure 8.27**.

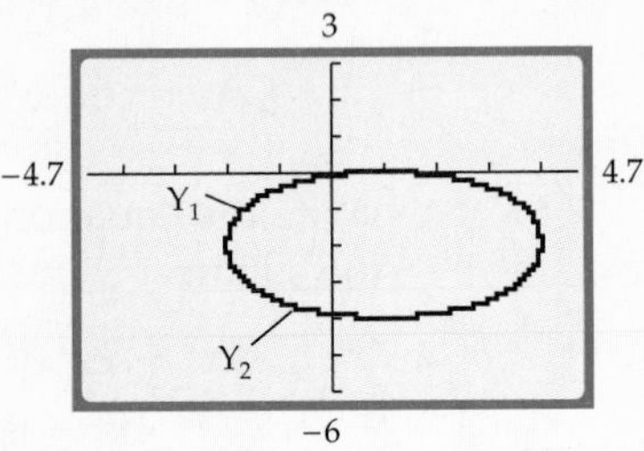

Figure 8.27

One advantage of this graphing procedure is that it does not require us to write the given equation in standard form. A disadvantage of the graphing procedure is that it does not indicate where the foci of the ellipse are located.

EXAMPLE 3 Find the Equation of an Ellipse

Find the standard form of the equation of the ellipse with center at $(4, -2)$, foci $F_2(4, 1)$ and $F_1(4, -5)$, and minor axis of length 10, as shown in **Figure 8.28.**

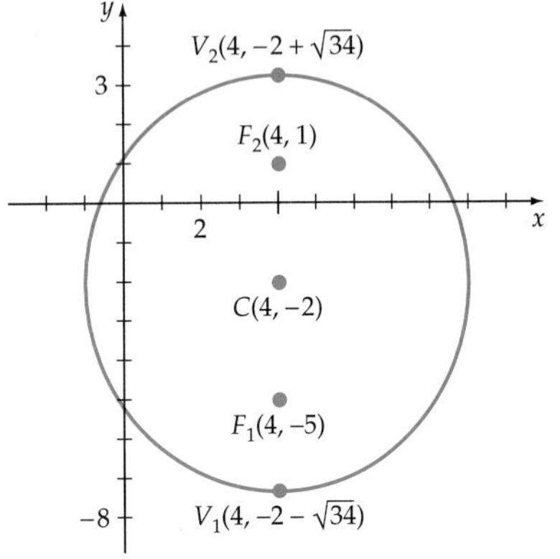

Figure 8.28

Solution

Because the foci are on the major axis, the major axis is parallel to the y-axis. The distance from the center of the ellipse to a focus is c. The distance between the center $(4, -2)$ and the focus $(4, 1)$ is 3. Therefore, $c = 3$.

The length of the minor axis is $2b$. Thus $2b = 10$ and $b = 5$.

To find a^2, use the equation $c^2 = a^2 - b^2$.

$$9 = a^2 - 25$$
$$a^2 = 34$$

Thus the equation in standard form is

$$\frac{(x-4)^2}{25} + \frac{(y+2)^2}{34} = 1$$

TRY EXERCISE 42, EXERCISE SET 8.2, PAGE 605

◆ ECCENTRICITY OF AN ELLIPSE

The graph of an ellipse can be very long and thin, or it can be much like a circle. The **eccentricity** of an ellipse is a measure of its "roundness."

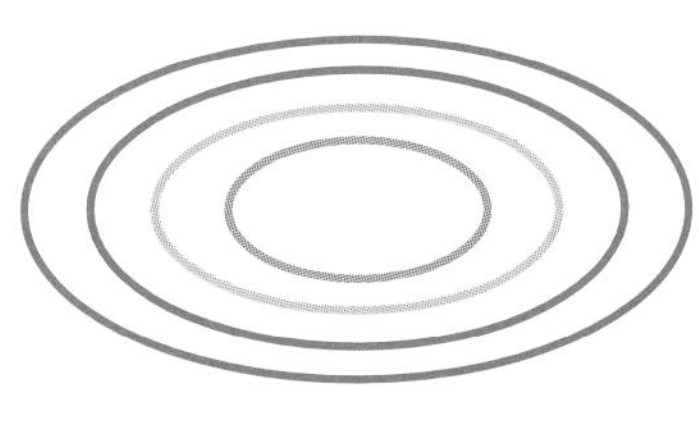

Eccentricity = 0.87

Figure 8.29

Eccentricity (e) of an Ellipse

The eccentricity e of an ellipse is the ratio of c to a, where c is the distance from the center to a focus and a is one-half the length of the major axis. (See **Figure 8.29.**) That is,

$$e = \frac{c}{a}$$

Because $c < a$, for an ellipse, $0 < e < 1$. When $e \approx 0$, the graph is almost a circle. When $e \approx 1$, the graph is long and thin. See **Figure 8.30.**

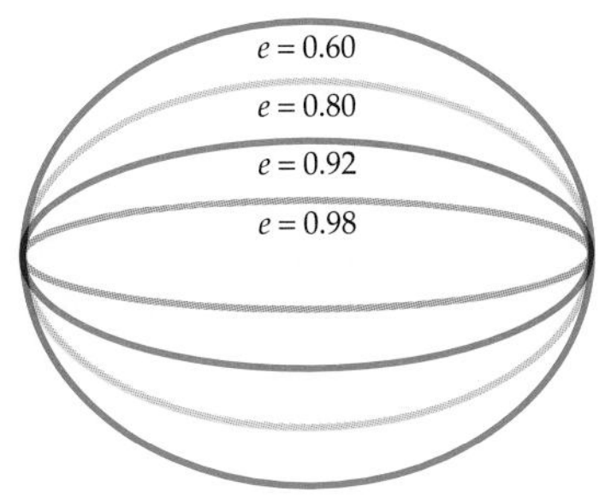

Figure 8.30

EXAMPLE 4 Find the Eccentricity of an Ellipse

Find the eccentricity of the ellipse given by $8x^2 + 9y^2 = 18$.

Continued ▸

Solution

First, write the equation of the ellipse in standard form. Divide each side of the equation by 18.

$$\frac{8x^2}{18} + \frac{9y^2}{18} = 1$$

$$\frac{4x^2}{9} + \frac{y^2}{2} = 1$$

$$\frac{x^2}{9/4} + \frac{y^2}{2} = 1 \qquad \bullet\ \frac{4}{9} = \frac{1}{9/4}$$

The last step is necessary because the standard form of the equation has coefficients of 1 in the numerator. Thus

$$a^2 = \frac{9}{4} \quad \text{and} \quad a = \frac{3}{2}$$

Use the equation $c^2 = a^2 - b^2$ to find c.

$$c^2 = \frac{9}{4} - 2 = \frac{1}{4} \quad \text{and} \quad c = \sqrt{\frac{1}{4}} = \frac{1}{2}$$

Now find the eccentricity.

$$e = \frac{c}{a} = \frac{1/2}{3/2} = \frac{1}{3}$$

The eccentricity of the ellipse is 1/3.

TRY EXERCISE 48, EXERCISE SET 8.2, PAGE 605

◆ APPLICATIONS

The planets travel around the sun in elliptical orbits. The sun is located at a focus of the orbit. The eccentricities of the orbits for the planets in our solar system are given in Table 8.1.

Table 8.1

Planet	Eccentricity
Mercury	0.206
Venus	0.007
Earth	0.017
Mars	0.093
Jupiter	0.049
Saturn	0.051
Uranus	0.046
Neptune	0.005
Pluto	0.250

QUESTION Which planet has the most nearly circular orbit?

The terms *perihelion* and *aphelion* are used to denote the position of a planet in its orbit around the sun. The perihelion is the point nearest the sun; the aphelion is the point farthest from the sun. See **Figure 8.31**. The length of the semimajor axis of a planet's elliptical orbit is called the *mean distance* of the planet from the sun.

ANSWER Neptune has the smallest eccentricity, so it is the planet with the most nearly circular orbit.

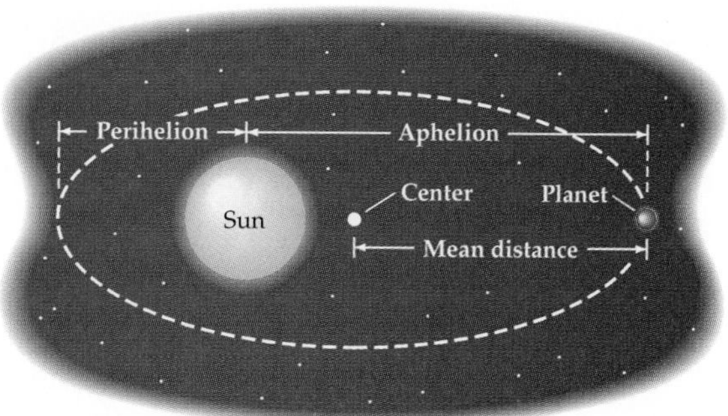

Figure 8.31

EXAMPLE 5 Determine an Equation for the Orbit of Earth

Earth has a mean distance of 93 million miles and a perihelion distance of 91.5 million miles. Find an equation for Earth's orbit.

Solution

A mean distance of 93 million miles implies that the length of the semimajor axis of the orbit is $a = 93$ million miles. Earth's aphelion distance is the length of the major axis less the length of the perihelion distance. Thus

$$\text{Aphelion distance} = 2(93) - 91.5 = 94.5 \text{ million miles}$$

The distance c from the sun to the center of Earth's orbit is

$$c = \text{aphelion distance} - 93 = 94.5 - 93 = 1.5 \text{ million miles}$$

The length b of the semiminor axis of the orbit is

$$b = \sqrt{a^2 - c^2} = \sqrt{93^2 - 1.5^2} = \sqrt{8646.75}$$

An equation of Earth's orbit is

$$\frac{x^2}{93^2} + \frac{y^2}{8646.75} = 1$$

TRY EXERCISE 54, EXERCISE SET 8.2, PAGE 605

◆ ACOUSTIC PROPERTY OF AN ELLIPSE

Sound waves, although different from light waves, have a similar reflective property. When sound is reflected from a point P on a surface, the angle of incidence equals the angle of reflection. Applying this principle to a room with an elliptical ceiling results in what are called whispering galleries. These galleries are based on the following theorem.

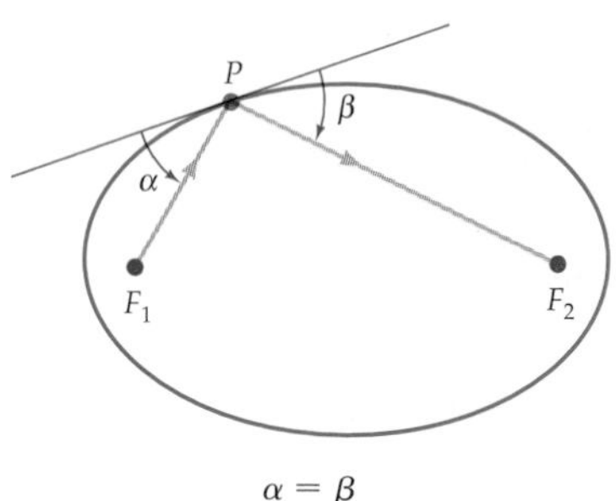

$\alpha = \beta$

Figure 8.32

The Reflective Property of an Ellipse

The lines from the foci to a point on an ellipse make equal angles with the tangent line at that point. See **Figure 8.32.**

The Statuary Hall in the Capital Building in Washington, D.C., is a whispering gallery. Two people standing at the foci of the elliptical ceiling can whisper and yet hear each other even though they are a considerable distance apart. The whisper from one person is reflected to the person standing at the other focus.

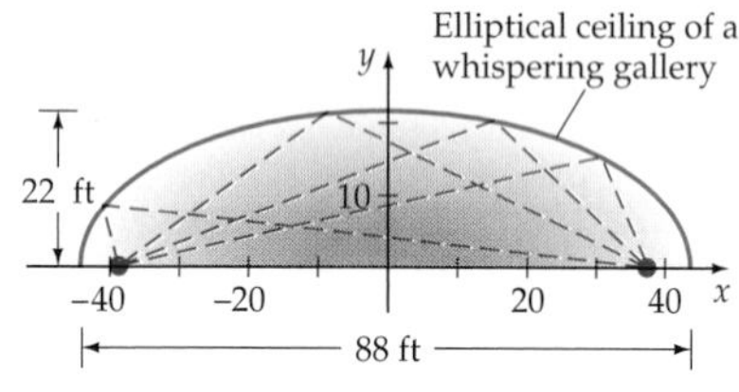

Figure 8.33

EXAMPLE 6 Locate the Foci of a Whispering Gallery

A room 88 feet long is constructed to be a whispering gallery. The room has an elliptical ceiling, as shown in **Figure 8.33.** If the maximum height of the ceiling is 22 feet, determine where the foci are located.

Solution

The length a of the semimajor axis of the elliptical ceiling is 44 feet. The height b of the semiminor axis is 22 feet. Thus

$$c^2 = a^2 - b^2$$
$$c^2 = 44^2 - 22^2$$
$$c = \sqrt{44^2 - 22^2} \approx 38.1 \text{ feet}$$

The foci are located about 38.1 feet from the center of the elliptical ceiling along its major axis.

TRY EXERCISE 56, EXERCISE SET 8.2, PAGE 606

TOPICS FOR DISCUSSION

1. In every ellipse, the length of the semimajor axis a is greater than the length of the semiminor axis b and greater than the distance c from a focus to the center of the ellipse. Do you agree? Explain.
2. How many vertices does an ellipse have?
3. Every ellipse has two y-intercepts. Do you agree? Explain.
4. Explain why the eccentricity of every ellipse is a number between 0 and 1.

EXERCISE SET 8.2

In Exercises 1 to 32, find the vertices and foci of the ellipse given by each equation. Sketch the graph.

1. $\frac{x^2}{16} + \frac{y^2}{25} = 1$
2. $\frac{x^2}{49} + \frac{y^2}{36} = 1$
3. $\frac{x^2}{9} + \frac{y^2}{4} = 1$
4. $\frac{x^2}{64} + \frac{y^2}{25} = 1$
5. $\frac{x^2}{7} + \frac{y^2}{9} = 1$
6. $\frac{x^2}{5} + \frac{y^2}{4} = 1$

7. $\frac{4x^2}{9} + \frac{y^2}{16} = 1$

8. $\frac{x^2}{9} + \frac{9y^2}{16} = 1$

9. $\frac{(x-3)^2}{25} + \frac{(y+2)^2}{16} = 1$

10. $\frac{(x+3)^2}{9} + \frac{(y+1)^2}{16} = 1$

11. $\frac{(x+2)^2}{9} + \frac{y^2}{25} = 1$

12. $\frac{x^2}{25} + \frac{(y-2)^2}{81} = 1$

13. $\frac{(x-1)^2}{21} + \frac{(y-3)^2}{4} = 1$

14. $\frac{(x+5)^2}{9} + \frac{(y-3)^2}{7} = 1$

15. $\frac{9(x-1)^2}{16} + \frac{(y+1)^2}{9} = 1$

16. $\frac{(x+6)^2}{25} + \frac{25y^2}{144} = 1$

17. $3x^2 + 4y^2 = 12$

18. $5x^2 + 4y^2 = 20$

19. $25x^2 + 16y^2 = 400$

20. $25x^2 + 12y^2 = 300$

21. $64x^2 + 25y^2 = 400$

22. $9x^2 + 64y^2 = 144$

23. $4x^2 + y^2 - 24x - 8y + 48 = 0$

24. $x^2 + 9y^2 + 6x - 36y + 36 = 0$

25. $5x^2 + 9y^2 - 20x + 54y + 56 = 0$

26. $9x^2 + 16y^2 + 36x - 16y - 104 = 0$

27. $16x^2 + 9y^2 - 64x - 80 = 0$

28. $16x^2 + 9y^2 + 36y - 108 = 0$

29. $25x^2 + 16y^2 + 50x - 32y - 359 = 0$

30. $16x^2 + 9y^2 - 64x - 54y + 1 = 0$

31. $8x^2 + 25y^2 - 48x + 50y + 47 = 0$

32. $4x^2 + 9y^2 + 24x + 18y + 44 = 0$

In Exercises 33 to 44, find the equation in standard form of each ellipse, given the information provided.

33. Center $(0, 0)$, major axis of length 10, foci at $(4, 0)$ and $(-4, 0)$

34. Center $(0, 0)$, minor axis of length 6, foci at $(0, 4)$ and $(0, -4)$

35. Vertices $(6, 0)$, $(-6, 0)$; ellipse passes through $(0, -4)$ and $(0, 4)$

36. Vertices $(7, 0)$, $(-7, 0)$; ellipse passes through $(0, 5)$ and $(0, -5)$

37. Major axis of length 12 on the x-axis, center at $(0, 0)$; ellipse passes through $(2, -3)$

38. Minor axis of length 8, center at $(0, 0)$; ellipse passes through $(-2, 2)$

39. Center $(-2, 4)$, vertices $(-6, 4)$ and $(2, 4)$, foci at $(-5, 4)$ and $(1, 4)$

40. Center $(0, 3)$, minor axis of length 4, foci at $(0, 0)$ and $(0, 6)$

41. Center $(2, 4)$, major axis parallel to the y-axis and of length 10; ellipse passes through the point $(3, 3)$

42. Center $(-4, 1)$, minor axis parallel to the y-axis and of length 8; ellipse passes through the point $(0, 4)$

43. Vertices $(5, 6)$ and $(5, -4)$, foci at $(5, 4)$ and $(5, -2)$

44. Vertices $(-7, -1)$ and $(5, -1)$, foci at $(-5, -1)$ and $(3, -1)$

In Exercises 45 to 52, use the eccentricity of each ellipse to find its equation in standard form.

45. Eccentricity $2/5$, major axis on the x-axis and of length 10, center at $(0, 0)$

46. Eccentricity $3/4$, foci at $(9, 0)$ and $(-9, 0)$

47. Foci at $(0, -4)$ and $(0, 4)$, eccentricity $2/3$

48. Foci at $(0, -3)$ and $(0, 3)$, eccentricity $1/4$

49. Eccentricity $2/5$, foci at $(-1, 3)$ and $(3, 3)$

50. Eccentricity $1/4$, foci at $(-2, 4)$ and $(-2, -2)$

51. Eccentricity $2/3$, major axis of length 24 on the y-axis, center at $(0, 0)$

52. Eccentricity $3/5$, major axis of length 15 on the x-axis, center at $(0, 0)$

53. THE ORBIT OF SATURN The distance from Saturn to the sun at Saturn's aphelion is 934.34 million miles, and the distance from Saturn to the sun at its perihelion is 835.14 million miles. Find an equation for the orbit of Saturn.

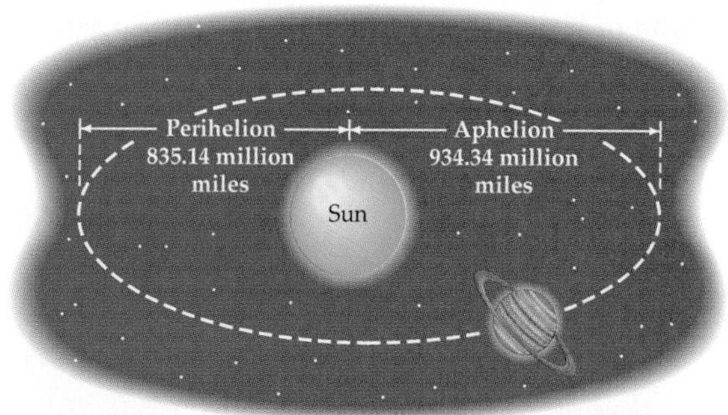

54. THE ORBIT OF VENUS Venus has a mean distance from the sun of 67.08 million miles, and the distance from Venus to the sun at its aphelion is 67.58 million miles. Find an equation for the orbit of Venus.

55. WHISPERING GALLERY An architect wishes to design a large room that will be a whispering gallery. See

Example 6. The ceiling of the room has a cross section that is an ellipse, as shown in the following figure.

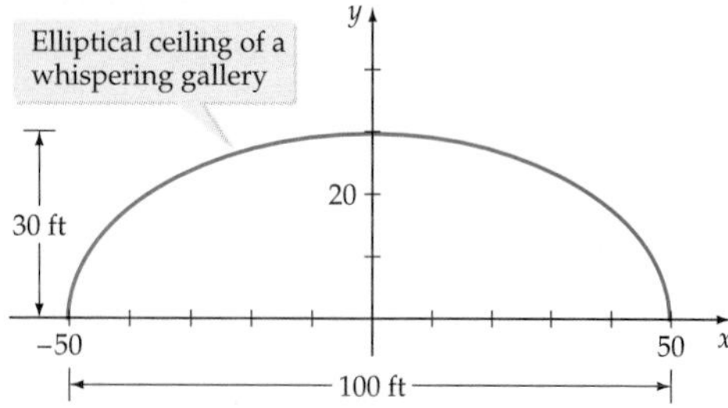

How far to the right and to the left of center are the foci located?

56. WHISPERING GALLERY An architect wishes to design a large room 100 feet long that will be a whispering gallery. The ceiling of the room has a cross section that is an ellipse, as shown in the following figure.

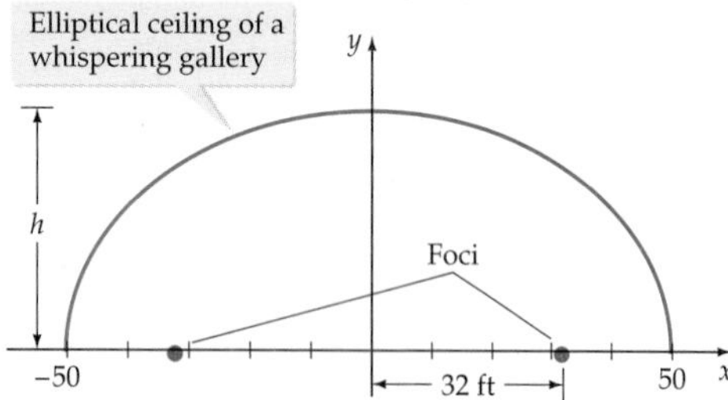

If the foci are to be located 32 feet to the right and to the left of center, find the height h of the elliptical ceiling (to the nearest 0.1 foot).

57. HALLEY'S COMET Find the equation of the path of Halley's comet in astronomical units by letting the sun (one focus) be at the origin and letting the other focus be on the positive x-axis. The length of the major axis of the orbit of Halley's comet is approximately 36 astronomical units (36 AU), and the length of the minor axis is 9 AU (1 AU = 92,960,000 miles).

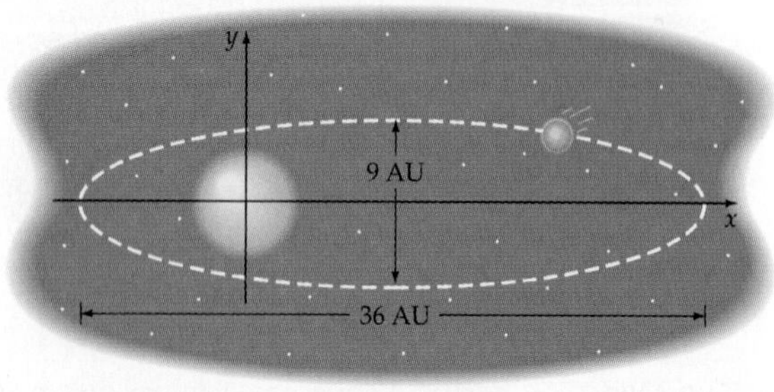

58. ELLIPTICAL RECEIVERS Some satellite receivers are made in an elliptical shape that enable the receiver to pick up signals from two satellites. The receiver shown at the top of the next column has a major axis of 24 inches and a minor axis of 18 inches.

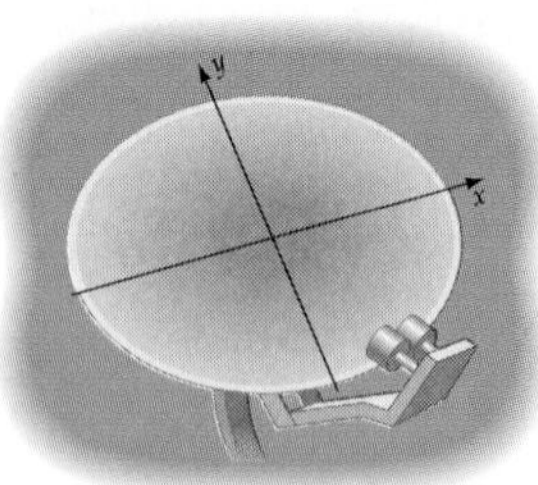

Determine, to the nearest 0.1 inch, the coordinates in the xy-plane of the foci of the ellipse. (*Note:* Because the receiver has only a slight curvature, we can estimate the location of the foci by assuming the receiver is flat.)

In Exercises 59 and 60, use the following formula for the perimeter p of an ellipse with semimajor axis a and semiminor axis b.

$$p \approx \pi\sqrt{2(a^2 + b^2)}$$

59. ELLIPTICAL EXERCISE EQUIPMENT Many exercise clubs have installed elliptical trainers. These machines are similar to step machines except that the motion of the feet follows an elliptical path. On one elliptical trainer, the path of a person's foot is elliptical with a major axis of 16 inches and a minor axis of 10 inches. How many revolutions must the left foot make to have completed a distance of 1 mile on this elliptical trainer?

60. ORBIT OF MARS Mars travels around the sun in an elliptical orbit. The orbit has a major axis of 3.04 AU and a minor axis of 2.99 AU. (1 AU is 1 astronomical unit, or approximately 92,960,000 miles, the average distance of Earth from the sun.) Estimate, to the nearest million miles, the perimeter of the orbit of Mars.

61. THE COLOSSEUM The base of the Colosseum in Rome has an elliptical shape.

a. Find an equation in standard form for the base of the Colosseum, which has a major axis of 615 feet and a minor axis of 510 feet.

b. The area of an ellipse with a semimajor axis of length a and a semiminor axis of length b is given by $A = \pi ab$. Find, to the nearest 100 square feet, the area of the base of the Colosseum.

In Exercises 62 to 67, use the quadratic formula to solve for y in terms of x. Then use a graphing utility to graph each equation.

62. $16x^2 + 9y^2 - 64x - 80 = 0$

63. $16x^2 + 9y^2 + 36y - 108 = 0$

64. $25x^2 + 16y^2 + 50x - 32y - 359 = 0$

65. $16x^2 + 9y^2 - 64x - 54y + 1 = 0$

66. $8x^2 + 25y^2 - 48x + 50y + 47 = 0$

67. $4x^2 + 9y^2 + 24x + 18y + 44 = 0$

SUPPLEMENTAL EXERCISES

68. Explain why the graph of $4x^2 + 9y - 16x - 2 = 0$ is or is not an ellipse. Sketch the graph of this equation.

In Exercises 69 to 72, find the equation in standard form of each ellipse by using the definition of an ellipse.

69. Find the equation of the ellipse with foci at $(-3, 0)$ and $(3, 0)$ that passes through the point $(3, 9/2)$.

70. Find the equation of the ellipse with foci at $(0, 4)$ and $(0, -4)$ that passes through the point $(9/5, 4)$.

71. Find the equation of the ellipse with foci at $(-1, 2)$ and $(3, 2)$ that passes through the point $(3, 5)$.

72. Find the equation of the ellipse with foci at $(-1, 1)$ and $(-1, 7)$ that passes through the point $(3/4, 1)$.

In Exercises 73 and 74, find the latus rectum of the given ellipse. A line segment with endpoints on the ellipse that is perpendicular to the major axis and passes through a focus is a *latus rectum* of the ellipse.

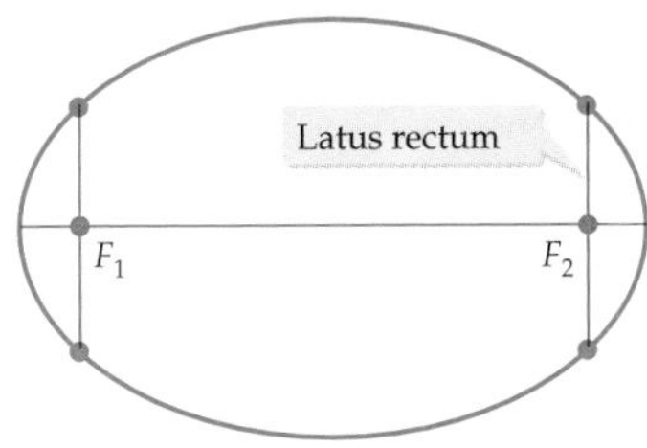

73. Find the length of a latus rectum of the ellipse given by

$$\frac{(x-1)^2}{9} + \frac{(y+1)^2}{16} = 1$$

74. Find the length of a latus rectum of the ellipse given by

$$9x^2 + 16y^2 - 36x + 96y + 36 = 0$$

75. Show that for any ellipse, the length of a latus rectum is $2b^2/a$.

76. Use the definition of an ellipse to find the equation of an ellipse with center at $(0, 0)$ and foci at $(0, c)$ and $(0, -c)$.

Recall that a parabola has a directrix that is a line perpendicular to the axis of symmetry. An ellipse has two directrices, both of which are perpendicular to the major axis and outside the ellipse. For an ellipse with center at the origin and whose major axis is the x-axis, the equations of the directrices are $x = a^2/c$ and $x = -a^2/c$.

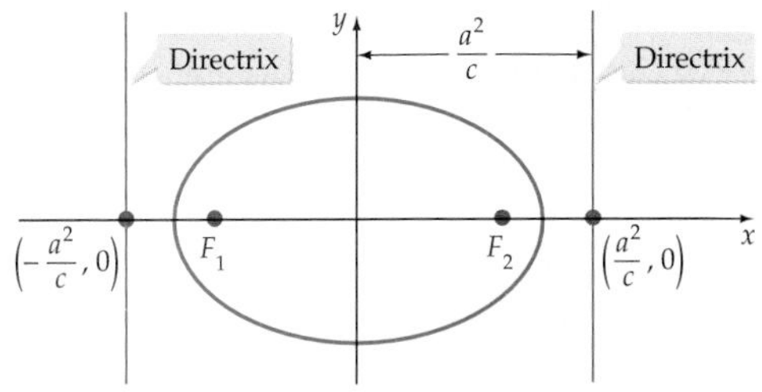

77. Find the directrices of the ellipse in Exercise 3.

78. Find the directrices of the ellipse in Exercise 4.

79. Let $P(x, y)$ be a point on the ellipse $\frac{x^2}{12} + \frac{y^2}{8} = 1$. Show that the distance from the point P to the focus $(2, 0)$ divided by the distance from the point P to the directrix $x = 6$ equals the eccentricity. (*Hint:* Solve the equation of the ellipse for y^2. Substitute this value for y^2 after applying the distance formula.)

80. Generalize the results of Exercise 79. That is, show that if $P(x, y)$ is a point on the ellipse $\frac{x^2}{a^2} + \frac{y^2}{b^2} = 1$, where $F(c, 0)$ is a focus and $x = a^2/c$ is a directrix, then the following equation is true: $e = d(P, F)/d(P, D)$. (*Hint:* Solve the equation of the ellipse for y^2. Substitute this value for y^2 after applying the distance formula.)

PROJECTS

1. **I. M. PEI'S OVAL** The poet and architect I. M. Pei suggested that the oval with the most appeal to the eye is given by the equation

$$\left(\frac{x}{a}\right)^{3/2} + \left(\frac{y}{b}\right)^{3/2} = 1$$

Use a graphing utility to graph this equation with $a = 5$ and $b = 3$. Then compare your graph with the graph of.

$$\left(\frac{x}{5}\right)^2 + \left(\frac{y}{3}\right)^2 = 1$$

2. **KEPLER'S LAWS** The German astronomer Johannes Kepler (1571–1630) derived three laws that describe how the planets orbit the sun. Write an essay that includes biographical information about Kepler and a statement of Kepler's Laws. In addition, use Kepler's Laws to answer the following questions.

   a. Where is a planet located in its orbit around the sun when it achieves its greatest velocity?

   b. What is the period of Mars if it has a mean distance from the sun of 1.52 astronomical units? (*Hint:* Use Earth as a reference with a period of 1 year and a mean distance from the sun of 1 astronomical unit.)

3. **NEPTUNE** The position of the planet Neptune was discovered by using celestial mechanics and mathematics. Write an essay that tells how, when, and by whom Neptune was discovered.

4. **GRAPH THE COLOSSEUM** Some of the Colosseum scenes in the movie *Gladiator* (Universal Studios, 2000) were computer-generated.

   a. You can create a simple but accurate scale image of the exterior of the Colosseum by using a computer and the mathematics software program *Maple*. Open a new *Maple* worksheet and enter the following two commands.

```
with(plots);
plots[implicitplot3d]((x^2)/(307.5^2)+(y^2)/(255^2)=1, x= –310..310, y= –260..260,
z=0..157, scaling=CONSTRAINED, style=PATCHNOGRID, axes=FRAMED);
```

   Execute each of the commands by placing the cursor in a command and pressing the *enter* key. After execution of the second command a 3-dimensional graph will appear. Click and drag on the graph to rotate the image.

   b. A graphing calculator can also be used to generate a simple "graph" of the exterior of the Colosseum. Here is a procedure for the TI-83 graphing calculator.

   Enter the following in the WINDOW menu.

   Xmin=–4.7 Xmax=4.7 Xscl=1 Ymin=–4 Ymax=9 Yscl=1

   Enter the following forumlas in the Y= menu.

   $Y_1=\sqrt{(9-X^2)}$ $Y_2=Y_1+4$ $Y_3=-Y_1$ $Y_4=Y_3+4$

   Press: QUIT (2nd MODE)

   Enter: Shade(Y_3,Y_4) and press **ENTER**. Note: "Shade(" is in the DRAW menu.

   Explain why this "Colosseum graph" appears to be constructed with ellipses even though the functions entered in the Y= menu are the equations of semicircles.

SECTION 8.3 HYPERBOLAS

- HYPERBOLAS WITH CENTER AT (0, 0)
- HYPERBOLAS WITH CENTER AT (h, k)
- ECCENTRICITY OF A HYPERBOLA
- APPLICATIONS

8

The hyperbola is a conic section formed when a plane intersects a right circular cone at a certain angle. If β is the angle at which the plane intersects the axis of the cone and α is the angle shown in **Figure 8.34,** a hyperbola is formed when $0° < \beta < \alpha$ or when the plane is parallel to the axis of the cone.

As with the other conic sections, there is a definition of a hyperbola in terms of a certain set of points in the plane.

take note

If the plane intersects the cone along the axis of the cone, the resulting curve is two intersecting straight lines. This is the *degenerate* form of a hyperbola. See the accompanying figure.

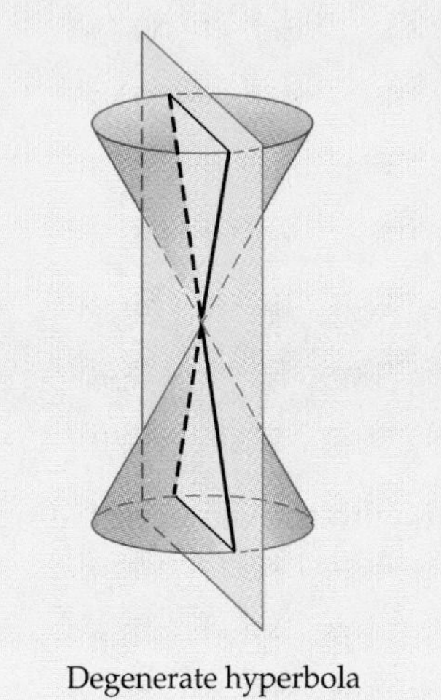

Degenerate hyperbola

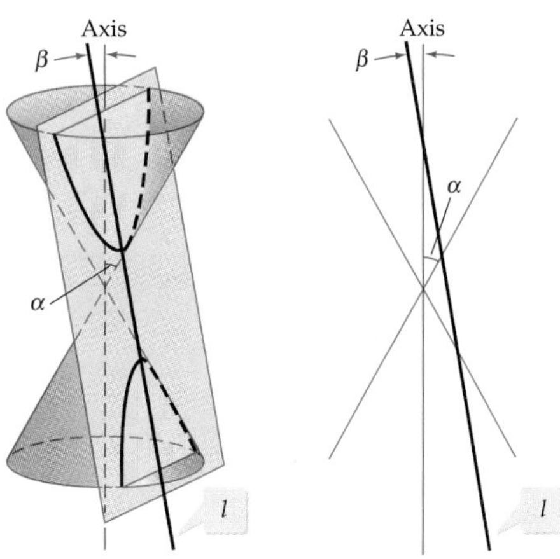

Figure 8.34

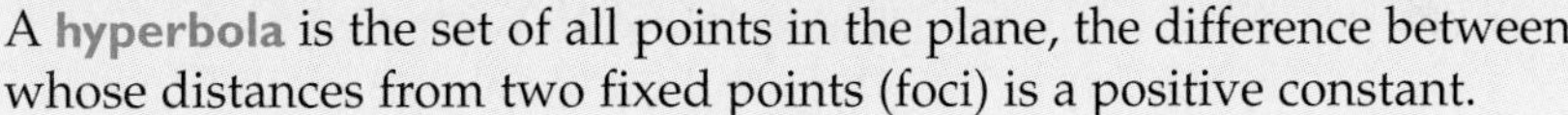

Definition of a Hyperbola

A **hyperbola** is the set of all points in the plane, the difference between whose distances from two fixed points (foci) is a positive constant.

This definition differs from that of an ellipse in that the ellipse was defined in terms of the *sum* of two distances, whereas the hyperbola is defined in terms of the *difference* of two distances.

take note

A web applet is available to experiment with hyperbolas by manipulating the foci and vertices. This applet, Hyperbola, can be found on our web site at

http://college.hmco.com

◆ HYPERBOLAS WITH CENTER AT (0, 0)

The **transverse axis** is the line segment joining the intercepts (see **Figure 8.35**). The midpoint of the transverse axis is called the **center** of the hyperbola. The **conjugate axis** passes through the center of the hyperbola and is perpendicular to the transverse axis.

The length of the transverse axis is customarily represented as $2a$, and the distance between the two foci is represented as $2c$. The length of the conjugate axis is represented as $2b$.

The **vertices** of a hyperbola are the points where the hyperbola intersects the transverse axis.

To determine the positive constant stated in the definition of a hyperbola, consider the point $V_1(a, 0)$, which is one vertex of a hyperbola, and the points $F_1(c, 0)$ and $F_2(-c, 0)$, which are the foci of the hyperbola (see **Figure 8.36**). The difference between the distance from $V_1(a, 0)$ to $F_1(c, 0)$, $c - a$, and the distance from $V_1(a, 0)$ to $F_2(-c, 0)$, $c + a$, must be a constant. By subtracting these distances, we find

$$|(c - a) - (c + a)| = |-2a| = 2a$$

Thus the constant is $2a$ and is the length of the transverse axis. The absolute value is used to ensure that the distance is a positive number.

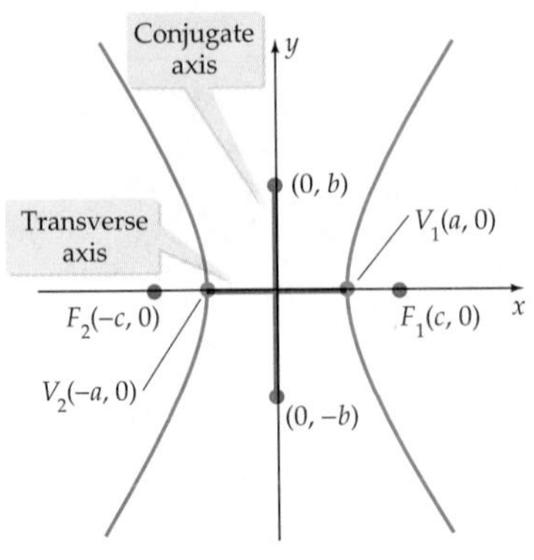

Figure 8.35

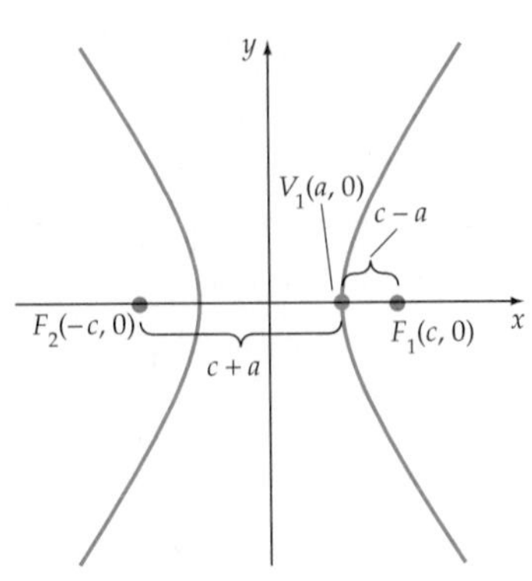

Figure 8.36

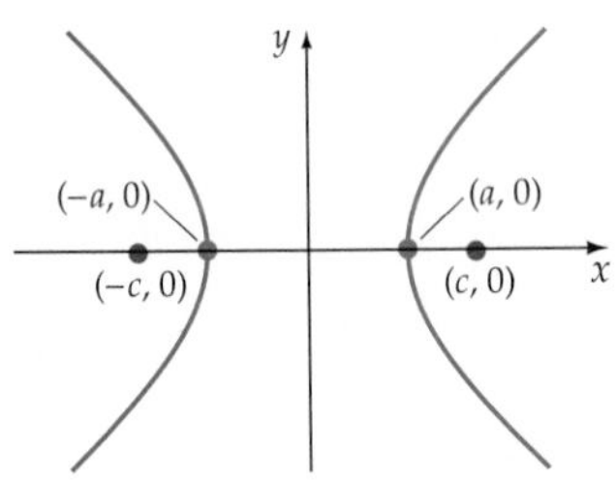

a. Transverse axis on the x-axis

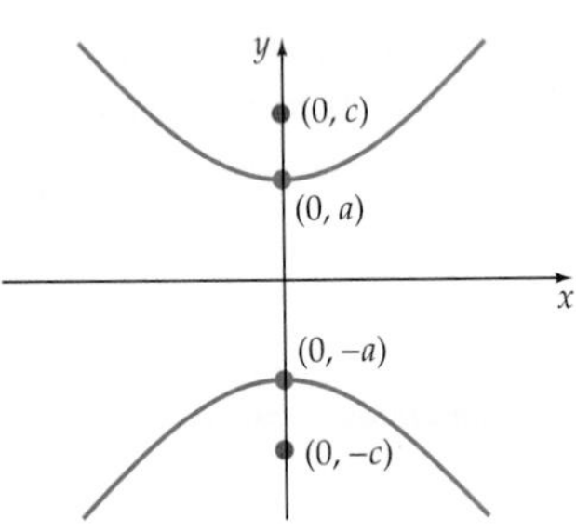

b. Transverse axis on the y-axis

Figure 8.37

Standard Forms of the Equation of a Hyperbola with Center at the Origin

Transverse Axis on the x-axis

The standard form of the equation of a hyperbola with the center at the origin and transverse axis on the x-axis (see **Figure 8.37a**) is given by

$$\frac{x^2}{a^2} - \frac{y^2}{b^2} = 1$$

The coordinates of the vertices are $(a, 0)$ and $(-a, 0)$, and the coordinates of the foci are $(c, 0)$ and $(-c, 0)$, where $c^2 = a^2 + b^2$.

Transverse Axis on the y-axis

The standard form of the equation of a hyperbola with the center at the origin and transverse axis on the y-axis (see **Figure 8.37b**) is given by

$$\frac{y^2}{a^2} - \frac{x^2}{b^2} = 1$$

The coordinates of the vertices are $(0, a)$ and $(0, -a)$, and the coordinates of the foci are $(0, c)$ and $(0, -c)$, where $c^2 = a^2 + b^2$.

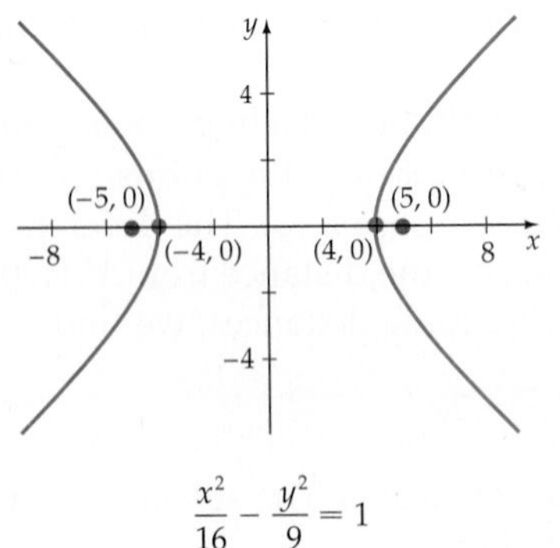

$\frac{x^2}{16} - \frac{y^2}{9} = 1$

Figure 8.38

By looking at the equations, it is possible to determine the location of the transverse axis by finding which term in the equation is positive. When the x^2 term is positive, the transverse axis is on the x-axis. When the y^2 term is positive, the transverse axis is on the y-axis.

Consider the hyperbola given by the equation $\frac{x^2}{16} - \frac{y^2}{9} = 1$. Because the x^2 term is positive, the transverse axis is on the x-axis, $a^2 = 16$, and thus $a = 4$. The vertices are $(4, 0)$ and $(-4, 0)$. To find the foci, we determine c.

$$c^2 = a^2 + b^2 = 16 + 9 = 25$$
$$c = \sqrt{25} = 5$$

The foci are $(5, 0)$ and $(-5, 0)$. The graph is shown in **Figure 8.38.**

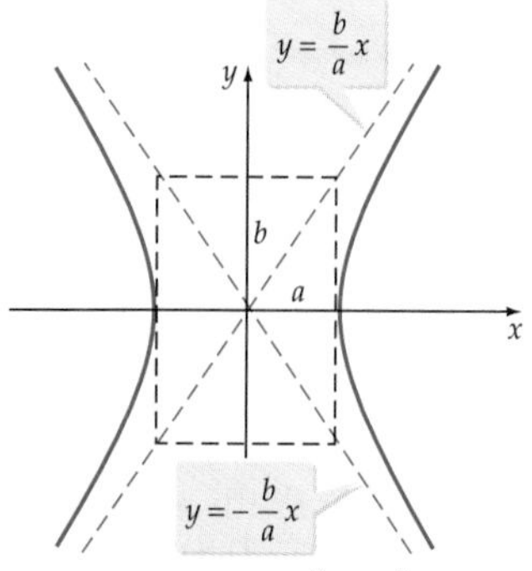

a. Asympyotes of $\frac{x^2}{a^2} - \frac{y^2}{b^2} = 1$

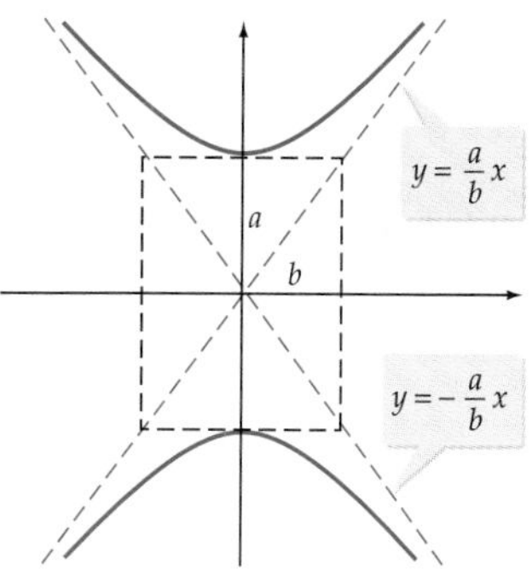

b. Asympyotes of $\frac{y^2}{a^2} - \frac{x^2}{b^2} = 1$

Figure 8.39

Each hyperbola has two asymptotes that pass through the center of the hyperbola. The asymptotes of the hyperbola are a useful guide to sketching the graph of the hyperbola.

Asymptotes of a Hyperbola with Center at the Origin

The **asymptotes** of the hyperbola defined by $\frac{x^2}{a^2} - \frac{y^2}{b^2} = 1$ are given by the equations $y = \frac{b}{a}x$ and $y = -\frac{b}{a}x$ (see **Figure 8.39a**).

The asymptotes of the hyperbola defined by $\frac{y^2}{a^2} - \frac{x^2}{b^2} = 1$ are given by the equations $y = \frac{a}{b}x$ and $y = -\frac{a}{b}x$ (see **Figure 8.39b**).

One method for remembering the equations of the asymptotes is to write the equation of a hyperbola in standard form but to replace 1 by 0 and then solve for y.

$$\frac{x^2}{a^2} - \frac{y^2}{b^2} = 0 \quad \text{so} \quad y^2 = \frac{b^2}{a^2}x^2, \text{ or } y = \pm\frac{b}{a}x$$

$$\frac{y^2}{a^2} - \frac{x^2}{b^2} = 0 \quad \text{so} \quad y^2 = \frac{a^2}{b^2}x^2, \text{ or } y = \pm\frac{a}{b}x$$

EXAMPLE 1 Find the Vertices, Foci, and Asymptotes of a Hyperbola

Find the vertices, foci, and asymptotes of the hyperbola given by the equation $\frac{y^2}{9} - \frac{x^2}{4} = 1$. Sketch the graph.

Solution

Because the y^2 term is positive, the transverse axis is on the y-axis. We know $a^2 = 9$; thus $a = 3$. The vertices are $V_1(0, 3)$ and $V_2(0, -3)$.

$$c^2 = a^2 + b^2 = 9 + 4$$
$$c = \sqrt{13}$$

The foci are $F_1(0, \sqrt{13})$ and $F_2(0, -\sqrt{13})$.

Because $a = 3$ and $b = 2$ ($b^2 = 4$), the equations of the asymptotes are $y = \frac{3}{2}x$ and $y = -\frac{3}{2}x$.

To sketch the graph, we draw a rectangle that has its center at the origin and has dimensions equal to the lengths of the transverse and conjugate axes. The asymptotes are extensions of the diagonals of the rectangle. See **Figure 8.40**.

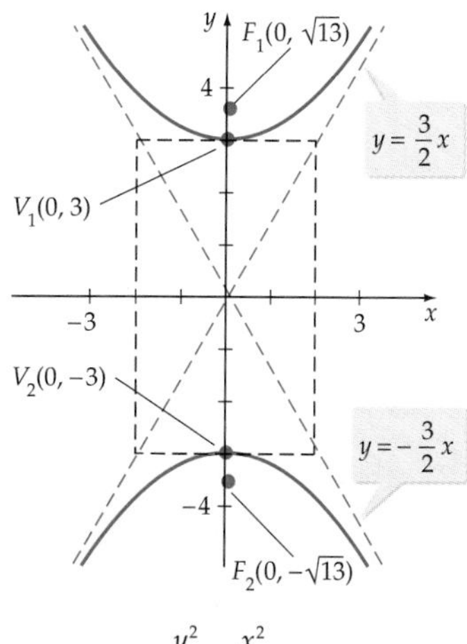

$\frac{y^2}{9} - \frac{x^2}{4} = 1$

Figure 8.40

TRY EXERCISE 4, EXERCISE SET 8.3, PAGE 618

◆ HYPERBOLAS WITH CENTER AT (h, k)

Using a translation of coordinates similar to that used for ellipses, we can write the equation of a hyperbola with its center at the point (h, k). Given coordinate axes labeled x' and y', an equation of a hyperbola with center at the origin is

$$\frac{(x')^2}{a^2} - \frac{(y')^2}{b^2} = 1 \qquad (1)$$

Figure 8.41

Now place the origin of this coordinate system at the point (h, k) of the xy-coordinate system, as shown in **Figure 8.41**. The relationship between an ordered pair in the $x'y'$-coordinate system and the xy-coordinate system is given by the transformation equations

$$x' = x - h$$
$$y' = y - k$$

Substitute the expressions for x' and y' into Equation (1). The equation of the hyperbola with center at (h, k) is

$$\frac{(x-h)^2}{a^2} - \frac{(y-k)^2}{b^2} = 1$$

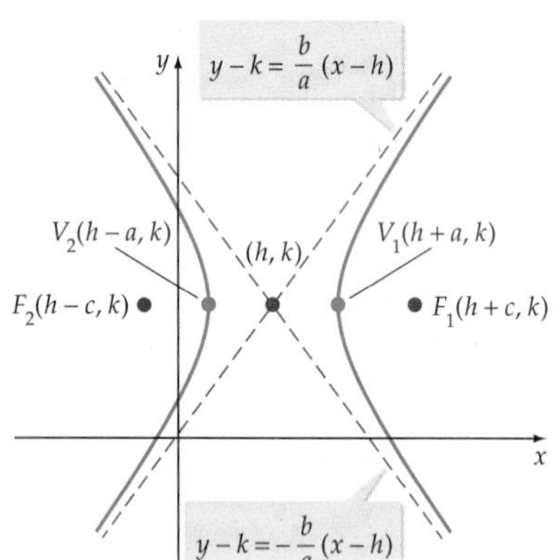

a. Transverse axis parallel to the x-axis

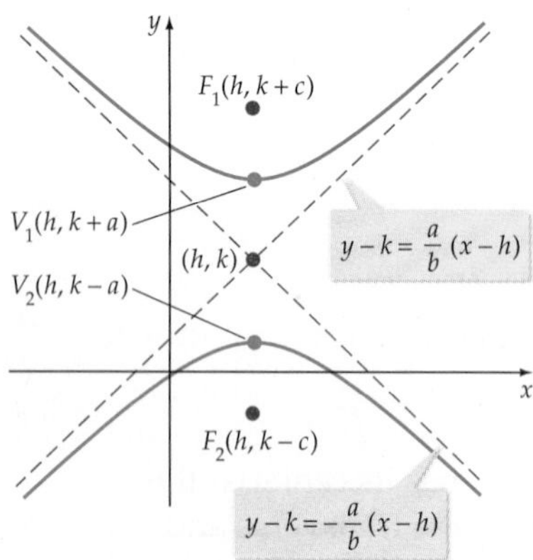

b. Transverse axis parallel to the y-axis

Figure 8.42

Standard Forms of the Equation of a Hyperbola with Center at (h, k)

Transverse Axis Parallel to the x-axis
The standard form of the equation of a hyperbola with center at (h, k) and transverse axis parallel to the x-axis (see **Figure 8.42a**) is given by

$$\frac{(x-h)^2}{a^2} - \frac{(y-k)^2}{b^2} = 1$$

The coordinates of the vertices are $V_1(h + a, k)$ and $V_2(h - a, k)$. The coordinates of the foci are $F_1(h + c, k)$ and $F_2(h - c, k)$, where $c^2 = a^2 + b^2$. The equations of the asymptotes are $y - k = \pm\frac{b}{a}(x - h)$.

Transverse Axis Parallel to the y-axis
The standard form of the equation of a hyperbola with center at (h, k) and transverse axis parallel to the y-axis (see **Figure 8.42b**) is given by

$$\frac{(y-k)^2}{a^2} - \frac{(x-h)^2}{b^2} = 1$$

The coordinates of the vertices are $V_1(h, k + a)$ and $V_2(h, k - a)$. The coordinates of the foci are $F_1(h, k + c)$ and $F_2(h, k - c)$, where $c^2 = a^2 + b^2$. The equations of the asymptotes are $y - k = \pm\frac{a}{b}(x - h)$.

EXAMPLE 2 Find the Vertices, Foci, and Asymptotes of a Hyperbola

Find the vertices, foci, and asymptotes of the hyperbola given by the equation $4x^2 - 9y^2 - 16x + 54y - 29 = 0$. Sketch the graph.

Solution

Write the equation of the hyperbola in standard form by completing the square.

$$4x^2 - 9y^2 - 16x + 54y - 29 = 0$$

$$4x^2 - 16x - 9y^2 + 54y = 29 \qquad \text{• Rearrange terms.}$$

$$4(x^2 - 4x) - 9(y^2 - 6y) = 29 \qquad \text{• Factor.}$$

$$4(x^2 - 4x + 4) - 9(y^2 - 6y + 9) = 29 + 16 - 81 \qquad \text{• Complete the square.}$$

$$4(x - 2)^2 - 9(y - 3)^2 = -36 \qquad \text{• Factor.}$$

$$\frac{(y-3)^2}{4} - \frac{(x-2)^2}{9} = 1 \qquad \text{• Divide by } -36.$$

The coordinates of the center are (2, 3). Because the term containing $(y - 3)^2$ is positive, the transverse axis is parallel to the y-axis. We know $a^2 = 4$; thus $a = 2$. The vertices are (2, 5) and (2, 1). See **Figure 8.43.**

$$c^2 = a^2 + b^2 = 4 + 9$$
$$c = \sqrt{13}$$

The foci are $(2, 3 + \sqrt{13})$ and $(2, 3 - \sqrt{13})$. We know $b^2 = 9$; thus $b = 3$. The equations of the asymptotes are $y - 3 = \pm(2/3)(x - 2)$ which simplifies to

$$y = \frac{2}{3}x + \frac{5}{3} \qquad \text{and} \qquad y = -\frac{2}{3}x + \frac{13}{3}$$

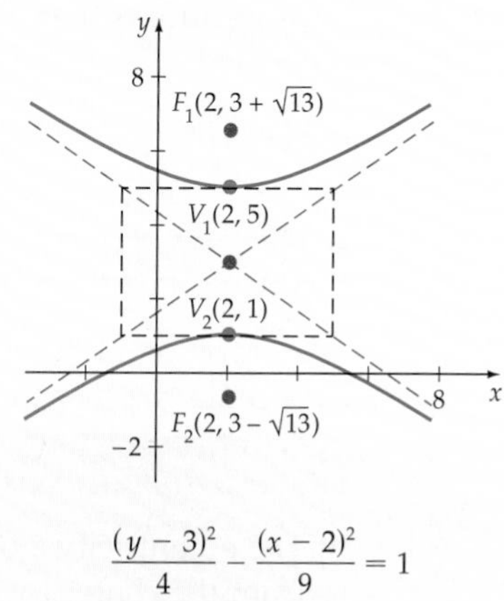

$$\frac{(y-3)^2}{4} - \frac{(x-2)^2}{9} = 1$$

Figure 8.43

TRY EXERCISE 26, EXERCISE SET 8.3, PAGE 618

A graphing utility can be used to graph a hyperbola. For instance, consider the equation $4x^2 - 9y^2 - 16x + 54y - 29 = 0$ from Example 2. Rewrite the equation as

$$-9y^2 + 54y + (4x^2 - 16x - 29) = 0$$

In this form, the equation is a quadratic equation in terms of the variable y with

$$A = -9, B = 54, \text{ and } C = 4x^2 - 16x - 29$$

Apply the quadratic formula to produce

$$y = \frac{-54 \pm \sqrt{2916 + 36(4x^2 - 16x - 29)}}{-18}$$

The graph of $Y_1 = \dfrac{-54 + \sqrt{2916 + 36(4x^2 - 16x - 29)}}{-18}$ is the upper branch of the hyperbola (see **Figure 8.44**). The graph of $Y_2 = \dfrac{-54 - \sqrt{2916 + 36(4x^2 - 16x - 29)}}{-18}$ is the lower branch of the hyperbola, as shown in **Figure 8.44.**

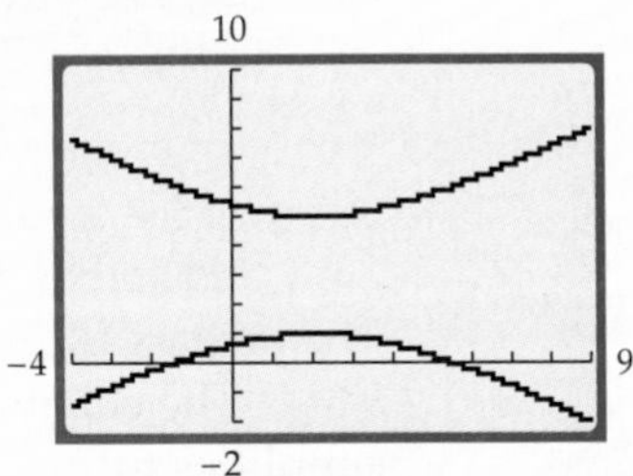

Figure 8.44

One advantage of this graphing procedure is that it does not require us to write the given equation in standard form. A disadvantage of the graphing procedure is that it does not indicate where the foci of the hyperbola are located.

◆ ECCENTRICITY OF A HYPERBOLA

The graph of a hyperbola can be very wide or very narrow. The **eccentricity** of a hyperbola is a measure of its "wideness."

Eccentricity (e) of a Hyperbola

The eccentricity e of a hyperbola is the ratio of c to a, where c is the distance from the center to a focus and a is the length of the semitransverse axis.

$$e = \frac{c}{a}$$

For a hyperbola, $c > a$ and therefore $e > 1$. As the eccentricity of the hyperbola increases, the graph becomes wider and wider, as shown in **Figure 8.45.**

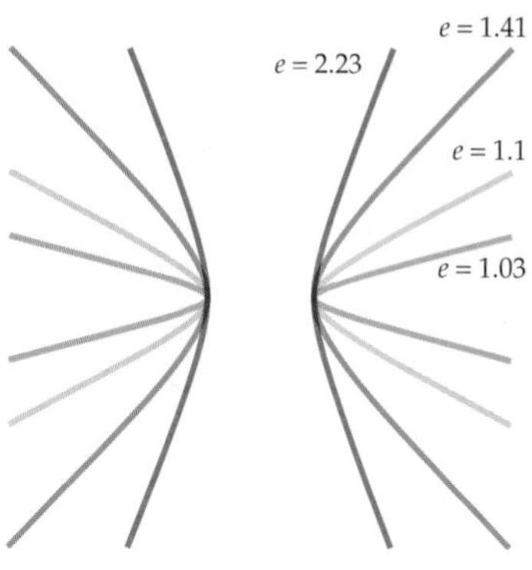

Figure 8.45

EXAMPLE 3 Find the Equation of a Hyperbola Given its Eccentricity

Find the standard form of the equation of the hyperbola that has eccentricity 3/2, center at the origin, and a focus (6, 0).

Solution

Because the focus is located at (6, 0) and the center is at the origin, $c = 6$. An extension of the transverse axis contains the foci, so the transverse axis is on the x-axis.

$$e = \frac{3}{2} = \frac{c}{a}$$

$$\frac{3}{2} = \frac{6}{a} \quad \text{• Substitute 6 for } c.$$

$$a = 4 \quad \text{• Solve for } a.$$

To find b^2, use the equation $c^2 = a^2 + b^2$ and the values for c and a.

$$c^2 = a^2 + b^2$$

$$36 = 16 + b^2$$

$$b^2 = 20$$

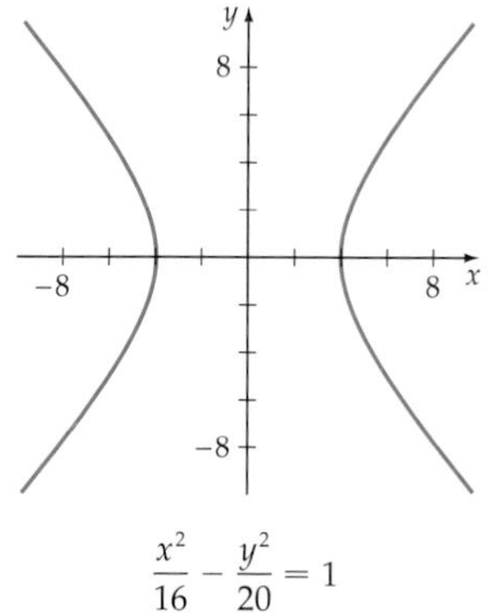

$\frac{x^2}{16} - \frac{y^2}{20} = 1$

Figure 8.46

The equation of the hyperbola is $\frac{x^2}{16} - \frac{y^2}{20} = 1$. See **Figure 8.46.**

TRY EXERCISE 48, EXERCISE SET 8.3, PAGE 618

◆ APPLICATIONS

Orbits of Comets In Section 8.2 we noted that the orbits of the planets are elliptical. Some comets have elliptical orbits also, the most notable being Halley's comet, whose eccentricity is 0.97.

MATH MATTERS

Caroline Herschel (1750–1848) became interested in mathematics and astronomy after her brother William discovered the planet Uranus. She was the first woman to receive credit for the discovery of a comet. In fact, between 1786 and 1797 she discovered eight comets. In 1828 she completed a catalog of over 2000 nebulae for which the Royal Astronomical Society of England presented her with its prestigious gold medal.

Other comets have hyperbolic orbits with the sun at a focus. These comets pass by the sun only once. The velocity of a comet determines whether its orbit is elliptical or hyperbolic. See **Figure 8.47.**

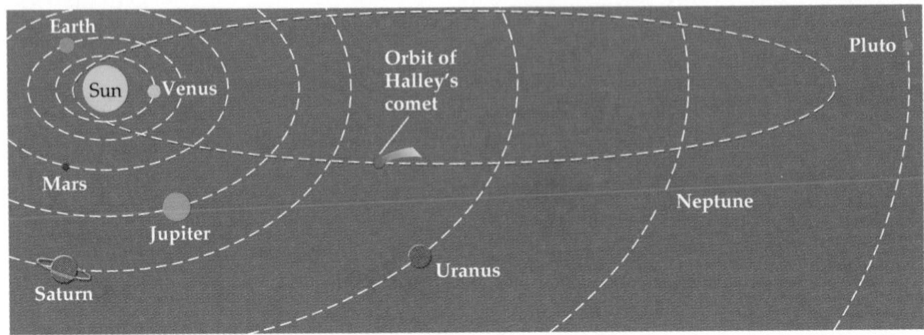

Figure 8.47

Hyperbolas as an Aid to Navigation Consider two radio transmitters, T_1 and T_2, placed some distance apart. A ship with electronic equipment measures the difference between the times it takes signals from the transmitters to reach the ship. Because the difference between the times is proportional to the difference between distances of the ship from the transmitters, the ship must be located on the hyperbola with foci at the two transmitters.

Using a third transmitter, T_3, we can find a second hyperbola with foci T_2 and T_3. The ship lies on the intersection of the two hyperbolas, as shown in **Figure 8.48.**

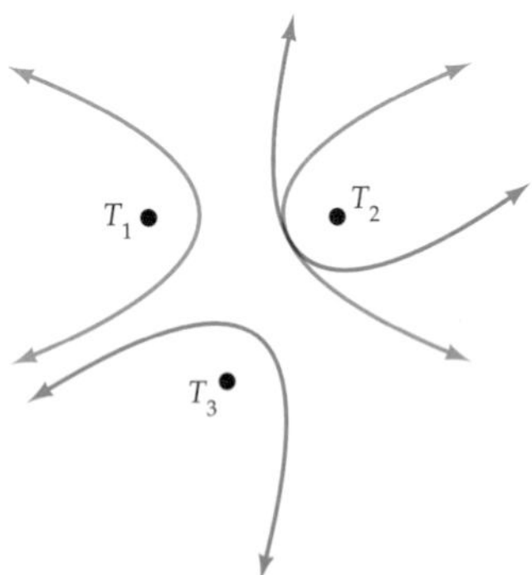

Figure 8.48

EXAMPLE 4 Determine the Position of a Ship

Two radio transmitters are positioned along a coastline, 500 miles apart. See **Figure 8.49.** Using a LORAN (LOng RAnge Navigation) system, a ship determines that a radio signal from transmitter T_1 reaches the ship 1600 microseconds before it receives a simultaneous signal from transmitter T_2.

a. Find an equation of a hyperbola (with foci located at T_1 and T_2) on which the ship lies. See **Figure 8.49.** (Assume the radio signals travel at 0.186 mile per microsecond.)

b. If the ship is directly north of transmitter T_1, determine how far (to the nearest mile) the ship is from the transmitter.

Solution

a. The ship lies on a hyperbola at point B, with foci at T_1 and T_2. The difference of the distances $d(T_2, B)$ and $d(T_1, B)$ is given by

$$\begin{aligned} \text{Distance} &= \text{rate} \times \text{time} \\ &= 0.186 \text{ mile/microsecond} \times 1600 \text{ microseconds} \\ &= 297.6 \text{ mile} \end{aligned}$$

Thus the ship is located on a hyperbola with transverse axis of 297.6 miles and semitransverse axis $a = 148.8$ miles. **Figure 8.49**

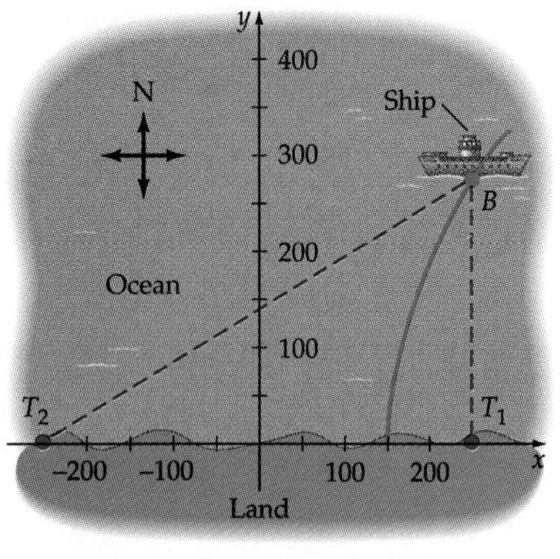

Figure 8.49

shows that the foci are located at (250, 0) and (−250, 0). Thus $c = 250$ miles, and

$$b = \sqrt{c^2 - a^2} = \sqrt{250^2 - 148.8^2} \approx 200.9 \text{ miles}$$

The ship is located on the hyperbola given by

$$\frac{x^2}{148.8^2} - \frac{y^2}{200.9^2} = 1$$

b. If the ship is directly north of T_1, then $x = 250$, and the distance from the ship to the transmitter T_1 is y, where

$$-\frac{y^2}{200.9^2} = 1 - \frac{250^2}{148.8^2}$$

$$y = \frac{200.9}{148.8}\sqrt{250^2 - 148.8^2} \approx 271 \text{ miles}$$

The ship is about 271 miles north of transmitter T_1.

TRY EXERCISE 54, EXERCISE SET 8.3, PAGE 619

Hyperbolas also have a reflective property that makes them useful in many applications.

Reflective Property of a Hyperbola

A ray of light directed toward one focus of a hyperbolic mirror is reflected toward the other focus. See **Figures 8.50** and **8.51.**

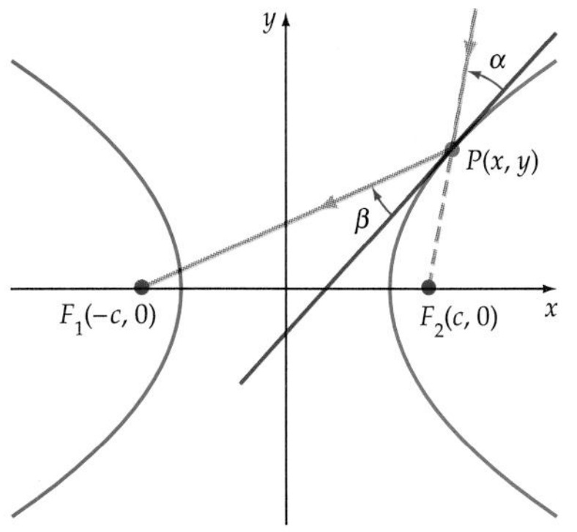

Figure 8.50

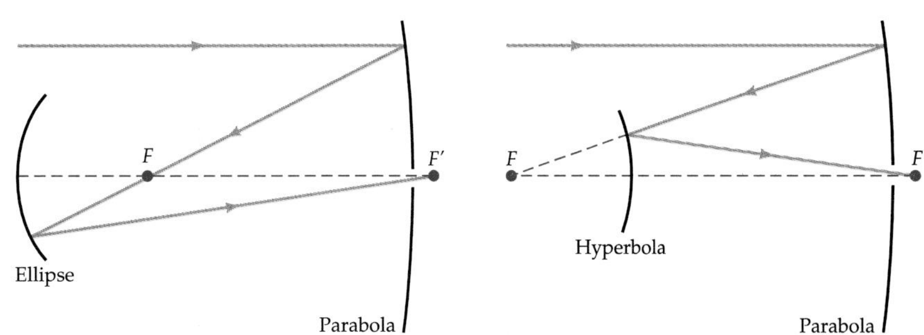

Figure 8.51

TOPICS FOR DISCUSSION

1. In every hyperbola, the distance c from a focus to the center of the hyperbola is greater than the length of the semi-transverse axis a. Do you agree? Explain.
2. How many vertices does a hyperbola have?

3. Explain why the eccentricity of every hyperbola is a number greater than 1.

4. Is the conjugate axis of a hyperbola perpendicular to the transverse axis of the hyperbola?

EXERCISE SET 8.3

In Exercises 1 to 26, find the center, vertices, foci, and asymptotes for the hyperbola given by each equation. Graph each equation.

1. $\frac{x^2}{16} - \frac{y^2}{25} = 1$

2. $\frac{x^2}{16} - \frac{y^2}{9} = 1$

3. $\frac{y^2}{4} - \frac{x^2}{25} = 1$

4. $\frac{y^2}{25} - \frac{x^2}{36} = 1$

5. $\frac{x^2}{7} - \frac{y^2}{9} = 1$

6. $\frac{x^2}{5} - \frac{y^2}{4} = 1$

7. $\frac{4x^2}{9} - \frac{y^2}{16} = 1$

8. $\frac{x^2}{9} - \frac{9y^2}{16} = 1$

9. $\frac{(x-3)^2}{16} - \frac{(y+4)^2}{9} = 1$

10. $\frac{(x+3)^2}{25} - \frac{y^2}{4} = 1$

11. $\frac{(y+2)^2}{4} - \frac{(x-1)^2}{16} = 1$

12. $\frac{(y-2)^2}{36} - \frac{(x+1)^2}{49} = 1$

13. $\frac{(x+2)^2}{9} - \frac{y^2}{25} = 1$

14. $\frac{x^2}{25} - \frac{(y-2)^2}{81} = 1$

15. $\frac{9(x-1)^2}{16} - \frac{(y+1)^2}{9} = 1$

16. $\frac{(x+6)^2}{25} - \frac{25y^2}{144} = 1$

17. $x^2 - y^2 = 9$

18. $4x^2 - y^2 = 16$

19. $16y^2 - 9x^2 = 144$

20. $9y^2 - 25x^2 = 225$

21. $9y^2 - 36x^2 = 4$

22. $16x^2 - 25y^2 = 9$

23. $x^2 - y^2 - 6x + 8y - 3 = 0$

24. $4x^2 - 25y^2 + 16x + 50y - 109 = 0$

25. $9x^2 - 4y^2 + 36x - 8y + 68 = 0$

26. $16x^2 - 9y^2 - 32x - 54y + 79 = 0$

In Exercises 27 to 32, use the quadratic formula to solve for y in terms of x. Then use a graphing utility to graph each equation.

27. $4x^2 - y^2 + 32x + 6y + 39 = 0$

28. $x^2 - 16y^2 + 8x - 64y + 16 = 0$

29. $9x^2 - 16y^2 - 36x - 64y + 116 = 0$

30. $2x^2 - 9y^2 + 12x - 18y + 18 = 0$

31. $4x^2 - 9y^2 + 8x - 18y - 6 = 0$

32. $2x^2 - 9y^2 - 8x + 36y - 46 = 0$

In Exercises 33 to 46, find the equation in standard form of the hyperbola that satisfies the stated conditions.

33. Vertices $(3, 0)$ and $(-3, 0)$, foci $(4, 0)$ and $(-4, 0)$

34. Vertices $(0, 2)$ and $(0, -2)$, foci $(0, 3)$ and $(0, -3)$

35. Foci $(0, 5)$ and $(0, -5)$, asymptotes $y = 2x$ and $y = -2x$

36. Foci $(4, 0)$ and $(-4, 0)$, asymptotes $y = x$ and $y = -x$

37. Vertices $(0, 3)$ and $(0, -3)$, passing through $(2, 4)$

38. Vertices $(5, 0)$ and $(-5, 0)$, passing through $(-1, 3)$

39. Asymptotes $y = \frac{1}{2}x$ and $y = -\frac{1}{2}x$, vertices $(0, 4)$ and $(0, -4)$

40. Asymptotes $y = \frac{2}{3}x$ and $y = -\frac{2}{3}x$, vertices $(6, 0)$ and $(-6, 0)$

41. Vertices $(6, 3)$ and $(2, 3)$, foci $(7, 3)$ and $(1, 3)$

42. Vertices $(-1, 5)$ and $(-1, -1)$, foci $(-1, 7)$ and $(-1, -3)$

43. Foci $(1, -2)$ and $(7, -2)$, slope of an asymptote $5/4$

44. Foci $(-3, -6)$ and $(-3, -2)$, slope of an asymptote 1

45. Passing through $(9, 4)$, slope of an asymptote $1/2$, center $(7, 2)$, transverse axis parallel to the y-axis

46. Passing through $(6, 1)$, slope of an asymptote 2, center $(3, 3)$, transverse axis parallel to the x-axis

In Exercises 47 to 52, use the eccentricity to find the equation in standard form of each hyperbola.

47. Vertices $(1, 6)$ and $(1, 8)$, eccentricity 2

48. Vertices $(2, 3)$ and $(-2, 3)$, eccentricity $5/2$

49. Eccentricity 2, foci $(4, 0)$ and $(-4, 0)$

50. Eccentricity $4/3$, foci $(0, 6)$ and $(0, -6)$

51. Center $(4, 1)$, conjugate axis of length 4, eccentricity $4/3$ (*Hint:* There are two answers.)

52. Center $(-3, -3)$, conjugate axis of length 6, eccentricity 2 (*Hint:* There are two answers.)

53. **LORAN** Two radio transmitters are positioned along the coast, 250 miles apart. A signal is sent simultaneously from each transmitter. The signal from transmitter T_2 is

received by a ship's LORAN 500 microseconds after it receives the signal from T_1. The radio signal travels 0.186 mile per microsecond.

a. Find an equation of a hyperbola, with foci at T_1 and T_2, on which the ship is located.

b. If the ship is 100 miles east of the y-axis, determine its distance from the coastline (to the nearest mile).

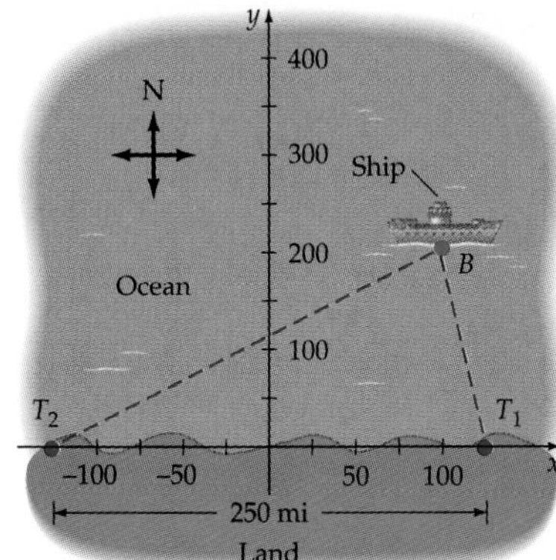

54. LORAN Two radio transmitters are positioned along the coast, 300 miles apart. A signal is sent simultaneously from each transmitter. The signal from transmitter T_1 is received by a ship's LORAN 800 microseconds after it receives the signal from T_2. The radio signal travels 0.186 mile per microsecond.

a. Find an equation of a hyperbola, with foci at T_1 and T_2, on which the ship is located.

b. If the ship continues to travel so that the difference of 800 microseconds is maintained, determine the point at which the ship will reach the coastline.

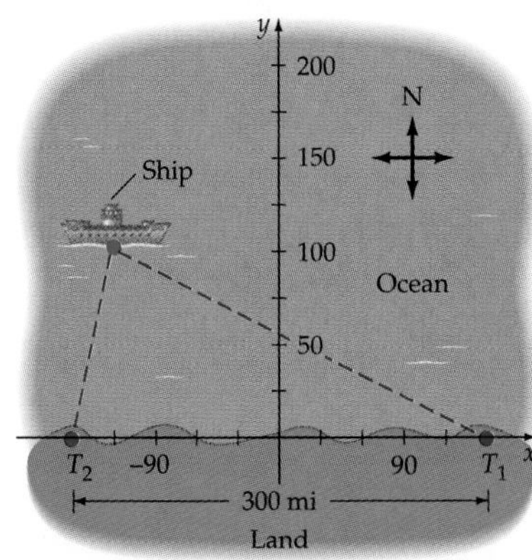

In Exercises 55 to 62, identify the graph of each equation as a parabola, ellipse, or hyperbola. Graph each equation.

55. $4x^2 + 9y^2 - 16x - 36y + 16 = 0$

56. $2x^2 + 3y - 8x + 2 = 0$

57. $5x - 4y^2 + 24y - 11 = 0$

58. $9x^2 - 25y^2 - 18x + 50y = 0$

59. $x^2 + 2y - 8x = 0$

60. $9x^2 + 16y^2 + 36x - 64y - 44 = 0$

61. $25x^2 + 9y^2 - 50x - 72y - 56 = 0$

62. $(x - 3)^2 + (y - 4)^2 = (x + 1)^2$

SUPPLEMENTAL EXERCISES

In Exercises 63 to 66, use the definition of a hyperbola to find the equation of the hyperbola in standard form.

63. Foci $(2, 0)$ and $(-2, 0)$; passes through the point $(2, 3)$

64. Foci $(0, 3)$ and $(0, -3)$; passes through the point $(5/2, 3)$

65. Foci $(0, 4)$ and $(0, -4)$; passes through the point $(7/3, 4)$

66. Foci $(5, 0)$ and $(-5, 0)$; passes through the point $(5, 9/4)$

Recall that an ellipse has two directrices that are lines perpendicular to the line containing the foci. A hyperbola also has two directrices; they are perpendicular to the transverse axis and outside the hyperbola. For a hyperbola with center at the origin and transverse axis on the x-axis, the equations of the directrices are $x = a^2/c$ and $x = -a^2/c$. In Exercises 67 to 70, use this information to solve each exercise.

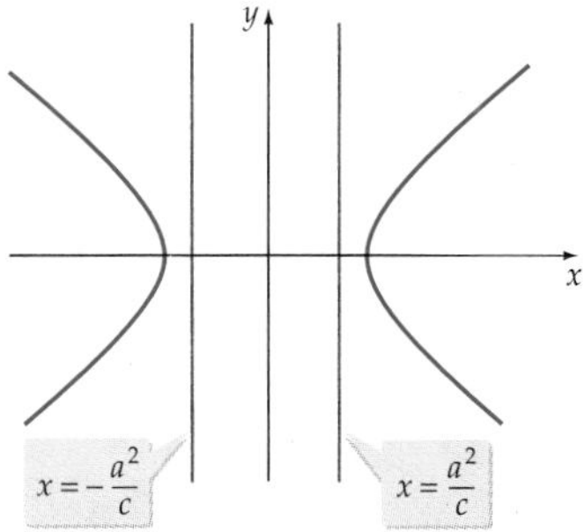

67. Find the directrices for the hyperbola in Exercise 1.

68. Find the directrices for the hyperbola in Exercise 2.

69. Let $P(x, y)$ be a point on the hyperbola $\frac{x^2}{9} - \frac{y^2}{16} = 1$. Show that the distance from the point P to the focus $(5, 0)$ divided by the distance from the point P to the directrix $x = 9/5$ equals the eccentricity.

70. Generalize the results of Exercise 69. That is, show that if $P(x, y)$ is a point on the hyperbola $\frac{x^2}{a^2} - \frac{y^2}{b^2} = 1$, $F(c, 0)$ is a focus, and $x = a^2/c$ is a directrix, then the following equation is true:

$$e = d(P, F)/d(P, D)$$

71. Sketch a graph of $\frac{x|x|}{16} - \frac{y|y|}{9} = 1$.

72. Sketch a graph of $\frac{x|x|}{16} + \frac{y|y|}{9} = 1$.

PROJECTS

1. **A HYPERBOLIC PARABOLOID** A *hyperbolic paraboloid* is a three-dimensional figure. Some of its cross sections are parabolas and some hyperbolas. Make a drawing of a hyperbolic paraboloid. Explain the relationship that exists between the equations of the parabolic cross sections and the relationship that exists between the equations of the hyperbolic cross sections.

2. **A HYPERBOLOID OF ONE SHEET** Make a sketch of a *hyperboloid of one sheet*. Explain the different cross sections of the hyperboloid of one sheet. Do some research on nuclear power plants, and explain why nuclear cooling towers are designed in the shape of hyperboloids of one sheet.

SECTION 8.4 ROTATION OF AXES

- THE ROTATION THEOREM FOR CONICS
- THE CONIC IDENTIFICATION THEOREM
- USE A GRAPHING UTILITY TO GRAPH SECOND-DEGREE EQUATIONS IN TWO VARIABLES

♦ THE ROTATION THEOREM FOR CONICS

The equation of a conic with axes parallel to the coordinate axes can be written in a general form.

General Equation of a Conic with Axes Parallel to Coordinate Axes

The **general equation of a conic** with axes parallel to the coordinate axes and not both A and C equal to zero is

$$Ax^2 + Cy^2 + Dx + Ey + F = 0$$

The graph of the equation is a parabola when $AC = 0$, an ellipse when $AC > 0$, and a hyperbola when $AC < 0$.

take note

Some choices of the constants A, B, C, D, E, and F may result in a degenerate conic or an equation that has no solutions.

The terms Dx, Ey, and F determine the translation of the conic from the origin. The general equation of a conic is a *second-degree equation* in two variables. A more general second-degree equation can be written that contains a Bxy term.

General Second-Degree Equation in Two Variables

The **general second-degree equation in two variables** is

$$Ax^2 + Bxy + Cy^2 + Dx + Ey + F = 0$$

The Bxy term ($B \neq 0$) determines a rotation of the conic so that its axes are no longer parallel to the coordinate axes.

A **rotation of axes** is a rotation of the x- and y-axes about the origin to another position denoted by x' and y'. We denote the measure of the **angle of rotation** by α.

Let P be some point in the plane, and let r represent the distance of P from the origin. The coordinates of P relative to the xy-coordinate system and the $x'y'$-coordinate system are $P(x, y)$ and $P(x', y')$, respectively.

Let $Q(x, 0)$ and $R(0, y)$ be the projections of P onto the x- and y-axis and let $Q'(x', 0')$ and $R'(0', y')$ be projections of P onto the x'- and the y'-axis. (See **Figure 8.52.**) The angle between the x'-axis and OP is denoted by θ. We can express the coordinates of P in each coordinate system in terms of α and θ.

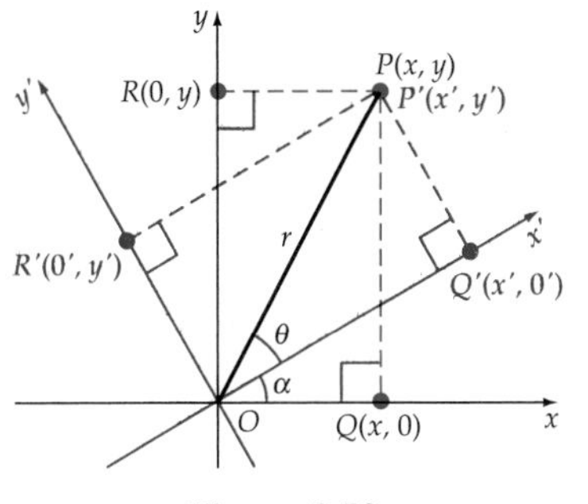

Figure 8.52

$$x = r\cos(\theta + \alpha) \qquad x' = r\cos\theta$$
$$y = r\sin(\theta + \alpha) \qquad y' = r\sin\theta$$

Applying the addition formulas for $\cos(\theta + \alpha)$ and $\sin(\theta + \alpha)$, we get

$$x = r\cos(\theta + \alpha) = r\cos\theta\cos\alpha - r\sin\theta\sin\alpha$$

$$y = r\sin(\theta + \alpha) = r\sin\theta\cos\alpha + r\cos\theta\sin\alpha$$

Now, substituting x' for $r\cos\theta$ and y' for $r\sin\theta$ into these equations yields

$$x = x'\cos\alpha - y'\sin\alpha \qquad \bullet\ x' = r\cos\theta,\ y' = r\sin\alpha$$
$$y = y'\cos\alpha + x'\sin\alpha$$

This proves the equations labeled (1) of the following theorem.

Rotation-of-Axes Formulas

Suppose an xy-coordinate system and an $x'y'$-coordinate system have the same origin and α is the angle between the positive x-axis and the positive x'-axis. If the coordinates of a point P are (x, y) in one system and (x', y') in the rotated system, then

$$\left.\begin{array}{l} x = x'\cos\alpha - y'\sin\alpha \\ y = y'\cos\alpha + x'\sin\alpha \end{array}\right\} \quad (1) \qquad \left.\begin{array}{l} x' = x\cos\alpha + y\sin\alpha \\ y' = y\cos\alpha - x\sin\alpha \end{array}\right\} \quad (2)$$

The derivations of the formulas for x' and y' are left as an exercise.

As we have noted, the appearance of the Bxy ($B \neq 0$) term in the general second-degree equation indicates that the graph of the conic has been rotated. The angle through which the axes have been rotated can be determined from the following theorem.

Rotation Theorem for Conics

Let $Ax^2 + Bxy + Cy^2 + Dx + Ey + F = 0$, $B \neq 0$, be the equation of a conic in an xy-coordinate system, and let α be an angle of rotation such that

$$\cot 2\alpha = \frac{A - C}{B}, \quad 0° < 2\alpha < 180°$$

Then the equation of the conic in the rotated coordinate system will be

$$A'x'^2 + C'y'^2 + D'x' + E'y' + F' = 0$$

where $0° < 2\alpha < 180°$ and

$$A' = A\cos^2\alpha + B\cos\alpha\sin\alpha + C\sin^2\alpha \tag{3}$$

$$C' = A\sin^2\alpha - B\cos\alpha\sin\alpha + C\cos^2\alpha \tag{4}$$

$$D' = D\cos\alpha + E\sin\alpha \tag{5}$$

$$E' = -D\sin\alpha + E\cos\alpha \tag{6}$$

$$F' = F \tag{7}$$

EXAMPLE 1 Use the Rotation Theorem to Sketch a Conic

Sketch the graph of $7x^2 - 6\sqrt{3}xy + 13y^2 - 16 = 0$.

Solution

We are given

$$A = 7, \quad B = -6\sqrt{3}, \quad C = 13, \quad D = 0, \quad E = 0, \quad \text{and} \quad F = -16$$

The angle of rotation α can be determined by solving

$$\cot 2\alpha = \frac{A - C}{B} = \frac{7 - 13}{-6\sqrt{3}} = \frac{-6}{-6\sqrt{3}} = \frac{1}{\sqrt{3}} = \frac{\sqrt{3}}{3}$$

This gives us $2\alpha = 60°$, or $\alpha = 30°$. Because $\alpha = 30°$, we have

$$\sin\alpha = \frac{1}{2} \quad \text{and} \quad \cos\alpha = \frac{\sqrt{3}}{2}$$

We determine the coefficients A', C', D', E', and F' by using Equations (3) to (7).

$$A' = 7\left(\frac{\sqrt{3}}{2}\right)^2 + (-6\sqrt{3})\left(\frac{\sqrt{3}}{2}\right)\left(\frac{1}{2}\right) + 13\left(\frac{1}{2}\right)^2 = 4$$

$$C' = 7\left(\frac{1}{2}\right)^2 - (-6\sqrt{3})\left(\frac{\sqrt{3}}{2}\right)\left(\frac{1}{2}\right) + 13\left(\frac{\sqrt{3}}{2}\right)^2 = 16$$

$$D' = 0\left(\frac{\sqrt{3}}{2}\right)^2 + 0\left(\frac{1}{2}\right) = 0$$

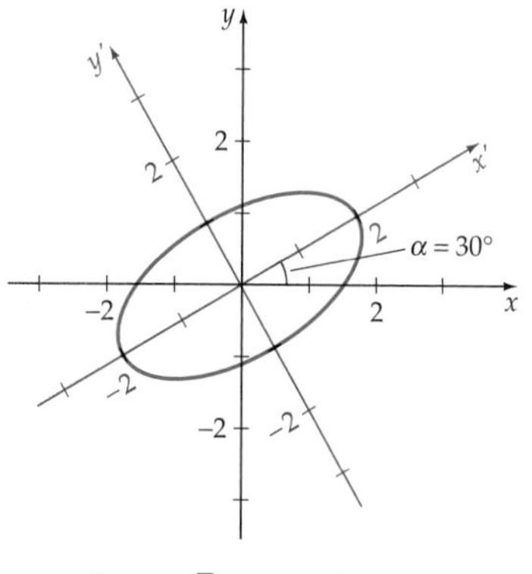

$7x^2 - 6\sqrt{3}xy + 13y^2 - 16 = 0$

Figure 8.53

$$E' = -0\left(\frac{1}{2}\right) + 0\left(\frac{\sqrt{3}}{2}\right) = 0$$

$$F' = F = -16$$

The equation of the conic in the $x'y'$-plane is $4(x')^2 + 16(y')^2 - 16 = 0$ or

$$\frac{(x')^2}{2^2} + \frac{(y')^2}{1^2} = 1$$

This is the equation of an ellipse that is centered at the origin of an $x'y'$-coordinate system. The ellipse has a semimajor axis $a = 2$ and a semiminor axis $b = 1$. See **Figure 8.53**.

TRY EXERCISE 8, EXERCISE SET 8.4, PAGE 627

In Example 1, the angle of rotation α was 30°, which is a special angle. In the next example, we demonstrate a technique that is often used when the angle of rotation is not a special angle.

EXAMPLE 2 Use the Rotation Theorem to Sketch a Conic

Sketch the graph of $32x^2 - 48xy + 18y^2 - 15x - 20y = 0$.

Solution

We are given

$$A = 32, \quad B = -48, \quad C = 18, \quad D = -15, \quad E = -20, \quad \text{and} \quad F = 0$$

Therefore,

$$\cot 2\alpha = \frac{A - C}{B} = \frac{32 - 18}{-48} = -\frac{7}{24}$$

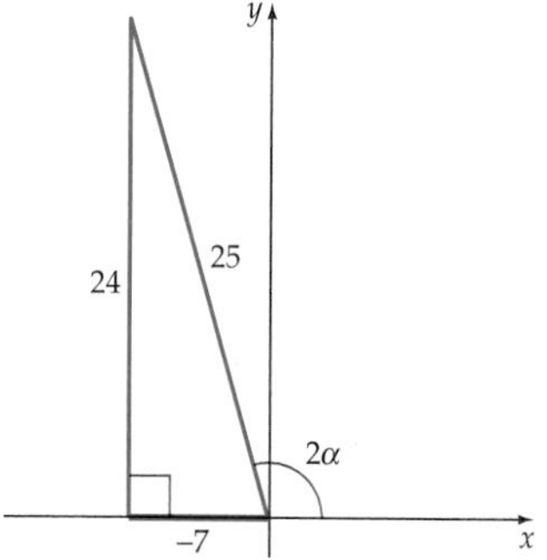

Figure 8.54

Figure 8.54 shows an angle 2α for which $\cot 2\alpha = -7/24$. From **Figure 8.54** we conclude that $\cos 2\alpha = -7/25$. The half-angle identities can be used to determine $\sin \alpha$ and $\cos \alpha$.

$$\sin \alpha = \sqrt{\frac{1 - (-7/25)}{2}} = \frac{4}{5} \quad \text{and} \quad \cos \alpha = \sqrt{\frac{1 + (-7/25)}{2}} = \frac{3}{5}$$

A calculator can be used to determine that $\alpha \approx 53.1°$.

Equations (3) to (7) give us

$$A' = 32\left(\frac{3}{5}\right)^2 + (-48)\left(\frac{3}{5}\right)\left(\frac{4}{5}\right) + 18\left(\frac{4}{5}\right)^2 = 0$$

$$C' = 32\left(\frac{4}{5}\right)^2 - (-48)\left(\frac{3}{5}\right)\left(\frac{4}{5}\right) + 18\left(\frac{3}{5}\right)^2 = 50$$

$$D' = (-15)\left(\frac{3}{5}\right) + (-20)\left(\frac{4}{5}\right) = -25$$

Continued

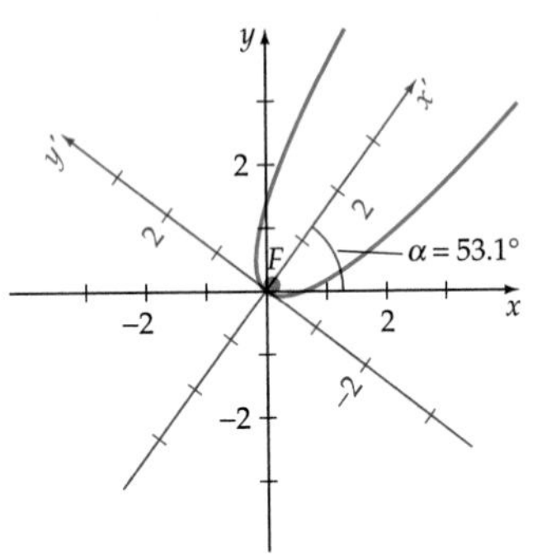

$32x^2 - 48xy + 18y^2 - 15x - 20y = 0$

Figure 8.55

$$E' = -(-15)\left(\frac{4}{5}\right) + (-20)\left(\frac{3}{5}\right) = 0$$

$$F' = F = 0$$

The equation of the conic in the $x'y'$-plane is $50(y')^2 - 25x' = 0$, or

$$(y')^2 = \frac{1}{2}x'$$

This is the equation of a parabola. Because $4p = 1/2$, we know $p = 1/8$, and the focus of the parabola is at $(1/8, 0)$ on the x'-axis. See **Figure 8.55.**

TRY EXERCISE 18, EXERCISE SET 8.4, PAGE 627

◆ THE CONIC IDENTIFICATION THEOREM

The following theorem provides us with a procedure that can be used to identify the type of conic that will be produced by graphing an equation that is in the form of the general second-degree equation in two variables.

Conic Identification Theorem

The graph of

$$Ax^2 + Bxy + Cy^2 + Dx + Ey + F = 0$$

is either a conic or a degenerate conic. If the graph is a conic, then the graph can be identified by its *discriminant* $B^2 - 4AC$. The graph is

- an ellipse or a circle, provided $B^2 - 4AC < 0$.
- a parabola, provided $B^2 - 4AC = 0$.
- a hyperbola, provided $B^2 - 4AC > 0$.

EXAMPLE 3 Identify Conic Sections

Each of the following equations has a graph that is a nondegenerate conic. Compute $B^2 - 4AC$ to identify the type of conic given by each equation.

a. $2x^2 - 4xy + 2y^2 - 6x - 10 = 0$ **b.** $-2xy + 11 = 0$

c. $3x^2 + 5xy + 4y^2 - 8x + 10y + 6 = 0$ **d.** $xy - 3y^2 + 2 = 0$

Solution

a. Because $B^2 - 4AC = (-4)^2 - 4(2)(2) = 0$, the graph is a parabola.

b. Because $B^2 - 4AC = (-2)^2 - 4(0)(0) > 0$, the graph is a hyperbola.

c. Because $B^2 - 4AC = 5^2 - 4(3)(4) < 0$, the graph is an ellipse or a circle.

d. Because $B^2 - 4AC = 1^2 - 4(0)(-3) > 0$, the graph is a hyperbola.

TRY EXERCISE 30, EXERCISE SET 8.4, PAGE 627

◆ USE A GRAPHING UTILITY TO GRAPH SECOND-DEGREE EQUATIONS IN TWO VARIABLES

To graph a general second-degree equation in two variables with a graphing utility requires that we first solve the general equation for y. Consider the general second-degree equation in two variables

$$Ax^2 + Bxy + Cy^2 + Dx + Ey + F = 0 \tag{8}$$

where A, B, C, D, E, and F are real constants, and $C \neq 0$. To solve Equation (8) for y, we first rewrite the equation as

$$Cy^2 + (Bx + E)y + (Ax^2 + Dx + F) = 0 \tag{9}$$

Applying the quadratic formula to Equation (9) yields

$$y = \frac{-(Bx + E) \pm \sqrt{(Bx + E)^2 - 4C(Ax^2 + Dx + F)}}{2C} \tag{10}$$

Thus the graph of Equation (8) can be constructed by graphing both

$$y_1 = \frac{-(Bx + E) + \sqrt{(Bx + E)^2 - 4C(Ax^2 + Dx + F)}}{2C} \tag{11}$$

and

$$y_2 = \frac{-(Bx + E) - \sqrt{(Bx + E)^2 - 4C(Ax^2 + Dx + F)}}{2C} \tag{12}$$

on the same grid.

take note

A TI graphing calculator program is available to graph a rotated conic section by entering its coefficients. This program, ROTATE, can be found on our web site at

http://college.hmco.com

EXAMPLE 4 Use a Graphing Utility to Graph a Conic

Use a graphing utility to graph each conic.

a. $7x^2 + 6xy + 2.5y^2 - 14x + 4y + 9 = 0$

b. $x^2 + 5xy + 3y^2 - 25x - 84y + 375 = 0$

c. $3x^2 - 6xy + 3y^2 - 15x - 12y - 8 = 0$

Continued ►

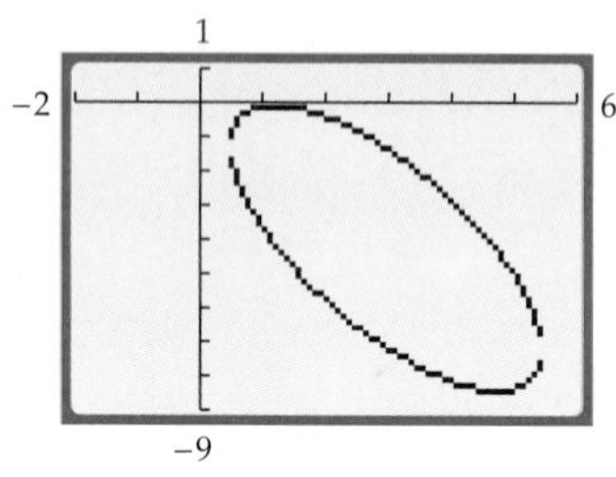

$7x^2 + 6xy + 2.5y^2 - 14x + 4y + 9 = 0$

Figure 8.56

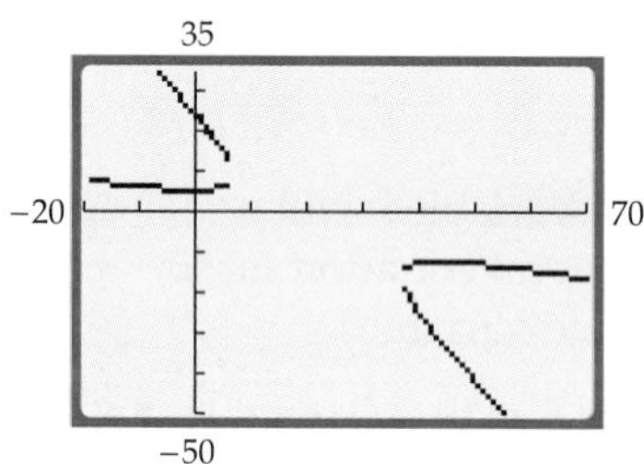

$x^2 + 5xy + 3y^2 - 25x - 84y + 375 = 0$

Figure 8.57

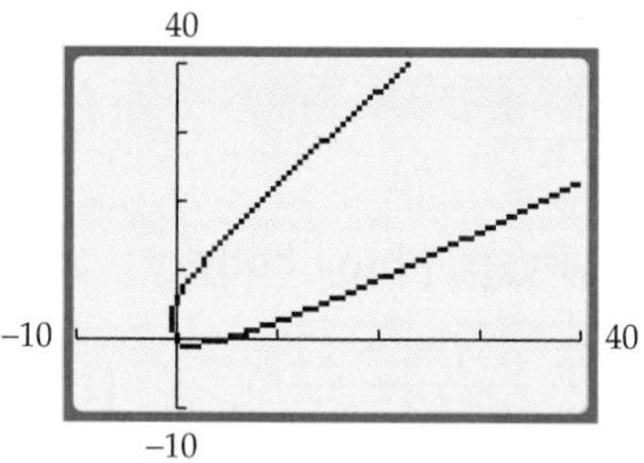

$3x^2 - 6xy + 3y^2 - 15x - 12y - 8 = 0$

Figure 8.58

Solution

Enter y_1 (Equation 11) and y_2 (Equation 12) into the function editing menu of a graphing utility.

a. Store the following constants in place of the indicated variables.

$$A = 7, \quad B = 6, \quad C = 2.5, \quad D = -14, \quad E = 4, \quad \text{and} \quad F = 9$$

Graph y_1 and y_2 on the same screen. The union of the two graphs is an ellipse. See **Figure 8.56.**

b. Store the following constants in place of the indicated variables.

$$A = 1, \quad B = 5, \quad C = 3, \quad D = -25, \quad E = -84, \quad \text{and} \quad F = 375$$

Graph y_1 and y_2 on the same screen. The union of the two graphs is a hyperbola. See **Figure 8.57.**

c. Store the following constants in place of the indicated variables.

$$A = 3, \quad B = -6, \quad C = 3, \quad D = -15, \quad E = -12, \quad \text{and} \quad F = -8$$

Graph y_1 and y_2 on the same screen. The union of the two graphs is a parabola. See **Figure 8.58.**

TRY EXERCISE 24, EXERCISE SET 8.4, PAGE 627

TOPICS FOR DISCUSSION

1. Two students disagree about the graph of $x^2 + y^2 = -4$. One student states that the equation graphs to be an ellipse. The other student claims that the equation does not have a graph. Which student is correct? Explain.
2. The graph of $4x^2 - y^2 = 0$ consists of two intersecting lines. What are the equations of the lines?
3. The graph of $xy = 12$ is a hyperbola. What are the equations of the asymptotes of the hyperbola?
4. Explain why the graph of the ellipse in **Figure 8.56** shows a gap at the left side and the right side of the ellipse.
5. Explain why any conic that is given by a general second-degree equation in which $D = 0$ and $E = 0$ will have a quadratic equation in the rotated coordinate system with $D' = 0$ and $E' = 0$.

EXERCISE SET 8.4

In Exercises 1 to 6, find an angle of rotation α that eliminates the xy term. State approximate solutions to the nearest 0.1°.

1. $xy = 3$
2. $5x^2 - 3xy - 5y^2 - 1 = 0$
3. $9x^2 - 24xy + 16y^2 - 320x - 240y = 0$
4. $x^2 + 4xy + 4y^2 - 6x - 5 = 0$
5. $5x^2 - 6\sqrt{3}xy - 11y^2 + 4x - 3y + 2 = 0$
6. $5x^2 + 4xy + 8y^2 - 6x + 3y - 12 = 0$

In Exercises 7 to 18, find an angle of rotation α that eliminates the xy term. Then find an equation in $x'y'$-coordinates. Graph the equation.

7. $xy = 4$

8. $xy = -10$

9. $6x^2 - 6xy + 14y^2 - 45 = 0$

10. $11x^2 - 10\sqrt{3}xy + y^2 - 20 = 0$

11. $x^2 + 4xy - 2y^2 - 1 = 0$

12. $9x^2 - 24xy + 16y^2 + 100 = 0$

13. $3x^2 + 2\sqrt{3}xy + y^2 + 2x - 2\sqrt{3}y + 16 = 0$

14. $x^2 + 2xy + y^2 + 2\sqrt{2}x - 2\sqrt{2}y = 0$

15. $9x^2 - 24xy + 16y^2 - 40x - 30y + 100 = 0$

16. $24x^2 + 16\sqrt{3}xy + 8y^2 - x + \sqrt{3}y - 8 = 0$

17. $6x^2 + 24xy - y^2 - 12x + 26y + 11 = 0$

18. $x^2 + 4xy + 4y^2 - 2\sqrt{5}x + \sqrt{5}y = 0$

In Exercises 19 to 24, use a graphing utility to graph each equation.

19. $6x^2 - xy + 2y^2 + 4x - 12y + 7 = 0$

20. $5x^2 - 2xy + 10y^2 - 6x - 9y - 20 = 0$

21. $x^2 - 6xy + y^2 - 2x - 5y + 4 = 0$

22. $2x^2 - 10xy + 3y^2 - x - 8y - 7 = 0$

23. $3x^2 - 6xy + 3y^2 + 10x - 8y - 2 = 0$

24. $2x^2 - 8xy + 8y^2 + 20x - 24y - 3 = 0$

25. Find the equations of the asymptotes, relative to an xy-coordinate system, for the hyperbola in Exercise 11. Assume that the xy-coordinate system has the same origin as the $x'y'$-coordinate system.

26. Find the coordinates of the foci and the equation of the directrix, relative to an xy-coordinate system, for the parabola in Exercise 14. Assume that the xy-coordinate system has the same origin as the $x'y'$-coordinate system.

27. Find the coordinates of the foci, relative to an xy-coordinate system, for the ellipse defined by the equation in Exercise 9. Assume that the xy-coordinate system has the same origin as the $x'y'$-coordinate system.

In Exercises 28 to 36, identify the graph of each equation as a parabola, ellipse (or circle), or hyperbola.

28. $xy = 4$

29. $x^2 + xy - y^2 - 40 = 0$

30. $11x^2 - 10\sqrt{3}xy + y^2 - 20 = 0$

31. $3x^2 + 2\sqrt{3}xy + y^2 - 3x + 2y + 20 = 0$

32. $9x^2 - 24xy + 16y^2 + 8x - 12y - 20 = 0$

33. $4x^2 - 4xy + y^2 - 12y + 20 = 0$

34. $5x^2 + 4xy + 8y^2 - 6x + 3y - 12 = 0$

35. $5x^2 - 6\sqrt{3}xy - 11y^2 + 4x - 3y + 2 = 0$

36. $6x^2 - 6xy + 14y^2 - 14x + 12y - 60 = 0$

SUPPLEMENTAL EXERCISES

37. By using the rotation-of-axes equations, show that for every choice of α, the equation $x^2 + y^2 = r^2$ becomes $x'^2 + y'^2 = r^2$.

38. The vertices of a hyperbola are $(1, 1)$ and $(-1, -1)$. The foci are $(\sqrt{2}, \sqrt{2})$ and $(-\sqrt{2}, -\sqrt{2})$. Find an equation of the hyperbola.

39. The vertices on the major axis of an ellipse are the points $(2, 4)$ and $(-2, -4)$. The foci are the points $(\sqrt{2}, 2\sqrt{2})$ and $(-\sqrt{2}, -2\sqrt{2})$. Find an equation of the ellipse.

40. The vertex of a parabola is the origin, and the focus is the point $(1, 3)$. Find an equation of the parabola.

41. **AN INVARIANT THEOREM** Let $Ax^2 + Bxy + Cy^2 + Dx + Ey + F = 0$ be an equation of a conic in an xy-coordinate system. Let the equation of the conic in the rotated $x'y'$-coordinate system be $A'x'^2 + B'x'y' + C'y'^2 + D'x' + E'y' + F' = 0$. Show that

$$A' + C' = A + C$$

42. **AN INVARIANT THEOREM** Let $Ax^2 + Bxy + Cy^2 + Dx + Ey + F = 0$ be an equation of a conic in an xy-coordinate system. Let the equation of the conic in the rotated $x'y'$-coordinate system be $A'x'^2 + B'x'y' + C'y'^2 + D'x' + E'y' + F' = 0$. Show that

$$B'^2 - 4A'C' = B^2 - 4AC$$

43. Using the result of Exercise 42, show that, except in degenerate cases,

$$B^2 - 4AC \begin{cases} < 0 & \text{for ellipses} \\ = 0 & \text{for parabolas} \\ > 0 & \text{for hyperbolas} \end{cases}$$

44. Derive Equation (2) of the rotation-of-axes formulas.

PROJECTS

1. **USE THE INVARIANT THEOREMS** The results of Exercises 41 and 42 illustrate that when the rotation theorem is used to transform equations of the form

$$Ax^2 + Bxy + Cy^2 - F = 0$$

to the form

$$A'(x')^2 + C'(y')^2 - F = 0$$

the following relationships hold:

$$A + C = A' + C' \quad \text{and} \quad B^2 - 4AC = (B')^2 - 4A'C'$$

Use these equations to transform $10x^2 + 24xy + 17y^2 - 26 = 0$ to the form

$$A'(x')^2 + C'(y')^2 - 26 = 0$$

without applying the rotation theorem.

SECTION 8.5 INTRODUCTION TO POLAR COORDINATES

- THE POLAR COORDINATE SYSTEM 
- GRAPHS OF EQUATIONS IN A POLAR COORDINATE SYSTEM 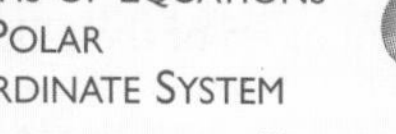
- TRANSFORMATIONS BETWEEN RECTANGULAR AND POLAR COORDINATES 
- WRITE POLAR COORDINATE EQUATIONS IN RECTANGULAR FORM, AND VICE VERSA 

Until now, we have used a *rectangular coordinate system* to locate a point in the coordinate plane. An alternative method is to use a *polar coordinate system,* wherein a point is located by giving a distance from a fixed point and an angle from some fixed direction.

◆ THE POLAR COORDINATE SYSTEM

A **polar coordinate system** is formed by drawing a horizontal ray. The ray is called the **polar axis,** and the endpoint of the ray is called the **pole.** A point $P(r, \theta)$ in the plane is located by specifying a distance r from the pole and an angle θ measured from the polar axis to the line segment OP. The angle can be measured in degrees or radians. See **Figure 8.59.**

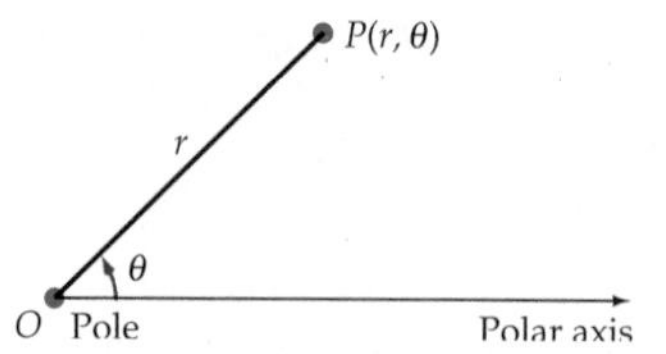

Figure 8.59

The coordinates of the pole are $(0, \theta)$, where θ is an arbitrary angle. Positive angles are measured counterclockwise from the polar axis. Negative angles are measured clockwise from the axis. Positive values of r are measured along the ray that makes an angle θ from the polar axis. Negative values of r are measured along the ray that makes an angle of $\theta + 180°$ from the polar axis. See **Figures 8.60** and **8.61.**

In a rectangular coordinate system, there is a one-to-one correspondence between the points in the plane and the ordered pairs (x, y). This is not true for a polar coordinate system. For polar coordinates, the relationship is one-to-many. Infinitely many ordered-pair descriptions correspond to each point $P(r, \theta)$ in a polar coordinate system.

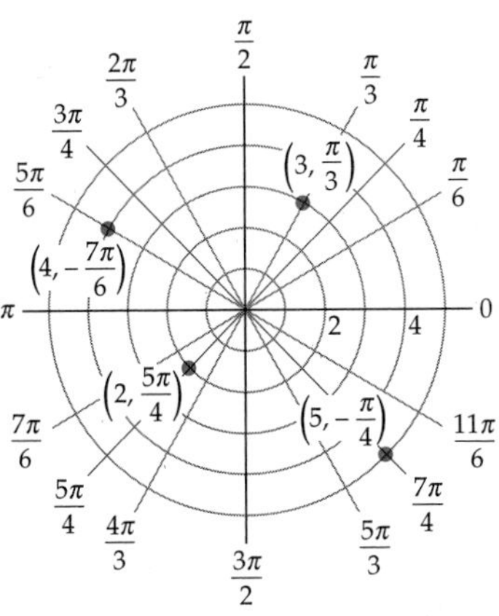

Figure 8.60

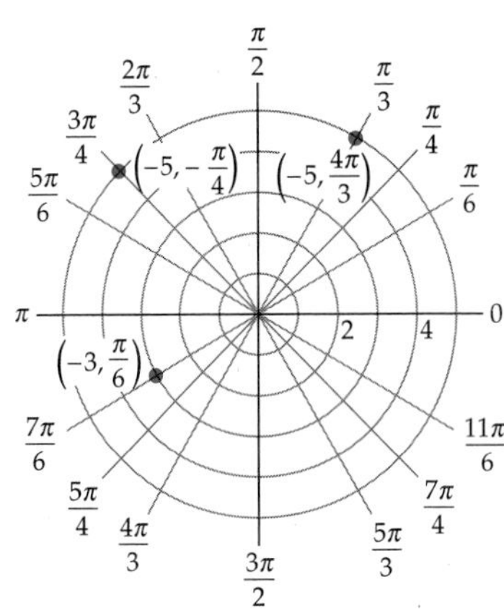

Figure 8.61

For example, consider a point whose coordinates are $P(3, 45°)$. Because there are 360° in one complete revolution around a circle, the point P could also be written as $(3, 405°)$, as $(3, 765°)$, as $(3, 1125°)$, and generally as $(3, 45° + n \cdot 360°)$, where n is an integer. It is also possible to describe the point $P(3, 45°)$ by $(-3, 225°)$, $(-3, -135°)$, and $(3, -315°)$, to name just a few.

The relationship between an ordered pair and a point is not one-to-many. That is, given an ordered pair (r, θ), there is exactly one point in the plane that corresponds to that ordered pair.

◆ GRAPHS OF EQUATIONS IN A POLAR COORDINATE SYSTEM

A **polar equation** is an equation in r and θ. A **solution** to a polar equation is an ordered pair (r, θ) that satisfies the equation. The **graph** of a polar equation is the set of all points whose ordered pairs are solutions of the equation.

The graph of the polar equation $\theta = \pi/6$ is a line. Because θ is independent of r, θ is $\pi/6$ radians from the polar axis for all values of r. The graph is a line that makes an angle of $\pi/6$ radians (30°) from the polar axis. See **Figure 8.62.**

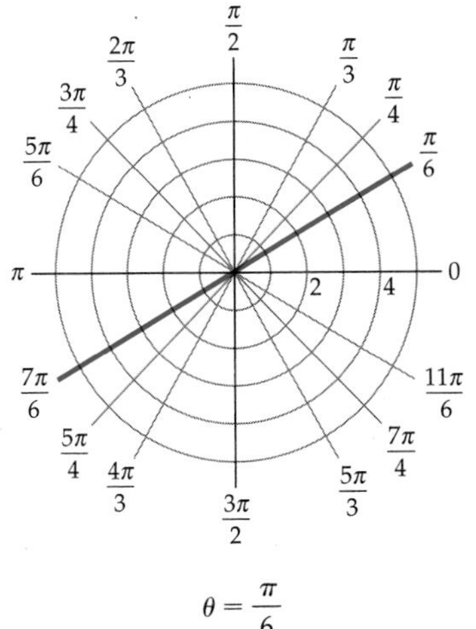

Figure 8.62

Polar Equations of a Line

The graph of $\theta = \alpha$ is a line through the pole at an angle of α from the polar axis. See **Figure 8.63a.**

The graph of $r \sin \theta = a$ is a horizontal line passing through the point $(a, \pi/2)$. See **Figure 8.63b.**

The graph of $r \cos \theta = a$ is a vertical line passing through the point $(a, 0)$. See **Figure 8.63c.**

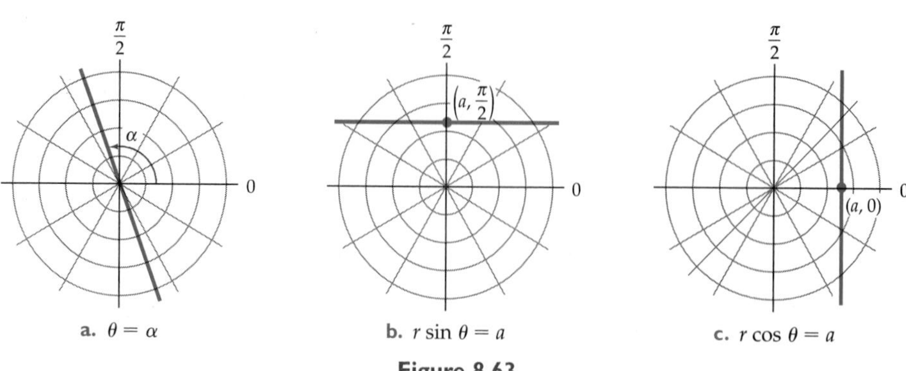

Figure 8.63

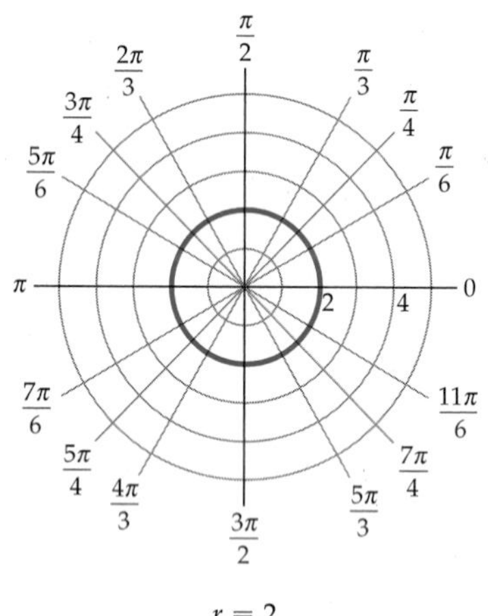

$r = 2$

Figure 8.64

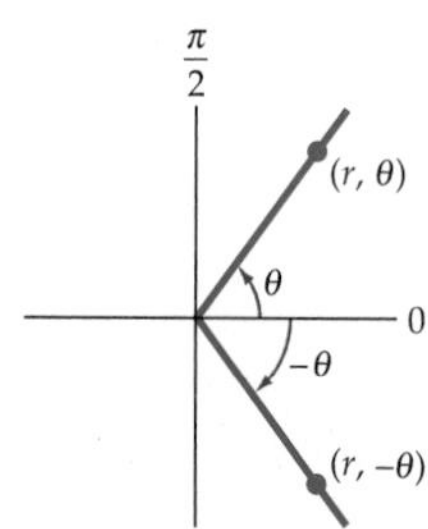

Symmetry with respect to the line $\theta = 0$

Figure 8.65

Figure 8.64 is the graph of the polar equation $r = 2$. Because r is independent of θ, r is 2 units from the pole for all values of θ. The graph is a circle of radius 2 with center at the pole.

The Graph of $r = a$

The graph of $r = a$ is a circle with center at the pole and radius a.

Suppose that whenever the ordered pair (r, θ) lies on the graph of a polar equation, $(r, -\theta)$ also lies on the graph. From **Figure 8.65,** the graph will have symmetry with respect to the polar axis $\theta = 0$. Thus one test for symmetry is to replace θ by $-\theta$ in the polar equation. If the resulting equation is equivalent to the original equation, the graph is symmetric with respect to the polar axis.

Table 8.2 shows the types of symmetry and their associated tests. For each type, if the recommended substitution results in an equivalent equation, the graph will have the indicated symmetry. **Figure 8.66** illustrates the tests for symmetry with respect to the line $\theta = \pi/2$ and for symmetry with respect to the pole.

Table 8.2 Tests for Symmetry

Substitution	Symmetry with respect to
$-\theta$ for θ	The line $\theta = 0$
$\pi - \theta$ for θ, $-r$ for r	The line $\theta = 0$
$\pi - \theta$ for θ	The line $\theta = \pi/2$
$-\theta$ for θ, $-r$ for r	The line $\theta = \pi/2$
$-r$ for r	The pole
$\pi + \theta$ for θ	The pole

The graph of a polar equation may have a symmetry even though a test for that symmetry fails. For example, as we will see later, the graph of $r = \sin 2\theta$ is symmetric with respect to the line $\theta = 0$. However, using the symmetry test of substituting $-\theta$ for θ, we have

$$\sin 2(-\theta) = -\sin 2\theta = -r \neq r$$

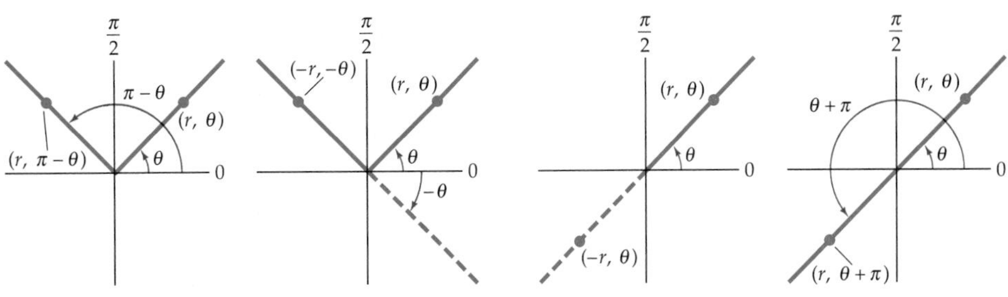

Symmetry with respect to the line $\theta = \frac{\pi}{2}$

Symmetry with respect to the pole

Figure 8.66

Thus this test fails to show symmetry with respect to the line $\theta = 0$. The symmetry test of substituting $\pi - \theta$ for θ and $-r$ for r establishes symmetry with respect to the line $\theta = 0$.

EXAMPLE 1 Graph a Polar Equation

Show that the graph of $r = 4\cos\theta$ is symmetric with respect to the polar axis. Graph the equation.

Solution

Test for symmetry with respect to the polar axis. Replace θ by $-\theta$.

$$r = 4\cos(-\theta) = 4\cos\theta \qquad \bullet \cos(-\theta) = \cos\theta$$

Because replacing θ by $-\theta$ results in the original equation $r = 4\cos\theta$, the graph is symmetric with respect to the polar axis.

To graph the equation, begin choosing various values of θ and finding the corresponding values of r. However, before doing so, consider two further observations that will reduce the number of points you must choose.

First, because the cosine function is a periodic function with period 2π, it is only necessary to choose points between 0 and 2π (0° and 360°). Second, when $\pi/2 < \theta < 3\pi/2$, $\cos\theta$ is negative, which means that any θ between these values will produce a negative r. Thus the point will be in the first or fourth quadrant. That is, we need consider only angles θ in the first or fourth quadrants. However, because the graph is symmetric with respect to the polar axis, it is only necessary to choose values of θ between 0 and $\pi/2$.

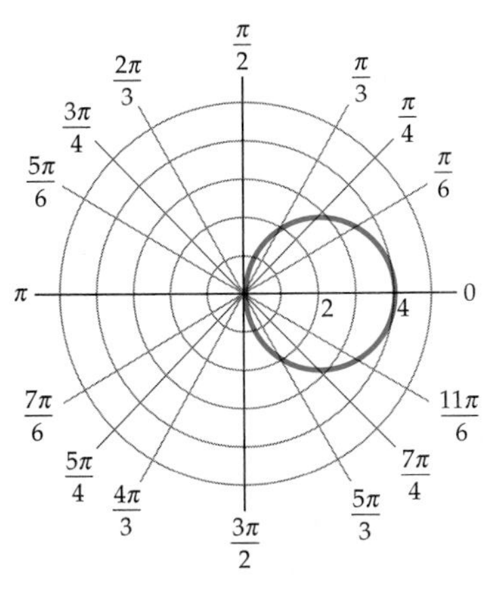

$r = 4\cos\theta$

Figure 8.67

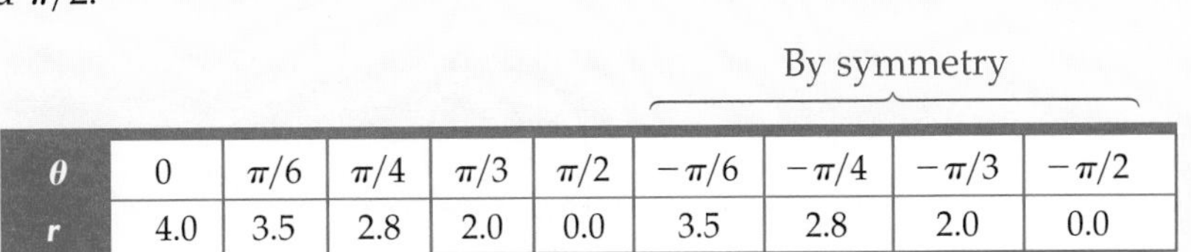

By symmetry (applies to columns $-\pi/6$ through $-\pi/2$)

θ	0	$\pi/6$	$\pi/4$	$\pi/3$	$\pi/2$	$-\pi/6$	$-\pi/4$	$-\pi/3$	$-\pi/2$
r	4.0	3.5	2.8	2.0	0.0	3.5	2.8	2.0	0.0

The graph of $r = 4\cos\theta$ is a circle with the center at (2, 0). See **Figure 8.67.**

TRY EXERCISE 14, EXERCISE SET 8.5, PAGE 638

Polar Equations of a Circle

The graph of the equation $r = a$ is a circle with center at the pole and radius a. See **Figure 8.68a.**

The graph of the equation $r = a \cos \theta$ is a circle that is symmetric with respect to the line $\theta = 0$. See **Figure 8.68b.**

The graph of $r = a \sin \theta$ is a circle that is symmetric with respect to the line $\theta = \pi/2$. See **Figure 8.68c.**

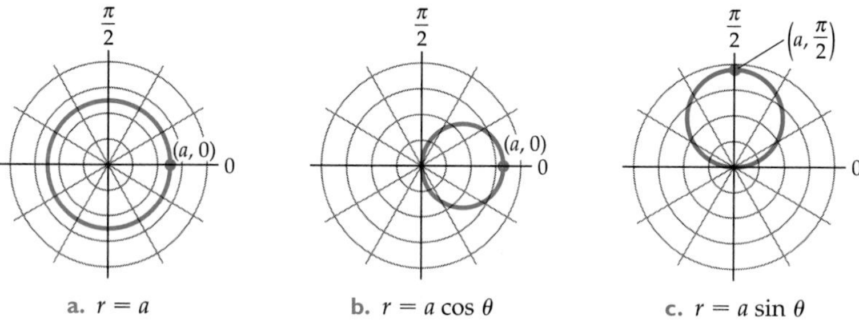

a. $r = a$ b. $r = a \cos \theta$ c. $r = a \sin \theta$

Figure 8.68

Just as there are specifically named curves in an xy-coordinate system (such as parabola and ellipse), there are named curves in an $r\theta$-coordinate system. Two of the many types are the *limaçon* and the *rose curve*.

Polar Equations of a Limaçon

The graph of the equation $r = a + b \cos \theta$ is a **limaçon** that is symmetric with respect to line $\theta = 0$.

The graph of the equation $r = a + b \sin \theta$ is a limaçon that is symmetric with respect to the line $\theta = \pi/2$.

In the special case where $|a| = |b|$, the graph is called a **cardioid.**

The graph of $r = a + b \cos \theta$ is shown in **Figure 8.69** for various values of a and b.

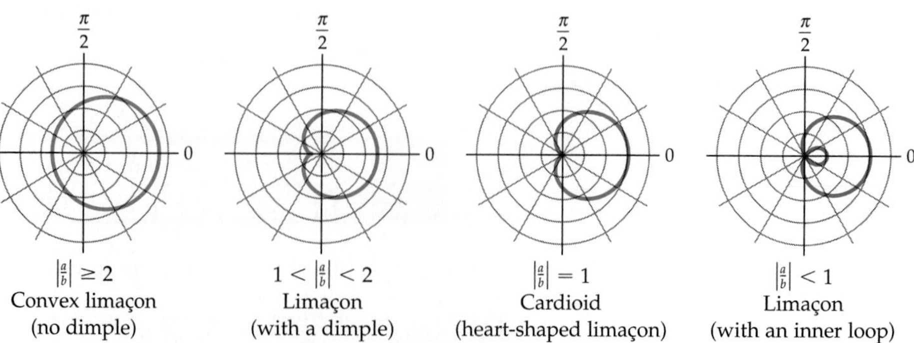

$\left|\frac{a}{b}\right| \geq 2$ Convex limaçon (no dimple)

$1 < \left|\frac{a}{b}\right| < 2$ Limaçon (with a dimple)

$\left|\frac{a}{b}\right| = 1$ Cardioid (heart-shaped limaçon)

$\left|\frac{a}{b}\right| < 1$ Limaçon (with an inner loop)

Figure 8.69

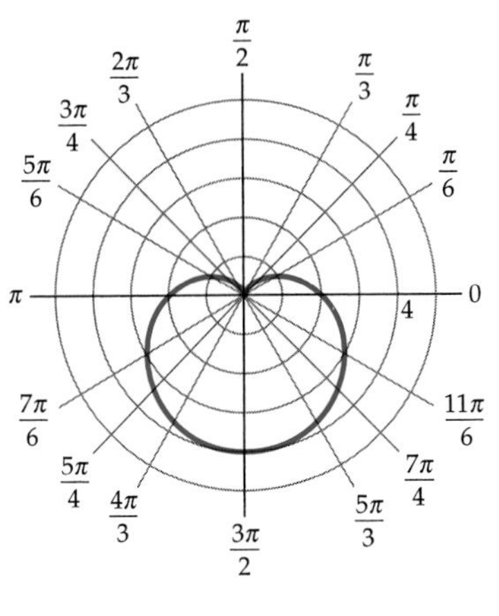

$r = 2 - 2\sin\theta$

Figure 8.70

EXAMPLE 2 Sketch the Graph of a Limaçon

Sketch the graph of $r = 2 - 2\sin\theta$.

Solution

From the general equation of a limaçon $r = a + b\sin\theta$ with $|a| = |b|$ ($|2| = |-2|$), the graph of $r = 2 - 2\sin\theta$ is a cardioid that is symmetric with respect to the line $\theta = \pi/2$.

Because we know that the graph is heart-shaped, we can sketch the graph by finding r for a few values of θ. When $\theta = 0$, $r = 2$. When $\theta = \pi/2$, $r = 0$. When $\theta = \pi$, $r = 2$. When $\theta = 3\pi/2$, $r = 4$. Sketching a heart-shaped curve through the four points

$$(0, 2), \quad \left(\frac{\pi}{2}, 0\right), \quad (\pi, 2), \quad \text{and} \quad \left(\frac{3\pi}{2}, 4\right)$$

produces the cardioid in **Figure 8.70.**

TRY EXERCISE 20, EXERCISE SET 8.5, PAGE 638

Example 3 gives the details necessary for using a graphing utility to construct a polar graph.

EXAMPLE 3 Use a Graphing Utility to Sketch the Graph of a Limaçon

Use a graphing utility to graph $r = 3 - 2\cos\theta$.

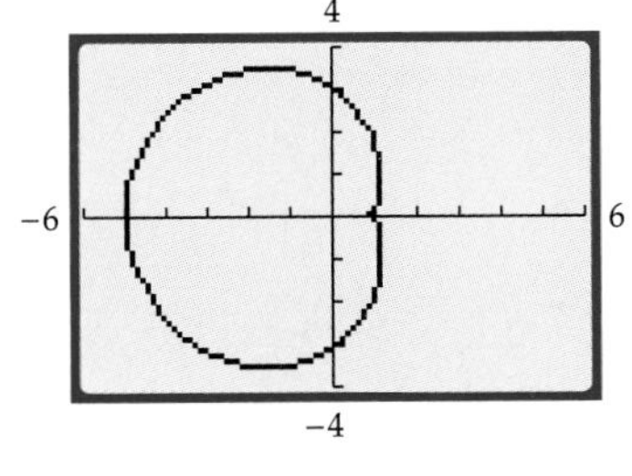

$r = 3 - 2\cos\theta$

Figure 8.71

Solution

From the general equation of a limaçon $r = a + b\cos\theta$, with $a = 3$ and $b = -2$, we know that the graph will be a limaçon with a dimple. The graph will be symmetric with respect to the line $\theta = 0$.

Use polar mode with angle measure in radians. Enter the equation $r = 3 - 2\cos\theta$ in the polar function editing menu. The graph in **Figure 8.71** was produced with a *TI-83* by using a window defined by the following:

θmin=0	Xmin=-6	Ymin=-4
θmax=2π	Xmax=6	Ymax=4
θstep=0.1	Xscl=1	Yscl=1

TRY EXERCISE 26, EXERCISE SET 8.5, PAGE 638

Polar Equations of Rose Curves

The graphs of the equations $r = a\cos n\theta$ and $r = a\sin n\theta$ are **rose curves.** When n is an even number, the number of petals is $2n$. See **Figure 8.72a.** When n is an odd number, the number of petals is n. See **Figure 8.72b.**

take note

When using a graphing utility in polar mode, choose the value of θstep carefully. If θstep is set too small, the graphing utility may require an excessively long period of time to complete the graph. If θstep is set too large, the resulting graph may give only a very rough approximation of the actual graph.

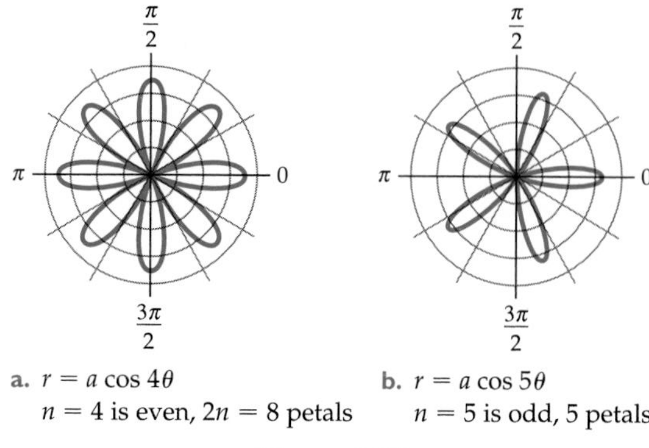

a. $r = a \cos 4\theta$
$n = 4$ is even, $2n = 8$ petals

b. $r = a \cos 5\theta$
$n = 5$ is odd, 5 petals

Figure 8.72

EXAMPLE 4 Sketch the Graph of a Rose Curve

Sketch the graph of $r = 2 \sin 3\theta$.

Solution

From the general equation of a rose curve $r = a \sin n\theta$, with $a = 2$ and $n = 3$, the graph of $r = 2 \sin 3\theta$ is a rose curve that is symmetric with respect to the line $\theta = \pi/2$. Because n is an odd number ($n = 3$), there will be three petals in the graph.

Choose some values for θ and find the corresponding values of r. Use symmetry to sketch the graph. See **Figure 8.73.**

θ	0	$\pi/18$	$\pi/6$	$5\pi/18$	$\pi/3$	$7\pi/18$	$\pi/2$
r	0.0	1.0	2.0	1.0	0.0	−1.0	−2.0

TRY EXERCISE 16, EXERCISE SET 8.5, PAGE 638

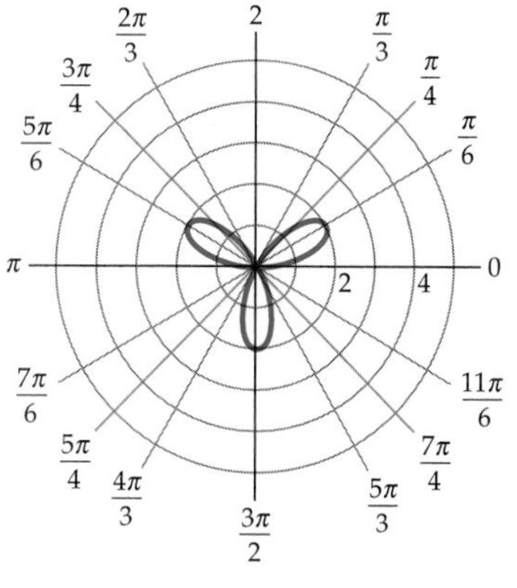

$r = 2 \sin 3\theta$

Figure 8.73

EXAMPLE 5 Use a Graphing Utility to Sketch the Graph of a Rose Curve

Use a graphing utility to graph $r = 4 \cos 2\theta$.

Solution

From the general equation of a rose curve $r = a \cos n\theta$, with $a = 4$ and $n = 2$, we know that the graph will be a rose curve with $2n = 4$ petals. The very tip of each petal will be $a = 4$ units away from the pole. Our symmetry tests also indicate that the graph is symmetric with respect to the line $\theta = 0$, the line $\theta = \pi/2$, and the pole.

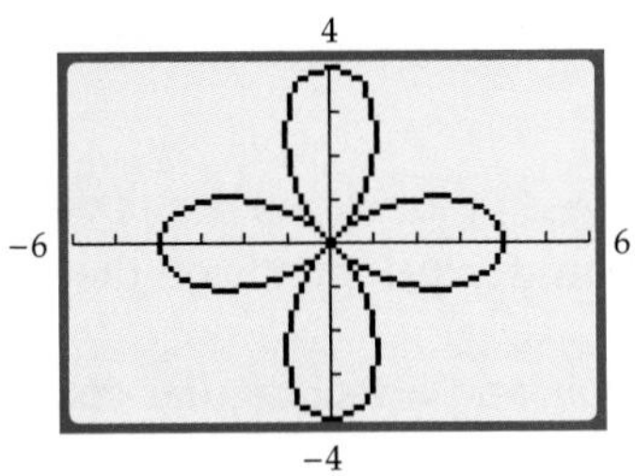

$r = 4 \cos 2\theta$

Figure 8.74

Use polar mode with angle measure in radians. Enter the equation $r = 4 \cos 2\theta$ in the polar function editing menu. The graph in **Figure 8.74** was produced with a *TI-83* by using a window defined by the following:

θmin=0	Xmin=-6	Ymin=-4
θmax=2π	Xmax=6	Ymax=4
θstep=0.1	Xscl=1	Yscl=1

TRY EXERCISE 32, EXERCISE SET 8.5, PAGE 638

◆ TRANSFORMATIONS BETWEEN RECTANGULAR AND POLAR COORDINATES

A transformation between coordinate systems is a set of equations that relate the coordinates of a point in one system with the coordinates in a second system. By superimposing a rectangular coordinate system on a polar system, we can derive the set of transformation equations.

Construct a polar coordinate system and a rectangular system so that the pole coincides with the origin and the polar axis coincides with the positive x-axis. Let a point P have coordinates (x, y) in one system and (r, θ) in the other $(r > 0)$.

From the definitions of $\sin \theta$ and $\cos \theta$, we have

$$\frac{x}{r} = \cos \theta \quad \text{or} \quad x = r \cos \theta$$

$$\frac{y}{r} = \sin \theta \quad \text{or} \quad y = r \sin \theta$$

It can be shown that these equations are also true when $r < 0$.

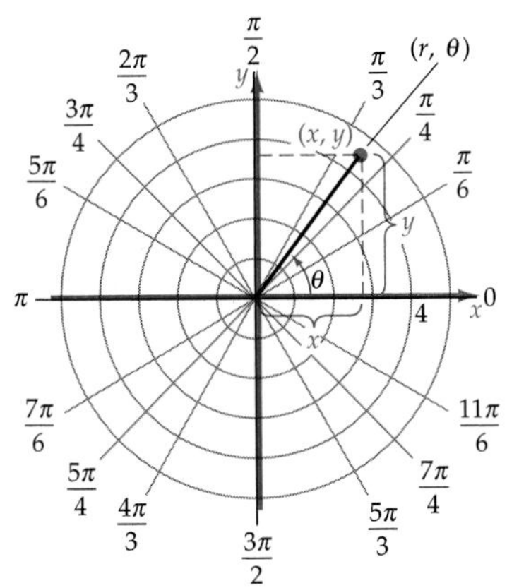

Figure 8.75

Thus, given the point (r, θ) in a polar coordinate system (see **Figure 8.75**), the coordinates of the point in the xy-coordinate system are given by

$$x = r \cos \theta \qquad y = r \sin \theta$$

For example, to find the point in the xy-coordinate system that corresponds to the point $(4, 2\pi/3)$ in the $r\theta$-coordinate system, substitute 4 for r and $\frac{2\pi}{3}$ for θ into the equations and simplify.

$$x = 4 \cos\left(\frac{2\pi}{3}\right) = 4\left(-\frac{1}{2}\right) = -2$$

$$y = 4 \sin\left(\frac{2\pi}{3}\right) = 4\left(\frac{\sqrt{3}}{2}\right) = 2\sqrt{3}$$

The point $(4, 2\pi/3)$ in the $r\theta$-coordinate system is $(-2, 2\sqrt{3})$ in the xy-coordinate system.

To find the polar coordinates of a given point in the xy-coordinate system, use the Pythagorean Theorem and the definition of the tangent function. Let $P(x, y)$ be a point in the plane, and let r be the distance from the origin to the point P. Then $r = \sqrt{x^2 + y^2}$.

From the definition of the tangent function of an angle in a right triangle,

$$\tan\theta = \frac{y}{x}$$

Thus θ is the angle whose tangent is y/x. The quadrant for θ depends on the sign of x and the sign of y.

The equations of transformations between a polar and a rectangular coordinate system are summarized as follows:

Transformations between Polar and Rectangular Coordinates

Given the point (r, θ) in the polar coordinate system, the transformation equations to change from polar to rectangular coordinates are

$$x = r\cos\theta \qquad y = r\sin\theta$$

Given the point (x, y) in the rectangular coordinate system, the transformation equations to change from rectangular to polar coordinates are

$$r = \sqrt{x^2 + y^2} \qquad \tan\theta = \frac{y}{x}, \quad x \neq 0$$

where $r \geq 0$, $0 \leq \theta < 2\pi$, and θ is chosen so that the point lies in the appropriate quadrant. If $x = 0$, then $\theta = \pi/2$ or $\theta = 3\pi/2$.

EXAMPLE 6 Transform from Polar to Rectangular Coordinates

Find the rectangular coordinates of the points whose polar coordinates are: **a.** $(6, 3\pi/4)$ **b.** $(-4, 30°)$

Solution

Use the equations $x = r\cos\theta$ and $y = r\sin\theta$.

a. $x = 6\cos\left(\frac{3\pi}{4}\right) = -3\sqrt{2} \qquad y = 6\sin\left(\frac{3\pi}{4}\right) = 3\sqrt{2}$

The rectangular coordinates of $(6, 3\pi/4)$ are $\left(-3\sqrt{2}, 3\sqrt{2}\right)$.

b. $x = -4\cos(30°) = -2\sqrt{3} \qquad y = -4\sin(30°) = -2$

The rectangular coordinates of $(-4, 30°)$ are $\left(-2\sqrt{3}, -2\right)$.

TRY EXERCISE 44, EXERCISE SET 8.5, PAGE 639

EXAMPLE 7 Transform from Rectangular to Polar Coordinates

Find the polar coordinates of the points whose rectangular coordinates are $\left(-2, -2\sqrt{3}\right)$

Solution

Use the equations $r = \sqrt{x^2 + y^2}$ and $\tan \theta = \dfrac{y}{x}$.

$$r = \sqrt{(-2)^2 + (-2\sqrt{3})^2} = \sqrt{4 + 12} = \sqrt{16} = 4$$

$$\tan \theta = \frac{-2\sqrt{3}}{-2} = \sqrt{3}$$

From this and the fact that $(-2, -2\sqrt{3})$ lies in the third quadrant, $\theta = 4\pi/3$. The polar coordinates of $(-2, -2\sqrt{3})$ are $(4, 4\pi/3)$.

TRY EXERCISE 48, EXERCISE SET 8.5, PAGE 639

◆ WRITE POLAR COORDINATE EQUATIONS IN RECTANGULAR FORM, AND VICE VERSA

Using the transformation equations, it is possible to write a polar coordinate equation in rectangular form or a rectangular coordinate equation in polar form.

EXAMPLE 8 Write a Polar Coordinate Equation in Rectangular Form

Find a rectangular form of the equation $r^2 \cos 2\theta = 3$.

Solution

$$r^2 \cos 2\theta = 3$$

$$r^2(1 - 2\sin^2 \theta) = 3 \qquad \bullet \ \cos 2\theta = 1 - 2\sin^2 \theta$$

$$r^2 - 2r^2 \sin^2 \theta = 3$$

$$r^2 - 2(r \sin \theta)^2 = 3$$

$$x^2 + y^2 - 2y^2 = 3 \qquad \bullet \ r^2 = x^2 + y^2;\ \sin \theta = \frac{y}{r}$$

$$x^2 - y^2 = 3$$

A rectangular form of $r^2 \cos 2\theta = 3$ is $x^2 - y^2 = 3$.

TRY EXERCISE 56, EXERCISE SET 8.5, PAGE 639

EXAMPLE 9 Write a Rectangular Coordinate Equation in Polar Form

Find a polar form of the equation $x^2 + y^2 - 2x = 3$.

Continued ▸

Solution

$$x^2 + y^2 - 2x = 3$$
$$(r\cos\theta)^2 + (r\sin\theta)^2 - 2r\cos\theta = 3$$
• Use the transformation equations $x = r\cos\theta$ and $y = r\sin\theta$.

$$r^2(\cos^2\theta + \sin^2\theta) - 2r\cos\theta = 3$$
• Simplify.
$$r^2 - 2r\cos\theta = 3$$

A polar form of $x^2 + y^2 - 2x = 3$ is $r^2 - 2r\cos\theta = 3$.

TRY EXERCISE 64, EXERCISE SET 8.5, PAGE 639

TOPICS FOR DISCUSSION

1. In what quadrant is the point $(-2, 150°)$ located?
2. To explain why the graph of $\theta = \pi/6$ is a line, a tutor rewrites the equation in the form $\theta = \pi/6 + 0 \cdot r$. In this form it is easy to see that regardless of the value of r, $\theta = \pi/6$. Use an analogous approach to explain to a classmate why the graph of $r = a$ is a circle.
3. Two students use a graphing calculator to graph the polar equation $r = 2\sin\theta$. One graph appears to be a circle, and the other graph appears to be an ellipse. Both graphs are correct. Explain.
4. A student reasons that the graph of $r^2 = 6\cos 2\theta$ should be a rose curve with four petals. Explain why the student is not correct.

EXERCISE SET 8.5

In Exercises 1 to 8, plot the point on a polar coordinate system.

1. $(2, 60°)$
2. $(3, -90°)$
3. $(1, 315°)$
4. $(2, 400°)$
5. $\left(-2, \dfrac{\pi}{4}\right)$
6. $\left(4, \dfrac{7\pi}{6}\right)$
7. $\left(-3, \dfrac{5\pi}{3}\right)$
8. $(-3, \pi)$

In Exercises 9 to 24, sketch the graph of each polar equation.

9. $r = 3$
10. $r = 5$
11. $\theta = 2$
12. $\theta = -\dfrac{\pi}{3}$
13. $r = 6\cos\theta$
14. $r = 4\sin\theta$
15. $r = 4\cos 2\theta$
16. $r = 5\cos 3\theta$
17. $r = 2\sin 5\theta$
18. $r = 3\cos 5\theta$
19. $r = 2 - 3\sin\theta$
20. $r = 2 - 2\cos\theta$
21. $r = 4 + 3\sin\theta$
22. $r = 2 + 4\sin\theta$
23. $r = 2(1 - 2\sin\theta)$
24. $r = 4(1 - \sin\theta)$

In Exercises 25 to 40, use a graphing utility to graph each equation.

25. $r = 3 + 3\cos\theta$
26. $r = 4 - 4\sin\theta$
27. $r = 4\cos 3\theta$
28. $r = 2\sin 4\theta$
29. $r = 3\sec\theta$
30. $r = 4\csc\theta$
31. $r = -5\csc\theta$
32. $r = -4\sec\theta$
33. $r = 4\sin(3.5\theta)$
34. $r = 6\cos(2.25\theta)$

35. $r = \theta, 0 \le \theta \le 6\pi$

36. $r = -\theta, 0 \le \theta \le 6\pi$

37. $r = 2^{\theta}, 0 \le \theta \le 2\pi$

38. $r = \dfrac{1}{\theta}, 0 \le \theta \le 4\pi$

39. $r = \dfrac{6 \cos 10\theta + \cos \theta}{\cos \theta}$

40. $r = \dfrac{4 \cos 3\theta + \cos 5\theta}{\cos \theta}$

In Exercises 41 to 48, transform the given coordinates to the indicated ordered pair.

41. $(1, -\sqrt{3})$ to (r, θ)

42. $(-2\sqrt{3}, 2)$ to (r, θ)

43. $\left(-3, \dfrac{2\pi}{3}\right)$ to (x, y)

44. $\left(2, -\dfrac{\pi}{3}\right)$ to (x, y)

45. $\left(0, -\dfrac{\pi}{2}\right)$ to (x, y)

46. $\left(3, \dfrac{5\pi}{6}\right)$ to (x, y)

47. $(3, 4)$ to (r, θ)

48. $(12, -5)$ to (r, θ)

In Exercises 49 to 60, find a rectangular form of each of the equations.

49. $r = 3 \cos \theta$

50. $r = 2 \sin \theta$

51. $r = 3 \sec \theta$

52. $r = 4 \csc \theta$

53. $r = 4$

54. $\theta = \dfrac{\pi}{4}$

55. $r = \tan \theta$

56. $r = \cot \theta$

57. $r = \dfrac{2}{1 + \cos \theta}$

58. $r = \dfrac{2}{1 - \sin \theta}$

59. $r(\sin \theta - 2 \cos \theta) = 6$

60. $r(2 \cos \theta + \sin \theta) = 3$

In Exercises 61 to 68, find a polar form of each of the equations.

61. $y = 2$

62. $x = -4$

63. $x^2 + y^2 = 4$

64. $2x - 3y = 6$

65. $x^2 = 8y$

66. $y^2 = 4y$

67. $x^2 - y^2 = 25$

68 $x^2 + 4y^2 = 16$

In Exercises 69 to 76, use a graphing utility to graph each equation.

69. $r = 3 \cos\left(\theta + \dfrac{\pi}{4}\right)$

70. $r = 2 \sin\left(\theta - \dfrac{\pi}{6}\right)$

71. $r = 2 \sin\left(2\theta - \dfrac{\pi}{3}\right)$

72. $r = 3 \cos\left(2\theta + \dfrac{\pi}{4}\right)$

73. $r = 2 + 2 \sin\left(\theta - \dfrac{\pi}{6}\right)$

74. $r = 3 - 2 \cos\left(\theta + \dfrac{\pi}{3}\right)$

75. $r = 1 + 3 \cos\left(\theta + \dfrac{\pi}{3}\right)$

76. $r = 2 - 4 \sin\left(\theta - \dfrac{\pi}{4}\right)$

SUPPLEMENTAL EXERCISES

77. Explain why the graph $r^2 = \cos^2 \theta$ and the graph of $r = \cos \theta$ are not the same.

78. Explain why the graph $r = \cos 2\theta$ and the graph of $r = 2 \cos^2 \theta - 1$ are identical.

In Exercises 79 to 86, use a graphing utility to graph each equation.

79. $r^2 = 4 \cos 2\theta$ (lemniscate)

80. $r^2 = -2 \sin 2\theta$ (lemniscate)

81. $r = 2(1 + \sec \theta)$ (conchoid)

82. $r = 2 \cos 2\theta \sec \theta$ (strophoid)

83. $r\theta = 2$ (spiral)

84. $r = 2 \sin \theta \cos^2 2\theta$ (bifolium)

85. $r = |\theta|$

86. $r = \ln \theta$

87. The graph of

$$r = 1.5^{\sin \theta} - 2.5 \cos 4\theta + \sin^7 \frac{\theta}{15}$$

is a *butterfly curve* similar to the one shown below.

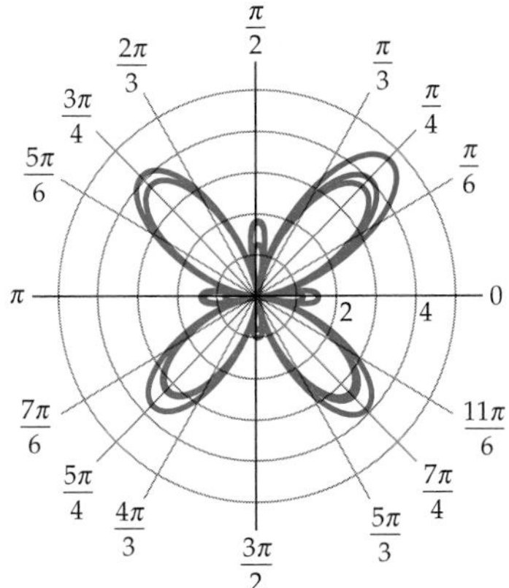

Use a graphing utility to graph the butterfly curve for

a. $0 \le \theta \le 5\pi$ **b.** $0 \le \theta \le 20\pi$

For additional information on butterfly curves, read "The Butterfly Curve" by Temple H. Fay, *The American Mathematical Monthly*, vol. 96, no. 5 (May 1989), p. 442.

PROJECTS

1. **A POLAR DISTANCE FORMULA** Let $P_1(r_1, \theta_1)$ and $P_2(r_2, \theta_2)$ be two distinct points in the $r\theta$-plane.
   a. Verify that the distance d between the points is
   $$d = \sqrt{r_1^2 + r_2^2 - 2r_1r_2\cos(\theta_2 - \theta_1)}$$
   b. Use the above formula to find the distance (to the nearest hundredth) between $(3, 60°)$ and $(5, 170°)$.
   c. Does the formula $d = \sqrt{r_1^2 + r_2^2 - 2r_1r_2\cos(\theta_1 - \theta_2)}$ also produce the correct distance between P_1 and P_2? Explain.
2. **ANOTHER POLAR FORM FOR A CIRCLE**
   a. Verify that the graph of the polar equation $r = a\sin\theta + b\cos\theta$ is a circle. Assume that a and b are not both 0.
   b. What are the center (in rectangular coordinates) and the radius of the circle?

SECTION 8.6 POLAR EQUATIONS OF THE CONICS

- POLAR EQUATIONS OF THE CONICS
- GRAPH A CONIC GIVEN IN POLAR FORM
- WRITE THE POLAR EQUATION OF A CONIC

8C

◆ POLAR EQUATIONS OF THE CONICS

The definition of a parabola was given in terms of a point (the focus) and a line (the directrix). The definitions of both the ellipse and the hyperbola were given in terms of two points (the foci). It is possible to define each conic in terms of a point and a line.

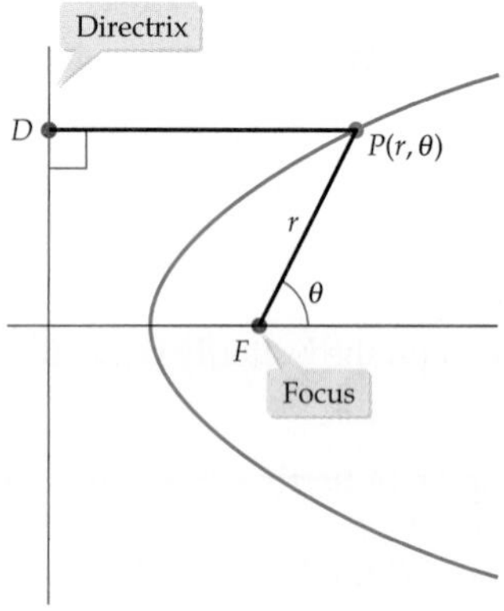

Focus-Directrix Definitions of the Conics

Let F be a fixed point and D a fixed line in a plane. Consider the set of all points P such that $\dfrac{d(P, F)}{d(P, D)} = e$, where e is a constant. The graph is a parabola for $e = 1$, an ellipse for $0 < e < 1$, and a hyperbola for $e > 1$. See **Figure 8.76.**

Figure 8.76

The fixed point is a focus of the conic, and the fixed line is a directrix. The constant e is the eccentricity of the conic. Using this definition, we can derive the polar equations of the conics.

Standard Form of the Polar Equations of the Conics

Let the pole be a focus of a conic section of eccentricity e with directrix d units from the focus. Then the equation of the conic is given by one of the following:

$$r = \frac{ed}{1 + e\cos\theta} \quad (1)$$

Vertical directrix to the right of the pole

$$r = \frac{ed}{1 - e\cos\theta} \quad (2)$$

Vertical directrix to the left of the pole

$$r = \frac{ed}{1 + e\sin\theta} \quad (3)$$

Horizontal directrix above the pole

$$r = \frac{ed}{1 - e\sin\theta} \quad (4)$$

Horizontal directrix below the pole

When the equation involves $\cos\theta$, the polar axis is an axis of symmetry. When the equation involves $\sin\theta$, the line $\theta = \pi/2$ is an axis of symmetry. Graphs of examples are shown in **Figure 8.77.**

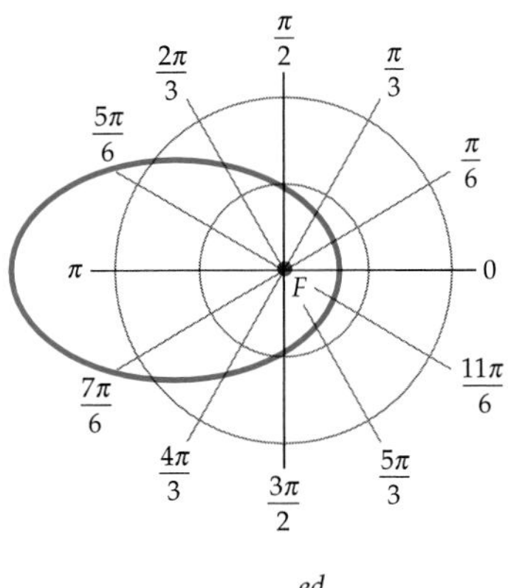

$r = \dfrac{ed}{1 + e\cos\theta}$

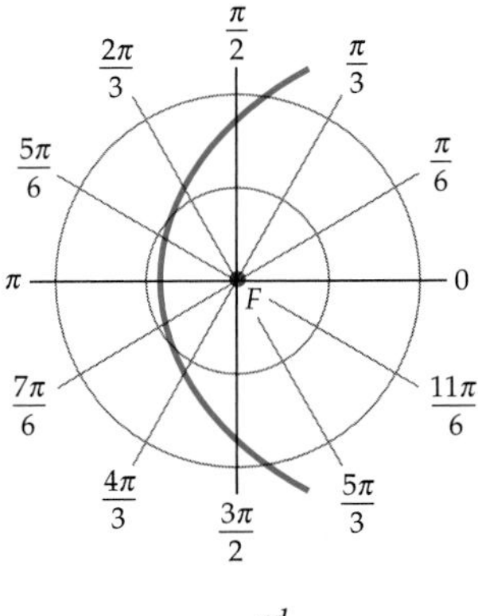

$r = \dfrac{ed}{1 - e\cos\theta}$

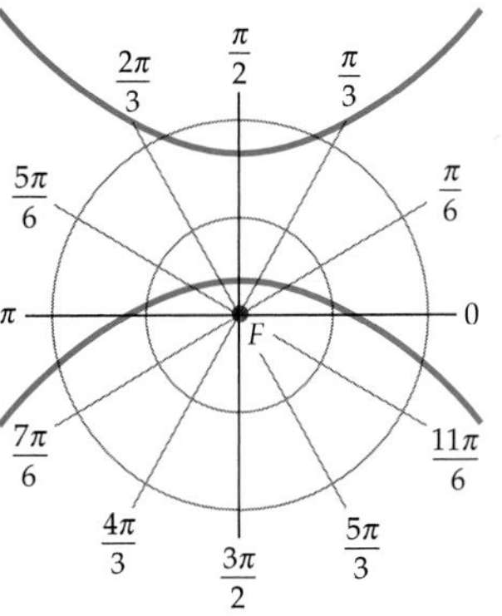

$r = \dfrac{ed}{1 + e\sin\theta}$

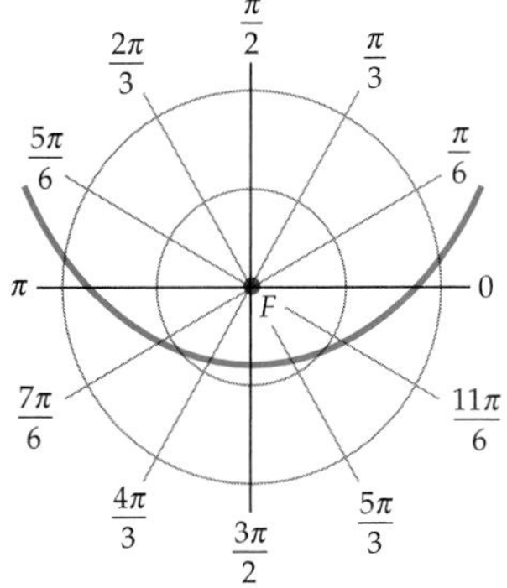

$r = \dfrac{ed}{1 - e\sin\theta}$

Figure 8.77

We will derive Equation (2). Let $P(r, \theta)$ be any point on a conic section. Then, by definition,

$$\frac{d(P, F)}{d(P, D)} = e \quad \text{or} \quad d(P, F) = e \cdot d(P, D)$$

From **Figure 8.78**, $d(P, F) = r$ and $d(P, D) = d(A, Q)$. But note that

$$d(A, Q) = d(A, F) + d(F, Q) = d + r\cos\theta$$

Thus

$$\begin{aligned} r &= e(d + r\cos\theta) && \bullet\ d(P, F) = e \cdot d(P, D) \\ &= ed + er\cos\theta \\ r - er\cos\theta &= ed && \bullet\ \text{Subtract } er\cos\theta. \\ r &= \frac{ed}{1 - e\cos\theta} && \bullet\ \text{Solve for } r. \end{aligned}$$

The remaining equations can be derived in a similar manner.

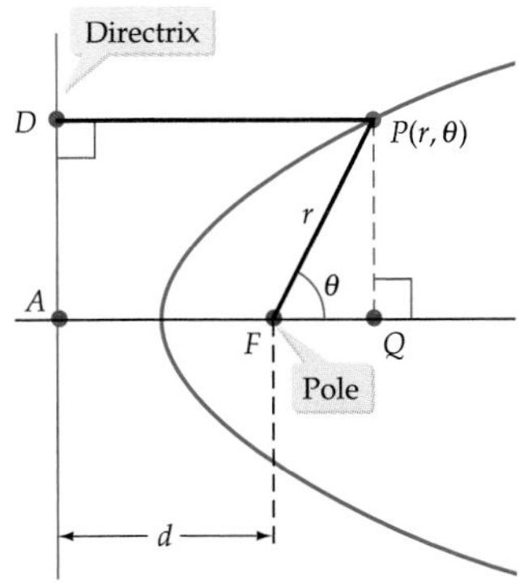

Figure 8.78

◆ GRAPH A CONIC GIVEN IN POLAR FORM

EXAMPLE 1 Sketch the Graph of a Hyperbola Given in Polar Form

Describe and sketch the graph of $r = \dfrac{8}{2 - 3\sin\theta}$.

Solution

Write the equation in standard form by dividing the numerator and denominator by 2, the constant term in the denominator.

$$r = \frac{4}{1 - \dfrac{3}{2}\sin\theta}$$

Because e is the coefficient of $\sin\theta$ and $e = 3/2 > 1$, the graph is a hyperbola with a focus at the pole. Because the equation contains the expression $\sin\theta$, the transverse axis is on the line $\theta = \pi/2$.

To find the vertices, choose θ equal to $\pi/2$ and $3\pi/2$. The corresponding values of r are -8 and $8/5$. The vertices are $(-8, \pi/2)$ and $(8/5, 3\pi/2)$. By choosing θ equal to 0 and π, we can determine the points (4, 0) and $(4, \pi)$ on the upper branch of the hyperbola. The lower branch can be determined by symmetry.

Plot some points (r, θ) for additional values of θ and corresponding values of r. See **Figure 8.79.**

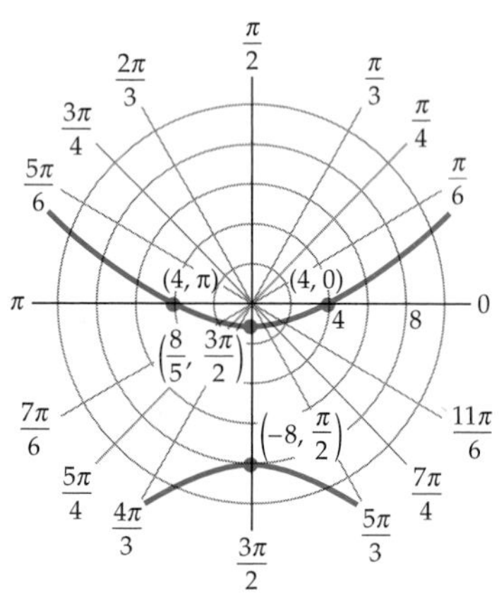

Figure 8.79

TRY EXERCISE 2, EXERCISE SET 8.6, PAGE 644

EXAMPLE 2 Sketch the Graph of an Ellipse Given in Polar Form

Describe and sketch the graph of $r = \dfrac{4}{2 + \cos\theta}$.

Solution

Write the equation in standard form by dividing the numerator and denominator by 2, which is the constant term in the denominator.

$$r = \frac{2}{1 + \dfrac{1}{2}\cos\theta}$$

Thus $e = 1/2$ and the graph is an ellipse with a focus at the pole. Because the equation contains the expression $\cos\theta$, the major axis is on the polar axis.

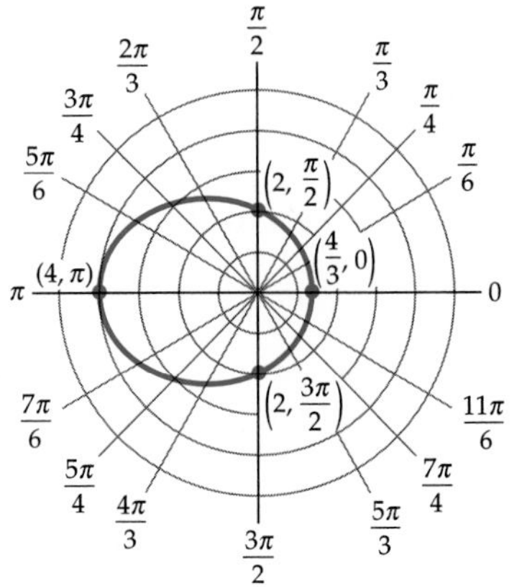

Figure 8.80

To find the vertices, choose θ equal to 0 and π. The corresponding values for r are 4/3 and 4. The vertices on the major axis are $(4/3, 0)$ and $(4, \pi)$. Plot some points (r, θ) for additional values of θ and the corresponding values of r. Two possible points are $(2, \pi/2)$ and $(2, 3\pi/2)$. See the graph of the ellipse in Figure 8.80.

TRY EXERCISE 4, EXERCISE SET 8.6, PAGE 644

◆ WRITE THE POLAR EQUATION OF A CONIC

EXAMPLE 3 Find the Equation of a Conic in Polar Form

Find the equation of the parabola, shown in Figure 8.81, with vertex at $(2, \pi/2)$ and focus at the pole.

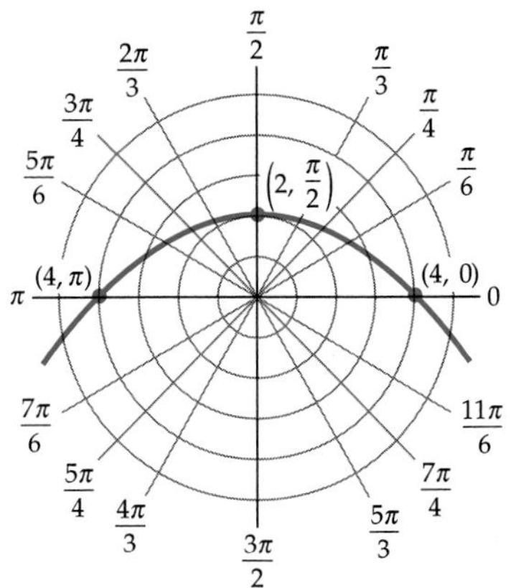

Figure 8.81

Solution

Because the vertex is on the line $\theta = \pi/2$ and the focus is at the pole, the axis of symmetry is the line $\theta = \pi/2$. Thus the equation of the parabola must involve $\sin \theta$. The parabola has a horizontal directrix above the pole, so the equation has the form

$$r = \frac{ed}{1 + e \sin \theta}$$

The distance from the vertex to the focus is 2 so the distance from the focus to the directrix is 4. Because the graph of the equation is a parabola, the eccentricity is 1. The equation is

$$r = \frac{(1)(4)}{1 + (1) \sin \theta} \qquad \bullet\ e = 1, d = 4$$

$$r = \frac{4}{1 + \sin \theta}$$

TRY EXERCISE 24, EXERCISE SET 8.6, PAGE 644

QUESTION In Example 3, why is there no point on the parabola that corresponds to $\theta = 3\pi/2$?

TOPICS FOR DISCUSSION

1. A student claims that the graph of $r = \dfrac{12}{2 + \cos \theta}$ is a parabola because the coefficient of the cosine term is 1. Explain why the student is wrong.

ANSWER When $\theta = 3\pi/2$, $\sin \theta = -1$. Thus $1 + \sin \theta = 0$, and $r = \dfrac{4}{1 + \sin \theta}$ is undefined.

2. The graph of $r = \dfrac{2}{2 + \sec\theta}$ is an ellipse except for the fact that it has two holes. Where are the holes located?

3. A tutor claims that there are two different ellipses that have a focus at the pole and a vertex at $\left(1, \dfrac{\pi}{2}\right)$. Do you agree? Explain.

4. Does the parabola given by $r = \dfrac{6}{1 + \sin\theta}$ have a horizontal axis of symmetry or a vertical axis of symmetry? Explain.

EXERCISE SET 8.6

In Exercises 1 to 14, describe and sketch the graph of each equation.

1. $r = \dfrac{12}{3 - 6\cos\theta}$
2. $r = \dfrac{8}{2 - 4\cos\theta}$
3. $r = \dfrac{8}{4 + 3\sin\theta}$
4. $r = \dfrac{6}{3 + 2\cos\theta}$
5. $r = \dfrac{9}{3 - 3\sin\theta}$
6. $r = \dfrac{5}{2 - 2\sin\theta}$
7. $r = \dfrac{10}{5 + 6\cos\theta}$
8. $r = \dfrac{8}{2 + 4\cos\theta}$
9. $r = \dfrac{4\sec\theta}{2\sec\theta - 1}$
10. $r = \dfrac{3\sec\theta}{2\sec\theta + 2}$
11. $r = \dfrac{12\csc\theta}{6\csc\theta - 2}$
12. $r = \dfrac{3\csc\theta}{2\csc\theta + 2}$
13. $r = \dfrac{3}{\cos\theta - 1}$
14. $r = \dfrac{2}{\sin\theta + 2}$

In Exercises 15 to 20, find a rectangular equation for the graphs in Exercises 1 to 6.

In Exercises 21 to 28, find a polar equation of the conic with the focus at the pole and the given eccentricity and directrix.

21. $e = 2, r\cos\theta = -1$
22. $e = 3/2, r\sin\theta = 1$
23. $e = 1, r\sin\theta = 2$
24. $e = 1, r\cos\theta = -2$
25. $e = 2/3, r\sin\theta = -4$
26. $e = 1/2, r\cos\theta = 2$
27. $e = 3/2, r = 2\sec\theta$
28. $e = 3/4, r = 2\csc\theta$

29. Find the polar equation of the parabola with a focus at the pole and vertex $(2, \pi)$.

30. Find the polar equation of the ellipse with a focus at the pole, vertex at $(4, 0)$, and eccentricity $1/2$.

31. Find the polar equation of the hyperbola with a focus at the pole, vertex at $(1, 3\pi/2)$, and eccentricity 2.

32. Find the polar equation of the ellipse with a focus at the pole, vertex at $(2, 3\pi/2)$, and eccentricity $2/3$.

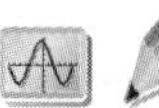

In Exercises 33 to 40, use a graphing utility to graph each equation. Write a sentence that explains how to obtain the graph from the graph of *r* as given in the exercise listed to the right of each equation.

33. $r = \dfrac{12}{3 - 6\cos\left(\theta - \dfrac{\pi}{6}\right)}$ (Compare with Exercise 1.)
34. $r = \dfrac{8}{2 - 4\cos\left(\theta - \dfrac{\pi}{2}\right)}$ (Compare with Exercise 2.)
35. $r = \dfrac{8}{4 + 3\sin(\theta - \pi)}$ (Compare with Exercise 3.)
36. $r = \dfrac{6}{3 + 2\cos\left(\theta - \dfrac{\pi}{3}\right)}$ (Compare with Exercise 4.)
37. $r = \dfrac{9}{3 - 3\sin\left(\theta + \dfrac{\pi}{6}\right)}$ (Compare with Exercise 5.)
38. $r = \dfrac{5}{2 - 2\sin\left(\theta + \dfrac{\pi}{2}\right)}$ (Compare with Exercise 6.)
39. $r = \dfrac{10}{5 + 6\cos(\theta + \pi)}$ (Compare with Exercise 7.)
40. $r = \dfrac{8}{2 + 4\cos\left(\theta + \dfrac{\pi}{3}\right)}$ (Compare with Exercise 8.)

SUPPLEMENTAL EXERCISES

In Exercises 41 to 46, use a graphing utility to graph each equation.

41. $r = \dfrac{3}{3 - \sec\theta}$

42. $r = \dfrac{5}{4 - 2\csc\theta}$

43. $r = \dfrac{3}{1 + 2\csc\theta}$

44. $r = \dfrac{4}{1 + 3\sec\theta}$

45. $r = 4\sin\sqrt{2}\theta, 0 \le \theta \le 12\pi$

46. $r = 4\cos\sqrt{3}\theta, 0 \le \theta \le 8\pi$

47. Let $P(r, \theta)$ satisfy the equation $r = \dfrac{ed}{1 - e\cos\theta}$. Show that $\dfrac{d(P, F)}{d(P, D)} = e$.

48. Show that the equation of a conic with a focus at the pole and directrix $r\sin\theta = d$ is given by $r = \dfrac{ed}{1 + e\sin\theta}$.

PROJECTS

1. **POLAR EQUATION OF A LINE** Verify that the polar equation of a line that is d units from the pole is given by

$$r = \frac{d}{\cos(\theta - \theta_p)}$$

where θ_p is the angle from the polar axis to a line segment that passes through the pole and is perpendicular to the line.

2. **POLAR EQUATION OF A CIRCLE THAT PASSES THROUGH THE POLE** Verify that the polar equation of a circle with center (a, θ_c) that passes through the pole is given by

$$r = 2a\cos(\theta - \theta_c)$$

SECTION 8.7 PARAMETRIC EQUATIONS

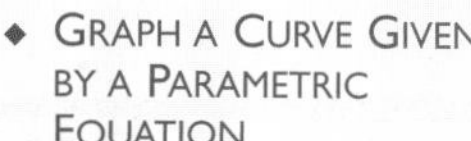

- PARAMETRIC EQUATIONS
- GRAPH A CURVE GIVEN BY A PARAMETRIC EQUATION
- ELIMINATE THE PARAMETER OF A PARAMETRIC EQUATION
- THE BRACHISTOCHRONE PROBLEM
- PARAMETRIC EQUATIONS AND PROJECTILE MOTION

◆ PARAMETRIC EQUATIONS

The graph of a function is a graph for which no vertical line can intersect the graph more than once. For a graph that is not the graph of a function (an ellipse or hyperbola, for example), it is frequently useful to describe the graph by *parametric equations.*

Curve and Parametric Equations

Let t be a number in an interval I. A **curve** is the set of ordered pairs (x, y), where

$$x = f(t), \qquad y = g(t) \quad \text{for } t \in I$$

The variable t is called a **parameter,** and the equations $x = f(t)$ and $y = g(t)$ are **parametric equations.**

◆ GRAPH A CURVE GIVEN BY A PARAMETRIC EQUATION

EXAMPLE 1 Sketch the Graph of a Curve Given in Parametric Form

Sketch the graph of the curve given by the parametric equations

$$x = t^2 + t, \qquad y = t - 1 \qquad \text{for } t \in (-\infty, \infty)$$

Solution

Begin by making a table of values of t and the corresponding values of x and y. Five values of t were arbitrarily chosen for the table that follows. Many more values might be necessary to determine an accurate graph.

t	$x = t^2 + t$	$y = t - 1$	(x, y)
-2	2	-3	$(2, -3)$
-1	0	-2	$(0, -2)$
0	0	-1	$(0, -1)$
1	2	0	$(2, 0)$
2	6	1	$(6, 1)$

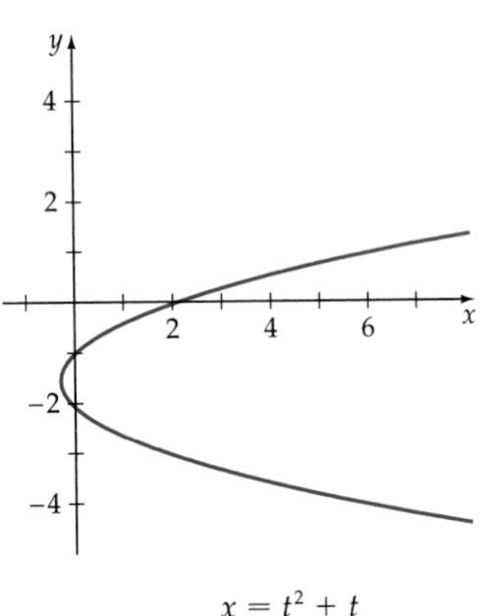

$x = t^2 + t$
$y = t - 1$

Figure 8.82

Graph the ordered pairs (x, y) and then draw a smooth curve through the points. See **Figure 8.82.**

TRY EXERCISE 6, EXERCISE SET 8.7, PAGE 650

◆ ELIMINATE THE PARAMETER OF A PARAMETRIC EQUATION

It may not be clear from Example 1 and the corresponding graph that the curve is a parabola. By **eliminating the parameter,** we can write one equation in x and y that is equivalent to the two parametric equations.

To eliminate the parameter, solve $y = t - 1$ for t.

$$y = t - 1 \quad \text{or} \quad t = y + 1$$

Substitute $y + 1$ for t in $x = t^2 + t$ and then simplify.

$$\begin{aligned} x &= (y + 1)^2 + (y + 1) \\ &= y^2 + 3y + 2 \end{aligned}$$

• **The equation of a parabola**

Complete the square and write the equation in standard form.

$$\left(x + \frac{1}{4}\right) = \left(y + \frac{3}{2}\right)^2$$

• **This is the equation of a parabola with vertex at $\left(-\frac{1}{4}, -\frac{3}{2}\right)$.**

EXAMPLE 2 Eliminate the Parameter and Sketch the Graph of a Curve

Eliminate the parameter and sketch the curve of the parametric equations

$$x = \sin t, \qquad y = \cos t \quad \text{for } 0 \le t < 2\pi$$

Solution

The process of eliminating the parameter sometimes involves trigonometric identities. To eliminate the parameter for the equations, square each side of each equation and then add.

$$x^2 = \sin^2 t$$
$$y^2 = \cos^2 t$$
$$x^2 + y^2 = \sin^2 t + \cos^2 t$$

Thus, using the trigonometric identity $\sin^2 t + \cos^2 t = 1$, we get

$$x^2 + y^2 = 1$$

This is the equation of a circle with center (0, 0) and radius equal to 1. See **Figure 8.83.**

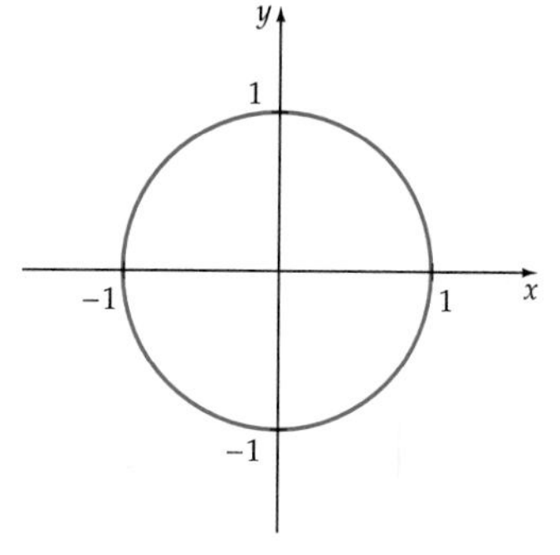

Figure 8.83

TRY EXERCISE 12, EXERCISE SET 8.7, PAGE 650

A parametric representation of a curve is not unique. That is, it is possible that a curve may be given by many different pairs of parametric equations. We will demonstrate this by using the equation of a line and providing two different parametric representations of the line.

Consider a line with slope m passing through the point (x_1, y_1). By the point-slope formula, the equation of the line is

$$y - y_1 = m(x - x_1)$$

Let $t = x - x_1$. Then $y - y_1 = mt$. A parametric representation is

$$x = x_1 + t, \qquad y = y_1 + mt \quad \text{for } t \text{ a real number} \tag{1}$$

Let $x - x_1 = \cot t$. Then $y - y_1 = m \cot t$. A parametric representation is

$$x = x_1 + \cot t, \qquad y = y_1 + m \cot t \quad \text{for } 0 < t < \pi \tag{2}$$

It can be verified that Equations (1) and (2) represent the original line.

Example 3 illustrates that the domain of the parameter t can be used to determine the domain and range of the function.

EXAMPLE 3 Sketch the Graph of a Curve Given by Parametric Equations

Eliminate the parameter and sketch the graph of the curve that is given by the parametric equations

$$x = 2 + 3\cos t, \qquad y = 3 + 2\sin t \quad \text{for } 0 \le t \le \pi$$

Continued

Solution

Rewrite each equation in terms of the trigonometric function.

$$\frac{x-2}{3} = \cos t \qquad \frac{y-3}{2} = \sin t \tag{3}$$

Using the trigonometric identity $\cos^2 t + \sin^2 t = 1$, we have

$$\cos^2 t + \sin^2 t = \left(\frac{x-2}{3}\right)^2 + \left(\frac{y-3}{2}\right)^2 = 1$$

$$\frac{(x-2)^2}{9} + \frac{(y-3)^2}{4} = 1$$

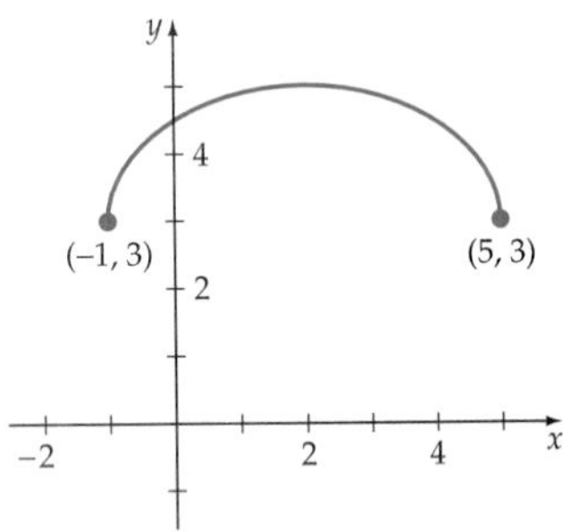

Figure 8.84

This is the equation of an ellipse with center at (2, 3) and major axis parallel to the x-axis. However, because $0 \le t \le \pi$, it follows that $-1 \le \cos t \le 1$ and $0 \le \sin t \le 1$. Therefore, we have

$$-1 \le \frac{x-2}{3} \le 1 \qquad 0 \le \frac{y-3}{2} \le 1$$ • **Using Equations (3)**

Solving these inequalities for x and y yields

$$-1 \le x \le 5 \quad \text{and} \quad 3 \le y \le 5$$

Because the values of y are between 3 and 5, the graph of the parametric equations is only the top half of the ellipse. See **Figure 8.84.**

TRY EXERCISE 14, EXERCISE SET 8.7, PAGE 650

◆ THE BRACHISTOCHRONE PROBLEM

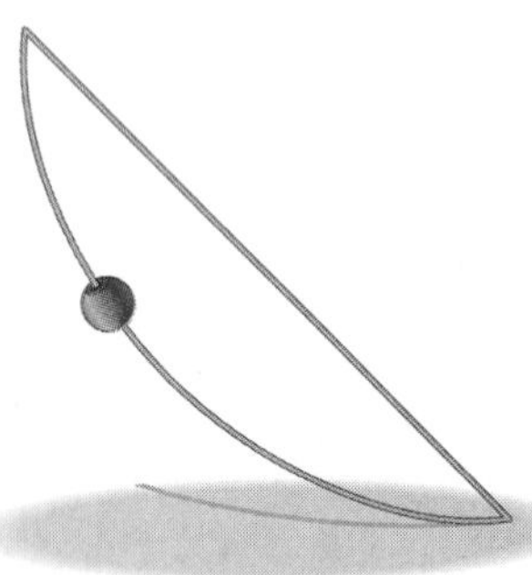
Figure 8.85

Parametric equations are useful in writing the equation of a moving point. One famous problem, involving a bead traveling down a frictionless wire, was posed in 1696 by the mathematician Johann Bernoulli. The problem was to determine the shape of a wire a bead could slide down so that the distance between two points was traveled in the shortest time. Problems that involve "shortest time" are called *brachistochrone problems.* They are very important in physics and form the basis for much of the classical theory of light propagation.

The answer to Bernoulli's problem is an arc of an inverted cycloid. See **Figure 8.85.** A **cycloid** is formed by letting a circle of radius a roll on a straight line L without slipping. See **Figure 8.86.** The curve traced by a point on the circumference of the circle is a cycloid. To find an equation for this curve, begin by placing a circle tangent to the x-axis with a point P on the circle and at the origin of a rectangular coordinate system.

Roll the circle along the x-axis. After the radius of the circle has rotated through an angle θ, the coordinates of the point $P(x, y)$ can be given by

$$x = h - a\sin\theta, \qquad y = k - a\cos\theta \tag{4}$$

where $C(h, k)$ is the current center of the circle.

Because the radius of the circle is a, $k = a$. See **Figure 8.86.** Because the circle rolls without slipping, the arc length subtended by θ equals h. Thus $h = a\theta$. Substituting for h and k in Equations (4), we have, after factoring,

$$x = a(\theta - \sin\theta), \qquad y = a(1 - \cos\theta) \quad \text{for } \theta \geq 0$$

See **Figure 8.87.**

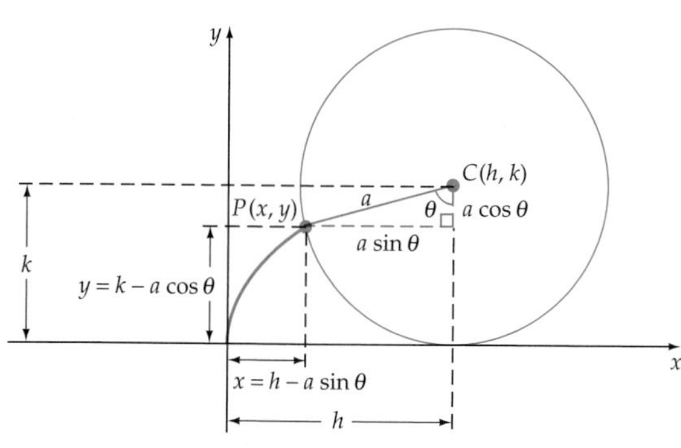

Figure 8.86

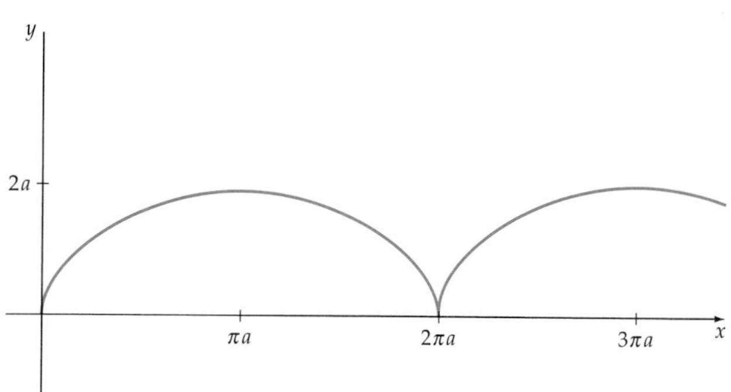

$x = a(\theta - \sin\theta), y = a(1 - \cos\theta)$

Figure 8.87
A cycloid

EXAMPLE 4 Graph a Cycloid

Use a graphing utility to graph the cycloid given by

$$x = 4(\theta - \sin\theta), \qquad y = 4(1 - \cos\theta) \quad \text{for } 0 \leq \theta \leq 4\pi$$

Solution

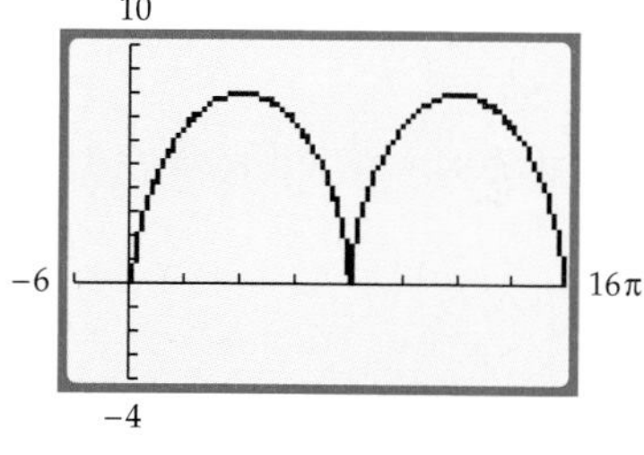

$x = 4(\theta - \sin\theta)$
$y = 4(1 - \cos\theta)$

Figure 8.88

Although θ is the parameter in the above equations, many graphing utilities use T as the parameter for parametric equations. Thus to graph the equations for $0 \leq \theta \leq 4\pi$, we use Tmin = 0 and Tmax = 4π as shown below. Use radian mode and parametric mode to produce the graph in **Figure 8.88.**

Tmin=0	Xmin=-6	Ymin=-4
Tmax=4π	Xmax=16π	Ymax=10
Tstep=0.5	Xscl=2π	Yscl=1

TRY EXERCISE 26, EXERCISE SET 8.7, PAGE 651

◆ PARAMETRIC EQUATIONS AND PROJECTILE MOTION

The path of a projectile (assume air resistance is negligible) that is launched at an angle θ from the horizon with an initial velocity of v_0 feet per second is given by the parametric equations

$$x = (v_0 \cos\theta)t, \qquad y = -16t^2 + (v_0 \sin\theta)t$$

where t is the time in seconds since the projectile was launched.

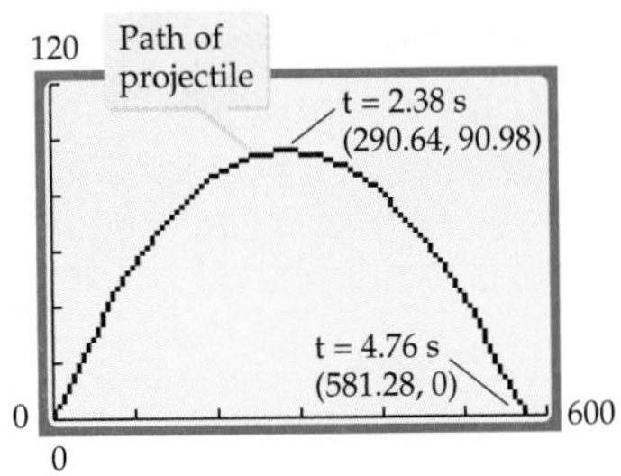

$x = (144 \cos 32°)t$
$y = -16t^2 + (144 \sin 32°)t$

Figure 8.89

take note

In **Figure 8.89**, the angle of launch does not appear to be 32° because 1 foot on the x-axis is smaller than 1 foot on the y-axis.

EXAMPLE 5 Sketch the Path of a Projectile

Use a graphing utility to sketch the path of a projectile that is launched at an angle of $\theta = 32°$ with an initial velocity of 144 feet per second. Use the graph to determine (to the nearest foot) the maximum height of the projectile and the range of the projectile. Assume the ground is level.

Solution

Use degree mode and parametric mode. Graph the parametric equations

$$x = (144 \cos 32°)t, \qquad y = -16t^2 + (144 \sin 32°)t \quad \text{for } 0 \le t \le 5$$

to produce the graph in **Figure 8.89** Use the TRACE feature to determine that the maximum height of 91 feet is attained when $t \approx 2.38$ seconds and that the projectile strikes the ground about 581 feet downrange when $t \approx 4.76$ seconds

TRY EXERCISE 32, EXERCISE SET 8.7, PAGE 651

TOPICS FOR DISCUSSION

1. It is always possible to eliminate the parameter of a pair of parametric equations. Do you agree? Explain.
2. The line $y = 3x + 5$ has more than one parametric representation. Do you agree? Explain.
3. Parametric equations are used only to graph functions. Do you agree? Explain.
4. Every function $y = f(x)$ can be written in parametric form by letting $x = t$ and $y = f(t)$. Do you agree? Explain.

EXERCISE SET 8.7

In Exercises 1 to 10, graph the parametric equations by plotting several points.

1. $x = 2t, y = -t$, for $t \in R$
2. $x = -3t, y = 6t$, for $t \in R$
3. $x = -t, y = t^2 - 1$, for $t \in R$
4. $x = 2t, y = 2t^2 - t + 1$, for $t \in R$
5. $x = t^2, y = t^3$, for $t \in R$
6. $x = t^2 + 1, y = t^2 - 1$, for $t \in R$
7. $x = 2 \cos t, y = 3 \sin t$, for $0 \le t < 2\pi$
8. $x = 1 - \sin t, y = 1 + \cos t$, for $0 \le t < 2\pi$
9. $x = 2^t, y = 2^{t+1}$, for $t \in R$
10. $x = t^2, y = 2 \log_2 t$, for $t \ge 1$

In Exercises 11 to 20, eliminate the parameter and graph the equation.

11. $x = \sec t, y = \tan t$, for $-\pi/2 < t < \pi/2$
12. $x = 3 + 2 \cos t, y = -1 - 3 \sin t$, for $0 \le t < 2\pi$
13. $x = 2 - t^2, y = 3 + 2t^2$, for $t \in R$
14. $x = 1 + t^2, y = 2 - t^2$, for $t \in R$
15. $x = \cos^3 t, y = \sin^3 t$, for $0 \le t < 2\pi$
16. $x = e^{-t}, y = e^t$, for $t \in R$
17. $x = \sqrt{t + 1}, y = t$, for $t \ge -1$
18. $x = \sqrt{t}, y = 2t - 1$, for $t \ge 0$
19. $x = t^3, y = 3 \ln t$, for $t > 0$
20. $x = e^t, y = e^{2t}$, for $t \in R$

21. Eliminate the parameter for the curves

$$C_1: \ x = 2 + t^2, \quad y = 1 - 2t^2$$

and $$C_2: \ x = 2 + t, \quad y = 1 - 2t$$

and then discuss the differences between their graphs.

22. Eliminate the parameter for the curves

$$C_1: \ x = \sec^2 t, \quad y = \tan^2 t$$

and $$C_2: \ x = 1 + t^2, \quad y = t^2$$

for $0 \le t < \pi/2$, and then discuss the differences between their graphs.

23. Sketch the graph of

$$x = \sin t, \qquad y = \csc t \quad \text{for } 0 < t \le \pi/2$$

Sketch another graph for the same pair of equations but choose the domain of t as $\pi < t \le 3\pi/2$.

24. Discuss the differences between

$$C_1: \ x = \cos t, \quad y = \cos^2 t$$

and $$C_2: \ x = \sin t, \quad y = \sin^2 t$$

for $0 \le t \le \pi$.

25. Use a graphing utility to graph the cycloid $x = 2(t - \sin t)$, $y = 2(1 - \cos t)$ for $0 \le t < 2\pi$.

26. Use a graphing utility to graph the cycloid $x = 3(t - \sin t)$, $y = 3(1 - \cos t)$ for $0 \le t \le 12\pi$.

Parametric equations of the form $x = a \sin \alpha t, y = b \cos \beta t$, for $t \ge 0$, are encountered in electrical circuit theory. The graphs of these equations are called *Lissajous figures*.

27. Graph: $x = 5 \sin 2t$, $y = 5 \cos t$

28. Graph: $x = 5 \sin 3t$, $y = 5 \cos 2t$

29. Graph: $x = 5 \sin 6t$, $y = 5 \cos 3t$

30. Graph: $x = 5 \sin 10t$, $y = 5 \cos 9t$

In Exercises 31 to 34, graph the path of the projectile that is launched at an angle θ with the horizon with an initial velocity of v_0. In each exercise, use the graph to determine the maximum height and the range of the projectile (to the nearest foot). Also state the time t at which the projectile reaches its maximum height and the time it hits the ground. Assume the ground is level and the only force acting on the projectile is gravity.

31. $\theta = 55°$, $v_0 = 210$ feet per second

32. $\theta = 35°$, $v_0 = 195$ feet per second

33. $\theta = 42°$, $v_0 = 315$ feet per second

34. $\theta = 52°$, $v_0 = 315$ feet per second

SUPPLEMENTAL EXERCISES

35. Let $P_1(x_1, y_1)$ and $P_2(x_2, y_2)$ be two distinct points in the plane, and consider the line L passing through those points. Choose a point $P(x, y)$ on the line L. Show that

$$\frac{x - x_1}{x_2 - x_1} = \frac{y - y_1}{y_2 - y_1}$$

Use this result to demonstrate that $x = (x_2 - x_1)t + x_1$, $y = (y_2 - y_1)t + y_1$ is a parametric representation of the line through the two points.

36. Show that $x = h + a \sin t$, $y = k + b \cos t$, for $a > 0$, $b > 0$, and $0 \le t < 2\pi$, are parametric equations for an ellipse with center at (h, k).

37. Suppose a string, held taut, is unwound from the circumference of a circle of radius a. The path traced by the end of the string is called the *involute* of a circle. Find parametric equations for the involute of a circle.

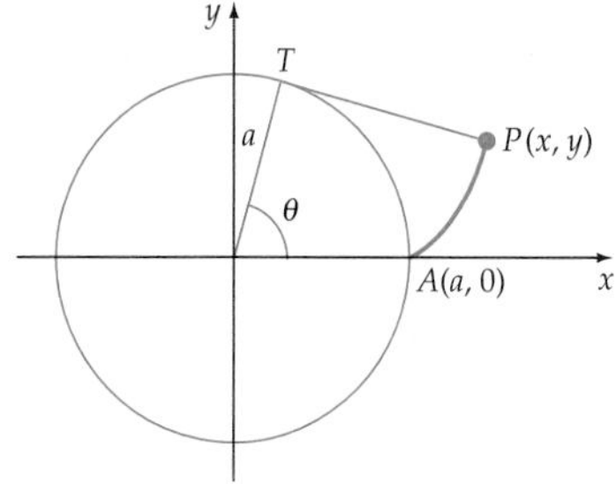

38. A circle of radius a rolls without slipping on the outside of a circle of radius $b > a$. Find the parametric equations of a point P on the smaller circle. The curve is called an *epicycloid*.

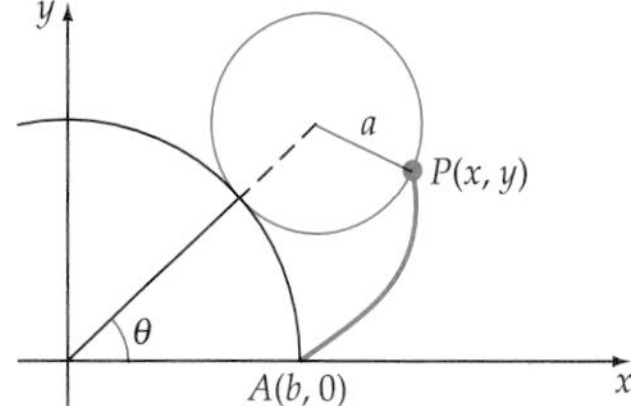

39. A circle of radius a rolls without slipping on the inside of a circle of radius $b > a$. Find the parametric equation of a point P on the smaller circle. The curve is called a *hypocycloid*.

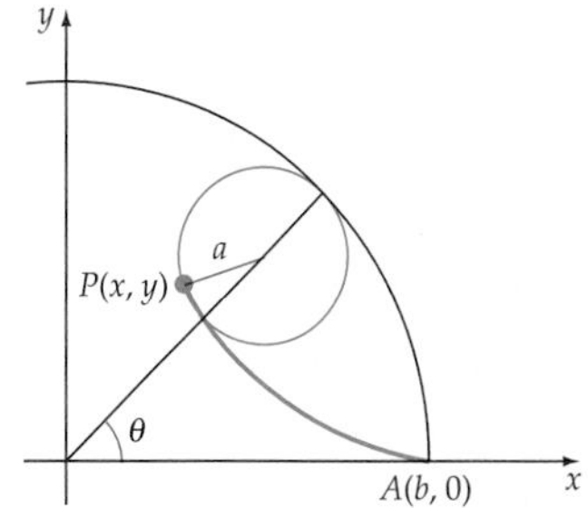

PROJECTS

1. **PARAMETRIC EQUATIONS IN AN xyz-COORDINATE SYSTEM**

   a. Graph the three-dimensional curve given by

   $$x = 3\cos t, \qquad y = 3\sin t, \qquad z = 0.5t$$

   b. Graph the three-dimensional curve given by

   $$x = 3\cos t, \qquad y = 6\sin t, \qquad z = 0.5t$$

   c. What is the main difference between these curves?

   d. What name is given to curves of this type?

EXPLORING CONCEPTS WITH TECHNOLOGY

Using a Graphing Calculator to Find the *n*th Roots of *z*

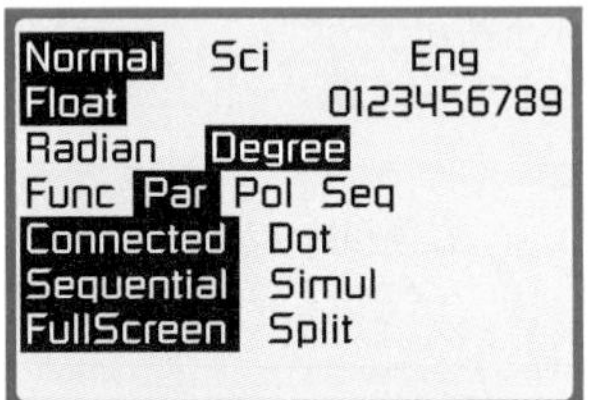

Figure 8.90

In Chapter 7 we used De Moivre's Theorem to find the nth roots of a number. The parametric feature of a graphing calculator can also be used to find and display the nth roots of $z = r(\cos\theta + i\sin\theta)$. Here is the procedure for a TI-83 graphing calculator. Put the calculator in parametric and degree mode. See **Figure 8.90.** To find the nth roots of $z = r(\cos\theta + i\sin\theta)$, enter in the [Y=] menu

X1T=r^(1/n)cos(θ/n+T) and Y1T=r^(1/n)sin(θ/n+T).

In the WINDOW menu, set Tmin=0, Tmax=360, and Tstep=360/n. Set Xmin, Xmax, Ymin, and Ymax to appropriate values that will allow the roots to be seen in the graph window. Press GRAPH to display a polygon. The x- and y-coordinates of each vertex of the polygon represent a root of z in the rectangular form $x + yi$. Here is a specific example that illustrates this procedure.

Example Find the fourth roots of $z = 16i$.

In trigonometric form, $z = 16(\cos 90° + i\sin 90°)$. Thus in this example, $r = 16$, $\theta = 90°$, and $n = 4$. In the [Y=] menu, enter

X1T=16^(1/4)cos(90/4+T) and Y1T=16^(1/4)sin(90/4+T)

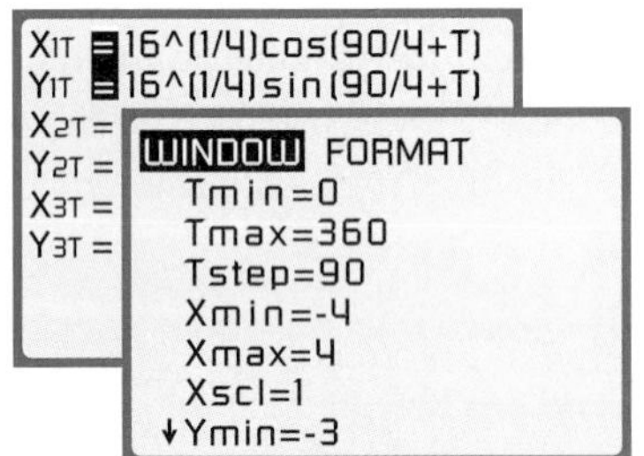

Figure 8.91

In the WINDOW menu, set

Tmin=0	Xmin=-4	Ymin=-3
Tmax=360	Xmax=4	Ymax=3
Tstep=360/4	Xscl=1	Yscl=1

See **Figure 8.91.** Press GRAPH to produce the quadrilateral in **Figure 8.92.** Use TRACE and the arrow key [▷] to move to each of the vertices of the quadrilateral. **Figure 8.92** shows that one of the roots of $z = 16i$ is $1.8477591 + 0.76536686i$. Continue to press the arrow key [▷] to find the other three roots, which are

$-0.7653669 + 1.8477591i$, $-1.847759 - 0.7653669i$, and $0.76536686 - 1.847759i$.

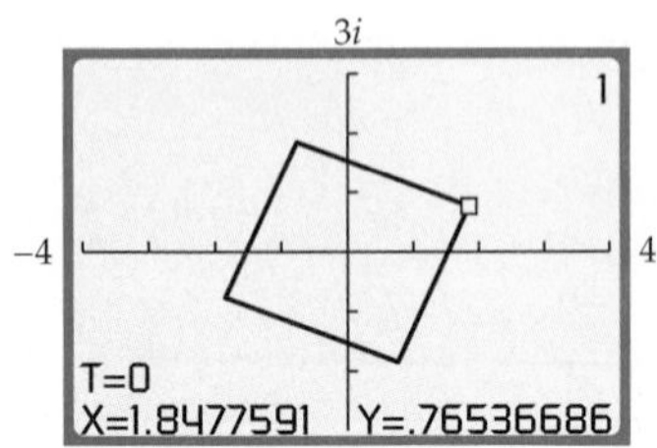

The vertices of the quadrilateral represent the fourth roots of $16i$ in the complex plane.

Figure 8.92

Use a graphing calculator to find, in rectangular form, each of the following.

1. The cube roots of -27.
2. The fifth roots of $32i$.
3. The fourth roots of $\sqrt{8} + \sqrt{8}i$.
4. The sixth roots of $-64i$.

CHAPTER 8 SUMMARY

8.1 Parabolas

- The equations of a parabola with vertex at (h, k) and axis of symmetry parallel to a coordinate axis are given by

$$(x - h)^2 = 4p(y - k);\ \text{focus } (h, k + p);\ \text{directrix } y = k - p$$
$$(y - k)^2 = 4p(x - h);\ \text{focus } (h + p, k);\ \text{directrix } x = h - p$$

8.2 Ellipses

- The equations of an ellipse with center at (h, k) and major axis parallel to a coordinate axis are given by

$$\frac{(x-h)^2}{a^2} + \frac{(y-k)^2}{b^2} = 1;\ \text{foci } (h \pm c, k);\ \text{vertices } (h \pm a, k)$$
$$\frac{(x-h)^2}{b^2} + \frac{(y-k)^2}{a^2} = 1;\ \text{foci } (h, k \pm c);\ \text{vertices } (h, k \pm a)$$

For each equation, $a > b$ and $c^2 = a^2 - b^2$.

- The eccentricity e of an ellipse is given by $e = c/a$.

8.3 Hyperbolas

- The equations of a hyperbola with center at (h, k) and transverse axis parallel to a coordinate axis are given by

$$\frac{(x-h)^2}{a^2} - \frac{(y-k)^2}{b^2} = 1;\ \text{foci } (h \pm c, k);\ \text{vertices } (h \pm a, k)$$
$$\frac{(y-k)^2}{a^2} - \frac{(x-h)^2}{b^2} = 1;\ \text{foci } (h, k \pm c);\ \text{vertices } (h, k \pm a)$$

For each equation, $c^2 = a^2 + b^2$.

- The eccentricity e of a hyperbola is given by $e = c/a$.

8.4 Rotation of Axes

- The rotation-of-axes formulas are

$$\begin{cases} x = x' \cos\alpha - y' \sin\alpha \\ y = y' \cos\alpha + x' \sin\alpha \end{cases} \qquad \begin{cases} x' = x\cos\alpha + y\sin\alpha \\ y' = y\cos\alpha - x\sin\alpha \end{cases}$$

- To eliminate the xy term from the general quadratic equation, rotate the coordinate axes through an angle α, where

$$\cot 2\alpha = \frac{A - C}{B},\quad B \neq 0,\quad 0° < 2\alpha < 180°$$

- The graph of $Ax^2 + Bxy + Cy^2 + Dx + Ey + F = 0$ is either a conic or a degenerate conic. If the graph is a conic, then the graph can be identified by its *discriminant* $B^2 - 4AC$. The graph is

  an ellipse or a circle, provided $B^2 - 4AC < 0$.

  a parabola, provided $B^2 - 4AC = 0$.

  a hyperbola, provided $B^2 - 4AC > 0$.

- The graph of $Ax^2 + Bxy + Cy^2 + Dx + Ey + F = 0$ can be constructed by using a graphing utility to graph both

$$y_1 = \frac{-(Bx + E) + \sqrt{(Bx + E)^2 - 4(C)(Ax^2 + Dx + F)}}{2(C)}$$

and

$$y_2 = \frac{-(Bx + E) - \sqrt{(Bx + E)^2 - 4(C)(Ax^2 + Dx + F)}}{2(C)}$$

8.5 Introduction to Polar Coordinates

- A polar coordinate system is formed by drawing a horizontal ray (*polar axis*). The *pole* is the origin of a polar coordinate system.
- A point is specified by coordinates (r, θ), where r is a directed distance from the pole and θ is an angle measured from the polar axis.

  The transformation equations between a polar coordinate system and a rectangular coordinate system are

  Polar to rectangular: $x = r\cos\theta \qquad y = r\sin\theta$

  Rectangular to polar: $r = \sqrt{x^2 + y^2} \qquad \tan\theta = y/x$

8.6 Polar Equations of the Conics

- The polar equations of the conics are given by

$$r = \frac{ed}{1 \pm e\cos\theta} \quad \text{or} \quad r = \frac{ed}{1 \pm e\sin\theta}$$

where e is the eccentricity and d is the distance of the directrix from the focus.

When

$0 < e < 1$, the graph is an ellipse.

$e = 1$, the graph is a parabola.

$e > 1$, the graph is a hyperbola.

8.7 Parametric Equations

- Let t be a number in an interval I. A *curve* is a set of ordered pairs (x, y), where

$$x = f(t), \qquad y = g(t) \quad \text{for } t \in I$$

The variable t is called a *parameter*, and the pair of equations are *parametric equations*.

- To *eliminate the parameter* is to find an equation in x and y that has the same graph as the given parametric equations.
- The path of a projectile (assume air resistance is negligible) that is launched at an angle θ from the horizon with an initial velocity of v_0 feet per second is given by

$$x = (v_0\cos\theta)t, \qquad y = -16t^2 + (v_0\sin\theta)t$$

where t is the time in seconds since the projectile was launched.

CHAPTER 8 TRUE/FALSE EXERCISES

In Exercises 1 to 10, answer true or false. If the statement is false, give an example to show that the statement is false.

1. The graph of a parabola is the same shape as that of one branch of a hyperbola.
2. For the two axes of an ellipse (which is not a circle), the major axis and the minor axis, the major axis is always the longer axis.
3. For the two axes of a hyperbola, the transverse axis and the conjugate axis, the transverse axis is always the longer axis.
4. If two ellipses have the same foci, they have the same graph.
5. A hyperbola is similar to a parabola in that both curves have asymptotes.
6. If a hyperbola with center at the origin and a parabola with vertex at the origin have the same focus, $(0, c)$, then the two graphs always intersect.
7. The graphs of all the conic sections are not the graphs of functions.
8. Only the graph of a function can be written using parametric equations.
9. The graph of $x = \sin t$, $y = \cos t$, for $0 \le t < 2\pi$, and the graph of $x = \cos t$, $y = \sin t$, for $0 \le t < 2\pi$, are exactly the same.
10. Each ordered pair (r, θ) in a polar coordinate system specifies exactly one point.

CHAPTER 8 REVIEW EXERCISES

In Exercises 1 to 12, find the foci and the vertices of each conic. If the conic is a hyperbola, find the asymptotes. Graph each equation.

1. $x^2 - y^2 = 4$
2. $y^2 = 16x$
3. $x^2 + 4y^2 - 6x + 8y - 3 = 0$
4. $3x^2 - 4y^2 + 12x - 24y - 36 = 0$
5. $3x - 4y^2 + 8y + 2 = 0$
6. $3x + 2y^2 - 4y - 7 = 0$
7. $9x^2 + 4y^2 + 36x - 8y + 4 = 0$
8. $11x^2 - 25y^2 - 44x - 50y - 256 = 0$
9. $4x^2 - 9y^2 - 8x + 12y - 144 = 0$
10. $9x^2 + 16y^2 + 36x - 16y - 104 = 0$
11. $4x^2 + 28x + 32y + 81 = 0$
12. $x^2 - 6x - 9y + 27 = 0$

In Exercises 13 to 20, find the equation of the conic that satisfies the given conditions.

13. Ellipse with vertices at $(7, 3)$ and $(-3, 3)$; length of minor axis is 8.
14. Hyperbola with vertices at $(4, 1)$ and $(-2, 1)$; eccentricity 4/3.
15. Hyperbola with foci $(-5, 2)$ and $(1, 2)$; length of transverse axis is 4.
16. Parabola with focus $(2, -3)$ and directrix $x = 6$.
17. Parabola with vertex $(0, -2)$ and passing through the point $(3, 4)$.
18. Ellipse with eccentricity 2/3 and foci $(-4, -1)$ and $(0, -1)$.
19. Hyperbola with vertices $(\pm 6, 0)$ and asymptotes whose equations are $y = \pm\frac{1}{9}x$.
20. Parabola passing through the points $(1, 0)$, $(2, 1)$, and $(0, 1)$ with axis of symmetry parallel to the y-axis.

In Exercises 21 to 24, write an equation without an *xy* term. Name the graph of the equation.

21. $11x^2 - 6xy + 19y^2 - 40 = 0$
22. $3x^2 + 6xy + 3y^2 - 4x + 5y - 12 = 0$
23. $x^2 + 2\sqrt{3}xy + 3y^2 + 8\sqrt{3}x - 8y + 32 = 0$
24. $xy - x - y - 1 = 0$

In Exercises 25 to 34, graph each polar equation.

25. $r = 4\cos 3\theta$
26. $r = 1 + \cos\theta$
27. $r = 2(1 - 2\sin\theta)$
28. $r = 4\sin 4\theta$
29. $r = 5\sin\theta$
30. $r = 3\sec\theta$
31. $r = 4\csc\theta$
32. $r = 4\cos\theta$
33. $r = 3 + 2\cos\theta$
34. $r = 4 + 2\sin\theta$

In Exercises 35 to 38, change each equation to a polar equation.

35. $y^2 = 16x$
36. $x^2 + y^2 + 4x + 3y = 0$
37. $3x - 2y = 6$
38. $xy = 4$

In Exercises 39 to 42, change each equation to a rectangular equation.

39. $r = \dfrac{4}{1 - \cos\theta}$

40. $r = 3\cos\theta - 4\sin\theta$

41. $r^2 = \cos 2\theta$

42. $\theta = 1$

In Exercises 43 to 46, graph the conic given by each polar equation.

43. $r = \dfrac{4}{3 - 6\sin\theta}$

44. $r = \dfrac{2}{1 + \cos\theta}$

45. $r = \dfrac{2}{2 - \cos\theta}$

46. $r = \dfrac{6}{4 + 3\sin\theta}$

In Exercises 47 to 53, eliminate the parameter and graph the curve given by the parametric equations.

47. $x = 4t - 2, y = 3t + 1$, for $t \in R$

48. $x = 1 - t^2, y = 3 - 2t^2$, for $t \in R$

49. $x = 4\sin t, y = 3\cos t$, for $0 \le t < 2\pi$

50. $x = \sec t, y = 4\tan t$, for $-\pi/2 < t < \pi/2$

51. $x = \dfrac{1}{t}, y = -\dfrac{2}{t}$, for $t > 0$

52. $x = 1 + \cos t, y = 2 - \sin t$, for $0 \le t < 2\pi$

53. $x = \sqrt{t}, y = 2^{-t}$, for $t \ge 0$

54. Use a graphing utility to graph the cycloid given by

$$x = 3(t - \sin t), \qquad y = 3(1 - \cos t) \quad \text{for } 0 \le t \le 18\pi$$

55. Use a graphing utility to graph the conic given by

$$x^2 + 4xy + 2y^2 - 2x + 5y + 1 = 0$$

56. Use a graphing utility to graph

$$r = \frac{6}{3 + \sin\left(\theta + \dfrac{\pi}{4}\right)}$$

57. The path of a projectile (assume air resistance is negligible) that is launched at an angle θ from the horizon with an initial velocity of v_0 feet per second is given by the parametric equations

$$x = (v_0\cos\theta)t, \qquad y = -16t^2 + (v_0\sin\theta)t$$

where t is the time in seconds since the projectile was launched. Use a graphing utility to graph the path of a projectile that is launched at an angle of 33° with an initial velocity of 245 feet per second. Use the graph to determine the maximum height of the projectile to the nearest foot.

CHAPTER 8 TEST

1. Find the vertex, focus, and directrix of the parabola given by the equation $y = \frac{1}{8}x^2$.

2. Graph: $\dfrac{x^2}{16} + \dfrac{y^2}{1} = 1$

3. Find the vertices and foci of the ellipse given by the equation $25x^2 - 150x + 9y^2 + 18y + 9 = 0$.

4. Find the equation in standard form of the ellipse with center $(0, -3)$, foci $(-6, -3)$ and $(6, -3)$, and minor axis of length 6.

5. Graph: $\dfrac{y^2}{25} - \dfrac{x^2}{16} = 1$

6. Find the vertices, foci, and asymptotes of the hyperbola given by the equation $\dfrac{x^2}{36} - \dfrac{y^2}{64} = 1$.

7. Graph: $16y^2 + 32y - 4x^2 - 24x = 84$

8. For the equation $x^2 - 4xy - 5y^2 + 3x - 5y - 20 = 0$, determine what angle of rotation (to the nearest 0.01°) would eliminate the xy term.

9. Determine whether the graph of the following equation is the graph of a parabola, an ellipse, or a hyperbola.

$$8x^2 + 5xy + 2y^2 - 10x + 5y + 4 = 0$$

10. $P(1, -\sqrt{3})$ are the coordinates of a point in an xy-coordinate system. Find the polar coordinates of P.

11. Graph: $r = 4\cos\theta$

12. Graph: $r = 3(1 - \sin\theta)$

13. Graph: $r = 2\sin 4\theta$

14. Find the rectangular coordinates of the point whose polar coordinates are $(5, 7\pi/3)$.

15. Find the rectangular form of $r - r\cos\theta = 4$.

16. Write $r = \dfrac{4}{1 + \sin\theta}$ as an equation in rectangular coordinates.

17. Eliminate the parameter and graph the curve given by the parametric equations $x = t - 3, y = 2t^2$.

18. Eliminate the parameter and graph the curve given by the parametric equations $x = 4\sin\theta,\ y = \cos\theta + 2$, where $0 \le \theta < 2\pi$.

19. Use a graphing utility to graph the cycloid given by

$$x = 2(t - \sin t), \qquad y = 2(1 - \cos t)$$

for $0 \le t \le 12\pi$.

20. The path of a projectile that is launched at an angle of 30° from the horizon with an initial velocity of 128 feet per second is given by

$$x = (128 \cos 30°)t, \qquad y = -16t^2 + (128 \sin 30°)t$$

where t is the time in seconds after the projectile is launched. Use a graphing utility to determine how far (to the nearest foot) the projectile will travel down-range if the ground is level.

CHAPTER 9

SYSTEMS OF EQUATIONS

Boardwalk? Park Place? How about New York?: A Winning Strategy for Monopoly

Monopoly, invented during the Depression, is still a very popular board game. In fact, Parker Brothers, the maker of Monopoly, has sponsored world championship Monopoly games in Atlantic City, home of Baltic and Mediterranean Avenues.

In the early 1980s, Stephen Heppe, at the time a student, became interested in winning Monopoly strategies. He wanted to know which properties on the Monopoly board paid greater rates of return for each dollar invested. Answering Heppe's question required solving a system of linear equations. Heppe's system of equations contained 123 equations with 123 variables. Here are some of the results obtained from solving this system of equations.

> A player is less likely to land on Mediterranean Avenue during the course of the game than on any other property. The chances of a player landing on Illinois Avenue are greater than those of landing on any other property.

◆ A modern Monopoly board.

Besides knowing which properties have the greatest chance of being occupied, Heppe also wanted to know which properties pay the greatest return for each dollar invested in houses or hotels. Some of his conclusions:

> New York with a hotel has the highest rate of return.
>
> The lowest rate of return for a property with a hotel is offered by Mediterranean. Assuming that all the railroads are owned, the B&O railroad has the greatest rate of return of all the railroads. This is because it is more likely that a player will land on this railroad than on the other railroads.

◆ The Boardwalk in Atlantic City, namesake of one of the most coveted Monopoly properties.

For more information on the mathematics of Monopoly, see "Matrix Mathematics: How to Win at Monopoly" by Dr. Crypton in the September 1985 issue of *Science Digest*.

SECTION 9.1 SYSTEMS OF LINEAR EQUATIONS IN TWO VARIABLES

- SUBSTITUTION METHOD FOR SOLVING A SYSTEM OF LINEAR EQUATIONS
- ELIMINATION METHOD FOR SOLVING A SYSTEM OF EQUATIONS
- APPLICATIONS OF SYSTEMS OF EQUATIONS

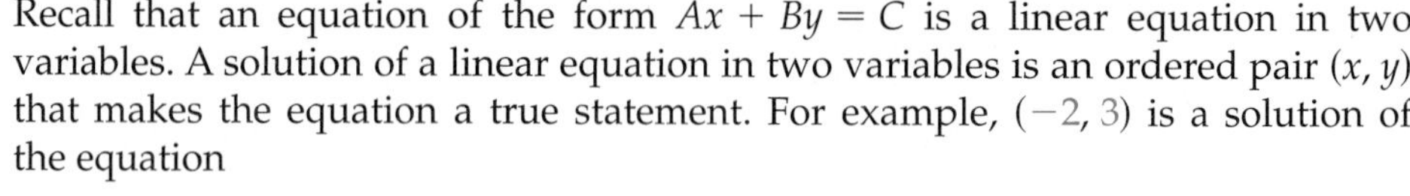

Recall that an equation of the form $Ax + By = C$ is a linear equation in two variables. A solution of a linear equation in two variables is an ordered pair (x, y) that makes the equation a true statement. For example, $(-2, 3)$ is a solution of the equation

$$2x + 3y = 5 \qquad \text{since} \qquad 2(-2) + 3(3) = 5$$

The graph of a linear equation, a straight line, is the set of points whose ordered pairs satisfy the equation. **Figure 9.1** is the graph of $2x + 3y = 5$.

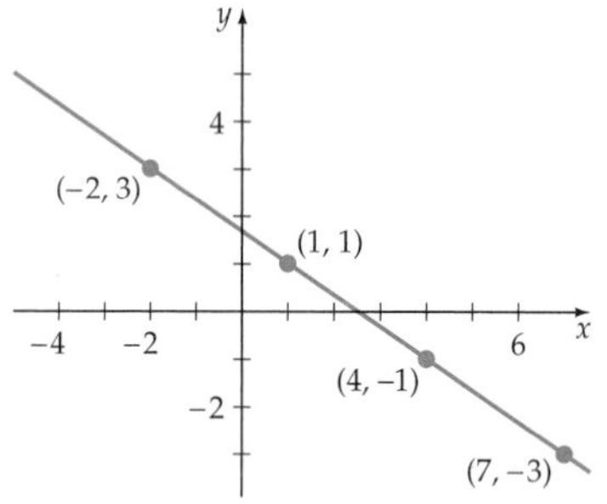

$2x + 3y = 5$

Figure 9.1

A **system of equations** is two or more equations considered together. The following system of equations is a **linear system of equations** in two variables.

$$\begin{cases} 2x + 3y = 4 \\ 3x - 2y = -7 \end{cases}$$

A **solution** of a system of equations in two variables is an ordered pair that is a solution of both equations.

In **Figure 9.2,** the graphs of the two equations in the system of equations above intersect at the point $(-1, 2)$. Because that point lies on both lines, $(-1, 2)$ is a solution of both equations and thus is a solution of the system of equations. The point $(5, -2)$ is a solution of the first equation but not the second equation. Therefore, $(5, -2)$ is not a solution of the system of equations.

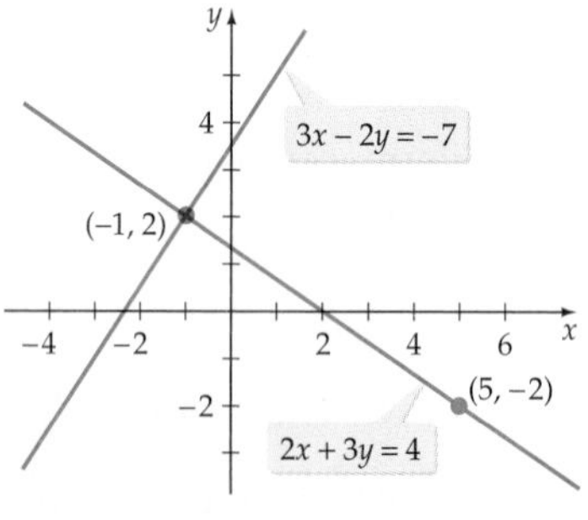

Figure 9.2

The graphs of two linear equations in two variables can intersect at a single point, be the same line, or be parallel. When the graphs intersect at a single point or are the same line, the system is called a **consistent** system of equations. The system is called an **independent** system of equations when the lines intersect at exactly one point. The system is called a **dependent** system of equations when the equations represent the same line. In this case, the system has an infinite number of solutions. When the graphs of the two equations are parallel lines, the system is called **inconsistent** and has no solution. See **Figure 9.3.**

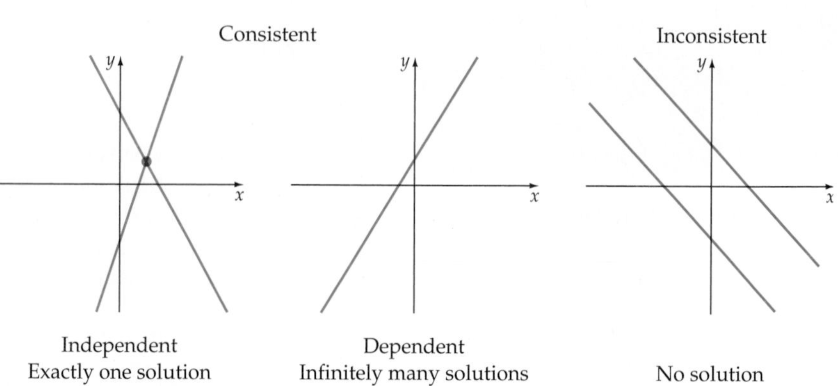

Figure 9.3

◆ SUBSTITUTION METHOD FOR SOLVING A SYSTEM OF LINEAR EQUATIONS

The **substitution method** is one procedure for solving a system of equations. This method is illustrated in Example 1.

EXAMPLE 1 Solve a System of Equations by the Substitution Method

Solve: $\begin{cases} 3x - 5y = 7 & (1) \\ y = 2x & (2) \end{cases}$

Solution

The solutions of $y = 2x$ are the ordered pairs $(x, 2x)$. For the system of equations to have a solution, ordered pairs of the form $(x, 2x)$ must also be solutions of $3x - 5y = 7$. To determine whether the ordered pairs $(x, 2x)$ are solutions of Equation (1), substitute $(x, 2x)$ into Equation (1) and solve for x. Think of this as *substituting* $2x$ for y.

$$3x - 5y = 7 \quad \text{• Equation (1)}$$
$$3x - 5(2x) = 7 \quad \text{• Substitute } 2x \text{ for } y.$$
$$3x - 10x = 7$$
$$-7x = 7$$
$$x = -1$$
$$y = 2x \quad \text{• Equation (2)}$$
$$= 2(-1) = -2 \quad \text{• Substitute } -1 \text{ for } x \text{ in Equation 2.}$$

The only ordered-pair solution of the system of equations is $(-1, -2)$. When a system of equations has a unique solution, the system of equations is independent.

Visualize the Solution

Graphing $3x - 5y = 7$ and $y = 2x$ shows that the ordered pair $(-1, -2)$ belongs to both lines. Therefore, $(-1, -2)$ is a solution of the system of equations. See **Figure 9.4**.

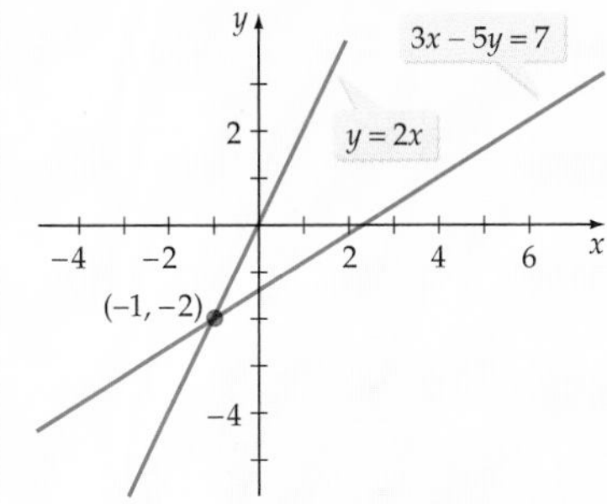

Figure 9.4
An independent system of equations

TRY EXERCISE 6, EXERCISE SET 9.1, PAGE 665

EXAMPLE 2 Identify an Inconsistent System of Equations

Solve: $\begin{cases} x + 3y = 6 & (1) \\ 2x + 6y = -18 & (2) \end{cases}$

Solution

Solve Equation (1) for y:

$$x + 3y = 6$$
$$y = -\frac{1}{3}x + 2$$

Visualize the Solution

Solving Equations (1) and (2) for y gives $y = -\frac{1}{3}x + 2$ and $y = -\frac{1}{3}x - 3$. Note that these

Continued ►

The solutions of $y = -\frac{1}{3}x + 2$ are the ordered pairs $\left(x, -\frac{1}{3}x + 2\right)$. For the system of equations to have a solution, ordered pairs of that form must also be solutions of $2x + 6y = -18$. To determine whether the ordered pairs $\left(x, -\frac{1}{3}x + 2\right)$ are solutions of Equation (2), substitute $\left(x, -\frac{1}{3}x + 2\right)$ into Equation (2) and solve for x.

$$2x + 6y = -18 \qquad \text{• Equation (2)}$$
$$2x + 6\left(-\frac{1}{3}x + 2\right) = -18 \qquad \text{• Substitute } -\frac{1}{3}x + 2 \text{ for } y.$$
$$2x - 2x + 12 = -18$$
$$12 = -18 \qquad \text{• A false statement}$$

The false statement $12 = -18$ means that no ordered pair that is a solution of Equation (1) is also a solution of Equation (2). The system of the equations have no ordered pairs in common and thus the system of equations has no solution. This is an inconsistent system of equations.

two equations have the same slope, $-\frac{1}{3}$, and different y-intercepts. Therefore, the graphs of the two lines are parallel and never intersect.

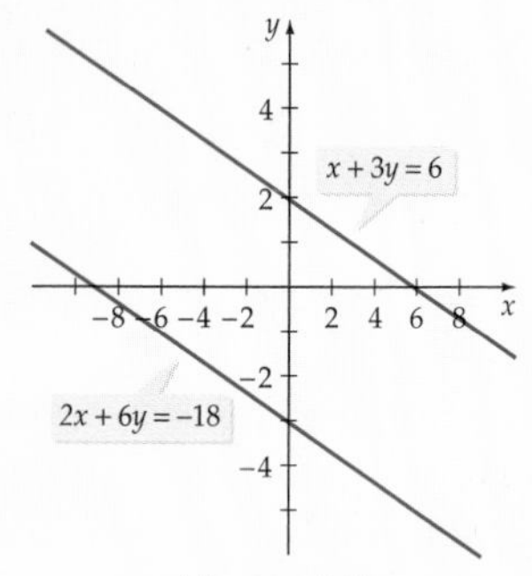

Figure 9.5
An inconsistent system of equations

TRY EXERCISE 18, EXERCISE SET 9.1, PAGE 666

EXAMPLE 3 Identify a Dependent System of Equations

Solve: $\begin{cases} 8x - 4y = 16 & (1) \\ 2x - y = 4 & (2) \end{cases}$

Solution

Solve Equation (2) for y:

$$2x - y = 4$$
$$y = 2x - 4$$

The solutions of $y = 2x - 4$ are the ordered pairs $(x, 2x - 4)$. For the system of equations to have a solution, ordered pairs of the form $(x, 2x - 4)$ must also be solutions of $8x - 4y = 16$. To determine whether the ordered pairs $(x, 2x - 4)$ are solutions of Equation (1), substitute $(x, 2x - 4)$ into Equation (1) and solve for x.

$$8x - 4y = 16 \qquad \text{• Equation (1)}$$
$$8x - 4(2x - 4) = 16 \qquad \text{• Substitute } 2x - 4 \text{ for } y.$$
$$8x - 8x + 16 = 16$$
$$16 = 16 \qquad \text{• A true statement}$$

The true statement $16 = 16$ means that the ordered pairs $(x, 2x - 4)$ that are solutions of Equation (2) are also solutions of Equation (1). Because x can be replaced by any real number c, the solution of the system of equations is the set of ordered pairs $(c, 2c - 4)$. This is a dependent system of equations.

Visualize the Solution

Solving Equations (1) and (2) for y gives $y = 2x - 4$ and $y = 2x - 4$. Note that these two equations have the same slope, 2, and the same y-intercept, $(0, -4)$. Therefore, the graphs of the two lines are exactly the same. One graph intersects the second graph infinitely often.

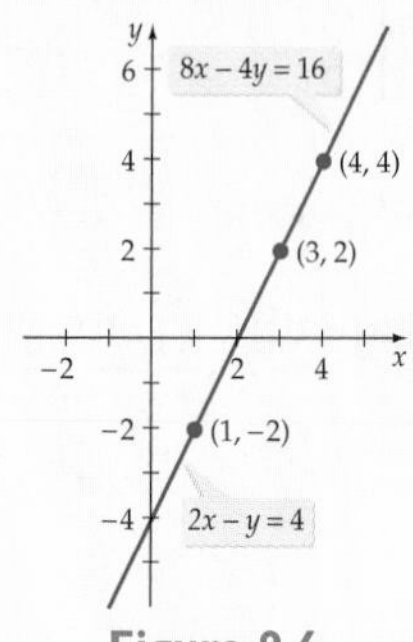

Figure 9.6

TRY EXERCISE 20, EXERCISE SET 9.1, PAGE 666

Some of the specific ordered-pair solutions in Example 3 can be found by choosing various values for c. The table below shows the ordered pairs that result from choosing c as 1, 3, and 4. The ordered pairs $(1, -2)$, $(3, 2)$, and $(4, 4)$ are specific solutions of the system of equations. These points are on the graphs of Equation (1) and Equation (2) as shown in **Figure 9.6.**

c	(c, 2c − 4)	(x, y)
1	$(1, 2(1) - 4)$	$(1, -2)$
3	$(3, 2(3) - 4)$	$(3, 2)$
4	$(4, 2(4) - 4)$	$(4, 4)$

take note

When a system of equations is dependent, there is more than one way to write the solutions of the solution set. The solution to Example 3 is the set of ordered pairs

$$(c, 2c - 4) \text{ or } \left(\frac{1}{2}b + 2, b\right)$$

However, there are infinitely more ways in which the ordered pairs could be expressed. For instance, let $b = 2w$. Then

$$\frac{1}{2}b + 2 = \frac{1}{2}(2w) + 2 = w + 2$$

The ordered-pair solutions, written in terms of w, are $(w + 2, 2w)$.

Before leaving Example 3, note that there is more than one way to represent the ordered-pair solutions. To illustrate this point, solve Equation (2) for x.

$$2x - y = 4 \qquad \text{• Equation (2)}$$

$$x = \frac{1}{2}y + 2 \qquad \text{• Solve for } x.$$

Because y can be replaced by any real number b, there are an infinite number of ordered pairs $\left(\frac{1}{2}b + 2, b\right)$ that are solutions of the system of equations. Choosing b as -2, 2, and 4 gives the same ordered pairs: $(1, -2)$, $(3, 2)$ and $(4, 4)$. There is always more than one way to describe the ordered pairs when writing the solution of a dependent system of equations. For Example 3, either the ordered pairs $(c, 2c - 4)$ or the ordered pairs $\left(\frac{1}{2}b + 2, b\right)$ would generate all the solutions of the system of equations.

◆ Elimination Method for Solving a System of Equations

Two systems of equations are **equivalent** if each system has exactly the same solutions. The systems

$$\begin{cases} 3x + 5y = 9 \\ 2x - 3y = -13 \end{cases} \quad \text{and} \quad \begin{cases} x = -2 \\ y = 3 \end{cases}$$

are equivalent systems of equations. Each system has the solution $(-2, 3)$, as shown in **Figure 9.7.**

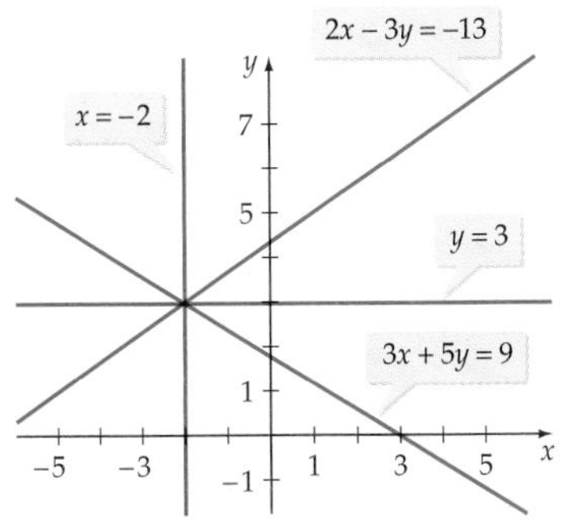

Figure 9.7

A second technique for solving a system of equations is similar to strategy for solving first-degree equations in one variable. The system of equations is replaced by a series of equivalent systems until the solution is obvious.

Operations That Produce Equivalent Systems of Equations

1. Interchange any two equations.
2. Replace an equation with a nonzero multiple of that equation.
3. Replace an equation with the sum of that equation and a nonzero constant multiple of another equation in the system.

Because the order in which the equations are written does not affect the system of equations, interchanging the equations does not affect its solution. The second operation restates the property that says that multiplying each side of an equation by the same nonzero constant does not change the solutions of the equation.

The third operation can be illustrated as follows. Consider the system of equations

$$\begin{cases} 3x + 2y = 10 & (1) \\ 2x - 3y = -2 & (2) \end{cases}$$

Multiply each side of Equation (2) by 2. (Any nonzero number would work.) Add the resulting equation to Equation (1).

$$\begin{aligned} 3x + 2y &= 10 && \text{• Equation (1)} \\ 4x - 6y &= -4 && \text{• 2 times Equation (2)} \\ \hline 7x - 4y &= 6 \quad (3) && \text{• Add the equations.} \end{aligned}$$

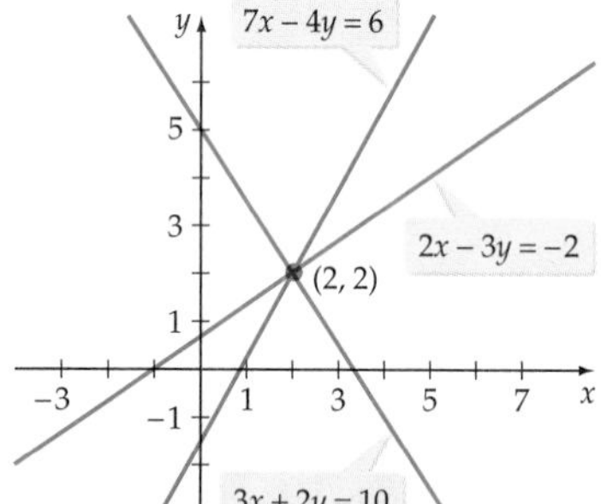

Figure 9.8

Replace Equation (1) with the new Equation (3) to produce the following equivalent system of equations.

$$\begin{cases} 7x - 4y = 6 & (3) \\ 2x - 3y = -2 & (2) \end{cases}$$

The third property states that the resulting system of equations has the same solutions as the original system and is therefore equivalent to the original system of equations. Figure 9.8 shows the graph of $7x - 4y = 6$. Note that the line passes through the same point at which the lines of the original system of equations intersect, the point (2, 2).

EXAMPLE 4 Solve a System of Equations by the Elimination Method

Solve: $\begin{cases} 3x - 4y = 10 & (1) \\ 2x + 5y = -1 & (2) \end{cases}$

Solution

Use the operations that produce equivalent equations to eliminate a variable from one of the equations. We will eliminate x from Equation (2) by multiplying each equation by a different constant so as to have a new system of equations in which the coefficients of x are additive inverses.

$$\begin{aligned} 6x - 8y &= 20 && \text{• 2 times Equation (1)} \\ -6x - 15y &= 3 && \text{• −3 times Equation (2)} \\ \hline -23y &= 23 && \text{• Add the equations.} \\ y &= -1 && \text{• Solve for } y. \end{aligned}$$

Solve Equation (1) for x by substituting -1 for y.

$$\begin{aligned} 3x - 4(-1) &= 10 \\ 3x &= 6 \\ x &= 2 \end{aligned}$$

The solution of the system of equations is $(2, -1)$.

Visualize the Solution

Graphing $3x - 4y = 10$ and $2x + 5y = -1$ shows that $(2, -1)$ belongs to both lines. Therefore, $(2, -1)$ is a solution of the system of equations. See Figure 9.9.

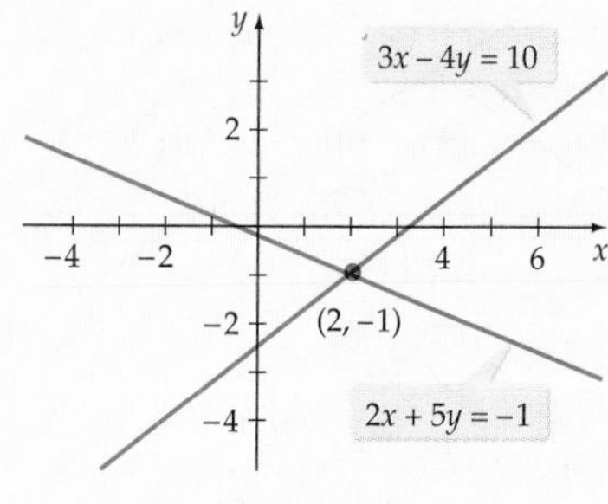

Figure 9.9

TRY EXERCISE 24, EXERCISE SET 9.1, PAGE 666

The method just described is called the **elimination method** for solving a system of equations, because it involves *eliminating* a variable from one of the equations.

You can use a graphing calculator to solve a system of equations in two variables. First, algebraically solve each equation for y.

Solve for y.

$$3x - 4y = 10 \quad \rightarrow \quad y = 0.75x - 2.5$$
$$2x + 5y = -1 \quad \rightarrow \quad y = -0.4x - 0.2$$

Now graph the equations. Enter 0.75x–2.5 into Y_1 and –0.4X–0.2 into Y_2 and graph the two equations in the standard viewing window. The sequence of steps shown in **Figure 9.10** can be used to find the point of intersection with a TI-83 calculator.

Press 2nd CALC
Select 5: intersect.
Press ENTER

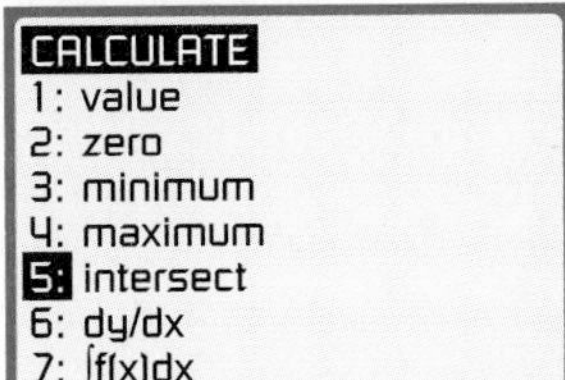

The "First curve?" shown on the bottom of the screen means to select the first of the two graphs that intersect. Just press ENTER

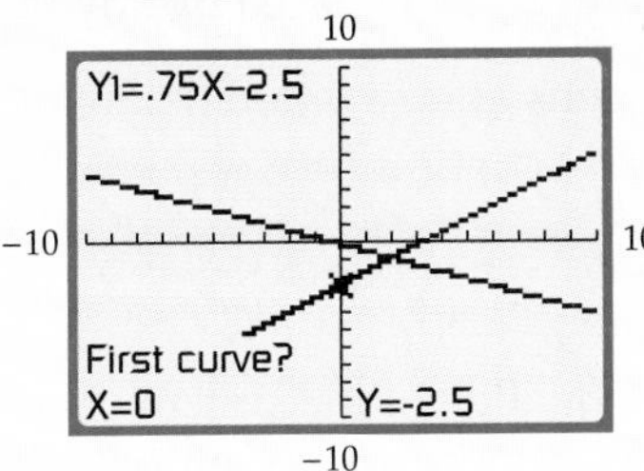

The "Second curve?" shown on the bottom of the screen means to select the second of the two graphs that intersect. Just press ENTER

10
Y2=-.4X-.2
–10
10
Second curve?
X=0
Y=-.2
–10

"Guess?" is shown on the bottom of the screen. Move the cursor until it is approximately on the point of intersection. Press ENTER

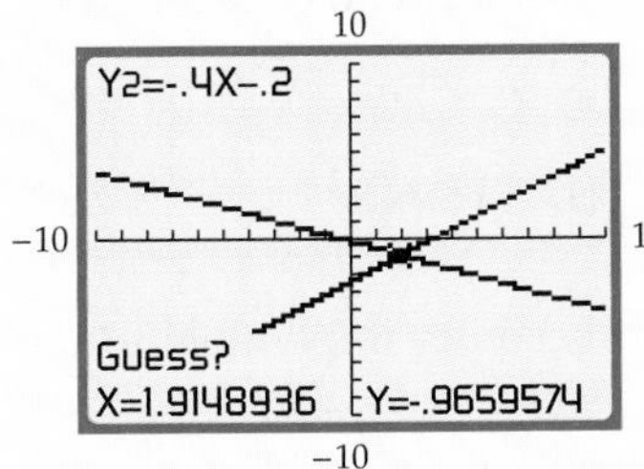

The coordinates of the point of intersection $(2, -1)$ are shown at the bottom of the screen.

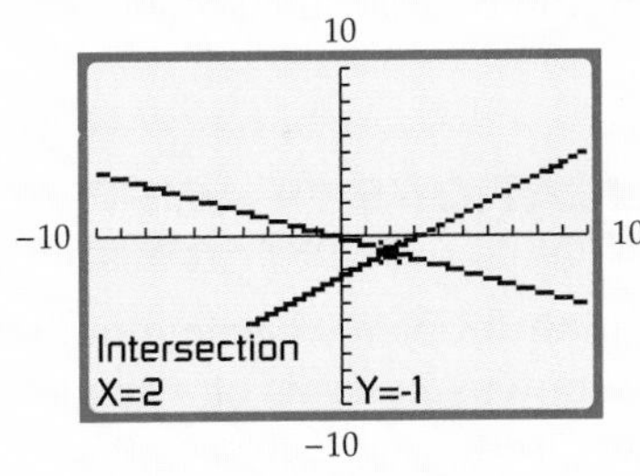

Figure 9.10

For the system of equations in Example 4, the intersection of the two graphs occurred at a point in the standard viewing window. If the point of intersection does not appear on the screen, you must adjust the viewing window so that the point of intersection is visible.

EXAMPLE 5 Solve a Dependent System of Equations

$$\text{Solve: } \begin{cases} x - 2y = 2 & (1) \\ 3x - 6y = 6 & (2) \end{cases}$$

Continued •➤

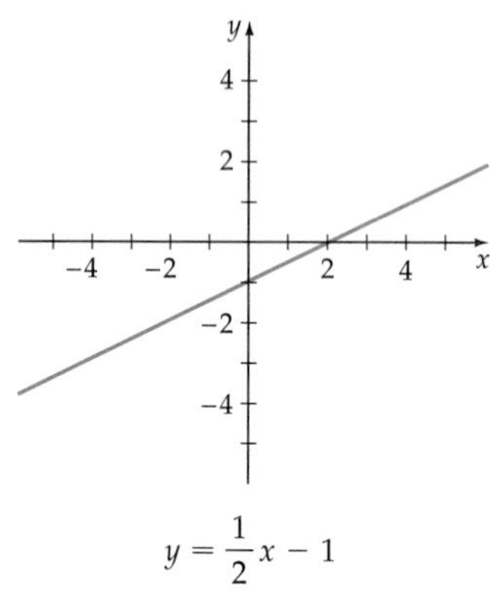

$y = \frac{1}{2}x - 1$

Figure 9.11

take note

Referring again to Example 5 and solving Equation (1) for x, we have $x = 2y + 2$. Because y can be any real number b, the ordered-pair solutions of the system of equations can be written also as $(2b + 2, b)$.

Solution

Eliminate x by multiplying Equation (2) by $-1/3$ and then adding the result to Equation (1).

$$\begin{aligned} x - 2y &= 2 \\ -x + 2y &= -2 \\ \hline 0 &= 0 \end{aligned}$$

• Equation (1)
• −1/3 times Equation (2)
• Add the two equations.

Replace Equation (2) by $0 = 0$.

$$\begin{cases} x - 2y = 2 \\ 0 = 0 \end{cases}$$

• This is an equivalent system of equations.

Because the equation $0 = 0$ is an identity, an ordered pair that is a solution of Equation (1) is also a solution of $0 = 0$. Thus the solutions are the solutions of $x - 2y = 2$. Solving for y, we find that $y = \frac{1}{2}x - 1$. Because x can be replaced by any real number c, the solutions of the system of equations are the ordered pairs $\left(c, \frac{1}{2}c - 1\right)$. See **Figure 9.11.**

TRY EXERCISE 28, EXERCISE SET 9.1, PAGE 666

If one equation of the system of equations is replaced by a false equation, the system of equations has no solution. For example, the system of equations

$$\begin{cases} x + y = 4 \\ 0 = 5 \end{cases}$$

has no solution because the second equation is false for any choice of x and y.

◆ APPLICATIONS OF SYSTEMS OF EQUATIONS

As application problems become more difficult, it becomes impossible to represent all unknowns in terms of a single variable. In some cases, a system of equations can be used.

EXAMPLE 6 Solve an Application

A rowing team rowing with the current traveled 18 miles in 2 hours. Against the current, the team rowed 10 miles in 2 hours. Find the rate of the boat in calm water and the rate of the current.

Solution

Let r_1 represent the rate of the boat in calm water, and let r_2 represent the rate of the current.

The rate of the boat *with the current* is $r_1 + r_2$.
The rate of the boat *against the current* is $r_1 - r_2$.

Because the rowing team traveled 18 miles in 2 hours with the current, we use the equation $d = rt$.

$$d = r \cdot t$$

$$18 = (r_1 + r_2) \cdot 2 \qquad \bullet\ d = 18, t = 2$$

$$9 = r_1 + r_2 \qquad \bullet\ \text{Divide each side by 2.}$$

Because the team rowed 10 miles in 2 hours against the current, we write

$$10 = (r_1 - r_2) \cdot 2 \qquad \bullet\ d = 10, t = 2$$

$$5 = r_1 - r_2 \qquad \bullet\ \text{Divide each side by 2.}$$

Thus we have a system of two linear equations in the variables r_1 and r_2.

$$\begin{cases} 9 = r_1 + r_2 \\ 5 = r_1 - r_2 \end{cases}$$

Solving the system by using the elimination method, we find that r_1 is 7 mph and r_2 is 2 mph. Thus the rate of the boat in calm water is 7 mph and the rate of the current is 2 mph. You should verify these solutions.

TRY EXERCISE 44, EXERCISE SET 9.1, PAGE 666

TOPICS FOR DISCUSSION

1. Explain how to use the substitution method to solve a system of equations.
2. Explain how to use the elimination method to solve a system of equations.
3. Give an example of a system of equations in two variables that is

   a. independent b. dependent c. inconsistent
4. If a linear system of equations in two variables has no solution, what does that mean about the graphs of the equations of the system?
5. If $A = \{(x, y)|x + y = 5\}$ and $B = \{(x, y)|x - y = 3\}$, explain the meaning of $A \cap B$.

EXERCISE SET 9.1

In Exercises 1 to 20, solve each system of equations by the substitution method.

1. $\begin{cases} 2x - 3y = 16 \\ x = 2 \end{cases}$

2. $\begin{cases} 3x - 2y = -11 \\ y = 1 \end{cases}$

3. $\begin{cases} 3x + 4y = 18 \\ y = -2x + 3 \end{cases}$

4. $\begin{cases} 5x - 4y = -22 \\ y = 5x - 2 \end{cases}$

5. $\begin{cases} -2x + 3y = 6 \\ x = 2y - 5 \end{cases}$

6. $\begin{cases} 8x + 3y = -7 \\ x = 3y + 15 \end{cases}$

7. $\begin{cases} 6x + 5y = 1 \\ x - 3y = 4 \end{cases}$

8. $\begin{cases} -3x + 7y = 14 \\ 2x - y = -13 \end{cases}$

9. $\begin{cases} 7x + 6y = -3 \\ y = \dfrac{2}{3}x - 6 \end{cases}$

10. $\begin{cases} 9x - 4y = 3 \\ x = \dfrac{4}{3}y + 3 \end{cases}$

11. $\begin{cases} y = 4x - 3 \\ y = 3x - 1 \end{cases}$

12. $\begin{cases} y = 5x + 1 \\ y = 4x - 2 \end{cases}$

13. $\begin{cases} y = 5x + 4 \\ x = -3y - 4 \end{cases}$

14. $\begin{cases} y = -2x - 6 \\ x = -2y - 2 \end{cases}$

15. $\begin{cases} 3x - 4y = 2 \\ 4x + 3y = 14 \end{cases}$

16. $\begin{cases} 6x + 7y = -4 \\ 2x + 5y = 4 \end{cases}$

17. $\begin{cases} 3x - 3y = 5 \\ 4x - 4y = 9 \end{cases}$

18. $\begin{cases} 3x - 4y = 8 \\ 6x - 8y = 9 \end{cases}$

19. $\begin{cases} 4x + 3y = 6 \\ y = -\dfrac{4}{3}x + 2 \end{cases}$

20. $\begin{cases} 5x + 2y = 2 \\ y = -\dfrac{5}{2}x + 1 \end{cases}$

In Exercises 21 to 40, solve each system of equations by the elimination method.

21. $\begin{cases} 3x - y = 10 \\ 4x + 3y = -4 \end{cases}$

22. $\begin{cases} 3x + 4y = -5 \\ x - 5y = -8 \end{cases}$

23. $\begin{cases} 4x + 7y = 21 \\ 5x - 4y = -12 \end{cases}$

24. $\begin{cases} 3x - 8y = -6 \\ -5x + 4y = 10 \end{cases}$

25. $\begin{cases} 5x - 3y = 0 \\ 10x - 6y = 0 \end{cases}$

26. $\begin{cases} 3x + 2y = 0 \\ 2x + 3y = 0 \end{cases}$

27. $\begin{cases} 6x + 6y = 1 \\ 4x + 9y = 4 \end{cases}$

28. $\begin{cases} 4x + 5y = 2 \\ 8x - 15y = 9 \end{cases}$

29. $\begin{cases} 3x + 6y = 11 \\ 2x + 4y = 9 \end{cases}$

30. $\begin{cases} 4x - 2y = 9 \\ 2x - y = 3 \end{cases}$

31. $\begin{cases} \dfrac{5}{6}x - \dfrac{1}{3}y = -6 \\ \dfrac{1}{6}x + \dfrac{2}{3}y = 1 \end{cases}$

32. $\begin{cases} \dfrac{3}{4}x + \dfrac{2}{5}y = 1 \\ \dfrac{1}{2}x - \dfrac{3}{5}y = -1 \end{cases}$

33. $\begin{cases} \dfrac{3}{4}x + \dfrac{1}{3}y = 1 \\ \dfrac{1}{2}x + \dfrac{2}{3}y = 0 \end{cases}$

34. $\begin{cases} \dfrac{3}{5}x - \dfrac{2}{3}y = 7 \\ \dfrac{2}{5}x - \dfrac{5}{6}y = 7 \end{cases}$

35. $\begin{cases} 2\sqrt{3}x - 3y = 3 \\ 3\sqrt{3}x + 2y = 24 \end{cases}$

36. $\begin{cases} 4x - 3\sqrt{5}y = -19 \\ 3x + 4\sqrt{5}y = 17 \end{cases}$

37. $\begin{cases} 3\pi x - 4y = 6 \\ 2\pi x + 3y = 5 \end{cases}$

38. $\begin{cases} 2x - 5\pi y = 3 \\ 3x + 4\pi y = 2 \end{cases}$

39. $\begin{cases} 3\sqrt{2}x - 4\sqrt{3}y = -6 \\ 2\sqrt{2}x + 3\sqrt{3}y = 13 \end{cases}$

40. $\begin{cases} 2\sqrt{2}x + 3\sqrt{5}y = 7 \\ 3\sqrt{2}x - \sqrt{5}y = -17 \end{cases}$

In Exercises 41 to 55, solve by using a system of equations.

41. **RATE OF WIND** Flying with the wind, a plane traveled 450 miles in 3 hours. Flying against the wind, the plane traveled the same distance in 5 hours. Find the rate of the plane in calm air and the rate of the wind.

42. **RATE OF WIND** A plane flew 800 miles in 4 hours while flying with the wind. Against the wind, it took the plane 5 hours to travel 800 miles. Find the rate of the plane in calm air and the rate of the wind.

43. **RATE OF CURRENT** A motorboat traveled a distance of 120 miles in 4 hours while traveling with the current. Against the current, the same trip took 6 hours. Find the rate of the boat in calm water and the rate of the current.

44. **RATE OF CURRENT** A canoeist can row 12 miles with the current in 2 hours. Rowing against the current, it takes the canoeist 4 hours to travel the same distance. Find the rate of the canoeist in calm water and the rate of the current.

45. **METALLURGY** A metallurgist made two purchases. The first purchase, which cost \$1080, included 30 kilograms of an iron alloy and 45 kilograms of a lead alloy. The second purchase, at the same prices, cost \$372 and included 15 kilograms of the iron alloy and 12 kilograms of the lead alloy. Find the cost per kilogram of the iron and lead alloys.

46. **CHEMISTRY** For \$14.10, a chemist purchased 10 liters of hydrochloric acid and 15 liters of silver nitrate. A second purchase, at the same prices, cost \$18.16 and included 12 liters of hydrochloric acid and 20 liters of silver nitrate. Find the cost per liter of each of the two chemicals.

47. **GEOMETRY** A right triangle in the first quadrant is bounded by the lines $y = 0$, $y = \dfrac{1}{2}x$, and $y = -2x + 6$. Find its area.

48. **GEOMETRY** The lines whose equations are $2x + 3y = 1$, $3x - 4y = 10$, and $4x + ky = 5$ all intersect at the same point. What is the value of k?

49. **NUMBER THEORY** Adding a three-digit number $5Z7$ to 256 gives $XY3$. If $XY3$ is divisible by 3, then what is the largest possible value of Z?

50. **NUMBER THEORY** Find the value of k if $2x + 5 = 6x + k = 4x - 7$.

51. **INVESTMENT** A broker invests \$25,000 of a client's money in two different municipal bonds. The annual rate of return on one bond is 6%, and the annual rate of return on the second bond is 6.5%. The investor receives a total annual interest payment from the two bonds of \$1555. Find the amount invested in each bond.

52. **INVESTMENT** An investment of \$3000 is placed in stocks and bonds. The annual rate of return on the stocks is 4.5%, and the rate of return on the bonds is 8%. The annual return from the stocks and bonds is \$177. Find the amount invested in bonds.

53. **CHEMISTRY** A goldsmith has two gold alloys. The first alloy is 40% gold; the second alloy is 60% gold. How many grams of each should be mixed to produce 20 grams of an alloy that is 52% gold?

54. **CHEMISTRY** One acetic acid solution is 70% water and another is 30% water. How many liters of each solution

should be mixed to produce 20 liters of a solution that is 40% water?

55. **CHEMISTRY** A chemist wants to make 50 milliliters of a 16% acid solution. How many milliliters each of a 13% acid solution and an 18% acid solution should be mixed to produce the desired solution?

SUPPLEMENTAL EXERCISES

In Exercises 56 to 65, solve for x and y. Use the fact that if $z_1 = a_1 + b_1 i$ and $z_2 = a_2 + b_2 i$ are two complex numbers, then $z_1 = z_2$ if and only if $a_1 = a_2$ and $b_1 = b_2$.

56. $(2 + i)x + (3 - i)y = 7$

57. $(3 + 2i)x + (4 - 3i)y = 2 - 16i$

58. $(4 - 3i)x + (5 + 2i)y = 11 + 9i$

59. $(2 + 6i)x + (4 - 5i)y = -8 - 7i$

60. $(-3 - i)x - (4 + 2i)y = 1 - i$

61. $(5 - 2i)x + (-3 - 4i)y = 12 - 35i$

62. $\begin{cases} 2x + 5y = 11 + 3i \\ 3x + y = 10 - 2i \end{cases}$

63. $\begin{cases} 4x + 3y = 11 + 6i \\ 3x - 5y = 1 + 19i \end{cases}$

64. $\begin{cases} 2x + 3y = 11 + 5i \\ 3x - 3y = 9 - 15i \end{cases}$

65. $\begin{cases} 5x - 4y = 15 - 41i \\ 3x + 5y = 9 + 5i \end{cases}$

PROJECTS

1. **INDEPENDENT AND DEPENDENT CONDITIONS** Consider the following problem: "Maria and Michael drove from Los Angeles to New York in 60 hours. How long did Maria drive?" It is difficult to answer this question. She may have driven all 60 hours while Micheal relaxed, or she may have relaxed while Michael drove all 60 hours. The difficulty is that there are two unknowns (how long each drove) and only one condition (the total driving time) relating the unknowns. If we added another condition, such as Michael drove 25 hours, then we could determine how long Maria drove, 35 hours. In most cases, an application problem will have a single answer only when there are as many *independent* conditions as there are variables. Conditions are independent if knowing one does *not* allow you to know the other.

Here is an example of conditions that are not independent "The perimeter of a rectangle is 50 meters. The sum of the width and length is 25 meters." To see that these conditions are dependent, write the perimeter equation and divide each side by 2.

$$2w + 2l = 50$$

$$w + l = 25 \qquad \text{• Divide each side by 2.}$$

Note that the resulting equation is the second condition: the sum of the width and length is 25. Thus knowing the first condition allows us to determine the second condition. The conditions are not independent, so there is no one solution to this problem.

For each of the problems below, determine whether the conditions are independent or dependent. For those problems that have independent conditions, find the solution (if possible). For those problems for which the conditions are dependent, find two solutions.

a. The sum of two numbers is 30. The difference between the two numbers is 10. Find the numbers.

b. The area of a square is 25 square meters. Find the length of each side.

c. The area of a rectangle is 25 square meters. Find the length of each side.

d. Emily spent $1000 for carpeting and tile. Carpeting cost $20 per square yard and tile cost $30 per square yard. How many square yards of each did she purchase?

e. The sum of two numbers is 20. Twice the smaller number is 10 minus twice the larger number. Find the two numbers.

f. Make up a word problem for which there are two independent conditions. Solve the problem.

g. Make up a word problem for which there are two dependent conditions. Find at least two solutions.

SECTION 9.2 SYSTEMS OF LINEAR EQUATIONS IN MORE THAN TWO VARIABLES

- SYSTEMS OF EQUATIONS IN THREE VARIABLES
- TRIANGULAR FORM
- NONSQUARE SYSTEMS OF EQUATIONS
- HOMOGENEOUS SYSTEMS OF EQUATIONS
- CURVE FITTING

9A

◆ SYSTEMS OF EQUATIONS IN THREE VARIABLES

An equation of the form $Ax + By + Cz = D$, with A, B, and C not all zero, is a linear equation in three variables. A solution of an equation in three variables is an **ordered triple** (x, y, z).

The ordered triple $(2, -1, -3)$ is one of the solutions of the equation $2x - 3y + z = 4$. The ordered triple $(3, 1, 1)$ is another solution. In fact, an infinite number of ordered triples are solutions of the equation.

Graphing an equation in three variables requires a third coordinate axis perpendicular to the xy-plane. This third axis is commonly called the **z-axis.** The result is a three-dimensional coordinate system called the xyz-coordinate system (**Figure 9.12**). To help visualize a three-dimensional coordinate system, think of a corner of a room: the floor is the xy-plane, one wall is the yz-plane, and the other wall is the xz-plane.

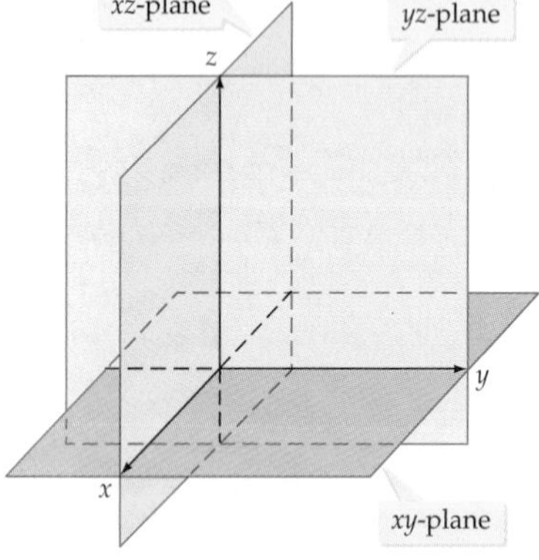

Figure 9.12

Graphing an ordered triple requires three moves, the first along the x-axis, the second along the y-axis, and the third along the z-axis. **Figure 9.13** is the graph of the points $(-5, -4, 3)$ and $(4, 5, -2)$.

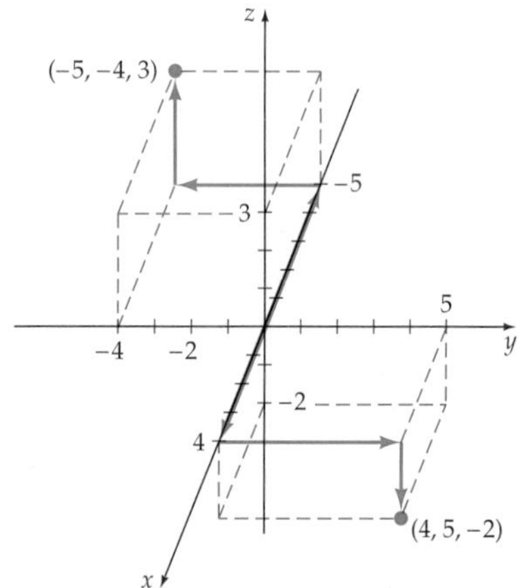

Figure 9.13

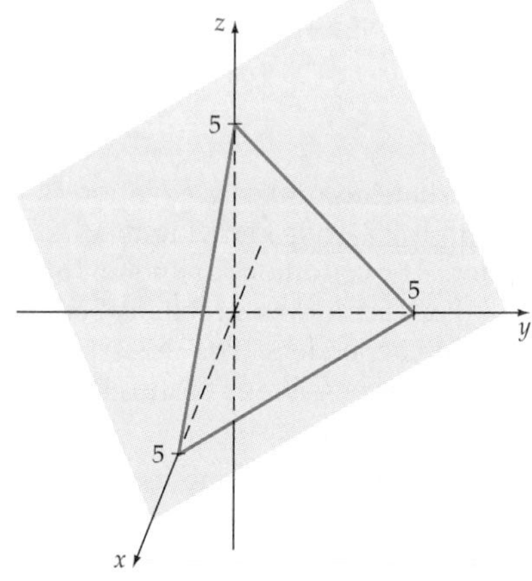

Figure 9.14

The graph of a linear equation in three variables is a plane. That is, if all the solutions of a linear equation in three variables were plotted in an xyz-coordinate system, the graph would look like a large piece of paper with infinite extent. **Figure 9.14** is a portion of the graph of $x + y + z = 5$.

There are different ways in which three planes can be oriented in an xyz-coordinate system. **Figure 9.15** illustrates several ways.

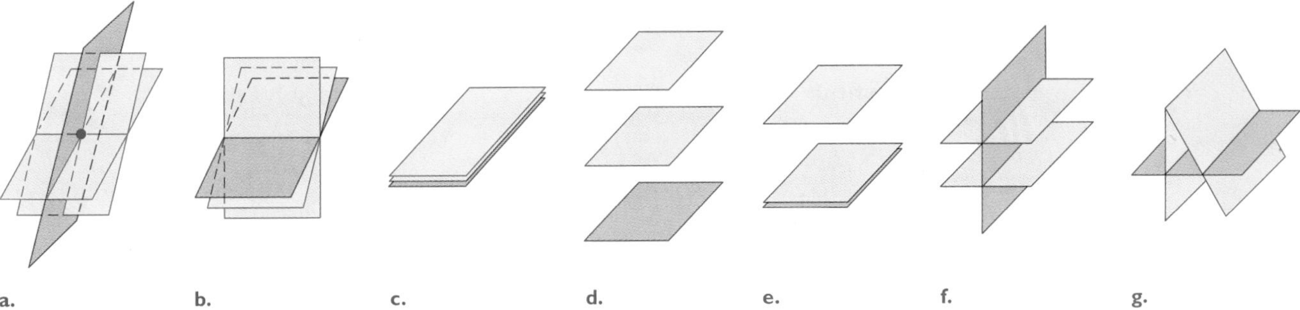

Figure 9.15

For a linear system of equations in three variables to have a solution, the graphs of the planes must intersect at a single point, they must intersect along a common line, or all equations must have a graph that is the same plane. In **Figure 9.15**, the graphs in (a), (b), and (c) represent systems of equations that have a solution. The system of equations represented in **Figure 9.15a** is a consistent system of equations. **Figures 9.15b** and **9.15c** are graphs of a dependent system of equations. The remaining graphs are examples of inconsistent systems of equations.

A system of equations in more than two variables can be solved by using the substitution method or the elimination method. To illustrate the substitution method, consider the system of equations

$$\begin{cases} x - 2y + z = 7 & (1) \\ 2x + y - z = 0 & (2) \\ 3x + 2y - 2z = -2 & (3) \end{cases}$$

Solve Equation (1) for x and substitute the result into Equations (2) and (3).

$$x = 2y - z + 7 \quad (4)$$

$$2(2y - z + 7) + y - z = 0$$
• **Substitute $2y - z + 7$ for x in Equation (2).**

$$4y - 2z + 14 + y - z = 0$$
• **Simplify.**

$$5y - 3z = -14 \quad (5)$$

$$3(2y - z + 7) + 2y - 2z = -2$$
• **Substitute $2y - z + 7$ for x in Equation (3).**

$$6y - 3z + 21 + 2y - 2z = -2$$
• **Simplify.**

$$8y - 5z = -23 \quad (6)$$

Now solve the system of equations formed from Equations (5) and (6).

$$\begin{cases} 5y - 3z = -14 & \textbf{multiply by 8} \longrightarrow & 40y - 24z = -112 \\ 8y - 5z = -23 & \textbf{multiply by } \mathbf{-5} \longrightarrow & -40y + 25z = 115 \end{cases}$$

$$z = 3$$

Substitute 3 for z into Equation (5) and solve for y.

$$\begin{aligned} 5y - 3z &= -14 && \bullet \text{ Equation (5)} \\ 5y - 3(3) &= -14 \\ 5y - 9 &= -14 \\ 5y &= -5 \\ y &= -1 \end{aligned}$$

Substitute -1 for y and 3 for z into Equation (4) and solve for x.

$$x = 2y - z + 7 = 2(-1) - (3) + 7 = 2$$

The ordered-triple solution is $(2, -1, 3)$. The graphs of the three planes intersect at a single point.

◆ Triangular Form

There are many approaches one can take to determine the solution of a system of equations by the elimination method. For consistency, we will always follow a plan that produces an equivalent system of equations in **triangular form.** Three examples of systems of equations in triangular form are

$$\begin{cases} 2x - 3y + z = -4 \\ 2y + 3z = 9 \\ -2z = -2 \end{cases} \qquad \begin{cases} w + 3x - 2y + 3z = 0 \\ 2x - y + 4z = 8 \\ -3y - 2z = -1 \\ 3z = 9 \end{cases} \qquad \begin{cases} 3x - 4y + z = 1 \\ 3y + 2z = 3 \end{cases}$$

Once a system of equations is written in triangular form, the solution can be found by *back substitution*—that is, by solving the last equation of the system and substituting *back* into the previous equation. This process is continued until the value of each variable has been found.

As an example of solving a system of equations by back substitution, consider the following system of equations in triangular form.

$$\begin{cases} 2x - 4y + z = -3 & (1) \\ 3y - 2z = 9 & (2) \\ 3z = -9 & (3) \end{cases}$$

Solve Equation (3) for z. Substitute the value of z into Equation (2) and solve for y.

$$\begin{aligned} 3z &= -9 && \bullet \text{ Equation (3)} \\ z &= -3 \end{aligned}$$

$$\begin{aligned} 3y - 2z &= 9 && \bullet \text{ Equation (2)} \\ 3y - 2(-3) &= 9 && \bullet\ z = -3 \\ 3y &= 3 \\ y &= 1 \end{aligned}$$

Replace z by -3 and y by 1 in Equation (1) and then solve for x.

$$\begin{aligned} 2x - 4y + z &= -3 && \bullet \text{ Equation (1)} \\ 2x - 4(1) + (-3) &= -3 \\ 2x - 7 &= -3 \\ x &= 2 \end{aligned}$$

The solution is the ordered triple $(2, 1, -3)$.

EXAMPLE 1 Solve an Independent System of Equations

Solve: $\begin{cases} x + 2y - z = 1 & (1) \\ 2x - y + z = 6 & (2) \\ 2x - y - z = 0 & (3) \end{cases}$

Solution

Eliminate x from Equation (2) by multiplying Equation (1) by -2 and then adding it to Equation (2). Replace Equation (2) by the new equation.

$$\begin{array}{rl} -2x - 4y + 2z = -2 & \text{• } -2 \text{ times Equation (1)} \\ 2x - y + z = 6 & \text{• Equation (2)} \\ \hline -5y + 3z = 4 & \text{• Add the equations.} \end{array}$$

$$\begin{cases} x + 2y - z = 1 & (1) \\ -5y + 3z = 4 & (4) \\ 2x - y - z = 0 & (3) \end{cases} \quad \text{• Replace Equation (2).}$$

Eliminate x from Equation (3) by multiplying Equation (1) by -2 and adding it to Equation (3). Replace Equation (3) by the new equation.

$$\begin{array}{rl} -2x - 4y + 2z = -2 & \text{• } -2 \text{ times Equation (1)} \\ 2x - y - z = 0 & \text{• Equation (3)} \\ \hline -5y + z = -2 & \text{• Add the equations.} \end{array}$$

$$\begin{cases} x + 2y - z = 1 & (1) \\ -5y + 3z = 4 & (4) \\ -5y + z = -2 & (5) \end{cases} \quad \text{• Replace Equation (3).}$$

Eliminate y from Equation (5) by multiplying Equation (4) by -1 and then adding it to Equation (5). Replace Equation (5) by the new equation.

$$\begin{array}{rl} 5y - 3z = -4 & \text{• } -1 \text{ times Equation (4)} \\ -5y + z = -2 & \text{• Equation (5)} \\ \hline -2z = -6 & \text{• Add the equations.} \end{array}$$

$$\begin{cases} x + 2y - z = 1 & (1) \\ -5y + 3z = 4 & (4) \\ -2z = -6 & (6) \end{cases} \quad \text{• Replace Equation (5).}$$

The system of equations is now in triangular form. Solve the system of equations by back substitution.

Solve Equation (6) for z. Substitute the value into Equation (4) and then solve for y.

$$\begin{array}{rl} -2z = -6 & \text{• Equation (6)} \\ z = 3 & \end{array}$$

$$\begin{array}{rl} -5y + 3z = 4 & \text{• Equation (4)} \\ -5y + 3(3) = 4 & \text{• Replace } z \text{ by 3.} \\ -5y = -5 & \text{• Solve for } y. \\ y = 1 & \end{array}$$

Continued •➤

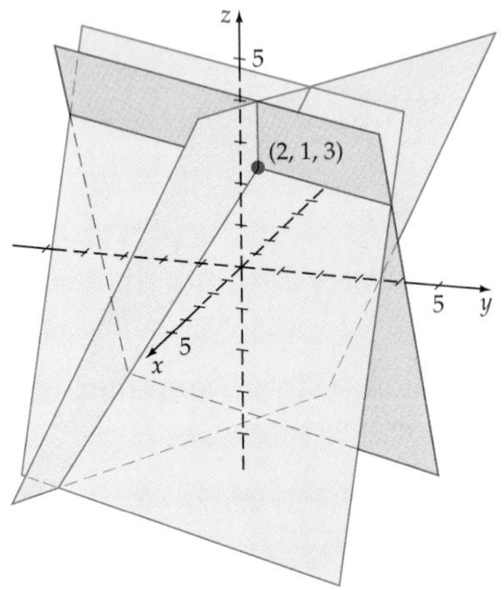

Figure 9.16

Replace z by 3 and y by 1 in Equation (1) and solve for x.

$$x + 2y - z = 1 \qquad \text{• Equation (1)}$$
$$x + 2(1) - 3 = 1 \qquad \text{• Replace } y \text{ by 1; replace } z \text{ by 3.}$$
$$x = 2$$

The system of equations is consistent. The solution is the ordered triple (2, 1, 3). See **Figure 9.16.**

TRY EXERCISE 12, EXERCISE SET 9.2, PAGE 678

EXAMPLE 2 Solve a Dependent System of Equations

Solve: $\begin{cases} 2x - y - z = -1 & (1) \\ -x + 3y - z = -3 & (2) \\ -5x + 5y + z = -1 & (3) \end{cases}$

Solution

Eliminate x from Equation (2) by multiplying Equation (2) by 2 and then adding it to Equation (1). Replace Equation (2) by the new equation.

$$\begin{array}{rl} 2x - y - z = -1 & \text{• Equation (1)} \\ -2x + 6y - 2z = -6 & \text{• 2 times Equation (2)} \\ \hline 5y - 3z = -7 & \text{• Add the equations.} \end{array}$$

$$\begin{cases} 2x - y - z = -1 & (1) \\ 5y - 3z = -7 & (4) \\ -5x + 5y + z = -1 & (3) \end{cases} \qquad \text{• Replace Equation (2).}$$

Eliminate x from Equation (3) by multiply Equation (1) by 5 and multiply Equation (3) by 2. Then add. Replace Equation (3) by the new equation.

$$\begin{array}{rl} 10x - 5y - 5z = -5 & \text{• 5 times Equation (1)} \\ -10x + 10y + 2z = -2 & \text{• 2 times Equation (3)} \\ \hline 5y - 3z = -7 & \text{• Add the equations.} \end{array}$$

$$\begin{cases} 2x - y - z = -1 & (1) \\ 5y - 3z = -7 & (4) \\ 5y - 3z = -7 & (5) \end{cases} \qquad \text{• Replace Equation (3).}$$

Eliminate y from Equation (5) by multiplying Equation (4) by -1 and then adding it to Equation (5). Replace Equation (5) by the new equation.

$$\begin{array}{rl} -5y + 3z = 7 & \text{• } -1 \text{ times Equation (4)} \\ 5y - 3z = -7 & \text{• Equation (5)} \\ \hline 0 = 0 & \text{• Add the equations.} \end{array}$$

$$\begin{cases} 2x - y - z = -1 & (1) \\ 5y - 3z = -7 & (4) \\ 0 = 0 & (6) \end{cases} \qquad \text{• Replace Equation (5).}$$

Because any ordered triple (x, y, z) is a solution of Equation (6), the solutions of the system of equations will be the ordered triples that are solutions of Equations (1) and (4).

Solve Equation (4) for y.

$$5y - 3z = -7$$
$$5y = 3z - 7$$
$$y = \frac{3}{5}z - \frac{7}{5}$$

Substitute $\frac{3}{5}z - \frac{7}{5}$ for y in Equation (1) and solve for x.

$$2x - y - z = -1 \quad \text{• Equation (1)}$$
$$2x - \left(\frac{3}{5}z - \frac{7}{5}\right) - z = -1 \quad \text{• Replace } y \text{ by } \frac{3}{5}z - \frac{7}{5}.$$
$$2x - \frac{8}{5}z + \frac{7}{5} = -1 \quad \text{• Simplify and solve for } x.$$
$$2x = \frac{8}{5}z - \frac{12}{5}$$
$$x = \frac{4}{5}z - \frac{6}{5}$$

By choosing any real number c for z, we have $y = \frac{3}{5}c - \frac{7}{5}$ and $x = \frac{4}{5}c - \frac{6}{5}$. For any real number c, the ordered-triple solutions of the system of equations are $\left(\frac{4}{5}c - \frac{6}{5}, \frac{3}{5}c - \frac{7}{5}, c\right)$. The solid red line shown in **Figure 9.17** is a graph of the solutions.

The three planes intersect along this line.
z
y
x

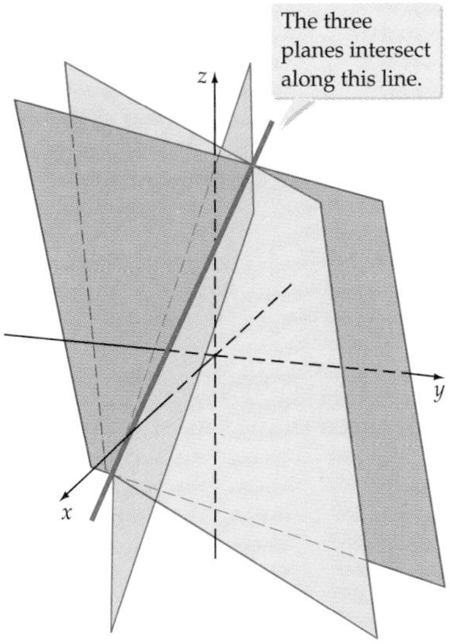

Figure 9.17

TRY EXERCISE 16, EXERCISE SET 9.2, PAGE 678

take note

Although the ordered triples

$$\left(\frac{4}{5}c - \frac{6}{5}, \frac{3}{5}c - \frac{7}{5}, c\right)$$

and

$$\left(a, \frac{3}{4}a - \frac{1}{2}, \frac{5}{4}a + \frac{3}{2}\right)$$

appear to be different, they represent exactly the same set of ordered triples. For instance, choosing $c = -1$, we have $(-2, -2, -1)$. Choosing $a = -2$ results in the same ordered triple, $(-2, -2, -1)$.

As in the case of a dependent system of equations in two variables, there is more than one way to represent the solutions of a dependent system of equations in three variables. For instance, from Example 2, let $a = \frac{4}{5}c - \frac{6}{5}$, the x-coordinate of the ordered triple $\left(\frac{4}{5}c - \frac{6}{5}, \frac{3}{5}c - \frac{7}{5}, c\right)$, and solve for c.

$$a = \frac{4}{5}c - \frac{6}{5} \longrightarrow c = \frac{5}{4}a + \frac{3}{2}$$

Substitute this value of c into each component of the ordered triple.

$$\left(\frac{4}{5}\left(\frac{5}{4}a + \frac{3}{2}\right) - \frac{6}{5}, \frac{3}{5}\left(\frac{5}{4}a + \frac{3}{2}\right) - \frac{7}{5}, \frac{5}{4}a + \frac{3}{2}\right) = \left(a, \frac{3}{4}a - \frac{1}{2}, \frac{5}{4}a + \frac{3}{2}\right)$$

Thus the solutions of the system of equation can also be written as

$$\left(a, \frac{3}{4}a - \frac{1}{2}, \frac{5}{4}a + \frac{3}{2}\right)$$

EXAMPLE 3 Identify an Inconsistent System of Equations

$$\text{Solve: } \begin{cases} x + 2y + 3z = 4 & (1) \\ 2x - y - z = 3 & (2) \\ 3x + y + 2z = 5 & (3) \end{cases}$$

Solution

Eliminate x from Equation (2) by multiplying Equation (1) by -2 and then adding it to Equation (2). Replace Equation (2). Eliminate x from Equation (3) by multiplying Equation (1) by -3 and adding it to Equation (3). Replace Equation (3). The equivalent system is

$$\begin{cases} x + 2y + 3z = 4 & (1) \\ -5y - 7z = -5 & (4) \\ -5y - 7z = -7 & (5) \end{cases}$$

Eliminate y from Equation (5) by multiplying Equation (4) by -1 and adding it to Equation (5). Replace Equation (5). The equivalent system is

$$\begin{cases} x + 2y + 3z = 4 & (1) \\ -5y - 7z = -5 & (4) \\ 0 = -2 & (6) \end{cases}$$

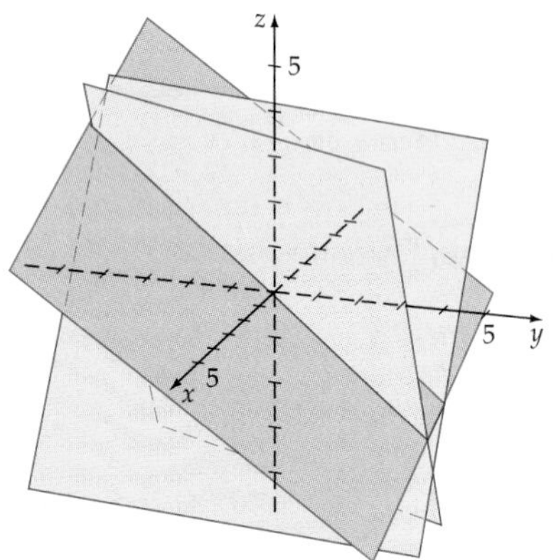

Figure 9.18

This system of equations contains a false equation. The system is inconsistent and has no solutions. There is no point on all three planes as shown in **Figure 9.18.**

TRY EXERCISE 18, EXERCISE SET 9.2, PAGE 678

◆ NONSQUARE SYSTEMS OF EQUATIONS

The linear systems of equations that we have solved so far contain the same number of variables as equations. These are *square systems of equations.* If there are fewer equations than variables—a *nonsquare system of equations*—the system has either no solution or an infinite number of solutions.

EXAMPLE 4 Solve a Nonsquare System of Equations

$$\text{Solve: } \begin{cases} x - 2y + 2z = 3 & (1) \\ 2x - y - 2z = 15 & (2) \end{cases}$$

Solution

Eliminate x from Equation (2) by multiplying Equation (1) by -2 and adding it to Equation (2). Replace Equation (2).

$$\begin{cases} x - 2y + 2z = 3 & (1) \\ 3y - 6z = 9 & (3) \end{cases}$$

Solve Equation (3) for y.

$$3y - 6z = 9$$
$$y = 2z + 3$$

Substitute $2z + 3$ for y into Equation (1) and solve for x.

$$x - 2y + 2z = 3$$
$$x - 2(2z + 3) + 2z = 3 \qquad \bullet\ y = 2z + 3$$
$$x = 2z + 9$$

For each value of z selected, there correspond values for x and y. If z is any real number c, then the solutions of the system are the ordered triples $(2c + 9, 2c + 3, c)$.

TRY EXERCISE 20, EXERCISE SET 9.2, PAGE 678

◆ HOMOGENEOUS SYSTEMS OF EQUATIONS

A linear system of equations for which the constant term is zero for all equations is called a **homogeneous system of equations.** Two examples of homogeneous systems of equations are

$$\begin{cases} 3x + 4y = 0 \\ 2x + 3y = 0 \end{cases} \qquad \begin{cases} 2x - 3y + 5z = 0 \\ 3x + 2y + z = 0 \\ x - 4y + 5z = 0 \end{cases}$$

The solution $(0, 0)$ is always a solution of a homogeneous system of equations in two variables, and $(0, 0, 0)$ is always a solution of a homogeneous system of equations in three variables. This solution is called the **trivial solution.**

Sometimes a homogeneous system of equations may have solutions other than the trivial solution. For example, $(1, -1, -1)$ is a solution to the homogeneous system of three equations in three variables above.

If a homogeneous system of equations has a unique solution, the graphs intersect only at the origin. Solutions to a homogeneous system of equations can be found by using the substitution method or the elimination method.

EXAMPLE 5 Solve a Homogeneous System of Equations

$$\text{Solve: } \begin{cases} x + 2y - 3z = 0 & (1) \\ 2x - y + z = 0 & (2) \\ 3x + y - 2z = 0 & (3) \end{cases}$$

Continued •➤

Solution

Eliminate x from Equations (2) and (3) and replace these equations by the new equations.

$$\begin{cases} x + 2y - 3z = 0 & (1) \\ \quad -5y + 7z = 0 & (4) \\ \quad -5y + 7z = 0 & (5) \end{cases}$$

Eliminate y from Equation (5). Replace Equation (5).

$$\begin{cases} x + 2y - 3z = 0 & (1) \\ \quad -5y + 7z = 0 & (4) \\ \qquad\quad 0 = 0 & (6) \end{cases}$$

Because Equation (6) is an identity, the solutions of the system are the solutions of Equations (1) and (4).

Solve Equation (4) for y.

$$y = \frac{7}{5}z$$

Substitute the expression for y into Equation (1) and solve for x.

$$x + 2y - 3z = 0 \qquad \bullet \text{ Equation (1)}$$

$$x + 2\left(\frac{7}{5}z\right) - 3z = 0 \qquad \bullet\ y = \frac{7}{5}z$$

$$x = \frac{1}{5}z$$

Letting z be any real number c, we find the solutions of the system are $\left(\frac{1}{5}c, \frac{7}{5}c, c\right)$.

TRY EXERCISE 32, EXERCISE SET 9.2, PAGE 678

◆ CURVE FITTING

One application of a system of equations is "curve fitting." Given a set of points in the plane, try to find an equation whose graph passes through those points, or "fits" those points.

EXAMPLE 6 Solve an Application of a System of Equations to Curve Fitting

Find an equation of the form $y = ax^2 + bx + c$ whose graph passes through the points whose coordinates are $(1, 4)$, $(-1, 6)$, and $(2, 9)$.

Solution

Substitute each of the given ordered pairs into the equation $y = ax^2 + bx + c$. Write the resulting system of equations.

$$\begin{cases} 4 = a(1)^2 + b(1) + c \\ 6 = a(-1)^2 + b(-1) + c \\ 9 = a(2)^2 + b(2) + c \end{cases} \quad \text{or} \quad \begin{cases} a + b + c = 4 & (1) \\ a - b + c = 6 & (2) \\ 4a + 2b + c = 9 & (3) \end{cases}$$

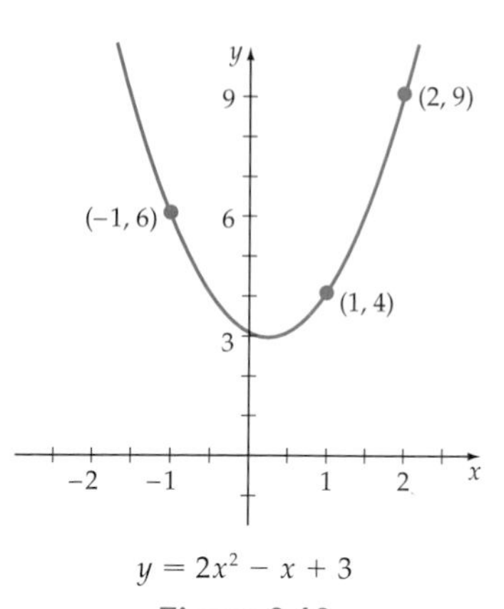

$y = 2x^2 - x + 3$

Figure 9.19

Solve the resulting system of equations for a, b, and c.

Eliminate a from Equation (2) by multiplying Equation (1) by -1 and then adding it to Equation (2). Now eliminate a from Equation (3) by multiplying Equation (1) by -4 and adding it to Equation (3). The result is

$$\begin{cases} a + b + c = 4 \\ -2b = 2 \\ -2b - 3c = -7 \end{cases}$$

Although this system of equations is not in triangular form, we can solve the second equation for b and use this value to find a and c.

Solving by substitution, we obtain $a = 2$, $b = -1$, $c = 3$. The equation of the form $y = ax^2 + bx + c$ whose graph passes through $(1, 4)$, $(-1, 6)$, and $(2, 9)$ is $y = 2x^2 - x + 3$. See **Figure 9.19.**

TRY EXERCISE 36, EXERCISE SET 9.2, PAGE 678

TOPICS FOR DISCUSSION

1. Can a system of equations contain more equations than variables? If not, explain why not. If so, give an example.

2. If a linear system of three equations in three variables is dependent, what does that mean about the graphs of the equations of the system?

3. If a linear system of three equations in three variables is inconsistent, what does that mean about the graphs of the equations of the system?

4. The equation of a circle centered at the origin with radius 5 is given by $x^2 + y^2 = 25$. Discuss the shape of $x^2 + y^2 + z^2 = 25$ in an xyz-coordinate system.

5. Consider the plane P given by $2x + 4y - 3z = 12$. The *trace* of the graph of P is obtained by letting one of the variables equal zero. For instance, the trace in the xy-plane is the graph of $2x + 4y = 12$ that is obtained by letting $z = 0$. Determine the traces of P in the xz- and yz-planes, and discuss how the traces can be used to visualize the graph of P.

EXERCISE SET 9.2

In Exercises 1 to 24, solve each system of equations.

1. $\begin{cases} 2x - y + z = 8 \\ 2y - 3z = -11 \\ 3y + 2z = 3 \end{cases}$

2. $\begin{cases} 3x + y + 2z = -4 \\ -3y - 2z = -5 \\ 2y + 5z = -4 \end{cases}$

3. $\begin{cases} x + 3y - 2z = 8 \\ 2x - y + z = 1 \\ 3x + 2y - 3z = 15 \end{cases}$

4. $\begin{cases} x - 2y + 3z = 5 \\ 3x - 3y + z = 9 \\ 5x + y - 3z = 3 \end{cases}$

5. $\begin{cases} 3x + 4y - z = -7 \\ x - 5y + 2z = 19 \\ 5x + y - 2z = 5 \end{cases}$

6. $\begin{cases} 2x - 3y - 2z = 12 \\ x + 4y + z = -9 \\ 4x + 2y - 3z = 6 \end{cases}$

7. $\begin{cases} 2x - 5y + 3z = -18 \\ 3x + 2y - z = -12 \\ x - 3y - 4z = -4 \end{cases}$

8. $\begin{cases} 4x - y + 2z = -1 \\ 2x + 3y - 3z = -13 \\ x + 5y + z = 7 \end{cases}$

9. $\begin{cases} x + 2y - 3z = -7 \\ 2x - y + 4z = 11 \\ 4x + 3y - 4z = -3 \end{cases}$

10. $\begin{cases} x - 3y + 2z = -11 \\ 3x + y + 4z = 4 \\ 5x - 5y + 8z = -18 \end{cases}$

11. $\begin{cases} 2x - 5y + 2z = -4 \\ 3x + 2y + 3z = 13 \\ 5x - 3y - 4z = -18 \end{cases}$

12. $\begin{cases} 3x + 2y - 5z = 6 \\ 5x - 4y + 3z = -12 \\ 4x + 5y - 2z = 15 \end{cases}$

13. $\begin{cases} 2x + y - z = -2 \\ 3x + 2y + 3z = 21 \\ 7x + 4y + z = 17 \end{cases}$

14. $\begin{cases} 3x + y + 2z = 2 \\ 4x - 2y + z = -4 \\ 11x - 3y + 4z = -6 \end{cases}$

15. $\begin{cases} 3x - 2y + 3z = 11 \\ 2x + 3y + z = 3 \\ 5x + 14y - z = 1 \end{cases}$

16. $\begin{cases} 2x + 3y + 2z = 14 \\ x - 3y + 4z = 4 \\ -x + 12y - 6z = 2 \end{cases}$

17. $\begin{cases} 2x - 3y + 6z = 3 \\ x + 2y - 4z = 5 \\ 3x + 4y - 8z = 7 \end{cases}$

18. $\begin{cases} 2x + 3y - 6z = 4 \\ 3x - 2y - 9z = -7 \\ 2x + 5y - 6z = 8 \end{cases}$

19. $\begin{cases} 2x - 3y + 5z = 14 \\ x + 4y - 3z = -2 \end{cases}$

20. $\begin{cases} x - 3y + 4z = 9 \\ 3x - 8y - 2z = 4 \end{cases}$

21. $\begin{cases} 6x - 9y + 6z = 7 \\ 4x - 6y + 4z = 9 \end{cases}$

22. $\begin{cases} 4x - 2y + 6z = 5 \\ 2x - y + 3z = 2 \end{cases}$

23. $\begin{cases} 5x + 3y + 2z = 10 \\ 3x - 4y - 4z = -5 \end{cases}$

24. $\begin{cases} 3x - 4y - 7z = -5 \\ 2x + 3y - 5z = 2 \end{cases}$

In Exercises 25 to 32, solve each homogeneous system of equations.

25. $\begin{cases} x + 3y - 4z = 0 \\ 2x + 7y + z = 0 \\ 3x - 5y - 2z = 0 \end{cases}$

26. $\begin{cases} x - 2y + 3z = 0 \\ 3x - 7y - 4z = 0 \\ 4x - 4y + z = 0 \end{cases}$

27. $\begin{cases} 2x - 3y + z = 0 \\ 2x + 4y - 3z = 0 \\ 6x - 2y - z = 0 \end{cases}$

28. $\begin{cases} 5x - 4y - 3z = 0 \\ 2x + y + 2z = 0 \\ x - 6y - 7z = 0 \end{cases}$

29. $\begin{cases} 3x - 5y + 3z = 0 \\ 2x - 3y + 4z = 0 \\ 7x - 11y + 11z = 0 \end{cases}$

30. $\begin{cases} 5x - 2y - 3z = 0 \\ 3x - y - 4z = 0 \\ 4x - y - 9z = 0 \end{cases}$

31. $\begin{cases} 4x - 7y - 2z = 0 \\ 2x + 4y + 3z = 0 \\ 3x - 2y - 5z = 0 \end{cases}$

32. $\begin{cases} 5x + 2y + 3z = 0 \\ 3x + y - 2z = 0 \\ 4x - 7y + 5z = 0 \end{cases}$

In Exercises 33 to 42, solve each exercise by solving a system of equations.

33. **CURVE FITTING** Find an equation of the form $y = ax^2 + bx + c$ whose graph passes through the points $(2, 3)$, $(-2, 7)$, and $(1, -2)$.

34. **CURVE FITTING** Find an equation of the form $y = ax^2 + bx + c$ whose graph passes through the points $(1, -2)$, $(3, -4)$, and $(2, -2)$.

35. **CURVE FITTING** Find the equation of the circle whose graph passes through the points $(5, 3)$, $(-1, -5)$, and $(-2, 2)$. (*Hint:* Use the equation $x^2 + y^2 + ax + by + c = 0$.)

36. **CURVE FITTING** Find the equation of the circle whose graph passes through the points $(0, 6)$, $(1, 5)$, and $(-7, -1)$. (*Hint:* See Exercise 35.)

37. **CURVE FITTING** Find the center and radius of the circle whose graph passes through the points $(-2, 10)$, $(-12, -14)$, and $(5, 3)$. (*Hint:* See Exercise 35.)

38. **CURVE FITTING** Find the center and radius of the circle whose graph passes through the points $(2, 5)$, $(-4, -3)$, and $(3, 4)$. (*Hint:* See Exercise 35.)

39. **COIN PROBLEM** A coin bank contains only nickels, dimes, and quarters. The value of the coins is \$2. There are twice as many nickels as dimes and one more dime than quarters. Find the number of each coin in the bank.

40. **COIN PROBLEM** A coin bank contains only nickels, dimes, and quarters. The value of the coins is \$5.50. The number of nickels is six more than twice the number of quarters. The number of dimes is one-third the number of nickels. Find the number of each coin in the bank.

41. **NUMBER THEORY** The sum of the digits of a positive three-digit number is 19. The tens digit is four less than twice the hundreds digit. The number is decreased by 99 when the digits are reversed. Find the number.

42. **NUMBER THEORY** The sum of the digits of a positive three-digit number is 10. The hundreds digit is one less than twice the ones digit. The number is decreased by 198 when the digits are reversed. Find the number.

SUPPLEMENTAL EXERCISES

In Exercises 43 to 48, solve each system of equations.

43. $\begin{cases} 2x + y - 3z + 2w = -1 \\ 2y - 5z - 3w = 9 \\ 3y - 8z + w = -4 \\ 2y - 2z + 3w = -3 \end{cases}$

44. $\begin{cases} 3x - y + 2z - 3w = 5 \\ 2y - 5z + 2w = -7 \\ 4y - 9z + w = -19 \\ 3y + z - 2w = -12 \end{cases}$

45. $\begin{cases} x - 3y + 2z - w = 2 \\ 2x - 5y - 3z + 2w = 21 \\ 3x - 8y - 2z - 3w = 12 \\ -2x + 8y + z + 2w = -13 \end{cases}$

46. $\begin{cases} x - 2y + 3z + 2w = 8 \\ 3x - 7y - 2z + 3w = 18 \\ 2x - 5y + 2z - w = 19 \\ 4x - 8y + 3z + 2w = 29 \end{cases}$

47. $\begin{cases} x + 2y - 2z + 3w = 2 \\ 2x + 5y + 2z + 4w = 9 \\ 4x + 9y - 2z + 10w = 13 \\ -x - y + 8z - 5w = 3 \end{cases}$

48. $\begin{cases} x - 2y + 3z - 2w = -1 \\ 3x - 7y - 2z - 3w = -19 \\ 2x - 5y + 2z - w = -11 \\ -x + 3y - 2z - w = 3 \end{cases}$

In Exercises 49 and 50, use the system of equations

$$\begin{cases} x - 3y - 2z = A^2 \\ 2x - 5y + Az = 9 \\ 2x - 8y + z = 18 \end{cases}$$

49. Find all values of A for which the system has no solutions.

50. Find all values of A for which the system has a unique solution.

In Exercises 51 to 53, use the system of equations

$$\begin{cases} x + 2y + z = A^2 \\ -2x - 3y + Az = 1 \\ 7x + 12y + A^2z = 4A^2 - 3 \end{cases}$$

51. Find all values of A for which the system has a unique solution.

52. Find all values of A for which the system has an infinite number of solutions.

53. Find all values of A for which the system has no solution.

54. Find an equation of a plane that contains the points $(2, 1, 1)$, $(-1, 2, 12)$, and $(3, 2, 0)$. (*Hint:* The equation of a plane can be written as $z = ax + by + c$.)

55. Find an equation of a plane that contains the points $(1, -1, 5)$, $(2, -2, 9)$, and $(-3, -1, -1)$. (*Hint:* The equation of a plane can be written as $z = ax + by + c$.)

PROJECTS

1. **CONCEPT OF DIMENSION** In this chapter we graphed first-degree equations in three variables. If we were to attempt to graph an equation in four variables, we would need a fourth axis perpendicular to the three axes of an *xyz*-coordinate system. It seems impossible to imagine a fourth dimension, but incorporating it is really a quite practical matter in mathematics. In fact, there are some systems that require an infinite-dimensional coordinate system. To gain some insight into the concept of dimension, read the book *Flatland* by Edwin A. Abbott, and then write an essay explaining what this book has to do with dimension.

2. **ABILITIES OF A FOUR-DIMENSIONAL HUMAN** There have been a number of attempts to describe the abilities of a four-dimensional human in a three-dimensional world. Read some of these accounts, and then write an essay on some of the actions a four-dimensional person could perform. Answer the following question in your essay. Can a four-dimensional person remove the money from a locked safe without first opening the safe?

SECTION 9.3 NONLINEAR SYSTEMS OF EQUATIONS

- SOLVING NONLINEAR SYSTEMS OF EQUATIONS

◆ SOLVING NONLINEAR SYSTEMS OF EQUATIONS

A **nonlinear system of equations** is one in which one or more equations of the system are not linear equations. **Figure 9.20** shows examples of nonlinear systems of equations and the corresponding graphs of the equations. Each point of intersection of the graphs is a solution of the system of equations. In the third example, the graphs do not intersect; therefore, the system of equations has no real number solution.

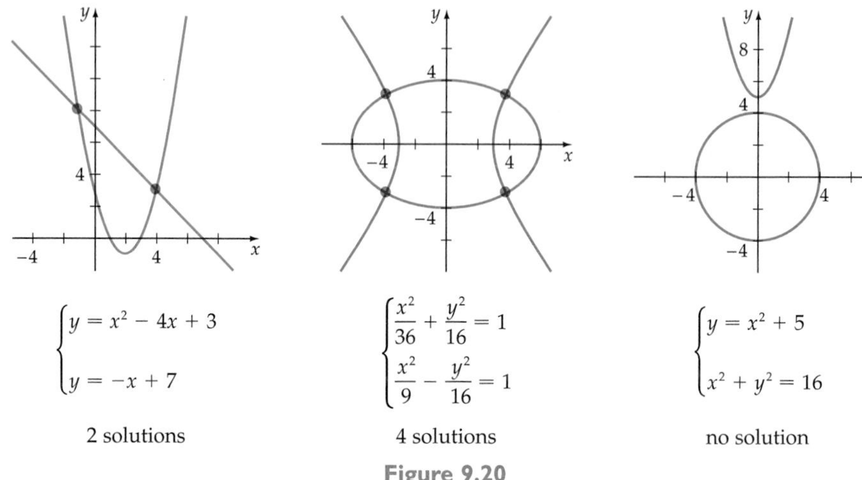

$$\begin{cases} y = x^2 - 4x + 3 \\ y = -x + 7 \end{cases}$$

2 solutions

$$\begin{cases} \dfrac{x^2}{36} + \dfrac{y^2}{16} = 1 \\ \dfrac{x^2}{9} - \dfrac{y^2}{16} = 1 \end{cases}$$

4 solutions

$$\begin{cases} y = x^2 + 5 \\ x^2 + y^2 = 16 \end{cases}$$

no solution

Figure 9.20

To solve a nonlinear system of equations, use the substitution method or the elimination method. The substitution method is usually easier for solving a nonlinear system that contains a linear equation.

EXAMPLE 1 Solve a Nonlinear System by the Substitution Method

Solve: $\begin{cases} y = x^2 - x - 1 & (1) \\ 3x - y = 4 & (2) \end{cases}$

Solution

We will use the substitution method. Using the equation $y = x^2 - x - 1$, substitute the expression for y into $3x - y = 4$.

$$3x - y = 4$$
$$3x - (x^2 - x - 1) = 4 \qquad \bullet\ y = x^2 - x - 1$$
$$-x^2 + 4x + 1 = 4 \qquad \bullet\ \text{Simplify.}$$
$$x^2 - 4x + 3 = 0 \qquad \bullet\ \text{Write the quadratic equation in standard form.}$$
$$(x - 3)(x - 1) = 0 \qquad \bullet\ \text{Solve for } x.$$
$$x - 3 = 0 \quad \text{or} \quad x - 1 = 0$$
$$x = 3 \quad \text{or} \quad x = 1$$

Substitute these values into Equation (1) and solve for y.

$$y = 3^2 - 3 - 1 = 5, \quad \text{or} \quad y = 1^2 - 1 - 1 = -1$$

The solutions are $(3, 5)$ and $(1, -1)$. See **Figure 9.21**.

Visualize the Solution

Graphing $y = x^2 - x - 1$ and $3x - y = 4$ shows that $(1, -1)$ and $(3, 5)$ belong to each graph. Therefore, these ordered pairs are solutions of the system of equations.

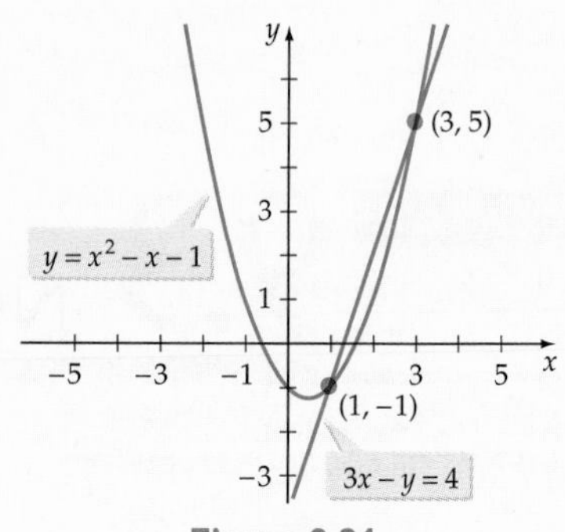

Figure 9.21

TRY EXERCISE 8, EXERCISE SET 9.3, PAGE 684

You can use a graphing calculator to solve some nonlinear systems of equations in two variables. For instance, to solve

$$\begin{cases} y = x^2 - 2x + 2 \\ y = x^3 + 2x^2 - 7x - 3 \end{cases}$$

enter X²–2X+2 into Y1 and X^3+2X²–7X–3 into Y2 and graph the two equations. Be sure to use a viewing window that will show all points of intersection. The sequence of steps shown in **Figure 9.22** can be used to find the points of intersection with a TI-83 calculator.

Press [2nd] CALC
Select 5: intersect.
Press [ENTER]

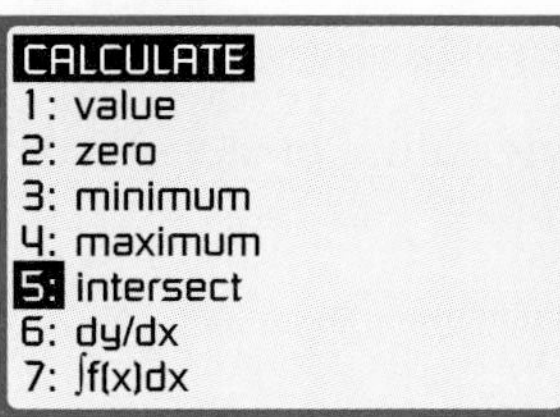

The "First curve?" shown on the bottom of the screen means to select the first of the two graphs that intersect. Just press [ENTER]

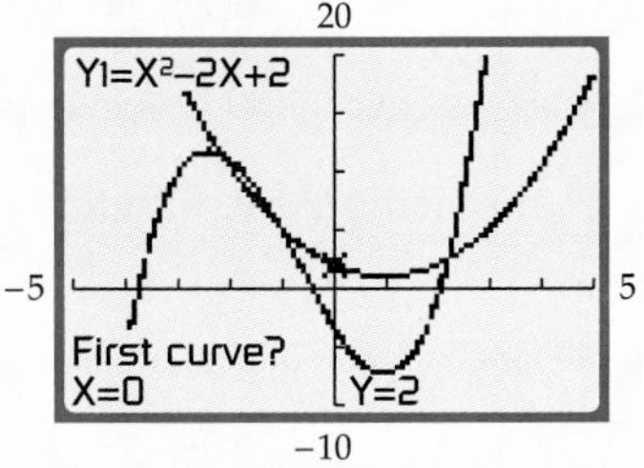

The "Second curve?" shown on the bottom of the screen means to select the second of the two graphs that intersect. Just press [ENTER]

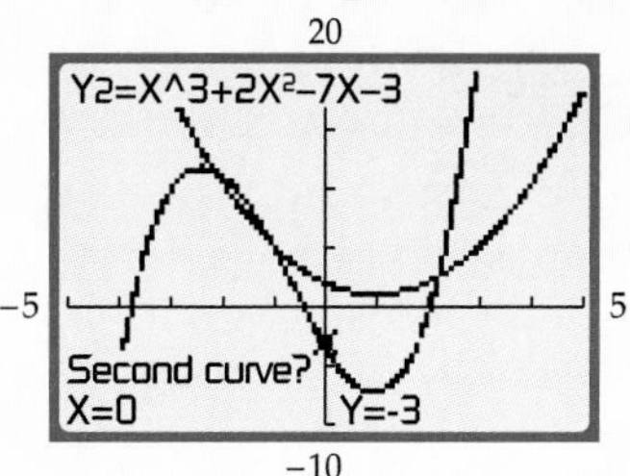

"Guess?" is shown on the bottom of the screen. Move the cursor until it is approximately at the first point of intersection. Press [ENTER]

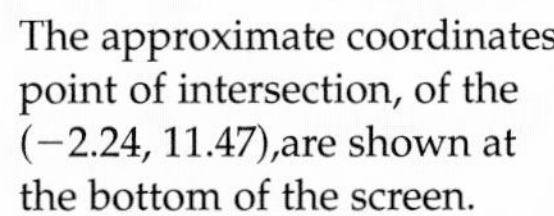

The approximate coordinates point of intersection, of the (−2.24, 11.47),are shown at the bottom of the screen.

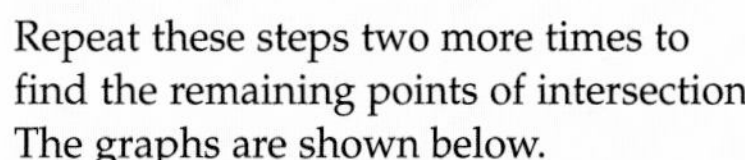

Repeat these steps two more times to find the remaining points of intersection. The graphs are shown below.

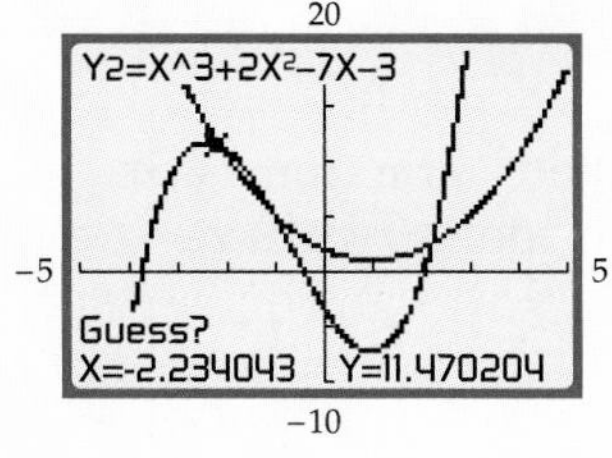

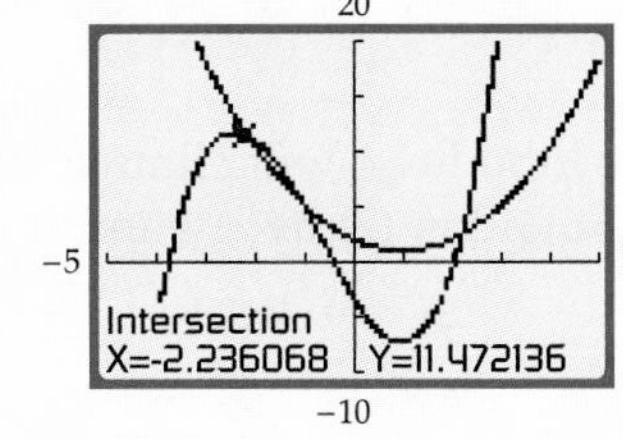

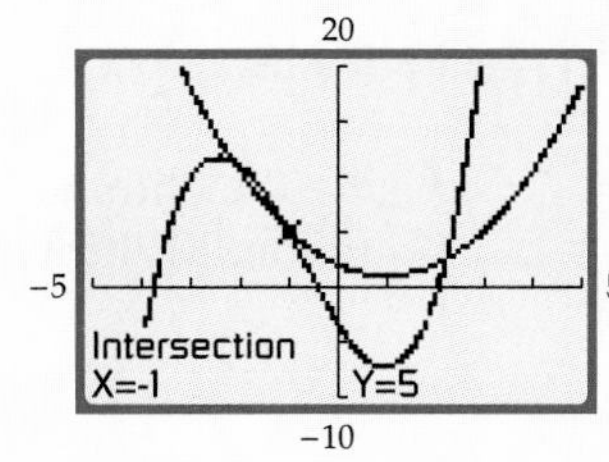

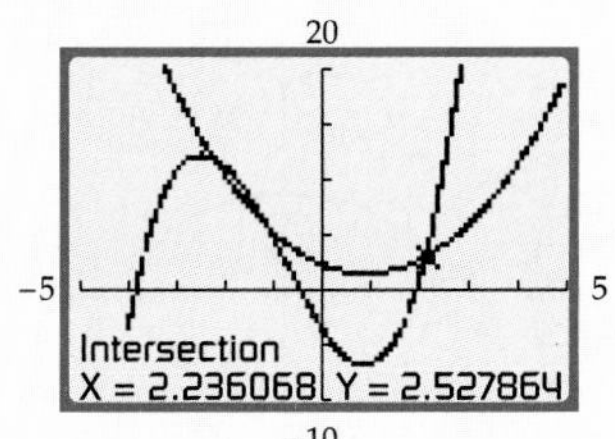

Figure 9.22

The coordinates of the points of intersection are $(-2.24, 11.47)$, $(-1, 5)$, and $(2.24, 2.53)$.

EXAMPLE 2 Solve a Nonlinear System by the Elimination Method

$$\text{Solve: } \begin{cases} 4x^2 + 3y^2 = 48 & (1) \\ 3x^2 + 2y^2 = 35 & (2) \end{cases}$$

Continued

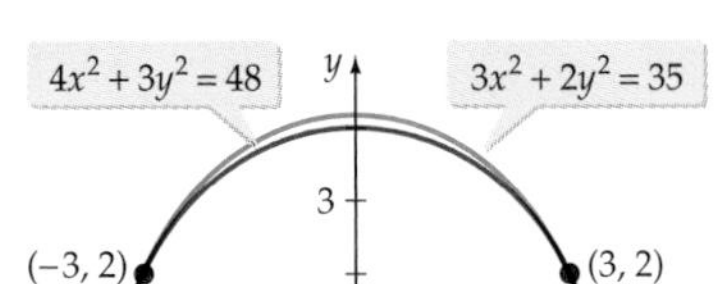
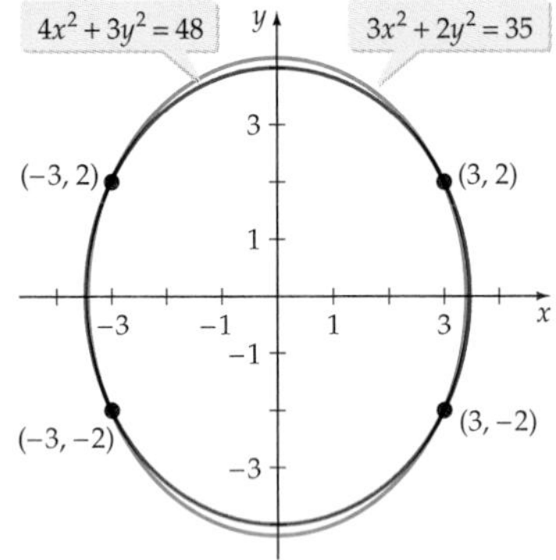

Figure 9.23

Solution

We will eliminate the x^2 term. Multiply Equation (1) by -3 and Equation (2) by 4. Then add the two equations.

$$\begin{aligned} -12x^2 - 9y^2 &= -144 \\ 12x^2 + 8y^2 &= 140 \\ \hline -y^2 &= -4 \\ y^2 &= 4 \\ y &= \pm 2 \end{aligned}$$

Substitute 2 for y into Equation (1) and solve for x.

$$\begin{aligned} 4x^2 + 3(2)^2 &= 48 \\ 4x^2 &= 36 \\ x^2 &= 9 \\ x &= \pm 3 \end{aligned}$$

Because $(-2)^2 = 2^2$, replacing y by -2 yields the same values of x: $x = 3$ or $x = -3$. The solutions are $(3, 2)$, $(3, -2)$, $(-3, 2)$, and $(-3, -2)$. See **Figure 9.23.**

TRY EXERCISE 16, EXERCISE SET 9.3, PAGE 684

EXAMPLE 3 Identify an Inconsistent System of Equations

$$\text{Solve: } \begin{cases} 4x^2 + 9y^2 = 36 & (1) \\ x^2 - y^2 = 25 & (2) \end{cases}$$

Solution

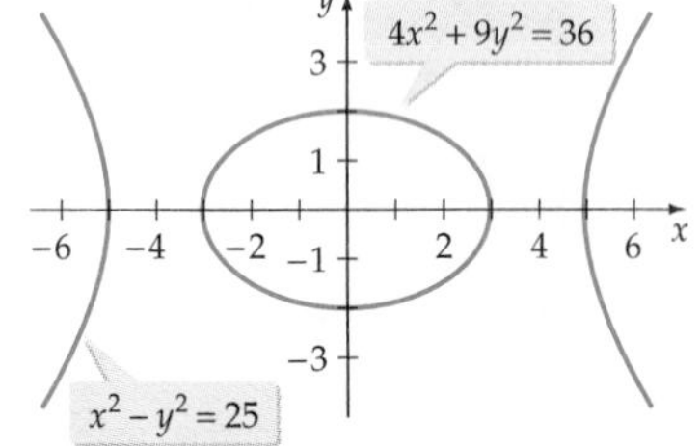

Figure 9.24

Using the elimination method, we will eliminate the x^2 term from each equation. Multiplying Equation (2) by -4 and then adding, we have

$$\begin{aligned} 4x^2 + 9y^2 &= 36 \\ -4x^2 + 4y^2 &= -100 \\ \hline 13y^2 &= -64 \end{aligned}$$

Because the equation $13y^2 = -64$ has no real number solutions, the system of equations has no real solutions. The graphs of the equations do not intersect. See **Figure 9.24.**

TRY EXERCISE 20, EXERCISE SET 9.3, PAGE 684

EXAMPLE 4 Solve a Nonlinear System of Equations

$$\text{Solve: } \begin{cases} (x + 3)^2 + (y - 4)^2 = 20 \\ (x + 4)^2 + (y - 3)^2 = 26 \end{cases}$$

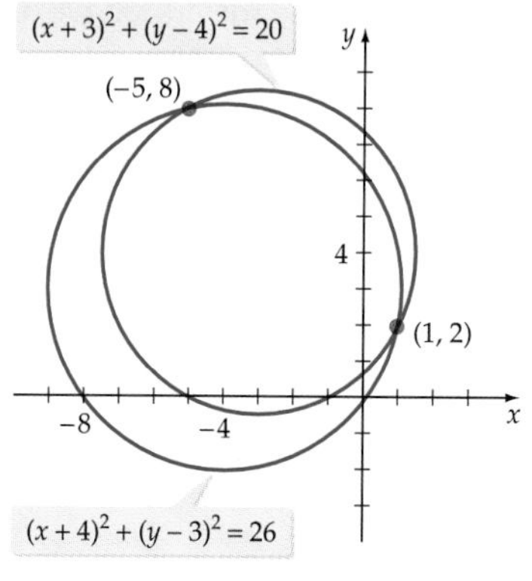

Figure 9.25

Solution

Expand the binomials in each equation. Then subtract the two equations and simplify.

$$\begin{array}{rl} x^2 + 6x + 9 + y^2 - 8y + 16 = 20 & (1) \\ x^2 + 8x + 16 + y^2 - 6y + 9 = 26 & (2) \\ \hline -2x - 7 \quad - 2y + 7 = -6 & \\ x + y = 3 & \end{array}$$

Now solve the resulting equation for y.

$$y = -x + 3$$

Substitute $-x + 3$ for y into Equation (1) and solve for x.

$$x^2 + 6x + 9 + (-x + 3)^2 - 8(-x + 3) + 16 = 20$$
$$2(x^2 + 4x - 5) = 0$$
$$2(x + 5)(x - 1) = 0$$
$$x = -5 \quad \text{or} \quad x = 1$$

Substitute -5 and 1 for x into the equation $y = -x + 3$ and solve for y. This yields $y = 8$ or $y = 2$. The solutions of the system of equations are $(-5, 8)$ and $(1, 2)$. See **Figure 9.25.**

TRY EXERCISE 28, EXERCISE SET 9.3, PAGE 684

TOPICS FOR DISCUSSION

1. What distinguishes a system of linear equations from a system of nonlinear equations? Give an example of both types of systems of equations.
2. Is the system of equations

$$\begin{cases} xy = 1 \\ x + y = 1 \end{cases}$$

a nonlinear system of equations? Why or why not?

3. Can a nonlinear system of equations have no solution? If so, give an example. If not, explain why not.
4. Make up a nonlinear system of equations in two variables that has at least $(2, -3)$ as a solution, contains one nonlinear equation, and contains one linear equation.

EXERCISE SET 9.3

In Exercises 1 to 32, solve the system of equations.

1. $\begin{cases} y = x^2 - x \\ y = 2x - 2 \end{cases}$

2. $\begin{cases} y = x^2 + 2x - 3 \\ y = x - 1 \end{cases}$

3. $\begin{cases} y = 2x^2 - 3x - 3 \\ y = x - 4 \end{cases}$

4. $\begin{cases} y = -x^2 + 2x - 4 \\ y = \dfrac{1}{2}x + 1 \end{cases}$

5. $\begin{cases} y = x^2 - 2x + 3 \\ y = x^2 - x - 2 \end{cases}$

6. $\begin{cases} y = 2x^2 - x + 1 \\ y = x^2 + 2x + 5 \end{cases}$

7. $\begin{cases} x + y = 10 \\ xy = 24 \end{cases}$

8. $\begin{cases} x - 2y = 3 \\ xy = -1 \end{cases}$

9. $\begin{cases} 2x - y = 1 \\ xy = 6 \end{cases}$

10. $\begin{cases} x - 3y = 7 \\ xy = -4 \end{cases}$

11. $\begin{cases} 3x^2 - 2y^2 = 1 \\ y = 4x - 3 \end{cases}$

12. $\begin{cases} x^2 + 3y^2 = 7 \\ x + 4y = 6 \end{cases}$

13. $\begin{cases} y = x^3 + 4x^2 - 3x - 5 \\ y = 2x^2 - 2x - 3 \end{cases}$

14. $\begin{cases} y = x^3 - 2x^2 + 5x + 1 \\ y = x^2 + 7x - 5 \end{cases}$

15. $\begin{cases} 2x^2 + y^2 = 9 \\ x^2 - y^2 = 3 \end{cases}$

16. $\begin{cases} 3x^2 - 2y^2 = 19 \\ x^2 - y^2 = 5 \end{cases}$

17. $\begin{cases} x^2 - 2y^2 = 8 \\ x^2 + 3y^2 = 28 \end{cases}$

18. $\begin{cases} 2x^2 + 3y^2 = 5 \\ x^2 - 3y^2 = 4 \end{cases}$

19. $\begin{cases} 2x^2 + 4y^2 = 5 \\ 3x^2 + 8y^2 = 14 \end{cases}$

20. $\begin{cases} 2x^2 + 3y^2 = 11 \\ 3x^2 + 2y^2 = 19 \end{cases}$

21. $\begin{cases} x^2 - 2x + y^2 = 1 \\ 2x + y = 5 \end{cases}$

22. $\begin{cases} x^2 + y^2 + 3y = 22 \\ 2x + y = -1 \end{cases}$

23. $\begin{cases} (x - 3)^2 + (y + 1)^2 = 5 \\ x - 3y = 7 \end{cases}$

24. $\begin{cases} (x + 2)^2 + (y - 2)^2 = 13 \\ 2x + y = 6 \end{cases}$

25. $\begin{cases} x^2 - 3x + y^2 = 4 \\ 3x + y = 11 \end{cases}$

26. $\begin{cases} x^2 + y^2 - 4y = 4 \\ 5x - 2y = 2 \end{cases}$

27. $\begin{cases} (x - 1)^2 + (y + 2)^2 = 14 \\ (x + 2)^2 + (y - 1)^2 = 2 \end{cases}$

28. $\begin{cases} (x + 2)^2 + (y - 3)^2 = 10 \\ (x - 3)^2 + (y + 1)^2 = 13 \end{cases}$

29. $\begin{cases} (x + 3)^2 + (y - 2)^2 = 20 \\ (x - 2)^2 + (y - 3)^2 = 2 \end{cases}$

30. $\begin{cases} (x - 4)^2 + (y - 5)^2 = 8 \\ (x + 1)^2 + (y + 2)^2 = 34 \end{cases}$

31. $\begin{cases} (x - 1)^2 + (y + 1)^2 = 2 \\ (x + 2)^2 + (y - 3)^2 = 3 \end{cases}$

32. $\begin{cases} (x + 1)^2 + (y - 3)^2 = 4 \\ (x - 3)^2 + (y + 2)^2 = 2 \end{cases}$

33. **GEOMETRY** Find the perimeter of the rectangle below.

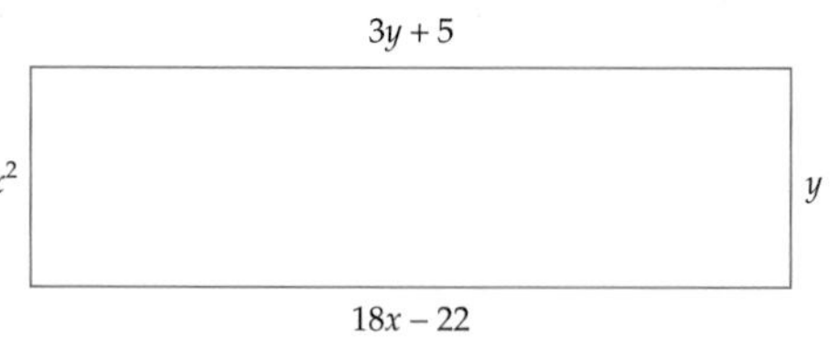

34. **CONSTRUCTION** A painter leans a ladder against a vertical wall. The top of the ladder is 7 meters above the ground. When the bottom of the ladder is moved 1 meter farther away from the wall, the top of the ladder is 5 meters above the ground. What is the length of the ladder? Round to the nearest hundredth of a meter.

35. **ANALYTIC GEOMETRY** For what values of the radius does the line $y = 2x + 1$ intersect (at one or more points) the circle whose equation is $x^2 + y^2 = r^2$?

36. **GEOMETRY** Three rectangles have exactly the same area. The dimensions of each rectangle (as length and width) are a and b; $a - 3$ and $b + 2$; and $a + 3$ and $b - 1$. Find the area of the rectangles.

In Exercises 37 to 44, approximate the real number solutions of each system of equations to the nearest ten-thousandth.

37. $\begin{cases} y = 2^x \\ y = x + 1 \end{cases}$

38. $\begin{cases} y = \log_2 x \\ y = x - 3 \end{cases}$

39. $\begin{cases} y = e^{-x} \\ y = x^2 \end{cases}$

40. $\begin{cases} y = \ln x \\ y = -x + 4 \end{cases}$

41. $\begin{cases} y = \sqrt{x} \\ y = \dfrac{1}{x - 1} \end{cases}$

42. $\begin{cases} y = \dfrac{6}{x + 1} \\ y = \dfrac{x}{x - 1} \end{cases}$

43. $\begin{cases} y = |x| \\ y = 2^{-x^2} \end{cases}$

44. $\begin{cases} y = \dfrac{2^x + 2^{-x}}{2} \\ y = \dfrac{2^x - 2^{-x}}{2} \end{cases}$

SUPPLEMENTAL EXERCISES

In Exercises 45 to 50, solve the system of equations for *rational number* ordered pairs.

45. $\begin{cases} y = x^2 + 4 \\ x = y^2 - 24 \end{cases}$

46. $\begin{cases} y = x^2 - 5 \\ x = y^2 - 13 \end{cases}$

47. $\begin{cases} x^2 - 3xy + y^2 = 5 \\ x^2 - xy - 2y^2 = 0 \end{cases}$

(*Hint:* Factor the second equation. Now use the principle of zero products and the substitution principle.)

48. $\begin{cases} x^2 + 2xy - y^2 = 1 \\ x^2 + 3xy + 2y^2 = 0 \end{cases}$

(*Hint:* See Exercise 47.)

49. $\begin{cases} 2x^2 - 4xy - y^2 = 6 \\ 4x^2 - 3xy - y^2 = 6 \end{cases}$

(*Hint:* Subtract the two equations.)

50. $\begin{cases} 3x^2 + 2xy - 5y^2 = 11 \\ x^2 + 3xy + y^2 = 11 \end{cases}$

(*Hint:* Subtract the two equations.)

51. Show that the line $y = mx$ intersects the hyperbola $\frac{x^2}{a^2} - \frac{y^2}{b^2} = 1$ if and only if $|m| < \left|\frac{b}{a}\right|$.

PROJECTS

1. **FINDING ZEROS OF A POLYNOMIAL** One zero of $P(x) = x^3 + 2x^2 + Cx - 6$ is the sum of the other two zeros of $P(x)$. Find C and the three zeros of $P(x)$.

2. **PROVING A GEOMETRY THEOREM** Consider the triangle that is shown at the right inscribed in a circle of radius a with one side along the diameter of the circle. Prove that the triangle is a right triangle by completing the following steps.

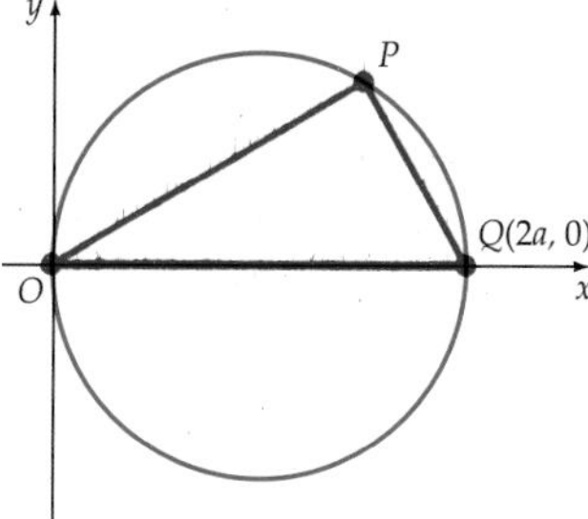

   a. Let $y = mx$, $m \geq 0$. Show that the graph of $y = mx$, $m \geq 0$, intersects the circle whose equation is $(x - a)^2 + y^2 = a^2$, $a > 0$, at $P\left(\frac{2a}{1 + m^2}, \frac{2ma}{1 + m^2}\right)$.

   b. Show that the slope of the line through P and $Q(2a, 0)$ is $-1/m$.

   c. What is the slope of the line between O and P?

   d. Prove that line segment OP is perpendicular to line segment PQ.

   e. How can you conclude from the foregoing that triangle OPQ is a right triangle?

SECTION 9.4 PARTIAL FRACTIONS

- PARTIAL FRACTION DECOMPOSITION

◆ PARTIAL FRACTION DECOMPOSITION

An algebraic application of systems of equations is a technique known as *partial fractions.* In Chapter 1, we reviewed the problem of adding two rational expressions. For example,

$$\frac{5}{x - 1} + \frac{1}{x + 2} = \frac{6x + 9}{(x - 1)(x + 2)}$$

Now we will take an opposite approach. That is, given a rational expression, we will find simpler rational expressions whose sum is the given expression. The method by which a more complicated rational expression is written as a sum of rational expressions is called **partial fraction decomposition.** This technique is based on the following theorem.

Partial Fraction Decomposition Theorem

If
$$f(x) = \frac{p(x)}{q(x)}$$
is a rational expression in which the degree of the numerator is less than the degree of the denominator, and $p(x)$ and $q(x)$ have no common factors, then $f(x)$ can be written as a partial fraction decomposition in the form
$$f(x) = f_1(x) + f_2(x) + \cdots + f_n(x)$$
where each $f_i(x)$ has one of the following forms:
$$\frac{A}{(px+q)^m} \quad \text{or} \quad \frac{Bx+C}{(ax^2+bx+c)^m}$$

The procedure for finding a partial fraction decomposition of a rational expression depends on factorization of the denominator of the rational expression. There are four cases.

Case 1 Nonrepeated Linear Factors

The partial fraction decomposition will contain an expression of the form $A/(x+a)$ for each nonrepeated linear factor of the denominator. Example:

$$\frac{3x-1}{x(3x+4)(x-2)}$$

• **Each linear factor of the denominator occurs only once.**

Partial fraction decomposition:

$$\frac{3x-1}{x(3x+4)(x-2)} = \frac{A}{x} + \frac{B}{3x+4} + \frac{C}{x-2}$$

Case 2 Repeated Linear Factors

The partial fraction decomposition will contain an expression of the form

$$\frac{A_1}{(x+a)} + \frac{A_2}{(x+a)^2} + \cdots + \frac{A_m}{(x+a)^m}$$

for each repeated linear factor of multiplicity m. Example:

$$\frac{4x+5}{(x-2)^2(2x+1)}$$

• **$(x-2)^2 = (x-2)(x-2)$, a repeated linear factor.**

Partial fraction decomposition:

$$\frac{4x+5}{(x-2)^2(2x+1)} = \frac{A_1}{x-2} + \frac{A_2}{(x-2)^2} + \frac{B}{2x+1}$$

Case 3 Nonrepeated Quadratic Factors

The partial fraction decomposition will contain an expression of the form

$$\frac{Ax+B}{ax^2+bx+c}$$

for each quadratic factor irreducible over the real numbers. Example:

$$\frac{x-4}{(x^2+x+1)(x-4)}$$ • $x^2 + x + 1$ is irreducible over the real numbers.

Partial fraction decomposition:

$$\frac{x-4}{(x^2+x+1)(x-4)} = \frac{Ax+B}{x^2+x+1} + \frac{C}{x-4}$$

Case 4 Repeated Quadratic Factors

The partial fraction decomposition will contain an expression of the form

$$\frac{A_1x+B_1}{ax^2+bx+c} + \frac{A_2x+B_2}{(ax^2+bx+c)^2} + \cdots + \frac{A_mx+B_m}{(ax^2+bx+c)^m}$$

for each quadratic factor irreducible over the real numbers. Example:

$$\frac{2x}{(x-2)(x^2+4)^2}$$ • $(x^2 + 4)^2$ is a repeated quadratic factor.

Partial fraction decomposition:

$$\frac{2x}{(x-2)(x^2+4)^2} = \frac{A_1x+B_1}{x^2+4} + \frac{A_2x+B_2}{(x^2+4)^2} + \frac{C}{x-2}$$

There are various methods for finding the constants of a partial fraction decomposition. One such method is based on a property of polynomials.

Equality of Polynomials

If the two polynomials $p(x) = a_nx^n + a_{n-1}x^{n-1} + \cdots + a_1x + a_0$ and $r(x) = b_nx^n + b_{n-1}x^{n-1} + \cdots + b_1x + b_0$ are of degree n, then $p(x) = r(x)$ if and only if $a_0 = b_0, a_1 = b_1, a_2 = b_2, \ldots, a_n = b_n$.

EXAMPLE 1 Find a Partial Fraction Decomposition Case 1: Nonrepeated Linear Factors

Find a partial fraction decomposition of $\dfrac{x+11}{x^2-2x-15}$.

Solution

First factor the denominator.

$$x^2 - 2x - 15 = (x+3)(x-5)$$

The factors are nonrepeated linear factors. Therefore, the partial fraction decomposition will have the form

$$\frac{x+11}{(x+3)(x-5)} = \frac{A}{x+3} + \frac{B}{x-5} \qquad (1)$$

Continued •

To solve for A and B, multiply each side of the equation by the least common multiple of the denominators, $(x + 3)(x - 5)$.

$$x + 11 = A(x - 5) + B(x + 3)$$
$$x + 11 = (A + B)x + (-5A + 3B) \qquad \text{• Combine like terms.}$$

Using the Equality of Polynomials Theorem, equate coefficients of like powers. The result will be the system of equations

$$\begin{cases} 1 = A + B \\ 11 = -5A + 3B \end{cases} \qquad \text{• Recall that } x = 1 \cdot x.$$

Solving the system of equations for A and B, we have $A = -1$ and $B = 2$. Substituting -1 for A and 2 for B into the form of the partial fraction decomposition (1), we obtain

$$\frac{x + 11}{(x + 3)(x - 5)} = \frac{-1}{x + 3} + \frac{2}{x - 5}$$

You should add the two expressions to verify the equality.

TRY EXERCISE 14, EXERCISE SET 9.4, PAGE 692

EXAMPLE 2 Find the Partial Fraction Decomposition Case 2: Repeated Linear Factors

Find a partial fraction decomposition of $\dfrac{x^2 + 2x + 7}{x(x - 1)^2}$.

Solution

The denominator has one nonrepeated factor and one repeated factor. The partial fraction decomposition will have the form

$$\frac{x^2 + 2x + 7}{x(x - 1)^2} = \frac{A}{x} + \frac{B}{x - 1} + \frac{C}{(x - 1)^2}$$

Multiplying each side by the LCD $x(x - 1)^2$, we have

$$x^2 + 2x + 7 = A(x - 1)^2 + B(x - 1)x + Cx$$

Expanding the right side and combining like terms give

$$x^2 + 2x + 7 = (A + B)x^2 + (-2A - B + C)x + A$$

Using the Equality of Polynomials Theorem, equate coefficients of like powers. This will result in the system of equations

$$\begin{cases} 1 = A + B \\ 2 = -2A - B + C \\ 7 = A \end{cases}$$

The solution is $A = 7$, $B = -6$, and $C = 10$. Thus the partial fraction decomposition is

$$\frac{x^2 + 2x + 7}{x(x-1)^2} = \frac{7}{x} + \frac{-6}{x-1} + \frac{10}{(x-1)^2}$$

TRY EXERCISE 22, EXERCISE SET 9.4, PAGE 692

EXAMPLE 3 Find the Partial Fraction Decomposition Case 3: Nonrepeated Quadratic Factor

Find the partial fraction decomposition of $\dfrac{3x + 16}{(x-2)(x^2+7)}$.

Solution

Because $(x - 2)$ is a nonrepeated linear factor and $x^2 + 7$ is an irreducible quadratic over the real numbers, the partial fraction decomposition will have the form

$$\frac{3x + 16}{(x-2)(x^2+7)} = \frac{A}{x-2} + \frac{Bx + C}{x^2 + 7}$$

Multiplying each side by the LCD $(x - 2)(x^2 + 7)$ yields

$$3x + 16 = A(x^2 + 7) + (Bx + C)(x - 2)$$

Expanding the right side and combining like terms, we have

$$3x + 16 = (A + B)x^2 + (-2B + C)x + (7A - 2C)$$

Using the Equality of Polynomials Theorem, equate coefficients of like powers. This will result in the system of equations

$$\begin{cases} 0 = A + B \\ 3 = \quad -2B + C \\ 16 = 7A \quad - 2C \end{cases}$$

• **Think of $3x + 16$ as $0x^2 + 3x + 16$.**

The solution is $A = 2$, $B = -2$, and $C = -1$. Thus the partial fraction decomposition is

$$\frac{3x + 16}{(x-2)(x^2+7)} = \frac{2}{x-2} + \frac{-2x - 1}{x^2 + 7}$$

TRY EXERCISE 24, EXERCISE SET 9.4, PAGE 692

EXAMPLE 4 Find a Partial Fraction Decomposition Case 4: Repeated Quadratic Factors

Find the partial fraction decomposition of $\dfrac{4x^3 + 5x^2 + 7x - 1}{(x^2 + x + 1)^2}$.

Continued ►

Solution

The quadratic factor $(x^2 + x + 1)$ is irreducible over the real numbers and is a repeated factor. The partial fraction decomposition will be of the form

$$\frac{4x^3 + 5x^2 + 7x - 1}{(x^2 + x + 1)^2} = \frac{Ax + B}{x^2 + x + 1} + \frac{Cx + D}{(x^2 + x + 1)^2}$$

Multiplying each side by the LCD $(x^2 + x + 1)^2$ and collecting like terms, we obtain

$$\begin{aligned} 4x^3 + 5x^2 + 7x - 1 &= (Ax + B)(x^2 + x + 1) + Cx + D \\ &= Ax^3 + Ax^2 + Ax + Bx^2 + Bx + B + Cx + D \\ &= Ax^3 + (A + B)x^2 + (A + B + C)x + (B + D) \end{aligned}$$

Equating coefficients of like powers gives the system of equations

$$\begin{cases} 4 = A \\ 5 = A + B \\ 7 = A + B + C \\ -1 = \quad B \quad + D \end{cases}$$

Solving this system, we have $A = 4$, $B = 1$, $C = 2$, and $D = -2$. Thus the partial fraction decomposition is

$$\frac{4x^3 + 5x^2 + 7x - 1}{(x^2 + x + 1)^2} = \frac{4x + 1}{x^2 + x + 1} + \frac{2x - 2}{(x^2 + x + 1)^2}$$

TRY EXERCISE 30, EXERCISE SET 9.4, PAGE 692

The Partial Fraction Decomposition Theorem requires that the degree of the numerator be less than the degree of the denominator. If this is *not* the case, use long division to first write the rational expression as a polynomial plus a remainder over the denominator.

EXAMPLE 5 Find a Partial Fraction Decomposition When the Degree of the Numerator Exceeds the Degree of the Denominator

Find the partial fraction decomposition of $F(x) = \dfrac{x^3 - 4x^2 - 19x - 35}{x^2 - 7x}$.

Solution

Because the degree of the denominator is less than the degree of the numerator, use long division first to obtain

$$F(x) = x + 3 + \frac{2x - 35}{x^2 - 7x}$$

The partial fraction decomposition of $\frac{2x - 35}{x^2 - 7x}$ will have the form

$$\frac{2x - 35}{x^2 - 7x} = \frac{2x - 35}{x(x - 7)} = \frac{A}{x} + \frac{B}{x - 7}$$

Multiplying each side by $x(x - 7)$ and combining like terms, we have

$$2x - 35 = (A + B)x + (-7A)$$

Equating coefficients of like powers yields

$$\begin{cases} 2 = A + B \\ -35 = -7A \end{cases}$$

The solution of this system is $A = 5$ and $B = -3$. The partial fraction decomposition is

$$\frac{x^3 - 4x^2 - 19x - 35}{x^2 - 7x} = x + 3 + \frac{5}{x} + \frac{-3}{x - 7}$$

TRY EXERCISE 34, EXERCISE SET 9.4, PAGE 692

TOPICS FOR DISCUSSION

1. What is the purpose of a partial fraction decomposition?
2. Discuss how the factors of the denominator of a rational expression dictate how a partial fraction decomposition is determined.
3. Discuss the Equality of Polynomials Theorem and how it is used in a partial fraction decomposition.
4. For the rational expression $\frac{3x - 1}{x^3 - 2x^2 - x + 2}$, what is the first step you perform to find a partial fraction decomposition? What equation or equations do you solve to find the partial fraction decomposition?

EXERCISE SET 9.4

In Exercises 1 to 10, determine the constants *A*, *B*, *C*, and *D*.

1. $\frac{x + 15}{x(x - 5)} = \frac{A}{x} + \frac{B}{x - 5}$
2. $\frac{5x - 6}{x(x + 3)} = \frac{A}{x} + \frac{B}{x + 3}$
3. $\frac{1}{(2x + 3)(x - 1)} = \frac{A}{2x + 3} + \frac{B}{x - 1}$
4. $\frac{6x - 5}{(x + 4)(3x + 2)} = \frac{A}{x + 4} + \frac{B}{3x + 2}$
5. $\frac{x + 9}{x(x - 3)^2} = \frac{A}{x} + \frac{B}{x - 3} + \frac{C}{(x - 3)^2}$
6. $\frac{2x - 7}{(x + 1)(x - 2)^2} = \frac{A}{x + 1} + \frac{B}{x - 2} + \frac{C}{(x - 2)^2}$
7. $\frac{4x^2 + 3}{(x - 1)(x^2 + x + 5)} = \frac{A}{x - 1} + \frac{Bx + C}{x^2 + x + 5}$
8. $\frac{x^2 + x + 3}{(x^2 + 7)(x - 3)} = \frac{Ax + B}{x^2 + 7} + \frac{C}{x - 3}$

9. $\dfrac{x^3 + 2x}{(x^2 + 1)^2} = \dfrac{Ax + B}{x^2 + 1} + \dfrac{Cx + D}{(x^2 + 1)^2}$

10. $\dfrac{3x^3 + x^2 - x - 5}{(x^2 + 2x + 5)^2} = \dfrac{Ax + B}{x^2 + 2x + 5} + \dfrac{Cx + D}{(x^2 + 2x + 5)^2}$

In Exercises 11 to 36, find the partial fraction decomposition of the given rational expression.

11. $\dfrac{8x + 12}{x(x + 4)}$

12. $\dfrac{x - 14}{x(x - 7)}$

13. $\dfrac{3x + 50}{x^2 - 7x - 18}$

14. $\dfrac{7x + 44}{x^2 + 10x + 24}$

15. $\dfrac{16x + 34}{4x^2 + 16x + 15}$

16. $\dfrac{-15x + 37}{9x^2 - 12x - 5}$

17. $\dfrac{x - 5}{(3x + 5)(x - 2)}$

18. $\dfrac{1}{(x + 7)(2x - 5)}$

19. $\dfrac{x^3 + 3x^2 - 4x - 8}{x^2 - 4}$

20. $\dfrac{x^3 - 13x - 9}{x^2 - x - 12}$

21. $\dfrac{3x^2 + 49}{x(x + 7)^2}$

22. $\dfrac{x - 18}{x(x - 3)^2}$

23. $\dfrac{5x^2 - 7x + 2}{x^3 - 3x^2 + x}$

24. $\dfrac{9x^2 - 3x + 49}{x^3 - x^2 + 10x - 10}$

25. $\dfrac{2x^3 + 9x^2 + 26x + 41}{(x + 3)^2(x^2 + 1)}$

26. $\dfrac{12x^3 - 37x^2 + 48x - 36}{(x - 2)^2(x^2 + 4)}$

27. $\dfrac{3x - 7}{(x - 4)^2}$

28. $\dfrac{5x - 53}{(x - 11)^2}$

29. $\dfrac{3x^3 - x^2 + 34x - 10}{(x^2 + 10)^2}$

30. $\dfrac{2x^3 + 9x + 1}{x^4 + 14x^2 + 49}$

31. $\dfrac{1}{k^2 - x^2}$, where k is a constant

32. $\dfrac{1}{x(k + lx)}$, where k and l are constants

33. $\dfrac{x^3 - x^2 - x - 1}{x^2 - x}$

34. $\dfrac{2x^3 + 5x^2 + 3x - 8}{2x^2 + 3x - 2}$

35. $\dfrac{2x^3 - 4x^2 + 5}{x^2 - x - 1}$

36. $\dfrac{x^4 - 2x^3 - 2x^2 - x + 3}{x^2(x - 3)}$

SUPPLEMENTAL EXERCISES

In Exercises 37 to 42, find the partial fraction decomposition of the given rational expression.

37. $\dfrac{x^2 - 1}{(x - 1)(x + 2)(x - 3)}$

38. $\dfrac{x^2 + x}{x^2(x - 4)}$

39. $\dfrac{-x^4 - 4x^2 + 3x - 6}{x^4(x - 2)}$

40. $\dfrac{3x^2 - 2x - 1}{(x^2 - 1)^2}$

41. $\dfrac{2x^2 + 3x - 1}{x^3 - 1}$

42. $\dfrac{x^3 - 2x^2 + x - 2}{x^4 - x^3 + x - 1}$

There is a short-cut for finding some partial fraction decompositions of quadratic polynomials that do not factor over the real numbers. Exercises 43 and 44 give one method and some examples.

43. Show that for real numbers a and b with $a \neq b$,

$$\frac{1}{(b - a)[p(x) + a]} + \frac{1}{(a - b)[p(x) + b]} = \frac{1}{[p(x) + a][p(x) + b]}$$

44. Use the result of Exercise 43 to find the partial fraction decomposition of

a. $\dfrac{1}{(x^2 + 4)(x^2 + 1)}$

b. $\dfrac{1}{(x^2 + 1)(x^2 + 9)}$

c. $\dfrac{1}{(x^2 + x + 1)(x^2 + x + 2)}$

d. $\dfrac{1}{(x^2 + 2x + 4)(x^2 + 2x + 9)}$

PROJECTS

1. Computer algebra systems (CAS) such as *Mathematica* and *Derive* provide computer assistance for partial fraction decompositions. The command in *Mathematica* is **Apart** and the command in *Derive* is **Expand**. Here is an example of using each of these programs to find the partial fraction decomposition of $\dfrac{x^3 - 4x^2 - 19x - 35}{x^2 - 7x}$.

Derive
Start up the *Derive* program. The menu bar on the bottom of the screen should begin with **Author**. If it does not, press the [Esc] key until it does. Now type

A((x^3–4x^2–19x–35)/(x^2–7x)) [Enter] E [Enter]

The **A** allows you to input the expression into the computer. The **E** after the first Enter is the command **Expand**, which performs the partial fraction decomposition. The result is displayed as $\frac{3}{7-x} + x + \frac{5}{x} + 3$.

Mathematica
Start up the *Mathematica* program. Now type

```
Apart[(x^3-4x^2-19x-35)/(x^2-7x)] Enter
```

The brackets, [and], are not interchangeable with parentheses, (and). The result is displayed as $3 - \frac{3}{-7+x} + \frac{5}{x} + x$.

Use a CAS program to find partial fraction decompositions for some of the exercises in this section. Include a printout of your work.

SECTION 9.5 INEQUALITIES IN TWO VARIABLES AND SYSTEMS OF INEQUALITIES

- GRAPH A NONLINEAR INEQUALITY
- SYSTEMS OF INEQUALITIES IN TWO VARIABLES
- NONLINEAR SYSTEMS OF INEQUALITIES

◆ GRAPH A NONLINEAR INEQUALITY

Two examples of inequalities in two variables are

$$2x + 3y > 6 \quad \text{and} \quad xy \le 1$$

A solution of an inequality in two variables is an ordered pair (x, y) that satisfies the inequality. For example, $(-2, 4)$ is a solution of the first inequality because $2(-2) + 3(4) > 6$. The ordered pair $(2, 1)$ is not a solution of the second inequality because $(2)(1) \not\le 1$.

The **solution set of an inequality** in two variables is the set of all ordered pairs that satisfy the inequality. The **graph** of an inequality is the graph of the solution set.

To sketch the graph of an inequality, first replace the inequality symbol by an equality sign and sketch the graph of the equation. Use a dashed graph for $<$ or $>$ to indicate that the curve is not part of the solution set. Use a solid graph for $\le$ or $\ge$ to show that the graph *is* part of the solution set.

It is important to test an ordered pair in each region of the plane defined by the graph. If the ordered pair satisfies the inequality, shade that entire region. Do this for each region into which the graph divides the plane. For example, consider the inequality $xy \ge 1$. **Figure 9.26** shows the three regions of the plane defined by this inequality. Because the inequality is $\ge$, a solid graph is used.

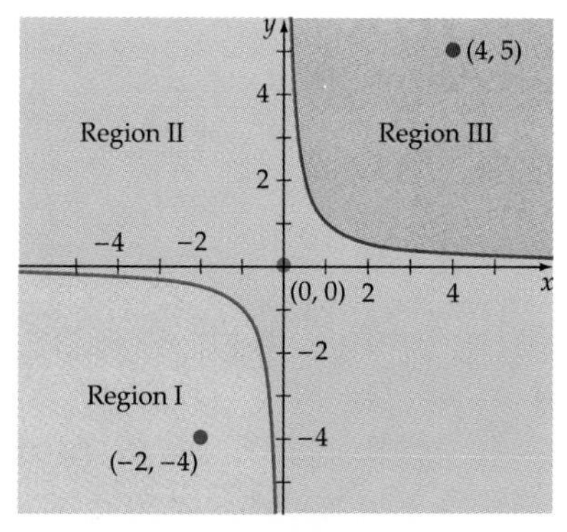

$xy \ge 1$

Figure 9.26

Choose an ordered pair in each of the three regions and determine whether that ordered pair satisfies the inequality. In Region I, choose a point, say $(-2, -4)$. Because $(-2)(-4) \ge 1$, Region I is part of the solution set. In Region II, choose a point, say $(0, 0)$. Because $0 \cdot 0 \not\ge 1$, Region II is not part of the solution set. In Region III, choose $(4, 5)$. Because $4 \cdot 5 \ge 1$, Region III is part of the solution set.

You may choose the coordinates of any point not on the graph of the equation as a test ordered pair; (0, 0) is usually a good choice.

EXAMPLE 1 Graph a Linear Inequality

Graph: $3x + 4y > 12$

Solution

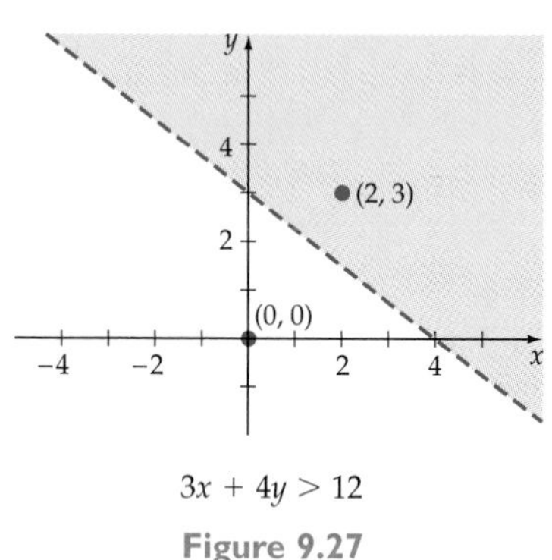

$3x + 4y > 12$

Figure 9.27

Graph the line $3x + 4y = 12$ using a dashed line.

$$\text{Test the ordered pair } (0, 0)\text{:} \quad 3(0) + 4(0) = 0 \not> 12$$

Because (0, 0) does not satisfy the inequality, do not shade this region.

$$\text{Test the ordered pair } (2, 3)\text{:} \quad 3(2) + 4(3) = 18 > 12$$

Because (2, 3) satisfies the inequality, the half-plane that includes (2, 3) is the solution set. See **Figure 9.27.**

TRY EXERCISE 6, EXERCISE SET 9.5, PAGE 698

In general, the solution set of a *linear inequality in two variables* will be one of the regions of the plane separated by a line. Each region is called a **half-plane.**

EXAMPLE 2 Graph a Nonlinear Inequality

Graph: $y \le x^2 + 2x - 3$

Solution

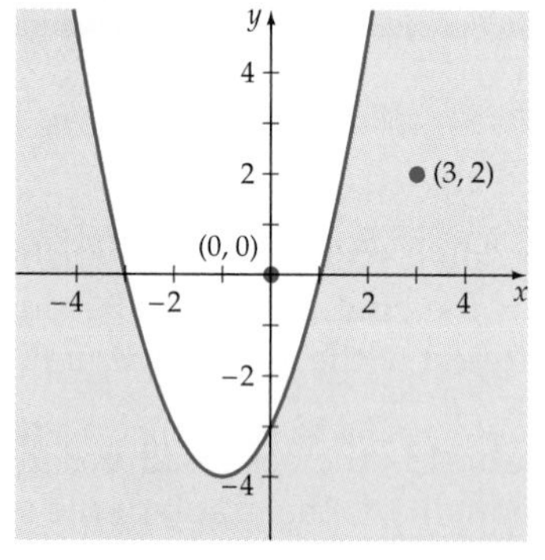

$y \le x^2 + 2x - 3$

Figure 9.28

Graph the parabola $y = x^2 + 2x - 3$ using a solid curve.

$$\text{Test the ordered pair } (0, 0)\text{:} \quad 0 \not\le 0^2 + 2(0) - 3$$

Because (0, 0) does not satisfy the inequality, do not shade this region.

$$\text{Test the ordered pair } (3, 2)\text{:} \quad 2 \le (3)^2 + 2(3) - 3$$

Because (3, 2) satisfies the inequality, shade this region of the plane. See **Figure 9.28.**

TRY EXERCISE 12, EXERCISE SET 9.5, PAGE 698

EXAMPLE 3 Graph an Absolute Value Inequality

Graph: $y \ge |x| + 1$

Solution

Graph the equation $y = |x| + 1$ using a solid graph.

$$\text{Test the ordered pair } (0, 0)\text{:} \quad 0 \not\ge |0| + 1$$

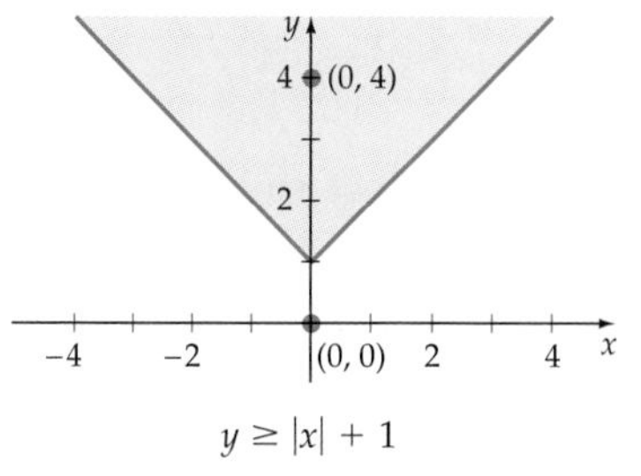

$y \geq |x| + 1$

Figure 9.29

Because $0 \ngeq 1$, (0, 0) does not belong to the solution set. Do not shade that portion of the plane that contains (0, 0).

$$\text{Test the ordered pair } (0, 4)\text{:}\quad 4 \geq |0| + 1$$

Because (0, 4) satisfies the inequality, shade this region. See **Figure 9.29.**

TRY EXERCISE 20, EXERCISE SET 9.5, PAGE 698

◆ SYSTEMS OF INEQUALITIES IN TWO VARIABLES

The **solution set of a system of inequalities** is the intersection of the solution sets of the individual inequalities. To graph the solution set of a system of inequalities, first graph the solution set of each inequality. The solution set of the system of inequalities is the region of the plane represented by the intersection of the shaded regions.

EXAMPLE 4 Graph a System of Linear Inequalities

Graph the solution set of the system of inequalities.

$$\begin{cases} 3x - 2y > 6 \\ 2x - 5y \leq 10 \end{cases}$$

Solution

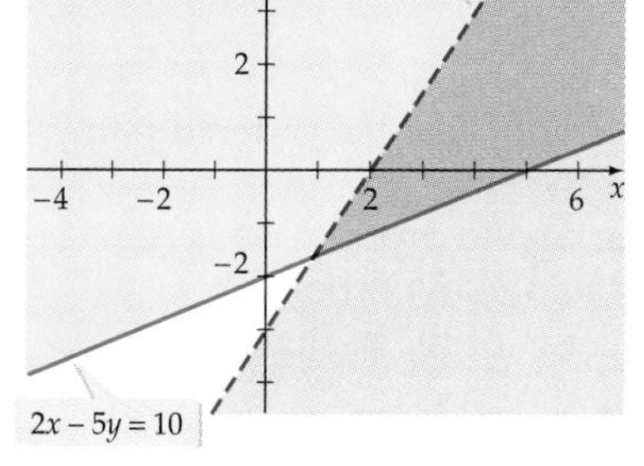

Figure 9.30

Graph the line $3x - 2y = 6$ using a dashed line. Test the ordered pair (0, 0). Because $3(0) - 2(0) \not> 6$, (0, 0) does not belong to the solution set. Do not shade the region that contains (0, 0). Instead, shade the region below and to the right of the graph of $3x - 2y = 6$, because any ordered pair from this region satisfies $3x - 2y > 6$.

Graph the line $2x - 5y = 10$ using a solid line. Test the ordered pair (0, 0). Because $2(0) - 5(0) \leq 10$, shade the region that contains (0, 0).

The solution set is the region of the plane represented by the intersection of the solution sets of the individual inequalities. See **Figure 9.30.**

TRY EXERCISE 28, EXERCISE SET 9.5, PAGE 698

◆ NONLINEAR SYSTEMS OF INEQUALITIES

EXAMPLE 5 Graph a Nonlinear System of Inequalities

Graph the solution set of the system of inequalities.

$$\begin{cases} x^2 - y^2 \leq 9 \\ 2x + 3y > 12 \end{cases}$$

Continued

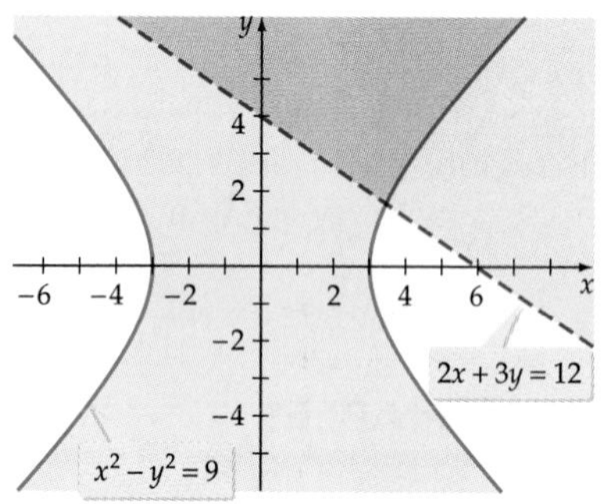

Figure 9.31

Solution

Graph the hyperbola $x^2 - y^2 = 9$ by using a solid graph. Test the ordered pair (0, 0). Because $0^2 - 0^2 \leq 9$, shade the region containing the origin. By choosing points in the other regions, you should show that those regions are not part of the solution set.

Graph the line $2x + 3y = 12$ by using a dashed graph. Test the ordered pair (0, 0). Because $2(0) + 3(0) \ngtr 12$, do not shade the half-plane below the line. Testing the ordered pair (4, 4) will show that we need to shade the half-plane above the line $2x + 3y = 12$.

The solution set is the region of the plane represented by the intersection of the solution sets of the individual inequalities. This intersection is shown by the dark color in **Figure 9.31.**

TRY EXERCISE 38, EXERCISE SET 9.5, PAGE 698

EXAMPLE 6 Identify a System of Inequalities with No Solution

Graph the solution set of the system of inequalities

$$\begin{cases} x^2 + y^2 \leq 16 \\ x^2 - y^2 \geq 36 \end{cases}$$

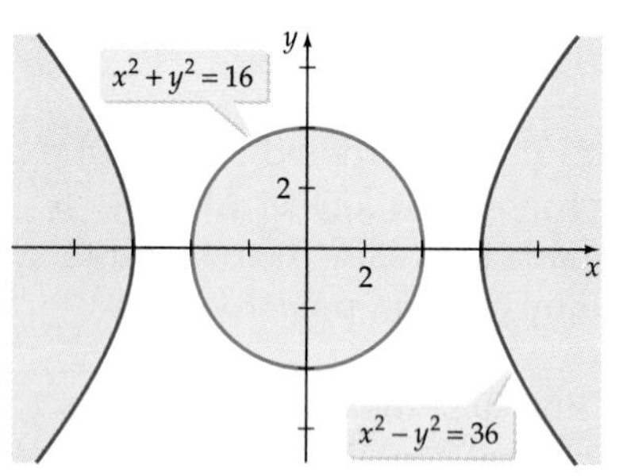

Figure 9.32

Solution

Graph the circle $x^2 + y^2 = 16$ by using a solid graph. Test the ordered pair (0, 0). Because $0^2 + 0^2 \leq 16$, shade the inside of the circle.

Graph the hyperbola $x^2 - y^2 = 36$ by using a solid graph. Use ordered pairs from each of the regions defined by the hyperbola to determine that the solution of $x^2 - y^2 > 36$ consists of the region to the right of the right branch of the hyperbola and the region to the left of the left branch.

Because the solution sets of the inequalities do not intersect, the system has no solution. The solution set is the empty set. See **Figure 9.32.**

TRY EXERCISE 42, EXERCISE SET 9.5, PAGE 698

EXAMPLE 7 Graph a System of Four Inequalities

Graph the solution set of the system of inequalities.

$$\begin{cases} 2x - 3y \leq 2 \\ 3x + 4y \geq 12 \\ x \geq -1, y \geq 2 \end{cases}$$

Solution

First graph the inequalities $x \geq -1$ and $y \geq 2$. Because $x \geq -1$ and $y \geq 2$, the solution set for this system will be on or above the line $y = 2$ and on or to the right of the line $x = -1$. See **Figure 9.33.**

Graph the solution set of $2x - 3y = 2$ by using a solid graph. Because $2(0) - 3(0) \leq 2$, shade the region above the line.

Graph the solution set of $3x + 4y = 12$ by using a solid graph. Test an ordered pair, say (3, 3), to determine that we need to shade above the line $3x + 4y = 12$.

The solution set of the system of inequalities is the region where the graphs of the solution sets of all four inequalities intersect. This intersection is indicated by the dark color in **Figure 9.34.**

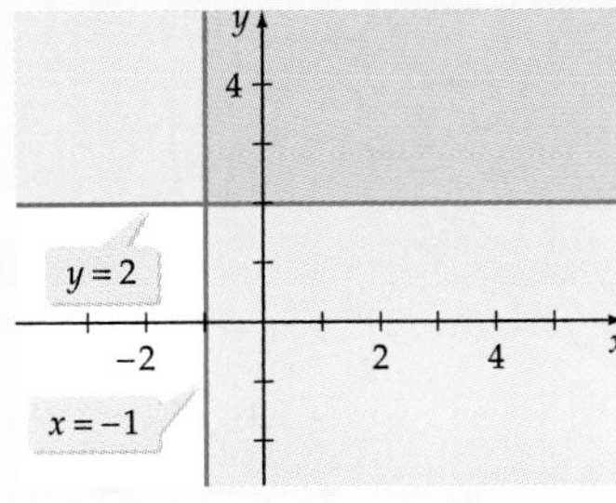

Figure 9.33

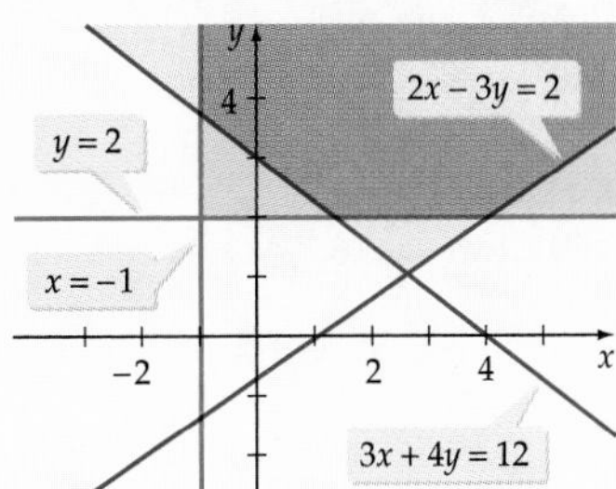

Figure 9.34

TRY EXERCISE 44, EXERCISE SET 9.5, PAGE 698

TOPICS FOR DISCUSSION

1. Does the graph of a linear inequality in two variables represent the graph of a function? Why or why not?
2. What is a half-plane?
3. Is it possible for a system of inequalities to have no solution? If so, give an example. If not, explain why not.
4. Let $A = \{(x, y) | x + y > 5\}$ and let $B = \{(x, y) | x - y < 3\}$. What is the significance of $A \cap B$?
5. Suppose a company makes two types of frying pans: regular and nonstick. Each week the company plans on making at least twice as many nonstick pans as regular. Production facilities are such that the company can make a maximum of 1500 pans per week. Letting x represent the number of nonstick pans and y represent the number of regular pans, the system of inequalities

$$\begin{cases} x \geq 2y \\ x + y \leq 1500 \\ x \geq 0, y \geq 0 \end{cases}$$

represents this situation. The graph of the solution set is shown in the accompanying figure. Explain the meaning of the solution set (shown shaded) in the context of this problem.

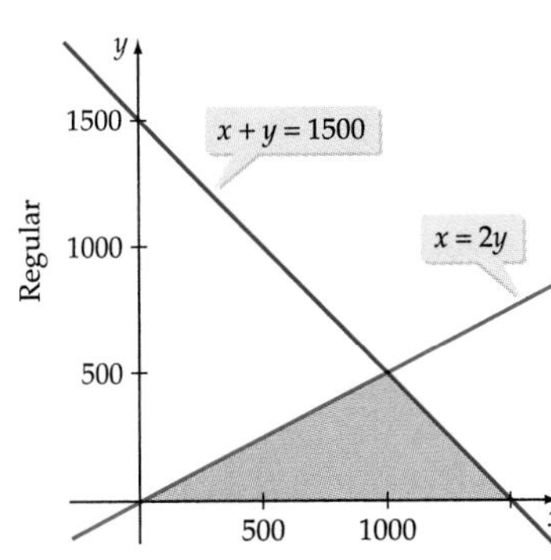

EXERCISE SET 9.5

In Exercises 1 to 22, sketch the graph of each inequality.

1. $y \le -2$
2. $x + y > -2$
3. $y \ge 2x + 3$
4. $y < -2x + 1$
5. $2x - 3y < 6$
6. $3x + 4y \le 4$
7. $4x + 3y \le 12$
8. $5x - 2y < 8$
9. $y < x^2$
10. $x > y^2$
11. $y \ge x^2 - 2x - 3$
12. $y < 2x^2 - x - 3$
13. $(x - 2)^2 + (y - 1)^2 < 16$
14. $(x + 2)^2 + (y - 3)^2 > 25$
15. $\dfrac{(x-3)^2}{9} - \dfrac{(y+1)^2}{16} > 1$
16. $\dfrac{(x+1)^2}{25} - \dfrac{(y-3)^2}{16} \le 1$
17. $4x^2 + 9y^2 - 8x + 18y \ge 23$
18. $25x^2 - 16y^2 - 100x - 64y < 64$
19. $y \ge |2x - 4|$
20. $y < |x|$
21. $y < 2^{x-1}$
22. $y > \log_3 x$
23. $y \le \log_2 (x - 1)$
24. $y > 3^x + 1$

In Exercises 25 to 48, sketch the graph of the solution set of each system of inequalities.

25. $\begin{cases} 1 \le x < 3 \\ -2 < y \le 4 \end{cases}$
26. $\begin{cases} -2 < x < 4 \\ y \ge -1 \end{cases}$
27. $\begin{cases} 3x + 2y \ge 1 \\ x + 2y < -1 \end{cases}$
28. $\begin{cases} 2x - 5y < -6 \\ 3x + y < 8 \end{cases}$
29. $\begin{cases} 2x - y \ge -4 \\ 4x - 2y \le -17 \end{cases}$
30. $\begin{cases} 4x + 2y > 5 \\ 6x + 3y > 10 \end{cases}$
31. $\begin{cases} 4x - 3y < 14 \\ 2x + 5y \le -6 \end{cases}$
32. $\begin{cases} 3x + 5y \ge -8 \\ 2x - 3y \ge 1 \end{cases}$
33. $\begin{cases} y < 2x + 3 \\ y > 2x - 2 \end{cases}$
34. $\begin{cases} y > 3x + 1 \\ y < 3x - 2 \end{cases}$
35. $\begin{cases} y < 2x - 1 \\ y \ge x^2 + 3x - 7 \end{cases}$
36. $\begin{cases} y \le 2x + 7 \\ y > x^2 + 3x + 1 \end{cases}$
37. $\begin{cases} x^2 + y^2 \le 49 \\ 9x^2 + 4y^2 \ge 36 \end{cases}$
38. $\begin{cases} y < 2x - 1 \\ y > x^2 - 2x + 2 \end{cases}$
39. $\begin{cases} (x - 1)^2 + (y + 1)^2 \le 16 \\ (x - 1)^2 + (y + 1)^2 \ge 4 \end{cases}$
40. $\begin{cases} (x + 2)^2 + (y - 3)^2 > 25 \\ (x + 2)^2 + (y - 3)^2 < 16 \end{cases}$
41. $\begin{cases} \dfrac{(x-4)^2}{16} - \dfrac{(y+2)^2}{9} > 1 \\ \dfrac{(x-4)^2}{25} + \dfrac{(y+2)^2}{9} < 1 \end{cases}$
42. $\begin{cases} \dfrac{(x+1)^2}{36} + \dfrac{(y-2)^2}{25} < 1 \\ \dfrac{(x+1)^2}{25} + \dfrac{(y-2)^2}{36} < 1 \end{cases}$
43. $\begin{cases} 2x - 3y \ge -5 \\ x + 2y \le 7 \\ x \ge -1, y \ge 0 \end{cases}$
44. $\begin{cases} 5x + y \le 9 \\ 2x + 3y \le 14 \\ x \ge -2, y \ge 2 \end{cases}$
45. $\begin{cases} 3x + 2y \ge 14 \\ x + 3y \ge 14 \\ x \le 10, y \le 8 \end{cases}$
46. $\begin{cases} 4x + y \ge 13 \\ 3x + 2y \ge 16 \\ x \le 15, y \le 12 \end{cases}$
47. $\begin{cases} 3x + 4y \le 12 \\ 2x + 5y \le 10 \\ x \ge 0, y \ge 0 \end{cases}$
48. $\begin{cases} 5x + 3y \le 15 \\ x + 4y \le 8 \\ x \ge 0, y \ge 0 \end{cases}$

SUPPLEMENTAL EXERCISES

In Exercises 49 to 58, sketch the graph of the inequality.

49. $|y| \ge |x|$
50. $|y| \le |x - 1|$
51. $|x + y| \le 1$
52. $|x - y| > 1$
53. $|x| + |y| \le 1$
54. $|x| - |y| > 1$
55. $y > [\![x]\!]$, where $[\![x]\!]$ is the greatest integer function
56. $y > x - [\![x]\!]$, where $[\![x]\!]$ is the greatest integer function
57. Sketch the graphs of $xy > 1$ and $y > 1/x$. Note that the two graphs are not the same, yet the second inequality can be derived from the first by dividing each side by x. Explain.
58. Sketch the graph of $x/y < 1$ and the graph of $x < y$. Note that the two graphs are not the same, yet the second inequality can be derived from the first by multiplying each side by y. Explain.

PROJECT

1. **A PARALLELOGRAM COORDINATE SYSTEM** The xy-coordinate system described in this chapter consisted of two coordinate lines that intersected at right angles. It is not necessary that coordinate lines intersect at right angles for a coordinate system to exist. Draw two coordinate lines that intersect at 0 for each line but for which the angle between the two axes is 45°. You now have a *parallelogram* coordinate system rather than a *rectangular* coordinate system. Explain the last sentence. Now experiment in this system. For example, is the graph of $3x + 4y = 12$ a straight line in the *parallelogram* coordinate system? In a parallelogram coordinate system, is the graph of $y = x^2$ a parabola?

SECTION 9.6 LINEAR PROGRAMMING

- INTRODUCTION TO LINEAR PROGRAMMING (9B)
- SOLVING OPTIMIZATION PROBLEMS

◆ INTRODUCTION TO LINEAR PROGRAMMING

Consider a business analyst who is trying to maximize the profit from the production of a product or an engineer who is trying to minimize the amount of energy an electrical circuit needs to operate. Generally, problems that seek to maximize or minimize a situation are called **optimization problems.** One strategy for solving certain of these problems was developed in the 1940s and is called **linear programming.**

A linear programming problem involves a **linear objective function,** which is the function that must be maximized or minimized. This objective function is subject to some **constraints,** which are inequalities or equations that restrict the values of the variables. To illustrate these concepts, suppose a manufacturer produces two types of computer monitors: monochrome and color. Past sales experience shows that at least twice as many monochrome monitors are sold as color monitors. Suppose further that the manufacturing plant is capable of producing 12 monitors per day. Let x represent the number of monochrome monitors produced, and let y represent the number of color monitors produced. Then

$$\begin{cases} x \geq 2y \\ x + y \leq 12 \end{cases}$$ • **These are the constraints.**

These two inequalities place a constraint, or restriction, on the manufacturer. For example, the manufacturer cannot produce 5 color monitors, because that would require producing at least 10 monochrome monitors, and $5 + 10 \nleq 12$.

Suppose a profit of \$50 is earned on each monochrome monitor sold and \$75 is earned on each color monitor sold. Then the manufacturer's profit, P in dollars, is given by the equation

$$P = 50x + 75y$$ • **Objective function**

The equation $P = 50x + 75y$ defines the objective function. The goal of this linear programming problem is to determine how many of each monitor should be produced to maximize the manufacturer's profit and at the same time satisfy the constraints.

Because the manufacturer cannot produce fewer than zero units of either monitor, there are two other implied constraints, $x \geq 0$ and $y \geq 0$. Our linear programming problem now looks like

$$\text{Objective function:} \quad P = 50x + 75y$$

$$\text{Constraints:} \quad \begin{cases} x - 2y \geq 0 \\ x + y \leq 12 \\ x \geq 0, y \geq 0 \end{cases}$$

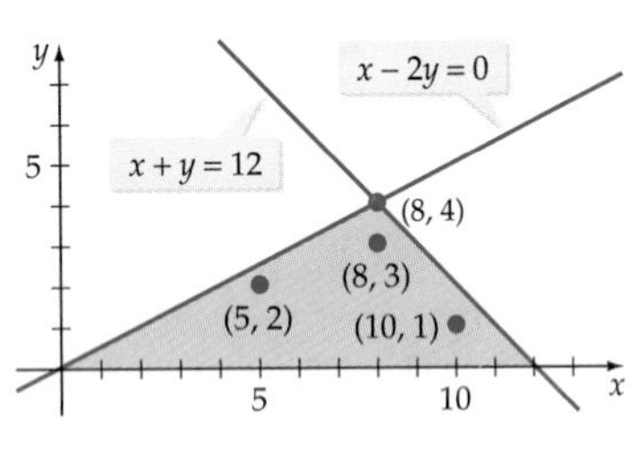

Figure 9.35

To solve this problem, graph the solution set of the constraints. The solution set of the constraints is called the **set of feasible solutions.** Ordered pairs in this set are used to evaluate the objective function to determine which ordered pair maximizes the profit. For example, (5, 2), (8, 3), and (10, 1) are three ordered pairs in the set. See **Figure 9.35.** For these ordered pairs, the profit would be

$$P = 50(5) + 75(2) = 400 \qquad \bullet\ x = 5, y = 2$$
$$P = 50(8) + 75(3) = 625 \qquad \bullet\ x = 8, y = 3$$
$$P = 50(10) + 75(1) = 575 \qquad \bullet\ x = 10, y = 1$$

take note

The set of feasible solutions includes ordered pairs with whole number coordinates as well as fractional coordinates. For instance, the ordered pair $(5, 2\frac{1}{2})$ is in the set of feasible solutions. During one day, the company could produce 5 monochrome monitors and $2\frac{1}{2}$ color monitors.

It would be impossible to check every ordered pair in the set of feasible solutions to find which maximizes profit. Fortunately, we can find that ordered pair by solving the objective function $P = 50x + 75y$ for y.

$$y = -\frac{2}{3}x + \frac{P}{75}$$

In this form, the objective function is a linear equation whose graph has slope $-2/3$ and y-intercept $P/75$. If P is as large as possible (P a maximum), then the y-intercept will be as large as possible. Thus the maximum profit will occur on the line that has a slope of $-2/3$ and has the largest possible y-intercept and intersects the set of feasible solutions.

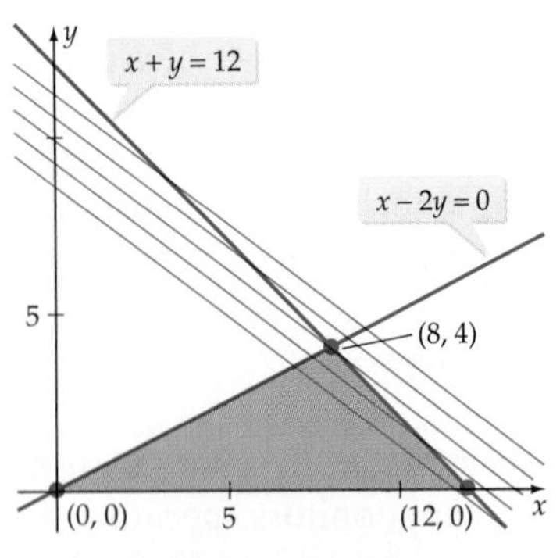

Figure 9.36

From **Figure 9.36,** the largest possible y-intercept occurs when the line passes through the point with coordinates (8, 4). At this point, the profit is

$$P = 50(8) + 75(4) = 700$$

The manufacturer will maximize profit by producing 8 monochrome monitors and 4 color monitors each day. The profit will be \$700 per day.

In general, the goal of any linear programming problem is to maximize or minimize the objective function, subject to the constraints. Minimization problems occur, for example, when a manufacturer wants to minimize the cost of operations.

Suppose that a cost minimization problem results in the following objective function and constraints.

$$\text{Objective function:} \quad C = 3x + 4y$$

$$\text{Constraints:} \quad \begin{cases} x + y \geq 1 \\ 2x - y \leq 5 \\ x + 2y \leq 10 \\ x \geq 0, y \geq 0 \end{cases}$$

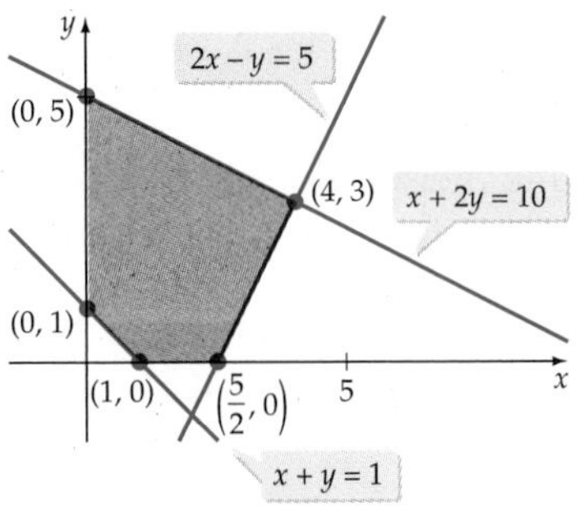

Figure 9.37

Figure 9.37 is the graph of the solution set of the constraints. The task is to find the ordered pair that satisfies all the constraints and that will give the smallest value of C. We again could solve the objective function for y and, because we want to minimize C, find the smallest y-intercept. However, a theorem from linear programming simplifies our task even more. The proof of this theorem, omitted here, is based on the techniques we used to solve our examples.

Fundamental Linear Programming Theorem

If an objective function has an optimal solution, then that solution will be at a vertex of the set of feasible solutions.

Following is a list of the values of C at the vertices. The minimum value of the objective function occurs at the point whose coordinates are (1, 0).

(x, y)	$C = 3x + 4y$	
$(1, 0)$	$C = 3(1) + 4(0) = 3$	• Minimum
$\left(\frac{5}{2}, 0\right)$	$C = 3\left(\frac{5}{2}\right) + 4(0) = 7.5$	
$(4, 3)$	$C = 3(4) + 4(3) = 24$	• Maximum
$(0, 5)$	$C = 3(0) + 4(5) = 20$	
$(0, 1)$	$C = 3(0) + 4(1) = 4$	

The maximum value of the objective function can also be determined from the list. It occurs at (4, 3).

It is important to realize that the maximum or minimum value of an objective function depends on the objective function and on the set of feasible solutions. For example, using the same set of feasible solutions as in **Figure 9.37** but changing the objective function to $C = 2x + 5y$ changes the maximum value of C to 25 at the ordered pair (0, 5). You should verify this result by making a list similar to the one shown above.

◆ Solving Optimization Problems

EXAMPLE 1 Solve a Minimization Problem

Minimize the objective function $C = 4x + 7y$ with the constraints

$$\begin{cases} 3x + y \geq 6 \\ x + y \geq 4 \\ x + 3y \geq 6 \\ x \geq 0, y \geq 0 \end{cases}$$

Continued •➤

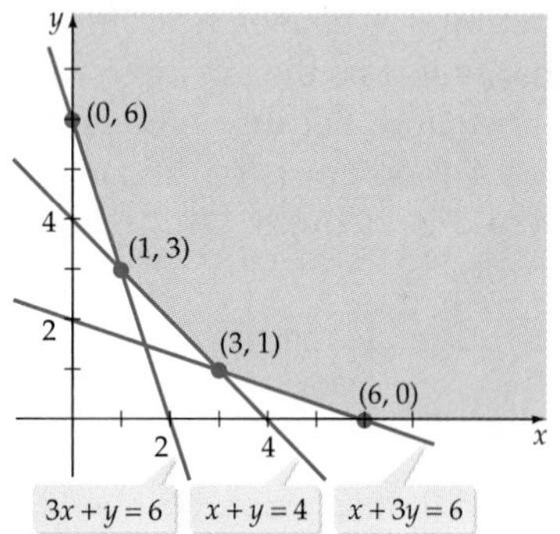

Figure 9.38

Solution

Determine the set of feasible solutions by graphing the solution set of the inequalities. See **Figure 9.38.** Note that in this instance the set of feasible solutions is an unbounded set.

Find the vertices of the region by solving the following systems of equations. These systems are formed by the equations of the lines that intersect to form a vertex of the set of feasible solutions.

$$\begin{cases} 3x + y = 6 \\ x + y = 4 \end{cases} \qquad \begin{cases} x + 3y = 6 \\ x + y = 4 \end{cases}$$

The solutions of the two systems are (1, 3) and (3, 1), respectively. The points (0, 6) and (6, 0) are the vertices on the y- and x-axes.

Evaluate the objective function at each of the four vertices of the set of feasible solutions.

(x, y)	$C = 4x + 7y$	
(0, 6)	$C = 4(0) + 7(6) = 42$	
(1, 3)	$C = 4(1) + 7(3) = 25$	
(3, 1)	$C = 4(3) + 7(1) = 19$	• **Minimum**
(6, 0)	$C = 4(6) + 7(0) = 24$	

The minimum value of the objective function is 19 at (3, 1).

TRY EXERCISE 12, EXERCISE SET 9.6, PAGE 705

Linear programming can be used to determine the best allocation of the resources available to a company. In fact, the word *programming* refers to a "program to allocate resources."

EXAMPLE 2 Solve an Applied Minimization Problem

A manufacturer of animal food makes two grain mixtures, G_1 and G_2. Each kilogram of G_1 contains 300 grams of vitamins, 400 grams of protein, and 100 grams of carbohydrate. Each kilogram of G_2 contains 100 grams of vitamins, 300 grams of protein, and 200 grams of carbohydrate. Minimum nutritional guidelines require that a feed mixture made from these grains contain at least 900 grams of vitamins, 2200 grams of protein, and 800 grams of carbohydrate. G_1 costs \$2.00 per kilogram to produce, and G_2 costs \$1.25 per kilogram to produce. Find the number of kilograms of each grain mixture that should be produced to minimize cost.

Solution

Let

$$x = \text{the number of kilograms of } G_1$$
$$y = \text{the number of kilograms of } G_2$$

The objective function is the cost function $C = 2x + 1.25y$.

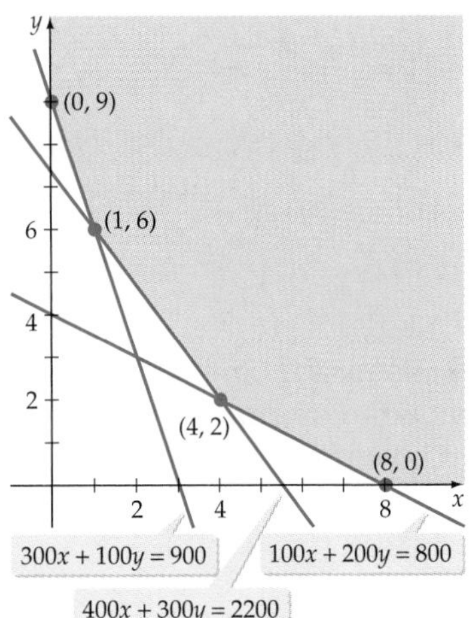

Figure 9.39

Because x kilograms of G_1 contain $300x$ grams of vitamins and y kilograms of G_2 contain $100y$ grams of vitamins, the total amount of vitamins contained in x kilograms of G_1 and y kilograms of G_2 is $300x + 100y$. At least 900 grams of vitamins are necessary, so $300x + 100y \geq 900$. Following similar reasoning, we have the constraints

$$\begin{cases} 300x + 100y \geq 900 \\ 400x + 300y \geq 2200 \\ 100x + 200y \geq 800 \\ x \geq 0, y \geq 0 \end{cases}$$

Two of the vertices of the set of feasible solutions (see **Figure 9.39**) can be found by solving two systems of equations. These systems are formed by the equations of the lines that intersect to form a vertex of the set of feasible solutions.

$$\begin{cases} 300x + 100y = 900 \\ 400x + 300y = 2200 \end{cases}$$ • **The vertex is (1, 6).**

$$\begin{cases} 100x + 200y = 800 \\ 400x + 300y = 2200 \end{cases}$$ • **The vertex is (4, 2).**

The vertices on the x- and y-axes are the x- and y-intercepts (8, 0) and (0, 9).

Substitute the coordinates of the vertices into the objective function.

(x, y)	$C = 2x + 1.25y$	
$(0, 9)$	$C = 2(0) + 1.25(9) = 11.25$	
$(1, 6)$	$C = 2(1) + 1.25(6) = 9.50$	• **Minimum**
$(4, 2)$	$C = 2(4) + 1.25(2) = 10.50$	
$(8, 0)$	$C = 2(8) + 1.25(0) = 16.00$	

The minimum value of the objective function is \$9.50. It occurs when the company produces a feed mixture that contains 1 kilogram of G_1 and 6 kilograms of G_2.

TRY EXERCISE 22, EXERCISE SET 9.6, PAGE 706

EXAMPLE 3 Solve an Applied Maximization Problem

A chemical firm produces two types of industrial solvents, S_1 and S_2. Each solvent is a mixture of three chemicals. Each kiloliter of S_1 requires 12 liters of chemical 1, 9 liters of chemical 2, and 30 liters of chemical 3. Each kiloliter of S_2 requires 24 liters of chemical 1, 5 liters of chemical 2, and 30 liters of chemical 3. The profit per kiloliter of S_1 is \$100, and the profit per kiloliter of S_2 is \$85. The inventory of the company shows 480 liters of chemical 1, 180 liters of chemical 2, and 720 liters of chemical 3. Assuming the company can sell all the solvent it makes, find the number of kiloliters of each solvent that the company should make to maximize profit.

Continued

Solution

Let

$$x = \text{the number of kiloliters of } S_1$$
$$y = \text{the number of kiloliters of } S_2$$

The objective function is the profit function $P = 100x + 85y$.

Because x kiloliters of S_1 require $12x$ liters of chemical 1, and y kiloliters of S_2 require $24y$ liters of chemical 1, the total amount of chemical 1 needed is $12x + 24y$. There are 480 liters of chemical 1 in inventory, so $12x + 24y \leq 480$. Following similar reasoning, we have the constraints

$$\begin{cases} 12x + 24y \leq 480 \\ 9x + 5y \leq 180 \\ 30x + 30y \leq 720 \\ x \geq 0, y \geq 0 \end{cases}$$

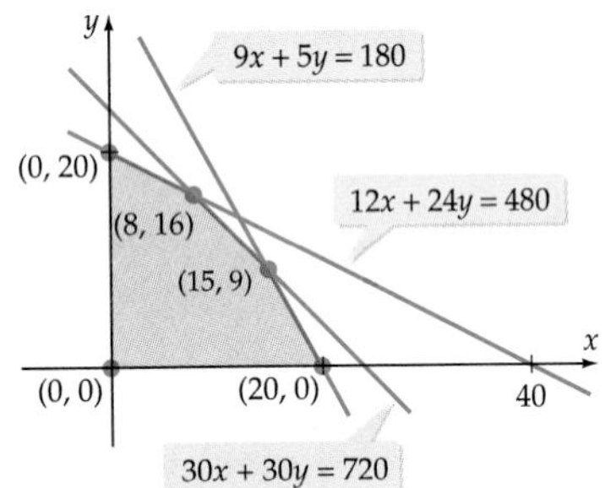

Figure 9.40

Two of the vertices of the set of feasible solutions (see **Figure 9.40**) can be found by solving two systems of equations. These systems are formed by the equations of the lines that intersect to form a vertex of the set of feasible solutions.

$$\begin{cases} 12x + 24y = 480 \\ 30x + 30y = 720 \end{cases}$$ • **The vertex is (8, 16).**

$$\begin{cases} 9x + 5y = 180 \\ 30x + 30y = 720 \end{cases}$$ • **The vertex is (15, 9).**

The vertices on the x- and y-axes are the x- and y-intercepts (20, 0) and (0, 20).

Substitute the coordinates of the vertices into the objective function.

(x, y)	$P = 100x + 85y$	
(0, 20)	$P = 100(0) + 85(20) = 1700$	
(8, 16)	$P = 100(8) + 85(16) = 2160$	
(15, 9)	$P = 100(15) + 85(9) = 2265$	• **Maximum**
(20, 0)	$P = 100(20) + 85(0) = 2000$	

The maximum value of the objective function is \$2265 when the company produces 15 kiloliters of S_1 and 9 kiloliters of S_2.

TRY EXERCISE 24, EXERCISE SET 9.6, PAGE 706

TOPICS FOR DISCUSSION

1. What is an optimization problem? Give an example of a situation in which optimization may be the goal.
2. What is a constraint for a linear programming problem? Explain what type of condition might be a constraint for the situation you gave in Exercise 1.

3. What is the objective function for a linear programming problem? Explain what the objective function might be for the situation you gave in Exercise 1.

4. What is the set of feasible solutions for a linear programming problem?

5. If a linear programming problem has an optimal solution, where in the set of feasible solutions must that optimal solution occur?

EXERCISE SET 9.6

In Exercises 1 to 20, solve the linear programming problem. Assume $x \geq 0$ and $y \geq 0$.

1. Minimize $C = 4x + 2y$ with the constraints

$$\begin{cases} x + y \geq 7 \\ 4x + 3y \geq 24 \\ x \leq 10, y \leq 10 \end{cases}$$

2. Minimize $C = 5x + 4y$ with the constraints

$$\begin{cases} 3x + 4y \geq 32 \\ x + 4y \geq 24 \\ x \leq 12, y \leq 15 \end{cases}$$

3. Maximize $C = 6x + 7y$ with the constraints

$$\begin{cases} x + 2y \leq 16 \\ 5x + 3y \leq 45 \end{cases}$$

4. Maximize $C = 6x + 5y$ with the constraints

$$\begin{cases} 2x + 3y \leq 27 \\ 7x + 3y \leq 42 \end{cases}$$

5. Minimize $C = 5x + 6y$ with the constraints

$$\begin{cases} 4x - 3y \leq 2 \\ 2x + 3y \geq 10 \end{cases}$$

6. Maximize $C = 4x + 5y$ with the constraints

$$\begin{cases} 2x - y \leq 0 \\ 0 \leq y \leq 10 \\ 0 \leq x \leq 10 \end{cases}$$

7. Maximize $C = x + 6y$ with the constraints

$$\begin{cases} 5x + 8y \leq 120 \\ 7x + 16y \leq 192 \end{cases}$$

8. Minimize $C = 4x + 5y$ with the constraints

$$\begin{cases} x + 3y \geq 30 \\ 3x + 4y \geq 60 \end{cases}$$

9. Minimize $C = 4x + y$ with the constraints

$$\begin{cases} 3x + 5y \geq 120 \\ x + y \geq 32 \end{cases}$$

10. Maximize $C = 7x + 2y$ with the constraints

$$\begin{cases} x + 3y \leq 108 \\ 7x + 4y \leq 280 \end{cases}$$

11. Maximize $C = 2x + 7y$ with the constraints

$$\begin{cases} x + y \leq 10 \\ x + 2y \leq 16 \\ 2x + y \leq 16 \end{cases}$$

12. Minimize $C = 4x + 3y$ with the constraints

$$\begin{cases} 2x + y \geq 8 \\ 2x + 3y \geq 16 \\ x + 3y \geq 11 \\ x \leq 20, y \leq 20 \end{cases}$$

13. Minimize $C = 3x + 2y$ with the constraints

$$\begin{cases} 3x + y \geq 12 \\ 2x + 7y \geq 21 \\ x + y \geq 8 \end{cases}$$

14. Maximize $C = 2x + 6y$ with the constraints

$$\begin{cases} x + y \leq 12 \\ 3x + 4y \leq 40 \\ x + 2y \leq 18 \end{cases}$$

15. Maximize $C = 3x + 4y$ with the constraints

$$\begin{cases} 2x + y \leq 10 \\ 2x + 3y \leq 18 \\ x - y \leq 2 \end{cases}$$

16. Minimize $C = 3x + 7y$ with the constraints

$$\begin{cases} x + y \geq 9 \\ 3x + 4y \geq 32 \\ x + 2y \geq 12 \end{cases}$$

17. Minimize $C = 3x + 2y$ with the constraints

$$\begin{cases} x + 2y \geq 8 \\ 3x + y \geq 9 \\ x + 4y \geq 12 \end{cases}$$

18. Maximize $C = 4x + 5y$ with the constraints

$$\begin{cases} 3x + 4y \leq 250 \\ x + y \leq 75 \\ 2x + 3y \leq 180 \end{cases}$$

19. Maximize $C = 6x + 7y$ with the constraints

$$\begin{cases} x + 2y \leq 900 \\ x + y \leq 500 \\ 3x + 2y \leq 1200 \end{cases}$$

20. Minimize $C = 11x + 16y$ with the constraints

$$\begin{cases} x + 2y \geq 45 \\ x + y \geq 40 \\ 2x + y \geq 45 \end{cases}$$

21. **MAXIMIZE PROFIT** A farmer is planning to raise wheat and barley. Each acre of wheat yields a profit of \$50, and each acre of barley yields a profit of \$70. To sow the crop, two machines, a tractor and a tiller, are rented. The tractor is available for 200 hours, and the tiller is available for 100 hours. Sowing an acre of barley requires 3 hours of tractor time and 2 hours of tilling. Sowing an acre of wheat requires 4 hours of tractor time and 1 hour of tilling. How many acres of each crop should be planted to maximize the farmer's profit?

22. **MINIMIZE COST** An ice cream supplier has two machines that produce vanilla and chocolate ice cream. To meet one of its contractual obligations, the company must produce at least 60 gallons of vanilla ice cream and 100 gallons of chocolate ice cream per day. One machine makes 4 gallons of vanilla and 5 gallons of chocolate ice cream per hour. The second machine makes 3 gallons of vanilla and 10 gallons of chocolate ice cream per hour. It costs \$28 per hour to run machine 1 and \$25 per hour to run machine 2. How many hours should each machine be operated to fulfill the contract at the least expense?

23. **MAXIMIZE PROFIT** A manufacturer makes two types of golf clubs: a starter model and a professional model. The starter model requires 4 hours in the assembly room and 1 hour in the finishing room. The professional model requires 6 hours in the assembly room and 1 hour in the finishing room. The total number of hours available in the assembly room is 108. There are 24 hours available in the finishing room. The profit for each starter model is \$35, and the profit for each professional model is \$55. Assuming all the sets produced can be sold, find how many of each set should be manufactured to maximize profit.

24. **MAXIMIZE PROFIT** A company makes two types of telephone answering machines: the standard model and the deluxe model. Each machine passes through three processes: P_1, P_2, and P_3. One standard answering machine requires 1 hour in P_1, 1 hour in P_2, and 2 hours in P_3. One deluxe answering machine requires 3 hours in P_1, 1 hour in P_2, and 1 hour in P_3. Because of employee work schedules, P_1 is available for 24 hours, P_2 is available for 10 hours, and P_3 is available for 16 hours. If the profit is \$25 for each standard model and \$35 for each deluxe model, how many units of each type should the company produce to maximize profit?

SUPPLEMENTAL EXERCISES

25. **MINIMIZE COST** A dietitian formulates a special diet from two food groups: *A* and *B*. Each ounce of food group *A* contains 3 units of vitamin A, 1 unit of vitamin C, and 1 unit of vitamin D. Each ounce of food group *B* contains 1 unit of vitamin A, 1 unit of vitamin C, and 3 units of vitamin D. Each ounce of food group *A* costs 40 cents, and each ounce of food group *B* costs 10 cents. The dietary constraints are such that at least 24 units of vitamin A, 16 units of vitamin C, and 30 units of vitamin D are required. Find the amount of each food group that should be used to minimize the cost. What is the minimum cost?

26. **MAXIMIZE PROFIT** Among the many products it produces, an oil refinery makes two specialized petroleum distillates: Pymex *A* and Pymex *B*. Each distillate passes through three stages: S_1, S_2, and S_3. Each liter of Pymex *A* requires 1 hour in S_1, 3 hours in S_2, and 3 hours in S_3. Each liter of Pymex *B* requires 1 hour in S_1, 4 hours in S_2, and 2 hours in S_3. There are 10 hours available for S_1, 36 hours available for S_2, and 27 hours available for S_3. The profit per liter of Pymex *A* is \$12, and the profit per liter of Pymex *B* is \$9. How many liters of each distillate should be produced to maximize profit? What is the maximum profit?

27. **MAXIMIZE PROFIT** An engine reconditioning company works on 4- and 6-cylinder engines. Each 4-cylinder engine requires 1 hour for cleaning, 5 hours for overhauling, and 3 hours for testing. Each 6-cylinder engine requires 1 hour for cleaning, 10 hours for overhauling, and 2 hours for testing. The cleaning station is available for at most 9 hours. The overhauling equipment is available for at most 80 hours, and the testing equipment is available for at most 24 hours. For each reconditioned 4-cylinder engine, the company makes a profit of \$150. A reconditioned 6-cylinder engine yields a profit of \$250. The company can sell all the reconditioned engines it produces. How many of each type should be produced to maximize profit? What is the maximum profit?

28. **MINIMIZE COST** A producer of animal feed makes two food products: F_1 and F_2. The products contain three major ingredients: M_1, M_2, and M_3. Each ton of F_1 requires 200 pounds of M_1, 100 pounds of M_2, and 100 pounds of M_3. Each ton of F_2 requires 100 pounds of M_1, 200 pounds of M_2, and 400 pounds of M_3. There are at least 5000 pounds of M_1 available, at least 7000 pounds of M_2 available, and at least 10,000 pounds of M_3 available. Each ton of F_1 costs \$450 to make, and each ton of F_2 costs \$300 to make. How many tons of each food product should the feed producer make to minimize cost? What is the minimum cost?

PROJECT

1. **HISTORY OF LINEAR PROGRAMMING** Linear programming has been used successfully to solve a wide range of problems in fields as diverse as providing health care and hardening nuclear silos. Write an essay on linear programming and some of the applications of this procedure in solving practical problems. Include in your essay the contributions of George Danzig, Narendra Karmarkar, and L.G. Khachian.

EXPLORING CONCEPTS WITH TECHNOLOGY

Ill-Conditioned Systems of Equations

Solving systems of equations algebraically as we did in this chapter is not practical for systems of equations that contain a large number of variables. In those cases, a computer solution is the only hope. Computer solutions are not without some problems, however.

Consider the system of equations

$$\begin{cases} 0.24567x + 0.49133y = 0.73700 \\ 0.84312x + 1.68623y = 2.52935 \end{cases}$$

It is easy to verify that the solution of this system of equations is (1, 1). However, change the constant 0.73700 to 0.73701 (add 0.00001) and the constant 2.52935 to 2.52936 (add 0.00001), and the solution is now (3, 0). Thus a very small change in the constant terms produced a dramatic change in the solution. A system of equations of this sort is said to be *ill-conditioned*.

These types of systems are important because computers generally cannot store numbers beyond a certain number of significant digits. Your calculator, for example, probably allows you to enter no more than 10 significant digits. If an exact number cannot be entered, then an approximation to that number is necessary. When a computer is solving an equation or system of equations, the hope is that approximations of the coefficients it uses will give reasonable approximations to the solutions. For ill-conditioned systems of equations, this is not always true.

In the system of equations above, small changes in the constant terms caused a large change in the solution. It is possible that small changes in the coefficients of the variables will also cause large changes in the solution.

In the two systems of equations that follow, examine the effects of approximating the fractional coefficients on the solutions. Try approximating each fraction to the nearest hundredth, to the nearest thousandth, to the nearest ten-thousandth, and then to the limits of your calculator. The exact solution of the first system of equations is (27, −192, 210). The exact solution of the second system of equations is (−64, 900, −2520, 1820).

$$\begin{cases} x + \frac{1}{2}y + \frac{1}{3}z = 1 \\ \frac{1}{2}x + \frac{1}{3}y + \frac{1}{4}z = 2 \\ \frac{1}{3}x + \frac{1}{4}y + \frac{1}{5}z = 3 \end{cases} \qquad \begin{cases} x + \frac{1}{2}y + \frac{1}{3}z + \frac{1}{4}w = 1 \\ \frac{1}{2}x + \frac{1}{3}y + \frac{1}{4}z + \frac{1}{5}w = 2 \\ \frac{1}{3}x + \frac{1}{4}y + \frac{1}{5}z + \frac{1}{6}w = 3 \\ \frac{1}{4}x + \frac{1}{5}y + \frac{1}{6}z + \frac{1}{7}w = 4 \end{cases}$$

Note how the solutions change as the approximations change and thus how important it is to know whether a system of equations is ill-conditioned. For systems that are not ill-conditioned, approximations of the coefficients yield reasonable approximations of the solution. For ill-conditioned systems of equations, that is not always true.

CHAPTER 9 SUMMARY

9.1 Systems of Linear Equations in Two Variables

- A system of equations is two or more equations considered together. A solution of a system of equations in two variables is an ordered pair that satisfies each equation of the system. Equivalent systems of equations have the same solution set.
- A system of equations is consistent if it has one or more solutions. A system of linear equations is independent if it has exactly one solution. A system is dependent if it has infinitely many solutions. An inconsistent system of equations has no solution.
- **Operations That Produce Equivalent Systems of Equations**
  1. Interchange any two equations.
  2. Replace an equation with a nonzero multiple of that equation.
  3. Replace an equation with the sum of that equation and a nonzero constant multiple of another equation in the system.

9.2 Systems of Linear Equations in More Than Two Variables

- An equation of the form $ax + by + cz = d$, with a, b, and c not all zero, is a linear equation in three variables. A solution of a system of equations in three variables is an ordered triple that satisfies each equation of the system.
- The graph of a linear equation in three variables is a plane.
- A linear system of equations for which the constant term is zero for all equations of the system is called a homogeneous system of equations.

9.3 Nonlinear Systems of Equations

- A nonlinear system of equations is a system in which one or more equations of the system are nonlinear.

9.4 Partial Fractions

- A rational expression can be written as the sum of terms whose denominators are factors of the denominator of the rational expression. This is called a partial fraction decomposition.

9.5 Inequalities in Two Variables and Systems of Inequalities

- The graph of an inequality in two variables frequently separates the plane into two or more regions.
- The solution set of a system of inequalities is the intersection of the solution sets of the individual inequalities.

9.6 Linear Programming

- A linear programming problem consists of a linear objective function and a number of constraints, which are inequalities or equations that restrict the values of the variables.
- The Fundamental Linear Programming Theorem states that if an objective function has an optimal solution, then that solution will be at a vertex of the set of feasible solutions.

CHAPTER 9 TRUE/FALSE EXERCISES

In Exercises 1 to 10, answer true or false. If the statement is false, give an example to show that the statement is false.

1. A system of equations will always have a solution as long as the number of equations is equal to the number of variables.
2. A system of two different quadratic equations can have at most four solutions.
3. A homogeneous system of equations is one in which all the variables have the same exponent.
4. In an xyz-coordinate system, the graph of the set of points formed by the intersection of two different planes is a straight line.
5. It is possible to find a partial fraction decomposition of a rational expression if the degree of the numerator is greater than the degree of the denominator.
6. Two systems of equations with the same solution set have the same equations in their respective systems.
7. The systems of equations
$$\begin{cases} x = 0 \\ y = 0 \end{cases} \quad \text{and} \quad \begin{cases} y = x \\ y = -x \end{cases}$$
are equivalent systems of equations.
8. For a linear programming problem, one or more constraints are used to define the set of feasible solutions.
9. A system of three linear equations in three variables for which two of the planes are parallel and the third plane intersects the first two is a dependent system of equations.
10. The inequality $xy < 1$ and the inequality $y < 1/x$ are equivalent inequalities.

CHAPTER 9 REVIEW EXERCISES

In Exercises 1 to 30, solve each system of equations.

1. $\begin{cases} 2x - 4y = -3 \\ 3x + 8y = -12 \end{cases}$

2. $\begin{cases} 4x - 3y = 15 \\ 2x + 5y = -12 \end{cases}$

3. $\begin{cases} 3x - 4y = -5 \\ y = \dfrac{2}{3}x + 1 \end{cases}$

4. $\begin{cases} 7x + 2y = -14 \\ y = -\dfrac{5}{2}x - 3 \end{cases}$

5. $\begin{cases} y = 2x - 5 \\ x = 4y - 1 \end{cases}$

6. $\begin{cases} y = 3x + 4 \\ x = 4y - 5 \end{cases}$

7. $\begin{cases} 6x + 9y = 15 \\ 10x + 15y = 25 \end{cases}$

8. $\begin{cases} 4x - 8y = 9 \\ 2x - 4y = 5 \end{cases}$

9. $\begin{cases} 2x - 3y + z = -9 \\ 2x + 5y - 2z = 18 \\ 4x - y + 3z = -4 \end{cases}$

10. $\begin{cases} x - 3y + 5z = 1 \\ 2x + 3y - 5z = 15 \\ 3x + 6y + 5z = 15 \end{cases}$

11. $\begin{cases} x + 3y - 5z = -12 \\ 3x - 2y + z = 7 \\ 5x + 4y - 9z = -17 \end{cases}$

12. $\begin{cases} 2x - y + 2z = 5 \\ x + 3y - 3z = 2 \\ 5x - 9y + 8z = 13 \end{cases}$

13. $\begin{cases} 3x + 4y - 6z = 10 \\ 2x + 2y - 3z = 6 \\ x - 6y + 9z = -4 \end{cases}$

14. $\begin{cases} x - 6y + 4z = 6 \\ 4x + 3y - 4z = 1 \\ 5x - 9y + 8z = 13 \end{cases}$

15. $\begin{cases} 2x + 3y - 2z = 0 \\ 3x - y - 4z = 0 \\ 5x + 13y - 4z = 0 \end{cases}$

16. $\begin{cases} 3x - 5y + z = 0 \\ x + 4y - 3z = 0 \\ 2x + y - 2z = 0 \end{cases}$

17. $\begin{cases} x - 2y + z = 1 \\ 3x + 2y - 3z = 1 \end{cases}$

18. $\begin{cases} 2x - 3y + z = 1 \\ 4x + 2y + 3z = 21 \end{cases}$

19. $\begin{cases} y = x^2 - 2x - 3 \\ y = 2x - 7 \end{cases}$

20. $\begin{cases} y = 2x^2 + x \\ y = 2x + 1 \end{cases}$

21. $\begin{cases} y = 3x^2 - x + 1 \\ y = x^2 + 2x - 1 \end{cases}$

22. $\begin{cases} y = 4x^2 - 2x - 3 \\ y = 2x^2 + 3x - 6 \end{cases}$

23. $\begin{cases} (x + 1)^2 + (y - 2)^2 = 4 \\ 2x + y = 4 \end{cases}$

24. $\begin{cases} (x - 1)^2 + (y + 1)^2 = 5 \\ y = 2x - 3 \end{cases}$

25. $\begin{cases} (x-2)^2+(y+2)^2 = 4 \\ (x+2)^2+(y+1)^2 = 17 \end{cases}$

26. $\begin{cases} (x+1)^2+(y-2)^2 = 1 \\ (x-2)^2+(y+2)^2 = 20 \end{cases}$

27. $\begin{cases} x^2-3xy+y^2 = -1 \\ 3x^2-5xy-2y^2 = 0 \end{cases}$

28. $\begin{cases} 2x^2+2xy-y^2 = -1 \\ 6x^2+xy-y^2 = 0 \end{cases}$

29. $\begin{cases} 2x^2-5xy+2y^2 = 56 \\ 14x^2-3xy-2y^2 = 56 \end{cases}$

30. $\begin{cases} 2x^2+7xy+6y^2 = 1 \\ 6x^2+7xy+2y^2 = 1 \end{cases}$

In Exercises 31 to 36, find the partial fraction decomposition.

31. $\dfrac{7x-5}{x^2-x-2}$

32. $\dfrac{x+1}{(x-1)^2}$

33. $\dfrac{2x-2}{(x^2+1)(x+2)}$

34. $\dfrac{5x^2-10x+9}{(x-2)^2(x+1)}$

35. $\dfrac{11x^2-x-2}{x^3-x}$

36. $\dfrac{x^4+x^3+4x^2+x+3}{(x^2+1)^2}$

In Exercises 37 to 48, graph the solution set of each inequality.

37. $4x-5y<20$

38. $2x+7y\geq -14$

39. $y\geq 2x^2-x-1$

40. $y<x^2-5x-6$

41. $(x-2)^2+(y-1)^2>4$

42. $(x+3)^2+(y+1)^2\leq 9$

43. $\dfrac{(x-3)^2}{16}-\dfrac{(y+2)^2}{25}\leq 1$

44. $\dfrac{(x+1)^2}{9}-\dfrac{(y-3)^2}{4}<-1$

45. $(2x-y+1)(x-2y-2)>0$

46. $(2x-3y-6)(x+2y-4)<0$

47. $x^2y^2<1$

48. $xy\geq 0$

In Exercises 49 to 60, graph the solution set of each system of inequalities.

49. $\begin{cases} 2x-5y<9 \\ 3x+4y\geq 2 \end{cases}$

50. $\begin{cases} 3x+y>7 \\ 2x+5y<9 \end{cases}$

51. $\begin{cases} 2x+3y>6 \\ 2x-y>-2 \\ x\leq 4 \end{cases}$

52. $\begin{cases} 2x+5y>10 \\ x-y>-2 \\ x\leq 4 \end{cases}$

53. $\begin{cases} 2x+3y\leq 18 \\ x+y\leq 7 \\ x\geq 0, y\geq 0 \end{cases}$

54. $\begin{cases} 3x+5y\geq 25 \\ 2x+3y\geq 16 \\ x\geq 0, y\geq 0 \end{cases}$

55. $\begin{cases} 3x+y\geq 6 \\ x+4y\geq 14 \\ 2x+3y\geq 16 \\ x\geq 0, y\geq 0 \end{cases}$

56. $\begin{cases} 3x+2y\geq 14 \\ x+y\geq 6 \\ 11x+4y\leq 48 \\ x\geq 0, y\geq 0 \end{cases}$

57. $\begin{cases} y<x^2-x-2 \\ y\geq 2x-4 \end{cases}$

58. $\begin{cases} y>2x^2+x-1 \\ y>x+3 \end{cases}$

59. $\begin{cases} x^2+y^2-2x+4y>4 \\ y<2x^2-1 \end{cases}$

60. $\begin{cases} x^2-y^2-4x-2y<-4 \\ x^2-y^2-4x+4y>8 \end{cases}$

In Exercises 61 to 66, solve the linear programming problem. In each problem, assume $x\geq 0$ and $y\geq 0$.

61. Objective function: $P=2x+2y$
Constraints: $\begin{cases} x+2y\leq 14 \\ 5x+2y\leq 30 \end{cases}$
Maximize the objective function.

62. Objective function: $P=4x+5y$
Constraints: $\begin{cases} 2x+3y\leq 24 \\ 4x+3y\leq 36 \end{cases}$
Maximize the objective function.

63. Objective function: $P=4x+y$
Constraints: $\begin{cases} 5x+2y\geq 16 \\ x+2y\geq 8 \\ x\leq 20, y\leq 20 \end{cases}$
Minimize the objective function.

64. Objective function: $P=2x+7y$
Constraints: $\begin{cases} 4x+3y\geq 24 \\ 4x+7y\geq 40 \\ x\leq 10, y\leq 10 \end{cases}$
Minimize the objective function.

65. Objective function: $P=6x+3y$
Constraints: $\begin{cases} 5x+2y\geq 20 \\ x+y\geq 7 \\ x+2y\geq 10 \\ x\leq 15, y\leq 15 \end{cases}$
Minimize the objective function.

66. Objective function: $P=5x+4y$
Constraints: $\begin{cases} x+y\leq 10 \\ 2x+y\leq 13 \\ 3x+y\leq 18 \end{cases}$
Maximize the objective function.

In Exercises 67 to 73, solve each exercise by solving a system of equations.

67. Find an equation of the form $y = ax^2 + bx + c$ whose graph passes through the points $(1, 0)$, $(-1, 5)$, and $(2, 3)$.

68. Find an equation of the circle that passes through the points $(4, 2)$, $(0, 1)$, and $(3, -1)$.

69. Find an equation of a plane that passes through the points $(2, 1, 2)$, $(3, 1, 0)$, and $(-2, -3, -2)$. Use the equation $z = ax + by + c$.

70. How many liters of a 20% acid solution should be mixed with 10 liters of a 10% acid solution so that the resulting solution is a 16% acid solution?

71. Flying with the wind, a small plane traveled 855 miles in 5 hours. Flying against the same wind, the plane traveled 575 miles in the same time. Find the rate of the wind and the rate of the plane in calm air.

72. A collection of ten coins has a value of \$1.25. The collection consists of only nickels, dimes, and quarters. How many of each coin are in the collection? (*Hint:* There is more than one solution.)

73. Consider the ordered triple (a, b, c). Find all real number values for a, b, and c so that the product of any two numbers equals the remaining number.

CHAPTER 9 TEST

In Exercises 1 to 8, solve each system of equations. If a system of equations is inconsistent, so state.

1. $\begin{cases} 3x + 2y = -5 \\ 2x - 5y = -16 \end{cases}$

2. $\begin{cases} x - \frac{1}{2}y = 3 \\ 2x - y = 6 \end{cases}$

3. $\begin{cases} x + 3y - z = 8 \\ 2x - 7y + 2z = 1 \\ 4x - y + 3z = 13 \end{cases}$

4. $\begin{cases} 3x - 2y + z = 2 \\ x + 2y - 2z = 1 \\ 4x - z = 3 \end{cases}$

5. $\begin{cases} 2x - 3y + z = -1 \\ x + 5y - 2z = 5 \end{cases}$

6. $\begin{cases} 4x + 2y + z = 0 \\ x - 3y - 2z = 0 \\ 3x + 5y + 3z = 0 \end{cases}$

7. $\begin{cases} y = x + 3 \\ y = x^2 + x - 1 \end{cases}$

8. $\begin{cases} y = x^2 - x - 3 \\ y = 2x^2 + 2x - 1 \end{cases}$

In Exercises 9 to 12, graph each inequality.

9. $3x - 4y > 8$

10. $y \le x^2 - 2x - 3$

11. $x^2 + 4y^2 \ge 16$

12. $x + y^2 < 0$

In Exercises 13 to 16, graph each system of inequalities. If the solution set is empty, so state.

13. $\begin{cases} 2x - 5y \le 16 \\ x + 3y \ge -3 \end{cases}$

14. $\begin{cases} x^2 + y^2 > 9 \\ x^2 + y^2 < 4 \end{cases}$

15. $\begin{cases} x + y \ge 8 \\ 2x + y \ge 11 \\ x \ge 0, y \ge 0 \end{cases}$

16. $\begin{cases} 2x + 3y \le 12 \\ x + y \le 5 \\ 3x + 2y \le 11 \\ x \ge 0, y \ge 0 \end{cases}$

In Exercises 17 and 18, find the partial fraction decomposition.

17. $\dfrac{3x - 5}{x^2 - 3x - 4}$

18. $\dfrac{2x + 1}{x(x^2 + 1)}$

19. A farmer has 160 acres available on which to plant oats and barley. It costs \$15 per acre for oat seed and \$13 per acre for barley seed. The labor cost is \$15 per acre for oats and \$20 per acre for barley. The farmer has \$2200 available to purchase seed and has set aside \$2600 for labor. The profit per acre for oats is \$120, and the profit per acre for barley is \$150. How many acres of oats should the farmer plant to maximize profit?

20. Find an equation of the circle that passes through the points $(3, 5)$, $(-3, -3)$, and $(4, 4)$. (*Hint:* Use $x^2 + y^2 + ax + by + c = 0$.)

CHAPTER 10

MATRICES

Matrices and Error-Correcting Codes

A matrix is a rectangular array of elements; an example is shown below. This particular matrix is a modification of what is called a *Hadamard matrix*. One application of Hadamard matrices is the enhancement of images sent to Earth by satellites or space probes such as the Mariner or Voyager.

$$\begin{bmatrix} 1 & 1 & 1 & 1 & 1 & 1 & 1 & 1 \\ 1 & 0 & 1 & 0 & 1 & 0 & 1 & 0 \\ 1 & 1 & 0 & 0 & 1 & 1 & 0 & 0 \\ 1 & 0 & 0 & 1 & 1 & 0 & 0 & 1 \\ 1 & 1 & 1 & 1 & 0 & 0 & 0 & 0 \\ 1 & 0 & 1 & 0 & 0 & 1 & 0 & 1 \\ 1 & 1 & 0 & 0 & 0 & 0 & 1 & 1 \\ 1 & 0 & 0 & 1 & 0 & 1 & 1 & 0 \end{bmatrix}$$

◆ Matrices make possible this detailed image of the surface of Mars.

◆ How a satellite "sees" the Nile River in Egypt.

An image to be transmitted from a space probe to a receiving station is first divided into very small squares called *pixels*. Each pixel is then assigned a number, the magnitude of which is a measure of the darkness of the square. For example, if pixels are assigned numbers from 0 to 63, a value of 0 corresponds to white and a value of 63 corresponds to black. These numbers are then represented as binary numbers, with $0 = 000000_{\text{two}}$ and $63 = 111111_{\text{two}}$. Using this conversion, the image is represented by a series of 0s and 1s (the 0s and 1s are called *bits*) that are sent to Earth.

To produce an accurate image on Earth, the bits transmitted by a space probe must be received accurately. However, because of what engineers call *noise* (it is similar to static on a radio), some of the 0s are changed to 1s and some of the 1s are changed to 0s. As a result, the image is blurred.

To minimize the effect of noise, *error-correcting codes* are used to help determine whether a bit has been changed. One way of establishing these codes is to use a Hadamard matrix.

SECTION 10.1 GAUSSIAN ELIMINATION METHOD

- INTRODUCTION TO MATRICES
- ELEMENTARY ROW OPERATIONS
- GAUSSIAN ELIMINATION METHOD
- APPLICATIONS: INTERPOLATING POLYNOMIALS

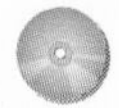

MATH MATTERS

The word *matrix* has the Latin word *mater* as its root. The mathematician James Sylvester (1814–1897) coined the term to indicate a place from which something else originates. Sylvester, born in London, spent the last part of his life at Johns Hopkins University, where he was influential in establishing graduate study in mathematics.

take note

When a term is missing from one of the equations of the system (as in the second equation), the coefficient of that term is 0, and a 0 is entered in the matrix. A vertical bar is frequently drawn in the matrix that separates the coefficients of the variables from the constants.

◆ INTRODUCTION TO MATRICES

A **matrix** is a rectangular array of numbers. Each number in a matrix is called an **element** of the matrix. The matrix below, with three rows and four columns, is called a 3×4 (read "3 by 4") matrix.

$$\begin{bmatrix} 2 & 5 & -2 & 5 \\ -3 & 6 & 4 & 0 \\ 1 & 3 & 7 & 2 \end{bmatrix}$$

A matrix of m rows and n columns is said to be of **order $m \times n$** or **dimension $m \times n$.** A **square matrix of order n** is a matrix with n rows and n columns. The matrix above has order 3×4. We will use the notation a_{ij} to refer to the element of a matrix in the ith row and jth column. For the matrix given above, $a_{23} = 4$, $a_{31} = 1$, and $a_{13} = -2$.

The elements $a_{11}, a_{22}, a_{33}, \ldots, a_{mm}$ form the **main diagonal** of a matrix. The elements 2, 6, and 7 form the main diagonal of the matrix shown above.

A matrix can be created from a system of linear equations. Consider the system of linear equations

$$\begin{cases} 2x - 3y + z = 2 \\ x - 3z = 4 \\ 4x - y + 4z = 3 \end{cases}$$

Using only the coefficients and constants of this system, we can write the 3×4 matrix

$$\left[\begin{array}{ccc|c} 2 & -3 & 1 & 2 \\ 1 & 0 & -3 & 4 \\ 4 & -1 & 4 & 3 \end{array}\right]$$

This matrix is called the **augmented matrix** of the system of equations. The matrix formed by the coefficients of the system is the **coefficient matrix.** The matrix formed from the constants is the **constant matrix** for the system. The coefficient matrix and constant matrix for the given system are

$$\text{Coefficient matrix: } \begin{bmatrix} 2 & -3 & 1 \\ 1 & 0 & -3 \\ 4 & -1 & 4 \end{bmatrix} \qquad \text{Constant matrix: } \begin{bmatrix} 2 \\ 4 \\ 3 \end{bmatrix}$$

We can write a system of equations from an augmented matrix.

$$\text{Augmented matrix: } \left[\begin{array}{ccc|c} 2 & -1 & 4 & 3 \\ 1 & 1 & 0 & 2 \\ 3 & -2 & -1 & 2 \end{array}\right] \quad \xrightarrow{\text{System:}} \quad \begin{cases} 2x - y + 4z = 3 \\ x + y = 2 \\ 3x - 2y - z = 2 \end{cases}$$

In certain cases, an augmented matrix represents a system of equations that we can solve by back substitution. Consider the following augmented matrix and the equivalent system of equations.

$$\left[\begin{array}{ccc|c} 1 & -3 & 4 & 5 \\ 0 & 1 & 2 & -4 \\ 0 & 0 & 1 & -1 \end{array}\right] \xrightarrow{\text{equivalent system}} \begin{cases} x - 3y + 4z = 5 \\ y + 2z = -4 \\ z = -1 \end{cases}$$

Solving this system by using back substitution, we find that the solution is $(3, -2, -1)$. The matrix above is in *echelon form*.

Echelon Form

A matrix is in **echelon form** if all the following conditions are satisfied.

1. The first nonzero number in any row is a 1.
2. Rows are arranged so that the column containing the first nonzero number in any row is to the left of the column containing the first nonzero number of the next row.
3. All rows consisting entirely of zeros appear at the bottom of the matrix.

Following are three examples of matrices in echelon form.

$$\begin{bmatrix} 1 & -3 & 4 & 2 \\ 0 & 1 & -2 & -1 \\ 0 & 0 & 0 & 0 \end{bmatrix} \quad \begin{bmatrix} 1 & 2 & -1 & 3 \\ 0 & 1 & 2 & -1 \end{bmatrix} \quad \begin{bmatrix} 1 & -1 & 3 & 2 \\ 0 & 1 & 2 & 5 \\ 0 & 0 & 1 & -2 \end{bmatrix}$$

◆ ELEMENTARY ROW OPERATIONS

We can write an augmented matrix in echelon form by using **elementary row operations.** These operations are a rewording, in matrix terminology, of the operations that produce equivalent equations.

take note

Many graphing calculators have the elementary row operations as built-in functions. See the Project at the end of this section for more details.

Elementary Row Operations

Given the augmented matrix for a system of linear equations, each of the following elementary row operations produces a matrix of an equivalent system of equations.

1. Interchanging any two rows
2. Multiplying all the elements in a row by the same nonzero number
3. Replacing a row by the sum of that row and a nonzero multiple of any other row

It is convenient to specify each operation symbolically as follows:

1. Interchanging the ith and jth rows: $R_i \longleftrightarrow R_j$
2. Multiplying the ith row by k, a nonzero constant: kR_i
3. Replacing the jth row by the sum of that row and a nonzero multiple of the ith row: $kR_i + R_j$

To demonstrate these operations, we will use the 3×3 matrix

$$\begin{bmatrix} 2 & 1 & -2 \\ 3 & -2 & 2 \\ 1 & -2 & 3 \end{bmatrix}$$

$$\begin{bmatrix} 2 & 1 & -2 \\ 3 & -2 & 2 \\ 1 & -2 & 3 \end{bmatrix} R_1 \longleftrightarrow R_3 \begin{bmatrix} 1 & -2 & 3 \\ 3 & -2 & 2 \\ 2 & 1 & -2 \end{bmatrix}$$

- **Interchange row 1 and row 3.**

$$\begin{bmatrix} 2 & 1 & -2 \\ 3 & -2 & 2 \\ 1 & -2 & 3 \end{bmatrix} \xrightarrow{-3R_2} \begin{bmatrix} 2 & 1 & -2 \\ -9 & 6 & -6 \\ 1 & -2 & 3 \end{bmatrix}$$

- **Multiply row 2 by −3.**

$$\begin{bmatrix} 2 & 1 & -2 \\ 3 & -2 & 2 \\ 1 & -2 & 3 \end{bmatrix} \xrightarrow{-2R_3 + R_1} \begin{bmatrix} 0 & 5 & -8 \\ 3 & -2 & 2 \\ 1 & -2 & 3 \end{bmatrix}$$

- **Multiply row 3 by −2 and add to row 1. Replace row 1 by the sum.**

◆ GAUSSIAN ELIMINATION METHOD

MATH MATTERS

The Gaussian elimination method is used in some computer programs as the procedure to solve a system of equations.

The **Gaussian elimination method** is an algorithm[1] that uses elementary row operations to solve a system of linear equations. The goal of this method is to rewrite an augmented matrix in echelon form.

We will now demonstrate how to solve a system of two equations in two variables by the Gaussian elimination method. Consider the system of equations

$$\begin{cases} 2x + 5y = -1 \\ 3x - 2y = 8 \end{cases} \qquad (1)$$

The augmented matrix for this system is

$$\left[\begin{array}{cc|c} 2 & 5 & -1 \\ 3 & -2 & 8 \end{array}\right]$$

The goal of the Gaussian elimination method is to rewrite the augmented matrix in echelon form by using elementary row operations. The row operations are chosen so that first, there is a 1 as a_{11}; second, there is a 0 as a_{21}; and third, there is a 1 as a_{22}.

Begin by multiplying row 1 by 1/2. The result is a 1 as a_{11}.

$$\left[\begin{array}{cc|c} 2 & 5 & -1 \\ 3 & -2 & 8 \end{array}\right] \xrightarrow{\frac{1}{2}R_1} \left[\begin{array}{cc|c} 1 & \frac{5}{2} & -\frac{1}{2} \\ 3 & -2 & 8 \end{array}\right]$$

Now multiply row 1 by −3 and add the result to row 2. Replace row 2. The result is a 0 as a_{21}.

$$\left[\begin{array}{cc|c} 1 & \frac{5}{2} & -\frac{1}{2} \\ 3 & -2 & 8 \end{array}\right] \xrightarrow{-3R_1 + R_2} \left[\begin{array}{cc|c} 1 & \frac{5}{2} & -\frac{1}{2} \\ 0 & -\frac{19}{2} & \frac{19}{2} \end{array}\right]$$

[1] An algorithm is a procedure used in calculations. The word is derived from Al-Khwarizmi, the name of the author of an Arabic algebra book written around A.D. 825.

Now multiply row 2 by $-2/19$. The result is a 1 as a_{22}. The matrix is now in row echelon form.

$$\begin{bmatrix} 1 & \frac{5}{2} & \bigg| & -\frac{1}{2} \\ 0 & -\frac{19}{2} & \bigg| & \frac{19}{2} \end{bmatrix} \xrightarrow{-\frac{2}{19}R_2} \begin{bmatrix} 1 & \frac{5}{2} & \bigg| & -\frac{1}{2} \\ 0 & 1 & \bigg| & -1 \end{bmatrix}$$

The system of equations written from the echelon form of the matrix is

$$\begin{cases} x + \dfrac{5}{2}y = -\dfrac{1}{2} \\ \qquad\quad y = -1 \end{cases} \qquad (2)$$

To solve by back substitution, replace y in the first equation by -1 and solve for x.

$$x + \left(\frac{5}{2}\right)(-1) = -\frac{1}{2}$$

$$x = 2$$

The solution of System (1) is $(2, -1)$.

To conserve space, we will occasionally perform more than one elementary row operation in one step. For example, the notation

$$\xrightarrow{\substack{3R_1 + R_2 \\ -2R_1 + R_3}}$$

means that two elementary row operations were performed. First multiply row 1 by 3 and add it to row 2. Replace row 2. Now multiply row 1 by -2 and add it to row 3. Replace row 3.

take note

Following a systematic procedure will help you reduce a matrix to echelon form. Using elementary row operations, change a_{11} to 1 and change the remaining elements in the first column to 0. Now change a_{22} to 1 and change the remaining elements below a_{22} to zero. Now move to a_{33} and repeat the procedure. Continue moving down the main diagonal until you reach a_{nn} or until all remaining elements on the main diagonal are zero.

EXAMPLE 1 Solve a System of Equations by the Gaussian Elimination Method

Solve by using the Gaussian elimination method.

$$\begin{cases} 3t - 8u + 8v + 7w = 41 \\ t - 2u + 2v + w = 9 \\ 2t - 2u + 6v - 4w = -1 \\ 2t - 2u + 3v - 3w = 3 \end{cases}$$

Solution

Write the augmented matrix and then use elementary row operations to rewrite the matrix in echelon form.

$$\left[\begin{array}{cccc|c} 3 & -8 & 8 & 7 & 41 \\ 1 & -2 & 2 & 1 & 9 \\ 2 & -2 & 6 & -4 & -1 \\ 2 & -2 & 3 & -3 & 3 \end{array}\right] \quad R_1 \longleftrightarrow R_2 \quad \left[\begin{array}{cccc|c} 1 & -2 & 2 & 1 & 9 \\ 3 & -8 & 8 & 7 & 41 \\ 2 & -2 & 6 & -4 & -1 \\ 2 & -2 & 3 & -3 & 3 \end{array}\right]$$

Continued •➤

$$\xrightarrow{\substack{-3R_1 + R_2 \\ -2R_1 + R_3 \\ -2R_1 + R_4}} \left[\begin{array}{rrrr|r} 1 & -2 & 2 & 1 & 9 \\ 0 & -2 & 2 & 4 & 14 \\ 0 & 2 & 2 & -6 & -19 \\ 0 & 2 & -1 & -5 & -15 \end{array}\right] \xrightarrow{-\frac{1}{2}R_2} \left[\begin{array}{rrrr|r} 1 & -2 & 2 & 1 & 9 \\ 0 & 1 & -1 & -2 & -7 \\ 0 & 2 & 2 & -6 & -19 \\ 0 & 2 & -1 & -5 & -15 \end{array}\right]$$

$$\xrightarrow{\substack{-2R_2 + R_3 \\ -2R_2 + R_4}} \left[\begin{array}{rrrr|r} 1 & -2 & 2 & 1 & 9 \\ 0 & 1 & -1 & -2 & -7 \\ 0 & 0 & 4 & -2 & -5 \\ 0 & 0 & 1 & -1 & -1 \end{array}\right] \quad R_4 \longleftrightarrow R_3 \quad \left[\begin{array}{rrrr|r} 1 & -2 & 2 & 1 & 9 \\ 0 & 1 & -1 & -2 & -7 \\ 0 & 0 & 1 & -1 & -1 \\ 0 & 0 & 4 & -2 & -5 \end{array}\right]$$

$$\xrightarrow{-4R_3 + R_4} \left[\begin{array}{rrrr|r} 1 & -2 & 2 & 1 & 9 \\ 0 & 1 & -1 & -2 & -7 \\ 0 & 0 & 1 & -1 & -1 \\ 0 & 0 & 0 & 2 & -1 \end{array}\right] \xrightarrow{\frac{1}{2}R_4} \left[\begin{array}{rrrr|r} 1 & -2 & 2 & 1 & 9 \\ 0 & 1 & -1 & -2 & -7 \\ 0 & 0 & 1 & -1 & -1 \\ 0 & 0 & 0 & 1 & -\frac{1}{2} \end{array}\right]$$

The last matrix is in echelon form. The system of equations written from the matrix is

$$\begin{cases} t - 2u + 2v + w = 9 \\ u - v - 2w = -7 \\ v - w = -1 \\ w = -\dfrac{1}{2} \end{cases}$$

Solve by back substitution. The solution is $(-13/2, -19/2, -3/2, -1/2)$.

TRY EXERCISE 14, EXERCISE SET 10.1, PAGE 723

EXAMPLE 2 Solve a Dependent System of Equations

Solve using the Gaussian elimination method.

$$\begin{cases} x - 3y + 4z = 1 \\ 2x - 5y + 3z = 6 \\ x - 2y - z = 5 \end{cases}$$

Solution

Write the augmented matrix and then use elementary row operations to rewrite the matrix in echelon form.

$$\left[\begin{array}{rrr|r} 1 & -3 & 4 & 1 \\ 2 & -5 & 3 & 6 \\ 1 & -2 & -1 & 5 \end{array}\right] \xrightarrow{\substack{-2R_1 + R_2 \\ -R_1 + R_3}} \left[\begin{array}{rrr|r} 1 & -3 & 4 & 1 \\ 0 & 1 & -5 & 4 \\ 0 & 1 & -5 & 4 \end{array}\right]$$

$$\xrightarrow{-R_2 + R_3} \left[\begin{array}{rrr|r} 1 & -3 & 4 & 1 \\ 0 & 1 & -5 & 4 \\ 0 & 0 & 0 & 0 \end{array}\right]$$

$$\begin{cases} x - 3y + 4z = 1 \\ \quad\;\; y - 5z = 4 \end{cases}$$ • Equivalent system

Any solution of the system of equations is a solution of $y - 5z = 4$. Solving this equation for y, we have $y = 5x + 4$.

$$x - 3y + 4z = 1$$

$$x - 3(5z + 4) + 4z = 1 \qquad \bullet\ y = 5z + 4$$

$$x = 11z + 13 \qquad \bullet\ \text{Solve for } x.$$

Both x and y are expressed in terms of z. Let z be any real number c. The solutions of the system of equations are $(11c + 13, 5c + 4, c)$.

TRY EXERCISE 18, EXERCISE SET 10.1, PAGE 723

EXAMPLE 3 Identify an Inconsistent System of Equations

Solve using the Gaussian elimination method.

$$\begin{cases} x - 3y + z = 5 \\ 3x - 7y + 2z = 12 \\ 2x - 4y + z = 3 \end{cases}$$

Solution

Write the augmented matrix and then use elementary row operations to rewrite the matrix in echelon form.

$$\left[\begin{array}{ccc|c} 1 & -3 & 1 & 5 \\ 3 & -7 & 2 & 12 \\ 2 & -4 & 1 & 3 \end{array}\right] \xrightarrow[-2R_1 + R_3]{-3R_1 + R_2} \left[\begin{array}{ccc|c} 1 & -3 & 1 & 5 \\ 0 & 2 & -1 & -3 \\ 0 & 2 & -1 & -7 \end{array}\right]$$

$$\xrightarrow{\frac{1}{2}R_2} \left[\begin{array}{ccc|c} 1 & -3 & 1 & 5 \\ 0 & 1 & -\frac{1}{2} & -\frac{3}{2} \\ 0 & 2 & -1 & -7 \end{array}\right] \xrightarrow{-2R_2 + R_3} \left[\begin{array}{ccc|c} 1 & -3 & 1 & 5 \\ 0 & 1 & -\frac{1}{2} & -\frac{3}{2} \\ 0 & 0 & 0 & -4 \end{array}\right]$$

$$\begin{cases} x - 3y + z = 5 \\ \quad\;\; y - \dfrac{1}{2}z = -\dfrac{3}{2} \\ \qquad\quad 0z = -4 \end{cases}$$ • Equivalent system

Because the equation $0z = -4$ has no solution, the system of equations has no solution.

TRY EXERCISE 20, EXERCISE SET 10.1, PAGE 723

EXAMPLE 4 Solve a Nonsquare System of Equations

take note

When there are fewer equations than variables (as in Example 4), the system of equations has either no solution or an infinite number of solutions. See the Project at the end of this section.

Solve the system of equations using the Gaussian elimination method.

$$\begin{cases} x_1 - 2x_2 - 3x_3 - 2x_4 = 1 \\ 2x_1 - 3x_2 - 4x_3 - 2x_4 = 3 \\ x_1 + x_2 + x_3 - 7x_4 = -7 \end{cases}$$

Solution

Write the augmented matrix and then use elementary row operations to rewrite the matrix in echelon form.

$$\left[\begin{array}{cccc|c} 1 & -2 & -3 & -2 & 1 \\ 2 & -3 & -4 & -2 & 3 \\ 1 & 1 & 1 & -7 & -7 \end{array}\right] \xrightarrow[-1R_1 + R_3]{-2R_1 + R_2} \left[\begin{array}{cccc|c} 1 & -2 & -3 & -2 & 1 \\ 0 & 1 & 2 & 2 & 1 \\ 0 & 3 & 4 & -5 & -8 \end{array}\right]$$

$$\xrightarrow[-\frac{1}{2}R_3]{-3R_2 + R_3} \left[\begin{array}{cccc|c} 1 & -2 & -3 & -2 & 1 \\ 0 & 1 & 2 & 2 & 1 \\ 0 & 0 & 1 & \frac{11}{2} & \frac{11}{2} \end{array}\right]$$

$$\begin{cases} x_1 - 2x_2 - 3x_3 - 2x_4 = 1 \\ x_2 + 2x_3 + 2x_4 = 1 \\ x_3 + \frac{11}{2}x_4 = \frac{11}{2} \end{cases}$$ • **Equivalent system**

Now express each of the variables in terms of x_4. Solve the third equation for x_3.

$$x_3 = -\frac{11}{2}x_4 + \frac{11}{2}$$

Substitute this value into the second equation and solve for x_2.

$$x_2 + 2\left(-\frac{11}{2}x_4 + \frac{11}{2}\right) + 2x_4 = 1$$

$$x_2 = 9x_4 - 10$$ • **Simplify**

Substitute the values for x_2 and x_3 into the first equation and solve for x_1.

$$x_1 - 2(9x_4 - 10) - 3\left(-\frac{11}{2}x_4 + \frac{11}{2}\right) - 2x_4 = 1$$

$$x_1 = \frac{7}{2}x_4 - \frac{5}{2}$$ • **Simplify**

If x_4 is any real number c, the solution is of the form

$$\left(\frac{7}{2}c - \frac{5}{2}, 9c - 10, -\frac{11}{2}c + \frac{11}{2}, c\right)$$

TRY EXERCISE 36, EXERCISE SET 10.1, PAGE 723

◆ APPLICATIONS: INTERPOLATING POLYNOMIALS

take note

Because the subscript on the first ordered pair is zero, there are $n + 1$ points. The degree of the interpolating polynomial, however, is n. Thus, if there were three ordered pairs, the degree of the interpolating polynomial would be at most 2. If there were seven ordered pairs, the degree of the interpolating polynomial would be at most 6.

One application of the Gaussian elimination method of solving a system of equations is in finding *interpolating polynomials*.

Interpolating Polynomial

Let $(x_0, y_0), (x_1, y_1), (x_2, y_2), \ldots, (x_n, y_n)$ be the coordinates of a set of points for which all the x_i are distinct. Then the **interpolating polynomial** is a unique polynomial of degree at most n that passes through the given points.

EXAMPLE 5 Find an Interpolating Polynomial

Find the interpolating polynomial that passes through the points whose coordinates are $(-2, 13)$, $(1, -2)$, and $(2, 1)$.

Solution

Because there are three given points, the degree of the interpolating polynomial will be at most 2. The form of the polynomial will be $p(x) = a_2x^2 + a_1x + a_0$. Use this polynomial to create a system of equations.

$$p(x) = a_2x^2 + a_1x + a_0$$

$$p(-2) = a_2(-2)^2 + a_1(-2) + a_0 = 4a_2 - 2a_1 + a_0 = 13 \qquad \bullet\ x = -2,\ p(x) = 13$$

$$p(1) = a_2(1)^2 + a_1(1) + a_0 = a_2 + a_1 + a_0 = -2 \qquad \bullet\ x = 1,\ p(x) = -2$$

$$p(2) = a_2(2)^2 + a_1(2) + a_0 = 4a_2 + 2a + a_0 = 1 \qquad \bullet\ x = 2,\ p(x) = 1$$

The system of equations and the associated augmented matrix are

$$\begin{cases} 4a_2 - 2a_1 + a_0 = 13 \\ a_2 + a_1 + a_0 = -2 \\ 4a_2 + 2a_1 + a_0 = 1 \end{cases} \qquad \left[\begin{array}{ccc|c} 4 & -2 & 1 & 13 \\ 1 & 1 & 1 & -2 \\ 4 & 2 & 1 & 1 \end{array}\right]$$

The augmented matrix in echelon form and the resulting system of equations are

$$\left[\begin{array}{ccc|c} 1 & -0.5 & 0.25 & 3.25 \\ 0 & 1 & 0 & -3 \\ 0 & 0 & 1 & -1 \end{array}\right] \qquad \begin{cases} a_2 - 0.5a_1 + 0.25a_0 = 3.25 \\ a_1 = -3 \\ a_0 = -1 \end{cases}$$

Solving the system of equations by back substitution yields $a_0 = -1$, $a_1 = -3$, and $a_2 = 2$. The interpolating polynomial is $p(x) = 2x^2 - 3x - 1$. See **Figure 10.1**.

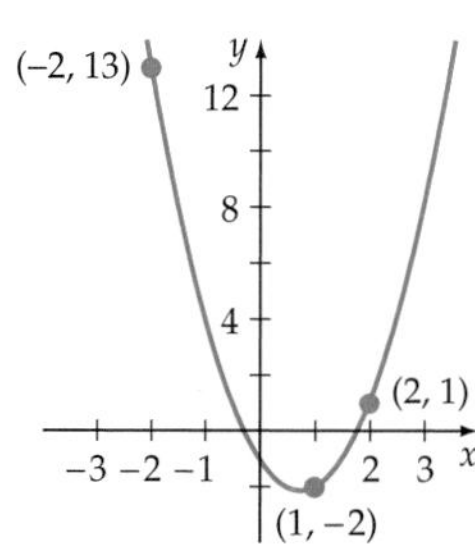

Figure 10.1

TRY EXERCISE 44, EXERCISE SET 10.1, PAGE 724

A graphing calculator can be used to write a matrix in echelon form. **Figure 10.2** offers some suggestions and screens for solving Example 5 with a TI-83. See Take Note below for the TI-83 PLUS.

Press [MATRX]. Then use the right-arrow key to highlight [EDIT]. [2nd QUIT]

```
NAMES MATH EDIT
1:[A] 3x4
2:[B]
3:[C]
4:[D]
5:[E]
6:[F]
7↓[G]
```

Enter the number of rows and columns and the elements of the matrix.

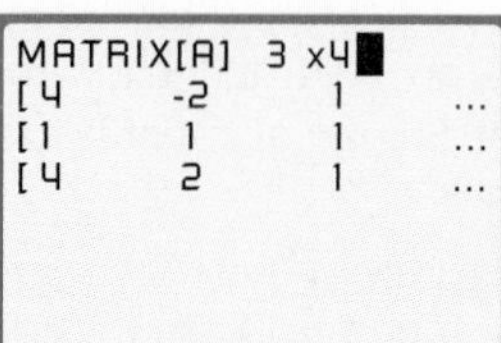

Press [MATRX]. Then use the right-arrow key to highlight MATH. Select A:ref(. Press [ENTER].

```
NAMES MATH EDIT
7↑augment(
8:Matr▶list(
9:List▶matr(
0:cumSum(
A:ref(
B:rref(
C↓rowSwap(
```

Press [MATRX]. Then select [A] and press [ENTER]. Press [ENTER] again. The echelon form of the matrix is shown.

```
ref([A])
[[1 -.5 .25 3.2 ...
 [0 1   0   -3  ...
 [0 0   1   -1  ...
```

Figure 10.2

take note

For matrix operations on a TI-83 PLUS, press [2nd] MATRX, then [x^{-1}] key

TOPICS FOR DISCUSSION

1. What is a matrix? What is the order of a matrix?
2. Explain how an augmented matrix differs from the coefficient matrix for a system of equations.
3. Give examples of matrices that are in echelon form and of matrices that are not in echelon form.
4. What are the elementary row operations? Give examples of each one.
5. After elementary row operations have been correctly performed on an augmented matrix, the result is $\begin{bmatrix} 1 & -2 & 3 & 0 \\ 0 & 1 & 2 & -1 \\ 0 & 0 & 0 & 3 \end{bmatrix}$. Does this result indicate that the system of equations has a unique solution, an infinite number of solutions, or no solution?

EXERCISE SET 10.1

In Exercises 1 to 4, write the augmented matrix, the coefficient matrix, and the constant matrix.

1. $\begin{cases} 2x - 3y + z = 1 \\ 3x - 2y + 3z = 0 \\ x + 5z = 4 \end{cases}$

2. $\begin{cases} -3y + 2z = 3 \\ 2x - y = -1 \\ 3x - 2y + 3z = 4 \end{cases}$

3. $\begin{cases} 2x - 3y - 4z + w = 2 \\ 2y + z = 2 \\ x - y + 2z = 4 \\ 3x - 3y - 2z = 1 \end{cases}$

4. $\begin{cases} x - y + 2z + 3w = -2 \\ 2x + z - 2w = 1 \\ 3x - 2w = 3 \\ -x + 3y - z = 3 \end{cases}$

In Exercises 5 to 12, use elementary row operations to write each matrix in echelon form.

5. $\begin{bmatrix} 2 & -1 & 3 & -2 \\ 1 & -1 & 2 & 2 \\ 3 & 2 & -1 & 3 \end{bmatrix}$

6. $\begin{bmatrix} 1 & 2 & 4 & 1 \\ 2 & 2 & 7 & 3 \\ 3 & 6 & 8 & -1 \end{bmatrix}$

7. $\begin{bmatrix} 4 & -5 & -1 & 2 \\ 3 & -4 & 1 & -2 \\ 1 & -2 & -1 & 3 \end{bmatrix}$

8. $\begin{bmatrix} -2 & 1 & -1 & 3 \\ 2 & 2 & 4 & 6 \\ 3 & 1 & -1 & 2 \end{bmatrix}$

9. $\begin{bmatrix} 1 & -2 & 3 & -4 \\ 3 & -6 & 10 & -14 \\ 5 & -8 & 19 & -21 \\ 2 & -4 & 7 & -10 \end{bmatrix}$

10. $\begin{bmatrix} 2 & -1 & 3 & 2 \\ 1 & 2 & -1 & 3 \\ 3 & 5 & -2 & 2 \\ 4 & 3 & 1 & 8 \end{bmatrix}$

11. $\begin{bmatrix} 1 & -3 & 4 & 2 & 1 \\ 2 & -3 & 5 & -2 & -1 \\ -1 & 2 & -3 & 1 & 3 \end{bmatrix}$

12. $\begin{bmatrix} 2 & -1 & 3 & 2 & 2 \\ 1 & -2 & 2 & 1 & -1 \\ 3 & -5 & -1 & -2 & 3 \end{bmatrix}$

In Exercises 13 to 38, solve each system of equations by the Gaussian elimination method.

13. $\begin{cases} x + 2y - 2z = -2 \\ 5x + 9y - 4z = -3 \\ 3x + 4y - 5z = -3 \end{cases}$

14. $\begin{cases} x - 3y + z = 8 \\ 2x - 5y - 3z = 2 \\ x + 4y + z = 1 \end{cases}$

15. $\begin{cases} 3x + 7y - 7z = -4 \\ x + 2y - 3z = 0 \\ 5x + 6y + z = -8 \end{cases}$

16. $\begin{cases} 2x - 3y + 2z = 13 \\ 3x - 4y - 3z = 1 \\ 3x + y - z = 2 \end{cases}$

17. $\begin{cases} x + 2y - 2z = 3 \\ 5x + 8y - 6z = 14 \\ 3x + 4y - 2z = 8 \end{cases}$

18. $\begin{cases} 3x - 5y + 2z = 4 \\ x - 3y + 2z = 4 \\ 5x - 11y + 6z = 12 \end{cases}$

19. $\begin{cases} 3x + 2y - z = 1 \\ 2x + 3y - z = 1 \\ x - y + 2z = 3 \end{cases}$

20. $\begin{cases} 2x + 5y + 2z = -1 \\ x + 2y - 3z = 5 \\ 5x + 12y + z = 10 \end{cases}$

21. $\begin{cases} x - 3y + 2z = 0 \\ 2x - 5y - 2z = 0 \\ 4x - 11y + 2z = 0 \end{cases}$

22. $\begin{cases} x + y - 2z = 0 \\ 3x + 4y - z = 0 \\ 5x + 6y - 5z = 0 \end{cases}$

23. $\begin{cases} 2x + y - 3z = 4 \\ 3x + 2y + z = 2 \end{cases}$

24. $\begin{cases} 3x - 6y + 2z = 2 \\ 2x + 5y - 3z = 2 \end{cases}$

25. $\begin{cases} 2x + 2y - 4z = 4 \\ 2x + 3y - 5z = 4 \\ 4x + 5y - 9z = 8 \end{cases}$

26. $\begin{cases} 3x - 10y + 2z = 34 \\ x - 4y + z = 13 \\ 5x - 2y + 7z = 31 \end{cases}$

27. $\begin{cases} x + 3y + 4z = 11 \\ 2x + 3y + 2z = 7 \\ 4x + 9y + 10z = 20 \\ 3x - 2y + z = 1 \end{cases}$

28. $\begin{cases} x - 4y + 3z = 4 \\ 3x - 10y + 3z = 4 \\ 5x - 18y + 9z = 10 \\ 2x + 2y - 3z = -11 \end{cases}$

29. $\begin{cases} t + 2u - 3v + w = -7 \\ 3t + 5u - 8v + 5w = -8 \\ 2t + 3u - 7v + 3w = -11 \\ 4t + 8u - 10v + 7w = -10 \end{cases}$

30. $\begin{cases} t + 4u + 2v - 3w = 11 \\ 2t + 10u + 3v - 5w = 17 \\ 4t + 16u + 7v - 9w = 34 \\ t + 4u + v - w = 4 \end{cases}$

31. $\begin{cases} 2t - u + 3v + 2w = 2 \\ t - u + 2v + w = 2 \\ 3t - 2v - 3w = 13 \\ 2t + 2u - 2w = 6 \end{cases}$

32. $\begin{cases} 4t + 7u - 10v + 3w = -29 \\ 3t + 5u - 7v + 2w = -20 \\ t + 2u - 3v + w = -9 \\ 2t - u + 2v - 4w = 15 \end{cases}$

33. $\begin{cases} 3t + 10u + 7v - 6w = 7 \\ 2t + 8u + 6v - 5w = 5 \\ t + 4u + 2v - 3w = 2 \\ 4t + 14u + 9v - 8w = 8 \end{cases}$

34. $\begin{cases} t - 3u + 2v + 4w = 13 \\ 3t - 8u + 4v + 13w = 35 \\ 2t - 7u + 8v + 5w = 28 \\ 4t - 11u + 6v + 17w = 56 \end{cases}$

35. $\begin{cases} t - u + 2v - 3w = 9 \\ 4t + 11v - 10w = 46 \\ 3t - u + 8v - 6w = 27 \end{cases}$

36. $\begin{cases} t - u + 3v - 5w = 10 \\ 2t - 3u + 4v + w = 7 \\ 3t + u - 2v - 2w = 6 \end{cases}$

37. $\begin{cases} 3t - 4u + v = 2 \\ t + u - 2v + 3w = 1 \end{cases}$

38. $\begin{cases} 2t + 3v - 4w = 2 \\ t + 2u - 4v + w = -3 \end{cases}$

Some graphing calculators and computer programs contain a program that will assist you in solving a system of linear equations by rewriting the system in echelon form. Try one of these programs for Exercises 39 to 50.

39. **INTERPOLATING POLYNOMIAL** Find a polynomial that passes through the points whose coordinates are $(-2, -7)$ and $(1, -1)$.

40. **INTERPOLATING POLYNOMIAL** Find a polynomial that passes through the points whose coordinates are $(-3, -8)$ and $(1, 4)$.

41. **INTERPOLATING POLYNOMIAL** Find a polynomial that passes through the points whose coordinates are $(-1, 6)$, $(1, 2)$, and $(2, 3)$.

42. **INTERPOLATING POLYNOMIAL** Find a polynomial that passes through the points whose coordinates are $(-2, -3)$, $(0, -1)$, and $(3, 17)$.

43. INTERPOLATING POLYNOMIAL Find a polynomial that passes through the points whose coordinates are $(-2, -12)$, $(0, 2)$, $(1, 0)$, and $(3, 8)$.

44. INTERPOLATING POLYNOMIAL Find a polynomial that passes through the points whose coordinates are $(-1, -5)$, $(0, 0)$, $(1, 1)$, and $(2, 4)$.

45. INTERPOLATING POLYNOMIAL Find a polynomial that passes through the points whose coordinates are $(-1, 3)$, $(1, 7)$, and $(2, 9)$. This exercise illustrates that the degree of the polynomial is at most one less than the number of points.

46. INTERPOLATING POLYNOMIAL Find a polynomial that passes through the points whose coordinates are $(-2, 7)$, $(1, -2)$, and $(2, -5)$. This exercise illustrates that the degree of the polynomial is at most one less than the number of points.

47. $\begin{cases} x_1 + 2x_2 - x_3 + 2x_4 + 3x_5 = 11 \\ x_1 - x_2 + 2x_3 - x_4 + 2x_5 = 0 \\ 2x_1 + x_2 - x_3 + 2x_4 - x_5 = 4 \\ 3x_1 + 2x_2 - x_3 + x_4 - 2x_5 = 2 \\ 2x_1 + x_2 - x_3 - 2x_4 + x_5 = 4 \end{cases}$

48. $\begin{cases} x_1 - 2x_2 + 2x_3 - 3x_4 + 2x_5 = 5 \\ x_1 - 3x_2 - x_3 + 2x_4 - x_5 = -4 \\ 3x_1 + x_2 - 2x_3 + x_4 + 3x_5 = 9 \\ 2x_1 - x_2 + 3x_3 - x_4 - 2x_5 = 2 \\ -x_1 + 2x_2 - 2x_3 + 3x_4 - x_5 = -4 \end{cases}$

49. $\begin{cases} x_1 + 2x_2 - 3x_3 - x_4 + 2x_5 = -10 \\ -x_1 - 3x_2 + x_3 + x_4 - x_5 = 4 \\ 2x_1 + 3x_2 - 5x_3 + 2x_4 + 3x_5 = -20 \\ 3x_1 + 4x_2 - 7x_3 + 3x_4 - 2x_5 = -16 \\ 2x_1 + x_2 - 6x_3 + 4x_4 - 3x_5 = -12 \end{cases}$

50. $\begin{cases} x_1 - 2x_2 + 2x_3 - 3x_4 + x_5 = 5 \\ 2x_1 - 3x_2 + 4x_3 - 5x_4 - x_5 = 13 \\ x_1 + x_2 - 2x_3 + 2x_4 + 2x_5 = -11 \\ 3x_1 - 2x_2 + 2x_3 - 2x_4 - 2x_5 = 7 \\ 4x_1 - 4x_2 + 4x_3 - 5x_4 - x_5 = 12 \end{cases}$

SUPPLEMENTAL EXERCISES

In Exercises 51 to 53, use the system of equations

$$\begin{cases} x + 3y - a^2z = a^2 \\ 2x + 3y + az = 2 \\ 3x + 4y + 2z = 3 \end{cases}$$

51. Find all values of a for which the system of equations has a unique solution.

52. Find all values of a for which the system of equations has infinitely many solutions.

53. Find all values of a for which the system of equations has no solutions.

54. Find an equation of the plane that passes through the points $(1, 2, 6)$, $(-1, 1, 7)$, and $(4, 2, 0)$. Use the equation $z = ax + by + c$.

55. Find an equation of the plane that passes through the points $(-1, 0, -4)$, $(2, 1, 5)$, and $(-1, 1, -1)$. Use the equation $z = ax + by + c$.

PROJECTS

1. ECHELON FORM BY USING A GRAPHING CALCULATOR Many graphing calculators have the elementary row operations as built-in functions. Complete this project using one of those calculators.

a. Enter into your calculator the augmented matrix for

$$\begin{cases} 2x - 3y + z = 4 \\ x + 2y - 2z = -2 \\ 3x + y - 3z = 4 \end{cases}$$

b. Complete the following steps to write the augmented matrix in echelon form. *Suggestion:* Suppose that you enter the augmented matrix as A. If you perform an elementary row operation on A, the new matrix will be displayed. However, matrix A has *not* been changed. The new matrix must be saved as another matrix, say B. Now perform the elementary row operations on B and save the result in B. When you have finished, the original matrix will still be in A and the echelon form of matrix A will be in B.

1. $R_1 \leftrightarrow R_2$ **2.** $-2R_1 + R_2 \to R_2$ **3.** $-3R_1 + R_3 \to R_3$

4. $-\frac{1}{7}R_2$ **5.** $5R_2 + R_3 \to R_3$ **6.** $-\frac{7}{4}R_3$

2.  For more on interpolating polynomials, see our web site at http://college.hmco.com.

SECTION 10.2 THE ALGEBRA OF MATRICES

- ADDITION AND SUBTRACTION OF MATRICES
- SCALAR MULTIPLICATION
- MATRIX MULTIPLICATION
- MATRIX PRODUCTS AND SYSTEMS OF EQUATIONS

◆ ADDITION AND SUBTRACTION OF MATRICES

Besides being convenient for solving systems of equations, matrices are useful tools to model problems in business and science. One very prevalent application of matrices is to spreadsheet programs.

The typical method used in spreadsheets is to number the rows 1, 2, 3, ... and to identify the columns as A, B, C, ... The partial spreadsheet below shows how a consumer's car loan is being repaid over a 5-year period. The elements in column A represent the loan amount at the beginning of a year; column B represents the amount owed after a year; and column C represents the amount of interest paid during the year.

	A	B	C
1	10,000.00	8,305.60	738.77
2	8,305.60	6,470.56	598.13
3	6,470.56	4,483.22	445.82
4	4,483.22	2,330.93	280.88
5	2,330.93	0.00	102.24

For instance, the element in 3C means that the consumer paid $445.82 in interest during the third year of the loan.

QUESTION What is the meaning of the element in 3A?

Matrices are effective for situations in which there are a number of items to be classified. For instance, suppose a music store has sales for January as shown in the matrix at the top of the next page.

ANSWER At the beginning of the third year of the loan, the consumer owed $6,470.56.

$$\begin{array}{l} \\ \text{CDs} \\ \text{Tapes} \\ \text{Videos} \end{array} \begin{array}{c} \begin{array}{ccccc} \text{Rock} & \text{R\&B} & \text{Rap} & \text{Classical} & \text{Other} \end{array} \\ \begin{bmatrix} 455 & 135 & 65 & 87 & 236 \\ 252 & 68 & 32 & 40 & 101 \\ 36 & 4 & 5 & 2 & 28 \end{bmatrix} \end{array}$$

This matrix indicates, for instance, that the music store sold 40 classical tapes in January.

Now consider a similar matrix for February.

$$\begin{array}{l} \\ \text{CDs} \\ \text{Tapes} \\ \text{Videos} \end{array} \begin{array}{c} \begin{array}{ccccc} \text{Rock} & \text{R\&B} & \text{Rap} & \text{Classical} & \text{Other} \end{array} \\ \begin{bmatrix} 402 & 128 & 68 & 101 & 255 \\ 259 & 35 & 28 & 51 & 115 \\ 28 & 7 & 3 & 5 & 33 \end{bmatrix} \end{array}$$

Looking at this matrix and the one for January reveals that the number of R&B tapes sold for the two months is $68 + 35 = 103$. By adding the elements in corresponding cells, we obtain the total sales for the two months. In matrix notation, this would be shown as

$$\begin{bmatrix} 455 & 135 & 65 & 87 & 236 \\ 252 & 68 & 32 & 40 & 101 \\ 36 & 4 & 5 & 2 & 28 \end{bmatrix} + \begin{bmatrix} 402 & 128 & 68 & 101 & 255 \\ 259 & 35 & 28 & 51 & 115 \\ 28 & 7 & 3 & 5 & 33 \end{bmatrix} = \begin{bmatrix} 857 & 263 & 133 & 188 & 491 \\ 511 & 103 & 60 & 91 & 216 \\ 64 & 11 & 8 & 7 & 61 \end{bmatrix}$$

In the matrix that represents the sum, 857 (in row 1, column 1) indicates that a total of 857 rock music CDs were sold in January and February. Similarly, a total of 91 (row 2, column 4) classical tapes were sold for the two months.

This example suggests that the addition of two matrices should be performed by adding the corresponding elements. Before we actually state this definition, we first introduce some notation and a definition of equality.

Throughout this book a matrix will be indicated by using a capital letter or by surrounding a lower-case letter with brackets. For instance, a matrix can be denoted as

$$A \quad \text{or} \quad [a_{ij}]$$

take note

The brackets around $[a_{ij}]$ indicate a matrix. If we write a_{ij} (no brackets), it refers to the element in the ith row and the jth column.

An important concept involving matrices is the principle of equality.

Definition of Equality of Two Matrices

Two matrices $A = [a_{ij}]$ and $B = [b_{ij}]$ are equal if and only if

$$a_{ij} = b_{ij}$$

for every i and j.

For example, if $A = \begin{bmatrix} a & -2 & b \\ 3 & c & 1 \end{bmatrix}$ and $B = \begin{bmatrix} 3 & x & -4 \\ 3 & -1 & y \end{bmatrix}$, then $A = B$ if and only if $a = 3$, $x = -2$, $b = -4$, $c = -1$, and $y = 1$.

QUESTION If two matrices A and B are equal, do they have the same order?

Definition of Addition of Matrices

If A and B are matrices of order $m \times n$, then the sum of the matrices is the $m \times n$ matrix given by

$$A + B = [a_{ij} + b_{ij}]$$

Here is an example. Let $A = \begin{bmatrix} 2 & -2 & 3 \\ 1 & 3 & -4 \end{bmatrix}$ and $B = \begin{bmatrix} 5 & -2 & 6 \\ -2 & 3 & 5 \end{bmatrix}$. Then

$$A + B = \begin{bmatrix} 2 & -2 & 3 \\ 1 & 3 & -4 \end{bmatrix} + \begin{bmatrix} 5 & -2 & 6 \\ -2 & 3 & 5 \end{bmatrix} = \begin{bmatrix} 2+5 & (-2)+(-2) & 3+6 \\ 1+(-2) & 3+3 & (-4)+5 \end{bmatrix}$$
$$= \begin{bmatrix} 7 & -4 & 9 \\ -1 & 6 & 1 \end{bmatrix}$$

Now let $C = \begin{bmatrix} 2 & -3 \\ 4 & 1 \end{bmatrix}$ and $D = \begin{bmatrix} 3 & 2 & 0 \\ 1 & -5 & 3 \end{bmatrix}$. Here $C + D$ is not defined because the matrices do not have the same order.

To define the subtraction of two matrices, we first define the additive inverse of a matrix.

Additive Inverse of a Matrix

Given the matrix $A = [a_{ij}]$, the additive inverse of A is $-A = [-a_{ij}]$.

For example, if $A = \begin{bmatrix} -2 & 3 & -1 \\ 0 & -1 & 4 \end{bmatrix}$, then the additive inverse of A is

$$-A = -\begin{bmatrix} -2 & 3 & -1 \\ 0 & -1 & 4 \end{bmatrix} = \begin{bmatrix} 2 & -3 & 1 \\ 0 & 1 & -4 \end{bmatrix}$$

Subtraction of two matrices is defined in terms of the additive inverse of a matrix.

Definition of Subtraction of Matrices

Given two matrices A and B of order $m \times n$, then $A - B$ is the sum of A and the additive inverse of B.

$$A - B = A + (-B)$$

ANSWER Yes. If they were of different order, there would be an element in one matrix for which there was no corresponding element in the second matrix.

As an example, let $A = \begin{bmatrix} 2 & -3 \\ -1 & 2 \\ 2 & 4 \end{bmatrix}$ and $B = \begin{bmatrix} -1 & 2 \\ -4 & 1 \\ 3 & -2 \end{bmatrix}$. Then

$$A - B = \begin{bmatrix} 2 & -3 \\ -1 & 2 \\ 2 & 4 \end{bmatrix} - \begin{bmatrix} -1 & 2 \\ -4 & 1 \\ 3 & -2 \end{bmatrix} = \begin{bmatrix} 2 & -3 \\ -1 & 2 \\ 2 & 4 \end{bmatrix} + \begin{bmatrix} 1 & -2 \\ 4 & -1 \\ -3 & 2 \end{bmatrix} = \begin{bmatrix} 3 & -5 \\ 3 & 1 \\ -1 & 6 \end{bmatrix}$$

Of special importance is the *zero matrix,* which is the matrix that consists of all zeros. The zero matrix is the additive identity for matrices.

Definition of the Zero Matrix

The $m \times n$ **zero matrix,** denoted by $\boldsymbol{O}$, is the matrix whose elements are all zeros.

Three examples of zero matrices are

$$\begin{bmatrix} 0 & 0 & 0 \\ 0 & 0 & 0 \end{bmatrix} \quad \begin{bmatrix} 0 & 0 & 0 & 0 \\ 0 & 0 & 0 & 0 \\ 0 & 0 & 0 & 0 \end{bmatrix} \quad \begin{bmatrix} 0 & 0 \\ 0 & 0 \end{bmatrix}$$

Properties of Matrix Addition

Given matrices A, B, C and the zero matrix $\boldsymbol{O}$, each of order $m \times n$, then the following properties hold.

Commutative	$A + B = B + A$
Associative	$A + (B + C) = (A + B) + C$
Additive inverse	$A + (-A) = \boldsymbol{O}$
Additive identity	$A + \boldsymbol{O} = \boldsymbol{O} + A = A$

◆ SCALAR MULTIPLICATION

Two types of products involve matrices. The first product we will discuss is the product of a real number and a matrix. Consider the matrix below, which shows the hourly wages for various job classifications in a construction firm before a 6% pay increase.

	Carpenter	Welder	Plumber	Electrician
Apprentice	12.75	15.86	14.76	16.87
Journeyman	15.60	18.07	16.89	19.05

After the pay increase, the pay in each job category will increase by 6%. This can be shown in matrix form as

$$1.06\begin{bmatrix} 12.75 & 15.86 & 14.76 & 16.87 \\ 15.60 & 18.07 & 16.89 & 19.05 \end{bmatrix} = \begin{bmatrix} 1.06 \cdot 12.75 & 1.06 \cdot 15.86 & 1.06 \cdot 14.76 & 1.06 \cdot 16.87 \\ 1.06 \cdot 15.60 & 1.06 \cdot 18.07 & 1.06 \cdot 16.89 & 1.06 \cdot 19.05 \end{bmatrix}$$

$$\approx \begin{bmatrix} 13.52 & 16.81 & 15.65 & 17.88 \\ 16.54 & 19.15 & 17.90 & 20.19 \end{bmatrix}$$

The element in row 1, column 4 indicates that an apprentice electrician will earn \$17.88 per hour after the pay increase.

This example suggests that to multiply a matrix by a constant, we multiply each entry in the matrix by the constant.

Definition of the Product of a Real Number and a Matrix

Given the $m \times n$ matrix $A = [a_{ij}]$ and the real number c, then $cA = [ca_{ij}]$.

Finding the product of a real number and a matrix is called **scalar multiplication.**

As an example of this definition, consider the matrix

$$A = \begin{bmatrix} 2 & -3 & 1 \\ 3 & 1 & -2 \\ 1 & -1 & 4 \end{bmatrix}$$

and the constant $c = -2$. Then

$$-2A = -2\begin{bmatrix} 2 & -3 & 1 \\ 3 & 1 & -2 \\ 1 & -1 & 4 \end{bmatrix} = \begin{bmatrix} -2(2) & -2(-3) & -2(1) \\ -2(3) & -2(1) & -2(-2) \\ -2(1) & -2(-1) & -2(4) \end{bmatrix} = \begin{bmatrix} -4 & 6 & -2 \\ -6 & -2 & 4 \\ -2 & 2 & -8 \end{bmatrix}$$

This definition is also used to factor a constant from a matrix.

$$\begin{bmatrix} \frac{3}{2} & -\frac{5}{4} & \frac{1}{4} \\ \frac{3}{4} & \frac{1}{2} & \frac{5}{2} \end{bmatrix} = \frac{1}{4}\begin{bmatrix} 6 & -5 & 1 \\ 3 & 2 & 10 \end{bmatrix}$$

Properties of Scalar Multiplication

Given real numbers a, b, and c and matrices $A = [a_{ij}]$ and $B = [b_{ij}]$ each of order $m \times n$, then

$$(b + c)A = bA + cA$$
$$c(A + B) = cA + cB$$
$$a(bA) = (ab)A$$

EXAMPLE 1 Find the Sum of Two Scalar Products

Given $A = \begin{bmatrix} -2 & 3 \\ 4 & -2 \\ 0 & 4 \end{bmatrix}$ and $B = \begin{bmatrix} 8 & -2 \\ -3 & 2 \\ -4 & 7 \end{bmatrix}$, find $2A + 5B$.

Continued

Solution

$$2A + 5B = 2\begin{bmatrix} -2 & 3 \\ 4 & -2 \\ 0 & 4 \end{bmatrix} + 5\begin{bmatrix} 8 & -2 \\ -3 & 2 \\ -4 & 7 \end{bmatrix}$$

$$= \begin{bmatrix} -4 & 6 \\ 8 & -4 \\ 0 & 8 \end{bmatrix} + \begin{bmatrix} 40 & -10 \\ -15 & 10 \\ -20 & 35 \end{bmatrix} = \begin{bmatrix} 36 & -4 \\ -7 & 6 \\ -20 & 43 \end{bmatrix}$$

TRY EXERCISE 6, EXERCISE SET 10.2, PAGE 736

◆ MATRIX MULTIPLICATION

Day	5:00 A.M.–5:00 P.M. $.23 per minute
Evening	5:00 P.M.–11:00 P.M. $.17 per minute
Night	11:00 P.M.–5:00 A.M. $.08 per minute

Now we turn to the product of two matrices. This product can be developed by considering long-distance telephone rates. The rates charged by a telephone company depend on the time of day a call is made. For this particular company, the schedule is shown in the table at the left. During one month, the number of long-distance minutes used by a customer of this telephone company was

Day	33 minutes
Evening	48 minutes
Night	15 minutes

The total cost for long-distance telephone service for that month is the sum of the products of the cost per minute and the number of minutes.

$$\text{Total cost} = 0.23(33) + 0.17(48) + 0.08(15) = \$16.95$$

In matrix terms, the cost per minute can be written as the *row* matrix $[0.23 \quad 0.17 \quad 0.08]$. The number of minutes of long-distance service used can be written as the *column* matrix $\begin{bmatrix} 33 \\ 48 \\ 15 \end{bmatrix}$. The product of the row matrix and the column matrix is

$$[0.23 \quad 0.17 \quad 0.08]\begin{bmatrix} 33 \\ 48 \\ 15 \end{bmatrix} = 0.23(33) + 0.17(48) + 0.08(15) = 16.95$$

In general, if A is a row matrix of order $1 \times n$,

$$A = [a_1 \quad a_2 \quad \cdots \quad a_n]$$

and B is a column matrix of order $n \times 1$,

$$B = \begin{bmatrix} b_1 \\ b_2 \\ \cdot \\ \cdot \\ \cdot \\ b_n \end{bmatrix}$$

then the product of A and B, written AB, is

$$AB = [a_1 \quad a_2 \quad a_3 \quad \cdots \quad a_n] \begin{bmatrix} b_1 \\ b_2 \\ \cdot \\ \cdot \\ \cdot \\ b_n \end{bmatrix} = a_1b_1 + a_2b_2 + a_3b_3 + \cdots + a_nb_n$$

take note

For the example at the right, the number of elements in the row matrix A equals the number of elements in the column matrix B. If this is not the case, AB is not defined. For instance, if

$A = [-2 \quad 3 \quad 1]$ and $B = \begin{bmatrix} 3 \\ 1 \end{bmatrix}$,

then AB is not defined.

For example, if $A = [2 \quad -1 \quad 4]$ and $B = \begin{bmatrix} -3 \\ 2 \\ 6 \end{bmatrix}$, then

$$AB = 2(-3) + (-1)(2) + 4(6) = 16$$

Now consider three phone companies (T_1, T_2, and T_3) with different rate structures and two customers (C_1 and C_2).

Telephone Company Rates (cost per minute)

	Day	Night	Evening
T_1	\$.23	\$.17	\$.08
T_2	\$.27	\$.12	\$.10
T_3	\$.26	\$.15	\$.09

Customer Time Chart (minutes)

	C_1	C_2
Day	45	52
Night	73	60
Evening	21	8

In terms of matrices, let the telephone companies' rate structure be denoted by T and the customers' time usage by C. Then

$$T = \begin{bmatrix} 0.23 & 0.17 & 0.08 \\ 0.27 & 0.12 & 0.10 \\ 0.26 & 0.15 & 0.09 \end{bmatrix} \quad \text{and} \quad C = \begin{bmatrix} 45 & 52 \\ 73 & 60 \\ 21 & 8 \end{bmatrix}$$

Let P denote the product TC. This product is determined by extending the concept of the product and a row-and-column matrix. Multiply each row of T and each column of C.

$$P = \begin{bmatrix} 0.23 & 0.17 & 0.08 \\ 0.27 & 0.12 & 0.10 \\ 0.26 & 0.15 & 0.09 \end{bmatrix} \begin{bmatrix} 45 & 52 \\ 73 & 60 \\ 21 & 8 \end{bmatrix}$$

$$= \begin{bmatrix} [0.23 \quad 0.17 \quad 0.08]\begin{bmatrix} 45 \\ 73 \\ 21 \end{bmatrix} & [0.23 \quad 0.17 \quad 0.08]\begin{bmatrix} 52 \\ 60 \\ 8 \end{bmatrix} \\ [0.27 \quad 0.12 \quad 0.10]\begin{bmatrix} 45 \\ 73 \\ 21 \end{bmatrix} & [0.27 \quad 0.12 \quad 0.10]\begin{bmatrix} 52 \\ 60 \\ 8 \end{bmatrix} \\ [0.26 \quad 0.15 \quad 0.09]\begin{bmatrix} 45 \\ 73 \\ 21 \end{bmatrix} & [0.26 \quad 0.15 \quad 0.09]\begin{bmatrix} 52 \\ 60 \\ 8 \end{bmatrix} \end{bmatrix}$$

$$= \begin{bmatrix} 0.23(45) + 0.17(73) + 0.08(21) & 0.23(52) + 0.17(60) + 0.08(8) \\ 0.27(45) + 0.12(73) + 0.10(21) & 0.27(52) + 0.12(60) + 0.10(8) \\ 0.26(45) + 0.15(73) + 0.09(21) & 0.26(52) + 0.15(60) + 0.09(8) \end{bmatrix}$$

$$= \begin{bmatrix} 24.44 & 22.80 \\ 23.01 & 22.04 \\ 24.54 & 23.24 \end{bmatrix}$$

Each entry in P is the total cost for long-distance service that each customer would incur for each of the three telephone companies. For example, $p_{11} = 24.44$ represents the amount company T_1 would charge customer C_1. The entry in row 3, column 2 ($p_{32} = 23.24$) represents the amount company T_3 would charge customer C_2. In each case, the subscripts on an element of P denote the company and the customer, respectively.

Using this application as a model, we now define the product of two matrices. The definition is an extension of the definition of the product of a row matrix and a column matrix.

take note

This definition may appear complicated, but basically, to multiply two matrices, multiply each row of the first matrix by each column of the second matrix.

Definition of the Product of Two Matrices

Let $A = [a_{ij}]$ be a matrix of order $m \times n$, and let $B = [b_{ij}]$ be a matrix of order $n \times p$. Then the product AB is the matrix of order $m \times p$ given by $AB = [c_{ij}]$, where each element c_{ij} is

$$c_{ij} = [a_{i1} \quad a_{i2} \quad a_{i3} \cdots a_{in}] \begin{bmatrix} b_{1j} \\ b_{2j} \\ b_{3j} \\ \vdots \\ b_{nj} \end{bmatrix} = a_{i1}b_{1j} + a_{i2}b_{2j} + a_{i3}b_{3j} + \cdots + a_{in}b_{nj}$$

For the product of two matrices to be possible, the number of columns of the first matrix must equal the number of rows of the second matrix.

$$\underset{m \times n}{A} \cdot \underset{n \times p}{B} = \underset{m \times p}{C}$$

Must be equal (the n of A and the n of B)

Order of product matrix (the m of A and the p of B)

The product matrix has as many rows as the first matrix and as many columns as the second matrix. For example, let

$$A = \begin{bmatrix} 2 & -3 & 0 \\ 1 & 4 & -1 \end{bmatrix} \quad \text{and} \quad \begin{bmatrix} 1 & 0 \\ 4 & -2 \\ 3 & 5 \end{bmatrix}$$

Then A has order 2×3 and B has order 3×2. Thus the order of AB is 2×2.

$$\begin{bmatrix} 2 & -3 & 0 \\ 1 & 4 & -1 \end{bmatrix}_{2\times 3} \begin{bmatrix} 1 & 0 \\ 4 & -2 \\ 3 & 5 \end{bmatrix}_{3\times 2} = \begin{bmatrix} [2 \quad -3 \quad 0]\begin{bmatrix} 1 \\ 4 \\ 3 \end{bmatrix} & [2 \quad -3 \quad 0]\begin{bmatrix} 0 \\ -2 \\ 5 \end{bmatrix} \\ [1 \quad 4 \quad -1]\begin{bmatrix} 1 \\ 4 \\ 3 \end{bmatrix} & [1 \quad 4 \quad -1]\begin{bmatrix} 0 \\ -2 \\ 5 \end{bmatrix} \end{bmatrix}_{2\times 2}$$

$$= \begin{bmatrix} 2(1) + (-3)(4) + 0(3) & 2(0) + (-3)(-2) + 0(5) \\ 1(1) + 4(4) \quad + (-1)(3) & 1(0) + 4(-2) \quad + (-1)(5) \end{bmatrix}_{2\times 2} = \begin{bmatrix} -10 & 6 \\ 14 & -13 \end{bmatrix}_{2\times 2}$$

EXAMPLE 2 Find the Product of Two Matrices

Find each product.

a. $\begin{bmatrix} 2 & 3 \\ -3 & 1 \\ 1 & -3 \end{bmatrix} \begin{bmatrix} 1 & 2 & -2 & 3 \\ -1 & 0 & 3 & -4 \end{bmatrix}$ **b.** $\begin{bmatrix} 1 & -1 & 3 \\ 2 & 2 & -1 \\ 0 & -2 & 3 \end{bmatrix} \begin{bmatrix} 4 & -2 & 0 \\ -1 & 3 & 1 \\ 2 & -3 & 1 \end{bmatrix}$

Solution

a. $\begin{bmatrix} 2 & 3 \\ -3 & 1 \\ 1 & -3 \end{bmatrix} \begin{bmatrix} 1 & 2 & -2 & 3 \\ -1 & 0 & 3 & -4 \end{bmatrix}$

$$= \begin{bmatrix} 2(1) + 3(-1) & 2(2) + 3(0) & 2(-2) + 3(3) & 2(3) + 3(-4) \\ (-3)(1) + 1(-1) & (-3)(2) + 1(0) & (-3)(-2) + 1(3) & (-3)3 + 1(-4) \\ 1(1) + (-3)(-1) & 1(2) + (-3)(0) & 1(-2) + (-3)(3) & 1(3) + (-3)(-4) \end{bmatrix}$$

$$= \begin{bmatrix} -1 & 4 & 5 & -6 \\ -4 & -6 & 9 & -13 \\ 4 & 2 & -11 & 15 \end{bmatrix}$$

b. $\begin{bmatrix} 1 & -1 & 3 \\ 2 & 2 & -1 \\ 0 & -2 & 3 \end{bmatrix} \begin{bmatrix} 4 & -2 & 0 \\ -1 & 3 & 1 \\ 2 & -3 & 1 \end{bmatrix}$

$$= \begin{bmatrix} 4 + 1 + 6 & -2 + (-3) + (-9) & 0 + (-1) + 3 \\ 8 + (-2) + (-2) & -4 + 6 + 3 & 0 + 2 + (-1) \\ 0 + 2 + 6 & 0 + (-6) + (-9) & 0 + (-2) + 3 \end{bmatrix}$$

$$= \begin{bmatrix} 11 & -14 & 2 \\ 4 & 5 & 1 \\ 8 & -15 & 1 \end{bmatrix}$$

TRY EXERCISE 16, EXERCISE SET 10.2, PAGE 737

Graphing calculators can be used to perform matrix operations. You first enter the dimension of the matrix and then its elements. The matrix is stored (usually) as an upper-case letter. Once you have entered the matrices, you can use the regular arithmetic operation keys and the variable names for the matrices to perform many operations. For instance, let $A = \begin{bmatrix} -2 & 3 & 4 \\ 1 & 0 & -2 \\ 3 & 1 & -1 \end{bmatrix}$ and $B = \begin{bmatrix} 1 & 2 & -1 \\ 3 & 1 & 0 \\ -1 & 1 & 2 \end{bmatrix}$. Typical calculator displays for addition and multiplication of matrices A and B are shown in **Figure 10.3.**

```
[A] + [B]
          [[-1  5  3]
           [4   1 -2]
           [2   2  1]]
```

```
[A] * [B]
          [[3  3  10]
           [3  0  -5]
           [7  6  -5]]
```

Figure 10.3

Generally, matrix multiplication is not commutative. That is, given two matrices A and B, $AB \neq BA$. In some cases, as in Example 2a, if the matrices were reversed, the product would not be defined.

$$\begin{bmatrix} 1 & 2 & -2 & 3 \\ -1 & 0 & 3 & -4 \end{bmatrix}_{2\times 4} \begin{bmatrix} 2 & 3 \\ -3 & 1 \\ 1 & -3 \end{bmatrix}_{3\times 2}$$

columns ≠ rows

Even in those cases where multiplication is defined, the products AB and BA may not be equal. Finding the product of Example 2b with the matrices reversed illustrates this point.

$$\begin{bmatrix} 4 & -2 & 0 \\ -1 & 3 & 1 \\ 2 & -3 & 1 \end{bmatrix} \begin{bmatrix} 1 & -1 & 3 \\ 2 & 2 & -1 \\ 0 & -2 & 3 \end{bmatrix} = \begin{bmatrix} 0 & -8 & 14 \\ 5 & 5 & -3 \\ -4 & -10 & 12 \end{bmatrix} \neq \begin{bmatrix} 11 & -14 & 2 \\ 4 & 5 & 1 \\ 8 & -15 & 1 \end{bmatrix}$$

Although matrix multiplication is not a commutative operation, the associative property of multiplication and the distributive property do hold for matrices.

Properties of Matrix Multiplication

Associative property Given matrices A, B, and C of orders $m \times n$, $n \times p$, and $p \times q$, respectively, then

$$A(BC) = (AB)C$$

Distributive property Given matrices A_1 and A_2 of order $m \times n$ and matrices B_1 and B_2 of order $n \times p$, then

$$A_1(B_1 + B_2) = A_1B_1 + A_1B_2 \quad \textbf{• Left distributive property}$$

$$(A_1 + A_2)B_1 = A_1B_1 + A_2B_1 \quad \textbf{• Right distributive property}$$

take note

You can verify that matrix multiplication is not commutative by using a graphing calculator. For instance, let $A = \begin{bmatrix} -1 & 2 & 4 \\ 3 & 0 & -2 \\ 3 & 1 & 5 \end{bmatrix}$ and $B = \begin{bmatrix} 4 & -2 & 0 \\ 1 & -3 & 6 \\ 2 & 1 & -1 \end{bmatrix}$. The matrix products AB and BA are shown in **Figure 10.4.**

```
[A] * [B]
          [[6   0  8]
           [8  -8  2]
           [23 -4  1]]
```

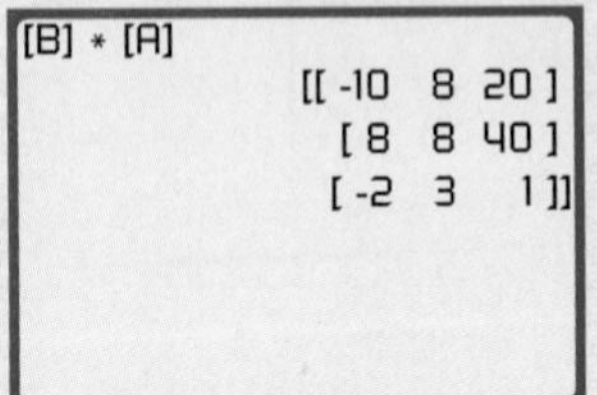

Figure 10.4

A square matrix that has a 1 for each element on the main diagonal and zeros elsewhere is called an *identity matrix*.

The **identity matrix** of order n, denoted I_n, is the $n \times n$ matrix

$$I_n = \begin{bmatrix} 1 & 0 & 0 & \cdots & 0 \\ 0 & 1 & 0 & \cdots & 0 \\ 0 & 0 & 1 & \cdots & 0 \\ \vdots & \vdots & \vdots & \ddots & \vdots \\ 0 & 0 & 0 & \cdots & 1 \end{bmatrix}_{n\times n}$$

The identity matrix has properties similar to those of the real number 1. For example, the product of a matrix A below and I_3 is A.

$$\begin{bmatrix} 2 & -3 & 0 \\ 4 & 7 & -5 \\ 9 & 8 & -6 \end{bmatrix}\begin{bmatrix} 1 & 0 & 0 \\ 0 & 1 & 0 \\ 0 & 0 & 1 \end{bmatrix} = \begin{bmatrix} 2 & -3 & 0 \\ 4 & 7 & -5 \\ 9 & 8 & -6 \end{bmatrix}$$

Multiplicative Identity Property for Matrices

If A is a square matrix of order n, and I_n is the identity matrix of order n, then $AI_n = I_nA = A$.

◆ MATRIX PRODUCTS AND SYSTEMS OF EQUATIONS

Consider the system of equations

$$\begin{cases} 2x + 3y - z = 5 \\ x - 2y + 2z = 6 \\ 4x + y - 3z = 5 \end{cases}$$

This system can be expressed as a product of matrices, as shown below.

$$\begin{bmatrix} 2x + 3y - z \\ x - 2y + 2z \\ 4x + y - 3z \end{bmatrix} = \begin{bmatrix} 5 \\ 6 \\ 5 \end{bmatrix}$$ • **Equality of matrices**

$$\begin{bmatrix} 2 & 3 & -1 \\ 1 & -2 & 2 \\ 4 & 1 & -3 \end{bmatrix}\begin{bmatrix} x \\ y \\ z \end{bmatrix} = \begin{bmatrix} 5 \\ 6 \\ 5 \end{bmatrix}$$ • **Definition of matrix multiplication**

Reversing this procedure, certain matrix products can represent systems of equations. Consider the matrix equation

$$\begin{bmatrix} 4 & 3 & -2 \\ 1 & -2 & 3 \\ 1 & 0 & 5 \end{bmatrix}_{3\times 3}\begin{bmatrix} x \\ y \\ z \end{bmatrix}_{3\times 1} = \begin{bmatrix} 2 \\ -1 \\ 3 \end{bmatrix}_{3\times 1}$$

Continued •►

$$\begin{bmatrix} 4x + 3y - 2z \\ x - 2y + 3z \\ x \qquad + 5z \end{bmatrix}_{3\times1} = \begin{bmatrix} 2 \\ -1 \\ 3 \end{bmatrix}_{3\times1}$$ • Definition of matrix multiplication

$$\begin{cases} 4x + 3y - 2z = 2 \\ x - 2y + 3z = -1 \\ x \qquad + 5z = 3 \end{cases}$$ • Equality of matrices

Performing operations on matrices that represent a system of equations is another method of solving systems of equations. This is discussed in the next section.

TOPICS FOR DISCUSSION

1. How are matrices related to spreadsheet programs?
2. Is it always possible to add two matrices? If so, explain why. If not, discuss what conditions must be met for two matrices to be added. Is matrix addition a commutative operation?
3. Is it always possible to multiply two matrices? If so, explain why. If not, discuss what conditions must be met for two matrices to be multiplied. Is matrix multiplication a commutative operation?
4. How does scalar multiplication differ from matrix multiplication?
5. Is it possible that two matrices could be added but not multiplied? Can two matrices be multiplied but not added? Discuss what type of conditions must be met for two matrices to be both added and multiplied.

EXERCISE SET 10.2

In Exercises 1 to 8, find *a.* **$A + B$,** *b.* **$A - B$,** *c.* **$2B$, and** *d.* **$2A - 3B$.**

1. $A = \begin{bmatrix} 2 & -1 \\ 3 & 3 \end{bmatrix} \quad B = \begin{bmatrix} -1 & 3 \\ 2 & 1 \end{bmatrix}$

2. $A = \begin{bmatrix} 0 & -2 \\ 2 & 3 \end{bmatrix} \quad B = \begin{bmatrix} 5 & -1 \\ 3 & 0 \end{bmatrix}$

3. $A = \begin{bmatrix} 0 & -1 & 3 \\ 1 & 0 & -2 \end{bmatrix} \quad B = \begin{bmatrix} -3 & 1 & 2 \\ 2 & 5 & -3 \end{bmatrix}$

4. $A = \begin{bmatrix} 2 & -2 & 4 \\ 0 & -3 & -4 \end{bmatrix} \quad B = \begin{bmatrix} 1 & -5 & 6 \\ 4 & -2 & -3 \end{bmatrix}$

5. $A = \begin{bmatrix} -3 & 4 \\ 2 & -3 \\ -1 & 0 \end{bmatrix} \quad B = \begin{bmatrix} 4 & 1 \\ 1 & -2 \\ 3 & -4 \end{bmatrix}$

6. $A = \begin{bmatrix} 2 & -2 \\ 3 & 4 \\ 1 & 0 \end{bmatrix} \quad B = \begin{bmatrix} -1 & 8 \\ 2 & -2 \\ -4 & 3 \end{bmatrix}$

7. $A = \begin{bmatrix} -2 & 3 & -1 \\ 0 & -1 & 2 \\ -4 & 3 & 3 \end{bmatrix} \quad B = \begin{bmatrix} 1 & -2 & 0 \\ 2 & 3 & -1 \\ 3 & -1 & 2 \end{bmatrix}$

8. $A = \begin{bmatrix} 0 & 2 & 0 \\ 1 & -3 & 3 \\ 5 & 4 & -2 \end{bmatrix} \quad B = \begin{bmatrix} -1 & 2 & 4 \\ 3 & 3 & -2 \\ -4 & 4 & 3 \end{bmatrix}$

In Exercises 9 to 16, find AB and BA if possible.

9. $A = \begin{bmatrix} 2 & -3 \\ 1 & 4 \end{bmatrix} \quad B = \begin{bmatrix} -2 & 4 \\ 2 & -3 \end{bmatrix}$

10. $A = \begin{bmatrix} 3 & -2 \\ 4 & 1 \end{bmatrix}$ $B = \begin{bmatrix} -1 & -1 \\ 0 & 4 \end{bmatrix}$

11. $A = \begin{bmatrix} 3 & -1 \\ 2 & 3 \end{bmatrix}$ $B = \begin{bmatrix} 4 & 1 \\ 2 & -3 \end{bmatrix}$

12. $A = \begin{bmatrix} -3 & 2 \\ 2 & -2 \end{bmatrix}$ $B = \begin{bmatrix} 0 & 2 \\ -2 & 4 \end{bmatrix}$

13. $A = \begin{bmatrix} 2 & -1 \\ 0 & 3 \\ 1 & -2 \end{bmatrix}$ $B = \begin{bmatrix} 1 & -2 & 3 \\ 2 & 0 & 1 \end{bmatrix}$

14. $A = \begin{bmatrix} -1 & 3 \\ 2 & 1 \\ -3 & -2 \end{bmatrix}$ $B = \begin{bmatrix} 0 & -1 & 2 \\ 1 & 2 & -4 \end{bmatrix}$

15. $A = \begin{bmatrix} 2 & -1 & 3 \\ 0 & 2 & -1 \\ 0 & 0 & 2 \end{bmatrix}$ $B = \begin{bmatrix} 2 & 0 & 0 \\ 1 & -1 & 0 \\ 2 & -1 & -2 \end{bmatrix}$

16. $A = \begin{bmatrix} -1 & 2 & 0 \\ 2 & -1 & 1 \\ -2 & 2 & -1 \end{bmatrix}$ $B = \begin{bmatrix} 2 & -1 & 0 \\ 1 & 5 & -1 \\ 0 & -1 & 3 \end{bmatrix}$

In Exercises 17 to 24, find *AB* if possible.

17. $A = [1 \quad -2 \quad 3]$ $B = \begin{bmatrix} 1 & 0 \\ 2 & -1 \\ 1 & 2 \end{bmatrix}$

18. $A = \begin{bmatrix} -2 & 3 \\ 1 & -2 \\ 0 & 2 \end{bmatrix}$ $B = \begin{bmatrix} 3 \\ -2 \end{bmatrix}$

19. $A = \begin{bmatrix} 2 & -1 \\ 3 & 3 \end{bmatrix}$ $B = \begin{bmatrix} 1 & -2 \\ 3 & 1 \\ 0 & -2 \end{bmatrix}$

20. $A = \begin{bmatrix} 2 & 0 & -1 \\ 3 & 4 & -3 \end{bmatrix}$ $B = \begin{bmatrix} 3 & -1 & 0 \\ 2 & 4 & 5 \end{bmatrix}$

21. $A = \begin{bmatrix} 2 & 3 \\ -4 & -6 \end{bmatrix}$ $B = \begin{bmatrix} 3 & 6 \\ -2 & -4 \end{bmatrix}$

22. $A = \begin{bmatrix} 2 & -1 & 3 \\ -1 & 2 & 1 \end{bmatrix}$ $B = \begin{bmatrix} 1 & 3 & 2 \\ 2 & -1 & 0 \\ 3 & 1 & 2 \end{bmatrix}$

23. $A = \begin{bmatrix} 1 & 2 & -2 & 3 \\ 0 & -2 & 1 & -3 \end{bmatrix}$ $B = \begin{bmatrix} -2 & 0 \\ 4 & -2 \end{bmatrix}$

24. $A = \begin{bmatrix} 2 & -2 & 4 \\ 1 & 0 & -1 \\ 2 & 1 & 3 \end{bmatrix}$ $B = \begin{bmatrix} 2 & 1 & -3 & 0 \\ 0 & -2 & 1 & -2 \\ 1 & -1 & 0 & 2 \end{bmatrix}$

In Exercises 25 to 28, given the matrices

$$A = \begin{bmatrix} -1 & 3 \\ 2 & -1 \\ 3 & 1 \end{bmatrix} \quad \text{and} \quad B = \begin{bmatrix} 0 & -2 \\ 1 & 3 \\ 4 & -3 \end{bmatrix}$$

find the 3 × 2 matrix *X* that is a solution of the equation.

25. $3X + A = B$

26. $2A - 3X = 5B$

27. $2X - A = X + B$

28. $3X + 2B = X - 2A$

In Exercises 29 to 32, use the matrices

$$A = \begin{bmatrix} 2 & -3 \\ 1 & -1 \end{bmatrix} \quad \text{and} \quad B = \begin{bmatrix} 3 & -1 & 0 \\ 2 & -2 & -1 \\ 1 & 0 & 2 \end{bmatrix}$$

If *A* is a square matrix, then $A^n = A \cdot A \cdot A \cdots A$, where the matrix *A* is repeated *n* times.

29. Find A^2.

30. Find A^3.

31. Find B^2.

32. Find B^3.

In Exercises 33 to 38, find the system of equations that is equivalent to the given matrix equation.

33. $\begin{bmatrix} 3 & -8 \\ 4 & 3 \end{bmatrix}\begin{bmatrix} x \\ y \end{bmatrix} = \begin{bmatrix} 11 \\ 1 \end{bmatrix}$

34. $\begin{bmatrix} 2 & 7 \\ 3 & -4 \end{bmatrix}\begin{bmatrix} x \\ y \end{bmatrix} = \begin{bmatrix} 1 \\ 16 \end{bmatrix}$

35. $\begin{bmatrix} 1 & -3 & -2 \\ 3 & 1 & 0 \\ 2 & -4 & 5 \end{bmatrix}\begin{bmatrix} x \\ y \\ z \end{bmatrix} = \begin{bmatrix} 6 \\ 2 \\ 1 \end{bmatrix}$

36. $\begin{bmatrix} 2 & 0 & 5 \\ 3 & -5 & 1 \\ 4 & -7 & 6 \end{bmatrix}\begin{bmatrix} x \\ y \\ z \end{bmatrix} = \begin{bmatrix} 9 \\ 7 \\ 14 \end{bmatrix}$

37. $\begin{bmatrix} 2 & -1 & 0 & 2 \\ 4 & 1 & 2 & -3 \\ 6 & 0 & 1 & -2 \\ 5 & 2 & -1 & -4 \end{bmatrix}\begin{bmatrix} x_1 \\ x_2 \\ x_3 \\ x_4 \end{bmatrix} = \begin{bmatrix} 5 \\ 6 \\ 10 \\ 8 \end{bmatrix}$

38. $\begin{bmatrix} 5 & -1 & 2 & -3 \\ 4 & 0 & 2 & 0 \\ 2 & -2 & 5 & -4 \\ 3 & 1 & -3 & 4 \end{bmatrix}\begin{bmatrix} x_1 \\ x_2 \\ x_3 \\ x_4 \end{bmatrix} = \begin{bmatrix} -2 \\ 2 \\ -1 \\ 2 \end{bmatrix}$

39. **LIFE SCIENCES** Biologists use capture-recapture models to estimate how many animals live in a certain area. A sample of, say, fish are caught and tagged. When subsequent samples of fish are caught, a biologist can use a capture history matrix to record (with a 1) which, if any, of the fish in the original sample are caught again. The rows of this matrix represent the particular fish (each has its own

identification number), and the columns represent the number of the sample in which the fish was caught. Here is a small capture history matrix.

Samples

	1	2	3	4
Fish A	1	0	0	1
Fish B	0	1	1	1
Fish C	0	0	1	1

a. What is the dimension of this matrix? Write a sentence that explains the meaning of dimension in this case.

b. What is the meaning of the 1 in row A, column 4?

c. Which fish was captured the most times?

40. **LIFE SCIENCE** Biologists can use a predator-prey matrix to study the relationships among animals in an ecosystem. Each row and each column represents an animal in the system. A 1 as an element in the matrix indicates that the animal represented by that row preys on the animal represented by that column. A 0 indicates that the animal in that row does not prey on the animal in that column. A simple predator-prey matrix is shown below. The abbreviations are H = hawk, R = rabbit, S = snake, C = coyote.

	H	R	S	C
H	0	1	1	0
R	0	0	0	0
S	1	1	0	0
C	0	1	1	0

a. What is the dimension of this matrix? Write a sentence that explains the meaning of dimension in this case.

b. What is the meaning of the 0 in row 2, column 1?

c. What is the meaning of there being all zeros in column C?

d. What is the meaning of all zeros in row R?

41. **BUSINESS** The matrix below shows the sales revenues, in millions of dollars, that a pharmaceutical company received from various divisions in different parts of the country. The abbreviations are W = western states, N = northern states, S = southern states, and E = eastern states.

	W	N	S	E
Patented drugs	2.0	1.4	3.0	1.4
Generic drugs	0.8	1.1	2.0	0.9
Nonprescription drugs	3.6	1.2	4.5	1.5

Suppose the business plan for this company indicates that it anticipates a 2% decrease in sales (because of competition) for each of its drug divisions for each region of the country. Express, to the nearest ten thousand dollars, this matrix as a scalar product and compute the anticipated sales matrix.

42. **SALARY SCHEDULES** The partial current-year salary matrix for an elementary school district is given below. Column A indicates a B.A. degree, column B a B.A. degree plus 15 graduate units, column C an M.A. degree, and column D an M.A. degree plus 30 additional graduate units. The rows give the numbers of years of teaching experience. Each entry is the annual salary in thousands of dollars.

Years	A	B	C	D
0 to 5	18.0	18.9	20.0	21.5
5 to 9	19.0	20.3	22.5	24.5
10 to 15	20.0	21.4	24.0	27.0

Express, as matrix scalar multiplication to the nearest hundred dollars, the result of the school board's approving a 6% salary increase for all teachers in this district, and compute the scalar product.

43. **SPORTS** The matrices for the number of wins and losses at home, H, and away, A, are shown for the top 3 finishers of the 2000 American League East division baseball teams.

$H =$

	W	L
New York	44	36
Boston	42	39
Toronto	45	36

$A =$

	W	L
New York	43	38
Boston	43	38
Toronto	38	43

a. Find $H + A$.

b. Write a sentence that explains the meaning of the sum of the two matrices.

c. Find $H - A$.

d. Write a sentence that explains the meaning of the difference of the two matrices.

44. **BUSINESS** Let A represent the number of televisions of various sizes in two stores of a company in one city, and let B represent the same situation for the company in a second city.

$A =$

	19-inch	25-inch	40-inch
Store 1	23	35	49
Store 2	32	41	24

$B =$

	19-inch	25-inch	40-inch
Store 1	19	28	36
Store 2	25	38	26

a. Find $A + B$.

b. Write a sentence that explains the meaning of the sum of the two matrices.

45. **GEOMETRIC TRANSFORMATION** Consider the rectangle shown below. Each pair of x- and y-coordinates of the points shown appears as a column of a matrix of dimension 2×4. This is matrix A below. Matrix T is called a *translation matrix*.

$$A = \begin{bmatrix} -2 & 4 & 2 & -4 \\ 5 & 2 & -2 & 1 \end{bmatrix}$$

$$T = \begin{bmatrix} 2 & 2 & 2 & 2 \\ -1 & -1 & -1 & -1 \end{bmatrix}$$

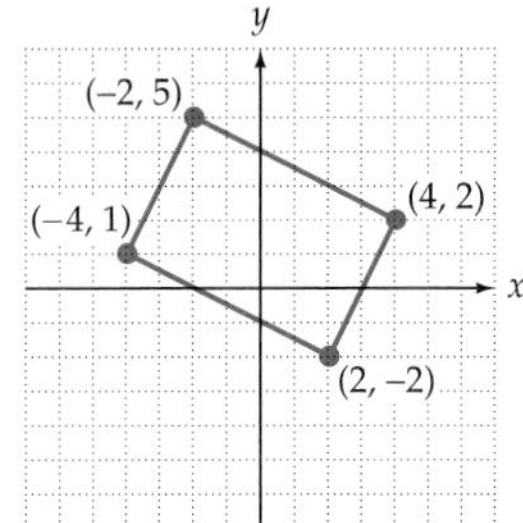

a. Find $A + T$.

b. Using the columns of $A + T$ as the x- and y-coordinates of four points, plot the points on the same coordinate grid as the original rectangle. Construct a polygon by connecting the points in the order of the columns. What polygon have you constructed?

c. Write a sentence that explains the relationship between the original polygon and the new one.

46. **GEOMETRIC TRANSFORMATION** Use the information in Exercise 45.

a. Find $A - T$.

b. Using the columns of $A - T$ as the x- and y-coordinates of four points, plot the points on the same coordinate grid as the original rectangle. Construct a polygon by connecting the points in the order of the columns. What polygon have you constructed?

c. Write a sentence that explains the relationship between the original polygon and the new one.

47. **GEOMETRIC TRANSFORMATION** Consider the information in Exercise 45 and the matrix $R = \begin{bmatrix} 0 & -1 \\ 1 & 0 \end{bmatrix}$.

a. Find $R \cdot A$.

b. Using the columns of $R \cdot A$ as the x- and y-coordinates of four points, plot the points on the same coordinate grid as the original rectangle. Construct a polygon by connecting the points in the order of the columns. What polygon have you constructed?

c. Write a sentence that explains the meaning of the product of these two matrices.

48. **GEOMETRIC TRANSFORMATION** Consider the information in Exercise 45 and the matrix $R = \begin{bmatrix} 0 & 1 \\ 1 & 0 \end{bmatrix}$.

a. Find $R \cdot A$.

b. Using the columns of $R \cdot A$ as the x- and y-coordinates of four points, plot the points on the same coordinate grid as the original rectangle. Construct a polygon by connecting the points in the order of the columns. What polygon have you constructed?

c. Write a sentence that explains the meaning of the product of these two matrices.

49. **BUSINESS INVENTORY** Matrix A gives the stock on hand of four products in a warehouse at the beginning of the week, and matrix B gives the stock on hand for the same four items at the end of the week. Find and interpret $A - B$.

$A =$

	Blue	Green	Red
Pens	530	650	815
Pencils	190	385	715
Ink	485	600	610
Colored Lead	150	210	305

$B =$

	Blue	Green	Red
Pens	480	500	675
Pencils	175	215	345
Ink	400	350	480
Colored Lead	70	95	280

50. **BUSINESS SERVICES** Matrix A gives the number of employees in the divisions of a company in the west coast branch, and matrix B gives the same information for the east coast branch. Find and interpret $A + B$.

$A =$

	Engineering	Administration	Data Processing
Division I	315	200	415
Division II	285	175	300
Division III	275	195	250

$B =$

	Engineering	Administration	Data Processing
Division I	200	175	350
Division II	150	90	180
Division III	105	50	175

51. **BUSINESS INVENTORY** The total unit sales matrix for three computer stores is given by

$S =$

	Monitors	Printers	Computers	Drives
Store A	25	31	35	12
Store B	20	12	30	15
Store C	16	19	25	18

The unit pricing matrix in dollars for the three stores is given by

$$P = \begin{bmatrix} 250 & 225 & 315 \\ 180 & 210 & 225 \\ 400 & 425 & 450 \\ 89 & 95 & 78 \end{bmatrix} \begin{matrix} \text{Monitor} \\ \text{Printer} \\ \text{Computer} \\ \text{Drive} \end{matrix}$$

(columns: Store A, Store B, Store C)

Find the gross income matrix.

52. **YOUTH SPORTS** The total unit sales matrix at three soccer games in a summer league for children is given by

$$S = \begin{bmatrix} 52 & 50 & 75 & 20 \\ 45 & 48 & 80 & 20 \\ 62 & 70 & 78 & 25 \end{bmatrix} \begin{matrix} \text{Game 1} \\ \text{Game 2} \\ \text{Game 3} \end{matrix}$$

(columns: Soft Drinks, Hot Dogs, Candy, Popcorn)

The unit pricing matrix in dollars for the wholesale cost of each item and the retail price of each item is given by

$$P = \begin{bmatrix} 0.25 & 0.50 \\ 0.30 & 0.75 \\ 0.15 & 0.45 \\ 0.10 & 0.50 \end{bmatrix} \begin{matrix} \text{Soft Drinks} \\ \text{Hot Dogs} \\ \text{Candy} \\ \text{Popcorn} \end{matrix}$$

(columns: Wholesale, Retail)

Use matrix multiplication to find the total cost and total revenue at each game.

In Exercises 53 to 58, use a graphing calculator to perform the indicated operations on matrices A and B.

$$A = \begin{bmatrix} 2 & -1 & 3 & 5 & -1 \\ 2 & 0 & 2 & -1 & 1 \\ -1 & -3 & 2 & 3 & 3 \\ 5 & -4 & 1 & 0 & 3 \\ 0 & 2 & -1 & 4 & 3 \end{bmatrix}$$

$$B = \begin{bmatrix} 0 & -2 & 1 & 7 & 2 \\ -3 & 0 & 2 & 3 & 1 \\ -2 & 1 & 1 & 4 & 5 \\ 6 & 4 & -4 & 2 & -3 \\ 3 & -2 & -5 & 1 & 3 \end{bmatrix}$$

53. AB
54. BA
55. A^3
56. B^3
57. $A^2 + B^2$
58. $AB - BA$

SUPPLEMENTAL EXERCISES

The elements of a matrix can be complex numbers. In Exercises 59 to 68, let

$$A = \begin{bmatrix} 2+3i & 1-2i \\ 1+i & 2-i \end{bmatrix} \quad \text{and} \quad B = \begin{bmatrix} 1-i & 2+3i \\ 3+2i & 4-i \end{bmatrix}$$

Perform the indicated operations.

59. $3A$
60. $-2B$
61. $2iB$
62. $3iA$
63. $A + B$
64. $A - B$
65. AB
66. BA
67. A^2
68. B^2

Matrices with complex number elements play a role in the theory of the atom. The following three matrices, called Pauli spin matrices, were used by Wolfgang Pauli in his early study of the electron. Use these matrices in Exercises 69 to 71.

$$\sigma_1 = \begin{bmatrix} 0 & 1 \\ 1 & 0 \end{bmatrix} \quad \sigma_2 = \begin{bmatrix} 0 & -i \\ i & 0 \end{bmatrix} \quad \sigma_3 = \begin{bmatrix} 1 & 0 \\ 0 & -1 \end{bmatrix}$$

69. Show that $(\sigma_i)^2 = I_2$ for $i = 1, 2$, and 3.
70. Show that $\sigma_1 \cdot \sigma_2 = i\sigma_3$.
71. Show that $\sigma_1 \cdot \sigma_2 + \sigma_2 \cdot \sigma_1 = O$.
72. Given two real numbers a and b and a matrix A of order 2×2, prove that $(a + b)A = aA + bA$.
73. Given two real numbers a and b and a matrix A of order 2×2, prove that $a(bA) = (ab)A$.

PROJECTS

1. **MATRICES IN GRAPHICS ART** Matrices can be used to translate and dilate geometric figures in the plane. These concepts are important in graphics design and art. Consider the kite shown at the right.

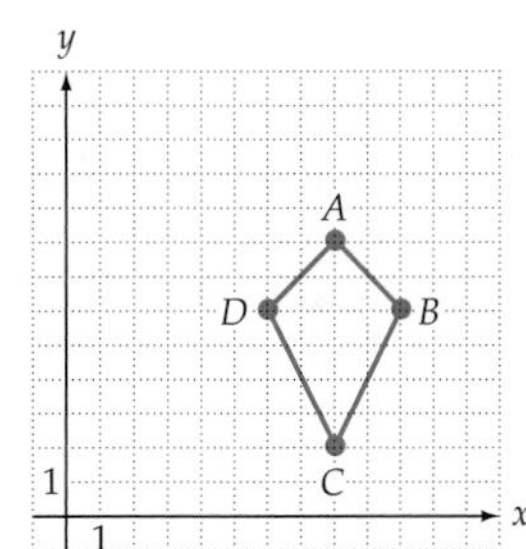

a. Find the coordinates of each vertex of the kite, and prepare a matrix K of dimension 2×4, where the columns are the x- and y-coordinates, respectively, of the vertices of the kite.

b. Find the product $\frac{1}{2}K = M$. Plot the points of M, using the columns as the x- and y-coordinates, respectively, of the vertices of a new kite.

c. Find the length of any two line segments of the original kite and the corresponding lengths for the new kite. Show that the **scale factor**, which is the ratio $\frac{\text{length of new}}{\text{length of original}}$, is $\frac{1}{2}$ for each pair of line segments. Observe that this is the coefficient of K in the product in **b**.

d. Find the product $2K = N$. Plot the points of N again, using the columns as the x- and y-coordinates, respectively, of the vertices of a new kite. Show that the new kite has a scale factor of 2.

e. Consider the vertices A, B, C, and D of the original kite and the corresponding vertices A', B', C', and D' of the kite from **c**. Show that the lines through AA', BB', CC', and DD' pass through the origin of the coordinate system. This point of intersection is called the **center of dilation**.

2. See our web site at http://college.hmco.com for a program that animates translation of geometric figures.

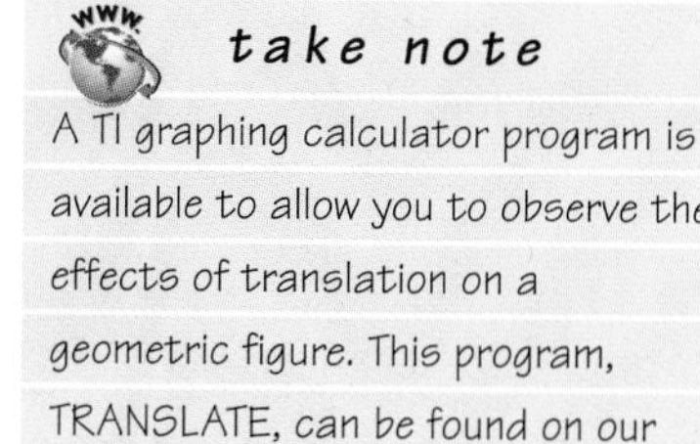

SECTION 10.3 THE INVERSE OF A MATRIX

◆ FINDING THE INVERSE OF A MATRIX

Recall that the multiplicative inverse of a nonzero real number c is $1/c$, the number whose product with c is 1. For example, the multiplicative inverse of $\frac{2}{3}$ is $\frac{3}{2}$ because $\frac{2}{3} \cdot \frac{3}{2} = 1$.

For some square matrices we can define a multiplicative inverse.

Multiplicative Inverse of a Matrix

If A is a square matrix of order n, then the **inverse** of matrix A, denoted by A^{-1}, has the property that

$$A \cdot A^{-1} = A^{-1} \cdot A = I_n$$

where I_n is the identity matrix of order n.

As we will see shortly, not all square matrices have a multiplicative inverse.

QUESTION Are there any real numbers that do not have a multiplicative inverse?

ANSWER The real number zero does not have a multiplicative inverse.

A procedure for finding the inverse (we will simply say *inverse* for *multiplicative inverse*) uses elementary row operations. The procedure will be illustrated by finding the inverse of a 2×2 matrix.

Let $A = \begin{bmatrix} 2 & 7 \\ 1 & 4 \end{bmatrix}$. To the matrix A we will merge the identity matrix I_2 to the right of A and denote this new matrix by $[A:I_2]$.

$$[A:I_2] = \left[\begin{array}{cc|cc} 2 & 7 & 1 & 0 \\ 1 & 4 & 0 & 1 \end{array}\right]$$

A ↑ ↑ I_2

Now we use elementary row operations in a manner similar to that of the Gaussian elimination method. The goal is to produce

$$[I_2:A^{-1}] = \left[\begin{array}{cc|cc} 1 & 0 & b_{11} & b_{12} \\ 0 & 1 & b_{21} & b_{22} \end{array}\right]$$

I_2 ↑ ↑ A^{-1}

In this form, the inverse matrix is the matrix that is to the right of the identity matrix. That is,

$$A^{-1} = \begin{bmatrix} b_{11} & b_{12} \\ b_{21} & b_{22} \end{bmatrix}$$

To find A^{-1}, we first use a series of elementary row operations that will result in a 1 in the first row and the first column.

$$\left[\begin{array}{cc|cc} 2 & 7 & 1 & 0 \\ 1 & 4 & 0 & 1 \end{array}\right] \xrightarrow{\frac{1}{2}R_1} \left[\begin{array}{cc|cc} 1 & \frac{7}{2} & \frac{1}{2} & 0 \\ 1 & 4 & 0 & 1 \end{array}\right] \xrightarrow{-1R_1 + R_2} \left[\begin{array}{cc|cc} 1 & \frac{7}{2} & \frac{1}{2} & 0 \\ 0 & \frac{1}{2} & -\frac{1}{2} & 1 \end{array}\right]$$

$$\xrightarrow{2R_2} \left[\begin{array}{cc|cc} 1 & \frac{7}{2} & \frac{1}{2} & 0 \\ 0 & 1 & -1 & 2 \end{array}\right] \xrightarrow{-\frac{7}{2}R_2 + R_1} \left[\begin{array}{cc|cc} 1 & 0 & 4 & -7 \\ 0 & 1 & -1 & 2 \end{array}\right]$$

The inverse matrix is the matrix to the right of the identity matrix. Therefore,

$$A^{-1} = \begin{bmatrix} 4 & -7 \\ -1 & 2 \end{bmatrix}$$

Each elementary row operation is chosen to advance the process of transforming the original matrix into the identity matrix.

EXAMPLE 1 Find the Inverse of a 3 × 3 Matrix

Find the inverse of the matrix $A = \begin{bmatrix} 1 & -1 & 2 \\ 2 & 0 & 6 \\ 3 & -5 & 7 \end{bmatrix}$.

Solution

$$\left[\begin{array}{ccc|ccc} 1 & -1 & 2 & 1 & 0 & 0 \\ 2 & 0 & 6 & 0 & 1 & 0 \\ 3 & -5 & 7 & 0 & 0 & 1 \end{array}\right]$$

• **Merge the given matrix with the identity matrix I_3.**

$$\xrightarrow{\substack{-2R_1 + R_2 \\ -3R_1 + R_3}} \left[\begin{array}{rrr|rrr} 1 & -1 & 2 & 1 & 0 & 0 \\ 0 & 2 & 2 & -2 & 1 & 0 \\ 0 & -2 & 1 & -3 & 0 & 1 \end{array}\right]$$

• Because a_{11} is already 1, we next produce zeros in a_{21} and a_{31}.

$$\xrightarrow{\frac{1}{2}R_2} \left[\begin{array}{rrr|rrr} 1 & -1 & 2 & 1 & 0 & 0 \\ 0 & 1 & 1 & -1 & \frac{1}{2} & 0 \\ 0 & -2 & 1 & -3 & 0 & 1 \end{array}\right]$$

• Produce a 1 in a_{22}.

$$\xrightarrow{2R_2 + R_3} \left[\begin{array}{rrr|rrr} 1 & -1 & 2 & 1 & 0 & 0 \\ 0 & 1 & 1 & -1 & \frac{1}{2} & 0 \\ 0 & 0 & 3 & -5 & 1 & 1 \end{array}\right]$$

• Produce a 0 in a_{32}.

$$\xrightarrow{\frac{1}{3}R_3} \left[\begin{array}{rrr|rrr} 1 & -1 & 2 & 1 & 0 & 0 \\ 0 & 1 & 1 & -1 & \frac{1}{2} & 0 \\ 0 & 0 & 1 & -\frac{5}{3} & \frac{1}{3} & \frac{1}{3} \end{array}\right]$$

• Produce a 1 in a_{33}.

$$\xrightarrow{\substack{-1R_3 + R_2 \\ -2R_3 + R_1}} \left[\begin{array}{rrr|rrr} 1 & -1 & 0 & \frac{13}{3} & -\frac{2}{3} & -\frac{2}{3} \\ 0 & 1 & 0 & \frac{2}{3} & \frac{1}{6} & -\frac{1}{3} \\ 0 & 0 & 1 & -\frac{5}{3} & \frac{1}{3} & \frac{1}{3} \end{array}\right]$$

• Now work upward. Produce a 0 in a_{23} and a_{13}.

$$\xrightarrow{R_2 + R_1} \left[\begin{array}{rrr|rrr} 1 & 0 & 0 & 5 & -\frac{1}{2} & -1 \\ 0 & 1 & 0 & \frac{2}{3} & \frac{1}{6} & -\frac{1}{3} \\ 0 & 0 & 1 & -\frac{5}{3} & \frac{1}{3} & \frac{1}{3} \end{array}\right]$$

• Produce a 0 in a_{12}.

The inverse matrix is $A^{-1} = \begin{bmatrix} 5 & -\frac{1}{2} & -1 \\ \frac{2}{3} & \frac{1}{6} & -\frac{1}{3} \\ -\frac{5}{3} & \frac{1}{3} & \frac{1}{3} \end{bmatrix}$.

You should verify that this matrix satisfies the condition of an inverse matrix. That is, show that $A^{-1} \cdot A = A \cdot A^{-1} = I_3$.

TRY EXERCISE 6, EXERCISE SET 10.3, PAGE 748

```
[A]-1
[[.5           ...
 [-.3333333333  ...
```

Figure 10.5

The inverse of a matrix can be found by using a graphing calculator. Enter and store the matrix in some variable, say A. To compute the inverse of A, use the $\boxed{x^{-1}}$ key to calculate the inverse. For instance, let $A = \begin{bmatrix} 4 & 3 \\ 2 & 3 \end{bmatrix}$. A typical calculator display of the inverse of A is shown in **Figure 10.5.** Because the elements of the matrix are decimals, it is possible to see only the first column of the inverse matrix. Use the arrow keys to see the remaining columns.

Another possibility for viewing the inverse of A is to use the function on your calculator that converts a decimal to a fraction. This will change the decimals to fractions, and you will be able to see more columns of the matrix.

A **singular matrix** is a matrix that does not have a multiplicative inverse. A matrix that has a multiplicative inverse is a **nonsingular matrix.** As you apply elementary row operation to a singular matrix, there will come a point where there are all zeros in a row of the *original* matrix. When that condition exists, the original matrix does not have an inverse.

EXAMPLE 2 Identify a Singular Matrix

Show that the matrix $\begin{bmatrix} 1 & -1 & -1 \\ 2 & -3 & 0 \\ 1 & -2 & 1 \end{bmatrix}$ is a singular matrix.

Solution

$$\left[\begin{array}{rrr|rrr} 1 & -1 & -1 & 1 & 0 & 0 \\ 2 & -3 & 0 & 0 & 1 & 0 \\ 1 & -2 & 1 & 0 & 0 & 1 \end{array}\right] \xrightarrow[-1R_1 + R_3]{-2R_1 + R_2} \left[\begin{array}{rrr|rrr} 1 & -1 & -1 & 1 & 0 & 0 \\ 0 & -1 & 2 & -2 & 1 & 0 \\ 0 & -1 & 2 & -1 & 0 & 1 \end{array}\right]$$

$$\xrightarrow{-1 \cdot R_2} \left[\begin{array}{rrr|rrr} 1 & -1 & -1 & 1 & 0 & 0 \\ 0 & 1 & -2 & 2 & -1 & 0 \\ 0 & -1 & 2 & -1 & 0 & 1 \end{array}\right] \xrightarrow{R_2 + R_3} \left[\begin{array}{rrr|rrr} 1 & -1 & -1 & -1 & 0 & 0 \\ 0 & 1 & -2 & 2 & -1 & 0 \\ 0 & 0 & 0 & 1 & -1 & 1 \end{array}\right]$$

There are zeros in a row of the original matrix. The original matrix does not have an inverse.

TRY EXERCISE 10, EXERCISE SET 10.3, PAGE 748

◆ SOLVING SYSTEMS OF EQUATIONS USING INVERSE MATRICES

Systems of linear equations can be solved by finding the inverse of the coefficient matrix. Consider the system of equations

$$\begin{cases} 3x_1 + 4x_2 = -1 \\ 3x_1 + 5x_2 = 1 \end{cases} \qquad (1)$$

Using matrix multiplication and the concept of equality of matrices, we can write this system as a matrix equation.

$$\begin{bmatrix} 3 & 4 \\ 3 & 5 \end{bmatrix} \begin{bmatrix} x_1 \\ x_2 \end{bmatrix} = \begin{bmatrix} -1 \\ 1 \end{bmatrix} \qquad (2)$$

If we let

$$A = \begin{bmatrix} 3 & 4 \\ 3 & 5 \end{bmatrix} \qquad X = \begin{bmatrix} x_1 \\ x_2 \end{bmatrix} \qquad B = \begin{bmatrix} -1 \\ 1 \end{bmatrix}$$

then Equation (2) can be written as $AX = B$. The inverse of the coefficient matrix A is $A^{-1} = \begin{bmatrix} \frac{5}{3} & -\frac{4}{3} \\ -1 & 1 \end{bmatrix}$.

To solve the system of equations, multiply each side of the equation $AX = B$ by the inverse A^{-1}.

$$\begin{bmatrix} \frac{5}{3} & -\frac{4}{3} \\ -1 & 1 \end{bmatrix}\begin{bmatrix} 3 & 4 \\ 3 & 5 \end{bmatrix}\begin{bmatrix} x_1 \\ x_2 \end{bmatrix} = \begin{bmatrix} \frac{5}{3} & -\frac{4}{3} \\ -1 & 1 \end{bmatrix}\begin{bmatrix} -1 \\ 1 \end{bmatrix}$$

$$\begin{bmatrix} x_1 \\ x_2 \end{bmatrix} = \begin{bmatrix} -3 \\ 2 \end{bmatrix}$$

Thus $x_1 = -3$ and $x_2 = 2$. The solution to System (1) is $(-3, 2)$.

take note

The disadvantage of using the inverse matrix method to solve a system of equations is that this method will not work if the system is dependent or inconsistent. In addition, this method cannot distinguish between inconsistent and dependent systems. However, in some applications this method is very efficient. See the material on input-output analysis later in this section.

EXAMPLE 3 Solve a System of Equations by Using the Inverse of the Coefficient Matrix

Find the solution of the system of equations by using the inverse of the coefficient matrix.

$$\begin{cases} x_1 + \qquad 7x_3 = 20 \\ 2x_1 + x_2 - x_3 = -3 \\ 7x_1 + 3x_2 + x_3 = 2 \end{cases} \qquad (1)$$

Solution

Write the system as a matrix equation.

$$\begin{bmatrix} 1 & 0 & 7 \\ 2 & 1 & -1 \\ 7 & 3 & 1 \end{bmatrix}\begin{bmatrix} x_1 \\ x_2 \\ x_3 \end{bmatrix} = \begin{bmatrix} 20 \\ -3 \\ 2 \end{bmatrix} \qquad (2)$$

The inverse of the coefficient matrix is $\begin{bmatrix} -\frac{4}{3} & -7 & \frac{7}{3} \\ 3 & 16 & -5 \\ \frac{1}{3} & 1 & -\frac{1}{3} \end{bmatrix}$.

Multiply each side of the matrix equation (2) by the inverse.

$$\begin{bmatrix} -\frac{4}{3} & -7 & \frac{7}{3} \\ 3 & 16 & -5 \\ \frac{1}{3} & 1 & -\frac{1}{3} \end{bmatrix}\begin{bmatrix} 1 & 0 & 7 \\ 2 & 1 & -1 \\ 7 & 3 & 1 \end{bmatrix}\begin{bmatrix} x_1 \\ x_2 \\ x_3 \end{bmatrix} = \begin{bmatrix} -\frac{4}{3} & -7 & \frac{7}{3} \\ 3 & 16 & -5 \\ \frac{1}{3} & 1 & -\frac{1}{3} \end{bmatrix}\begin{bmatrix} 20 \\ -3 \\ 2 \end{bmatrix}$$

$$\begin{bmatrix} x_1 \\ x_2 \\ x_3 \end{bmatrix} = \begin{bmatrix} -1 \\ 2 \\ 3 \end{bmatrix}$$

Thus $x_1 = -1$, $x_2 = 2$, and $x_3 = 3$. The solution to System (1) is $(-1, 2, 3)$.

TRY EXERCISE 20, EXERCISE SET 10.3, PAGE 749

The advantage of using the inverse matrix to solve a system of equations is not apparent unless it is necessary to solve repeatedly a system of equations with the same coefficient matrix but different constant matrices. *Input-output analysis* is one such application of this method.

◆ INPUT-OUTPUT ANALYSIS

In an economy, some of the output of an industry is used by the industry to produce its own product. For example, an electric company uses water and electricity to produce electricity, and a water company uses water and electricity to produce drinking water. **Input-output analysis** attempts to determine the necessary output of industries to satisfy each other's demands plus the demands of consumers. Wassily Leontief, a Harvard economist, was awarded the Nobel prize for his work in this field.

An **input-output matrix** is used to express the interdependence among industries in an economy. Each column of this matrix gives the dollar values of the inputs an industry needs to produce $1 worth of output.

To illustrate the concepts, we will assume an economy with only three industries: agriculture, transportation, and oil. Suppose that to produce $1 worth of agricultural products requires $.05 worth of agriculture, $.02 worth of transportation, and $.05 worth of oil. To produce $1 worth of transportation requires $.10 worth of agriculture, $.08 worth of transportation, and $.10 worth of oil. To produce $1 worth of oil requires $.10 worth of agriculture, $.15 worth of transportation, and $.13 worth of oil. The input-output matrix A is

		Input requirements of		
		Agriculture	Transportation	Oil
	Agriculture	0.05	0.10	0.10
from	Transportation	0.02	0.08	0.15
	Oil	0.05	0.10	0.13

Consumers (other than the industries themselves) want to purchase some of the output from these industries. The amount of output that the consumer will want is called the **final demand** on the economy. This is represented by a column matrix.

Suppose in our example that the final demand is $3 billion worth of agriculture, $1 billion worth of transportation, and $2 billion worth of oil. The final demand matrix is

$$\begin{bmatrix} 3 \\ 1 \\ 2 \end{bmatrix} = D$$

We represent the total output of each industry (in billions of dollars) as follows:

$$\begin{aligned} x &= \text{total output of agriculture} \\ y &= \text{total output of transportation} \\ z &= \text{total output of oil} \end{aligned}$$

The object of input-output analysis is to determine the values of x, y, and z that will satisfy the amount the consumer demands. To find these values, consider agriculture. The amount of agriculture left for the consumer (demand d) is

$$d = x - (\text{amount of agriculture used by industries}) \qquad (1)$$

To find the amount of agriculture used by the three industries in our economy, refer to the input-output matrix. Production of x billion dollars worth of agriculture takes $0.05x$ of agriculture, production of y billion dollars worth of transportation takes $0.10y$ of agriculture, and production of z billion dollars worth of oil takes $0.10z$ of agriculture. Thus,

$$\text{Amount of agriculture used by industries} = 0.05x + 0.10y + 0.10z \qquad (2)$$

Combining Equations (1) and (2), we have

$$d = x - (0.05x + 0.10y + 0.10z)$$

$$3 = 0.95x - 0.10y - 0.10z$$

• *d* is $3 billion for agriculture.

We could continue this way for each of the other industries. The result would be a system of equations. Instead, however, we will use a matrix approach.

If X = total output of the three industries of the economy, then

$$X = \begin{bmatrix} x \\ y \\ z \end{bmatrix}$$

The product of A, the input-output matrix, and X is

$$AX = \begin{bmatrix} 0.05 & 0.10 & 0.10 \\ 0.02 & 0.08 & 0.15 \\ 0.05 & 0.10 & 0.13 \end{bmatrix} \begin{bmatrix} x \\ y \\ z \end{bmatrix}$$

This matrix represents the dollar amount of products used in production for all three industries. Thus the amount available for consumer demand is $X - AX$. As a matrix equation, we can write

$$X - AX = D$$

Solving this equation for X, we determine the output necessary to meet the needs of our industries and the consumer.

$$IX - AX = D$$ **• *I* is the identity matrix. Thus *IX* = *X*.**

$$(I - A)X = D$$ **• Right distributive property**

$$X = (I - A)^{-1}D$$ **• Assuming the inverse of (*I* − *A*) exists**

The last equation states that the solution to an input-output problem can be found by multiplying the demand matrix D by the inverse of $(I - A)$. In our example, we have

$$I - A = \begin{bmatrix} 1 & 0 & 0 \\ 0 & 1 & 0 \\ 0 & 0 & 1 \end{bmatrix} - \begin{bmatrix} 0.05 & 0.10 & 0.10 \\ 0.02 & 0.08 & 0.15 \\ 0.05 & 0.10 & 0.13 \end{bmatrix} = \begin{bmatrix} 0.95 & -0.10 & -0.10 \\ -0.02 & 0.92 & -0.15 \\ -0.05 & -0.10 & 0.87 \end{bmatrix}$$

$$(I - A)^{-1} \approx \begin{bmatrix} 1.063 & 0.131 & 0.145 \\ 0.034 & 1.112 & 0.196 \\ 0.065 & 0.135 & 1.180 \end{bmatrix}$$

The consumer demand is

$$X = (I - A)^{-1}D$$

$$X \approx \begin{bmatrix} 1.063 & 0.131 & 0.145 \\ 0.034 & 1.112 & 0.196 \\ 0.065 & 0.135 & 1.180 \end{bmatrix} \begin{bmatrix} 3 \\ 1 \\ 2 \end{bmatrix} \approx \begin{bmatrix} 3.61 \\ 1.61 \\ 2.69 \end{bmatrix}$$

This matrix indicates that \$3.61 billion worth of agriculture, \$1.61 billion worth of transportation, and \$2.69 billion worth of oil must be produced by the industries to satisfy consumers' demands and the industries' internal requirements.

If we change the final demand matrix,

$$D = \begin{bmatrix} 2 \\ 2 \\ 3 \end{bmatrix}$$

then the total output of the economy can be found as

$$X \approx \begin{bmatrix} 1.063 & 0.131 & 0.145 \\ 0.034 & 1.112 & 0.196 \\ 0.065 & 0.135 & 1.180 \end{bmatrix} \begin{bmatrix} 2 \\ 2 \\ 3 \end{bmatrix} \approx \begin{bmatrix} 2.82 \\ 2.88 \\ 3.94 \end{bmatrix}$$

Thus agriculture must produce output worth \$2.82 billion, transportation must produce output worth \$2.88 billion, and oil must produce output worth \$3.94 billion to satisfy the given consumer demand and the industries' internal requirements.

TOPICS FOR DISCUSSION

1. Explain how to find the inverse of a matrix.
2. Explain the difference between a singular matrix and a nonsingular matrix.
3. Discuss the advantages and disadvantages of solving a system of equations by using an inverse matrix.
4. Do all square matrices have an inverse? If not, give an example of a square matrix that does not have an inverse.

EXERCISE SET 10.3

In Exercises 1 to 10, find the inverse of the given matrix.

1. $\begin{bmatrix} 1 & -3 \\ -2 & 5 \end{bmatrix}$

2. $\begin{bmatrix} 1 & 2 \\ -2 & -3 \end{bmatrix}$

3. $\begin{bmatrix} 1 & 4 \\ 2 & 10 \end{bmatrix}$

4. $\begin{bmatrix} -2 & 3 \\ -6 & -8 \end{bmatrix}$

5. $\begin{bmatrix} 1 & 2 & -1 \\ 2 & 5 & 1 \\ 3 & 6 & -2 \end{bmatrix}$

6. $\begin{bmatrix} 1 & 3 & -2 \\ -1 & -5 & 6 \\ 2 & 6 & -3 \end{bmatrix}$

7. $\begin{bmatrix} 1 & 2 & -1 \\ 2 & 6 & 1 \\ 3 & 6 & -4 \end{bmatrix}$

8. $\begin{bmatrix} 2 & 1 & -1 \\ 6 & 4 & -1 \\ 4 & 2 & -3 \end{bmatrix}$

9. $\begin{bmatrix} 2 & 4 & -4 \\ 1 & 3 & -4 \\ 2 & 4 & -3 \end{bmatrix}$

10. $\begin{bmatrix} 1 & -2 & 2 \\ 2 & -3 & 1 \\ 3 & -6 & 6 \end{bmatrix}$

In Exercises 11 to 14, use a graphing calculator to find the inverse of the given matrix.

11. $\begin{bmatrix} 1 & -1 & 2 & 1 \\ 2 & -1 & 5 & 1 \\ 3 & -3 & 7 & 5 \\ -2 & 3 & -4 & -1 \end{bmatrix}$

12. $\begin{bmatrix} 1 & 1 & -1 & 2 \\ 3 & 2 & -1 & 5 \\ 2 & 2 & -1 & 5 \\ 4 & 4 & -4 & 7 \end{bmatrix}$

13. $\begin{bmatrix} 1 & -1 & 1 & 3 \\ 2 & -1 & 4 & 8 \\ 1 & 1 & 6 & 10 \\ -1 & 5 & 5 & 4 \end{bmatrix}$

14. $\begin{bmatrix} 1 & -1 & 1 & 2 \\ 2 & -1 & 6 & 6 \\ 3 & -1 & 12 & 12 \\ -2 & -1 & -14 & -10 \end{bmatrix}$

In Exercises 15 to 24, solve each system of equations by using inverse matrix methods.

15. $\begin{cases} x + 4y = 6 \\ 2x + 7y = 11 \end{cases}$

16. $\begin{cases} 2x + 3y = 5 \\ x + 2y = 4 \end{cases}$

17. $\begin{cases} x - 2y = 8 \\ 3x + 2y = -1 \end{cases}$

18. $\begin{cases} 3x - 5y = -18 \\ 2x - 3y = -11 \end{cases}$

19. $\begin{cases} x + y + 2z = 4 \\ 2x + 3y + 3z = 5 \\ 3x + 3y + 7z = 14 \end{cases}$

20. $\begin{cases} x + 2y - z = 5 \\ 2x + 3y - z = 8 \\ 3x + 6y - 2z = 14 \end{cases}$

21. $\begin{cases} x + 2y + 2z = 5 \\ -2x - 5y - 2z = 8 \\ 2x + 4y + 7z = 19 \end{cases}$

22. $\begin{cases} x - y + 3z = 5 \\ 3x - y + 10z = 16 \\ 2x - 2y + 5z = 9 \end{cases}$

23. $\begin{cases} w + 2x + z = 6 \\ 2w + 5x + y + 2z = 10 \\ 2w + 4x + y + z = 8 \\ 3w + 6x + 4z = 16 \end{cases}$

24. $\begin{cases} w - x + 2y = 5 \\ 2w - x + 6y + 2z = 16 \\ 3w - 2x + 9y + 4z = 28 \\ w - 2x - z = 2 \end{cases}$

In Exercises 25–28, solve each application by writing a system of equations that models the conditions and then applying inverse matrix methods.

25. **BUSINESS REVENUE** A vacation resort offers a helicopter tour of an island. The price for an adult ticket is \$20; the price for a children's ticket is \$15. The records of the tour operator show that 100 people took the tour on Saturday and 120 people took the tour on Sunday. The total receipts for Saturday were \$1900, and on Sunday the receipts were \$2275. Find the number of adults and the number of children who took the tour on Saturday and on Sunday.

26. **BUSINESS REVENUE** A company sells a standard and a deluxe model tape recorder. Each standard tape recorder costs \$45 to manufacture, and each deluxe model costs \$60 to manufacture. The January manufacturing budget for 90 of these recorders was \$4650; the February budget for 100 recorders was \$5250. Find the number of each type of recorder manufactured in January and in February.

27. **SOIL SCIENCE** The following table shows the active chemical content of three different soil additives.

	Grams per 100 Grams		
Additive	**Ammonium Nitrate**	**Phosphorus**	**Iron**
1	30	10	10
2	40	15	10
3	50	5	5

A soil chemist wants to prepare two chemical samples. The first sample contains 380 grams of ammonium nitrate, 95 grams of phosphorus, and 85 grams of iron. The second sample requires 380 grams of ammonium nitrate, 110 grams of phosphorus, and 90 grams of iron. How many grams of each additive are required for sample 1, and how many grams of each additive are required for sample 2?

28. **NUTRITION** The following table shows the carbohydrate, fat, and protein content of three food types.

	Grams per 100 Grams		
Food Type	**Carbohydrate**	**Fat**	**Protein**
I	13	10	13
II	4	4	3
III	1	0	10

A nutritionist must prepare two diets from these three food groups. The first diet must contain 23 grams of carbohydrate, 18 grams of fat, and 39 grams of protein. The second diet must contain 35 grams of carbohydrate, 28 grams of fat, and 42 grams of protein. How many grams of each food type are required for the first diet, and how many grams of each food type are required for the second diet?

In Exercises 29 to 32, use a graphing calculator to find the inverse of each matrix. Where necessary, round values to the nearest thousandth.

29. $\begin{bmatrix} 2 & -2 & 3 & 1 \\ 5 & 2 & -2 & 3 \\ 6 & -1 & 2 & 3 \\ 2 & 3 & -1 & 5 \end{bmatrix}$

30. $\begin{bmatrix} 3 & -1 & 0 & 1 \\ 2 & -2 & 3 & 0 \\ -1 & -3 & 5 & 3 \\ 5 & 3 & -2 & 1 \end{bmatrix}$

31. $\begin{bmatrix} -\frac{2}{7} & 4 & -\frac{1}{6} \\ -2 & \sqrt{2} & -3 \\ \sqrt{3} & 3 & -\sqrt{5} \end{bmatrix}$

32. $\begin{bmatrix} 6 & \pi & -\frac{4}{7} \\ -5 & \sqrt{7} & 2 \\ \frac{5}{6} & -\sqrt{3} & \sqrt{10} \end{bmatrix}$

33. **INPUT-OUTPUT ANALYSIS** A simplified economy has three industries: manufacturing, transportation, and service. The input-output matrix for this economy is

$$\begin{bmatrix} 0.20 & 0.15 & 0.10 \\ 0.10 & 0.30 & 0.25 \\ 0.20 & 0.10 & 0.10 \end{bmatrix}$$

Find the gross output needed to satisfy the consumer demand of \$120 million worth of manufacturing, \$60 million worth of transportation, and \$55 million worth of service.

34. **INPUT-OUTPUT ANALYSIS** A four-sector economy consists of manufacturing, agriculture, service, and transportation. The input-output matrix for this economy is

$$\begin{bmatrix} 0.10 & 0.05 & 0.20 & 0.15 \\ 0.20 & 0.10 & 0.30 & 0.10 \\ 0.05 & 0.30 & 0.20 & 0.40 \\ 0.10 & 0.20 & 0.15 & 0.20 \end{bmatrix}$$

Find the gross output needed to satisfy the consumer demand of \$80 million worth of manufacturing, \$100 million worth of agriculture, \$50 million worth of service, and \$80 million worth of transportation.

35. **INPUT-OUTPUT ANALYSIS** A conglomerate is composed of three industries: coal, iron, and steel. To produce \$1 worth of coal requires \$.05 worth of coal, \$.02 worth of iron, and \$.10 worth of steel. To produce \$1 worth of iron requires \$.20 worth of coal, \$.03 worth of iron, and \$.12 worth of steel. To produce \$1 worth of steel requires \$.15 worth of coal, \$.25 worth of iron, and \$.05 worth of steel. How much should each industry produce to allow for a consumer demand of \$30 million worth of coal, \$5 million worth of iron, and \$25 million worth of steel?

36. **INPUT-OUTPUT ANALYSIS** A conglomerate has three divisions: plastics, semiconductors, and computers. For each \$1 worth of output, the plastics division needs \$.01 worth of plastics, \$.03 worth of semiconductors, and \$.10 worth of computers. Each \$1 worth of output from the semiconductor division requires \$.08 worth from plastics, \$.05 worth from semiconductors, and \$.15 worth from computers. For each \$1 worth of output, the computer division needs \$.20 worth from plastics, \$.20 worth from semiconductors, and \$.10 worth from computers. The conglomerate estimates consumer demand of \$100 million worth from the plastics division, \$75 million worth from the semiconductor division, and \$150 million worth from the computer division. At what level should each division produce to satisfy this demand?

SUPPLEMENTAL EXERCISES

37. Let $A = \begin{bmatrix} 2 & -3 \\ -6 & 9 \end{bmatrix}$ and $B = \begin{bmatrix} -3 & 15 \\ -2 & 10 \end{bmatrix}$. Show that $AB = O$, the 2×2 zero matrix. This illustrates that for matrices, if $AB = O$, it is not necessarily so that $A = O$ or $B = O$.

38. Show that if a matrix A has an inverse and $AB = O$, then $B = O$.

39. Let $A = \begin{bmatrix} 2 & -1 \\ -4 & 2 \end{bmatrix}$, $B = \begin{bmatrix} 3 & 4 \\ 1 & 5 \end{bmatrix}$, and $C = \begin{bmatrix} 4 & 7 \\ 3 & 11 \end{bmatrix}$. Show that $AB = AC$ but that $B \neq C$. This illustrates that the cancellation rule of real numbers may not apply to matrices.

40. (Continuation of Exercise 39.) Show that if A is a matrix that has an inverse and $AB = AC$, then $B = C$.

41. Show that if $A = \begin{bmatrix} a & b \\ c & d \end{bmatrix}$ and $ad - bc \neq 0$ then

$$A^{-1} = \frac{1}{ad - bc}\begin{bmatrix} d & -b \\ -c & a \end{bmatrix}$$

42. Use the result of Exercise 41 to show that a square matrix of order 2 has an inverse if and only if $ad - bc \neq 0$.

43. Use the result of Exercise 41 to find the inverse of each matrix.

   a. $\begin{bmatrix} 2 & -3 \\ 4 & -5 \end{bmatrix}$ b. $\begin{bmatrix} 5 & 6 \\ 3 & 4 \end{bmatrix}$ c. $\begin{bmatrix} 0 & -1 \\ 4 & 4 \end{bmatrix}$

44. Let $A = \begin{bmatrix} 3 & -2 \\ 1 & 1 \end{bmatrix}$ and $B = \begin{bmatrix} 2 & -1 \\ 2 & 3 \end{bmatrix}$. Use Exercise 41 to show that

$$A^{-1} = \frac{1}{5}\begin{bmatrix} 1 & 2 \\ -1 & 3 \end{bmatrix} \quad \text{and} \quad B^{-1} = \frac{1}{8}\begin{bmatrix} 3 & 1 \\ -2 & 2 \end{bmatrix}$$

Now show that $(AB)^{-1} = B^{-1} \cdot A^{-1}$.

45. Generalize the last result in Exercise 44. That is, show that if A and B are square matrices of order n and each has an inverse matrix, then $(AB)^{-1} = B^{-1} \cdot A^{-1}$. (*Hint:* Begin with the equation $(AB)(AB)^{-1} = I$, where I is the identity matrix. Now multiply each side of the equation by A^{-1} and then by B^{-1}.)

PROJECTS

1. Cryptography is the study of the techniques of concealing the meaning of a message. The message that is to be concealed is called **plaintext**. The concealed message is called **ciphertext**. One way to change plaintext to ciphertext is to give each letter of the alphabet a numerical equivalent. Then matrices are used to scramble the numbers so that it is difficult to determine which number is associated with which letter.

   a. One way to assign each letter a number is to use the ASCII coding system. Check the Internet to determine how this system assigns a number to each letter and punctuation mark.

   b. Now write a short sentence, such as "THE BUCK STOPS HERE." Group the letters of the sentence into packets of, say, 3, using 0 (zero) for a space. Our sentence would look like

(THE)(0BU)(CK0)(STO)(PS0)(HER)(E.0)

Replace each letter and punctuation mark by its numerical ASCII equivalent. For our message, the first three groups would be

$$(84 \quad 72 \quad 69)(48 \quad 66 \quad 85)(67 \quad 75 \quad 48)\ldots$$

Place these numbers in a matrix, using the set of three numbers as a column. For our example, the first three columns are

$$W = \begin{bmatrix} 84 & 48 & 67 & \ldots \\ 72 & 66 & 75 & \ldots \\ 69 & 85 & 48 & \ldots \end{bmatrix}$$

c. Now construct a 3×3 matrix E with integer elements that has an inverse. You can use any 3×3 matrix as long as you can find the inverse. (A graphing calculator may be useful here.)

d. Find the product $E \cdot W = M$. The numbers in the matrix M would be sent as the coded message. Do this for your message.

e. The person who receives this message would multiply the matrix M by E^{-1} to restore the message to its original form. Do this for your message.

SECTION 10.4 DETERMINANTS

- DETERMINANT OF A 2 × 2 MATRIX
- MINORS AND COFACTORS
- EVALUATE A DETERMINANT USING EXPANDING BY COFACTORS
- EVALUATE A DETERMINANT USING ELEMENTARY ROW OPERATIONS
- CONDITION FOR A SQUARE MATRIX TO HAVE A MULTIPLICATIVE INVERSE

◆ DETERMINANT OF A 2 × 2 MATRIX

Associated with each square matrix A is a number called the *determinant* of A. We will denote the determinant of the matrix A by det(A) or by $|A|$. For the remainder of this chapter, we assume that all matrices are square matrices.

The Determinant of a 2 × 2 Matrix

The **determinant** of the matrix $A = [a_{ij}]$ of order 2 is

$$|A| = \begin{vmatrix} a_{11} & a_{12} \\ a_{21} & a_{22} \end{vmatrix} = a_{11}a_{22} - a_{21}a_{12}$$

Caution Be careful not to confuse the notation for a matrix and that for a determinant. The symbol [ ] (brackets) is used for a matrix; the symbol | | (vertical bars) is used for the determinant of a matrix.

An easy way to remember the formula for the determinant of a 2×2 matrix is to recognize that the determinant is the difference between the products of the diagonal elements. That is,

$$\begin{vmatrix} a_{11} & a_{12} \\ a_{21} & a_{22} \end{vmatrix} = a_{11}a_{22} - a_{21}a_{12}$$

EXAMPLE 1 Find the Value of a Determinant

Find the value of the determinant of the matrix $A = \begin{bmatrix} 5 & 3 \\ 2 & -3 \end{bmatrix}$.

Solution

$$|A| = \begin{vmatrix} 5 & 3 \\ 2 & -3 \end{vmatrix} = 5(-3) - 2(3) = -15 - 6 = -21$$

TRY EXERCISE 2, EXERCISE SET 10.4, PAGE 758

◆ MINORS AND COFACTORS

To define the determinant of a matrix of order greater than 2, we first need two other definitions.

The Minor of a Matrix

The **minor** M_{ij} of the element a_{ij} of a square matrix A of order $n \geq 3$ is the determinant of the matrix of order $n - 1$ obtained by deleting the ith row and the jth column of A.

Consider the matrix $A = \begin{bmatrix} 2 & -1 & 5 \\ 4 & 3 & -7 \\ 8 & -7 & 6 \end{bmatrix}$. The minor M_{23} is the determinant of the matrix A formed by deleting row 2 and column 3 from A.

$$M_{23} = \begin{vmatrix} 2 & -1 \\ 8 & -7 \end{vmatrix} \qquad \bullet \begin{vmatrix} 2 & -1 & 5 \\ 4 & 3 & -7 \\ 8 & -7 & 6 \end{vmatrix}$$

$$= 2(-7) - 8(-1) = -14 + 8 = -6$$

The minor M_{31} is the determinant of the matrix A formed by deleting row 3 and column 1 from A.

$$M_{31} = \begin{vmatrix} -1 & 5 \\ 3 & -7 \end{vmatrix} \qquad \bullet \begin{vmatrix} 2 & -1 & 5 \\ 4 & 3 & -7 \\ 8 & -7 & 6 \end{vmatrix}$$

$$= (-1)(-7) - 3(5) = 7 - 15 = -8$$

The second definition we need is that of the *cofactor* of a matrix.

Cofactor of a Matrix

The **cofactor** C_{ij} of the element a_{ij} of a square matrix A is given by $C_{ij} = (-1)^{i+j}M_{ij}$, where M_{ij} is the minor of a_{ij}.

When $i + j$ is an even integer, $(-1)^{i+j} = 1$. When $i + j$ is an odd integer, $(-1)^{i+j} = -1$. Thus

$$C_{ij} = \begin{cases} M_{ij}, & i + j \text{ is an even integer} \\ -M_{ij}, & i + j \text{ is an odd integer} \end{cases}$$

EXAMPLE 2 Find the Minor and Cofactor of a Matrix

Given $A = \begin{bmatrix} 4 & 3 & -2 \\ 5 & -2 & 4 \\ 3 & -2 & -6 \end{bmatrix}$, find M_{32} and C_{12}.

Solution

$$M_{32} = \begin{vmatrix} 4 & -2 \\ 5 & 4 \end{vmatrix} = 4(4) - 5(-2) = 16 + 10 = 26$$

$$C_{12} = (-1)^{1+2}M_{12} = -M_{12} = -\begin{vmatrix} 5 & 4 \\ 3 & -6 \end{vmatrix} = -(-30 - 12) = 42$$

TRY EXERCISE 14, EXERCISE SET 10.4, PAGE 759

◆ EVALUATE A DETERMINANT USING EXPANDING BY COFACTORS

Cofactors are used to evaluate the determinant of a matrix of order 3 or greater. The technique used to evaluate a determinant by using cofactors is called *expanding by cofactors.*

Determinants by Expanding by Cofactors

Given the square matrix A of order 3 or greater, the value of the determinant of A is the sum of the products of the elements of any row or column and their cofactors. For the rth row of A, the value of the determinant of A is

$$|A| = a_{r1}C_{r1} + a_{r2}C_{r2} + a_{r3}C_{r3} + \cdots + a_{rn}C_{rn}$$

For the cth column of A, the determinant of A is

$$|A| = a_{1c}C_{1c} + a_{2c}C_{2c} + a_{3c}C_{3c} + \cdots + a_{nc}C_{nc}$$

This theorem states that the value of a determinant can be found by expanding by cofactors of *any* row or column. The value of the determinant is the same in each case. To illustrate the method, consider the matrix $A = \begin{bmatrix} 2 & 3 & -1 \\ 4 & -2 & 3 \\ 1 & -3 & 4 \end{bmatrix}$.

Expanding the determinant of A by some row, say row 2, gives

$$\begin{aligned} |A| = \begin{vmatrix} 2 & 3 & -1 \\ 4 & -2 & 3 \\ 1 & -3 & 4 \end{vmatrix} &= 4C_{21} + (-2)C_{22} + 3C_{23} \\ &= 4(-1)^{2+1}M_{21} + (-2)(-1)^{2+2}M_{22} + 3(-1)^{2+3}M_{23} \\ &= (-4)\begin{vmatrix} 3 & -1 \\ -3 & 4 \end{vmatrix} + (-2)\begin{vmatrix} 2 & -1 \\ 1 & 4 \end{vmatrix} + (-3)\begin{vmatrix} 2 & 3 \\ 1 & -3 \end{vmatrix} \\ &= (-4)9 + (-2)9 + (-3)(-9) = -27 \end{aligned}$$

Expanding the determinant of A by some column, say column 3, gives

$$\begin{aligned} |A| = \begin{vmatrix} 2 & 3 & -1 \\ 4 & -2 & 3 \\ 1 & -3 & 4 \end{vmatrix} &= (-1)C_{13} + 3C_{23} + 4C_{33} \\ &= (-1)(-1)^{1+3}M_{13} + 3(-1)^{2+3}M_{23} + 4(-1)^{3+3}M_{33} \\ &= (-1)\begin{vmatrix} 4 & -2 \\ 1 & -3 \end{vmatrix} + (-3)\begin{vmatrix} 2 & 3 \\ 1 & -3 \end{vmatrix} + 4\begin{vmatrix} 2 & 3 \\ 4 & -2 \end{vmatrix} \\ &= (-1)(-10) + (-3)(-9) + 4(-16) = -27 \end{aligned}$$

The value of the determinant of A is the same whether we expanded by cofactors of the elements of a row or by cofactors of the elements of a column. When evaluating a determinant, choose the most convenient row or column, which usually is the row or column containing the most zeros.

take note

Example 3 illustrates that choosing a row or column with the most zeros and then expanding about that row or column will reduce the number of calculations you must perform. For Example 3 we have $0 \cdot C_{32} = 0$, and it is not necessary to compute C_{32}.

EXAMPLE 3 Evaluate a Determinant by Cofactors

Evaluate the determinant of $A = \begin{bmatrix} 5 & -3 & -1 \\ -2 & 1 & -1 \\ 1 & 0 & 2 \end{bmatrix}$ by expanding by cofactors.

Solution

Because $a_{32} = 0$, expand using row 3 or column 2. Row 3 will be used here.

$$\begin{aligned} |A| &= 1C_{31} + 0C_{32} + 2C_{33} = 1(-1)^{3+1}M_{31} + 0(-1)^{3+2}M_{32} + 2(-1)^{3+3}M_{33} \\ &= 1\begin{vmatrix} -3 & -1 \\ 1 & -1 \end{vmatrix} + 0 + 2\begin{vmatrix} 5 & -3 \\ -2 & 1 \end{vmatrix} = 1[3 - (-1)] + 0 + 2[5 - 6] \\ &= 4 - 2 = 2 \end{aligned}$$

TRY EXERCISE 20, EXERCISE SET 10.4, PAGE 759

The determinant of a matrix can be found by using a graphing calculator. Many of these calculators use *det* as the operation that produces the value of the determinant. For instance, if $A = \begin{bmatrix} -2 & 3 & 4 \\ 1 & 0 & -2 \\ 3 & 1 & 1 \end{bmatrix}$, then a typical calculator display of the determinant of A is shown in **Figure 10.6.**

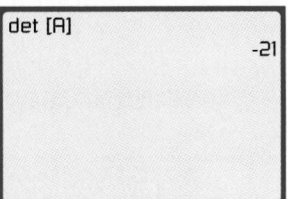

Figure 10.6

◆ EVALUATE A DETERMINANT USING ELEMENTARY ROW OPERATIONS

take note

The properties of determinants stated at the right remain true when the word "row" is replaced by "column." In that case, we would have elementary *column* operations.

MATH MATTERS

There is a vast difference between computing the determinant of a matrix by cofactors and doing so by row reduction. For instance, approximately 4×10^{12} operations are necessary to evaluate a determinant of order 15 by cofactors. At 4 million operations per second, a computer would need approximately 11 days to compute the determinant. On the other hand, approximately 2000 operations are needed to find the determinant by row reduction. Using the same computer, it would take less than 0.001 second to find the determinant.

Effects of Elementary Row Operations on the Value of a Determinant of a Matrix

If A is a square matrix of order n, then the following elementary row operations produce the indicated changes in the determinant of A.

1. Interchanging any two rows of A changes the sign of $|A|$.
2. Multiplying a row of A by a constant k multiplies the determinant of A by k.
3. Adding a multiple of a row of A to another row does not change the value of the determinant of A.

To illustrate these properties, consider the matrix $A = \begin{bmatrix} 2 & 3 \\ 1 & -2 \end{bmatrix}$. The determinant of A is $|A| = 2(-2) - 1(3) = -7$. Now consider each of the elementary row operations.

Interchange the rows of A and evaluate the determinant.

$$\begin{vmatrix} 1 & -2 \\ 2 & 3 \end{vmatrix} = 1(3) - 2(-2) = 3 + 4 = 7 = -|A|$$

Multiply row 2 of A by -3 and evaluate the determinant.

$$\begin{vmatrix} 2 & 3 \\ -3 & 6 \end{vmatrix} = 2(6) - (-3)3 = 12 + 9 = 21 = -3|A|$$

Multiply row 1 of A by -2 and add it to row 2. Then evaluate the determinant.

$$\begin{vmatrix} 2 & 3 \\ -3 & -8 \end{vmatrix} = 2(-8) - (-3)(3) = -16 + 9 = -7 = |A|.$$

These elementary row operations are often used to rewrite a matrix in *triangular form.* A matrix is in **triangular form** if all elements below or above the main diagonal are zero. The matrices

$$A = \begin{bmatrix} 2 & -2 & 3 & 1 \\ 0 & -2 & 4 & 2 \\ 0 & 0 & 6 & 9 \\ 0 & 0 & 0 & -5 \end{bmatrix} \quad \text{and} \quad B = \begin{bmatrix} 3 & 0 & 0 & 0 \\ 2 & -3 & 0 & 0 \\ 6 & 4 & -2 & 0 \\ 8 & 3 & 4 & 2 \end{bmatrix}$$

are in triangular form.

Determinant of a Matrix in Triangular Form

Let A be a square matrix of order n in triangular form. The determinant of A is the product of the elements on the main diagonal.

$$|A| = a_{11}a_{22}a_{33} \cdots a_{nn}$$

For the matrices A and B given above,

$$|A| = 2(-2)(6)(-5) = 120$$
$$|B| = 3(-3)(-2)(2) = 36$$

EXAMPLE 4 Evaluate a Determinant by Elementary Row Operations

Evaluate the determinant by rewriting in triangular form.

$$\begin{vmatrix} 2 & 1 & -1 & 3 \\ 2 & 2 & 0 & 1 \\ 4 & 5 & 4 & -3 \\ 2 & 2 & 7 & -3 \end{vmatrix}$$

Solution

Rewrite the determinant in triangular form by using elementary row operations.

$$\begin{vmatrix} 2 & 1 & -1 & 3 \\ 2 & 2 & 0 & 1 \\ 4 & 5 & 4 & -3 \\ 2 & 2 & 7 & -3 \end{vmatrix} \begin{matrix} -1R_1 + R_2 \\ -2R_1 + R_3 \\ -1R_1 + R_4 \\ = \end{matrix} \begin{vmatrix} 2 & 1 & -1 & 3 \\ 0 & 1 & 1 & -2 \\ 0 & 3 & 6 & -9 \\ 0 & 1 & 8 & -6 \end{vmatrix}$$

$$\begin{matrix} \text{Factor 3,} \\ \text{from row 3.} \\ = \end{matrix} \; 3\begin{vmatrix} 2 & 1 & -1 & 3 \\ 0 & 1 & 1 & -2 \\ 0 & 1 & 2 & -3 \\ 0 & 1 & 8 & -6 \end{vmatrix} \begin{matrix} -1R_2 + R_3 \\ -1R_2 + R_4 \\ = \end{matrix} \; 3\begin{vmatrix} 2 & 1 & -1 & 3 \\ 0 & 1 & 1 & -2 \\ 0 & 0 & 1 & -1 \\ 0 & 0 & 7 & -4 \end{vmatrix}$$

$$\overset{-7R_3 + R_4}{=} 3\begin{vmatrix} 2 & 1 & -1 & 3 \\ 0 & 1 & 1 & -2 \\ 0 & 0 & 1 & -1 \\ 0 & 0 & 0 & 3 \end{vmatrix} = 3(2)(1)(1)(3) = 18$$

TRY EXERCISE 42, EXERCISE SET 10.4, PAGE 759

In some cases it is possible to recognize when the determinant of a matrix is zero.

Conditions for a Zero Determinant

If A is a square matrix, then $|A| = 0$ when any one of the following is true.

1. A row (column) consists entirely of zeros.
2. Two rows (columns) are identical.
3. One row (column) is a constant multiple of a second row (column).

Proof To prove part 2 of this theorem, let A be the given matrix and let $D = |A|$. Now interchange the two identical rows. Then $|A| = -D$. Thus

$$D = -D$$

Zero is the only real number that is its own additive inverse, and hence $D = |A| = 0$. ◆

The proofs of the other two properties are left as exercises.

QUESTION If I is the identity matrix of order n, what is the value of $|I|$?

The last property of determinants that we will discuss is a product property.

Product Property of Determinants

If A and B are square matrices of order n, then

$$|AB| = |A||B|$$

◆ CONDITION FOR A SQUARE MATRIX TO HAVE A MULTIPLICATIVE INVERSE

Recall that a singular matrix is one that does not have a multiplicative inverse. The Product Property of Determinants can be used to determine whether a matrix has an inverse.

ANSWER The identity matrix is in diagonal form with 1s on the main diagonal. Thus $|I|$ is a product of 1s, or $|I| = 1$.

Consider a matrix A with an inverse A^{-1}. Then, by the last theorem,

$$|A \cdot A^{-1}| = |A||A^{-1}|$$

But $A \cdot A^{-1} = I$, the identity matrix, and $|I| = 1$. Therefore,

$$1 = |A||A^{-1}|$$

From the last equation, $|A| \neq 0$. And, in particular,

$$|A^{-1}| = \frac{1}{|A|}$$

These results are summarized in the following theorem.

Existence of the Inverse of a Square Matrix

If A is a square matrix of order n, then A has a multiplicative inverse if and only if $|A| \neq 0$. Furthermore,

$$|A^{-1}| = \frac{1}{|A|}$$

We proved only part of this theorem. It remains to show that given $|A| \neq 0$, then A has an inverse. This proof can be found in most texts on linear algebra.

TOPICS FOR DISCUSSION

1. Discuss the difference between a matrix and a determinant.
2. Explain the difference between the minor and the cofactor of an element of a matrix.
3. Discuss how determinants are used to discover whether a matrix has an inverse.
4. Explain how to calculate the value of a determinant by expanding by cofactors.
5. Discuss how elementary row operations are used to find the determinant of a matrix.

EXERCISE SET 10.4

In Exercises 1 to 8, evaluate the determinants.

1. $\begin{vmatrix} 2 & -1 \\ 3 & 5 \end{vmatrix}$ **2.** $\begin{vmatrix} 2 & 9 \\ -6 & 2 \end{vmatrix}$ **3.** $\begin{vmatrix} 5 & 0 \\ 2 & -3 \end{vmatrix}$

4. $\begin{vmatrix} 0 & -8 \\ 3 & 4 \end{vmatrix}$ **5.** $\begin{vmatrix} 4 & 6 \\ 2 & 3 \end{vmatrix}$ **6.** $\begin{vmatrix} -3 & 6 \\ 4 & -8 \end{vmatrix}$

7. $\begin{vmatrix} 0 & 9 \\ 0 & -2 \end{vmatrix}$ **8.** $\begin{vmatrix} -3 & 9 \\ 0 & 0 \end{vmatrix}$

In Exercises 9 to 12, evaluate the indicated minor and cofactor for the determinant

$$\begin{vmatrix} 5 & -2 & -3 \\ 2 & 4 & -1 \\ 4 & -5 & 6 \end{vmatrix}$$

9. M_{11}, C_{11} **10.** M_{21}, C_{21}

11. M_{32}, C_{32} **12.** M_{33}, C_{33}

In Exercises 13 to 16, evaluate the indicated minor and cofactor for the determinant

$$\begin{vmatrix} 3 & -2 & 3 \\ 1 & 3 & 0 \\ 6 & -2 & 3 \end{vmatrix}$$

13. M_{22}, C_{22} **14.** M_{13}, C_{13} **15.** M_{31}, C_{31} **16.** M_{23}, C_{23}

In Exercises 17 to 26, evaluate the determinant by expanding by cofactors.

17. $\begin{vmatrix} 2 & -3 & 1 \\ 2 & 0 & 2 \\ 3 & -2 & 4 \end{vmatrix}$ **18.** $\begin{vmatrix} 3 & 1 & -2 \\ 2 & -5 & 4 \\ 3 & 2 & 1 \end{vmatrix}$

19. $\begin{vmatrix} -2 & 3 & 2 \\ 1 & 2 & -3 \\ -4 & -2 & 1 \end{vmatrix}$ **20.** $\begin{vmatrix} 3 & -2 & 0 \\ 2 & -3 & 2 \\ 8 & -2 & 5 \end{vmatrix}$

21. $\begin{vmatrix} 2 & -3 & 10 \\ 0 & 2 & -3 \\ 0 & 0 & 5 \end{vmatrix}$ **22.** $\begin{vmatrix} 6 & 0 & 0 \\ 2 & -3 & 0 \\ 7 & -8 & 2 \end{vmatrix}$

23. $\begin{vmatrix} 0 & -2 & 4 \\ 1 & 0 & -7 \\ 5 & -6 & 0 \end{vmatrix}$ **24.** $\begin{vmatrix} 5 & -8 & 0 \\ 2 & 0 & -7 \\ 0 & -2 & -1 \end{vmatrix}$

25. $\begin{vmatrix} 4 & -3 & 3 \\ 2 & 1 & -4 \\ 6 & -2 & -1 \end{vmatrix}$ **26.** $\begin{vmatrix} -2 & 3 & 9 \\ 4 & -2 & -6 \\ 0 & -8 & -24 \end{vmatrix}$

In Exercises 27 to 40, without expanding, give a reason for each equality.

27. $\begin{vmatrix} 2 & -1 & 3 \\ 0 & 0 & 0 \\ 3 & 4 & 1 \end{vmatrix} = 0$ **28.** $\begin{vmatrix} 2 & 3 & 0 \\ 1 & -2 & 0 \\ 4 & 1 & 0 \end{vmatrix} = 0$

29. $\begin{vmatrix} 1 & 4 & -1 \\ 2 & 4 & 12 \\ 3 & 1 & 4 \end{vmatrix} = 2\begin{vmatrix} 1 & 4 & -1 \\ 1 & 2 & 6 \\ 3 & 1 & 4 \end{vmatrix}$

30. $\begin{vmatrix} 1 & -3 & 4 \\ 4 & 6 & 1 \\ 0 & -9 & 3 \end{vmatrix} = -3\begin{vmatrix} 1 & 1 & 4 \\ 4 & -2 & 1 \\ 0 & 3 & 3 \end{vmatrix}$

31. $\begin{vmatrix} 1 & 5 & -2 \\ 2 & -1 & 4 \\ 3 & 0 & -2 \end{vmatrix} = \begin{vmatrix} 1 & 5 & -2 \\ 0 & -11 & 8 \\ 3 & 0 & -2 \end{vmatrix}$

32. $\begin{vmatrix} 1 & 1 & -3 \\ 2 & 2 & 5 \\ 1 & -2 & 4 \end{vmatrix} = \begin{vmatrix} 1 & 1 & -3 \\ 2 & 2 & 5 \\ 0 & -3 & 7 \end{vmatrix}$

33. $\begin{vmatrix} 4 & -3 & 2 \\ 6 & 2 & 1 \\ -2 & 2 & 4 \end{vmatrix} = 2\begin{vmatrix} 2 & -3 & 2 \\ 3 & 2 & 1 \\ -1 & 2 & 4 \end{vmatrix}$

34. $\begin{vmatrix} 2 & -1 & 3 \\ 3 & 0 & 1 \\ -4 & 2 & -6 \end{vmatrix} = 0$

35. $\begin{vmatrix} 2 & -4 & 5 \\ 0 & 3 & 4 \\ 0 & 0 & -2 \end{vmatrix} = -12$ **36.** $\begin{vmatrix} 3 & 0 & 0 \\ 2 & -1 & 0 \\ 3 & 4 & 5 \end{vmatrix} = -15$

37. $\begin{vmatrix} 3 & 5 & -2 \\ 2 & 1 & 0 \\ 9 & -2 & -3 \end{vmatrix} = -\begin{vmatrix} 9 & -2 & -3 \\ 2 & 1 & 0 \\ 3 & 5 & -2 \end{vmatrix}$

38. $\begin{vmatrix} 6 & 0 & -2 \\ 2 & -1 & -3 \\ 1 & 5 & -7 \end{vmatrix} = -\begin{vmatrix} 0 & 6 & -2 \\ -1 & 2 & -3 \\ 5 & 1 & -7 \end{vmatrix}$

39. $a^3\begin{vmatrix} 1 & 1 & 1 \\ a & a & a \\ a^2 & a^2 & a^2 \end{vmatrix} = \begin{vmatrix} a & a & a \\ a^2 & a^2 & a^2 \\ a^3 & a^3 & a^3 \end{vmatrix}$

40. $\begin{vmatrix} 1 & 1 & 1 \\ 2 & 2 & 2 \\ 3 & 3 & 3 \end{vmatrix} = 0$

In Exercises 41 to 50, evaluate the determinant by first rewriting the determinant in triangular form.

41. $\begin{vmatrix} 2 & 4 & 1 \\ 1 & 2 & -1 \\ 1 & 2 & 2 \end{vmatrix}$ **42.** $\begin{vmatrix} 3 & -2 & -1 \\ 1 & 2 & 4 \\ 2 & -2 & 3 \end{vmatrix}$

43. $\begin{vmatrix} 1 & 2 & -1 \\ 2 & 3 & 1 \\ 3 & 4 & 3 \end{vmatrix}$ **44.** $\begin{vmatrix} 1 & 2 & 5 \\ -1 & 1 & -2 \\ 3 & 1 & 10 \end{vmatrix}$

45. $\begin{vmatrix} 0 & -1 & 1 \\ 1 & 0 & -2 \\ 2 & 2 & 0 \end{vmatrix}$ **46.** $\begin{vmatrix} 2 & -1 & 3 \\ 1 & 1 & 1 \\ 3 & -4 & 5 \end{vmatrix}$

47. $\begin{vmatrix} 1 & 2 & -1 & 2 \\ 1 & -2 & 0 & 3 \\ 3 & 0 & 1 & 5 \\ -2 & -4 & 1 & 6 \end{vmatrix}$ **48.** $\begin{vmatrix} 1 & -1 & -1 & 2 \\ 0 & 2 & 4 & 6 \\ 1 & 1 & 4 & 12 \\ 1 & -1 & 0 & 8 \end{vmatrix}$

49. $\begin{vmatrix} 1 & 2 & 3 & -1 \\ 6 & 5 & 9 & 8 \\ 2 & 4 & 12 & -1 \\ 1 & 2 & 6 & -1 \end{vmatrix}$ **50.** $\begin{vmatrix} 1 & 2 & 0 & -2 \\ -1 & 1 & 3 & 5 \\ 2 & 1 & 4 & 0 \\ -2 & 5 & 2 & 6 \end{vmatrix}$

In Exercises 51 to 54, use a graphing calculator to find the value of the determinant of the matrix. Where necessary, round answer to the nearest thousandth.

51. $\begin{bmatrix} 2 & -2 & 3 & 1 \\ 5 & 2 & -2 & 3 \\ 6 & -1 & 2 & 3 \\ 2 & 3 & -1 & 5 \end{bmatrix}$

52. $\begin{bmatrix} 3 & -1 & 0 & 1 \\ 2 & -2 & 3 & 0 \\ -1 & -3 & 5 & 3 \\ 5 & 3 & -2 & 1 \end{bmatrix}$

53. $\begin{bmatrix} -\frac{2}{7} & 4 & -\frac{1}{6} \\ -2 & \sqrt{2} & -3 \\ \sqrt{3} & 3 & -\sqrt{5} \end{bmatrix}$

54. $\begin{bmatrix} 6 & \pi & -\frac{4}{7} \\ -5 & \sqrt{7} & 2 \\ \frac{5}{6} & -\sqrt{3} & \sqrt{10} \end{bmatrix}$

SUPPLEMENTAL EXERCISES

The area of a triangle with vertices (x_1, y_1), (x_2, y_2), and (x_3, y_3) can be given as the absolute value of the determinant

$$\frac{1}{2}\begin{vmatrix} x_1 & y_1 & 1 \\ x_2 & y_2 & 1 \\ x_3 & y_3 & 1 \end{vmatrix}$$

Use this formula to find the area of each triangle whose coordinates are given in Exercises 55 to 58.

55. (2, 3), (−1, 0), (4, 8)

56. (−3, 4), (1, 5), (5, −2)

57. (4, 9), (8, 2), (−3, −2)

58. (0, 4), (−5, 7), (2, 9)

59. Given a square matrix of order 3 where one row is a constant multiple of a second row, show that the determinant of the matrix is zero. (*Hint:* Use an elementary row operation and part 2 of the theorem for conditions for a zero determinant.)

60. Given a square matrix of order 3 with a zero as every element in a column, show that the determinant of the matrix is zero. (*Hint:* Expand the determinant by cofactors using the column of zeros.)

61. Show that the determinant $\begin{vmatrix} x & y & 1 \\ x_1 & y_1 & 1 \\ x_2 & y_2 & 1 \end{vmatrix} = 0$ is the equation of a line through the points (x_1, y_1) and (x_2, y_2).

62. Use Exercise 61 to find the equation of the line passing through the points (2, 3) and (−1, 4).

63. Use Exercise 61 to find the equation of the line passing through the points (−3, 4) and (2, −3).

64. Show that $\begin{vmatrix} a_1 & b_1 \\ a_2 & b_2 \end{vmatrix} = \begin{vmatrix} a_1 & b_1 \\ ka_1 + a_2 & kb_1 + b_2 \end{vmatrix}$.
What property of determinants does this illustrate?

65. Surveyors use a formula to find the area of a plot of land. *Surveyor's Area Formula:* If the vertices (x_1, y_1), (x_2, y_2), $(x_3, y_3), \ldots, (x_n, y_n)$ of a simple polygon are listed counterclockwise around the perimeter, the area of the polygon is

$$A = \frac{1}{2}\left\{\begin{vmatrix} x_1 & x_2 \\ y_1 & y_2 \end{vmatrix} + \begin{vmatrix} x_2 & x_3 \\ y_2 & y_3 \end{vmatrix} + \begin{vmatrix} x_3 & x_4 \\ y_3 & y_4 \end{vmatrix} + \cdots + \begin{vmatrix} x_n & x_1 \\ y_n & y_1 \end{vmatrix}\right\}$$

Use the Surveyor's Area Formula to find the area of the polygon with vertices (8, −4), (25, 5), (15, 9), (17, 20), and (0, 10).

PROJECTS

1. Consider the rectangle in the accompanying diagram. Construct a matrix M of dimension 2×4 where the columns are the x- and y-coordinates, respectively, of successive vertices of the rectangle.

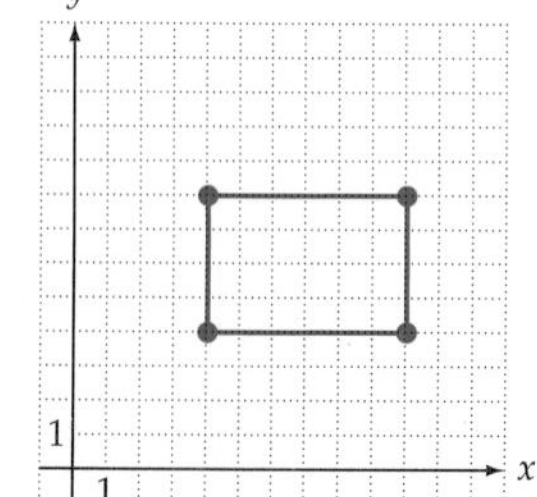

a. Consider the matrix $A = \begin{bmatrix} 2 & 1 \\ 3 & 2 \end{bmatrix}$. Find the product AM. Plot the points of AM again, using the columns as x- and y-coordinates, respectively, of successive vertices of a new figure. Show that the area of the new figure is the same as the area of the original rectangle and that $\det(A) = 1$.

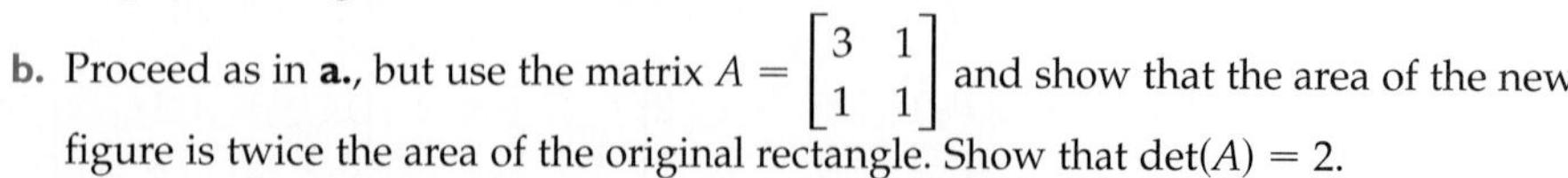

b. Proceed as in **a.**, but use the matrix $A = \begin{bmatrix} 3 & 1 \\ 1 & 1 \end{bmatrix}$ and show that the area of the new figure is twice the area of the original rectangle. Show that $\det(A) = 2$.

c. Proceed as in **a.**, but use the matrix $A = \begin{bmatrix} 1 & 2 \\ 0.5 & 1 \end{bmatrix}$ and show that the new figure is a line segment and that, therefore, the figure has no area. Show that $\det(A) = 0$.

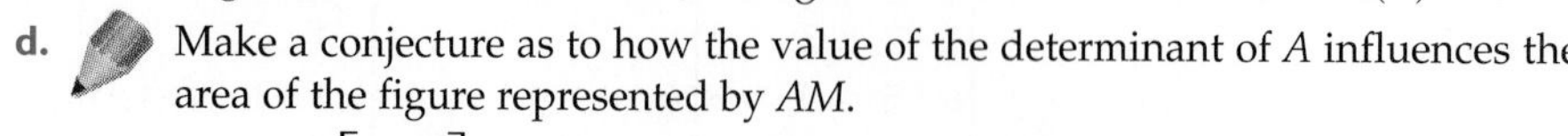

d. Make a conjecture as to how the value of the determinant of A influences the area of the figure represented by AM.

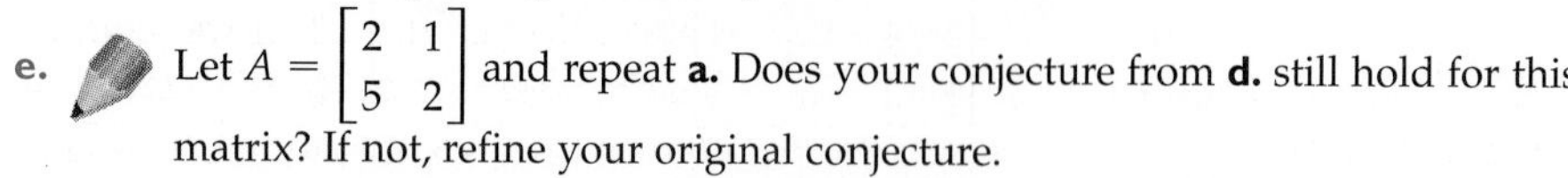

e. Let $A = \begin{bmatrix} 2 & 1 \\ 5 & 2 \end{bmatrix}$ and repeat **a.** Does your conjecture from **d.** still hold for this matrix? If not, refine your original conjecture.

SECTION 10.5 CRAMER'S RULE

◆ SOLVING A SYSTEM OF EQUATIONS USING CRAMER'S RULE

◆ SOLVING A SYSTEM OF EQUATIONS USING CRAMER'S RULE

An application of determinants is to solve a system of linear equations. Consider the system

$$\begin{cases} a_{11}x_1 + a_{12}x_2 = b_1 \\ a_{21}x_1 + a_{22}x_2 = b_2 \end{cases}$$

To eliminate x_2 from this system, we first multiply the top equation by a_{22} and the bottom equation by a_{12}. Then we subtract.

$$\begin{aligned} a_{22}a_{11}x_1 + a_{22}a_{12}x_2 &= a_{22}b_1 \\ a_{12}a_{21}x_1 + a_{12}a_{22}x_2 &= a_{12}b_2 \\ \hline (a_{22}a_{11} - a_{12}a_{21})x_1 \qquad &= a_{22}b_1 - a_{12}b_2 \\ x_1 &= \frac{a_{22}b_1 - a_{12}b_2}{a_{22}a_{11} - a_{12}a_{21}} \\ \text{or} \quad x_1 &= \frac{\begin{vmatrix} b_1 & a_{12} \\ b_2 & a_{22} \end{vmatrix}}{\begin{vmatrix} a_{11} & a_{12} \\ a_{21} & a_{22} \end{vmatrix}}, \quad \begin{vmatrix} a_{11} & a_{12} \\ a_{21} & a_{22} \end{vmatrix} \neq 0 \end{aligned}$$

We can find x_2 in a similar manner. The results are given in Cramer's Rule for a System of Two Linear Equations.

Cramer's Rule for a System of Two Linear Equations

Let
$$\begin{cases} a_{11}x_1 + a_{12}x_2 = b_1 \\ a_{21}x_1 + a_{22}x_2 = b_2 \end{cases}$$

be the system of equations for which the determinant of the coefficient matrix is not zero. The solution of the system of equations is the ordered pair whose coordinates are

$$x_1 = \frac{\begin{vmatrix} b_1 & a_{12} \\ b_2 & a_{22} \end{vmatrix}}{\begin{vmatrix} a_{11} & a_{12} \\ a_{21} & a_{22} \end{vmatrix}} \quad \text{and} \quad x_2 = \frac{\begin{vmatrix} a_{11} & b_1 \\ a_{21} & b_2 \end{vmatrix}}{\begin{vmatrix} a_{11} & a_{12} \\ a_{21} & a_{22} \end{vmatrix}}$$

Note that the denominator is the determinant of the coefficient matrix of the variables. The numerator of x_1 is formed by replacing column 1 of the coefficient determinant with the constants b_1 and b_2. The numerator of x_2 is formed by replacing column 2 of the coefficient determinant with the constants b_1 and b_2.

EXAMPLE 1 **Solve a System of Equations by Using Cramer's Rule**

Solve the following system of equations using Cramer's Rule.

$$\begin{cases} 5x_1 - 3x_2 = 6 \\ 2x_1 + 4x_2 = -7 \end{cases}$$

Solution

$$x_1 = \frac{\begin{vmatrix} 6 & -3 \\ -7 & 4 \end{vmatrix}}{\begin{vmatrix} 5 & -3 \\ 2 & 4 \end{vmatrix}} = \frac{3}{26} \qquad x_2 = \frac{\begin{vmatrix} 5 & 6 \\ 2 & -7 \end{vmatrix}}{\begin{vmatrix} 5 & -3 \\ 2 & 4 \end{vmatrix}} = -\frac{47}{26}$$

The solution is (3/26, −47/26).

TRY EXERCISE 4, EXERCISE SET 10.5, PAGE 764

Cramer's Rule can be used for a system of three linear equations in three variables. For example, consider the system of equations

$$\begin{cases} 2x - 3y + z = 2 \\ 4x + 2z = -3 \\ 3x + y - 2z = 1 \end{cases} \qquad (1)$$

To solve this system of equations, we extend the concepts behind the solution for a system of two linear equations. The solution of the system has the form (x, y, z), where

$$x = \frac{D_x}{D} \qquad y = \frac{D_y}{D} \qquad z = \frac{D_z}{D}$$

The determinant D is the determinant of the coefficient matrix. The determinants D_x, D_y, and D_z are the determinants of the matrices formed by replacing the first, second, and third columns, respectively, by the constants. For System (1),

$$x = \frac{D_x}{D} \qquad y = \frac{D_y}{D} \qquad z = \frac{D_z}{D}$$

where $D = \begin{vmatrix} 2 & -3 & 1 \\ 4 & 0 & 2 \\ 3 & 1 & -2 \end{vmatrix} = -42 \qquad D_x = \begin{vmatrix} 2 & -3 & 1 \\ -3 & 0 & 2 \\ 1 & 1 & -2 \end{vmatrix} = 5$

$$D_y = \begin{vmatrix} 2 & 2 & 1 \\ 4 & -3 & 2 \\ 3 & 1 & -2 \end{vmatrix} = 49 \qquad D_z = \begin{vmatrix} 2 & -3 & 2 \\ 4 & 0 & -3 \\ 3 & 1 & 1 \end{vmatrix} = 53$$

Thus

$$x = -\frac{5}{42} \qquad y = -\frac{7}{6} \qquad z = -\frac{53}{42}$$

The solution of System (1) is

$$\left(-\frac{5}{42}, -\frac{7}{6}, -\frac{53}{42}\right)$$

Cramer's Rule can be extended to a system of n linear equations in n variables.

Cramer's Rule

Let
$$\begin{cases} a_{11}x_1 + a_{12}x_2 + a_{13}x_3 + \cdots + a_{1n}x_n = b_1 \\ a_{21}x_1 + a_{22}x_2 + a_{23}x_3 + \cdots + a_{2n}x_n = b_2 \\ a_{31}x_1 + a_{32}x_2 + a_{33}x_3 + \cdots + a_{3n}x_n = b_3 \\ \vdots \\ a_{n1}x_1 + a_{n2}x_2 + a_{n3}x_3 + \cdots + a_{nn}x_n = b_n \end{cases}$$

be a system of n equations in n variables. The solution of the system is given by $(x_1, x_2, x_3, \ldots, x_n)$, where

$$x_1 = \frac{D_1}{D} \qquad x_2 = \frac{D_2}{D} \qquad \cdots \qquad x_i = \frac{D_i}{D} \qquad \cdots \qquad x_n = \frac{D_n}{D}$$

and D is the determinant of the coefficient matrix, $D \neq 0$. D_i is the determinant formed by replacing the ith column of the coefficient matrix with the column of constants $b_1, b_2, b_3, \ldots, b_n$.

Because the determinant of the coefficient matrix must be nonzero for us to use Cramer's Rule, this method is not appropriate for systems of linear equations with no solution or infinitely many solutions. In fact, the only time a system of linear equations has a unique solution is when the coefficient determinant is not zero, a fact summarized in the following theorem.

Systems of Linear Equations with Unique Solutions

A system of n linear equations in n variables has a unique solution if and only if the determinant of the coefficient matrix is not zero.

Cramer's Rule is also useful when we want to determine the value of only a single variable in a system of equations.

EXAMPLE 2 Determine the Value of a Single Variable in a System of Linear Equations

Find x_3 for the system of equations

$$\begin{cases} 4x_1 + 3x_3 - 2x_4 = 2 \\ 3x_1 + x_2 + 2x_3 - x_4 = 4 \\ x_1 - 6x_2 - 2x_3 + 2x_4 = 0 \\ 2x_1 + 2x_2 - x_4 = -1 \end{cases}$$

Continued

Solution

Find D and D_3.

$$D = \begin{vmatrix} 4 & 0 & 3 & -2 \\ 3 & 1 & 2 & -1 \\ 1 & -6 & -2 & 2 \\ 2 & 2 & 0 & -1 \end{vmatrix} = 39 \qquad D_3 = \begin{vmatrix} 4 & 0 & 2 & -2 \\ 3 & 1 & 4 & -1 \\ 1 & -6 & 0 & 2 \\ 2 & 2 & -1 & -1 \end{vmatrix} = 96$$

Thus $x_3 = 96/39 = 32/13$.

TRY EXERCISE 24, EXERCISE SET 10.5, PAGE 765

TOPICS FOR DISCUSSION

1. Discuss the advantages and disadvantages of using Cramer's Rule to solve a system of equations.
2. Can Cramer's Rule be used to solve any system of linear equations? If not, explain when Cramer's Rule will lead to a solution and when it will not.

EXERCISE SET 10.5

In Exercises 1 to 20, solve each system of equations by using Cramer's Rule.

1. $\begin{cases} 3x_1 + 4x_2 = 8 \\ 4x_1 - 5x_2 = 1 \end{cases}$

2. $\begin{cases} x_1 - 3x_2 = 9 \\ 2x_1 - 4x_2 = -3 \end{cases}$

3. $\begin{cases} 5x_1 + 4x_2 = -1 \\ 3x_1 - 6x_2 = 5 \end{cases}$

4. $\begin{cases} 2x_1 + 5x_2 = 9 \\ 5x_1 + 7x_2 = 8 \end{cases}$

5. $\begin{cases} 7x_1 + 2x_2 = 0 \\ 2x_1 + x_2 = -3 \end{cases}$

6. $\begin{cases} 3x_1 - 8x_2 = 1 \\ 4x_1 + 5x_2 = -2 \end{cases}$

7. $\begin{cases} 3x_1 - 7x_2 = 0 \\ 2x_1 + 4x_2 = 0 \end{cases}$

8. $\begin{cases} 5x_1 + 4x_2 = -3 \\ 2x_1 - x_2 = 0 \end{cases}$

9. $\begin{cases} 1.2x_1 + 0.3x_2 = 2.1 \\ 0.8x_1 - 1.4x_2 = -1.6 \end{cases}$

10. $\begin{cases} 3.2x_1 - 4.2x_2 = 1.1 \\ 0.7x_1 + 3.2x_2 = -3.4 \end{cases}$

11. $\begin{cases} 3x_1 - 4x_2 + 2x_3 = 1 \\ x_1 - x_2 + 2x_3 = -2 \\ 2x_1 + 2x_2 + 3x_3 = -3 \end{cases}$

12. $\begin{cases} 5x_1 - 2x_2 + 3x_3 = -2 \\ 3x_1 + x_2 - 2x_3 = 3 \\ x_1 - 2x_2 + 3x_3 = -1 \end{cases}$

13. $\begin{cases} x_1 + 4x_2 - 2x_3 = 0 \\ 3x_1 - 2x_2 + 3x_3 = 4 \\ 2x_1 + x_2 - 3x_3 = -1 \end{cases}$

14. $\begin{cases} 4x_1 - x_2 + 2x_3 = 6 \\ x_1 + 3x_2 - x_3 = -1 \\ 2x_1 + 3x_2 - 2x_3 = 5 \end{cases}$

15. $\begin{cases} 2x_2 - 3x_3 = 1 \\ 3x_1 - 5x_2 + x_3 = 0 \\ 4x_1 + 2x_3 = -3 \end{cases}$

16. $\begin{cases} 2x_1 + 5x_2 = 1 \\ x_1 - 3x_3 = -2 \\ 2x_1 - x_2 + 2x_3 = 4 \end{cases}$

17. $\begin{cases} 4x_1 - 5x_2 + x_3 = -2 \\ 3x_1 + x_2 = 4 \\ x_1 - x_2 + 3x_3 = 0 \end{cases}$

18. $\begin{cases} 3x_1 - x_2 + x_3 = 5 \\ x_1 + 3x_3 = -2 \\ 2x_1 + 2x_2 - 5x_3 = 0 \end{cases}$

19. $\begin{cases} 2x_1 + 2x_2 - 3x_3 = 0 \\ x_1 - 3x_2 + 2x_3 = 0 \\ 4x_1 - x_2 + 3x_3 = 0 \end{cases}$

20. $\begin{cases} x_1 + 3x_2 = -2 \\ 2x_1 - 3x_2 + x_3 = 1 \\ 4x_1 + 5x_2 - 2x_3 = 0 \end{cases}$

In Exercises 21 to 26, solve for the indicated variable.

21. Solve for x_2: $\begin{cases} 2x_1 - 3x_2 + 4x_3 - x_4 = 1 \\ x_1 + 2x_2 + 2x_4 = -1 \\ 3x_1 + x_2 - 2x_4 = 2 \\ x_1 - 3x_2 + 2x_3 - x_4 = 3 \end{cases}$

22. Solve for x_4: $\begin{cases} 3x_1 + x_2 - 2x_3 + 3x_4 = 4 \\ 2x_1 - 3x_2 + 2x_3 = -2 \\ x_1 + x_2 - 2x_3 + 2x_4 = 3 \\ 2x_1 + 3x_3 - 2x_4 = 4 \end{cases}$

23. Solve for x_1: $\begin{cases} x_1 - 3x_2 + 2x_3 + 4x_4 = 0 \\ 3x_1 + 5x_2 - 6x_3 + 2x_4 = -2 \\ 2x_1 - x_2 + 9x_3 + 8x_4 = 0 \\ x_1 + x_2 + x_3 - 8x_4 = -3 \end{cases}$

24. Solve for x_3: $\begin{cases} 2x_1 + 5x_2 - 5x_3 - 3x_4 = -3 \\ x_1 + 7x_2 + 8x_3 - x_4 = 4 \\ 4x_1 + x_3 + x_4 = 3 \\ 3x_1 + 2x_2 - x_3 = 0 \end{cases}$

25. Solve for x_4: $\begin{cases} 3x_2 - x_3 + 2x_4 = 1 \\ 5x_1 + x_2 + 3x_3 - x_4 = -4 \\ x_1 - 2x_2 + 9x_4 = 5 \\ 2x_1 + 2x_3 = 3 \end{cases}$

26. Solve for x_1: $\begin{cases} 4x_1 + x_2 - 3x_4 = 4 \\ 5x_1 + 2x_2 - 2x_3 + x_4 = 7 \\ x_1 - 3x_2 + 2x_3 - 2x_4 = -6 \\ 3x_3 + 4x_4 = -7 \end{cases}$

SUPPLEMENTAL EXERCISES

27. A solution of the system of equations

$$\begin{cases} 2x_1 - 3x_2 + x_3 = 9 \\ x_1 + x_2 - 2x_3 = -3 \\ 4x_1 - x_2 - 3x_3 = 3 \end{cases}$$

is $(1, -2, 1)$. However, this solution cannot be found by using Cramer's Rule. Explain.

28. Verify the solution for x_2 given in Cramer's Rule for a System of Two Equations by solving the system of equations

$$\begin{cases} a_{11}x_1 + a_{12}x_2 = b_1 \\ a_{21}x_1 + a_{22}x_2 = b_2 \end{cases}$$

for x_2 by using the elimination method.

29. For what values of k does the system of equations

$$\begin{cases} kx + 3y = 7 \\ kx - 2y = 5 \end{cases}$$

have a unique solution?

30. For what values of k does the system of equations

$$\begin{cases} kx + 4y = 5 \\ 9x - ky = 2 \end{cases}$$

have a unique solution?

31. For what values of k does the system of equations

$$\begin{cases} x + 2y - 3z = 4 \\ 2x + ky - 4z = 5 \\ x - 2y + z = 6 \end{cases}$$

have a unique solution?

32. For what values of k does the system of equations

$$\begin{cases} kx_1 + x_2 = 1 \\ x_2 - 4x_3 = 1 \\ x_1 + kx_3 = 1 \end{cases}$$

have a unique solution?

33. Find real values for r and s so that $ru + sv = w$, where u, v, and w are complex numbers and $u = 2 + 3i$, $v = 4 - 2i$, and $w = -6 + 15i$.

34. Find real values for r and s such that $ru + sv = w$, where $u = 3 - 4i$, $v = 1 + 2i$, and $w = 4 - 22i$.

PROJECT

1. Prove Cramer's Rule for a system of three linear equations in three variables.

EXPLORING CONCEPTS WITH TECHNOLOGY

Stochastic Matrices

Matrices can be used to predict how percents of populations will change over time. Consider two neighborhood supermarkets, Super A and Super B. Each week Super A loses 5% of its customers to Super B, and each week Super B loses 8% of its customers to Super A. If this trend continues and if Super A currently has 40% of the neighborhood customers and Super B the remaining 60% of the neighborhood customers, what percent of the neighborhood will each have after n weeks?

We will approach this problem by examining the changes on a week-by-week basis. Because Super A loses 5% of its customers each week, it retains 95% of its customers. It has 40% of the neighborhood customers now, so after 1 week it will have 95% of its 40% share, or 38% $(0.95 \cdot 0.40)$ of the customers. In that same week, it gains 8% of the customers of Super B. Because Super B has 60% of the neighborhood customers, Super A's gain is 4.8% $(0.08 \cdot 0.60)$. After 1 week, Super A has 38% + 4.8% = 42.8% of the neighborhood customers. Super B has the remaining 57.2% of the customers.

The changes for the second week are calculated similarly. Super A retains 95% of its 42.8% and gains 8% of Super B's 57.2%. After week 2, Super A has

$$0.95 \cdot 0.428 + 0.08 \cdot 0.572 \approx 0.452$$

or approximately 45.2%, of the neighborhood customers. Super B has the remaining 54.8%.

We could continue in this way, but using matrices is a more convenient way to proceed. Let $T = \begin{bmatrix} 0.95 & 0.05 \\ 0.08 & 0.92 \end{bmatrix}$ where column 1 represents the percent retained by Super A and column 2 represents the percent retained by Super B. Let $X = [0.40 \quad 0.60]$ be the current market shares of Super A and Super B, respectively. Now form the product XT.

$$[0.40 \quad 0.60]\begin{bmatrix} 0.95 & 0.05 \\ 0.08 & 0.92 \end{bmatrix} = [0.428 \quad 0.572]$$

For the second week, multiply the market share after week 1 by T.

$$[0.428 \quad 0.572]\begin{bmatrix} 0.95 & 0.05 \\ 0.08 & 0.92 \end{bmatrix} \approx [0.452 \quad 0.548]$$

The last product can also be expressed as

$$[0.452 \quad 0.548] = [0.428 \quad 0.572]\begin{bmatrix} 0.95 & 0.05 \\ 0.08 & 0.92 \end{bmatrix} = \overbrace{[0.40 \quad 0.60]\begin{bmatrix} 0.95 & 0.05 \\ 0.08 & 0.92 \end{bmatrix}}^{[0.428 \quad 0.572]}\begin{bmatrix} 0.95 & 0.05 \\ 0.08 & 0.92 \end{bmatrix}$$

$$= [0.40 \quad 0.60]\begin{bmatrix} 0.95 & 0.05 \\ 0.08 & 0.92 \end{bmatrix}^2 = XT^2$$

Note that the exponent on T corresponds to the fact that 2 weeks have passed. In general, the market share after n weeks is XT^n. The matrix T is called a **stochastic matrix.** A stochastic matrix is characterized by the fact that each element of the matrix is nonnegative and the sum of the elements in each row is 1.

Use a calculator to calculate the market share of Super A and Super B after 20 weeks, 40 weeks, 60 weeks, and 100 weeks. What observations do you draw from your calculations? We started this problem with the assumption that Super A had 40% of the market and Super B had 60% of the market. Suppose, however, that originally Super A had 99% of the market and Super B 1%. Does this affect the market share each will have after 100 weeks? If Super A had 1% of the market and Super B had 99% of the market, what will the market share of each be after 100 weeks?

As another example, suppose each of three department stores is vying for the business of the other two stores. In one month, Store A loses 15% of its customers to Store B and 8% of its customers to Store C. Store B loses 10% of its customers to Store A and 12% to Store C. Store C loses 5% to Store A and 9% to Store B. Assuming these three stores have 100% of the market and the trend continues, determine what market share each will have after 100 months.

CHAPTER 10 SUMMARY

10.1 Gaussian Elimination Method

- A matrix is a rectangular array of numbers. A matrix with m rows and n columns is of order $m \times n$ or dimension $m \times n$.
- For a system of equations, it is possible to form a coefficient matrix, an augmented matrix, and a constant matrix.
- A matrix is an echelon form if all of the following conditions are satisfied:
  1. The first nonzero in any row is a 1.
  2. Rows are arranged so that the column containing the first nonzero number is to the left of the column containing the first nonzero number of the next row.
  3. All rows consisting entirely of zeros appear at the bottom of the matrix.
- The Gaussian elimination method uses elementary row operations to solve a system of linear equations.
- **Elementary Row Operations**
  The elementary row operations for a matrix are
  1. Interchanging two rows
  2. Multiplying all the elements in a row by the same nonzero number
  3. Replacing a row by the sum of that row and a nonzero multiple of any other row
- **Interpolating Polynomial**
  Let $(x_0, y_0), (x_1, y_1), (x_2, y_2), \ldots, (x_n, y_n)$ be the coordinates of a set of points for which all the x_i are distinct. Then the interpolating polynomial is a unique polynomial of degree at most n that passes through the given points.

10.2 The Algebra of Matrices

- Two matrices $A = [a_{ij}]$ and $B = [b_{ij}]$ are equal if and only if $a_{ij} = b_{ij}$ for every i and j.
- The sum of two matrices of the same order is the matrix whose elements are the sum of the corresponding elements of the two matrices.
- The $m \times n$ zero matrix is the matrix whose elements are all zeros.
- Taking the product of a real number and a matrix is called scalar multiplication.
- In order for us to multiply two matrices, the number of columns of the first matrix must equal the number of rows of the second matrix.
- In general, matrix multiplication is not commutative.
- The multiplicative identity matrix is the matrix with 1s on the main diagonal and zeros everywhere else.

10.3 The Inverse of a Matrix

- The multiplicative inverse of a square matrix A, denoted by A^{-1}, has the property that
$$A \cdot A^{-1} = A^{-1} \cdot A = I_n$$
where I_n is the multiplicative identity matrix.
- A singular matrix is one that does not have a multiplicative inverse.
- Input-output analysis attempts to determine the necessary output of industries to satisfy each other's demands plus the demands of consumers.

10.4 Determinants

- Associated with each square matrix is a number called the determinant of the matrix.
- The minor of the element a_{ij} of a square matrix A is the determinant of the matrix obtained by deleting the ith row and the jth column of A.
- The cofactor of the element a_{ij} of a square matrix A is $(-1)^{i+j}M_{ij}$, where M_{ij} is the minor of a_{ij}.
- The value of a determinant can be found by multiplying the elements of any row or column by their respective cofactors and then adding the results. This is called expanding by cofactors.

10.5 Cramer's Rule

- Cramer's Rule is a method of solving a system of n equations in n variables by using determinants.

CHAPTER 10 TRUE/FALSE EXERCISES

In Exercises 1 to 15, answer true or false. If the statement is false, give an example to show that the statement is false.

1. If $A = \begin{bmatrix} 2 & 3 \\ 1 & 4 \end{bmatrix}$, then $A^2 = \begin{bmatrix} 4 & 9 \\ 1 & 16 \end{bmatrix}$.
2. Every matrix has an additive inverse.
3. Every square matrix has a multiplicative inverse.
4. Let the matrices A, B, and C be square matrices of order n. If $AB = AC$, then $B = C$.

5. It is possible to find the determinant of every square matrix.

6. If A and B are square matrices of order n, then

$$\det(A + B) = \det(A) + \det(B)$$

7. Cramer's Rule can be used to solve any system of three equations in three variables.

8. If A and B are matrices of order n, then $AB - BA = O$.

9. A nonsingular matrix has a multiplicative inverse.

10. If A, B, and C are square matrices of order n, then the product ABC depends on which two matrices are multiplied first. That is, $(AB)C$ produces a different result from $A(BC)$.

11. The Gaussian elimination method for solving a system of linear equations can be applied only to systems of equations that have the same number of variables as equations.

12. If A is a square matrix of order n, then $\det(2A) = 2\det(A)$.

13. If A and B are matrices, then the product AB is defined when the number of columns of A equals the number of rows of B.

14. If A and B are square matrices of order n and $AB = O$ (the zero matrix), then $A = O$ or $B = O$.

15. If $A = \begin{bmatrix} 3 & 6 \\ -1 & -2 \end{bmatrix}$, then $A^5 = A$.

CHAPTER 10 REVIEW EXERCISES

In Exercises 1 to 18, perform the indicated operations. Let

$A = \begin{bmatrix} 2 & -1 & 3 \\ 3 & 2 & -1 \end{bmatrix}$, $B = \begin{bmatrix} 0 & -2 \\ 4 & 2 \\ 1 & -3 \end{bmatrix}$, $C = \begin{bmatrix} 2 & 6 & 1 \\ 1 & 2 & -1 \\ 2 & 4 & -1 \end{bmatrix}$, **and**

$D = \begin{bmatrix} -3 & 4 & 2 \\ 4 & -2 & 5 \end{bmatrix}$.

1. $3A$
2. $-2B$
3. $-A + D$
4. $2A - 3D$
5. AB
6. DB
7. BA
8. BD
9. C^2
10. C^3
11. BAC
12. ADB
13. $AB - BA$
14. $DB - BD$
15. $(A - D)C$
16. $AC - DC$
17. C^{-1}
18. $|C|$

In Exercises 19 to 34, solve the system of equations by using the Gaussian elimination method.

19. $\begin{cases} 2x - 3y = 7 \\ 3x - 4y = 10 \end{cases}$

20. $\begin{cases} 3x + 4y = -9 \\ 2x + 3y = -7 \end{cases}$

21. $\begin{cases} 4x - 5y = 12 \\ 3x + y = 9 \end{cases}$

22. $\begin{cases} 2x - 5y = 10 \\ 5x + 2y = 4 \end{cases}$

23. $\begin{cases} x + 2y + 3z = 5 \\ 3x + 8y + 11z = 17 \\ 2x + 6y + 7z = 12 \end{cases}$

24. $\begin{cases} x - y + 3z = 10 \\ 2x - y + 7z = 24 \\ 3x - 6y + 7z = 21 \end{cases}$

25. $\begin{cases} 2x - y - z = 4 \\ x - 2y - 2z = 5 \\ 3x - 3y - 8z = 19 \end{cases}$

26. $\begin{cases} 3x - 7y + 8z = 10 \\ x - 3y + 2z = 0 \\ 2x - 8y + 7z = 5 \end{cases}$

27. $\begin{cases} 4x - 9y + 6z = 54 \\ 3x - 8y + 8z = 49 \\ x - 3y + 2z = 17 \end{cases}$

28. $\begin{cases} 3x + 8y - 5z = 6 \\ 2x + 9y - z = -8 \\ x - 4y - 2z = 16 \end{cases}$

29. $\begin{cases} x + y + 2z = -5 \\ 2x + 3y + 5z = -13 \\ 2x + 5y + 7z = -19 \end{cases}$

30. $\begin{cases} x - 2y + 3z = 9 \\ 3x - 5y + 8z = 25 \\ x - z = 5 \end{cases}$

31. $\begin{cases} w + 2x - y + 2z = 1 \\ 3w + 8x + y + 4z = 1 \\ 2w + 7x + 3y + 2z = 0 \\ w + 3x - 2y + 5z = 6 \end{cases}$

32. $\begin{cases} w - 3x - 2y + z = -1 \\ 2w - 5x + 3z = 1 \\ 3w - 7x + 3y = -18 \\ 2w - 3x - 5y - 2z = -8 \end{cases}$

33. $\begin{cases} w + 3x + y - 4z = 3 \\ w + 4x + 3y - 6z = 5 \\ 2w + 8x + 7y - 5z = 11 \\ 2w + 5x - 6z = 4 \end{cases}$

34. $\begin{cases} w + 4x - 2y + 3z = 6 \\ 2w + 9x - y + 5z = 13 \\ w + 7x + 6y + 5z = 9 \\ 3w + 14x + 7z = 20 \end{cases}$

35. **INTERPOLATING POLYNOMIAL** Find a polynomial that passes through the points whose coordinates are $(-1, -4)$, $(2, 8)$, and $(3, 16)$.

36. **INTERPOLATING POLYNOMIAL** Find a polynomial that passes through the points whose coordinates are $(-1, 4)$, $(1, 0)$, and $(2, -5)$.

In Exercises 37 to 48, find the inverse, if it exists, of the given matrix.

37. $\begin{bmatrix} 2 & -2 \\ 3 & -2 \end{bmatrix}$

38. $\begin{bmatrix} 3 & 4 \\ 2 & 3 \end{bmatrix}$

39. $\begin{bmatrix} -2 & 3 \\ 2 & 4 \end{bmatrix}$

40. $\begin{bmatrix} 5 & -4 \\ 3 & 2 \end{bmatrix}$

41. $\begin{bmatrix} 1 & 2 & 1 \\ 2 & 6 & 4 \\ 3 & 8 & 6 \end{bmatrix}$

42. $\begin{bmatrix} 1 & -3 & 2 \\ 3 & -8 & 7 \\ 2 & -3 & 6 \end{bmatrix}$

43. $\begin{bmatrix} 3 & -2 & 7 \\ 2 & -1 & 5 \\ 3 & 0 & 10 \end{bmatrix}$

44. $\begin{bmatrix} 4 & 9 & -11 \\ 3 & 7 & -8 \\ 2 & 6 & -3 \end{bmatrix}$

45. $\begin{bmatrix} 1 & -1 & 2 & 3 \\ 2 & -1 & 6 & 5 \\ 3 & -1 & 9 & 6 \\ 2 & -2 & 4 & 7 \end{bmatrix}$

46. $\begin{bmatrix} 1 & 2 & -2 & 1 \\ 3 & 7 & -3 & 1 \\ 2 & 7 & 4 & 3 \\ 1 & 4 & 2 & 4 \end{bmatrix}$

47. $\begin{bmatrix} 3 & 7 & -1 & 8 \\ 2 & 5 & 0 & 5 \\ 3 & 6 & -4 & 8 \\ 2 & 4 & -4 & 4 \end{bmatrix}$

48. $\begin{bmatrix} 3 & 1 & 5 & -5 \\ 2 & 1 & 4 & -3 \\ 3 & 0 & 4 & -3 \\ 4 & 1 & 8 & 1 \end{bmatrix}$

In Exercises 49 to 52, solve the given system of equations for each set of constants. Use the inverse matrix method.

49. $\begin{cases} 3x + 4y = b_1 \\ 2x + 3y = b_2 \end{cases}$

a. $b_1 = 2, b_2 = -3$

b. $b_1 = -2, b_2 = 4$

50. $\begin{cases} 2x - 5y = b_1 \\ 3x - 7y = b_2 \end{cases}$

a. $b_1 = -3, b_2 = 4$

b. $b_1 = 2, b_2 = -5$

51. $\begin{cases} 2x + y - z = b_1 \\ 4x + 4y + z = b_2 \\ 2x + 2y - 3z = b_3 \end{cases}$

a. $b_1 = -1, b_2 = 2, b_3 = 4$

b. $b_1 = -2, b_2 = 3, b_3 = 0$

52. $\begin{cases} 3x - 2y + z = b_1 \\ 3x - y + 3z = b_2 \\ 6x - 4y + z = b_3 \end{cases}$

a. $b_1 = 0, b_2 = 3, b_3 = -2$

b. $b_1 = 1, b_2 = 2, b_3 = -4$

In Exercises 53 to 60, evaluate each determinant by using elementary row or column operations.

53. $\begin{vmatrix} 2 & 6 & 4 \\ 1 & 2 & 1 \\ 3 & 8 & 6 \end{vmatrix}$

54. $\begin{vmatrix} 3 & 0 & 10 \\ 3 & -2 & 7 \\ 2 & -1 & 5 \end{vmatrix}$

55. $\begin{vmatrix} 3 & -8 & 7 \\ 2 & -3 & 6 \\ 1 & -3 & 2 \end{vmatrix}$

56. $\begin{vmatrix} 4 & 9 & -11 \\ 2 & 6 & -3 \\ 3 & 7 & -8 \end{vmatrix}$

57. $\begin{vmatrix} 1 & -1 & 2 & 1 \\ 2 & -1 & 6 & 3 \\ 3 & -1 & 8 & 7 \\ 3 & 0 & 9 & 9 \end{vmatrix}$

58. $\begin{vmatrix} 1 & 2 & -2 & 3 \\ 3 & 7 & -3 & 11 \\ 2 & 3 & -5 & 11 \\ 2 & 6 & 1 & 8 \end{vmatrix}$

59. $\begin{vmatrix} 1 & 2 & -2 & 1 \\ 2 & 5 & -3 & 1 \\ 2 & 0 & -10 & 1 \\ 3 & 8 & -4 & 1 \end{vmatrix}$

60. $\begin{vmatrix} 1 & 3 & -2 & 0 \\ 3 & 11 & -4 & 4 \\ 2 & 9 & -8 & 2 \\ 3 & 12 & -10 & 2 \end{vmatrix}$

In Exercises 61 to 66, solve each system of equations by using Cramer's Rule.

61. $\begin{cases} 2x_1 - 3x_2 = 2 \\ 3x_1 + 5x_2 = 2 \end{cases}$

62. $\begin{cases} 3x_1 + 4x_2 = -3 \\ 5x_1 - 2x_2 = 2 \end{cases}$

63. $\begin{cases} 2x_1 + x_2 - 3x_3 = 2 \\ 3x_1 + 2x_2 + x_3 = 1 \\ x_1 - 3x_2 + 4x_3 = -2 \end{cases}$

64. $\begin{cases} 3x_1 + 2x_2 - x_3 = 0 \\ x_1 + 3x_2 - 2x_3 = 3 \\ 4x_1 - x_2 - 5x_3 = -1 \end{cases}$

65. $\begin{cases} 2x_2 + 5x_3 = 2 \\ 2x_1 - 5x_2 + x_3 = 4 \\ 4x_1 + 3x_2 = 2 \end{cases}$

66. $\begin{cases} 2x_1 - 3x_2 - 4x_3 = 2 \\ x_1 - 2x_2 + 2x_3 = -1 \\ 2x_1 + 7x_2 - x_3 = 2 \end{cases}$

In Exercises 67 and 68, use Cramer's Rule to solve for the indicated variable.

67. Solve for x_3: $\begin{cases} x_1 - 3x_2 + x_3 + 2x_4 = 3 \\ 2x_1 + 7x_2 - 3x_3 + x_4 = 2 \\ -x_1 + 4x_2 + 2x_3 - 3x_4 = -1 \\ 3x_1 + x_2 - x_3 - 2x_4 = 0 \end{cases}$

68. Solve for x_2: $\begin{cases} 2x_1 + 3x_2 - 2x_3 + x_4 = -2 \\ x_1 - x_2 - 3x_3 + 2x_4 = 2 \\ 3x_1 + 3x_2 - 4x_3 - x_4 = 4 \\ 5x_1 - 5x_2 - x_3 + 2x_4 = 7 \end{cases}$

In Exercises 69 and 70, solve the input-output problem.

69. **BUSINESS RESOURCE ALLOCATION** An electronics conglomerate has three divisions, which produce computers, monitors, and disk drives. For each \$1 worth of output, the computer division needs \$.05 worth of computers, \$.02 worth of monitors, and \$.03 worth of disk drives. For each \$1 worth of output, the monitor division needs \$.06 worth of computers, \$.04 worth of monitors, and \$.03 worth of disk drives. For each \$1 worth of output, the disk drive division requires \$.08 worth of computers, \$.04 worth of monitors, and \$.05 worth of disk drives. Sales estimates are \$30 million for the computer division, \$12 million for the monitor division, and \$21 million for the disk drive division. At what level should each division produce to satisfy this demand?

70. **BUSINESS RESOURCE ALLOCATION** A manufacturing conglomerate has three divisions, which produce paper,

lumber, and prefabricated walls. For each \$1 worth of output, the lumber division needs \$.07 worth of lumber, \$.03 worth of paper, and \$.03 worth of prefabricated walls. For each \$1 worth of output, the paper division needs \$.04 worth of lumber, \$.07 worth of paper, and \$.03 worth of prefabricated walls. For each \$1 worth of output, the prefabricated walls division requires \$.07 worth of lumber, \$.04 worth of paper, and \$.02 worth of prefabricated walls. Sales estimates are \$27 million for the lumber division, \$18 million for the paper division, and \$10 million for the prefabricated walls division. At what level should each division produce to satisfy this demand?

CHAPTER 10 TEST

1. Write the augmented matrix, the coefficient matrix, and the constant matrix, for the system of equations

$$\begin{cases} 2x + 3y - 3z = 4 \\ 3x \qquad + 2z = -1 \\ 4x - 4y + 2z = 3 \end{cases}$$

2. Write a system of equations that is equivalent to the augmented matrix $\begin{bmatrix} 3 & -2 & 5 & -1 & 9 \\ 2 & 3 & -1 & 4 & 8 \\ 1 & 0 & 3 & 2 & -1 \end{bmatrix}$.

In Exercises 3 to 5, solve the system of equations by using the Gaussian elimination method.

3. $\begin{cases} x - 2y + 3z = 10 \\ 2x - 3y + 8z = 23 \\ -x + 3y - 2z = -9 \end{cases}$

4. $\begin{cases} 2x + 6y - z = 1 \\ x + 3y - z = 1 \\ 3x + 10y - 2z = 1 \end{cases}$

5. $\begin{cases} w + 2x - 3y + 2z = 11 \\ 2w + 5x - 8y + 5z = 28 \\ -2w - 4x + 7y - z = -18 \end{cases}$

In Exercises 6 to 18, let $A = \begin{bmatrix} -1 & 3 & 2 \\ 1 & 4 & -1 \end{bmatrix}$, $B = \begin{bmatrix} 2 & -1 & 3 \\ 4 & -2 & -1 \\ 3 & 2 & 2 \end{bmatrix}$, **and** $C = \begin{bmatrix} 1 & -2 & 3 \\ 2 & -3 & 8 \\ -1 & 3 & -2 \end{bmatrix}$. **Perform each possible operation. If an operation is not possible, so state.**

6. $-3A$

7. $A + B$

8. $3B - 2C$

9. AB

10. $AB - A$

11. CA

12. $BC - CB$

13. A^2

14. B^2

15. C^{-1}

16. Find the minor and cofactor of b_{21} for matrix B.

17. Find the determinant of B by expanding by cofactors of row 3.

18. Find the determinant of C by using elementary row operations.

19. Find the value of z for the following system of equations by using Cramer's Rule.

$$\begin{cases} 3x + 2y - z = 12 \\ 2x - 3y + 2z = -1 \\ 5x + 6y + 3z = 4 \end{cases}$$

20. A simplified economy has three major industries: mining, manufacturing, and transportation. The input-output matrix for this economy is

$$\begin{bmatrix} 0.15 & 0.23 & 0.11 \\ 0.08 & 0.10 & 0.05 \\ 0.16 & 0.11 & 0.07 \end{bmatrix}$$

Set up, but do not solve, a matrix equation that, when solved, will determine the gross output needed to satisfy consumer demand for \$50 million worth of mining, \$32 million worth of manufacturing, and \$8 million worth of transportation.

CHAPTER 11

SEQUENCES, SERIES, AND PROBABILITY

A Snowflake with Infinite Perimeter

We started this book with a discussion of infinities. Now that we have reached the last chapter, it seems appropriate to conclude with a discussion of infinity as well.

We begin with an equilateral triangle each side of which is 1 unit long. The perimeter, then, is 3 units. Now construct an identical but smaller triangle onto the middle third of each side. See the following figures.

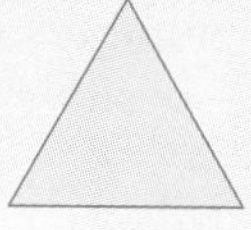

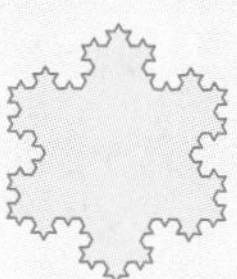

The "snowflake" now has 12 line segments each of length 1/3 unit. The perimeter of the snowflake is 12(1/3) or 4 units. Repeat the procedure and construct identical but smaller triangles on the middle third of each of the 12 sides of the snowflake. The perimeter is $48(1/9) = 16/3$ units. Continuing this procedure, we can show that the perimeter of each succeeding snowflake is 4/3 the perimeter of the preceding one. Thus the perimeter continues to grow without bound and becomes infinite.

Now examine the area of each snowflake. Let A be the area of the original triangle. In the second stage, each new triangle has an area that is 1/9 of A. The area of this snowflake is the original area plus the area of these 3 triangles: $A + 3(1/9)A = (4/3)A$.

For the next snowflake, each new triangle has an area equal to $(1/81)A$. The area of this snowflake is now $(40/27)A$. Continuing in this way, it is possible to show that the area of the succeeding snowflakes approaches the finite number $(8/5)A$. Thus we have a sequence of snowflakes whose perimeter increases without bound but whose area remains finite.

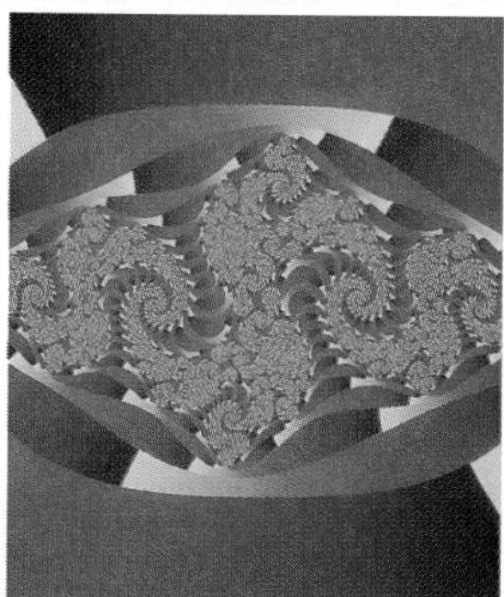

◆ A fractal is a geometric figure consisting of a basic pattern that keeps repeating on a smaller and smaller scale. Here are two other examples.

SECTION 11.1 INFINITE SEQUENCES AND SUMMATION NOTATION

- INFINITE SEQUENCES
- FACTORIALS
- PARTIAL SUMS AND SUMMATION NOTATION

◆ INFINITE SEQUENCES

The *ordered* list of numbers 2, 4, 8, 16, 32, ... is called an infinite sequence. The list is ordered simply because order makes a difference. The sequence 2, 8, 4, 16, 32, ... contains the same numbers, but in a different order. Therefore, it is a different infinite sequence.

An infinite sequence can be thought of as a pairing between positive numbers and real numbers. For example, 1, 4, 9, 16, 25, 36, ..., n^2, ... pairs a natural number with its square.

$$\begin{array}{cccccccccc} 1 & 2 & 3 & 4 & 5 & 6 & \ldots & n & \ldots \\ \downarrow & \downarrow & \downarrow & \downarrow & \downarrow & \downarrow & & \downarrow & \\ 1 & 4 & 9 & 16 & 25 & 36 & \ldots & n^2 & \ldots \end{array}$$

This pairing of numbers enables us to define an infinite sequence as a function with domain the positive integers.

Infinite Sequence

An **infinite sequence** is a function whose domain is the positive integers and whose range is a set of real numbers.

Although the positive integers do not include zero, it is occasionally convenient to include zero in the domain of an infinite sequence. Also, we will frequently use the word *sequence* instead of the phrase *infinite sequence.*

As an example of a sequence, let $f(n) = 2n - 1$. The range of this function is

$$\begin{array}{cccccccc} f(1), & f(2), & f(3), & f(4), & \ldots, & f(n), & \ldots \\ 1, & 3, & 5, & 7, & \ldots, & 2n-1, & \ldots \end{array}$$

The elements in the range of a sequence are called the **terms** of the sequence. For our example, the terms are 1, 3, 5, 7, ..., $2n - 1$, ... The **first term** of the sequence is 1, the **second term** is 3, and so on. The ***n*th term,** or the **general term,** is $2n - 1$.

Rather than use functional notation for sequences, it is customary to use a subscript notation. Thus a_n represents the nth term of a sequence. Using this notation, we would write

$$a_n = 2n - 1$$

Thus $a_1 = 1, a_2 = 3, a_3 = 5, a_4 = 7$.

EXAMPLE 1 Find the Terms of a Sequence

a. Find the first three terms of the sequence $a_n = \dfrac{1}{n(n+1)}$.

b. Find the eighth term of the sequence $a_n = \dfrac{2^n}{n^2}$.

Solution

a. $a_1 = \dfrac{1}{1(1+1)} = \dfrac{1}{2}, a_2 = \dfrac{1}{2(2+1)} = \dfrac{1}{6}, a_3 = \dfrac{1}{3(3+1)} = \dfrac{1}{12}$

b. $a_8 = \dfrac{2^8}{8^2} = \dfrac{256}{64} = 4$

TRY EXERCISE 6, EXERCISE SET 11.1, PAGE 777

An **alternating sequence** is one in which the signs of the terms *alternate* between positive and negative values. The sequence defined by $a_n = (-1)^{n+1} \cdot 1/n$ is an alternating sequence.

$$a_1 = (-1)^{1+1} \cdot \frac{1}{1} = 1 \qquad a_2 = (-1)^{2+1} \cdot \frac{1}{2} = -\frac{1}{2} \qquad a_3 = (-1)^{3+1} \cdot \frac{1}{3} = \frac{1}{3}$$

The first six terms of the sequence are

$$1, -\frac{1}{2}, \frac{1}{3}, -\frac{1}{4}, \frac{1}{5}, -\frac{1}{6}$$

MATH MATTERS

Recursive subroutines are a part of many computer programming languages. One of the most influential people in the development of computer languages (especially COBOL) was Grace Murray Cooper. She was the first woman to receive a doctorate in mathematics from Yale University. She went on to become an admiral in the U.S. Navy.

A **recursively defined sequence** is one in which each succeeding term of the sequence is defined by using some of the preceding terms. For example, let $a_1 = 1$, $a_2 = 1$, and $a_{n+1} = a_{n-1} + a_n$.

$$a_3 = a_1 + a_2 = 1 + 1 = 2 \qquad \bullet\ n = 2$$
$$a_4 = a_2 + a_3 = 1 + 2 = 3 \qquad \bullet\ n = 3$$
$$a_5 = a_3 + a_4 = 2 + 3 = 5 \qquad \bullet\ n = 4$$
$$a_6 = a_4 + a_5 = 3 + 5 = 8 \qquad \bullet\ n = 5$$

This recursive sequence 1, 1, 2, 3, 5, 8, ... is called the **Fibonacci sequence,** named after Leonardo Fibonacci (1180?–?1250), an Italian mathematician.

EXAMPLE 2 Find Terms of a Sequence Defined Recursively

Let $a_1 = 1$ and $a_n = na_{n-1}$. Find a_2, a_3, and a_4.

Solution

$$a_2 = 2a_1 = 2 \cdot 1 = 2 \qquad a_3 = 3a_2 = 3 \cdot 2 = 6 \qquad a_4 = 4a_3 = 4 \cdot 6 = 24$$

TRY EXERCISE 28, EXERCISE SET 11.1, PAGE 777

◆ FACTORIALS

It is possible to find an nth term formula for the sequence defined recursively in Example 2 by

$$a_1 = 1 \qquad a_n = na_{n-1}$$

Consider the term a_5 of that sequence.

$$\begin{aligned} a_5 &= 5a_4 \\ &= 5 \cdot 4a_3 && \bullet\ a_4 = 4a_3 \\ &= 5 \cdot 4 \cdot 3a_2 && \bullet\ a_3 = 3a_2 \\ &= 5 \cdot 4 \cdot 3 \cdot 2a_1 && \bullet\ a_2 = 2a_1 \\ &= 5 \cdot 4 \cdot 3 \cdot 2 \cdot 1 && \bullet\ a_1 = 1 \end{aligned}$$

Continuing in this manner for a_n, we have

$$\begin{aligned} a_n &= na_{n-1} \\ &= n(n-1)a_{n-2} \\ &= n(n-1)(n-2)a_{n-3} \\ &\ \vdots \\ &= n(n-1)(n-2)(n-3)\cdots 2 \cdot 1 \end{aligned}$$

The number $n \cdot (n-1)\cdots 3 \cdot 2 \cdot 1$ is called **n factorial** and is written $n!$.

The Factorial of a Number

If n is a positive integer, then $n!$, which is read "n factorial," is

$$n! = n \cdot (n-1) \cdots 3 \cdot 2 \cdot 1$$

We also define

$$0! = 1$$

MATH MATTERS

The factorial symbol as an exclamation point was first introduced in 1808 so that printers setting the type in a book would not have to write out, for example, 5 · 4 · 3 · 2 · 1 but instead could write 5!. This notation was not accepted by everyone, and alternative suggestions were made. One other notation was $\underline{n|}$. Although this angle bracket saved the printer from writing a product, it had to be constructed, whereas the exclamation point was already in the printer's tray.

It may seem strange to define $0! = 1$, but we shall see later that it is a reasonable definition.

Examples of factorials include

$$5! = 5 \cdot 4 \cdot 3 \cdot 2 \cdot 1 = 120$$
$$10! = 10 \cdot 9 \cdot 8 \cdot 7 \cdot 6 \cdot 5 \cdot 4 \cdot 3 \cdot 2 \cdot 1 = 3{,}628{,}800$$

Note that we can write 12! as

$$12! = 12 \cdot 11! = 12 \cdot 11 \cdot 10! = 12 \cdot 11 \cdot 10 \cdot 9!$$

In general,

$$n! = n \cdot (n-1)!$$

EXAMPLE 3 Evaluate Factorial Expressions

Evaluate each factorial expression. **a.** $\dfrac{8!}{5!}$ **b.** $6! - 4!$

Solution

a. $\dfrac{8!}{5!} = \dfrac{8 \cdot 7 \cdot 6 \cdot 5!}{5!} = 8 \cdot 7 \cdot 6 = 336$

b. $6! - 4! = (6 \cdot 5 \cdot 4 \cdot 3 \cdot 2 \cdot 1) - (4 \cdot 3 \cdot 2 \cdot 1) = 720 - 24 = 696$

TRY EXERCISE 42, EXERCISE SET 11.1, PAGE 777

◆ PARTIAL SUMS AND SUMMATION NOTATION

Another important way of obtaining a sequence is by adding the terms of a given sequence. For example, consider the sequence whose general term is given by $a_n = 1/2^n$. The terms of this sequence are

$$\frac{1}{2}, \frac{1}{4}, \frac{1}{8}, \frac{1}{16}, \frac{1}{32}, \ldots, \frac{1}{2^n}, \ldots$$

From this sequence we can generate a new sequence that is the sum of the terms of $1/2^n$.

$$S_1 = \frac{1}{2}$$

$$S_2 = \frac{1}{2} + \frac{1}{4} = \frac{3}{4}$$

$$S_3 = \frac{1}{2} + \frac{1}{4} + \frac{1}{8} = \frac{7}{8}$$

$$S_4 = \frac{1}{2} + \frac{1}{4} + \frac{1}{8} + \frac{1}{16} = \frac{15}{16}$$

and, in general, $$S_n = \frac{1}{2} + \frac{1}{4} + \frac{1}{8} + \frac{1}{16} + \cdots + \frac{1}{2^n}$$

The term S_n is called the **nth partial sum** of the infinite sequence, and the sequence $S_1, S_2, S_3, \ldots, S_n$ is called the **sequence of partial sums.**

A convenient notation used for partial sums is called **summation notation.** The sum of the first n terms of a sequence a_n is represented by using the Greek letter Σ (sigma).

$$\sum_{i=1}^{n} a_i = a_1 + a_2 + a_3 + \cdots + a_n$$

This sum is called a **series.** The letter i is called the **index of the summation;** n is the **upper limit** of the summation; 1 is the **lower limit** of the summation.

MATH MATTERS

Leonard Euler (1707–1783) found that the sequence of terms given by

$$S_n = 1 - \frac{1}{3} + \frac{1}{5} - \frac{1}{7} + \cdots + \frac{(1)^{n-1}}{2n-1}$$

became closer and closer to $\pi/4$ as n increased. In summation notation, we would write each term as

$$S_n = \sum_{k=1}^{n} \frac{(-1)^{k-1}}{2k-1}$$

EXAMPLE 4 Evaluating Series

Evaluate each series. **a.** $\sum_{i=1}^{4} \frac{i}{i+1}$ **b.** $\sum_{j=2}^{5} (-1)^j j^2$

Solution

a. $\sum_{i=1}^{4} \frac{i}{i+1} = \frac{1}{2} + \frac{2}{3} + \frac{3}{4} + \frac{4}{5} = \frac{163}{60}$

b. $\sum_{j=2}^{5} (-1)^j j^2 = (-1)^2 2^2 + (-1)^3 3^2 + (-1)^4 4^2 + (-1)^5 5^2$

$= 4 - 9 + 16 - 25 = -14$

take note

Example 4b illustrates that it is not necessary for a summation to begin at 1. The index of the summation can be any letter.

TRY EXERCISE 52, EXERCISE SET 11.1, PAGE 777

Properties of Summation Notation

If a_n and b_n are sequences and c is a real number, then

1. $\sum_{i=1}^{n} (a_i \pm b_i) = \sum_{i=1}^{n} a_i \pm \sum_{i=1}^{n} b_i$
2. $\sum_{i=1}^{n} ca_i = c\sum_{i=1}^{n} a_i$
3. $\sum_{i=1}^{n} c = nc$

The proof of property (1) depends on the commutative and associative properties of real numbers.

$$\begin{aligned}\sum_{i=1}^{n} (a_i \pm b_i) &= (a_1 \pm b_1) + (a_2 \pm b_2) + \cdots + (a_n \pm b_n) \\ &= (a_1 + a_2 + \cdots + a_n) \pm (b_1 + b_2 + \cdots + b_n) \\ &= \sum_{i=1}^{n} a_i \pm \sum_{i=1}^{n} b_i\end{aligned}$$

Property (2) is proved by using the distributive property; this is left as an exercise.

To prove property (3), let $a_n = c$. That is, each a_n is equal to the same constant c. (This is called a **constant sequence.**) Then

$$\sum_{i=1}^{n} a_n = a_1 + a_2 + \cdots + a_n = \underbrace{c + c + \cdots + c}_{n \text{ terms}} = nc$$

TOPICS FOR DISCUSSION

1. Discuss the difference between a finite sequence and an infinite sequence. Give an example of each type.
2. Discuss the difference between a sequence and a series.
3. What is a recursive sequence? Give an example of a recursive sequence.
4. What is an alternating sequence? Give an example of an alternating sequence.

EXERCISE SET 11.1

In Exercises 1 to 24, find the first three terms and the eighth term of the sequence that has the given *n*th term.

1. $a_n = n(n-1)$
2. $a_n = 2n$
3. $a_n = 1 - \frac{1}{n}$
4. $a_n = \frac{n+1}{n}$
5. $a_n = \frac{(-1)^{n+1}}{n^2}$
6. $a_n = \frac{(-1)^{n+1}}{n(n+1)}$
7. $a_n = \frac{(-1)^{2n-1}}{3n}$
8. $a_n = \frac{(-1)^n}{2n-1}$
9. $a_n = \left(\frac{2}{3}\right)^n$
10. $a_n = \left(\frac{-1}{2}\right)^n$
11. $a_n = 1 + (-1)^n$
12. $a_n = 1 + (-0.1)^n$
13. $a_n = (1.1)^n$
14. $a_n = \frac{n}{n^2+1}$
15. $a_n = \frac{(-1)^{n+1}}{\sqrt{n}}$
16. $a_n = \frac{3^{n-1}}{2^n}$
17. $a_n = n!$
18. $a_n = \frac{n!}{(n-1)!}$
19. $a_n = \log n$
20. $a_n = \ln n$ (natural logarithm)
21. a_n is the digit in the nth place in the decimal expansion of 1/7.
22. a_n is the digit in the nth place in the decimal expansion of 1/13.
23. $a_n = 3$
24. $a_n = -2$

In Exercises 25 to 34, find the first three terms of each recursively defined sequence.

25. $a_1 = 5, a_n = 2a_{n-1}$
26. $a_1 = 2, a_n = 3a_{n-1}$
27. $a_1 = 2, a_n = na_{n-1}$
28. $a_1 = 1, a_n = n^2a_{n-1}$
29. $a_1 = 2, a_n = (a_{n-1})^2$
30. $a_1 = 4, a_n = \frac{1}{a_{n-1}}$
31. $a_1 = 2, a_n = 2na_{n-1}$
32. $a_1 = 2, a_n = (-3)na_{n-1}$
33. $a_1 = 3, a_n = (a_{n-1})^{1/n}$
34. $a_1 = 2, a_n = (a_{n-1})^n$
35. $a_1 = 1, a_2 = 3, a_n = \frac{1}{2}(a_{n-1} + a_{n-2})$. Find a_3, a_4, and a_5.
36. $a_1 = 1, a_2 = 4, a_n = (a_{n-1})(a_{n-2})$. Find a_3, a_4, and a_5.

In Exercises 37 to 44, evaluate the factorial expression.

37. $7! - 6!$
38. $(4!)^2$
39. $\frac{9!}{7!}$
40. $\frac{10!}{5!}$
41. $\frac{8!}{3!\,5!}$
42. $\frac{12!}{4!\,8!}$
43. $\frac{100!}{99!}$
44. $\frac{100!}{98!\,2!}$

In Exercises 45 to 58, evaluate the series.

45. $\sum_{i=1}^{5} i$
46. $\sum_{i=1}^{4} i^2$
47. $\sum_{i=1}^{5} i(i-1)$
48. $\sum_{i=1}^{7} (2i+1)$
49. $\sum_{k=1}^{4} \frac{1}{k}$
50. $\sum_{k=1}^{6} \frac{1}{k(k+1)}$
51. $\sum_{j=1}^{8} 2j$
52. $\sum_{i=1}^{6} (2i+1)(2i-1)$
53. $\sum_{i=3}^{5} (-1)^i 2^i$
54. $\sum_{i=3}^{5} \frac{(-1)^i}{2^i}$
55. $\sum_{n=1}^{7} \log \frac{n+1}{n}$
56. $\sum_{n=2}^{8} \ln \frac{n}{n+1}$

57. $\displaystyle\sum_{k=0}^{8} \frac{8!}{k!\,(8-k)!}$

58. $\displaystyle\sum_{k=0}^{7} \frac{1}{k!}$

In Exercises 59 to 66, write the given series in summation notation.

59. $\frac{1}{1} + \frac{1}{4} + \frac{1}{9} + \frac{1}{16} + \frac{1}{25} + \frac{1}{36}$

60. $2 + 4 + 6 + 8 + 10 + 12 + 14$

61. $2 - 4 + 8 - 16 + 32 - 64 + 128$

62. $1 - 8 + 27 - 64 + 125$

63. $7 + 10 + 13 + 16 + 19$

64. $30 + 26 + 22 + 18 + 14 + 10$

65. $\frac{1}{2} + \frac{1}{4} + \frac{1}{8} + \frac{1}{16}$

66. $1 - \frac{2}{3} + \frac{4}{9} - \frac{8}{27} + \frac{16}{81} - \frac{32}{243}$

SUPPLEMENTAL EXERCISES

67. **NEWTON'S METHOD** Newton's approximation to the square root of a number is given by the recursive sequence

$$a_1 = \frac{N}{2} \qquad a_n = \frac{1}{2}\left(a_{n-1} + \frac{N}{a_{n-1}}\right)$$

Approximate $\sqrt{7}$ by computing a_4. Compare this result with the calculator value of $\sqrt{7} \approx 2.6457513$.

68. Use the formula in Exercise 67 to approximate $\sqrt{10}$ by finding a_5.

69. Let $a_1 = N$ and $a_n = \sqrt{a_{n-1}}$. Find a_{20} when $N = 7$. (*Hint:* Enter 7 into your calculator and then press the $\boxed{\sqrt{\ }}$ key nineteen times.) Make a conjecture as to the value of a_n as n increases without bound.

70. Let $a_n = i^n$, where i is the imaginary unit. Find the first eight terms of the sequence defined by a_n. Find a_{237}.

71. Let $a_n = \left[\frac{1}{2}\left(-1 + i\sqrt{3}\right)\right]^n$. Find the first six terms of the sequence defined by a_n. Find a_{99}.

72. **STIRLING'S FORMULA** By using a calculator, evaluate $\sqrt{2\pi n}(n/e)^n$, where e is the base of the natural logarithms for $n = 10$, 20, and 30. This formula is called Stirling's formula and is used as an approximation for $n!$. For $n > 20$, the error in the approximation is less than 0.1%.

73. Prove that $\sum_{i=1}^{n} ca_i = c\sum_{i=1}^{n} a_i$, where c is a constant.

PROJECTS

1. **FORMULAS FOR INFINITE SEQUENCES** It is not possible to define an infinite sequence by giving a finite number of terms of the sequence. For instance, the question "What is the next term in the sequence 2, 4, 6, 8, ...?" does not have a unique answer.

   a. Verify this statement by finding a formula for a_n such that the first four terms of the sequence are 2, 4, 6, 8 and the next term is 43. *Suggestion:* The formula

   $$a_n = \frac{n(n-1)(n-2)(n-3)(n-4)}{4!} + 2n$$

   generates the sequence 2, 4, 6, 8, 15 for $n = 1, 2, 3, 4, 5$.

   b. Extend the result in **a.** by finding a formula for a_n that will give the first four terms as 2, 4, 6, 8 and the fifth term as x, where x is any real number.

SECTION 11.2 ARITHMETIC SEQUENCES AND SERIES

- ARITHMETIC SEQUENCES
- ARITHMETIC SERIES
- ARITHMETIC MEANS

11A

ARITHMETIC SEQUENCES

Note that in the sequence

$$2, 5, 8, 11, 14, \ldots, 3n - 1, \ldots$$

the difference between successive terms is always 3. Such a sequence is an *arithmetic sequence* or an *arithmetic progression.* These sequences have the following property: The difference between successive terms is the same constant. This constant is called the *common difference.* For the sequence above, the common difference is 3.

In general, an arithmetic sequence can be defined as follows:

Arithmetic Sequence

Let d be a real number. A sequence a_n is an **arithmetic sequence** if

$$a_{i+1} - a_i = d \quad \text{for all } i$$

The number d is the **common difference** for the sequence.

Further examples of arithmetic sequences include

$$3, 8, 13, 18, \ldots, 5n - 2, \ldots$$
$$11, 7, 3, -1, \ldots, -4n + 15, \ldots$$
$$1, 2, 3, 4, \ldots, n, \ldots$$

Consider an arithmetic sequence in which the first term is a_1 and the common difference is d. By adding the common difference to each successive term of the arithmetic sequence, we can find a formula for the nth term.

$$a_1 = a_1$$
$$a_2 = a_1 + d$$
$$a_3 = a_2 + d = a_1 + d + d = a_1 + 2d$$
$$a_4 = a_3 + d = a_1 + 2d + d = a_1 + 3d$$

Note the relationship between the term number and the multiplier of d. The multiplier is 1 less than the term number.

Formula for the *n*th Term of an Arithmetic Sequence

The ***n*th term of an arithmetic sequence** with common difference of d is given by

$$a_n = a_1 + (n - 1)d$$

EXAMPLE 1 Find the *n*th Term of an Arithmetic Sequence

a. Find the twenty-fifth term of the arithmetic sequence whose first three terms are $-12, -6, 0$.

b. The fifteenth term of an arithmetic sequence is -3 and the first term is 25. Find the tenth term.

Solution

a. Find the common difference: $d = a_2 - a_1 = -6 - (-12) = 6$. Use the formula $a_n = a_1 + (n - 1)d$ with $n = 25$.

$$a_{25} = -12 + (25 - 1)(6) = -12 + 24(6) = -12 + 144 = 132$$

b. Solve the equation $a_n = a_1 + (n - 1)d$ for d, given that $n = 15$, $a_1 = 25$, and $a_{15} = -3$.

$$-3 = 25 + (14)d$$
$$d = -2$$

Now find the tenth term.

$$a_n = a_1 + (n - 1)d$$
$$a_{10} = 25 + (9)(-2) = 7 \qquad \bullet\ n = 10,\ a_1 = 25,\ d = -2$$

TRY EXERCISE 16, EXERCISE SET 11.2, PAGE 784

MATH MATTERS

Galileo (1564–1642), using the fact that the sum of a set of odd integers is the square of the number of integers being added (as is shown below under Arithmetic Sequences), was able to show that objects of different weights fall at the same rate. By constructing inclines of various slopes similar to the one shown below, with a track in the center and equal intervals marked along the incline, he measured the distance various balls of different weights traveled in equal intervals of time. He concluded from his observations that the distance an object falls is proportional to the square of the time it takes to fall and does not depend on its weight. Galileo's views were contrary to the prevailing (Aristotelian) theory on this subject, and he lost his post at the University of Pisa because of them.

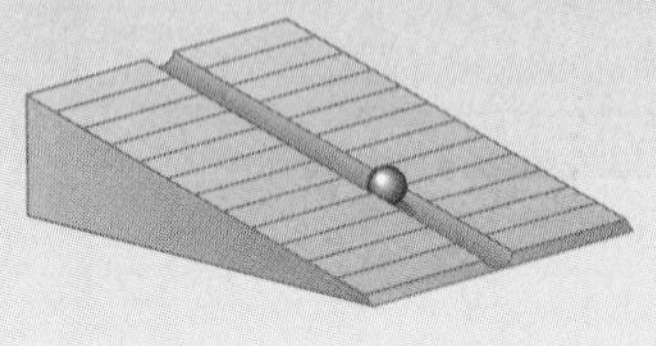

◆ ARITHMETIC SERIES

Consider the arithmetic sequence given by

$$1, 3, 5, \ldots, 2n - 1, \ldots$$

Adding successive terms of this sequence, we generate a sequence of partial sums. The sum of the terms of an arithmetic sequence is called an **arithmetic series.**

$$S_1 = 1$$
$$S_2 = 1 + 3 = 4$$
$$S_3 = 1 + 3 + 5 = 9$$
$$S_4 = 1 + 3 + 5 + 7 = 16$$
$$S_5 = 1 + 3 + 5 + 7 + 9 = 25$$
$$\vdots \qquad \vdots$$
$$S_n = 1 + 3 + \cdots + (2n - 1) \stackrel{?}{=} n^2$$

The first five terms of this sequence are 1, 4, 9, 16, 25. It appears from this example that the sum of the first n odd integers is n^2. Shortly, we will be able to prove this result by using the following formula.

Formula for the *n*th Partial Sum of an Arithmetic Sequence

The **nth partial sum S_n of an arithmetic sequence a_n** with common difference d is

$$S_n = \frac{n}{2}(a_1 + a_n)$$

Proof We write S_n in both forward and reverse order.

$$\begin{aligned} S_n &= a_1 + a_2 + a_3 + \cdots + a_{n-2} + a_{n-1} + a_n \\ S_n &= a_n + a_{n-1} + a_{n-2} + \cdots + a_3 + a_2 + a_1 \end{aligned}$$

Add the two partial sums.

$$\begin{aligned} 2S_n = (a_1 + a_n) + (a_2 + a_{n-1}) + (a_3 + a_{n-2}) + \cdots \\ + (a_{n-2} + a_3) + (a_{n-1} + a_2) + (a_n + a_1) \end{aligned} \qquad (1)$$

Consider the term $(a_3 + a_{n-2})$. Using the formula for the nth term of an arithmetic sequence, we have

$$\begin{aligned} a_3 &= a_1 + (3 - 1)d = a_1 + 2d \\ a_{n-2} &= a_1 + [(n - 2) - 1]d = a_1 + nd - 3d \end{aligned}$$

Thus

$$\begin{aligned} a_3 + a_{n-2} &= (a_1 + 2d) + (a_1 + nd - 3d) \\ &= a_1 + (a_1 + nd - d) = a_1 + [a_1 + (n - 1)d] \\ &= a_1 + a_n \end{aligned}$$

In a similar manner, we can show that each term in parentheses in Equation (1) equals $(a_1 + a_n)$. Because there are n such terms, we have

$$\begin{aligned} 2S_n &= n(a_1 + a_n) \\ S_n &= \frac{n}{2}(a_1 + a_n) \end{aligned}$$

◆

There is an alternative form of the formula for the sum of n terms of an arithmetic sequence.

Alternative Formula for the Sum of an Arithmetic Series

The nth partial sum S_n of an arithmetic sequence with common difference d is

$$S_n = \frac{n[2a_1 + (n - 1)d]}{2}$$

The proof of this theorem is left as a project.

EXAMPLE 2 **Find a Partial Sum of an Arithmetic Sequence**

a. Find the sum of the first 100 positive odd integers.

b. Find the sum of the first 50 terms of the arithmetic sequence whose first three terms are 2, 13/4, and 9/2.

Solution

Use the formula $S_n = \frac{n}{2}[2a_1 + (n - 1)d]$.

a. We have $a_1 = 1$, $d = 2$, and $n = 100$. Thus

$$S_{100} = \frac{100}{2}[2(1) + (100 - 1)2] = 10{,}000$$

b. We have $a_1 = 2$, $d = \frac{5}{4}$, and $n = 50$. Thus

$$S_{50} = \frac{50}{2}\left[2(2) + (50 - 1)\frac{5}{4}\right] = \frac{6525}{4}$$

TRY EXERCISE 22, EXERCISE SET 11.2, PAGE 784

The first n positive integers $1, 2, 3, 4, \ldots, n$ are part of an arithmetic sequence with a common difference of 1, $a_1 = 1$, and $a_n = n$. A formula for the sum of the first n positive integers can be found by using the formula for the nth partial sum of an arithmetic sequence.

$$S_n = \frac{n}{2}(a_1 + a_n)$$

Replacing a_1 by 1 and a_n by n yields

$$S_n = \frac{n}{2}(1 + n) = \frac{n(n + 1)}{2}$$

This proves the following theorem.

Sum of the First n Positive Integers

The sum of the first n positive integers is given by

$$S_n = \frac{n(n + 1)}{2}$$

To find the sum of the first 85 positive integers, use $n = 85$.

$$S_{85} = \frac{85(85 + 1)}{2} = 3655$$

◆ ARITHMETIC MEANS

The **arithmetic mean** of two numbers a and b is $(a + b)/2$. The three numbers a, $(a + b)/2$, b form an arithmetic sequence. In general, given two numbers a and b, it is possible to insert k numbers $c_1, c_2, \ldots, c_k$ in such a way that the sequence

$$a, c_1, c_2, \ldots, c_k, b$$

is an arithmetic sequence. This is called *inserting k arithmetic means between a and b.*

EXAMPLE 3 Insert Arithmetic Means

Insert three arithmetic means between 3 and 13.

Solution

After we insert the three terms, the sequence will be

$$a = 3, c_1, c_2, c_3, b = 13$$

The first term of the sequence is 3, the fifth term is 13, and n is 5. Thus

$$a_n = a_1 + (n - 1)d$$
$$13 = 3 + 4d$$
$$d = \frac{5}{2}$$

The three arithmetic means are

$$c_1 = a + d = 3 + \frac{5}{2} = \frac{11}{2}$$
$$c_2 = a + 2d = 3 + 2\left(\frac{5}{2}\right) = 8$$
$$c_3 = a + 3d = 3 + 3\left(\frac{5}{2}\right) = \frac{21}{2}$$

TRY EXERCISE 34, EXERCISE SET 11.2, PAGE 784

TOPICS FOR DISCUSSION

1. Discuss what distinguishes an arithmetic sequence from all other types of sequences.
2. Is $a_n = 2/n$ a possible formula for the nth term of an arithmetic sequence? Why or why not?
3. Discuss the characteristics of an arithmetic series. Give an example of an arithmetic series.
4. Consider the series $\sum_{k=1}^{n} f(k)$. Discuss how you can determine whether this is an arithmetic series.

EXERCISE SET 11.2

In Exercises 1 to 14, find the ninth, twenty-fourth, and *n*th terms of the arithmetic sequence.

1. $6, 10, 14, \ldots$
2. $7, 12, 17, \ldots$
3. $6, 4, 2, \ldots$
4. $11, 4, -3, \ldots$
5. $-8, -5, -2, \ldots$
6. $-15, -9, -3, \ldots$
7. $1, 4, 7, \ldots$
8. $-4, 1, 6, \ldots$
9. $a, a + 2, a + 4, \ldots$
10. $a - 3, a + 1, a + 5, \ldots$
11. $\log 7, \log 14, \log 28, \ldots$
12. $\ln 4, \ln 16, \ln 64, \ldots$
13. $\log a, \log a^2, \log a^3, \ldots$
14. $\log_2 5, \log_2 5a, \log_2 5a^2, \ldots$
15. The fourth and fifth terms of an arithmetic sequence are 13 and 15. Find the twentieth term.
16. The sixth and eighth terms of an arithmetic sequence are -14 and -20. Find the fifteenth term.
17. The fifth and seventh terms of an arithmetic sequence are -19 and -29. Find the seventeenth term.
18. The fourth and seventh terms of an arithmetic sequence are 22 and 34. Find the twenty-third term.

In Exercises 19 to 32, find the *n*th partial sum of the arithmetic sequence.

19. $a_n = 3n + 2; n = 10$
20. $a_n = 4n - 3; n = 12$
21. $a_n = 3 - 5n; n = 15$
22. $a_n = 1 - 2n; n = 20$
23. $a_n = 6n; n = 12$
24. $a_n = 7n; n = 14$
25. $a_n = n + 8; n = 25$
26. $a_n = n - 4; n = 25$
27. $a_n = -n; n = 30$
28. $a_n = 4 - n; n = 40$
29. $a_n = n + x; n = 12$
30. $a_n = 2n - x; n = 15$
31. $a_n = nx; n = 20$
32. $a_n = -nx; n = 14$

In Exercises 33 to 36, insert *k* arithmetic means between the given numbers.

33. -1 and 23; $k = 5$
34. 7 and 19; $k = 5$
35. 3 and $\frac{1}{2}$; $k = 4$
36. $\frac{11}{3}$ and 6; $k = 4$
37. Show that the sum of the first n positive odd integers is n^2.
38. Show that the sum of the first n positive even integers is $n^2 + n$.
39. **STACKING LOGS** Logs are stacked so that there are 25 logs in the bottom row, 24 logs in the second row, and so on, decreasing by 1 log each row. How many logs are stacked in the sixth row? How many logs are there in all six rows?
40. **THEATER SEATING** The seating section in a theater has 27 seats in the first row, 29 seats in the second row, and so on, increasing by 2 seats each row for a total of 10 rows. How many seats are in the tenth row, and how many seats are there in the section?
41. **CONTEST PRIZES** A contest offers 15 prizes. The first prize is \$5000, and each successive prize is \$250 less than the preceding prize. What is the value of the fifteenth prize? What is the total amount of money distributed in prizes?
42. **PHYSICAL FITNESS** An exercise program calls for walking 15 minutes each day for a week. Each week thereafter, the amount of time spent walking increases by 5 minutes per day. In how many weeks will a person be walking 60 minutes each day?
43. **PHYSICS** An object dropped from a cliff will fall 16 feet the first second, 48 feet the second second, 80 feet the third second, and so on, increasing by 32 feet each second. What is the total distance the object will fall in 7 seconds?
44. **PHYSICS** The distance a ball rolls down a ramp each second is given by the arithmetic sequence whose nth term is $2n - 1$ feet. Find the distance the ball rolls during the tenth second and the total distance the ball travels in 10 seconds.

SUPPLEMENTAL EXERCISES

45. If $f(x)$ is a linear polynomial, show that $f(n)$, where n is a positive integer, is an arithmetic sequence.
46. Find the formula for a_n in terms of a_1 and n for the sequence that is defined recursively by $a_1 = 3, a_n = a_{n-1} + 5$.
47. Find a formula for a_n in terms of a_1 and n for the sequence that is defined recursively by $a_1 = 4, a_n = a_{n-1} - 3$.
48. Suppose a_n and b_n are two sequences such that $a_1 = 4$, $a_n = b_{n-1} + 5$ and $b_1 = 2$, $b_n = a_{n-1} + 1$. Show that a_n and b_n are arithmetic sequences. Find a_{100}.
49. Suppose a_n and b_n are two sequences such that $a_1 = 1$, $a_n = b_{n-1} + 7$ and $b_1 = -2$, $b_n = a_{n-1} + 1$. Show that a_n and b_n are arithmetic sequences. Find a_{50}.

PROJECTS

1. **ANGLES OF A TRIANGLE** The sum of the interior angles of a triangle is 180°.
   a. Using this fact, what is the sum of the interior angles of a quadrilateral?
   b. What is the sum of the interior angles of a pentagon?
   c. What is the sum of the interior angles of a hexagon?
   d. On the basis of your previous results, what is the apparent formula for the sum of the interior angles of a polygon of n sides?
2. **PROVE A FORMULA** Prove the Alternative Formula for the Sum of an Arithmetic Series.

SECTION 11.3 GEOMETRIC SEQUENCES AND SERIES

- GEOMETRIC SEQUENCES
- FINITE GEOMETRIC SERIES
- INFINITE GEOMETRIC SERIES
- FUTURE VALUE OF AN ANNUITY

11A

◆ GEOMETRIC SEQUENCES

Arithmetic sequences are characterized by a common *difference* between successive terms. A *geometric sequence* is characterized by a common *ratio* between successive terms.

The sequence

$$3, 6, 12, 24, \ldots, 3(2^{n-1}), \ldots$$

is a geometric sequence. Note that the ratio of any two successive terms is 2.

$$\frac{6}{3} = 2 \qquad \frac{12}{6} = 2 \qquad \frac{24}{12} = 2$$

Geometric Sequence

Let r be a nonzero constant real number. A sequence is a **geometric sequence** if

$$\frac{a_{i+1}}{a_i} = r \quad \text{for all positive integers } i.$$

The number r is called the **common ratio.**

EXAMPLE 1 Determine Whether a Sequence Is a Geometric Sequence

Which of the following are geometric sequences?

a. $4, -2, 1, \ldots, 4\left(-\frac{1}{2}\right)^{n-1}, \ldots$ b. $1, 4, 9, \ldots, n^2, \ldots$

Continued ▶

Solution

To determine whether the sequence is a geometric sequence, calculate the ratio of successive terms.

a. $\dfrac{a_{i+1}}{a_i} = \dfrac{4(-1/2)^i}{4(-1/2)^{i-1}} = -\dfrac{1}{2}.$

Because the ratio of successive terms is a constant, the sequence is a geometric sequence.

b. $\dfrac{a_{i+1}}{a_i} = \dfrac{(i+1)^2}{i^2} = \left(1 + \dfrac{1}{i}\right)^2$

Because the ratio of successive terms is not a constant, the sequence is not a geometric sequence.

TRY EXERCISE 6, EXERCISE SET 11.3, PAGE 792

Consider a geometric sequence in which the first term is a_1 and the common ratio is r. By multiplying each successive term of the geometric sequence by the common ratio, we can derive a formula for the nth term.

$$a_1 = a_1$$
$$a_2 = a_1 r$$
$$a_3 = a_2 r = (a_1 r)r = a_1 r^2$$
$$a_4 = a_3 r = (a_1 r^2)r = a_1 r^3$$

Note the relationship between the number of the term and the number that is the exponent on r. The exponent on r is 1 less than the number of the term. With this observation, we can write a formula for the nth term of a geometric sequence.

nth Term of a Geometric Sequence

The **nth term of a geometric sequence** with first term a_1 and common ratio r is

$$a_n = a_1 r^{n-1}$$

EXAMPLE 2 Find the nth Term of a Geometric Sequence

Find the nth term of the geometric sequence whose first three terms are

a. $4, 8/3, 16/9, \ldots$ b. $5, -10, 20, \ldots$

Solution

a. $r = \dfrac{8/3}{4} = \dfrac{2}{3}$ and $a_1 = 4$. Thus $a_n = 4\left(\dfrac{2}{3}\right)^{n-1}$.

b. $r = \dfrac{-10}{5} = -2$ and $a_1 = 5$. Thus $a_n = 5(-2)^{n-1}$.

TRY EXERCISE 18, EXERCISE SET 11.3, PAGE 792

◆ FINITE GEOMETRIC SERIES

Adding the terms of a geometric sequence, we can define the nth partial sum of a geometric sequence in a manner similar to that of an arithmetic sequence. Consider the geometric sequence $1, 2, 4, 8, \ldots, 2^{n-1}, \ldots$.

$$\begin{aligned} S_1 &= 1 \\ S_2 &= 1 + 2 = 3 \\ S_3 &= 1 + 2 + 4 = 7 \\ S_4 &= 1 + 2 + 4 + 8 = 15 \\ &\vdots \\ S_n &= 1 + 2 + 4 + 8 + \cdots + 2^{n-1} \end{aligned}$$

The first four terms of the sequence of partial sums are 1, 3, 7, 15.

To find a general formula for S_n, the nth term of the sequence of partial sums of a geometric sequence, let

$$S_n = a_1 + a_1 r + a_1 r^2 + \cdots + a_1 r^{n-1}$$

Multiply each side of this equation by r.

$$\begin{aligned} S_n &= a_1 + a_1 r + a_1 r^2 + \cdots + a_1 r^{n-2} + a_1 r^{n-1} \\ rS_n &= \qquad a_1 r + a_1 r^2 + \cdots + a_1 r^{n-2} + a_1 r^{n-1} + a_1 r^n \end{aligned}$$

Subtract the two equations.

$$\begin{aligned} S_n - rS_n &= a_1 - a_1 r^n \\ S_n(1 - r) &= a_1(1 - r^n) && \text{• Factor out the common factors.} \\ S_n &= \frac{a_1(1 - r^n)}{1 - r} && \text{• } r \neq 1 \end{aligned}$$

This proves the following theorem.

The nth Partial Sum of a Geometric Sequence

The **nth partial sum of a geometric sequence** with first term a_1 and common ratio r is

$$S_n = \frac{a_1(1 - r^n)}{1 - r} \qquad r \neq 1$$

QUESTION If $r = 1$, what is the nth partial sum of a geometric sequence?

EXAMPLE 3 Find the nth Partial Sum of a Geometric Sequence

Find the partial sum of each geometric sequence.

a. $5, 15, 45, \ldots, 5(3)^{n-1}, \ldots; n = 4$ **b.** $\sum_{n=1}^{17} 3\left(\frac{3}{4}\right)^{n-1}$

Solution

a. We have $a_1 = 5$, $r = 3$, and $n = 4$. Thus

$$S_4 = \frac{5[1 - 3^4]}{1 - 3} = \frac{5(-80)}{-2} = 200$$

b. When $n = 1$, $a_1 = 3$. The first term is 3. The second term is $9/4$. Therefore, the common ratio is $r = 3/4$. Thus

$$S_{17} = \frac{3[1 - (3/4)^{17}]}{1 - (3/4)} \approx 11.909797$$

TRY EXERCISE 40, EXERCISE SET 11.3, PAGE 792

◆ INFINITE GEOMETRIC SERIES

Following are two examples of geometric sequences for which $|r| < 1$.

$$3, \frac{3}{4}, \frac{3}{16}, \frac{3}{64}, \frac{3}{256}, \frac{3}{1024}, \ldots \qquad \bullet\ r = \frac{1}{4}$$

$$2, -1, \frac{1}{2}, -\frac{1}{4}, \frac{1}{8}, -\frac{1}{16}, \frac{1}{32}, \ldots \qquad \bullet\ r = -\frac{1}{2}$$

Note that when the absolute value of the common ratio of a geometric sequence is less than 1, the terms of the geometric sequence approach zero as n increases. We write, for $|r| < 1$, $|r|^n \to 0$ as $n \to \infty$.

Consider again the geometric sequence

$$3, \frac{3}{4}, \frac{3}{16}, \frac{3}{64}, \frac{3}{256}, \frac{3}{1024}, \ldots$$

The nth partial sums for $n = 3, 6, 9$, and 12 are given in Table 11.1, along with the value of r^n. As n increases, S_n is closer to 4 and r^n is closer to zero. By finding more values of S_n for larger values of n, we would find that $S_n \to 4$ as $n \to \infty$. As n becomes larger and larger, S_n is the nth partial sum of more and more terms of the sequence. The sum of *all* the terms of a sequence is called an **infinite series.** If the sequence is a geometric sequence, we have an **infinite geometric series.**

Table 11.1

n	S_n	r^n
3	3.93750000	0.01562500
6	3.99902344	0.00024414
9	3.99998474	0.00000381
12	3.99999976	0.00000006

ANSWER When $r = 1$, the sequence is the constant sequence a_1. The nth partial sum of a constant sequence is na_1.

Sum of an Infinite Geometric Sequence

If a_n is a geometric sequence with $|r| < 1$ and first term a_1, then the sum of the infinite geometric series is

$$S = \frac{a_1}{1 - r}$$

A formal proof of this formula requires topics that typically are studied in calculus. We can, however, give an intuitive argument. Start with the formula for the nth partial sum of a geometric sequence.

$$S_n = \frac{a_1(1 - r^n)}{1 - r}$$

When $|r| < 1$, $|r|^n \approx 0$ when n is large. Thus

$$S_n = \frac{a_1(1 - r^n)}{1 - r} \approx \frac{a_1(1 - 0)}{1 - r} = \frac{a_1}{1 - r}$$

An infinite series is represented by $\sum_{n=1}^{\infty} a_n$. One application of infinite geometric series concerns repeating decimals. Consider the repeating decimal

$$0.\overline{6} = \frac{6}{10} + \frac{6}{100} + \frac{6}{1000} + \frac{6}{10,000} + \cdots$$

The right-hand side is a geometric series with $a_1 = 6/10$ and common ratio $r = 1/10$. Thus

$$S = \frac{6/10}{1 - (1/10)} = \frac{6/10}{9/10} = \frac{2}{3}$$

The repeating decimal $0.\overline{6} = 2/3$. We can write any repeating decimal as a ratio of two integers by using the formula for the sum of an infinite geometric series.

> **take note**
>
> The sum of an infinite geometric series is not defined when $|r| \geq 1$. For instance, the infinite geometric series
>
> $$2 + 4 + 8 + \cdots + 2^n + \cdots$$
>
> with $r = 2$ increases without bound. However, applying the formula
>
> $S = \frac{a_1}{1 - r}$ with $r = 2$ and $a_1 = 2$
>
> gives $S = -2$, which is not correct.

EXAMPLE 4 Find the Value of an Infinite Geometric Series

a. Evaluate the infinite geometric series $\sum_{n=1}^{\infty} \left(-\frac{2}{3}\right)^{n-1}$.

b. Write $0.3\overline{45}$ as the ratio of two integers in lowest terms.

Solution

a. To find the first term, we let $n = 1$. Then $a_1 = (-2/3)^{1-1} = (-2/3)^0 = 1$. The common ratio $r = -2/3$. Thus

$$S = \frac{1}{1 - (-2/3)} = \frac{1}{(5/3)} = \frac{3}{5}$$

Continued

b. $0.3\overline{45} = \dfrac{3}{10} + \left[\dfrac{45}{1000} + \dfrac{45}{100{,}000} + \dfrac{45}{10{,}000{,}000} + \cdots\right]$

The terms in the brackets form an infinite geometric series. Evaluate that series with $a_1 = 45/1000$ and $r = 1/100$, and then add the term 3/10.

$$\frac{45}{1000} + \frac{45}{100{,}000} + \frac{45}{10{,}000{,}000} + \cdots = \frac{45/1000}{1 - (1/100)} = \frac{1}{22}$$

Thus $0.3\overline{45} = \dfrac{3}{10} + \dfrac{1}{22} = \dfrac{19}{55}$

TRY EXERCISE 62, EXERCISE SET 11.3, PAGE 792

◆ FUTURE VALUE OF AN ANNUITY

In an earlier chapter we discussed compound interest by using exponential functions. As an extension of this idea, suppose that for each of the next 5 years, P dollars are deposited on December 31 into an account earning i% annual interest compounded annually. Using the compound interest formula, we can find the total value of all the deposits. Table 11.2 shows the growth of the investment.

Table 11.2

Deposit number	Value of each deposit	
1	$P(1 + i)^4$	Value of first deposit after 4 years
2	$P(1 + i)^3$	Value of second deposit after 3 years
3	$P(1 + i)^2$	Value of third deposit after 2 years
4	$P(1 + i)$	Value of fourth deposit after 1 year
5	P	Value of fifth deposit

The total value of the investment after the last deposit, called the **future value** of the investment, is the sum of the values of all the deposits.

$$A = P + P(1 + i) + P(1 + i)^2 + P(1 + i)^3 + P(1 + i)^4$$

This is a geometric series with first term P and common ratio $1 + i$. Thus, using the formula for the nth partial sum of a geometric sequence

$$S = \frac{a_1(1 - r^n)}{1 - r}$$

we have

$$A = \frac{P[1 - (1 + i)^5]}{1 - (1 + i)} = \frac{P[(1 + i)^5 - 1]}{i}$$

Deposits of equal amounts at equal intervals of time are called **annuities.** When the amounts are deposited at the end of a compounding period (as in our example), we have an **ordinary annuity.**

Future Value of an Ordinary Annuity

Let $r = i/n$ and $m = nt$, where i is the annual interest rate, n is the number of compounding periods per year, and t is the number of years. Then the future value A of an ordinary annuity after m compounding periods is given by

$$A = \frac{P[(1 + r)^m - 1]}{r}$$

where P is the amount of each deposit.

EXAMPLE 5 Find the Future Value of an Ordinary Annuity

An employee savings plan allows any employee to deposit $25 at the end of each month into a savings account earning 6% annual interest compounded monthly. Find the future value of this savings plan if an employee makes the deposits for 10 years.

Solution

We are given $P = 25$, $i = 0.06$, $n = 12$, and $t = 10$. Thus,

$$r = \frac{i}{n} = \frac{0.06}{12} = 0.005 \quad \text{and} \quad m = nt = 12(10) = 120$$

$$A = \frac{25[(1 + 0.005)^{120} - 1]}{0.005} \approx 4096.9837$$

The future value after 10 years is $4096.98.

TRY EXERCISE 70, EXERCISE SET 11.3, PAGE 793

TOPICS FOR DISCUSSION

1. Discuss what distinguishes a geometric sequence from all other types of sequences.

2. Is $a_n = n^2$ a possible formula for the nth term of a geometric sequence? Why or why not?

3. Discuss the characteristics of a geometric series.

4. Consider the series $\sum_{k=1}^{n} f(k)$. Explain how you can determine whether this is a geometric series.

EXERCISE SET 11.3

In Exercises 1 to 12, determine which sequences are geometric. For geometric sequences, find the common ratio.

1. $4, 16, 64, \ldots, 4^n, \ldots$
2. $1, 6, 36, \ldots, 6^{n-1}, \ldots$
3. $1, \frac{1}{2}, \frac{1}{3}, \ldots, \frac{1}{n}, \ldots$
4. $\frac{1}{2}, \frac{1}{4}, \frac{1}{8}, \ldots, \frac{1}{2^n}, \ldots$
5. $2^x, 2^{2x}, 2^{3x}, \ldots, 2^{nx}, \ldots$
6. $e^x, -e^{2x}, e^{3x}, \ldots, (-1)^{n-1}e^{nx}, \ldots$
7. $3, 6, 12, \ldots, 3(2^{n-1}), \ldots$
8. $5, -10, 20, \ldots, 5(-2)^{n-1}, \ldots$
9. $x^2, x^4, x^6, \ldots, x^{2n}, \ldots$
10. $3x, 6x^2, 9x^3, \ldots, 3nx^n, \ldots$
11. $\ln 5, \ln 10, \ln 15, \ldots, \ln 5n, \ldots$
12. $\log x, \log x^2, \log x^4, \ldots, \log x^{2n-1}, \ldots$

In Exercises 13 to 32, find the *n*th term of the geometric sequence.

13. $2, 8, 32, \ldots$
14. $1, 5, 25, \ldots$
15. $-4, 12, -36, \ldots$
16. $-3, 6, -12, \ldots$
17. $6, 4, \frac{8}{3}, \ldots$
18. $8, 6, \frac{9}{2}, \ldots$
19. $-6, 5, -\frac{25}{6}, \ldots$
20. $-2, \frac{4}{3}, -\frac{8}{9}, \ldots$
21. $9, -3, 1, \ldots$
22. $8, -\frac{4}{3}, \frac{2}{9}, \ldots$
23. $1, -x, x^2, \ldots$
24. $2, 2a, 2a^2, \ldots$
25. $c^2, c^5, c^8, \ldots$
26. $-x^2, x^4, -x^6, \ldots$
27. $\frac{3}{100}, \frac{3}{10{,}000}, \frac{3}{1{,}000{,}000}, \ldots$
28. $\frac{7}{10}, \frac{7}{10{,}000}, \frac{7}{10{,}000{,}000}, \ldots$
29. $0.5, 0.05, 0.005, \ldots$
30. $0.4, 0.004, 0.00004, \ldots$
31. $0.45, 0.0045, 0.000045, \ldots$
32. $0.234, 0.000234, 0.000000234, \ldots$

33. Find the third term of a geometric sequence whose first term is 2 and whose fifth term is 162.
34. Find the fourth term of a geometric sequence whose third term is 1 and whose eighth term is 1/32.
35. Find the second term of a geometric sequence whose third term is 4/3 and whose sixth term is −32/81.
36. Find the fifth term of a geometric sequence whose fourth term is 8/9 and whose seventh term is 64/243.

In Exercises 37 to 46, find the sum of the geometric series.

37. $\sum_{n=1}^{5} 3^n$
38. $\sum_{n=1}^{7} 2^n$
39. $\sum_{n=1}^{6} \left(\frac{2}{3}\right)^n$
40. $\sum_{n=1}^{14} \left(\frac{4}{3}\right)^n$
41. $\sum_{n=0}^{8} \left(-\frac{2}{5}\right)^n$
42. $\sum_{n=0}^{7} \left(-\frac{1}{3}\right)^n$
43. $\sum_{n=1}^{10} (-2)^{n-1}$
44. $\sum_{n=0}^{7} 2(5)^n$
45. $\sum_{n=0}^{9} 5(3)^n$
46. $\sum_{n=0}^{10} 2(-4)^n$

In Exercises 47 to 56, find the sum of the infinite geometric series.

47. $\sum_{n=1}^{\infty} \left(\frac{1}{3}\right)^n$
48. $\sum_{n=1}^{\infty} \left(\frac{3}{4}\right)^n$
49. $\sum_{n=1}^{\infty} \left(-\frac{2}{3}\right)^n$
50. $\sum_{n=1}^{\infty} \left(-\frac{3}{5}\right)^n$
51. $\sum_{n=1}^{\infty} \left(\frac{9}{100}\right)^n$
52. $\sum_{n=1}^{\infty} \left(\frac{7}{10}\right)^n$
53. $\sum_{n=1}^{\infty} (0.1)^n$
54. $\sum_{n=1}^{\infty} (0.5)^n$
55. $\sum_{n=0}^{\infty} (-0.4)^n$
56. $\sum_{n=0}^{\infty} (-0.8)^n$

In Exercises 57 to 68, write each rational number as the quotient of two integers in simplest form.

57. $0.\overline{3}$
58. $0.\overline{5}$
59. $0.\overline{45}$
60. $0.\overline{63}$
61. $0.\overline{123}$
62. $0.\overline{395}$
63. $0.\overline{422}$
64. $0.\overline{355}$
65. $0.25\overline{4}$
66. $0.37\overline{2}$
67. $1.2\overline{084}$
68. $2.2\overline{590}$
69. **TIME VALUE OF MONEY** Find the future value of an ordinary annuity that calls for depositing \$100 at the end of every 6 months for 8 years into an account that earns 9% interest compounded semiannually.

70. **TIME VALUE OF MONEY** To save for the replacement of a computer, a business deposits \$250 at the end of each month into an account that earns 8% annual interest compounded monthly. Find the future value of the ordinary annuity in 4 years.

SUPPLEMENTAL EXERCISES

71. If the sequence a_n is a geometric sequence, make a conjecture about the sequence $\log a_n$ and give a proof.

72. If the sequence a_n is an arithmetic sequence, make a conjecture about the sequence 2^{a_n} and give a proof.

73. Does $\sum_{i=0}^{\infty} x^i \quad (x \neq 0)$ represent an infinite geometric series? Why or why not?

74. Consider a square with a side of length 1. Construct another square inside the first one by connecting the midpoints of the first square. What is the area of the inscribed square? Continue constructing squares in the same way. Find the area of the nth inscribed square.

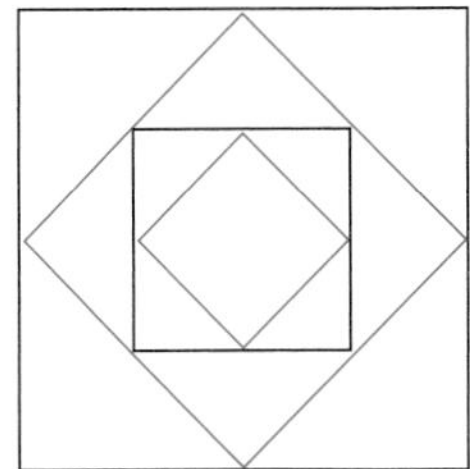

75. The product $P_n = a_1 \cdot a_2 \cdot a_3 \cdots a_n$ is called the nth partial product of a sequence. Find a formula for the nth partial product of the geometric sequence whose nth term is ar^{n-1}.

76. Let $f(x) = ab^x$, $a, b > 0$. Show that if x is restricted to positive integers n, then $f(n)$ is a geometric sequence.

77. **PHYSICS** A ball is dropped from a height of 5 feet. The ball rebounds 80% of the distance after each fall. Use an infinite geometric series to find the total distance the ball will travel.

78. **PENDULUM** The bob of a pendulum swings through an arc of 30 inches on its first swing. Each successive swing is 90% of the length of the previous swing. Find the total distance the bob will travel.

79. **GENEALOGY** Some people can trace their ancestry back ten generations, which means two parents, four grandparents, eight great-grandparents, and so on. How many grandparents does such a family tree include?

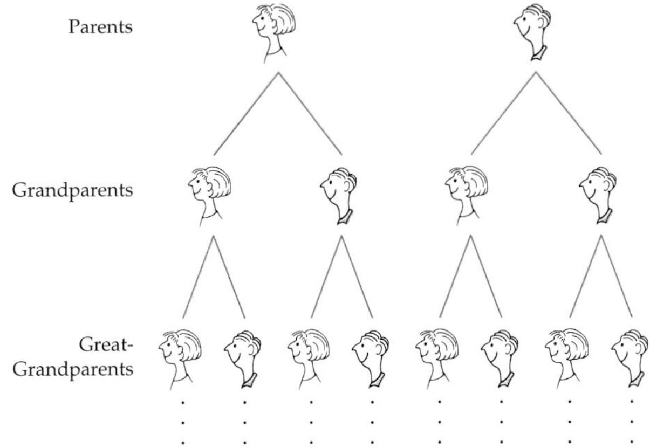

PROJECTS

1. **FRACTALS** The snowflake that was created at the beginning of this chapter is an example of a fractal. Here is another example of a fractal. Begin with a square with each side 1 unit long. Construct another, smaller square onto the middle third of each side. Continue this procedure of constructing similar but smaller squares on each of the line segments. The figure at the right shows the result after the process has been completed twice.

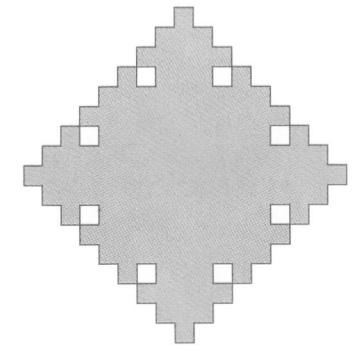

a. What is the perimeter of the figure after the process has been completed n times?

b. As n approaches infinity, what value does the perimeter approach?

c. What is the area of the figure after the process has been completed n times?

d. As n approaches infinity, what value does the area approach? *Suggestion:* The series in **c.** is a geometric series after the first term.

SECTION 11.4 MATHEMATICAL INDUCTION

- PRINCIPLE OF MATHEMATICAL INDUCTION
- EXTENDED PRINCIPLE OF MATHEMATICAL INDUCTION

11B

Consider the sequence

$$\frac{1}{1 \cdot 2}, \frac{1}{2 \cdot 3}, \frac{1}{3 \cdot 4}, \ldots, \frac{1}{n(n+1)}, \ldots$$

and the sequence of partial sums for this sequence:

$$S_1 = \frac{1}{1 \cdot 2} = \frac{1}{2}$$

$$S_2 = \frac{1}{1 \cdot 2} + \frac{1}{2 \cdot 3} = \frac{2}{3}$$

$$S_3 = \frac{1}{1 \cdot 2} + \frac{1}{2 \cdot 3} + \frac{1}{3 \cdot 4} = \frac{3}{4}$$

$$S_4 = \frac{1}{1 \cdot 2} + \frac{1}{2 \cdot 3} + \frac{1}{3 \cdot 4} + \frac{1}{4 \cdot 5} = \frac{4}{5}$$

This pattern suggests the conjecture that

$$S_n = \frac{1}{1 \cdot 2} + \frac{1}{2 \cdot 3} + \frac{1}{3 \cdot 4} + \cdots + \frac{1}{n(n+1)} = \frac{n}{n+1}$$

How can we be sure that the pattern does not break down when $n = 50$ or maybe $n = 2000$ or some other large number? As we will show, this conjecture is true for all values of n.

As a second example, consider the conjecture that the expression $n^2 - n + 41$ is a prime number for all positive integers. To test this conjecture, we will try various values of n. See Table 11.3. The results suggest that the conjecture is true. But again, how can we be sure? In fact, this conjecture is false when $n = 41$. In that case we have

Table 11.3

n	$n^2 - n + 41$	
1	41	Prime
2	43	Prime
3	47	Prime
4	53	Prime
5	61	Prime

$$n^2 - n + 41 = (41)^2 - 41 + 41 = (41)^2$$

and $(41)^2$ is not a prime.

The last example illustrates that just verifying a conjecture for a few values of n does not constitute a proof of the conjecture. To prove theorems about statements involving positive integers, a process called *mathematical induction* is used. This process is based on an axiom called the *induction axiom*.

Induction Axiom

Suppose S is a set of positive integers with the following two properties:

1. 1 is an element of S.
2. If the positive integer k is in S, then $k + 1$ is in S.

Then S contains all the positive integers.

Part 2 of this axiom states that if some positive integer, say 8, is in S, then $8 + 1$, or 9, is in S. But because 9 is in S, part 2 says that $9 + 1$, or 10, is in S, and so on. Part 1 states that 1 is in S. Thus 2 is in S; thus 3 is in S; thus 4 is in S;.... Therefore all the positive integers are in S.

◆ PRINCIPLE OF MATHEMATICAL INDUCTION

The induction axiom is used to prove the *Principle of Mathematical Induction.*

Principle of Mathematical Induction

Let P_n be a statement about a positive integer n. If

1. P_1 is true, and

2. The truth of P_k implies the truth of P_{k+1}

then P_n is true for all positive integers.

Part 2 of the Principle of Mathematical Induction is referred to as the **induction hypothesis.** When applying this step, we assume the statement P_k is true and then try to prove that P_{k+1} is also true.

As an example, we will prove that the first conjecture we made in this section is true for all positive integers. Every induction proof has the two distinct parts stated in the theorem. First we must show that the result is true for $n = 1$. Second, we assume the statement is true for some positive integer k and, using that assumption, prove the statement is true for $n = k + 1$.

Prove that

$$S_n = \frac{1}{1 \cdot 2} + \frac{1}{2 \cdot 3} + \frac{1}{3 \cdot 4} + \cdots + \frac{1}{n(n+1)} = \frac{n}{n+1}$$

for all positive integers n.

Proof

1. For $n = 1$,

$$S_1 = \frac{1}{1(1+1)} = \frac{1}{2}, \text{ and } \frac{n}{n+1} = \frac{1}{1+1} = \frac{1}{2}$$

The statement is true for $n = 1$.

2. Assume the statement is true for some positive integer k.

$$S_k = \frac{1}{1 \cdot 2} + \frac{1}{2 \cdot 3} + \frac{1}{3 \cdot 4} + \cdots + \frac{1}{k(k+1)} = \frac{k}{k+1}$$ • Induction hypothesis

Now verify that the formula is true when $n = k + 1$. That is, verify that

$$S_{k+1} = \frac{k+1}{(k+1)+1} = \frac{k+1}{k+2}$$ • This is the goal of the induction proof.

It is helpful, when proving a theorem about sums, to note that

$$S_{k+1} = S_k + a_{k+1}$$

Begin by noting that $a_k = \dfrac{1}{k(k+1)}$; thus, $a_{k+1} = \dfrac{1}{(k+1)(k+2)}$.

$$\begin{aligned} S_{k+1} &= S_k + a_{k+1} \\ &= \frac{k}{k+1} + \frac{1}{(k+1)(k+2)} \\ &= \frac{k(k+2)}{(k+1)(k+2)} + \frac{1}{(k+1)(k+2)} \\ &= \frac{k(k+2)+1}{(k+1)(k+2)} = \frac{k^2+2k+1}{(k+1)(k+2)} = \frac{(k+1)^2}{(k+1)(k+2)} \\ S_{k+1} &= \frac{k+1}{k+2} \end{aligned}$$

• **By the induction hypothesis and substituting for a_{k+1}**

Because we have verified the two parts of the Principle of Mathematical Induction, we can conclude that the statement is true for all positive integers. ◆

EXAMPLE 1 Prove by Mathematical Induction

Prove that $1^2 + 2^2 + 3^2 + \cdots + n^2 = \dfrac{n(n+1)(2n+1)}{6}$.

Solution

Verify the two parts of the Principle of Mathematical Induction.

1. Let $n = 1$.

$$S_1 = 1^2 = 1 = \frac{1(1+1)(2 \cdot 1 + 1)}{6}$$

2. Assume the statement is true for some positive integer k.

$$S_k = 1^2 + 2^2 + 3^2 + \cdots + k^2 = \frac{k(k+1)(2k+1)}{6}$$

• **Induction hypothesis**

Verify that the statement is true when $n = k + 1$. Show that

$$S_{k+1} = \frac{(k+1)(k+2)(2k+3)}{6}$$

Because $a_k = k^2$, $a_{k+1} = (k+1)^2$.

$$\begin{aligned} S_{k+1} &= S_k + a_{k+1} \\ &= \frac{k(k+1)(2k+1)}{6} + (k+1)^2 \\ &= \frac{k(k+1)(2k+1)}{6} + \frac{6(k+1)^2}{6} = \frac{k(k+1)(2k+1) + 6(k+1)^2}{6} \end{aligned}$$

$$= \frac{(k+1)[k(2k+1)+6(k+1)]}{6} = \frac{(k+1)(2k^2+7k+6)}{6}$$

$$S_{k+1} = \frac{(k+1)(k+2)(2k+3)}{6}$$

By the Principle of Mathematical Induction, the statement is true for all positive integers.

TRY EXERCISE 8, EXERCISE SET 11.4, PAGE 799

Mathematical induction can also be used to prove statements about sequences, products, and inequalities.

EXAMPLE 2 Prove a Product Formula by Mathematical Induction

Prove that

$$\left(1+\frac{1}{1}\right)\left(1+\frac{1}{2}\right)\left(1+\frac{1}{3}\right)\cdots\left(1+\frac{1}{n}\right) = n+1$$

Solution

1. Verify for $n = 1$.

$$\left(1+\frac{1}{1}\right) = 2, \text{ and } 1+1=2$$

2. Assume the statement is true for some positive integer k.

$$P_k = \left(1+\frac{1}{1}\right)\left(1+\frac{1}{2}\right)\left(1+\frac{1}{3}\right)\cdots\left(1+\frac{1}{k}\right) = k+1$$

• Induction hypothesis

Verify that the statement is true when $n = k + 1$. That is, prove $P_{k+1} = k + 2$.

$$P_{k+1} = \left(1+\frac{1}{1}\right)\left(1+\frac{1}{2}\right)\left(1+\frac{1}{3}\right)\cdots\left(1+\frac{1}{k}\right)\left(1+\frac{1}{k+1}\right)$$

$$= P_k\left(1+\frac{1}{k+1}\right) = (k+1)\left(1+\frac{1}{k+1}\right) = k+1+1$$

$$P_{k+1} = k+2$$

By the Principle of Mathematical Induction, the statement is true for all positive integers.

TRY EXERCISE 12, EXERCISE SET 11.4, PAGE 800

EXAMPLE 3 **Prove an Inequality by Mathematical Induction**

Prove that $1 + 2n \leq 3^n$ for all positive integers.

Solution

1. Let $n = 1$. Then $1 + 2(1) = 3 \leq 3^1$. The statement is true when n is 1.
2. Assume the statement is true for some positive integer k.

$$1 + 2k \leq 3^k \qquad \text{• Induction hypothesis}$$

Now prove the statement is true for $n = k + 1$. That is, prove that $1 + 2(k + 1) \leq 3^{k+1}$.

$$\begin{aligned} 3^{k+1} &= 3^k(3) \\ &\geq (1 + 2k)(3) && \text{• Because by the induction hypothesis, } 1 + 2k \leq 3^k. \\ &= 6k + 3 \\ &> 2k + 2 + 1 && \text{• } 6k > 2k, \text{ and } 3 = 2 + 1. \\ &= 2(k + 1) + 1 \end{aligned}$$

Thus $1 + 2(k + 1) \leq 3^{k+1}$.

By the Principle of Mathematical Induction, $1 + 2k \leq 3^k$ for all positive integers.

TRY EXERCISE 16, EXERCISE SET 11.4, PAGE 800

◆ EXTENDED PRINCIPLE OF MATHEMATICAL INDUCTION

The Principle of Mathematical Induction can be extended to cases where the beginning index is greater than 1.

Extended Principle of Mathematical Induction

Let P_n be a statement about a positive integer n. If

1. P_j is true for some positive integer j, and
2. For $k \geq j$ the truth of P_k implies the truth of P_{k+1} then P_n is true for all positive integers $n \geq j$.

EXAMPLE 4 **Prove an Inequality by Mathematical Induction**

For $n \geq 3$, prove that $n^2 > 2n + 1$.

Solution

1. Let $n = 3$. Then $3^2 = 9$; $2(3) + 1 = 7$. Thus $n^2 > 2n + 1$ for $n = 3$.
2. Assume the statement is true for some positive integer $k \geq 3$.

$$k^2 > 2k + 1 \qquad \text{• Induction hypothesis}$$

Verify that the statement is true when $n = k + 1$. That is, show that

$$(k + 1)^2 > 2(k + 1) + 1 = 2k + 3$$

$$\begin{aligned} (k + 1)^2 &= k^2 + 2k + 1 \\ &> (2k + 1) + 2k + 1 && \text{• Induction hypothesis} \\ &> 2k + 1 + 1 + 1 && \text{• } 2k > 1 \\ &= 2k + 3 \end{aligned}$$

Thus $(k + 1)^2 > 2k + 3$.

By the Extended Principle of Mathematical Induction, $n^2 > 2n + 1$ for all $n \geq 3$.

TRY EXERCISE 20, EXERCISE SET 11.4, PAGE 800

TOPICS FOR DISCUSSION

1. Discuss the purpose of mathematical induction.
2. What is an induction hypothesis?
3. What is the Extended Principle of Mathematical Induction and how is it used?
4. Mathematical induction can be used to prove that $x^m \cdot x^n = x^{m+n}$ when m and n are natural numbers. Explain why mathematical induction cannot be used if we are attempting to prove the result when m and n are real numbers.

EXERCISE SET 11.4

In Exercises 1 to 12, use mathematical induction to prove each statement.

1. $\sum_{i=1}^{n} (3i - 2) = 1 + 4 + 7 + \cdots + 3n - 2 = \dfrac{n(3n - 1)}{2}$

2. $\sum_{i=1}^{n} 2i = 2 + 4 + 6 + \cdots + 2n = n(n + 1)$

3. $\sum_{i=1}^{n} i^3 = 1 + 8 + 27 + \cdots + n^3 = \dfrac{n^2(n + 1)^2}{4}$

4. $\sum_{i=1}^{n} 2^i = 2 + 4 + 8 + \cdots + 2^n = 2(2^n - 1)$

5. $\sum_{i=1}^{n} (4i - 1) = 3 + 7 + 11 + \cdots + 4n - 1 = n(2n + 1)$

6. $\sum_{i=1}^{n} 3^i = 3 + 9 + 27 + \cdots + 3^n = \dfrac{3(3^n - 1)}{2}$

7. $\sum_{i=1}^{n} (2i - 1)^3 = 1 + 27 + 125 + \cdots + (2n - 1)^3$
$= n^2(2n^2 - 1)$

8. $\sum_{i=1}^{n} i(i + 1) = 2 + 6 + 12 + \cdots + n(n + 1)$
$= \dfrac{n(n + 1)(n + 2)}{3}$

9. $\displaystyle\sum_{i=1}^{n} \frac{1}{(2i-1)(2i+1)} = \frac{1}{1\cdot 3} + \frac{1}{3\cdot 5} + \frac{1}{5\cdot 7} + \cdots + \frac{1}{(2n-1)(2n+1)} = \frac{n}{2n+1}$

10. $\displaystyle\sum_{i=1}^{n} \frac{1}{2i(2i+2)} = \frac{1}{2\cdot 4} + \frac{1}{4\cdot 6} + \frac{1}{6\cdot 8} + \cdots + \frac{1}{2n(2n+2)} = \frac{n}{4(n+1)}$

11. $\displaystyle\sum_{i=1}^{n} i^4 = 1 + 16 + 81 + \cdots + n^4 = \frac{n(n+1)(2n+1)(3n^2+3n-1)}{30}$

12. $\displaystyle P_n = \left(1-\frac{1}{2}\right)\left(1-\frac{1}{3}\right)\left(1-\frac{1}{4}\right)\cdots\left(1-\frac{1}{n+1}\right) = \frac{1}{n+1}$

In Exercises 13 to 20, use mathematical induction to prove each inequality.

13. $\left(\frac{3}{2}\right)^n > n + 1, n \geq 4$

14. $\left(\frac{4}{3}\right)^n > n, n \geq 7$

15. If $0 < a < 1$, show that $a^{n+1} < a^n$ for all positive integers n.

16. If $a > 1$, show that $a^{n+1} > a^n$ for all positive integers n.

17. $1 \cdot 2 \cdot 3 \cdot \cdots \cdot n > 2^n, n \geq 4$

18. $\frac{1}{\sqrt{1}} + \frac{1}{\sqrt{2}} + \frac{1}{\sqrt{3}} + \cdots + \frac{1}{\sqrt{n}} \geq \sqrt{n}$

19. For $a > 0$, show that $(1+a)^n \geq 1 + na$ for all positive integers.

20. $\log_{10} n < n$ for all positive integers. (*Hint:* Because $\log_{10} x$ is an increasing function, $\log_{10}(n+1) \leq \log_{10}(n+n)$.)

In Exercises 21 to 30, use mathematical induction to prove each statement.

21. 2 is a factor of $n^2 + n$ for all positive integers n.

22. 3 is a factor of $n^3 - n$ for all positive integers n.

23. 4 is a factor of $5^n - 1$ for all positive integers n. (*Hint:* $5^{k+1} - 1 = 5 \cdot 5^k - 5 + 4$.)

24. 5 is a factor of $6^n - 1$ for all positive integers n.

25. $(xy)^n = x^n y^n$ for all positive integers n.

26. $\left(\frac{x}{y}\right)^n = \frac{x^n}{y^n}$ for all positive integers n.

27. For $a \neq b$, show that $(a - b)$ is a factor of $a^n - b^n$, where n is a positive integer. *Hint:*

$$a^{k+1} - b^{k+1} = (a \cdot a^k - ab^k) + (ab^k - b \cdot b^k)$$

28. For $a \neq -b$, show that $(a + b)$ is a factor of $a^{2n+1} + b^{2n+1}$, where n is a positive integer. *Hint:*

$$a^{2k+3} + b^{2k+3} = (a^{2k+2} + b^{2k+2})(a+b) - ab(a^{2k+1} + b^{2k+1})$$

29. $\displaystyle\sum_{k=1}^{n} ar^{k-1} = \frac{a(1-r^n)}{1-r}$ for $r \neq 1$

30. $\displaystyle\sum_{k=1}^{n} (ak + b) = \frac{n[(n+1)a + 2b]}{2}$

SUPPLEMENTAL EXERCISES

In Exercises 31 to 35, use mathematical induction to prove each statement.

31. Using a calculator, find the smallest integer N for which $\log N! > N$. Now prove that $\log n! > n$ for all $n > N$.

32. Let a_n be a sequence for which there is a number r and an integer N for which $a_{n+1}/a_n < r$ for $n \geq N$. Show that $a_{N+k} < a_N r^k$ for each positive integer k.

33. For constant positive integers m and n, show that $(x^m)^n = x^{mn}$.

34. Prove that $\displaystyle\sum_{i=0}^{n} \frac{1}{i!} \leq 3 - \frac{1}{n}$ for all positive integers n.

35. Prove that $\left(\frac{n+1}{n}\right)^n < n$ for all integers $n \geq 3$.

PROJECTS

1. **STEPS IN A MATHEMATICAL INDUCTION PROOF** In every proof by mathematical induction, it is important that both parts of the Principle of Mathematical Induction be verified. For instance, consider the formula

$$2 + 4 + 8 + \cdots + 2^n \stackrel{?}{=} 2^{n+1} + 1$$

   a. Show that if we assume the formula is true for some positive integer k, then the formula is true for $k + 1$.

b. Show that the formula is not true for $n = 1$.

c. Show that the formula is not valid for any value of n by showing that the left side is always an even number and the right side is always an odd number.

d. Explain how this shows that both steps of the Principle of Mathematical Induction must be verified.

2. **THE TOWER OF HANOI** The Tower of Hanoi is a game that consists of three pegs and n disks of distinct diameter arranged on one of the pegs such that the largest disk is on the bottom, then the next largest, and so on. The object of the game is to move all the disks from one peg to a second peg. The rules require that only one disk be moved at a time and that a larger disk may not be placed on a smaller disk. All pegs may be used.

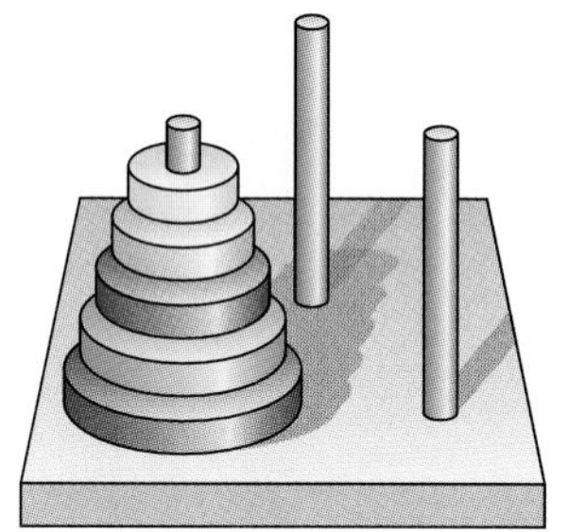

a. Show that it is possible to complete the game in $2^n - 1$ moves.

b. A legend says that in the center of the universe, high priests have the task of moving 64 golden disks from one of three diamond needles by using the rules of the Tower of Hanoi game. When they have completed the transfer, the universe will cease to exist. If one move is made every second, and the priests started 5 billion years ago (the approximate age of Earth), how many more years does the legend predict the universe will continue to exist?

SECTION 11.5 THE BINOMIAL THEOREM

- BINOMIAL THEOREM
- *i*th TERM OF A BINOMIAL EXPANSION
- PASCAL'S TRIANGLE

In certain situations in mathematics, it is necessary to write $(a + b)^n$ as the sum of its terms. Because $(a + b)$ is a binomial, this process is called **expanding the binomial.** For small values of n, it is relatively easy to write the expansion by using multiplication.

Earlier in the text, we found

$$(a + b)^1 = a + b$$
$$(a + b)^2 = a^2 + 2ab + b^2$$
$$(a + b)^3 = a^3 + 3a^2b + 3ab^2 + b^3$$

Building on these expansions, we can write a few more.

$$(a + b)^4 = a^4 + 4a^3b + 6a^2b^2 + 4ab^3 + b^4$$
$$(a + b)^5 = a^5 + 5a^4b + 10a^3b^2 + 10a^2b^3 + 5ab^4 + b^5$$

We could continue to build on previous expansions and eventually have quite a comprehensive list of binomial expansions. Instead, however, we will look for a theorem that will enable us to expand $(a + b)^n$ directly without multiplying.

Look at the variable parts of each expansion above. Note that for each $n = 1, 2, 3, 4, 5$

- The first term is a^n. The exponent on a decreases by 1 for each successive term.
- The exponent on b increases by 1 for each successive term. The last term is b^n.
- The degree of each term is n.

To find a pattern for the coefficients in each expansion, first note that there are $n + 1$ terms and that the coefficient of the first and last term is 1. To find the remaining coefficients, consider the expansion of $(a + b)^5$.

$$(a + b)^5 = a^5 + 5a^4b + 10a^3b^2 + 10a^2b^3 + 5ab^4 + b^5$$

$$\frac{5}{1} = 5 \qquad \frac{5 \cdot 4}{2 \cdot 1} = 10 \qquad \frac{5 \cdot 4 \cdot 3}{3 \cdot 2 \cdot 1} = 10 \qquad \frac{5 \cdot 4 \cdot 3 \cdot 2}{4 \cdot 3 \cdot 2 \cdot 1} = 5$$

Observe from these patterns that there is a strong relationship to factorials. In fact, we can express each coefficient by using factorial notation.

$$\frac{5!}{1!\,4!} = 5 \qquad \frac{5!}{2!\,3!} = 10 \qquad \frac{5!}{3!\,2!} = 10 \qquad \frac{5!}{4!\,1!} = 5$$

In each denominator, the first factorial is the exponent of b and the second factorial is the exponent of a.

In general, we will conjecture that the coefficient of the term $a^{n-k}b^k$ in the expansion of $(a + b)^n$ is $\frac{n!}{k!\,(n - k)!}$. Each coefficient of a term of a binomial expansion is called a **binomial coefficient** and is denoted by $\binom{n}{k}$.

Formula for a Binomial Coefficient

The coefficient of the term whose variable part is $a^{n-k}b^k$ in the expansion of $(a + b)^n$ is

$$\binom{n}{k} = \frac{n!}{k!\,(n - k)!}$$

The first term of the expansion of $(a + b)^n$ can be thought of as a^nb^0. In that case, we can calculate the coefficient of that term as

$$\binom{n}{0} = \frac{n!}{0!\,(n - 0)!} = \frac{n!}{1 \cdot n!} = 1$$

EXAMPLE 1 Evaluate a Binomial Coefficient

Evaluate each binomial coefficient. a. $\binom{9}{6}$ b. $\binom{10}{10}$

Solution

a. $\binom{9}{6} = \frac{9!}{6!\,(9 - 6)!} = \frac{9!}{6!\,3!} = \frac{9 \cdot 8 \cdot 7 \cdot 6!}{6! \cdot 3 \cdot 2 \cdot 1} = 84$

b. $\binom{10}{10} = \frac{10!}{10!\,(10 - 10)!} = \frac{10!}{10!\,0!} = 1.$ • Remember that $0! = 1$.

TRY EXERCISE 4, EXERCISE SET 11.5, PAGE 805

◆ BINOMIAL THEOREM

We are now ready to state the Binomial Theorem for positive integers.

Binomial Theorem for Positive Integers

If n is a positive integer, then

$$(a + b)^n = \sum_{i=0}^{n} \binom{n}{i} a^{n-i} b^i$$
$$= \binom{n}{0} a^n + \binom{n}{1} a^{n-1} b + \binom{n}{2} a^{n-2} b^2 + \cdots + \binom{n}{n} b^n$$

EXAMPLE 2 Expand the Sum of Two Terms

Expand: $(2x^2 + 3)^4$

Solution

$$(2x^2 + 3)^4 = \binom{4}{0}(2x^2)^4 + \binom{4}{1}(2x^2)^3(3) + \binom{4}{2}(2x^2)^2(3)^2 + \binom{4}{3}(2x^2)(3)^3 + \binom{4}{4}(3)^4$$
$$= 16x^8 + 96x^6 + 216x^4 + 216x^2 + 81$$

TRY EXERCISE 18, EXERCISE SET 11.5, PAGE 805

EXAMPLE 3 Expand a Difference of Two Terms

Expand: $(\sqrt{x} - 2y)^5$

Solution

$$(\sqrt{x} - 2y)^5 = \binom{5}{0}(\sqrt{x})^5 + \binom{5}{1}(\sqrt{x})^4(-2y) + \binom{5}{2}(\sqrt{x})^3(-2y)^2 + \binom{5}{3}(\sqrt{x})^2(-2y)^3 + \binom{5}{4}(\sqrt{x})(-2y)^4 + \binom{5}{5}(-2y)^5$$
$$= x^{5/2} - 10x^2y + 40x^{3/2}y^2 - 80xy^3 + 80x^{1/2}y^4 - 32y^5$$

take note

If exactly one of the terms a and b is negative, the terms of the expansion alternate in sign.

TRY EXERCISE 20, EXERCISE SET 11.5, PAGE 805

◆ *i*th TERM OF A BINOMIAL EXPANSION

The Binomial Theorem can also be used to find a specific term in the expansion of $(a + b)^n$.

take note

The exponent on b is 1 less than the term number.

Formula for the ith Term of a Binomial Expansion

The ith term of the expansion of $(a + b)^n$ is given by

$$\binom{n}{i-1}a^{n-i+1}b^{i-1}$$

EXAMPLE 4 Find the ith Term of a Binomial Expansion

Find the fourth term in the expansion of $(2x^3 - 3y^2)^5$.

Solution

With $a = 2x^3$ and $b = -3y^2$, and using the last theorem with $i = 4$ and $n = 5$, we have

$$\binom{5}{3}(2x^3)^2(-3y^2)^3 = -1080x^6y^6$$

The fourth term is $-1080x^6y^6$.

TRY EXERCISE 34, EXERCISE SET 11.5, PAGE 805

◆ PASCAL'S TRIANGLE

A pattern for the coefficients of the terms of an expanded binomial can be found by writing the coefficients in a triangular array known as **Pascal's Triangle.** See **Figure 11.1.**

Each row begins and ends with the number 1. Any other number in a row is the sum of the two closest numbers above it. For example, $4 + 6 = 10$. Thus each succeeding row can be found from the preceding row.

$(a + b)^1$:	1 1
$(a + b)^2$:	1 2 1
$(a + b)^3$:	1 3 3 1
$(a + b)^4$:	1 4 6 4 1
$(a + b)^5$:	1 5 10 10 5 1
$(a + b)^6$:	1 6 15 20 15 6 1

Figure 11.1

Pascal's triangle can be used to expand a binomial for small values of n. For instance, the seventh row of Pascal's Triangle is

$$1 \quad 7 \quad 21 \quad 35 \quad 35 \quad 21 \quad 7 \quad 1$$

Therefore,

$$(a + b)^7 = a^7 + 7a^6b + 21a^5b^2 + 35a^4b^3 + 35a^3b^4 + 21a^2b^5 + 7ab^6 + b^7$$

TOPICS FOR DISCUSSION

1. Discuss the use of the Binomial Theorem.
2. Can the Binomial Theorem be used to expand $(a+b)^n$, n a natural number, for any expressions for a and b? Why or why not?
3. What is Pascal's Triangle and how is it related to expanding a binomial?
4. Explain how Pascal's Triangle suggests that $\binom{n-1}{k-1} + \binom{n-1}{k} = \binom{n}{k}$.

EXERCISE SET 11.5

In Exercises 1 to 8, evaluate the binomial coefficients.

1. $\binom{7}{4}$
2. $\binom{8}{6}$
3. $\binom{9}{2}$
4. $\binom{10}{5}$
5. $\binom{12}{9}$
6. $\binom{6}{5}$
7. $\binom{11}{0}$
8. $\binom{14}{14}$

In Exercises 9 to 28, expand the binomial.

9. $(x-y)^6$
10. $(a-b)^5$
11. $(x+3)^5$
12. $(x-5)^4$
13. $(2x-1)^7$
14. $(2x+y)^6$
15. $(x+3y)^6$
16. $(x-4y)^5$
17. $(2x-5y)^4$
18. $(3x+2y)^4$
19. $\left(x+\frac{1}{x}\right)^6$
20. $(2x-\sqrt{y})^7$
21. $(x^2-4)^7$
22. $(x-y^3)^6$
23. $(2x^2+y^3)^5$
24. $(2x-y^3)^6$
25. $\left(\frac{2}{x}-\frac{x}{2}\right)^4$
26. $\left(\frac{a}{b}+\frac{b}{a}\right)^3$
27. $(s^{-2}+s^2)^6$
28. $(2r^{-1}+s^{-1})^5$

In Exercises 29 to 36, find the indicated term without expanding.

29. $(3x-y)^{10}$; eighth term
30. $(x+2y)^{12}$; fourth term
31. $(x+4y)^{12}$; third term
32. $(2x-1)^{14}$; thirteenth term
33. $(\sqrt{x}-\sqrt{y})^9$; fifth term
34. $(x^{-1/2}+x^{1/2})^{10}$; sixth term
35. $\left(\frac{a}{b}+\frac{b}{a}\right)^{11}$; ninth term
36. $\left(\frac{3}{x}-\frac{x}{3}\right)^{13}$; seventh term
37. Find the term that contains b^8 in the expansion of $(2a-b)^{10}$.
38. Find the term that contains s^7 in the expansion of $(3r+2s)^9$.
39. Find the term that contains y^8 in the expansion of $(2x+y^2)^6$.
40. Find the term that contains b^9 in the expansion of $(a-b^3)^8$.
41. Find the middle term of $(3a-b)^{10}$.
42. Find the middle term of $(a+b^2)^8$.
43. Find the two middle terms of $(s^{-1}+s)^9$.
44. Find the two middle terms of $(x^{1/2}-y^{1/2})^7$.

In Exercises 45 to 50, use the Binomial Theorem to simplify the powers of the complex numbers.

45. $(2-1)^4$
46. $(3+2i)^3$
47. $(1+2i)^5$
48. $(1-3i)^5$
49. $\left(\frac{\sqrt{2}}{2}+i\frac{\sqrt{2}}{2}\right)^8$
50. $\left(\frac{1}{2}+i\frac{\sqrt{3}}{2}\right)^6$

SUPPLEMENTAL EXERCISES

51. Let n be a positive integer. Expand and simplify $\frac{(x+h)^n-x^n}{h}$, where x is any real number and $h \neq 0$.
52. Show that $\binom{n}{k} = \binom{n}{n-k}$ for all positive integers n and k with $0 \le k \le n$.
53. Show that $\sum_{k=0}^{n}\binom{n}{k} = 2^n$. (*Hint:* Use the Binomial Theorem with $x=1$, $y=1$.)

54. Prove that $\binom{n}{k} + \binom{n}{k+1} = \binom{n+1}{k+1}$, n and k integers, $0 \le k \le n$.

55. Prove that $\sum_{i=0}^{n} (-1)^i \binom{n}{i} = 0$.

56. Approximate $(0.98)^8$ by evaluating the first three terms of $(1 - 0.02)^8$.

57. Approximate $(1.02)^8$ by evaluating the first three terms of $(1 + 0.02)^8$.

There is an extension of the Binomial Theorem called the *Multinomial Theorem*. This theorem is used in determining probabilities. *Multinomial Theorem:* If n, r, and k are positive integers, then the coefficient of $a^r b^k c^{n-r-k}$ in the expansion of $(a + b + c)^n$ is

$$\frac{n!}{r!\,k!\,(n - r - k)!}$$

In Exercises 58 to 61, use the Multinomial Theorem to find the indicated coefficient.

58. Find the coefficient of $a^2b^3c^5$ in the expansion of $(a + b + c)^{10}$.

59. Find the coefficient of $a^5b^2c^2$ in the expansion of $(a + b + c)^9$.

60. Find the coefficient of a^4b^5 in the expansion of $(a + b + c)^9$.

61. Find the coefficient of a^3c^5 in the expansion of $(a + b + c)^8$.

PROJECTS

1. PASCAL'S TRIANGLE Write an essay on Pascal's Triangle. Include some of the earliest known examples of the triangle and some of its applications.

2. SOME OTHER FUNCTIONS Do some research and determine a definition for positive integers for each of the following types of numbers. Give examples of calculations using each type of number.

   a. Pochammer (m, n) b. double factorial $(n!!)$

SECTION 11.6 PERMUTATIONS AND COMBINATIONS

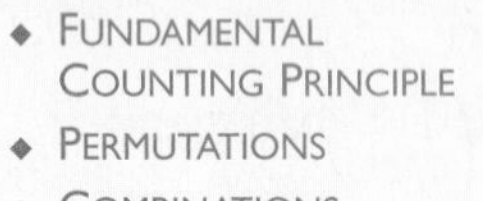

- FUNDAMENTAL COUNTING PRINCIPLE
- PERMUTATIONS
- COMBINATIONS

◆ FUNDAMENTAL COUNTING PRINCIPLE

Suppose an electronics store offers a three-component stereo system for \$250. A buyer must choose one amplifier, one tuner, and one pair of speakers. If the store has two models of amplifiers, three models of tuners, and two speaker models, how many different stereo systems could a consumer purchase?

This problem belongs to a class of problems called *counting problems.* The problem is to determine the number of ways in which the conditions of the problem can be satisfied. One way to do this is to make a tree diagram and then count the items on the list. We will organize the list in a table using A_1 and A_2 for the amplifiers; T_1, T_2, and T_3 for the tuners; and S_1 and S_2 for the speakers. See Figure 11.2.

By counting the possible systems that can be purchased, we find there are 12 different systems. Another way to arrive at this result is to find the product of the number of options available.

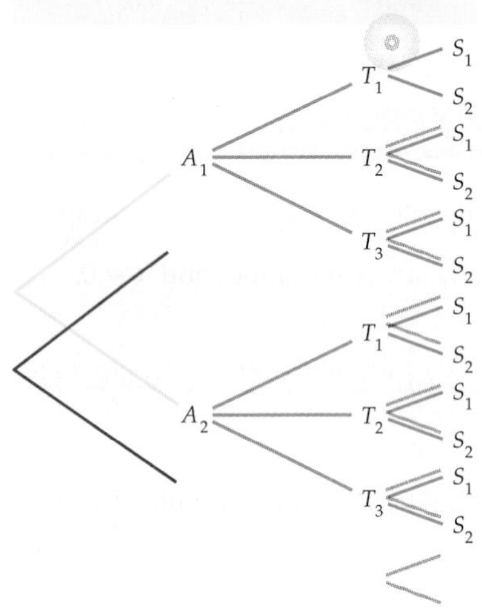

Figure 11.2

$$\begin{array}{ccccccc} \text{Number of} & \times & \text{number of} & \times & \text{number of} & = & \text{number of} \\ \text{amplifiers} & & \text{tuners} & & \text{speakers} & & \text{systems} \\ 2 & \times & 3 & \times & 2 & = & 12 \end{array}$$

In some states, a standard car license plate consists of a nonzero digit, followed by three letters, followed by three more digits. What is the maximum number of car license plates of this type that could be issued? If we begin a list of the possible license plates, it soon becomes apparent that listing them all would be very time-consuming and impractical.

1AAA000, 1AAA001, 1AAA002, 1AAA003,...

Instead, the following counting principle is used. This principle forms the basis for all counting problems.

Fundamental Counting Principle

Let $T_1, T_2, T_3, \ldots, T_n$ be a sequence of n conditions. Suppose that T_1 can occur in m_1 ways, T_2 can occur in m_2 ways, T_3 can occur in m_3 ways, and so on until finally T_n can occur in m_n ways. Then the number of ways of satisfying the conditions, $T_1, T_2, T_3, \ldots, T_n$ in succession is given by the product

$$m_1 m_2 m_3 \cdots m_n$$

Table 11.4

Condition	Number of ways
T_1: a nonzero digit	$m_1 = 9$
T_2: a letter	$m_2 = 26$
T_3: a letter	$m_3 = 26$
T_4: a letter	$m_4 = 26$
T_5: a digit	$m_5 = 10$
T_6: a digit	$m_6 = 10$
T_7: a digit	$m_7 = 10$

To apply the counting principle to the license plate problem first find the number of ways each condition can be satisfied, as shown in Table 11.4. Thus, we have

$$\begin{matrix}\text{Number of car} \\ \text{license plates}\end{matrix} = 9 \cdot 26 \cdot 26 \cdot 26 \cdot 10 \cdot 10 \cdot 10 = 158{,}184{,}000$$

EXAMPLE 1 Apply the Fundamental Counting Principle

An automobile dealer offers three mid-size cars. A customer selecting one of these cars must choose one of three different engines, one of five different colors, and one of four different interior packages. How many different selections can the customer make?

Solution

T_1: mid-size car $\quad m_1 = 3$

T_2: engine $\quad m_2 = 3$

T_3: color $\quad m_3 = 5$

T_4: interior $\quad m_4 = 4$

Number of different selections $= 3 \cdot 3 \cdot 5 \cdot 4 = 180.$

TRY EXERCISE 12, EXERCISE SET 11.6, PAGE 811

◆ PERMUTATIONS

An application of the Fundamental Counting Principle is to determine the number of arrangements of distinct elements in a definite order.

Permutation

A **permutation** is an arrangement of distinct objects in a definite order.

For example, *abc* and *bca* are two of the possible permutations of the three elements *a*, *b*, *c*.

Consider a race with 10 runners. In how many different orders can the runners finish first, second, and third (assuming no ties)?

Any one of the 10 runners could finish first: $m_1 = 10$

Any one of the remaining 9 runners could be second: $m_2 = 9$

Any one of the remaining 8 runners could be third: $m_3 = 8$

By the Fundamental Counting Principle, there are $10 \cdot 9 \cdot 8 = 720$ possible first-, second-, and third-place finishes for the 10 runners. Using the language of permutations, we would say, "There are 720 permutations of 10 objects (the runners) taken 3 (the possible finishes) at a time."

Permutations occur so frequently in counting problems that a formula, rather than the counting principle, is often used.

take note

Some graphing calculators use the notation *nPr* to represent the number of permutations of *n* objects taken *r* at a time. The calculation of the number of permutations of 15 objects taken 4 at a time is shown below.

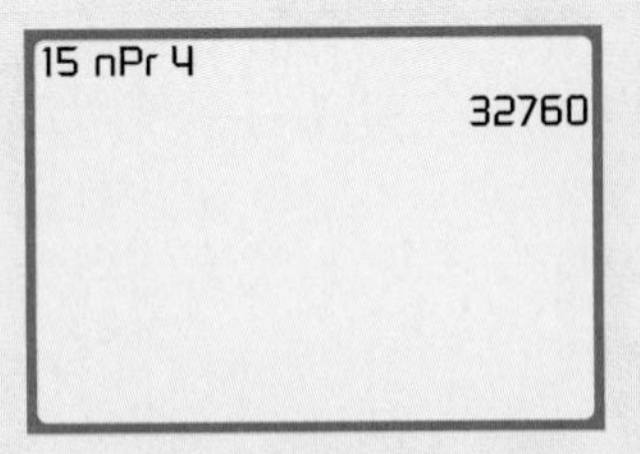

Formula for a Permutation of *n* Distinct Objects Taken *r* at a Time

The number of permutations of *n* distinct objects taken *r* at a time is

$$P(n, r) = \frac{n!}{(n - r)!}$$

EXAMPLE 2 Find the Number of Permutations

In how many ways can a president, vice president, secretary, and treasurer be selected from a committee of fifteen people?

Solution

There are fifteen distinct people to place in four positions. Thus $n = 15$ and $r = 4$.

$$P(15, 4) = \frac{15!}{(15 - 4)!} = \frac{15!}{11!} = \frac{15 \cdot 14 \cdot 13 \cdot 12 \cdot 11!}{11!} = 32{,}760$$

TRY EXERCISE 16, EXERCISE SET 11.6, PAGE 811

EXAMPLE 3 Find the Number of Seating Permutations

Six people attend a movie and all sit in the same row with six seats.

a. Find the number of ways the group can sit together.

b. Find the number of ways the group can sit together if two people in the group must sit side-by-side.

c. Find the number of ways the group can sit together if two people in the group refuse to sit side-by-side.

Solution

a. There are six distinct people to place in six distinct positions. Thus $n = 6$ and $r = 6$.

$$P(6, 6) = \frac{6!}{(6-6)!} = \frac{6!}{0!} = \frac{6!}{1} = 720$$

b. Think of the two people who must sit together as a single object and count the number of arrangements of the *five* objects $(AB), C, D, E, F$. Thus $n = 5$ and $r = 5$.

$$P(5, 5) = \frac{5!}{(5-5)!} = \frac{5!}{0!} = \frac{5!}{1} = 120$$

There are also 120 arrangements with A and B reversed $(BA), C, D, E, F$. Thus the total number of arrangements is $120 + 120 = 240$.

c. From **a.**, there are 720 possible seating arrangements. From **b.**, there are 240 arrangements with two specific people next to each other. Thus there are $720 - 240 = 480$ arrangements where two specific people are not seated together.

TRY EXERCISE 22, EXERCISE SET 11.6, PAGE 811

◆ COMBINATIONS

Up to this point, we have been counting the number of distinct arrangements of objects. In some cases we may be interested in determining the number of ways of selecting objects without regard to the order of the selection. For example, suppose we want to select a committee of three people from five candidates denoted by A, B, C, D, and E. One possible committee is A, C, D. If we select D, C, A, we still have the same committee because the order of the selection is not important. An arrangement of objects for which the order of the selection is not important is a **combination.**

take note

Recall that a binomial coefficient is given by $\binom{n}{r} = \frac{n!}{r!\,(n-r)!}$ which is the same as $C(n, r)$.

Formula for the Combination of n Objects Taken r at a Time

The number of combinations of n objects taken r at a time is

$$C(n, r) = \frac{n!}{r!\,(n-r)!}$$

take note

Some calculators use the notation *nCr* to represent a combination of *n* objects taken *r* at a time.

EXAMPLE 4 Find the Number of Combinations

A standard deck of playing cards consists of fifty-two cards. How many five-card hands can be chosen from this deck?

Solution

We have $n = 52$ and $r = 5$. Thus

$$C(52, 5) = \frac{52!}{5!\,(52-5)!} = \frac{52!}{5!\,47!} = \frac{52 \cdot 51 \cdot 50 \cdot 49 \cdot 48 \cdot 47!}{5 \cdot 4 \cdot 3 \cdot 2 \cdot 1 \cdot 47!} = 2{,}598{,}960$$

TRY EXERCISE 20, EXERCISE SET 11.6, PAGE 811

EXAMPLE 5 Find the Number of Combinations

A chemist has nine samples of a solution, of which four are type A and five are type B. If the chemist chooses three of the solutions at random, determine in how many ways the chemist can have exactly one type A solution.

Solution

The chemist has chosen three solutions, one of which is type A. If one is type A, then two are type B. The number of ways of choosing one type A solution from four type A solutions is $C(4, 1)$.

$$C(4, 1) = \frac{4!}{1!\,(4-1)!} = \frac{4!}{1!\,3!} = 4$$

The number of ways of choosing two type B solutions from five type B solutions is $C(5, 2)$.

$$C(5, 2) = \frac{5!}{2!\,(5-2)!} = \frac{5!}{2!\,3!} = 10$$

By the counting principle, there are

$$C(4, 1) \cdot C(5, 2) = 4 \cdot 10 = 40$$

ways to have one type A and two type B solutions.

TRY EXERCISE 30, EXERCISE SET 11.6, PAGE 812

The difficult part of counting is determining whether to use the counting principle, the permutation formula, or the combination formula. Following is a summary of guidelines.

Guidelines for Solving Counting Problems

1. The counting principle will always work but is not always the easiest method to apply.
2. When reading a problem, ask yourself, "Is the order of the selection process important?" If the answer is yes, the arrangements are permutations. If the answer is no, the arrangements are combinations.

TOPICS FOR DISCUSSION

1. Discuss the Fundamental Counting Principle and how it is used.
2. Discuss the difference between a permutation and a combination.
3. Explain why $\binom{n}{k}$ occurs as a coefficient in the binomial formula.

EXERCISE SET 11.6

In Exercises 1 to 10, evaluate each quantity.

1. $P(6, 2)$
2. $P(8, 7)$
3. $C(8, 4)$
4. $C(9, 2)$
5. $P(8, 0)$
6. $P(9, 9)$
7. $C(7, 7)$
8. $C(6, 0)$
9. $C(10, 4)$
10. $P(10, 4)$

11. **COMPUTER SYSTEMS** A computer manufacturer offers a computer system with three different disk drives, two different monitors, and two different keyboards. How many different computer systems could a consumer purchase from this manufacturer?

12. **COLOR MONITORS** A computer monitor produces color by blending colors on *palettes*. If a computer monitor has four palettes and each palette has four colors, how many blended colors can be formed? Assume each palette must be used each time.

13. **LIGHT SWITCHES** A large conference room has four doors. At the entrance to each door there is a single light switch. How many different configurations of "on" and "off" are possible for the light switches?

14. **COMPUTER MEMORY** An integer is stored in a computer's memory as a series of zeros and ones. Each memory unit contains 8 spaces for a zero or a one. The first space is used for the sign of the number, and the remaining 7 spaces are used for the integer. How many positive integers can be stored in one memory unit of this computer?

15. **SCHEDULING** In how many different ways can six employees be assigned to six different jobs?

16. **CONTEST WINNERS** First-, second-, and third-place prizes are to be awarded in a dance contest in which twelve contestants are entered. In how many ways can the prizes be awarded?

17. **MAIL BOXES** There are five mailboxes outside a post office. In how many ways can three letters be deposited into the five boxes?

18. **COMMITTEE MEMBERSHIP** How many different committees of three people can be selected from nine people?

19. **TEST QUESTIONS** A professor provides to a class 25 possible essay questions for an upcoming test. Of the 25 questions, the professor will ask 5 of the questions on the exam. How many different tests can the professor prepare?

20. **TENNIS MATCHES** Twenty-six people enter a tennis tournament. How many different first-round matches are possible if each player can be matched with any other player?

21. **EMPLOYEE INITIALS** A company has more than 676 employees. Explain why there must be at least 2 employees who have the same first and last initials.

22. **SEATING ARRANGEMENTS** A car holds six passengers, three in the front seat and three in the back seat. How many different seating arrangements of six people are possible if one person refuses to sit in front and one person refuses to sit in back?

23. **COMMITTEE MEMBERSHIP** A committee of six people is chosen from six senators and eight representatives. How many committees are possible if there are to be three senators and three representatives on the committee?

24. **ARRANGING NUMBERS** The numbers 1, 2, 3, 4, 5, 6 are to be arranged. How many different arrangements are possible under each of the following conditions?
    a. All the even numbers come first.
    b. The arrangements are such that the numbers alternate between even and odd.

25. **TEST QUESTIONS** A true-false examination contains ten questions. In how many ways can a person answer the questions on this test by just guessing? Assume that all questions are answered.

26. **TEST QUESTIONS** A twenty-question, four-option multiple-choice examination is given as a pre-employment test. In how many ways could a prospective employee answer the questions on this test by just guessing? Assume that all questions are answered.

27. **STATE LOTTERY** A state lottery game requires a person to select six different numbers from forty numbers. The order of the selection is not important. In how many ways can this be done?

28. **TEST QUESTIONS** A student must answer eight of ten questions on an exam. How many different choices can the student make?

29. **ACCEPTANCE SAMPLING** A warehouse receives a shipment of ten computers, of which three are defective. Five computers are then randomly selected from the ten and delivered to a store.
    a. In how many ways can the store receive no defective computers?
    b. In how many ways can the store receive one defective computer?
    c. In how many ways can the store receive all three defective computers?

30. **CONTEST** Fifteen students, of whom seven are seniors, are selected as semifinalists for a literary award. Of the fifteen students, ten finalists will be selected.
    a. In how many ways can ten finalists be selected from the fifteen students?
    b. In how many ways can the ten finalists contain three seniors?
    c. In how many ways can the ten finalists contain at least five seniors?

31. **SERIAL NUMBERS** A television manufacturer uses a code for the serial number of a television set. The first symbol is the letter *A*, *B*, or *C* and represents the location of the manufacturing plant. The next two symbols are 01, 02, . . . , 12 and represent the month in which the set was manufactured. The next symbol is a 5, 6, 7, 8, or 9 and represents the year the set was manufactured. The last seven symbols are digits. How many serial numbers are possible?

32. **CARD GAMES** Five cards are chosen at random from an ordinary deck of playing cards. In how many ways can the cards be chosen under each of the following conditions?
    a. All are hearts.
    b. All are the same suit.
    c. Exactly three are kings.
    d. Two or more are aces.

33. **ACCEPTANCE SAMPLING** A quality control inspector receives a shipment of ten computer disk drives and randomly selects three of the drives for testing. If two of the disk drives in the shipment are defective, find the number of ways in which the inspector could select at most one defective drive.

34. **BASKETBALL TEAMS** A basketball team has twelve members. In how many ways can five players be chosen under each of the following conditions?
    a. The selection is random.
    b. The two tallest players are always among the five selected.

35. **ARRANGING NUMBERS** The numbers 1, 2, 3, 4, 5, 6 are arranged in random order. In how many ways can the numbers 1 and 2 appear next to one another in the order 1, 2?

36. **OCCUPANCY PROBLEM** Seven identical balls are randomly placed in seven available containers in such a way that two balls are in one container. Of the remaining six containers, each receives at most one ball. Find the number of ways in which this can be accomplished.

37. **LINES IN A PLANE** Seven points lie in a plane in such a way that no three points lie on the same line. How many lines are determined by seven points?

38. **CHESS MATCHES** A chess tournament has twelve participants. How many games must be scheduled if every player must play every other player exactly once?

39. **CONTEST WINNERS** Eight couples attend a benefit at which two prizes are given. In how many ways can two names be randomly drawn so that the prizes are not awarded to the same couple?

40. **GEOMETRY** Suppose there are twelve distinct points on a circle. How many different triangles can be formed with vertices at the given points?

41. **TEST QUESTIONS** In how many ways can a student answer a twenty-question true-false test if the student marks ten of the questions true and ten of the questions false?

42. **COMMITTEE MEMBERSHIP** From a group of fifteen people a committee of eight is formed. From the committee a president, secretary, and treasurer are selected. Find the number of ways in which the two consecutive operations can be carried out.

43. **COMMITTEE MEMBERSHIP** From a group of twenty people a committee of twelve is formed. From the committee of twelve, a subcommittee of four people is chosen. Find the number of ways in which the two consecutive operations can be carried out.

44. **CHECKERBOARDS** A checkerboard consists of 8 rows and 8 columns of squares. Starting at the top left square of a checkerboard, how many possible paths will end at the bottom right square if the only way a player can legally move is right one square or down one square from the current position?

45. **ICE CREAM CONES** An ice cream store offers 31 flavors of ice cream. How many different triple-decker cones are possible? *Note:* Assume that different orders of the same flavors are *not* different cones. Thus a scoop of rocky road

followed by two scoops of mint chocolate is the same as one scoop of mint chocolate followed by one scoop of rocky road followed by a second scoop of mint chocolate.

46. **COMPUTER SCREENS** A typical computer monitor consists of pixels each of which can, in some cases, be assigned any one of 2^{16} different colors. If a computer screen has a resolution of 1024 pixels by 768 pixels, how many different images can be displayed on the screen? [*Suggestion:* Write your answer as a power of 2.]

47. **DART BOARDS** How many different arrangements are there of the integers 1 through 20 on a typical dart board assuming that 20 is always at the top?

SUPPLEMENTAL EXERCISES

48. **LINES IN A PLANE** Generalize Exercise 37. That is, given n points in a plane, no three of which lie on the same line, how many lines are determined by n points?

49. **BIRTHDAYS** Seven people are asked the month of their birth. In how many ways can each of the following conditions exist?

    a. No two people have a birthday in the same month.

    b. At least two people have a birthday in the same month.

50. **SUMS OF COINS** From a penny, nickel, dime, and quarter, how many different sums of money can be formed using one or more of the coins?

51. **BIOLOGY** Five sticks of equal length are broken into a short piece and a long piece. The ten pieces are randomly arranged in five pairs. In how many ways will each pair consist of a long stick and a short stick? (This exercise actually has a practical side. When cells are exposed to harmful radiation, some chromosomes break. If two long sides unite or two short sides unite, the cell dies.)

52. **ARRANGING NUMBERS** Four random digits are drawn (repetitions are allowed). Among the four digits, in how many ways can two or more repetitions occur?

53. **RANDOM WALK** An aimless tourist, standing on a street corner, tosses a coin. If the result is heads, the tourist walks one block north. If the result is tails, the tourist walks one block south. At the new corner, the coin is tossed again and the same rule applied. If the coin is tossed ten times, in how many ways will the tourist be back at the original corner? This problem is an elementary example of what is called a *random walk*. Random walk problems have many applications in physics, chemistry, and economics.

PROJECTS

1. **EXPLAIN PERMUTATIONS AND COMBINATIONS** Write an outline of a lesson that you could use to teach permutations and combinations. Include at least five examples of permutations and five examples of combinations.

2. **APPLICATION OF COUNTING** Calculating the number of ways in which balls can be distributed in boxes has a variety of applications. For instance, a traffic engineer may want to know how traffic accidents (the balls) are distributed throughout the days of the week (the boxes). Or a physicist may want to know how electrons (the balls) can be distributed in the energy orbits (the boxes) of an atom. The formula for counting the number of ways in which n distinguishable balls can be placed in k distinguishable boxes, where each box must have at least 1 ball, is given by

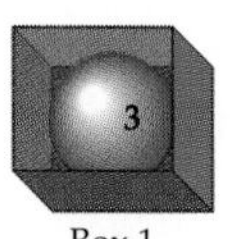

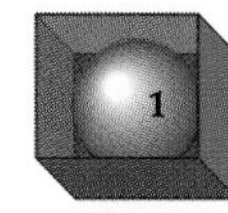

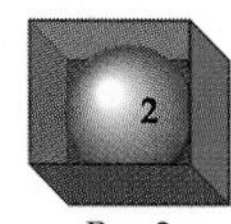

Box 1 Box 2 Box 3

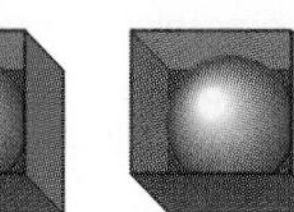

Box 1 Box 2 Box 3

$$\binom{k}{0}k^n - \binom{k}{1}(k-1)^n + \binom{k}{2}(k-2)^n + \cdots + (-1)^{k-1}\binom{k}{k-1}$$

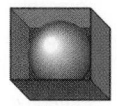

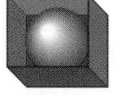

The word *distinguishable* is important. This formula refers to counting under circumstances similar to the first figure at the right, where the boxes are numbered and the balls are numbered. The second figure shows a situation that is not covered by this formula. Although the boxes are numbered, the balls are not and are therefore *indistinguishable*.

a. A computer network consists of five computers and three printers. How many possible connections can be made if each computer must be hooked to a printer and all printers are used?

b. A supermarket has four checkout lanes. Assuming shoppers are efficient and will not leave a checkout lane empty, in how many ways can ten shoppers line up for the checkout lanes?

SECTION 11.7 INTRODUCTION TO PROBABILITY

- SAMPLE SPACES
- EVENTS
- PROBABILITY OF AN EVENT
- BINOMIAL PROBABILITIES

MATH MATTERS

The beginning of probability theory is frequently associated with letters sent between Pascal (of Pascal's Triangle) and Fermat (Fermat's Last Theorem) in which they discuss the solution of a problem posed to them by Antoine Gombaud, Chevalier de Mere, a French aristocrat who liked to gamble. The basic question was "How many tosses of two dice are necessary to have a better than 50–50 chance of rolling 2 sixes?" Although the correct analysis of this problem by Fermat and Pascal presaged probability theory, historical records indicate that commerce and the need to insure ships, cargo, and lives was another motivating factor to calculating probabilities. The first merchant insurance companies were established in the fourteenth century to insure ships. Life insurance companies were established at the end of the seventeenth century. Lloyd's of London was established sometime before 1690.

Many events in the world around us have random character, such as the chances of an accident occurring on a certain freeway, the chances of winning a state lottery, and the chances that the nucleus of an atom will undergo fission. By repeatedly observing such events, it is often possible to recognize certain patterns. **Probability** is the mathematical study of random patterns.

When a weather reporter predicts a 30% chance of rain, the forecaster is saying that similar weather conditions have led to rain 30 times out of 100. When a fair coin is tossed, we expect heads to occur 1/2, or 50%, of the time. The numbers 30% (or 0.3) and 1/2 are the probabilities of the events.

◆ SAMPLE SPACES

An activity with an observable outcome is called an **experiment.** Examples of experiments include

1. Flipping a coin and observing the side facing upward
2. Observing the incidence of a disease in a certain population
3. Observing the length of time a person waits in a checkout line in a grocery store

The **sample space** of an experiment is the set of *all possible* outcomes of that experiment.

Consider the experiment of tossing one coin three times and recording the number of occurrences of the upward side of the coin. The sample space is

$$S = \{\text{HHH, HHT, HTH, THH, HTT, THT, TTH, TTT}\}$$

EXAMPLE 1 List the Elements of a Sample Space

Suppose that among five batteries, two are defective. Two batteries are randomly drawn from the five and tested for defects. List the elements in the sample space.

Solution

Label the nondefective batteries, N_1, N_2, N_3 and the defective batteries D_1, D_2. The sample space is

$$S = \{N_1D_1, N_2D_1, N_3D_1, N_1D_2, N_2D_2, N_3D_2, N_1N_2, N_1N_3, N_2N_3, D_1D_2\}$$

TRY EXERCISE 6, EXERCISE SET 11.7, PAGE 821

◆ EVENTS

An **event** E is any subset of a sample space. For the sample space defined in Example 1, several of the events we could define are

E_1: There are no defective batteries.

E_2: At least one battery is defective.

E_3: Both batteries are defective.

Because an event is a subset of the sample space, each of these events can be expressed as a set.

$$E_1 = \{N_1N_2, N_1N_3, N_2N_3\}$$
$$E_2 = \{N_1D_1, N_2D_1, N_3D_1, N_1D_2, N_2D_2, N_3D_2, D_1D_2\}$$
$$E_3 = \{D_1D_2\}$$

There are two methods by which elements are drawn from a sample space: with replacement and without replacement. *With replacement* means that after the element is drawn, it is returned to the sample space. The same element could be selected on the next drawing. When elements are drawn *without replacement*, an element drawn is not returned to the sample space and therefore is not available for any subsequent drawing.

EXAMPLE 2 List the Elements of an Event

A two-digit number is formed by choosing from the digits 1, 2, 3, 4, both with replacement and without replacement. Express each event as a set.

a. E_1: The second digit is greater than or equal to the first digit.

b. E_2: Both digits are less than zero.

Solution

a. With replacement: $E_1 = \{11, 12, 13, 14, 22, 23, 24, 33, 34, 44\}$

Without replacement: $E_1 = \{12, 13, 14, 23, 24, 34\}$

b. $E_2 = \varnothing$

Choosing from the digits 1, 2, 3, 4, this event is impossible. The impossible event is denoted by the empty set or null set.

TRY EXERCISE 14, EXERCISE SET 11.7, PAGE 821

◆ PROBABILITY OF AN EVENT

The probability of an event is defined in terms of the concepts of sample space and event.

Probability of an Event

Let $n(S)$ and $n(E)$ represent the number of elements in the sample space S and the event E, respectively. The probability of event E, $P(E)$, is

$$P(E) = \frac{n(E)}{n(S)}$$

Because E is a subset of S, $n(E) \leq n(S)$. Thus $P(E) \leq 1$. If E is an impossible event, then $E = \varnothing$ and $n(E) = 0$. Thus $P(E) = 0$. If E is the event that *always* occurs, then $E = S$ and $n(E) = n(S)$. Thus $P(E) = 1$. Thus, we have, for any event E,

$$0 \leq P(E) \leq 1$$

EXAMPLE 3 Calculate the Probability of an Event

A coin is tossed three times. What is the probability of each outcome?

a. E_1: Two or more heads will appear.

b. E_2: At least one tail will appear.

Solution

First determine the number of elements in the sample space. The sample space for this experiment is

$$S = \{\text{HHH, HHT, HTH, THH, HTT, THT, TTH, TTT}\}$$

Therefore $n(S) = 8$. Now determine the number of elements in each event. Then calculate the probability of the event by using $P(E) = n(E)/n(S)$.

a. $E_1 = \{\text{HHH, HHT, HTH, THH}\}$

$$P(E_1) = \frac{n(E_1)}{n(S)} = \frac{4}{8} = \frac{1}{2}$$

b. $E_2 = \{\text{HHT, HTH, THH, HTT, THT, TTH, TTT}\}$

$$P(E_2) = \frac{n(E_2)}{n(S)} = \frac{7}{8}$$

TRY EXERCISE 22, EXERCISE SET 11.7, PAGE 821

Calculating probabilities by listing and then counting the elements of a sample space is not always practical. Instead, we will use the counting principles developed in the last section to determine the number of elements in the sample space and in an event.

EXAMPLE 4 Use the Counting Principles to Calculate a Probability

A state lottery game allows a person to choose five numbers from the integers 1 to 40. Repetitions of numbers are not allowed. If three or more numbers match the numbers chosen by the lottery, the player wins a prize. Find the probability that a player will match

a. exactly three numbers

b. exactly four numbers

Solution

The sample space S is the number of ways in which five numbers can be chosen from forty numbers. This is a combination because the order of the drawing is not important.

$$n(S) = C(40, 5) = \frac{40!}{5!\,35!} = 658{,}008$$

We will call the five numbers chosen by the state lottery "lucky" and the remaining thirty-five numbers "unlucky."

a. Let E_1 be the event a player has three lucky and therefore two unlucky numbers. The three lucky numbers are chosen from the five lucky numbers. There are $C(5, 3)$ ways to do this. The two unlucky numbers are chosen from the thirty-five unlucky numbers. There are $C(35, 2)$ ways to do this. By the counting principle, the number of ways the event E_1 can occur is

$$N(E_1) = C(5, 3) \cdot C(35, 2) = 10 \cdot 595 = 5950$$

$$P(E_1) = \frac{n(E_1)}{n(S)} = \frac{C(5, 3) \cdot C(35, 2)}{C(40, 5)} = \frac{5950}{658{,}008} \approx 0.009042$$

b. Let E_2 be the event a player has four lucky numbers and one unlucky number. The number of ways a person can select four lucky numbers and one unlucky number is $C(5, 4) \cdot C(35, 1)$.

$$P(E_2) = \frac{n(E_2)}{n(S)} = \frac{C(5, 4) \cdot C(35, 1)}{C(40, 5)} = \frac{175}{658{,}008} \approx 0.000266$$

TRY EXERCISE 32, EXERCISE SET 11.7, PAGE 822

The expression "one or the other of two events occurs" is written as the union of the two sets. For example, suppose an experiment leads to the sample space $S = \{1, 2, 3, 4, 5, 6\}$ and the events are

Draw a number less than four, $E_1 = \{1, 2, 3\}$

Draw an even number, $E_2 = \{2, 4, 6\}$

Then the event $E_1 \cup E_2$ is described by drawing a number less than four *or* an even number. Thus

$$E_1 \cup E_2 = \{1, 2, 3\} \cup \{2, 4, 6\} = \{1, 2, 3, 4, 6\}$$

Two events E_1 and E_2 that cannot occur at the same time are **mutually exclusive** events. Using set notation, if $E_1 \cap E_2 = \emptyset$, then E_1 and E_2 are mutually exclusive.

For example, using the same sample space $\{1, 2, 3, 4, 5, 6\}$, a third event is

$$\text{Draw an odd number, } E_3 = \{1, 3, 5\}$$

Then $E_2 \cap E_3 = \emptyset$ and the events E_2 and E_3 are mutually exclusive. On the other hand,

$$E_1 \cap E_2 = \{2\}$$

so the events E_1 and E_2 are not mutually exclusive.

One of the axioms of probability involves the union of mutually exclusive events.

A Probability Axiom

If E_1 and E_2 are mutually exclusive events, then

$$P(E_1 \cup E_2) = P(E_1) + P(E_2)$$

If the events are not mutually exclusive, the addition rule for probabilities can be used.

Addition Rule for Probabilities

If E_1 and E_2 are two events, then

$$P(E_1 \cup E_2) = P(E_1) + P(E_2) - P(E_1 \cap E_2)$$

The probability axiom and addition rule are useful when calculating probabilities of events connected by the word *or*.

Using the calculations of Example 4, we can find the probability that a player will have three or four lucky numbers in the lottery. Because the events E_1 and E_2 as defined in Example 4 are mutually exclusive,

$$P(E_1 \cup E_2) = P(E_1) + P(E_2) = 0.009042 + 0.000266 = 0.009308$$

As an example of nonmutually exclusive events, draw a card at random from a deck of ordinary playing cards. Find the probability of drawing an ace or a heart.

$$S = \{\text{52 ordinary playing cards}\}$$

Let $E_1 = \{\text{an ace}\}$ and $E_2 = \{\text{a heart}\}$. Then

$$P(E_1) = \frac{n(E_1)}{n(S)} = \frac{4}{52} = \frac{1}{13} \qquad P(E_2) = \frac{n(E_2)}{n(S)} = \frac{13}{52} = \frac{1}{4}$$

We have $E_1 \cup E_2 = \{\text{an ace } or \text{ a heart}\}$ and $E_1 \cap E_2 = \{\text{ace of hearts}\}$. First, we find $P(E_1 \cap E_2)$.

$$P(E_1 \cap E_2) = \frac{n(E_1 \cap E_2)}{n(S)} = \frac{1}{52}$$

Now we can find $P(E_1 \cup E_2)$.

$$P(E_1 \cup E_2) = P(E_1) + P(E_2) - P(E_1 \cap E_2) = \frac{1}{13} + \frac{1}{4} - \frac{1}{52} = \frac{16}{52} = \frac{4}{13}$$

Two events are **independent** if the outcome of the first event does not influence the outcome of the second event. As an example, consider tossing a fair coin twice. The outcome of the first toss has no bearing on the outcome of the second toss. The two events are independent.

Now consider drawing two cards in succession, without replacement, from a regular deck of playing cards. The probability that the second card drawn will be an ace depends on the card drawn first.

Probability Rule for Independent Events

If E_1 and E_2 are two independent events, then the probability that both E_1 *and* E_2 will occur is

$$P(E_1) \cdot P(E_2)$$

EXAMPLE 5 Calculate a Probability for Independent Events

A coin is tossed and then a die is rolled. What is the probability that the coin will show a head and that the die will show a six?

Solution

The events are independent because the result of one does not influence the probability of the other. $P(\text{head}) = \frac{1}{2}$ and $P(\text{six}) = \frac{1}{6}$. Thus the probability of tossing a head and rolling a six is

$$P(\text{head}) \cdot P(\text{six}) = \frac{1}{2} \cdot \frac{1}{6} = \frac{1}{12}$$

TRY EXERCISE 34, EXERCISE SET 11.7, PAGE 822

◆ BINOMIAL PROBABILITIES

Some probabilities can be calculated from formulas. One of the most important of those formulas is the *Binomial Probability Formula*. This formula is used to calculate probabilities for *independent* events.

MATH MATTERS

Airlines "overbook" flights. That is, they sell more tickets than there are seats on the plane. An airline company can determine the probability that some number of passengers will be "bumped" on a certain flight by using the Binomial Probability Formula. For instance, suppose a plane has 200 seats and the airline sells 240 tickets. If the probability that a person will show up for this flight is 0.8, then the probability that one or more people will have to be bumped is given by

$$\sum_{k=1}^{40}\binom{240}{200+k}0.8^{200+k}0.2^{40-k}$$

Using a computer reveals that this sum is approximately 0.08. That is, there is only an 8% chance that someone will have to be bumped, even though the number of tickets sold exceeded the plane's capacity by 20%.

Binomial Probability Formula

Let an experiment consist of n trials for which the probability of success on a single trial is p and the probability of failure is $q = 1 - p$. Then the probability of k successes in n trials is given by

$$\binom{n}{k}p^k q^{n-k}$$

EXAMPLE 6 Use the Binomial Formula

A multiple-choice exam consists of ten questions. For each question there are four possible choices, of which only one is correct. If someone randomly guesses at the answers, what is the probability of guessing six answers correctly?

Solution

Selecting an answer is one trial of the experiment. Because there are ten questions, $n = 10$. There are four possible choices for each question, of which only one is correct. Therefore,

$$p = \frac{1}{4} \quad \text{and} \quad q = 1 - p = 1 - \frac{1}{4} = \frac{3}{4}$$

A success for this experiment occurs each time a correct answer is guessed. Thus $k = 6$. By the Binomial Probability Formula,

$$P = \binom{10}{6}\left(\frac{1}{4}\right)^6\left(\frac{3}{4}\right)^4 \approx 0.016222$$

The probability of guessing six answers correctly is approximately 0.0162.

TRY EXERCISE 40, EXERCISE SET 11.7, PAGE 822

Following are five guidelines for calculating probabilities.

Guidelines for Calculating a Probability

1. The word "or" usually means to add the probabilities of each event.
2. The word "and" usually means to multiply the probabilities of each event.
3. The phrase "at least n" means n *or* more. At least 5 is 5 or more.
4. The phrase "at most n" means n *or* less. At most 5 is 5 or less.
5. "Exactly n" means just that. Exactly 5 heads in 7 tosses of a coin means 5 heads *and therefore* 2 tails.

TOPICS FOR DISCUSSION

1. What is the meaning of probability and what are the possible values of a probability?
2. What is the sample space of an experiment? What are events and how are they related to the sample space?
3. Discuss the difference between mutually exclusive events and events that are not mutually exclusive. Give examples of each type.
4. Discuss the Addition Rule for Probabilities and how it is used.
5. What is the Binomial Probability Formula and how is it used?

EXERCISE SET 11.7

In Exercises 1 to 10, list the elements in the sample space defined by the given experiment.

1. Two people are selected from two senators and three representatives.
2. A letter is chosen at random from the word "Tennessee."
3. A fair coin is tossed and then a random integer between 1 and 4, inclusive, is selected.
4. A fair coin is tossed four times.
5. Two identical tennis balls are randomly placed in three tennis ball cans.
6. Two people are selected from among one Republican, one Democrat, and one Independent.
7. Three cards are randomly chosen from the ace of hearts, ace of spades, ace of clubs, and ace of diamonds.
8. Three letters addressed to A, B, and C, respectively, are randomly put into three envelopes addressed to A, B, and C, respectively.
9. Two vowels are randomly chosen from a, e, i, o, and u.
10. Three computer disks are randomly chosen from one defective disk and three nondefective disks.

In Exercises 11 to 15, use the sample space defined by the experiment of tossing a fair coin four times. Express each event as a subset of the sample space.

11. There are no tails.
12. There are exactly two heads.
13. There are at most two heads.
14. There are more than two heads.
15. There are twelve tails.

In Exercises 16 to 20, use the sample space defined by the experiment of choosing two random numbers, in succession, from the integers 1, 2, 3, 4, 5, and 6. The numbers are chosen with replacement. Express each event as a subset of the sample space.

16. The sum of the numbers is 7.
17. The two numbers are the same.
18. The first number is greater than the second number.
19. The second number is a 4.
20. The sum of the two numbers is greater than 1.

In Exercises 21 through 44, calculate the probabilities of the events.

21. **CARD GAMES** From a deck of regular playing cards, one card is chosen at random. Find the probability of each event.

    a. The card is a king. b. The card is a spade.

22. **NUMBER THEORY** A single number is chosen from the digits 1, 2, 3, 4, 5, and 6. Find the probability that the number is an even number or a number divisible by 3.
23. **ECONOMICS** An economist predicts that the probability of an increase in gross domestic product (GDP) is 0.64 and that the probability of an increase in inflation is 0.55. The economist also predicts that the probability of an increase in GDP *and* inflation is 0.22. Find the probability of an increase in GDP *or* an increase in inflation.
24. **NUMBER THEORY** Four digits are selected from the digits 1, 2, 3, and 4, and a number is formed. Find the probability that the number is greater than 3000, assuming digits can be repeated.

25. **BUILDING INDUSTRY** An owner of a construction company has bid for the contracts on two buildings. If the contractor estimates that the probability of getting the first contract is $1/2$, that of getting the second contract is $1/5$, and that of getting both contracts is $1/10$, find the probability that the contractor will get at least one of the two building contracts.

26. **ACCEPTANCE SAMPLING** A shipment of ten calculators contains two defective calculators. Two calculators are chosen from the shipment. Find the probability of each event.

    a. Both are defective.

    b. At least one is defective.

27. **NUMBER THEORY** Five random digits are selected from 0 to 9 with replacement. What is the probability (to the nearest hundredth) that 0 does not occur?

28. **QUEUING THEORY** Six persons are arranged in a line. What is the probability that two specific people, say A and B, are standing next to each other?

29. **LOTTERY** A box contains 500 envelopes, of which 50 have \$100 in cash, 75 have \$50 in cash, and 125 have \$25 in cash. If an envelope is selected at random from this box, what is the probability that it will contain at least \$50?

30. **JURY SELECTION** A jury of twelve people is selected from thirty people: fifteen women and fifteen men. What is the probability that the jury will have six men and six women?

31. **QUEUING THEORY** Three girls and three boys are randomly placed in six adjacent seats. What is the probability that the boys and girls will be in alternating seats?

32. **COMMITTEE MEMBERSHIP** A committee of four is chosen from three accountants and five actuaries. Find the probability that the committee consists of two accountants and two actuaries.

33. **EXTRA-SENSORY PERCEPTION** A magician claims to be able to read minds. To test this claim, five cards numbered 1 to 5 are used. A subject selects two cards from the five and concentrates on the numbers. What is the probability that the magician can correctly identify the two cards by just guessing?

34. **CARD GAMES** One card is randomly drawn from a regular deck of playing cards. The card is replaced and another card is drawn. Are the events independent? What is the probability that both cards drawn are aces?

35. **SCHEDULING** A meeting is scheduled by randomly choosing a weekday and then randomly choosing an hour between 8:00 A.M. and 4:00 P.M. What is the probability that Monday at 8:00 A.M. is chosen?

36. **NATIONAL DEFENSE** A missile radar detection system consists of two radar screens. The probability that any one of the radar screens will detect an incoming missile is 0.95. If radar detections are assumed to be independent events, what is the probability that a missile that enters the detection space of the radar will be detected?

37. **OIL INDUSTRY** An oil drilling venture involves drilling four wells in different parts of the country. For each well, the probability that it will be profitable is 0.10, and the probability that it will be unprofitable is 0.90. If these events are independent, what is the probability of drilling at least one unprofitable well?

38. **MANUFACTURING** A manufacturer of CD-ROMs claims that only 1 of every 1000 CD-ROMs manufactured is defective. If this claim is correct and if defective CD-ROMs are independent events, what is the probability that of the next three CD-ROMs produced, all are not defective?

39. **PREFERENCE TESTING** A software firm is considering marketing two newly designed spreadsheet programs, A and B. To test the appeal of the programs, the firm installs them in four corporations. After 2 weeks, the firm asks each corporation to evaluate each program. If the corporations have no preference, what is the probability that all four will choose product A?

40. **AGRICULTURE** A fruit grower claims that one-fourth of the orange trees in a grove crop have suffered frost damage. Find the probability that among eight orange trees, exactly three have frost damage.

41. **QUALITY CONTROL** A quality control inspector receives a shipment of 20 computer monitors. From the 20 monitors, the inspector randomly chooses 5 for inspection. If the probability of a monitor being defective is 0.05, what is the probability that at least one of the monitors chosen by the inspector is defective?

42. **LOTTERY** Consider a lottery that sells 1000 tickets and awards two prizes. If you purchase 10 tickets, what is the probability that you will win a prize?

43. **AIRLINE SCHEDULING** An airline estimates that 75% of the people who make a reservation for a certain flight will actually show up for the flight. Suppose the airline sells 25 tickets on this flight and the plane has room for 20 passengers. What is the probability that 21 or more people with tickets will show up for the flight?

44. **AIRLINE SCHEDULING** Suppose that an airplane's engines operate independently and that the probability that any

one engine will fail is 0.03. A plane can make a safe landing if at least one-half of its engines operate. Is a safe flight more likely to occur in a two-engine or a four-engine plane? Why?

SUPPLEMENTAL EXERCISES

45. **SPREAD OF A RUMOR** A club has nine members. One member starts a rumor by telling it to a second club member, who repeats the rumor to a third person, and so on. At each stage, the recipient of the rumor is chosen at random from the nine club members. What is the probability that the rumor will be told three times without returning to the originator?

46. **EXTRASENSORY PERCEPTION** As a test for extrasensory perception (ESP), ten cards, five black and five white, are shuffled, and then a person looks at each card. In another room, the ESP subject attempts to guess whether the card is black or white. The ESP subject must guess black five times and white five times. If the ESP subject has no extrasensory perception, what is the probability that the subject will correctly name eight of the ten cards?

47. **TELEPHONE NUMBER EXTENSIONS** The telephone extensions at a university are four-digit numbers chosen from the digits 1–9. If two telephone numbers are randomly chosen from the telephone book, what is the probability that the first three digits are different and the fourth digit matches the third digit?

48. **ARRANGING LETTERS OF A WORD** Each arrangement of the letters of the word "Tennessee" is written on a piece of paper, and all the pieces of paper are placed in a bowl. One piece of paper is selected at random. What is the probability that the first letter in the arrangement is a T?

PROJECTS

1. **MONTE HALL PROBLEM** The grand prize in a game show is behind one of three curtains. A contestant selects one of the three curtains, say curtain A. To add drama to the show, the game show host reveals a prize behind one of the other curtains. This prize is not the grand prize. Now the contestant has an opportunity to cancel the original choice (curtain A) and choose the remaining closed curtain or stay with the original choice. What is the probability that the contestant will now choose the grand prize? *Note:* This problem is sometimes referred to as the *Monte Hall* problem after the game show *Let's Make a Deal*. For more information on this problem, read the "Ask Marilyn" column (by syndicated columnist Marilyn vos Savant) in which this problem was discussed (see *Parade* magazine, September 9, 1990, p. 15).

2. **PROBABILITY AND AUTOMATIC GARAGE DOOR OPENERS** A home construction company builds 500 homes in a planned community, and each home has one garage with an automatic garage door opener. The garage door opener has six switches that can be set to either 0 or 1. Suppose a homeowner chooses some sequence, say 011101. Assuming that all the homes in the development are sold, what is the probability that at least two of the homeowners have chosen the same sequence and could therefore open their neighbor's garage?

3. WWW See our web site at http://college.hmco.com for a project dealing with overbooking by airlines.

EXPLORING CONCEPTS WITH TECHNOLOGY

Mathematical Expectation

Expectation E is a number used to determine the fairness of a gambling game. It is defined as the probability of winning a bet times the amount A available to win.

$$E = P \cdot A$$

A game is called fair if the expectation of the game equals the amount bet. For example, if you and a friend each bet \$1 on who can guess the side facing up on the flip of a coin, then the expectation is $E = \frac{1}{2} \cdot \$2 = \$1$. Because the amount of your bet equals the expectation, the game is fair.

When a game is unfair, it benefits one of the players. If you bet \$1 and your friend bets \$2 on who can guess the flip of a coin, your expectation is $E = \frac{1}{2} \cdot \$3 = \$1.50$. Because your expectation is greater than the amount you bet, the game is advantageous to you. Your friend's expectation is also \$1.50, which is less than the amount your friend bet. This is a disadvantage to your friend.

Keno is a game of chance played in many casinos. In this game, a large basket contains 80 balls numbered from 1 to 80. From these balls, the casino randomly chooses 20 balls. The number of ways in which the dealer can choose 20 balls from 80 is the number of combinations of 80 things chosen 20 at a time, or $C(80, 20)$.

In one particular game, a gambler can bet \$1 and mark five numbers. The gambler will win a prize if three of the five numbers marked are included in the 20 numbered balls chosen by the dealer. By the counting principle, there are $C(20, 3) \cdot C(60, 2) = 2{,}017{,}800$ ways the gambler can do this. The probability of this event is $\frac{C(20, 3) \cdot C(60, 2)}{C(80, 5)} \approx 0.0839$. The amount the gambler wins for this event is \$2 (the \$2 bet plus \$1 from the casino), so the expectation of the gambler is approximately \$.17 ($0.0839 \cdot \$2$).

Each casino has different rules and different methods of awarding prizes. The tables below give the prizes for a \$2 bet for some of the possible choices a gambler can make at four casinos. Complete the Expectation column. In each case, the Mark column indicates how many numbers the gambler marked, and the Catch column shows how many of the numbers marked by the gambler were also chosen by the dealer.

Casino 1

Mark	Catch	Win	Expectation
6	4	\$8	
6	5	\$176	
6	6	\$2960	

Casino 2

Mark	Catch	Win	Expectation
6	4	\$6	
6	5	\$160	
6	6	\$3900	

Casino 3

Mark	Catch	Win	Expectation
6	4	\$8	
6	5	\$180	
6	6	\$3000	

Casino 4

Mark	Catch	Win	Expectation
6	4	\$6	
6	5	\$176	
6	6	\$3000	

Adding the expectations in each column gives you the total expectation for marking six numbers. Find the total expectation for each casino. Which casino offers the gambler the greatest expectation?

CHAPTER 11 SUMMARY

11.1 Infinite Sequences and Summation Notation

- An infinite sequence is a function whose domain is the positive integers and whose range is a set of real numbers.
- An alternating sequence is one in which the signs of the terms alternate between positive and negative values.
- A recursively defined sequence is one in which each succeeding term of the sequence is defined by using some of the preceding terms.
- If n is a positive integer, then n factorial, $n!$, is the product of the first n positive integers.

$$n! = n(n-1)(n-2)\cdots 3\cdot 2\cdot 1$$

- If a_n is a sequence, then $S_n = \sum_{i=1}^{n} a_i$ is the nth partial sum of the sequence.

11.2 Arithmetic Sequences and Series

- Given that d is a real number, the sequence a_n is an arithmetic sequence if $a_{i+1} - a_i = d$ for all i. The number d is called the common difference for the sequence.
- The nth term of an arithmetic sequence with common difference of d is $a_n = a_1 + (n-1)d$.
- If a_n is an arithmetic sequence, then the nth partial sum S_n of the sequence is given by

$$S_n = \frac{n}{2}(a_1 + a_n)$$

11.3 Geometric Sequences and Series

- Given that $r \neq 0$ is a constant real number, the sequence a_n is a geometric sequence if $a_{i+1}/a_i = r$ for all positive integers i. The ratio r is called the common ratio for the geometric sequence.
- The nth term of the geometric sequence is $a_n = a_1 r^{n-1}$, where a_1 is the first term of the sequence and r is the common ratio.
- If a_n is a geometric sequence, then the nth partial sum of the sequence is given by

$$S_n = \frac{a_1(1-r^n)}{1-r} \qquad r \neq 1$$

- If $|r| < 1$, then the sum of an infinite geometric series is given by

$$S = \frac{a_1}{1-r}$$

11.4 Mathematical Induction

- **Principle of Mathematical Induction**
  Let P_n be a statement that involves positive integers. If
  1. P_1 is true, and
  2. The truth of P_k implies the truth of P_{k+1} then P_n is true for all positive integers.

11.5 The Binomial Theorem

- **Binomial Theorem for Positive Integers**
  If n is a positive integer, then

$$(a+b)^n = \sum_{i=0}^{n} \binom{n}{i} a^{n-i} b^i$$

- The ith term of the expansion of $(a+b)^n$ is

$$\binom{n}{i-1} a^{n-i+1} b^{i-1}$$

11.6 Permutations and Combinations

- The Fundamental Counting Principle is used to count the number of ways in which a sequence of n conditions can occur.
- A permutation is an arrangement of distinct objects in a definite order. The formula for the permutations of n distinct objects taken r at a time is

$$P(n, r) = \frac{n!}{(n-r)!}$$

- A combination is an arrangement of objects for which the order of the selection is not important. The formula for the number of combinations of n objects taken r at a time is

$$C(n, r) = \frac{n!}{r!\,(n-r)!}$$

11.7 Introduction to Probability

- Probability is the mathematical study of random patterns. The sample space of an experiment is the set of all possible outcomes of that experiment. An event is any subset of a sample space.
- If S is the sample space of an experiment and E is an event in the sample space, then the probability of the event is given by

$$P(E) = \frac{n(E)}{n(S)}$$

where $n(E)$ and $n(S)$ are the number of elements in E and S, respectively.

- **Addition Rule for Probabilities**
  If E_1 and E_2 are two events, then
  $$P(E_1 \cup E_2) = P(E_1) + P(E_2) - P(E_1 \cap E_2)$$
- **Probability Rule for Independent Events**
  If E_1 and E_2 are two independent events, then the probability that both E_1 and E_2 will occur is
  $$P(E_1) \cdot P(E_2)$$
- **Binomial Probability Formula**
  Let an experiment consist of n trials for which the probability of success on a single trial is p and the probability of failure is $q = 1 - p$. Then the probability of k successes in n trials is given by
  $$\binom{n}{k} p^k q^{n-k}$$

CHAPTER 11 TRUE/FALSE EXERCISES

In Exercises 1 to 15, answer true or false. If the statement is false, give an example to show that the statement is false.

1. $0! \cdot 4! = 0$
2. $\left(\sum_{i=1}^{3} a_i\right)\left(\sum_{i=1}^{3} b_i\right) = \sum_{i=1}^{3} a_i b_i$
3. $\dfrac{n(n-1)(n-2)\cdots(n-k+1)}{k!} = C(n, k)$
4. No two terms of a sequence can be equal.
5. $1, 8, 27, 64, \ldots, k^3, \ldots$ is a geometric sequence.
6. $a_1 = 2, a_{n+1} = a_n - 3$ defines an arithmetic sequence.
7. $0.\overline{9} = 1$
8. Adding all the terms of an infinite sequence produces an infinite sum.
9. Because the first step of an induction proof is normally easy, this step can be omitted.
10. In the expansion of $(a + b)^8$, the exponent of a for the fifth term is 5.
11. The counting principle states that if there are n ways to satisfy one condition and m ways to satisfy a second condition, then there are $n + m$ ways to satisfy both conditions.
12. The number of permutations of n things taken r at a time is given by $n!/r!$.
13. If E is an event in a sample space, then $0 \le P(E) \le 1$, where $P(E)$ is the probability of E.
14. If A and B are mutually exclusive events, then $P(A \cap B) = 1$.
15. If a fair coin is tossed five times, then the probability of observing HHHHH is the same as the probability of observing HTHHT.

CHAPTER 11 REVIEW EXERCISES

In Exercises 1 to 20, find the third and seventh terms of the sequence defined by a_n.

1. $a_n = n^2$
2. $a_n = n!$
3. $a_n = 3n + 2$
4. $a_n = 1 - 2n$
5. $a_n = 2^{-n}$
6. $a_n = 3^n$
7. $a_n = \dfrac{1}{n!}$
8. $a_n = \dfrac{1}{n}$
9. $a_n = \left(\dfrac{2}{3}\right)^n$
10. $a_n = \left(-\dfrac{4}{3}\right)^n$
11. $a_1 = 2, a_n = 3a_{n-1}$
12. $a_1 = -1, a_n = 2a_{n-1}$
13. $a_1 = 1, a_n = -na_{n-1}$
14. $a_1 = 2, a_n = n^2 a_{n-1}$
15. $a_1 = 4, a_n = a_{n-1} + 2$
16. $a_1 = 3, a_n = a_{n-1} - 3$
17. $a_1 = 1, a_2 = 2, a_n = a_{n-1}a_{n-2}$
18. $a_1 = 1, a_2 = 2, a_n = a_{n-1}/a_{n-2}$
19. $a_1 = -1, a_n = 3na_{n-1}$
20. $a_1 = 2, a_n = -2na_{n-1}$

21–40. Classify each sequence defined in Exercises 1 to 20 as arithmetic, geometric, or neither.

In Exercises 41 to 56, find the indicated sum of the series.

41. $\sum_{n=1}^{9} (2n - 3)$
42. $\sum_{i=1}^{11} (1 - 3i)$
43. $\sum_{k=1}^{8} (4k + 1)$
44. $\sum_{i=1}^{10} (i^2 + 3)$
45. $\sum_{n=1}^{6} 3 \cdot 2^n$
46. $\sum_{i=1}^{5} 2 \cdot 4^{i-1}$
47. $\sum_{k=1}^{9} (-1)^k 3^k$
48. $\sum_{i=1}^{8} (-1)^{i+1} 2^i$
49. $\sum_{i=1}^{10} \left(\dfrac{2}{3}\right)^i$
50. $\sum_{i=1}^{11} \left(\dfrac{3}{2}\right)^i$
51. $\sum_{n=1}^{9} \dfrac{(-1)^{n+1}}{n^2}$
52. $\sum_{k=1}^{5} \dfrac{(-1)^{k+1}}{k!}$

53. $\sum_{n=1}^{\infty}\left(\frac{1}{4}\right)^n$

54. $\sum_{i=1}^{\infty}\left(-\frac{5}{6}\right)^i$

55. $\sum_{k=1}^{\infty}\left(-\frac{4}{5}\right)^k$

56. $\sum_{j=0}^{\infty}\left(\frac{1}{5}\right)^j$

In Exercises 57 to 64, prove each statement by mathematical induction.

57. $\sum_{i=1}^{n}(5i + 1) = \frac{n(5n + 7)}{2}$

58. $\sum_{i=1}^{n}(3 - 4i) = n(1 - 2n)$

59. $\sum_{i=0}^{n}\left(-\frac{1}{2}\right)^i = \frac{2[1 - (-1/2)^{n+1}]}{3}$

60. $\sum_{i=0}^{n}(-1)^i = \frac{1 - (-1)^{n+1}}{2}$

61. $n^n \geq n!$

62. $n! > 4^n, \quad n \geq 9$

63. 3 is a factor of $n^3 + 2n$ for all positive integers n.

64. Let $a_1 = \sqrt{2}$ and $a_n = (\sqrt{2})^{a_{n-1}}$. Prove that $a_n < 2$ for all positive integers n.

In Exercises 65 to 68, use the Binomial Theorem to expand each binomial.

65. $(4a - b)^5$

66. $(x + 3y)^6$

67. $(\sqrt{a} + 2\sqrt{b})^8$

68. $\left(2x - \frac{1}{2x}\right)^7$

69. Find the fifth term in the expansion of $(3x - 4y)^7$.

70. Find the eighth term in the expansion of $(1 - 3x)^9$.

71. **COMPUTER PASSWORDS** A computer password consists of eight letters. How many passwords are possible? Assume there is no difference between lower-case and upper-case letters.

72. **SERIAL NUMBERS** The serial number on an airplane consists of the letter N, followed by six numerals, followed by one letter. How many serial numbers are possible?

73. **COMMITTEE MEMBERSHIP** From a committee of fifteen members, a president, vice president, and treasurer are elected. In how many ways can this be accomplished?

74. **SCHEDULING** The emergency staff for a hospital consists of four supervisors and twelve regular employees. How many shifts of four people can be formed if each shift must contain exactly one supervisor?

75. **COMMITTEE MEMBERSHIP** From twelve people, a committee of five people is formed. In how many ways can this be accomplished if there are two people among the twelve who refuse to serve together on the committee?

76. **ACCEPTANCE SAMPLING** A shipment of ten calculators contains two defective ones. A quality control inspector randomly chooses four of the calculators for testing. What is the probability that the inspector will choose one defective calculator?

77. **SUMS OF COINS** A nickel, dime, and quarter are tossed. What is the probability that the nickel and dime will show heads and the quarter will show tails? What is the probability that only one of the coins will show tails?

78. **ARRANGEMENTS OF CARDS** A deck of ten cards contains five red and five black cards. If four cards are drawn from the deck, what is the probability that two are red and two are black?

79. **NUMBER THEORY** For the 1000 numbers 000 to 999, what is the probability that the middle digit is greater than the other two digits?

80. **NUMBER THEORY** Two numbers are chosen, with replacement, from the digits 1, 2, 3, 4, 5, and 6, and their sum is recorded. Now two more digits are selected and their sum noted. This process continues until the sum is 7 or the original sum is obtained. If the original sum was 9, what is the probability of having another sum of 9 before having a sum of 7? (*Hint:* Assume the events are independent. The probability can be found by summing an infinite geometric series.)

81. **CARD GAMES** Which of the following has the greater probability: drawing an ace and a ten-card (ten, jack, queen, or king) from a regular deck of fifty-two playing cards or drawing an ace and a ten-card from two decks of regular playing cards?

82. **NUMBER THEORY** From the digits 1, 2, 3, 4, and 5, two numbers are chosen without replacement. What is the probability that the second number is greater than the first number?

83. **EMPLOYEE BADGES** A room contains twelve people who are wearing badges numbered 1 to 12. If three people are randomly selected, what is the probability that the person wearing badge 6 will be included?

CHAPTER 11 TEST

In Exercises 1 to 3, find the third and fifth terms of the sequence defined by a_n.

1. $a_n = \frac{2^n}{n!}$

2. $a_n = \frac{(-1)^{n+1}}{2n}$

3. $a_1 = 3, a_n = 2a_{n-1}$

In Exercises 4 to 6, classify each sequence as an arithmetic sequence, a geometric sequence, or neither.

4. $a_n = -2n + 3$

5. $a_n = 2n^2$

6. $a_n = \frac{(-1)^{n-1}}{3^n}$

In Exercises 7 to 9, find the indicated sum of the series.

7. $\sum_{i=1}^{6} \frac{1}{i}$

8. $\sum_{j=1}^{10} \frac{1}{2^j}$

9. $\sum_{k=1}^{20} (3k - 2)$

10. The third term of an arithmetic sequence is 7 and the eighth term is 22. Find the twentieth term.

11. Find the sum of the infinite geometric series given by $\sum_{k=1}^{\infty} \left(\frac{3}{8}\right)^k$.

12. Write $0.\overline{15}$ as the quotient of integers in simplest form.

In Exercises 13 and 14, prove the statement by mathematical induction.

13. $\sum_{i=1}^{n} (2 - 3i) = \frac{n(1 - 3n)}{2}$

14. $n! > 3^n, \quad n \geq 7$

15. Write the binomial expansion of $(x - 2y)^5$.

16. Write the binomial expansion of $\left(x + \frac{1}{x}\right)^6$.

17. Find the sixth term in the expansion of $(3x + 2y)^8$.

18. Three cards are randomly chosen from a regular deck of playing cards. In how many ways can the cards be chosen?

19. A serial number consists of seven characters. The first three characters are upper-case letters of the alphabet. The next two characters are selected from the digits 1 through 9. The last two characters are upper-case letters of the alphabet. How many serial numbers are possible if no letter or number can be used twice in the same serial number?

20. Five cards are randomly selected from a deck of cards containing eight black cards and ten red cards. What is the probability that three black cards and two red cards are selected?

SOLUTIONS TO SELECTED EXERCISES

Exercise Set P.1, page 8

2. a. Integers: 21, 53

b. Rational numbers: $5.\overline{17}$, -4.25, $1/4$, 21, 53, 0.45454545...

c. Irrational numbers: π

d. Real numbers: All of the given numbers are real numbers.

e. Prime numbers: 53

f. Composite numbers: 21

14. $\{0, 1, 2, 3, 4\} \cap \{1, 3, 6, 10\} = \{1, 3\}$

26. Commutative property of addition

28. Symmetric property of equality

40. $\dfrac{2a}{5} + \dfrac{3a}{7} = \dfrac{2a \cdot 7}{5 \cdot 7} + \dfrac{3a \cdot 5}{7 \cdot 5} = \dfrac{29a}{35}$

Exercise Set P.2, page 15

16. 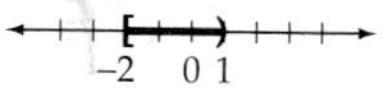 $[-2, 1)$

56. $|x + 6| + |x - 2| = x + 6 - (x - 2)$
$= x + 6 - x + 2$
$= 8$

64. $|-5 - 8| = 13$

80. $|y + 3| > 6$

Exercise Set P.3, page 28

24. $\left(\dfrac{2ab^2c^3}{5ab^2}\right)^3 = \left(\dfrac{2c^3}{5}\right)^3 = \dfrac{8c^9}{125}$

36. $\dfrac{x^{1/3}y^{5/6}}{x^{3/2}y^{1/6}} = x^{1/3-3/2}y^{5/6-1/6} = x^{2/6-9/6}y^{4/6} = \dfrac{y^{2/3}}{x^{7/6}}$

76. $\sqrt{18x^2y^5} = \sqrt{9x^2y^4}\,\sqrt{2y} = 3|x|y^2\sqrt{2y}$

84. $-3x\sqrt[3]{54x^4} + 2\sqrt[3]{16x^7} = -3x\sqrt[3]{3^3 \cdot 2x^4} + 2\sqrt[3]{2^4x^7}$
$= -3x\sqrt[3]{3^3x^3}\,\sqrt[3]{2x} + 2\sqrt[3]{2^3x^6}\,\sqrt[3]{2x}$
$= -3x\left(3x\sqrt[3]{2x}\right) + 2\left(2x^2\sqrt[3]{2x}\right)$
$= -9x^2\sqrt[3]{2x} + 4x^2\sqrt[3]{2x}$
$= -5x^2\sqrt[3]{2x}$

92. $(3\sqrt{5y} - 4)^2 = (3\sqrt{5y} - 4)(3\sqrt{5y} - 4)$
$= 9 \cdot 5y - 12\sqrt{5y} - 12\sqrt{5y} + 16$
$= 45y - 24\sqrt{5y} + 16$

104. $\dfrac{2}{\sqrt[4]{4y}} = \dfrac{2}{\sqrt[4]{4y}} \cdot \dfrac{\sqrt[4]{4y^3}}{\sqrt[4]{4y^3}} = \dfrac{2\sqrt[4]{4y^3}}{2y} = \dfrac{\sqrt[4]{4y^3}}{y}$

110. $\dfrac{5}{\sqrt{y} - \sqrt{3}} \cdot \dfrac{(\sqrt{y} + \sqrt{3})}{\sqrt{y} + \sqrt{3}} = \dfrac{5\sqrt{y} + 5\sqrt{3}}{y - 3}$

Exercise Set P.4, page 36

24. $(5y^2 - 7y + 3) + (2y^2 + 8y + 1) = 7y^2 + y + 4$

32. $(5x - 7)(3x^2 - 8x - 5)$
$= 15x^3 - 40x^2 - 25x - 21x^2 + 56x + 35$
$= 15x^3 - 61x^2 + 31x + 35$

56. $(4x^2 - 3y)(4x^2 + 3y) = (4x^2)^2 - (3y)^2 = 16x^4 - 9y^2$

66. $-x^2 - 5x + 4 = -(-5)^2 - 5(-5) + 4$
$= -25 + 25 + 4 = 4$

76. $\dfrac{1}{6}n^3 - \dfrac{1}{2}n^2 + \dfrac{1}{3}n = \dfrac{1}{6}(21)^3 - \dfrac{1}{2}(21)^2 + \dfrac{1}{3}(21)$
$= 1330$ committees

78. a. $4.3 \times 10^{-6}(1000)^2 - 2.1 \times 10^{-4}(1000)$
$= 4.09$ seconds

b. $4.3 \times 10^{-6}(5000)^2 - 2.1 \times 10^{-4}(5000)$
$= 106.45$ seconds

c. $4.3 \times 10^{-6}(10{,}000)^2 - 2.1 \times 10^{-4}(10{,}000)$
$= 427.9$ seconds

Exercise Set P.5, page 47

6. $6a^3b^2 - 12a^2b + 72ab^3 = 6ab(a^2b - 2a + 12b^2)$

12. $b^2 + 12b - 28 = (b + 14)(b - 2)$

16. $57y^2 + y - 6 = (19y - 6)(3y + 1)$

24. $b^2 - 4ac = 8^2 - 4(16)(-35) = 2304 = 48^2$
The trinomial is factorable over the integers.

32. $81b^2 - 16c^2 = (9b - 4c)(9b + 4c)$

42. $b^2 - 24b + 144 = (b - 12)^2$

48. $b^3 + 64 = (b + 4)(b^2 - 4b + 16)$

58. $a^2y^2 - ay^3 + ac - cy = ay^2(a - y) + c(a - y)$
$= (a - y)(ay^2 + c)$

64. $81y^4 - 16 = (9y^2 - 4)(9y^2 + 4)$
$= (3y - 2)(3y + 2)(9y^2 + 4)$

Exercise Set P.6, page 56

2. $\dfrac{2x^2 - 5x - 12}{2x^2 + 5x + 3} = \dfrac{(2x + 3)(x - 4)}{(2x + 3)(x + 1)} = \dfrac{x - 4}{x + 1}$

16. $\dfrac{x^2 - 16}{x^2 + 7x + 12} \cdot \dfrac{x^2 - 4x - 21}{x^2 - 4x}$

$= \dfrac{(x - 4)(x + 4)(x + 3)(x - 7)}{(x + 3)(x + 4)x(x - 4)} = \dfrac{x - 7}{x}$

30. $\dfrac{3y - 1}{3y + 1} - \dfrac{2y - 5}{y - 3} = \dfrac{(3y - 1)(y - 3)}{(3y + 1)(y - 3)} - \dfrac{(2y - 5)(3y + 1)}{(y - 3)(3y + 1)}$

$= \dfrac{(3y^2 - 10y + 3) - (6y^2 - 13y - 5)}{(3y + 1)(y - 3)}$

$= \dfrac{-3y^2 + 3y + 8}{(3y + 1)(y - 3)}$

42. $\dfrac{3 - \frac{2}{a}}{5 + \frac{3}{a}} = \dfrac{\left(3 - \frac{2}{a}\right)a}{\left(5 + \frac{3}{a}\right)a} = \dfrac{3a - 2}{5a + 3}$

60. $\dfrac{e^{-2} - f^{-1}}{ef} = \dfrac{\frac{1}{e^2} - \frac{1}{f}}{ef} = \dfrac{f - e^2}{e^2 f} \div \dfrac{ef}{1}$

$= \dfrac{f - e^2}{e^2 f} \cdot \dfrac{1}{ef} = \dfrac{f - e^2}{e^3 f^2}$

64. a. $\dfrac{v_1 + v_2}{1 + \frac{v_1 v_2}{c^2}} = \dfrac{1.2 \times 10^8 + 2.4 \times 10^8}{1 + \frac{(1.2 \times 10^8)(2.4 \times 10^8)}{(6.7 \times 10^8)^2}} \approx 3.4 \times 10^8 \text{ mph}$

b. $\dfrac{v_1 + v_2}{1 + \frac{v_1 \cdot v_2}{c^2}} = \dfrac{c^2(v_1 + v_2)}{c^2\left(1 + \frac{v_1 \cdot v_2}{c^2}\right)} = \dfrac{c^2(v_1 + v_2)}{c^2 + v_1 \cdot v_2}$

Exercise Set 1.1, page 69

2. $-3y + 20 = 2$

$-3y = -18$

$y = 6$

12. $\dfrac{1}{2}x + 7 - \dfrac{1}{4}x = \dfrac{19}{2}$

$4\left(\dfrac{1}{2}x + 7 - \dfrac{1}{4}x\right) = 4\left(\dfrac{19}{2}\right)$

$2x + 28 - x = 38$

$x = 38 - 28$

$x = 10$

18. $5(x + 4)(x - 4) = (x - 3)(5x + 4)$

$5(x^2 - 16) = 5x^2 - 11x - 12$

$5x^2 - 80 = 5x^2 - 11x - 12$

$-80 + 12 = -11x$

$-68 = -11x$

$\dfrac{68}{11} = x$

24. $2x + \dfrac{1}{3} = \dfrac{6x + 1}{3}$ **• Rewrite the left side.**

$\dfrac{6x + 1}{3} = \dfrac{6x + 1}{3}$

This equation is an identity.

30. $\dfrac{4}{y + 2} = \dfrac{7}{y - 4} \qquad y \neq 2, y \neq 4$

$4(y - 4) = 7(y + 2)$

$4y - 16 = 7y + 14$

$4y - 7y = 14 + 16$

$-3y = 30$

$y = -10$

50. $|2x - 3| = 21$

$2x - 3 = 21$ or $2x - 3 = -21$

$2x = 24 \qquad 2x = -18$

$x = 12 \qquad x = -9$

62. Patents (in thousands) $= 5.4x + 110$

$150 = 5.4x + 110$ **• Substitute 150 for "patents."**

$40 = 5.4x$

$x = \dfrac{40}{5.4}$

$x \approx 7.4$

Because x is approximately 7.4, the year in $1993 + 7.4 = 2000.4$. The number of patents will first exceed 150,000 in 2000.

Exercise Set 1.2, page 79

4. $A = P + Prt$

$A = P(1 + rt)$ **• Factor.**

$P = \dfrac{A}{(1 + rt)}$

22. $P = 2l + 2w, \qquad w = \dfrac{1}{2}l + 1$

$110 = 2l + 2\left(\dfrac{1}{2}l + 1\right)$ **• Substitute for *w*.**

$$110 = 2l + l + 2 \quad \text{• Simplify.}$$
$$108 = 3l$$
$$36 = l$$
$$l = 36 \text{ meters}$$
$$w = \frac{1}{2}l + 1 = \frac{1}{2}(36) + 1 = 19 \text{ meters}$$

28. Let t_1 = the time it takes to travel to the island.
Let t_2 = the time it takes to make the return trip.

$$t_1 + t_2 = 7.5$$
$$t_2 = 7.5 - t_1$$
$$15t_1 = 10t_2$$
$$15t_1 = 10(7.5 - t_1) \quad \text{• Substitute for } t_2.$$
$$15t_1 = 75 - 10t_1$$
$$25t_1 = 75$$
$$t_1 = 3 \text{ hours}$$
$$D = 15t_1 = 15(3) = 45 \text{ nautical miles}$$

36. Let x = the number of glasses of orange juice.

$$\text{Profit} = \text{revenue} - \text{cost}$$
$$\$2337 = 0.75x - 0.18x$$
$$2337 = 0.57x$$
$$x = \frac{2337}{0.57}$$
$$x = 4100$$

40. Let x = the amount of money invested at 5%.

5%	x
7%	$7500 - x$

$$0.05x + 0.07(7500 - x) = 405$$
$$0.05x + 525 - 0.07x = 405$$
$$-0.02x = -120$$
$$x = 6000$$
$$7500 - x = 1500$$

\$6000 was invested at 5%. \$1500 was invested at 7%.

44. Let x = the number of liters of the 40% solution to be mixed with the 24% solution.

0.40	x
0.24	4
0.30	$4 + x$

$$0.40x + 0.24(4) = 0.30(4 + x)$$
$$0.40x + 0.96 = 1.2 + 0.30x$$
$$0.10x = 0.24$$
$$x = 2.4$$

Thus 2.4 liters of 40% sulfuric acid should be mixed with 4 liters of a 24% sulfuric acid solution, to produce the 30% solution.

54. Let x = the number of hours needed to print the report if both the printers are used.
Printer A prints 1/3 of the report every hour.
Printer B prints 1/4 of the report every hour.
Thus

$$\frac{1}{3}x + \frac{1}{4}x = 1$$
$$4x + 3x = 12 \cdot 1$$
$$7x = 12$$
$$x = \frac{12}{7} \approx 1.71$$

It would take approximately 1.71 hours to print the report.

Exercise Set 1.3, page 96

6. $12w^2 - 41w + 24 = 0$

$$(4w - 3)(3w - 8) = 0$$
$$4w - 3 = 0 \quad \text{or} \quad 3w - 8 = 0$$
$$w = \frac{3}{4} \qquad w = \frac{8}{3}$$

14. $y^2 = 225$

$$y = \pm\sqrt{225}$$
$$y = \pm 15$$

22. $(x + 2)^2 + 28 = 0$

$$(x + 2)^2 = -28$$
$$\sqrt{(x + 2)^2} = \sqrt{-28}$$
$$x + 2 = \pm i\sqrt{28} = \pm 2i\sqrt{7}$$
$$x = -2 \pm 2i\sqrt{7}$$

The solutions are $-2 - 2i\sqrt{7}$ and $-2 + 2i\sqrt{7}$.

32. $(5 + 3i)(-2 - 4i) = -10 - 20i - 6i - 12i^2$

$$= -10 - 26i - 12(-1)$$
$$= -10 - 26i + 12 = 2 - 26i$$

40. $\dfrac{5 - 7i}{5 + 7i} = \dfrac{(5 - 7i)(5 - 7i)}{(5 + 7i)(5 - 7i)} = \dfrac{-24 - 70i}{74}$

$$= -\frac{12}{37} - \frac{35}{37}i$$

56. $x^2 - 6x + 10 = 0$

$$x^2 - 6x = -10$$
$$x^2 - 6x + 9 = -10 + 9$$
$$(x - 3)^2 = -1$$
$$\sqrt{(x - 3)^2} = \sqrt{-1}$$
$$x - 3 = \pm i$$
$$x = 3 \pm i$$

The solutions are $3 - i$ and $3 + i$.

60. $2x^2 + 10x - 3 = 0$

$$2x^2 + 10x = 3$$
$$2(x^2 + 5x) = 3$$
$$x^2 + 5x = \frac{3}{2}$$

$$x^2 + 5x + \frac{25}{4} = \frac{3}{2} + \frac{25}{4}$$
$$\left(x + \frac{5}{2}\right)^2 = \frac{31}{4}$$
$$x + \frac{5}{2} = \pm\sqrt{\frac{31}{4}}$$
$$x = -\frac{5}{2} \pm \frac{\sqrt{31}}{2}$$
$$x = \frac{-5 + \sqrt{31}}{2} \quad \text{or} \quad x = \frac{-5 - \sqrt{31}}{2}$$

70. $2x^2 + 4x - 1 = 0$
$$x = \frac{-4 \pm \sqrt{4^2 - 4(2)(-1)}}{4}$$
$$x = \frac{-4 \pm \sqrt{16 + 8}}{4} = \frac{-4 \pm \sqrt{24}}{4}$$
$$x = \frac{-4 \pm 2\sqrt{6}}{4} = \frac{-2 \pm \sqrt{6}}{2}$$
$$x = \frac{-2 + \sqrt{6}}{2} \quad \text{or} \quad x = \frac{-2 - \sqrt{6}}{2}$$

78. $x^2 + 3x - 11 = 0$

$b^2 - 4ac = 3^2 - 4(1)(-11) = 9 + 44 = 53 > 0$

Thus the equation has two distinct real roots.

84. $\left(10 \text{ feet} + \frac{1}{4} \text{ inch}\right)^2 = (10 \text{ feet})^2 + x^2$
$$120.25^2 = 120^2 + x^2 \quad \text{• Change feet to inches.}$$
$$120.25^2 - 120^2 = x^2$$
$$\sqrt{120.25^2 - 120^2} = x$$
$$7.75 = x$$

To the nearest inch, the concrete will rise 8 inches.

90. Let P = perimeter and A = area.

$A = 4800 = lw$
$$l = \frac{4800}{w}$$
$$P = 4w + 2l = 400$$
$$2w + l = 200$$
$$2w + \frac{4800}{w} = 200 \quad \text{• Substitute for } l.$$
$$2w^2 + 4800 = 200w$$
$$w^2 - 100w + 2400 = 0$$
$$(w - 60)(w - 40) = 0$$
$$w = 60 \quad \text{or} \quad w = 40$$
$$l = \frac{4800}{60} = 80 \quad \text{or} \quad l = \frac{4800}{40} = 120$$

There are two solutions: 60 yards × 80 yards or 40 yards × 120 yards.

Exercise Set 1.4, page 105

6.
$$x^4 - 36x^2 = 0$$
$$x^2(x^2 - 36) = 0$$
$$x^2(x - 6)(x + 6) = 0$$
$$x = 0, x = 6, x = -6$$

14. $\sqrt{10 - x} = 4$ *Check:* $\sqrt{10 - (-6)} = 4$
$$10 - x = 16 \qquad \sqrt{16} = 4$$
$$-x = 6 \qquad 4 = 4$$
$$x = -6$$

The solution is -6.

16.
$$x = \sqrt{5 - x} + 5$$
$$(x - 5)^2 = \left(\sqrt{5 - x}\right)^2$$
$$x^2 - 10x + 25 = 5 - x$$
$$x^2 - 9x + 20 = 0$$
$$(x - 5)(x - 4) = 0$$
$$x = 5 \quad \text{or} \quad x = 4$$

Check: $5 = \sqrt{5 - 5} + 5$ $4 = \sqrt{5 - 4} + 5$
$$5 = 0 + 5 \qquad 4 = 1 + 5$$
$$5 = 5 \qquad 4 = 6 \quad \text{False}$$

The solution is 5.

20.
$$\sqrt{x + 7} - 2 = \sqrt{x - 9}$$
$$\left(\sqrt{x + 7} - 2\right)^2 = \left(\sqrt{x - 9}\right)^2$$
$$x + 7 - 4\sqrt{x + 7} + 4 = x - 9$$
$$-4\sqrt{x + 7} = -20$$
$$\left(\sqrt{x + 7}\right)^2 = (5)^2$$
$$x + 7 = 25$$
$$x = 18$$

Check: $\sqrt{18 + 7} - 2 = \sqrt{18 - 9}$
$$\sqrt{25} - 2 = \sqrt{9}$$
$$5 - 2 = 3$$
$$3 = 3$$

The solution is 18.

32. $(4z + 7)^{1/3} = 2$ *Check:* $\left[4\left(\frac{1}{4}\right) + 7\right]^{1/3} = 2$
$$[(4z + 7)^{1/3}]^3 = 2^3 \qquad 8^{1/3} = 2$$
$$4z + 7 = 8 \qquad 2 = 2$$
$$4z = 1$$
$$z = \frac{1}{4}$$

The solution is $\frac{1}{4}$.

42. $x^4 - 10x^2 + 9 = 0$ • **Let $u = x^2$.**

$u^2 - 10u + 9 = 0$

$(u - 9)(u - 1) = 0$

$u = 9$ or $u = 1$

$x^2 = 9$ $\quad x^2 = 1$

$x = \pm 3$ $\quad x = \pm 1$

The solutions are 3, −3, 1, and −1.

52. $6x^{2/3} - 7x^{1/3} - 20 = 0$ • **Let $u = x^{1/3}$.**

$6u^2 - 7u - 20 = 0$

$(3u + 4)(2u - 5) = 0$

$u = -\frac{4}{3}$ or $u = \frac{5}{2}$

$x^{1/3} = -\frac{4}{3}$ $\quad x^{1/3} = \frac{5}{2}$

$(x^{1/3})^3 = \left(-\frac{4}{3}\right)^3$ $\quad (x^{1/3})^3 = \left(\frac{5}{2}\right)^3$

$x = -\frac{64}{27}$ $\quad x = \frac{125}{8}$

The solutions are $-64/27$ and $125/8$.

Exercise Set 1.5, page 117

8. $-4(x - 5) \ge 2x + 15$

$-4x + 20 \ge 2x + 15$

$-6x \ge -5$

$x \le \frac{5}{6}$

The solution set is $\{x \mid x \le 5/6\}$.

12. $2x + 5 > -16$ and $2x + 5 < 9$

$2x > -21$ and $2x < 4$

$x > -\frac{21}{2}$ and $x < 2$

$$\left\{x \mid x > -\frac{21}{2}\right\} \cap \{x \mid x < 2\} = \left\{x \,\middle|\, -\frac{21}{2} < x < 2\right\}$$

The solution set is $\{x \mid -21/2 < x < 2\}$.

22. $|2x - 9| < 7$

$-7 < 2x - 9 < 7$

$2 < 2x < 16$

$1 < x < 8$

In interval notation, the solution set is $(1, 8)$.

26. $|2x - 5| \ge 1$

$2x - 5 \le -1$ or $2x - 5 \ge 1$

$2x \le 4$ $\quad 2x \ge 6$

$x \le 2$ $\quad x \ge 3$

In interval notation, the solution set is $(-\infty, 2] \cup [3, \infty)$.

42. $x^2 + 5x + 6 < 0$

$(x + 2)(x + 3) = 0$

$x = -2$ and $x = -3$ • **Critical values**

Use a test number from each of the intervals $(-\infty, -3)$, $(-3, -2)$, and $(-2, \infty)$ to determine where $x^2 + 5x + 6$ is negative.

+ + + + + + + | − | + + + + +

−3 −2 0

In interval notation the solution set is $(-3, -2)$.

56. $\frac{3x + 1}{x - 2} \ge 4$

$\frac{3x + 1}{x - 2} - 4 \ge 0$

$\frac{3x + 1 - 4(x - 2)}{x - 2} \ge 0$

$\frac{-x + 9}{x - 2} \ge 0$

$x = 2$ and $x = 9$ • **Critical values**

Use a test number from each of the intervals $(-\infty, 2)$, $(2, 9)$, and $(9, \infty)$ to determine where $(-x + 9)/(x - 2)$ is positive. The solution set is $(2, 9]$.

74. Let m = the number of miles driven.
Company A: $29 + 0.12m$
Company B: $22 + 0.21m$

$29 + 0.12m < 22 + 0.21m$

$77.\overline{7} < m$

Company A is less expensive if you drive at least 78 miles.

78. $41 \le F \le 68$

$41 \le \frac{9}{5}C + 32 \le 68$

$9 \le \frac{9}{5}C \le 36$

$\frac{5}{9}(9) \le \left(\frac{5}{9}\right)\left(\frac{9}{5}\right)C \le \frac{5}{9}(36)$

$5 \le C \le 20$

The Celsius temperature is between 5°C and 20°C.

Exercise Set 1.6, page 125

22. $d = kw$

$6 = k \cdot 80$

$\frac{6}{80} = k$

$$k = \frac{3}{40}$$

Thus $d = \frac{3}{40} \cdot 100 = 7.5$ inches.

24. $r = kv^2$

$140 = k \cdot 60^2$

$\frac{140}{60^2} = k$

$\frac{7}{180} = k$

Thus $r = \frac{7}{180} \cdot 65^2 \approx 164$ feet.

28. $I = \frac{k}{d^2}$

$50 = \frac{k}{10^2}$

$5000 = k$

Thus $I = \frac{5000}{d^2} = \frac{5000}{15^2} = \frac{5000}{225} \approx 22.2$ footcandles.

30. $L = kwd^2$

$200 = k \cdot 2 \cdot 6^2$

$k = \frac{200}{2 \cdot 6^2} = \frac{25}{9}$

Thus $L = \frac{25}{9} \cdot 4 \cdot 4^2 = \frac{1600}{9} \approx 177.8$ pounds.

34. $L = k\frac{wd^2}{l}$

$800 = k\frac{4 \cdot 8^2}{12}$

$\frac{12 \cdot 800}{4 \cdot 8^2} = k$

$37.5 = k$

Thus $L = 37.5\frac{3.5 \cdot 6^2}{16} = 295.3125 \approx 295$ pounds.

Exercise Set 2.1, page 144

26.

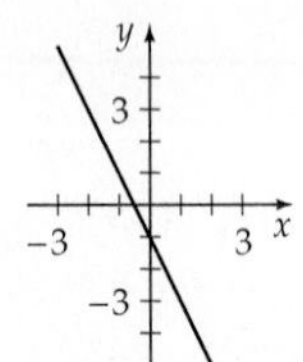

30.

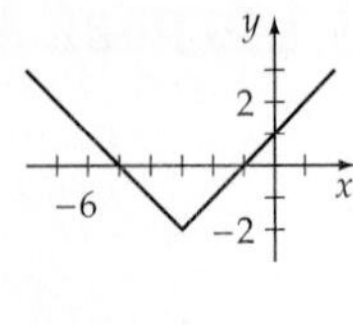

32.

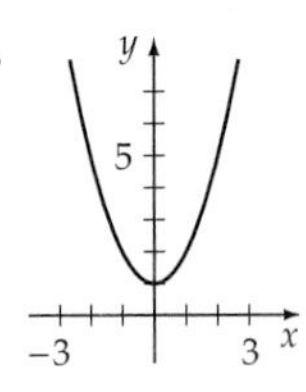

40. y-intercept: $(0, -15/4)$
x-intercept: $(5, 0)$

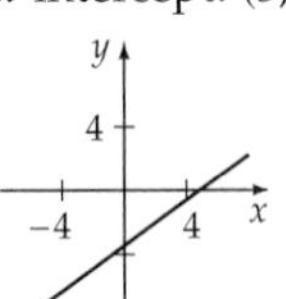

66. $r = \sqrt{(1 - (-2))^2 + (7 - 5)^2}$
$= \sqrt{9 + 4} = \sqrt{13}$

Using the standard form

$(x - h)^2 + (y - k)^2 = r^2$

with $h = -2$, $k = 5$, and $r = \sqrt{13}$ yields

$(x + 2)^2 + (y - 5)^2 = (\sqrt{13})^2$

68. $x^2 + y^2 - 6x - 4y + 12 = 0$

$x^2 - 6x + y^2 - 4y = -12$

$x^2 - 6x + 9 + y^2 - 4y + 4 = -12 + 9 + 4$

$(x - 3)^2 + (y - 2)^2 = 1^2$

center $(3, 2)$, radius 1

Exercise Set 2.2, page 158

2. Given $g(x) = 2x^2 + 3$

a. $g(3) = 2(3)^2 + 3 = 18 + 3 = 21$

b. $g(-1) = 2(-1)^2 + 3 = 2 + 3 = 5$

c. $g(0) = 2(0)^2 + 3 = 0 + 3 = 3$

d. $g(1/2) = 2(1/2)^2 + 3 = 1/2 + 3 = 7/2$

e. $g(c) = 2(c)^2 + 3 = 2c^2 + 3$

f. $g(c + 5) = 2(c + 5)^2 + 3 = 2c^2 + 20c + 50 + 3$
$= 2c^2 + 20c + 53$

10. a. Because $0 \le 0 \le 5$, $Q(0) = 4$.

b. Because $6 < e < 7$, $Q(e) = -e + 9$.

c. Because $1 < n < 2$, $Q(n) = 4$.

d. Because $1 < m \le 2$, $8 < m^2 + 7 \le 11$. Thus
$Q(m^2 + 7) = \sqrt{(m^2 + 7) - 7} = \sqrt{m^2} = m$

14. $x^2 - 2y = 2$ **• Solve for y.**

$-2y = -x^2 + 2$

$y = \frac{1}{2}x^2 - 1$

y is a function of x because each x value will yield one and only one y value.

28. The domain is the set of all real numbers.

40. Domain all real numbers

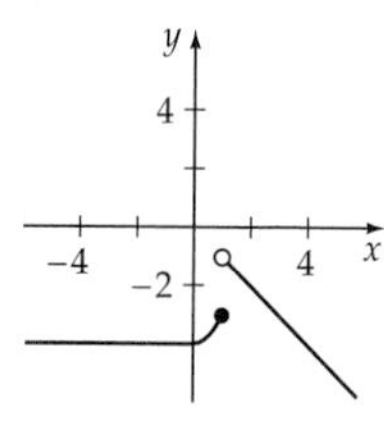

48. $C(4.75) = 0.85 - 0.50\ \text{int}(1 - 4.75)$
$= 0.85 - 0.50(-4) = 2.85$

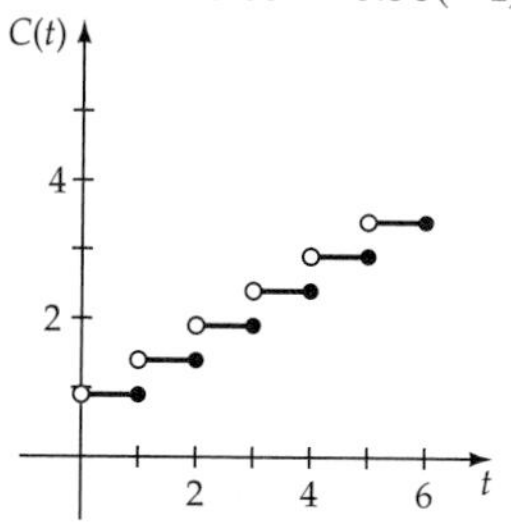

50. a. This is the graph of a function. Every vertical line intersects the graph in at most one point.

b. This is not the graph of a function. Some vertical lines intersect the graph at two points.

c. This is not the graph of a function. The vertical line at $x = -2$ intersects the graph at more than one point.

d. This is the graph of a function. Every vertical line intersects the graph at exactly one point.

66. $V(t) = 44{,}000 - 4200t, 0 \le t \le 8$

68. a. $V(x) = (30 - 2x)^2 x$
$= (900 - 120x + 4x^2)x$
$= 900x - 120x^2 + 4x^3$

b. Domain $\{x \mid 0 < x < 15\}$

72. $d(A, B) = \sqrt{1 + x^2}$. The time required to swim from A to B at 2 mph is $\sqrt{1 + x^2}/2$.

$d(B, C) = 3 - x$. The time required to run from B to C at 8 mph is $(3 - x)/8$.

Thus the total time to reach point C is

$$t = \frac{\sqrt{1 + x^2}}{2} + \frac{3 - x}{8} \text{ hours}$$

Exercise Set 2.3, page 175

2. $m = \dfrac{1 - 4}{5 - (-2)} = -\dfrac{3}{7}$

16. $m = -1$
$b = 1$

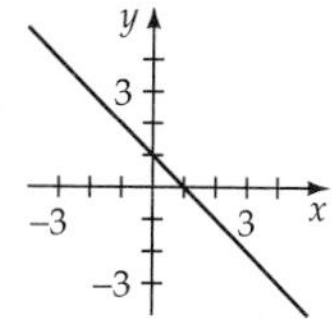

28. $y - 5 = -2(x - 0)$
$y = -2x + 5$

42. $f(x) = \dfrac{2x}{3} + 2$

$4 = \dfrac{2x}{3} + 2$

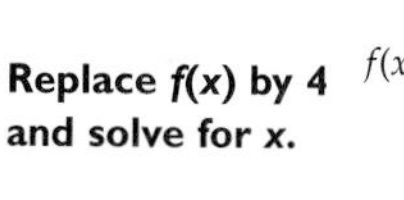

$2 = \dfrac{2x}{3}$

$3 = x$

When $x = 3, f(x) = 4$.

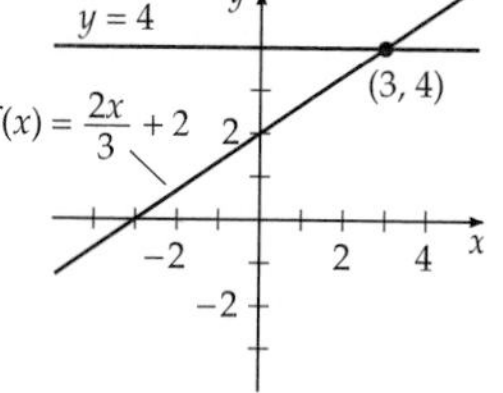

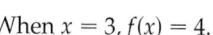

46. $f(x) = 0$
$-2x - 4 = 0$
$-2x = 4$
$x = -2$

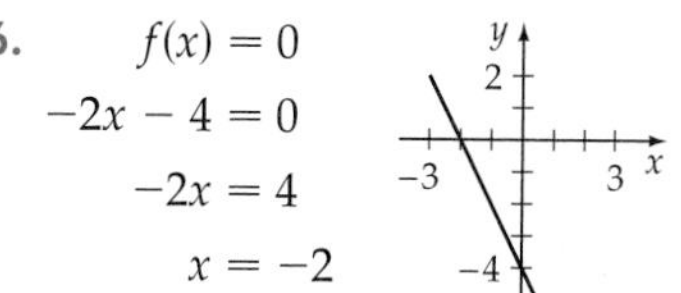

50. $f_1(x) = f_2(x)$
$-2x - 11 = 3x + 7$
$-5x - 11 = 7$
$-5x = 18$
$x = -\dfrac{18}{5} = -3.6$

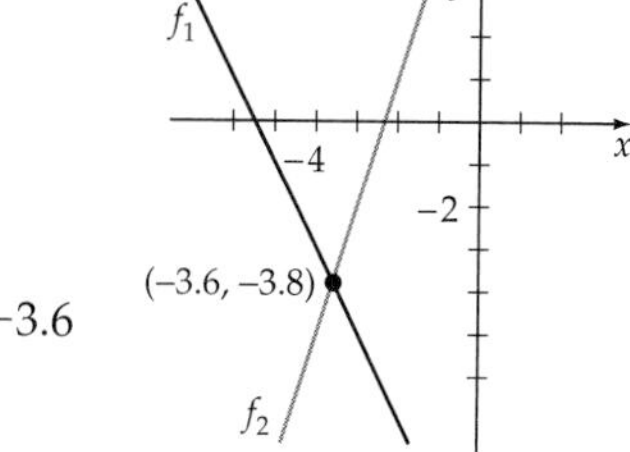

54. Use $f(x) = 0.46175x - 17.84843$.

a. Evaluate f when $x = 1025$.

$f(x) = 0.46175x - 17.84843$
$f(1025) = 0.46175(1025) - 17.84843$
$= 455.44532$

The airline's planes would fly approximately 455 million miles.

b. Evaluate f when $x = 2100$.

$f(x) = 0.46175x - 17.84843$
$f(2100) = 0.46175(2100) - 17.84843$
$= 951.82657$

The airline's planes would fly approximately 952 million miles.

56. $P(x) = R(x) - C(x)$
$P(x) = 124x - (78.5x + 5005)$
$P(x) = 45.5x - 5005$
$45.5x - 5005 = 0$
$45.5x = 5005$
$x = 110$ **• The break-even point**

66. a. Let x represent a person's current age, and let y represent the average remaining lifetime of that person.

Find the slope of the line through the given points $P_1(0, 76.5)$ and $P_2(75, 11.2)$. Then use the point-slope formula to find the equation of the line through the given points.

$$m = \frac{y_2 - y_1}{x_2 - x_1} = \frac{76.5 - 11.2}{0 - 75} \approx -0.87$$

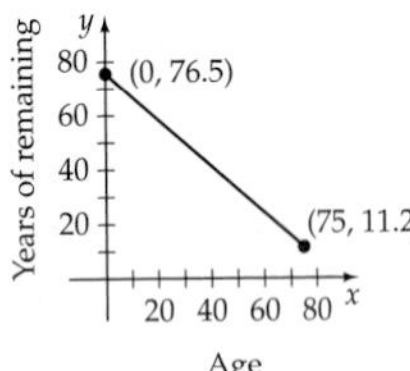

$$y - y_1 = m(x - x_1)$$
$$y - 76.5 = -0.87(x - 0)$$
$$y - 76.5 = -0.87x$$
$$y = -0.87x + 76.5$$

b. To find the average remaining lifetime for a person whose age is 25, replace x by 25 and solve for y.

$$y = -0.87x + 76.5$$
$$= -0.87(25) + 76.5 = 54.75$$

According to the model, a person whose age is 25 has an average remaining lifetime of approximately 55 years.

72. a. The slope of the radius from $(0, 0)$ to $(\sqrt{15}, 1)$ is $1/\sqrt{15}$. The slope of the linear path of the rock is $-\sqrt{15}$. The path of the rock is given by

$$y - 1 = -\sqrt{15}(x - \sqrt{15})$$
$$y - 1 = -\sqrt{15}x + 15$$
$$y = -\sqrt{15}x + 16$$

Every point on the wall has a y value of 14. Thus

$$14 = -\sqrt{15}x + 16$$
$$-2 = -\sqrt{15}x$$
$$x = \frac{2}{\sqrt{15}} \approx 0.52$$

The rock hits the wall at $(0.52, 14)$.

Exercise Set 2.4, page 188

10. $f(x) = x^2 + 6x - 1$

$$= x^2 + 6x + 9 + (-1 - 9)$$
$$= (x + 3)^2 - 10$$

vertex $(-3, -10)$
axis of symmetry $x = -3$

20. $h = -\frac{b}{2a} = -\frac{-6}{2(1)} = 3$

$k = f(3) = 3^2 - 6(3) = -9$

vertex $(3, -9)$
$f(x) = (x - 3)^2 - 9$

32. Determine the y-coordinate of the vertex of the graph of $f(x) = 2x^2 + 6x - 5$.

$f(x) = 2x^2 + 6x - 5$ • **$a = 2$, $b = 6$, $c = -5$.**

$h = -\frac{b}{2a} = -\frac{6}{2(2)} = -\frac{3}{2}$ • **Find the x-coordinate of the vertex.**

$k = f\left(-\frac{3}{2}\right) = 2\left(-\frac{3}{2}\right)^2 + 6\left(-\frac{3}{2}\right) - 5 = -\frac{19}{2}$ • **Find the y-coordinate of the vertex.**

The vertex is $\left(-\frac{3}{2}, -\frac{19}{2}\right)$. Because the parabola opens up, $-\frac{19}{2}$ is the minimum value of f. Therefore, the range of f is $\left\{y \,\middle|\, y \geq -\frac{19}{2}\right\}$.

To determine the values of x for which $f(x) = 15$, replace $f(x)$ by $2x^2 + 6x - 5$ and solve for x.

$f(x) = 15$

$2x^2 + 6x - 5 = 15$ • **Replace $f(x)$ by $2x^2 + 6x - 5$.**

$2x^2 + 6x - 20 = 0$ • **Solve for x.**

$2(x - 2)(x + 5) = 0$ • **Factor**

$x - 2 = 0 \quad x + 5 = 0$ • **Use the Principle of Zero Products to solve for x.**

$x = 2 \quad x = -5$

The values of x for which $f(x) = 15$ are 2 and -5.

36. $f(x) = -x^2 - 6x$

$$= -(x^2 + 6x)$$
$$= -(x^2 + 6x + 9) + 9$$
$$= -(x + 3)^2 + 9$$

Maximum value of f is 9 when $x = -3$.

46. a. $l + w = 240$, so $w = 240 - l$.

b. $A = lw = l(240 - l) = 240l - l^2$.

c. The l value of the vertex point of the graph of $A = 240l - l^2$ is

$$\frac{-b}{2a} = \frac{-240}{2(-1)} = 120$$

Thus $l = 120$ meters and $w = 240 - 120 = 120$ meters are the dimensions that produce the greatest area.

64. Let $x =$ the number of parcels.

a. $R(x) = xp = x(22 - 0.01x) = -0.01x^2 + 22x$

b. $P(x) = R(x) - C(x)$

$$= (-0.01x^2 + 22x) - (2025 + 7x)$$
$$= -0.01x^2 + 15x - 2025$$

c. $-\frac{b}{2a} = -\frac{15}{2(-0.01)} = 750$

The maximum profit is

$P(750) = -0.01(750)^2 + 15(750) - 2025 = \$3600$

d. The price per parcel that yields the maximum profit is

$p(750) = 22 - 0.01(750) = \$14.50$

e. The break-even point(s) occur when $R(x) = C(x)$.

$$-0.01x^2 + 22x = 2025 + 7x$$
$$0 = 0.01x^2 - 15x + 2025$$
$$x = \frac{-(-15) \pm \sqrt{(-15)^2 - 4(0.01)(2025)}}{2(0.01)}$$

$x = 150$ and $x = 1350$ are the breakeven points.

Thus the minimum number of parcels the air freight company must ship to break even is 150.

66. $h(t) = -16t^2 + 64t + 80$

$$t = \frac{-b}{2a} = \frac{-64}{2(-16)} = 2$$

$$h(2) = -16(2)^2 + 64(2) + 80$$
$$= -64 + 128 + 80 = 144$$

a. The vertex $(2, 144)$ gives us the maximum height of 144 feet.

b. The vertex of the graph of h is $(2, 144)$, so the time when it achieves this maximum height is at time $t = 2$ seconds.

c. $-16t^2 + 64t + 80 = 0$ • **Solve for *t* with *h* = 0.**

$$-16(t^2 - 4t - 5) = 0$$
$$-16(t + 1)(t - 5) = 0$$
$$t = -1 \qquad t - 5 = 0$$
$$\text{no} \qquad t = 5$$

The projectile will have a height of 0 feet at time $t = 5$ seconds.

Exercise Set 2.5, page 201

14. Symmetric with respect to the x-axis, because replacing y with $-y$ leaves the equation unaltered. The graph is not symmetric with respect to the y-axis, because replacing x with $-x$ alters the equation.

24. The graph is symmetric with respect to the origin because $(-y) = (-x)^3 - (-x)$ simplifies to $-y = -x^3 + x$, which is equivalent to the original equation $y = x^3 - x$.

44. Even, because $h(-x) = (-x)^2 + 1 = x^2 + 1 = h(x)$.

58.

60.

62.

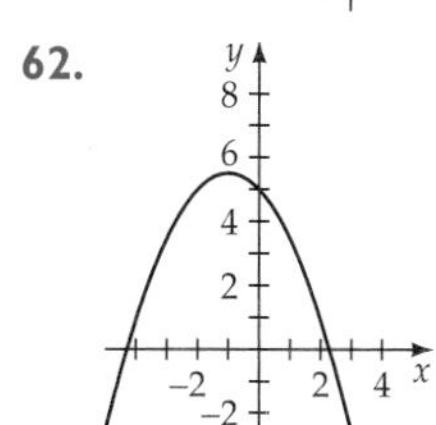

64. a.

b.

Exercise Set 2.6, page 213

10. $f(x) + g(x) = \sqrt{x - 4} - x$ domain $\{x \mid x \geq 4\}$

$f(x) - g(x) = \sqrt{x - 4} + x$ domain $\{x \mid x \geq 4\}$

$f(x)g(x) = -x\sqrt{x - 4}$ domain $\{x \mid x \geq 4\}$

$f(x)/g(x) = -\dfrac{\sqrt{x - 4}}{x}$ domain $\{x \mid x \geq 4\}$

14. $(f + g)(x) = (x^2 - 3x + 2) + (2x - 4) = x^2 - x - 2$

$(f + g)(-7) = (-7)^2 - (-7) - 2 = 49 + 7 - 2 = 54$

30. $$\frac{f(x + h) - f(x)}{h} = \frac{[4(x + h) - 5] - (4x - 5)}{h} = \frac{4x + 4(h) - 5 - 4x + 5}{h} = \frac{4(h)}{h} = 4$$

38. $(g \circ f)(x) = g[f(x)] = g[2x - 7]$

$= 3[2x - 7] + 2 = 6x - 19$

$(f \circ g)(x) = f[g(x)] = f[3x + 2]$

$= 2[3x + 2] - 7 = 6x - 3$

50. $(f \circ g)(4) = f[g(4)]$

$= f[4^2 - 5(4)]$

$= f[-4] = 2(-4) + 3 = -5$

66. a. $l = 3 - 0.5t$ for $0 \leq t \leq 6$. $l = -3 + 0.5t$ for $t > 6$. In either case, $l = |3 - 0.5t|$. $w = |2 - 0.2t|$ as in Example 7.

b. $A(t) = |3 - 0.5t||2 - 0.2t|$

c. A is decreasing on $[0, 6]$ and on $[8, 10]$. A is increasing on $[6, 8]$ and on $[10, 14]$.

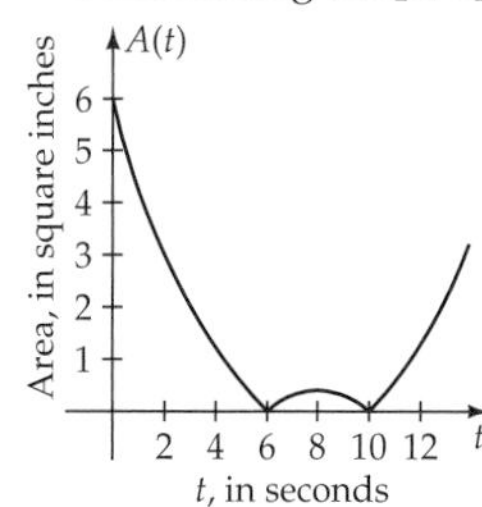

d. The highest point on the graph of A for $0 \leq t \leq 14$ occurs when $t = 0$ seconds.

72. a. On $[2, 3]$,

$a = 2$

$\Delta t = 3 - 2 = 1$

$f(a + \Delta t) = f(3) = 6 \cdot 3^2 = 54$

$f(a) = f(2) = 6 \cdot 2^2 = 24$

$$\text{Average velocity} = \frac{f(a + \Delta t) - f(a)}{\Delta t} = \frac{f(3) - f(2)}{1} = 54 - 24 = 30 \text{ feet per second}$$

This is identical to the slope of the line through $(2, f(2))$ and $(3, f(3))$ because

$$m = \frac{f(3) - f(2)}{3 - 2} = f(3) - f(2)$$

b. On $[2, 2.5]$,

$a = 2$

$\Delta t = 2.5 - 2 = 0.5$

$f(a + \Delta t) = f(2.5) = 6(2.5)^2 = 37.5$

$$\text{Average velocity} = \frac{f(2.5) - f(2)}{0.5} = \frac{37.5 - 24}{0.5} = \frac{13.5}{0.5} = 27 \text{ feet per second}$$

c. On $[2, 2.1]$,

$a = 2$

$\Delta t = 2.1 - 2 = 0.1$

$f(a + \Delta t) = f(2.1) = 6(2.1)^2 = 26.46$

$$\text{Average velocity} = \frac{f(2.1) - f(2)}{0.1} = \frac{26.46 - 24}{0.1} = \frac{2.46}{0.1} = 24.6 \text{ feet per second}$$

d. On $[2, 2.01]$,

$a = 2$

$\Delta t = 2.01 - 2 = 0.01$

$f(a + \Delta t) = f(2.01) = 6(2.01)^2 = 24.2406$

$$\text{Average velocity} = \frac{f(2.01) - f(2)}{0.01} = \frac{24.2406 - 24}{0.01} = \frac{0.2406}{0.01} = 24.06 \text{ feet per second}$$

e. On $[2, 2.001]$,

$a = 2$

$\Delta t = 2.001 - 2 = 0.001$

$f(a + \Delta t) = f(2.001) = 6(2.001)^2 = 24.024006$

$$\text{Average velocity} = \frac{f(2.001) - f(2)}{0.001} = \frac{24.024006 - 24}{0.001} = \frac{0.024006}{0.001} = 24.006 \text{ feet per second}$$

f. On $[2, 2 + \Delta t]$,

$$\frac{f(2 + \Delta t) - f(2)}{\Delta t} = \frac{6(2 + \Delta t)^2 - 24}{\Delta t}$$

$$= \frac{6(4 + 4(\Delta t) + (\Delta t)^2) - 24}{\Delta t}$$

$$= \frac{24 + 24(\Delta t) + 6(\Delta t)^2 - 24}{\Delta t}$$

$$= \frac{24\Delta t + 6(\Delta t)^2}{\Delta t} = 24 + 6(\Delta t)$$

As Δt approaches zero, the average velocity seems to approach 24 feet per second.

Exercise Set 2.7, page 223

18. Enter the data in the table. Then use your calculator to find the linear regression equation.

a. The linear regression equation is

$y = -72.06131724x + 14926.16191$

b. Evaluate the linear regression equation when $x = 55$.

$y = -72.06131724(55) + 14926.16191$

$= 10{,}962.79$

The approximate trade-in value of the car is \$10,462.

32. Enter the data in the table. Then use your calculator to find the quadratic regression model.

a. $y = 0.05208x^2 - 3.56026x + 82.32999$

b. The speed at which the bird has minimum oxygen consumption is the x-coordinate of the vertex of the graph of the regression equation. Recall that the x-coordinate of the vertex is given by $x = -\dfrac{b}{2a}$.

$$x = -\frac{b}{2a} = -\frac{-3.56026}{2(0.05208)}$$

≈ 34

The speed that minimizes oxygen consumption is 34 kilometers per hour.

Exercise Set 3.1, page 245

6.

$$\begin{array}{r} x^2 - 9 \\ 2x^2 - x - 5\overline{)2x^4 - x^3 - 23x^2 + 9x + 45} \\ \underline{2x^4 - x^3 - 5x^2} \\ -18x^2 + 9x + 45 \\ \underline{-18x^2 + 9x + 45} \\ 0 \end{array}$$

12.

$$\begin{array}{r|rrrr} 5 & 5 & 6 & -8 & 1 \\ & & 25 & 155 & 735 \\ \hline & 5 & 31 & 147 & 736 \end{array}$$

$$\frac{5x^3 + 6x^2 - 8x + 1}{x - 5} = 5x^2 + 31x + 147 + \frac{736}{x - 5}$$

30. $$\begin{array}{r|rrrr} 3 & 2 & -1 & 3 & -1 \\ & & 6 & 15 & 54 \\ \hline & 2 & 5 & 18 & 53 \end{array}$$

$P(c) = P(3) = 53$

40. $$\begin{array}{r|rrrr} -6 & 1 & 4 & -27 & -90 \\ & & -6 & 12 & 90 \\ \hline & 1 & -2 & -15 & 0 \end{array}$$

A remainder of 0 implies that $x + 6$ is a factor of $P(x)$.

62. $$\begin{array}{r|rrrrr} -1 & 1 & 5 & 3 & -5 & -4 \\ & & -1 & -4 & 1 & 4 \\ \hline & 1 & 4 & -1 & -4 & 0 \end{array}$$

The reduced polynomial is $x^3 + 4x^2 - x - 4$.

$x^4 + 5x^3 + 3x^2 - 5x - 4 = (x + 1)(x^3 + 4x^2 - x - 4)$

Exercise Set 3.2, page 254

2. Because $a_n = -2$ is negative and $n = 3$ is odd, the graph of P goes up to the far left and down to the far right.

26. The volume of the box is $V = lwh$, with $h = x$, $l = 18 - 2x$, and $w = \dfrac{42 - 3x}{2}$. Therefore, the volume is

$$V(x) = (18 - 2x)\left(\frac{42 - 3x}{2}\right)x$$
$$= 3x^3 - 69x^2 + 378x$$

Use a graphing utility to graph $V(x)$. The graph is shown below. The value of x that produces the maximum volume is 3.571 inches (to the nearest 0.001 inch). The maximum volume is approximately 606.6 cubic inches.

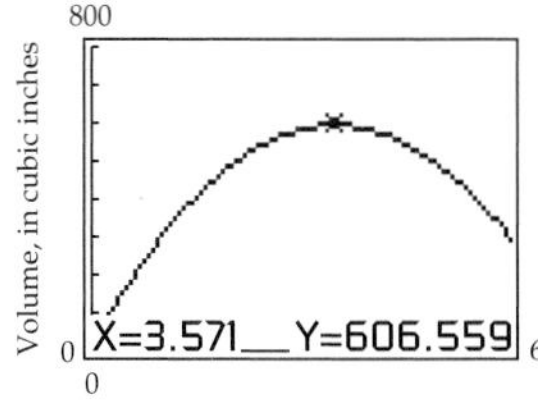

34. $$\begin{array}{r|rrrr} 0 & 4 & -1 & -6 & 1 \\ & & 0 & 0 & 0 \\ \hline & 4 & -1 & -6 & 1 \end{array}$$ • $P(0) = 1$

$$\begin{array}{r|rrrr} 1 & 4 & -1 & -6 & 1 \\ & & 4 & 3 & -3 \\ \hline & 4 & 3 & -3 & -2 \end{array}$$ • $P(1) = -2$

Because $P(0)$ and $P(1)$ have opposite signs, P must have a real zero between 0 and 1.

52. The exponent of $(x + 2)^3$ is odd. Thus P crosses the x-axis at the x-intercept $(-2, 0)$. The exponent of $(x - 6)^{10}$ is even. Thus P intersects but does not cross the x-axis at the x-intercept $(6, 0)$.

Exercise Set 3.3, page 267

14. $p = \pm 1, \pm 2, \pm 4, \pm 8$

$q = \pm 1, \pm 3$

$\dfrac{p}{q} = \pm 1, \pm 2, \pm 4, \pm 8, \pm\frac{1}{3}, \pm\frac{2}{3}, \pm\frac{4}{3}, \pm\frac{8}{3}$

22. $$\begin{array}{r|rrrr} 3 & 1 & 0 & -19 & -28 \\ & & 3 & 9 & -30 \\ \hline & 1 & 3 & -10 & -58 \end{array}$$

$$\begin{array}{r|rrrr} 4 & 1 & 0 & -19 & -28 \\ & & 4 & 16 & -12 \\ \hline & 1 & 4 & -3 & -40 \end{array}$$

$$\begin{array}{r|rrrr} 5 & 1 & 0 & -19 & -28 \\ & & 5 & 25 & 30 \\ \hline & 1 & 5 & 6 & 2 \end{array}$$

All numbers are positive, so 5 is an upper bound.

$$\begin{array}{r|rrrr} -2 & 1 & 0 & -19 & -28 \\ & & -2 & 4 & 30 \\ \hline & 1 & -2 & -15 & 2 \end{array}$$

$$\begin{array}{r|rrrr} -3 & 1 & 0 & -19 & -28 \\ & & -3 & 9 & 30 \\ \hline & 1 & -3 & -10 & 2 \end{array}$$

$$\begin{array}{r|rrrr} -4 & 1 & 0 & -19 & -28 \\ & & -4 & 16 & 12 \\ \hline & 1 & -4 & -3 & -16 \end{array}$$

$$\begin{array}{r|rrrr} -5 & 1 & 0 & -19 & -28 \\ & & -5 & 25 & -30 \\ \hline & 1 & -5 & 6 & -58 \end{array}$$

These numbers alternate signs, so -5 is a lower bound.

32. One positive real zero because the polynomial P has one variation in sign.

$P(-x) = (-x)^3 - 19(-x) - 30 = -x^3 + 19x - 30$

2 or no negative real zeros because $-x^3 + 19x - 30$ has two variations in sign.

42. One positive and two or no negative real zeros (see Exercise 32).

$$\begin{array}{r|rrrr} 5 & 1 & 0 & -19 & -30 \\ & & 5 & 25 & 30 \\ \hline & 1 & 5 & 6 & 0 \end{array}$$

The reduced polynomial is

$x^2 + 5x + 6 = (x + 3)(x + 2)$

which has -3 and -2 as zeros. The zeros of $x^3 - 19x - 30$ are 5, -2, and -3.

54. One positive zero; four, two, or no negative zeros

$\dfrac{p}{q} = \pm 1, \pm 2, \pm 7, \pm 14, \pm\frac{1}{3}, \pm\frac{2}{3}, \pm\frac{7}{3}, \pm\frac{14}{3}$

$$\begin{array}{r|rrrrrr} 2 & 3 & 16 & 2 & -58 & -61 & -14 \\ & & 6 & 44 & 92 & 68 & 14 \\ \hline & 3 & 22 & 46 & 34 & 7 & 0 \end{array}$$

$\dfrac{p}{q} = \pm 1, \pm 7, \pm\frac{1}{3}, \pm\frac{7}{3}$

$$\begin{array}{r|rrrrr} -1 & 3 & 22 & 46 & 34 & 7 \\ & & -3 & -19 & -27 & -7 \\ \hline -\frac{1}{3} & 3 & 19 & 27 & 7 & 0 \\ & & -1 & -6 & -7 & \\ \hline & 3 & 18 & 21 & 0 & \end{array}$$

Divide each term of $3x^2 + 18x + 21 = 0$ by 3 to produce $x^2 + 6x + 7 = 0$. Solve by the quadratic formula.

$$x = \frac{-6 \pm \sqrt{6^2 - 4(1)(7)}}{2(1)} = \frac{-6 \pm \sqrt{8}}{2} = \frac{-6 \pm 2\sqrt{2}}{2}$$
$$= -3 \pm \sqrt{2}$$

The zeros of $3x^5 + 16x^4 + 2x^3 - 58x^2 - 61x - 14$ are $2, -1, -\frac{1}{3}, -3 + \sqrt{2}$, and $-3 - \sqrt{2}$.

64. The resulting solid has dimensions of n by $n - 1$ by $n - 3$. The volume of the solid is $V = n(n - 1)(n - 3) = n^3 - 4n^2 + 3n$. We need to solve $n^3 - 4n^2 + 3n = 1560$. This is equivalent to finding the zeros of

$$n^3 - 4n^2 + 3n - 1560 = 0$$

The following division shows that 15 is an upper bound

$$\begin{array}{r|rrrr} 15 & 1 & -4 & 3 & -1560 \\ & & 15 & 165 & 2520 \\ \hline & 1 & 11 & 168 & 960 \end{array}$$

The positive divisors of 1560 that are less than 15 are 1, 2, 4, 5, 6, 8, 10, 12, and 13.

$$\begin{array}{r|rrrr} 13 & 1 & -4 & 3 & -1560 \\ & & 13 & 117 & 1560 \\ \hline & 1 & 9 & 120 & 0 \end{array}$$

Thus $n = 13$ inches. The original cube had 13-inch sides.

66. The volume of space inside the shell is $\frac{4}{3}\pi(r - 8)^3$, The volume of the shell itself is $\frac{4}{3}\pi r^3 - \frac{4}{3}\pi(r - 8)^3$.

$$\frac{4}{3}\pi(r - 8)^3 = \frac{1}{10}\left[\frac{4}{3}\pi r^3 - \frac{4}{3}\pi(r - 8)^3\right]$$
$$\frac{4}{3}\pi(r - 8)^3 = \frac{2}{15}\pi r^3 - \frac{2}{15}\pi(r - 8)^3$$
$$10(r - 8)^3 = r^3 - (r - 8)^3$$
$$0 = r^3 - 11(r - 8)^3$$
$$0 = r^3 - 11(r^3 - 24r^2 + 192r - 512)$$
$$0 = r^3 - 11r^3 + 264r^2 - 2112r + 5632$$
$$0 = -10r^3 + 264r^2 - 2112r + 5632$$

Now use a graphing utility to show that $r = 14.54$ mm (nearest 0.01).

Exercise Set 3.4, page 276

2.
$$\begin{array}{r|rrrr} 5 + 3i & 3 & -29 & 92 & 34 \\ & & 15 + 9i & -97 + 3i & -34 \\ \hline & 3 & -14 + 9i & -5 + 3i & 0 \end{array}$$

$$\begin{array}{r|rrr} 5 - 3i & 3 & -14 + 9i & -5 + 3i \\ & & 15 - 9i & 5 - 3i \\ \hline & 3 & 1 & 0 \end{array}$$

The reduced polynomial $3x + 1$ has $-1/3$ as a zero. The zeros of $3x^3 - 29x^2 + 92x + 34$ are $5 + 3i$, $5 - 3i$, and $-1/3$.

14.
$$\begin{array}{r|rrrrrr} 3i & 1 & -6 + 0i & 22 + 0i & -64 + 0i & 117 + 0i & -90 \\ & & 0 + 3i & -9 - 18i & 54 + 39i & -117 - 30i & 90 \\ \hline -3i & 1 & -6 + 3i & 13 - 18i & -10 + 39i & 0 - 30i & 0 \\ & & 0 - 3i & 0 + 18i & 0 - 39i & 30i & \\ \hline & 1 & -6 & 13 & -10 & 0 & \end{array}$$

$$\frac{p}{q} = \pm 1, \pm 2, \pm 5, \pm 10$$

$$\begin{array}{r|rrrr} 2 & 1 & -6 & 13 & -10 \\ & & 2 & -8 & 10 \\ \hline & 1 & -4 & 5 & 0 \end{array}$$

Use the quadratic formula to solve $x^2 - 4x + 5 = 0$.

$$x = \frac{-(-4) \pm \sqrt{(-4)^2 - 4(1)(5)}}{2(1)} = \frac{4 \pm \sqrt{-4}}{2}$$
$$= \frac{4 \pm 2i}{2} = 2 \pm i$$

The zeros of $x^5 - 6x^4 + 22x^3 - 64x^2 + 117x - 90$ are $3i$, $-3i$, 2, $2 + i$, and $2 - i$.

28. The graph of $P(x) = 4x^3 + 3x^2 + 16x + 12$ is shown below. Applying Descartes' Rule of Signs, we find that the real zeros are all negative numbers. From the Upper- and Lower-Bound Theorem there is no real zero less than -1, and from the Rational Zero Theorem the possible rational zeros (that are negative and greater than -1) are $p/q = -1/2, -1/4, -3/4$. From the graph, it appears that $-3/4$ is a zero.

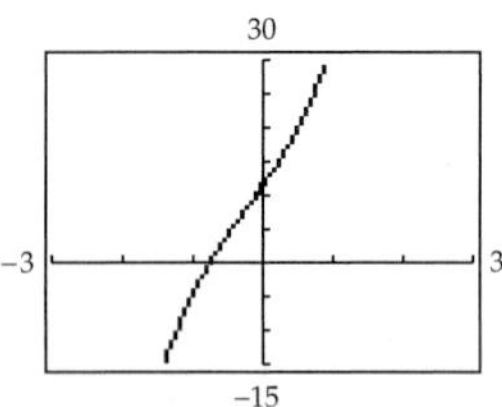

Using synthetic division, we have

$$\begin{array}{r|rrrr} -\frac{3}{4} & 4 & 3 & 16 & 12 \\ & & -3 & 0 & -12 \\ \hline & 4 & 0 & 16 & 0 \end{array}$$

Thus $-3/4$ is a zero, and by the Factor Theorem,

$$4x^3 + 3x^2 + 16x + 12 = \left(x + \frac{3}{4}\right)(4x^2 + 16) = 0$$

Solving $4x^2 + 16 = 0$, we have $x = -2i$ and $x = 2i$. The solutions of the original equation are $-3/4$, $-2i$, and $2i$.

36. $6x^3 - 23x^2 - 4x = x(6x^2 - 23x - 4)$
$$= x(6x + 1)(x - 4)$$

56. Because P has real coefficients, use the Conjugate Pair Theorem.

$$\begin{aligned} P &= (x - [3 + 2i])(x - [3 - 2i])(x - 7) \\ &= (x - 3 - 2i)(x - 3 + 2i)(x - 7) \\ &= (x^2 - 6x + 13)(x - 7) \\ &= x^3 - 13x^2 + 55x - 91 \end{aligned}$$

Exercise Set 3.5, page 291

2.
$$\begin{aligned} x^2 - 4 &= 0 \\ (x - 2)(x + 2) &= 0 \\ x = 2 \quad \text{or} \quad x &= -2 \end{aligned}$$

The vertical asymptotes are $x = 2$ and $x = -2$.

6. The horizontal asymptote is $y = 0$ (x-axis) because the degree of the denominator is larger than the degree of the numerator.

10. Vertical asymptote: $x - 2 = 0$
$$x = 2$$
Horizontal asymptote: $y = 0$

26. Vertical asymptote:
$$\begin{aligned} x^2 - 6x + 9 &= 0 \\ (x - 3)(x - 3) &= 0 \\ x &= 3 \end{aligned}$$

The horizontal asymptote is $y = 1/1 = 1$ (the Theorem on Horizontal Asymptotes) because numerator and denominator both have degree 2.

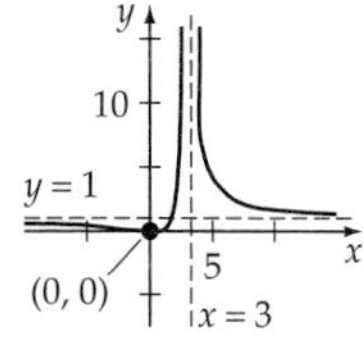

34.
$$\begin{array}{r} x + 1 \\ x^2 - 3x + 5\overline{)\,x^3 - 2x^2 + 3x + 4} \\ \underline{x^3 - 3x^2 + 5x} \\ x^2 - 2x + 4 \\ \underline{x^2 - 3x + 5} \\ x - 1 \end{array}$$

$$F(x) = x + 1 + \frac{x - 1}{x^2 - 3x + 5}$$

Slant asymptote: $y = x + 1$

40. Vertical asymptote:
$$\begin{aligned} 2x + 5 &= 0 \\ 2x &= -5 \\ x &= -\frac{5}{2} \end{aligned}$$

The vertical asymptote is $x = -5/2$.
Slant asymptote:

$$\begin{array}{r} \frac{1}{2}x - \frac{13}{4} \\ 2x + 5\overline{)\,x^2 - 4x - 5} \\ \underline{x^2 + \frac{5}{2}x} \\ -\frac{13}{2}x - 5 \\ \underline{-\frac{13}{2}x - \frac{65}{4}} \\ \frac{45}{4} \end{array}$$

$$F(x) = \frac{1}{2}x - \frac{13}{4} + \frac{45/4}{2x + 5}$$

The slant asymptote is $y = \frac{1}{2}x - \frac{13}{4}$.

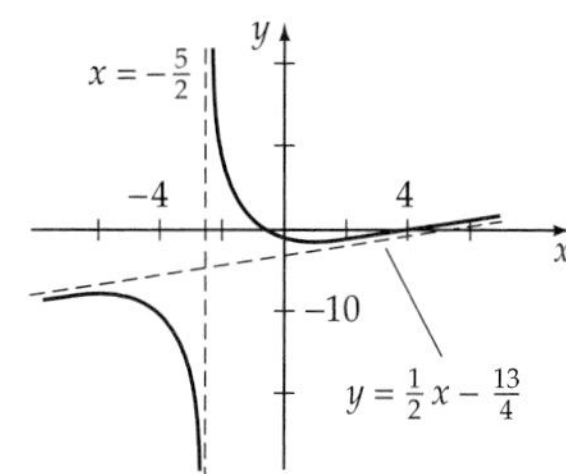

50. $F(x) = \dfrac{x^2 - x - 12}{x^2 - 2x - 8} = \dfrac{(x - 4)(x + 3)}{(x - 4)(x + 2)} = \dfrac{x + 3}{x + 2}, x \neq 4$

The function F is undefined at $x = 4$. Thus the graph of F is the graph of $y = \dfrac{x + 3}{x + 2}$ with an open circle at $(4, 7/6)$.

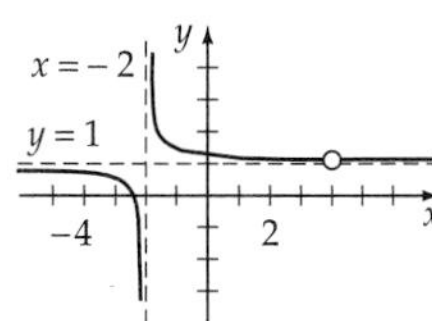

64. $A(x) = \dfrac{40{,}000 + 20x + 0.0001x^2}{x}$

a. $A(5000) = \dfrac{40{,}000 + 20(5000) + 0.0001(5000)^2}{5000}$
$= \$28.50$

b. $A(10{,}000) = \$25$

c. $y = 0.0001x + 20$

d. Use a graphing utility to graph $y = A(x)$.

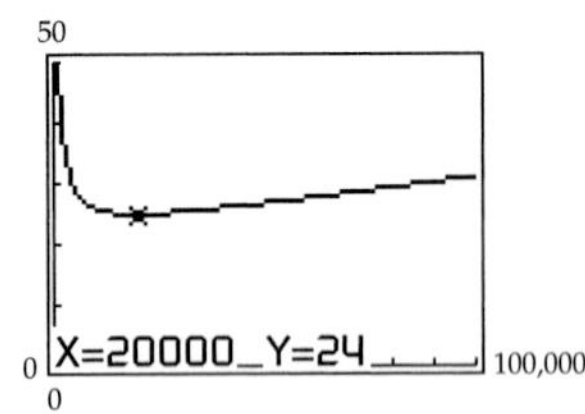

To minimize cost, 20,000 books should be published.

Exercise Set 4.1, page 306

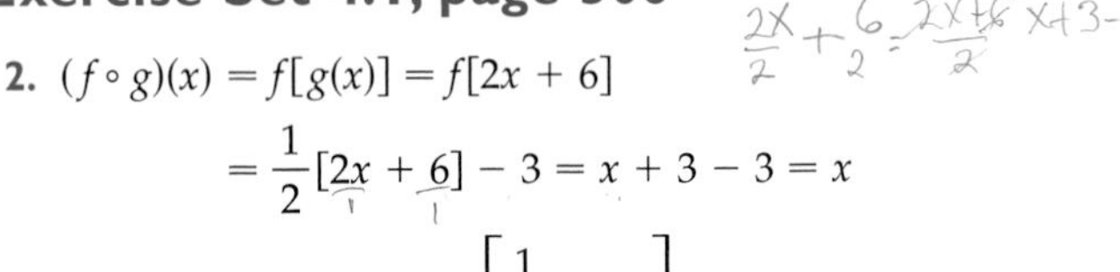

2. $(f \circ g)(x) = f[g(x)] = f[2x + 6]$

$$= \frac{1}{2}[2x + 6] - 3 = x + 3 - 3 = x$$

$$(g \circ f)(x) = g[f(x)] = g\left[\frac{1}{2}x - 3\right]$$

$$= 2\left[\frac{1}{2}x - 3\right] + 6 = x - 6 + 6 = x$$

14. $g(x) = \frac{2}{3}x + 4$

$$y = \frac{2}{3}x + 4$$

$$x = \frac{2}{3}y + 4 \quad \text{• Interchange } x \text{ and } y.$$

$$x - 4 = \frac{2}{3}y \quad \text{• Solve for } y.$$

$$\frac{3}{2}x - 6 = y$$

Thus $g^{-1}(x) = \frac{3}{2}x - 6$.

22. $G(x) = \frac{3x}{x - 5}, \quad x \neq 5$

$$y = \frac{3x}{x - 5}$$

$$x = \frac{3y}{y - 5} \quad \text{• Interchange } x \text{ and } y.$$

$$xy - 5x = 3y \quad \text{• Solve for } y.$$

$$xy - 3y = 5x$$

$$y = \frac{5x}{x - 3}$$

Thus $G^{-1}(x) = \frac{5x}{x - 3}, \quad x \neq 3$.

36. $f(x) = x^2 + 6x - 6, x \geq -3$

Domain f is $\{x | x \geq -3\}$, range f is $\{y | y \geq -15\}$.

$$y = x^2 + 6x - 6$$

$$x = y^2 + 6y - 6 \quad \text{• Interchange } x \text{ and } y.$$

$$x + 6 = y^2 + 6y$$

$$x + 15 = y^2 + 6y + 9 \quad \text{• Complete the square.}$$

$$x + 15 = (y + 3)^2$$

Choose the positive root, because the range of f^{-1} is $\{y | y \geq -3\}$.

$$\sqrt{x + 15} = y + 3$$

$$-3 + \sqrt{x + 15} = y$$

Thus $f^{-1}(x) = -3 + \sqrt{x + 15}$, domain f^{-1} is $\{x | x \geq -15\}$, and range f^{-1} is $\{y | y \geq -3\}$.

40.

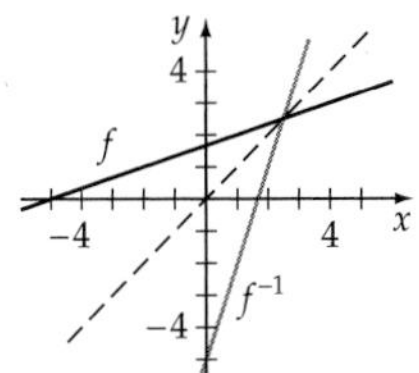

46. a. $UK(9) = 1.3(9) - 4.7 = 7$

b.

$$y = 1.3x - 4.7$$

$$x = 1.3y - 4.7$$

$$x + 4.7 = 1.3y$$

$$y = \frac{x + 4.7}{1.3}$$

$$UK^{-1} = \frac{x + 4.7}{1.3}$$

$$UK^{-1}(5) = \frac{5 + 4.7}{1.3} \approx 7.5$$

Exercise Set 4.2, page 317

36.

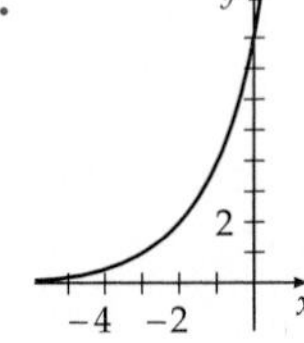

40.

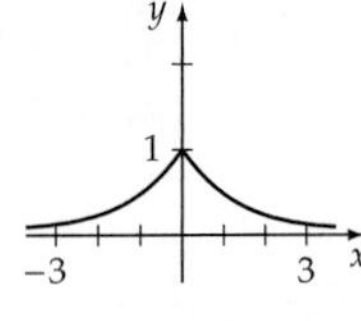

44.

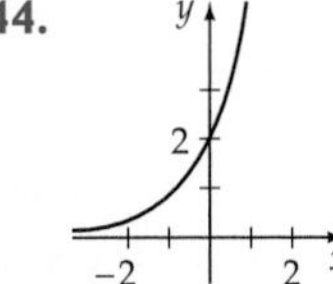

48.

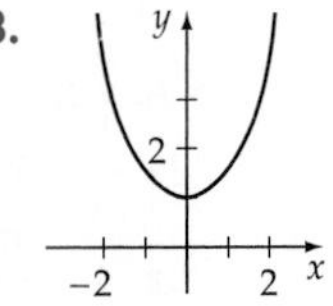

50. Graph $f(x) = 3^{-x} - 4$. Then use the features of a graphing utility to locate the zero. The graph is shown below. To the nearest hundredth, the zero of f is -1.26.

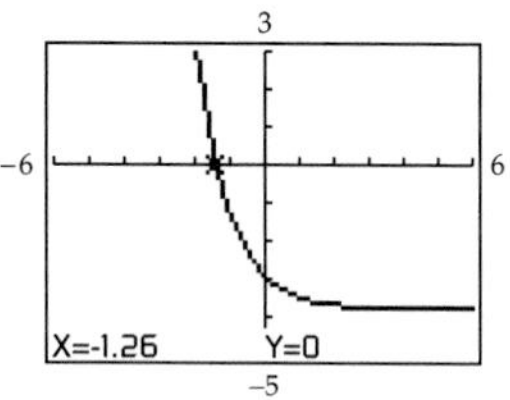

58. $$\begin{aligned} I(1) &= 100e^{-1.5 \cdot 1} \\ &= 100e^{-1.5} \\ &\approx 22.3 \end{aligned}$$

Thus 22.3% of the radiation will penetrate a lead shield that is 1 mm thick.

Exercises Set 4.3, page 327

2. $$\begin{aligned} \log_{10} 1000 &= 3 \\ 1000 &= 10^3 \end{aligned}$$

12. $$\begin{aligned} 3^5 &= 243 \\ \log_3 243 &= 5 \end{aligned}$$

22. $\log_b b = 1$ because $b^1 = b$

30. $$\begin{aligned} \log_{10} \frac{1}{1000} &= n \\ 10^n &= \frac{1}{1000} \\ 10^n &= 10^{-3} \\ n &= -3 \end{aligned}$$

32. The exponential form of $f(x) = \log_5 x$ is $x = 5^y$. Choose values of y and calculate the corresponding values of x to produce a table as shown below.

$x = 5^y$	$\frac{1}{5}$	1	5	25
y	-1	0	1	2

Plot the ordered pairs and connect the points with a smooth curve.

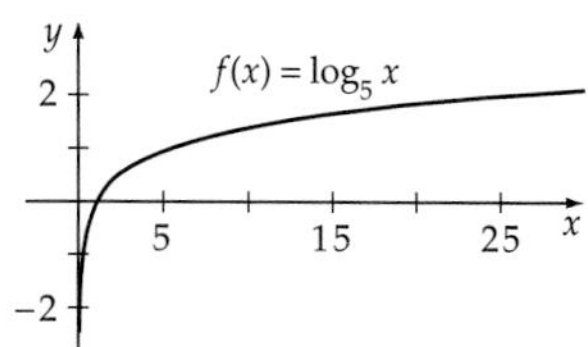

42. $$\begin{aligned} 3x - 1 &> 0 \\ 3x &> 1 \\ x &> \frac{1}{3} \end{aligned}$$

The domain of g is the set of all real numbers greater than 1/3. In interval notation the domain is $(1/3, \infty)$.

52. Translate the graph of $f(x) = \log_5 x$ to the left 3 units.

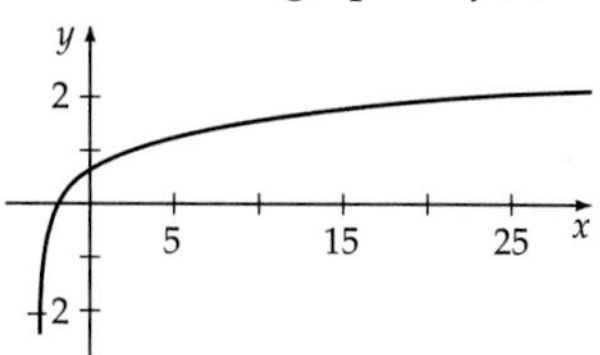

80. Find the value of A for which $N = 6$ by finding the intersection of $N = 1.6 + 2.3A$ and $N = 6$.
When $N = 6$, $A \approx 6774$.
Approximately $6774 was spent on advertising.

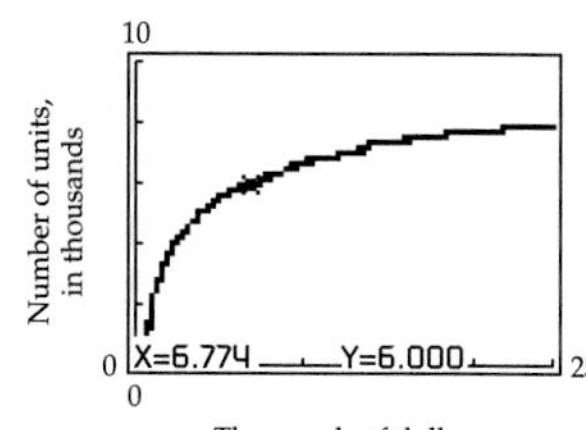

Note: The x-value shown is rounded.

Exercise Set 4.4, page 339

2. $\log_b(x^2y^3) = \log_b x^2 + \log_b y^3 = 2\log_b x + 3\log_b y$

12. $$\begin{aligned} &5\log_3 x - 4\log_3 y + 2\log_3 z \\ &= \log_3 x^5 - \log_3 y^4 + \log_3 z^2 \\ &= \log_3 \frac{x^5z^2}{y^4} \end{aligned}$$

22. $$\begin{aligned} \log_7 20 &= \log_7 2^2 \cdot 5 \\ &= 2\log_7 2 + \log_7 5 \\ &= 2(0.3562) + (0.8271) \\ &= 1.5395 \end{aligned}$$

32. $\log_5 37 = \dfrac{\log 37}{\log 5} \approx 2.2436$

42.

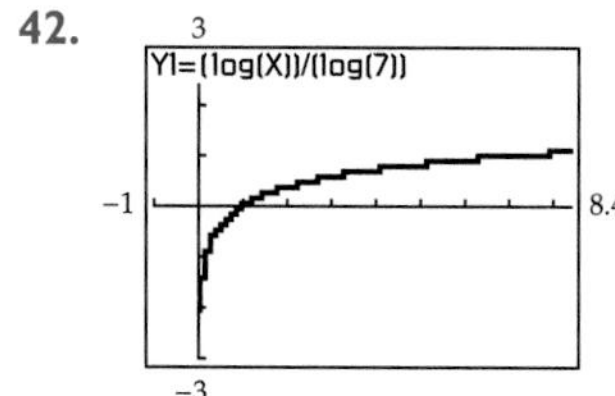

54. $M = \log\left(\dfrac{398107000 I_0}{I_0}\right) = \log 398107000 \approx 8.6$

56. $9.5 = \log\left(\frac{I}{I_0}\right)$

$10^{9.5} = \frac{I}{I_0}$

$10^{9.5}I_0 = I$

$3{,}162{,}277{,}660I_0 \approx I$

58. $\frac{I_1}{I_2} = \frac{10^{9.5}I_0}{10^{8.3}I_0} = 10^{1.2} \approx 15.8$

62. $M = \log 26 + 3\log(8 \cdot 17) - 2.92 \approx 4.9$

64. $\text{pH} = -\log[\text{H}^+] = -\log[1.26 \times 10^{-3}] \approx 2.9$; acid

66. $\text{H}^+ = 10^{-5.6} \approx 2.51 \times 10^{-6}$

Exercise Set 4.5, page 348

2. $3^x = 243$

$3^x = 3^5$

$x = 5$

10. $6^x = 50$

$\log(6^x) = \log 50$

$x\log 6 = \log 50$

$x = \frac{\log 50}{\log 6} \approx 2.18$

18. $3^{x-2} = 4^{2x+1}$

$\log 3^{x-2} = \log 4^{2x+1}$

$(x-2)\log 3 = (2x+1)\log 4$

$x\log 3 - 2\log 3 = 2x\log 4 + \log 4$

$x\log 3 - 2\log 3 - 2x\log 4 = \log 4$

$x\log 3 - 2x\log 4 = \log 4 + 2\log 3$

$x(\log 3 - 2\log 4) = \log 4 + 2\log 3$

$x = \frac{\log 4 + 2\log 3}{\log 3 - 2\log 4}$

$x \approx -2.141$

22. $\log(x^2 + 19) = 2$

$x^2 + 19 = 10^2$

$x^2 + 19 = 100$

$x^2 = 81$

$x = \pm 9$

A check shows that 9 and -9 are both solutions of the original equation.

26. $\log_3 x + \log_3(x+6) = 3$

$\log_3[x(x+6)] = 3$

$3^3 = x(x+6)$

$27 = x^2 + 6x$

$x^2 + 6x - 27 = 0$

$(x+9)(x-3) = 0$

$x = -9$ or $x = 3$

Because $\log_3 x$ is defined only for $x > 0$, the only solution is $x = 3$.

36. $\ln x = \frac{1}{2}\ln\left(2x + \frac{5}{2}\right) + \frac{1}{2}\ln 2$

$= \frac{1}{2}\left[\ln\left(2x + \frac{5}{2}\right) + \ln 2\right]$

$\ln x = \frac{1}{2}\ln\left[2\left(2x + \frac{5}{2}\right)\right]$

$\ln x = \frac{1}{2}\ln(4x + 5)$

$\ln x = \ln(4x+5)^{1/2}$

$x = \sqrt{4x+5}$

$x^2 = 4x + 5$

$0 = x^2 - 4x - 5$

$0 = (x-5)(x+1)$

$x = 5$ or $x = -1$

Check: $\ln 5 = \frac{1}{2}\ln\left(10 + \frac{5}{2}\right) + \frac{1}{2}\ln 2$

$1.6094 \approx 1.2629 + 0.3466$

Because $\ln(-1)$ is not defined, -1 is not a solution. Thus the only solution is $x = 5$.

40. $\frac{10^x + 10^{-x}}{2} = 8$

$10^x + 10^{-x} = 16$

$10^x(10^x + 10^{-x}) = (16)10^x$ • **Multiply each side by 10^x.**

$10^{2x} + 1 = 16(10^x)$

$10^{2x} - 16(10^x) + 1 = 0$

$u^2 - 16u + 1 = 0$ • **Let $u = 10^x$.**

$u = \frac{16 \pm \sqrt{16^2 - 4(1)(1)}}{2} = 8 \pm 3\sqrt{7}$

$10^x = 8 \pm 3\sqrt{7}$ • **Replace u with 10^x.**

$\log 10^x = \log(8 \pm 3\sqrt{7})$

$x = \log(8 \pm 3\sqrt{7}) \approx \pm 1.20241$

70. a. $t = \frac{9}{24}\ln\frac{24+v}{24-v}$

$1.5 = \frac{9}{24}\ln\frac{24+v}{24-v}$

$4 = \ln\frac{24+v}{24-v}$

$e^4 = \frac{24+v}{24-v}$ • **$N = \ln M$ means $e^N = M$.**

$(24 - v)e^4 = 24 + v$

$-v - ve^4 = 24 - 24e^4$

$$v(-1 - e^4) = 24 - 24e^4$$

$$v = \frac{24 - 24e^4}{-1 - e^4} \approx 23.14$$

The velocity is 23.14 feet per second.

b. The vertical asymptote is $v = 24$.

c. Because of the air resistance, the object cannot exceed a velocity of 24 feet per second.

Exercise Set 4.6, page 361

4. a. $P = 12{,}500$, $r = 0.08$, $t = 10$, $n = 1$.

$$A = 12{,}500\left(1 + \frac{0.08}{1}\right)^{10} \approx \$26{,}986.56$$

b. $n = 365$

$$A = 12{,}500\left(1 + \frac{0.08}{365}\right)^{3650} \approx \$27{,}816.82$$

c. $n = 8760$

$$A = 12{,}500\left(1 + \frac{0.08}{8760}\right)^{87600} \approx \$27{,}819.16$$

6. $P = 32{,}000$, $r = 0.08$, $t = 3$.

$$A = Pe^{rt} = 32{,}000e^{3(0.08)} \approx \$40{,}679.97$$

10. $t = \dfrac{\ln 3}{r}$ $\quad r = 0.055$

$$t = \frac{\ln 3}{0.055}$$

$t \approx 20$ years (to the nearest year)

18. a. $P(12) = 20{,}899(1.027)^{12} \approx 28{,}772$ thousands, or 28,772,000.

b. P is in thousands, so

$$35{,}000 = 20{,}899(1.027)^t$$

$$\frac{35{,}000}{20{,}899} = 1.027^t$$

$$\ln\left(\frac{35{,}000}{20{,}899}\right) = t \ln 1.027$$

$$\frac{\ln\left(\frac{35{,}000}{20{,}899}\right)}{\ln 1.027} = t$$

$$19.35 \approx t$$

According to the growth function, the population will first exceed 35 million in 19.35 years—that is, in the year 1991 + 19 = 2010.

20.

$$N(t) = N_0e^{kt}$$

$$N(138) = N_0e^{138k}$$

$$0.5N_0 = N_0e^{138k}$$

$$0.5 = e^{138k}$$

$$\ln 0.5 = 138k$$

$$k = \frac{\ln 0.5}{138} \approx -0.005023$$

$$N(t) = N_0(0.5)^{t/138} \approx N_0e^{-0.005023t}$$

24.

$$N(t) = N_0(0.5)^{t/5730}$$

$$0.65N_0 = N_0(0.5)^{t/5730}$$

$$0.65 = (0.5)^{t/5700}$$

$$\ln 0.65 = \ln (0.5)^{t/5730}$$

$$t = 5730\frac{\ln 0.65}{\ln 0.5} \approx 3600$$

The bone is approximately 3600 years old.

32. a.

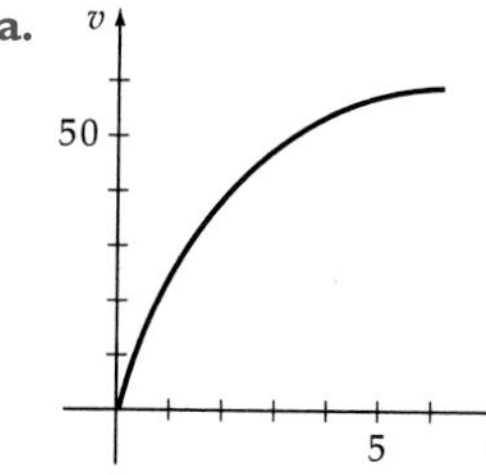

b. Here is an algebraic solution. An approximate solution can be obtained from the graph.

$$v = 64(1 - e^{-1/2})$$

$$50 = 64(1 - e^{-t/2})$$

$$\frac{50}{64} = (1 - e^{-t/2})$$

$$1 - \frac{50}{64} = e^{-t/2}$$

$$\ln\left(1 - \frac{50}{64}\right) = -\frac{t}{2}$$

$$t = -2\ln\left(1 - \frac{50}{64}\right) \approx 3.0$$

The velocity is 50 feet per second in approximately 3.0 seconds.

c. As $t \to \infty$, $e^{-t/2} \to 0$. Therefore, $64(1 - e^{-t/2}) \to 64$. The horizontal asymptote is $v = 64$.

d. Because of the air resistance, the velocity of the object will never exceed 64 feet per second.

36. a.

$$P(t) = \frac{mP_0}{P_0 + (m - P_0)e^{-kt}}$$

$$900 = \frac{(5500)800}{800 + 4700e^{-k}}$$ **• $t = 1$, $P(1) = 900$, $m = 5500$, $P_0 = 800$**

$$800 + 4700e^{-k} = \frac{(5500)800}{900}$$

$$4700e^{-k} = \frac{(5500)800}{900} - 800$$

$$e^{-k} = \frac{\frac{(5500)800}{900} - 800}{4700}$$

$$k = -\ln\left(\frac{\frac{(5500)800}{900} - 800}{4700}\right)$$

$$k \approx 0.14$$

b.
$$P(t) = \frac{5500(800)}{800 + 4700e^{-0.14t}}$$

$$2000 = \frac{5500(800)}{800 + 4700e^{-0.14t}}$$

$$800 + 4700e^{-0.14t} = \frac{5500(800)}{2000}$$

$$e^{-0.14t} = \frac{1400}{4700}$$

$$-0.14t = \ln\frac{14}{47}$$

$$t \approx 8.65$$

The population will first exceed 2000 in the eighth year after 1998, or in 2006.

Exercise Set 4.7, page 375

2. The data can be effectively modeled by a linear function, a quadratic function, a decreasing exponential function, or a decreasing logarithmic function.

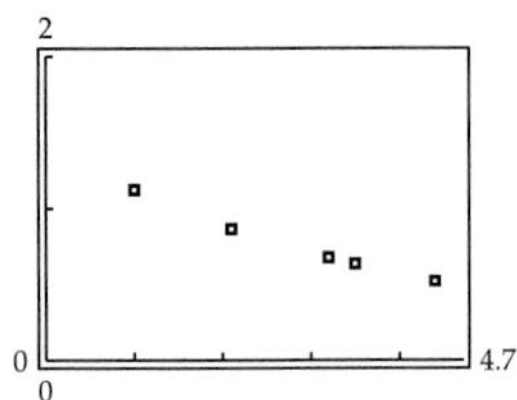

8. $y \approx 1.48874 \cdot 2.50469^x \qquad r \approx 0.999985$

16. $y \approx \dfrac{78.02635}{1 + 2.09585e^{-1.19988x}}$

20. a. *Logistic model:*

$$\text{Distance} \approx \frac{71.84158}{1 + 13.77825e^{-0.07915t}}$$

Logarithmic model:

$$\text{Distance} \approx -22.58293 + 20.91655 \ln t$$

b. A graph shows that the logistic model provides a better fit to the data.

c. The year 2008 is represented by $t = 108$.

$$\text{Distance}(108) \approx \frac{71.84158}{1 + 13.77825e^{-0.07915 \cdot 108}}$$

$$\approx 71.65 \text{ feet (nearest 0.01 foot)}$$

Exercise Set 5.1, page 399

2. The measure of the complement of an angle of 87° is

$$(90° - 87°) = 3°$$

The measure of the supplement of an angle of 87° is

$$(180° - 87°) = 93°$$

14. Because $765° = 2 \cdot 360° + 45°$, α is coterminal with an angle that has a measure of 45°. α is a Quadrant I angle.

32. $-45° = -45°\left(\dfrac{\pi \text{ radians}}{180°}\right) = -\dfrac{\pi}{4} \text{ radians}$

40. $\dfrac{\pi}{4} \text{ radians} = \dfrac{\pi}{4} \text{ radians}\left(\dfrac{180°}{\pi \text{ radians}}\right) = 45°$

62. $s = r\theta = 5\left(144° \cdot \dfrac{\pi}{180°}\right) \approx 12.57 \text{ meters}$

66. Let θ_2 be the angle through which the pulley with a diameter of 0.8 meter turns. Let θ_1 be the angle through which the pulley with a diameter of 1.2 meters turns. Let $r_2 = 0.4$ meter be the radius of the smaller pulley, and let $r_1 = 0.6$ meter be the radius of the larger pulley.

$$\theta_1 = 240° = \frac{4}{3}\pi \text{ radians}$$

Thus $r_2\theta_2 = r_1\theta_1$

$$0.4\theta_2 = 0.6\left(\frac{4}{3}\pi\right)$$

$$\theta_2 = \frac{0.6}{0.4}\left(\frac{4}{3}\pi\right) = 2\pi \text{ radians or } 360°$$

68. The earth makes 1 revolution ($\theta = 2\pi$) in 1 day.

$$t = 24 \cdot 3600 = 86{,}400 \text{ seconds}$$

$$\omega = \frac{\theta}{t} = \frac{2\pi}{86{,}400} \approx 7.27 \times 10^{-5} \text{ radian/second}$$

74. $C = 2\pi r = 2\pi(18 \text{ inches}) = 36\pi \text{ inches}$
Thus one conversion factor is $(36\pi \text{ inches}/1 \text{ rev})$.

$$\frac{500 \text{ rev}}{1 \text{ minute}} = \frac{500 \text{ rev}}{1 \text{ minute}}\left(\frac{36\pi \text{ inches}}{1 \text{ rev}}\right) = \frac{18{,}000\pi \text{ inches}}{1 \text{ minute}}$$

Now convert inches to miles and minutes to hours.

$$\frac{18{,}000\pi \text{ inches}}{1 \text{ minute}} = \frac{18{,}000\pi \text{ inches}}{1 \text{ minute}}\left(\frac{1 \text{ foot}}{12 \text{ inches}}\right)\left(\frac{1 \text{ mile}}{5280 \text{ feet}}\right)\left(\frac{60 \text{ minutes}}{1 \text{ hour}}\right)$$

$$\approx 54 \text{ mph}$$

Exercise Set 5.2, page 410

6.

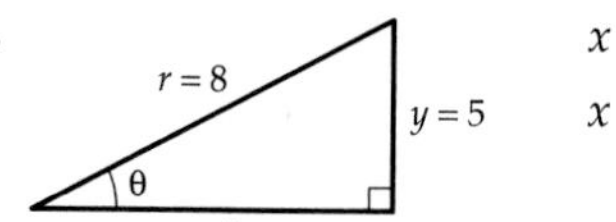

$$x = \sqrt{8^2 - 5^2}$$
$$x = \sqrt{64 - 25} = \sqrt{39}$$

$$\sin\theta = \frac{y}{r} = \frac{5}{8} \qquad \csc\theta = \frac{r}{y} = \frac{8}{5}$$

$$\cos\theta = \frac{x}{r} = \frac{\sqrt{39}}{8} \qquad \sec\theta = \frac{r}{x} = \frac{8}{\sqrt{39}} = \frac{8\sqrt{39}}{39}$$

$$\tan\theta = \frac{y}{x} = \frac{5}{\sqrt{39}} = \frac{5\sqrt{39}}{39} \qquad \cot\theta = \frac{x}{y} = \frac{\sqrt{39}}{5}$$

20. Because $\tan\theta = \frac{y}{x} = \frac{4}{3}$, let $y = 4$ and $x = 3$.

$$r = \sqrt{3^2 + 4^2} = 5$$

$$\sec\theta = \frac{r}{x} = \frac{5}{3}$$

38. $$\sin\frac{\pi}{3}\cos\frac{\pi}{4} - \tan\frac{\pi}{4} = \frac{\sqrt{3}}{2}\cdot\frac{\sqrt{2}}{2} - 1$$
$$= \frac{\sqrt{6}}{4} - 1 = \frac{\sqrt{6} - 4}{4}$$

66.

$$\tan 68.9° = \frac{h}{116}$$
$$h = 116 \tan 68.9°$$
$$h \approx 301 \text{ m (3 significant digits)}$$

68.

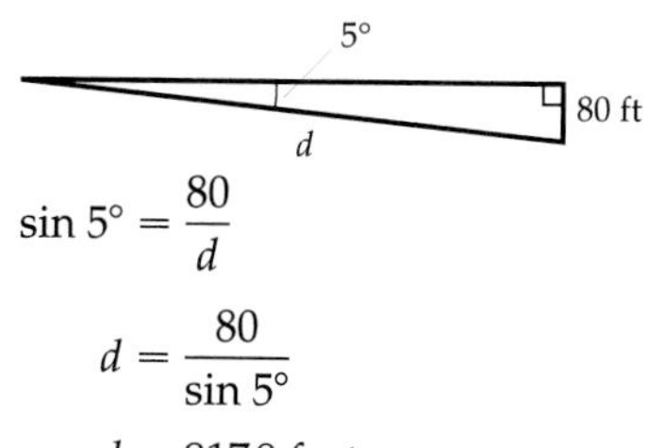

$$\sin 5° = \frac{80}{d}$$
$$d = \frac{80}{\sin 5°}$$
$$d \approx 917.9 \text{ feet}$$

Change 9 mph to feet per minute.

$$r = 9\frac{\text{miles}}{\text{hour}} = \frac{9 \text{ miles}}{1 \text{ hour}}\cdot\frac{5280 \text{ feet}}{1 \text{ mile}}\cdot\frac{1 \text{ hour}}{60 \text{ minutes}}$$
$$= \frac{9(5280)}{60}\,\frac{\text{feet}}{\text{minute}} = 792\,\frac{\text{feet}}{\text{minute}}$$

$$t = \frac{d}{r}$$

$$t \approx \frac{917.9 \text{ feet}}{792 \text{ feet per minute}} \approx 1.16 \text{ minutes}$$

72.

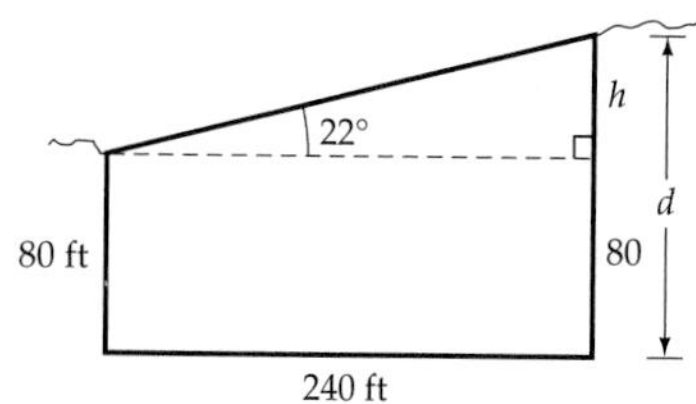

$$\tan 22° = \frac{h}{240}$$
$$h = 240 \tan 22°$$
$$d = 80 + h$$
$$d = 80 + 240 \tan 22°$$
$$d \approx 180 \text{ feet (2 significant digits)}$$

Exercise Set 5.3, page 421

6. $x = -6, y = -9, r = \sqrt{(-6)^2 + (-9)^2} = \sqrt{117} = 3\sqrt{13}$

$$\sin\theta = \frac{y}{r} = \frac{-9}{3\sqrt{13}} = -\frac{3}{\sqrt{13}} = -\frac{3\sqrt{13}}{13} \qquad \csc\theta = -\frac{\sqrt{13}}{3}$$

$$\cos\theta = \frac{x}{r} = \frac{-6}{3\sqrt{13}} = -\frac{2}{\sqrt{13}} = -\frac{2\sqrt{13}}{13} \qquad \sec\theta = -\frac{\sqrt{13}}{2}$$

$$\tan\theta = \frac{y}{x} = \frac{-9}{-6} = \frac{3}{2} \qquad \cot\theta = \frac{2}{3}$$

18. $\sec\theta = \frac{2\sqrt{3}}{3} = \frac{r}{x}$ **• Let $r = 2\sqrt{3}$ and $x = 3$.**

$$y = \pm\sqrt{(2\sqrt{3})^2 - 3^2} = \pm\sqrt{3}$$
$y = -\sqrt{3}$ because $y < 0$ in Quadrant IV
$$\sin\theta = \frac{-\sqrt{3}}{2\sqrt{3}} = -\frac{1}{2}$$

26. $\theta' = 255° - 180° = 75°$

38. $\cos 300° > 0$, $\theta' = 360° - 300° = 60°$

Thus $\cos 300° = \cos 60° = \frac{1}{2}$.

Exercise Set 5.4, page 431

10. $t = -\frac{7\pi}{4}$; $W(t) = P(x, y)$ where

$$y = \sin t \qquad x = \cos t$$
$$= \sin\left(-\frac{7\pi}{4}\right) \qquad = \cos\left(-\frac{7\pi}{4}\right)$$
$$= \sin\frac{\pi}{4} \qquad = \cos\frac{\pi}{4}$$
$$= \frac{\sqrt{2}}{2} \qquad = \frac{\sqrt{2}}{2}$$

$$W\left(-\frac{7\pi}{4}\right) = \left(\frac{\sqrt{2}}{2}, \frac{\sqrt{2}}{2}\right)$$

16. The reference angle for $-\dfrac{5\pi}{6}$ is $\dfrac{\pi}{6}$.

$$\sec\left(-\frac{5\pi}{6}\right) = -\sec\frac{\pi}{6} \qquad \text{• sec } t < 0 \text{ for } t \text{ in Quadrant III}$$
$$= -\frac{2\sqrt{3}}{3}$$

36. $F(-x) = \tan(-x) + \sin(-x)$

$= -\tan x - \sin x$ **• tan *x* and sin *x* are odd functions.**

$= -(\tan x + \sin x)$

$= -F(x)$

Because $F(-x) = -F(x)$, the function defined by $F(x) = \tan x + \sin x$ is an odd function.

42.

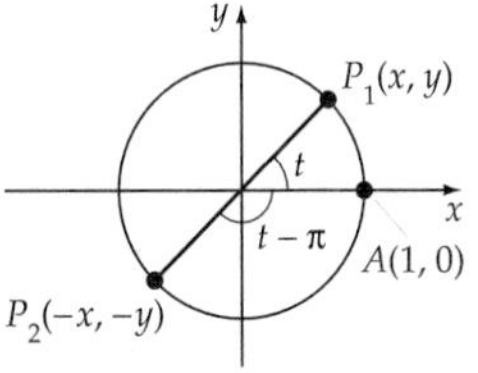

$$\tan t = \frac{y}{x}$$

$$\tan(t - \pi) = \frac{-y}{-x} = \frac{y}{x} \qquad \text{• From the unit circle}$$

Therefore, $\tan t = \tan(t - \pi)$.

60. $$\frac{1}{1 - \sin t} + \frac{1}{1 + \sin t} = \frac{1 + \sin t + 1 - \sin t}{(1 - \sin t)(1 + \sin t)}$$
$$= \frac{2}{1 - \sin^2 t}$$
$$= \frac{2}{\cos^2 t}$$
$$= 2\sec^2 t$$

66. $1 + \tan^2 t = \sec^2 t$

$\tan^2 t = \sec^2 t - 1$

$\tan t = \pm\sqrt{\sec^2 t - 1}$

Because $3\pi/2 < t < 2\pi$, $\tan t$ is negative. Thus $\tan t = -\sqrt{\sec^2 t - 1}$.

70. March 5 is represented by $t = 2$.

$$T(2) = -41\cos\left(\frac{\pi}{6}\cdot 2\right) + 36$$
$$= -41\cos\left(\frac{\pi}{3}\right) + 36$$
$$= -41(0.5) + 36$$
$$= 15.5°\text{F}$$

July 20 is represented by $t = 6.5$.

$$T(6.5) = -41\cos\left(\frac{\pi}{6}\cdot 6.5\right) + 36$$
$$\approx -41(-0.9659258263) + 36$$
$$\approx 75.6°\text{F}$$

Exercise Set 5.5, page 440

20. $y = -\dfrac{3}{2}\sin x$

$a = \left|-\dfrac{3}{2}\right| = \dfrac{3}{2}$

period $= 2\pi$

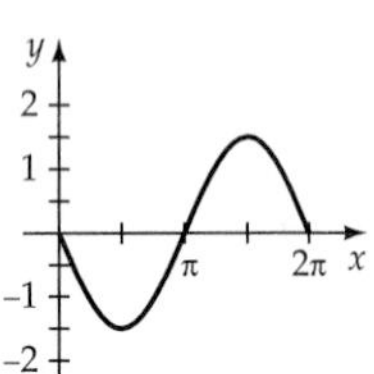

30. $y = \sin\dfrac{3\pi}{4}x$

$a = 1$

period $= \dfrac{2\pi}{b} = \dfrac{2\pi}{\frac{3\pi}{4}} = \dfrac{8}{3}$

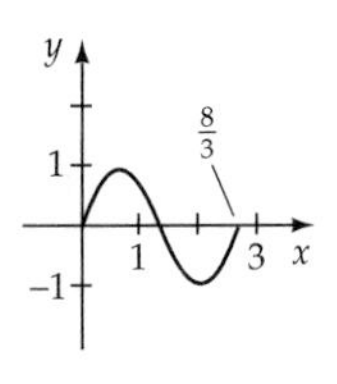

32. $y = \cos 3\pi x$

$a = 1$

period $= \dfrac{2\pi}{b} = \dfrac{2\pi}{3\pi} = \dfrac{2}{3}$

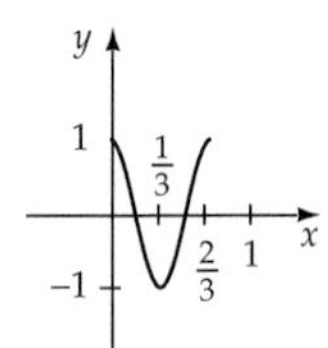

38. $y = \dfrac{1}{2}\sin\dfrac{\pi x}{3}$

$a = \dfrac{1}{2}$

period $= \dfrac{2\pi}{b} = \dfrac{2\pi}{\pi/3} = 6$

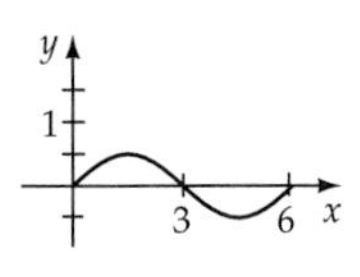

46. $y = -\dfrac{3}{4}\cos 5x$

$a = \left|-\dfrac{3}{4}\right| = \dfrac{3}{4}$

period $= \dfrac{2\pi}{b} = \dfrac{2\pi}{5}$

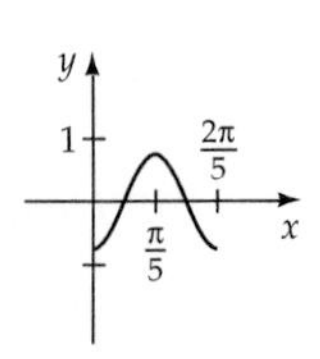

52. $y = -\left|3\sin\dfrac{2}{3}x\right|$

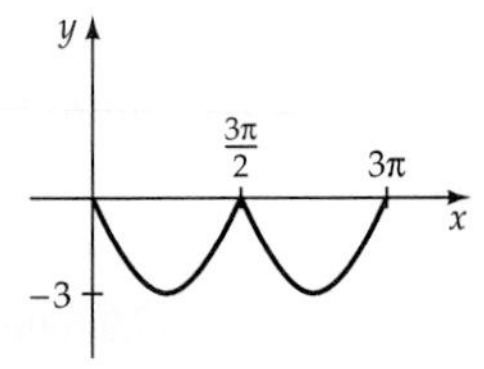

Exercise Set 5.6, page 448

22. $y = \frac{1}{3}\tan x$

period $= \frac{\pi}{b} = \pi$

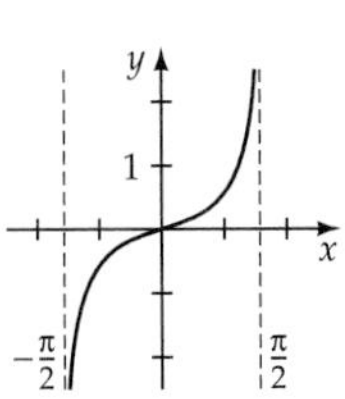

30. $y = -3\tan 3x$

period $= \frac{\pi}{b} = \frac{\pi}{3}$

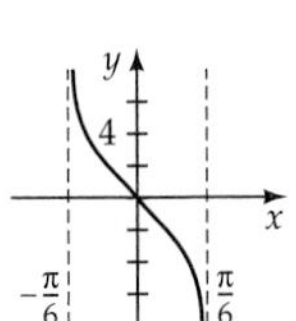

32. $y = \frac{1}{2}\cot 2x$

period $= \frac{\pi}{b} = \frac{\pi}{2}$

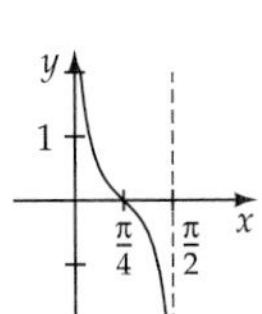

38. $y = 3\csc\frac{\pi x}{2}$

period $= \frac{2\pi}{b} = \frac{2\pi}{\pi/2} = 4$

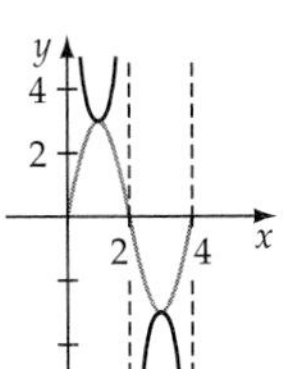

42. $y = \sec\frac{x}{2}$

period $= \frac{2\pi}{b} = \frac{2\pi}{1/2} = 4\pi$

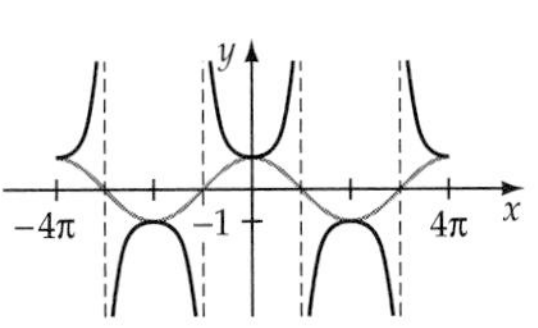

Exercise Set 5.7, page 456

20. $y = \cos\left(2x - \frac{\pi}{3}\right)$

$a = 1$

period $= \pi$

phase shift $= -\frac{c}{b}$

$= -\frac{-\pi/3}{2} = \frac{\pi}{6}$

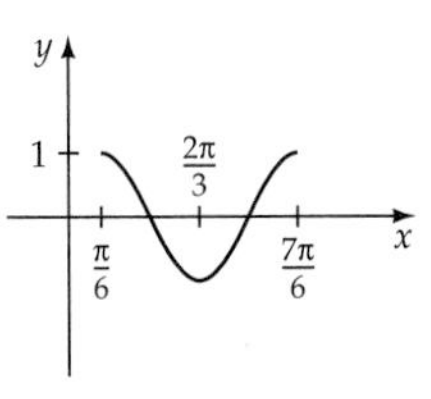

22. $y = \tan(x - \pi)$

period $= \pi$

phase shift $= -\frac{c}{b}$

$= -\frac{-\pi}{1} = \pi$

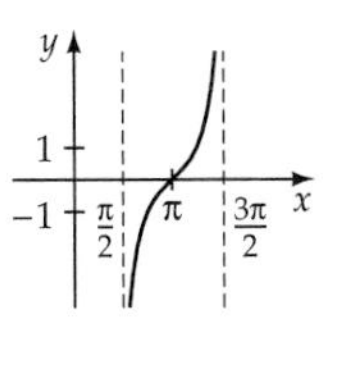

40. $y = 2\sin\left(\frac{\pi x}{2} + 1\right) - 2$

$a = 2$

period $= 4$

phase shift $= -\frac{c}{b}$

$= -\frac{1}{\pi/2} = -\frac{2}{\pi}$

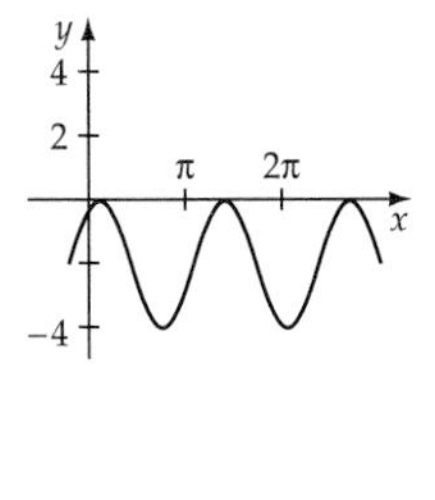

42. $y = -3\cos(2\pi x - 3) + 1$

$a = 3$

period $= 1$

phase shift $= -\frac{c}{b} = \frac{3}{2\pi}$

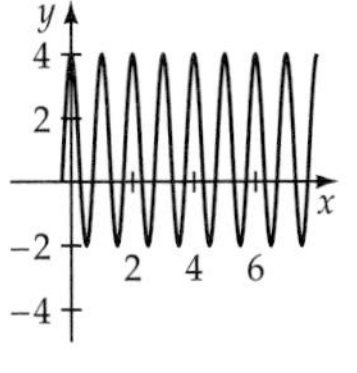

48. $y = \csc\frac{x}{3} + 4$

period $= 6\pi$

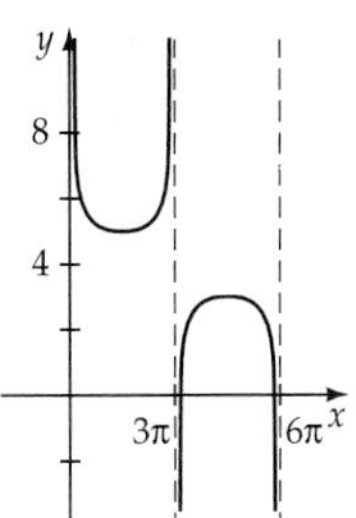

52. a. Phase shift: $-\frac{c}{b} = -\frac{\left(-\frac{7}{12}\pi\right)}{\left(\frac{\pi}{6}\right)} = 3.5$ months,

period: $\frac{2\pi}{b} = \frac{2\pi}{\left(\frac{\pi}{6}\right)} = 12$ months

b. First graph $y_1 = 2.7\cos\left(\frac{\pi}{6}t\right)$. Because the phase shift is 3.5 months, shift the graph of y_1 3.5 units to the right to produce the graph of y_2. Now shift the graph of y_2 upward 4 units to produce the graph of S.

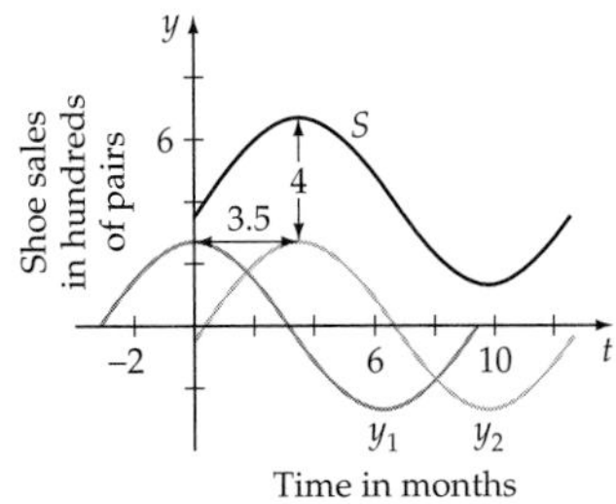

c. 3.5 months after January 1 is the middle of April.

54. $y = \dfrac{x}{2} + \cos x$

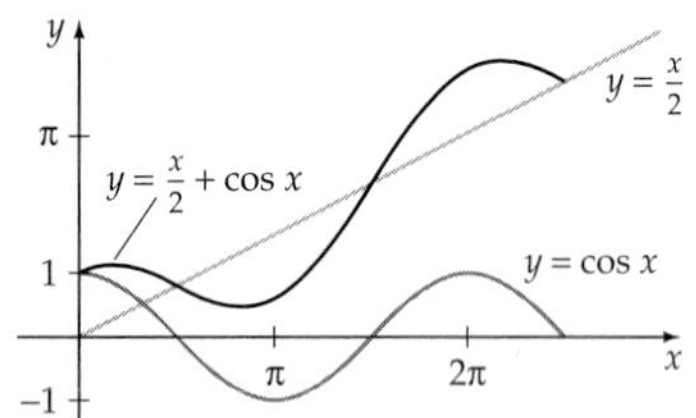

58. $y = -\sin x + \cos x$

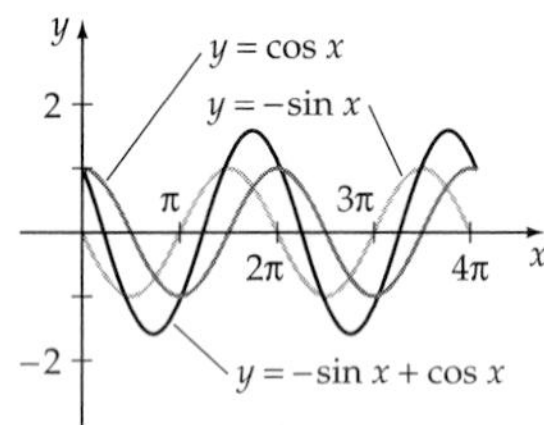

78. $y = x \cos x$

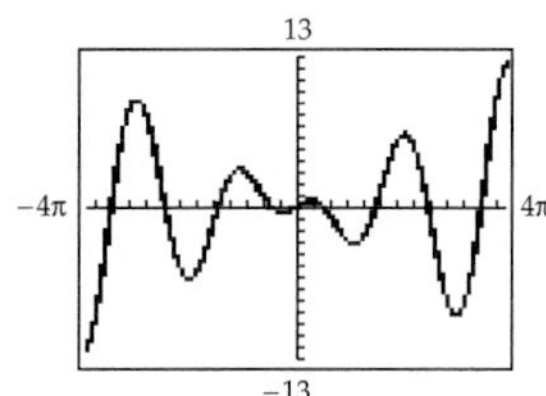

Exercise Set 5.8, page 464

20. Amplitude $= 3$, frequency $= 1/\pi$, period $= \pi$.

Because $2\pi/b = \pi$, we have $b = 2$. Thus $y = 3\cos 2t$.

28. Amplitude $= |-1.5| = 1.5$

$$f = \frac{1}{2\pi}\sqrt{\frac{k}{m}} = \frac{1}{2\pi}\sqrt{\frac{3}{27}} = \frac{1}{2\pi}\cdot\frac{1}{3} = \frac{1}{6\pi}, \text{ period} = 6\pi$$

$$y = a\cos 2\pi f t = -1.5\cos\left[2\pi\left(\frac{1}{6\pi}\right)t\right] = -1.5\cos\frac{1}{3}t$$

30.

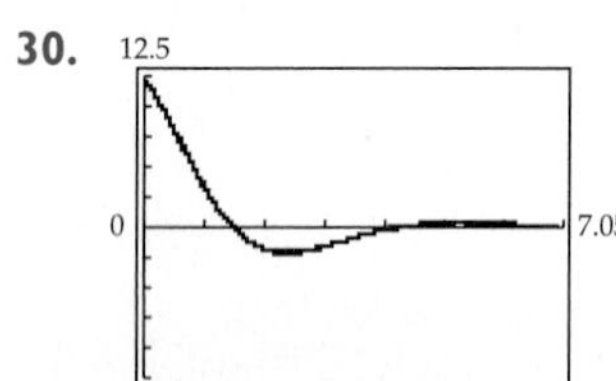

a. f has pseudoperiod $\dfrac{2\pi}{1} = 2\pi$.

$10 \div (2\pi) \approx 1.59$

Thus f completes only one full oscillation on $0 \le t \le 10$.

b. The following graph of f shows that $|f(t)| < 0.01$ for $t > 10.5$.

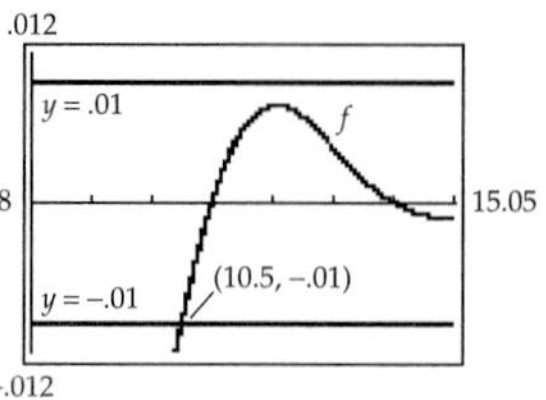

Exercise Set 6.1, page 476

2. The equation $\tan 2x = 2\tan x$ is not an identity. To verify that it is not an identity, let $x = \pi/6$. Then

$$\tan 2x = \tan 2\left(\frac{\pi}{6}\right) = \tan\frac{\pi}{3} = \sqrt{3}$$

whereas $2\tan x = 2\tan \pi/6 = 2\sqrt{3}/3$. Because $\sqrt{3} \neq 2(\sqrt{3}/3)$, we have shown that the equation is not an identity.

24.
$$\begin{aligned}\sin^4 x - \cos^4 x &= (\sin^2 x + \cos^2 x)(\sin^2 x - \cos^2 x)\\ &= 1(\sin^2 x - \cos^2 x) = \sin^2 x - \cos^2 x\end{aligned}$$

36.
$$\begin{aligned}&\frac{2\sin x\cot x + \sin x - 4\cot x - 2}{2\cot x + 1}\\ &= \frac{(\sin x)(2\cot x + 1) - 2(2\cot x + 1)}{2\cot x + 1}\\ &= \frac{(2\cot x + 1)(\sin x - 2)}{2\cot x + 1} = \sin x - 2\end{aligned}$$

46.
$$\begin{aligned}\frac{\frac{1}{\sin x} + \frac{1}{\cos x}}{\frac{1}{\sin x} - \frac{1}{\cos x}} &= \frac{\frac{1}{\sin x} + \frac{1}{\cos x}}{\frac{1}{\sin x} - \frac{1}{\cos x}}\cdot\frac{\sin x\cos x}{\sin x\cos x}\\ &= \frac{\cos x + \sin x}{\cos x - \sin x}\\ &= \frac{\cos x + \sin x}{\cos x - \sin x}\cdot\frac{\cos x - \sin x}{\cos x - \sin x}\\ &= \frac{\cos^2 x - \sin^2 x}{\cos^2 x - 2\sin x\cos x + \sin^2 x}\\ &= \frac{\cos^2 x - \sin^2 x}{1 - 2\sin x\cos x}\end{aligned}$$

56.
$$\begin{aligned}\frac{\frac{1}{\tan x} + \cot x}{\frac{1}{\tan x} + \tan x} &= \frac{\frac{1}{\tan x} + \cot x}{\frac{1}{\tan x} + \tan x}\cdot\frac{\tan x}{\tan x}\\ &= \frac{1 + 1}{1 + \tan^2 x} = \frac{2}{\sec^2 x}\end{aligned}$$

Exercise Set 6.2, page 484

20. $\sin x \cos 3x + \cos x \sin 3x = \sin(x + 3x) = \sin 4x$

32. $\tan \alpha = 24/7$, with $0° < \alpha < 90°$, $\sin \alpha = 24/25$,

$\cos \alpha = 7/25$

$\sin \beta = -8/17$, with $180° < \beta < 270°$

$\cos \beta = -15/17$, $\tan \beta = 8/15$

a. $$\sin(\alpha + \beta) = \sin \alpha \cos \beta + \cos \alpha \sin \beta = \left(\frac{24}{25}\right)\left(-\frac{15}{17}\right) + \left(\frac{7}{25}\right)\left(-\frac{8}{17}\right) = -\frac{360}{425} - \frac{56}{425} = -\frac{416}{425}$$

b. $$\cos(\alpha + \beta) = \cos \alpha \cos \beta - \sin \alpha \sin \beta = \left(\frac{7}{25}\right)\left(-\frac{15}{17}\right) - \left(\frac{24}{25}\right)\left(-\frac{8}{17}\right) = -\frac{105}{425} + \frac{192}{425} = \frac{87}{425}$$

c. $$\tan(\alpha - \beta) = \frac{\tan \alpha - \tan \beta}{1 + \tan \alpha \tan \beta} = \frac{\frac{24}{7} - \frac{8}{15}}{1 + \left(\frac{24}{7}\right)\left(\frac{8}{15}\right)} = \frac{\frac{24}{7} - \frac{8}{15}}{1 + \frac{192}{105}} \cdot \frac{105}{105} = \frac{360 - 56}{105 + 192} = \frac{304}{297}$$

44. $$\cos(\theta + \pi) = \cos \theta \cos \pi - \sin \theta \sin \pi = (\cos \theta)(-1) - (\sin \theta)(0) = -\cos \theta$$

56. $$\cos 5x \cos 3x + \sin 5x \sin 3x = \cos(5x - 3x) = \cos 2x = \cos(x + x) = \cos x \cos x - \sin x \sin x = \cos^2 x - \sin^2 x$$

68. $$\sin(\theta + 2\pi) = \sin \theta \cos 2\pi + \cos \theta \sin 2\pi = (\sin \theta)(1) + (\cos \theta)(0) = \sin \theta$$

Exercise Set 6.3, page 491

2. $2 \sin 3\theta \cos 3\theta = \sin[2(3\theta)] = \sin 6\theta$

26. $\cos \theta = 24/25$ with $270° < \theta < 360°$

$$\sin \theta = -\sqrt{1 - \left(\frac{24}{25}\right)^2} = -\frac{7}{25} \qquad \tan \theta = \frac{-7/25}{24/25} = -\frac{7}{24}$$

$$\sin 2\theta = 2 \sin \theta \cos \theta = 2\left(\frac{-7}{25}\right)\left(\frac{24}{25}\right) = -\frac{336}{625} \qquad \cos 2\theta = \cos^2 \theta - \sin^2 \theta = \left(\frac{24}{25}\right)^2 - \left(-\frac{7}{25}\right)^2 = \frac{527}{625}$$

$$\tan 2\theta = \frac{2 \tan \theta}{1 - \tan^2 \theta} = \frac{2(-7/24)}{1 - (-7/24)^2} = \frac{-7/12}{1 - 49/576} \cdot \frac{576}{576} = -\frac{336}{527}$$

54. $$\frac{1}{1 - \cos 2x} = \frac{1}{1 - 1 + 2\sin^2 x} = \frac{1}{2\sin^2 x} = \frac{1}{2}\csc^2 x$$

72. $$\cos^2 \frac{x}{2} = \left[\pm\sqrt{\frac{1 + \cos x}{2}}\right]^2 = \frac{1 + \cos x}{2} = \frac{1 + \cos x}{2} \cdot \frac{\sec x}{\sec x} = \frac{\sec x + 1}{2 \sec x}$$

78. $$\tan^2 \frac{x}{2} = \left(\frac{1 - \cos x}{\sin x}\right)^2 = \frac{(1 - \cos x)^2}{\sin^2 x} = \frac{(1 - \cos x)^2}{1 - \cos^2 x} = \frac{(1 - \cos x)^2}{(1 - \cos x)(1 + \cos x)} = \frac{1 - \cos x}{1 + \cos x} = \frac{\frac{1}{\cos x} - \frac{\cos x}{\cos x}}{\frac{1}{\cos x} + \frac{\cos x}{\cos x}} = \frac{\sec x - 1}{\sec x + 1}$$

Exercise Set 6.4, page 500

22. $$\cos 3\theta + \cos 5\theta = 2 \cos \frac{3\theta + 5\theta}{2} \cos \frac{3\theta - 5\theta}{2} = 2 \cos 4\theta \cos(-\theta) = 2 \cos 4\theta \cos \theta$$

36. $$\sin 5x \cos 3x = \frac{1}{2}[\sin(5x + 3x) + \sin(5x - 3x)] = \frac{1}{2}(\sin 8x + \sin 2x) = \frac{1}{2}(2 \sin 4x \cos 4x + 2 \sin x \cos x) = \sin 4x \cos 4x + \sin x \cos x$$

44. $$\frac{\cos 5x - \cos 3x}{\sin 5x + \sin 3x} = \frac{-2 \sin \frac{5x + 3x}{2} \sin \frac{5x - 3x}{2}}{2 \sin \frac{5x + 3x}{2} \cos \frac{5x - 3x}{2}} = -\frac{\sin 4x \sin x}{\sin 4x \cos x} = -\tan x$$

62. $a = 1, b = \sqrt{3}, k = \sqrt{(\sqrt{3})^2 + (1)^2} = 2$. Thus α is a first-quadrant angle.

$$\sin \alpha = \frac{\sqrt{3}}{2} \quad \text{and} \quad \cos \alpha = \frac{1}{2}$$

Thus $\alpha = \pi/3$.

$$y = k \sin (x + \alpha).$$

$$y = 2 \sin \left(x + \frac{\pi}{3}\right)$$

70. From Exercise 62, we know that

$$y = \sin x + \sqrt{3} \cos x = 2 \sin \left(x + \frac{\pi}{3}\right)$$

Thus the graph is the graph of $y = 2 \sin x$ shifted $\pi/3$ units to the left.

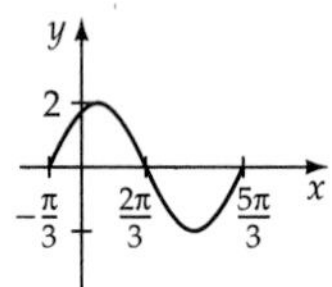

Exercise Set 6.5, page 513

2. $y = \sin^{-1} \dfrac{\sqrt{2}}{2}$ implies

$$\sin y = \frac{\sqrt{2}}{2} \quad \text{for } -\frac{\pi}{2} \le y \le \frac{\pi}{2}$$

Thus $y = \pi/4$.

24. Because $\tan (\tan^{-1} x) = x$ for all real numbers, we have $\tan [\tan^{-1} (1/2)] = 1/2$.

46. Let $x = \cos^{-1} 3/5$. Thus

$$\cos x = \frac{3}{5} \quad \text{and} \quad \sin x = \sqrt{1 - \left(\frac{3}{5}\right)^2} = \frac{4}{5}$$

$$y = \tan \left(\cos^{-1} \frac{3}{5}\right) = \tan x = \frac{\sin x}{\cos x} = \frac{4/5}{3/5} = \frac{4}{3}$$

54. $y = \cos \left(\sin^{-1} \dfrac{3}{4} + \cos^{-1} \dfrac{5}{13}\right)$

Let $\alpha = \sin^{-1} \dfrac{3}{4}$, $\sin \alpha = \dfrac{3}{4}$,

$$\cos \alpha = \sqrt{1 - \left(\frac{3}{4}\right)^2} = \frac{\sqrt{7}}{4}.$$

$$\beta = \cos^{-1} \frac{5}{13}, \cos \beta = \frac{5}{13}, \sin \beta = \sqrt{1 - \left(\frac{5}{13}\right)^2} = \frac{12}{13}.$$

$$y = \cos (\alpha + \beta)$$
$$= \cos \alpha \cos \beta - \sin \alpha \sin \beta$$
$$= \frac{\sqrt{7}}{4} \cdot \frac{5}{13} - \frac{3}{4} \cdot \frac{12}{13} = \frac{5\sqrt{7}}{52} - \frac{36}{52} = \frac{5\sqrt{7} - 36}{52}$$

64. $\sin^{-1} x + \cos^{-1} \dfrac{4}{5} = \dfrac{\pi}{6}$

$$\sin^{-1} x = \frac{\pi}{6} - \cos^{-1} \frac{4}{5}$$

$$\sin (\sin^{-1} x) = \sin \left(\frac{\pi}{6} - \cos^{-1} \frac{4}{5}\right)$$

$$x = \sin \frac{\pi}{6} \cos \left(\cos^{-1} \frac{4}{5}\right) - \cos \frac{\pi}{6} \sin \left(\cos^{-1} \frac{4}{5}\right)$$
$$= \frac{1}{2} \cdot \frac{4}{5} - \frac{\sqrt{3}}{2} \cdot \frac{3}{5} = \frac{4 - 3\sqrt{3}}{10}$$

72. Let $\alpha = \cos^{-1} x$ and $\beta = \cos^{-1} (-x)$. Thus $\cos \alpha = x$ and $\cos \beta = -x$. We have $\sin \alpha = \sqrt{1 - x^2}$ and we have $\sin \beta = \sqrt{1 - x^2}$ because α is in Quadrant I and β is in Quadrant II.

$$\cos^{-1} x + \cos^{-1} (-x)$$
$$= \alpha + \beta$$
$$= \cos^{-1} [\cos (\alpha + \beta)]$$
$$= \cos^{-1} (\cos \alpha \cos \beta - \sin \alpha \sin \beta)$$
$$= \cos^{-1} [x(-x) - \sqrt{1 - x^2} \cdot \sqrt{1 - x^2}]$$
$$= \cos^{-1} (-x^2 - 1 + x^2)$$
$$= \cos^{-1} (-1) = \pi$$

76. Recall that the graph of $y = f(x - a)$ is a horizontal shift of the graph of $y = f(x)$. Therefore, the graph of $y = \cos^{-1} (x - 1)$ is the graph of $y = \cos^{-1} x$ shifted 1 unit to the right.

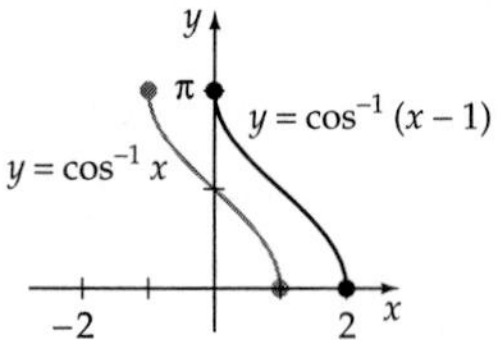

84. **a.**

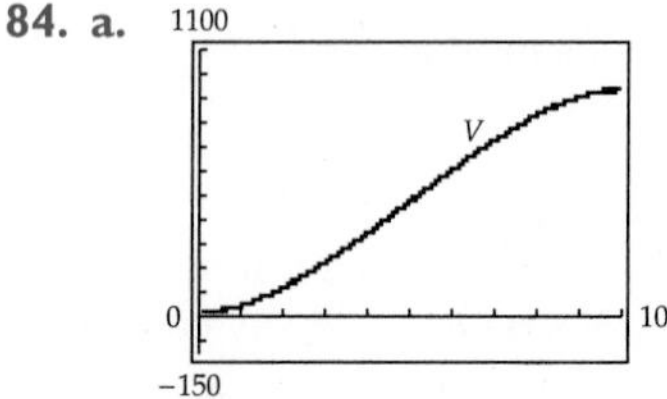

b. Although the water rises 0.1 foot in each case, there is more surface area (and thus more volume of water) at the 4.9-to-5.0-foot level near the diameter of the cylinder than at the 0.1-to-0.2-foot level near the bottom.

c.

$$V(4) = 12\left\{25\cos^{-1}\left(\frac{5-(4)}{5}\right) - [5-(4)]\sqrt{10(4)-(4)^2}\right\}$$

$$= 12\left\{25\cos^{-1}\left(\frac{1}{5}\right) - \sqrt{24}\right\}$$

$$\approx 352.04 \text{ cubic feet}$$

d.

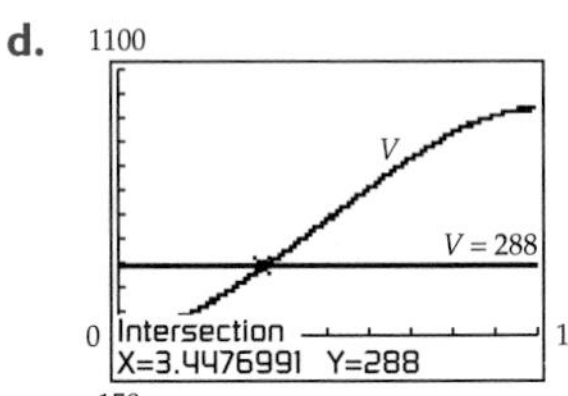

When $V = 288$ cubic feet, $x \approx 3.45$ feet.

Exercise Set 6.6, page 524

14.

$$2\cos^2 x + 1 = -3\cos x$$

$$2\cos^2 x + 3\cos x + 1 = 0$$

$$(2\cos x + 1)(\cos x + 1) = 0$$

$2\cos x + 1 = 0$ or $\cos x + 1 = 0$

$\cos x = -\dfrac{1}{2}$ $\qquad$ $\cos x = -1$

$x = \dfrac{2\pi}{3}, \dfrac{4\pi}{3}$ $\qquad$ $x = \pi$

The solutions in the interval $0 \le x < 2\pi$ are $2\pi/3$, π, and $4\pi/3$.

52. $\sin x + 2\cos x = 1$

$$\sin x = 1 - 2\cos x$$

$$(\sin x)^2 = (1 - 2\cos x)^2$$

$$\sin^2 x = 1 - 4\cos x + 4\cos^2 x$$

$$1 - \cos^2 x = 1 - 4\cos x + 4\cos^2 x$$

$$0 = \cos x(5\cos x - 4)$$

$\cos x = 0$ or $5\cos x - 4 = 0$

$x = 90°, 270°$ $\qquad$ $\cos x = \dfrac{4}{5}$

$x \approx 36.9°, 323.1°$

The solutions in the interval $0° \le x < 360°$ are $90°$ and $323.1°$. (*Note:* $x = 270°$ and $x = 36.9°$ are extraneous solutions. Neither of these satisfies the original equation.)

56. $2\cos^2 x - 5\cos x - 5 = 0$

$$\cos x = \frac{5 \pm \sqrt{(-5)^2 - 4(2)(-5)}}{2(2)} = \frac{5 \pm \sqrt{65}}{4}$$

$\cos x \approx 3.27$ or $\cos x \approx -0.7656$

no solution $\qquad$ $x \approx 140.0°, 220.0°$

The solutions in the interval $0° \le x < 360°$ are $140°$ and $220°$.

66. $\cos 2x = -\dfrac{\sqrt{3}}{2}$

$2x = \dfrac{5\pi}{6} + 2k\pi$ or $2x = \dfrac{7\pi}{6} + 2k\pi$, k an integer

$x = \dfrac{5\pi}{12} + k\pi$ or $x = \dfrac{7\pi}{12} + k\pi$, k an integer

84. $2\sin x\cos x - 2\sqrt{2}\sin x - \sqrt{3}\cos x + \sqrt{6} = 0$

$$2\sin x(\cos x - \sqrt{2}) - \sqrt{3}(\cos x - \sqrt{2}) = 0$$

$$(\cos x - \sqrt{2})(2\sin x - \sqrt{3}) = 0$$

$\cos x = \sqrt{2}$ or $\sin x = \dfrac{\sqrt{3}}{2}$

no solution $\qquad$ $x = \dfrac{\pi}{3}, \dfrac{2\pi}{3}$

The solutions in the interval $0 \le x < 2\pi$ are $\pi/3$ and $2\pi/3$.

86. The following graph shows that the solutions in the interval $[0, 2\pi)$ are $x = 0$ and $x = 1.895$.

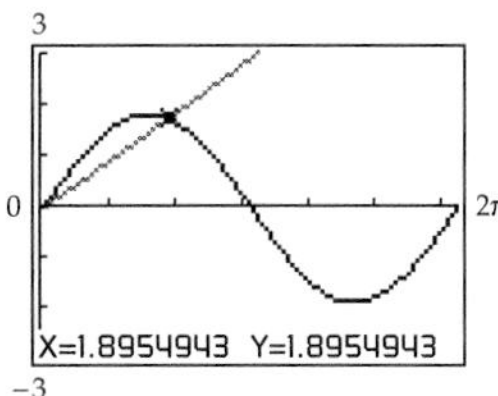

92. When $\theta = 45°$, d attains its maximum of 4394.5 feet.

94. a. The following work shows the sine regression procedure for a TI-83 calculator. Note that each time (given in hours:minutes) must be converted to hours. Thus 17:03 is $17 + 3/60 = 17.05$ hours.

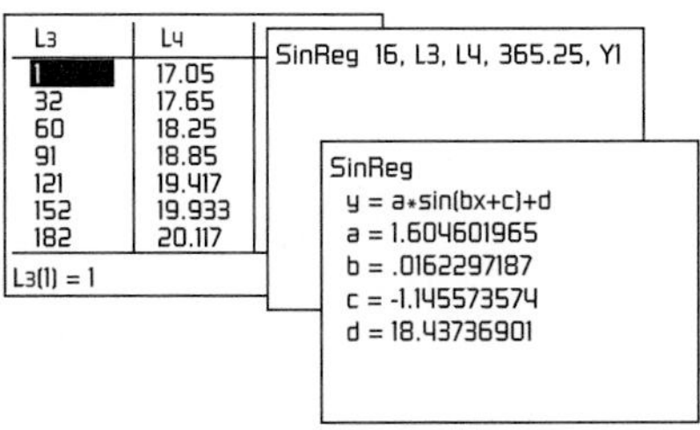

The sine regression function is
$Y_1 \approx 1.6046\sin(0.01623x - 1.1456) + 18.437$

b. Using the "value" command in the CALC menu gives

$$\text{Y1}(141) \approx 19.897246$$
$$\approx 19{:}54 \text{ (to the nearest minute)}$$

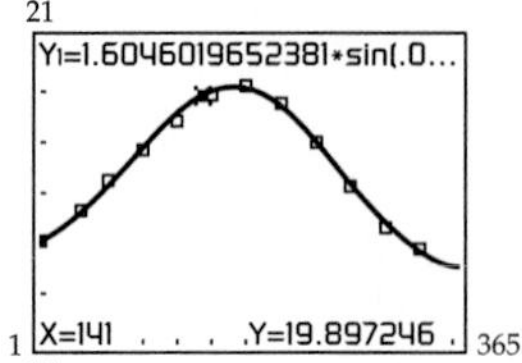

Exercise Set 7.1, page 541

4.

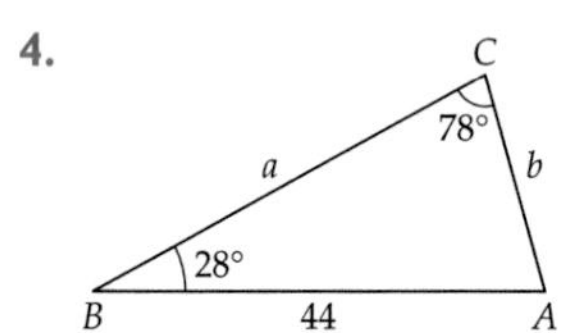

$A = 180° - 78° - 28° = 74°$

$$\frac{b}{\sin B} = \frac{c}{\sin C} \qquad \frac{a}{\sin A} = \frac{c}{\sin C}$$

$$\frac{b}{\sin 28°} = \frac{44}{\sin 78°} \qquad \frac{a}{\sin 74°} = \frac{44}{\sin 78°}$$

$$b = \frac{44 \sin 28°}{\sin 78°} \approx 21 \qquad a = \frac{44 \sin 74°}{\sin 78°} \approx 43$$

18. $\dfrac{a}{\sin A} = \dfrac{b}{\sin B}$

$$\frac{13.8}{\sin A} = \frac{5.55}{\sin 22.6}$$

$\sin A = 0.9555$

$A \approx 72.9°$ or $107.1°$

If $A = 72.9°$, $C \approx 180° - 72.9° - 22.6° = 84.5°$

$$\frac{c}{\sin 84.5°} = \frac{5.55}{\sin 22.6°}$$

$$c = \frac{5.55 \sin 84.5°}{\sin 22.6°} \approx 14.4$$

If $A = 107.1°$, $C \approx 180° - 107.1° - 22.6° = 50.3°$

$$\frac{c}{\sin 50.3°} = \frac{5.55}{\sin 22.6°}$$

$$c = \frac{5.55 \sin 50.3°}{\sin 22.6°} \approx 11.1$$

Case 1: $A = 72.9°$, $C = 84.5°$, and $c = 14.4$

Case 2: $A = 107.1°$, $C = 50.3°$, and $c = 11.1$

26. The angle with its vertex at the helicopter measures $180° - (59.0° + 77.2°) = 43.8°$. Let the distance from the helicopter to the carrier be x. Using the Law of Sines, we have

$$\frac{x}{\sin 77.2°} = \frac{7620}{\sin 43.8°}$$

$$x = \frac{7620 \sin 77.2°}{\sin 43.8°}$$

$$\approx 10{,}700 \text{ feet} \quad \text{(3 significant digits)}$$

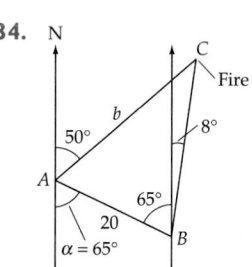

34.

$\alpha = 65°$

$B = 65° + 8° = 73°$

$A = 180° - 50° - 65° = 65°$

$C = 180° - 65° - 73° = 42°$

$$\frac{b}{\sin B} = \frac{c}{\sin C}$$

$$\frac{b}{\sin 73°} = \frac{20}{\sin 42°}$$

$$b = \frac{20 \sin 73°}{\sin 42°}$$

$b \approx 29$ miles

Exercise Set 7.2, page 550

12. $c^2 = a^2 + b^2 - 2ab \cos C$

$c^2 = 14.2^2 + 9.30^2 - 2(14.2)(9.30) \cos 9.20°$

$c = \sqrt{14.2^2 + 9.30^2 - 2(14.2)(9.30) \cos 9.20°}$

$c \approx 5.24$

18. $\cos A = \dfrac{b^2 + c^2 + a^2}{2bc}$

$$\cos A = \frac{132^2 + 160^2 - 108^2}{2(132)(160)} \approx 0.7424$$

$A \approx \cos^{-1}(0.7424) \approx 42.1°$

26. $K = \dfrac{1}{2} ac \sin B$

$K = \dfrac{1}{2}(32)(25) \sin 127° \approx 320$ square units

28. $A = 180° - 102° - 27° = 51°$

$K = \dfrac{a^2 \sin B \sin C}{2 \sin A}$

$K = \dfrac{8.5^2 \sin 102° \sin 27°}{2 \sin 51°} \approx 21$ square units

36. $s = \dfrac{1}{2}(a + b + c)$

$= \dfrac{1}{2}(10.2 + 13.3 + 15.4) = 19.45$

$K = \sqrt{s(s - a)(s - b)(s - c)}$

$= \sqrt{19.45(19.45 - 10.2)(19.45 - 13.3)(19.45 - 15.4)}$

≈ 66.9 square units

52. $\alpha = 270° - 254° = 16°$

$A = 16° + 90° + 32° = 138°$

$b = 4 \cdot 16 = 64$ miles

$c = 3 \cdot 22 = 66$ miles

$a^2 = b^2 + c^2 - 2bc \cos A$

$a^2 = 64^2 + 66^2 - 2(64)(66) \cos 138°$

$a = \sqrt{64^2 + 66^2 - 2(64)(66) \cos 138°}$

$a \approx 120$ miles

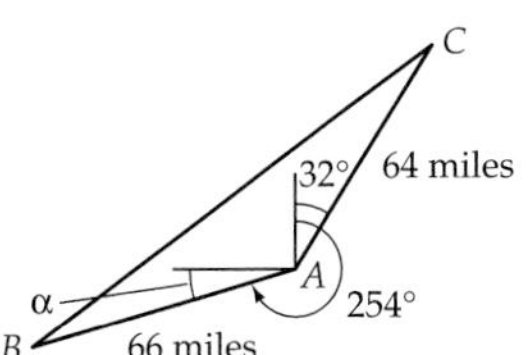

60. $S = \dfrac{1}{2}(324 + 412 + 516) = 626$

$K = \sqrt{626(626 - 324)(626 - 412)(626 - 516)}$

$= \sqrt{4{,}450{,}284{,}080}$

cost $= 4.15(\sqrt{4{,}450{,}284{,}080}) \approx \$276{,}848$

Exercise Set 7.3, page 566

6. $a = 3 - 3 = 0$

$b = 0 - (-2) = 2$

A vector equivalent to P_1P_2 is $\mathbf{v} = \langle 0, 2 \rangle$.

8. $\|\mathbf{v}\| = \sqrt{6^2 + 10^2}$

$= \sqrt{36 + 100} = \sqrt{136} = 2\sqrt{34}$

$\theta = \tan^{-1} \dfrac{10}{6} = \tan^{-1} \dfrac{5}{3} \approx 59.0°$

Thus $\mathbf{v}$ has a direction of about 59° as measured from the positive x-axis.

A unit vector in the direction of $\mathbf{v}$ is

$\mathbf{u} = \left\langle \dfrac{6}{2\sqrt{34}}, \dfrac{10}{2\sqrt{34}} \right\rangle = \left\langle \dfrac{3\sqrt{34}}{34}, \dfrac{5\sqrt{34}}{34} \right\rangle$

20. $\dfrac{3}{4}\mathbf{u} - 2\mathbf{v} = \dfrac{3}{4}\langle -2, 4 \rangle - 2\langle -3, -2 \rangle$

$= \left\langle -\dfrac{3}{2}, 3 \right\rangle - \langle -6, -4 \rangle$

$= \left\langle \dfrac{9}{2}, 7 \right\rangle$

26. $3\mathbf{u} + 2\mathbf{v} = 3(3\mathbf{i} - 2\mathbf{j}) + 2(-2\mathbf{i} + 3\mathbf{j})$

$= (9\mathbf{i} - 6\mathbf{j}) + (-4\mathbf{i} + 6\mathbf{j})$

$= (9 - 4)\mathbf{i} + (-6 + 6)\mathbf{j}$

$= 5\mathbf{i} + 0\mathbf{j}$

$= 5\mathbf{i}$

36. $a_1 = 2 \cos 8\pi/7 \approx -1.8$

$a_2 = 2 \sin 8\pi/7 \approx -0.9$

$\mathbf{v} = a_1\mathbf{i} + a_2\mathbf{j} \approx -1.8\mathbf{i} - 0.9\mathbf{j}$

40.

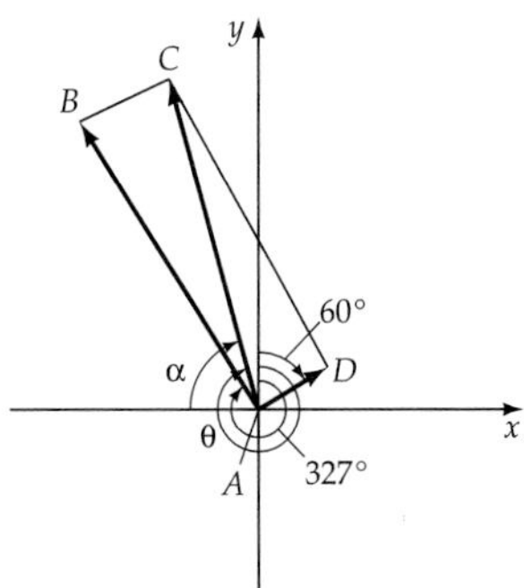

$\mathbf{AB} = 18 \cos 123°\mathbf{i} + 18 \sin 123°\mathbf{j} \approx -9.8\mathbf{i} + 15.1\mathbf{j}$

$\mathbf{AD} = 4 \cos 30°\mathbf{i} + 4 \sin 30°\mathbf{j} \approx 3.5\mathbf{i} + 2\mathbf{j}$

$\mathbf{AC} = \mathbf{AB} + \mathbf{AD} = -9.8\mathbf{i} + 15.1\mathbf{j} + 3.5\mathbf{i} + 2\mathbf{j}$

$= -6.3\mathbf{i} + 17.1\mathbf{j}$

$\|\mathbf{AC}\| = \sqrt{(-6.3)^2 + (17.1)^2} \approx 18$

$\alpha = \tan^{-1}\left|\dfrac{17.1}{-6.3}\right| = \tan^{-1}\dfrac{17.1}{6.3} \approx 70°$

$\theta \approx 270° + 70° = 340°$

The course of the boat is about 18 mph at an approximate heading of 340°.

42.

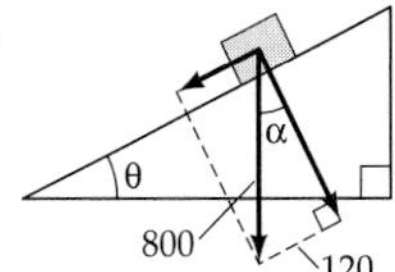

$\alpha = \theta$

$\sin \alpha = \dfrac{120}{800}$

$\alpha \approx 8.6°$

50. $\mathbf{v} \cdot \mathbf{w} = (5\mathbf{i} + 3\mathbf{j}) \cdot (4\mathbf{i} - 2\mathbf{j})$
$= 5(4) + 3(-2) = 20 - 6 = 14$

60. $\cos \theta = \dfrac{\mathbf{v} \cdot \mathbf{w}}{\|\mathbf{v}\|\,\|\mathbf{w}\|}$

$\cos \theta = \dfrac{(3\mathbf{i} - 4\mathbf{j}) \cdot (6\mathbf{i} - 12\mathbf{j})}{\sqrt{3^2 + (-4)^2}\sqrt{6^2 + (-12)^2}}$

$\cos \theta = \dfrac{3(6) + (-4)(-12)}{\sqrt{25}\sqrt{180}}$

$\cos \theta = \dfrac{66}{5\sqrt{180}} \approx 0.9839$

$\theta \approx 10.3°$

62. $\text{proj}_{\mathbf{w}}\,\mathbf{v} = \dfrac{\mathbf{v} \cdot \mathbf{w}}{\|\mathbf{w}\|}$

$\text{proj}_{\mathbf{w}}\,\mathbf{v} = \dfrac{\langle -7, 5\rangle \cdot \langle -4, 1\rangle}{\sqrt{(-4)^2 + 1^2}} = \dfrac{33}{\sqrt{17}} = \dfrac{33\sqrt{17}}{17} \approx 8.0$

70. $W = \|\mathbf{F}\|\,\|\mathbf{S}\| \cos \alpha$

$W = 100 \cdot 25 \cdot \cos 42°$

$W \approx 1858$ foot-pounds

Exercise Set 7.4, page 574

12. $r = \sqrt{1^2 + (\sqrt{3})^2} = 2$

$\alpha = \tan^{-1}\left|\dfrac{\sqrt{3}}{1}\right| = \tan^{-1}\sqrt{3} = 60°$

$\theta = \alpha = 60°, z = 2 \text{ cis } 60°$

26. $z = 4\left(\cos\dfrac{5\pi}{3} + i \sin\dfrac{5\pi}{3}\right) = 4\left(\dfrac{1}{2} - \dfrac{\sqrt{3}}{2}i\right) = 2 - 2i\sqrt{3}$

52. $\sqrt{3} - i = 2 \text{ cis } (-30°); \quad 1 + i\sqrt{3} = 2 \text{ cis } 60°$

$(\sqrt{3} - i)(1 + i\sqrt{3}) = 2 \text{ cis } (-30°) \cdot 2 \text{ cis } 60°$
$= 2 \cdot 2 \text{ cis } (-30° + 60°)$
$= 4 \text{ cis } 30°$
$= 4(\cos 30° + i \sin 30°)$
$= 4\left(\dfrac{\sqrt{3}}{2} + i\dfrac{1}{2}\right)$
$= 2\sqrt{3} + 2i$

56. $z_1 = 1 + i$

$r_1 = \sqrt{1^2 + 1^2} = \sqrt{2}$

$\alpha_1 = \tan^{-1}\left|\dfrac{1}{1}\right| = 45°; \theta_1 = 45°$

$z_1 = \sqrt{2}(\cos 45° + i \sin 45°) = \sqrt{2} \text{ cis } 45°$

$z_2 = 1 - i$

$r_2 = \sqrt{1^2 + (-1)^2} = \sqrt{2}$

$\alpha_2 = \tan^{-1}\left|\dfrac{-1}{1}\right| = 45°; \theta_2 = 315°$

$z_2 = \sqrt{2}(\cos 315° + i \sin 315°) = \sqrt{2} \text{ cis } 315°$

$\dfrac{z_1}{z_2} = \dfrac{\sqrt{2} \text{ cis } 45°}{\sqrt{2} \text{ cis } 315°}$
$= \text{cis } (-270°)$
$= \cos 270° - i \sin 270° = 0 - (-i) = i$

Exercise Set 7.5, page 579

6. $(2 \text{ cis } 330°)^4 = 2^4 \text{ cis } (4 \cdot 330°)$
$= 16 \text{ cis } 1320°$
$= 16 \text{ cis } 240°$
$= 16(\cos 240° + i \sin 240°)$
$= 16\left[-\dfrac{1}{2} + i\left(-\dfrac{\sqrt{3}}{2}\right)\right]$
$= -8 - 8i\sqrt{3}$

14. $\left(-\dfrac{\sqrt{2}}{2} + i\dfrac{\sqrt{2}}{2}\right)^{12} = [1(\cos 135° + i \sin 135°)]^{12}$
$= 1^{12}[\cos (12 \cdot 135°) + i \sin (12 \cdot 135°)]$
$= \cos 1620° + i \sin 1620°$
$= -1 + 0i$, or -1

26. $-2 + 2i\sqrt{3} = 4(\cos 120° + i \sin 120°)$

$w_k = 4^{1/3}\left(\cos\dfrac{120° + 360°k}{3} + i \sin\dfrac{120° + 360°k}{3}\right)$
$k = 0, 1, 2$

$w_0 = 4^{1/3}\left(\cos\dfrac{120°}{3} + i \sin\dfrac{120°}{3}\right) \approx 1.216 + 1.020i$

$w_1 = 4^{1/3}\left(\cos\dfrac{120° + 360°}{3} + i \sin\dfrac{120° + 360°}{3}\right)$
$\approx -1.492 + 0.543i$

$w_2 = 4^{1/3}\left(\cos\dfrac{120° + 360° \cdot 2}{3} + i \sin\dfrac{120° + 360° \cdot 2}{3}\right)$
$\approx 0.276 - 1.563i$

28. $-1 + i\sqrt{3} = 2(\cos 120° + i \sin 120°)$

$w_k = 2^{1/2}\left(\cos\dfrac{120° + 360°k}{2} + i \sin\dfrac{120° + 360°k}{2}\right), k = 0, 1$

$w_0 = 2^{1/2}(\cos 60° + i \sin 60°) = \dfrac{\sqrt{2}}{2} + \dfrac{\sqrt{6}}{2}i \quad (k = 0)$

$w_1 = 2^{1/2}\left(\cos\dfrac{120° + 360°}{2} + i \sin\dfrac{120° + 360°}{2}\right) \quad (k = 1)$
$= 2^{1/2}(\cos 240° + i \sin 240°)$
$= -\dfrac{\sqrt{2}}{2} - \dfrac{\sqrt{6}}{2}i$

Exercise Set 8.1, page 592

4. Comparing $x^2 = 4py$ with $x^2 = -\frac{1}{4}y$, we have

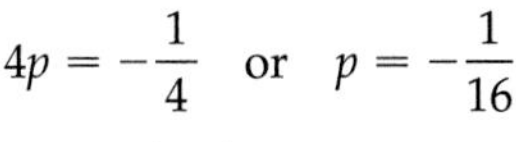

$$4p = -\frac{1}{4} \quad \text{or} \quad p = -\frac{1}{16}$$

vertex $(0, 0)$
focus $(0, -1/16)$
directrix $y = 1/16$

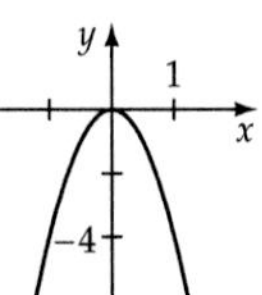

20. $x^2 + 5x - 4y - 1 = 0$

$$x^2 + 5x = 4y + 1$$

$$x^2 + 5x + \frac{25}{4} = 4y + 1 + \frac{25}{4} \qquad \text{• Complete the square.}$$

$$\left(x + \frac{5}{2}\right)^2 = 4\left(y + \frac{29}{16}\right) \qquad \text{• } h = -\frac{5}{2}, k = -\frac{29}{16}$$

$$4p = 4 \qquad \text{• Compare to } (x - h)^2 = 4p(y - k)^2.$$

$$p = 1$$

vertex $(-5/2, -29/16)$
foxus $(h, k + p) = (-5/2, -13/16)$
directrix $y = k - p = -45/16$

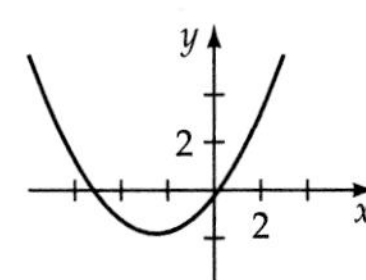

28. vertex $(0, 0)$, focus $(5, 0)$, $p = 5$ because focus is $(p, 0)$

$$y^2 = 4px$$
$$y^2 = 4(5)x$$
$$y^2 = 20x$$

30. vertex $(2, -3)$, focus $(0, -3)$

$(h, k) = (2, -3)$, so $h = 2$ and $k = -3$.

Focus is $(h + p, k) = (2 + p, -3) = (0, -3)$.

Therefore, $2 + p = 0$ and $p = -2$.

$$(y - k)^2 = 4p(x - h)$$
$$(y + 3)^2 = 4(-2)(x - 2)$$
$$(y + 3)^2 = -8(x - 2)$$

36.
$$x^2 = 4py$$
$$40.5^2 = 4p(16)$$
$$p = \frac{40.5^2}{64}$$
$$p \approx 25.6 \text{ feet}$$

Exercise Set 8.2, page 604

20. $25x^2 + 12y^2 = 300$

$$\frac{x^2}{12} + \frac{y^2}{25} = 1 \qquad \text{• } a^2 = 25, b^2 = 12, \quad c^2 = 25 - 12$$
$$a = 5, \quad b = 2\sqrt{3}, \quad c = \sqrt{13}$$

center $(0, 0)$
vertices $(0, 5)$ and $(0, -5)$
foci $\left(0, \sqrt{13}\right)$ and $\left(0, -\sqrt{13}\right)$

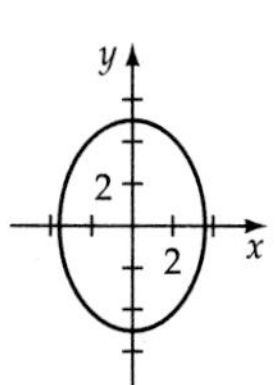

26.
$$9x^2 + 16y^2 + 36x - 16y - 104 = 0$$
$$9x^2 + 36x + 16y^2 - 16y - 104 = 0$$
$$9(x^2 + 4x) + 16(y^2 - y) = 104$$
$$9(x^2 + 4x + 4) + 16\left(y^2 - y + \frac{1}{4}\right) = 104 + 36 + 4$$
$$9(x + 2)^2 + 16\left(y - \frac{1}{2}\right)^2 = 144$$
$$\frac{(x + 2)^2}{16} + \frac{\left(y - \frac{1}{2}\right)^2}{9} = 1$$

center $(-2, 1/2)$
$a = 4, b = 3,$
$c = \sqrt{4^2 - 3^2} = \sqrt{7}$
vertices $(2, 1/2)$ and $(-6, 1/2)$,
foci $\left(-2 + \sqrt{7}, 1/2\right)$ and
$\left(-2 - \sqrt{7}, 1/2\right)$

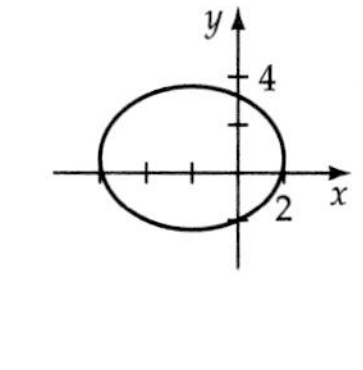

42. Center $(-4, 1) = (h, k)$. Therefore, $h = -4$ and $k = 1$. Length of minor axis is 8, so $2b = 8$ or $b = 4$. The equation of the ellipse is of the form

$$\frac{(x - h)^2}{a^2} + \frac{(y - k)^2}{b^2} = 1$$

$$\frac{(x + 4)^2}{a^2} + \frac{(y - 1)^2}{16} = 1 \qquad \text{• } h = -4, k = 1, b = 4$$

$$\frac{(0 + 4)^2}{a^2} + \frac{(4 - 1)^2}{16} = 1 \qquad \text{• The point } (0, 4) \text{ is on the graph. Thus } x = 0 \text{ and } y = 4 \text{ satisfy the equation.}$$

$$\frac{16}{a^2} + \frac{9}{16} = 1 \qquad \text{• Solve for } a^2.$$

$$\frac{16}{a^2} = \frac{7}{16}$$

$$a^2 = \frac{256}{7}$$

$$\frac{(x + 4)^2}{256/7} + \frac{(y - 1)^2}{16} = 1$$

48. Because the foci are $(0, -3)$ and $(0, 3)$, $c = 3$ and center is $(0, 0)$, the midpoint of the line segment between $(0, -3)$ and $(0, 3)$.

$$e = \frac{c}{a}$$

$\frac{1}{4} = \frac{3}{a}$ **• e = $\frac{1}{4}$**

$a = 12$

$3^2 = 12^2 - b^2$ **• $c^2 = a^2 - b^2$**

$b^2 = 144 - 9 = 135$ **• Solve for b^2.**

The equation of the ellipse is $\frac{x^2}{135} + \frac{y^2}{144} = 1$.

54. The mean distance is $a = 67.08$ million miles.

Aphelion $= a + c = 67.58$ million miles

Thus $c = 67.58 - a = 0.50$ million miles.

$b = \sqrt{a^2 - c^2} = \sqrt{67.08^2 - 0.50^2} \approx 67.078$

An equation of the orbit of Venus is

$$\frac{x^2}{67.08^2} + \frac{y^2}{67.078^2} = 1$$

56. The length of the semimajor axis is 50 feet. Thus

$c^2 = a^2 - b^2$

$32^2 = 50^2 - b^2$

$b^2 = 50^2 - 32^2$

$b = \sqrt{50^2 - 32^2}$

$b \approx 38.4$ feet

Exercise Set 8.3, page 618

4. $\frac{y^2}{25} - \frac{x^2}{36} = 1$

$a^2 = 25$ $\quad b^2 = 36$ $\quad c^2 = a^2 + b^2 = 25 + 36 = 61$

$a = 5$ $\quad b = 6$ $\quad c = \sqrt{61}$

Transverse axis is on y-axis because y^2 term is positive.

center $(0, 0)$

foci $(0, \sqrt{61})$ and $(0, -\sqrt{61})$

asymptotes $y = \frac{5}{6}x$ and $y = -\frac{5}{6}x$

vertices $(0, 5)$ and $(0, -5)$

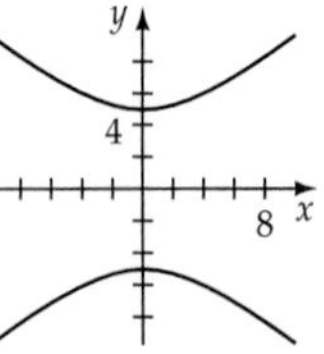

26. $16x^2 - 9y^2 - 32x - 54y + 79 = 0$

$16(x^2 - 2x + 1) - 9(y^2 + 6y + 9) = -79 + 16 - 81$

$= -144$

$$\frac{(y+3)^2}{16} - \frac{(x-1)^2}{9} = 1$$

Transverse axis is parallel to y-axis because y^2 term is positive. Center is at $(1, -3)$; $a^2 = 16$ so $a = 4$.

vertices $(h, k + a) = (1, 1)$

$(h, k - a) = (1, -7)$

$c^2 = a^2 + b^2 = 16 + 9 = 25$

$c = \sqrt{25} = 5$

foci $(h, k + c) = (1, 2)$

$(h, k - c) = (1, -8)$

Because $b^2 = 9$ and $b = 3$, the asymptotes are

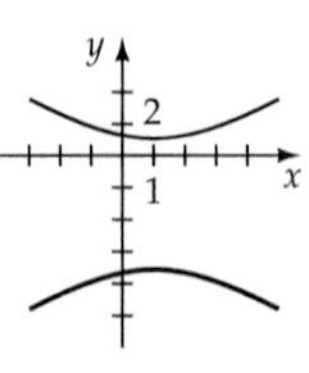

$y + 3 = \frac{4}{3}(x - 1)$ and

$y + 3 = -\frac{4}{3}(x - 1)$.

48. Because the vertices are $(2, 3)$ and $(-2, 3)$, $a = 2$ and center is $(0, 3)$.

$e = \frac{c}{a}$ $\quad c^2 = a^2 + b^2$

$\quad 5^2 = 2^2 + b^2$

$\frac{5}{2} = \frac{c}{2}$ $\quad b^2 = 25 - 4 = 21$

$c = 5$

Substituting into the standard equation yields

$$\frac{x^2}{4} - \frac{(y-3)^2}{21} = 1.$$

54. a. Because the transmitters are 300 miles apart, $2c = 300$ and $c = 150$.

$2a = \text{rate} \times \text{time}$

$2a = 0.186 \times 800 = 148.8$ miles

Thus $a = 74.4$ miles.

$b = \sqrt{c^2 - a^2}$

$= \sqrt{150^2 - 74.4^2} \approx 130.25$ miles

The ship is located on the hyperbola given by

$$\frac{x^2}{74.4^2} - \frac{y^2}{130.25^2} = 1$$

b. The ship will reach the coastline when $x < 0$ and $y = 0$. Thus

$\frac{x^2}{74.4^2} - \frac{0^2}{130.25^2} = 1$

$\frac{x^2}{74.4^2} = 1$

$x^2 = 74.4^2$

$x = -74.4$

The ship reaches the coastline 74.4 miles to the left of the origin, which is 75.6 miles to the right of transmitter T_2.

Exercise Set 8.4, page 626

8. $xy = -10$

$xy + 10 = 0$

$A = 0, B = 1, C = 0, F = 10$

$$\cot 2\alpha = \frac{A - C}{B} = \frac{0 - 0}{1} = 0$$

Thus $2\alpha = 90°$ and $\alpha = 45°$.

$$A' = A\cos^2\alpha + B\cos\alpha\sin\alpha + C\sin^2\alpha$$
$$= 0\left(\frac{\sqrt{2}}{2}\right)^2 + 1\left(\frac{\sqrt{2}}{2}\right)\left(\frac{\sqrt{2}}{2}\right) + 0\left(\frac{\sqrt{2}}{2}\right)^2 = \frac{1}{2}$$

$$C' = A\sin^2\alpha - B\cos\alpha\sin\alpha + C\cos^2\alpha$$
$$= 0\left(\frac{\sqrt{2}}{2}\right)^2 - 1\left(\frac{\sqrt{2}}{2}\right)\left(\frac{\sqrt{2}}{2}\right) + 0\left(\frac{\sqrt{2}}{2}\right)^2 = -\frac{1}{2}$$

$$F' = F = 10$$

$$\frac{1}{2}x'^2 - \frac{1}{2}y'^2 + 10 = 0 \text{ or } \frac{y'^2}{20} - \frac{x'^2}{20} = 1$$

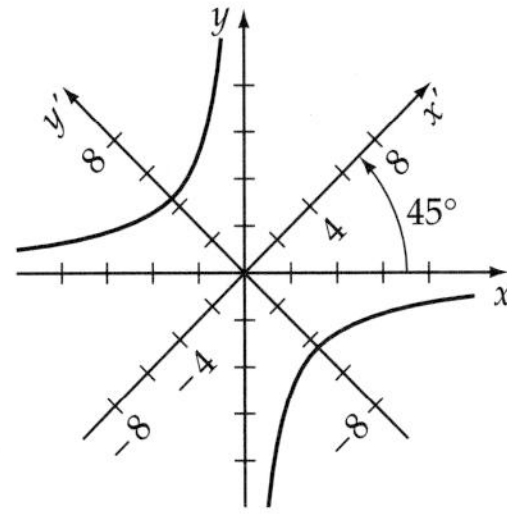

18. $x^2 + 4xy + 4y^2 - 2\sqrt{5}x + \sqrt{5}y = 0$

$A = 1, B = 4, C = 4, D = -2\sqrt{5}, E = \sqrt{5}, F = 0$

$$\cot 2\alpha = \frac{A - C}{B} = \frac{1 - 4}{4} = -\frac{3}{4}$$

$$\csc^2 2\alpha = \cot^2 2\alpha + 1$$

$$\csc^2 2\alpha = \left(-\frac{3}{4}\right)^2 + 1 = \frac{25}{16}$$

$$\csc 2\alpha = +\sqrt{\frac{25}{16}} = \frac{5}{4} \text{ (}2\alpha\text{ is in Quadrant II)}$$

$$\sin 2\alpha = \frac{1}{\csc 2\alpha} = \frac{4}{5}$$

$$\sin^2 2\alpha + \cos^2 2\alpha = 1$$

$$\cos^2 2\alpha = 1 - \sin^2 2\alpha$$

$$\cos^2 2\alpha = 1 - \left(\frac{4}{5}\right)^2 = \frac{9}{25}$$

$$\cos 2\alpha = -\sqrt{\frac{9}{25}} = -\frac{3}{5} \text{ (}2\alpha\text{ is in Quadrant II)}$$

$$\sin\alpha = \sqrt{\frac{1 - \left(-\frac{3}{5}\right)}{2}} = \frac{2\sqrt{5}}{5}$$

$$\cos\alpha = \sqrt{\frac{1 + \left(-\frac{3}{5}\right)}{2}} = \frac{\sqrt{5}}{5}$$

$$\alpha \approx 63.4°$$

$$A' = A\cos^2\alpha + B\cos\alpha\sin\alpha + C\sin^2\alpha$$
$$= 1\left(\frac{\sqrt{5}}{5}\right)^2 + 4\left(\frac{\sqrt{5}}{5}\right)\left(\frac{2\sqrt{5}}{5}\right) + 4\left(\frac{2\sqrt{5}}{5}\right)^2 = 5$$

$$C' = A\sin^2\alpha - B\cos\alpha\sin\alpha + C\cos^2\alpha$$
$$= 1\left(\frac{2\sqrt{5}}{5}\right)^2 - 4\left(\frac{\sqrt{5}}{5}\right)\left(\frac{2\sqrt{5}}{5}\right) + 4\left(\frac{\sqrt{5}}{5}\right)^2 = 0$$

$$D' = D\cos\alpha + E\sin\alpha$$
$$= -2\sqrt{5}\left(\frac{\sqrt{5}}{5}\right) + \sqrt{5}\left(\frac{2\sqrt{5}}{5}\right) = 0$$

$$E' = -D\sin\alpha + E\cos\alpha$$
$$= 2\sqrt{5}\left(\frac{2\sqrt{5}}{5}\right) + \sqrt{5}\left(\frac{\sqrt{5}}{5}\right) = 5$$

$$5x'^2 + 5y' = 0 \text{ or } y' = -x'^2$$

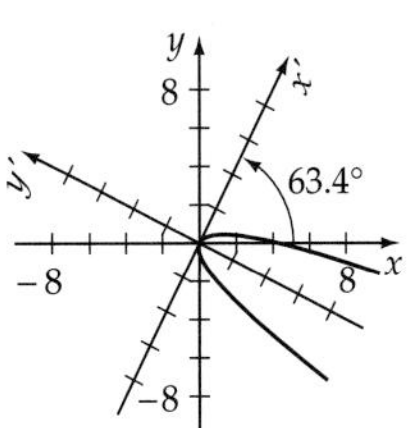

24. Use y_1 and y_2 as in Equations (11) and (12) that precede Example 4. Store the following constants.

$A = 2, B = -8, C = 8, D = 20, E = -24, F = -3$

Graph y_1 and y_2 on the same screen to produce the following parabola.

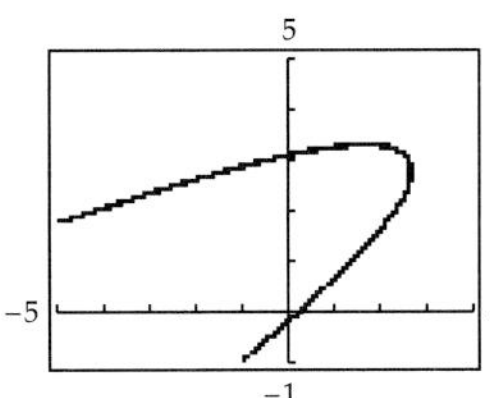

30. Because

$$B^2 - 4AC = (-10\sqrt{3})^2 - 4(11)(1)$$
$$= 300 - 44 > 0$$

the graph is a hyperbola.

Exercise Set 8.5, page 638

14.

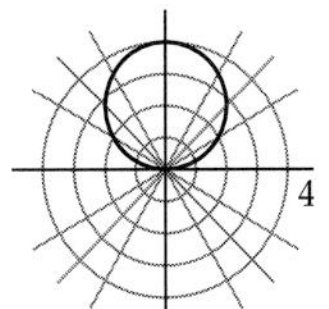

16. $r = 5\cos 3\theta$

Because 3 is odd, this is a rose with 3 petals.

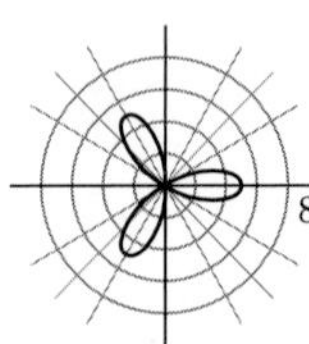

20. Because $|a| = |b| = 2$, the graph of $r = 2 - 2\csc\theta$, $-\pi \le \theta \le \pi$, is a cardioid.

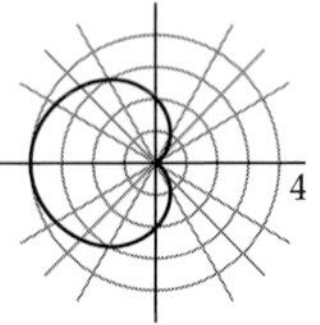

26. $r = 4 - 4\sin\theta$

θ min $= 0$
θ max $= 2\pi$
θ step $= 0.1$

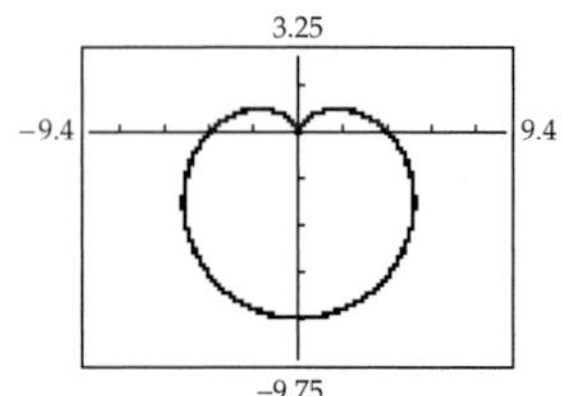

32. $r = -4\sec\theta$

θ min $= 0$
θ max $= 2\pi$
θ step $= 0.1$

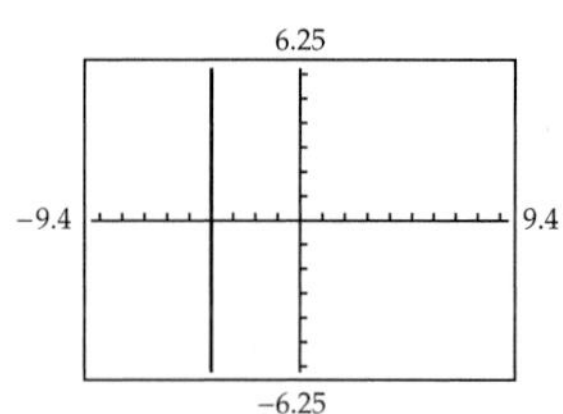

44. $x = r\cos\theta$

$$= (2)\left[\cos\left(-\frac{\pi}{3}\right)\right]$$

$$= (2)\left(\frac{1}{2}\right) = 1$$

$y = r\sin\theta$

$$= (2)\left[\sin\left(-\frac{\pi}{3}\right)\right]$$

$$= (2)\left(-\frac{\sqrt{3}}{2}\right) = -\sqrt{3}$$

The rectangular coordinates of the point are $(1, -\sqrt{3})$.

48. $r = \sqrt{x^2 + y^2}$

$$= \sqrt{(12)^2 + (-5)^2}$$

$$= \sqrt{144 + 25}$$

$$= \sqrt{169}$$

$$= 13$$

$\alpha = \tan^{-1}\left|\dfrac{y}{x}\right|$ • **α is the reference angle for θ.**

$$= \tan^{-1}\left|\frac{-5}{12}\right| \approx 22.6^\circ$$

θ is a Quadrant IV angle. Thus $\theta \approx 360^\circ - 22.6^\circ = 337.4^\circ$

The approximate polar coordinates of the point are $(13, 337.4^\circ)$.

56.

$$r = \cot\theta$$

$$r = \frac{\cos\theta}{\sin\theta} \qquad \text{• } \mathbf{\cot\theta = \frac{\cos\theta}{\sin\theta}}$$

$$r\sin\theta = \cos\theta$$

$$r(r\sin\theta) = r\cos\theta \qquad \text{• Multiply both sides by } r.$$

$$(\sqrt{x^2 + y^2})y = x \qquad \text{• } y = r\sin\theta;\ x = r\cos\theta$$

$$(x^2 + y^2)y^2 = x^2 \qquad \text{• Square each side.}$$

$$y^4 + x^2y^2 - x^2 = 0$$

64.

$$2x - 3y = 6$$

$$2r\cos\theta - 3r\sin\theta = 6 \qquad \text{• } x = r\cos\theta;\ y = r\sin\theta$$

$$r(2\cos\theta - 3\sin\theta) = 6$$

$$r = \frac{6}{2\cos\theta - 3\sin\theta}$$

Exercise Set 8.6, page 644

2. $r = \dfrac{8}{2 - 4\cos\theta}$ • **Divide numerator and denominator by 2.**

$$r = \frac{4}{1 - 2\cos\theta}$$

$e = 2$, so the graph is a hyperbola. The transverse axis is on the polar axis because the equation involves $\cos\theta$. Let $\theta = 0$.

$$r = \frac{8}{2 - 4\cos 0} = \frac{8}{2 - 4} = -4$$

Let $\theta = \pi$.

$$r = \frac{8}{2 - 4\cos\pi} = \frac{8}{2 + 4} = \frac{4}{3}$$

The vertices are $(-4, 0)$ and $(4/3, \pi)$.

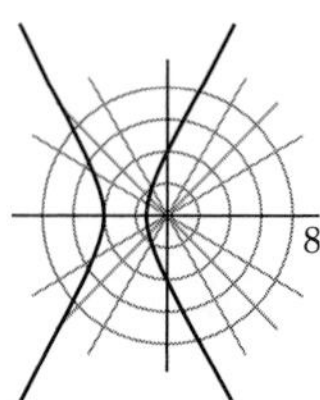

4. $r = \dfrac{6}{3 + 2\cos\theta}$

$$r = \frac{2}{1 + (2/3)\cos\theta} \qquad \text{• Divide numerator and denominator by 3.}$$

$e = 2/3$, so the graph is an ellipse. The major axis is on the polar axis because the equation involves $\cos\theta$. Let $\theta = 0$.

$$r = \frac{6}{3 + 2\cos 0} = \frac{6}{3 + 2} = \frac{6}{5}$$

Let $\theta = \pi$.

$$r = \frac{6}{3 + 2\cos\pi} = \frac{6}{3 - 2} = 6$$

The vertices on the major axis are $(6/5, 0)$ and $(6, \pi)$. Let $\theta = \pi/2$.

$$r = \frac{6}{3 + 2\cos(\pi/2)} = \frac{6}{3 + 0} = 2$$

Let $\theta = 3\pi/2$.

$$r = \frac{6}{3 + 2\cos(3\pi/2)} = \frac{6}{3 + 0} = 2$$

Two additional points on the ellipse are $(2, \pi/2)$ and $(2, 3\pi/2)$.

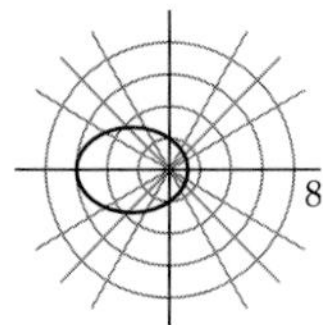

24. $e = 1, r\cos\theta = -2$

Thus the directrix is $x = -2$. The distance from the focus (pole) to the directrix is $d = 2$.

$$r = \frac{ed}{1 - e\cos\theta}$$ **• Use Eq (2) since the vertical directrix is to the left to the pole.**

$$r = \frac{(1)(2)}{1 - (1)\cos\theta}$$ **• $e = 1, d = 2$**

$$r = \frac{2}{1 - \cos\theta}$$

Exercise Set 8.7, page 650

6. Plotting points for several values of t yields the following graph.

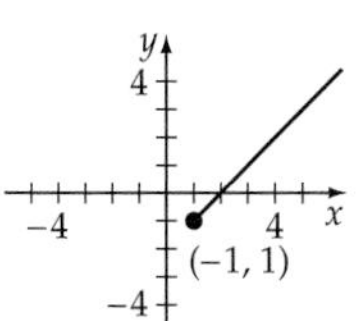

12. $x = 3 + 2\cos t, y = -1 - 3\sin t, 0 \le t < 2\pi$

$$\cos t = \frac{x - 3}{2}, \sin t = -\frac{y + 1}{3}$$

$$\cos^2 t + \sin^2 t = 1$$

$$\left(\frac{x - 3}{2}\right)^2 + \left(-\frac{y + 1}{3}\right)^2 = 1$$

$$\frac{(x - 3)^2}{4} + \frac{(y + 1)^2}{9} = 1$$

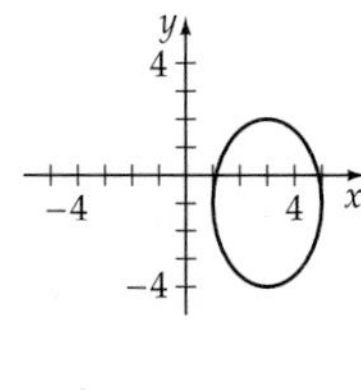

14. $x = 1 + t^2, y = 2 - t^2, t \in R$

$$x = 1 + t^2$$
$$t^2 = x - 1$$
$$y = 2 - (x - 1)$$
$$y = -x + 3$$

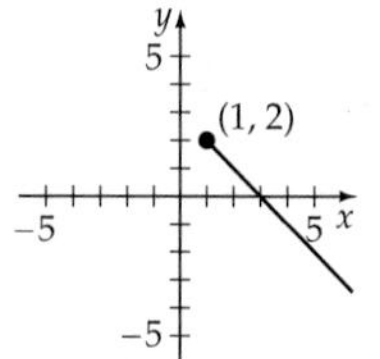

Because $x = 1 + t^2$ and $t^2 \ge 0$ for all real numbers t, $x \ge 1$ for all t. Similarly, $y \le 2$ for all t.

26. The maximum height of the cycloid is $2a = 2(3) = 6$. The cycloid intersects the x-axis every $2\pi a = 2\pi(3) = 6\pi$ units.

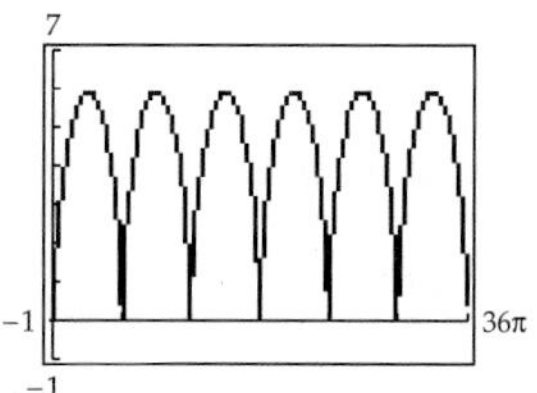

32.

The maximum height is about 195 feet when $t \approx 3.50$ seconds. The range is 1117 feet when $t \approx 6.99$ seconds.

Exercise Set 9.1, page 665

6. $\begin{cases} 8x + 3y = -7 & (1) \\ x = 3y + 15 & (2) \end{cases}$

$$8(3y + 15) + 3y = -7$$ **• Replace x in Eq. (1).**

$$24y + 120 + 3y = -7$$ **• Simplify.**

$$27y = -127$$

$$y = -\frac{127}{27}$$

$$x = 3\left(-\frac{127}{27}\right) + 15 = \frac{8}{9}$$ **• Substitute $-\frac{127}{27}$ for y in Eq. (2).**

The solution is $(8/9, -127/27)$.

18. $\begin{cases} 3x - 4y = 8 & (1) \\ 6x - 8y = 9 & (2) \end{cases}$

$$8y = 6x - 9$$ **• Solve Eq. (2) for y.**

$$y = \frac{3}{4}x - \frac{9}{8}$$

$3x - 4\left(\frac{3}{4}x - \frac{9}{8}\right) = 8$ • **Replace y in Eq. (1).**

$3x - 3x + \frac{9}{2} = 8$ • **Simplify.**

$\frac{9}{2} = 8$

This is a false equation. Therefore, the system of equations is inconsistent and has no solution.

20. $\begin{cases} 5x + 2y = 2 & (1) \\ y = -\frac{5}{2}x + 1 & (2) \end{cases}$

$5x + 2\left(-\frac{5}{2}x + 1\right) = 2$ • **Replace y in Eq. (1).**

$5x - 5x + 2 = 2$ • **Simplify.**

$2 = 2$

This is a true statement, therefore the system of equations is dependent. Let $x = c$. Then $y = -\frac{5}{2}c + 1$. Thus the solutions are $\left(c, -\frac{5}{2}c + 1\right)$.

24. $\begin{cases} 3x - 8y = -6 & (1) \\ -5x + 4y = 10 & (2) \end{cases}$

$3x - 8y = -6$

$-10x + 8y = 20$ • **2 times Eq. (2)**

$-7x = 14$

$x = -2$

$3(-2) - 8y = -6$ • **Substitute −2 for x in Eq. (1). Solve for y.**

$-8y = 0$

$y = 0$

The solution is $(-2, 0)$.

28. $\begin{cases} 4x + 5y = 2 & (1) \\ 8x - 15y = 9 & (2) \end{cases}$

$12x + 15y = 6$ • **3 times Eq. (1)**

$8x - 15y = 9$

$20x = 15$

$x = \frac{3}{4}$

$4\left(\frac{3}{4}\right) + 5y = 2$ • **Substitute $\frac{3}{4}$ for x in Eq. (1).**

$3 + 5y = 2$ • **Solve for y.**

$y = -\frac{1}{5}$

The solution is $(3/4, -1/5)$.

44. Let r = the rate of the canoeist.
Let w = the rate of the current.
Rate of canoeist with the current: $r + w$
Rate of canoeist against the current: $r - w$

$r \cdot t = d$

$(r + w) \cdot 2 = 12$ (1)

$(r - w) \cdot 4 = 12$ (2)

$r + w = 6$ • **Divide Eq. (1) by 2.**

$r - w = 3$ • **Divide Eq. (2) by 4.**

$2r = 9$

$r = 4.5$

$4.5 + w = 6$

$w = 1.5$

Rate of canoeist = 4.5 mph
Rate of current = 1.5 mph

Exercise Set 9.2, page 677

12. $\begin{cases} 3x + 2y - 5z = 6 & (1) \\ 5x - 4y + 3z = -12 & (2) \\ 4x + 5y - 2z = 15 & (3) \end{cases}$

$15x + 10y - 25z = 30$ • **5 times Eq. (1)**

$-15x + 12y - 9z = 36$ • **−3 times Eq. (2)**

$22y - 34z = 66$ • **Divide by 2.**

$11y - 17z = 33$ (4)

$12x + 8y - 20z = 24$ • **4 times Eq. (1)**

$-12x - 15y + 6z = -45$ • **−3 times Eq. (3)**

$-7y - 14z = -21$ • **Divide by −7.**

$y + 2z = 3$ (5)

$11y - 17z = 33$ (4)

$-11y - 22z = -33$ • **−11 times Eq. (5)**

$-39z = 0$

$z = 0$ (6)

$11y - 17(0) = 33$

$y = 3$

$3x + 2(3) - 5(0) = 6$

$x = 0$

The solution is $(0, 3, 0)$.

16. $\begin{cases} 2x + 3y + 2z = 14 & (1) \\ x - 3y + 4z = 4 & (2) \\ -x + 12y - 6z = 2 & (3) \end{cases}$

$2x + 3y + 2z = 14$ (1)

$-2x + 6y - 8z = -8$ • **−2 times Eq. (2)**

$9y - 6z = 6$ • **Divide by 3.**

$3y - 2z = 2$ (4)

$$\begin{array}{rl} 2x + 3y + 2z = 14 & (1) \\ -2x + 24y - 12z = 4 & \\ \hline 27y - 10z = 18 & (5) \end{array}$$ • **2 times Eq. (3)**

$$\begin{array}{rl} -27y + 18z = -18 & \\ 27y - 10z = 18 & (5) \\ \hline 8z = 0 & \\ z = 0 & (6) \end{array}$$ • **−9 times Eq. (4)**

$$3y - 2(0) = 2$$
$$y = \frac{2}{3}$$
• **Substitute z = 0 in Eq. (4).**

$$2x + 3\left(\frac{2}{3}\right) + 2(0) = 14$$
$$x = 6$$
• **Substitute y = 2/3 and z = 0 in Eq. (1).**

The solution is $(6, 2/3, 0)$.

18. $\begin{cases} 2x + 3y - 6z = 4 & (1) \\ 3x - 2y - 9z = -7 & (2) \\ 2x + 5y - 6z = 8 & (3) \end{cases}$

$$\begin{array}{rl} 6x + 9y - 18z = 12 & \\ -6x + 4y + 18z = 14 & \\ \hline 13y = 26 & \\ y = 2 & (4) \end{array}$$ • **3 times Eq. (1)** • **−2 times Eq. (2)**

$$\begin{array}{rl} 2x + 3y - 6z = 4 & (1) \\ -2x = 5y + 6z = -8 & \\ \hline -2y = -4 & \\ y = 2 & (5) \end{array}$$ • **−1 times Eq. (3)**

$$\begin{array}{rl} y = 2 & (4) \\ -y = -2 & \\ \hline 0 = 0 & (6) \end{array}$$ • **−1 times Eq. (5)**

The equations are dependent. Let $z = c$.

$$2x + 3(2) - 6c = 4$$
$$x = 3c - 1$$
• **Substitute y = 2 and z = c in Eq. (1).**

The solutions are $(3c - 1, 2, c)$.

20. $\begin{cases} x - 3y + 4z = 9 & (1) \\ 3x - 8y - 2x = 4 & (2) \end{cases}$

$$\begin{array}{rl} -3x + 9y - 12z = -27 & \\ 3x - 8y - 2z = 4 & (2) \\ \hline y - 14z = -23 & (3) \end{array}$$ • **−3 times Eq. (1)**

$$y = 14z - 23$$ • **Solve Eq. (3) for y.**

$$x - 3(14z - 23) + 4z = 9$$ • **Substitute 14z − 23 for y in Eq. (1).**

$$x = 38z - 60$$ • **Solve for x.**

Let $z = c$. The solutions are $(38c - 60, 14c - 23, c)$.

32. $\begin{cases} 5x + 2y + 3z = 0 & (1) \\ 3x + y - 2z = 0 & (2) \\ 4x - 7y + 5z = 0 & (3) \end{cases}$

$$\begin{array}{rl} 15x + 6y + 9z = 0 & \\ -15x - 5y + 10z = 0 & \\ \hline y + 19z = 0 & (4) \end{array}$$ • **3 times Eq. (1)** • **−5 times Eq. (2)**

$$\begin{array}{rl} 20x + 8y + 12z = 0 & \\ -20x + 35y - 25z = 0 & \\ \hline 43y - 13z = 0 & (5) \end{array}$$ • **4 times Eq. (1)** • **−5 times Eq. (3)**

$$\begin{array}{rl} -43y - 817z = 0 & \\ 43y - 13z = 0 & (5) \\ \hline -830z = 0 & \\ z = 0 & (6) \end{array}$$ • **−43 times Eq. (4)**

Solving by back substitution, the only solution is $(0, 0, 0)$.

36. $x^2 + y^2 + ax + by + c = 0$

$\begin{cases} 0 + 36 + a(0) + b(6) + c = 0 \\ 1 + 25 + a(1) + b(5) + c = 0 \\ 49 + 1 + a(-7) + b(-1) + c = 0 \end{cases}$ • **Let x = 0, y = 6.** • **Let x = 1, y = 5.** • **Let x = −7, y = −1.**

$\begin{cases} 6b + c = -36 & (1) \\ a + 5b + c = -26 & (2) \\ -7a - b + c = -50 & (3) \end{cases}$

$$\begin{array}{rl} 7a + 35b + 7c = -182 & \\ -7a - b + c = -50 & (3) \\ \hline 34b + 8c = -232 & \\ 17b + 4c = -116 & (4) \end{array}$$ • **7 times Eq. (2)**

$$\begin{array}{rl} -24b - 4c = 144 & \\ 17b + 4c = -116 & (4) \\ \hline -7b = 28 & \\ b = -4 & \end{array}$$ • **−4 times Eq. (1)**

$$17(-4) + 4c = -116$$
$$c = -12$$
• **Substitute −4 for b in Eq. (4).**

$$-7a - (-4) - 12 = -50$$
$$a = 6$$
• **Substitute −4 for b and −12 for c in Eq. (3)**

An equation of a circle whose graph passes through the three given points is $x^2 + y^2 + 6x - 4y - 12 = 0$.

Exercise Set 9.3, page 683

8. $\begin{cases} x - 2y = 3 & (1) \\ xy = -1 & (2) \end{cases}$

$$x = 2y + 3$$ • **Solve Eq. (1) for x.**

$$(2y + 3)y = -1$$ • **Replace x by 2y + 3 in Eq. (2).**

$$2y^2 + 3y + 1 = 0$$ • **Solve for y.**

$$(2y + 1)(y + 1) = 0$$

$$y = -\frac{1}{2} \quad \text{or} \quad y = -1$$

$$x - 2\left(-\frac{1}{2}\right) = 3 \qquad x - 2(-1) = 3$$ **• Substitute for y in Eq. (1).**

$$x = 2 \qquad x = 1$$

The solutions are $(2, -1/2)$ and $(1, -1)$.

16. $\begin{cases} 3x^2 - 2y^2 = 19 & (1) \\ x^2 - y^2 = 5 & (2) \end{cases}$

$$3x^2 - 2y^2 = 19 \quad (1)$$
$$-3x^2 + 3y^2 = -15$$ **• Multiply Eq. (2) by −3.**
$$y^2 = 4$$ **• Add the equations.**
$$y = \pm 2$$ **• Solve for y.**
$$x^2 - (-2)^2 = 5$$ **• Substitute −2 for y in Eq. (2).**
$$x^2 - 4 = 5$$
$$x^2 = 9$$
$$x = \pm 3$$
$$x^2 - 2^2 = 5$$ **• Substitute 2 for y in Eq. (2).**
$$x^2 - 4 = 5$$
$$x^2 = 9$$
$$x = \pm 3$$

The solutions are $(3, -2)$, $(-3, -2)$, $(3, 2)$, $(-3, 2)$.

20. $\begin{cases} 2x^2 + 3y^2 = 11 & (1) \\ 3x^2 + 2y^2 = 19 & (2) \end{cases}$

Use the elimination method to eliminate y^2.

$$4x^2 + 6y^2 = 22$$ **• 2 times Eq. (1)**
$$-9x^2 - 6y^2 = -57$$ **• −3 times Eq. (2)**
$$-5x^2 = -35$$
$$x^2 = 7$$
$$2(7) + 3y^2 = 11$$ **• Substitute for x in Eq. (1).**
$$3y^2 = -3$$
$$y^2 = -1$$

$y^2 = -1$ has no real number solutions. The graphs of the equations do not intersect. The system is inconsistent and has no solution.

28. $\begin{cases} (x + 2)^2 + (y - 3)^2 = 10 \\ (x - 3)^2 + (y + 1)^2 = 13 \end{cases}$

$$x^2 + 4x + 4 + y^2 - 6y + 9 = 10 \quad (1)$$
$$x^2 - 6x + 9 + y^2 + 2y + 1 = 13 \quad (2)$$
$$10x - 5 \quad - 8y + 8 = -3$$ **• Subtract**
$$10x - 8y = -6$$
$$y = \frac{5x + 3}{4} \quad (3)$$ **• Solve for y.**
$$(x + 2)^2 + \left(\frac{5x - 9}{4}\right)^2 = 10$$ **• Substitute for y.**
$$x^2 + 4x + 4 + \frac{25x^2 - 90x + 81}{16} = 10$$ **• Solve for x.**
$$16x^2 + 64x + 64 + 25x^2 - 90x + 81 = 160$$
$$41x^2 - 26x - 15 = 0$$
$$(41x + 15)(x - 1) = 0$$
$$x = -\frac{15}{41} \quad \text{or} \quad x = 1$$
$$y = \frac{5}{4}\left(-\frac{15}{41}\right) + \frac{3}{4} \quad \text{or} \quad y = \frac{5(1) + 3}{4}$$ **• Substitute for x into Eq. (3).**
$$y = \frac{12}{41} \qquad y = 2$$

The solutions are $(-15/41, 12/41)$ and $(1, 2)$.

Exercise Set 9.4, page 691

14. $\dfrac{7x + 44}{x^2 + 10x + 24} = \dfrac{7x + 44}{(x + 4)(x + 6)} = \dfrac{A}{x + 4} + \dfrac{B}{x + 6}$

$$7x + 44 = A(x + 6) + B(x + 4)$$
$$7x + 44 = (A + B)x + (6A + 4B)$$
$$\begin{cases} 7 = A + B \\ 44 = 6A + 4B \end{cases}$$

The solution is $A = 8$, $B = -1$.

$$\frac{7x + 44}{x^2 + 10x + 24} = \frac{8}{x + 4} + \frac{-1}{x + 6}$$

22. $\dfrac{x - 18}{x(x - 3)^2} = \dfrac{A}{x} + \dfrac{B}{x - 3} + \dfrac{C}{(x - 3)^2}$

$$x - 18 = A(x - 3)^2 + Bx(x - 3) + Cx$$
$$x - 18 = Ax^2 - 6Ax + 9A + Bx^2 - 3Bx + Cx$$
$$x - 18 = (A + B)x^2 + (-6A - 3B + C)x + 9A$$
$$\begin{cases} 0 = A + B \\ 1 = -6A - 3B + C \\ -18 = 9A \end{cases}$$

The solution is $A = -2$, $B = 2$, $C = -5$.

$$\frac{x - 18}{x(x - 3)^2} = \frac{-2}{x} + \frac{2}{x - 3} + \frac{-5}{(x - 3)^2}$$

24. $x^3 - x^2 + 10x - 10 = (x - 1)(x^2 + 10)$

$$\frac{9x^2 - 3x + 49}{(x - 1)(x^2 + 10)} = \frac{A}{x - 1} + \frac{Bx + C}{x^2 + 10}$$
$$9x^2 - 3x + 49 = A(x^2 + 10) + (Bx + C)(x - 1)$$
$$9x^2 - 3x + 49 = (A + B)x^2 + (-B + C)x + (10A - C)$$
$$\begin{cases} 9 = A + B \\ -3 = -B + C \\ 49 = 10A - C \end{cases}$$

The solution is $A = 5$, $B = 4$, $C = 1$.

$$\frac{9x^2 - 3x + 49}{x^3 - x^2 + 10x - 10} = \frac{5}{x - 1} + \frac{4x + 1}{x^2 + 10}$$

30. $\dfrac{2x^3 + 9x + 1}{(x^2 + 7)^2} = \dfrac{Ax + B}{x^2 + 7} + \dfrac{Cx + D}{(x^2 + 7)^2}$

$2x^3 + 9x + 1 = (Ax + B)(x^2 + 7) + Cx + D$

$2x^3 + 9x + 1 = Ax^3 + Bx^2 + (7A + C)x + (7B + D)$

$$\begin{cases} 2 = A \\ 0 = B \\ 9 = 7A + C \\ 1 = 7B + D \end{cases}$$

The solutions are $A = 2, B = 0, C = -5, D = 1$.

$\dfrac{2x^3 + 9x + 1}{x^4 + 14x^2 + 49} = \dfrac{2x}{x^2 + 7} + \dfrac{-5x + 1}{(x^2 + 7)^2}$

34. $2x^2 + 3x - 2 \overline{\smash{)}\,2x^3 + 5x^2 + 3x - 8}$ with quotient $x + 1$

$$\begin{array}{r} 2x^3 + 5x^2 + 3x - 8 \\ \underline{2x^3 + 3x^2 - 2x} \\ 2x^2 + 5x - 8 \\ \underline{2x^2 + 3x - 2} \\ 2x - 6 \end{array}$$

$\dfrac{2x^3 + 5x^2 + 3x - 8}{2x^2 + 3x - 2} = x + 1 + \dfrac{2x - 6}{2x^2 + 3x - 2}$

$\dfrac{2x - 6}{(2x - 1)(x + 2)} = \dfrac{A}{2x - 1} + \dfrac{B}{x + 2}$

$2x - 6 = A(x + 2) + B(2x - 1)$

$2x - 6 = Ax + 2A + 2Bx - B$

$2x - 6 = (A + 2B)x + (2A - B)$

$$\begin{cases} 2 = A + 2B \\ -6 = 2A - B \end{cases}$$

The solutions are $A = -2, B = 2$.

$\dfrac{2x^3 + 5x^2 + 3x - 8}{2x^2 + 3x - 2} = x + 1 + \dfrac{-2}{2x - 1} + \dfrac{2}{x + 2}$

Exercise Set 9.5, page 698

6.
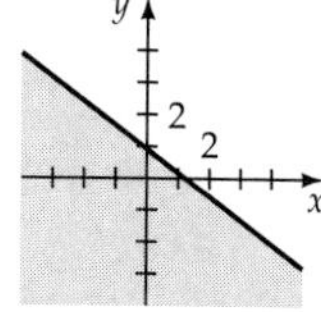

12.
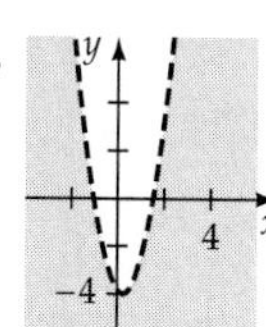

20.
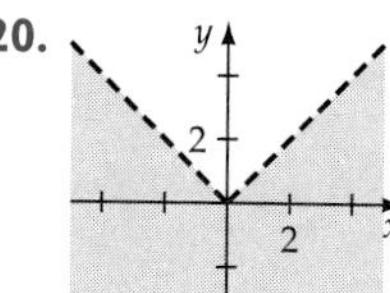

28.
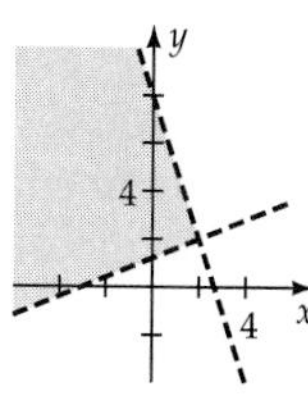

38.
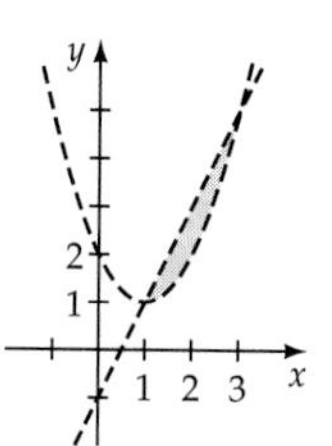

42.
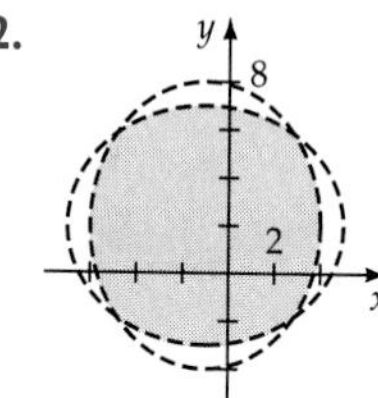

44.
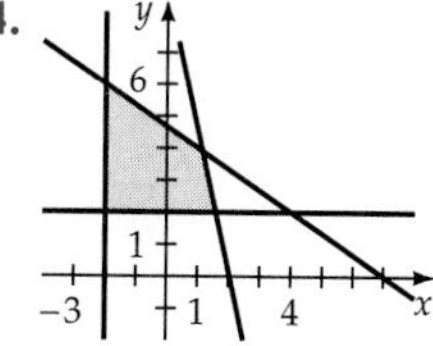

Exercise Set 9.6, page 705

12. $C = 4x + 3y$

(x, y)	C	
(0, 8)	24	
(2, 4)	20	• Minimum
(5, 2)	26	
(11, 0)	44	
(20, 0)	80	
(20, 20)	140	
(0, 20)	60	

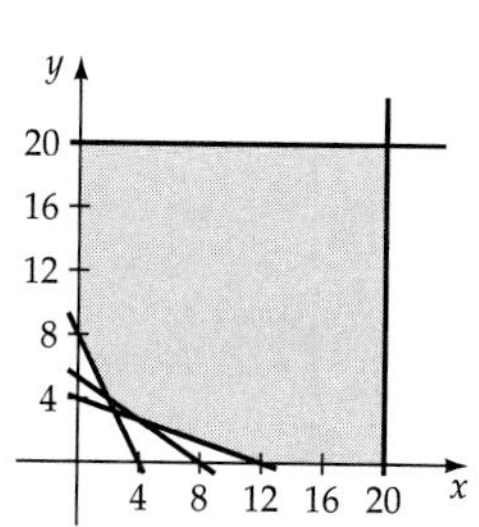

22. x = hours of machine 1 use
y = hours of machine 2 use
Cost $= 28x + 25y$

Constraints: $\begin{cases} 4x + 3y \geq 60 \\ 5x + 10y \geq 100 \\ x \geq 0, y \geq 0 \end{cases}$

(x, y)	Cost	
(0,20)	500	
(12,4)	436	• Minimum
(20,0)	560	

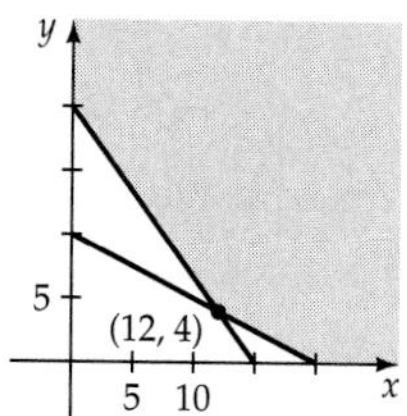

To achieve the minimum cost, use machine 1 for 12 hours and machine 2 for 4 hours.

24. Let x = number of standard models.
Let y = number of deluxe models.
Profit $= 25x + 35y$

Constraints: $\begin{cases} x + 3y \le 24 \\ x + \ \ y \le 10 \\ 2x + \ \ y \le 16 \\ x \ge 0, y \ge 0 \end{cases}$

(x, y)	Profit	
(0, 0)	0	
(0, 8)	280	
(6, 4)	290	
(3, 7)	320	• **Maximum**
(8, 0)	200	

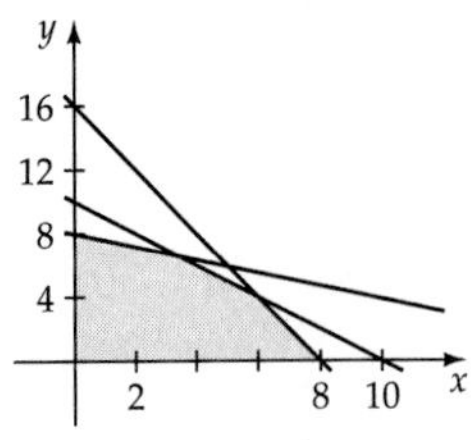

To maximize profits, produce 3 standard models and 7 deluxe models.

Exercise Set 10.1, page 722

14. $\left[\begin{array}{rrr|r} 1 & -3 & 1 & 8 \\ 2 & -5 & -3 & 2 \\ 1 & 4 & 1 & 1 \end{array}\right] \xrightarrow[-1R_1 + R_3]{-2R_1 + R_2} \left[\begin{array}{rrr|r} 1 & -3 & 1 & 8 \\ 0 & 1 & -5 & -14 \\ 0 & 7 & 0 & -7 \end{array}\right]$

$\xrightarrow{-7R_2 + R_3} \left[\begin{array}{rrr|r} 1 & -3 & 1 & 8 \\ 0 & 1 & -5 & -14 \\ 0 & 0 & 35 & 91 \end{array}\right]$

$\xrightarrow{\frac{1}{35}R_3} \left[\begin{array}{rrr|r} 1 & -3 & 1 & 8 \\ 0 & 1 & -5 & -14 \\ 0 & 0 & 1 & \frac{13}{5} \end{array}\right]$

$$\begin{cases} x - 3y + z = 8 \\ \qquad y - 5z = -14 \\ \qquad\qquad z = \dfrac{13}{5} \end{cases}$$

By back substitution, the solution is (12/5, −1, 13/5).

18. $\left[\begin{array}{rrr|r} 3 & -5 & 2 & 4 \\ 1 & -3 & 2 & 4 \\ 5 & -11 & 6 & 12 \end{array}\right] \xrightarrow{R_2 \leftrightarrow R_1} \left[\begin{array}{rrr|r} 1 & -3 & 2 & 4 \\ 3 & -5 & 2 & 4 \\ 5 & -11 & 6 & 12 \end{array}\right]$

$\xrightarrow[-5R_1 + R_3]{-3R_1 + R_2} \left[\begin{array}{rrr|r} 1 & -3 & 2 & 4 \\ 0 & 4 & -4 & -8 \\ 0 & 4 & -4 & -8 \end{array}\right] \xrightarrow{\frac{1}{4}R_2} \left[\begin{array}{rrr|r} 1 & -3 & 2 & 4 \\ 0 & 1 & -1 & -2 \\ 0 & 4 & -4 & -8 \end{array}\right]$

$\xrightarrow{-4R_2 + R_3} \left[\begin{array}{rrr|r} 1 & -3 & 2 & 4 \\ 0 & 1 & -1 & -2 \\ 0 & 0 & 0 & 0 \end{array}\right]$

$$\begin{cases} x - 3y + 2z = 4 & (1) \\ \qquad y - z = -2 & (2) \\ \qquad\quad 0 = 0 & (3) \end{cases}$$

$y - z = -2$ or $y = z - 2$

$x - 3(z - 2) + 2z = 4$ • **Substitute $z - 2$ for y in Eq. (1).**

$x - 3z + 6 + 2z = 4$

$x = z - 2$

Let $z = c$. The solutions are $(c - 2, c - 2, c)$.

20. $\left[\begin{array}{rrr|r} 2 & 5 & 2 & -1 \\ 1 & 2 & -3 & 5 \\ 5 & 12 & 1 & 10 \end{array}\right] \xrightarrow{R_2 \leftrightarrow R_1} \left[\begin{array}{rrr|r} 1 & 2 & -3 & 5 \\ 2 & 5 & 2 & -1 \\ 5 & 12 & 1 & 10 \end{array}\right]$

$\xrightarrow[-5R_1 + R_3]{-2R_1 + R_2} \left[\begin{array}{rrr|r} 1 & 2 & -3 & 5 \\ 0 & 1 & 8 & -11 \\ 0 & 2 & 16 & -15 \end{array}\right]$

$\xrightarrow{-2R_2 + R_3} \left[\begin{array}{rrr|r} 1 & 2 & -3 & 5 \\ 0 & 1 & 8 & -11 \\ 0 & 0 & 0 & 7 \end{array}\right]$

$$\begin{cases} x + 2y - 3z = 5 \\ \qquad y + 8z = -11 \\ \qquad\qquad 0 = 7 \end{cases}$$

Because $0 = 7$ is a false equation, the system of equations has no solution.

36. $\left[\begin{array}{rrrr|r} 1 & -1 & 3 & -5 & 10 \\ 2 & -3 & 4 & 1 & 7 \\ 3 & 1 & -2 & -2 & 6 \end{array}\right]$

$\xrightarrow[-3R_1 + R_3]{-2R_1 + R_2} \left[\begin{array}{rrrr|r} 1 & -1 & 3 & -5 & 10 \\ 0 & -1 & -2 & 11 & -13 \\ 0 & 4 & -11 & 13 & -24 \end{array}\right]$

$\xrightarrow{-1R_2} \left[\begin{array}{rrrr|r} 1 & -1 & 3 & -5 & 10 \\ 0 & 1 & 2 & -11 & 13 \\ 0 & 4 & -11 & 13 & -24 \end{array}\right]$

$\xrightarrow{-4R_2 + R_3} \left[\begin{array}{rrrr|r} 1 & -1 & 3 & -5 & 10 \\ 0 & 1 & 2 & -11 & 13 \\ 0 & 0 & -19 & 57 & -76 \end{array}\right]$

$\xrightarrow{-\frac{1}{19}R_3} \left[\begin{array}{rrrrr} 1 & -1 & 3 & -5 & 10 \\ 0 & 1 & 2 & -11 & 13 \\ 0 & 0 & 1 & -3 & 4 \end{array}\right]$

$$\begin{cases} t - u + 3v - 5w = 10 & (1) \\ \quad u + 2v - 11w = 13 & (2) \\ \qquad\quad v - 3w = 4 & (3) \end{cases}$$

$v = 3w + 4$

$u + 2(3w + 4) - 11w = 13$ • **Substitute $3w + 4$ for v in Eq. (2).**

$u = 5w + 5$

$t - (5w + 5) + 3(3w + 4) - 5w = 10$ • **Substitute $5w + 5$ for u and $3w + 4$ for u in Eq. (1).**

$t = w + 3$

Let w by any real number c. The solution of the system of equations is $(c + 3, 5c + 5, 3c + 4, c)$.

44. Because there are four given points, the degree of the interpolating polynomial will be at most 3. The form of the polynomial will be $p(x) = a_3x^3 + a_2x^2 + a_1x + a_0$. Use this polynomial to create a system of equations.

$$
\begin{aligned}
p(x) &= a_3x^3 + a_2x^2 + a_1x + a_0\\
p(-1) &= a_3(-1)^3 + a_2(-1)^2 + a_1(-1) + a_0\\
&= -a_3 + a_2 - a_1 + a_0 = -5\\
p(0) &= a_3(0)^3 + a_2(0)^2 + a_1(0) + a_0 = a_0 = 0\\
p(1) &= a_3(1)^3 + a_2(1)^2 + a_1(1) + a_0\\
&= a_3 + a_2 + a_1 + a_0 = 1\\
p(2) &= a_3(2)^3 + a_2(2)^2 + a_1(2) + a_0\\
&= 8a_3 + 4a_2 + 2a_1 + a_0 = 4
\end{aligned}
$$

The system of equations and the associated augmented matrix are

$$
\begin{cases}
-a_3 + a_2 - a_1 + a_0 = -5\\
a_0 = 0\\
a_3 + a_2 + a_1 + a_0 = 1\\
8a_3 + 4a_2 + 2a_1 + a_0 = 4
\end{cases}
\qquad
\left[\begin{array}{cccc|c}
-1 & 1 & -1 & 1 & -5\\
0 & 0 & 0 & 1 & 0\\
1 & 1 & 1 & 1 & 1\\
8 & 4 & 2 & 1 & 4
\end{array}\right]
$$

The augmented matrix in echelon form and the resulting system of equations are

$$
\left[\begin{array}{cccc|c}
1 & 0.5 & 0.25 & 0.125 & 0.5\\
0 & 1 & -0.5 & 0.75 & -3\\
0 & 0 & 1 & 0.5 & 2\\
0 & 0 & 0 & 1 & 0
\end{array}\right]
$$

$$
\begin{cases}
a_1 \quad 0.5a_2 \quad 0.25a_3 \quad 0.125a_4 = 0.5\\
a_2 \quad -0.5a_3 \quad 0.75a_4 = -3\\
a_3 \quad 0.5a_4 = 2\\
a_4 = 0
\end{cases}
$$

Solving the system of equations by back substitution yields $a_0 = 0$, $a_1 = 2$, $a_2 = -2$, and $a_3 = 1$. The interpolating polynomial is $p(x) = x^3 - 2x^2 + 2x$.

Exercise Set 10.2, page 736

6. a. $A + B = \begin{bmatrix} 2 & -2\\ 3 & 4\\ 1 & 0 \end{bmatrix} + \begin{bmatrix} -1 & 8\\ 2 & -2\\ -4 & 3 \end{bmatrix} = \begin{bmatrix} 1 & 6\\ 5 & 2\\ -3 & 3 \end{bmatrix}$

b. $A - B = \begin{bmatrix} 2 & -2\\ 3 & 4\\ 1 & 0 \end{bmatrix} - \begin{bmatrix} -1 & 8\\ 2 & -2\\ -4 & 3 \end{bmatrix} = \begin{bmatrix} 3 & -10\\ 1 & 6\\ 5 & -3 \end{bmatrix}$

c. $2B = 2\begin{bmatrix} -1 & 8\\ 2 & -2\\ -4 & 3 \end{bmatrix} = \begin{bmatrix} -2 & 16\\ 4 & -4\\ -8 & 6 \end{bmatrix}$

d. $2A - 3B = 2\begin{bmatrix} 2 & -2\\ 3 & 4\\ 1 & 0 \end{bmatrix} - 3\begin{bmatrix} -1 & 8\\ 2 & -2\\ -4 & 3 \end{bmatrix} = \begin{bmatrix} 7 & -28\\ 0 & 14\\ 14 & -9 \end{bmatrix}$

16. $AB = \begin{bmatrix} -1 & 2 & 0\\ 2 & -1 & 1\\ -2 & 2 & -1 \end{bmatrix}\begin{bmatrix} 2 & -1 & 0\\ 1 & 5 & -1\\ 0 & -1 & 3 \end{bmatrix}$

$$
= \begin{bmatrix}
(-1)(2) + (2)(1) + (0)(0) & (-1)(-1) + (2)(5) + (0)(-1) & (-1)(0) + (2)(-1) + (0)(3)\\
(2)(2) + (-1)(1) + (1)(0) & (2)(-1) + (-1)(5) + (1)(-1) & (2)(0) + (-1)(-1) + (1)(3)\\
(-2)(2) + (2)(1) + (-1)(0) & (-2)(-1) + (2)(5) + (-1)(-1) & (-2)(0) + (2)(-1) + (-1)(3)
\end{bmatrix}
$$

$$
= \begin{bmatrix} 0 & 11 & -2\\ 3 & -8 & 4\\ -2 & 13 & -5 \end{bmatrix}
$$

$$
BA = \begin{bmatrix} 2 & -1 & 0\\ 1 & 5 & -1\\ 0 & -1 & 3 \end{bmatrix}\begin{bmatrix} -1 & 2 & 0\\ 2 & -1 & 1\\ -2 & 2 & -1 \end{bmatrix}
$$

$$
= \begin{bmatrix}
(2)(-1) + (-1)(-2) + (0)(-2) & (2)(2) + (-1)(-1) + (0)(2) & (2)(0) + (-1)(1) + (0)(-1)\\
(1)(-1) + (5)(2) + (-1)(-2) & (1)(2) + (5)(-1) + (-1)(2) & (1)(0) + (5)(1) + (-1)(-1)\\
(0)(-1) + (-1)(2) + (3)(-2) & (0)(2) + (-1)(-1) + (3)(2) & (0)(0) + (-1)(1) + (3)(-1)
\end{bmatrix}
$$

$$
= \begin{bmatrix} -4 & 5 & -1\\ 11 & -5 & 6\\ -8 & 7 & -4 \end{bmatrix}
$$

Exercise Set 10.3, page 748

6. $\begin{bmatrix} 1 & 3 & -2 & 1 & 0 & 0\\ -1 & -5 & 6 & 0 & 1 & 0\\ 2 & 6 & -3 & 0 & 0 & 1 \end{bmatrix}$

$$
\xrightarrow[-2R_1 + R_3]{R_1 + R_2}
\begin{bmatrix} 1 & 3 & -2 & 1 & 0 & 0\\ 0 & -2 & 4 & 1 & 1 & 0\\ 0 & 0 & 1 & -2 & 0 & 1 \end{bmatrix}
$$

$$
\xrightarrow{-\frac{1}{2}R_2}
\begin{bmatrix} 1 & 3 & -2 & 1 & 0 & 0\\ 0 & 1 & -2 & -\frac{1}{2} & -\frac{1}{2} & 0\\ 0 & 0 & 1 & -2 & 0 & 1 \end{bmatrix}
$$

$$
\xrightarrow[2R_3 + R_1]{2R_3 + R_2}
\begin{bmatrix} 1 & 3 & 0 & -3 & 0 & 2\\ 0 & 1 & 0 & -\frac{9}{2} & -\frac{1}{2} & 2\\ 0 & 0 & 1 & -2 & 0 & 1 \end{bmatrix}
$$

$$
\xrightarrow{-3R_2 + R_1}
\begin{bmatrix} 1 & 0 & 0 & \frac{21}{2} & \frac{3}{2} & -4\\ 0 & 1 & 0 & -\frac{9}{2} & -\frac{1}{2} & 2\\ 0 & 0 & 1 & -2 & 0 & 1 \end{bmatrix}
$$

The inverse matrix is $\begin{bmatrix} \frac{21}{2} & \frac{3}{2} & -4 \\ -\frac{9}{2} & -\frac{1}{2} & 2 \\ -2 & 0 & 1 \end{bmatrix}$.

10. $\left[\begin{array}{ccc|ccc} 1 & -2 & 2 & 1 & 0 & 0 \\ 2 & -3 & 1 & 0 & 1 & 0 \\ 3 & -6 & 6 & 0 & 0 & 1 \end{array}\right]$

$\xrightarrow[-3R_1+R_3]{-2R_1+R_2} \left[\begin{array}{ccc|ccc} 1 & -2 & 2 & 1 & 0 & 0 \\ 0 & 1 & -3 & -2 & 1 & 0 \\ 0 & 0 & 0 & -3 & 0 & 1 \end{array}\right]$

Because there are zeros in a row of the original matrix, the matrix does not have an inverse.

20. $\begin{bmatrix} 1 & 2 & -1 \\ 2 & 3 & -1 \\ 3 & 6 & -2 \end{bmatrix}\begin{bmatrix} x \\ y \\ z \end{bmatrix} = \begin{bmatrix} 5 \\ 8 \\ 14 \end{bmatrix}$

The inverse of the coefficient matrix is

$$\begin{bmatrix} 0 & 2 & -1 \\ -1 & -1 & 1 \\ -3 & 0 & 1 \end{bmatrix}$$

Multiplying each side of the equation by the inverse, we have

$$\begin{bmatrix} x \\ y \\ z \end{bmatrix} = \begin{bmatrix} 0 & 2 & -1 \\ -1 & -1 & 1 \\ -3 & 0 & 1 \end{bmatrix}\begin{bmatrix} 5 \\ 8 \\ 14 \end{bmatrix} = \begin{bmatrix} 2 \\ 1 \\ -1 \end{bmatrix}$$

The solution is $(2, 1, -1)$.

Exercise Set 10.4, page 758

2. $\begin{vmatrix} 2 & 9 \\ -6 & 2 \end{vmatrix} = 2 \cdot 2 - (-6)(9) = 4 + 54 = 58$

14. $M_{13} = \begin{vmatrix} 1 & 3 \\ 6 & -2 \end{vmatrix} = 1(-2) - 6(3) = -2 - 18 = -20$

$C_{13} = (-1)^{1+3} \cdot M_{13} = 1 \cdot M_{13} = 1(-20) = -20$

20. Expanding with cofactors of row 1 yields

$$\begin{vmatrix} 3 & -2 & 0 \\ 2 & -3 & 2 \\ 8 & -2 & 5 \end{vmatrix} = 3C_{11} + (-2)C_{12} + 0 \cdot C_{13}$$

$$= 3\begin{vmatrix} -3 & 2 \\ -2 & 5 \end{vmatrix} + 2\begin{vmatrix} 2 & 2 \\ 8 & 5 \end{vmatrix} + 0\begin{vmatrix} 2 & -3 \\ 8 & -2 \end{vmatrix}$$

$$= 3(-15 + 4) + 2(10 - 16) + 0$$

$$= 3(-11) + 2(-6) = -33 + (-12)$$

$$= -45$$

42. Let $D = \begin{vmatrix} 3 & -2 & -1 \\ 1 & 2 & 4 \\ 2 & -2 & 3 \end{vmatrix}$. Then

$$D \overset{R_1 \leftrightarrow R_2}{=} -\begin{vmatrix} 1 & 2 & 4 \\ 3 & -2 & -1 \\ 2 & -2 & 3 \end{vmatrix} \overset{\substack{-3R_1+R_2 \\ -2R_1+R_3}}{=} -\begin{vmatrix} 1 & 2 & 4 \\ 0 & -8 & -13 \\ 0 & -6 & -5 \end{vmatrix}$$

$$\overset{-\frac{1}{8}R_2}{=} 8\begin{vmatrix} 1 & 2 & 4 \\ 0 & 1 & \frac{13}{8} \\ 0 & -6 & -5 \end{vmatrix} \overset{6R_2+R_3}{=} 8\begin{vmatrix} 1 & 2 & 4 \\ 0 & 1 & \frac{13}{8} \\ 0 & 0 & \frac{19}{4} \end{vmatrix}$$

$$= 8(1)(1)\left(\tfrac{19}{4}\right) = 38$$

Exercise Set 10.5, page 764

4. $x_1 = \dfrac{\begin{vmatrix} 9 & 5 \\ 8 & 7 \end{vmatrix}}{\begin{vmatrix} 2 & 5 \\ 5 & 7 \end{vmatrix}} = \dfrac{63 - 40}{14 - 25} = \dfrac{23}{-11} = -\dfrac{23}{11}$

$x_2 = \dfrac{\begin{vmatrix} 2 & 9 \\ 5 & 8 \end{vmatrix}}{\begin{vmatrix} 2 & 5 \\ 5 & 7 \end{vmatrix}} = \dfrac{16 - 45}{14 - 25} = \dfrac{-29}{-11} = \dfrac{29}{11}$

The solution is $(-23/11, 29/11)$.

24. $x_3 = \dfrac{\begin{vmatrix} 2 & 5 & -3 & -3 \\ 1 & 7 & 4 & -1 \\ 4 & 0 & 3 & 1 \\ 3 & 2 & 0 & 0 \end{vmatrix}}{\begin{vmatrix} 2 & 5 & -5 & -3 \\ 1 & 7 & 8 & -1 \\ 4 & 0 & 1 & 1 \\ 3 & 2 & -1 & 0 \end{vmatrix}} = \dfrac{157}{168}$

Exercise Set 11.1, page 777

6. $a_n = \dfrac{(-1)^{n+1}}{n(n+1)}$, $a_1 = \dfrac{(-1)^{1+1}}{1(1+1)} = \dfrac{1}{2}$,

$a_2 = \dfrac{(-1)^{2+1}}{2(2+1)} = -\dfrac{1}{6}$, $a_3 = \dfrac{(-1)^{3+1}}{3(3+1)} = \dfrac{1}{12}$,

$a_8 = \dfrac{(-1)^{8+1}}{8(8+1)} = -\dfrac{1}{72}$

28. $a_1 = 1, a_2 = 2^2 \cdot a_1 = 4 \cdot 1 = 4, a_3 = 3^2 \cdot a_2 = 9 \cdot 4 = 36$

42. $\dfrac{12!}{4!\,8!} = \dfrac{12 \cdot 11 \cdot 10 \cdot 9 \cdot 8!}{4!\,8!} = \dfrac{12 \cdot 11 \cdot 10 \cdot 9}{4 \cdot 3 \cdot 2 \cdot 1} = 495$

52. $\displaystyle\sum_{i=1}^{6} (2i + 1)(2i - 1) = \sum_{i=1}^{6} (4i^2 - 1)$

$$= (4 \cdot 1^2 - 1) + (4 \cdot 2^2 - 1) + (4 \cdot 3^2 - 1) + (4 \cdot 4^2 - 1) + (4 \cdot 5^2 - 1) + (4 \cdot 6^2 - 1)$$

$$= 3 + 15 + 35 + 63 + 99 + 143$$

$$= 358$$

Exercise Set 11.2, page 784

16. $a_6 = -14, a_8 = -20$

$$a_8 = a_6 + 2d$$

$$\frac{a_8 - a_6}{2} = d \quad \text{• Solve for } d.$$

$$\frac{-20 - (-14)}{2} = d$$

$$-3 = d$$

$$a_n = a_1 + (n-1)d$$

$$a_6 = a_1 + (6-1)(-3)$$

$$-14 = a_1 + (-15)$$

$$a_1 = 1$$

$$a_{15} = 1 + (15-1)(-3) = 1 + (14)(-3) = -41$$

22. $S_{20} = \dfrac{20}{2}(a_1 + a_{20})$

$$a_1 = 1 - 2(1) = -1$$

$$a_{20} = 1 - 2(20) = -39$$

$$S_{20} = 10[-1 + (-39)] = 10(-40) = -400$$

34. $a = 7, c_1, c_2, c_3, c_4, c_5, b = 19$

$$a_n = a_1 + (n-1)d$$

$$19 = 7 + (7-1)d \quad \text{• There are 7 terms, so } n = 7.$$

$$19 = 7 + 6d$$

$$d = 2$$

$$c_1 = a_1 + d = 7 + 2 = 9$$

$$c_2 = a_1 + 2d = 7 + 4 = 11$$

$$c_3 = a_1 + 3d = 7 + 6 = 13$$

$$c_4 = a_1 + 4d = 7 + 8 = 15$$

$$c_5 = a_1 + 5d = 7 + 10 = 17$$

Exercise Set 11.3, page 792

6. $\dfrac{a_{i+1}}{a_i} = \dfrac{(-1)^i e^{(i+1)x}}{(-1)^{i-1}e^{ix}} = -e^{(i+1)x-ix} = -e^x$

Because x is a constant, $-e^x$ is a constant and the sequence is a geometric sequence.

18. $\dfrac{a_2}{a_1} = \dfrac{6}{8} = \dfrac{3}{4} = r$

$$a_n = a_1 r^{n-1}$$

$$a_n = 8\left(\frac{3}{4}\right)^{n-1}$$

40. $r = \dfrac{4}{3}, a_1 = \dfrac{4}{3}, n = 14$

$$S_n = \frac{a_1(1-r^n)}{1-r}$$

$$S_{14} = \frac{\frac{4}{3}\left[1 - \left(\frac{4}{3}\right)^{14}\right]}{1 - \frac{4}{3}} = \frac{\frac{4}{3}\left[\frac{-263{,}652{,}487}{4{,}782{,}969}\right]}{-\frac{1}{3}} \approx 220.49$$

62. $0.3\overline{95} = \dfrac{3}{10} + \dfrac{95}{1000} + \dfrac{95}{100{,}000} + \cdots = \dfrac{3}{10} + \dfrac{95/1000}{1 - 1/100}$

$$= \frac{3}{10} + \frac{95}{990} = \frac{392}{990} = \frac{196}{495}$$

70. $A = \dfrac{P[(1+r)^m - 1]}{r}$; $P = 250, r = \dfrac{0.08}{12}, m = 12(4)$

$$A = \frac{250[(1 + 0.08/12)^{48} - 1]}{0.08/12} \approx 14087.48$$

Exercise Set 11.4, page 799

8. $S_n = 2 + 6 + 12 + \cdots + n(n+1) = \dfrac{n(n+1)(n+2)}{3}$

1. When $n = 1$, $S_1 = 1(1+1) = 2$; $\dfrac{1(1+1)(1+2)}{3} = 2$

   Therefore, the statement is true for $n = 1$.

2. Assume the statement is true for $n = k$.

   $$S_k = 2 + 6 + 12 + \cdots + k(k+1)$$

   $$= \frac{k(k+1)(k+2)}{3} \quad \text{• Induction hypothesis}$$

   Prove the statement is true for $n = k + 1$. That is, prove

   $$S_{k+1} = \frac{(k+1)(k+2)(k+3)}{3}.$$

   Because $a_k = k(k+1)$ and $a_{k+1} = (k+1)(k+2)$,

   $$S_{k+1} = S_k + a_{k+1} = \frac{k(k+1)(k+2)}{3} + (k+1)(k+2)$$

   $$= \frac{k(k+1)(k+2) + 3(k+1)(k+2)}{3}$$

   $$= \frac{(k+1)(k+2)(k+3)}{3} \quad \text{• Factor out } (k+1) \text{ and } (k+2) \text{ from each term.}$$

   By the Principle of Mathematical Induction, the statement is true for all positive integers n.

12. $P_n = \left(1 - \dfrac{1}{2}\right)\left(1 - \dfrac{1}{3}\right)\cdots\left(1 - \dfrac{1}{n+1}\right) = \dfrac{1}{n+1}$

1. Let $n = 1$; then $P_1 = \left(1 - \dfrac{1}{2}\right) = \dfrac{1}{2}$; $\dfrac{1}{1+1} = \dfrac{1}{2}$

   The statement is true for $n = 1$.

2. Assume the statement is true for $n = k$.

   $$P_k = \left(1 - \frac{1}{2}\right)\left(1 - \frac{1}{3}\right)\cdots\left(1 - \frac{1}{k+1}\right) = \frac{1}{k+1}$$

Prove the statement is true for $n = k + 1$. That is, prove

$$P_{k+1} = \left(1 - \frac{1}{2}\right)\left(1 - \frac{1}{3}\right)\cdots\left(1 - \frac{1}{k+1}\right)\left(1 - \frac{1}{k+2}\right)$$
$$= \frac{1}{k+2}$$

Because $a_k = \left(1 - \frac{1}{k+1}\right)$ and $a_{k+1} = \left(1 - \frac{1}{k+2}\right)$,

$$P_{k+1} = P_k \cdot a_{k+1} = \frac{1}{k+1} \cdot \left(1 - \frac{1}{k+2}\right)$$
$$= \frac{1}{k+1} \cdot \frac{k+1}{k+2} = \frac{1}{k+2}$$

By the Principle of Mathematical Induction, the statement is true for all positive integers n.

16. If $a > 1$, show that $a^{n+1} > a^n$ for all positive integers n.

1. Because $a > 1$, $a \cdot a > a \cdot 1$ or $a^2 > a$. Thus the statement is true when $n = 1$.
2. Assume the statement is true for $n = k$.

$a^{k+1} > a^k$ • **Induction hypothesis**

Prove the statement is true for $n = k + 1$. That is, prove

$a^{k+2} > a^{k+1}$

Because $a^{k+1} > a^k$ and $a > 0$,

$a(a^{k+1}) > a(a^k)$

$a^{k+2} > a^{k+1}$

By the Principle of Mathematical Induction, the statement is true for all positive integers n.

20. 1. Let $n = 1$. Because $\log_{10} 1 = 0$,

$\log_{10} 1 < 1$

The inequality is true for $n = 1$.

2. Assume $\log_{10} k < k$ is true for some positive integer k (induction hypothesis). Prove the inequality is true for $n = k + 1$. That is, prove $\log_{10}(k + 1) < k + 1$ is true when $n = k + 1$.

$$\log_{10}(k+1) \le \log_{10}(k+k)$$
$$= \log_{10} 2k = \log_{10} 2 + \log_{10} k < 1 + k$$

Thus $\log_{10}(k + 1) < k + 1$. By the Principle of Mathematical Induction, $\log_{10} n < n$ for all positive integers n.

Exercise Set 11.5, page 805

4. $\binom{10}{5} = \frac{10!}{5!\,5!} = \frac{10 \cdot 9 \cdot 8 \cdot 7 \cdot 6 \cdot 5!}{5!\,5!} = \frac{10 \cdot 9 \cdot 8 \cdot 7 \cdot 6}{5 \cdot 4 \cdot 3 \cdot 2 \cdot 1}$

$= 252$

18. $(3x + 2y)^4$

$$= (3x)^4 + 4(3x)^3(2y) + 6(3x)^2(2y)^2 + 4(3x)(2y)^3 + (2y)^4$$
$$= 81x^4 + 216x^3y + 216x^2y^2 + 96xy^3 + 16y^4$$

20. $(2x - \sqrt{y})^7 = \binom{7}{0}(2x)^7 + \binom{7}{1}(2x)^6(-\sqrt{y})$

$$+ \binom{7}{2}(2x)^5(-\sqrt{y})^2 + \binom{7}{3}(2x)^4(-\sqrt{y})^3$$
$$+ \binom{7}{4}(2x)^3(-\sqrt{y})^4 + \binom{7}{5}(2x)^2(-\sqrt{y})^5$$
$$+ \binom{7}{6}(2x)(-\sqrt{y})^6 + \binom{7}{7}(-\sqrt{y})^7$$
$$= 128x^7 - 448x^6\sqrt{y} + 672x^5y - 560x^4y\sqrt{y}$$
$$+ 280x^3y^2 - 84x^2y^2\sqrt{y}$$
$$+ 14xy^3 - y^3\sqrt{y}$$

34. $\binom{10}{6-1}(x^{-1/2})^{10-6+1}(x^{1/2})^{6-1} = \binom{10}{5}(x^{-1/2})^5(x^{1/2})^5 = 252$

Exercise Set 11.6, page 811

12. Because there are 4 palettes and each palette contains 4 colors, by the counting principle there are $4 \cdot 4 \cdot 4 \cdot 4 = 256$ possible colors.

16. There are three possible finishes (first, second, and third) for the 12 contestants. Because the order of finish is important, these are the permutations of the 12 contestants selected 3 at a time.

$$P(12,3) = \frac{12!}{(12-3)!} = \frac{12!}{9!} = 12 \cdot 11 \cdot 10 = 1320$$

There are 1320 possible finishes.

20. Player A matched against Player B in the same tennis match as Player B matched against Player A. Therefore, this is a combination of 26 players selected 2 at a time.

$$C(26, 2) = \frac{26!}{2!(26-2)!} = \frac{26!}{2!\,24!} = \frac{26 \cdot 25 \cdot 24!}{2 \cdot 1 \cdot 24!} = 325$$

There are 325 possible first-round matches.

22. The person who refuses to sit in the back seat can be placed in any one of the 3 front seats. Similarly, the person who refuses to sit in the front can be placed in any of the 3 back seats. The remaining 4 people can sit in any of the remaining seats. The number of seating arrangements is

$3 \cdot 3 \cdot 4 \cdot 3 \cdot 2 \cdot 1 = 216$

30. a. The number of ways in which 10 finalists can be selected from 15 semifinalists is the combination of 15 students selected 10 at a time.

$C(15, 10) = 3003$

There are 3003 ways in which the finalists can be chosen.

b. The number of ways in which the 10 finalists can include 3 seniors is the product of the combination of 7 seniors selected 3 at a time and the combination of 8 remaining students selected 7 at a time.

$C(7,3)C(8,7) = 35 \cdot 8 = 280$

There are 280 ways in which the finalists can include 3 seniors.

c. "At least five seniors" means 5 or 6 or 7 seniors are finalists (there are only 7 seniors). Because the events are related by "or," sum the number of ways each event can occur.

$$C(7,5)C(8,5) + C(7,6)C(8,4) + C(7,7)C(8,3)$$
$$= 21 \cdot 56 + 7 \cdot 70 + 1 \cdot 56 = 1176 + 490 + 56 = 1722$$

There are 1722 ways in which the finalists can include at least 5 seniors.

Exercise Set 11.7, page 821

6. Let R represent the Republican, D the Democrat, and I the Independent. The sample space is

$\{(R, D), (R, I), (D, I)\}$

14. {HHHT, HHTH, HTHH, THHH, HHHH}

22. Let $E = \{2, 4, 6\}$, $T = \{3, 6\}$, $S = \{1, 2, 3, 4, 5, 6\}$.

$$E \cup T = \{2, 3, 4, 6\}$$
$$P(E \cup T) = \frac{N(E \cup T)}{N(S)} = \frac{4}{6} = \frac{2}{3}$$

32. $\dfrac{C(3,2) \cdot C(5,2)}{C(8,4)} = \dfrac{3 \cdot 10}{70} = \dfrac{3}{7}$

34. Yes, because the card was replaced. The probability of an ace on each draw is $4/52 = 1/13$.

$$P(2\text{ aces}) = \frac{1}{13} \cdot \frac{1}{13} = \frac{1}{169}$$

40. This is a binomial experiment; $p = 1/4$; $q = 3/4$, $n = 8$, $k = 3$.

$$\binom{8}{3}\left(\frac{1}{4}\right)^3\left(\frac{3}{4}\right)^5 = 56\left(\frac{1}{64}\right)\left(\frac{243}{1024}\right) \approx 0.2076$$

ANSWERS TO ODD-NUMBERED EXERCISES

Exercise Set P.1, page 8

1. a. $-3, 4, 11, 57$ are integers. **b.** $-3, 4, 1/5, 11, 3.14, 57$ are rational numbers. **c.** $0.25225222522225\ldots$ is an irrational number. **d.** All are real numbers. **e.** 11 is a prime number. **f.** 4, 57 are composite numbers. **3.** $A = \{4, 6, 8, 9, 10\}$ **5.** $C = \{53, 59\}$ **7.** $\{2, 4, 6, 8\}$ **9.** $\{3, 5, 7, 9\}$ **11.** $\{0, 1, 2, 3\}$ **13.** $\{1, 3\}$ **15.** $\{1, 3\}$ **17.** $\{0, 2, 4\}$ **19.** $\{0, 1, 2, 3, 4, 5, 11\}$ **21.** $\{1, 3\}$ **23.** $\varnothing$ **25.** associative property of addition **27.** identity property of multiplication **29.** reflexive property of equality **31.** transitive property of equality **33.** inverse property of addition **35.** commutative property of addition **37.** transitive property of equality **39.** $-3a/7$ **41.** $-7a/20$ **43.** $-10/21$ **45.** $-18/5$ **47.** $-2a/15$ **49. a.** 26/55 of the pool **b.** $26x/165$ of the pool **51.** 8/59 **53.** $2 - 1 \neq 1 - 2$ **55.** $(3 - 1) - 5 = -3; 3 - (1 - 5) = 7$ **57.** False **59.** True **61.** False **63.** True **65.** True **67. a.** $0.\overline{72}$ **b.** 0.825 **c.** $0.\overline{285714}$ **d.** $0.1\overline{35}$ **69.** 0.16620626, 0.16662040, 0.16666204, 0.16667; $\dfrac{\sqrt{x+9}-3}{x}$ is approaching $0.1\overline{6} = 1/6$. **71.** all the properties except for the identity property of addition and the inverse properties **73.** all the properties **75. a.** $5 + 7$ **b.** $23 + 7$

Exercise Set P.2, page 15

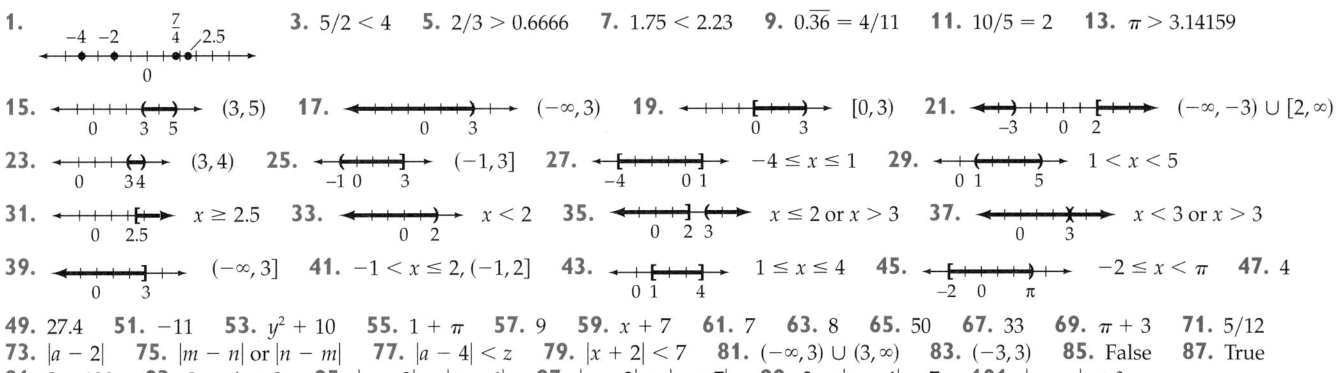

1. (number line: -4, -2, $\frac{7}{4}$, 2.5, 0) **3.** $5/2 < 4$ **5.** $2/3 > 0.6666$ **7.** $1.75 < 2.23$ **9.** $0.\overline{36} = 4/11$ **11.** $10/5 = 2$ **13.** $\pi > 3.14159$

15. $(3, 5)$ **17.** $(-\infty, 3)$ **19.** $[0, 3)$ **21.** $(-\infty, -3) \cup [2, \infty)$ **23.** $(3, 4)$ **25.** $(-1, 3]$ **27.** $-4 \le x \le 1$ **29.** $1 < x < 5$ **31.** $x \ge 2.5$ **33.** $x < 2$ **35.** $x \le 2$ or $x > 3$ **37.** $x < 3$ or $x > 3$ **39.** $(-\infty, 3]$ **41.** $-1 < x \le 2, (-1, 2]$ **43.** $1 \le x \le 4$ **45.** $-2 \le x < \pi$ **47.** 4

49. 27.4 **51.** -11 **53.** $y^2 + 10$ **55.** $1 + \pi$ **57.** 9 **59.** $x + 7$ **61.** 7 **63.** 8 **65.** 50 **67.** 33 **69.** $\pi + 3$ **71.** 5/12 **73.** $|a - 2|$ **75.** $|m - n|$ or $|n - m|$ **77.** $|a - 4| < z$ **79.** $|x + 2| < 7$ **81.** $(-\infty, 3) \cup (3, \infty)$ **83.** $(-3, 3)$ **85.** False **87.** True **91.** $I \le 120$ **93.** $2 \le A < 3$ **95.** $|x - 2| < |x - 6|$ **97.** $|x - 3| > |x + 7|$ **99.** $2 < |x - 4| < 7$ **101.** $|x - a| < \delta$

Exercise Set P.3, page 28

1. -256 **3.** 1 **5.** 81/16 **7.** 8 **9.** -3 **11.** -16 **13.** 16 **15.** 1/3 **17.** 1/9 **19.** 3/4 **21.** 9/4 **23.** $6x^7y^4$ **25.** $\dfrac{36x^4}{y^4}$ **27.** $\dfrac{y}{3x}$ **29.** $\dfrac{a+b}{ab}$ **31.** $\dfrac{16}{b^2}$ **33.** $3|xy^3|$ **35.** $a^{1/2}b^{3/10}$ **37.** $a^2 + 7a$ **39.** $p - q$ **41.** $m^2n^{3/2}$ **43.** 1 **45.** $r^{(m-n)/(mn)}$ **47.** 2.1×10^7 **49.** 9.5×10^{-4} **51.** 6500 **53.** 0.000217 **55.** $\sqrt{3x}$ **57.** $5\sqrt[4]{xy}$ **59.** $\sqrt[3]{(5w)^2}$ **61.** $(17k)^{1/3}$ **63.** $a^{2/5}$ **65.** $\left(\dfrac{7a}{3}\right)^{1/2}$ **67.** $3\sqrt{5}$ **69.** $2\sqrt[3]{3}$ **71.** $-3\sqrt[3]{3}$ **73.** $-2\sqrt[3]{4}$ **75.** $2|xy|\sqrt{6x}$ **77.** $-2ay^2\sqrt[3]{2y}$ **79.** $-13\sqrt{2}$ **81.** $-10\sqrt[4]{3}$ **83.** $17y\sqrt[3]{4y}$ **85.** $22x^2y\sqrt[3]{y}$ **87.** $29 + 11\sqrt{5}$ **89.** $2x - 9$ **91.** $50y + 10\sqrt{6yz} + 3z$ **93.** $x + 10\sqrt{x-3} + 22$ **95.** $2x + 14\sqrt{2x+5} + 54$ **97.** $\sqrt{2}$ **99.** $\dfrac{\sqrt{10}}{6}$ **101.** $\dfrac{3\sqrt[3]{4}}{2}$ **103.** $\dfrac{2\sqrt[3]{x}}{x}$ **105.** $\dfrac{\sqrt{5}}{3}$ **107.** $\dfrac{\sqrt{6xy}}{9y}$ **109.** $\dfrac{3(\sqrt{5}-\sqrt{x})}{5-x}$ **111.** $-\dfrac{2\sqrt{7}+7}{3}$ **113. a.** $2^{16} = 65{,}536$ **b.** $2^{32} = 4{,}294{,}967{,}296$ **115.** 1.97×10^4 seconds **117.** \$4873.50 **119. a.** 81% **b.** 66% **121.** x^3y^{n+1} **123.** $\dfrac{y^{n+1}}{x^{5n}}$ **125.** 8/5 **127.** $-19/12$ **129.** $3^{(3^3)}$ **131.** ?? **133.** $\dfrac{1}{\sqrt{4+h}+2}$ **135.** $\dfrac{1}{\sqrt{a+h}+a}$ **137.** $\dfrac{1}{\sqrt{n^2+1}+n}$

Exercise Set P.4, page 36

1. D **3.** H **5.** G **7.** B **9.** J **11. a.** $x^2 + 2x - 7$ **b.** 2 **c.** $1, 2, -7$ **d.** 1 **e.** $x^2, 2x, -7$ **13. a.** $x^3 - 1$ **b.** 3 **c.** $1, -1$ **d.** 1 **e.** $x^3, -1$ **15. a.** $2x^4 + 3x^3 + 4x^2 + 5$ **b.** 4 **c.** $2, 3, 4, 5$ **d.** 2 **e.** $2x^4, 3x^3, 4x^2, 5$ **17.** 3 **19.** 5 **21.** 2 **23.** $5x^2 + 11x + 3$ **25.** $9w^3 + 8w^2 - 2w + 6$ **27.** $-2r^2 + 3r - 12$ **29.** $-3u^2 - 2u + 4$ **31.** $8x^3 + 18x^2 - 67x + 40$ **33.** $6x^4 - 19x^3 + 26x^2 - 29x + 10$ **35.** $10x^2 + 22x + 4$ **37.** $y^2 + 3y + 2$ **39.** $4z^2 - 19z + 12$ **41.** $a^2 + 3a - 18$ **43.** $10x^2 - 57xy + 77y^2$ **45.** $18x^2 + 55xy + 25y^2$ **47.** $6p^2 - 11pq - 35q^2$ **49.** $12d^2 + 4d - 8$ **51.** $r^3 + s^3$ **53.** $60c^3 - 49c^2 + 4$

55. $9x^2 - 25$ **57.** $9x^4 - 6x^2y + y^2$ **59.** $16w^2 + 8wz + z^2$ **61.** $x^2 + 10x + 25 - y^2$ **63.** 29 **65.** −17 **67.** −1 **69.** 33 **71. a.** 1.6 pounds **b.** 3.6 pounds **73. a.** 72π in.^{3} **b.** 300π cm^3 **75. a.** 0.076 seconds **b.** 0.085 seconds **77.** 11,175 matches **79.** 14.8 sec; 90.4 sec **81.** $a^3 - 3a^2b + 3ab^2 - b^3$ **83.** $y^3 + 6y^3 + 12y + 8$ **85.** $27x^3 + 135x^2y + 225xy^2 + 125y^3$

Exercise Set P.5, page 47

1. $5(x + 4)$ **3.** $-3x(5x + 4)$ **5.** $2xy(5x + 3 - 7y)$ **7.** $(x - 3)(2a + 4b)$ **9.** $(x + 3)(x + 4)$ **11.** $(a - 12)(a + 2)$ **13.** $(6x + 1)(x + 4)$ **15.** $(17x + 4)(3x - 1)$ **17.** $(3x + 8y)(2x - 5y)$ **19.** $(x^2 + 5)(x^2 + 1)$ **21.** $(6x^2 + 5)(x^2 + 3)$ **23.** factorable over the integers **25.** not factorable over the integers **27.** not factorable over the integers **29.** $(x - 3)(x + 3)$ **31.** $(2a - 7)(2a + 7)$ **33.** $(1 - 10x)(1 + 10x)$ **35.** $(x^2 - 3)(x^2 + 3)$ **37.** $(x + 3)(x + 7)$ **39.** $(x + 5)^2$ **41.** $(a - 7)^2$ **43.** $(2x + 3)^2$ **45.** $(z^2 + 2w^2)^2$ **47.** $(x - 2)(x^2 + 2x + 4)$ **49.** $(2x - 3y)(4x^2 + 6xy + 9y^2)$ **51.** $(2 - x^2)(4 + 2x^2 + x^4)$ **53.** $(x - 3)(x^2 - 3x + 3)$ **55.** $(3x + 1)(x^2 + 2)$ **57.** $(x - 1)(ax + b)$ **59.** $(3w + 2)(2w^2 - 5)$ **61.** $2(3x - 1)(3x + 1)$ **63.** $(2x - 1)(2x + 1)(4x^2 + 1)$ **65.** $a(3x - 2y)(4x - 5y)$ **67.** $b(3x + 4)(x - 1)(x + 1)$ **69.** $2b(6x + y)^2$ **71.** $(w - 3)(w^2 - 12w + 39)$ **73.** $(x + 3y - 1)(x + 3y + 1)$ **75.** not factorable over the integers **77.** $(2x - 5)^2(3x + 5)$ **79.** $(2x - y)(2x + y + 1)$ **81.** 8 **83.** 64 **85.** $(x^n - 1)(x^n + 1)(x^{2n} + 1)$ **87.** $\pi(R - r)(R + r)$ **89.** $r^2(4 - \pi)$

Exercise Set P.6, page 56

1. $\frac{x + 4}{3}$ **3.** $\frac{x - 3}{x - 2}$ **5.** $\frac{a^2 - 2a + 4}{a - 2}$ **7.** $-\frac{x + 8}{x + 2}$ **9.** $-\frac{4y^2 + 7}{y + 7}$ **11.** $-\frac{8}{a^3b}$ **13.** $\frac{10}{27q^2}$ **15.** $\frac{x(3x + 7)}{2x + 3}$ **17.** $\frac{x + 3}{2x + 3}$ **19.** $\frac{(2y + 3)(3y - 4)}{(2y - 3)(y + 1)}$ **21.** $\frac{1}{a - 8}$ **23.** $\frac{3p - 2}{r}$ **25.** $\frac{8x(x - 4)}{(x - 5)(x + 3)}$ **27.** $\frac{3y - 4}{y + 4}$ **29.** $\frac{7z(2z - 5)}{(2z - 3)(z - 5)}$ **31.** $\frac{-2x^2 + 14x - 3}{(x - 3)(x + 3)(x + 4)}$ **33.** $\frac{(2x - 1)(x + 5)}{x(x - 5)}$ **35.** $\frac{-q^2 + 12q + 5}{(q - 3)(q + 5)}$ **37.** $\frac{3x^2 - 7x - 13}{(x + 3)(x + 4)(x - 3)(x - 4)}$ **39.** $\frac{(x + 2)(3x - 1)}{x^2}$ **41.** $\frac{4x + 1}{x - 1}$ **43.** $\frac{x - 2y}{y(y - x)}$ **45.** $\frac{(5x + 9)(x + 3)}{(x + 2)(4x + 3)}$ **47.** $\frac{(b + 3)(b - 1)}{(b - 2)(b + 2)}$ **49.** $\frac{x - 1}{x}$ **51.** $2 - m^2$ **53.** $\frac{-x^2 + 5x + 1}{x^2}$ **55.** $\frac{-x - 7}{x^2 + 6x - 3}$ **57.** $\frac{2x - 3}{x + 3}$ **59.** $\frac{a + b}{ab(a - b)}$ **61.** $\frac{(b - a)(b + a)}{ab(a^2 + b^2)}$ **63. a.** 136.55 mph **b.** $\frac{2v_1v_2}{v_1 + v_2}$ **65.** $\frac{2x + 1}{x(x + 1)}$ **67.** $\frac{3x^2 - 4}{x(x - 2)(x + 2)}$ **69.** $\frac{x^2 + 9x + 25}{(x + 5)^2}$ **71.** $\frac{x(1 - 4xy)}{(1 - 2xy)(1 + 2xy)}$ **73.** $R\left[\frac{(1 + i)^n - 1}{i(1 + i)^n}\right]$

Chapter P True/False Exercises, page 60

1. True **2.** False; if $a = 1/2$, then $(1/2)^2 = 1/4 < 1/2$. **3.** True **4.** False; $\sqrt{2} + (-\sqrt{2}) = 0$, which is a rational number. **5.** False; $(2 \oplus 4) \oplus 6 \neq 2 \oplus (4 \oplus 6)$. **6.** False; $x > a$ is written as (a, ∞). **7.** False; $\sqrt{(-2)^2} \neq -2$. **8.** False **9.** False

Chapter P Review Exercises, page 61

1. integer, rational number, real number, prime number **3.** rational number, real number **5.** $\{1, 2, 3, 5, 7, 11\}$ **7.** distributive property **9.** associative property of multiplication **11.** identity property of addition **13.** symmetric property of equality **15.**  $(-4, 2]$ **17.** 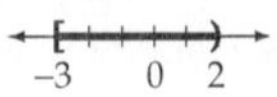 $-3 \le x < 2$ **19.** 7 **21.** $4 - \pi$ **23.** 17 **25.** −36 **27.** $12x^8y^3$ **29.** 6.2×10^5 **31.** 35,000 **33.** $-a^2 - 2a - 1$ **35.** $6x^4 + 5x^3 - 13x^2 + 22x - 20$ **37.** $3(x + 5)^2$ **39.** $4(5a^2 - b^2)$ **41.** $\frac{3x - 2}{x + 4}$ **43.** $\frac{2x + 3}{2x - 5}$ **45.** $\frac{x(3x + 10)}{(x + 3)(x - 3)(x + 4)}$ **47.** $\frac{2x - 9}{3x - 17}$ **49.** 5 **51.** $x^{17/12}$ **53.** $x^{3/4}y^2$ **55.** $4ab^3\sqrt{3b}$ **57.** $6x\sqrt{2y}$ **59.** $\frac{3y\sqrt{15y}}{5}$ **61.** $\frac{7\sqrt[3]{4x}}{2}$ **63.** $-3y^2\sqrt[3]{5x^2y}$

Chapter P Test, page 62

1. distributive property **2.** $\{0, 1, 2, 3, 4, 5, 6, 7, 8, 9\}$ **3.** 7 **4.** $\frac{4}{9x^4y^2}$ **5.** $\frac{96bc^2}{a^5}$ **6.** 1.37×10^{-3} **7.** $x^3 - 2x^2 + 5xy - 2x^2y - 2y^2$ **8.** −94 **9.** $(7x - 1)(x + 5)$ **10.** $(a - 4b)(3x - 2)$ **11.** $2x(2x - y)(4x^2 + 2xy + y^2)$ **12.** $-\frac{x + 3}{x + 5}$ **13.** $\frac{(x - 6)(x + 1)}{(x + 3)(x - 2)(x - 3)}$ **14.** $\frac{x(x + 2)}{x - 3}$ **15.** $\frac{3a^2 - 3ab - 10a + 5b}{a(2a - b)}$ **16.** $\frac{x(2x - 1)}{2x + 1}$ **17.** $\frac{x^{5/6}}{y^{9/4}}$ **18.** $7xy\sqrt[3]{3xy}$ **19.** $\frac{\sqrt[4]{8x}}{2}$ **20.** $\frac{3\sqrt{x} - 6}{x - 4}$

Exercise Set 1.1, page 69

1. 15 **3.** −4 **5.** 9/2 **7.** 108/23 **9.** 2/9 **11.** 12 **13.** 16 **15.** 9 **17.** 1/2 **19.** 22/13 **21.** 95/18 **23.** identity **25.** conditional equation **27.** contradiction **29.** 31 **31.** 2 **33.** no solution **35.** no solution **37.** 7/2 **39.** 6 **41.** −12 **43.** 1 **45.** −4, 4 **47.** 7, 3 **49.** 8, −3 **51.** 2, −8 **53.** 20, −12 **55.** no solution **57.** 12, −18 **59.** $(a+b)/2, (a-b)/2$ **61.** between 2002 and 2003 **63.** 15 min **65.** maximum 166 beats per minute, minimum 127 beats per minute **67.** no **69.** yes **73.** $\{x \mid x \geq -4\}$ **75.** $\{x \mid x \leq -7\}$ **77.** $\{x \mid x \geq -\frac{7}{2}\}$ **79.** $\{3, -5\}$

Exercise Set 1.2, page 79

1. $h = \dfrac{3V}{\pi r^2}$ **3.** $t = \dfrac{I}{Pr}$ **5.** $m_1 = \dfrac{Fd^2}{Gm_2}$ **7.** $v_0 = \dfrac{s + 16t^2}{t}$ **9.** $T_w = \dfrac{-Q_w + m_w c_w T_f}{m_w c_w}$ **11.** $d = \dfrac{a_n - a_1}{n - 1}$ **13.** $r = \dfrac{S - a_1}{S}$ **15.** $f_1 = \dfrac{w_1 f - w_2 f_2 + w_2 f}{w_1}$ **17.** $v_{LC} = \dfrac{f_{LC}v - f_v v}{f_v}$ **19.** 100 **21.** 30 feet by 57 feet **23.** 12 centimeters, 36 centimeters, 36 centimeters **25.** $(x+5)(x-1)$ **27.** 240 meters **29.** 2 hours **31.** 3 miles **33.** 98 **35.** 850 **37.** $937.50 **39.** $7600 invested at 8%, $6400 invested at 6.5% **41.** $3750 **43.** $18\frac{2}{11}$ grams **45.** 64 liters **47.** 1200 at $14 and 1800 at $25 **49.** $6\frac{2}{3}$ pounds of the $12 coffee and $13\frac{1}{3}$ pounds of the $9 coffee **51.** 10 grams **53.** 7.875 hours **55.** $13\frac{1}{3}$ hours **57.** $10.05 for book, $0.05 for bookmark **59.** 6.25 feet **61.** 40 pounds **63.** 1384 feet **65.** 84 years old

Exercise Set 1.3, page 96

1. 5, −3 **3.** $-\frac{1}{2}, 1$ **5.** −24, 3/8 **7.** 0, 7/3 **9.** 8, 2 **11.** 3, 8/3 **13** ±9 **15.** $\pm 2\sqrt{6}$ **17.** $\pm 2i$ **19.** 11, −1 **21.** $3 \pm 4i$ **23.** $5 + 12i$ **25.** $-2 - 5i$ **27.** $12 - 2i$ **29.** $16 + 16i$ **31.** $23 + 2i$ **33.** 74 **35.** $\dfrac{1}{2} - \dfrac{1}{2}i$ **37.** $\dfrac{7}{58} + \dfrac{3}{58}i$ **39.** $\dfrac{5}{13} + \dfrac{12}{13}i$ **41.** $-16 - 30i$ **43.** −1 **45.** −1 **47.** $-i$ **49.** i **51.** $-3 \pm 2\sqrt{2}$ **53.** 5, −3 **55.** $-2 \pm i$ **57.** $\dfrac{-3 \pm \sqrt{13}}{2}$ **59.** $\dfrac{-2 \pm \sqrt{6}}{2}$ **61.** $\dfrac{4 \pm \sqrt{13}}{3}$ **63.** $3 \pm 2i\sqrt{6}$ **65.** −3, 5 **67.** $\dfrac{-1 \pm \sqrt{5}}{2}$ **69.** $\dfrac{-2 \pm \sqrt{2}}{2}$ **71.** $\dfrac{5 \pm i\sqrt{11}}{6}$ **73.** $\dfrac{-3 \pm \sqrt{41}}{4}$ **75.** $-\dfrac{\sqrt{2}}{2}, -\sqrt{2}$ **77.** 81, two distinct real numbers **79.** −116, two distinct nonreal complex numbers **81.** 0, one real number **83.** 76.4 inches **85.** 26.8 centimeters **87.** 10 centimeters by 3.5 centimeters **89.** 100 feet by 150 feet **91.** 6:56 A.M. and 8:22 A.M. **93.** 7:11 A.M. **95.** 35 mph for the first part and 45 mph for the last part **97.** 12 hours **99.** 1.66 seconds **101.** yes **103.** yes **105.** yes **107.** $\dfrac{v_0 \pm \sqrt{v_0^2 - 2gs_0}}{-g}$ **109.** $\dfrac{-1 \pm \sqrt{33 + 4x}}{2}$ **111.** 7/2 **113.** 0 **117.** 5 **119.** $\sqrt{29}$ **121.** $\sqrt{65}$ **123.** 3

Exercise Set 1.4, page 105

1. 0, ±5 **3.** 2, ±1 **5.** 0, ±3 **7.** 0, −5, 8 **9.** 0, ±4 **11.** $2, -1 \pm i\sqrt{3}$ **13.** 40 **15.** 3 **17.** 7 **19.** 7 **21.** 9 **23.** 5/2 **25.** 1, −6 **27.** 4 **29.** 1, 5 **31.** 2 **33.** 23, −31 **35.** 2, −1/8 **37.** 0, 1/256 **39.** −1, −59/3 **41.** $\pm\sqrt{7}, \pm\sqrt{2}$ **43.** $\pm 2, \pm\sqrt{6}/2$ **45.** $\sqrt[3]{2}, -\sqrt[3]{3}$ **47.** $-\sqrt[3]{36}/3, \sqrt[3]{98}/7$ **49.** 1, 16 **51.** −1/27, 64 **53.** $\pm\sqrt{15}/3$ **55.** ±1 **57.** 1/2, −1/5 **59.** 256/81, 16 **61.** ±0.62, ±1.62 **63.** ±0.34, ±2.98 **65.** $x = \pm\sqrt{9 - y^2}$ **67.** $x = y + 2\sqrt{yz} + z$ **69.** $x = (7 - 2y^2)/(2y)$ **71.** 9, 36 **73.** 3 inches **75.** 10.5 mm **77.** 87 feet **79. a.** 8.93 inches **b.** $5\sqrt{3}$ inches **81.** $s = \left(\dfrac{-275 + 5\sqrt{3025 + 176T}}{2}\right)^2$

Exercise Set 1.5, page 117

1. $\{x \mid x < 4\}$ **3.** $\{x \mid x < -6\}$ **5.** $\{x \mid x \leq -3\}$ **7.** $\{x \mid x \geq -13/8\}$ **9.** $\{x \mid x < 2\}$ **11.** $\{x \mid -3/4 < x \leq 4\}$ **13.** $\{x \mid 1/3 \leq x \leq 11/3\}$ **15.** $\{x \mid -3/8 \leq x < 11/4\}$ **17.** $\{x \mid x < 1\}$ **19.** $\{x \mid x > -1\}$ **21.** $(-\infty, -3/2) \cup (5/2, \infty)$ **23.** $(-\infty, -8] \cup [2, \infty)$ **25.** $[-4/3, 8]$ **27.** $(-\infty, -4] \cup [28/5, \infty)$ **29.** $(-\infty, \infty)$ **31.** $\{4\}$ **33.** no solution **35.** $(-\infty, \infty)$ **37.** $(-\infty, -7) \cup (0, \infty)$ **39.** $[-4, 4]$ **41.** $(-5, -2)$ **43.** $(-\infty, -4] \cup [7, \infty)$ **45.** $\left[-\frac{1}{2}, 1\right]$ **47.** $(-4, 1)$ **49.** $[-29/2, -8)$ **51.** $[-4, -7/2)$ **53.** $(-\infty, -1) \cup (2, 4)$ **55.** $(-\infty, 5) \cup [12, \infty)$ **57.** $(-2/3, 0) \cup (5/2, \infty)$ **59.** $(-\infty, 5) \cup \{3\}$ **61.** $[4, \infty)$ **63.** $[-9, \infty)$ **65.** $[-3, 3]$ **67.** $(-\infty, -4] \cup [4, \infty)$ **69.** $(-\infty, -3] \cup [5, \infty)$ **71.** $(-\infty, \infty)$ **73.** if you write more than 57 checks a month **75.** at least 34 sales **77.** $20° \leq C \leq 40°$ **79.** $\{12, 14, 16\}, \{14, 16, 18\}$ **81.** $(0, 210)$ **83.** maximum radius 4.480 inches, minimum radius 4.432 inches **85.** $(-\infty, 3) \cup (3, 6) \cup (6, \infty)$ **87.** $(-3, \infty)$ **89.** $\left[-\sqrt{3}, 0\right] \cup \left[\sqrt{3}, \infty\right)$ **91.** $(-\infty, \infty)$ **93.** $\left(-\infty, -2\sqrt{6}\right] \cup \left[2\sqrt{6}, \infty\right)$ **95.** $\left(-\infty, -2\sqrt{14}\right] \cup \left[2\sqrt{14}, \infty\right)$ **97.** $(-5, -1) \cup (1, 5)$ **99.** $(-7, -3] \cup [3, 7)$ **101.** $(a - \delta, a) \cup (a, a + \delta)$ **103.** $(2, 4) \cup (8, 10)$ **105.** $(1/2, \infty)$ **107.** $(-4, -2) \cup (2, 4)$ **109.** 1 second $< t <$ 3 seconds

Exercise Set 1.6, page 125

1. $d = kt$ **3.** $y = k/x$ **5.** $m = knp$ **7.** $V = klwh$ **9.** $A = ks^2$ **11.** $F = km_1m_2/d^2$ **13.** $y = kx, k = 4/3$ **15.** $r = kt^2, k = 1/81$ **17.** $T = krs^2, k = 7/25$ **19.** $V = klwh, k = 1$ **21.** 1.02 liters **23.** 437.5 pounds per square foot **25. a.** approximately 3.3 seconds **b.** approximately 3.7 feet **27.** 112 decibels **29. a.** 9 times larger **b.** 3 times larger **c.** 27 times larger **31.** 6 times larger **33.** approximately 3.2 miles per second **35.** approximately 3950 pounds **37.** $d \approx 142$ million miles

Chapter 1 True/False Exercises, page 130

1. False; $(-3)^2 = 9$. **2.** False; one has solution set $\{3\}$, and the other has solution set $\{3, -4\}$. **3.** True **4.** True **5.** False; $100 > 1$ but $1/100 \ngtr 1/1$. **6.** False; the discriminant is $b^2 - 4ac$. **7.** False; $\sqrt{1} + \sqrt{1} = 1 + 1 = 2$ but $1 + 1 = 2 \neq 2^2$. **8.** True **9.** False; $3x^2 - 48 = 0$ has roots of 4 and -4. **10.** True

Chapter 1 Review Exercises, page 130

1. $3/2$ **3.** $1/2$ **5.** $-38/15$ **7.** $3, 2$ **9.** $(1 \pm \sqrt{13})/6$ **11.** $0, 5/3$ **13.** $\pm 2\sqrt{3}/3, \pm\sqrt{10}/2$ **15.** $-5, 3$ **17.** 4 **19.** -4 **21.** $-2, -4$ **23.** $5, 1$ **25.** $2, -3$ **27.** $-2, -1$ **29.** $14, -31\frac{1}{2}$ **31.** $(-\infty, 2]$ **33.** $[-5, 2]$ **35.** $\left[\frac{145}{9}, 35\right]$ **37.** $(-\infty, 0] \cup [3, 4]$ **39.** $(-\infty, -3) \cup (4, \infty)$ **41.** $(-\infty, 5/2] \cup (3, \infty)$ **43.** $(2/3, 2)$ **45.** $(-2, 0) \cup (0, 2)$ **47.** $(1, 2) \cup (2, 3)$ **49.** $h = \frac{V}{\pi r^2}$ **51.** $b_1 = \frac{2A - hb_2}{h}$ **53.** $m = \frac{e}{c^2}$ **55.** $3 - 8i$, conjugate $3 + 8i$ **57.** $5 + 2i$ **59.** $25 - 19i$ **61.** 1 **63.** 1 **65.** 80 **67.** 24 nautical miles **69.** \$1750 in the 4% account, \$3750 in the 6% account **71.** \$864 **73.** 18 hours **75.** 13 feet **77.** 1.64 meters/second^2

Chapter 1 Test, page 132

1. 9.6 or 48/5 **2.** $-14/3$ **3.** $x = \frac{c - cd}{a - c}, a \neq c$ **4.** $-2 \pm \sqrt{5}$ **5.** $\frac{-1 \pm 2\sqrt{7}}{3}$ **6.** $-4, 1, -\frac{1}{2} - \frac{\sqrt{3}}{2}i, -\frac{1}{2} + \frac{\sqrt{3}}{2}i$ **7.** 3 **8.** $8/27, -64$ **9.** $67, -61$ **10.** $x \leq 5/2$ **11.** $[-4, -1) \cup [3, \infty)$ **12.** $-1, -6$ **13.** $(-7, -1)$ **14.** $(-\infty, -5/3] \cup [3, \infty)$ **15.** $\frac{4}{13} + \frac{7}{13}i$ **16.** 2 mph **17.** 2.25 liters **18.** 15 hours **19.** more than 100 miles **20.** 4.4 miles/second

Exercise Set 2.1, page 144

1.

(2, 4) (−2, 1) (−5, −3) (0, −3)

3. a.

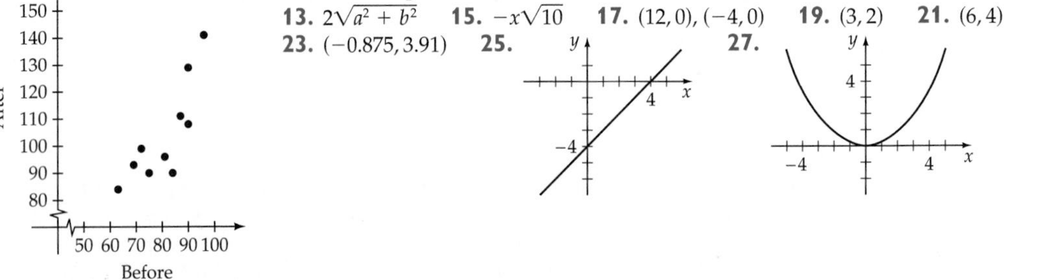

b. 23.4 beats per minute **5.** $7\sqrt{5}$ **7.** $\sqrt{1261}$ **9.** $\sqrt{89}$ **11.** $\sqrt{38 - 12\sqrt{6}}$ **13.** $2\sqrt{a^2 + b^2}$ **15.** $-x\sqrt{10}$ **17.** $(12, 0), (-4, 0)$ **19.** $(3, 2)$ **21.** $(6, 4)$ **23.** $(-0.875, 3.91)$ **25.** **27.**

29.

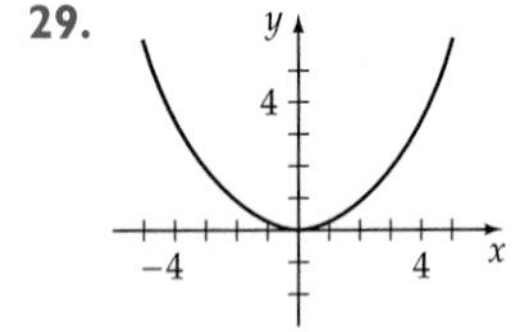

31. **33.**

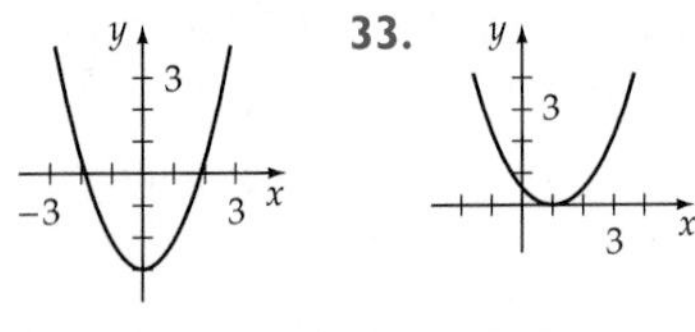

35. **37.**

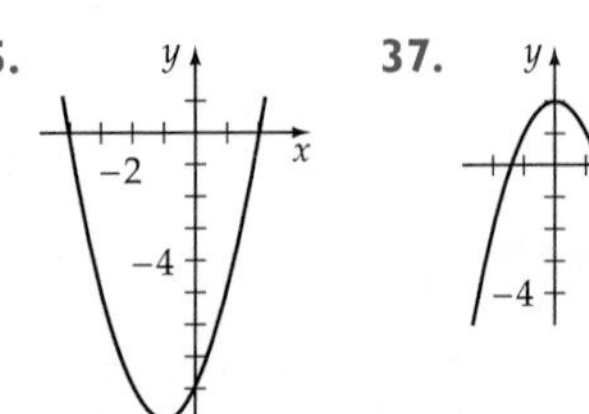

39. $(0, 12/5), (6, 0)$

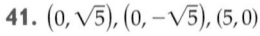

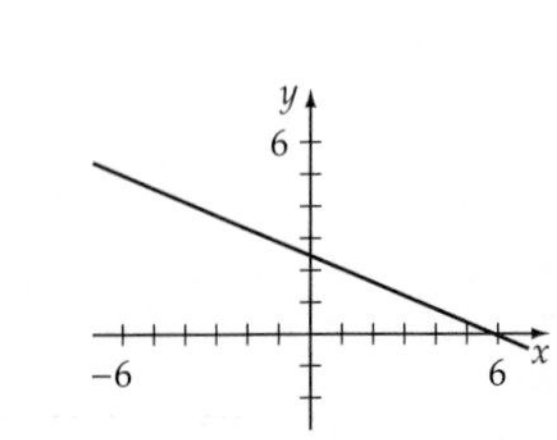

41. $(0, \sqrt{5}), (0, -\sqrt{5}), (5, 0)$

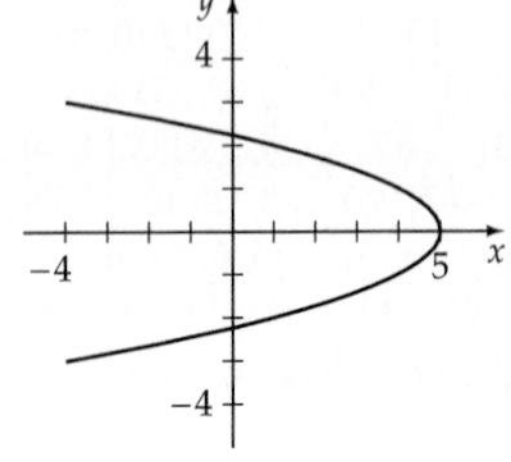

43. (0, 4), (0, −4), (−4, 0) **45.** (0, ±2), (±2, 0) **47.** (0, ±4), (±4, 0)

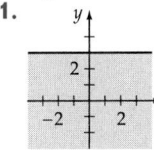

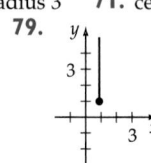

49. center (0, 0), radius 6
51. center (0, 0), radius 10
53. center (1, 3), radius 7
55. center (−2, −5), radius 5
57. center (8, 0), radius 1/2
59. $(x - 4)^2 + (y - 1)^2 = 2^2$
61. $(x - 1/2)^2 + (y - 1/4)^2 = (\sqrt{5})^2$
63. $(x - 0)^2 + (y - 0)^2 = 5^2$
65. $(x - 1)^2 + (y - 3)^2 = 5^2$
67. center (3, 0), radius 2

69. center (2, 5), radius 3 **71.** center (7, −4), radius 3 **73.** center (−1/2, 0), radius 4 **75.** center (1/2, −1/3), radius 1/6

77. 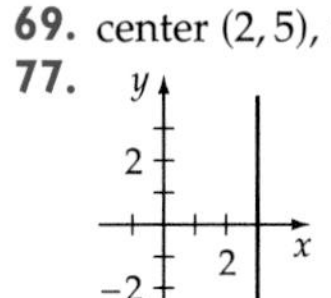 **79.** **81.** **83.** **85.** **87.**

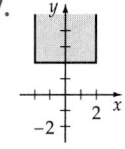

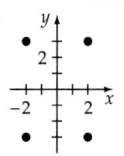

89. (13, 5) **91.** (7, −6) **93.** (5/2, 71/4) **95.** (5/2, 17/8) **97.** yes **99.** $x^2 - 6x + y^2 - 8y = 0$ **101.** $9x^2 + 25y^2 = 225$
103. $(x + 1)^2 + (y - 7)^2 = 5^2$ **105.** $(x - 7)^2 + (y - 11)^2 = 11^2$ **107.** $(x + 3)^2 + (y - 3)^2 = 3^2$

Exercise Set 2.2, page 158

1. a. 5 **b.** −4 **c.** −1 **d.** 1 **e.** $3k - 1$ **f.** $3k + 5$ **3. a.** $\sqrt{5}$ **b.** 3 **c.** 3 **d.** $\sqrt{21}$ **e.** $\sqrt{r^2 + 2r + 6}$ **f.** $\sqrt{c^2 + 5}$
5. a. 1/2 **b.** 1/2 **c.** 5/3 **d.** 1 **e.** $\dfrac{1}{c^2 + 4}$ **f.** $\dfrac{1}{|2 + h|}$ **7. a.** 1 **b.** 1 **c.** −1 **d.** −1 **e.** 1 **f.** −1 **9. a.** −11 **b.** 6
c. $3c + 1$ **d.** $-k^2 - 2k + 10$ **11.** yes **13.** no **15.** no **17.** yes **19.** no **21.** yes **23.** yes **25.** yes **27.** all real numbers
29. all real numbers **31.** $\{x \mid x \neq -2\}$ **33.** $\{x \mid x \geq -7\}$ **35.** $\{x \mid -2 \leq x \leq 2\}$ **37.** $\{x \mid x > -4\}$

39. 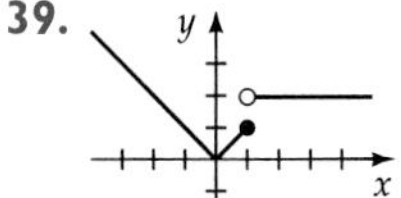

domain: all real numbers

41. 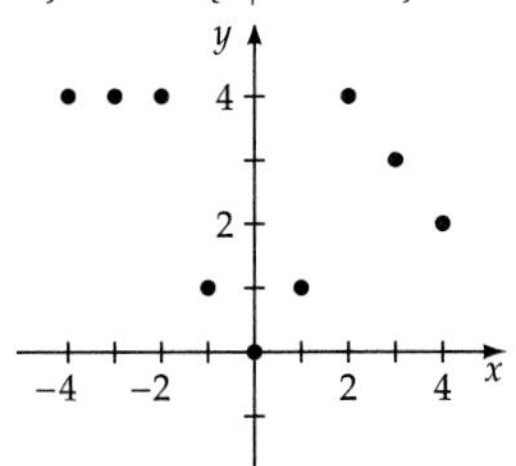

domain: $\{-4, -3, -2, -1, 0, 1, 2, 3, 4\}$

43.

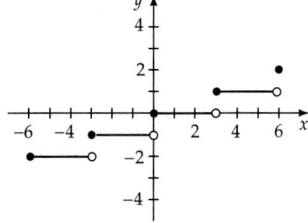

domain: $\{x \mid -6 \leq x \leq 6\}$

45. 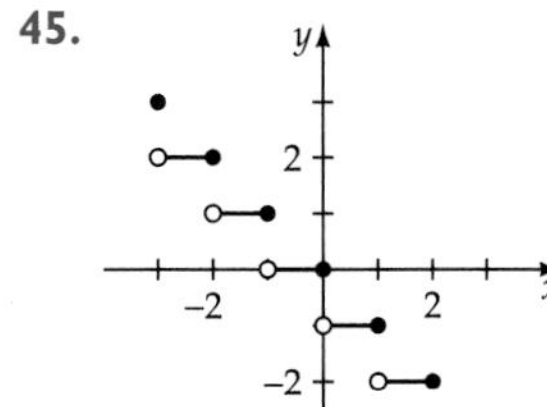

domain: $\{x \mid -3 \leq x \leq 3\}$

47. a. $C(3.97) = 1.16$ **b.** $c(w)$

49. **a**, **b**, and **d**.
51. decreasing on $(-\infty, 0]$; increasing on $[0, \infty)$
53. increasing on $(-\infty, \infty)$
55. decreasing on $(-\infty, -3]$; increasing on $[-3, 0]$; decreasing on $[0, 3]$; increasing on $[3, \infty)$
57. constant on $(-\infty, 0]$; increasing on $[0, \infty)$
59. decreasing on $(-\infty, 0]$; constant on $[0, 1]$; increasing on $[1, \infty)$
61. g and F
63. a. $w = 25 - l$ **b.** $A = 25l - l^2$
65. $v(t) = 80{,}000 - 6500t, 0 \leq t \leq 10$

67. a. $C(x) = 2000 + 22.80x$ **b.** $R(x) = 37.00x$ **c.** $P(x) = 14.20x - 2000$ **69.** $h = 15 - 5r$ **71.** $d = \sqrt{(3t)^2 + 50^2}$
73. $d = \sqrt{(45 - 8t)^2 + (6t)^2}$ **75.** 275, 375, 385, 390, 394 **77.** $c = -2$ or $c = 3$ **79.** 1 is not in the range of f.

81. 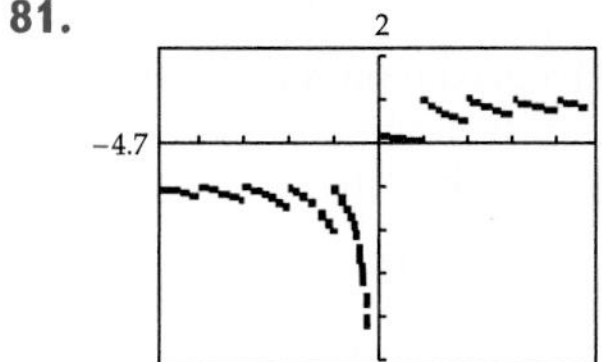

83.

85.

87. 4
89. 2
91. a. 36
b. 13
c. 12
d. 30
e. $13k - 2$
f. $8k - 11$

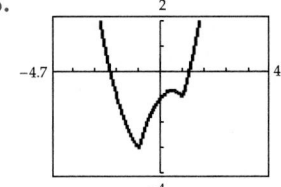

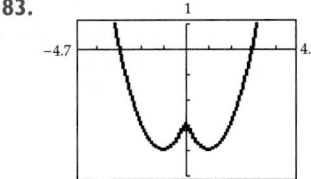

93. $4\sqrt{21}$ **95.** 1, −3 **97.**

Exercise Set 2.3, page 175

1. −3/2 **3.** −1/2 **5.** The line does not have slope. **7.** 6 **9.** 9/19 **11.** $\dfrac{f(3+h)-f(3)}{h}$ **13.** $\dfrac{f(h)-f(0)}{h}$

15. 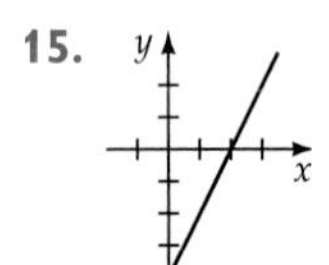 **17.** 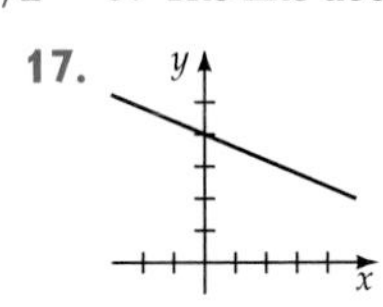 **19.** 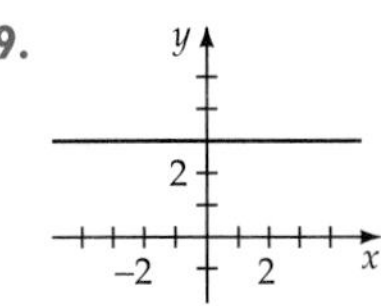 **21.** 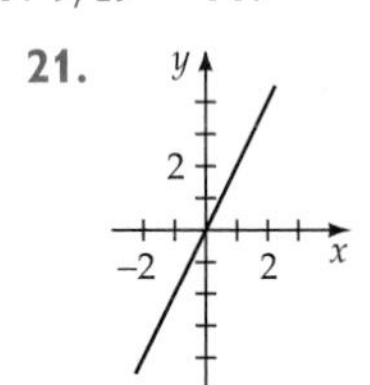 **23.** 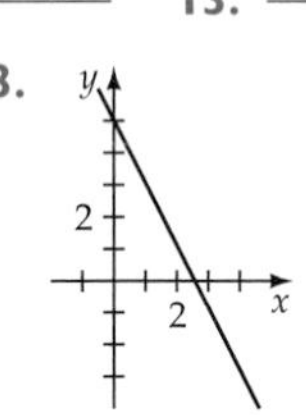 **25.** 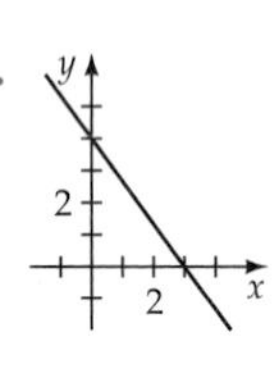

27. $y = x + 3$ **29.** $y = \dfrac{3}{4}x + \dfrac{1}{2}$ **31.** $y = (0)x + 4 = 4$ **33.** $y = -4x - 10$ **35.** $y = -\dfrac{3}{4}x + \dfrac{13}{4}$ **37.** $y = \dfrac{12}{5}x - \dfrac{29}{5}$ **39.** −2 **41.** −1/2 **43.** −4 **45.** 4 **47.** −20 **49.** 1/3 **51.** 16/3 **53. a.** approx. 26 mpg **b.** approx. 31 mpg **55.** $P(x) = 40.50x - 1782$, $x = 44$, the break-even point **57.** $P(x) = 79x - 10{,}270$, $x = 130$, the break-even point **59. a.** \$275 **b.** \$283 **c.** \$355 **d.** \$8 **61. a.** $C(t) = 19{,}500.00 + 6.75t$ **b.** $R(t) = 55.00t$ **c.** $P(t) = 48.25t - 19{,}500.00$ **d.** approximately 405 days **63. a.** $y = 0.087x + 1.773$ **b.** 2.8 million **65. a.** $y = 1.842x - 18.947$ **b.** 147 **67.** $y = -\dfrac{3}{4}x + \dfrac{15}{4}$ **69.** $y = x + 1$ **71.** −5 ft **73. a.** $Q = (3, 10)$, $m = 5$ **b.** $Q = (2.1, 5.41)$, $m = 4.1$ **c.** $Q = (2.01, 5.0401)$, $m = 4.01$ **d.** 4 **79.** $y = -2x + 11$ **81.** $5x + 3y = 15$ **83.** $3x + y = 17$

Exercise Set 2.4, page 188

1. d **3.** b **5.** g **7.** c

9. $f(x) = (x + 2)^2 - 3$
vertex: $(-2, -3)$
axis of symmetry: $x = -2$

11. $f(x) = (x - 4)^2 - 11$
vertex: $(4, -11)$
axis of symmetry: $x = 4$

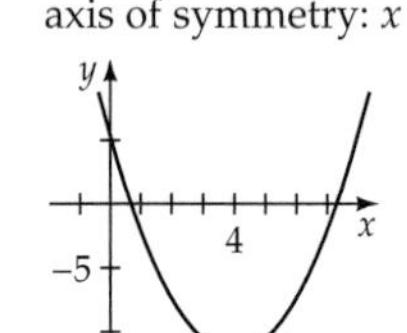

13. $f(x) = (x - (-3/2))^2 - 5/4$
vertex: $(-3/2, -5/4)$
axis of symmetry: $x = -3/2$

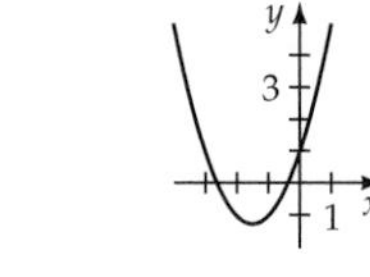

15. $f(x) = -(x - 2)^2 + 6$
vertex: $(2, 6)$
axis of symmetry: $x = 2$

17. $f(x) = -3(x - 1/2)^2 + 31/4$
vertex: $(1/2, 31/4)$
axis of symmetry: $x = 1/2$

19. vertex: $(5, -25)$, $f(x) = (x - 5)^2 - 25$
21. vertex: $(0, -10)$, $f(x) = x^2 - 10$
23. vertex: $(3, 10)$, $f(x) = -(x - 3)^2 + 10$
25. vertex: $(3/4, 47/8)$, $f(x) = 2(x - 3/4)^2 + 47/8$
27. vertex: $(1/8, 17/16)$, $f(x) = -4(x - 1/8)^2 + 17/16$
29. $\{y \mid y \geq -2\}$, −1 and 3
31. $\left\{y \mid y \leq \frac{17}{8}\right\}$, 1 and $\frac{3}{2}$
33. No, $3 \notin \left\{y \geq \frac{15}{4}\right\}$
35. −16, minimum
37. 11, maximum

39. −1/8, minimum **41.** −11, minimum **43.** 35, maximum **45. a.** 27 feet **b.** $22\frac{5}{16}$ feet **c.** 20.1 feet from the center **47. a.** $w = \dfrac{600 - 2l}{3}$ **b.** $A = 200l - \dfrac{2}{3}l^2$ **c.** $w = 100$ feet, $l = 150$ feet **49. a.** 12:43 PM **b.** 91°F **51. a.** 41 mph **b.** 34 mpg **53.** y-intercept $(0, 0)$; x-intercepts $(0, 0)$ and $(-6, 0)$ **55.** y-intercept $(0, -6)$; no x-intercepts **57.** 740 units yield a maximum revenue of \$109,520. **59.** 85 units yield a maximum profit of \$24.25. **61.** $P(x) = -0.1x^2 + 50x - 1840$, break-even points: $x = 40$ and $x = 460$ **63. a.** $R(x) = -0.25x^2 + 30.00x$ **b.** $P(x) = -0.25x^2 + 27.50x - 180$ **c.** \$576.25 **d.** 55 **65. a.** $t = 4$ seconds **b.** 256 feet **c.** $t = 8$ seconds **67.** 30 feet **69.** $r = \dfrac{48}{4 + \pi} \approx 6.72$ feet, $h = r \approx 6.72$ feet **73.** $f(x) = \dfrac{3}{4}x^2 - 3x + 4$ **75. a.** $w = 16 - x$ **b.** $A = 16x - x^2$ **77.** The discriminant is $b^2 - 4(1)(-1) = b^2 + 4$, which is positive for all b. **79.** increases the height of each point on the graph by c units. **81.** 4, 4

Exercise Set 2.5, page 201

1. $C(-5, 3)$, $B(5, 3)$, $A(-5, -3)$, $P(5, -3)$ **3.** $R(-2, 3)$, $A(2, 3)$, $B(-2, -3)$, $C(2, -3)$ **5.** $B(-4, 5)$, $C(4, 5)$, $T(-4, -5)$, $A(4, -5)$ **7.** **9.** **11.**

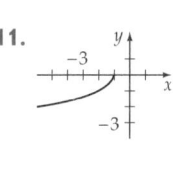

13. a. no **b.** yes **15. a.** no **b.** no **17. a.** yes **b.** yes **19. a.** yes **b.** yes **21. a.** yes **b.** yes **23.** no **25.** yes **27.** yes **29.** yes **31.**

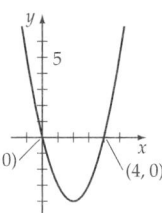

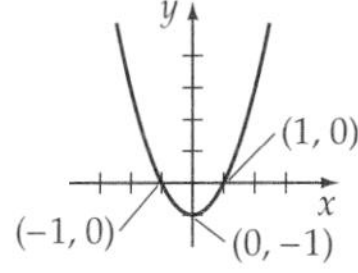

33. **35.**

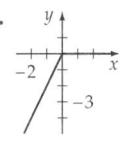

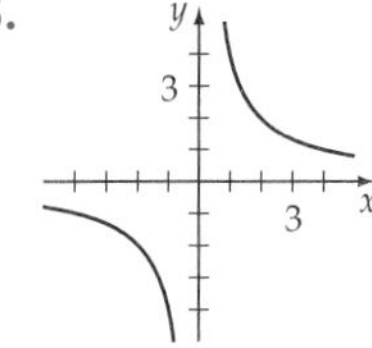

37.

39. **41.**

43. even
45. odd
47. even
49. even
51. even
53. even
55. neither

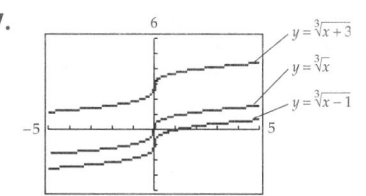

57. a., b.

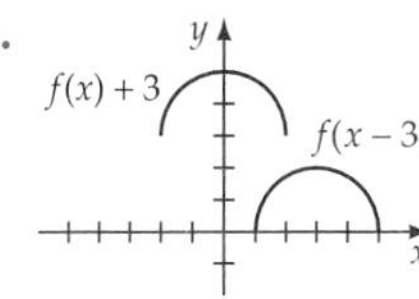

59. a., b.

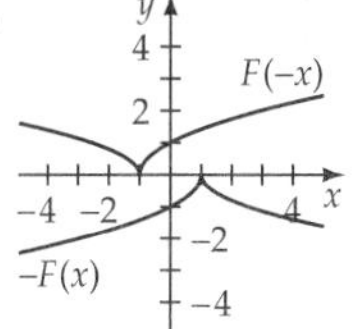

61.

63. a.

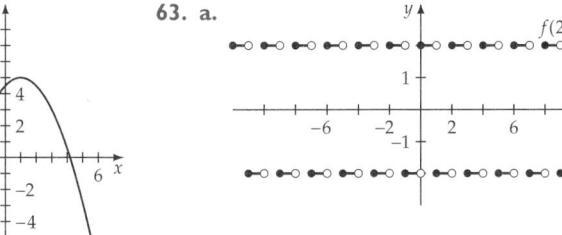

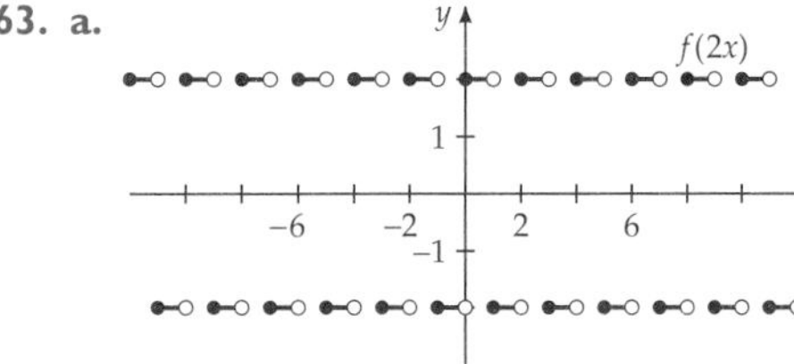

b.

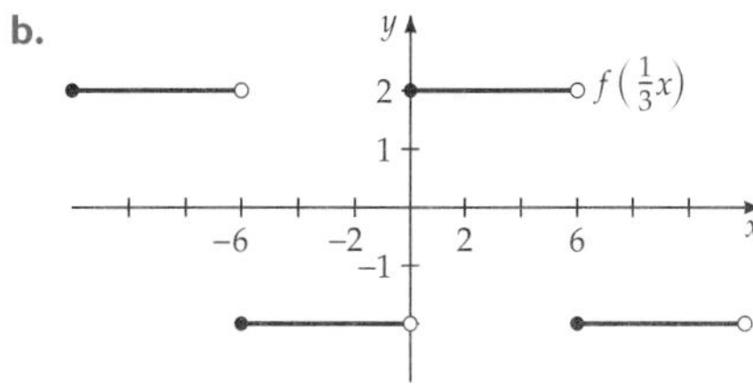

65. a.

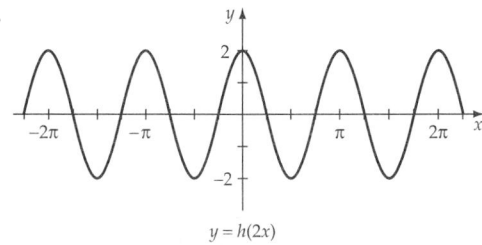

$y = h(2x)$

b.

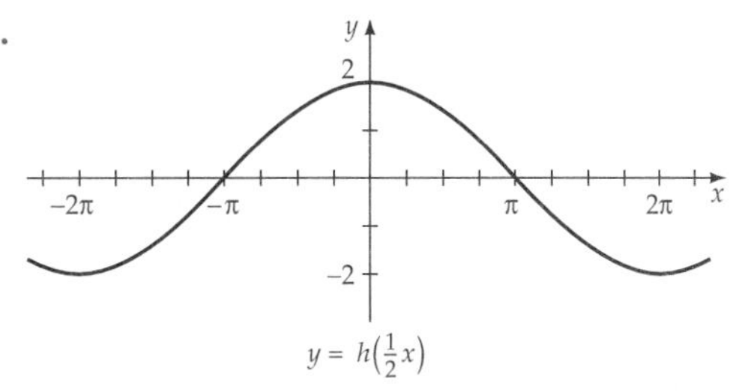

$y = h\left(\frac{1}{2}x\right)$

67. $y = \sqrt[3]{x+3}$, $y = \sqrt[3]{x}$, $y = \sqrt[3]{x-1}$

69.

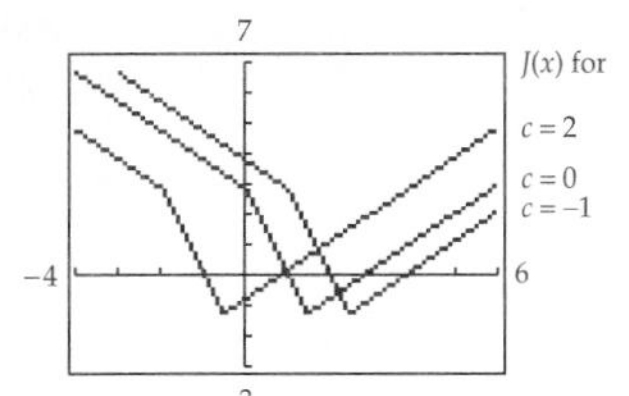

71.

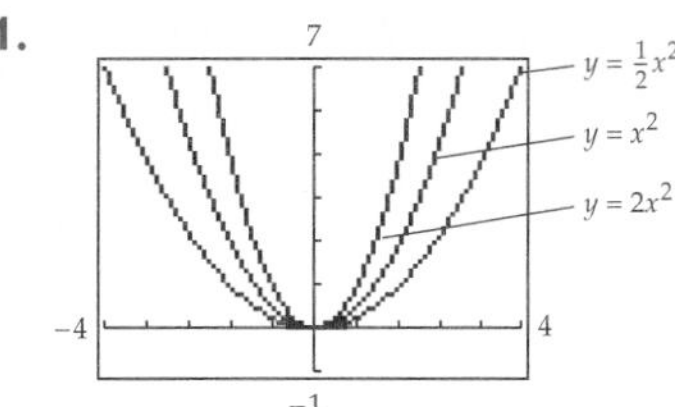

73.

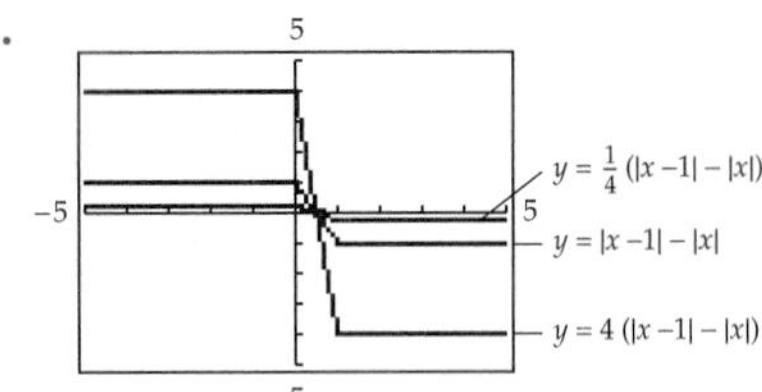

75. a.

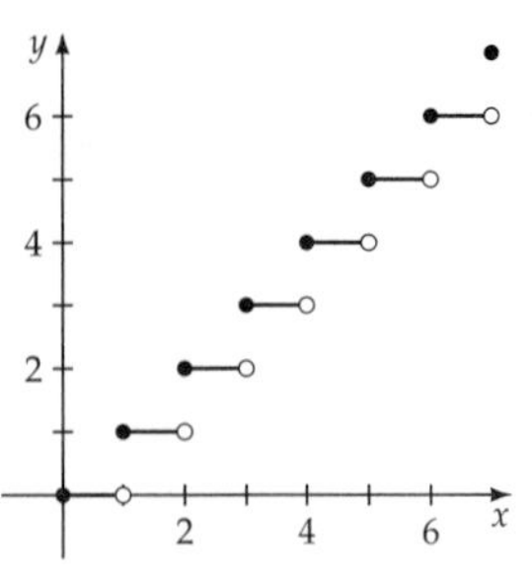

b.

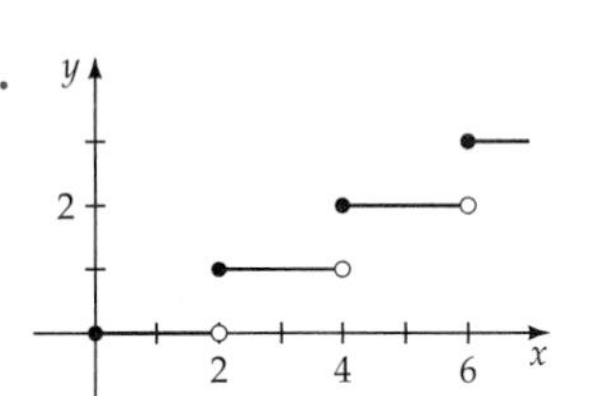

c.

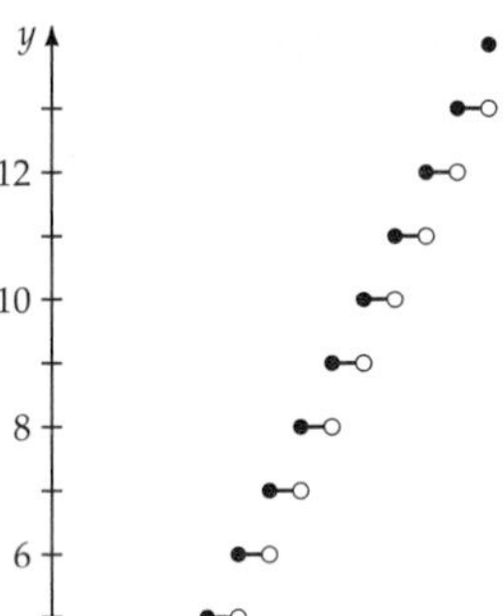

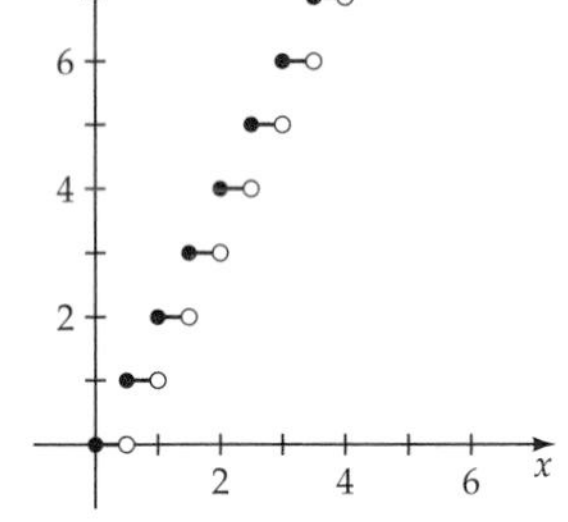

77. a. $f(x) = \dfrac{2}{(x+1)^2 + 1} + 1$

b. $f(x) = -\dfrac{2}{(x-2)^2 + 1}$

Exercise Set 2.6, page 213

1. $f(x) + g(x) = x^2 - x - 12$, domain all real numbers
$f(x) - g(x) = x^2 - 3x - 18$, domain all real numbers
$f(x) \cdot g(x) = x^3 + x^2 - 21x - 45$, domain all real numbers
$f(x)/g(x) = x - 5$, domain $\{x \mid x \neq -3\}$

3. $f(x) + g(x) = 3x + 12$, domain all real numbers
$f(x) - g(x) = x + 4$, domain all real numbers
$f(x) \cdot g(x) = 2x^2 + 16x + 32$, domain all real numbers
$f(x)/g(x) = 2$, domain $\{x \mid x \neq -4\}$

5. $f(x) + g(x) = x^3 - 2x^2 + 8x$, domain all real numbers
$f(x) - g(x) = x^3 - 2x^2 + 6x$, domain all real numbers
$f(x) \cdot g(x) = x^4 - 2x^3 + 7x^2$, domain all real numbers
$f(x)/g(x) = x^2 - 2x + 7$, domain $\{x \mid x \neq 0\}$

7. $f(x) + g(x) = 4x^2 + 7x - 12$, domain all real numbers
$f(x) - g(x) = x - 2$, domain all real numbers
$f(x) \cdot g(x) = 4x^4 + 14x^3 - 12x^2 - 41x + 35$, domain all real numbers
$f(x)/g(x) = 1 + \dfrac{x-2}{2x^2 + 3x - 5}$, domain $\{x \mid x \neq 1, x \neq -5/2\}$

9. $f(x) + g(x) = \sqrt{x-3} + x$, domain $\{x \mid x \geq 3\}$
$f(x) - g(x) = \sqrt{x-3} - x$, domain $\{x \mid x \geq 3\}$
$f(x) \cdot g(x) = x\sqrt{x-3}$, domain $\{x \mid x \geq 3\}$
$f(x)/g(x) = \dfrac{\sqrt{x-3}}{x}$, domain $\{x \mid x \geq 3\}$

11. $f(x) + g(x) = \sqrt{4-x^2} + 2 + x$, domain $\{x \mid -2 \leq x \leq 2\}$
$f(x) - g(x) = \sqrt{4-x^2} - 2 - x$, domain $\{x \mid -2 \leq x \leq 2\}$
$f(x) \cdot g(x) = (\sqrt{4-x^2})(2 + x)$, domain $\{x \mid -2 \leq x \leq 2\}$
$f(x)/g(x) = \dfrac{\sqrt{4-x^2}}{2+x}$, domain $\{x \mid -2 < x \leq 2\}$

13. 18 **15.** −9/4 **17.** 30 **19.** 12 **21.** 300 **23.** −384/125 **25.** −5/2 **27.** −1/4 **29.** 2 **31.** $2x + h$ **33.** $4x + 2h + 4$
35. $-8x - 4h$ **37.** $(g \circ f)(x) = 6x + 3$, $(f \circ g)(x) = 6x - 16$ **39.** $(g \circ f)(x) = x^2 + 4x + 1$, $(f \circ g)(x) = x^2 + 8x + 11$ **41.** $(g \circ f)(x) = -5x^3 - 10x$, $(f \circ g)(x) = -125x^3 - 10x$

43. $(g \circ f)(x) = \dfrac{1 - 5x}{x + 1}$, $(f \circ g)(x) = \dfrac{2}{3x - 4}$ **45.** $(g \circ f)(x) = \dfrac{\sqrt{1 - x^2}}{|x|}$, $(f \circ g)(x) = \dfrac{1}{x - 1}$ **47.** $(g \circ f)(x) = -\dfrac{2|5 - x|}{3}$, $(f \circ g)(x) = \dfrac{3|x|}{|5x + 2|}$ **49.** 66 **51.** 51 **53.** −4 **55.** 41 **57.** −3848/625

59. $6 + 2\sqrt{3}$ **61.** $16c^2 + 4c - 6$ **63.** $9k^4 + 36k^3 + 45k^2 + 18k - 4$ **65. a.** $A(t) = \pi(1.5t)^2$, $A(2) = 9\pi$ square feet ≈ 28.27 square feet
b. $V(t) = 2.25\pi t^3$, $V(3) = 60.75\pi$ cubic feet ≈ 190.85 cubic feet **67. a.** $d(t) = \sqrt{(48 - t)^2 - 4^2}$ **b.** $s(35) = 13$ feet, $d(35) \approx 12.37$ feet
69. $(Y \circ F)(x)$ converts x inches to yards. **71. a.** 99.8; this is identical to the slope of the line through $(0, C(0))$ and $(1, C(1))$.
b. 156.2 **c.** −49.7 **d.** −30.8 **e.** −16.4 **f.** 0 **79. a.** $(s \circ m)(x) = 87 + 49{,}300/x$ **b.** \$89

Section 2.7, page 223

1. No relationship **3.** Linear **5.** Figure A **7.** $y = 2.00862069x + 0.5603448276$ **9.** $y = -0.7231182796x + 9.233870968$
11. $y = 2.222641509x - 7.364150943$ **13.** $y = 1.095779221x^2 - 2.69642857x + 1.136363636$
15. $y = -0.2987274717x^2 - 3.20998141x + 3.416463667$ **17. a.** $y = 23.55706665x - 24.4271215$ **b.** 1247.7 cm
19. a. $y = 1.671510024x + 16.32830605$ **b.** 46.4 cm **21. a.** $y = 0.1628623408x - 0.6875682232$ **b.** 25 **23.** No, because the linear correlation coefficient is close to 0. **25.** Yes, there is a strong linear correlation. **a.** $-0.9033088235x + 78.62573529$ **b.** 56 years
27. No. The linear correlation coefficient is close to 0. **29.** $y = -0.6328671329x^2 + 33.6160839x - 379.4405594$
31. a. $y = -0.0165034965x^2 + 1.366713287x + 5.685314685$ **b.** 32.8 mpg
33. a. 5 pound—$y = 0.6130952381t^2 - 0.0714285714t + 0.1071428571$
10 pound—$y = 0.6091269841t^2 - 0.0011904762t - 0.3$
15 pound—$y = 0.5922619048t^2 + 0.3571428571t - 1.520833333$
b. All the regression equations are approximately the same. Therefore, the equation of motion of the three masses are the same.
35. quadratic; r^2 is close to 1 for a quadratic model.

Chapter 2 True/False Exercises, page 231

1. False. Let $f(x) = x^2$. Then $f(3) = f(-3) = 9$, but $3 \neq -3$. **2.** False. Consider $f(x) = x + 1$ and $g(x) = x^2 - 2$. **3.** True **4.** True **5.** False. Let $f(x) = 3x$. $[f(x)]^2 = 9x^2$, whereas $f[f(x)] = f(3x) = 3(3x) = 9x$. **6.** False. Let $f(x) = x^2$. Then $f(1) = 1$, $f(2) = 4$. Thus $f(2)/f(1) = 4 \neq 2/1$. **7.** True **8.** False. Let $f(x) = |x|$. Then $f(-1 + 3) = f(2) = 2$. $f(-1) + f(3) = 1 + 3 = 4$. **9.** True **10.** True **11.** True **12.** True **13.** True. $f(x) = x^2 + 1, c = i$ **14.** False. The coefficient of determination is r^2 and therefore non-negative.

Chapter 2 Review Exercises, page 232

1. $\sqrt{181}$ **3.** $(-1/2, 10)$ **5.** center $(3, -4)$, radius 9 **7.** $(x - 2)^2 + (y + 3)^2 = 5^2$ **9. a.** 2 **b.** 10 **c.** $3t^2 + 4t - 5$ **d.** $3x^2 + 6xh + 3h^2 + 4x + 4h - 5$ **e.** $9t^2 + 12t - 15$ **f.** $27t^2 + 12t - 5$ **11. a.** 5 **b.** -11 **c.** $x^2 - 12x + 32$ **d.** $x^2 + 4x - 8$ **13.** $8x + 4h - 3$

15.

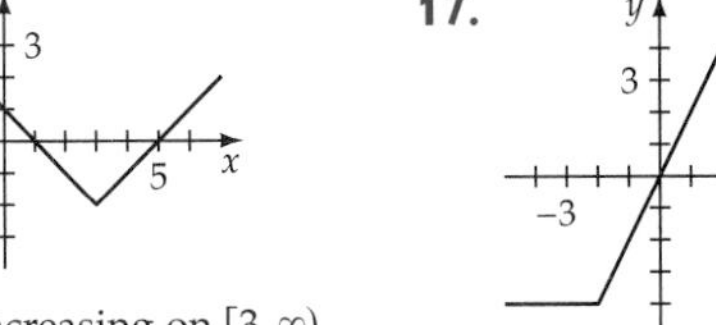

increasing on $[3, \infty)$
decreasing on $(-\infty, 3]$

17.

increasing on $[-2, 2]$
constant on $(-\infty, -2] \cup [2, \infty)$

19. increasing on $(-\infty, \infty)$

21. domain $\{x \mid x \text{ is a real number}\}$ **23.** domain $\{x \mid -5 \leq x \leq 5\}$ **25.** $y = -2x + 1$ **27.** $y = \frac{3}{4}x + \frac{19}{2}$ **29.** $f(x) = (x + 3)^2 + 1$ **31.** $f(x) = -(x + 4)^2 + 19$ **33.** $f(x) = -3(x - 2/3)^2 - 11/3$ **35.** $(1, 8)$ **37.** $(5, 161)$ **39.** $4\sqrt{5}/5$ **41.**

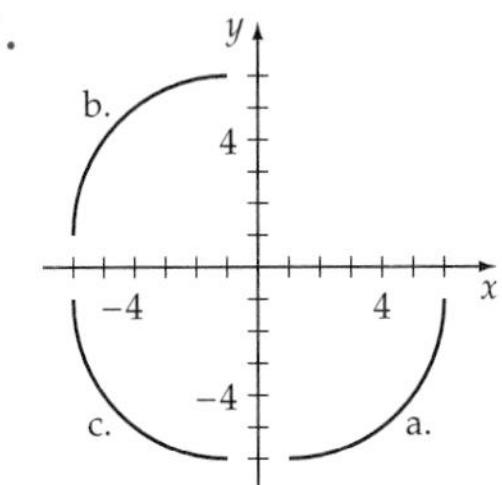

43. symmetric to the y-axis **45.** symmetric to the origin **47.** symmetric to the x-axis, the y-axis, and the origin
49. symmetric to the x-axis, the y-axis, and the origin

51.

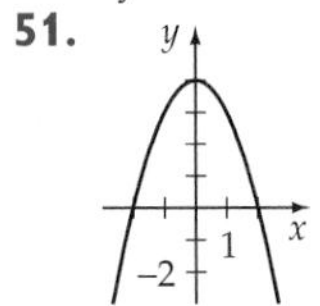

a. domain all real numbers
range $\{y \mid y \leq 4\}$
b. even

53.

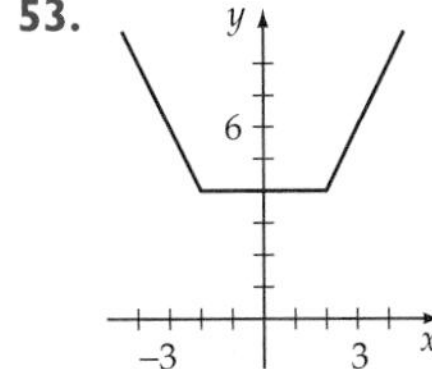

a. domain all real numbers
range $\{y \mid y \geq 4\}$
b. even

55.

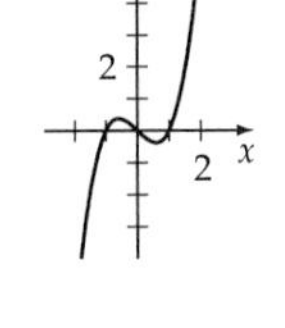

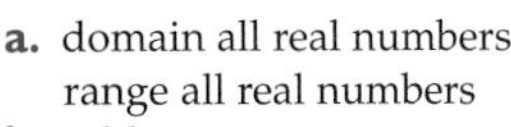

a. domain all real numbers
range all real numbers

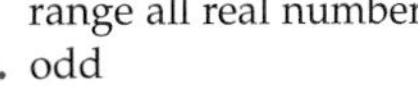

b. odd

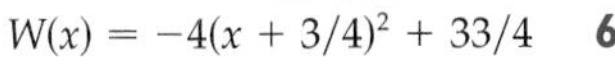

57. $F(x) = (x + 2)^2 - 11$

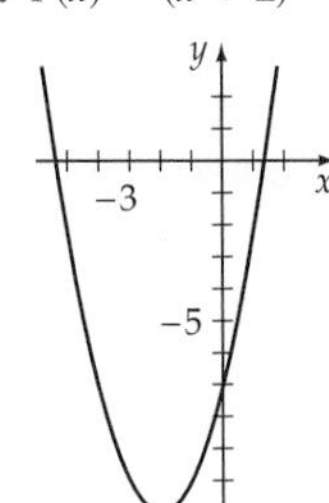

59. $P(x) = 3(x - 0)^2 - 4$

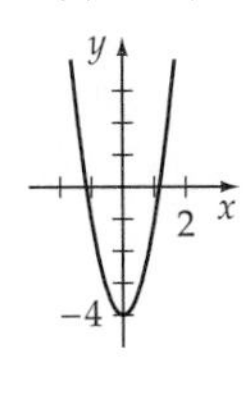

61. $W(x) = -4(x + 3/4)^2 + 33/4$

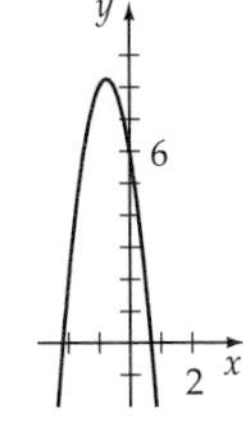

63.

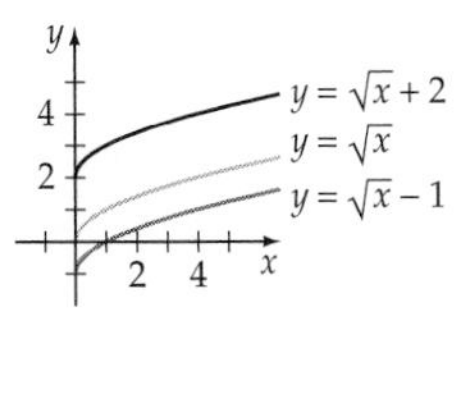

65.

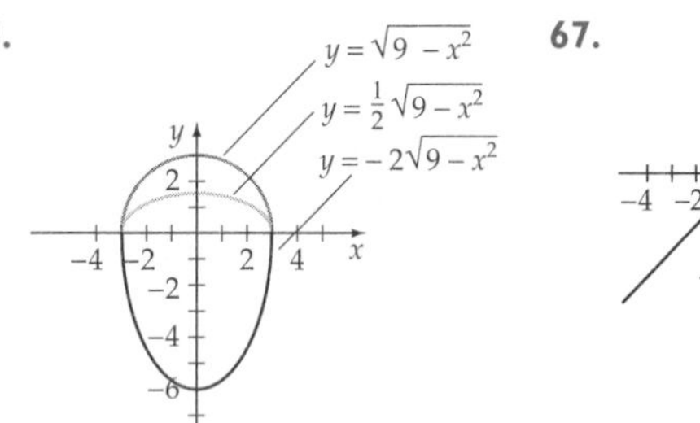

67.

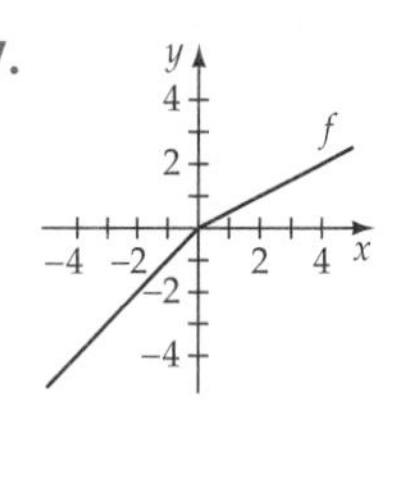

69. $f(x) + g(x) = x^2 + x - 6$, domain all real numbers
$f(x) - g(x) = x^2 - x - 12$, domain all real numbers
$f(x) \cdot g(x) = x^3 + 3x^2 - 9x - 27$, domain all real numbers
$f(x)/g(x) = x - 3$, domain $\{x \mid x \neq -3\}$
71. 25, 25
73. a. 18 feet per second **b.** 15 feet per second **c.** 13.5 feet per second **d.** 12.03 feet per second **e.** 12 feet per second
75. a. $y = 0.0180247x + 0.0005005$ **b.** Yes. $r = 0.999$, which is very close to 1. **c.** 1.8 seconds

Chapter 2 Test, page 234

1. midpoint $(1, 1)$; length $2\sqrt{13}$ **2.** $(0, \sqrt{2}), (0, -\sqrt{2}), (-4, 0)$

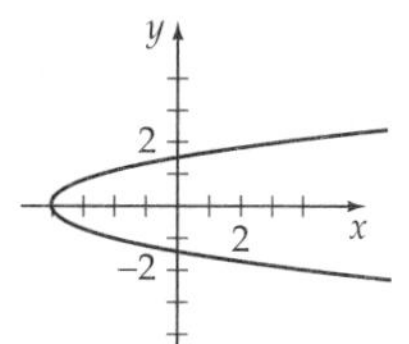

3. **4.** center $(2, -1)$; radius 3

5. domain $\{x \mid x \geq 4 \text{ or } x \leq -4\}$ **6.** $3\sqrt{10}/5$ **7.**

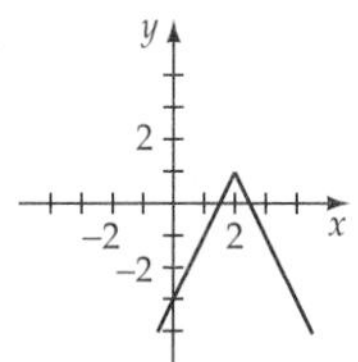

increasing on $(-\infty, 2]$
decreasing on $[2, \infty)$

8.

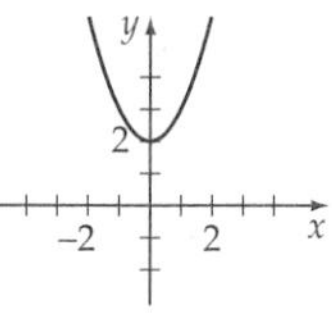

domain all real numbers; range $\{y \mid y \geq 2\}$

9. a. $R = 12.00x$ **b.** $P = 11.25x - 875$ **c.** $x = 78$ **10.**

11. a. even **b.** odd **c.** neither
12. $y = -\dfrac{2}{3}x + \dfrac{2}{3}$
13. -12, minimum
14. $x^2 + x - 3; \dfrac{x^2 - 1}{x - 2}, x \neq 2$
15. $2x + h$
16. $x - 2\sqrt{x - 2} - 1$ **17. a.** 25 feet per second **b.** 22.5 feet per second **c.** 20.05 feet per second
18. a. $y = -7.98245614x + 767.122807$ **b.** 57 calories

Exercise Set 3.1, page 245

1. $5x^2 - 9x + 10 + \dfrac{-10}{x + 3}$ **3.** $x^3 + 5x^2 - 9x - 45$ **5.** $x^2 - 33 + \dfrac{33x + 3}{3x^2 + x + 1}$ **7.** $4x^2 + 1 + \dfrac{11}{5x^2 - 2}$
9. $x + 4 + \dfrac{6x - 3}{x^2 + x - 4}$ **11.** $4x^2 + 3x + 12 + \dfrac{17}{x - 2}$ **13.** $4x^2 - 4x + 2 + \dfrac{1}{x + 1}$ **15.** $x^4 + 4x^3 + 6x^2 + 24x + 101 + \dfrac{403}{x - 4}$
17. $x^4 + x^3 + x^2 + x + 1$ **19.** $8x^2 + 6$ **21.** $x^7 + 2x^6 + 5x^5 + 10x^4 + 21x^3 + 42x^2 + 85x + 170 + \dfrac{344}{x - 2}$
23. $x^5 - 3x^4 + 9x^3 - 27x^2 + 81x - 242 + \dfrac{716}{x + 3}$ **25.** $3x - 3.1 + \dfrac{4.07}{x - 0.3}$ **27.** $2x^2 - 11x - 17 + \dfrac{3}{x}$ **29.** 25
31. 45 **33.** -2230 **35.** -80 **37.** -187 **39.** yes **41.** yes **43.** yes **45.** yes **47.** yes **49.** yes **61.** $x^2 + 3x + 7$, $(x - 2)(x^2 + 3x + 7)$ **63.** $x^3 + 3x^2 + 3x + 1$, $(x - 4)(x^3 + 3x^2 + 3x + 1)$ **65.** $(x + 3)(x - 1)(x - 2)$ **67.** $(x + 3)(x + 2)(x + 1)(x - 4)$ **71.** 13 **73.** By the Factor Theroem, $P(x)$ has a factor of $x - c$ if and only if $P(c) = 0$. However, $P(c) = 4c^4 + 7c^2 + 12$, which is greater than 0 for any real number c.

Exercise Set 3.2, page 254

1. up to far left, up to far right **3.** down to far left, up to far right **5.** down to far left, down to far right
7. down to far left, up to far right **9.** up to far left, down to far right **11.** $(-2, -5)$, minimum is -5 **13.** $(-4, 17)$, maximum is 17
15. 62.5 feet by 125 feet **17.** 6

19. $(-2.1, 5.0)$, maximum $(1.4, -16.9)$, minimum **21.** $(4, -77)$, minimum $(-2, 31)$, maximum **23.** $(-1, -14)$ and $(3, -14)$, minimum $(1, 2)$, maximum

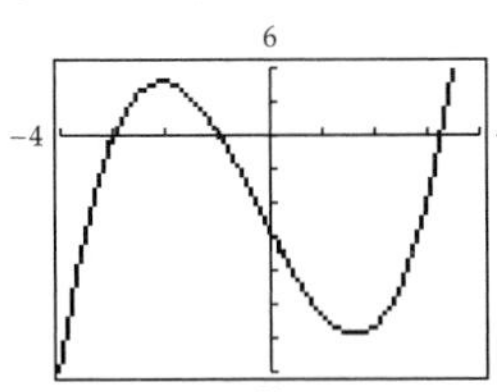

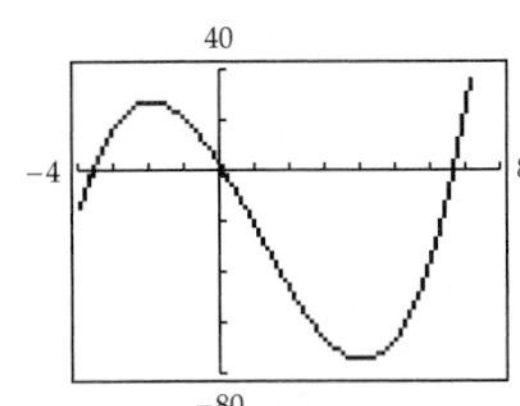

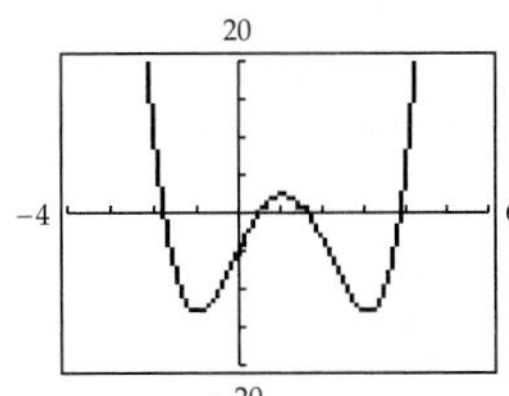

25. 2.137 inches, 337.1 cubic inches **27.** 103.3 kilometers/hour; 1:05.11 P.M. **29.** 55°, 1:44 P.M. **31.** 0.4 cm **39.** $(-7/2, 0)$, $(-2, 0)$, $(3, 0)$ **41.** $(0, 0)$, $(2/5, 0)$, $(1, 0)$ **43.** $(-5, 0)$, $(-7/3, 0)$, $(11/2, 0)$ **45.** $(-3, 0)$, $(0, 0)$, $(2, 0)$ **47.** $(1, 0)$ **49.** crosses at $(-1, 0)$, $(1, 0)$, and $(3, 0)$ **51.** crosses at $(7, 0)$; intersects but does not cross at $(3, 0)$ **53.** crosses at $(1, 0)$; intersects but does not cross at $(3/2, 0)$ **55.** crosses at $(2/3, 0)$; intersects but does not cross at $(0, 0)$ and $(201, 0)$ **57.** crosses at $(0, 0)$; intersects but does not cross at $(3, 0)$ **59.** 3 and 4 **61.** False. As one example, let $P(x) = x^3 - 5x^2 + 6x$, $a = 1$, and $b = 4$. In this case, $P(a) > 0$ and $P(b) > 0$; however, $P(x)$ has a zero when $x = 2$.

63.

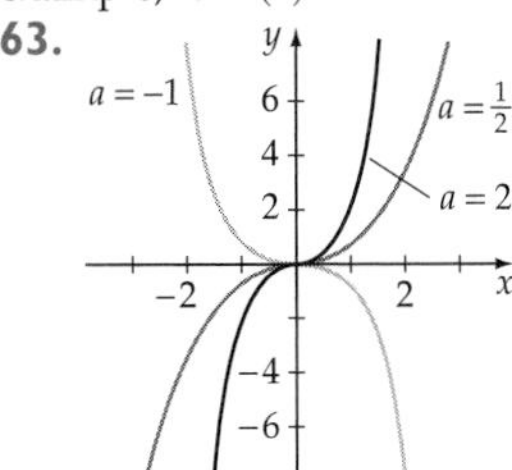

Exercise Set 3.3, page 267

1. 3 (multiplicity 2), −5 (multiplicity 1) **3.** 0 (multiplicity 2), −5/3 (multiplicity 2) **5.** 2 (multiplicity 1), −2 (multiplicity 1), −3 (multiplicity 2) **7.** 5 (multiplicity 2), −2 (multiplicity 2) **9.** −3 (multiplicity 1), 3 (multiplicity 1), −1 (multiplicity 1), 1 (multiplicity 1) **11.** $\pm1, \pm2, \pm4, \pm8$ **13.** $\pm1, \pm2, \pm3, \pm4, \pm6, \pm12, \pm1/2, \pm3/2$ **15.** $\pm1, \pm2, \pm4, \pm1/2, \pm1/3, \pm2/3, \pm4/3, \pm1/6$ **17.** $\pm1, \pm7, \pm1/2, \pm7/2, \pm1/4, \pm7/4$ **19.** $\pm1, \pm2, \pm4, \pm8, \pm16, \pm32$ **21.** upper bound 2, lower bound −5 **23.** upper bound 4, lower bound −4 **25.** upper bound 1, lower bound −4 **27.** upper bound 4, lower bound −2 **29.** upper bound 2, lower bound −1 **31.** one positive, two or no negative **33.** two or no positive, one negative **35.** one positive, three or one negative **37.** three or one positive, one negative **39.** one positive, no negative **41.** $2, -1, -4$ **43.** $3, -4, 1/2$ **45.** $1/2, -1/3, -2$ (multiplicity 2) **47.** $1/2, 4, \sqrt{3}, -\sqrt{3}$ **49.** $6, 1 + \sqrt{5}, 1 - \sqrt{5}$ **51.** $5, 1/2, 2 + \sqrt{3}, 2 - \sqrt{3}$ **53.** $1, -1, -2, -2/3, 3 + \sqrt{3}, 3 - \sqrt{3}$ **55.** $2, -1$ (multiplicity 2) **57.** $0, -2, 1 + \sqrt{2}, 1 - \sqrt{2}$ **59.** -1 (multiplicity 3), 2 **61.** $-3/2, 1$ (multiplicity 2), 8 **63.** $n = 9$ inches **65.** $x = 3.85$ cm **67.** $-5, -1, 1, 4$ **69.** $-2.5, -1, 2, 3.5$ **71.** $-0.5, 2$ **79.** yes **81.** yes

Exercise Set 3.4, page 276

1. $1 - i, 1/2$ **3.** $i, -3$ **5.** $-i\sqrt{2}, 1, \sqrt{5}, -\sqrt{5}$ **7.** $1 + \frac{1}{2}i, 1 - \frac{1}{2}i$ **9.** $1 - 3i, 1 + 2i, 1 - 2i$ **11.** $-i, 3, -1$ (multiplicity 2) **13.** $-i, 4 + i, 4 - i$ **15.** $5 - 2i, \frac{7}{2} + \frac{\sqrt{3}}{2}i, \frac{7}{2} - \frac{\sqrt{3}}{2}i$ **17.** $2, -3, 2i, -2i$ **19.** $1/2, -3, 1 + 5i, 1 - 5i$ **21.** 1 (multiplicity 3), $3 + 2i, 3 - 2i$ **23.** $-3, -1/2, 2 + i, 2 - i$ **25.** $4, 2, \frac{1}{2} + \frac{3}{2}i, \frac{1}{2} - \frac{3}{2}i$ **27.** $\frac{3}{2}, \frac{-1 + i\sqrt{7}}{2}, \frac{-1 - i\sqrt{7}}{2}$ **29.** $-2/3, 3/4, 5/2$ **31.** $-i, i, 2$ (multiplicity 2) **33.** -3 (multiplicity 2), 1 (multiplicity 2) **35.** $x(x - 2)(x + 1)$ **37.** $x(x^2 + 9)$ **39.** $(x^2 + 6)(x + 2)(x - 2)$ **41.** $(x^2 + 2)(x^2 + 1)$ **43.** $(x + 1)(x - 3)(x^2 + 4)$ **45.** $x^3 - 3x^2 - 10x + 24$ **47.** $x^3 - 3x^2 + 4x - 12$ **49.** $x^4 - 10x^3 + 63x^2 - 214x + 290$ **51.** $x^5 - 22x^4 + 212x^3 - 1012x^2 + 2251x - 1830$ **53.** $4x^3 - 19x^2 + 224x - 159$ **55.** $x^3 + 13x + 116$ **57.** $x^4 - 18x^3 + 131x^2 - 458x + 650$ **59.** $3x^3 - 12x^2 + 3x + 18$ **61.** $-2x^4 + 4x^3 + 36x^2 - 140x + 150$ **63.** Because $x^3 - x^2 - ix^2 - 9x + 9 + 9i$ does not have real coefficients, the theorem does not apply. **65.** $P(x) = (x - 2)^3(x^2 + 9)$ **67.** $P(x) = \frac{1}{2}x^5 - 4x^4 + \frac{25}{2}x^3 - 19x^2 + 14x - 4$

Exercise Set 3.5, page 291

1. vertical asymptotes: $x = 0, x = -3$ **3.** vertical asymptotes: $x = 4/3, x = -1/2$ **5.** horizontal asymptote: $y = 4$

7. horizontal asymptote: $y = 30$

9. vertical asymptote: $x = -4$
horizontal asymptote: $y = 0$

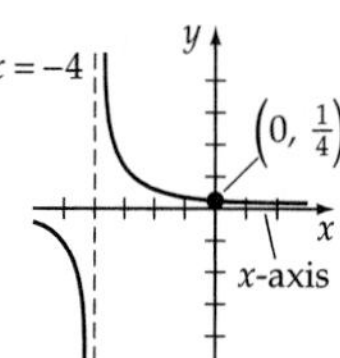

11. vertical asymptote: $x = 3$
horizontal asymptote: $y = 0$

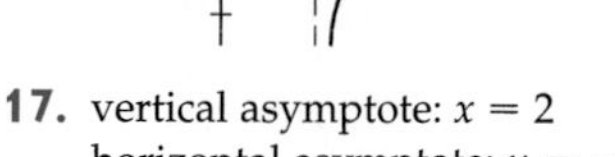

13. vertical asymptote: $x = 0$
horizontal asymptote: $y = 0$

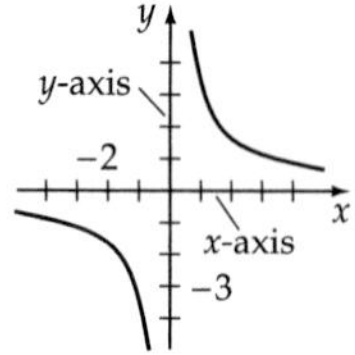

15. vertical asymptote: $x = -4$
horizontal asymptote: $y = 1$

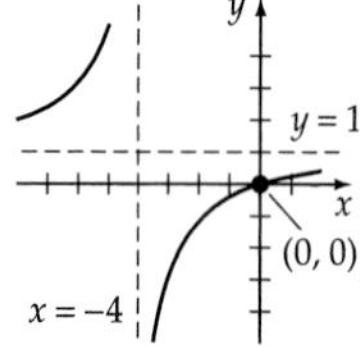

17. vertical asymptote: $x = 2$
horizontal asymptote: $y = -1$

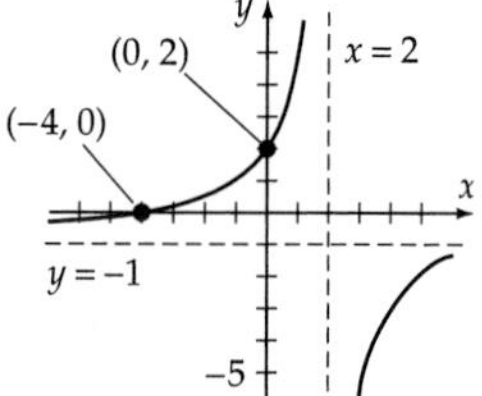

19. vertical asymptotes: $x = 3, x = -3$
horizontal asymptote: $y = 0$

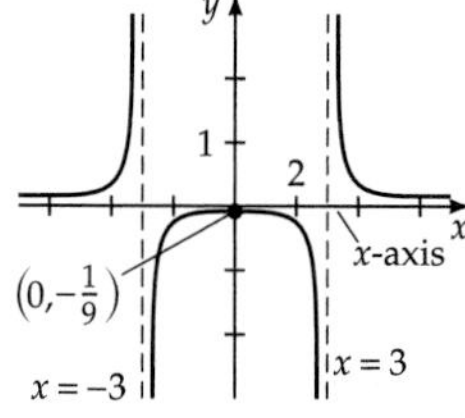

21. vertical asymptotes: $x = -3, x = 1$
horizontal asymptote: $y = 0$

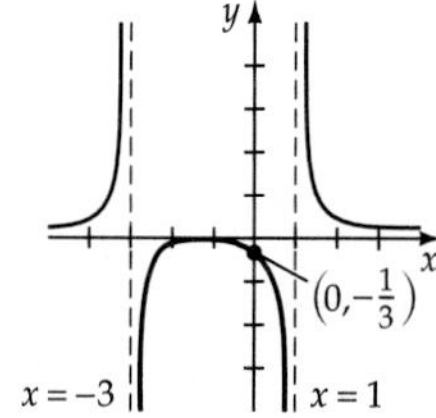

23. vertical asymptotes: $x = 3, x = -3$
horizontal asymptote: $y = 0$

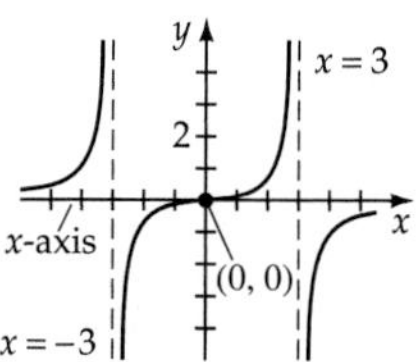

25. vertical asymptote: $x = -2$
horizontal asymptote: $y = 1$

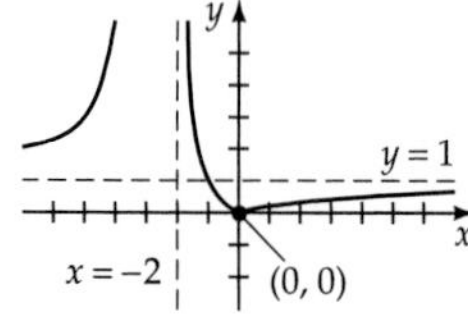

27. vertical asymptote: none
horizontal asymptote: $y = 0$

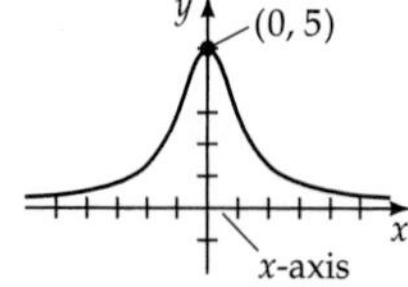

29. vertical asymptotes: $x = 3, x = -3$
horizontal asymptote: $y = 2$

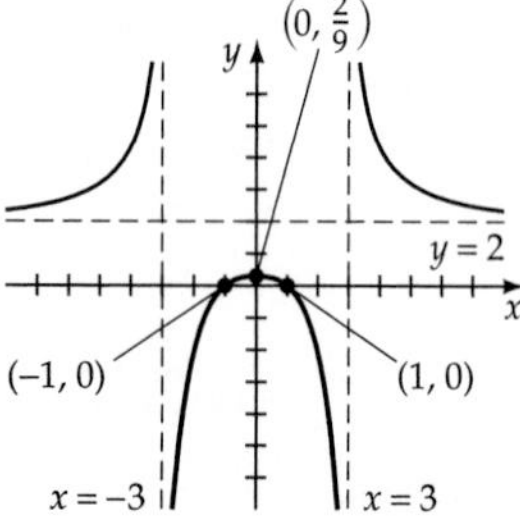

31. vertical asymptotes: $x = -1 + \sqrt{2}, x = -1 - \sqrt{2}$
horizontal asymptote: $y = 1$

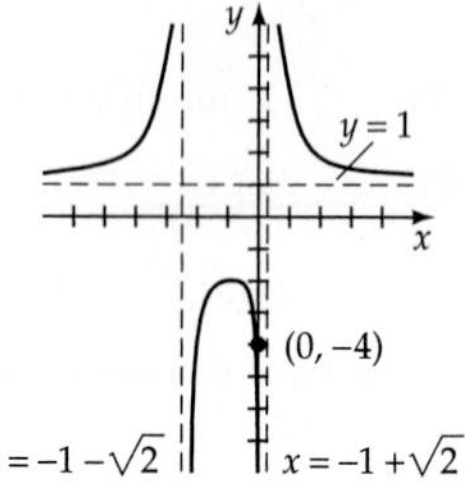

33. $y = 3x - 7$ **35.** $y = x$

37. vertical asymptote: $x = 0$
slant asymptote: $y = x$

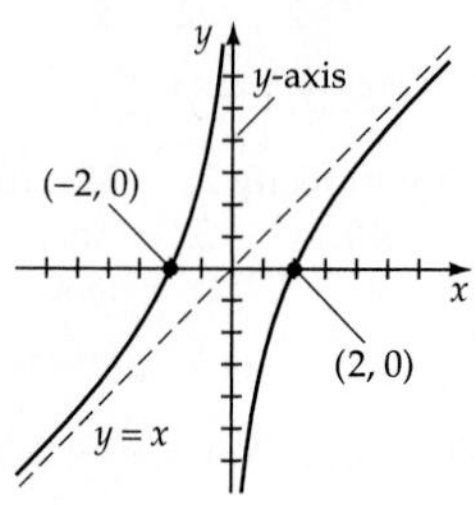

39. vertical asymptote: $x = -3$
slant asymptote: $y = x - 6$

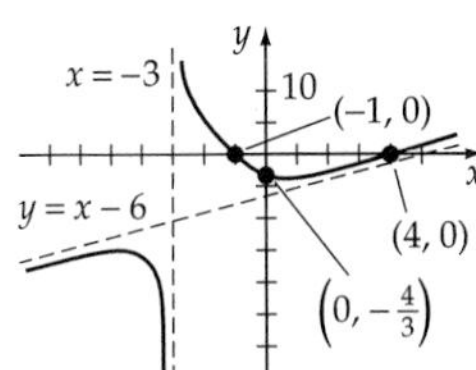

41. vertical asymptote: $x = 4$
slant asymptote: $y = 2x + 13$

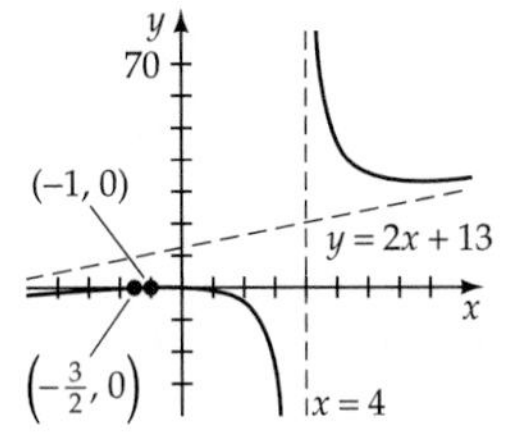

43. vertical asymptote: $x = -2$
slant asymptote: $y = x - 3$

45. vertical asymptotes: $x = 2, x = -2$
slant asymptote: $y = x$

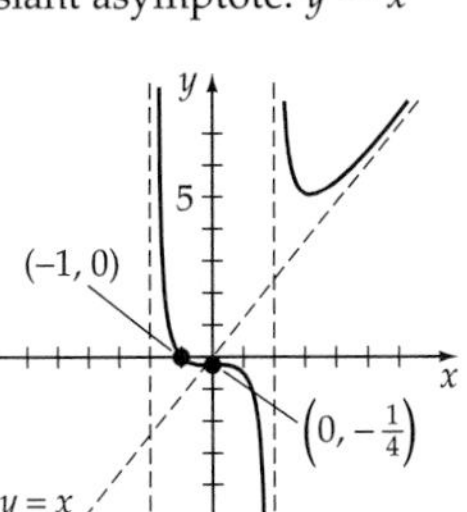

47.

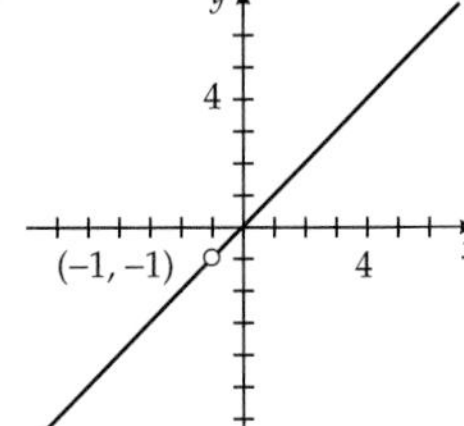

49.

51.

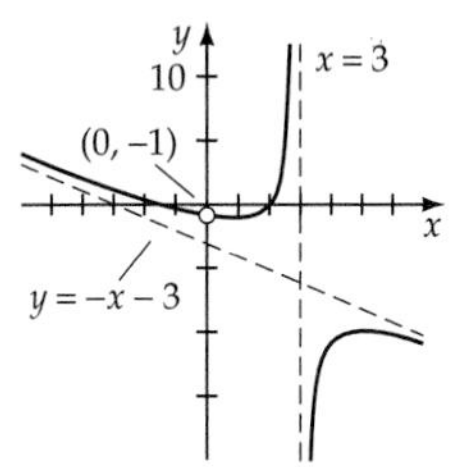

53.

55.

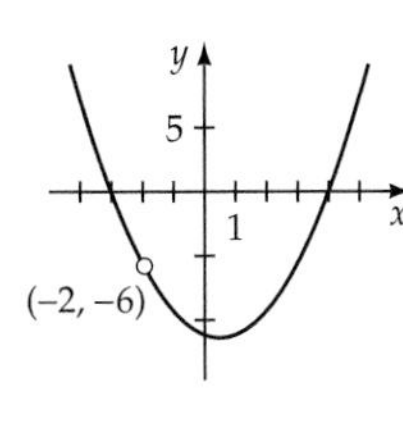

57. $x = -1.3, x = 2.3$
59. $x = -2, x = 1, x = 3$

61. a. \$1333.33 **b.** \$8000
c.

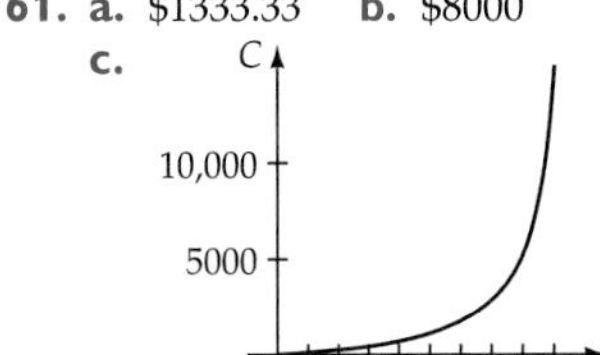

63. a. 611
b. 1777
c. $y = 2000$
d.

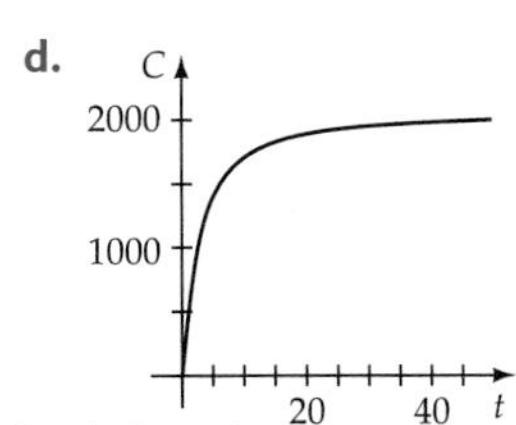

17.4 weeks

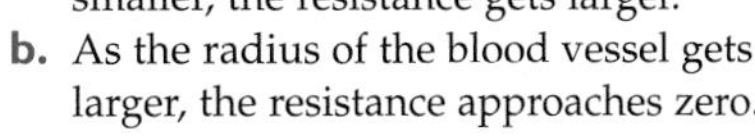
65. a. As the radius of the blood vessel gets smaller, the resistance gets larger.
b. As the radius of the blood vessel gets larger, the resistance approaches zero.
c.

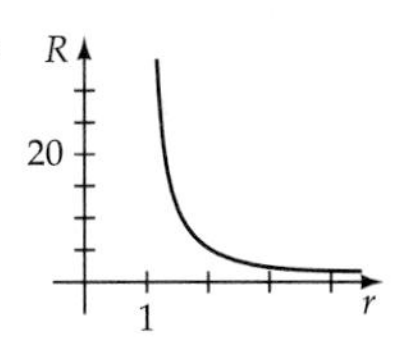

67. $(-2, 2)$ **69.** $(0, 1), (-4, 1)$

Chapter 3 True/False Exercises, page 296

1. False; $x - i$ has a zero of i, but it does not have a zero of $-i$. **2.** False: Descartes' Rule of Signs indicates that $x^3 - x^2 + x - 1$ has 3 or 1 positive zeros. **3.** True **4.** True **5.** False; $f(x) = \dfrac{x}{x^2 + 1}$ does not have a vertical asymptote.
6. False; $f(x) = \dfrac{(x - 2)^2}{(x - 3)(x - 2)} = \dfrac{x - 2}{x - 3}, x \neq 2$. **7.** True **8.** True **9.** True **10.** True **11.** True **12.** True **13.** True
14. False; $x^2 + 1$ does not have a real zero.

Chapter 3 Review Exercises, page 297

1. $x + 4 + \dfrac{-5x - 29}{x^2 + x + 3}$ **3.** $-x + \dfrac{3x^2 - 12x - 3}{x^3 + x}$ **5.** $3x^2 - 5x - 1$ **7.** $4x^2 + x + 8 + \dfrac{22}{x - 3}$ **9.** $3x^2 - 6x + 7 + \dfrac{-13}{x + 2}$

11. $3x^2 + 5x - 11$ **13.** 77 **15.** 33 **21.** **23.** **25.**

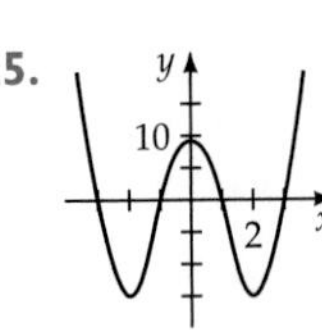

27. $\pm1, \pm2, \pm3, \pm6$

29. $\pm1, \pm2, \pm3, \pm4, \pm6, \pm12, \pm1/3, \pm2/3, \pm4/3, \pm1/5, \pm2/5, \pm3/5, \pm4/5, \pm6/5, \pm12/5, \pm1/15, \pm2/15, \pm4/15$ **31.** ±1 **33.** no positive and three or one negative **35.** one positive and one negative **37.** $1, -2, -5$ **39.** -2 (multiplicity 2), $-1/2, -4/3$ **41.** 1 (multiplicity 4) **43.** $2x^3 - 3x^2 - 23x + 12$ **45.** $x^4 - 3x^3 + 27x^2 - 75x + 50$ **47.** vertical asymptote: $x = -2$, horizontal asymptote: $y = 3$ **49.** vertical asymptote: $x = -1$, slant asymptote: $y = 2x + 3$

51. **53.**

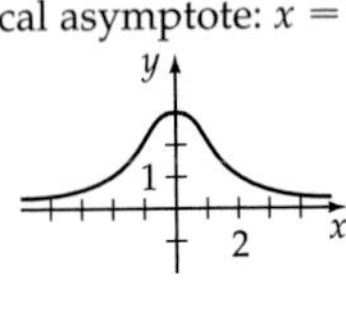

55.

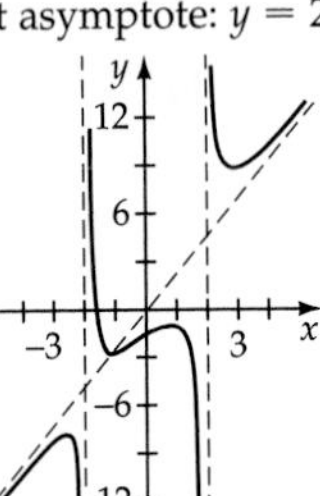

57.

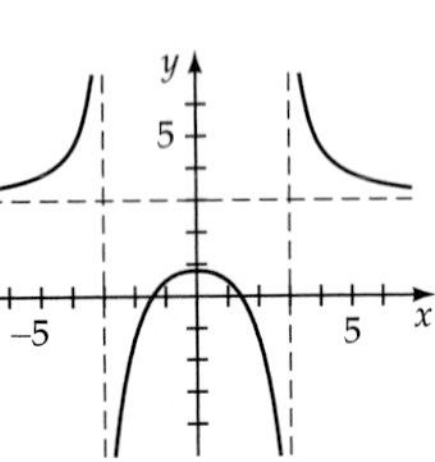

59. 1.325 **61.** 0.786

Chapter 3 Test, page 298

1. $3x^2 - x + 6 - \dfrac{13}{x+2}$ **2.** 43 **4.** up to the far left and down to the far right **5.** $0, 2/3, -3$ **6.** $P(1) < 0$, $P(2) > 0$. Therefore P has a zero between 1 and 2. **7.** 2 (multiplicity 2), -2 (multiplicity 2), $3/2$ (multiplicity 1), -1 (multiplicity 3) **8.** $\pm1, \pm3, \pm1/2, \pm3/2, \pm1/3, \pm1/6$ **9.** upper bound 4, lower bound -5 **10.** 4, 2, or 0 positive zeros, no negative zero. **11.** $1/2, 3, -2$ **12.** $2 + 3i, 2 - 3i, -2/3, -5/2$ **13.** 0, 1 (multiplicity 2), $2 + i, 2 - i$ **14.** $x^4 - 5x^3 + 8x^2 - 6x$ **15.** vertical asymptotes: $x = 3, x = 2$ **16.** horizontal aymptote: $y = 3/2$

17. **18.**

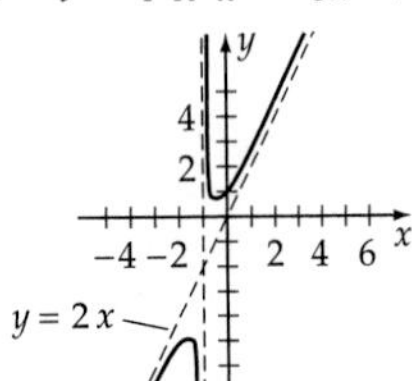

19.

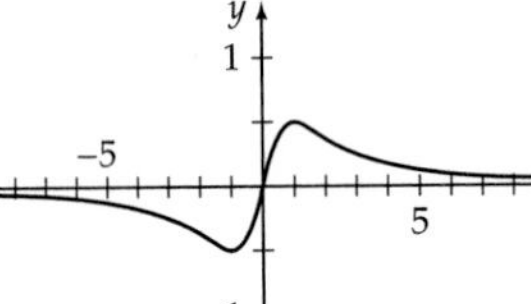

20. 1.8

Exercise Set 4.1, page 306

9. $\{(1, -3), (2, -2), (5, 1), (-7, 4)\}$ **11.** $\{(1, 0), (2, 1), (4, 2), (8, 3), (16, 4)\}$ **13.** $f^{-1}(x) = \dfrac{x-1}{4}$ **15.** $F^{-1}(x) = \dfrac{1-x}{6}$ **17.** $j^{-1}(t) = \dfrac{t-1}{2}$ **19.** $f^{-1}(v) = \sqrt[3]{1-v}$ **21.** $f^{-1}(x) = \dfrac{-4x}{x+3}, x \neq -3$ **23.** $M^{-1}(t) = \dfrac{5}{1-t}, t \neq 1$ **25.** $r^{-1}(t) = -\sqrt{\dfrac{1}{t}}, t > 0$ **27.** $J^{-1}(x) = \sqrt{x-4}, x \geq 4$ **29.** $f^{-1}(x) = \sqrt{x-3}$, domain $\{x | x \geq 3\}$, range $\{y | y \geq 0\}$; f has domain $\{x | x \geq 0\}$, range $\{y | y \geq 3\}$. **31.** $f^{-1}(x) = x^2$, domain $\{x | x \geq 0\}$, range $\{y | y \geq 0\}$; f has domain $\{x | x \geq 0\}$, range $\{y | y \geq 0\}$. **33.** $f^{-1}(x) = \sqrt{9-x^2}$, domain $\{x | 0 \leq x \leq 3\}$, range $\{y | 0 \leq y \leq 3\}$; f has domain $\{x | 0 \leq x \leq 3\}$, range $\{y | 0 \leq y \leq 3\}$. **35.** $f^{-1}(x) = 2 + \sqrt{x+3}$, domain $\{x | x \geq -3\}$, range $\{y | y \geq 2\}$; f has domain $\{x | x \geq 2\}$, range $\{y | y \geq -3\}$. **37.** $f^{-1}(x) = -4 - \sqrt{x+25}$, domain $\{x | x \geq -25\}$, range $\{y | y \leq -4\}$. f has domain $\{x | x \leq -4\}$, range $\{y | y \geq -25\}$.

39.

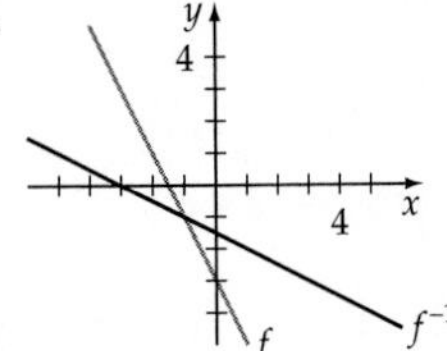

41.

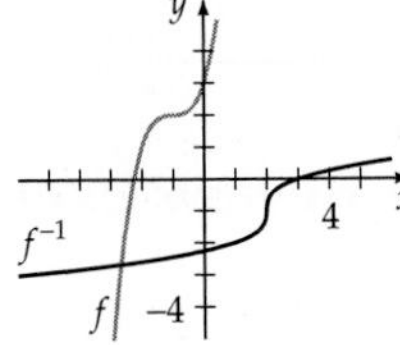

43.
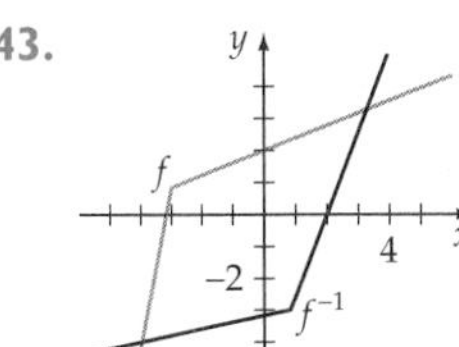

45. a. $s(x) = x + 32$ **b.** $s^{-1} = x - 32$; this means a woman's U.S. shoe size is her Italian shoe size less 32

47.
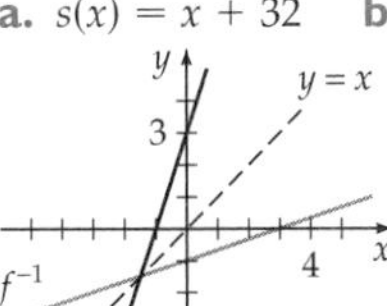

49.
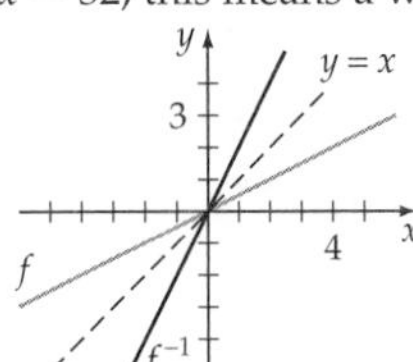

51.
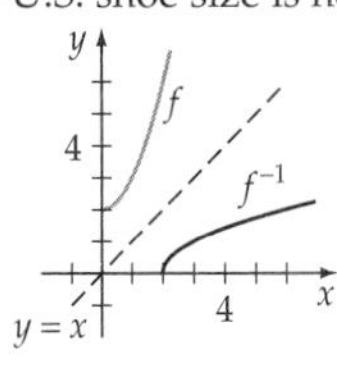

53.
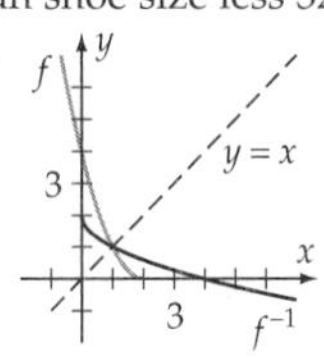

55. $f^{-1}(x) = \dfrac{x - b}{a}, a \neq 0$ **57.** $f^{-1}(x) = \dfrac{x + 1}{1 - x}, x \neq 1$ **59.** no **61.** yes **63.** yes **65.** no **67.** 5 **69.** 4

71. The reflection of f about the line $y = x$ yields f. Thus f is its own inverse.

Exercise Set 4.2, page 317

1. 4.7288 **3.** 442.3350 **5.** 2.1746 **7.** 164.0219 **9.** 5.6522 **11.** 0.9695 **13.** 70.4503 **15.** 19.8130 **17.** 14.0940
19. 15.1543 **21.** 3353.3255 **23.** 8103.0839

25.
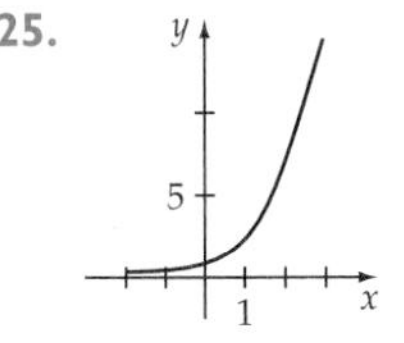

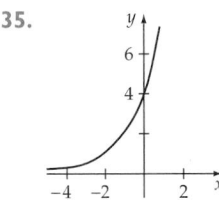

27.
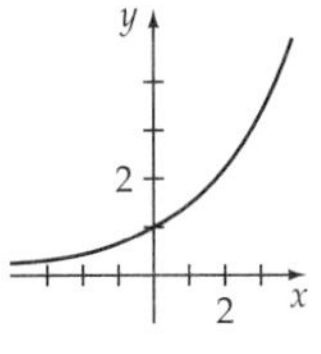

29.
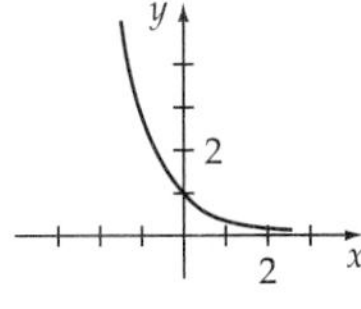

31.
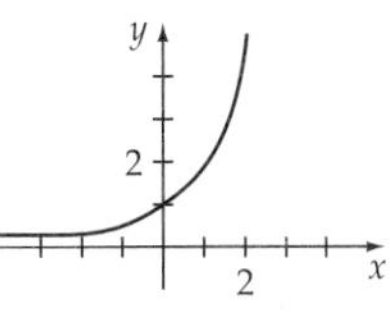

33.
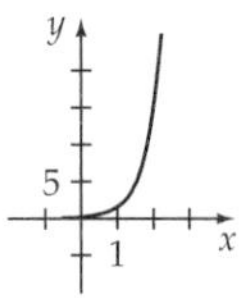

35.
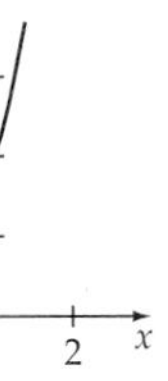

37.
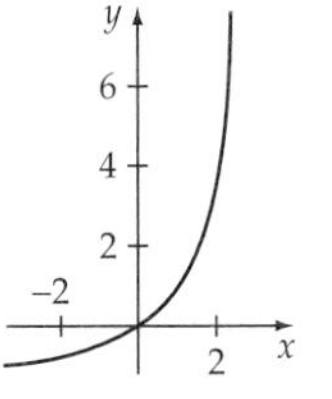

39.
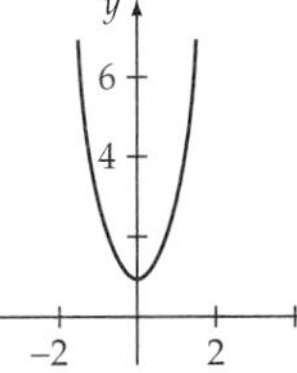

41.
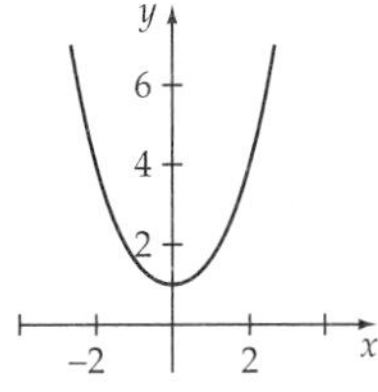

43.
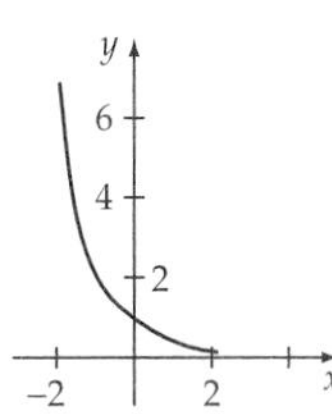

45.
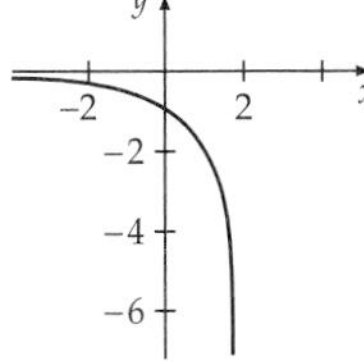

47.
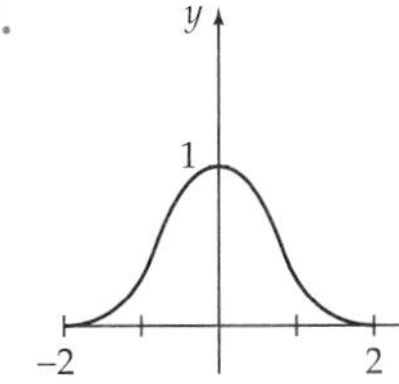

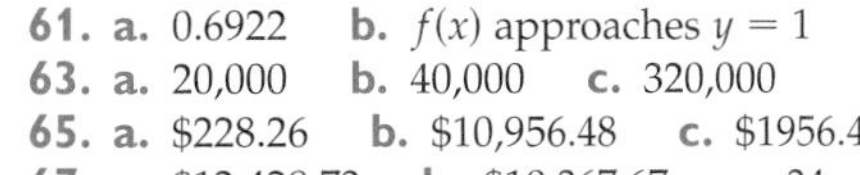

49. 1.58 **51.** 0.69 **53.** 0.79 **55.** 0.80 **57.** 15.0% **59.** 64 million
61. a. 0.6922 **b.** $f(x)$ approaches $y = 1$
63. a. 20,000 **b.** 40,000 **c.** 320,000
65. a. \$228.26 **b.** \$10,956.48 **c.** \$1956.48
67. a. \$12,428.73 **b.** \$10,367.67 **c.** 34

69.
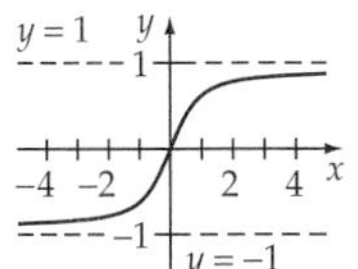

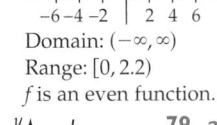

Domain: $(-\infty, \infty)$
Range: $(-1, 1)$
f is an odd function.

71.
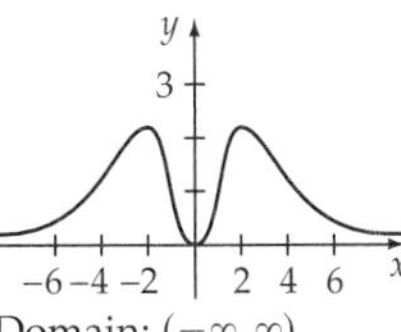
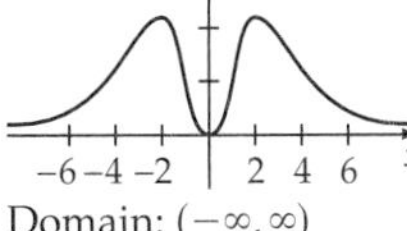

Domain: $(-\infty, \infty)$
Range: $[0, 2.2)$
f is an even function.

73.
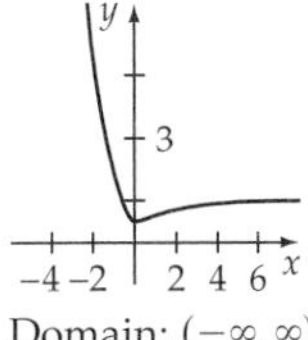
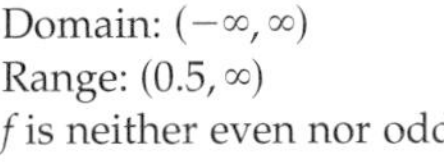

Domain: $(-\infty, \infty)$
Range: $(0.5, \infty)$
f is neither even nor odd.

75.
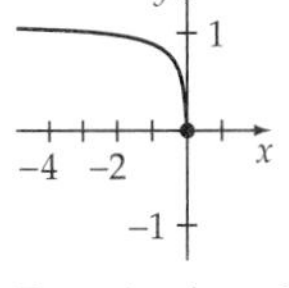
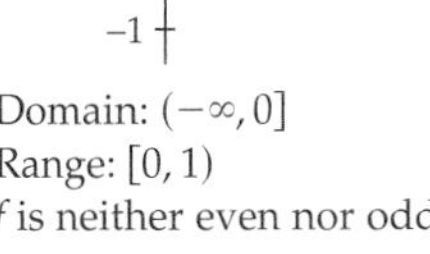

Domain: $(-\infty, 0]$
Range: $[0, 1)$
f is neither even nor odd.

77. a.
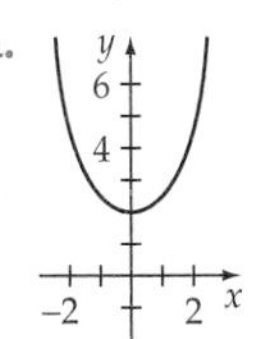

Domain: $(-\infty, \infty)$
Range: $[2, \infty)$

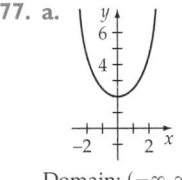

b.
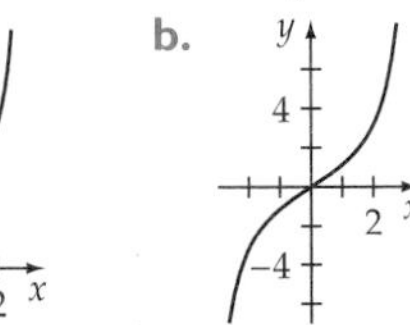

Domain: $(-\infty, \infty)$
Range: $(-\infty, \infty)$

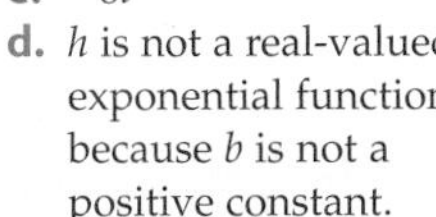

79. a. -4 **b.** 16
c. $-8i$
d. h is not a real-valued exponential function because b is not a positive constant.

81.
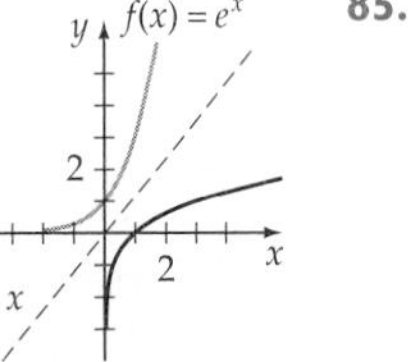

85. e^{π}

87. a. Vertical asymptote: none
Horizontal asymptote: $y = 0$
$y = 2$

b. none

c.
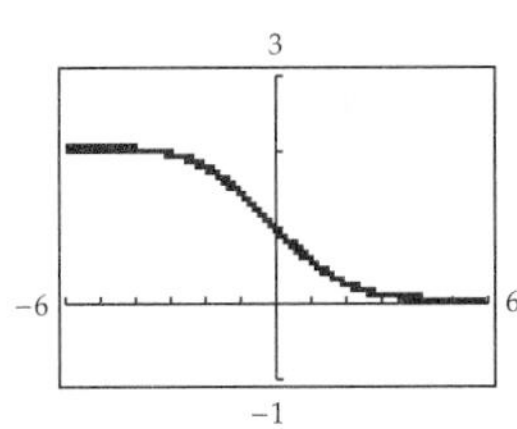

89. a. Vertical asymptote: $x = 0$
Horizontal asymptote: $y = 0$
b. $(-1, 0)$
c.

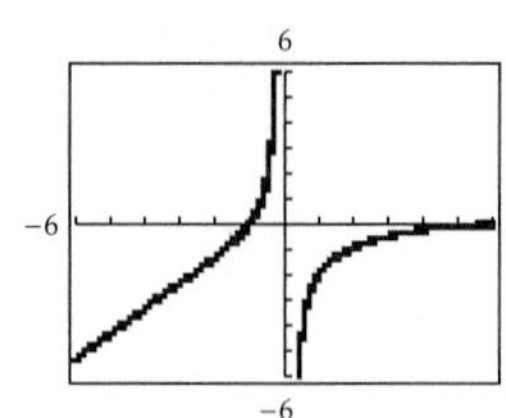

91. a. Vertical asymptote: $x = 0$
Horizontal asymptote: $y = 0$
b. 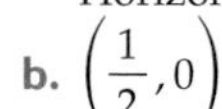 $\left(\frac{1}{2}, 0\right)$
c.

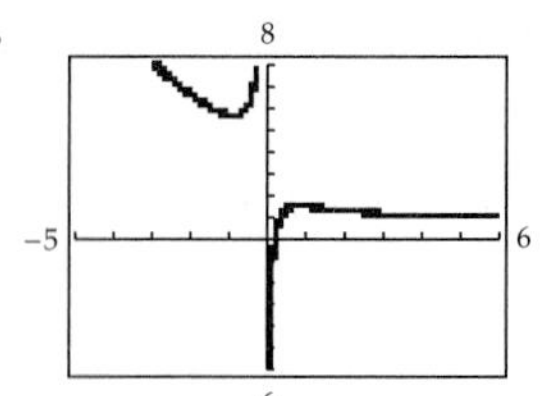

Exercise Set 4.3, page 327

1. $10^2 = 100$ **3.** $5^3 = 125$ **5.** $3^4 = 81$ **7.** $b^t = r$ **9.** $3^{-3} = 1/27$ **11.** $\log_2 16 = 4$ **13.** $\log_7 343 = 3$ **15.** $\log_{10} 10{,}000 = 4$ **17.** $\log_b j = k$ **19.** $\log_b b = 1$ **21.** 6 **23.** 5 **25.** 3 **27.** −2 **29.** 0 **31.**

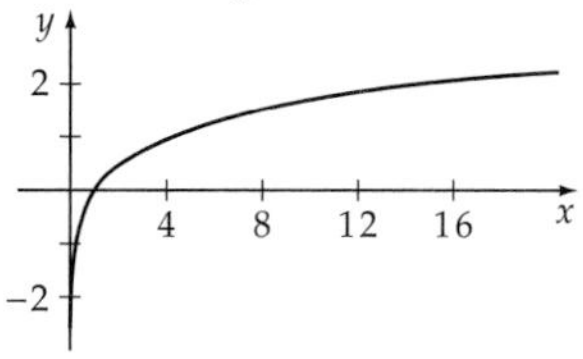

33.

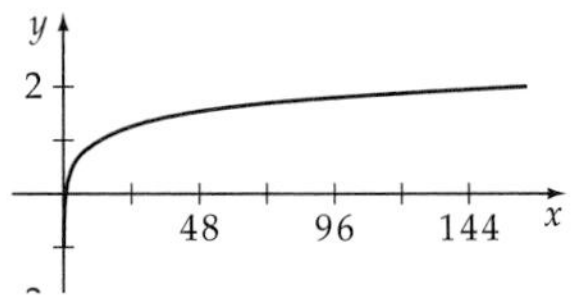

35.

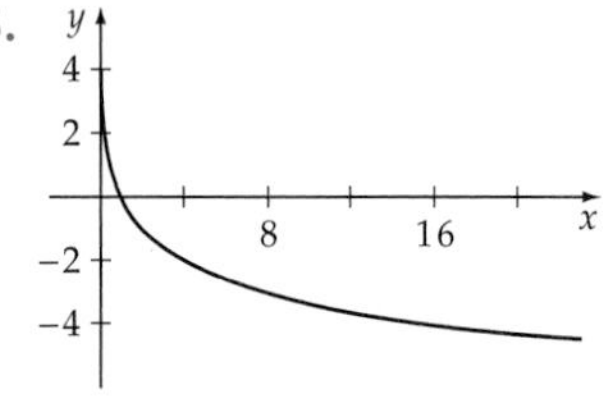

37.

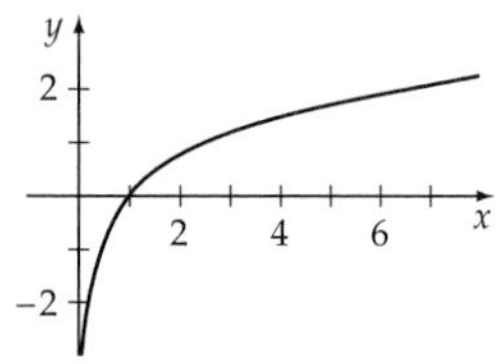

39.

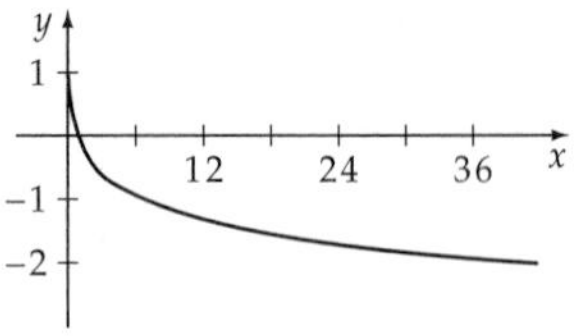

41. $(3, \infty)$ **43.** $(-\infty, 11)$ **45.** $(-\infty, -2) \cup (2, \infty)$ **47.** $(4, \infty)$ **49.** $(-1, 0) \cup (1, \infty)$

51.

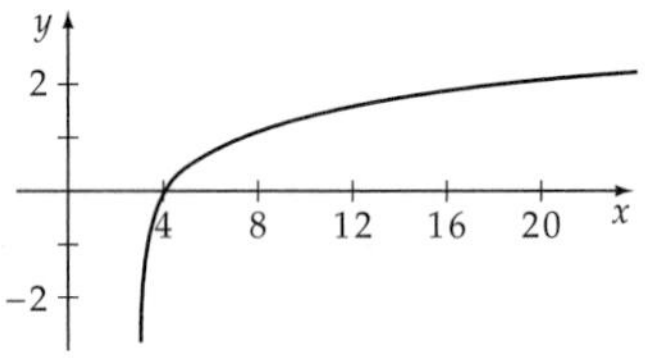

53.

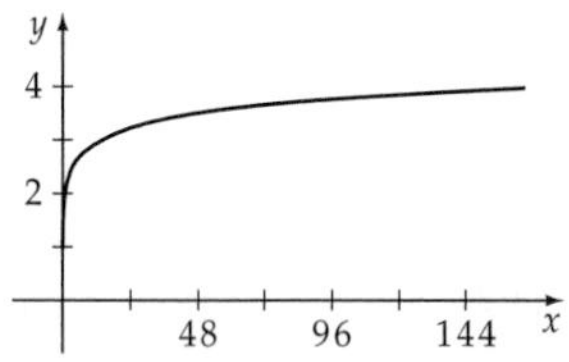

55.

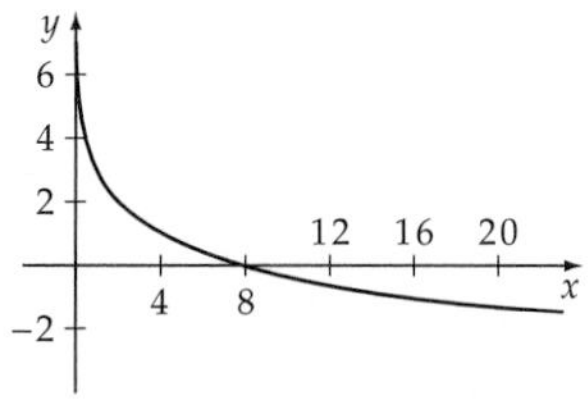

57.

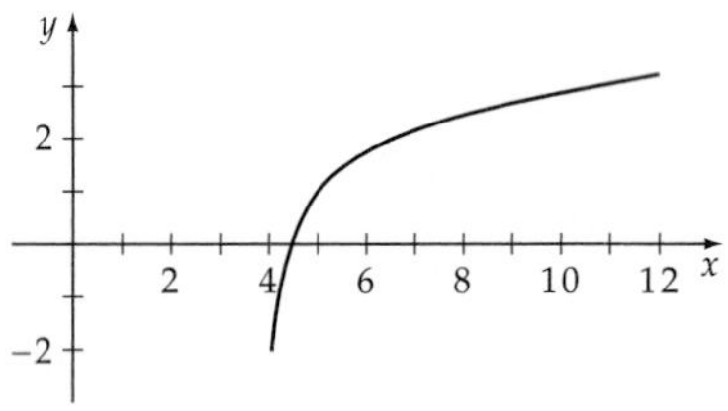

59.

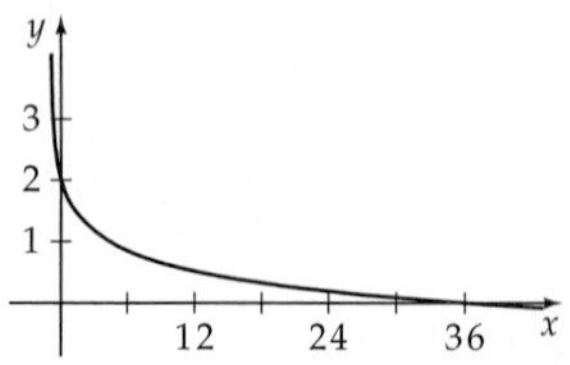

61. Y1=-2ln(X)

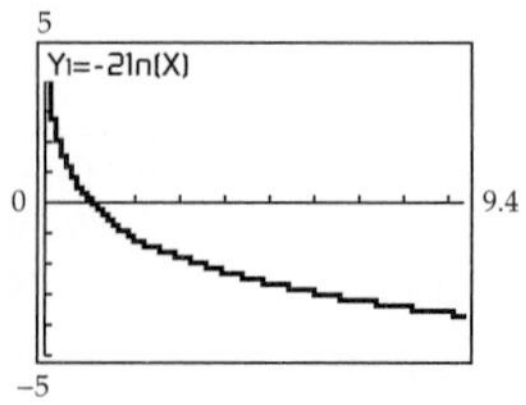

63. Y1=abs(ln(X))

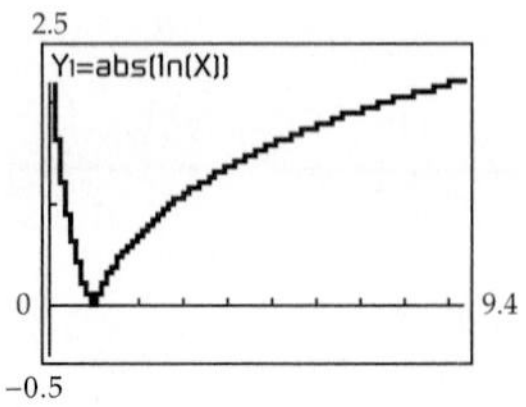

65. Y1=log(X^(1/3))

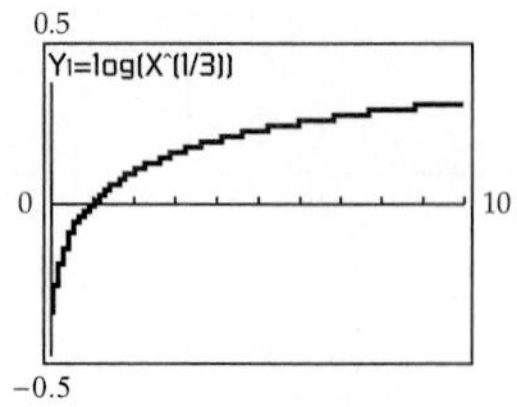

67. Y1=log(X+10)

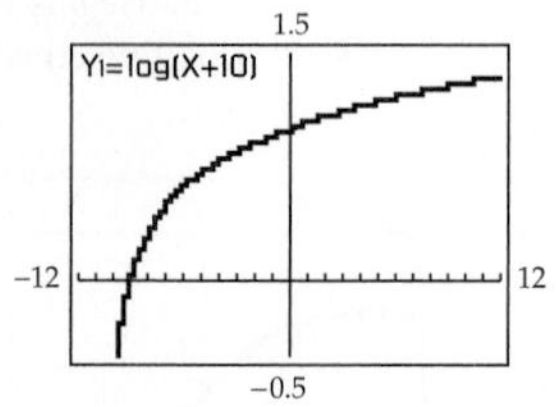

69. Y1=ln(X-5)

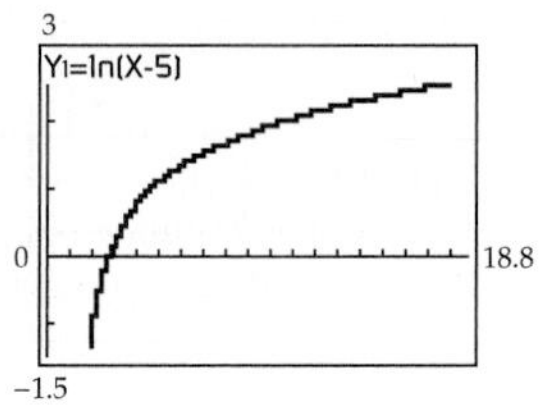

71.

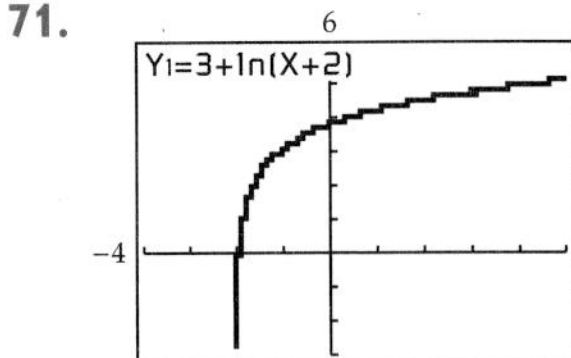

73.

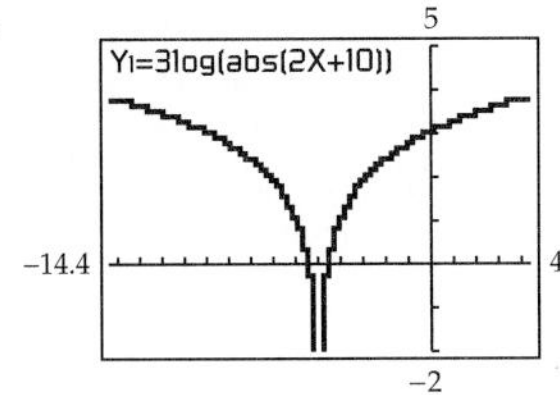

75. $f(x) \to 0$

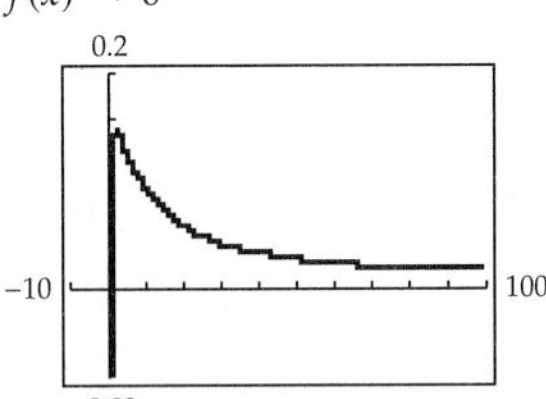

77. Yes

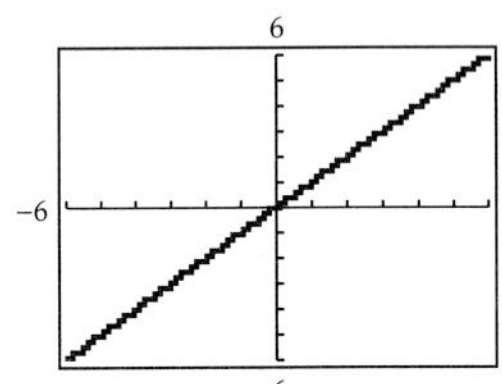

79. a. 49 b. 2.2 81. a. 1.50 m^2 b. 2.05 m^2 c. 0.67 m^2 83. a. 4 b. 96 c. 3385 d. 2,098,960 Both numbers have the same number of digits, because $2^{6972593}$ does not equal a power of 10.

85.

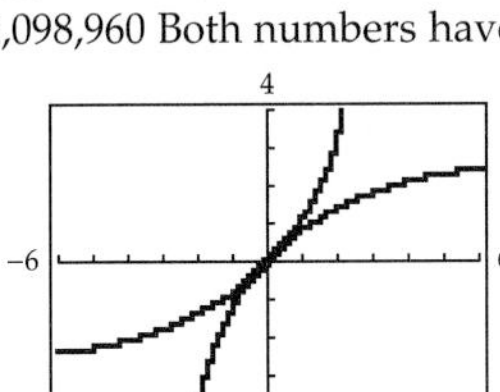

$f(x)$ and $g(x)$ are inverse functions.

87.

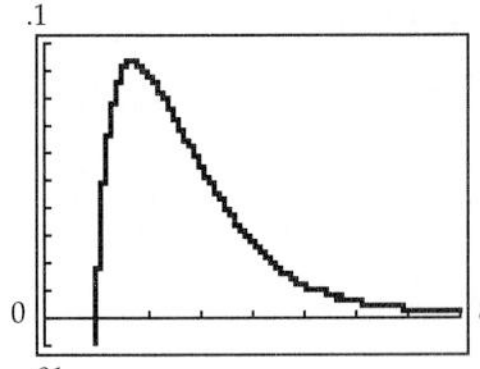

89.

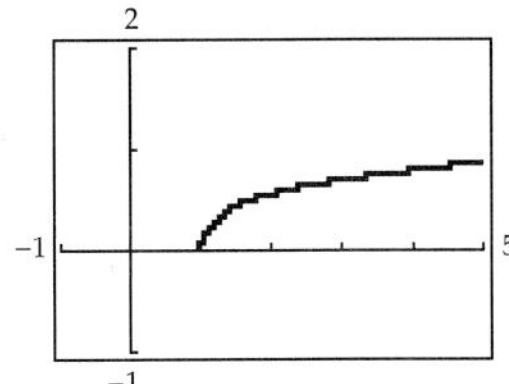

Domain $\{x \mid x \ge 1\}$, Range $\{y \mid y \ge 0\}$

91.

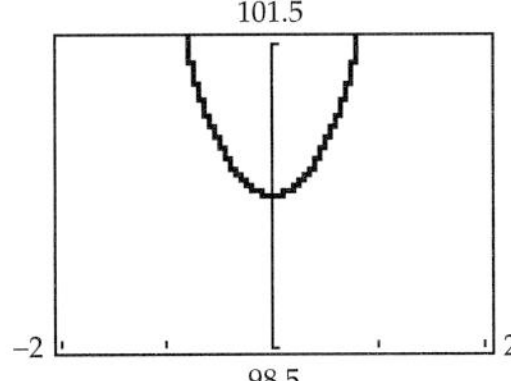

Domain $\{x \mid -1 < x < 1\}$, Range $\{y \mid y \ge 100\}$

93.

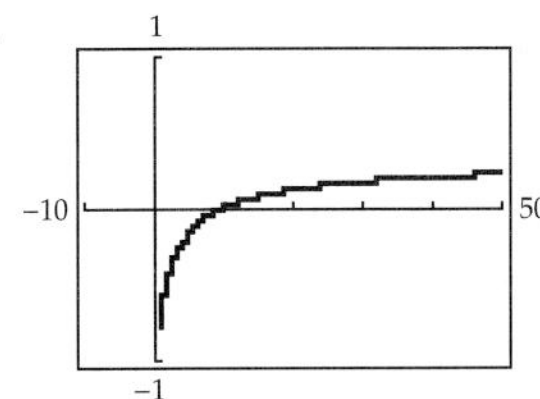

Domain $\{x \mid x > 1\}$, Range all real numbers

Exercise Set 4.4, page 339

1. $\log_b x + \log_b y + \log_b z$ 3. $\log_3 x - 4 \log_3 z$ 5. $(1/2) \log_b x - 3 \log_b y$ 7. $\log_b x + (2/3) \log_b y - (1/3) \log_b z$

9. $(1/2) \log_7 x + (1/2) \log_7 z - 2 \log_7 y$ 11. $\log_{10} [x^2(x + 5)]$ 13. $\log_b \sqrt{\dfrac{(x - y)^3(x + y)}{z}}$ 15. $\log_8 (x + y)$ 17. $\ln [x^2(x - 3)^4]$

19. $\ln (xz/y)$ 21. 0.9208 23. 1.1292 25. −0.4709 27. 1.7479 29. 1.3562 31. 1.5395 33. 0.86719 35. −1.7308

37. −2.3219 39. 0.87357 41.

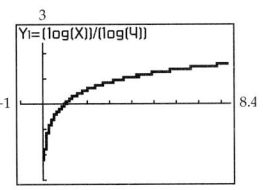

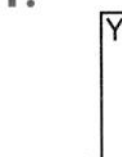

43.

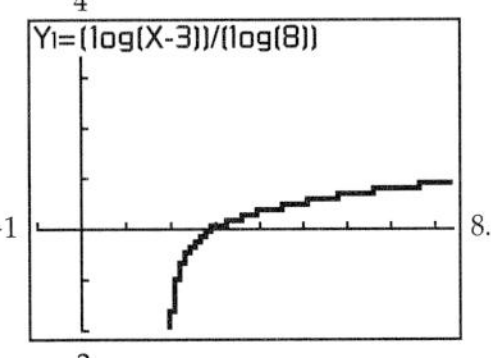

45.

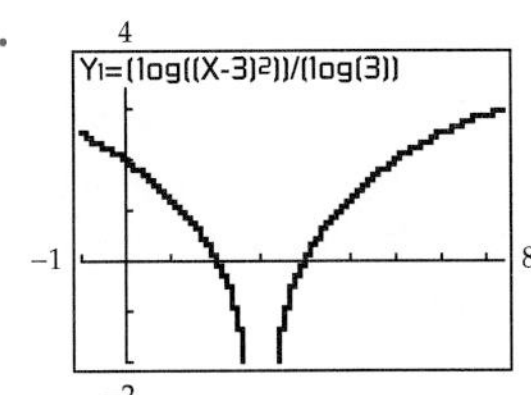

47.

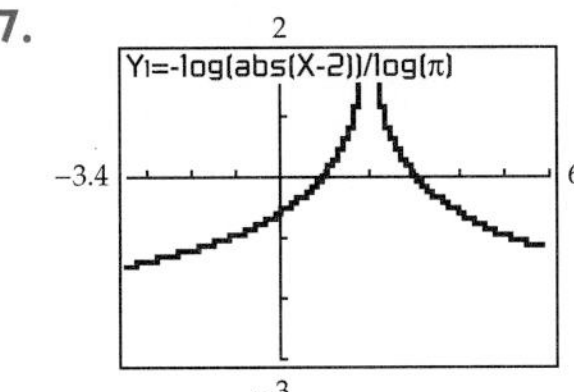

49.

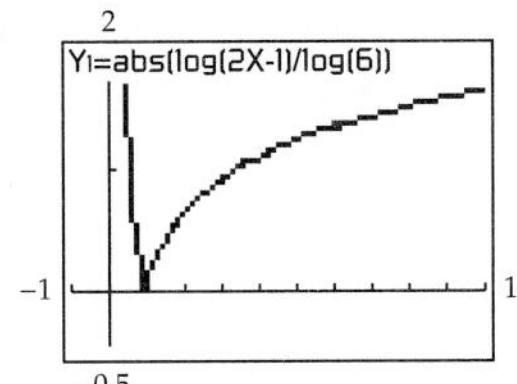

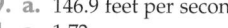

51.

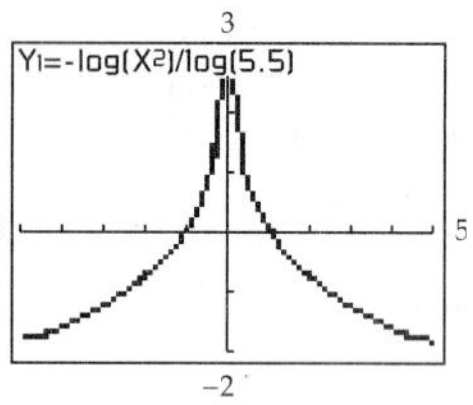

53. 5 **55.** $10^{6.5}I_0$ or $3{,}162{,}277.7I_0$ **57.** 100 to 1 **59.** $10^{1.8}$ to 1 or 63 to 1 **61.** 5.5 **63.** 11.9; base **65.** 3.16×10^{-10} **67. a.** 82.0 **b.** 40.3 **c.** 115.0 **d.** 152.0 **69.** 10 **71.** 2 **73.** Definition of a logarithm. (Write in exponential form.)
Multiply each side of $M = b^x$ by the same quantity.
The product property of exponents.
The logarithm of each side property.
The $\log_b b^p = p$ property.
Substitution.
75. $[1, 10^{1000}]$

77. $\left[e^e, e^{(e^e)}\right]$ **79.** $(0, 1)$ **81. a.** $M \approx 6.1$ **b.** $M \approx 3.8$ These results are close to the magnitudes produced by the amplitude–time-difference formula.

Exercise Set 4.5, page 348

1. 6 **3.** $-3/2$ **5.** $-6/5$ **7.** 3 **9.** $\dfrac{\log 70}{\log 5}$ **11.** $-\dfrac{\log 120}{\log 3}$ **13.** $\dfrac{\log 315 - 3}{2}$ **15.** ln 10 **17.** $\dfrac{\ln 2 - \ln 3}{\ln 6}$ **19.** $\dfrac{3 \log 2 - \log 5}{2 \log 2 + \log 5}$ **21.** 7 **23.** 4 **25.** $2 + 2\sqrt{2}$ **27.** 199/95 **29.** -1 **31.** 3 **33.** 10^{10} **35.** 2 **37.** 5 **39.** $\log\left(20 + \sqrt{401}\right)$ **41.** $(1/2) \log (3/2)$ **43.** $\ln\left(15 \pm 4\sqrt{14}\right)$ **45.** $\ln\left(1 + \sqrt{65}\right) - \ln 8$ **47.** 1.61 **49.** 0.96 **51.** 2.20 **53.** -1.93 **55.** -1.34 **57. a.** 8500, 10,285 **b.** in 6 years **59. a.** 60°F **b.** 27 minutes

61. a.

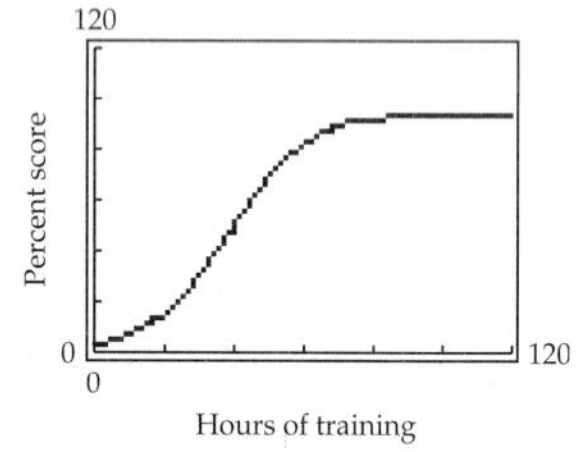

b. 48 hours **c.** $P = 100$ **d.** As the number of hours of training increases, the test scores approach 100%.

63. a.

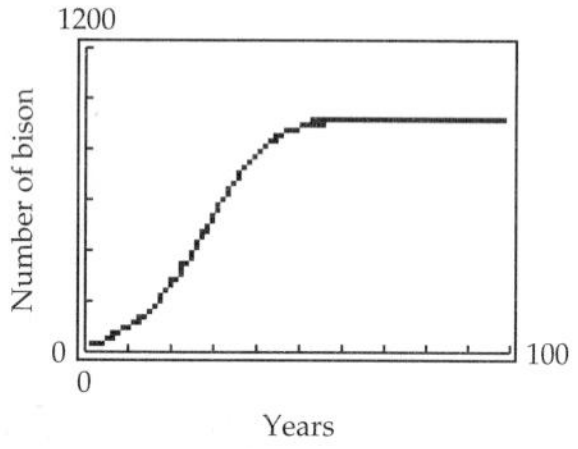

b. in 27 years or 2026 **c.** $B = 1000$ **d.** As the number of years increases, the bison population approaches but never exceeds 1000.

65. a.

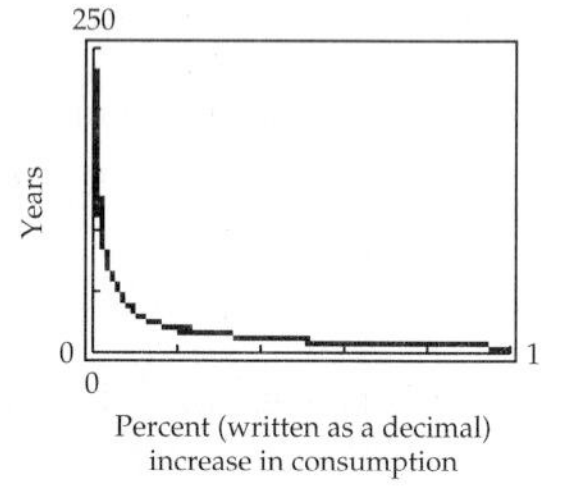

b. 77 years **c.** 1.9%

67. a.

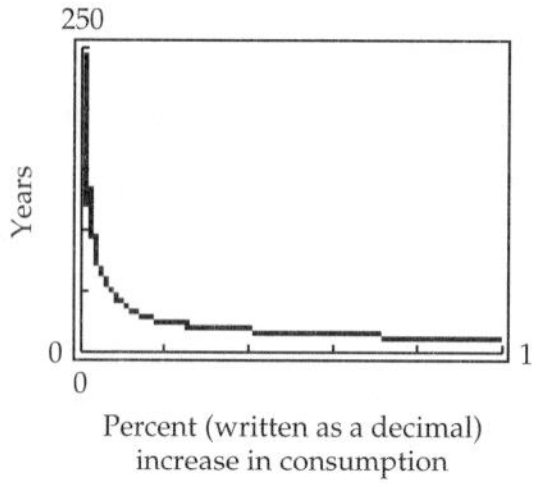

b. 78 years **c.** 1.9% **d.** More; For any $r > 0$, T in Exercise 67 is greater than T in Exercise 65. Thus this model predicts more time before the coal is depleted.

69. a. 146.9 feet per second **b.** $v = 175$ **c.** The velocity of the sky diver will never exceed 175 feet per second. **71. a.** 1.72 **b.** $v = 100$ **c.** The object cannot fall faster than 100 feet per second.

73. a.

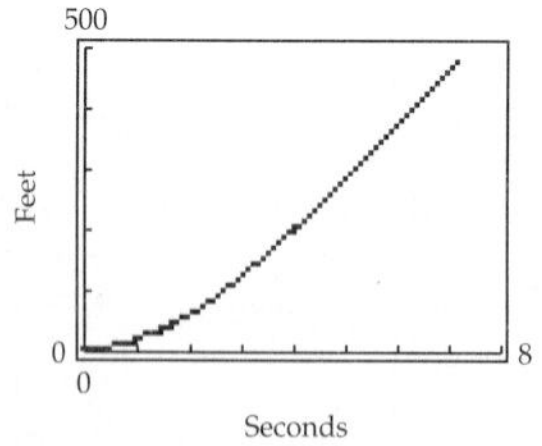

b. 2.6 seconds **75. a.** the whole numbers 0, 1, 2, 3, ..., 196 **b.** 138

77. The second step; because $\log 0.5 < 0$, the inequality sign must be reversed. **79.** $x = y/(y - 1)$ **81.** $e^{0.336} \approx 1.4$ **83.** $(0, 1/\ln 2)$

Exercise Set 4.6, page 361

1. a. \$9724.05 **b.** \$11,256.80 **3. a.** \$48,885.72 **b.** \$49,282.20 **c.** \$49,283.30 **5.** \$24,730.82 **7.** 8.8 years **9.** $t = (\ln 3)/r$ **11.** 14 years **13. a.** 2200 bacteria **b.** 17,600 bacteria **15. a.** $N(t) \approx 22{,}600e^{0.01368t}$ **b.** 27,700 **17. a.** 10,755,000 **b.** 2042 **19. a.**

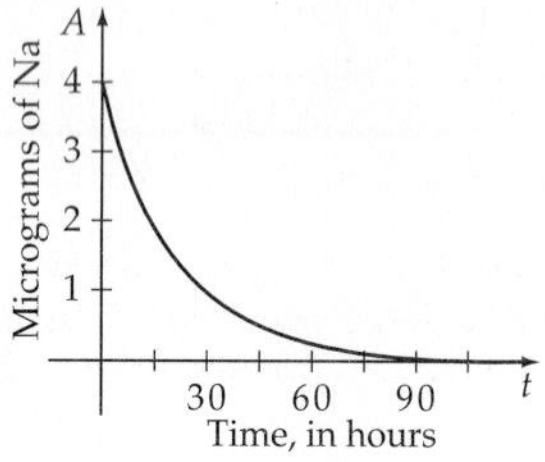

b. 3.18 micrograms **c.** 15.07 hours **d.** 30.14 hours **21.** 6601 years ago **23.** 2378 years old **25. a.** 0.056 **b.** 42° **c.** 54 minutes **27. a.** 211 hours **b.** 1386 hours **29.** 3.1 years

31. a.

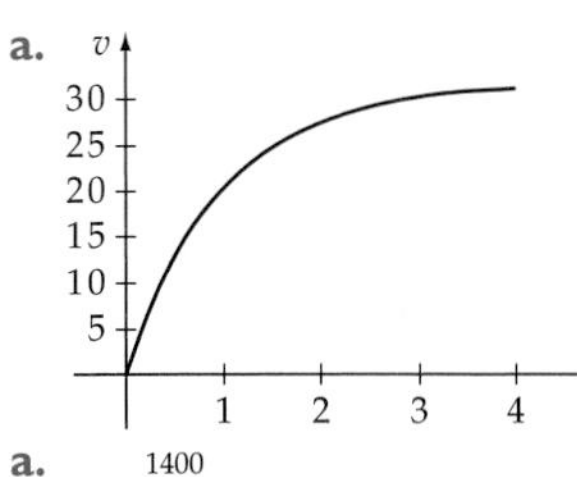

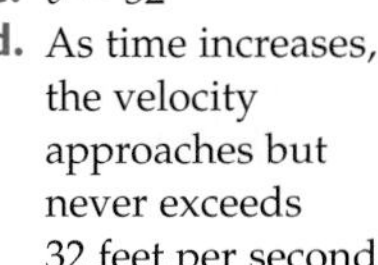

b. 0.98 seconds **c.** $v = 32$ **d.** As time increases, the velocity approaches but never exceeds 32 feet per second.

33. a.

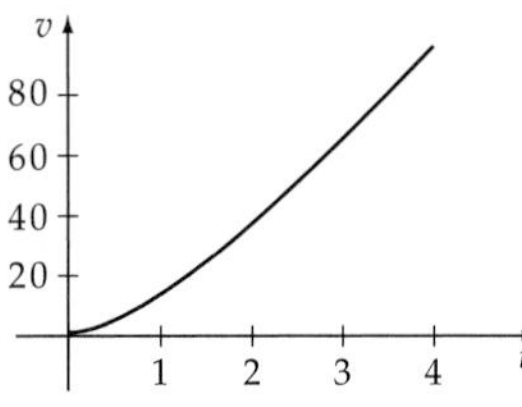

b. 2.5 seconds **c.** 24.56 feet per second **d.** The average speed of the object was 24.56 feet per second between $t = 1$ and $t = 2$.

35. a. 0.29 **b.** 2004

37. a.

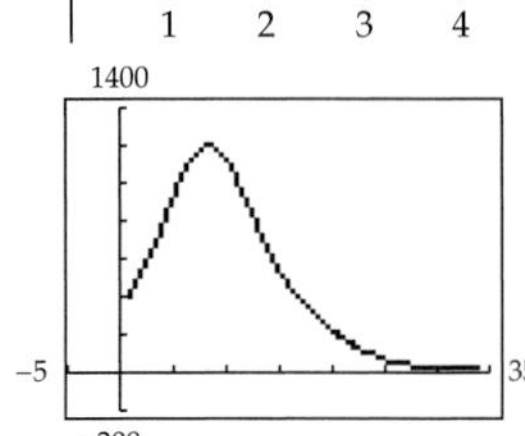

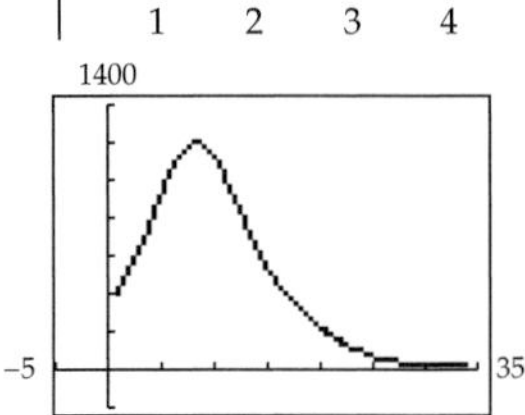

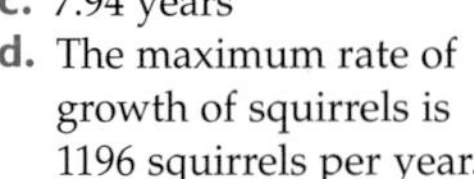

b. 1196 squirrels per year **c.** 7.94 years **d.** The maximum rate of growth of squirrels is 1196 squirrels per year.

39. a.

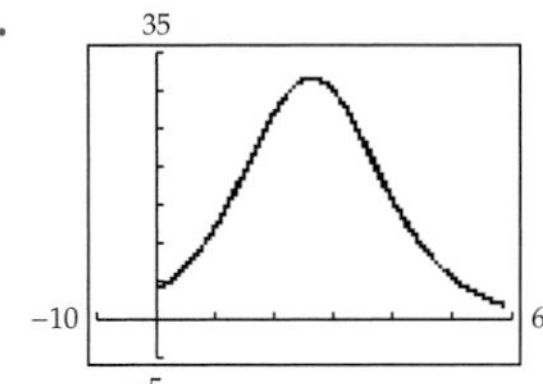

b. $\approx$ 32 bison per year **c.** 26.78 years **d.** The maximum rate of growth of bison is about 32 bison per year.

41. 45 hours **43. a.**

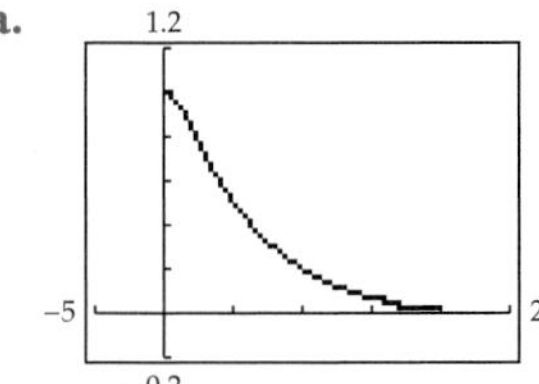

b. 0.504 **c.** 8.08 million liters **d.** As the number of liters of water per day becomes very large ($x \to \infty$), the probability of providing the water approaches 0.

45. a. 2.2 seconds **b.** The velocity approaches but never exceeds 100 feet per second. **47. a.** 0.71 **b.** 0.96 **c.** 0.52 **49. a.** 1.7% **b.** 13.9% **c.** 19.0% **d.** 19.5% **e.** 1.5% $P \to 0$ **51.** 13,715,120,270 centuries **53. a.** 286,471 gallons **b.** 250,662 gallons **c.** 34 hours **55. a.** 1.7747×10^{-4} g/m^3 **b.** 4.7510×10^{-5} g/m^3 **c.** 2.1608×10^{-5} g/m^3; 0.2 km

Exercise Set 4.7, page 375

1. Quadratic function; increasing exponential function

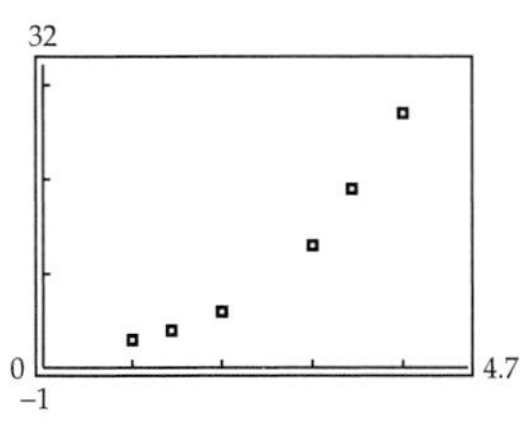

3. Quadratic function.

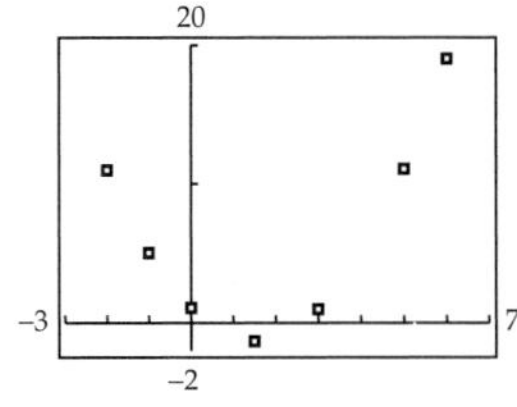

5. Linear function; quadratic function; decreasing logarithmic function or decreasing exponential function

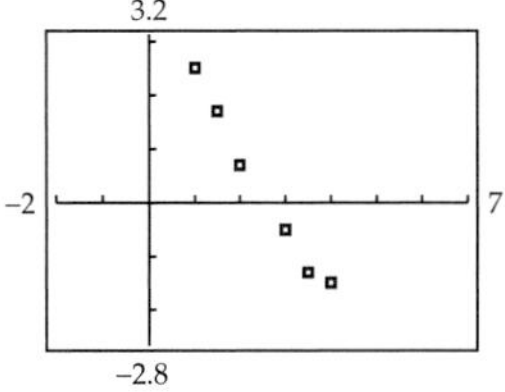

7. $y \approx 0.99628 \cdot 1.20052^x$; $r \approx 0.85705$ **9.** $y \approx 1.81505 \cdot 0.519793^x$; $r \approx -0.99978$ **11.** $y \approx 4.89060 - 1.35073 \ln x$; $r \approx -0.99921$ **13.** $y \approx 14.05858 + 1.76393 \ln x$; $r \approx 0.99983$ **15.** $y \approx \dfrac{26.01827}{1 + 2.99372e^{-0.79650x}}$ **17.** $y \approx \dfrac{799.76773}{1 + 14.33048e^{-0.75471x}}$
19. a. Linear: height $\approx 0.22448t + 58.87986$, Logarithmic: height $\approx -7.07160 + 19.17358 \ln t$ **b.** Linear model: $r \approx 0.86012$. Logarithmic model: $r \approx 0.88386$. The logarithmic model provides a slightly better fit. **c.** 83.4 inches or 6 feet 11.4 inches
21. a. Linear: Time $\approx 1.04035x + 154.96491$ Logarithmic: Time $\approx 149.56876 + 7.63077 \ln x$ **b.** Linear model: $r \approx 0.88458$. Logarithmic model: $r \approx 0.93101$. The logarithmic model provides a better fit. **c.** 2008 **23. a.** $T - 70° \approx 96.16776667 \cdot 0.93786526^t$ **b.** 35 minutes
25. a. Linear: $p \approx 0.0401t + 0.36$, Logarithmic: $p \approx -10.23519 + 3.161541 \ln t$ **b.** Linear model: $r \approx 0.99096$. Logarithmic model: $r \approx 0.99738$. The logarithmic model provides a slightly better fit. **c.** 4.48 pounds **27. a.** $p \approx 1.05258 \cdot 1.01791^t$; 2037
b. $p \approx \dfrac{11.26828}{1 + 15.89102e^{-0.029238t}}$; 11.26828 billion **c.** The logistic model is more realistic. **29. a.** $p \approx 3200 \cdot 0.91894^t$; 2012 **b.** No. The model fits the data perfectly because there are only two data points. **31.** (1.4, 13.0) **33.** (7.8, 1982.7) **35. a.** $c = m$ **b.** $b = k$ **c.** $a = \dfrac{m - P_0}{P_0}$
37. a. The x-coordinate of the first ordered pair is 0, and 0 is not in the domain of $y = \ln x$. **b.** Use a horizontal translation of the data. For instance, add 1 to each of the x-coordinates. Find the logarithmic regression function for this new data set. Remember that each x-value in the regression represents $x - 1$ in the original data set. **39. a.** Exponential: $y \approx 1.8112 \cdot 1.6174^x$; $r \approx 0.96793$ Power: $y \approx 2.0939x^{1.4025}$; $r \approx 0.99999$ **b.** The power regression provides the best fit. **41. a.** $t \approx 1.1109l^{0.50113}$ **b.** 115.4 feet

Chapter 4 True/False Exercises, page 382

1. False. $f(x) = x^2$ does not have an inverse function. **2.** False. Let $f(x) = 2x$, $g(x) = 3x$. Then $f(g(0)) = 0$ and $g(f(0))$, but f and g are not inverse functions. **3.** True **4.** True **5.** True **6.** False; because f is not defined for negative values of x, and thus $g(f(x))$ is undefined for negative values of x. **7.** False; $h(x)$ is not an increasing function for $0 < b < 1$. **8.** False; $j(x)$ is not an increasing function for $0 < b < 1$. **9.** True **10.** True **11.** True **12.** True **13.** False; $\log x + \log y = \log(xy)$. **14.** True **15.** True **16.** True

Chapter 4 Review Exercises, page 383

1. yes **3.** yes **5.** $f^{-1}(x) = \dfrac{x+4}{3}$

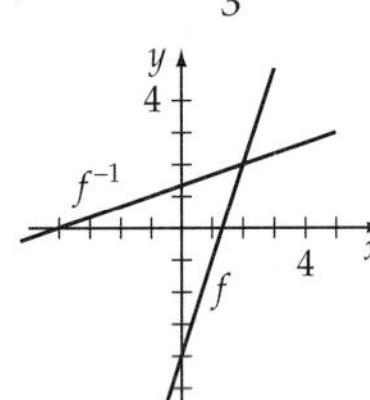

7. $h^{-1}(x) = -2x - 4$

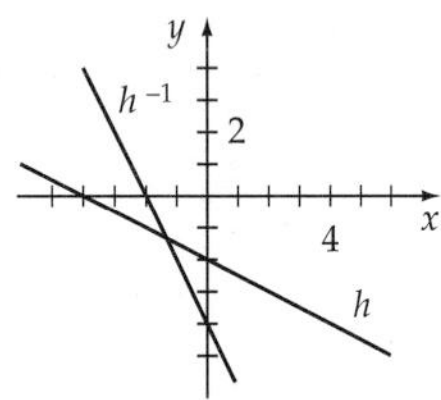

9. 2 **11.** 3 **13.** -2 **15.** -3 **17.** ± 1000 **19.** 7

21.

23.

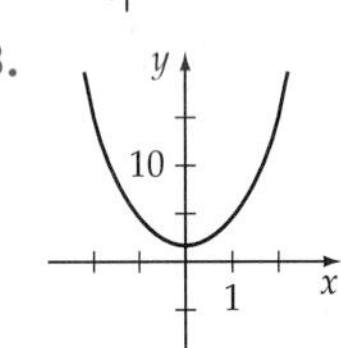

25.

27.

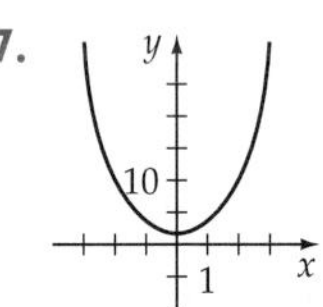

29.

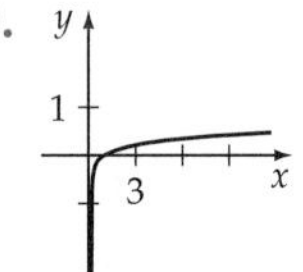

31.

33.

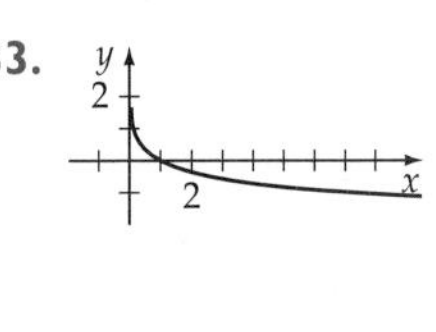

35. $4^3 = 64$ **37.** $(\sqrt{2})^4 = 4$ **39.** $\log_5 125 = 3$ **41.** $\log_{10} 1 = 0$ **43.** $2\log_b x + 3\log_b y - \log_b z$ **45.** $\ln x + 3\ln y$ **47.** $\log\left(x^2\sqrt[3]{x+1}\right)$ **49.** $\ln\dfrac{\sqrt{2xy}}{z^3}$ **51.** 2.86754 **53.** -0.117233 **55.** $\ln 30/\ln 4$ **57.** 4 **59.** 4 **61.** $\ln 3/(2\ln 4)$ **63.** 10^{1000} **65.** 1,000,005 **67.** 81 **69.** 4 **71.** 7.7 **73.** 3162 to 1 **75.** 4.2 **77. a.** \$20,323.79 **b.** \$20,339.99 **79.** \$4,438.10 **81.** $N(t) \approx e^{0.8047t}$ **83.** $N(t) \approx 3.783e^{0.0558t}$ **85. a.** $P(t) \approx 25{,}200e^{0.06155789t}$ **b.** 38,800

87. a. Linear: $P \approx -76{,}740t + 8{,}888{,}642$; $r \approx -0.96553$. Exponential: $P \approx 125{,}169{,}617(0.95504879)^t$; $r \approx -0.97604$ Logarithmic: $P \approx 34{,}707{,}491 - 7{,}271{,}279 \ln t$; $r \approx -0.96872$ **b.** The exponential regression provides the best fit to the data. **c.** 1,050,000

89. a. $P(t) \approx \dfrac{294{,}000}{210 + (1190)e^{-0.22457636 \cdot t}}$ **b.** 1070

Chapter 4 Test, page 385

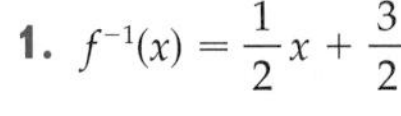

1. $f^{-1}(x) = \dfrac{1}{2}x + \dfrac{3}{2}$

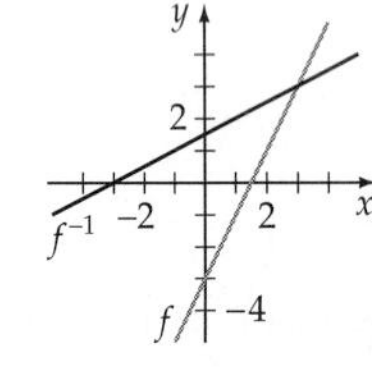

2. $f^{-1}(x) = \dfrac{8x}{4x-1}$

Domain f^{-1}: all real numbers except $\dfrac{1}{4}$

Range f^{-1}: all real numbers except 2

3.

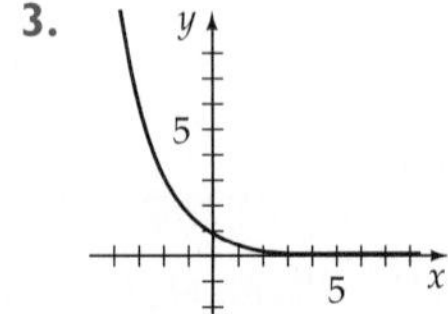

4.

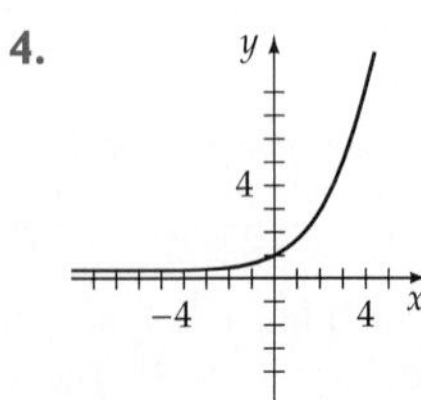

5. $b^c = 5x - 3$ **6.** $\log_3 y = x/2$ **7.** $2\log_b z - 3\log_b y - (1/2)\log_b x$ **8.** $\log\dfrac{2x+3}{(x-2)^3}$ **9.** 1.7925 **10.**

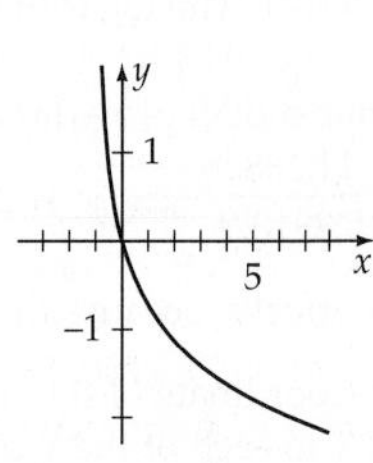

11. 1.9206 **12.** $\dfrac{\ln\left(21 \pm 2\sqrt{110}\right)}{\ln 3}$ **13.** 1 **14.** -3 **15. a.** \$29,502.36 **b.** \$29,539.62 **16. a.** 7.6 **b.** 63 to 1

17. a. $P(t) \approx 34{,}600e^{(0.04667108)t}$ **b.** 55,000 **18.** 690 years **19. a.** $y \approx 1.67199(2.47188)^x$ **b.** 1945
20. a. Interest rate $\approx 0.03939 + 0.00249 \ln t$. The predicted interest rate for 3.5 years is 4.25%. **b.** 6.4 years

Exercise Set 5.1, page 399

1. 75°, 165° **3.** 19°45′, 109°45′ **5.** 33°26′45″, 123°26′45″ **7.** $\pi/2 - 1, \pi - 1$ **9.** $\pi/4, 3\pi/4$ **11.** $0, \pi/2$ **13.** Quadrant III, 250° **15.** Quadrant II, 105° **17.** Quadrant IV, 296° **19.** 24°33′36″ **21.** 64°9′28.8″ **23.** 3°24′7.2″ **25.** 25.42° **27.** 183.56° **29.** 211.78° **31.** $\pi/6$ **33.** $\pi/2$ **35.** $11\pi/12$ **37.** $7\pi/3$ **39.** $13\pi/4$ **41.** 36° **43.** 30° **45.** 67.5° **47.** 660° **49.** 85.94° **51.** 2.32 **53.** 472.69° **55.** 4, 229.18° **57.** 2.38, 136.63° **59.** 6.28 inches **61.** 18.33 centimeters **63.** 3π **65.** $5\pi/12$ radians or 75° **67.** $\pi/30$ radians per second **69.** $5\pi/3$ radians per second **71.** $10\pi/9$ radians per second $\approx$ 3.49 radians per second **73.** 40 mph **75.** 840,000 miles **77. a.** 3.9 radians per hour **b.** 27,300 kilometers per hour **79. a.** B **b.** Both points have the same linear velocity. **81. a.** 1.15 statute miles **b.** 10% **83.** 13 square inches **85.** 4680 square centimeters **87.** 436 square meters **89. a.** 1039 meters **b.** 3.5 meters/second **91.** 18.8 feet/second **93.** $\approx$23.1 square inches **95.** 1780 miles

Exercise Set 5.2, page 410

1. $\sin\theta = 12/13$ $\csc\theta = 13/12$ $\cos\theta = 5/13$ $\sec\theta = 13/5$ $\tan\theta = 12/5$ $\cot\theta = 5/12$

3. $\sin\theta = 4/7$ $\csc\theta = 7/4$ $\cos\theta = \sqrt{33}/7$ $\sec\theta = 7\sqrt{33}/33$ $\tan\theta = 4\sqrt{33}/33$ $\cot\theta = \sqrt{33}/4$

5. $\sin\theta = 5\sqrt{29}/29$ $\csc\theta = \sqrt{29}/5$ $\cos\theta = 2\sqrt{29}/29$ $\sec\theta = \sqrt{29}/2$ $\tan\theta = 5/2$ $\cot\theta = 2/5$

7. $\sin\theta = \sqrt{21}/7$ $\csc\theta = \sqrt{21}/3$ $\cos\theta = 2\sqrt{7}/7$ $\sec\theta = \sqrt{7}/2$ $\tan\theta = \sqrt{3}/2$ $\cot\theta = 2\sqrt{3}/3$

9. $\sin\theta = 2\sqrt{30}/15$ $\csc\theta = \sqrt{30}/4$ $\cos\theta = \sqrt{105}/15$ $\sec\theta = \sqrt{105}/7$ $\tan\theta = 2\sqrt{14}/7$ $\cot\theta = \sqrt{14}/4$

11. $\sin\theta = \dfrac{\sqrt{3}}{2}$ $\csc\theta = \dfrac{2\sqrt{3}}{3}$ $\cos\theta = \dfrac{1}{2}$ $\sec\theta = 2$ $\tan\theta = \sqrt{3}$ $\cot\theta = \dfrac{\sqrt{3}}{3}$

13. $\sin\theta = \dfrac{6\sqrt{61}}{61}$ $\csc\theta = \dfrac{\sqrt{61}}{6}$ $\cos\theta = \dfrac{5\sqrt{61}}{61}$ $\sec\theta = \dfrac{\sqrt{61}}{5}$ $\tan\theta = \dfrac{6}{5}$ $\cot\theta = \dfrac{5}{6}$

15. 3/4 **17.** 4/5 **19.** 3/4 **21.** 12/13 **23.** 13/5 **25.** 3/2 **27.** $\sqrt{2}$ **29.** $-3/4$ **31.** 5/4

33. $\sqrt{3} - \sqrt{6}$ **35.** $\sqrt{3}$ **37.** $\dfrac{3\sqrt{2} + 2\sqrt{3}}{6}$ **39.** $\dfrac{3 - \sqrt{3}}{3}$ **41.** $2\sqrt{2} - \sqrt{3}$ **43.** 0.6249 **45.** 0.4488 **47.** 0.8221 **49.** 1.0053 **51.** 0.4816 **53.** 1.0729 **55.** 0.3153 **57.** 1.2331 **59.** 9.5 feet **63.** 4:28 P.M. **65.** 5.1 feet **67.** 1.7 miles **69.** 1400 feet **71.** 612 feet **73.** 5.60×10^2 feet **75. a.** 559 feet **b.** 193 feet **77.** 5.2 meters **79.** 49.9 meters $\le h \le$ 52.1 meters **81.** 8.5 feet

Exercise Set 5.3, page 421

1. $\sin\theta = 3\sqrt{13}/13$ $\csc\theta = \sqrt{13}/3$ $\cos\theta = 2\sqrt{13}/13$ $\sec\theta = \sqrt{13}/2$ $\tan\theta = 3/2$ $\cot\theta = 2/3$

3. $\sin\theta = 3\sqrt{13}/13$ $\csc\theta = \sqrt{13}/3$ $\cos\theta = -2\sqrt{13}/13$ $\sec\theta = -\sqrt{13}/2$ $\tan\theta = -3/2$ $\cot\theta = -2/3$

5. $\sin\theta = -5\sqrt{89}/89$ $\csc\theta = -\sqrt{89}/5$ $\cos\theta = -8\sqrt{89}/89$ $\sec\theta = -\sqrt{89}/8$ $\tan\theta = 5/8$ $\cot\theta = 8/5$

7. $\sin\theta = 0$ $\csc\theta$ is undefined $\cos\theta = -1$ $\sec\theta = -1$ $\tan\theta = 0$ $\cot\theta$ is undefined

9. Quadrant I **11.** Quadrant IV **13.** Quadrant III **15.** $\sqrt{3}/3$ **17.** -1 **19.** $-\sqrt{3}/3$

21. $2\sqrt{3}/3$ **23.** $-\sqrt{3}/3$ **25.** 20° **27.** 9° **29.** $\pi/5$ **31.** $\pi - 8/3$ **33.** 34° **35.** 65° **37.** $-\sqrt{2}/2$ **39.** 1 **41.** $-2\sqrt{3}/3$ **43.** $\sqrt{2}/2$ **45.** $\sqrt{2}$ **47.** cot 540° is undefined. **49.** 0.798636 **51.** -0.438371 **53.** -1.26902 **55.** -0.587785 **57.** -1.70130 **59.** -3.85522 **61.** 0 **63.** 1 **65.** $-3/2$ **67.** 1 **69.** 30°, 150° **71.** 150°, 210° **73.** 225°, 315° **75.** $3\pi/4, 7\pi/4$ **77.** $5\pi/6, 11\pi/6$ **79.** $\pi/3, 2\pi/3$ **91.** (0.2079, 0.9781) **93.** $(-0.9900, 0.1411)$ **95.** $(0.3746, -0.9272)$ **97. a.** 1 **b.** increases from 0 to 1 **c.** decreases from 1 to 0

Exercise Set 5.4, page 431

1. $(\sqrt{3}/2, 1/2)$ **3.** $(-\sqrt{3}/2, -1/2)$ **5.** $(1/2, -\sqrt{3}/2)$ **7.** $(\sqrt{3}/2, -1/2)$ **9.** $(-1, 0)$ **11.** $(-1/2, -\sqrt{3}/2)$ **13.** $-\sqrt{3}/3$ **15.** $-1/2$ **17.** $-2\sqrt{3}/3$ **19.** -1 **21.** $-2\sqrt{3}/3$ **23.** 0.9391 **25.** -1.1528 **27.** -0.2679 **29.** 0.8090 **31.** 48.0889 **33.** odd **35.** neither **37.** even **39.** odd **49.** $\sin t$ **51.** $\sec t$ **53.** $-\tan^2 t$ **55.** $-\cot t$ **57.** $\cos^2 t$ **59.** $2\csc^2 t$ **61.** $\csc^2 t$ **63.** 1 **65.** $\sqrt{1 - \cos^2 t}$ **67.** $\sqrt{1 + \cot^2 t}$ **69.** 750 miles **71.** $-\sin^2 t/\cos t$ **73.** $\csc t \sec t$ **75.** $1 - 2\sin t + \sin^2 t$ **77.** $1 - 2\sin t\cos t$ **79.** $\cos^2 t$ **81.** $2\csc t$ **83.** $(\cos t - \sin t)(\cos t + \sin t)$ **85.** $(\tan t + 2)(\tan t - 3)$ **87.** $(2\sin t + 1)(\sin t - 1)$ **89.** $(\cos t - \sin t)(\cos t + \sin t)$ **91.** $\sqrt{2}/2$ **93.** $-\sqrt{3}/3$

Exercise Set 5.5, page 440

1. $2, 2\pi$ **3.** $1, \pi$ **5.** $1/2, 1$ **7.** $2, 4\pi$ **9.** $1/2, 2\pi$ **11.** $1, 8\pi$ **13.** $2, 6$ **15.** $3, 3\pi$ **17.** **19.**

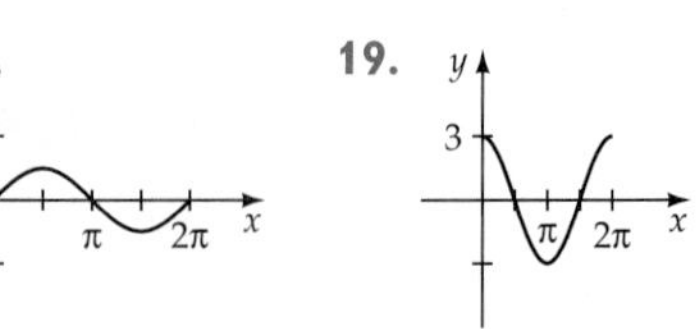

21.

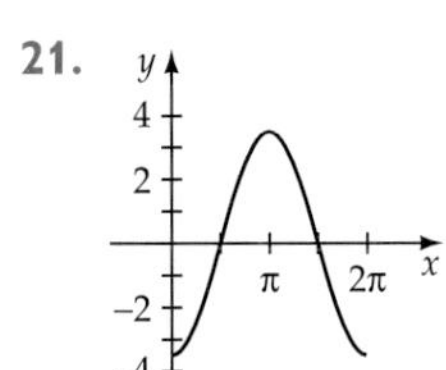

23.

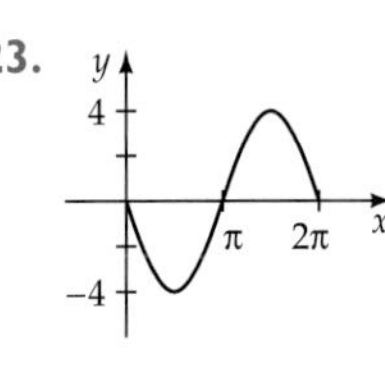

25.

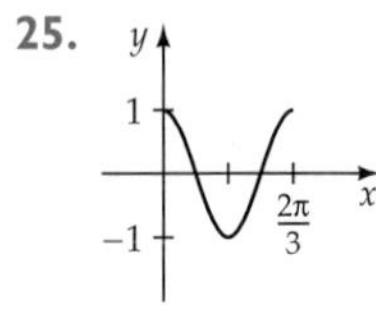

27.

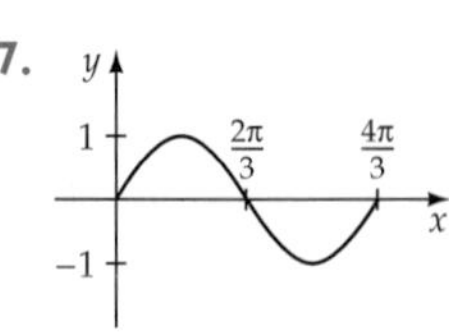

29.

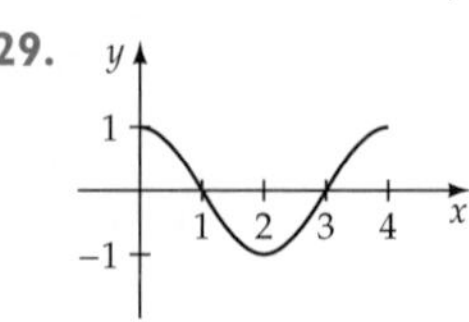

31.

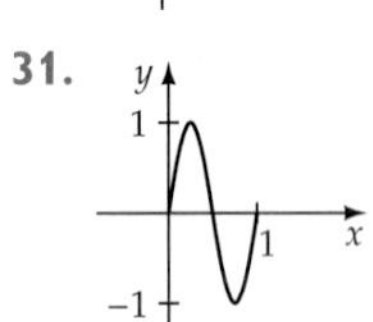

33.

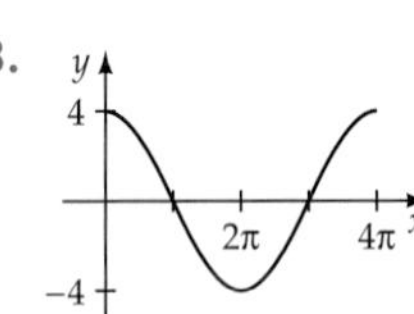

35.

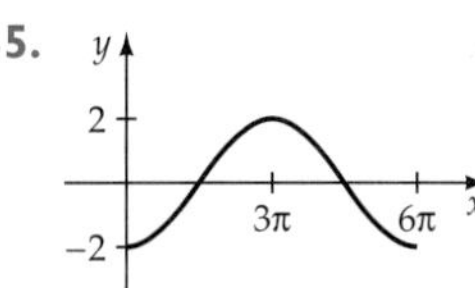

37.

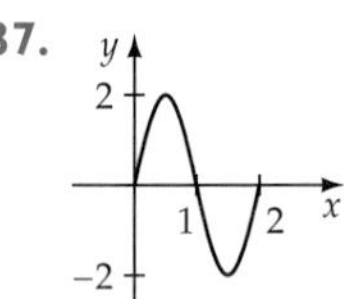

39.

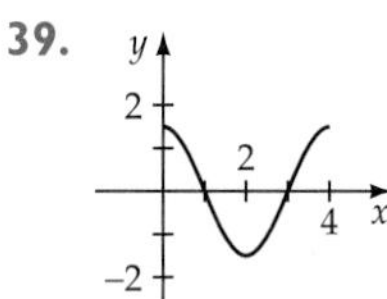

41.

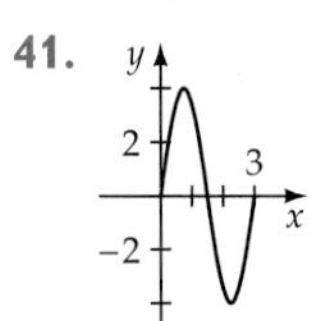

43.

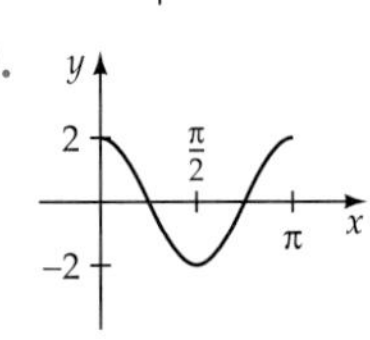

45.

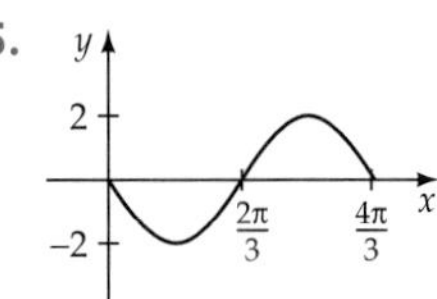

47.

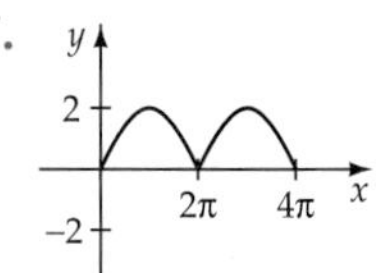

49.

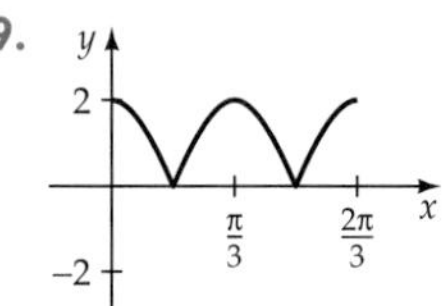

51.

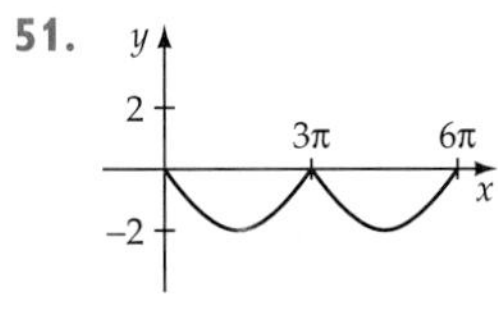

53.

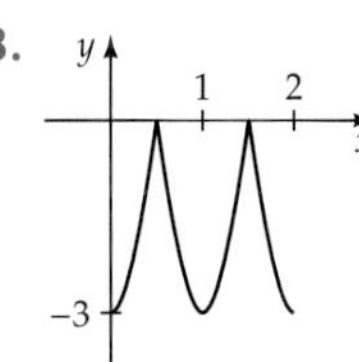

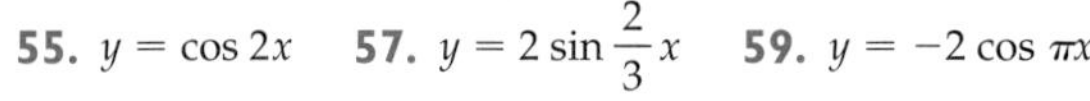

55. $y = \cos 2x$ **57.** $y = 2 \sin \frac{2}{3}x$ **59.** $y = -2 \cos \pi x$

61.

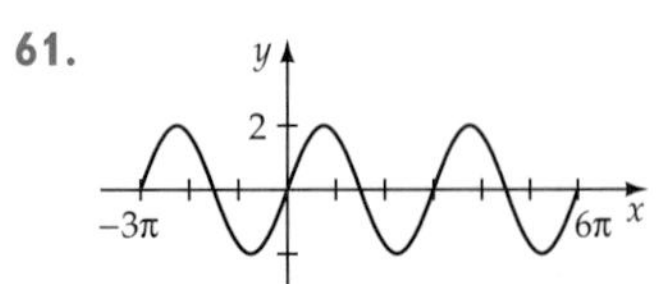

63.

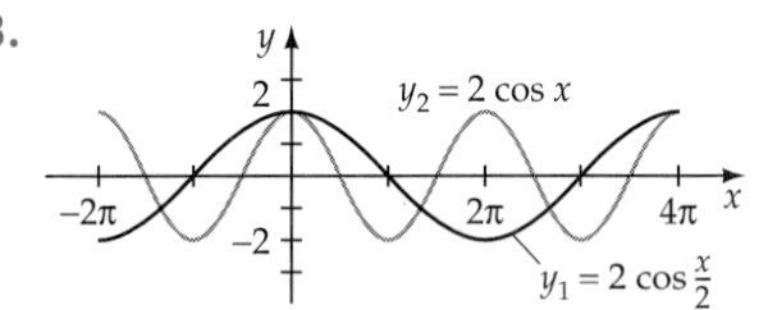

65.

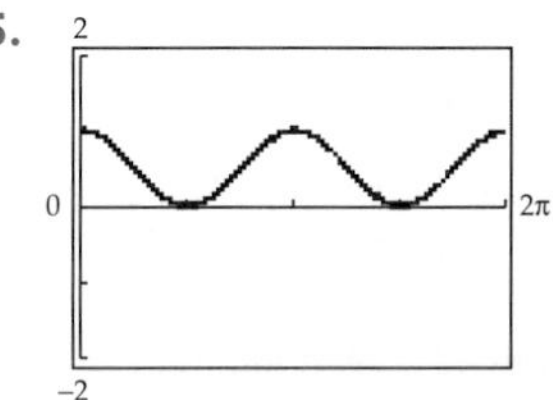

67.

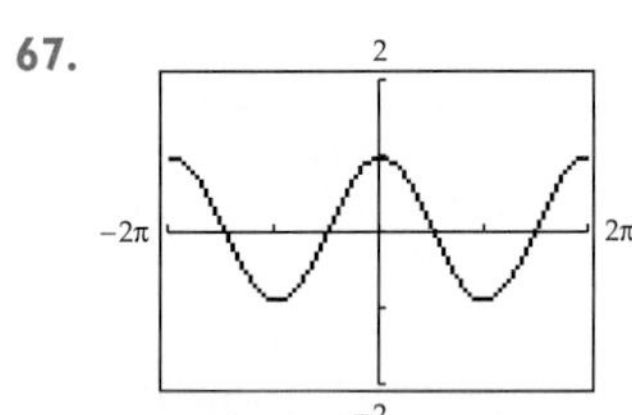

69.

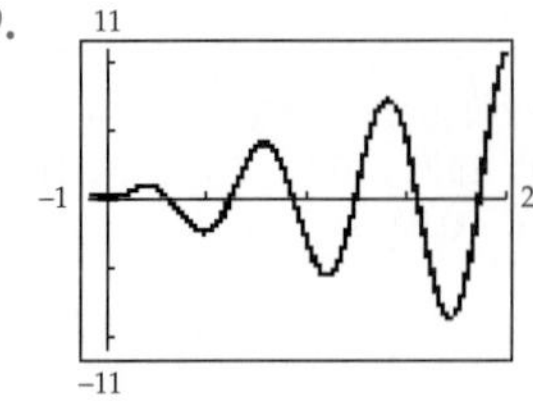

71.

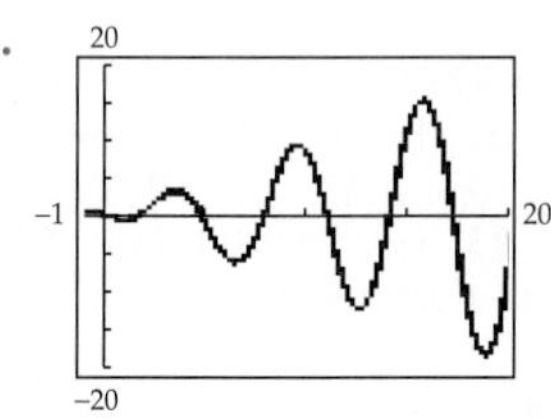

73.

x	-0.1	-0.05	-0.01	-0.001	0.001	0.01	0.05	0.1
$\frac{\sin x}{x}$	0.99833	0.99958	0.99998	0.99999	0.99999	0.99998	0.99958	0.99833

The graph of f approaches 1 as x approaches 0. The graph does not have a vertical asymptote at $x = 0$.

75.

max $= e$ min $= 1/e \approx 0.3679$
period $= 2\pi$

77. $y = 2\sin\frac{2}{3}x$ **79.** $y = 4\sin\pi x$ **81.** $y = 3\cos 4x$ **83.** $y = 3\cos\frac{4\pi}{5}x$

85. $a = 60, p = 20, b = \frac{\pi}{10}$

$f(t) = 60\cos\frac{\pi}{10}t$

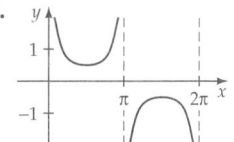

87. $S(t) = 4\sin 3t$, 3.1 feet

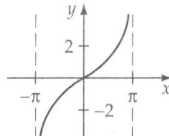

Exercise Set 5.6, page 448

1. $\pi/2 + k\pi$, k an integer **3.** $\pi/2 + k\pi$, k an integer **5.** 2π **7.** π **9.** 2π **11.** $2\pi/3$ **13.** $\pi/3$ **15.** 8π **17.** 1 **19.** 4

21. **23.** **25.** **27.** **29.**

31.

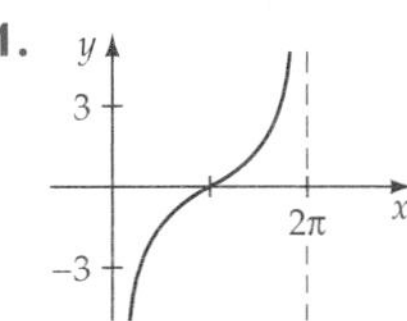

33.

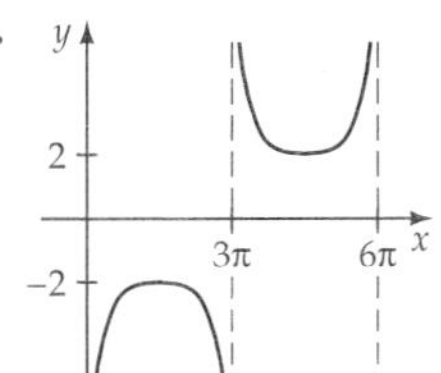

35.

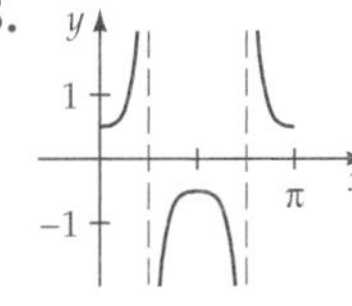

37.

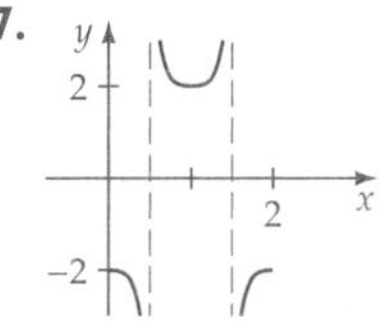

39.

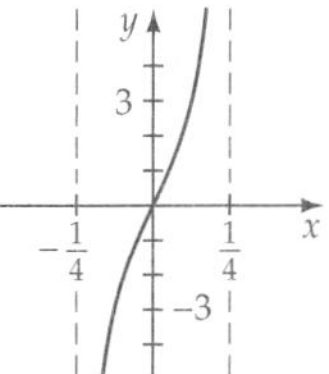

41.

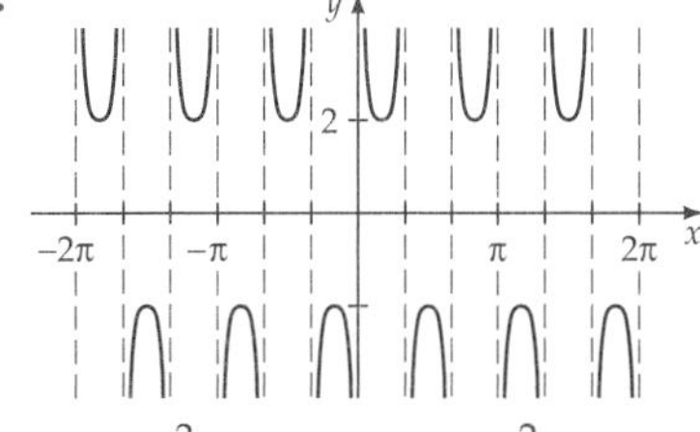

43.

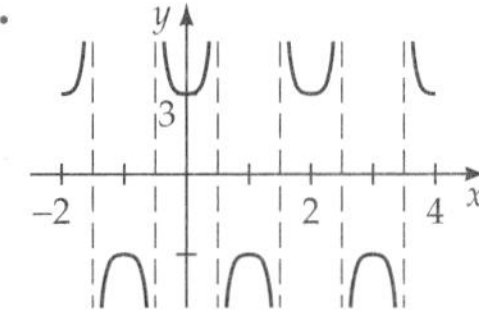

45.

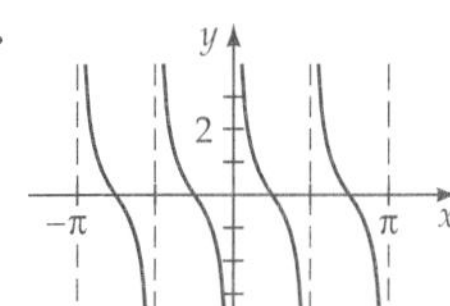

47.

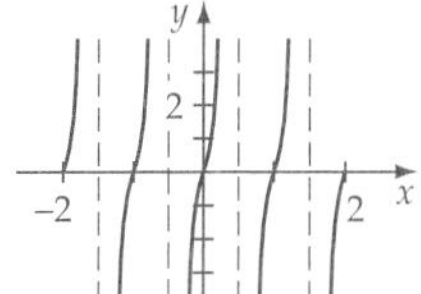

49. $y = \cot\frac{3}{2}x$ **51.** $y = \csc\frac{2}{3}x$ **53.** $y = \sec\frac{3}{4}x$ **55.**

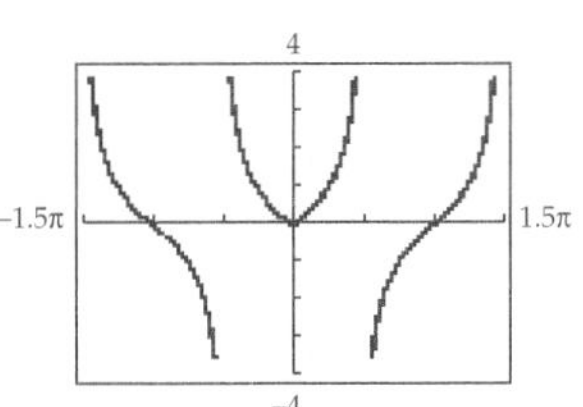

57.

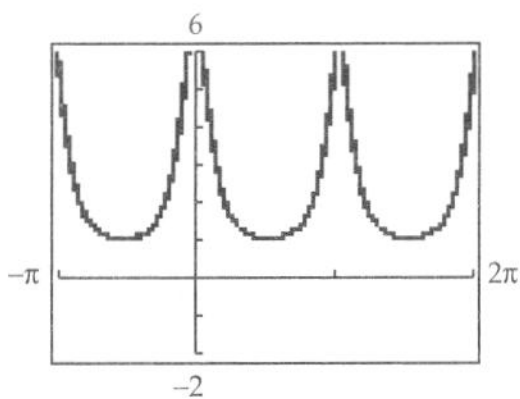

59.

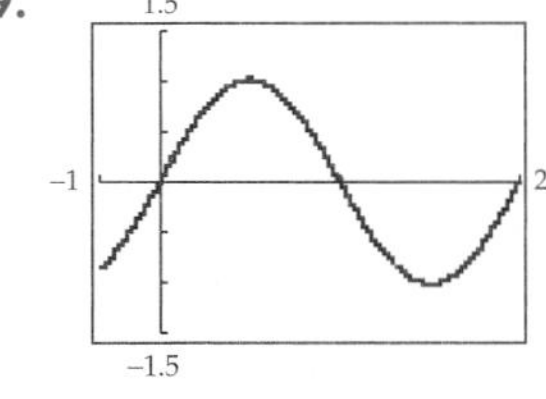

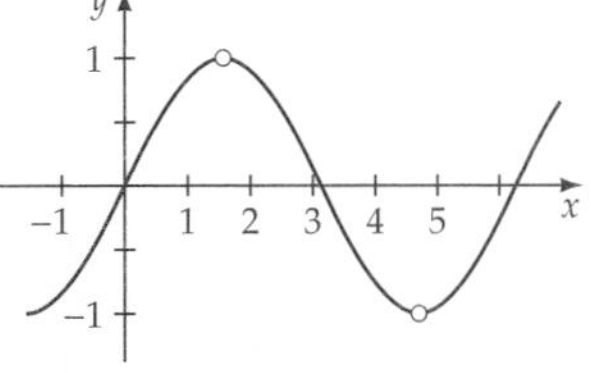

Note: The display screen at the left fails to show that on $[-1, 2\pi]$ the function is undefined at $x = \frac{\pi}{2}$ and $x = \frac{3\pi}{2}$.

61.

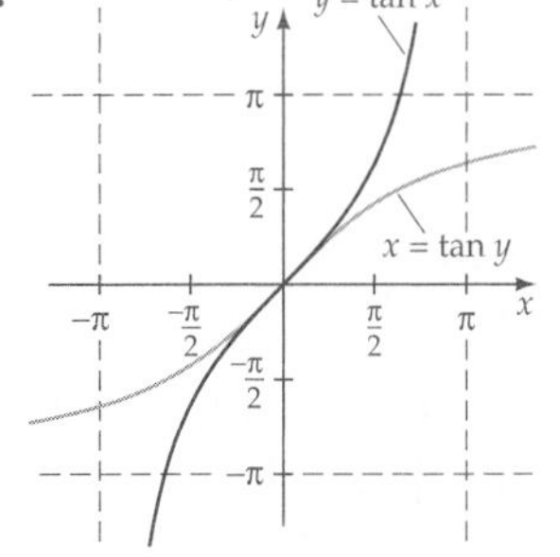

63. $y = \tan 3x$

65. $y = \sec\frac{8}{3}x$ **67.** $y = \cot\frac{\pi}{2}x$ **69.** $y = \csc\frac{4\pi}{3}x$

Exercise 5.7, page 456

1. 2, $\pi/2$, 2π **3.** 1, $\pi/8$, π **5.** 4, $-\pi/4$, 3π **7.** 5/4, $2\pi/3$, $2\pi/3$ **9.** $\pi/8$, $\pi/2$ **11.** -3π, 6π **13.** $\pi/16$, π **15.** -12π, 4π

17. **19.** **21.** **23.**

25. **27.** **29.** **31.**

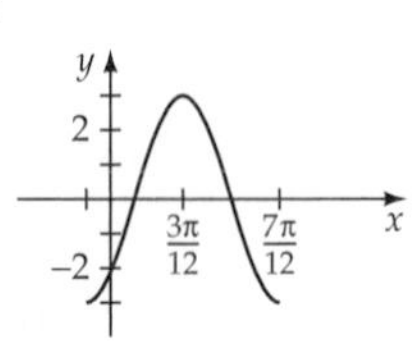

33. **35.** **37.** **39.**

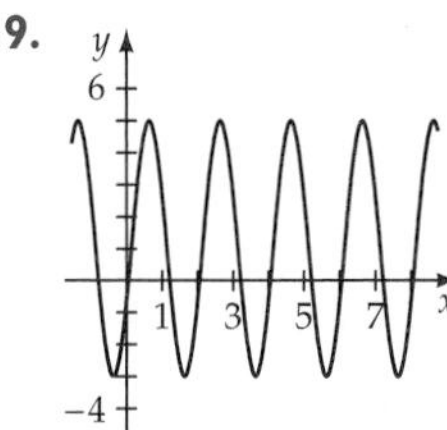

41.

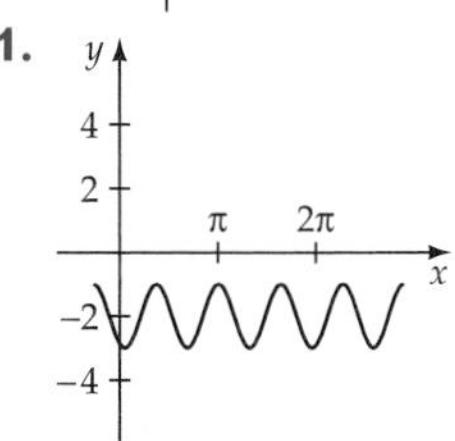

43. **45.**

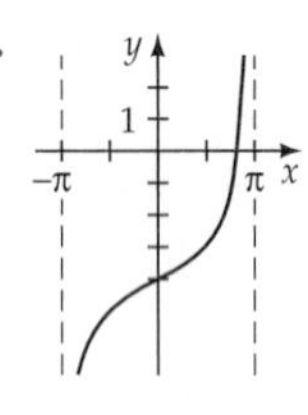

47.

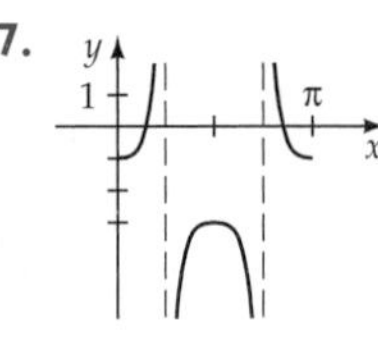

49.

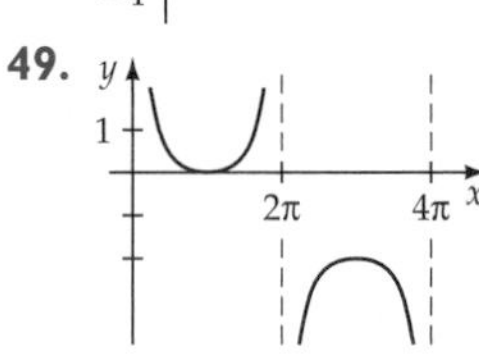

51. a. 7.5 months, 12 months

b.

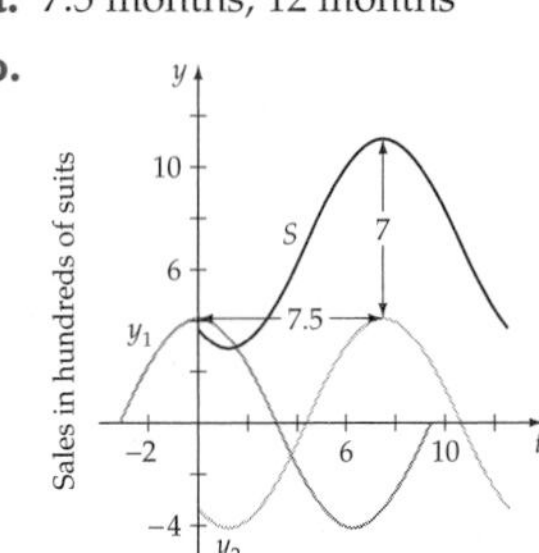

c. August

53. **55.**

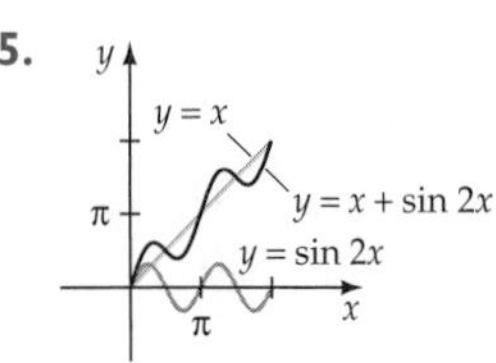

57.

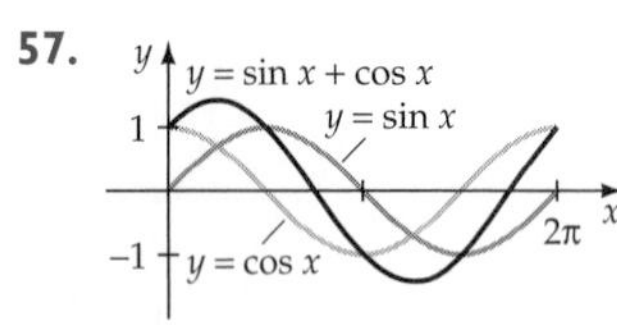

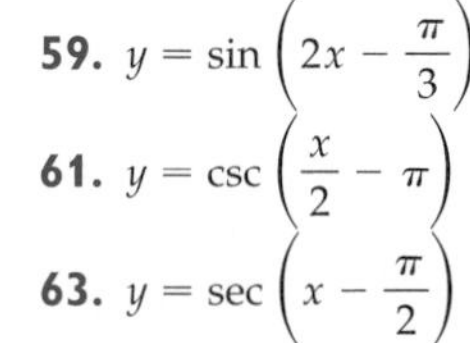

59. $y = \sin\left(2x - \dfrac{\pi}{3}\right)$

61. $y = \csc\left(\dfrac{x}{2} - \pi\right)$

63. $y = \sec\left(x - \dfrac{\pi}{2}\right)$

65. $\approx$ 25 ppm **67.** $s = 7\cos 10\pi t + 5$ **69.** $s = 400\tan\dfrac{\pi}{5}t$, t in seconds **71.** $y = 3\cos\dfrac{\pi}{6}t + 9$, $y = 12$ at 6:00 P.M.

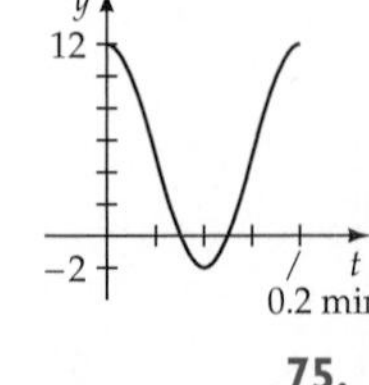

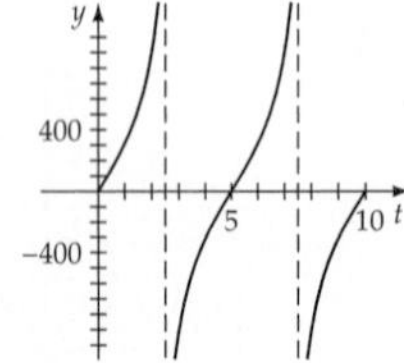

73.

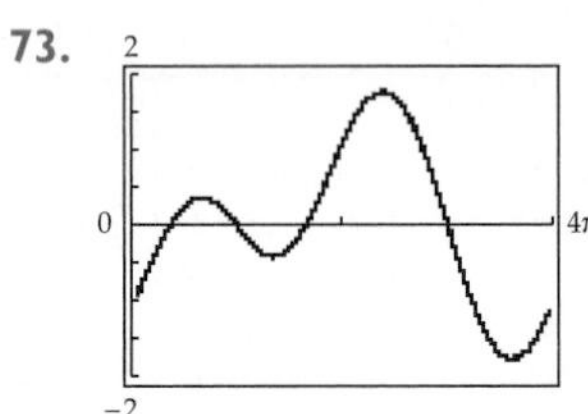

75.

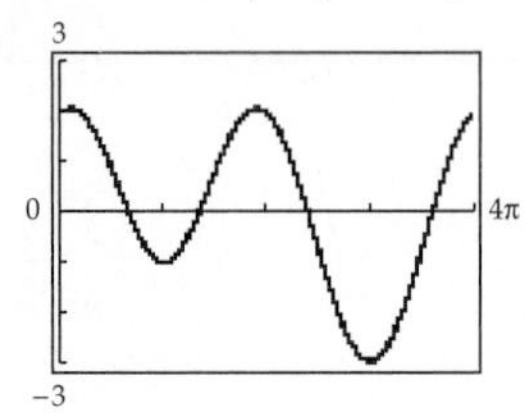

77.

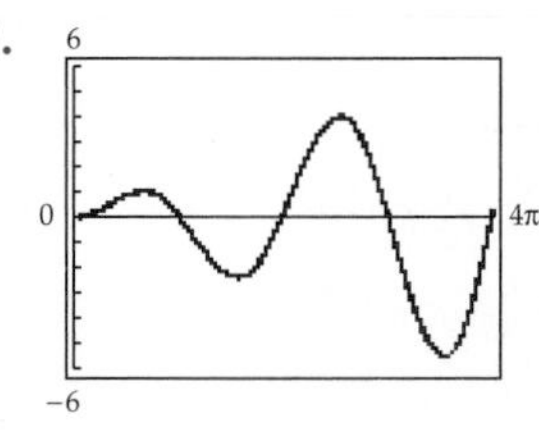

79.

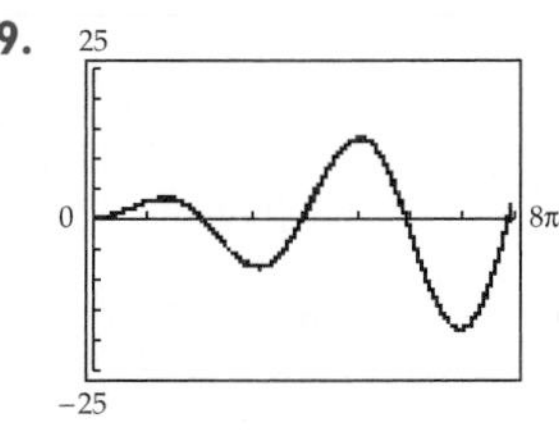

81. 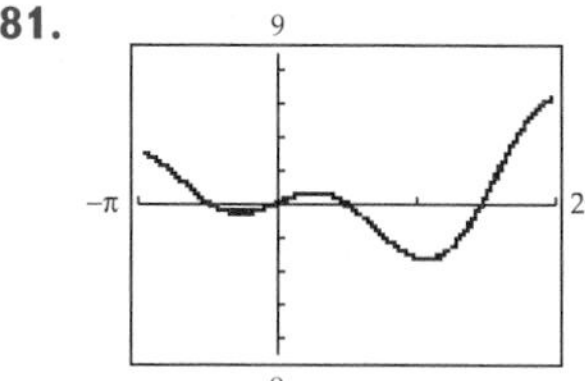

83. 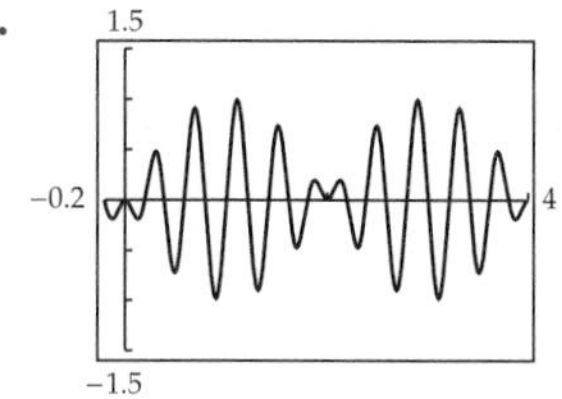

85. 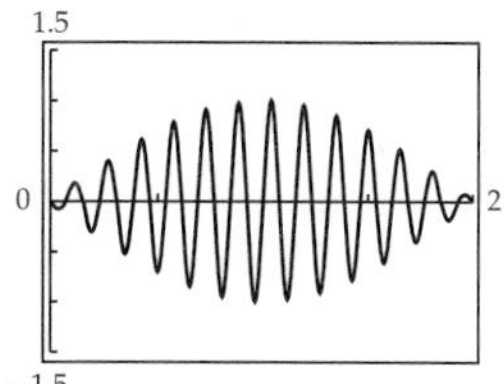

87. $y = 2\sin(2x - 2\pi/3)$
89. $y = \tan(x/2 - \pi/4)$
91. $y = \sec(x/2 - 3\pi/8)$
93. 1
95. $\cos^2 x + 2$

97. 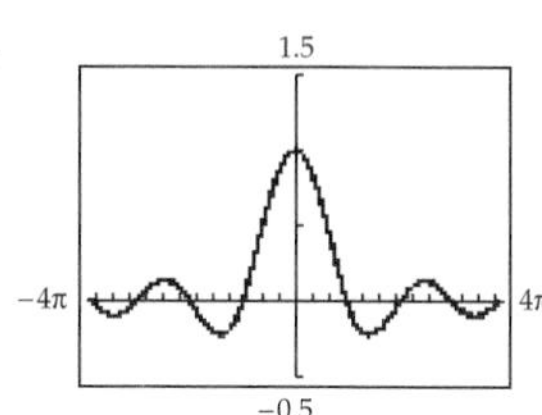

The graph above does not show that the function is undefined at $x = 0$.

99. 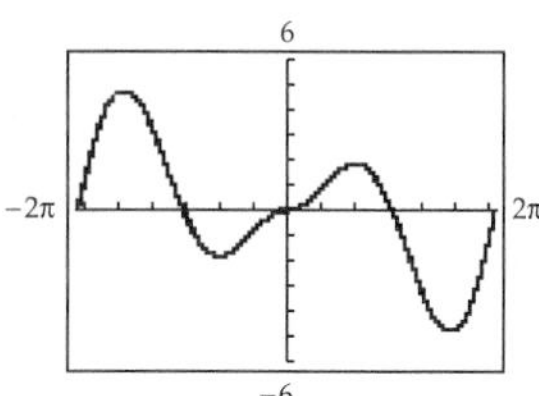

Exercise Set 5.8, page 464

1. $2, \pi, 1/\pi$ **3.** $3, 3\pi, 1/(3\pi)$ **5.** $4, 2, 1/2$ **7.** $3/4, 4, 1/4$

9. $y = 4\cos 3\pi t$

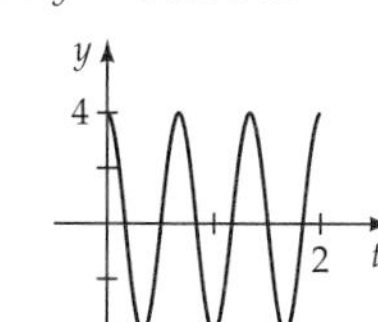

11. $y = \dfrac{3}{2}\cos\dfrac{4\pi}{3}t$

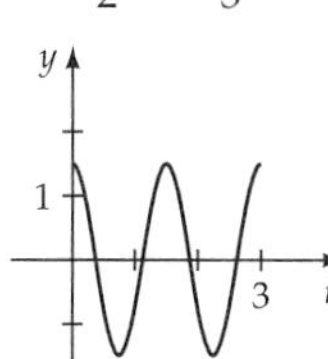

13. $y = 2\sin 2t$

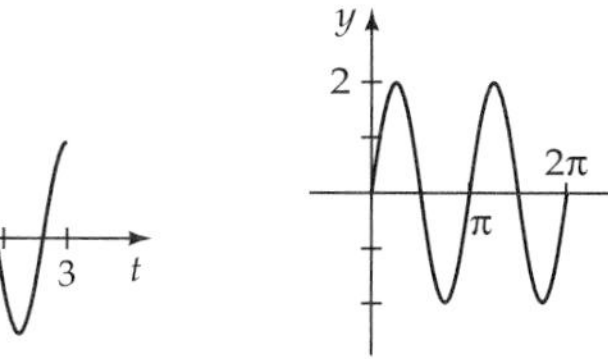

15. $y = \sin \pi t$

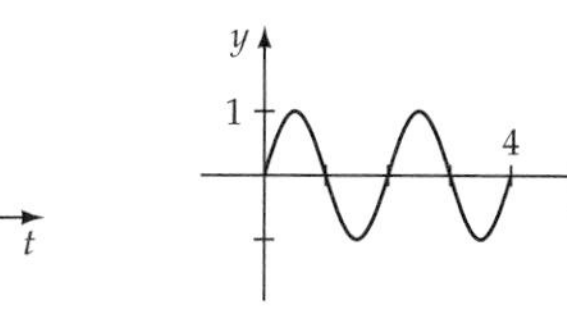

17. $y = 2\sin 2\pi t$

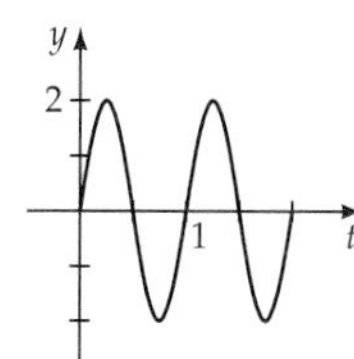

19. $y = \dfrac{1}{2}\cos 4t$ **21.** $y = 2.5\cos \pi t$ **23.** $y = \dfrac{1}{2}\cos\dfrac{2\pi}{3}t$ **25.** $y = 4\cos 4t$ **27.** $4\pi, 1/(4\pi)$, 2 feet; $y = -2\cos\dfrac{t}{2}$
29. **a.** 3 **b.** 59.8 seconds **31.** **a.** 10 **b.** 71.0 seconds **33.** **a.** 10 **b.** 9.1 seconds **35.** **a.** 10 **b.** 6.1 seconds
37. The new period is 3 times the original period. **39.** yes **41.** yes

Chapter 5 True/False Exercises, page 467

1. False; the initial side must be along the positive x-axis. **2.** True **3.** True **4.** False; in the third quadrant $\cos\theta < 0$ and $\tan\theta > 0$. **5.** False; $\sec^2\theta - \tan^2\theta = 1$ is an identity. **6.** False; the tangent function has no amplitude. **7.** False; the period is 2π. **8.** True **9.** False; $\sin 45° + \cos(90° - 45°) = \sin 45° + \cos 45° = \sqrt{2}/2 + \sqrt{2}/2 = \sqrt{2}$. **10.** False; $\sin(\pi/2 + \pi/2) = \sin\pi = 0$, and $\sin\pi/2 + \sin\pi/2 = 1 + 1 = 2$. **11.** False; $\sin^2\pi/6 = (1/2)^2 = 1/4$, and $\sin(\pi/6)^2 = \sin\pi^2/36 \approx 0.2707$. **12.** False; the phase shift is $\dfrac{\pi/3}{2} = \dfrac{\pi}{6}$. **13.** True **14.** False; 1 radian $\approx 57.3°$. **15.** False; the graph lies on or between the graphs of $y = 2^{-x}$ and $y = -2^{-x}$. **16.** False; $|f(t)| \to 0$ as $t \to \infty$.

Chapter 5 Review Exercises, page 467

1. complement measures 25°; supplement measures 115° **3.** 114.59° **5.** 3.93 meters **7.** 55 radians per second **9.** $\sqrt{5}/2$ **11.** $3\sqrt{5}/5$ **13.** **a.** $-2\sqrt{3}/3$ **b.** 1 **c.** -1 **d.** $-1/2$ **15.** **a.** $-1/2$ **b.** $\sqrt{3}/3$ **17.** **a.** $\sqrt{2}/2$ **b.** -1 **19.** even **23.** $\tan\phi$ **25.** $\tan^2\phi$ **27.** 0 **29.** no amplitude, period $\pi/3$, phase shift 0 **31.** amplitude 1, period π, phase shift $\pi/3$ **33.** no amplitude, period 2π, phase shift $\pi/4$

35. 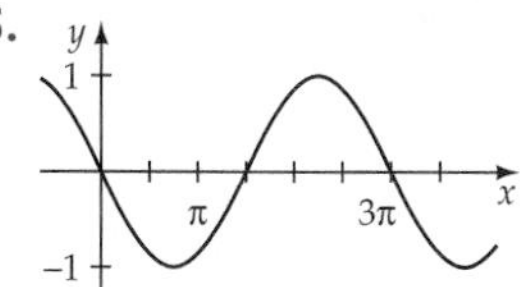

37. 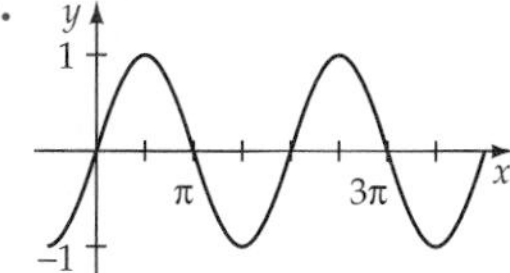

39. 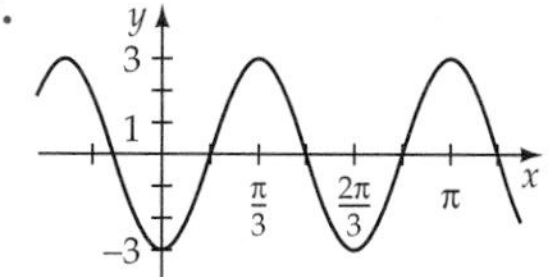

41. 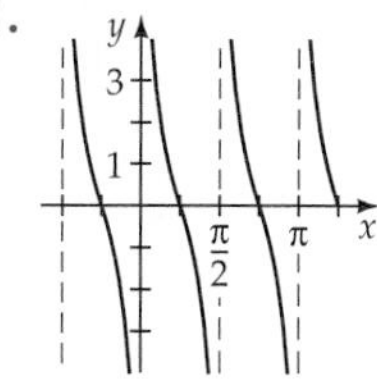

43. 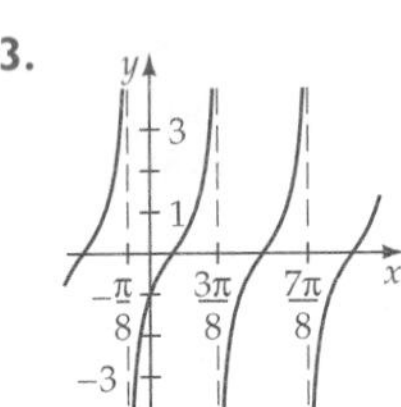 **45.** 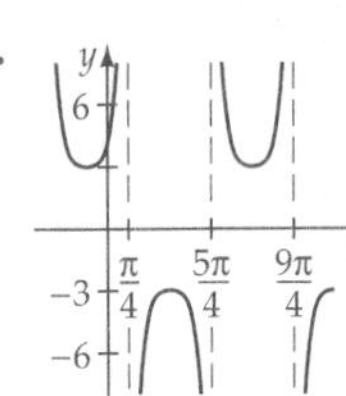 **47.** 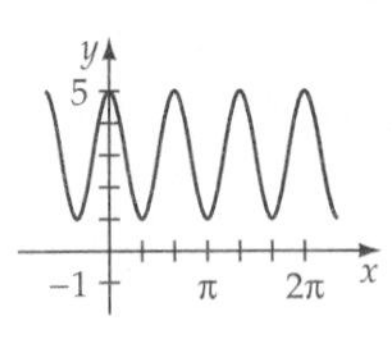 **49.** 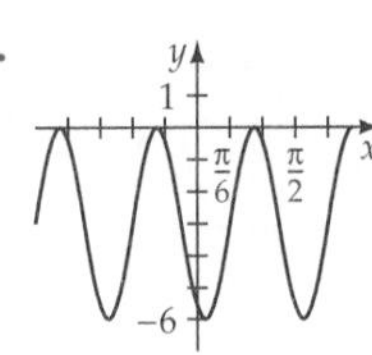 **51.** 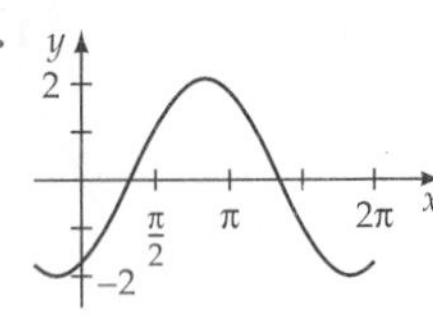

53. 12.3 feet **55.** 46 feet **57.** amplitude = 0.5, $f = \frac{1}{\pi}$, $p = \pi$, $y = -0.5 \cos 2t$

Chapter 5 Test, page 469

1. $5\pi/6$ **2.** $\pi/12$ **3.** 13.1 centimeters **4.** 12π radians/second **5.** 80 centimeters/second **6.** $\sqrt{58}/7$ **7.** 1.0864 **8.** $(\sqrt{3} - 6)/6$ **9.** $(\sqrt{3}/2, -1/2)$ **10.** $\sin^2 t$ **11.** $\pi/3$ **12.** amplitude 3, period π, phase shift $-\pi/4$ **13.** period 3, phase shift $-\frac{1}{2}$

14. 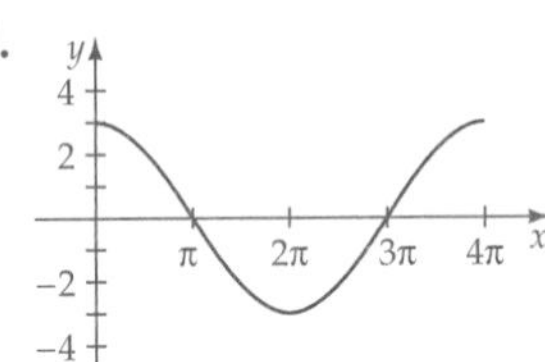 **15.** 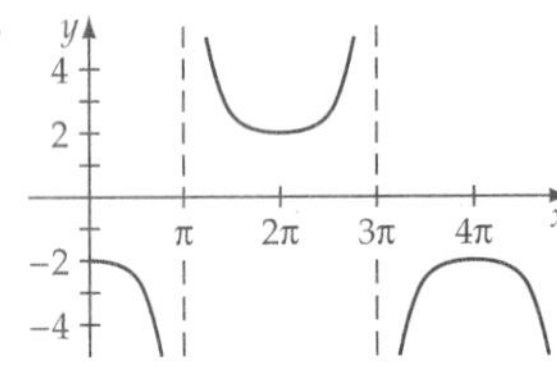

16. Shift the graph [of $y = 2 \sin (2x)$] $\frac{\pi}{4}$ units to the right and down 1 unit.

17. 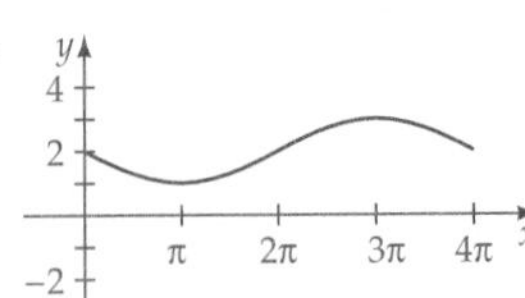 **18.** 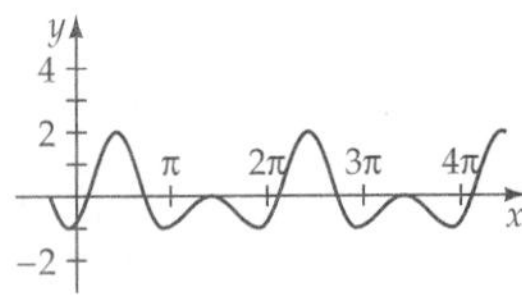

19. 25.5 meters **20.** $y = 13 \sin \frac{2\pi}{5} t$

Exercise Set 6.1, page 476

1. Not an identity. If $x = \pi/4$, the left side is 2 and the right side is 1. **3.** Not an identity. If $x = 0°$, the left side is $\sqrt{3}/2$ and the right side is $(2 + \sqrt{3})/2$. **5.** Not an identity. If $x = 0$, the left side is -1 and the right side is 1. **7.** Not an identity. If $x = 0$, the left side is -1 and the right side is 1. **9.** Not an identity. The left side is 1 and the right side is $\sqrt{3}/2$. **11.** Not an identity. If $x = \pi/2$, the left side is 1 and the right side is 2. **67.** $\cos x = \pm\sqrt{1 - \sin^2 x}$ **69.** $\sec x = \pm \frac{\sqrt{1 - \sin^2 x}}{1 - \sin^2 x}$

Each of the graphs shown in Exercises 71, 73, 75, and 77 is the graph of both the left and the right side of the respective equation.

71. 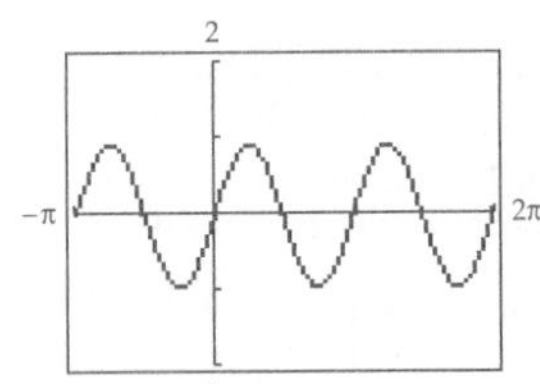 **73.** 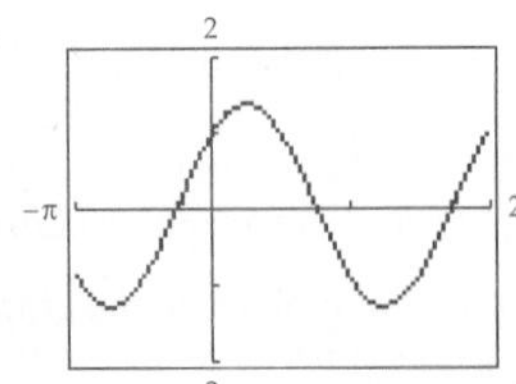 **75.** 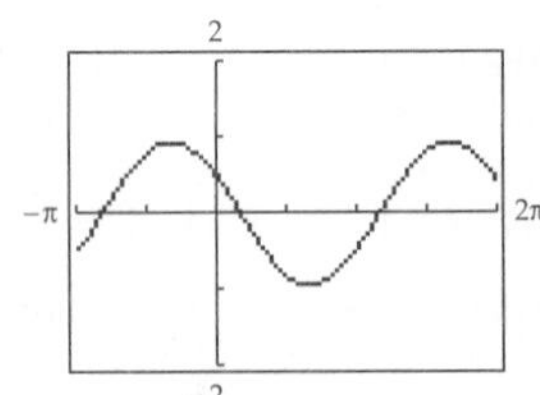 **77.** 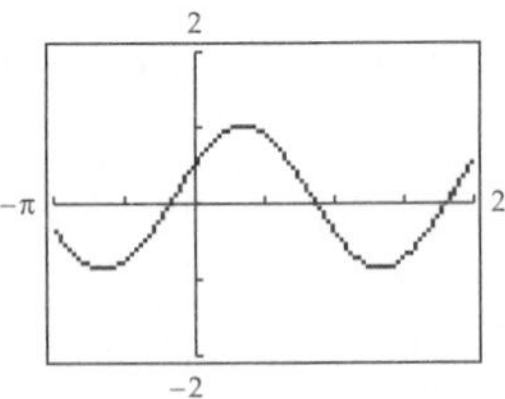

Exercise Set 6.2, page 484

1. $\frac{\sqrt{6} + \sqrt{2}}{4}$ **3.** $\frac{\sqrt{6} + \sqrt{2}}{4}$ **5.** $2 - \sqrt{3}$ **7.** $\frac{-\sqrt{6} + \sqrt{2}}{4}$ **9.** $-\frac{\sqrt{6} + \sqrt{2}}{4}$ **11.** $2 + \sqrt{3}$ **13.** 0 **15.** 1/2 **17.** $\sqrt{3}$ **19.** $\sin 5x$ **21.** $\cos x$ **23.** $\sin 4x$ **25.** $\cos 2x$ **27.** $\sin x$ **29.** $\tan 7x$ **31. a.** $-77/85$ **b.** 84/85 **c.** 77/36 **33. a.** $-63/65$ **b.** $-56/65$ **c.** $-63/16$ **35. a.** 63/65 **b.** 56/65 **c.** 33/56 **37. a.** $-77/85$ **b.** $-84/85$ **c.** $-13/84$ **39. a.** $-33/65$ **b.** $-16/65$ **c.** 63/16 **41. a.** $-56/65$ **b.** $-63/65$ **c.** 16/63 **67.** $\cos(\theta + 3\pi) = -\cos\theta$ **69.** $\tan(\theta + \pi) = \tan\theta$ **71.** $\sin(\theta + 2k\pi) = \sin\theta$

73. $y = \sin\left(\frac{\pi}{2} - x\right)$ and $y = \cos x$ both have the following graph.

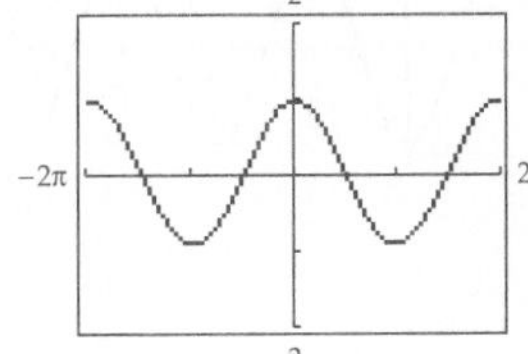

75. $y = \sin 7x \cos 2x - \cos 7x \sin 2x$ and $y = \sin 5x$ both have the following graph.

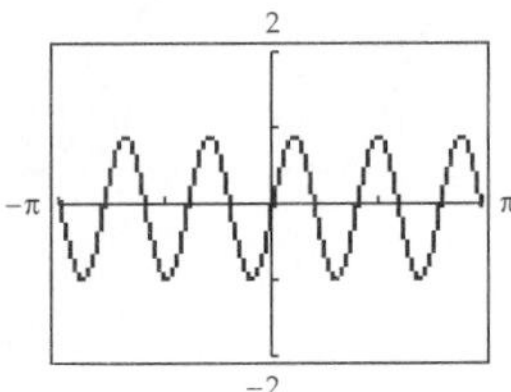

Exercise Set 6.3, page 491

1. $\sin 4\alpha$ **3.** $\cos 10\beta$ **5.** $\cos 6\alpha$ **7.** $\tan 6\alpha$ **9.** $\dfrac{\sqrt{2+\sqrt{3}}}{2}$ **11.** $\sqrt{2}+1$ **13.** $-\dfrac{\sqrt{2+\sqrt{2}}}{2}$ **15.** $\dfrac{\sqrt{2-\sqrt{2}}}{2}$ **17.** $\dfrac{\sqrt{2-\sqrt{2}}}{2}$ **19.** $\dfrac{\sqrt{2-\sqrt{3}}}{2}$ **21.** $-2-\sqrt{3}$ **23.** $\dfrac{\sqrt{2+\sqrt{3}}}{2}$ **25.** $\sin 2\theta = -24/25$, $\cos 2\theta = 7/25$, $\tan 2\theta = -24/7$
27. $\sin 2\theta = -240/289$, $\cos 2\theta = 161/289$, $\tan 2\theta = -240/161$ **29.** $\sin 2\theta = -336/625$, $\cos 2\theta = -527/625$, $\tan 2\theta = 336/527$
31. $\sin 2\theta = 240/289$, $\cos 2\theta = -161/289$, $\tan 2\theta = -240/161$ **33.** $\sin 2\theta = -720/1681$, $\cos 2\theta = 1519/1681$, $\tan 2\theta = -720/1519$
35. $\sin 2\theta = 240/289$, $\cos 2\theta = -161/289$, $\tan 2\theta = -240/161$ **37.** $\sin \alpha/2 = 5\sqrt{26}/26$, $\cos \alpha/2 = \sqrt{26}/26$, $\tan \alpha/2 = 5$
39. $\sin \alpha/2 = 5\sqrt{34}/34$, $\cos \alpha/2 = -3\sqrt{34}/34$, $\tan \alpha/2 = -5/3$ **41.** $\sin \alpha/2 = \sqrt{5}/5$, $\cos \alpha/2 = 2\sqrt{5}/5$, $\tan \alpha/2 = 1/2$
43. $\sin \alpha/2 = \sqrt{2}/10$, $\cos \alpha/2 = -7\sqrt{2}/10$, $\tan \alpha/2 = -1/7$ **45.** $\sin \alpha/2 = \sqrt{17}/17$, $\cos \alpha/2 = 4\sqrt{17}/17$, $\tan \alpha/2 = 1/4$
47. $\sin \alpha/2 = 5\sqrt{34}/34$, $\cos \alpha/2 = -3\sqrt{34}/34$ $\tan \alpha/2 = -5/3$

95. $y = \sin^2 x + \cos 2x$ and $y = \cos^2 x$ both have the following graph.

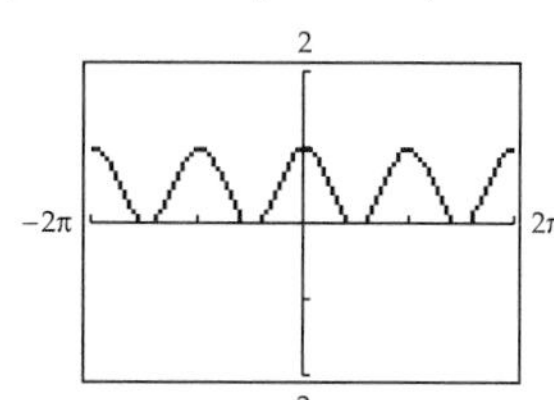

97. $y = 2\sin\dfrac{x}{2}\cos\dfrac{x}{2}$ and $y = \sin x$ both have the following graph.

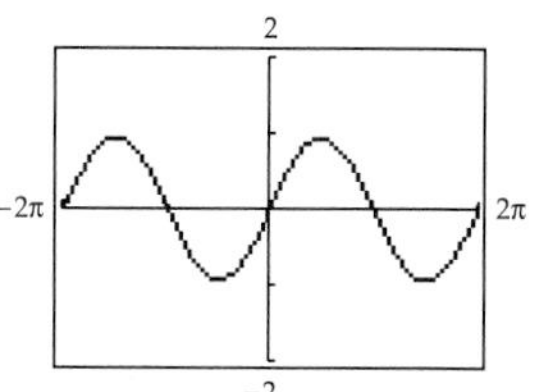

Exercise Set 6.4, page 500

1. $\sin 3x - \sin x$ **3.** $\dfrac{1}{2}[\sin 8x - \sin 4x]$ **5.** $\sin 8x + \sin 2x$ **7.** $\dfrac{1}{2}[\cos 4x - \cos 6x]$ **9.** $1/4$ **11.** $-\sqrt{2}/4$ **13.** $-1/4$ **15.** $\dfrac{\sqrt{3}-2}{4}$
17. $2\sin 3\theta\cos\theta$ **19.** $2\cos 2\theta\cos\theta$ **21.** $-2\sin 4\theta\sin 2\theta$ **23.** $2\cos 4\theta\cos 3\theta$ **25.** $2\sin 7\theta\cos 2\theta$ **27.** $-2\sin\dfrac{3}{2}\theta\sin\dfrac{1}{2}\theta$
29. $2\sin\dfrac{3}{4}\theta\sin\dfrac{\theta}{4}$ **31.** $2\cos\dfrac{5}{12}\theta\sin\dfrac{1}{12}\theta$ **49.** $y = \sqrt{2}\sin(x - 135°)$ **51.** $y = \sin(x - 60°)$ **53.** $y = \dfrac{\sqrt{2}}{2}\sin(x - 45°)$
55. $y = 3\sqrt{2}\sin(x + 135°)$ **57.** $y = \pi\sqrt{2}\sin(x - 45°)$ **59.** $y = \sqrt{2}\sin(x + 3\pi/4)$ **61.** $y = \sin(x + \pi/6)$ **63.** $y = 20\sin\left(x + \dfrac{2\pi}{3}\right)$
65. $y = 5\sqrt{2}\sin(x + 3\pi/4)$

67. y; 2; $\frac{8\pi}{3}$; $\frac{2\pi}{3}$; x **69.** y; 2; $\frac{7\pi}{4}$; $-\frac{\pi}{4}$; x **71.** y; 2; $\frac{17\pi}{6}$; $\frac{5\pi}{6}$; x **73.** y; 10; $-\frac{2\pi}{3}$; $\frac{4\pi}{3}$; −10; x **75.** y; 12; −6; $\frac{\pi}{6}$; $\frac{13\pi}{6}$; x

Each of the graphs shown in Exercises 77, 79, and 81 is the graph of both the left and the right sides of the respective equation.

77.

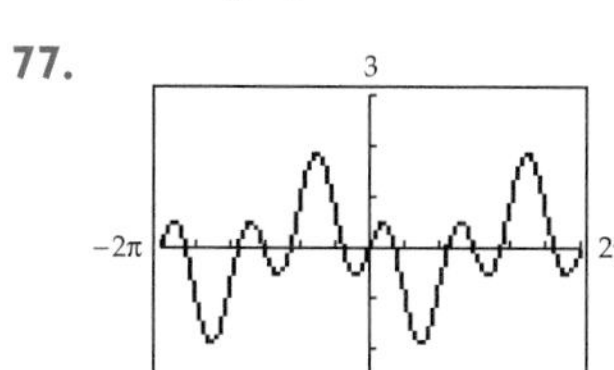

79.

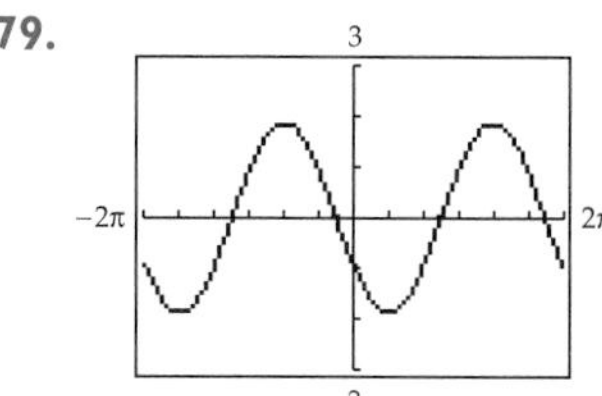

81.

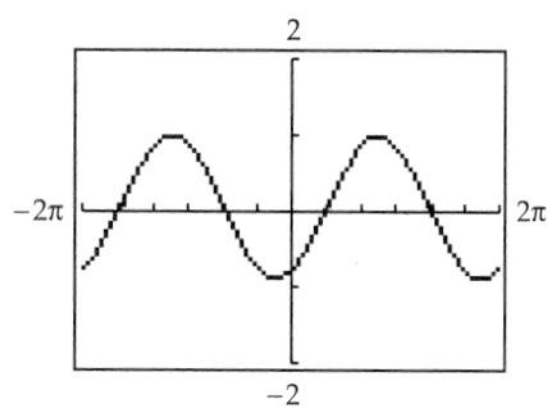

97. $\sqrt{2}, \dfrac{\pi}{2}, 4\pi$

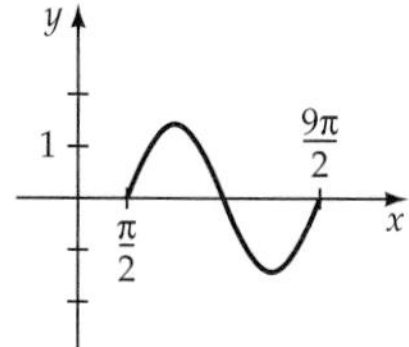

99. $2, \pi/12, \pi$

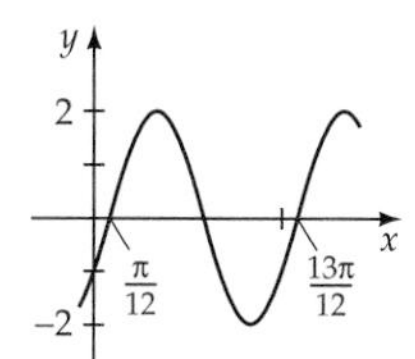

101. $2, -1/3, 2$

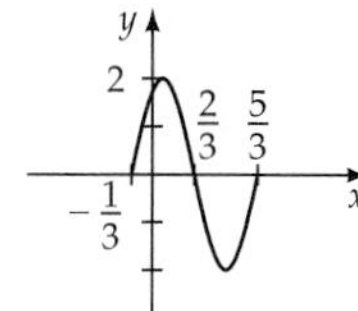

Exercise Set 6.5, page 513

1. $\pi/2$ **3.** $5\pi/6$ **5.** $-\pi/4$ **7.** $\pi/3$ **9.** $\pi/3$ **11.** $-\pi/4$ **13.** $-\pi/3$ **15.** $2\pi/3$ **17.** $\pi/6$ **19.** $\pi/6$ **21.** 1/2 **23.** 2 **25.** 3/5 **27.** 1 **29.** 1/2 **31.** $\pi/6$ **33.** $\pi/4$ **35.** not defined **37.** 0.4636 **39.** $-\pi/6$ **41.** $\sqrt{3}/3$ **43.** $4\sqrt{15}/15$ **45.** 24/25 **47.** 13/5 **49.** 0 **51.** 24/25 **53.** $\dfrac{2+\sqrt{15}}{6}$ **55.** $\dfrac{1}{5}(3\sqrt{7}-4\sqrt{3})$ **57.** 12/13 **59.** 2 **61.** $\dfrac{2-\sqrt{2}}{2}$ **63.** $7\sqrt{2}/10$ **65.** $\cos\dfrac{5\pi}{12}\approx 0.2588$ **67.** $\sqrt{1-x^2}$ **69.** $\dfrac{\sqrt{x^2-1}}{|x|}$

75.

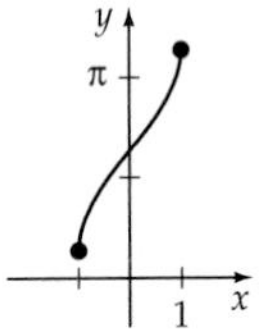

77.

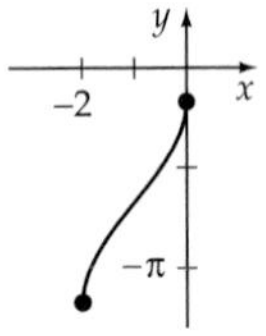

79.

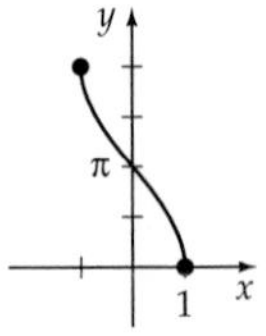

81.

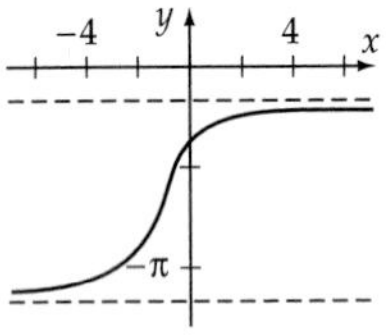

83. a. 0.1014 **b.** 0.1552

85.

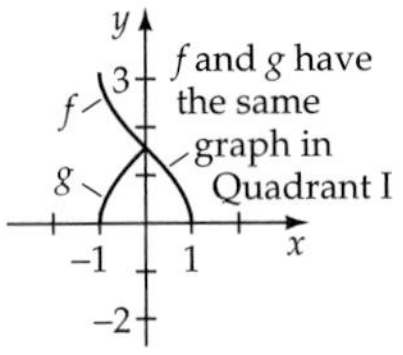

No. $f(x)$ is neither odd nor even. $g(x)$ is an even function.

87.

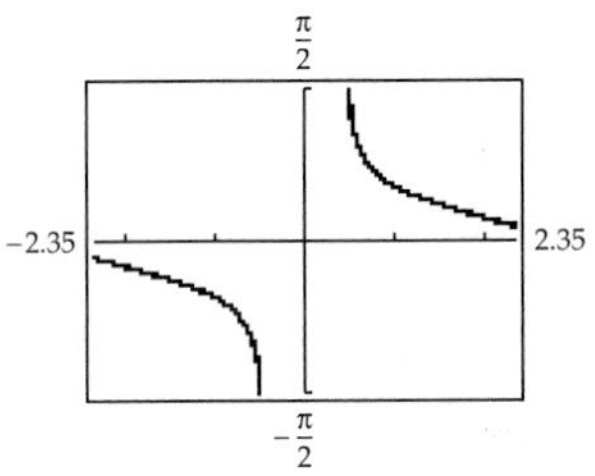

89.

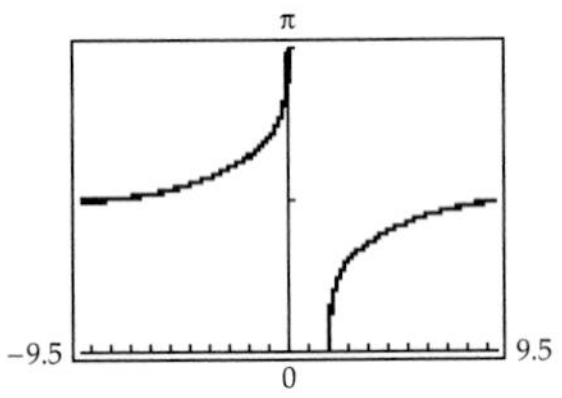

91.

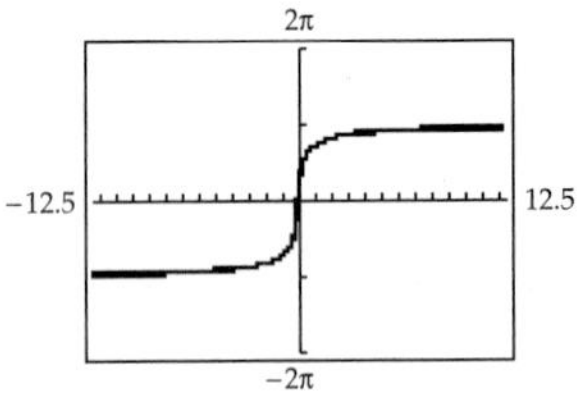

93.

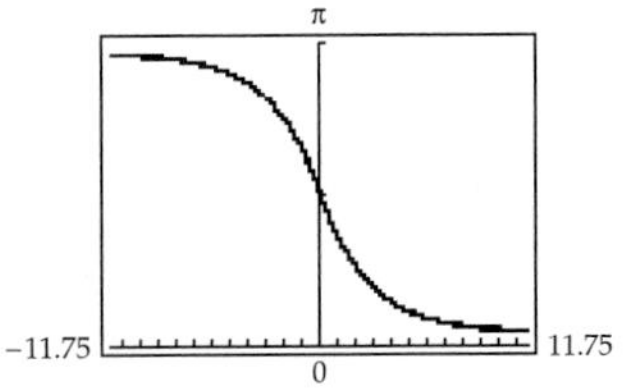

99. $y=\dfrac{1}{3}\tan 5x$

101. $y=3+\cos\left(x-\dfrac{\pi}{3}\right)$

Exercise Set 6.6, page 524

1. $\pi/4, 7\pi/4$ **3.** $\pi/3, 4\pi/3$ **5.** $\pi/4, \pi/2, 3\pi/4, 3\pi/2$ **7.** $\pi/2, 3\pi/2$ **9.** $\pi/6, \pi/4, 3\pi/4, 11\pi/6$ **11.** $\pi/4, 3\pi/4$ **13.** $\pi/6, \pi/2, 5\pi/6$ **15.** $\pi/6, 5\pi/6, 7\pi/6, 11\pi/6$ **17.** $0, \pi/4, 3\pi/4, \pi, 5\pi/4, 7\pi/4$ **19.** $\pi/6, 5\pi/6, 4\pi/3, 5\pi/3$ **21.** $0, \pi/2, \pi, 3\pi/2$ **23.** 41.4°, 318.6° **25.** no solution **27.** 68.0°, 292° **29.** no solution **31.** 12.8°, 167.2° **33.** 15.5°, 164.5° **35.** 0°, 33.7°, 180°, 213.7° **37.** no solution **39.** no solution **41.** 0°, 120°, 240° **43.** 70.5°, 289.5° **45.** 68.2°, 116.6°, 248.2°, 296.6° **47.** 19.5°, 90°, 160.5°, 270° **49.** 60°, 90°, 300° **51.** 53.1°, 180° **53.** 72.4°, 220.2° **55.** 50.1°, 129.9°, 205.7°, 334.3° **57.** no solution **59.** 22.5°, 157.5° **61.** $\pi/8 + k\pi/2$, where k is an integer **63.** $\pi/10 + 2k\pi/5$, where k is an integer **65.** $0 + 2k\pi, \pi/3 + 2k\pi, \pi + 2k\pi, 5\pi/3 + 2k\pi$, where k is an integer **67.** $\pi/2 + k\pi, 5\pi/6 + k\pi$, where k is an integer **69.** $0 + 2k\pi$, where k is an integer **71.** $0, \pi$ **73.** $0, \pi/6, \pi/2, 5\pi/6, \pi, 7\pi/6, 3\pi/2, 11\pi/6$ **75.** $0, \pi/2, 3\pi/2$ **77.** $0, \pi/3, 2\pi/3, \pi, 4\pi/3, 5\pi/3$ **79.** $4\pi/3, 5\pi/3$ **81.** $0, \pi/4, 3\pi/4, \pi, 5\pi/4, 7\pi/4$ **83.** $\pi/6, 5\pi/6, \pi$ **85.** 0.7391 **87.** −3.2957, 3.2957 **89.** 1.16 **91.** 14.99° and 75.01°

The sine regression functions in Exercises 93, 95, and 97 were obtained on a TI-83 calculator by using an iteration factor of 16. The use of a different iteration factor may produce a sine regression function that varies from the regression functions listed below.

93. a. $y \approx 1.1213\sin(0.01595x + 1.8362) + 6.6257$ **b.** 6:49 **95. a.** $y \approx 49.515\sin(0.21274x - 2.9120) + 53.185$ **b.** 28% **97. a.** $y \approx 35.323\sin(0.30236x - 2.1426) + 1.7201$ **b.** 24.6° **99.** day 74 to day 268 = 195 days **101.** $\pi/6, \pi/2$ **103.** $5\pi/3, 0$ **105.** $0, \pi/4, \pi/2, 3\pi/4, \pi, 5\pi/4, 3\pi/2, 7\pi/4$ **107.** $0, \pi/2, \pi, 3\pi/2$ **109.** $\pi/6, \pi/2, 5\pi/6, 7\pi/6, 3\pi/2, 11\pi/6$ **111.** 0.93 foot, 1.39 feet

Chapter 6 True/False Exercises, page 531

1. False; $\dfrac{\tan 45°}{\tan 60°} \neq \dfrac{45°}{60°}$. **2.** False, if $y = 0$, $\tan\dfrac{x}{0}$ is not defined. **3.** False. Let $x = 1$. Then $\sin^{-1} 1 = \dfrac{\pi}{2}$ and $\csc 1^{-1} \approx 1.18$. **4.** False; if $\alpha = \dfrac{\pi}{2}$, then $\sin 2\alpha = \sin \pi = 0$ but $2\sin\dfrac{\pi}{2} = 2$. **5.** False; $\sin(30° + 60°) \neq \sin 30° + \sin 60°$. **6.** False; $\sin x = 0$ has an infinite number of solutions $x = k\pi$, but $\sin x = 0$ is not an identity. **7.** False; $\tan 45° = \tan 225°$ but $45° \neq 225°$.

8. False; $\cos^{-1}\left(\cos\frac{3\pi}{2}\right) = \cos^{-1}(0) = \frac{\pi}{2} \neq \frac{3\pi}{2}$. **9.** False; $\cos(\cos^{-1} 2) \neq 2$, because $\cos^{-1} 2$ is undefined. **10.** False; if $\alpha = 1$, then we get $\frac{\pi}{2} \neq \frac{1}{\csc 1} \approx 0.8415$. **11.** False; $\sin(180° - \theta) = \sin\theta$. **12.** False; because $\sin^2\theta \geq 0$ for all θ but $\sin\theta^2$ can be < 0.

Chapter 6 Review Exercises, page 531

1. $\frac{\sqrt{6}-\sqrt{2}}{4}$ **3.** $\frac{\sqrt{6}-\sqrt{2}}{4}$ **5.** $-\frac{\sqrt{6}+\sqrt{2}}{4}$ **7.** $\frac{\sqrt{2-\sqrt{2}}}{2}$ **9.** $\sqrt{2}+1$ **11. a.** 0 **b.** $\sqrt{3}$ **c.** 1/2 **13. a.** $\sqrt{3}/2$ **b.** $-\sqrt{3}$ **c.** $-\frac{\sqrt{2-\sqrt{3}}}{2}$ **15.** $\sin 6x$ **17.** $\sin 3x$ **19.** $\cos\beta$ **21.** 0.2740 **23.** $-\sqrt{3}/2$ **25.** $2\sin 3\theta \sin\theta$ **27.** $2\sin 4\theta\cos 2\theta$ **47.** 13/5 **49.** 3/2 **51.** 4/5 **53.** 30°, 150°, 240°, 300° **55.** $\pi/2 + 2k\pi$, $3.8713 + 2k\pi$, $5.553 + 2k\pi$, where k is an integer **57.** $\pi/12, 5\pi/12, 13\pi/12, 17\pi/12$ **59.** $2, -\pi/6$ **61.** $2, -4\pi/3$ **63.** **65.**

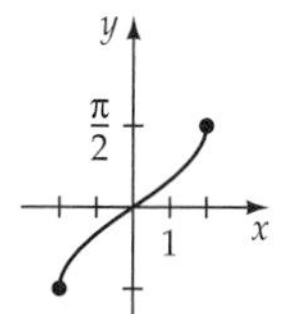

67. a. $y \approx 1.5682\sin(0.01632x + 1.8135) + 5.9374$ **b.** 5:22

Chapter 6 Test, page 533

5. $\frac{-\sqrt{6}+\sqrt{2}}{4}$ **6.** $-\sqrt{2}/10$ **8.** $\sin 9x$ **9.** $-7/25$ **12.** $\frac{2-\sqrt{3}}{4}$ **13.** $y = \sin\left(x + \frac{5\pi}{6}\right)$ **14.** 0.701 **15.** 5/13 **16.** **17.** 41.8°, 138.2° **18.** $0, \frac{\pi}{6}, \pi, \frac{11\pi}{6}$ **19.** $\frac{\pi}{2}, \frac{2\pi}{3}, \frac{4\pi}{3}$ **20. a.** $y \approx 38.961\sin(0.40033x + 2.9206) + 33.053$ **b.** 63.8°

Exercise Set 7.1, page 541

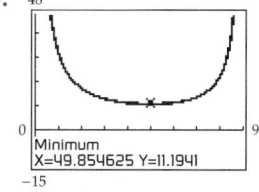

1. $C = 77°, b \approx 16, c \approx 17$ **3.** $B = 38°, a \approx 18, c \approx 10$ **5.** $C \approx 15°, B \approx 33°, c \approx 7.8$ **7.** $C = 32.6°, c \approx 21.6, a \approx 39.8$ **9.** $B = 47.7°, a \approx 57.4, b \approx 76.3$ **11.** $A \approx 58.5°, B \approx 7.3°, a \approx 81.5$ **13.** $C = 59°, B = 84°, b \approx 46$ or $C = 121°, B = 22°, b \approx 17$ **15.** No triangle is formed **17.** No triangle is formed **19.** $C = 19.8°, B = 145.4°, b \approx 10.7$ or $C = 160.2°, B = 5.0°, b \approx 1.64$ **21.** No triangle is formed **23.** $C = 51.21°, A = 11.47°, c \approx 59.00$ **25.** ≈ 68.8 miles **27.** ≈ 110 feet **29.** ≈ 130 yards **31.** ≈ 96 feet **33.** ≈ 33 feet **35.** ≈ 8.1 miles **37.** ≈ 1200 miles **39.** ≈ 260 meters **43.** minimum value of $L \approx 11.19$ meters

Exercise Set 7.2, page 550

1. ≈ 13 **3.** ≈ 150 **5.** ≈ 29 **7.** ≈ 9.5 **9.** ≈ 10 **11.** ≈ 40.1 **13.** ≈ 90.7 **15.** ≈ 39° **17.** ≈ 90° **19.** ≈ 47.9° **21.** ≈ 116.67° **23.** ≈ 80.3° **25.** ≈ 140 square units **27.** ≈ 53 square units **29.** ≈ 81 square units **31.** ≈ 299 square units **33.** ≈36 square units **35.** ≈ 7.3 square units **37.** ≈ 710 miles **39.** ≈ 74 feet **41.** ≈60.9° **43.** ≈ 350 miles **45.** 40 centimeters **47.** ≈ 9.7 inches, 25 inches **49.** ≈ 55 centimeters **51.** ≈ 2800 feet **53.** ≈ 47,500 square meters **55.** ≈ 203 square meters **57.** 162 square inches **59.** ≈ $41,000 **61.** ≈ 6.23 acres **63. a.** 64 square units **b.** 55 square units **65.** ≈ 12.5° **67.** ≈ 52.0 centimeters **73.** ≈ 140 cubic inches

Exercise Set 7.3, page 566

1. $a = 7, b = -1; \langle 7, -1 \rangle$ **3.** $a = -7, b = -5; \langle -7, -5 \rangle$ **5.** $a = 0, b = 8; \langle 0, 8 \rangle$ **7.** $5, \approx 126.9°, \left\langle -\frac{3}{5}, \frac{4}{5} \right\rangle$
9. $\approx 44.7, \approx 296.6°, \left\langle \frac{\sqrt{5}}{5}, \frac{-2\sqrt{5}}{5} \right\rangle$ **11.** $\approx 4.5, \approx 296.6°, \left\langle \frac{\sqrt{5}}{5}, \frac{-2\sqrt{5}}{5} \right\rangle$ **13.** $\approx 45.7, \approx 336.8°, \left\langle \frac{7\sqrt{58}}{58}, \frac{-3\sqrt{58}}{58} \right\rangle$ **15.** $\langle -6, 12 \rangle$
17. $\langle -1, 10 \rangle$ **19.** $\left\langle -\frac{11}{6}, \frac{7}{3} \right\rangle$ **21.** $2\sqrt{5}$ **23.** $2\sqrt{109}$ **25.** $-8\mathbf{i} + 12\mathbf{j}$ **27.** $14\mathbf{i} - 6\mathbf{j}$ **29.** $\frac{11}{12}\mathbf{i} + \frac{1}{2}\mathbf{j}$ **31.** $\sqrt{113}$
33. $a_1 \approx 4.5, a_2 \approx 2.3, 4.5\mathbf{i} + 2.3\mathbf{j}$ **35.** $a_1 \approx 2.8, a_2 \approx 2.8, 2.8\mathbf{i} + 2.8\mathbf{j}$ **37.** ≈ 380 mph **39.** ≈ 250 mph at a heading of 86°
41. ≈ 293 pounds **43.** ≈ 24.7 pounds **45.** -3 **47.** 0 **49.** 1 **51.** 0 **53.** $\approx 79.7°$ **55.** 45° **57.** 90°, orthogonal **59.** 180°
61. $\frac{46}{5}$ **63.** $\frac{14\sqrt{29}}{29} \approx 2.6$ **65.** $\sqrt{5} \approx 2.2$ **67.** $-\frac{11\sqrt{5}}{5} \approx -4.9$ **69.** ≈ 954 foot-pounds **71.** ≈ 779 foot-pounds
73.

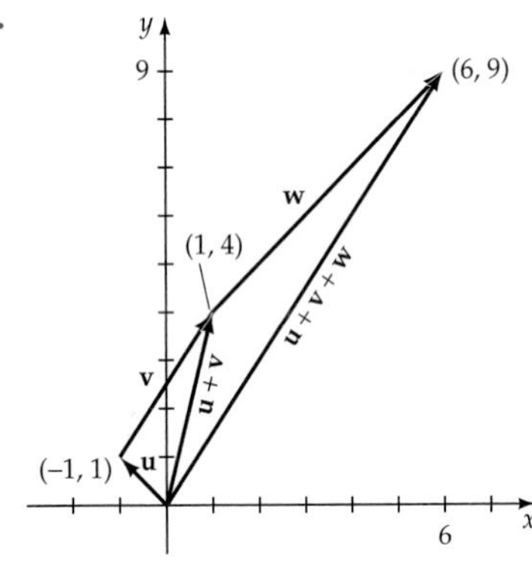

$\langle 6, 9 \rangle$ **75.** The vector from $P_1(3, -1)$ to $P_2(5, -4)$ is equivalent to $2\mathbf{i} - 3\mathbf{j}$.
77. $\mathbf{v} \cdot \mathbf{w} = 0$; the vectors are perpendicular. **79.** $\langle 7, 2 \rangle$ is one example **81.** 6.4 pounds **83.** No
87. The same amount of work is done.

Exercise Set 7.4, page 574

1–7.
$|-2 - 2i| = 2\sqrt{2}$
$|\sqrt{3} - i| = 2$
$|-2i| = 2$
$|3 - 5i| = \sqrt{34}$

9. $\sqrt{2}$ cis 315° **11.** 2 cis 330° **13.** 3 cis 90° **15.** 5 cos 180° **17.** $\sqrt{2} + i\sqrt{2}$
19. $\frac{\sqrt{2}}{2} - \frac{\sqrt{2}}{2}i$ **21.** $-3\sqrt{2} + 3i\sqrt{2}$ **23.** 8 **25.** $-\sqrt{3} + i$ **27.** $-3i$ **29.** $-4\sqrt{2} + 4i\sqrt{2}$ **31.** $\frac{9\sqrt{3}}{2} - \frac{9}{2}i$
33. $\approx -0.832 + 1.819i$ **35.** 6 cis 255° **37.** 12 cis 335° **39.** 10 cis $\frac{16\pi}{15}$ **41.** 24 cis 6.5 **43.** $-4 - 4i\sqrt{3}$ **45.** 3i **47.** $-\frac{3\sqrt{3}}{2} + \frac{3i}{2}$
49. $\approx -2.081 + 4.546i$ **51.** $\approx 2.732 - 0.732i$ **53.** $6 + 0i = 6$ **55.** $-\frac{1}{2} + \frac{\sqrt{3}}{2}i$ **57.** $0 - \sqrt{2}i = -\sqrt{2}i$ **59.** $16 - 16i$
61. $-\frac{3}{8} + \frac{\sqrt{3}}{8}i$ **63.** $\approx 59.1 + 42.9i$ **65.** r^2 or $a^2 + b^2$

Exercise Set 7.5, page 579

1. $-128 - 128i\sqrt{3}$ **3.** $-16 + 16i\sqrt{3}$ **5.** $16\sqrt{2} + 16i\sqrt{2}$ **7.** $64 + 0i = 64$ **9.** $0 - 32i = -32i$ **11.** $1024 - 1024i$ **13.** $0 - 1i = -i$

15. $3 + 0i = 3$; $-3 + 0i = -3$

17. $2 + 0i = 2$; $1 + i\sqrt{3}$; $-1 + i\sqrt{3}$; $-2 + 0i = -2$; $-1 - i\sqrt{3}$; $1 - i\sqrt{3}$

19. $0.809 + 0.588i$; $-0.309 + 0.951i$; $-1 + 0i = -1$; $-0.309 - 0.951i$; $0.809 - 0.588i$

21. $1 + 0i = 1$; $-\frac{1}{2} + \frac{i\sqrt{3}}{2}$; $-\frac{1}{2} - \frac{i\sqrt{3}}{2}$

23. $1.070 + 0.213i$; $-0.213 + 1.070i$; $-1.070 - 0.213i$; $0.213 - 1.070i$

25. $-0.276 + 1.563i$; $-1.216 - 1.020i$; $1.492 - 0.543i$

27. $2\sqrt{2} + 2i\sqrt{6}$; $-2\sqrt{2} - 2i\sqrt{6}$

29. 2 cis 60°; 2 cis 180°; 2 cis 300°

31. cis 67.5°; cis 157.5°; cis 247.5°; cis 337.5°

33. 3 cis 0°; 3 cis 120°; 3 cis 240°

35. 3 cis 45°; 3 cis 135°; 3 cis 225°; 3 cis 315°

37. $\sqrt[4]{2}$ cis 75°; $\sqrt[4]{2}$ cis 165°; $\sqrt[4]{2}$ cis 255°; $\sqrt[4]{2}$ cis 345°

39. $\sqrt[3]{2}$ cis 80°; $\sqrt[3]{2}$ cis 200°; $\sqrt[3]{2}$ cis 320°

Chapter 7 True/False Exercises, page 581

1. False; we cannot solve a triangle using the Law of Cosines given the angle opposite one of the given sides. **2.** True **3.** True **4.** False; $2\mathbf{i} \neq 2\mathbf{j}$ **5.** True **6.** True **7.** True **8.** True **9.** True **10.** False; $\mathbf{v} \cdot \mathbf{v} = a^2 + b^2$. **11.** False; let $\mathbf{v} = \mathbf{i} + \mathbf{j}$, and $\mathbf{w} = \mathbf{i} - \mathbf{j}$. Then $\mathbf{v} \cdot \mathbf{w} = 1 - 1 = 0$. **12.** True **13.** False; $z^2 = r^2(\cos 2\theta + i \sin 2\theta)$. **14.** True **15.** False; $i = \cos\frac{\pi}{2} + i \sin\frac{\pi}{2}$. **16.** True

Chapter 7 Review Exercises, page 582

1. $B = 53°, a \approx 11, c \approx 18$ **3.** $B \approx 48°, C \approx 95°, A \approx 37°$ **5.** $c \approx 13, A \approx 55°, B \approx 90°$ **7.** No triangle is formed. **9.** $C = 45°, a \approx 29, b \approx 35$ **11.** ≈ 360 square units **13.** ≈ 920 square units **15.** ≈ 790 square units **17.** ≈ 170 square units **19.** $a_1 = 5, a_2 = 3, \langle 5, 3\rangle$ **21.** $\approx 4.5, 153.4°$ **23.** $\approx 3.6, 123.7°$ **25.** $\left\langle -\frac{8\sqrt{89}}{89}, \frac{5\sqrt{89}}{89}\right\rangle$ **27.** $\frac{5\sqrt{26}}{26}\mathbf{i} + \frac{\sqrt{26}}{26}\mathbf{j}$ **29.** $\langle -7, -3\rangle$ **31.** $-6\mathbf{i} - (17/2)\mathbf{j}$ **33.** ≈ 420 mph **35.** 18 **37.** -9 **39.** $\approx 86°$ **41.** $\approx 125°$ **43.** $\frac{10\sqrt{41}}{41}$ **45.** ≈ 662 foot-pounds **47.** $\approx 5.29, 161°$ **49.** $2\sqrt{3}$ cis 120° **51.** $-3 - 3i\sqrt{3}$ **53.** $\approx -8.918 + 8.030i$ **55.** $\approx -6.012 - 13.742i$ **57.** 3 cis 110° **59.** $\sqrt{2}$ cis 285° **61.** $-\frac{1}{2} + \frac{i\sqrt{3}}{2}$ **63.** $32{,}768i$ **65.** $\sqrt[4]{8}$ cis 22.5°, $\sqrt[4]{8}$ cis 112.5°, $\sqrt[4]{8}$ cis 202.5°, $\sqrt[4]{8}$ cis 292.5° **67.** $\sqrt[10]{2}$ cis 45°, $\sqrt[10]{2}$ cis 117°, $\sqrt[10]{2}$ cis 189°, $\sqrt[10]{2}$ cis 261°, $\sqrt[10]{2}$ cis 333°

Chapter 7 Test, page 583

1. $B = 94°, a \approx 48, b \approx 51$ **2.** $\approx 11°$ **3.** ≈ 14 **4.** $\approx 48°$ **5.** $K \approx 39$ square units **6.** $K \approx 93$ square units **7.** $K \approx 260$ square units **8.** $-9.2\mathbf{i} - 7.7\mathbf{j}$ **9.** $-19\mathbf{i} - 29\mathbf{j}$ **10.** -1 **11.** $\approx 103°$ **12.** $\approx 3\sqrt{3}$ cis 145° **13.** $\frac{5\sqrt{2}}{2} - \frac{5\sqrt{2}}{2}i$ **14.** $\approx 8.208 - 22.553i$ **15.** 2.5 cis (−95°) or 2.5 cis 265° **16.** $\approx -15.556 - 1.000i$ **17.** $\frac{3\sqrt{3}}{2} + \frac{3}{2}i, -\frac{3\sqrt{3}}{2} + \frac{3}{2}i, 0 - 3i$ or $-3i$ **18.** ≈ 27 miles **19.** ≈ 21 miles **20.** $\approx \$66{,}000$

Exercise Set 8.1, page 592

1. vertex: $(0, 0)$; focus: $(0, -1)$; directrix: $y = 1$

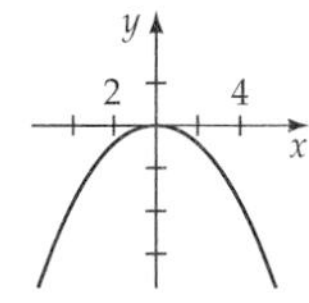

3. vertex: $(0, 0)$; focus: $(1/12, 0)$; directrix: $x = -1/12$

5. vertex: $(2, -3)$; focus: $(2, -1)$; directrix: $y = -5$

7. vertex: $(2, -4)$; focus: $(1, -4)$; directrix: $x = 3$

9. vertex: $(-4, 1)$; focus: $(-7/2, 1)$; directrix: $x = -9/2$

11. vertex: $(2, 2)$; focus: $(2, 5/2)$; directrix: $y = 3/2$

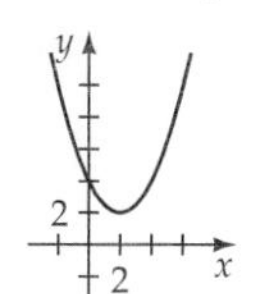

13. vertex: $(-4, -10)$; focus: $(-4, -39/4)$; directrix: $y = -41/4$

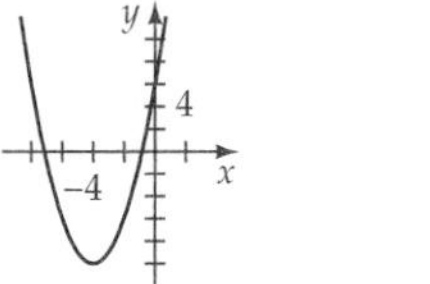

15. vertex: $(-7/4, 3/2)$; focus: $(-2, 3/2)$; directrix: $x = -3/2$

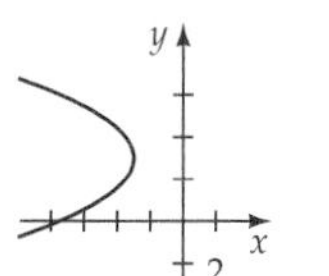

17. Vertex: $(-5, -3)$; focus: $(-9/2, -3)$; directrix: $x = -11/2$

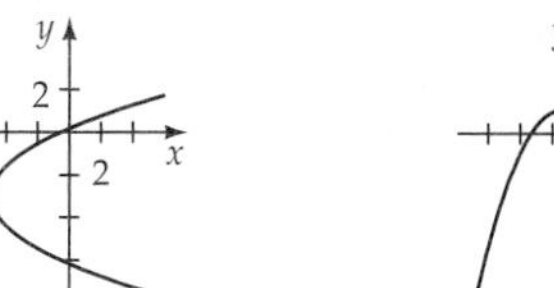

19. vertex: $(-3/2, 13/12)$; focus: $(-3/2, 1/3)$; directrix: $y = 11/6$

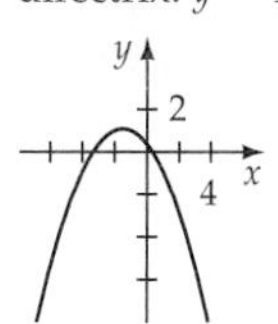

21. vertex: $(2, -5/4)$
focus: $(2, -3/4)$
directrix: $y = -7/4$

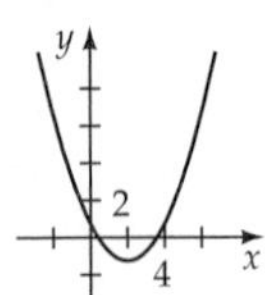

23. vertex: $(9/2, -1)$
focus: $(35/8, -1)$
directrix: $x = 37/8$

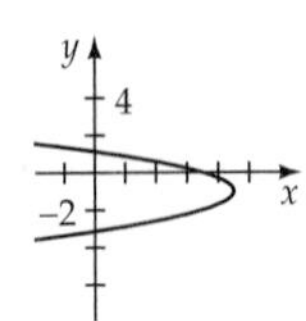

25. vertex: $(1, 1/9)$
focus: $(1, 31/36)$
directrix: $y = -23/36$

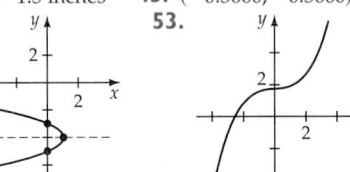

27. $x^2 = -16y$ **29.** $(x + 1)^2 = 4(y - 2)$ **31.** $(x - 3)^2 = 4(y + 4)$ **33.** $(x + 4)^2 = 4(y - 1)$ **35.** on axis 4 feet above vertex
37. 6.0 inches **39. a.** 5900 square feet **b.** 56,800 square feet **41.** $a = 1.5$ inches **43.** $(-0.3660, -0.3660)$ and $(1.3660, 1.3660)$
45. $(-1.5616, 3.8769)$ and $(2.5616, 12.1231)$ **47.** 4 **49.** $4|p|$ **51.**

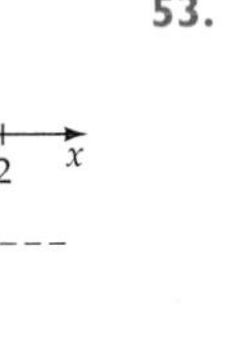

53.

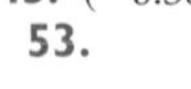

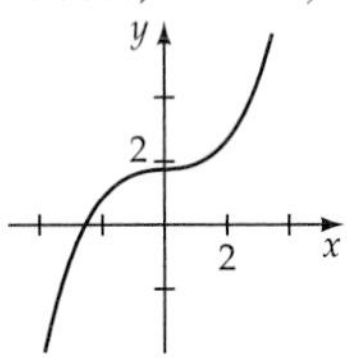

55. $x^2 + y^2 - 8x - 8y - 2xy = 0$

Exercise Set 8.2, page 604

1. vertices: $(0, 5)$, $(0, -5)$
foci: $(0, 3)$, $(0, -3)$

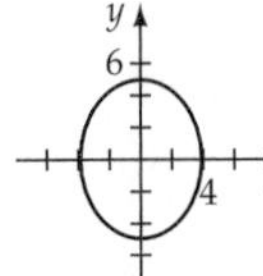

3. vertices: $(3, 0)$, $(-3, 0)$
foci: $(\sqrt{5}, 0)$, $(-\sqrt{5}, 0)$

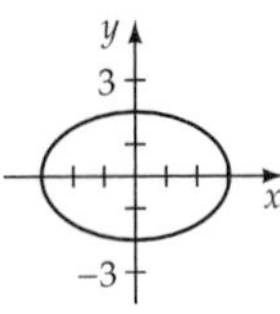

5. vertices: $(0, 3)$, $(0, -3)$
foci: $(0, \sqrt{2})$, $(0, -\sqrt{2})$

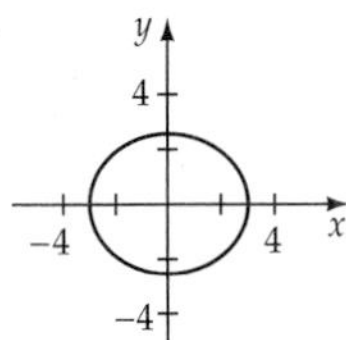

7. vertices: $(0, 4)$, $(0, -4)$
foci: $(0, \sqrt{55}/2)$, $(0, -\sqrt{55}/2)$

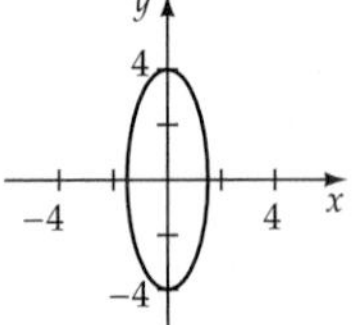

9. vertices: $(8, -2)$, $(-2, -2)$
foci: $(6, -2)$, $(0, -2)$

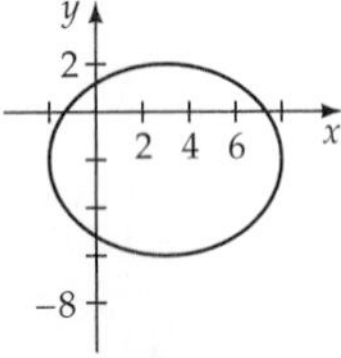

11. vertices: $(-2, 5)$, $(-2, -5)$
foci: $(-2, 4)$, $(-2, -4)$

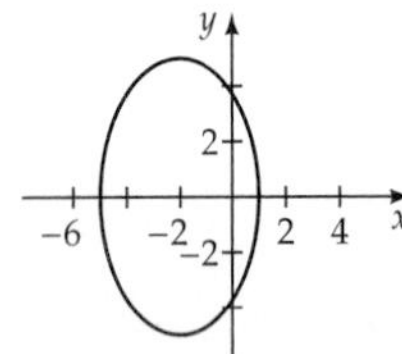

13. vertices: $(1 + \sqrt{21}, 3)$, $(1 - \sqrt{21}, 3)$
foci: $(1 + \sqrt{17}, 3)$, $(1 - \sqrt{17}, 3)$

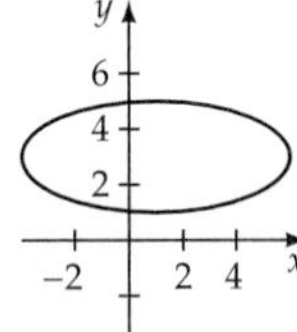

15. vertices: $(1, 2)$, $(1, -4)$
foci: $(1, -1 + \sqrt{65}/3)$, $(1, -1 - \sqrt{65}/3)$

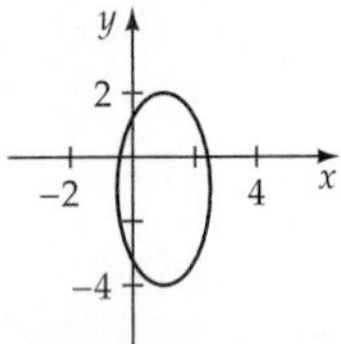

17. vertices $(2, 0)$, $(-2, 0)$
foci: $(1, 0)$, $(-1, 0)$

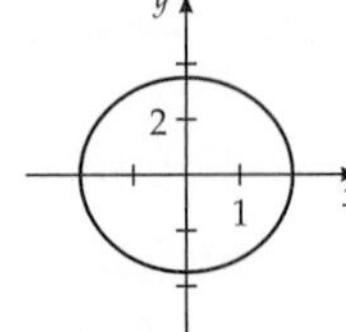

19. vertices: $(0, 5)$, $(0, -5)$
foci: $(0, 3)$, $(0, -3)$

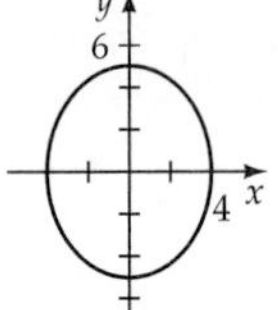

21. vertices: $(0, 4)$, $(0, -4)$
foci: $(0, \sqrt{39}/2)$, $(0, -\sqrt{39}/2)$

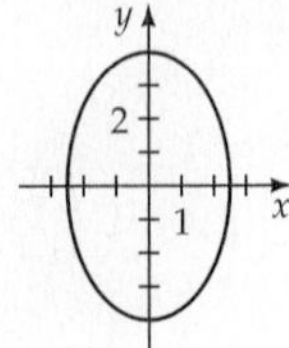

23. vertices: $(3, 6)$, $(3, 2)$
foci: $(3, 4 + \sqrt{3})$, $(3, 4 - \sqrt{3})$

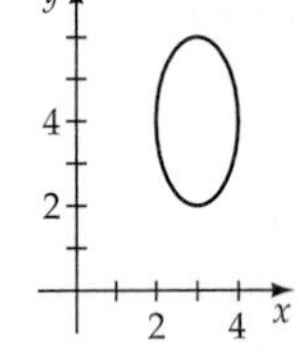

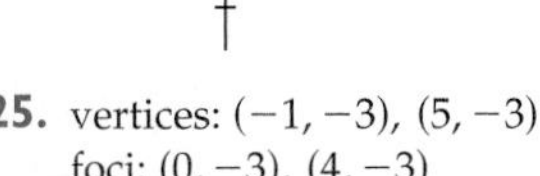

25. vertices: $(-1, -3)$, $(5, -3)$
foci: $(0, -3)$, $(4, -3)$

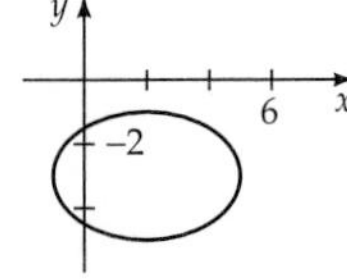

27. vertices: $(2, 4)$, $(2, -4)$
foci: $(2, \sqrt{7})$, $(2, -\sqrt{7})$

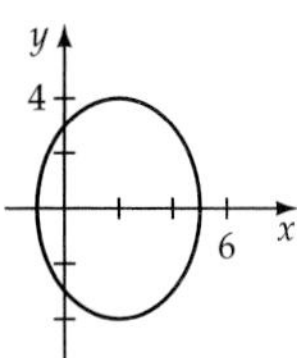

29. vertices: $(-1, 6)$, $(-1, -4)$
foci: $(-1, 4)$, $(-1, -2)$

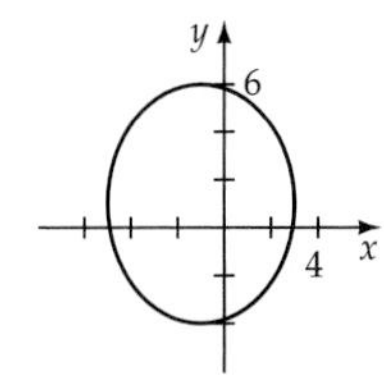

31. vertices: $(11/2, -1)$, $(1/2, -1)$
foci: $(3 + \sqrt{17}/2, -1)$, $(3 - \sqrt{17}/2, -1)$

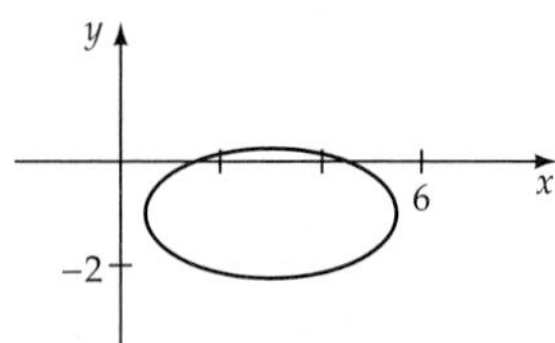

33. $\dfrac{x^2}{25} + \dfrac{y^2}{9} = 1$ **35.** $\dfrac{x^2}{36} + \dfrac{y^2}{16} = 1$ **37.** $\dfrac{x^2}{36} + \dfrac{y^2}{81/8} = 1$ **39.** $\dfrac{(x+2)^2}{16} + \dfrac{(y-4)^2}{7} = 1$ **41.** $\dfrac{(x-2)^2}{25/24} + \dfrac{(y-4)^2}{25} = 1$

43. $\dfrac{(x-5)^2}{16} + \dfrac{(y-1)^2}{25} = 1$ **45.** $\dfrac{x^2}{25} + \dfrac{y^2}{21} = 1$ **47.** $\dfrac{x^2}{20} + \dfrac{y^2}{36} = 1$ **49.** $\dfrac{(x-1)^2}{25} + \dfrac{(y-3)^2}{21} = 1$ **51.** $\dfrac{x^2}{80} + \dfrac{y^2}{144} = 1$

53. $\dfrac{x^2}{884.74^2} + \dfrac{y^2}{883.35^2} = 1$ **55.** 40 feet **57.** $\dfrac{\left(x - 9\sqrt{15}/2\right)^2}{324} + \dfrac{y^2}{81/4} = 1$ **59.** 1512 **61. a.** $\dfrac{x^2}{307.5^2} + \dfrac{y^2}{255^2} = 1$ **b.** 246,300 square feet

63. $y = \dfrac{-36 \pm \sqrt{1296 - 36(16x^2 - 108)}}{18}$

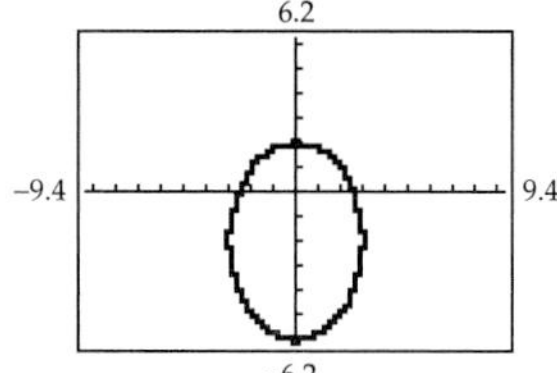

65. $y = \dfrac{54 \pm \sqrt{2916 - 36(16x^2 - 64x + 1)}}{18}$

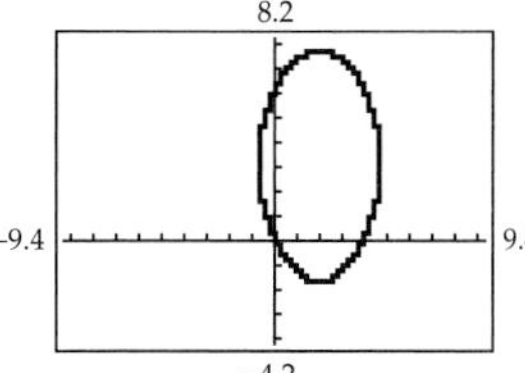

67. $y = \dfrac{-18 \pm \sqrt{324 - 36(4x^2 + 24x + 44)}}{18}$

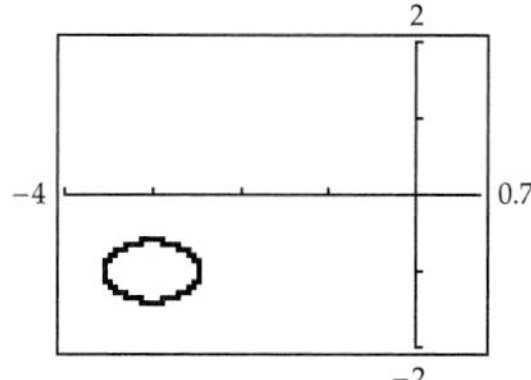

69. $\dfrac{x^2}{36} + \dfrac{y^2}{27} = 1$ **71.** $\dfrac{(x-1)^2}{16} + \dfrac{(y-2)^2}{12} = 1$ **73.** $9/2$ **77.** $x = \pm\dfrac{9\sqrt{5}}{5}$

Exercise Set 8.3, page 618

1. center: $(0, 0)$
vertices: $(\pm 4, 0)$
foci: $(\pm\sqrt{41}, 0)$
asymptotes: $y = \pm 5x/4$

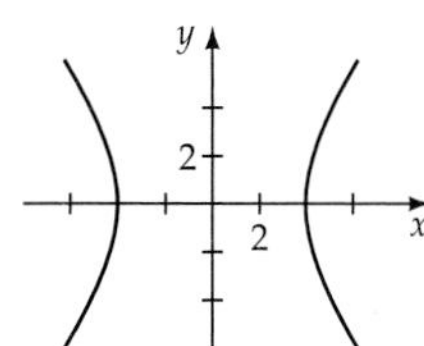

3. center: $(0, 0)$
vertices: $(0, \pm 2)$
foci: $(0, \pm\sqrt{29})$
asymptotes: $y = \pm 2x/5$

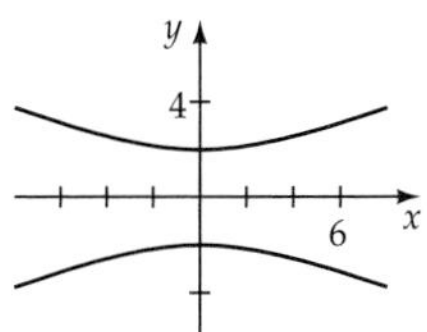

5. center: $(0, 0)$
vertices: $(\pm\sqrt{7}, 0)$
foci: $(\pm 4, 0)$
asymptotes: $y = \pm 3\sqrt{7}x/7$

7. center: $(0, 0)$
vertices: $(\pm 3/2, 0)$
foci: $(\pm\sqrt{73}/2, 0)$
asymptotes: $y = \pm 8x/3$

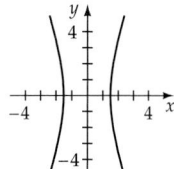

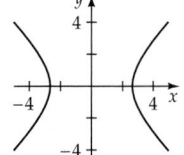

9. center: $(3, -4)$
vertices: $(7, -4)$, $(-1, -4)$
foci: $(8, -4)$, $(-2, -4)$
asymptotes: $y + 4 = \pm 3(x - 3)/4$

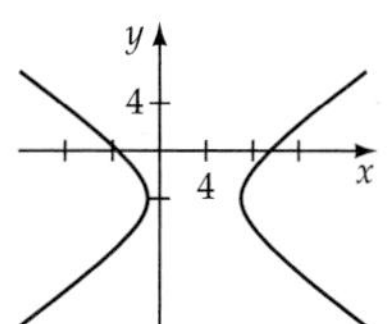

11. center: $(1, -2)$
vertices: $(1, 0)$, $(1, -4)$
foci: $(1, -2 \pm 2\sqrt{5})$
asymptotes: $y + 2 = \pm(x - 1)/2$

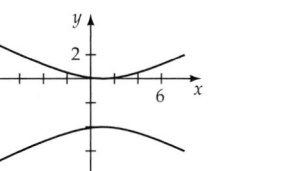

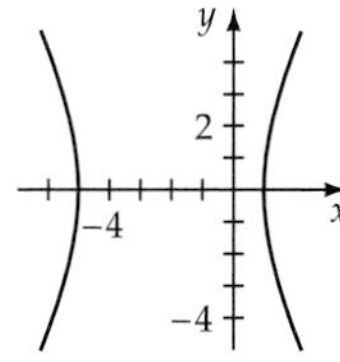

13. center: $(-2, 0)$
vertices: $(1, 0)$, $(-5, 0)$
foci: $(-2 \pm \sqrt{34}, 0)$
asymptotes: $y = \pm 5(x + 2)/3$

15. center: $(1, -1)$
vertices: $(7/3, -1)$, $(-1/3, -1)$
foci: $(1 \pm \sqrt{97}/3, -1)$
asymptotes: $y + 1 = \pm 9(x - 1)/4$

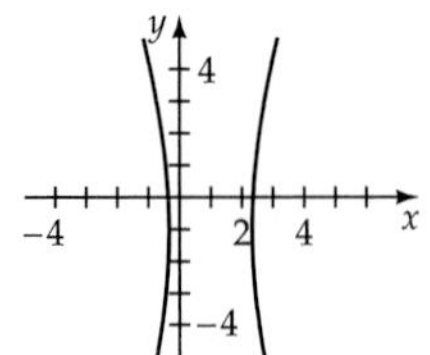

17. center: $(0, 0)$
vertices: $(\pm 3, 0)$
foci: $(\pm 3\sqrt{2}, 0)$
asymptotes: $y = \pm x$

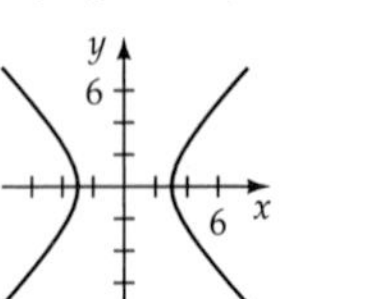

19. center: $(0, 0)$
vertices: $(0, \pm 3)$
foci: $(0, \pm 5)$
asymptotes: $y = \pm 3x/4$

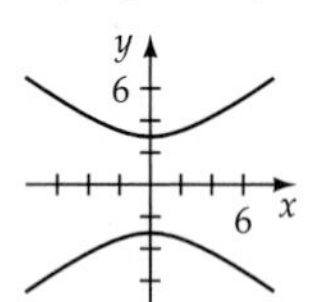

21. center: $(0, 0)$
vertices: $(0, \pm 2/3)$
foci: $(0, \pm\sqrt{5}/3)$
asymptotes: $y = \pm 2x$

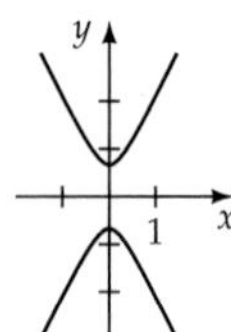

23. center: $(3, 4)$
vertices: $(3, 6)$, $(3, 2)$
foci: $(3, 4 \pm 2\sqrt{2})$
asymptotes: $y - 4 = \pm(x - 3)$

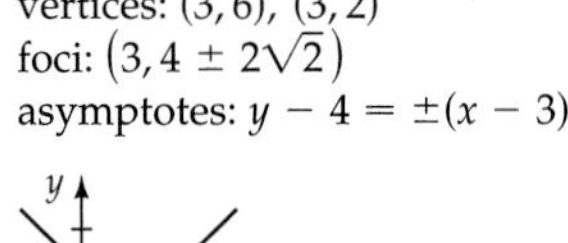

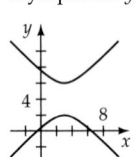

25. center: $(-2, -1)$
vertices: $(-2, 2)$, $(-2, -4)$
foci: $(-2, -1 \pm \sqrt{13})$
asymptotes: $y + 1 = \pm 3(x + 2)/2$

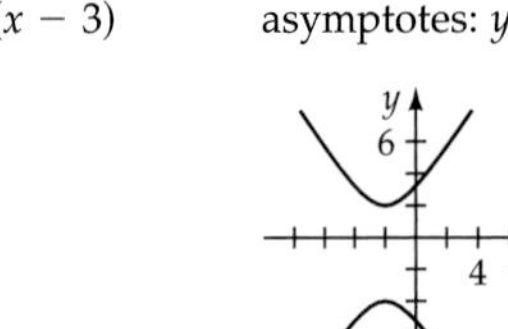

27. $y = \dfrac{-6 \pm \sqrt{36 + 4(4x^2 + 32x + 39)}}{-2}$

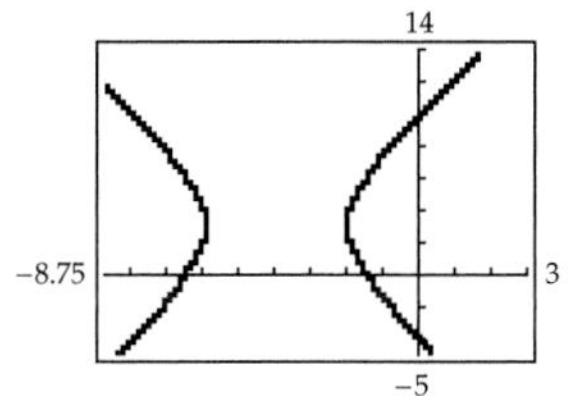

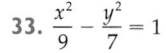

29. $y = \dfrac{64 \pm \sqrt{4096 + 64(9x^2 - 36x + 116)}}{-32}$

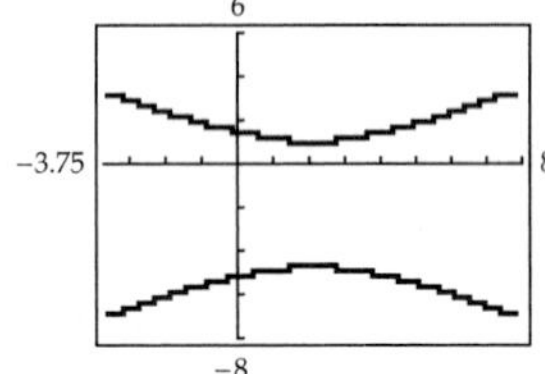

31. $y = \dfrac{18 \pm \sqrt{324 + 36(4x^2 + 8x - 6)}}{-18}$

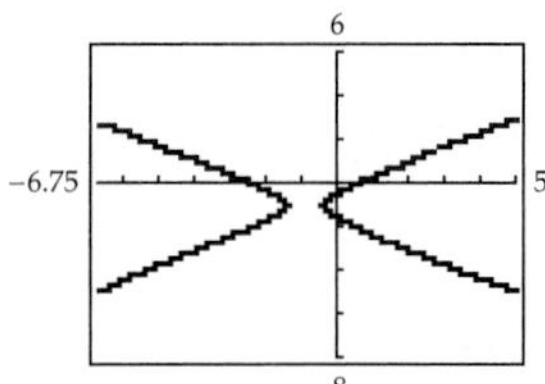

33. $\dfrac{x^2}{9} - \dfrac{y^2}{7} = 1$ **35.** $\dfrac{y^2}{20} - \dfrac{x^2}{5} = 1$

37. $\dfrac{y^2}{9} - \dfrac{x^2}{36/7} = 1$ **39.** $\dfrac{y^2}{16} - \dfrac{x^2}{64} = 1$ **41.** $\dfrac{(x-4)^2}{4} - \dfrac{(y-3)^2}{5} = 1$ **43.** $\dfrac{(x-4)^2}{144/41} - \dfrac{(y+2)^2}{225/41} = 1$ **45.** $\dfrac{(y-2)^2}{3} - \dfrac{(x-7)^2}{12} = 1$

47. $\dfrac{(y-7)^2}{1} - \dfrac{(x-1)^2}{3} = 1$ **49.** $\dfrac{x^2}{4} - \dfrac{y^2}{12} = 1$ **51.** $\dfrac{(x-4)^2}{36/7} - \dfrac{(y-1)^2}{4} = 1$ and $\dfrac{(y-1)^2}{36/7} - \dfrac{(x-4)^2}{4} = 1$

53. **a.** $\dfrac{x^2}{2162.25} - \dfrac{y^2}{13{,}462.75} = 1$
b. 221 miles

55. ellipse

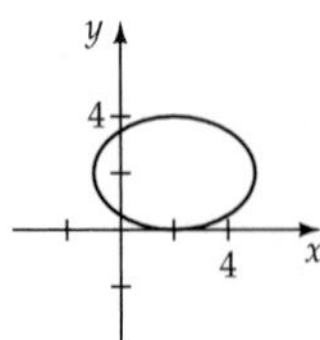

57. parabola

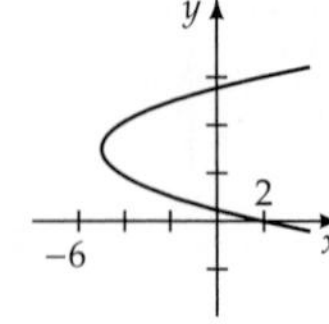

59. parabola

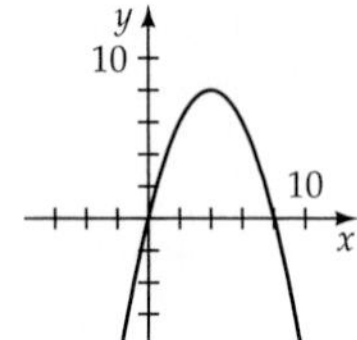

61. ellipse

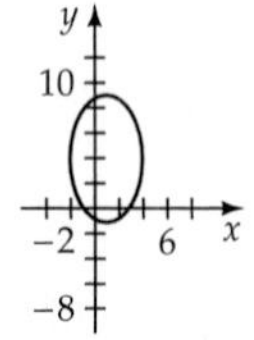

63. $\dfrac{x^2}{1} - \dfrac{y^2}{3} = 1$ **65.** $\dfrac{y^2}{9} - \dfrac{x^2}{7} = 1$ **67.** $x = \pm\dfrac{16\sqrt{41}}{41}$ **71.**

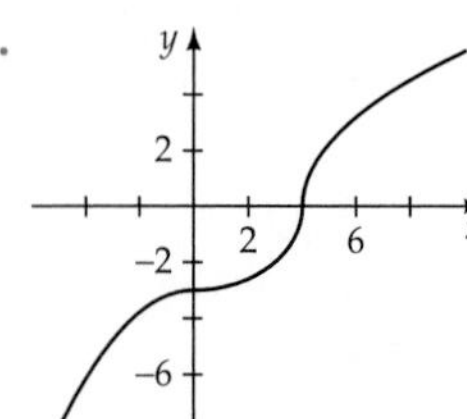

Exercise Set 8.4, page 626

1. 45° **3.** 36.9° **5.** 73.5° **7.** 45°, $\frac{x'^2}{8} - \frac{y'^2}{8} = 1$ **9.** 18.4°, $\frac{x'^2}{9} + \frac{y'^2}{3} = 1$ **11.** 26.6°, $\frac{x'^2}{1/2} - \frac{y'^2}{1/3} = 1$

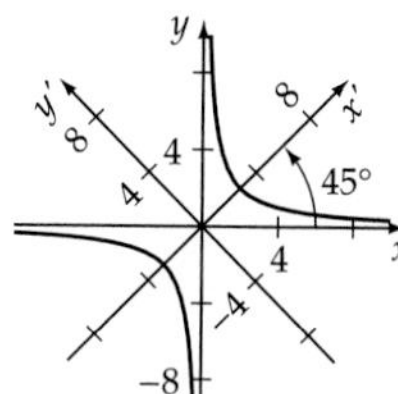

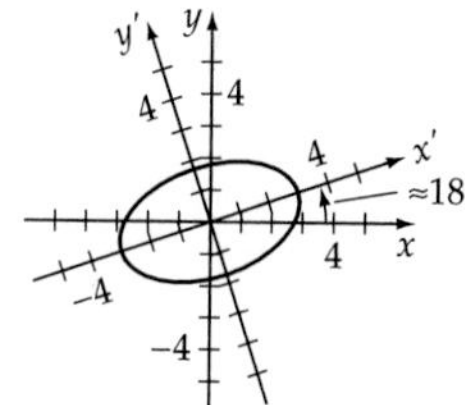

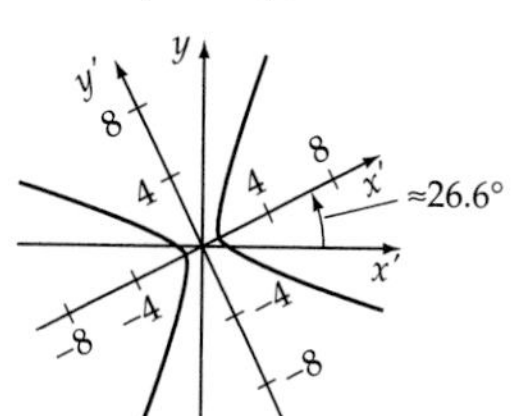

13. 30°, $y' = x'^2 + 4$ **15.** 36.9°, $y'^2 = 2(x' - 2)$ **17.** 36.9°, $15x'^2 - 10y'^2 + 6x' + 28y' + 11 = 0$

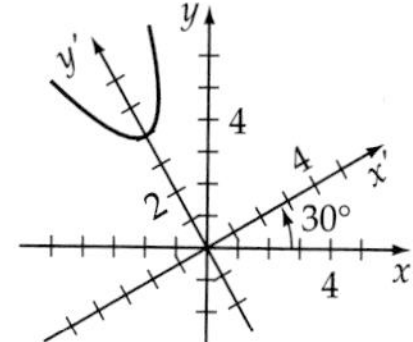

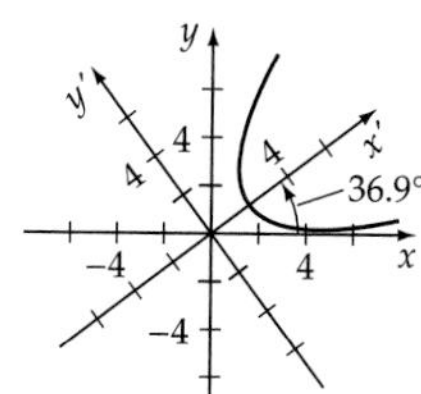

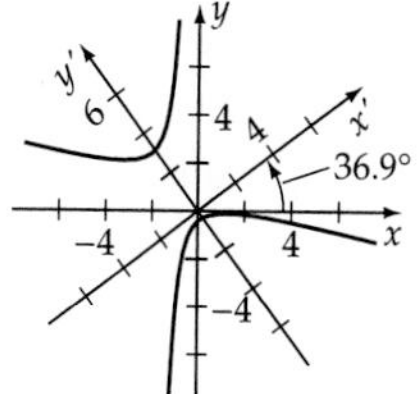

19.

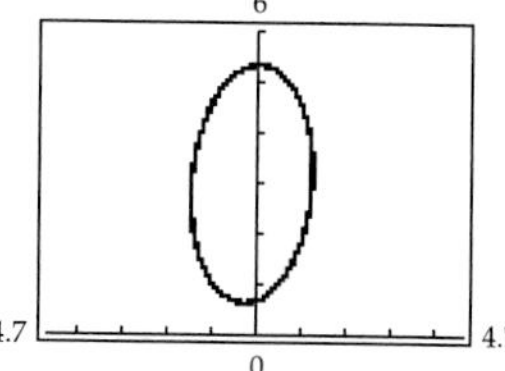

21.

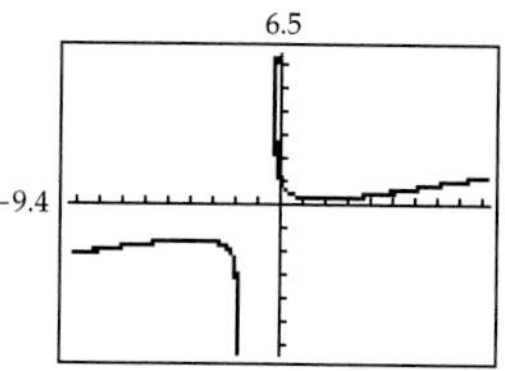

23.

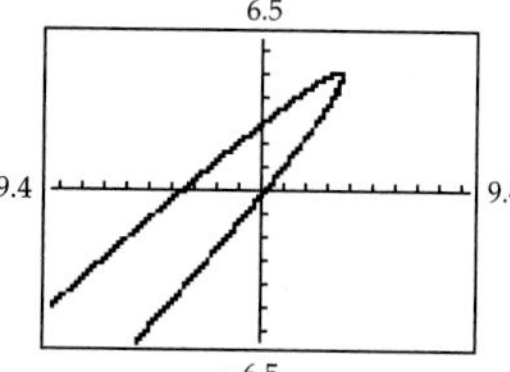

25. $y = \frac{\sqrt{3} + 2\sqrt{2}}{2\sqrt{3} - \sqrt{2}}x$ and $y = \frac{\sqrt{3} - 2\sqrt{2}}{2\sqrt{3} + \sqrt{2}}x$ **27.** $\left(\frac{3\sqrt{15}}{5}, \frac{\sqrt{15}}{5}\right)$ and $\left(-\frac{3\sqrt{15}}{5}, -\frac{\sqrt{15}}{5}\right)$ **29.** hyperbola **31.** parabola **33.** parabola **35.** hyperbola **39.** $9x^2 - 4xy + 6y^2 = 100$

Exercise Set 8.5, page 638

1–7.

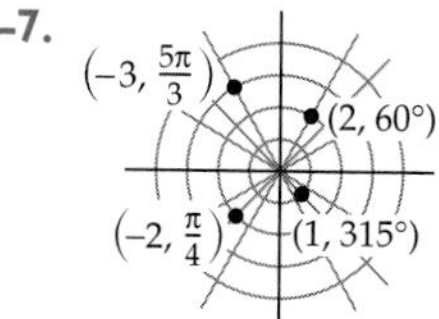

9.

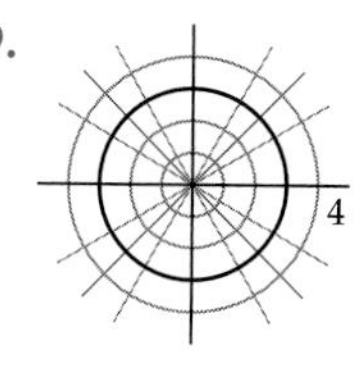

$0 \le \theta \le 2\pi$

11.

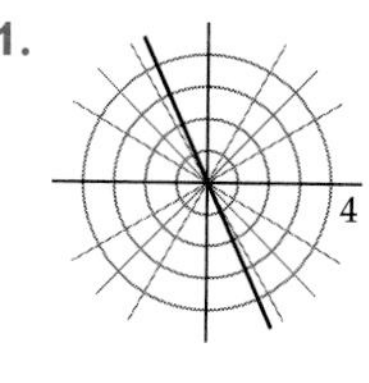

13.

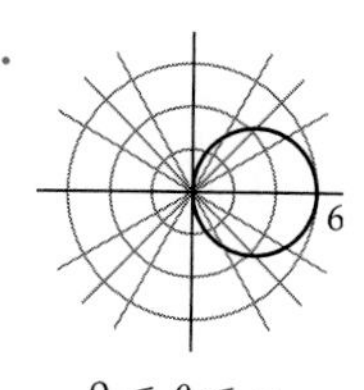

$0 \le \theta \le \pi$

15.

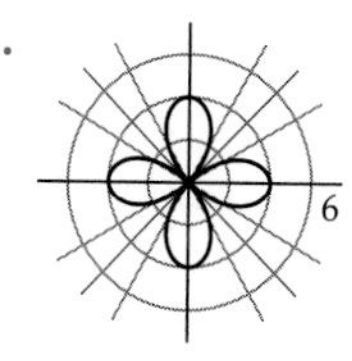

$0 \le \theta \le 2\pi$

17.

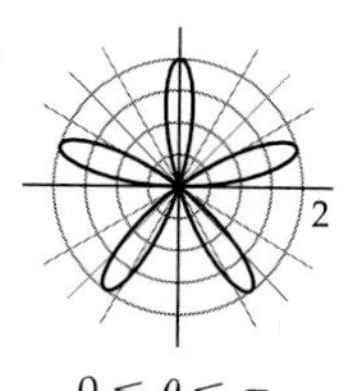

$0 \le \theta \le \pi$

19.

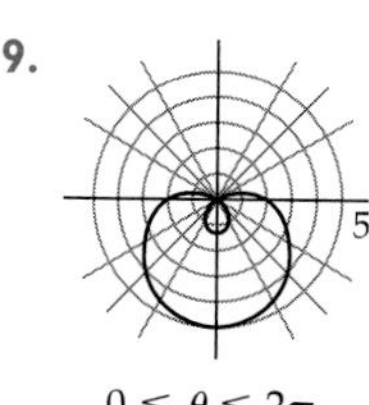

$0 \le \theta \le 2\pi$

21.

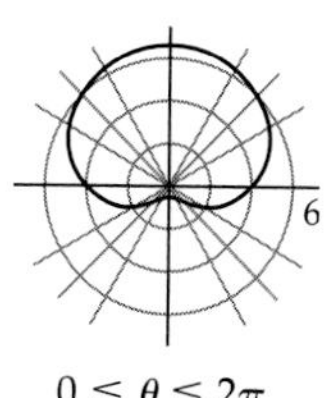

$0 \le \theta \le 2\pi$

23.

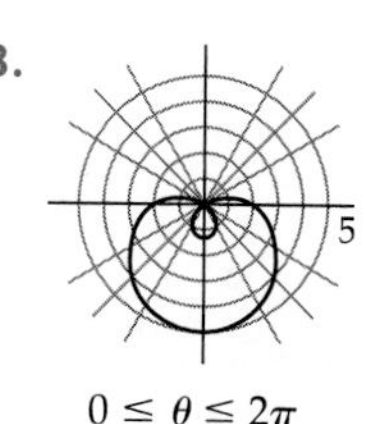

$0 \le \theta \le 2\pi$

25.

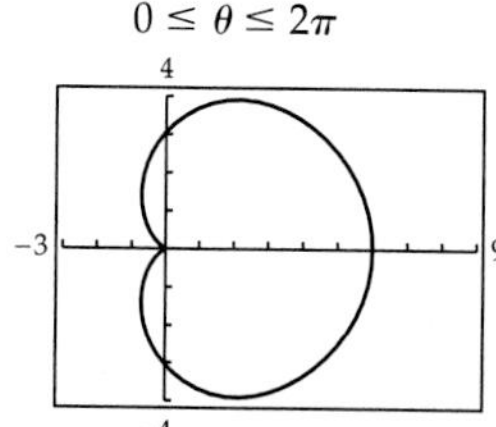

$0 \le \theta \le 2\pi$

27.

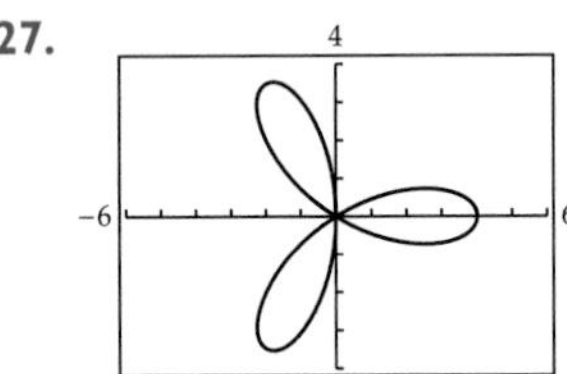

$0 \le \theta \le \pi$

29.

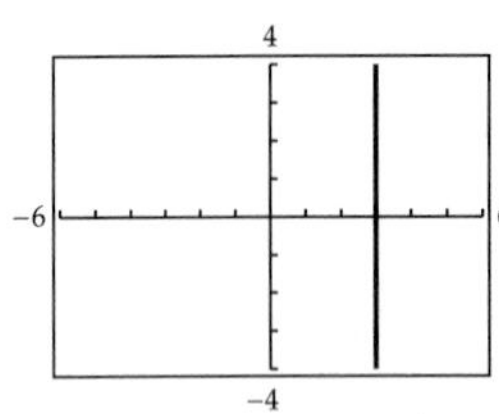

$0 \le \theta \le \pi$

31.

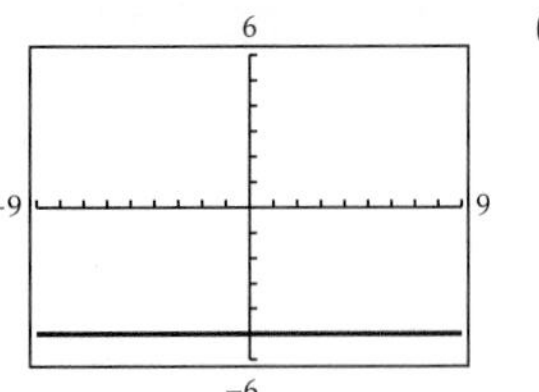

$0 \le \theta \le \pi$

33.

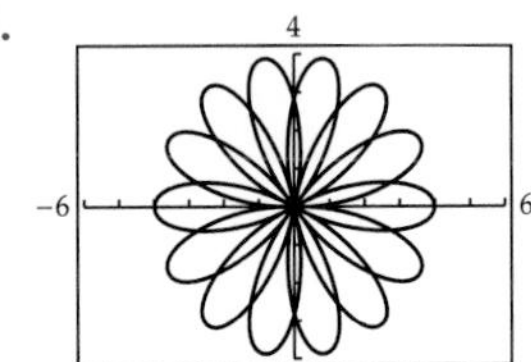

$0 \le \theta \le 4\pi$

35.

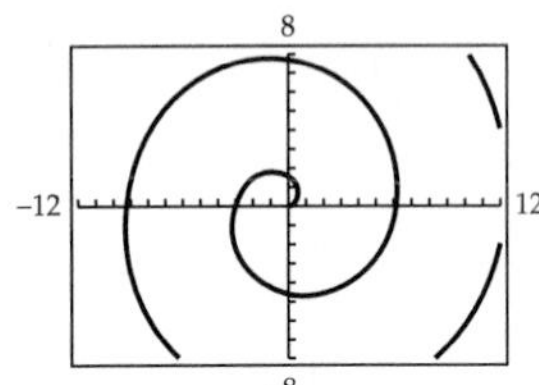

$0 \le \theta \le 6\pi$

37.

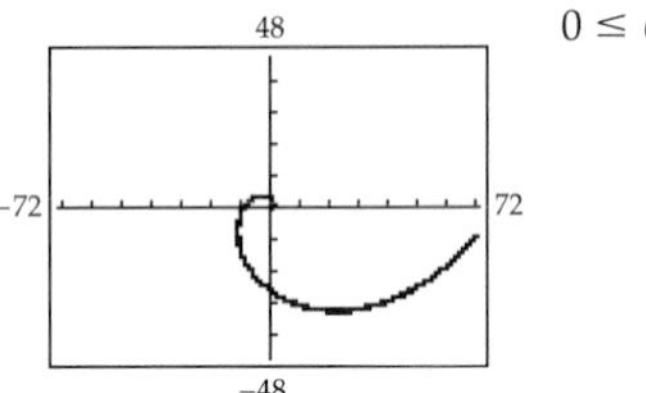

$0 \le \theta \le 2\pi$

39.

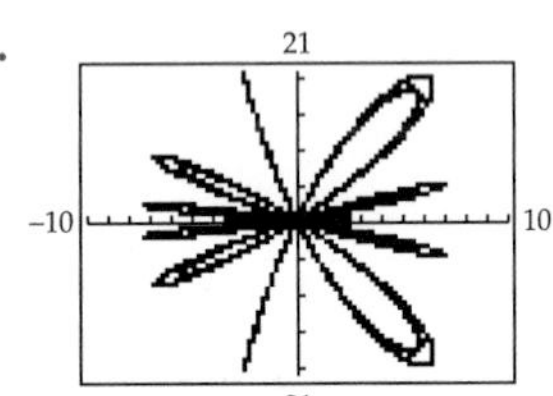

$0 \le \theta \le 2\pi$

41. $(2, -60°)$ **43.** $(3/2, -3\sqrt{3}/2)$ **45.** $(0, 0)$ **47.** $(5, 53.1°)$ **49.** $x^2 + y^2 - 3x = 0$

51. $x = 3$ **53.** $x^2 + y^2 = 16$ **55.** $x^4 - y^2 + x^2y^2 = 0$ **57.** $y^2 + 4x - 4 = 0$ **59.** $y = 2x + 6$ **61.** $r = 2 \csc \theta$ **63.** $r = 2$

65. $r \cos^2 \theta = 8 \sin \theta$ **67.** $r^2(\cos 2\theta) = 25$

69.

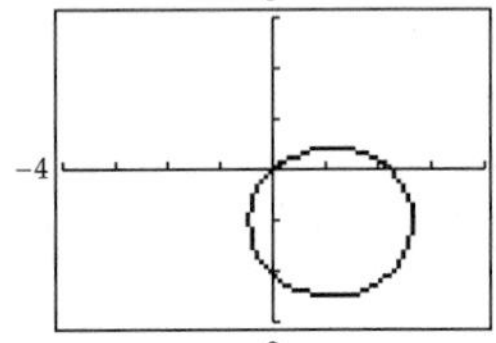

71.

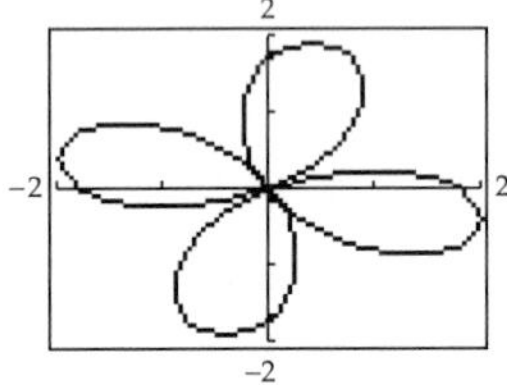

73.

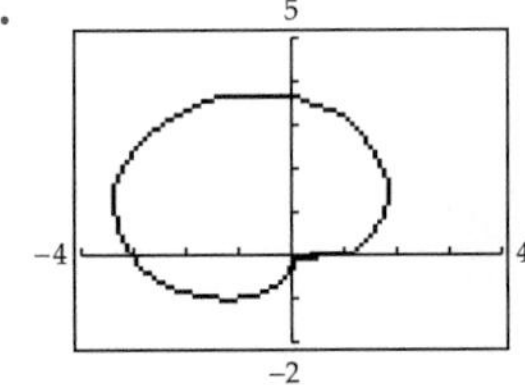

75.

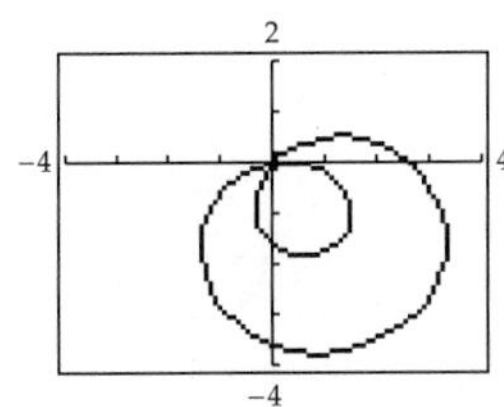

79.

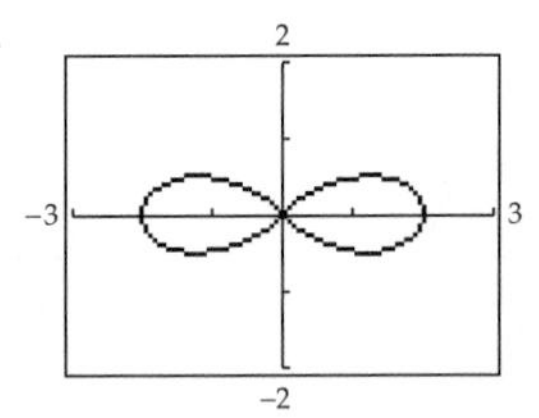

$0 \le \theta \le 2\pi$

81.

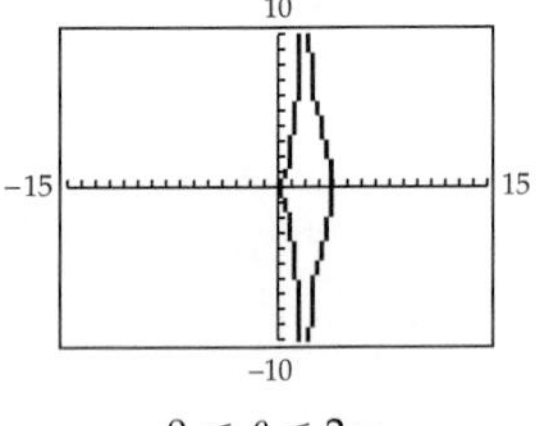

$0 \le \theta \le 2\pi$

83.

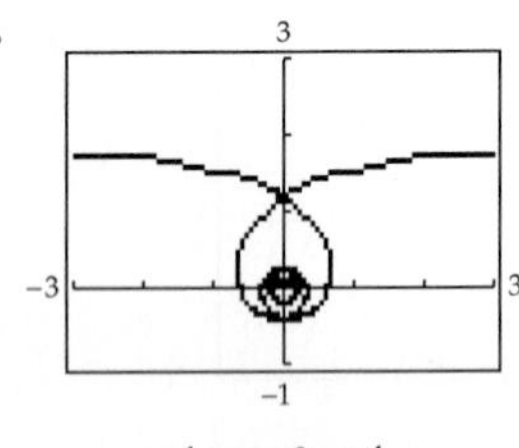

$-4\pi \le \theta \le 4\pi$

85.

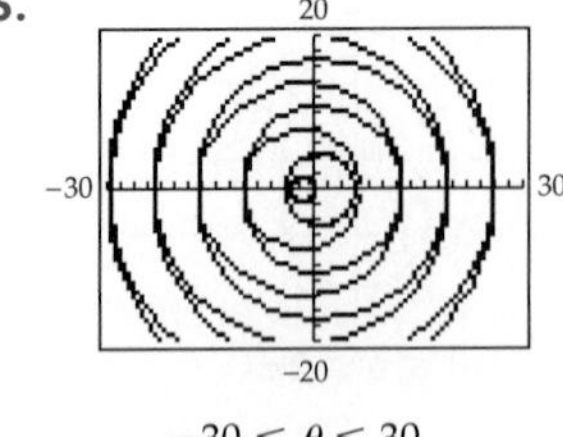

$-30 \le \theta \le 30$

87. a.

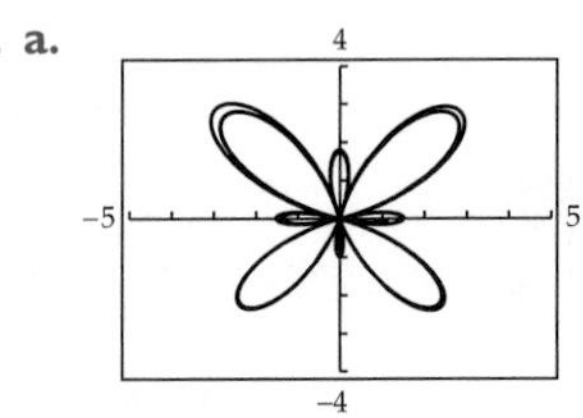

$0 \le \theta \le 5\pi$

b.

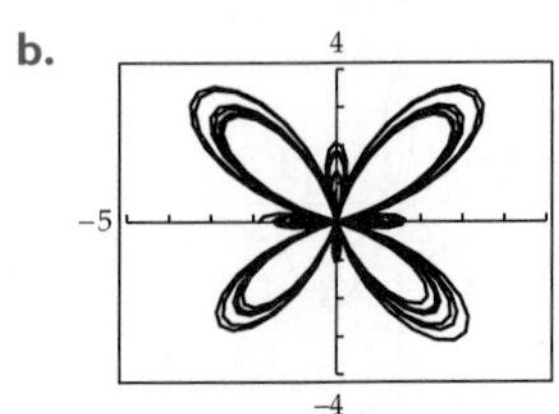

$0 \le \theta \le 20\pi$

Exercise Set 8.6, page 644

1. Hyperbola

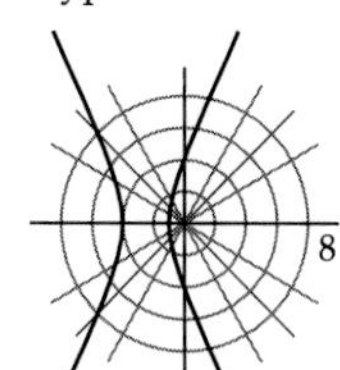

3. Ellipse

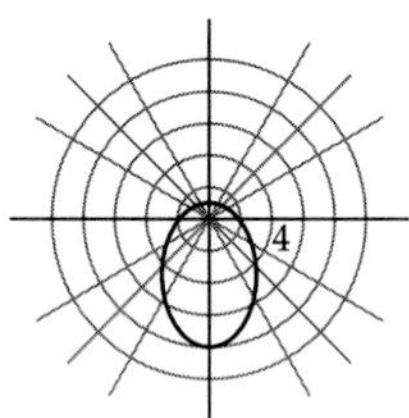

5. Parabola

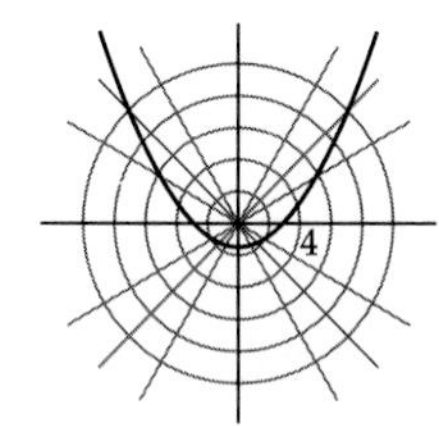

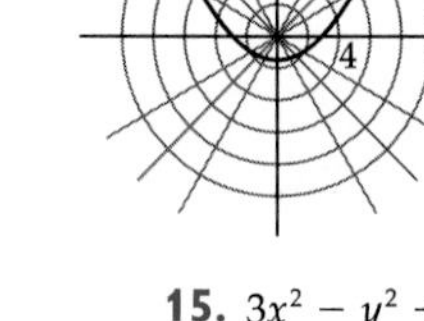

7. Hyperbola

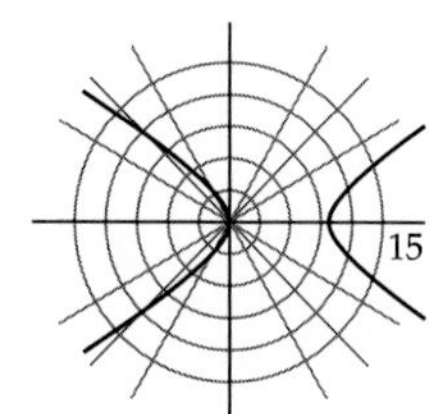

9. Ellipse with 2 holes

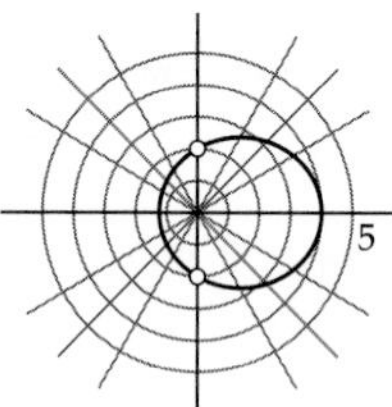

11. Ellipse with 2 holes

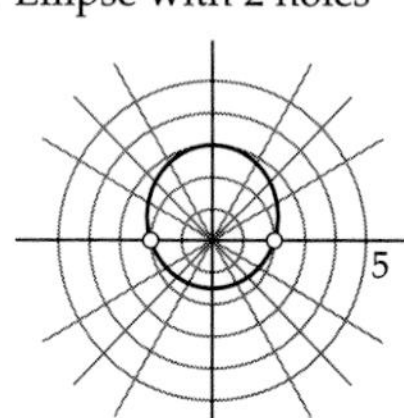

13. Parabola

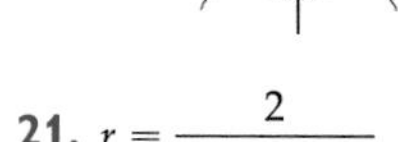

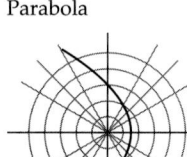

15. $3x^2 - y^2 + 16x + 16 = 0$ **17.** $16x^2 + 7y^2 + 48y - 64 = 0$

19. $x^2 - 6y - 9 = 0$

21. $r = \dfrac{2}{1 - 2\cos\theta}$ **23.** $r = \dfrac{2}{1 + \sin\theta}$ **25.** $r = \dfrac{8}{3 - 2\sin\theta}$ **27.** $r = \dfrac{6}{2 + 3\cos\theta}$ **29.** $r = \dfrac{4}{1 - \cos\theta}$ **31.** $r = \dfrac{3}{1 - 2\sin\theta}$

33.

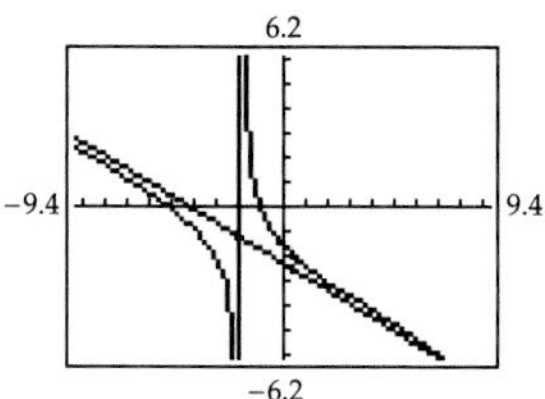

Rotate the graph in Exercise 1 counterclockwise $\pi/6$ radians.

35.

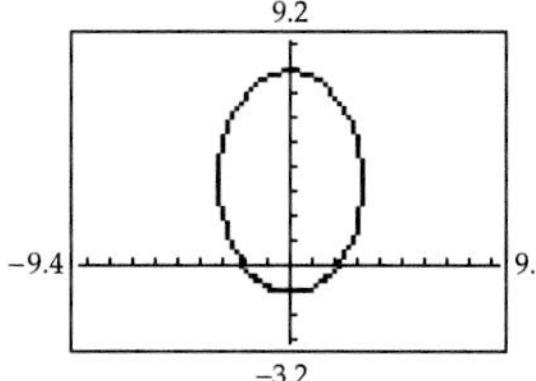

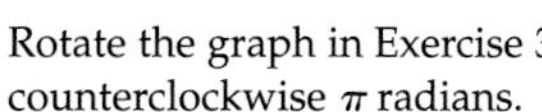

Rotate the graph in Exercise 3 counterclockwise π radians.

37.

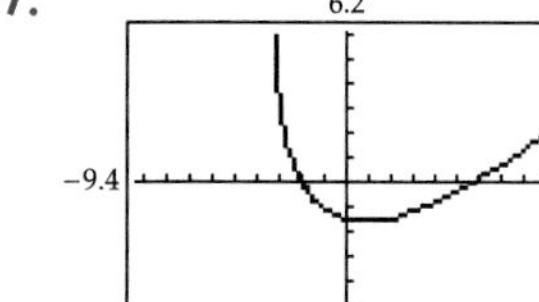

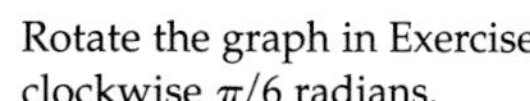

Rotate the graph in Exercise 5 clockwise $\pi/6$ radians.

39.

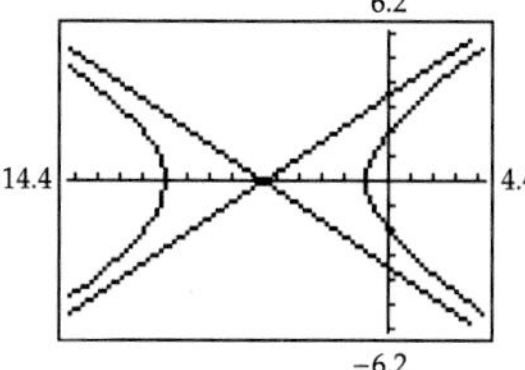

Rotate the graph in Exercise 7 clockwise π radians.

41.

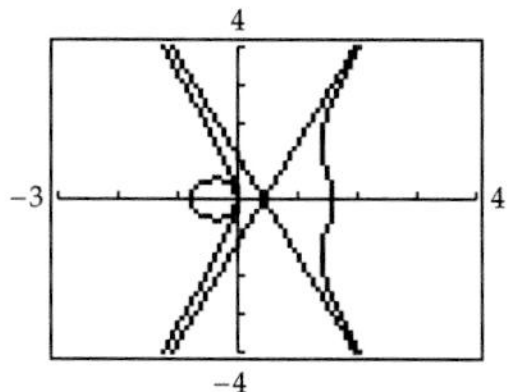

43.

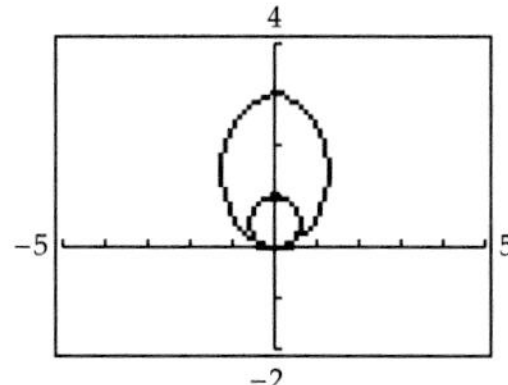

45.

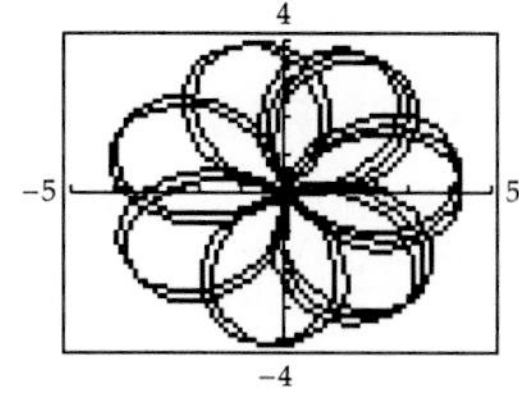

$0 \le \theta \le 12\pi$

Exercise Set 8.7, page 650

1.

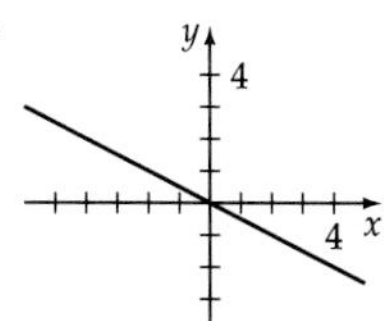

3.

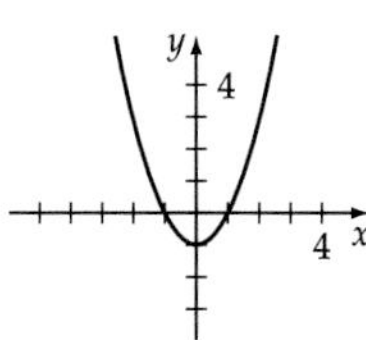

5.

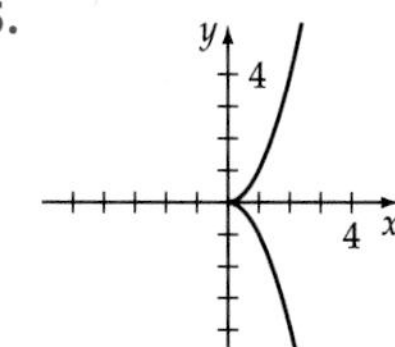

7.

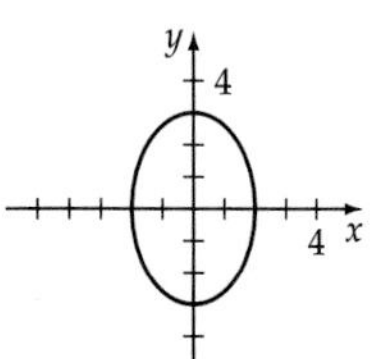

9.

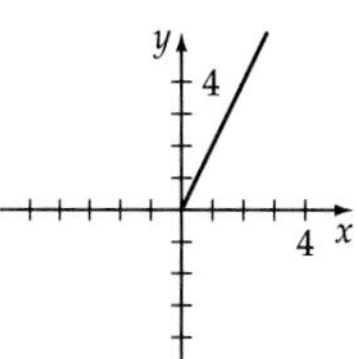

11. $x^2 - y^2 - 1 = 0$
$x \ge 1$
$y \in R$

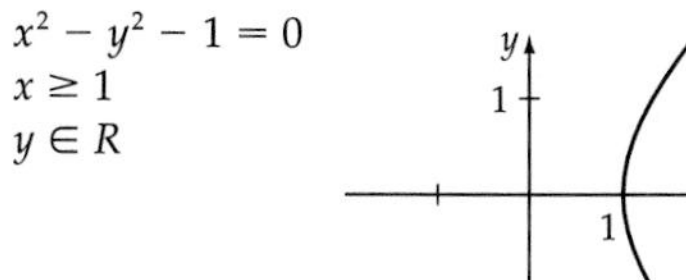

13. $y = -2x + 7$
$x \le 2$
$y \ge 3$

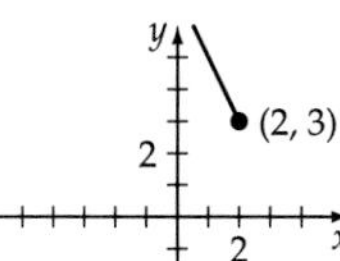

15. $x^{2/3} + y^{2/3} = 1$
$-1 \le x \le 1$
$-1 \le y \le 1$

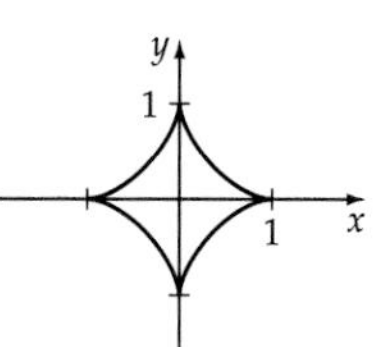

17. $y = x^2 - 1$
$x \geq 0$
$y \geq -1$

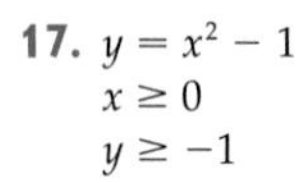

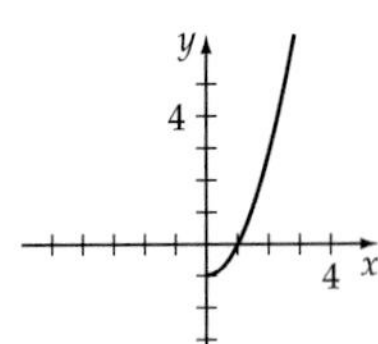

19. $y = \ln x$
$x > 0$
$y \in R$

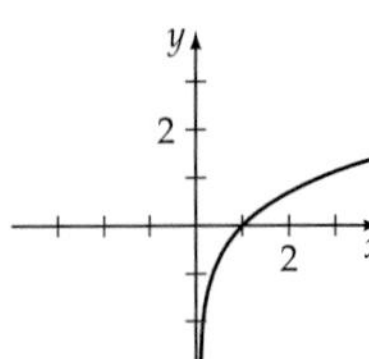

21. C_1: $y = -2x + 5, x \geq 2$; C_2: $y = -2x + 5, x \in R$.
C_2 is a line. C_1 is a ray.

23.

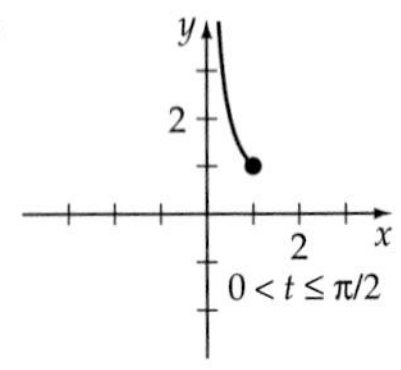

25.

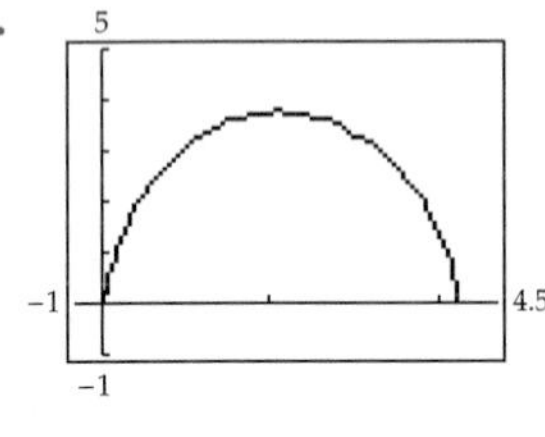

Xscl $= 2\pi$

27.

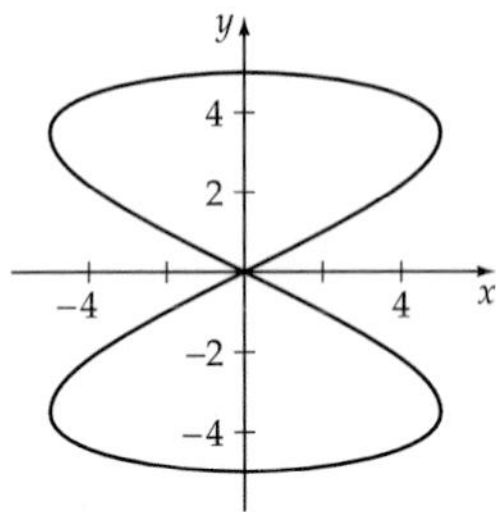

29.

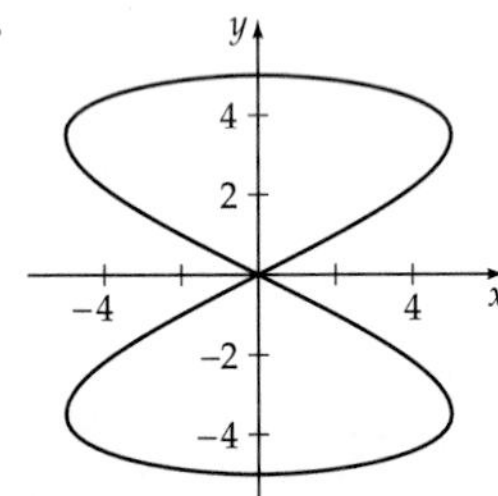

31.

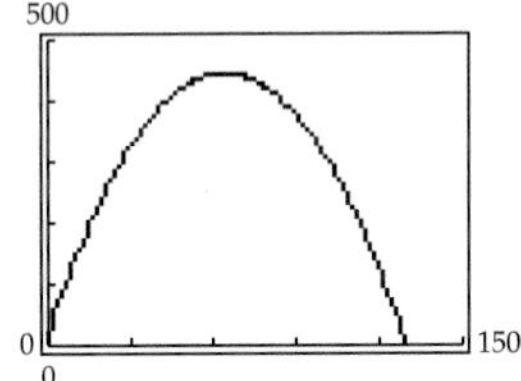

Max height (nearest foot) of 462 feet is attained when $t \approx 5.38$ seconds.

Range (nearest foot) of 1295 feet is attained when $t \approx 10.75$ seconds.

33.

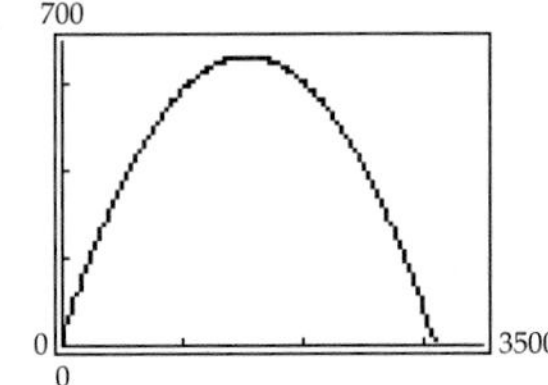

Max height (nearest foot) of 694 feet is attained when $t \approx 6.59$ seconds.

Range (nearest foot) of 3084 feet is attained when $t \approx 13.17$ seconds.

37. $x = a\cos\theta + a\theta\sin\theta$
$y = a\sin\theta - a\theta\cos\theta$

39. $x = (b - a)\cos\theta + a\cos\left(\frac{b-a}{a}\theta\right)$
$y = (b - a)\sin\theta - a\sin\left(\frac{b-a}{a}\theta\right)$

Chapter 8 True/False Exercises, page 654

1. False; a parabola has no asymptotes. **2.** True **3.** False; by keeping foci fixed and varying asymptotes, we can make conjugate axis any size needed. **4.** False; $\frac{x^2}{25} + \frac{y^2}{9} = 1$ and $\frac{x^2}{36} + \frac{y^2}{20} = 1$ have the same c's but different a's. **5.** False; parabolas have no asymptotes. **6.** True **7.** False; the graph of a parabola can be a function. **8.** False; $x = \cos t$, $y = \sin t$ graphs to be a circle. **9.** True **10.** True

Chapter 8 Review Exercises, page 654

1. vertices: $(\pm 2, 0)$
foci: $(\pm 2\sqrt{2}, 0)$
asymptotes: $y = \pm x$

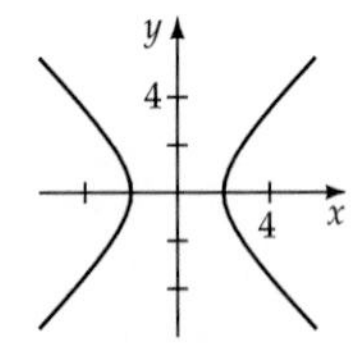

3. vertices: $(-1, -1)$, $(7, -1)$
foci: $(3 \pm 2\sqrt{3}, -1)$

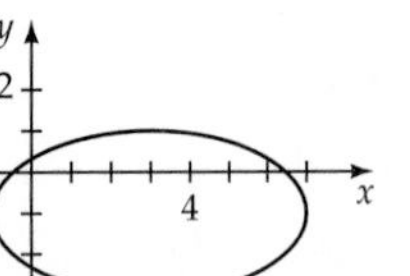

5. vertex: $(-2, 1)$
foci: $(-29/16, 1)$

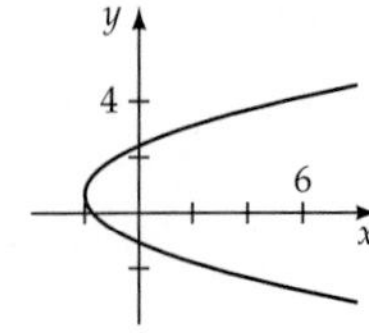

7. vertices: $(-2, -2)$, $(-2, 4)$
foci: $(-2, 1 \pm \sqrt{5})$

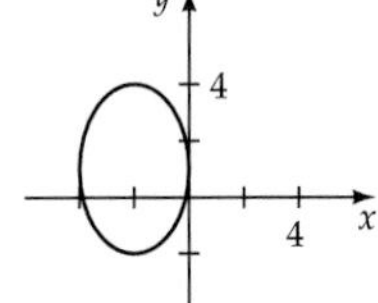

9. vertices: $(-5, 2/3)$, $(7, 2/3)$
foci: $(1 \pm 2\sqrt{13}, 2/3)$
asymptotes: $y - 2/3 = \pm 2(x - 1)/3$

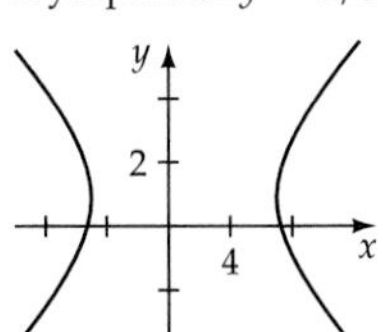

11. vertex: $(-7/2, -1)$
focus: $(-7/2, -3)$

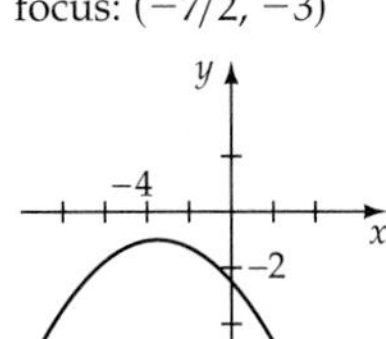

13. $\dfrac{(x-2)^2}{25} + \dfrac{(y-3)^2}{16} = 1$

15. $\dfrac{(x+2)^2}{4} - \dfrac{(y-2)^2}{5} = 1$

17. $x^2 = 3(y+2)/2$ or $(y+2)^2 = 12x$

19. $\dfrac{x^2}{36} - \dfrac{y^2}{4/9} = 1$

21. $x'^2 + 2y'^2 - 4 = 0$, ellipse

23. $x'^2 - 4y' + 8 = 0$, parabola

25. 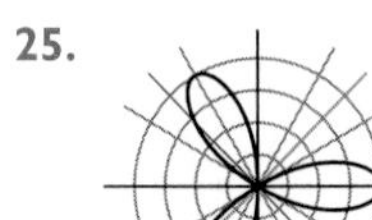
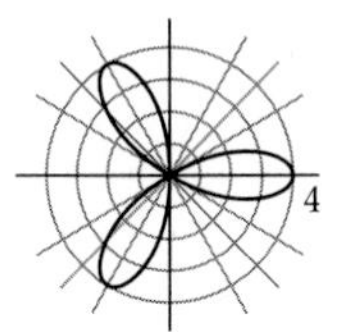

27.

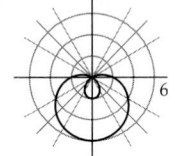

29.

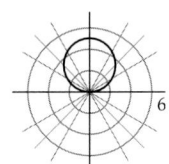

31. 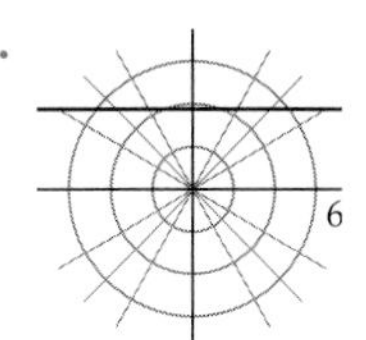

33. 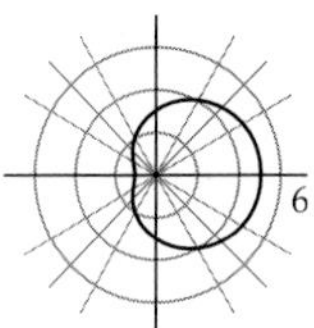

35. $r \sin^2 \theta = 16 \cos \theta$ 37. $3r \cos \theta - 2r \sin \theta = 6$ 39. $y^2 = 8x + 16$ 41. $x^4 + y^4 + 2x^2y^2 - x^2 + y^2 = 0$

43. 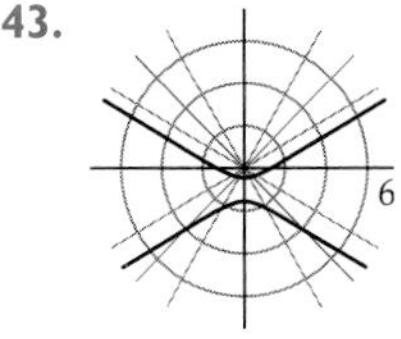

45. 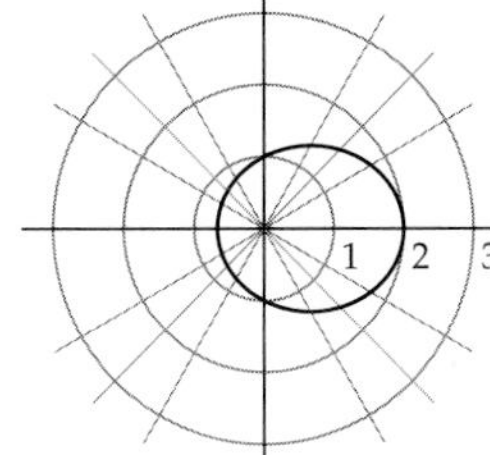

47. $y = \dfrac{3}{4}x + \dfrac{5}{2}$

49. $\dfrac{x^2}{16} + \dfrac{y^2}{9} = 1$

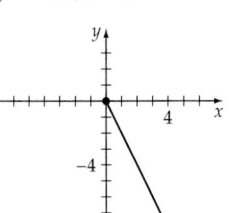

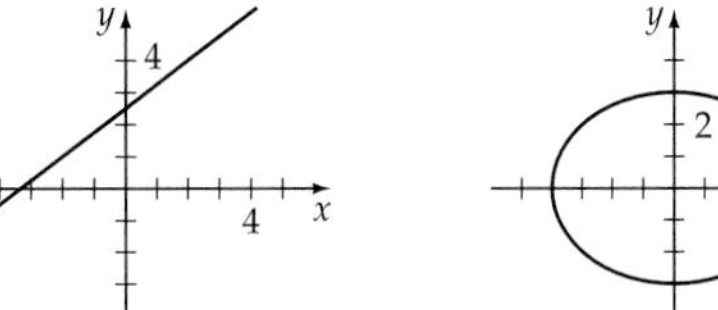

51. $y = -2x, x > 0$

53. $y = 2^{-x^2}, x \geq 0$

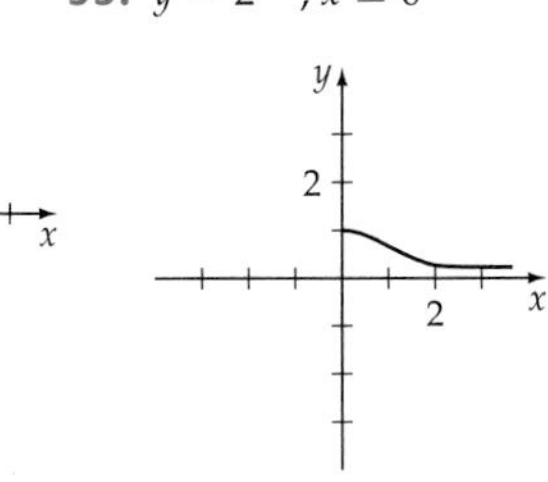

55. 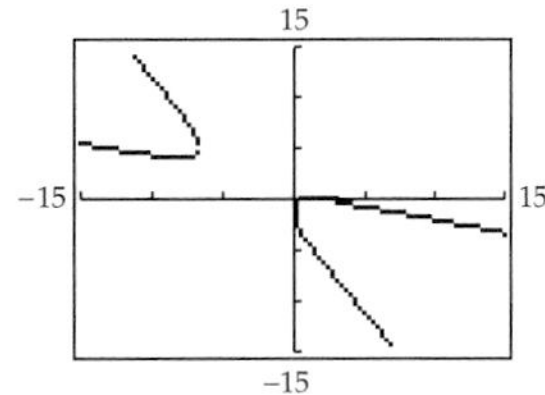

57. 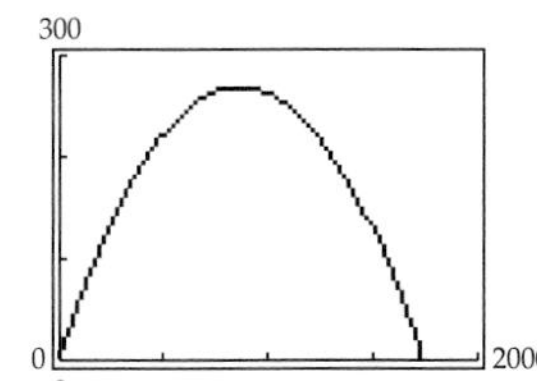

Max. height (nearest foot) of 278 feet is attained when $t \approx 4.17$ seconds.

Chapter 8 Test, page 655

1. focus: $(0, 2)$
vertex: $(0, 0)$
directrix: $y = -2$

2. 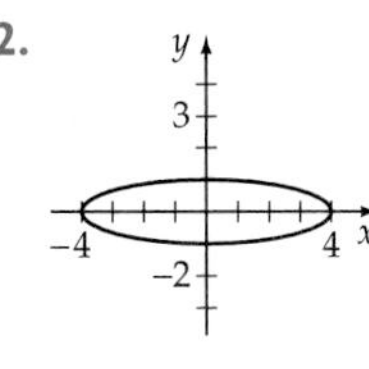

3. vertices: $(3, 4)$, $(3, -6)$
foci: $(3, 3)$, $(3, -5)$

4. $\dfrac{x^2}{45} + \dfrac{(y+3)^2}{9} = 1$

5.

6. vertices: $(6, 0)$, $(-6, 0)$
foci: $(-10, 0)$, $(10, 0)$
asymptotes: $y = \pm 4x/3$

7. 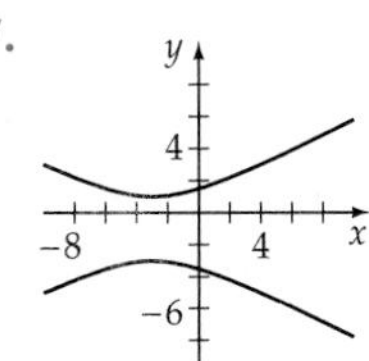

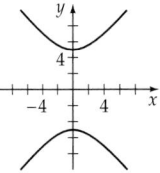

8. 73.15° 9. ellipse 10. $P(2, 300°)$

11. 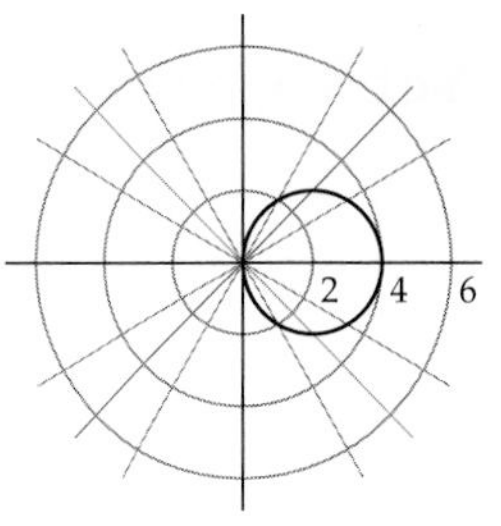

12.

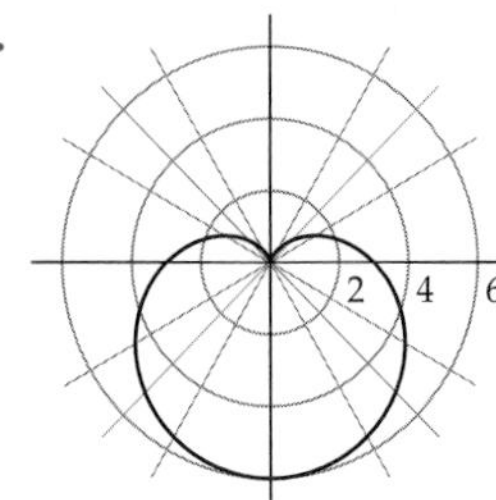

13.

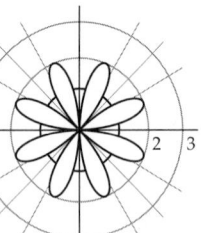

14. $(5/2, 5\sqrt{3}/2)$ **15.** $y^2 - 8x - 16 = 0$ **16.** $x^2 + 8y - 16 = 0$

17. $(x + 3)^2 = \frac{1}{2}y$

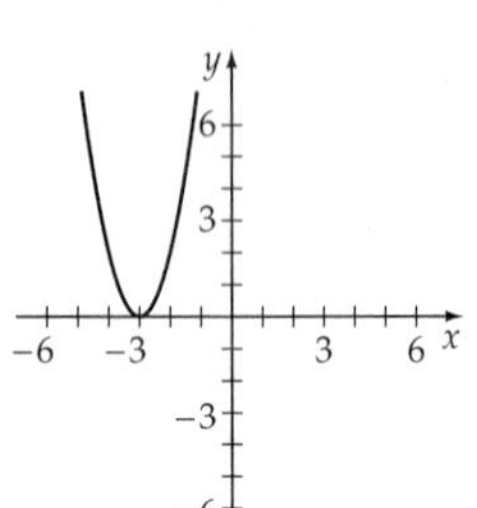

18. $\frac{x^2}{16} + \frac{(y - 2)^2}{1} = 1$

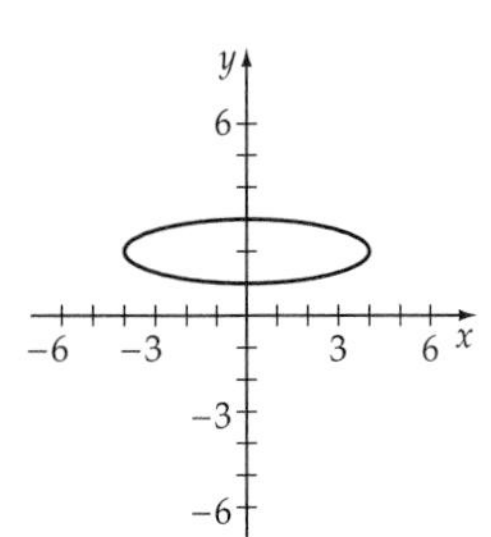

19.

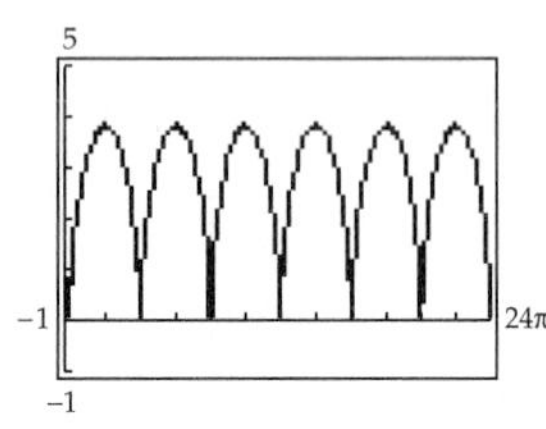

Xscl = 2π

20. $256\sqrt{3}$ feet $\approx$ 443 feet

Exercise Set 9.1, page 665

1. $(2, -4)$ **3.** $(-6/5, 27/5)$ **5.** $(3, 4)$ **7.** $(1, -1)$ **9.** $(3, -4)$ **11.** $(2, 5)$ **13.** $(-1, -1)$ **15.** $(62/25, 34/25)$ **17.** no solution **19.** $(c, -4c/3 + 2)$ **21.** $(2, -4)$ **23.** $(0, 3)$ **25.** $(3c/5, c)$ **27.** $(-1/2, 2/3)$ **29.** no solution **31.** $(-6, 3)$ **33.** $(2, -3/2)$ **35.** $(2\sqrt{3}, 3)$ **37.** $(38/(17\pi), 3/17)$ **39.** $(\sqrt{2}, \sqrt{3})$ **41.** plane: 120 mph, wind: 30 mph **43.** boat: 25 mph, current: 5 mph **45.** \$12 per kilogram for iron, \$16 per kilogram for lead **47.** 9/5 square units **49.** 8 **51.** \$14,000 at 6%, \$11,000 at 6.5% **53.** 8 gm of 40% gold, 12 gm of 60% gold **55.** 20 ml of 13% solution, 30 ml of 18% solution **57.** $x = -58/17, y = 52/17$ **59.** $x = -2, y = -1$ **61.** $x = 153/26, y = 151/26$ **63.** $x = 2 + 3i, y = 1 - 2i$ **65.** $x = 3 - 5i, y = 4i$

Exercise Set 9.2, page 677

1. $(2, -1, 3)$ **3.** $(2, 0, -3)$ **5.** $(2, -3, 1)$ **7.** $(-5, 1, -1)$ **9.** $(3, -5, 0)$ **11.** $(0, 2, 3)$ **13.** $(5c - 25, 48 - 9c, c)$ **15.** $(3, -1, 0)$ **17.** no solution **19.** $((50 - 11c)/11, (11c - 18)/11, c)$ **21.** no solution **23.** $((25 + 4c)/29, (55 - 26c)/29, c)$ **25.** $(0, 0, 0)$ **27.** $(5c/14, 4c/7, c)$ **29.** $(-11c, -6c, c)$ **31.** $(0, 0, 0)$ **33.** $y = 2x^2 - x - 3$ **35.** $x^2 + y^2 - 4x + 2y - 20 = 0$ **37.** center $(-7, -2)$, radius 13 **39.** 5 dimes, 10 nickels, 4 quarters **41.** 685 **43.** $(3, 5, 2, -3)$ **45.** $(1, -2, -1, 3)$ **47.** $(14a - 7b - 8, -6a + 2b + 5, a, b)$ **49.** $A = -13/2$ **51.** $A \neq -3, A \neq 1$ **53.** $A = -3$ **55.** $3x - 5y - 2z = -2$

Exercise Set 9.3, page 683

1. $(1, 0), (2, 2)$ **3.** $((2 + \sqrt{2})/2, (-6 + \sqrt{2})/2, ((2 - \sqrt{2})/2, (-6 - \sqrt{2})/2)$ **5.** $(5, 18)$ **7.** $(4, 6), (6, 4)$ **9.** $(-3/2, -4), (2, 3)$ **11.** $(19/29, -11/29), (1, 1)$ **13.** $(-2, 9), (1, -3), (-1, 1)$ **15.** $(-2, 1), (-2, -1), (2, 1), (2, -1)$ **17.** $(4, 2), (-4, 2), (4, -2), (-4, -2)$ **19.** no real number solution **21.** $(12/5, 1/5), (2, 1)$ **23.** $(26/5, -3/5), (1, -2)$ **25.** $(39/10, -7/10), (3, 2)$ **27.** $((-3 + \sqrt{3})/2, (1 + \sqrt{3})/2), ((-3 - \sqrt{3})/2, (1 - \sqrt{3})/2)$ **29.** $(19/13, 22/13), (1, 4)$ **31.** no real number solution **33.** 82 units **35.** $r \geq \sqrt{\frac{1}{5}}$ or $\frac{\sqrt{5}}{5}$ **37.** $(0, 1), (1, 2)$ **39.** $(0.7035, 0.4949)$ **41.** $(1.7549, 1.3247)$ **43.** $(-0.7071, 0.7071), (0.7071, 0.7071)$ **45.** $(1, 5)$ **47.** $(-1, 1), (1, -1)$ **49.** $(1, -2), (-1, 2)$

Exercise Set 9.4, page 691

1. $A = -3, B = 4$ **3.** $A = -2/5, B = 1/5$ **5.** $A = 1, B = -1, C = 4$ **7.** $A = 1, B = 3, C = 2$ **9.** $A = 1, B = 0, C = 1, D = 0$ **11.** $\frac{3}{x} + \frac{5}{x + 4}$ **13.** $\frac{7}{x - 9} + \frac{-4}{x + 2}$ **15.** $\frac{5}{2x + 3} + \frac{3}{2x + 5}$ **17.** $\frac{20}{11(3x + 5)} + \frac{-3}{11(x - 2)}$ **19.** $x + 3 + \frac{1}{x - 2} + \frac{-1}{x + 2}$ **21.** $\frac{1}{x} + \frac{2}{x + 7} + \frac{-28}{(x + 7)^2}$ **23.** $\frac{2}{x} + \frac{3x - 1}{x^2 - 3x + 1}$ **25.** $\frac{2}{x + 3} + \frac{-1}{(x + 3)^2} + \frac{4}{x^2 + 1}$ **27.** $\frac{3}{x - 4} + \frac{5}{(x - 4)^2}$ **29.** $\frac{3x - 1}{x^2 + 10} + \frac{4x}{(x^2 + 10)^2}$

31. $\dfrac{1}{2k(k-x)} + \dfrac{1}{2k(k+x)}$ **33.** $x + \dfrac{1}{x} + \dfrac{-2}{x-1}$ **35.** $2x - 2 + \dfrac{3}{x^2 - x - 1}$ **37.** $\dfrac{1}{5(x+2)} + \dfrac{4}{5(x-3)}$

39. $\dfrac{1}{x} + \dfrac{2}{x^2} + \dfrac{3}{x^4} + \dfrac{-2}{x-2}$ **41.** $\dfrac{4}{3(x-1)} + \dfrac{2x+7}{3(x^2+x+1)}$

Exercise Set 9.5, page 698

1. 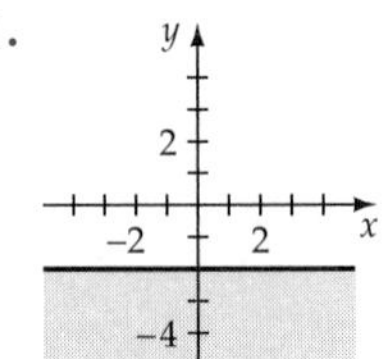

3. 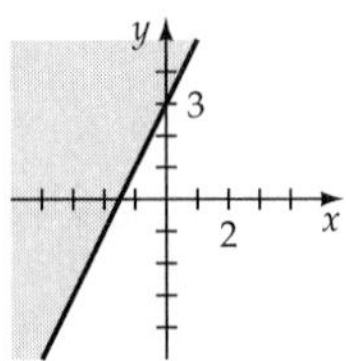

5. 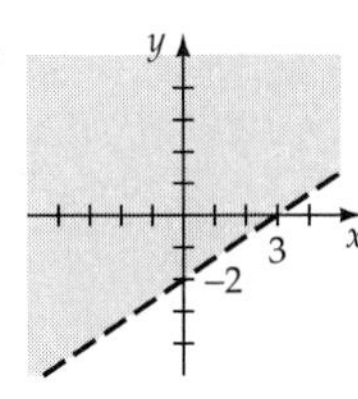

7. 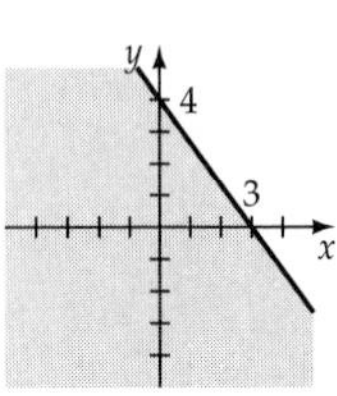

9. 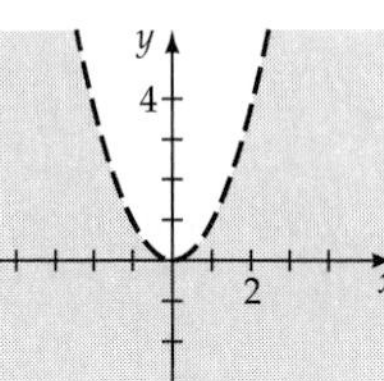

11. 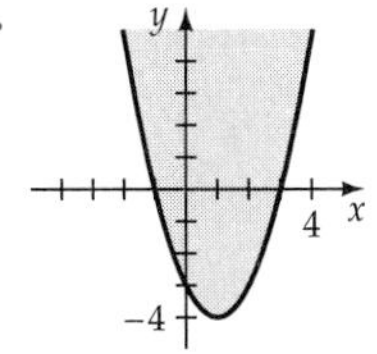

13. 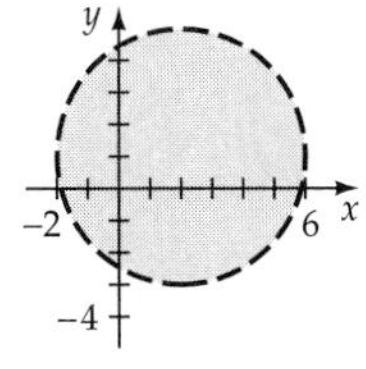

15. 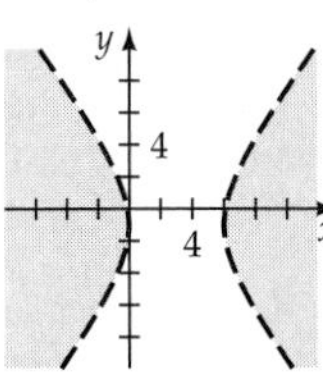

17. 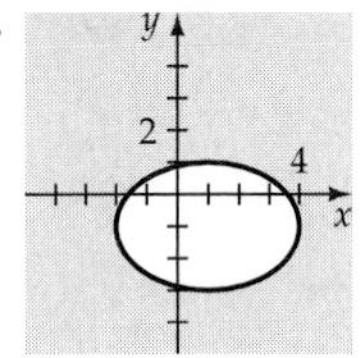

19. 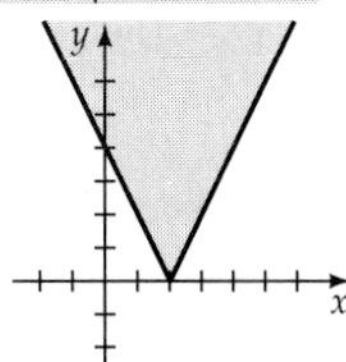

21. 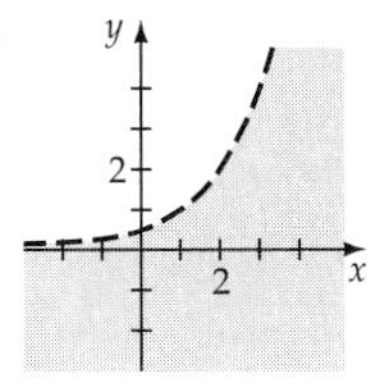

23. 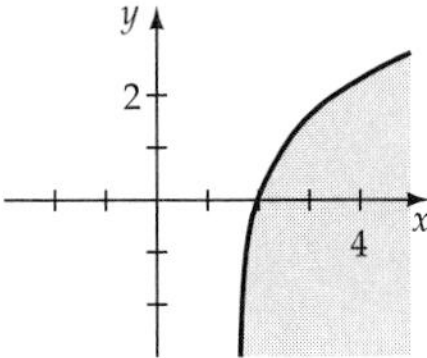

25. 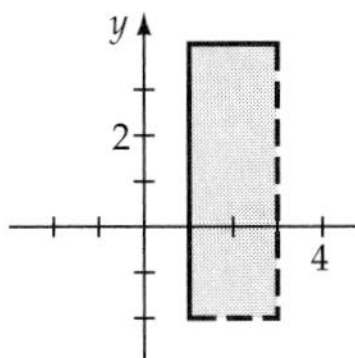

27. 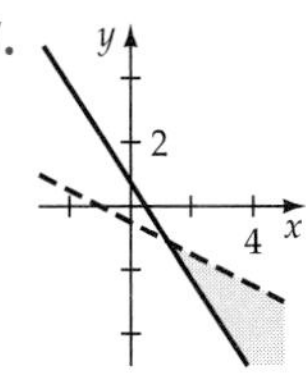

29. no solution

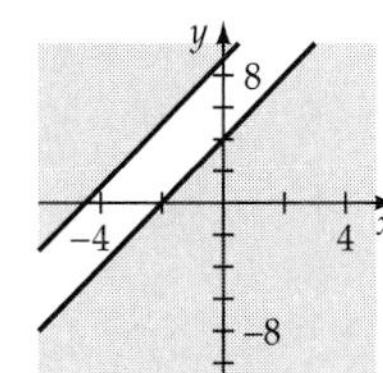

31. 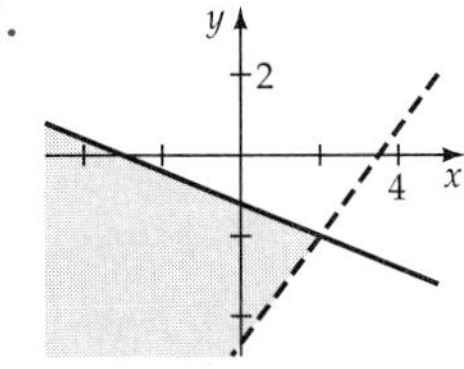

33. 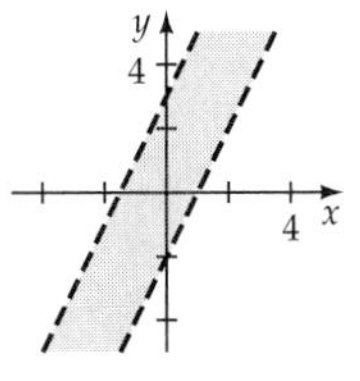

35. 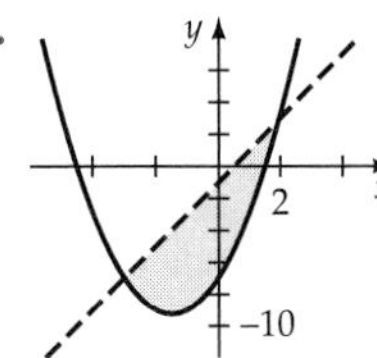

37. 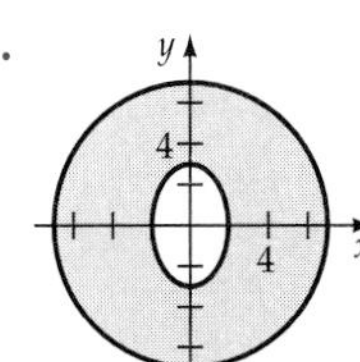

39. 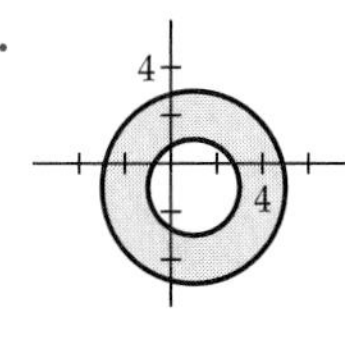

41. 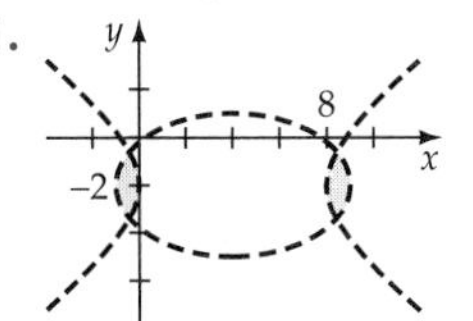

43. 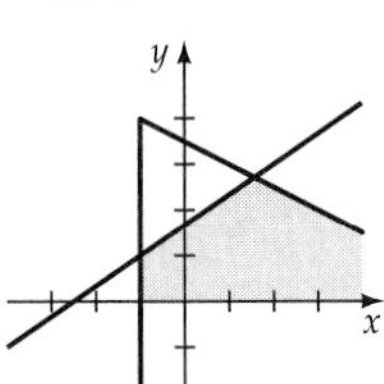

45. 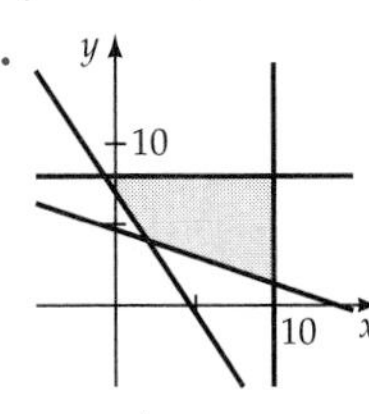

47. 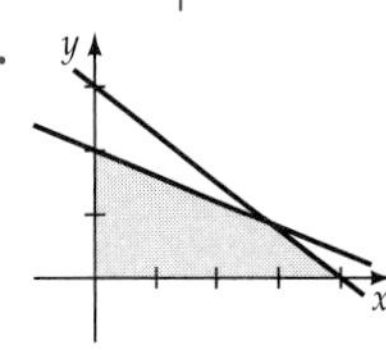

49. 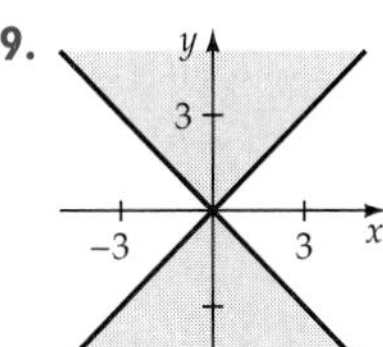

51. 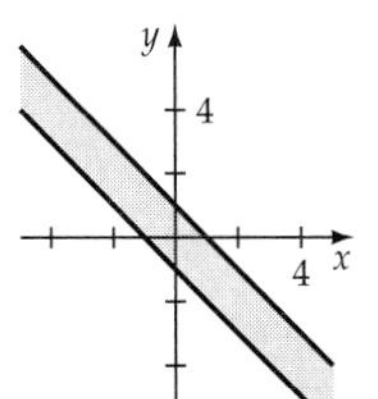

53. 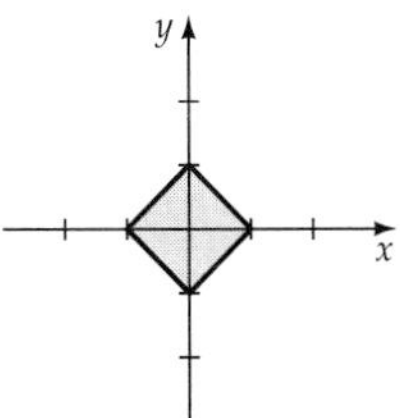

55. 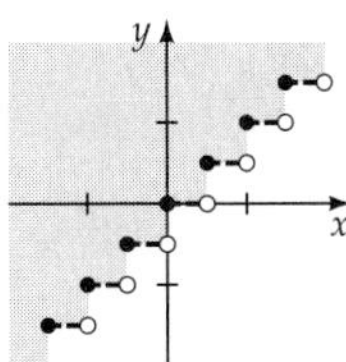

57. a. 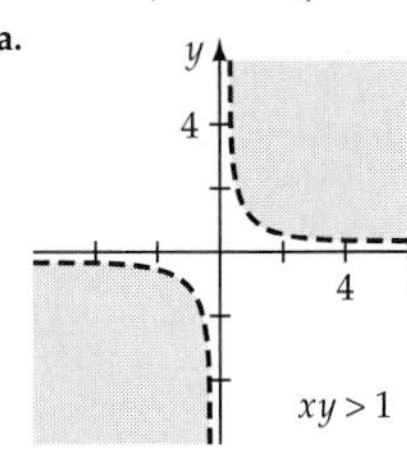

b. 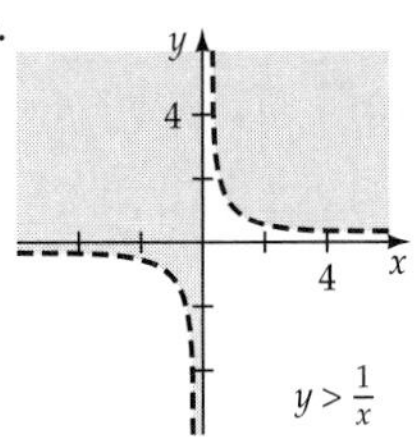

If x is a negative number, then the inequality is reversed when both sides of the inequality are multiplied by a negative number.

Exercise Set 9.6, page 705

1. minimum at $(0, 8)$: 16 **3.** maximum at $(6, 5)$: 71 **5.** minimum at $(0, 10/3)$: 20 **7.** maximum at $(0, 12)$: 72
9. minimum at $(0, 32)$: 32 **11.** maximum at $(0, 8)$: 56 **13.** minimum at $(2, 6)$: 18 **15.** maximum at $(3, 4)$: 25
17. minimum at $(2, 3)$: 12 **19.** maximum at $(100, 400)$: 3400 **21.** 20 acres of wheat and 40 acres of barley
23. 0 starter sets and 18 pro sets **25.** 24 ounces of group B and 0 ounces of group A; yields a minimum cost of \$2.40.
27. two 4-cylinder engines and seven 6-cylinder engines; yields a maximum profit of \$2050.

Chapter 9 True/False Exercises, page 709

1. False; $\begin{cases} x + y = 1 \\ x + y = 2 \end{cases}$ has no solution **2.** True **3.** False; a homogenous system is one where the constant term in each equation is zero.
4. True **5.** True **6.** False; $\begin{cases} x + y = 2 \\ x + 2y = 3 \end{cases}$ and $\begin{cases} 2x + 3y = 5 \\ 2x - 2y = 0 \end{cases}$ are two systems with the same solution but not common equations. **7.** True
8. True **9.** False; it is inconsistent **10.** False; $(-1, 1)$ satisfies the first equation but not the second, and $(-2, -1)$ satisfies the second but not the first.

Chapter 9 Review Exercises, page 709

1. $(-18/7, -15/28)$ **3.** $(-3, -1)$ **5.** $(3, 1)$ **7.** $((5 - 3c)/2, c)$ **9.** $(1/2, 3, -1)$ **11.** $((7c - 3)/11, (16c - 43)/11, c)$
13. $(2, (3c + 2)/2, c)$ **15.** $(14c/11, -2c/11, c)$ **17.** $((c + 1)/2, (3c - 1)/4, c)$ **19.** $(2, -3)$ **21.** no real solution **23.** $(1/5, 18/5)$, $(1, 2)$
25. $(2, 0)$, $(18/17, -64/17)$ **27.** $(2, 1)$, $(-2, -1)$ **29.** $(2, -3)$, $(-2, 3)$ **31.** $\frac{3}{x - 2} + \frac{4}{x + 1}$ **33.** $\frac{6x - 2}{5(x^2 + 1)} + \frac{-6}{5(x + 2)}$

35. $\frac{2}{x} + \frac{4}{x - 1} + \frac{5}{x + 1}$ **37.**

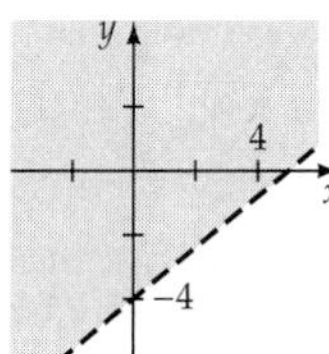

39. **41.**

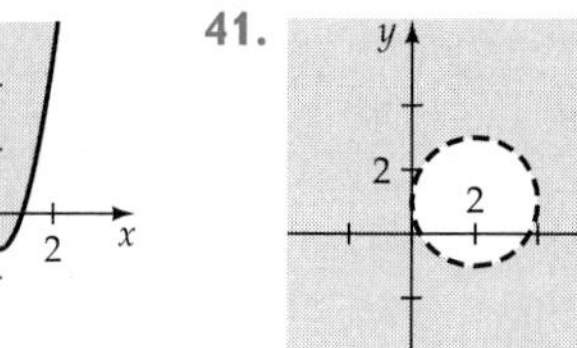

43.

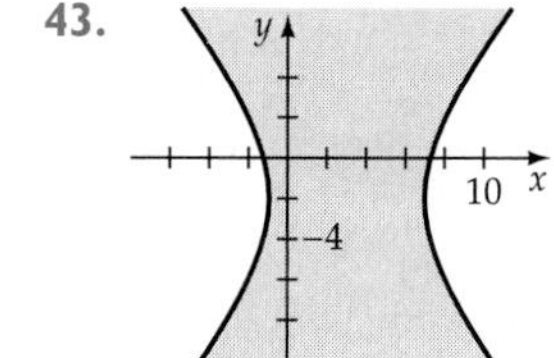

45.

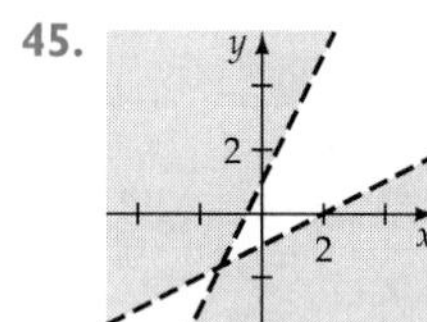

47.

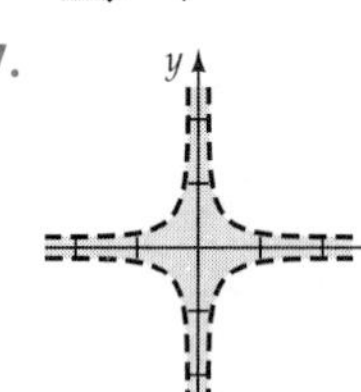

49.

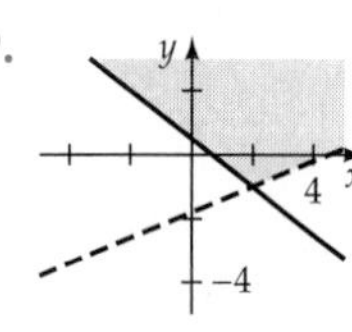

51.

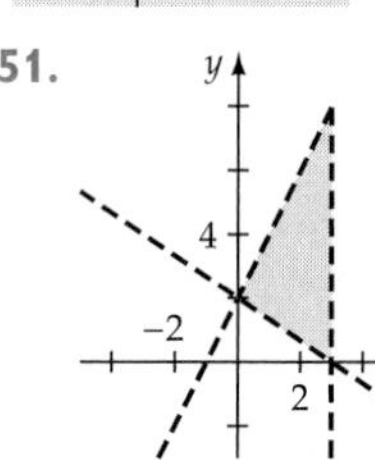

53.

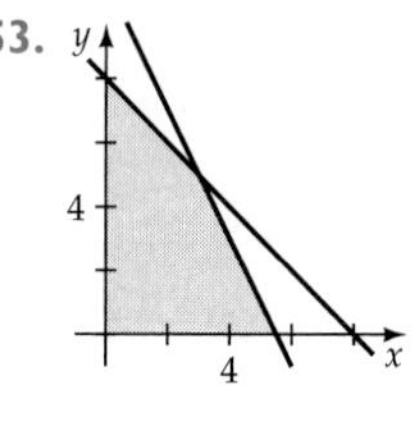

55.

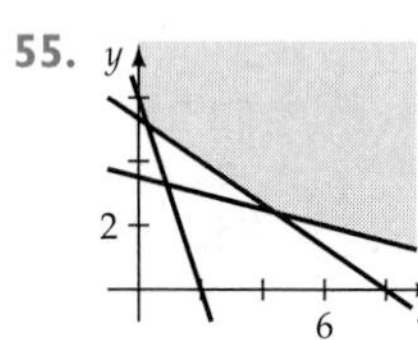

57.

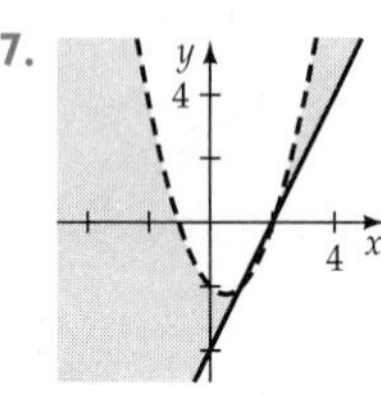

59.

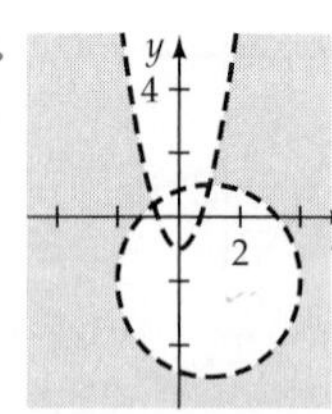

61. maximum at $(4, 5)$: 18 **63.** minimum at $(0, 8)$: 8

65. minimum at $(2, 5)$: 27 **67.** $y = \frac{11}{6}x^2 - \frac{5}{2}x + \frac{2}{3}$ **69.** $z = -2x + 3y + 3$ **71.** wind: 28 mph, plane: 143 mph
73. $(0, 0, 0)$, $(1, 1, 1)$, $(1, -1, -1)$, $(-1, -1, 1)$, $(-1, 1, -1)$

Chapter 9 Test, page 711

1. $(-3, 2)$ **2.** $((6 + c)/2, c)$ **3.** $(173/39, 29/39, -4/3)$ **4.** $((c + 3)/4, (7c + 1)/8, c)$ **5.** $((c + 10)/13, (5c + 11)/13, c)$
6. $(c/14, -9c/14, c)$ **7.** $(2, 5)$, $(-2, 1)$ **8.** $(-2, 3)$, $(-1, -1)$ **9.** **10.**

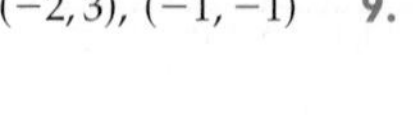

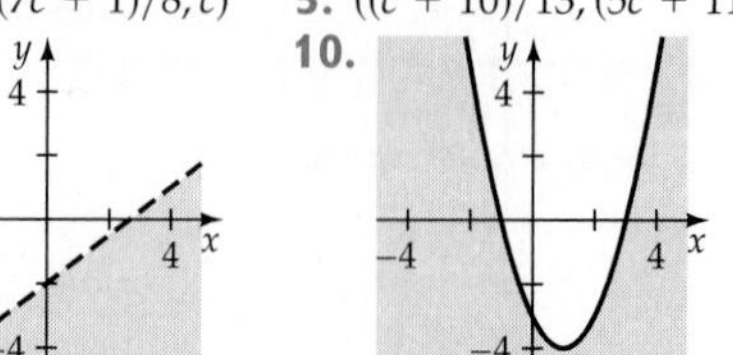

11.

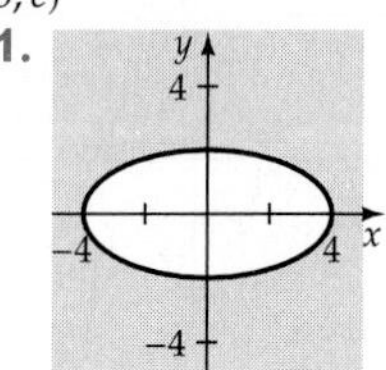

12.

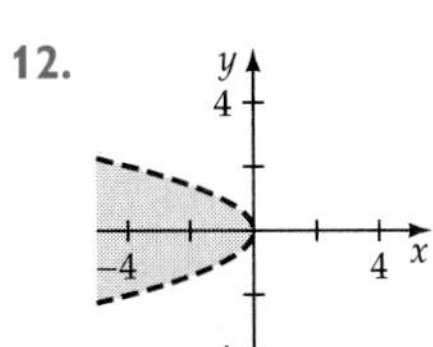

13.

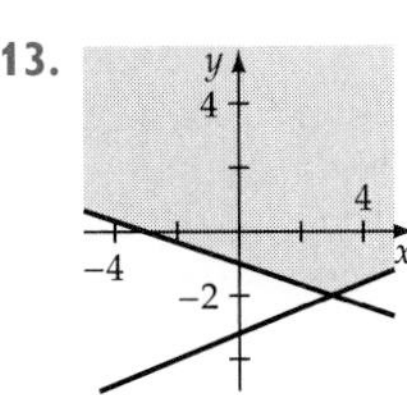

14. no graph; the solution set is the empty set.

15.

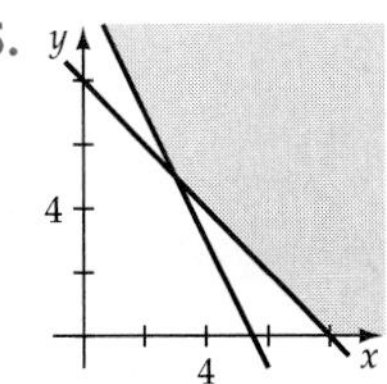

16.

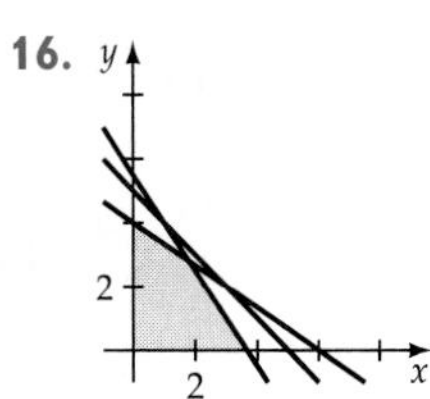

17. $\dfrac{7}{5(x-4)} + \dfrac{8}{5(x+1)}$ **18.** $\dfrac{1}{x} + \dfrac{-x+2}{x^2+1}$ **19.** 680/7 acres of oats and 400/7 acres of barley **20.** $x^2 + y^2 - 2y - 24 = 0$

Exercise Set 10.1, page 722

1. $\begin{bmatrix} 2 & -3 & 1 & 1 \\ 3 & -2 & 3 & 0 \\ 1 & 0 & 5 & 4 \end{bmatrix} \begin{bmatrix} 2 & -3 & 1 \\ 3 & -2 & 3 \\ 1 & 0 & 5 \end{bmatrix} \begin{bmatrix} 1 \\ 0 \\ 4 \end{bmatrix}$ **3.** $\begin{bmatrix} 2 & -3 & -4 & 1 & 2 \\ 0 & 2 & 1 & 0 & 2 \\ 1 & -1 & 2 & 0 & 4 \\ 3 & -3 & -2 & 0 & 1 \end{bmatrix} \begin{bmatrix} 2 & -3 & -4 & 1 \\ 0 & 2 & 1 & 0 \\ 1 & -1 & 2 & 0 \\ 3 & -3 & -2 & 0 \end{bmatrix} \begin{bmatrix} 2 \\ 2 \\ 4 \\ 1 \end{bmatrix}$ **5.** $\begin{bmatrix} 1 & -1 & 2 & 2 \\ 0 & 1 & -1 & -6 \\ 0 & 0 & 1 & -27/2 \end{bmatrix}$

7. $\begin{bmatrix} 1 & -2 & -1 & 3 \\ 0 & 1 & 2 & -11/2 \\ 0 & 0 & 1 & -13/6 \end{bmatrix}$ **9.** $\begin{bmatrix} 1 & -2 & 3 & -4 \\ 0 & 1 & 2 & -1/2 \\ 0 & 0 & 1 & -2 \\ 0 & 0 & 0 & 0 \end{bmatrix}$ **11.** $\begin{bmatrix} 1 & -3 & 4 & 2 & 1 \\ 0 & 1 & -1 & -2 & -1 \\ 0 & 0 & 0 & 1 & 3 \end{bmatrix}$ **13.** $(2, -1, 1)$ **15.** $(1, -2, -1)$

17. $(2 - 2c, 2c + 1/2, c)$ **19.** $(1/2, 1/2, 3/2)$ **21.** $(16c, 6c, c)$ **23.** $(7c + 6, -11c - 8, c)$ **25.** $(c + 2, c, c)$ **27.** no solution
29. $(2, -2, 3, 4)$ **31.** $(21/10, -8/5, 2/5, -5/2)$ **33.** $(3, -3/2, 1, -1)$ **35.** $(27c/2 + 39, 5c/2 + 10, -4c - 10, c)$
37. $(c_1 - 12c_2/7 + 6/7, c_1 - 9c_2/7 + 1/7, c_1, c_2)$ **39.** $p(x) = 2x - 3$ **41.** $p(x) = x^2 - 2x + 3$ **43.** $p(x) = x^3 - 2x^2 - x + 2$
45. $p(x) = 2x + 5$ **47.** $(1, 0, -2, 1, 2)$ **49.** $((77c + 151)/3, (-25c - 50)/3, (14c + 34)/3, (-3c - 7, c)$
51. all values of a except $a \neq 1$ and $a \neq -6$ **53.** $a = -6$ **55.** $z = 2x + 3y - 2$

Exercise Set 10.2, page 736

1. a. $\begin{bmatrix} 1 & 2 \\ 5 & 4 \end{bmatrix}$ **b.** $\begin{bmatrix} 3 & -4 \\ 1 & 2 \end{bmatrix}$ **c.** $\begin{bmatrix} -2 & 6 \\ 4 & 2 \end{bmatrix}$ **d.** $\begin{bmatrix} 7 & -11 \\ 0 & 3 \end{bmatrix}$ **3. a.** $\begin{bmatrix} -3 & 0 & 5 \\ 3 & 5 & -5 \end{bmatrix}$ **b.** $\begin{bmatrix} 3 & -2 & 1 \\ -1 & -5 & 1 \end{bmatrix}$ **c.** $\begin{bmatrix} -6 & 2 & 4 \\ 4 & 10 & -6 \end{bmatrix}$

d. $\begin{bmatrix} 9 & -5 & 0 \\ -4 & -15 & 5 \end{bmatrix}$ **5. a.** $\begin{bmatrix} 1 & 5 \\ 3 & -5 \\ 2 & -4 \end{bmatrix}$ **b.** $\begin{bmatrix} -7 & 3 \\ 1 & -1 \\ -4 & 4 \end{bmatrix}$ **c.** $\begin{bmatrix} 8 & 2 \\ 2 & -4 \\ 6 & -8 \end{bmatrix}$ **d.** $\begin{bmatrix} -18 & 5 \\ 1 & 0 \\ -11 & 12 \end{bmatrix}$ **7. a.** $\begin{bmatrix} -1 & 1 & -1 \\ 2 & 2 & 1 \\ -1 & 2 & 5 \end{bmatrix}$

b. $\begin{bmatrix} -3 & 5 & -1 \\ -2 & -4 & 3 \\ -7 & 4 & 1 \end{bmatrix}$ **c.** $\begin{bmatrix} 2 & -4 & 0 \\ 4 & 6 & -2 \\ 6 & -2 & 4 \end{bmatrix}$ **d.** $\begin{bmatrix} -7 & 12 & -2 \\ -6 & -11 & 7 \\ -17 & 9 & 0 \end{bmatrix}$ **9.** $\begin{bmatrix} -10 & 17 \\ 6 & -8 \end{bmatrix} \begin{bmatrix} 0 & 22 \\ 1 & -18 \end{bmatrix}$ **11.** $\begin{bmatrix} 10 & 6 \\ 14 & -7 \end{bmatrix} \begin{bmatrix} 14 & -1 \\ 0 & -11 \end{bmatrix}$

13. $\begin{bmatrix} 0 & -4 & 5 \\ 6 & 0 & 3 \\ -3 & -2 & 1 \end{bmatrix} \begin{bmatrix} 5 & -13 \\ 5 & -4 \end{bmatrix}$ **15.** $\begin{bmatrix} 9 & -2 & -6 \\ 0 & -1 & 2 \\ 4 & -2 & -4 \end{bmatrix} \begin{bmatrix} 4 & -2 & 6 \\ 2 & -3 & 4 \\ 4 & -4 & 3 \end{bmatrix}$ **17.** $[0, 8]$ **19.** The product is not possible. **21.** $\begin{bmatrix} 0 & 0 \\ 0 & 0 \end{bmatrix}$

23. The product is not possible. **25.** $\begin{bmatrix} 1/3 & -5/3 \\ -1/3 & 4/3 \\ 1/3 & -4/3 \end{bmatrix}$ **27.** $\begin{bmatrix} -1 & 1 \\ 3 & 2 \\ 7 & -2 \end{bmatrix}$ **29.** $\begin{bmatrix} 1 & -3 \\ 1 & -2 \end{bmatrix}$ **31.** $\begin{bmatrix} 7 & -1 & 1 \\ 1 & 2 & 0 \\ 5 & -1 & 4 \end{bmatrix}$

33. $\begin{cases} 3x - 8y = 11 \\ 4x + 3y = 1 \end{cases}$ **35.** $\begin{cases} x - 3y - 2z = 6 \\ 3x + y = 2 \\ 2x - 4y + 5z = 1 \end{cases}$ **37.** $\begin{cases} 2x_1 - x_2 + 2x_4 = 5 \\ 4x_1 + x_2 + 2x_3 - 3x_4 = 6 \\ 6x_1 + x_3 - 2x_4 = 10 \\ 5x_1 + 2x_2 - x_3 - 4x_4 = 8 \end{cases}$

39. a. 3×4, Three different fish were caught in 4 different samples. **b.** Fish A was caught in sample 4. **c.** Fish B

41. $\begin{bmatrix} 1.96 & 1.37 & 2.94 & 1.37 \\ 0.78 & 1.08 & 1.96 & 0.88 \\ 3.53 & 1.18 & 4.41 & 1.47 \end{bmatrix}$ **43. a.** $\begin{bmatrix} 87 & 74 \\ 85 & 77 \\ 83 & 79 \end{bmatrix}$ **b.** The matrix represents the total number of wins and losses for each team.

c. $\begin{bmatrix} 1 & -2 \\ -1 & 1 \\ 7 & -7 \end{bmatrix}$ **d.** The matrix represents the difference between performance at home and performance away.

45. **a.** $\begin{bmatrix} 0 & 6 & 4 & -2 \\ 4 & 1 & -3 & 0 \end{bmatrix}$ **b.**

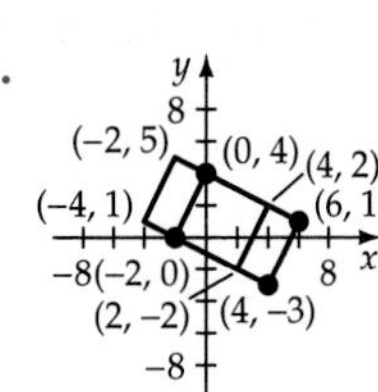

c. The new rectangle is shifted 2 units right and 1 unit down from the original rectangle.

47. **a.** $\begin{bmatrix} -5 & -2 & 2 & -1 \\ -2 & 4 & 2 & -4 \end{bmatrix}$ **b.**

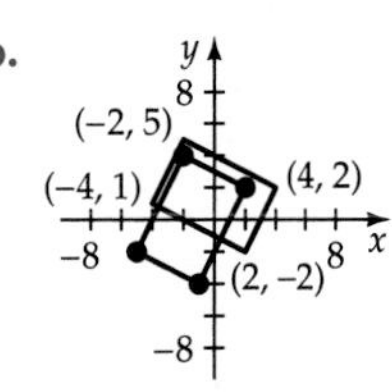

c. Second rectangle obtained by reflecting first about $y = x$ and then reflecting the result about $x = 0$.

49. $A - B = \begin{bmatrix} 50 & 150 & 140 \\ 15 & 170 & 370 \\ 85 & 250 & 130 \\ 80 & 115 & 25 \end{bmatrix}$

$A - B$ is the number sold of each item during the week.

51. $\begin{bmatrix} 26{,}898 & 28{,}150 & 31{,}536 \\ 20{,}495 & 21{,}195 & 23{,}670 \\ 19{,}022 & 19{,}925 & 21{,}969 \end{bmatrix}$

53. $\begin{bmatrix} 24 & 21 & -12 & 32 & 0 \\ -7 & -8 & 3 & 21 & 20 \\ 32 & 10 & -32 & 1 & 5 \\ 19 & -15 & -17 & 30 & 20 \\ 29 & 9 & -28 & 13 & -6 \end{bmatrix}$

55. $\begin{bmatrix} 46 & -100 & 36 & 273 & 93 \\ 82 & -93 & 19 & 27 & 97 \\ 73 & -10 & -23 & 109 & 83 \\ 212 & -189 & 52 & 37 & 156 \\ 68 & -22 & 54 & 221 & 58 \end{bmatrix}$

57. $\begin{bmatrix} 76 & -8 & -25 & 30 & 6 \\ 14 & 16 & -10 & 14 & 2 \\ 39 & 0 & -45 & 22 & 27 \\ 0 & -4 & 23 & 83 & -16 \\ 56 & -20 & -22 & 7 & 5 \end{bmatrix}$

59. $\begin{bmatrix} 6 + 9i & 3 - 6i \\ 3 + 3i & 6 - 3i \end{bmatrix}$ **61.** $\begin{bmatrix} 2 + 2i & -6 + 4i \\ -4 + 6i & 2 + 8i \end{bmatrix}$ **63.** $\begin{bmatrix} 3 + 2i & 3 + i \\ 4 + 3i & 6 - 2i \end{bmatrix}$

65. $\begin{bmatrix} 12 - 3i & -3 + 3i \\ 10 + i & 6 - i \end{bmatrix}$ **67.** $\begin{bmatrix} -2 + 11i & 8 - 6i \\ 2 + 6i & 6 - 5i \end{bmatrix}$

Exercise Set 10.3, page 748

1. $\begin{bmatrix} -5 & -3 \\ -2 & -1 \end{bmatrix}$ **3.** $\begin{bmatrix} 5 & -2 \\ -1 & 1/2 \end{bmatrix}$ **5.** $\begin{bmatrix} -16 & -2 & 7 \\ 7 & 1 & -3 \\ -3 & 0 & 1 \end{bmatrix}$ **7.** $\begin{bmatrix} 15 & -1 & -4 \\ -11/2 & 1/2 & 3/2 \\ 3 & 0 & -1 \end{bmatrix}$ **9.** $\begin{bmatrix} 7/2 & -2 & -2 \\ -5/2 & 1 & 2 \\ -1 & 0 & 1 \end{bmatrix}$

11. $\begin{bmatrix} 19/2 & -1/2 & -3/2 & 3/2 \\ 7/4 & 1/4 & -1/4 & 3/4 \\ -7/2 & 1/2 & 1/2 & -1/2 \\ 1/4 & -1/4 & 1/4 & 1/4 \end{bmatrix}$ **13.** $\begin{bmatrix} 2 & 3/5 & -7/5 & 4/5 \\ 4 & -7/5 & -2/5 & 4/5 \\ -6 & 14/5 & -1/5 & -3/5 \\ 3 & -8/5 & 2/5 & 1/5 \end{bmatrix}$ **15.** $(2, 1)$ **17.** $(7/4, -25/8)$ **19.** $(1, -1, 2)$ **21.** $(23, -12, 3)$

23. $(0, 4, -6, -2)$ **25.** on Saturday 80 adults, 20 children on Sunday 95 adults, 25 children **27.** Sample 1: 500 g of additive 1, 200 g of additive 2, 300 g of additive 3 Sample 2: 400 g of additive 1, 400 g of additive 2, 200 g of additive 3

29. $\begin{bmatrix} -5.667 & -3.667 & 5 & 0.333 \\ -27.667 & -18.667 & 24 & 2.333 \\ -19.333 & -13.333 & 17 & 1.667 \\ 15 & 10 & -13 & -1 \end{bmatrix}$ **31.** $\begin{bmatrix} -0.150 & -0.217 & 0.302 \\ 0.248 & -0.024 & 0.013 \\ 0.217 & -0.200 & -0.195 \end{bmatrix}$

33. \$194.67 million worth of manufacturing, \$156.03 million worth of transportation, \$212.82 million worth of services

35. \$39.69 million worth of coal, \$14.30 million worth of iron, \$32.30 million worth of steel

43. **a.** $\begin{bmatrix} -5/2 & 3/2 \\ -2 & 1 \end{bmatrix}$ **b.** $\begin{bmatrix} 2 & -3 \\ -3/2 & 5/2 \end{bmatrix}$ **c.** $\begin{bmatrix} 1 & 1/4 \\ -1 & 0 \end{bmatrix}$

Exercise Set 10.4, page 758

1. 13 **3.** −15 **5.** 0 **7.** 0 **9.** 19, 19 **11.** 1, −1 **13.** −9, −9 **15.** −9, −9 **17.** 10 **19.** 53 **21.** 20 **23.** 46
25. 0 **27.** Row 2 consists of zeros, so the determinant is zero. **29.** 2 was factored from row 2.
31. Row 1 was multiplied by −2 and added to row 2. **33.** 2 was factored from column 1.

35. The matrix is in diagonal form. The value of the determinant is the product of the terms on the main diagonal.
37. Row 1 and row 3 were interchanged, so the sign of the determinant was changed.
39. Each row of the determinant was multiplied by a. **41.** 0 **43.** 0 **45.** 6 **47.** −90 **49.** 21 **51.** 3 **53.** −38.933
55. 9/2 square units **57.** $46\frac{1}{2}$ square units **63.** $7x + 5y = -1$ **65.** 263.5

Exercise Set 10.5, page 764

1. $x_1 = 44/31, x_2 = 29/31$ **3.** $x_1 = 1/3, x_2 = -2/3$ **5.** $x_1 = 2, x_2 = -7$ **7.** $x_1 = 0, x_2 = 0$ **9.** $x_1 = 1.28125, x_2 = 1.875$
11. $x_1 = 21/17, x_2 = -3/17, x_3 = -29/17$ **13.** $x_1 = 32/49, x_2 = 13/49, x_3 = 6/7$ **15.** $x_1 = -29/64, x_2 = -25/64, x_3 = -19/32$
17. $x_1 = 50/53, x_2 = 62/53, x_3 = 4/53$ **19.** $x_1 = 0, x_2 = 0, x_3 = 0$ **21.** $x_2 = -35/19$ **23.** $x_1 = -121/131$ **25.** $x_4 = 4/3$
27. The determinant of the coefficient matrix is zero, so Cramer's Rule cannot be used. The system of equations has infinitely many solutions.
29. all values of k except $k = 0$ **31.** all values of k except $k = 2$ **33.** $r = 3, s = -3$

Chapter 10 True/False Exercises, page 767

1. False; $A^2 = A \cdot A = \begin{bmatrix} 7 & 18 \\ 6 & 19 \end{bmatrix}$. **2.** True **3.** False; a singular matrix does not have a multiplicative inverse.
4. False; as an example, $A = \begin{bmatrix} 2 & -1 \\ -4 & 2 \end{bmatrix}$, $B = \begin{bmatrix} 3 & 4 \\ 1 & 5 \end{bmatrix}$, and $C = \begin{bmatrix} 4 & 7 \\ 3 & 11 \end{bmatrix}$. $AB = AC$ but $B \neq C$. **5.** True **6.** False; for example, if $A = \begin{bmatrix} 1 & 4 \\ -2 & 3 \end{bmatrix}$ and $B = \begin{bmatrix} 2 & 0 \\ -1 & 5 \end{bmatrix}$, then $\det(A) + \det(B) \neq \det(A + B)$. **7.** False; if the determinant of the coefficient matrix is zero, Cramer's rule cannot be used to solve the system of equations. **8.** False; matrix multiplication is not commutative—that is, $AB \neq BA$, $AB - BA \neq 0$. **9.** True **10.** False; by the Associative Property of Matrix Multiplication, given A, B, and C square matrices of order n, $(AB)C = A(BC)$. **11.** False; if the number of equations is less than the number of variables, the Gaussian elimination method can be used to solve the system of linear equations. If the system of equations has a solution, the solutions will be given in terms of one or more of the variables. **12.** False; for example, for a 2 × 2 matrix, $\det(2A) = 2 \cdot 2\det(A)$, and for a 3 × 3 matrix, $\det(2A) = 4 \cdot 2\det(A)$. **13.** True
14. False, for example, given $A = \begin{bmatrix} -3 & 2 \\ -6 & 4 \end{bmatrix}$ and $B = \begin{bmatrix} 2 & 4 \\ 3 & 6 \end{bmatrix}$, then $AB = \begin{bmatrix} 0 & 0 \\ 0 & 0 \end{bmatrix} = \mathbf{O}$, but $A \neq \mathbf{O}$ and $B \neq \mathbf{O}$. **15.** True

Chapter 10 Review Exercises, page 768

1. $\begin{bmatrix} 6 & -3 & 9 \\ 9 & 6 & -3 \end{bmatrix}$ **3.** $\begin{bmatrix} -5 & 5 & -1 \\ 1 & -4 & 6 \end{bmatrix}$ **5.** $\begin{bmatrix} -1 & -15 \\ 7 & 1 \end{bmatrix}$ **7.** $\begin{bmatrix} -6 & -4 & 2 \\ 14 & 0 & 10 \\ -7 & -7 & 6 \end{bmatrix}$ **9.** $\begin{bmatrix} 12 & 28 & -5 \\ 2 & 6 & 0 \\ 6 & 16 & -1 \end{bmatrix}$ **11.** $\begin{bmatrix} -12 & -36 & -4 \\ 48 & 124 & 4 \\ -9 & -32 & -6 \end{bmatrix}$
13. not possible **15.** $\begin{bmatrix} 7 & 24 & 9 \\ -10 & -22 & 1 \end{bmatrix}$ **17.** $\begin{bmatrix} -1 & -5 & 4 \\ 1/2 & 2 & -3/2 \\ 0 & -2 & 1 \end{bmatrix}$ **19.** $(2, -1)$ **21.** $(3, 0)$ **23.** $(3, 1, 0)$ **25.** $(1, 0, -2)$
27. $(3, -4, 1)$ **29.** $(-c - 2, -c - 3, c)$ **31.** $(1, -2, 2, 3)$ **33.** $(-37c + 2, 16c, -7c + 1, c)$ **35.** $y = x^2 + 3x - 2$ **37.** $\begin{bmatrix} -1 & 1 \\ -3/2 & 1 \end{bmatrix}$
39. $\begin{bmatrix} -2/7 & 3/14 \\ 1/7 & 1/7 \end{bmatrix}$ **41.** $\begin{bmatrix} 2 & -2 & 1 \\ 0 & 3/2 & -1 \\ -1 & -1 & 1 \end{bmatrix}$ **43.** $\begin{bmatrix} -10 & 20 & -3 \\ -5 & 9 & -1 \\ 3 & -6 & 1 \end{bmatrix}$ **45.** $\begin{bmatrix} -1 & -7 & 4 & 2 \\ -6 & -3 & 2 & 3 \\ 1 & 2 & -1 & -1 \\ -2 & 0 & 0 & 1 \end{bmatrix}$ **47.** The matrix does not have an inverse.
49. a. $(18, -13)$ **b.** $(-22, 16)$ **51. a.** $(-18/7, 23/7, -6/7)$ **b.** $(-31/14, 20/7, 3/7)$ **53.** −2 **55.** −1 **57.** 0 **59.** 0
61. $x_1 = 16/19, x_2 = -2/19$ **63.** $x_1 = 13/44, x_2 = 1/4, x_3 = -17/44$ **65.** $x_1 = 18/23, x_2 = -26/69, x_3 = 38/69$ **67.** $x_3 = 115/126$
69. \$34.47 million computer division, \$14.20 million monitor division, \$23.64 million disk drive division.

Chapter 10 Test, page 770

1. $\begin{bmatrix} 2 & 3 & -3 & 4 \\ 3 & 0 & 2 & -1 \\ 4 & -4 & 2 & 3 \end{bmatrix}$, $\begin{bmatrix} 2 & 3 & -3 \\ 3 & 0 & 2 \\ 4 & -4 & 2 \end{bmatrix}$, $\begin{bmatrix} 4 \\ -1 \\ 3 \end{bmatrix}$ **2.** $\begin{cases} 3x - 2y + 5z - w = 9 \\ 2x + 3y - z + 4w = 8 \\ x + 3z + 2w = -1 \end{cases}$ **3.** $(2, -1, 2)$ **4.** $(3, -1, -1)$
5. $(3c - 5, -7c + 14, 4 - 3c, c)$ **6.** $\begin{bmatrix} 3 & -9 & -6 \\ -3 & -12 & 3 \end{bmatrix}$ **7.** $A + B$ is not defined. **8.** $\begin{bmatrix} 4 & 1 & 3 \\ 8 & 0 & -19 \\ 11 & 0 & 10 \end{bmatrix}$ **9.** $\begin{bmatrix} 16 & -1 & -2 \\ 15 & -11 & -3 \end{bmatrix}$

10. $\begin{bmatrix} 17 & -4 & -4 \\ 14 & -15 & -2 \end{bmatrix}$ **11.** CA is not defined. **12.** $\begin{bmatrix} -6 & -1 & -19 \\ -15 & -25 & -27 \\ 1 & 3 & 31 \end{bmatrix}$ **13.** A^2 is not defined. **14.** $\begin{bmatrix} 9 & 6 & 13 \\ -3 & -2 & 12 \\ 20 & -3 & 11 \end{bmatrix}$

15. $\begin{bmatrix} 18 & -5 & 7 \\ 4 & -1 & 2 \\ -3 & 1 & -1 \end{bmatrix}$ **16.** $M_{21} = -8, C_{21} = 8$ **17.** 49 **18.** -1 **19.** $-140/41$ **20.** $\left(\begin{bmatrix} 1 & 0 & 0 \\ 0 & 1 & 0 \\ 0 & 0 & 1 \end{bmatrix} - \begin{bmatrix} 0.15 & 0.23 & 0.11 \\ 0.08 & 0.10 & 0.05 \\ 0.16 & 0.11 & 0.07 \end{bmatrix}\right)^{-1} \begin{bmatrix} 50 \\ 32 \\ 8 \end{bmatrix}$

Exercise Set 11.1, page 777

1. $0, 2, 6, a_8 = 56$ **3.** $0, 1/2, 2/3, a_8 = 7/8$ **5.** $1, -1/4, 1/9, a_8 = -1/64$ **7.** $-1/3, -1/6, -1/9, a_8 = -1/24$
9. $2/3, 4/9, 8/27, a_8 = 256/6561$ **11.** $0, 2, 0, a_8 = 2$ **13.** $1.1, 1.21, 1.331, a_8 = 2.14358881$ **15.** $1, -\sqrt{2}/2, \sqrt{3}/3, a_8 = -\sqrt{2}/4$
17. $1, 2, 6, a_8 = 40320$ **19.** $0, 0.3010, 0.4771, a_8 = 0.9031$ **21.** $1, 4, 2, a_8 = 4$ **23.** $3, 3, 3, a_8 = 3$ **25.** 5, 10, 20 **27.** 2, 4, 12
29. 2, 4, 16 **31.** 2, 8, 48 **33.** $3, \sqrt{3}, \sqrt[6]{3}$ **35.** 2, 5/2, 9/4 **37.** 4320 **39.** 72 **41.** 56 **43.** 100 **45.** 15 **47.** 40 **49.** 25/12
51. 72 **53.** -24 **55.** 3 log 2 **57.** 256 **59.** $\sum_{i=1}^{6} \frac{1}{i^2}$ **61.** $\sum_{i=1}^{7} 2^i(-1)^{i+1}$ **63.** $\sum_{i=0}^{4} (7 + 3i)$ **65.** $\sum_{i=1}^{4} \frac{1}{2^i}$ **67.** 2.6457520
69. $a_{20} \approx 1.0000037, a_{100} \approx 1$ **71.** $\frac{1}{2}(-1 + i\sqrt{3}), \frac{1}{2}(-1 - i\sqrt{3}), 1, \frac{1}{2}(-1 + i\sqrt{3}), \frac{1}{2}(-1 - i\sqrt{3}), 1, a_{99} = 1$

Exercise Set 11.2, page 784

1. $a_9 = 38, a_{24} = 98, a_n = 4n + 2$ **3.** $a_9 = -10, a_{24} = -40, a_n = 8 - 2n$ **5.** $a_9 = 16, a_{24} = 61, a_n = 3n - 11$
7. $a_9 = 25, a_{24} = 70, a_n = 3n - 2$ **9.** $a_9 = a + 16, a_{24} = a + 46, a_n = a + 2n - 2$
11. $a_9 = \log 7 + 8 \log 2, a_{24} = \log 7 + 23 \log 2, a_n = \log 7 + (n - 1) \log 2$ **13.** $a_9 = 9 \log a, a_{24} = 24 \log a, a_n = n \log a$
15. 45 **17.** -79 **19.** 185 **21.** -555 **23.** 468 **25.** 525 **27.** -465 **29.** $78 + 12x$ **31.** $210x$ **33.** 3, 7, 11, 15, 19
35. 5/2, 2, 3/2, 1 **39.** 20 on 6th row, 135 in the 6 rows **41.** \$1500, \$48,750 **43.** 784 feet **47.** $a_n = 7 - 3n$ **49.** $a_{50} = 197$

Exercise Set 11.3, page 792

1. geometric; $r = 4$ **3.** not geometric **5.** geometric; $r = 2^x$ **7.** geometric; $r = 2$ **9.** geometric; $r = x^2$ **11.** not geometric
13. 2^{2n-1} **15.** $-4(-3)^{n-1}$ **17.** $6(2/3)^{n-1}$ **19.** $-6(-5/6)^{n-1}$ **21.** $(-1/3)^{n-3}$ **23.** $(-x)^{n-1}$ **25.** c^{3n-1} **27.** $3(1/100)^n$ **29.** $5(0.1)^n$
31. $45(0.01)^n$ **33.** 18 **35.** -2 **37.** 363 **39.** 1330/729 **41.** $\frac{279{,}091}{390{,}625}$ **43.** -341 **45.** 147,620 **47.** 1/2 **49.** $-2/5$ **51.** 9/91
53. 1/9 **55.** 5/7 **57.** 1/3 **59.** 5/11 **61.** 41/333 **63.** 422/999 **65.** 229/900 **67.** 997/825 **69.** \$2271.93
71. Because log r is a constant, the sequence log a_n is an arithmetic sequence. **73.** Yes. The common ratio is x. **75.** $a^n r^{[(n-1)n]/2}$
77. 45 feet **79.** 2044

Exercise Set 11.4, page 799

No answers are provided because each exercise is a verification.

Exercise Set 11.5, page 805

1. 35 **3.** 36 **5.** 220 **7.** 1 **9.** $x^6 - 6x^5y + 15x^4y^2 - 20x^3y^3 + 15x^2y^4 - 6xy^5 + y^6$ **11.** $x^5 + 15x^4 + 90x^3 + 270x^2 + 405x + 243$
13. $128x^7 - 448x^6 + 672x^5 - 560x^4 + 280x^3 - 84x^2 + 14x - 1$ **15.** $x^6 + 18x^5y + 135x^4y^2 + 540x^3y^3 + 1215x^2y^4 + 1458xy^5 + 729y^6$
17. $16x^4 - 160x^3y + 600x^2y^2 - 1000xy^3 + 625y^4$ **19.** $x^6 + 6x^4 + 15x^2 + 20 + 15/x^2 + 6/x^4 + 1/x^6$
21. $x^{14} - 28x^{12} + 336x^{10} - 2240x^8 + 8960x^6 - 21{,}540x^4 + 28{,}672x^2 - 16{,}384$ **23.** $32x^{10} + 80x^8y^3 + 80x^6y^6 + 40x^4y^9 + 10x^2y^{12} + y^{15}$
25. $16/x^4 - 16/x^2 + 6 - x^2 + x^4/16$ **27.** $s^{-12} + 6s^{-8} + 15s^{-4} + 20 + 15s^4 + 6s^8 + s^{12}$ **29.** $-3240x^3y^7$ **31.** $1056x^{10}y^2$ **33.** $126x^2y^2\sqrt{x}$
35. $165b^5/a^5$ **37.** $180a^2b^8$ **39.** $60x^2y^8$ **41.** $-61{,}236a^5b^5$ **43.** $126s^{-1}, 126s$ **45.** $-7 - 24i$ **47.** $41 - 38i$ **49.** 1
51. $nx^{n-1} + \frac{n(n-1)x^{n-2}h}{2} + \frac{n(n-1)(n-2)x^{n-3}h^2}{6} + \cdots + h^{n-1}$ **57.** 1.1712 **59.** 756 **61.** 56

Exercise Set 11.6, page 811

1. 30 **3.** 70 **5.** 1 **7.** 1 **9.** 210 **11.** 12 **13.** 16 **15.** 720 **17.** 125 **19.** 53,130
21. There are 676 ways to arrange 26 letters taken 2 at a time. Now if there are more than 676 employees, then at least 2 employees will have the same first and last initials **23.** 1120 **25.** 1024 **27.** 3,838,380 **29. a.** 21 **b.** 105 **c.** 21 **31.** 1.8×10^9
33. 112 **35.** 120 **37.** 21 **39.** 112 **41.** 184,756 **43.** 62,355,150 **45.** 5456 **47.** 19! **49. a.** 3,991,680 **b.** 31,840,128
51. 120 **53.** 252

Exercise Set 11.7, page 821

1. $\{S_1R_1, S_1R_2, S_1R_3, S_2R_1, S_2R_2, S_2R_3, R_1R_2, R_1R_3, R_2R_3, S_1S_2\}$ **3.** {H1, H2, H3, H4, T1, T2, T3, T4} **5.** Let the three cans be represented by A, B, and C and let (x, y) represent the cans that balls 1 and 2 are placed in; e.g., (A, B) means ball 1 in can A and ball 2 in B. $S = \{(A, A), (A, B), (A, C), (B, B), (B, C), (B, A), (C, C), (C, A), (C, B)\}$ **7.** {HSC, HSD, HCD, SCD} **9.** {ae, ai, ao, au, ei, eo, eu, io, iu, ou} **11.** {HHHH} **13.** {TTTT, HTTT, THTT, TTHT, TTTH, TTHH, THTH, HTHT, THHT, HTTH, HHTT} **15.** $\varnothing$ **17.** {(1, 1), (2, 2), (3, 3), (4, 4), (5, 5), (6, 6)} **19.** {(1, 4), (2, 4), (3, 4), (4, 4), (5, 4), (6, 4)} **21. a.** 1/13 **b.** 1/4 **23.** 0.97 **25.** 3/5 **27.** 0.59 **29.** 0.25 **31.** 0.1 **33.** 0.1 **35.** 0.025 **37.** 0.9999 **39.** 1/16 **41.** 0.2262 **43.** 0.2137 **45.** $(7/8)^2$ **47.** $\dfrac{56}{729}$

Chapter 11 True/False Exercises, page 826

1. False; $0! \cdot 4! = 1 \cdot 4 \cdot 3 \cdot 2 \cdot 1 = 24$. **2.** False; $\left(\sum_{i=1}^{3} i\right)\left(\sum_{i=1}^{3} i\right) \neq \sum_{i=1}^{3} i^2$. **3.** True **4.** False; the constant sequence has all terms equal. **5.** False; $\dfrac{(k+1)^3}{k^3} = (1 + 1/k)^3$ is not a constant. **6.** True **7.** True **8.** False; $\sum_{i=1}^{\infty} \dfrac{1}{2^i} = 1$. **9.** False; see Project 1, Section 12.4. **10.** False; the exponent is 4. **11.** False; there are $m \cdot n$ ways. **12.** False; $P(n, r) = \dfrac{n!}{(n-r)!}$. **13.** True **14.** False; $P(A \cap B) = P(\varnothing) = 0$. **15.** True

Chapter 11 Review Exercises, page 826

1. $a_3 = 9, a_7 = 49$ **3.** $a_3 = 11, a_7 = 23$ **5.** $a_3 = 1/8, a_7 = 1/128$ **7.** $a_3 = 1/6, a_7 = 1/5040$ **9.** $a_3 = 8/27, a_7 = 128/2187$ **11.** $a_3 = 18, a_7 = 1458$ **13.** $a_3 = 6, a_7 = 5040$ **15.** $a_3 = 8, a_7 = 16$ **17.** $a_3 = 2, a_7 = 256$ **19.** $a_3 = -54, a_7 = -3{,}674{,}160$ **21.** neither **23.** arithmetic **25.** geometric **27.** neither **29.** geometric **31.** geometric **33.** neither **35.** arithmetic **37.** neither **39.** neither **41.** 63 **43.** 152 **45.** 378 **47.** −14,763 **49.** $\dfrac{116{,}050}{59{,}049} \approx 1.9653$ **51.** 0.8280 **53.** 1/3 **55.** −4/9 **65.** $1024a^5 - 1280a^4b + 640a^3b^2 - 160a^2b^3 + 20ab^4 - b^5$ **67.** $a^4 + 16a^{7/2}b^{1/2} + 112a^3b + 448a^{5/2}b^{3/2} + 1120a^2b^2 + 1792a^{3/2}b^{5/2} + 1792ab^3 + 1024a^{1/2}b^{7/2} + 256b^4$ **69.** $241{,}920x^3y^4$ **71.** 26^8 **73.** 2730 **75.** 672 **77.** 1/8, 3/8 **79.** 0.285 **81.** drawing an ace and a ten-card from one deck **83.** 1/4

Chapter 11 Test, page 827

1. $a_3 = 4/3, a_5 = 4/15$ **2.** $a_3 = 1/6, a_5 = 1/10$ **3.** $a_3 = 12, a_5 = 48$ **4.** arithmetic **5.** neither **6.** geometric **7.** 49/20 **8.** 1023/1024 **9.** 590 **10.** 58 **11.** 3/5 **12.** 5/33 **15.** $x^5 - 10x^4y + 40x^3y^2 - 80x^2y^3 + 80xy^4 - 32y^5$ **16.** $x^6 + 6x^4 + 15x^2 + 20 + 15/x^2 + 6/x^4 + 1/x^6$ **17.** $48{,}384x^3y^5$ **18.** 132,600 **19.** 568,339,200 **20.** $\dfrac{5}{17} \approx 0.294118$

GLOSSARY

abscissa The x-coordinate of an ordered pair. (Section 2.1)
absolute minimum A minimum value of a function f that is also the smallest range value of f. (Section 3.2)
absolute value The absolute value of the real number a, denoted $|a|$, equals a when $a \geq 0$ and equals $-a$ when $a < 0$. (Section P.2)
acute angle An angle that has a measure greater than 0° but less than 90°. (Section 5.1)
addition Addition of the two real numbers a and b is designated by $a + b = c$, where c is the sum and the real numbers a and b are called terms. (Section P.1)
additive inverse The number $-b$ is called the additive inverse of b. (Section P.1)
additive inverse of a polynomial If $P(x)$ is a polynomial, then $-P(x)$ is the additive inverse of $P(x)$. (Section P.4)
airspeed The speed of a plane if there were no wind. (Section 7.3)
alternating sequence A sequence in which the signs of the terms alternate between positive and negative values. (Section 11.1)
angle An angle is formed by rotating a given ray about its endpoint to some terminal position. (Section 5.1)
angle of depression An angle measured below the line of sight. (Section 5.2)
angle of elevation An angle measured above the line of sight. (Section 5.2)
angular speed The angle through which a point on a circle moves per unit time. (Section 5.1)
annuities Deposits of equal amounts at equal intervals of time. (Section 11.3)
antilogarithm In $\log_a M = N$, the number M. (Section 5.2)
arc A portion of a circle. (Section 5.1)
argument The independent variable of a function. (Section 5.2)
argument of a complex number For a complex number written in the form $z = r \text{ cis } \theta$, the angle θ. (Section 7.4)
arithmetic mean The arithmetic mean of two numbers a and b is $(a + b)/2$. (Section 11.2)
arithmetic sequence A sequence in which the difference between any two successive terms is constant. (Section 11.2)
arithmetic series The sum of the terms of an arithmetic sequence. (Section 11.2)
asymptotes A line (or curve) approached by another curve in the sense that the perpendicular distance from a point on the curve to the asymptote approaches zero as the point moves an infinite distance from the origin of the coordinate system. (Section 3.5)
augmented matrix A matrix consisting of the coefficients and constants of a system of equations. (Section 10.2)
average velocity The ratio of the change in distance to the change in time. (Section 2.6)
axis of symmetry of a parabola The line that passes through the focus and is perpendicular to the directrix. (Section 8.1)
base In the expression b^x, b is the base. (Section P.3)
bearing The angular direction used to locate one object in relation to another object. (Section 7.1)
binomial A simplified polynomial that has two terms. (Section P.4)
binomial coefficient The coefficient of a term of a binomial expansion. (Section 11.5)
Boyle's Law The volume V of a sample of gas (at a constant temperature) varies inversely as the pressure P. (Section 1.6)
break-even point The value of x for which $R(x) = C(x)$ (revenue equals cost). (Section 2.3)
cardioid A graph of an equation of the form $r = a(1 + \cos \theta)$ or $r = a(1 + \sin \theta)$. (Section 8.5)
Cartesian coordinate system A two-dimensional coordinate system formed by the intersection of two perpendicular number lines. (Section 2.1)
center of a hyperbola The midpoint of the transverse axis. (Section 8.3)
center of an ellipse The midpoint of the major axis. (Section 8.2)
central angle The angle formed by two radii of a circle. (Section 5.1)
circle The set of points in a plane that are a fixed distance from a specified point. (Section 2.1)
closed interval $[a, b]$ represents all real numbers between a and b, including a and including b. (Section P.2)
coefficient The constant of a monomial. (Section P.4)
coefficient matrix The matrix formed by the coefficients of a system of equations. (Section 10.1)
coefficient of determination r^2 A measure of the percent of the total variation in the dependent variable that is explained by the regression line. (Section 2.7)
cofactor $(-1)^{i+j}M_{ij}$, where M_{ij} is the minor of the matrix. (Section 10.4)
cofunctions Any pair of trigonometric functions f and g for which $f(x) = g(90° - x)$ and $g(x) = f(90° - x)$. (Section 6.2)
combination An arrangement of objects for which the order of the selection is not important. (Section 11.6)
combined variation A variation that involves more than one type of variation. (Section 1.6)
common difference In an arithmetic sequence, the difference between any two successive terms. (Section 11.2)
common logarithm A logarithm with a base of 10. (Section 5.2)
complementary angles Two positive angles for which the sum of the measures of the angles is 90°. (Section 5.1)
complex conjugates The complex numbers $a + bi$ and $a - bi$. (Section 1.3)
complex number A number in the form $a + bi$, where a and b are real numbers and i is the imaginary unit. (Section 1.3)
complex plane The coordinate system with the real axis along the x-axis and the imaginary axis along the y-axis. (Section 7.4)
composite function A function formed from the composition of two functions f and g, given by $(f \circ g)(x) = f(g(x))$. (Section 2.6)
composite number A composite number is an integer greater than 1 that is not a prime number. (Section P.1)
compound continuously In compounding interest, to increase the number of compounding periods without bound. (Section 4.6)
compound inequality An inequality formed by joining two inequalities with the connective word *and* or *or*. (Section 1.5)
compound interest Interest that is added to principal at regular intervals so that interest is paid on interest as well as on principal. (Section 4.6)

conditional equation Any equation that is true for some values of the variable but is not true for other values of the variable. (Section 1.1)

conjugate axis of a hyperbola The axis that passes through the center of the hyperbola and is perpendicular to the transverse axis. (Section 8.3)

conjugates The complex numbers $a + bi$ and $a - bi$. (Section 1.3)

consistent A system of equations for which the graphs intersect at a single point or are the same line. (Section 9.1)

constant function A function of the form $f(x) = a$, where a is a real number. (Section 2.2)

constant matrix The matrix formed from the constants of a system of equations. (Section 10.1)

constant of proportionality In the direct variation equation $y = kx$, the value of k. (Section 2.8)

constant polynomial A nonzero constant, such as 5. (Section P.4)

constant sequence A sequence in which each term is the same. (Section 11.1)

constant term A monomial with no variable part. (Section P.4)

constraints Equations or inequalities that force the solution of a linear programming problem to lie within a particular set. (Section 9.6)

contradiction An equation that has no solutions. (Section 1.1)

conversion factor A rate that provides a conversion between two different units. (Section 5.1)

coordinate The number associated with a particular point on a real number line. (Section P.2)

coordinate axis A line on which each real number can be designated by a point. (Section P.1)

coordinate plane The set of all points on a flat, two-dimensional surface. (Section 2.1)

coordinates An ordered pair of numbers. (Section 2.1)

cosecant In a right triangle, the function of an acute angle that is the ratio of the hypotenuse to the opposite side. (Section 5.2)

cosine In a right triangle, the function of an acute angle that is the ratio of the adjacent side to the hypotenuse. (Section 5.2)

cost function The function, C, that gives a manufacturer's cost to produce x units of a product. (Section 2.3)

cotangent In a right triangle, the function of an acute angle that is the ratio of the adjacent side to the opposite side. (Section 5.2)

coterminal angles Angles in standard position that have the same terminal side. (Section 5.1)

critical values of a rational expression The numbers that cause the numerator or the denominator of the rational expression to equal zero. (Section 1.5)

cube The product of the same three factors. (Section P.5)

cube root One of the three equal factors of a cube. (Section P.5)

cubic equation An equation of the form $ax^3 + bx^2 + cx + d = 0$, where $a \neq 0$. (Section 1.4)

cycloid A curve traced by a point on the circumference of a circle as the circle rolls on a straight line without slipping. (Section 8.7)

decreasing function A function f for which, for all x_1 and x_2 in the domain of f, $f(x_1) > f(x_2)$ whenever $x_1 < x_2$. (Section 2.2)

degree The measure of an angle formed by rotating a ray 1/360 of a complete revolution. (Section 5.1)

degree of a monomial The sum of the exponents of the variables in the monomial. (Section P.4)

degree of a polynomial The largest degree of the terms in the polynomial. (Section P.4)

denominator The nonzero real number b in the fraction a/b. (Section P.1)

dependent A system of equations that has an infinite number of solutions. (Section 9.1)

dependent variable For a function defined by an equation, the variable that represents elements of the range. (Section 2.2)

depressed polynomial See *reduced polynomial.* (Section 3.1)

Descartes' Rule of Signs A theorem that describes the number of positive or negative zeros a polynomial function may have. (Section 3.3)

determinant A square array of elements having a value determined by a rule involving the sum of the products of certain elements. (Section 10.4)

diagonal form A matrix is in diagonal form if all elements below and above the main diagonal are zero. (Section 10.4)

difference If $a - b = c$, then c is called the difference of a and b. (Section P.1)

difference of two squares An expression of the form $a^2 - b^2$. (Section P.5)

difference quotient The quotient defined by $[f(x + h) - f(x)]/h$. (Section 2.6)

dimension of a matrix A matrix of m rows and n columns has dimension $m \times n$ (read "m by n"). (Section 10.1)

direction angle The angle between a vector and the positive x-axis. (Section 7.3)

directly proportional If $y = kx$, the variable y varies directly as the variable x, or y is directly proportional to x. (Section 1.6)

directrix A line perpendicular to the line containing the foci of a conic section. (Section 8.1)

discriminant For $ax^2 + bx + c$ where $a \neq 0$, the discriminant is $b^2 - 4ac$. (Section 1.3)

dividend A quantity to be divided. (Section 3.1)

division The division of a and b, designated by $a \div b$. (Section P.1)

divisor The quantity by which another quantity, the dividend, is to be divided. (Section 3.1)

domain The set of all the first coordinates of the ordered pairs of a function. (Section 2.2)

domain of a rational expression The set of all real numbers that can be used as replacements for the variable in the rational expression. (Section P.6)

double root A root of an equation that is repeated twice. (Section 1.3)

eccentricity A measure used to describe a characteristic of a conic section. The value of the eccentricity is c/a, where c is the distance from the center to a focus and a is the distance from the center to a vertex. (Sections 8.2, 8.3)

echelon form A form of a matrix in which the first nonzero element in any row is a 1, the rows are arranged so that the column containing the first nonzero number in any row is to the left of the column containing the first nonzero number of the next row, and all rows consisting entirely of zeros appear at the bottom of the matrix. (Section 10.1)

element of a set Each member of the set. (Section P.1)

element of a matrix Each member in the matrix. (Section 10.1)

elementary row operation An operation performed on the rows of a matrix. (Section 10.1)

elimination method A method of solving a system of equations. (Section 9.1)

ellipse The set of all points in a plane, the sum of whose distances from two fixed points (foci) is a positive constant. (Section 8.2)

empty set The set without any elements. (Section P.1)
equals a equals b (denoted by $a = b$) if $a - b = 0$. (Section P.2)
equation A statement of equality between two numbers or two expressions. (Sections P.1, 1.1)
equivalent systems of equations Systems of equations that have exactly the same solution(s). (Section 9.1)
equivalent equations Equations that have exactly the same solution(s). (Section 1.1)
equivalent inequalities Inequalities that have the same solution set. (Section 1.5)
equivalent vectors Vectors that have the same magnitude and the same direction. (Section 7.3)
evaluate a polynomial To substitute the given value(s) for the variable(s) and then perform the indicated operations using the Order of Operations Agreement. (Section P.4)
even function A function f for which $f(-x) = f(x)$ for all x in the domain of f. (Section 2.5)
event Any subset of a sample space. (Section 11.7)
experiment An activity with an observable outcome. (Section 11.7)
exponent In the expression b^n, n is the exponent. (Section P.3)
exponential decay function A function of the form $A(x) = Ae^{kt}$, where $k < 0$ and $t \geq 0$. (Section 4.6)
exponential equation An equation in which a variable appears as an exponent in a term of the equation. (Section 4.5)
exponential function A function defined by $f(x) = b^x$, where $b > 0, b \neq 1$, and x is any real number. (Section 4.2)
exponential growth function A function of the form $A(x) = Ae^{kt}$, where $k > 0$ and $t \geq 0$. (Section 4.6)
exponential notation An expression written in the form b^x. (Section P.3)
extraneous solution An apparent solution of an equation that is not a solution of the original equation. (Section 1.4)
factor by grouping To factor by first grouping together pairs of terms that have a common factor. (Section P.5)
factoring Writing a polynomial as a product of polynomials of lower degree. (Section P.5)
factoring over the integers Factoring by using only polynomial factors that have integer coefficients. (Section P.5)
factors If $ab = c$, then a and b are called factors of c. (Section P.1)
final demand The amount of output that a consumer will want. (Section 10.3)
formula An equation that expresses known relationships between two or more variables. (Section 1.2)
function A set of ordered pairs in which no two ordered pairs that have the same first coordinate have different second coordinates. (Section 2.2)
Fundamental Theorem of Algebra If $P(x)$ is a polynomial with complex number coefficients and is of degree greater than or equal to 1, then $P(x)$ has at least one complex zero. (Section 3.4)
future value The total value of an investment after the last deposit. (Section 11.3)
Gaussian elimination method An algorithm that uses elementary row operations to solve a system of linear equations. (Section 10.2)
general form of the equation of a circle An equation of the form $x^2 + y^2 + ax + by + c = 0$. (Section 2.1)
general form of the equation of a line An equation of the form $Ax + By + C = 0$, where A, B, and C are real numbers and both A and B are not 0. (Section 2.3)
geometric sequence A sequence in which the ratio of any two successive terms is a constant. (Section 11.3)
graph of a polar equation The set of all points whose coordinates are solutions of the equation. (Section 8.5)
graph of a function The graph of all the ordered pairs that belong to the function. (Section 2.2)
graph of an equation The set of all points whose coordinates satisfy the equation. (Section 2.1)
greater than a is greater than b (denoted by $a > b$) if $a - b$ is positive. (Section P.2)
greatest integer function The function, denoted by $f(x) = [\![x]\!]$, for which the value of the function is the greatest integer less than or equal to x. (Section 2.2)
ground speed The magnitude of the actual velocity of a plane. (Section 7.3)
half-life The time required for the disintegration of half of the atoms in a sample of a radioactive substance. (Section 4.6)
half-line One of the two parts into which a point P on a line L separates the line. (Section 5.1)
half-open interval $(a, b]$ represents all real numbers between a and b, not including a, but including b; $[a, b)$ represents all real numbers between a and b, including a, but not including b. (Section P.2)
half-plane Each region in a plane separated by a line. (Section 9.5)
heading The angular direction in which a craft is pointed. (Section 7.1)
homogeneous system of equations A linear system of equations for which the constant term of each equation is 0. (Section 9.2)
horizontal asymptote If $\lim_{x \to \pm\infty} f(x) = A$, then $y = A$ is a horizontal asymptote of the graph of f. (Section 3.5)
hyperbola The set of all points in a plane, the difference between whose distances from two fixed points (foci) is a positive constant. (Section 8.3)
hypotenuse In a right triangle, the side opposite the 90° angle. (Section 1.3)
identity An equation that is true for *every* real number for which all terms of the equation are defined. (Sections 1.1, 6.1)
identity matrix An $n \times n$ matrix that has 1s on the main diagonal and 0s as the remaining elements. (Section 10.2)
imaginary axis The vertical axis of the complex plane. (Section 7.4)
imaginary number A number in the form ai, where i is the imaginary unit and a is a real number. (Section 1.3)
imaginary part of a complex number The real number b for the complex number $a + bi$. (Section 1.3)
imaginary unit The number i, defined so that $i^2 = -1$. (Section 1.3)
inconsistent system of equations A system of equations that has no solution. (Section 9.1)
increasing function A function f for which, for all elements x_1 and x_2 in the domain of f, $f(x_1) < f(x_2)$ whenever $x_1 < x_2$. (Section 2.2)
independent A system of equations for which the graphs intersect at exactly one point. (Section 9.1)
independent events Two events for which the outcome of the first event does not influence the outcome of the second event. (Section 11.7)
independent variable For a function defined by an equation, the variable that represents elements of the domain. (Section 2.2)

index of a radical In the expression $\sqrt[n]{a}$, the positive integer n is the index of the radical. (Section P.3)

infinite sequence A funtion whose domain is the positive integers and whose range is a set of real numbers. (Section 11.1)

infinite series The sum of all the terms of an infinite sequence. (Section 11.3)

initial side The beginning position of the ray that rotates to form an angle. (Section 5.1)

integers The numbers $\ldots -4, -3, -2, -1, 0, 1, 2, 3, \ldots$. (Section P.1)

intercept Any point on a graph that has an x- or a y-coordinate of 0; a point where the graph intersects the x- or the y-axis. (Section 2.1)

interest Money paid for the use of money. (Section 4.6)

intersection of sets The intersection of sets A and B, denoted by $A \cap B$, is the set of all elements that belong to both set A and set B. (Section P.1)

interval notation A compact notation used to represent subsets of real numbers. (Section P.2)

inverse of a matrix The inverse of matrix A, denoted by A^{-1}, is the matrix with the property that $AA^{-1} = I$, the identity matrix. (Section 10.3)

inverse function The function, denoted by f^{-1}, that is formed by interchanging the x and y coordinates of a function f. (Section 4.1)

inversely proportional If $y = k/x$, the variable y varies inversely as the variable x, or y is inversely proportional to x. (Section 1.6)

irrational numbers The set of all nonterminating, nonrepeating decimals. (Section P.1)

irreducible over the reals A quadratic factor with no real zeros. (Section 3.4)

leading coefficient The coefficient a_n of a polynomial of degree n. (Section P.4)

least-squares regression line The line that minimizes the sum of the squares of the vertical deviations of all data points from the line. (Section 2.7)

legs In a right triangle, the two sides other than the hypotenuse. (Section 1.3)

less than a is less than b (denoted by $a < b$) if $b - a$ is positive. (Section P.2)

like radicals Radicals that have the same radicand and the same index. (Section P.1)

like terms Terms that have exactly the same variables raised to the same powers. (Section P.4)

limacon A graph of an equation of the form $r = a + b\cos\theta$ or $r = a + b\sin\theta$. (Section 8.5)

line of best fit See *least-squares regression line*. (Section 2.7)

linear correlation coefficient r A measure of how closely the points of a data set can be modeled by a straight line. (Section 2.7)

linear equation An equation that can be written in the form $ax + b = 0$, where a and b are real numbers and $a \neq 0$. (Section 1.1)

linear extrapolation A linear approximation of data beyond the given values. (Section 2.3)

linear function A function of the form $f(x) = ax + b$. (Section 2.3)

linear interpolation A linear approximation of data between the given values. (Section 2.3)

linear objective function The function to be maximized or minimized in a linear programming problem. (Section 9.6)

linear programming A technique of solving some types of maximization or minimization problems. (Section 9.6)

linear speed Distance traveled per unit time. (Section 5.1)

linear system of equations A system of equations in which each equation is a linear equation. (Section 9.1)

logarithmic function f with base b $y = \log_b x$ if and only if $b^x = y$. (Section 4.3)

logarithmic equation An equation that involves logarithms. (Section 4.5)

lower bound A real number a for which no zero of the polynomial function P is less than a. (Section 3.3)

Mach number The speed of an object divided by the speed of sound. (Section 6.3)

major axis The longer axis of the graph of an ellipse. (Section 8.2)

matrix A rectangular array of numbers. (Section 10.1)

maximum value of a quadratic function If $a < 0$, then the vertex (h, k) is the highest point on the graph of $f(x) = a(x - h)^2 + k$, and k is the maximum value of the function f. (Section 2.4)

maximum value of a function The largest range element of the function. (Section 2.2)

measure The measure of an angle is determined by the amount of rotation of the initial ray. (Section 5.1)

midpoint The point on a line segment that is equidistant from the endpoints of the segment. (Section 2.1)

minimum value of a function The smallest range element of the function. (Section 2.2)

minimum value (of a quadratic function) If $a > 0$, then the vertex (h, k) is the lowest point on the graph of $f(x) = a(x - h)^2 + k$, and k is the minimum value of the function f. (Section 2.4)

minor of a matrix The determinant formed by removing the ith row and jth column of the determinant of the matrix; denoted by M_{ij}. (Section 10.4)

minor axis The shorter axis of the graph of an ellipse. (Section 8.2)

modulus of a complex number The number r for a complex number written in the form $z = r\operatorname{cis}\theta$. (Section 7.4)

monomial A constant, a variable, or a product of a constant and one or more variables, with the variables having only nonnegative integer exponents. (Section P.4)

multiplication Multiplication of the real numbers a and b is designated by ab. (Section P.1)

multiplicative inverse The multiplicative inverse or reciprocal of the nonzero number b is $1/b$. (Section P.1)

mutually exclusive events Two events A and B for which $A \cap B = \varnothing$. (Section 11.7)

n factorial ($n!$) $n! = n(n-1)(n-2)\cdots 3 \cdot 2 \cdot 1$, n a natural number. $0! = 1$. (Section 11.1)

natural exponential function The function defined by $f(x) = e^x$ for all real numbers x. (Section 4.2)

natural logarithm A logarithm with base e. (Section 4.3)

natural number A positive integer. (Section P.1)

negative angles An angle formed by a clockwise rotation. (Section 5.1)

negative integer An integer less than 0. (Section P.1)

negative real number A real number less than 0. (Section P.2)

nonfactorable over the integers A polynomial that cannot be factored into the product of two polynomials having integer coefficients. (Section P.5)

nonlinear system of equations A system of equations in which one or more equations are not linear equations. (Section 9.3)
nonsingular matrix A matrix that has a multiplicative inverse. (Section 10.3)
***n*th partial sum** The sum of the first n terms of a sequence. (Section 11.1)
null set The set without any elements; denoted by $\varnothing$. (Section P.1)
numerator The real number a in the fraction a/b. (Section P.1)
numerical coefficient The constant in a monomial. (Section P.4)
oblique triangle A triangle that does not contain a right angle. (Section 7.1)
obtuse angle An angle that has a measure greater than 90° but less than 180°. (Section 5.1)
odd function A function for which $f(-x) = -f(x)$. (Section 2.5)
one-to-one function A function that satisfies the additional condition that given any y value, there is only one x value paired with that given y value. (Section 2.2)
open interval (a, b) represents all real numbers between a and b, not including a and not including b. (Section P.2)
optimization problem A problem that requires a situation to be maximized or minimized. (Section 9.6)
order of a matrix A matrix of m rows and n columns has order $m \times n$ (read "m by n"). (Section 10.1)
ordinary annuity An annuity for which the amounts are deposited at the end of a compounding period. (Section 11.3)
ordinate The y-coordinate of an ordered pair. (Section 2.1)
origin The point, (0, 0), where x- and y-axes intersect. (Section 2.1)
parabola The set of all points in a plane that are equidistant from a fixed line (directrix) and a fixed point (focus) not on the directrix. (Section 8.1)
parallel lines Two nonintersecting lines in a plane. (Section 2.3)
partial fraction decomposition The method by which a more complicated rational expression is written as a sum of simpler rational expressions. (Section 9.4)
Pascal's Triangle A triangular array of the coefficients of the terms of expanded binomials. (Section 11.5)
perfect-square trinomial A trinomial that is the square of a binomial. (Section P.5)
permutation An arrangement of distinct objects in a definite order. (Section 11.6)
perpendicular lines Two lines that intersect to form adjacent angles each of which measures 90°. (Section 2.3)
pH The negative of the common logarithm of the molar hydronium-ion concentration. (Section 4.5)
phase shift The horizontal shift of the graph of a trigonometric function. (Section 5.7)
piecewise-defined function A function represented by more than one equation. (Section 2.2)
plot a point To draw a dot at the point's location in the coordinate plane. (Section 2.1)
point-slope form The equation of a straight line written in the form $y - y_1 = m(x - x_1)$. (Section 2.3)
polar axis The horizontal ray of a polar coordinate system. (Section 8.5)
polar coordinate system A coordinate system formed by rays and concentric circles; the rays emanate from the center of the concentric circles. (Section 8.5)
polar equation An equation of the form $r = f(\theta)$. (Section 8.5)
polar form The trigonometric form of a complex number. (Section 7.4)
pole The origin of a polar coordinate system. (Section 8.5)
polynomial A sum of a finite number of monomials. (Section P.4)
positive angle An angle formed by a counterclockwise rotation. (Section 5.1)
positive integer An integer greater than zero. (Section P.1)
positive real number A number to the right of the origin. (Section P.2)
power The expression b^n is the nth power of b. (Section P.3)
prime number A positive integer greater than 1 that has no positive-integer factors other than itself and 1. (Section P.1)
principal An amount of money invested. (Section 4.6)
principal square root The positive square root of a number. (Section P.3)
probability The mathematical study of random patterns. (Section 11.7)
product If $ab = c$, then c is the product. (Section P.1)
profit function The function, P, that gives a manufacturer's profit from selling x units of a product. (Section 2.3)
Pythagorean identities The Pythagorean identities are based on the equation of a unit circle and on the definitions of the sine and cosine functions. They are $\cos^2 t + \sin^2 t = 1$, $1 + \tan^2 t = \sec^2 t$, and $1 + \cot^2 t = \csc^2 t$. (Section 5.4)
quadrantal angles An angle in standard position whose terminal side lies on a coordinate axis. (Section 5.1)
quadrants The four regions formed by the axes. (Section 2.1)
quadratic equation An equation that can be written in the standard quadratic form $ax^2 + bx + c = 0$, where $a \neq 0$. (Section 1.3)
quadratic formula If $ax^2 + bx + c = 0, a \neq 0$, then $x = \dfrac{-b \pm \sqrt{b^2 - 4ac}}{2a}$. (Section 1.3)
quadratic function A function that can be represented by an equation of the form $f(x) = ax^2 + bx + c, a \neq 0$. (Section 2.4)
quadratic in form A polynomial that can be written in the form $au^2 + bu + c = 0$, where $a \neq 0$. (Section 1.4)
quotient If $a \div b = c$, then c is the quotient of a and b; the number obtained when dividing one quantity by another. (Sections P.1, 3.1)
radian The measure of the central angle subtended by an arc of length r on a circle of radius r. (Section 5.1)
radicals Expressions using the notation $\sqrt[n]{b}$, also used to denote roots. (Section P.3)
radicand In the expression $\sqrt[n]{b}$, the number b is the radicand. (Section P.3)
radius The distance from the center of a circle or sphere to a point on the circle or sphere. (Section 2.1)
range The set of all the second coordinates of the ordered pairs of a function. (Section 2.2)
ratio identities The ratio identities are obtained by writing the tangent and cotangent functions in terms of the sine and cosine functions. They are $\tan t = \sin t/\cos t$ and $\cot t = \cos t/\sin t$. (Section 5.4)
rational expression A fraction in which the numerator and the denominator are polynomials. (Section P.6)
rational function A function that can be expressed as a quotient of polynomials. (Section 3.5)
rational inequalities An inequality that involves rational expressions. (Section 1.5)

rational numbers The set of all terminating or repeating decimals. (Section P.1)

rationalize the denominator To write a fraction in an equivalent form that does not involve any radicals in the denominator. (Section P.1)

ray The union of a point P and a half-line formed by P. (Section 5.1)

real axis The horizontal axis of the coordinate plane. (Section 7.4)

real number line A coordinate axis is used to represent the real numbers geometrically. (Section P.1)

real numbers The set of all rational or irrational numbers. (Section P.1)

real part of a complex number The real number a for the complex number $a + bi$. (Section 1.3)

reciprocal The multiplicative inverse or reciprocal of the nonzero number b is $1/b$. (Section P.1)

reciprocal function The function denoted by $1/f$. (Sections 4.1, 5.2)

rectangular form of a complex number A complex number written in the form $z = a + bi$. (Section 7.4)

recursively defined sequence A sequence in which each succeeding term of the sequence is defined using one or more of the preceding terms. (Section 11.1)

reduced polynomial The polynomial formed from $P(x)/(x - a)$, where a is a zero of $P(x)$. (Section 3.1)

reference angle For an angle θ in standard position, the positive acute angle formed by the terminal side of θ and the x-axis. (Section 5.3)

relation Any set of ordered pairs. (Section 2.2)

remainder The number left over when one integer is divided by another. (Section 3.1)

resultant See *resultant vector.* (Section 7.3)

resultant vector The sum of two vectors. (Section 7.3)

revenue function The function, R, that gives a manufacturer's revenue from the sale of x units of a product. (Section 2.3)

right angle An angle that measures 90°. (Section 5.1)

right triangle A triangle that contains one 90° angle. (Section 1.3)

roots of an equation The values of the variable that satisfy an equation. (Section 1.1)

roots of a polynomial The values of x for which a polynomial $P(x)$ is equal to 0. (Section 3.1)

rose curve A polar graph of an equation in the form $r = a \sin n\theta$ or $r = a \cos n\theta$, where n is an integer, $n \geq 2$. (Section 8.4)

sample space The set of all possible outcomes of an experiment. (Section 11.7)

scalar The number used to indicate the magnitude of a measurement. (Section 7.3)

scalar multiplication The product of a real number and a vector or a matrix. (Sections 7.3, 11.2)

scalar quantities Measurements such as area, mass, distance, speed, and time. (Section 7.3)

scientific notation A number in the form $a \times 10^n$, where n is an integer and $1 \leq a < 10$. (Section P.3)

secant In a right triangle, the function of an acute angle that is the ratio of the hypotenuse to the adjacent side. (Section 5.2)

semi-perimeter One-half the perimeter of a triangle. (Section 7.2)

sequence of partial sums A sequence formed from the partial sums of another sequence. (Section 11.1)

series The indicated sum of a sequence. (Section 11.1)

set of feasible solutions The solution set of the constraints of a linear programming problem. (Section 9.6)

set-builder notation Makes use of a variable and a characteristic property that the elements of the set alone possess. (Section P.1)

simple interest Interest that is a fixed percent r, per time period t, of the amount of money invested. (Section 4.6)

simple zero A zero of multiplicity 1. (Section 3.3)

simplified A rational expression is simplified when 1 is the only common polynomial factor of both the numerator and the denominator. (Section P.6)

simplify a rational expression To factor the numerator and the denominator of the rational expression. (Section P.6)

sine In a right triangle, the function of an acute angle that is the ratio of the opposite side to the hypotenuse. (Section 5.2)

singular matrix A matrix that does not have a multiplicative inverse. (Section 10.3)

slant asymptote A linear asymptote that is not a vertical or horizontal line. (Section 3.5)

slope of a line The ratio of the change in y to the change in x between two points on the line. (Section 2.3)

slope-intercept form The equation of a line written in the form $f(x) = mx + b$; the slope is m and the y-intercept is $(0, b)$. (Section 2.3)

smooth, continuous curve A curve that does not have sharp corners, a break, or a hole. (Section 3.2)

solution of a system of equations in two variables An ordered pair that is a solution of both equations of the system. (Section 9.1)

solution of a polar equation An ordered pair (r, θ) that satisfies the polar equation. (Section 8.5)

solution set of a system of inequalities The intersection of the solution sets of the individual inequalities. (Section 9.5)

solution set of an inequality in one variable The set of all solutions of the inequality. (Section 1.5)

solution set of an inequality in two variables The set of all ordered pairs that satisfy the inequality. (Section 9.5)

solutions of an equation The values of the variable that satisfy the equation. (Section 1.1)

solve an equation To find all values of the variable that satisfy the equation. (Section 1.1)

square matrix of order n A matrix with n rows and n columns. (Section 10.1)

square root If the index n equals 2, then the radical $\sqrt[n]{b}$ is written as simply $\sqrt{b}$, and it is referred to as the principal square root of b or simply the square root of b. (Section P.3)

standard form of a complex number A complex number written in the form $z = a + bi$. (Section 1.3)

standard form of a quadratic function A quadratic function $f(x) = ax^2 + bx + c$ written in the form $f(x) = a(x - h)^2 + k$. (Section 2.4)

standard form of the equation of a circle An equation of a circle written in the form $(x - h)^2 + (y - k)^2 = r^2$, where (h, k) is the center and r is the radius. (Section 2.1)

standard form of a polynomial A polynomial in the variable x written with decreasing powers of x. (Section P.4)

standard position An angle superimposed in a Cartesian coordinate system is in standard position if its vertex is at the origin and its initial side is on the positive x-axis. (Section 5.1)

standard quadratic form A quadratic equation written in the form $ax^2 + bx + c = 0$, where $a \neq 0$. (Section 1.3)

step function A function of the form $f(x) = [\![x]\!]$, where $[\![x]\!]$ is the greatest integer function. (Section 2.2)
subset Set A is a subset of set B if every element of set A is also an element of set B. (Section P.1)
substitution method A method of solving a system of equations. (Section 9.1)
subtraction Subtraction of the real numbers a and b is designated by $a - b$. (Section P.1)
sum If $a + b = c$, then c is the sum. (Section P.1)
summation notation A convenient notation used for partial sums. (Section 11.1)
supplementary angles Two positive angles for which the sum of the measures of the angles is 180°. (Section 5.1)
symmetric with respect to a line A graph is symmetric with respect to a line L if, for each point P on the graph, there is a point P' on the graph such that the line L is the perpendicular bisector of the line segment PP'. (Section 2.4)
symmetric with respect to a point A graph is symmetric with respect to a point Q if, for each point P on the graph, there is a point P' on the graph such that Q is the midpoint of the line segment PP'. (Section 2.5)
symmetric with respect to the x-axis A graph is symmetric with respect to the x-axis if, whenever the point given by (x, y) is on the graph, then $(x, -y)$ is also on the graph. (Section 2.5)
symmetric with respect to the y-axis A graph is symmetric with respect to the y-axis if, whenever the point given by (x, y) is on the graph, then $(-x, y)$ is also on the graph. (Section 2.5)
synthetic division A procedure for dividing a polynomial by a binomial of the form $x - c$. (Section 3.1)
system of equations Two or more equations considered together. (Section 9.1)
tangent function In a right triangle, the function of an acute angle that is the ratio of the opposite side to the adjacent side. (Section 5.2)
term Each monomial of a polynomial. (Section P.4)
terminal side The ending position of the ray that rotates to form an angle. (Section 5.1)
terms If $a + b = c$, then the real numbers a and b are called the terms. (Section P.1)
terms of a sequence The elements in the range of the sequence. (Section 11.1)
third-degree equation A cubic equation. (Section 1.4)
tolerance The acceptable amount by which a dimension may differ from a given standard. (Section 1.6)
transverse axis of a hyperbola The line segment joining the vertices. (Section 8.3)
triangular form of a matrix A matrix in which the entries above or below the main diagonal are zero. (Section 10.2)
trinomial A simplified polynomial that has three terms. (Section P.4)
trivial solution For a homogeneous system of equations, the solution that contains all zeros. (Section 9.2)
turning point A point where a function changes from an increasing function to a decreasing function, or vice versa. (Section 3.2)
union The union of sets A and B, denoted by $A \cup B$, is the set of all elements belonging to set A, to set B, or to both. (Section P.1)
unit circle A circle given by the equation $x^2 + y^2 = 1$. (Section 5.4)
unit fraction See *conversion factor.* (Section 5.1)
unit vector A vector whose magnitude is 1. (Section 7.3)
upper bound A real number b for which no zero of the polynomial function P is greater than b. (Section 3.3)
variation Many real-life situations involve variables that are related by a type of function called a variation. (Section 1.6)
variation constant In the direct variation equation $y = kx$, the value of k. (Section 1.6)
varies directly If $y = kx$, the variable y varies directly as the variable x. (Section 1.6)
varies directly as the nth power If y varies directly as the nth power of x, then $t = kx^n$, where k is a constant. (Section 1.6)
varies inversely If $y = k/x$, the variable y varies inversely as the variable x. (Section 1.6)
varies inversely as the nth power If y varies inversely as the nth power of x, then $y = k/x^n$, where k is a constant. (Section 1.6)
varies jointly The variable z varies jointly as the variables x and y if and only if $z = kxy$, where k is a constant. (Section 1.6)
vector A directed line segment. (Section 7.3)
vector quantity A quantity that has a magnitude (numerical and unit description) and a direction. (Section 7.3)
vertex of an angle The common endpoint of the sides of the angle. (Section 5.1)
vertex of a parabola The lowest point of a parabola that opens up or the highest point on a parabola that opens down; the midpoint of the line segment joining the focus and directrix of the parabola. (Sections 2.4, 8.1)
vertical asymptote If $\lim_{x \to a} f(x) = \pm\infty$, then $x = a$ is a vertical asymptote of the graph of f. (Section 3.5)
vertices of a hyperbola The points where the hyperbola intersects the transverse axis. (Section 8.3)
vertices of an ellipse The endpoints of the major axis of the ellipse. (Section 8.2)
x-axis A horizontal coordinate axis. (Section 2.1)
x-coordinate In the ordered pair (a, b), the real number a. (Section 2.1)
x-intercept A point at which a graph crosses the x-axis. (Section 2.1)
y-axis A vertical coordinate axis. (Section 2.1)
y-coordinate In the ordered pair (a, b), the real number b. (Section 2.1)
y-intercept A point at which a graph crosses the y-axis. (Section 2.1)
z-axis A third coordinate axis perpendicular to the xy plane. (Section 9.2)
zero of a polynomial Any value of x that causes a polynomial in x to equal 0 is called a zero of the polynomial. (Sections 1.5, 3.1)
zero matrix A matrix in which all the elements are 0. (Section 10.2)
zero of multiplicity k If a polynomial $P(x)$ has $(x - r)$ as a factor exactly k times, then r is a zero of multiplicity k of $P(x)$. (Section 3.3)
zero product property The zero product property states that if the product of two factors equals 0, then at least one of the factors is 0. (Section 1.3)
zeros of a function For a function f, the values of x for which $f(x) = 0$. (Section 3.1)

INDEX

Properties of Exponents

$a^m a^n = a^{m+n}$ $\qquad \dfrac{a^m}{a^n} = a^{m-n}$ $\qquad (a^m)^n = a^{mn}$

$(a^m b^n)^p = a^{mp} b^{np}$ $\qquad \left(\dfrac{a^m}{b^n}\right)^p = \dfrac{a^{mp}}{b^{np}}$ $\qquad b^{-p} = \dfrac{1}{b^p}$

Properties of Logarithms

$y = \log_b x$ if and only if $b^y = x$

$\log_b b = 1$ $\qquad \log_b 1 = 0$ $\qquad \log_b (b)^p = p$

$b^{\log_b p} = p$ $\qquad \log x = \log_{10} x$ $\qquad \ln x = \log_e x$

$\log_b (MN) = \log_b M + \log_b N$

$\log_b (M/N) = \log_b M - \log_b N$

$\log_b M^p = p \log_b M$

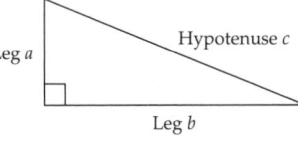

Properties of Radicals

$(\sqrt[n]{b})^m = \sqrt[n]{b^m} = b^{m/n}$ $\qquad \dfrac{\sqrt[n]{a}}{\sqrt[n]{b}} = \sqrt[n]{\dfrac{a}{b}}$

$\sqrt[n]{a}\sqrt[n]{b} = \sqrt[n]{ab}$ $\qquad \sqrt[m]{\sqrt[n]{b}} = \sqrt[mn]{b}$

Properties of Absolute Value Inequalities

$|x| < c \ (c \geq 0)$ if and only if $-c < x < c$.

$|x| > c \ (c \geq 0)$ if and only if either $x > c$ or $x < -c$.

Important Theorems

Pythagorean Theorem
$c^2 = a^2 + b^2$

Leg a Hypotenuse c Leg b

Remainder Theorem
If a polynomial $P(x)$ is divided by $x - c$, then the remainder is $P(c)$.

Factor Theorem
A polynomial $P(x)$ has a factor $(x - c)$ if and only if $P(c) = 0$.

Fundamental Theorem of Algebra
If P is a polynomial of degree $n \geq 1$ with complex coefficients, then P has at least one complex zero.

Binomial Theorem

$$(a + b)^n = a^n + \binom{n}{1}a^{n-1}b + \binom{n}{2}a^{n-2}b^2 + \cdots + \binom{n}{k}a^{n-k}b^k + \cdots + b^n$$

Important Formulas

The *distance* between $P_1(x_1, y_1)$ and $P_2(x_2, y_2)$ is

$$d(P_1, P_2) = \sqrt{(x_1 - x_2)^2 + (y_1 - y_2)^2}$$

The *slope* m of a line through $P_1(x_1, y_1)$ and $P_2(x_2, y_2)$ is

$$m = \frac{y_2 - y_1}{x_2 - x_1}, \quad x_1 \neq x_2$$

The *slope-intercept form* of a line with slope m and y-intercept b is $y = mx + b$

The *point-slope formula* for a line with slope m passing through $P_1(x_1, y_1)$ is

$$y - y_1 = m(x - x_1)$$

Quadratic Formula
If $a \neq 0$, the solutions of $ax^2 + bx + c = 0$ are

$$x = \frac{-b \pm \sqrt{b^2 - 4ac}}{2a}$$

Properties of Functions

A *function* is a set of ordered pairs in which no two ordered pairs that have the same first coordinate have different second coordinates.

If a and b are elements of an interval I that is a subset of the domain of a function f, then

- f is an *increasing* function on I if $f(a) < f(b)$ whenever $a < b$.
- f is a *decreasing* function on I if $f(a) > f(b)$ whenever $a < b$.
- f is a *constant* function on I if $f(a) = f(b)$ for all a and b.

A *one-to-one* function satisfies the additional condition that given any y, there is one and only one x that can be paired with that given y.

Graphing Concepts

Odd Functions
A function f is an odd function if $f(-x) = -f(x)$ for all x in the domain of f. The graph of an odd function is symmetric with respect to the origin.

Even Functions
A function is an even function if $f(-x) = f(x)$ for all x in the domain of f. The graph of an even function is symmetric with respect to the y-axis.

Vertical and Horizontal Translations
If f is a function and c is a positive constant, then the graph of

- $y = f(x) + c$ is the graph of $y = f(x)$ shifted up *vertically* c units.
- $y = f(x) - c$ is the graph of $y = f(x)$ shifted down *vertically* c units.
- $y = f(x + c)$ is the graph of $y = f(x)$ shifted left *horizontally* c units.
- $y = f(x - c)$ is the graph of $y = f(x)$ shifted right *horizontally* c units.

Reflections
If f is a function then the graph of

- $y = -f(x)$ is the graph of $y = f(x)$ reflected across the x-axis.
- $y = f(-x)$ is the graph of $y = f(x)$ reflected across the y-axis.

Vertical Shrinking and Stretching

- If $c > 0$ and the graph of $y = f(x)$ contains the point (x, y), then the graph of $y = c \cdot f(x)$ contains the point (x, cy).
- If $c > 1$, the graph of $y = cf(x)$ is obtained by stretching the graph of $y = f(x)$ away from the x-axis by a factor of c.
- If $0 < c < 1$, the graph of $y = cf(x)$ is obtained by shrinking the graph of $y = f(x)$ toward the x-axis by a factor of c.

Horizontal Shrinking and Stretching

- If $a > 0$ and the graph of $y = f(x)$ contains the point (x, y), then the graph of $y = f(ax)$ contains the point $\left(\frac{1}{a}x, y\right)$.
- If $a > 1$, the graph of $y = f(ax)$ is a *horizontal shrinking* of the graph of $y = f(x)$.
- If $0 < a < 1$, the graph of $y = f(ax)$ is a *horizontal stretching* of the graph of $y = f(x)$.

Definitions of Trigonometric Functions

$\sin \theta = \frac{b}{r}$ $\csc \theta = \frac{r}{b}$

$\cos \theta = \frac{a}{r}$ $\sec \theta = \frac{r}{a}$

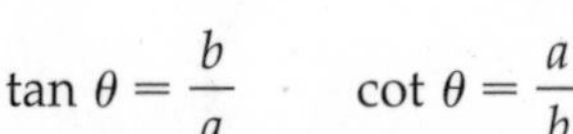

$\tan \theta = \frac{b}{a}$ $\cot \theta = \frac{a}{b}$

where $r = \sqrt{a^2 + b^2}$

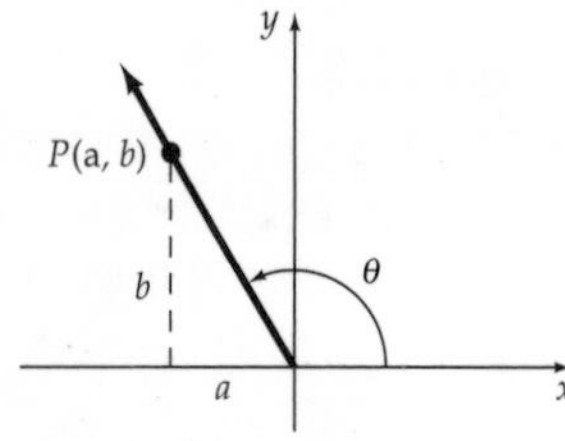

Definitions of Circular Functions

$\sin t = y$ $\csc t = \frac{1}{y}$

$\cos t = x$ $\sec t = \frac{1}{x}$

$\tan t = \frac{y}{x}$ $\cot t = \frac{x}{y}$

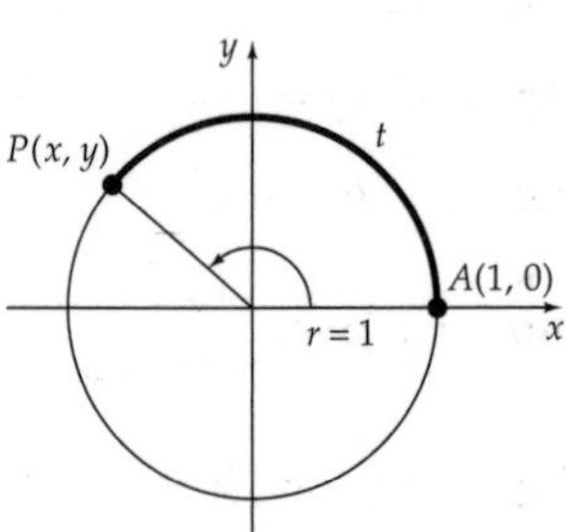

Formulas for Triangles

For any triangle ABC, the following formula can be used.

Law of Sines

$$\frac{a}{\sin A} = \frac{b}{\sin B} = \frac{c}{\sin C}$$

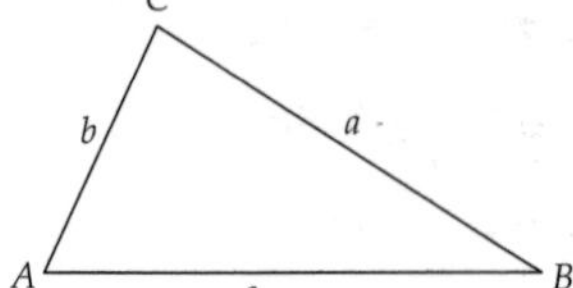

Law of Cosines

$$c^2 = a^2 + b^2 - 2ab \cos C$$

Area of a Triangle

$$K = \frac{1}{2}ab \sin C \qquad K = \frac{a^2 \sin B \sin C}{2 \sin A}$$

$$K = \sqrt{s(s-a)(s-b)(s-c)}, \text{ where } s = \frac{a+b+c}{2}$$

Fundamental Identities

$\tan \theta = \frac{\sin \theta}{\cos \theta}$ $\cot \theta = \frac{\cos \theta}{\sin \theta}$

$\sin^2 \theta + \cos^2 \theta = 1$ $1 + \tan^2 \theta = \sec^2 \theta$

$1 + \cot^2 \theta = \csc^2 \theta$

Formulas for Negatives

$\sin(-\theta) = -\sin \theta$ $\cos(-\theta) = \cos \theta$

$\tan(-\theta) = -\tan \theta$

Reciprocal Identities

$\csc \theta = \frac{1}{\sin \theta}$ $\sec \theta = \frac{1}{\cos \theta}$ $\cot \theta = \frac{1}{\tan \theta}$